QUICK ARITHMETIC

Short-Cut Methods of Solving the Problems
(Fully Solved)

An ideal book for :

- SSC (CGL Tier I and Tier II, CHSL, FCI, MTS.)
- Bank PO, SBI PO, RBI, Bank Clerical Exams.
- AAO Exam – LIC, GIC
- Railway Recruitment Board Exams.
- IGNOU, MBA, Hotel Management Exams.
- Sub-Inspectors of Police, CBI, CPO Exam.
- CDS, Management Exams, IGNOU

Ashish Aggarwal

S. CHAND PUBLISHING
empowering minds

S. CHAND & COMPANY PVT. LTD.
(ISO 9001 Certified Company)

S Chand And Company Limited
(ISO 9001 Certified Company)
Head Office: Block B-1, House No. D-1, Ground Floor, Mohan Co-operative Industrial Estate, New Delhi – 110 044 | Phone: 011-66672000
Registered Office: A-27, 2nd Floor, Mohan Co-operative Industrial Estate, New Delhi – 110 044
Phone: 011-49731800
www.**schandpublishing.com**; e-mail: **info@schandpublishing.com**

Branches

Chennai : Ph: 044-2363 2120, chennai@schandpublishing.com
Guwahati : Ph: 0361-2738 811, 2735 640, guwahati@schandpublishing.com
Hyderabad : Ph: 040-4018 6018, hyderabad@schandpublishing.com
Jalandhar : Ph: 0181-4645 630, jalandhar@schandpublishing.com
Kolkata : Ph: 033-2335 7458, 2335 3914, kolkata@schandpublishing.com
Lucknow : Ph: 0522-4003 633, lucknow@schandpublishing.com
Mumbai : Ph: 022-2500 0297, mumbai@schandpublishing.com
Patna : Ph: 0612-2260 011, patna@schandpublishing.com

First Edition 2004
Subsequent Editions and Reprints 2005, 2006, 2007 (Twice), 2009, 2010, 2011, 2012, 2013, 2014
Sixth Revised Edition 2014; Reprint 2015
Seventh Revised Edition 2016
Reprint 2022 (Thrice)

ISBN : 978-93-525-3195-0 **Product Code :** H4ARI49MATH40ENAG16O

PRINTED IN INDIA

By Vikas Publishing House Private Limited, Plot 20/4, Site-IV, Industrial Area Sahibabad, Ghaziabad – 201 010 and Published by S Chand And Company Limited, A-27, 2nd Floor, Mohan Co-operative Industrial Estate, New Delhi – 110 044

PREFACE TO THE SEVENTH REVISED EDITION

Mathematics has always been a very important part of various competitive examinations and with the time, it's importance has increased, hence there is a need for learning the short-cut methods of solving the mathematical problems in the competitive exams where time taken is of prime essence.

This edition of the book is thoroughly revised and enlarged according to the latest syllabi and trend of the various competitive exams. Taking into consideration the feedback from our readers, new and better short-cut methods, based on logics, to solve the questions have been explained in this edition. Important questions from various competitive exams have been included. This book will also be useful for many competitive exams having quantitative aptitude as one of the paper.

It gives me a great pleasure to note the appreciation shown by the students of the previous editions of this book. I hope the readers would get as much pleasure in reading this book as I get in writing the same. I am thankful to my publisher S. Chand & Company Pvt. Ltd., New Delhi for all the help, support and guidance.

The salient features of this edition are as follows :

- Written strictly as per the new syllabi of various exams. including SSC CGL Tier I and Tier II Exams
- Each chapter starts with important rules and formulae followed by fully solved examples
- Contains solved previous year question papers of popular examinations
- Inclusion of same important topics such as *Vedic Maths, Geometry* and *Trigonometry* makes this book apt for various competitive examinations
- New and improved short-cut methods with tricks of solving problems in almost no time.
- Rated ***Five Stars*** by most readers

This book is very useful to candidates appearing in the following exams. :

- SSC : Combined Graduate Level (Tiers I & II), Combined Higher Secondary Level, Food Corporation of India, Multi-Tasking, Constables, Sub-Inspector in Delhi Police, CISF, etc.
- Bank : PO, Clerical, RBI
- Railway Recruitment Board
- LIC/GIC : AAO, Clerical
- Management Exams., IGNOU, CDS

Any suggestions and feedback for the improvement of the book are most welcome. I wish you success in the exams.

ASHISH AGGARWAL
ashishmea@yahoo.com

PREFACE TO THE FIRST EDITION

I am happy to present this book to the students who wish to appear in the objective type competitive examinations. In such examinations, the students are required to solve questions in a limited time. The conventional method of solving the mathematical questions is not only time-consuming but also prone to mistakes because of the lengthy methods involved. This book explains the short-cut methods of solving the sums in a faster way. For the convenience of the students and easy understanding, the questions are divided into various chapters. The book covers basic terms, short-cut methods of multiplication, ratio and proportion, percentage, profit and loss, time and distance, time and work, simple and compound interest, mensuration, etc. Each chapter contains rules and formulae followed by fully solved exercises. Questions are also taken from previous papers of SSC, Banks and other exams. The important rules are given in the boxes so that one can locate them easily for future reference. In the chapters, logic of formulae is explained so that the students learn the logic of formulae, instead of memorising the formulae directly.

While teaching, I have experienced students facing difficulty in memorising methods, and also making wrong application of rules, leading to wrong answers. I hope that this book will be of immense help to those who appear in objective type mathematical papers. I am very thankful to my family and friends for their encouragement and support in writing this book. Any suggestions for the improvement of quality of the book and rectification of errors are most welcome.

ASHISH AGGARWAL
ashishmea@yahoo.com

Contents

CHAPTER 1

NUMBER SYSTEM

TYPE OF NUMBERS

Natural Numbers (N): (Counting numbers, Positive Integers) 1, 2, 3, 4, ...∞

Whole Numbers (W): (Set of all natural numbers and 0) 0, 1, 2, 3, ...∞

Integers (Z): – ∞........... – 3, – 2, – 1, 0, 1, 2, 3, 4, 5,...........∞

Even Numbers: 2, 4, 6, 8, ... ∞

Numbers that are exactly divisible by 2. These numbers end in even number or zero.

Odd Numbers: 1, 3, 5, 7, 9, ... ∞

Numbers that are not exactly divisible by 2. These numbers end in odd number.

Note: Sum of any two odd numbers is always an even number.

Example: 3 + 5 = 8 (an even number) 7 + 23 = 30 (an even number)

Note: Product of any two odd numbers is always an odd number.

Example: 3 × 7 = 21 (an odd number) 5 × 13 = 65 (an odd number)

Rational Numbers (Q): Numbers which can be expressed in the form of $\frac{p}{q}$, where p and q both are integers, and q is a non-zero integer. A number expressed in the form of $\frac{p}{q}$ is also called a **fraction**, where p is numerator and q is denominator. A fraction when converted into decimal, gives finite or recurring digits after decimal sign. Rational numbers include integers. Every rational number can be expressed in decimal (terminating or non-terminating).

Example: $\frac{1}{3}, \frac{1}{2}, \frac{7}{4}$ **Note:** $\frac{x}{1} = x, \frac{0}{x} = 0$ and $\frac{x}{0}$ = Not defined

Mixed Numbers: Number which consists of a whole number and a fraction.

Example: $3\frac{1}{4}, 5\frac{2}{7}, 4\frac{6}{17}$, etc.

Irrational Numbers: Numbers that cannot be expressed in the form of $\frac{p}{q}$ where p and q both are integers *i.e.,* the numbers which are not rational. These numbers when converted into decimal, give infinite and non-recurring digits after decimal sign.

Example: $\pi, \sqrt{2}, \sqrt{3}, \sqrt{27}$ etc. (Remember: 22/7 or 3.14 is approximate value of π)

Real numbers(R): Real numbers are combination of rational and irrational numbers.

Complex or imaginary numbers: These numbers are written as $a + bi$ where $i^2 = -1$.

Prime Numbers: A number which is not divisible by any number other than 1 and itself. Such numbers have only two factors, the number itself and 1.

Example: 2, 3, 5, 7, 11, 13, 17, 19, 23.

Note: (*i*) 1 is neither a prime number nor a composite number.

(*ii*) 2 is the only even prime number. Every prime number, except 2 is odd. But every odd number is not necessarily a prime number. No even number (other than 2) is prime number.

(*iii*) Product of two prime numbers is always a composite number, since the product is divisible by both the prime numbers. **Example:** $7 \times 11 = 77$ (a composite number)

(*iv*) All prime numbers (other than 2 and 3) can be expressed in the form $(6N - 1)$ or $(6N + 1)$ where N is a natural number. The converse is not always true, i.e. every number in the form of $6N \pm 1$ is not a prime number.

(*v*) $p^2 - 1$ always divisible by 24, for every $p \geq 5$, where p is a prime number.

(*vi*) There are 15 prime numbers between 1 and 50, 10 prime numbers between 50 and 100, 21 prime numbers between 100 and 200.

Twin-prime Numbers: Pairs of such prime numbers whose difference is 2.

Example: 3 and 5, 11 and 13, 17 and 19.

How to check whether a given number is prime or not ?

Steps: (*i*) Find approximate square root of the given number.

(*ii*) Divide the given number by every prime number less than the approximate square root.

(*iii*) If the given number is exactly divisible by atleast one of the prime numbers, the number is a composite number otherwise it is a prime number.

Example: Is 401 a prime number?

Sol. Approximate square root of 401 is 20.

Prime numbers less than 20 are 2, 3, 5, 7, 11, 13, 17 and 19.

401 is not divisible by 2, 3, 5, 7, 11, 13, 17 or 19. $\Rightarrow$ 401 is a prime number.

(**Hint:** Next prime number after 19 is 23, which is greater than 20, so we need not check further.)

Composite Number: The number that is not a prime number *i.e.*, the number which has more than two different factors.

Example: 4, 6, 8, 9, 10, 12, 14, 15, 16, 18, 20, 21, 22, 24.

Co-prime Numbers: Co-prime numbers are those numbers which are prime to each other *i.e.*, they don't have any common factor other than 1.

Since these numbers do not have any common factor, their HCF is 1 and their LCM is equal to product of the numbers.

Note: Co-prime numbers can be prime or composite numbers. Any two prime numbers are always co-prime numbers.

Example 1 : 3 and 5: Both numbers are prime numbers.

Example 2 : 8 and 15: Both numbers are composite numbers but they are prime to each other *i.e.,* they don't have any common factor.

Face value and Place value: Face Value is absolute value of a digit in a number.

Place Value (or Local Value) is value of a digit in relation to its position in the number.

Example: Face value and Place value of 9 in 14921 is 9 and 900 respectively.

RULES

1. Product of two positive numbers is always a positive number.

$(+a) \times (+b) = +(a \times b)$ **Example:** $(+4) \times (+3) = (+)\ 12 = 12$

2. Product of two negative numbers is always a positive number.
$(-a) \times (-b) = +(a \times b)$ **Example:** $(-4) \times (-3) = (+)\ 12 = 12$
3. Product of a negative number and a positive number is always a negative number.
$(-a) \times (+b) = (+a) \times (-b) = -(a \times b)$
Example: $(-4) \times (3) = (-)\ 12 = -12$, $(4) \times (-3) = (-)\ 12 = -12$
4. Any number multiplied by 0 gives 0. *i.e.*, $a \times 0 = 0 \times a = 0$.
Example: $0 \times 127 = 0$, $92567 \times 0 = 0$.
5. **Modulus or Absolute value of a number:**
Modulus of a number x is expressed as $|x|$ and its modulus value is x if $x \geq 0$ and $-x$ if $x < 0$.
Exampe: $|-3| = 3\ (\because -3 < 0)$; $|7| = 7\ (\because 7 > 0)$
6. **If $x + y = c$ (where c is a constant number). Then xy is maximum when $x = y = \frac{c}{2}$.**
For xy = constant, $x + y$ is minimum when $x = y$
Example: Let us take a number, say 14. Half of 14 is 7.
$7 \times 7 = 49$ is the maximum product value for any two numbers whose sum is 14.
Similarly, if $x + y + z = c$. Then xyz is maximum when $x = y = z = \frac{c}{3}$
Example: Let us take $c = 15$. Then $5 \times 5 \times 5 = 125$. $6 \times 5 \times 4 = 120$.
7. **Rule of BODMAS.** The simplification of an expression may involve various operations, these operations must be performed in the following order of preference:
B : Bracket:
If more than one type of brackets are involved, they are operated in the following order:
First: (‾) line or bar bracket or vinculum bracket
Second: ( ) small bracket
Third: { } Mid bracket
Fourth : [ ] Large Bracket
O : Of ('Of' means multiplication but is operated before division)
D : Division
M : Multiplication
A : Addition
S : Subtraction

Example: $4 \times 8 \div 2$ of $2 + (15 - 3) = ?$
(*a*) 12 (*b*) 20 (*c*) 40 (*d*) 44
Sol. (*b*) $4 \times 8 \div 2$ of $2 + (15 - 3) = 4 \times 8 \div 2$ of $2 + 12 = 4 \times 8 \div 4 + 12$
$$= 4 \times 8 \times \frac{1}{4} + 12 = 8 + 12 = 20$$

8. If sum of two numbers a and b is x and difference between the numbers is y.
Then $a = \frac{x + y}{2}$ and $b = \frac{x - y}{2}$

Example: If the sum of two numbers is 33 and their difference is 15, the numbers are:
(*a*) 24, 9 (*b*) 33, 15 (*c*) 17, 16 (*d*) 48, 18
Sol. (*a*) Bigger number $\frac{33 + 15}{2} = \frac{48}{2} = 24$. Smaller number $= \frac{33 - 15}{2} = \frac{18}{2} = 9$.

9. If on adding 9 to a two-digit number (say *xy*), its digits are reversed. Then unit's digit of the number exceeds its ten's digit by 1. Similarly, if on adding 9*a* to a two-digit number, its digits are reversed, then unit's digit of the number exceeds its ten's digit by *a i.e.* $x + a = y$

Note: *If digits are reversed on subtracting 9a from the number, then* $x - a = y$.

Note: Difference between two digits of the number is ***Number added/subtracted ÷ 9.***

Proof: Let the original number is *xy. Then on adding 9 to it, the number becomes yx.*

xy can be written as $10x + y$, and *yx* can be written as $10y + x$

$\therefore\ (10x + y) + 9 = (10y + x) \Rightarrow (10x - x) + 9 = (10y - y) \Rightarrow 9x + 9 = 9y \Rightarrow x + 1 = y$

Example: The sum of the digits of a two-digit number is 9. If 9 is added to the number, its digits are reversed. What is the number ?

(*a*) 18 (*b*) 36 (*c*) 45 (*d*) 81

Sol. (*c*) On adding 9 to the number, its digits are reversed.

$\therefore$ Unit's digit is $9 \div 9 = 1$ more than the ten's digit.

Ten's digit $= \dfrac{9-1}{2} = 4 \Rightarrow$ Unit's digit is $4 + 1 = 5 \Rightarrow$ The required number is 45.

10. If on adding 99*a* to a three-digit number (say *xyz*), its digits are reversed, then $x + a = z$.

Note: If digits are reversed on subtracting 99*a* from the number, then $x - a = z$.

Example: $923 - 329 = 594$ (*i.e.* 99×6). (**Note:** $9 - 3 = 6$)

11. If a composite number *C*, can be expressed as $a^m \times b^n \times c^p$ ...., where *a*, *b*, *c*....... are prime factors and *m*, *n*, *p*.... are natural numbers, then total number of factors of *C*

$= (m + 1) \times (n + 1) \times (p + 1)$.....

Note: Total number of factors includes unity and the number itself.

Example: Find the number of factors of 12.

Sol. $12 = 2^2 \times 3^1$.

$\therefore$ Total number of factors of $12 = (2+1) \times (1+1) = 3 \times 2 = 6$

Note: The power of 2 can be any of (0, 1, 2) and power of 3 can be one of (0, 1).

The following factors of 12 are possible :

$(2^0 \times 3^0), (2^0 \times 3^1), (2^1 \times 3^0), (2^1 \times 3^1), (2^2 \times 3^0), (2^2 \times 3^1)$.

Note: Total number of factors excluding unity and the number itself $= 6 - 2 = 4$

12. The highest power a prime number *p*, which divides *x*! exactly is :

$$\left[\frac{x}{p}\right] + \left[\frac{x}{p^2}\right] + \left[\frac{x}{p^3}\right] + \ldots + \left[\frac{x}{p^n}\right] \text{ where } \frac{x}{p^n} \nless 1$$

Note: [*a*] denotes greatest integer less than or equal to *a*.

Example : Find the highest power of 5 that can divide 620!

Sol: $\dfrac{620}{5} + \dfrac{620}{5^2} + \dfrac{620}{5^3} = 124 + 24 + 4 = 152$ (Note $620 < 5^4$)

Trick: First divide 620 by 5 and then divide successive quotients by 5, till we get final quotient 0.

13. If *x* and *y* are two positive rational numbers such that $\sqrt{x \times y}$ is irrational number, then $\sqrt{xy}$ is an irrational number between *x* and *y*. $\dfrac{x+y}{2}$ is a rational number between them.

Example 1: Irrational number between 2 and 3 is $\sqrt{2 \times 3} = \sqrt{6}$

Example 2: Rational number between 2 and 3 is $\frac{2+3}{2} = 2.5$

Example 3: Irrational number between 2 and 2.5 = $\sqrt{2 \times 2.5} = \sqrt{5}$

Example 4: Irrational number between $\sqrt{2}$ and $\sqrt{3} = \sqrt{\sqrt{2} \times \sqrt{3}} = \sqrt{\sqrt{6}} = 6^{1/4}$

MORE PROPERTIES OF NUMBERS

1. A three-digit number formed by repeating a one-digit number thrice, is always divisible by 111. **(SSC CGL Tier I 2015)**
   Since $111 = 3 \times 37$, such numbers are divisible by 3 and 37 also.
   Example: 222, 777, 888, etc. are divisible by 3, 37 and 111.
2. A four-digit number formed by repeating a two-digit number, is always divisible by 101. **(CDS 1997, SSC CGL Tier I 2005, 2010, SSC CHSL 2010)**
   Example: 2525, 3636, 8383, etc. are divisible by 101.
3. The number zero is surrounded by the same two-digit number on both sides. Any such five-digit number is always divisible by 1001.
   Since $1001 = 7 \times 11 \times 13$. Such numbers are divisible by 7, 11 and 13 also.
   Example: 52052, 68068, 93093, etc. are divisible by 7, 11, 13 and 1001.
4. A six-digit number formed by repeating a three-digit number is always divisible by 1001. **(SSC CGL Tier I 1994, 2000)**
   Since $1001 = 7 \times 11 \times 13$. Such numbers are divisible by 7, 11 and 13 also.
   Example: 436436, 619619, 851851, etc. are divisible by 7, 11, 13 and 1001.
5. If sum of digits of a natural number is subtracted from the number itself, the resulting number is always divisible by 9.
   Example: 143 : Sum of digits = $1 + 4 + 3 = 8$
   Now, $143 - 8 = 135$, which is divisible by 9.
6. The difference of a number consisting of two digits and the number formed by interchanging the digits is always divisible by 9. **(CDS 2000)**
   Example: $41 - 14 = 27$; $83 - 38 = 45$
7. Sum of a two-digit number and number formed by interchanging its digits, is always divisible by 11. **(CDS 2003, SSC CGL Tier I 2008)**
   Proof: Let the number is xy. Then new number on interchanging digits is yx.
   $(10x + y) + (10y + x) = 11(x + y)$
   Example: $38 + 83 = 121$ which is divisible by 11.
8. Digits of a two digit number are reversed. Difference between the original number and new number is always divisible by 9.
   Example: $71 - 17 = 54$, which is divisible by 9.
9. Digits of a three digit number are reversed. Difference between the original number and new number is always divisible by 99.
   Proof: Let the number is $xyz = 100x + 10y + z$
   Then new number on reversing the digits = $zyx = 100z + 10y + x$
   Difference between two number = $(100x + 10y + z) - (100z + 10y + x) = 99(x - z)$
   Example: $731 - 137 = 594$, which is divisible by 99.
10. Difference between square of a two-digit number (xy) and square of a number formed by interchanging the digits (yx), is always divisible by 99.

Proof: $(10x + y)^2 - (10y + x)^2 = (100x^2 + y^2 + 20xy) - (100y^2 + x^2 + 20xy)$
$= 99x^2 - 99y^2 = 99(x^2 - y^2) = 99(x - y)(x + y)$

Example : $21^2 - 12^2 = 441 - 144 = 297 = 99 \times 3$

Logic: Sum is divisible by 11 and the difference by 9.

11. The number of 'n' digits numbers are $9 \times 10^{(n-1)}$.

Example: One digit numbers are (from 1 to 9) $= 9 = 9 \times 10^{(1-1)}$.
Two digits numbers are (from 10 to 99) $= 90 = 9 \times 10^{(2-1)}$.
Three digits numbers are (from 100 to 999) $= 900 = 9 \times 10^{(3-1)}$.

12. **Fermat's Little Theorem** : If p is a prime number and N is a natural number, then $N^p - N$ is a multiple of p.

Example : $2^7 - 2 = 126 = 7 \times 18$ is a multiple of 7.
$2^5 - 2 = 30 = 5 \times 6$ is a multiple of 5. $3^7 - 3 = 2184 = 7 \times 312$ is a multiple of 7.
$5^{11} - 5 = 48828120 = 11 \times 4438920$ is a multiple of 11.

SOLVED EXERCISE

1. If we divide a positive integer by another integer, what is the resulting number ? **(CDS 2009)**
   (*a*) It is always a natural number (*b*) It is always an integer
   (*c*) It is always a rational number (*d*) It is always an irrational number

Sol. (*c*) It is always a rational number (except when denominator is zero)

2. Consider the following statements : **(CDS 2012)**
   1. Every composite number is a natural number.
   2. Every whole number is a natural number.
   Which one of the statements given above is/are correct ?
   (*a*) 1 only (*b*) 2 only (*c*) Both 1 and 2 (*d*) Neither 1 nor 2

Sol. (*a*) Statement 2 is incorrect as zero is a whole number but not a natural number.

3. The product of a rational number and an irrational number is: **(CDS 2010)**
   (*a*) a natural number (*b*) an irrational number
   (*c*) a composite number (*d*) a rational number

Sol. (*b*) Always an irrational number.

4. A piece of cloth measured with a metre stick, one cm short, is 100 metres long. Reckoning the metre stick as being right, the actual length of the cloth (in cm) is **(SSC CGL Tier I 2014)**
   (*a*) 3900 (*b*) 9900 (*c*) 8000 (*d*) 6100

Sol. (*b*) 1 metre stick = (100 – 1) cm = 99 cm ⇒ 100 metres = 100 × 99 = 9900 cm (actual)

5. $15 - 15 \div 15 \times 6 = ?$
   (*a*) 6 (*b*) 84 (*c*) 9 (*d*) 0

Sol. (*c*) $15 - 15 \div 15 \times 6 = 15 - 15 \times \frac{1}{15} \times 6 = 15 - 1 \times 6 = 15 - 6 = 9$

6. $5 \times 16 \div 8$ of $2 = ?$
   (*a*) 5 (*b*) 6 (*c*) 20 (*d*) 30

Sol. (*a*) $5 \times 16 \div 8$ of $2 = 5 \times 16 \div 16 = 5 \times 16 \times \frac{1}{16} = 5$

7. $13800 \div 230 \div 30 = ?$
   (*a*) 1 (*b*) 2 (*c*) 3 (*d*) None of these

Sol. (*b*) $? = 13800 \div 30 \div 230 = 460 \div 230 = 2$

8. $100 \times 10 - 100 + 2000 \div 100 = ?$ **(SSC CGL Tier I 2012)**

(*a*) 29 (*b*) 920 (*c*) 980 (*d*) 1000

Sol. (*b*) $100 \times 10 - 100 + 2000 \div 100 = 100 \times 10 - 100 + 20 = 1000 - 100 + 20 = 920$

9. $\dfrac{8 + 2 \text{ of } 5 - 2}{8 - 2 \text{ of } (5 - 2)} = ?$

(*a*) 2 (*b*) 6 (*c*) 8 (*d*) 24

Sol. (*c*) $\dfrac{8 + 2 \text{ of } 5 - 2}{8 - 2 \text{ of } (5 - 2)} = \dfrac{8 + 10 - 2}{8 - 2 \text{ of } 3} = \dfrac{18 - 2}{8 - 6} = \dfrac{16}{2} = 8$

10. $8.7 - [7.6 - \{6.5 - (5.4 - \overline{4.3 - 2})\}]$ is simplified to: **(SSC CGL Tier I 2004)**

(*a*) 2.5 (*b*) 3.5 (*c*) 4.5 (*d*) 5.5

Sol. (*c*) Given expression $= 8.7 - [7.6 - \{6.5 - (5.4 - 2.3)\}] = 8.7 - [7.6 - \{6.5 - 3.1\}]$
$= 8.7 - [7.6 - \{3.4\}] = 8.7 - 4.2 = 4.5$

11. A number N is divisible by 6 but not divisible by 4. Which of the following will not be an integer ? **(SSC CGL Tier I 2013)**

(*a*) $\dfrac{N}{3}$ (*b*) $\dfrac{N}{2}$ (*c*) $\dfrac{N}{6}$ (*d*) $\dfrac{N}{12}$

Sol. (*d*) $\dfrac{N}{12} = \dfrac{N}{3 \times 4}$

12. If $7n + 9 > 100$ and n is an integer, the smallest possible value of n is **(SSC CHSL 2012)**

(*a*) 13 (*b*) 12 (*c*) 14 (*d*) 15

Sol. (*c*) $7n + 9 > 100 \Rightarrow 7n > 100 - 9 \Rightarrow 7n > 91 \Rightarrow n > 13 \Rightarrow$ Required number = 14

13. $a^c \times b^a \times ? = 0$ **(Bank PO 2005)**

(*a*) 1 (*b*) – 1 (*c*) 0 (*d*) c^b

Sol. (*c*) Any number multiplied by 0 yields 0.

14. The irrational number between 3 and 4 is

(*a*) 3.5 (*b*) $\sqrt{7}$ (*c*) $\sqrt{15}$ (*d*) $\sqrt{17}$

Sol. (*c*) $3^2 = 9$ and $4^2 = 16$; $\sqrt{9} < \sqrt{15} < \sqrt{16}$

15. A canteen requires 119 kg of wheat for a week. How many kg of wheat will it require for 79 days? **(SBI Clerical 2008)**

(*a*) 357 (*b*) 1343 (*c*) 9401 (*d*) None of these

Sol. (*b*) Daily requirement $= 119 \div 7 = 17$ kg $\Rightarrow$ Requirement for 79 days $= 79 \times 17 = 1343$ kg
Note: $79 \times 17 = (80 - 1) \times 17 = 80 \times 17 - 17 = 1360 - 17$ kg $= 1343$ kg

16. There are 7 dozen candles kept in a box. If there are 14 such boxes, how many candles are there in all boxes together ? **(SSC CGL Tier I 2008)**

(*a*) 1176 (*b*) 98 (*c*) 1216 (*d*) 168

Sol. (*a*) Total candles $= 7 \times 12 \times 14 = 1176$

17. The product of two numbers is 192 and the sum of these two numbers is 28. What is the smaller of these two numbers ? **(Bank PO 1999)**

(*a*) 12 (*b*) 14 (*c*) 16 (*d*) 18

Sol. (*a*) In option (*a*): Smaller number = 12 $\Rightarrow$ Bigger number = 28 – 12 = 16
Product $= 12 \times 16 = 192$
Note: Option (*b*), (*c*) and (*d*) are not possible as bigger numbers in these options = 28 – 14 = 14 (same), 28 – 16 = 12 (smaller) and 28 – 18 = 10 (smaller)

18. Find the difference between the place value and the face value of 7 in the number 897346.

(*a*) 0 (*b*) 6993 (*c*) 7337 (*d*) None of these

Sol. (*b*) Place value of 7 = 7000 Face value of 7 = 7 Difference = 7000 – 7 = 6993

19. If a number is as much greater than 31 as it is less than 75, then the number is **(SSC CHSL 2013)**

(*a*) 53 (*b*) 106 (*c*) 44 (*d*) 74

Sol. (*a*) Let the required number be x. Then $75 - x = x - 31 \Rightarrow 2x = 106 \Rightarrow x = 53$

20. Between 100 and 300, how many numbers are divisible by 7?

(*a*) 14 (*b*) 21 (*c*) 28 (*d*) 35

Sol. (*c*) $100 = 14 \times 7 + 2 \Rightarrow$ Between 1 and 100, 14 numbers are divisible by 7.

$300 = 42 \times 7 + 6 \Rightarrow$ Between 1 and 300, 42 numbers are divisible by 7.

$\therefore$ Between 100 and 300, 42 – 14 = 28 numbers are divisible by 7.

Hint: $100 = 14 \times 7 + 2$. $\therefore 300 = 3 \times (14 \times 7 + 2) = 42 \times 7 + 6$

21. The smallest 3-digit prime number is: **(Bank P.O. Exam)**

(*a*) 100 (*b*) 101 (*c*) 103 (*d*) None of these

Sol. (*b*) 101 is not divisible by 2, 3, 5 and 7.

22. In a two digit number, the digit in unit's place is three times the digit in ten's place and sum of the digits is equal to 12. Find the number.

(*a*) 39 (*b*) 93 (*c*) Date inadequate (*d*) None of these

Sol. (*a*) Let ten's place digit be x. Then unit's place digit be $3x$.

Now $x + 3x = 12 \Rightarrow 4x = 12 \Rightarrow x = 3$

$\therefore$ Ten's place digit = 3. Unit's place digit = $3x = 9$ $\therefore$ The reqiured number is 39.

23. Sarika is 19th from either end of a row of girls. How many girls are there in the row? **(SSC CHSL 2013)**

(*a*) 20 (*b*) 37 (*c*) 38 (*d*) 39

Sol. (*b*) Sarika is 19th from both the sides. $\therefore$ There are 18 students on her either side.

$\therefore$ Total students in the queue = $18 \times 2 + 1 = 37$

Alternative method: $19 \times 2 - 1 = 37$

24. Atul is twelfth from the right and fourth from the left in a queue of boys. The number of boys to be added to the queue such that there are 28 boys in the queue is : **(SSC CGL Tier I)**

(*a*) 12 (*b*) 13 (*c*) 14 (*d*) 16

Sol. (*b*) Boys in the queue = 12 + 4 – 1 =15 $\Rightarrow$ Boys to be added = 28 – 15 = 13

25. In a row, A is 5th from the left side and B is 8th from the right side. They both interchange their positions and then A becomes 12th from the left side. How many persons are there in the row? **(SSC CHSL)**

(*a*) 16 (*b*) 17 (*c*) 19 (*d*) 20

Sol. (*c*) B is 8th from the right side. When they interchange their places, A is at B's place and his place becomes 12th from the left side.

$\therefore$ Same place is 8th from one side and 12th from the other side.

$\therefore$ Total persons in the row = 8 + 12 – 1 = 19

26. Find the maximum number of trees which can be planted, 20 metres apart, on the two sides of a straight road 1760 metres long. **(SSC CGL Tier I 2013)**

(*a*) 176 (*b*) 180 (*c*) 178 (*d*) 174

Sol. (*c*) Number of trees = $2 \times [(1760 \div 20) + 1] = 178$

27. If the sum of the digits of any integer lying between 100 and 1000 is subtracted from the number, the result always is divisible by **(SSC CHSL 2013)**

(*a*) 5 (*b*) 6 (*c*) 2 (*d*) 9

Sol. (*d*) (Number – Sum of digits) is always divisible by 9.

28. On multiplying a number by 7, all the digits in the product appear as 3's. The smallest such number is: **(SSC CGL 2006)**

(*a*) 47719 (*b*) 47619 (*c*) 48619 (*d*) 47649

Sol. (*b*) All given options have 5 digits ⇒ the required number is of 6 digits.

$333333 \div 7 = 47619$

29. If x and y be two real numbers and $x + y = 8$, then the greatest value of xy is **(SSC CPO SI 2012)**

(*a*) 16 (*b*) 18 (*c*) 12 (*d*) 15

Sol. (*a*) xy is maximum when $x = y = 8 \div 2 = 4, \Rightarrow xy = x^2 = 4^2 = 16$

Trick: xy is square number. Out of given options, only 16 is a square number.

30. If $a + b + c + d = 1$, then the maximum value of $(1 + a)(1 + b)(1 + c)(1 + d)$ is **(SSC CGL Tier I 2012)**

(*a*) 1 (*b*) $\left(\frac{1}{2}\right)^3$ (*c*) $\left(\frac{3}{4}\right)^3$ (*d*) $\left(\frac{5}{4}\right)^4$

Sol. (*d*) $(1 + a)(1 + b)(1 + c)(1 + d)$ is maximum when $a = b = c = d = 0.25$

$\therefore$ Maximum value $= (1 + 0.25)^4 = \left(\frac{5}{4}\right)^4$

31. In a group of 40 persons, 15 drink tea but not coffee and 18 drink coffee but not tea. If they drink atleast one of the drinks, how many persons in the group drink tea ?

(*a*) 15 (*b*) 18 (*c*) 22 (*d*) 25

Sol. (*c*) Persons who drink tea = Total persons – Persons who don't drink tea = 40 – 18 = 22

32. A sum of ₹ 100 is divided among A, B and C. If A and B together receive ₹ 70, B and C together receive ₹ 50. Find the amount received by B.

(*a*) ₹ 10 (*b*) ₹ 20 (*c*) ₹ 40 (*d*) ₹ 60

Sol. (*b*) (A + B) + (B + C) = ₹70 + ₹50 = ₹120 ∴ B's share = ₹ 120 – ₹ 100 = ₹ 20

33. A and B together have ₹ 158, C has ₹ 101 less than what A and B together have and B has ₹ 23 more than C. Then amount of A is: **(SSC CGL Tier I 2002)**

(*a*) ₹ 80 (*b*) ₹ 78 (*c*) ₹ 57 (*d*) ₹ 88

Sol. (*b*) A + B = 158 C = (A + B) – 101 = 158 – 101 = 57

B = C + 23 = 57 + 23 = 80 A = (A + B) – B = 158 – 80 = ₹ 78

34. In a class of 50 students, 25 students can speak English and 40 students Hindi. If each student can speak atleast one of the two languages, how many students can speak both the languages?

(*a*) 10 (*b*) 15 (*c*) 20 (*d*) 25

Sol. (*b*) Students who can speak both the languages = 25 + 40 – 50 = 15

35. Find the number which when added to itself 13 times, gives 182 as result. **(Bank Clerical)**

(*a*) 12 (*b*) 13 (*c*) 14 (*d*) None of these

Sol. (*b*) Let x be the required number. Then $x + 13x = 14x = 182 \Rightarrow x = 182 \div 14 = 13$

36. A boy was asked to multiply a certain number by 53. By mistake, he multiplied it by 35 and got his answer less than the correct one by 1206. The number was : **(SSC CGL Tier I 2006, CDS 2015 M)**

(*a*) 67 (*b*) 76 (*c*) 77 (*d*) 66

Sol. (*a*) Let x be the required number. Then $53x - 35x = 1206 \Rightarrow 18x = 1206 \Rightarrow x = 67$

37. What is the greater of two numbers whose product is 1092 and the sum of the two numbers exceeds their difference by 42 ? **(SBI Bank PO 2008)**

(*a*) 48 (*b*) 44 (*c*) 52 (*d*) 54

Sol. (*c*) Let the numbers be x and y.

Then $(x + y) - (x - y) = 42 \Rightarrow 2y = 42 \Rightarrow y = 42 \div 2 = 21$

Now $xy = 1092 \Rightarrow 21x = 1092 \Rightarrow x = 1092 \div 21 = 52$

38. In a division sum, the divisor is 10 times the quotient and 5 times the remainder. If the remainder is 24, the dividend is: **(SSC CHSL 2013)**

(*a*) 1440 (*b*) 1464 (*c*) 2892 (*d*) None of these

Sol. (*b*) Divisor = 5 × 24 = 120, Quotient = 120 ÷ 10 = 12

∴ Dividend = (Divisor × Quotient) + Remainder = 120 × 12 + 24 = 1440 + 24 = 1464

39. In a question of division, the divisor is 7 times the quotient and 3 times the remainder. If the remainder is 28, then the dividend is: **(SSC CGL Tier I 2006, 2003M)**

(*a*) 588 (*b*) 784 (*c*) 823 (*d*) 1036

Sol. (*d*) Divisor = 28 × 3 = 84; Quotient = 84 ÷ 7 = 12

∴ Dividend = (Divisor × Quotient) + Remainder = (84 × 12) + 28 = 1008 + 28 = 1036

40. A divisor is 25 times the quotient and 5 times the remainder. If the quotient is 16, the dividend is : **(SSC CGL Tier I 2005)**

(*a*) 6400 (*b*) 6480 (*c*) 400 (*d*) 480

Sol. (*b*) Divisor = 25 × Quotient = 25 × 16 = 400; Remainder = 400 ÷ 5 = 80

∴ Dividend = (Divisor × Quotient) + Remainder = 400 × 16 + 80 = 6400 + 80 = 6480

41. In a question of division, the dividend is 37693, quotient is 52 and remainder lies between 52 and 104. Divisor will be: **(SSC CGL Tier I 2007)**

(*a*) 724 (*b*) 723 (*c*) 722 (*d*) 721

Sol. (*b*) 37693 ÷ 52 = 724 + (remainder 45)

∴ 37693 = 724 × 52 + 45 = 723 × 52 + (45 + 52) = 723 × 52 + 97

∴ 723 is the required divisor.

42. A single person takes 6 minutes to fill a bottle. If from 11 a.m. to 12.30 p.m., 1845 bottles are to be filled, how many persons should be employed on this job?

(*a*) 121 (*b*) 123 (*c*) 131 (*d*) 137

Sol. (*b*) Bottles filled by a single person from 11 a.m. to 12.30 p.m. = 90 ÷ 6 = 15

∴ Persons required to fill 1845 bottles = 1845 ÷ 15 = 123

43. The cost of 7 tables and 12 chairs is ₹ 48,250. What is the cost of 21 tables and 36 chairs ? **(SBI Bank PO 2008)**

(*a*) ₹ 96,500 (*b*) ₹ 1,25,500 (*c*) ₹ 1,44,750 (*d*) Data insufficient

Sol. (*c*) 21 tables + 36 chairs = 3 × (7 tables + 12 chairs) = 3 × ₹ 48,250 = ₹ 144,750

44. The cost of 12 pens and 16 pencils is ₹ 844. What will be the cost of 9 pens and 12 pencils?

(*a*) ₹ 366 (*b*) ₹ 636 (*c*) ₹ 633 (*d*) None of these

Sol. (*c*) Cost of 12 pens + 16 pencils = ₹ 844 ⇒ 4 (Cost of 3 pens + 4 pencils) = ₹ 844

∴ Cost of 3 pens + 4 pencils = ₹ 844 ÷ 4 = ₹ 211

∴ Cost of 9 pens + 12 pencils = 3 × ₹ 211 = 633

45. Total cost of 3 tables and 2 chairs is ₹ 1300 and that of 2 tables and 3 chairs is ₹ 1200. Find the cost of one chair.

(*a*) ₹ 100 (*b*) ₹ 150 (*c*) ₹ 200 (*d*) ₹ 250

Sol. (*c*) In the second case, 1 table is replaced by 1 chair.

As a result of it, total cost is decreased by ₹ 1300 – ₹ 1200 = ₹ 100

∴ Cost of 1 table = Cost of 1 chair + ₹ 100

Cost of 3 tables + cost of 2 chairs = ₹ 1300

∴ 3 × (Cost of 1 chair + ₹ 100) + Cost of 2 chairs = ₹ 1300

∴ Cost of 5 chairs = ₹ 1300 – ₹ 300 = ₹ 1000 ⇒ Cost of 1 chair = ₹ 1000 ÷ 5 = ₹ 200

46. Total cost of 5 pencils and 7 pens is ₹ 45. Total cost of 7 pencils and 6 pens is ₹ 44. Find the cost of one pencil.

(*a*) ₹ 2 (*b*) ₹ 3 (*c*) ₹ 4 (*d*) ₹ 5

Sol. (*a*) In the second case, 1 pen is replaced by 2 pencils.

Decrease in total cost = ₹ 45 – ₹ 44 = ₹ 1 ⇒ Cost of 1 pen = Cost of 2 pencils + ₹ 1

Cost of 5 pencils + cost of 7 pens = ₹ 45

∴ Cost of 5 pencils + 7 × (cost of 2 pencils + ₹ 1) = ₹ 45

∴ Cost of 19 pencils = ₹ 45 – ₹ 7 = ₹ 38 ⇒ Cost of 1 pencil = ₹ $\frac{38}{19}$ = ₹ 2

47. Two pens and three pencils cost ₹ 89. Four pens and one pencil cost ₹ 103. What is the difference between the cost of a pen and a pencil ?

(*a*) ₹ 14 (*b*) ₹ 7 (*c*) ₹ 4 (*d*) ₹ 5

Sol. (*b*) 2 pens + 3 pencils = ₹ 89 ...(*i*) 4 pens + 1 pencil = ₹ 103 ...(*ii*)

(*ii*) – (*i*) : 2 pens – 2 pencils = ₹ 14 ∴ Cost of 1 pen – Cost of 1 pencil = ₹ 7

48. A student was asked to multiply a number by 3. But he divided the number by 3 and got 6.9 as answer. What is the correct answer?

(*a*) 2.3 (*b*) 20.7 (*c*) 62.1 (*d*) None of these

Sol. (c) Let x be the given number. Then student's answer = $x \div 3$

Correct answer = $3x = 9 \times (x \div 3) = 9 \times 6.9 = 62.1$

49. Pooja started her job with certain monthly salary and gets a fixed increment every year. If her salary was ₹ 4200 after 3 years and ₹ 6800 after 8 years of service, then what are her initial salary and the annual increment respectively ? **(CDS 2009)**

(*a*) 2640, 320 (*b*) 2460, 320 (*c*) 2460, 520 (*d*) 2640, 520

Sol. (*d*) Five increments = ₹ 6800 – ₹ 4200 = ₹ 2600 ⇒ One increment = 2600 ÷ 5 = ₹ 520

Three increments = ₹ 520 × 3 = ₹ 1560

Salary after 3 years = Initial salary + 3 increments

⇒ Initial salary = Salary after 3 years – 3 increments = ₹4200 – ₹ 1560 = ₹ 2640

50. The number of times 2 is used while writing the numbers from 1 to 100 is **(SSC CPO SI 2012)**

(*a*) 18 (*b*) 19 (*c*) 20 (*d*) 21

Sol. (*c*) 2 used at unit's place : 2, 12, 22, 32, ..., 92 = 10 times

2 used at ten's place : 20, 21, 22, 23, ..., 29 = 10 times

∴ Total times = 10 + 10 = 20 times.

51 What is the total number of digits printed if a book containing 150 pages is to be numbered from 1 to 150 ?

(*a*) 450 (*b*) 360 (*c*) 342 (*d*) 262

Sol. (*c*) Single digit number are 1 to 9 $\Rightarrow$ Digits = $9 \times 1 = 9$

Two digit numbers are 10 to 99 $\Rightarrow$ Digits = $90 \times 2 = 180$

Three digit numbers are 100 to 150 $\Rightarrow$ Digits = $51 \times 3 = 153$

Total number of digits = $9 + 180 + 153 = 342$

52 How many times does the digit 2 apper in all 4-digit numbers ?

(*a*) 2000 (*b*) 4000 (*c*) 3700 (*d*) 4700

Sol. (*c*) Total numbers = 1000 to 9999 = 9000 $\Rightarrow$ Answer = $\frac{9000}{9} + \frac{9000}{10} \times 3 = 3{,}700$

53. How many numbers less than 1000 are multiples of both 10 and 13? **(SSC CGL Tier I 2006)**

(*a*) 9 *b*) 8 (*c*) 6 (*d*) 7

Sol. (*d*) 10 and 13 are co-prime numbers. Product of 10 and 13 = $10 \times 13 = 130$

$1000 \div 130 = 7$ (Quotient) + 90 (Remainder)

54. A vessel contains 12 litres of water. If water is drained from the vessel until it holds 6 litres more than the quantity drained away. Find the quantity of the water (in litres) drained away.

(*a*) 3 (*b*) 4 (*c*) 6 (*d*) 9

Sol. (*a*) Let x litres of water is drained away. Then water remaining in the vessel is $x + 6$

$\therefore x + (x + 6) = 12 \Rightarrow 2x + 6 = 12 \Rightarrow 2x = 12 - 6 = 6 \Rightarrow x = 3$ litres

55. A number N is a positive three digit number. If x is in its hundred's place and y is in its unit's place, then the numebr $N - 100x - y$ is always divisible by **(SSC FCI Asstt. 2012)**

(*a*) 8 (*b*) 9 (*c*) 10 (*d*) 11

Sol. (*c*) Let ten's digit = a. Then, $N = 100x + 10a + y$

$\therefore N - 100x - y = (100x + 10a + y) - (100x + y) = 10a$

56. The difference between a two-digit number and the number obtained by interchanging the two digits of the number is 36. What is the difference between the two digits of the number ? **(SBI Associate Bank PO 2007)**

(*a*) 6 (*b*) 4 (*c*) 3 (*d*) None of these

Sol. (*b*) Difference between the digits = $36 \div 9 = 4$

57. The sum of digits of a two digit number is 8 and the difference between the number and that formed by reversing the digits is 18. What is the difference between the digits of the number? **(CDS 2011)**

(*a*) 1 (*b*) 2 (*c*) 3 (*d*) 4

Sol. (*b*) Difference between the digits = $18 \div 9 = 2$

58. The sum of the digits of a two-digit number is 15. On subtracting 27 from the given number, its digits are reversed. Find the number. **(Bank Clerical)**

(*a*) 69 (*b*) 78 (*c*) 87 (*d*) 96

Sol. (*d*) Ten's digit – Unit's digit = $27 \div 9 = 3$ $\therefore$ Unit's digit = $\frac{15 - 3}{2} = 6$

Ten's digit = $6 + 3 = 9$ $\Rightarrow$ The required number is 96.

59. A number consists of two digits, whose sum is 7. If the digits are reversed, the number is increased by 27. The product of digits of the number is **(CDS 2015)**

(*a*) 6 (*b*) 8 (*c*) 10 (*d*) 12

Sol. (*c*) Unit's digit – Ten's digit = $27 \div 9 = 3$ $\therefore$ Ten's digit = $\frac{7-3}{2} = 2$

Unit's digit = $2 + 3 = 5$ $\Rightarrow$ The required number is 25. $\Rightarrow$ Product = $2 \times 5 = 10$

60. The sum of the digits of a two-digit number is 12. On adding 36 to the number, its digits are reversed. What is the number ? **(Bank P.O. Exam)**

(*a*) 39 (*b*) 48 (*c*) 57 (*d*) 75

Sol. (*b*) Unit's digit – ten's digit = $36 \div 9 = 4$

Of the given options 48 is the only number that satisfies this condition.

61. In a two-digit number, the unit digit is equal to the square of tens digit. The difference between the original number and the number formed by interchanging the digits is 54. What is the original number ? **(Bank P.O. Exam)**

(*a*) 24 (*b*) 39 (*c*) 54 (*d*) 93

Sol. (*b*) Unit's digit – ten's digit = $54 \div 9 = 6$

39 is the only number that satisfies this condition.

62. A number consists of two digits. The sum of the digits is 10. On reversing the digits of the number, the number decreases by 36. What is the product of the two digits ? **(CDS 2012)**

(*a*) 21 (*b*) 24 (*c*) 36 (*d*) 42

Sol. (*a*) Difference between two digits = $36 \div 9 = 4$, Sum of digits = 10 (given)

The digits are = $\frac{10+4}{2}$ and $\frac{10-4}{2}$ = 7 and 3 $\Rightarrow$ Multiplication of digits = $7 \times 3 = 21$

63. A number consists of two digits such that the digit in the ten's place is less by 2 than the digit in the unit's place. Three times the number added to $\frac{6}{7}$ times the number obtained by reversing the digits equals 108. The sum of digits in the number is: **(SSC CGL Tier I 2003)**

(*a*) 8 (*b*) 9 (*c*) 6 (*d*) 7

Sol. (*c*) Going through the options:

Option (*a*) If sum is 8, difference is 2, then the number is 35. Reverse = 53 (not divisible by 7)

Option (*b*) If sum is 9, difference is 2, one number = $\frac{9+2}{2}$ = 5.5 (not a natural number)

Option (*c*) If sum is 6, difference is 2, unit's digit = $\frac{6+2}{2}$ = 4. $\Rightarrow$ The number is 24.

Verification = $3 \times 24 + \frac{6}{7} \times 42 = 72 + 36 = 108$.

64. A number consisting of two digits is four times the sum of its digits. On adding 18 to the number, the digits of the given number are reversed. Find the original number.

(*a*) 24 (*b*) 36 (*c*) 46 (*d*) 68

Sol. (*a*) Unit's digit – ten's digit = $18 \div 9 = 2$ $\Rightarrow$ 36 is not the required number.

Since, the number is four times the sum of its digits, whatever is the sum of digits, the number is divisible by 4. $\Rightarrow$ 46 is not the required number.

Now we check the remaining options *i.e.*, 24 and 68.

Sum of the digits should be $\frac{24}{4}$ and $\frac{68}{4}$ *i.e.*, 6 and 17 respectively.

68 : Sum of digits = $6 + 8 = 14 \neq 17$ $\Rightarrow$ 68 is not the required number.

24 : Sum of digits = $2 + 4 = 6 = 6$ $\Rightarrow$ 24 is the required number.

65. A two-digit number is three times the sum of its digits. If 45 is added to the number, its digits are reversed. The sum of digits of the number is:

(*a*) 5 (*b*) 7 (*c*) 9 (*d*) 11

Sol. (*c*) Let the number is $10x + y$. Then $10x + y = 3(x + y) \Rightarrow 7x - 2y = 0 \Rightarrow 7x = 2y$

Since the digits can be natural numbers only, the possible values of $x = 2$ and $y = 7$.

$\therefore$ The number is 27 and the sum of digits is $2 + 7 = 9$

Alternative Method: Unit's digit – ten's digit $= 45 \div 9 = 5$

Going through the given options, we can proceed as follows:

Option (*a*) : Sum 5 is not possible. Option (*b*) : If sum is 7, number is $16 \neq 3 \times 7$

Option (*c*) : If sum is 9, number is $27 = 3 \times 9$.

Option (*d*) : If sum is 11, number is $38 \neq 3 \times 11$

66. A two digit number is five times the sum of its digits. If 9 is added to the number, the digits interchange their positions. The sum of digits of the number is: **(SSC CGL Tier I 2004)**

(*a*) 11 (*b*) 9 (*c*) 7 (*d*) 6

Sol. (*b*) Let the number be $10x + y$. Then $10x + y = 5 \times (x + y) \Rightarrow 5x = 4y$

Since the digits can be natural number only, possible values of $x = 4$ and $y = 5$.

The number is 45. [Note: $45 + 9 = 54$] $\Rightarrow$ Sum of digits $= 4 + 5 = 9$

Trick: Unit's digit – ten's digit $= 9 \div 9 = 1$

$\therefore$ Possible numbers from given options 56, 45, 34. Option (*d*) not possible

Out of given options, only 45 is divisible by 5.

67. With a two digit prime number, if 18 is added, we get another prime number with digits reversed. How many such numbers are possible ? **(SSC CGL Tier I 2012)**

(*a*) 2 (*b*) 3 (*c*) 0 (*d*) 1

Sol. (*a*) On adding 18, the digits are reversed. $\Rightarrow$ Unit's digit – ten's digit $= 18 \div 9 = 2$.

Prime numbers with difference of digits 2 = 13, 79

Prime numbers on reversing the digits = 31, 97

68. In a three digit number, the digit in the unit's place is 75% of the digit in the ten's place. The digit in ten's place is greater than the digit in the hundred's place by 1. If the sum of the digits in the ten's place and the hundred's place is 15, what is the number ? **(Dena Bank PO 2006)**

(*a*) 795 (*b*) 786 (*c*) 687 (*d*) None of these

Sol. (*b*) Unit's digit is 75% of ten's digit. Therefore, ten's digit may be either 4 or 8 so that unit's digit (*i.e.* its 75%) is a natural number i.e. 3 or 6.

Sum of hundred's digit and ten's digit is 15. Therefore, hundred's digit may be either 11 (which is not possible) or 7 (possible).

$\therefore$ Required number is 786.

Trick: In option (*b*) : $7 + 8 = 15$

69. A three digit number is divisible by 11 and has its digit in the unit's place equal to 1. The number is 297 more than the number obtained by reversing the digits. What is the number ? **(CDS 2010)**

(*a*) 121 (*b*) 231 (*c*) 561 (*d*) 451

Sol. (*d*) Hundred's digit – unit's digit $= 297 \div 99 = 3$

Only option (*d*) satisfies this condition.

70. In a three-digit number, the digit in the unit's place is 4 times the digit in the hundred's place. If the digit in the unit's place and ten's place are interchanged, the new number such formed is 18 more than the original number. If the digit in the hundred's place is one-third of the digit in the ten's place, what is 25% of the original number ? **(LIC AAO 2005)**

(*a*) 31 (*b*) 67 (*c*) 107 (*d*) None of these

Sol. (*b*) Number may have 1 or 2 at hundred's place and 4 or 8 at unit's place.

unit's digit – ten's digit = 18 ÷ 9 = 2 ∴ Ten's digit may be 4 – 2 = 2 or 8 – 2 = 6.

Given that hundred's place digit is one-third of the ten's place digit.

∴ Hundred's place digit = 6 ÷ 3 = 2 ⇒ The given number is 268.

25% of the number is 25% × 268 = 67

71. If a two digit number is three times the sum of its digits, the number formed by interchanging the digits is how many times of their sum ?

(*a*) 7 (*b*) 8 (*c*) 9 (*d*) 11

Sol. (*b*) We know that sum of two such numbers is 11 times the sum of its digits.

∴ Number on interchanging = 11 – 3 = 8 times of sum of the digits.

72. Four prime numbers are written in an ascending order. The product of the first three numbers is 385 and that of the last three numbers is 1001. What is the largest number? **(CBI 1995, MBA)**

(*a*) 7 (*b*) 11 (*c*) 13 (*d*) 17

Sol. (*c*) $a \times b \times c = 385$ and $b \times c \times d = 1001$

HCF of 385 and 1001 is $b \times c = 77$

∴ The largest number $= d = (b \times c \times d) \div (b \times c) = 1001 \div 77 = 13.$

Alternative Method: $\dfrac{abc}{bcd} = \dfrac{385}{1001} \Rightarrow \dfrac{a}{d} = \dfrac{5}{13} \Rightarrow d = 13$

73. Three numbers which are co-prime to one another are such that the product of the first two is 551 and that of the last two numbers is 1073. The sum of the three numbers is : **(SSC CGL Tier I 2000)**

(*a*) 75 (*b*) 81 (*c*) 85 (*d*) 89

Sol. (*c*) Let the number are *x*, *y* and *z*.

$x \times y = 551 = 19 \times 29$ and $y \times z = 1073 = 29 \times 37$

∴ The numbers are 19, 29 and 37. ⇒ Sum of the numbers = 19 + 29 + 37 = 85

74. The difference between two numbers is 1375. On dividing the bigger number by the smaller number, the quotient is 6 and the remainder is 25. Find the smaller number.

(*a*) 250 (*b*) 270 (*c*) 275 (*d*) 300

Sol. (*b*) Let the smaller number = *x*. Then bigger number = $x + 1375$

∴ $6x + 25 = x + 1375 \Rightarrow 5x = 1350 \Rightarrow x = 270$

75. The numbers 1, 3, 5, 7, ... 99 and 128 are multiplied together. The number of zeros at the end of the product must be: **(SSC CGL Tier I 2000)**

(*a*) 19 (*b*) 22 (*c*) 7 (*d*) Nil

Sol. (*c*) There are 10 numbers up to 100 *viz.* 5, 15, 25, 35,...........95 which have 5 at the unit's place and therefore are divisible by 5 but 25 and 75 are divisible by 25.

∴ The product of odd numbers up to 100 can be divided 10 + 2 = 12 times by 5.

And $128 = 2^7$

We know that 2 × 5 = 10 i.e. product of one pair of 2 and 5 gives one zero.

In the given question, 7 such pairs are possible.

∴ Total zeros at the end of the product are 7.

76. All the even numbers up to 100 *i.e.*, 2, 4, 6, ... 100 are multiplied together. The number of zeros at the end of the product will be: **(SSC CHSL & SSC CGL Tier I 2000)**

(*a*) 13 (*b*) 12 (*c*) 11 (*d*) 10

Sol. (*b*) There are 10 multiples of 10 up to 100 viz. 10, 20, 30, ... 100.

These numbers have 0 at the unit place but 100 has two zeros.

∴ Zeros in the product of these numbers = 10 + 1 = 11. Now 50 = 5 × 10

∴ Product of 5 and an even number is also divisible by 10.

∴ Total number of zeros in the product = 11 + 1 = 12.

Alt: $2 \times 4 \times 6 \times .... \times 100 = 2^{50} \times (1 \times 2 \times 3 .... \times 50) \Rightarrow \text{Zeros} = \frac{50}{5} + \frac{50}{5^2} = 10 + 2 = 12$

77. The numbers 1, 2, 3, 4, ... 1000 are multiplied together. The number of zeros at the end (on the right) of the product must be: **(SSC CGL Tier I 2005)**

(*a*) 30 (*b*) 200 (*c*) 211 (*d*) 249

Sol. (*d*) 2 × 5 = 10 *i.e.*, product of 2 and 5 gives one zero. Hence, number of zeros dependes on the number of 5's and 2's. Here, number of 5's is the constraint.

Numbers divisible by 5 = 1000 ÷ 5 = 200

Numbers divisible by 5^2 = 1000 ÷ 5^2 = 40

Numbers divisible by 5^3 = 1000 ÷ 5^3 = 8

Numbers divisible by 5^4 = 1000 ÷ 5^4 = 1

Total multiples of 5 = 200 + 40 + 8 + 1 = 249

∴ Total number of zeros is 249 (minimum of numbers divisible by 2 or 5)

Alt: $\frac{200}{5} = 40; \frac{40}{5} = 8; \frac{8}{5} = 1$ $200 + 40 + 8 + 1 = 249$

78. What is the highest power of 2 that divides 20! completely?

(*a*) 10 (*b*) 12 (*c*) 18 (*d*) 20

Sol. (*c*) $\frac{20}{2^1} + \frac{20}{2^2} + \frac{20}{2^3} + \frac{20}{2^4} = 10 + 5 + 2 + 1 = 18$

79. First 100 multiples of 10 are multipied together. The number of zeros at the end of product must be: **(SSC CGL Tier I 2000)**

(*a*) 110 (*b*) 121 (*c*) 124 (*d*) 125

Sol. (*c*) $(10 \times 1) \times (10 \times 2) \times (10 \times 3) \times .... (10 \times 100) = 10^{100} \times (1 \times 2 \times 3 \times .... 100)$

Zeros 10^{100} = 100 and zeros in $(1 \times 2 \times 3 \times .... 100) = (100 \div 5) + (100 \div 5^2) = 20 + 4 = 24$

∴ Total number of zeros = 100 + 24 = 124

80. Ramesh and Umesh had equal shares, in a flock of sheep. In distribution of sheep, Ramesh took 32 sheep from the flock while Umesh took 40 sheep and paid Ramesh ₹2800. The cost of each sheep was: **(SSC CGL Tier I 2007)**

(*a*) ₹ 600 (*b*) ₹ 700 (*c*) ₹ 800 (*d*) ₹ 200

Sol. (*b*) Total number of sheep = 32 + 40 = 72. Share of each = 72 ÷ 2 = 36

∴ Cost of (36 – 32) sheep = 4 sheep = ₹ 2800

∴ Cost of each sheep = ₹ 2800 ÷ 4 = ₹ 700

81. Find the total number of factors of 1400, divisible by 5.

(*a*) 24 (*b*) 16 (*c*) 8 (*d*) 12

Sol. (*b*) $1400 = 2^3 \times 5^2 \times 7$

Total factors, divisible by 5 = $(2^0, 2^1, 2^2, 2^3) \times (5^1, 5^2) \times (7^0, 7^1) = 4 \times 2 \times 2 = 16$

82. Find the total number of factors of 1080, divisible by 15.

(*a*) 24 (*b*) 16 (*c*) 8 (*d*) 12

Sol. (*d*) $1080 = 2^3 \times 3^3 \times 5^1$

Total factors, divisible by 15 = $(2^0, 2^1, 2^2, 2^3) \times (3^1, 3^2, 3^3) \times (5^1) = 4 \times 3 \times 1 = 12$

83. Find the total number of even factors of a number: $2^4 \times 3^5 \times 5^3$

(*a*) 60 (*b*) 80 (*c*) 96 (*d*) 120

Sol. (*c*) Total number of factors = $(4 + 1) \times (5 + 1) \times (3 + 1) = 120$

Odd factors = $(5 + 1) \times (3 + 1) = 24$ ⇒ Even factors = $120 - 24 = 96$

Note: The odd factors are formed only with the combination of powers of 3 and 5 only.

Alt: Even factors = $4 \times (5 + 1) \times (3 + 1) = 96$ (Eliminating 2^0 from the factors)

84. A bucket full of water weighs 50 kg, while half-filled bucket weighs 30 kg. Find the weight of the empty bucket.

(*a*) 10 kg (*b*) 15 kg (*c*) 20 kg (*d*) 25 kg

Sol. (*a*) Weight of bucket + full water = 50 kg ...(1)

And weight of bucket + half water = 30 kg ...(2)

(1) – (2) : Weight of half water = (50 – 30) kg = 20 kg.

∴ Weight of empty bucket = 30 kg – 20 kg = 10 kg.

85. The weight of a container full of water, is 28 kg. When it is 1/4 full, it weighs 19 kg. Its weight, when it is 2/3 full of water, will be: **(SSC CGL Tier I 2006)**

(*a*) 8 kg (*b*) 20 kg (*c*) 24 kg (*d*) 18.67 kg

Sol. (*c*) Weight of container + full water = 28 kg ... (1)

And weight of container + 1/4 water = 19 kg ... (2)

Weight of 4 containers + 1 water = 19 × 4 = 76 kg ... (3)

(3) – (1) : Weight of 3 empty containers = (76 – 28) kg = 48 kg.

∴ Weight of one empty container = 48 ÷ 3 = 16 kg.

Weight of 1 full water = (28 – 16) kg = 12 kg.

Weight of $\frac{2}{3}$ water = $\frac{2}{3} \times 12 = 8$ kg. ⇒ Required weight = 16 + 8 = 24 kg.

86. A man engaged a servant on the condition that he would pay him ₹ 3000 and one uniform after one year of service. He served only for 9 months and got ₹ 2200 and a uniform. What is the price of uniform ?

(*a*) ₹ 100 (*b*) ₹ 200 (*c*) ₹ 250 (*d*) ₹ 400

Sol. (*b*) 12 months wages = ₹ 3000 + One uniform ...(1)

9 months wages = ₹ 2200 + One uniform ...(2)

(1) – (2) ⇒ 3 months wages = ₹ 800 ⇒ 12 months wages = 4 × ₹ 800 = ₹ 3200

From (1) : Cost of uniform = ₹ 3200 – ₹ 3000 = ₹ 200

87. A labourer was engaged for 20 days on the condition that he will receive ₹ 60 for each day he works and he will be fined ₹ 5 for each day he is absent. If he received ₹ 745 in all, then the number of days he remained absent is : **(CDS 2003)**

(*a*) 3 (*b*) 5 (*c*) 7 (*d*) 9

Sol. (*c*) Let days of absence = x. Then $60(20 - x) - 5x = 745$

$\Rightarrow 1200 - 60x - 5x = 745 \Rightarrow 65x = 455 \Rightarrow x = 7$

Alt : Total wages for 20 days = $20 \times 60 = 1200 \Rightarrow$ Difference = $1200 - 745 = 455$

Loss for each day of absence = $60 + 5 = 65 \Rightarrow$ Total days of absence = $455 \div 65 = 7$

88. A labourer was appointed by a contractor on the condition that he would be paid ₹ 75 for each day of his work but would be fined at the rate of ₹ 15 per day for his absence, apart from losing his wages. After 20 days, the contractor paid the labourer ₹ 1140. The number of days the labourer absented from the work was: **(SSC CGL Tier I 2007)**

(*a*) 3 days (*b*) 5 days (*c*) 4 days (*d*) 2 days

Sol. (*c*) Wages for 20 days @ ₹ 75 per day = ₹ 1500. Wages actually earned = ₹ 1140

Total wages deducted for absence = ₹ 1500 – ₹ 1140 = ₹ 360

Wages deducted for one day absence = ₹ 75 + ₹ 15 = ₹ 90

∴ Number of days he was absent for = $360 \div 90 = 4$ days

89. In an examination, a student is awarded 4 marks for every correct answer and loses 2 marks for every wrong answer. If a student attempted all 75 questions and secured 150 marks, the number of questions he attempted correct is: **(SSC CGL Tier I 2006)**

(*a*) 45 (*b*) 50 (*c*) 55 (*d*) 48

Sol. (*b*) Let he attempted x sums correctly.

Then his total marks = $4x - 2(75 - x) = 150 \Rightarrow 6x = 300 \Rightarrow x = 50$

Alternative method: Let he attempted all the questions correctly.

Then his total marks = $4 \times 75 = 300$ marks

Excess over actual marks = $300 - 150 = 150$ marks

For every wrong answers he loses = $4 + 2 = 6$ marks

∴ Wrong answers = $150 \div 6 = 25$ $\Rightarrow$ Correct answers = $75 - 25 = 50$

90. Mohan gets 3 marks for each correct sum and loses 2 marks for each wrong sum. He attempted 30 sums and obtains 40 marks. The number of sums solved correctly is: **(SSC CGL Tier I 2013)**

(*a*) 10 (*b*) 15 (*c*) 20 (*d*) 25

Sol. (*c*) Let he attempted x sums correctly.

Then his total marks = $3x - 2(30 - x) = 40 \Rightarrow 5x = 100 \Rightarrow x = 20$

Alternative method: Let he attempted all the questions correctly.

Then his total marks = $3 \times 30 = 90$ marks

Excess over actual marks = $90 - 40 = 50$ marks

For every wrong answers he loses = $3 + 2 = 5$ marks

∴ Wrong answers = $50 \div 5 = 10$ $\Rightarrow$ Correct answers = $30 - 10 = 20$

91. The taxi charges in a city contain fixed charges and additional charge/km. The fixed charge is for a distance of upto 5 km and additional charge/km thereafter. The charge for a distance of 10 km is ₹ 350 and for 25 km is ₹ 800. The charge for a distance of 30 km is: **(SSC CGL Tier II 2012)**

(*a*) ₹ 900 (*b*) ₹ 950 (*c*) ₹ 800 (*d*) ₹ 750

Sol. (*b*) Additional km charges for 15 km (i.e. 25 – 10) km = ₹ 800 – ₹ 350 = ₹ 450

Additional km charges for 1 km = ₹ 450 ÷ 15 = ₹ 30

Additional km charges for 5 km = ₹ 30 × 5 = ₹ 150

∴ Total charges for 30 km = Total charges for 25 km + Additional km charges for 5 km

= ₹ 800 + ₹ 150 = ₹ 950

92. A worker earns ₹ 25 on the first day and spends ₹ 20 on the second day. earns ₹ 25 on the third day and spends ₹ 20 on the fourth day and so on. On which day would he had ₹ 100?

(*a*) 20th day (*b*) 30th day (*c*) 31st day (*d*) 40th day

Sol. (*c*) He earns ₹ 25 on first day and spends ₹ 20 on the next day.

∴ His savings in first 2 days = ₹ 25 – ₹ 20 = ₹ 5

He will save ₹ 100 – ₹ 25 = ₹ 75 in $75 \times \frac{2}{5} = 30$ days

On 31st day, he will have = ₹ 75 + ₹ 25 = ₹ 100

93. A monkey wanted to climb on a smooth vertical pole of height 35 meters. In the first one minute, he climbs up 5 meters, in the next one minute, he slips down by 2 metres. Further, he repeated the same process till he reached on the top of the pole. How many times it has to go upward to reach the apex of the pole ?

(*a*) 11 (*b*) 12 (*c*) 27 (*d*) 35

Sol. (*a*) Monkey climbs in first two minutes = 5 – 2 = 3 meters

Time taken to ascend first 30 meters = $\frac{30}{3} \times 2 = 20$ minutes

Time taken to ascend remaining 5 meters = 5 ÷ 5 = 1 minute

Total time taken = 20 + 1 = 21 minutes and total climbs = 10 + 1 = 11

94. A monkey ascends a greased pole 30 metres high. He ascends 3 metres in the first minute and slips down one metre in the next minute, if he keeps on doing so, in what time will he reach the top? **(SS CHSL 2006)**

(*a*) 15 minutes (*b*) $28\frac{2}{3}$ minutes (*c*) 29 minutes (*d*) 30 minutes

Sol. (*b*) Ascending 3 metres in a minute and slipping down 1 metre in the next minute means ascending 2 metres in 2 minutes.

Hence the monkey can ascend a distance of 28 metres in 28 minutes.

Remaining 2 metres he will ascend in $2 \times \frac{1}{3}$ minute = $\frac{2}{3}$ minute

∴ Total time taken = $28\frac{2}{3}$ minutes

95. The sum of three different even numbers, each greater than 15, is 58. One of them must be: **(SSC CGL Tier I 2005)**

(*a*) 16 (*b*) 18 (*c*) 20 (*d*) 22

Sol. (*a*) Two possible combinations : 16 + 18 + 24 = 58 and 16 + 20 + 22 = 58

16 is the common number in both possible combinations of numbers.

Note: 18 + 20 + 22 = 60 > 58 and 16 + 18 + 20 = 54 < 58

96. The sum of two consecutive multiples of a certain number is 120. Find the number.

(*a*) 4 (*b*) 7 (*c*) 8 (*d*) 15

Sol. (*c*) 120 = 15 × 8 = (7 + 8) × 8 = 7 × 8 + 8 × 8 ⇒ The required number is 8.

Hint: Let the number is x. Then sum = $x(n) + x(n + 1) = x(2n + 1)$

Note: $(2n + 1)$ is always an odd number.

97. A man has some hens and some cows. If the number of heads is 50 and number of feet is 142, the number of cows is: **(Bank Clerical 2008)**

(*a*) 21 (*b*) 25 (*c*) 27 (*d*) 29

Sol. (*a*) Let the man has hens only. Then total heads = 50 × 1 = 50
And legs = 50 × 2 = 100 which is short by 42 from the actual legs *i.e.*, 142.
Replacement of one hen with one cow means same number of heads and two more legs.
∴ Hens replaced with cows = 42 ÷ 2 = 21 ⇒ Cows = 21

98. A man has some hens and cows. If the number of heads : number of feet = 12 : 35, find out the number of hens, if the number of heads alone is 48 **(SSC Constable GD 2012)**

(*a*) 28 (*b*) 26 (*c*) 24 (*d*) 22

Sol. (*b*) Heads = 12x = 48 ⇒ x = 4 ⇒ Feet = 35x = 35 × 4 = 140
Let all are cows. Then feet = 48 × 4 = 192
Excess feet = 192 – 140 = 52 ⇒ Hens = 52 ÷ 2 = 26

99. 380 mangoes are distributed among some boys and girls who are 85 in number. Each boy gets four mangoes and each girl gets five. The number of boys is: **(SSC CGL Tier I 2001)**

(*a*) 15 (*b*) 38 (*c*) 40 (*d*) 45

Sol. (*d*) Let all are boys. Then total mangoes required = 4 × 85 = 340
Mangoes left = 380 – 340 = 40
For replacement of one boy with one girl one extra mango is required.
∴ Number of girls = 40 ÷ 1 = 40 ⇒ Number of boys = 85 – 40 = 45

100. A sum of ₹ 350 made up of 110 coins, which are of either ₹ 1 or ₹ 5 denomination. How many coins are of ₹ 5 ?

(*a*) 50 (*b*) 55 (*c*) 60 (*d*) 175

Sol. (*c*) Let all the coins are of ₹ 1 denomination.
Then total value of 110 coins = 110 × 1 = ₹ 110 *i.e.* ₹ 240 less than ₹ 350.
Replacing 1 one-rupee coin with five-rupee coin gives ₹ 4 extra with same number of coins.
∴ Five-rupee coins = 240 ÷ 4 = 60 coins.

101. A person has only ₹ 1 and ₹ 2 coins with her. If the total number of coins that she has is 50 and the amount of money with her is ₹ 75, then the number of ₹ 1 and ₹ 2 coins are, respectively. **(UPSC Civil Services (Pre.) 2011)**

(*a*) 15 and 35 (*b*) 35 and 15 (*c*) 30 and 20 (*d*) 25 and 25

Sol. (*d*) 1 × 50 = 100; 75 – 50 = 25; Required coins = 25, 25
Alt : Go through the options, 25 × 1 + 25 + 2 = 75

102. 24 is divided into two parts such that 7 times the first part added to 5 times the second part makes 146. The first part is **(SSC CHSL 2013)**

(*a*) 11 (*b*) 15 (*c*) 13 (*d*) 17

Sol. (*c*) If we take 5 times 24, we get 120 which is 26 less than 146.
We can increase the total by 2 (i.e. 7 – 5) by moving one number from second part to first.
⇒ First part = 26 ÷ 2 = 13

103. A sum of ₹ 80 made up of 100 coins, which are either of ₹ 1 or 50 paise denomination. How many coins are of 50 paise?

(*a*) 80 (*b*) 75 (*c*) 60 (*d*) 40

Sol. (*d*) Let all the coins are of 50 paise denomination.

Then total value of 100 coins = $100 \times \frac{1}{2}$ = ₹ 50 *i.e.* ₹ 30 less than ₹ 80.

Replacement of one fifty paise coin with one-rupee coin means 50 paise extra.

∴ One-rupee coins = $\frac{30}{0.50} = 30 \times 2 = 60$ coins ⇒ 50 paise coins = 100 – 60 = 40

Alternative method: Let all the coins are of ₹ 1 denomination.

Then total value of 100 coins = 100 × 1 = ₹ 100 *i.e.* ₹ 20 more than ₹ 80.

Replacement of one one-rupee coin with one fifty paise coin means 50 paise less.

∴ 50 paise coins = 20 ÷ 0.50 = 40 coins

104. Mohan bought two varieties of wheat – first at ₹ 9.50 per kg and second at ₹ 10 per kg. If he bought 50 kg at a total cost of ₹ 480, what is the quantity of wheat bought at ₹ 10 per kg. **(MBA Entrance Exam)**

(*a*) 10 kg (*b*) 20 kg (*c*) 30 kg (*d*) 40 kg

Sol. (*a*) Let whole quantity bought at ₹ 10 per kg. Then total cost = ₹ 10 × 50 = ₹ 500

Actual cost is short by ₹ 500 – ₹ 480 = ₹ 20

Difference between two rates = ₹ 10 – ₹ 9.50 = ₹ 0.50 per kg

∴ Quantity bought at ₹ 9.50 per kg = 20 ÷ 0.50 = 40 kg

∴ Quantity bought at ₹ 10 per kg = (50 – 40) kg = 10 kg

105. A wire when cut into pieces of 14 metre each, leaves a piece of 11 metre. When another wire of same length is cut into pieces of 15 metre each, a piece of 3 metre is left. The minimum length of the wire is:

(*a*) 18 metre (*b*) 25 metre (*c*) 123 metre (*d*) 221 metre

Sol. (*c*) Difference between the pieces left = 11 – 3 = 8 metre.

Difference in length of two types of pieces = 15 – 14 = 1 metre each.

∴ Number of pieces cut from the wire = $\frac{8}{1} = 8$

∴ Length of the wire = 14 × 8 + 11 = 123 metre

Or, length of the wire = 15 × 8 + 3 = 123 metre

106. A mass of gold and silver weighting 20 grams costs ₹ 780. Had the weights of gold and silver are exchanged it would have cost ₹ 920. Find the price of 1 gram of gold, if one gram of silver costs ₹ 25.

(*a*) ₹ 25 (*b*) ₹ 40 (*c*) ₹ 50 (*d*) ₹ 60

Sol. (*d*) Let the mass contains x and $(20 - x)$ grams of gold and silver respectively.

Then new mass contains $(20 - x)$ and x grams of gold and silver respectively.

Total weight of gold in two masses = $x + (20 - x) = 20$ grams

Total weight of silver in two masses = $(20 - x) + x = 20$ grams

∴ Total weight of gold and silver in two masses is 20 grams each.

Total cost of two masses = ₹ 780 + ₹ 920 = ₹ 1700

Cost of 20 grams of silver = 20 × ₹ 25 = ₹ 500

∴ Cost of 20 grams of gold = ₹ 1700 – ₹ 500 = ₹ 1200

∴ Cost of 1 gram of gold = ₹ $\frac{1200}{20}$ = ₹ 60

107. A mass of copper and brass weighs 25 kg. and is worth ₹ 137. Had the weights of copper and brass are interchanged it would worth ₹ 163. If 3 kg. of copper and 5 kg. of brass costs ₹ 50, find the cost of brass per kg. **(MBA Ent. Exam.)**

(*a*) ₹ 5 (*b*) ₹ 6 (*c*) ₹ 7 (*d*) ₹ 8

Sol. (*c*) Cost of 25 kg of copper + Cost of 25 kg. of brass = ₹ 137 + ₹ 163 = ₹ 300

∴ Cost of 1 kg. of copper + Cost of 1 kg. of brass = ₹ 300 ÷ 25 = ₹ 12

∴ Cost of 3 kg. of copper + Cost of 3 kg. of brass = ₹ 12 × 3 = ₹ 36

Cost of 3 kg. of copper + Cost of 5 kg. of brass = ₹ 50 (Given)

∴ Cost of 2 kg. of brass = ₹ 50 – ₹ 36 = ₹ 14

∴ Cost of 1 kg. of brass = ₹ 14 ÷ 2 = ₹ 7

108. A man buys 3 cows and 8 goats in ₹ 47,200. Instead, if he would have bought 8 cows and 3 goats, he had to pay ₹ 53,000 more. Cost of one cow is : **(SSC CGL Tier I 2013)**

(*a*) ₹ 12,000 (*b*) ₹ 13,000 (*c*) ₹ 10,000 (*d*) ₹ 11,000

Sol. (*a*) 3 Cows + 8 Goats = ₹ 47,200 ⇒ 8 Cows + 3 Goats = ₹ 47,200 + ₹ 53,000

∴ 5 Cows – 5 Goats = ₹ 53,000 ⇒ Cow – Goat = ₹ 53,000 ÷ 5 = ₹ 10,600

Now, 3 Cows + 8 Goats = ₹ 47,200 ⇒ 3 × (Goat + ₹ 10,600) + 8 Goats = ₹ 47,200

⇒ 11 Goats = ₹ 47,200 – 3 × ₹ 10,600 = ₹ 47,200 – ₹ 31,800 = ₹ 15,400

1 Goat = ₹ 15,400 ÷ 11 = ₹ 1,400 ⇒ Cow = ₹ 1,400 + ₹ 10,600 = ₹ 12,000

109. If 5 + 3 + 2 = 151022, 9 + 2 + 4 = 183652, 8 + 6 + 3 = 482466, and 5 + 4 + 5 = 202541, then 7 + 2 + 5 = ? **(MBA Ent. Exam.)**

Sol. The pattern followed, for pairs of two digits starting from left hand side, is:

First number × Second number; First number × Third number;

First number × (Second number + Third number) – Second number

∴ ? = 7 × 2 7 × 5 7 × (2 + 5) – 2 = 143547

CHAPTER

2

FRACTIONS AND DECIMALS

Fraction or Rational number: A number which can be expressed in the form of p/q, where p and q both are integers, and q is a non-zero integer. In the fraction, p is called numerator and q is called denominator. Like the natural numbers, fractions can also be added, subtracted, multiplied and divided.

Proper Fraction: Numerator is smaller than the denominator.

Example: $\frac{2}{3}, \frac{1}{5}, \frac{5}{7}$ **Note:** The value of a proper fraction lies between 0 and 1.

Improper Fraction: Numerator is greater than the denominator.

Example: $\frac{4}{3}, \frac{8}{5}, \frac{10}{7}$ **Note:** The value of an improper fraction is greater than 1.

Simplest/Standard/Lowest form/terms of a Rational number: The simplest form of a rational number is obtained by dividing its numerator and the denominator by their greatest common factor. When a fraction is in lowest terms, its numerator and the denominator are co-prime numbers. By multiplying (or dividing) numerator and denominator of a fraction by same non-zero number we obtain an equivalent fraction.

Example: $\frac{12}{18} = \frac{2}{3}$ (Obtained by dividing N and D by 6, *i.e.* common factor)

Like Fractions: The fractions whose denominators are equal. **Example:** $\frac{1}{7}, \frac{3}{7}, \frac{6}{7}$

Unlike Fractions: The fractions whose denominators are not equal.

Mixed numbers: A number which consists of an integer and a proper fraction.

Example: $3\frac{1}{2}, 5\frac{3}{4}$.

Mixed numbers can be converted into improper fractions and vice-versa.

How to convert a mixed number into improper fraction.

Multiply the whole number with the denominator of the fraction and then add the result to the numerator of the fraction. The final result is the numerator and the denominator remains the same.

Examples: $3\frac{1}{2} = \frac{3 \times 2 + 1}{2} = \frac{7}{2}$; $5\frac{3}{4} = \frac{5 \times 4 + 3}{4} = \frac{23}{4}$

How to convert an improper fraction into a mixed number.

Divide the numerator by the denominator. The quotient is the whole number, the remainder is the numerator of the fraction of the mixed number and the denominator remains the same.

Examples: $\frac{5}{3} = 1\frac{2}{3}$; $\frac{9}{2} = 4\frac{1}{2}$

Addition/Subtraction of fractions:

1. Find LCM of the denominators of the fractions.
2. Divide the LCM by each denominator of the fractions and multiply the quotient with the respective numerators.
3. Add the multiplication results. The sum is the numerator and the LCM is the denominator.

Example 1: Find the value of $\frac{1}{2}+\frac{2}{3}+\frac{2}{5}$.

Sol. LCM of denominators 2, 3 and 5 is 30.

On dividing 30 by 2, 3 and 5, the quotients are 15, 10 and 6 respectively.

Multiply the numerators 1, 2 and 2 with 15, 10 and 6 respectively.

$$\therefore \frac{1}{2}+\frac{2}{3}+\frac{2}{5}=\frac{1\times15+2\times10+2\times6}{30}=\frac{15+20+12}{30}=\frac{47}{30}=1\frac{17}{30}$$

Example 2. $2\frac{1}{3}+1\frac{1}{5}+2\frac{1}{4}=?$ **(UCO Bank PO 2009)**

(*a*) $8\frac{23}{60}$ (*b*) $4\frac{53}{60}$ (*c*) $5\frac{47}{60}$ (*d*) $6\frac{37}{60}$

Sol. (*c*) $2\frac{1}{3}+1\frac{1}{5}+2\frac{1}{4}=(2+1+2)+\left(\frac{1}{3}+\frac{1}{5}+\frac{1}{4}\right)=5+\frac{20+12+15}{60}=5\frac{47}{60}$

Example 3. $3\frac{1}{6}+8\frac{1}{4}-5\frac{7}{8}=?$ **(SBI Clerical)**

(*a*) $4\frac{13}{24}$ (*b*) $5\frac{11}{24}$ (*c*) $5\frac{13}{24}$ (*d*) $6\frac{13}{24}$

Sol. (*c*) $3\frac{1}{6}+8\frac{1}{4}-5\frac{7}{8}=(3+8-5)+\left(\frac{1}{6}+\frac{1}{4}-\frac{7}{8}\right)=6+\frac{4+6-21}{24}=6-\frac{11}{24}=5\frac{13}{24}$

Note: While adding (or subtracting) two or more mixed numbers, we can add (or subtract) whole numbers and fractions separately.

Another method of adding/subtracting the fractions : Multiply the denominator and numerator of each fraction by a number so as to make the denominators of all the fractions equal to their LCM. Then add or subtract the numerators and afterwards divide the result by the LCM. Example 1 above can be solved by using this method as follows:

L.C.M. of 2, 3 and 5 is 30.

∴ The denominators of these fractions are made equal to 30.

$\frac{1}{2}=\frac{1}{2}\times\frac{15}{15}=\frac{15}{30}$ $\frac{2}{3}=\frac{2}{3}\times\frac{10}{10}=\frac{20}{30}$ $\frac{2}{5}=\frac{2}{5}\times\frac{6}{6}=\frac{12}{30}$

$$\frac{1}{2}+\frac{2}{3}+\frac{2}{5}=\frac{15}{30}+\frac{20}{30}+\frac{12}{30}=\frac{47}{30}=1\frac{17}{30}$$

Multiplication of fractions: Change the mixed numbers, if any, into improper fractions. The product of all the numerators is written as numerator and similarly the product of all the denominators is written as denominator. Reduce the result by common factors, if any, in numerator and denominator.

$\frac{2}{3}\times\frac{5}{7}=\frac{2\times5}{3\times7}=\frac{10}{21}$ $1\frac{2}{3}\times\frac{4}{5}=\frac{1\times3+2}{3}\times\frac{4}{5}=\frac{5}{3}\times\frac{4}{5}=\frac{4}{3}$ $\frac{8}{15}\times\frac{25}{32}=\frac{1}{3}\times\frac{5}{4}=\frac{5}{12}$

(8 and 32 have common factor 8) (15 and 25 have common factor 5)

Multiplicative inverse or Reciprocal: The multiplicative inverse of a non-zero rational number is a number such that its product with the given number gives 1. Reciprocal of a rational number is obtained by interchanging its numerator and the denominator.

Example: $\frac{a}{b} \times \frac{b}{a} = 1 \Rightarrow \frac{b}{a}$ is reciprocal of $\frac{a}{b}$ and vice-versa

Division of fractions: Multiply the dividend by the multiplicative inverse of the divisor.

Example: $\frac{5}{4} \div \frac{3}{2} = \frac{5}{4} \times \frac{2}{3} = \frac{5}{6}$

FRACTIONS AND THEIR EQUIVALENT IN DECIMALS

$\frac{1}{2} = 0.5$ $\quad \frac{1}{3} = 0.\overline{3}$ $\quad \frac{1}{4} = 0.25$ $\quad \frac{1}{5} = 0.2$

$\frac{1}{6} = 0.1\overline{6} = 0.167$ (approx.) $\quad \frac{1}{8} = 0.125$ $\quad \frac{1}{9} = 0.\overline{1}$

$\frac{1}{10} = 0.1$ $\quad \frac{1}{20} = 0.5$ $\quad \frac{2}{3} = 0.\overline{6} = 0.67$ (approx.)

MULTIPLICATION OF NUMBERS WITH DECIMALS

Steps: 1. Multiply the numbers as if there is no decimal sign in the given numbers.

2. Put the decimal after n digits of the solution counting from right hand side where n is equal to number of total digits to the right hand side of the decimal points in the numbers multiplied.

Example: $1.41 \times 1.1 = ?$

Sol. $141 \times 11 = 1551$

After the decimal point, there are 2 digits in the first number and 1 digit in the second number.

∴ The multiplication will have 2 + 1 = 3 digits after the decimal point.

∴ $1.41 \times 1.1 = 1.551$

COMPARISON OF FRACTIONS FOR POSITIVE NUMBERS

Two or more fractions can be compared by using the following methods:

(*a*) If $a > b$, then $\frac{1}{a} < \frac{1}{b}$ **Example:** $\frac{1}{7} > \frac{1}{8} > \frac{1}{9} > \frac{1}{10}$

(*b*) $\frac{a}{x} > \frac{b}{x}$, if $a > b$ **Example:** $\frac{13}{7} > \frac{11}{7} > \frac{5}{7}$

⇒ If denominators of two or more fractions are same.

Greater the numerator, greater the value of fraction is.

(*c*) $\frac{a}{x} < \frac{a}{y}$, if $x > y$ **Example:** $\frac{4}{5} > \frac{4}{7} > \frac{4}{13}$

⇒ If numerators of two or more fractions are same.

Greater the denominator, smaller the value of fraction is.

(*d*) $\frac{x}{y} < \frac{x+k}{y+k}$ where k is a constant number and $x < y$ (*i.e.* Proper Fraction).

⇒ If a constant number is added to both the numerator and the denominator of a proper fraction, the value of fraction with higher numerator is more.

Example: $\frac{1}{3} < \frac{1+2}{3+2} = \frac{3}{5} < \frac{3+6}{5+6} = \frac{9}{11}$

Note: Difference between numerator and denominator of the above fractions is same. Value of fraction increases and approaches to 1.

(e) $\frac{x}{y} > \frac{x+k}{y+k}$ where '*k*' is a constant number and $x > y$ (*i.e.* Improper Fraction)

⇒ If a constant number is added to both the numerator and the denominator of a improper fraction, the value of fraction with smaller numerator is more.

Example: $\frac{4}{3} > \frac{4+1}{3+1} = \frac{5}{4} > \frac{5+7}{4+7} = \frac{12}{11}$

Note: Difference between numerator and denominator of each fraction is same. Value of fraction decreases and approaches to 1.

(*f*) In a Proper Fraction $\frac{x}{nx+k} < \frac{x+c}{n(x+c)+k}$ where *c* and *k* are constant numbers

Example: $\frac{2}{3\times 2+1} < \frac{3}{3\times 3+1} < \frac{7}{3\times 7+1}$

(*g*) Cross Multiplication Method: $\frac{a}{b} > \frac{c}{d}$ if $ad > bc$. Multiply the numerator of one fraction with the denominator of the other fraction.

⇒ The fraction with greater numerator after cross-multiplication, is bigger.

Example: Which of the two fractions $\frac{5}{6}$ or $\frac{2}{5}$ is bigger ?

Sol. Using cross multiplication method, we get the following values:

5 × 5 and 2 × 6 or 25 and 12. Clearly 25 is bigger than 12. $\therefore \frac{5}{6} > \frac{2}{5}$

(*h*) LCM Method: Find LCM of the denominators. Divide the LCM by each denominator. Multiply the quotients by the respective numerators. The fraction with the largest numerator is the largest fraction.

Example: Compare $\frac{5}{6}, \frac{7}{4}$ and $\frac{9}{10}$.

Solution: LCM of the denominators *i.e.* LCM of 6, 4 and 10 is 60.

∴ Denominators of all the fractions are made equal to 60.

Numerators become 5 × 10, 7 × 15, 9 × 6 or 50, 105 and 54 respectively.

$\therefore\ 105 > 54 > 50 \quad \Rightarrow \quad \frac{7}{4} > \frac{9}{10} > \frac{5}{6}$

(*i*) Division Method: (*i*) Fractions are converted into decimals by dividing the numerators of the given fractions by their respective denominators.

Example: Compare $\frac{14}{3}$ and $\frac{17}{5}$

Sol. $\frac{14}{3} = 4.67$ and $\frac{17}{5} = 3.4$ $\quad \therefore 4.67 > 3.4 \quad \therefore \frac{14}{3} > \frac{17}{5}$

(*ii*) Fractions can be converted into decimals by dividing the denominators of the fractions by their respective numerators.

Note: This method can be used conveniently when the denominator is greater than numerator.

Example: Compare $\frac{3}{11}, \frac{2}{9}$ and $\frac{1}{5}$

Sol. $\frac{3}{11} = \frac{1}{3.67}, \frac{2}{9} = \frac{1}{4.5}$ and $\frac{1}{5} = \frac{1}{5}$ $\because \frac{1}{3.67} > \frac{1}{4.5} > \frac{1}{5}$ $\therefore \frac{3}{11} > \frac{2}{9} > \frac{1}{5}$

CONVERSION OF DECIMALS INTO FRACTIONS

(*a*) Finite or terminating decimals: In this case, denominator of the fraction, in its lowest terms, can be expressed as product of powers of 2 and 5 only.

Method:

Write the given number, after removing the decimal sign, as numerator of the fraction.

Denominator consists of 1 followed by as many 0's as many are the digits after the decimal point in the given number.

Example: $0.7 = \frac{7}{10}, 0.23 = \frac{23}{100}, 0.47 = \frac{47}{100}, 5.18 = 5\frac{18}{100}$

Addition (or subtraction) of terminating decimals:

Non-recurring decimal numbers can be added as natural numbers after placing decimals in one column. Zeros may be suffixed after decimal portion to get equal number of digits after the decimal point.

Example 1. 0.03 + 0.7 = ?

```
 0.03
 0.70
-----
 0.73
```

Example 2 : 0.27 + 31.35 + 10.738 + 118.56 = ?

```
  0.270
 31.350
 10.738
118.560
-------
160.918
```

(*b*) Recurring and non-terminating Decimals: A digit or set of digits repeats itself and division process never comes to end. It can also be expressed by putting a bar over the set of repeating digits or by putting dots over the first and the last digits which are repeated.

(*i*) Pure recurring decimals: Where all the digits after decimal point are repeated again and again. In this case, denominator is divisible neither by 2 nor by 5.

Method:

The given number (taking repeated digits only once) after removing the decimal sign is written as numerator of the fraction.

Denominator consists of as many 9's as there are digits repeating in the given number after the decimal point.

Examples: $0.\overline{1} = \frac{1}{9}, 0.\overline{2} = \frac{2}{9}, 0.\overline{3} = \frac{3}{9}, 15.\overline{7} = 15\frac{7}{9}$

$0.\overline{01} = \frac{1}{99}, 0.\overline{17} = \frac{17}{99}, 0.\overline{376} = \frac{376}{999}, 78.\overline{08} = 78\frac{8}{99}$

Addition (or subtraction) of pure recurring decimals:

Add number of digits equal to LCM of recurring digits in the given numbers.

Example 1: $2.\overline{07}+3.\overline{19}+13.\overline{08}=?$

(*a*) $18.\overline{34}$ (*b*) $18.\overline{35}$ (*c*) $19.\overline{34}$ (*d*) $19.\overline{35}$

Sol. (*a*) In this example, there are two recurring digits in each number. Therefore, we may add two recurring digits from each number. However, we may add two extra digits to get the correct carry over.

$$\begin{array}{r} 2.07\ 07 \\ 3.19\ 19 \\ +\ 13.08\ 08 \\ \hline 18.\overline{34}\ 34 \\ \hline \end{array}$$

Alternative Method:

$2.\overline{07}+3.\overline{19}+13.\overline{08} = 2+3+13+0.\overline{07}+0.\overline{19}+0.\overline{08}$

$= 18+\frac{7}{99}+\frac{19}{99}+\frac{8}{99} = 18+\frac{34}{99} = 18.\overline{34}$

Direct Method: $(2+3+13)+(0.\overline{07}+0.\overline{19}+0.\overline{08}) = 18+0.\overline{34} = 18.\overline{34}$

Note: Since there is no carry over and number of recurring digits are same. The given numbers are simply added.

Example 2: $0.\overline{4}+0.\overline{5}+0.\overline{9}+0.\overline{13}=?$

(*a*) $0.\overline{31}$ (*b*) $1.\overline{93}$ (*c*) $2.\overline{13}$ (*d*) $2.\overline{31}$

Sol. (*c*)

$$\begin{array}{r} 0.44\ \ 44 \\ 0.55\ \ 55 \\ 0.99\ \ 99 \\ +\ 0.13\ \ 13 \\ \hline 2.\overline{13}\ \ 11 \\ \hline \end{array}$$

Alternative Method :

$0.\overline{4}+0.\overline{5}=0.\overline{9}=1$

$\therefore$ Given expression $=1+1+0.\overline{13} = 2.\overline{13}$

(*ii*) Mixed recurring decimals: One or more digits after decimal point are not repeated which are followed by a set of one or more digits that are repeated. In such case, denominator contains factors of 2 and / or 5 in addition to other factors.

Method: From the given number (taking repeated digits only once) deduct number formed by non-recurring digits and write the result as the numerator.

Denominator consists of as many 9's as there are recurring digits followed by as many 0's as there are non-recurring digits after the decimal point.

$$\frac{\text{Full number} - \text{Number formed by non-recurring digits}}{\text{9's equal to number of recurring digits followed by 0's equal to number of non-recurring digits after the decimal point}}$$

Examples:

$0.1\overline{6}=\frac{16-1}{90}=\frac{15}{90}$; $0.3\overline{7}=\frac{37-3}{90}=\frac{34}{90}$; $0.12\overline{43}=\frac{1243-12}{9900}=\frac{1231}{9900}$

$2.\overline{3}=\frac{23-2}{9}=\frac{21}{9}$; $2.\overline{3}=2+0.\overline{3}=2+\frac{3}{9}=\frac{21}{9}$

$5.0\overline{17}=5+\frac{17-0}{990}=5\frac{17}{990}$ $\quad$ $1.2\overline{57}=1+\frac{257-2}{990}=1\frac{255}{990}=1\frac{51}{198}=\frac{249}{198}$

Addition (or subtraction) of mixed recurring decimals:

Take digits equal to number of maximum non-recurring digits out of the given numbers. This would be followed by number of digits equal to LCM of number of recurring digits in the given numbers. Add two more digits from each number to correct the error due to carry over.

Example. $0.11\overline{62} + 0.07\overline{21} = ?$

(*a*) $0.18\overline{83}$ (*b*) $0.18\overline{84}$ (*c*) $0.19\overline{83}$ (*d*) $0.19\overline{84}$

Sol. (*a*)
$$\begin{array}{r} 0.11\ 62\ 62 \\ +0.07\ 21\ 21 \\ \hline 0.18\ \overline{83}\ 83 \\ \hline \end{array}$$

Alternative Method: $0.11\overline{62} + 0.07\overline{21} = (0.11 + 0.07) + (0.00\overline{62} + 0.00\overline{21}) = 0.18\overline{83}$

SOLVED EXERCISE

1. $31.2 \times 56.8 = ?$

(*a*) 177216 (*b*) 17.7216 (*c*) 1772.16 (*d*) 177.216

Sol. (*c*) Ignoring the decimal point, all the options are same *i.e.* 177216.

Decimal to be placed after two digits from right-hand side as both the numbers have 1 digit each on the right-hand side of the decimal from right side and $1 + 1 = 2$.

$\therefore$ The answer is 1772.16.

2. If $154 \times 18 = 2772$, then $27.72 \div 1.8 = ?$

(*a*) 1.54 (*b*) 15.4 (*c*) 154 (*d*) 1540

Sol. (*b*) $2772 \div 18 = 154$, what we need to find is the place of decimal.

Decimal is to be placed after 1 digit *i.e.* 2 digits (of the dividend) – 1 digit (of the divisor).

$\therefore$ The answer is 15.4.

3. $\frac{5}{6} \times \frac{2}{9} \div \frac{4}{9} \div \frac{5}{18} = ?$ **(Bank P.O. Exam)**

(*a*) 1 (*b*) 1.5 (*c*) 2 (*d*) None of these

Sol. (*b*) $\frac{5}{6} \times \frac{2}{9} \div \frac{4}{9} \div \frac{5}{18} = \frac{5}{6} \times \frac{2}{9} \times \frac{9}{4} \times \frac{18}{5} = \frac{3}{2} = 1.5$

4. $5\frac{1}{7} \times 8\frac{1}{6} \div 5\frac{1}{4} = ?$ **(Bank Clerical)**

(*a*) $8\frac{2}{21}$ (*b*) $8\frac{1}{21}$ (*c*) 8 (*d*) None of these

Sol. (*c*) Given expression $= \frac{36}{7} \times \frac{49}{6} \div \frac{21}{4} = \frac{36}{7} \times \frac{49}{6} \times \frac{4}{21} = 8$

5. $\dfrac{1\frac{1}{4} \div 1\frac{1}{2}}{\frac{1}{15} + 1 - \frac{9}{10}}$ is equal to: **(SSC CGL Tier I 2004)**

(*a*) 3 (*b*) 6 (*c*) $\frac{2}{5}$ (*d*) 5

Sol. (*d*) Given expression $= \dfrac{\frac{5}{4}\times\frac{2}{3}}{\frac{1}{15}+\frac{1}{10}} = \dfrac{\frac{5}{6}}{\frac{1}{6}} = \frac{5}{6}\times\frac{6}{1} = 5$

6. $\left(\frac{2}{3}+\frac{2}{3}+\frac{2}{3}+\frac{2}{3}+\frac{2}{3}\right)\div\frac{2}{3} = ?$

(*a*) $\frac{2}{3}$ (*b*) $\frac{3}{2}$ (*c*) 4 (*d*) 5

Sol. (*d*) $\left(\frac{2}{3}+\frac{2}{3}+\frac{2}{3}+\frac{2}{3}+\frac{2}{3}\right)\div\frac{2}{3} = 5\times\frac{2}{3}\div\frac{2}{3} = 5$

7. One-third of one-seventh of a number is 200. What is the one-fourth of the same number ?

(*a*) 50 (*b*) 800 (*c*) 1050 (*d*) 4200

Sol. (*c*) The original number is 200 × 3 × 7. ⇒ Required number $= \frac{200\times3\times7}{4} = 1050$

8. The sum of a number and its reciprocal is 10/3, then the numbers are : **(CDS 2012)**

(*a*) 3, 1/3 (*b*) 3, –1/3 (*c*) –3, 1/3 (*d*) –3, –1/3

Sol. (*a*) $\frac{10}{3} = 3\frac{1}{3} = 3+\frac{1}{3}$

9. If $(a + 2b) = 6$ and $ab = 4$, then $\left(\frac{2}{a}+\frac{1}{b}\right)$ **(MBA Entrance)**

(*a*) 1 (*b*) $\frac{2}{3}$ (*c*) $\frac{3}{2}$ (*d*) 2

Sol. (*c*) $\frac{2}{a}+\frac{1}{b} = \frac{2b+a}{ab} = \frac{6}{4} = \frac{3}{2}$

10. Find the value of * in the following: $1\frac{2}{3}\div\frac{2}{7}\times\frac{*}{7} = 1\frac{1}{4}\times\frac{2}{3}\div\frac{1}{6}$. **(SSC CGL Tier I 2002)**

(*a*) $\frac{1}{6}$ (*b*) 0.6 (*c*) 0.006 (*d*) 6

Sol. (*d*) $\frac{5}{3}\times\frac{7}{2}\times\frac{*}{7} = \frac{5}{4}\times\frac{2}{3}\times\frac{6}{1} \Rightarrow \frac{*}{7} = \frac{5}{4}\times\frac{2}{3}\times\frac{6}{1}\times\frac{3}{5}\times\frac{2}{7} = \frac{6}{7} \Rightarrow * = 6.$

11. If * represents a number, then the value of * in $5\frac{3}{*}\times3\frac{1}{2} = 19$ is:**(SSC CGL Tier I 06)**

(*a*) 7 (*b*) 4 (*c*) 6 (*d*) 2

Sol. (*a*) $5\frac{3}{*}\times3\frac{1}{2} = 19 \Rightarrow 5\frac{3}{*} = 19\div3\frac{1}{2} = 19\times\frac{2}{7} = \frac{38}{7} = 5\frac{3}{7} \Rightarrow x = 7$

12. If $4\frac{1}{a}\times b\frac{2}{3} = 7$, find the values of a and b.

(*a*) 1, 5 (*b*) 2, 3 (*c*) 3, 2 (*d*) 5, 1

Sol. (*d*) $4b = 7 \Rightarrow b = \left[\frac{7}{4}\right] = 1 \Rightarrow b = 1$

Now the expression becomes $4\frac{1}{a}\times1\frac{2}{3} = 7 \Rightarrow 4\frac{1}{a}\times\frac{5}{3} = 7 \Rightarrow 4\frac{1}{a} = 7\times\frac{3}{5} = \frac{21}{5} = 4\frac{1}{5}$

∴ $a = 5$ and $b = 1$

13. The difference between a number and its two-fifth is 270 more than two-fifth of the number. The number is: **(Bank PO Exam)**

(*a*) 54 (*b*) 450 (*c*) 1350 (*d*) None of these

Sol. (*c*) Difference between a number and its two-fifth = $1 - \frac{2}{5} = \frac{3}{5}$ of the number.

Difference between $\frac{3}{5}$ and $\frac{2}{5}$ of the number = $\frac{1}{5}$ of the number = 270

∴ The number = 270 × 5 = 1350

14. If $2 = x + \cfrac{1}{1+\cfrac{1}{3+\cfrac{1}{4}}}$, then value of x is: **(SSC CGL Tier I 2002)**

(*a*) $\frac{18}{17}$ (*b*) $\frac{21}{17}$ (*c*) $\frac{13}{17}$ (*d*) $\frac{12}{17}$

Sol. (*b*) $2 = x + \cfrac{1}{1+\cfrac{1}{3+\cfrac{1}{4}}} = x + \cfrac{1}{1+\cfrac{4}{13}} = x + \cfrac{1}{\cfrac{17}{13}} = x + \frac{13}{17} \Rightarrow 2 = x + \frac{13}{17} \Rightarrow x = 2 - \frac{13}{17} = \frac{21}{17}$

15. Given that $3.718 = \frac{1}{0.2689}$, then $\frac{1}{0.0003718} = ?$ **(Railways 1995, SSC CGL Tier I 2007)**

(*a*) 2689 (*b*) 2.689 (*c*) 26890 (*d*) 0.2689

Sol. (*a*) $\frac{1}{0.0003718} = \frac{10000}{3.718} = 10000 \times 0.2689 = 2689$

16. Find the smallest fraction that should be deducted from the sum of $1\frac{1}{2} + 3\frac{1}{4} + 5\frac{1}{8}$ to make the result a whole number.

(*a*) $\frac{3}{8}$ (*b*) $\frac{1}{8}$ (*c*) $\frac{7}{8}$ (*d*) $\frac{1}{2}$

Sol. (*c*) Sum of the fractions = $\frac{1}{2}+\frac{1}{4}+\frac{1}{8} = \frac{7}{8}$ (Proper fraction)

∴ Deducting $\frac{7}{8}$ from the sum of the numbers will make the result a whole number.

17. Find the smallest fraction that should be added to the sum of $1\frac{1}{2}+3\frac{3}{4}+5\frac{5}{8}$ to make the result a whole number.

(*a*) $\frac{3}{8}$ (*b*) $\frac{1}{8}$ (*c*) $\frac{7}{8}$ (*d*) $\frac{1}{2}$

Sol. (*b*) Sum of the fractions = $\frac{1}{2}+\frac{3}{4}+\frac{5}{8} = \frac{4+6+5}{8} = \frac{15}{8}$

∴ $\frac{1}{8}$ is the required fraction as $\frac{1}{8}+\frac{15}{8} = \frac{16}{8}$ = 2, which is a whole number.

18. $\frac{6}{7}$ of a number exceeds its $\frac{4}{5}$ by 4. The number is:

(*a*) 28 (*b*) 35 (*c*) 70 (*d*) 140

Sol. (*c*) Let *x* be the number. Then $\frac{6x}{7}-\frac{4x}{5}=\frac{(30-28)x}{35}=\frac{2x}{35}=4 \Rightarrow x=\frac{4\times35}{2}=70$

19. A farmer divides his herd of *n* cows among his four sons, so that the first son gets one-half the herd, the second one-fourth, the third son $\frac{1}{5}$ and the fourth son 7 cows. Then the value of *n* is: **(SSC CGL Tier I 2012)**

(*a*) 180 (*b*) 140 (*c*) 240 (*d*) 100

Sol. (*b*) $\frac{n}{2}+\frac{n}{4}+\frac{n}{5}+7=n \Rightarrow \frac{19}{20}n+7=n \Rightarrow \frac{n}{20}=7 \Rightarrow n=140$

20. The denominator of a rational number is 3 more than its numerator. If the numerator is increased by 7 and the denominator is decreased by 2, the result is 2. The fraction is:

(*a*) $\frac{3}{5}$ (*b*) $\frac{5}{8}$ (*c*) $\frac{7}{10}$ (*d*) $\frac{8}{11}$

Sol. (*b*) Let the numerator be *x*. Then the fraction is $\frac{x}{x+3}$.

New number $=\frac{x+7}{(x+3)-2}=2 \Rightarrow x+7=2x+2 \Rightarrow x=5$

$\therefore$ The fraction is $\frac{x}{x+3}=\frac{5}{5+3}=\frac{5}{8}$

21. If 1 is added to denominator of a fraction it becomes $\frac{1}{2}$ and if 1 is added to the numerator, the fraction becomes 1. What is the fraction ? **(CDS 2010)**

(*a*) $\frac{5}{9}$ (*b*) $\frac{2}{3}$ (*c*) $\frac{4}{7}$ (*d*) $\frac{10}{11}$

Sol. (*b*) Let the fraction $=\frac{x}{y}$. Then $\frac{x+1}{y}=1 \Rightarrow y=x+1$ .... (1)

And $\frac{x}{y+1}=\frac{1}{2} \Rightarrow 2x=y+1$ .... (2)

Solving (1) and (2), we get: $x=2$ and $y=3 \Rightarrow$ The fraction is $\frac{2}{3}$.

Alternatively we can check the options instead of solving the equations. On checking the given options, we find $\frac{2}{3}$ that satisfies the two given conditions.

22. A girl was asked to multiply a number by $\frac{7}{8}$, instead she divided the number by $\frac{7}{8}$ and got the result 15 more than the correct result. The sum of the digits of the number is: **(SSC CHSL 2011)**

(*a*) 4 (*b*) 6 (*c*) 8 (*d*) 11

Sol. (*d*) Let x be the number. The girl was required to find $\frac{7x}{8}$ instead of $\frac{8x}{7}$.

$\therefore$ Difference between the two results $= \frac{8x}{7} - \frac{7x}{8} = \frac{15x}{56} = 15$

$\therefore x = \frac{15 \times 56}{15} = 56 \Rightarrow$ The sum of digits = 5 + 6 = 11

Note: $\frac{x}{y} - \frac{y}{x} = \frac{x^2 - y^2}{xy} = \frac{(x+y) \times (x-y)}{xy}$

23. In an examination, a student was asked to find 3/14 of a certain number. By mistake, he found 3/4 of it. His answer was 150 more than the correct answer. The given number is: **(SSC CGL Tier I 2006)**

(*a*) 500 (*b*) 280 (*c*) 240 (*d*) 180

Sol. (*b*) $\left(\frac{3}{4} - \frac{3}{14}\right)$ of a number = 150 $\Rightarrow \frac{30}{56}$ or $\frac{15}{28}$ of the number = 150

$\Rightarrow$ The number $= 150 \times \frac{28}{15} = 280$

24. A student was asked to multiply a number by $\frac{3}{2}$ but he divided that number by $\frac{3}{2}$. His result was 10 less than the correct answer. The number was: **(SSC CGL Tier I 2002)**

(*a*) 10 (*b*) 12 (*c*) 15 (*d*) 20

Sol. (*b*) Let x be the number. Then $\frac{3}{2}x - \frac{2}{3}x = 10 \Rightarrow \frac{5}{6}x = 10 \Rightarrow x = 10 \times \frac{6}{5} = 12$

25. $\frac{1}{5}$ of a number exceeds $\frac{1}{7}$ of the same number by 10. The number is: **(SSC CGL 1999)**

(*a*) 125 (*b*) 150 (*c*) 175 (*d*) 200

Sol. (*c*) $\frac{1}{5} - \frac{1}{7} = \frac{2}{35}$ of the number = 10 $\Rightarrow$ Number $= 10 \times \frac{35}{2} = 175$

26. The difference between the $\frac{3}{4}$ th of the $\frac{4}{5}$ th of a number and $\frac{1}{6}$ th of $\frac{2}{5}$ th of the same number is 648. What is the number ? **(Bank PO 2009)**

(*a*) 1110 (*b*) 1215 (*c*) 1325 (*d*) 1440

Sol. (*b*) Difference $= \frac{3}{4} \times \frac{4}{5} - \frac{1}{6} \times \frac{2}{5} = \frac{3}{5} - \frac{1}{15} = \frac{8}{15}$ of the number = 648

$\therefore$ The number $= 648 \times \frac{15}{8} = 1215$

27. The sum and product of two numbers are 12 and 35 respectively. What will be the sum of their reciprocals ? **(SSC CGL Tier I 2000, SSC CHSL 2010 modified)**

(*a*) $\frac{1}{3}$ (*b*) $\frac{1}{5}$ (*c*) $\frac{12}{35}$ (*d*) $\frac{35}{12}$

Sol. (*c*) $x + y = 12$ and $xy = 35$. Sum of reciprocals $= \frac{1}{x} + \frac{1}{y} = \frac{x+y}{xy} = \frac{12}{35}$

28. Divide 50 into two parts so that the sum of their reciprocals is 1/12. **(SSC CHSL 2013)**

(*a*) 28, 22 (*b*) 35, 15 (*c*) 20, 30 (*d*) 24, 36

Sol. (*c*) Let the two parts be x and y. Given, $x + y = 50$. Then $\frac{1}{x} + \frac{1}{y} = \frac{x+y}{xy} = \frac{1}{12} = \frac{50}{600}$.

Out of given options, $20 \times 30 = 600$ and $20 + 30 = 50$.

29. If difference between two numbers is 7 and their product is 78, find the difference between their reciprocals?

(*a*) $\frac{6}{13}$ (*b*) $\frac{7}{15}$ (*c*) $\frac{13}{78}$ (*d*) $\frac{7}{78}$

Sol. (*d*) $x - y = 7$ and $xy = 78$ $\Rightarrow$ Difference between reciprocals $= \frac{1}{x} - \frac{1}{y} = \frac{x-y}{xy} = \frac{7}{78}$

30. Of the following, the largest fraction is: $\frac{6}{7}, \frac{5}{6}, \frac{7}{8}, \frac{4}{5}$ **(SSC CGL Tier I 2003, CHSL 2010 M)**

(*a*) $\frac{6}{7}$ (*b*) $\frac{4}{5}$ (*c*) $\frac{4}{5}$ (*d*) $\frac{7}{8}$

Sol. (*d*) Numerator = Denominator – 1 in each case. $\Rightarrow$ $\frac{7}{8}$ is the largest fraction.

31. If $p = 3/5$, $q = 7/9$, $r = 5/7$, then which of the following inequality is true?

(SSC CHSL 2012, SSC CGL Tier I 2014)

(*a*) $p < q < r$ (*b*) $q < r < p$ (*c*) $p < r < q$ (*d*) $r < q < p$

Sol. (*c*) $\frac{3}{5} < \frac{5}{7} < \frac{7}{9}$ $\Rightarrow$ $p < r < q$ ($\because$ N – D = 2 in each case)

32. Which of the following numbers is greatest?

(*a*) $\frac{11}{16}$ (*b*) $\frac{13}{18}$ (*c*) $\frac{13}{16}$ (*d*) $\frac{11}{17}$

Sol. (*c*) $\frac{11}{16} > \frac{11}{17}$ and $\frac{13}{16} > \frac{13}{18}$ (smaller the denominator, bigger the number)

$\frac{13}{16} > \frac{11}{16}$ (bigger the numerator, bigger the number)

$\therefore$ $\frac{13}{16}$ is the greatest number.

33. Which of the following fractions is smallest?

(*a*) $\frac{5}{8}$ (*b*) $\frac{7}{10}$ (*c*) $\frac{3}{4}$ (*d*) $\frac{9}{16}$

Sol. (*d*) $\frac{5}{8} < \frac{7}{10} \Rightarrow \frac{7}{10}$ can't be the smallest number. Now we compare $\frac{5}{8}, \frac{3}{4}$ and $\frac{9}{16}$.

LCM of denominators *i.e.* LCM of 8, 4 and 16 is 16.

$\frac{5}{8} = \frac{10}{16}, \frac{3}{4} = \frac{12}{16}, \frac{9}{16} = \frac{9}{16}$ $\Rightarrow$ $\frac{9}{16} < \frac{10}{16} < \frac{12}{16} \Rightarrow \frac{9}{16} < \frac{5}{8} < \frac{3}{4}$

Note: $\frac{3}{4} > \frac{5}{8}$ and $\frac{5}{8} > \frac{9}{16}$ **(Since denominator is double but numerator is less than double)**

34. Which of the following numbers is greatest?

(*a*) $\frac{14}{3}$ (*b*) $\frac{19}{4}$ (*c*) $\frac{23}{5}$ (*d*) $\frac{29}{6}$

Sol. (*d*) $\frac{14}{3} = 4.67; \frac{19}{4} = 4.75; \frac{23}{5} = 4.6; \frac{29}{6} = 4.83 \Rightarrow 4.83$ or $\frac{29}{6}$ is the biggest.

35. Find the greatest fraction among following **(SSC FCI 2012)**

(*a*) $\frac{7}{9}$ (*b*) $\frac{4}{5}$ (*c*) $\frac{11}{12}$ (*d*) $\frac{17}{19}$

Sol. (*c*) $\frac{7}{9}$ $: 9 - 7 = 2 \Rightarrow 7 \div 2 = 3.5;$ $\frac{4}{5} : 5 - 4 = 1 \Rightarrow 4 \div 1 = 4;$

$\frac{11}{12} : 12 - 11 = 1 \Rightarrow 12 \div 1 = 12;$ $\frac{17}{19} : 19 - 17 = 2 \Rightarrow 17 \div 2 = 8.5$

$12 > 8.5 > 4 > 3.5 \Rightarrow \frac{11}{12}$ is largest fraction.

36. Which of the following fractions lies between $\frac{3}{5}$ and $\frac{2}{3}$?

(*a*) $\frac{2}{5}$ (*b*) $\frac{1}{3}$ (*c*) $\frac{1}{15}$ (*d*) $\frac{31}{50}$

Sol. (*d*) $\frac{3}{5} = 0.6$ and $\frac{2}{3} = 0.67$ (approx.)

$\frac{2}{5} = 0.4, \frac{1}{3} = 0.33$ (approx), $\frac{1}{15} = 0.067$ (approx) and $\frac{31}{50} = 0.62$

0.62 lies between 0.6 and 0.67. $\Rightarrow \frac{31}{50}$ lies between $\frac{3}{5}$ and $\frac{2}{3}$.

Alternative method: $\frac{3}{5} < \frac{2}{3}$ $\frac{2}{5}, \frac{1}{3}$ and $\frac{1}{15}$ are smaller than $\frac{3}{5}$.

37. A rational number between $\frac{3}{4}$ and $\frac{3}{8}$ is **(SSC CGL Tier I 2013)**

(*a*) $\frac{16}{9}$ (*b*) $\frac{9}{16}$ (*c*) $\frac{12}{7}$ (*d*) $\frac{7}{3}$

Sol. (*b*) $\frac{3}{4} = \frac{3}{4} \times \frac{4}{4} = \frac{12}{16}$ and $\frac{3}{8} = \frac{3}{8} \times \frac{2}{2} = \frac{6}{16}$. $\frac{6}{16} < \frac{9}{16} < \frac{12}{16} \Rightarrow \frac{9}{16}$ lies between $\frac{3}{4}$ and $\frac{3}{8}$.

Trick: $\frac{3}{4} < 1, \frac{3}{8} < 1$ but $\frac{16}{9}, \frac{12}{7}, \frac{7}{3} > 1$; only $\frac{9}{16} < 1$

38. Arrange the fractions in ascending order: $\frac{4}{15}, \frac{5}{17}, \frac{10}{33}$ and $\frac{8}{27}$. **(EPFO SSA)**

(*a*) $\frac{4}{15}, \frac{5}{17}, \frac{10}{33}, \frac{8}{27}$ (*b*) $\frac{5}{17}, \frac{10}{33}, \frac{8}{27}, \frac{4}{15}$ (*c*) $\frac{4}{15}, \frac{5}{17}, \frac{8}{27}, \frac{10}{33}$ (*d*) $\frac{10}{33}, \frac{8}{27}, \frac{5}{17}, \frac{4}{15}$

Sol. (*c*) On Dividing the denominators by the numerators, we get:

$$\frac{4}{15}=\frac{1}{3.75},\ \frac{5}{17}=\frac{1}{3.4},\ \frac{10}{33}=\frac{1}{3.3},\ \frac{8}{27}=\frac{1}{3.375}$$

Now $\frac{1}{3.75}<\frac{1}{3.4}<\frac{1}{3.375}<\frac{1}{3.3} \Rightarrow \frac{4}{15}<\frac{5}{17}<\frac{8}{27}<\frac{10}{33}$

39. The number 1.232323......, when written as a fraction in lowest terms, the difference of numerator and denominator is **(SSC FCI 2012)**

(*a*) 24 (*b*) 7 (*c*) 23 (*d*) 868

Sol. (*c*) $1.232323...... = \frac{123-1}{99}=\frac{122}{99}$ $\Rightarrow$ Required difference = 122 – 99 = 23

40. $\bar{3}.72+\bar{2}.48 = ?$

(*a*) $\bar{6}.2$ (*b*) $\bar{3}.8$ (*c*) $\bar{4}.02$ (*d*) None of these

Sol. (*d*) $\bar{3}.72+\bar{2}.48 = (-3 + 0.72) + (-2 + 0.48) = -5 + 1.2 = -4 + 0.2 = \bar{4}.2$

41. $0.\bar{6}+0.\bar{4}=?$

(*a*) 1.0 (*b*) 0.10 (*c*) 1.1 (*d*) $1.\bar{1}$

Sol. (*d*) $0.\bar{6}+0.\bar{4}=\frac{6}{9}+\frac{4}{9}=\frac{10}{9}=1\frac{1}{9}=1.\bar{1}$

$$\begin{array}{r} 0.6\,66 \\ +\ 0.4\,44 \\ \hline 1.\bar{1}\,10 \\ \hline \end{array}$$

42. What is the value of $2.\bar{6}-1.\bar{9}$? **(CDS 2011)**

(*a*) $0.\bar{6}$ (*b*) $0.\bar{9}$ (*c*) $0.\bar{7}$ (*d*) 0.7

Sol. (*a*) $2.\bar{6}-1.\bar{9}=2.\bar{6}-2=0.\bar{6}$

43. What is the value of $1.\overline{34}+4.1\bar{2}$ = ?

(*a*) $\frac{133}{90}$ (*b*) $\frac{371}{90}$ (*c*) $5\frac{219}{990}$ (*d*) $5\frac{461}{990}$

Sol. (*d*) $1.\overline{34}+4.1\bar{2}=(1+4)+\frac{34}{99}+\frac{11}{90}=5\frac{461}{990}$

44. $1.\overline{19}+5.\bar{3}+0.\overline{07}$ = ?

(*a*) $6.\overline{56}$ (*b*) $6.\overline{59}$ (*c*) $7.\overline{56}$ (*d*) $7.\overline{59}$

Sol. (*b*) $1.\overline{19}+5.\bar{3}+0.\overline{07}==(1+5+0)+\frac{19}{99}+\frac{33}{99}+\frac{7}{99}=6+\frac{59}{99}=6\cdot\overline{59}$

Direct Method: $1.\overline{19}+5.\overline{33}+0.\overline{07}=(1+5+0)+\left(0.\overline{19}+0.\overline{33}+0.\overline{07}\right)=6.\overline{59}$

(Note: This method can be used when number of recurring digits are same and there is no carry over.

45. The value of $0.\bar{2}+0.\bar{3}+0.3\bar{2}$ is: **(SSC CGL Tier I 2005)**

(*a*) $0.\overline{87}$ (*b*) $0.\overline{77}$ (*c*) $0.\overline{82}$ (*d*) $0.\overline{86}$

Sol. (*a*) Given expression = $0.\overline{22}+0.\overline{33}+0.\overline{32}=\frac{22}{99}+\frac{33}{99}+\frac{32}{99}=\frac{87}{99}=0.\overline{87}$

Alternative method:

$$\begin{array}{r} 0.22\ \ 22 \\ 0.33\ \ 33 \\ +\ 0.32\ \ 32 \\ \hline 0.\overline{87}\ \ 87 \\ \hline \end{array}$$

Alternative method: $0.\overline{22}+0.\overline{33}+0.\overline{32}=0.\overline{87}$

Note: Since there is no carry over and number of recurring digits are same. The given numbers are simply added.

46. The simplification of $(0.\overline{63}+0.\overline{37}+0.\overline{80})$ yields the result: **(SSC CGL 2004, CDS 2015)**

(*a*) $1.\overline{80}$ (*b*) $1.\overline{81}$ (*c*) $1.\overline{79}$ (*d*) 1.80

Sol. (*b*) G.E. = $\frac{63}{99}+\frac{37}{99}+\frac{80}{99}=\frac{180}{99}=1\frac{81}{99}=1.\overline{81}$

Alternative Method:

$$\begin{array}{r} 0.63\ 63 \\ 0.37\ 37 \\ +\ \ 0.80\ 80 \\ \hline 1.\overline{81}\ 80 \\ \hline \end{array}$$

47. The simplification of $3.\overline{36}-2.\overline{05}+1.\overline{33}$ equals: **(SSC CGL Tier I 2003)**

(*a*) 2.60 (*b*) $2.\overline{61}$ (*c*) 2.64 (*d*) $2.\overline{64}$

Sol. (*d*) Given expression = $3\frac{36}{99}-2\frac{5}{99}+1\frac{33}{99}=(3-2+1)+\left(\frac{36}{99}+\frac{33}{99}-\frac{5}{99}\right)=2+\frac{64}{99}=2.\overline{64}$

Direct Method: Given expression $=(3.\overline{36}+1.\overline{33})-2.\overline{05}=+\,4.\overline{69}-2.\overline{05}=2.\overline{64}$

48. $\left(5.\overline{88}-4.\overline{58}\right)-\left(0.\overline{64}+0.\overline{36}\right)$ is equal to: **(SSC CGL Tier I 2006)**

(*a*) $1.\overline{01}$ (*b*) $1.\overline{30}$ (*c*) $1.\overline{19}$ (*d*) $0.\overline{29}$

Sol. (*d*) Given expression $=\left[5\frac{88}{99}-4\frac{58}{99}\right]-\left[\frac{64}{99}+\frac{36}{99}\right]=\left(1+\frac{30}{99}\right)-\left(1+\frac{1}{99}\right)=\frac{29}{99}=0.\overline{29}$

Alternative Method: $\left(5.\overline{88}-4.\overline{58}\right)-\left(0.\overline{64}+0.\overline{36}\right)=1.\overline{30}-1.\overline{01}=0.\overline{29}$

49. $0.34\overline{67}+0.1\overline{33}\ =?$ **(SSC CGL 1993, CGL Tier I 2002, LIC AAO 2009)**

(*a*) $0.35\overline{7}$ (*b*) $0.4\overline{801}$ (*c*) $0.48\overline{01}$ (*d*) $0.35\overline{97}$

Sol. (*c*)

$$\begin{array}{r} 0.34\ 67\ 67 \\ +\ 0.13\ 33\ 33 \\ \hline 0.\,48\ \overline{01}\ 00 \\ \hline \end{array}$$

50. $8.3\overline{1}+0.\overline{6}+0.00\overline{2}\ =?$ **(SSC CGL Tier I 2005)**

(*a*) $8.\overline{912}$ (*b*) $8.9\overline{12}$ (*c*) 8.98 (*d*) $8.9\overline{79}$

Sol. (*c*)

$$\begin{array}{r} 8.31\ 1\ 11 \\ 0.66\ 6\ 66 \\ +\ 0.00\ 2\ 22 \\ \hline 8.97\ \bar{9} \\ \hline \end{array}$$

and $8.97\bar{9} = 8.98$

51. $1.\overline{73}+5.\overline{006}=?$

(*a*) $6.\overline{739}$ (*b*) $6.\overline{743}$ (*c*) $6.\overline{743379}$ (*d*) $6.\overline{744}$

Sol. (*c*) First number has two recurring digits and the second number has 3 recurring digits. So we took 6 digits (*i.e.* LCM of 2 and 3). Extra digits are not required to be added as there is no carry over.

$$\begin{array}{l} 1.737373 \\ \underline{5.006006} \\ \underline{6.743379} = 6.\overline{743379} \end{array}$$

52. $2.\overline{57}+3.7\overline{183}=?$

(*a*) $6.2\overline{941}$ (*b*) $6.2\overline{940758}$ (*c*) $6.\overline{2941}$ (*d*) None of these

Sol. (*b*) After adding non-recurring decimal (one digit), we add 6 recurring digits (*i.e.* LCM of 2 and 3, number of recurring digits). Two extra digits are added to correct the carry over.

$$\begin{array}{l} \ \ 2.5\ 757575\ 75 \\ +3.7\ 183183\ 18 \\ \hline \ \ 6.2\ \overline{940758}\ 93 = 6.2\overline{940758} \\ \hline \end{array}$$

53. What is $3.\overline{76}-1.4\overline{576}$ equal to ? **(CDS 2011)**

(*a*) $2.3\overline{100191}$ (*b*) $2.3\overline{101091}$ (*c*) $2.3\overline{110091}$ (*d*) $2.3\overline{110901}$

Sol. (*a*)

$$\begin{array}{r|ccc|c} 3.7 & 67 & 67 & 67 & 67 \\ -1.4 & 57 & 65 & 76 & 57 \\ \hline 2.3 & 10 & 01 & 91 & 10 \\ \hline \end{array} = 2.3\overline{100191}$$

54. Find the value of (0.333......) × (0.444......). **(SSC CGL Tier I 1995, 2003)**

(*a*) $0.\overline{012}$ (*b*) $0.\overline{12}$ (*c*) $0.\overline{148}$ (*d*) $0.\overline{13}$

Sol. (*c*) $(0.333......) \times (0.444......) = \frac{3}{9}\times\frac{4}{9}=\frac{4}{27}=\frac{4}{27}\times\frac{37}{37}=\frac{148}{999}=0.\overline{148}$

CHAPTER

3

DIVISIBILITY TEST AND REMAINDER RULES

A number is exactly divisible by a certain divisor if on dividing the number by that divisor, the remainder is zero. On dividing a number by the divisor, we can tell whether the number is exactly divisible and if not, what is the remainder. In this chapter, we will learn the short-cut methods of finding whether a number is divisible by the given divisor, if not what will be the remainder on dividing the number, without actually dividing the numbers.

RULES OF DIVISIBILITY TEST

A number is divisible by:

2: If unit's digit of the number is either 0 or a multiple of 2 *i.e.,* unit's digit is 0 or an even number.

Example: 18, 110, 98514 are divisible by 2. But 23, 147, 1521 are not divisible by 2.

3: If sum of the digits of the number is exactly divisible by 3.

Note: *While totalling the digits, any digit or groups of digits totalling to 9 may be ignored and the final sum is reduced to single digit number.*

Example: 132 is divisible by 3 as sum of its digits *i.e.* $1 + 3 + 2 = 6$ is exactly divisible by 3.

But 275 is not divisible by 3 as sum of the digits *i.e.* $2 + 7 + 5 = 14$ is not divisible by 3.

In this example, we can ignore $2 + 7 = 9$. Ignoring 9, we get 5 as sum of digits, which is not divisible by 3.

4: If the number formed by the last two digits are either zeros or a multiple of 4.

Example: 24, 564, 9900, 97632 and 859196 are divisible by 4.

But 918, 547622, 6853 and 35994 are not divisible by 4.

Note: A number having odd number at unit's place is never divisible by any even number.

Note: In 859196, the last two digits are 96. $100 - 96 = 4$ which is divisible by 4.

In 35994, the last two digits are 94. $100 - 94 = 6$ which is not divisible by 4.

5: If unit's digit of the number is 0 or 5.

Example: 135, 93215, 75290 are divisible by 5. But 1019, 50852 are not divisible by 5.

7: Deduct double of unit's digit from the remaining number and check whether the result is divisible by 7.

If the result is divisible by 7, the number is also divisible by 7.

Note: If the given number is large enough, then the process can be repeated till such a small number is obtained that can be easily checked whether it is divisible by 7.

Example: 161 : $16 - 2 \times 1 = 14 \Rightarrow$ The number is divisible by 7.

147 : $14 - 2 \times 7 = 0 \Rightarrow$ The number is divisible by 7.

8: If the number formed by the last three digits are either zeros or a multiple of 8.

Example: 721000, 1345144 are divisible by 8.

But 172433 and 258996 are not divisible by 8.

Note: In 258996, 1000 – 996 = 4, which is not divisible by 8.

9: If sum of digits of the number is exactly divisible by 9.

Note: While totalling the digits, any digit or groups of digits totalling to 9 may be ignored and the final sum is reduced to single digit number.

Example: 216 is divisible by 9 as sum of digits *i.e.* 2 + 1 + 6 = 9 is divisible by 9.

3294 is divisible by 9 as sum of digits *i.e.* 3 + 2 + 4 = 9 (ignoring 9) is divisible by 9.

Note: If we ignore both 9 while totalling digits of 3294, then we get 0 as the sum of digits.

⇒ If we strike out all 9 while totalling the digits of a number, sum of digits is always 0 when the number is exactly divisible by 9.

10: If unit digit of the number is 0.

Example: 2760, 38610 are divisible by 10. But 1531, 2376 are not divisible by 10.

11: If difference of the sum of digits in the even places and sum of digits in odd places is either 0 or a multiple of 11 i.e. exactly divisible by 11.

Examples:
1. 3256 : Difference between (3 + 5) and (2 + 6) is 0.
   ∴ The number is divisible by 11.
2. 291918 : Difference between (2 + 1 + 1) and (9 + 9 + 8) is 22, which is a multiple of 11. ⇒ The number is divisible by 11.
3. 517 : Difference between (5 + 7) and 1 is 11. ⇒ 517 is divisible by 11.
4. 10901 : Difference between (1 + 9 + 1) and (0 + 0) is 11.
   ∴ The number is divisible by 11.
5. 123 : Difference between (1 + 3) and (2) is 2.
   ∴ The number is not divisible by 11.

25: If last two digits of the number are either zeros or are divisible by 25.

∴ Numbers divisible by 25 must end in 00, 25, 50 or 75.

Example: 2125, 4350, 1375, 7800 are divisible by 25.

But 165, 4327, 7681 are not divisible by 25.

125: If last three digits of the number are either zeros or are divisible by 125.

99: Starting from right-hand side, make pairs of two digits each. Add all the pairs and divide the sum by 99. If the sum is exactly divisible by 99, the number is also divisible by 99.

Examples: 297 : 2 + 97 = 99 (99 is divisible by 99) ⇒ 297 is divisible by 99.

73854 : 7 + 38 + 54 = 99 (99 is divisible by 99) ⇒ 73854 is divisible by 99.

598 : 5 + 98 = 103 (103 is not divisible by 99) ⇒ 598 is not divisible by 99.

999: Starting from right-hand side, make groups of three digits each. Add all the groups and divide the sum by 999. If the sum is exactly divisible by 999, the number is also divisible by 999.

Examples: 4995 : 4 + 995 = 999 (999 is divisible by 999) ⇒ 4995 is divisible by 999.

175824 : 175 + 824 = 999 (999 is divisible by 999)

∴ 175824 is divisible by 999.

17424 : 17 + 424 = 441 (441 is not divisible by 999)

∴ 17424 is not divisible by 999.

7, 11 or 13: Starting from right-hand side, make groups of three digits each. Find the difference of the sum of groups at odd places and the sum of groups at even places. If the difference is divisible by 7, 11 or 13, the given number is divisible by that divisor.

Examples: 442,778 : 778 – 442 = 336 is divisible by 7 ⇒ 442778 is divisible by 7
6,895,938 : (6 + 938) – 895 = 49 is divisible by 7.
106,238 : 238 – 106 = 132 is disible by 11 ⇒ 106238 is divisible by 11.
18,084 : 084 – 18 = 66 is divisible by 11 ⇒ 18084 is divisible by 11
345,033 : 345 – 033 = 312 is divisible 13.
120,211 : 211 – 120 = 91 is divisible by 13
87,933, 703 : 933 – (703 + 87) = 143 is divisible by 11 and 13 both.

Note: If a number is divisible by two or more co-prime numbers, then the given number is also divisible by the product of the those co-prime numbers.

6: If the number is divisible by both 2 and 3 (Co-prime factors)

Examples:

1. Is 204 divisible by 6 ?

Sol. 204 is divisible by 2. 204 is divisible by 3. ⇒ 204 is divisible by 6.

2. Is 1431 divisible by 6?

Sol. 1431 is not divisible by 2. ⇒ 1431 is not divisible by 6.

3. Is 2156 divisible by 6?

Sol. 2156 is divisible by 2. 2156 is not divisible by 3.

∴ 2156 is not divisible by 6.

12: If the number is divisible by both 3 and 4 (Co-prime factors)

Example: Is 1068 divisible by 12 ?

Sol. 1068 is divisible by 3. 1068 is divisible by 4. ⇒ 1068 is divisible by 12.

14: If the number is divisible by both 2 and 7 (Co-prime factors)

Example: Is 938 divisible by 14 ?

Sol. 938 is divisible by 2. 938 is divisible by 7. ⇒ 938 is divisible by 14.

15: If the number is divisible by both 3 and 5 (Co-prime factors)

Example: Is 345 divisible by 15 ?

Sol. 345 is divisible by 3. 345 is divisible by 5. ⇒ 345 is divisible by 15.

18: If the number is divisible by both 2 and 9 (Co-prime factors)

Example: Is 1134 divisible by 18 ?

Sol. 1134 is divisible by 2. 1134 is divisible by 9. ⇒ 1134 is divisible by 18.

General Rule: If a number is divisible by two or more co-prime numbers say *a*, *b*, *c*......, then the given number is always divisible by $a \times b \times c \times$ ....

Example: 2160 is divisible by 2, 3 and 5 (co-prime numbers)

∴ 2160 is divisible by 2 × 3 × 5 = 30

REMAINDER RULES

If a number is exactly divisible by a divisor, remainder is '0' and in the other cases remainder may be any number lying between one and one less than the divisor. Generally, remainders are obtained by actually dividing the given number by the divisor. Using the Rules of Divisibility Test, we can also find the remainder without actually dividing the number :

2: Remainder is always 1, if the given number is not divisible by 2 *i.e.*, when the number ends in an odd number.

3: Remainder is obtained by dividing the sum of digits of the given number by 3.

Note: While totalling the digits, any digit or groups of digits totalling to 9 may be ignored.

Example: 275 is not divisible by 3 as sum of the digits *i.e.*, 2 + 7 + 5 = 5 (ignoring 9) is not divisible by 3.

On dividing 5 by 3, we get remainder 2. ∴ Remainder on dividing 275 by 3 is 2.

4: Remainder is obtained by dividing the last two digits of the number by 4.

Example: Remainders on dividing 918, 547621, 6855 by 4 are equal to remainder on dividing 18, 21, 55 by 4 *i.e.*, 2, 1 and 3 respectively.

5: Remainder is obtained by dividing the unit's digit of the given number by 5.

Example: Remainders on dividing 1019, 50852 by 5 are 4 and 2 respectively *i.e.*, remainders obtained on dividing 9 and 2 by 5.

8: Remainder is obtained on dividing the last three digits of the given number by 8.

Example: 172433: On dividing 433 by 8, we get remainder 1.

∴ Remainder on dividing 172433 by 8 is 1.

9: Remainder is obtained on dividing the sum of digits of the given number by 9.

Note: While totalling the digits, a digit or groups of digits totalling to 9 may be ignored.

Example: 2183: Sum of digits = 2 + 1 + 8 + 3 = 2 + 3 = 5 (ignoring 9)

∴ Remainder is 5.

10: Unit's digit of the given number is the remainder.

Example: Find the remainder on dividing 1276 by 10.

Sol. Remainder is unit's digit *i.e.*, 6.

99: Starting from right-hand side, make pairs of two digits each. Add all the pairs and divide the sum by 99. Remainder obtained is the answer.

Example: What is the remainder when 175394 is divided by 99?

Sol. Sum of pairs is 94 + 53 + 17 = 164. Remainder on dividing 164 by 99 is 65.

∴ Remainder on dividing 175394 by 99 is 65. **[Note:** 1 + 64 = 65**]**

999: Starting from right-hand side, make pairs of three digits each. Add all the pairs and divide the sum by 999. Remainder obtained is the answer.

Example: What is the remainder when 13528 is divided by 999?

Sol. Sum of pairs is 13 + 528 = 541. ∴ Remainder on dividing 13528 by 999 is 541.

MORE REMAINDER RULES

1. A certain number when successively divided by two different numbers leaves some remainders. The same number when divided by the product of the two divisors will leave remainder equal to:

First Divisor × Second Remainder + First Remainder.

Proof: Let the number is 'N' and it is divided successively by 'x' and 'y' leaving remainders 'a' and 'b' respectively.

We know that, Divided = (Divisor × Quotient) + Remainder

Let $N = mx + a$ (where 'm' is assumed quotient)

And $m = ny + b$ (where 'n' is assumed quotient)

$\therefore N = mx + a = (ny + b)x + a = nxy + bx + a$

Clearly, if we divide this number by 'xy', the remainder = $bx + a$.

i.e. First Divisor × Second Remainder + First Remainder

Example : A certain number when successively divided by 4 and 7 leaves the remainders 2 and 3 respectively. What is the remainder if the same number is divided by 28?

(*a*) 5 (*b*) 14 (*c*) 17 (*d*) 26

Sol. (*b*) Remainder = First Divisor × Second Remainder + First Remainder = 4 × 3 + 2 = 14

2. Two numbers *A* and *B* when separately divided by *x*, leaves remainders *a* and *b* respectively. (*A* + *B*) when divided by *x*, will leave remainder (*a* + *b*).

Logic: $A = mx + a, B = nx + b;\ A + B = mx + a + nx + b = (m + n)x + (a + b)$

Example: When two numbers are separately divided by 33, the remainders are 21 and 28 respectively. If the sum of the two numbers is divided by 33, the remainder will be :

(SSC CHSL 2010)

(*a*) 10 (*b*) 12 (*c*) 14 (*d*) 16

Sol. (*d*) The remainder should be 21 + 28 = 49. Remainder is reduced by 33.

⇒ Actual remainder = 49 – 33 = 16.

3. Two numbers *A* and *B* when separately divided by *x*, leaves remainders *a* and *b* respectively. (*A* × *B*) when divided by *x*, will leave remainder (*a* × *b*).

Logic: $(x + a)(x + b) = x^2 + (a + b)x + ab$

Example 1: Find the remainder on dividing 25 × 26 by 24.

Sol. 25 and 26 when divided by 24, leave remainders 1 and 2 respectively.

∴ 25 × 26 when divided by 24, leaves remainder 1× 2 = 2

Example 2: Find the remainder on dividing 153 × 799 by 5.

Solution: 153 and 799 when divided by 5, leave remainders 3 and 4 respectively.

∴ 153 × 799 when divided by 5, leaves remainder 3 × 4 = 12

We know that remainder cannot exceed the divisor.

∴ Remainder is 2 *i.e.*, remainder obtained on dividing 12 by 5.

Alternative Method: Remainder = 3 × (– 1) = – 3 or 5 – 3 = 2

Example 3: Find the remainder on dividing 17 × 27 × 95 × 36 by 8.

Sol. 17, 27, 95 and 36 when divided by 8, leave remainders 1, 3, 7 and 4 respectively.

∴ 17 × 27 × 95 × 36 when divided by 8, leaves remainder 1× 3 × 7 × 4 = 84.

∵ Remainder cannot exceed the divisor.

∴ Remainder is 4 *i.e.*, remainder obtained on dividing 84 by 8.

Alternative Method: Remainder = 1× 3 × (– 1) × 4 = – 12 or 16 – 12 = 4

4. If two numbers *A* and *B* when separately divided by *x*, leave remainders *a* and *b* respectively. Then $A^m \times B^n$, when divided by *x*, will leave remainder $a^m \times b^n$.

Example 1: Find the remainder on dividing $76^2 \times 49^3$ by 24.

Sol. 76 and 49 when divided by 24, leave remainders 4 and 1 respectively.

∴ $76^2 \times 49^3$ when divided by 24, leaves remainder $4^2 \times 1^3 = 16$.

Example 2: Find the remainder on dividing $71^2 \times 36^3 \times 21^2$ by 17.

Sol. 71, 36 and 21 when divided by 17, leave remainders 3, 2 and 4 respectively.

∴ $71^2 \times 36^3 \times 21^2$ when divided by 17, leaves remainder $3^2 \times 2^3 \times 4^2 = 1152$.

∵ Remainder cannot exceed the divisor.

∴ Remainder is 13 *i.e.*, remainder obtained on dividing 1152 by 17.

Alternatively, remainder = 9 × 8 × (– 1) = –72 or 85 – 72 = 13

5. How to check whether $(x^n - a)$ is divisible by y?

Method: 1. (x^n) is divided into some parts in such a way that value of each part is slightly more (or less) than the multiple of y.

2. Each part is divided by y.

3. Remainders obtained in each part are multiplied together.

4. Subtract a from the product of remainders.

5. If result is 0 or a multiple of y, then $(x^n - a)$ is divisible by y.

Example 1: Is $7^6 - 1$ divisible by 16 ?

Sol. $7^6 - 1 = 7^2 \times 7^2 \times 7^2 - 1 = 49 \times 49 \times 49 - 1$

Remainder obtained on dividing each part by $16 = 1 \times 1 \times 1 - 1 = 0$

$\therefore$ $7^6 - 1$ is divisible by 16.

Example 2: Is $3^9 - 1$ divisible by 26 ?

Sol. $3^9 - 1 = 3^3 \times 3^3 \times 3^3 - 1 = 27 \times 27 \times 27 - 1$

Remainder obtained on dividing each part by $26 = 1 \times 1 \times 1 - 1 = 0$

$\therefore$ $3^9 - 1$ is divisible by 26.

Example 3: Is $5^5 - 1$ divisible by 11 ?

Sol. $5^5 - 1 = 5^2 \times 5^3 - 1 = 25 \times 125 - 1$

Remainder obtained on dividing each part by 11 is $3 \times 4 - 1 = 12 - 1 = 11$, which is divisible by 11.

$\therefore$ $5^5 - 1$ is divisible by 11.

Example 4: Is $2^{12} - 4$ divisible by 11 ?

Sol. $2^{12} - 4 = 2^6 \times 2^6 - 4 = 64 \times 64 - 4$

Remainder obtained on dividing each part by 11 is $9 \times 9 - 4 = 81 - 4 = 77$, which is divisible by 11.

$\therefore$ $2^{12} - 4$ is divisible by 11.

Alternatively, $2^{12} - 4 = 2^5 \times 2^5 \times 2^2 - 4 = (-1) \times (-1) \times (4) - 4 = 0$.

SOLVED EXERCISE

1. Consider the following statements: **(CDS 2009)**
A number $a_1a_2a_3a_4a_5$ is divisible by 9 if :
1. $a_1 + a_2 + a_3 + a_4 + a_5$ is divisible by 9. 2. $a_1 - a_2 + a_3 - a_4 + a_5$ is divisible by 9.
Which of the above statements is/are correct ?
(*a*) 1 only (*b*) 2 only (*c*) both 1 & 2 (*d*) Neither 1 nor 2

Sol. (*a*) A number is divisible by 9 if and only if sum of its digits is divisible by 9.

2. If 72*72 is divisible by 9, the missing * digit will be: **(SSC CGL Tier I 2013)**
(*a*) 0 or 9 (*b*) 3 or 6 (*c*) 2 or 4 (*d*) 5 or 8

Sol. (*a*) Sum of given digits = 9.

3. Which one of the number is divisible by 25 ? **(SSC CGL Tier II 2013)**
(*a*) 303310 (*b*) 373355 (*c*) 303375 (*d*) 22040

Sol. (*c*) 303375 ends in 75, which is divisible by 25.

4. Which of the following numbers is NOT divisible by 18:
(*a*) 54036 (*b*) 50436 (*c*) 34056 (*d*) 65043

Sol. (*d*) A number is exactly divisible by 18 if it is divisible by 2 and 9 both.
65043 is an odd number and hence not divisible by 2.

5. Which of the following numbers is exactly divisible by 45?
(*a*) 2283 (*b*) 3920 (*c*) 4275 (*d*) 5184

Sol. (*c*) A number is divisible by 45 if it is divisible by both 5 and 9.

We can easily rule out 2283 and 5184 for not being divisible by 5 as these numbers don't have 5 or 0 at unit's place.

Both the remaining numbers *i.e.* 3920, 4275 are divisible by 5.

3920: Sum of digits of 3920 is 3 + 2 = 5. ∴ 3920 is not divisible by 9.

4275 : It being the only option left, is the answer. However, we can check 4275 also.

Sum of digits = 4 + 2 + 7 + 5 = 0 (ignoring both 9's) ∴ 4275 is divisible by 9 also.

6. Which among the following is the largest four digit number that is divisible by 88 ? **(SSC CGL Tier I 2015 CDS 2011)**

(*a*) 9988 (*b*) 9966 (*c*) 9944 (*d*) 8888

Sol. (*c*) A number is divisible by 88 if it is divisible by 8 and 11.

9988: Not divisible by 8 as the number formed by its last three digits is not divisible by 8.

Hint: 1000 is exactly divisible by 8. 988 is 12 less than 1000.

Since, 12 is not divisible by 8, 9988 is also not divisible by 8.

9966: Not divisible by 8 as the number formed by its last three digits is not divisible by 8.

Hint: 1000 is exactly divisible by 8. 966 is 34 less than 1000.

Since, 34 is not divisible by 8, 9966 is also not divisible by 8.

8888: Divisible by 8 and 11.

9944: Last three digits *i.e.* 944 is 56 less than 1000 and 56 is divisible by 8.

∴ 9944 is divisible by 8.

9944 is divisible by 11 as (9 + 4) – (9 + 4) = 0.

∴ 8888 and 9944 both are divisible by 8 and 11.

But 9944 being the greater number, is the answer.

7. Which of the followings numbers is exactly divisible by 99 ? **(SSC CGL Tier I 2006)**

(*a*) 114543 (*b*) 1592307 (*c*) 111437 (*d*) 931741

Sol. (*a*) A number is divisible by 99 if it is divisible by 9 and 11.

114543 is the only number which is divisible by both 9 and 11.

Alternative method: We check sum of pairs for each option:

(*a*) 43 + 45 + 11 = 99 which is divisible by 99.

(*b*) 07 + 23 + 59 + 1 = 90 which is not divisible by 99

(*c*) 37 + 14 + 11 = 62 which is not divisible by 99

(*d*) 41 + 17 + 93 = 151 which is not divisible by 99.

8. The least number, which must be added to 6709 to make it exactly divisible by 9, is **(SSC CGL Tier I 2004)**

(*a*) 5 (*b*) 4 (*c*) 7 (*d*) 2

Sol. (*a*) Sum of digits = 6 + 7 + 0 + 9 = 4 (ignoring 9s). ⇒ Required number = 9 – 4 = 5

9. The smallest number to be added to 1000 so that 45 divides the sum exactly, is: **(SSC CGL Tier I 2000)**

(*a*) 35 (*b*) 80 (*c*) 20 (*d*) 10

Sol. (*a*) On dividing 1000 by 45, the remainder is 10. ∴ Required number = 45 – 10 = 35

Alternative method: 45 = 5 × 9. 1000 is divisible by 5.

Sum of digits = 1 + 0 + 0 + 0 = 1. ∴ Remainder on dividing 1000 by 9 is 1.

∴ Smallest number to be added to 1000, must be a multiple of 5 and sum of digits be 8. 35 is the smallest such number.

10. The number nearest to 75070 which is divisible by 65, is **(SSC CGL Tier I 2012)**

(*a*) 75010 (*b*) 75065 (*c*) 75070 (*d*) 75075

Sol. (*d*) $75070 \div 65$ = Remainder 60 ⇒ Required number = $75070 + 65 - 60 = 75075$

11. Find the greatest four-digit number which is exactly divisible by 5, 6, 8 and 9.

(*a*) 8640 (*b*) 9999 (*c*) 9800 (*d*) 9720

Sol. (*d*) 9999 : Not divisible by 5 as its unit's digit is neither 0 nor 5.

9800 : Not divisible by 9 as sum of digits i.e. 8 is not divisible by 9.

8640 and 9720 both are divisible by 5, 6, 8 and 9, but 9720 being the greater number, is the answer.

12. The greatest 4-digit number exactly divisible by 10, 15, 20 is **(SSC CGL Tier II 2013)**

(*a*) 9990 (*b*) 9960 (*c*) 9980 (*d*) 9995

Sol. (*b*) 9995 : Not divisible by 10. 9990 : Not divisible by 20

9980 : $9 + 9 + 8 = 26$ ⇒ Not divisible by 15.

9960 : Divisible by 10, 15, 20.

13. An even number, sum of whose digits is 36, is always divisible by: **(SSC)**

(*a*) 9 (*b*) 18 (*c*) 36 (*d*) 72

Sol. (*b*) An even number is always divisible by 2.

Further any number sum of whose digits is 36 is always divisible by 9.

∴ The number is divisible by both 2 and 9.

∴ The number is divisible by $2 \times 9 = 18$.

14. Find the least number which when divided by 13 leaves remainder 3 and when divided by 5, leaves remainder 2. **(MBA)**

(*a*) 16 (*b*) 22 (*c*) 29 (*d*) 42

Sol. (*d*) Out of given options, 42 satisfies both the conditions.

Note: Next higher numbers = (LCM of 13 and 5) $\times x + 42 = 65x + 42$

15. When an integer K is divided by 3, the remainder is 1, and when K + 1 is divided by 5, the remainder is 0. Of the following, a possible value of K is: **(SSC CHSL 2011)**

(*a*) 62 (*b*) 63 (*c*) 64 (*d*) 65

Sol. (*c*) 64 when divided by 3, leaves remainder 1. 64 + 1 is fully divisible by 5.

16. What value must be given to * so that the number 9781*4 is exactly divisible by 6:

(*a*) 0 (*b*) 1 (*c*) 2 (*d*) 3

Sol. (*b*) A number is divisible by 6 if it is divisible by 2 and 3.

The given number is divisible by 2 as it has an even number in the unit's place.

Sum of the digits of the given number is 11 or 2.

On adding 1 to it, sum of digits will become 3 and hence divisible by 3.

∴ * is to be replaced by 1.

17. What must be added to 583171 to make it exactly divisible by 11: **(Bank Clerical)**

(*a*) 5 (*b*) 6 (*c*) 7 (*d*) 8

Sol. (*a*) Sum of digits at odd places (from left side) = $5 + 3 + 7 = 15$.

Sum of digits at even places (from left side) = $8 + 1 + 1 = 10$.

Whatever we add to the given number, will increase the sum of digits which includes the unit's digit *i.e.* to 10. If we add 5 to given number difference between the sum of digits becomes $15 - 15 = 0$. $\Rightarrow$ 5 is to be added to the given number.

18. If * is a digit such that 5824* is divisible by 11, then * equals **(SSC CGL Tier I 2000)**

(*a*) 5 (*b*) 4 (*c*) 7 (*d*) 2

Sol. (*a*) Sum of digits at odd places (from left side) $= 5 + 2 + * = 7 + *$

Sum of digits at even places (from left side) $= 8 + 4 = 12$

$7 + * = 12 \Rightarrow * = 12 - 7 = 5$

19. What value must be given to * to make 8597*65 exactly divisible by 11?

(*a*) 0 (*b*) 2 (*c*) 4 (*d*) 7

Sol. (*d*) Sum of digits at odd places (from left side) $= 8 + 9 + * + 5 = 22 + *$

Sum of digits at even places (from left side) $= 5 + 7 + 6 = 18$

Difference between the two sums $= (22 + *) - 18 = 4 + *$

We know that any number is divisible by 11, if the difference is 0 or a multiple of 11.

The difference between $22 + *$ and 18 will be 11 when $* = 7$.

20. What is the remainder when 823442 is divided by 11 ?

(*a*) 0 (*b*) 2 (*c*) 4 (*d*) 7

Sol. (*c*) Difference of sums of digits $= (8 + 3 + 4) - (2 + 4 + 2) = 7$

$\Rightarrow$ If unit digit of the given number is more by 7, the number is divisible by 11.

$\Rightarrow$ Required remainder $= 11 - 7 = 4$

21. What is the remainder when 579658 is divided by 11 ?

(*a*) 0 (*b*) 2 (*c*) 5 (*d*) 9

Sol. (*b*) Difference of sums of digits $= (5 + 9 + 5) - (7 + 6 + 8) = 19 - 21 = -2$

$\Rightarrow$ If unit digit is reduced by 2, the number is divisible by 11.

$\Rightarrow$ Required remainder = 2

22. A three-digit number 4a3 is added to another three-digit number 984 to give the four digit number 13b7 which is divisible by 11. Then the value of $(a + b)$ is:

(CBI 1993, SSC CGL Tier I 2012)

(*a*) 11 (*b*) 12 (*c*) 9 (*d*) 10

Sol. (*d*) Sum of digits at odd places (from left side) $= 1 + b$

Sum of digits at even places (from left side) $= 3 + 7 = 10$

Difference between the two sums $= (1 + b) - 10 = b - 9 = 0 \quad \Rightarrow \quad b = 9$

$\Rightarrow \quad a + 8 = b = 9. \quad \Rightarrow \quad a = 9 - 8 = 1 \quad \Rightarrow \quad a + b = 9 + 1 = 10$

23. Both the end digits of a 99 digit number *N* are 2. *N* is divisible by 11, then all the middle digits are **(SSC FCI 2012)**

(*a*) 1 (*b*) 2 (*c*) 3 (*d*) 4

Sol. (*d*) 242, 24442, $2\underbrace{44....}_{97\text{ times}}2$ are divisible by 11.

24. What should be added to 15762 to make it divisible by 18?

(*a*) 4 (*b*) 6 (*c*) 9 (*d*) 12

Sol. (*b*) A number is divisible by 18, if it is divisible by 2 and 9 both.

Unit's digit of the given number is an even number and hence divisible by 2.

$\therefore$ The number to be added must be an even number so that sum of two numbers is even number.

A number is divisible by 9, if sum of its digits is divisible by 9.

Sum of digits of 15762 is 3. $3 + 6 = 9$ *i.e.* sum of digits is divisible by 9.

$\therefore$ 6 is the required number.

25. What should be subtracted from 331683 to make it divisible by 36 ?

(*a*) 6 (*b*) 11 (*c*) 15 (*d*) 24

Sol. (*c*) A number is divisible by 36, if it is divisible by 4 and 9 both.

A number is divisible by 4, if its last two digits are divisible by 4.

In 331683, last two digits are 83, which is an odd number.

Any number divisible by 4 must be an even number.

$\therefore$ The number to be deducted from 331683 must be odd number.

$\therefore$ 6 and 24 are not possible.

11 : $83 - 11 = 72$ is divisible by 4. (Possible)

15 : $83 - 15 = 68$ is divisible by 4. (Possible)

Now we check divisibility test rule for 9: Sum of digits of 331683 is 6.

$\therefore$ Sum of digits of the number deducted must be 6 so that the result is divisible by 9.

11: Sum of digits is $1 + 1 = 2$ (not possible)

15: Sum of digits is $1 + 5 = 6$ (possible) $\Rightarrow$ 15 is the required number.

26. When 335 is added to 5A7, the result is 8B2. 8B2 is divisible by 3. What is the largest possible value of A? **(SSC CGL Tier II 2013)**

(*a*) 8 (*b*) 2 (*c*) 1 (*d*) 4

Sol. (*d*) $335 + 5A7 = 8B2 \Rightarrow (3 + 1) + A = B \Rightarrow$ when $A = 4$, $B = 8$

$8B2 = 882$, sum of digits $= 8 + 8 + 2 = 18$, which is divisible by 3.

27. By which least number 325 must be multiplied to get a multiple of 35 ?

(*a*) 3 (*b*) 5 (*c*) 7 (*d*) 11

Sol. (*c*) $35 = 5 \times 7 \Rightarrow$ Any multiple of 35 must be divisible by 5 and 7.

$325 = 5 \times 5 \times 13 \Rightarrow$ 325 is divisible by 5, but not by 7.

$\therefore$ 325 must be multiplied by 7 to make it a multiple of 35.

28. In a question on division, if four times of the divisor is added to the dividend, then the new remainder, in comparison with the original remainder, will : **(SSC CGL Tier I 2006)**

(*a*) Be increased by 4 (*b*) Be decreased by 4 (*c*) Remain unchanged (*d*) Be 4 times

Sol. (*c*) 4 times of the divisor is fully divisible by the divisor itself.

$\therefore$ The remainder will remain unchanged.

29. When a number is divided by 56, the remainder obtained is 29. What will be the remainder when the number is divided by 8 ? **(SSC CGL Tier I 2002)**

(*a*) 4 (*b*) 5 (*c*) 3 (*d*) 7

Sol. (*b*) Let the number be $56x + 29 = 8 \times 7x + 8 \times 3 + 5 = 8 \times (7x + 3) + 5$

Required remainder is 5.

Trick : 56 is fully divisible by 8, therefore the required remainder may be obtained by dividing the original remainder by 8. We get remainder 5 on dividing 29 by 8.

30. A number when divided by 119 leaves remainder 19. The same number when divided by 17, will leave remainder: **(SSC CGL 1993, SSC CGL Tier I 2008)**

(*a*) 2 (*b*) 7 (*c*) 10 (*d*) 12

Sol. (*a*) Since 119 is exactly divisible by 17, the remainder may be obtained by dividing the 19 by 17.
$\therefore$ Required Remainder = 2

31. A certain number when divided by 175 leaves a remainder 132. When the same number is divided by 25, the remainder is: **(SSC CHSL 2011)**

(*a*) 6 (*b*) 7 (*c*) 8 (*d*) 9

Sol. (*b*) Since 175 is exactly divisible by 25, the remainder may be obtained by dividing the 132 by 25. $\therefore$ Required Remainder = 7

32. When a natural number *n* is divided by 6, the remainder is 4. When 2*n* is divided by 6, the remainder is ? **(SSC CHSL 2015)**

(*a*) 1 (*b*) 2 (*c*) 4 (*d*) 0

Sol. (*b*) The remainder should be $2 \times 4 = 8$. But a remainder can't exceed the divisor.
Actual remainder is reduced by the divisor. $\Rightarrow$ Actual remainder = $8 - 6 = 2$

33. What is the remainder obtained when $1421 \times 1423 \times 1425$ is divided by 12 ? **(CDS 2015)**

(*a*) 1 (*b*) 2 (*c*) 3 (*d*) 4

Sol. (*c*) The remainder should be $5 \times 7 \times 9 = 315$. But a remainder can't exceed the divisor.
$\therefore$ Remainder is 3 *i.e.*, remainder obtained on dividing 315 by 12.

Alternative Method: Remainder = $5 \times (-5) \times (-3) = 75$ or $75 - 72 = 3$

34. When a natural number *n* is divided by 4, the remainder is 3. What is the remainder when 2*n* is divided by 4 ? **(CDS 2009)**

(*a*) 1 (*b*) 2 (*c*) 3 (*d*) 6

Sol. (*b*) The remainder should be $2 \times 3 = 6$. But a remainder can't exceed the divisor.
Actual remainder is reduced by the divisor. $\Rightarrow$ Actual remainder = $6 - 4 = 2$

35. When a positive integer *n* is divided by 5, the remainder is 2. What is the remainder when number 3*n* is divided by 5 ? **(CDS 2011)**

(*a*) 1 (*b*) 2 (*c*) 3 (*d*) 4

Sol. (*a*) Remainder = $3 \times 2 = 6$. But a remainder can't exceed the divisor.
$\therefore$ Actual remainder = $6 - 5 \times 1 = 1$

36. If two numbers are each divided by the same divisor, the remainders are respectively 3 and 4. If the sum of the numbers be divided by the same divisor, the remainder is 2. The divisor is : **(SSC CGL Tier I 2007)**

(*a*) 9 (*b*) 7 (*c*) 5 (*d*) 3

Sol. (*c*) The remainder should be $3 + 4 = 7$. Actual remainder = 2 $\Rightarrow$ Divisor = $7 - 2 = 5$

37. A number when divided by a certain divisor leaves remainder 50. When double of that number is divided by the same divisor, the remainder is 8. Find the divisor.

(*a*) 42 (*b*) 58 (*c*) 92 (*d*) 108

Sol. (*c*) When the number is doubled, the remainder should have also be doubled.
$\therefore$ Remainder should be $50 \times 2 = 100$. But a remainder can't exceed the divisor.
$\therefore$ Actual remainder is 8 which is reduced by the divisor. $\therefore$ Divisor = $100 - 8 = 92$

38. The remainder on dividing given integers *a* and *b* by 7 are respectively 5 and 4. What is the remainder when *ab* is divided by 7 ? **(CDS 2010)**

(*a*) 3 (*b*) 4 (*c*) 5 (*d*) 6

Sol. (*d*) Remainder = $5 \times 4 = 20$. But a remainder can't exceed the divisor.
$\therefore$ Actual remainder = $20 - 7 \times 2 = 6$

39. A number when divided by 5 leaves a remainder 3. What is the remainder when the square of the same number is divided by 5 ? **(SSC CGL Tier I 2008, 2011 M)**

(*a*) 2 (*b*) 3 (*c*) 4 (*d*) 9

Sol. (*c*) Since the number (dividend) is squared, the remainder will also be squared.

But a remainder can't exceed the divisor. ∴ Required remainder = $3^2 - 5 = 9 - 5 = 4$

Alt: Let the number be $5x + 3$. Then square of the number = $(5x + 3)^2 = 25x^2 + 30x + 9$

∴ Required remainder = $9 - 5 = 4$

40. A number when divided by 6 leaves a remainder 3. When the square of the same number is divided by 6, the remainder is: **(SSC CGL Tier I 2005)**

(*a*) 0 (*b*) 1 (*c*) 2 (*d*) 3

Sol. (*d*) Since the number (dividend) is squared, the remainder will also be squared.

But a remainder can't exceed the divisor. ∴ Required remainder = $3^2 - 6 = 9 - 6 = 3$

Alt: Let the number be $6x + 3$. Then square of the number = $(6x + 3)^2 = 36x^2 + 36x + 9$

∴ Required remainder = $9 - 6 = 3$

41. Find the remainder on dividing 26^{16} by 25.

(*a*) 0 (*b*) 1 (*c*) 15 (*d*) 24

Sol. (*b*) Remainder on dividing 26 by 25 is 1. ⇒ Remainder on dividing (26^{16}) by 25 is $1^{16} = 1$

42. Find the remainder on dividing 17^{200} by 18. **(SSC CGL Tier I 2000, 2011)**

(*a*) 1 (*b*) 2 (*c*) 16 (*d*) 17

Sol (*a*) Remainder on dividing 17 by 18 is – 1.

∴ Remainder on dividing 17^{200} by 18 is $(-1)^{200} = 1$.

43. What will be the remainder when 19^{100} is divided by 20 ? **(SSC CGL Tier II 2012)**

(*a*) 3 (*b*) 1 (*c*) 19 (*d*) 20

Sol. (*b*) Remainder on dividing 19 by 20 is – 1.

∴ Remainder on dividing 19^{100} by 20 is $(-1)^{100} = 1$.

44. If $(67^{67} + 67)$ is divided by 68, the remainder is: **(SSC CGL Tier I 2000)**

(*a*) 1 (*b*) 67 (*c*) 63 (*d*) 66

Sol. (*d*) Remainder on dividing 67 by 68 is –1. Actual remainder = $(-1)^{67} + 67 = -1 + 67 = 66$.

45. Find the remainder on dividing $(9^{16} + 6)$ by 8. **(SSC CGL Tier II 2012)**

(*a*) 5 (*b*) 7 (*c*) 2 (*d*) 3

Sol. (*b*) Remainder on dividing 9 by 8 is 1.

∴ $(9^{16} + 6)$ when divided by 8 leaves remainder $(1)^{16} + 6 = 7$.

46. 25^{25} is divided by 26, the remainder is: **(SSC CGL Tier I 2000)**

(*a*) 1 (*b*) 2 (*c*) 24 (*d*) 25

Sol. (*c*) Remainder on dividing 25 by 26 is –1. Actual remainder = $(-1)^{25} = -1$ or $25 - 1 = 24$.

47. Find the remainder on dividing 17^{31} by 18.

(*a*) 0 (*b*) 1 (*c*) 16 (*d*) 17

Sol. (*d*) Remainder on dividing 17 by 18 is – 1.

∴ Remainder on dividing 17^{31} by 18 = $(-1)^{31} = -1$ or $18 - 1 = 17$

48. $2^{16} - 1$ is divisible by **(SSC CGL Tier I 2011)**

(*a*) 11 (*b*) 13 (*c*) 17 (*d*) 19

Sol. (*c*) $2^{16} - 1 = (2^8)^2 - 1^2 = (2^8 + 1)(2^8 - 1) = 257 \times 255 = 257 \times (15 \times 17)$

49. A number divided by 68 gives the quotient 269 and remainder zero. If the same number is divided by 67, the remainder is: **(SSC CGL Tier I 2005)**

(*a*) 0 (*b*) 1 (*c*) 2 (*d*) 3

Sol. (*b*) Number $= 68 \times 269 = (67 + 1) \times 269 = 67 \times 269 + (1 \times 269) = 67 \times 269 + (67 \times 4 + 1)$

$\therefore$ Required remainder = 1

Alt. Number $= 68 \times 269 \Rightarrow$ Remainder $= 1 \times 1 = 1$.

50. A number when divided by 3 leaves a remainder 1. When the quotient is divided by 2, it leaves a remainder 1. What will be the remainder when the number is divided by 6 ?

(SSC CGL Tier I 2004)

(*a*) 3 (*b*) 4 (*c*) 5 (*d*) 2

Sol. (*b*) Required remainder = First Divisor × Second Remainder + First Remainder

$= 3 \times 1 + 1 = 3 + 1 = 4$

Alternative method: Let final quotient be 1. Then

$$\begin{array}{r|ll} 3 & y & \\ \hline 2 & x & 1 \\ \hline & 1 & 1 \end{array}$$

$x = 2 \times 1 + 1 = 3$

$y = 3x + 1 = 3 \times 3 + 1 = 10$

$\therefore$ Required remainder $= 10 \div 6 = 1$ (Quotient) + 4 (Remainder)

51. A number divided by 13 leaves a remainder 1 and if the quotient, thus obtained, is divided by 5, we get a remainder of 3. What will be the remainder if the number is divided by 65 ?

(SSC CGL Tier I 2005)

(*a*) 28 (*b*) 16 (*c*) 18 (*d*) 40

Sol. (*d*) Required remainder = First Divisor × Second Remainder + First Remainder

$= 13 \times 3 + 1 = 40$

Alternative method: Let final quotient be 1. Then,

$$\begin{array}{r|ll} 13 & y & \\ \hline 5 & x & 1 \\ \hline & 1 & 3 \end{array}$$

$x = 5 \times 1 + 3 = 8$

$y = 13x + 1 = 13 \times 8 + 1 = 105$

$\therefore$ Required remainder $= 105 \div 65 = 1$ (Quotient) + 40 (Remainder)

52. A number being successively divided by 3, 5 and 7 leaves remainders 1, 2 and 3 respectively. What is the number ?

(*a*) 111 (*b*) 146 (*c*) 157 (*d*) 183

Sol. (*c*) Let the final quotient be 1. Then,

$$\begin{array}{r|ll} 3 & z & \\ \hline 5 & y & 1 \\ \hline 7 & x & 2 \\ \hline & 1 & 3 \end{array}$$

$x = 7 \times 1 + 3 = 10$

$y = 5x + 2 = 5 \times 10 + 2 = 52$

$z = 3y + 1 = 3 \times 52 + 1 = 157$

53. **A number being successively divided by 2, 3 and 5 leaves remainders 1, 2 and 3 respectively. What is the remainder when the same number is divided by 30 ?**

(*a*) 3 (*b*) 5 (*c*) 6 (*d*) 23

Sol. (*d*) Let the final quotient be 1. Then,

2	z	
3	y	1
5	x	2
	1	3

$x = 5 \times 1 + 3 = 8$

$y = 3x + 2 = 3 \times 8 + 2 = 26$

$z = 2y + 1 = 2 \times 26 + 1 = 53$

Remainder on dividing 53 by 30 is 23.

CHAPTER

4

MULTIPLICATION AND DIVISION OF NUMBERS – VEDIC METHODS

Multiplication of numbers is extensively used in solving numerous problems in the examinations. Conventional method of multiplication of numbers is very lengthy especially when the numbers to be multiplied are larger. As the conventional method involves a lot of calculations, the method is time consuming and also prone to mistakes. By using the short-cut methods explained in this chapter, even the multiplication of large numbers can be done in few seconds, simply by adding and subtracting the numbers. In this chapter we will also learn the method of checking the correctness of product of two or more numbers.

SHORT-CUT METHODS OF MULTIPLICATION

1. Multiplication of numbers – one or more of which have 0's at the end:

Steps: 1. Multiply the numbers ignoring 0's at the end of the given numbers.

2. At the end of multiplication result, put as many 0's as many 0's are at the end of the given numbers.

Example: $257 \times 200 = ?$

Sol. Second number has two 0's at the end of it.

Ignoring 0's at the end, we get $257 \times 2 = 514$.

Now we put two 0's at the end of the result. $\therefore \ 257 \times 200 = 51400$

2. Multiplication by 5^n *i.e.* 5, 25 and 125, etc. :

Logic $5^n = (10 \div 2)^n = 10^n \div 2^n$

(*a*) Multiply a number by 5:

Method: Divide the given number by 2 and multiply the result by 10. **Note:** $10 \div 2 = 5$

Example: $112 \times 5 = \frac{112}{2} \times 10 = 56 \times 10 = 560$

$576 \times 5 = \frac{576}{2} \times 10 = 288 \times 10 = 2880$

$175 \times 5 = \frac{175}{2} \times 10 = 87.5 \times 10 = 875$

(*b*) Multiply a number by 25:

Method: Divide the number by 4 and multiply the result by 100. **Note:** $\frac{100}{4} = 25$

Example: $16 \times 25 = \frac{16}{4} \times 100 = 4 \times 100 = 400$

$128 \times 25 = \frac{128}{4} \times 100 = 32 \times 100 = 3200$

$73 \times 25 = \frac{73}{4} \times 100 = 18\frac{1}{4} \times 100 = 1825$

$$127 \times 25 = \frac{127}{4} \times 100 = 31\frac{3}{4} \times 100 = 3175$$

(c) Multiply a number by 125:

Method: Divide the number by 8 and Multiply the result by 1000. **Note:** $\frac{1000}{8} = 125$

Example: $24 \times 125 = \frac{24}{8} \times 1000 = 3 \times 1000 = 3000$

$$45 \times 125 = \frac{45}{8} \times 1000 = 5\frac{5}{8} \times 1000 = 5 \times 1000 + \frac{5}{8} \times 1000 = 5000 + 5 \times 125 = 5625$$

3. Multiplication by 9, 99, 999, 9999 i.e., $(10^n - 1)$

Logic = Number × 9999.... *n* times = Number × $(10^n - 1)$ = Number × 10^n – Number

Steps for multiplication:

1. At the end of the given number put '*n*' zeros.
2. From the new number such formed subtract the Multilicant (i.e. original number).

Examples:

238×9	217 × 99	258 × 999	1295 × 9999
2380	21700	258000	12950000
–238	– 217	– 258	– 1295
2142	21483	257742	12948705

4. Multiplication by 11, 101, 1001, ... or $(10^n + 1)$

Logic = Number × $(10^n + 1)$ = Number × 10^n + Number

Steps:

1. At the end of given number, put '*n*' zeros.
2. Add the original number to the number such formed.

Example:

259×11	1293×101	517×1001	21931×10001
2590	129300	517000	219310000
259	1293	517	21931
2849	130593	517517	219331931

5. Multiplication by a number consisting of 1's, i.e. 11, 111, 1111 ....

(*a*) Multiply a number by 11:

Steps: *(Solution from right to left)*

1. The unit's digit of the number is written down as the first part of the solution.
2. Last two digits are added to get the next part and successively 2 digits are added by moving to the left side by 1 digit each time to get the next parts.
3. First digit of the number becomes last part of the solution.

Note: One digit (right-hand side) is taken from each part and extra digit, if any, is to carry over to the immediate left column.

Example 1: 23 × 11 = ?

Sol. Part I: 3

Part II: 2 + 3 = 5

Example 2: 124 × 11 = ?

Sol. Part I: 4

Part II: 2 + 4 = 6

Part III: 2

∴ 23 × 11 = 253

Part III: 1 + 2 = 3

Part IV: 1

∴ 124 × 11 = 1364

Example 3: 2475 × 11 = ?

Sol. Part I: 5

Part II: 7 + 5 = 12

Part III: (4 + 7) + 1 = 12

Part IV: (2 + 4) + 1 = 7

Part V: 2

∴ 2475 × 11 = 27225

(b) Multiplication of a number by 111:

Steps: *(Start calculation from right hand side and proceed to left)*

1. The unit's digit of the number is written as first part of the solution.
2. Unit's and ten's digits are added to get the next part of the solution.
3. Last three digits are added to get next part and successively 3 digits are added moving to left side by 1 digit each time to get next parts.
4. First 2 digits are added to get second last part of the solution.
5. First digit of the given number becomes last part of the solution.

Note: One digit (right-hand side) is taken from each part and extra digit, if any, is to carry over to the left part.

Example 1: 413 × 111 = ?

Sol. Part I: 3

Part II: 1 + 3 = 4

Part III: 4 + 1 + 3 = 8

Part IV: 4 + 1 = 5

Part V: 4

∴ 413 × 111 = 45843

Example 2: 2643 × 111 = ?

Sol. Part I: 3

Part II: 4 + 3 = 7

Part III: 6 + 4 + 3 = 13

Part IV: (2 + 6 + 4) + 1 = 13

Part V: (2 + 6) + 1 = 9

Part VI: 2

∴ 2643 × 111 = 293373

Example 3: 27 × 111 = ?

Sol. We write the problem as 027 × 111 to convert 27 into 3-digit number.

Part I: 7

Part II: 2 + 7 = 9

Part III: 0 + 2 + 7 = 9

Part IV: 0 + 2 = 2

∴ 27 × 111 = 2997

(c) Multiplication of a number by 1111:

Steps: (*Start calculation from right hand side and proceed to left*)

1. The unit's digit of the number is written as first part of the solution.
2. Unit's and ten's digits are added to get the next part of the solution.
3. Last three digits are added to get the next part of the solution.
4. Last four digits are added to get the next part and successively digits are added in the group of four digits each time shifting to left side by one digit to get the next parts.

5. First 3 digits are added to get the next part of the solution.
6. First 2 digits are added to get second last part of the solution.
7. First digit of the given number becomes last part of the solution.

Note: One digit (right-hand side) is taken from each part and extra digit, if any, is to carry over to the left part.

Example 1: 24587 × 1111 = ?

Sol. Part I: 7 = 7
Part II: 8 + 7 = 15
Part III: (5 + 8 + 7) + 1 = 21
Part IV: (4 + 5 + 8 + 7) + 2 = 26
Part V: (2 + 4 + 5 + 8) + 2 = 21
Part VI: (2 + 4 + 5) + 2 = 13
Part VII: (2 + 4) + 1 = 7
Part VIII: 2 = 2
∴ 24587 × 1111 = 27316157

Example 2: 432 × 1111 = ?

Sol. We write the question as 0432 × 1111 to convert 432 into 4-digit number.
Part I: 2 = 2
Part II: 3 + 2 = 5
Part III: 4 + 3 + 2 = 9
Part IV: 0 + 4 + 3 + 2 = 9
Part V: 0 + 4 + 3 = 7
Part VI: 0 + 4 = 4
∴ 432 × 1111 = 479952

5. Multiplication with a multiple of 11, 111, ... *i.e.*, 22, 33, 222, etc.

Example 1: Find the value of 42 × 22.

Sol. 42 × 22 = (42 × 11) × 2 = 462 × 2 = 924

Example 2: Find the value of 132 × 333.

Sol. 132 × 333 = (132 × 111) × 3 = 14652 × 3 = 43956

6. Multiplication of two numbers when both numbers have 5 at unit's place:

Sol. *(from right to left)*:

1. First two digits will always be 5 × 5 = 25.
2. To get remaining digits: Multiply the remaining digits of the two numbers and add half of their sum to it.

Example:

$$25 \times 65 = (2 \times 6) + \frac{1}{2}\ (2 + 6)\ |\ 25 = 16\ 25$$

$$45 \times 65 = (4 \times 6) + \frac{1}{2}\ (4 + 6)\ |\ 25 = 29\ 25$$

$$125 \times 105 = (12 \times 10) + \frac{1}{2}\ (12 + 10)\ |\ 25 = 131\ 25$$

Note: *When in second part, we get 0.5 on solving, we write 75 instead of 25 in the first part.*

Example: $35 \times 65 = (3 \times 6) + \frac{1}{2}(3+6)|25 = 22.5|25 = 22\ 75$

$115 \times 85 = (11 \times 8) + \frac{1}{2}(11+8)|25 = 97.5|25 = 97\ 75$

7. Multiplication of two numbers when both numbers are either less than or greater than base *i.e.*, less than or greater than 10^n

Method: Let the two numbers are a_1 and a_2.

Where $a_1 = 10^n \pm b_1$ and $a_2 = 10^n \pm b_2$. Then $a_1 \times a_2 = (a_1 + a_2) | b_1 \times b_2$

Note: (*i*) Subtract 1 from the left-most digit of $(a_1 + a_2)$ or subtract the base from the sum of given numbers or subtract the sum of two deficient numbers from 10^n.

(*ii*) Right hand side part contains '*n*' digits.

(*a*) Both numbers smaller than the base (Deficit Method) :

Example 1: Find the value of 98 × 94.

Sol.

$98 = 100 - 2$

$94 = 100 - 6$

$b_1 \times b_2 = 2 \times 6 = 12$

$a_1 + a_2 = 98 + 94 = 192$

$\therefore a_1 \times a_2 = 192 | 12 = 9212$

Example 2: Find the value of 97 × 99.

Sol.

$97 = 100 - 3$

$99 = 100 - 1$

$b_1 \times b_2 = 3 \times 1 = 3$

$a_1 + a_2 = 97 + 99 = 196$

$\therefore a_1 \times a_2 = 196 | 03 = 9603$

More Examples:

$$\begin{array}{r} 995 \rightarrow 5 \\ \times\ 999 \rightarrow 1 \\ \hline 994 \rightarrow 5 \\ \hline = 994\ 005 \end{array} \qquad \begin{array}{r} 9991 \rightarrow 9 \\ \times\ 9997 \rightarrow 3 \\ \hline 9988 \rightarrow 27 \\ \hline = 9988\ 0027 \end{array} \qquad \begin{array}{r} 99981 \rightarrow 19 \\ \times\ 99993 \rightarrow 7 \\ \hline 99974 \rightarrow 133 \\ \hline = 99974\ 00133 \end{array}$$

Note: When the numbers to be multiplied have equal number of digits, left part can also be calculated by adding diagonally one of the number and difference of the other number from the base.

$999 - 5 = 994 \qquad 9997 - 9 = 9988 \qquad 99981 - 7 = 99974$

$995 - 1 = 994 \qquad 9991 - 3 = 9988 \qquad 99993 - 19 = 99974$

When the two numbers don't have equal digits:

Steps for multiplication:

(*i*) Add first digit (left-hand side) of the first number to first digit (left-hand side) of the second number and so on.

(*ii*) Right hand side part contains digits equal to zeros in the base of the smaller number and left-hand side part is simply the sum of the two numbers.

Examples:

$$\begin{array}{r} 999 \rightarrow 1 \\ \times\ 99 \rightarrow 1 \\ \hline 989 \rightarrow 1 \\ \hline = 989\ 01 \end{array} \qquad \begin{array}{r} 9994 \rightarrow 6 \\ \times\ 85 \rightarrow 15 \\ \hline 8494 \rightarrow 90 \\ \hline = 8494\ 90 \end{array} \qquad \begin{array}{r} 9989 \rightarrow 11 \\ \times\ 997 \rightarrow 3 \\ \hline 9959 \rightarrow 33 \\ \hline = 9959\ 033 \end{array}$$

(*b*) Both numbers greater than the base (Surplus Method):

Example 1: Find the value of 102 × 106.

Sol.

$102 = 100 + 2$

$106 = 100 + 6$

$b_1 \times b_2 = 2 \times 6 = 12$

$$a_1 + a_2 = 102 + 106 = 208$$
$$\therefore\ a_1 \times a_2 = 208 \mid 12 = 10812$$

Note: 102 + 106 = 208. We subtract 1 from '2' of 208, to get the left part.

More Examples:

$$\begin{array}{r} 102 \rightarrow 2 \\ \times\ 107 \rightarrow 7 \\ \hline 109 \rightarrow 14 \\ \hline = 109\ 14 \end{array} \qquad \begin{array}{r} 105 \rightarrow 5 \\ \times\ 103 \rightarrow 3 \\ \hline 108 \rightarrow 15 \\ \hline = 108\ 15 \end{array} \qquad \begin{array}{r} 107 \rightarrow 7 \\ \times\ 101 \rightarrow 1 \\ \hline 108 \rightarrow 7 \\ \hline = 108\ 07 \end{array} \qquad \begin{array}{r} 112 \rightarrow 12 \\ \times\ 114 \rightarrow 14 \\ \hline 126 \rightarrow 168 \\ \hline = 12768 \end{array}$$

Note: When the numbers to be multiplied have equal number of digits, left part can also be obtained by adding one of the number and difference of other number from the base.

$$\begin{array}{l} 102 + 7 = 109 \\ 107 + 2 = 109 \end{array} \qquad \begin{array}{l} 105 + 3 = 108 \\ 103 + 5 = 108 \end{array} \qquad \begin{array}{l} 107 + 1 = 108 \\ 101 + 7 = 108 \end{array} \qquad \begin{array}{l} 112 + 14 = 126 \\ 114 + 12 = 126 \end{array}$$

When the two numbers don't have equal number of digits:

$$\begin{array}{r} 1001 \rightarrow 1 \\ \times\ 12 \quad \rightarrow 2 \\ \hline 1201 \rightarrow 2 \\ \hline = 1201\ 2 \end{array} \qquad \begin{array}{r} 1009 \rightarrow 9 \\ \times 108 \quad \rightarrow 8 \\ \hline 1089 \rightarrow 72 \\ \hline = 1089\ 72 \end{array} \qquad \begin{array}{r} 1009 \quad \rightarrow 9 \\ \times 12 \qquad \rightarrow 2 \\ \hline 1209 \rightarrow 18 \\ \hline = 1210\ 8 \end{array}$$

Note: (*i*) Right-hand side part consists of digits one less than the digits in the smaller number, *i.e.* digits equal to the zeros in the base of the smaller number.

8. Multiplication of two numbers – one of which is greater than a base and the other is less than the same base:

Method: Let the two numbers are a_1 and a_2. Where $a_1 = 10^n + b_1$ and $a_2 = 10^n - b_2$.

Then $a_1 \times a_2 = (a_1 + a_2) \mid -(b_1 \times b_2)$

Note: (*i*) Subtract 1 from the extreme left digit of $(a_1 + a_2)$.

(*ii*) Put 0's at the end of the sum of the numbers so that zeros in the base of the sum (before subtraction) are equal to sum of zeros of the bases of the numbers multiplied.

(*iii*) The first digit (left-hand side) of the number less than base is added to the second digit of the number bigger than base and so on.

Example:

$$\begin{array}{r} 102 \rightarrow 2\,(+) \\ \times 8 \rightarrow 2\,(-) \\ \hline 82 \rightarrow 4\,(-) \\ \hline = 820 - 4 \\ = 816 \end{array} \qquad \begin{array}{r} 1001 \rightarrow 1\,(+) \\ \times\ 9 \rightarrow 1\,(-) \\ \hline 901 \rightarrow 1\,(-) \\ \hline = 9010 - 1 \\ = 9009 \end{array} \qquad \begin{array}{r} 105 \rightarrow 5\,(+) \\ \times 989 \rightarrow 11\,(-) \\ \hline 1039 \rightarrow 55\,(-) \\ \hline = 1039\ 00 - 55 \\ = 1038\ 45 \end{array}$$

$$\begin{array}{r} 1002 \rightarrow 2\,(+) \\ \times 999 \quad \rightarrow 1\,(-) \\ \hline 1001 \rightarrow 2\,(-) \\ \hline = 100100\ 0 - 2 \\ = 1000998 \end{array} \qquad \begin{array}{r} 103 \rightarrow 3\,(+) \\ \times\ 98 \rightarrow 2\,(-) \\ \hline 101 \rightarrow 6\,(-) \\ \hline = 10100 - 6 \\ = 10094 \end{array}$$

Note:

$$\begin{array}{l} 999 + 2 = 1001 \\ 1002 - 1 = 1001 \end{array} \qquad \begin{array}{l} 103 - 2 = 101 \\ 98 + 3 = 101 \end{array}$$

When the numbers to be multiplied have the same base, left part can be obtained by adding one number to the difference of the other number from the common base.

9. Multiplication of two numbers which are either more than or less than half of base, *i.e.* more than or less than 5, 50, 500, 5000, etc.:

Sol. *(from right to left)*:

1. Multiply the differences and write as right-hand side part.

2. Add the two numbers and deduct 5 from extreme left digit. Half the result to obtain the left-hand side part.

Note: Digits in the solution are equal to total digits of two numbers multiplied.

Example:

$$\begin{array}{r} 51 \to 1 \\ \times\, 53 \to 3 \\ \hline 54 \to 3 \\ \hline = 27\ 03 \end{array} \qquad \begin{array}{r} 56 \to 6 \\ \times\, 58 \to 8 \\ \hline 64 \to 48 \\ \hline = 32\ 48 \end{array} \qquad \begin{array}{r} 513 \to 13 \\ \times\, 511 \to 11 \\ \hline 524 \to 143 \\ \hline = 262\ 143 \end{array} \qquad \begin{array}{r} 52 \to 2 \\ \times\, 57 \to 7 \\ \hline 59 \to 14 \\ \hline = 29\ 64 \end{array}$$

Note: $\frac{1}{2} \times 54 = 27$ $\quad \frac{1}{2} \times 64 = 32$ $\quad \frac{1}{2} \times 524 = 262$ $\quad \frac{1}{2} \times 59 = 29.5$

Note: When the numbers to be multiplied have equal digits, we can use the following method also to get the left-hand side part:

51 + 3 = 54	56 + 8 = 64	513 + 11 = 524	52 + 7 = 59
53 + 1 = 54	58 + 6 = 64	511 + 13 = 524	57 + 2 = 59

When the numbers have unequal digits:

$$\begin{array}{r} 501 \to 1 \\ \times\ 51 \to 1 \\ \hline 511 \to 1 \\ \hline = 255.5 \mid 01 \\ = 255\ 51 \end{array} \qquad \begin{array}{r} 5004 \to 4 \\ \times\, 53 \to 3 \\ \hline 5304 \to 12 \\ \hline = 2652\ 12 \end{array} \qquad \begin{array}{r} 50003 \to 3 \\ \times\quad 502 \quad \to 2 \\ \hline 50203 \to 6 \\ \hline = 25101506 \end{array}$$

$\frac{1}{2} \times 511 = 255.5$ $\quad \frac{1}{2} \times 5304 = 2652$ $\quad \frac{1}{2} \times 50203 = 25101.5$

Example:

$$\begin{array}{r} 47 \to 3 \\ \times\, 49 \to 1 \\ \hline 46 \to 3 \\ \hline = 23\ 03 \end{array} \qquad \begin{array}{r} 474 \to 26 \\ \times\, 492 \to 8 \\ \hline 466 \to 208 \\ \hline = 233\ 208 \end{array} \qquad \begin{array}{r} 4997 \to 3 \\ \times\, 4996 \to 4 \\ \hline 4993 \to 12 \\ \hline = 2496\ 5012 \end{array}$$

Note: $\frac{1}{2} \times 46 = 23$ $\quad \frac{1}{2} \times 466 = 233$ $\quad \frac{1}{2} \times 4993 = 2496.5$

Note: When the numbers to be multiplied have equal number of digits, we can use the following method also to get the first part:

47 – 1 = 46	474 – 8 = 466	4997 – 4 = 4993
49 – 3 = 46	492 – 26 = 466	4996 – 3 = 4993

When the two numbers don't have equal digits:

$$\begin{array}{r} 498 \to 2 \\ \times\, 49 \to 1 \\ \hline 488 \to 2 \\ \hline = 244\ 02 \end{array} \qquad \begin{array}{r} 4993 \to 7 \\ \times\ 489 \to 11 \\ \hline 4883 \to 77 \\ \hline = 2441.5 \mid 077 \\ = 2441\ 577 \end{array}$$

Note: $\frac{1}{2} \times 488 = 244$ $\qquad$ $\frac{1}{2} \times 4883 = 2441.5$

10. Multiplication of two numbers – one of which is more than and the other is less than half of a base:

Steps:

1. Add the two numbers. Put 0's at the end of sum of the numbers so that digits of the sum are equal to total digits of the numbers multiplied. Deduct 5 from extreme left digit.
2. Half the result obtained in Step 1.
3. Multiply the differences and deduct it from the result obtained in Step 2.

Example:

$51 \to 1\,(+)$	$53 \to 3\,(+)$	$507 \to 7\,(+)$
$\times\, 47 \to 3\,(-)$	$\times\, 46 \to 4\,(-)$	$\times\, 498 \to 2\,(-)$
$48 \to 3\,(-)$	$49 \to 12\,(-)$	$505 \to 14\,(-)$
$= 24\ 00 - 3$	$= 24\ 50 - 12$	$= 252\ 500 - 14$
$= 23\ 97$	$= 24\ 38$	$= 252\ 486$

Note: $\frac{1}{2} \times 48 = 24$ $\qquad$ $\frac{1}{2} \times 49 = 24.5$ $\qquad$ $\frac{1}{2} \times 505 = 252.5$

Note: When the numbers to be multiplied have equal number of digits, we can use the following method also to get the left hand side part:

$51 - 3 = 48$	$53 - 4 = 49$	$507 - 2 = 505$
$47 + 1 = 48$	$46 + 3 = 49$	$498 + 7 = 505$

When the numbers have unequal digits:

$503 \to 3\,(+)$	$49988 \to 12\,(-)$	$481 \to 19\,(-)$
$\times\, 49 \to 1\,(-)$	$\times\, 511 \to 11\,(+)$	$\times\, 52 \to 2\,(+)$
$493 \to 3\,(-)$	$51088 \to 132\,(-)$	$501 \to 38\,(-)$
$= 246\ 50 - 3$	$= 25544\ 000 - 132$	$= 250\ 50 - 38$
$= 246\ 47$	$= 25543\ 868$	$= 250\ 12$

11. Multiplication of two numbers when their base is x times of 10^n:

Solution (from right to left):

1. Multiply the difference from base and write down as the right-hand side part.
2. Add first number and difference from base for the second number, or second number and difference from base for first number, to get the left-hand side part. Multiply the result with 'x'.

Example:

Base $300 = 3 \times 100$	Base $40 = 4 \times 10$	Base $70 = 7 \times 10$
$304 \to 4$	$39 \to 1\,(-)$	$72 \to 2$
$\times\, 305 \to 5$	$\times\, 37 \to 3\,(-)$	$\times\, 76 \to 6$
$309 \to 20$	$36 \to 3$	$78 \to 12$
$= 927\ 20$	$= 144\ 3$	$= 546 \mid 12 = 5472$

Note:

$304 + 5 = 309$	$39 - 3 = 36$	$72 + 6 = 78$
$305 + 4 = 309$	$37 - 1 = 36$	$76 + 2 = 78$
$3 \times 309 = 927$	$4 \times 36 = 144$	$7 \times 78 = 546$

Base $20 = 2 \times 10$	Base $30 = 3 \times 10$	Base $60 = 6 \times 10$
$21 \to 1\,(+)$	$32 \to 2\,(+)$	$63 \to 3\,(+)$
$\times\ 22 \to 2\,(+)$	$\times\ 34 \to 4\,(+)$	$\times\ 58 \to 2\,(-)$
$22 + 1 = 23$	$32 + 4 = 36$	$63 - 2 = 61$
$20 \times 23 = 460$	$30 \times 36 = 1080$	$60 \times 61 = 3660$
$1 \times 2 = 2$	$2 \times 4 = 8$	$(+3) \times (-2) = (-6)$
$= 460 + 2 = 462$	$= 1080 + 8 = 1088$	$= 3660 - 6 = 3654$

12. Multiplication of two numbers – one of which is more (or less) than a base and the other is more (or less) than half of same base:

Method: Let the numbers are a_1 and a_2. Where $a_1 = 10^n \pm b_1$ and $a_2 = \frac{10^n}{2} \pm b_2$

Then $a_1 \times a_2 = a_2 \pm \frac{1}{2} b_1 \mid b_1 \times b_2$ **Note :- Right-hand side part contains *n* digits.**

Example:

More than base:

$102 \to 2$	$103 \to 3$	$1012 \to 12$
$\times\ 51 \to 1$	$\times\ 57 \to 7$	$\times\ 508 \to 8$
$51 + \frac{1}{2} \times 2 \to 02$	$57 + \frac{1}{2} \times 3 \to 21$	$508 + \frac{1}{2} \times 12 \to 096$
$= 52\ 02$	$= 58.5 \mid 21 = 58\ 71$	$= 514\ 096$

Less than base:

$96 \to 4$	$999 \to 1$	$9986 \to 14$
$\times\ 48 \to 2$	$\times\ 496 \to 4$	$\times\ 4994 \to 6$
$48 - \frac{1}{2} \times 4 \to 8$	$496 - \frac{1}{2} \times 1 \to 4$	$4994 - \frac{1}{2} \times 14 \to 84$
$= 46\ 08$	$= 495.5 \mid 004 = 495\ 504$	$= 4987\ 0084$

13. Multiplication of two numbers – one of which is more (or less) than a base and other is less (or more) than half of same base:

Method:

Let the numbers are a_1 and a_2. Where $a_1 = 10^n \pm b_1$ and $a_2 = \frac{10^n}{2} \mp b_2$

Then, $a_1 \times a_2 = a_2 \pm \frac{1}{2} b_1 \mid (-)\, b_1 \times b_2$ **Note:** Right-hand side part contains *n* digits.

Example:

$104 \to 4\,(+)$	$1018 \to 18\,(+)$
$\times\ 47 \to 3\,(-)$	$\times\ 497 \to 3\,(-)$
$47 + \frac{1}{2} \times 4 \to 12\,(-)$	$497 + \frac{1}{2} \times 18 \to 54\,(-)$
$= 49\ 00 - 12$	$= 506\ 000 - 54$
$= 48\ 88$	$= 505\ 946$

$$\begin{array}{r} 98 \rightarrow 2\,(-) \\ \times\ 53 \rightarrow 3\,(+) \\ \hline 53 - \frac{1}{2} \times 2 \rightarrow 6\,(-) \\ \hline = 52\,00 - 6 \\ = 51\,94 \end{array} \qquad \begin{array}{r} 993 \rightarrow 7\,(-) \\ \times\ 512 \rightarrow 12\,(+) \\ \hline 512 - \frac{1}{2} \times 7 \rightarrow 84\,(-) \\ \hline = 508\,500 - 84 \\ = 508\,416 \end{array}$$

14. Multiplication of two numbers whose unit's digits total to 10 and the remaining digits are same:

Steps: 1. Multiply the unit's digits of the given numbers to get last 2 digits of the solution.

2. Multiply the common digit in the two numbers with the next higher number to get the remaining digits of the solution.

Example: $62 \times 68 = 6 \times 7 \mid 2 \times 8 = 42\ 16$

$94 \times 96 = 9 \times 10 \mid 4 \times 6 = 90\ 24$

$123 \times 127 = 12 \times 13 \mid 3 \times 7 = 156\ 21$

$31 \times 39 = 3 \times 4 \mid 1 \times 9 = 12\ 09$

Note: $44 \times 48 = 44 \times 46 + 44 \times 2 = 2024 + 88 = 2112$

$64 \times 63 = 64 \times 66 - 64 \times 3 = 4224 - 192 = 4032$

15. Multiplication of two numbers whose last two digits total to 100 and remaining digits are same:

Steps: 1. Multiply the last two digits of both the numbers to get last 4 digits of the solution.

2. Multiply the common digits in the two numbers with the next higher number to get the remaining digits of the solution.

Examples: $301 \times 399 = 3 \times 4 \mid 01 \times 99 = 12\ 0099$

$204 \times 296 = 2 \times 3 \mid 04 \times 96 = 6\ 0384$

$411 \times 489 = 4 \times 5 \mid 11 \times 89 = 20\ 0979$

$1103 \times 1197 = 11 \times 12 \mid 03 \times 97 = 132\ 0291$

$1235 \times 1265 = 12 \times 13 \mid 35 \times 65 = 156\ 2275$

16. Multiplication of two numbers whose last three digits total to 1000 and remaining digits are same:

Steps: 1. Multiply the last three digits of both the numbers to get last 6 digits of the solution.

2. Multiply the common digits in the two numbers with the next higher number to get remaining digits of the solution.

Examples: $2002 \times 2998 = 2 \times 3 \mid 002 \times 998 = 6\ 001996$

$9001 \times 9999 = 9 \times 10 \mid 001 \times 999 = 90\ 000999$

$13006 \times 13994 = 13 \times 14 \mid 006 \times 994 = 182\ 005964$

17. Multiplication of two numbers whose ten's digits total to 10 and the unit's digits are same:

Steps: 1. Multiply the unit's digits of the given numbers to get last two digits of the solution.

2. Multiply the ten's digits of the two numbers and add common unit's digit to it to get remaining digits of the solution.

Examples: $43 \times 63 = 4 \times 6 + 3 \mid 3^2 = 27\ 09$

$27 \times 87 = 2 \times 8 + 7 \mid 7^2 = 23\ 49$

$98 \times 18 = 9 \times 1 + 8 \mid 8^2 = 17\ 64$

18. Multiplication of two numbers sum of whose units' digits is 5 and the remaining digits are same:

Steps: 1. Multiply the unit's digits of the given numbers to get last 2 digits of the solution.

2. Multiply the common digits in the two numbers and add half of the common unit's digit to get the remaining digits.

Examples: $22 \times 23 = 2 \times 2 + \frac{1}{2} \times 2 \mid 2 \times 3 = 5\ 06$

$803 \times 802 = 80 \times 80 + \frac{1}{2} \times 80 \mid 3 \times 2 = 6440\ 06$

$11 \times 14 = 1 \times 1 + \frac{1}{2} \times 1 \mid 1 \times 4 = 1.5 \mid 04 = 1\ 54$

$52 \times 53 = 5 \times 5 + \frac{1}{2} \times 5 \mid 2 \times 3 = 27.5 \mid 06 = 27\ 56$

19. Multiplication of two numbers whose last two digits total to 20, 25, 40, 50, 60, 75, 80, 90, 110, 120, 125, 130, 140, 150, etc. and the remaining digits are same:

Steps: 1. To obtain the last 4 digits of the solution, multiply the last two digits of the given numbers.

2. To obtain remaining digits of the solution:

(*a*) Find the sum of the numbers formed by last 2 digits of the given numbers.

(*b*) Divide the sum obtained in step (*a*) by 100.

(*c*) Multiply the common digits of the two numbers.

(*d*) Multiply the common digit with the fraction obtained in step (*b*).

(*e*) Add the results obtained at Steps (*c*) and (*d*) to get the left-hand side part of the solution.

Examples:

1. $512 \times 508 = 5 \times 5 + \frac{1}{5} \times 5 \mid 12 \times 08 = 26\ 0096$ (**Note:** $12 + 8 = 20$; And $\frac{20}{100} = \frac{1}{5}$)

2. $811 \times 814 = 8 \times 8 + \frac{1}{4} \times 8 \mid 11 \times 14 = 66\ 0154$ (**Note:** $11 + 14 = 25$; And $\frac{25}{100} = \frac{1}{4}$)

3. $1016 \times 1024 = 10 \times 10 + \frac{2}{5} \times 10 \mid 16 \times 24 = 104\ 0384$

   (**Note:** $16 + 24 = 40$; And $\frac{40}{100} = \frac{2}{5}$)

4. $604 \times 646 = 6 \times 6 + \frac{1}{2} \times 6 \mid 04 \times 46 = 39\ 0184$ (**Note:** $4 + 46 = 50$; And $= \frac{50}{100} = \frac{1}{2}$)

5. $1011 \times 1099 = 10 \times 10 + \frac{11}{10} \times 10 \mid 11 \times 99 = 111\ 1089$

   (**Note:** $11 + 99 = 110$; And $= \frac{110}{100} = \frac{11}{10}$)

6. $851 \times 899 = 8 \times 8 + \frac{3}{2} \times 8 \mid 51 \times 99 = 76\ 5049$ (**Note:** $51 + 99 = 150$; And $\frac{150}{100} = \frac{3}{2}$)

Note: When the result in left part is in decimals:

1. $102 \times 108 = 1 \times 1 + \frac{1}{10} \times 1 \mid 02 \times 08 = 1.1 \mid 0016 = 1\ 1016$

   (**Note:** $2 + 8 = 10$; And $\frac{10}{100} = \frac{1}{10}$)

2. $303 \times 327 = 3 \times 3 + \frac{3}{10} \times 3 \mid 03 \times 27 = 9.9 \mid 0081 = 9\ 9081$

(**Note:** $3 + 27 = 30$; And $= \frac{30}{100} = \frac{3}{10}$)

20. Multiplication of two numbers when both numbers are either more or less than 50:

Method: $(50 \pm a) \times (50 \pm b) = 25 \pm \frac{1}{2}(a + b) \mid ab$

Note: 2 digits are taken from each part.

Examples: 1. Find the value of 53×57.

Sol. Here $a = 3, b = 7, a + b = 10$ $\therefore 53 \times 57 = 25 + \frac{1}{2} \times 10 \mid 3 \times 7 = 30\ 21$

2. Find the value of 58×62.

Sol. Here $a = 8, b = 12, a + b = 20$ $\therefore 58 \times 62 = 25 + \frac{1}{2} \times 20 \mid 8 \times 12 = 35\ 96$

3. Find the value of 51×52.

Sol. Here $a = 1, b = 2, a + b = 3$ $\therefore 51 \times 52 = 25 + \frac{1}{2} \times 3 \mid 1 \times 2 = 26.5 \mid 02 = 26\ 52$

4. Find the value of 48×46.

Sol. Here $a = 2, b = 4$ and $a + b = 6$ $\therefore 48 \times 46 = 25 - \frac{1}{2} \times 6 \mid 2 \times 4 = 22\ 08$

5. Find the value of 41×47.

Sol. Here $a = 9, b = 3$ and $a + b = 12$ $\therefore 41 \times 47 = 25 - \frac{1}{2} \times 12 \mid 9 \times 3 = 19\ 27$

6. Find the value of 42×43.

Sol. Here $a = 8, b = 7$ and $a + b = 15$ $\therefore 42 \times 43 = 25 - \frac{1}{2} \times 15 \mid 8 \times 7 = 17.5 \mid 56 = 18\ 06$

21. Multiplication of two numbers – one of which is bigger than 50 and the other is less than 50:

Method: If the numbers are $(50 + a)$ and $(50 - b)$.

Then $(50 + a) \times (50 - b) = 25 + \frac{1}{2}(a - b) \mid (-ab)$

Note: 2 digits are taken from each part.

Example 1. Find the value of 53×49.

Sol. Here $a = 3, b = 1$, and $a - b = 2$

$\therefore 53 \times 49 = 25 + \frac{1}{2} \times (3 - 1) \mid -3 = 2600 - 3 = 25\ 97$

2. Find the value of 52×49.

Sol. Here $a = 2, b = 1$, and $a - b = 1$

$\therefore 52 \times 49 = 25 + \frac{1}{2} \times (2 - 1) \mid -2 = 2550 - 2 = 25\ 48$

3. Find the value of 51×47.

Sol. Here $a = 1, b = 3$, and $a - b = -2$

$\therefore 51 \times 47 = 25 + \frac{1}{2} \times (1 - 3) \mid -3 = 2400 - 3 = 23\ 97$

4. Find the value of 52 × 43.

Solution: Here $a = 2$, $b = 7$, and $a - b = -5$

$\therefore 52 \times 43 = 25 + \frac{1}{2} \times (2 - 7) \mid - 14 = 2250 - 14 = 22\ 36$

22. Multiplication of two numbers one of which is greater than and other number is less than 50 and difference from 50 is same in both the cases:

Method: Let the numbers are $(50 + a)$ and $(50 - a)$. Then $(50 - a) \times (50 + a) = 2500 - a^2$

Examples: 1. Find the value of 51 × 49.

Sol. Here common number is 50 and difference is 1.

$\therefore 51 \times 49 = 2500 - 1^2 = 2500 - 1 = 2499$

2. Find the value of 55 × 45.

Sol. Here common number is 50 and difference is 5.

$\therefore 55 \times 45 = 2500 - 5^2 = 2500 - 25 = 2475$

3. Find the value of 58 × 42.

Sol. Here common number is 50 and difference is 8.

$\therefore 58 \times 42 = 2500 - 8^2 = 2500 - 64 = 2436$

23. Multiplication of two numbers one of which is greater than and the other is less than a common number and difference from that common number is same in both the cases:

Method: Let the numbers are $(x + a)$ and $(x - a)$. Then $(x + a) \times (x - a) = x^2 - a^2$

Note: Common number must be the number whose square can be calculated easily.

Example 1. Find the value of 42 × 38.

Sol. Here common number is 40 and difference is 2. $\therefore 42 \times 38 = 40^2 - 2^2 = 1600 - 4 = 1596$

2. Find the value of 53 × 47.

Sol. Here common number is 50 and difference is 3. $\therefore 53 \times 47 = 50^2 - 3^2 = 2500 - 9 = 2491$

3. Find the value of 62 × 58.

Sol. Here common number is 60 and difference is 2. $\therefore 62 \times 58 = 60^2 - 2^2 = 3600 - 4 = 3596$

4. Find the value of 94 × 86.

Sol. Here common number is 90 and difference is 4. $\therefore 94 \times 86 = 90^2 - 4^2 = 8100 - 16 = 8084$

5. Find the value of 224 × 216.

Sol. Here common number is 220 and difference is 4. $\therefore 224 \times 216 = 220^2 - 4^2 = 48400 - 16 = 48384$

6. Find the value of 42 × 39.

Sol. $42 \times 39 = 42 \times (38 + 1) = 42 \times 38 + 42 = 40^2 - 2^2 + 42 = 1600 - 4 + 42 = 1638$

Alt: $42 \times 39 = (41 + 1) \times 39 = 41 \times 39 + 39 = 40^2 - 1^2 + 39 = 1600 - 1 + 39 = 1638$

7. Find the value of 77 × 82.

Sol. $77 \times 82 = (78 - 1) \times 82 = 78 \times 82 - 82 = 80^2 - 2^2 - 82 = 6400 - 4 - 82 = 6314$

Alt: $77 \times 82 = 77 \times (83 - 1) = 77 \times 83 - 77 = 80^2 - 3^2 - 77 = 6400 - 9 - 77 = 6314$

24. $(x - 1)(y - 1) = xy - (x + y) + 1$

Example: $79 \times 59 = (80 - 1) \times (60 - 1) = 80 \times 60 - (80 + 60) + 1 = 4800 - 140 + 1 = 4661$

25. Multiplication of two numbers with any number of digits:

$$\begin{array}{cccccc} & A & B & C & D & \\ \times & P & Q & R & S & \\ \hline \end{array}$$

$$AP \mid AQ + BP \mid AR + BQ + CP \mid AS + BR + CQ + DP \mid BS + CR + DQ \mid CS + DR \mid DS$$

Note: 1. We start calculation from right hand side and move to left hand side.

2. One digit is taken from each part of the result. Extra digits (left hand side), if any, is carried over and added to the result obtained in immediate left part.

Example 1: 2538
$\underline{\times 3176}$

First part = 8 × 6 = 48 (4 is carry over)

Second part = 3 × 6 + 8 × 7 = 18 + 56 = 74   74 + 4 = 78 (7 is carry over)

Third part = 5 × 6 + 3 × 7 + 8 × 1 = 30 + 21 + 8 = 59   59 + 7 = 66 (6 is carry over)

Fourth part = 2 × 6 + 5 × 7 + 3 × 1 + 8 × 3 = 12 + 35 + 3 + 24 = 74;   74 + 6 = 80 (8 is carry over)

Fifth part = 2 × 7 + 5 × 1 + 3 × 3 = 14 + 5 + 9 = 28   28 + 8 = 36 (3 is carry over)

Sixth part = 2 × 1 + 5 × 3 = 2 + 15 = 17   17 + 3 = 20 (2 is carry over)

Seventh part = 2 × 3 = 6   6 + 2 = 8

∴ 2538 × 3176 = 8060688

Example 2: 427
$\underline{\times 46}$

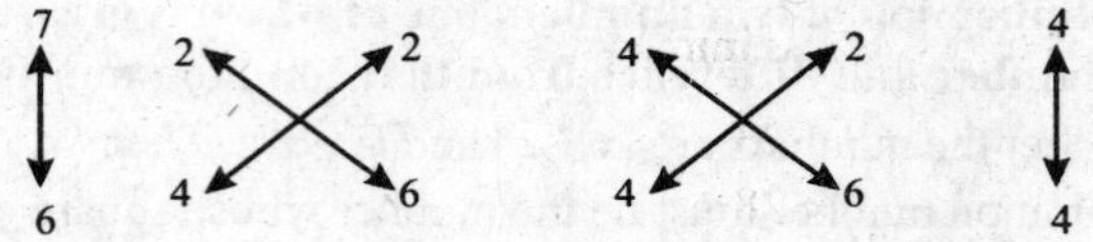

First part = 7 × 6 = 42 (4 is carry over)

Second part = 2 × 6 + 7 × 4 = 12 + 28 = 40   40 + 4 = 44 (4 is carry over)

Third part = 4 × 6 + 2 × 4 = 24 + 8 = 32   32 + 4 = 36 (3 is carry over)

Fourth part = 4 × 4 = 16   16 + 3 = 19

∴ 427 × 26 = 19642

Example 3: 3562
$\underline{\times 123}$

First part = 2 × 3 = 6

Second part = 6 × 3 + 2 × 2 = 18 + 4 = 22 (2 is carry over)

Third part = 5 × 3 + 6 × 2 + 2 × 1 = 15 + 12 + 2 = 29   29 + 2 = 31 (3 is carry over)

Fourth part = 3 × 3 + 5 × 2 + 6 × 1 = 9 + 10 + 6 = 25   25 + 3 = 28 (2 is carry over)

Fifth part = 3 × 2 + 5 × 1 = 6 + 5 = 11   11 + 2 = 13 (1 is carry over)

Sixth part = 3 × 1 = 3   3 + 1 = 4

∴ 3562 × 123 = 438126

26. Multiplication of two 2-digit numbers both having same unit place digit.

$$\begin{array}{ccc} & A & C \\ \times & B & C \\ \hline \multicolumn{3}{c}{A \times B \mid (A+B) \times C \mid C \times C} \\ \hline \end{array}$$

Note: 1. We start calculation from right hand side and move to left hand side.

2. One digit is taken from each part. Extra digit (left hand side), if any, is carry over and added to the result obtained in the immediate left part.

Examples:

$$\begin{array}{r} 13 \\ \times 23 \\ \hline 1\times2\,|\,(1+2)\,3\,|\,3\times3 \\ \hline =299 \end{array} \qquad \begin{array}{r} 24 \\ \times 34 \\ \hline 2\times3\,|\,(2+3)\,4\,|\,4\times4 \\ \hline =6\,|\,20\,|\,16=816 \end{array} \qquad \begin{array}{r} 58 \\ \times 98 \\ \hline 5\times9\,|\,(5+9)\,8\,|\,8\times8 \\ \hline =45\,|\,112\,|\,64=5684 \end{array}$$

Note: The above method can be used to find multiplication of two 3-digit numbers where last two digits are same in both the numbers.

Example: 1. 312 × 512 = 3 × 5 | (3 + 5) × 12 | 12 × 12 = 15 | 96 | 144 = 15 97 44

2. 213 × 813 = 2 × 8 | (2 + 8) × 13 | 13 × 13 = 16 | 130 | 169 = 17 31 69

3. 102 × 202 = 1 × 2 | (1 + 2) × 2 | 2 × 2 = 2 | 6 | 4 = 2 06 04

Note: In these examples two digits are taken from each part.

27. Multiplication of two 2-digit numbers both having same digit in ten's place.

$$\begin{array}{r} A \quad B \\ \times \quad A \quad C \\ \hline A\times A\,|\,A\times(B+C)\,|\,B\times C \\ \hline \end{array}$$

Note: 1. We start calculation from right hand side and move to left hand side.

2. One digit is taken from each part. Extra digit (left hand side), if any, is carry over and added to the result obtained in the immediate left part.

Example:

$$\begin{array}{r} 23 \\ \times 28 \\ \hline 2\times2\,|\,2\,(3+8)\,|\,3\times8 \\ \hline =4\,|\,22\,|\,24=644 \end{array} \qquad \begin{array}{r} 57 \\ \times 56 \\ \hline 5\times5\,|\,5\,(7+6)\,|\,7\times6 \\ \hline =25\,|\,65\,|\,42=3192 \end{array}$$

Note: The above method can be used to multiply two 3-digit numbers where first two digits in both the numbers are same.

Example 1. 113 × 115 = 11 × 11 | (3 + 5) × 11 | 3 × 5 = 121 | 88 | 15 = 129 9 5

2. 254 × 256 = 25 × 25 | (4 + 6) × 25 | 4 × 6 = 625 | 250 | 24 = 650 2 4

Note: One digit is taken from each part.

28. Verification of the correctness of the multiplication of two or more numbers:

Digit sum method or casting out nines Method:

1. Find sum of digits of each of the number to be multiplied. If sum of digits consists of more than one digit, keep on adding the digits of the results till we get the sum in single digit.

2. Multiply the sum of digits of the respective numbers and then find sum of digits of the product.

3. Find sum of the digits of the multiplication result of the given numbers.

4. Compare the results obtained in Step 2 and Step 3. If the two results are same the product is correct.

Example 1:

$$\begin{array}{r} 25 \\ \times 35 \\ \hline 875 \\ \hline \end{array}$$

Step 1: 2 + 5 = 7 and 3 + 5 = 8

Step 2: 7 × 8 = 56 and sum of digits = 5 + 6 = 11; 1 + 1 = 2

Step 3: 8 + 7 + 5 = 20; 2 + 0 = 2

Step 4: Since the two results are same, *i.e.* 2, the solution is correct.

Example 2:

$$\begin{array}{r} 87 \\ \times 71 \\ \hline 6177 \\ \hline \end{array}$$

Step 1: 8 + 7 = 15; 1 + 5 = 6; and 7 + 1 = 8

Step 2: 6 × 8 = 48 and sum of digits = 4 + 8 = 12; 1 + 2 = 3

Step 3: 6 + 1 + 7 + 7 = 21; 2 + 1 = 3

Step 4: Since the two results are same i.e. 3, the solution is correct.

Important Notes: (*i*) If the results obtained in Step 2 and Step 3 are not same, the product is definitely incorrect but opposite of this is not always true. In other words, equality of two results does not necessarily mean that the product is correct. For example, if in the above example, we change the product as 6267 instead of 6177, then the sum of digits of the product will remain same and hence will tally but 6267 is the wrong answer.

(*ii*) If two or more numbers are multiplied together, sum of digits of any one or more of the numbers is 9, then sum of the digits of the product of the numbers will always be 9.

Example 3:

$$\begin{array}{r} 18 \\ \times 41 \\ \hline 738 \\ \hline \end{array}$$

Step 1: 1 + 8 = 9 and 4 + 1 = 5

Step 2: Since sum of digits of one of the numbers is 9. This result will always be 9. However, we can verify this result also. 9 × 5 = 45 and sum of digits = 4 + 5 = 9

Step 3: 7 + 3 + 8 = 18; 1 + 8 = 9

Step 4: Since the two results are same i.e. 9, the solution is correct.

29. Multiplication of two mixed numbers whose whole numbers are same and sum of the fractions is 1, $\frac{1}{2}, \frac{1}{4}$, etc.

Steps: 1. Multiply the common whole number with the next higher number, if sum of the fractions is 1.

(If the sum is $\frac{1}{2}$ or $\frac{1}{4}$, etc., square the common number and add to it the product of common whole number and sum of the fractions.)

2. Multiply the fractions.

Examples: $8\frac{1}{4} \times 8\frac{3}{4} = 8 \times 9 \,|\, \frac{1}{4} \times \frac{3}{4} = 72\frac{3}{16}$ (**Note:** sum of the fractions is 1)

$13\frac{2}{11} \times 13\frac{9}{11} = 13 \times 14 \,|\, \frac{2}{11} \times \frac{9}{11} = 182\frac{18}{121}$ (**Note:** sum of the fractions is 1)

$6\frac{3}{8} \times 6\frac{1}{8} = 6 \times 6 + \frac{1}{2} \times 6 \,|\, \frac{3}{8} \times \frac{1}{8} = 39\frac{3}{64}$ (**Note:** sum of the fractions is $\frac{3}{8} + \frac{1}{8} = \frac{1}{2}$)

$4\frac{1}{6} \times 4\frac{1}{12} = 4 \times 4 + \frac{1}{4} \times 4 \,|\, \frac{1}{6} \times \frac{1}{12} = 17\frac{1}{72}$ (**Note:** sum of the fractions is $\frac{1}{6} + \frac{1}{12} = \frac{1}{4}$)

30. Multiplication of Mixed Numbers:

Base: $(x + a) \times (y + b) = xy + ab + (xb + ay)$

i.e. Product of whole numbers

+ Product of fractions + Sum of cross multiplication of whole numbers and fractions

Example 1: $13\frac{1}{4}\times 12\frac{1}{5}$

Sol. $13\times 12+\frac{1}{4}\times\frac{1}{5}+12\times\frac{1}{4}+13\times\frac{1}{5}=156+\frac{1}{20}+3+2\frac{3}{5}$

$=156+3+2+\frac{1}{20}+\frac{3}{5}=161+\frac{1}{20}+\frac{12}{20}=161\frac{13}{20}$

Example 2: $25\frac{1}{6}\times 13\frac{1}{4}$

Sol. $25\times 13+\frac{1}{6}\times\frac{1}{4}+25\times\frac{1}{4}+13\times\frac{1}{6}=325+\frac{1}{24}+6\frac{1}{4}+2\frac{1}{6}=325+6+2+\frac{1}{24}+\frac{1}{4}+\frac{1}{6}=333\frac{11}{24}$

31. Multiplication using algebra:

(i) $x\times y+x\times z=x\times(y+z)$

Example 1: $657\times 618+657\times 382=657\times(618+382)=657\times 1000=657000$

(ii) $x\times y-x\times z=x\times(y-z)$

Example 1: $349\times 649-349\times 549=349\times(649-549)=349\times 100=34900$

(iii) $x\times x-y\times y=(x+y)\times(x-y)$

Example 1: $67\times 67-33\times 33=(67+33)\times(67-33)=100\times 34=3400$

(iv) $x\times x+y\times y=\frac{1}{2}[(x+y)^2+(x-y)^2]$

Example 1: $112\times 112+88\times 88=\frac{1}{2}[(112+88)^2+(112-88)^2]=\frac{1}{2}[200^2+24^2]$

$=\frac{1}{2}\times(40000+576)=\frac{1}{2}\times 40576=20288$

32. By re-arranging the numbers:

Example 1 : $125\times 43\times 8=?$

Sol. $125\times 43\times 8=(125\times 8)\times 43=1000\times 43=43000$

Example 2 : $28\times 225=?$

Sol. $(28\times 25)\times 9=700\times 9=6300$

Example 3 : $4\frac{1}{3}\times 1.80=?$

Sol. $4\frac{1}{3}\times 1.80=4\times 1.80+\frac{1}{3}\times 1.80=7.20+0.60=7.80$

Example 4 : $10.25\times 4.80=?$

Sol. We know that $0.25=\frac{1}{4}$

$\therefore 10.25\times 4.80=10\frac{1}{4}\times 4.80=10\times 4.80+\frac{1}{4}\times 4.80=48+1.20=49.20$

DIVISION OF NUMBERS

Division by 5:

Method: Multiply the number by 2 and divide the result by 10. (Note: $10\div 2=5$)

Example: $135\div 5=(135\times 2)\div 10=270\div 10=27$

$570\div 5=(570\times 2)\div 10=57\times 2=114$

$713\div 5=(713\times 2)\div 10=1426\div 10=142.6$

Short-cut Method:

1. Divide last digit of the number by 5.
2. Multiply the number formed by the remaining digits by 2.
3. Add the results obtained in Step 1 and 2.

Example: $135 \div 5 = (13 \times 2) + (5 \div 5) = 26 + 1 = 27$

$570 \div 5 = (57 \times 2) + (0 \div 5) = 114 + 0 = 114$

$713 \div 5 = (71 \times 2) + (3 \div 5) = 142 + 0.6 = 142.6$

Division by 25:

Method: Multiply the number by 4 and divide the result by 100. (Note: $100 \div 4 = 25$)

Example: $725 \div 25 = (725 \times 4) \div 100 = 2900 \div 100 = 29$

Short-cut Method:

1. Divide last 2 digits of the number by 25.
2. Multiply the number formed by the remaining digits by 4.
3. Add the results obtained in Step 1 and 2.

Example: $725 \div 25 = 7 \times 4 + (25 \div 25) = 28 + 1 = 29$

$3275 \div 25 = 32 \times 4 + (75 \div 25) = 128 + 3 = 131$

Division by 125:

***Method:* Multiply the number by 8 and divide the result by 1000. (Note: $1000 \div 8 = 125$)**

Example: $8375 \div 125 = (8375 \times 8) \div 1000 = 67000 \div 1000 = 67$

$9250 \div 125 = (9250 \times 8) \div 1000 = 74000 \div 1000 = 74$

Short-cut Method:

1. Divide last 3 digits of the number by 125.
2. Multiply the number formed by the remaining digits by 8.
3. Add the results obtained in Step 1 and 2.

Example: $8375 \div 125 = (8 \times 8) + (375 \div 125) = 64 + 3 = 67$

$9250 \div 125 = (9 \times 8) + (250 \div 125) = 72 + 2 = 74$

Division by 9:

Method: Starting from left-hand side, the first digit of the dividend is written as quotient. Adding first digit of quotient and second digit of dividend, we get second digit of quotient. Adding second digit of quotient and third digit of dividend, we get third digit of the quotient and so on.

Note: If remainder at any stage is equal to or more than 9 (*i.e.* the divisor), divide the remainder by 9, add this quotient to the previous quotient and the new remainder is the revised remainder.

Example: 21032 ÷ 9 = 2 | 2 + 1 | 3 + 0 | 3 + 3 | 6 + 2 = 2336 | 8 (Remainder)

203101 ÷ 9 = 2 | 2 + 0 | 2 + 3 | 5 + 1 | 6 + 0 | 6 + 1 = 22566 | 7 (Remainder)

11036343 ÷ 9 = 1 | 1 + 1 | 2 + 0 | 2 + 3 (+1) | 5 + 6 (–9) | 2 + 3 (+1)| 5 + 4 (–9) | 0 + 3 = 1226260 | 3 (Remainder)

Division by 99, 999, 9999, ..., 99999....n, etc.:

Method:

1. Divide the dividend into two parts, right hand side part containing digits equal to digits in the divisor. Write remaining digits as quotient.

2. Divide the quotient such obtained into two parts in the same manner. Repeat this process till the final quotient is 0.

3. Add all quotients and the remainders separately.

4. If the sum of remainders is greater than the divisor, divide the same into two parts, as explained in step 1.

5. Add the quotients in Step 3 and 4. Remainder in Step 4 is final remainder.

Example: $687931 \div 99$ | $67878 \div 99$ | $584478 \div 999$

Solution:

	Quotient	Remainder	Quotient	Remainder	Quotient	Remainder
	6879	31	678	78	584	478
	68	79	6	78	0	584
	0	68	0	6		
Step 3. Total	6947	178	684	162	584	1062
Step 4.	1	78	1	62	1	062
	0	1	0	1	0	1
Final	6948	79	685	63	585	063

DIVISION OF MIXED NUMBERS

The number is divided into two parts and then divided by the divisor.

Example 1: $42\frac{2}{3} \div 6 = 42\frac{2}{3} \times \frac{1}{6} = 42 \times \frac{1}{6} + \frac{2}{3} \times \frac{1}{6} = 7 + \frac{1}{9} = 7\frac{1}{9}$.

Example 2: $25\frac{1}{3} \div 11 = 25\frac{1}{3} \times \frac{1}{11} = 25 \times \frac{1}{11} + \frac{1}{3} \times \frac{1}{11} = 2\frac{3}{11} + \frac{1}{33} = 2 + \frac{3 \times 3 + 1}{33} = 2\frac{10}{33}$.

SOLVED EXERCISE

1. Find unit's digit in $217 \times 818 \times 193$.

(*a*) 0 (*b*) 1 (*c*) 7 (*d*) 8

Sol. (*d*) Since $7 \times 8 = 56$ $\Rightarrow$ Unit's digit in 217×818 is 6.

Now $6 \times 3 = 18$ $\Rightarrow$ Unit's digit in $(217 \times 818) \times 193$ is 8.

2. The unit digit in $3 \times 38 \times 537 \times 1256$ is **(SSC CGL Tier II 2013)**

(*a*) 4 (*b*) 2 (*c*) 6 (*d*) 8

Sol. (*d*) $(3 \times 8) \times 7 \times 6 = (4 \times 7) \times 6 = 8 \times 6 = 8$

3. Find unit's digit in $200 \times 217 \times 818 \times 193$.

(*a*) 0 (*b*) 1 (*c*) 7 (*d*) 8

Sol. (*a*) Since one of the number has two 0's at the end.

∴ Multiplication of these numbers will also have two 0's at the end.

∴ Unit's digit in the product is 0.

4. The value of $999\frac{995}{999} \times 999$ is: **(SSC CGL Tier I 2003, SSC CHSL 2010)**

(*a*) 990809 (*b*) 998996 (*c*) 999824 (*d*) 998999

Sol. (*b*) $999\frac{995}{999} \times 999 = 999 \times 999 + \frac{995}{999} \times 999 = 998001 + 995 = 998996$

5. Find the value of $\frac{1}{5} + 999\frac{494}{495} \times 99$ **(SSC CGL Tier I 2003)**

(*a*) 90000 (*b*) 99000 (*c*) 90900 (*d*) 99990

Sol. (*b*) Given expression $= \frac{1}{5} + 999 \times 99 + \frac{494}{495} \times 99 = \frac{1}{5} + 98901 + \frac{494}{5} = 98901 + 99 = 99000$

Alt: $\frac{1}{5} + 999 \times 99 + \frac{494}{495} \times 99 = \frac{1}{5} + \left(1000 - \frac{1}{495}\right) \times 99 = \frac{1}{5} + 99000 - \frac{1}{5} = 99000$

6. If $a = 4011$ and $b = 3989$, then value of ab = ? **(SSC CGL Tier I 2013)**

(*a*) 15999879 (*b*) 15899879 (*c*) 15989979 (*d*) 15998879

Sol. (*a*) $4011 \times 3989 = (4000 + 11)(4000 - 11) = 16000000 - 121 = 15999879$

7. $12345679 \times 9 = ?$

(*a*) 111111111 (*b*) 21212121 (*c*) 111111112 (*d*) 222222221

Sol. (*a*) Unit's digit of the solution must be 1 *i.e.*, unit's digit in 9×9. ⇒ (*c*) is ruled out.

Since $12 \times 9 = 108$, (*b*) & (*d*) are ruled out as first three digits (left-hand side) are much greater than 108.

Sum of digits of solution must be 9 as the sum of digits of one multiplier is 9.

∴ (*a*) is the correct answer.

8. On multiplying a number by 153, 102325 is obtained as result. It is found that both the 2's are wrong. Find the correct result.

(*a*) 104345 (*b*) 107375 (*c*) 108385 (*d*) 109395

Sol. (*d*) 153 is a multiple of 9 as sum of its digits is 9. ⇒ The result must also be a multiple of 9.

Sum of digits (other than the wrong digits) = 1 + 0 + 3 + 5 = 9.

∴ The number 10*3*5 is divisible by 9 (ignoring both *).

∴ The number will be divisible by 9 when * + * is divisible by 9.

Sum of replaced digits = 4 + 4, 7 + 7, 8 + 8, 9 + 9 or 8, 14, 16, 18.

Only 18 is divisible by 9. ∴ * is to be replaced by 9. ∴ 109395 is divisible by 153.

9. $16.3 \times 5.8 = ?$ **(EPFO SSA)**

(*a*) 80.24 (*b*) 90.12 (*c*) 94.54 (*d*) None of these

Sol. (*c*) $16.3 \times 5.8 = (16 + 0.3) \times (5 + 0.8) = 16 \times 5 + 16 \times 0.8 + 0.3 \times 5 + 0.3 \times 0.8$

$= 80 + 12.8 + 1.5 + 0.24 = 94.54$

10. The product of two consecutive even numbers is 12768. What is the greater number ? **(Bank Clerical 2008)**

(*a*) 112 (*b*) 114 (*c*) 116 (*d*) 118

Sol. (*b*) $11^2 < 127 < 12^2$ ⇒ First two digits must be 11.

$2 \times 4 = 8$ suggests that numbers are 112 and 114. However, we can check 112×114.

$$\begin{array}{l} 112 \rightarrow 12 \\ \times\ 114 \rightarrow 14 \\ \hline 126 \rightarrow 168 = 12768 \end{array}$$

11. The product of two successive numbers is 9506. Which is the smaller of the two numbers ? **(Canara Bank PO 2009)**

(*a*) 96 (*b*) 97 (*c*) 98 (*d*) 99

Sol. (*b*) If the smaller number is 96. Then $6 \times 7 = 42$ (unit's digit is 2)

If the smaller number is 97. Then $7 \times 8 = 56$ (unit's digit is 6)

We can check 97×98

$$\begin{array}{l} 97 \rightarrow 3 \\ \times 98 \rightarrow 2 \\ \hline 95 \rightarrow 6 = 9506 \end{array}$$

CHAPTER

5

HCF AND LCM

Lowest Common Multiple (LCM): LCM of two or more numbers is the smallest number, that is exactly divisible by each of the given numbers.

Highest Common Factor (HCF): HCF of two or more numbers is the greatest number that divides each of them exactly.

Notes: (*i*) For any set of given numbers LCM is always a multiple of HCF. In other words HCF always exactly divides LCM.

Example. The HCF of two number is 8. Which one of the following can never be their LCM ? **(SSC CGL Tier I 2000)**

(*a*) 24 (*b*) 48 (*c*) 56 (*d*) 60

Sol. (*d*) LCM is always a multiple of HCF. 60 is not a multiple of 8.

(*ii*) If two numbers are separately divisible by a given number, then their sum is also divisible by that number. If HCF of x and y is a, then a divides $(x + y)$ also. In other words, on adding or subtracting any two multiples of an integer, we get another multiple of the same integer.

Example: 56 and 21 are divisible by 7. (56 + 21) is also divisible by 7.

(*iii*) If two numbers are separately divisible by a given number, then difference between them is also divisible by that number. If HCF of x and y is a, then a divides $(x - y)$ also.

Example: 56 and 21 are divisible by 7. (56 – 21) is also divisible by 7.

(*iv*) If two or more numbers when separately divided by a certain divisor leave the same remainder in each case, then difference between any of two numbers is exactly divisible by the divisor.

(*v*) The greatest number, that divides the numbers x, y and z, leaving remainders a, b, and c, respectively = HCF of $(x - a)$, $(y - b)$, $(z - c)$

Example 1: Find the greatest number that divides 132 and 77 leaving remainder 2 in each case.

(*a*) 5 (*b*) 11 (*c*) 22 (*d*) 25

Sol. (*a*) Remainder is 2 on dividing 132 by the divisor.

$\therefore$ 132 – 2 = 130 is exactly divisible by that divisor.

Similarly 77 – 2 = 75 is exactly divisible by the divisor.

$\therefore$ The required number is HCF of 130 and 75, *i.e.* 5.

Example 2: The greatest number, that divides 122 and 243 leaving respectively 2 and 3 as remainder, is: **(SSC CGL Tier I 2004)**

(*a*) 12 (*b*) 24 (*c*) 30 (*d*) 120

Sol. (*d*) On dividing 122 by HCF, the remainder is 2.

$\Rightarrow$ 122 – 2 = 120 is exactly divisible by the HCF.

Similarly on dividing 243 by HCF, the remainder is 3.

$\Rightarrow$ 243 – 3 = 240 is exactly divisible by the HCF.

Required number is HCF of (122 – 2) and (243 – 3) = HCF of 120 and 240 = 120

(*vi*) For any pair of two numbers, product of the numbers is equal to the product of their LCM and HCF *i.e.*, Product of numbers = HCF × LCM

Example: The HCF of two numbers is 15 and their LCM is 300. If one of the numbers is 60, the other is: **(SSC CGL Tier I 2004)**

(*a*) 50 (*b*) 75 (*c*) 65 (*d*) 100

Sol. (*b*) Second number $= \dfrac{\text{HCF} \times \text{LCM}}{\text{One number}} = \dfrac{15 \times 300}{60} = 75$

(*vii*) If the numbers are co-prime, their HCF is 1, then LCM of the numbers = Product of the numbers.

HCF/LCM of fractions:

$$\text{HCF of fractions} = \frac{\text{HCF of Numerators}}{\text{LCM of Denomiantors}} \qquad \text{LCM of fractions} = \frac{\text{LCM of Numerators}}{\text{HCF of Denominators}}$$

Example 1: The HCF of $\frac{1}{2}, \frac{3}{4}, \frac{5}{6}, \frac{7}{8}, \frac{9}{10}$ **(SSC CHSL 2013)**

(*a*) 1/10 (*b*) 9/120 (*c*) 1/120 (*d*) 1/2

Sol. (*c*) HCF $= \dfrac{\text{HCF of } 1,3,5,7,9}{\text{LCM of } 2,4,6,8,10} = \dfrac{1}{120}$

Example 2: The LCM of $\frac{2}{5}, \frac{3}{5}$ and $\frac{1}{2}$ is:

(*a*) $\frac{1}{10}$ (*b*) $\frac{6}{10}$ (*c*) $\frac{3}{25}$ (*d*) 6

Sol. (*d*) LCM $= \dfrac{\text{LCM of } 2, 3, 1}{\text{HCF of } 5, 5, 2} = \dfrac{6}{1} = 6$

HCF/LCM of Decimals:

Method:

1. Put zeros in the end of the numbers, if required, to get the same number of digits after decimal point in each number.
2. Find the HCF/LCM of the given number without considering the decimals.
3. In the result, put decimal at as many places as there are in the each of the number after step 1.

Example: Find HCF of 12.1 and 33.

(*a*) 0.11 (*b*) 1.1 (*c*) 11 (*d*) 3.3

Sol. (*b*) We change the numbers as 12.1 and 33.0. HCF of 121 and 330 is 11.

∴ The required number is 1.1.

Finding HCF of two numbers by Division Method:

Steps:

1. Divide the greater number by the smaller number and find the remainder.
2. Divide the previous divisor by the remainder.
3. Repeat this process, till final remainder is 0. The last divisor is HCF of the given numbers.

Example: Find HCF of 15 and 21.

Sol.

```
15)21(1
   15
    6)15(2
      12
       3)6(2
         6
         ×
```

∴ HCF = 3

Finding HCF of more than two numbers by Division Method:

Steps: 1.Find HCF of any two numbers. It is preferable to take two numbers whose difference is the least, to make the calculation easy.

2. Find HCF of the third number and of HCF calculated in first Step.

3. Find HCF of next number and HCF calculated at step 2 and so on.

Example: Find HCF of 12, 18 and 33.

Sol. First we find HCF of 12 and 18. HCF of 12 and 18 is 6.

HCF of first HCF *i.e.,* 6 and the third number *i.e.,* 33 is 3.

Finding HCF of two or more numbers by Factor Method:

Steps: 1. Express each of the given numbers as product of prime factors.

2. Find the factor or set of factors which are common in all the numbers.

3. The multiplication of the common factors is HCF.

Note: If no factor is common in all the numbers, then the HCF is 1.

Example 1: Find HCF of 15 and 21.

Sol. $15 = 3 \times 5$ $\quad 21 = 3 \times 7$ $\Rightarrow$ 3 being the only common factor is HCF.

Example 2: Find HCF of 20 and 50.

Sol. $20 = 2 \times 2 \times 5$ $\quad 50 = 2 \times 5 \times 5$

Common factors are 2 and 5. $\quad \therefore$ HCF $= 2 \times 5 = 10$

Example 3: Find HCF of 8 and 15.

Sol. $8 = 2 \times 2 \times 2$ $\quad 15 = 3 \times 5$

There is no common factor in above two numbers. $\Rightarrow$ Their HCF is 1.

Note: 8 and 15 are co-prime numbers.

Quicker Method of finding HCF:

1. Instead of finding factors of each number, we find factors of one number and then check with the help of Divisibility Test Method if these factors divide the other numbers or not.

2. To get HCF, multiply the factors of the first number which divide each of the remaining numbers.

Example 1: Find the HCF of 42, 63 and 70.

Sol. $70 = 2 \times 5 \times 7$

Now we check whether 42 and 63 are divisible by 2, 5 and 7.

2: 63 is not divisible by 2. (no need to check for 42)

5: 42 is not divisible by 5 (no need to check for 63).

7: 42 and 63 both are divisible by 7.

7 is the only factor which divides each of the given numbers.

$\therefore$ HCF of 42, 63 and 70 is 7.

Factor method for finding LCM of two or more numbers:

Steps:

1. Write factors of one of the numbers.

2. Write factors of the second number. Strike out the factors that have already occured in the first number. Multiply the remaining factors of the second number with the factors of the first number to get LCM of the first and the second number.

3. Write factors of the third number. Strike out the factors that have already occured in LCM of first two numbers. Multiply the remaining factors of the third number with the LCM of two first two numbers.

4. Repeat process mentioned in Step 3 for the remaining numbers till all the numbers are covered.

Example 1: Find LCM of 15 and 21.

Sol. $15 = 3 \times 5$

$21 = 3 \times 7$ (3 in 3×7 has already occurred in 3×5 and therefore ignored)

$\therefore$ LCM of 15 and 21 is $3 \times 5 \times 7 = 105$.

Example 2: Find LCM of 12, 15 and 18.

Sol. $12 = 2 \times 2 \times 3$

$15 = 3 \times 5$ (3 in 3×5 has already occurred in $2 \times 2 \times 3$ and therefore ignored)

$\therefore$ LCM of 12 and 15 is $2 \times 2 \times 3 \times 5$.

$18 = 2 \times 3 \times 3$

(2×3 in $2 \times 3 \times 3$ has already occurred in $2 \times 2 \times 3 \times 5$ and therefore ignored)

$\therefore$ LCM of 12, 15, 18 is $2 \times 2 \times 3 \times 5 \times 3 = 180$.

SOLVED EXERCISE

1. HCF of 18, 27 and 81 is:

(*a*) 9 (*b*) 18 (*c*) 27 (*d*) 45

Sol. (*a*) $18 = 2 \times 3 \times 3$ $27 = 3 \times 3 \times 3$ $81 = 3 \times 3 \times 3 \times 3$

Here 3×3 is common in all the three numbers. $\Rightarrow$ HCF $= 3 \times 3 = 9$

2. LCM of 25, 30 and 45 is:

(*a*) 5 (*b*) 450 (*c*) 900 (*d*) 2250

Sol. (*b*) $25 = 5 \times 5$

$30 = 2 \times 3 \times 5$ (But 5 has already occured in factors of 25)

$\therefore$ LCM of 25 and 30 is $5 \times 5 \times 2 \times 3$.

$45 = 3 \times 3 \times 5$ (But 3×5 has already occured in $5 \times 5 \times 2 \times 3$)

$\therefore$ LCM of 25, 30 and 45 $= 5 \times 5 \times 2 \times 3 \times 3 = 450$

3. The LCM of 26, 56, 104, 182 is **(Hotel Management)**

(*a*) 364 (*b*) 728 (*c*) 758 (*d*) 1092

Sol. (*b*) $26 = 2 \times 13$, $56 = 8 \times 7$, $104 = 8 \times 13$, $182 = 2 \times 7 \times 13$

$\therefore$ LCM $= 2 \times 13 \times 4 \times 7 = 728$

4. Find HCF of 2268, 2058 and 2100.

(*a*) 6 (*b*) 12 (*c*) 14 (*d*) 42

Sol. (*d*) $2268 - 2058 = 210$ and $2100 - 2058 = 42$. HCF of 210 and 42 is 42.

5. There are five hobby clubs in a college viz., photography, yachting, chess, electronics and gardening. The gardening group meets every second day, the electronics group meets every third day, the chess group meets every fourth day, the yachting group meets every fifth day and the photography meets every sixth day. How many times do all the five groups meet on the same day within 180 days? **[UPSC Civil Services (Pre.) 2013]**

(*a*) 3 (*b*) 5 (*c*) 10 (*d*) 18

Sol. (*a*) LCM of 2, 3, 4, 5, 6 $= 2 \times 3 \times 2 \times 5 = 60$ $\therefore$ Required number $= 180 \div 60 = 3$

6. Product of two co-prime numbers is 117. Then their L.C.M. is **(SSC CGL Tier I 2013)**

(*a*) 13 (*b*) 117 (*c*) 39 (*d*) 3

Sol. (*b*) LCM of co-prime numbers = Product of numbers = 117.

7. A milkman has 75 litres milk in one can and 45 litres in another. The maximum capacity of container which can measure milk of either container exact number of times, is:

(SSC CGL Tier I 2002)

(*a*) 1 litre (*b*) 5 litres (*c*) 15 litres (*d*) 25 litres

Sol. (*c*) HCF of 75 and 45 = 15 ∴ Maximum capacity = 15 litres

8. If the students of 9th class are arranged in rows of 6, 8, 12 or 16, no student is left behind. Then the posible number of students in the class is **(SSC CGL Tier II 2013)**

(*a*) 60 (*b*) 72 (*c*) 80 (*d*) 96

Sol. (*d*) LCM of 6, 8, 12, 16 = 6 × 4 × 2 = 48. 96 is a multiple of 48.

9. Three persons start walking together and their steps measure 40 cm, 42 cm and 45 cm respectively. What is the minimum distance each should walk so that each can cover the same distance in complete steps ? **(UPSC Civil Services Pre. 2011)**

(*a*) 25 m 20 cm (*b*) 50 m 40 cm (*c*) 75 m 60 cm (*d*) 100 m 80 cm

Sol. (*b*) Required distance = LCM of 40, 42, 45 = 4 × 10 × 3 × 7 × 3 = 2520cm = 25m20cm

10. A bell rings every 18 minutes. A second bell rings every 24 minutes. A third bell rings every 32 minutes. If all the three bells ring at the same time at 8 o'clock in the morning, at what other time will they all ring together ? **(UPSC Civil Services (Pre.) 2014)**

(*a*) 12:40 hrs (*b*) 12:48 hrs (*c*) 12:56 hrs (*d*) 13:04 hrs

Sol. (*b*) LCM of 18, 24, 32 = 2 × 3 × 3 × 4 × 4 = 288 minutes

⇒ Required time = 8:00 + 288 minutes = 12:48 hrs

11. The sum of two numbers is 36 and their HCF and LCM are 3 and 105 respectively. The sum of the reciprocals of the two numbers will be: **(SSC CHSL 2010)**

(*a*) 13 (*b*) $\frac{9}{11}$ (*c*) $\frac{7}{35}$ (*d*) $\frac{4}{35}$

Sol. (*d*) Sum of reciprocals $= \frac{\text{Sum}}{\text{Product}} = \frac{\text{Sum}}{\text{HCF}\times\text{LCM}} = \frac{36}{3\times105} = \frac{4}{35}$

12. The HCF of two numbers is 15 and their LCM is 225. If one of the numbers is 75, then the other number is: **(SSC CHSL 2010)**

(*a*) 105 (*b*) 90 (*c*) 60 (*d*) 45

Sol. (*d*) Other number $= \frac{15\times225}{75} = 45$

13. The LCM of two numbers is 1920 and their HCF is 16. If one of the numbers is 128, find the other number. **(SI 1997, SSC CGL Tier I 2002)**

(*a*) 204 (*b*) 240 (*c*) 260 (*d*) 320

Sol. (*b*) Other number $= \frac{1920\times16}{128} = 240$

14. The LCM of two numbers is 864 and their HCF is 144. If one of the numbers is 288, the other number is: **(SSC CGL Tier I 1999)**

(*a*) 576 (*b*) 1296 (*c*) 432 (*d*) 144

Sol. (*c*) Other number $= \frac{864\times144}{288} = 432$

Trick: If we compare the unit's digit:

Unit's digit in the product of HCF and LCM is 4 × 4 = 6.

The number 288 has 8 in the unit's place.

Out of the given options only 8 × 2 has 6 in unit's place.

∴ 432 is the required number.

15. The LCM of two multiples of 12 is 1056. If one of the numbers is 132, the other number is: **(SSC CPO Sub-Inspector 2009)**

(*a*) 12 (*b*) 72 (*c*) 96 (*d*) 132

Sol. (*c*) The number are multiple of 12 $\Rightarrow$ HCF of numbers = 12.

Required number = $\frac{1056 \times 12}{132} = 96$

16. The product of two numbers is 4107. If the HCF of the numbers is 37, the greater number is: **(SSC CGL Tier I 2003)**

(*a*) 185 (*b*) 111 (*c*) 107 (*d*) 101

Sol. (*b*) LCM = $\frac{\text{Product of numbers}}{\text{HCF}} = \frac{4107}{37} = 111 = 37 \times 3$ $\Rightarrow$ The numbers are 37 and 111.

17. The product of the LCM and HCF of two numbers is 24. The difference of the two numbers is 2. Find the numbers. **(SSC CGL Tier I 1999)**

(*a*) 8 and 6 (*b*) 8 and 10 (*c*) 2 and 4 (*d*) 6 and 4

Sol. (*d*) Product of numbers = Product of LCM and HCF = 24 $\Rightarrow$ The numbers are 6 and 4.

18. The ratio of two numbers is 3 : 5 and their LCM is 300. Then one of the numbers is: **(SSC CHSL 2006)**

(*a*) 30 (*b*) 50 (*c*) 60 (*d*) 75

Sol. (*c*) Let the numbers are $3x$ and $5x$. Then LCM = $15x = 300 \Rightarrow x = 20$

$\therefore$ The numbers are 60 and 100 respectively.

19. The least number which when divided by 5, 6, 7 and 8 leaves a remainder 3 is **(CDS 2011)**

(*a*) 423 (*b*) 843 (*c*) 1683 (*d*) 2523

Sol. (*b*) LCM of 5, 6, 7, 8 is $5 \times 6 \times 7 \times 4 = 840$. $\Rightarrow$ Required number = 840 + 3 = 843.

20. The least number which when divided by 6, 9, 12, 15, 18 leaves the same remainder 2 in each case is **(SSC CGL Tier I 2015)**

(*a*) 178 (*b*) 182 (*c*) 180 (*d*) 176

Sol. (*b*) LCM of 6, 9, 12, 15, 18 is multiple of 9. $\Rightarrow$ Sum of digits of the LCM must be 9.

Required number = Multiple of 9 + 2 $\Rightarrow$ Option (*b*) only possible here.

21. Which is the greatest three-digit number which when divided by 3, 4, 5 and 6 leaves remainder 2 in each case?

(*a*) 060 (*b*) 062 (*c*) 960 (*d*) 962

Sol. (*d*) LCM of 3, 4, 5 and 6 is $3 \times 4 \times 5 = 60$.

Greatest 3-digit number is 999. On dividing 999 by 60, remainder is 39.

$\therefore$ 999 – 39 = 960 is exactly divisible by 60.

$\therefore$ 960 + 2 = 962 is the number which when divided by 60 will leave remainder 2.

Alt: 60 and 62, being two-digit numbers can't be the required number.

960 and 962 when reduced by 2 becomes 958 and 960 respectively.

958 is not divisible by 5.

Using Divisibility Test Method, we find 960 is divisible by 3, 4, 5 and 6.

22. Let the least number of six digits which when divided by 4, 6, 10, 15 leaves in each case same remainder 2 be *N*. The sum of digits in *N* is: **(SSC CGL Tier I 2003)**

(*a*) 3 (*b*) 5 (*c*) 4 (*d*) 6

Sol. (*b*) LCM of 4, 6, 10, 15 = $2 \times 2 \times 3 \times 5 = 60$. Remainder on dividing 100000 by 60 is 40.

Least number of 6 digits divisible by 60 = 100000 + 60 – 40 = 100020.

$\therefore$ N = 100020 + 2 = 100022 $\Rightarrow$ Sum of digits = 1 + 2 + 2 = 5

23. Find the greatest number of five digits which when divided by 3, 5, 8, 12 have 2 as remainder. **(SSC CGL Tier I 2001)**

(*a*) 99999 (*b*) 99958 (*c*) 99960 (*d*) 99962

Sol. (*d*) LCM of 3, 5, 8 and 12 is $3 \times 5 \times 8 = 120$. Remainder on dividing 99999 by 120 is 39.

$\therefore$ Required number = $(99999 - 39) + 2 = 99962$

Alt: LCM of 3, 5, 8 and 12 is $3 \times 5 \times 8 = 120$ which is a multiple of 10.

$99962 - 2$ is the only option which is multiple of 10.

24. What least number must be subtracted from 1936 so that the resulting number when divided by 9, 10 and 15 will leave in each case the same remainder 7 ? **(SSC CGL Tier I 2008)**

(*a*) 37 (*b*) 36 (*c*) 39 (*d*) 30

Sol. (*c*) LCM of 9, 10 and 15 is 90. $1936 = 21 \times 90 + 46$ $\therefore$ The required number = $46 - 7 = 39$

25. Which of the followings is the least square number by which 12, 20 and 25 are exactly divisible:

(*a*) 100 (*b*) 300 (*c*) 900 (*d*) 3600

Sol. (*c*) As we are to find the least number, we start with the smallest number.

100 is not divisible by 12. 300 is not a square number.

900 is a square number and is also divisible by 12, 20 and 25.

26. The least five-digit perfect square number which is divisible by 3, 4, 5 and 8 is: **(SSC CGL Tier II 2012)**

(*a*) 10800 (*b*) 10201 (*c*) 14400 (*d*) 32400

Sol. (*c*) We may check the options, starting from the smallest number:

10201 : A square number but not divisible by 3, 4 or 5.

10800 : Not a square number

14400 : Square number and also divisible by 3, 4, 5 and 8.

27. HCF of two different numbers is 24, which one of the following can be their LCM: **(SSC CGL Tier I 1997)**

(*a*) 12 (*b*) 18 (*c*) 36 (*d*) 48

Sol. (*d*) We know that LCM is always exactly divisible by HCF of the numbers.

Here 48 is the only number which is exactly divisible by 24.

Note: The possible numbers are 24 and 48.

28. If the highest common factor of two positive integers is 24, then their least common multiple ***cannot*** be **(CDS 2010)**

(*a*) 2352 (*b*) 3840 (*c*) 3728 (*d*) 3744

Sol. (*c*) Here 3728 is the only number which is not exactly divisible by 24.

29. LCM of three different numbers is 120. Which one of the following cannot be their HCF ? **(SSC CGL Tier I 2000, 2011)**

(*a*) 8 (*b*) 12 (*c*) 24 (*d*) 35

Sol. (*d*) 35 out of the given options does not exactly divide 120, the LCM.

30. If the HCF of three numbers 144, x and 192 is 12, then the number x cannot be: **(CDS 2011)**

(*a*) 180 (*b*) 84 (*c*) 60 (*d*) 48

Sol. (*d*) HCF of 48, 144 and 192 is 48.

31. The greater number which divides 1657 and 2037 leaving remainders 6 and 5 respectively is: **(SSC CGL Tier I 2006)**

(*a*) 123 (*b*) 127 (*c*) 235 (*d*) 305

Sol. (*b*) Required number is HCF of $(1657 - 6)$ and $(2037 - 5)$ = HCF of 1651 and 2032 = 127

32. What is the greatest number that divides 307 and 330 leaving remainder 3 and 7 respectively?
(SSC CGL Tier I 2005, 2002M)

(*a*) 19 (*b*) 16 (*c*) 17 (*d*) 23

Sol. (*a*) Required number is HCF of (307 – 3) and (330 – 7) = HCF of 304 and 323 = 19

33. What is the sum of the digits of the least number which, when divided by 52, leaves 33 as remainder; when divided by 78, leaves 59 and when divided by 117, leaves 98 as remainder ? **(CDS 2010)**

(*a*) 17 (*b*) 18 (*c*) 19 (*d*) 21

Sol. (*a*) LCM of 52, 78 and 117 = 13 × 2 × 2 × 3 × 3 = 468

Difference between divisor and respective remainder = 52 – 33 = 78 – 59 = 117 – 98 = 19

The number = 468 – 19 = 449 ⇒ Sum of digits = 4 + 4 + 9 = 17

34. What is the smallest positive integer which when divided by 4, 5, 8, 9 leaves remainder 3, 4, 7, 8 respectively ? **(CDS 2010)**

(*a*) 119 (*b*) 319 (*c*) 359 (*d*) 719

Sol. (*c*) LCM of 4, 5, 8, 9 = 4 × 5 × 2 × 9 = 360

Difference = 4 – 3 = 5 – 4 = 8 – 7 = 9 – 8 =1

∴ The required number = 360 – 1 = 359

35. The largest number which divides 25, 73, and 97 to leave the same remainder in each case, is
(SSC CGL Tier I 2007)

(*a*) 24 (*b*) 23 (*c*) 21 (*d*) 16

Sol. (*a*) Required number is HCF of (97 – 73) and (73 – 25) = HCF of 24 and 48 = 24

36. The greatest number that will divide 19, 35 and 59 to leave the same remainder in each case is:
(SSC CGL Tier I 2012)

(*a*) 9 (*b*) 6 (*c*) 7 (*d*) 8

Sol. (*d*) Required number is HCF of (59 – 35) and (35 – 19) = HCF of 24 and 16 = 8

37. Let *N* be the greatest number that will divide 1305, 4665 and 6905 leaving the same remainder in each case. The sum of the digits in *N* is: **(SSC CGL Tier I 2004)**

(*a*) 4 (*b*) 5 (*c*) 6 (*d*) 8

Sol. (*a*) Required number is HCF of (4665 – 1305) and (6905 – 4665).
= HCF of 3360 and 2240 = 1120 ⇒ Sum of the digits = 1 + 1 + 2 + 0 = 4

38. What is the greatest number which divides 392, 486 and 627 so as to leave the same remainder in each case ? **(CDS 2009)**

(*a*) 47 (*b*) 43 (*c*) 37 (*d*) 34

Sol. (*a*) Required number is HCF of (486 – 392) and (627 – 486) = HCF of 94 and 141 = 47

39. The numbers 11284 and 7655, when divided by a certain 3-digit number leave the same remainder. The 3-digit number is: **(SSC CGL Tier I 2006, SSC CHSL 2013)**

(*a*) 199 (*b*) 197 (*c*) 193 (*d*) 191

Sol. (*d*) Difference between the two numbers = 11284 – 7655 = 3629 is fully divisible the divisor.

3629 = 19 × 191 ∴ The required number is 191.

40. Which one of the following three-digit number divides 9238 and 7091 with the same remainder in each case ? **(CDS 2011)**

(*a*) 113 (*b*) 209 (*c*) 317 (*d*) 191

Sol. (*a*) 9238 – 7091 = 2147 = 113 × 19 ⇒ The required number is 113.

41. Two numbers 2184 and 4038, when divided by a certain divisor of three digits, leave the same remainder in each case, find the divisor and the remainder:

(*a*) 101, 63 (*b*) 103, 21 (*c*) 105, 84 (*d*) 111, 75

Sol. (*b*) Difference between the two numbers = 4038 – 2184 = 1854 is divisible the divisor.
Out of given options, only 103 divides 1854 exactly.
Remainder on dividing the given numbers by 103 is 21 in each case.

42. What is the least numbers which, when divided by 5, 6, 7, 8 gives remainder 3 but is divisible by 9? **(SSC CGL Tier I 2011)**

(*a*) 1463 (*b*) 1573 (*c*) 1683 (*d*) 1793

Sol. (*c*) **Trick:** Out of given options, only 1683 is disible by 9.

43. Find the least multiple of 23, which when divided by 18, 21 and 24 leaves the remainder 7, 10 and 13 respectively. **(SSC CGL Tier I 2001)**

(*a*) 3013 (*b*) 3024 (*c*) 3002 (*d*) 3036

Sol. (*a*) LCM of 18, 21 and 24 = $2 \times 3 \times 3 \times 7 \times 4 = 504$
Remainder in each case is 11 less than the respective divisor.
$\therefore$ Required number is $(504x - 11) = (23 \times 21x) + (21x - 11)$
Remainder on dividing the number by 23 is $(21x - 11)$
$(21x - 11)$ is divisible by 23, when $x = 6$
The required number is $504x - 11 = 504 \times 6 - 11 = 3024 - 11 = 3013$
Trick : LCM of 18, 21 and 24 = $2 \times 3 \times 3 \times 7 \times 4 = 504$
Given options are around 3000 which is approximately 6 times 504.
$\therefore$ Required number = $504 \times 6 - 11 = 3024 - 11 = 3013$

44. The least multiple of 13, which on dividing by 4, 5, 6, 7 and 8 leaves the remainder 2 in each case, is: **(SSC CGL Tier I 2002, CGL Tier I 2012M)**

(*a*) 2520 (*b*) 842 (*c*) 2522 (*d*) 840

Sol. (*c*) LCM of 4, 5, 6, 7 and 8 = $4 \times 5 \times 3 \times 7 \times 2 = 840$. Remainder is 2 in each case.
$\therefore$ Required number is $(840x + 2) = (64x \times 13) + (8x + 2)$
On dividing $(840x + 2)$ by 13, remainder is $(8x + 2)$.
$(8x + 2)$ is divisible by 13, when $x = 3$.
$\therefore$ Required number = $840x + 2 = 840 \times 3 + 2 = 2522$
Alternative method:
2520 – 2 = 2518 and 840 – 2 = 838 are not divisible by 5. 842 is not divisible by 13.
2522 – 2 = 2520 is divisible by 4, 5, 6, 7 and 8. 2522 is divisible by 13.

45. The LCM and the HCF of the numbers 28 and 42 are in the ratio **(SSC CGL Tier I 2003)**

(*a*) 6 : 1 (*b*) 2 : 3 (*c*) 3 : 2 (*d*) 7 : 2

Sol. (*a*) $28 = 2 \times 2 \times 7$ $42 = 2 \times 3 \times 7$
LCM of 28 and 42 is $2 \times 2 \times 7 \times 3$. HCF of 28 and 42 is 2×7.
$\therefore$ LCM : HCF = $(2 \times 2 \times 7 \times 3) : (2 \times 7) = (2 \times 3) : 1 = 6 : 1$

46. The ratio of two numbers is 3 : 4 and their HCF is 4. Their LCM is: **(SSC CGL Tier I 2001)**

(*a*) 12 (*b*) 16 (*c*) 24 (*d*) 48

Sol. (*d*) The number are 3×4 and 4×4 = 12 and 16. LCM of 12 and 16 is 48.
Alternative method: LCM of 3 and 4 is 12. LCM of numbers = $4 \times 12 = 48$.

47. Three numbers are in the ratio 1 : 2 : 3 and their HCF is 12. The numbers are **(SSC CGL Tier I 2014)**

(*a*) 12, 24, 36 (*b*) 5, 10, 15 (*c*) 4, 8, 12 (*d*) 10, 20, 30

Sol. (*a*) Common factor of 12, 24, 36 is 12 and 12 : 24 : 36 = 1 : 2 : 3

48. Three numbers are in the ratio 2 : 3 : 4 and their HCF is 12. The LCM of the numbers is **(SSC CGL Tier I 2007)**

(*a*) 144 (*b*) 192 (*c*) 96 (*d*) 72

Sol. (*a*) LCM of 2, 3 and 4 is 12.

∴ LCM of the numbers = HCF × LCM of ratios = 12 × 12 = 144

49. The sum of two numbers is 45. Their difference is 1/9 of their sum. Their LCM is: **(SSC CGL 2007)**

(*a*) 100 (*b*) 150 (*c*) 200 (*d*) 250

Sol. (*a*) Let the numbers be x and y. Then $x + y = 45$, and $x - y = \frac{1}{9} \times 45 = 5$

$\Rightarrow x = \frac{1}{2}(45 + 5) = 25$ and $y = \frac{1}{2}(45 - 5) = 20$. LCM of 25 and 20 = 100

50. The HCF and LCM of two numbers are 13 and 455 respectively. If one of the numbers lies between 75 and 125, then, that number is: **(SSC CGL Tier I 1999)**

(*a*) 78 (*b*) 91 (*c*) 104 (*d*) 117

Sol. (*b*) 455 = 13 × 7 × 5 ⇒ Required number 13 × 7 = 91

51. The HCF and LCM of two numbers are 21 and 4641 respectively. If one number lies between 200 and 300, then the numbers are: **(MBA 2006)**

(*a*) 210, 357 (*b*) 273, 357 (*c*) 253, 351 (*d*) 273, 361

Sol. (*b*) Let the numbers be $21x$ and $21y$ respectively.

Then $21x \times 21y = 21 \times 4641 \Rightarrow x \times y = 4641 \div 21 = 221 = 13 \times 17$

The numbers are 21 × 13 and 21× 17 = 273, 357.

52. If the HCF of two numbers (each greater than 13) be 13 and LCM 273, then the sum of the numbers is: **(SSC 2006)**

(*a*) 130 (*b*) 290 (*c*) 34 (*d*) 286

Sol. (*a*) Let the numbers be $13x$ and $13y$. Then $13xy = 273 \Rightarrow xy = 21 \Rightarrow x = 3, y = 7$

∴ The numbers are 13 × 3, 13 × 7. Sum of numbers = 13 × 3 + 13 × 7 = 13 × 10 = 130

53. Product of two numbers is 1440 and their HCF is 12. How many pairs of such numbers are possible?

(*a*) 2 (*b*) 3 (*c*) 4 (*d*) 6

Sol. (*a*) Let the numbers are $12x$ and $12y$. Then $(12x) \times (12y) = 1440 \Rightarrow xy = \frac{1440}{12 \times 12} = 10$

∴ The possible values of x and y for which $x \times y$ is 10 are (1, 10), (2, 5).

∴ Only two pairs are valid.

54. HCF and LCM of two numbers are 7 and 140 respectively. If the numbers are between 20 and 45, the sum of the numbers, is: **(SSC CGL Tier I 2003)**

(*a*) 70 (*b*) 77 (*c*) 63 (*d*) 56

Sol. (*c*) Let the numbers be $7x$ and $7y$. Then $7xy = 140 \Rightarrow x \times y = 20 = 4 \times 5$

∴ The numbers are 7 × 4 and 7 × 5 = 28 and 35.

Sum of numbers = 28 + 35 = 63

55. The HCF of two numbers is 98 and their LCM is 2352. The sum of the numbers may be **(CDS 2013)**

(*a*) 1372 (*b*) 1398 (*c*) 1426 (*d*) 1484

Sol. (*a*) 1372 is divisible by 98. (**Hint:** $98 \times 14 = 100 \times 14 - 2 \times 14 = 1400 - 28 = 1372$)

56. LCM and HCF of two different numbers are 288 and 24 respectively. How many such pairs are possible?

(*a*) 1 (*b*) 2 (*c*) 3 (*d*) 4

Sol. (*b*) Let the numbers are $24x$ and $24y$. LCM $= 24xy = 288 \Rightarrow xy = 288 \div 24 = 12$

The possible values of x and y for which xy is 12 are (1, 12), (2, 6), (3, 4).

But (2, 6) is not valid pair as they are not co-prime numbers. $\therefore$ Two pairs are valid.

Note: Quotients on dividing the numbers by the HCF must be co-prime numbers.

For example, if $x = 2$ and $y = 6$, Then the numbers are $24 \times 2 = 48$ and $24 \times 6 = 144$.

HCF and LCM of 48 and 144 are 48 and 144 respectively.

57. The sum of two numbers is 36 and their HCF is 4. How many pairs of such numbers are possible ? **(SSC CGL Tier I 2004)**

(*a*) 1 (*b*) 2 (*c*) 3 (*d*) 4

Sol. (*c*) Let the numbers are $4x$ and $4y$. Then $(4x) + (4y) = 36 \Rightarrow x + y = 36 \div 4 = 9$

The possible values of x and y for which $(x + y) = 9$ are (1, 8), (2, 7), (3, 6), (4, 5).

But (3, 6) is not valid as they are not co-prime numbers $\Rightarrow$ Three pairs are valid.

58. Sum of two numbers is 384. HCF of the numbers is 48. The difference of the numbers is: **(SSC CPO Sub-Inspector Exam. 2009)**

(*a*) 100 (*b*) 192 (*c*) 288 (*d*) 336

Sol. (*c*) Let the numbers be $48x$ and $48y$. Then $48(x + y) = 384 \Rightarrow x + y = 384 \div 48 = 8$

Valid pairs for $x + y = 8$ are (1, 7) and (3, 5)

Possible values of difference $= 48 \times (7 - 1) = 288$ or $48 \times (5 - 3) = 96$

59. The LCM of two numbers is 495 and their HCF is 5. If the sum of the numbers is 100, then their difference is: **(SSC CGL Tier I 1999)**

(*a*) 10 (*b*) 46 (*c*) 70 (*d*) 90

Sol. (*a*) Let the numbers be $5x$ and $5y$. Then LCM $= 5xy = 495 = 5 \times 9 \times 11 \Rightarrow (x, y) = (9, 11)$

Difference of numbers $= 5 \times 11 - 5 \times 9 = 10$

60. The sum of two numbers is 528 and their HCF is 33. The number of pairs of such numbers is: **(SSC CGL Tier I 2006, SSC CHSL 2013)**

(*a*) 1 (*b*) 2 (*c*) 3 (*d*) 4

Sol. (*d*) Let the numbers be $33x$ and $33y$. Then $33x + 33y = 528 \Rightarrow x + y = 528 \div 33 = 16$

$\therefore$ Valid pairs are (1, 15), (3, 13), (5, 11), (7, 9) $\Rightarrow$ 4 pairs are possible.

61. Two 3-digit numbers have their HCF 29 and LCM 4147. The sum of the numbers is: **(SSC CGL Tier I 2005 and 2002 modified)**

(*a*) 966 (*b*) 696 (*c*) 669 (*d*) 666

Sol. (*b*) Let the number are $29x$ and $29y$. Then $29x \times 29y = 4147 \times 29$

$$\therefore\ x \times y = \frac{4147 \times 29}{29 \times 29} = 143 = 11 \times 13$$

$\therefore$ Sum of numbers $= 29 \times 11 + 29 \times 13 = 29 \times (11 + 13) = 29 \times 24 = 696.$

62. The sum of the HCF and LCM of two numbers is 680 and LCM is 84 times the HCF. If one of the numbers is 56, the other is: **(SSC CGL Tier I 2005)**

(*a*) 84 (*b*) 12 (*c*) 8 (*d*) 96

Sol. (*d*) Let the HCF be x. Then LCM = $84x$.

$x + 84x = 85x = 680 \Rightarrow x = 680 \div 85 = 8 \Rightarrow$ One number = 56 = $7x$

$\therefore$ The other number = $12x$ (as LCM of two numbers is $84x$) = $12 \times 8 = 96$

63. The LCM of two numbers is 12 times their HCF. The sum of HCF and LCM is 403. If one number is 93, the other number is: **(NDA 2000, SSC CGL Tier I 2008 and 2015, CDS 2015)**

(*a*) 124 (*b*) 248 (*c*) 372 (*d*) 403

Sol. (*a*) Let HCF = x, then LCM = $12x$

Given $x + 12x = 403 \Rightarrow x = 403 \div 13 = 31 \Rightarrow$ HCF = 31, LCM = 12×31

One number = 93 $\Rightarrow$ Other number = $\dfrac{31\times12\times31}{93} = 124$

64. The HCF and LCM of two numbers are respectively 12 and 2448. If the difference of the numbers is 60, their sum is: **(SSC CGL Tier I 2008)**

(*a*) 348 (*b*) 284 (*c*) 248 (*d*) 204

Sol. (*a*) Let the numbers be $12x$ and $12y$. Then $12x - 12y = 60 \Rightarrow x - y = 60 \div 12 = 5$

$12xy = 2448 \Rightarrow xy = 2448 \div 12 = 204$

$(x + y)^2 = (x - y)^2 + 4xy = 5^2 + 4 \times 204 = 25 + 816 = 841 = 29^2$

$\therefore x + y = 29 \Rightarrow$ Sum of numbers = $12x + 12y = 12(x + y) = 12 \times 29 = 348$

65. HCF and LCM of 77, 99 and x are 11 and 3465 respectively. The least value of x is: **(SSC CGL Tier I 2006)**

(*a*) 11 (*b*) 35 (*c*) 45 (*d*) 55

Sol. (*d*) Let $x = 11a$. LCM of 77, 99 and $11a = 7 \times 11 \times 9 \times a = 3465$

$\therefore a = \dfrac{3465}{7\times11\times9} = 5 \Rightarrow$ The required number is $11a = 11 \times 5 = 55$

66. In finding HCF of two numbers by the division method the quotients are 1, 2 and 3 and the last divisor is 15, what are the numbers?

(*a*) 105, 120 (*b*) 90, 120 (*c*) 105, 150 (*d*) 120, 150

Sol. (*c*) In such questions, we start calculation in the reverse order.

$15 \times 3 = 45$, $45 \times 2 + 15 = 105$ and $105 \times 1 + 45 = 150$

67. A worker was engaged for a certain number of days. But he remained absent for some days and was paid ₹ 1387. Had he worked all the days, he would have earned ₹ 1752. How many days did he work for?

(*a*) 19 (*b*) 21 (*c*) 24 (*d*) 29

Sol. (*a*) Daily wage is HCF of ₹ 1387 and ₹ 1752.

$1387 = 73 \times 19$ and $1752 = 73 \times 24 \Rightarrow$ Daily wages = ₹ 73 (*i.e.* HCF)

$\therefore$ Number of days, he worked for is 19 (*i.e.* $1387 \div 73$)

CHAPTER

6

SQUARE AND CUBE–VEDIC MATHS

The square of a number is equal to product of the number by itself. Square of x is denoted by x^2. Therefore, a number is a perfect square number if it can be expressed as product of a natural number by itself. Square of a number of n digits, may contain $2n$ or $2n - 1$ digits.

Example: Square of 7 = $7^2 = 7 \times 7 = 49$

Square of 18 = $18^2 = 18 \times 18 = 324$

The cube of a number is equal to product of the number by itself twice. Cube of x is denoted by x^3. Therefore, a number is a perfect cube number if it can be expressed as product of a natural number by itself and itself again. Cube of a number of n digits, may contain $3n$, $(3n - 1)$ or $(3n - 2)$ digits.

Example: Cube of 2 = $2^3 = 2 \times 2 \times 2 = 8$. Cube of 4 = $4^3 = 4 \times 4 \times 4 = 64$

Cube of 5 = $5^3 = 5 \times 5 \times 5 = 125$

COMMON SQUARE NUMBERS

$11^2 = 121$	$12^2 = 144$	$13^2 = 169$	$14^2 = 196$	$15^2 = 225$
$16^2 = 256$	$17^2 = 289$	$18^2 = 324$	$19^2 = 361$	$21^2 = 441$
$22^2 = 484$	$23^2 = 529$	$24^2 = 576$	$25^2 = 625$	$26^2 = 676$
$27^2 = 729$	$28^2 = 784$	$29^2 = 841$	$31^2 = 961$	$32^2 = 1024$
$41^2 = 1681$				

Note: Difference between square of two consecutive numbers is equal to sum of the numbers themselves.

1681: It is a four digit perfect square number. The numbr formed by the first two digits and the number formed by last two digits are also perfect squares.

Example: $996^2 - 995^2 = ?$

(*a*) 0 (*b*) 1 (*c*) 100 (*d*) 1991

Sol. (*d*) We know that, difference between the squares of two consecutive numbers is equal to the sum of the given numbers.

$\therefore 996^2 - 995^2 = 996 + 995 = 1991$

COMMON CUBE NUMBERS

$2^3 = 8$	$3^3 = 27$	$4^3 = 64$	$5^3 = 125$	$6^3 = 216$
$7^3 = 343$	$8^3 = 512$	$9^3 = 729$	$11^3 = 1331$	$12^3 = 1728$
$13^3 = 2197$	$21^3 = 9261$			

PROPERTIES OF SQUARE NUMBERS

1. Unit's digit of a square number can only be one of the following numbers:
   1, 4, 5, 6, 9, 0 (Even number of 0's)

This implies a square number can never have the following digits at unit's place: 2, 3, 7, 8

2. Square numbers ending in an even number is always divisible by 4.
3. Square number ending in an odd number when reduced by 1 is exactly divisible by 8.

Proof: Let the number is $(2x + 1)$.
Then $(2x + 1)^2 - 1 = (4x^2 + 4x + 1) - 1 = 4x^2 + 4x = 4(x^2 + x) = 4(x)(x + 1)$
Ether x or $(x + 1)$ is an even number.
Hence, $4(x)(x + 1)$ is divisible by $4 \times 2 = 8$ for all avalues of x.

Note: If a square number ends in an odd number, it is always square of an odd number.

4. Square number ending in 5, always ends in 25 and is always preceded by an even number.

Proof: A number ending is 5 can be written as $(10x + 5)$.
Now, $(10x + 5)^2 = 100x^2 + 100x + 25 = 100(x^2 + x) + 25 = 100x(x + 1) + 25$

5. If a square number is multiple of 3, it is divisible by 9 also.

Proof: $(3x)^2 = 9x^2$

6. If a square number is not a multiple of 3, it exceeds multiple of 3 by 1.

Proof: If a number is not a multiple of 3, if may be 1 or 2 more than the multiple of 3.
Let the numbers are $(3x + 1)$ and $(3x + 2)$.
$(3x + 1)^2 = 9x^2 + 6x + 1 = (9x^2 + 6x) + 1 = 3 \times (3x^2 + 2x) + 1$
$(3x + 2)^2 = 9x^2 + 12x + 4 = (9x^2 + 12x + 3) + 1 = 3 \times (3x^2 + 4x + 1) + 1$
Clearly, square of both numbers exceed multiple of 3 by 1.

UNIT'S AND TEN'S DIGITS OF A SQUARE NUMBER

Unit's digit	*unit's digit of square number*	*Preceded by*
1 or 9	1	An even number or zero
2 or 8	4	An even number or zero
3 or 7	9	An even number or zero
5	25 (last two digits)	An even number or zero
4 or 6	6	An odd number.

UNIT'S DIGIT OF A NUMBER AND ITS CUBE

0 – 0   1 – 1   4 – 4   5 – 5   6 – 6   9 – 9

i.e., same digit in unit's place

2 – 8   3 – 7   7 – 3   8 – 2

i.e., sum of unit digits of the number and its cube is 10.

Note: Numbers in the second category are the numbers that are not possible at unit's place in a square number.

MORE PROPERTIES OF SQUARE AND CUBE NUMBERS

1. Difference between squares of two even numbers is always divisible by 4.

Proof: Let the even number are $2m$ and $2n$. The difference between their squares

$$= (2m)^2 - (2n)^2 = 4(m^2 - n^2) = 4 \times (m + n) \times (m - n)$$

Note: Square of an even number is always divisible by 4. Therefore, the difference between any two such numbers is also always divisible by 4.

2. Difference between squares of any two odd numbers is always divisible by 8. **(CDS 2015)**

Proof: Let the odd numbers are $(2m - 1)$ and $(2n - 1)$.

Then difference between their squares $= (2m - 1)^2 - (2n - 1)^2$

$= (4m^2 - 4m + 1) - (4n^2 - 4n + 1) = 4m^2 - 4m + 1 - 4n^2 + 4n - 1$

$= 4(m^2 - m - n^2 + n) = 4[(m^2 - n^2) - (m - n)] = 4[(m - n)(m + n) - (m - n)]$

$= 4[(m - n)\{(m + n) - 1\}]$

Either $(m - n)$ or $\{(m + n) - 1\}$ is an even number.

$\therefore$ The result is divisible by $4 \times 2 = 8$

3. Product of two consecutive odd or even numbers increased by 1 is equal to square of average of the numbers. $(x - 1)(x + 1) + 1 = x^2 - 1 + 1 = x^2$ [Note average of number $(x - 1)$ and $(x + 1)$ is x.]

Examples:

1. $3 \times 5 + 1 = 16 = 4^2$ (**Note:** 4 is average of 3 and 5)
2. $4 \times 6 + 1 = 25 = 5^2$ (**Note:** 5 is average of 4 and 6)
3. $7 \times 9 + 1 = 64 = 8^2$ (**Note:** 8 is average of 7 and 9)

4. Sum of cubes of any three consecutive numbers is always divisible by the sum of the given numbers and by 9 separately.

Proof: Let the numbers are $(n - 1)$, (n) and $(n + 1)$, then:

Sum of the cubes of the number $= (n - 1)^3 + (n)^3 + (n + 1)^3$

$= (n^3 - 3n^2 + 3n - 1) + n^3 + (n^3 + 3n^2 + 3n + 1) = 3n^3 + 6n = 3n(n^2 + 2)$

(*a*) If 'n' is multiple of 3, then $3n$ is divisible by $3 \times 3 = 9$

It 'n' is not a multiple of 3, then n^2 exceeds multiple of 3 by 1, therefore, $(n^2 + 2)$ is divisible by 3.

(*b*) Sum of the mumbers $= (n - 1) + (n) + (n + 1) = 3n$.

Clearly, $3n(n^2 + 2)$ is always divisible by $3n$.

5. Cube of a natural number is either of form $9n$, $9n + 1$, $9n + 8$ where n is a natural number

PYTHAGOREAN TRIPLES

$3^2 + 4^2 = 5^2$	$5^2 + 12^2 = 13^2$	$8^2 + 15^2 = 17^2$	$7^2 + 24^2 = 25^2$
$6^2 + 8^2 = 10^2$	$10^2 + 24^2 = 26^2$	$16^2 + 30^2 = 34^2$	$20^2 + 21^2 = 29^2$
$9^2 + 12^2 = 15^2$	$9^2 + 40^2 = 41^2$	$11^2 + 60^2 = 61^2$	$12^2 + 35^2 = 37^2$
$16^2 + 63^2 = 65^2$	$28^2 + 45^2 = 53^2$	$33^2 + 56^2 = 65^2$	

SHORT-CUT METHODS OF FINDING SQUARE

1. Number consisting of 1's only:

Steps: 1. Count number of 1's in the given number. Let there are n number of digits.

2. Write down counting from 1 to n in the ascending order. Thereafter write counting in descending order upto 1.

Example: $11^2 = 121$ $\quad 111^2 = 12321$ $\quad 1111^2 = 1234321$

$101^2 = 10201$ $\quad 10101^2 = 102030201$ $\quad 1010101^2 = 1020304030201$

$1001^2 = 1002001$ $\quad 1001001^2 = 1002003002001$

2. Number ending in 5:

Method: A number ending in 5 can be written as $(10x + 5)$.

$\mathbf{(10x + 5)^2 = 100x^2 + 100x + 25 = 100x(x + 1) + 25 = x(x + 1) \mid 25}$

Example: $25^2 = 2 \times 3 \mid 25 = 625$

$35^2 = 3 \times 4 \mid 25 = 1225$

$85^2 = 8 \times 9 \mid 25 = 7225$

$195^2 = 19 \times 20 \mid 25 = 38025$

$1005^2 = 100 \times 101 \mid 25 = 1010025$

Note: Square of these numbers ends at 25 and are preceded by the even numbers.

3. Number ending in 25:

Method: A number ending in 25 can be written as $(100x + 25)$.

$$(100x + 25)^2 = 10000x^2 + 5000x + 625 = 10000x\left(x + \frac{1}{2}\right) + 625 = x\left(x + \frac{1}{2}\right) \mid 0625$$

Example: $825^2 = 8 \times 8.5 \mid 0625 = 68\ 0625$

$725^2 = 7 \times 7.5 \mid 0625 = 52.5 \mid 0625 = 52\ 5625$

$1425^2 = 14 \times 14.5 \mid 0625 = 203\ 0625$

4. Numbers less than base:

Base: $x^2 = x^2 - y^2 + y^2 = (x + y) \times (x - y) + y^2$ **[Note:** $x + y$ is equal to 10^n]

Rule: If $a = 10^n - b$, then $a^2 = a - b \mid b^2$

Note: Squares of the 'differences from the respective bases' are converted into n digit number where base is 10^n.

Example 1: Find square of 9.

Sol. $9 = 10^1 - 1 \Rightarrow 9^2 = 9 - 1 \mid 1^2 = 8 \mid 1 = 81$

Example 2: Find square of 97.

Sol. $97 = 10^2 - 3 \Rightarrow 97^2 = 97 - 3 \mid 3^2 = 94 \mid 09 = 9409$

Example 3: Find square of 999.

Sol. $999 = 10^3 - 1 \Rightarrow 999^2 = 999 - 1 \mid 1^2 = 998 \mid 001 = 998001$

Example 4: Find square of 99987.

Sol. $99987 = 10^5 - 13 \Rightarrow 99987^2 = 99987 - 13 \mid 13^2 = 99974 \mid 00169 = 9997400169$

Example 5: Find square of 999973.

Sol. $999973 = 10^6 - 27 \Rightarrow 999973^2 = 999973 - 27 \mid 27^2 = 999946 \mid 000729 = 999946000729$

Example 6: Find square of 89.

Sol. $89 = 10^2 - 11 \Rightarrow 89^2 = 89 - 11 \mid 11^2 = 78 \mid 121 = 79\ 21$

Note: Here square of 'difference from base' is a 3-digit number whereas, we need two digits from this part. ∴ We write down 21 and carry over 1 and add it to the next part.

5. Numbers greater than base:

Base: $x^2 = x^2 - y^2 + y^2 = (x + y) \times (x - y) + y^2$ **[Note:** $x - y$ is equal to 10^n]

Rule: If $a = 10^n + b$, then $a^2 = a + b \mid b^2$

Note: Square of the 'difference from the base' is converted into n digit number where base is 10^n.

Example 1: Find square of 12.

Sol. $12 = 10^1 + 2 \Rightarrow 12^2 = 12 + 2 \mid 2^2 = 14 \mid 4 = 144$

Example 2: Find square of 101.

Sol. $101 = 10^2 + 1 \Rightarrow 101^2 = 101 + 1 \mid 1^2 = 102 \mid 01 = 10201$

Example 3: Find square of 1009.

Sol. $1009 = 10^3 + 9 \Rightarrow 1009^2 = 1009 + 9 \mid 9^2 = 1018 \mid 081 = 1018081$

Example 4: Find square of 10018.

Sol. $10018 = 10^4 + 18 \Rightarrow 10018^2 = 10018 + 18 \mid 18^2 = 10036 \mid 0324 = 100360324$

Example 5: Find square of 119.

Sol. $119 = 10^2 + 19 \Rightarrow 119^2 = 119 + 19 \mid 19^2 = 138 \mid 361 = 14161$

Note: Here square of 'difference from base' is a 3-digit number whereas we require two digits from this part. $\therefore$ We write down 61 and carry over 3 and add it to the next part.

6. Two digit-number (General Formula)

Base: $(x + y)^2 = x^2 + 2xy + y^2 \quad \therefore \; (xy)^2 = x^2 \mid 2xy \mid y^2$

Note: Starting from right-hand side, write down one digit from each part and extra digit, if any, is carried over and added to the next part.

Example: $26^2 = 2^2 \mid 2 \times 2 \times 6 \mid 6^2 = 4 \mid 24 \mid 36 = 676$

$32^2 = 3^2 \mid 2 \times 3 \times 2 \mid 2^2 = 9 \mid 12 \mid 4 = 1024$

$51^2 = 5^2 \mid 2 \times 5 \times 1 \mid 1^2 = 25 \mid 10 \mid 1 = 2601$

Note: Square of **three-digit number** when **middle digit is 0** (2 digits are taken from each part):

Example: $209^2 = 2^2 \mid 2 \times 2 \times 9 \mid 9^2 = 4 \mid 36 \mid 81 = 43681$

$503^2 = 5^2 \mid 2 \times 5 \times 3 \mid 3^2 = 25 \mid 30 \mid 09 = 253009$

Note the following examples (2 digits are taken from each part):

$413^2 = 4^2 \mid 2 \times 4 \times 13 \mid 13^2 = 16 \mid 104 \mid 169 = 170569$

$712^2 = 7^2 \mid 2 \times 7 \times 12 \mid 12^2 = 49 \mid 168 \mid 144 = 50 \mid 69 \mid 44 = 506944$

Note the following examples (1 digit are taken from each part):

$113^2 = 11^2 \mid 2 \times 11 \times 3 \mid 3^2 = 121 \mid 66 \mid 9 = 12769$

$127^2 = 12^2 \mid 2 \times 12 \times 7 \mid 7^2 = 144 \mid 168 \mid 49 = 16129$

Note. Four-digit number when **middle two digits are 0** (3 digits are taken from each part):

Example: $3002^2 = 3^2 \mid 2 \times 3 \times 2 \mid 2^2 = 9 \mid 012 \mid 004 = 9012004$

$7005^2 = 7^2 \mid 2 \times 7 \times 5 \mid 5^2 = 49 \mid 070 \mid 025 = 49070025$

Note the following examples (2 digits are taken from each part).

$1013^2 = 10^2 \mid 2 \times 10 \times 13 \mid 13^2 = 100 \mid 260 \mid 169 = 1026169$

$2302^2 = 23^2 \mid 2 \times 23 \times 2 \mid 2^2 = 529 \mid 92 \mid 04 = 5299204$

$1113^2 = 11^2 \mid 2 \times 11 \times 13 \mid 13^2 = 121 \mid 286 \mid 169 = 1238769$

7. Any number of digits (Duplex Method)

Steps: 1. Sets of digits are made in the following manner:

First set consists of unit's digit only, second set of last 2 digits, third set of last 3 digits and similarly one digit is added each time, till all the digits are included.

Next set consists of all digits except unit's digit, next set of all digits except last 2 digits and so on.

2. Duplex of each pair is calculated in the following manner:

(*a*) Pairs of two numbers equidistant from both sides: Find twice the product of such numbers.

(*b*) Single middle number: Find square of that number.

(*c*) Add the results obtained in (*a*) and (*b*).

3. Write the Duplex of each pair starting from right-hand side.

4. One digit (right-hand side) is written down from each part and extra digit, if any is carried over and added to its immediate left part.

Note: This method can be used for squaring a number having any number of digits.

Example 1: Find square of 543.

Sol. Step 1: The sets are 3, 43, 543, 54, 5

Step 2: Duplex of 3 is $3^2 = 9$

Duplex of 43 is $2 \times (4 \times 3) = 24$

Duplex of 543 is $4^2 + 2 \times (5 \times 3) = 16 + 30 = 46$

Duplex of 54 is $2 \times (5 \times 4) = 40$

Duplex of 5 is $5^2 = 25$

$\therefore 543^2 = 25 \mid 40 \mid 46 \mid 24 \mid 9 = 294849$

Example 2: Find square of 1427.

Sol. Step 1: The sets are 7, 27, 427, 1427, 142, 14, 1

Step 2: Duplex of 7 is $7^2 = 49$

Duplex of 27 is $2 \times (2 \times 7) = 28$

Duplex of 427 is $2^2 + 2 \times (4 \times 7) = 60$

Duplex of 1427 is $2 \times (1 \times 7) + 2 \times (4 \times 2) = 30$

Duplex of 142 is $4^2 + 2 \times (1 \times 2) = 20$

Duplex of 14 is $2 \times (1 \times 4) = 8$

Duplex of 1 is $1^2 = 1$

$\therefore 1427^2 = 1 \mid 8 \mid 20 \mid 30 \mid 60 \mid 28 \mid 49 = 2036329.$

8. Numbers around 50:

Base: $(50 \pm a)^2 = (25 \pm a) \times 100 + a^2$

Rule: If $x = 50 \pm a$, then $x^2 = 25 \pm a \mid a^2$ **[Note:** a^2 **is written as two-digit number.]**

Example 1: Find square of 53.

Sol. $53 = 50 + 3 \Rightarrow 53^2 = 25 + 3 \mid 3^2 = 28 \mid 09 = 2809$

Example 2: Find square of 63.

Sol. $63 = 50 + 13 \Rightarrow 63^2 = 25 + 13 \mid 13^2 = 38 \mid 169 = 3969$

Example 3: Find square of 49.

Sol. $49 = 50 - 1 \Rightarrow 49^2 = 25 - 1 \mid 1^2 = 24 \mid 01 = 2401$

Example 4: Find square of 42.

Sol. $42 = 50 - 8 \Rightarrow 42^2 = 25 - 8 \mid 8^2 = 17 \mid 64 = 1764$

9. Numbers around 500:

Base: $(500 \pm a)^2 = (250 \pm a) \times 1000 + a^2$

Rule: If $x = 500 \pm a$, then $x^2 = 250 \pm a \mid a^2$ **[Note:** a^2 **is written as three-digit number.]**

Example 1: Find square of 501.

Sol. $501 = 500 + 1 \Rightarrow 501^2 = 250 + 1 \mid 1^2 = 251 \mid 001 = 251001$

Example 2: Find square of 511.

Sol. $511 = 500 + 11 \Rightarrow 511^2 = 250 + 11 \mid 11^2 = 261 \mid 121 = 261121$

Example 3: Find square of 498.

Sol. $498 = 500 - 2 \Rightarrow 498^2 = 250 - 2 \mid 2^2 = 248 \mid 004 = 248004$

10. Base : $x^2 = x^2 - a^2 + a^2 = (x + a)(x - a) + a^2$

Method: Add and subtract the same number so as to make one number round off and then solve as shown below.

Examples: $31^2 = (31 + 1) \times (31 - 1) + 1^2 = 32 \times 30 + 1^2 = 960 + 1 = 961$

$38^2 = (38 + 2) \times (38 - 2) + 2^2 = 40 \times 36 + 2^2 = 1440 + 4 = 1444$

$83^2 = (83 + 3) \times (83 - 3) + 3^2 = 86 \times 80 + 3^2 = 6880 + 9 = 6889$

11. Numbers ending in 15, 25, 35, 45, etc.

Note: Four digits are taken from the right-hand side part.

Numbers ending in 15 (Note: Double of last two digits is 30):

$$215^2 = 2^2 + 2 \times \frac{3}{10} \mid 15^2 = 4.6 \mid 0225 = 4\ 6225$$

$$515^2 = 5^2 + 5 \times \frac{3}{10} \mid 15^2 = 26.5 \mid 0225 = 26\ 5225$$

Numbers ending in 25 (Note: Double of last two digits is 50):

$$225^2 = 2^2 + \frac{1}{2} \times 2 \mid 25^2 = 5 \mid 0625 = 5\ 0625$$

$$825^2 = 8^2 + \frac{1}{2} \times 8 \mid 25^2 = 68 \mid 0625 = 68\ 0625$$

$$1025^2 = 10^2 + \frac{1}{2} \times 10 \mid 25^2 = 105 \mid 0625 = 105\ 0625$$

$$725^2 = 7^2 + \frac{1}{2} \times 7 \mid 25^2 = 52.5 \mid 0625 = 52\ 5625$$

Numbers ending in 35 (Note: Double of last two digits is 70):

$$135^2 = 1^2 + 1 \times \frac{7}{10} \mid 35^2 = 1.7 \mid 1225 = 1\ 8225$$

$$235^2 = 2^2 + 2 \times \frac{7}{10} \mid 35^2 = 5.4 \mid 1225 = 5\ 5225$$

Numbers ending in 45 (Note: Double of last two digits is 90):

$$245^2 = 2^2 + 2 \times \frac{9}{10} \mid 45^2 = 5.8 \mid 2025 = 6\ 0025$$

Numbers ending in 55, 65, 75 etc. :

$$455^2 = 4^2 + 4 \times \frac{11}{10} \mid 55^2 = 20.4 \mid 3025 = 20\ 7025$$

$$1065^2 = 10^2 + 10 \times \frac{13}{10} \mid 65^2 = 113 \mid 4225 = 113\ 4225$$

$$675^2 = 6^2 + 6 \times \frac{15}{10} \mid 75^2 = 45 \mid 5625 = 45\ 5625$$

12. When square of previous or next number is easy to calculate.

(*i*) $(x + 1)^2 = x^2 + 2x + 1 = x^2 + [x + (x + 1)]$

Examples: $31^2 = 30^2 + (30 + 31) = 900 + 61 = 961$

$91^2 = 90^2 + (90 + 91) = 8100 + 181 = 8281$

$231^2 = 230^2 + (230 + 231) = 52900 + 461 = 53361$

$2501^2 = 2500^2 + (2500 + 2501) = 6250000 + 5001 = 6255001$

(ii) $(x-1)^2 = x^2 - 2x + 1 = x^2 - [x + (x-1)]$

Examples: $49^2 = 50^2 - (50 + 49) = 2500 - 99 = 2401$

$179^2 = 180^2 - (179 + 180) = 32400 - 359 = 32041$

$3499^2 = 3500^2 - (3499 + 3500) = 12250000 - 6999 = 12243001$

(**Note:** $3499 + 3500 = 2 \times 3500 - 1$)

13. Square of mixed number which can be expressed as $x\frac{1}{2}$ is $x(x+1) \mid \frac{1}{4}$

Example: $\left(2\frac{1}{2}\right)^2 = 2 \times 3 \mid \frac{1}{4} = 6\frac{1}{4}$ $\left(5\frac{1}{2}\right)^2 = 5 \times 6 \mid \frac{1}{4} = 30\frac{1}{4}$

FINDING CUBE

1. Cube of a number consisting of 9s only:

$$(999 \ldots\ldots n \text{ times})^3 = \underbrace{9\ldots\ldots}_{(n-1)\text{ times}}7\underbrace{0\ldots\ldots}_{(n-1)\text{ times}}2\underbrace{9\ldots\ldots}_{n\text{ times}}$$

Example: $9^3 = 729$

$99^3 = 970299$

$999^3 = 997002999$

$9999^3 = 999700029999$

2. Cube of a number ending in 5 $= n(n+1) \times \frac{n+(n+1)}{2} \,\Big|\, 125 \times [n \times (n+1)]$

[Note: $125 \times 8 = 1000$]

Example 1: $35^3 = 3 \times 4 \times \frac{7}{2} \mid 125 \times 7 = 42 \mid 875 = 42875$

Example 2: $45^3 = 4 \times 5 \times \frac{9}{2} \mid 125 \times 9 = 90 \mid 1125 = 91125$

Example 3: $95^3 = 9 \times 10 \times \frac{19}{2} \mid 125 \times 19 = 855 \mid 2375 = 857375$

3. Cube of a number (General Formula):

Base: $(x+y)^3 = x^3 + 3x^2y + 3xy^2 + y^3$ (**Note:** One digit is taken from each part.)

Examples: 12^3 *i.e.* $x = 1$ and $y = 2$

$= 1^3 \mid 3 \times 1^2 \times 2 \mid 3 \times 1 \times 2^2 \mid 2^3 = 1 \mid 6 \mid 12 \mid 8 = 1728$

23^3 *i.e.* $x = 2$ and $y = 3$

$= 2^3 \mid 3 \times 2^2 \times 3 \mid 3 \times 2 \times 3^2 \mid 3^3 = 8 \mid 36 \mid 54 \mid 27 = 12167$

In the following examples, two digits are taken from each part:

104^3 *i.e.* $x = 1$ and $y = 4$

$= 1^3 \mid 3 \times 1^2 \times 4 \mid 3 \times 1 \times 4^2 \mid 4^3 = 1 \mid 12 \mid 48 \mid 64 = 1124864$

205^3 *i.e.* $x = 2$ and $y = 5$

$= 2^3 \mid 3 \times 2^2 \times 5 \mid 3 \times 2 \times 5^2 \mid 5^3 = 8 \mid 60 \mid 150 \mid 125 = 8615125$

Note: Three digits are taken from each part in the following example:

1005^3 *i.e.* $x = 1$ and $y = 5$

$= 1^3 \mid 3 \times 1^2 \times 5 \mid 3 \times 1 \times 5^2 \mid 5^3 = 1 \mid 015 \mid 075 \mid 125 = 1015075125$

Note: One digit is taken from each part in the following example:

$114^3 = 11^3 \mid 3 \times 11^2 \times 4 \mid 3 \times 11 \times 4^2 \mid 4^3 = 1331 \mid 1452 \mid 528 \mid 64 = 1481544$

Note: Two digits are taken from each part in the following example:

$512^3 = 5^3 \mid 3 \times 5^2 \times 12 \mid 3 \times 5 \times 12^2 \mid 12^3 = 125 \mid 900 \mid 2160 \mid 1728 = 134217728$

4. Another method of finding cube of a number using identity:

Base: $(x + y)^3 = x^3 + (x^2y + 2x^2y) + (xy^2 + 2xy^2) + y^3$. This can also be written as

$$\begin{array}{rcccc} & x^3 & x^2y & xy^2 & y^3 \\ + & & 2x^2y & 2xy^2 & \\ \hline \end{array}$$

Note: In each successive term power of x is reduced by 1 and power of y is increased by 1 *i.e.* each successive term is y/x times of the previous term.

Examples:

$12^3 =$

$$\begin{array}{cccc} 1 & 2 & 4 & 8 \\ & 4 & 8 & \\ \hline 1 & 6 & 12 & 8 \\ \hline \end{array}$$

$= 1728$

$26^3 =$

$$\begin{array}{cccc} 8 & 24 & 72 & 216 \\ & 48 & 144 & \\ \hline 8 & 72 & 216 & 216 \\ \hline \end{array}$$

$= 17576$

$27^3 =$

$$\begin{array}{cccc} 8 & 28 & 98 & 343 \\ & 56 & 196 & \\ \hline 8 & 84 & 294 & 343 \\ \hline \end{array}$$

$= 19683$

$52^3 =$

$$\begin{array}{cccc} 125 & 50 & 20 & 8 \\ & 100 & 40 & \\ \hline 125 & 150 & 60 & 8 \\ \hline \end{array}$$

$= 140608$

5. Another method of finding cube of a number, using Algebric Identity:

Base: $(x + y)^3 = x^3 + y^3 + 3xy\,(x + y)$ and $(x - y)^3 = x^3 - y^3 - 3xy\,(x - y)$

Example: $1001^3 = (1000 + 1)^3 = 1000^3 + 1^3 + 3 \times 1000 \times 1 \times (1000 + 1)$ **(SSC CGL I 2014)**

$= 1000000000 + 1 + 3003000 = 1003003001$

Example: $999^3 = (1000 - 1)^3 = 1000^3 - 1^3 - 3 \times 1000 \times 1 \times (1000 - 1)$

$= 1000000000 - 1 - 3000000 + 3000 = 997002999$

$1^2 + 2^2 + 2^2 = 3^2$ (Note: $1^2 + 2^2 = 5 = 2 + 3$)

$2^2 + 3^2 + 6^2 = 7^2$ (Note: $2^2 + 3^2 = 13 = 6 + 7$)

$3^2 + 4^2 + 12^2 = 13^2$ (Note: $3^2 + 4^2 = 25 = 12 + 13$)

$4^2 + 5^2 + 20^2 = 21^2$ (Note: $4^2 + 5^2 = 41 = 20 + 21$)

$5^2 + 6^2 + 30^2 = 31^2$ (Note: $5^2 + 6^2 = 61 = 30 + 31$)

$6^2 + 7^2 + 42^2 = 43^2$ (Note: $6^2 + 7^2 = 85 = 42 + 43$)

$1^3 + 6^3 + 8^3 = 9^3$

$3^3 + 4^3 + 5^3 = 6^3 \Rightarrow 6^3 + 8^3 + 10^3 = 12^3$

$8^2 + 9^2 + 12^2 = 17^2$

$1^3 = 1 = 1^2$

$1^3 + 2^3 = 9 = 3^2$

$1^3 + 2^3 + 3^3 = 36 = 6^2$

$1^3 + 2^3 + 3^3 + 4^3 = 100 = 10^2$

$1^3 + 2^3 + 3^3 + 4^3 + 5^3 = 225 = 15^2$

$1^3 + 2^3 + 3^3 + 4^3 + 5^3 + 6^3 = 441 = 21^2$

$1^3 + 2^3 + 3^3 + 4^3 + 5^3 + 6^3 + 7^3 = 784 = 28^2$

SOLVED EXERCISE

1. The number of perfect cubes between 1 and 1000009 which are exactly divisible by 9 is **(SSC CHSL 2012)**

(*a*) 33 (*b*) 101 (*c*) 35 (*d*) 99

Sol. (*a*) $100^3 < 1000009 < 101^3$

Cube of a number which is multiple of 3, is divisible by 9.

Between 1 and 100, there are $100 \div 3 = 33$ perfrect cubes which are multiples of 3.

2. Which one of the following numbers is not a square of any natural number ? **(CDS 2010)**

(*a*) 5041 (*b*) 9852 (*c*) 1936 (*d*) 4624

Sol. (*b*) Unit digit 2 is not possible in a square number.

3. If $59^2 + 12^2 = x^2 + 600$, find the value of x? **(SBI Clerical 2009)**

(*a*) 30 (*b*) 45 (*c*) 53 (*d*) None of these

Sol. (*d*) $59^2 = 25 + 9 \mid 9^2 = 3481$ $x^2 = 3481 + 144 - 600 = 3025$

$\Rightarrow x = 55$ (Note: $55^2 = 5 \times 6 \mid 25 = 3025$)

4. If the expression $x + 809436 \times 809438$ be a perfect square, then the value of x is **(SSC CGL Tier II 2012)**

(*a*) 809436 (*b*) 809438 (*c*) 0 (*d*) 1

Sol. (*d*) $809436 \times 809438 = (809437 - 1) \times (809437 + 1) = 809437^2 - 1$

Note: Product of two consecutive odd or even numbers increased by unity is equal to square of average of the numbers.

5. If ($x + 7954 \times 7956$) be a square number, then the value of 'x' is **(SSC CGL Tier I 2013)**

(*a*) 1 (*b*) 16 (*c*) 9 (*d*) 4

Sol. (*a*) $(7954 \times 7956) = (7955 - 1) \times (7955 + 1) = 7955^2 - 1$

Note: Product of two consecutive odd or even numbers increased by unity is equal to square of average of the numbers.

6. If $x = (0.08)^2$, $y = \dfrac{1}{(0.08)^2}$ and $z = (1 - 0.08^2) - 1$, then out of the following the true relation is **(SSC CGL Tier I 2012)**

(*a*) $y < z < x$ (*b*) $z < x < y$ (*c*) $y < x$ and $x = z$ (*d*) $x < y$ and $x = z$

Sol. (*b*) $x = (0.08)^2 \Rightarrow 0 < x < 1$ $y = \dfrac{1}{(0.08)^2} = (12.5)^2 > 1$

$z = (1 - 0.08^2) - 1 = -0.08^2 < 0$ $\therefore z < x < y$

7. What should be added to 1752 to make it a perfect square number?

(*a*) 6 (*b*) 10 (*c*) 12 (*d*) 18

Sol. (*c*) $41^2 = 1681$, which is smaller than 1752.

And $42^2 = 1764$, which is slightly bigger than 1752.

$\therefore$ Required number is $1764 - 1752 = 12$.

Trick: We check all the options one by one, concentrating on unit's digit.

6 : Unit's digit is $2 + 6 = 8$ (8 is not possible at unit's place).

10 : Unit's digit is $2 + 0 = 2$ (2 is not possible at unit's place).

12 : Unit's digit is $2 + 2 = 4$ (4 is possible at unit's digit).

18 : Last two digits are $52 + 18 = 70$ (Single '0' is not possible at the end of a square number).

$\therefore$ 12 is the only possible number out of the given options.

8. A gardener has 1000 plants. He wants to plant them in such a way that the number of rows and the number of columns remains the same. What is the minimum number of plants that he needs more for this purpose ? **[UPSC Civil Services (Pre.) 2013]**

(*a*) 14 (*b*) 24 (*c*) 32 (*d*) 34

Sol. (*b*) $32^2 = 1024 \Rightarrow$ Required number $= 1024 - 1000 = 24$

9. Find the least number to be added to 1425 to make it perfect square. **(Bank Clerical)**

(*a*) 9 (*b*) 19 (*c*) 96 (*d*) None of these

Sol. (*b*) $38^2 = 1444$ and $37^2 = 1369 \quad \Rightarrow \quad 37^2 < 1425 < 38^2$

$\therefore$ Required number $= 1444 - 1425 = 19$

Alternative Method: 1425 is closer to 1600 *i.e.* 40^2 $\quad 40^2 = 1600$

$39^2 = 1600 - 79$ (*i.e.* $40 + 39$) $= 1521$ $\quad 38^2 = 1521 - 77$ (*i.e.* $39 + 38$) $= 1444$

$\Rightarrow$ Required number $= 1444 - 1425 = 19$

10. What should be added to 16259 to make it a perfect square number?

(*a*) 107 (*b*) 125 (*c*) 212 (*d*) 423

Sol. (*b*) $127^2 = 16129$, which is smaller than 16259. And $128^2 = 16384$, which is bigger than 16259.

$\therefore$ Required number is $16384 - 16259 = 125$.

Trick: We check the given options one by one:

423 : Unit's digit is $9 + 3 = 2$ (not possible).

107 : Unit's digit is $9 + 7 = 6$ (possible). But $59 + 07 = 66$ (The number is an even number but not divisible by 4 and hence not possible). [also if unit's digit is 6, ten's digist must be an odd number.]

212 : Unit's digit is $9 + 2 = 1$ (possible). Last two digits are $59 + 12 = 71$ *i.e.* 1 at unit's place (this implies square of a number ending in 1 or 9), preceded by an odd number but it should have been either an even number or 0 only and hence not possible.

125 : Unit's digit is $9 + 5 = 4$ (possible). Last two digits are $59 + 25 = 84$. 84 is an even number which is divisible by 4. It being the only option left, is the answer.

11. What should be subtracted from 83908 to make it a perfect square number?

(*a*) 103 (*b*) 387 (*c*) 404 (*d*) 501

Sol. (*b*) $290^2 = 84100$, which is bigger than 83908.

And $289^2 = 83521$, which is smaller than 83908

$\therefore$ Required number is $83908 - 83521 = 387$

Trick: We check the given options one by one:

501: Unit's digit is $8 - 1 = 7$ (not possible).

103: Unit's digit is $8 - 3 = 5$ (possible). Ten's digit must be 2.

Here $08 - 03 = 05$ (Not possible).

404: Unit's digit is 8 – 4 = 4. (possible)
Last two digits = 08 – 04 = 04 (Possible as the number is even and is divisible by 4).
Sum of digits of (83908 – 404) = 1 – 8 = 2 (ignoring 9).
∴ The number when reduced by 1 is not divisible by 3. (Not possible).

387: 83908 – 387 = 83521. Unit's digit is 1 (Possible).
Ten's digit is 2, an even number (Possible).
Sum of digits of 83521 is 1 *i.e.* the number which when divided by 3 will leave remainder 1. (Possible)
∴ 387 being the only possible number, is the answer.

12. The least number that must be subtracted from 63520 to make the result a perfect square, is: **(SSC CGL 2002)**

(*a*) 16 (*b*) 20 (*c*) 24 (*d*) 30

Sol. (*a*) $252^2 = 63504 \Rightarrow$ Required number = 63520 – 63504 = 16

13. Find the least number to be subtracted from 16800 to make it a perfect square. **(SSC CGL Tier I 2014)**

(a) 169 (b) 159 (c) 249 (d) 219

Sol. (*b*) $130^2 = 16900 \Rightarrow 129^2 = 16900 - (130 + 129) = 16641$

$\Rightarrow$ Required number = 16800 – 16641 = 159

14. $(23)^3 = ?$ **(Vijaya Bank PO 2008)**

(*a*) 12167 (*b*) 17576 (*c*) 2197 (*d*) 5832

Sol. (*a*) $23^3 =$

8 12 18 27
  24 36
8 36 54 27
= 12167

15. $(1515)^2 = ? \div 1515$ **(Bank PO 2008)**

(*a*) 3030 (*b*) 2295225 (*c*) 4485 (*d*) None of these

Sol. (*d*) $(1515)^2 = ? \div 1515 \Rightarrow ? = (1515)^3$

$15^3 =$
1 5 25 125
  10 50
1 15 75 125
= 3375

$(15\,15)^3 =$
3375 3375 3375 3375
  6750 6750
3375 10125 10125 3375
= 3477 26 58 75 (2 digits taken from each part)

16. The product of two consecutive numbers is 3192. What is the smaller number ? **(EPFO SSA 2007)**

(*a*) 56 (*b*) 57 (*c*) 58 (*d*) None of these

Sol. (*a*) 31 = 25 + 6 $\Rightarrow$ 3192 is closer to square of 50 + 6 = 56

Now 56 × 57 = 3192

17. The product of two consecutive natural numbers is 6162, what is the smaller number ? **(SBI Clerical 2008)**

(*a*) 72 (*b*) 76 (*c*) 78 (*d*) None of these

Sol. (*c*) 6162 is smaller than 80^2 or 6400. 2 is the unit digit in 6162 and $8 \times 9 = 72$

$78 \times 79 = 6162$ $\Rightarrow$ Required number is 78.

18. If 103^2 is added to the square of a number, the answer so obtained is 11138, what is the number? **(SBI Clerical 2008)**

(*a*) 19 (*b*) 23 (*c*) 27 (*d*) None of these

Sol. (*b*) Let the required number be x. Then $x^2 + 103^2 = 11138$

$\Rightarrow x^2 = 11138 - 103^2 = 11138 - 10609 = 529 \Rightarrow x = \sqrt{529} = 23$

19. $91^2 + 41^2 - \sqrt{?} = 9858.$ **(SBI Clerical 2009)**

(*a*) 72 (*b*) 10816 (*c*) 78 (*d*) None of these

Sol. (*b*) $8281 + 1681 - \sqrt{?} = 9858 \Rightarrow \sqrt{?} = 8281 + 1681 - 9858 = 104 \Rightarrow ? = 104^2 = 10816$

20. $(19.89)^2 - \sqrt{625} = ?$ **(Bank PO Exam.)**

(*a*) 395.6121 (*b*) 359.6121 (*c*) 370.1621 (*d*) 370.6121

Sol. (*d*) $(1989)^2$ = Duplex of 1 | 19 | 198 | 1989 | 989 | 89 | 9

= 1 | 18 | 97 | 162 | 226 | 144 | 81 = 3956121

$\therefore (19.89)^2 = 395.6121 \Rightarrow (19.89)^2 - \sqrt{625} = 395.6121 - 25 = 370.6121$

Alternative method: $(19.89)^2 = (20 - 0.11)^2 = 400 + 0.0121 - 4.4 = 395.6121$

$\therefore (19.89)^2 - \sqrt{625} = 395.6121 - 25 = 370.6121$

21. $(?)^2 + (123)^2 = (246)^2 - (99)^2 - 2462$ **(Canara Bank PO 2009)**

(*a*) 184 (*b*) 186 (*c*) 182 (*d*) 180

Sol. (*c*) 123^2 = Duplex of 1 | 12 | 123 | 23 | 3 = 1 | 4 | 10 | 12 | 9 = 15129

$(246)^2 = (123 \times 2)^2 = 123^2 \times 4 = 15129 \times 4 = 60516$

$(?)^2 = (246)^2 - (99)^2 - 2462 - (123)^2 = 60516 - 9801 - 2462 - 15129 = 33124 = 182^2$

22. A toy factory manufactured a batch of electronic toys. If the toys were packed in boxes of 115 each, 13 boxes would not be filled completely. If the toys were packed in boxes of 65 each, 22 boxes would not be enough to pack all of them. Coincidentally, in the end, the toys were packed in n boxes containing n toys each, without any remainder. The total number of toys was: **(SSC CGL Tier II 2012)**

(*a*) 1444 (*b*) 1454 (*c*) 1424 (*d*) 1434

Sol. (*a*) n boxes containing n toys each $\Rightarrow$ Total number of toys $= n \times n = n^2$

Out of given options, only 1444 is a square number as $38^2 = 1444$.

23. If $N = p^3 \times q^4 \times r^7$, then number of perfect squares which are factors of N.

(*a*) 14 (*b*) 84 (*c*) 7 (*d*) 24

Sol. (*d*) Powers of p can be 0 or 2, *i.e.*, 2 possiblities.

Powers of q can be 0, 2, or 4 *i.e.*, 3 possibilities.

Powers of r can be 0, 2, 4 or 6, *i.e.*, 4 possiblities

$\therefore$ Total factors (square numbers) $= 2 \times 3 \times 4 = 24$

CHAPTER 7

SQUARE ROOT AND CUBE ROOT

Square root: Square root of a number is the number, product of which by itself is equal to the given number. Square root is also known as under-root or 2nd root. Square root is denoted by $\sqrt{\ }$.

If $x^2 = y$, then $\sqrt{y} = x$

Examples: $3^2 = 9 \Rightarrow \sqrt{9} = 3$ $\quad 11^2 = 121 \Rightarrow \sqrt{121} = 11$

How to find number of digits in square-root of a number having *n* number of digits:

1. Make pairs of two digits each. Single digit left, if any, would make one pair.
2. Digits in the square-root of the number will be equal to number of pairs of the square number.

Alternatively, divide n by 2. If quotient is a natural number, quotient is the number of digits in the square root. If quotient is a mixed number, round it up to the next higher natural number.

Cube root: Cube root of a number is the number, product of which by itself and itself again is equal to the given number. In other words, cube root of a number is the number whose cube is equal to the given number itself. Cube root is denoted by $\sqrt[3]{\ }$. It is also called 3rd root.

If $x^3 = y$, then $\sqrt[3]{y} = x$

Examples: $2^3 = 8 \Rightarrow \sqrt[3]{8} = 2$

$5^3 = 125 \Rightarrow \sqrt[3]{125} = 5$

How to find number of digits in cube-root of a number with *n* number of digits:

1. Make groups of three digits each. Digits left (single or two), if any, would make one group.
2. Digits in the cube-root of the number will be equal to number of groups of the cube number.

Division Method of finding square root:

Method:

1. Make pairs of two digits each starting from unit digit side.
2. Find the largest whole number whose square is equal to or less than the left most pair.
3. Find the remainder and bring down the next pair.
4. Double the quotient obtained and write it down to left side of the pair brought down.
5. Write one digit number to the right of number of step 4.
   Multiply the number so obtained with same single digit so that the product is equal to or less than the number of step 3.
6. Repeat the processes from 3 to 5 till all the pairs are taken.

Example:

Divisor	234	
2	5, 47, 56	Make pairs of 2 digits each starting from RHS
	4	$2^2 \le 5$
43	147	$2 \times 2 = 4$ is next divisor; $43 \times 3 \le 147$
	129	
464	1856	$23 \times 2 = 46$ or $43 + 3 = 46$; $464 \times 4 \le 1856$
	1856	
	×	

Divisor	489	
4	23, 91, 21	$4^2 \le 23$;
	16	
88	791	$4 \times 2 = 8$; $88 \times 8 \le 791$
	704	
969	8721	$48 \times 2 = 96$ or $88 + 8 = 96$; $969 \times 9 \le 8721$
	8721	
	×	

Another Method of finding square root of 3 or 4 digit square numbers:

Example: Find the square root of 3136.

Sol. Starting from right hand side, make pairs of two digits each. The digits in the square root is equal to the number of pairs in the number. Therefore, solution in this case will have two digits.

Find square root of first pair (LHS). Square root of 31 is 5. This is the first digit of our answer. Write this digit on top of the number and write double of this digit as our next divisor on the left side of the number.

$$\begin{array}{r|l} & \underline{\;\;5\;\;\;\;} \\ 10 & 31,36 \end{array}$$

Subtract square of the first digit from the first pair: $31 - 5^2 = 6$

Carry the remainder 6 to the next digit of the number. Divide the new number 63 by the new divisor 10. This gives 6 as next digit of answer and 3 remainder.

$$\begin{array}{r|l} & \underline{\;56\;\;\;} \\ 10 & 31,36 \\ & \;\;63 \end{array}$$

36 is our new working number. Duplex of 6 is $6^2 = 36$.

There is no remainder. 3136 is perfect square and 56 is its square root.

PRIME FACTORISATION METHOD FOR FINDING SQUARE ROOT AND CUBE ROOT

Steps for finding square root

1. Write prime factors of the given number.
2. Select the factors that are in pair.
3. Take one number from each pair and multiply them. The product of such numbers followed by the product (written under $\sqrt{\;}$ sign) of the numbers which are not in pair.

Example: Find square root of 11025 (By Prime factor Method).

Sol. Prime factors of 11025 are $3 \times 3 \times 5 \times 5 \times 7 \times 7$.

We want to find square-root *i.e.* 2nd root. $\Rightarrow$ Pairs are made with two same factors.

$11025 = (3 \times 3) \times (5 \times 5) \times (7 \times 7)$ $\Rightarrow$ $\sqrt{11025} = 3 \times 5 \times 7 = 105$

Note: This method can be used to find out any root of a given number.

Example: Find cube-root of 216.

Sol. $216 = 2 \times 2 \times 2 \times 3 \times 3 \times 3$

As we want to find cube-root *i.e.* 3rd root. $\therefore$ Groups are made with three same factors.

$216 = (2 \times 2 \times 2) \times (3 \times 3 \times 3)$ $\therefore$ Cube-root of 216 is $2 \times 3 = 6$.

SQUARE ROOT OF 3 OR 4-DIGIT SQUARE NUMBER

Base: Square of *(a b)* by Duplex Method is $a^2 \mid 2ab \mid b^2$

Steps: 1. Starting from left side, divide the given number in three parts. First part contains 1 digit (in case of 3-digit number) or 2 digits (in case of 4-digit number) and second and third parts contain one digit each.

2. Find maximum value of one-digit number (say a) whose square does not exceed first part from left-hand side.

3. Find one-digit number (say b) so that value of $(2ab)$ does not exceed second part.

4. If the third part is equal to b^2, then ab is square-root of the given number.

Note: (*i*) Remainders, if any, in the first and second parts are carried over to second and third parts respectively.

(*ii*) Square-root of any three or four-digit number is always a two-digit number.

(*iii*) By using this method, we can find out the square root of a perfect square number only.

Example 1: Find square root of 676.

Sol. 676 is divided into 3 parts: 6 | 7 | 6.

Part I : $6 \geq a^2 \Rightarrow a = 2$, Remainder = 2.

Part II : $27 \geq 2ab \Rightarrow b = 27 \div 2a = 27 \div 4 = 6$, Remainder = 3.

Part III : $36 = b^2 = 6^2$, Remainder = 0

$\therefore a = 2$ and $b = 6$ $\therefore$ Square-root of 676 is 26.

Example 2: Find square root of 9604.

Sol. 9604 is divided into 3 parts : 96 | 0 | 4

Part I : $96 \geq a^2 \Rightarrow a = 9$, Remainder = 15

Part II : $150 \geq 2ab \Rightarrow b = 150 \div 2a = 150 \div 18 = 8$, Remainder = 6

Part III : $64 = b^2 = 8^2$, Remainder = 0

$\therefore a = 9$ and $b = 8$ $\therefore$ Square-root of 9604 is 98.

Example 3: Find square root of 841.

Sol. The number is divided into three parts: 8 | 4 | 1

Part I : $8 \geq a^2 \Rightarrow a = 2$, Remainder = 4

Part II : $44 \geq 2ab \Rightarrow b = 44 \div 2a = 44 \div 4 = 9$, Remainder = 8

Part III : $81 = b^2 = 9^2$, Remainder = 0

$\therefore a = 2$ and $b = 9$ $\therefore$ The square-root of 841 is 29.

Note: Maximum value of 'b' in Part II is 11. But, we can't take $b = 11$, for two reasons – firstly 11 is a two-digit number, secondly, higher the value of 'b' the more remainder is required for 3rd part.

Example: Find the square root of 15129.

The number is divided into three parts: 151 | 2 | 9

Part I : $151 \geq a^2 \Rightarrow a = 12$, Remainder = 7

Part II : $72 \geq 2ab \Rightarrow b = 72 \div 2a = 72 \div 24 = 3$, Remainder = 0

Part III : $9 = b^2 = 3^2$, Remainder = 0

$\therefore a = 12$ and $b = 3$ $\therefore$ Square-root of 15129 is 123.

EASIER METHOD OF FINDING SQUARE-ROOT OF 3 OR 4 DIGIT SQUARE NUMBER

Method:

1. Make pairs of two digits each starting from right hand side.
2. Find the largest number whose square is less than or equal to the number in left-most pair. This is the tens digit of the square root.
3. Based on unit digit of the given number, guess the unit digit of the square root.

Example:

1. Find square root of 9801.

Sol. Two pairs are 98 and 01.

$9^2 < 98 < 10^2 \Rightarrow$ Ten's digit of square root is 9.

Second part is 01. Since the number ends in 1, square root must ends in 1 or 9.

91 is closer to 90 and 99 is closer to 100. $90^2 = 8100$ and $100^2 = 10000$

9801 is closer to 10000.

$\therefore$ The square root is 99. (we can verify $99^2 = 9801$)

2. Find square root of 5184.

Sol. Two pairs are 51 and 84.

$7^2 < 51 < 8^2 \Rightarrow$ Ten's digit of square root is 7.

Second part is 84. Since the number ends in 4, square root must ends in 2 or 8.

72 is closer to 70 and 78 is closer to 80. $70^2 = 4900$ and $80^2 = 6400$

5184 is closer to 4900.

$\therefore$ The square root is 72. (we can verify $72^2 = 5184$)

FINDING SQUARE ROOT BY APPROXIMATION

Method: Divide the given number by a divisor around approximate square root of the number. Take average of the divisor and quotient. This is approximate square root of the number.

Example 1: Find approximate square root of 310.

Sol. Take 15 as divisor. $310 \div 15$ or $620 \div 30 = 20.6$ (rounded off downwards)

Average of 15 and 20.6 is 17.8

Alternative method:

Take 20 as divisor. $310 \div 20 = 15.5$

Average of 20 and 15.5 is 17.75

Example 2: Find approximate square root of 950.

Sol. Take 30 as divisor $950 \div 30 = 31.66$ (rounded off downward)

Average of 30 and 31.66 is 30.83

Example 3: Find approximate square root of 2809.

Sol. Take 50 as divisor $2809 \div 50 = 56.18$

Average of 50 and 56.18 is 53.09

Example 4: Find approximate square root of 1200.

Sol. Take 30 as divisor $1200 \div 30 = 40$

Average of 30 and 40 is 35

Note: We can divide 1200 by 35 (above average) and average again for better approximation.

FINDING CUBE ROOT BY APPROXIMATION

Method: Divide the given number by a divisor around approximate cube root of the number. Divide the quotient by the same divisor again. Take average of the two divisors and final quotient. This is approximate cube root of the number.

Example 1: Find cube root of 800.

Sol. Take 10 as divisor $800 \div 10 = 80$ and $80 \div 10 = 8$

Average of 10, 10, 8 is $28 \div 3 = 9.3$

Example 2: Find cube root of 230.

Sol. Take 6 as divisor. $230 \div 6 = 38.3$ and $38.3 \div 6 = 6.38$ (rounded off downwards)

Average of 6, 6 and 6.38 is 6.126.

FINDING CUBE ROOT OF EXACT CUBES

Method: Finding cube root of exact cubes consisting of atmost six digits:

1. Determine the number of digits in the cube-root: Starting from right hand side, make pairs of three digits. A perfect cube number having 1 to 3 digits will have 1 digit in cube root and a cube number having 4 to 6 digits will have 2 digits in cube root. ($1^3 = 1$, $10^3 = 1000$, $100^3 = 1000000$)

2. Consider the last three digits of the cube to determine the unit digit of the cube root. Cube of numbers ending in 0, 1, 4, 5, 6, 9 ends in 0, 1, 4, 5, 6, 9 respectively. Cube of numbers ending in 2, 3, 7, 8 ends in 8, 7, 3, 2 respectively.

3. Determine the tens digit: Consider the remaining digits on left hand side. Find the largest digit whose cube is less than or equal to this number. This is the tens digit of the cube root.

Example 1: Find cube root of 314,432.

(*a*) 62 (*b*) 52 (*c*) 58 (*d*) 68

Sol. (*d*) $6^3 < 314 < 7^3$ $\Rightarrow$ Ten's digit is 6. Unit's digit is $10 - 2 = 8$

The required number is 68.

Example 1: $\sqrt[3]{148877} = ? = ?$ **(SBI PO 2009)**

(*a*) 51 (*b*) 59 (*c*) 53 (*d*) 49

Sol. (*c*) $5^3 < 148 < 6^3$ $\Rightarrow$ Ten's digit is 5. Unit's digit is $10 - 7 = 3$

The required number is 53. However, we may check the cube of 53.

We check : $53^3 =$

125	75	45	27
	150	90	
125	225	135	27 = 148 8 7 7

COMMON SQUARE ROOTS

$\sqrt{2} = 1.4142$ $\sqrt{3} = 1.7321$ $\sqrt{5} = 2.2361$

SOLVED EXERCISE

1. Number of digits in the square root of 62478078 is: **(SSC CGL Tier I 2007M, 2013)**

(*a*) 3 (*b*) 4 (*c*) 5 (*d*) 6

Sol. (*b*) 8 digits will make 4 pairs.

2. The number of digits in the square root of a 17 digit perfect square number is:

(*a*) 8 (*b*) 9 (*c*) 10 (*d*) 11

Sol. (*b*) 17 digits will make (8 pairs + 1 digit) $\therefore$ Digits in the square root = 8 + 1 = 9

3. Find the approximate value (not required to calculate exact value) of $\sqrt[3]{1500}$. **(UCO Bank PO 2009)**

(*a*) 11 (*b*) 16 (*c*) 15 (*d*) 19

Sol. (*a*) $11^3 = 1331 \Rightarrow$ 11 is approximate value of $\sqrt[3]{1500}$.

4. $\sqrt{64^2} = ?$ **(SBI Clerical 2009)**

(*a*) 8 (*b*) 16 (*c*) 64 (*d*) None of these

Sol. (*c*) Square root of square of a number is the number itself.

5. If $\frac{x}{372} = \frac{93}{x}$, find the value of *x*. **(SBI Clerical 2008)**

(*a*) 31 (*b*) 62 (*c*) 93 (*d*) None of these

Sol. (*d*) $x^2 = 93 \times 372 = 93 \times 93 \times 4 \quad \Rightarrow \quad x = 93 \times 2 = 186$

6. Which number should replace both the question marks in the following equation ?

$\frac{?}{576} = \frac{256}{?}$ **(Bank PO 2009)**

(*a*) 384 (*b*) 398 (*c*) 404 (*d*) 416

Sol. (*a*) $\frac{?}{576} = \frac{256}{?} \Rightarrow ? \times ? = 256 \times 576 \Rightarrow ? = \sqrt{256 \times 576} = \sqrt{16 \times 16 \times 24 \times 24}$

$= 16 \times 24 = 384$

7. $\sqrt{12321} = ?$ **(SBI Clerical 2008)**

(*a*) 91 (*b*) 101 (*c*) 111 (*d*) 121

Sol. (*c*) $111^2 = 12321$

8. The value of $\sqrt[3]{1372} \times \sqrt[3]{1458} = ?$ **(CPO SI Exam 2012)**

(*a*) 116 (*b*) 126 (*c*) 106 (*d*) 136

Sol. (*b*) $\sqrt[3]{1372} \times \sqrt[3]{1458} = \sqrt[3]{2 \times 2 \times 7 \times 7 \times 7} \times \sqrt[3]{2 \times 9 \times 9 \times 9} = 2 \times 7 \times 9 = 126$

9. The square root of $(272^2 - 128^2)$ is: **(SSC CGL Tier I 2000)**

(*a*) 200 (*b*) 240 (*c*) 144 (*d*) 400

Sol. (*b*) $\sqrt{272^2 - 128^2} = \sqrt{(272+128) \times (272-128)} = \sqrt{400 \times 144} = 20 \times 12 = 240$

10. $\sqrt[3]{\frac{0.000729}{0.085184}} = ?$ **(SSC CGL Tier II 2012)**

(*a*) $\frac{44}{9}$ (*b*) $\frac{27}{42}$ (*c*) $\frac{27}{44}$ (*d*) $\frac{9}{44}$

Sol. (*d*) $9^3 = 729 \Rightarrow$ Only possible option (*d*).

11. $\frac{(6.25)^{1/2} \times (0.0144)^{1/2} + 1}{(0.027)^{1/3} \times (81)^{1/4}}$ is simplified to **(SSC CGL Tier I 2005)**

(*a*) 0.14 (*b*) 1.4 (*c*) 1 (*d*) $1.\overline{4}$

Sol. (*d*) Given expression $= \frac{2.5 \times 0.12 + 1}{0.3 \times 3} = \frac{0.3 + 1}{0.9} = \frac{13}{9} = 1.\overline{4}$

12. 1008 divided by which single digit number gives a perfect square ? **(SSC CGL Tier I 2000)**

(a) 9 (b) 4 (c) 8 (d) 7

Sol. (d) $1008 = 2 \times 2 \times 2 \times 2 \times 3 \times 3 \times 7$. Here 7 is not in pair.

13. By what least number should 675 be multiplied to obtain a number which is a perfect cube ? **(CBI 1993, SSC C GL Tier I 2010 and 2012)**

(*a*) 5 (*b*) 6 (*c*) 7 (*d*) 8

Sol. (*a*) 675 = (3 × 3 × 3) × (5 × 5)

14. By what least number 252 must be multiplied to make it a perfect square number?

(*a*) 7 (*b*) 6 (*c*) 5 (*d*) 4

Sol. (*a*) 252 = 2 × 2 × 3 × 3 × 7. Since 7 occurs once in the factors of 252.

∴ 252 must be multiplied by 7 to get a square number.

Trick: We check unit's digit for all the options one by one:

7 : 252 × 7 = 4 (Possible)

6 : 252 × 6 = 2 (Not possible)

5 : 252 × 5 = 10 (Single 0 is not possible at the end)

4 : 252 × 4 = 8 (Not possible)

Hence out of the given options, only 7 is possible. ∴ 7 is the answer.

15. The smallest positive integer which when multiplied by 392, gives a perfect square is **(SSC CGL Tier I 2013)**

(*a*) 2 (*b*) 3 (*c*) 5 (*d*) 7

Sol. (*a*) 392 = 2 × 2 × 2 × 7 × 7 ⇒ Required number = 2

16. 1008 divided by which single digit number gives a perfect square ? **(SSC CGL Tier I 2000)**

(*a*) 9 (*b*) 4 (*c*) 8 (*d*) 7

Sol. (*d*) $1008 = 2^4 \times 3^2 \times 7$ ∴ Required number is 7.

17. Find the least number which when multiplied by 588 gives a perfect square number.

(*a*) 1 (*b*) 2 (*c*) 3 (*d*) 7

Sol. (*c*) 588 = 2 × 2 × 3 × 7 × 7. Here except for 3 all other factors are in pairs.

∴ We have to multiply 588 by 3 to make it a perfect square number.

Note: The number 588 can also be divided by 3 to get a square number.

Trick: We note that 588 is an even number and is divisible by 4. (Possible)

Sum of digits is 21, *i.e.* the number is divisible by 3 (but not by 9)

∴ The number must be multiplied by 3 to make it a square number.

18. A general wishing to draw up his 2350 men army in the form of a solid square found that he had 46 men left over. Find the number of men in each row. **(SSC)**

(*a*) 23 (*b*) 46 (*c*) 48 (*d*) None of these

Sol. (*c*) The number of men who formed the solid square = 2350 – 46 = 2304

The number of men in each row = $\sqrt{2304} = 48$

19. The least number by which 1944 must be multiplied so as to make the result a perfect cube, is:

(*a*) 2 (*b*) 3 (*c*) 6 (*d*) 13

Sol. (*b*) $1944 = 2^3 \times 3^5$ ∴ 1944 must be multiplied by 3.

20. The sum of the digits of the smallest number which, when multiplied by 1800, gives a perfect cube, is: **(SSC CGL Tier I 2004)**

(*a*) 2 (*b*) 3 (*c*) 6 (*d*) 8

Sol. (*c*) $1800 = 2^3 \times 3^2 \times 5^2$

Smallest number to multiply = $3^1 \times 5^1 = 15$ ∴ Required sum = 1 + 5 = 6

21. Sum of the digits of the smallest number by which 1440 be multiplied so that it becomes a perfect cube, is: **(SSC CGL Tier I 2003)**

(*a*) 4 (*b*) 6 (*c*) 7 (*d*) 8

Sol. (*b*) $1440 = 2^5 \times 3^2 \times 5^1$

Smallest number to multiply $= 2^1 \times 3^1 \times 5^2 = 150$ $\quad\therefore$ Required sum $= 1 + 5 + 0 = 6$

22. The largest number of 5 digits, which is a perfect square is:

(*a*) 99999 (*b*) 99976 (*c*) 99856 (*d*) 99764

Sol. (*c*) The largest number of 5-digit is 99999. $\quad 99999 = 316^2 + 143$

$\therefore$ The required number is $99999 - 143 = 99856$

23. What is cube root of 2197 ?

(*a*) 11 (*b*) 13 (*c*) 17 (*d*) 23

Sol. (*b*) $1^3 < 2 < 2^3 \Rightarrow$ Ten's digit of cube root is 1.

Other part is 197. Since the number ends in 7, therefore, cube root of the number ends in 3.

$\therefore$ The cube root is 13.

24. What is cube root of 29791 ?

(*a*) 21 (*b*) 31 (*c*) 37 (*d*) 39

Sol. (*b*) $3^3 < 29 < 4^3 \Rightarrow$ Ten's digit of cube root is 3.

Other part is 791. Since the number ends in 1, cube root of the number ends in 1.

$\therefore$ The cube root is 31.

25. The square root of 180625 is:

(*a*) 425 (*b*) 435 (*c*) 475 (*d*) 525

Sol. (*a*) $4^2 < 18 < 5^2 \Rightarrow$ The first digit of the square root is 4.

Last 4 digits of the given numbers are 0625 $\Rightarrow$ Square-root has last two digits 25.

$\therefore$ 425 may be the required number.

Verification : $425^2 = 4 \times 4.5 \mid 0625 = 180625$

26. Find the cube root of 185193. **(Bank Clerical)**

(*a*) 47 (*b*) 53 (*c*) 57 (*d*) 63

Sol. (*c*) $5^3 < 185 < 6^3 \Rightarrow$ Ten's digit is 5

Unit's digit of the given number is 3. $\Rightarrow$ Its cube root will end in 7. $\Rightarrow$ Cube root = 57.

27. $\sqrt[3]{1906624} = ?$ **(RBI Note Printing Asstt. Manager 2006)**

(*a*) 126 (*b*) 124 (*c*) 128 (*d*) 119

Sol. (*b*) 3 pairs are 1, 906 and 624. $\quad 12^3 < 1906 < 13^3$

Unit's digit of the given number is 4. $\Rightarrow$ Its cube root will end in 4. $\Rightarrow$ Cube root = 124.

28. $2432 \div ? = \sqrt{23104}$ **(Canara Bank PO 2009)**

(*a*) 12 (*b*) 14 (*c*) 18 (*d*) 16

Sol. (*d*) $2432 \div ? = \sqrt{23104} = 152 \Rightarrow ? = 2432 \div 152 = 16$

29. $\sqrt[3]{13824} \times \sqrt{?} = 864.$ **(SBI Clerical 2009)**

(*a*) 1024 (*b*) 1296 (*c*) 1156 (*d*) None of these

Sol. (*b*) $\sqrt{?} = 864 \div \sqrt[3]{13824} = 864 \div 24 = 36 \quad \Rightarrow \quad ? = 36^2 = 1296$

30. $\sqrt{?} \times \sqrt{3025} = 2695$ **(SBI Bank PO 2008)**

(*a*) 2401 (*b*) 2209 (*c*) 2601 (*d*) 2304

Sol. (*a*) $\sqrt{?} \times \sqrt{3025} = 2695 \Rightarrow \sqrt{?} \times 55 = 2695 \Rightarrow \sqrt{?} = 2695 \div 55 = 49 \Rightarrow ? = 49^2 = 2401$

31. $\sqrt[3]{4096} = ?$ **(UCO Bank PO 2009)**

(*a*) 16 (*b*) 26 (*c*) 18 (*d*) 24

Sol. (*a*) $\sqrt[3]{4096} = \sqrt[3]{2^{12}} = 2^4 = 16$

32. $\sqrt[3]{681472} = ?$ **(PNB Officers Exam. 2009)**

(*a*) 88 (*b*) 98 (*c*) 96 (*d*) 76

Sol. (*a*) The cube root of a number ending in 2, will end in 8.

98 is closer to 100. $100^3 = 1000000$

88 is closer to 90. $90^3 = 729000$, which is closer to 681472

We check: $88^3 =$

512	512	512	512
	1024	1024	
512	1536	1536	512 = 681 4 7 2

33. $\sqrt[3]{?} = (28 \times 24) \div 14$ **(SBI PO 2008)**

(*a*) 85184 (*b*) 140608 (*c*) 97336 (*d*) 110592

Sol. (*d*) $\sqrt[3]{?} = \dfrac{28 \times 24}{14} = 48 \Rightarrow ? = 48^3$

$48^3 =$

64	128	256	512
	256	512	
64	384	768	512 = 110 5 9 2

34. $\sqrt[3]{19683} = ? \times 3$ **(Bank PO 2006)**

(*a*) 90 (*b*) 27 (*c*) 3 (*d*) None of these

Sol. (*d*) Pairs are 19 and 683. Two pairs suggests that cube-root is a two-digit number.

$2^3 < 19 < 3^3 \Rightarrow$ Ten's digit is 2. Unit's digit 3 $\Rightarrow$ Unit digit of the cube-root is 7.

$\therefore$ Cube root of 19683 is 27. $\therefore$ $27 = ? \times 3 \Rightarrow ? = 27 \div 3 = 9$

35. Find the approximate value of $\sqrt{624.98} + \sqrt{729.25}$ **(Bank PO 2009)**

(*a*) 58 (*b*) 56 (*c*) 52 (*d*) 63

Sol. (*c*) $\sqrt{624.98} + \sqrt{729.25} \approx \sqrt{625} + \sqrt{729} = 25 + 27 = 52$

36. $(42 \times 221) \div (9261)^{1/3} = ?$ **(Bank PO)**

(*a*) 424 (*b*) 442 (*c*) 414 (*d*) 84

Sol. (*b*) $(42 \times 221) \div (9261)^{1/3} = \dfrac{42 \times 221}{(21^3)^{1/3}} = \dfrac{42 \times 221}{21} = 442$

37. The digit in the unit's place in the square-root of 15876 is: **(SSC CGL Tier I 2000)**

(*a*) 8 (*b*) 6 (*c*) 4 (*d*) 2

Sol. (*b*) $\sqrt{15876} = 126$ $\therefore$ Unit's digit is 6.

Alternative method: $12^2 < 158 < 13^2$

Unit's digit 6 in 15876 suggests that unit digit of the square-root may be either 4 or 6.

$\therefore$ The square-root may be 124 or 126.

Sum of digits of 124 and 126 and 7 and 9 respectively.

Sum of digits of squares of 7 and 9 and 4 + 9 = 13 and 8 + 1 = 9 respectively.

Sum of digits of 15876 is 9. $\therefore$ The square-root is 126.

38. What is the unit's digit in the square root of 23409 ?

(*a*) 1 (*b*) 3 (*c*) 5 (*d*) 7

Sol. (*b*) The number is divided into three parts: 234 | 0 | 9

Part I : $234 \geq a^2 \Rightarrow a = 15$, Remainder = 9

Part II : $90 \geq 2ab \Rightarrow b = 90 \div 2a = 90 \div 30 = 3$, Remainder = 0

Part III: $9 = b^2 = 3^2$, Remainder = 0

$\therefore a = 15$ and $b = 3$ $\quad\therefore$ Square-root of 23409 is 153.

Trick: Root of first 3 digits of the given number = 15, Remainder = 9.

The unit's digit of the square root of 15129 may be either 3 or 7.

Since the number has 5 digits, square-root of 15129 will have 3 digits.

$\therefore$ The number is either 153 or 157.

The sum of the digits of 153 and 157 are 9 and 4 respectively.

Sum of the digits of square of 153 must be 9 and sum of digits of square of 157 must exceed multiple of 3 by 1.

Sum of the digits of the number 23409 is 9.

$\therefore$ The square root is 153. $\quad\therefore$ The unit's digit is 3.

39. Find the square root of $1\frac{9}{16}$

(*a*) $1\frac{3}{4}$ (*b*) $1\frac{3}{16}$ (*c*) $1\frac{1}{4}$ (*d*) None of these

Sol. (*c*) $\sqrt{1\frac{9}{16}} = \sqrt{\frac{25}{16}} = \sqrt{\frac{5}{4}} = 1\frac{1}{4}$.

40. Find the square root of $33\frac{1}{16}$. **(Bank Clerical)**

(*a*) $33\frac{1}{4}$ (*b*) $\frac{23}{16}$ (*c*) $5\frac{3}{4}$ (*d*) None of these

Sol. (*c*) $\sqrt{33\frac{1}{16}} = \sqrt{\frac{529}{16}} = \frac{23}{4} = 5\frac{3}{4}$

41. Find the square root of $0.\bar{4}$. **(SSC CGL Tier I 2004)**

(*a*) $0.\bar{2}$ (*b*) $0.\bar{4}$ (*c*) $0.\bar{6}$ (*d*) $0.\bar{8}$

Sol. (*c*) $\sqrt{0.\bar{4}} = \sqrt{\frac{4}{9}} = \frac{2}{3} = \frac{6}{9} = 0.\bar{6}$ **Note:** If $x < 1$, then $x < \sqrt{x}$

42. Find the cube root of $0.\overline{037}$. **(SSC FCI 2012)**

(*a*) $0.\bar{3}$ (*b*) $0.\overline{27}$ (*c*) $0.\overline{37}$ (*d*) $0.\bar{6}$

Sol. (*a*) $0.\overline{037} = \frac{37}{999} = \frac{1}{27} \Rightarrow$ cube-root $= \frac{1}{3} = \frac{3}{9} = 0.\bar{3}$

43. Find the square root of 12.5 × 1.5. **(SSC)**

(*a*) 2.5 (*b*) $\frac{5\sqrt{3}}{2}$ (*c*) 15 (*d*) $\frac{3\sqrt{3}}{2}$

Sol. (*b*) $\sqrt{12.5 \times 1.5} = \sqrt{\frac{25 \times 5}{10} \times \frac{5 \times 3}{10}} = \frac{5 \times 5}{10}\sqrt{3} = \frac{5}{2}\sqrt{3}$

CHAPTER

8

SERIES

Arithmetic Progression (A.P.) : A constant number is added to each term to get the next term. Hence difference between each pair of two adjacent terms remains constant.

$\therefore$ A.P. = $a, a+d, a+2d, a+3d, \ldots\ldots, a+(n-1)d$

n^{th} term, *i.e.* $a_n = [a+(n-1)d]$

Where a is the first term of the series and d is the common difference between two adjacent terms.

If a, b, c are in A.P., then $b = \dfrac{a+c}{2}$.

Sum of first n terms is $S_n = \dfrac{n}{2}[2a+(n-1)d] = \dfrac{n}{2}[a+\{a+(n-1)d\}] = \dfrac{n}{2}(a_1+a_n)$

$$\therefore S_n = \text{Number of terms} \times \frac{\text{First term + Last term}}{2}$$

i.e. Number of terms × Average of first and last term.

Second Degree Arithmetic Series : Difference between two consecutive terms is in A.P.

Example : 1, 2, 4, 7, 11, 16, ...

Here difference between terms of series is : 1, 2, 3, 4, 5, ...

Geometric Progression (G.P.) : Each term is multiplied by a constant number to get the next term. Hence ratio between any two consecutive terms remains constant.

G.P. = $b, br, br^2, br^3, br^4, \ldots\ldots$

Note: n^{th} term = $b(r^{n-1})$, where b is the first term and r is the common ratio.

Sum of first n terms in a Geometric Progression is

$$S_n = \frac{b(r^n-1)}{r-1} \text{ if } r > 1, \text{ or } S_n = \frac{b(1-r^n)}{1-r} \text{ if } r < 1$$

Sum of infinite series in a G.P. is $S_\infty = \dfrac{b}{1-r}$ [If $|r| < 1$]

Combination of A.P. and G.P.: Difference between successive terms are in G.P.

Example : 1, 2, 6, 22, 86 ....

Difference between terms of series is : 1, 4, 16, 64, ... (a G.P.)

Combination of two or more series:

Example : 2, 3, 4, 5, 8, 7, 16, 9, 32, 11, 64, ...

First series : 2, 4, 8, 16, 32, 64, ... (a G.P.)

Second series : 3, 5, 7, 9, 11, ... (an A.P.)

OTHER SERIES

1. Sum of Natural numbers: $1 + 2 + 3 + 4 + \ldots\ldots\ldots\ldots\ldots + n = \frac{n(n+1)}{2}$

2. Sum of odd numbers:
$1 + 3 + 5 + 7 + \ldots\ldots\ldots\ldots\ldots + (2n-1) = n^2 = \left(\frac{\text{Last Number} + 1}{2}\right)^2$ **(NDA 2000)**

3. Sum of even umbers:
$2 + 4 + 6 + 8 + \ldots\ldots\ldots\ldots\ldots + 2n = n \times (n+1) = \left(\frac{\text{Last Number}}{2}\right) \times \left(\frac{\text{Last number}}{2} + 1\right)$

4. Sum of squares of natural numbers:
$1^2 + 2^2 + 3^2 + \ldots\ldots\ldots\ldots\ldots + n^2 = \frac{n(n+1)(2n+1)}{6}$

5. Sum of squares of odd numbers:
$1^2 + 3^2 + 5^2 + \ldots\ldots\ldots\ldots\ldots + (2n-1)^2 = \frac{n(2n-1)(2n+1)}{3}$

6. Sum of cubes of natural numbers:
$1^3 + 2^3 + 3^3 + \ldots\ldots\ldots\ldots\ldots + n^3 = \left[\frac{n(n+1)}{2}\right]^2$ **(RBI 2001)**

7. $\frac{1}{2^1} + \frac{1}{2^2} + \frac{1}{2^3} + \ldots\ldots\ldots\ldots\ldots + \frac{1}{2^n} = 1 - \frac{1}{2^n}$

8. $(1 \times 2) + (2 \times 3) + (3 \times 4) + \ldots\ldots\ldots\ldots\ldots + n(n+1) = \frac{n(n+1) \times (n+2)}{3}$

9. $\frac{1}{1 \times 2} + \frac{1}{2 \times 3} + \frac{1}{3 \times 4} + \ldots\ldots\ldots\ldots\ldots + \frac{1}{n(n+1)} = \frac{n}{(n+1)}$

10. $2^0 + 2^1 + 2^2 + 2^3 + 2^4 + \ldots\ldots\ldots\ldots\ldots + 2^n = \frac{2^0 \times (2^{n+1} - 1)}{2 - 1} = 2^{n+1} - 1$

SOLVED EXERCISE

1. Find 307th term of the sequence:
1, 2, 5, 8, 1, 2, 5, 8, 1, 2, 5, 8,..................

(*a*) 1 (*b*) 2 (*c*) 5 (*d*) 8

Sol. (*c*) Four terms make one set. Remainder on dividing 307 by 4 is 3.
307th term is same as third term in the series. Third term in the series is 5.

2. Find the sum of first 50 terms in the series: 5 + 3 − 7 + 5 + 3 − 7 ....

(*a*) 1 (*b*) 8 (*c*) 24 (*d*) 40

Sol. (*c*) Three terms make one set and sum of one set = 5 + 3 − 7 = 1
Since 50 = 3 × 16 + 2. 50 terms make 16 sets and 2 terms.
∴ Sum of 50 terms = 1 × 16 + (5 + 3) = 24

3. If a clock strikes appropriate number of times at each hour, how many times will it strike in a day ? **(SSC CGL Tier II 2013)**

(*a*) 300 (*b*) 156 (*c*) 68 (*d*) 78

Sol. (*b*) Strikes from 1 to 12 = 1 + 2 + 3 + ... + 12 = $\frac{12 \times 13}{2} = 78$

Total strikes in 24 hours = 2 × 78 = 156.

4. The sum of first 50 odd natural numbers is: **(SSC CGL Tier I 2008)**

(*a*) 1250 (*b*) 2500 (*c*) 2601 (*d*) 5202

Sol. (*b*) Sum of first *n* odd natural numbers = $n^2 = 50^2 = 2500$

5. (1 + 3 + 5 + 7 + ... + 99) = ? **(SSC CGL Tier I 2007)**

(*a*) 2050 (*b*) 2500 (*c*) 2005 (*d*) 2002

Sol. (*b*) Sum of given series = $\left(\frac{1+99}{2}\right)^2 = 50^2 = 2500$

6. (1 × 2 + 2 × 3 + 3 × 4 + 4 × 5 + 5 × 6 + 6 × 7 + 7 × 8 + 8 × 9 + 9 × 10 + 10 × 11) is equal to: **(SSC CHSL 2007)**

(*a*) 770 (*b*) 660 (*c*) 440 (*d*) 330

Sol. (*c*) Here *n* = 10. Required sum = $\frac{n(n+1) \times (n+2)}{3} = \frac{10 \times 11 \times 12}{3} = 440$

7. The sum (5 + 6 + 7 + 8 + ... + 19) is equal to **(SSC CGL Tier I 2005)**

(*a*) 150 (*b*) 170 (*c*) 180 (*d* 190

Sol. (*c*) (5 + 6 + 7 + 8 + ... + 19) = (4 + 1) + (4 + 2) + (4 + 3) + (4 + 4) + ... (4 + 15)

$= 4 \times 15 + (1 + 2 + 3 + 4 + \ldots\ldots + 15) = 60 + \frac{15 \times 16}{2} = 60 + 120 = 180$

Alt : Number of terms = (19 – 5) + 1 = 15 ⇒ Required sum = $15 \times \frac{5+19}{2} = 180$

8. The sum (101 + 102 + 103 + ... + 200) is equal to: **(SSC CGL Tier I 2000)**

(*a*) 15000 (*b*) 15025 (*c*) 15050 (*d*) 25000

Sol. (*c*) Number of terms = 100 ⇒ Sum of the series = $100 \times \frac{101+200}{2} = 100 \times \frac{301}{2} = 15050$

Alt: G.E. = $100 \times 100 + \frac{1+100}{2} \times 100 = 10000 + 5050 = 15050$

9. 45 + 46 + 47 + ... + 114 + 115 = ? **(SSC CGL Tier I 2007)**

(*a*) 5600 (*b*) 5680 (*c*) 5760 (*d*) None of these

Sol. (*b*) Number of terms = (115 – 45) + 1 = 71 ⇒ Sum = $\frac{45+115}{2} \times 71 = 80 \times 71 = 5680$

10. The numbers 1 to 19 are written side by side as follows: 123456789 ... 19. Find the remainder on dividing the number such formed by 9.

(*a*) 0 (*b*) 1 (*c*) 3 (*d*) 5

Sol. (*b*) Sum of numbers from 1 to 19 = $\frac{19 \times 20}{2} = 190$

Sum of digits = 1 + 9 + 0 = 10 ⇒ 1 + 0 = 1 ⇒ Required remainder = 1.

11. The sum of all the three-digit numbers is: **(SSC CGL Tier I)**

(*a*) 98901 (*b*) 494550 (*c*) 8991 (*d*) 899

Sol. (*b*) The sum of three-digit numbers is 100 + 101 + 102 + ... + 999

Number of terms = (999 – 100) + 1 = 900

$\therefore$ Sum $= 900 \times \dfrac{100+999}{2} = 450 \times 1099 = 494550$

12. Given $1 + 2 + 3 + 4 + \ldots\ldots\ldots\ldots + 10 = 55$, then the sum $6 + 12 + 18 + 24 + \ldots\ldots\ldots\ldots + 60$ is equal to: **(SSC CGL Tier I 2004)**

(*a*) 300 (*b*) 655 (*c*) 330 (*d*) 455

Sol. (*c*) $6 + 12 + 18 + 24 + \ldots\ldots\ldots\ldots + 60 = 6 \times (1 + 2 + 3 + 4 + \ldots\ldots\ldots\ldots + 10) = 6 \times 55 = 330$

13. The sum of the series $(1 + 0.6 + 0.06 + 0.006 + 0.0006 + \ldots)$ **(SSC CGL Tier I 2010)**

(*a*) $1\frac{2}{3}$ (*b*) $1\frac{1}{3}$ (*c*) $2\frac{1}{3}$ (*d*) $2\frac{2}{3}$

Sol. (*a*) G.E. $= 1.6666666\ldots = 1\frac{6}{9} = 1\frac{2}{3}$

14. $1 + \dfrac{1}{10} + \dfrac{1}{10^2} + \dfrac{1}{10^3} + \dfrac{1}{10^4} = ?$

(*a*) 1.1 (*b*) 1.11 (*c*) 1.111 (*d*) 1.1111

Sol. (*d*) $1 + \dfrac{1}{10} + \dfrac{1}{10^2} + \dfrac{1}{10^3} + \dfrac{1}{10^4} = 1 + 0.1 + 0.01 + 0.001 + 0.0001 = 1.1111$

15. Given that $(1^2 + 2^2 + 3^2 + 4^2 + 5^2) = 55$; $(2^2 + 4^2 + 6^2 + 8^2 + 10^2)$ is equal to: **(SSC CGL Tier I 2006, SSC CHSL 2011 M)**

(*a*) 110 (*b*) 165 (*c*) 220 (*d*) 55^2

Sol. (*c*) $2^2 + 4^2 + 6^2 + 8^2 + 10^2 = 2^2 \times (1^2 + 2^2 + 3^2 + 4^2 + 5^2) = 4 \times 55 = 220$

16. Given that $1^2 + 2^2 + 3^2 + \ldots\ldots\ldots\ldots + 20^2 = 2870$, the value of $2^2 + 4^2 + 6^2 + \ldots\ldots\ldots\ldots + 40^2$ is: **(SSC CGL Tier I 2005)**

(*a*) 11480 (*b*) 5740 (*c*) 28700 (*d*) 2870

Sol. (*a*) $(2^2 + 4^2 + 6^2 + \ldots\ldots\ldots + 40^2) = 2^2 \times (1^2 + 2^2 + 3^2 + \ldots\ldots\ldots + 20^2) = 4 \times 2870 = 11480$

17. $(2^2 + 3^2 + 4^2 + 5^2 + \ldots\ldots\ldots + 10^2)$ is equal to: **(SSC CHSL 2008)**

(*a*) 385 (*b*) 2916 (*c*) 540 (*d*) 384

Sol. (*d*) Given expression $= (1^2 + 2^2 + 3^2 + \ldots\ldots\ldots + 10^2) - 1^2 = \dfrac{10 \times 11 \times 21}{6} - 1 = 385 - 1 = 384$

18. If $1^2 + 2^2 + 3^2 + \ldots\ldots\ldots + x^2 = \dfrac{x(x+1)(2x+1)}{6}$, then $1^2 + 3^2 + 5^2 + \ldots\ldots\ldots + 19^2$ is equal to: **(SSC CGL Tier I 2003)**

(*a*) 1330 (*b*) 2100 (*c*) 2485 (*d*) 2500

Sol. (*a*) $(1^2 + 3^2 + 5^2 + \ldots\ldots\ldots + 19^2) = (1^2 + 2^2 + 3^2 + \ldots\ldots\ldots + 20^2) - (2^2 + 4^2 + 6^2 + \ldots\ldots\ldots + 20^2)$

$= (1^2 + 2^2 + 3^2 + \ldots\ldots\ldots + 20^2) - 2^2 \times (1^2 + 2^2 + 3^2 + \ldots\ldots\ldots + 10^2)$

$= \dfrac{20 \times 21 \times 41}{6} - 4 \times \dfrac{10 \times 11 \times 21}{6} = 2870 - 4 \times 385 = 2870 - 1540 = 1330$

Alt: $1^2 + 3^2 + 5^2 + \ldots\ldots\ldots + (2n-1)^2 = \dfrac{n(2n-1)(2n+1)}{3} = \dfrac{10 \times 19 \times 21}{3} = 1330$

(Hint : $19 = 2 \times 10 - 1 \Rightarrow n = 10$**)**

19. If $1^3 + 2^3 + 3^3 + ... + 10^3 = 3025$, then $4 + 32 + 108 + ... + 4000$ is equal to: **(SSC CGL Tier I 2002)**

(*a*) 12000 (*b*) 12100 (*c*) 12200 (*d*) 12400

Sol. (*b*) $4 + 32 + 108 + ... + 4000 = 4 \times (1 + 8 + 27 + ... + 1000)$

$= 4 \times (1^3 + 2^3 + 3^3 + ... + 10^3) = 4 \times 3025 = 12100$

20. The value of (13 + 23 + 33 + ... + 153) – (1 + 2 + 3 + ... + 15) is

(*a*) 975 (*b*) 1125 (*c*) 1325 (*d*) 1475

Sol. (*b*) $= \frac{13+153}{2} \times 15 - \frac{15 \times 16}{2} = \frac{15}{2}(166-16) = \frac{15}{2} \times 150 = \frac{2250}{2} = 1125$

21. If $1^3 + 2^3 + 3^3 + ... + 9^3 = 2025$, then $(0.11)^3 + (0.22)^3 + ... + (0.99)^3$ is close to : **(SSC CGL Tier I 2003, 2012)**

(*a*) 0.2695 (*b*) 2.695 (*c*) 3.695 (*d*) 0.3695

Sol. (*b*) $(0.11)^3 + (0.22)^3 + ... + (0.99)^3$

$= (0.11)^3 \times (1^3 + 2^3 + 3^3 + ... + 9^3) = 0.11^3 \times 2025 = 2.695275$

22. Find the value of $5^3 + 6^3 + ... + 10^3$. **(SSC CGL Tier I 2000)**

(*a*) 2295 (*b*) 2425 (*c*) 2495 (*d*) 2925

Sol. (*d*) Required Sum $= (1^3 + 2^3 + 3^3 + ... + 10^3) - (1^3 + 2^3 + 3^3 + 4^3)$

$= \left[\frac{10 \times 11}{2}\right]^2 - \left[\frac{4 \times 5}{2}\right]^2 = 55^2 - 10^2 = 3025 - 100 = 2925$

23. Find the sum of even numbers between 11 and 31.

(*a*) 200 (*b*) 210 (*c*) 242 (*d*) 262

Sol. (*b*) Sum $= \frac{12 + 30}{2} \times 10 = 210$

24. Find the sum of odd numbers between 20 and 50.

(*a*) 400 (*b*) 500 (*c*) 525 (*d* 625

Sol. (*c*) Required sum $= \left(\frac{49+1}{2}\right)^2 - \left(\frac{19+1}{2}\right)^2 = 25^2 - 10^2 = 525$

Alternative method: $\frac{21 + 49}{2} \times 15 = 525$

25. (6 + 12 + 24 + ……… + 768) is equal to :

(*a*) 768 (*b*) 1268 (*c*) 1500 (*d*) 1530

Sol. (*d*) Given expression $= 6 \times (1 + 2 + 4 + ... + 128) = 6 \times (2^0 + 2^1 + 2^2 + ... + 2^7)$

$= 6 \times \frac{2^0 (2^8 - 1)}{2 - 1} = 6 \times (2^8 - 1) = 6 \times (256 - 1) = 1530$

26. In a book, containing 20 pages, one sheet is missing. Sum of the page numbers of the remaining pages is 195. The numbers written on both the sides of the missing sheet must be: **(SSC CGL Tier I 2006)**

(*a*) 5, 6 (*b*) 6, 7 (*c*) 7, 8 (*d*) 8, 9

Sol. (*c*) Sum of all page numbers (including missing pages) $= \frac{20 \times 21}{2} = 210$

Sum of page numbers (excluding missing pages) = 195

∴ Sum of page numbers of missing sheet = 210 – 195 = 15

∴ The required sheet numbers are 7 and 8.

27. $999\frac{1}{7}+999\frac{2}{7}+999\frac{3}{7}+999\frac{4}{7}+999\frac{5}{7}+999\frac{6}{7}$ is simplified to :

(SSC CGL 2004, SSC CHSL 2010)

(*a*) 5997 (*b*) 5979 (*c*) 5994 (*d*) 2997

Sol. (*a*) $999\times 6+\left(\frac{1}{7}+\frac{2}{7}+\frac{3}{7}+\ldots\ldots\ldots+\frac{6}{7}\right)=5994+\frac{6\times 7}{2\times 7}=5994+3=5997$

28. $\left(1-\frac{1}{2}\right)\times\left(1-\frac{1}{3}\right)\times\left(1-\frac{1}{4}\right)\ldots\ldots\ldots\ldots\times\left(1-\frac{1}{n}\right)=?$

(SSC CGL Tier I 2008, CDS 2004, SSC CHSL 2014)

(*a*) 0 (*b*) 1 (*c*) $\frac{1}{n}$ (*d*) $\frac{1}{n-1}$

Sol. (*c*) Given expression = $\frac{1}{2}\times\frac{2}{3}\times\frac{3}{4}\times\ldots\ldots\ldots\ldots\times\frac{n-1}{n}=\frac{1}{n}$

29. The simplified value of $\left(1+\frac{1}{x}\right)\left(1+\frac{1}{x+1}\right)\left(1+\frac{1}{x+2}\right)\left(1+\frac{1}{x+3}\right)$ **(SSC CGL Tier I 2000)**

(*a*) 0 (*b*) 1 (*c*) $\frac{x+4}{x}$ (*d*) $\frac{x}{x+4}$

Sol. (*c*) Given expression = $\left(\frac{x+1}{x}\right)\left(\frac{x+2}{x+1}\right)\left(\frac{x+3}{x+2}\right)\left(\frac{x+4}{x+3}\right)=\frac{x+4}{x}$

30. The simplified value of $\left(1-\frac{1}{3}\right)\left(1-\frac{1}{4}\right)\left(1-\frac{1}{5}\right)\ldots\ldots\ldots\ldots\left(1-\frac{1}{99}\right)\left(1-\frac{1}{100}\right)$.

(SSC CGL Tier I 2003)

(*a*) $\frac{2}{99}$ (*b*) $\frac{1}{25}$ (*c*) $\frac{1}{50}$ (*d*) $\frac{1}{100}$

Sol. (*c*) Given expression = $\frac{2}{3}\times\frac{3}{4}\times\frac{4}{5}\times\ldots\ldots\ldots\ldots\times\frac{98}{99}\times\frac{99}{100}=\frac{2}{100}=\frac{1}{50}$

31. $\left(2-\frac{1}{3}\right)\left(2-\frac{3}{5}\right)\left(2-\frac{5}{7}\right)\ldots\ldots\left(2-\frac{997}{999}\right)$ is equal to **(SAS 2010)**

(*a*) $\frac{1001}{999}$ (*b*) $\frac{999}{1001}$ (*c*) $\frac{1001}{3}$ (*d*) $\frac{5}{1001}$

Sol. (*c*) G.E. = $\frac{5}{3}\times\frac{7}{5}\times\frac{9}{7}\times\ldots\ldots\times\frac{1001}{999}=\frac{1001}{3}$

32. $\left(1\frac{1}{2}\right)\times\left(1\frac{1}{3}\right)\times\left(1\frac{1}{4}\right)\times\ldots\ldots\ldots\ldots\times\left(1\frac{1}{120}\right)=?$

(*a*) 30 (*b*) 40.5 (*c*) 60.5 (*d*) 121

Sol. (*c*) Given Expression $=\frac{3}{2}\times\frac{4}{3}\times\frac{5}{4}\times\ldots\ldots\ldots\ldots\times\frac{121}{120}=\frac{121}{2}=60.5$

33. Find the sum of $\left(1-\frac{1}{n+1}\right)+\left(1-\frac{2}{n+1}\right)+\left(1-\frac{3}{n+1}\right)+....\left(1-\frac{n}{n+1}\right)$

(SSC CGL Tier II 2013)

(*a*) n (*b*) $\frac{1}{2}n$ (*c*) $(n+1)$ (*d*) $\frac{1}{2}(n+1)$

Sol. (*b*) G.E. = $\frac{n}{n+1}+\frac{n-1}{n+1}+\frac{n-2}{n+1}+......\frac{1}{n+1}=\frac{n+(n-1)(n-2)+...1}{n+1}=\frac{n(n+1)}{2(n+1)}=\frac{n}{2}$

34. $\frac{1}{1.2}+\frac{1}{2.3}+\frac{1}{3.4}+..............+\frac{1}{50.51}=?$ **(SSC CGL Tier I Exam.)**

(*a*) $\frac{1}{50}$ (*b*) $\frac{1}{51}$ (*c*) $\frac{50}{51}$ (*d*) $\frac{51}{50}$

Sol. (*c*) $\frac{1}{1\times 2}+\frac{1}{2\times 3}+\frac{1}{3\times 4}+..............+\frac{1}{n(n+1)}=\frac{n}{(n+1)}\Rightarrow$ Given expression = $\frac{50}{51}$

Alt: G.E. = $\left(1-\frac{1}{2}\right)+\left(\frac{1}{2}-\frac{1}{3}\right)+\left(\frac{1}{3}-\frac{1}{4}\right)+..............+\left(\frac{1}{50}-\frac{1}{51}\right)=1-\frac{1}{51}=\frac{50}{51}$

35. $\frac{1}{20}+\frac{1}{30}+\frac{1}{42}+\frac{1}{56}+\frac{1}{72}+\frac{1}{90}+\frac{1}{110}+\frac{1}{132}$ is equal to: **(SSC CGL Tier I 2002, CHSL 2013)**

(*a*) $\frac{1}{8}$ (*b*) $\frac{1}{7}$ (*c*) $\frac{1}{6}$ (*d*) $\frac{1}{10}$

Sol. (*c*) Given Expression = $\frac{1}{4\times 5}+\frac{1}{5\times 6}+\frac{1}{6\times 7}+..............+\frac{1}{11\times 12}$

$=\left(\frac{1}{4}-\frac{1}{5}\right)+\left(\frac{1}{5}-\frac{1}{6}\right)+\left(\frac{1}{6}-\frac{1}{7}\right)+..............+\left(\frac{1}{11}-\frac{1}{12}\right)=\frac{1}{4}-\frac{1}{12}=\frac{2}{12}=\frac{1}{6}$

36. $\frac{1}{1\times 4}+\frac{1}{4\times 7}+\frac{1}{7\times 10}+\frac{1}{10\times 13}+\frac{1}{13\times 16}=?$ **(SSC CHSL 2014, CDS 2015)**

(*a*) $\frac{1}{3}$ (*b*) $\frac{5}{16}$ (*c*) $\frac{3}{8}$ (*d*) $\frac{41}{7280}$

Sol. (*b*) Given expression = $\frac{1}{3}\times\left(\frac{3}{1\times 4}\right)+\frac{1}{3}\times\left(\frac{3}{4\times 7}\right)+\frac{1}{3}\times\left(\frac{3}{7\times 10}\right)+\frac{1}{3}\times\left(\frac{3}{10\times 13}\right)+\frac{1}{3}\times\left(\frac{3}{13\times 16}\right)$

$=\frac{1}{3}\times\left[\left(1-\frac{1}{4}\right)+\left(\frac{1}{4}-\frac{1}{7}\right)+\left(\frac{1}{7}-\frac{1}{10}\right)+\left(\frac{1}{10}-\frac{1}{13}\right)+\left(\frac{1}{13}-\frac{1}{16}\right)\right]=\frac{1}{3}\times\left(1-\frac{1}{16}\right)=\frac{5}{16}$

37. $\frac{1}{3}+\frac{1}{15}+\frac{1}{35}+\frac{1}{63}+\frac{1}{99}=?$ **(SSC FCI 2012)**

(*a*) $\frac{10}{11}$ (*b*) $\frac{5}{11}$ (*c*) $\frac{1}{11}$ (*d*) $\frac{7}{11}$

Sol. (*b*) G.E. $= \frac{1}{2}\left(\frac{2}{1\times3}+\frac{2}{3\times5}+\frac{2}{5\times7}+\frac{2}{7\times9}+\frac{2}{9\times11}\right)$

$= \frac{1}{2}\left(1-\frac{1}{3}+\frac{1}{3}-\frac{1}{5}+\frac{1}{5}-\frac{1}{7}+\frac{1}{7}-\frac{1}{9}+\frac{1}{9}-\frac{1}{11}\right) = \frac{1}{2}\left(1-\frac{1}{11}\right) = \frac{1}{2}\times\frac{10}{11} = \frac{5}{11}$

Alt: G.E. = 0.33 + 0.07 + 0.03 + 0.01 + 0.01 = 0.45 (approx).

Only 5/11 is close to this value. 5/11 = 0.45

38. $\left(1-\frac{1}{2^2}\right)\left(1-\frac{1}{3^2}\right)$ .... $\left(1-\frac{1}{9^2}\right)\left(1-\frac{1}{10^2}\right) = ?$ **(SSC CGL Tier I 2007)**

(*a*) $\frac{11}{20}$ (*b*) $\frac{1}{20}$ (*c*) $\frac{1}{2}$ (*d*) $\frac{10}{21}$

Sol. (*a*) G.E. $= \left(1-\frac{1}{2}\right)\left(1+\frac{1}{2}\right)\left(1-\frac{1}{3}\right)\left(1+\frac{1}{3}\right)$ .... $\left(1-\frac{1}{9}\right)\left(1+\frac{1}{9}\right)\left(1-\frac{1}{10}\right)\left(1+\frac{1}{10}\right)$

$= \left(1-\frac{1}{2}\right)\left(1-\frac{1}{3}\right)$ .... $\left(1-\frac{1}{9}\right)\left(1-\frac{1}{10}\right) \times \left(1+\frac{1}{2}\right)\left(1+\frac{1}{3}\right)$ .... $\left(1+\frac{1}{9}\right)\left(1+\frac{1}{10}\right)$

$= \left(\frac{1}{2}\times\frac{2}{3}\times \text{..........} \times\frac{8}{9}\times\frac{9}{10}\right)\times\left(\frac{3}{2}\times\frac{4}{3}\times \text{..........} \times\frac{10}{9}\times\frac{11}{10}\right) = \frac{1}{10}\times\frac{11}{2} = \frac{11}{20}$

39. The value of $\frac{3}{1^2.2^2}+\frac{5}{2^2.3^2}+\frac{7}{3^2.4^2}+\frac{9}{4^2.5^2}+\frac{11}{5^2.6^2}+\frac{13}{6^2.7^2}+\frac{15}{7^2.8^2}+\frac{17}{8^2.9^2}+\frac{19}{9^2.10^2}$ is : **(SSC CGL Tier I 2004)**

(*a*) $\frac{1}{100}$ (*b*) $\frac{99}{100}$ (*c*) $\frac{101}{100}$ (*d*) 1

Sol. (*b*) Given expression $= \frac{2^2-1^2}{1^2.2^2}+\frac{3^2-2^2}{2^2.3^2}+\frac{4^2-3^2}{3^2.4^2}+\frac{5^2-4^2}{4^2.5^2}+\text{.........}+\frac{10^2-9^2}{9^2.10^2}$

$= \left(1-\frac{1}{4}\right)+\left(\frac{1}{4}-\frac{1}{9}\right)+\left(\frac{1}{9}-\frac{1}{16}\right)+\left(\frac{1}{16}-\frac{1}{25}\right)+\text{..........}+\left(\frac{1}{81}-\frac{1}{100}\right) = 1-\frac{1}{100} = \frac{99}{100}$

40. $(1^2-2^2+3^2-4^2+5^2-6^2+7^2-8^2+9^2-10)$ is equal to

(*a*) –5 (*b*) 5 (*c*) 55 (*d*) –55

Sol. (*d*) G.E. $= (1^2-2^2)+(3^2-4^2)+(5^2-6^2)+(7^2-8^2)+(9^2-10^2)$

$= (1-2)(1+2)+(3-4)(3+4)+ \text{....} + (9-10)(9+10)$

$= -(1+2+3+4+\text{....}+9+10) = -\frac{10\times11}{2} = -55$

41. The sum of $2n$ terms of series : $1^2-2^2+3^2-4^2+$ .... **(SSC CGL Tier I 2006)**

(*a*) $n^2(2n+1)$ (*b*) $-n^2(2n+1)$ (*c*) $n(2n+1)$ (*d*) $-n(2n+1)$

Sol. (*d*) G.E $= (1^2-2^2)+(3^2-4^2)+$ .... $[(2n-1)^2-(2n)^2]$

$= (1-2)(1+2)+(3-4)(3+4)+$ .... $[(2n-1)-(2n)][(2n-1)+(2n)]$

$= (-1)(1+2)+(-1)(3+4)+$ .... $(-1)[(2n-1)+(2n)]$

$= -[1+2+3+4+$ .... $(2n-1)+(2n)] = -\frac{2n(2n+1)}{2} = -n(2n+1)$

42. If p, q, r are in Geometric Progression, then which is true among the following ? **(SSC CGL Tier I 2012)**

(*a*) $q = \frac{p+r}{2}$ (*b*) $p^2 = qr$ (*c*) $q = \sqrt{pr}$ (*d*) $\frac{p}{r} = \frac{r}{q}$

Sol. (*c*) $\frac{p}{q} = \frac{q}{r} \Rightarrow q^2 = pr \Rightarrow q = \sqrt{pr}$

43. Find the sum of infinite series: $\frac{1}{2} + \frac{1}{4} + \frac{1}{8} + \frac{1}{16} + \frac{1}{32}$ ....

(*a*) 1 (*b*) 2 (*c*) $\frac{3}{4}$ (*d*) $\frac{5}{4}$

Sol. (*a*) Given series is a G.P. with $b = \frac{1}{2}$ and $r = \frac{1}{2}$ $\Rightarrow$ Sum $= \frac{b}{1-r} = \frac{\frac{1}{2}}{1-\frac{1}{2}} = 1$

44. A man starts going for morning walk every day. The distance walked by him on the first day was 2 km. Every day he walks half of the distance walked on the previous day. What is the total distance walked by him in his life time ? **(SBI Bank PO 2001)**

(*a*) 4 km (*b*) 6 km (*c*) 8 km (*d*) 12 km

Sol. (*a*) Total distance covered $= 2 + 1 + \frac{1}{2} + \frac{1}{4} + \ldots\ldots\ldots$ (G.P. with $b = 2$ and $r = \frac{1}{2}$)

$$S_\infty = \frac{b}{1-r} = \frac{2}{1-\frac{1}{2}} = \frac{2}{\frac{1}{2}} = 4$$

45. The value of $1 - \frac{1}{20} + \frac{1}{20^2} - \frac{1}{20^3} + \ldots\ldots\ldots$ correct up to 5 places of decimals is :

(SSC CGL Tier I 2004)

(*a*) 1.05 (*b*) 0.95238 (*c*) 0.95239 (*d*) 10.5

Sol. (*b*) Given series is a Geometric Progression with $b = 1$ and $r = -\frac{1}{20}$

$$\text{Sum} = \frac{b}{1-r} = \frac{1}{1-\left(-\frac{1}{20}\right)} = \frac{20}{21} = 0.95238$$

46. Shyam earns ₹100 on the first day. In the next days, he earns 90% of what he earned on the previous day. Find the maximum total amount he will earn.

(*a*) ₹900 (*b*) ₹1000 (*c*) ₹2000 (*d*) ₹200

Sol. (*b*) Here, $b = 100$, $r = 0.9 \Rightarrow S_\alpha = \frac{100}{1-0.9} = \frac{100}{0.1} =$ ₹1000

47. A contract on construction job specifies a penalty for delay in completion of the work beyond a certain date is as follows : ₹ 200 for the first day, ₹ 250 for the second day, ₹ 300 for the third day etc., the penalty for each succeeding day being ₹ 50 more than that of the preceding day. How much penalty should the contractor pay if he delays the work by 10 days ?

(UPSC Civil Services (Pre.) 2011)

(*a*) ₹4950 (*b*) ₹4250 (*c*) ₹3600 (*d*) ₹650

Sol. (*b*) Fine for 10 days = 200 + 250 + 300 + ... + 650 = $\frac{200+650}{2} \times 10 =$ ₹ 4250

Alt : $S_{10} = \frac{n}{2}[2a+(n-1)d] = 5 \times (400+450) = 4250$

48. What is the sum of all the numbers between 200 and 600 which are divisible by 16 ?

(*a*) 5000 (*b*) 10000 (*c*) 15000 (*d*) 16000

Sol. (*b*) $200 = 16 \times 12 + 8$

$600 = 3 \times (16 \times 12 + 8) = 16 \times 36 + 24 = 16 \times 37 + 8$

Required sum = $(16 \times 13) + (16 \times 14) + (16 \times 15) + \ldots\ldots + (16 \times 37)$

$= 16 \times (13 + 14 + 15 + \ldots\ldots + 37) = 16 \times \frac{13+37}{2} \times 25 = 10000$

49. The sum of all the three-digit numbers each of which on division by 5 leaves remainder 3, is : **(SSC CGL Tier I 2008)**

(*a*) 180 (*b*) 1550 (*c*) 6995 (*d*) 99090

Sol. (*d*) First term = 103 (*i.e.* 100 + 3) and the last term = 998 (i.e. 995 + 3)

Number of terms = $\frac{998-103}{5}+1 = \frac{895}{5}+1 = 180$

$\therefore$ Required sum = $\frac{103+998}{2} \times 180 = 99090$

50. If third term of a A.P. is 3, then the sum of first five terms of the A.P. is

(*a*) 15 (*b*) 27 (*c*) 243 (*d*) Date insufficient

Sol. (*a*) Let the third term of the AP is a and the common difference is d.

Then the first five terms are $(a - 2d), (a - d), (a), (a + d), (a + 2d)$

Sum of these terms = $5a = 5 \times 3 = 15$

51. If third term of a G.P. is 3, then the product of first five terms of the G.P. is **(MBA)**

(*a*) 15 (*b*) 27 (*c*) 243 (*d*) Date insufficient

Sol. (*c*) Let the third term of the GP is a and the common ratio is r.

Then the first five terms are $\frac{a}{r^2}, \frac{a}{r}, a, ar, ar^2$ $\Rightarrow$ Product = $a^5 = 3^5 = 243$

52. Mohan earns ₹ 10 profit on the first day, ₹ 20 profit on the second day, ₹ 40 profit on the third day, and so on. What will he earn in the first 6 days ?

(*a*) ₹ 60 (*b*) ₹ 180 (*c*) ₹ 240 (*d*) ₹ 630

Sol. (*d*) Here $b = 10, r = 2, n = 6$ $\Rightarrow$ $S_6 = \frac{b(r^n-1)}{r-1} = \frac{10(2^6-1)}{2-1} = 630$

53. If the sum of n terms of an A.P. is $an + bn^2$, find the common difference of the A.P. ?

(*a*) $a - b$ (*b*) $a + b$ (*c*) $2a$ (*d*) $2b$ **(MBA)**

Sol. (*d*) $S_1 = a + b = a_1$ $S_2 = 2a + 4b = a_1 + a_2$

$d = a_2 - a_1 = S_2 - 2a_1 = (2a + 4b) - 2(a + b) = 2b$

54. If $a_1 + a_5 + a_{10} + a_{15} + a_{20} + a_{24} = 45$, find the value of $a_1 + a_2 + a_3 + \ldots + a_{24}$. **(MBA)**

(*a*) 72 (*b*) 96 (*c*) 180 (*d*) 225

Sol. (*c*) $a_1 + a_5 + a_{10} + a_{15} + a_{20} + a_{24} = 45$

$\Rightarrow (a_1 + a_{24}) + (a_5 + a_{20}) + (a_{10} + a_{15}) = 45$

$\Rightarrow\ 3(a_1 + a_{24}) = 45 \Rightarrow\ (a_1 + a_{24}) = 15$

$\therefore\ a_1 + a_2 + a_3 + ... + a_{24} = 12(a_1 + a_{24}) = 12 \times 15 = 180.$

What should come in place of question mark (?) in the following number series:

55. 841, 961, 1089, 1225, 1369, 1521, ? **(Canara Bank PO 2009)**

(*a*) 1581 (*b*) 1681 (*c*) 1781 (*d*) 1881

Sol. (*b*) $841 = 29^2$, $961 = 31^2$, $1089 = 33^2$, $1225 = 35^2$, $1369 = 37^2$, $1521 = 39^2$

$\therefore\ ? = 41^2 = 1681$

56. 1548, 516, 129, 43, ? **(Bank of Maharashtra 2008)**

(*a*) 11 (*b*) 10.75 (*c*) 9.5 (*d*) 12

Sol. (*b*) 1548 516 129 43

÷ 3 ÷ 4 ÷ 3

$\therefore\ ? = 43 \div 4 = 10.75$

57. 1, 4, 27, 256, 3125, 46656, ? **(Bank PO 2007)**

(*a*) 117649 (*b*) 279936 (*c*) 705894 (*d*) 823543

Sol. (*d*) $1 = 1^1, 4 = 2^2,\ 27 = 3^3,\ 256 = 4^4,\ 3125 = 5^5,\ 46656 = 6^6 \Rightarrow\ ? = 7^7 = 823543$

Trick: Unit digit in 7^7 is same as in 7^3 *i.e.* 3. 823543 is the only option with unit's digit 3.

58. Complete the series 7, 26, 63, 124, 215, 342, ? **(SSC CGL Tier II 2012)**

(*a*) 391 (*b*) 421 (*c*) 481 (*d*) 511

Sol. (*d*) $7 = 2^3 - 1, 26 = 3^3 - 1, 63 = 4^3 - 1, 124 = 5^3 - 1, 215 = 6^3 - 1, 342 = 7^3 - 1$

$\therefore\ ? = 8^3 - 1 = 511$

59. 30, 46, 78, 126, 190, 270, ? **(Bank PO 2007)**

(*a*) 356 (*b*) 366 (*c*) 382 (*d*) 398

Sol. (*b*) 30 46 78 126 190 270 ?

Difference = 16 32 48 64 80 (96)

$\therefore\ ? = 270 + 96 = 366$

60. 104, 153, 189, 214, ? **(Andhra Bank IT Officers Exam.)**

(*a*) 239 (*b*) 225 (*c*) 264 (*d*) 230

Sol. (*d*) 104 153 189 214 ?

Diff = $49 = 7^2$ $36 = 6^2$ $25 = 5^2$ $(16 = 4^2)$

$\therefore\ ? = 214 + 4^2 = 214 + 16 = 230$

61. 1250, 500, 200, 80, 32, 12.8, ? **(Bank PO 2007)**

(*a*) 5.12 (*b*) 6.4 (*c*) 4.3 (*d*) 6.02

Sol. (*a*) 1250 500 200 80 32 12.8 ?

÷ 2.5 ÷ 2.5 ÷ 2.5 ÷ 2.5 ÷ 2.5 (÷ 2.5)

$\therefore\ ? = 12.8 \div 2.5 = 5.12$

62. 13, 14, 30, 93, 376, 1885, ? **(Bank PO)**

(*a*) 10818 (*b*) 10316 (*c*) 11316 (*d*) 11318

Sol. (*c*) 13   14   30   93   376   1885   ?

× 1 + 1   × 2 + 2   × 3 + 3   × 4 + 4   × 5 + 5   (× 6 + 6)

$\therefore$? = 1885 × 6 + 6 = 11310 + 6 = 11316

63. 18, 20, 44, 138, 560, 2810, ? **(Canara Bank PO 2009)**

(*a*) 16818 (*b*) 16836 (*c*) 16854 (*d*) 16872

Sol. (*d*) 18 × 1 + 2 = 18 + 2 = 20   20 × 2 + 4 = 40 + 4 = 44

44 × 3 + 6 = 132 + 6 = 138   138 × 4 + 8 = 552 + 8 = 560

560 × 5 + 10 = 2800 + 10 = 2810 $\therefore$? = 2810 × 6 + 12 = 16860 + 12 = 16872

64. 19, 25, 20, 24, 22, 23, ? **(Bank PO)**

(*a*) 22 (*b*) 24 (*c*) 20 (*d*) 25

Sol. (*d*) In the given series, there are two series :

− 1   − 1

19   25   20   24   22   23   ?

+ 1   + 2   (+ 3)

$\therefore$? = 22 + 3 = 25

65. 286, 142, 70, ?, 16, 7 **(Bank PO 2009)**

(*a*) 36 (*b*) 30 (*c*) 32 (*d*) 34

Sol. (*d*) (286 ÷ 2) − 1 = 143 − 1 = 142 $\therefore$? = (70 ÷ 2) − 1 = 35 − 1 = 34

66. 9, 4.5, 4.5, 6.75, 13.5, 33.75, ? **(Bank PO 2007)**

(*a*) 101.25 (*b*) 103.75 (*c*) 99.75 (*d*) 105.50

Sol. (*a*) 9   4.5   4.5   6.75   13.5   33.75   ?

× 0.5   × 1   × 1.5   × 2   × 2.5   (× 3)

$\therefore$? = 33.75 × 3 = 101.25

67. 705, 728, 774, 843, 935, 1050, ? **(Bank PO 2007)**

(*a*) 1190 (*b*) 1180 (*c*) 1185 (*d*) 1188

Sol. (*d*) 705   728   774   843   935   1050   ?

Diff = 23 × 1   23 × 2   23 × 3   23 × 4   23 × 5   23 × 6

$\therefore$? = 1050 + (23 × 6) = 1050 + 138 = 1188

68. 3, 10, 32, 100, ? **(Syndicate Bank PO 2004)**

(*a*) 345 (*b*) 460 (*c*) 308 (*d*) 440

Sol. (*c*) 3   10   32   100   ?

× 3 + 1   × 3 + 2   × 3 + 4   (× 3 + 8)

$\therefore$? = 100 × 3 + 8 = 308

69. 5, 3, 4, ?, 38 **(Syndicate Bank PO 2004)**

(*a*) 8.5 (*b*) 10 (*c*) 7.5 (*d*) 8

Sol. (*b*) 5   3   4   ?   38

× 1 − 2   × 2 − 2   × 3 − 2   × 4 − 2

$\therefore$? = 4 × 3 − 2 = 12 − 2 = 10   10 × 4 − 2 = 40 − 2 = 38

70. 5, 9, 18, 34, 59, 95, ? **(Canara Bank PO 2009)**

(*a*) 272 (*b*) 168 (*c*) 116 (*d*) 144

Sol. (*d*) 5 9 18 34 59 95 ?

Diff = $4 = 2^2$, $9 = 3^2$, $16 = 4^2$, $25 = 5^2$, $36 = 6^2$, $(49 = 7^2)$

$\therefore ? = 95 + 7^2 = 95 + 49 = 144$

71. 12, 22, 69, 272, 1365, ? **(Andhra Bank IT Officers Exam.)**

(*a*) 8196 (*b*) 8184 (*c*) 8195 (*d*) 6830

Sol. (*b*) 12 22 69 272 1365 ?

$\times 2 - 2$, $\times 3 + 3$, $\times 4 - 4$, $\times 5 + 5$, $\times 6 - 6$

$\therefore ? = 1365 \times 6 - 6 = 8190 - 6 = 8184$

72. 14, 16, 35, 109, 441, ? **(RBI Note Printing Asstt. Manager 2006)**

(*a*) 2651 (*b*) 2205 (*c*) 2315 (*d*) 2211

Sol. (*d*) 14 16 35 109 441 ?

$\times 1 + 2$, $\times 2 + 3$, $\times 3 + 4$, $\times 4 + 5$, $(\times 5 + 6)$

$\therefore ? = 441 \times 5 + 6 = 2211$

73. 3600, 725, 150, 35, 12, ? **(RBI Note Printing Asstt. Manager 2006)**

(*a*) 8 (*b*) 7.4 (*c*) 10.5 (*d*) 10

Sol. (*b*) 3600 725 150 35 12 ?

$\div 5 + 5$, $\div 5 + 5$, $\div 5 + 5$, $\div 5 + 5$, $\div 5 + 5$

$\therefore ? = 12 \div 5 + 5 = 2.4 + 5 = 7.4$

In the following number series only one number is wrong. Find out the wrong number.

74. 1, 4, 27, 256, 3125, 46658 **(Bank PO)**

(*a*) 46658 (*b*) 4 (*c*) 27 (*d*) 3125

Sol. (*a*) $1 = 1^1, 4 = 2^2, 27 = 3^3, 256 = 4^4, 3125 = 5^5, 46658 \neq 6^6$

$\therefore$ 46658 is the wrong number.

75. 12, 237, 406, 527, 604, 657 **(Bank PO)**

(*a*) 237 (*b*) 406 (*c*) 527 (*d*) 604

Sol. (*d*) $237 - 12 = 225 = 15^2, 406 - 237 = 169 = 13^2, 527 - 406 = 121 = 11^2, 604 - 527 = 77 \neq 9^2$

$\therefore$ 604 is the wrong number.

76. The odd term in the sequence 0, 7, 26, 63, 124, 217 is **(SSC CGL Tier II 2013)**

(*a*) 217 (*b*) 7 (*c*) 26 (*d*) 63

Sol. (*a*) $217 \neq n^3 - 1$

CHAPTER

9

LINEAR EQUATIONS

Linear Equations in one variable

1. In the general or standard form the linear equation in one variable is written as $ax + b = c$ where a is not equal to zero and x is a variable. A value of the variable that satisfies the equation is known as a solution or root of the equation.

2. The equality sign of an equation is not affected, if we :

(*i*) add the same number to both the sides of the equation.
(*ii*) subtract the same number from both the sides of the equation.
(*iii*) multiply both sides of the equation by same non-zero number.
(*iv*) divide both sides of the equation by same non-zero number.

Linear Equations in two variables

1. $ax + by + c = 0$ or $ax + by = c$ (where a, b, c, are real numbers, and $a \neq 0, b \neq 0$) is a linear equation in two variables. The graph of a linear equation is a straight line.

$ax = b$ (straight line parallel to y-axis)

$ay = c$ (straight line parallel to x-axis)

2. $a_1x + b_1y + c_1 = 0$ and $a_2x + b_2y + c_2 = 0$ are a pair of linear equations in two variables. A pair of linear equations will have either a unique solution or infinitely many solutions or no solution.

(*i*) Lines intersect at one point if $\frac{a_1}{a_2} \neq \frac{b_1}{b_2}$ (one solution)

(*ii*) Lines are parallel (no solution) *i.e.*, they do not intersect if $\frac{a_1}{a_2} = \frac{b_1}{b_2} \neq \frac{c_1}{c_2}$

(*iii*) Coincident lines (infinitely many solutions) if $\frac{a_1}{a_2} = \frac{b_1}{b_2} = \frac{c_1}{c_2}$

3. Cross multiplication method of solving a pair of two linear equations

$$x = \frac{b_1 c_2 - b_2 c_1}{a_1b_2 - a_2 b_1} \text{ and } y = \frac{c_1a_2 - c_2a_1}{a_1b_2 - a_2b_1}$$

$$\left[\frac{x}{b_1 c_1 \atop b_2 c_2} = \frac{-y}{a_1 c_1 \atop a_2 c_2} = \frac{1}{a_1 b_1 \atop a_2 b_2}\right]$$

4. Any pair of values of x and y which satisfies the given equation, is called its solution.
Example: $x = 2, y = 1$ is a solution of equation $2x - 3y = 1$
When $x = 2, y = 1$, LHS $= 4 - 3 = 1 =$ RHS.

Note: Every linear equation in two variables has infinitely many solutions.

SOLVED EXERCISE

1. Find the value of k for which the given equations has infinitely many solutions.
(*i*) $5x + 2y = k$ and (*ii*) $10x + 4y = 3$
(*a*) 3 (*b*) $\frac{3}{2}$ (*c*) $\frac{1}{2}$ (*d*) 2

Sol. (*b*) $\frac{5}{10} = \frac{2}{4} = \frac{k}{3} \Rightarrow \frac{k}{3} = \frac{1}{2} \Rightarrow k = \frac{3}{2}$

2. Find the value of k for which the given equation has infinitely many solutions
(*i*) $(k-3)x + 3y = k$ and (*ii*) $kx + ky = 12$
(*a*) 0 (*b*) 3 (*c*) 6 (*d*) 9

Sol. (*c*) $\frac{k-3}{k} = \frac{3}{k}$ and $\frac{3}{k} = \frac{k}{12} \Rightarrow k - 3 = 3$ and $k^2 = 36$

$\Rightarrow k = 6$ and $k^2 = 36 \Rightarrow k = 6$ and $k = \pm 6 \Rightarrow k = 6$

3. The system of equations $x + 2y = 3$ and $3x + 6y = 9$ has **(CDS 2011)**
(*a*) unique solution (*b*) no solution
(*c*) infinitely many solutions (*d*) finite number of solutions

Sol. (*c*) Here, $\frac{1}{3} = \frac{2}{6} = \frac{3}{9}$ $\Rightarrow$ Infinitely many solutions.

4. Number of solutions of the two equations $4x - y = 2$ and $2y - 8x + 4 = 0$ is **(SCC CHSL 2012)**
(*a*) infinitely many (*b*) zero (*c*) one (*d*) two

Sol. (*a*) $2y - 8x + 4 = 0 \Rightarrow 8x - 2y = 4$

Here, $\frac{4}{8} = \frac{-1}{-2} = \frac{2}{4} = \frac{1}{2} \Rightarrow \frac{a_1}{a_2} = \frac{b_1}{b_2} = \frac{c_1}{c_2}$ $\Rightarrow$ Equations have infinite solutions

Alt: $2y - 8x + 4 = 0 \Rightarrow 8x - 2y = 4 \Rightarrow 4x - y = 2$

$\Rightarrow$ Two equations are same *i.e.*, infinite solutions.

5. The value of k for which the system of equations $x + 2y = 5$, $3x + ky + 15 = 0$ has no solution, is **(SCC CPO SI 2012)**
(*a*) 2 (*b*) – 2 (*c*) 6 (*d*) – 6

Sol. (*c*) No solution $\Rightarrow \frac{a_1}{a_2} = \frac{b_1}{b_2} \neq \frac{c_1}{c_2} \Rightarrow \frac{1}{3} = \frac{2}{k} \neq \frac{5}{-15} \Rightarrow k = 6.$

6. In the *xy*-coordinate system, if (a, b) and $(a + 3, b + k)$ are two points on the line defined by the equation $x = 3y - 7$, then $k =$ **(SCC CGL Tier 2012)**
(*a*) 9 (*b*) 3 (*c*) 7/3 (*d*) 1

Sol. (*d*) $x = 3y - 7 \Rightarrow x + 7 = 3y$
Any point on the line will satisfy this equation.
$\therefore$ $a + 7 = 3b$ ... (1) $(a + 3) + 7 = 3(b + k) \Rightarrow a + 10 = 3b + 3k$ ... (2)
(2) – (1) : $3 = 3k \Rightarrow k = 1.$

7. The graphs of $x + y = 4$ and $3x + 3y = 6$ **(SCC CHSL 2013)**
(*a*) instersect at the point (1,3) (*b*) instersect at the point (3,1)
(*c*) instersect at the point (2,2) (*d*) do not instersect

Sol. (*d*) $1/3 = 1/3 \neq 4/6 \Rightarrow$ Do not intersect

CHAPTER

10

ALGEBRA

ALGEBRAIC FORMULAE

$$x(y+z) = xy + xz$$

$$x(y-z) = xy - xz$$

$$(x+y)^2 = x^2 + y^2 + 2xy = (x-y)^2 + 4xy$$

$$(x-y)^2 = x^2 + y^2 - 2xy = (x+y)^2 - 4xy$$

$$(x-y)(x+y) = x^2 - y^2$$

$$x^2 + y^2 = (x-y)^2 + 2xy = (x+y)^2 - 2xy$$

$$(x+y)^2 - (x-y)^2 = 4xy$$

$$(x+y)^2 + (x-y)^2 = 2(x^2+y^2) \Rightarrow x^2 + y^2 = \frac{1}{2}[(x+y)^2 + (x-y)^2]$$

$$(x+y)^3 = x^3 + y^3 + 3xy(x+y) = x^3 + y^3 + 3x^2y + 3xy^2$$

$$(x-y)^3 = x^3 - y^3 - 3xy(x-y) = x^3 - y^3 - 3x^2y + 3xy^2$$

$$(x+y)^3 + (x-y)^3 = 2x^3 + 6xy^2 = 2(x^3 + 3xy^2)$$

$$(x+y)^3 - (x-y)^3 = 2y^3 + 6x^2y = 2(y^3 + 3x^2y)$$

$$x^3 + y^3 = (x+y)(x^2 - xy + y^2) \text{ or } (x+y)^3 - 3xy(x+y)$$

$$x^3 - y^3 = (x-y)(x^2 + xy + y^2) \text{ or } (x-y)^3 + 3xy(x-y)$$

$$a^3 + 1 = (a+1)(a^2 - a + 1)$$

$$a^3 - 1 = (a-1)(a^2 + a + 1)$$

$$(x+y+z)^2 = x^2 + y^2 + z^2 + 2xy + 2yz + 2xz$$

$$= x^2 + y^2 + z^2 + 2(xy + yz + xz)$$

$$\Rightarrow \text{If } (x+y+z) = 0, \text{ then } x^2 + y^2 + z^2 = -2(xy + yz + xz)$$

$$(x+y)(y+z)(x+z) = x^2(y+z) + y^2(x+z) + z^2(x+y) + 2xyz$$

$$= xy(x+y) + yz(y+z) + xz(x+z) + 2xyz$$

$$(x-y)(y-z)(z-x) = x^2(z-y) + y^2(x-z) + z^2(y-x)$$

$$x^3 + y^3 + z^3 - 3xyz = \frac{1}{2}(x+y+z)[(x-y)^2 + (y-z)^2 + (z-x)^2]$$

$$= (x+y+z)[(x+y+z)^2 - 3(xy+yz+xz)]$$

$$= (x+y+z) \times (x^2 + y^2 + z^2 - xy - yz - xz)$$

$\Rightarrow$ If $(x+y+z) = 0$, then $x^3 + y^3 + z^3 = 3xyz$ **(SSC CHSL 2013)**

$(x-y)^3 + (y-z)^3 + (z-x)^3 = 3(x-y)(y-z)(z-x)$ **[Hint:** $(x-y) + (y-z) + (z-x) = 0$]

$$(x+y+z)^3 = x^3 + y^3 + z^3 + 3(x+y)(y+z)(x+z)$$

$$(x^4 + x^2 + 1) = (x^2 + x + 1)(x^2 - x + 1)$$

$$x^4 + y^4 + x^2y^2 = (x^2 + y^2 + xy)(x^2 + y^2 - xy)$$

$$(x+a)(x+b) = x^2 + (a+b)x + ab$$

$$(x+a)(x+b)(x+c) = x^3+(a+b+c)x^2+(ab+bc+ca)x+abc$$
$$(x-a)(x-b)(x-c) = x^3-(a+b+c)x^2+(ab+bc+ca)x-abc$$
$$(x+y+z)(xy+yz+xz) = x^2(y+z)+y^2(z+x)+z^2(x+y)+3xyz$$
$$\left(x+\frac{1}{x}\right)^2 = x^2+\frac{1}{x^2}+2 \quad \Rightarrow x^2+\frac{1}{x^2} = \left(x+\frac{1}{x}\right)^2-2$$
$$\left(x-\frac{1}{x}\right)^2 = x^2+\frac{1}{x^2}-2 \quad \Rightarrow x^2+\frac{1}{x^2} = \left(x-\frac{1}{x}\right)^2+2$$
$$\left(x+\frac{1}{x}\right)^3 = x^3+\frac{1}{x^3}+3\left(x+\frac{1}{x}\right) \quad \Rightarrow x^3+\frac{1}{x^3} = \left(x+\frac{1}{x}\right)^3-3\left(x+\frac{1}{x}\right)$$
$$\left(x-\frac{1}{x}\right)^3 = x^3-\frac{1}{x^3}-3\left(x-\frac{1}{x}\right) \quad \Rightarrow x^3-\frac{1}{x^3} = \left(x-\frac{1}{x}\right)^3+3\left(x-\frac{1}{x}\right)$$

RULES

1. $(x-a)$ is a factor of $f(x)$ if $f(a)=0$
   $(x+a)$ is a factor of $f(x)$ if $f(-a)=0$
   $(ax-b)$ is a factor of $f(x)$ if $f\left(\frac{b}{a}\right)=0$
   $(ax+b)$ is a factor of $f(x)$ if $f\left(-\frac{b}{a}\right)=0$

   Hint: We put the divisor equal to zero and find the value of x. For example, if divisor is $(x-a)$, then $x-a=0 \Rightarrow x=a$
   If $f(x)=0$ when $x=a$ then $f(x)$ is exactly divisible by $(x-a)$
2. If $f(x)$ is divided by $(x-a)$, the remainder is $f(a)$.
   If $f(x)$ is divided by $(x+a)$, the remainder is $f(-a)$.
   If $f(x)$ is divided by $(ax-b)$, the remainder is $f\left(\frac{b}{a}\right)$.
   If $f(x)$ is divided by $(ax+b)$, the remainder is $f\left(-\frac{b}{a}\right)$.
3. Remainder on dividing (x^n-1) by $(x-1)$ is $1^n-1=0$
   $\therefore$ (x^n-1) is always divisible by $(x-1)$.
   (x^n-1) is divisible by $(x+1)$ for even values of n.
   $\therefore$ (x^n-1) is not divisible by $(x+1)$ for odd values of n.
   Example: $(10^{21}-1)$ is not divisible by:
   (*a*) 3 (*b*) 9 (*c*) 11 (*d*) 27
   Sol. (*c*) We know that (x^n-1) is not divisible by $(x+1)$ for odd values of n.
   $\therefore$ $(10^{21}-1)$ is not divisible by $(10+1)=11$ since 21 is an odd number.
   Note: $(10^{3n}-1)$ is always divisible by $9\times 3=27$. $(10^{9n}-1)$ is always divisible by $9\times 9=81$
4. (x^n+1) is divisible by $(x+1)$ for all odd values of n.
   (x^n+1) is not divisible by $(x+1)$ for all even values of n.
   (x^n+1) is not divisible by $(x-1)$ for all values of n.
5. $(x^{ab}-1)$ is always divisible by (x^a-1) and (x^b-1). [Logic: $(x^{ab}-1)=(x^a)^b-1)$]
   $(x^{abc}-1)$ is always divisible by (x^a-1), (x^b-1) and (x^c-1).

6. $(x^n - y^n)$ is always divisible by $(x - y)$.
$(x^n - y^n)$ is divisible by $(x + y)$ for even values of n.
$\therefore$ $(x^n - y^n)$ is not divisible by $(x + y)$ for odd values of n.
7. $(x^n + y^n)$ is divisible by $(x + y)$ for odd values of n.
$(x^n + y^n)$ is not divisible by $(x + y)$ for even values of n.
$(x^n + y^n)$ is not divisible by $(x - y)$ for all values of n.
8. The common factor of $(x^m - 1)$ and $(x^n - 1)$ is $(x^{\text{HCF of } m \text{ and } n} - 1)$.

Example 1. Find the HCF of $(2^{15} - 1)$ and $(2^{25} - 1)$

Sol. HCF of 15 and 25 is 5. $\Rightarrow$ Required HCF $= 2^5 - 1 = 31$

Example 2: Find the HCF of $(3^{75} - 1)$ and $(3^{35} - 1)$.

Sol. HCF of 75 and 35 = 5 $\Rightarrow$ Required HCF $= 3^5 - 1 = 242$.

9. **Important Trick:** If $x + \frac{1}{x} = 2 \Rightarrow x^2 - 2x + 1 = 0 \Rightarrow (x-1)^2 = 0 \Rightarrow x = 1$.
$\Rightarrow$ When sum of one number and its reciprocal is 2. Then the number is 1.
When $x + \frac{1}{x} = 2 \Rightarrow x^n + \frac{1}{x^n} = 1 + 1 = 2$

Example. If $x + \frac{1}{x} = 2$, $x \neq 0$, then value of $x^2 + \frac{1}{x^3}$ is equal to **(SSC FCI Asstt. 2012)**

(*a*) 1 (*b*) 2 (*c*) 3 (*d*) 4

Sol. (*b*) $x + \frac{1}{x} = 2 \Rightarrow x = 1 \Rightarrow x^2 + \frac{1}{x^3} = 1 + 1 = 2$

10. If all the coefficients of a polynomial equation are intergers and coefficient of the highest power is 1, then any factor of the constant term is root of the given equation.
Example: Possible roots of equation $x^2 - 6x + 9 = 0$ are factors of 9. *i.e.*, $\pm 1, \pm 3, \pm 9$.
11. If $px^n + ... + q = 0$ is an equation where p is the coefficient of the highest power and q is the constant term, then possible roots of the equation are $\frac{\text{Factors of } q}{\text{Factors of } p}$.

Example: Find roots of the equation $2x^3 + 3x^2 - 11x - 6 = 0$
Factors of $6 = \pm 1, \pm 2, \pm 3, \pm 6$, Factors of $2 = \pm 1, \pm 2$

$\therefore$ Possible roots $= \pm 1, \pm 2, \pm 3, \pm 6, \pm \frac{1}{2}, \pm \frac{3}{2}$

Note: $f(2)$, $f(-3)$ and $f\left(-\frac{1}{2}\right) = 0 \Rightarrow 2, -3, -\frac{1}{2}$ are the rational roots of the given equation.

12. A nth degree equation can have maximum n real roots.
Hence, a linear equation has only one root, a quadratic equation has maximum two roots and so on.
13. How to find minimum value of an algebric expression using arithmetic mean, geometric mean method ?
Method: Arithmetic mean $\geq$ Geometric Mean where
A.M. $= \frac{x_1 + x_2 + x_3 + ... x_n}{n}$ and G.M. $= \sqrt[n]{x_1 x_2 x_3 ... x_n}$.
$\Rightarrow$ For a set of two numbers, A.M. $= \frac{x+y}{2}$ and G.M. $= \sqrt{xy}$

A.M. ≥ G.M. ⇒ $\frac{x+y}{2} \geq \sqrt{xy} \Rightarrow (x+y) \geq 2\sqrt{xy}$

Example 1: If a be a positive number, then the least value of $a+\frac{1}{a}$ is **(SSC CGL Tier I 2013)**

(*a*) 1 (*b*) 0 (*c*) 2 (*d*) $\frac{1}{2}$

Sol. (*c*) $a+\frac{1}{a} \geq 2\sqrt{a \times \frac{1}{a}} \Rightarrow a+\frac{1}{a} \geq 2$

Example 2: The least value of $3^x + 3^{-x}$ is

(*a*) 2 (*b*) 1 (*c*) 2/3 (*d*) $\frac{1}{3}$

Sol. (*a*) $(3^x + 3^{-x}) \geq 2\sqrt{3^x \times 3^{-x}} \Rightarrow (3^x + 3^{-x}) \geq 2$

Remember: The minimum value of sum of a number and its reciprocal is 2.

QUADRATIC EQUATIONS

A quadratic equation is an equation in the form of $ax^2 + bx + c = 0$, where x is a variable and a, b, c are constants with $a \neq 0$. It is also called equation of second order. a, b, c are called coefficients.

Roots of quadratic equations : The values of x that satisfies the given equation are called its roots or solutions. Every quadratic equation can have at most two solutions. Some quadratic equations may not have any real solution. Different methods of finding roots :

1. Factorisation Method : Write b as sum of two numbers whose product is equal to ac. Let the factors are α and β. Then $b = \alpha + \beta$. $ac = \alpha \times \beta$. $ax^2 + bx + c = 0$ is expressed as $(x - \alpha)(x - \beta = 0)$. Then, put $(x - \alpha) = 0$ and $(x - \beta) = 0 \Rightarrow (x = \alpha)$ or $(x = \beta)$

Example : $x^2 + 7x + 12 = 0$.

Sol. Factors of 12 (*i.e.*, constant term) are ± 1, ± 2, ± 3, ± 4, ± 6, ± 12

With *trial-and-error method,* we can tell the values of α and β are 3 and 4.

Note: 3 + 4 = 7 (i.e., the coefficient of x) and 3 × 4 = 12 (i.e. the constant term).

$\Rightarrow x^2 + 3x + 4x + 12 = 0$

$\Rightarrow x(x + 3) + 4(x + 3) = 0$ $\Rightarrow (x + 4) \times (x + 3) = 0$

$\Rightarrow x + 4 = 0$ or $x + 3 = 0$ $\Rightarrow x = -4$ or -3

2. Completion of square method:

$x^2 - 7x + 12 = 0$ $\Rightarrow x^2 - 7x = -12 \Rightarrow \left(x - \frac{7}{2}\right)^2 = -12 + \left(\frac{7}{2}\right)^2$

$$\Rightarrow \left(x - \frac{7}{2}\right)^2 = \frac{1}{4} \Rightarrow x - \frac{7}{2} = \pm\frac{1}{2} \Rightarrow x = \frac{7}{2} + \frac{1}{2} \text{ or } \frac{7}{2} - \frac{1}{2} \Rightarrow x = 4 \text{ or } 3$$

3. Using Formula : $x = \frac{-b \pm \sqrt{b^2 - 4ac}}{2a}$

Note: If $b^2 - 4ac > 0$, then the equation has two real and distinct solutions.

If $b^2 - 4ac < 0$, then the equation has no real solution.

If $b^2 - 4ac = 0$, then the equation has one real solution.

$b^2 - 4ac$ is called the discriminant. (denoted by D)

Example : $x^2 - 7x + 12 = 0$.

Sol. $x = \dfrac{-(-7) \pm \sqrt{(-7)^2 - 4 \times 1 \times 12}}{2 \times 1} = \dfrac{7 \pm 1}{2} = 4 \text{ or } 3$

Note: Sum of roots $\alpha + \beta = -\dfrac{b}{a}$ and product of roots $\alpha\beta = \dfrac{c}{a}$

If roots are reciprocals to each other, then $\dfrac{c}{a} = 1$ or $c = a$

If roots are negative reciprocals to each other, then $\dfrac{c}{a} = -1$ or $c = -a$ **(CDS 2010)**

If roots are opposite in signs, then sum = 0 or $b = 0$
If one root = 0, then $c = 0$
A quadratic equation with given roots α and β is $x^2 - (\alpha + \beta)\, x + \alpha\, \beta = 0$

15. If $f(x) = ax^2 + bx + c$ where $a \neq 0$, then Maximum/minimum value of $f(x)$ is at $x = -\dfrac{b}{2a}$

Minimum value = $\dfrac{4ac - b^2}{4a}$, if $a > 0$ Maximum value = $\dfrac{4ac - b^2}{4a}$, if $a < 0$

Example: If x be real, the minimum value of $4x^2 + 4x + 9$ is **(SSC FCI 2012)**

(*a*) $\dfrac{3}{4}$ (*b*) 8 (*c*) 9 (*d*) $-\dfrac{1}{2}$

Sol. (*b*) $4x^2 + 4x + 9 = (2x + 1)^2 + 8 \Rightarrow$ Minimum value = 0 + 8 = 8 when $2x + 1 = 0$

Alt: Minimum value = $\dfrac{4ac - b^2}{4a} = c - \dfrac{b^2}{4a} = 9 - \dfrac{16}{16} = 8$ when $x = \dfrac{-b}{2a} = -\dfrac{4}{8} = -\dfrac{1}{2}$

SOLVED EXERCISE

1. A number exceeds its two fifth by 75. The number is **(SSCCGLTierI2015)**

(*a*) 112 (*b*) 150 (*c*) 125 (*d*) 100

Sol. (*c*) $x - \dfrac{2}{5}x = 75 \Rightarrow \dfrac{3}{5}x = 75 \Rightarrow x = 75 \times \dfrac{5}{3} = 125$

2. $(53 \times 87 + 159 \times 21 + 106 \times 25)$ is equal to : **(SSC CGL Tier I 2007)**

(*a*) 16000 (*b*) 1060 (*c*) 10600 (*d*) 60100

Sol. (*c*) Given expression $= 53 \times 87 + (53 \times 3) \times 21 + (53 \times 2) \times 25$
$= 53 \times 87 + 53 \times 63 + 53 \times 50 = 53 \times (87 + 63 + 50) = 53 \times 200 = 10600$

3. $55^2 \times 12^2 - 160^2 = ?$ **(Bank Clerical 2008)**

(*a*) 500 (*b*) 2,50,000 (*c*) 4,10,000 (*d*) None of these

Sol. (*c*) $? = (55 \times 12)^2 - 160^2 = 660^2 - 160^2 = 4{,}35{,}600 - 25{,}600 = 4{,}10{,}000$
Or $= (660 + 160)(660 - 160) = 820 \times 500 = 4{,}10{,}000$

4. If $a^3b = abc = 180$. a, b, c are positive integers, then the value of c is

(Hotel Management 1993, SSC CGL Tier I 2012)

(*a*) 110 (*b*) 1 (*c*) 4 (*d*) 25

Sol. (*b*) $a^3b = abc = 180 = 2 \times 2 \times 3 \times 3 \times 5 = 2^2 \times 3^2 \times 5$
$\Rightarrow a^3b = 1^3 \times 2^2 \times 3^2 \times 5 \Rightarrow a^3 = 1^3 \Rightarrow a = 1$
$a^3b = 180 \Rightarrow b = 180 \div 1 = 180$. 180 is not divisible by 2^3, 3^3 or 5^3.
Now, $abc = 180 \Rightarrow 1 \times 180 \times c = 180 \Rightarrow c = 1$

Trick: No option, other than 1 (option *b*) exactly divides 180.

5. The product of two positive numbers is 2500. If one number is four times the other, the sum of the two numbers is : **(SSC CGL Tier I 2002)**

(*a*) 25 (*b*) 125 (*c*) 225 (*d*) 250

Sol. (*b*) Let the numbers be x and $4x$.

Then $x \times 4x = 4x^2 = 2500 \Rightarrow x^2 = 2500 \div 4 = 625 \Rightarrow x = \sqrt{625} = 25$

Second number $= 4x = 4 \times 25 = 100$. Sum of the numbers $= 25 + 100 = 125$

6. If difference between the squares of two consecutive numbers is 47. Find the smaller number.

(*a*) 21 (*b*) 22 (*c*) 23 (*d*) 24

Sol. (*c*) Let the numbers be x and $(x + 1)$.

Difference between square of two consecutive numbers is equal to the sum of the numbers.

$\therefore x + (x + 1) = 2x + 1 = 47 \Rightarrow x = \dfrac{47-1}{2} = 23$

7. If $x + y = 9$ and $xy = 8$, then what is one of the values of $(x^4 - y^4)$ **(CDS 2012)**

(*a*) 4025 (*b*) 4065 (*c*) 4095 (*d*) 4905

Sol. (*c*) $x + y = 9$ and $xy = 8 \Rightarrow$ The numbers are 1 and 8.

$x^4 - y^4 = (x + y)(x - y)(x^2 + y^2) = 9 \times 7 \times 65 = 4095$

8. A student was asked to divide a number by 6 and add 12 to the quotient. He, however, first added 12 to the number and then divided it by 6, getting 112 as the answer. The correct answer should have been **(SSC CGL Tier I 2011)**

(*a*) 124 (*b*) 122 (*c*) 118 (*d*) 114

Sol. (*b*) $\dfrac{x+12}{6} = 112 \Rightarrow \dfrac{x}{6} + 2 = 112 \Rightarrow \dfrac{x}{6} + 12 = \left(\dfrac{x}{6} + 2\right) + 10 = 112 + 10 = 122$

9. If $x = 1.75$ and $y = 0.5$, then find the value of $(4x^2 + 4xy + y^2)$ **(SSC CGL Tier I 2013)**

(*a*) 15.75 (*b*) 16.00 (*c*) 16.25 (*d*) 16.75

Sol. (*b*) $(4x^2 + 4xy + y^2) = (2x + y)^2 = (3.5 + 0.5)^2 = 4^2 = 16$

10. If $x^2 + y^2 = 45$ and $xy = 18$, then find the value of $\dfrac{1}{x} + \dfrac{1}{y}$

(*a*) 1/2 (*b*) 1/3 (*c*) 1/4 (*d*) 2/3

Sol. (*a*) $(x + y)^2 = x^2 + y^2 + 2xy = 45 + 2 \times 18 = 81 \Rightarrow (x + y) = 9$

$\Rightarrow \dfrac{1}{x} + \dfrac{1}{y} = \dfrac{x+y}{xy} = \dfrac{9}{18} = \dfrac{1}{2}$

11. If $p + q = 10$ and $pq = 5$, then the numerical value of $\dfrac{p}{q} + \dfrac{q}{p}$ will be **(SSC CHSL 2012)**

(*a*) 22 (*b*) 18 (*c*) 16 (*d*) 20

Sol. (*b*) $\dfrac{p}{q} + \dfrac{q}{p} = \dfrac{p^2+q^2}{pq} = \dfrac{(p+q)^2 - 2pq}{pq} = \dfrac{10^2 - 10}{5} = \dfrac{90}{5} = 18$

12. If $m - 5n = 2$ then the value of $(m^3 - 125n^3 - 30mn)$ is **(SSC CGL Tier I 2015)**

(*a*) 8 (*b*) 7 (*c*) 6 (*d*) 9

Sol. (*a*) $(m^3 - 125n^3 - 30mn) = m^3 - (5n)^3 - 3(m)(5n)(2)$

$= m^3 - (5n)^3 - 3(m)(5n)(m - n) = (m - 5n)^3 = 2^3 = 8$

13. Two numbers are such that the sum of twice the first number and thrice the second number is 36 and the sum of thrice the first number and twice the second number is 39. Which is the smaller number ? **(SBI Bank PO 2008)**

(*a*) 9 (*b*) 7 (*c*) 6 (*d*) 5

Sol. (*c*) Let the numbers be x and y. Then $2x + 3y = 36$ and $3x + 2y = 39$

$\Rightarrow (3x + 2y) - (2x + 3y) = 39 - 36 \Rightarrow x - y = 3$

$(2x + 3y) + (3x + 2y) = 36 + 39 \Rightarrow 5(x + y) = 75 \Rightarrow (x + y) = 75 \div 5 = 15$

$\therefore$ Smaller number $= \dfrac{15-3}{2} = 6$

14. Out of three consecutive odd numbers, the sum of the first two is 33 more than the third number. The second number is:

(*a*) 33 (*b*) 35 (*c*) 37 (*d*) 39

Sol. (*c*) Let the three numbers be x, $x + 2$, $x + 4$. Then $(x) + (x + 2) - (x + 4) = 33 \Rightarrow x = 35$

$\therefore$ Second number $= x + 2 = 35 + 2 = 37$.

15. $54 \times 54 - 46 \times 46 = ?$

(*a*) 800 (*b*) 144 (*c*) 64 (*d*) 8

Sol. (*a*) $54 \times 54 - 46 \times 46 = (54 + 46) \times (54 - 46) = 100 \times 8 = 800$

Trick : Unit's digit is 6 in both the cases i.e. 54×54 and 46×46.

$\therefore$ Unit's digit in $54 \times 54 - 46 \times 46$ is $6 - 6 = 0 \Rightarrow$ (*a*) is the answer.

16. If $49^2 - 25^2 = 37x$, then what is x equal to ? **(CDS 2013)**

(*a*) 64 (*b*) 74 (*c*) 48 (*d*) 42

Sol. (*c*) $37x = 49^2 - 25^2 = (49 + 25)(49 - 25) = 74 \times 24 \Rightarrow x = \dfrac{74 \times 24}{37} = 48$

Trick : Unit's digit $= 1 - 5 = 6 = 7 \times 8$

17. $10^{30} - 10^{29} = ?$

(*a*) 10 (*b*) 10^{28} (*c*) 10^{29} (*d*) 9×10^{29}

Sol. (*d*) $10^{30} - 10^{29} = 10 \times 10^{29} - 10^{29} = 10^{29} \times (10 - 1) = 9 \times 10^{29}$

18. $106 \times 106 + 94 \times 94 = ?$

(*a*) 2400 (*b*) 4800 (*c*) 10036 (*d*) 20072

Sol. (*d*) $106^2 + 94^2 = (100 + 6)^2 + (100 - 6)^2 = 2 \times (100^2 + 6^2) = 2 \times (10000 + 36) = 20072$

Trick : Unit's digit $= 6 \times 6 + 4 \times 4 = 2$ (unit's digit)

19. If $a - b = 4$ and $a^2 + b^2 = 40$, where a and b are positive integers, then $a^3 + b^3$ is equal to **(CDS 2015)**

(*a*) 264 (*b*) 280 (*c*) 300 (*d*) 324

Sol. (*b*) $a = 6$ and $b = 2$ satisfies the given equations $\Rightarrow a^3 + b^3 = 6^3 + 2^3 = 216 + 64 = 280$

20. If $x - y = 2$ and $xy = 24$, then the value of $(x^2 + y^2)$ is **(SSC CHSL 2012)**

(*a*) 52 (*b*) 25 (*c*) 36 (*d*) 63

Sol. (*a*) $x - y = 2$ and $xy = 24 \Rightarrow$ Possible value of $(x, y) = (6, 4) \Rightarrow x^2 + y^2 = 6^2 + 4^2 = 36 + 16 = 52$

21. Each member of a picnic party contributed twice as many rupees as the total number of members and the total collection was ₹ 3042. The number of members present in the party was : **(SSC CGL Tier I 2002)**

(*a*) 2 (*b*) 32 (*c*) 40 (*d*) 39

Sol. (*d*) Let number of members in the party be x.
Then $x(2x) = 3042 \Rightarrow 2x^2 = 3042 \Rightarrow x^2 = 3042 \div 2 = 1521 \Rightarrow x = \sqrt{1521} = 39$

22. 800 chocolates were distributed among the students of a class. Each student got twice as many chocolates as the number of studetns in the class. Find the number of students in the class. **(SSC CGL Tier I 2000)**

(*a*) 25 (*b*) 30 (*c*) 35 (*d*) 20

Sol. (*d*) Let students = x. Then each student get = $2x$ chocolates $\Rightarrow 2x^2 = 800 \Rightarrow x = \sqrt{400} = 20$

23. If x is real, then the minimum value of $(x^2 - x + 1)$ is **(SSC CGL Tier I 2007)**

(*a*) $\frac{3}{4}$ (*b*) $\frac{1}{4}$ (*c*) 0 (*d*) 1

Sol. (*a*) $x^2 - x + 1 = \left(x - \frac{1}{2}\right)^2 + 1 - \frac{1}{4} = \left(x - \frac{1}{2}\right)^2 + \frac{3}{4}$

Minimum value of $\left(x - \frac{1}{2}\right)^2$ is 0 $\Rightarrow$ Required minimum value $= 0 + \frac{3}{4} = \frac{3}{4}$

Alt: Maximum/Minimum of Parabola is at vertex i.e. at $x = \frac{-b}{2a} = \frac{1}{2} \Rightarrow y = \frac{1}{4} - \frac{1}{2} + 1 = \frac{3}{4}$

24. The minimum value of $(x - 2)(x - 9)$ is **(SSC CGL Tier I 2013)**

(*a*) $-\frac{11}{4}$ (*b*) $\frac{49}{4}$ (*c*) 0 (*d*) $-\frac{49}{4}$

Sol. (*d*) The expression will have minimum value when the product of two factors is maximum and the product sign is negative.

$\therefore (x - 2) = -(x - 9) \Rightarrow x - 2 = -x + 9 \Rightarrow 2x = 11 \Rightarrow x = \frac{11}{2}$

$\therefore (x-2)(x-9) = \left(\frac{11}{2} - 2\right)\left(\frac{11}{2} - 9\right) = \frac{7}{2}\left(-\frac{7}{2}\right) = -\frac{49}{4}$

Alt: $(x - 2)(x - 9) = x^2 - 11x + 18 = \left(x - \frac{11}{2}\right)^2 + 18 - \frac{121}{4} = \left(x - \frac{11}{2}\right)^2 - \frac{49}{4}$

Its minimum value is $\frac{-49}{4}$ when $\left(x - \frac{11}{2}\right) = 0$

Alt: Derivative method : $dy/dx = (x - 2) + (x - 9) = 0 \Rightarrow x = 11/2$

25. $\frac{15^2 - 3^2}{9^2 - 3^2} = ?$ **(SBI Clerical 2008)**

(*a*) $\frac{5}{3}$ (*b*) $\frac{3}{2}$ (*c*) 2 (*d*) 3

Sol. (*d*) $\frac{15^2 - 3^2}{9^2 - 3^2} = \frac{(15+3)\times(15-3)}{(9+3)\times(9-3)} = \frac{18 \times 12}{12 \times 6} = 3$

26. Find the value of $\left(\frac{5^3 + 5^4 + 5^5 + 5^6 + 5^7}{5^1 + 5^2 + 5^3 + 5^4 + 5^5}\right)^{\frac{1}{2}}$.

(*a*) 5 (*b*) 25 (*c*) 125 (*d*) 625

Sol. (*a*) $\left(\frac{5^3+5^4+5^5+5^6+5^7}{5^1+5^2+5^3+5^4+5^5}\right)^{\frac{1}{2}} = \left(\frac{5^2 \times (5^1+5^2+5^3+5^4+5^5)}{(5^1+5^2+5^3+5^4+5^5)}\right)^{\frac{1}{2}} = (5^2)^{½} = 5.$

27. If *a*, *b*, *c* are real and $a^3 + b^3 + c^3 = 3abc$ and $a + b + c \neq 0$, then the relation with *a*, *b*, *c* will be **(SSC CGL Tier I 2012)**

(*a*) $a + b = c$ (*b*) $a + c = b$ (*c*) $a = b = c$ (*d*) $b + c = a$

Sol. (*c*) We check option (*c*): LHS = $a^3 + a^3 + a^3 = 3a^3$; RHS = $3aaa = 3a^3$.

28. If $|x - 4| < 5$, then

(*a*) $-1 < x < 9$ (*b*) $-9 < x < 1$ (*c*) $1 < x < 9$ (*d*) $5 < x < 9$

Sol. (*a*) $|x - 4| < 5 \Rightarrow x - 4 < 5$ or $4 - x < 5 \Rightarrow x < 9$ or $x > -1 \Rightarrow -1 < x < 9$.

29. If $a^2 + b^2 + c^2 + d^2 = 1$, then the maximum value of *abcd* is:

(*a*) 1 (*b*) $\frac{1}{2}$ (*c*) $\frac{1}{4}$ (*d*) $\frac{1}{16}$

Sol. (*d*) *abcd* is maximum when $a = b = c = d = k$ (say)

$\Rightarrow a^2 + b^2 + c^2 + d^2 = 4k^2 = 1 \Rightarrow k^2 = \frac{1}{4} \Rightarrow k = \frac{1}{2} \Rightarrow abcd = \left(\frac{1}{2}\right)^4 = \frac{1}{16}$

30. If $a = 23$ and $b = -29$, then the value of $25a^2 + 40ab + 16b^2$ is: **(SSC FCI 2012)**

(*a*) 1 (*b*) –1 (*c*) 0 (*d*) 2

Sol. (*a*) $25a^2 + 40ab + 16b^2 = (5a + 4b)^2 = (5 \times 23 - 4 \times 29)^2 = (115 - 116)^2 = (-1)^2 = 1$

31. Three numbers are in the ratio 2 : 3 : 4. If the sum of their squares is 1856, then the numbers are **(SSC CGL Tier II 2013)**

(*a*) 8,12,16 (*b*) 16, 24, 32 (*c*) 12, 18, 24 (*d*) None of these

Sol. (*b*) $(2x)^2 + (3x)^2 + (4x)^2 = 1856 \Rightarrow (4 + 9 + 16)x^2 = 1856$

$\Rightarrow x^2 = 1856 \div 29 = 64 \Rightarrow x = 8 \Rightarrow$ The numbers are 16, 24, 32.

32. If *a* and *b* are positive integers such that $a^2 - b^2 = 19$, then the value of *a* is **(SSC CGL Tier I 2010)**

(*a*) 19 (*b*) 20 (*c*) 9 (*d*) 10

Sol. (*d*) $a^2 - b^2 = (a + b)(a - b) = 19 \times 1 \Rightarrow a + b = 19$ and $a - b = 1 \Rightarrow a = \frac{19+1}{2} = 10$

33. If the difference of the squares of two natural numbers is 37, then the sum of the squares of the numbers will be **(SAS 2010)**

(*a*) 690 (*b*) 687 (*c*) 685 (*d*) 672

Sol. (*c*) $x^2 - y^2 = 37 \Rightarrow (x + y)(x - 1) = 37 \times 1 \Rightarrow (x + y) = 37$ and $(x - y) = 1$

$\Rightarrow x = 19, y = 18 \Rightarrow x^2 + y^2 = 19^2 + 18^2 = 361 + 324 = 685$

34. If *a*, *b*, *c* are natural numbers such that $a^2 + b^2 = 52$ and $b^2 + c^2 = 45$, find the value of $a + b + c$.

(*a*) 9 (*b*) 11 (*c*) 12 (*d*) 13

Sol. (*d*) $(a^2 + b^2) - (b^2 + c^2) = 52 - 45 \Rightarrow a^2 - c^2 = 7 \Rightarrow (a + c)(a - c) = 7 \times 1$

$\Rightarrow a + c = 7$ and $a - c = 1 \Rightarrow a = 4, c = 3$

$b = \sqrt{52 - 16} = \sqrt{36} = 6 \Rightarrow a + b + c = 4 + 6 + 3 = 13$

35. If $-2 \leq x \leq 6$ and $-5 \leq y \leq 4$, what is the minimum possible value of $x - y$?

(*a*) –7 (*b*) –6 (*c*) –2 (*d*) 2

Sol. (*b*) The value of $x - y$ is minimum when x is smallest and y is greatest.
Smallest possible value of $x = -2$ and greatest possible value of $y = 4$
$\therefore$ Smallest possible value of $x - y = -2 - 4 = -6$

36. Find the number which gives the same result when added to 3/2 or when multiplied by 3/2. **(SSC CHSL 2004)**

(*a*) 2/3 (*b*) 1 (*c*) 2 (*d*) 3

Sol. (*d*) Let the number be x. Then $\frac{3}{2} + x = \frac{3}{2}x \Rightarrow \frac{1}{2}x = \frac{3}{2} \Rightarrow x = 3$

37. On subtracting a number from its square, we get 4 times the number. What is the number? **(EPF SSA 2007)**

(*a*) 3 (*b*) 4 (*c*) 5 (*d*) None of these

Sol. (*c*) $x^2 - x = 4x \Rightarrow x^2 = 5x \Rightarrow x = 5$

38. If $(a-3)^2 + (b-4)^2 + (c-9)^2 = 0$, then the value of $\sqrt{a+b+c}$ is : **(SSC CHSL 2011)**

(*a*) –4 (*b*) 4 (*c*) ± 4 (*d*) ± 2

Sol. (*c*) $(a-3)^2 + (b-4)^2 + (c-9)^2 = 0 \Rightarrow a-3=0, b-4=0, c-9=0 \Rightarrow a=3, b=4, c=9$
(Hint : Square of number cannot have negative sign)
$\therefore \sqrt{a+b+c} = \sqrt{3+4+9} = \sqrt{16} = \pm 4$

39. If $\frac{1}{10} + \frac{1}{20} = \frac{1}{x}$, then $x = ?$ **(MBA Entrance)**

(*a*) $\frac{1}{30}$ (*b*) $\frac{3}{20}$ (*c*) $\frac{20}{3}$ (*d*) 30

Sol. (*c*) $\frac{1}{10} + \frac{1}{20} = \frac{3}{20} \Rightarrow \frac{1}{x} = \frac{3}{20} \Rightarrow x = \frac{20}{3}$

40. If the number p is 5 more than q and the sum of the squares of p and q is 55, then the product of p and q is **(SSC CGL Tier I 2013)**

(*a*) 10 (*b*) –10 (*c*) 15 (*d*) –15

Sol. (*c*) $p = q + 5 \Rightarrow p - q = 5$; $p^2 + q^2 = 55$
$(p-q)^2 = (p^2 + q^2) - 2pq \Rightarrow 5^2 = 55 - 2pq \Rightarrow 2pq = 30 \Rightarrow pq = 15$

41. The product of two numbers is 24 times the difference of these two numbers. If the sum of these numbers is 14, the larger number is : **(SSC CPO Sub-Inspector Exam 2009)**

(*a*) 9 (*b*) 8 (*c*) 7 (*d*) 10

Sol. (*b*) Let the numbers be x and y. Then $xy = 24(x - y)$.
From the given options, the possible pairs are (9, 5), (8, 6), (7, 7), (10, 4)
8×6 is the only number, which is divisible by 24.

42. If $x + \frac{1}{x} = 2$ then the value of $\sqrt{x} + \frac{1}{\sqrt{x}}$ is **(SAS 2010)**

(*a*) $\sqrt{2}$ (*b*) 2 (*c*) $\sqrt{2} + 1$ (*d*) 1

Sol. (*b*) $x + \frac{1}{x} = 2 \Rightarrow x = 1 \Rightarrow \sqrt{x} + \frac{1}{\sqrt{x}} = 1 + 1 = 2.$

43. If $x + \frac{1}{x} = 2$, then the value of $x^{12} - \frac{1}{x^{12}}$ is **(SSC CGL Tier I 2015)**

(*a*) 0 (*b*) 2 (*c*) 4 (*d*) – 4

Sol. (*a*) $x+\frac{1}{x}=1 \Rightarrow x=1 \Rightarrow x^{12}-\frac{1}{x^{12}}=1-1=0$

44. If $a+\frac{1}{a+2}=0$, then value of $(a+2)^3+\frac{1}{(a+2)^3}$ is **(SSC CGL Tier I 2012)**

(*a*) 4 (*b*) 6 (*c*) 2 (*d*) 3

Sol. (*c*) $a+\frac{1}{a+2}=0 \Rightarrow (a+2)+\frac{1}{a+2}=2 \Rightarrow (a+2)=1 \Rightarrow$ G.E. $=1^3+1^3=2$

45. If $x+\frac{1}{x}=2$, then the value of $\frac{2x^2+2}{3x^2+5x+3}$ is **(SAS 2010)**

(*a*) $\frac{4}{11}$ (*b*) $\frac{1}{2}$ (*c*) $1\frac{3}{4}$ (*d*) $2\frac{1}{3}$

Sol. (*a*) $x+\frac{1}{x}=2 \Rightarrow x=1 \Rightarrow \frac{2x^2+2}{3x^2+5x+3}=\frac{2+2}{3+5+3}=\frac{4}{11}$

46. If $x+\frac{1}{x}=1$, then the value of $\frac{2}{x^2-x+2}$ is **(SSC CGL Tier I 2015)**

(*a*) 2/3 (*b*) 2 (*c*) 1 (*d*) 4

Sol. (*b*) $x+\frac{1}{x}=1 \Rightarrow x^2+1=x \Rightarrow x^2-x=-1 \Rightarrow \frac{2}{x^2-x+2}=\frac{2}{-1+2}=2$

47. If $x+\frac{1}{x}=-2$, then find the value of $x^{2n+1}+\frac{1}{x^{2n+1}}$, where n is a natural number

(SSC CGL Tier I 2008)

(*a*) 0 (*b*) 2 (*c*) –2 (*d*) –5

Sol. (*c*) $x+\frac{1}{x}=-2 \Rightarrow x^2+1=-2x \Rightarrow x^2+2x+1=0 \Rightarrow (x+1)^2=0 \Rightarrow x=-1$

$\therefore x^{2n+1}+\frac{1}{x^{2n+1}}=(-1)+(-1)=-2$

(Remember : If sum of a number and its reciprocal is –2, then the number is –1)

48. If a and b two odd positive integers, by which of the following integers is (a^4-b^4) always divisible ? **(SSC CGL Tier I 2010)**

(*a*) 3 (*b*) 6 (*c*) 8 (*d*) 12

Sol. (*c*) $a^4-b^4=(a-b)(a+b)(a^2+b^2)$. Each of the factor is divisible by 2 (even number)

$\therefore$ a^4-b^4 is always divisible by $2\times2\times2=8$.

49. The product of two numbers is 9375 and the quotient, when the larger one is divided by smaller, is 15. The sum of the numbers is : **(SSC CGL Tier I 2004)**

(*a*) 395 (*b*) 380 (*c*) 400 (*d*) 425

Sol. (*c*) Let the numbers be x and $15x$.

Then $x\times(15x)=9375 \Rightarrow 15x^2=9375 \Rightarrow x^2=9375\div15=625 \Rightarrow x=\sqrt{625}=25$

$15x=15\times25=375$ $\therefore$ Required sum $=x+15x=25+375=400$

50. The product of two positive numbers is 11520 and their quotient is $\frac{9}{5}$. Find the difference of two numbers. **(SSC CGL Tier I 2002)**

(*a*) 60 (*b*) 64 (*c*) 74 (*d*) 70

Sol. (*b*) Let the numbers be *x* and *y*.

Then $xy = 11520$ and $\frac{x}{y} = \frac{9}{5} \Rightarrow xy \times \frac{x}{y} = 11520 \times \frac{9}{5} \Rightarrow x^2 = 2304 \times 9$

$\Rightarrow x = \sqrt{2304 \times 9} = 48 \times 3 = 144 \Rightarrow \frac{x}{y} = \frac{9}{5} \Rightarrow y = \frac{5}{9}x \Rightarrow y = \frac{5}{9} \times 144 = 80$

Difference between the numbers $= x - y = 144 - 80 = 64$

51. The product of two fractions is $\frac{14}{15}$ and their quotient is $\frac{35}{24}$. The greater of the fractions is **(SSC CGL Tier I 2002)**

(*a*) $\frac{7}{4}$ (*b*) $\frac{7}{6}$ (*c*) $\frac{7}{3}$ (*d*) $\frac{4}{5}$

Sol. (*b*) Let the fractions be *x* and *y*. Then $xy = \frac{14}{15}$ and $\frac{x}{y} = \frac{35}{24}$

Multiplying the two equations, we get : $(xy)\left(\frac{x}{y}\right) = \frac{14}{15} \times \frac{35}{24}$

$\Rightarrow x^2 = \frac{49}{36} \Rightarrow x = \sqrt{\frac{49}{36}} = \frac{7}{6} \Rightarrow xy = \frac{14}{15} \Rightarrow \left(\frac{7}{6}\right)y = \frac{14}{15} \Rightarrow y = \frac{14}{15} \times \frac{6}{7} = \frac{4}{5}$

As $\frac{7}{6} > \frac{4}{5} \Rightarrow \frac{7}{6}$ is the required fraction.

52. $\left[\left(7\frac{1}{2}\right)^2 - \left(3\frac{3}{4}\right)^2\right] \div 11\frac{1}{4} = ?$

(*a*) $7\frac{1}{2}$ (*b*) $3\frac{3}{4}$ (*c*) $11\frac{1}{4}$ (*d*) $11\frac{1}{2}$

Sol. (*b*) Given expression $= \left(7\frac{1}{2} + 3\frac{3}{4}\right)\left(7\frac{1}{2} - 3\frac{3}{4}\right) \div 11\frac{1}{4} = \left(11\frac{1}{4}\right)\left(3\frac{3}{4}\right) \div 11\frac{1}{4} = 3\frac{3}{4}$

53. What is the sum of the squares of the roots of the equation $x^2 + 2x - 143 = 0$ **(NDA 2012)**

(*a*) 170 (*b*) 180 (*c*) 190 (*d*) 290

Sol. (*d*) $\alpha + \beta = -2$ and $\alpha\beta = -143$

$\alpha^2 + \beta^2 = (\alpha + \beta)^2 - 2\alpha\beta = (-2)^2 - 2(-143) = 4 + 286 = 290$

54. A and B solved a quadratic equation. In solving it, A made a mistake in the constant term and got the roots as 6 and 2, while B made a mistake in the coefficient of *x* only and obtained roots as –7 and –1. The correct roots of the equation are:

(*a*) – 6, –2 (*b*) 7, 1 (*c*) –7, 2 (*d*) – 6, –1

Sol. (*b*) Correct sum of roots = 6 + 2 = 8; Correct product of roots = (–7) (–1) = 7

$\Rightarrow$ Correct roots are 7 and 1 (Note: 7 + 1 = 8; 7 × 1 = 7).

55. The sum of the squares of three consecutive natural numbers is 2030. Then, what is the middle number ? **(SSC CGL Tier I 2000)**

(*a*) 25 (*b*) 26 (*c*) 27 (*d*) 28

Sol. (*b*) Let the numbers are $(n-1), (n), (n+1)$. Then sum of their squares $= 3n^2 + 2 = 2030$

$\therefore 3n^2 = 2030 - 2 = 2028 \Rightarrow n^2 = \dfrac{2028}{3} = 676 \Rightarrow n = \sqrt{676} = 26$ (middle-number)

56. The sum of the squares of three consecutive positive numbers is 365. The sum of the number is **(SSC MTS 2011)**

(*a*) 30 (*b*) 33 (*c*) 36 (*d*) 45

Sol. (*b*) $3x^2 + 2 = 365 \Rightarrow 3x^2 = 363 \Rightarrow x^2 = 121 \Rightarrow x = 11 \Rightarrow \text{sum} = 3x = 33$

57. Three-fifth of the square of a certain number is 126.15. What is the number ? **(SSC CGL Tier I 2001)**

(*a*) 201.25 (*b*) 75.69 (*c*) 14.5 (*d*) 145

Sol. (*c*) Let the number be x. Then $\dfrac{3}{5}x^2 = 126.15$

$\Rightarrow x^2 = 126.15 \times \dfrac{5}{3} = 42.05 \times 5 = 8.41 \times 5 \times 5 \Rightarrow x = \sqrt{8.41 \times 5 \times 5} = 2.9 \times 5 = 14.5$

58. Thrice the square of a natural number decreased by four times the number is equal to 50 more than the number. The number is : **(SSC CGL Tier I 2003)**

(*a*) 4 (*b*) 5 (*c*) 10 (*d*) 6

Sol. (*b*) $3x^2 - 4x = x + 50 \Rightarrow 3x^2 - 5x - 50 = 0 \Rightarrow 3x^2 + 10x - 15x - 50 = 0$

$\Rightarrow x(3x + 10) - 5(3x + 10) = 0 \Rightarrow (x - 5) \times (3x + 10) = 0 \Rightarrow x = 5$

59. $x = \dfrac{a}{a-1}$ and $y = \dfrac{1}{a-1}$ **(SBI Associate Bank PO 2007)**

(*a*) $x > y$ (*b*) $y < x$ (*c*) $x = y$ (*d*) Can't say

Sol. (*a*) $x - y = \dfrac{a}{a-1} - \dfrac{1}{a-1} = \dfrac{a-1}{a-1} = 1 \Rightarrow x = y + 1 \Rightarrow x > y$

60. If $3x - 5y = 5$ and $\dfrac{x}{x+y} = \dfrac{5}{7}$, then what is the value of $x - y$?

(*a*) 9 (*b*) 6 (*c*) 4 (*d*) 3

Sol. (*d*) $\dfrac{x}{x+y} = \dfrac{5}{7} \Rightarrow 7x = 5x + 5y \Rightarrow 2x = 5y$. Now $3x - 5y = 5 \Rightarrow 3x - 2x = 5 \Rightarrow x = 5$

Now $2x = 5y \Rightarrow 5y = 2 \times 5 = 10 \Rightarrow y = 2$. $\therefore x - y = 5 - 2 = 3$

61. A person bought 5 tickets from a station P to a station Q and 10 tickets from the station P to a station R. He paid ₹ 350. If the sum of a ticket from P to Q and P to R is ₹ 42, then what is the fare from P to Q ? **(CDS 2009)**

(*a*) 12 (*b*) 14 (*c*) 16 (*d*) 18

Sol. (*b*) Let the fare from P to Q and P to R be ₹ x and ₹ y respectively.

Then $5x + 10y = 350$..............(1)

and $x + y = 42 \Rightarrow 10x + 10y = 420$ ....(2)

(2) – (1) : $5x =$ ₹ $70 \Rightarrow x = 70 \div 5 = 14$

62. The product of two consecutive odd numbers is 4899. What is the smaller number ? **(Bank Clerical 2008)**

(*a*) 67 (*b*) 69 (*c*) 71 (*d*) 73

Sol. (*b*) Let the numbers be $(x - 1)$ and $(x + 1)$.

Then $(x - 1) \times (x + 1) = 4899 \Rightarrow x^2 = 4899 + 1 = 4900 \Rightarrow x = 70$

$\therefore$ The numbers are 69 and 71.

63. The number, whose square is equal to the difference of the squares of 75.15 and 60.12, is : **(SSC CGL Tier I 2004)**

(*a*) 46.09 (*b*) 48.09 (*c*) 45.09 (*d*) 47.09

Sol. (*c*) Let x be the required number.

Then $x^2 = 75.15^2 - 60.12^2 = (75.15 + 60.12) \times (75.15 - 60.12) = 135.27 \times 15.03$

$= (9 \times 15.03) \times 15.03 = 3^2 \times 15.03^2 \Rightarrow x = 3 \times 15.03 = 45.09$.

64. Sum of two numbers is 20 and difference between the numbers is 4. Find the difference between their squares. **(SSC CGL Tier I 2005)**

(*a*) 5 (*b*) 16 (*c*) 24 (*d*) 80

Sol. (*d*) Let the numbers be x and y. Then $x + y = 20$ and $x - y = 4$

$\therefore\ x^2 - y^2 = (x + y) \times (x - y) = 20 \times 4 = 80$

65. If the sum of two numbers is 3 and the sum of their squares is 12, then their product is equal to : **(SSC CGL Tier I 2000)**

(*a*) $\frac{3}{2}$ (*b*) $\frac{2}{3}$ (*c*) $-\frac{3}{2}$ (*d*) $-\frac{2}{3}$

Sol. (*c*) Let the numbers be x and y. Then $x + y = 3$ and $x^2 + y^2 = 12$

$(x + y)^2 = x^2 + y^2 + 2xy \Rightarrow 3^2 = 12 + 2xy \Rightarrow 2xy = 9 - 12 = -3 \Rightarrow xy = -\frac{3}{2}$

66. If the sum of the two numbers is 22 and the sum of their squares is 404, then the product of the numbers is : **(SSC CGL Tier I 2000)**

(*a*) 40 (*b*) 44 (*c*) 80 (*d*) 88

Sol. (*a*) Let the numbers be x and y. Then $x + y = 22$ and $x^2 + y^2 = 404$

$(x + y)^2 = (x^2 + y^2) + 2xy \Rightarrow 2xy = (x + y)^2 - (x^2 + y^2) = 22^2 - 404 = 484 - 404 = 80$

$\Rightarrow xy = 80 \div 2 = 40$

67. The sum of the squares of two positive integers is 100 and difference of their square is 28. The sum of the numbers is : **(SSC CGL Tier I 2006 & 2001)**

(*a*) 12 (*b*) 13 (*c*) 14 (*d*) 15

Sol. (*c*) $x^2 + y^2 = 100$ .... (1) And $x^2 - y^2 = 28$ .... (2)

Adding (1) and (2) $\Rightarrow 2x^2 = 128 \Rightarrow x = \sqrt{64} = 8$

From (1) : $y^2 = 100 - x^2 = 100 - 8^2 = 36 \Rightarrow y = \sqrt{36} = 6$

$\therefore$ The sum of the numbers $= 8 + 6 = 14$

68. If difference between two positive numbers is 3 and sum of squares of the numbers is 369, then the sum of the numbers is : **(SSC CGL Tier I 2003)**

(*a*) 23 (*b*) 27 (*c*) 33 (*d*) 37

Sol. (*b*) $2xy = (x^2 + y^2) - (x - y)^2 = 369 - 9 = 360$

$(x + y)^2 = (x^2 + y^2) + 2xy = 369 + 360 = 729 \Rightarrow x + y = \sqrt{729} = 27$

69. If difference between two numbers is 3 and sum of squares of the numbers is 117. Find the product of the numbers.

(*a*) 21 (*b*) 27 (*c*) 31 (*d*) 54

Sol. (*d*) $2xy = (x^2 + y^2) - (x - y)^2 = 117 - 3^2 = 117 - 9 = 108 \Rightarrow xy = \frac{108}{2} = 54$

70. The product of two numbers is 120. The sum of their squares is 289. The difference of these two numbers is **(SSC CGL Tier I 2001)**

(*a*) 9 (*b*) 7 (*c*) 8 (*d*) 6

Sol. (*b*) $(x - y)^2 = (x^2 + y^2) - 2xy = 289 - 2 \times 120 = 49 = 7^2$

71. The product of two numbers is 120 and the sum of their squares is 289. The sum of the two numbers is : **(SSC Data Entry Operator 2008)**

(*a*) 23 (*b*) 7 (*c*) 13 (*d*) 169

Sol. (*a*) Let the numbers be x and y. Then $xy = 120$ and $x^2 + y^2 = 289$

$\therefore (x + y)^2 = x^2 + y^2 + 2xy = 289 + 2 \times 120 = 289 + 240 = 529 = 23^2 \Rightarrow (x + y) = 23$

72. If sum of two numbers is 25 and difference between them is 11. What is the product of the numbers.

(*a*) 275 (*b*) 252 (*c*) 250 (*d*) 126

Sol. (*d*) $(x + y)^2 - (x - y)^2 = 4xy \Rightarrow 25^2 - 11^2 = 4xy \Rightarrow 4xy = (25 + 11) \times (25 - 11)$

$\Rightarrow xy = \frac{36 \times 14}{4} = 126$

Trick: Bigger number = $\frac{25 + 11}{2} = 18$ and Smaller number = $\frac{25 - 11}{2} = 7$

$\therefore$ Multiplication of the numbers = $18 \times 7 = 126$

73. $(986 + 937)^2 - 4 \times 986 \times 937 = ?$

(*a*) 1923 (*b*) 49 (*c*) 2401 (*d*) 2104

Sol. (*c*) Given expression = $(x + y)^2 - 4xy = (x - y)^2 = (986 - 937)^2 = 49^2 = 2401$

74. $\frac{1 + 876542 \times 876544}{876543 \times 876543}$ is equal to **(SSC CHSL 2012)**

(*a*) 3 (*b*) 0 (*c*) 1 (*d*) 2

Sol. (*c*) Given expression = $\frac{1 + x(x + 2)}{(x + 1) \times (x + 1)} = \frac{1 + x^2 + 2x}{(x + 1)^2} = \frac{(x + 1)^2}{(x + 1)^2} = 1$

75. The sum of two numbers is 42 and their product is 437. Then the absolute difference between the numbers is : **(SSC CGL Tier I 2003)**

(*a*) 3 (*b*) 4 (*c*) 5 (*d*) 7

Sol. (*b*) $(x - y)^2 = (x + y)^2 - 4xy = 42^2 - 4 \times 437 = 1764 - 1748 = 16 \Rightarrow x - y = \sqrt{16} = 4$

76. If $(a - b)$ is 6 more than $(c + d)$ and $(a + b)$ is 3 less than $(c - d)$, then the value of $(a - c)$ is : **(SSC CGL Tier I 2002, CDS 2004)**

(*a*) 0.5 (*b*) 1.0 (*c*) 1.5 (*d*) 2.0

Sol. (*c*) $(a - b) = (c + d) + 6 \Rightarrow a - b - c - d = 6$ .... (1)

$(a + b) = (c - d) - 3 \Rightarrow a + b - c + d = -3$ .... (2)

Adding equations (1) and (2): $2(a - c) = 3 \Rightarrow (a - c) = \frac{3}{2} = 1.5$

77. What is the sum of two consecutive even numbers, the difference of whose squares is 84 ? **(SSC CGL Tier I 2002)**

(*a*) 38 (*b*) 34 (*c*) 42 (*d*) 46

Sol. (*c*) Let the required number be x and $(x + 2)$. Then $(x + 2)^2 - x^2 = 84$

$\Rightarrow x^2 + 4x + 4 - x^2 = 84 \Rightarrow 4x = 84 - 4 = 80 \Rightarrow x = 80 \div 4 = 20$

Required sum = 20 + (20 + 2) = 42

78. If the sum of two numbers is multiplied by the each of the number separately, the result is 574 and 1107 respectively. Find the smaller number.

(*a*) 14 (*b*) 27 (*c*) 41 (*d*) None of these

Sol. (*a*) Let the numbers be x and y. Then $x(x + y) + y(x + y) = 574 + 1107$

$\Rightarrow (x + y)(x + y) = 1681 \Rightarrow (x + y) = \sqrt{1681} = 41$

Now $x(x + y) = x \times 41 = 574 \Rightarrow x = 574 \div 41 = 14$

79. If $x + y + z = 0$, then $x^3 + y^3 + z^3 + 3xyz$ is **(CDS 2015)**

(*a*) 0 (*b*) $6xyz$ (*c*) $12xyz$ (*d*) xyz

Sol. (*b*) If $x + y + z = 0$, then $x^3 + y^3 + z^3 + 3xyz \Rightarrow$ G.E. $= 3xyz + 3xyz = 6xyz$

80. If $x + y + z = 13$, then the maximum value of $(x - 2)(y + 1)(z - 3)$ is **(SSC FCI 2012)**

(*a*) 25 (*b*) 30 (*c*) 54 (*d*) 27

Sol. (*d*) The value of G.E. is maximum when $(x - 2) = (y + 1) = (z - 3) = \dfrac{x-2+y+1+z-3}{3}$

$= \dfrac{x+y+z-4}{3} = \dfrac{13-4}{3} = \dfrac{9}{3} = 3$ $\therefore$ Required value $= 3 \times 3 \times 3 = 27$.

Trick: 27 is the only cube number.

81. What is the solution of the equations $x - y = 0.9$ and $11(x + y)^{-1} = 2$? **(CDS 2010)**

(*a*) $x = 3.2, y = 2.3$ (*b*) $x = 1, y = 0.1$ (*c*) $x = 2, y = 1.1$ (*d*) $x = 1.2, y = 0.3$

Sol. (*a*) $x - y = 0.9$ $11(x + y)^{-1} = 2 \Rightarrow (x + y) = \dfrac{11}{2} = 5.5$

The numbers are $\dfrac{5.5 + 0.9}{2}$ and $\dfrac{5.5 - 0.9}{2} = 3.2$ and 2.3

Trick : $11(x + y)^{-1} = 2 \Rightarrow (x + y) = \dfrac{11}{2} = 5.5$

Out of given options, only (*a*) satisfies this condition.

82. If $2a + b = 8$ and $5a - 2b = 11$, then $(5a - b)$ is equal to :

(*a*) 3 (*b*) 13 (*c*) 17 (*d*) 19

Sol. (*b*) $2a + b = 8$ .... (1)

$5a - 2b = 11$ .... (2)

Multiplying (1) by 2, we get : $4a + 2b = 16$ ...... (3)

Adding (2) and (3), we get : $9a = 27 \Rightarrow a = 3$

$2a + b = 8 \Rightarrow 2 \times 3 + b = 8 \Rightarrow b = 8 - 6 = 2$. $\therefore 5a - b = 5 \times 3 - 2 = 13$

83. The largest number that exactly divides each number of the form $(n^3 - n)$, where n is a natural number is : **(SSC CGL Tier I 2005, SSC CHSL 2010)**

(*a*) 12 (*b*) 2 (*c*) 3 (*d*) 6

Sol. (*d*) $(n^3 - n) = n(n^2 - 1) = n(n - 1)(n + 1)$

$= (n - 1)\, n\, (n + 1)$ i.e. product of three consecutive numbers.

For every $n > 1$, the given expression is always divisible by $1 \times 2 \times 3 = 6$

84. In a classroom there are certain number of benches. If 6 students are made to sit on a bench, then to accommodate all of them, one more bench is needed. However, if 7 students are made to sit on a bench, then after accommodating all of them, space for 5 students is left. What is the total number of students in the class ? **(SSC CGL Tier I 2005)**

(*a*) 30 (*b*) 42 (*c*) 72 (*d*) None of these

Sol. (*c*) Let the number of benches be x.

Then, students $= 6(x + 1) = 7x - 5 \Rightarrow 6x + 6 = 7x - 5 \Rightarrow 7x - 6x = 6 + 5 \Rightarrow x = 11$

$\therefore$ Number of students $= 6(x + 1) = 6 \times 12 = 72$

85. A librarian purchased 50 story-books for his library. But he saw that he could have get 14 more books by spending ₹76 more and the average price per book would be reduced by ₹1. The average price (in ₹) of each book he bought, was ? **(SSC CGL Tier I 2015)**

(*a*) 20 (*b*) 25 (*c*) 15 (*d*) 10

Sol. (*d*) Let the original price per book be x.

Then $50x + 76 = 64(x - 1) \Rightarrow 50x + 76 = 64x - 64 \Rightarrow 14x = 140 \Rightarrow x = 10$

86. The numerator and denominator of a fraction are in the ratio of 2 : 3. If 6 is subtracted from the numerator, the result is a fraction that has a value 2/3 of the original fraction, the numerator of the original fraction is:

(*a*) 16 (*b*) 18 (*c*) 27 (*d*) 36

Sol. (*b*) $\frac{2x-6}{3x} = \frac{2}{3} \times \frac{2}{3} \Rightarrow 18x - 54 = 12x \Rightarrow 6x = 54 \Rightarrow x = 9 \Rightarrow$ Numerator $= 2x = 18$.

87. For which of the following values of x the $2x^2 + 5x < 3$ is satisfied ? **(Bank PO 2004)**

(*a*) $x < -3$ and $x > \frac{1}{2}$ (*b*) $-3 < x < \frac{1}{2}$ (*c*) $-\frac{1}{2} < x < 3$ (*d*) $\frac{1}{2} < x < 3$

Sol. (*b*) $2x^2 + 5x < 3 \Rightarrow 2x^2 + 5x - 3 < 0 \Rightarrow 2x^2 + 6x - x - 3 < 0$

$\Rightarrow 2x(x + 3) - 1(x + 3) < 0 \Rightarrow (2x - 1)(x + 3) < 0$

This is possible only if one number is smaller than zero and other is greater than zero.

$\Rightarrow (2x - 1) < 0$ and $(x + 3) > 0$ or $(2x - 1) > 0$ and $(x + 3) < 0$

$\Rightarrow x < \frac{1}{2}$ and $x > -3$ (possible) or $x > \frac{1}{2}$ and $x < -3$ (not possible)

$\Rightarrow -3 < x < \frac{1}{2}$ satisfies the equation.

88. If $x^2 + 7x + 12 = 0$ and $y^2 + 9y + 20 = 0$, then what is the relation between x and y ? **(Vijaya Bank PO 2008)**

(*a*) $x > y$ (*b*) $x < y$ (*c*) $x \geq y$ (*d*) $x \leq y$

Sol. (*c*) $x^2 + 7x + 12 = 0 \Rightarrow x^2 + 3x + 4x + 12 = 0 \Rightarrow (x + 3)(x + 4) = 0 \Rightarrow x = -3$ or $x = -4$

$y^2 + 9y + 20 = 0 \Rightarrow y^2 + 4y + 5y + 20 = 0 \Rightarrow (y + 4)(y + 5) = 0 \Rightarrow y = -4$ or $y = -5$

Hence $x \geq y$

89. If $5x^2 - 4xy + y^2 - 2x + 1 = 0$, then the values of x and y are **(SSC CGL Tier II 2012)**

(*a*) $x = 1, y = 2$ (*b*) $x = 2, y = 1$ (*c*) $x = 1, y = -1$ (*d*) $x = -1, y = 1$

Sol. (*a*) $5x^2 - 4xy + y^2 - 2x + 1 = 0 \Rightarrow (x^2 - 2x + 1) + (4x^2 - 4xy + y^2) = 0$

$\Rightarrow (x - 1)^2 + (2x - y)^2 = 0 \Rightarrow x = 1$ and $y = 2x = 2$

90. If $2x^2 - 7xy + 3y^2 = 0$, then $x : y$ = ?

(*a*) 3 : 1 (*b*) 1 : 2 (*c*) Both (*d*) None of these

Sol. (*c*) $2x^2 - 7xy + 3y^2 = 0 \Rightarrow 2x^2 - 6xy - xy + 3y^2 = 0$

$\Rightarrow 2x(x-3y) - y(x-3y) = 0 \Rightarrow (2x-y)(x-3y) = 0$

$\Rightarrow 2x - y = 0, x - 3y = 0 \Rightarrow 2x = y, x = 3y \Rightarrow x : y = 1 : 2$ or $3 : 1$

91. If $x = 2$ and $y = 3$, find the value of $8x^3 + 27y^3 + 36x^2y + 54xy^2$.

(*a*) 125 (*b*) 1728 (*c*) 2791 (*d*) 2197

Sol. (*d*) $8x^3 + 27y^3 + 36x^2y + 54xy^2 = (2x)^3 + (3y)^3 + 3(2x)^2(3y) + 3(2x)(3y)^2$

$= (2x + 3y)^3 = (4 + 9)^3 = 13^3 = 2197.$

92. The value of $(0.98)^3 + (0.02)^3 + 3 \times 0.98 \times 0.02 - 1$ is : **(SSC CGL Tier I 2005)**

(*a*) 1.98 (*b*) 1.09 (*c*) 1 (*d*) 0

Sol. (*d*) Given expression $= x^3 + y^3 + 3xy - 1 = x^3 + y^3 + 3xy(x + y) - 1$

$= (x + y)^3 - 1 = 1^3 - 1 = 0$

93. If $3y + 9x = 54$ and $\frac{28x}{13y} = \frac{140}{39}$ then what is the value of $y - x$? **(Bank PO 2009)**

(*a*) – 1 (*b*) – 2 (*c*) 2 (*d*) 1

Sol. (*b*) $3y + 9x = 54 \Rightarrow y + 3x = 18 \Rightarrow 3x + y = 18$ .... (i)

$\frac{28x}{13y} = \frac{140}{39} \Rightarrow \frac{x}{y} = \frac{140}{39} \times \frac{13}{28} = \frac{5}{3} \Rightarrow 3x = 5y \Rightarrow 3x - 5y = 0$ .... (ii)

(*i*) – (*ii*) : $6y = 18 \Rightarrow y = 3$

$3x = 5y = 5 \times 3 = 15 \Rightarrow x = 15 \div 3 = 5 \Rightarrow y - x = 3 - 5 = -2$

94. Two candles of the same height are lighted at the same time. The first is consumed in 4 hours and the second in 3 hours. Assuming that each candle burns at its own constant rate, find the number of hours, after being lighted, after which the first candle will be twice the height of the second candle. **(SSC CGL Tier II 2000)**

(*a*) $2\frac{2}{5}$ hours (*b*) 2 hours (*c*) $2\frac{1}{4}$ hours (*d*) $1\frac{3}{4}$ hours

Sol. (*a*) Let required time = *t* hours. Then $\left(1-\frac{t}{4}\right) = 2\left(1-\frac{t}{3}\right) \Rightarrow \frac{4-t}{4} = \frac{6-2t}{3}$

$\Rightarrow 12 - 3t = 24 - 8t \Rightarrow 5t = 12 \Rightarrow t = 2\frac{2}{5}$ hours

95. If $(x^{13} + 1)$ is divided by $(x - 1)$, the remainder is :

(*a*) 1 (*b*) – 1 (*c*) 0 (*d*) 2

Sol. (*d*) On putting the value of divisor equal to 0, we get $x - 1 = 0$ or $x = 1$

$\therefore$ Remainder $= f(1) = 1^{13} + 1 = 2.$

96. If $x^{51} + 51$ is divided by $(x + 1)$, the remainder is **(SSC CGL Tier I 2013)**

(*a*) 1 (*b*) 49 (*c*) 0 (*d*) 50

Sol. (*d*) Remainder $= (-1)^{51} + 51 = 50$

97. What is the remainder when $(x^{11} + 1)$ is divided by $(x + 1)$? **(CDS 2011)**

(*a*) 0 (*b*) 2 (*c*) 11 (*d*) 12

Sol. (*a*) Remainder $= (-1)^{11} + 1 = -1 + 1 = 0$

98. If $x - 2$ is a factor of $x^3 + kx^2 - 2x - 24$ then what is the value of k ? **(CDS 2012)**

(*a*) 6 (*b*) 5 (*c*) 1 (*d*) – 2

Sol. (*b*) Put $x - 2 = 0 \Rightarrow x = 2$ in the given equation then $f(x) = 0$

Then $2^3 + k2^2 - 2 \times 2 - 24 = 0 \Rightarrow 8 + 4k - 4 - 24 = 0 \Rightarrow 4k = 20 \Rightarrow k = 5$

99. The HCF of $x^6 - 1$ and $x^4 + 2x^3 - 2x - 1$ is: **(SSC CGL Tier I 2015)**

(a) $x^2 - 1$ (b) $x^2 + 1$ (c) $x + 1$ (d) $x - 1$

Sol. (d) If we put $x - 1 = 0 \Rightarrow x = 1$ in the given equations then $f(x) = 0$ for both equations.

100. If $x^3 - 3x^2 + 5x + 1$ is divided by $x - 2$, the remainder is :

(a) 0 (b) 2 (c) 5 (d) 7

Sol. (d) On putting the value of divisor equal to 0, we get $x - 2 = 0$ or $x = 2$

$\therefore$ Remainder $= f(2) = 2^3 - 3(2)^2 + 5(2) + 1 = 8 - 12 + 10 + 1 = 7$

101. If $(x - 3)$ is a factor of $(x^2 + 4px - 11p)$, then what is the value of p ? **(CDS 2011)**

(a) –9 (b) –3 (c) –1 (d) 1

Sol. (a) $3^2 + 4(p)3 - 11p = 0 \Rightarrow 9 + 12p - 11p = 0 \Rightarrow p = -9$

102. The expression $x^4 - 2x^2 + k$ will be a perfect square when the value of k is **(SSC CGL Tier I 2013)**

(a) 2 (b) 1 (c) – 1 (d) – 2

Sol. (b) $x^4 - 2x^2 + k = y^2 - 2y + k$ (on replacing $x^2 = y$) $= (y - 1)^2$ when $k = 1$

103. The value of k for which the expression $9x^2 - kx + 16$ be a perfect square is **(SSC CHSL 2012)**

(a) ± 12 (b) ± 24 (c) ± 2 (d) ± 6

Sol. (b) $(3x \pm 4)^2 = 9x^2 \pm 24x + 16 \Rightarrow k = \pm 24$ $[(3x)^2 = 9x^2; 4^2 = 16]$

104. What is the value of a for which the equations $x^3 + ax + 1 = 0$ and $x^4 + ax^2 + 1 = 0$ have a common root ? **(UPSC Civil Services (Pre) 2009)**

(a) 2 (b) –2 (c) 1 (d) –1

Sol. (c) Possible values of x = factors of 1 = 1 and – 1

If $x = 1$, then $1 + a + 1 = 0$ and $1 + a + 1 = 0 \Rightarrow a = -2$ and $a = -2$ (Possible)

If $x = -1$, then $-1 - a + 1 = 0$ and $1 + a + 1 = 0 \Rightarrow a = 0$ and $a = -2$ (not possible)

105. If $(x + 2)$ is HCF of $x^2 + ax + b$ and $x^2 + cx + d$ ($a \neq c$ and $b \neq d$), then which one of the following is correct ? **(CDS 2009)**

(a) $a + c = b + d$ (b) $2a + b = 2c + d$ (c) $b + 2c = 2a + d$ (d) $b - 2c = 2a - d$

Sol. (c) $(x + 2)$ is the factor of $x^2 + ax + b$ and $x^2 + cx + d$

$(x + 2) = 0 \Rightarrow x = -2$ is the solution of both functions.

$x^2 + ax + b = 0 \Rightarrow (-2)^2 - 2a + b = 0 \Rightarrow 4 - 2a + b = 0$

$x^2 + cx + d = 0 \Rightarrow (-2)^2 - 2c + d = 0 \Rightarrow 4 - 2c + d = 0$

$\Rightarrow 4 - 2a + b = 4 - 2c + d \Rightarrow -2a + b = -2c + d \Rightarrow b + 2c = 2a + d$

106. If α and β are the roots of the equation $ax^2 + bx + c = 0$, then the roots of the equation $cx^2 + bx + a = 0$ are:

(a) $\alpha, \frac{1}{\beta}$ (b) $\frac{1}{\alpha}, \beta$ (c) $\frac{1}{\alpha}, \frac{1}{\beta}$ (d) $-\alpha, -\beta$

Sol. (c) $ax^2 + bx + c = 0 \Rightarrow$ sum of roots $= \frac{-b}{a}$, product of roots $= \frac{c}{a}$

In $cx^2 + bx + a = 0$, product of roots $= \frac{a}{c}$ = reciprocal of $\frac{c}{a}$

Sum of roots $= \frac{-b}{c} = \frac{-b}{a} \times \frac{a}{c} = \frac{\alpha + \beta}{\alpha b} \Rightarrow$ Required roots are $\frac{1}{\alpha}$ and $\frac{1}{\beta}$

107. If α, β are the roots of quadratic equation $4x^2 - 4x + 1 = 0$, then $\alpha^3 + \beta^3 = ?$ **(SSC CHSL 2012)**

(*a*) $\frac{1}{4}$ (*b*) $\frac{1}{8}$ (*c*) 8 (*d*) 16

Sol. (*a*) $\alpha + \beta = \frac{-b}{a} = \frac{4}{4} = 1; \quad \alpha\beta = \frac{c}{a} = \frac{1}{4}$

$$\alpha^3 + \beta^3 = (\alpha + \beta)^3 - 3\alpha\beta(\alpha + \beta) = 1^3 - 3 \times \frac{1}{4} \times 1 = 1 - \frac{3}{4} = \frac{1}{4}$$

Alt: $\alpha + \beta = 1$ and $\alpha\beta = \frac{1}{4} \Rightarrow \alpha = \beta = \frac{1}{2}$ as $\frac{1}{2} + \frac{1}{2} = 1$ and $\frac{1}{2} \times \frac{1}{2} = \frac{1}{4}$

$$\therefore \quad \alpha^3 + \beta^3 = \left(\frac{1}{2}\right)^3 + \left(\frac{1}{2}\right)^3 = \frac{1}{8} + \frac{1}{8} = \frac{1}{4}$$

Alt: $4x^2 - 4x + 1 = 0, \Rightarrow (2x - 1)^2 = 0 \Rightarrow x = 1/2$

108. Find the value of $\frac{23 \times 23 + 17 \times 17 + 2 \times 23 \times 17}{23 \times 23 - 17 \times 17}$

(*a*) $\frac{40}{3}$ (*b*) $\frac{17}{23}$ (*c*) $\frac{23}{17}$ (*d*) $\frac{20}{3}$

Sol. (*d*) Substituting $x = 23$ and $y = 17$, the given expression becomes

$$\frac{x^2 + y^2 + 2xy}{x^2 - y^2} = \frac{(x + y)^2}{x^2 - y^2} = \frac{(x + y)(x + y)}{(x + y)(x - y)} = \frac{(x + y)}{(x - y)} = \frac{40}{6} = \frac{20}{3}$$

109. $\frac{(63 + 36)^2 + (63 - 36)^2}{63^2 + 36^2} = ?$ **(ITI 1983)**

(*a*) 1 (*b*) 2 (*c*) 405 (*d*) 831

Sol. (*b*) Given expression $= \frac{(x + y)^2 + (x - y)^2}{x^2 + y^2} = \frac{2(x^2 + y^2)}{(x^2 + y^2)} = 2$

110. $\left(\frac{4}{3} + \frac{3}{4}\right)^2 - \left(\frac{4}{3} - \frac{3}{4}\right)^2 = ?$

(*a*) 1 (*b*) 2 (*c*) 3 (*d*) 4

Sol. (*d*) $(x + y)^2 - (x - y)^2 = 4xy \Rightarrow$ The given expression $= 4 \times \frac{4}{3} \times \frac{3}{4} = 4$

111. $\frac{998^2 - 997^2 - 45}{98^2 - 97^2} = ?$ **(SSC CGL Tier I 2008)**

(*a*) 1995 (*b*) 195 (*c*) 95 (*d*) 10

Sol. (*d*) Given expression $= \frac{(998 + 997) \times (998 - 997) - 45}{(98 + 97) \times (98 - 97)} = \frac{1995 - 45}{195} = \frac{1950}{195} = 10$

112. $\frac{1.49 \times 14.9 - 0.51 \times 5.1}{14.9 - 5.1} = ?$ **(SSC CGL Tier I 2004)**

(*a*) 0.20 (*b*) 20.00 (*c*) 2.00 (*d*) 22.00

Sol. (*c*) Given expression = $\dfrac{1.49\times1.49\times10-0.51\times0.51\times10}{1.49\times10-0.51\times10}$

$=\dfrac{10\times(1.49\times1.49-0.51\times0.51)}{10\times(1.49-0.51)}=1.49+0.51=2.00$

113. $\dfrac{(5.9)^2-(4.9)^2}{(6.8)^2+(6.8\times6.4)+(3.2)^2}=?$

(*a*) 0.108 (*b*) 1.08 (*c*) 10.8 (*d*) 108

Sol. (*a*) $\dfrac{(5.9)^2-(4.9)^2}{(6.8)^2+(6.8\times64)+(3.2)^2}=\dfrac{(5.9+4.9)(5.9-4.9)}{(6.8+3.2)^2}=\dfrac{10.8\times1}{100}=0.108$

114. Find the square root of $\dfrac{(0.064-0.008)(0.16-0.04)}{(0.16+0.08+0.04)(0.4+0.2)^3}$ **(SSC CGL Tier I 2015)**

(*a*) 1/3 (*b*) 3 (*c*) 2/3 (*d*) 3/2

Sol. (*a*) Substituting $x = 0.4$ and $y = 0.2$, the given expression becomes

$\dfrac{(x^3-y^3)(x^2-y^2)}{(x^2+xy+y^2)(x+y)^3}=\dfrac{(x-y)(x^2+xy+y^2)(x-y)(x+y)}{(x^2+xy+y^2)(x+y)^3}=\dfrac{(x-y)^2}{(x+y)^2}$

Square root $=\dfrac{(x-y)}{(x+y)}=\dfrac{0.4-0.2}{0.4+0.2}=\dfrac{0.2}{0.6}=\dfrac{1}{3}$

115. If $\dfrac{a^3-b^3}{a^3+b^3}=\dfrac{13}{14}$, then find the value of $\dfrac{a+b}{a-b}$

(*a*) 1 (*b*) 2 (*c*) 1/2 (*d*) 3/2

Sol. (*b*) $\dfrac{a^3-b^3}{a^3+b^3}=\dfrac{13}{14}\Rightarrow 14a^3-14b^3=13a^3+13b^3\Rightarrow a^3=27b^3=(3b)^3$

$\Rightarrow a=(3b)\Rightarrow a:b=3:1\Rightarrow\dfrac{a+b}{a-b}=\dfrac{3+1}{3-1}=\dfrac{4}{2}=2$

116. If $a+\dfrac{1}{a}+1=0\ (a\neq0)$ then the value of (a^4-a) is : **(SSC CHSL 2010)**

(*a*) 0 (*b*) 1 (*c*) 2 (*d*) – 1

Sol. (*a*) $a+\dfrac{1}{a}+1=0\Rightarrow a^2+1+a=0$

$\Rightarrow (a^4-a)=a(a^3-1)=a(a-1)(a^2+1+a)=a(a-1)\times0=0$

117. If $\dfrac{1}{x}+\dfrac{1}{y}=0$, then $x^3+y^3=?$

(*a*) –1 (*b*) 0 (*c*) 1 (*d*) 3

Sol. (*b*) $\dfrac{1}{x}+\dfrac{1}{y}=0\Rightarrow\dfrac{x+y}{xy}=0\Rightarrow x+y=0$

$\Rightarrow x^3+y^3=(x+y)(x^2-xy+y^2)=0\times(x^2-xy+y^2)=0$

118. If $\dfrac{a}{b}+\dfrac{b}{a}=1$, then the value of is a^3+b^3 is

(SSC CGL Tier I 2007, 2012, SSC CPO SI 2011, SSC FCI 2012)

(*a*) 0 (*b*) 1 (*c*) 2 (*d*) 3

Sol. (*a*) $\frac{a}{b}+\frac{b}{a}=1 \Rightarrow a^2+b^2=ab \Rightarrow a^2+b^2-ab=0$

$\Rightarrow a^3+b^3=(a+b)(a^2+b^2-ab)=(a+b)\times 0=0$

119. If $a=\frac{b^2}{b-a}$, then the value of a^3+b^3 is **(SSC CHSL 2013)**

(*a*) 2 (*b*) 6 *ab* (*c*) 0 (*d*) 1

Sol. (*c*) $a=\frac{b^2}{b-a} \Rightarrow ab-a^2=b^2 \Rightarrow a^2+b^2-ab=0$

$\Rightarrow a^3+b^3=(a+b)(a^2+b^2-ab)=(a+b)\times 0=0$

120. If $x-\frac{1}{x}=4$, then $x+\frac{1}{x}=?$ **(SSC CGL Tier I 2008)**

(*a*) $5\sqrt{2}$ (*b*) $2\sqrt{5}$ (*c*) $4\sqrt{2}$ (*d*) $4\sqrt{5}$

Sol. (*b*) $\left(x+\frac{1}{x}\right)^2=\left(x-\frac{1}{x}\right)^2+4=4^2+4=20 \quad \therefore x+\frac{1}{x}=\sqrt{20}=2\sqrt{5}$

121. The value of $\left(\sqrt[3]{3.5}+\sqrt[3]{2.5}\right)\left\{\left(\sqrt[3]{3.5}\right)^2-\sqrt[3]{8.75}+\left(\sqrt[3]{2.5}\right)^2\right\}$ is : **(SSC CGL Tier I 2004)**

(*a*) 5.375 (*b*) 1 (*c*) 6 (*d*) 5

Sol. (*c*) Substituting $x=\sqrt[3]{3.5}$ and $y=\sqrt[3]{2.5}$, the given expression becomes

$(x+y)(x^2-xy+y^2)=x^3+y^3=\left(\sqrt[3]{3.5}\right)^3+\left(\sqrt[3]{2.5}\right)^3=3.5+2.5=6$

122. $\frac{(0.73)^3+(0.27)^3}{(0.73)^2+(0.27)^2-(0.73\times 0.27)}$ simplifies to : **(SSC CGL Tier I 2005, 2002 modified)**

(*a*) 1 (*b*) 0.4087 (*c*) 0.73 (*d*) 0.27

Sol. (*a*) Substituting $x = 0.73$ and $y = 0.27$, the given expression becomes

$$\frac{x^3+y^3}{x^2+y^2-xy}=\frac{(x+y)(x^2+y^2-xy)}{x^2+y^2-xy}=x+y=0.73+0.27=1$$

123. Find the value of $\frac{1.\bar{3}\times 1.\bar{3}\times 1.\bar{3}-1}{1.\bar{3}+1.\bar{3}\times 1.\bar{3}+1}$ **(SSC CGL Tier I 2006)**

(*a*) $\frac{1}{3}$ (*b*) $1\frac{1}{3}$ (*c*) $\frac{37}{91}$ (*d*) $\frac{27}{91}$

Sol. (*a*) Let $x=1.\bar{3} \Rightarrow$ Given expression $=\frac{x^3-1}{x+x^2+1}=x-1=1.\bar{3}-1=0.\bar{3}=\frac{3}{9}=\frac{1}{3}$

124. The value of $\frac{0.125+0.027}{0.25-0.15+0.09}$ **(SSC CGL Tier I 2008)**

(*a*) 0.2 (*b*) 0.25 (*c*) 0.3 (*d*) 0.8

Sol. (*d*) Given expression $=\frac{x^3+y^3}{x^2-xy+y^2}=x+y=0.5+0.3=0.8$

125. The value of $\dfrac{0.9\times0.9\times0.9+0.2\times0.2\times0.2+0.3\times0.3\times0.3-3\times0.9\times0.2\times0.3}{0.9\times0.9+0.2\times0.2+0.3\times0.3-0.9\times0.2-0.2\times0.3-0.3\times0.9}$ **(SSC CGL Tier I 2003)**

(*a*) 1.4 (*b*) 0.054 (*c*) 0.8 (*d*) 1.0

Sol. (*a*) G.E. $=\dfrac{x^3+y^3+z^3-3xyz}{x^2+y^2+z^2-xy-yz-zx}=x+y+z=0.9+0.2+0.3=1.4$

126. $\dfrac{8\times(3.75)^3+1}{(7.5)^2-6.5}=?$ **(SSC CGL Tier I 2008, 2003)**

(*a*) 2.75 (*b*) 1.8 (*c*) 4.75 (*d*) 8.5

Sol. (*d*) Given expression $=\dfrac{(2\times3.75)^3+1^3}{(7.5)^2-(7.5\times1)+(1)^2}=\dfrac{x^3+y^3}{x^2-xy+y^2}=x+y=7.5+1=8.5$

127. If $P=99$, then $P(P^2+3P+3)=?$ **(SSC CGL Tier I 2008, 2014 ; SSC CHSL 2012)**

(*a*) 999 (*b*) 9999 (*c*) 99999 (*d*) 999999

Sol. (*d*) G.E. $=(P^3+3P^2+3P+1)-1=(P+1)^3-1=(99+1)^3-1=1000000-1=999999$

128. If $P=999$, then $\sqrt[3]{P(P^2+3P+3)+1}=?$ **(SSC CGL Tier I 2008)**

(*a*) 998 (*b*) 999 (*c*) 1000 (*d*) 1002

Sol. (*c*) Given expression $=\sqrt[3]{P^3+3P^2+3P+1}=\sqrt[3]{(P+1)^3}=P+1=999+1=1000$

129. If $a+b+c=2$ and $ab+bc+ca=-1$, then value of $(a+b)^2+(b+c)^2+(c+a)^2$ is **(SSC FCI 2012)**

(*a*) 5 (*b*) 10 (*c*) 15 (*d*) 25

Sol. (*b*) $(a+b)^2+(b+c)^2+(c+a)^2=2[a^2+b^2+c^2+ab+bc+ca]$

$=2[(a+b+c)^2-(ab+bc+ca)]=2[(2)^2-(-1)]=2\times5=10$

130. If $x+y=7$, then the value of x^3+y^3+21xy is **(SSC CGL Tier I 2007)**

(*a*) 243 (*b*) 143 (*c*) 343 (*d*) 443

Sol. (*c*) $x^3+y^3+21xy=x^3+y^3+3xy\times(7)=x^3+y^3+3xy\times(x+y)=(x+y)^3=7^3=343.$

131. If $x+y=1$, then the value of x^3+y^3+3xy is **(SSC CHSL 2012)**

(*a*) – 2 (*b*) 1 (*c*) 2 (*d*) 3

Sol. (*b*) $(x+y)^3=1^3\Rightarrow x^3+y^3+3xy(x+y)=1\Rightarrow x^3+y^3+3xy=1$

132. If $a+b+1=0$, then the value of $a^3+b^3+1-3ab+1$ is **(SSC CGL Tier I 2012)**

(*a*) –1 (*b*) 1 (*c*) 3 (*d*) 0

Sol. (*d*) $a+b=-1\Rightarrow(a+b)^3=(-1)^3\Rightarrow a^3+b^3+3ab(a+b)=-1$

$\Rightarrow a^3+b^3+3ab(-1)=-1\Rightarrow a^3+b^3-3ab+1=0$

133. If $x-y=1$ then $x^3-y^3-3(xy-2)$ is equal to **(SSC CHSL 2013)**

(*a*) 7 (*b*) 6 (*c*) 10 (*d*) 4

Sol. (*a*) $(x-y)=1\Rightarrow(x-y)^3=1^3\Rightarrow x^3-y^3-3xy(x-y)=1\Rightarrow x^3-y^3-3xy=1$

$\therefore\ x^3-y^3-3(xy-2)=x^3-y^3-3xy+6=1+6=7$

134. If $5x+9y=5$, then the value of $125x^3+729y^3=120$, then the value of product of x and y is **(SSC CGL Tier I 2015)**

(*a*) 135 (*b*) 1/135 (*c*) 1/9 (*d*) 45

Sol. (*b*) $(5x + 9y)^3 = (5)^3 \Rightarrow 125x^3 + 729y^3 + 3(5x)(9y)(5x + 9y) = 125$
$\Rightarrow 125x^3 + 729y^3 + 135xy(5) = 125 \Rightarrow 675xy = 125 - 120 = 5 \Rightarrow xy = 1/135$

135. If $a^4 + a^2b^2 + b^4 = 8$ and $a^2 + ab + b^2 = 4$, then $ab = ?$

(*a*) 0 (*b*) 1 (*c*) –1 (*d*) 2

Sol. (*b*) $\dfrac{a^4 + a^2b^2 + b^4}{a^2 + ab + b^2} = \dfrac{8}{4} \Rightarrow a^2 - ab + b^2 = 2$

$(a^2 + ab + b^2) - (a^2 - ab + b^2) = 4 - 2 \Rightarrow 2ab = 2 \Rightarrow ab = 1$

Note: $a^4 + a^2b^2 + b^4 = a^4 + 2a^2b^2 + b^4 - a^2b^2 = (a^2 + b^2)^2 - (ab)^2 = (a^2 + b^2 + ab)(a^2 + b^2 - ab)$

136. If $\dfrac{x}{a} = \dfrac{1}{a} - \dfrac{1}{x}$, then the value of $x - x^2$ is : **(SSC CGL Tier I 2013)**

(*a*) a (*b*) $-a$ (*c*) $\dfrac{1}{a}$ (*d*) $-\dfrac{1}{a}$

Sol. (*a*) $\dfrac{x}{a} = \dfrac{1}{a} - \dfrac{1}{x} \Rightarrow x^2 = x - a$ (on multiplying each term by ax) $\Rightarrow x - x^2 = a$

137. If $a + b + c = 0$, then the value of $\dfrac{a^2 + b^2 + c^2}{a^2 - bc}$ is : **(SSC CGL Tier II 2012)**

(*a*) 1 (*b*) 2 (*c*) 3 (*d*) 4

Sol. (*b*) $a + b + c = 0 \Rightarrow b + c = -a \Rightarrow (b + c)^2 = (-a)^2 \Rightarrow b^2 + c^2 + 2bc = a^2 \Rightarrow b^2 + c^2 = a^2 - 2bc$

$\Rightarrow \dfrac{a^2 + b^2 + c^2}{a^2 - bc} = \dfrac{a^2 + a^2 - 2bc}{a^2 - bc} = \dfrac{2(a^2 - bc)}{a^2 - bc} = 2$

138. If $a^3 - b^3 = 56$ and $a - b = 2$, then value of $a^2 + b^2$ will be **(SSC CHSL 2012)**

(*a*) 5 (*b*) 48 (*c*) 20 (*d*) 22

Sol. (*c*) $\dfrac{a^3 - b^3}{a - b} = \dfrac{56}{2} \Rightarrow a^2 + b^2 + ab = 28$ ... (1)

$(a - b)^2 = 2^2 \Rightarrow a^2 + b^2 - 2ab = 4$ ... (2)

(1) – (2) gives, $3ab = 24 \Rightarrow ab = 8$

From (1), $a^2 + b^2 = 28 - ab = 28 - 8 = 20$

139. If $a + b + c = 2s$, then $\dfrac{(s-a)^2 + (s-b)^2 + (s-c)^2 + s^2}{a^2 + b^2 + c^2}$ is equal to **(SSC CGL Tier I 2013)**

(*a*) $a^2 + b^2 + c^2$ (*b*) 0 (*c*) 1 (*d*) 2

Sol. (*c*) G.E. $= \dfrac{s^2 + a^2 - 2as + s^2 + b^2 - 2bs + s^2 + c^2 - 2cs + s^2}{a^2 + b^2 + c^2}$

$= \dfrac{4s^2 - 2s(a+b+c) + a^2 + b^2 + c^2}{a^2 + b^2 + c^2} = \dfrac{4s^2 - 2s(2s) + a^2 + b^2 + c^2}{a^2 + b^2 + c^2} = \dfrac{a^2 + b^2 + c^2}{a^2 + b^2 + c^2} = 1$

140. If $x - y = \dfrac{x + y}{7} = \dfrac{xy}{4}$, then the numerical value of xy is **(SSC)**

(*a*) 1/3 (*b*) 4/3 (*c*) 1/4 (*d*) 3/4

Sol. (*b*) $x - y = \dfrac{x + y}{7} = \dfrac{xy}{4} = k$ (say) $\Rightarrow x - y = k$ and $x + y = 7k \Rightarrow x = 4k$ and $y = 3k$

$\Rightarrow xy = (4k) \Rightarrow (4k)(3k) = (4k) \Rightarrow (3k) = 1 \Rightarrow k = 1/3 \Rightarrow xy = (4k) = 4 \times \dfrac{1}{3} = \dfrac{4}{3}$

141. If $a = xb$ and $b = ay$, then the value of $\frac{1}{x+1}+\frac{1}{y+1}$: **(SSC CGL Tier I 2013)**

(*b*) $2ab$ (*b*) 1 (*c*) a (*d*) b

Sol. (*b*) $a = xb \Rightarrow x = \frac{a}{b}$, $b = ay \Rightarrow y = \frac{b}{a} \Rightarrow \frac{1}{x+1}+\frac{1}{y+1} = \frac{1}{\frac{a}{b}+1}+\frac{1}{\frac{b}{a}+1} = \frac{b}{a+b}+\frac{a}{a+b} = 1$

Alt. $a = xb \Rightarrow x = \frac{a}{b}, b = ay \Rightarrow y = \frac{b}{a} \Rightarrow xy = \frac{a}{b}\times\frac{b}{a} = 1$

$\therefore \frac{1}{x+1}+\frac{1}{y+1} = \frac{y+1+x+1}{xy+x+y+1} = \frac{x+y+2}{1+x+y+1} = \frac{x+y+2}{x+y+2} = 1$

142. If $\frac{1}{x+y} = \frac{1}{x}+\frac{1}{y}$, find the value of $x^3 - y^3$. **(SSC CHSL 2011)**

(*a*) 0 (*b*) 1 (*c*) 2 (*d*) – 1

Sol. (*a*) $\frac{1}{x+y} = \frac{1}{x}+\frac{1}{y} \Rightarrow \frac{1}{x+y} = \frac{x+y}{xy} \Rightarrow xy = (x+y)^2 \Rightarrow x^2+y^2+xy = 0$

$\Rightarrow x^3 - y^3 = (x-y)\times(x^2+y^2+xy) = (x-y)\times 0 = 0$

143. If $p - 2q = 4$, then the value of $p^3 - 8q^3 - 24pq - 64$ is **(SSC CGL Tier I 2013)**

(*a*) 0 (*b*) 3 (*c*) –1 (*d*) 2

Sol. (*a*) $(p-2q)^3 = p^3 - 8q^3 - 3p(2q)(p-2q) = p^3 - 8q^3 - 3p(2q)(4) = p^3 - 8q^3 - 24pq = 4^3$
$\Rightarrow p^3 - 8q^3 - 24pq - 64 = 4^3 - 64 = 0$

144. If $x = 2+\sqrt{3}$, find the value of $x^2 - 4x + 2$. **(SSC FCI 2014)**

(a) 1 (b) 2 (c) 3 (d) 4

Sol. (*a*) $x^2 - 4x + 2 = (x-2)^2 - 2 = (\sqrt{3})^2 - 2 = 3 - 2 = 1$

145. If $x + y + z = 0$, then what is $\frac{xyz}{(x+y)(y+z)(z+x)}$ equal to **(CDS 2011)**

(*a*) – 1 (*b*) 1 (*c*) $xy + yz + zx$ (*d*) None of these

Sol. (*a*) $x+y+z = 0 \Rightarrow x+y = -z, x+z = -y, x+z = -y$

$\therefore \frac{xyz}{(x+y)(y+z)(z+x)} = \frac{xyz}{(-z)(-x)(-y)} = -1$

146. If $a + b + c = 0$, then the value of $\left(\frac{a+b}{c}+\frac{b+c}{a}+\frac{c+a}{b}\right)\left(\frac{a}{b+c}+\frac{b}{c+a}+\frac{c}{a+b}\right)$ is : **(SSC CGL Tier I 2013)**

(*a*) 0 (*b*) 8 (*c*) –3 (*d*) 9

Sol. (*d*) $a+b+c = 0 \Rightarrow a+b = -c, b+c = -a, a+c = -b,$

$\therefore \text{GE} = \left(\frac{-c}{c}+\frac{-a}{a}+\frac{-b}{b}\right)\left(\frac{a}{-a}+\frac{b}{-b}+\frac{c}{-c}\right) = (-3)(-3) = 9$

147. If $a + b + c = 0$, then value of $\frac{1}{(a+b)(b+c)}+\frac{1}{(a+c)(b+a)}+\frac{1}{(c+a)(c+b)}$ is : **(SSC CHSL 2011)**

(*a*) 1 (*b*) 0 (*c*) – 1 (*d*) – 2

Sol. (*b*) G.E. $= \dfrac{(c+a)+(b+c)+(a+b)}{(a+b)(b+c)(c+a)} = \dfrac{2(a+b+c)}{(a+b)(b+c)(c+a)} = 0$

148. If $x+y+z=0$, then $\dfrac{x^2}{yz}+\dfrac{y^2}{xz}+\dfrac{z^2}{xy}=?$ **(SSC CGL Tier I 2012, 2013, CDS 2010)**

(*a*) 0 (*b*) 1 (*c*) 2 (*d*) 3

Sol. (*d*) $\dfrac{x^2}{yz}+\dfrac{y^2}{xz}+\dfrac{z^2}{xy}=\dfrac{x^3+y^3+z^3}{xyz}=\dfrac{3xyz}{xyz}=3$

149. If $x+y=z$, then $x^3+y^3-z^3+3xyz=?$ **(SSC CHSL 2011)**

(*a*) 0 (*b*) $3xyz$ (*c*) $-3xyz$ (*d*) z^3

Sol. (*a*) $x+y=z \Rightarrow x+y+(-z)=0$

$\Rightarrow x^3+y^3+(-z)^3=3xy(-z) \Rightarrow x^3+y^3-z^3+3xyz=0$

150. If $a+\dfrac{1}{a}=1$, then the value of a^3 is : **(SSC CGL Tier I 2012)**

(*a*) -2 (*b*) 2 (*c*) -1 (*d*) 4

Sol. (*c*) $a+\dfrac{1}{a}=1 \Rightarrow a^2-a+1=0$

$\Rightarrow a^3+1=(a+1)(a^2-a+1)=(a+1)\times 0 \Rightarrow a^3+1=0 \Rightarrow a^3=-1$

151. If $a^2+a+1=0$, then the value of a^9 is : **(SSC CGL Tier I 2013)**

(*a*) 2 (*b*) 3 (*c*) 1 (*d*) 0

Sol. (*c*) $a^3-1=(a-1)(a^2+a+1)=0 \Rightarrow a^3=1 \Rightarrow a^9=(a^3)^3=1^3=1$

152. If $x+\dfrac{a}{x}=1$, then the value of $\dfrac{x^2+x+a}{x^3-x^2}$ is **(SSC CGL Tier II 2012)**

(*a*) $\dfrac{2}{a}$ (*b*) $-\dfrac{2}{a}$ (*c*) -2 (*d*) $-\dfrac{a}{2}$

Sol. (*b*) $x+\dfrac{a}{x}=1 \Rightarrow x^2+a=x \Rightarrow \dfrac{x^2+x+a}{x^3-x^2}=\dfrac{x+x}{x(x^2-x)}=\dfrac{2}{x^2-x}=\dfrac{2}{-a}=\dfrac{-2}{a}$

153. If x, y are real and $x^2+y^2+2=2(y-x)$, then the value of x is : **(SSC CGL Tier I 2013)**

(*a*) 0 (*b*) -2 (*c*) 1 (*d*) -1

Sol. (*d*) $x^2+y^2+2=2(y-x) \Rightarrow x^2+y^2+2-2y+2x=0 \Rightarrow (x^2+2x+1)+(y^2-2y+1)=0$

$\Rightarrow (x+1)^2+(y-1)^2=0 \Rightarrow (x+1)^2=0 \Rightarrow (x+1)=0 \Rightarrow x=-1$

154. If $x^2+y^2+z^2=2(x+z-1)$, then the value of $x^3+y^3+z^3$ is : **(SSC CGL Tier I 2015)**

(*a*) 1 (*b*) 0 (*c*) -1 (*d*) 2

Sol. (*d*) $x^2+y^2+z^2=2(x+z-1) \Rightarrow x^2+y^2+z^2=2x+2z-2$

$\Rightarrow x^2+y^2+z^2-2x-2z+2=0 \Rightarrow (x^2-2x+1)+(y^2)+(z^2-2z+1)=0$

$\Rightarrow (x-1)^2+(y)^2+(z-1)^2=0 \Rightarrow (x-1)=0, (y)=0, (z-1)=0$

$\Rightarrow x=1, y=0, z=1 \Rightarrow x^3+y^3+z^3=1+0+1=2$

155. If $a^2+b^2+c^2+3=2(a-b-c)$, then the value of $2a-b+c$ is : **(SSC CGL Tier I 2013)**

(*a*) 4 (*b*) 0 (*c*) 2 (*d*) 3

Sol. (*c*) $a^2+b^2+c^2+3=2(a-b-c) \Rightarrow a^2+b^2+c^2+3=2a-2b-2c$

$\Rightarrow a^2+b^2+c^2+3-2a+2b+2c=0$

$\Rightarrow (a^2-2a+1)+(b^2+2b+1)+(c^2+2c+1)=0$

$\Rightarrow (a-1)^2+(b+1)^2+(c+1)^2=0 \Rightarrow (a-1)=0, (b+1)=0, (c+1)=0$

$\Rightarrow a=1, b=-1, c=-1 \Rightarrow 2a-b+c=2+1-1=2$

156. If $a^2+b^2+c^2=2(a-b-c)-3$, then the value of $4a-3b+5c$ is **(SSC CGL Tier I 2012M, SSC CHSL 2011M, SSC CGL Tier II 2013)**

(*a*) 2 (*b*) 3 (*c*) 5 (*d*) 6

Sol. (*a*) $(a^2-2a+1)+(b^2+2b+1)+(c^2+2c+1)=0$

$\Rightarrow (a-1)^2+(b+1)^2+(c+1)^2=0 \Rightarrow a=1, b=-1, c=-1$

$4a-3b+5c=4(1)-3(-1)+5(-1)=4+3-5=2.$

157. If $a^2+b^2+c^2-ab-bc-ca=0$, then

(*a*) $a+b+c=0$ (*b*) $a+b+c=1$ (*c*) $a=b=c$ (*d*) $a \neq b \neq c$

Sol. (*c*) G.E. $\Rightarrow 2a^2+2b^2+2c^2-2ab-2bc-2ca=0$

$\Rightarrow (a^2-2ab+b^2)+(b^2-2bc+c^2)+(c^2-2ca+a^2)=0$

$\Rightarrow (a-b)^2+(b-c)^2+(c-a)^2=0$

$\Rightarrow a-b=0, b-c=0, c-a=0 \Rightarrow a=b, b=c, c=a \Rightarrow a=b=c$

158. If $a^2+b^2+4c^2=2(a+b-2c)-3$ and a, b, c are real, then the value of $(a^2+b^2+c^2)$ is **(SSC CGL Tier I 2013)**

(*a*) 2 (*b*) $2\frac{1}{4}$ (*c*) 3 (*d*) $3\frac{1}{4}$

Sol. (*b*) $a^2+b^2+4c^2-2a-2b+4c+3=0$

$\Rightarrow (a^2-2a+1)+(b^2-2b+1)+(4c^2+4c+1)=0$

$\Rightarrow (a-1)^2+(b-1)^2+(2c+1)^2=0 \Rightarrow a=1, b=1, c=-\frac{1}{2}$

$\Rightarrow a^2+b^2+c^2=1^2+1^2+\left(-\frac{1}{2}\right)^2=2\frac{1}{4}$

159. If $a^2+b^2+2b+4a+5=0$, then the value of $\frac{a-b}{a+b}$ is **(SSC)**

(*a*) 3 (*b*) -3 (*c*) $\frac{1}{3}$ (*d*) $\frac{-1}{3}$

Sol. (*c*) $(a^2+4a+4)+(b^2+2b+1)=0 \Rightarrow (a+2)^2+(b+1)^2=0$

$\Rightarrow a+2=0$ and $b+1=0 \Rightarrow a=-2$ and $b=-1$

$\Rightarrow \frac{a-b}{a+b}=\frac{-2-(-1)}{-2-1}=\frac{-1}{-3}=\frac{1}{3}$

160. If $x^2+y^2+z^2=xy+yz+zx$, $(x \neq 0)$, then the value of $\frac{4x+2y-3z}{2x}$ is **(FCI Asstt. 2012)**

(*a*) 0 (*b*) 1 (*c*) $\frac{3}{2}$ (*d*) $\frac{1}{2}$

Sol. (*c*) $x^2+y^2+z^2=xy+yz+zx$ is possible when $x=y=z=k$ (say)

$\therefore \frac{4x+2y-3z}{2x}=\frac{4k+2k-3k}{2k}=\frac{3k}{2k}=\frac{3}{2}$

161. If $a=4$, $b=5$ and $c=6$, find the value of $2a^2+2b^2+2c^2-2ab-2bc-2ca$

(*a*) -1 (*b*) -2 (*c*) -6 (*d*) 6

Sol. (*d*) $2a^2 + 2b^2 + 2c^2 - 2ab - 2bc - 2ca$
$= (a^2 - 2ab + b^2) + (b^2 - 2bc + c^2) + (c^2 - 2ca + b^2)$
$= (a-b)^2 + (b-c)^2 + (c-a)^2 = (4-5)^2 + (5-6)^2 + (6-4)^2 = 1 + 1 + 4 = 6$

162. If $x \neq 0, y \neq 0$ and $z \neq 0$ and $\frac{1}{x^2}+\frac{1}{y^2}+\frac{1}{z^2}=\frac{1}{xy}+\frac{1}{yz}+\frac{1}{zx}$, then the relation among x, y, z is **(SSC CGL Tier I 2013)**

(*a*) $x = y = z$ (*b*) $x + y + z = 0$ (*c*) $x + y = z$ (*d*) $\frac{1}{x}+\frac{1}{y}+\frac{1}{z}=0$

Sol. (*a*) $\frac{2}{x^2}+\frac{2}{y^2}+\frac{2}{z^2}=\frac{2}{xy}+\frac{2}{yz}+\frac{2}{zx}$

$\Rightarrow \left(\frac{1}{x^2}-\frac{2}{xy}+\frac{1}{y^2}\right)+\left(\frac{1}{y^2}-\frac{2}{yz}+\frac{1}{z^2}\right)+\left(\frac{1}{z^2}-\frac{2}{zx}+\frac{1}{x^2}\right)=0$

$\Rightarrow \left(\frac{1}{x}-\frac{1}{y}\right)^2+\left(\frac{1}{y}-\frac{1}{z}\right)^2+\left(\frac{1}{z}-\frac{1}{x}\right)^2=0 \Rightarrow \frac{1}{x}=\frac{1}{y}, \frac{1}{y}=\frac{1}{z}, \frac{1}{z}=\frac{1}{x} \Rightarrow x=y=z$

Alt: If $x = y = z$, then $\frac{1}{xy}+\frac{1}{yz}+\frac{1}{zx}=\frac{1}{x^2}+\frac{1}{y^2}+\frac{1}{z^2}$

163. If $x = b + c - 2a, y = c + a - 2b, z = a + b - 2c$, then the value of $x^2 + y^2 - z^2 + 2xy$ is **(SSC CHSL 2011)**

(*a*) 0 (*b*) $a + b + c$ (*c*) $a - b + c$ (*d*) $a + b - c$

Sol. (*a*) Here, $x + y + z = (b + c - 2a) + (c + a - 2b) + (a + b - 2c) = 0$
$x^2 + y^2 - z^2 + 2xy = x^2 + y^2 + 2xy - z^2 = (x + y)^2 - z^2 = (x + y + z)(x + y - z) = 0 \times (x + y - z) = 0$

164. If $xy(x + y) = 1$, then the value of $\frac{1}{x^3y^3} - x^3 - y^3$ **(SSC CGL Tier I 2013)**

(*a*) 3 (*b*) – 3 (*c*) 1 (*d*) – 1

Sol. (*a*) $xy(x + y) = 1 \Rightarrow x + y = \frac{1}{xy} \Rightarrow (x + y)^3 = \frac{1}{x^3y^3}$

$\Rightarrow x^3 + y^3 + 3xy(x + y) = \frac{1}{x^3y^3} \Rightarrow \frac{1}{x^3y^3} - x^3 - y^3 = 3xy(x + y) = 3 \times 1 = 3$

165. If $a + \frac{1}{b} = 1$, and $b + \frac{1}{c} = 1$, then $c + \frac{1}{a}$ is equal to : **(SSC CHSL 2013, SSC CGL Tier I 2008)**

(*a*) 0 (*b*) $\frac{1}{2}$ (*c*) 1 (*d*) 2

Sol. (*c*) $a+\frac{1}{b}=1 \Rightarrow a=1-\frac{1}{b}=\frac{b-1}{b} \Rightarrow \frac{1}{a}=\frac{b}{b-1}$ $b+\frac{1}{c}=1 \Rightarrow \frac{1}{c}=1-b \Rightarrow c=\frac{1}{1-b}$

$\therefore c+\frac{1}{a}=\frac{1}{1-b}+\frac{b}{b-1}=\frac{1}{1-b}-\frac{b}{1-b}=\frac{1-b}{1-b}=1$

166. If a, b, c are non-zero, $a + \frac{1}{b} = 1$, and $b + \frac{1}{c} = 1$, then the value of abc is: **(SSC CGL Tier I 2013)**

(*a*) 1 (*b*) –1 (*c*) 3 (*d*) –3

Sol. (*b*) $a+\frac{1}{b}=1 \Rightarrow a=1-\frac{1}{b}=\frac{b-1}{b}$ $\qquad b+\frac{1}{c}=1 \Rightarrow \frac{1}{c}=1-b \Rightarrow c=\frac{1}{1-b}$

$\therefore abc=\frac{b-1}{b}\times b\times\frac{1}{1-b}=\frac{b-1}{b}\times b\times\frac{-1}{b-1}=-1$

167. If $a + b + c = 8$, then the value of $(a-4)^3 + (b-3)^3 + (c-1)^3 - 3(a-4)(b-3)(c-1)$ is **(SSC CGL Tier I 2012)**

(*a*) 2 (*b*) 4 (*c*) 1 (*d*) 0

Sol. (*d*) $a + b + c = 8 \Rightarrow a + b + c - 8 = 0 \Rightarrow (a-4) + (b-3) + (c-1) = a + b + c - 8 = 0$
If $(x + y + z) = 0$, then $x^3 + y^3 + z^3 - 3xyz = 0$

168. $(a + b - 2c)^3 + (b + c - 2a)^3 + (c + a - 2b)^3$ is equal to **(SSC CHSL 2013)**

(*a*) $(a + b - 2c)(b + c - 2a)(c + a - 2b)$ (*b*) $2(a + b - 2c)(b + c - 2a)(c + a - 2b)$
(*c*) $3(a + b - 2c)(b + c - 2a)(c + a - 2b)$ (*d*) $-3(a + b - 2c)(b + c - 2a)(c + a - 2b)$

Sol. (*c*) We know that $x^3 + y^3 + z^3 = 3xyz$ when $x + y + z = 0$
Here, $(a + b - 2c) + (b + c - 2a) + (c + a - 2b) = 0$
$\Rightarrow (a + b - 2c)^3 + (b + c - 2a)^3 + (c + a - 2b)^3 = 3(a + b - 2c)(b + c - 2a)(c + a - 2b)$

169. Out of the given responses, one of the factors of $(a^2 - b^2)^3 + (b^2 - c^2)^3 + (c^2 - a^2)^3$ is **(SSC CGL Tier II 2013)**

(*a*) $(a + b)(a - b)$ (*b*) $(a + b)(a + b)$ (*c*) $(a - b)(a - b)$ (*d*) $(b - c)(b - c)$

Sol. (*a*) G.E. $= 3(a^2 - b^2)(b^2 - c^2)(c^2 - a^2) = 3(a - b)(a + b)(b^2 - c^2)(c^2 - a^2)$

170. If $a^3 - b^3 - c^3 = 0$, then the value of $a^9 - b^9 - c^9 - 3a^3b^3c^3$ is. **(SSC CGL Tier I 2013)**

(*a*) 0 (*b*) 1 (*c*) $2xyz$ (*d*) $3xyz$

Sol. (*a*) $a^9 - b^9 - c^9 - 3a^3b^3c^3 = (a^3)^3 - (b^3)^3 - (c^3)^3 - 3a^3b^3c^3 = 0$

171. $a^3(b - c)^3 + b^3(c - a)^3 + c^3(a - b)^3 = ?$

(*a*) $3abc$ (*b*) $3(a - b)(b - c)(c - a)$
(*c*) $3abc(a - b)(b - c)(a - c)$ (*d*) $3abc(a - b)(b - c)(c - a)$

Sol. (*d*) $a(b - c) + b(c - a) + c(a - b) = ab - ac + bc - ab + ac - bc = 0$
$\therefore$ G.E. $= 3abc(a - b)(b - c)(c - a)$

172. Find the value of $15^3 - 6^3 - 9^3 - 5^3$.

(*a*) 0 (*b*) -125 (*c*) 2305 (*d*) 2430

Sol. (*c*) $15^3 - 6^3 - 9^3 = 3 \times 15 \times (-6) \times (-9) = 2430$ ($\because 15 - 6 - 9 = 0$)
$\therefore$ Given Expression $= 2430 - 5^3 = 2430 - 125 = 2305$

173. Find the value of $50^3 + (-30)^3 + (-20)^3$. **(SSC CGL Tier I 2008)**

(*a*) 0 (*b*) -9000 (*c*) 90000 (*d*) None of these

Sol. (*c*) $50^3 + (-30)^3 + (-20)^3 = 3 \times 50 \times (-30) \times (-20) = 90000$ [$\because 50 + (-30) + (-20) = 0$]

174. If $a = 1.21$, $b = 2.12$ and $c = -3.33$, then the value of $a^3 + b^3 + c^3 - 3abc$ is : **(SSC CGL Tier I 2001)**

(*a*) 0 (*b*) 1 (*c*) 2 (*d*) 3

Sol. (*a*) Here $a + b + c = 1.21 + 2.12 - 3.33 = 0 \Rightarrow a^3 + b^3 + c^3 = 3abc \Rightarrow a^3 + b^3 + c^3 - 3abc = 0$

175. If $a = 4.36$, $b = 2.39$ and $c = 1.97$, then the value of $a^3 - b^3 - c^3 - 3abc$ is : **(SSC CGL Tier I 2005, 2002 M)**

(*a*) 3.94 (*b*) 2.39 (*c*) 0 (*d*) 1

Sol. (*c*) Here $a + (-b) + (-c) = 4.36 - 2.39 - 1.97 = 0$

$\therefore\ a^3 + (-b)^3 + (-c)^3 = 3a(-b)(-c) \Rightarrow a^3 - b^3 - c^3 - 3abc = 0$

176. Simplified value of $[(0.111)^3 + (0.222)^3 - (0.333)^3 + (0.333)^2 (0.222)]^3$ is : **(SSC CGL Tier I 2004)**

(*a*) 0.999 (*b*) 0 (*c*) 0.888 (*d*) 0.111

Sol. (*b*) If $a + b + c = 0$, then $a^3 + b^3 + c^3 = 3abc$

$\therefore$ Given expression $= [3 \times 0.111 \times 0.222 \times (-0.333) + (0.333)^2 (0.222)]^3$

$= [-(0.333)^2 \times (0.222) + (0.333)^2 (0.222)]^3 = (0)^3 = 0$

177. The value of the expression $x^4 - 17x^3 + 17x^2 - 17x + 17$ at $x = 16$ is **(SSC FCI Asstt 2012)**

(*a*) 1 (*b*) 3 (*c*) 2 (*d*) 0

Sol. (*a*) G.E. $= (x^4 - 16x^3) + (-x^3 + 16x^2) + (x^2 - 16x) + (-x + 17) = 0 + 0 + 0 - 16 + 17 = 1$

178. If $x + y = \sqrt{3}$ and $x - y = \sqrt{2}$, then $8xy(x^2 + y^2)$ is : **(SSC CGL 2006, CHSL 2012)**

(*a*) 1 (*b*) 5 (*c*) $\sqrt{6}$ (*d*) $8\sqrt{6}$

Sol. (*b*) $(x + y)^2 + (x - y)^2 = \left(\sqrt{3}\right)^2 + \left(\sqrt{2}\right)^2 \Rightarrow 2(x^2 + y^2) = 3 + 2 = 5$

$(x + y)^2 - (x - y)^2 = \left(\sqrt{3}\right)^2 - \left(\sqrt{2}\right)^2 \Rightarrow 4xy = 3 - 2 = 1$

$\therefore\ 8xy(x^2 + y^2) = 4xy \times 2(x^2 + y^2) = 1 \times 5 = 5$

179. If $x = \sqrt{3} + \sqrt{2}$ and $y = \sqrt{3} - \sqrt{2}$, then the value of $8xy(x^2 + y^2)$ is **(SSC CHSL 2013)**

(*a*) $16\sqrt{6}$ (*b*) 32 (*c*) $48\sqrt{2}$ (*d*) 80

Sol. (*d*) $(x + y)^2 + (x - y)^2 = \left(\sqrt{3} + \sqrt{2} + \sqrt{3} - \sqrt{2}\right)^2 + \left(\sqrt{3} + \sqrt{2} - \sqrt{3} + \sqrt{2}\right)^2$

$\Rightarrow\ 2(x^2 + y^2) = \left(2\sqrt{3}\right)^2 + \left(2\sqrt{2}\right)^2 = 12 + 8 = 20$

$(x + y)^2 - (x - y)^2 = (2\sqrt{3})^2 - (2\sqrt{2})^2 = 12 - 8 \Rightarrow 4xy = 4$

$\therefore\ 8xy(x^2 + y^2) = 4xy[2(x^2 + y^2)] = 4 \times 20 = 80$

180. If $x^{1/3} + y^{1/3} + z^{1/3} = 0$, then $(x + y + z)^3 = ?$ **(CDS 2002, SSC CHSL 2012)**

(*a*) 0 (*b*) xyz (*c*) $\sqrt[3]{xyz}$ (*d*) $27xyz$

Sol. (*d*) $x^{1/3} + y^{1/3} = -z^{1/3} \Rightarrow (x^{1/3} + y^{1/3})^3 = (-z^{1/3})^3$

$\Rightarrow x + y + 3x^{1/3}y^{1/3}(x^{1/3} + y^{1/3})^3 = -z \Rightarrow x + y + z + 3x^{1/3}y^{1/3}(-z)^{1/3} = 0$

$\Rightarrow x + y + z = 3x^{1/3}y^{1/3}z^{1/3} \Rightarrow (x + y + z)^3 = (3x^{1/3}y^{1/3}z^{1/3})^3 = 27xyz$

Alt: If $a + b + c = 0 \Rightarrow a^3 + b^3 + c^3 = 3abc$

$\Rightarrow$ Here, $x + y + z = 3x^{1/3}y^{1/3}z^{1/3} \Rightarrow (x^3 + y^3 + z^3)^3 = (3x^{1/3}y^{1/3}z^{1/3})^3 = 27xyz$

181. When $a = \dfrac{4}{3}$, the value of $27a^3 - 108a^2 + 144a - 317$ is : **(SSC CHSL 2012)**

(*a*) 261 (*b*) − 253 (*c*) − 245 (*d*) 0

Sol. (*b*) $a = \dfrac{4}{3} \Rightarrow 3a = 4 \Rightarrow 3a - 4 = 0 \Rightarrow (3a - 4)^3 = 27a^3 - 108a^2 + 144a - 64 = 0^3$

$\Rightarrow 27a^3 - 108a^2 + 144a - 317 = (27a^3 - 108a^2 + 144a - 64) - 253 = 0 - 253 = -253$

182. If $a^2 + b^2 + c^2 = 19$ and $ab - bc - ca = 15$, find the value of $a + b - c$.

(*a*) 2 (*b*) 4 (*c*) 5 (*d*) 7

Sol. (*d*) $(a + b - c)^2 = a^2 + b^2 + c^2 + 2 \times (ab - bc - ca) = 19 + 2 \times 15 = 49 \Rightarrow a + b - c = 7$

183. Find the value of $(x^3+y^3+z^3-3xyz)$, if $(x+y+z)=9$ and $(xy+yz+xz)=26$.
(*a*) 27 (*b*) 52 (*c*) 81 (*d*) 576

Sol. (*a*) $x^3+y^3+z^3-3xyz=(x+y+z)[(x+y+z)^2-3(xy+yz+xz)]=9\times(9^2-3\times26)=9\times3=27$

184. If $x+y+z=6$ and $x^2+y^2+z^2=20$ then the value of $x^3+y^3+z^3-3xyz$ is **(SSC CGL Tier I 2013)**
(*a*) 64 (*b*) 70 (*c*) 72 (*d*) 76

Sol. (*c*) $(x+y+z)^2=x^2+y^2+z^2+2(xy+yz+xz)=6^2=36$
$\Rightarrow 2(xy+yz+xz)=36-20=16 \Rightarrow (xy+yz+xz)=8$
Now $(x^3+y^3+z^3-3xyz)=(x+y+z)[(x^2+y^2+z^2)-(xy+yz+xz)]$
$=6\times(20-8)=6\times12=72$

185. If $x+y+z=1$, $xy+yz+zx=-1$, $xyz=-1$, then $x^3+y^3+z^3$ is **(SSC FCI Asstt. 2012)**
(*a*) –2 (*b*) –1 (*c*) 0 (*d*) 1

Sol. (*d*) $x+y+z=1 \Rightarrow (x+y+z)^2=1^2 \Rightarrow x^2+y^2+z^2+2(xy+yz+zx)=1$
$\Rightarrow x^2+y^2+z^2+2(-1)=1 \Rightarrow x^2+y^2+z^2=1+2=3$
$x^3+y^3+z^3-3xyz=(x+y+z)[(x^2+y^2+z^2)-(xy+yz+zx)]$
$\Rightarrow x^3+y^3+z^3-3(-1)=1[(3)-(-1)] \Rightarrow x^3+y^3+z^3+3=3+1=4$
$\Rightarrow x^3+y^3+z^3=4-3=1.$

186. If $x=997$, $y=998$, $z=999$, then the value of $x^2+y^2+z^2-xy-yz-zx$ will be **(SSC CGL Tier I 2012, 2014, SSC CHSL 2014)**
(*a*) 3 (*b*) 9 (*c*) 16 (*d*) 4

Sol. (*a*) $x^2+y^2+z^2-xy-yz-zx=\frac{1}{2}[(x-y)^2+(y-z)^2+(z-x)^2]$
$=\frac{1}{2}(1^2+1^2+2^2)=\frac{1}{2}\times6=3$

Note: If *x*, *y* and *z* are three consecutive natural numbers. Then the value of $x^2+y^2+z^2-xy-yz-zx$ is always 3.

187. Find the value of $a^3+b^3+c^3-3abc$ when $a=225$, $b=226$, $c=227$ **(SSC CGL Tier II 2012)**
(*a*) 2034 (*b*) 2340 (*c*) 2304 (*d*) 2430

Sol. (*a*) $a^3+b^3+c^3-3abc=\frac{1}{2}(a+b+c)[(a-b)^2+(b-c)^2+(c-a)^2]$
$=\frac{1}{2}\times(225+226+227)\times[(225-226)^2+(226-227)^2+(227-225)^2]$
$=\frac{1}{2}\times(678)\times[(1)^2+(1)^2+(2)^2]=\frac{1}{2}\times(678)\times6=2034$

188. If $x=y=333$ and $z=334$, then the value of $x^3+y^3+z^3-3xyz$ is **(SSC CGL Tier II 2013)**
(*a*) 0 (*b*) 667 (*c*) 1000 (*d*) 2334

Sol. (*c*) $x^3+y^3+z^3-3xyz=\frac{1}{2}(x+y+z)[(x-y)^2+(y-z)^2+(z-x)^2]$
$=\frac{1}{2}\times(333+333+334)(0^2+1^2+1^2)=\frac{1}{2}\times1000\times2=1000$

189. If $x=z=225$ and $y=226$, then the value of $x^3+y^3+z^3-3xyz$ is **(SSC CGL Tier I 2015)**
(*a*) 576 (*b*) 674 (*c*) 765 (*d*) 676

Sol. (*d*) $x^3 + y^3 + z^3 - 3xyz = \frac{1}{2}(x + y + z)[(x - y)^2 + (y - z)^2 + (z - x)^2]$

$= \frac{1}{2} \times (225 + 226 + 225)[(1)^2 + (1)^2 + (0)^2] = \frac{1}{2} \times (676) \times 2 = 676$

190. If $(a - b) = 3$, $(b - c) = 5$ and $(c - a) = 1$, then the value of $\frac{a^3 + b^3 + c^3 - 3abc}{a + b + c}$ is

(SSC CGL Tier I 2012)

(*a*) 17.5 (*b*) 20.5 (*c*) 10.5 (*d*) 35

Sol. (*a*) G.E. $= \frac{\frac{1}{2}(a+b+c)[(a-b)^2 + (b-c)^2 + (c-a)^2]}{(a+b+c)}$

$= \frac{1}{2}[(a-b)^2 + (b-c)^2 + (c-a)^2] = \frac{1}{2}[3^2 + 5^2 + 1^2] = \frac{1}{2}(9 + 25 + 1) = \frac{35}{2} = 17.5$

191. What would be the minimum value of $\frac{x^2(y+z) + y^2(x+z) + z^2(x+y)}{xyz}$, where *x*, *y*, z are distinct positive real numbers. **(CAT 2013)**

(*a*) 3 (*b*) 4 (*c*) 5 (*d*) 6

Sol. (*d*) G.E. $= \frac{x^2(y+z)}{xyz} + \frac{y^2(x+z)}{xyz} + \frac{z^2(x+y)}{xyz} = \left(\frac{x}{z} + \frac{x}{y}\right) + \left(\frac{y}{z} + \frac{y}{x}\right) + \left(\frac{z}{y} + \frac{z}{x}\right)$

$= \left(\frac{x}{y} + \frac{y}{x}\right) + \left(\frac{y}{z} + \frac{z}{y}\right) + \left(\frac{z}{x} + \frac{x}{z}\right) \geq 2 + 2 + 2$ or 6

192. If $x + \frac{1}{x} = 8$, then $x^2 + \frac{1}{x^2} = ?$

(*a*) 62 (*b*) 64 (*c*) 66 (*d*) 72

Sol. (*a*) $x^2 + \frac{1}{x^2} = \left(x + \frac{1}{x}\right)^2 - 2 = 8^2 - 2 = 64 - 2 = 62$

193. If $x - \frac{1}{x} = 5$, find the value of $x^2 + \frac{1}{x^2} = ?$

(*a*) 25 (*b*) 27 (*c*) 30 (*d*) 32

Sol. (*b*) $x^2 + \frac{1}{x^2} = \left(x - \frac{1}{x}\right)^2 + 2 = 5^2 + 2 = 27$

194. If $x - \frac{1}{x} = 7$, then $x^3 - \frac{1}{x^3} = ?$

(*a*) 309 (*b*) 322 (*c*) 364 (*d*) 401

Sol. (*c*) $x^3 - \frac{1}{x^3} = \left(x - \frac{1}{x}\right)^3 + 3\left(x - \frac{1}{x}\right) = 7^3 + 3 \times 7 = 343 + 21 = 364$

195. If $x + \frac{1}{x} = \sqrt{13}$, then the value of $x^3 - \frac{1}{x^3}$ is

(*a*) 9 (*b*) 18 (*c*) 27 (*d*) 36

Sol. (*d*) $\left(x - \frac{1}{x}\right)^2 = \left(x + \frac{1}{x}\right)^2 - 4 = 13 - 4 = 9 = 3^2$

$$\Rightarrow x^3 - \frac{1}{x^3} = \left(x - \frac{1}{x}\right)^3 + 3\left(x - \frac{1}{x}\right) = 3^3 + 3 \times 3 = 27 + 9 = 36$$

196. If $x + \frac{1}{x} = 5$, then $x^3 + \frac{1}{x^3} = ?$

(*a*) 125 (*b*) 120 (*c*) 115 (*d*) 110

Sol. (*d*) $x^3 + \frac{1}{x^3} = \left(x + \frac{1}{x}\right)^3 - 3\left(x + \frac{1}{x}\right) = 5^3 - 3 \times 5 = 125 - 15 = 110$

197. If $\left(x + \frac{1}{x}\right) = 4$, then the value of $x^4 + \frac{1}{x^4}$ is **(SSC CGL Tier I 2013)**

(*a*) 124 (*b*) 64 (*c*) 194 (*d*) 81

Sol. (*c*) $x^2 + \frac{1}{x^2} = \left(x + \frac{1}{x}\right)^2 - 2 = 4^2 - 2 = 14 \Rightarrow x^4 + \frac{1}{x^4} = \left(x^2 + \frac{1}{x^2}\right)^2 - 2 = 14^2 - 2 = 194$

198. If $x - \frac{1}{x} = 2$, then $x^4 + \frac{1}{x^4} = ?$ **(SSC CHSL 2013)**

(*a*) 8 (*b*) 4 (*c*) 12 (*d*) 34

Sol. (*d*) $\left(x - \frac{1}{x}\right)^2 = 2^2 \Rightarrow x^2 + \frac{1}{x^2} = 2^2 + 2 = 6$

$$\Rightarrow \left(x^2 + \frac{1}{x^2}\right)^2 = 6^2 \Rightarrow x^4 + \frac{1}{x^4} = 36 - 2 = 34$$

199. If $x + \frac{1}{x} = 3$, then $x^8 + \frac{1}{x^8} = ?$

(*a*) 27 (*b*) 2207 (*c*) 47 (*d*) 2209

Sol. (*b*) $x^2 + \frac{1}{x^2} = \left(x + \frac{1}{x}\right)^2 - 2 = 3^2 - 2 = 7 \Rightarrow x^4 + \frac{1}{x^4} = \left(x^2 + \frac{1}{x^2}\right)^2 - 2 = 7^2 - 2 = 47$

$$\Rightarrow x^8 + \frac{1}{x^8} = \left(x^4 + \frac{1}{x^4}\right)^2 - 2 = 47^2 - 2 = 2207$$

200 If $x^2 + \frac{1}{x^2} = 38$, then $x - \frac{1}{x} = ?$

(*a*) 3 (*b*) 4 (*c*) 5 (*d*) 6

Sol. (*d*) $\left(x - \frac{1}{x}\right)^2 = x^2 + \frac{1}{x^2} - 2 = 38 - 2 = 36 \Rightarrow x - \frac{1}{x} = \sqrt{36} = 6$

201. If $\frac{x}{y} + \frac{y}{x} = 5$, then $\frac{x^2}{y^2} + \frac{y^2}{x^2} = ?$

(*a*) 23 (*b*) 25 (*c*) 27 (*d*) None of these

Sol. (*a*) $\frac{x^2}{y^2} + \frac{y^2}{x^2} = \left(\frac{x}{y} + \frac{y}{x}\right)^2 - 2 = 5^2 - 2 = 23$

202. If $\frac{a}{b} + \frac{b}{a} = 4$, then $\frac{a^4}{b^4} + \frac{b^4}{a^4} = ?$

(*a*) 64 (*b*) 194 (*c*) 196 (*d*) 326

Sol. (*b*) $\frac{a^2}{b^2}+\frac{b^2}{a^2}=\left(\frac{a}{b}+\frac{b}{a}\right)^2-2=4^2-2=14 \Rightarrow \frac{a^4}{b^4}+\frac{b^4}{a^4}=\left(\frac{a^2}{b^2}+\frac{b^2}{a^2}\right)^2-2=14^2-2=194$

203. If $x-\frac{1}{x}=\frac{1}{3}$, then what $9x^2+\frac{9}{x^2}$ is equal to ? **(CDS 2011)**

(*a*) 18 (*b*) 19 (*c*) 20 (*d*) 21

Sol. (*b*) $x-\frac{1}{x}=\frac{1}{3}\Rightarrow\left(x-\frac{1}{x}\right)^2=\left(\frac{1}{3}\right)^2\Rightarrow x^2+\frac{1}{x^2}-2=\frac{1}{9}\Rightarrow 9x^2+\frac{9}{x^2}-18=1$

$\Rightarrow 9x^2+\frac{9}{x^2}=1+18=19$

204. If $a^2+1=a$, then the value of $a^{12}+a^6+1$ is : **(SSC CGL Tier I 2013)**

(*a*) 3 (*b*) – 3 (*c*) 1 (*d*) 2

Sol. (*a*) $a^2+1=a \Rightarrow a^2-a+1=0 \Rightarrow (a+1)(a^2-a+1)=0$

$\Rightarrow a^3+1=0 \Rightarrow a^3=-1$

$\therefore a^{12}+a^6+1=(-1)^{12}+(-1)^6+1=1+1+1=3$

205. If $\left(x+\frac{1}{x}\right)^2=3$, then the value of $x^{206}+x^{200}+x^{90}+x^{84}+x^{12}+x^6+1$ is

(SSC CGL Tier II 2012)

(*a*) 84 (*b*) 206 (*c*) 0 (*d*) 1

Sol. (*c*) $\left(x+\frac{1}{x}\right)^2=3 \Rightarrow \left(x+\frac{1}{x}\right)=\sqrt{3} \Rightarrow \left(x+\frac{1}{x}\right)^3=\left(\sqrt{3}\right)^3$

$\Rightarrow x^3+\frac{1}{x^3}=\left(\sqrt{3}\right)^3-3\left(x+\frac{1}{x}\right)=3\sqrt{3}-3\sqrt{3}=0 \Rightarrow x^6+1=0$

$\therefore x^{206}+x^{200}+x^{90}+x^{84}+x^{12}+x^6+1$

$=x^{200}(x^6+1)+x^{84}(x^6+1)+x^6(x^6+1)+1(x^6+1)=0$

Remember the rule : If $\left(x+\frac{1}{x}\right)^2=3$, then $(x^6+1)=0$

206. If $a+\frac{1}{a}=\sqrt{3}$, then the value of $a^6-\frac{1}{a^6}+2$ will be **(SSC CHSL 2012)**

(*a*) $3\sqrt{3}$ (*b*) 5 (*c*) 1 (*d*) 2

Sol. (*d*) $a+\frac{1}{a}=\sqrt{3} \Rightarrow a^3+\frac{1}{a^3}=\left(\sqrt{3}\right)^2-3\sqrt{3}=0 \Rightarrow a^6=-1 \Rightarrow a^6-\frac{1}{a^6}+2=1-1+2=2$

207. If $x^2-3x+1=0$, then the value of $x+\frac{1}{x}$ is : **(SSC CGL Tier I 2007)**

(*a*) 0 (*b*) 1 (*c*) 2 (*d*) 3

Sol. (*d*) $x+\frac{1}{x}=\frac{x^2+1}{x}$. Given $x^2-3x+1=0 \Rightarrow x^2+1=3x \Rightarrow$ Given expression $=\frac{3x}{x}=3$

208. If $x^2-3x+1=0$, then the value of $\frac{x^6+x^4+x^2+1}{x^3}$ will be **(SSC CGL Tier I 2013, 2014)**

(*a*) 18 (*b*) 15 (*c*) 21 (*d*) 30

Sol. (*c*) $x^2 - 3x + 1 = 0 \Rightarrow x^2 + 1 = 3x \Rightarrow x + \frac{1}{x} = 3$

$$\frac{x^6 + x^4 + x^2 + 1}{x^3} = x^3 + x + \frac{1}{x} + \frac{1}{x^3} = \left(x^3 + \frac{1}{x^3}\right) + \left(x + \frac{1}{x}\right) = (3^3 - 3 \times 3) + 3 = 27 - 9 + 3 = 21$$

209. If $\frac{x^{24} + 1}{x^{12}} = 7$, then the value of $\frac{x^{72} + 1}{x^{36}}$ will be **(SSC CGL Tier I 2015)**

(*a*) 432 (*b*) 433 (*c*) 343 (*d*) 322

Sol. (*d*) $\frac{x^{24} + 1}{x^{12}} = 7 \Rightarrow x^{12} + \frac{1}{x^{12}} = 7$

$$\Rightarrow \frac{x^{72} + 1}{x^{36}} = x^{36} + \frac{1}{x^{36}} = \left(x^{12} + \frac{1}{x^{12}}\right)^3 - 3\left(x^{12} + \frac{1}{x^{12}}\right) = 7^3 - 3 \times 7 = 343 - 21 = 322$$

210. If $x^4 + \frac{1}{x^4} = 23$, then the value of $\left(x - \frac{1}{x}\right)^2$ will be **(SSC CGL Tier I 2013)**

(*a*) – 3 (*b*) 3 (*c*) 7 (*d*) –7

Sol. (*b*) $\left(x^2 + \frac{1}{x^2}\right)^2 = x^4 + \frac{1}{x^4} + 2 = 23 + 2 = 25 = 5^2 \Rightarrow \left(x - \frac{1}{x}\right)^2 = x^2 + \frac{1}{x^2} - 2 = 5 - 2 = 3.$

211. If $x^3 + \frac{1}{x^3} = 52$, find the value of $x + \frac{1}{x}$.

(*a*) 3 (*b*) 5 (*c*) 4 (*d*) 6

Sol. (*c*) $\left(x + \frac{1}{x}\right)^3 = x^3 + \frac{1}{x^3} + 3\left(x + \frac{1}{x}\right)$.

Substituting $\left(x + \frac{1}{x}\right) = y$, we get: $y^3 = 52 + 3y \Rightarrow y = 4$

212. If $x^4 + \frac{1}{x^4} = 119$, then $x^3 - \frac{1}{x^3} = ?$

(*a*) 36 (*b*) 72 (*c*) 18 (*d*) 11

Sol. (*a*) $\left(x^2 + \frac{1}{x^2}\right)^2 = x^4 + \frac{1}{x^4} + 2 = 119 + 2 = 121 = 11^2$

$$\left(x - \frac{1}{x}\right)^2 = x^2 + \frac{1}{x^2} - 2 = 11 - 2 = 9 = 3^2$$

$$\left(x^3 - \frac{1}{x^3}\right) = \left(x - \frac{1}{x}\right)^3 + 3\left(x - \frac{1}{x}\right) = 3^3 + 3 \times 3 = 27 + 9 = 36$$

213. If $x + \frac{1}{x} = 1$, then the value of $\frac{x^2 + 3x + 1}{x^2 + 7x + 1}$ is : **(SSC CGL Tier I 2015)**

(*a*) 1 (*b*) 2 (*c*) 1/2 (*d*) 3/7

Sol. (*c*) $\frac{x^2 + 3x + 1}{x^2 + 7x + 1} = \frac{x + 3 + \frac{1}{x}}{x + 7 + \frac{1}{x}} = \frac{1 + 3}{1 + 7} = \frac{4}{8} = \frac{1}{2}$

214. If $\frac{2p}{p^2-2p+1}=\frac{1}{4}$, $p\neq 0$, then the value of $p+\frac{1}{p}$ is **(SSC FCI Asstt. 2012)**

(*a*) 4 (*b*) 5 (*c*) 10 (*d*) 12

Sol. (*c*) G.E. $\Rightarrow 8p=p^2-2p+1 \Rightarrow p^2-10p+1=0 \Rightarrow p^2+1=10p \Rightarrow p+\frac{1}{p}=10$

Alt: $\frac{p^2-2p+1}{p}=8 \Rightarrow p+\frac{1}{p}=8+2=10$

215. If $\frac{x}{x^2-2x+1}=\frac{1}{3}$, then the value of $x^3+\frac{1}{x^3}$ is **(SSC CGL Tier I 2013)**

(*a*) 110 (*b*) 125 (*c*) 27 (*d*) 81

Sol. (*a*) $\frac{x}{x^2-2x+1}=\frac{1}{3} \Rightarrow \frac{x^2-2x+1}{x}=3 \Rightarrow x-2+\frac{1}{x}=3 \Rightarrow x+\frac{1}{x}=5$

$\Rightarrow x^3+\frac{1}{x^3}=\left(x+\frac{1}{x}\right)^3-3\left(x+\frac{1}{x}\right)=5^3-3\times 5=125-15=110$

216. If $x+\frac{1}{x}=5$, then the value of $\frac{x^4+\frac{1}{x^2}}{x^2-3x+1}$ is **(SSC CGL Tier I 2012)**

(*a*) 70 (*b*) 50 (*c*) 110 (*d*) 55

Sol. (*d*) $x+\frac{1}{x}=5 \Rightarrow x^2+1=5x \Rightarrow x^2-3x+1=2x$

$$\therefore \frac{x^4+\frac{1}{x^2}}{x^2-3x+1}=\frac{x\left(x^3+\frac{1}{x^3}\right)}{2x}=\frac{\left(x+\frac{1}{x}\right)^3-3\left(x+\frac{1}{x}\right)}{2}=\frac{5^3-3\times 5}{2}=\frac{125-15}{2}=55$$

Alt: $$\text{G.E.}=\frac{x^3+\frac{1}{x^3}}{x-3+\frac{1}{x}}=\frac{\left(x+\frac{1}{x}\right)^3-3\left(x+\frac{1}{x}\right)}{\left(x+\frac{1}{x}\right)-3}=\frac{5^3-3\times 5}{5-3}=\frac{125-15}{2}=\frac{110}{2}=55$$

217. If $x+\frac{1}{x}=5$, then the value of $\frac{2x}{3x^2-5x+3}$ is **(SSC CHSL 2012, SSC FCI 2012)**

(*a*) 2 (*b*) $1\frac{1}{5}$ (*c*) $\frac{1}{5}$ (*d*) $\frac{3}{5}$

Sol. (*c*) $$\frac{2x}{3x^2-5x+3}=\frac{2}{3x-5+\frac{3}{x}}=\frac{2}{3\left(x+\frac{1}{x}\right)-5}=\frac{2}{3\times 5-5}=\frac{2}{10}=\frac{1}{5}$$

218. If $x^2+y^2+\frac{1}{x^2}+\frac{1}{y^2}=4$, then the value of x^2+y^2 is **(SSC CPO SI 2012)**

(*a*) 2 (*b*) 4 (*c*) 8 (*d*) 16

Sol. (*a*) $x^2+y^2+\frac{1}{x^2}+\frac{1}{y^2}=4 \Rightarrow \left(x^2+\frac{1}{x^2}-2\right)+\left(y^2+\frac{1}{y^2}-2\right)=0$

$$\left(x-\frac{1}{x}\right)^2+\left(y-\frac{1}{y}\right)^2=0 \Rightarrow x=\frac{1}{x}, y=\frac{1}{y} \Rightarrow x=1 \text{ and } y=1 \Rightarrow x^2+y^2=1^2+1^2=2$$

219. If $a^2+\frac{1}{a^2}=98\ (a>0)$, then the value of $a^3+\frac{1}{a^3}$ will be **(SSC CGL Tier I 2012)**

(*a*) 535 (*b*) 790 (*c*) 970 (*d*) 1030

Sol. (*c*) $a^2+\frac{1}{a^2}=98 \Rightarrow \left(a+\frac{1}{a}\right)^2=98+2=100=10^2$

$$\Rightarrow a^3+\frac{1}{a^3}=\left(a+\frac{1}{a}\right)^3-3\left(a+\frac{1}{a}\right)=10^3-3(10)=1000-30=970$$

220. If $x^2+\frac{1}{x^2}=66$, then the value of $\frac{x^2-1+2x}{x}=?$ **(SSC CGL Tier I 2013)**

(*a*) ±8 (*b*) 10, –6 (*c*) 6, –10 (*d*) ±4

Sol. (*b*) $x^2+\frac{1}{x^2}=66 \Rightarrow \left(x-\frac{1}{x}\right)^2=66-2=64 \Rightarrow \left(x-\frac{1}{x}\right)=\pm 8$

$$\therefore \frac{x^2-1+2x}{x}=x-\frac{1}{x}+2=\pm 8+2=+8+2 \text{ or } -8+2=10,-6$$

221. If $x+\frac{9}{x}=6$, then the value of $\left(x^2+\frac{9}{x^2}\right)$ is : **(SSC CHSL 2010)**

(*a*) 8 (*b*) 9 (*c*) 10 (*d*) 12

Sol. (*c*) $x+\frac{9}{x}=6 \Rightarrow x^2+9=6x \Rightarrow x^2-6x+9=0 \Rightarrow (x-3)^2=0 \Rightarrow x=3$

$$\therefore \left(x^2+\frac{9}{x^2}\right)=\left(3^2+\frac{9}{3^2}\right)=9+1=10$$

222. If $x+\frac{4}{x}=4$, find the value of $x^3+\frac{4}{x^3}$ **(SSC CGL Tier I 2013, CGL 2011M)**

(*a*) 8 (*b*) $8\frac{1}{2}$ (*c*) 16 (*d*) $16\frac{1}{2}$

Sol. (*b*) $x+\frac{4}{x}=4 \Rightarrow x^2+4=4x \Rightarrow x^2-4x+4=0 \Rightarrow (x-2)^2=0 \Rightarrow x=2$

$$\Rightarrow x^3+\frac{4}{x^3}=2^3+\frac{4}{2^3}=8+\frac{4}{8}=8+\frac{1}{2}=8\frac{1}{2}$$

Alt: $x+\frac{4}{x}=4 \Rightarrow \frac{x}{2}+\frac{2}{x}=2 \Rightarrow \frac{x}{2}=\frac{2}{x}=1 \Rightarrow x=2$

Alt: When $x=2$, $x+\frac{4}{x}=2+2=4$

223. If $2k-\frac{1}{2k}=6$, then $k^2+\frac{1}{16k^2}$ is equal to : **(SSC CGL Tier I 2006)**

(*a*) 7 (*b*) $8\frac{1}{2}$ (*c*) $9\frac{1}{2}$ (*d*) 11

Sol. (*c*) $2k-\frac{1}{2k}=6 \Rightarrow k-\frac{1}{4k}=3 \Rightarrow \left(k-\frac{1}{4k}\right)^2=3^2=9$

$$\Rightarrow k^2 + \frac{1}{16k^2} - 2 \times k \times \frac{1}{4k} = 9 \Rightarrow k^2 + \frac{1}{16k^2} = 9 + \frac{1}{2} = 9\frac{1}{2}$$

224. If $2x + \frac{2}{x} = 3$, then the value of $x^3 + \frac{1}{x^3} + 2$ is **(SSC CGL Tier II 2013)**

(*a*) $-\frac{9}{8}$ (*b*) $-\frac{25}{8}$ (*c*) $\frac{7}{8}$ (*d*) 11

Sol. (*c*) $x + \frac{1}{x} = \frac{3}{2} \Rightarrow \left(x^3 + \frac{1}{x^3}\right) + 2 = \left(x + \frac{1}{x}\right)^3 - 3\left(x + \frac{1}{x}\right) + 2$

$$= \left(\frac{3}{2}\right)^3 - 3\left(\frac{3}{2}\right) + 2 = \frac{27}{8} - \frac{9}{2} + 2 = \frac{-9}{8} + 2 = \frac{7}{8}$$

225. If $y + \frac{1}{4y} = 1$, then the value of $8y^3 + \frac{1}{8y^3}$ is **(SSC FCI 2012)**

(*a*) 2 (*b*) 4 (*c*) 6 (*d*) 8

Sol. (*a*) $y + \frac{1}{4y} = 1 \Rightarrow 2y + \frac{1}{2y} = 2$

$$8y^3 + \frac{1}{8y^3} = \left(2y + \frac{1}{2y}\right)^3 - 3\left(2y + \frac{1}{2y}\right) = 2^3 - 3 \times 2 = 8 - 6 = 2$$

Trick : $2y + \frac{1}{2y} = 2 \Rightarrow 2y = 1 \quad \Rightarrow$ G.E. $= 1^3 + 1^3 = 2$

226. If $a + \frac{1}{a-2} = 4$, then the value of $(a-2)^2 + \left(\frac{1}{a-2}\right)^2$ is **(SSC CGL Tier I 2013)**

(*a*) 0 (*b*) 2 (*c*) -2 (*d*) 4

Sol. (*b*) $a + \frac{1}{a-2} = 4 \Rightarrow (a-2) + \frac{1}{a-2} = 2 \Rightarrow (a-2)^2 + \left(\frac{1}{a-2}\right)^2 = 2^2 - 2 = 2$

Trick : $(a-2) + \frac{1}{a-2} = 2 \Rightarrow (a-2) = 1 \Rightarrow$ G.E. $= 1^2 + 1^2 = 2$

227. If $x\left(3 - \frac{2}{x}\right) = \frac{3}{x}, x \neq 0$, then the value of $x^2 + \frac{1}{x^2}$ is **(SSC FCI Asstt 2012)**

(*a*) $2\frac{1}{3}$ (*b*) $2\frac{2}{3}$ (*c*) $2\frac{4}{9}$ (*d*) $2\frac{5}{9}$

Sol. (*c*) $x\left(3 - \frac{2}{x}\right) = \frac{3}{x} \Rightarrow 3x - \frac{3}{x} = 2 \Rightarrow x - \frac{1}{x} = \frac{2}{3}$

$$\left(x - \frac{1}{x}\right)^2 = \frac{4}{9} \Rightarrow x^2 + \frac{1}{x^2} - 2 = \frac{4}{9} \Rightarrow x^2 + \frac{1}{x^2} = 2 + \frac{4}{9} = 2\frac{4}{9}$$

228. If x is real, $x + \frac{1}{x} \neq 0$ and $x^3 + \frac{1}{x^3} = 0$, then the value of $\left(x + \frac{1}{x}\right)^4$ is **(SSC CGL Tier I 2012)**

(*a*) 4 (*b*) 9 (*c*) 16 (*d*) 25

Sol. (*b*) $\left(x^3+\frac{1}{x^3}\right)=\left(x+\frac{1}{x}\right)\left(x^2+\frac{1}{x^2}-1\right)=0 \Rightarrow \left(x^2+\frac{1}{x^2}-1\right)=0$

$$\Rightarrow x^2+\frac{1}{x^2}=1 \Rightarrow \left(x+\frac{1}{x}\right)^2=1+2=3 \Rightarrow \left(x+\frac{1}{x}\right)^4=3^2=9$$

229. If $4a-\frac{4}{a}+3=0$, then the value of $a^3-\frac{1}{a^3}+3=?$ **(SSC CGL Tier I 2015)**

(*a*) 3/8 (*b*) 3/4 (*c*) 7/8 (*d*) 21/64

Sol. (*d*) $4a-\frac{4}{a}+3=0 \Rightarrow a-\frac{1}{a}=-\frac{3}{4} \Rightarrow a^3-\frac{1}{a^3}+3=\left(a-\frac{1}{a}\right)^3+3\left(a-\frac{1}{a}\right)+3$

$$=\left(-\frac{3}{4}\right)^3+3\left(-\frac{3}{4}\right)+3=-\frac{27}{64}-\frac{9}{4}+3=\frac{3}{4}-\frac{27}{64}=\frac{21}{64}$$

230. If $\left(2x-\frac{3}{x}\right)=5$, find the value of $\left(4x^2-\frac{9}{x^2}\right)$.

(*a*) 25 (*b*) 30 (*c*) 35 (*d*) 49

Sol. (*c*) $\left(2x-\frac{3}{x}\right)^2=5^2 \Rightarrow \left(4x^2+\frac{9}{x^2}\right)-12=25 \Rightarrow \left(4x^2+\frac{9}{x^2}\right)=25+12=37$

$$\left(2x+\frac{3}{x}\right)^2=\left(4x^2+\frac{9}{x^2}\right)+12=37+12=49 \Rightarrow \left(2x+\frac{3}{x}\right)=\sqrt{49}=7$$

$$\left(4x^2-\frac{9}{x^2}\right)=\left(2x+\frac{3}{x}\right)\times\left(2x-\frac{3}{x}\right)=7\times 5=35$$

231. If $5a+\frac{1}{3a}=5$, the value of $9a^2+\frac{1}{25a^a}$ is : **(SSC CGL Tier I 2012)**

(*a*) $\frac{34}{5}$ (*b*) $\frac{39}{5}$ (*c*) $\frac{42}{5}$ (*d*) $\frac{52}{5}$

Sol. (*b*) $5a+\frac{1}{3a}=5 \Rightarrow \frac{5}{3}\left(3a+\frac{1}{5a}\right)=5 \Rightarrow 3a+\frac{1}{5a}=5\times\frac{3}{5}=3$

$$\Rightarrow \left(3a+\frac{1}{5a}\right)^2=3^2 \Rightarrow 9a^2+\frac{1}{25a^2}+2\,(3a)\left(\frac{1}{5a}\right)=9 \Rightarrow 9a^2+\frac{1}{25a^2}=9-\frac{6}{5}=\frac{39}{5}$$

232. If $3x+\frac{1}{2x}=5$, then the value of $8x^3+\frac{1}{27x^3}$ is **(SSC CHSL 2011)**

(*a*) $118\frac{1}{2}$ (*b*) $30\frac{10}{27}$ (*c*) 0 (*d*) 1

Sol. (*b*) $3x+\frac{1}{2x}=5 \Rightarrow \frac{3}{2}\left(2x+\frac{1}{3x}\right)=5 \Rightarrow 2x+\frac{1}{3x}=\frac{10}{3}$

$$\left(2x+\frac{1}{3x}\right)^3=\left(\frac{10}{3}\right)^3 \Rightarrow 8x^3+\frac{1}{27x^3}+3\times 2x\times\frac{1}{3x}\left(2x+\frac{1}{3x}\right)=\frac{1000}{27}$$

$$\Rightarrow 8x^3+\frac{1}{27x^3}=\frac{1000}{27}-2\times\frac{10}{3}=\frac{1000}{27}-\frac{20}{3}=\frac{820}{27}=30\frac{10}{27}$$

233. If $x + \frac{1}{x} = 3$, the value of $\left(x^5 + \frac{1}{x^5}\right)$ is **(SSC CGL Tier I 2012)**

(*a*) 322 (*b*) 126 (*c*) 123 (*d*) 113

Sol. (*c*) $x + \frac{1}{x} = 3 \Rightarrow x^2 + \frac{1}{x^2} = \left(x+\frac{1}{x}\right)^2 - 2 = 3^2 - 2 = 9 - 2 = 7$

$$x^3 + \frac{1}{x^3} = \left(x+\frac{1}{x}\right)^3 - 3\left(x+\frac{1}{x}\right) = 3^3 - 3 \times 3 = 27 - 9 = 18$$

$$\left(x^2 + \frac{1}{x^2}\right) \times \left(x^3 + \frac{1}{x^3}\right) = x^5 + \frac{1}{x^5} + x + \frac{1}{x}$$

$$\Rightarrow \quad x^5 + \frac{1}{x^5} = \left(x^2 + \frac{1}{x^2}\right) \times \left(x^3 + \frac{1}{x^3}\right) - \left(x+\frac{1}{x}\right) = 7 \times 18 - 3 = 123.$$

234. If $\frac{3-5x}{x} + \frac{3-5y}{y} + \frac{3-5z}{z} = 0$, the value of $\frac{1}{x} + \frac{1}{y} + \frac{1}{z}$ is **(SSC CGL Tier I 2012)**

(*a*) 2 (*b*) 3 (*c*) 5 (*d*) –5

Sol. (*c*) $\frac{3}{x} - \frac{5x}{x} + \frac{3}{y} - \frac{5y}{y} + \frac{3}{z} - \frac{5z}{z} = 0$

$$\Rightarrow \frac{3}{x} + \frac{3}{y} + \frac{3}{z} = \frac{5x}{x} + \frac{5y}{y} + \frac{5z}{z} \Rightarrow 3\left(\frac{1}{x} + \frac{1}{y} + \frac{1}{z}\right) = 3 \times 5 \Rightarrow \frac{1}{x} + \frac{1}{y} + \frac{1}{z} = 5$$

235. If $\frac{4x-3}{x} + \frac{4y-3}{y} + \frac{4z-3}{z} = 0$, then the value of $\frac{1}{x} + \frac{1}{y} + \frac{1}{z}$ is **(SSC CGL Tier I 2013)**

(*a*) 4 (*b*) 6 (*c*) 9 (*d*) 3

Sol. (*a*) $\frac{4x-3}{x} + \frac{4y-3}{y} + \frac{4z-3}{2} = 0 \Rightarrow \left(4 - \frac{3}{x}\right) + \left(4 - \frac{3}{y}\right) + \left(4 - \frac{3}{z}\right) = 0$

$$\Rightarrow \quad 4+4+4 = \frac{3}{x} + \frac{3}{y} + \frac{3}{z} \Rightarrow 3\left(\frac{1}{x} + \frac{1}{y} + \frac{1}{z}\right) = 12 \Rightarrow \frac{1}{x} + \frac{1}{y} + \frac{1}{z} = \frac{12}{3} = 4$$

236. If $(x - a)(x - b) = 1$ and $a - b + 5 = 0$, then the value of $(x-a)^3 - \frac{1}{(x-a)^3}$ is **(SSC CGL Tier II 2013)**

(*a*) –125 (*b*) 1 (*c*) 125 (*d*) 140

Sol. (*d*) $a - b + 5 = 0 \quad \Rightarrow \quad b = a + 5$

$$(x-a)(x-b) = 1 \Rightarrow \frac{1}{(x-a)} = x - b = (x - a - 5) \Rightarrow (x-a) - \frac{1}{(x-a)} = 5$$

$$\Rightarrow \quad (x-a)^3 - \frac{1}{(x-a)^3} = (5)^3 + 3 \times 5 = 125 + 15 = 140$$

237. Which one is one of the factors of $x^2 + \frac{1}{x^2} + 8\left(x + \frac{1}{x}\right) + 14$? **(CDS 2013)**

(*a*) $x + \frac{1}{x} + 1$ (*b*) $x + \frac{1}{x} + 3$ (*c*) $x + \frac{1}{x} + 6$ (*d*) $x + \frac{1}{x} + 7$

Sol. (*c*) G.E. = $\left(x+\frac{1}{x}\right)^2+8\left(x+\frac{1}{x}\right)+12=y^2+8y+12=(y+2)(y+6)$

$$=\left[\left(x+\frac{1}{x}\right)+2\right]\left[\left(x+\frac{1}{x}\right)+6\right]$$

238. If $a^2+b^2=2$ and $c^2+d^2=1$, then the value of $(ad-bc)^2+(ac+bd)^2$ is **(SSC CPO SI 2011)**

(*a*) $\frac{4}{9}$ (*b*) $\frac{1}{2}$ (*c*) 1 (*d*) 2

Sol. (*d*) $(ad-bc)^2+(ac+bd)^2=a^2d^2+b^2c^2+a^2c^2+b^2d^2$

$=a^2(c^2+d^2)+b^2(c^2+d^2)=(a^2+b^2)(c^2+d^2)=2\times 1=2$

239. If $ax+by=6$, $bx-ay=2$ and $x^2+y^2=4$, then the value of (a^2+b^2) would be: **(SSC CGL Tier I 2012)**

(*a*) 2 (*b*) 4 (*c*) 5 (*d*) 10

Sol. (*d*) $(ax+by)^2+(bx-ay)^2=6^2+2^2 \Rightarrow a^2x^2+b^2y^2+b^2x^2+a^2y^2=36+4$

$\Rightarrow a^2(x^2+y^2)+b^2(x^2+y^2)=40$

$\Rightarrow (a^2+b^2)(x^2+y^2)=40 \Rightarrow (a^2+b^2)=40\div(x^2+y^2)=40\div 4=10$

240. If $a^3-b^3=56$ and $a-b=2$, then the value of (a^2+b^2) is: **(SSC CGL Tier I 2012)**

(*a*) –10 (*b*) –12 (*c*) 20 (*d*) 18

Sol. (*c*) $a^3-b^3=(a-b)(a^2+b^2+ab) \Rightarrow 56=2(a^2+b^2+ab)$

$\Rightarrow a^2+b^2+ab=28 \Rightarrow (a-b)^2+3ab=28 \Rightarrow 2^2+3ab=28$

$\Rightarrow 3ab=28-4=24 \Rightarrow ab=8$

$a^2+b^2+ab=28 \Rightarrow a^2+b^2=28-ab=28-8=20$

241. If $\frac{x-a}{b+c}+\frac{x-b}{c+a}+\frac{x-c}{a+b}=3$, then x is :

(*a*) 0 (*b*) 1 (*c*) $a+b+c$ (*d*) abc

Sol. (*c*) $\frac{x-a}{b+c}+\frac{x-b}{c+a}+\frac{x-c}{a+b}=3 \Rightarrow \frac{x-a}{b+c}-1+\frac{x-b}{c+a}-1+\frac{x-c}{a+b}-1=0$

$\Rightarrow \frac{x-a-b-c}{b+c}+\frac{x-b-c-a}{c+a}+\frac{x-c-a-b}{a+b}=0$

$\Rightarrow (x-a-b-c)\left(\frac{1}{b+c}+\frac{1}{c+a}+\frac{1}{a+b}\right)=0$

$\Rightarrow (x-a-b-c)=0 \Rightarrow x=a+b+c$

Trick: If we put $x=a+b+c$ Then G.E. = $\frac{b+c}{b+c}+\frac{c+a}{c+a}+\frac{a+b}{a+b}=1+1+1=3$

242. If $x=\frac{a-b}{a+b}$, $y=\frac{b-c}{b+c}$, $z=\frac{c-a}{c+a}$, then find the value of $\frac{1+x}{1-x}\times\frac{1+y}{1-y}\times\frac{1+z}{1-z}$.

(*a*) 0 (*b*) 1 (*c*) –1 (*d*) 2

Sol. (*b*) G.E. = $\frac{1+\frac{a-b}{a+b}}{1-\frac{a-b}{a+b}}\times\frac{1+\frac{b-c}{b+c}}{1-\frac{b-c}{b+c}}\times\frac{1+\frac{c-a}{c+a}}{1-\frac{c-a}{c+a}}=\frac{2a}{2b}\times\frac{2b}{2c}\times\frac{2c}{2a}=1$

243. If $\frac{x}{2}+\frac{y}{3}=4$ and $\frac{2}{x}+\frac{3}{y}=1$ then what is $x+y$ equal to ? **(CDS 2013)**

(*a*) 11 (*b*) 10 (*c*) 9 (*d*) 8

Sol. (*b*) $\frac{x}{2}+\frac{y}{3}=4 \Rightarrow 3x+2y=24;\quad \frac{2}{x}+\frac{3}{y}=1 \Rightarrow 3x+2y=xy$

$\Rightarrow xy=24 \Rightarrow (x, y)=(4, 6) \Rightarrow x+y=4+6=10$

Trick: $\frac{x}{2}+\frac{y}{3}=\frac{4}{2}+\frac{6}{3}=4$ and $\frac{2}{4}+\frac{3}{6}=\frac{1}{2}+\frac{1}{2}=1$

244. $\frac{(a+b)^2-c^2}{a^2-(b+c)^2}=?$

(*a*) 1 (*b*) $\frac{a+b+c}{a-b-c}$ (*c*) $\frac{a+b-c}{a-b-c}$ (*d*) $\frac{a-b+c}{a-b-c}$

Sol. (*c*) Given expression $=\frac{(a+b+c)\times(a+b-c)}{(a+b+c)\times(a-b-c)}=\frac{a+b-c}{a-b-c}$

245. $\frac{a^2-b^2-c^2-2bc}{a^2+b^2-c^2+2ab}=?$

(*a*) $\frac{a-b-c}{a+b-c}$ (*b*) $\frac{a+b+c}{a+b-c}$ (*c*) $\frac{a-b-c}{a-b+c}$ (*d*) $\frac{a-b-c}{a+b+c}$

Sol. (*a*) $\frac{a^2-b^2-c^2-2bc}{a^2+b^2-c^2+2ab}=\frac{a^2-(b+c)^2}{(a+b)^2-c^2}=\frac{(a+b+c)(a-b-c)}{(a+b+c)(a+b-c)}=\frac{a-b-c}{a+b-c}$

246. If $a+b+c=0$, then $\frac{(a^2+b^2+c^2)^2}{a^2b^2+b^2c^2+c^2a^2}=?$

(*a*) 0 (*b*) 1 (*c*) 2 (*d*) 4

Sol. (*d*) $a+b+c=0 \Rightarrow (a+b+c)^2=0 \Rightarrow a^2+b^2+c^2+2(ab+bc+ca)=0$

$\Rightarrow a^2+b^2+c^2=-2(ab+bc+ca) \Rightarrow (a^2+b^2+c^2)^2=4(ab+bc+ca)^2$

$\Rightarrow (a^2+b^2+c^2)^2=4[a^2b^2+b^2c^2+c^2a^2+2(ab^2c+abc^2+a^2bc)]$

$=4[a^2b^2+b^2c^2+c^2a^2+2abc(a+b+c)]=4(a^2b^2+b^2c^2+c^2a^2)$

$\therefore \frac{(a^2+b^2+c^2)^2}{a^2b^2+b^2c^2+c^2a^2}=\frac{4(a^2b^2+b^2c^2+c^2a^2)}{a^2b^2+b^2c^2+c^2a^2}=4$

247. $\frac{1}{1-a+a^2}-\frac{1}{1+a^2+a}-\frac{2a}{1+a^2+a^4}=?$ **(SSC CHSL 2012)**

(*a*) –1 (*b*) 0 (*c*) 1 (*d*) 2

Sol. (*b*) G.E. $=\frac{1+a^2+a-1-a^2+a}{(1+a^2)^2-(a)^2}-\frac{2a}{1+a^2+a^4}=\frac{2a}{1+a^2+a^4}-\frac{2a}{1+a^2+a^4}=0$

248. If $(ab-b+1)=0$ and $(bc-c+1)=0$, then $(a-ac)=?$

(*a*) abc (*b*) 0 (*c*) –1 (*d*) 1

Sol. (*d*) $ab-b+1 \Rightarrow b(1-a)=1 \Rightarrow b=\frac{1}{1-a}$; $bc-c+1=0 \Rightarrow bc=c-1 \Rightarrow b=\frac{c-1}{c}$

$$\Rightarrow \frac{1}{1-a} = \frac{c-1}{c} \Rightarrow c = (1-a)(c-1) \Rightarrow c = c-1-ac+a \Rightarrow a-ac=1$$

249. Find unit's digit in $(26^{2n} - 11^{3n})$, if n is a natural number.

(*a*) 1 (*b*) 3 (*c*) 5 (*d*) 6

Sol. (*c*) Unit digit in 26^{2n} is always 6 for all values of $2n$.

Unit digit in 11^{3n} is always 1 for all values of $3n$.

$\therefore$ Unit digit in $(26^{2n} - 11^{3n})$ is $6 - 1 = 5$

250. $7^{12} - 4^{12}$ is exactly divisible by which of the following numbers? **(SSC FCI Asstt. 2012)**

(*a*) 34 (*b*) 33 (*c*) 36 (*d*) 35

Sol. (*b*) $7^{12} - 4^{12} = 49^6 - 16^6$ is always divisible by $49 - 16 = 33$

251. The expression $(5^{2n} - 2^{3n})$ is always divisible by:

(*a*) 3 (*b*) 8 (*c*) 15 (*d*) 17

Sol. (*d*) $5^{2n} - 2^{3n} = (5^2)^n - (2^3)^n = (25^n - 8^n) \Rightarrow (25^n - 8^n)$ is always divisible by $25 - 8 = 17$.

252. The expression $(2^{6n} - 4^{2n})$ is always divisible by:

(*a*) 10 (*b*) 12 (*c*) 32 (*d*) 48

Sol. (*d*) $2^{6n} - 4^{2n} = 64^n - 16^n \Rightarrow 2^{6n} - 4^{2n}$ is always divisible by $64 - 16 = 48$.

252. $49^{15} - 1$ is exactly divisible by **(SSC CGL 1999)**

(*a*) 50 (*b*) 51 (*c*) 29 (*d*) 8

Sol. (*d*) $49^{15} - 1$ is divisible by $49 - 1 = 48$

254. A common factor of $(13^7 + 11^7)$ and $(13^5 + 11^5)$ is : **(SSC CGL 2000)**

(*a*) 24 (*b*) $13^5 + 11^5$ (*c*) $13^2 + 11^2$ (*d*) None of these

Sol. (*a*) $(x^n + y^n)$ is divisible by $(x + y)$ for all odd values of n.

$\therefore$ Both $(13^7 + 11^7)$ and $(13^5 + 11^5)$ are divisible by $(13 + 11) = 24$ as 7 and 5 are odd numbers.

255. If $(2^{32} + 1)$ is exactly divisible by a certain number, which one of the following is also definitely divisible by the same number ? **(SSC CGL Tier I 2007)**

(*a*) $2^{16} - 1$ (*b*) $2^{16} + 1$ (*c*) $2^{96} + 1$ (*d*) 7×2^{33}

Sol. (*c*) $x^3 + 1 = (x + 1)(x^2 - x + 1) \Rightarrow$ The required number is $(2^{32})^3 + 1 = 2^{96} + 1$

256. The greatest common divisor of $3^{3^{333}} + 1$ and $3^{3^{334}} + 1$ is : **(SSC CGL Tier I 2013)**

(*a*) 20 (*b*) 2 (*c*) 1 (*d*) $3^{3^{333}} + 1$

Sol. (*d*) $a^{3^1} = a^3, a^{3^2} = a^9 = (a^3)^3, a^{3^3} = a^{27} = (a^9)^3 \Rightarrow$ If $a^{3^n} = x, a^{3^{n+1}} = x^3$

Common divisor of $(x + 1)$ and $(x^3 + 1)$ is $(x + 1)$

257. $(x + 1) \times (1 - x) \times (x^2 - x + 1) \times (x^2 + x + 1)$.

(*a*) $x^6 - 1$ (*b*) $x^6 + 1$ (*c*) $1 - x^6$ (*d*) None of these

Sol. (*c*) G.E. $= [(x + 1) \times (x^2 - x + 1)] \times [(1 - x) \times (x^2 + x + 1)] = (x^3 + 1) \times (1 - x^3) = (1 - x^6)$

258. Find the value of $\left(2^{1/4} - 1\right)\left(2^{3/4} + 2^{1/2} + 2^{1/4} + 1\right)$.

(*a*) –1 (*b*) 0 (*c*) 1 (*d*) 2

Sol. (*c*) Putting $2^{1/4} = x$, G.E. $= (x - 1)(x^3 + x^2 + x + 1) = (x - 1)[x^2(x + 1) + (x + 1)]$

$= (x - 1)(x + 1)(x^2 + 1) = (x^2 - 1)(x^2 + 1) = x^4 - 1 = 2^1 - 1 = 1$

259. $2^{32} - (2 + 1)(2^2 + 1)(2^4 + 1)(2^8 + 1)(2^{16} + 1)$ **(SSC CGL Tier II 2012)**

(*a*) 2 (*b*) 2^{16} (*c*) 0 (*d*) 1

Sol. (*d*) Given expression $= 2^{32} - (2^2 - 1)(2^2 + 1)(2^4 + 1)(2^8 + 1)(2^{16} + 1)$

$= 2^{32} - (2^4 - 1)(2^4 + 1)(2^8 + 1)(2^{16} + 1)$

$= 2^{32} - (2^8 - 1)(2^8 + 1)(2^{16} + 1) = 2^{32} - (2^{16} - 1)(2^{16} + 1) = 2^{32} - (2^{32} - 1) = 1$

260. If $x(x + y + z) = 9$, $y(x + y + z) = 16$ and $z(x + y + z) = 144$, then what is x equal to? **(CDS 2011)**

(*a*) $\frac{9}{5}$ (*b*) $\frac{9}{7}$ (*c*) $\frac{9}{13}$ (*d*) $\frac{16}{13}$

Sol. (*c*) $x(x + y + z) + y(x + y + z) + z(x + y + z) = 9 + 16 + 144 = 169$

$(x + y + z)(x + y + z) = 169 \Rightarrow (x + y + z)^2 = 169 = 13^2 \Rightarrow (x + y + z) = 13$

$x(x + y + z) = 9 \Rightarrow x \times 13 = 9 \Rightarrow x = \frac{9}{13}$

261. If $(x + y)^2 - z^2 = 4$, $(y + z)^2 - x^2 = 9$, $(z + x)^2 - y^2 = 36$, then $(x + y + z) = ?$ **(CDS)**

(*a*) 0 (*b*) ± 1 (*c*) ± 3 (*d*) ± 7

Sol. (*d*) $(x + y + z)(x + y - z) = 4 \Rightarrow (x + y - z) = \frac{4}{x+y+z}$ .... (1)

$(y + z + x)(y + z - x) = 9 \Rightarrow (y + z - x) = \frac{9}{x+y+z}$ .... (2)

$(z + x + y)(z + x - y) = 36 \Rightarrow (z + x - y) = \frac{36}{x+y+z}$ .... (3)

Adding equations (1), (2) and (3) : $(x + y + z) = \frac{49}{x+y+z}$

$\therefore (x + y + z)^2 = 49 \Rightarrow (x + y + z) = \sqrt{49} = \pm 7$

262. If $x + y + z = 0$ and $xyz \neq 0$, find the value of $\frac{1}{x^2 + y^2 - z^2} + \frac{1}{y^2 + z^2 - x^2} + \frac{1}{z^2 + x^2 - y^2}$

(*a*) 0 (*b*) 1 (*c*) –1 (*d*) 2

Sol. (*a*) G.E. $= \frac{1}{x^2 + y^2 - (-x-y)^2} + \frac{1}{y^2 + z^2 - (-y-z)^2} + \frac{1}{z^2 + x^2 - (-x-z)^2}$

$= \frac{1}{-2xy} + \frac{1}{-2yz} + \frac{1}{-2xz} = \frac{z+x+y}{-2xyz} = 0$

263. If $\frac{x-a^2}{b+c} + \frac{x-b^2}{c+a} + \frac{x-c^2}{a+b} = 4(a+b+c)$, then x is equal to **(SSC CGL Tier II 2013)**

(*a*) $(a + b + c)^2$ (*b*) $a^2 + b^2 + c^2$ (*c*) $ab + bc + ca$ (*d*) $a^2 + b^2 + c^2 - ab - bc - ca$

Sol. (*a*) If $x = (a + b + c)^2$, then $\frac{x-a^2}{b+c} = \frac{(a+b+c)^2 - a^2}{b+c}$

$= \frac{(a+b+c-a)(a+b+c+a)}{(b+c)} = \frac{(b+c)(2a+b+c)}{b+c} = 2a+b+c$

Similarly $\frac{x-b^2}{c+a} = a+2b+c$; and $\frac{x-c^2}{a+b} = a+b+2c$

$(2a + b + c) + (a + 2b + c) + (c + b + 2c) = 4(a + b + c) = \text{R.H.S.}$

264. If $x^2 = y + z$, $y^2 = z + x$ and $z^2 = x + y$, then the value of $\frac{1}{1+x}+\frac{1}{1+y}+\frac{1}{1+z}$ is :

(SSC CPO SI 2011, SSC CGL Tier I 2013, CDS 2002)

(*a*) 2 (*b*) 0 (*c*) –1 (*d*) 1

Sol. (*d*) $x^2 = y + z \Rightarrow x + x^2 = x + y + z \Rightarrow x(1 + x) = x + y + z \Rightarrow (1+x) = \frac{x+y+z}{x}$

Similarly, $(1+y) = \frac{x+y+z}{y}$ and $(1+z) = \frac{x+y+z}{z}$

$$\therefore \frac{1}{1+x}+\frac{1}{1+y}+\frac{1}{1+z} = \frac{x}{x+y+z}+\frac{y}{x+y+z}+\frac{z}{x+y+z} = \frac{x+y+z}{x+y+z} = 1$$

Alt: $x^2 = y+z \Rightarrow x = \frac{y+z}{x} \Rightarrow x+1 = \frac{y+z}{x}+1 \Rightarrow x+1 = \frac{x+y+z}{x} \Rightarrow \frac{1}{x+1} = \frac{x}{x+y+z}$

$$\text{G.E.} = \frac{x}{x+y+z}+\frac{y}{y+x+z}+\frac{z}{z+x+y} = \frac{x+y+z}{x+y+z} = 1$$

265. The roots of the equation $\frac{1}{p+q+x} = \frac{1}{p}+\frac{1}{q}+\frac{1}{x}$ are

(*a*) $p+q, p-q$ (*b*) $-p, -q$ (*c*) $\frac{1}{p+q}, \frac{1}{p-q}$ (*d*) $\frac{pq}{p+q}, \frac{pq}{p-q}$

Sol. (*b*) $\frac{1}{p+q+x} - \frac{1}{x} = \frac{1}{p}+\frac{1}{q} \Rightarrow \frac{x-(p+q+x)}{x(p+q+x)} = \frac{p+q}{pq}$

$$\Rightarrow \frac{-(p+q)}{x(p+q+x)} = \frac{(p+q)}{pq} \Rightarrow x(p+q+x) = -pq \Rightarrow x^2 + (p+q)x + pq = 0$$

$$\Rightarrow (x+p)(x+q) = 0 \Rightarrow x = -p, -q$$

266. If $xy + yz + zx = 0$, then $\left(\frac{1}{x^2-yz}+\frac{1}{y^2-zx}+\frac{1}{z^2-xy}\right)$ $(x, y, z \neq 0)$ is equal to

(SSC CHSL 2013)

(*a*) 0 (*b*) 3 (*c*) 1 (*d*) $x + y + z$

Sol. (*a*) $xy + yz + zx = 0 \Rightarrow x(y+z) = -yz \Rightarrow x^2 - yz = x^2 + x(y+z) = x(x+y+z)$

Similarly $y^2 - zx = y(x+y+z)$ and $z^2 - xy = z(x+y+z)$

$$\therefore \text{G.E.} = \frac{1}{x(x+y+z)}+\frac{1}{y(x+y+z)}+\frac{1}{z(z+y+z)} = \frac{yz+xz+xy}{xyz(x+y+z)} = \frac{0}{xyz(x+y+z)} = 0$$

267. If $a + b + c = 0$, then the value of $\frac{a^2}{a^2-bc}+\frac{b^2}{b^2-ca}+\frac{c^2}{c^2-ab}$ is

(*a*) 1 (*b*) –1 (*c*) 2 (*d*) 0

Sol. (*c*) $a + b + c = 0 \Rightarrow a = -(b+c)$

$$\therefore \frac{a^2}{a^2-bc}+\frac{b^2}{b^2-ca}+\frac{c^2}{c^2-ab} = \frac{(b+c)^2}{(b+c)^2-bc}+\frac{b^2}{b^2+c(b+c)}+\frac{c^2}{c^2+(b+c)b}$$

$$= \frac{b^2+c^2+2bc}{b^2+c^2+bc}+\frac{b^2}{b^2+c^2+bc}+\frac{c^2}{b^2+c^2+bc} = \frac{2(b^2+c^2+bc)}{b^2+c^2+bc} = 2$$

268. The solution(s) of $\frac{x+\sqrt{x^2-1}}{x-\sqrt{x^2-1}}+\frac{x-\sqrt{x^2-1}}{x+\sqrt{x^2-1}}=14$ is/are: **(SSC CHSL 2012)**

(*a*) ±8 (*b*) ±6 (*c*) ±2 (*d*) ±4

Sol. (*c*) $\frac{(x+\sqrt{x^2-1})^2+(x-\sqrt{x^2-1})^2}{x^2-(x^2-1)}=14$ **[Apply rule $(a+b)^2+(a-b)^2=2(a^2+b^2)$]**

$\Rightarrow 2(x^2+x^2-1)=14 \Rightarrow 2x^2-1=7 \Rightarrow 2x^2=8 \Rightarrow x^2=4 \Rightarrow x=\pm 2$

269. If $(x-2a)(x-5a)(x-8a)(x-11a)+ka^4$ is a perfect square, then the value of k is **(SSC FCI Asstt. 2012)**

(*a*) 81 (*b*) 64 (*c*) 49 (*d*) 72

Sol. (*a*) $(x-2a)(x-11a)(x-5a)(x-8a)+ka^4$

$= (x^2-13a+22a^2)(x^2-13a+40a^2)+ka^4$

$= (y+22a^2)(y+40a^2)+ka^4$ [let $x^2-13a=y$]

$= y^2+22a^2y+40a^2y+880a^4+ka^4=y^2+62a^2y+(880+k)a^4$

$= (y+31a)^2$ when $31^2=880+k \Rightarrow k=961-880=81$

270. If $\frac{\sqrt{x}+\sqrt{y}}{\sqrt{x}-\sqrt{y}}=2$, find the value of $\frac{x^2-xy+y^2}{x^2+xy+y^2}$

(*a*) 4 (*b*) 3 (*c*) 73/91 (*d*) 91/73

Sol. (*c*) $\sqrt{x}+\sqrt{y}=2\left(\sqrt{x}-\sqrt{y}\right) \Rightarrow \sqrt{x}=3\sqrt{y} \Rightarrow x=9y$

$$\left[\textbf{Alt:}\ \frac{\left(\sqrt{x}+\sqrt{y}\right)+\left(\sqrt{x}-\sqrt{y}\right)}{\left(\sqrt{x}+\sqrt{y}\right)-\left(\sqrt{x}-\sqrt{y}\right)}=\frac{2+1}{1} \Rightarrow \frac{\sqrt{x}}{\sqrt{y}}=3 \Rightarrow \frac{x}{y}=9\right]$$

$$\therefore \frac{x^2-xy+y^2}{x^2+xy+y^2}=\frac{(x/y)^2-(x/y)+1}{(x/y)^2+(x/y)+1}=\frac{9^2-9+1}{9^2+9+1}=\frac{73}{91}$$

Note: (Rule of componendo and Dividendo) : If $\frac{a}{b}=\frac{c}{d}$, then $\frac{a+b}{a-b}=\frac{c+d}{c-d}$

271. If $\frac{\sqrt{x+7}+\sqrt{x-1}}{\sqrt{x+7}-\sqrt{x-1}}=\frac{2}{1}$, find the value of x.

(*a*) 0 (*b*) 1 (*c*) 2 (*d*) –1

Sol. (*c*) $\frac{\left(\sqrt{x+7}+\sqrt{x-1}\right)+\left(\sqrt{x+7}-\sqrt{x-1}\right)}{\left(\sqrt{x+7}+\sqrt{x-1}\right)-\left(\sqrt{x+7}-\sqrt{x-1}\right)}=\frac{2+1}{2-1}$ (Applying componendo and dividendo)

$\Rightarrow \frac{2\sqrt{x+7}}{2\sqrt{x-1}}=\frac{3}{1} \Rightarrow \frac{x+7}{x-1}=9 \Rightarrow x+7=9x-9 \Rightarrow 8x=16 \Rightarrow x=2$

Alternatively, $1\left(\sqrt{x+7}+\sqrt{x-1}\right)=2\left(\sqrt{x+7}-\sqrt{x-1}\right)$ (By cross-multiplication)

$\Rightarrow \sqrt{x+7}=3\sqrt{x-1} \Rightarrow x+7=9x-9 \Rightarrow 8x=16 \Rightarrow x=2$

Alternatively, if we check through the options:

$\sqrt{0+7}$ and $\sqrt{1+7}$ are not rational number. $\sqrt{2+7}=3$ is a rational number.

272. If $x = \frac{4ab}{a+b}$ $(a \neq b)$, the value of $\frac{x+2a}{x-2a} + \frac{x+2b}{x-2b}$ is **(SSC CHSL 2011)**

(*a*) a (*b*) b (*c*) $2ab$ (*d*) 2

Sol. (*d*) $x = \frac{4ab}{a+b} \Rightarrow \frac{x}{2a} = \frac{2b}{a+b} \Rightarrow \frac{x+2a}{x-2a} = \frac{2b+(a+b)}{2b-(a+b)} = \frac{a+3b}{b-a}$

$x = \frac{4ab}{a+b} \Rightarrow \frac{x}{2b} = \frac{2a}{a+b} \Rightarrow \frac{x+2b}{x-2b} = \frac{2a+(a+b)}{2a-(a+b)} = \frac{3a+b}{a-b}$

$\therefore \frac{x+2a}{x-2a} + \frac{x+2b}{x-2b} = \frac{a+3b}{b-a} + \frac{3a+b}{a-b} = \frac{(a+3b)-(3a+b)}{b-a} = \frac{2(b-a)}{b-a} = 2$

273. If $x = \frac{2\sqrt{6}}{\sqrt{3}+\sqrt{2}}$, then the value of $\frac{x+\sqrt{2}}{x-\sqrt{2}} + \frac{x+\sqrt{3}}{x-\sqrt{3}}$ is

(SSC Constable GD 2012, SSC CGL Tier I 2012)

(*a*) $\sqrt{3}$ (*b*) $\sqrt{6}$ (*c*) 2 (*d*) $\sqrt{2}$

Sol. (*c*) $x = \frac{2\sqrt{6}}{\sqrt{3}+\sqrt{2}} \Rightarrow \frac{x}{\sqrt{2}} = \frac{2\sqrt{3}}{\sqrt{3}+\sqrt{2}} \Rightarrow \frac{x+\sqrt{2}}{x-\sqrt{2}} = \frac{2\sqrt{3}+\sqrt{3}+\sqrt{2}}{2\sqrt{3}-\sqrt{3}-\sqrt{2}} = \frac{3\sqrt{3}+\sqrt{2}}{\sqrt{3}-\sqrt{2}}$

$x = \frac{2\sqrt{6}}{\sqrt{3}+\sqrt{2}} \Rightarrow \frac{x}{\sqrt{3}} = \frac{2\sqrt{2}}{\sqrt{3}+\sqrt{2}} \Rightarrow \frac{x+\sqrt{3}}{x-\sqrt{3}} = \frac{2\sqrt{2}+\sqrt{3}+\sqrt{2}}{2\sqrt{2}-\sqrt{3}-\sqrt{2}} = \frac{3\sqrt{2}+\sqrt{3}}{-\sqrt{3}+\sqrt{2}} = \frac{-(3\sqrt{2}+\sqrt{3})}{\sqrt{3}-\sqrt{2}}$

$\therefore \frac{x+\sqrt{2}}{x-\sqrt{2}} + \frac{x+\sqrt{3}}{x-\sqrt{3}} = \frac{3\sqrt{3}+\sqrt{2}}{\sqrt{3}-\sqrt{2}} - \frac{3\sqrt{2}+\sqrt{3}}{\sqrt{3}-\sqrt{2}} = \frac{2\sqrt{3}-2\sqrt{2}}{\sqrt{3}-\sqrt{2}} = \frac{2(\sqrt{3}-\sqrt{2})}{\sqrt{3}-\sqrt{2}} = 2$

274. If $x = \frac{2\sqrt{24}}{\sqrt{3}+\sqrt{2}}$, then the value of $\frac{x+\sqrt{8}}{x-\sqrt{8}} + \frac{x+\sqrt{12}}{x-\sqrt{12}}$ is **(SSC CGL Tier I 2012)**

(*a*) 1 (*b*) 2 (*c*) –1 (*d*) –2

Sol. (*b*) $x = \frac{2\sqrt{24}}{\sqrt{3}+\sqrt{2}} \Rightarrow \frac{x}{\sqrt{8}} = \frac{2\sqrt{3}}{\sqrt{3}+\sqrt{2}} \Rightarrow \frac{x+\sqrt{8}}{x-\sqrt{8}} = \frac{2\sqrt{3}+\sqrt{3}+\sqrt{2}}{2\sqrt{3}-\sqrt{3}-\sqrt{2}} = \frac{3\sqrt{3}+\sqrt{2}}{\sqrt{3}-\sqrt{2}}$

$x = \frac{2\sqrt{24}}{\sqrt{3}+\sqrt{2}} \Rightarrow \frac{x}{\sqrt{12}} = \frac{2\sqrt{2}}{\sqrt{3}+\sqrt{2}} \Rightarrow \frac{x+\sqrt{12}}{x-\sqrt{12}} = \frac{2\sqrt{2}+\sqrt{3}+\sqrt{2}}{2\sqrt{2}-\sqrt{3}-\sqrt{2}} = -\frac{\sqrt{3}+3\sqrt{2}}{\sqrt{3}-\sqrt{2}}$

$\Rightarrow \frac{x+\sqrt{8}}{x-\sqrt{8}} + \frac{x+\sqrt{12}}{x-\sqrt{12}} = \frac{3\sqrt{3}+\sqrt{2}}{\sqrt{3}-\sqrt{2}} - \frac{\sqrt{3}+3\sqrt{2}}{\sqrt{3}-\sqrt{2}} = \frac{2(\sqrt{3}-\sqrt{2})}{\sqrt{3}-\sqrt{2}} = 2$

275. Find the value of $3 + \cfrac{1}{3 + \cfrac{1}{3 + \cfrac{1}{3 + \ldots\ldots}}}$ **(MBA)**

(*a*) $\frac{3 \pm \sqrt{13}}{2}$ (*b*) $\frac{3+\sqrt{13}}{2}$ (*c*) $\frac{3-\sqrt{13}}{2}$ (*d*) 4

Sol. (*b*) Let given expression = x

Then $x = 3 + \frac{1}{x} \Rightarrow x^2 - 3x - 1 = 0 \Rightarrow x = \frac{3 \pm \sqrt{9+4}}{2} = \frac{3 \pm \sqrt{13}}{2}$

As $x > 3$, $x = \frac{3-\sqrt{13}}{2}$ is not possible $\therefore x = \frac{3+\sqrt{13}}{2}$

CHAPTER

11

POWERS AND ROOTS

a^n is read as a raised to the power n, where a is the base and n is the power or index. a^n is equal to the multiplication of a with a for n times, where a is a real number and n is a natural number.

$$a^n = a \times a \times a \times a \times \ldots\ldots\ldots\ldots\ldots\ldots \times a \text{ (} n \text{ times)}$$

EXPONENTS

1. $a^m \times a^n = a^{m+n}$ **Example** : $3^{13} \times 3^2 = ?$ **(SBI Clerical 2008)**

   (*a*) 3^{15} (*b*) 9^{15} (*c*) 3^{26} (*d*) 9^{26}

   Sol. (*a*) $3^{13} \times 3^2 = 3^{13+2} = 3^{15}$

2. $\dfrac{a^m}{a^n} = a^m \div a^n = a^{m-n}$ **Example :** $11^7 \div 11^3 = ?$

   (*a*) 11^{10} (*b*) 11^{21} (*c*) 11^4 (*d*) 11^0

   Sol. (*c*) $11^7 \div 11^3 = 11^{7-3} = 11^4$

3. $\dfrac{a^m}{a^m} = a^{m-m} = a^0 = 1$ (for non-zero values of a)

4. $(a^m)^n = a^{m \times n}$ **Example** : $(5^2)^3 = 5^{2 \times 3} = 5^6$

5. $a^{\frac{m}{n}} = \left(a^{\frac{1}{n}}\right)^m = \sqrt[n]{a^m}$ **Example :** $8^{\frac{2}{3}} = \left(\sqrt[3]{8}\right)^2 = 2^2 = 4$

6. $(ab)^n = a^n \times b^n$ **Example:** $2^5 \times 3^5 = ?$

   (*a*) 5^5 (*b*) 6^5 (*c*) 5^{10} (*d*) 6^{10}

   Sol. (*b*) $2^5 \times 3^5 = (2 \times 3)^5 = 6^5$

7. $\left(\dfrac{a}{b}\right)^m = \dfrac{a^m}{b^m}$ **Example :** $\left(\dfrac{2}{3}\right)^5 = \dfrac{2^5}{3^5} = \dfrac{32}{243}$

8. $a^{-m} = \dfrac{1}{a^m}$ **Example :** $2^{-3} = \dfrac{1}{2^3} = \dfrac{1}{8}$

9. $\dfrac{1}{a^{-m}} = a^m$ **Example :** $\dfrac{1}{2^{-3}} = 2^3 = 8$

10. $\left(\dfrac{a}{b}\right)^{-\frac{p}{q}} = \left(\dfrac{b}{a}\right)^{\frac{p}{q}}$ **Example :** $\left(\dfrac{3}{2}\right)^{-3} = ?$

    (*a*) $\dfrac{4}{9}$ (*b*) $\dfrac{9}{4}$ (*c*) $\dfrac{8}{27}$ (*d*) $\dfrac{27}{8}$

    Sol. (*c*) $\left(\dfrac{3}{2}\right)^{-3} = \left(\dfrac{2}{3}\right)^3 = \dfrac{8}{27}$

11. If $a^x = b^x \Rightarrow a = b$ (for positive numbers)
12. If $a^x = a^y \Rightarrow x = y$ (for positive numbers)

UNIT PLACE DIGIT RULE

$(4n + 1)^{th}$ power of a number always has the same digit at the unit's place which is unit's digit of the given number. $(x)^{4n+a}$ has the same unit's digit as is the unit's digit in x^a.

Note: If a (i.e. remainder) = 0, then we can find unit digit of x^4.

Remember the following points:

(*a*) A number with 0, 1, 5 and 6 at unit's place, raised to any power always has 0, 1, 5 and 6 respectively at unit's place.

(*b*) Unit's digit in $(\ldots 1)^n$, $(\ldots 3)^{4n}$, $(\ldots 7)^{4n}$ and $(\ldots 9)^{2n}$ is 1 for all values of n.

(*c*) Unit's digit in $(\ldots 2)^{4n}$, $(\ldots 4)^{2n}$, $(\ldots 6)^n$ and $(\ldots 8)^{4n}$ is 6 for all values of n.

$\Rightarrow$ $(4n)^{th}$ power of an even number (other than numbers ending in 0) always has 6 at unit's place.

Examples :

1. Find unit digit in 25^{1136}.

(*a*) 0 (*b*) 1 (*c*) 5 (*d*) 6

Sol. (*c*) We know that a number with 5 at unit place raised to any power always ends in 5.

$\therefore$ Unit digit in 25^{1136} is 5.

2. Find unit digit in 34^{325}.

(*a*) 1 (*b*) 4 (*c*) 6 (*d*) 8

Sol. (*b*) On dividing 325 or 25 by 4, remainder is 1.

$\therefore$ 34^{325} have the same unit digit as is the unit digit in 4^1 *i.e.* 4.

POWERS OF 2

$2^1 = 2$ $2^2 = 4$ $2^3 = 8$ $2^4 = 16$ $2^5 = 32$ $2^6 = 64$

$2^7 = 128$ $2^8 = 256$ $2^9 = 512$ $2^{10} = 1024$

POWERS OF 3

$3^1 = 3$ $3^2 = 9$ $3^3 = 27$ $3^4 = 81$ $3^5 = 243$ $3^6 = 729$

POWERS OF 5

$5^1 = 5$ $5^2 = 25$ $5^3 = 125$ $5^4 = 625$ $5^5 = 3125$

ROOTS

Root is inverse of Power. If $x^n = y$. Then $x = \sqrt[n]{y} = y^{\frac{1}{n}}$; $y^{\frac{1}{n}}$ is called the surd of nth order.

Example : $2^3 = 8 \Rightarrow \sqrt[3]{8} = 2$ $3^4 = 81 \Rightarrow \sqrt[4]{81} = 3$

COMMON SQUARE ROOTS

$\sqrt{2} = 1.4142$ $\sqrt{3} = 1.7321$ $\sqrt{5} = 2.2361$

RADICALS

1. $x\sqrt{a}+y\sqrt{a}=(x+y)\sqrt{a}$ **2.** $x\sqrt{a}-y\sqrt{a}=(x-y)\sqrt{a}$ **3.** $\sqrt[n]{x}\times\sqrt[n]{y}=\sqrt[n]{xy}$

4. $\sqrt{x}\times\sqrt{x}=x$ **Example :** $\sqrt{2}\times\sqrt{2}=2$

5. $\sqrt{x}\times\sqrt{y}=\sqrt{xy}$ **Example :** $\sqrt{2}\times\sqrt{3}=\sqrt{6}$

6. $x\sqrt{z}+y\sqrt{z}=(x+y)\sqrt{z}$ **Example:** $2\sqrt{2}+3\sqrt{2}=(2+3)\sqrt{2}=5\sqrt{2}$

7. $x\sqrt{z}-y\sqrt{z}=(x-y)\sqrt{z}$ **Example:** $7\sqrt{2}-4\sqrt{2}=(7-4)\sqrt{2}=3\sqrt{2}$

8. $\sqrt{x}\div\sqrt{y}=\sqrt{\dfrac{x}{y}}$ **Example :** $\sqrt{100}\div\sqrt{25}=\sqrt{\dfrac{100}{25}}=\sqrt{4}=2$

9. $\dfrac{\sqrt[n]{x}}{\sqrt[n]{y}}=\sqrt[n]{\dfrac{x}{y}}$ **Example :** $\sqrt[3]{64}\div\sqrt[3]{8}=\sqrt[3]{\dfrac{64}{8}}=\sqrt[3]{8}=2$

10. $\left(\sqrt{x}+\sqrt{y}\right)^2=x+y+2\sqrt{xy}$ **Example:** $(\sqrt{5}+\sqrt{3})^2=5+3+2\sqrt{15}=8+2\sqrt{15}$

11. $\left(\sqrt{x}-\sqrt{y}\right)^2=x+y-2\sqrt{xy}$ **Example:** $(\sqrt{5}-\sqrt{3})^2=5+3-2\sqrt{15}=8-2\sqrt{15}$

12. $\left(\sqrt{x}+\sqrt{y}\right)\times\left(\sqrt{x}-\sqrt{y}\right)=x-y$ **Example:** $(\sqrt{5}+\sqrt{3})(\sqrt{5}\times\sqrt{3})=5-3=2$

13. $\dfrac{1}{\sqrt{x}}=\dfrac{1}{\sqrt{x}}\times\dfrac{\sqrt{x}}{\sqrt{x}}=\dfrac{\sqrt{x}}{x}$ **Example :** $\dfrac{1}{\sqrt{3}}=?$

(*a*) 0.577 (*b*) 1.732 (*c*) 0.575 (*d*) 1.734

Sol. (*a*) $\dfrac{1}{\sqrt{3}}=\dfrac{\sqrt{3}}{3}=\dfrac{1.732}{3}=0.577$

14. $\dfrac{1}{\sqrt{x}+\sqrt{y}}=\dfrac{1}{\sqrt{x}+\sqrt{y}}\times\dfrac{\sqrt{x}-\sqrt{y}}{\sqrt{x}-\sqrt{y}}=\dfrac{\sqrt{x}-\sqrt{y}}{x-y}$ **Example :** $\dfrac{1}{\sqrt{6}+\sqrt{5}}=?$

(*a*) 1 (*b*) 11 (*c*) $\sqrt{6}+\sqrt{5}$ (*d*) $\sqrt{6}-\sqrt{5}$

Sol. (*d*) $\dfrac{1}{\sqrt{6}+\sqrt{5}}=\sqrt{6}-\sqrt{5}$

15. $\dfrac{1}{\sqrt{x}-\sqrt{y}}=\dfrac{1}{\sqrt{x}-\sqrt{y}}\times\dfrac{\sqrt{x}+\sqrt{y}}{\sqrt{x}+\sqrt{y}}=\dfrac{\sqrt{x}+\sqrt{y}}{x-y}$ **Example :** $\dfrac{1}{\sqrt{8}-\sqrt{7}}=?$

(*a*) 1 (*b*) 15 (*c*) $\sqrt{8}+\sqrt{7}$ (*d*) $\sqrt{8}-\sqrt{7}$

Sol. (*c*) $\dfrac{1}{\sqrt{8}-\sqrt{7}}=\sqrt{8}+\sqrt{7}$

16. $\dfrac{\sqrt{x}+\sqrt{y}}{\sqrt{x}-\sqrt{y}}=\dfrac{\sqrt{x}+\sqrt{y}}{\sqrt{x}-\sqrt{y}}\times\dfrac{\sqrt{x}+\sqrt{y}}{\sqrt{x}+\sqrt{y}}=\dfrac{\left(\sqrt{x}+\sqrt{y}\right)^2}{x-y}=\dfrac{x+y+2\sqrt{xy}}{x-y}$

Example : $\dfrac{\sqrt{8}+\sqrt{7}}{\sqrt{8}-\sqrt{7}}=?$

(*a*) 1 (*b*) 15 (*c*) $15+2\sqrt{56}$ (*d*) $15-2\sqrt{56}$

Sol. (*c*) $\dfrac{\sqrt{8}+\sqrt{7}}{\sqrt{8}-\sqrt{7}}=\dfrac{\left(\sqrt{8}+\sqrt{7}\right)^2}{8-7}=\dfrac{8+7+2\sqrt{56}}{1}=15+2\sqrt{56}$

17. $\frac{\sqrt{x}-\sqrt{y}}{\sqrt{x}+\sqrt{y}}=\frac{\sqrt{x}-\sqrt{y}}{\sqrt{x}+\sqrt{y}}\times\frac{\sqrt{x}-\sqrt{y}}{\sqrt{x}-\sqrt{y}}=\frac{\left(\sqrt{x}-\sqrt{y}\right)^2}{x-y}=\frac{x+y-2\sqrt{xy}}{x-y}$

Example : $\frac{\sqrt{6}-\sqrt{5}}{\sqrt{6}+\sqrt{5}}=?$

(*a*) 1 (*b*) $\frac{1}{11}$ (*c*) $11+2\sqrt{30}$ (*d*) $11-2\sqrt{30}$

Sol. (*d*) $\frac{\sqrt{6}-\sqrt{5}}{\sqrt{6}+\sqrt{5}}=\frac{\left(\sqrt{6}-\sqrt{5}\right)^2}{6-5}=\frac{6+5-2\sqrt{30}}{1}=11-2\sqrt{30}$

18. $\left(\sqrt{a}+\sqrt{b}\right)\left(\sqrt{c}-\sqrt{d}\right)+\left(\sqrt{a}-\sqrt{b}\right)\left(\sqrt{c}+\sqrt{d}\right)=2\left(\sqrt{ac}-\sqrt{bd}\right)$

19. (*a*) $\sqrt{n+\sqrt{n+\sqrt{n+...\infty}}}=?$

Sol. Let $x=\sqrt{n+\sqrt{n+\sqrt{n+...\infty}}}\Rightarrow x^2=n+\sqrt{n+\sqrt{n+\sqrt{n+...\infty}}}$

$\Rightarrow x^2=n+x\Rightarrow x^2-x=n\Rightarrow x(x-1)=n$

$\Rightarrow$ We express n as product of two consecutive natural numbers, Bigger number is the answer.

Example : $\sqrt{6+\sqrt{6+\sqrt{6+...\infty}}}=?$ **(SSC CGL Tier I 2000, CDS 1999)**

Sol. $6=2\times 3\Rightarrow$ G . E. = 3 (Bigger number)

(*b*) $\sqrt{n-\sqrt{n-\sqrt{n-...\infty}}}=?$

Sol. Let $x=\sqrt{n-\sqrt{n-\sqrt{n-...\infty}}}\Rightarrow x^2=n-\sqrt{n-\sqrt{n-\sqrt{n-...\infty}}}$

$\Rightarrow x^2=n-x\Rightarrow x^2+x=n\Rightarrow x\,(x+1)=n$

$\Rightarrow$ We express n as product of two consecutive natural numbers, the smaller number is the answer.

Example : $\sqrt{6-\sqrt{6-\sqrt{6-...\infty}}}=?$

Sol. $6=2\times 3\Rightarrow$ G . E. = 2 (smaller number)

Important : If $x=n\times(n+1)$, then

$\sqrt{x-\sqrt{x-\sqrt{x-\sqrt{x-........\infty}}}}=n$ and $\sqrt{x+\sqrt{x+\sqrt{x+\sqrt{x+......\infty}}}}=n+1$

SOLVED EXERCISE

1. $\sqrt[4]{500}$ lies between :

(*a*) 2 and 3 (*b*) 3 and 4 (*c*) 4 and 5 (*d*) 5 and 6

Sol. (*c*) $2^4=16, 3^4=81, 4^4=256, 5^4=625\Rightarrow\sqrt[4]{500}$ lies between 4 and 5.

2. Square root of 5^{2^4} is **(SSC FCI Asstt. 2012)**

(*a*) 5^{2^2} (*b*) 5^{2^3} (*c*) 5^2 (*d*) 5^3

Sol. (*a*) $\sqrt{5^{2^4}} = 5^{2^2}$

3. $9^3 \times 27^2 = ?$

(*a*) 3^{12} (*b*) 243^5 (*c*) 9^5 (*d*) 9^6

Sol. (*a*) $9^3 \times 27^2 = (3^2)^3 \times (3^3)^2 = 3^6 \times 3^6 = 3^{6+6} = 3^{12}$

4. $21^? \times 21^{6.5} = 21^{12.4}$ **(UCO Bank PO 2009)**

(*a*) 18.9 (*b*) 4.4 (*c*) 6.9 (*d*) None of these

Sol. (*d*) $21^? = 21^{12.4} \div 21^{6.5} \Rightarrow ? = 12.4 - 6.5 = 5.9$

5. $3^8 \times 3^4 = ?$ **(Canara Bank PO 2009)**

(*a*) 27^3 (*b*) 729^2 (*c*) 27^5 (*d*) 729^3

Sol. (*b*) $3^8 \times 3^4 = 3^{8+4} = 3^{12} = (3^6)^2 = (729)^2$

6. $(4)^{0.5} \times (0.5)^4$ is equal to : **(SSC CGL Tier I 2000)**

(*a*) 1 (*b*) 4 (*c*) $\frac{1}{8}$ (*d*) $\frac{1}{32}$

Sol. (*c*) $(4)^{0.5} \times (0.5)^4 = 2 \times \left(\frac{1}{2}\right)^4 = \frac{1}{8}$

7. $(16^{0.16} \times 2^{0.36})$ is equal to **(SSC CGL Tier I 2000)**

(*a*) 2 (*b*) 16 (*c*) 32 (*d*) 64

Sol. (*a*) $(16^{0.16} \times 2^{0.36}) = (2^4)^{0.16} \times 2^{0.36} = 2^{0.64+0.36} = 2^1 = 2$

8. $16^{0.16} \times 16^{0.04} \times 2^{0.2}$ is equal to : **(SSC CGL Tier I 2005)**

(*a*) 1 (*b*) 2 (*c*) 4 (*d*) 16

Sol. (*b*) G.E. $= (16)^{0.16+0.04} \times 2^{0.2} = (2^4)^{0.2} \times 2^{0.2} = (2)^{0.8} \times 2^{0.2} = 2^1 = 2$

9. The value of $(256)^{0.16} \times (16)^{0.18}$ is : **(SSC CGL Tier I 2002)**

(*a*) 4 (*b*) – 4 (*c*) 16 (*d*) 256

Sol. (*a*) G.E. $= (4^4)^{0.16} \times (4^2)^{0.18} = (4)^{0.64} \times 4^{0.36} = (4)^{0.64+0.36} = 4^1 = 4$

10. $1000^{20} \div 10^{15} = ?$ **(Bank PO Exam.)**

(*a*) 10^5 (*b*) 10^{35} (*c*) 10^{65} (*d*) None of these

Sol. (*d*) $1000^{20} \div 10^{15} = (10^3)^{20} \div 10^{15} = 10^{60} \div 10^{15} = 10^{60-15} = 10^{45}$

11. $8^{0.4} \times 4^{1.6} \times 2^{1.6} = ?$ **(SBI Bank PO 2008)**

(*a*) 52 (*b*) 48 (*c*) 64 (*d*) 76

Sol. (*c*) $(2^3)^{0.4} \times (2^2)^{1.6} \times (2)^{1.6} = 2^{1.2} \times 2^{3.2} \times 2^{1.6} = 2^{1.2+3.2+1.6} = 2^6 = 64$

12. $6^7 \times 3^7 \times 2^7 = 6^?$ **(Bank PO Exam.)**

(*a*) 7 (*b*) 6^7 (*c*) 14 (*d*) 6^{14}

Sol. (*c*) $6^7 \times 3^7 \times 2^7 = 6^7 \times (3 \times 2)^7 = 6^7 \times 6^7 = 6^{14} \Rightarrow ? = 14$

13. $\dfrac{4^2 \times 9^4}{8 \times 27^2} = ?$

(*a*) $\frac{1}{6}$ (*b*) 1 (*c*) 6 (*d*) 18

Sol. (*d*) $\dfrac{4^2 \times 9^4}{8 \times 27^2} = \dfrac{(2^2)^2 \times (3^2)^4}{(2^3) \times (3^3)^2} = \dfrac{2^4 \times 3^8}{2^3 \times 3^6} = (2^{4-3}) \times (3^{8-6}) = (2^1) \times (3^2) = 2 \times 9 = 18$

14. $[12^2 \times 14^2] \div 16^2 = ?$ **(SBI Bank PO 2008)**

(*a*) 282.24 (*b*) 1764 (*c*) 110.25 (*d*) 104.25

Sol. (*c*) $? = \dfrac{12 \times 12 \times 14 \times 14}{16 \times 16} = \dfrac{441}{4} = 110.25$

15. $81^{0.25} \times 9^{0.5} \times 27^{1.5} \div 243^{0.5} = ?$ **(SSC CGL Tier I 2006)**

(*a*) 81 (*b*) 83 (*c*) 85 (*d*) 87

Sol. (*a*) Given expression $= (3^4)^{0.25} \times (3^2)^{0.5} \times (3^3)^{1.5} \div (3^5)^{0.5} = 3^{1+1+4.5-2.5} = 3^4 = 81$

16. $(a^2)^? = a^6$ **(Bank PO 2005)**

(*a*) 6 (*b*) 5 (*c*) 3 (*d*) None of these

Sol. (*c*) $2 \times ? = 6 \Rightarrow ? = 6 \div 2 = 3$

17. $c^0 \times b = ?$ **(Bank PO 2005)**

(*a*) 1 (*b*) 2 (*c*) *cb* (*d*) None of these

Sol. (*d*) $c^0 \times b = 1 \times b = b$

18. $a^b \times b^c \times ? = 0$ **(Bank PO 2005)**

(*a*) 0 (*b*) 1 (*c*) a^c (*d*) c^a

Sol. (*a*) $x \times 0 = 0$

19. $(a^b)^? = 1$ **(SBI Bank PO 2005)**

(*a*) 1 (*b*) 0 (*c*) *a* (*d*) *b*

Sol. (*b*) $(a^b)^? = 1 = a^0 = (a^b)^0$

20. The largest among the numbers 0.9, $(0.9)^2$, $\sqrt{0.9}$, $0.\overline{9}$ **(SSC DEO 2010)**

(*a*) 0.9 (*b*) $(0.9)^2$ (*c*) $\sqrt{0.9}$ (*d*) $0.\overline{9}$

Sol. (*d*) $0.\overline{9} = \dfrac{9}{9} = 1$ whereas value of all other numbers is less than 1.

21. Which one of the following will completely divide $5^{71} + 5^{72} + 5^{73}$? **(SSC CGL 2011)**

(*a*) 150 (*b*) 160 (*c*) 155 (*d*) 30

Sol. (*c*) $5^{71} + 5^{72} + 5^{73} = 5^{71}(1 + 5 + 25) = 5^{71} \times 31 = 5^{70} \times 5 \times 31 = 5^{70} \times 155$

22. The expression $(4^{61} + 4^{62} + 4^{63} + 4^{64})$ is divisible by **(SSC CGL 1996)**

(*a*) 3 (*b*) 11 (*c*) 13 (*d*) 17

Sol. (d) $4^{61} \times (1 + 4 + 16 + 64) = 461 \times 85$ is divisible by 17.

23. The unit digit in $(264)^{102} + (264)^{103}$ is **(SSC CGL 1996)**

(*a*) 0 (*b*) 4 (*c*) 5 (*d*) 8

Sol. (*a*) $(264)^{102} + (264)^{103} = (264)^{102} \times (1 + 264) = (264)^{102} \times (265)$

= Even number × Multiple of 5 ⇒ The given expression will end in 0.

24. Find the largest natural number for which $n^{200} < 5^{300}$

(*a*) 9 (*b*) 10 (*c*) 11 (*d*) 12

Sol. (*c*) $n^{200} < 5^{300} \Rightarrow (n^2)^{100} < (5^3)^{100} \Rightarrow n^2 < 5^3 \Rightarrow n^2 < 125 \Rightarrow n \le 11$

25. Which one of the followings is greatest : **(SSC CGL Tier I 2010)**

(*a*) 2^{60} (*b*) 3^{48} (*c*) 4^{36} (*d*) 5^{24}

Sol. (*b*) HCF of powers = HCF of 60, 48, 36 and 24 = 12

$2^{60} = (2^5)^{12} = 32^{12}$ $3^{48} = (3^4)^{12} = 81^{12}$

$4^{36} = (4^3)^{12} = 64^{12}$ $5^{24} = (5^2)^{12} = 25^{12}$

⇒ 81^{12} or 3^{48} is the greatest of the given numbers.

26. Arrange the following in ascending order : $625^6, 125^7, 25^{10}$

(*a*) $625^6 < 25^{10} < 125^7$ (*b*) $625^6 > 125^7 > 25^{10}$

(*c*) $625^6 > 125^7 < 25^{10}$ (*d*) None of these

Sol. (*b*) $625^6 = (5^4)^6 = 5^{24}$; $125^7 = (5^3)^7 = 5^{21}$; $25^{10} = (5^2)^{10} = 5^{20}$;

$\therefore 5^{24} > 5^{21} > 5^{20} \Rightarrow 625^6 > 125^7 > 25^{10}$

27. If $(125)^x = 3125$, then $x = ?$ **(CBI 1995, SSC CGL Tier I 2008)**

(*a*) $\frac{4}{3}$ (*b*) $\frac{5}{3}$ (*c*) $\frac{1}{5}$ (*d*) 5

Sol. (*b*) $(125)^x = 3125 \Rightarrow (5^3)^x = 5^5 \Rightarrow 3x = 5 \Rightarrow x = \frac{5}{3}$

28. If $3^x \times 27^x = 9^{x+4}$, then what is x equal to ? **(CDS 2011)**

(*a*) 4 (*b*) 5 (*c*) 6 (*d*) 7

Sol. (*a*) $3^x \times 3^{3x} = 3^{2(x+4)} \Rightarrow x + 3x = 2(x+4) \Rightarrow 4x = 2x + 8 \Rightarrow 2x = 8 \Rightarrow x = 4$

29. If $27^{2x-1} = 243^3$, then the value of x is : **(SSC CGL Tier I 1999)**

(*a*) 3 (*b*) 6 (*c*) 7 (*d*) 9

Sol. (*a*) $27^{2x-1} = 243^3 \Rightarrow (3^3)^{2x-1} = (3^5)^3 \Rightarrow 3(2x-1) = 15 \Rightarrow 2x - 1 = 5 \Rightarrow x = 3$

30. If $(2^x)(2^y) = 8$ and $(9^x)(3^y) = 81$, then (x, y) is : **(SSC FCI 2012)**

(*a*) (1, 2) (*b*) (2, 1) (*c*) (1, 1) (*d*) (2, 2)

Sol. (*a*) $2^{x+y} = 8 = 2^3 \Rightarrow x + y = 3 \Rightarrow (3^{2x})(3^y) = 81 = 3^4 \Rightarrow 2x + y = 4 \Rightarrow (x, y) = (1, 2)$

31. If $2^m + 2^{1+m} = 24$, then what is the value of m ? **(CDS 2010)**

(*a*) 0 (*b*) 1/3 (*c*) 3 (*d*) 6

Sol. (*c*) $2^m + 2^{1+m} = 24 \Rightarrow 2^m(1+2) = 24 \Rightarrow 2^m = 24 \div 3 = 8 = 2^3$

32. Solve for x : $3^x - 3^{x-1} = 486$ **(SSC CGL Tier I 2014)**

(*a*) 7 (*b*) 9 (*c*) 5 (*d*) 6

Sol. (*d*) $3^{x-1}(3-1) = 486 \Rightarrow 3^{x-1} = 243 = 3^5 \Rightarrow x - 1 = 5 \Rightarrow x = 6$

33. If $3^x - 3^{x-1} = 18$, then the value of x^x is **(SSC FCI 2012)**

(*a*) 3 (*b*) 8 (*c*) 27 (*d*) 216

Sol. (*c*) $3^{x-1}(3-1) = 18 \Rightarrow 3^{x-1} = 18 \div 2 = 9 = 3^2 \Rightarrow x - 1 = 2 \Rightarrow x = 3 \Rightarrow x^x = 3^3 = 27$

34. If $\left(\frac{a}{b}\right)^{2x-1} = \left(\frac{b}{a}\right)^{x-8}$, find the value of x. **(MBA Entrance Exam.)**

(*a*) –7 (*b*) –3 (*c*) 3 (*d*) 7

Sol. (*c*) $\left(\frac{a}{b}\right)^{2x-1} = \left(\frac{b}{a}\right)^{x-8} = \left(\frac{a}{b}\right)^{8-x} \Rightarrow 2x - 1 = 8 - x \Rightarrow 3x = 9 \Rightarrow x = 3$

35. If $\left(\frac{3}{4}\right)^3\left(\frac{4}{3}\right)^{-7} = \left(\frac{3}{4}\right)^{2x}$, then x is : **(SSC CGL Tier I 2013)**

(*a*) 2½ (*b*) – 2 (*c*) 2 (*d*) 5

Sol. (*d*) $\left(\frac{3}{4}\right)^3\left(\frac{4}{3}\right)^{-7} = \left(\frac{3}{4}\right)^{2x} \Rightarrow \left(\frac{3}{4}\right)^3\left(\frac{3}{4}\right)^7 = \left(\frac{3}{4}\right)^{2x} \Rightarrow 2x = 3 + 7 \Rightarrow x = 5$

36. If $5^{2x} - 5^{x+3} + 125 = 5^x$, find the value(s) of x.

(*a*) 0, 1 (*b*) 0, 2 (*c*) 0, 3 (*d*) 0, – 3

Sol. (*c*) G.E. $\Rightarrow 5^x(5^x - 5^3) - 5^x + 5^3 = 0 \Rightarrow 5^x(5^x - 5^3) - 1(5^x - 5^3) = 0$

$\Rightarrow (5^x - 1)(5^x - 5^3) = 0 \Rightarrow 5^x = 1 = 5^0$ or $5^x = 5^3 \Rightarrow x = 0, 3$

37. Find the value of $\dfrac{2^x + 2^{x-1}}{2^{x+1} - 2^x}$

(*a*) 1/2 (*b*) 3/2 (*c*) 1 (*d*) 2

Sol. (*b*) $\dfrac{2^x + 2^{x-1}}{2^{x+1} - 2^x} = \dfrac{2^{x-1}(2^1 + 1)}{2^{x-1}(2^2 - 2^1)} = \dfrac{2+1}{4-2} = \dfrac{3}{2}$

38. The real value of x, that satisfies the equation $\sqrt{4x-9} + \sqrt{4x+9} = 5 + \sqrt{7}$ is **(SSC CGL Tier II 2012)**

(*a*) $\dfrac{3}{\sqrt{7}}$ (*b*) 4 (*c*) $\sqrt{5}$ (*d*) $2\sqrt{3}$

Sol. (*b*) Comparing the two sides of the equation, we get:

$\sqrt{4x+9} = 5$ and $\sqrt{4x-9} = \sqrt{7} \Rightarrow 4x + 9 = 25$ and $4x - 9 = 7$

$\Rightarrow 4x = 16$ and $4x = 16 \Rightarrow x = 16 \div 4 = 4$

39. Which one of the following is the greatest :

(*a*) 0.999 (*b*) $(0.999)^2$ (*c*) $\sqrt{0.999}$ (*d*) $\dfrac{1}{0.999}$

Sol. (*d*) $\dfrac{1}{0.999} > 1$ whereas 0.999, $(0.999)^2$ and $\sqrt{0.999} < 1$

40. $5^{\frac{1}{2}} \times 5^{\frac{1}{4}} \times 5^{\frac{1}{8}} \times 5^{\frac{1}{16}}$ .... = ? **(SSC CGL Tier I 2008)**

(*a*) 0 (*b*) 1 (*c*) 5 (*d*) 25

Sol. (*c*) Given expression = $5^{\frac{1}{2}+\frac{1}{4}+\frac{1}{8}+\frac{1}{16}+\ldots\ldots\ldots}$

Now $\dfrac{1}{2} + \dfrac{1}{4} + \dfrac{1}{8} + \dfrac{1}{16}$ .... is a Geometric Progression with $b = \dfrac{1}{2}, r = \dfrac{1}{2}$

$\Rightarrow$ Sum of G.P. = $\dfrac{b}{1-r} = \dfrac{1/2}{1 - 1/2} = 1 \Rightarrow$ Given expression = $5^1 = 5$

41. If $5^{\sqrt{x}} + 12^{\sqrt{x}} = 13^{\sqrt{x}}$ then x is equal to : **(MBA 1987, SSC CGL Tier I 2008)**

(*a*) 25/4 (*b*) 4 (*c*) 9 (*d*) 16

Sol. (*b*) We know that $5^2 + 12^2 = 13^2 \Rightarrow \sqrt{x} = 2 \Rightarrow x = 2^2 = 4$

42. Find the difference between 2^{2^3} and $(2^2)^3$. **(SSC CHSL)**

(*a*) 0 (*b*) 8 (*c*) 64 (*d*) 192

Sol. (*d*) $2^{2^3} = 2^{(2^3)} = 2^8 = 256$ $(2^2)^3 = 2^{2\times3} = 2^6 = 64$ $\therefore 2^{2^3} - (2^2)^3 = 256 - 64 = 192$

43. $\{(-2)^{(-2)}\}^{(-2)} = ?$ **(SSC CGL Tier I 2005)**

(*a*) 16 (*b*) 8 (*c*) – 8 (*d*) – 16

Sol. (*a*) G.E = $\left\{\dfrac{1}{(-2)^2}\right\}^{(-2)} = \left(\dfrac{1}{4}\right)^{(-2)} = 4^2 = 16$

Alternative method : Given expression = $(-2)^{(-2)(-2)} = (-2)^{(4)} = 16$

44. $(0.04)^{-1.5} = ?$ **(SSC CGL Tier I 2008, 2004)**

(*a*) 0.08 (*b*) 25 (*c*) 125 (*d*) 625

Sol. (*c*) $(0.04)^{-1.5} = (0.04)^{-\frac{3}{2}} = \left(\frac{1}{0.04}\right)^{\frac{3}{2}} = 25^{\frac{3}{2}} = (5^2)^{\frac{3}{2}} = 5^3 = 125$

45. $\frac{1}{(64)^{-\frac{1}{2}}} + \frac{1}{(729)^{-\frac{2}{3}}} + \frac{1}{(256)^{-\frac{3}{4}}} + \frac{1}{243^{-\frac{4}{5}}} = ?$

(*a*) 243 (*b*) 234 (*c*) 256 (*d*) None of these

Sol. (*b*) Given expression $= (64)^{\frac{1}{2}} + (729)^{\frac{2}{3}} + (256)^{\frac{3}{4}} + (243)^{\frac{4}{5}}$

$= (8^2)^{\frac{1}{2}} + (9^3)^{\frac{2}{3}} + (4^4)^{\frac{3}{4}} + (3^5)^{\frac{4}{5}} = 8 + 9^2 + 4^3 + 3^4 = 8 + 81 + 64 + 81 = 234$

46. $\frac{5^{0.25} \times 125^{0.25}}{256^{0.10} \times 256^{0.15}} = ?$ **(SSC CHSL)**

(*a*) $\frac{4}{5}$ (*b*) $\frac{5}{4}$ (*c*) 4 (*d*) 5

Sol. (*b*) $\frac{5^{0.25} \times 125^{0.25}}{256^{0.10} \times 256^{0.15}} = \frac{5^{0.25} \times (5^3)^{0.25}}{256^{0.25}} = \frac{5^{0.25} \times 5^{0.75}}{\sqrt[4]{256}} = \frac{5}{4}$

47. How many prime factors are in $2^8 \times 3^5 \times 5^9$?

(*a*) 10 (*b*) 18 (*c*) 22 (*d*) 25

Sol. (*c*) $2^8 = 2 \times 2 \times 2 \times 2 \times 2 \times 2 \times 2 \times 2 \Rightarrow 2^8$ has 8 prime factors.

Similarly 3^5 and 5^9 have 5 and 9 prime factors respectively.

$\therefore$ Total prime factors $= 8 + 5 + 9 = 22$

48. The number of prime factors in the expression $6^{10} \times 7^{17} \times 11^{27}$ is **(SSC FCI 2012)**

(*a*) 54 (*b*) 64 (*c*) 71 (*d*) 81

Sol. (*b*) $6^{10} \times 7^{17} \times 11^{27} = (2 \times 3)^{10} \times 7^{17} \times 11^{27} = 2^{10} \times 3^{10} \times 7^{17} \times 11^{27}$

$\therefore$ Total prime factors $= 10 + 10 + 17 + 27 = 64$

49. What is the last digit in the expansion of 3^{4798}. **(CDS 2010)**

(*a*) 1 (*b*) 3 (*c*) 7 (*d*) 9

Sol. (*d*) On dividing 4798 or 98 by 4, remainder is 2.

$\therefore$ Unit's digit in 3^{4798} will be same as in 3^2 *i.e.* 9.

50. The digit in one's place of the product $(2153)^{167}$ is : **(SSC CGL Tier I 2004)**

(*a*) 1 (*b*) 3 (*c*) 7 (*d*) 9

Sol. (*c*) On dividing 167 or 67 by 4, remainder is 3.

$\therefore$ 2153^{167} has the same unit digit as is the unit digit in 3^3 *i.e.* 7.

51. The last digit in the expansion of 17^{256} is : **(CDS 2015)**

(*a*) 1 (*b*) 3 (*c*) 7 (*d*) 9

Sol. (*a*) On dividing 256 or 56 by 4, remainder is 0.

$\therefore$ 17^{256} has the same unit digit as is the unit digit in 7^4 *i.e.* 1.

52. What is the last digit in the expansion of $(2457)^{754}$? **(CDS 2009)**

(*a*) 3 (*b*) 7 (*c*) 8 (*d*) 9

Sol. (*d*) On dividing 754 or 54 by 4, remainder is 2.

$\therefore$ 2457^{754} has the same unit digit as is the unit digit in 7^2 *i.e.* 9.

53. Find unit's digit in 42^{148}.

(*a*) 2 (*b*) 4 (*c*) 6 (*d*) 8

Sol. (*c*) 148 or 48 is fully divisible by 4.

$4n^{th}$ power of an even number always has 6 at unit's place. ⇒ Required unit's digit = 6.

54. What is the remainder when $(27^{10} + 7^{51})$ is divided by 10 :

(Special Class Railway Apprentics 2009)

(*a*) 1 (*b*) 2 (*c*) 3 (*d*) 5

Sol. (*b*) Unit digit in $(27^{10} + 7^{51})$ = Unit digit in $(7^2 + 7^3) = 9 + 3 = 2$

∴ Required remainder = Unit digit of given expression = 2

55. When 2^{33} is divided by 10, the remainder will be **(SSC CGLTier I 2011)**

(*a*) 2 (*b*) 3 (*c*) 4 (*d*) 8

Sol. (*a*) Remainder on dividing a number by 10 will be same as unit digit of the dividend.

$2^{33} = (2^4)^8 \times 2^1 \Rightarrow$ unit digit $= 2^1 = 2$

56. What will be the remainder on dividing $(72)^{795}$ by 5?

(*a*) 3 (*b*) 7 (*c*) 8 (*d*) 9

Sol. (*a*) On dividing 795 or 95 by 4, remainder is 3.

∴ 72^{795} has the same unit digit as is the unit digit in 2^3 *i.e.* 8.

Required remainder is = 8 ÷ 5 = 3 (remainder)

Note: Unit's digit method can be used to find remainder on dividing by numbers 2, 5 and 10 only i.e. the factors of 10.

57. Find unit digit in $35^{15} \times 19^{29} \times 21^{25}$.

(*a*) 0 (*b*) 1 (*c*) 5 (*d*) 9

Sol. (*c*) Numbers ending in 5 and 1 raised to any power always have 5 and 1 respectively at the unit place.

19^{29} will have same unit digit as in 19^1 *i.e.* 9.

∴ Unit digit in the given expression is 5 × 9 × 1 = 5.

Trick : Here one number is multiple of 5 and all other numbers are odd numbers. Any odd multiple of 5 always ends in 5.

58. The unit digit in $3^{65} \times 6^{53} \times 7^{51}$ is : **(LIC Exam.)**

(*a*) 3 (*b*) 4 (*c*) 6 (*d*) 7

Sol. (*b*) The unit digit in $3^{65} \times 6^{53} \times 7^{51}$ is same as unit digit in $3^1 \times 6^1 \times 7^3 = 3 \times 6 \times 3 = 4$

59. Find unit digit in $31^{197} + 25^{303} + 50^{55} - 46^{67}$

(*a*) 0 (*b*) 1 (*c*) 5 (*d*) 6

Sol. (*a*) Numbers ending in 0, 1, 5, 6 raised to any power always have same digit in the unit place.

Unit digit in the given expression is 1 + 5 + 0 – 6 = 0.

60. The unit digit in the sum of $(124)^{372} + (124)^{373}$ is **(SSC CGL Tier I 2011)**

(*a*) 2 (*b*) 4 (*c*) 5 (*d*) 0

Sol. (*d*) $(124)^{372} + (124)^{373} = (124)^{372} \times (1 + 124)$ = Even number × Multiple of 5 = unit digit 0

61. The digit in the units place of the resulting number of the expression $(234)^{100} + (234)^{101}$ is

(CDS 2015)

(*a*) 6 (*b*) 4 (*c*) 2 (*d*) 0

Sol. (*d*) $(234)^{100} + (234)^{101} = (124)^{100} \times (1 + 234)$ = Even number × Multiple of 5 = unit digit 0

62. Find the digit in unit place in $15^{28} + 11^{22} - 9^{27}$. **(MBA Entrance Exam.)**

(*a*) 2 (*b*) 3 (*c*) 7 (*d*) 9

Sol. (*c*) Required unit's digit = $5 + 1 - 9^3 = 5 + 1 - 9 = -3$.
But a digit can't have negative value. ⇒ Unit digit in the given expression is $10 - 3 = 7$.

63. Find the unit's digit of $(11353)^{178!}$ **(MBA)**

(*a*) 3 (*b*) 9 (*c*) 1 (*d*) 8

Sol. (*c*) 178! is multiple of 4 and therefore can be written as $4n$. Unit's digit in $3^4 = 1$

64. Find the last two digits of $101 \times 102 \times 103 \times 197 \times 198 \times 199$. **(MBA)**

(*a*) 36 (*b*) 24 (*c*) 32 (*d*) 64

Sol. (*d*) Remainders on dividing by $100 = 1 \times 2 \times 3 \times (-3) \times (-2) \times (-1) = -36 = 100 - 36 = 64$

65. What is the remainder when 4^{96} is divided by 6 ? **(CAT 2003, CDS 2015)**

(*a*) 1 (*b*) 2 (*c*) 4 (*d*) 3

Sol. (*c*)

Number ÷ 6	4^1	4^2	4^3	4^4	4^5
Remainder	4	4	4	4	4

We note that the remainder for $4^n \div 6$ is always 4, for all values of n.
Hint: Remainder $(4^3 \div 6)$ = Remainder $(4^2 \div 6)$ × Remainder $(4^1 \div 6)$ = Rem. $(4 \times 4) \div 6 = 4$

66. What is the remainder when 4^{500} is divided by 7 ?

(*a*) 1 (*b*) 2 (*c*) 4 (*d*) 5

Sol. (*b*)

Number ÷ 7	4^1	4^2	4^3	4^4	4^5	4^6
Remainder	4	2	1	4	2	1

We note that the remainder for $4^n \div 7$ has cyclicity of 3, i.e. remainder repeats in the set of 3.
On dividing 500 by 3, Remainder is 2 ⇒ Required remainder = $4^2 \div 7 = 2$

67. The number of digits in $(48^4 \times 5^{12})$ is : **(SSC CGL Tier I 2006)**

(*a*) 18 (*b*) 16 (*c*) 14 (*d*) 12

Sol. (*b*) $48^4 \times 5^{12} = (3 \times 2^4)^4 \times 5^{12} = 3^4 \times 2^{16} \times 5^{12} = 3^4 \times (2 \times 5)^{12} \times 2^4 = 3^4 \times 2^4 \times 10^{12} = 1296 \times 10^{12}$
∴ Total number of digits = 4 + 12 = 16

68. If $a^{1/3} = 11$, then the value of $a^2 - 331a$ is **(SSC CGL Tier I 2011)**

(*a*) 1331331 (*b*) 1331000 (*c*) 1334331 (*d*) 1330030

Sol. (*b*) $a^{1/3} = 11 \Rightarrow a = 11^3 = 1331$

$\therefore a^2 - 331a = a(a - 331) = 1331(1331 - 331) = 1331 \times 1000 = 1331000$

69. The value of x when 5% of $\sqrt{2x}$ is 0.01 will be **(SSC CGL Tier I 2012)**

(*a*) 0.02 (*b*) 0.01 (*c*) 0.05 (*d*) 0.03

Sol. (*a*) $\frac{1}{20} \times \sqrt{2x} = \frac{1}{100} \Rightarrow \sqrt{2x} = \frac{1}{5} \Rightarrow 2x = \frac{1}{25} \Rightarrow x = \frac{1}{50} = 0.02$

70. $\sqrt{3\sqrt{3\sqrt{3 \ldots\ldots\ldots\ldots}}}$ is equal to **(CDS 2010)**

(*a*) $\sqrt{3}$ (*b*) 3 (*c*) $2\sqrt{3}$ (*d*) $3\sqrt{3}$

Sol. (*b*) Let $x = \sqrt{3\sqrt{3\sqrt{3 \ldots\ldots\ldots\ldots}}} \Rightarrow x^2 = 3x \Rightarrow x = 3$

71. $\sqrt{1 + \sqrt{1 + \sqrt{1 + \ldots\ldots\ldots\ldots}}}$ **(SSC CGL Tier I 2002)**

(*a*) Equals 1 (*b*) Lies between 0 and 1

(*c*) Lies between 1 and 2 (*d*) Is greater than 2

Sol. (*c*) Let $x = \sqrt{1+\sqrt{1+\sqrt{1+..........}}}$ On squaring both sides, $x^2 = 1 + x \Rightarrow x^2 - x - 1 = 0$

$$\therefore\ x = \frac{1 \pm \sqrt{1+4}}{2} = \frac{1 \pm \sqrt{5}}{2} = \frac{1 \pm 2.23}{2} = \frac{3.23}{2} = 1.615 \text{ (Taking +ve value)}$$

∴ The value lies between 1 and 2.

72. $\sqrt{2-\sqrt{2-\sqrt{2-\sqrt{2..........}}}} = ?$

(*a*) 0 (*b*) 1 (*c*) 2 (*d*) None of these

Sol. (*b*) $2 = 1 \times 2 \Rightarrow x = 1$ (smaller number)

73. $\sqrt{2+\sqrt{2+\sqrt{2+\sqrt{2..........}}}} = ?$ **(SSC CGL Tier I 2008, 2002)**

(*a*) 1.42 (*b*) 4 (*c*) 2 (*d*) 2.414

Sol. (*c*) $2 = 1 \times 2 \Rightarrow x = 2$ (Bigger number)

74. $\sqrt{12-\sqrt{12-\sqrt{12-\sqrt{12..........}}}} = ?$

(*a*) 3 (*b*) 4 (*c*) 6 (*d*) 12

Sol. (*a*) $12 = 3 \times 4 \Rightarrow x = 3$ (smaller number)

75. $\sqrt{12+\sqrt{12+\sqrt{12+\sqrt{12..........}}}} = ?$ **(CBI 1990, SAS 2010)**

(*a*) 3 (*b*) 4 (*c*) 6 (*d*) 12

Sol. (*b*) $12 = 3 \times 4 \Rightarrow x = 4$ (Bigger number)

76. The value of $\sqrt{20+\sqrt{20+\sqrt{20+....}}}$ is **(CDS 2000)**

(*a*) 4 (*b*) 5 (*c*) 6 (*d*) greater than 6

Sol. (*b*) $20 = 4 \times 5 \Rightarrow x = 5$ (Bigger number)

77. Find the value of $\sqrt{30+\sqrt{30+\sqrt{30+....}}}$ **(SSC CGL Tier II 2013)**

(*a*) 5 (*b*) $3\sqrt{10}$ (*c*) 6 (*d*) 7

Sol. (*c*) $30 = 5 \times 6 \Rightarrow x = 6$ (Bigger number)

78. If the square root of *x* is equal to cube root of *y*, then the relation between *x* and *y* is **(SSC FCI Asstt 2012)**

(*a*) $x^3 = y^2$ (*b*) $x^2 = y^3$ (*c*) $x = y$ (*d*) $x^6 = y^5$

Sol. (*a*) $x^{1/2} = y^{1/3} \Rightarrow (x^{1/2})^6 = (y^{1/3})^6 \Rightarrow x^3 = y^2$

79. The value of $\sqrt{248+\sqrt{52+\sqrt{144}}}$ is : **(SSC CGL Tier I 2004 , 2001)**

(*a*) 14 (*b*) 16 (*c*) 16.6 (*d*) 18.8

Sol. (*b*) $\sqrt{248+\sqrt{52+12}} = \sqrt{248+8} = \sqrt{256} = 16$

80. $\left(\sqrt{8}-\sqrt{4}-\sqrt{2}\right)$ equals : **(SSC CGL Tier I 2000)**

(*a*) $2 - \sqrt{2}$ (*b*) $\sqrt{2} - 2$ (*c*) 2 (*d*) –2

Sol. (*b*) $\left(\sqrt{8}-\sqrt{4}-\sqrt{2}\right) = 2\sqrt{2} - 2 - \sqrt{2} = \sqrt{2}\,(2-1) - 2 = \sqrt{2} - 2$

81. The value of $\frac{\sqrt{0.441}}{\sqrt{0.625}}$ is equal to : **(SSC CGL Tier I 2002)**

(*a*) 0.048 (*b*) 0.84 (*c*) 0.48 (*d*) 0.084

Sol. (*b*) Given expression $= \frac{\sqrt{441}}{\sqrt{625}} = \frac{21}{25} = 0.84$

82. $\sqrt{\frac{48.4}{0.289}}$ is equal to : **(SSC CGL Tier I 2004)**

(*a*) $129\frac{7}{17}$ (*b*) $1\frac{5}{17}$ (*c*) $12\frac{16}{17}$ (*d*) $12\frac{1}{17}$

Sol. (*c*) Given expression $= \sqrt{\frac{48400}{289}} = \frac{220}{17} = 12\frac{16}{17}$

83. If $\sqrt{1+\frac{x}{9}} = \frac{13}{3}$, then the value of x is : **(SSC CGL Tier I 2011)**

(*a*) $\frac{1439}{9}$ (*b*) 160 (*c*) $\frac{1443}{9}$ (*d*) 169

Sol. (*b*) $\sqrt{1+\frac{x}{9}} = \frac{13}{3} \Rightarrow 1+\frac{x}{9} = \frac{169}{9} \Rightarrow \frac{x}{9} = \frac{169}{9} - 1 = \frac{160}{9} \Rightarrow x = 160$

84. $\sqrt{27} + \sqrt{75} = ?$

(*a*) 8 (*b*) $8\sqrt{3}$ (*c*) 16 (*d*) $16\sqrt{3}$

Sol. (*b*) $\sqrt{27} + \sqrt{75} = \sqrt{3\times3\times3} + \sqrt{3\times5\times5} = 3\sqrt{3} + 5\sqrt{3} = (3+5)\sqrt{3} = 8\sqrt{3}$

85. $\sqrt{60} + \sqrt{15} - \sqrt{135} = ?$

(*a*) 0 (*b*) 1 (*c*) $\sqrt{3}$ (*d*) $\sqrt{60}$

Sol. (*a*) $\sqrt{60} + \sqrt{15} - \sqrt{135} = \sqrt{2\times2\times3\times5} + \sqrt{3\times5} - \sqrt{3\times3\times3\times5}$

$= 2\sqrt{15} + \sqrt{15} - 3\sqrt{15} = (2+1-3)\sqrt{15} = 0$

86. The value of $\sqrt{32} - \sqrt{128} + \sqrt{50}$ correct to 3 places of decimals is : **(SSC CGL T-I 2004)**

(*a*) 1.732 (*b*) 1.141 (*c*) 1.414 (*d*) 1.441

Sol. (*c*) $\sqrt{32} - \sqrt{128} + \sqrt{50} = 4\sqrt{2} - 8\sqrt{2} + 5\sqrt{2} = (4-8+5)\sqrt{2} = \sqrt{2} = 1.414$

87. If $\sqrt{6}\times\sqrt{15} = x\sqrt{10}$, then the value of x is **(SSC CGL Tier I 2012)**

(*a*) $\sqrt{3}$ (*b*) $\sqrt{6}$ (*c*) 3 (*d*) ± 3

Sol. (*d*) $\sqrt{6}\times\sqrt{15} = x\sqrt{10} \Rightarrow \sqrt{90} = x\sqrt{10} \Rightarrow x = \sqrt{\frac{90}{10}} = \sqrt{9} = \pm 3$

88. If $6\sqrt{5} + \sqrt{80} = 22.35$, find the value of $5\sqrt{5} + \sqrt{45}$.

(*a*) 22.35 (*b*) 15.645 (*c*) 17.88 (*d*) None of these

Sol. (*c*) $6\sqrt{5} + \sqrt{80} = 6\sqrt{5} + 4\sqrt{5} = 10\sqrt{5} = 22.35 \Rightarrow \sqrt{5} = 2.235$

$5\sqrt{5} + \sqrt{45} = 5\sqrt{5} + 3\sqrt{5} = 8\sqrt{5} = 8\times2.235 = 17.88$

89. $\sqrt{\sqrt[3]{0.004096}}$ is equal to : **(SSC CGL Tier I 2003)**

(*a*) 4 (*b*) 0.4 (*c*) 0.04 (*d*) 0.004

Sol. (*b*) $\sqrt{\sqrt[3]{0.004096}} = \sqrt{\sqrt[3]{2^{12} \times 10^{-6}}} = \sqrt{2^4 \times 10^{-2}} = 2^2 \times 10^{-1} = 4 \times 0.1 = 0.4$

90. The value of $\sqrt[3]{2^4\sqrt{2^{-5}\sqrt{2^6}}}$ is : **(SSC CGL Tier I 2005)**

(*a*) 2 (*b*) $2^{\frac{5}{3}}$ (*c*) 2^5 (*d*) 1

Sol. (*a*) Given expression $= \sqrt[3]{2^4\sqrt{2^{-5} \times 2^3}} = \sqrt[3]{2^4\sqrt{2^{-2}}} = \sqrt[3]{2^4 \times 2^{-1}} = \sqrt[3]{2^3} = 2$

91. If $\dfrac{\sqrt{3+x}+\sqrt{3-x}}{\sqrt{3+x}-\sqrt{3-x}} = 2$, then x is equal to **(SSC CGL Tier I 2010)**

(*a*) $\dfrac{5}{12}$ (*b*) $\dfrac{12}{5}$ (*c*) $\dfrac{5}{7}$ (*d*) $\dfrac{7}{5}$

Sol. (*b*) G.E. $\Rightarrow \sqrt{3+x}+\sqrt{3-x} = 2\sqrt{3+x}-2\sqrt{3-x} \Rightarrow \sqrt{3+x} = 3\sqrt{3-x}$

$\Rightarrow 3 + x = 9\,(3 - x) \Rightarrow 3 + x = 27 - 9x \Rightarrow 10x = 24 \Rightarrow x = \dfrac{24}{10} = \dfrac{12}{5}$

92. The square root of $\left(7+3\sqrt{5}\right)\left(7-3\sqrt{5}\right)$ is : **(SSC CGL Tier I 2004)**

(*a*) 4 (*b*) $\sqrt{5}$ (*c*) $3\sqrt{5}$ (*d*) 2

Sol. (*d*) $\left(7+3\sqrt{5}\right)\left(7-3\sqrt{5}\right) = 49 - 45 = 4 \Rightarrow$ Required number $= \sqrt{4} = 2$

93. $\sqrt[3]{64} + 3\sqrt{64} = ?$

(*a*) 28 (*b*) $\sqrt{128}$ (*c*) 48 (*d*) $3\sqrt{128}$

Sol. (*a*) $\sqrt[3]{64} + 3\sqrt{64} = \sqrt[3]{4^3} + 3 \times 8 = 4 + 24 = 28$

94. If $\sqrt{256 \times \sqrt{16 \div x}} = 16$, find the value of x.

(*a*) 4 (*b*) 8 (*c*) 12 (*d*) 16

Sol. (*d*) Given $\sqrt{256 \times \sqrt{16 \div x}} = 16$

Squaring both sides, we get: $256 \times \sqrt{16 \div x} = 256$

$\therefore\ \sqrt{16 \div x} = 1 \Rightarrow 16 \div x = 1$ (Squaring both the sides) $\Rightarrow x = 16$

95. If $48\sqrt{x} + 32\sqrt{x} = 320$, find the value of x. **(Railways Exam.)**

(*a*) 2 (*b*) 4 (*c*) 8 (*d*) 16

Sol. (*d*) $48\sqrt{x} + 32\sqrt{x} = 80\sqrt{x} = 320 \Rightarrow \sqrt{x} = 320 \div 80 = 4 \Rightarrow x = 4^2 = 16$

96. $\sqrt[3]{333 + \sqrt[3]{987 + \sqrt[3]{2197}}} = ?$ **(SSC CGL Tier I 2007)**

(*a*) 21 (*b*) 18 (*c*) 7 (*d*) 3

Sol. (*c*) Given expression $= \sqrt[3]{333 + \sqrt[3]{987 + 13}} = \sqrt[3]{333 + 10} = \sqrt[3]{343} = 7$

97. $\sqrt[3]{(4.032)^2-(3.968)^2}=?$ **(SSC CGL Tier I 2007)**

(*a*) 1 (*b*) 1.6 (*c*) 0.08 (*d*) $\frac{4}{5}$

Sol. (*d*) Given expression $=\sqrt[3]{(4.032+3.968)\times(4.032-3.968)}=\sqrt[3]{8\times 0.064}$

$$=\sqrt[3]{2^3\times(0.4)^3}=2\times 0.4=0.8=\frac{4}{5}$$

98. Given that $\sqrt{3}=1.732$, then $\sqrt{147}-\frac{1}{4}\sqrt{48}-\sqrt{75}$ is equal to : **(SSC CGL Tier I 2006, 2004M)**

(*a*) 5.196 (*b*) 3.464 (*c*) 1.732 (*d*) 0.866

Sol. (*c*) G.E. $=7\sqrt{3}-\frac{1}{4}\times 4\sqrt{3}-5\sqrt{3}=7\sqrt{3}-\sqrt{3}-5\sqrt{3}=(7-1-5)\sqrt{3}=\sqrt{3}=1.732$

99. If $\sqrt{4096}=64$, then the value of $\sqrt{40.96}+\sqrt{0.4096}+\sqrt{0.004096}+\sqrt{0.00004096}$ up to two places of decimals is : **(SSC CGL Tier I 2005 & 2001)**

(*a*) 7.09 (*b*) 7.10 (*c*) 7.11 (*d*) 7.12

Sol. (*c*) Given expression = 6.4 + 0.64 + 0.064 + 0.0064 = 7.1104 = 7.11 (approx)

100. If cube root of 175616 is 56, then the value of $\sqrt[3]{175.616}+\sqrt[3]{0.175616}+\sqrt[3]{0.000175616}$ is equat to : **(SSC CGL Tier I 2002, 2015 M)**

(*a*) 0.168 (*b*) 62.16 (*c*) 6.216 (*d*) 6.116

Sol. (*c*) Given expression = 5.6 + 0.56 + 0.056 = 6.216

101. What is the value of $\sqrt{29.16}+\sqrt{0.2916}+\sqrt{0.002916}+\sqrt{0.00002916}$ **(CDS 2010)**

(*a*) 5.9949 (*b*) 5.9894 (*c*) 5.9984 (*d*) 5.9994

Sol. (*d*) $\sqrt{2916}=54\Rightarrow$ Given expression = 5.4 + 0.54 + 0.054 + 0.0054 = 5.9994

102. Given $\sqrt{13}=3.605$ and $\sqrt{130}=11.40$, then the value of $\sqrt{1.3}+\sqrt{1300}+\sqrt{0.013}$ is equal to: **(SSC CGL Tier I 1999)**

(*a*) 36.164 (*b*) 37.304 (*c*) 36.304 (*d*) 37.164

Sol. (*b*) Given expression $=\sqrt{\frac{130}{100}}+\sqrt{13\times 100}+\sqrt{\frac{130}{10000}}=\frac{\sqrt{130}}{10}+10\sqrt{13}+\frac{\sqrt{130}}{100}$

$$=1.14+36.05+0.114=37.304$$

103. If a and b are two distinct natural numbers, which one of the following is true? **(SSC CPO SI 2007)**

(*a*) $\sqrt{a+b}<\sqrt{a}+\sqrt{b}$ (*b*) $\sqrt{a+b}>\sqrt{a}+\sqrt{b}$

(*c*) $\sqrt{a+b}=\sqrt{a}+\sqrt{b}$ (*d*) None of these

Sol. (*a*) $(\sqrt{a}+\sqrt{b})^2=a+b+2\sqrt{ab}$ $(\sqrt{a+b})^2=a+b$

Clearly, $(\sqrt{a+b})^2<(\sqrt{a}+\sqrt{b})^2\Rightarrow\sqrt{a+b}<\sqrt{a}+\sqrt{b}$

104. If $\sqrt{6}=2.450$, then the value of $\frac{3\times\sqrt{2}}{2\times\sqrt{3}}=?$ **(NDA 1994)**

(*a*) 1.225 (*b*) 0.817 (*c*) 0.408 (*d*) 0.490

Sol. (*a*) $\frac{3\times\sqrt{2}}{2\times\sqrt{3}}=\frac{3}{\sqrt{3}}\times\frac{\sqrt{2}}{2}=\sqrt{3}\times\frac{\sqrt{2}}{2}=\frac{\sqrt{6}}{2}=\frac{2.450}{2}=1.225$

105. If $\sqrt{15} = 3.88$, then what is the value of $\sqrt{\frac{5}{3}}$ **(SSC CGL Tier I 2003)**

(*a*) $1.29\overline{3}$ (*b*) 1.2934 (*c*) 1.29 (*d*) 1.295

Sol. (*a*) $\sqrt{\frac{5}{3}} = \sqrt{\frac{5\times3}{3\times3}} = \frac{\sqrt{15}}{3} = \frac{3.88}{3} = 1.29\overline{3}$

106. $\frac{\sqrt{72}-\sqrt{18}}{\sqrt{12}} = ?$ **(SSC CGL Tier I 2000)**

(*a*) $\sqrt{6}$ (*b*) $\frac{\sqrt{3}}{2}$ (*c*) $\frac{\sqrt{2}}{3}$ (*d*) $\frac{\sqrt{6}}{2}$

Sol. (*d*) $\frac{\sqrt{72}-\sqrt{18}}{\sqrt{12}} = \frac{6\sqrt{2}-3\sqrt{2}}{2\sqrt{3}} = \frac{3\sqrt{2}}{2\sqrt{3}} = \sqrt{3}\times\frac{\sqrt{2}}{2} = \frac{\sqrt{6}}{2}$

107. $\frac{\sqrt{80}-\sqrt{112}}{\sqrt{45}-\sqrt{63}} = ?$ **(SSC CGL Tier I 2000)**

(*a*) $\frac{3}{4}$ (*b*) $1\frac{3}{4}$ (*c*) $1\frac{1}{3}$ (*d*) $1\frac{7}{9}$

Sol. (*c*) $\frac{\sqrt{80}-\sqrt{112}}{\sqrt{45}-\sqrt{63}} = \frac{4\sqrt{5}-4\sqrt{7}}{3\sqrt{5}-3\sqrt{7}} = \frac{4(\sqrt{5}-\sqrt{7})}{3(\sqrt{5}-\sqrt{7})} = \frac{4}{3} = 1\frac{1}{3}$

108. $\frac{2+\sqrt{6}}{\sqrt{2}+\sqrt{3}} = ?$

(*a*) $\sqrt{2}$ (*b*) $\sqrt{3}$ (*c*) 2 (*d*) 3

Sol. (*a*) $\frac{2+\sqrt{6}}{\sqrt{2}+\sqrt{3}} = \frac{\sqrt{2\times2}+\sqrt{2\times3}}{\sqrt{2}+\sqrt{3}} = \frac{\sqrt{2}\left(\sqrt{2}+\sqrt{3}\right)}{\sqrt{2}+\sqrt{3}} = \sqrt{2}$

109. $\frac{1}{\sqrt{7}-\sqrt{5}} = ?$

(*a*) 2 (*b*) 12 (*c*) $\frac{\sqrt{7}+\sqrt{5}}{2}$ (*d*) $\frac{\sqrt{7}-\sqrt{5}}{2}$

Sol. (*c*) $\frac{1}{\sqrt{7}-\sqrt{5}} = \frac{\sqrt{7}+\sqrt{5}}{7-5} = \frac{\sqrt{7}+\sqrt{5}}{2}$

110. If $x = \frac{1}{\sqrt{2}+1}$ then $(x+1)$ equals to **(SSC CGL Tier I 2015)**

(*a*) 2 (*b*) $\sqrt{2}+1$ (*c*) $\sqrt{2}-1$ (*d*) $\sqrt{2}$

Sol. (*d*) $x = \frac{1}{\sqrt{2}+1} = \sqrt{2}-1 \Rightarrow (x+1) = \sqrt{2}-1+1 = \sqrt{2}$

111. Find the value of $\frac{1}{\sqrt{9}-\sqrt{8}} - \frac{1}{\sqrt{8}-\sqrt{7}} + \frac{1}{\sqrt{7}-\sqrt{6}} - \frac{1}{\sqrt{6}-\sqrt{5}} + \frac{1}{\sqrt{5}-\sqrt{4}}$

(SSC CGL 2008, CDS 2009 & 2010, SSC CPO SI 2009, SSC MTS 2011, CHSL 2011)

(*a*) 0 (*b*) 1 (*c*) 5 (*d*) 6

Sol. (*c*) Given expression $= \left(\sqrt{9}+\sqrt{8}\right) - \left(\sqrt{8}+\sqrt{7}\right) + \left(\sqrt{7}+\sqrt{6}\right) - \left(\sqrt{6}+\sqrt{5}\right) + \left(\sqrt{5}+\sqrt{4}\right)$

$= \sqrt{9}+\sqrt{8}-\sqrt{8}-\sqrt{7}+\sqrt{7}+\sqrt{6}-\sqrt{6}-\sqrt{5}+\sqrt{5}+\sqrt{4} = \sqrt{9}+\sqrt{4} = 3+2 = 5$

112. If $x = \frac{1}{2+\sqrt{3}}$ and $y = \frac{1}{2-\sqrt{3}}$, then the value of $\left(\frac{1}{x+1}+\frac{1}{y+1}\right)$ is **(SAS 2010, SSC CGL Tier I 2012)**

(*a*) $\frac{1}{\sqrt{3}}$ (*b*) $\sqrt{3}$ (*c*) 2 (*d*) 1

Sol. (*d*) $\frac{1}{x+1}+\frac{1}{y+1} = \frac{1}{3+\sqrt{3}}+\frac{1}{3-\sqrt{3}} = \frac{3-\sqrt{3}+3+\sqrt{3}}{9-3} = \frac{6}{6} = 1$

113. $\frac{1}{\sqrt{8}+\sqrt{5}} = ?$

(*a*) 3 (*b*) 13 (*c*) $\frac{\sqrt{8}+\sqrt{5}}{3}$ (*d*) $\frac{\sqrt{8}-\sqrt{5}}{3}$

Sol. (*d*) $\frac{1}{\sqrt{8}+\sqrt{5}} = \frac{\sqrt{8}-\sqrt{5}}{8-5} = \frac{\sqrt{8}-\sqrt{5}}{3}$

114. $\frac{2}{\sqrt{5}+\sqrt{3}} = ?$

(*a*) 4 (*b*) $\sqrt{5}-\sqrt{3}$ (*c*) $2\left(\sqrt{5}-\sqrt{3}\right)$ (*d*) $\sqrt{5}+\sqrt{3}$

Sol. (*b*) $\frac{2}{\sqrt{5}+\sqrt{3}} = \frac{2\left(\sqrt{5}-\sqrt{3}\right)}{5-3} = \sqrt{5}-\sqrt{3}$

115. $\frac{\sqrt{5}-\sqrt{3}}{\sqrt{5}+\sqrt{3}} = ?$ **(RRB 2008, CGL 1993 M)**

(*a*) $4-\sqrt{15}$ (*b*) $4+\sqrt{15}$ (*c*) $3+\sqrt{15}$ (*d*) $\sqrt{15}-4$

Sol. (*a*) $\frac{\sqrt{5}-\sqrt{3}}{\sqrt{5}+\sqrt{3}} = \frac{\left(\sqrt{5}-\sqrt{3}\right)^2}{5-3} = \frac{5+3-2\sqrt{15}}{2} = 4-\sqrt{15}$

116. If $x = \frac{\sqrt{5}+\sqrt{3}}{\sqrt{5}-\sqrt{3}}$ and $y = \frac{\sqrt{5}-\sqrt{3}}{\sqrt{5}+\sqrt{3}}$, then $(x + y)$ equals **(SSC CGL Tier I 2005)**

(*a*) 8 (*b*) 16 (*c*) $2\sqrt{15}$ (*d*) $2(\sqrt{5}+\sqrt{3})$

Sol. (*a*) $x = 4+\sqrt{15}$, $y = 4-\sqrt{15} \Rightarrow x+y = 8$

117. Given that $\sqrt{3} = 1.732$, find the value of $\frac{\sqrt{6}+\sqrt{2}}{\sqrt{6}-\sqrt{2}}$. **(SSC CGL Tier I 2007)**

(*a*) 3.713 (*b*) 3.721 (*c*) 3.732 *d*) 3.752

Sol. (*c*) $\frac{\sqrt{6}+\sqrt{2}}{\sqrt{6}-\sqrt{2}} = \frac{\sqrt{6}+\sqrt{2}}{\sqrt{6}-\sqrt{2}} \times \frac{\sqrt{6}+\sqrt{2}}{\sqrt{6}+\sqrt{2}} = \frac{6+2+2\sqrt{12}}{6-2} = \frac{8+4\sqrt{3}}{4}$

$= 2+\sqrt{3} = 2+1.732 = 3.732$

118. If $\sqrt{3} = 1.732$, the value of $\frac{2+\sqrt{3}}{2-\sqrt{3}}$ is **(SSC CHSL 2008)**

(*a*) 11.732 (*b*) 13.928 (*c*) 12.928 (*d*) 13.925

Sol. (*b*) $\frac{2+\sqrt{3}}{2-\sqrt{3}} = \frac{\sqrt{4}+\sqrt{3}}{\sqrt{4}-\sqrt{3}} = \frac{\left(\sqrt{4}+\sqrt{3}\right)^2}{4-3} = \frac{4+3+4\sqrt{3}}{1} = 7+4\sqrt{3}$

$= 7 + (4 \times 1.732) = 7 + 6.928 = 13.928$

119. If $x = 7 - 4\sqrt{3}$, then $x + \frac{1}{x} = ?$ **(SSC CGL Tier I 2000)**

(*a*) $3\sqrt{3}$ (*b*) $8\sqrt{3}$ (*c*) $14+8\sqrt{3}$ (*d*) 14

Sol. (*d*) $\frac{1}{x} = \frac{1}{7-4\sqrt{3}} = 7+4\sqrt{3}$ (**Hint**: $7 = \sqrt{49}$ and $4\sqrt{3} = \sqrt{48}$)

$\therefore\ x + \frac{1}{x} = (7-4\sqrt{3}) + (7+4\sqrt{3}) = 14$

120. $\sqrt{\frac{\sqrt{12}+\sqrt{3}}{\sqrt{12}-\sqrt{3}}} = ?$

(*a*) $\sqrt{2}$ (*b*) $\sqrt{3}$ (*c*) $\sqrt{6}$ (*d*) 3

Sol. (*b*) $\sqrt{\frac{\sqrt{12}+\sqrt{3}}{\sqrt{12}-\sqrt{3}}} = \sqrt{\frac{(\sqrt{12}+\sqrt{3})^2}{12-3}} = \frac{2\sqrt{3}+\sqrt{3}}{3} = \frac{3\sqrt{3}}{3} = \sqrt{3}$

121. If $\sqrt{2} = 1.4142$, find the value of $\sqrt{\frac{\sqrt{2}-1}{\sqrt{2}+1}}$ **(I Tax 1990)**

(*a*) 2 (*b*) $\sqrt{2}$ (*c*) 1 (*d*) 0.414

Sol. (*d*) $\sqrt{\frac{\sqrt{2}-1}{\sqrt{2}+1}} = \sqrt{\frac{\sqrt{2}-1}{\sqrt{2}+1} \times \frac{\sqrt{2}-1}{\sqrt{2}-1}} = \sqrt{\frac{(\sqrt{2}-1)^2}{2-1}} = \sqrt{2}-1 = 1.414 - 1 = 0.414$

122. The value of $\sqrt{\frac{(0.03)^2+(0.21)^2+(0.065)^2}{(0.003)^2+(0.021)^2+(0.0065)^2}}$ is : **(SSC CGL Tier I 2002)**

(*a*) 0.1 (*b*) 10 (*c*) 10^2 (*d*) 10^3

Sol. (*b*) Given expression $= \sqrt{\frac{10^2 \times \{(0.003)^2+(0.021)^2+(0.0065)^2\}}{(0.003)^2+(0.021)^2+(0.0065)^2}} = \sqrt{10^2} = 10$

123. Given that $\sqrt{2} = 1.4142$, $\frac{4+\sqrt{2}}{\sqrt{2}\,(1+\sqrt{2})} = ?$ **(SSC CGL Tier I 2007)**

(*a*) 2.5858 (*b*) 1.5858 (*c*) 2.4142 (*d*) 1.4142

Sol. (*b*) $\frac{4+\sqrt{2}}{\sqrt{2}\,(1+\sqrt{2})} = \frac{2\sqrt{2}+1}{(\sqrt{2}+1)} = \frac{2\sqrt{2}+1}{(\sqrt{2}+1)} \times \frac{\sqrt{2}-1}{\sqrt{2}-1} = \frac{4-2\sqrt{2}+\sqrt{2}-1}{2-1}$

$= 3 - \sqrt{2} = 3 - 1.4142 = 1.5858$

124. $\frac{3+\sqrt{6}}{\sqrt{75}-\sqrt{48}-\sqrt{32}+\sqrt{50}} = ?$

(*a*) 1 (*b*) $\sqrt{2}$ (*c*) $\sqrt{3}$ (*d*) $\sqrt{5}$

Sol. (*c*) G.E. = $\frac{\sqrt{9}+\sqrt{6}}{5\sqrt{3}-4\sqrt{3}-4\sqrt{2}+5\sqrt{2}} = \frac{\sqrt{3}(\sqrt{3}+\sqrt{2})}{5(\sqrt{3}+\sqrt{2})-4(\sqrt{3}+\sqrt{2})} = \frac{\sqrt{3}(\sqrt{3}+\sqrt{2})}{(5-4)(\sqrt{3}+\sqrt{2})} = \sqrt{3}$

125. $3-\frac{3+\sqrt{5}}{4}-\frac{1}{3+\sqrt{5}}$ is equal to **(SSC CGL Tier I 2012)**

(*a*) $\frac{\sqrt{5}}{2}$ (*b*) $\sqrt{5}$ (*c*) 0 (*d*) $\frac{3}{2}$

Sol. (*d*) G.E. $=3-\frac{3+\sqrt{5}}{4}-\frac{3-\sqrt{5}}{4}=3-\frac{1}{4}\left(3+3+\sqrt{5}-\sqrt{5}\right)=3-\frac{1}{4}(6)=3-\frac{3}{2}=\frac{3}{2}$

126. If $x=1+\sqrt{2}+\sqrt{3}$, then $\left(x+\frac{1}{x-1}\right)=?$ **(SSC FCI 2012)**

(*a*) $1+2\sqrt{3}$ (*b*) $2+\sqrt{3}$ (*c*) $3+\sqrt{2}$ (*d*) $2\sqrt{3}-1$

Sol. (*a*) $\frac{1}{x-1}=\frac{1}{\sqrt{3}+\sqrt{2}}=\sqrt{3}-\sqrt{2}$

$\therefore\ x+\frac{1}{x-1}=1+\sqrt{2}+\sqrt{3}+\sqrt{3}-\sqrt{2}=1+2\sqrt{3}$

127. If $\frac{5+2\sqrt{3}}{7+4\sqrt{3}}$ can be expressed as $a+b\sqrt{3}$, find the value of b.

(*a*) 6 (*b*) – 6 (*c*) $6\sqrt{3}$ (*d*) $-6\sqrt{3}$

Sol. (*b*) Given expression = $\frac{5+2\sqrt{3}}{7+4\sqrt{3}}\times\frac{7-4\sqrt{3}}{7-4\sqrt{3}}=\frac{35-20\sqrt{3}+14\sqrt{3}-24}{49-48}=11-6\sqrt{3}$

$\therefore\ 11-6\sqrt{3}=a+b\sqrt{3}\Rightarrow b=-6$

128. $\frac{\sqrt{5}}{\sqrt{3}+\sqrt{2}}-\frac{3\sqrt{3}}{\sqrt{5}+\sqrt{2}}+\frac{2\sqrt{2}}{\sqrt{5}+\sqrt{3}}=?$ **(SSC CGL Tier I 2004)**

(*a*) 0 (*b*) $2\sqrt{15}$ (*c*) $2\sqrt{10}$ (*d*) $2\sqrt{6}$

Sol. (*a*) G.E. = $\frac{\sqrt{5}(\sqrt{3}-\sqrt{2})}{3-2}-\frac{3\sqrt{3}(\sqrt{5}-\sqrt{2})}{5-2}+\frac{2\sqrt{2}(\sqrt{5}-\sqrt{3})}{5-3}$

$=\sqrt{5}\times(\sqrt{3}-\sqrt{2})-\sqrt{3}\times(\sqrt{5}-\sqrt{2})+\sqrt{2}\times(\sqrt{5}-\sqrt{3})$

$=\sqrt{15}-\sqrt{10}-\sqrt{15}+\sqrt{6}+\sqrt{10}-\sqrt{6}=0$

129. $\frac{3\sqrt{2}}{\sqrt{3}+\sqrt{6}}-\frac{4\sqrt{3}}{\sqrt{6}+\sqrt{2}}+\frac{\sqrt{6}}{\sqrt{2}+\sqrt{3}}=?$

(SSC CGL Tier I 2008 & 2003, Tier II 2013, CPO SI 2007)

(*a*) 0 (*b*) $\sqrt{6}$ (*c*) 1 (*d*) $\sqrt{2}$

Sol. (*a*) $\frac{3\sqrt{2}}{\sqrt{3}+\sqrt{6}}=\frac{3\sqrt{2}}{\sqrt{6}+\sqrt{3}}\times\frac{\sqrt{6}-\sqrt{3}}{\sqrt{6}-\sqrt{3}}=\frac{3\sqrt{2}(\sqrt{6}-\sqrt{3})}{3}=\sqrt{12}-\sqrt{6}$

$\frac{4\sqrt{3}}{\sqrt{6}+\sqrt{2}}=\frac{4\sqrt{3}}{\sqrt{6}+\sqrt{2}}\times\frac{\sqrt{6}-\sqrt{2}}{\sqrt{6}-\sqrt{2}}=\frac{4\sqrt{3}(\sqrt{6}-\sqrt{2})}{4}=\sqrt{18}-\sqrt{6}$

$$\frac{\sqrt{6}}{\sqrt{2}+\sqrt{3}}=\frac{\sqrt{6}}{\sqrt{3}+\sqrt{2}}\times\frac{\sqrt{3}-\sqrt{2}}{\sqrt{3}-\sqrt{2}}=\frac{\sqrt{6}\left(\sqrt{3}-\sqrt{2}\right)}{3-2}=\sqrt{18}-\sqrt{12}$$

$$\text{GE}=(\sqrt{12}-\sqrt{6})-(\sqrt{18}-\sqrt{6})+(\sqrt{18}-\sqrt{12})=\sqrt{12}-\sqrt{6}-\sqrt{18}+\sqrt{6}+\sqrt{18}-\sqrt{12}=0$$

130. $1+\frac{4\sqrt{3}}{2-\sqrt{2}}-\frac{30}{4\sqrt{3}-\sqrt{18}}-\frac{\sqrt{18}}{3+2\sqrt{3}}=?$ **(SSC CGL Tier I 2006)**

(*a*) 0 (*b*) 1 (*c*) $\sqrt{3}$ (*d*) $\sqrt{2}$

Sol. (*b*) $\frac{4\sqrt{3}}{2-\sqrt{2}}=\frac{4\sqrt{3}}{2-\sqrt{2}}\times\frac{2+\sqrt{2}}{2+\sqrt{2}}=\frac{8\sqrt{3}+4\sqrt{6}}{2}=4\sqrt{3}+2\sqrt{6}$

$$\frac{30}{4\sqrt{3}-\sqrt{18}}=\frac{30}{4\sqrt{3}-\sqrt{18}}\times\frac{4\sqrt{3}+\sqrt{18}}{4\sqrt{3}+\sqrt{18}}=\frac{120\sqrt{3}+90\sqrt{2}}{48-18}=4\sqrt{3}+3\sqrt{2}$$

$$\frac{\sqrt{18}}{3+2\sqrt{3}}=\frac{\sqrt{18}}{3+2\sqrt{3}}\times\frac{3-2\sqrt{3}}{3-2\sqrt{3}}=\frac{9\sqrt{2}-6\sqrt{6}}{9-12}=-3\sqrt{2}+2\sqrt{6}$$

Given expression $=1+(4\sqrt{3}+2\sqrt{6})-(4\sqrt{3}+3\sqrt{2})-(-3\sqrt{2}+2\sqrt{6})$

$=1+4\sqrt{3}+2\sqrt{6}-4\sqrt{3}-3\sqrt{2}+3\sqrt{2}-2\sqrt{6}=1$

131. $3+\frac{1}{\sqrt{3}}+\frac{1}{3+\sqrt{3}}+\frac{1}{\sqrt{3}-3}$ is equal to : **(SSC CHSL 2013, SSC CGL Tier I 2007, 2011)**

(*a*) 1 (*b*) 3 (*c*) $3+\sqrt{3}$ (*d*) $3-\sqrt{3}$

Sol. (*b*) Given expression $=3+\frac{1}{\sqrt{3}}+\frac{1}{3+\sqrt{3}}-\frac{1}{3-\sqrt{3}}$

$$=3+\frac{\sqrt{3}}{3}+\frac{3-\sqrt{3}}{6}-\frac{3+\sqrt{3}}{6}=3+\frac{\sqrt{3}}{3}-\frac{2\sqrt{3}}{6}=3$$

132. If $x=\frac{1}{1+\sqrt{2}}$, then the value of x^2+2x+3 is : **(SSC CHSL 2013)**

(*a*) 0 (*b*) 1 (*c*) 4 (*d*) 3

Sol. (*c*) $\frac{1}{1+\sqrt{2}}=\sqrt{2}-1 \Rightarrow$ G.E. $=x(x+2)+3=\left(\sqrt{2}-1\right)\left(\sqrt{2}+1\right)+3=(2-1)+3=4$

133. $\left(\frac{2+\sqrt{3}}{2-\sqrt{3}}+\frac{2-\sqrt{3}}{2+\sqrt{3}}+\frac{\sqrt{3}-1}{\sqrt{3}+1}\right)$ simplifies to : **(SSC CGL Tier I 2000)**

(*a*) $2-\sqrt{3}$ (*b*) $2+\sqrt{3}$ (*c*) $16-\sqrt{3}$ (*d*) $40-\sqrt{3}$

Sol. (*c*) $\frac{2+\sqrt{3}}{2-\sqrt{3}}=4+3+4\sqrt{3}=7+4\sqrt{3}$; $\frac{2-\sqrt{3}}{2+\sqrt{3}}=4+3-4\sqrt{3}=7-4\sqrt{3}$

$$\frac{\sqrt{3}-1}{\sqrt{3}+1}=\frac{3+1-2\sqrt{3}}{3-1}=2-\sqrt{3}$$

$\therefore$ Given expression $=7+4\sqrt{3}+7-4\sqrt{3}+2-\sqrt{3}=16-\sqrt{3}$

134. If $x=\sqrt{3}+\sqrt{2}$ then the value of $\left(x^2+\frac{1}{x^2}\right)$ is : **(SSC CHSL 2010)**

(*a*) 4 (*b*) 6 (*c*) 9 (*d*) 10

Sol. (*d*) $\frac{1}{x} = \frac{1}{\sqrt{3}+\sqrt{2}} = \sqrt{3} - \sqrt{2}$

$$\left(x^2 + \frac{1}{x^2}\right) = \left(x + \frac{1}{x}\right)^2 - 2 = [(\sqrt{3}+\sqrt{2}) + (\sqrt{3}-\sqrt{2})]^2 - 2 = (2\sqrt{3})^2 - 2 = 12 - 2 = 10$$

135. If $x = \sqrt{a} + \frac{1}{\sqrt{a}}$, $y = \sqrt{a} - \frac{1}{\sqrt{a}}$, then the value of $x^4 + y^4 - 2x^2y^2$ is **(SSC CGL Tier I 2012)**

(*a*) 16 (*b*) 20 (*c*) 10 (*d*) 5

Sol. (*a*) $x^4 + y^4 - 2x^2y^2 = (x^2 - y^2)^2 = \left[\left(\sqrt{a}+\frac{1}{\sqrt{a}}\right)^2 - \left(\sqrt{a}-\frac{1}{\sqrt{a}}\right)^2\right]^2 = \left(4\times\sqrt{a}\times\frac{1}{\sqrt{a}}\right)^2 = 16$

136. If $x = \frac{\sqrt{3}+1}{\sqrt{3}-1}$ and $y = \frac{\sqrt{3}-1}{\sqrt{3}+1}$ the value of $x^2 + y^2$ is : **(SSC CGL Tier I 2003)**

(*a*) 14 (*b*) 13 (*c*) 15 (*d*) 10

Sol. (*a*) $x = \frac{\sqrt{3}+1}{\sqrt{3}-1} \times \frac{\sqrt{3}+1}{\sqrt{3}+1} = \frac{(\sqrt{3}+1)^2}{2} = \frac{3+1+2\sqrt{3}}{2} = 2 + \sqrt{3}$

$$y = \frac{\sqrt{3}-1}{\sqrt{3}+1} \times \frac{\sqrt{3}-1}{\sqrt{3}-1} = \frac{(\sqrt{3}-1)^2}{2} = \frac{3+1-2\sqrt{3}}{2} = 2 - \sqrt{3}$$

$$x^2 + y^2 = \left(2+\sqrt{3}\right)^2 + \left(2-\sqrt{3}\right)^2 = 2\left(2^2 + \sqrt{3}^2\right) = 2 \times (4+3) = 14$$

137. If $x = \frac{\sqrt{5}+\sqrt{3}}{\sqrt{5}-\sqrt{3}}$ and $y = \frac{\sqrt{5}-\sqrt{3}}{\sqrt{5}+\sqrt{3}}$, find the value of $3x^2 + 4xy - 3y^2$.

(*a*) $4 + 48\sqrt{15}$ (*b*) $4 - 48\sqrt{15}$ (*c*) $2\sqrt{15}$ (*d*) $48\sqrt{15}$

Sol. (*a*) $x = \frac{5+3+2\sqrt{15}}{2} = 4+\sqrt{15}$; $y = \frac{5+3-2\sqrt{15}}{2} = 4-\sqrt{15}$

$$xy = \frac{\sqrt{5}+\sqrt{3}}{\sqrt{5}-\sqrt{3}} \times \frac{\sqrt{5}-\sqrt{3}}{\sqrt{5}+\sqrt{3}} = 1$$

$$3x^2 - 3y^2 + 4xy = 3(x+y)(x-y) + 4xy = 3 \times 8 \times 2\sqrt{15} + 4 \times 1 = 48\sqrt{15} + 4$$

138. If $x = 2 + \sqrt{3}$, $y = 2 - \sqrt{3}$, then the value of $\frac{x^2+y^2}{x^3+y^3}$ is **(SSC CGL Tier I 2012)**

(*a*) $\frac{7}{38}$ (*b*) $\frac{7}{40}$ (*c*) $\frac{7}{19}$ (*d*) $\frac{7}{26}$

Sol. (*d*) $x + y = (2+\sqrt{3}) + (2-\sqrt{3}) = 4$; $xy = (2+\sqrt{3})(2-\sqrt{3}) = 4 - 3 = 1$

$$\therefore \frac{x^2+y^2}{x^3+y^3} = \frac{(x+y)^2 - 2xy}{(x+y)^3 - 3xy(x+y)} = \frac{4^2 - 2\times1}{4^3 - 3\times1\times4} = \frac{16-2}{64-12} = \frac{14}{52} = \frac{7}{26}$$

139. If $a = \frac{\sqrt{3}-\sqrt{2}}{\sqrt{3}+\sqrt{2}}$, $b = \frac{\sqrt{3}+\sqrt{2}}{\sqrt{3}-\sqrt{2}}$, then the value of $\frac{a^2}{b}+\frac{b^2}{a}$ is : **(SSC CGL Tier I 2012)**

(*a*) 900 (*b*) 970 (*c*) 930 (*d*) 1030

Sol. (*b*) $a + b = \frac{\sqrt{3}-\sqrt{2}}{\sqrt{3}+\sqrt{2}} + \frac{\sqrt{3}+\sqrt{2}}{\sqrt{3}-\sqrt{2}} = \frac{\left(\sqrt{3}-\sqrt{2}\right)^2 + \left(\sqrt{3}+\sqrt{2}\right)^2}{3-2} = 2(3+2) = 10;$

$$ab = \frac{\sqrt{3}-\sqrt{2}}{\sqrt{3}+\sqrt{2}} \times \frac{\sqrt{3}+\sqrt{2}}{\sqrt{3}-\sqrt{2}} = 1$$

$$\therefore \frac{a^2}{b}+\frac{b^2}{a} = \frac{a^3+b^3}{ab} = \frac{(a+b)^3 - 3ab(a+b)}{ab} = \frac{10^3 - 3(10)}{1} = 1000 - 30 = 970$$

140. $\left(\frac{\sqrt{5}+\sqrt{3}}{\sqrt{5}-\sqrt{3}}\right)^2 + \left(\frac{\sqrt{5}-\sqrt{3}}{\sqrt{5}+\sqrt{3}}\right)^2$ is equal to **(SSC CGL Tier I 2000)**

(*a*) 64 (*b*) 62 (*c*) 66 (*d*) 68

Sol. (*b*) $\left(\frac{\sqrt{5}+\sqrt{3}}{\sqrt{5}-\sqrt{3}}\right) = \frac{5+3+2\sqrt{15}}{2} = 4+\sqrt{15}$; $\left(\frac{\sqrt{5}-\sqrt{3}}{\sqrt{5}+\sqrt{3}}\right) = \frac{5+3-2\sqrt{15}}{2} = 4-\sqrt{15}$

$$= \left(\frac{\sqrt{5}+\sqrt{3}}{\sqrt{5}-\sqrt{3}}\right)^2 + \left(\frac{\sqrt{5}-\sqrt{3}}{\sqrt{5}+\sqrt{3}}\right)^2 = \left(4+\sqrt{15}\right)^2 + \left(4-\sqrt{15}\right)^2$$

$$= 2\left[4^2 + \left(\sqrt{15}\right)^2\right] = 2(16+15) = 62$$

Alt: $x^2 + y^2 = (x + y)^2 - 2$

141. If $x = \frac{\sqrt{2}+1}{\sqrt{2}-1}$ and $xy = 1$, the value of $\frac{2x^2+3xy+2y^2}{2x^2-3xy+2y^2}$ is **(SAS 2010)**

(*a*) $\frac{71}{65}$ (*b*) $3 + 2\sqrt{2}$ (*c*) $\frac{81}{65}$ (*d*) $3 - 2\sqrt{2}$

Sol. (*a*) $x = \frac{\sqrt{2}+1}{\sqrt{2}-1} = \left(\sqrt{2}+1\right)^2 = 3+2\sqrt{2}$; $y = \frac{1}{x} = \frac{1}{3+2\sqrt{2}} = 3-2\sqrt{2}$

$$\text{G.E.} = \frac{2(x+y)^2 - xy}{2(x+y)^2 - 7xy} = \frac{2(6)^2 - 1}{2(6)^2 - 7} = \frac{72-1}{72-7} = \frac{71}{65}$$

142. What is the value of $\frac{1}{1+\sqrt{2}+\sqrt{3}} + \frac{1}{1-\sqrt{2}+\sqrt{3}}$? **(CDS 2011)**

(*a*) 1 (*b*) $\sqrt{2}$ (*c*) $\sqrt{3}$ (*d*) 2

Sol. (*a*) G.E. $= \frac{1}{(1+\sqrt{3})+(\sqrt{2})} + \frac{1}{(1+\sqrt{3})-(\sqrt{2})} = \frac{(1+\sqrt{3})-(\sqrt{2})+(1+\sqrt{3})+(\sqrt{2})}{(1+\sqrt{3})^2 - (\sqrt{2})^2}$

$$= \frac{2+2\sqrt{3}}{1+3+2\sqrt{3}-2} = \frac{2+2\sqrt{3}}{2+2\sqrt{3}} = 1$$

143. $\frac{1}{\sqrt{2}+\sqrt{3}-\sqrt{5}} + \frac{1}{\sqrt{2}-\sqrt{3}-\sqrt{5}}$ in simplified form equals to : **(SSC CGL Tier I 2005)**

(*a*) 1 (*b*) $\sqrt{2}$ (*c*) $\frac{1}{\sqrt{2}}$ (*d*) 0

Sol. (*c*) G.E. $= \frac{1}{(\sqrt{2}-\sqrt{5})+\sqrt{3}} + \frac{1}{(\sqrt{2}-\sqrt{5})-\sqrt{3}} = \frac{(\sqrt{2}-\sqrt{5})-\sqrt{3}+(\sqrt{2}-\sqrt{5})+\sqrt{3}}{(\sqrt{2}-\sqrt{5})^2-(\sqrt{3})^2}$

$= \frac{2(\sqrt{2}-\sqrt{5})}{2+5-2\sqrt{10}-3} = \frac{2(\sqrt{2}-\sqrt{5})}{4-2\sqrt{10}} = \frac{2(\sqrt{2}-\sqrt{5})}{2\sqrt{2}(\sqrt{2}-\sqrt{5})} = \frac{1}{\sqrt{2}}$

144. $\frac{1+\sqrt{2}}{\sqrt{5}+\sqrt{3}} + \frac{1-\sqrt{2}}{\sqrt{5}-\sqrt{3}} = ?$ **(SSC CGL Tier I 2000)**

(*a*) $\sqrt{5}+\sqrt{6}$ (*b*) $2\sqrt{5}+\sqrt{6}$ (*c*) $\sqrt{5}-\sqrt{6}$ (*d*) $2\sqrt{5}-3\sqrt{6}$

Sol. (*c*) Substituting $a = 1$, $b = 2$, $c = 5$ and $d = 3$, the given expression becomes

$\frac{\sqrt{a}+\sqrt{b}}{\sqrt{c}+\sqrt{d}} + \frac{\sqrt{a}-\sqrt{b}}{\sqrt{c}-\sqrt{d}} = \frac{(\sqrt{a}+\sqrt{b})(\sqrt{c}-\sqrt{d})+(\sqrt{a}-\sqrt{b})(\sqrt{c}+\sqrt{d})}{(\sqrt{c}+\sqrt{d})(\sqrt{c}-\sqrt{d})}$

$= \frac{2(\sqrt{ac}-\sqrt{bd})}{c-d} = \frac{2(\sqrt{5}-\sqrt{6})}{2} = \sqrt{5}-\sqrt{6}$

145. Find the value of $\sqrt{\frac{(\sqrt{12}-\sqrt{8})(\sqrt{3}+\sqrt{2})}{5+\sqrt{24}}}$. **(SSC CGL Tier I 2002)**

(*a*) $\sqrt{6}-\sqrt{2}$ (*b*) $\sqrt{6}+\sqrt{2}$ (*c*) $\sqrt{6}-2$ (*d*) $2-\sqrt{6}$

Sol. (*c*) GE $= \sqrt{\frac{2(\sqrt{3}-\sqrt{2})(\sqrt{3}+\sqrt{2})}{(\sqrt{3}+\sqrt{2})^2}} = \sqrt{\frac{2(3-2)}{(\sqrt{3}+\sqrt{2})^2}} = \frac{\sqrt{2}}{\sqrt{3}+\sqrt{2}} \times \frac{\sqrt{3}-\sqrt{2}}{\sqrt{3}-\sqrt{2}} = \frac{\sqrt{6}-2}{3-2} = \sqrt{6}-2$

146. If $a(2-\sqrt{3}) = b(2+\sqrt{3}) = 1$, then the value of (a^2-b^2) is **(SAS 2010)**

(*a*) $8\sqrt{3}$ (*b*) $3\sqrt{8}$ (*c*) $6\sqrt{3}$ (*d*) $2\sqrt{6}$

Sol. (*a*) $a(2-\sqrt{3}) = 1 \Rightarrow a = \frac{1}{2-\sqrt{3}} = 2+\sqrt{3}$; $b(2+\sqrt{3}) = 1 \Rightarrow b = \frac{1}{2+\sqrt{3}} = 2-\sqrt{3}$

$\therefore (a^2-b^2) = (a+b)(a-b) = (2+\sqrt{3}+2-\sqrt{3})(2+\sqrt{3}-2+\sqrt{3}) = 4(2\sqrt{3}) = 8\sqrt{3}$

147. If $\sqrt{\frac{x-a}{x-b}} + \frac{a}{x} = \sqrt{\frac{x-b}{x-a}} + \frac{b}{x}$, $b \neq a$, then the value of x is **(SSC CGL Tier I 2012)**

(*a*) $\frac{ab}{a+b}$ (*b*) $\frac{a}{a+b}$ (*c*) $\frac{b}{a+b}$ (*d*) 1

Sol. (*a*) $\sqrt{\frac{x-a}{x-b}} - \sqrt{\frac{x-b}{x-a}} = \frac{b}{x} - \frac{a}{x} \Rightarrow \frac{x-a-x+b}{\sqrt{(x-a)(x-b)}} = \frac{b-a}{x} \Rightarrow \frac{b-a}{\sqrt{(x-a)(x-b)}} = \frac{b-a}{x}$

$\Rightarrow x = \sqrt{(x-a)(x-b)} \Rightarrow x^2 = (x-a)(x-b) \Rightarrow x^2 = x^2-(a+b)x+ab$

$\Rightarrow (a+b)x = ab \Rightarrow x = \frac{ab}{a+b}$

148. If $x = \sqrt{3}+\sqrt{2}$, then the value of $x^3 + \frac{1}{x^3} = ?$ **(SSC CGL Tier I 2007)**

(*a*) $6\sqrt{3}$ (*b*) $12\sqrt{3}$ (*c*) $18\sqrt{3}$ (*d*) $24\sqrt{3}$

Sol. (*c*) $\frac{1}{x} = \frac{1}{\sqrt{3}+\sqrt{2}} = \frac{\sqrt{3}-\sqrt{2}}{3-2} = \sqrt{3}-\sqrt{2} \Rightarrow x + \frac{1}{x} = (\sqrt{3}+\sqrt{2}) + (\sqrt{3}-\sqrt{2}) = 2\sqrt{3}$

$$x^3 + \frac{1}{x^3} = \left(x + \frac{1}{x}\right)^3 - 3\left(x + \frac{1}{x}\right) = (2\sqrt{3})^3 - 3\,(2\sqrt{3}) = 24\sqrt{3} - 6\sqrt{3} = 18\sqrt{3}$$

149. The value of $\sqrt{5+2\sqrt{6}} - \frac{1}{\sqrt{5+2\sqrt{6}}}$ is : **(SSC CGL Tier I 1999)**

(*a*) $2\sqrt{2}$ (*b*) $2\sqrt{3}$ (*c*) $1+\sqrt{5}$ (*d*) $\sqrt{5}-1$

Sol. (*a*) $\sqrt{5+2\sqrt{6}} = \sqrt{(\sqrt{3})^2 + (\sqrt{2})^2 + 2\,(\sqrt{3})\,(\sqrt{2})} = \sqrt{3}+\sqrt{2}$

Given expression = $(\sqrt{3}+\sqrt{2}) - \frac{1}{\sqrt{3}+\sqrt{2}} = (\sqrt{3}+\sqrt{2}) - (\sqrt{3}-\sqrt{2}) = 2\sqrt{2}$

150. The value of $\sqrt{11+2\sqrt{30}} - \frac{1}{\sqrt{11+2\sqrt{30}}}$ is : **(SSC CGL Tier I 2002)**

(*a*) $2\sqrt{5}$ (*b*) $2\sqrt{3}$ (*c*) $1+\sqrt{5}$ (*d*) $\sqrt{5}-1$

Sol. (*a*) $\sqrt{11+2\sqrt{30}} = \sqrt{6+5+2\sqrt{30}} = \sqrt{(\sqrt{6})^2 + (\sqrt{5})^2 + 2\,(\sqrt{6})\,(\sqrt{5})} = \sqrt{6}+\sqrt{5}$

$\frac{1}{\sqrt{11+2\sqrt{30}}} = \frac{1}{\sqrt{6}+\sqrt{5}} = \sqrt{6}-\sqrt{5} \Rightarrow$ Given expression $= (\sqrt{6}+\sqrt{5}) - (\sqrt{6}-\sqrt{5}) = 2\sqrt{5}$

151. What is one of square roots of $9 - 2\sqrt{14}$: **(CDS 2011)**

(*a*) $\sqrt{7}-\sqrt{3}$ (*b*) $\sqrt{6}-\sqrt{3}$ (*c*) $\sqrt{7}-\sqrt{5}$ (*d*) $\sqrt{7}-\sqrt{2}$

Sol. (*d*) $9 - 2\sqrt{14} = 7 + 2 - 2\sqrt{14} = \left(\sqrt{7}-\sqrt{2}\right)^2 \Rightarrow$ The square root = $\left(\sqrt{7}-\sqrt{2}\right)$

152. The square root of $33 - 4\sqrt{35}$ is **(SSC CGL Tier I 2013)**

(*a*) $\pm\left(2\sqrt{7}-\sqrt{5}\right)$ (*b*) $\pm\left(2\sqrt{7}+\sqrt{5}\right)$ (*c*) $\pm\left(\sqrt{7}+2\sqrt{5}\right)$ (*d*) $\pm\left(\sqrt{7}-2\sqrt{5}\right)$

Sol. (*a*) $\left(2\sqrt{7}\right)^2 + \left(\sqrt{5}\right)^2 = 28 + 5 = 33$ and $2\left(2\sqrt{7}\right)\left(\sqrt{5}\right) = 4\sqrt{35}$

153. The value $\sqrt{-\sqrt{3} + \sqrt{3 + 8\sqrt{7+4\sqrt{3}}}}$ of is : **(SSC CGL Tier I 2003, 2012)**

(*a*) 1 (*b*) 2 (*c*) 3 (*d*) 8

Sol. (*b*) $\sqrt{-\sqrt{3} + \sqrt{3 + 8\sqrt{7+4\sqrt{3}}}} = \sqrt{-\sqrt{3} + \sqrt{3 + 8\sqrt{4+3+4\sqrt{3}}}}$

$$= \sqrt{-\sqrt{3} + \sqrt{3 + 8\sqrt{2^2 + \sqrt{3}^2 + 2\left(2\sqrt{3}\right)}}} = \sqrt{-\sqrt{3} + \sqrt{3 + 8\left(2+\sqrt{3}\right)}} = \sqrt{-\sqrt{3} + \sqrt{19 + 8\sqrt{3}}}$$

$$= \sqrt{-\sqrt{3} + \sqrt{16 + 3 + 8\sqrt{3}}} = \sqrt{-\sqrt{3} + \sqrt{(4+\sqrt{3})^2}} = \sqrt{-\sqrt{3} + 4 + \sqrt{3}} = \sqrt{4} = 2$$

154. $\sqrt{30+12\sqrt{6}} - \sqrt{4+2\sqrt{3}} - \sqrt{19-6\sqrt{2}}$ is simplified to **(SAS 2010)**

(*a*) $3\sqrt{2}$ (*b*) $\sqrt{2}$ (*c*) $2\sqrt{3}$ (*d*) $\sqrt{3}$

Sol. (*d*) G.E. $= \sqrt{18+12+12\sqrt{6}} - \sqrt{3+1+2\sqrt{3}} - \sqrt{18+1-6\sqrt{2}}$

$$= \sqrt{(3\sqrt{2})^2+(2\sqrt{3})^2+2(3\sqrt{2})(2\sqrt{3})} - \sqrt{(\sqrt{3})^2+(1)^2+2\sqrt{3}} - \sqrt{(3\sqrt{2})^2+(1)^2-2(3\sqrt{2})}$$

$$= \sqrt{(3\sqrt{2}+2\sqrt{3})^2} - \sqrt{(\sqrt{3}+1)^2} - \sqrt{(3\sqrt{2}-1)^2} = 3\sqrt{2}+2\sqrt{3}-\sqrt{3}-1-3\sqrt{2}+1=\sqrt{3}$$

155. The square root of $3-\sqrt{5}$ is : **(SSC CGL Tier I 2006)**

(*a*) $\sqrt{3}-\sqrt[4]{5}$ (*b*) $\frac{1}{2}(\sqrt{5}-\sqrt{3})$ (*c*) $\frac{1}{2}(\sqrt{5}-1)$ (*d*) $\frac{1}{\sqrt{2}}(\sqrt{5}-1)$

Sol. (*d*) $3-\sqrt{5}=\frac{1}{2}(6-2\sqrt{5})=\frac{1}{2}(5+1-2\sqrt{5})=\frac{1}{2}(\sqrt{5}-\sqrt{1})^2=\left[\frac{1}{\sqrt{2}}(\sqrt{5}-1)\right]^2$

$\therefore$ The square root is $\frac{1}{\sqrt{2}}(\sqrt{5}-1)$

156. The square root of $\sqrt[3]{4}+\sqrt[3]{16}+1$ is

(*a*) $\sqrt[3]{2}+1$ (*b*) $\sqrt[3]{4}+1$ (*c*) $\sqrt[3]{2}-1$ (*d*) $\sqrt[3]{4}-1$

Sol. (*a*) $\sqrt[3]{4}+\sqrt[3]{16}+1 = \sqrt[3]{2\times2}+\sqrt[3]{2\times8}+1=\sqrt[3]{2}\times\sqrt[3]{2}+\sqrt[3]{2}\times\sqrt[3]{8}+1$

$= (\sqrt[3]{2})^2+2\cdot\sqrt[3]{2}+(1)^2=\left(\sqrt[3]{2}+1\right)^2 \Rightarrow$ square-root $= (\sqrt[3]{2}+1)$

157. If $\sqrt{28-6\sqrt{3}}=\sqrt{3}a-b$ (where *a*, *b* are rationals), value of $(a-b)$ is **(SSC CGL Tier II 2012)**

(*a*) 1 (*b*) – 1 (*c*) – 2 (*d*) 2

Sol. (*d*) $\sqrt{28-6\sqrt{3}}=\sqrt{(3\sqrt{3}-1)^2}=(3\sqrt{3}-1)=\sqrt{3}a-b \Rightarrow a=3, b=1 \Rightarrow (a-b)=3-1=2$

158. If $x=7-4\sqrt{3}$, then $\sqrt{x}+\frac{1}{\sqrt{x}}=?$

(*a*) 1 (*b*) 2 (*c*) 4 (*d*) 16

Sol. (*c*) $\frac{1}{x}=\frac{1}{7-4\sqrt{3}}=7+4\sqrt{3}$ $\qquad x+\frac{1}{x}=(7-4\sqrt{3})+(7+4\sqrt{3})=14$

$\therefore \left(\sqrt{x}+\frac{1}{\sqrt{x}}\right)^2=x+\frac{1}{x}+2=14+2=16 \Rightarrow \left(\sqrt{x}+\frac{1}{\sqrt{x}}\right)=\sqrt{16}=4$

159. If $x=8+2\sqrt{15}$, then the value of $\sqrt{x}+\frac{1}{\sqrt{x}}$ is **(SSC CGL Tier I 2006)**

(*a*) $2\sqrt{3}$ (*b*) $2\sqrt{5}$ (*c*) $\frac{3}{2}\sqrt{5}+\frac{\sqrt{3}}{2}$ (*d*) $\frac{\sqrt{5}}{2}+\frac{3}{2}\sqrt{3}$

Sol. (*c*) $x=8+2\sqrt{15}=5+3+2\sqrt{15}=(\sqrt{5})^2+(\sqrt{3})^2+2(\sqrt{5})(\sqrt{3})=(\sqrt{5}+\sqrt{3})^2$

$\sqrt{x}=\sqrt{5}+\sqrt{3} \Rightarrow \frac{1}{\sqrt{x}}=\frac{1}{\sqrt{5}+\sqrt{3}}=\frac{\sqrt{5}-\sqrt{3}}{2}$

$\sqrt{x}+\frac{1}{\sqrt{x}}=\sqrt{5}+\sqrt{3}+\frac{\sqrt{5}-\sqrt{3}}{2}=\frac{3}{2}\sqrt{5}+\frac{\sqrt{3}}{2}$

160. If $x=2+\sqrt{3}$, then the value of $\sqrt{x}+\frac{1}{\sqrt{x}}$ is : **(SSC CGL Tier I 2012)**

(*a*) $\sqrt{3}$ (*b*) $\sqrt{6}$ (*c*) $2\sqrt{6}$ (*d*) 6

Sol. (*b*) $\frac{1}{x} = \frac{1}{2+\sqrt{3}} = 2-\sqrt{3} \Rightarrow x+\frac{1}{x} = (2+\sqrt{3})+(2-\sqrt{3}) = 4$

$$\Rightarrow \left(\sqrt{x}+\frac{1}{\sqrt{x}}\right)^2 = x+\frac{1}{x}+2 = 4+2 = 6 \Rightarrow \sqrt{x}+\frac{1}{\sqrt{x}} = \sqrt{6}$$

161. The value of $\frac{1}{\sqrt{(12-\sqrt{140})}} - \frac{1}{\sqrt{(8-\sqrt{60})}} - \frac{2}{\sqrt{(10+\sqrt{84})}}$ is

(SAS 2010, SSC CGL Tier I 2003)

(*a*) 0 (*b*) 1 (*c*) 2 (*d*) 3

Sol. (*a*) G.E. = $\frac{1}{\sqrt{(7+5-2\sqrt{35})}} - \frac{1}{\sqrt{(5+3-2\sqrt{15})}} - \frac{2}{\sqrt{(7+3+2\sqrt{21})}}$

$$= \frac{1}{\sqrt{(\sqrt{7}-\sqrt{5})^2}} - \frac{1}{\sqrt{(\sqrt{5}-\sqrt{3})^2}} - \frac{2}{\sqrt{(\sqrt{7}+\sqrt{3})^2}} = \frac{1}{\sqrt{7}-\sqrt{5}} - \frac{1}{\sqrt{5}-\sqrt{3}} - \frac{2}{\sqrt{7}+\sqrt{3}}$$

$$= \frac{\sqrt{7}+\sqrt{5}}{2} - \frac{\sqrt{5}+\sqrt{3}}{2} - \frac{\sqrt{7}-\sqrt{3}}{2} = \frac{\sqrt{7}+\sqrt{5}-\sqrt{5}-\sqrt{3}-\sqrt{7}+\sqrt{3}}{2} = 0$$

162. The value of $\sqrt{10+2\sqrt{6}+2\sqrt{10}+2\sqrt{15}}$

(SSC CGL Tier I 2007)

(*a*) $\sqrt{2}+\sqrt{3}-\sqrt{5}$ (*b*) $\sqrt{2}+\sqrt{3}+\sqrt{5}$ (*c*) $\sqrt{2}-\sqrt{3}+\sqrt{5}$ (*d*) $\sqrt{3}+\sqrt{5}-\sqrt{2}$

Sol. (*b*) G.E. = $\sqrt{2+3+5+2\sqrt{6}+2\sqrt{10}+2\sqrt{15}}$

$$= \sqrt{(\sqrt{2})^2+(\sqrt{3})^2+(\sqrt{5})^2+2(\sqrt{2}.\sqrt{3}+2\sqrt{2}.\sqrt{5}+2\sqrt{3}.\sqrt{5})} = \sqrt{(\sqrt{2}+\sqrt{3}+\sqrt{5})^2} = \sqrt{2}+\sqrt{3}+\sqrt{5}$$

Trick: G.E. ~ $\sqrt{10+5+6+7.5} = \sqrt{28.5} > 5$ Note: $2\sqrt{6} = \sqrt{24} \sim 5$

$\sqrt{2}+\sqrt{3}+\sqrt{5} = 1.4+1.7+2.2 = 5.3$

163. If $x = \sqrt{\frac{\sqrt{5}+1}{\sqrt{5}-1}}$, then $x^2 - x - 1$, is equal to

(SSC CGL Tier II 2012)

(*a*) 2 (*b*) 5 (*c*) 0 (*d*) 1

Sol. (*c*) $x = \sqrt{\frac{\sqrt{5}+1}{\sqrt{5}-1}} = \sqrt{\frac{\sqrt{5}+1}{\sqrt{5}-1} \times \frac{\sqrt{5}+1}{\sqrt{5}+1}} = \sqrt{\frac{(\sqrt{5}+1)^2}{5-1}} = \frac{\sqrt{5}+1}{2}$

$$x^2 - x - 1 = x(x-1) - 1 = \frac{\sqrt{5}+1}{2}\left(\frac{\sqrt{5}+1}{2}-1\right)-1 = \frac{\sqrt{5}+1}{2}\left(\frac{\sqrt{5}-1}{2}\right)-1 = \frac{5-1}{4} = 1-1 = 0$$

164. If $x = \sqrt{\frac{\sqrt{5}+1}{\sqrt{5}-1}}$, then the value of $5x^2 - 5x - 1$ is

(SSC CGL Tier I 2011)

(*a*) 0 (*b*) 3 (*c*) 4 (*d*) 5

Sol. (*c*) $x = \sqrt{\frac{\sqrt{5}+1}{\sqrt{5}-1}} \times \sqrt{\frac{\sqrt{5}+1}{\sqrt{5}+1}} = \sqrt{\frac{(\sqrt{5}+1)^2}{5-1}} = \frac{\sqrt{5}+1}{2} \Rightarrow 5x^2 - 5x - 1 = 5x(x-1) - 1$

$$= 5\times\frac{\sqrt{5}+1}{2}\left(\frac{\sqrt{5}-1}{2}\right)-1 = 5\times\frac{5-1}{4} = 5-1 = 4$$

165. If $x = \sqrt[3]{5} + 2$, then the value of $x^3 - 6x^2 + 12x - 13$ is **(SSC CGL Tier II 2013)**

(*a*) – 1 (*b*) 1 (*c*) 2 (*d*) 0

Sol. (*d*) $x - 2 = \sqrt[3]{5} \Rightarrow (x-2)^3 = 5 \Rightarrow x^3 - 8 - 3(x)(2)(x-2) = 5$

$\Rightarrow x^3 - 8 - 6x^2 + 12x - 5 = 0 \Rightarrow x^3 - 6x^2 + 12x - 13 = 0$

166. If $x = \dfrac{1}{2-\sqrt{3}}$, find the value of $x^3 - 2x^2 - 7x + 10$

(*a*) 0 (*b*) 1 (*c*) 5 (*d*) 8

Sol. (*d*) $x = \dfrac{1}{2-\sqrt{3}} = 2 + \sqrt{3} \Rightarrow x - 2 = \sqrt{3}$

$\Rightarrow (x-2)^2 = 3 \Rightarrow x^2 - 4x + 4 = 3 \Rightarrow x^2 - 4x + 1 = 0$

Now, $x^3 - 2x^2 - 7x + 10 = x(x^2 - 4x + 1) + 2(x^2 - 4x + 1) + 8 = x \times 0 + 2 \times 0 + 8 = 8$

167. If $\sqrt{x} = \sqrt{3} - \sqrt{5}$, then the value of $x^2 - 16x + 6$ is **(SSC CGL Tier II 2013)**

(*a*) 0 (*b*) – 2 (*c*) 2 (*d*) 4

Sol. (*c*) $x = 3 + 5 - 2\sqrt{15} = 8 - 2\sqrt{15}$; $x^2 = 64 + 60 - 32\sqrt{15} = 124 - 32\sqrt{15}$

$x^2 - 16x + 6 = 124 - 32\sqrt{15} - 16(8 - 2\sqrt{15}) + 6 = 124 - 32\sqrt{15} - 128 + 32\sqrt{15} + 6 = 2$

168. If $x = 1 + \sqrt{2} + \sqrt{3}$, then the value of $(2x^4 - 8x^3 - 5x^2 + 26x - 28)$ is **(SSC CGL Tier I 2012)**

(*a*) $6\sqrt{6}$ (*b*) 0 (*c*) $3\sqrt{6}$ (*d*) $2\sqrt{6}$

Sol. (*a*) $x = 1 + \sqrt{2} + \sqrt{3} \Rightarrow (x-1)^2 = (\sqrt{2} + \sqrt{3})^2$

$\Rightarrow x^2 - 2x + 1 = 5 + 2\sqrt{6} \Rightarrow x^2 - 2x = 4 + 2\sqrt{6} \Rightarrow x^2 - 2x - 4 = 2\sqrt{6}$

$\Rightarrow (x^2 - 2x - 4)^2 = (2\sqrt{6})^2 \Rightarrow x^4 + 4x^2 + 16 - 4x^3 + 16x - 8x^2 = 24$

$\Rightarrow x^4 - 4x^3 - 4x^2 + 16x - 8 = 0 \Rightarrow 2x^4 - 8x^3 - 8x^2 + 32x - 16 = 0$

$\Rightarrow 2x^4 - 8x^3 - 5x^2 + 26x - 28 = 3x^2 - 6x - 12 = 3(x^2 - 2x) - 12$

$= 3(4 + 2\sqrt{6}) - 12 = 6\sqrt{6}$

169. If $a = (\sqrt{3} + \sqrt{2})$ and $b = (\sqrt{3} - \sqrt{2})$, find the value of $(a+1)^{-1} + (b+1)^{-1}$. **(SSC CHSL 2012)**

(*a*) 0 (*b*) 1 (*c*) 2 (*d*) –1

Sol. (*b*) $\dfrac{1}{a} = \dfrac{1}{\sqrt{3}+\sqrt{2}} = \dfrac{\sqrt{3}-\sqrt{2}}{1} = b$

$(a+1)^{-1} + (b+1)^{-1} = (a+1)^{-1} + \left(\dfrac{1}{a} + 1\right)^{-1} = \dfrac{1}{a+1} + \dfrac{a}{a+1} = 1$

170. If $a = (\sqrt{2} - 1)^{1/3}$, then the value of $(a - a^{-1})^3 + 3(a - a^{-1})$ is **(SSC CHSL 2012)**

(*a*) –2 (*b*) 2 (*c*) $\sqrt{2}$ (*d*) $2\sqrt{2}$

Sol. (*b*) $a^3 = \sqrt{2} - 1 \Rightarrow \dfrac{1}{a^3} = \dfrac{1}{\sqrt{2}-1} = \sqrt{2} + 1$

$\therefore$ G.E. $= \left(a-\frac{1}{a}\right)^3 + 3\left(a-\frac{1}{a}\right) = a^3 - \frac{1}{a^3} = (\sqrt{2}-1)-(\sqrt{2}+1) = -2$

171. $x = 3+2\sqrt{2}$, find the value of $\sqrt{x} - \frac{1}{\sqrt{x}} = ?$ **(MBA 1988, SSC CGL Tier I 2012, 2013)**

(*a*) $\pm 2\sqrt{2}$ (*b*) ± 2 (*c*) $\pm\sqrt{2}$ (*d*) $\pm\frac{1}{2}$

Sol. (*b*) $\frac{1}{x} = 3-2\sqrt{2} \Rightarrow \left(\sqrt{x}-\frac{1}{\sqrt{x}}\right)^2 = x+\frac{1}{x}-2 = (3+2\sqrt{2})+(3-2\sqrt{2})-2 = 6-2 = 4$

$\left(\sqrt{x}-\frac{1}{\sqrt{x}}\right) = \pm 2$

Alt: $x = 3+2\sqrt{2} = \left(\sqrt{2}+1\right)^2 \Rightarrow \frac{1}{\sqrt{x}} = \frac{1}{\sqrt{2}+1} = \sqrt{2}-1 \Rightarrow \sqrt{x}-\frac{1}{\sqrt{x}} = \sqrt{2}+1-\left(\sqrt{2}-1\right) = 2$

172. If $x = 3 + 2\sqrt{2}$, then the value of $x^3 + \frac{1}{x^3}$ is **(SSC FCI Asstt. 2012)**

173. The value of $(3+2\sqrt{2})^{-3} + (3-2\sqrt{2})^{-3}$ is : **(RRB 2008, SSC CGL Tier I 2004)**

(*a*) 216 (*b*) 201 (*c*) 198 (*d*) 162

Sol. (*c*) $\frac{1}{x} = \frac{1}{3+2\sqrt{2}} = 3-2\sqrt{2} \Rightarrow x+\frac{1}{x} = (3+2\sqrt{2})+(3-2\sqrt{2}) = 6$

$\Rightarrow x^3+\frac{1}{x^3} = \left(x+\frac{1}{x}\right)^3 - 3\left(x+\frac{1}{x}\right) = 6^3 - 3\times 6 = 216-18 = 198$

174. If $x = 3+2\sqrt{2}$, find the value of $x^4+\frac{1}{x^4} = ?$

(*a*) 17 (*b*) 32 (*c*) 34 (*d*) 1154

Sol. (*d*) $\frac{1}{x} = 3-2\sqrt{2} \Rightarrow x+\frac{1}{x} = 6 \Rightarrow \left(x^2+\frac{1}{x^2}\right) = 6^2 - 2 = 34$

$\Rightarrow \left(x^4+\frac{1}{x^4}\right) = 34^2 - 2 = 1156 - 2 = 1154$

175. If $a = \frac{\sqrt{5}+1}{\sqrt{5}-1}$ and $b = \frac{\sqrt{5}-1}{\sqrt{5}+1}$, find the value of $\frac{a^2+ab+b^2}{a^2-ab+b^2}$ **(CBI 90, SSC CGL 2005)**

(*a*) $\frac{3}{4}$ (*b*) $\frac{4}{3}$ (*c*) $\frac{3}{5}$ (*d*) $\frac{5}{3}$

Sol. (*b*) $(a+b) = \frac{(\sqrt{5}+1)^2+(\sqrt{5}-1)^2}{(\sqrt{5}-1)\times(\sqrt{5}+1)} = \frac{2\times(5+1)}{5-1} = 3$

$ab = 1$ and $a^2+b^2 = (a+b)^2 - 2ab = 9-2 = 7 \Rightarrow$ GE $= \frac{7+1}{7-1} = \frac{8}{6} = \frac{4}{3}$

176. Find the value of $(4^{\frac{3}{4}} + 4^{\frac{1}{2}} + 4^{\frac{1}{4}} + 1)\times(4^{\frac{1}{4}} - 1)$.

(*a*) 1 (*b*) 3 (*c*) 4 (*d*) 6

Sol. (*b*) Let $4^{\frac{1}{4}} = x$, then given expression becomes $(x^3 + x^2 + x + 1) \times (x - 1)$

$$= x^4 - 1 = 4^{\frac{4}{4}} - 1 = 4 - 1 = 3$$

177. If $x = 2^{1/3} + 2^{2/3}$, then $x^3 - 6x - 6 = ?$

(*a*) 0 (*b*) 1 (*c*) 2 (*d*) 3

Sol. (*a*) $x^3 = (2^{1/3} + 2^{2/3})^3 = (2^{1/3})^3 + (2^{2/3})^3 + 3 \times 2^{1/3} \times 2^{2/3} \times (2^{1/3} + 2^{2/3})$

$$= 2 + 2^2 + 3 \times 2 \times (2^{1/3} + 2^{2/3}) = 6 + 6x \Rightarrow x^3 = 6 + 6x \Rightarrow x^3 - 6x - 6 = 0$$

178. If $x = 2 + 2^{1/3} + 2^{2/3}$, then find the value of $x^3 - 6x^2 + 6x$ **(MAT)**

(*a*) 0 (*b*) 1 (*c*) 2 (*d*) – 1

Sol. (*c*) $x = 2 + 2^{1/3} + 2^{2/3} \Rightarrow x - 2 = 2^{1/3} + 2^{2/3}$

$\Rightarrow (x - 2)^3 = (2^{1/3} + 2^{2/3})^3 \Rightarrow x^3 - 8 - 3(x)(2)(x - 2) = 2 + 2^2 + 3(2^{1/3})(2^{2/3})(2^{1/3} + 2^{2/3})$

$\Rightarrow x^3 - 8 - 6x^2 + 12x = 2 + 4 + 6(x - 2) \Rightarrow x^3 - 8 - 6x^2 + 12x = 6 + 6x - 12$

$\Rightarrow x^3 - 6x^2 + 6x = 6 - 12 + 8 = 2$

179. If $x = \frac{\sqrt{3}}{2}$ then $\sqrt{1+a} + \sqrt{1-a} = ?$ **(SSC CGL Tier I 2007, SSC CGL Tier II 2012)**

(*a*) $\sqrt{3}$ (*b*) $-\sqrt{3}$ (*c*) 3 (*d*) 9

Sol. (*a*) Let $y = \sqrt{1+a} + \sqrt{1-a}$. Then $y^2 = (1 + a) + (1 - a) + 2\sqrt{1-a^2} = 2 + 2\sqrt{1-a^2}$

$$\Rightarrow y^2 = 2 + 2\sqrt{1 - \left(\frac{\sqrt{3}}{2}\right)^2} = 2 + 2\sqrt{\frac{1}{4}} = 2 + 2 \times \frac{1}{2} = 3 \Rightarrow y = \sqrt{3}$$

180. If $x = \frac{\sqrt{3}}{2}$, then $\frac{\sqrt{1+x} + \sqrt{1-x}}{\sqrt{1+x} - \sqrt{1-x}} = ?$ **(SSC CGL Tier I 2010, SSC CGL Tier II 2012)**

(*a*) $\sqrt{3}$ (*b*) $-\sqrt{3}$ (*c*) 3 (*d*) 9

Sol. (*a*) $\frac{\sqrt{1+x} + \sqrt{1-x}}{\sqrt{1+x} - \sqrt{1-x}} = \frac{\sqrt{1+x} + \sqrt{1-x}}{\sqrt{1+x} - \sqrt{1-x}} \times \frac{\sqrt{1+x} + \sqrt{1-x}}{\sqrt{1+x} + \sqrt{1-x}} = \frac{(1+x) + (1-x) + 2\sqrt{1-x^2}}{(1+x) - (1-x)}$

$$= \frac{2 + 2\sqrt{1-x^2}}{2x} = \frac{1 + \sqrt{1-x^2}}{x} = \frac{1 + \sqrt{1 - \frac{3}{4}}}{\frac{\sqrt{3}}{2}} = \frac{1 + \frac{1}{2}}{\frac{\sqrt{3}}{2}} = \frac{3}{2} \times \frac{2}{\sqrt{3}} = \sqrt{3}$$

Alt: $x = \frac{\sqrt{3}}{2} \Rightarrow \frac{1}{x} = \frac{2}{\sqrt{3}} \Rightarrow \frac{1+x}{1-x} = \frac{2+\sqrt{3}}{2-\sqrt{3}} = \frac{(2+\sqrt{3})^2}{4-3} \Rightarrow \frac{\sqrt{1+x}}{\sqrt{1-x}} = \frac{2+\sqrt{3}}{1}$

$$\Rightarrow \frac{\sqrt{1+x} + \sqrt{1-x}}{\sqrt{1+x} - \sqrt{1-x}} = \frac{2+\sqrt{3}+1}{2+\sqrt{3}-1} = \frac{3+\sqrt{3}}{\sqrt{3}+1} = \frac{\sqrt{3}(\sqrt{3}+1)}{\sqrt{3}+1} = \sqrt{3}$$

181. The greatest among the numbers $3\sqrt{2}$, $3\sqrt{7}$, $6\sqrt{5}$, $2\sqrt{20}$ is **(SSC CHSL 2011)**

(*a*) $3\sqrt{2}$ (*b*) $3\sqrt{7}$ (*c*) $6\sqrt{5}$ (*d*) $2\sqrt{20}$

Sol. (*c*) $3\sqrt{2} = \sqrt{18}$, $3\sqrt{7} = \sqrt{63}$, $6\sqrt{5} = \sqrt{180}$, $2\sqrt{20} = 80$

$\Rightarrow \sqrt{180}$ or $6\sqrt{5}$ is the greatest

182. Among $\sqrt{2}, \sqrt[3]{3}, \sqrt[4]{5}, \sqrt[3]{2}$, which one is the greatest : **(SSC CGL Tier I 2002)**

(*a*) $\sqrt[4]{5}$ (*b*) $\sqrt{2}$ (*c*) $\sqrt[3]{3}$ (*d*) $\sqrt[3]{2}$

Sol. (*a*) LCM of root indices 2, 3, 4 is 12. $\therefore$ We reduce the given numbers to the 12th order.

$\sqrt{2} = 2^{1/2} = (2^6)^{1/12} = 64^{1/12}$ $\sqrt[3]{3} = 3^{1/3} = (3^4)^{1/12} = 81^{1/12}$

$\sqrt[4]{5} = 5^{1/4} = (5^3)^{1/12} = 125^{1/12}$ $\sqrt[3]{2} = 2^{1/3} = (2^4)^{1/12} = 16^{1/12}$

$\therefore$ $125^{1/2}$ or $\sqrt[4]{5}$ is the greatest.

183. Which of the following is the greatest ? **(CAT 2006, RRB 2000, SSC CGL Tier I 2008)**

(*a*) $\sqrt{2}$ (*b*) $\sqrt[3]{3}$ (*c*) $\sqrt[4]{4}$ (*d*) $\sqrt[6]{6}$

Sol. (*b*) LCM of the root indices *i.e.* LCM of 2, 3, 4 and 6, is 12.

$\therefore$ We reduce the given numbers to the 12th order.

$\sqrt{2} = 2^{\frac{1}{2}} = (2^6)^{\frac{1}{12}} = 64^{\frac{1}{12}}$ $\sqrt[3]{3} = 3^{\frac{1}{3}} = (3^4)^{\frac{1}{12}} = 81^{\frac{1}{12}}$

$\sqrt[4]{4} = 4^{\frac{1}{4}} = (4^3)^{\frac{1}{12}} = 64^{\frac{1}{12}}$ $\sqrt[6]{6} = 6^{\frac{1}{6}} = (6^2)^{\frac{1}{12}} = 36^{\frac{1}{12}}$

$\therefore$ $81^{\frac{1}{12}}$ or $\sqrt[3]{3}$ is the greatest of the given numbers.

184. The greatest number among $\sqrt[3]{2}, \sqrt{3}, \sqrt[3]{5}$, 1.5 is : **(SSC CGL Tier I 2006)**

(*a*) $\sqrt[3]{2}$ (*b*) $\sqrt{3}$ (*c*) $\sqrt[3]{5}$ (*d*) 1.5

Sol. (*b*) LCM of the root indices *i.e.* LCM of 3, 2 and 3 is 6.

$\therefore$ We reduce the given number to the 6th order.

$\sqrt[3]{2} = 2^{1/3} = (2^2)^{1/6} = 4^{1/6}$ $\sqrt{3} = 3^{1/2} = (3^3)^{1/6} = 27^{1/6}$

$\sqrt[3]{5} = 5^{1/3} = (5^2)^{1/6} = 25^{1/6}$ $1.5 = \left[\left(\frac{3}{2}\right)^6\right]^{1/6} = \left(\frac{729}{64}\right)^{1/6}$

$\therefore$ $27^{1/6}$ or $\sqrt{3}$ is the greatest of the given numbers.

185. $\sqrt[3]{4} \times \sqrt[4]{8} = ?$

(*a*) 2 (*b*) $\sqrt[12]{32}$ (*c*) $2 \times \sqrt[12]{32}$ (*d*) $\sqrt[12]{32^7}$

Sol. (*c*) $\sqrt[3]{4} \times \sqrt[4]{8} = \sqrt[3]{2^2} \times \sqrt[4]{2^3} = 2^{\frac{2}{3}} \times 2^{\frac{3}{4}} = 2^{\frac{2}{3}+\frac{3}{4}} = 2^{\frac{17}{12}} = 2 \times 2^{\frac{5}{12}} = 2 \times \sqrt[12]{2^5} = 2 \times \sqrt[12]{32}$

186. $16\sqrt[3]{9} \div 8\sqrt{3} = ?$

(*a*) $2\sqrt{3}$ (*b*) $2\sqrt[4]{3}$ (*c*) $2\sqrt[6]{3}$ (*d*) $2\sqrt{27}$

Sol. (*c*) $\frac{16\sqrt[3]{9}}{8\sqrt{3}} = \frac{16}{8} \times \frac{9^{\frac{1}{3}}}{3^{\frac{1}{2}}} = 2 \times \frac{(9^2)^{\frac{1}{6}}}{(3^3)^{\frac{1}{6}}} = 2 \times \left(\frac{81}{27}\right)^{\frac{1}{6}} = 2 \times 3^{\frac{1}{6}} = 2\sqrt[6]{3}$

187. Find the value of $\sqrt[3]{20+14\sqrt{2}} + \sqrt[3]{20-14\sqrt{2}}$.

(*a*) 3 (*b*) 4 (*c*) 5 (*d*) 6

Sol. (*b*) Let G.E. $= x \Rightarrow x^3 = (20 + 14\sqrt{2}) + (20 - 14\sqrt{2}) + 3(\sqrt[3]{20+14\sqrt{2}})(\sqrt[3]{20-14\sqrt{2}})(x)$

$\Rightarrow x^3 = 40 + 3(\sqrt[3]{400-392})x = 40 + 3(\sqrt[3]{8})x = 40 + 3(2)x$

$\Rightarrow x^3 = 40 + 6x$. Out of given options $x = 4$ satisfies this equation.

188. $\sqrt{2\sqrt{2\sqrt{2\sqrt{2}}}} = ?$

(*a*) $\sqrt{2}$ (*b*) 4 (*c*) $2^{\frac{7}{8}}$ (*d*) $2^{\frac{15}{16}}$

Sol. (*d*) $\sqrt{2\sqrt{2\sqrt{2\sqrt{2}}}} = \sqrt{2\sqrt{2\sqrt{2 \times 2^{\frac{1}{2}}}}} = \sqrt{2\sqrt{2\sqrt{2^{\frac{3}{2}}}}}$

$= \sqrt{2\sqrt{2 \times 2^{\frac{3}{4}}}} = \sqrt{2\sqrt{2^{\frac{7}{4}}}} = \sqrt{2 \times 2^{\frac{7}{8}}} = \sqrt{2^{\frac{15}{8}}} = 2^{\frac{15}{16}}$

189. The value of $\sqrt{2\sqrt[3]{4\sqrt{2\sqrt[3]{4\sqrt{2\sqrt[3]{4...}}}}}}$ is **(SSC CGL Tier II 2013)**

(*a*) 2 (*b*) 2^2 (*c*) 2^3 (*d*) 2^5

Sol. (*a*) Let G.E. $= x$. Then $x^2 = 2\sqrt[3]{4\sqrt{2\sqrt[3]{4\sqrt{2\sqrt[3]{4...}}}}} \Rightarrow (x^2)^3 = 2^3 \times 4(x)$

$\Rightarrow \quad x^6 = 32x \Rightarrow x^5 = 32 = 2^5 \Rightarrow x = 2$

190. Number of solutions of the equation $\sqrt{x^2 - x + 1} + \dfrac{1}{\sqrt{x^2 - x + 1}} = 2 - x^2$ **(SSC CGL Tier II 2012)**

(*a*) 2 (*b*) 4 (*c*) 0 (*d*) 1

Sol. (*d*) We know that $a + \dfrac{1}{a} \geq 2 \Rightarrow \sqrt{x^2 - x + 1} + \dfrac{1}{\sqrt{x^2 - x + 1}} \geq 2$

$\Rightarrow 2 - x^2 \geq 2 \Rightarrow x^2 \leq 0 \Rightarrow x = 0$ (only one solution)

191. Which of the following is the biggest in value: **(SSC CGL Tier I 2007)**

(*a*) $\sqrt{19} - \sqrt{17}$ (*b*) $\sqrt{13} - \sqrt{11}$ (*c*) $\sqrt{7} - \sqrt{5}$ (*d*) $\sqrt{5} - \sqrt{3}$

Sol. (*d*) $\sqrt{19} - \sqrt{17} = \dfrac{19 - 17}{\sqrt{19} + \sqrt{17}} = \dfrac{2}{\sqrt{19} + \sqrt{17}}$ $\quad \sqrt{13} - \sqrt{11} = \dfrac{13 - 11}{\sqrt{13} + \sqrt{11}} = \dfrac{2}{\sqrt{13} + \sqrt{11}}$

$\sqrt{7} - \sqrt{5} = \dfrac{7 - 5}{\sqrt{7} + \sqrt{5}} = \dfrac{2}{\sqrt{7} + \sqrt{5}}$ $\quad \sqrt{5} - \sqrt{3} = \dfrac{5 - 3}{\sqrt{5} + \sqrt{3}} = \dfrac{2}{\sqrt{5} + \sqrt{3}}$

In the above fractions, all have same numerator while denominators are different. The fraction with smallest denominator is biggest and $\sqrt{5} + \sqrt{3}$ is the smallest of the four denominators.

Therefore, $\dfrac{2}{\sqrt{5} + \sqrt{3}}$ is the biggest fraction.

Note: If $x > y$, then $\sqrt{x + k} - \sqrt{x} < \sqrt{y + k} - \sqrt{x}$

192. Let $a = \sqrt{6} - \sqrt{5}$, $b = \sqrt{5} - 2$, $c = 2 - \sqrt{3}$. Then point out the correct alternative among the four alternatives given below. **(SSC CHSL 2013)**

(*a*) $a < b < c$ (*b*) $b < a < c$ (*c*) $a < c < b$ (*d*) $b < c < a$

Sol. (*a*) $a = \sqrt{6}-\sqrt{5} = \dfrac{1}{\sqrt{6}+\sqrt{5}}$, $b = \sqrt{5}-\sqrt{4} = \dfrac{1}{\sqrt{5}+\sqrt{4}}$, $c = \sqrt{4}-\sqrt{3} = \dfrac{1}{\sqrt{4}+\sqrt{3}}$

$\therefore a < b < c$

193. If $x^{m^n} = x^{mn}$ find the value of m. **(MBA Entrance Exam.)**

(*a*) 0 (*b*) 1 (*c*) $n^{\frac{1}{n}}$ (*d*) $n^{\frac{1}{(n-1)}}$

Sol. (*d*) $x^{m^n} = x^{mn} \Rightarrow m^n = mn \Rightarrow m \times m^{n-1} = mn \Rightarrow m^{n-1} = n \Rightarrow m = n^{\frac{1}{(n-1)}}$

194. $\dfrac{1}{1+a^{m-n}} + \dfrac{1}{1+a^{n-m}} = ?$ **(MBA Entrance Exam.)**

(*a*) 0 (*b*) 1 (*c*) a^{m-n} (*d*) a^{m+n}

Sol. (*b*) Given expression $= \dfrac{1}{1+\dfrac{a^m}{a^n}} + \dfrac{1}{1+\dfrac{a^n}{a^m}} = \dfrac{a^n}{a^m+a^n} + \dfrac{a^m}{a^m+a^n} = 1$

195. $\left(\dfrac{x^a}{x^b}\right)^{a+b} \times \left(\dfrac{x^b}{x^c}\right)^{b+c} \times \left(\dfrac{x^c}{x^a}\right)^{c+a} = ?$

(*a*) 0 (*b*) 1 (*c*) x (*d*) $\dfrac{1}{x}$

Sol. (*b*) Given expression $= x^{(a-b)(a+b)} \times x^{(b-c)(b+c)} \times x^{(c-a)(c+a)}$

$= x^{a^2-b^2} \times x^{b^2-c^2} \times x^{c^2-a^2} = x^{a^2-b^2+b^2-c^2+c^2-a^2} = x^0 = 1$

196. $\left(\dfrac{x^a}{x^b}\right)^{(a+b-c)} \left(\dfrac{x^b}{x^c}\right)^{(b+c-a)} \left(\dfrac{x^c}{x^a}\right)^{(c+a-b)} = ?$ **(MBA Entrance Exam.)**

(*a*) –1 (*b*) 0 (*c*) 1 (*d*) $a+b+c$

Sol. (*c*) Given expression $= (x)^{(a-b)(a+b-c)} (x)^{(b-c)(b+c-a)} (x)^{(c-a)(c+a-b)}$

$= (x)^{(a-b)(a+b-c)+(b-c)(b+c-a)+(c-a)(c+a-b)}$

$= (x)^{(a-b)(a+b)+(b-c)(b+c)+(c-a)(c+a)} \times (x)^{(a-b)(-c)+(b-c)(-a)+(c-a)(-b)}$

$= (x)^{(a^2-b^2)+(b^2-c^2)+(c^2-a^2)} \times (x)^{(-ac+bc)+(-ab+ac)+(-bc+ab)} = x^0 \times x^0 = 1$

197. If $x + \dfrac{1}{x} = \sqrt{3}$, then the value of $x^{18}+x^{12}+x^6+1$ is **(SSC CPO SI 2011)**

(*a*) 0 (*b*) 1 (*c*) 2 (*d*) 3

Sol. (*a*) $x^3 + \dfrac{1}{x^3} = \left(x+\dfrac{1}{x}\right)^3 - 3\left(x+\dfrac{1}{x}\right) = (\sqrt{3})^3 - 3\sqrt{3} = 0 \Rightarrow x^6+1 = 0$

$\Rightarrow x^{18}+x^{12}+x^6+1 = x^{12}(x^6+1)+(x^6+1) = 0+0 = 0$

Alt: $x^6+1 = 0 \Rightarrow x^6 = -1 \Rightarrow$ G.E. $= -1+1-1+1 = 0$

198. If $a^x = (x+y+z)^y$, $a^y = (x+y+z)^z$ and $a^z = (x+y+z)^x$, then $x+y+z = ?$ **(SSC CGL Tier I 2012)**

(*a*) 0 (*b*) 1 (*c*) a (*d*) $3a$

Sol. (*c*) $a^x.a^y.a^z = (x+y+z)^y \times (x+y+z)^z \times (x+y+z)^x$

$\Rightarrow a^{x+y+z} = (x+y+z)^{x+y+z} \Rightarrow x+y+z = a$

190. If $a^x = b$, $b^y = c$ and $c^z = a$, find the value of xyz. **(SSC CHSL 2013)**

(*a*) 1 (*b*) 2 (*c*) 0 (*d*) abc

Sol. (*a*) $a = c^z = (b^y)^z = b^{yz} = (a^x)^{yz} = a^{xyz} \Rightarrow xyz = 1$

199. If $\sqrt[x]{a} = \sqrt[y]{b} = \sqrt[z]{c}$ and $abc = 1$, then $x + y + z = ?$

(*a*) –1 (*b*) 0 (*c*) 1 (*d*) 3

Sol. (*b*) Let $\sqrt[x]{a} = \sqrt[y]{b} = \sqrt[z]{c} = k \Rightarrow a = k^x, b = k^y, c = k^z$

Given $abc = 1 \Rightarrow k^x \times k^y \times k^z = 1 \Rightarrow k^{x+y+z} = 1 = k^0 \Rightarrow x + y + z = 0$

200. If $2^x = 3^y = 6^{-z}$ then $\frac{1}{x} + \frac{1}{y} + \frac{1}{z} = ?$ **(SSC CHSL 2011)**

(*a*) 0 (*b*) 1 (*c*) $\frac{3}{2}$ (*d*) $-\frac{1}{2}$

Sol. (*a*) Let $2^x = 3^y = 6^{-z} = k \Rightarrow 2 = k^{\frac{1}{x}}, 3 = k^{\frac{1}{y}}, 6 = k^{-\frac{1}{z}}$

$$2 \times 3 = 6 \Rightarrow k^{\frac{1}{x}} \times k^{\frac{1}{y}} = k^{-\frac{1}{z}} \Rightarrow \frac{1}{x} + \frac{1}{y} = -\frac{1}{z} \Rightarrow \frac{1}{x} + \frac{1}{y} + \frac{1}{z} = 0$$

201. If $4^x = 5^y = 20^z$ then $z = ?$ **(MBA 1984)**

(*a*) xy (*b*) $\frac{1}{xy}$ (*c*) $\frac{x+y}{xy}$ (*d*) $\frac{xy}{x+y}$

Sol. (*d*) Let $4^x = 5^y = 20^z = k \Rightarrow 4 = k^{\frac{1}{x}}, 5 = k^{\frac{1}{y}}, 20 = k^{\frac{1}{z}}$

$$\Rightarrow 4 \times 5 = 20 \Rightarrow k^{\frac{1}{x}} \times k^{\frac{1}{y}} = k^{\frac{1}{z}} \Rightarrow \frac{1}{x} + \frac{1}{y} = \frac{1}{z} \Rightarrow \frac{x+y}{xy} = \frac{1}{z} \Rightarrow \frac{xy}{x+y} = z$$

202. If $2^x = 3^y = 12^z$, then :

(*a*) $\frac{1}{x} + \frac{1}{y} + \frac{1}{z} = 0$ (*b*) $xyz = 0$ (*c*) $\frac{1}{z} = \frac{1}{y} + \frac{2}{x}$ (*d*) $\frac{1}{z} = \frac{1}{x} + \frac{1}{y}$

Sol. (*c*) Let $2^x = 3^y = 12^z = k$. Then $2 = k^{1/x}, 3 = k^{1/y}, 12 = k^{1/z}$

$$12 = 2^2 \times 3 \Rightarrow k^{1/z} = k^{2/x} \times k^{1/y} \Rightarrow \frac{1}{z} = \frac{2}{x} + \frac{1}{y}$$

203. If $2^x = 4^y = 8^z$ and $xyz = 288$ then the value of $\frac{1}{2x} + \frac{1}{4y} + \frac{1}{8z}$ is : **(SSC CGL Tier I 2014)**

(*a*) 11/12 (*b*) 11/96 (*c*) 29/96 (*d*) 27/96

Sol. (*b*) $2^x = 4^y \Rightarrow x = 2y;\ 2^x = 8^z \Rightarrow x = 3z \Rightarrow x : y : z = 6 : 3 : 2$

$xyz = 288 \Rightarrow (6k \times 3k \times 2k) = 288 \Rightarrow (36k^3) = 288 \Rightarrow k^3 = 8 \Rightarrow k = 2$

$$\Rightarrow \frac{1}{2x} + \frac{1}{4y} + \frac{1}{8z} = \frac{1}{2 \times 12} + \frac{1}{4 \times 6} + \frac{1}{8 \times 4} = \frac{1}{24} + \frac{1}{24} + \frac{1}{32} = \frac{11}{96}$$

204. If $a^{\frac{1}{x}} = b^{\frac{1}{y}} = c^{\frac{1}{z}}$, and a, b, c are in GP, then $2y = ?$

(*a*) $x - z$ (*b*) $x + z$ (*c*) xz (*d*) $\frac{x}{z}$

Sol. (*b*) Let $a^{\frac{1}{x}} = b^{\frac{1}{y}} = c^{\frac{1}{z}} = k \Rightarrow a = k^x, b = k^y, c = k^z$

a, b, c are in G.P. $\Rightarrow b^2 = ac \Rightarrow (k^y)^2 = k^x \cdot k^z \Rightarrow 2y = x + z.$

CHAPTER

12

RATIO, PROPORTION AND PARTNERSHIP

Ratio between two numbers x and y is expressed as $x : y$. It can also be expressed as x/y. The first term in a ratio is called **antecedent** and the second term is called **consequent**. Ratio has no unit. Ratio between ₹ 300 and ₹ 500 is 3 : 5 and not ₹ 3 and ₹ 5.

Note : Multiplication or division of each term of a ratio by the same non-zero number does not affect the ratio. But the same is not true for Addition or Subtraction.

$x : y = ax : ay = \frac{x}{a} : \frac{y}{a}$, where $a \neq 0$. But $x : y \neq (x \pm a) : (y \pm a)$

Example: 3 : 5 is same as 6 : 10 or 30 : 50.

Proportion: The equality of two ratios is called **proportion**.

a, b, c and d are in proportion, if $a : b = c : d$. It can also be expressed as $a : b :: c : d$

In any proportion, the first and fourth terms are called the **extreme terms** while the second and the third terms are called **mean terms**.

In a proportion, Product of Extremes = Product of Means *i.e.* $a \times d = b \times c$

Example. Which of the following represents a correct proportion ? **(SSC CGL Tier I 2014)**

(*a*) 12 : 9 = 16 : 12 (*b*) 13 : 11 = 5 : 4 (*c*) 30 : 45 = 13 : 24 (*d*) 3 : 5 = 2 : 5

Sol. (*a*) $12 \times 12 = 9 \times 16$

Inverse or Reciprocal Ratio : If ratio between two numbers is $a : b$. Then the inverse ratio is $\frac{1}{a} : \frac{1}{b}$ or $b : a$. **Example**: Inverse ratio of 2 : 3 is 3 : 2.

Direct Proportion : Two quantities are said to be in direct proportion if on increase (or decrease) in one quantity, the other quantity also increases (or decreases) proportionately *i.e.* in the same ratio.

Examples: Number of articles bought and total cost of articles *i.e.* more articles, more cost and less articles, less cost.

Cost per article and total cost of certain number of articles.

Indirect or Inverse proportion : Two quantities are said to be in indirect or inverse proportion, if on increase in one quantity, the other quantity decreases in the same ratio and vice-versa. Remember that the product of the two quantities will remain the same.

Examples: Number of workers and days taken to do a certain work.

Speed of a train and time taken to cover a certain distance. (More speed, less time)

Price of a product and quantity of the product that can be purchased for a certain amount.

Note : If ratio of prices of two commodities change in the ratio $x : y$, then with the same amount of money, quantity purchased changes to ratio $y : x$.

Continued proportion : Three numbers are said to be in continued proportion when the ratio of the first number to the second is equal to ratio of the second number to the third.

x, y and z are said to be in proportion if $x : y :: y : z$. **Example** : 4 : 6 :: 6 : 9

Mean Proportion: If x is Mean or Second Proportion of a and b.

Then $a : x = x : b \Rightarrow x^2 = ab \Rightarrow x = \sqrt{ab}$

Example: Mean proportional of 32 and 2 is:

(*a*) 1 (*b*) 8 (*c*) 16 (*d*) 64

Sol. (*b*) Let mean proportional to 32 and 2 is x. Then $32 : x = x : 2 \Rightarrow x = \sqrt{32 \times 2} = 8$

Third Proportion: If x is Third Proportion of a and b. Then $a : b = b : x \Rightarrow x = \frac{b^2}{a}$

Example 1. Third proportional to 16 and 4 is:

(*a*) 1 (*b*) 4 (*c*) 12 (*d*) 64

Sol. (*a*) Let third proportional to 16 and 4 is x. Then $16 : 4 = 4 : x \Rightarrow x = \frac{4 \times 4}{16} = 1$

Fourth Proportion: If x is Fourth Proportion of a, b and c. Then $a : b = c : x \Rightarrow x = \frac{bc}{a}$

Example 1. Fourth proportional to 6, 9 and 20 is: **(SSC CGL Tier I)**

(*a*) 5 (*b*) 23 (*c*) 30 (*d*) 45

Sol. (*c*) Let fourth proportional to 6, 9 and 20 is x. Then $6 : 9 = 20 : x \Rightarrow x = \frac{9 \times 20}{6} = 30$

Example 2. The fourth proportional to 7, 11 and 14 is: **(CDS 2011)**

(*a*) 16 (*b*) 18 (*c*) 20 (*d*) 22

Sol. (*d*) Let fourth proportional to 7, 11 and 14 is x. Then $7 : 11 = 14 : x \Rightarrow x = \frac{11 \times 14}{7} = 22$

RULES

1. If $a : b = m : n$ and $b : c = p : q$. Then $a : c = \frac{m}{n} \times \frac{p}{q}$

Note: This rule can be extended for any number of ratios.

Example: If $a : b = 2 : 3$, $b : c = 4 : 5$ and $c : d = 6 : 7$, then $a : d = ?$ **(SSC CGL Tier I 2012M)**

(*a*) 2 : 7 (*b*) 16 : 35 (*c*) 12 : 21 (*d*) None of these

Sol. (*b*) $\frac{a}{d} = \frac{2}{3} \times \frac{4}{5} \times \frac{6}{7} = \frac{16}{35} \Rightarrow a : d = 16 : 35$

2. If x times of a number is equal to y times of other number

Then ratio between two numbers $= \frac{1}{x} : \frac{1}{y} = y : x$

Example: If 5 times a number is equal to 3 times another number. Find ratio between two numbers.

(*a*) 2 : 3 (*b*) 4 : 5 (*c*) 3 : 5 (*d*) 5 : 3

Sol. (*c*) Ratio between two numbers $= \frac{1}{5} : \frac{1}{3} = 3 : 5$

3. If the sum of two numbers is x and their difference is y, ratio of two numbers is $(x + y) : (x - y)$.

Note: The two numbers are $\frac{x+y}{2}$ and $\frac{x-y}{2}$ respectively.

Proof: Let the numbers are a and b. Then $a + b = x$, and $a - b = y$

Now $(a + b) + (a - b) = x + y \Rightarrow a = \frac{x+y}{2}$ and $(a + b) - (a - b) = x - y \Rightarrow b = \frac{x-y}{2}$

$\therefore a : b = \frac{x+y}{2} : \frac{x-y}{2} = (x+y) : (x-y)$

Example: The ratio of the sum and difference of two numbers is 5 : 2. What is the ratio of two numbers?

(*a*) 7 : 3 (*b*) 3 : 7 (*c*) 1 : 10 (*d*) 5 : 2

Sol. (*a*) Ratio of the numbers $= \frac{5+2}{5-2} = 7 : 3$

4. If $\frac{am}{bn} = \frac{x}{y} \Rightarrow \frac{a}{b} = \frac{x}{y} : \frac{m}{n} = \frac{x}{m} : \frac{y}{m}$

Example: Product of two sets of two numbers are in the ratio of 3 : 5. If the ratio of two numbers one from each set is 2 : 3, find the ratio between the two other numbers.

Sol. Required ratio $= \frac{3}{2} : \frac{5}{3} = 9 : 10$

5. What must be added to each of the four numbers *a*, *b*, *c* and *d* so that the resultant numbers are in proportion ?

Sol. The number to be added is *x* where

$$x = \frac{bc - ad}{(a+d)-(b+c)} = \frac{\text{Product of means} - \text{Product of extremes}}{\text{Sum of extremes} - \text{Sum of means}}$$

Alternative Method: 1. Find the difference between the first term and the second term and similarly difference between the third and the fourth term, separately.

2. Divide bigger difference by the smaller difference and get the quotient.

3. Let us assume that the second difference is bigger, then :

Third term – (First Term × Quotient) will give the surplus value.

4. Divide Surplus by (Quotient – 1)

Example 1: What must be added to each of the numbers 7, 10, 19 and 25 so that the resultant numbers are in proportion ?

(*a*) 2 (*b*) 3 (*c*) 4 (*d*) 5

Sol. (*d*) Required number $= x = \frac{10\times19 - 7\times25}{(7+25)-(10+19)} = \frac{15}{3} = 5$

Alt: Differences = 10 – 7 = 3 and 25 – 19 = 6

Quotient = 6 ÷ 3 = 2   19 – 7 × 2 = 5   5 ÷ (2 – 1) = 5

New numbers = (7 + 5), (10 + 5), (19 + 5), (25 + 5) = 12, 15, 24, 30 and 12 : 15 = 24 : 30

Note: 24 ÷ 12 = 2 and 30 ÷ 15 = 2 (equal to our quotient)

Example 2: The number to be added to each of the numbers 7, 16, 43, 79 to make the numbers in proportion is **(SSC CGL Tier I 2012)**

(*a*) 2 (*b*) 3 (*c*) 5 (*d*) 1

Sol. (*c*) Required number $= x = \frac{16\times43 - 7\times79}{(7+79)-(16+43)} = \frac{135}{27} = 5$

Alt: Differences = 16 – 7 = 9 and 79 – 43 = 36

Quotient = 36 ÷ 9 = 4   43 – 7 × 4 = 15   15 ÷ (4 – 1) = 5

New numbers = (7 + 5), (16 + 5), (43 + 5), (75 + 5) = 12 : 21 = 48 : 84
Note: 48 ÷ 12 = 4 and 84 ÷ 21 = 4 (equal to our quotient)

6. If $\frac{a}{b}=\frac{c}{d}$ then

(*i*) $\frac{b}{a}=\frac{d}{c}$ (Invertendo Rule: Obtained by dividing 1 by the given ratios)

(*ii*) $\frac{a}{c}=\frac{b}{d}$ (Alternendo Rule)

(*iii*) $\frac{a+b}{b}=\frac{c+d}{d}$ (Rule of Componendo. Arrived at by adding 1 to each side)

(*iv*) $\frac{a-b}{b}=\frac{c-d}{d}$ (Rule of Dividendo. Arrived at by subtracting 1 from each side)

(*v*) $\frac{a+b}{a-b}=\frac{c+d}{c-d}$ (Rule of Componendo and Dividendo: Arrived at by dividing (*iii*) by (*iv*)

(*vi*) $\frac{a}{b}=\frac{c}{d}=\frac{a+c}{b+d}$ (Rule of Addendo)

(*vii*) $\frac{a}{b}=\frac{c}{d}=\frac{a-c}{b-d}$ (Rule of Subtranendo)

(*viii*) $\frac{a}{a-b}=\frac{c}{c-d}$ (Rule of Convertendo)

7. (*i*) If $a : b$ and $c : d$ are two ratios, ratio compunded of two ratios is $ac : bd$
Similarly compounded ratio of $a : b$, $c : d$ and $e : f$ is $ace : bdf$

(*ii*) A ratio compounded to itself is called its duplicate ratio. $x^2 : y^2$ is duplicate ratio of $x : y$.
Example : Duplicate ratio of 2 : 3 is $2^2 : 3^2 = 4 : 9$.

(*iii*) Triplicate ratio of $x : y$ is $xxx : yyy = x^3 : y^3$.
Example : Triplicate ratio of 2 : 3 is $2^3 : 3^3 = 8 : 27$.

(*iv*) Sub duplicate ratio of $x : y$ is $\sqrt{x} : \sqrt{y}$.
Example : Sub-duplicate ratio of 4 : 9 is $\sqrt{4} : \sqrt{9} = 2 : 3$.

(*v*) Sub triplicate ratio of $x : y$ is $\sqrt[3]{x} : \sqrt[3]{y}$.
Example : Sub-triplicate ratio of 8 : 27 is $\sqrt[3]{8} : \sqrt[3]{27} = 2 : 3$.

8. If $a : b = c : d = e : f = $ ...., then each of these ratios is equal to the followings :

(*i*) $(a + c + e + \text{..........}) : (b + d + f + \text{..........})$

(*ii*) $(a - c - e - \text{..........}) : (b - d - f - \text{..........})$

(*iii*) $(ra + sc + te + \text{..........}) : (rb + sd + tf + \text{..........})$

Example: If $a : b = c : d = e : f = 1 : 2$, then $(3a + 5c + 7e) : (3b + 5d + 7f)$ is equal to : **(SSC CGL Tier I 2005)**

(*a*) 8 : 7 (*b*) 2 : 1 (*c*) 1 : 4 (*d*) 1 : 2

Sol. (*d*) Required ratio = $a : b = 1 : 2$

9. If two numbers are in the ratio $x : y$ and on adding a constant number a to both the numbers, the new ratio becomes $(x + 1) : (y + 1)$, Or On subtracting a from both the numbers, the ratio becomes $(x - 1) : (y - 1)$. Then the original numbers are ax and ay respectively.

Example 1. If two numbers are in the ratio 2 : 3 and the ratio becomes 3 : 4 when 8 is added to both the numbers, then the sum of the two numbers is : **(SSC CGL Tier I 2015)**

(*a*) 80 (*b*) 40 (*c*) 10 (*d*) 100

Sol. (*b*) Here on adding 8 to both the numbers, both ratios are increased by one.

∴ The original numbers are 2 × 8 and 3 × 8 i.e. 16 and 24. Sum of 16 + 24 = 40

Example 2. Two numbers are in the ratio 8 : 11. If we subtract 6 from each number, then new ratio becomes 7 : 10. The original numbers are:

(*a*) 48, 66 (*b*) 42, 60 (*c*) 40, 55 (*d*) 56, 77

Sol. (*a*) On subtracting 6 from both the numbers, the new ratio is decreased by same ratio i.e. 1 in each case. ∴ The original numbers are 8 × 6 and 11 × 6 = 48 and 66.

(*a*) 48, 66 (*b*) 42, 60 (*c*) 40, 55 (*d*) 56, 77

10. If two numbers are in the ratio $x : y$ and on adding a constant number a to both the numbers, the new ratio becomes $(x + k) : (y + k)$, Or On subtracting a from both the numbers, the ratio becomes $(x - k) : (y - k)$. Then the original numbers are $\frac{a}{k}x$ and $\frac{a}{k}y$ respectively.

Example. The ratio of monthly incomes of A, B is 6 : 5 and their monthly expenditures are in the ratio 4 : 3. If each of them saves ₹ 400 per month, find the sum of their monthly incomes. **(SSC CGL Tier I 2013)**

(*a*) ₹ 2300 (*b*) ₹ 2400 (*c*) ₹ 2200 (*d*) ₹ 2500

Sol. (*c*)	A		B
Income	6	:	5
Expenditure	4	:	3
Savings	2	:	2

Hence savings of each = 2 ratio = ₹ 400 ⇒ 1 ratio = ₹ 200

∴ Sum of monthly income of A and B = 11 ratio = 11 × ₹ 200 = ₹ 2200

11. If two numbers are in the ratio $x : y$. On adding a constant number a to the first number, the new ratio becomes $(x + 1) : (y)$, Or On subtracting a from the first number, the ratio becomes $(x - 1) : (y)$. Then the original numbers are ax and ay respectively.

Example 1. Ratio of milk and water in a mixture is 4 : 1. If 5 litre of milk is added in the mixture, new ratio becomes 5 : 1. The quantity of milk in the mixture originally was:

(*a*) 10 litres (*b*) 15 litres (*c*) 20 litres (*d*) 25 litres

Sol. (*c*) On adding 5 litres of milk, its ratio in the mixture is increased by 1.

∴ Original quantity of milk in the mixture was 4 × 5 = 20 litres.

12. If two numbers are in the ratio $a : b$ and on adding k, a constant number to one of the numbers, new ratio becomes $m : n$, i.e. both the ratios change instead of one ratio only.

Method:

1. Multiply the ratios (old, new or both) so that the old and new ratios become the same for the number, to which k is not added to or subtracted from.
2. Find increase or decrease in the revised new ratio from revised old ratio.
3. Number added to get the desired change in the ratio is :

$$\frac{\text{Difference in revised old and new ratios}}{\text{Sum of the revised old ratio}} \times \text{Total quantity (initial)}$$

Example. 132 litres of a mixture contains milk and water in the ratio 8 : 3. How much water must be added to the mixture so as to make the ratio of milk and water 2 : 1?

(*a*) 11 litres (*b*) 12 litres (*c*) 13 litres (*d*) 15 litres

Sol. (*b*) LCM of ratios of milk *i.e.* LCM of 8 and 2 is 8.

∴ We make ratio of milk equal to 8 in the two ratios.

	Milk	:	Water
New ratio = 2 : 1 =	8	:	4
Old ratio =	8	:	3
Difference =	0	:	1

$$\therefore \text{ Water to be added} = \frac{1}{8+3} \times 132 = 12 \text{ litres}$$

13. Two numbers are in the ratio *a* : *b*. If *k*, a constant number is **added to both the numbers, new ratio of the numbers becomes *m* : *n*.**

Then the original numbers are *ax* and *bx*, where $x = \frac{(m-n)k}{an-bm}$

$$= \frac{\text{Difference between new ratios} \times \text{Number added to both numbers}}{\text{Difference between cross multiplication of the two ratios}}$$

Note: Constant number *k* added to both the numbers is $k = \frac{an - bm}{m - n} \times x$

Proof: Let the original numbers are '*ax*' and '*bx*'. Then $\frac{ax+k}{bx+k} = \frac{m}{n}$

$$\therefore anx + nk = bmx + mk \Rightarrow (an - bm)x = (m - n)k \Rightarrow x = \frac{(m-n)k}{an-bm}$$

Example 1. The ratio between two numbers is 5 : 3. If 3 is added to both the numbers, the ratio becomes 14 : 9. Find the smaller number.

(*a*) 15 (*b*) 18 (*c*) 25 (*d*) 28

Sol. (*a*) $x = \frac{(14-9) \times 3}{5 \times 9 - 3 \times 14} = \frac{5 \times 3}{45 - 42} = \frac{5 \times 3}{3} = 5 \Rightarrow$ Smaller number = 3*x* = 3 × 5 = 15

Alt : Difference in origianl old ratios = 5 – 3 = 2

Difference in origianl new ratios = 14 – 9 = 5

∴ We multiply the old ratios by 5 and new ratios by 2.

Old ratio = 5 : 3 =	25	:	15
New ratio = 14 : 9 =	28	:	18
Difference =	3	:	3

3 ratio = 3 ⇒ 1 ratio = 3 ÷ 3 = 1 ⇒ Smaller number = 15 ratio = 15

Example 2. Two numbers are in the ratio 5 : 7. On diminishing each of them by 40, they become in the ratio 17 : 27. The difference of the number is : **(SSC CGL Tier I 2011)**

(*a*) 18 (*b*) 52 (*c*) 137 (*d*) 50

Sol. (*d*) $x = \dfrac{(17-27)\times(-40)}{5\times 27 - 7\times 17} = \dfrac{400}{16} = 25$

$\therefore$ Difference between numbers $= 7x - 5x = 2x = 2 \times 25 = 50$

Alternative method :

Old ratio = 5 : 7 =	25	:	35	(Difference = (7 – 5) ratio = 2 ratio)
New ratio =	17	:	27	(Difference = (27 – 17) ratio = 10 ratio)
Difference =	8	:	8	

$\therefore$ Difference in each case = 8 ratio = 40 $\Rightarrow$ 1 ratio = 5

Difference between the numbers = (35 – 25) ratio = 10 ratio = 10 × 5 = 50

14. The incomes of two persons are in the ratio $a : b$ and their expenditures are in the ratio $m : n$. If both saves ₹ S, then their incomes are ax and bx, where $x = \dfrac{S(n-m)}{an-bm}$

Example. The monthly incomes of two persons are in the ratio 2 : 3 and their monthly expenses are in the ratio 5 : 9. If each of them saves ₹ 600 per month, then their monthly incomes are : **(SSC CGL Tier I 2005)**

(*a*) ₹ 1500; ₹ 2250 (*b*) ₹ 1200; ₹ 1800 (*c*) ₹ 1600; ₹ 2400 (*d*) ₹ 1400; ₹ 2100

Sol. (*c*) $x = \dfrac{S(n-m)}{an-bm} = \dfrac{600\times(9-5)}{2\times 9 - 3\times 5} = \dfrac{600\times 4}{3} = ₹\ 800$

$\therefore$ A's income = 2 × ₹ 800 = ₹ 1600. B's income = 3 × ₹ 800 = ₹ 2400

Alt : Let their incomes be $2x$ and $3x$ and expenditures be $5y$ and $9y$.

$2x - 5y = 3x - 9y = ₹\ 600 \Rightarrow x = 4y$

$2x - 5y = ₹\ 600 \Rightarrow 8y - 5y = ₹\ 600 \Rightarrow 3y = ₹\ 600 \Rightarrow y = ₹\ 200$

$x = 4y = 4 \times ₹\ 200 = ₹\ 800$

$\therefore$ A's income = 2 × ₹ 800 = ₹ 1600 B's income = 3 × ₹ 800 = ₹ 2400

15. When two or more persons start a business by investing some money, profit or loss is distributed among the partners in the ratio of their equivalent monthly investment, unless stated otherwise.

$$\frac{\text{Profit of A}}{\text{Profit of B}} = \frac{\text{Investment by A} \times \text{Period of Investment by A}}{\text{Investment by B} \times \text{Period of Investment by B}}$$

SOLVED EXERCISE

1. A man's expenditure on food is 3/5 of his expenditure on clothes, find ratio between his expenditure on the two items.

(*a*) 2 : 5 (*b*) 3 : 5 (*c*) 2 : 3 (*d*) 5 : 3

Sol. (*b*) Let his expenditure on clothes is ₹ 5. Then expenditure on food is ₹ 3.

$\therefore$ Ratio of expenditure on food and clothes = 3 : 5

2. A mixture of milk and water is such that the quantity of milk is 3/5 that of water. The proportion of milk in the mixture is : **(SSC CHSL 2012)**

(*a*) $\frac{1}{8}$ (*b*) $\frac{1}{2}$ (*c*) $\frac{3}{8}$ (*d*) $\frac{5}{8}$

Sol. (*c*) Let quantity of water = 5. Then quantity of milk = 3. ⇒ Required ratio = $\frac{3}{8}$

3. 16 litres of mixture contains milk and water in the ratio 5 : 3. If 4 litres of milk is added to this mixture, the ratio of milk to water in the mixture would be **(CDS 2015)**

(*a*) 2 : 1 (*b*) 7 : 3 (*c*) 4 : 3 (*d*) 8 : 3

Sol. (*b*) Total mixture = (5 + 3) ratio = 16 litres ⇒ 1 ratio = 2 litres

Milk in original mixture = 5 × 2 = 10 litres ⇒ water = 16 – 10 = 6 litres

⇒ New ratio = (10 + 4) : 6 = 14 : 6 = 7 : 3

4. Out of 120 applications for a post, 70 are male and 80 have a driver's license. What is the ratio between the minimum to maximum number of males having driver's license? **(UPSC Civil Services (Pre.) 2013)**

(*a*) 1 to 2 (*b*) 2 to 3 (*c*) 3 to 7 (*d*) 5 to 7

Sol. (*c*) Number of females = 120 – 70 = 50

Minimum male license holder = 80 – 50 = 30. Maximum male license holder = 70

⇒ Required ratio = 30 : 70 = 3 : 7

5. 0.6 of a number is equal to 0.08 of another number. The ratio of the numbers will be : **(SSC CGL Tier I 1999)**

(*a*) 3 : 4 (*b*) 4 : 3 (*c*) 2 : 15 (*d*) 2 : 9

Sol. (*c*) Required ratio = $\frac{1}{0.6} : \frac{1}{0.08}$ = 0.08 : 0.6 = 8 : 60 = 2 : 15

6. In a class there are '*z*' students. Out of them '*x*' are boys. What part of the class is composed of girls? **(SSC CGL Tier II 2013)**

(*a*) $\frac{x}{z}$ (*b*) $\frac{z}{x}$ (*c*) $1-\frac{x}{z}$ (*d*) $\frac{x}{z}-1$

Sol. (*c*) Boys = $\frac{x}{z}$ ⇒ Girls = $1-\frac{x}{z}$

7. What is the mean proportional between $(15+\sqrt{200})$ and $(27-\sqrt{648})$? **(CDS 2012)**

(*a*) 4 (*b*) $14\sqrt{7}$ (*c*) $3\sqrt{5}$ (*d*) $5\sqrt{3}$

Sol. (*c*) Mean proportional = $\sqrt{(15+\sqrt{200})(27-\sqrt{648})}$

$$= \sqrt{(15+10\sqrt{2})(27-18\sqrt{2})} = \sqrt{405+270\sqrt{2}-270\sqrt{2}-360} = \sqrt{45} = 3\sqrt{5}$$

8. The third proportional of 12 and 18 is **(SSC CGL Tier II 2013)**

(*a*) 3 (*b*) 6 (*c*) 27 (*d*) 144

Sol. (*c*) 12 : 18 = 18 : *x* ⇒ $x = \frac{18\times18}{12} = 27$

9. The third proportional to $\left(\frac{x}{y}+\frac{y}{x}\right)$ and $\sqrt{x^2+y^2}$ is **(SSC CGL Tier I 2013)**

(*a*) xy (*b*) $\sqrt{xy}$ (*c*) $\sqrt[3]{xy}$ (*d*) $\sqrt[4]{xy}$

Sol. (*a*) Third proportional = $\frac{\left(\sqrt{x^2+y^2}\right)^2}{\left(\frac{x}{y}+\frac{y}{x}\right)} = \frac{x^2+y^2}{\frac{x^2+y^2}{xy}} = xy$

10. If $x : y :: y : z$, then $x : z$ is equal to

(*a*) 1 : 1 (*b*) $z : x$ (*c*) $x^2 : y^2$ (*d*) $y^2 : x^2$

Sol. (*c*) $x : y = y : z \Rightarrow xz = y^2 \Rightarrow z = \frac{y^2}{x} \Rightarrow \frac{1}{z} = \frac{x}{y^2} \Rightarrow \frac{x}{z} = x \times \frac{x}{y^2} = \frac{x^2}{y^2}$

11. Ratio between sum and difference of two numbers is 5 : 1. Find ratio between the numbers.

(*a*) 5 : 1 (*b*) 1 : 5 (*c*) 2 : 3 (*d*) 3 : 2

Sol. (*d*) Ratio between the numbers is (5 + 1) : (5 – 1) = 6 : 4 = 3 : 2

12. The sum of two numbers is 40 and their difference is 4. The ratio of the numbers is : **(SSC CGL Tier I 2000)**

(*a*) 21 : 19 (*b*) 22 : 9 (*c*) 11 : 9 (*d*) 11 : 18

Sol. (*c*) Required ratio = (40 + 4) : (40 – 4) = 44 : 36 = 11 : 9

13. The sum of two numbers is twice the difference between them. If the smaller number is 20, find the bigger number.

(*a*) 5 (*b*) 40 (*c*) 60 (*d*) 80

Sol. (*c*) Ratio between sum and difference of the numbers is 2 : 1.

∴ Ratio between the numbers = (2 + 1) : (2 – 1) = 3 : 1.

∴ Smaller number = 1 ratio = 20. ⇒ Bigger number = 3 ratio = 3 × 20 = 60.

14. Two numbers are such that their difference, their sum and their product are in the ratio 1 : 7 : 24. Find the product of the numbers. **(SSC CGL Tier I 2006)**

(*a*) 42 (*b*) 48 (*c*) 52 (*d*) 60

Sol. (*b*) Out of given options 48 is divisible by 24. (The numbers are 6 and 8)

Alternative method : $(x + y) : (x - y) = 7 : 1 \Rightarrow x : y = 8 : 6$

15. Ratio between two numbers is 3 : 4 and sum of their squares is 625. The numbers are:

(*a*) 6, 8 (*b*) 9, 12 (*c*) 12, 16 (*d*) 15, 20

Sol. (*d*) Let the numbers be $3x$ and $4x$. Then $(3x)^2 + (4x)^2 = 625$

$\Rightarrow 9x^2 + 16x^2 = 625 \Rightarrow 25x^2 = 625 \Rightarrow x^2 = 25 \Rightarrow x = \sqrt{25} = 5$

∴ The numbers are 3 × 5 and 4 × 5 i.e. 15 and 20.

16. A and B are in the ratio 4 : 5 and the difference of their squares is 81, what is the value of A? **(SSC CGL Tier I 2015)**

(*a*) 36 (*b*) 12 (*c*) 45 (*d*) 15

Sol. (*b*) Let the numbers be $4x$ and $5x$. Then $(5x)^2 - (4x)^2 = 81 \Rightarrow 25x^2 - 16x^2 = 81$

$\Rightarrow 9x^2 = 81 \Rightarrow x^2 = 9 \Rightarrow x = \sqrt{9} = 3$ ∴ $A = 4x = 4 \times 3 = 12$

17. The cost of a piece of diamond varies with the square of its weight. A diamond of ₹ 5,184 value is cut into 3 pieces whose weights are in the ratio 1 : 2 : 3. Find the loss involved in the cutting. **(SSC CGL Tier I 2012)**

(*a*) ₹ 3175 (*b*) ₹ 3168 (*c*) ₹ 3068 (*d*) ₹ 3088

Sol. (*b*) Ratio of values of bigger and smaller pieces

$= (1 + 2 + 3)^2 : (1^2 + 2^2 + 3^2) = 6^2 : (1 + 4 + 9) = 36 : 14$

Loss = (36 – 14) ratio = 22 ratio = $\frac{22}{36} \times$ ₹ 5184 = ₹ 3168

18. If $(5x^2 - 3y^2) : xy = 11 : 2$, then the positive value of x/y is : **(SSC CGL Tier I 2012)**

(*a*) $\frac{7}{2}$ (*b*) $\frac{5}{2}$ (*c*) $\frac{3}{2}$ (*d*) $\frac{5}{3}$

Sol. (*c*) $\frac{5x^2 - 3y^2}{xy} = \frac{11}{2} \Rightarrow 2(5x^2 - 3y^2) = 11xy \Rightarrow 10x^2 - 11xy - 6y^2 = 0$

$\Rightarrow 10x^2 - 15xy + 4xy - 6y^2 = 0 \Rightarrow 5x(2x - 3y) + 2y(2x - 3y) = 0$

$\Rightarrow (5x + 2y)(2x - 3y) = 0 \Rightarrow 5x = -2y$ and $2x = 3y \Rightarrow x : y = 3 : 2$

19. What number should be added to each 6, 14, 18 and 38 so that the resulting numbers make a proportion ? **(SSC CHSL 2010)**

(*a*) 1 (*b*) 2 (*c*) 3 (*d*) 4

Sol. (*b*) $\frac{\text{Product of means - Product of extremes}}{\text{Sum of extremes - Sum of means}} = \frac{(14\times18)-(6\times38)}{(6+38)-(14+18)} = \frac{252-228}{44-32} = \frac{24}{12} = 2$

Alt: Differences = 14 – 6 = 8 and 38 – 18 = 20

20 ÷ 8 = 2.5 18 – (6 × 2.5) = 3 3 ÷ (2.5 – 1) = 2

20. What must be deducted from each of the number 7, 10, 12 and 18 so that the resultant number are in proportion ? **(Bank Clerical)**

(*a*) 1 (*b*) 2 (*c*) 3 (*d*) 4

Sol. (*b*) The required number $= x = \frac{10\times12 - 7\times18}{(7+18)-(10+12)} = \frac{-6}{3} = -2$

Alt: Differences = 10 – 7 = 3 and 18 – 12 = 6

6 ÷ 3 = 2 12 – (7 × 2) = – 2 –2 ÷ (2 – 1) = – 2

Alternatively we can check through the given options.

(*a*) 6, 9, 11, 17 (Not in proportion) (*b*) 5, 8, 10, 16 (In proportion)

21. What number must be added to each of 4, 10, 12, 24, so that the resulting numbers are in proportion ? **(SSC MTS 2014)**

(*a*) 3 (*b*) 4 (*c*) 6 (*d*) 9

Sol. (*b*) $\frac{4+x}{10+x} = \frac{12+x}{24+x} \Rightarrow$ Solving, $x = 4$

Alt: Differences = 10 – 4 = 6 and 24 – 12 = 12

12 ÷ 6 = 2 12 – (4 × 2) = 4 4 ÷ (2 – 1) = 4

22. Same quantity of rice is required for each member of a family of 15 members. On a particular day due to the absence of some members of the family, the consumption of rice was reduced in the ratio 5 : 3. The number of members absent on that day was : **(SSC CGL 2005, NDA)**

(*a*) 3 (*b*) 6 (*c*) 8 (*d*) 9

Sol. (*b*) Ratio of family members = 5 : 3 $\Rightarrow$ Total members = $5x = 15 \Rightarrow x = 3$

Members absent = $2x$ = 6 members

23. A mixture contains spirit and water in the ratio 3 : 2. If it contains 3 litres more spirit than water, the quantity of spirit in the mixture is : **(SSC CGL Tier I 2003)**

(*a*) 10 litres (*b*) 12 litres (*c*) 8 litres (*d*) 9 litres

Sol. (*d*) Difference of quantities = (3 – 2) ratio = 1 ratio = 3 litres

$\therefore$ Quantity of spirit = 3 ratio = 3 × 3 litres = 9 litres

24. The ratio between male population and female population of a village is 17 : 13. If the number of males is 280 more than that of the females, find total size of population in the village. **(Bank Clerical 2006)**

(*a*) 1190 (*b*) 910 (*c*) 1120 (*d*) None of these

Sol. (*d*) Difference in male and female population = (17 – 13) ratio = 4 ratio = 280

$\therefore$ 1 ratio = 280 ÷ 4 = 70 $\Rightarrow$ Total population = (17 + 13) ratio = 30 ratio = 30 × 70 = 2100

25. One-fifth of Ram's expenses is equal to one-third of his savings. If his monthly income is ₹ 3,200, what is his savings ?

(*a*) ₹ 1200 (*b*) ₹ 1600 (*c*) ₹ 1800 (*d*) ₹ 2000

Sol. (*a*) Expenses : Savings = 5 : 3 $\Rightarrow$ Savings = $\frac{3}{8}$ × ₹ 3200 = ₹ 1200

26. The ratio of two numbers is 10 : 7 and their difference is 105. The sum of the numbers is: **(SSC CGL Tier I 2006)**

(*a*) 595 (*b*) 805 (*c*) 1190 (*d*) 1610

Sol. (*a*) Difference between numbers = (10 – 7) ratio = 3 ratio = 105 $\Rightarrow$ 1 ratio = 105 ÷ 3 = 35

Sum of numbers = (10 + 7) = 17 ratio = 17 × 35 = 595

27. A fraction in its lowest terms can be expressed as $\frac{4}{17}$. If its numerator and denominator differ by 91, find its numerator.

(*a*) 7 (*b*) 13 (*c*) 52 (*d*) 28

Sol. (*d*) Difference in numerator and denominator = (17 – 4) ratio = 13 ratio = 91

$\Rightarrow$ 1 ratio = 91 ÷ 13 = 7 $\Rightarrow$ Numerator = 4 ratio = 4 × 7 = 28

28. The prices of a school bag and a shoe are in the ratio 7 : 5. The price of a school bag is ₹ 200 more than the price of a shoe. Then the price of a shoe is:

(*a*) ₹ 1,200 (*b*) ₹ 500 (*c*) ₹ 700 (*d*) ₹ 200

Sol. (*b*) Difference in prices = (7 – 5) ratio = 2 ratio = ₹ 200 $\Rightarrow$ 1 ratio = ₹ 100

$\therefore$ Price of shoe = 5 ratio = ₹ 500

29. A certain amount of money has to be divided between two persons P and Q in the ratio 3 : 5. But it was divided in the ratio 2 : 3 and thereby Q loses ₹ 10. What was the amount? **(CDS 2011)**

(*a*) ₹ 250 (*b*) ₹ 300 (*c*) ₹ 350 (*d*) ₹ 400

Sol. (*d*) Q's loss = $\frac{5}{8} - \frac{3}{5} = \frac{1}{40}$ of the total amount = ₹ 10 $\Rightarrow$ Total amount = ₹ 10 × 40 = ₹ 400

30. A certain amount of money is distributed among A, B and C. A gets $\frac{3}{16}$ and B gets $\frac{1}{4}$ of the whole amount. If C gets ₹ 81, then B gets : **(SSC CGL Tier I 2002)**

(*a*) ₹ 30 (*b*) ₹ 36 (*c*) ₹ 32 (*d*) ₹ 40

Sol. (*b*) A's share = $\frac{3}{16}$ and B's share = $\frac{1}{4}=\frac{4}{16}$ $\Rightarrow$ C's share = $1-\left(\frac{3}{16}+\frac{4}{16}\right)=\frac{9}{16}$ ratio = ₹ 81

$\therefore$ B's share = $\frac{4}{16}$ ratio = $81\times\frac{16}{9}\times\frac{4}{16}$ = ₹36

31. If ratio of boys and girls in a class is 7 : 5, which of the following cannot be the total number of students in the class?

(*a*) 36 (*b*) 50 (*c*) 60 (*d*) 120

Sol. (*b*) Sum of the ratios = 7 + 5 = 12

50 is the only number which is not divisible by 12 (*i.e.* sum of the ratios) and number of students cannot be in fractions.

32. An employer reduces the number of employees in the ratio 8 : 5 and increases their wages in the ratio 7 : 9. As a result, the overall wages bill is **(SSC CGL Tier I 2012)**

(*a*) increased in the ratio of 56 : 69 (*b*) decreased in the ratio 56 : 45

(*c*) increased in the ratio 13 : 17 (*d*) decreased in the ratio 17 : 13

Sol. (*b*) Ratio of wages = (8 × 7) : (5 × 9) = 56 : 45

33. The wages of labourers in a factory has increased in the ratio 22 : 25 and their number decreased in the ratio 3 : 2. What was the original wage bill of the factory if the present bill is ₹ 5,000 ? **(CDS 2010)**

(*a*) ₹ 4,000 (*b*) ₹ 6,000 (*c*) ₹ 8,000 (*d*) None of these

Sol. (*d*) Ratio of wage bill = 22 × 3 : 25 × 2 = 33 : 25 $\Rightarrow$ Original bill = $\frac{33}{25}\times 5000$ = ₹ 6,600

34. Find the ratio between $2^{1.5}$ and $2^{0.5}$. **(SSC CGL Tier I 2000)**

(*a*) 2 : 1 (*b*) 3 : 1 (*c*) 6 : 1 (*d*) 3 : 2

Sol. (*a*) $2^{1.5} : 2^{0.5} = 2^1 \times 2^{0.5} : 2^{0.5} = 2^1 : 1 = 2 : 1$

35. If 21 pens cost ₹ 35, find the cost of 30 pens. **(Bank Clerical)**

(*a*) ₹ 18 (*b*) ₹ 26 (*c*) ₹ 44 (*d*) ₹ 50

Sol. (*d*) Let cost of 30 pens be ₹ *x*. Then 21 : 35 = 30 : *x* (Direct Proportion)

$\therefore x = \frac{35\times 30}{21}$ = ₹ 50

36. The shadow of a pole 6 metre high is 15 m long and at the same time the shadow of a tree is 25 metre long. What is the height of the tree ? **(CDS 2010)**

(*a*) 21 metre (*b*) 10 metre (*c*) 35 metre (*d*) None of these

Sol. (*b*) Let the length of second pole = *x* metre. Then 6 : 15 = *x* : 25 (Direct Proportion)

$\therefore x = \frac{6\times 25}{15}$ = 10 metre

37. A vertical stick, 12 cm long, casts a shodow 8 cm long on the ground. At the same time a tower casts a shadow 40 cm long on the ground. Determine the height of the tower. **(SSC CHSL 2013)**

(*a*) 80 cm (*b*) 60 cm (*c*) 40 cm (*d*) 50 cm

Sol. (*b*) $12 : 8 = x : 40 \Rightarrow x = \frac{12\times 40}{8}$ = 60 cm

38. A man having height 169 cm is standing near a pole. He casts a shadow of 130 cm long. What is the length of the pole if it gives a shadow 420 cm long? **(SSC CGL Tier I 2014)**

(*a*) 550 cm (*b*) 589 cm (*c*) 323 cm (*d*) 546 cm

Sol. (*d*) 169 : 130 = *x* : 420 $\Rightarrow$ Required length = $\frac{169\times 420}{130}$ = 546 cm

39. A photograph of size 18 cm × 12 cm is enlarged, so that its new length becomes 27 cm. The breadth of the enlarged photograph will be **(SSC MTS 2014)**

(*a*) 16 cm (*b*) 14 cm (*c*) 18 cm (*d*) 20 cm

Sol. (c) $18 : 12 = 27 : x \Rightarrow$ New breadth $= \frac{27}{18} \times 12 = 18$ cm

40. *A* and *B* have together three times what *B* and *C* have, while *A*, *B*, *C* toghter have thirty rupees more than that of *A*. If *B* has 5 times that of *C*, then *A* has **(SSC CGL Tier I 2013)**

(*a*) ₹60 (*b*) ₹65 (*c*) ₹75 (*d*) ₹45

Sol. (*b*) $A + B + C = A + 30 \Rightarrow B + C = 30;\quad B = 5C \Rightarrow B : C = 5 : 1 \Rightarrow B = \frac{5}{6} \times 30 = 25$

$A + B = 3(B + C) = 3 \times 30 = 90 \Rightarrow A =$ ₹$(90 - 25) =$ ₹65

41. The marks of 3 students A, B and C are in the ratio 10 : 12 : 15. If the maximum marks of the paper are 100, then the marks of B cannot be in the range of **(SSC CGL Tier II 2012)**

(*a*) 70 - 80 (*b*) 80 - 90 (*c*) 20 - 30 (*d*) 40 - 50

Sol. (*b*) Let marks obtained by A, B and C be $10x$, $12x$ and $15x$ respectively.

Then C's maximum marks = $15 \times 6 = 90$ (Note: $15 \times 7 = 105$ is not possible)

Then B's maximum marks = $12x = 12 \times 6 = 72$

42. A vessel contains milk and water in the ratio 5 : 7. Had it contained 2 litres more of milk and 2 litres less of water, the ratio would have become 4 : 5. What is the quantity of milk in the vessel ? **(GMAT)**

(*a*) 28 litres (*b*) 30 litres (*c*) 32 litres (*d*) 42 litres

Sol. (*b*) $\frac{5x+2}{7x-2} = \frac{4}{5} \Rightarrow 25x + 10 = 28x - 8 \Rightarrow 3x = 18 \Rightarrow x = 6 \Rightarrow$ Milk $= 5x = 30$ litres

43. Two numbers are in the ratio 3 : 5. If 2 is added to the first number and 5 to the second, the ratio between the numbers becomes 4 : 7. Find the smaller number.

(*a*) 12 (*b*) 18 (*c*) 30 (*d*) 25

Sol. (*b*) $\frac{3x+2}{5x+5} = \frac{4}{7} \Rightarrow 21x + 14 = 20x + 20 \Rightarrow x = 6 \Rightarrow 3x = 18$

44. If the cost of 9 mangoes and 5 apples is equal to cost of 7 mangoes and 8 apples, find the ratio between the cost of 1 mango and the cost of 1 apple.

(*a*) 2 : 3 (*b*) 3 : 2 (*c*) 5 : 8 (*d*) 9 : 7

Sol. (*b*) In the second case, 2 mangoes (*i.e.* 9 – 7) are replaced by 3 apples (*i.e.* 8 – 5), total cost remains the same in both the cases. ∴ Ratio of their costs = 3 : 2

45. A man encashes a cheque for ₹600 from a bank. The bank pays him money in 10-rupee notes and 5-rupee notes only, totalling 72. The ratio of the number of 10-rupess notes to that of 5-rupee notes is : **(SSC CGL Tier I 2006)**

(*a*) 1 : 2 (*b*) 2 : 1 (*c*) 2 : 3 (*d*) 3 : 2

Sol. (*b*) Let all the notes are of ₹5 only. Then total amount = 72 × ₹5 = ₹360

Short from actual amount = ₹600 – ₹360 = ₹240

∴ 10-rupee notes = 240 ÷ 5 = 48 ⇒ 5-rupee notes = 72 – 48 = 24

∴ Ratio of 10-rupee and 5-rupee notes = 48 : 24 = 2 : 1

46. 94 is divided into two parts in such a way that the fifth part of the first and the eighth part of the second are in the ratio 3 : 4. The first part is **(SSC CHSL 2012)**

(*a*) 28 (*b*) 30 (*c*) 36 (*d*) 40

Sol. (*b*) Let the two parts be *x* and *y*. Then $\frac{x}{5}:\frac{y}{8} = 3:4$

$\Rightarrow \frac{x}{5}\times 4 = \frac{y}{8}\times 3 \Rightarrow \frac{x}{y} = \frac{3}{8}\times\frac{5}{4} = 15:32. \Rightarrow$ First part $= \frac{15}{47}\times 94 = 30$

47. The ratio of 2-digit number to the sum of its digits is 7 : 1. If the digit in the ten's place is 1 more than the digit in the one's place, then the number is : **(SSC CGL Tier I 2006)**

(*a*) 65 (*b*) 43 (*c*) 32 (*d*) 21

Sol. (*d*) Given that the number is 7 times the sum of its digits.

∴ The number must be a multiple of 7.

Out of the given options, 21 is the only number which is multiple of 7 and also satisfies the other condition.

48. Rate of income tax is increased from 4% to 5%. However, the total tax liability of a person reamains the same as was in the last year. If his income for the last year was ₹ 10000, find his present inccome.

(*a*) ₹ 12,5000 (*b*) ₹ 9,000 (*c*) ₹ 8,000 (*d*) ₹ 7,500

Sol. (*c*) Ratio of tax = 4 : 5 ⇒ Ratio of income = 5 : 4

New income = ₹ 10,000 × $\frac{4}{5}$ = ₹ 8,000

49. If $a : b = 7 : 9$ and $b : c = 15 : 7$, then $a : c$ is **(SSC CGL Tier I 2006)**

(*a*) 3 : 5 (*b*) 5 : 3 (*c*) 7 : 15 (*d*) 7 : 21

Sol. (*b*) $\frac{a}{c} = \frac{a}{b}\times\frac{b}{c} = \frac{7}{9}\times\frac{15}{7} = \frac{5}{3} = 5:3$

50. If $(a + b) : (b + c) : (c + a) = 7 : 8 : 9$, find $a : b : c$. **(SSC CGL Tier I 2006)**

(*a*) 7 : 8 : 9 (*b*) 9 : 8 : 7 (*c*) 5 : 3 : 4 (*d*) 4 : 3 : 5

Sol. (*d*) $(a + b) + (b + c) + (c + a) = 2(a + b + c) = 7 + 8 + 9 = 24 \Rightarrow (a + b + c) = 24 \div 2 = 12$

$a = (a + b + c) - (b + c) = 12 - 8 = 4.$ $b = (a + b + c) - (c + a) = 12 - 9 = 3$

$c = (a + b + c) - (a + b) = 12 - 7 = 5 \Rightarrow a : b : c = 4 : 3 : 5$

51. If $a : b = 2 : 3$ and $b : c = 4 : 5$, then $a : b : c$ = ? **(SSC CHSL)**

(*a*) 8 : 12 : 15 (*b*) 2 : 3 : 5 (*c*) 2 : 4 : 5 (*d*) 8 : 12 : 20

Sol. (*a*) In such questions, values of common ratio is made equal to their LCM.

Here *b* is common in $a : b$ and $b : c$. The values of *b* are 3 and 4 in the given ratios. LCM of 3 and 4 is 12.

We multiply $(a : b)$ by 4 and $(b : c)$ by 3 to get value of *b* equal to 12 in both the ratios.

$a : b = 2 : 3 = (2 : 3) \times 4 = 8 : 12$ $b : c = 4 : 5 = (4 : 5) \times 3 = 12 : 15$

∴ $a : b : c = 8 : 12 : 15$

52. If $a : b = 3 : 4$ and $b : c = 6 : 7$, then $a : b : c$ = ? **(SAS 2010)**

(*a*) 9 : 12 : 14 (*b*) 6 : 8 : 21 (*c*) 3 : 6 : 7 (*d*) 3 : 4 : 7

Sol. (*a*) LCM of 4 and 6 is 12.

We multiply $(a : b)$ by 3 and $(b : c)$ by 2 to get values of *b* equal to 12 in both the ratios.

$a : b = 3 : 4 = (3 : 4) \times 3 = 9 : 12$ $b : c = 6 : 7 = (6 : 7) \times 2 = 12 : 14$

∴ $a : b : c = 9 : 12 : 14$

53. If $a : b = 3 : 4$ and $b : c = 8 : 9$, then $a : b : c$ = ?

(*a*) 3 : 8 : 9 (*b*) 6 : 8 : 9 (*c*) 3 : 6 : 9 (*d*) 3 : 4 : 9

Sol. (*b*) Here b is common in $a : b$ and $b : c$.
The values of b are 4 and 8 in the given ratios. LCM of 4 and 8 is 8.
We multiply $(a : b)$ by 2 to get value of b equal to 8 and $(b : c)$ remains same.
$a : b = 3 : 4 = (3 : 4) \times 2 = 6 : 8 \quad b : c = 8 : 9 \quad \therefore a : b : c = 6 : 8 : 9$

54. A : B = 4 : 9 and A : C = 2 : 3, then (A + B) : (A + C) is **(SSC CHSL 2011)**
(*a*) 10 : 13 (*b*) 13 : 10 (*c*) 13 : 15 (*d*) 15 : 13

Sol. (*b*) $A : B = 4 : 9;\ A : C = 2 : 3 = 4 : 6 \Rightarrow A : B : C = 4 : 9 : 6$
$\Rightarrow (A + B) : (A + C) = (4 + 9) : (4 + 6) = 13 : 10$

55. If $a : b = 1\frac{1}{2} : 2\frac{1}{4}$ and $b : c = 2 : 3\frac{1}{2}$, then what is $a : b : c$ equal to ? **(CDS 2009)**
(*a*) 12 : 8 : 21 (*b*) 8 : 21 : 12 (*c*) 8 : 12 : 21 (*d*) 21 : 8 : 12

Sol. (*c*) $a : b = 1\frac{1}{2} : 2\frac{1}{4} = \frac{3}{2} : \frac{9}{4} = 6 : 9 = 2 : 3 = (2 : 3) \times 4 = 8 : 12$

$b : c = 2 : 3\frac{1}{2} = 2 : \frac{7}{2} = 4 : 7 = (4 : 7) \times 3 = 12 : 21 \Rightarrow a : b : c = 8 : 12 : 21$

56. If $a : c = 2 : 3$, $b : c = 8 : 9$, and $a + b + c = 161$, find the value of b.
(*a*) 49 (*b*) 42 (*c*) 56 (*d*) 63

Sol. (*c*) $a : c = 2 : 3 = (2 : 3) \times 3 = 6 : 9.\ \ b : c = 8 : 9 \quad \therefore a : b : c = 6 : 8 : 9$
$a + b + c = (6 + 8 + 9)$ ratio = 23 ratio = 161 $\quad \therefore \quad b = \frac{8}{23} \times 161 = 56$

57. If A : B = 1 : 2 and B : C = 3 : 4 and C : D = 5 : 6, find D : C : B : A = ? **(SSC CGL Tier I 2002)**
(*a*) 6 : 5 : 4 : 2 (*b*) 6 : 3 : 2 : 1 (*c*) 6 : 4 : 2 : 1 (*d*) 48 : 40 : 30 : 15

Sol. (*d*) $A : B = 1 : 2 = (1 : 2) \times 3 = 3 : 6. \quad B : C = 3 : 4 = (3 : 4) \times 2 = 6 : 8$
$\therefore A : B : C = 3 : 6 : 8 = (3 : 6 : 8) \times 5 = 15 : 30 : 40$
$C : D = 5 : 6 = (5 : 6) \times 8 = 40 : 48$
$\therefore A : B : C : D = 15 : 30 : 40 : 48 \quad \therefore D : C : B : A = 48 : 40 : 30 : 15$

58. If x runs are scored by A, y runs by B and z runs by C, then $x : y = y : z = 3 : 2$. If total number of runs scored by A, B and C is 342, the runs scored by each would be respectively **(SSC CGL Tier II 2013)**
(*a*) 144, 96, 64 (*b*) 162, 108, 72 (*c*) 180, 120, 80 (*d*) 189, 126, 84

Sol. (*b*) $x : y : z = 9 : 6 : 4 \Rightarrow x = \frac{9}{19} \times 342 = 162; y = \frac{6}{19} \times 342 = 108; z = \frac{4}{19} \times 342 = 72$

Trick: Total runs = 162 + 108 + 72 = 342 only in option (*b*)

59. Three students A, B and C play cricket. The runs scored by A and B respectively are in the ratio 3 : 2. B's runs to C's runs are also in the same ratio. Together they score 342 runs. Then the runs scored by B are **(SSC CGL Tier I 2012)**
(*a*) 90 (*b*) 171 (*c*) 216 (*d*) 108

Sol. (*d*) $A : B = 3 : 2$ and $B : C = 3 : 2 \Rightarrow A : B : C = 9 : 6 : 4 \Rightarrow$ B's runs $= \frac{6}{19} \times 342 = 108$

60. ₹ 900 is divided among *A*, *B*, *C*; the division is such that $\frac{1}{2}$ of A's money = $\frac{1}{3}$rd of B's money = $\frac{1}{4}$th of C's money. Find the amount (in ₹) received by A, B, C.
(SSC CGL Tier II 2013)
(*a*) 300, 400, 200 (*b*) 350, 450, 100 (*c*) 200, 300, 400 (*d*) 400, 150, 350

Sol. (*c*) Let $\frac{A}{2}=\frac{B}{3}=\frac{C}{4}=x$. Then A = 2*x*, B = 3*x*, C = 4*x*

$\therefore$ A : B : C = 2*x* : 3*x* : 4*x* = 2 : 3 : 4

Their shares = $\frac{2}{9}\times 900, \frac{3}{9}\times 900, \frac{4}{9}\times 900 = 200, 300, 400$

61. Production of wheat is $2\frac{1}{4}$ times that of rice, but cost of rice is $1\frac{1}{4}$ times that of wheat. If a farmer produces wheat in place of rice, then his income in terms of previous income is : **(SSC CGL Tier I 2006)**

(*a*) 2 times (*b*) $1\frac{1}{4}$ times (*c*) $2\frac{1}{4}$ times (*d*) $1\frac{4}{5}$ times

Sol. (*d*)

	Wheat : Rice
Ratio of production	9 : 4
Ratio of cost	4 : 5
Ratio of income	36 : 20

$\therefore$ Required ratio = $\frac{36}{20}=\frac{9}{5}=1\frac{4}{5}$ times

62. If ₹ 2,600 is divided among three persons A, B and C in the ratio $\frac{1}{2}:\frac{1}{3}:\frac{1}{4}$, how much does A get ? **(CDS 2011)**

(*a*) ₹ 600 (*b*) ₹ 800 (*c*) ₹ 1000 (*d*) ₹ 1200

Sol. (*d*) LCM of denominators of the fractions i.e. LCM of 2, 3 and 4 is 12.

A : B : C = $\frac{1}{2}\times 12 : \frac{1}{3}\times 12 : \frac{1}{4}\times 12$ = 6 : 4 : 3 $\Rightarrow$ A's share = $\frac{6}{13}$ × ₹ 2,600 = ₹ 1200

63. If A : B : C = 2 : 3 : 4, then $\frac{A}{B}:\frac{B}{C}:\frac{C}{A}$ is equal to : **(SSC CGL Tier I 2002)**

(*a*) 8 : 9 : 16 (*b*) 8 : 9 : 12 (*c*) 8 : 9 : 24 (*d*) 4 : 9 : 16

Sol. (*c*) $\frac{A}{B}:\frac{B}{C}:\frac{C}{A}=\frac{2}{3}:\frac{3}{4}:\frac{4}{2}=\frac{2}{3}:\frac{3}{4}:2$ = 8 : 9 : 24 (on multiplying each ratio by 12)

64. If $\frac{1}{x}:\frac{1}{y}:\frac{1}{z}$ = 2 : 3 : 5, then *x* : *y* : *z* = ?

(*a*) 2 : 3 : 5 (*b*) 5 : 3 : 2 (*c*) 15 : 10 : 6 (*d*) 6 : 10 : 15

Sol. (*c*) $\frac{1}{x}:\frac{1}{y}:\frac{1}{z}$ = 2 : 3 : 5 $\Rightarrow$ *x* : *y* : *z* = $\frac{1}{2}:\frac{1}{3}:\frac{1}{5}$ = 15 : 10 : 6

65. A person distributes his pens among four friends A, B, C, D in the ratio $\frac{1}{3}:\frac{1}{4}:\frac{1}{5}:\frac{1}{6}$. What is the minimum number of pens that the person should have ? **(SSC CGL Tier I 2013)**

(*a*) 45 (*b*) 57 (*c*) 65 (*d*) 75

Sol. (*b*) $\frac{1}{3}:\frac{1}{4}:\frac{1}{5}:\frac{1}{6}=\frac{60}{3}:\frac{60}{4}:\frac{60}{5}:\frac{60}{6}$ = 20 : 15 : 12 : 10

$\therefore$ Minimum number of pens required = 20 + 15 + 12 + 10 = 57

66. A's share is twice B's. Three times B's share is equal to 4 times C's share. Find A : B : C.

(*a*) 1 : 2 : 4 (*b*) 1 : 3 : 4 (*c*) 6 : 3 : 4 (*d*) 8 : 4 : 3

Sol. (*d*) A = 2B ⇒ A : B = 2 : 1 = (2 : 1) × 4 = 8 : 4

3B = 4C ⇒ B : C = 4 : 3 ⇒ A : B : C = 8 : 4 : 3

67. If 2A = 3B = 4C, then A : B : C = ? **(SSC CHSL 2010)**

(*a*) 3 : 4 : 5 (*b*) 2 : 3 : 4 (*c*) 4 : 3 : 2 (*d*) 6 : 4 : 3

Sol. (*d*) Given 2A = 3B = 4C. LCM of 2, 3, 4 is 12.

∴ We make each part equal to the LCM *i.e.* 12.

Then $2A = 12 \Rightarrow A = \frac{12}{2} = 6$ And $3B = 12 \Rightarrow B = \frac{12}{3} = 4$ And $4C = 12 \Rightarrow C = \frac{12}{4} = 3$

∴ A : B : C = 6 : 4 : 3

68. If $A = \frac{1}{4}B$ and $B = \frac{1}{2}C$, then A : B : C is **(SSC CHSL 2010)**

(*a*) 8 : 4 : 1 (*b*) 4 : 2 : 1 (*c*) 1 : 4 : 8 (*d*) 1 : 2 : 4

Sol. (*c*) $A = \frac{1}{4}B \Rightarrow A : B = 1 : 4$ and $B = \frac{1}{2}C \Rightarrow B : C = 1 : 2 \Rightarrow A : B : C = 1 : 4 : 8$

69. If $A : B = \frac{1}{2} : \frac{3}{8}$, $B : C = \frac{1}{3} : \frac{5}{9}$, and $C : D = \frac{5}{6} : \frac{3}{4}$, then the ratio $A : B : C : D$ is :

(SSC CGL Tier I 2005 and 2001)

(*a*) 6 : 4 : 8 : 10 (*b*) 6 : 8 : 9 : 10 (*c*) 8 : 6 : 10 : 9 (*d*) 4 : 8 : 6 : 10

Sol. (*c*) $A : B = \frac{1}{2} : \frac{3}{8} = 4 : 3$, $B : C = \frac{1}{3} : \frac{5}{9} = 3 : 5$, and $C : D = \frac{5}{6} : \frac{3}{4} = 10 : 9$

∴ $A : B : C = 4 : 3 : 5 = 8 : 6 : 10$ ∴ $A : B : C : D = 8 : 6 : 10 : 9$

70. In a class $\frac{3}{5}$ of the students are girls and rest are boys. If $\frac{2}{9}$ of the girls and $\frac{1}{4}$ of the boys are absent. What part of the total number of students are present ? **(SSC CGL Tier I 1999)**

(*a*) $\frac{23}{30}$ (*b*) $\frac{23}{36}$ (*c*) $\frac{18}{49}$ (*d*) $\frac{17}{25}$

Sol. (*a*) Girls = $\frac{3}{5}$ ⇒ Boys = $\frac{2}{5}$

Students absent = $\frac{3}{5} \times \frac{2}{9} + \frac{2}{5} \times \frac{1}{4} = \frac{2}{15} + \frac{1}{10} = \frac{7}{30}$ ⇒ Students present = $1 - \frac{7}{30} = \frac{23}{30}$

71. A watermelon is cut into two pieces in the ratio of 3 : 5 by weight. The bigger of the two is further cut in the ratio of 5 : 7 by weight. Find the ratio of each of the three pieces.

(SSC CHSL 2012)

(*a*) 15 : 25 : 26 (*b*) 5 : 7 : 9 (*c*) 3 : 5 : 7 (*d*) 36 : 25 : 35

Sol. (*d*) Ratio of first two pieces = $\frac{3}{8} : \frac{5}{8}$

Ratio of three pieces = $\frac{3}{8} : \frac{5}{8} \times \frac{5}{12} : \frac{5}{8} \times \frac{7}{12} = \frac{3}{8} : \frac{25}{96} : \frac{35}{96}$ = 36 : 25 : 35

72. Ram got twice as many marks in English as in science. His total marks in English, Science and Maths are 180. If the ratio of his marks in English and Maths is 2 : 3, what are his marks in science? **(SSC CGL Tier II 2013)**

(*a*) 30 (*b*) 60 (*c*) 72 (*d*) 90

Sol. (*a*) English : Science = 2 : 1. English : Maths = 2 : 3

⇒ English : Science : Maths = 2 : 1 : 3 ⇒ Marks in science = $\frac{1}{6}\times 180 = 30$

73. ₹ 750 are divided among A, B and C in such a manner that A : B = 5 : 2 and B : C = 7 : 13. What is A's share? **(SSC CGL Tier I 2004, 2002)**

(*a*) ₹ 350 (*b*) ₹ 260 (*c*) ₹ 140 (*d*) ₹ 250

Sol. (*a*) A : B = 5 : 2 = (5 : 2) × 7 = 35 : 14

B : C = 7 : 13 = (7 : 13) × 2 = 14 : 26 ∴ A : B : C = 35 : 14 : 26

∴ A's share = $\frac{35}{35+14+26}\times 750 = \frac{35}{75}\times 750$ = ₹ 350

74. ₹ 700 is divided among A, B and C in such a way that the ratio of the amounts of A and B is 2 : 3 and that of B and C is 4 : 5. Find the amounts in ₹ each received, in the order A, B, C. **(SSC CGL Tier I 2013)**

(*a*) 150, 250, 300 (*b*) 160, 240, 300 (*c*) 150, 250, 290 (*d*) 150, 240, 310

Sol. (*b*) A : B = 2 : 3 = (2 : 3) × 4 = 8 : 12 B : C = 4 : 5 = (4 : 5) × 3 = 12 : 15

∴ A : B : C = 8 : 12 : 15

Sum of ratios = (8 + 12 + 15) ratio = 35 ratio = ₹ 700 ⇒ 1 ratio = ₹ 700 ÷ 35 = ₹ 20

∴ Amounts received by A, B, C = (8, 12, 15) ratio = 160, 240, 300

Trick : Only in option (*b*), ratio of shares of A and B is in the ratio of 2 : 3.

75. A, B, C and D purchase a gift worth ₹ 60. A pays 1/2 of what others are paying, B pays 1/3rd of what others are paying and C pays 1/4th of what others are paying. What is the amount paid by D ? **(SSC CGL Tier I 2013)**

(*a*) ₹15 (*b*) ₹16 (*c*) ₹13 (*d*) ₹14

Sol. (*c*) A : (B + C + D) = 1 : 2; B : (A + C + D) = 1 : 3; C : (A + B + D) = 1 : 4

A = $\frac{1}{3}$ × ₹ 60 = ₹ 20, B = $\frac{1}{4}$ × ₹ 60 = ₹ 15, C = $\frac{1}{5}$ × ₹ 60 = ₹ 12

∴ D = ₹ 60 – (₹ 20 + ₹ 15 + ₹ 12) = ₹ 13

76. A sum of ₹ 9000 is to be distributed among A, B and C in the ratio 4 : 5 : 6. What will the difference between A's and C's shares ? **(SSC CGL 2001)**

(*a*) ₹ 600 (*b*) ₹ 1000 (*c*) ₹ 900 (*d*) ₹ 1200

Sol. (*d*) Sum of ratios = (4 + 5 + 6) ratio = 15 ratios

Difference between A and C shares = (6 – 4) ratio = 2 ratios

∴ Required difference = $\frac{2}{15}\times 9000$ = ₹ 1200

77. If ₹126.50 is divided among A, B and C in the ratio of 2 : 5 : 4, the share of B exceeds that of A by **(SSC CGL Tier II 2013)**

(*a*) ₹ 36.50 (*b*) ₹ 35.50 (*c*) ₹ 34.50 (*d*) ₹ 33.50

Sol. (*c*) $\frac{5-2}{11}\times 126\cdot 50 = \frac{3}{11}\times 126\cdot 50$ = ₹ 34.50

78. A sum of money is divided among A, B, C and D in the ratio of 3 : 4 : 9 : 10 respectively. If the share of C is ₹ 2,580 more than the share of B, then what is the total amount of money A and D receive together ? **(Canara Bank PO 2009)**

(*a*) ₹ 5,676 (*b*) ₹ 6,192 (*c*) ₹ 6,708 (*d*) ₹ 7,224

Sol. (*c*) C – B = (9 – 4) ratio = 5 ratio = ₹ 2580 ⇒ 1 ratio = 2580 ÷ 5 = ₹ 516

A + D = (3 + 10) ratio = 13 ratio = 13 × ₹ 516 = ₹ 6708

79. A sum of ₹ 300 is divided among P, Q and R in such a way that Q gets ₹ 30 more than P and R gets ₹ 60 more than Q. The ratio of their share is: **(SSC CGL Tier I 2013)**

(*a*) 2 : 5 : 3 (*b*) 2 : 3 : 5 (*c*) 5 : 3 : 2 (*d*) 3 : 2 : 5

Sol. (*b*) Q = P + 30; R = Q + 60 = P + 90

P + Q + R = 300 ⇒ P + (P + 30) + (P + 90) = 300 ⇒ 3P = 180 ⇒ P = 60

∴ P : Q : R = 60 : (60 + 30) : (60 + 90) = 60 : 90 : 150 = 2 : 3 : 5

80. A, B, C subscribe together ₹ 50,000 for business. A subscribes ₹4,000 more than B and B ₹5,000 more than C. Out of a total profit of ₹35,000, A receives **(SSC CGL Tier I 2012)**

(*a*) ₹8,500 (*b*) ₹11,998 (*c*) ₹12,600 (*d*) ₹14,700

Sol. (*d*) Let C subscribe = x, then B = $(x + 5000)$ and A = $(x + 5000) + 4000$

Total = $x + (x + 5000) + (x + 5000) + 4000 = 50,000$

⇒ $3x + 14000 = 50,000 \Rightarrow 3x = 36,000 \Rightarrow x = 12,000$

⇒ Ratio of shares of A : B : C = 21000 : 17000 : 12000 = 21 : 17 : 12

∴ A's share = $\frac{21}{50} \times 35000$ = ₹ 14700

81. Three numbers are in the ratio $\frac{1}{2}:\frac{2}{3}:\frac{3}{4}$. The difference between the greatest and the smallest numbers is 36. The numbers are : **(SSC CGL Tier I 2005)**

(*a*) 72, 84, 108 (*b*) 60, 72, 96 (*c*) 72, 84, 96 (*d*) 72, 96, 108

Sol. (*d*) LCM of denominators of the fractions *i.e.* LCM of 2, 3 and 4 is 12.

∴ We multiply each ratio by 12. The ratio becomes $\frac{1}{2} \times 12 : \frac{2}{3} \times 12 : \frac{3}{4} \times 12 = 6 : 8 : 9$

Difference between greatest and smallest numbers = (9 – 6) ratio = 3 ratio = 36

∴ 1 ratio = 36 ÷ 3 = 12

∴ The numbers are (6, 8, 9) ratio = 6 × 12, 8 × 12, 9 × 12 = 72, 96, 108

82. Instead of dividing ₹117 among P, Q, R in the ratio $\frac{1}{2}:\frac{1}{3}:\frac{1}{4}$, by mistake it was divided in the ratio 2 : 3 : 4. Who gained in the transaction ? **(SSC CGL Tier I 2000, 2014M)**

(*a*) Only P (*b*) Only Q (*c*) Only R (*d*) Both Q and R

Sol. (*d*) Correct ratio = $\frac{1}{2}:\frac{1}{3}:\frac{1}{4}$ = 6 : 4 : 3 ⇒ Their shares = ₹54, ₹36, ₹27

Incorrect ratio = 2 : 3 : 4 ⇒ Their shares = ₹26, ₹39, ₹52. Hence both Q and R gain.

83. Mohan distributed his assets to his wife, three sons, two daughters and five grand children in such a way that each grand child got one-eighth of each son and one-tenth of each daughter. His wife got 40% of the total share of his sons and daughter together. If each daughter receives assets of worth ₹ 1.25 lakh, what is the share of his wife ? **(SBI PO 2005)**

(*a*) ₹ 12,500 (*b*) ₹ 2,20,000 (*c*) ₹ 2,50,000 (*d*) None of these

Sol. (*b*) Share of 1 grand child = $\frac{1}{10}$ × ₹ 1.25 lakhs = ₹ 0.125 lakhs

Share of 1 son = 8 × ₹ 0.125 lakh = ₹ 1 lakh ⇒ Share of 3 sons = 3 × ₹ 1 lakh = ₹ 3 lakh

Share of 2 daughters = 2 × ₹ 1.25 lakh = ₹ 2.5 lakh

Total share of sons and daughters = ₹ (3 + 2.5) lakhs = ₹ 5.5 lakhs

Share of wife = $\frac{4}{10}$ × ₹ 5.5 lakh = ₹ 2.2 lakh

84. The ratio of monthly incomes of A and B is 4 : 5 and that of B and C is 2 : 3. If A's monthly income is ₹ 800, find C's income.

(*a*) ₹1000 (*b*) ₹1500 (*c*) ₹1800 (*d*) ₹2400

Sol. (*b*) A's income = $4x$ = ₹ 800 ⇒ x = ₹ 200

⇒ B's income = $5x$ = 5 × ₹ 200 = ₹ 1000 = $2y$ (say) ⇒ y = ₹ 500

⇒ C's income = $3y$ = 3 × ₹ 500 = ₹ 1500

Alt: A : B = 4 : 5 = 8 : 10; B : C = 2 : 3 = 10 : 15 ⇒ A : B : C = 8 : 10 : 15

⇒ A's income = $8x$ = ₹ 800 ⇒ x = 100 ⇒ C's income = $15x$ = 15 × ₹ 100 = ₹1500

85. The ratio of the first and second class train fares between two stations is 3 : 1 and that of the number of passengers travelling between the two stations by first and second classes is 1 : 50. If on a particular day, ₹1325 are collected from passengers travelling between the two stations, then the amount collected from the second class passengers is : **(SSC CGL 2005)**

(*a*) ₹ 1250 (*b*) ₹ 1000 (*c*) ₹ 850 (*d*) ₹ 750

Sol. (*a*)

	First	Second
Ratio of Fares	= 3	1
Ratio of Passengers	= 1	50
Ratio of Money	= 3	50

∴ Required amount = $\frac{50}{53}$ × ₹1325 = ₹1250

86. In a particular type of a fertilizer, the ratio of two chemicals A and B is 2 : 5. In 21 kg of this fertilizer, if 3 kg of A is added, the ratio of A to B in the new fertilizer will be : **(SSC CGL 2005)**

(*a*) 1 : 1 (*b*) 2 : 3 (*c*) 3 : 5 (*d*) 4 : 5

Sol. (*c*) Chemical A = $\frac{2}{7} \times 21$ = 6 kg. Chemical B = 21 – 6 = 15 kg.

∴ Required ratio = (6 + 3) : 15 = 9 : 15 = 3 : 5

Alternative method : (2 + 5) ratio = 7 ratio = 21 kg ⇒ 1 ratio = 21 ÷ 7 = 3 kg

Quantity of A added = 3 kg = 1 ratio ∴ New ratio = (2 + 1) : 5 = 3 : 5

87. The ratio of successfull and unsuccessful examinees in an examination in a school is 6 : 1. The ratio would have been 9 : 1, if 6 more examinees had been successful. The total number of examinees is **(SSC Constable GD 2012)**

(*a*) 140 (*b*) 120 (*c*) 200 (*d*) 160

Sol. (*a*) $\frac{6x+6}{x-6} = \frac{9}{1} \Rightarrow 6x + 6 = 9x - 54 \Rightarrow 3x = 60 \Rightarrow x = 20 \Rightarrow$ Examinees = $7x$ = 7 × 20 = 140

88. If $\frac{a}{3}=\frac{b}{4}=\frac{c}{7}$ than $\frac{a+b+c}{c}$ is equal to **(SSC CGL Tier I 2001)**

(*a*) 0 (*b*) 1 (*c*) 2 (*d*) 3

Sol. (*c*) $\frac{a}{3}=\frac{b}{4}=\frac{c}{7} \Rightarrow a:b:c=3:4:7 \Rightarrow \frac{a+b+c}{c}=\frac{3+4+7}{7}=2$

89. If $m : n = 3 : 2$ then $(4m + 5n) : (4m - 5n)$ is equal to **(SSC CGL Tier I 2000)**

(*a*) 4 : 9 (*b*) 9 : 4 (*c*) 11 : 1 (*d*) 9 : 1

Sol. (*c*) $(4m + 5n) : (4m - 5n) = (4 \times 3 + 5 \times 2) : (4 \times 3 - 5 \times 2) = 22 : 2 = 11 : 1$

90. If $x : y = 3 : 4$, then $(2x + 3y) : (3y - 2x)$ is **(SSC CGL Tier I 2015)**

(*a*) 1 : 1 (*b*) 2 : 1 (*c*) 3 : 1 (*d*) 3 : 2

Sol. (*c*) $(2x + 3y) : (3y - 2x) = (2 \times 3 + 3 \times 4) : (3 \times 4 - 2 \times 3) = 18 : 6 = 3 : 1$

91. If $\frac{a}{b} = 0.25$, then $\frac{2b-a}{2b+a}+\frac{2}{9} = ?$ **(Railways Exam.)**

(*a*) 0 (*b*) 1 (*c*) $\frac{4}{9}$ (*d*) 2

Sol. (*b*) $\frac{a}{b} = 0.25 = \frac{1}{4} \Rightarrow \frac{2b-a}{2b+a}+\frac{2}{9}=\frac{2\times4-1}{2\times4+1}+\frac{2}{9}=\frac{7}{9}+\frac{2}{9}=1$

92. If $a : b = 5 : 7$ and $c : d = 2a : 3b$, then $ac : bd = ?$ **(SSC CGL Tier I 2005)**

(*a*) 20 : 38 (*b*) 50 : 147 (*c*) 10 : 21 (*d*) 50 : 151

Sol. (*b*) $ac : bd = \frac{ac}{bd}=\frac{a}{b}\times\frac{c}{d}=\frac{5}{7}\times\frac{2a}{3b}=\frac{5}{7}\times\frac{2}{3}\times\frac{5}{7}=\frac{50}{147}$

93. If $x : y = 2 : 1$, then $(x^2 - y^2) : (x^2 + y^2) = ?$ **(Railways 2010)**

(*a*) 3 : 5 (*b*) 5 : 3 (*c*) 1 : 3 (*d*) 3 : 1

Sol. (*a*) $(x^2 - y^2) : (x^2 + y^2) = (2^2 - 1^2) : (2^2 + 1^2) = (4 - 1) : (4 + 1) = 3 : 5$

94. If $x : y = 7 : 3$, then $\frac{xy + y^2}{x^2 - y^2} = ?$ **(SSC CGL Tier I 2009)**

(*a*) 3 : 4 (*b*) 4 : 3 (*c*) 3 : 7 (*d*) 7 : 3

Sol. (*a*) $\frac{xy + y^2}{x^2 - y^2}=\frac{7\times3+3^2}{7^2-3^2}=\frac{21+9}{49-9}=\frac{30}{40}=\frac{3}{4}$

95. If $a : b = b : c$, then $a^4 : b^4$ is equal to : **(SSC CGL Tier I 2005, SAS 2010)**

(*a*) $ac : b^2$ (*b*) $a^2 : c^2$ (*c*) $c^2 : a^2$ (*d*) $b^2 : ac$

Sol. (*b*) $a : b = b : c \Rightarrow b^2 = ac \Rightarrow \frac{a^4}{b^4}=\frac{a^4}{(ac)^2}=\frac{a^2}{c^2}$

96. If $a : b = 2 : 3$ and $b : c = 4 : 5$, find $a^2 : b^2 : bc$. **(SSC CGL Tier I 2001)**

(*a*) 4 : 9 : 45 (*b*) 16 : 36 : 45 (*c*) 16 : 36 : 20 (*d*) 4 : 36 : 20

Sol. (*b*) $a : b = 2 : 3 = (2 : 3) \times 4 = 8 : 12$ $\quad b : c = 4 : 5 = (4 : 5) \times 3 = 12 : 15$

$\therefore a : b : c = 8 : 12 : 15$

$a^2 : b^2 : bc = 8^2 : 12^2 : 12 \times 15 = 64 : 144 : 180 = 16 : 36 : 45$

97. If $\frac{2a}{3b}=\frac{1}{2}$, then $\frac{a+b}{a-b}=?$ **(MBA Entrance Exam.)**

(a) –3 (b) 3 (c) 5 (d) 7

Sol. (d) $\frac{2a}{3b}=\frac{1}{2}\Rightarrow\frac{a}{b}=\frac{1}{2}\times\frac{3}{2}=\frac{3}{4}\Rightarrow\frac{a+b}{a-b}=\frac{4+3}{4-3}=7$

98. If $5a+3b:4a+7b=3:4$, then $a:b$ is equal to **(SSC CHSL 2012)**

(a) 9 : 8 (b) 7 : 11 (c) 8 : 9 (d) 11 : 9

Sol. (a) $5a+3b:4a+7b=3:4\Rightarrow 4(5a+3b)=3(4a+7b)\Rightarrow 20a+12b=12a+21b$

$\Rightarrow 8a=9b\Rightarrow a:b=9:8$

99. If $\frac{2a+b}{a+4b}=3$, then find the value of $\frac{a+b}{a+2b}$: **(SSC CGL Tier I 2001)**

(a) $\frac{5}{9}$ (b) $\frac{2}{7}$ (c) $\frac{10}{9}$ (d) $\frac{10}{7}$

Sol. (c) $\frac{2a+b}{a+4b}=3\Rightarrow 2a+b=3a+12b\Rightarrow -a=11b\Rightarrow a:b=-11:1$

$\therefore\ \frac{a+b}{a+2b}=\frac{-11+1}{-11+2}=\frac{-10}{-9}=\frac{10}{9}$

100. If $\frac{a}{b+c}=\frac{b}{c+a}=\frac{c}{a+b}=x$, what is the value of x?

(a) $\frac{1}{2}$ (b) 1 (c) $\frac{3}{2}$ (d) 2

Sol. (a) $x=\frac{a+b+c}{b+c+c+a+a+b}$ (Rule of Addendo) $\Rightarrow x=\frac{a+b+c}{2(a+b+c)}=\frac{1}{2}$

101. If $\frac{x}{(2x+y+z)}=\frac{y}{(x+2y+z)}=\frac{z}{(x+y+2z)}=a$, find the value of a, where $(x+y+z)\neq 0$

(a) $\frac{1}{2}$ (b) $\frac{1}{3}$ (c) $\frac{1}{4}$ (d) $\frac{1}{6}$

Sol. (c) $a=\frac{x+y+z}{(2x+y+z)+(x+2y+z)+(x+y+2z)}=\frac{x+y+z}{4(x+y+z)}=\frac{1}{4}$

102. If $\frac{x}{a-b}=\frac{y}{b-c}=\frac{z}{c-a}$, then $x+y+z=?$

(a) –1 (b) 0 (c) 1 (d) $a+b+c$

Sol. (b) Let $\frac{x}{a-b}=\frac{y}{b-c}=\frac{z}{c-a}=k\ \Rightarrow\ x=(a-b)k, y=(b-c)k, z=(c-a)k$

$\therefore x+y+z=(a-b+b-c+c-a)\,k=0\times k=0$

103. If $x:y=3:2$, then the ratio $(2x^2+3y^2):(3x^2-2y^2)$ is equal to : **(SSC CGL Tier I 2002)**

(a) 12 : 5 (b) 6 : 5 (c) 30 : 19 (d) 5 : 3

Sol. (c) $\frac{2x^2+3y^2}{3x^2-2y^2}=\frac{2(3)^2+3(2)^2}{3(3)^2-2(2)^2}=\frac{18+12}{27-8}=\frac{30}{19}$

104. If $\frac{a}{b}=\frac{b}{c}=\frac{c}{d}$, then $\frac{b^3+c^3+d^3}{a^3+b^3+c^3}$ will be equal to : **(CDS 2002)**

(*a*) $\frac{a}{b}$ (*b*) $\frac{b}{c}$ (*c*) $\frac{c}{d}$ (*d*) $\frac{d}{a}$

Sol. (*d*) Let $\frac{a}{b}=\frac{b}{c}=\frac{c}{d}=k \Rightarrow c=dk,\ b=ck=dk^2,\ a=bk=ck^2=dk^3$

$$\therefore \frac{b^3+c^3+d^3}{a^3+b^3+c^3}=\frac{(dk^2)^3+(dk)^3+d^3}{(dk^3)^3+(dk^2)^3+(dk)^3}=\frac{d^3(k^6+k^3+1)}{d^3k^3(k^6+k^3+1)}=\frac{1}{k^3}=1\div\frac{a}{d}=\frac{d}{a}$$

105. If $x+y+z=15$, $xy+yz+zx=75$, then $\frac{x+4y+z}{3z}$ is equal to **(SSC CHSL 2013)**

(*a*) 1 (*b*) 0 (*c*) 2 (*d*) –1

Sol. (*c*) $x=y=z=5$ satisfies both the equations $\Rightarrow x:y:z=1:1:1$

$$\therefore \frac{x+4y+z}{3z}=\frac{1+4+1}{3}=\frac{6}{3}=2$$

106. If $a:(b+c)=1:3$ and $c:(a+b)=5:7$, then $b:(a+c)=?$ **(SSC CPO SI 2006)**

(*a*) 1 : 5 (*b*) 3 : 5 (*c*) 1 : 2 (*d*) 1 : 3

Sol. (*c*) $\frac{a}{b+c}=\frac{1}{3}\Rightarrow\frac{a}{a+b+c}=\frac{1}{4}$ $\frac{c}{a+b}=\frac{5}{7}\Rightarrow\frac{c}{a+b+c}=\frac{5}{12}$

Let $a+b+c$ = LCM of 4, 12 = 12

Then $a=\frac{1}{4}\times 12=3$ $c=\frac{5}{12}\times 12=5 \Rightarrow b=12-(3+5)=4$

$\Rightarrow b:(a+c)=4:(3+5)=4:8=1:2$

107. If $yz:zx:xy=1:2:3$ and $(xyz\neq 0)$, then $\frac{x}{yz}\div\frac{y}{zx}$ is? **(SSC CHSL 2013)**

(*a*) 1/2 (*b*) 2 (*c*) 4 (*d*) 1/4

Sol. (*c*) $\frac{x}{yz}\div\frac{y}{zx}=\frac{x}{yz}\times\frac{zx}{y}=\frac{(xz)^2}{(yz)^2}=\left(\frac{2}{1}\right)^2=4$

108. From each of two given numbers, half the smaller number is subtracted. After such subtraction, the larger number is 4 times as large as the smaller number. What is the ratio of the numbers? **(SSC CGL Tier II 2012)**

(*a*) 4 : 1 (*b*) 4 : 5 (*c*) 5 : 2 (*d*) 1 : 4

Sol. (*c*) Option (c) satisfies the given condition. [5 – (2 ÷ 2)] : [2 – (2 ÷ 2)] = 4 : 1

109. Divide ₹ 1570 between A and B so that ₹ 25 being deducted from A's share and ₹ 45 from B's share, their share becomes 2 : 3. Find the amount received by A.

(*a*) ₹ 625 (*b*) ₹ 628 (*c*) ₹ 942 (*d*) ₹ 945

Sol. (*a*) Total amount to be deducted = ₹ 25 + ₹ 45 = ₹ 70

Remaining amount = ₹ 1570 – ₹ 70 = ₹ 1500 is distributed between A and B in the ratio 2 : 3.

∴ A's share in ₹ 1500 = $\frac{2}{5}$ × ₹ 1500 = ₹ 600 ⇒ A's share in ₹ 1570 = ₹ 600 + ₹ 25 = ₹ 625

Alt: $(2x+25)+(3x+45)=1570\Rightarrow 5x+70=1570\Rightarrow 5x=1570-70=1500$

$\Rightarrow x=300\Rightarrow$ A's share $=2x+25=600+25=625$

110. A sum of ₹ 2550 was divided among A, B and C in such a way that if ₹10, ₹20 and ₹20 respectively deducted from their shares, the balances are in the ratio 5 : 7 : 13. Find the share received by A.

(*a*) ₹ 450 (*b*) ₹ 510 (*c*) ₹ 580 (*d*) ₹ 670

Sol. (*b*) Divide ₹ 2550 – ₹(10 + 20 + 20) = ₹ 2500 among A, B and C in ratio 5 : 7 : 13.

$\therefore$ A's share in ₹ 2500 = ₹ 2500 × $\frac{5}{5+7+13}$ = ₹ 2500 × $\frac{5}{25}$ = ₹ 500

$\therefore$ A's total share = ₹ 500 + ₹ 10 = ₹ 510

111. The cost of 3 horses is same as the cost of 5 cows. If total cost of 4 horses and 6 cows is ₹1900, find the cost of one horse.

(*a*) ₹ 50 (*b*) ₹ 150 (*c*) ₹ 200 (*d*) ₹ 250

Sol. (*d*) Ratio of cost of horse and cow $= \frac{1}{3} : \frac{1}{5} = 5 : 3$

Let cost of horse and cow are $5x$ and $3x$ respectively.

Then $4 \times 5x + 6 \times 3x$ = ₹ 1900 $\Rightarrow 38x = 1900 \Rightarrow x = \frac{1900}{38}$ = ₹ 50

$\therefore$ Cost of one horse = 5 × ₹ 50 = ₹ 250

112. 6 pens and 14 exercise books cost ₹162. 5 pens and 8 exercise books cost ₹102. The ratio of cost of 1 pen to that of 1 exercise book is : **(SSC CGL Tier I 2006)**

(*a*) 10 : 3 (*b*) 2 : 3 (*c*) 3 : 2 (*d*) 5 : 6

Sol. (*b*) Let cost of 1 pen be ₹x and that of 1 exercise book be ₹y.

Then $6x + 14y = 162 \Rightarrow 3x + 7y = 81$ .... (1)

$5x + 8y = 102$ .... (2)

(1) × 5 – (2) × 3 $\Rightarrow 11y = 99 \Rightarrow y = 9$. Putting $y = 9$ in equation (1), we get : $x = 6$

$\therefore x : y = 6 : 9 = 2 : 3$

113. The price of 10 chairs is equal to that of 4 tables. The price of 15 chairs and 2 tables together is ₹ 4000. The total price of 12 chairs and 3 tables is : **(SSC CGL Tier I 2001)**

(*a*) ₹ 3750 (*b*) ₹ 3840 (*c*) ₹ 3500 (*d*) ₹ 3900

Sol. (*d*) Cost of 4 tables = Cost of 10 chairs ⇒ Cost of 1 table = 10 ÷ 4 = 2.5 chairs

⇒ 15 chairs + 2 tables = ₹ 4000 ⇒ 15 chairs + 2 × 2.5 chairs = ₹ 4000

⇒ 20 chairs = ₹ 4000 ⇒ 1 chair = ₹ 4000 ÷ 20 = ₹ 200

⇒ Cost of 1 table = 2.5 × ₹ 200 = ₹ 500

$\therefore$ Cost of 12 chairs + cost of 3 tables = 12 × ₹ 200 + 3 × ₹ 500 = ₹ 2400 + ₹ 1500 = ₹ 3900

114. If $6x^2 - 17xy + 12y^2 = 0$, find $x : y$ **(MBA Entrance Exam.)**

(*a*) 3 : 2 (*b*) 4 : 3 (*c*) Both (*d*) None of these

Sol. (*c*) $6x^2 - 17xy + 12y^2 = 0 \Rightarrow 6x^2 - 8xy - 9xy + 12y^2 = 0 \Rightarrow 2x(3x - 4y) - 3y(3x - 4y) = 0$

$\Rightarrow (2x - 3y)(3x - 4y) = 0 \Rightarrow 2x - 3y = 0$ or $3x - 4y = 0 \Rightarrow 2x = 3y$ or $3x = 4y$

$\Rightarrow x : y = 3 : 2$ or $4 : 3$

115. A man plays 100 games with cards. He gets ₹ 5 if he wins and pays ₹ 2 if he loses. If he wins ₹ 3 on the whole, in how many games did he win?

(*a*) 25 (*b*) 28 (*c*) 29 (*d*) 32

Sol. (*c*) Suppose he wins the first game and loses the second. Then his gain = ₹ 5 – ₹ 2 = ₹ 3
Out of remaining 98 games, amount won = amount lost ⇒ Games won: Games lost = 2 : 5

∴ In last 98 games, he wins = $\frac{2}{7} \times 98 = 28$ games ⇒ Total games he won = 28 + 1 = 29

116. There are 200 questions on a 3 hour examination. Among 200 questions, 50 are from Maths, 100 are in G.K. and 50 are in Science. He spent twice as much as time on each Mathematics question as for each other question. How many minutes did he spend on Mathematics questions ? **(SSC CGL Tier II 2012)**

(*a*) 100 (*b*) 60 (*c*) 36 (*d*) 72

Sol. (*d*) Ratio of questions on Mathematics and others = 50 : 150 = 1 : 3
Ratio of time spent on Mathematics and other questions = (1 × 2) : (3 × 1) = 2 : 3

∴ Time spent on Mathematics = $\frac{2}{5} \times 180$ minutes = 72 minutes

117. A milkman mixes 20 litres of water with 80 litres of milk. After selling one-fourth of this mixture, he adds water to replenish the quantity he has sold, what is the current proportion of water to milk ? **(CAT 2003)**

(*a*) 1 : 3 (*b*) 2 : 3 (*c*) 3 : 4 (*d*) 4 : 3

Sol. (*b*) Milk remaining after selling $\frac{1}{4}$ of mixture = $80 \times \left(1-\frac{1}{4}\right)$ = 60 litres

⇒ Water = 100 – 60 = 40 litres ⇒ Required ratio = 40 : 60 = 2 : 3

118. When a ball bounces, it rises to 2/3 of the height from which it fell. If the ball is dropped from a height of 36 m, how high will it rise at the third bounce ? **(CGL Tier I 2015, 2000 M)**

(*a*) $10\frac{2}{3}$ m (*b*) $13\frac{2}{3}$ m (*c*) $12\frac{1}{3}$ m (*d*) $12\frac{2}{3}$

Sol. (*a*) Height at third bounce = $36 \times \left(\frac{2}{3}\right)^3 = 10\frac{2}{3}$ metre

119. A vessel contains 100 litres of milk. From this, 10 litres of milk was taken out and replaced by water. The process was repeated further two times. How much milk is now contained by the vessel ? **(SSC CGL Tier I 2012)**

(*a*) 72.9 lites (*b*) 60.4 litres (*c*) 71.6 litres (*d*) 81 litres

Sol. (*a*) Milk after three replacements = $100 \times \left(\frac{90}{100}\right)^3$ = 72.9 litres

120. From a vessel containing 100 *l* of wine, 10 *l* are drawn out and an equal amount of water is added. From the mixture, 10 *l* is again drawn out and same quantity of water is added. What is the final ratio of wine and water ? **(SSC CGL Tier II 2012)**

(*a*) 91 : 9 (*b*) 81 : 19 (*c*) 80 : 20 (*d*) 90 : 10

Sol. (*b*) Quantity of wine after two replacements = $100\ l \times \left(\frac{90}{100}\right)^2 = 81\ l$

∴ Required ratio = 81 : (100 – 81) = 81 : 19

121. A vessel contains 100 litres of pure milk. 10 litres of milk is taken out from the vessel and is replaced by 10 litres of water. Now, 20 litres of the mixture is taken from the vessel and is replaced by water. What is the quantity of milk left in the vessel?

(*a*) 67 litres (*b*) 70 litres (*c*) 72 litres (*d*) 75 litres

Sol. (*c*) Quantity of milk left in the vessel

$$= 100 \times \frac{100-10}{100} \times \frac{100-20}{100} = 100 \times \frac{9}{10} \times \frac{8}{10} = 72 \text{ litres}$$

122. 5 litres of milk taken out from a container full of milk, then filled with water. Again 5 litre of the mixture is drawn and container filled with water. If the ratio of milk and water now left in the container is 16 : 9, find the capacity of container? **(MBA)**

(*a*) 9 litres (*b*) 16 litres (*c*) 25 litres (*d*) 125 litres

Sol. (*c*) Milk : Water = 16 : 9 ⇒ Milk : Total capacity = 16 : (16 + 9) = 16 : 25

∴ Ratio of milk and water after each replacement = $\sqrt{\frac{16}{25}} = \frac{4}{5}$

∴ (5 – 4) ratio = 1 ratio ≡ 5 litres ∴ Capacity of container = 5 ratio ≡ 5 × 5 = 25 litre

123. Equal amounts of water were poured into two empty jars of different capacities, which made one jar $\frac{1}{4}$ full and the other jar $\frac{1}{3}$ full. If the water in the jar with lesser capacity is then poured into the jar with greater capacity, then the part of the larger jar filled with water is : **(SSC CPO SI 2012)**

(*a*) $\frac{1}{2}$ (*b*) $\frac{7}{12}$ (*c*) $\frac{1}{4}$ (*d*) $\frac{1}{3}$

Sol. (*a*) $\frac{1}{4} < \frac{1}{3} \Rightarrow \frac{1}{4}$ filled jar is with greater capacity.

Greater capacity jar is again filled with $\frac{1}{4}$ of its capacity. ⇒ New quantity = $\frac{1}{4} + \frac{1}{4} = \frac{1}{2}$

124. Some sweets were to be distributed equally among 175 students of a school. But due to absence of 35 students, each child got 4 more sweets. How many sweets were distributed? **(Bank PO 2003, Bank PO 2005 M)**

(*a*) 2400 (*b*) 2480 (*c*) 2680 (*d*) 2800

Sol. (*d*) Let each students gets x sweets. Then $175x = (175 - 35)(x + 4)$

$175x = 140x + 140 \times 4 \Rightarrow x = \frac{140 \times 4}{35} = 16$

∴ Number of sweets = $140 \times (x + 4) = 140 \times 20 = 2800$

Alternative Method : Ratio of number of students = 175 : (175 – 35) = 175 : 140 = 5 : 4

∴ Ratio of sweets distributed to each student = 4 : 5

Difference in ratios = (5 – 4) ratio = 1 ratio = 4 sweets

Sweets distributed originally to each student = 5 ratio = 5 × 4 = 20

∴ Number of sweets = 140 × 20 = 2800

125. One year ago, the ratio between A's and B's salaries was 2 : 3. The ratio of their own salaries of last year and present year was 3 : 4 and 4 : 5 respectively. If sum of their present salaries is ₹ 1140, find the salary of A.

(*a*) ₹ 640 (*b*) ₹ 800 (*c*) ₹ 900 (*d*) None of these

Sol. (*a*) Ratio of present salaries of A and B = $2 \times \frac{4}{3} : 3 \times \frac{5}{4} = \frac{8}{3} : \frac{15}{4} = 32 : 45$

A's salary = $\frac{32}{77} \times 1140$ = ₹ 640

126. If $p : q = r : s = t : u = 2 : 3$, then $(mp + nr + ot) : (mq + ns + ou) = ?$ **(SSC CGL Tier I 2000)**

(*a*) 1 : 3 (*b*) 1 : 2 (*c*) 2 : 3 (*d*) 3 : 2

Sol. (*c*) $\frac{p}{q} = \frac{r}{s} = \frac{t}{u} \Rightarrow \frac{mp + nr + ot}{mq + ns + ou} = \frac{(m + n + o) \times 2}{(m + n + o) \times 3} = \frac{2}{3}$

127. If $a : b = c : d$; then $\frac{ma + nc}{mb + nd} = ?$ **(SSC CGL Tier I 2006)**

(*a*) $m : n$ (*b*) $a : b$ (*c*) $b : a$ (*d*) $n : m$

Sol. (*b*) Let $\frac{a}{b} = \frac{c}{d} = k$. Then $a = bk$, and $c = dk$ $\Rightarrow \frac{ma + nc}{mb + nd} = \frac{mbk + ndk}{mb + nd} = k = \frac{a}{b} = a : b$

Alt: $\frac{ma + nc}{mb + nd} = \frac{a}{b}$

128. A man spends $\frac{3}{5}$ of his income and still saves ₹ 600. His income is:

(*a*) ₹ 240 (*b*) ₹ 360 (*c*) ₹ 1000 (*d*) ₹ 1500

Sol. (*d*) His savings $= 1 - \frac{3}{5} = \frac{2}{5}$ of income. His actual savings = ₹ 600

$\therefore$ Income = ₹ 600 $\times \frac{5}{2}$ = ₹ 1500

129. A man spends $\frac{1}{5}$ of his income on food and $\frac{3}{5}$ on clothing. If he still saves ₹ 100, what is his total income?

(*a*) ₹ 20 (*b*) ₹ 80 (*c*) ₹ 400 (*d*) ₹ 500

Sol. (*d*) Total expenditure $= \frac{1}{5} + \frac{3}{5} = \frac{4}{5} \Rightarrow$ His savings $= 1 - \frac{4}{5} = \frac{1}{5}$ of his income

His actual savings = ₹ 100 $\Rightarrow$ His income = ₹ 100 $\times \frac{5}{1}$ = ₹ 500

130. A man spends $\frac{2}{5}$ of his income on food and $\frac{4}{5}$ of the remaining income on clothing. If he still saves ₹ 600, what is his total income?

(*a*) ₹ 480 (*b*) ₹ 500 (*c*) ₹ 1920 (*d*) ₹ 5000

Sol. (*d*) Expenditure on food $= \frac{2}{5}$. Balance $= 1 - \frac{2}{5} = \frac{3}{5}$

Expenditure on clothes $= \frac{4}{5} \times \frac{3}{5} = \frac{12}{25}$. Savings $= \frac{3}{5} - \frac{12}{25} = \frac{3}{25}$

$\therefore$ Income $= \frac{25}{3} \times$ ₹ 600 = ₹ 5000

Alternative Method : Expenditure on food $= \frac{2}{5}$. Balance $= 1 - \frac{2}{5} = \frac{3}{5}$

Expenditure on clothing is $\frac{4}{5}$ of the balance amount. Savings $= 1 - \frac{4}{5} = \frac{1}{5}$

$\therefore$ Income $= 600 \div \frac{1}{5} \div \frac{3}{5}$ = ₹ 600 $\times \frac{5}{1} \times \frac{5}{3}$ = ₹ 5000

131. A man gives $\frac{1}{3}$ of his money to his wife, $\frac{1}{4}$ of the remaining to his son and $\frac{2}{5}$ of the remaining to his daughter. If he still have a balance of ₹ 3000, how much money does he had originally? **(EPFO SSA Exam.)**

(*a*) ₹ 100 (*b*) ₹ 2250 (*c*) ₹ 6000 (*d*) ₹ 10000

Sol. (*d*) Table showing distribution of amount

	Share	Balance	Inverse
Wife	$\frac{1}{3}$	$\frac{2}{3}$	$\frac{3}{2}$
Son	$\frac{1}{4}$	$\frac{3}{4}$	$\frac{4}{3}$
Daughter	$\frac{2}{5}$	$\frac{3}{5}$	$\frac{5}{3}$

Money originally he had = ₹ 3,000 $\times \frac{5}{3} \times \frac{4}{3} \times \frac{3}{2}$ = ₹ 10,000

132. A man won a lottery of ₹ 20,00,000. He gave $\frac{2}{5}$ of the amount to his wife and $\frac{1}{4}$ of the balance to his son. He donated $\frac{2}{9}$ of the balance amount. How much money was donated ?

(*a*) ₹ 1,00,000 (*b*) ₹ 1,50,000 (*c*) ₹ 2,00,000 (*d*) ₹ 2,50,000

Sol. (*c*) Donation = ₹ 2000000 $\times \left(1-\frac{2}{5}\right) \times \left(1-\frac{1}{4}\right) \times \frac{2}{9}$ = ₹20,00,000 $\times \frac{3}{5} \times \frac{3}{4} \times \frac{2}{9}$ = ₹2,00,000

133. A father gives 1% of his monthly salary to his two sons as pocket money. The elder son gets 80% of the total amount given to the two sons and spends 80% of his share. If he saves ₹20 every month, then the monthly salary of the father is: **(SSC CGL Tier I 2008)**

(*a*) ₹ 10,000 (*b*) ₹ 11,500 (*c*) ₹ 12,000 (*d*) ₹ 12,500

Sol. (*d*) Monthly income = ₹ $20 \times \frac{100}{1} \times \frac{100}{80} \times \frac{100}{20}$ = ₹12,500

134. A man deposits certain amount in his bank account. After a few days, he withdraws half of the money deposited and deposits ₹ 500 more. If he has a balance of ₹ 2000 in his bank account, find the amount deposited initially.

(*a*) ₹ 2000 (*b*) ₹ 2500 (*c*) ₹ 3000 (*d*) ₹ 4000

Sol. (*c*) Balance left in his account on the last day = ₹ 2000

Amount deposited on the last day = ₹ 500

∴ Balance before that deposit = ₹ 2000 – ₹ 500 = ₹ 1500

₹ 1500 is half of the amount deposited initially.

∴ Amount deposited initially = 2 × ₹ 1500 = ₹ 3000

135. A train at the first station drops one-third of the passengers and takes 50 more. At the second station, it drops half of the remaining passengers and takes 25 more. If there were 150 passengers left in the train now, how many passengers were there in the train at the starting point?

(*a*) 200 (*b*) 250 (*c*) 300 (*d*) 350

Sol. (*c*) Passengers in the train at the third station = 150

Passengers in the train at the second station (before taking 25 more passengers)
= 150 – 25 = 125

Passengers in the train at the time of leaving the first station = 2 × 125 = 250

Passengers in the train at the first station (before taking 50 more passengers)
= 250 – 50 = 200

Passengers at starting point = $\frac{3}{2} \times 200 = 300$

Hint: At the first station, trains drops 1/3 passengers.

∴ Remaining passengers = $1-\frac{1}{3}=\frac{2}{3}$. Inverse of $\frac{2}{3}=\frac{3}{2}$

136. A bus starts with some passengers. At the first stop, one fifth of the passengers gets down and 40 passengers get in. At the second stop, half of the passengers gets down and 30 get in. The number of passengers now is 70. The number of passengers with which the bus started was **(CDS 2013)**

(*a*) 40 (*b*) 50 (*c*) 60 (*d*) 70

Sol. (*b*) Passengers at second stop = 2 × (70 – 30) = 80

Passengers at first stop = $\frac{5}{4} \times (80 - 40) = 50$

137. A shopkeeper sold half of his stock of apples and 1 more to his first customer, half of the remaining stock and 2 more to the second customer and half of the remaining stock and 3 more to the third customer. If he still has 12 apples with him, how many apples had he in the beginning?

(*a*) 56 (*b*) 91 (*c*) 110 (*d*) 130

Sol. (*d*) Stock left at the end = 12

Apples left before selling to the 3rd customer = 2 (12 + 3) = 30

Apples left before selling to the 2nd customer = 2 (30 + 2) = 64

Apples left before selling to the 1st customer = 2 (64 + 1) = 130

∴ He had 130 apples at the beginning.

138. A shopkeeper sold half of his stock of apples and 1 more to his first customer, half of the remaining stock and 1 more to the second customer and half of the remaining stock and 1 more to the third customer and so on to fourth customer. If he exhausted the stock, how many apples had he in the beginning?

(*a*) 4 (*b*) 12 (*c*) 30 (*d*) 40

Sol. (*c*) Stock left at the end = 0

Apples left before selling to the 4th customer = 2 (0 + 1) = 2

Apples left before selling to the 3rd customer = 2 (2 + 1) = 6

Apples left before selling to the 2nd customer = 2 (6 + 1) = 14

Apples left before selling to the 1st customer = 2 (14 + 1) = 30

∴ He had 30 apples at the beginning.

139. A women sells to the first customer half her stock of apples and half an apple, to the second customer half an apple and half of her remaining stock and so also to a third and to a fourth customer. She finds that she has now 15 apples left. How many had she at first ? **(SSC CGL Tier II 2012)**

(*a*) 125 (*b*) 255 (*c*) 250 (*d*) 155

Sol. (*c*) Stock left at the end = 15

Apples left before selling to the 4th customer = 2 × 15 + 0.5 = 30.5

Apples left before selling to the 3rd customer = 2 × (30.5 + 0.5) = 62

Apples left before selling to the 2nd customer = 2 × 62 + 0.5 = 124.5

Apples left before selling to the 1st customer = 2 × (124.5 + 0.5) = 250

∴ She had 250 apples at the beginning.

140. A and B joined a partnership business by investing ₹ 30,000 and ₹ 50,000 respectively. If they earn a profit of ₹ 4000, find A's share in the profit.

(*a*) ₹ 1500 (*b*) ₹ 2000 (*c*) ₹ 2500 (*d*) ₹ 1000

Sol. (*a*) Since periods for which the two amounts are invested, are same.

∴ Ratio in which profit is to distributed between A and B is 30,000 : 50,000 = 3 : 5

∴ A's share in profit = $\frac{3}{8} \times 4000$ = ₹ 1500

141. A, B and C rent a pasture. A puts in 10 cows for 7 months, B 12 cows for 5 months and C 15 cows for 3 months for grazing. If the rent of the pasture is ₹ 175, how much must C pay as his share of rent ? (SSC CGL Tier I 2000)

(*a*) ₹ 45 (*b*) ₹ 50 (*c*) ₹ 55 (*d*) ₹ 60

Sol. (*a*) Ratio of shares of A, B and C in the rent = 10 × 7 : 12 × 5 : 15 × 3 = 14 : 12 : 9

∴ C's share in the rent = ₹ 175 × $\frac{9}{35}$ = ₹ 45

142. A and B joined a partnership and invested ₹ 50,000 and ₹ 60,000 respectively. After 8 months, B leaves and C joins with capital of ₹ 90,000. If their profit of one year is ₹ 36,000, find A's share in the profit.

(*a*) ₹ 9,000 (*b*) ₹ 12,000 (*c*) ₹ 15,000 (*d*) ₹ 21,000

Sol. (*c*) Capitals of A, B and C are invested for 12, 8 and 4 months respectively.

Profit sharing ratio = (50,000 × 12) : (60,000 × 8) : (90,000 × 4) = 5 : 4 : 3

A's share in profit = $\frac{5}{12} \times 36,000$ = ₹ 15,000

143. A and B started a partnership investing amounts in the ratio 2 : 3. After 6 months, C joined the partnership, with an amount equal to that of B. The profit at the end of one year should be distributed among A, B and C in the ratio :

(*a*) 2 : 3 : 3 (*b*) 4 : 6 : 3 (*c*) 2 : 6 : 3 (*d*) None of these

Sol. (*b*) Equivalent capital of A, B and C for one month = (2 × 12) : (3 × 12) : (3 × 6) = 4 : 6 : 3

Alt: Equivalent capital of A, B and C for six months = (2 × 2) : (3 × 2) : (3 × 1) = 4 : 6 : 3

144. A, B and C started a business with investment in the ratio 5 : 6 : 8 respectively. After one year C withdrew 50% of his capital and A increased his capital by 60% of his investment. After two years in what ratio should the earned profit be distributed among A, B and C respectively? (Syndicate Bank PO 2004)

(*a*) 2 : 3 : 3 (*b*) 4 : 3 : 2 (*c*) 13 : 12 : 12 (*d*) None of these

Sol. (*c*) Investment for first year = 5 : 6 : 8

A's capital for second year = 5 + 60% of 5 = 5 + 3 = 8

C's capital for second year = 8 – 50% of 8 = 8 – 4 = 4

∴ Required ratio = (5 + 8) : (6 + 6) : (8 + 4) = 13 : 12 : 12

145. Raksha invested an amount of ₹ 60,000 to start a software business. After six months, Kamal joined her with an amount of ₹ 90,000. After one year from the commencement of the business, Raksha put in an additional amount of ₹ 20,000. At the end of three years, they earned profit of ₹ 71,20,000. What is Raksha's share in the profit ? **(Bank PO 2006)**

(*a*) ₹ 35,20,000 (*b*) ₹ 26,40,000 (*c*) ₹ 27,00,000 (*d*) ₹ 38,40,000

Sol. (*a*) Profit sharing ratio = (60,000 × 1 + 80,000 × 2) : (90,000 × 2½)
= 2,20,000 : 2,25,000 = 44 : 45

Raksha's share = $\frac{44}{89}$ × 71,20,000 = ₹ 35,20,000

146. A started a business with a capital of ₹1,00,000. One year later, B joined him with a capital of ₹2,00,000. At the end of 3 years from the start of the business, the profit earned was ₹84,000. The share of B in the profit exceeded the share of A by : **(SSC CGL Tier I 2008)**

(*a*) ₹ 10,000 (*b*) ₹ 12,000 (*c*) ₹ 14,000 (*d*) ₹ 15,000

Sol. (*b*) Ratio in which profit is to distributed between A and B = 1,00,000 × 3 : 2,00,000 × 2 = 3 : 4

∴ Difference in their share in profit = $\frac{4-3}{3+4}$ × ₹84,000 = ₹ 12,000

147. A and B entered into a partnership investing ₹16000 and ₹12000 respectively. After 3 months, A withdrew ₹5000 while B invested ₹5000 more. After 3 more months, C joins the business with a Capital of ₹21000. The share of B exceeds that of C, out of total profit of ₹26400 after one year by **(SSC CGL Tier I 2015)**

(*a*) ₹ 2400 (*b*) ₹ 1200 (*c*) ₹ 4800 (*d*) ₹ 3600

Sol. (*d*) Ratio of capitals = (16 × 3 + 11 × 9):(12 × 3 + 17 × 9):(21 × 6) = 147:189:126 = 7 : 9 : 6

∴ B – C = $\frac{3}{22}$ × 26400 = 3600

148. A, B and C started a business with capitals in the ratio 5 : 6 : 8. If at the end of one year, they shared the profit in the ratio 5 : 3 : 12, find the ratio of time for which they had contributed their capitals?

(*a*) 25 : 18 : 96 (*b*) 1 : 2 : 3 (*c*) 2 : 1 : 3 (*d*) 3 : 6 : 2

Sol. (*c*) Ratio of capitals = 5 : 6 : 8. Ratio of share in profit = 5 : 3 : 12

∴ Ratio of periods = $\frac{\text{Ratio of profits}}{\text{Ratio of capitals}} = \frac{5}{5}:\frac{3}{6}:\frac{12}{8} = 1:\frac{1}{2}:\frac{3}{2}$ = 2 : 1 : 3

149. A began business with ₹ 45,000 and was joined afterwards by B with ₹54,000. After how many months did B join if the profits at the end of the year were divided in the ratio 2 : 1 ? **(SSC CGL Tier I 2005)**

(*a*) 4 (*b*) 5 (*c*) 6 (*d*) 7

Sol. (*d*) 45000 × 12 : 54000 × *x* = 2 : 1 ⇒ *x* = $\frac{45000 \times 12 \times 1}{54000 \times 2}$ = 5 ⇒ B joined after 7 months.

Alternative Method: Ratio of capitals = 45000 : 54000 = 5 : 6. Ratio of profits = 2 : 1

∴ Ratio of periods = $\frac{\text{Ratio of profits}}{\text{Ratio of capitals}} = \frac{2}{5}:\frac{1}{6}$ = 12 : 5 ⇒ B joined after 7 months.

150. A and B are partners in a business, A contributes 1/4 of the capital for 15 months and B received 2/3 of the profit. Find how long B's money was used. **(SSC CGL Tier I 2012)**

(*a*) 6 months (*b*) 8 months (*c*) 10 months (*d*) 12 months

Sol. (*c*) Ratio of capitals = 1 : 3. Profit sharing ratio = 1 : 2.

Ratio of periods = $\frac{1}{1}:\frac{2}{3}=3:2 \Rightarrow$ Required period = $\frac{2}{3}\times 15$ = 10 months

151. A starts business with ₹ 7000 and after 5 months, B joined as a partner. After a year, the profit is divided in the ratio 2 : 3. The capital of B is **(SSC CGL Tier I 2012, 2000M)**
(*a*) ₹ 10,000 (*b*) ₹ 6,500 (*c*) ₹ 18,000 (*d*) ₹ 9,000

Sol. (*c*) $\frac{7000\times 12}{x\times 7}=\frac{2}{3} \Rightarrow x=\frac{7000\times 3\times 12}{7\times 2}=18,000$

Alt: Profit-sharing ratio = 2 : 3. Ratio of time = 12 : 7

$\therefore$ Ratio of capitals = $\frac{2}{12}:\frac{3}{7}=\frac{1}{6}:\frac{3}{7}$ = 7 : 18

$\therefore$ A's capital = 7 ratio = ₹ 7000 $\Rightarrow$ B's capital = 18 ratio = ₹ 18000

152. A and B have 3 and 5 loaves respectively. They share their food with C who pays them ₹ 16. If each of them got equal food, find A's share in ₹ 16.
(*a*) ₹ 2 (*b*) ₹ 3 (*c*) ₹ 5 (*d*) ₹ 6

Sol. (*a*) Total loaves = 3 + 5 = 8

Share of each = $\frac{8}{3}$. A's sacrifice = $3-\frac{8}{3}=\frac{1}{3}$. B's sacrifice = $5-\frac{8}{3}=\frac{7}{3}$

$\therefore$ Ratio of their sacrifice = $\frac{1}{3}:\frac{7}{3}$ = 1 : 7 $\Rightarrow$ A's share in ₹ 16 = $\frac{1}{1+7}\times$ ₹ 16 = ₹ 2

153. A and B started a business with ₹ 20,000 and ₹ 35,000 respectively. They agreed to share the profit in the ratio of their capital. C joins the partnership with the condition that A, B and C will share profit equally and pays ₹ 2,20,000 as premium for this, to be shared between A and B. This is to be divided between A and B in the ratio of **(SSC CGL Tier II 2012)**
(*a*) 9 : 10 (*b*) 10 : 9 (*c*) 10 : 1 (*d*) 1 : 10

Sol. (*d*) Old ratio of A and B = 20000 : 35000 = 4 : 7 $\Rightarrow$ New ratio of A, B, C = 1 : 1 : 1
A's share is reduced by $\frac{4}{11}-\frac{1}{3}=\frac{1}{33}$ B's share is reduced by $\frac{7}{11}-\frac{1}{3}=\frac{10}{33}$

Required ratio = $\frac{1}{33}:\frac{10}{33}=1:10$

154. A group of 15 persons spends ₹ 4500 in 6 months, find total expenditure of a group of 25 persons in 4 months, if expenditure per person is same in both the cases.
(*a*) ₹ 2000 (*b*) ₹ 3000 (*c*) ₹ 4000 (*d*) ₹ 5000

Sol. (*d*) Let expenditure of 25 persons is x, then

$\left.\begin{matrix}15:25 \text{ Persons}\\ 6:4 \text{ Months}\end{matrix}\right] - 4500 : x \text{ Expenditure}$

More persons, more expenditure i.e. direct proportion
Less months, less expenditure i.e. direct proportion

$\therefore x = 4500\times\frac{25}{15}\times\frac{4}{6}$ = ₹ 5000

155. A sum of ₹ 4800 was divided among A, B and C in the ratio 9 : 8 : 7. If an amount of ₹ 200 is added to the share of each, what will be their new ratio? **(SBI Clerical 2008)**
(*a*) 7 : 8 : 9 (*b*) 1 : 1 : 1 (*c*) 10 : 9 : 8 (*d*) None of these

Sol. (*c*) 1 ratio = $\frac{4800}{9+8+7}$ = ₹ 200 ⇒ New ratio = (9 + 1) : (8 + 1) : (7 + 1) = 10 : 9 : 8

156. Two numbers are in the ratio 2 : 3. If 3 be added to both of them, then their ratio becomes 3 : 4. Find the sum of the numbers. **(SSC CGL Tier I 2003)**

(*a*) 10 (*b*) 15 (*c*) 20 (*d*) 25

Sol. (*b*) New ratio 3 : 4
Old ratio 2 : 3
Difference 1 : 1
⇒ 1 ratio = 3 ⇒ Sum = (2 + 3) ratio = 5 ratio = 5 × 3 = 15

157. Three numbers are in the ratio 1 : 2 : 3. By adding 5 to each of them, the new numbers are in the ratio 2 : 3 : 4. The numbers are : **(SSC CGL Tier I 2013)**

(*a*) 5, 10, 15 (*b*) 10, 20, 30 (*c*) 15, 30, 45 (*d*) 1, 2, 3

Sol. (*a*) New ratio 2 : 3 : 4
Old ratio 1 : 2 : 3
Difference 1 : 1 : 1
⇒ 1 ratio = 5 ⇒ Numbers are (1, 2, 3) ratio = 5, 10, 15

158. The students in three classes are in the ratio 2 : 3 : 5. If 20 students are increased in each class, the ratio changes to 4 : 5 : 7. Originally the total number of students was : **(SSC CGL Tier I 2005 & 2002 modified)**

(*a*) 50 (*b*) 90 (*c*) 100 (*d*) 150

Sol. (*c*) Increase in ratio for each of the 3 classes is 4 – 2 = 5 – 3 = 7 – 5 = 2 ratio
20 more students are admitted to each class. ⇒ 2 ratio = 20 ⇒ 1 ratio = 10
∴ Original number of students = (2 + 3 + 5) ratio = 10 ratio = 10 × 10 = 100

159. An alloy contains copper, zinc and nickel in the ratio 5 : 3 : 2. The quantity of nickel in kilogram that must be added to 100 kg of this alloy to have the new ratio 5 : 3 : 3, is **(SSC CGL Tier I 2011)**

(*a*) 8 (*b*) 10 (*c*) 12 (*d*) 15

Sol. (*b*)

	Copper	*Zinc*	*Nickel*
New ratio	5	3	3
Old ratio	5	3	2
Difference	0	0	1

Original quantity = (5 + 3 + 2) ratio = 10 ratio = 100 kg ⇒ 1 ratio = 10 kg
⇒ Nickel added = 1 ratio = 10 kg.

160. The number of students in three classes are in the ratio 2 : 3 : 4. If 12 students are increased in each class, this ratio changes to 8 : 11 : 14. The total number of students in three classes in the beginning was : **(SSC CGL Tier I 2008)**

(*a*) 162 (*b*) 108 (*c*) 96 (*d*) 54

Sol. (*a*) New ratio 8 : 11 : 14
Old ratio = 2 : 3 : 4 or 6 : 9 : 12
Difference 2 2 2
(Note : difference in old ratios = 1, difference in new ratios = 3 ∴ old ratio is multiplied by 3)
2 ratio = 12 student ⇒ 1 ratio = 6 students
Total students in the beginning = (6 + 9 + 12) ratio = 27 ratio = 27 × 6 = 162

161. A sum of money is distributed among A, B and C in the ratio 6 : 11 : 9. If C gives ₹ 500 from his share to B, the ratio of incomes of A, B and C becomes 3 : 6 : 4. Find the total sum.

(*a*) ₹ 6,500 (*b*) ₹ 13,000 (*c*) ₹ 15,000 (*d*) ₹ 26,000

Sol. (*b*) Old ratio 6 : 11 : 9

New ratio = 3 : 6 : 4 = 6 : 12 : 8 (multiplied by 2 to make A's share equal to 6 ratio)

Difference 0 +1 –1

₹ 500 = 1 ratio ⇒ Total sum = (6 + 11 + 9) ratio = 26 ratio = 26 × ₹ 500 = ₹ 13,000

162. A mixture contains milk and water in the ratio 5 : 1. On adding 5 litres of water, the ratio of milk and water becomes 5 : 2. What is the quantity of milk in the original mixute ?

(CDS 2010)

(*a*) 5 litres (*b*) 25 litres (*c*) 27.5 litres (*d*) 32.5 litres

Sol. (*b*)

	Milk		Water
New mixture	5	:	2
Old mixture	5	:	1
Difference	0		1

Water added = 1 ratio = 5 litres

Milk in original mixture = 5 ratio = 5 × 5 = 25 litres

163. In a mixture of 63 litres, the milk and water are in the ratio 7 : 2. If 7 litres of water is added to the mixture, find the ratio in the resulting mixture.

(*a*) 3 : 1 (*b*) 4 : 1 (*c*) 7 : 3 (*d*) 7 : 5

Sol. (*c*) Quantity of water in the mixture $= \frac{2}{9} \times 63 = 14$ litres

Water added = 7 litres ⇒ Increased quantity of water = 14 + 7 = 21 litres

∴ Ratio of milk and water = (63 – 14) : 21 = 49 : 21 = 7 : 3

Alternative Method: Sum of ratios = 7 + 2 = 9 ⇒ 1 ratio = 63 ÷ 9 = 7 litres

∴ New ratio = 7 : (2 + 1) = 7 : 3

164. In a 729 litres mixture of milk and water, the ratio of milk to water is 7 : 2. To get a new mixture containing milk and water in the ratio 7 : 3, the amount of water to be added is :

(SSC CGL Tier I 2008)

(*a*) 56 litres (*b*) 71 litres (*c*) 81 litres (*d*) 91 litres

Sol. (*c*) Sum of ratios in the original mixture = 7 + 2 = 9 ratios

∴ 9 ratio = 729 litres ⇒ 1 ratio = 729 ÷ 9 = 81 litres

Change in ratio of water = 3 – 2 = 1 ratio ⇒ Water to be added = 1 ratio = 81 litres

165. The ratio of milk and water in a mixture of 84 litres is 3 : 4. What will be the new ratio if 3 litres each of milk and water is added to the mixture ?

(*a*) 7 : 9 (*b*) 4 : 5 (*c*) 13 : 17 (*d*) 9 : 16

Sol. (*c*) Quantity of milk in the mixture $= \frac{3}{7} \times 84 = 36$ litres

Water in the mixture = 84 – 36 = 48 litres

∴ New ratio of milk and water = (36 + 3) : (48 + 3) = 39 : 51 = 13 : 17

Alt: Total mixture = (3 + 4) ratio = 84 lites ⇒ 1 ratio = 84 ÷ 7 = 12 litres

3 litres $= \frac{3}{12} = \frac{1}{4}$ ratio ⇒ New ratio $= \left(3+\frac{1}{4}\right):\left(4+\frac{1}{4}\right) = 13:17$

166. In a class, the number of girls is 20% more than that of the boys. The strength of the class is 66. If 4 more girls are admitted to the class, the ratio of the number of boys to that of the girls is : **(SSC CGL Tier I 2005)**

(*a*) 1 : 2 (*b*) 3 : 4 (*c*) 1 : 4 (*d*) 3 : 5

Sol. (*b*) Let the number of boys be 100. Then number of girls = 120

Boys : Girls = 100 : 120 = 5 : 6 $\Rightarrow$ Boys = $\frac{5}{11} \times 66 = 30 \Rightarrow$ Girls = 66 – 30 = 36

New ratio = 30 : (36 + 4) = 30 : 40 = 3 : 4

Alternative method : Let the number of boys be 100. Then number of girls = 120

$\therefore$ Boys : Girls = 100 : 120 = 5 : 6

Total students = (5 + 6) ratio = 11 ratio = 66 $\Rightarrow$ 1 ratio = 66 ÷ 11 = 6

Girls admitted = $\frac{4}{6} = \frac{2}{3}$ New ratio = $5:\left(6+\frac{2}{3}\right) = 5:\frac{20}{3}$ = 15 : 20 = 3 : 4

Alternative method : Old ratio of boys = $\frac{5}{11}$. New ratio = $\frac{5}{11}\times\frac{66}{70}=\frac{3}{7}$

$\therefore$ Boys : Girls = 3 : (7 – 3) = 3 : 4

167. The ratio of numbers of boys and girls of a school with 504 students is 13 : 11. What will be the new ratio if 12 more girls are admitted ? **(SSC CGL Tier I 2002)**

(*a*) 91 : 81 (*b*) 81 : 91 (*c*) 9 : 10 (*d*) 10 : 9

Sol. (*a*) Boys = $\frac{13}{24} \times 504 = 273$ $\Rightarrow$ Girls = 504 – 273 = 231

New ratio = 273 : (231 + 12) = 273 : 243 = 91 : 81

Alternative Method: $x = \frac{504}{13+11} = \frac{504}{24} = 21$ $\Rightarrow$ Girls admitted = $\frac{12}{21}$ or $\frac{4}{7}$ of x

$\therefore$ New ratio $= 13:\left(11+\frac{4}{7}\right) = 91:81$

168. The ratio of incomes of P and Q is 3 : 4 and the ratio of their expenditures is 2 : 3. If both of them save ₹ 6000, the income of P is :

(*a*) ₹ 20000 (*b*) ₹ 12000 (*c*) ₹ 18000 (*d*) ₹ 24000

Sol. (*c*)

	P		Q
Income	3	:	4
Expenditure	2	:	3
Savings	1	:	1

Hence savings of each = 1 ratio = ₹ 6000 $\therefore$ Income of P = 3 ratio = 3 × ₹ 6000 = ₹ 18000

169. A mixture of 70 litres contains milk and water in the ratio 3 : 4. How many litres of milk must be added to the mixture so as to make the ratio 5 : 4?

(*a*) 10 litres (*b*) 15 litres (*c*) 20 litres (*d*) 25 litres

Sol. (*c*)

	Milk	:	Water
New ratio	5	:	4
Old ratio =	3	:	4
Difference =	2	:	0

Sum of old ratios = (3 + 4) ratio = 7 ratio = 70 litres $\Rightarrow$ 1 ratio = 10 litres

$\therefore$ Milk added = 2 ratio = 2 × 10 = 20 litres

170. In 30 litres of mixture of acid, the ratio of acid and water is 2 : 3. What amount of water should be added to the mixture so that the ratio of acid and water becomes 2 : 5? **(SSC CGL 2002)**

(*a*) 10 litres (*b*) 15 litres (*c*) 18 litres (*d*) 12 litres

Sol. (*d*)

	Acid	:	Water
New ratio	2	:	5
Old ratio =	2	:	3
Difference =	0	:	2

Sum of old ratios = (2 + 3) ratio = 5 ratio = 30 litres ⇒ 1 ratio = 6 litres

∴ Water to be added = 2 ratio = 2 × 6 litres = 12 litres

171. The ratio of boys and girls in a class of 80 students is 5 : 3. How many more girls should join the class to make the ratio 1 : 1 ?

(*a*) 10 (*b*) 15 (*c*) 20 (*d*) 25

Sol. (*c*)

	Boys	:	Girls
New ratio = 1 : 1 =	5	:	5
Old ratio =	5	:	3
Difference =	0	:	2

Sum of old ratios = 5 + 3 = 8 ⇒ Girls should join = $\frac{2}{8} \times 80 = 20$

Trick: Girls should join = $\frac{5-3}{5+3} \times 80 = 20$

Note: This formula can be used only when new ratio is 1 : 1.

172. The ratio of the quantities of an acid and water in a mixture is 1 : 3. If 5 litres of acid is further added to the mixture, the new ratio becomes 1 : 2. The quantity of new mixture in litres is : **(SSC CGL Tier I 2011)**

(*a*) 32 (*b*) 40 (*c*) 42 (*d*) 45

Sol. (*d*)

	Acid	:	water
Old ratio = 1 : 3 =	2	:	6
New ratio = 1 : 2 =	3	:	6
Difference =	1	:	0

1 ratio = 5 litres ⇒ New mixture = (3 + 6) ratio = 9 ratio = 9 × 5 litres = 45 litres

173. In a rare collection, there is one gold coin for every three non-gold coins. 10 more gold coins are added to the collection and the ratio of gold coins to non-gold coins would be 1 : 2. Based on the information, the total number of coins in the collection now becomes **(UPSC Civil Services (Pre.) 2013)**

(*a*) 90 (*b*) 80 (*c*) 60 (*d*) 50

Sol. (*a*)

	Gold	:	Non-Gold
Old ratio = 1 : 3 =	2	:	6
New ratio = 1 : 2 =	3	:	6
Difference =	1	:	0

1 ratio = 10 coins ⇒ Coins now = (3 + 6) ratio = 9 ratio = 9 × 10 coins = 90 coins

173. In 40 litres mixture of milk and water, the ratio of milk to water is 7 : 1. In order to make the ratio of milk and water 3 : 1, the quantity of water added to the mixture is :

(SSC CGL Tier I 2008)

(*a*) 6 litres (*b*) $6\frac{2}{3}$ litres (*c*) 10 litres (*d*) 20 litres

Sol. (*b*) LCM of ratios of milk *i.e.* LCM of 7 and 3 is 21.

∴ We make ratio of milk equal to 21 in the two ratios.

	Milk	:	Water
New ratio = 3 : 1 =	21	:	7
Old ratio = 7 : 1 =	21	:	3
Difference =	0	:	4

∴ Water added = $\frac{4}{21+3}\times 40 = \frac{20}{3} = 6\frac{2}{3}$ litres

Alt : Milk in original mixture = $\frac{7}{8}\times 40$ = 35 litres ⇒ Water = 40 – 35 = 5 litres

Ratio of milk and water in new mixture = 3 : 1

∴ Quantity of water in the new mixture = $\frac{1}{3}\times 35$ litres = $11\frac{2}{3}$ litres

Water to be added = $11\frac{2}{3}-5=6\frac{2}{3}$ litres

175. The total number of students of a school was 660.The ratio between boys and girls was 13 : 9. After some days, 30 girls joined the school and some boys left the school and new ratio between boys and girls became 6 : 5. The number of boys who left the school is

(SSC CHSL 2012)

(*a*) 60 (*b*) 50 (*c*) 40 (*d*) 30

Sol. (*d*) Boys = $\frac{13}{22}\times 660$ = 390 ⇒ Girls = 660 – 390 = 270

Girls joined = 30 ⇒ Girls = 270 + 30 = 300 = 5 ratio ⇒ 1 ratio = 60

Boys = 6 ratio (new) = 6 × 60 = 360 ⇒ Boys left = 390 – 360 = 30

Trick: Old ratio = 13 : 9, New ratio = 6 : 5 = 12 : 10 i.e. the Boy's ratio is decreased by 1 ratio and Girl's ratio in increased by 1 ratio.

∴ Boys Left = Girls joined = 30

176. Two types of oil A and B are mixed in the ratio of 3 : 2 and the total mixture is 90 litres. When one-third of the mixture was consumed, some more quantity of oil A was mixed and the new ratio becomes 2 : 1. How much quantity of A was added later on.

(*a*) 10 litres (*b*) 12 litres (*c*) 15 litres (*d*) 20 litres

Sol. (*b*) Mixture after one-third consumption = $\frac{2}{3}\times 90$ = 60 litres

Quantity of type A oil in remaining mixture = $\frac{3}{5}\times 60$ = 36 litres

B in remaining mixture = 60 – 36 = 24 litres = 1 ratio (in new mixture)

Quantity of type A oil in new mixture = 2 ratio = 2 × 24 = 48 litres

∴ Quantity of type A oil added = 48 – 36 = 12 litres

177. Zinc and copper are in the ratio 5 : 3 in 200 gm of an alloy. How much grams of copper be added to make the ratio as 3 : 5? **(SSC CGL Tier I 2001)**

(*a*) $133\frac{1}{3}$ (*b*) $\frac{1}{200}$ (*c*) 72 (*d*) 66

Sol. (*a*) LCM of ratios of copper *i.e.* LCM of 3 and 5 is 15.

	Zinc	:	Copper
New ratio = 3 : 5 =	15	:	25
Old ratio = 5 : 3 =	15	:	9
Difference =	0	:	16

$\therefore$ Copper to be added $= \frac{16}{15+9} \times 200 = 133\frac{1}{3}$ grams

Alternative Method: Copper in the given mixture $= \frac{3}{8} \times 200 = 75$ grams

Zinc in the given mixture $= 200 - 75 = 125$ grams

Copper in the new mixture $= \frac{5}{3} \times 125 = \frac{625}{3} = 208\frac{1}{3}$ grams

$\therefore$ Copper to be added $= 208\frac{1}{3} - 75 = 133\frac{1}{3}$ grams

178. In a mixture of 75 litres, the ratio of milk to water is 2 : 1. The amount of water to be further added to the mixture so as to make the ratio of milk to water 1 : 2 : **(SSC CGL Tier I 2008)**

(*a*) 45 litres (*b*) 60 litres (*c*) 75 litres (*d*) 80 litres

Sol. (*c*) LCM of ratios of milk *i.e.* 2 and 1 is 2

	Milk	:	Water
New ratio = 1 : 2 =	2	:	4
Old ratio = 2 : 1 =	2	:	1
Difference =	0	:	3

$\therefore$ Water to be added $= \frac{3}{3} \times 75$ litres $= 75$ litres

Alt: Milk in original mixture $= \frac{2}{3} \times 75 = 50$ litres $\Rightarrow$ Water $= 75 - 50 = 25$ litres

$\therefore$ Water in new mixture $= 2 \times 50$ litres $= 100$ litres $\Rightarrow$ Water added $= 100 - 25 = 75$ litres

179. The ratio of males and females in a party is 3 : 2. When 20 more females joined, the ratio was reversed. How many males were there at the party?

(*a*) 16 (*b*) 24 (*c*) 32 (*d*) 40

Sol. (*b*) Old ratio of males and females is 3 : 2. $\Rightarrow$ New ratio of males and females is 2 : 3.

LCM of ratios of male *i.e.* LCM of 3 and 2 is 6.

	Male	:	Female
New ratio = 2 : 3 =	6	:	9
Old ratio = 3 : 2 =	6	:	4
Difference =	0	:	5

Female joined = 5 ratio = 20 $\Rightarrow$ 1 ratio = 4 $\Rightarrow$ Males in the beginning = 6 ratio = $6 \times 4 = 24$

178. Incomes of A and B are in the ratio 3 : 4. Had A's income more by ₹ 15,000, the ratio of their incomes would have been 3 : 2. What is the income of B ?

(*a*) ₹ 15,000 (*b*) ₹ 20,000 (*c*) ₹ 25,000 (*d*) ₹ 30,000

Sol. (*b*)

Old ratio =	3	:	4
New ratio = 3 : 2 =	6	:	4
Difference =	3	:	0

3 ratio = ₹ 15,000 ⇒ 1 ratio = ₹ 5,000 ⇒ B's income = 4 ratio = 4 × ₹ 5,000 = ₹ 20,000

179. The ratio between two numbers is 2 : 3. On adding 3 to both the numbers, the ratio becomes 5 : 7. Find the smaller number.

(*a*) 10 (*b*) 12 (*c*) 18 (*d*) 21

Sol. (*b*) $\frac{2x+3}{3x+3}=\frac{5}{7} \Rightarrow 14x+21=15x+15 \Rightarrow x=6 \Rightarrow 2x=12$

180. A's and B's incomes are in the ratio 3 : 2 and their expenses are in the ratio 5 : 3. If each of them saves ₹ 2000, then the income of A is. **(SSC CHSL 2013)**

(*a*) ₹ 8000 (*b*) ₹ 10000 (*c*) ₹ 12000 (*d*) ₹ 15000

Sol. (*c*)

	A		B	
Income ratio = 3 : 2 =	6	:	4	(Difference of ratio = 3 – 2 = 1)
Expenditure ratio =	5	:	3	(Difference of ratio = 5 – 3 = 2)
Difference =	1	:	1	

Savings = 1 ratio = ₹ 2000 ⇒ A's income = 6 ratio = 6 × ₹ 2000 = ₹ 12000

181. Two numbers are in the ratio 5 : 7. On diminishing each of them by 40, they become in the ratio 17 : 27. The difference of the numbers is : **(SSC CGL Tier I 2001)**

(*a*) 18 (*b*) 52 (*c*) 137 (*d*) 50

Sol. (*d*)

Old ratio = 5 : 7 =	25	:	35	(Difference = 7 – 5 = 2 ratio)
New ratio =	17	:	27	(Difference = 27 – 17 = 10 ratio)
Difference =	8	:	8	

∴ Difference in each case = 8 ratio = 40 ⇒ 1 ratio = 5

Difference between the numbers = (35 – 25) ratio = 10 ratio = 10 × 5 = 50

182. Two numbers are in the ratio 3 : 5. If 9 is subtracted from each number, then they are in the ratio of 12 : 23. What is the second number ? **(CDS 2011)**

(*a*) 44 (*b*) 55 (*c*) 66 (*d*) 77

Sol. (*b*)

Numbers	I	:	II	
Old ratio = 3 : 5 =	33	:	55	(Difference of ratio = 5 – 3 = 2)
New ratio = 12 : 23 =	24	:	46	(Difference of ratio = 23 – 12 = 11)
Difference =	9	:	9	

⇒ 9 ratio = 9 ⇒1 ratio = 1 ⇒ Second original number = 55 ratio = 55

Alt: $\frac{3x-9}{5x-9}=\frac{12}{23} \Rightarrow 69x - 23 \times 9 = 60x - 12 \times 9$

$\Rightarrow 9x = (23 - 12) \times 9 \Rightarrow x = 11 \Rightarrow 5x = 5 \times 11 = 55$

183. A sum of money is to be distributed among P, Q and R in the ratio 6 : 19 : 7. If R gives ₹200 from his share to Q, the ratio of P, Q and R becomes 3 : 10 : 3. What is the total sum ? **(SBI PO 2000)**

(*a*) ₹3200 (*b*) ₹6400 (*c*) ₹12800 (*d*) None of these

Sol. (*b*)

	P	:	Q	:	R
Old ratio	6	:	19	:	7
Change			+ ₹200	:	–₹200
New ratio	3	:	10	:	3
Revised New ratio	6	:	20	:	6
Difference =	0	:	+1	:	–1

⇒ 1 ratio = ₹200 ⇒ Total = (6 + 19 + 7) ratio = 32 × ₹200 = ₹6400

184. Two numbers are in the ratio $1\frac{1}{2}:2\frac{2}{3}$. When each of these is increased by 15, they become in the ratio $1\frac{2}{3}:2\frac{1}{2}$. The greater of the numbers is : **(SSC CGL 2002)**

(*a*) 27 (*b*) 36 (*c*) 48 (*d*) 64

Sol. (*c*) Old ratio = $1\frac{1}{2}:2\frac{2}{3}=\frac{3}{2}:\frac{8}{3}=9:16$ New ratio = $1\frac{2}{3}:2\frac{1}{2}=\frac{5}{3}:\frac{5}{2}=2:3$

$\therefore\ x=\frac{(2-3)\times 15}{9\times 3-16\times 2}=\frac{15}{5}=3$ ⇒ Greater number = $16x = 16 \times 3 = 48$

185. A mixture of 55 litres contains milk and water in the ratio 7 : 4. How many litres of milk and water each must be added to the mixture to make the ratio 3 : 2 ?

(*a*) 2 litres (*b*) 5 litres (*c*) 8 litres (*d*) 10 litres

Sol. (*d*)

Numbers	Milk	:	Water	
Old ratio	7	:	4	(Difference of ratios = 7 – 4 = 3)
New ratio = 3 : 2 =	9	:	6	(Difference of ratios = 3 – 2 = 1)
Difference =	2	:	2	

Total quantity = (7 + 4) ratio = 11 ratio = 55 litres ⇒1 ratio = 5 litres

∴ Required quantity = 2 ratio = 10 litres

186. What number should be subtracted from both the terms of the ratio 15 : 19 in order to make it 3 : 4? **(SSC CGL Tier I 2000)**

(*a*) 9 (*b*) 2 (*c*) 5 (*d*) 3

Sol. (*d*) Old ratio = 15 : 19 ⇒ Difference = 19 – 15 = 4 ⇒ Revised ratio = Ratio × 1 = 15 : 19

New ratio = 3 : 4 ⇒ Difference = 4 – 3 = 1 ⇒ Revised ratio = Ratio × 4 = 12 : 16

Difference = 3 : 3 ⇒ Required number = 3

187. What number should be subtracted from both terms of the ratio 11 : 15 so as to make it as 2 : 3? **(SSC CGL Tier I 2013)**

(*a*) 2 (*b*) 3 (*c*) 4 (*d*) 5

Sol. (*b*) $\frac{11-x}{15-x}=\frac{2}{3}\Rightarrow 33-3x=30-2x\Rightarrow x=3$

Alt: Old rato = 11 : 15. New ratio = 4 × (2 : 3) = 8 : 12

⇒ Difference = 11 – 8 = 15 – 12 = 3.

188. What must be added to each term of the ratio 2 : 5 so that it may equal to 5 : 6 ? **(SSC CGL Tier I 2015)**

(*a*) 12 (*b*) 78 (*c*) 65 (*d*) 13

Sol. (*d*) Let k be the required number. Then $x = \frac{(2-5)\times k}{2\times 6-5\times 5} = \frac{3}{13}k$

The smallest value of k is 13 for which x is a natural number.

Alternative method :

New ratio = 5 : 6 = 15 : 18 (Difference = 6 – 5 = 1)

Old ratio = 2 : 5 = 2 : 5 (Difference = 5 – 2 = 3)

Difference = 13 : 13

189. What must be added to each term of the ratio 7 : 11 so as to make it equal to 3 : 4? **(SSC CGL Tier I 2006, 2011)**

(*a*) 1 (*b*) 2 (*c*) 3 (*d*) 5

Sol. (*d*) Let k be the required number. Then $x = \frac{(3-4)\times k}{7\times 4-11\times 3} = \frac{1}{5}k$

The smallest value of k is 5 for which x is a natural number.

Alternative method :

New ratio = 3 : 4 = 12 : 16 (Difference = 4 – 3 = 1)

Old ratio = 7 : 11 = 7 : 11 (Difference =11 –7 = 4)

Difference = 5 : 5

190. Two numbers are in the ratio 2 : 3. If 2 is subtracted from the first and 2 is added to the second, the ratio becomes 1 : 2. The sum of the numbers is **(SSC CGL Tier I 2005)**

(*a*) 30 (*b*) 28 (*c*) 24 (*d*) 10

Sol. (*a*) $\frac{2x-2}{3x+2} = \frac{1}{2} \Rightarrow 4x-4 = 3x+2 \Rightarrow x = 6 \Rightarrow \text{sum} = 2x+3x = 5x = 30$

Alternatively, we can check from the options. (*a*) 30 is divisible by 5 i.e. the sum of ratios.

⇒ Original numbers are 12 and 18. ⇒ New numbers are 10 and 20. ⇒ New ratio = 1 : 2

191. The ratio of incomes of two persons A and B is 5 : 3 and that of their expenditure is 9 : 5. Find the income of each person, if they save ₹ 1300 and ₹ 900 respectively. **(SSC CGL Tier I 2002, 2005 M)**

(*a*) ₹ 4000; ₹ 2400 (*b*) ₹ 3000; ₹ 1800 (*c*) ₹ 5000; ₹ 3000 (*d*) ₹ 4500; ₹ 2700

Sol. (*a*) Let their incomes are $5x$ and $3x$ and expenditures are $9y$ and $5y$.

$5x - 9y =$ ₹ 1300 ⇒ $15x - 27y =$ ₹ 3900

$3x - 5y =$ ₹ 900 ⇒ $15x - 25y =$ ₹ 4500

Solving the equations, $y = 300$ and $x =$ ₹ 800

A's income = 5 × ₹ 800 = ₹ 4000 and B's income = 3 × ₹ 800 = ₹ 2400

192. x varies inversely as the square root of y and $y = 81$ when $x = 9$. The value of x when $y = 9$ is **(CDS 2004)**

(*a*) 3 (*b*) 27 (*c*) 81 (*d*) 29

Sol. (*b*) $x \propto \frac{1}{\sqrt{y}} \Rightarrow x = k\frac{1}{\sqrt{y}} \Rightarrow 9 = k\frac{1}{\sqrt{81}} = \frac{k}{9} \Rightarrow k = 81$

Now, $x = 81 \times \frac{1}{\sqrt{9}} = 81 \times \frac{1}{3} = 27$

193. x varies directly as y and inversely as square of z. When $y = 4$ and $z = 14$, then $x = 10$. If $y = 13$ and $z = 7$, then find the value of x.

(*a*) 91 (*b*) 130 (*c*) 169 (*d*) 260

Sol. (*b*) $x \propto y$ and $x \propto \frac{1}{z^2} \Rightarrow x \propto \frac{y}{z^2} \Rightarrow x = k\frac{y}{z^2}$

$\Rightarrow \quad 10 = k\frac{4}{14^2} \Rightarrow k = \frac{14 \times 14 \times 10}{4} = 490$. Now, $x = k\frac{y}{z^2} = 490 \times \frac{13}{7^2} = 130$

194. An alloy contains zinc, copper and tin in the ratio 2 : 3 : 1 and another alloy contains copper, tin and lead in the ratio 5 : 4 : 3. If equal weights of both the alloys are melted together to form a third alloy, then the weight of lead per kg in the new alloy is : **(SSC CGL 2008)**

(*a*) $\frac{5}{24}$ kg (*b*) $\frac{1}{8}$ kg (*c*) $\frac{1}{4}$ kg (*d*) $\frac{1}{3}$ kg

Sol. (*b*) Let 1 kg of both alloys are taken. Lead in 1 kg of first alloy = 0 kg

Lead in 1 kg of second alloy = $\frac{3}{12} = \frac{1}{4}$ kg. Total lead in 2 kg of mixture = $\frac{1}{4}$ kg

$\therefore$ Lead per kg in third alloy = $\frac{1}{4} \div 2 = \frac{1}{8}$ kg

195. Two vessels contain mixture of milk and water in the ratio 5 : 2 and 3 : 1 respectively. Find the ratio of milk and water in the new solution, if two mixtures are mixed in equal amount.

(*a*) 5 : 2 (*b*) 3 : 1 (*c*) 8 : 3 (*d*) 41 : 15

Sol. (*d*) Sum of the ratios are 5 + 2 and 3 + 1 = 7 and 4 respectively.

LCM of sum of the ratios *i.e.* LCM of 7 and 4 is 28.

$\therefore$ We assume that 28 litres of mixture is taken from each vessel.

Milk in first vessel = $\frac{5}{7} \times 28 = 20$ litres $\Rightarrow$ Water in first vessel = 28 – 20 = 8 litres

Milk in second vessel = $\frac{3}{4} \times 28 = 21$ litres $\Rightarrow$ Water in second vessel = 28 – 21 = 7 litres

Total quantity of milk in the resultant mixture = 20 + 21 = 41

Total quantity of water in the resultant mixture = 8 + 7 = 15

$\therefore$ Ratio of milk and water in the new solution = 41 : 15

Alternative Method : Milk in the resultant mixture = $\frac{5}{7} + \frac{3}{4} = \frac{41}{28}$

Water in the resultant mixture = $\frac{2}{7} + \frac{1}{4} = \frac{15}{28}$, or $2 - \frac{41}{28} = \frac{15}{28}$

$\therefore$ Ratio of milk and water = $\frac{41}{28} : \frac{15}{28} = 41 : 15$

196. There are three containers of equal capacity. The ratio of sulphuric acid to water in the first container is 3 : 2, that in the second container is 7 : 3 and in the third container it is 11 : 4. If all the liquids are mixed together, then the ratio of sulphuric acid to water in the mixture will be: **(SSC CGL Tier I 2004)**

(*a*) 61 : 29 (*b*) 61 : 28 (*c*) 60 : 29 (*d*) 59 : 29

Sol. (*a*) Sum of the ratios are 3 + 2, 7 + 3, 11 + 4 = 5, 10 and 15.

LCM of the sum of ratios i.e. LCM of 5, 10 and 15 is 30.

$\therefore$ We assume that 30 litre of mixture is taken from each vessel.

Vessel	Milk	:	Water
I	18	:	12
II	21	:	9
III	22	:	8
Total	61	:	29

∴ Ratio in the resultant mixture = 61 : 29

Alt: Sulphuric acid in resultant mixture = $\frac{3}{5}+\frac{7}{10}+\frac{11}{15}=\frac{18+21+22}{30}=\frac{61}{30}$

Water in the resultant mixture = $3-\frac{61}{30}=\frac{90-61}{30}=\frac{29}{30}$

∴ Required ratio = $\frac{61}{30}:\frac{29}{30}=61:29$

197. The ratio of milk and water in four solutions are 1 : 2, 2 : 3, 3 : 2 and 7 : 8 respectively. If their equal quantities are mixed together, then the ratio of milk and water in the new solution will be: **(SSC CGL Tier I 2006)**

(*a*) 13 : 15 (*b*) 7 : 9 (*c*) 9 : 11 (*d*) 11 : 9

Sol. (*c*) Milk in the resultant mixture $\frac{1}{3}+\frac{2}{5}+\frac{3}{5}+\frac{7}{15}=\frac{27}{15}$

Water in the resultant mixture = $4-\frac{27}{15}=\frac{33}{15}$

∴ Ratio of milk and water = $\frac{27}{15}:\frac{33}{15}$ = 27 : 33 = 9 : 11

198. Two vessels are full of milk with milk-water ratio 1 : 3 and 3 : 5 respectively. If both are mixed in the ratio 3 : 2, what is the ratio of milk and water in the new mixutre ? **(CDS 2011)**

(*a*) 4 : 15 (*b*) 3 : 7 (*c*) 6 : 7 (*d*) None of these

Sol. (*b*) Quantity of milk in two mixtures = $\frac{1}{4}$ and $\frac{3}{8}$.

Quantity of mixture taken from two vessels $\frac{3}{5}$ and $\frac{2}{5}$ respectively.

Quantity of milk in resultant mixture = $\frac{3}{5}\times\frac{1}{4}+\frac{2}{5}\times\frac{3}{8}=\frac{3}{20}+\frac{3}{20}=\frac{3}{10}$

Quantity of water in resultant mixture = $1-\frac{3}{10}=\frac{7}{10}$

Ratio of milk and water = $\frac{3}{10}:\frac{7}{10}$ = 3 : 7

Trick: Let quantities taken from two vessels be 12 litres and 8 litres (*i.e.* in the ratio 3 : 2)

	Milk		Water	Total quanity
First vessel = 1 : 3 =	3	:	9	12
Second Vessel = 3 : 5 =	3	:	5	8
Total	6	:	14	

∴ Required ratio = 6 : 14 = 3 : 7

199. Two alloys contain tin and iron in the ratio of 1 : 2 and 2 : 3. If the two alloys are mixed in the proportion of 3 : 4 respectively (by weight), the ratio of tin and iron in the newly formed allow is **(SSC CGL Tier I 2015)**

(*a*) 12 : 23 (*b*) 13 : 32 (*c*) 14 : 25 (*d*) 10 : 21

Sol. (*b*) Tin = $\frac{1}{3}\times\frac{3}{7}+\frac{2}{5}\times\frac{4}{7}=\frac{1}{7}+\frac{8}{35}=\frac{13}{35}$ $\Rightarrow$ Required ratio = 13 : (35 – 13) = 13 : 22

200. Three vessels contain milk and water in the ratio 1 : 3, 3 : 5 and 5 : 7 respectively. The quantities of the solution in the vessels are in the ratio 2 : 3 : 4 respectively. If the contents of the three vessels are mixed together, find the ratio of milk and water in the resultant mixture.

(*a*) 137 : 79 (*b*) 79 : 137 (*c*) 47 : 25 (*d*) 25 : 47

Sol. (*b*) Milk in new mixture = $2\times\frac{1}{4}+3\times\frac{3}{8}+4\times\frac{5}{12}=\frac{1}{2}+\frac{9}{8}+\frac{5}{3}=\frac{1}{2}+1\frac{1}{8}+1\frac{2}{3}=2\frac{31}{24}=3\frac{7}{24}$

$\Rightarrow$ Water = $(2+3+4)-3\frac{7}{24}=5\frac{17}{24}$ $\therefore$ Required ratio = $3\frac{7}{24}:5\frac{17}{24}$ = 79 : 137

201. A bag contains ₹ 90 in coins of denominations of 50 paise, 25 paise and 10 paise. If coins of 50 paise, 25 paise and 10 paise are in the ratio 2 : 3 : 5, then the number of 25 paise coins in the bag is : **(SSC CGL Tier I 2003)**

(*a*) 80 (*b*) 120 (*c*) 100 (*d*) 135

Sol. (*b*)

Denomination of coins	50p	25p	10p
Number of coins	2	3	5
Value of coins	100	75	50

Total value = 100 + 75 + 50 = 225 paise $\Rightarrow$ Number of 25 paise coins = $3\times\frac{9000}{225}=120$

202. A bag contains ₹ 112 in the form of 1-rupee, 50-paise and 10-paise coins in the ratio 3 : 8 : 10. What is the number of 50-paise coins ? **(CDS 2011, 2009 M, SSC CGL Tier I 2008 M)**

(*a*) 112 (*b*) 108 (*c*) 96 (*d*) 84

Sol. (*a*)

Denomination of coins	₹ 1	50p	10p
Number of coins	3	8	10
Value of coins	₹ 3	₹ 4	₹ 1

Total value = ₹ (3 + 4 + 1) = ₹ 8 $\Rightarrow$ Value of 50-paise coins = $\frac{4}{8}\times$ ₹ 112 = ₹ 56

$\Rightarrow$ Number of 50 paise coins = $56\div\frac{1}{2}=56\times2=112$ coins

203. There are ₹ 225 consisting of one rupee, 50 paise and 25 paise coins. The ratio of their numbers in that order is 8 : 5 : 3. The number of one-rupee coins is: **(SSC CGL 2008, 2004)**

(*a*) 80 (*b*) 112 (*c*) 160 (*d*) 172

Sol. (*c*)

Denomination of coins	100p	50p	25p
Ratio of coins	8	5	3
Value of coins	800	250	75

Total value = 800 + 250 + 75 = 1125 paise = ₹ 11.25

Number of one-rupee coins = $8\times\frac{225}{11.25}=160$

Alt : Denomination of coins 100p 50p 25p

Denomination of coins	100p	50p	25p
Ratio of coins	8	5	3
Value of coins	800	250	75

Ratio of values = 800 : 250 : 75 = 32 : 10 : 3

Value of 1 rupee coins = $\frac{32}{45} \times 225 =$ ₹ 160

Number of one-rupee coins = 160 ÷ 1 = 160 coins

Alt: Ratio of denomination of coins 4 2 1

Ratio of denomination of coins	4	2	1
Ratio of coins	8	5	3
Ratio of Value of coins	32	10	3

Value of 1 rupee coins = $\frac{32}{45} \times 225 =$ ₹ 160

Number of one-rupee coins = 160 ÷ 1 = 160 coins

204. A box contains 280 coins of one rupee, 50 paisa and 25 paisa. The values of the three kinds of coins are in the ratio 8 : 4 : 3. The number of 25 paisa coins is **(SSC CGL Tier I 2013)**

(*a*) 80 (*b*) 120 (*c*) 100 (*d*) 30

Sol. (*b*)

Denomination of coins	100p	50p	25p
Value of coins	800	400	300
Number of coins	8	8	12
Ratio of number of coins	2	2	3

⇒ Number of 25 paise coins = $\frac{3}{7} \times 280 = 120$

205. If 378 coins consists of rupees, 50 paise and 25 paise coins whose values are in the ratio of 13 : 11 : 7, the number of 50 paise coins will be : **(SSC CGL Tier I 2003)**

(*a*) 132 (*b*) 128 (*c*) 136 (*d*) 133

Sol. (*a*) Let value of rupee, 50 paise and 25 paise coins be $13x$, $11x$ and $7x$ respectively.

Then number of rupee coins = $13x$

Number of 50 paise coins = $11x \div 0.50 = 22x$

Number of 25 paise coins = $7x \div 0.25 = 28x$

Total number of coins = $13x + 22x + 28x = 63x = 378 \Rightarrow x = 378 \div 63 = 6$

∴ Number of 50 paise coins = $22x = 22 \times 6 = 132$ coins

206. A box has 210 coins of denomination one-rupee and fifty paise only. The ratio of their respective value is 13 : 11. The number of one-rupee coin is : **(SSC CGL Tier I 2008)**

(*a*) 65 (*b*) 66 (*c*) 77 (*d*) 78

Sol. (*d*) Ratio of number of coins of ₹ 1 and 50 paise = 13 : (11 × 2) = 13 : 22

∴ Number of 1 rupee coins = $\frac{13}{35} \times 210 = 78$ coins

207. 465 coins consists of 1 rupee, 50 paise and 25 paise coins. Their values are in the ratio of 5 : 3 : 1. The number of each type of coins respectively is **(SSC CGL Tier I 2012)**

(*a*) 155, 186, 124 (*b*) 154, 187, 124 (*c*) 154, 185, 126 (*d*) 150, 140, 175

Sol. (*a*) Ratio of number of coins of ₹ 1, 50 paise and 25 paise

= (5 ÷ 1) : (3 ÷ 0.50) : (1 ÷ 0.25) = 5 : 6 : 4

Total coins = (5 + 6 + 4) ratio = 15 ratio = 465 coins ⇒ 1 ratio = 31 coins

∴ Number of coins = (5, 6, 4) ratio = 155, 186, 124 coins

208. A sum of ₹ 600 is divided among 8 men, 9 women and 12 boys. If share of one man, one woman and one boy are in the ratio 6 : 4 : 3, find share of each boy.

(*a*) ₹ 15 (*b*) ₹ 50 (*c*) ₹ 100 (*d*) ₹ 180

Sol. (*a*) Ratio of money received by men, women and boys

$= (8 \times 6) : (9 \times 4) : (12 \times 3) = 48 : 36 : 36 = 4 : 3 : 3$

Sum of ratios of money received = 4 + 3 + 3 = 10

$\therefore$ Share of 12 boys $= \frac{3}{10} \times 600 =$ ₹ 180 $\Rightarrow$ Share of 1 boy $= \frac{180}{12} =$ ₹ 15

Alternative Method: Value of money received by men, women and boys

$= (8 \times 6) + (9 \times 4) + (12 \times 3) = 48 + 36 + 36 = 120$

Share of one boy in ₹120 = ₹3 $\Rightarrow$ Share of 1 boy in ₹ 600 $= \frac{3}{120} \times 600 =$ ₹ 15

209. Men, women and children are employed to do a work in the proportion of 3 : 2 : 1 and their wages as 5 : 3 : 2. When 90 men are employed, total daily wages of all amounts to ₹ 10,350. Find the daily wage of a man. **(SSC CGL Tier II 2012)**

(*a*) ₹ 115 (*b*) ₹ 75 (*c*) ₹ 45 (*d*) ₹ 57.50

Sol. (*b*) Ratio of men, women and children = 3 : 2 : 1. Ratio of their wages = 5 : 3 : 2

Sum of ratio of their wages $= (3 \times 5) + (2 \times 3) + (1 \times 2) = 15 + 6 + 2 = 23$ ratio

Given, men = 90 i.e. 30 times of the given ratio $\Rightarrow$ 1 ratio = ₹ 10,350 ÷ 23 ÷ 30 = ₹ 15

$\therefore$ Wages of one man = 5 ratio = 5 × ₹ 15 = ₹ 75

210. A man leaves ₹ 12,600 to be divided among 7 sons, 3 daughters and 5 nephews. If each daughter receives three times as much as each nephew and each son seven times as much as each nephew, then each daughter's share is : **(SSC CPO SI 2012)**

(*a*) ₹ 700 (*b*) ₹ 650 (*c*) ₹ 600 (*d*) ₹ 750

Sol. (*c*)

	Sons	Daughters	Nephews
Number	7	3	5
Ratio of share of each	7	3	1
Ratio of total shares	49	9	5

Sum of total of their shares = (49 + 9 + 5) ratio = 63 ratio = ₹ 12,600 $\Rightarrow$ 1 ratio = ₹ 200

Share of each daughter = 3 ratio = 3 × ₹ 200 = ₹ 600

211. The income of A and B are in the ratio 5 : 3. The expenses of A, B and C are in the ratio 8 : 5 : 2. If C spends ₹2000 and B saves ₹700, then A saves : **(MBA)**

(*a*) ₹1500 (*b*) ₹1000 (*c*) ₹500 (*d*) ₹250

Sol. (*a*)

	A	B	C
Income	5	3	
Expenditure	8	5	2

C's expenditure = 2 ratio = ₹2000 $\Rightarrow$ 1 ratio (expenditure) = 2000 ÷ 2 = ₹ 1000

B's expenditure = 5 ratio = 5 × ₹ 1000 = ₹5000

B's income = ₹5000 + ₹700 = ₹5700 = 3 ratio $\Rightarrow$ 1 ratio (income) = 5700 ÷ 3 = ₹ 1900

A's income = 5 ratio = 5 × ₹ 1900 = ₹9500 A's expenditure = 8 ratio = ₹8000

A's savings = ₹9500 – ₹8000 = ₹1500

212. Two numbers are in the ratio 6 : 11. If the first number is increased by 50% and the second number by 200, new ratio becomes 3 : 5, find the original numbers. **(MBA Entrance Exam.)**

(*a*) 15, 24 (*b*) 20, 32 (*c*) 25, 40 (*d*) 300, 550

Sol. (*d*) Let the original numbers are $6x$ and $11x$.

The first number after 50% increase $= 6x \times 1.5 = 9x$

New ratio = 3 : 5 $\Rightarrow$ Second number $= 9x \times \frac{5}{3} = 15x$

$\therefore$ Increase in the second ratio $= 15x - 11x = 4x = 200 \Rightarrow x = 50$

$\therefore$ Original numbers are 50×6 and $50 \times 11 = 300$ and 550

Alternative method: $\frac{9x}{11x+200} = \frac{3}{5} \Rightarrow 45x = 33x + 600 \Rightarrow 12x = 600 \Rightarrow x = 50$

Original numbers $= 6x$ and $11x = 300$ and 550.

213. The ratio of the monthly incomes of A and B is 4 : 3. The ratio of their monthly expenditures is 5 : 4. If A saves one fourth of his monthly income, find the ratio of their monthly savings.

(*a*) 2 : 3 (*b*) 3 : 2 (*c*) 3 : 5 (*d*) 5 : 3

Sol. (*d*) Let monthly incomes of A and B be $4x$ and $3x$ and monthly expenditures be $5y$ and $4y$.

$\therefore$ A's savings $= 4x - 5y = \frac{1}{4} \times 4x \Rightarrow 3x = 5y \Rightarrow x : y = 5 : 3$

B's savings $= 3x - 4y = 5y - 4y = y \Rightarrow$ Required ratio $= x : y = 5 : 3$.

214. The incomes of A, B and C are in the ratio 7 : 9 : 12 and their spendings are in the ratio 8 : 9 : 15. If A save 1/4th of his income, then the saving of A, B and C are in the ratio of : **(SSC CGL Tier I 2003, 2011)**

(*a*) 56 : 99 : 69 (*b*) 69 : 56 : 99 (*c*) 99 : 56 : 69 (*d*) 99 : 69 : 56

Sol. (*a*) Let their incomes be $7x$, $9x$ and $12x$ and expenditures be $8y$, $9y$ and $15y$ respectively.

A's expenditure $= 7x - \frac{7}{4}x = \frac{21}{4}x = 8y \Rightarrow x = 8 \times \frac{4}{21} = \frac{32}{21}y$

Savings of A $= \frac{7}{4}x = \frac{7}{4} \times \frac{32}{21}y = \frac{56}{21}y$

Income of B $= 9x = 9 \times \frac{32}{21}y = \frac{288}{21}y$. Savings of B $= \frac{288}{21}y - 9y = \frac{99}{21}y$

Income of C $= 12x = 12 \times \frac{32}{21}y = \frac{384}{21}y$. Savings of C $= \frac{384}{21}y - 15y = \frac{69}{21}y$

$\therefore$ Required ratio $= \frac{56}{21}y : \frac{99}{21}y : \frac{69}{21}y = 56 : 99 : 69$

215. Brothers A and B had some savings in the ratio 4 : 5. They decided to buy a gift for their sister, sharing the cost in the ratio 3 : 4. After they bought, A spent one-third of his amount while B is left with ₹145. Then the value of the gift is : **(SSC CGL Tier II 2012)**

(*a*) ₹140 (*b*) ₹175 (*c*) ₹70 (*d*) ₹105

Sol. (*a*) Let their savings be $4x$ and $5x$; Amount spent by them for the gift be $3y$ and $4y$.

Then A's expenditure $= \frac{4x}{3} = 3y \Rightarrow 4x = 9y \Rightarrow x = 9/4y$

B's final savings = $5x - 4y = 145 \Rightarrow \frac{45}{4}y - 4y = 145 \Rightarrow \frac{29}{4}y = 145 \Rightarrow y = 145 \times \frac{4}{29} = 20$

$\therefore$ Value of gift = $3y + 4y = 7y = 7 \times$ ₹ 20 = ₹ 140.

216. **A sum of ₹ 216 is divided among A, B, C and D in such a way that if ₹ 5 added to A's share, ₹ 5 deducted from B's share, C's share multiplied by 5 and D's share divided by 5, all results will be equal. Find A's share in the sum.** **(MBA Ent. Exam.)**

(*a*) ₹ 10 (*b*) ₹ 15 (*c*) ₹ 20 (*d*) ₹ 25

Sol. (*d*) Let $A + 5 = B - 5 = C \times 5 = D \div 5 = x$. Then $A = x - 5$, $B = x + 5$, $C = x \div 5$, $D = x \times 5$

$\therefore (x - 5) + (x + 5) + (x \div 5) + (x \times 5) =$ ₹ 216

$\therefore \frac{36}{5}x =$ ₹ 216 $\Rightarrow x =$ ₹ 30 $\Rightarrow$ A's share = $x - 5$ = ₹ 30 – ₹ 5 = ₹ 25

CHAPTER

13

PROBLEMS ON AGES

FORMULAE

1. If the ratio of present ages of A and B is $x : y$. After a years from now, the ratio of their ages becomes $(x + 1) : (y + 1)$. Then their present ages are ax and ay respectively.

Example : Ratio of present ages of A and B is 2 : 3. After 6 years, ratio of their ages is 3 : 4. Find the present age of A. **(Bank Clerical)**

(*a*) 6 years (*b*) 12 years (*c*) 18 years (*d*) 24 years

Sol. (*b*) New Ratio 3 : 4

Old Ratio 2 : 3

Difference 1 : 1

Difference = 1 ratio = 6 years ⇒ Present age of A = 2 ratio = 2 × 6 years = 12 years

2. The ratio of present ages of A and B is $x : y$. After a years from now, the ratio of ages becomes $(x + k) : (y + k)$. Then their present ages are $\frac{a}{k}x$ and $\frac{a}{k}y$ respectively.

Example : The ratio of the ages of Ram and Sohan is 5 : 8. The ratio of their ages 10 years hence would be 7 : 10. What is the present age of Ram? **(Bank Clerical)**

(*a*) 25 years (*b*) 35 years (*c*) 40 years (*d*) 50 years

Sol. (*a*) New Ratio 7 : 10

Old Ratio 5 : 8

Difference 2 : 2

Difference = 2 ratio = 10 years ⇒ 1 ratio = 5 years

Ram's present age = 5 ratio = 25 years

3. Let ages of A and B are in the ratio $a : b$. After y years, the ratio of their ages becomes $m : n$. Then their present ages are ax and bx years, where

$$x = \frac{y(m-n)}{an - bm} = \frac{\text{Difference in time} \times \text{Difference in new ratio}}{\text{Difference in cross multiplication of new and old ratios}}$$

Example : Present age of Ram's father is 4 times Ram's age. 5 years from now, father's age will become 3 times of Ram's age, find present age of Ram's father.

(*a*) 40 years (*b*) 45 years (*c*) 50 years (*d*) 60 years

Sol. (*a*)

	Father		Son
Old Ratio =	4	:	1
New Ratio =	3	:	1

$$\therefore\ x = \frac{5 \times (3-1)}{4 \times 1 - 3 \times 1} = \frac{5 \times 2}{1} = 10 \quad \Rightarrow \quad \text{Father's present age} = 4 \times 10 = 40 \text{ years}$$

Alternative method :

New ratio = 3 : 1 = 9 : 3 (Difference in ratios = 3 – 1 = 2)

Old ratio = 4 : 1 = 8 : 2 (Difference in ratios = 4 – 1 = 3)

Difference = 1 : 1

⇒ 1 ratio = 5 years ⇒ Father's present age = 8 ratio = 8 × 5 years = 40 years

SOLVED EXERCISE

1. The present ages of A, B and C are in the ratio 8 : 14 : 22 respectively. The present ages of B, C and D are in the ratio 21 : 33 : 44 respectively. Which of the the following represents the ratio of the present ages of A, B, C and D respectively ? **(SBI Bank PO 2005)**

   (*a*) 12 : 21 : 33 : 44 (*b*) 12 : 22 : 31 : 44 (*c*) 12 : 21 : 36 : 44 (*d*) None of these

Sol. (*a*) $A : B : C = 8 : 14 : 22 = (8 : 14 : 22) \times \frac{3}{2} = 12 : 21 : 33$

$B : C : D = 21 : 33 : 44$ $\therefore$ $A : B : C : D = 12 : 21 : 33 : 44$

2. The difference between the present ages of Arun and Deepak is 14 years. Seven years ago, the ratio of their ages was 5 : 7 respectively. What is Deepak's present age ? **(SBI Bank PO 2008)**

   (*a*) 49 years (*b*) 42 years (*c*) 63 years (*d*) None of these

Sol. (*d*) Difference between ages = (7 – 5) ratio = 2 ratio = 14 years ⇒ 1 ratio = 7 years

7 years ago, Deepak's age = 7 ratio = 7 × 7 = 49 years

Deepak's present age = 49 + 7 = 56 years

3. The present ages of A and B are in the ratio 4 : 5 and after 5 years, their ages will be in the ratio 5 : 6. The present age of A is : **(SSC CGL Tier I 2008)**

   (*a*) 15 years (*b*) 20 years (*c*) 25 years (*d*) 30 years

Sol. (*b*) New Ratio 5 : 6

Old Ratio 4 : 5

Difference 1 : 1

Difference = 1 ratio = 5 years ⇒ Present age of A = 4 ratio = 4 × 5 years = 20 years

4. Present ages of Ram and Sita are in the ratio 13 : 12 respectively. 8 years later, their ages will be in the ratio 15 : 14. What is Sita's present age ? **(Corp. Bank Officers Exam. 2004)**

   (*a*) 50 years (*b*) 54 years (*c*) 48 years (*d*) None of these

Sol. (*c*)

	Ram	Sita
New Ratio	15	14
Old Ratio	13	12
Difference	2	2

Difference = 2 ratio = 8 years ⇒ 1 ratio = 8 ÷ 2 = 4 years

Present age of Sita = 12 ratio = 12 × 4 years = 48 years

5. The ratio between the ages of A and B is 2 : 5. After 8 years, their ages will be in the ratio 1 : 2. What is the difference between their present ages ? **(CDS 2010)**

   (*a*) 20 years (*b*) 22 years (*c*) 24 years (*d*) 25 years

Sol. (*c*) New Ratio = 1 : 2 = 3 : 6

Old Ratio 2 : 5

Difference 1 : 1

Difference = 1 ratio = 8 years.

Difference between present ages = (5 – 2) ratio = 3 ratio = 3 × 8 = 24 years

6. The ages of Samir and Tanuj are in the ratio 8 : 15. After 9 years the ratio of their ages will be 11 : 18. What is the difference in years between their ages ? **(Bank of Baroda PO 2007)**

(*a*) 24 years (*b*) 20 years (*c*) 33 years (*d*) 21 years

Sol. (*d*)

New Ratio	11 : 18
Old Ratio	8 : 15
Difference	3 : 3

Difference = 3 ratio = 9 years ⇒ 1 ratio = 3 years

Difference in ages = (15 – 8) ratio = 7 ratio = 7 × 3 = 21 years

7. The ratio between the present ages of A and B is 7 : 5. The ratio between A's age 5 years ago and B's age 5 years hence is 1 : 1. What is the ratio of age of A after 5 years and that of B 5 years ago ?

(*a*) 4 : 3 (*b*) 2 : 1 (*c*) 1 : 2 (*d*) 3 : 4

Sol. (*b*)

Present ages	7	:	5
	– 5 years	:	+ 5 years
New ratio	6	:	6 = 1 : 1

Difference = 1 ratio = 5 years ⇒ Required ratio = (7 + 1) : (5 – 1) = 8 : 4 = 2 : 1

8. The average age of a man and his son is 28 years. The ratio of their ages is 5 : 2. What will be the ratio of their ages after 8 years? **(SBI Clerical)**

(*a*) 7 : 4 (*b*) 2 : 1 (*c*) 3 : 1 (*d*) None of these

Sol. (*b*) Let their ages be $5x$ and $2x$. Sum of ages = $7x = 28 \times 2 \Rightarrow x = 8$

Their present ages = $5x$ and $2x = 5 \times 8$ and 2×8 = 40 and 16 years respectively.

Ratio of ages after 8 years = (40 + 8) : (16 + 8) = 48 : 24 = 2 : 1

Alt: Sum of ratio = (5 + 2) ratio = 7 ratio = 28 × 2 years ⇒ 1 ratio = 8 years

∴ Ratio after 8 years = (5 + 1) : (2 + 1) = 6 : 3 = 2 : 1

9. If ratio of Mohan's age to that of his father is 1 : 5. After 10 years, their ages will be in the ratio 1 : 3. What is the father's present age? **(SSC CGL Tier I Exam.)**

(*a*) 25 years (*b*) 30 years (*c*) 40 years (*d*) 50 years

Sol. (*d*)

	Father		Mohan
Old Ratio =	5	:	1
New Ratio =	3	:	1

$$\therefore\ x = \frac{10 \times (3 - 1)}{5 \times 1 - 3 \times 1} = \frac{10 \times 2}{2} = 10 \Rightarrow \text{Father's present age} = 5x = 5 \times 10 \text{ years} = 50 \text{ years}$$

Alt: New ratio = 3 : 1 = 6 : 2 (Difference in ratios = 3 – 1 = 2)

Old ratio = 5 : 1 (Difference in ratios = 5 – 1 = 4)

Difference = 1 : 1

1 ratio = 10 years ⇒ Father's present age = 5 ratio = 5 × 10 years = 50 years.

10. The ratio of present ages of Sita and her son is 5 : 2 respectively. Seven years hence, the ratio will be 2 : 1. What was the age of Sita when her son was born ? **(Bank PO 2004)**

(*a*) 28 years (*b*) 35 years (*c*) 21 years (*d*) Data insufficient

Sol. (*c*)

	Sita		Son	
New Ratio = 2 : 1 =	6	:	3	(Difference in ratios = 2 – 1 = 1)
Old Ratio	5	:	2	(Difference in ratios = 5 – 2 = 3)
Difference	1	:	1	

Difference = 1 ratio = 7 years

Difference in their ages = (5 – 2) ratio = 3 ratio = 3 × 7 = 21 years

Alt: $(5x + 7) = 2(2x + 7) \Rightarrow 5x + 7 = 4x + 14 \Rightarrow x = 7$

11. 3 years ago, the age of a father was 7 times of his son's age. If the father's present age is 5 times his son's age. Find the present age of the father.

(*a*) 35 years (*b*) 39 years (*c*) 42 years (*d*) 45 years

Sol. (*d*)

	Father		Son
Old Ratio =	7	:	1
New Ratio =	5	:	1

Let 3 years ago, the father's age was $7x$ years, where $x = \dfrac{3 \times (5-1)}{7-5} = \dfrac{3 \times 4}{2} = 6$

$\therefore$ 3 years ago, father's age = $7x = 7 \times 6 = 42$ years $\Rightarrow$ His present age = 42 + 3 = 45 years

12. 5 years ago, the father's age was 3 times of his son's age. 5 years hence the ratio between the ages of father and the son becomes 11 : 5. Find the father's present age.

(*a*) 40 years (*b*) 45 years (*c*) 50 years (*d*) 55 years

Sol. (*c*)

	Father		Son
Old ratio =	3	:	1
New ratio =	11	:	5

Let 5 years ago father's age was $3x$ years, where $x = \dfrac{(5+5) \times (11-5)}{3 \times 5 - 1 \times 11} = \dfrac{10 \times 6}{4} = 15$

$\therefore$ Father's age (5 years ago) = 3 × 15 = 45 years $\Rightarrow$ His present age = 45 + 5 = 50 years

Alt: Old ratio = 3 : 1 =	9	:	3	(Difference in original ratio = 3 – 1 = 2)
New ratio =	11	:	5	(Difference in ratio = 11 – 5 = 6)
Difference =	2	:	2	

Difference = 2 ratio = 10 years $\Rightarrow$ 1 ratio = 5 years

$\therefore$ Father's age (5 years ago) = 9 ratio = 5 × 9 = 45 $\Rightarrow$ His present age = 45 + 5 = 50 years

Alt: $11(x + 10) = 5(3x + 10) \Rightarrow 11x + 110 = 15x + 50 \Rightarrow 4x = 60 \Rightarrow x = 15$

13. A father's age was five times his son's age five years ago and will be three times the son's age after two years. The ratio of the present age of the father to that of the son is **(SSC CGL Tier I 2013)**

(*a*) 33 : 5 (*b*) 11 : 5 (*c*) 3 : 1 (*d*) 10 : 3

Sol. (*d*)

	Father		Son
Old ratio =	5	:	1
New ratio =	3	:	1

Let 5 years ago, age of son and father were x and $5x$ respectively.

$$\therefore\ x = \frac{(5+2) \times (3-1)}{5 \times 1 - 3 \times 1} = \frac{7 \times 2}{2} = 7 \text{ years}$$

$\therefore$ 5 years ago, father's age was $5x = 5 \times 7 = 35$ years and son's age = $1x = 1 \times 7 = 7$ years

$\therefore$ Ratio of their present ages = (35 + 5) : (7 + 5) = 40 : 12 = 10 : 3

14. The ratio of present ages two brothers is 1 : 2 and 5 years back the ratio was 1 : 3. What will be the ratio of their ages after 5 years? **(SSC CGL Tier I 2006)**

(*a*) 1 : 4 (*b*) 2 : 3 (*c*) 3 : 5 (*d*) 5 : 6

Sol. (*c*)

	I		II
Old Ratio =	1	:	3
New Ratio =	1	:	2

$\therefore\ x = \dfrac{5 \times (2-1)}{3 \times 1 - 2 \times 1} = 5$ years $\Rightarrow$ Their ages (5 years ago) = x and $3x$ = 5 and 15 years

Their ages (5 years later) = (5 + 10) and (15 + 10) years = 15 and 25 years respectively.

$\therefore$ Required ratio = 15 : 25 = 3 : 5

Alt: New ratio = 1 : 2 = 2 : 4 (Difference in ratios = 2 – 1 = 1)

Old ratio = 1 : 3 (Difference in ratios = 3 – 1 = 2)

Difference = 1 : 1

Difference = 1 ratio = 5 years

$\therefore$ Required ratio = (2 + 1) : (4 + 1) = 3 : 5

15. The present age of Suresh is twice his daughter's present age. Six years hence, the ratio between the ages of Suresh and his daughter will be respectively 23 : 13. What is the present age of Suresh ? **(Vijaya Bank PO 2004)**

(*a*) 40 years (*b*) 36 years (*c*) 46 years (*d*) None of these

Sol. (*a*)

	Suresh		Daughter
Old Ratio =	2	:	1
New Ratio =	23	:	13

$x = \dfrac{6 \times (23-13)}{2 \times 13 - 1 \times 23} = \dfrac{6 \times 10}{26-23} = \dfrac{6 \times 10}{3} = 20$ years

Suresh's present age = $2x = 2 \times 20 = 40$ years

Alt: Old Ratio = 2 : 1 = 20 : 10 (Difference in ratios = 2 – 1 = 1)

New Ratio 23 : 13 (Difference in ratio = 23 – 13 = 10)

Difference 3 : 3

Difference = 3 ratio = 6 years $\Rightarrow$ 1 ratio = 2 years

Suresh's present age = 20 ratio = 20 × 2 = 40 years

16. 4 years ago the ratio of the ages of A and B was 2 : 3 and 4 years hence the ratio will be 5 : 7. Their present ages (in years) are: **(SSC CGL Tier I 2006)**

(*a*) 36, 40 (*b*) 32, 48 (*c*) 40, 56 (*d*) 36, 52

Sol. (*d*) Difference of time = 4 + 4 = 8 years $\Rightarrow x = \dfrac{8 \times (5-7)}{2 \times 7 - 3 \times 5} = 16$ years

Ages 4 years ago, $2x$ and $3x$ = 32 years and 48 years

Their present ages = (32 + 4) and (48 + 4) = 36 and 52 years

17. Ratio between present ages of A and B is 3 : 4. What is the present age of A, if 4 years ago, the ratio was 5 : 7?

(*a*) 20 years (*b*) 24 years (*c*) 28 years (*d*) 32 years

Sol. (*b*)

	A		B
Old ratio =	5	:	7
New ratio =	3	:	4

Difference in time = 4 years $\Rightarrow x = \frac{4\times(3-4)}{5\times4-7\times3} = \frac{4\times(-1)}{20-21} = \frac{-4}{-1} = 4$

$\therefore$ A's age (4 years ago) = 5 × 4 = 20 years ⇒ A's present age = 20 + 4 = 24 years.

Alt : Old ratio = 5 : 7

New ratio = 3 : 4 = 6 : 8

Difference = 1 : 1 ⇒ Difference in time = 1 ratio = 4 years

A's present age = 6 ratio = 6 × 4 = 24 years.

18. The ratio of the present ages of a son and his father is 1 : 5 and that of his mother and father is 4 : 5. After 2 years the ratio of the ages of the son to that of his mother becomes 3 : 10. What is the present age of the father ? **(BSRB 2001)**

(*a*) 30 years (*b*) 34 years (*c*) 35 years (*d*) 37 years

Sol. (*c*)

	Son	:	Father	:	Mother
Old ratio	1	:	5	:	4
New ratio	3	:		:	10

$\Rightarrow x = \frac{(10-3)\times2}{4\times3-1\times10} = 7 \Rightarrow$ Father's age = 5*x* = 5 × 7 years = 35 years

19. Harsha is 40 years old and Ritu is 60 years old. How many years ago was the ratio of their ages 3 : 5 ? **(SSC CGL Tier I 2001)**

(*a*) 10 years (*b*) 20 years (*c*) 37 years (*d*) 5 years

Sol. (*a*) $\frac{40-x}{60-x} = \frac{3}{5} \Rightarrow 200 - 5x = 180 - 3x \Rightarrow 2x = 20 \Rightarrow x = 10$ years

Alt: Ratio of present ages = 40 : 60 = 4 : 6 ⇒ 4 ratio = 40 years ⇒ 1 ratio = 10 years

Old ratio = 3 : 5

Difference in ratio = (4 – 3) = (6 – 5) = 1 ratio ⇒ Required time = 1 ratio = 10 years

20. Mohan and Sohan are 35 and 40 years old respectively. After how many years would the ratio of their ages become 9 : 10 ? **(Corporation Bank Clerical 2004)**

(*a*) 5 years (*b*) 10 years (*c*) 15 years (*d*) Data insufficient

Sol. (*b*) $\frac{35+x}{40+x} = \frac{9}{10} \Rightarrow 350 + 10x = 360 + 9x \Rightarrow x = 10$

Alt: Ratio of ages = 35 : 40 = 7 : 8 ⇒ Difference = (8 – 7) ratio = 1 ratio = 40 – 35 = 5 years

⇒ Required time = (9 – 7) ratio or (10 – 8) ratio = 2 ratio = 2 × 5 = 10 years

21. In 30 years, age of a person becomes 3 times of his present age. What is his present age?

(*a*) 7.5 years (*b*) 10 years (*c*) 12.5 years (*d*) 15 years

Sol. (*d*) Let his present age be *x* years. Then his age after 30 years will be 3*x* years.

Given $x + 30 = 3x \Rightarrow 2x = 30$ $\therefore$ Present age = $x = \frac{30}{2} = 15$ years

22. Mohan got married 9 years ago. Today his age is $1\frac{1}{3}$ times of his age at the time of his marriage. Find his present age.

(*a*) 12 years (*b*) 27 years (*c*) 36 years (*d*) 45 years

Sol. (*c*) 9 years = $\frac{1}{3}$ of his age at the time of marriage

$\therefore$ Mohan's age at the time of marriage = 3 × 9 years = 27 years

$\therefore$ His present age = 27 + 9 = 36 years

23. Sum of present ages of A and B is 60 years. 5 years hence, their ages will be in the ratio 3 : 4. Find A's present age. **(SSC CHSL)**

(*a*) 25 years (*b*) 30 years (*c*) 35 years (*d*) 40 years

Sol. (*a*) Sum of their present ages = 60 years.

$\therefore$ Sum of their ages, 5 years hence = 60 + (5 × 2) = 70 years.

After 5 years, ratio of ages of A and B will be 3 : 4.

$\therefore$ A's age after 5 years = $\frac{3}{7} \times 70 = 30$ years $\Rightarrow$ A's present age = 30 – 5 = 25 years

24. The sum of ages of Sunil and his father is 46 years. 3 years back, father's age was 4 times Sunil's age. Find present age of Sunil. **(SSC CGL Tier I 2007)**

(*a*) 7 years (*b*) 8 years (*c*) 11 years (*d*) 12 years

Sol. (*c*) Sum of their ages, 3 years ago = 46 – (2 × 3) = 40 years

$\therefore$ Sunil's age, 3 years ago $= \frac{1}{5} \times 40 = 8$ years $\Rightarrow$ Sunil's present age = 8 + 3 = 11 years

25. Sum of present ages of A, B and C is 72 years. If 4 years ago, their ages were in the ratio 1 : 2 : 3, find A's present age.

(*a*) 7 years (*b*) 10 years (*c*) 12 years (*d*) 14 years

Sol. (*d*) Sum of present ages of A, B and C is 72 years.

$\therefore$ Sum of their ages (4 years ago) = 72 – (3 × 4) = 60 years

4 years ago, ratio of ages of A, B and C was 1 : 2 : 3.

$\therefore$ A's age, 4 years ago = $\frac{1}{6} \times 60 = 10$ years $\Rightarrow$ A's present age = 10 + 4 = 14 years

26. Present ages of A and B are 50 years and 18 years respectively. In how many years will A be twice as old as B ?

(*a*) 14 years (*b*) 15 years (*c*) 16 years (*d*) 18 years

Sol. (*a*) Let the required time be *x* years.

Then $(50 + x) = 2(18 + x) \Rightarrow 50 + x = 36 + 2x \Rightarrow x = 14$ years

Alternative method : Double of B's age = 2 × 18 = 36 years

Difference between A's present age and double of B's age = 50 – 36 = 14 years

$\therefore$ A will be twice of B's age after 14 years.

27. The sum of ages of Mohan and his father is 35 years. When Mohan's age will be equal to present age of his father, then sum of their ages will be 85 years. Find present age of the father.

(*a*) 20 years (*b*) 25 years (*c*) 30 years (*d*) 32 years

Sol. (*c*) Sum of present ages = 35 years

Sum of their ages will be 85 years after $\frac{85-35}{2} = 25$ years from now.

$\therefore$ The father is 25 years older than his son. $\therefore$ Present age of father $= \frac{35+25}{2} = 30$ years

28. The sum of present ages of a father and his son is 36 years. When the son reaches father's present age, the sum of their ages will be 80 years. What is the present age of the son ?

(*a*) 4 years (*b*) 7 years (*c*) 9 years (*d*) 12 years

Sol. (*b*) Difference in two sums = 80 – 36 = 44 years

$\therefore$ Age of both are increased by 44 ÷ 2 = 22 years

$\Rightarrow$ Difference in ages of father and son = 22 years

Sum of present ages of father and son = 36 years

$\therefore$ Son's present age = $\frac{1}{2} \times (36 - 22) = 7$ years

29. The ages of A and B are in the ratio 8 : 5. If the sum of their ages is 39 years, what will be the ratio of their ages after 9 years ? **(SBI Clerical)**

(*a*) 3 : 2 (*b*) 8 : 7 (*c*) 10 : 7 (*d*) 11 : 8

Sol. (*d*) Sum of ratios = 8 + 5 = 13. Sum of ages = 39 years

$\therefore$ 13 ratio = 39 years $\Rightarrow$ 1 ratio = 3 years $\Rightarrow$ 9 years = 3 ratio

New ratio = (8 + 3) : (5 + 3) = 11 : 8

30. A is 2 years older than B who is twice as old as C. If sum of ages of A, B and C is 37 years, find the age of A. **(Hotel Management)**

(*a*) 7 years (*b*) 9 years (*c*) 14 years (*d*) 16 years

Sol. (*d*) Let the age of C be x years. Then age of B = $2x$ and age of A = $2x + 2$

Sum of ages of A, B and C = $(2x + 2) + (2x) + (x) = 37 \Rightarrow (5x + 2) = 37 \Rightarrow x = 7$

$\therefore$ Age of A = $2x + 2 = 16$ years

31. At the time of marriage, a man was 6 years older to his wife. 12 years after their marriage, his age is $\frac{6}{5}$ times the age of his wife. What was wife's age at the time of marriage ?

(*a*) 18 years (*b*) 24 years (*c*) 30 years (*d*) 36 years

Sol. (*a*) Ratio of their ages, 12 years hence = 6 : 5 $\Rightarrow$ Difference in ages = 1 ratio = 6 years

$\therefore$ Wife's age (12 years hence) = 5 ratio = 5 × 6 years = 30 years

$\therefore$ Wife's age at the time of marriage = 30 – 12 = 18 years

32. A and B are 3 years and 2 years old respectively. Their father is 40 years. After how many years, father's age would be twice of combined age of A and B ?

(*a*) 5 years (*b*) 10 years (*c*) 15 years (*d*) 20 years

Sol. (*b*) Father's age = 40 years. Childern's combined age = 3 + 2 = 5 years

Double of their ages = 2 × 5 = 10 years

Differenc between father's age and combined age of childern = 40 – 10 = 30 years

$\therefore$ Required time = $\frac{1}{2+1} \times 30 = 10$ years

33. 3 years ago, a father's age was twice the sum of ages of his 4 sons. In 3 years time, the father's age will be equal to sum of ages of the sons. find the present age of father.

(*a*) 24 years (*b*) 36 years (*c*) 39 years (*d*) 45 years

Sol. (*c*) 3 years ago, let sum of age of four sons = x. Then father's age (3 years ago) = $2x$

Father's age (after 6 years) = $2x + 6$

Sum of the ages of 4 sons (6 years hence) = $x + (6 \times 4) = x + 24$

$\therefore 2x + 6 = x + 24 \Rightarrow x = 18$

Father's present age = $2x + 3 = 18 \times 2 + 3 = 39$ years.

34. Ratio of father's age and sum of ages of his three children is 3 : 1. After 3 years, father's age will be twice of children's age. Find the present age of the father.

(*a*) 45 years (*b*) 54 years (*c*) 24 years (*d*) 30 years

Sol. (*a*) $\frac{3x+3}{x+(3 \times 3)} = \frac{2}{1} \Rightarrow 3x + 3 = 2x + 18 \Rightarrow x = 15$

$\therefore$ Father's age = $3x = 3 \times 15 = 45$ years

CHAPTER

14

PERCENTAGE

Per cent: Per cent means 'per hundred' or 'out of every hundred'. It is denoted by % (percentage) sign. If we say $x\%$, it implies x out of 100. In fraction, $x\%$ can be written as $\frac{x}{100}$.

Example: 27% means 27 out of 100. In fraction, 27% can be written as $\frac{27}{100}$ and in decimal it can be written as 0.27.

To convert per cent into a fraction or decimal, divide the given number by 100.

Example: $20\% = \frac{20}{100} = \frac{1}{5}$ (or 0.20 in decimal)

To convert a fraction or decimal into percentage, multiply the given fraction or decimal by 100.

Example: $\frac{1}{4} = \frac{1}{4} \times 100\% = 25\%$, $0.38 = 0.38 \times 100 = 38\%$

How to calculate percentage?

Method: To convert x out of y into % form, divide x by y and multiply the result by 100.

x out of $y = \frac{x}{y} \times 100\%$. Note: Both x and y should be in the same unit.

Example :

A student secured 240 marks out of 300 in an examination. Find his percentage of marks.

Sol : $\% \text{ Marks} = \frac{\text{Marks secured}}{\text{Total Marks}} = \frac{240}{300} \times 100 = 80\%$

How to find certain percentage of a given number?

If we need to find $x\%$ of y. We calculate $\frac{x}{100} \times y$.

Example : Find 80% of 300. **Sol**: 80% of 300 = $\frac{80}{100} \times 300 = 240$.

COMMON FRACTIONS AND EQUIVALENT PERCENTAGE

$1 = 100\%$	$\frac{1}{2} = 50\%$	$\frac{1}{3} = 33.33\%$	$\frac{1}{4} = 25\%$	$\frac{1}{5} = 20\%$
$\frac{1}{6} = 16.67\%$	$\frac{1}{8} = 12.5\%$	$\frac{1}{9} = 11.11\%$	$\frac{1}{10} = 10\%$	$\frac{1}{11} = 9.09\%$
$\frac{1}{12} = 8.33\%$	$\frac{1}{16} = 6.25\%$	$\frac{1}{20} = 5\%$	$\frac{1}{25} = 4\%$	$\frac{1}{50} = 2\%$
$\frac{2}{3} = 66.67\%$	$\frac{2}{5} = 40\%$	$\frac{3}{5} = 60\%$	$\frac{3}{4} = 75\%$	

FORMULAE

1. If a number is increased by x%, then new number = Original number × $\frac{100+x}{100}$.

If a number is increased by x%, then new number = Original number × $(100+x)$%

2. If a number is decreased by x%, then new number = Original number × $\frac{100-x}{100}$

If a person spends x% of his income. Then his savings is $(100-x)$% of the income.

Example. A man spends 20% of his income on food and 70% on clothing. If he still has a balance of ₹ 200, what is his total income? **(SSC CGL Tier I Exam.)**

(*a*) ₹ 100 (*b*) ₹ 200 (*c*) ₹ 1000 (*d*) ₹ 2000

Sol. (*d*) Total expenditure = 20% + 70% = 90% of income.

∴ Savings = (100 – 90)% = 10% of income = ₹ 200 (*i.e.* 20 times 10)

∴ His income = ₹ 20 × 100 = ₹ 2000

3. If x% of a number is y, then the number is $y \times \frac{100}{x}$

Note: If y can be expressed as ax, then the number = $a \times 100$.

Example: If 25% of a number is 200. Find the number.

Sol. Here x is 25% and y is 200. ⇒ The number is $200 \times \frac{100}{25} = 800$.

Alternative Method : 200 is 8 times of 25. ⇒ The number is 8 × 100 = 800.

4. If x% of a number is equal to y% of second number.

Then ratio between first and second number = $\frac{1}{x}:\frac{1}{y} = y:x$

Example: If 20% of a number is equal to 25% of the other number, find ratio between the two numbers.

(*a*) 2 : 3 (*b*) 3 : 4 (*c*) 4 : 5 (*d*) 5 : 4

Sol. (*d*) $x:y = \frac{1}{20}:\frac{1}{25} = 25:20 = 5:4$

5. Two numbers are in the ratio $x : y$. If the first number is increased by a% and the second number is increased by b%, then

(*a*) Percentage increase in the sum of the numbers $= \frac{ax+by}{x+y} \times 100$

(*b*) Percentage change in difference of the numbers $= \frac{ax-by}{x-y} \times 100$

Proof: Sum of original numbers = $x + y$. ⇒ Sum of new numbers = $x + \frac{ax}{100} + y + \frac{by}{100}$

∴ Increase in sum = $\frac{ax}{100} + \frac{by}{100} = \frac{ax+by}{100}$ ⇒ % Increase = $\frac{ax+by}{x+y} \times 100$

Example 1 : Income and expenditure of a person are in the ratio $x : y$. If his income increases by a% and his expenditure increases by b%, then increase in his savings = $\frac{ax-by}{x-y} \times 100$

(**Note:** Income – Expenditure = Savings)

Example 2 : Ram's expenditure and savings are in the ratio 5 : 3. If his income increases by 12% and expenditure by 15%, then how much per cent do his savings increase ? **(SSC CGL Tier I 2008)**

(*a*) 3% (*b*) 5% (*c*) 7% (*d*) 27%

Sol. (*c*)

Expenditure	Savings	Income
5	3	8
15%	?	12%

Savings = Income – Expenditure ⇒ Increase in savings $= \frac{8\times12-5\times15}{8-5} = \frac{21}{3} = 7\%$

6. The product of two or more numbers is increased by x% if one of the numbers is increased by x%. The product of two or more numbers is decreased by x% if one of the numbers is decreased by x%.

7. If price of a commodity is changed from ₹ 1 to ₹ $\frac{x}{y}$, then to maintain his expenditure, a consumer must buy $\frac{y}{x}$ units of the commodity instead of 1 unit. $\left(\textbf{Hint}: \frac{x}{y}\times\frac{y}{x}=1\right)$

8. If we have two numbers – one of them is **increased by x%** and the other is **decreased by x%**. Then **product of the numbers will be decreased** by $\frac{x^2}{100}$%.

Note: This formula can be used if a number is increased by x% and then the resultant number is decreased by x% and vice-versa.

Example 1: Salary of a person is increased by 20%, then it is decreased by 20%. Change in his salary is: **(SSC CGL Tier I 2001)**

(*a*) 4% decreased (*b*) 4% increased (*c*) 8% decreased (*d*) remains same

Sol. (*a*) Change in salary $= \frac{20\times20}{100} = 4\%$ decrease

Example 2: A number is increased by 10% and then resultant number is decreased by 10%. Find net increase or decrease in the number. **(MAT 2001)**

(*a*) 1% decrease (*b*) 1% increase (*c*) No effect (*d*) 5% decrease

Sol. (*a*) Net decrease in the new price from original price $= \frac{10\times10}{100} = 1\%$ decrease

9. If out of two numbers **first number is increased by x%** and **second number is increased by y%**, then **per cent increase in the product of two numbers** is: $x + y + \frac{xy}{100}$.

Base of formula : $\left(1+\frac{x}{100}\right)\left(1+\frac{y}{100}\right) = 1+\frac{x+y}{100}+\frac{xy}{100^2}$

Note: Decrease in a number can be written as negative increase *e.g.* if the first number is decreased by x% then we can write it as $(-x)$% increase.

Note: This formula **can be used in numerous situations**, such as:

⇒ If a number is increased by x% and then the increased number is again increased by y%, then total increase (in percentage) $= x + y + \frac{xy}{100}$

⇒ If a number is increased by x% and the other number is decreased by y%, then per cent increase in the product of two numbers $= x + (-y) + \frac{x(-y)}{100} = x - y - \frac{xy}{100}$

Alternative Method:

1. Assume that the original numbers are 10 each.
2. Find new numbers after adding /subtracting per cent increase/decrease in original numbers.
3. Find product of the new numbers.
4. Deduct 100 from the product of the new numbers. Balance is net increase or decrease in the product of two numbers.

Example 1: If out of two numbers, one number is increased by 10% and the other number by 20%, find per cent increase in the product of the two numbers.

Sol. Method 1: Net increase $= 10 + 20 + \frac{10 \times 20}{100} = 32\%$

Method 2: Let both the numbers are 10 each. Then new numbers are 11 and 12.

$11 \times 12 - 100 = 32 \Rightarrow$ Increase $= 32\%$.

Example 2. The salary of an employee is first increased by 20% and then again increased by 30%. Find per cent increase in his salary from the original salary.

(*a*) 6% (*b*) 44% (*c*) 50% (*d*) 56%

Sol. (*d*) Net effect on his salary $= 20 + 30 + \frac{20 \times 30}{100} = 20 + 30 + 6 = 56\%$ increase

Alternative Method : $12 \times 13 - 100 = 56\%$

Example 3. A number is increased by 20% and then it is decreased by 10%. Find the net increase or decrease per cent : **(SSC CGL Tier I 2002)**

(*a*) 10% increase (*b*) 10% decrease (*c*) 8% increase (*d*) 8% decrease

Sol. (*c*) Net effect $= 20 - 10 - \frac{20 \times 10}{100} = 20 - 10 - 2 = 8\%$ increase

Alternative Method : $12 \times 9 - 100 = 8\%$

10. (*a*) If A is $\frac{1}{x}$ more than B, then B is less than A by $\frac{1}{x+1}$.

(*b*) If A is $\frac{1}{x}$ less than B, then B is more than A by $\frac{1}{x-1}$.

Note: What happens **if the fraction is $\frac{a}{x}$?**

(*a*) If A is $\frac{a}{x}$ more than B, then B is less than A by $\frac{a}{x+a}$.

(*b*) If A is $\frac{a}{x}$ less than B, then B is more than A by $\frac{a}{x-a}$.

Alternative method :

(*a*) If A is x% or $\frac{x}{100}$ more than B, then B is less than A by $\frac{x}{100+x}$.

(*b*) If A is x% or $\frac{x}{100}$ less than B, then B is more than A by $\frac{x}{100-x}$.

Important to note : The increase in percentage / fraction value is always more than the decrease in percentage/fraction.

Formula stated above can be used in numerous situations, such as :

1. If a number is increased and then by how much the new number is to be decreased to get the original number and vice-versa.

2. Price of a commodity is first increased and then by how much new price is to be reduced to restore to the original price and vice-versa.
3. If price of a commodity is increased then by how much quantity purchased is to be reduced so that total expenditure remains constant and vice-versa.
4. If first number is more than the second number then by how much the second number is less than the first number and vice-versa.

Examples:

1. A's salary is 25% more than B's salary, B's salary is how much less than A's salary ? **(SAS 2010, SSC CHSL 2010, MAT)**

(*a*) 20% (*b*) 24% (*c*) 25% (*d*) $33\frac{1}{3}$% %

Sol. (*a*) Let B = 100, then A = 125 ⇒ B – A = 25 ⇒ Required percentage = $\frac{25}{125} \times 100 = 20\%$

Alternative Method : A's salary is more than B's salary by 25% = $\frac{1}{4}$

∴ B's salary is less than A's salary by $\frac{1}{4+1} = \frac{1}{5} = 20\%$

2. B got 20% marks less than A. What per cent marks did A got more than B ? **(SSC CGL Tier I 2008)**

(*a*) 20% (*b*) 30% (*c*) 25% (*d*) $33\frac{1}{3}$%

Sol. (*c*) Let A's marks = 100. Then B's marks = 100 – 20 = 80

Difference in marks = 100 – 80 = 20 ⇒ Required percentage = $\frac{20}{80} \times 100 = 25\%$

Alt : B is less than A by 20% = $\frac{20}{100} = \frac{1}{5}$ ⇒ A is more than B by $\frac{20}{100-20} = \frac{1}{4} = 25\%$

3. If A's income is 40% less than that of B, how much per cent is B's income more than that of A? **(SSC CGL Tier I 2006)**

(*a*) 60% (*b*) 40% (*c*) $66\frac{2}{3}$% (*d*) $33\frac{1}{3}$%

Sol. (*c*) Let B = 100, then A = 100 – 40 = 60, ⇒ B – A = 40

Required percentage = $\frac{40}{60} \times 100 = 66\frac{2}{3}\%$

Alternative Method : A's income is less than B's income by 40% = $\frac{2}{5}$

∴ B's income is more than A's income by $\frac{2}{5-2} = \frac{2}{3} = 66\frac{2}{3}\%$

SOLVED EXERCISE

1. Half of 1 per cent written as a decimal is : **(SSC CGL Tier I 2000)**

(*a*) 0.2 (*b*) 0.002 (*c*) 0.005 (*d*) 0.0005

Sol. (*c*) 1% = 0.01 ⇒ Half of 1% = 0.01 ÷ 2 = 0.005

2. What is 20% of 25% of 300 ? **(SSC CGL Tier I 2002)**

(*a*) 150 (*b*) 60 (*c*) 45 (*d*) 15

Sol. (*d*) 20% of 25% of 300 = $\frac{20}{100} \times \frac{25}{100} \times 300 = 15$

Alt: 20% = $\frac{1}{5}$ and 25% = $\frac{1}{4}$ ⇒ 20% of 25% of 300 = $\frac{1}{5} \times \frac{1}{4} \times 300 = 15$

3. 18% of which number is equal to 12% of 75 ? **(SSC CGL Tier I 2002)**

(*a*) 50 (*b*) 100 (*c*) 2 (*d*) $\frac{3}{2}$

Sol. (*a*) Required number = $\frac{12 \times 75}{18} = 50$

4. If 50% of P = 25% of Q, then P = x% of Q. Find x. **(SSC CGL Tier I 2002)**

(*a*) 0.5 (*b*) 2 (*c*) 50 (*d*) 0.005

Sol. (*c*) 50% of P = 25% of Q ⇒ P : Q = 25: 50 = 1 : 2 ⇒ P = 50% of Q

5. What is the number whose 20% is 30% of 40 ? **(CDS 2011)**

(*a*) 90 (*b*) 80 (*c*) 60 (*d*) 50

Sol. (*c*) 30% of 40 = 12. ⇒ Required number = 12 ÷ 20% = 60

6. If 36% of 36 = 54% of x, find the value of x ? **(SBI Clerical 2008)**

(*a*) 24 (*b*) 36 (*c*) 54 (*d*) None of these

Sol. (*a*) $x = \frac{36 \times 36}{54} = 24$

7. 60 = ?% of 400. **(Bank PO 2006)**

(*a*) 6 (*b*) 12 (*c*) 20 (*d*) 15

Sol. (*d*) ?% = $\frac{60}{400} \Rightarrow ? = \frac{60}{400} \times 100 = 15$

8. The difference between 57% of a number and its 41% is 240. What is the number ? **(Bank Clerical 2008)**

(*a*) 1000 (*b*) 1200 (*c*) 1500 (*d*) 1800

Sol. (*c*) 57% – 41% = 16% of the number = 240 ⇒ The number = $\frac{240}{16} \times 100 = 1500$

9. 38 litres of milk was poured into a tub and the tub was found to be 5% empty. To completely fill the tub, what amount of additional milk must be poured : **(CDS 2011)**

(*a*) 1 litre (*b*) 2 litres (*c*) 3 litres (*d*) 4 litres

Sol. (*b*) (100 – 5)% of capacity = 38 litres ⇒ 5% of capacity = $38 \times \frac{5}{95} = 2$ litres

10. The total number of boys in a school is 15% more than the total number of girls in the school. What is the ratio of number of boys to the number of girls in the school ? **(Bank Clerical 2006)**

(*a*) 13 : 7 (*b*) 7 : 13 (*c*) 23 : 20 (*d*) None of these

Sol. (*c*) Ratio of boys : girls = 115 : 100 = 23 : 20

11. If the difference between 73% of a number and 48% of the same number is 615, what is 15% of the number ? **(SBI Clerical)**

(*a*) 123 (*b*) 92.75 (*c*) 369 (*d*) None of these

Sol. (*c*) (73 – 48)% of x = 615 ⇒ 25% of x = 615 ⇒ 15% of $x = \frac{615}{25} \times 15 = 369$

12. (7.9% of 134) – (3.4% of 79) = ? **(Canara Bank P.O. 2009)**

(*a*) 8.1 (*b*) 7.8 (*c*) 7.3 (*d*) None of these

Sol. (*d*) (7.9% of 134) – (3.4% of 79) = (7.9% of 134) – (34 of 7.9%)

= 7.9% of (134 – 34) = 7.9% of 100 = 7.9

13. 47% of 583 – 41% of 473 = ? **(SBI Clerical)**

(*a*) 34.98 (*b*) 28.38 (*c*) 80.08 (*d*) None of these

Sol. (*c*) ? = 274.01 – 193.93 = 80.08

Alternative method : GE = (41% + 6%) of 583 – 41% of 473

= 41% of (583 – 473) + 6% of 583 = 45.1 + 34.98 = 80.08

14. If 120 is 20% of a number, then 120% of that number will be **(SSC CGL Tier I 1999)**

(*a*) 20 (*b*) 120 (*c*) 480 (*d*) 720

Sol. (*d*) 20% of number = 120 $\Rightarrow$ 120% (i.e. 6 times 20%) = 6 × 120 = 720

15. If x% of $\frac{25}{2}$ is 150, then the value of x is : **(SSC CGL Tier I 2002)**

(*a*) 1000 (*b*) 1200 (*c*) 1400 (*d*) 1500

Sol. (*b*) $\frac{x}{100} \times \frac{25}{2} = 150 \Rightarrow x = 150 \times \frac{2}{25} \times 100 = 1200$

16. If p% of p is 36, then p is equal to : **(SSC CGL Tier I 2000)**

(*a*) 3600 (*b*) 600 (*c*) 60 (*d*) 15

Sol. (*c*) p% of $p = \frac{p^2}{100} = 36 \Rightarrow p^2 = 3600 \Rightarrow p = 60$

17. The ratio 5 : 4 expressed as a per cent equals : **(SSC CGL Tier I 2000)**

(*a*) 125% (*b*) 80% (*c*) 40% (*d*) 12.5%

Sol. (*a*) $5 : 4 = \frac{5}{4} \times 100 = 125\%$

18. 20% of 30% of a number is 12, the number is:

(*a*) 24 (*b*) 40 (*c*) 100 (*d*) 200

Sol. (*d*) Let the number be x.

Then 20% of 30% of $x = 12 \Rightarrow \frac{20}{100} \times \frac{30}{100} \times x = 12 \Rightarrow x = 12 \times \frac{100}{20} \times \frac{100}{30} = 200$

19. 60% of 2300 = 30% of ?

(*a*) 6000 (*b*) 4600 (*c*) 2300 (*d*) 3000

Sol. (*b*) Let 60% of 2300 = 30% of $x \Rightarrow x = \frac{60}{100} \times 2300 \times \frac{100}{30} = 4600$

Trick: Here first part is halved *i.e.* from 60% to 30%.

$\Rightarrow$ Second part is to be doubled to get the same product.

$\therefore$ The second number is 2300 × 2 = 4600

20. Find the amount for which a car worth ₹ 4,50,000 should be insured at 10% so that in case of loss of car, cost of the car as well as the amount paid towards insurance premium is recovered.

(*a*) ₹ 45,000 (*b*) ₹ 4,95,000 (*c*) ₹ 5,00,000 (*d*) ₹ 6,00,000

Sol. (*c*) Premium is paid on the insured value of the goods.

∴ Car worth ₹ 90 should be insured for ₹ 100.

∴ Car worth ₹ 4,50,000 should be insured for ₹ 4,50,000 × $\frac{100}{90}$ = ₹ 5,00,000

21. A saves 20% of his monthly salary. If his monthly expenditure is ₹ 6,000, then his monthly savings is **(SSC CHSL 2012)**

(*a*) ₹ 1,200 (*b*) ₹ 1,500 (*c*) ₹ 1,800 (*d*) ₹ 2,400

Sol. (*b*) Savings : Expenditure = 20 : 80 = 1 : 4 ⇒ The expenditure is 4 times savings

⇒ Savings = ₹ 6,000 ÷ 4 = ₹ 1,500

22. Instead of multiplying a number by 5, it has been mistakenly divided by 5. Find the percentage error due to this mistake.

(*a*) 25% (*b*) 500% (*c*) 125% (*d*) 96%

Sol. (*d*) Let number = 20 ⇒ Then error = 20 × 5 – 20 ÷ 5 = 100 – 4 = 96 or 96%.

23. If 70% of the students in a school are boys and the number of girls be 504, the number of boys is : **(SSCCGLTierI2002)**

(*a*) 1176 (*b*) 1008 (*c*) 1208 (*d*) 3024

Sol. (*a*) Girls = 30% of students = 504 ⇒ Boys = 70% of students = $\frac{504}{30}$ × 70 = 1176

24. The weight of the container alone is 25% of the container filled with a certain fluid. When some fluid is removed, the weight of the container and remaining fluid is 50% of the original total weight. What fractional part of the liquid has been removed ? **(DMRC 2003)**

(*a*) $\frac{1}{3}$ (*b*) $\frac{1}{2}$ (*c*) $\frac{2}{3}$ (*d*) $\frac{3}{4}$

Sol. (c) Weight of water in full container = 100% – 25% = 75%.

Weight of water after removal = 50% – 25% = 25%

∴ Water removed = 75% – 25% = 50% ⇒ Fraction $\frac{50}{75} = \frac{2}{3}$

25. If rate of income tax is increased from 20% to 25%. Find the additional amount of tax, a person has to pay, if his total income is ₹ 40,000.

(*a*) ₹ 2,000 (*b*) ₹ 2,500 (*c*) ₹ 8,000 (*d*) ₹ 10,000

Sol. (*a*) Increase in tax-rate = (25 – 20)% = 5% = $\frac{1}{20}$

∴ Increase in tax amount on ₹ 40,000 = $\frac{1}{20}$ × 40,000 = ₹ 2,000

26. Salaries of A and B are in the ratio of 3 : 4. Both of them spend 75% of their salaries and save rest of the money. Find the ratio of their savings.

(*a*) 3 : 4 (*b*) 4 : 3 (*c*) 1: 1 (*d*) Can't say

Sol. (*a*) Ratio of savings = 25% of 3 : 25% of 4 = 3 : 4

27. A plot of 640 sq. ft. is available at the rate of ₹ 4,600 per sq. ft. If 30% of the total cost of the plot is payable at booking of the plot, how much is the booking amount ? **(SBI Clerical 2008)**

(*a*) ₹ 8,83,200 (*b*) ₹ 8,38,200 (*c*) ₹ 9,83,200 (*d*) None of these

Sol. (*a*) Cost of the plot = 640 × 4600

Booking amount = 640 × 4600 × $\frac{30}{100}$ = ₹ 8,83,200

28. To pass an examination, a candidate needs 40% marks. All questions carry equal marks. A candidate just passed by getting 10 answers correct by attempting 15 of the total questions. How many questions are there in the examination ? **(CDS 2011)**

(*a*) 25 (*b*) 30 (*c*) 40 (*d*) 45

Sol. (*a*) 40% of questions = 10 $\Rightarrow$ Total questions = $10 \times \dfrac{100}{40} = 25$

29. 1 minute 48 seconds in percentage of one hour is represented by : **(SSC CHSL 2013)**

(*a*) 5% (*b*) 2% (*c*) 3% (*d*) 4%

Sol. (*c*) 1 minute 48 seconds = (60 + 48) seconds = 108 seconds. 1 hour = 3600 seconds

$\therefore$ Required rate per cent = $\dfrac{108}{3600} \times 100 = 3\%$

30. The ratio between two numbers is 5 : 4. If 40% of the first number is 12, then what would be 50% of the second number ? **(Bank of Baroda PO 1999)**

(*a*) 6 (*b*) 12 (*c*) 18 (*d*) 24

Sol. (*b*) Let the numbers be $5x$ and $4x$. Then 40% of $5x = 2x = 12$

50% of second number = 50% of $4x = 2x = 12$

31. The difference of the squares of two numbers is 80% of the sum of their squares. The ratio of the larger number to the smaller number is: **(SSC CGL Tier I 2006)**

(*a*) 1 : 3 (*b*) 3 : 1 (*c*) 5 : 2 (*d*) 2 : 5

Sol. (*b*) $(x^2 - y^2) = 80\% \times (x^2 + y^2) \Rightarrow (x^2 - y^2) = \dfrac{4}{5} \times (x^2 + y^2)$

$\Rightarrow 5x^2 - 5y^2 = 4x^2 + 4y^2 \Rightarrow x^2 = 9y^2 \Rightarrow x = 3y \Rightarrow x : y = 3 : 1$

32. One dozen pairs of socks quoted at ₹160 are available at a discount of 10%. Find how many pairs of socks can be bought for ₹60. **(SSC CGL Tier I 2014)**

(a) 5 pairs (*b*) 7 pairs (*c*) 6 pairs (d) 8 pairs

Sol. (*a*) Price of 12 socks = 160 – 16 = ₹144 $\Rightarrow$ Price of 1 sock = 144 ÷ 12 = ₹12

Socks bought for ₹60 = 60 ÷ 12 = 5

33. A fruit seller had some apples. He sells 40% apples and still has 420 apples. Originally, he had : **(SSC CGL Tier I 2003)**

(*a*) 588 apples (*b*) 600 apples (*c*) 672 apples (*d*) 700 apples

Sol. (*d*) Apples left = (100 – 40)% = 60% = 420 apples (i.e. 7 times 60)

$\Rightarrow$ Total apples 7 × 100 = 700 apples.

34. A person spends 30% of monthly salary on rent, 25% on food, 20% on children's education and 12% on electricity, and the balance of ₹ 1,040 on the remaining items. What is the monthly salary of the person ? **(CDS 2010)**

(*a*) ₹ 8,000 (*b*) ₹ 9,000 (*c*) ₹ 9,600 (*d*) ₹ 10,600

Sol. (*a*) Expenditure on remaining items = 100% – (30 + 25 + 20 + 12)% = 13% = ₹ 1,040

Monthly salary = $\dfrac{1040}{13} \times 100$ = ₹ 8,000

35. Radha spends 40% of her salary on food, 20% on house rent, 10% on entertainment and 10% on conveyance. If her savings at the end of a month are ₹ 1500, then her salary per month (in ₹) is : **(SSC CGL 2003)**

(*a*) 8000 (*b*) 7500 (*c*) 6000 (*d*) 10000

Sol. (*b*) Total expenditure = (40 + 20 + 10 + 10)% = 80%

$\therefore$ Savings = 100% – 80% = 20% of income = ₹ 1500 ⇒ Income = $1500 \times \frac{100}{20}$ = ₹ 7500

36. Ms Sujata invests ₹ 2170 which is 7% of her monthly salary in mutual funds. Later she invests 18% of her monthly salary in recurring deposit also, she invests 6% of her salary in NSCs. What is the total annual amount invested by Ms. Sujata ? **(Bank of Baroda PO 2007)**

(*a*) ₹ 125320 (*b*) ₹ 113520 (*c*) ₹ 135120 (*d*) ₹ 115320

Sol. (*d*) 7% of monthly salary = ₹ 2170

Total monthly investment = 7% + 18% + 6% = 31% = $2170 \times \frac{31}{7}$ = 9610

$\therefore$ Annual investment = 9610 × 12 = ₹ 115320

37. One-fifth of a number is 18. What will be 30% of the same number ? **(SBI Clerical)**

(*a*) 5.4 (*b*) 24 (*c*) 60 (*d*) None of these

Sol. (*d*) One-fifth of the number = 18. ⇒ The number = 18 × 5

$\therefore$ 30% of the number = $18 \times 5 \times \frac{30}{100} = 27$

38. 605 sweets were distributed equally among children in such a way that the number of sweets received by each child is 20% of the total number of children. How many sweets did each child receive ? **(Dena Bank PO 2006)**

(*a*) 24 (*b*) 11 (*c*) 45 (*d*) None of these

Sol. (*b*) Let number of students = x. Then each child got = 20% of x

20% of $x \times x = 605 \Rightarrow x^2 = 605 \times 5 = 3025 \Rightarrow x = 55$

Each child got = 20% of 55 = 11 sweets

39. In a class of 65 students and 4 teachers, each student got sweets that are 20% of total number of students and each teacher got sweets that are 40% of the total number of students. How many sweets were there ? **(Canara Bank PO 2009)**

(*a*) 845 (*b*) 897 (*c*) 949 (*d*) 104,

Sol. (*c*) Each student got = 20% of 65 = 13 sweets. Each teacher got = 40% of 65 = 26 sweets

Total number of sweets = 65 × 13 + 4 × 26 = 845 + 104 = 949

40. A reduction in the price of apples enables a person to purchase 3 apples for ₹ 1 instead of ₹1.25. What is the % of reduction in price (approximately) **(SSC CGL Tier I 2013)**

(*a*) 20 (*b*) 25 (*c*) 30 (*d*) $33\frac{1}{3}$

Sol. (*a*) Reduction in price = ₹ 1.25 – ₹ 1 = ₹ 0.25 ⇒ % Reduction = $\frac{25}{125} \times 100 = 20\%$

41. Price of an article is increased by 3%. Had it been ₹ 16 less, price would have been decreased by 1% of the original price. Find original price of the article. **(Bank Clerical)**

(*a*) ₹ 400 (*b*) ₹ 500 (*c*) ₹ 800 (*d*) ₹ 1600

Sol. (*a*) Difference in two prices = 3% + 1% = 4% of original price

$\therefore$ 4% of the original price = 16 (*i.e.* 4 times 4) ⇒ Original price = 4 × ₹ 100 = ₹ 400

42. A salesman receives salary of ₹ 4000 per month and 10% commission on total sales. What is his earnings, if sales is ₹ 20000?

(*a*) ₹ 3000 (*b*) ₹ 5000 (*c*) ₹ 6000 (*d*) ₹ 8000

Sol. (*c*) Salary = ₹ 4000. Commission = 10% of ₹ 20000 = ₹ 2000

$\therefore$ Total earnings = ₹ 4000 + ₹ 2000 = ₹ 6000

43. A salesman receives salary of ₹ 5000 per month and 20% commission on total sales. What is the amount of sales, if his total earnings in a month is ₹ 9,000?

(*a*) ₹ 15,000 (*b*) ₹ 20,000 (*c*) ₹ 50,000 (*d*) ₹ 60,000

Sol. (*b*) Total earnings = ₹ 9000. Salary = ₹ 5000

Commission = Total earnings – Salary = ₹ 9000 – ₹ 5000 = ₹ 4000

Commission = 20% of Sales = ₹ 4000 ∴ Sales = ₹ 4000 × $\frac{100}{20}$ = ₹ 20000

44. The total number of students in two schools A and B is 990. If the number of students of school A is decreased by 5% and the number of students of school B is increased by 14%, the number of students of the two schools will become equal. The ratio of the number of students of schools A and B is : **(SSC CGL Tier I 2006)**

(*a*) 5 : 6 (*b*) 6 : 5 (*c*) 115 : 84 (*d*) 84 : 115

Sol. (*b*) 95% of A = 114% of B ⇒ $\frac{A}{B} = \frac{114}{95} = \frac{6}{5}$

45. A salesman receives a salary of ₹ 1000 per month and 10% commission on sales in excess of ₹ 5000. What is the amount of sales, if he earns ₹ 2000 in a particular month?

(*a*) ₹ 10000 (*b*) ₹ 15000 (*c*) ₹ 20000 (*d*) ₹ 25000

Sol. (*b*) Total earnings = ₹ 2000. Salary = ₹ 1000 ⇒ Commission = ₹ 2000 – ₹ 1000 = ₹ 1000

Sales in excess of ₹ 5000 (to earn ₹ 1000 as commission) = 1000 × 10 = ₹ 10000

∴ Total sales = ₹ 10000 + ₹ 5000 = ₹ 15000

46. If 39% of a number is equal to 52% of the other number. What is the ratio between the two numbers ? **(SBI Clerical 2008)**

(*a*) 2 : 3 (*b*) 3 : 4 (*c*) 4 : 3 (*d*) None of these

Sol. (*c*) Ratio between two numbers = 52 : 39 = 4 : 3

47. If population of women in a village is 80% of population of men, what is the population of men as a percentage of population of women ? **(CISF Head Constable 2011)**

(*a*) 80% (*b*) 100% (*c*) 112.5% (*d*) 125%

Sol. (*d*) Required % = $\frac{100}{80} \times 100 = 125\%$

48. The tank-full petrol in Arun's motor-cycle lasts for 10 days. If he starts using 25% more everyday, how many days will the tank-full petrol last ? **(UPSC Civil Services (Pre.) 2013)**

(*a*) 5 (*b*) 6 (*c*) 7 (*d*) 8

Sol. (*d*) Required number of days = $10 \times \frac{100}{125}$ = 8 days

49. A spends 90% of his salary and B spends 85% of his salary. But savings of both are equal. Find the income of B, if sum of their incomes is ₹ 5000.

(*a*) ₹ 2000 (*b*) ₹ 2500 (*c*) ₹ 3000 (*d*) ₹ 3500

Sol. (*a*) Savings of A = (100 – 90)% = 10% of his salary.

Savings of B = (100 – 85)% = 15% of his salary.

But both saves equal amount. ⇒ 10% of A's salary = 15% of B's salary

∴ A's salary : B's salary = 15 : 10 = 3 : 2 ⇒ B's salary = $\frac{2}{5}$ × ₹ 5000 = ₹ 2000

50. Three-fourth of one number is equal to 60% of another number. What is the ratio of the first number to the second number ? **(Corporation Bank Special Officers Exam. 2004)**

(*a*) 3 : 4 (*b*) 5 : 4 (*c*) 6 : 7 (*d*) None of these

Sol. (*d*) $\frac{3}{4}$ of x = 60% of $y \Rightarrow \frac{3}{4}$ of $x = \frac{3}{5}$ of $y \Rightarrow \frac{x}{y} = \frac{3}{5} \div \frac{3}{4} = \frac{3}{5} \times \frac{4}{3} = \frac{4}{5} = 4:5$

Alt: $\frac{60}{100} : \frac{3}{4} = \frac{3}{5} : \frac{3}{4} = 4:5$

51. One-fourth of sixty per cent of a number is equal to two-fifth of twenty per cent of another number. What is the respective ratio of the first number to the second number ? **(Vijaya Bank PO 2008)**

(*a*) 4 : 7 (*b*) 8 : 13 (*c*) 5 : 9 (*d*) None of these

Sol. (*d*) $\frac{1}{4} \times 60\% \times x = \frac{2}{5} \times 20\% \times y \Rightarrow 15\%$ of $x = 8\%$ of $y \Rightarrow x : y = 8 : 15$

Alternative Method : Required ratio $= \frac{2}{5} \times 20 : \frac{1}{4} \times 60 = 8 : 15$

52. If 30% of a number is equal to four-fifth of the another number. What is the ratio between the two numbers ? **(Bank Clerical 2006)**

(*a*) 3 : 4 (*b*) 3 : 8 (*c*) 8 : 3 (*d*) None of these

Sol. (*c*) Required ratio $= \frac{4}{5} : \frac{30}{100} = 400 : 150 = 8 : 3$

53. If 30% of A = 0.25 of B = $\frac{1}{5}$ of C, then A : B : C is equal to: **(SSC CGL Tier I 2004, 2007 M)**

(*a*) 5 : 6 : 4 (*b*) 5 : 24 : 5 (*c*) 6 : 5 : 4 (*d*) 10 : 12 : 15

Sol. (*d*) $\frac{30}{100}A = \frac{25}{100}B = \frac{1}{5}C \Rightarrow \frac{3}{10}A = \frac{1}{4}B = \frac{1}{5}C = k$ (say)

$\Rightarrow A = \frac{10}{3}k$, B = 4*k*, C = 5*k* $\Rightarrow$ A : B : C = $\frac{10}{3} : 4 : 5 = 10 : 12 : 15$

54. If $\frac{2}{3}$ of A = 75% of B=0.6 of C, then $A : B : C$ is: **(SSC CGL Tier I 2006, 2008, LDC 2012)**

(*a*) 2 : 3 : 4 (*b*) 3 : 4 : 5 (*c*) 4 : 5 : 6 (*d*) 9 : 8 : 10

Sol. (*d*) $\frac{2}{3}A = \frac{3}{4}B = \frac{3}{5}C = k \Rightarrow A : B : C = \frac{3}{2} : \frac{4}{3} : \frac{5}{3} = 9 : 8 : 10$

55. Given that 10% of A's income = 15% of B's income = 20% of C's income. If sum of their income is ₹ 7800, then B's income is **(SSC CGL Tier I 2005)**

(*a*) ₹ 3600 (*b*) ₹ 3000 (*c*) ₹ 2400 (*d*) ₹ 1800

Sol. (*c*) 10% of A = 15% of B = 20% of C $\Rightarrow$ 2A = 3B = 4C

Dividing the ratio by LCM of 2, 3 and 4 i.e. by 12, we get : $\frac{A}{6} = \frac{B}{4} = \frac{C}{3}$

$\therefore$ A : B : C = 6 : 4 : 3 $\therefore$ B's income $= \frac{4}{13} \times 7800 =$ ₹ 2400

56. If 40% of a number is equal to $\frac{3}{7}$ of other number and sum of these numbers is 145, what is the larger number ? **(Bank PO 2004)**

(*a*) 85 (*b*) 70 (*c*) 75 (*d*) None of these

Sol. (*c*) Ratio = $\frac{3}{7}:\frac{2}{5} = 15:14 \Rightarrow$ Larger number = $\frac{15}{29}\times 145 = 75$

57. Sum of 5% of a number and 9% of other number is equal to sum of 8% of first number and 7% of the second number. Find ratio between the numbers.

(*a*) 2 : 3 (*b*) 3 : 2 (*c*) 5 : 7 (*d*) 8 : 9

Sol. (*a*) Let the numbers are x and y. Then 5% of x + 9% of y = 8% of x + 7% of y

$\therefore$ 3% of x = 2% of $y \Rightarrow 3x = 2y \Rightarrow x : y = 2 : 3$

Trick: First number is increased by 3% and second number is decreased by 2% but the result remains the same

$\therefore$ Ratio between numbers is 2 : 3 *i.e.* inverse ratio of 3% and 2%

58. The sum of 5% of a number and 4% of the other number is equal to $\frac{2}{3}$ of sum of 6% of the first number and 8% of the second number. The ratio of the first number to the second is : **(SSC CGL Tier I 1999)**

(*a*) 2 : 3 (*b*) 3 : 2 (*c*) 3 : 4 (*d*) 4 : 3

Sol. (*d*) Let the numbers are x and y. Then 5% of x + 4% of y = $\frac{2}{3}$ × (6% of x + 8% of y)

$\therefore$ 15% of x + 12% of y = 12% of x + 16% of $y \Rightarrow$ 3% of x = 4% of $y \Rightarrow x : y = 4 : 3$

59. In a group of persons, 70% of the persons are male and 30% of the persons are married. If two-sevenths of the males are married, what fraction of the female is single ? **(UPSC Civil Services Pre. 2011)**

(*a*) 3/7 (*b*) 2/7 (*c*) 1/3 (*d*) 2/3

Sol. (*d*) Females = 100% – 70% = 30% $\Rightarrow$ Married males = 70% × 2/7 = 20% of total population

$\Rightarrow$ Married females = 20/30 = 2/3

60. In an examination, there were 1100 boys and 900 girls, 50% of boys and 40% of girls passed. The percentage of candidates failed is **(MAT)**

(*a*) 45% (*b*) 54.5% (*c*) 50% (*d*) 60%

Sol. (*b*) Candidates passed = 50% of 1100 + 40% of 900 = 550 + 360 = 910

$\Rightarrow$ Candidates failed = (1100 + 900) – 910 = 1090 $\Rightarrow$ Required % = $\frac{1090}{2000}\times 100 = 54.5\%$

61. A number, on subtracting 15 from it, reduces to its 80%. What is 40% of the number? **(SSC Constable GD 2012)**

(*a*) 75 (*b*) 60 (*c*) 30 (*d*) 90

Sol. (*c*) Number is reduced by 15 or 20% of original number.

$\therefore$ 40% of number = 2 × 20% of number = 2 × 15 = 30

62. When 60 is subtracted from 60% of a number, the result is 60. The number is: **(SSC CGL Tier I 2000)**

(*a*) 120 (*b*) 150 (*c*) 180 (*d*) 200

Sol. (*d*) Let the number is x. Then (60% of x) – 60 = 60

$\therefore$ 60% of x = 120 (*i.e.* 2 times 60) $\Rightarrow x = 2 \times 100 = 200$

63. 75% of a number when added to 75 is equal to the number. The number is **(SSC CGL Tier I 2000, SSC CHSL 2012, SSC CGL Tier 2013M)**

(*a*) 225 (*b*) 270 (*c*) 300 (*d*) 325

Sol. (*c*) Let the number is x. Then 75% of $x + 75 = x \Rightarrow (100 - 75)\%$ of $x = 75$

$\Rightarrow$ 25% of $x = 75$ (*i.e.* 3 times 25) $\Rightarrow x = 3 \times 100 = 300$

64. When 60% of a number is added to 48, the result is the number itself. What is the number?

(*a*) 80 (*b*) 90 (*c*) 120 (*d*) None of these

Sol. (*c*) 48 = (100 – 60)% = 40% of the number $\Rightarrow$ The number $= 48 \times \frac{100}{40} = 120$

65. If 40% of one number is added to another number then it becomes 125% of itself. What will be the ratio of first and second numbers? **(BSRBBankPO2000)**

(*a*) 12 : 25 (*b*) 8 : 25 (*c*) 5 : 7 (*d*) None of these

Sol. (*d*) Let the numbers and x and y. Then 40% of $x + y$ = 125% of x

$\Rightarrow y = 85\%$ of $x \Rightarrow y : x = 85 : 100 = 17 : 20 \Rightarrow x : y = 20 : 17$

66. A number increased by $22\frac{1}{2}\%$ gives 98. The number is **(SSC CGL Tier II 2013)**

(*a*) 45 (*b*) 18 (*c*) 80 (*d*) 81

Sol. (*c*) $100 + 22\frac{1}{2}\% = 122\frac{1}{2}\% = \frac{245}{2}\% \Rightarrow$ Number $= 98 \times \frac{200}{245} = 80$

67. If two numbers are respectively 20% and 50% of a third number, what is the ratio between the two numbers ? **(SSC CGL Tier I 2001)**

(*a*) 5 : 2 (*b*) 2 : 5 (*c*) 1 : 5 (*d*) 1 : 2

Sol. (*b*) Let the third number be 100. Then the first two numbers are 20 and 50.

$\therefore$ Required ratio = 20 : 50 = 2 : 5

68. Two numbers are less than the third number by 50% and 54% respectively. By how much per cent is the second number less than the first number ? **(Bank of Maharashtra 2008)**

(*a*) 13 (*b*) 10 (*c*) 12 (*d*) None of these

Sol. (*d*) Let the third number be 100.

Then first number = 100 – 50 = 50 and second number = 100 – 54 = 46

Difference between two numbers = 50 – 46 = 4

$\therefore$ Required percentage $= \frac{4}{50} \times 100 = 8\%$

69. Two numbers are more than the third number by 20% and 50% respectively. First number is what per cent of the second number ? **(SSC CGL Tier I 2002)**

(*a*) 100% (*b*) 150% (*c*) 80% (*d*) 120%

Sol. (*c*) Let the third number be 100. Then the first two numbers are 120 and 150.

$\therefore$ Required percentage $= \frac{120}{150} \times 100 = 80\%$

70. Two numbers are respectively 25% and 40% less than a third number. What per cent of the first number is the second number ? **(SSC CHSL 2013)**

(*a*) 40% (*b*) 60% (*c*) 70% (*d*) 80%

Sol. (*d*) Let the third number be 100. Then the first two numbers are 75 and 60.

$\therefore$ Required percentage $= \frac{60}{75} \times 100 = 80\%$

71. Two numbers X and Y are respectively 20% and 28% less than a third number Z. By what percentage is the number Y less than the number X ? **(UPSC Civil Services Pre. 2011)**

(*a*) 7% (*b*) 8% (*c*) 9% (*d*) 10%

Sol. (*d*) Let Z = 100, then X = 100 – 20 = 80 and Y = 100 – 28 = 72; X – Y = 80 – 72 = 8

$\Rightarrow$ Required percentage = 8/80 × 100 = 10%

72. There are some coins and rings of either gold or silver in a box. 60% of the objects are coins. 40% of the rings are of gold and 30% of the coins are of silver. What is the percentage of gold articles ? **(Bank PO)**

(*a*) 16 (*b*) 27 (*c*) 58 (*d*) 70

Sol. (*c*) Let total articles be 100. Then number of coins = 60 and number of rings = 100 – 60 = 40

Gold articles = 60 × (100 – 30)% + 40 × 40% = 42 + 16 = 58%

73. Given A is 50% larger than C and B is 25% larger than C, then A is what per cent larger than B ? **(SSC CGL Tier I 2013)**

(*a*) 20% (*b*) 25% (*c*) 50% (*d*) 75%

Sol. (*a*) Let C = 100. Then A and B are 150 and 125.

A – B = 150 – 125 = 25 $\therefore$ Required percentage $= \frac{25}{125} \times 100 = 20\%$

74. Rakesh got 273 marks in an examination and scored 5% more than the pass%. If Lokesh got 312 marks, then by what % above the pass mark did he pass the examination? **(SSC CGL Tier I 2013)**

(*a*) 20% (*b*) 15% (*c*) 25% (*d*) 27%

Sol. (*a*) 273 = 105% of passing marks $\Rightarrow 312 = \frac{312}{273} \times 105\%$ = 120% of passing marks

75. The cost of an apple is twice that of banana and the cost of a banana is 25% less that of guava. If the cost of each type of fruit increases by 10%, then the percentage increase in cost of 4 bananas, 2 apples and 3 guavas is **(SSC CGL Tier I 2011)**

(*a*) 10% (*b*) 8% (*c*) 12% (*d*) 16%

Sol. (*a*) Cost of each fruit is increased by 10%. Therefore the required precentage is 10%.

76. If present price of sugar after 10% reduction is ₹ 18 per kg., find the original price of sugar per kg.

(*a*) ₹ 1.80 (*b*) ₹ 10.80 (*c*) ₹ 16.20 (*d*) ₹ 20

Sol. (*d*) Let original price of sugar = ₹ 100 per kg. Then reduced price = ₹ 100 – ₹ 10 = ₹ 90

But reduced price = ₹ 18 $\Rightarrow$ Original price = ₹ 18 × $\frac{100}{90}$ = ₹ 20

77. The difference of two numbers is 45% of their sum. The ratio of the larger number to the smaller number is : **(SSC CGL Tier I 2004)**

(*a*) 20 : 9 (*b*) 9 : 20 (*c*) 29 : 11 (*d*) 11 : 29

Sol. (*c*) Let sum of the numbers be 100. Then difference between the numbers = 45

$\therefore$ Required ratio = (100 + 45) : (100 – 45) = 145 : 55 = 29 : 11

78. The difference of two numbers is 15% of their sum. The ratio of the larger number to the smaller number is : **(SSC CGL Tier I 2006)**

(*a*) 23 : 17 (*b*) 11 : 9 (*c*) 17 : 11 (*d*) 23 : 11

Sol. (*a*) Let the sum of the numbers be 100.

Then difference between the numbers = 15% of 100 = 15

∴ Required ratio = (100 + 15) : (100 – 15) = 115 : 85 = 23 : 17

79. If 60% of A's income is equal to 75% of B's income, then B's income is equal to x% of A's income. The value of x is : **(SSC CGL Tier I 2003)**

(*a*) 70 (*b*) 60 (*c*) 80 (*d*) 90

Sol. (*c*) 60% of A's income = 75% of B's income ⇒ A's income : B's income = 75 : 60 = 5 : 4

∴ Required percentage = $\frac{4}{5} \times 100 = 80\%$

80. Out of two numbers, 40% of the greater number is equal to 60% of the smaller. If the sum of the numbers is 150, then the greater number is : **(SSC CGL Tier I 2005)**

(*a*) 70 (*b*) 80 (*c*) 90 (*d*) 60

Sol. (*c*) 40% of greater number = 60% of smaller number

∴ Greater number : Smaller number = 60 : 40 = 3 : 2 ⇒ Greater number = $\frac{3}{5} \times 150 = 90$

81. If 20% of $(P + Q)$ = 50% of $(P - Q)$, then find P : Q. **(SSC CGL Tier I 2005)**

(*a*) 7 : 8 (*b*) 7 : 3 (*c*) 7 : 5 (*d*) 5 : 7

Sol. (*b*) $\frac{P+Q}{P-Q} = \frac{50}{20} = \frac{5}{2} \Rightarrow P : Q = (5 + 2) : (5 - 2) = 7 : 3$

82. If 15% of $(A + B)$ = 25% of $(A - B)$, then what per cent of B is equal to A ? **(SSC CGL Tier I 2002)**

(*a*) 10% (*b*) 60% (*c*) 200% (*d*) 400%

Sol. (*d*) $\frac{A+B}{A-B} = \frac{25}{15} = \frac{5}{3} \Rightarrow A : B = (5 + 3) : (5 - 3) = 8 : 2 = 4 : 1 \Rightarrow A$ is 4 times or 400% of B.

83. If 50% of $(x - y)$ = 30% of $(x + y)$, then what per cent of x is y ? **(SSC CGL Tier I 2002, SSC FCI 2012, CDS 2010 M)**

(*a*) 25% (*b*) $33\frac{1}{3}\%$ (*c*) 40% (*d*) 400%

Sol. (*a*) $\frac{x+y}{x-y} = \frac{5}{3} \Rightarrow x : y = (5 + 3) : (5 - 3) = 8 : 2 = 4 : 1 \Rightarrow y$ is 1/4 or 25% of x

84. If the numerator of a fraction is increased by 20% and its denominator by 25%, then the fraction so obtained is $\frac{3}{5}$. What is the original fraction ? **(Bank PO 2004)**

(*a*) $\frac{3}{5}$ (*b*) $\frac{2}{8}$ (*c*) $\frac{5}{8}$ (*d*) None of these

Sol. (*c*) Let the original fraction is $\frac{x}{y}$. Then $\frac{120\% \text{ of } x}{125\% \text{ of } y} = \frac{24x}{25y} = \frac{3}{5} \Rightarrow \frac{x}{y} = \frac{3}{5} \times \frac{25}{24} = \frac{5}{8}$

85. If numerator of fraction is increased by 20% and the denominator is decreased by 5%, the value of the new fraction becomes 5/2. The original fraction is : **(SSC CGL Tier I 2005)**

(*a*) $\frac{24}{19}$ (*b*) $\frac{3}{18}$ (*c*) $\frac{95}{48}$ (*d*) $\frac{48}{95}$

Sol. (*c*) Let the original fraction be $\frac{x}{y}$. Then $\frac{120\% \text{ of } x}{95\% \text{ of } y} = \frac{5}{2} \Rightarrow \frac{x}{y} = \frac{5}{2} \times \frac{95}{120} = \frac{95}{48}$

86. Difference of two numbers is 1660. If $6\frac{1}{2}$% of one number is $8\frac{1}{2}$% of the other number, the smaller number is :

(SSC CGL Tier I 2003)

(*a*) 7055 (*b*) 5395 (*c*) 3735 (*d*) 2075

Sol. (*b*) $x : y = 8\frac{1}{2} : 6\frac{1}{2} = 17 : 13$

Difference of numbers $= 17x - 13x = 1660 \Rightarrow 4x = 1660 \Rightarrow x = 415$

$\therefore$ The smaller number $= 13x = 13 \times 415 = 5395$

87. In a test consisting of 75 questions carrying one mark each, Samir answered 75% of the first 40 questions correctly. What approximate per cent of the other 35 questions does he need to answer correctly to score 80% on the entire test ?

(Nabard PO 2008)

(*a*) 90 (*b*) 75 (*c*) 86 (*d*) 60

Sol. (*c*) Total marks obtained = 75 × 80% = 60

Marks obtained in first 40 questions = 40 × 75% = 30

Remaining marks to secure = 60 – 30 = 30 out of remaining 75 – 40 = 35 questions

$\therefore$ Required percentage $= \frac{30}{35} \times 100 \approx 86\%$

88. 1300 boys and 700 girls appeared in an examination, 80% of the boys and 70% of the girls passed in the exam, find the total percentage of failed students.

(*a*) 12.5% (*b*) 23.5% (*c*) 47% (*d*) 76.5%

Sol. (*b*) Boys failed = 1300 × (100 – 80)% = 260. Girls failed = 700 × (100 – 70)% = 210

$\therefore$ % of Students failed $= \frac{260 + 210}{2000} \times 100 = 23.5\%$

89. In an examination, 70% students qualified from school A. There are 20% more students in school B than in school A and number of students qualified in the exam are 50% more than the students qualified from school A. What is the percentage of students qualified to the number of students appeared from school B ?

(*a*) 30% (*b*) 70% (*c*) 87.5% (*d*) 78.5%

Sol. (*c*) Let students in school A = 100

Then students qualified from school A = 70% of 100 = 70

Total students in school B = 100 + 20% of 100 = 120

Students qualified from school B = 70 + 50% of 70 = 105

Required percentage $= \frac{105}{120} \times 100 = \frac{175}{2} = 87.5\%$

90. The prices of sugar and rice are in the ratio 4 : 5. If the price of sugar is increased by 10% and that of rice by 20%, find the ratio between increased prices of sugar and rice.

(Bank Clerical 2007)

(*a*) 1 : 2 (*b*) 5 : 6 (*c*) 11 : 15 (*d*) 11 : 12

Sol. (*c*) After 10% increase, the price of sugar will become $\frac{110}{100}$ or 1.1 of the original price.

Similarly the price of rice will become 1.2 of the original price.

$\therefore$ The ratio of new prices = (4 × 1.1) : (5 × 1.2) = 4.4 : 6.0 = 11 : 15

91. The ratio of the number of students studying in schools A, B and C is 6 : 8 : 7 respectively. If the number of student studying in each of the schools is increased by 20%, 15% and 20% respectively, what will be the new ratio of the number of students in schools A, B and C ?

(Nabard Officers Exam 2008, SBI Associate Bank PO 2007 M)

(*a*) 18 : 23 : 21 (*b*) 12 : 18 : 17 (*c*) 18 : 21 : 17 (*d*) None of these

Sol. (*a*)

	A	B	C
Old ratio	6	8	7
Increase :	20% of 6	15% of 8	20% of 7
Total increase	1.2	1.2	1.4
New numbers	7.2	9.2	8.4

$\therefore$ New ratio = 7.2 : 9.2 : 8.4 = 18 : 23 : 21

92. Total number of employees in companies A, B and C are in the ratio 4 : 5 : 6. If the number of employees in these companies are increased by 25%, 30% and 50% respectively, what will be the new ratio of employees working in these companies? **(SBI Clerical 2009)**

(*a*) 5 : 6 : 9 (*b*) 10 : 19 : 18 (*c*) 10 : 13 : 15 (*d*) None of these

Sol. (*d*) Original ratio = 4 : 5 : 6

New ratio = $(4 \times 1.25) : (5 \times 1.3) : (6 \times 1.5) = 5 : 6.5 : 9 = 10 : 13 : 18$

93. Fresh fruit contains 68% water and dry fruit contains 20% of water. How much dry fruit can be obtained from 100 kg of fresh fruit? **(MBA 1991 M, SSC CGL Tier I 2004)**

(*a*) 32 kg (*b*) 40 kg (*c*) 52 kg (*d*) 80 kg

Sol. (*b*) Non-water content from 100 kg of fresh fruit

= (100 – 68)% = 32 kg = (100 – 20)% of dry fruit

$\Rightarrow$ Dry fruit = $32 \div 80\% = 32 \times \frac{100}{80} = 40$ kg

94. 15 litres of a mixture contains alcohol and water in the ratio 1 : 4. If 3 litres of water is mixed in it, the percentage of alcohol in the new mixture will be **(SSC CGL Tier I 2013)**

(*a*) 15 (*b*) $16\frac{2}{3}$ (*c*) 17 (*d*) $18\frac{1}{2}$

Sol. (*b*) Alcohol in original mixture = $\frac{1}{5} \times 15 = 3$ litres

$\therefore$ Required ratio = $\frac{3}{15+3} \times 100 = 16\frac{2}{3}\%$

Alt: Alcohal in original mixture = $\frac{1}{5}$; Quantity of new mixture = 15 + 3 = 18 litres

$\Rightarrow$ Alcohal in new mixture = $\frac{1}{5} \times \frac{15}{18} = \frac{1}{6} = 16\frac{2}{3}\%$

95. 15 litres of a mixture contains 20% alcohol and rest water. If 3 litres of water is added in it percentage of alcohol in the new mixture will be?

(*a*) 16.67% (*b*) 20% (*c*) 25% (*d*) 33.33%

Sol. (*a*) Required percentage = $20 \times \frac{15}{15+3} = 16.67\%$

96. 5 litres of a solution of salt and water contains 8% salt. If 1 kg of water is evaporated, the percentage of salt in the remaining mixture will be :

(*a*) 7% (*b*) 8% (*c*) 9% (*d*) 10%

Sol. (*d*) Required percentage = $8\% \times \frac{5}{4} = 10\%$

97. From 5 litres of 20% solution of alcohol in water, 2 litres of solution is taken out and 2 litres water is added to it. The strength of the alcohol in the new solution is : **(SSC CGL Tier I 06)**

(*a*) 12% (*b*) 15% (*c*) 16% (*d*) 18%

Sol. (*a*) Total solution = 5 litres. Solution taken out = 2 litres.

Remaining solution = 5 – 2 = 3 litres

Strength of solution = 20%. Water added = 2 litres

Strength of new mixture = $20\% \times \frac{3}{3+2} = 12\%$

98. A solution of salt and water contains 20% salt by weight. If 20 kg of water is evaporated from the solution, the solution contains 25% salt. What is the original quantity (in kg) of the solution ?

(*a*) 80 kg (*b*) 100 kg (*c*) 150 kg (*d*) 200 kg

Sol. (*b*) Ratio of salt = 20 : 25 ⇒ Ratio of original and remaining solution = 25 : 20 = 5 : 4

∴ Difference = (5 – 4) ratio = 1 ratio = 20 kg.

∴ Original solution = 5 ratio = 5 × 20 kg. = 100 kg.

99. A solution of salt and water contains 15% salt. If 4 kg of water is evaporated, solution contains 18% salt. Find the original quantity of solution.

(*a*) 12 kg. (*b*) 18 kg. (*c*) 24 kg. (*d*) 36 kg.

Sol. (*c*) Ratio of salt = 15% : 18% = 5 : 6 ⇒ Ratio of solution = 6 : 5

Difference = (6 – 5) ratio = 1 ratio ≡ 4 kg ⇒ Original quantity = 6 ratio = 6 × 4 = 24 kg

100. How much water must be added to 16 litres of a solution containing 45% of acid to reduce it to 20% ?

(*a*) 36 litres (*b*) 30 litres (*c*) 24 litres (*d*) 20 litres

Sol. (*d*) Acid in the solution = $16 \times \frac{45}{100}$ litres

Quantity of acid remains unchanged in the new solution.

Ratio of acid and total quantity of the new mixture = 20 : 100

Total quantity of new solution having 20% acid = $16 \times \frac{45}{100} \times \frac{100}{20}$ = 36 litres

∴ Water to be added = 36 – 16 = 20 litres

Alt: Ratio of acid in two mixtures = 45 : 20 = 9 : 4 ⇒ Ratio of quantities = 4 : 9

First quantity = 4 ratio = 16 litres ⇒ 1 ratio = 4 litres

Second quantity = 9 ratio = 9 × 4 litres = 36 litres ⇒ Water to be added = 36 – 16 = 20 litres

101. 45 litres of a mixture of milk and water contains 15% water. How much water must be added so as to make water 25% in the resultant mixture?

(*a*) 10 litres (*b*) 6 litres (*c*) 4.5 litres (*d*) 9 litres

Sol. (*b*) Milk in mixture = $45 \times \frac{85}{100}$ litres

Quantity of milk remains unchanged in the new mixture.

Ratio of milk and water in the new mixture = 75 : 25

Quantity of new mixture having 25% water = $45 \times \frac{85}{100} \times \frac{100}{75} = 51$ litres

∴ Water to be added = 51 – 45 = 6 litres

Alt: Ratio of milk in two mixtures = 85 : 75 = 17 : 15 ⇒ Ratio of quantities = 15 : 17

First mixture = 15 ratio = 45 litres ⇒ 1 ratio = 3 litres

New mixture = 17 ratio = 17 × 3 litres = 51 litres ⇒ Water to be added = 51 – 45 = 6 litres

102. In a mixture of milk and water, the proportion of water by weight was 75%. If in the 60 grams of mixture 15 grams water was added, what would be the percentage of water in the new mixture?

(*a*) 75% (*b*) 80% (*c*) 90% (*d*) 100%

Sol. (*b*) Water in original mixture = 75% of 60 grams = 45 grams

Water in new mixture = 45 + 15 = 60 grams

New mixture = 60 + 15 = 75 grams ⇒ Required percentage = $\frac{60}{75} \times 100 = 80\%$

Alternative method : Milk in original mixture = (100 – 75)% = 25%

Milk in new mixture = $25\% \times \frac{60}{60+15} = 20\%$

Water in new mixture = (100 – 20)% = 80%

103. A and B are two fixed points 5 cm apart and C is a point on AB such that AC = 3 cm. If the length of AC is increased by 6%, the length of BC is decreased by : **(SSC CGL Tier I 2007)**

(*a*) 6% (*b*) 7% (*c*) 8% (*d*) 9%

Sol. (*d*) BC = AB – AC = 5 – 3 = 2 cm

Increase in AB = 5 × 6% = Decrease in BC

∴ Decrease in BC = $6\% \times \frac{AC}{BC} = 6\% \times \frac{3}{2} = 9\%$

104. If a number x is 10% less than another number y and y is 10% more than 125, then x is equal to **(SSC CGL Tier I 2005)**

(*a*) 150 (*b*) 143 (*c*) 140.55 (*d*) 123.75

Sol. (*d*) y is 10% more than 125 ⇒ $y = 125 \times \frac{110}{100}$

x is 10% less than y ⇒ $x = 125 \times \frac{110}{100} \times \frac{90}{100} = 123.75$

105. A sum of ₹ 2236 is divided among A, B and C such that A receives 25% more than C and C receives 25% less than B. What is A's share in the amount ? **(Indian Overseas Bank PO 2009)**

(*a*) ₹ 460 (*b*) ₹ 890 (*c*) ₹ 780 (*d*) ₹ 1280

Sol. (*c*) C = B – 25% = 75% of B A = C + 25% = 125% of C = 75% of 125% of B

∴ A : B : C = 75% of 125% of B : B : 75% of B = $\frac{3}{4} \times \frac{5}{4} : 1 : \frac{3}{4}$ = 15 : 16 : 12

∴ A = $\frac{15}{43} \times 2236$ = ₹ 780

106. In an examination a candidate must secure 40% marks to pass. A candidate who gets 220 marks, fails by 20 marks. What are the maximum marks for the examination ? **(SSC CGL Tier I 2000)**

(*a*) 1200 (*b*) 800 (*c*) 600 (*d*) 450

Sol. (*c*) Minimum passing marks = 40% of the maximum marks = 220 + 20 = 240 (*i.e.* 6 times 40)

$\therefore$ Maximum marks = 6 × 100 = 600

107. In an examination, a student must get 36% marks to pass. A student who gets 190 marks failed by 35 marks. The total marks in that examination is **(SSC CGL Tier I 2015)**

(*a*) 500 (*b*) 810 (*c*) 450 (*d*) 625

Sol. (*d*) Minimum passing marks = 36% of the maximum marks = 190 + 35 = 225

$\therefore$ Maximum marks = $\frac{225}{36} \times 100 = 625$

108. A student is required to obtain 288 marks to pass an exam. He gets 33% of the maximum marks and fails by 24 marks. The maximum marks are: **(SBI Clerical)**

(*a*) 650 (*b*) 700 (*c*) 800 (*d*) 900

Sol. (*c*) Marks obtained = 33% of maximum marks = 288 – 24 = 264

Maximum marks = $\frac{264}{33} \times 100 = 800$

109. A student is required to get 230 marks to pass an exam. He gets 184 marks and fails by 8% of the maximum marks. The maximum marks are: **(Bank Clerical 2008)**

(*a*) 2875 (*b*) 2300 (*c*) 575 (*d*) None of these

Sol. (*c*) 8% of maximum marks = 230 – 184 = 46 ⇒ Maximum marks = $\frac{46}{8} \times 100 = 575$

110. A student got 42% marks and has secured 12 marks more than the minimum passing marks. Another student who got 45% has obtained 30 marks more than the minimum passing marks. The maximum marks are :

(*a*) 250 (*b*) 300 (*c*) 450 (*d*) 600

Sol. (*d*) Difference in percentage of marks = 45% – 42% = 3%

Difference in marks = 30 – 12 = 18 (*i.e.* 6 times 3) ⇒ Maximum marks = 6 × 100 = 600

111. Two students A and B appeared in the same examination. A secures 35% of marks and fails by 15 marks. B secures 165 marks and fails by 25 marks. Find the maximum marks.

(*a*) 600 (*b*) 750 (*c*) 500 (*d*) 900

Sol. (*c*) Passing marks = 35% of maximum marks + 15 marks = (165 + 25) marks

35% of maximum marks = 165 + 25 – 15 = 175 marks (*i.e.* 5 times 35)

$\therefore$ Maximum marks = 5 × 100 = 500

112. A student gets 35% marks and fails by 10 marks and another student who gets 31% marks fails by 30 marks. The maximum marks are:

(*a*) 400 (*b*) 500 (*c*) 600 (*d*) 1000

Sol. (*b*) Difference in percentage of marks = 35% – 31% = 4%

Difference in marks = – 10 – (– 30) = – 10 + 30 = 20 (*i.e.* 5 times 4)

$\therefore$ Maximum marks = 5 × 100 = 500

113. A candidate secured 30% marks in an examination and failed by 6 marks. Another secured 40% marks and got 6 marks more than the bare minimum to pass. The maximum marks are: **(SSC CGL Tier I 2003)**

(*a*) 150 (*b*) 120 (*c*) 100 (*d*) 180

Sol. (*b*) Difference in percentage of marks = 40% – 30% = 10%

Difference in actual marks = 6 – (– 6) = 6 + 6 = 12 ⇒ Total marks = $12 \times \frac{100}{10} = 120$

114. Two students A and B appeared at an examination. If one of them secured 9 marks more than the other and his marks were 56% of sum of their marks. The marks obtained by them are: **(SSC CGL Tier I 2004)**

(*a*) 42, 33 (*b*) 43, 34 (*c*) 41, 32 (*d*) 39, 30

Sol. (*a*) Marks secured by B = (100 – 56)% = 44%

∴ Difference in marks = (56 – 44)% = 12% = 9 marks

∴ Sum of marks $= \frac{9}{12} \times 100 = 75$ marks

A's marks $= \frac{75+9}{2} = 42$ marks B's marks = 75 – 42 = 33

115. A candidate who get 20% marks in an examination fails by 30 marks but another candidate who gets 32% gets 42 marks more than the pass mark. Then the percentage of pass mark is: **(SSC CGL Tier I 2003)**

(*a*) 52% (*b*) 50% (*c*) 33% (*d*) 25%

Sol. (*d*) Difference in marks = (32% – 20%) of maximum marks = 42 – (– 30) = 72 marks

∴ 12% of maximum marks = 72 marks ⇒ 1% of maximum marks = 6 marks

Failed by 30 marks = 5% of maximum marks

∴ Pass percentage = 20% + 5% = 25% of maximum marks

116. In an examination, a student who get 20% of the maximum marks fails by 5 marks. Another student who scores 30% of the maximum marks get 20 marks more than the pass marks. The necessary percentage required for passing is : **(SSC CGL Tier I 2000)**

(*a*) 32% (*b*) 23% (*c*) 22% (*d*) 20%

Sol. (*c*) Difference in marks = (30% – 20%) of maximum marks = 20 – (– 5) = 25 marks

∴ 10% of maximum marks = 25 marks ⇒ 1% of maximum marks = 2.5 marks

Failed by 5 marks = 2% of maximum marks

∴ Pass percentage = 20% + 2% = 22% of maximum marks

117. In an examination, a student got 30% marks and failed by 12 marks. In the same examination, another student got 40% marks and secured 28 marks more than bare minimum marks to pass. The pass marks is: **(SSC CGL Tier I 2006)**

(*a*) 112 (*b*) 120 (*c*) 132 (*d*) 150

Sol. (*c*) (40% – 30%) of maximum marks = 28 – (–12) = 28 + 12 = 40 marks

⇒ 10% of marks = 40 (i.e. 4 times 10) ⇒ Maximum marks = 4 × 100 = 400

Minimum marks to pass = 30% of 400 + 12 = 120 + 12 = 132

118. A person saves 10% of his income. If his income is increased by 20% and he saves 15% of the new income, by what per cent his actual savings will increase?

(*a*) 5% (*b*) 35% (*c*) 75% (*d*) 80%

Sol. (*d*) Let initial income = ₹ 100. Initial savings = 10% of ₹ 100 = ₹ 10

Increased income = ₹ 100 + 20% of ₹ 100 = ₹ 120

New savings = 15% of ₹ 120 = ₹ 18 ⇒ Increase in savings = ₹ 18 – ₹ 10 = ₹ 8

∴ Per cent increase in savings $= \frac{8}{10} \times 100 = 80\%$

Alternative Method: Increase in savings = 5% of ₹ 100 + 15% of ₹ 20 = ₹ 8

$\therefore$ Per cent increase in savings = $\frac{8}{10} \times 100 = 80\%$

119. In an election between two candidates, the candidate getting 60% of the votes polled, is elected by a majority of 14000 votes. The number of votes polled for the winning candidate is : **(SSC CGL Tier I 2004)**

(*a*) 28000 (*b*) 32000 (*c*) 42000 (*d*) 46000

Sol. (*c*) Votes casted in favour of winning candidate = 60%

Votes casted in favour of other candidate = (100 – 60)% = 40%

$\therefore$ Winning candidate wins by (60% – 40%) = 20% of the votes = 14000 votes

Votes in favour of winning candidate = 60% of votes = $14000 \times \frac{60}{20} = 42000$

120. In an election, a candidate who gets 84% of the votes is elected by a majority of 476 votes. What is the total number of votes polled? **(SSC CGL Tier I 2014)**

(*a*) 900 (*b*) 810 (*c*) 600 (*d*) 700

Sol. (*d*) Votes casted in favour of defeated candidate = 100 – 84 = 16%

$\therefore$ Winning candidate wins by (84% – 16%) = 68% of the votes = 476 votes (i.e. 7 times 68)

Total voters = 100 % = 100 × 7 = 700

121. In an election there were only two candidates. One of the candidates secured 40% of the votes and is defeated by the other candidates by 298 votes. The total number of votes polled is **(SSC CGL Tier II 2012)**

(*a*) 1490 (*b*) 1500 (*c*) 745 (*d*) 1460

Sol. (*a*) Votes casted in favour of defeated candidate = 40%

Votes casted in favour of other candidate = (100 – 40)% = 60%

$\therefore$ Winning candidate wins by (60% – 40%) = 20% of the votes = 298 votes

Total number of voters = 298 × 5 = 1490

122. 8% of the voters in an election did not cast their votes. In this election, there were only two candidates. The winner by obtaining 48% of the total votes defeated his contestant by 1100 votes. The total number of voters in the election was : **(SSC CGL Tier I 2003)**

(*a*) 21000 (*b*) 23500 (*c*) 22000 (*d*) 27500

Sol. (*d*) Votes polled = 100% – 8% = 92%

Votes in favour of the winning candidate = 48%

Votes in favour of other candidate = (92 – 48)% = 44%

Difference in votes = 48% – 44% = 4% = 1100 votes

$\therefore$ Total Voters = $1100 \times \frac{100}{4} = 27500$

123. In an election between two candidates, one got 52% of total valid votes. 25% of the total votes were invalid. The total number of votes were 8400. How many valid votes did the other person get ? **(Bank of Maharashtra PO 2008)**

(*a*) 3276 (*b*) 3196 (*c*) 3024 (*d*) None of these

Sol. (*c*) Valid votes = 100% – 25% = 75%

Votes in favour of other person = 100% – 52% = 48% of valid votes

Required votes = 8400 × 75% × 48% = $8400 \times \frac{75}{100} \times \frac{48}{100}$ = 3024 votes

124. In an election contest between two candidates, 10% of total voters did not take part. The candidate who was elected got 60% of the votes casted. If he got 180 votes more than his opponent, find total number of voters.

(*a*) 300 (*b*) 900 (*c*) 1000 (*d*) 1800

Sol. (*c*) Votes casted = 100% – 10% = 90%

Votes in favour of winning candidates = 60% of 90% of total votes

Votes in favour of opponent = 40% of 90% of total votes

Difference = (60 – 40)% of 90% of total votes = 180 ⇒ 20% of 90% of total votes = 180

$\therefore$ Total voters $= 180 \times \frac{100}{90} \times \frac{100}{20} = 1000$

125. In a town 40% of the male population and 45% of the female population are married. Find ratio between male and female population in the town assuming that one man marries one woman and vice-versa.

(*a*) 8 : 9 (*b*) 9 : 8 (*c*) 5 : 7 (*d*) 7 : 5

Sol. (*b*) Number of married males = Number of married females

$\therefore$ 40% of the male population = 45% of the female population

$\therefore$ Male : Female = 45 : 40 = 9 : 8

126. Population of a village is 90000, $\frac{5}{9}$th of them are males and rest are females. If 40% of the males are married, what is the percentage of married females?

(*a*) 40% (*b*) 45% (*c*) 50% (*d*) 60%

Sol. (*c*) Male population $= 90{,}000 \times \frac{5}{9} = 50{,}000$

Female population = 90,000 – 50,000 = 40,000

Married males = 50,000 × 40% = 20,000 ⇒ Married females = 20,000

Percentage of married females $= \frac{20{,}000}{40{,}000} \times 100 = 50\%$

Alternative Method: Let male population = $5x$. Then female population = $4x$

Let a% of females are married. Then 40% of $5x$ = a% of $4x$ ⇒ $a = 40\% \times \frac{5x}{4x} = 50\%$

127. The population of a village is 3400. $\frac{10}{17}$ of them are males and rest are females. If 42% of males are married, find the percentage of married females.

(*a*) 36% (*b*) 42% (*c*) 60% (*d*) None of these

Sol. (*c*) Males : Females = 10 : 7 ⇒ % of married females $= 42 \times \frac{10}{7} = 60\%$

128. In a town, 45% of male population and 25% of female population are married. Assuming that one man marries one woman and vice-versa, the perecentage of total population who are married, is **(SSC CGL Tier II 2012)**

(*a*) 30% (*b*) 35% (*c*) $33\frac{1}{3}\%$ (*d*) $32\frac{1}{7}\%$

Sol. (*d*) 45% of male = 25% of female ⇒ Male : Female = 25 : 45 = 5 : 9

% of married population $= \frac{5 \times 45\% + 9 \times 25\%}{5+9} = \frac{225\% + 225\%}{14} = \frac{450}{14}\% = \frac{225}{7}\% = 32\frac{1}{7}\%$

129. The population of a town is 15000. If the number of males increase by 8% and that of the females by 10%, then the population would increases to 16300. Find the number of females in the town. **(SSC CGL Tier I 2012)**

(*a*) 3000 (*b*) 5000 (*c*) 4000 (*d*) 6000

Sol. (*b*) Let the population consists of females only.

Then increase in the population is 10% of 15000 = 1500

∴ Increased population = 15000 + 1500 = 16500

But actual increased population = 16300 ⇒ Difference = 16500 – 16300 = 200

200 is 2% (*i.e.* 10% – 8%) of the male population.

$\Rightarrow$ Male population $= \frac{200}{2} \times 100 = 10000$

$\Rightarrow$ Female population = 15000 – 10000 = 5000

130. A gave 50% of his money to his wife and distributed the remaining amount equally among his 4 children. If each child got ₹ 1250, what was the total amount he had? **(Bank Clerical 2006)**

(*a*) ₹ 2,500 (*b*) ₹ 5,000 (*c*) ₹ 7,500 (*d*) None of these

Sol. (*d*) Each child got = ₹ 1250 ⇒ Amount given to 4 children = 4 × ₹ 1250 = ₹ 5000

Total amount he had = 2 × ₹ 5000 = ₹ 10,000

131. Rubina decided to donate 16% of her monthly salary to an NGO. On the day of donation she changed her mind and donated ₹ 6,567 which was 75% of what she had decided earlier. How much is Rubina's monthly salary ? **(Canara Bank PO 2009)**

(*a*) ₹ 8,756 (*b*) ₹ 54,725 (*c*) ₹ 6,56,700 (*d*) ₹ 45,696

Sol. (*b*) Let monthly income of Rubina be x. Then 16% of 75% of x = ₹ 6,567

$\Rightarrow x = 6567 \times \frac{100}{16} \times \frac{100}{75} =$ ₹ 54,725

132. A man had some eggs. Out of them 10% had to be thrown away as they were broken. 80% of the remaining eggs were sold and now he is left with only 90 eggs. How many eggs had he in the beginning?

(*a*) 500 (*b*) 800 (*c*) 900 (*d*) 1000

Sol. (*a*) Eggs in the beginning $= 90 \times \frac{100}{100-10} \times \frac{100}{100-80} = 90 \times \frac{100}{90} \times \frac{100}{20} = 500$ eggs

133. A man loses 20% of his money. After spending 25% of the remainder, he has ₹ 480 left. What is amount of money he originally had? **(CDS 2012)**

(*a*) ₹ 600 (*b*) ₹ 720 (*c*) ₹ 800 (*d*) ₹ 840

Sol. (*c*) $480 \times \frac{100}{100-20} \times \frac{100}{100-25} = 480 \times \frac{100}{80} \times \frac{100}{75} =$ ₹ 800

134. Mr. Giridhar spends 50% of his monthly income on household items and out of the remaining he spends 50% on transport, 25% on entertainment, 10% on sports and remaining amount of ₹900 is saved. What is Mr. Giridhar's monthly income ? **(Bank PO 2009)**

(*a*) ₹ 6000 (*b*) ₹ 12000 (*c*) ₹ 9000 (*d*) None of these

Sol. (*b*) Expenses on household items = 50%

Expenses on other items = (50 + 25 + 10)% = 85%

∴ Income $= 900 \times \frac{100}{100-50} \times \frac{100}{100-85} = 900 \times \frac{100}{50} \times \frac{100}{15} =$ ₹ 12000

135. A fruit-seller bought some mangoes of which 10% were rotten. He sold $46\frac{2}{3}$% of the rest. If he still has 192 mangoes, the total number of mangoes, originally bought by him, was : **(SSC CGL Tier I 2006)**

(*a*) 375 (*b*) 400 (*c*) 425 (*d*) 450

Sol. (*b*) Mangoes sold = $46\frac{2}{3}\% = \frac{140}{3}\%$ $\quad\therefore$ Remaining mangoes = $100 - \frac{140}{3} = \frac{160}{300}$

$\therefore$ Total mangoes = $192 \times \frac{100}{90} \times \frac{300}{160} = 400$

136. A person gave 20% of his income to his elder son, 30% of the remaining to the younger son and 10% of the balance to a trust. If he is left with ₹ 10,080, his income was : **(Bank Clerical 2007, SSC CGL Tier I 2003)**

(*a*) ₹ 20000 (*b*) ₹ 25000 (*c*) ₹ 30000 (*d*) ₹ 40000

Sol. (*a*) Income $= 10080 \times \frac{100}{100-20} \times \frac{100}{100-30} \times \frac{100}{100-10} = 10080 \times \frac{10}{8} \times \frac{10}{7} \times \frac{10}{9} =$ ₹ 20000

137. A man spends $12\frac{1}{2}$% of his salary on items of daily use and 30% of the remainder on house rent. After that he is left with ₹ 2940. How much is his salary ? **(SSC CGL Tier I 2004)**

(*a*) ₹ 4800 (*b*) ₹ 5200 (*c*) ₹ 4500 (*d*) ₹ 4000

Sol. (*a*) Expenditure on items of daily use = $12\frac{1}{2}\% = \frac{1}{8}$

Expenditure on house rent = $30\% = \frac{30}{100} = \frac{3}{10}$

$\therefore$ Salary = $2940 \div \left(1-\frac{1}{8}\right) \div \left(1-\frac{3}{10}\right) = 2940 \times \frac{8}{7} \times \frac{10}{7} =$ ₹ 4800

138. A man had a certain amount with him. He spent 20% of that to buy an article and 5% of the remaining on transport. Then he gifted ₹ 120. If he is left with ₹ 1,400, the amount he spent on transport is **(SSC CGL Tier II 2012)**

(*a*) ₹ 95 (*b*) ₹ 80 (*c*) ₹ 76 (*d*) ₹ 61

Sol. (*b*) After spending 5% of the remaining amount on transport, he is left with ₹ 1,520.

$\therefore$ Expenditure on transport is 5% and amount left thereafter = 95%.

$\therefore$ Expenditure on transport = $\frac{5}{95}$ × ₹ 1,520 = ₹ 80

139. A spent 25% of his money on food and 20% of the remaining on clothes. If he is still left with ₹ 1,280, how much money did he spend on clothes?

(*a*) ₹ 256 (*b*) ₹ 300 (*c*) ₹ 320 (*d*) ₹ 350

Sol. (*c*) After spending 20% on clothes, he is left with 80%.

If balance is ₹1280, then expenditure on clothes = ₹ 1280 × $\frac{20}{80}$ = ₹ 320

140. If A exceeds B by 40%, B is less than C by 20%, then A : C is : **(SSC CGL Tier I 1999)**

(*a*) 28 : 25 (*b*) 26 : 25 (*c*) 3 : 2 (*d*) 3 : 1

Sol. (*a*) A : B = 140 : 100 and B : C = 80 : 100 $\Rightarrow \frac{A}{C} = \frac{A}{B} \times \frac{B}{C} = \frac{140}{100} \times \frac{80}{100} = \frac{28}{25}$

141. If A's salary is 110% of B's, B's salary is 80% of C's. Find C's salary, if A's salary is ₹ 440.

(*a*) ₹ 450 (*b*) ₹ 500 (*c*) ₹ 550 (*d*) ₹ 600

Sol. (*b*) If B's salary is ₹ 100, then A's salary = ₹ 110% of ₹ 100 = ₹ 110.

$\therefore$ If A's salary is ₹ 440, then B's salary = ₹ $440 \times \frac{100}{110}$

If C's salary is ₹ 100, then B's salary = 80% of ₹ 100 = ₹ 80.

$\therefore$ C's salary = ₹ $440 \times \frac{100}{110} \times \frac{100}{80}$ = ₹ 500

142. A sum of ₹ 817 is divided among A, B and C such that A receives 25% more than B and B received 25% less than C. What is the A share in the amount ? **(Bank of Baroda PO 2007)**

(*a*) ₹ 228 (*b*) ₹ 247 (*c*) ₹ 285 (*d*) ₹ 304

Sol. (*c*) If B receives ₹ 100, then A receives = 100 + 25 = 125

If C receives ₹ 100, then B receives = 100 – 25 = 75

A : B = 125 : 100 = 5 : 4, B : C = 75 : 100 = 3 : 4

$\therefore$ A : B : C = 15 : 12 : 16 $\Rightarrow$ A's share = $\frac{15}{43} \times 817$ = ₹ 285

143. In an examination A got 25% marks more than B, B got 10% less than C and C got 25% more than D. If D got 320 marks out of 500, the marks obtained by A were **(SSC CGL Tier II 2013)**

(*a*) 405 (*b*) 450 (*c*) 360 (*d*) 400

Sol. (*b*) A : B = 125 : 100; B : C = 90 : 100; C : D = 125 : 100

A's marks = $320 \times \frac{125}{100} \times \frac{90}{100} \times \frac{125}{100} = 450$

144. In an examination every candidate took Maths or English or both. If 65% took Maths and 60% English, how many took Maths and English both?

(*a*) 5% (*b*) 25% (*c*) 62.5% (*d*) 125%

Sol. (*b*) Let total students = 100. Students who took Maths = 65. Students who took English = 60

$\therefore$ Students who took both the subjects = 65 + 60 – 100 = 25%

Note: (65 – 25) + (60 – 25) + 25 = 100

145. In a group of students, 70% can speak English and 65% can speak Hindi. If 27% of the students can speak none of the two languages, then what per cent of group can speak both the languages ? **(SSC CGL Tier I 2004)**

(*a*) 38% (*b*) 62% (*c*) 28% (*d*) 23%

Sol. (*b*) Let total students = 100. Students who know none of the two languages = 27

Remaining students = 100 – 27 = 73

$\therefore$ Students who know both languages = 70 + 65 – 73 = 62%

Note: (70 – 62) + (65 – 62) + 62 + 27 = 100

146. In an examination, 34% failed in Mathematics and 42% failed in English. If 20% failed in both the subjects, the percentage of students who passed in both subjects was : **(SSC CGL Tier I 2003)**

(*a*) 54% (*b*) 50% (*c*) 44% (*d*) 56%

Sol. (*c*) Students failed in Maths only = 34% – 20% = 14%

Students failed in English only = 42% – 20% = 22%

Students failed in both subjects = 20%

$\therefore$ Students failed in either or both subjects = (14 + 22 + 20)% = 56%

$\therefore$ Students passed in both subjects = (100 – 56)% = 44%

147. In an examination, 52% of the candidates failed in English and 42% failed in Mathematics. If 17% failed in both the subjects, then the percentage of candidates, who passed in both the subjects, was ? **(SSC CGL Tier II 2012, SSC CGL Tier I 2004, CDS 2011)**

(*a*) 25 (*b*) 22 (*c*) 23 (*d*) 21

Sol. (*c*) Total students failed in either or both subjects = 52% + 42% – 17% = 77%

Students passed in both subjects = 100% – 77% = 23%

148. There are 100 students in a particular class. 60% students play cricket, 30% students play football and 10% students play both the games. What is the number of students who play neither cricket nor football ? **(UPSC Civil Services Pre. 2011)**

(*a*) 25 (*b*) 20 (*c*) 18 (*d*) 15

Sol. (*b*) Students play either game = 60% + 30% – 10% = 80%

Students who play neither cricket nor football = 100% – 80% = 20%

149. 90% of the students in school passed in English, 85% passed in Mathematics and 150 students passed in both the subjects. If no student failed in both the subjects, find the total number of students. **(SSC CGL Tier I 2012)**

(*a*) 120 (*b*) 220 (*c*) 200 (*d*) 300

Sol. (*c*) Total students failed in either subjects = 10% + 15% = 25%

Students passed in both subjects = 100% – 25% = 75% = 150 (i.e. 2 times 75)

Total number of students = 2 × 100 = 200 students

150. In an examination 70% of candidates passed in English. 80% passed in Mathematics. 10% failed in both subjects. If 144 candidates passed in both, the total number of candidates was **(SSC CGL Tier I 2003)**

(*a*) 125 (*b*) 200 (*c*) 240 (*d*) 375

Sol. (*c*) Candidates failed in English = (100 – 70)% = 30%

Candidates failed in Mathematics = (100 – 80)% = 20%

Candidates failed in either or both subjects = (30 + 20 – 10)% = 40%

∴ Candidates passed in both subjects = (100 – 40)% = 60%

$$\therefore \text{ Total number of candidates} = \frac{100}{60} \times 144 = 240$$

151. In an examination, 80% candidates passed in English and 85% candidates passed in Maths. If 73% candidates passed in both these subjects, what per cent of students failed in both the subjects : **(SSC CGL Tier I 2008)**

(*a*) 8% (*b*) 19% (*c*) 27% (*d*) 35%

Sol. (*a*) Percentage of candidates passed in one or both subjects

$= n(E \cup M) = n(E) + n(M) - n(E \cap M) = 80 + 85 - 73 = 92$

∴ Candidates failed in both subjects $= n(\cup) - n(E \cup M) = 100 - 92 = 8$

152. In an examination 60% of the students pass in English, 70% pass in Hindi and 40% pass in both. What per cent of students fails in both English and Hindi ? **(SSC CGL Tier I 2000)**

(*a*) 10 (*b*) 20 (*c*) 25 (*d*) 30

Sol. (*a*) Percentage of candidates passed in one or both subjects

$= n(E \cup H) = n(E) + n(H) - n(E \cap H) = 60 + 70 - 40 = 90$

∴ Candidates failed in both subjects $= n(\cup) - n(E \cup H) = 100 - 90 = 10$

153. In an examination, 35% of the candidates failed in Maths and 25% in English while 15% failed in both subjects. How many candidates passed in one subject but not in the other?

(*a*) 30% (*b*) 45% (*c*) 55% (*d*) 70%

Sol. (*a*) Students failed in Maths only = (35 – 15)% = 20%
∴ Students passed in English only = 20%
Students failed in English only = (25 – 15)% = 10%
∴ Students passed in Maths only = 10%
∴ Total candidates passed in one subject but not in the other = 20% + 10% = 30%

154. In the certain examination, 77% candidates passed in English and 34% failed in Mathematics. If 13% failed in both the subjects and 784 candidates passed in both the subjects, then the total number of candidates was **(SSCCGLTierI2014)**
(*a*) 1200 (*b*) 1400 (*c*) 1600 (*d*) 1800

Sol. (*b*) Students failed in English = (100 – 77)% = 23%
Students failed in English only = (23 – 13) = 10%
Students failed in Maths only = (34 – 13)% = 21%
∴ Total candidates failed in either or both subjects = (10 + 21 + 13)% = 44%
Candidates passed in both the subjects = (100 – 44)% = 56% = 784
∴ Total number of candidates = 784 ÷ 56% = 1400

155. In a class 60% of the students pass in Hindi and 45% pass in Sanskrit. If 25% of them pass in both subjects, what percentage of the students fail in both the subjects ? **(SSC CGL Tier I 2000)**
(*a*) 80% (*b*) 75% (*c*) 20% (*d*) 25%

Sol. (*c*) Students passed in atleast one subject = (60 – 25) + (45 – 25) + 25 = 80%
∴ Students failed in both subjects = (100 – 80) = 20%

156. A reduction of ₹ 2 per kg in the price of sugar enables a man to buy 4 kg more of sugar for ₹16. Find the original price (per kg.) of sugar.
(*a*) ₹ 2 (*b*) ₹ 4 (*c*) ₹ 8 (*d*) ₹ 16

Sol. (*b*) Original price × Original quantity = ₹ 16 = New price × New quantity
Option (*a*) $2 \times 8 = 16$ and $(2 - 2) \times (8 + 4) = 0 \times 12 \neq 16$
Option (*b*) $4 \times 4 = 16$ and $(4 - 2) \times (4 + 4) = 2 \times 8 = 16$, which satisfies the equation.

157. When the price of eggs is reduced by 20%, it enables a man to buy 20 more eggs for ₹ 40. The reduced price per egg is : **(SSC CGL Tier I 2004)**
(*a*) 35 paise (*b*) 40 paise (*c*) 50 paise (*d*) 56 paise

Sol. (*b*) Saving due to reduction in price = 20% of ₹ 40 = ₹ 8
With this amount he can buy 20 more eggs.
∴ Reduced price per egg = $\frac{8}{20} \times 100$ paise = 40 paise.

158. If the price of rice is reduced by 20%, one can buy 2 kg more for ₹ 100. The reduced price of rice (per kg.) is : **(SSC CGL Tier I 2001)**
(*a*) ₹ 50 (*b*) ₹ 10 (*c*) ₹ 40 (*d*) ₹5

Sol. (*b*) Saving due to reduction in price = 20% of ₹ 100 = ₹ 20 = Reduced price of 2 kg rice.
∴ Reduced price of rice = 20 ÷ 2 = ₹ 10 per kg.
Alternative Method: Price = 100 : 80 = 5 : 4 ⇒ Quantity = 4 : 5
Difference = (5 – 4) ratio = 1 ratio = 2 kg ⇒ New quantity = 5 ratio = 10 kg
New price = ₹ 100 ÷ 10 = ₹10

159. A reduction of 20% in the price of sugar enables Mr. Bhadra to buy an extra 5 kg. of it for ₹320. The original rate of the sugar is: **(SSC CHSL 2013)**
(*a*) ₹ 12 (*b*) ₹ 15 (*c*) ₹ 16 (*d*) ₹ 20

Sol. (*c*) Saving due to reduction in price = 20% of ₹ 320 = ₹ 64 = Reduced price of 5 kg

∴ Reduced price per kg. of sugar = $\frac{64}{5} \times \frac{100}{80}$ = ₹ 16

Alt: Let original price be ₹100. Then reduced price = ₹100 – ₹20 = ₹80
Ratio of prices = 100 : 80 = 5 : 4 ⇒ Ratio of quantities = 4 : 5
Difference = (5 – 4) ratio = 1 ratio = 5 kg ⇒ Original quantity = 4 × 5 kg. = 20 kg.
Original price = ₹ 320 ÷ 20 = ₹ 16

160. A reduction of 20% in the price of rice enables a person to buy 3.5 kg more rice for ₹ 385. The original price of rice per kg is : **(SSC CGL Tier I 2003)**

(*a*) ₹ 20 (*b*) ₹ 22.50 (*c*) ₹ 25 (*d*) ₹ 27.50

Sol. (*d*) Saving due to reduction in price = 20% of ₹ 385 = ₹ 77

∴ Reduced price per kg. = 77 ÷ 3.5 = ₹ 22

∴ Original price per kg = ₹ 22 × $\frac{100}{80}$ = ₹ 27.50.

161. The price of sugar having done down by 10%, a consumer can buy 5 kg more sugar for ₹ 270. The difference between the original and reduced price per kg is **(SSC CGL Tier I 2013)**

(*a*) 60 paise (*b*) 75 paise (*c*) 53 paise (*d*) 62 paise

Sol. (*a*) Saving due to reduction in price = 10% of ₹ 270 = ₹ 27

∴ Reduced price per kg. = 27 ÷ 5 = ₹ 5.40

∴ Original price per kg = ₹ 5.40 × $\frac{100}{90}$ = ₹ 6.00

∴ Difference in prices = ₹ 6.00 – ₹ 5.40 = ₹ 0.60

162. A reduction of $33\frac{1}{3}$% in the price of an item would enable a purchaser to get 4 more a rupee. The price before reduction was: **(SSC CGL Tier I 2005)**

(*a*) 2 per rupee (*b*) 4 per rupee (*c*) 12 per rupee (*d*) 8 per rupee

Sol. (*d*) Reduction in price = $33\frac{1}{3}\% = \frac{1}{3}$ ⇒ Reduced price is $1 - \frac{1}{3} = \frac{2}{3}$ of original price

∴ Ratio of quantities = 2 : 3 ⇒ Difference = (3 – 2) ratio = 1 ratio = 4 items

∴ Original quantity = 2 ratio = 8 items.

163. When the price of oranges is reduced by 20%, it enables a man to buy 20 more oranges for ₹ 120. Then original price per orange is : **(SSC CGL Tier I 2006)**

(*a*) ₹ 1.50 (*b*) ₹ 1.20 (*c*) ₹ 1.60 (*d*) ₹ 1.80

Sol. (*a*) Saving due to reduction = 20% × ₹ 120 = ₹ 24

∴ Reduced price = $\frac{24}{20} = \frac{6}{5}$ ⇒ Original price = $\frac{6}{5} \times \frac{100}{80} = \frac{3}{2}$ = ₹ 1.50

164. The price of sugar is reduced by 20%. Now a person can buy 500 g more sugar for ₹36. The original price of the sugar per kg was **(SSC CGL Tier I 2008)**

(*a*) ₹ 14.40 (*b*) ₹ 16.20 (*c*) ₹ 16.50 (*d*) ₹ 18

Sol. (*d*) Saving due to reduction in price = 20% of ₹ 36 = ₹ 7.20

∴ Reduced price per kg. of sugar = ₹ 7.20 ÷ 1/2 = ₹ 14.40

∴ Original price = 14.40 × $\frac{10}{8}$ = ₹ 18

165. When the price of rice increases by 10%, a man can get 5 kg less rice for ₹ 500. The present cost of the rice per kg is:

(*a*) ₹ 9 (*b*) ₹ 12 (*c*) ₹ 10 (*d*) ₹ 15

Sol. (*c*) New price of rice = (500 × 10%) ÷ 5 = ₹ 10 per kg

Alt: Ratio of prices = 100 : 110 = 10 : 11 $\Rightarrow$ Ratio of quantities = 11 : 10

Difference = 1 ratio = 5 kg $\Rightarrow$ New quantity = 10 Ratio = 50 kg

New price = 500 ÷ 50 = ₹ 10 per kg

166. Due to an increase of 50% in the price of eggs, 4 eggs less are available for ₹ 24. The present rate of eggs per dozen is : **(SSC Data Entry Operator 2010)**

(*a*) ₹ 24 (*b*) ₹ 27 (*c*) ₹ 36 (*d*) ₹ 42

Sol. (*c*) New price = (24 × 50%) ÷ 4 = ₹ 3 per egg $\Rightarrow$ Price per dozen = 12 × ₹ 3 = ₹ 36

167. A mixture of 3 litres contains 20% alcohol and another mixture of 7 litres contains 30% alcohol. Find the strength of alcohol in the resulting mixture, if two mixtures are mixed together.

(*a*) 25% (*b*) 27% (*c*) 27.5% (*d*) 28%

Sol. (*b*) Total mixture = 3 + 7 = 10 litres

Alcohol in the mixture = 3 × 0.2 + 7 × 0.3 = 0.6 + 2.1 = 2.7 litres

$\therefore$ Percentage of Alcohol = $\frac{2.7}{10} \times 100 = 27\%$

Alternative method : Percentage of Alcohol = $\frac{3 \times 20 + 7 \times 30}{3 + 7} = \frac{270}{10} = 27\%$

168. Three sets of 40, 50 and 60 students appeared for an examination and the pass percentage was 100, 90 and 80 respectively. The pass percentage of the whole set is **(SSC CGL Tier II 2013)**

(*a*) $88\frac{2}{3}$ (*b*) $84\frac{2}{3}$ (*c*) $88\frac{1}{3}$ (*d*) $84\frac{1}{3}$

Sol. (*a*) $\frac{40 \times 100\% + 50 \times 90\% + 60 \times 80\%}{40 + 50 + 60} = \frac{40 + 45 + 48}{150} = \frac{133}{150} \times 100 = \frac{266}{3} = 88\frac{2}{3}\%$

169. The ratio of number of boys and girls in a school is 3 : 2. If 20% of boys and 30% of girls are scholarship holders, then the percentage of students, who do not get scholarship, is **(SSC CGL Tier I 2007, SSC, CHSL 2010 modified)**

(*a*) 50% (*b*) 72% (*c*) 75% (*d*) 76%

Sol. (*d*) Let the number of students be 500.

Scholarship holders = 300 × 20% + 200 × 30% = 120

$\therefore$ Percentage of scholarship holders = $\frac{120}{500} \times 100 = 24\%$

$\therefore$ Required percentage = (100 – 24)% = 76%

Alternative Method : Boys in class = $\frac{3}{5} \Rightarrow$ Girls in the class = $1 - \frac{3}{5} = \frac{2}{5}$

$\therefore$ Percentage of scholarship holders = $\frac{3}{5} \times 20\% + \frac{2}{5} \times 30\% = 12\% + 12\% = 24\%$

$\therefore$ Required percentage = (100 – 24)% = 76%

170. The ratio of number of boys to that of girls in a school is 4 : 1. If 75% of boys and 70% of girls are scholarship-holders, then the percentage of students who do not get scholarship is **(SSC CGL Tier I 2003)**

(*a*) 50% (*b*) 28% (*c*) 75% (*d*) 26%

Sol. (*d*) Boys in school = $\frac{4}{5}$ and girls in school = $\frac{1}{5}$

Scholarship-holders = $\frac{4}{5} \times 75\% + \frac{1}{5} \times 70\% = 60\% + 14\% = 74\%$

$\therefore$ Required percentage = (100 – 74)% = 26%

171. 500 students of a school appeared in an examination, 40% of the students are girls and 5% of the girls failed in the exam. Find percentage of boys failed if the total pass percentage of the school is 92%.

(*a*) 3% (*b*) 6% (*c*) 8% (*d*) 10%

Sol. (*d*) Girls = 200, Boys = 300. Total students failed = 8% of 500 = 40

Girls failed = 5% of 200 = 10 ⇒ Boys failed = 40 – 10 = 30

Required percentage = $\frac{30}{300} \times 100 = 10\%$

Alternative method : Total students failed = (100 – 92)% = 8%

Percent girls failed = 5% of 40% = 2% (of total students)

$\therefore$ Percentage of boys failed = 8% – 2% = 6% (of total students)

Percentage of boys in the school = 100% – 40% = 60%

$\therefore$ Per cent boys failed (of boys) = $\frac{6}{60} \times 100 = 10\%$

172. A man spends 75% of his income. His income is increased by 20% and he increases his expenditure by 10%. His savings are increased by:

(SSC CGL Tier I 2007, 2004 and 2015, SSC CHSL 2013)

(*a*) 10% (*b*) 25% (*c*) 37% (*d*) 50%

Sol. (*d*) Ratio of income and expenditure = 100 : 75 = 4 : 3

Required percentage = $\frac{ax - by}{x - y} = \frac{4 \times 20 - 3 \times 10}{4 - 3} = 50\%$

173. The monthly income of a person was ₹13500 and his monthly expenditure was ₹ 9000. Next year his income increased by 14% and his expenditure increased by 7%. The percentage of increase in his savings is : **(SSC CGL Tier I 2004)**

(*a*) 7% (*b*) 21% (*c*) 28% (*d*) 35%

Sol. (*c*) Ratio of income and expenditure = 13500 : 9000 = 3 : 2

$\therefore$ Required percentage = $\frac{ax - by}{x - y} = \frac{3 \times 14 - 2 \times 7}{3 - 2} = 28\%$

174. Sohan's expenditure and savings are in the ratio 3 : 2. His income increases by 12% and savings by 15%. By what per cent would his expenditure increase ?

(*a*) 3% (*b*) 5% (*c*) 8% (*d*) 10%

Sol. (*d*) Ratio of his income and savings = (3 + 2) : 2 = 5 : 2

Per cent increase in expenditure = $\frac{5 \times 12 - 2 \times 15}{5 - 2} = 10\%$

175. Mohan's expenditure and savings are in the ratio 5 : 3. In the next year, his income increases by 25% but his expenses is increased by 60%. If savings has been decreased by ₹ 2000, find his income in the previous year.

(*a*) ₹ 12000 (*b*) ₹ 16000 (*c*) ₹ 24000 (*d*) ₹ 20000

Sol. (*b*)

Income	Expenditure	Savings
$8x$	$5x$	$3x$
$+25\% = 2x$	$+60\% = 3x$	
$10x$	$8x$	

Decrease in savings = $3x - 2x = x =$ ₹2000 ⇒ Original income = $8x = 8 \times 2000 =$ ₹16000

176. The marked price of a shirt and trousers are in the ratio 1 : 2. The shopkeeper gives 40% discount on the shirt. If the total discount on the set of the shirt and trousers is 30%, the discount offered on the trousers is : **(SSC CGL Tier I 2007)**

(*a*) 15% (*b*) 20% (*c*) 25% (*d*) 40%

Sol. (*c*) Let the cost prices of the shirt and trousers be ₹ 100 and ₹ 200 respectively.

Then total discount = ₹ 300 × 30% = ₹ 90

Discount on shirt = ₹ 100 × 40% = ₹ 40. Discount on trousers = ₹ 90 – ₹ 40 = ₹ 50

Required discount rate = $\frac{50}{200} \times 100 = 25\%$

177. A man usually buys goods from market A. In market B, he can get the same goods 10% cheaper. If he spends extra amount of ₹ 55 on transportation in market B and still saves ₹ 145 by buying in market B. What does he pay for the goods ?

(*a*) ₹ 550 (*b*) ₹ 1450 (*c*) ₹ 1800 (*d*) ₹ 2000

Sol. (*c*) His total savings in buying from market B = ₹ 55 + ₹ 145 = ₹ 200

He saves 10% of the bill amount by buying in market B.

∴ Cost of goods purchased to save ₹ 200 = $200 \times \frac{100}{10}$ = ₹ 2000

∴ Amount paid by him = 2000 – 10% of 2000 = ₹ 1800

178. On increasing price of an article by 20%, number of articles sold fell by 20%. What is effect of this on total sales of the article ?

(*a*) 4% decrease (*b*) 4% increase (*c*) 20% decrease (*d*) No effect

Sol. (*a*) Effect on sales amount = $\frac{20 \times 20}{100} = 4\%$ Decrease

179. The cost of an article was ₹ 75. The cost was first increased by 20% and later on it was reduced by 20%. The present cost of the article is **(SSC CGL 2008)**

(*a*) ₹ 72 (*b*) ₹ 75 (*c*) ₹ 78 (*d*) ₹ 76

Sol. (*a*) Net per cent decrease = $\frac{20 \times 20}{100} = 4\%$

∴ Decrease in price = 4% of ₹ 75 = ₹ 3 ∴ New price = ₹ 75 – ₹ 3 = ₹ 72

Alternative method : New price = ₹ $75 \times \frac{120}{100} \times \frac{80}{100}$ = ₹ 72

180. Selling price of a commodity is increased by 15% in a year and then the new price is decreased by 15% in the next year. Find net increase or decrease in the selling price.

(*a*) No effect (*b*) 2% decrease (*c*) 2.25% decrease (*d*) 2.25% increase

Sol. (*c*) Net decrease in the selling price of the commodity = $\frac{15 \times 15}{100} = 2.25\%$ decrease

181. The number of employees working in a farm is increased by 25% and the wages per head are decreased by 25%. If it results in x% decrease in total wages, then the value of x is : **(SSC CGL Tier I 2004)**

(*a*) 0 (*b*) 25 (*c*) 20 (*d*) $\frac{25}{4}$

Sol. (*d*) Net decrease in total wages = $\frac{25 \times 25}{100}\% = \frac{25}{4}\%$

182. A number is first decreased by 10% and then increased by 10%. The number so obtained is 50 less than the original number. The original number is : **(SSC CGL Tier I 2005)**

(*a*) 5900 (*b*) 5000 (*c*) 5500 (*d*) 5050

Sol. (*b*) Net decrease = $\frac{10 \times 10}{100}$ = 1% of original number = 50

Original number = $50 \times \frac{100}{1} = 5000$

183. The price of an article was increased by *r*%. Later the new price was decreased by *r*%. If the latest price was ₹ 1, then the original price was : **(SSC CGL Tier I 2004)**

(*a*) 1 (*b*) $\frac{1-r^2}{100}$ (*c*) $\frac{\sqrt{1-r^2}}{100}$ (*d*) $\frac{10000}{10000-r^2}$

Sol. (*d*) Let original price be ₹ 1. Then net decrease in price = $\frac{r^2}{100^2}$

New price = $1 - \frac{r^2}{100^2} = \frac{10000 - r^2}{10000}$

If new price is ₹ 1, then the original price = $\frac{10000}{10000 - r^2}$

184. A's salary was increased by 40% and then decreased by 20%. On the whole A's salary is increased by **(CDS 2015)**

(*a*) 60% (*b*) 40% (*c*) 20% (*d*) 12%

Sol. (*d*) Net effect = $40 - 20 - \frac{40 \times 20}{100} = 40 - 20 - 8 = 12\%$ increase

Alternative Method: 14 × 8 – 100 = 12%

185. Selling price of a commodity is increased by 20%. As a result of it, quantity sold is dereased by 30%. What is effect of this on total sales?

(*a*) 16% decrease (*b*) 10% increase (*c*) 10% decrease (*d*) 50% increase

Sol. (*a*) Net effect on sales = $20 - 30 - \frac{20 \times 30}{100} = 20 - 30 - 6 = (-)\ 16\%$ i.e. 16% decrease

Alternative Method: 12 × 7 – 100 = – 16%

186. Due to increase in the price of sugar by 40%, a person reduces the consumption by 20%. What is the effect of this on the expenditure? **(RRB Kolkata (TC) 2003)**

(*a*) 12% increase (*b*) 12% decrease (*c*) 20% decrease (*d*) 20% increase

Sol. (*a*) Effect = $40 - 20 - \frac{40 \times 20}{100} = 40 - 20 - 8 = 12\%$ increase

Alternative Method: 14 × 8 = 112; 112 – 100 = 12% (increase)

187. If price of a book is first decreased by 25% and then increased by 20%, the net change in the price of the book will be : **(SSC CGL Tier I 2003)**

(*a*) 10% decrease (*b*) 5% decrease (*c*) No change (*d*) 5% increase

Sol. (*a*) Net effect on price $= -25 + 20 - \frac{25 \times 20}{100} = -25 + 20 - 5 = (-)$ 10% i.e. 10% decrease

Alternative Method: $7.5 \times 12 - 100 = -10\%$

188. The population of a city is decreased by 20% in a year. In the next year, the population is again decreased by 30%. Find per cent decrease in population of the city from the original population.

(*a*) 44% (*b*) 50% (*c*) 55% (*d*) 56%

Sol. (*a*) Net decrease in population $= -20 - 30 + \frac{20 \times 30}{100} = -20 - 30 + 6 = 44\%$ decrease

Alternative Method: $8 \times 7 - 100 = -44\%$

189. The price of an article is reduced by 25% but the daily sale of the article is increased by 30%. The net effect on the daily sales receipts is : **(SSC CGL Tier I 2008)**

(*a*) 2.5% increase (*b*) 2.5% decrease (*c*) 2% increase (*d*) 2% decrease

Sol. (*b*) Net effect $= -25 + 30 - \frac{25 \times 30}{100} = -2.5$ decrease

Alternative method : $7.5 \times 13 - 100 = 97.5 - 100 = -2.5\%$

190. The number of seats in a cinema hall is increased by 25%. The cost of a ticket is also increased by 10%. The overall percentage increase in the revenue is: **(SSC CGL Tier I 2006, 2012M)**

(*a*) 10.5% (*b*) 27.5% (*c*) 37.5% (*d*) 40.5%

Sol. (*c*) Net effect $= 25 + 10 + \frac{25 \times 10}{100} = 37.5$

Alternative method : $12.5 \times 11 - 100 = 137.5 - 100 = 37.5\%$

191. B's income is 20% more than A's and C's income is 25% more than B's. How many per cent is C's income more than A's?

(*a*) 5% (*b*) 40% (*c*) 45% (*d*) 50%

Sol. (*d*) C's income is more than A's by $= 20 + 25 + \frac{20 \times 25}{100} = 20 + 25 + 5 = 50\%$

Alternative Method: $12 \times 12.5 - 100 = 50\%$

192. If B's salary is 30% more than A's and C's salary is 20% less than B's. How many per cent C's salary is more than A's?

(*a*) 4% (*b*) 6% (*c*) 8% (*d*) 10%

Sol. (*a*) C's salary is more than A's by $= 30 - 20 - \frac{30 \times 20}{100} = 30 - 20 - 6 = 4\%$

Alternative Method: $13 \times 8 - 100 = 4\%$

193. Annual income of a person is increased by 30% but rate of income tax is decreased by 20%. Find the effect on this on his tax liability.

(*a*) 6% decrease (*b*) 4% increase (*c*) 4% decrease (*d*) 10% increase

Sol. (*b*) Effect on tax liability $= 30 - 20 - \frac{30 \times 20}{100} = 4\%$ increase

Alternative Method: $13 \times 8 - 100 = 4\%$

194. The population of a town is 1,50,000. It increases by 15% in the first year and decreases by 20% in the second year. What will be the size of the population at the end of second year ? **(Bank Clerical)**

(*a*) 1,38,000 (*b*) 1,42,500 (*c*) 2,07,000 (*d*) None of these

Sol. (*a*) Population at the end of second year = $1,50,000 \times \frac{115}{100} \times \frac{80}{100} = 1,38,000$

Alternative Method: Net Increase = $15 - 20 - \frac{15 \times 20}{100} = -8\%$

$\Rightarrow$ Total decrease = 8% of 1,50,000 = 12,000

$\therefore$ Population at the end of second year = 1,50,000 – 12,000 = 1,38,000

195. A cow is given rich diet and gains 15% in weight. She is then put on a poor diet and loses 20% in weight. On the whole the cow **(NIFT)**

(*a*) loses 8% weight (*b*) gains 6% weight (*c*) loses 5% weight (*d*) loses 10% weight

Sol. (*a*) $+15 - 20 - \frac{15 \times 20}{100} = -8\%$

196. The price of a commodity was increased in the ratio 4 : 5. If a customer reduces its consumption in the ratio 4 : 3, how much per cent his expenditure on the commodity will increase or decrease?

(*a*) 1% increase (*b*) 1% decrease (*c*) 6.25% increase (*d*) 6.25% decrease

Sol. (*d*) Expenditure in first case = 4 × 4 = 16 ⇒ Expenditure in second case = 5 × 3 = 15

Decrease in expenditure = (16 – 15) ratio = 1 ratio

$\therefore$ Percentage decrease in expenditure $= \frac{1}{16} \times 100 = \frac{25}{4} = 6.25\%$

197. A man's income is increased by ₹ 2000 in a year. Rate of income-tax having reduced from 25% to 20%, his tax-liability remains the same as in the last year. Find his income in the current year

(*a*) ₹ 4000 (*b*) ₹ 5000 (*c*) ₹ 8000 (*d*) ₹ 10000

Sol. (*d*) Ratio of tax rates = 25% : 20% = 5 : 4 ⇒ Ratio of incomes = 4 : 5

Difference = (5 – 4) ratio = 1 ratio = ₹ 2000

Income in current year = 5 ratio = 5 × ₹ 2000 = ₹ 10000

198. Rate of commission is increased from 9% to 10%, but due to decrease in sale amount by ₹1000, a selling agent got the same commission as in the last year. Find the amount of sale achieved by him in the last year.

(*a*) ₹ 8000 (*b*) ₹ 9000 (*c*) ₹ 10000 (*d*) ₹ 12500

Sol. (*c*) Ratio of commission = 9 : 10 ⇒ Ratio of sales = 10 : 9

Difference = (10 – 9) ratio = 1 ratio = ₹1000 ⇒ Sales in first year = 10 × ₹1000 = ₹10000

199. A person could save 10% of his income. But 2 years later, when his income increased by 20%, he could save the same amount only as before. By how much percentage has his expenditure increased ? **(CDS 2015)**

(*a*) $22\frac{2}{9}\%$ (*b*) $23\frac{1}{3}\%$ (*c*) $24\frac{2}{9}\%$ (*d*) $25\frac{2}{9}\%$

Sol. (*a*) Let original income = ₹100. Then savings = ₹10 ⇒ Expenditure = 100 – 10 = ₹90

Increase in expenditure = Increase in income = 20% of ₹100 = ₹20

$\therefore$ Required % $= \frac{20}{90} \times 100 = 22\frac{2}{9}\%$

200. The sum of the number of boys and girls in a school is 150. If the number of boys is x, the number of girls becomes $x\%$ of the total number of students. The number of boys is : **(SSC CGL Tier I 2002)**

(*a*) 90 (*b*) 40 (*c*) 45 (*d*) 60

Sol. (*d*) $x + \frac{x}{100} \times 150 = 150 \Rightarrow \frac{5}{2}x = 150 \Rightarrow x = \frac{150 \times 2}{5} = 60$

201. A man saves 20% of his salary. If on increase in prices, he has to increase his expenses by 20%, and hence he could save ₹ 30 only. What is his salary?

(*a*) ₹ 500 (*b*) ₹ 750 (*c*) ₹ 900 (*d*) None of these

Sol. (*b*) Original expenses = 100 – 20 = 80% ⇒ Increased expenses = $80 \times \frac{120}{100} = 96\%$

∴ Savings = 4% of salary = ₹ 30 ⇒ Salary = $30 \times \frac{100}{4}$ = ₹ 750

202. Price of an article increases by 20%. As a result turnover increases by 12%. Find the derease in quantity sold.

(*a*) 5% (*b*) 6.67% (*c*) 8% (*d*) 16%

Sol. (*b*) Let original price = ₹100 and turnover = ₹100

Increased price = 100 + 20 = 120%. New turnover = 100 + 12 = 112%

Quantity sold after increase in price $= \frac{112}{120} \times 100 = \frac{280}{3}$

∴ Decrease in quantity sold $= 100 - \frac{280}{3} = \frac{20}{3} = 6.67\%$

203. A sells his goods 20% cheaper than B but 20% dearer than C. How much a customer of B saves, if he buys goods from C for ₹ 100?

(*a*) ₹ 4 (*b*) ₹ 40 (*c*) ₹ 44 (*d*) ₹ 50

Sol. (*d*) If C sells goods for ₹ 100, then A charges = ₹ 100 + ₹ 20 = ₹ 120

If B sells goods for ₹ 100, then A charges = ₹ 100 – ₹ 20 = ₹ 80

∴ Amount charged by B (for goods worth ₹ 100 of C) = $100 \times \frac{120}{100} \times \frac{100}{80}$ = ₹ 150

∴ Savings = ₹ 150 – ₹ 100 = ₹ 50

204. If A's income is $\frac{5}{6}$th of B's income, then B's income is more than A's income by **(SSC CHSL 2012)**

(*a*) $\frac{25}{2}\%$ (*b*) 15% (*c*) 20% (*d*) 22%

Sol. (*c*) A : B = 5 : 6 ∴ Required percentage = $\frac{1}{5} \times 100 = 20\%$

205. The price of sugar is increased by 25%. If a family wants to keep its expenses on sugar unaltered, then the family will have to reduce the consumption of sugar by : **(SSC CHSL 2010, SSC, FCI Assit 2012, SAS 2010)**

(*a*) 20% (*b*) 21% (*c*) 22% (*d*) 25%

Sol. (*a*) Increase in price = $25\% = \frac{1}{4}$ ⇒ Reduction in consumption = $\frac{1}{4+1} = \frac{1}{5} = 20\%$

206. X's income is 20% more than that of Y. What per cent is Y's income less than X ?

(SSC CGL Tier I 2002, CDS 2010 modified, CHSL 2013 M)

(*a*) $83\frac{1}{3}\%$ (*b*) $16\frac{2}{3}\%$ (*c*) $83\frac{2}{3}\%$ (*d*) $16\frac{1}{3}\%$

Sol. (*b*) Let Y's income = 100, then X's income = 120 ⇒ X – Y = 20

$\therefore$ Required percentage = $\frac{20}{120} \times 100 = 16\frac{2}{3}\%$

Alternative Method : X's income is more than Y's income by 20% = $\frac{1}{5}$

$\therefore$ Y's income is less than X's income by $\frac{1}{5+1} = \frac{1}{6} = 16\frac{2}{3}\%$

207. Price of a commodity is reduced by 25%. By how much new price must be increased by to restore it to the original price?

(*a*) 20% (*b*) 25% (*c*) $33\frac{1}{3}\%$ (*d*) 50%

Sol. (*c*) Decrease in price = 25% = $\frac{25}{100} = \frac{1}{4} \Rightarrow$ Price is to be increased by $\frac{1}{4-1} = \frac{1}{3} = 33\frac{1}{3}\%$

Alt: Reduction in price = 25% = $\frac{25}{100} \Rightarrow$ Increase in price = $\frac{25}{100-25} = \frac{1}{3} = 33\frac{1}{3}\%$

208. From 2008 to 2009, the sales of a book decreased by 80%. If the sales in 2010 were the same as in 2008, by what per cent did it increase from 2009 to 2010? **(SSC CHSL 2012)**

(*a*) 80% (*b*) 100% (*c*) 120% (*d*) 400%

Sol. (*d*) Let sales in 2008 = 100. Then sales in 2009 = 100 – 80 = 20

Sales in 2010 = 100. Increase in sale = 100 – 20 = 80

Per cent increase = $\frac{80}{20} \times 100 = 400\%$

209. The price of cooking oil is increased by 25%. By what per cent a family should reduce the consumption of cooking oil so as not to increase the expenditure on this account ?

(FMS Delhi 2003, SSC CGL Tier I 2002)

(*a*) 20 (*b*) 25 (*c*) 50 (*d*) 40

Sol. (*a*) Increase in price = 25% = $\frac{25}{100} = \frac{1}{4} \Rightarrow$ Decrease in consumption = $\frac{1}{4+1} = \frac{1}{5} = 20\%$

Alternative method : Let original price = ₹ 100. Then new price = 100 + 25 = ₹ 125

Decrease in consumption = $\frac{25}{125} = \frac{1}{5} = 20\%$

Alternative Method: New price = $1 + \frac{1}{4} = \frac{5}{4}$ of original price

$\therefore$ New quantity with the same amount = $\frac{4}{5}$ of the original quantity

$\therefore$ Decrease in quantity = $1 - \frac{4}{5} = \frac{1}{5} = 20\%$

210. The price of coal is increased by 20%. By what per cent a family should decrease its consumption so that expenditure remains same ? **(SSC CGL Tier I 2001)**

(*a*) 40% (*b*) $23\frac{1}{3}\%$ (*c*) 20% (*d*) $16\frac{2}{3}\%$

Sol. (*d*) Increase in price = 20% = $\frac{20}{100} = \frac{1}{5}$ ⇒ Decrease in consumption = $\frac{1}{5+1} = \frac{1}{6} = 16\frac{2}{3}\%$

Alternative method : Let original price = ₹ 100. Then new price = 100 + 20 = ₹ 120

Decrease in consumption = $\frac{20}{120} = \frac{1}{6} = 16\frac{2}{3}\%$

211. If food prices go up by 10%, by how much should a man reduce his consumption so as not to increase his expenditure ? **(SSC CGL Tier I 2000)**

(*a*) $9\frac{1}{11}\%$ (*b*) 10% (*c*) $11\frac{1}{9}\%$ (*d*) Data insufficient

Sol. (*a*) Increase in price = 10% = $\frac{1}{10}$ ⇒ Decrease in consumption = $\frac{1}{10+1} = \frac{1}{11} = 9\frac{1}{11}\%$

212. If price of oil is increased by $33\frac{1}{3}\%$, by how much a housewife must reduce consumption of oil so that her expenditure on oil remains the same?

(*a*) $33\frac{1}{3}\%$ (*b*) 25% (*c*) 40% (*d*) 50%

Sol. (*b*) New price = $1 + \frac{1}{3} = \frac{4}{3}$ of original price

∴ New quantity with the same amount = $\frac{3}{4}$ of the original quantity

∴ Reduction in quantity = $1 - \frac{3}{4} = \frac{1}{4}$ = 25%

Alt: Price increased by $33\frac{1}{3}\% = \frac{1}{3}$ ⇒ Consumption decreased by $\frac{1}{3+1} = \frac{1}{4}$ = 25%

213. The price of a commodity rises from ₹ 6 per kg to ₹ 7.50 per kg. If the expenditure cannot increase, then percentage of reduction in consumption is : **(SSC CGL Tier I 2011)**

(*a*) 15 (*b*) 20 (*c*) 25 (*d*) 30

Sol. (*b*) New price = $\frac{7.5}{6} = \frac{5}{4}$ of original price ⇒ New quantity = $\frac{4}{5}$ of original quantity

⇒ Reduction in consumption = $\frac{1}{5}$ = 20%

214. The salary of a person was reduced by 10%. By what per cent should his reduced salary be raised so as to bring it at par with his original salary ? **(SSC CGL Tier I 2004)**

(*a*) 9% (*b*) $11\frac{1}{9}\%$ (*c*) $9\frac{1}{11}\%$ (*d*) 11%

Sol. (*b*) Reduction in salary = 10% = $\frac{1}{10}$ ⇒Increase in salary to restore = $\frac{1}{10-1} = \frac{1}{9} = 11\frac{1}{9}\%$

Alt: Let original salary = ₹ 100. Then reduced salary = ₹ 90

⇒ Required increase = ₹10 on ₹90 ⇒ Required % = $\frac{10}{90} \times 100 = 11\frac{1}{9}\%$

215. When water is frozen to ice, its volume increases by 10%. What per cent of the volume of ice decreases when it melts to water ? **(SSC CGL Tier I 2012)**

(*a*) $10\frac{1}{11}\%$ (*b*) $6\frac{1}{5}\%$ (*c*) $9\frac{1}{11}\%$ (*d*) 11%

Sol. (*c*) Increase in volume $= 10\% = \frac{1}{10} \Rightarrow$ Decrease in volume $= \frac{1}{10+1} = \frac{1}{11} = 9\frac{1}{11}\%$

216. The difference of two numbers is 20% of the larger number. If the smaller number is 20, the larger number is: **(SSC CGL Tier I 2000)**

(*a*) 25 (*b*) 45 (*c*) 50 (*d*) 80

Sol. (*a*) Difference between the numbers is $20\% = \frac{1}{5}$ of the larger number.

$\therefore$ Difference $= \frac{1}{5-1} = \frac{1}{4}$ of smaller number $= \frac{1}{4} \times 20 = 5 \Rightarrow$ Larger number $= 20 + 5 = 25$

217. The difference of two numbers is 25% of the larger number. If the smaller number is 30, then the larger number is: **(SAS 2010)**

(*a*) 40 (*b*) 45 (*c*) 48 (*d*) 50

Sol. (*a*) Ratio of two numbers = 75 : 100 = 3 : 4

Smaller number = 3 ratio = 30 $\Rightarrow$ Larger number = 4 ratio = 40

218. Price of cloth having been raised by 75%, by how much per cent a householder must reduce his consumption of cloth so as not to increase his expenditure?

(*a*) $42\frac{6}{7}\%$ (*b*) $57\frac{1}{7}\%$ (*c*) 75% (*d*) 50%

Sol. (*a*) Increased in price $= 75\% = \frac{3}{4} \Rightarrow$ Decrease in consumption $= \frac{3}{4+3} = \frac{3}{7} = 42\frac{6}{7}\%$

Alt: $\frac{75}{100+75} \times 100 = \frac{3}{7} \times 100 = 42\frac{6}{7}\%$

219. Price of cloth having been reduced by 75%, how much per cent a householder can increase his consumption of cloth so that his expenditure on the cloth remains the same?

(*a*) 50% (*b*) 75% (*c*) 200% (*d*) 300%

Sol. (*d*) Decrease in price $= 75\% = \frac{3}{4} \Rightarrow$ Increase in consumption $= \frac{3}{4-3} = \frac{3}{1} = 300\%$

Alt: $\frac{75}{100-75} \times 100 = \frac{75}{25} \times 100 = 300\%$

220. A number is increased by 20% and then again by 20%. By what per cent should the increased number be reduced so as to get back the original number ? **(SSC CGL Tier I 2004)**

(*a*) $30\frac{5}{9}\%$ (*b*) $19\frac{11}{13}\%$ (*c*) 40% (*d*) 44%

Sol. (*a*) Net increase $= 20 + 20 + \frac{20 \times 20}{100} = 44\%$

Let the original number be 100. Then new number = 144

Required percentage $= \frac{44}{144} \times 100 = \frac{275}{9} = 30\frac{5}{9}\%$

221. If the price of sugar is increased by 50%, by how much per cent a housewife must reduce the consumption of sugar so that her expenditure on sugar is increased by 20% only.

(*a*) 10% (*b*) 12% (*c*) 20% (*d*) 25%

Sol. (*c*) New Price = 150% = $\frac{3}{2}$ of original price

∴ Consumption with same amount = $\frac{2}{3}$ of original consumption

∴ New Expenditure = 100 + 20 = 120% of original expenditure

∴ New consumption = $\frac{2}{3}$ × 120% = 80%

⇒ Decrease in consumption = 100% – 80% = 20%

222. If the price of sugar is increased by 25%, by how much per cent a housewife must reduce the consumption of sugar so that her expenditure on sugar is decreased by 10%?

(*a*) 35% (*b*) 30% (*c*) 28% (*d*) 25%

Sol. (*c*) New price = 125% = $\frac{5}{4}$ of original price

New consumption with same amount = $\frac{4}{5}$ of original consumption

∴ New consumption = $\frac{4}{5}$ × 90% = 72%

∴ Decrease in consumption = 100% – 72% = 28%

223. If the price of an article is reduced by 10%, by how much per cent will the consumption of the article be increased so that total expenditure increases by 8%?

(*a*) 2% (*b*) 9% (*c*) 18% (*d*) 20%

Sol. (*d*) Let original price be = ₹ 100. Then reduced price = ₹ 90

∴ New consumption = $\frac{100}{90}$ of original consumption

∴ Consumption with 8% extra expenditure = $\frac{100}{90}$ × 108% = 120%

∴ Increase in consumption = (120 – 100)% = 20%

224. Price of admission ticket is reduced by 20%. As a result, total collection amount is increased by 60%. Find per cent increase in the sale of tickets.

(*a*) 24% (*b*) 40% (*c*) 80% (*d*) 100%

Sol. (*d*) New Price = 1 – $\frac{1}{5}$ = $\frac{4}{5}$ of original price

∴ Tickets sold for same collection = $\frac{5}{4}$ of original sale of tickets

But new collection = (100 + 60)% = 160% of original collection

∴ New sale = $\frac{5}{4}$ × 160 = 200% of original sale of tickets

∴ Increase in sale of tickets = 200% – 100% = 100% of original

Alternative method : New price = 80%. New collection = 160%. New sales = $\frac{160}{80}$ = 2 times

225. If the price of a commodity is reduced by 10%. How much quantity of the commodity can be bought with the same money which was sufficient to buy 90 kg. at the original price?

(*a*) 80 kg (*b*) 81 kg (*c*) 99 kg (*d*) 100 kg

Sol. (*d*) Reduction in the price = 10% = $\frac{1}{10}$

$\Rightarrow$ Increase in quantity = $\frac{1}{9}$ of original quantity = $\frac{1}{9} \times 90$ kg. = 10 kg.

$\therefore$ Increased quantity = 90 + 10 = 100 kg.

Alt: Let original price = ₹100 $\Rightarrow$ new price = ₹90 $\Rightarrow$ New quantity = $90 \times \frac{100}{90}$ = 100 kg.

Alternative Method: Ratio of prices = 100 : 90 = 10 : 9 $\Rightarrow$ Ratio of quantities = 9 : 10
Original quantity = 9 ratio = 90 kg $\Rightarrow$ 1 ratio = 10 kg
New quantity = 10 ratio = 10 × 10 kg = 100 kg

226. Price of sugar having been fallen by 10%, a consumer can buy 10 kg. more of sugar than what he was buying originally. Find the quantity he was buying originally.

(*a*) 80 kg (*b*) 90 kg (*c*) 100 kg (*d*) 110 kg

Sol. (*b*) Fall in price = 10% = $\frac{1}{10}$ $\Rightarrow$ Quantity bought is increased by $\frac{1}{9}$ of original quantity.

$\Rightarrow$Original quantity is 9 times of increase in quantity $\Rightarrow$ Original quantity = 9×10kg = 90kg
Alternative Method: Ratio of prices = 100 : 90 = 10 : 9 $\Rightarrow$ Ratio of quantities = 9 : 10
Difference = (10 – 9)ratio = 1 ratio = 10 kg $\Rightarrow$ Original quantity = 9 ratio = 9×10 kg = 90 kg

227. Price of a commodity having been increased by 20%, a consumer can buy 10 kg. less than the quantity he was buying originally. Find the quantity he was buying originally.

(*a*) 60 kg (*b*) 80 kg (*c*) 90 kg (*d*) 100 kg

Sol. (*a*) Increase in price = 20% = $\frac{1}{5}$

$\therefore$ Quantity bought is to be reduced by $\frac{1}{6}$ of original quantity.

$\therefore$ Original quantity = 6 times of decrease in quantity$\Rightarrow$Original quantity = 6×10 kg= 60 kg
Alternative Method: Ratio of prices = 100 : 120 = 5 : 6 $\Rightarrow$ Ratio of quantities = 6 : 5
Difference = (6 – 5) ratio = 1 ratio = 10 kg.
Original quantity = 6 ratio = 6 × 10 kg = 60 kg

228. If price of sugar is fallen by 5%, a man can buy 220 kg sugar. Find the quantity of sugar he was buying originally.

(*a*) 209 kg (*b*) 210 kg (*c*) 240 kg (*d*) 242 kg

Sol. (*a*) Fall in price = 5% = $\frac{1}{20}$ $\Rightarrow$ Increase in quantity = $\frac{1}{19}$ of the original quantity

= $\frac{1}{20}$ of the increased quantity = $\frac{1}{20} \times 220$ kg = 11 kg.

$\therefore$ Original quantity of sugar = 220 – 11 = 209 kg.

Alt: Ratio of prices = 100 : 95 $\Rightarrow$ Ratio of quantities = 95 : 100 = 19 : 20

New quantity = 20 ratio = 220 kg $\Rightarrow$ 1 ratio = 11 kg

Original quantity = 19 ratio = 19 × 11 kg = 209 kg

229. The price of rice is reduced by 2%. How many kilograms of rice can now be bought for the money which was sufficient to buy 49 kg of rice earlier? **(SSC CGL Tier I 2004)**

(*a*) 48 kg (*b*) 49 kg (*c*) 50 kg (*d*) 51 kg

Sol. (*c*) Ratio of prices = 100 : 98 ⇒ Ratio of quantities = 98 : 100 = 49 : 50

If original quantity was 49 kg, then new quantity = $\frac{50}{49} \times 49 = 50$ kg.

230. The price of sugar falls by 20%. How much sugar can be bought now with the money that was sufficient to buy 20 kg of sugar originally?

(*a*) 25 kg (*b*) 15 kg (*c*) 20 kg (*d*) 30 kg

Sol. (*a*) Ratio of prices = 100 : 80 = 5 : 4 ⇒ Ratio of quantities = 4 : 5
⇒ Original quantity = 4 ratio = 20 kg ⇒ 1 ratio = 5 kg
⇒ New quantity = 5 ratio = 5 × 5 kg = 25 kg

231. If price of rice is increased by 10%, a housewife can buy 110 kg. rice for a certain amount. Find quantity of rice she was buying at original price.

(*a*) 99 kg (*b*) 100 kg (*c*) 110 kg (*d*) 121 kg

Sol. (*d*) Ratio of prices = 10 : 11 ⇒ Ratio of quantities = 11 : 10

If new quantity is 110 kg, then original quantity = $\frac{11}{10} \times 110 = 121$ kg.

232. Price of sugar having been fallen by 10%, a consumer can buy 22 kg. more than before. Had the price been increased by 10%, how much quantity of sugar could he have bought for the same sum?

(*a*) 180 kg (*b*) 200 kg (*c*) 220 kg (*d*) 250 kg

Sol. (*a*) **Situation I:** Decrease in price = 10% = $\frac{1}{10}$

∴ Quantity bought is increased by $\frac{1}{9}$ of the original quantity

∴ Original quantity is 9 times of increase in quantity. ⇒ Original quantity = (9 × 22) kg.

Situation II: Increase in price = 10% = $\frac{1}{10}$ of original price.

∴ Quantity bought is to be decreased by $\frac{1}{11}$ of the original quantity

∴ Decreased quantity = $1 - \frac{1}{11} = \frac{10}{11}$ of the original quantity = $(9 \times 22) \times \frac{10}{11} = 180$ kg.

Alt : Situation I: Ratio of prices = 100 : 90 ⇒ Ratio of quantities = 90 : 100 = 9 : 10
Difference = (10 – 9) ratio = 1 ratio = 22 kg. ∴ Original quantity = 9 × 22 kg
Situation II : Ratio of prices = 100 : 110 = 10 : 11 ⇒ Ratio of quantities = 11 : 10

New quantity = $(9 \times 22) \times \frac{10}{11} = 180$ kg.

233. A, B and C spend 70%, 75% and 80% respectively of their salaries. Their actual savings are in the ratio of 4 : 5 : 6. What is ratio of their salaries?

(*a*) 4 : 5 : 6 (*b*) 4 : 6 : 9 (*c*) 6 : 5 : 4 (*d*) 9 : 6 : 4

Sol. (*b*) Let their savings are $4x$, $5x$ and $6x$ respectively.

Then (100 – 70)% = 30% of A's salary = $4x$ ⇒ A's salary = $4x \times \frac{100}{30} = \frac{40x}{3}$

$(100 - 75)\% = 25\%$ of B's salary $= 5x \Rightarrow$ B's salary $= 5x \times \frac{100}{25} = 20x$

$(100 - 80)\% = 20\%$ of C's salary $= 6x \Rightarrow$ C's salary $= 6x \times \frac{100}{20} = 30x$

$\therefore$ Ratio of their salaries $= \frac{40}{3} : 20 : 30 = 4 : 6 : 9$

Short-cut Method: Ratio of their actual savings = 4 : 5 : 6

Ratio of their savings in terms of theirs salaries = 30 : 25 : 20 = 6 : 5 : 4

$$\text{Ratio of salaries} = \frac{\text{Ratio of actual savings}}{\text{Ratio of their savings in terms of salaries}}$$

$$= \frac{4}{6} : \frac{5}{5} : \frac{6}{4} = \frac{2}{3} : 1 : \frac{3}{2} = 4 : 6 : 9$$

234. Three persons A, B, C whose salaries together amount to ₹ 72,000 spend 80, 85 and 75 percent of their salaries respectively. If their savings are in the ratio 8 : 9 : 20, then A's salary is: **(SSC CHSL 2011)**

(*a*) ₹ 20,000 (*b*) ₹ 16,000 (*c*) ₹ 22,000 (*d*) ₹ 18,000

Sol. (*b*) Ratio of actual savings = 8 : 9 : 20

Ratio of savings to respective salary = 20 : 15 : 25 = 4 : 3 : 5

Ratio of salaries $= \frac{8}{4} : \frac{9}{3} : \frac{20}{5} = 2 : 3 : 4$ $\therefore$ A's salary $= \frac{2}{9} \times$ ₹ 72,000 = ₹ 16,000

235. Income-tax rate is 20% in a particular year. If in the next year, a person's income is increased by 25% but total tax liability remaining the same, find tax-rate during the next year.

(*a*) 15% (*b*) 16% (*c*) 20% (*d*) 25%

Sol. (*b*) Let his income in first year = ₹ 100

Then income in the next year = ₹ 100 + 25% of ₹ 100 = ₹ 125

Let tax-rate in the second year be x%. Then 20% of ₹ 100 = x% of ₹ 125

$\therefore x = 20\% \times \frac{100}{125} = 16\%$

Alternative Method: Increase in income in second year $= 25\% = \frac{1}{4}$

$\therefore$ Decrease in tax-rate so that tax liability remains same $= \frac{1}{4+1} = \frac{1}{5}$

$\therefore$ Decrease in tax-rate in second year $= \frac{1}{5} \times 20\% = 4\%$

$\therefore$ Tax-rate in second year = 20% – 4% = 16%

Alternative Method: Ratio of salary = 100 : 125 = 4 : 5 ⇒ Ratio of tax rate = 5 : 4

Previous tax rate = 5 ratio = 20% ⇒ 1 ratio = 20% ÷ 5 = 4%

New tax rate = 4 ratio = 4 × 4% = 16%.

236. The income of a sales agent remains the same even though the rate of commission is increased from 4% to 5%. Find the per cent decrease in the sales.

(*a*) 1% (*b*) 5% (*c*) 20% (*d*) 25%

Sol. (*c*) Increase in the rate of commission = 5% – 4% = 1%

$\therefore$ Relative increase in the rate of commission $= \frac{5-4}{4} = \frac{1}{4}$

$\therefore$ Decrease in the sales $= \dfrac{1}{4+1} = \dfrac{1}{5} = 20\%$

Alt: Ratio of commission = 4:5$\Rightarrow$Ratio of sales = 5:4 $\Rightarrow$ Decrease in sales $= \dfrac{5-4}{5} = 20\%$

237. The income-tax rate is increased from 4% to 5% but the tax liability of a person is increased by 10% only. Find the per cent decrease in his taxable income.

(*a*) 10% (*b*) 12% (*c*) 15% (*d*) 18%

Sol. (*b*) Let tax in the first case = ₹ 100. Then new tax $= 100 \times \dfrac{5}{4}$ = ₹ 125

But actual revised tax = ₹ 110 $\Rightarrow$ Reduced income $= \dfrac{110}{125} \times 100 = 88\%$

$\therefore$ Decrease in income = 100 – 88 = 12%

238. If the income tax is increased by 19%, the net income is reduced by 1%. The rate of income tax is : **(SSC CGL Tier I 2007)**

(*a*) 4% (*b*) 5% (*c*) 6% (*d*) 7.2%

Sol. (*b*) Let total income be ₹100 and income-tax rate be x%.

Total tax = 100 × x% = ₹ $x \Rightarrow$ Net income = ₹ $(100 - x)$

Given 19% of tax = 1% of net income

$\therefore$ 19% of x = 1% of $(100 - x) \Rightarrow 19x = (100 - x) \Rightarrow 20x = 100 \Rightarrow x = 5$

239. The population of a village is increased by 5% in one year and in the next year it is decreased by 5%. If at the end of second year, the population is 39,900, what was the population at the beginning of first year?

(*a*) 39,000 (*b*) 39,500 (*c*) 39,800 (*d*) 40,000

Sol. (*d*) Population in the beginning of first year

$$= 39{,}900 \times \frac{100}{100+5} \times \frac{100}{100-5} = 39{,}900 \times \frac{20}{21} \times \frac{20}{19} = 40{,}000$$

Alt : Decrease in population in 2 years $= \dfrac{5}{100} \times \dfrac{5}{100} = \dfrac{1}{400}$ of original population

$= \dfrac{1}{400-1} = \dfrac{1}{399}$ of the population at the end of two years

$\therefore$ Decrease in population $= \dfrac{1}{399} \times 39{,}900 = 100$

$\therefore$ Population at the beginning = 39,900 + 100 = 40,000.

240. The present population of a village is 5,500. If the number of males increases by 11% and the number of females increases by 20%, then the population will become 6,330. What is the present population of females in the village ? **(UPSC Civil Services Pre. 2011)**

(*a*) 2,000 (*b*) 2,500 (*c*) 3,000 (*d*) 3,500

Sol. (*b*) 11% of 5,500 = 605 $\Rightarrow$ Increased population = 5500 + 605 = 6105

Short from actual = 6330 – 6105 = 225 which is equal to (20 – 11)% = 9% of female population

Female population = 225 ÷ 9% = 2500

Alt: Going through the options, Let females = 2500, then males = 5500 – 2500 = 3000

Then increased population = 3000 + 11%×3000 + 2500 + 20%×2500 = 3330 + 3000 = 6330

CHAPTER

15

DISCOUNT

Discount: Selling an article at a price lower than its marked or printed price. If a shopkeeper allows x% discount on the marked price of an article, then selling price of the article = $(100 - x)$% of the marked price. Discount is offered on the Marked Price.

$$\text{Discount} = \frac{\text{Rate of Discount}}{\text{Marked price}} \times 100 \quad or \quad \text{Discount} = \text{Marked price} - \text{Selling price}$$

Selling price = Marked price – Discount

$$\text{Selling price} = \text{Marked price} \times \frac{100 - \text{Discount }\%}{100}$$

$$\text{Marked price} = \text{Selling price} \times \frac{100}{100 - \text{Discount }\%}$$

Example: A man bought an article for ₹ 21. What was the marked price of the article if he had bought the article at 30% discount?

(*a*) ₹ 30 (*b*) ₹ 6.30 (*c*) ₹ 14.70 (*d*) ₹ 70

Sol. (*a*) Marked price of the article $= 21 \times \frac{100}{100 - 30} = 21 \times \frac{100}{70} =$ ₹ 30

Series of discount: More than one discount may be allowed on an article. In such case, the first discount is allowed on the total price of the article and then the second discount is allowed on the amount payable after the first discount.

Amount payable after a series of two discounts

$$= \text{Marked Price} \times \frac{100 - \text{First Discount}}{100} \times \frac{100 - \text{Second Discount}}{100}$$

Note: This formula can be used for any number of discount series.

Finding single discount rate equivalent to the series of two discounts

$$= \text{First Discount} + \text{Second Discount} - \frac{\text{First Discount} \times \text{Second Discount}}{100}$$

Example: Successive discounts of 10% and 20% are equivalent to a single discount of :

(CDS 2004, SAS 2010, SSC CGL Tier I 2002, SSC MTS 2014 M)

(*a*) 30% (*b*) 15% (*c*) 28% (*d*) 72%

Sol: (*c*) Single equivalent discount rate $= 10 + 20 - \frac{10 \times 20}{100} = 30 - 2 = 28\%$

How to find discount rate equivalent to a series of more than two discount rates?

Method: 1. Find discount rate equivalent to series of any two given discount rates.

2. Find discount rate equivalent to third discount and discount equivalent to first two discounts.

Example: Find a simple discount equivalent to a discount series of 10%, 20% and 25%. **(SSC CGL Tier I 2015)**

(*a*) 45% (*b*) 55% (*c*) 52% (*d*) 46%

Sol (*d*): Discount rate equivalent to 10% and 20% $= 10 + 20 - \frac{10 \times 20}{100} = 10 + 20 - 2 = 28\%$

Discount equivalent to 25% and 28% $= 25 + 28 - \frac{25 \times 28}{100} = 25 + 28 - 7 = 46\%$

∴ Discount equivalent to 10%, 20% and 25% is 46%.

Alternative Method: Let Marked Price = ₹ 100

Selling price $= 100 \times \frac{100-10}{100} \times \frac{100-20}{100} \times \frac{100-25}{100} = 100 \times \frac{90}{100} \times \frac{80}{100} \times \frac{75}{100} =$ ₹ 54

∴ Single equivalent discount = 100% – 54% = 46%

SOLVED EXERCISE

1. If A man allows 20% discount on all his articles, the number of the articles sold is increased by 20%. What is the effect on total sale?

(*a*) 40% Increase (*b*) 4% Increase (*c*) 4% Decrease (*d*) No effect

Sol. (*c*) Effect on sales $= \frac{20 \times 20}{100} = 4\%$ decrease

2. A cloth store is offering "Buy 3, get 1 free". What is the net percentage discount being offered by the store? **(CDS 2012)**

(*a*) 20% (*b*) 25% (*c*) 30% (*d*) $33\frac{1}{3}\%$

Sol. (*b*) Let MP of 1 cloth = 100 ⇒ MP of 4 cloths = 400 ⇒ Discount = 400 – 300 = 100

⇒ Rate of discount $= \frac{100}{400} \times 100 = 25\%$

3. List price of a book is ₹100. A dealer sells three such books for ₹274.50 after allowing discount at a certain rate Find the rate of discount. **(SSC CGL Tier I 2015)**

(*a*) 8.16% (*b*) 8.5% (*c*) 8.33% (*d*) 8.34%

Sol. (*b*) SP of one book = ₹ 274.50 ÷ 3 = ₹ 91.50 ⇒ Rate of discount = 100 – 91.50 = 8.5%

4. Rahim bought a TV with 20% discount on list price. Had he bought it with 25% discount, he would have saved ₹500. At what price did he buy the T.V? **(SSC CHSL 2011)**

(*a*) ₹16,000 (*b*) ₹12,000 (*c*) ₹10,000 (*d*) ₹8000

Sol. (*d*) Difference = 25% – 20% = 5% = ₹500 (i.e. 100 times)

Price paid = 80% = 80 × 100 = ₹8000

5. Deepa bought a calculator with 30% discount on the listed price. Had she not got the discount, she would have paid ₹ 82.50 extra. At what price did she buy the calculator ? **(SBI Associate Bank PO 2007, Canara Bank PO 2003)**

(*a*) ₹ 192.50 (*b*) ₹ 275 (*c*) ₹ 117.85 (*d*) None of these

Sol. (*a*) Discount = 30% of listed price = ₹ 82.50

Price paid = 70% of listed price = ₹ 82.50 × $\frac{70}{30}$ = ₹ 192.50

6. Two successive discounts of $a\%$ and $b\%$ on the marked price of an article are equivalent to the single discount of **(SSC CGL Tier II 2013, CHSL 2011)**

(*a*) $(a + b)\%$ (*b*) $\left(a+b-\frac{ab}{100}\right)\%$ (*c*) $\frac{a+b}{100}\%$ (*d*) $\frac{a+b}{2}\%$

Sol. (*b*) $\left(a+b-\frac{ab}{100}\right)\%$

7. The equivalent single discount for two successive discounts of 15% and 10% is : **(SSC CGL Tier I 2003)**

(*a*) 25% (*b*) 20% (*c*) 23.5% (*d*) 20.5%

Sol. (*c*) Equivalent Discount $= 15 + 10 - \frac{15\times10}{100} = 25 - 1.5 = 23.5\%$

8. Successive discounts of 50% and 50% is equivalent to **(SSC CHSL 2013)**

(*a*) 100% (*b*) 75% (*c*) 50% (*d*) 25%

Sol. (*b*) Equivalent discount $= 50 + 50 - \frac{50\times50}{100} = 100 - 25 = 75\%$

9. A single discount equivalent to a discount series of 10%, 20% and 30%. **(SSC CHSL 2008)**

(*a*) 60% (*b*) 49.6% (*c*) 44% (*d*) 18%

Sol. (*b*) Discount equivalent to 10% and 20% $= 10 + 20 - \frac{10\times20}{100} = 30 - 2 = 28\%$

Discount equivalent to 28% and 30% $= 28 + 30 - \frac{28\times30}{100} = 49.6\%$

Alt: SP $= 100 \times \frac{90}{100}\times\frac{80}{100}\times\frac{70}{100} = 50.4 \Rightarrow$ Equivalent Discount $= 100 - 50.4 = 49.6\%$

10. The difference between successive discounts of 40% followed by 30% and 45% followed by 20% on the marked price of an article is ₹12. The marked price of the article is : **(SSC CGL Tier I 2015)**

(*a*) ₹ 400 (*b*) ₹ 200 (*c*) ₹ 800 (*d*) ₹ 600

Sol. (*d*) Equivalent discount of 40% and 30% = (40 + 30 – 12)% = 58%

Equivalent discount of 45% and 20% = (45 + 20 – 9)% = 56%

∴ Difference = (58 – 56)% = 2% of marked price = ₹12 (i.e. 6 times 2)

∴ Marked price = ₹ 100 × 6 = ₹ 600

11. The marked price of an article is ₹ 500. It is sold at successive discounts of 20% and 10%. The selling price of the article is : **(SSC CGL Tier I 2003)**

(*a*) ₹ 350 (*b*) ₹ 375 (*c*) ₹ 360 (*d*) ₹ 400

Sol. (*c*) Selling price $= 500 \times \frac{100-20}{100}\times\frac{100-10}{100} = 500\times\frac{80}{100}\times\frac{90}{100} =$ ₹ 360

12. A shopkeeper marks the price of an article at ₹ 80. What will be the selling price, if he allows two successive discounts at 5% each ? **(SSC CGL Tier I 2013)**

(*a*) ₹ 7.2 (*b*) ₹ 72.2 (*c*) ₹ 72 (*d*) ₹ 85

Sol. (*b*) Selling price = ₹ 80 × $\frac{95}{100}\times\frac{95}{100}$ = ₹ 72.2

13. List price of an article at a show room is ₹ 2000 and it being sold at successive discounts of 20% and 10%. Its net selling price will be : **(SSC CGL Tier I 2004)**

(*a*) ₹ 1900 (*b*) ₹ 1700 (*c*) ₹ 1440 (*d*) ₹ 1400

Sol. (*c*) Net price $= 2000 \times \frac{100-20}{100} \times \frac{100-10}{100} = 2000 \times \frac{80}{100} \times \frac{90}{100} =$ ₹ 1440

14. An article is sold at discount of 20% and an additional discount of 30% is allowed on cash payment. If Vidya purchased the article by paying ₹ 2240 in cash, the marked price of the article was : **(SSC CGL Tier I 2008)**

(*a*) ₹ 4000 (*b*) ₹ 4368 (*c*) ₹ 4400 (*d*) ₹ 4480

Sol. (*a*) Marked price $= 2240 \times \frac{100}{100-20} \times \frac{100}{100-30} = 2240 \times \frac{100}{80} \times \frac{100}{70} =$ ₹ 4000

15. The marked price of a piano was ₹ 15,000. At the time of sale, there were successive discounts of 20%, 10% and 10% respectively on it. The sale price was **(SSC CGL Tier II 2013)**

(*a*) ₹ 9,720 (*b*) ₹ 9,750 (*c*) ₹ 9,760 (*d*) ₹ 9,780

Sol. (*a*) $15{,}000 \times \frac{80}{100} \times \frac{90}{100} \times \frac{90}{100} =$ ₹ 9,720

15. A sofa-set is marked at ₹ 20,000. The shopkeeper allows successive discounts of 10%, 5% and 2% on it. What is the net selling price ? **(SSC CHSL 2012)**

(*a*) ₹ 17598 (*b*) ₹ 16758 (*c*) ₹ 16648 (*d*) ₹ 17768

Sol. (*b*) ₹ $20{,}000 \times \frac{90}{100} \times \frac{95}{100} \times \frac{98}{100} =$ ₹ 16758

17. After 3 successive discounts of 20%, 10% and 10%, a washing machine was sold for ₹ 6,480. Then the original marked price of washing machine was **(SSC CHSL 2012)**

(*d*) ₹ 10,800 (*b*) ₹ 12,000 (*c*) ₹ 8,960 (*d*) ₹ 10,000

Sol. (*d*) MP $= 6480 \times \frac{100}{100-20} \times \frac{100}{100-10} \times \frac{100}{100-10} = 6480 \times \frac{100}{80} \times \frac{100}{90} \times \frac{100}{90} =$ ₹ 10000

18. A trader gives two successive discounts of 15% and 10% on the marked price of an article. If the selling price is ₹765, the marked price is **(SSC CGL Tier I 2013)**

(*a*) ₹ 850 (*b*) ₹ 1,000 (*c*) ₹ 800 (*d*) ₹ 900

Sol. (b) Marked price = ₹$765 \times \frac{100}{85} \times \frac{100}{90} = 1{,}000$

19. The price of an article is raised by 30% and then two successive discounts of 10% each are allowed. Ultimately the price of the article is : **(SSC CGL Tier I 2003)**

(*a*) Increased by 10% (*b*) Increased by 5.3% (*c*) Decreased by 3% (*d*) Decreased by 5.3%

Sol. (*b*) Let the original price be ₹ 100. Then raised price = ₹ 130

Price after two successive discounts = ₹ $130 \times \frac{90}{100} \times \frac{90}{100} =$ ₹ 105.30 ⇒ Net increase = 5.3%

Alt: $-10-10+\frac{10\times10}{100} = -19\%$ $\quad 30-19-\frac{30\times19}{100} = +5.3\%$

20. The difference between a discount of 35% and two successive discounts of 20% on a certain bill was ₹ 22. The amount of the bill was **(SSC MTS 2011)**

(*a*) ₹ 200 (*b*) ₹ 220 (*c*) ₹ 1100 (*d*) ₹ 2200

Sol. (*d*) Equivalent discount of 20% each $= 20 + 20 - \frac{20 \times 20}{100} = 36\%$

Difference in two discounts = (36% – 35%) of Bill amount = ₹ 22

⇒ Amount of the Bill = ₹ 22 ÷ 1% = ₹ 2200

21. A shopkeeper allows two successive discounts of 3% and 7% while another allows two successive discounts of 2% and 8%. From which shopkeeper should a customer buy if marked price is same at both shops ?

(*a*) First (*b*) Second (*c*) Both are same (*d*) Data Inadequate

Sol. (*b*) Discount at the first shop $= 3 + 7 - \frac{3 \times 7}{100} = 10 - 2.1 = 7.9$

Discount at the second shop $= 2 + 8 - \frac{2 \times 8}{100} = 10 - 1.6 = 8.4$

Hence offer at second shop is beneficial to the customer.

Trick : In both the cases, sum of discounts rates is same and the customer saves more on buying from the shop where product of the discount rates is less. $(3 \times 7 > 2 \times 8)$

22. A company offers three types of successive discounts (*i*) 25% and 15%, (*ii*) 30% and 10%, (*iii*) 35% and 5%. Which officer is the best for a customer ? **(SSC CGL Tier I 2007)**

(*a*) First (*b*) Second (*c*) Third (*d*) All are equal

Sol. (*c*) Sum of rates of discounts are same in each case. Therefore we will compare the product of the rates of discounts in the three cases :

(*i*) $25 \times 15 = 375$ (*ii*) $30 \times 10 = 300$ (*iii*) $35 \times 5 = 175$

Since the product of rates is minimum in the third case, it is best offer for a customer.

23. Which of the following successive discounts better to a customer

(*i*) 20%, 15%, 10% (*ii*) 25%, 12%, 8% **(SSC CGL Tier I 2013)**

(*a*) (*i*) is better (*b*) (*ii*) is better (*c*) Both are same (*d*) None of these

Sol. (*b*) Case (*i*): Equivalent to 20% and 15% $= 20\% + 15\% - \frac{20 \times 15}{100}\% = 32\%$

Equivalent to 32% and 10% $= 32\% + 10\% - \frac{32 \times 10}{100}\% = 38.8\%$

Case (*ii*): Equivalent to 25% and 12% $= 25\% + 12\% - \frac{25 \times 12}{100}\% = 34\%$

Equivalent to 34% and 8% $= 34\% + 8\% - \frac{34 \times 8}{100}\% = 39.28\%$

24. An article is listed at ₹ 900 and two successive discounts of 8% and 8% are given on it. How much would the seller gain or lose, if he gives a single discount of 16%, instead of two successive discounts ? **(SSC CGL Tier I 2007, SSC CPO SI 2012 M)**

(*a*) Gain of ₹ 4.76 (*b*) Loss of ₹ 5.76 (*c*) Gain of ₹ 5.76 (*d*) Loss of ₹ 4.76

Sol. (*b*) Discount rate in the first case $= 8 + 8 - \frac{8 \times 8}{100} = (16 - 0.64)\%$

Discount rate in the second case = 16%

∴ Loss in the second case = 16 – (16 – 0.64)% = 0.64% ⇒ Loss = 900 × 0.64% = ₹ 5.76

25. The marked price of an article is ₹ 1050. A customer pays ₹ 798 for it with two successive discounts. If the rate of first discount is 20%, the rate of second discount is:

(SSC CGL Tier I 2006)

(*a*) 5% (*b*) 6% (*c*) 8% (*d*) 10%

Sol. (*a*) Price after first discount = 1050 – 20% of 1050 = 1050 – 210 = ₹ 840

Price after second discount = ₹ 798 ⇒ Second discount = ₹ 840 – ₹ 798 = ₹ 42

∴ Rate of second discount = $\frac{42}{840} \times 100 = 5\%$

26. The market price of a watch is ₹800. A shopkeeper gives two successive discounts and sells the watch for ₹612. If the first discount is 10%, then the second discount is

(SSC CGL Tier I 2002)

(*a*) 10% (*b*) 12% (*c*) 15% (*d*) 20%

Sol. (*c*) First discount = 10% of ₹ 800 = ₹80 ⇒ Price after first discount = ₹800 – ₹80 = ₹720

Second discount = ₹720 – ₹612 = ₹108 ⇒ % of second discount = 108/720 × 100 = 15%

27. The marked price of a mixie is ₹ 1600. The shopkeeper gives successive discount of 10% and x% to the customer. If the customer pays ₹ 1224 for the mixie, find the value of x :

(SSC CGL Tier I 2013)

(*a*) 8% (*b*) 10% (*c*) 12% (*d*) 15%

Sol. (*d*) Price after first discount = 1600 – 10% of 1600 = ₹ 1440

Price after second discount = ₹ 1224 ⇒ Second discount = ₹ 1440 – ₹ 1224 = ₹ 216

∴ Rate of second discount = $\frac{216}{1440} \times 100 = 15\%$

28. A fan is listed at ₹ 1500 and a discount of 20% is offered on the list price. What additional discount must be offered to the customer to bring the net price to ₹ 1104?

(SSC CGL Tier I 2002)

(*a*) 8% (*b*) 10% (*c*) 12% (*d*) 15%

Sol. (*a*) First discount = 20% of ₹ 1500 = ₹ 300

Price after first discount = ₹ 1500 – ₹ 300 = ₹ 1200

Second discount = ₹ 1200 – ₹ 1104 = ₹ 96

∴ Rate of the second discount = $\frac{96}{1200} \times 100 = 8\%$

29. The price of an article is ₹ 25. After two successive cuts by the same percentage, the price becomes ₹ 20.25. If each time the cut was x%, then :

(CDS 2012)

(*a*) $x = 9$ (*b*) $x = 10$ (*c*) $x = 11$ (*d*) $x = 11.5$

Sol. (*b*) Total discount = ₹ 25 – ₹ 20.25 = ₹ 4.75. ⇒ Rate of discount = $\frac{4.75}{25} \times 100 = 19\%$

Single discount equivalent to two discounts of 10% each = $10 + 10 - \frac{10 \times 10}{100} = 19\%$

CHAPTER 16

PROFIT AND LOSS

BASICS TERMS

Cost Price (C.P.): The price at which the article is purchased is called its cost price.

Selling Price(S.P.) : The price at which the article is sold is called its selling price.

Marked price (M.P.): The price marked on the goods. Also called list price or catalogue price.

Selling Price = Marked price – Discount

Profit: An article is said to be sold at profit, if its selling price is more than its cost price.

Profit = Selling Price – Cost Price. Selling Price = Cost Price + Profit

Loss: An article is said to be sold at loss, if its selling price is less than its cost price.

Loss = Cost Price – Selling Price. Selling Price = Cost Price – Loss

PROFIT AND LOSS PER CENT

Profit/Loss per cent is relation between Profit/Loss and cost price of an article. Therefore, profit or loss per cent is always calculated on cost price, unless mentioned otherwise.

$$\% \text{ Profit} = \frac{\text{Profit}}{\text{Cost price}} \times 100 \qquad \% \text{ Loss} = \frac{\text{Loss}}{\text{Cost price}} \times 100$$

Example1: An article costing ₹ 10 is sold for a profit of ₹ 2. Find profit per cent.

Sol. Profit per cent = $\frac{2}{10} \times 100 = 20\%$

2. A person buys an article for ₹120 and sells at ₹150. Find gain percent. **(SSC MTS 2014)**

(a) 15% (b) 20% (c) 25% (d) 30%

Sol. (*c*) Profit = ₹150 – ₹120 = ₹ 30 ⇒ Profit % = 30/120 × 100 = 25%

Margin: Profit as a percentage of selling price.

RELATION BETWEEN SELLING PRICE AND COST PRICE

When the article is sold at Profit:

$$\text{Selling Price} = \text{Cost Price} \times \frac{100 + \%\text{ Profit}}{100}$$

$$\Rightarrow \text{Cost Price} = \text{Selling Price} \times \frac{100}{100 + \%\text{ Profit}}$$ **(SSC MTS 2014)**

Example 1: An article costing ₹ 400 is sold at 20% profit. Find its selling price.

(*a*) ₹ 320 (*b*) ₹ 380 (*c*) ₹ 420 (*d*) ₹ 480

Sol. (*d*) Selling price = ₹ 400 × $\frac{100 + 20}{100}$ = ₹ 400 × $\frac{120}{100}$ = ₹ 480

Alt: Selling Price = Cost price + Profit = ₹ 400 + 20% of ₹ 400 = ₹ 400 + ₹ 80 = ₹ 480

Example 2 : A shopkeeper sells his articles at 13% profit. Find the cost price of the article sold at ₹14,916. **(SBI Clerical 2008)**

(*a*) ₹ 12,200 (*b*) ₹ 13,200 (*c*) ₹ 14,100 (*d*) None of these

Sol. (*b*) Cost price = ₹ 14,916 × $\frac{100}{113}$ = ₹ 13,200

When the article is sold at Loss:

Selling Price = Cost Price × $\frac{100 - \%\text{ Loss}}{100}$ ⇒ Cost Price = Selling Price × $\frac{100}{100 - \%\text{ Loss}}$

Example 1: An article costing ₹ 500 is sold at 10% loss. Find its selling price.

(*a*) ₹ 450 (*b*) ₹ 550 (*c*) ₹ 600 (*d*) ₹ 400

Sol. (*a*) Selling price = ₹ 500 × $\frac{100 - 10}{100}$ = ₹ 500 × $\frac{90}{100}$ = ₹ 450

Alt: Selling Price = Cost price – Loss = ₹ 500 – 10% of ₹ 500 = ₹ 500 – ₹ 50 = ₹ 450

CALCULATING PROFIT PER CENT WHEN GAIN IN TERMS OF ARTICLES IS GIVEN

If on selling *x* articles, gain is equal to the cost price of *y* articles, then % Profit = $\frac{y}{x} \times 100$

If on selling *x* articles, gain is equal to the selling price of *y* articles, Profit % = $\frac{y}{x-y} \times 100$

If cost price of *x* articles = selling price of *y* articles. Then Profit % = $\frac{x-y}{y} \times 100$

Example 1. By selling 4 articles, a shopkeeper gains equal to the cost price of 1 article. Find his gain per cent.

(*a*) 10% (*b*) 20% (*c*) 25% (*d*) 30%

Sol. (*c*) Profit % = $\frac{1}{4}$ = 25%

Example 2. If on selling 12 notebooks, a seller makes a profit equal to the selling price of 4 notebooks, what is his per cent profit ? **(SBI PO 2006)**

(*a*) 50% (*b*) 25% (*c*) $16\frac{2}{3}$% (*d*) None of these

Sol. (*a*) Profit % = $\frac{4}{12-4} \times 100 = \frac{4}{8} \times 100 = 50\%$

Example 3. If the cost price of 18 articles is equal to selling price of 15 articles. The gain is **(SSC CHSL 2013)**

(*a*) 15% (*b*) 18% (*c*) 20% (*d*) 25%

Sol. (*c*) Gain % = $\frac{18 - 15}{15} \times 100 = 20\%$

Alt : Profit on selling 15 articles = CP of 18 articles – CP of 15 articles = CP of 3 articles

$\therefore$ Profit % = $\frac{3}{15} \times 100 = 20\%$

Alternative Method: Cost price of 18 articles = Selling price of 15 articles

Cost price : Selling price = 15 : 18 $\Rightarrow$ Profit = $\frac{3}{15}\times100 = 20\%$

MORE FORMULAE

1. A shopkeeper **marks** his goods at **x% above** the cost price and then **allows y% discount** on the marked price, then his profit percent = $x - y - \frac{xy}{100}$

Example 1. A merchant marks price of his articles 20% above cost price and allows 10% discount on the marked price. What per cent profit does he make?

(*a*) 8% (*b*) 10% (*c*) 12% (*d*) 30%

Sol. (*a*) Profit = $20 - 10 - \frac{20\times10}{100} = 10 - 2 = 8\%$

Alternative Method: $120\times\frac{90}{100} = 108$ $\Rightarrow$ Profit = 108 – 100 = ₹ 8 on 100 = 8%

Example 2. A shopkeeper marks his goods 20% above cost price, but allows 30% discount for cash. His net loss is : **(SSC CGL Tier I 2000)**

(*a*) 8% (*b*) 10% (*c*) 16% (*d*) 20%

Sol. (*c*) Profit = $20 - 30 - \frac{20\times30}{100} = -10 - 6 = -16\%$

Alt: Let cost price = ₹ 100. Then its marked price = ₹ 100 + ₹ 20 = ₹ 120

Discount = 30% of 120 = ₹ 36 $\Rightarrow$ Selling price = ₹ 120 – ₹ 36 = ₹ 84

$\therefore$ Net loss = ₹ 100 – ₹ 84 = ₹ 16 on ₹ 100 = 16%

Alternative method : 12 × 7 = 84 ; $\Rightarrow$ Loss = 84 – 100 = – 16%

2. If price of an article is **marked x% above** the cost price **and x% discount** is allowed on its marked price, then **loss** per cent = $\frac{x^2}{100}$

Example. A shopkeeper marks price of his articles 12% above the cost price and then allows 12% discount on the marked price. What is his loss per cent?

(*a*) 0% (*b*) 1% (*c*) 1.44% (*d*) 2.4%

Sol. (*c*) Loss % = $\frac{12^2}{100} = 1.44\%$

3. If **two articles** are sold for the **same selling price. On selling first article, a man gains x%** and on selling the **other article he loses x%,** in such cases **there will be always loss**, and

Loss % = $\frac{x^2}{100}$ *i.e.* $\frac{(\text{Common Gain and Loss }\%)^2}{100}$

Example. A T.V. manufacturer sells two televisions at ₹ 9,900 each. He earned a profit of 10% on the first one and suffered 10% loss on another. His net profit or loss percent is **(SSC CGL Tier I 2013)**

(*a*) 1% loss (*b*) 1% gain (*c*) 2% loss (*d*) No profit/loss

Sol. (*a*) Loss % = $\frac{(\text{Common Gain and Loss})^2}{100} = \frac{10^2}{100} = 1\%$

4. If **two articles** are sold for the **same selling price**. **On selling the first, the gain is** $x\%$ and on selling the **second the gain is** $y\%$, then net % profit = $\frac{2\times(100+x)(100+y)}{(100+x)+(100+y)} - 100$

Note: Negative value of 'numerator' denotes loss.

Example. A man sold two articles for ₹ 600 each. On selling first, he gains 20% and on the other, he gains 30%. What is profit per cent in the transaction ? **(MBA Exam.)**

(*a*) 24.8% (*b*) 25% (*c*) 50% (*d*) 56%

Sol. (*a*) Profit% = $\frac{2\times 120\times 130}{120+130} - 100 = 124.8 - 100 = 24.8\%$

Alt: Let SP = LCM of 12 and 13 = ₹156 of each article ⇒ Total SP = 2 × 156 = ₹312

Total cost price = $156\times\frac{100}{120} + 156\times\frac{100}{130} = 130 + 120 =$ ₹ 250

Total profit = 312 – 250 = ₹62 % Profit = $\frac{62}{250} \times 100 = 24.8\%$

5. Relation between Profit/Loss and Cost Price/Selling price :

WHEN THE ARTICLE IS SOLD AT PROFIT

Cost price + Profit = Selling Price ⇒ Cost Price < Selling Price

∴ If we shift base from cost price to selling price (*i.e.* to bigger base), smaller fraction is required and vice-versa.

(*a*) If profit is $\frac{1}{x}$ of cost price, then profit is $\frac{1}{x+1}$ of the selling price.

Example: If profit on selling a commodity is $\frac{1}{3}$ of C.P., then profit is $\frac{1}{3+1} = \frac{1}{4}$ of S.P.

(*b*) If profit is $\frac{1}{x}$ of selling price, then profit is $\frac{1}{x-1}$ of the cost price.

Example: If profit on selling a commodity is $\frac{1}{3}$ of S.P., then profit is $\frac{1}{3-1} = \frac{1}{2}$ of C.P.

WHEN THE ARTICLE IS SOLD AT LOSS

Cost price – Loss = Selling price ⇒ Cost Price > Selling Price

∴ If we shift base from cost price to selling price (i.e. to smaller base), bigger fraction is required and vice-versa.

(*a*) If loss is $\frac{1}{x}$ of cost price, then loss is $\frac{1}{x-1}$ of the selling price

Example: If loss on selling a commodity is $\frac{1}{3}$ of C.P., then loss is $\frac{1}{3-1} = \frac{1}{2}$ of S.P.

(*b*) If loss is $\frac{1}{x}$ of selling price, then loss is $\frac{1}{x+1}$ of the cost price

Example 1: If loss on selling a commodity is $\frac{1}{3}$ of S.P., then loss is $\frac{1}{3+1} = \frac{1}{4}$ of C.P.

Example 2: On the basis of selling price of an article, the loss is calculated to be 25%. The percentage of loss on the basis of cost price is: **(SSC CGL Tier I 2008)**

(*a*) 20% (*b*) 25% (*c*) 33 1/3% (*d*) 40%

Sol. (*a*) Loss = 25% = $\frac{1}{4}$ of selling price = $\frac{1}{5}$ of cost price (As CP > SP)

Alternative Method: Loss = $\frac{25}{100}$ of SP = $\frac{25}{125}$ of CP = $\frac{1}{5}$ of CP = 20%

7. On selling an article for ₹ *x*, a person earns *a*% profit. In order to earn *b*% profit, he must sell the article for ₹ $x \times \frac{100+b}{100+a}$,

***i.e.* New Selling price = Old Selling Price × $\frac{\textbf{100 + Desired Profit (in \%)}}{\textbf{100 + Initial Profit (in \%)}}$**

Example 1: On selling an article for ₹ 550, a man gains 10%. What should be the selling price, if desired profit is 20%? **(SSC CGL Tier I Exam.)**

(*a*) ₹ 500 (*b*) ₹ 600 (*c*) ₹ 660 (*d*) ₹ 715

Sol. (*b*) Selling price = $550 \times \frac{100+20}{100+10}$ = ₹ $550 \times \frac{120}{110}$ = ₹ 600

Example 2 : Ram sold a cow for ₹ 136 and thus lost 15%. At what price he should have sold it to gain 15% ? **(SSC CHSL 2012)**

(*a*) ₹ 180 (*b*) ₹ 150 (*c*) ₹ 200 (*d*) ₹ 184

Sol. (*d*) New Selling Price = $136 \times \frac{100+15}{100-15} = 136 \times \frac{115}{85}$ = ₹ 184

8. If **two articles are sold** – first article at ***x*% profit** and the second article at ***y*% loss,** thus **making no profit, no loss in the transaction,**

Then cost price of the two articles are in the ratio *y* : *x*.

Proof: Let cost price of two articles is ₹ *a* and ₹ *b* respectively.

Then profit on the first article = $\frac{ax}{100}$ and loss on the second article = $\frac{by}{100}$

There is no profit or no loss in the bargain. $\Rightarrow \frac{ax}{100} = \frac{by}{100} \Rightarrow ax = by \Rightarrow \frac{a}{b} = \frac{y}{x}$

Example. A man purchased two articles for total cost of ₹ 9000. He sold the first article at 15% profit and the second at 12% loss. In the bargain, he neither gained nor lost anything. Find cost price of the first article.

(*a*) ₹ 4000 (*b*) ₹ 4500 (*c*) ₹ 5000 (*d*) ₹ 5500

Sol. (*a*) Ratio of cost price of the first and the second article = 12 : 15 = 4 : 5

∴ Cost price of the first article = $\frac{4}{9} \times 9000$ = ₹ 4000

SOLVED EXERCISE

1. A trader sells 145 metres of cloth for ₹ 12,325 at the profit of ₹ 10 per metre of cloth. What is the cost price of 1 metre of cloth ? **(SBI Bank PO 2008)**

(*a*) ₹ 65 (*b*) ₹ 75 (*c*) ₹ 95 (*d*) ₹ 85

Sol. (*b*) Selling price per 1 metre of cloth = 12325 ÷ 145 = ₹ 85

Cost price per 1 metre of cloth = 85 – 10 = ₹ 75

2. A man bought an old typewriter for ₹ 1200 and spent ₹ 200 on its repair. He sold it for ₹ 1680. His profit per cent is : **(SSC CGL Tier I 2003)**

(*a*) 20% (*b*) 10% (*c*) 8% (*d*) 15%

Sol. (*a*) Total cost price = ₹ 1200 + ₹ 200 = ₹ 1400 ⇒ Profit = ₹ (1680 – 1400) = ₹ 280

$$\text{Profit \%} = \frac{\text{Profit}}{\text{Cost price}} \times 100 = \frac{280}{1400} \times 100 = 20\%$$

3. A man purchased a car for ₹1,35,000 and spent ₹25,000 on repairs. At what price was the car sold if he suffered 10% loss on it ? **(SSC CGL Tier I 2014)**

(*a*) ₹ 1,21,500 (*b*) ₹ 1,50,000 (*c*) ₹ 1,76,000 (*d*) ₹ 1,44,000

Sol. (*d*) CP = 1,35,000 + 25,000 = 1,60,000 ⇒ SP = 1,60,000 – 16,000 = 1,44,000

4. If Suman buys an article at 75% of its value and sells it for 20% more than its value, his profit will be **(SSC CGL Tier I 2014)**

(*a*) 20% (*b*) 75% (*c*) 45% (*d*) 60%

Sol. (*d*) Let value = ₹100, then CP = ₹75 and SP = ₹120 ⇒ Profit = 120 – 75 = ₹45

⇒ Profit percentage = 45/75 × 100 = 60%

5. A man buys a cycle for ₹ 1400 and sells it at a loss of 15%. What is the selling price of the cycle? **(SSC CGL Tier I 2002)**

(*a*) ₹ 1202 (*b*) ₹ 1190 (*c*) ₹ 1160 (*d*) ₹ 1000

Sol. (*b*) Selling price = ₹ 1400 × $\frac{100-15}{100}$ = ₹ 1400 × $\frac{85}{100}$ = ₹ 1190

Alt: Selling Price = Cost price – Loss = ₹ 1400 – 15% of ₹ 1400 = ₹ 1400 – ₹ 210 = ₹ 1190

6. If the ratio of cost price and the selling price is 5 : 6, the gain per cent is : **(SSC CGL 2003)**

(*a*) 20% (*b*) $33\frac{1}{3}\%$ (*c*) 25% (*d*) 30%

Sol. (*a*) CP : SP = 5 : 6 ⇒ Profit = 6 – 5 = 1 ⇒ Gain% = $\frac{1}{5} \times 100 = 20\%$

7. If the ratio of cost price and selling price be 9 : 10, then the percentage in profit is **(SSC CGL Tier I 2012)**

(*a*) $11\frac{1}{9}\%$ (*b*) 9% (*c*) 11% (*d*) 10%

Sol. (*a*) CP : SP = 9 : 10 ⇒ Profit = 10 – 9 = 1 ⇒ Profit per cent = $\frac{1}{9} \times 100 = 11\frac{1}{9}\%$

8. If the cost price and selling price of an article are in the ratio 10 : 11, then the percentage of profit is : **(SSC CGL Tier I 2011, SSC CHSL 2010)**

(*a*) 10 (*b*) 9 (*c*) 3 (*d*) 1

Sol. (*a*) CP : SP = 10 : 11 ⇒ Profit = 11 – 10 = 1 ⇒ Profit % = $\frac{1}{10} \times 100 = 10\%$

9. If 3 toys are sold at the cost price of 4 toys of the same kind, the profit will be : **(SSC CGL Tier I 2000)**

(*a*) 25% (*b*) $33\frac{1}{3}$ (*c*) $62\frac{2}{3}\%$ (*d*) 50%

Sol. (*b*) CP : SP = 3 : 4 ⇒ Profit = 4 – 3 = 1 ⇒ Profit % = $\frac{1}{3} = 33\frac{1}{3}\%$

10. The cost price of 100 books is equal to the selling price of 60 books. The gain percentage/loss percentage is : **(SSC CGL Tier I 2015)**

(*a*) 66% (*b*) 67% (*c*) $66\frac{2}{3}\%$ (*d*) 66%

Sol. (*c*) CP : SP = 60 : 100 = 3 : 5 ⇒ Profit % = $\frac{2}{3} \times 100 = 66\frac{2}{3}\%$

11. The ratio of cost price and selling price is 5 : 4, the loss per cent is : **(SSC CGL Tier I 2002)**

(*a*) 20% (*b*) 25% (*c*) 40% (*d*) 50%

Sol. (*a*) Cost price : Selling price = 5 : 4 ⇒ Loss = 5 – 4 = 1 ⇒ Loss % = $\frac{1}{5} \times 100 = 20\%$

12. A person buys some pencils at 5 for a rupee and sells them at 3 for a rupee. His gain per cent will be : **(SSC CGL Tier I 2002)**

(*a*) $66\frac{2}{3}\%$ (*b*) $76\frac{2}{3}\%$ (*c*) $56\frac{2}{3}\%$ (*d*) $46\frac{2}{3}\%$

Sol. (*a*) CP : SP = $\frac{1}{5} : \frac{1}{3} = 3 : 5$ ⇒ Profit per cent = $\frac{5-3}{3} \times 100 = \frac{2}{3} \times 100 = 66\frac{2}{3}\%$

13. If selling price of an article is $\frac{8}{5}$ times its cost price, the profit per cent on it is : **(SSC CGL Tier I 2002)**

(*a*) 120% (*b*) 160% (*c*) 40% (*d*) 60%

Sol. (*d*) Profit = $\frac{8}{5} - 1 = \frac{3}{5} = \frac{3}{5} \times 100\% = 60\%$

14. If the cost price of an article is 80% of its selling price, the profit per cent is : **(SSC CHSL 2010)**

(*a*) 20% (*b*) $22\frac{1}{2}\%$ (*c*) 24% (*d*) 25%

Sol. (*d*) Cost price : Selling price = 80 : 100 ⇒ Profit % = $\frac{20}{80} \times 100 = 25\%$

15. The cost price of an article is equal to 50% of its selling price. How much is the percentage of the selling price to cost price ?

(*a*) 50 (*b*) 75 (*c*) 100 (*d*) 200

Sol. (d) CP : SP = 50 : 100 = 1 : 2 ⇒ SP = 2 CP = 200% of CP

16. The difference between CP and SP of an article is ₹ 240. If the profit is 20%, the SP is :

(*a*) ₹ 1200 (*b*) ₹ 1240 (*c*) ₹ 1400 (*d*) ₹ 1440

Sol. (*d*) CP : SP = 5 : 6 ⇒ Profit = 1 ratio = ₹ 240 ⇒ SP = 6 ratio = ₹ 240 × 6 = ₹ 1440.

17. A shopkeeper purchased a chair marked at ₹ 600 at two successive discounts of 15% and 20% respectively. He spent ₹ 28 on transportation and sold the chair for ₹545. His gain per cent was **(SSC CGL Tier II 2013, CGL 2000 M)**

(*a*) 25% (*b*) 30% (*c*) 35% (*d*) 20%

Sol. (*a*) Cost price = ₹ $600 \times \frac{85}{100} \times \frac{80}{100}$ + ₹ 28 = ₹ (408 + 28) = ₹ 436

Profit = ₹ 545 – ₹ 436 = ₹ 109 ⇒ Profit % = $\frac{109}{436} \times 100 = 25\%$

18. While selling a cooler, a shopkeeper gives a discount of 10% on the marked price. If he gives a discount of 12%, he earns ₹35 less as profit. The marked price of the cooler is : **(SSC CGL Tier I 2008)**

(*a*) ₹1650 (*b*) ₹1625 (*c*) ₹1725 (*d*) ₹1750

Sol. (*d*) Let MP = ₹ 100 ⇒ Selling price in first case = ₹ 100 – ₹ 10 = ₹ 90

SP in second case = ₹ 100 – ₹ 12 = ₹ 88 ⇒ Difference between two SP = ₹90 – ₹88 = ₹2

If difference is ₹2, MP = ₹100 ⇒ If difference is ₹35, MP = ₹ $35 \times \frac{100}{2}$ = ₹1750

Alternative method: 12% – 10% = 2% of MP = ₹ 35 ⇒ MP = $35 \times \frac{100}{2}$ = ₹ 1750

19. Ramesh purchased two types of wheat, one costing ₹ 20 per kg and the other ₹ 25 per kg in quantities having ratio 5 : 4. He mixed the two and sold the mixture at ₹ 24 per kg. In the transaction, he got : **(SSC CGL Tier I 2006)**

(*a*) 4% loss (*b*) 4% gain (*c*) 8% loss (*d*) 8% gain

Sol. (*d*) Cost price of 9 kg of mixture = 5 × ₹ 20 + 4 × ₹ 25 = ₹ 200

Selling price of 9 kg of mixture = 9 × ₹ 24 = ₹ 216

Profit = ₹ 216 – ₹ 200 = ₹ 16 ⇒ Profit per cent = $\frac{16}{200} \times 100 = 8\%$

20. Sourav purchased 30 kg of rice at the rate of ₹ 10 per kg and 35 kg at the rate of ₹ 11 per kg. He mixed the two. At what price per kg (in ₹) should he sell the mixture to make a 30% profit in the transaction? **(SSC CGL Tier II 2013)**

(*a*) 12.5 (*b*) 13 (*c*) 13.7 (*d*) 14.25

Sol. (*c*) Total cost price (of 65 kg) = 30 × 10 + 35 × 11 = 300 + 385 = 685

Total SP = 685 + 30% of 685 = 890.50 ⇒ SP per kg = 890.50 ÷ 65 = ₹ 13.70

21. A grocer purchased 2 kg of rice at the rate of ₹ 15 per kg. and 3 kg. of rice at the rate of ₹ 13 per kg. At what price per kg. should he sell the mixture to earn $33\frac{1}{3}\%$ profit on the cost price? **(SBI Bank PO 2006)**

(a) ₹ 28 (b) ₹ 20 (c) ₹ 18.40 (d) ₹ 17.40

Sol. (c) Total cost price (of 5 kg. rice) = 2 × ₹ 15 + 3 × ₹ 13 = ₹ 30 + ₹ 39 = ₹ 69

Desired profit = $33\frac{1}{3}\%$ of ₹ 69 = $\frac{1}{3}$ of ₹ 69 = ₹ 23

Total selling price = ₹ 69 + ₹ 23 = ₹ 92 ⇒ Selling price per kg. = ₹ 92 ÷ 5 = ₹ 18.40

22. If cost price of an article is $\frac{3}{2}$ of its selling price, the profit or loss per cent is:

(a) 33.33 % loss (b) 50 % profit (c) 50% loss (d) 33.33% profit

Sol. (a) Let selling price of the article = ₹ 2. Then cost price = ₹ 3

∴ Loss = ₹ 3 – ₹ 2 = ₹1 on ₹ 3 ⇒ Loss = $\frac{1}{3}$ = 33.33%

Alt: $\frac{CP}{SP}=\frac{3}{2}\Rightarrow\frac{SP}{CP}=\frac{2}{3}\Rightarrow \text{Loss}=1-\frac{2}{3}=\frac{1}{3}=33.33\%$

23. An article is sold for ₹ 900 and hence gaining 10% of the selling price. Find its cost price.

(a) ₹ 800 (b) ₹ 810 (c) ₹ 990 (d) ₹ 1000

Sol. (b) Cost Price = Selling Price – Profit = ₹ 900 – 10% of ₹ 900 = ₹ 900 – ₹ 90 = ₹ 810

24. If selling an article for ₹990 causes 10% loss on the selling price, find its cost price.

(a) ₹ 891 (b) ₹ 900 (c) ₹ 1089 (d) ₹ 1100

Sol. (c) Cost Price = Selling Price + Loss = ₹ 990 + 10% of ₹ 990 = ₹ 990 + ₹ 99 = ₹ 1089

25. By selling an umbrella for ₹30, a shopkeeper gains 20%. During a clearance sale, the shopkeeper allows a discount of 10% on marked price. His gain during the sale season is

(a) 18% (b) 8% (c) 12% (d) 10%

Sol. (b) Let cost price = ₹ 100. Then marked price = 100 + 20% = ₹120.

New SP = 120 – 10% = 120 – 12 = 108 ⇒ New gain = 108 – 100 = ₹ 8 on ₹ 100 = 8%

Alt: $20-10-\frac{20\times10}{100}=8\%$

26. Two items A and B are sold at a profit of 10% and 15% respectively. If the amount of profit received is the same, then the cost price of A and B may be **(SSC CGL Tier II 2013)**

(a) ₹ 1000, ₹ 1500 (b) ₹ 5000, ₹ 2000 (c) ₹ 3000, ₹ 2000 (d) ₹ 3000, ₹ 5000

Sol. (c) 10% of A = 15% of B ⇒ A : B = 15 : 10 = 3 : 2

27. There is a loss of 4% if an article is sold at ₹ x and gain of 12% if it is sold at y. Then $x : y$ is **(SSC CHSL 2012)**

(a) 2 : 3 (b) 6 : 7 (c) 4 : 5 (d) 8 : 9

Sol. (b) Let CP = ₹ 100. Then x = ₹ 100 – ₹ 4 = ₹ 96 and y = ₹ 100 + ₹ 12 = ₹ 112

⇒ $x : y$ = 96 : 112 = 6 : 7

28. A man gains 20% by selling an article for a certain price. If he sells it at double the price, then percentage of profit will be : **(SSC CGL Tier I 2004)**

(a) 40% (b) 140% (c) 100% (d) 120%

Sol. (b) Selling price = 100% + 20% = 120% of cost price

New selling price = 2 × 120% of cost price = 240% of cost price

∴ New Profit = 240% – 100% = 140% of cost price

29. A man gains 10% by selling an article for a certain price. If he sells it at double the price, then the profit made is **(MBA 2001, SSC MTS 2013)**

(*a*) 120% (*b*) 20% (*c*) 40% (*d*) 100%

Sol. (*a*) SP = 110% of CP $\Rightarrow$ New SP = 220% of CP $\Rightarrow$ Profit = 120% of CP

30. On selling an article for a certain price, a man loses 30%. What is his loss/profit per cent, if he sells the article for double the price?

(*a*) 60% loss (*b*) 15% loss (*c*) 40% profit (*d*) 30% profit

Sol. (*c*) SP = (100 – 30)% = 70% of CP $\Rightarrow$ New SP = 2 × 70% = 140% of CP

$\therefore$ Profit = 140% – 100% = 40% of cost price

31. If by selling an article at 2/3 of its normal price, there is a loss of 10%, find the profit percentage on selling the articles at normal selling price. **(MAT 2001)**

(a) 33.33% (b) 35% (c) 50% (d) 60%

Sol. (b) 2/3 of SP = 90% of CP $\Rightarrow$ Normal SP = 3/2 × 90% = 135% of CP

$\Rightarrow$ Profit = 135% – 100% = 35% of CP

32. Find the profit per cent made on selling an article for a certain amount, if on selling the article for $\frac{3}{4}$ of the amount causes 10% loss?

(*a*) 2.5% (*b*) 7.5% (*c*) 15% (*d*) 20%

Sol. (*d*) At 10% loss means 90% of the cost price, which is equal to $\frac{3}{4}$ of the selling price.

$\therefore$ Selling price = $\frac{4}{3}$ × 90% = 120% of cost price $\therefore$ Profit = 120% – 100% = 20%

33. A man gains 10% by selling an article for a certain price. What is profit/loss if the article is sold for half the price?

(*a*) 45% loss (*b*) 5% loss (*c*) 5% profit (*d*) 55% profit

Sol. (*a*) SP = (100 + 10)% = 110% of CP $\Rightarrow$ New SP = $\frac{1}{2}$ × 110% = 55% of CP

$\therefore$ Loss = 100% – 55% = 45% of cost price

34. A cloth merchant has announced 25% rebate on marked price. If one needs to have a rebate of ₹40, then how many shirts, each with a marked price of ₹32, should be purchased ?

(SSC CGL Tier I 2014)

(*a*) 5 (*b*) 6 (*c*) 7 (*d*) 8

Sol. (*a*) Shirts purchased = (40 ÷ 25%) ÷ 32 = 160 ÷ 32 = 5

35. A man buys milk at 45 p per litre, add one-fourth of water to it and sells the mixture at 54 p per litre. The profit per cent is :

(*a*) 20% (*b*) 30% (*c*) 40% (*d*) 50%

Sol. (*d*) Cost price of one litre of mixture = 45 × $\frac{4}{5}$ = 36 p

Profit per litre = (54 – 36) p = 18 p $\Rightarrow$ Profit % = $\frac{18}{36}$ × 100 = 50%

36. A dishonest dealer professes to sell his goods at cost price, but uses a false weight of 800 grams for one kg. What is his profit per cent? **(SSC MTS 2014)**

(*a*) 15% (*b*) 20% (*c*) 25% (*d*) 30%

Sol. (*c*) The merchant gives 800 grams charging the price of 1000 grams.

His gain is 200 grams on every 800 grams. ⇒ Profit per cent = $\frac{200}{800} \times 100 = 25\%$

37. A dishonest dealer professes to sell his goods at cost price, but uses a false weight of 900 grams for one kg. What is his profit per cent?

(*a*) 10% (*b*) $11\frac{1}{9}\%$ (*c*) $9\frac{1}{11}\%$ (*d*) 11%

Sol. (*b*) The merchant gives 900 grams charging the price of 1000 grams.

His gain is 100 grams on every 900 grams. ⇒ Profit per cent = $\frac{100}{900} \times 100 = 11\frac{1}{9}\%$

38. A merchant professes to sell goods at the loss of 5% but uses weight of 900 grams in place of one kilogram. What is his profit per cent?

(*a*) 5% (*b*) $5\frac{5}{9}\%$ (*c*) 6% (*d*) 10%

Sol. (*b*) The merchant gives 900 grams charging the price of 950 grams.

∴ His gain is 50 grams on every 900 grams. ⇒ Profit per cent = $\frac{50}{900} \times 100 = 5\frac{5}{9}\%$

39. A merchant professes to sell goods at 20% profit but uses weight of 800 grams in place of a kilogram. What is his actual profit per cent?

(*a*) 20% (*b*) 25% (*c*) 50% (*d*) 33.33%

Sol. (*c*) The merchant gives 800 grams charging the price of 1200 grams.

∴ His gain is 400 grams on every 800 grams. ⇒ Profit per cent = $\frac{400}{800} \times 100 = 50\%$

40. A trader has a weighing balance that shows 1,200 gm for a kilogram. He further marks up his cost price by 10%. Then the net profit percentage is **(SSC CGL Tier I 2012)**

(*a*) 32% (*b*) 23% (*c*) 31.75% (*d*) 23.5%

Sol. (*a*) The merchant gives 1000 grams charging the price of 1320 (=1200 + 10%) gm.

∴ Net profit = $\frac{320}{1000} \times 100 = 32\%$

41. A retailer profess to sell his goods at cost price. If using a false weight, he still gains 25%, find the weight he uses in place of 1 kg.

(*a*) 200 g (*b*) 600 g (*c*) 750 g (*d*) 800 g

Sol. (*d*) Let defraud = x grams per kg. Then $\frac{x}{1000 - x} \times 100 = 25 \Rightarrow 4x = 1000 - x$

$\Rightarrow 5x = 1000 \Rightarrow x = 200$ ⇒ He uses 1000 – 200 = 800 grams for one kg.

Trick : $\frac{x}{1000 - x} \times 100 = 25 \Rightarrow \frac{x}{1000 - x} = \frac{25}{100} = \frac{1}{4} \Rightarrow \frac{x}{1000} = \frac{1}{5} = 200$ g

42. A milkman claims to sell milk at its cost price only, but he is making a profit of 20% since he has mixed some amount of water in the milk. What is the percentage of milk in the mixture? **(CDS 2015)**

(*a*) 80% (*b*) $\frac{250}{3}\%$ (*c*) 75% (*d*) $\frac{200}{3}\%$

Sol. (*c*) $\frac{\text{Milk}}{\text{Water}} = \frac{1}{5} \Rightarrow \text{Milk}\,\% = \frac{5}{6} \times 100 = \frac{250}{3}$

43. A shopkeeper by means of his false balance defrauds his customers to the extent of 20% by purchasing the goods as well as by selling the goods. Find his net profit. **(SSC)**

(*a*) 36% (*b*) 40% (*c*) 44% (*d*) 50%

Sol. (*c*) $20 + 20 + \frac{20 \times 20}{100} = 44\%$

44. A shopkeeper by means of his false balance defrauds his customers to the extent of 10% by purchasing the goods as well as by selling the goods. Find his net profit. **(SSC)**

(*a*) 22% (*b*) 20% (*c*) 21% (*d*) 22 2/9%

Sol. (*d*)

$10 + 10 + \frac{10 \times 10}{100} = 21\%$

45. A shopkeeper buys some pens. If he sells them @ ₹ 13 per pen, his total loss is ₹150 but on selling them @ ₹ 15 per pen, his total gain is ₹ 100. How many pens did he sell?

(*a*) 100 (*b*) 125 (*c*) 150 (*d*) 250

Sol. (*b*) Difference in sales amount due to change in selling price = ₹ 150 + ₹ 100 = ₹ 250

Difference in selling price per pen = ₹ 15 – ₹ 13 = ₹ 2

∴ On selling 1 pen, sales amount is increased by ₹ 2 in second case.

∴ Total pens sold = $\frac{250}{2}$ = 125 pens

46. Ram buys 4 horses and 9 cows for ₹ 13,400. If he sells the horses at 10% profit and the cows at 20% profit, then he earns a total profit of ₹ 1880. The cost price of a horse is: **(CDS 2003)**

(*a*) ₹ 2500 (*b*) ₹ 3000 (*c*) ₹ 2000 (*d*) ₹ 1000

Sol. (*c*) 10% of ₹ 13,400 = ₹ 1340 ⇒ Difference = ₹1880 – ₹ 1340 = ₹ 540

∴ Total cost of cows = ₹ 540 ÷ 10% = ₹ 5,400

⇒ Total cost of 4 horses = ₹ 13400 – ₹ 5400 = ₹ 8000

∴ Cost price of 1 horse = ₹ 8000 ÷ 4 = ₹ 2000

47. A shopkeeper sells an article at 15% gain. Had he sold it for ₹ 18 more, he would have gained 18%. The cost price (in ₹)of the article is **(SSC CHSL 2013)**

(*a*) 540 (*b*) 600 (*c*) 350 (*d*) 318

Sol. (*b*) Difference of two selling prices = (18 – 15)% of CP = 3% of CP = ₹ 18 (6 times 3)

Cost price 6 × 100 = ₹ 600

48. An article is sold at a profit of 20%. If it had been sold at a profit of 25%, it would have fetched ₹ 35 more. The cost price of the article is : **(SSC CGL Tier I 2004)**

(*a*) ₹ 650 (*b*) ₹ 700 (*c*) ₹ 750 (*d*) ₹ 800

Sol. (*b*) Difference in two selling prices = 25% – 20% = 5% of cost price

Actual difference in two selling prices = ₹ 35 (*i.e.* 7 times 5)

∴ Cost Price = 7 × ₹ 100 = ₹ 700

49. An article is sold at a gain of 15%. Had it been sold for ₹ 27 more, the profit would have been 20%. The cost price of the article is **(SSC CGL Tier II 2013)**

(*a*) ₹ 500 (*b*) ₹ 700 (*c*) ₹ 540 (*d*) ₹ 545

Sol. (*c*) Difference = (20% – 15%) of CP = 5% of CP = ₹ 27 ⇒ CP = ₹ 27 × 20 = ₹ 540

50. A man sold an article at a loss of 20%. If he sells the article for ₹12 more, he would have gained 10%. The cost price of the article is **(SSCCGLTierI2012)**

(*a*) ₹ 60 (*b*) ₹ 40 (*c*) ₹ 30 (*d*) ₹ 22

Sol. (*b*) ₹ 12 = (20 + 10)% of CP ⇒ CP = 12 × 100/30 = ₹ 40

51. By selling a table for ₹ 350 instead of ₹ 400, loss per cent increases by 5%. The cost price of the table is:

(*a*) ₹ 1050 (*b*) ₹ 417.50 (*c*) ₹ 435 (*d*) ₹ 1000

Sol. (*d*) Difference in two selling prices = ₹ 400 – ₹ 350 = ₹ 50

Difference in two selling prices = 5% of cost price = ₹ 50 (*i.e.* 10 times 5)

∴ Cost Price = 10 × ₹ 100 = ₹ 1000

52. An article is sold at 10% loss. Had it been sold for ₹ 30 more, the loss would have been 5% only. What is the cost price of the article?

(*a*) ₹ 200 (*b*) ₹ 300 (*c*) ₹ 500 (*d*) ₹ 600

Sol. (*d*) Difference in two selling prices = 10% – 5% = 5% of cost price

Actual difference in two selling prices = ₹ 30 (i.e. 6 times 5)

∴ Cost Price = 6 × ₹ 100 = ₹ 600

53. A tradesman sold an article at a loss of 20%. If the selling price had been increased by ₹ 100, there would have been a gain of 5%. The cost price of the article was : **(SSC CGL 2004)**

(*a*) ₹ 200 (*b*) ₹ 25 (*c*) ₹ 400 (*d*) ₹ 250

Sol. (*c*) Difference in two selling prices = 5% – (– 20%) = 25% of cost price

Actual difference in two selling prices = ₹ 100 (*i.e.* 4 times 25)

∴ Cost Price = 4 × ₹ 100 = ₹ 400

54. A loss of 19% gets converted into a profit of 17% when the selling price is increased by ₹ 162. The cost price of the article is **(SSC CGL Tier II 2012)**

(*a*) ₹ 360 (*b*) ₹ 540 (*c*) ₹ 450 (*d*) ₹ 600

Sol. (*c*) Difference in two selling prices = 17% – (– 19%) = 36% of cost price

Actual difference in two selling prices = ₹ 162 ⇒ Cost Price = $\frac{162}{36} \times 100$ = ₹ 450

55. On selling an article for ₹ 250, a shopkeeper gains ₹ 50. His gain per cent is:

(*a*) 16.67% (*b*) 20% (*c*) 25% (*d*) 33.33%

Sol. (*c*) CP = SP – Profit = ₹ 250 – ₹ 50 = ₹ 200 ⇒ Profit % = $\frac{50}{200} = \frac{1}{4} = 25\%$

Alternative Method: Profit = $\frac{50}{250} = \frac{1}{5}$ of SP = $\frac{1}{5-1}$ of CP = $\frac{1}{4}$ of CP = 25%

56. On selling an article for ₹ 200, a man loses ₹ 20. Find his loss per cent.

(*a*) 10% (*b*) $9\frac{1}{11}\%$ (*c*) $11\frac{1}{9}\%$ (*d*) 20%

Sol. (*b*) CP = SP + loss = ₹ 200 + ₹ 20 = ₹ 220 ⇒ Loss % = $\frac{20}{220} = \frac{1}{11} = 9\frac{1}{11}\%$

Alternative Method : Loss = $\frac{20}{200} = \frac{1}{10}$ of SP = $\frac{1}{10+1}$ of CP = $\frac{1}{11}$ of CP = $9\frac{1}{11}\%$

57. 20% loss on selling price is what per cent loss on the cost price ? **(SSC CGL Tier I 2011)**

(*a*) 25% (*b*) 15% (*c*) $16\frac{2}{3}\%$ (*d*) $16\frac{1}{3}\%$

Sol. (*c*) Loss = 20% of S.P. = $\frac{1}{5}$ of SP = $\frac{1}{6}$ of CP = $16\frac{2}{3}\%$

Alternative method : SP : CP = 100 : 120 ⇒ Loss % = $\frac{20}{120} \times 100 = 16\frac{2}{3}\%$

58. On selling an article for ₹ 300, a shopkeeper gains 20% of the cost price. Find cost price of the article.

(*a*) ₹ 240 (*b*) ₹ 250 (*c*) ₹ 350 (*d*) ₹ 360

Sol. (*b*) Cost price = $300 \times \frac{100}{100+20}$ = ₹ $300 \times \frac{100}{120}$ = ₹ 250

Alternative Method: Profit = $\frac{1}{5}$ of cost price = $\frac{1}{5+1}$ of SP = $\frac{1}{6} \times 300$ = ₹ 50

∴ Cost Price = ₹ 300 – ₹ 50 = ₹ 250

59. On selling an article for ₹ 900, a shopkeeper loses 10% of the cost price. Find cost price of the article.

(*a*) ₹1000 (*b*) ₹ 810 (*c*) ₹ 990 (*d*) ₹ 800

Sol. (*a*) Cost price = ₹900 × $\frac{100}{100-10}$ = ₹ $900 \times \frac{100}{90}$ = ₹ 1000

Alternative Method: Loss = 10% of C.P. = $\frac{1}{10}$ of C.P. = $\frac{1}{10-1}$ of S.P. = $\frac{1}{9} \times 900$ = ₹ 100

∴ Cost price = 900 + 100 = ₹ 1000

60. If an article costing ₹ 1000 is sold gaining 20% of its selling price. Find selling price of the article.

(*a*) ₹ 1250 (*b*) ₹ 1200 (*c*) ₹ 980 (*d*) ₹ 975

Sol. (*a*) Profit = $\frac{1}{5}$ of selling price = $\frac{1}{5-1}$ of cost price = $\frac{1}{4} \times 1000$ = ₹ 250

∴ Selling Price = ₹ 1000 + ₹ 250 = ₹ 1250

61. A retailer got a discount of 10% on the marked price of a watch from the wholesaler. In turn he sells it at its marked price, then his gain per cent is **(SSC FCI 2012)**

(*a*) 10 (*b*) $11\frac{1}{9}$ (*c*) $27\frac{7}{9}$ (*d*) 26

Sol. (*b*) MP = 100 ⇒ CP = ₹ 90, Profit = ₹ 10 ⇒ Profit% = $\frac{10}{90} \times 100 = \frac{100}{9} = 11\frac{1}{9}\%$

62. Arvind purchased a wrist watch with 30% discount on the labelled price. He sold it with 40% profit on the price he bought. What was his per cent loss on the labelled price ? **(SSC CGL Tier I 2013)**

(*a*) 8 (*b*) 2 (*c*) 6 (*d*) 10

Sol. (*b*) Let labelled price = ₹ 100. Then cost price = ₹ 70.

Selling price = ₹ 70 + 40% of ₹ 70 = ₹ 98 $\Rightarrow$ Loss on labelled price = ₹ 2

Alternative method: $-30 + 40 - \frac{30 \times 40}{100} = 10 - 12 = -2\%$ (Loss)

63. What is the maximum percentage discount (approximately) that a merchant can offer on his marked price so that he ends up selling at no profit or loss, if he initially marked his goods up by 40% ? **(SSC CGL Tier II 2012)**

(*a*) 33.5% (*b*) 28.5% (*c*) 60% (*d*) No discount

Sol. (*b*) Increase in price = 40% = $\frac{40}{100} = \frac{2}{5}$

Discount / Decrease in price = $\frac{2}{5+2} = \frac{2}{7}$ = 28.5% (approximately)

64. A person bought an article on 40% discount and sold it at 50% more than the marked price. What profit did he get? **(SSC CGL Tier I 2006)**

(*a*) 50% (*b*) 90% (*c*) 150% (*d*) 250%

Sol. (*c*) Let MP = ₹ 100. Then CP = ₹(100 – 40) = ₹ 60 $\Rightarrow$ SP = ₹(100 + 50) = ₹ 150

Profit = ₹(150 – 60) = ₹ 90 $\Rightarrow$ Profit % = $\frac{90}{60} \times 100$ = 150%

65. Mahesh purchased a radio at $\frac{9}{10}$ of its selling price and sold it at 8% more than its original selling price. His gain per cent is : **(SSC Data Entry Operator 2010)**

(*a*) 20% (*b*) 18% (*c*) 10% (*d*) 8%

Sol. (*a*) Let marked selling price = ₹ 100. Then cost price = ₹100 × $\frac{9}{10}$ = ₹ 90

Actual selling price = ₹ 100 + 8% of ₹ 100 = ₹ 108

$\therefore$ Profit = ₹ 108 – ₹ 90 = ₹ 18 on ₹ 90 $\Rightarrow$ Profit % = $\frac{18}{90} = \frac{1}{5}$ = 20%

66. Raghavan purchased a scooter at $\frac{13}{15}$ of its selling price and sold it at 12% more than its selling price. His gain is : **(SSC CGL Tier I 2005)**

(*a*) 20% (*b*) 30% (*c*) $38\frac{1}{13}\%$ (*d*) $29\frac{3}{13}\%$

Sol. (*d*) Let marked price be ₹ x. Then cost price = $\frac{13}{15}x$. Selling price = $\frac{112}{100}x = \frac{28}{25}x$

$$\frac{SP}{CP} \times 100 = \frac{28}{25}x \div \frac{13}{15}x \times 100 = \frac{28}{25} \times \frac{15}{13} \times 100 = \frac{1680}{13} = 129\frac{3}{13}\%$$

$\therefore$ Profit% = $129\frac{3}{13}\% - 100\% = 29\frac{3}{13}\%$

67. A retailer buys 40 pens at the marked price of 36 pens from a wholesaler. If he sells these pens giving a discount of 1%, what is the profit per cent ? **(SSC CGL Tier I 2000)**

(*a*) 9% (*b*) 10% (*c*) $10\frac{1}{9}\%$ (*d*) 11%

Sol. (*b*) Marked price : Cost price = 40 : 36 = 10 : 9. Selling price = 10 – 1% of 10 = ₹ 9.90

$\therefore$ Profit = 9.90 – 9.00 = ₹ 0.90 on ₹ 9.00 = $\frac{0.90}{9.00}\times 100\% = 10\%$

68. Ram bought a bat with 20% discount on the marked price. He sold the bat for ₹ 1200 with a profit of ₹ 80. The marked price of the bat was: **(SSC CGL Tier I 2006)**

(*a*) ₹ 2400 (*b*) ₹ 1600 (*c*) ₹ 1400 (*d*) ₹ 1250

Sol. (*c*) Cost price of bat = Selling price – Profit = 1200 – 80 = ₹ 1120

$\therefore$ 80% of marked price = ₹ 1120 $\Rightarrow$ Marked price = $1120\times\frac{100}{80}$ = ₹ 1400

69. 12 copies of a book were sold for ₹ 1800 thereby gaining cost price of 3 copies. The cost price of a copy is : **(SSC CGL Tier I 2000)**

(*a*) ₹ 120 (*b*) ₹ 150 (*c*) ₹ 1200 (*d*) ₹ 1500

Sol. (*a*) ₹ 1800 = SP of 12 = CP of 12 + Profit = CP of 12 + CP of 3 = CP of 15

$\therefore$ Cost price of 1 book = ₹ 1800 ÷ 15 = ₹ 120

70. On selling 17 balls at ₹ 720, there is a loss equal to the cost price of 5 balls. The cost price (in ₹) of a ball is **(SSC CGL Tier II 2013, Tier I 2004)**

(*a*) 45 (*b*) 50 (*c*) 55 (*d*) 60

Sol. (*d*) SP of 17 = CP of 17 – CP of 5 = CP of 12 = ₹ 720 $\Rightarrow$ CP of 1 ball = 720 ÷ 12 = ₹ 60

71. A man bought 25 kg of rice for ₹ 750. On selling them for a certain price, he suffered loss equal to the selling price of 5 kg. Find the selling price of 1 kg of rice.

(*a*) ₹ 25 (*b*) ₹ 30 (*c*) ₹ 35 (*d*) ₹ 37.50

Sol. (*a*) SP of 25 kg + SP of 5 kg = ₹ 750 $\Rightarrow$ Selling price of 1 kg = ₹ $\frac{750}{30}$ = ₹ 25

72. A man bought 25 pens for ₹ 4800 and sold them at a profit equal to the selling price of one pen. What is the selling price of a pen?

(*a*) ₹ 192 (*b*) ₹ 196 (*c*) ₹ 200 (*d*) Can't say

Sol. (*c*) Selling price of 25 pens = Cost price of 25 pens + Selling price of 1 pen

$\therefore$ Selling price of 24 pens = ₹ 4800 $\Rightarrow$ Selling price of 1 pen = ₹ 4800 ÷ 24 = ₹ 200

73. Some quantity of tea is sold at ₹ 22 per kg, making 10% profit. If total gain is ₹ 88, what is the quantity of tea sold?

(*a*) 4 kg (*b*) 40 kg (*c*) 44 kg (*d*) 50 kg

Sol. (*c*) Profit = 10% of cost price = $\frac{1}{10}$ of cost price = $\frac{1}{11}$ of selling price = $\frac{1}{11}$ × ₹ 22 = ₹ 2 per kg.

But actual gain = ₹ 88 $\Rightarrow$ Quantity sold = 88 ÷ 2 = 44 kg.

74. On selling an article for ₹ 170, a shopkeeper loses 15%. In order to gain 20%, he must sell that article at rupees : **(SSC CGL Tier I 2013)**

(*a*) 210 (*b*) 215.50 (*c*) 212.50 (*d*) 240

Sol. (*d*) Selling price = $170\times\frac{100+20}{100-15}$ = ₹ $170\times\frac{120}{85}$ = ₹ 240

75. If an article is sold for ₹ 178 at a loss of 11%, what should be its selling price in order to earn a profit of 11% ? **(SSC CGL Tier I 2002 and 2001 modified, CDS 2010 M)**

(*a*) ₹ 222.50 (*b*) ₹ 267 (*c*) ₹ 222 (*d*) ₹ 220

Sol. (*c*) Required selling price = ₹ $178 \times \frac{100+11}{100-11}$ = ₹ $178 \times \frac{111}{89}$ = ₹ 222

Alternative method : (100 – 11)% of cost price = 89% of cost price = ₹ 178

(100 + 11)% of cost price = 111% of cost price = ₹ $178 \times \frac{111}{89}$ = ₹ 222

76. If a man were to sell his chair for ₹720, he would lose 25%. To gain 25%, he should sell it for: **(SSC CGL Tier I 2007, 2001M)**

(*a*) ₹ 1200 (*b*) ₹ 1000 (*c*) ₹ 960 (*d*) ₹ 900

Sol. (*a*) Required selling price = $720 \times \frac{100+25}{100-25}$ = ₹ $720 \times \frac{125}{75}$ = ₹1200

77. A merchant fixed selling price of his articles at ₹ 700 after adding 40% profit to the cost price. As the sale was very low at this price level, he decided to fix the selling price at 10% profit. Find the new selling price. **(SSC CHSL 2012)**

(*a*) ₹ 450 (*b*) ₹ 490 (*c*) ₹ 500 (*d*) ₹ 550

Sol. (*d*) New Selling price = ₹ $700 \times \frac{100+10}{100+40}$ = ₹ $700 \times \frac{110}{140}$ = ₹ 550

78. A shopkeeper bought some eggs at the rate of ₹ 16 per dozen. Due to mishandling 20% of the eggs bought were broken during the transportation. At what rate of per dozen should he sell the remaining eggs so as to gain 30% on the total cost price?

(*a*) ₹ 16 (*b*) ₹ 18 (*c*) ₹ 24 (*d*) ₹ 26

Sol. (*d*) Since 20% of the quantity is spoiled, selling the eggs at cost price will result in 20% loss.

∴ We are to find the selling price which gives him 30% profit instead of 20% loss.

∴ Selling Price = ₹ $16 \times \frac{130}{80}$ = ₹ 26 per dozen

79. A man purchases 5 pens for a rupee. How many pens should he sell for a rupee if he wants to earn 25% profit?

(*a*) 4 (*b*) 5 (*c*) 6 (*d*) 8

Sol. (*a*) Cost price of 1 pen = ₹ $\frac{1}{5}$ ⇒ Selling price of 1 pen = $\frac{1}{5} \times \frac{125}{100} = \frac{1}{4}$

80. A man sells eggs @ 10 eggs for 7 rupee, gaining 30%. How many eggs did he buy for 7 rupee?

(*a*) 7 (*b*) 11 (*c*) 13 (*d*) 15

Sol. (*c*) SP of 1 egg = $\frac{7}{10}$ ⇒ CP of 1 egg = $\frac{7}{10} \times \frac{100}{130} = \frac{7}{13}$ ⇒ He bought 13 eggs for ₹ 7.

81. A vendor sells lemons at the rate of 5 for ₹ 14, gaining thereby 40%. For how much did he buy a dozen lemons ? **(SSC CHSL 2010)**

(*a*) ₹ 20 (*b*) ₹ 21 (*c*) ₹ 24 (*d*) ₹ 28

Sol. (*c*) Selling price of 1 lemon = $\frac{14}{5}$ ⇒ Cost price of 1 lemon = $\frac{14}{5} \times \frac{100}{140}$

Cost price of 12 lemons = $\frac{14}{5} \times \frac{100}{140} \times 12$ = ₹ 24

82. By selling 9 articles for a rupee, a man incurred a loss of 4%. To make a gain of 44%, the number of articles to be sold for a rupee is **(SSC CGL Tier I 2012)**

(*a*) 5 (*b*) 3 (*c*) 4 (*d*) 6

Sol. (*d*) $\frac{1}{9}\times\frac{144}{96} = \frac{1}{6} \Rightarrow$ 6 articles sold for a rupee.

83. On selling 7 pens for ₹ 44, a man gains 10% of the cost price, what should be the selling price of 10 pens, if desired profit is 40%?

(*a*) ₹ 50 (*b*) ₹ 56 (*c*) ₹ 80 (*d*) ₹ 100

Sol. (*c*) Selling price of 1 pen = $\frac{44}{7} \Rightarrow$ Selling price of 10 pens = $10\times\frac{44}{7}$

$\therefore$ Selling price of 10 pens at new rate = $10\times\frac{44}{7}\times\frac{140}{110}$ = ₹ 80

84. Ravi buys some toffees at 2 for a rupee and sells them at 5 for a rupee. His loss per cent is: **(SSC CGL Tier I 2005)**

(*a*) 120 (*b*) 90 (*c*) 30 (*d*) 60

Sol. (*d*) Let toffees bought = LCM of 2 and 5 = 10

Then C.P. = ₹ 5, S.P. = ₹ 2, Loss = 5 – 2 = ₹ 3 $\Rightarrow$ Loss % = $\frac{3}{5}\times 100 = 60\%$

Alternative method : CP : SP = $\frac{1}{2}:\frac{1}{5} = 5:2 \Rightarrow$ Loss = $\frac{3}{5}\times 100 = 60\%$

85. A fruit-seller buys lemons at 2 for a rupee and sells them at 5 for three rupees. What is his gain per cent? **(CDS 2011)**

(*a*) 10% (*b*) 15% (*c*) 20% (*d*) 25%

Sol. (*c*) CP = $\frac{1}{2}$, SP = $\frac{3}{5} \Rightarrow \frac{SP}{CP} = \frac{3}{5}\times 2 = \frac{6}{5} \Rightarrow$ Profit = $\frac{1}{5} = 20\%$

86. By selling an article for ₹ 69, there is a loss of 8%, when the article is sold for ₹ 78, the gain or loss per cent is : **(SSC CGL Tier I 2004)**

(*a*) Neither gain nor loss (*b*) 40% gain (*c*) 4% gain (*d*) 4% loss

Sol. (*c*) Selling price = ₹ 69 = 100% – 8% = 92% of the cost price

New selling price = ₹ 78 = $92\%\times\frac{78}{69}$ = 104% of cost price

$\therefore$ Profit = 104% – 100% = 4% of the cost price

87. An article is sold for ₹ 300 at a profit of 20%. Had it been sold for ₹ 235, the loss percentage would have been **(SSC CGL Tier I 2013)**

(*a*) 6 (*b*) 3 (*c*) 4 (*d*) 9

Sol. (*a*) Cost price = ₹ $300\times\frac{100}{120}$ = ₹ 250

New S.P. = ₹ 235 $\Rightarrow$ Loss = ₹ 250 – ₹ 235 = ₹ 15 $\Rightarrow$ Loss% = $\frac{15}{250}\times 100 = 6\%$

Alt: ₹ 300 = 120% of C.P. $\Rightarrow$ ₹ 235 = $120\times\frac{235}{300}$ = 94% of CP

$\Rightarrow$ Loss (100 – 94)% = 6% of C.P.

88. By selling a tape-recorder for ₹ 950, I lose 5%. What per cent shall I gain by selling it for ₹1040? **(SSC CGL Tier I 2003)**

(*a*) 5 (*b*) 4 (*c*) 4.5 (*d*) 9

Sol. (*b*) Selling price = ₹ 950 = 100% – 5% = 95% of the cost price

New selling price = ₹ 1040 = $95\% \times \frac{1040}{950}$ = 104% of cost price

∴ Profit = 104% – 100% = 4% of the cost price

89. A tape recorder is sold for ₹ 935 at a profit of 10%. What would have been the actual loss if it had been sold for ₹835 ? **(SSC CGL Tier I 2014)**

(*a*) ₹85 (*b*) ₹60 (*c*) ₹15 (*d*) ₹50

Sol. (*c*) Selling price = ₹ 935 = 100% + 10% = 110% of the cost price

Cost price = ₹ 935 × $\frac{100}{110}$ = ₹ 850 ⇒ Loss = ₹ 850 – ₹ 835 = ₹ 15

90. By selling an article at 80% of its marked price, a trader makes a loss of 10%. What will be the profit percentage if he sells it at 95% of its marked price ? **(SSC CGL Tier II 2012)**

(*a*) 5.9 (*b*) 12.5 (*c*) 6.9 (*d*) 5

Sol. (*c*) 80% of marked price = (100 – 10)% = 90% of the cost price

∴ 95% of the marked price = $\frac{95}{80} \times 90$ = 106.9% (approx) of the cost price

Profit = (106.9 – 100)% = 6.9% of the cost price

91. On selling an article for ₹ 64, a man gains $\frac{1}{7}$ of the cost price. Find his profit percentage, if the article is sold for ₹ 63.

(*a*) 12.5% (*b*) 20% (*c*) 25% (*d*) 33.33%

Sol. (*a*) Gain = $\frac{1}{7}$ of Cost Price ⇒ Selling price = $1+\frac{1}{7}=\frac{8}{7}$ of cost price

∴ ₹ 64 = $\frac{8}{7}$ of cost price ⇒ ₹ 63 = $\frac{8}{7} \times \frac{63}{64} = \frac{9}{8}$ of cost price

∴ Profit = $\frac{9}{8}-1=\frac{1}{8}$ = 12.5% of cost price

92. A shopkeeper buys articles at 20% discount on marked price and sells them at 50% profit. How much percent the selling price is above the marked price?

(*a*) 20% (*b*) 30% (*c*) 60% (*d*) 70%

Sol. (*a*) Let marked price = ₹ 100. Cost price = ₹ 100 – ₹ 20 = ₹ 80

Selling price = ₹ $80 \times \frac{150}{100}$ = ₹ 120 ⇒ Selling price is 20% above the marked price.

93. On selling an article for ₹ 50, a man loses $\frac{1}{6}$ of his cost price. Find his profit percentage, if the article is sold for ₹ 66.

(*a*) 5% (*b*) 10% (*c*) 20% (*d*) 25%

Sol. (*b*) Loss = $\frac{1}{6}$ of CP = $\frac{1}{5}$ of SP = $\frac{1}{5}$ × 50 = 10 ⇒ CP = 50 + 10 = 60

New SP = 66 ⇒ Gain = 66 – 60 = 6 Profit % = $\frac{6}{60}$ × 100 = 10%

94. A sells an article to B at a profit of 10%. B sells the article back to A at a loss of 10%. In the transaction: **(SSC CGL Tier I 2008)**

(*a*) A neither gains nor loses (*b*) A gains 11% (*c*) A gains 20% (*d*) A gains 21%

Sol. (*b*) Let cost price of the article be ₹100. ⇒ A sells to B for = ₹110

B sold it to A for ₹ 110 – 10% of ₹110 = ₹ 99 ⇒A gains = ₹ 110 – ₹ 99 = ₹ 11

95. A house worth ₹ 1,50,000 is sold by X at 5% profit to Y. Y sells the house back to X at 2% loss. Then in the entire transaction : **(SSC CGL Tier I 2000)**

(*a*) X gains ₹ 4,350 (*b*) X loses ₹ 4,350 (*c*) Y gains ₹ 3,150 (*d*) Y loses ₹ 3,150

Sol. (*d*) X sold house to Y for ₹1,50,000 × $\frac{105}{100}$ = ₹ 1,57,500

∴ Y's loss = 2% of ₹ 1,57,500 = ₹ 3,150

96. The profit earned after selling an article for ₹ 1754 is same as loss incurred after selling the article for ₹ 1492. What is the cost price of the article ?

(Canara Bank PO 2009, Bank of Baroda PO 2007 (modified))

(*a*) ₹ 1623 (*b*) ₹ 1523 (*c*) ₹ 1689 (*d*) ₹ 1589

Sol. (*a*) Let Loss = Gain = x ⇒ CP = 1754 – x = 1492 + x ⇒ $2x$ = 1754 – 1492 = 262 ⇒ x = 131

∴ Cost Price = Selling Price (second) + Loss = ₹ 1492 + ₹ 131 = ₹ 1623

Or Cost Price = Selling Price (first) – Profit = ₹ 1754 – ₹ 131 = ₹ 1623

Alternative Method: Let the loss and gain be ₹ x in each case.

Then difference in two S.P. = $2x$ = ₹ 1754 – ₹ 1492 ⇒ $2x$ = ₹ 262 ⇒ x = 262 ÷ 2 = ₹ 131

∴ Cost Price = Selling Price (second) + Loss = ₹ 1492 + ₹ 131 = ₹ 1623

Alternative method : Cost price = ₹ (1754 + 1492) ÷ 2 = ₹ 1623

97. Profit after selling a commodity for ₹ 524 is the same as loss after selling it for ₹ 452. The cost price of the commodity is : **(SSC CGL Tier I 2003)**

(*a*) ₹ 480 (*b*) ₹ 500 (*c*) ₹ 488 (*d*) ₹ 485

Sol. (*c*) Let loss = profit = x. Then CP = 524 – x = 452 + x ⇒ $2x$ = 524 – 452 = 72 ⇒ x = 36

Cost price = 452 + 36 = ₹ 488

Alternative method : Cost price = ₹ (524 + 452) ÷ 2 = ₹ 488

98. The percentage of loss when an article is sold at ₹50 is the same as that of the profit when it is sold at ₹70. The above-mentioned percentage of profit or loss on the article is :

(SSC CGL Tier I 2008)

(*a*) 10% (*b*) $16\frac{2}{3}\%$ (*c*) 20% (*d*) $22\frac{2}{3}\%$

Sol. (*b*) Let cost price of the article be ₹ x. Then x – 50 = 70 – x ⇒ x = 60

Profit = ₹ (70 – 60) = ₹10 on ₹60 ⇒ Profit % = $\frac{10}{60}\times 100 = 16\frac{2}{3}\%$

99. The per cent of profit, when an article is sold for ₹ 78, is twice than when it is sold for ₹ 69, the cost of the article is : **(SSC CHSL 2010)**

(*a*) ₹ 49 (*b*) ₹ 51 (*c*) ₹ 57 (*d*) ₹ 60

Sol. (*d*) Let profit = x. Then CP + x = 69 and CP + $2x$ = 78 ⇒ x = 78 – 69 = 9

Cost price = 69 – 9 = ₹ 60

100. Profit on selling an article for ₹ 84 is thrice of profit earned on selling it for ₹ 76. Find the cost price of the article.

(*a*) ₹ 80 (*b*) ₹ 78 (*c*) ₹ 72 (*d*) ₹ 76

Sol. (*c*) Cost price = ₹ 84 – $3x$ = 76 – $x \Rightarrow 2x = 8 \Rightarrow x = 4$

$\Rightarrow$ Cost price = ₹ 84 – $3x$ = 84 – 3 × 4 = 72 or 76 – x = 76 – 4 = 72

101. The percentage profit earned by selling an article for ₹ 1730 is equal to the percentage loss incurred by selling the same article for ₹ 1270. At what price the article should be sold to earn 20% profit ?

(*a*) ₹ 1500 (*b*) ₹ 1650 (*c*) ₹ 1800 (*d*) Data inadequate

Sol. (*c*) Let CP = ₹ $x \Rightarrow$ CP = 1730 – x = x – 1270 $\Rightarrow 2x$ = 1730 + 1270 = 3000 $\Rightarrow x$ = ₹ 1500

Profit = 20% of ₹ 1500 = ₹ 300 $\Rightarrow$ Required selling price = 1500 + 300 = ₹ 1800

102. On selling an article for ₹ 270, the amount of loss suffered is as much as profit earned on selling the article at 10% profit. What is the selling price if the desired profit is 20%?

(*a*) ₹ 290 (*b*) ₹ 300 (*c*) ₹ 330 (*d*) ₹ 360

Sol. (*d*) Loss suffered = Profit made = 10% of the CP $\Rightarrow$ SP = ₹ 270 = ₹ 90% of CP

$\therefore$ Selling price at 20% profit = 120% of cost price = $270 \times \frac{120}{90}$ = ₹ 360

103. An article is sold for ₹ 500 and hence lost something. Had the article been sold for ₹ 700, the merchant would have gained three times the former loss. Find the cost price of the article.

(*a*) ₹ 525 (*b*) ₹ 550 (*c*) ₹ 600 (*d*) ₹ 650

Sol. (*b*) Let loss in the first case = ₹ x. Then profit in second case = ₹ $3x$.

$\therefore$ ₹ 500 = cost price – ₹ x .... (1)

And ₹ 700 = cost price + ₹ $3x$ .... (2)

Solving equations (1) and (2), we get: $4x$ = ₹ 200 $\Rightarrow x$ = ₹ 50

$\therefore$ Cost price = selling price (first) + loss = ₹ 500 + ₹ 50 = ₹ 550

Or, Cost price = Selling Price (second) – Profit = ₹ 700 – 3 × ₹ 50 = ₹ 550

Alt: Let loss in the first case = ₹ x. Then profit in the second case = ₹ $3x$

Difference in two selling prices = $x + 3x$ = ₹ 700 – ₹ 500 $\Rightarrow 4x$ = ₹ 200 $\Rightarrow x$ = ₹ 50

$\therefore$ Cost price = selling price (first) + loss = ₹ 500 + ₹ 50 = ₹ 550

104. On selling an article for ₹ 440, the amount of loss suffered is double of profit made on selling the article at 6% profit. What is the cost price of the article?

(*a*) ₹ 400 (*b*) ₹ 450 (*c*) ₹ 500 (*d*) ₹ 550

Sol. (*c*) Loss suffered = 2 × 6% of the cost price = 12% of cost price

$\therefore$ Selling price = ₹ 440 = 88% of the cost price $\Rightarrow$ Cost price = ₹ $440 \times \frac{100}{88}$ = ₹ 500

105. An article is sold for ₹ 190, hence gaining a certain amount. Had the article been sold for ₹175, he would have suffered loss equal to 50% of the gain in the first case. Find cost price of the article.

(*a*) ₹ 178 (*b*) ₹ 180 (*c*) ₹ 185 (*d*) ₹ 188

Sol. (*b*) Let gain in the first case = ₹ $2x$. Then loss in the second case = ₹ x.

Difference between two selling prices = ₹ $2x$ – (– ₹ x) = ₹ $2x$ + ₹ x = ₹ $3x$

Actual difference between the two SP = ₹ 190 – ₹ 175 = ₹ 15 $\Rightarrow 3x$ = ₹ 15 $\Rightarrow x$ = ₹ 5

$\therefore$ CP = ₹ 190 – $2x$ = ₹ 190 – ₹ 10 = ₹ 180 *or* cost price = ₹ 175 + x = ₹ 175 + ₹ 5 = ₹ 180

106. If 7% of selling price of an article is equal to 8% of its cost price and 9% of its selling price exceeds 10% of its cost price by ₹1, then find cost price of the article. **(MAT)**

(*a*) ₹ 300 (*b*) ₹ 350 (*c*) ₹ 400 (*d*) ₹ 500

Sol. (*b*) 7% of SP = 8% of CP $\Rightarrow$ SP : CP = 8 : 7 = $8x$, $7x$ (assume)

Then 9% of $8x$ – 10% of $7x = 1 \Rightarrow (0.72 - 0.7)x = 1 \Rightarrow 0.02x = 1 \Rightarrow x = 50$

$\Rightarrow$ CP = $7x = 7 \times 50 =$ ₹ 350

107. 5 litres of water is added to a certain quantity of pure milk costing ₹ 3 per litre. If the mixture is sold at the cost price of pure milk, a profit of 20% is made, what is the quantity of pure milk in the mixture?

(*a*) 1 litre (*b*) 20 litre (*c*) 25 litre (*d*) 30 litre

Sol. (*c*) Let pure milk = x litre. Total profit = CP of 5 litres = ₹ 3 × 5 = ₹ 15

$\Rightarrow$ Cost of milk = $3x \Rightarrow$ Profit = 20% of $3x$ = ₹ 15 $\Rightarrow x = \frac{15 \times 5}{3} = 25$ litre

108. The cost price of a book is ₹ 300. The shopkeeper wants to gain 20% after allowing a discount of 10% on the marked price. Then, the marked price of the book must be **(SSC MTS 2013)**

(*a*) ₹ 360 (*b*) ₹ 400 (*c*) ₹ 336 (*d*) ₹ 396

Sol. (*b*) If CP = ₹100, SP = ₹120; If MP = ₹100, SP = ₹90 $\Rightarrow$ MP = ₹$300 \times \frac{120}{100} \times \frac{100}{90}$ = ₹400

109. A sells a suitcase to B at 10% profit. B sells it to C at 30% profit. If C pays ₹ 2860 for it, then the price at which A bought it is **(SSC CGL Tier II 2013)**

(*a*) ₹ 1000 (*b*) ₹ 1600 (*c*) ₹ 2000 (*d*) ₹ 2500

Sol. (*c*) Cost price for A = $2860 \times \frac{100}{110} \times \frac{100}{130}$ = ₹ 2000

110. A sells a bicycle to B at a profit of 20%. B sells it to C at a profit of 25%. If C pays ₹ 225 for it, the cost price (in ₹) of the bicycle for A is : **(SSC CGL Tier I 2013 & 2002)**

(*a*) 110 (*b*) 120 (*c*) 125 (*d*) 150

Sol. (*d*) Cost price for A = ₹ $225 \times \frac{100}{125} \times \frac{100}{120}$ = ₹ 150

111. A shopkeeper allows 25% discount on the marked price of his articles and hence gains 25% of the cost price. What is the marked price of the article on selling which he gains ₹ 120?

(*a*) ₹ 400 (*b*) ₹ 600 (*c*) ₹ 800 (*d*) ₹ 1000

Sol. (*c*) If profit is ₹ 25, then selling price = ₹ 100 + ₹ 25 = ₹125

If marked price is ₹100, then selling price = ₹100 – ₹25 = ₹75

Marked price of the article = ₹ $120 \times \frac{125}{25} \times \frac{100}{75}$ = ₹ 800

112. Kabir buys an article with 25% discount on its marked price. He makes a profit of 10% by selling it at ₹ 660. The marked price is **(SSC CGL Tier I 2013)**

(*a*) ₹ 600 (*b*) ₹ 685 (*c*) ₹ 700 (*d*) ₹ 800

Sol. (*d*) Selling price = ₹ 660 $\Rightarrow$ Cost price = ₹$660 \times \frac{100}{110}$

Marked price = cost price × $\frac{100}{75} = 660 \times \frac{100}{110} \times \frac{100}{75}$ = ₹ 800

113. The cost price of a table is ₹ 3,200. A merchant wants to make 25% profit by selling it. At the time of sale he declares a discount of 20% on the marked price. The marked price is : **(SSC CGL Tier I 2014)**

(*a*) ₹ 5000 (*b*) ₹ 4000 (*c*) ₹ 6000 (*d*) ₹ 4500

Sol. (*a*) Marked price = $3200\times\frac{125}{100}\times\frac{100}{80}$ = ₹ 5000

114. A shopkeeper allows a discount of 12.5% on the marked price of a certain article and makes a profit of 20%. If the article costs the shopkeeper ₹210, then the marked price of the article will be **(SSC CGL Tier I 2014)**

(*a*) ₹ 387 (*b*) ₹ 350 (*c*) ₹ 386 (*d*) ₹ 288

Sol. (*d*) Marked price = $210\times\frac{120}{100}\times\frac{100}{87.5}$ = ₹ 288

115. Charging 30% above its production cost a radio maker puts a label of ₹286 on a radio as its price. But at the time of selling it, he allows 10% discount on the labelled price. What will his gain be ? **(SSC CGL Tier I 2014)**

(*a*) ₹ 257.40 (*b*) ₹ 254.40 (*c*) ₹ 198 (*d*) ₹ 37.40

Sol. (*d*) CP = 286×100/130 = ₹220; SP = 286×90/100 = ₹257.40 ⇒ Profit = 257.40 – 220 = ₹37.40

116. A man bought a horse for ₹ 10000 and sold it to B at 10% profit and B sold it to C at 10% loss. Find the amount paid by C.

(*a*) ₹ 10000 (*b*) ₹ 9900 (*c*) ₹ 9999 (*d*) ₹ 11000

Sol. (*b*) Difference = $\frac{10\times10}{100}$ = 1% decrease

∴ Required amount = 10000 – 1% of 10000 = 10000 – 100 = 9900

Alt : If A buys the horse for ₹ 100, B pays ₹ 110. If B buys the horse for ₹ 100, C pays ₹ 90.

∴ Amount paid by C = $10000\times\frac{110}{100}\times\frac{90}{100}$ = 9900

117. A sells a table costing ₹ 2000 to B and earns a profit of 6%. B sells it to C at a loss of 5%. At what price did B sell the table ?

(*a*) ₹ 2054 (*b*) ₹ 2050 (*c*) ₹ 2024 (*d*) ₹ 2014

Sol. (*d*) Difference = $6-5-\frac{6\times5}{100}$ = 0.7%

∴ Required price = 2000 + 0.7% of 2000 = 2000 + 14 = ₹ 2014

Alt: Required price = $2000\times\frac{100+6}{100}\times\frac{100-5}{100}=2000\times\frac{106}{100}\times\frac{95}{100}$ = ₹ 2014

Alternative mehod : Selling price for A = 2000 + 6% of 2000 = 2120

Selling price for B = 2120 – 5% of 2120 = 2120 – 106 = ₹ 2014

118. A manufacturer sells an article to a wholesale dealer at a profit of 10%. The wholesale dealer sells it to a shopkeeper at 20% profit. The shopkeeper sells it to a customer for ₹ 56,100 at a loss of 15%. Then the cost price of the article to the manufacturer is **(SSC CGL Tier II 2012)**

(*a*) ₹ 50,000 (*b*) ₹ 55,000 (*c*) ₹ 25,000 (*d*) ₹ 10,000

Sol. (*a*) Cost price to the manufacturer = ₹ $56{,}100\times\frac{100}{85}\times\frac{100}{120}\times\frac{100}{110}$ = ₹ 50,000

119. A sells an article to B making a profit of 1/5 of his outlay. B sells it to C, gaining 20%. If C sells it for ₹ 600 and incurs a loss of 1/6 of his outlay, the cost price of A is **(SSC CGL Tier II 2012)**

(*a*) ₹ 720 (*b*) ₹ 800 (*c*) ₹ 600 (*d*) ₹ 500

Sol. (*d*) Cost price of A = ₹ $600\times\frac{6}{5}\times\frac{100}{120}\times\frac{5}{6}$ = ₹ 500

120. A man purchased two articles for ₹ 10,000 each. On selling first, he gains 20% and on the other, he loses 20%. What is profit/loss in the transaction? **(Bank Clerical)**

(*a*) 4% loss (*b*) 4% gain (*c*) 2% loss (*d*) No profit/loss

Sol. (*d*) The cost price of both the articles are same.

∴ Profit made on one item is exactly equal to loss suffered on the other.

∴ No profit, no loss.

121. If Sohan, while selling two goats at the same price, makes a profit of 10% on one goat and suffers a loss of 10% on the other **(UPSC Civil Services (Pre.) 2014)**

(*a*) he makes no profit no loss (*b*) he makes a profit of 1%

(*c*) he suffers a loss of 1% (*d*) he suffers a loss of 1%

Sol. (c) Loss in such cases $= \frac{x^2}{100}$ *i.e.* $\frac{(\text{Common Gain and Loss \%})^2}{100} = \frac{10 \times 10}{100} = 1\%$

122. A dealer sold two types of goods for 10,000 each. On one of them, he lost 20% and on the other he gained 20%. His gain or loss percent in the entire transaction was **(SSC CGL Tier II 2012, SSC CGL Tier I 2001)**

(*a*) 4% gain (*b*) 4% loss (*c*) 2% loss (*d*) 2% gain

Sol. (*b*) Loss % $= \frac{(\text{Common Gain and Loss})^2}{100} = \frac{20^2}{100} = 4\%$

123. A trader sells two computers at the same price. On one he makes a profit of 15% and on the other he suffers a loss of 15%. Find his gain or loss per cent. **(SSC CGL Tier I 2013, 2005 M)**

(*a*) Loss 2.25% (*b*) Gain 2% (*c*) No loss no gain (*d*) Loss 2%

Sol. (*a*) Loss $= \frac{(15)^2}{100} = 2.25\%$

124. A man sold two horses for ₹ 9900 each. On selling first, he gains 10% and on the other, he loses 10%. Find his total gain or loss.

(*a*) ₹ 99 (*b*) ₹ 100 (*c*) ₹ 198 (*d*) ₹ 200

Sol. (*d*) Loss % $= \frac{10^2}{100} = 1\%$ of CP $= \frac{1}{100}$ of CP $= \frac{1}{100-1} = \frac{1}{99}$ of SP

∴ Total loss on two articles $= 2 \times 9900 \times \frac{1}{99} =$ ₹ 200

Alt: CP of first article $= 9900 \times \frac{100}{110} = 9000$ and CP of second article $= 9900 \times \frac{100}{90} = 10000$

Total CP = 9000 + 11000 = 20000. Total SP = 9900 × 2 = 19800

Total Loss = 20000 – 19800 = 200

125. A man sold two articles for ₹120000 each. On selling first, he gains 20% and on the other, he loses 20%. Find his total gain or loss. **(SSC CGL Tier I 2007, 2005M, CHSL 2013M)**

(*a*) ₹ 7500 (*b*) ₹ 5000 (*c*) ₹ 9600 (*d*) ₹ 10000

Sol. (*d*) Loss % $= \frac{20^2}{100} = 4\%$ of the CP $= \frac{1}{25}$ of the CP $= \frac{1}{25-1} = \frac{1}{24}$ of SP

∴ Total loss on two articles $= 2 \times 120000 \times \frac{1}{24} =$ ₹ 10000

126. A trader marks his goods at 20% above the cost price. If he allows a discount of 5% on the marked price, what profit per cent does he make ? **(SSCCGLTier I 2003)**

(a) 14% (b) 16% (c) 18% (d) 20%

Sol. (a) Let CP = 100, then MP = 120 $\Rightarrow$ Discount = 5% of 120 = 6

$\Rightarrow$ SP = 120 – 6 = 114 $\Rightarrow$ Profit = 14%

Alt : $20 - 5 - \frac{20\times5}{100} = 14$

127. A trade man marks his goods at 25% above the cost price. If he allows his customer 10% discount how much percent profit does he make ? **(SSC MTS 2014)**

(*a*) 7.5% (*b*) 8.5% (*c*) 10.5% (*d*) 12.5%

Sol. (*d*) Let CP = ₹ 100. Then MP = ₹ 125. $\Rightarrow$ Discount = ₹ 125 × 10% = 12.50

Selling price = ₹ 125 – ₹ 12.50 = ₹ 112.50 $\Rightarrow$ Profit = 12.5%

Alt: $25 - 10 - \frac{25\times10}{100} = 25 - 10 - 2.5 = 12.5\%$

128. A shopkeeper marks his goods 30% above the cost price and allows 10% discount on the marked price. His profit per cent is : **(SSC CGL Tier I 2008, 2012)**

(*a*) 10% (*b*) 12% (*c*) 17% (*d*) 20%

Sol. (*c*) Marked price = 130% of cost price

Sale price = 130% – 10% of 130 = 117% of cost price $\Rightarrow$ Profit per cent = 17%

Alternative method : Profit per cent = $30 - 10 - \frac{30\times10}{100} = 30 - 10 - 3 = 17\%$

Alternative method : 13 × 9 = 117 ; 117 – 100 = 17

129. By selling an umbrella for ₹ 30, a shopkeeper gains 20%. During a clearance sale, the shopkeeper allows a discount of 10% of the marked price. His gain percent during the sale season is **(SSC CGL Tier II 2012)**

(*a*) 8 (*b*) 9 (*c*) 7 (*d*) 7.5

Sol. (*a*) Required profit per cent = $20 - 10 - \frac{20\times10}{100} = 20 - 10 - 2 = 8\%$

Alternative method : $120\times\frac{90}{100} = 108$

130. A trader marked the price of his commodity so as to include a profit of 25%. He allowed discount of 16% on the marked price. His actual profit was : **(SSC CGL Tier I 2004)**

(*a*) 5% (*b*) 9% (*c*) 16% (*d*) 25%

Sol. (*a*) Profit per cent = $25 - 16 - \frac{25\times16}{100} = 25 - 16 - 4 = 5\%$

131. A shopkeeper marks the prices at 15% higher than the original price. Due to increase in demand, he further increase the price by 10%. How much percentage profit will he get ? **(SSC CGL Tier II 2012)**

(*a*) 20 (*b*) 24.5 (*c*) 25 (*d*) 26.5

Sol. (*d*) Required percentage = $15 + 10 + \frac{15\times10}{100} = 15 + 10 + 1.5 = 26.5\%$

132. A merchant fixes the sale price of his goods at 15% above the cost price. He sells his goods at 12% less than the fixed price. His percentage of profit is : **(SSC CGL Tier I 2003)**

(*a*) $2\frac{1}{2}\%$ (*b*) $1\frac{1}{5}\%$ (*c*) $1\frac{1}{2}\%$ (*d*) 2%

Sol. (*b*) Profit % = $15 - 12 - \frac{15\times12}{100} = 3 - 1.8 = 1.2\% = 1\frac{1}{5}\%$

133. A retailer purchased radiosets at the rate of ₹ 400 each from a wholesaler. He raised the price by 30% and then allowed a discount of 8% on each set. His profit will be **(SSC CGL Tier I 2013)**

(*a*) 19% (*b*) 78.4% (*c*) 22% (*d*) 19.6%

Sol. (*d*) Let CP = ₹ 100. Then MP = ₹ 130. ⇒ Discount = ₹ 130 × 8% = 10.4
Selling price = ₹ 130 – ₹ 10.4 = ₹ 119.6 ⇒ Profit = 19.6%

Alternative method : $30 - 8 - \frac{30\times8}{100} = 22 - 2.4 = 19.6\%$

134. The marked price of an article is 10% higher than cost price. A discount of 10% is given on marked price. In this kind of sale, the seller bears : **(SSC CGL Tier I 2004, CDS 2011)**

(*a*) no gain, no loss (*b*) a loss of 5% (*c*) a gain of 1% (*d*) a loss of 1%

Sol. (*d*) Marked price = ₹ 110% of cost price ⇒ SP = 110% – 10% of 110 = 99% of CP
∴ Loss per cent = 100% – 99% = 1%

Alternative method : Net loss = $\frac{10\times10}{100} = 1\%$

135. An item costing ₹ 300 is being sold at 10% gain. If the selling price is reduced by 15%, what will be the gain or loss per cent? **(SSC FCI 2012)**

(*a*) $6\frac{1}{2}\%$ gain (*b*) $6\frac{1}{2}\%$ loss (*c*) 5% loss (*d*) 5% gain

Sol. (*b*) $+10 - 15 - \frac{10\times15}{100} = -5 - 1.5 = -6.5\%$ Loss

136. A shopkeeper marks his goods 20% above cost price, but allows 30% discount for cash. His net loss is : **(SSC CGL Tier I 2000)**

(*a*) 8% (*b*) 10% (*c*) 16% (*d*) 20%

Sol. (*c*) Net loss = $20 - 30 - \frac{20\times30}{100} = 20 - 30 - 6 = -16\%$

137. A tradesman marks his goods 30% more than the cost price. If he allows a discount of $6\frac{1}{4}\%$, then his gain per cent is **(SSC CGL Tier II 2013)**

(*a*) $23\frac{3}{4}\%$ (*b*) 22% (*c*) $21\frac{7}{8}\%$ (*d*) 30%

Sol. (*c*) $30 - 6\frac{1}{4} - \frac{30\times25}{400} = 23\frac{3}{4} - \frac{15}{8} = 21\frac{3}{4} + \frac{1}{8} = 21\frac{7}{8}\%$

138. A sells an article to B at a profit of 20%. B sells it to C at a profit of 10%. How many per cent will C pay more than what A pays? **(SSC CHSL 2012)**

(*a*) 28% (*b*) 30% (*c*) 32% (*d*) 35%

Sol. (*c*) $20 + 10 + \frac{20\times10}{100} = 20 + 10 + 2 = 32\%$

Alt: Let CP = ₹ 100 ⇒ Amount paid by C = ₹ $100 \times \frac{120}{100} \times \frac{110}{100}$ = ₹ 132

∴ C pays more than A = ₹ 132 – ₹ 100 = ₹ 32 or 32%

139. Two tables are purchased for the total cost of ₹5000. First table is sold at 40% profit and second at 40% loss. If selling price is same for both the tables, what is the cost price of the table that was sold at profit?

(*a*) ₹ 1500 (*b*) ₹ 2000 (*c*) ₹ 3500 (*d*) ₹ 4000

Sol. (*a*) 140% of cost price of the first table = 60% of cost price of the second table

∴ Cost price of the first table : Cost price of the second table = 60 : 140 = 3 : 7

∴ Cost price of the first table = $\frac{3}{10}$ × ₹ 5000 = ₹ 1500

140. A shopkeeper bought 2 pens for ₹ 480. He sold the first at a loss of 15% and the second at a profit of 19%. Find cost price of the second pen, if selling price of both the pens are same.

(*a*) ₹ 195 (*b*) ₹ 200 (*c*) ₹ 250 (*d*) ₹ 280

Sol. (*b*) 85% of cost price of first pen = 119% of cost price of second pen

∴ Cost price of first pen : Cost price of second pen = 119 : 85 = 7 : 5

∴ Cost price of the second pen = $\frac{5}{12}$ × 480 = ₹ 200

141. A man bought two goats for ₹ 1008. He sold one at a loss of 20% and other at a profit of 44%. If each goat was sold for the same price, the cost price of the goat, which was sold at loss was: **(SSC CGL Tier I 2004)**

(*a*) ₹ 648 (*b*) ₹ 360 (*c*) ₹ 568 (*d*) ₹ 440

Sol. (*a*) 80% of cost price of first goat = 144% of cost price of second goat

∴ Cost price of first goat : Cost price of second goat = 144 : 80 = 9 : 5

∴ Cost price of the first goat = $\frac{9}{14}$ × 1008 = ₹ 648

142. The marked price of an article is ₹ 200. A discount of $12\frac{1}{2}$% is allowed on the marked price and a profit of 25% is made. The cost price of the article is : **(SSC CGL Tier I 2002)**

(*a*) ₹ 200 (*b*) ₹ 175 (*c*) ₹ 120 (*d*) ₹ 140

Sol. (*d*) Rate of discount = $12\frac{1}{2}\% = \frac{1}{8}$ ⇒ Selling price = $1-\frac{1}{8}=\frac{7}{8}$ of the marked price

Profit = 25% = $\frac{1}{4}$ of CP ⇒ CP = $\frac{4}{5}$ of SP ⇒ CP = ₹ $200\times\frac{7}{8}\times\frac{4}{5}$ = ₹ 140

143. A man sold two articles for ₹ 2500 each. On selling first, he gains 20% and on the other he loses 10%. What is his per cent profit in the transaction? **(MBA Exam.)**

(*a*) 2.18% (*b*) 2.86% (*c*) 5% (*d*) 10%

Sol. (*b*) Let SP = LCM of (100 + 20) and (100 – 10) = 360 ⇒ Total SP = 2 × 360 = 720

Total CP = $\frac{100}{120}\times360+\frac{100}{90}\times360$ = 300 + 400 = 700

Profit = 720 – 700 = 20 ⇒ Profit % = $\frac{20}{700}\times100$ = 2.86% (approx.)

Direct Method : $\frac{2\times120\times90}{210}$ – 100 = 102.86 – 100 = 2.86%

144. If I purchased 11 books for ₹ 100 and sold 10 books for ₹ 110, the percentage of profit per book sold is **(SSC MTS 2011, CGL Tier I 2001 M)**

(*a*) 10 (*b*) 11.5 (*c*) 17.3 (*d*) 21

Sol. (*d*) SP : CP = $\frac{110}{10}:\frac{100}{11}$ = 121 : 100 ⇒ Profit % = $\frac{21}{100}\times 100$ = 21%

145. Oranges are bought at the rate of 10 for ₹ 25 and sold at the rate of 9 for ₹ 25. The profit is : **(SSC CGL Tier I 2003)**

(*a*) $9\frac{1}{11}$% (*b*) 10% (*c*) $11\frac{1}{9}$% (*d*) $12\frac{1}{2}$%

Sol. (*c*) Cost price : Selling price = $\frac{25}{10}:\frac{25}{9}=\frac{1}{10}:\frac{1}{9}$ = 9 : 10 ⇒ Profit = $\frac{1}{9}\times 100 = 11\frac{1}{9}$%

146. A fruit seller buys oranges @ 8 for ₹ 10 and sells them @ 10 for ₹ 8. His profit/loss percentage is: **(SSC CGL Tier I Exam.)**

(*a*) 36% loss (*b*) 36% gain (*c*) 45% gain (*d*) 45% loss

Sol. (*a*) SP : CP = $\frac{8}{10}:\frac{10}{8}$ = 64 : 100 ⇒ Loss = 100% – 64% = 36%

147. Some articles were bought at 6 for ₹ 5, and sold at 5 for ₹ 6. Gain is: **(SSC CGL 2004)**

(*a*) 5% (*b*) 6% (*c*) 30% (*d*) 44%

Sol. (*d*) SP : CP = $\frac{6}{5}:\frac{5}{6}$ = 36 : 25 ⇒ Profit = $\frac{36}{25}-1=\frac{11}{25}=\frac{11}{25}\times 100\%$ = 44%

148. A fruit seller buys lemons at 2 for a rupee and sells them at 5 for three rupees. His profit per cent is :

(*a*) 10% (*b*) 15% (*c*) 20% (*d*) 25%

Sol. (*c*) $\frac{\text{Selling price}}{\text{Cost price}}=\frac{3}{5}\div\frac{1}{2}=\frac{3}{5}\times 2=\frac{6}{5}$ ⇒ Profit = $\frac{6}{5}-1=\frac{1}{5}$ = 20%

149. A retailer bought some apples @ 7 apples for ₹ 4 and sold them @ 8 apples for ₹ 5. If he gains ₹ 30 on that day, find the quantity of apples sold by him on that day.

(*a*) 56 (*b*) 70 (*c*) 80 (*d*) 560

Sol. (*d*) LCM of quantities purchased and sold *i.e.* LCM of 7 and 8 is 56.

Cost price of 56 apples = $56\times\frac{4}{7}$ = ₹ 32 Selling price of 56 apples = $56\times\frac{5}{8}$ = ₹ 35

Profit on selling 56 apples = ₹ 35 – ₹ 32 = ₹ 3

But actual profit is ₹ 30 (*i.e.* 10 times 3) ⇒ Apples sold = 56 × 10 = 560 apples

150. A man purchased some eggs at 3 for ₹ 5 and sold them at 5 for ₹ 12. Thus he gained ₹ 143 in all. The number of eggs he bought is **(SSC CGL Tier I 2011)**

(*a*) 210 (*b*) 200 (*c*) 195 (*d*) 190

Sol. (*c*) Profit on each egg = $\frac{12}{5}-\frac{5}{3}=\frac{11}{15}$ ∴ Number of eggs bought = $\frac{15}{11}\times 143$ = 195

151. Nisha bought a number of oranges at 2 for a rupee and an equal number at 3 for a rupee. To make a profit of 20% she should sell a dozen for **(SSC CGL Tier I 2014, CDS 2010)**

(*a*) ₹ 10 (*b*) ₹ 8 (*c*) ₹ 6 (*d*) ₹ 12

Sol. (*c*) Let oranges bought at two rates = LCM of 2 and 3 = 6

Cost price of 6 oranges @ 2 oranges for a rupee = $\frac{1}{2}\times 6$ = ₹ 3

Cost price of 6 oranges @ 3 oranges for a rupee = $\frac{1}{3}\times 6$ = ₹ 2

$\therefore$ Cost price of 12 (*i.e.* 6 + 6) oranges = ₹ 3 + ₹ 2 = ₹ 5

$\therefore$ Desired selling price = $5 \times \frac{120}{100}$ = ₹ 6

152. A fruit seller buys some oranges at the rate of 4 for ₹ 10 and an equal number more at 5 for ₹10. He sells the whole lot at 9 for ₹ 20. What is his loss or gain per cent?

(SSC CGL Tier I 2013)

(*a*) Loss $1\frac{19}{81}$% (*b*) Gain $1\frac{19}{81}$% (*c*) 2% loss (*d*) No loss no profit

Sol. (*a*) Let he buys 18 oranges each of two types. Total CP = $\frac{10}{4} \times 18 + \frac{10}{5} \times 18 = 45 + 36 =$ ₹81

Total SP = $\frac{20}{9} \times 36$ = ₹ 80 $\Rightarrow$ Loss = ₹ 1 on ₹ 81 = $\frac{1}{81} \times 100\% = 1\frac{19}{81}\%$

153. A shopkeeper sold one-third of his articles at 16% profit and the remaining at 10% profit. Find his overall profit per cent.

(*a*) 12% (*b*) 15% (*c*) 26% (*d*) 36%

Sol. (*a*) Let total articles sold are 3 of ₹ 100 (cost price) each.

Then total gain = 1 × 16 + 2 × 10 = ₹ 36 $\Rightarrow$ Profit per cent = $\frac{36}{3} = 12\%$

Alternative Method: Overall profit = $\frac{1}{3} \times 16\% + \frac{2}{3} \times 10\% = \frac{36}{3} = 12\%$

154. A cloth merchant sold half of his cloth at 40% profit, half of remaining at 40% loss and the rest was sold at the cost price. In the total transaction, his gain or loss will be

(*a*) 20% gain (*b*) 25% loss (*c*) 10% gain (*d*) 15% loss

Sol. (*c*) Profit % = $\frac{1}{2} \times 40 + \frac{1}{4} \times (-40) + \frac{1}{4} \times 0 = 20 - 10 + 0 = 10\%$

155. A dealer sold 3/4 of his articles at a gain of 24% and the remaining at the cost price. Percentage of gain in the whole transaction is **(SSC MTS 2011)**

(*a*) 15 (*b*) 18 (*c*) 24 (*d*) 32

Sol. (*b*) Profit % = $\frac{3}{4} \times 24\% + \frac{1}{4} \times 0 = 18\%$

156. A shopkeeper sold one-third of his articles at 5% profit. At what profit per cent should he sell the remaining articles so that his overall gain is 15%.

(*a*) 10% (*b*) 20% (*c*) 25% (*d*) 30%

Sol. (*b*) Let total articles sold are 3 of ₹ 100 (cost price) each.

Then total desired gain = 3 × 15 = ₹ 45. Gain on the first item = ₹ 5

Desired gain = 45 – 5 = ₹ 40 on the remaining two articles.

$\therefore$ Profit percent on remaining stock = $\frac{40}{2} = 20\%$

157. A merchant sold one-third of his goods at 10% loss. At what profit should he sell the remaining goods so as to gain 20% on selling the whole goods?

(*a*) 5% (*b*) 10% (*c*) 30% (*d*) 35%

Sol. (*d*) Let articles bought = 3 of ₹ 100 each. Then total profit made = 3 × 20 = ₹60

Profit on first article = – 10 ⇒ Desired profit on remaining 2 articles = 60 – (–10) = ₹70

∴ Profit desired on each of the two articles = 70 ÷ 2 = 35

158. Two-third of a consignment was sold at a profit of 6% and the rest at a loss of 3%. If there was an overall profit of ₹ 540, the value of the consignment was :

(*a*) ₹ 15,000 (*b*) ₹ 16,000 (*c*) ₹ 18,000 (*d*) ₹ 20,000

Sol. (*c*) Total profit = $\frac{2}{3}\times 6\% - \frac{1}{3}\times 3\%$ = (4 – 1)% = 3% = ₹ 540

Value of consignment = ₹ 540 × $\frac{100}{3}$ = ₹ 18000

159. A man purchased 150 pens at the rate of ₹ 12 per pen. He sold 50 pens at a gain of 10%. The percentage gain at which he must sell the remaining pens so as to gain 15% of the whole outlay is **(SSC CGL Tier II 2012)**

(*a*) 17% (*b*) $17\frac{1}{2}$ % (*c*) $21\frac{1}{2}$ % (*d*) 20%

Sol. (*b*) Ratio of pens sold and remaining pens = 50 : (150 – 50) = 1 : 2

Let pens bought = 3 of ₹ 100 each. Then total profit = 3 × 15 = ₹45

Profit on first article = ₹10 ⇒ Profit on remaining 2 articles = 45 – 10 = ₹35

∴ Profit desired on each of the remaining two articles = 35 ÷ 2 = 17.5%

160. A shopkeeper sold one-fourth of his articles at 15% loss. At what profit percent should he sell the remaining articles so that his overall gain is 6% ?

(*a*) 13% (*b*) 16% (*c*) 17% (*d*) 21 %

Sol. (*a*) Let total articles sold are 4 of ₹ 100 (cost price) each.

Then total desired gain = 4 × 6 = ₹ 24. Loss on the first article = ₹ 15

Required gain = 24 + 15 = ₹ 39 on the remaining three articles.

∴ Profit per cent on the remaining articles = 39 ÷ 3 = 13%

161. Gita buys a plot of land for ₹ 96000. She sells 2/5 of it at a loss of 6%. She wants to make a profit of 10% on the whole transaction by selling the remaining land. The gain % on the remaining land is **(SSC CGL Tier II 2013)**

(*a*) 20 (*b*) $20\frac{2}{3}$ (*c*) 14 (*d*) 7

Sol. (*b*) $\frac{5\times 10\% + 2\times 6\%}{5-2} = \frac{62}{3}\% = 20\frac{2}{3}\%$

162. A man purchases three television sets of ₹ 7500 each and sold them at average price of ₹ 7800. The first was sold at a profit of 8%, the second was sold at loss of 5%. Find the gain on selling the third television.

(*a*) 4% (*b*) 6% (*c*) 9% (*d*) 12%

Sol. (*c*) Average gain on each set = 7800 – 7500 = ₹ 300

Average gain% on each set = $\frac{300}{7500}\times 100$ = 4%

Let cost price of one set be ₹ 100. Then average profit on one set = ₹ 4

Total profit on three sets = 4 × 3 = ₹ 12

Profit desired on third set = 12 – (8 – 5) = ₹ 9 or 9%

163. A person bought 50 pens for ₹ 50 each. He sold 40 of them at a loss of 5%. He wants to gain 10% on the whole. Then his gain percent on the remaining pens should be **(SSC CPO SI 2011)**

(*a*) 15 (*b*) 40 (*c*) 50 (*d*) 70

Sol. (*d*) Let cost price of each pen = ₹ 100. Then profit required on 50 pens = 50 × ₹ 10 = ₹ 500

Loss suffered on 40 pens = 40 × ₹ 5 = ₹ 200 ⇒ Required profit % = $\frac{500+200}{10} = 70\%$

164. A merchant sells his goods at 10% profit but he mixes low quality of goods equal to one-fourth of good quality of goods. If the price of low quality of goods is four-fifth of that of good quality of goods, what is his net profit percentage?

(*a*) 8% (*b*) 12% (*c*) $14\frac{7}{12}\%$ (*d*) $16\frac{3}{7}\%$

Sol. (*c*) Good quality goods : Low quality goods = 4 : 1

Total CP of goods whose SP is ₹ 110 = $100\times\frac{4}{5}+100\times\frac{1}{5}\times\frac{4}{5}$ = ₹(80 + 16) = ₹ 96

∴ Profit = ₹ 14 on ₹ 96 = $\frac{14}{96}\times100\% = \frac{175}{12}\% = 14\frac{7}{12}\%$

165. A trader marked his goods at 20% above the cost price. He sold half the stock at the marked price, one quarter at a discount of 20% on the marked price and the rest at a discount of 40% on the marked price. His total gain is : **(SSC CGL Tier I 2004)**

(*a*) 2% (*b*) 4.5% (*c*) 13.5% (*d*) 15%

Sol. (*a*) Let cost price of each article be ₹ 100. Then marked price of each article = ₹ 120

Total selling price = $\frac{1}{2}\times120+\frac{1}{4}\times(120-20\% \text{ of } 120)+\left(1-\frac{1}{2}-\frac{1}{4}\right)\times(120-40\% \text{ of } 120)$

$= \frac{1}{2}\times120+\frac{1}{4}\times96+\frac{1}{4}\times72 = 60+24+18 = 102$

∴ Profit = 102 – 100 = 2 on 100 = 2%

166. An article passing through two hands, is sold at a profit of 38% of the original cost price. If the first person makes a profit of 20%, then the profit per cent made by the second person is:

(*a*) 5% (*b*) 10% (*c*) 12.5% (*d*) 15%

Sol. (*d*) Let original price = ₹ 100. Then final price = ₹ 138

Price paid by second person = ₹ 120 ⇒ His profit = ₹ 138 – ₹ 120 = ₹ 18

∴ Required profit % = $\frac{18}{120}\times100 = 15\%$

167. The price of jewel, passing through three hands, rises on the whole by 65%. If the first and the second sellers earned 20% and 25% profit respectively, the profit earned by the third seller is **(SSC CPO SI 2003)**

(*a*) 20% (*b*) 15% (*c*) 10% (*d*) 5%

Sol. (*c*) Selling price of second person = $100\times\frac{120}{100}\times\frac{125}{100} = 150$

⇒ Profit earned by third seller = 165 – 150 = ₹ 15

Required rate of profit = $\frac{15}{150}\times100 = 10\%$

168. On selling 10 articles, a merchant loses equal to cost price of 2 articles. Find his loss per cent.

(*a*) 12.5% (*b*) 20% (*c*) 25% (*d*) 33.33%

Sol. (*b*) Loss % = $\frac{2}{10} = \frac{1}{5} = 20\%$

169. By selling 25 metres of cloth, trader gains the selling price of 5 metres of cloth. The gain of the trader in % is **(SSC CGL Tier II 2013)**

(*a*) 25 (*b*) 20 (*c*) 28 (*d*) 29

Sol. (*a*) Profit % = $\frac{5}{25-5} \times 100 = 25\%$

170. A cloth merchant on selling 33 metre of cloth obtains a profit equal to the selling price of 11 metre of cloth. The profit is **(SSC CHSL 2013, MAT)**

(*a*) 22% (*b*) 11% (*c*) 50% (*d*) 40%

Sol. (*c*) Gain % = $\frac{11}{33-11} \times 100 = \frac{11}{22} \times 100 = 50\%$

171. By selling 8 dozen pencils, a shopkeeper gains the selling price of 1 dozen pencils. What is the gain ? **(CDS 2010)**

(*a*) $12\frac{1}{2}\%$ (*b*) $13\frac{1}{7}\%$ (*c*) $14\frac{2}{7}\%$ (*d*) $87\frac{1}{2}\%$

Sol. (*c*) Gain % = $\frac{1}{8-1} = \frac{1}{7} = 14\frac{2}{7}\%$

172. By selling 18 articles, a shopkeeper loses the selling price of 2 articles. Find his loss per cent.

(*a*) 10% (*b*) 11.11% (*c*) 12.5% (*d*) 25%

Sol. (*a*) Profit % = $\frac{-2}{18-(-2)} = \frac{-2}{20} = \frac{-1}{10}$ = (–) 10% = 10% Loss

173. If the cost price of 12 oranges is equal to selling price of 10 oranges, then the percentage of profit is : **(SSC CGL Tier I 2003)**

(*a*) $16\frac{2}{3}\%$ (*b*) 20% (*c*) 18% (*d*) 25%

Sol. (*b*) Profit % = $\frac{12-10}{10} \times 100 = 20\%$

Alt: Profit on selling 10 oranges = CP of 12 oranges – CP of 10 oranges = CP of 2 oranges

$\therefore$ Profit % = $\frac{2}{10} \times 100 = 20\%$

Alternative Method: CP of 12 oranges = SP of 10 oranges ⇒ CP : SP = 10 : 12

$\therefore$ Profit = $\frac{2}{10} \times 100 = 20\%$

174. By selling 15 mangoes, a fruit vendor recovers the cost price of 20 mangoes. What is the profit percentage of the fruit vender ? **(UPSC Civil Services (Pre.) 2011)**

(a) 20.5% (b) 30.67% (c) 33.33% (d) 35.4%

Sol. (c) Profit % = $\frac{20-15}{15} = \frac{5}{15} = 33.33\%$

175. The selling price of 20 articles is equal to the cost price of 22 articles. The gain percentage is **(SSC MTS 2013)**

(*a*) 12% (*b*) 9% (*c*) 10% (*d*) 11%

Sol. (*c*) Profit % = $\frac{22-20}{20}\times 100 = 10\%$

Alternative Method: SP : CP = 22 : 20 = 11 : 10 ⇒ Profit % = $\frac{1}{10}\times 100 = 10\%$

176. The cost price of 40 articles is the same as the selling price of 25 articles. Find the gain %. **(SSC CGL Tier II 2012)**

(*a*) 15% (*b*) 75% (*c*) 65% (*d*) 60%

Sol. (*d*) Profit % = $\frac{40-25}{25}\times 100 = 60\%$

Alternative Method: CP of 40 articles = SP of 25 articles ⇒ CP : SP = 25 : 40 = 5 : 8

∴ Profit = $\frac{3}{5}\times 100 = 60\%$

Alt : Profit on selling 25 articles = CP of 40 articles – CP of 25 pens = C.P. of 15 pens

∴ Profit = $\frac{15}{25}\times 100 = 60\%$

177. If the cost price of 12 pens is equal to the selling price of 8 pens, the gain per cent is : **(SSC CGL Tier I 2004)**

(*a*) $33\frac{1}{3}\%$ (*b*) $62\frac{2}{3}\%$ (*c*) 25% (*d*) 50%

Sol. (*d*) Profit % = $\frac{12-8}{8}\times 100 = 50\%$

Alt: CP of 12 pens = SP of 8 pens ⇒ CP : SP = 8 : 12 = 2 : 3 ⇒ Profit = $\frac{1}{2}\times 100 = 50\%$

Alt : Profit on selling 8 pens = CP of 12 pens – CP of 8 pens = C.P. of 4 pens

∴ Profit = $\frac{4}{8}\times 100 = 50\%$

178. The cost price of 20 articles is same as selling price of x articles. If the profit is 25%, find the value of x. **(MAT 2004)**

(*a*) 15 (*b*) 16 (*c*) 24 (*d*) 25

Sol. (*b*) In Option (b) : Profit = $\frac{20-16}{16} = \frac{4}{16} = \frac{1}{4} = 25\%$

179. If the cost price of 15 tables be equal to the selling price of 20 tables, the loss per cent is : **(SSC CGL Tier I 2001, SSC CHGL 2011)**

(*a*) 20% (*b*) 30% (*c*) 25% (*d*) 37.5%

Sol. (*c*) Profit % = $\frac{15-20}{20} = \frac{-5}{20} = \frac{-1}{4} = -25\% = 25\%$ Loss

Alternative method : Cost price : Selling price = 20 : 15 = 4 : 3 ⇒ Loss = $\frac{1}{4} = 25\%$

180. If selling price of an article is doubled, the profit triples. Find the profit percent. **(MAT 2001)**

(*a*) 100% (*b*) 200% (*c*) 300% (*d*) 400%

Sol. (*a*) 2SP – CP = 3(SP – CP) $\Rightarrow$ SP = 2 CP $\Rightarrow$ SP : CP = 2 : 1 $\Rightarrow$ Profit = 1 = 100%

181. If the selling price of an article is doubled, then its loss percent is converted into equal profit per cent. The loss per cent on the article is : **(SSC CGL Tier I 2008)**

(*a*) $26\frac{2}{3}\%$ (*b*) $33\frac{1}{3}\%$ (*c*) 50% (*d*) 20%

Sol. (*b*) Let cost price of the article be ₹100 and profit / loss percent be x%.

Then $(100 + x) = 2(100 - x) \Rightarrow 100 + x = 200 - 2x \Rightarrow 3x = 100$, or $x = 33\frac{1}{3}\%$

182. The cost price of an article is 64% of the marked price. The gain percentage after allowing a discount of 12% on the marked price is **(SSC CGL Tier I 2011)**

(*a*) 37.5% (*b*) 48% (*c*) 50.5% (*d*) 52%

Sol. (*a*) Let marked price = ₹ 100. Then cost price = ₹ 64. $\Rightarrow$ SP = ₹ 100 – 12% = ₹ 88.

$\Rightarrow$ Profit = ₹ 88 – ₹ 64 = ₹ 24 $\Rightarrow$ Profit % = $\frac{24}{64} \times 100 = 37.5\%$

183. A man sells a cycle for ₹ 600, gaining thereby 1/5 of its cost price. What will be the selling price if he had suffered loss 1/4 of the selling price?

(*a*) ₹ 400 (*b*) ₹ 500 (*c*) ₹ 600 (*d*) None of these

Sol. (*a*) Profit = 1/5 of cost price = 1/6 of selling price = 1/6 × ₹ 600 = ₹ 100

Cost price = ₹ 600 – ₹ 100 = ₹ 500

In the second situation : Loss = 1/4 of the SP = 1/5 of the CP = 1/5 × ₹ 500 = ₹ 100

New selling price = ₹ 500 – ₹ 100 = ₹ 400

184. X sells two articles for ₹ 4,000 each with no loss or no gain in the transaction. If one was sold at a gain of 25%, the other is sold at a loss of **(SSC CGL Tier I 2011)**

(*a*) 25% (*b*) $18\frac{2}{9}\%$ (*c*) $16\frac{2}{3}\%$ (*d*) 20%

Sol. (*c*) Profit on first article = $\frac{1}{4}$ of CP = $\frac{1}{5}$ of SP = $\frac{1}{5} \times 4000$ = ₹ 800

Loss on second article = ₹ 800. $\Rightarrow$ Its CP = ₹ 4000 + ₹ 800 = ₹ 4800

Required percentage = $\frac{800}{4800} \times 100 = 16\frac{2}{3}\%$

Alt: Profit = $\frac{1}{4}$ of CP = $\frac{1}{5}$ of SP $\Rightarrow$ Loss = $\frac{1}{5}$ of SP = $\frac{1}{6}$ of CP = $16\frac{2}{3}\%$

185. One trader calculates the percentage profit on the buying price and another calculates on the selling price. When their selling prices are the same, then the difference of their actual profits is ₹ 85 and both claim to have made 20% profit. What is the selling price of each ? **(SSC CGL Tier I 2008)**

(*a*) ₹1700 (*b*) ₹2100 (*c*) ₹2550 (*d*) ₹2750

Sol. (*c*) Let the CP in first case = ₹100 $\Rightarrow$ Profit = 20% of CP = ₹20 $\Rightarrow$ SP = ₹100 + ₹20 = ₹120

Profit in the second case, on selling price of ₹ 120 = 20% of ₹ 120 = ₹ 24

Difference between profits in two cases = ₹ 24 – ₹ 20 = ₹ 4

If difference is ₹4, SP = ₹120 (*i.e.* 30 times 4) ⇒ If difference is ₹85, SP = ₹85 × 30 = ₹2550

Alternative Method: Profit in the first case = 1/5 of cost price = 1/6 of selling price

Profit in the second case = 1/5 of selling price

$\therefore$ Difference in SP = $\frac{SP}{5} - \frac{SP}{6}$ = ₹ 85 ⇒ Selling price = ₹ 85 × 30 = ₹ 2550

186. A milk seller purchased milk at the rate of ₹10 per litre and adds one fifth of water to it. What per cent profit does he make by selling the mixture at ₹ 12 per litre?

(*a*) 20% (*b*) 25% (*c*) 30% (*d*) 50%

Sol. (*d*) The mixture contains $\frac{1}{5}$ water in it.

$\therefore$ Quantity of milk in 1 litre of mixture = $1 - \frac{1}{5} = \frac{4}{5}$ litre

Actual cost price per litre = ₹ 10 × $\frac{4}{5}$ = ₹ 8 per litre

Profit per litre = ₹ 12 – ₹ 8 = ₹ 4 on ₹ 8 ⇒ Profit = $\frac{4}{8} = \frac{1}{2}$ = 50%

187. The marked price of an article is ₹200. A discount of $12\frac{1}{2}$% is allowed on the marked price and a profit of 25% is made. The cost price of the article is. **(SSC CGL Tier I 2002)**

(*a*) ₹ 200 (*b*) ₹175 (*c*) ₹ 120 (*d*) ₹140

Sol. (*d*) Discount = $12\frac{1}{2}\% = \frac{1}{8}$ of marked price = $\frac{1}{8} \times 200$ = ₹ 25

Selling price = ₹200 – ₹ 25 = ₹175 ⇒ Cost price = ₹ $175 \times \frac{100}{125}$ = ₹140

188. Jasmine allows 4% discount on the marked price of her goods and still earns a profit of 20%. What is the cost price of a shirt if its marked price is ₹ 850 ? **(SSC CGL Tier I 2013)**

(*a*) ₹ 650 (*b*) ₹ 720 (*c*) ₹ 700 (*d*) ₹ 680

Sol. (*d*) CP = ₹ $850 \times \frac{96}{100} \times \frac{100}{120}$ = ₹ 680

189. A shopkeeper offers 10% discount on the marked price of his article and still makes a profit of 20%. What is the actual cost of the article marked ₹ 500 for him ? **(SSC CGL 2005)**

(*a*) ₹440 (*b*) ₹425 (*c*) ₹400 (*d*) ₹375

Sol. (*d*) SP = ₹ $500 \times \frac{100-10}{100}$ ⇒ CP = ₹ $500 \times \frac{100-10}{100} \times \frac{100}{100+20} = 500 \times \frac{90}{100} \times \frac{100}{120}$ = ₹ 375

190. By giving a discount of 10% on the marked price of ₹ 1100 of a cycle, a dealer gains 10%. The cost price of the cycle is : **(SSC CGL Tier I 2003)**

(*a*) ₹ 11000 (*b*) ₹ 900 (*c*) ₹ 1089 (*d*) ₹ 891

Sol. (*b*) Cost price = $1100 \times \frac{100-10}{100} \times \frac{100}{100+10} = 1100 \times \frac{90}{100} \times \frac{100}{110}$ = ₹ 900

191. A dealer buys an article listed at ₹ 100 and gets successive discounts of 10% and 20%. He spends 10% of the Cost Price on transporation. At what price should he sell the article to earn a profit of 15% **(SSC CGL Tier I 2015)**

(*a*) ₹ 92.00 (*b*) ₹ 91.08 (*c*) ₹ 90.80 (*d*) ₹ 91.20

Sol. (*b*) Selling price $= 100 \times \frac{90}{100} \times \frac{80}{100} \times \frac{110}{100} \times \frac{115}{100} =$ ₹ 91.08

192. A shopkeeper allows a discount of 10% on the marked price of a camera. Marked price of the camera, which cost him ₹ 600, to make a profit of 20% should be **(SSC CGL Tier I 2015)**

(*a*) ₹ 700 (*b*) ₹ 800 (*c*) ₹ 750 (*d*) ₹ 650

Sol. (*b*) Marked price $= 600 \times \frac{120}{100} \times \frac{100}{90} =$ ₹ 800

193. By selling an article at 80% of the marked price, there is a loss of 10%. If the article is sold at marked price, the profit percent will be: **(SSC CGL Tier I 2006)**

(*a*) 20 (*b*) 18 (*c*) 15 (*d*) 12.5

Sol. (*d*) 80% of marked price = 90% of cost price

$\therefore$ MP $= \frac{100}{80} \times 90\% = \frac{225}{2}\% = 112.50$ of CP. $\Rightarrow$ Profit = 112.50 – 100 = 12.5%

194. A tradesman gives 4% discount on the marked price and gives 1 article free for buying every 15 articles and thus gains 35%. The marked price is increased above the cost price by : **(SSC CGL Tier I 2003)**

(*a*) 40% (*b*) 39% (*c*) 50% (*d*) 20%

Sol. (*c*) Let cost price = ₹ 100. Then marked price = ₹ $100 \times \frac{100}{96} \times \frac{16}{15} \times \frac{135}{100} =$ ₹ 150

Per cent increase = 150% – 100% = 50%

195. A retailer buys a sewing machine at a discount of 15% and sells it for ₹ 1955. Thus he makes a profit of 15%. The discount is **(SSC CGL Tier I 2013, 2008)**

(*a*) ₹ 270 (*b*) ₹ 290 (*c*) ₹ 300 (*d*) ₹ 330

Sol. (*c*) CP = SP $\times \frac{100}{115} =$ ₹ $1955 \times \frac{100}{115} =$ ₹ 1700 $\Rightarrow$ MP = CP $\times \frac{100}{85} =$ ₹ $1700 \times \frac{100}{85} =$ ₹ 2000

$\therefore$ Discount = ₹ 2000 – ₹ 1700 = ₹ 300

Alt: SP of the article = ₹ 1955 $\Rightarrow$ C.P. of the article = ₹ $1955 \times \frac{100}{100+15} =$ ₹ 1700

Discount $= \frac{15}{100}$ of marked price $= \frac{15}{85}$ of cost price $= \frac{15}{85} \times 1700 =$ ₹ 300

196. A shopkeeper sells his goods at 10% discount on the marked price. What price should he mark on an article that costs him ₹ 900 to gain 10% ? **(SSC CGL Tier I 2003)**

(*a*) ₹ 1275 (*b*) ₹1250 (*c*) ₹ 1175 (*d*) ₹1100

Sol. (*d*) SP $= 900 \times \frac{110}{100} \Rightarrow$ Discount = 10% on MP $\Rightarrow$ MP $= 900 \times \frac{110}{100} \times \frac{100}{90} =$ ₹ 1100

197. A dealer allows his customers a discount of 25% and still gains 25%. If the cost price of the article is ₹ 720, then the marked price is **(SSC CHSL 2012)**

(*a*) ₹1200 (*b*) ₹1100 (*c*) ₹1300 (*d*) ₹1400

Sol. (*a*) Marked price = $720 \times \frac{100+25}{100} \times \frac{100}{100-25} = 720 \times \frac{125}{100} \times \frac{100}{75}$ = ₹1200

198. A shopkeeper allows discount of 15% discount on marked price and still gains 19%. How much percent is the marked price above the cost price?

(*a*) 40% (*b*) 30% (*c*) 60% (*d*) 70%

Sol. (*a*) Let CP = ₹ 100. SP = ₹ 100 + ₹ 19 = ₹ 119

MP = 119 × 100/85 = 140 ⇒ MP is 40% above the CP.

199. The cost of manufacture of a tape recorder is ₹ 1500. The manufacturer fixes the marked price 20% above the cost of manufacture and allows a discount in such a way as to get a profit of 8%. The rate of discount is **(SSC CGL Tier I 2012)**

(*a*) 12% (*b*) 8% (*c*) 20% (*d*) 10%

Sol. (*d*) Let CP = ₹ 100 ⇒ MP = ₹ 100 + ₹ 20 = ₹ 120 ⇒ SP = ₹ 100 + ₹ 8 = ₹ 108.

⇒ Discount = ₹ 120 – ₹ 108 = ₹ 12 ⇒ Rate of discount = $\frac{12}{120} = \frac{1}{10}$ = 10%

200. A merchant marks price of the articles 25% above the cost price and allows some discount on the marked price and makes a profit of 10%. Find the rate of discount. **(SSC CGL 2008)**

(*a*) 10% (*b*) 12% (*c*) 12.5% (*d*) 15%

Sol. (*b*) Let CP = ₹ 100 ⇒ MP = ₹ 100 + ₹ 25 = ₹ 125 ⇒ SP = ₹ 100 + ₹ 10 = ₹ 110.

⇒ Discount = ₹ 125 – ₹ 110 = ₹ 15 ⇒ Rate of discount = $\frac{15}{125} \times 100$ = 12%

201. At what per cent above the cost price, must a shopkeeper mark his goods so that he gains 20% even after giving a discount of 10% on the marked price ? **(SSC CGL 2004)**

(*a*) 25% (*b*) 30% (*c*) $33\frac{1}{3}\%$ (*d*) $37\frac{1}{2}\%$

Sol. (*c*) Let CP = ₹ 100. SP = ₹ 100 + ₹ 20 = ₹ 120 ⇒ MP = ₹ 120 × $\frac{100}{90} = \frac{400}{3} = 133\frac{1}{3}\%$

∴ Required % = $133\frac{1}{3}\% - 100\% = 33\frac{1}{3}\%$

202. The printed price of an article is 40% higher than its copst price. Then the rate of discount such that he gain 12% profit is : **(SSC CGL Tier I 2015)**

(*a*) 21% (*b*) 20% (*c*) 18% (*d*) 15%

Sol. (*b*) Let CP = ₹ 100. SP = ₹ 100 + ₹ 12 = ₹ 112. MP = ₹ 100 + ₹ 40 = ₹ 140

Discount = 140 – 112 = 28 ⇒ Rate of discount = $\frac{28}{140} \times 100$ = 20%

203. How much percent more than the cost price should a shopkeeper mark his goods so that after allowing a discount of 12.5%, he still gains 5% ? **(SSC CGL Tier I 2006)**

(*a*) 15 (*b*) 20 (*c*) 25 (*d*) 30

Sol. (*b*) Let CP = ₹ 100. Then SP = ₹ 105 ⇒ MP = $105 \times \frac{100}{100-12.5} = 105 \times \frac{100}{87.5}$ = ₹ 120

204. A shopkeeper makes a profit of 20% even after allowing a discount of 10% on the marked price. If he makes a profit of ₹ 720, find the marked price.

(*a*) ₹ 3600 (*b*) ₹ 4500 (*c*) ₹ 4800 (*d*) ₹ 5400

Sol. (*c*) If the cost price is ₹ 100, selling price at 20% profit = ₹ (100 + 20) = ₹ 120

If marked price is ₹ 100, selling price at 10% discount = ₹ (100 – 10) = ₹ 90

If profit is ₹ 720, marked price = $720\times\frac{120}{20}\times\frac{100}{90}$ = ₹ 4800

205. A shopkeeper mark up his goods at x% above the cost price. He cheats his customers by giving 880 g only instead of 1 kg. If his total profit is 25%, find the value of x.

(*a*) 10% (*b*) 12% (*c*) 12.5% (*d*) 15%

Sol. (*a*) Let CP = ₹ 880 ⇒ Profit = $\frac{1}{4}$ × 880 = 220 ⇒ SP = ₹ 880 + ₹ 220 = ₹ 1100.

$\Rightarrow x = \frac{1100-1000}{1000} \times 100 = 10\%$

206. A shopkeeper sells a badmintion racket whose marked price is ₹ 30, at a discount of 15% and gives a shuttle cock costing ₹ 1.50 free with each racket. Even then he makes a profit of 20%. His cost price, per racket is : **(SSC CGL Tier I 2004)**

(*a*) ₹ 21.00 (*b*) ₹ 21.25 (*c*) ₹ 20.00 (*d*) ₹ 19.75

Sol. (*d*) Marked price = ₹ 30. Selling price = ₹ 30 × $\frac{100-15}{100}$ = ₹ 25.50

Cost price (total) = ₹ 25.50 × $\frac{100}{100+20}$ = ₹ 21.25 = Cost of racket + Cost of Shuttle cock

∴ Cost of racket = ₹ 21.25 – ₹ 1.50 = ₹ 19.75

207. A trader gains 15% after selling an item at 10% discount on the printed price. The ratio of the cost price and printed price of the item is : **(SSC CGL Tier I 2008)**

(*a*) 18 : 23 (*b*) 17 : 18 (*c*) 17 : 23 (*d*) 18 : 25

Sol. (*a*) Let the printed price be ₹100. Selling price = ₹100 – ₹10 = ₹90

$CP = 90\times\frac{100}{115} = 90\times\frac{20}{23} \Rightarrow CP : MP = 90\times\frac{20}{23} : 100 = 18 : 23$

208. A shopkeeper earns a profit of 12% on selling a book at 10% discount on the printed price. The ratio of the cost price and the printed price of the book is : **(SSC CGL Tier I 2007, 2010)**

(*a*) 99 : 125 (*b*) 25 : 37 (*c*) 50 : 61 (*d*) 45 : 56

Sol. (*d*) Let printed price = ₹100 ⇒ SP = $100\times\frac{90}{100}$ = ₹90 ⇒ CP = ₹ $90\times\frac{100}{100+12} = \frac{45\times25}{14}$

∴ Required ratio = $\frac{45\times25}{14} : 100 = 45\times25 : 100\times14 = 45 : 56$

Alternative method : Let CP = 100, Then SP = 112 MP = $112\times\frac{100}{90}$

Required ratio = $100 : 112\times\frac{100}{90} = 90 : 112 = 45 : 56$

209. A shopkeeper gains 17% after allowing a discount of 10% on the marked price of an article. Find his profit percent if the articles are sold at marked price allowing no discount.

(SSC CGL Tier I 2012, 2013 M, 2014, 2015)

(*a*) 23% (*b*) 27% (*c*) 30% (*d*) 37%

Sol. (*c*) Let CP = ₹ 100 ⇒ SP = CP + Profit = ₹ 100 + ₹17 = ₹117 ⇒ MP = ₹ 117 × $\frac{100}{90}$ = ₹ 130

∴ Profit = ₹ 130 – ₹ 100 = ₹ 30 on ₹ 100 *i.e.* 30%.

210. At what percentage above the cost price must an article be marked so as to gain 33% after allowing a customer a discount of 5% ? **(MBA 2001)**

(*a*) 38% (*b*) 40% (*c*) 43% (*d*) 48%

Sol. (*b*) Let CP = 100, then SP = 133. Then MP = 133 × 100/95 = 140

⇒ Required % = 140 – 100 = 40 or 40%

211. A shopkeeper allows 10% discount on the marked price and earns 50% on the cost price. What will be his profit percentage, if 25% discount is allowed?

(*a*) 20% (*b*) 25% (*c*) 30% (*d*) 50%

Sol. (*b*) In first case, SP = 90% of MP In second case, SP = 75% of MP

90% of MP = 150% of CP ⇒ 75% of MP = $150\times\frac{75}{90}$ = 125% of CP ⇒ Profit% = 125–100 = 25%

Alt: Let MP = 100, then SP = 90 ⇒ CP = $90\times\frac{100}{150}$ = ₹ 60 ⇒ New SP = ₹100 – ₹25 = ₹75

Profit = ₹ 75 – ₹ 60 = ₹ 15 ⇒ Profit% = $\frac{15}{60}\times 100 = 25\%$

212. If the discount of 10% is given on the marked price of radio, the gain is 20%. If the discount is increased to 20%, the gain is : **(SSC CGL Tier I 2012 & 2006, SSC DEO 2010)**

(*a*) 5% (*b*) $6\frac{2}{3}\%$ (*c*) 8% (*d*) 10%

Sol. (*b*) Let CP = ₹ 100 ⇒ SP = 100 + 20% = ₹ 120

⇒ MP = $120\times\frac{100}{90}=\frac{400}{3}$ ⇒ New discount = 20% of $\frac{400}{3}$ = ₹ $\frac{80}{3}$

⇒ New SP = $\frac{400}{3}-\frac{80}{3}=\frac{320}{3}$ ⇒ Profit = $\frac{320}{3}-100=\frac{20}{3}=6\frac{2}{3}\%$

Alt: New SP = $120\times\frac{10}{9}\times\frac{8}{10}=\frac{320}{3}$ ⇒ New profit = $\frac{320}{3}-100=\frac{20}{3}=6\frac{2}{3}\%$

213. If a commission of 10% is given on the marked price of an article, then the gain is 20%. If the commission is increased to 20%, then the gain per cent is **(SSC CHSL 2012)**

(*a*) 10% (*b*) $16\frac{2}{3}\%$ (*c*) $8\frac{2}{3}\%$ (*d*) $6\frac{2}{3}\%$

Sol. (*d*) Let marked price = ₹ 100. Then MP – Commission = ₹ 100 – ₹ 10 = ₹ 90

Cost price of the article = ₹ 90 × $\frac{100}{120}$ = ₹ 75

New selling price in second case : MP – Commission = ₹ 100 – ₹ 20 = ₹ 80

Required profit percent = $\frac{5}{75}\times 100 = 6\frac{2}{3}\%$

214. A shopkeeper allows 4% discount on the marked price of an article and earns 20% profit. Find his profit per cent, if he allows 10% discount.

(*a*) 12% (*b*) 12.5% (*c*) 14% (*d*) 16%

Sol. (*b*) Let marked price = ₹ 100 ⇒ Selling price after 4% discount = ₹ 100 – ₹ 4 = ₹ 96

Cost price when profit is 20% $= 96 \times \frac{100}{120} =$ ₹ 80

SP after 10% discount = 100 – 10 = ₹ 90 ⇒ Profit per cent $= \frac{10}{80} \times 100 = 12.5\%$

Alt: Let cost price of the article = ₹ 100. Then MP $= 100 \times \frac{100}{96} \times \frac{120}{100} =$ ₹ 125

SP after 10% discount = 125 – 10% of 125 = ₹112.50 ⇒ Profit % = 112.50 – 100 = 12.5%

215. If a commission of 10% is given on the marked price of a work, the publisher gains 20%. If the commission is increased to 15%, the gain per cent is: **(SSC CGL 2012)**

(*a*) 15% (*b*) $16\frac{2}{3}\%$ (*c*) $13\frac{1}{3}\%$ (*d*) $15\frac{1}{6}\%$

Sol. (*c*) 90% of MP = 120% of CP ⇒ 85% of MP $= \frac{120}{90} \times 85 = \frac{340}{3}$ of CP

Required profit per cent $= \frac{340}{3} - 100 = \frac{40}{3} = 13\frac{1}{3}\%$

216. For a certain article, if discount is 25%, the profit is 25%. If the discount is 10%, then the profit is **(SSC FCI Asstt. 2012)**

(*a*) 50% (*b*) 40% (*c*) 30% (*d*) $33\frac{1}{3}\%$

Sol. (*a*) Let cost price = ₹ 100. Then MP $= 100 \times \frac{125}{100} \times \frac{100}{75} = \frac{500}{3}$

Selling price at 10% discount $= \frac{500}{3} \times \frac{90}{100} =$ ₹ 150 ⇒ Profit % = 50%

215. A shopkeeper earns a profit of 25%, after allowing 10% discount on the list price. What is the cost price of an article whose list price is ₹ 500 ?

(*a*) ₹ 360 (*b*) ₹ 400 (*c*) ₹ 450 (*d*) ₹ 540

Sol. (*a*) List price = ₹ 500 ⇒ Selling price after 10% discount = ₹ $500 \times \frac{90}{100}$

SP is 125% or $\frac{5}{4}$ of the CP ⇒ Cost price = ₹ $500 \times \frac{90}{100} \times \frac{4}{5} =$ ₹ 360

218. A shopkeeper marks his articles 20% above the cost price and gives a discount of 10% on the marked price. Find the cost price of the article, if the selling price is ₹ 540.

(*a*) ₹ 450 (*b*) ₹ 500 (*c*) ₹ 550 (*d*) ₹ 600

Sol. (*b*) Selling price of the article = ₹ 540 ⇒ Marked price = ₹ $540 \times \frac{100}{90}$

Cost price = Marked price × $\frac{100}{120} =$ ₹ $540 \times \frac{100}{90} \times \frac{100}{120} =$ ₹ 500

219. The cost of manufacture of an article was ₹ 900. The trader wants to gain 25% after giving a discount of 10%. The marked price should be : **(SSC CGL Tier I 2005, 2013)**

(*a*) ₹ 1500 (*b*) ₹ 1250 (*c*) ₹ 1200 (*d*) ₹ 1000

Sol. (*b*) Cost price of the article = ₹ 900. Selling price = ₹ 900 × $\frac{125}{100}$

Marked price = Selling price × $\frac{100}{90}$ = ₹ 900 × $\frac{125}{100} \times \frac{100}{90}$ = ₹ 1250

220. A profit of 20% is earned even after allowing 4% discount on the marked price. How much the marked price is above the cost price?

(*a*) 16% (*b*) 24% (*c*) 25% (*d*) 30%

Sol. (*c*) Let CP = ₹ 100. Then SP = CP + Profit = ₹ 100 + ₹ 20 = ₹ 120

∴ MP = ₹ 120 × $\frac{100}{96}$ = ₹ 125 ⇒ M.P. is 25% higher than C.P.

221. Selling an item for ₹ 1800 at a discount of 10% a shopkeeper had a gain of ₹ 200. Had he sold the item without the discount the percentage of profit would have been :

(SSC CGL Tier I 2013)

(*a*) 10% (*b*) 20% (*c*) 25% (*d*) 30%

Sol. (*c*) Marked Price = ₹ 1800 × $\frac{100}{90}$ = ₹ 2000 ⇒ CP = SP – Profit = 1800 – 200 = ₹ 1600

New profit = 2000 – 1600 = ₹ 400 ⇒ Profit% = $\frac{400}{1600} \times 100 = 25\%$

222. An article listed at ₹ 800 is sold at successive discounts of 25% and 15%. The buyer desires to sell it off at a profit of 20% after allowing 10% discount. What would be his list price ?

(SSC CGL Tier I 2007)

(*a*) ₹ 600 (*b*) ₹ 620 (*c*) ₹ 640 (*d*) ₹ 680

Sol. (*d*) Cost price for the buyer = ₹ 800 × $\frac{75}{100} \times \frac{85}{100}$ = ₹ 510

∴ Selling price = ₹ 510 × $\frac{120}{100}$ ⇒ Marked Price = ₹ 510 × $\frac{120}{100} \times \frac{100}{90}$ = ₹ 680

223. On allowing 20% discount, 12% loss is suffered. How much the marked price is above the cost price?

(*a*) 8% (*b*) 10% (*c*) 12% (*d*) 15%

Sol. (*b*) Let CP = ₹ 100. Then SP = Cost price – Loss = ₹ 100 – ₹ 12 = ₹ 88

∴ MP = ₹ 88 × $\frac{100}{80}$ = ₹ 110 ⇒ MP is 10% higher than the CP.

224. A shopkeeper allows 23% commission on his advertised price and still makes a profit of 10%. If he gains ₹ 56 on one item, his advertised price of the item, (in ₹) is

(SSC CGL Tier I 2011, SSC CHSL 2011)

(*a*) 820 (*b*) 780 (*c*) 790 (*d*) 800

Sol. (*d*) Profit = ₹ 56 ⇒ CP (at 10% profit) = ₹ 56 × 10 = ₹ 560

⇒ Selling price = $560 \times \frac{11}{10}$ ⇒ Advertised price = ₹$560 \times \frac{11}{10} \times \frac{100}{77}$ = ₹ 800

225. A man sells two articles for ₹ 5000 each neither losing nor gaining in the deal. If he sold one of them at a gain of 25%, the other article is sold at a loss of **(SSC CGL Tier I 2011)**

(*a*) $15\frac{2}{3}\%$ (*b*) $16\frac{2}{3}\%$ (*c*) $17\frac{1}{3}\%$ (*d*) $18\frac{1}{3}\%$

Sol. (*b*) Profit on first article = ₹ $5000 \times \frac{25}{125}$ = ₹ 1000 = Loss on second article

⇒ Cost price of second article = ₹ (5000 + 1000) = ₹ 6000

∴ Required Loss% = $\frac{1000}{6000} \times 100 = \frac{50}{3} = 16\frac{2}{3}\%$

226. A person bought two bicycles for ₹ 1600 and sold the first at 10% profit and the second at 20% profit. If he sold the first at 20% profit and the second at 10% profit, he would get ₹ 5 more. The difference of the cost price of the two bicycle was: **(SSC CGL Tier I 2013)**

(*a*) ₹ 75 (*b*) ₹ 50 (*c*) ₹ 40 (*d*) ₹ 25

Sol. (*b*) Let the cost price of two bicycle be ₹ x and ₹ y

Then (20% of x + 10% of y) – (10% of x + 20% of y) = ₹ 5

∴ 10% of $(x - y)$ = ₹ 5 ⇒ $(x - y)$ = ₹ 50

227. A man sold one watch and one pen for ₹ 492 and ₹ 168 respectively, gaining 10% of the total cost price of the two articles. Had he sold the watch for ₹ 435 and the pen at its cost price, he would have lost 5% on the total cost price. Find the cost price of the watch.

(*a*) ₹ 135 (*b*) ₹ 225 (*c*) ₹ 375 (*d*) ₹ 465

Sol. (*d*) In the first case: Total selling price = ₹ 492 + ₹ 168 = ₹ 660

Profit earned = 10% ∴ Cost price of both the articles = ₹ $660 \times \frac{100}{110}$ = ₹ 600

In the second case, total loss = 5%

Total selling price at 5% loss = ₹ 600 – 5% of ₹ 600 = ₹ 570

∴ Selling price of watch + Cost price of pen = ₹ 570

∴ ₹ 435 + Cost of pen = ₹ 570 ∴ Cost of pen = ₹ 570 – ₹ 435 = ₹ 135

∴ Cost of watch = Total Cost Price – Cost Price of pen = ₹ 600 – ₹ 135 = ₹ 465

228. A clock was sold for ₹ 144. If the percentage of profit was numerically equal to the cost price, the cost of the clock was :

(SSC CGL Tier I 2005, SSC CPO SI 1999, SSC Constable GD 2011)

(*a*) ₹ 72 (*b*) ₹ 80 (*c*) ₹ 90 (*d*) ₹ 100

Sol. (*b*) Let cost price of the article = ₹ x. Then profit = x%

∴ Selling price = cost price × $\frac{100 + \text{Profit}\%}{100} = \frac{x(100 + x)}{100}$

∴ If profit is 10%, then cost price is ₹ 10 and selling price = $\frac{10 \times 110}{100} = 1 \times 11 = 11$

If profit is 20%, then cost price is ₹ 20 and selling price = $\frac{20 \times 120}{100} = 2 \times 12 = 24$

Similarly for 30% profit, selling price = 3 × 13 = 39

On checking we find that 8 × 18 =144 ⇒ Cost price = ₹ 80.

229. A merchant sold an article for ₹ 75, at a profit percent equal to its cost price. The cost price of the article was : **(SSC Data Entry Operator 2010)**

(*a*) ₹ 45 (*b*) ₹ 50 (*c*) ₹ 54 (*d*) ₹ 60

Sol. (*b*) Going through the options: $\frac{50 \times 150}{100} = 75 \Rightarrow$ Cost price = ₹ 50

230. The selling price of an article is ₹ 39. If its cost price is numerically equal to its profit per cent, then its cost price is : **(SSC CHSL 2013)**

(*a*) ₹ 25 (*b*) ₹ 30 (*c*) ₹ 35 (*d*) ₹ 40

Sol. (*b*) Going through the options : $\frac{130 \times 30}{100} = 39 \Rightarrow$ Cost price = ₹ 30

231. An article was sold for ₹96. If percentage of profit was numerically equal to the cost price, the cost price of the article is **(SSC CGL Tier I 2014)**

(*a*) ₹ 90 (*b*) ₹ 100 (*c*) ₹ 60 (*d*) ₹ 75

Sol. (*c*) From options: $\frac{60 \times 160}{100} = 96 \Rightarrow$ Cost price = ₹ 60

232. A shopkeeper bought 200 articles, each costing the same. He sold 30% of the articles at 20% profit and the remaining at 10% profit. If total profit made by him is ₹ 2600, find cost price of one article. **(SSC CHSL 2012)**

(*a*) ₹ 100 (*b*) ₹ 200 (*c*) ₹ 1300 (*d*) ₹ 2600

Sol. (*a*) Profit in the transaction = 30% × 20% + 70% × 10% = 6% + 7% = 13% = ₹ 2600

∴ Cost Price of 200 articles = 2600 ÷ 13% = $2600 \times \frac{100}{13}$

∴ Cost price of 1 article = $2600 \times \frac{100}{13} \times \frac{1}{200}$ = ₹ 100

233. A man purchased 10 horses @ ₹ 3,000 each. 1 horse died and he sold 2 horses at 5% loss. At what rate he should sell the remaining horses so as to gain a profit of 10% on the total cost?

(*a*) ₹ 3,100 (*b*) ₹ 3,500 (*c*) ₹ 3,900 (*d*) ₹ 4,200

Sol. (*c*) Cost price of 10 horses = ₹ 3000 × 10 = ₹ 30,000

Total profit desired = 10% of ₹ 30,000 = ₹ 3,000

Total desired selling price = ₹ 30,000 + ₹ 3,000 = ₹ 33,000

Selling price of one horse (died) = ₹ 0

Selling price of two other horses = 2 × (3000 – 5% of ₹ 3000) = ₹ 5,700

Balance desired from selling of 7 horses = ₹ 33,000 – ₹ 5,700 = ₹ 27,300

∴ Selling price of one horse = ₹ 27,300 ÷ 7 = ₹ 3,900

Alt :

Total profit desired =	₹ 3,000
Loss on 1 horse died =	₹ 3,000
Loss on 2 horses (sold at 5% loss) =	₹ 300
Total	₹ 6,300

Profit desired on selling each of remaining 7 horses = ₹ 6,300 ÷ 7 = ₹ 900

Selling price of each horse = ₹ 3,000 + ₹ 900 = ₹ 3,900

234. A man purchased 10 horses @ ₹3,000 each. 1 horse died and he sold 2 horses at 5% loss. At what profit per cent should he sell the remaining horses so as to gain a profit of 10% on the total cost?

(*a*) 12% (*b*) 15% (*c*) 25% (*d*) 30%

Sol. (*d*) Let CP of each horse = ₹ 100. Then total profit desired = 10 × (10% of ₹ 100) = ₹ 100

Loss on one horse (died) = ₹ 100 ⇒ Loss on selling two horses = 2 × ₹ 5 = ₹ 10

Total profit desired on remaining 7 horses = ₹ (100 + 100 + 10) = ₹ 210

∴ Profit desired on 1 horse = ₹ 210 ÷ 7 = ₹30 or 30%

235. A man purchased 10 horses and 6 cows for ₹10,500. He sold the horse at 10% profit and cows at 15% loss. If his net profit is ₹ 300, find the cost of 1 horse.

(*a*) ₹ 450 (*b*) ₹ 500 (*c*) ₹ 900 (*d*) ₹ 750

Sol. (*d*) Cost price to earn ₹300 as profit = $300\times\frac{100}{10}$ = ₹ 3000

Balance cost price = ₹ 10500 – ₹ 3000 = ₹ 7500

Ratio of cost prices to set off the loss with gain = $\frac{1}{10}:\frac{1}{15} = 3:2$

∴ Remaining cost price towards horses = $\frac{3}{3+2}\times 7500$ = ₹ 4500

Total cost of 5 horses = ₹ 3000 + ₹ 4500 = ₹ 7500 ⇒ CP of 1 horse = ₹ 7500 ÷ 10 = ₹ 750

Alternative Method : Let total cost price of 10 horses be ₹ x.

Then 10% of x – 15% of (10500 – x) = ₹300 ⇒ 10% of x – 1575 + 15% of x = ₹ 300

⇒ 25% of x = ₹ 1875 ⇒ x = ₹ 7500 ⇒ CP of 1 horse = ₹ 7500 ÷ 10 = ₹ 750

236. A man buys a chair and a table for ₹6000. He sells the chair at a loss of 10% and the table at a gain of 10%. He still gains ₹100 on the whole. The cost price of the chair is **(SAS 2010)**

(*a*) ₹ 2500 (*b*) ₹ 2850 (*c*) ₹ 3050 (*d*) ₹ 3500

Sol. (*a*) Cost price to earn profit of ₹100 (at 10% profit) = $\frac{100}{10}\times 100 = 1000$

Cost price of chair : Remaining cost of table = 10 : 10 = 1 : 1

∴ Cost price of chair = $\frac{1}{2}\times$ ₹ (6000 – 1000) = ₹ 2500

237. A man buys one table and one chair for ₹ 500. He sells the table at a loss of 10% and the chair at a gain of 10%. He still gains ₹ 10 on the whole. The cost price of the chair is **(SSC CGL Tier I 2012)**

(*a*) ₹ 300 (*b*) ₹ 350 (*c*) ₹ 200 (*d*) ₹ 250

Sol. (*a*) Cost price to earn ₹ 10 profit at 10% = 10 ÷ 10% = ₹ 100

Remaining Cost price = 500 – 100 = ₹ 400

Ratio of CP of table : Remaining CP of chair = 10 : 10 = 1 : 1

∴ CP of table = Remaining CP of chair = $\frac{1}{2}\times$ ₹ 400 = ₹ 200

∴ CP of chair = ₹ 100 + ₹ 200 = ₹ 300

238. A man sold an article at a loss of 10%. Had he bought it for 20% less and sold it for ₹ 55 more, he would have made a profit of 40%. Original cost price of the article is:

(*a*) ₹ 225 (*b*) ₹ 250 (*c*) ₹ 300 (*d*) ₹ 500

Sol. (*b*) Let cost price of the article = ₹ 100. Then its selling price = ₹ 100 – ₹ 10 = ₹ 90

In the second situation: CP = ₹ 100 – ₹ 20 = ₹ 80. SP = ₹ 80 × $\frac{140}{100}$ = ₹ 112

Difference between two selling prices = ₹ 112 – ₹ 90 = ₹ 22

But actual difference = ₹ 55 ⇒ Cost price = $\frac{55}{22}\times$ ₹ 100 = ₹ 250

Alternative Method: 90% of cost price + ₹ 55 = 140% of 80% of cost price

∴ 90% of cost price + ₹ 55 = 112% of cost price ⇒ 22% of cost price = ₹ 55

∴ Cost price = $55 \times \frac{100}{22}$ = ₹ 250

239. A person sold a horse at a gain of 15%. Had he bought it for 25% less and sold it for ₹600 less, he would have made a profit of 32%. The cost price of the horse was :

(SSC CGL Tier I 2008)

(*a*) ₹ 3750 (*b*) ₹ 3250 (*c*) ₹ 2750 (*d*) ₹ 2250

Sol. (*a*) Let cost price of the horse be ₹ 100. Then its selling price = ₹ 100 + ₹ 15 = ₹ 115

In the second case: Cost price = ₹ 100 – ₹ 25 = ₹ 75.

Profit = ₹ 75 × $\frac{32}{100}$ = ₹24. Selling price = ₹ 75 + ₹24 = ₹ 99

Difference between two selling prices = ₹ 115 – ₹ 99 = ₹ 16

But actual difference = ₹ 600 ⇒ Cost price = $\frac{600}{16}$ × ₹ 100 = ₹ 3750

Alternative Method: 115% of cost price – ₹ 600 = 75% of 132% of cost price

∴ 115% of cost price – ₹ 600 = 99% of cost price

∴ 16% of cost price = ₹ 600 ⇒ Cost price = $\frac{600}{16}$ × ₹ 100 = ₹ 3750

240. A dealer sold a bicycle at a profit of 10%. Had he bought the bicycle at 10% less price and sold it at a price ₹ 60 more, he would have gained 25%. The cost price of the bicycle was:

(SSC CGL Tier I 2015)

(*a*) ₹ 2600 (*b*) ₹ 2000 (*c*) ₹ 2200 (*d*) ₹ 2400

Sol. (*d*) Let cost price of the article = ₹ 100. Then its selling price = ₹ 110

In the second case: CP = ₹ 100 – ₹ 10 = ₹ 90. ⇒ SP = ₹ 90 × $\frac{125}{100}$ = ₹ 112.50

Difference between two selling prices = ₹ 112.50 – ₹ 110 = ₹ 2.50

But actual difference = ₹ 60 (*i.e.* 24 times ₹ 2.5) ⇒ Cost price = 24 × ₹ 100 = ₹ 2400

Alternative Method: 110% of CP + ₹ 60 = 90% of 125% of CP = 112.5% of CP

∴ 2.5% of CP = ₹ 60 (*i.e.* 24 times 2.5) ⇒ CP = 24 × ₹ 100 = ₹ 2400

241. A man sold an article at 10% loss. Had he bought it for 10% more and sold it for ₹ 20 less, he would have lost 20%. Original cost price of the article is:

(*a*) ₹ 200 (*b*) ₹ 500 (*c*) ₹ 1000 (*d*) ₹ 2000

Sol. (*c*) Let cost price of the article = ₹ 100. Then its selling price = ₹ 90

In the second case: CP = ₹ 100 + ₹ 10 = ₹ 110. ⇒ SP = ₹ 110 × $\frac{80}{100}$ = ₹ 88

Difference between two selling prices = ₹ 90 – ₹ 88 = ₹ 2

But actual difference = ₹ 20 (*i.e.* 10 times ₹ 2) ⇒ Cost price = 10 × ₹ 100 = ₹ 1000.

Alternative Method: 90% of CP – ₹ 20 = 80% of 110% of CP = 88% of cost price

∴ 2% of cost price = ₹ 20 (*i.e.* 10 times 2) ⇒ Cost price = 10 × ₹ 100 = ₹ 1000.

242. An article is sold at 10% profit. Had it bought for 10% discount and sold for ₹ 2 less, the profit would have $16\frac{2}{3}$%. The original cost price of the article is : **(SSC CGL 2008)**

(*a*) ₹ 32 (*b*) ₹ 36 (*c*) ₹ 40 (*d*) ₹ 48

Sol. (*c*) Let CP = ₹100 ⇒ SP = ₹100 + ₹10 = ₹110 ⇒ In second case: CP = ₹100 – ₹10 = ₹90

Profit = $16\frac{2}{3}\% = \frac{50}{3}\%$. Selling price = ₹ 90 × $\frac{350}{300}$ = ₹ 105

Difference between two selling prices = ₹ 110 – ₹ 105 = ₹ 5

If difference is ₹ 5, CP = ₹100 ⇒ If difference is ₹ 2, CP = ₹100 × $\frac{2}{5}$ = ₹ 40

Alternative Method: In the first situation, selling price = 110% of original cost price

In the second situation, cost price = 90% of original cost price

∴ New selling price = $\frac{350}{3}$% of 90% of original cost price = 105% of original cost price

∴ (110% – 105%) of original cost price = ₹ 2 ⇒ 5% of cost price = ₹ 2

∴ Original cost price = ₹100 × $\frac{2}{5}$ = ₹ 40

243. An article is sold at a profit of 20%. If both the cost price and selling price are ₹ 100 less, the profit would be 25%. The cost price of the article is :

(*a*) ₹ 100 (*b*) ₹ 200 (*c*) ₹ 250 (*d*) ₹ 500

Sol. (*d*) Let CP = ₹ x. ⇒ SP = $1.2x$ ⇒ New CP = $x - 100$ and new selling price = $1.2x - 100$

Now $\frac{5}{4}(x - 100) = 1.2x - 100 \Rightarrow 5x - 500 = 4.8x - 400 \Rightarrow 0.2x = 100 \Rightarrow x =$ ₹ 500

Trick: C.P. = $\frac{\text{New profit \%}}{\text{Increase in profit \%}} \times$ Decrease in amount $= \frac{25}{5} \times 100$ = ₹ 500

Note: Above formula can be used when CP and SP are reduced by the same amount.

Alt (From options) : When CP = ₹ 500, SP = 500 + 20% of 500 = 600

New CP = ₹ 400, New SP = ₹ 500, Profit % = 25%

244. An article is sold at 25% profit. If its cost price and selling price are increased by ₹ 20 and ₹4 respectively, the profit would be 10%, find the original cost price.

(*a*) ₹ 120 (*b*) ₹ 150 (*c*) ₹ 175 (*d*) ₹ 200

Sol. (*a*) 125% of CP + ₹ 4 = 110% of (CP + ₹ 20) ⇒ 125% of CP + ₹ 4 = 110% of CP + ₹ 22

⇒ 15% of cost price = ₹ 18 ⇒ Cost price = $\frac{18}{15} \times 100$ = ₹ 120

245. If a shopkeeper sells a table at a profit of 10% and a chair at a loss of 15%, he neither gain nor loss on the whole. If he sells the table at 15% profit and chair at a loss of 10%, he would have gained ₹150 on the whole. Find the cost price of the table.

(*a*) ₹ 1200 (*b*) ₹ 1800 (*c*) ₹ 2400 (*d*) ₹ 3000

Sol. (*b*) Cost price of the table : cost price of chair = $\frac{1}{10} : \frac{1}{15}$ = 3 : 2

Let cost price of a table = ₹ 300. Then cost price of a chair = ₹ 200

Total profit (in second case) = 3 × 15 – 2 × 10 = ₹ 45 – ₹ 20 = ₹ 25

But actual profit = ₹ 150 (*i.e.* 6 times ₹ 25) ⇒ Cost of a table = 6 × ₹ 300 = ₹ 1800.

246. A horse and a cow were sold for ₹ 580, making a profit of 20% on the horse and 10% profit on the cow. If profit on horse is 10% and on cow is 20%, total selling price is ₹ 570. Find the cost price of the cow. **(MBA)**

(*a*) ₹ 200 (*b*) ₹ 240 (*c*) ₹ 150 (*d*) ₹ 120

Sol. (*a*) Let cost price of horse and cow are ₹ x and ₹ y respectively.

Then 120% of x + 110% of y = ₹ 580 ....(1)

And, 110% of x + 120% of y = ₹ 570 ....(2)

Subtracting (2) from (1), 10% of $(x - y)$ = ₹ 10 ⇒ $x - y$ = ₹ 10 × 10 = ₹ 100

Selling price of ₹ 100 at 120% = ₹ 120

Remaining selling price in the first equation is ₹ 580 – ₹ 120 = ₹ 460

Cost price of the remaining amount = ₹ 460 × $\frac{100+100}{120+110}$ = ₹ 400

Cost price of cow = ₹ 400 ÷ 2 = ₹ 200

Alternative method : Let CP of horse and cow are ₹ x and ₹ y respectively.

Then 120% of x + 110% of y = ₹ 580 ....(1)

And, 110% of x + 120% of y = ₹ 570 ....(2)

Subtracting (2) from (1), 10% of $(x - y)$ = ₹ 10 ⇒ $x - y$ = ₹ 10 × 10 = ₹ 100

Total selling price = ₹ 580 + ₹ 570 = ₹ 1150. Total CP = ₹ 1150 × $\frac{100}{110+120}$ = ₹ 500

Cost price of cow = $\frac{500-100}{2}$ = ₹ 200

CHAPTER

17

STOCK AND SHARES

1. **Stock or shares**: The capital of the company is divided into small units usually of ₹100. These small units are called shares.
2. **Face value**: The value of the share printed on the share certificate.
3. **Market value** : The pirce at which the shares are offered in the market. A share can be offered:
   (*i*) **At Par**: When its market value is same as its Face value.
   (*ii*) **At Premium or Above Par**: If its market value is more than its Face Value.
   Example. Shares having face value of ₹100 is sold at ₹110.
   (*iii*) **At Discount or Below Par**: If its market value is less than its face value.
   Example. Share having face value of ₹100 is sold at ₹95.

$$\text{Investment} = \frac{\text{Stock} \times \text{M.V}}{100} \Rightarrow \text{Stock} = \frac{\text{Investment} \times 100}{\text{M.V.}}$$

4. **Dividend**: The annual profit of the company, distributed among its share holders, is called dividend. Dividend is fixed per share or as a percentage of Face Value of the share.

$$\text{Dividend} = \frac{\text{Face value} \times \text{Rate}}{100} = \frac{\text{Investment} \times \text{Rate}}{\text{M.V.}}$$

5. **Brokerage.** It is commission paid on selling and buying the shares.
   (*i*) When the shares are purchased, brokerage is added to the purchase price.
   (*ii*) When the shares are sold, brokerage is deducted from the selling price.

SOLVED EXERCISE

1. Find the face value of 5% stock at 10 above par, purchased for ₹5500.

Sol. M.V. of one share = 100 + 10 = ₹110 $\Rightarrow$ Face value $= 5500 \times \frac{100}{110} =$ ₹5000

2. How much money is required to buy ₹ 8000, 9% stock at 104 (brokerage 1%).

Sol. Gross market value of 1 share = 104 + 1 = ₹105

$\Rightarrow$ Total money required $= 8000 \times \frac{105}{100} =$ ₹ 8400

3. Find the money realised on sale of ₹ 5500, 8.5% stock at 8 below per (brokerage 2%).

Sol. Net market value (selling price) of 1 share = 100 – (8 + 2) = ₹ 90

$\Rightarrow$ Total money realised $= 5500 \times \frac{90}{100} =$ ₹ 4950

4. Find the dividend income on ₹ 5,700, 12% stock at 5% discount.

Sol. M.V. of each share = 100 – 5 = ₹ 95 ⇒ Dividend = $\frac{5700}{95} \times 12$ = ₹ 720

5. Find the cost price of ₹ 1000 shares available at 4 discount and brokerage $\frac{1}{2}$%

Sol. CP of one share = $100 - 4 + \frac{1}{2}$ = ₹ 96.5

⇒ CP of 1000 shares = ₹ 96.5 × 1000 = ₹ 96500

6. Find the annual income obtained by investing ₹ 6300 in 12% stock at ₹ 126.

Sol. Income = $\frac{12}{126}$ × ₹ 6300 = ₹ 600

7. A company paid 20% dividend but a person calculated 16% on his investment. At what price did he purchase the share of ₹ 100?

Sol. Market value = $\frac{20}{16} \times 100$ = ₹ 125

8. Which one of the following is better investment?

(*i*) 8% stock at 160 (*ii*) 5% stock at 95

Sol. (*ii*) Amount of investment to earn dividend of ₹ 1, is

(*i*) 160 ÷ 8 = ₹ 20 (*ii*) 95 ÷ 5 = ₹ 19

Alt. Let investment in each case = ₹ (160 × 95), then income:

(*i*) 8 × 95 = ₹ 760 (*ii*) 5 × 160 = ₹ 800

Alt. Effective rate on ₹ 100:

(*i*) $\frac{100}{160} \times 8 = 5\%$ (*ii*) $\frac{100}{95} \times 5 = \frac{100}{19} = 5\frac{5}{19}\%$

9. Find the cost of 96 shares of ₹ 10 each at $\frac{3}{4}$ discount, brokerage being $\frac{1}{4}$ per share.

Sol. Cost of one share = $\left(10 - \frac{3}{4}\right) + \frac{1}{4} = 9\frac{1}{2}$

Cost of 96 shares = $96 \times 9\frac{1}{2}$ = 960 – 48 = ₹ 912

10. Find the amount of investment to earn an income of ₹ 600 from 10% stock at ₹ 96.

Sol. If income = ₹ 10, Investment = ₹ 96

⇒ If income = ₹ 600, Investment = $\frac{₹ 96 \times 600}{10}$ = ₹ 5760

11. An investment of ₹ 9000 is made in 9% stock at 90. This stock was sold at 96 and sale proceeds invested in 12% stock. By doing so, annual income rose by ₹ 540. Find the market price of the second stock purchased.

Sol. Face value of first stock = $\frac{9000}{90} \times 100$ = ₹ 10,000

Sale proceeds of the first stock = $\frac{96}{100} \times 10{,}000$ = ₹ 9,600

Annual income from first stock = 9% of 10,000 = ₹ 900

Annual income of second stock = ₹ (900 + 540) = ₹ 1, 440

$$\text{Face value of second stock} = \frac{1440 \times 100}{12} = ₹\ 12000$$

$$\therefore \quad \text{M.V. of second stock} = \frac{9600 \times 100}{12000} = ₹\ 80$$

12. What investment should be made in 5% stock at 114 to earn a net income of ₹ 150 if 10% income tax is charged on income from dividend?

Sol. Net income on ₹ 114 = ₹ 5 – 10% of ₹ 5 = ₹ 4.50

$$\text{Investment} = \frac{114}{4.5} \times 150 = ₹\ 3,800$$

$$\textbf{Alt:}\ \text{Investment} = 150 \times \frac{100}{90} \times \frac{114}{5} = ₹\ 3,800$$

CHAPTER

18

AVERAGE

1. Average or Arithmetic Mean is calculated by adding all the observations and dividing their sum by number of observations.

$\therefore$ Average of n terms $= \dfrac{\text{sum of } n \text{ terms}}{n} = \dfrac{x_1 + x_2 + x_3 \ldots\ldots + x_n}{n} = \dfrac{\sum x_i}{n}$

$\therefore$ If average of n terms is x, then sum of n terms $= nx$

2. When average of n_1 items is x_1, average of n_2 items is x_2, average of n_3 items is x_3, and so on.

Then combined average of all items $= \dfrac{\sum n_i x_i}{\sum n_i}$

Example. The average age of 20 boys in a class is 12 years. What will be average age of the class if 5 new boys take admission whose average age is 7 years ? **(SSC CGL Tier I 2006)**

(*a*) $11\frac{1}{2}$ years (*b*) 11 years (*c*) $10\frac{1}{2}$ years (*d*) 10 years

Sol. (*b*) New average $= \dfrac{20 \times 12 + 5 \times 7}{20 + 5} = \dfrac{240 + 35}{25} = \dfrac{275}{25} = 11$ years

Alternative method : Short age of 5 new boys $= 5 \times (12 - 7) = 25$ years

Decrease in average of all $= 25 \div 25 = 1$ year $\Rightarrow$ New average age $= 12 - 1 = 11$ years

3. If average of some numbers is x and a constant number $\boldsymbol{a}$

(*a*) **added to each term** of the series, then new average $= x + a$

(*b*) **subtracted from each term** of the series, then new average $= x - a$

(*c*) **multiplied with each term** of the series, then new average $= ax$

(*d*) **divides each term** of the series, then new average $= x \div a$

Examples 1: The average of 99 items is 27. If 3 is added to each number of the series, the new average is:

(*a*) 24 (*b*) 27 (*c*) 30 (*d*) 33

Sol. (*c*) New average $= 27 + 3 = 30$

Examples 2: The average of 18 numbers is 24. If each item of the series is multiplied by 3, the new average is:

(*a*) 8 (*b*) 24 (*c*) 27 (*d*) 72

Sol. (*d*) New average $= 24 \times 3 = 72$

4. Average of *n* terms of an Arithmetic Progression is :

(*a*) Mid-term *i.e.* $\left(\dfrac{n+1}{2}\right)^{\text{th}}$ term (if n is an odd number).

(*b*) Average of two mid-terms *i.e.* average of $\dfrac{n}{2}$ and $\left(\dfrac{n+1}{2}\right)^{\text{th}}$ term (if n is an even number).

Example 1: Find average of 12, 14, 16, 18, 20. (an A.P.)

Sol. Number of terms is 5, an odd number. $\Rightarrow$ Average = $\left(\frac{5+1}{2}\right)^{th}$ or 3rd term = 16

Example 2: Find average of 4, 6, 8, 10. (an A.P.)

Sol. Number of terms is 4, an even number. $\Rightarrow$ Average = Average of two mid-terms *i.e.*

Average of $\left(\frac{4}{2}\right)^{th}$ and $\left(\frac{4}{2}+1\right)^{th}$ term = Average of 2nd and 3rd term *i.e.* $\frac{6+8}{2} = 7$

5. Average of an Arithmetic Progession with given first and last terms

$= \frac{1}{2} \times$ (First term + Last term)

Example. The average of first 47 natural numbers is:

(*a*) 23 (*b*) 24 (*c*) 25 (*d*) 47

Sol. (*b*) Average = $\frac{\text{Sum of the numbers}}{\text{Number of terms}} = \frac{47 \times 48}{2} \times \frac{1}{47} = 24$

Alternative Method: Average = Mid-term = $\left(\frac{47+1}{2}\right)^{th}$ term = 24th term = 24

Alt: Average = $\frac{1+47}{2} = 24$

6. If a number is increased while the other numbers remain constant in a series, the average of the series will increase and vice-versa.

This implies if one or more numbers are increased, then one or more of the remaining numbers must be decreased to keep the same average for the series and vice-versa.

Example 1: Average of 8, 9, 11 and 12 is 40 ÷ 4 = 10

Average of 5, 9, 11 and 15 is 40 ÷ 4 = 10

We note that in the second series, the first term is decreased by 3 and the last term is increased by 3 to get the same average in both the series.

Example 2: Average of 4, 5 and 6 is 15 ÷ 3 = 5.

If we decrease first and second numbers by 3 each, then we have to increase the third term by 2 × 3 = 6 to get the same average. ∴ New series is 1, 2, 12 and average is $\frac{15}{3} = 5$.

7. Addition Case: When one or more terms are included in the given series.

Let average of n terms is x.

On including one more term to the series, average of $(n + 1)$ terms becomes y.

Then value of new term included = $y + n \times (y - x)$ or $x + (n + 1)(y - x)$

***i.e.* New average + Number of terms (previous) × Increase in average**

Or Old average + Number of terms (all) × Increase in average

Note : What happens if more than one term (say m) have been included ?

Average of new items $= y + \frac{n}{m} \times (y - x)$

$=$ New Average + $\frac{\text{Number of items (previous)}}{\text{Number of items (new)}}$ × Increase in Average

Example 1. The average age of 40 students in a class is 15 years. If the age of teacher is also included, the average becomes 16 years, find the age of the teacher.

(*a*) 15 years (*b*) 16 years (*c*) 55 years (*d*) 56 years

Sol. (*d*) Age of teacher = Average age of all + Total increase in age = 16 + (1 × 40) = 56 years

Example 2. The average weight of a class of 24 students is 35 kg. If the weight of the teacher be included, the average rises by 400 g. The weight of the teacher is :

(SSC CHSL 2012M, SSC CGL Tier I 2003)

(*a*) 50 kg (*b*) 55 kg (*c*) 45 kg (*d*) 53 kg

Sol. (*c*) Weight of teacher = Average weight + Increase in average

= 35 kg + (24 + 1) × 400 g = 35 kg + 10 kg = 45 kg

Example 3. The average of 20 numbers is 15. Average on inclusion of 5 more numbers becomes 18. Find the average of new numbers included.

(*a*) 21 (*b*) 24 (*c*) 27 (*d*) 30

Sol. (*d*) $18+\frac{20}{5}\times(18-15) = 18 + 12 = 30$

Alternative method: $\frac{25\times18-20\times15}{5} = \frac{450-300}{5} = \frac{150}{5} = 30$

8. Replacement Case:

When one term is excluded and one new term is included in a series of *n* terms and the average of the series is increased by *a* by doing so, then

New term = Replaced term + $a \times n$

i.e. New term = Replaced term + (Increase in Average × Number of terms)

Note: Decrease means negative increase.

Example. Average age of 12 men is increased by 2 years when one man whose age is 43 years is replaced by a new man. Find the age of the new man.

(*a*) 19 years (*b*) 41 years (*c*) 45 years (*d*) 67 years

Sol. (*d*) Age of new man = Age of man replaced + Total increase = 43 + 2 × 12 = 67 years

9. When terms are divided into two groups and one of the terms is excluded from both the groups, then: Excluded term = Average of all terms

+ (Combined average – Average of first group) × Number of terms in the first group

+ (Combined average – Average of second group) × Number of terms in the second group.

Example 1. The average of 25 results is 18. The average of first 12 of them is 14 and that of last 12 is 17. The thirteenth result is: **(SSC CGL Tier I 2006)**

(*a*) 28 (*b*) 72 (*c*) 78 (*d*) 85

Sol. (*c*) 13th result = 18 + 12 × (18 – 14) + 12 × (18 – 17) = 18 + 48 + 12 = 78

Explanation:

13th result = (25 × 18) – (12 × 14) – (12 × 17) = (12 + 12 + 1) × 18 – 12 × 14 – 12 × 17

= 12 × (18 – 14) + 12 × (18 – 17) + 1 × 18 = (12 × 4) + (12 × 1) + 18 = 78.

Example 2. The average of 11 observations is 72. If average of first 5 observations is 70 and that of last 5 observations is 71, then the 6th observation is:

(*a*) 72 (*b*) 75 (*c*) 82 (*d*) 87

Sol. (*d*) 6th Observation = 72 + 5 × (72 – 70) + 5 × (72 – 71) = 72 + 10 + 5 = 87

Example 3. The average of 20 numbers is 12. The average of the first 12 numbers is 11 and that of the next 7 numbers is 10. The last number is : **(SSC CGL Tier I 2004)**

(*a*) 40 (*b*) 38 (*c*) 48 (*d*) 50

Sol. (*b*) Last observation = 12 + 12 × (12 – 11) + 7 × (12 – 10) = 12 + 12 + 14 = 38

10. When terms are divided into two groups and one of the terms is included in both the groups, then: Common-term = Average of all terms

+ (Average of first group – Combined Average) × Number of terms in the first group

+ (Average of second group – Combined Average) × Number of terms in the second group

Example 1 : The average of 11 observations is 50. If average of first 6 observations is 49 and that of last 6 observations is 52, then the 6th observation is: **(SSC CHSL 2006)**

(*a*) 44 (*b*) 56 (*c*) 64 (*d*) 70

Sol. (*b*) 6th observation = 50 + 6 × (49 – 50) + 6 × (52 – 50) = 50 – 6 + 12 = 56

Example 2 : The average of 9 numbers is 25. The average of first 6 numbers is 22 and that of last 4 numbers is 27. Find the 6th number.

(*a*) 15 (*b*) 25 (*c*) 30 (*d*) 35

Sol. (*a*) 6th number = 25 + 6 × (22 – 25) + 4 × (27 – 25) = 25 – 18 + 8 = 15

11. Range of Data = Maximum value – Minimum Value

SOLVED EXERCISE

1. The average of 7, 11, 15, x, 14, 21, 25 is 15, then the value of x is **(SSC CGL Tier I 2015)**

(*a*) 3 (*b*) 12 (*c*) 13.3 (*d*) 14.5

Sol. (*b*) $7 + 11 + 15 + x + 14 + 21 + 25 = 15 \times 7 \Rightarrow 93 + x = 105 \Rightarrow x = 12$

2. The ratio of ducks and frogs in a pond is 37 : 41. The average of ducks and frogs is 156. What is the number of frogs in the pond? **(SBI Clerical 2009)**

(*a*) 78 (*b*) 74 (*c*) 82 (*d*) 164

Sol. (*d*) Total number = 156 × 2 ⇒ Number of frogs = $156 \times 2 \times \frac{41}{78} = 164$

3. The average age of three boys is 15 years. If their ages are in the ratio 3 : 5 : 7, then the age of the youngest boy is **(SSC CGL Tier I 2014)**

(*a*) 9 years (*b*) 15 years (*c*) 18 years (*d*) 21 years

Sol. (*a*) Sum of ages = 3 × 15 = 45 years ⇒ Age of the youngest boy = $\frac{3}{15} \times 45 = 9$ years

4. The average of ten numbers is 30. If 5 is added to every number, the new average is **(SSC CHSL 2012)**

(*a*) 30 (*b*) 30.5 (*c*) 35 (*d*) 53

Sol. (*c*) New average = 30 + 5 = 35

5. If $\overline{x}$ is the mean of n observations $x_1, x_2 ..., x_n$, then mean of $\frac{x_1}{a}, \frac{x_2}{a}, ..., \frac{x_n}{a}$ is: **(SSC CHSL 2012)**

(*a*) $\overline{x}$ (*b*) $a\overline{x}$ (*c*) $\frac{\overline{x}}{a}$ (*d*) $\overline{x}+a$

Sol. (*c*) New mean = $\overline{x} \div a = \frac{\overline{x}}{a}$

6. The arithmetic mean of a set of 10 numbers is 20. If each number is first multiplied by 2 and then increased by 5, then what is the mean of new numbers ? **(CDS 2009)**
(*a*) 20 (*b*) 25 (*c*) 40 (*d*) 45

Sol. (*d*) New mean = $(20 \times 2) + 5 = 45$

7. The mean of 10 observations is 5. If 2 is added to each observation and then multiplied by 3, then that will be the new mean? **(NDA 2012)**
(*a*) 5 (*b*) 7 (*c*) 15 (*d*) 21

Sol. (*d*) New mean = $(5 + 2) \times 3 = 21$

8. The arithmetic mean of 100 observations is 24. 6 is added to each of the observations and then each of them is multiplied by 2.5. Find the new arithmetic mean. **(SSC CGL Tier II 2013)**
(*a*) 30 (*b*) 75 (*c*) 35 (*d*) 60

Sol. (*b*) New mean = $(24 + 6) \times 2.5 = 30 \times 2.5 = 75$

9. What is the arithmetic mean of first 20 odd natural numbers? **(SSC CGL Tier I 2015)**
(*a*) 17 (*b*) 19 (*c*) 22 (*d*) 20

Sol. (*d*) First term = 1; and 20th term = $20 \times 2 - 1 = 39 \Rightarrow$ Average = $\frac{1+39}{2} = 20$

10. The average of all prime numbers and composite numbers upto 100 is:
(*a*) 50 (*b*) 50.5 (*c*) 51 (*d*) 51.5

Sol. (*c*) First term = 2 ($\because$ 1 is neither prime nor composite number); Last term = 100

$\therefore$ Average = $\frac{2+100}{2} = 51$

11. The average of all non-negative integers upto 100 is:
(*a*) 50 (*b*) 50.5 (*c*) 49 (*d*) 49.5

Sol. (*a*) First term = 0, Last term = 100 $\Rightarrow$ Required average = $\frac{0+100}{2} = 50$

12. The average age of a man and his son is 18 years. The ratio of their ages is 5 : 1. What will be the ratio of their ages after 6 years? **(SBI Clerical 2008)**
(*a*) 5 : 2 (*b*) 3 : 1 (*c*) 6 : 1 (*d*) None of these

Sol. (*b*) Sum of ages = (5 + 1) ratio = 6 ratio = $18 \times 2 = 36$ years $\Rightarrow$ 1 Ratio = $36 \div 6 = 6$ years
$\therefore$ Ratio after 6 years = (5 + 1) : (1 + 1) = 6 : 2 = 3 : 1

13. If $16a + 16b = 48$, what is the average of a and b ? **(Bank P.O. 2006)**
(*a*) 1.5 (*b*) 2 (*c*) 2.5 (*d*) 3

Sol. (*a*) $16a + 16b = 48 \Rightarrow a + b = 48 \div 16 = 3 \Rightarrow$ Average of a and $b = (a + b) \div 2 = 1.5$

14. If the average of x and $\frac{1}{x}$ is M, then the average of x^2 and $\frac{1}{x^2}$ is **(SSC CHSL 2011)**
(*a*) $1 - M^2$ (*b*) $1 - 2M$ (*c*) $2M^2 - 1$ (*d*) $2M^2 + 1$

Sol. (*c*) $x + \frac{1}{x} = M \times 2 = 2M \Rightarrow x^2 + \frac{1}{x^2} = \left(x + \frac{1}{x}\right)^2 - 2 = (2M)^2 - 2 = 4M^2 - 2$

Average of x^2 and $\frac{1}{x^2} = (4M^2 - 2) \div 2 = 2M^2 - 1$

15. The average of eight successive numbers is 6.5. The average of the smallest and the greatest numbers among them will be : **(SSC CHSL 2010)**
(*a*) 4 (*b*) 6.5 (*c*) 7.5 (*d*) 9

Sol. (*b*) Average of smallest and greatest number = Average of all numbers = 6.5

16. Eight consecutive numbers are given. If the average of the two numbers that appear in the middle is 6, then the sum of the eight given numbers is **(SSC CHSL 2012)**

(*a*) 36 (*b*) 48 (*c*) 54 (*d*) 64

Sol. (*b*) Average of all numbers = Average of middle two numbers = 6

Sum of eight numbers = 8 × 6 = 48

17. The sum of three consecutive odd natural numbers is 147. Then, the middle number is : **(SSC CGL Tier I 1999)**

(*a*) 47 (*b*) 48 (*c*) 49 (*d*) 51

Sol. (*c*) Average = Mid-term *i.e.* 2nd term = 147 ÷ 3 = 49

18. The average of 7 consecutive numbers is 20. The largest of these numbers is : **(SSC CGL Tier I 2000, SSC constable (GD) 2012)**

(*a*) 24 (*b*) 23 (*c*) 22 (*d*) 20

Sol. (*b*) Average = Mid-term *i.e.* 4th term = 20 ⇒ Largest number = 7th term = 20 + 3 = 23

19. If the sum of five consecutive integers is S, then the largest of those integers in terms of S is **(SSC CHSL 2011)**

(*a*) $\frac{S-10}{5}$ (*b*) $\frac{S+4}{4}$ (*c*) $\frac{S+5}{4}$ (*d*) $\frac{S+10}{5}$

Sol. (*d*) Average $= \frac{\text{Sum}}{5} \Rightarrow 3^{rd}\text{ term} = \frac{S}{5} \Rightarrow \text{Fifth term} = \frac{S}{5} + 2 = \frac{S+10}{5}$

20. Sum of three consecutive even integers is 54. Find the least among them. **(SSC CGL 2002)**

(*a*) 18 (*b*) 15 (*c*) 14 (*d*) 16

Sol. (*d*) Average = Mid-term = Second term = 54 ÷ 3 = 18 ⇒ Least number = 18 – 2 = 16

21. The sum of three consecutive odd natural numbers is 87. The smallest of these numbers is: **(SSC CGL Tier I 2001)**

(*a*) 29 (*b*) 31 (*c*) 23 (*d*) 27

Sol. (*d*) Average = Mid-term = Second term = 87 ÷ 3 = 29 ⇒ Smallest number = 29 – 2 = 27

22. In a Mathematics examination, the number scored by 5 candidates are 5 succeessive odd integers. If their marks is 185, the highest score is **(SSC CGL Tier I 2014)**

(*a*) 39 (*b*) 43 (*c*) 41 (*d*) 37

Sol. (*c*) Average = Mid-term = 3rd term = 185 ÷ 5 = 37⇒Largest number = 5th term = 37 + 4 = 41

23. The average of seven consecutive positive integers is 26. The smallest of these integers is: **(SSC CHSL 2010)**

(*a*) 21 (*b*) 23 (*c*) 25 (*d*) 26

Sol. (*b*) Average = Mid-term = Fourth term = 26 ⇒ First term = 26 – 3 = 23

24. Sum of three consecutive even numbers is 32 more than the average of the numbers. Find middle number in the series. **(Bank Clerical)**

(*a*) 14 (*b*) 16 (*c*) 18 (*d*) 20

Sol. (*b*) Let average = Middle term = x and sum of three numbers = 3 × x = $3x$

Then $3x - x = 32 \Rightarrow x = 16$

25. The average of two numbers is 62. If 2 is added to the smallest number, the ratio between the numbers becomes 1 : 2. The difference of the numbers is **(SSC FCI Asstt. 2012)**

(*a*) 62 (*b*) 40 (*c*) 84 (*d*) 44

Sol. (*d*) Sum of numbers = 2 × 62 = 124 ⇒ New sum = 124 + 2 = 126

∴ Smaller number = $\frac{1}{3} \times 126 = 42$ ⇒ Bigger number = 126 – 42 = 84

∴ Required difference = 84 – (42 – 2) = 84 – 40 = 44

26. Of the three numbers whose average is 60, the first is one fourth of the sum of the others. The first number is : **(SSC CGL Tier I 1999)**

(*a*) 30 (*b*) 36 (*c*) 42 (*d*) 45

Sol. (*b*) Let the first number be 1. Then sum of second and third number = 4.

Sum of three numbers = 3 × 60 = 180 ⇒ First number = $180 \times \frac{1}{5} = 36$

27. Out of 4 numbers, whose average is 60, the first one is one-fourth of the sum of the last three. The first number is **(SSC CGL Tier I 2011)**

(*a*) 15 (*b*) 45 (*c*) 48 (*d*) 60

Sol. (*c*) Sum of 4 numbers = 4 × 60 = 240 ⇒ First number = $\frac{1}{5} \times 240 = 48$

28. A sum of ₹ 700 has to be used to give seven cash prizes to the students of a school for their overall academic performance. If each prize is ₹ 20 less than its preceding prize, what is the least value of the prize ? **[UPSC Civil Services (Pre.) 2013]**

(*a*) ₹ 30 (*b*) ₹ 40 (*c*) ₹ 60 (*d*) ₹ 80

Sol. (*b*) Average = 4th prize = ₹ 700 ÷ 7 = ₹ 100 ⇒ Least prize = ₹ 100 – 3 × ₹ 20 = ₹ 40

29. *a*, *b*, *c* and *d* are four consecutive numbers. If the sum of *a* and *c* is 124, what is product of *b* and *d*? **[RBI Note Printing Asstt. Manager 2006]**

(*a*) 4032 (*b*) 3782 (*c*) 3906 (*d*) 3968

Sol. (*d*) $a + c = 124 \Rightarrow b = (a + c) \div 2 = 124 \div 2 = 62 \Rightarrow d = 62 + 2 = 64$

$\Rightarrow b \times d = 62 \times 64 = 3968$

30. A, B, C and D are four consecutive even numbers and their average is 65. What is the product of A and D ? **(Bank of Baroda PO Exam 2007)**

(*a*) 3698 (*b*) 4216 (*c*) 4092 (*d*) 4352

Sol. (*b*) Average of even number of terms = Average of two mid-terms

$\Rightarrow \frac{B + C}{2} = 65$ ⇒ B = 64 and C = 66 (Average of 64 and 66 is 65)

∴ A = 62 and D = 68 ⇒ A × D = 62 × 68 = 4216

31. *a*, *b*, *c*, *d* and *e* are five consecutive even numbers. If the sum of *a* and *d* is 162, what is the sum of all the numbers ? **(SBI Bank PO 2008)**

(*a*) 400 (*b*) 380 (*c*) 420 (*d*) None of these

Sol. (*d*) $a + d = 162 \Rightarrow b + d = 162 + 2 = 164 \Rightarrow c =$ Average of *b* and *d* = 164 ÷ 2 = 82

$a + b + c + d + e = 82 \times 5 = 410$

32. The average of five consecutive numbers A, B, C, D and E is 105. What is the product of B and E? **(Bank Clerical 2008)**

(*a*) 11025 (*b*) 12024 (*c*) 11128 (*d*) Can't say

Sol. (*c*) Average = Middle number = C = 105 ⇒ B = 105 – 1 = 104 and E = 105 + 2 = 107

B × E = 104 × 107 = 11,128

33. Of the three numbers, the first is twice the second and the second is 3 times the third. If their average is 100, the largest of the three numbers is : **(SSC CGL Tier I 2004)**

(*a*) 120 (*b*) 150 (*c*) 180 (*d*) 300

Sol. (*c*) Let the numbers are $6x$, $3x$ and x. Then sum of numbers $= 10x = 100 \times 3 = 300$

$\therefore x = 300 \div 10 = 30 \Rightarrow$ Largest number $= 6x = 6 \times 30 = 180$

34. Of the three numbers, second is twice the first and is also thrice the third. If the average of the three numbers is 44, the largest number is : **(SSC CGL Tier I 2002)**

(*a*) 24 (*b*) 72 (*c*) 36 (*d*) 108

Sol. (*b*) Let the numbers are $3x$, $6x$ and $2x$. $\Rightarrow$ Sum of the numbers $= 11x = 44 \times 3$

$\Rightarrow x = 44 \times 3 \div 11 = 12 \Rightarrow$ Largest number $= 6x = 6 \times 12 = 72$

35. The average of first three numbers is double of the fourth number. If the average of all the four numbers is 12, find the 4th number. **(SSC CGL Tier II 2013)**

(*a*) 16 (*b*) $\frac{48}{7}$ (*c*) 20 (*d*) $\frac{18}{7}$

Sol. (*b*) Sum of 4 numbers $= 4 \times 12 = 48$. Let the fourth number $= x$.

Then sum of first three numbers $= 3 \times 2x = 6x \Rightarrow x + 6x = 48 \Rightarrow x = \frac{48}{7}$

36. Out of four numbers, the average of first three numbers is thrice the fourth number. If the average of the four numbers is 5, the fourth number is : **(SSC CGL Tier I 2006)**

(*a*) 4.5 (*b*) 5 (*c*) 2 (*d*) 4

Sol. (*c*) Sum of four numbers $= 4 \times 5 = 20$. Let the fourth number be x.

Average of first three numbers $= 3x \Rightarrow$ Sum of first three numbers $= 3 \times 3x = 9x$

Sum of four numbers $= 9x + x = 10x = 20 \Rightarrow$ Fourth number $= 20 \div 10 = 2$

37. Out of three numbers, the first is twice the second and is half of the third. If the average of the three numbers is 56, then difference of first and third numbers is : **(SSC CGL Tier I 2001)**

(*a*) 12 (*b*) 20 (*c*) 24 (*d*) 48

Sol. (*d*) Let the third number be 4. Then first number $= 4 \div 2 = 2$

And second number $= 2 \div 2 = 1$

Sum of the numbers $= (2 + 1 + 4)$ ratio $= 7$ ratio $= 56 \times 3$

Difference of first and third numbers $= (4 - 2)$ ratio $= 2$ ratio $= 56 \times 3 \times \frac{2}{7} = 48$

38. Of three numbers, the first is twice the second and second is thrice the third. If the average of the three numbers is 10, then the largest number is : **(SSC CGL Tier I 2014)**

(*a*) 12 (*b*) 30 (*c*) 18 (*d*) 15

Sol. (*c*) Let third number $= 1 \Rightarrow$ Second number $= 3 \times 1 = 3 \Rightarrow$ First number $= 2 \times 3 = 6$

Sum of the numbers $= (6 + 3 + 1)$ ratio $= 10$ ratio $= 10 \times 3 = 30$

$\Rightarrow$ Largest number $= \frac{6}{10} \times 30 = 18$

39. A milkman bought 15 kg of milk and mixed 3 kg of water in it. If the price per kg of the mixture becomes ₹ 22, what is the cost price of the milk per kg? **(CDS 2012)**

(*a*) ₹ 28.00 (*b*) ₹ 26.40 (*c*) ₹ 24.00 (*d*) ₹ 22.60

Sol. (*b*) Cost price of mixture $= 22 \times 18 \Rightarrow$ Cost price of milk $= \frac{22 \times 18}{15} =$ ₹ 26.40

40. Sum and average of some numbers are 400 and 40 respectively. If two numbers with average of 100 are also included, find new average of all the numbers.

(*a*) 50 (*b*) 60 (*c*) 70 (*d*) Data inadequate

Sol. (*a*) Number of terms = 400 ÷ 40 = 10. Sum of new numbers = 2 × 100 = 200

New average = $\frac{400+200}{10+2} = 50$

41. The average of the first nine integral multiples of 3 is **(SSC CGL Tier I 2013)**

(*a*) 15 (*b*) 18 (*c*) 21 (*d*) 12

Sol. (*a*) Average = Mid-term = $\left(\frac{9+1}{2}\right)^{th}$ term = 5th term = 3 × 5 = 15

Alt: Average = $\frac{3+27}{2} = 15$

42. Find the average of first eight multiples of 6.

(*a*) 18 (*b*) 24 (*c*) 27 (*d*) 30

Sol. (*c*) Sum of first eight multiples of 6 = (6 × 1) + (6 × 2) + (6 × 3) + …….. + (6 × 8)

$= 6 \times (1+2+3+ \ldots\ldots + 8) = 6 \times \frac{8 \times 9}{2} \Rightarrow$ Average $= 6 \times \frac{8 \times 9}{2 \times 8} = 27$

Alternative Method: Average = Average of $\left(\frac{8}{2}\right)^{th}$ term and $\left(\frac{8}{2}+1\right)^{th}$ term

= Average of 4th and 5th term $= \frac{(6 \times 4) + (6 \times 5)}{2} = \frac{6 \times (4+5)}{2} = \frac{6 \times 9}{2} = 27$

Alt: Average = $\frac{6+48}{2} = 27$

43. Average of first five odd multiples of 3 is:

(*a*) 12 (*b*) 15 (*c*) 18 (*d*) 21

Sol. (*b*) Sum of first 5 odd multiples of 3 = (3 × 1) + (3 × 3) + (3 × 5) + (3 × 7) + (3 × 9)

$= 3 \times (1+3+5+7+9) = 3 \times 5^2 \Rightarrow$ Average $= 3 \times 5^2 \times \frac{1}{5} = 15$

Direct Method: Average = Mid-term = $\left(\frac{5+1}{2}\right)^{th}$ *i.e.* 3rd term = 3 × 5 = 15

Alt: Average = $\frac{3+27}{2} = 15$

44. The average of $1^2, 2^2, 3^2, 4^2, 5^2, \ldots, 10^2$ is **(SSC CGL Tier I 2012, CDS 2015)**

(*a*) 38.5 (*b*) 50.5 (*c*) 42.5 (*d*) 46.5

Sol. (*a*) Average of square numbers $= \frac{n(n+1)(2n+1)}{6} \times \frac{1}{n} = \frac{11 \times 21}{6} = 38.5$

45. What is the arithmetic mean of first 16 natural numbers with weights being the number itself? **(NDA 2012)**

(*a*) 17/2 (*b*) 33/2 (*c*) 11 (*d*) 187/2

Sol. (*d*) Average $= \dfrac{1^2+2^2+3^2+...+16^2}{16} = \dfrac{1}{16}\times\dfrac{16\times17\times33}{6} = \dfrac{187}{2}$

46. If average of 5 observations is 46 and that of 5 other observations is 50, find the average of all the 10 observations.

(*a*) 96 (*b*) 50 (*c*) 48 (*d*) 46

Sol. (*c*) Since number of terms are equal in both the cases.

Average of all the terms is equal to average of two averages.

$\therefore$ Average of 10 observations $= \dfrac{46+50}{2} = 48$

47. Average of n observations is 35, average of n other observations is 37 and average of remaining n observations is 42. Average of all the observations is:

(*a*) 36 (*b*) 38 (*c*) 40 (*d*) 57

Sol. (*b*) Since number of terms are equal in all the three cases.

Average of all the terms is equal to average of three averages.

$\therefore$ Average of all observations $= \dfrac{35+37+42}{3} = \dfrac{114}{3} = 38$

48. The average of 5 consecutive numbers is n. If the next two numbers are also included, the average of the 7 numbers will **(SSC CGL Tier I 2012, MAT 2005, UPSC Civil Serv. 2011)**

(*a*) increase by 2 (*b*) increase by 1 (*c*) remains the same (*d*) increase by 1.4

Sol. (*b*) Average of 5 consecutive numbers = 3rd number = n

Average of 7 consecutive numbers = 4th number = $n + 1$

49. The average of 21 consecutive odd numbers is n. If next two numbers are also included, the average of all the 23 numbers will be :

(*a*) 23 (*b*) $n - 1$ (*c*) $n + 1$ (*d*) $n + 2$

Sol. (*d*) Average of 21 consecutive odd numbers = Middle term $= \dfrac{21+1}{2}$ = 11th term = n (Given)

Average of 23 consecutive odd numbers = Middle term $= \dfrac{23+1}{2}$ = 12th term = $n + 2$

50. The average of 5 consecutive integers starting with 'm' is n. What is the average of 6 consecutive integers starting with $(m + 2)$? **(SSC CGL Tier II 2012)**

(*a*) $(n + 3)$ (*b*) $\dfrac{2n+9}{2}$ (*c*) $\dfrac{2n+5}{2}$ (*d*) $(n + 2)$

Sol. (*c*) Average of 5 consecutive integers starting with $m = m + 2 = n$ (given)

Average of 6 consecutive integers starting with $(m + 2)$ = Average of 3rd and 4th term

$= \dfrac{(m+4)+(m+5)}{2} = \dfrac{2m+9}{2} = \dfrac{2(m+2)+5}{2} = \dfrac{2n+5}{2}$

51. One year ago, the population of a town was 12000. In one year, male population is increased by 8% and female population by 6% but the total population is increased by 7%. What was the size of the male population one year ago.

(*a*) 5000 (*b*) 5500 (*c*) 6000 (*d*) 7000

Sol. (*c*) Total increase in population is 7% which is equal to average of 6% and 8%.

$\therefore$ Male population = Female population $\Rightarrow$ Male population $= \dfrac{1}{2} \times 12000 = 6000$

52. If the average of m numbers is n^2 and that of n numbers is m^2, find the average of $(m + n)$ numbers **(SSC CHSL 2013)**

(a) $m - n$ (b) $m + n$ (c) $m \div n$ (d) mn

Sol. (d) Sum of $(m + n)$ numbers $= mn^2 + m^2n = mn(n + m)$

Average of $(m + n)$ numbers $= \dfrac{mn(m+n)}{(m+n)} = mn$

53. The sum of five numbers is 555. The average of first two numbers is 75 and the third number is 115. What is the average of last two numbers ? **(Bank of Baroda PO Exam 2007)**

(a) 145 (b) 290 (c) 265 (d) 150

Sol. (a) Sum of first three numbers $= 75 \times 2 + 115 = 265$

Sum of last two numbers $= 555 - 265 = 290$

$\therefore$ Average of last two numbers $= 290 \div 2 = 145$

54. The average of eight numbers is 20. The average of first two numbers is $15\frac{1}{2}$, and that of next three numbers is $21\frac{1}{3}$. Seventh and eighth number exceed the sixth number by 4 and 7 respectively. What is the eighth number ? **(SSC CGL Tier I 2008)**

(a) 11 (b) 18 (c) 22 (d) 25

Sol. (d) Sum of eight numbers $= 8 \times 20 = 160$ Sum of first 5 numbers $= 2\times\frac{31}{2}+3\times\frac{64}{3} = 95$

Let the sixth number be x. Then $x + (x + 4) + (x + 7) = (160 - 95) \Rightarrow 3x + 11 = 65 \Rightarrow x = 18$

$\therefore$ Eighth number $= x + 7 = 18 + 7 = 25$

55. The average of n numbers is x. When 36 is subracted from two of the numbers, the new average becomes $(x - 8)$. The value of n is : **(SSC CGL Tier I 2006)**

(a) 6 (b) 8 (c) 9 (d) 72

Sol. (c) Decrease in sum of numbers $= 2 \times 36 = 72$

Decrease in average $= 8$ $\therefore$ Number of terms $= 72 \div 8 = 9$

Alternative method: $nx = n(x - 8) + 72 \Rightarrow 8n = 72 \Rightarrow n = 9$

56. The sum of five consecutive integers is a and the sum of next five consecutive integers is b. Then, $\dfrac{(b-a)}{100}$ is equal to **(SSC FCI Asstt. 2012)**

(a) $\frac{1}{4}$ (b) $\frac{1}{2}$ (c) 4 (d) 2

Sol. (a) Difference in average of two series $= 5 \Rightarrow b - a = 5 \times 5 \Rightarrow \dfrac{b-a}{100} = \dfrac{5\times5}{100} = \dfrac{1}{4}$

57. If the average of a, b, c be M and $ab + bc + ca = 0$, then the average of a^2, b^2, c^2 is :

(a) M^2 (b) $3M^2$ (c) $9M^2$ (d) $27M^2$

Sol. (b) $(a + b + c)/3 = M \Rightarrow a + b + c = 3M$

$\Rightarrow (a + b + c)^2 = a + b + c + 2(ab + bc + ca) \Rightarrow (3M)^2 = a^2 + b^2 + c^2$

$\Rightarrow$ Required average $= \dfrac{(a+b+c)^2}{3} = \dfrac{(3M)^2}{3} = 3M^2$

58. If the average of 6 consecutive even numbers is 25, the difference between the largest and the smallest number is **(SSC CGL Tier II 2013)**

(a) 8 (b) 10 (c) 12 (d) 14

Sol. (b) Difference $= (6 - 1) \times 2 = 10$ [**Note:** Average is immaterial here]

59. The sum of all natural numbers between 100 and 200, which are multiples of 3 is: **(SSC CGL Tier I 2000)**

(*a*) 5000 (*b*) 4950 (*c*) 4980 (*d*) 4900

Sol. (*b*) $100 = 33 \times 3 + 1 \Rightarrow$ Terms (divisible by 3) up to 100 = 33

$200 = 66 \times 3 + 2 \Rightarrow$ Terms (divisible by 3) up to 200 = 66

$\therefore$ Total terms (divisible by 3) between 100 and 200 are 66 – 33 = 33

Average = Mid-term = (33 + 17) or 50th term = 150 $\Rightarrow$ Sum of 33 terms is $150 \times 33 = 4950$

Note: Sum of the terms must be divisible by 3 as each of the term is divisible by 3.

Alternative method: $\frac{102+198}{2} \times 33 = 4950$

60. The average weight of A, B, C is 45 kg. If the average weight of A and B be 40 kg and that of B and C 43 kg, then the weight of B is : **(SSC CGL Tier I 2004)**

(*a*) 17 kg (*b*) 20 kg (*c*) 26 kg (*d*) 31 kg

Sol. (*d*) $A + B + C = 3 \times 45 = 135$; $A + B = 2 \times 40 = 80$ kg; $B + C = 2 \times 43 = 86$ kg

$\Rightarrow (A + B) + (B + C) - (A + B + C) = 80 + 86 - 135 \Rightarrow B = 31$ kg

61. The average monthly income of X and Y is ₹ 5050. The average monthly income of Y and Z is ₹ 6250 and the average monthly income of X and Z is ₹ 5200. The monthly income of X is : **(SSC CGL Tier I 2004)**

(*a*) ₹ 4050 (*b*) ₹ 3050 (*c*) ₹ 4000 (*d*) ₹ 5000

Sol. (*c*) $X + Y = 2 \times 5050 = 10100$, $Y + Z = 2 \times 6250 = 12500$, $X + Z = 2 \times 5200 = 10400$

$\therefore (X + Y) + (X + Z) - (Y + Z) = 10100 + 10400 - 12500 \Rightarrow 2X = 8000 \Rightarrow X =$ ₹ 4000

62. The average weight of three men A, B and C is 84 kg. Another man D joins the group and the average now becomes 80 kg. If another man E whose weight is 3 kg more than that of D, replaces A, then the average weight of B, C, D and E becomes 79 kg. The weight of A is : **(SSC CGL Tier I 2003 & 2015, Bank PO 1989, MBA 2000)**

(*a*) 72 kg (*b*) 70 kg (*c*) 75 kg (*d*) 80 kg

Sol. (*c*) $A + B + C = 84 \times 3 = 252$ kg. $A + B + C + D = 80 \times 4 = 320$ kg

$\therefore D = 320 - 252 = 68$ kg. $E = D + 3 = 68 + 3 = 71$ kg

$B + C + D + E = 79 \times 4 = 316$ kg

$A = (A + B + C + D + E) - (B + C + D + E) = (320 + 71) - 316 = 75$ kg

63. The average of 4 numbers is 67.5. The highest and the smallest numbers are 108 and 23 respectively. If difference between other two numbers is 31, find the bigger number of the two. **(Bank PO 2007)**

(*a*) 139 (*b*) 108 (*c*) 85 (*d*) 54

Sol. (*c*) Sum of 4 numbers = $4 \times 67.5 = 270$. Sum of two given numbers = 108 + 23 = 131

Sum of remaining two numbers = 270 – 131 = 139

Difference of the two numbers = 31 $\Rightarrow$ The bigger number = $\frac{139+31}{2} = 85$

64. In a group of 135 persons, 50% people contributed ₹ 40 each, 30% contributed ₹ 60 each and remaining persons ₹ 80 each. Find average contribution for the group.

(*a*) ₹ 50 (*b*) ₹ 54 (*c*) ₹ 60 (*d*) ₹ 62

Sol. (*b*) Average contribution = 50% of 40 + 30% of 60 + 20% of 80 = 20 + 18 + 16 = ₹ 54

65. The average age of A, B and C is 26 years. If the average age of A and C is 29 years, what is the age of B in years ? **(SBI Bank Officers Exam. 2007)**

(*a*) 26 years (*b*) 20 years (*c*) 29 years (*d*) 23 years

Sol. (*b*) $A + B + C = 3 \times 26$ years = 78 years. $A + C = 2 \times 29$ years = 58 years

$\therefore B = (A + B + C) - (A + C) = 78 - 58 = 20$ years

Alternative method : Age of B = Average of all + Increase in average

$= 26 + (26 - 29) \times 2 = 26 - 3 \times 2 = 26 - 6 = 20$ years

66. The average of 11 numbers is 70 and that of first 10 numbers is 69. The 11th number is:

(*a*) 59 (*b*) 79 (*c*) 80 (*d*) 81

Sol. (*c*) 11th number = Average of all + Total increase = $70 + (1 \times 10) = 80$

Alt: $70 \times 11 - 69 \times 10 = 70 \times 1 + 70 \times 10 - 69 \times 10 = 70 + 10(70 - 69) = 80$

67. The average age of 14 girls and their teacher's age is 15 years. If the teacher's age is excluded, the average reduces by 1. What is teacher's age? **(SSC CGL Tier I 2000, 2013)**

(*a*) 35 years (*b*) 32 years (*c*) 30 years (*d*) 29 years

Sol. (*d*) Teacher's age = $15 + 14 \times (15 - 14) = 29$ years

Alt: $15 \times 15 - 14 \times 14 = 15 + 14 = 29$

68. A cricketer has a mean score of 60 runs in 10 innings. Find out how many runs are to be scored in the eleventh innings to raise the mean score to 62. **(SSC CHSL 2012)**

(*a*) 80 (*b*) 81 (*c*) 82 (*d*) 83

Sol. (c) Required number = $11 \times 62 - 10 \times 60 = 682 - 600 = 82$

Alt: $11 \times 62 - 10 \times 60 = 62 + 10 \times (62 - 60) = 62 + 20 = 82$

69. The average age of four brothers is 12 years. If the age of their mother is also included, the average is increased by 5 years. The age of the mother is **(SSC CGL Tier I 2000)**

(*a*) 37 years (*b*) 43 years (*c*) 48 years (*d*) 53 years

Sol. (*a*) Average age of all = 12 + 5 = 17 years

Total increase on inclusion of mother's age = $4 \times 5 = 20$ years

$\therefore$ Mother's age = 17 + 20 = 37 years

Alt: $5 \times 17 - 4 \times 12 = 85 - 48 = 37$.

70. The average of 7 numbers is 8. If one number is added, their average is 9. Then the added number is **(SSC CGL Tier I 2012)**

(*a*) 16 (*b*) 14 (*c*) 12 (*d*) 11

Sol. (*a*) Required number = $8 \times 9 - 7 \times 8 = 16$ **Alt:** $9 + 7 \times (9 - 8) = 16$

71. Four years ago the average age of A, B and C was 25 years. Five years ago the average age of B and C was 20 years. A's present age is : **(SSC CGL Tier I 2006)**

(*a*) 60 years (*b*) 62 years (*c*) 15 years (*d*) 37 years

Sol. (*d*) Present average age of A, B and C = 25 + 4 = 29 years

Present average age of B and C = 20 + 5 = 25 years

$\therefore$ A's present age = $3 \times 29 - 2 \times 25 = 37$ years

Alternative method : A's present age = $29 + 2 \times (29 - 25) = 37$ years

72. A cricketer played 80 innings and scored an average of 99 runs. His score in the last inning is zero run. To have an average of 100 at the end, his score in the last innings should have been **(SSC CGL Tier II 2012)**

(*a*) 60 runs (*b*) 80 runs (*c*) 10 runs (*d*) 1 run

Sol. (*b*) His average would have been 100 if he had scored $80 \times (100 - 99) = 80$ runs

73. The average of runs of a cricket player of 10 innings was 32. How many runs must he make in his next innings so as to increase his average of runs by 4 ? **(SSC CGL Tier I 2004)**

(*a*) 76 (*b*) 70 (*c*) 4 (*d*) 2

Sol. (*a*) New average of all the innings = 32 + 4 = 36 runs
Increase in runs = 10 × 4 = 40 ⇒ Required runs = 36 + 40 = 76 runs

74. Average age of 6 persons is decreased by 1 year when one new person is included in the group. Find the age of new man, if average age of 6 persons was 39 years.
(*a*) 32 years (*b*) 33 years (*c*) 38 years (*d*) 40 years

Sol. (*a*) New average = 39 – 1 = 38 years
Age of new man = Average age of 7 persons – Total decrease in age = 38 – (1 × 6) = 32 years

75. Out of 10 terms, the average of first 5 terms is 34 and that of last 5 terms is 38. If the average of first 9 terms is 35, find the 10th term.
(*a*) 26 (*b*) 35 (*c*) 44 (*d*) 45

Sol. (*d*) Average of 10 terms = $\frac{34+38}{2}$ = 36. Average of 9 terms = 35
∴ 10th term = Average of all + Total increase = 36 + (1 × 9) = 36 + 9 = 45

76. A batsman makes a score of 58 runs in the 15th innings and thus increases his average by 3 runs. What is the average after 15th inning ? **(SSC CGL Tier I 2012)**
(*a*) 12 (*b*) 14 (*c*) 16 (*d*) 18

Sol. (c) Increase in score of 14 innings = 14 × 3 = 42
Average after 15 innings = 58 – 42 = 16 runs
Alt: $14x + 58 = 15(x + 3) \Rightarrow x = 13 \Rightarrow x + 3 = 16$ runs.

77. The average marks obtained by a group of 10 students is 41 marks. Find the new average if a new student who scored 63 marks is also included in the group.
(*a*) 39 (*b*) 40 (*c*) 43 (*d*) 45

Sol. (*c*) Excess marks obtained by 11th student over the average of 10 students = 63 – 41 = 22
Increase in average of 11 students = $\frac{22}{11}$ = 2 marks
∴ Average marks of 11 students = 41 + 2 = 43 marks.
Alt: $\frac{(10 \times 41) + 63}{11} = \frac{473}{11} = 43$ marks

78. The average of 9 terms is 79. If one new term whose value is 59 is also included in the series, find average of all the 10 terms.
(*a*) 77 (*b*) 80 (*c*) 81 (*d*) 83

Sol. (*a*) 10th term is short from the average of 9 terms by 79 – 59 = 20
∴ Decrease in average of 10 terms due to its inclusion = $\frac{20}{10}$ = 2
∴ Average of 10 terms = 79 – 2 = 77

79. A student on her first 3 tests received an average score on N points. If she exceeds her previous average score by 20 points on her fourth test, then what is the average score for the first 4 tests ? **(UPSC Civil Services Pre. 2011)**

Sol. (c) Total of 4 tests = 3 × N + (N + 20) = 4 N + 20 = (4 N + 20) ÷ 4 = N + 5

80. The average score of three boys in Mathematics is 30. When the score of a fourth boy is added to the total score of the three boys, the average is lowered to 25. What is the score of the fourth boy ? **(UPSC Civil Services Pre. 2011)**
(*a*) 10 (*b*) 15 (*c*) 20 (*d*) 25

Sol. (c) Total score of 4 boys = 4 × 25 = 100

Total score of 3 boys = 3 × 30 = 90 ⇒ Score of fourth boy = 100 – 90 = 10

Alt : New average – Total decrease on inclusion = 25 + 3 × (25 – 30) = 10

81. Average age of 9 members of a club is 29 years. If 2 more persons with the average age of 40 years become the members of the club, find average age of all the 11 members. **(GMAT)**

(*a*) 30 years (*b*) 31 years (*c*) 32 years (*d*) 35 years

Sol. (*b*) Average age of 9 members = 29 years

Excess age of 2 new members = (40 – 29) × 2 = 22 years

∴ Increase in the average age on inclusion of 2 new members = $\frac{22}{11}$ = 2 years

∴ Average age of 11 members = 29 + 2 = 31 years

82. The average of runs of a cricket player in 15 matches is 33. If the average of the first 10 matches is 45, then the average of the last 5 matches is **(SSC CHSL 2012)**

(*a*) 15 (*b*) 23 (*c*) 9 (*d*) 13.5

Sol. (*c*) Average of last 5 matches $\frac{15 \times 33 - 10 \times 45}{15 - 10} = \frac{495 - 450}{5} = \frac{45}{5} = 9$

83. The average of 50 numbers is 38. If two numbers namely 45 and 55 are discarded, the average of the remaining numbers is : **(CBI 1990, SSC CGL Tier I 2013, 2014)**

(*a*) 36 (*b*) 35 (*c*) 32.5 (*d*) 37.5

Sol. (*d*) New average = $\frac{50 \times 38 - (45 + 55)}{48} = \frac{1800}{48} = 37.5$

Alt: Average of 50 numbers = 38. Sum of two numbers discarded = 45 + 55 = 100

Excess of 2 numbers over average = 100 – 2 × 38 = 24

Decrease in average of remaining numbers = 24 ÷ 48 = 0.5

∴ New average of remaining 48 numbers = 38 – 0.5 = 37.5

84. The average of marks of 28 students in Mathematics was 50; 8 students left the school, then this average increased by 5. What is the average of marks obtained by the students who left the school ? **(SAS 2010, SSC CGL Tier I 2005)**

(*a*) 50.5 (*b*) 37.5 (*c*) 42.5 (*d*) 45

Sol. (*b*) Total marks of 28 students = 28 × 50 = 1400

Total marks of remaining 20 students = 20 × (50 + 5) = 1100

∴ Marks of 8 students = 1400 – 1100 = 300 ⇒ Average of 8 students = 300 ÷ 8 = 37.5 marks

Alternative Method : Net marks of 8 students = 8 × 50 – 5 × (28 – 8) = 400 – 100 = 300.

∴ Average of 8 students = 300 ÷ 8 = 37.5 marks

85. The average age of 40 students of a class is 18 years. When 20 new students are admitted to the same class, the average age of the students of the class is increased by six months. The average age of newly admitted students is : **(SSC CGL Tier I 2008)**

(*a*) 19 years (*b*) 19 years 6 months (*c*) 20 years (*d*) 20 years 6 months

Sol. (*b*) Total age of 40 students = 40 × 18 = 720 years

Total age of 60 students = 60 × 18.5 = 1110 years

Total age of 20 new students = 1110 – 720 = 390 years

Average age of 20 new students = 390 ÷ 20 = 19 years 6 months

Alternative Method : Average age of 60 students = 18 years 6 months

Total increase in age of 40 students = 40 × 6 months = 20 years

Increase in age of new students = 20 ÷ 20 = 1 year

∴ Average of new 20 students = 18 years 6 months + 1 year = 19 years 6 months

86. A cricketer had a certain average of runs for his 64 innings. In his 65th inning, he is bowled out for no score on his part. This brings down his average by 2 runs. His new average of runs is: **(SSC CGL Tier I 2008)**

(*a*) 130 (*b*) 128 (*c*) 70 (*d*) 68

Sol. (*b*) Let original average be x. Then $64x = 65(x-2) \Rightarrow x = 130$

∴ New average = 130 – 2 = 128 runs

Alternative Method : Decrease in average = 2 runs

Total deacrease in 64 innings = 64 × 2 = 128 runs ⇒ New average = 0 + 128 = 128 runs

87. A cricket player after playing 10 tests scored 100 runs in the 11th test. As a result, the average of his runs is increased by 5. The present average of runs is **(SSC CGL Tier I 2013, ICWA)**

(*a*) 45 (*b*) 40 (*c*) 50 (*d*) 55

Sol. (*c*) Let original average be x. Then $10x + 100 = 11(x+5) \Rightarrow x = 45$

∴ New average = 45 + 5 = 50 runs

Alternative Method : Increase in average = 5 runs

Total increase in 10 innings = 10 × 5 = 50 runs ⇒ New average = 100 – 50 = 50 runs

88. A cricket batsman had a certain average of runs for 11 innings. In the 12th innings, he made a score of 90 runs and thereby his average of runs was decreased by 5. His average of runs after 12th innings is: **(SSC CHSL 2011)**

(*a*) 155 (*b*) 150 (*c*) 145 (*d*) 140

Sol. (*c*) Let average of first 11 innings be x runs.

Then $11x + 90 = 12(x-5) \Rightarrow 11x + 90 = 12x - 60 \Rightarrow x = 90 + 60 = 150$ runs

∴ Average of 12 innings = $x - 5$ = 150 – 5 = 145 runs

Alterative Method : Decrease in average = 5 runs

∴ Decrease in total of 11 innings = 11 × 5 = 55 runs

∴ Average of 12 innings = 90 + 55 = 145 runs

89. Sachin Tendulkar has a certain average for 11 innings. In the 12th innings he scores 120 runs and thereby increases his average by 5 runs. His new average is **(SSC CGL Tier II 2013)**

(*a*) 60 (*b*) 62 (*c*) 65 (*d*) 66

Sol. (*c*) New average = 120 – (5 × 11) = 120 – 55 = 65 runs

Alt: $11x + 120 = 12(x+5) \Rightarrow x = 60 \Rightarrow (x+5) = 65$

90. The batting average of a cricket player for 64 innings is 62 runs. His highest score exceeds his lowest score by 180 runs. Excluding these two innings, the average of the remaining innings becomes 60 runs. His highest score is **(SSC CHSL 2011)**

(*a*) 180 runs (*b*) 214 runs (*c*) 209 runs (*d*) 212 runs

Sol. (*b*) Total runs of the two innings = 2 × 62 + 62 × (64 – 62) = 124 + 124 = 248

Highest score – Lowest score = 180 runs ⇒ Highest score = $\frac{248+180}{2}$ = 214 runs

91. The batting average for 40 innings of a cricket player is 50 runs. His highest score exceeds his lowest score by 172 runs. If these two innings are not included, the average of the remaining 38 innings is 48 runs. The highest score of the player is : **(SSC CGL Tier I 2006)**

(*a*) 165 runs (*b*) 170 runs (*c*) 172 runs (*d*) 174 runs

Sol. (*d*) Total runs of the two innings = 2 × 50 + (50 – 48) × 38 = 100 + 76 = 176 runs

Highest score – lowest score = 172 runs ⇒ Highest score $\frac{176+172}{2}$ = 174 runs

Alt: (40 × 50) – (38 × 48) = 2000 – 1824 = 176

92. The highest score in an innings was $\frac{3}{11}$ of the total and the next highest was $\frac{3}{11}$ of the remainder. If the scores differed by 9, then the total score is **(SSC CGL Tier I 2013)**

(*a*) 110 (*b*) 121 (*c*) 132 (*d*) 99

Sol. (*b*) Highest score = $\frac{3}{11}$ ⇒ Remaining score = $1-\frac{3}{11}=\frac{8}{11}$

⇒ Next highest = $\frac{3}{11}\times\frac{8}{11}=\frac{24}{121}$ ⇒ Difference = $\left(\frac{3}{11}-\frac{24}{121}\right)$ of total score = 9

⇒ $\frac{9}{121}$ of score = 9 ⇒ Total score = $9\times\frac{121}{9}$ = 121 runs

93. Average age of 6 sons of a family is 8 years. Average age of sons together with their parents is 22 years. If father is older than mother by 8 years, the age of mother is : **(SSC CGL Tier I 2003)**

(*a*) 44 years (*b*) 52 years (*c*) 60 years (*d*) 68 years

Sol. (*c*) Combined age of parents = 2 × 22 + 6 × (22 – 8) = 44 + 84 = 128

∴ Age of mother = $\frac{128-8}{2}$ = 60 years

Alt : Age of parents = 8 × 22 – 6 × 8 = 128 years ⇒ Age of mother = $\frac{128-8}{2}$ = 60 years

94. The average of 100 numbers is 44. The average of these 100 numbers and 4 other numbers is 50. The average of 4 new numbers will be : **(SSC CGL Tier I 2007)**

(*a*) 800 (*b*) 200 (*c*) 176 (*d*) 24

Sol. (*b*) Sum of 4 numbers = 4 × 50 + 100 × (50 – 44) = 200 + 600 = 800

∴ Average of 4 numbers = 800 ÷ 4 = 200

Alt: Sum of 100 numbers = 44 × 100 = 4400 Sum of 104 numbers = 50 × 104 = 5200

Sum of 4 numbers = 5200 – 4400 = 800 ∴ Average of 4 numbers = 800 ÷ 4 = 200

95. Average weight of 25 persons is increased by 1 kg when one man weighing 60 kg is replaced by a new person. Weight of new person is : **(SSC CGL Tier I 2004)**

(*a*) 50 kg (*b*) 61 kg (*c*) 86 kg (*d*) 85 kg

Sol. (*d*) Weight of new person = Weight of man replaced + Total increase in weight = 60 + 25 × 1 = 85 kg

96. The average age of 8 persons is increased by 2 years, when one of them, whose age is 24 years, is replaced by a new person. The age of the new person is : **(SSC CGL Tier I 2001)**

(*a*) 42 years (*b*) 40 years (*c*) 38 years (*d*) 45 years

Sol. (*b*) Age of new person = Age of man replaced + Total increase in age = 24 + 8 × 2 = 40 years

97. The average weight of 12 persons is 50 kg. On replacing a man whose weight is 53 kg. with a new man, new average becomes 49 kg. The weight of the new man is:

(*a*) 41 kg (*b*) 46 kg (*c*) 54 kg (*d*) 65 kg

Sol. (*a*) Weight of new man = Weight of man replaced – Total decrease in weight
$= 53 - 1 \times 12 = 41$ kg

98. On replacing an old member of a group with a younger member, the average age of 10 members of the group remains same as it was 2 years ago. How much is the new member younger than the old one?
(*a*) 10 years (*b*) 12 years (*c*) 15 years (*d*) 20 years

Sol. (*d*) Difference between the ages of old and new members
= Total members × Time Difference $= 10 \times 2 = 20$ years

99. Average age of 6 persons in a group is same as it was 5 years ago, when an old man in the group is replaced by a young man, what is the age of the new person if the man replaced is of 70 years.
(*a*) 30 years (*b*) 35 years (*c*) 40 years (*d*) 45 years

Sol. (*c*) Difference between the age of two persons $= 6 \times 5$ years $= 30$ years
$\therefore$ Age of new man $= 70 - 30 = 40$ years

100. Average weight of 30 students in a class is increased by 200 grams when one student whose weight is 20 kg. is replaced by a new student. Find the weight of the new student admitted.
(*a*) 10 kg (*b*) 25 kg (*c*) 26 kg (*d*) 30 kg

Sol. (*c*) 200 grams $= \frac{1}{5}$ kg. $\Rightarrow$ Total increase in weight $= \frac{1}{5} \times 30 = 6$ kg.

Weight of the new student = Weight of student replaced + Total increase in weight
$= (20 + 6)$ kg $= 26$ kg

101. Average age of 15 men is increased by 1 year when one man in the group is replaced with other man whose age is 49 years. Find the age of the man replaced.
(*a*) 34 years (*b*) 48 years (*c*) 50 years (*d*) 64 years

Sol. (*a*) Age of man replaced = Age of new man – Total increase in age $= 49 - 1 \times 15 = 34$ years

102. Out of 10 teachers of a school, one teacher retires and in his place, a new teacher of age 25 years joins. As a result, average age of teachers is reduced by 3 years. The age (in years) of the retired teacher is : **(SSC CGL Tier I 2013)**
(*a*) 50 (*b*) 58 (*c*) 60 (*d*) 55

Sol. (*d*) Age of retired man = Age of new man + Total decrease in age $= 25 + 3 \times 10 = 55$ years

103. The average age of 8 men is increased by 2 years when two of them whose ages are 21 and 23 years are replaced by two new men. The average age of the two new men is :
(SSC CGL Tier I 2002)
(*a*) 22 years (*b*) 24 years (*c*) 28 years (*d*) 30 years

Sol. (*d*) Total age of old (replaced) members $= 21 + 23 = 44$ years
Total increase in age $= 8 \times 2$ years $= 16$ years
$\therefore$ Total age of new members $= 44 + 16 = 60$ years
$\therefore$ Average age of new members $= 60 \div 2 = 30$ years

104. The average age of 11 players of a cricket team is increased by 2 months when two of them aged 18 years and 20 years are replaced by two new players. The average age of the new players is: **(SSC CGL Tier I 2005, 2011)**
(*a*) 19 years 1 month (*b*) 19 years 9 months (*c*) 19 years 11 months (*d*) 19 years 5 months

Sol. (*c*) Increase in average = 2 months $\Rightarrow$ Total increase = 11 × 2 months = 22 months

Total age of new players = (18 + 20) years + 22 months

$\therefore$ Average age of new players = (38 years + 22 months) ÷ 2 = 19 years 11 months

105. The average age of 11 players of a cricket team is decreased by 2 months when two of them aged 17 years and 20 years are replaced by two new players. The average age of the new players is: **(SSC CGL Tier I 2005)**

(*a*) 17 years 1 month (*b*) 17 years 7 months (*c*) 17 years 11 months (*d*) 18 years 3 months

Sol. (*b*) Decrease in average = 2 months $\Rightarrow$ Total decrease = 11 × 2 months = 22 months

Total age of new players = (17 + 20) years – 22 months = 35 years 2 months

$\therefore$ Average age of new players = (35 years 2 months) ÷ 2 = 17 years 7 months

106. The average age of 30 boys in a class is 15 years. One boy aged 20 years left the class, but two new boys came in his place whose ages differ by 5 years. If the average age of all the boys now in the class still remains 15 years, the age of the younger new comer is : **(SSC CGL 2001)**

(*a*) 20 years (*b*) 15 years (*c*) 10 years (*d*) 8 years

Sol. (*b*) Sum of ages of two new students = (20 + 15) years = 35 years

Difference of ages = 5 years $\Rightarrow$ Age of younger = $\frac{35-5}{2}$ = 15 years

107. The arithmetic mean of 10 numbers was computed as 7.6. It was later discovered that a number 8 was wrongly read as 3 during the computation. What should be the correct mean? **(CDS 2011)**

(*a*) 7.1 (*b*) 7.6 (*c*) 8.1 (*d*) 8.6

Sol. (*c*) Sum of numbers = 10 × 7.6 = 76

Correct sum of numbers = 76 + 8 – 3 = 81 $\Rightarrow$ Correct mean = 81 ÷ 10 = 8.1

108. The mean of 50 observations was 36. It was found later that an observation 48 was wrongly taken as 23. The corrected (new) mean is : **(SSC CGL Tier I 2003)**

(*a*) 35.2 (*b*) 36.1 (c) 36.5 (*d*) 39.1

Sol. (*c*) Sum of the observations was wrongly decreased by = 48 – 23 = 25

$\therefore$ Average was wrongly decreased by 25 ÷ 50 = 0.5

$\therefore$ Correct average = 36 + 0.5 = 36.5

109. The average marks obtained by a student in 6 subjects is 88. On subsequent verification it was found that the marks obtained by him in a subject was wrongly copied as 86 instead of 68. The correct average of the marks obtained by him is **(SSC CGL Tier I 2015)**

(*a*) 85 (*b*) 84 (c) 86 (*d*) 87

Sol. (*a*) Sum of marks wrongly increased by = 86 – 68 = 18

$\therefore$ Average was wrongly increased by 18 ÷ 6 = 3 $\Rightarrow$ Correct average = 88 – 3 = 85

110. The arithmetic mean of 100 numbers was computed as 89.05. It was later found that two numbers 92 and 83 have been misread s 192 and 33 respectively. What is the correct arithmetic mean of the numbers ? **(CDS 2010)**

(*a*) 88.55 (*b*) 87.55 (*c*) 89.55 (*d*) Data insufficient

Sol. (*a*) Sum of marks were wrongly increased by = (192 + 33) – (92 + 83) = 50

Average was wrongly increased by 50 ÷ 100 = 0.5

$\therefore$ Correct mean = 89.05 – 0.5 = 88.55

111. In an examination, the average marks of all the students calculated to be 58 marks. It was later found that marks of 60 students were wrongly written as 70 instead of 50. If the corrected average is 55, find the total number of students who took the exam.

(*a*) 100 (*b*) 250 (*c*) 400 (*d*) 500

Sol. (*c*) Total decrease in marks = 60 × (70 – 50) = 1200

Decrease in average = 58 – 55 = 3 ∴ Number of students = 1200 ÷ 3 = 400

112. Four years ago, the average age of 4 persons A, B, C and D was 38 years. With joining of E now, the present average age of 5 persons becomes 36 years. Find E's present age.

(*a*) 26 years (*b*) 12 years (*c*) 30 years (*d*) 32 years

Sol. (*b*) Average age (present) of A, B, C and D = 38 + 4 = 42 years.

Average age (present) of A, B, C, D and E = 36 years.

Total decrease in age on including E = 4 × (42 – 36) = 4 × 6 = 24 years

∴ E's age = 36 – 24 = 12 years

Alt: E's age = 5 × 36 – 4 × 42 = 180 – 168 = 12

113. Five numbers are arranged in ascending order. If average of first two numbers is 8 less than the average of last two numbers and sum of last two numbers is 56, what is the sum of the first two numbers ?

(*a*) 40 (*b*) 28 (*c*) 20 (*d*) None of these

Sol. (*a*) Sum of first two numbers = 56 – 2 × 8 = 40

114. The present average age of a family of four members is 36 years. If the present age of the youngest member of the family be 12 years, the average age of the family at the birth of the youngest member was : **(SSC CGL Tier I 2008)**

(*a*) 48 years (*b*) 40 years (*c*) 32 years (*d*) 24 years

Sol. (*c*) Sum of age 12 years ago = 4 × (36 – 12) = 96 years

The youngest member was not born 12 years ago ⇒ Required average = 96 ÷ 3 = 32 years

Alt: Average age, 12 years ago = 36 – 12 = 24 years ⇒ Required average = $\frac{4 \times 24}{3}$ = 32 years

115. The average age of a family of 6 members is 22 years. If the age of the youngest member be 7 years, what was the average age of the family at the time of birth of the youngest member?

(*a*) 16 years (*b*) 17 years (*c*) 18 years (*d*) 20 years

Sol. (*c*) Total age of 6 members = 6 × 22 = 132; Total age of 5 members = 132 – 7 = 125

Average age of 5 members = 125 ÷ 5 = 25 years ⇒ Required average = 25 – 7 = 18 years.

116. 3 years ago, the average age of a family of 5 members was 17 years. A baby having born, the average age of the family is the same today. The present age of the baby is :

(SSC FCI 2012, SSC CGL Tier I 1992, 2004, 2013, SSC CHSL 2014)

(*a*) 3 years (*b*) 2 years (*c*) $1\frac{1}{2}$ years (*d*) 1 year

Sol. (*b*) Present average age of 5 members = 17 + 3 = 20 years

Total present age of 5 members = 5 × 20 = 100 years

Present average age of 6 members = 17 years

Total present age of 6 members = 6 × 17 = 102 years

∴ Present age of baby = 102 – 100 = 2 years

Alternative method : Present average age of all members = 17 years

Decrease in average on inclusion of baby = 3 years

Total decrease in age = 5 × 3 years = 15 years ⇒ Age of baby = 17 – 15 = 2 years

117. 3 years ago, the average age of a family of 6 members was 19 years. A baby having been born, the average age of the family is same today. The present age of the baby is :

(*a*) 1 year (*b*) 2 years (*c*) 3 years (*d*) None of these

Sol. (*a*) Present age of 7 members = 19 years ⇒ Present age of 6 members = 19 + 3 = 22 years

Age of baby = 19 – 6 × (22 – 19) = 1 year

118. The average age of a family of 6 members is 25 years. The average age of the family after the demise of a 40 years old member will be: **(SSC CHSL 2006)**

(*a*) 20 years (*b*) 21 years (*c*) 22 years (*d*) 23 years

Sol. (*c*) Total age of 6 members = 6 × 25 = 150 years

Total age of remaning 5 members = 150 – 40 = 110 years

Average age of remaning 5 members = 110 ÷ 5 = 22 years

Alt: Excess age of demised member over average of all = 40 – 25 = 15 years

His exclusion will decrease average of the remaning 5 members by = 15 ÷ 5 = 3 years

∴ Average of remaning 5 members = 25 – 3 = 22 years

119. The average monthly income of a family of 4 earning members was ₹ 7350. One of the earning members died. Thereby the average came down to ₹ 6500. The monthly income of the deceased was : **(SSC CGL Tier I 2006)**

(*a*) ₹850 (*b*) ₹8200 (*c*) ₹9900 (*d*) ₹10750

Sol. (*c*) Total income of 4 members = 4 × ₹7350 = ₹29400

Total income of 3 members = 3 × ₹6500 = ₹19500

∴ Income of the deceased member = 29400 – 19500 = ₹9900

Alt: Decrease in average = 7350 – 6500 = ₹850 ⇒ Total decrease = 3 × ₹850 = ₹2550

∴ Income of the deceased member = 7350 + 2550 = ₹9900

120. Average weight of 8 persons is 48 kg. If one man weighing 34 kg. die, what is average weight the remaining 7 persons.

(*a*) 46 kg (*b*) 50 kg (*c*) 55 kg (*d*) 60 kg

Sol. (*b*) Excess of average weight over the weight of man died = 48 kg. – 34 kg. = 14 kg.

∴ Increase in weight of the remaining 7 persons = $\frac{14}{7}$ = 2 kg.

∴ Average weight of remaining 7 persons = 48 + 2 = 50 kg.

Alternative method : $\frac{8 \times 48 - 34}{7} = \frac{350}{7} = 50$ kg

121. Average runs scored by 11 players of a cricket team is 23 runs. If the first player scored 113 runs. Find the average runs of the remaining players.

(*a*) 8 runs (*b*) 12 runs (*c*) 14 runs (*d*) 27 runs

Sol. (*c*) Excess runs scored by 1st player over average runs = 113 – 23 = 90 runs

∴ Decrease in average on excluding his runs = $\frac{90}{10}$ = 9 runs

∴ Average of remaining 10 players = 23 – 9 = 14 runs

Alternative method : $\frac{11 \times 23 - 113}{10} = \frac{140}{10} = 14$ runs

122. Kamlesh bought 65 books for ₹ 1050 from one shop and 50 books for ₹ 1020 from another. What is the average price he paid per book? **(Bank of Maharashtra PO 2008)**

(*a*) ₹ 36.40 (*b*) ₹ 18.20 (*c*) ₹ 24 (*d*) ₹ 18

Sol. (*d*) Total number of books = 65 + 50 = 115. Total cost = 1050 + 1020 = ₹ 2070

∴ Average price = 2070 ÷ 115 = ₹ 18

123. The average age of 5 boys is 12 years. The average age of 3 others is 16 years. The average age of all the 8 boys is : **(SSC CGL Tier I 2004)**

(*a*) $13\frac{1}{2}$ years (*b*) 14 years (*c*) $12\frac{1}{2}$ years (*d*) 13 years

Sol. (*a*) Average age of all = $\frac{5\times12+3\times16}{5+3}=\frac{60+48}{8}=\frac{108}{8}=\frac{27}{2}=13\frac{1}{2}$ years

Alternative method : Assumed average age = 12 years

Total excess over assumed mean = 3 × (16 – 12) = 12 years

∴ Increase in average of all = $\frac{12}{8}=1\frac{1}{2}$ years ⇒ Average of all = $12+1\frac{1}{2}=13\frac{1}{2}$ years

124. Average expenditure of a person for the first 3 days of a week is ₹ 350 and for the next 4 days is ₹ 420. Average expenditure of the man for the whole week is:

(*a*) ₹ 410 (*b*) ₹ 390 (*c*) ₹ 385 (*d*) ₹ 375

Sol. (*b*) Assumed mean = ₹ 350

Total excess over assumed mean = 4 × (₹ 420 – ₹ 350) = ₹ 280

∴ Increase in average expenditure = ₹ $\frac{280}{7}$ = ₹ 40

∴ Average expenditure for 7 days = ₹ 350 + ₹ 40 = ₹ 390

125. A company produces an average of 4000 items per month for the first 3 months. How many items, it must produce on an average per month over the next 9 months to average 4375 items per month over the whole year ? **(SSC CGL Tier I 1999)**

(*a*) 4500 (*b*) 4600 (*c*) 4680 (*d*) 4710

Sol. (*a*) Excess average required during last 9 months = $\frac{375\times12}{9}$ = 500

∴ Monthly production required (during last 9 months) = 4000 + 500 = 4500

Alt: 4375 × 12 – 4000 × 3 = (52500 – 12000) = 40500 40500 ÷ 9 = 4500

126. On a school's Annual Day sweets were to be equally distributed amongst 112 children. But on that particular day, 32 children were absent. Thus the remaining children got 6 extra sweets. How many sweets was each child originally supposed to get ? **(Indian Overseas Bank PO 2009)**

(*a*) 24 (*b*) 18 (*c*) 15 (*d*) Data insufficient

Sol. (*c*) Children attended the function = 112 – 32 = 80

Total extra sweets = 80 × 6 = 480 due to absence of 32 children.

Original share of each child = 480 ÷ 32 = 15

Alternative method : $112x = (112 - 32)\times(x+6) \Rightarrow 32x = 6\times80 \Rightarrow x=\frac{6\times80}{32}=15$

127. A man's average expenditure for first 11 days of a month was ₹ 104 per day and for the next 19 days, it was ₹ 98 per day. If he could save ₹ 54 during that month, find his income for the month.

(*a*) ₹ 2940 (*b*) ₹ 3000 (*c*) ₹ 3060 (*d*) ₹ 3120

Sol. (*c*) Assumed mean = ₹ 100

Total excess income = Saving + Deviation from assumed mean

= ₹ 54 + 11 × (₹ 104 – ₹ 100) + 19 × (₹ 98 – ₹ 100) = ₹ 54 + ₹ 44 – ₹ 38 = ₹ 60

∴ Average daily income = ₹ 100 + ₹ $\frac{60}{30}$ = ₹ 102 ⇒ Monthly income = ₹ 102 × 30 = ₹ 3060

128. A man spends an average of ₹ 1,694.70 per month for the first 7 months and ₹ 1810.50 per month for the next 5 months. His monthly salary, if he saves ₹ 3084.60 during the whole year is: **(SSC CGL Tier II 2012)**

(*a*) ₹ 1000 (*b*) ₹ 2000 (*c*) ₹ 2,400 (*d*) ₹ 3000

Sol. (*b*) Average expenditure per month = (1694 + 1810) ÷ 2 = 1750 (roughly)

Average savings per month = 3000 ÷ 12 = ₹ 250 (roughly)

∴ Average salary per month = 1750 + 250 = ₹ 2000 (roughly)

129. Average age of 12 persons in a group is 25 years. On including 3 more persons in the group, the average age is decreased by 1 year. Find the average age of the three new members.

(*a*) 19 years (*b*) 20 years (*c*) 23 years (*d*) 24 years

Sol. (*b*) Total decrease in the age of 12 persons = 1 × 12 = 12 years

∴ Average age of 3 new members is less by $\frac{12}{3}$ = 4 years

∴ Average age of new members = 24 – 4 = 20 years

Alternative method: $24 + \frac{12}{3} \times (-1) = 24 - 4 = 20$ years

Alt: 15 × 24 – 12 × 25 = 360 – 300 = 60; 60 ÷ 3 = 20 years.

130. The average weight of the students of a class is 40 kg. 5 new students with the average weight of 46 kg. having joined the class, the average weight of the class is increased by 2 kg. Find the number of students in the class originally?

(*a*) 10 (*b*) 15 (*c*) 20 (*d*) 25

Sol. (*a*) Total increase in weight on including 5 more students = 5 × (46 – 40) = 30 kg.

Increase in average weight = 2 kg. ⇒ Total students (at present) = $\frac{30}{2}$ = 15 students

∴ Students (originally) = 15 – 5 = 10 students

Alternative Method : Let original no. of students be x.

Then $40x + 230 = 42(x + 5) \Rightarrow 2x = 230 - 210 = 20 \Rightarrow x = 10$

131. The average score of class is A. The ratio of boys and girls in the class is 3 : 1. If the average score of the boys is (A + 1), the average score of the girls is : **(SSC 2006)**

(*a*) A – 1 (*b*) A + 1 (*c*) A – 3 (*d*) A + 3

Sol. (*c*) Let total students be 3 + 1 = 4. Then total marks = 4A

Marks obtained by boys = 3 (A + 1) ⇒ Marks of girls = 4A – 3 (A + 1) = A – 3

132. When the average age of a husband, wife and their son was 42 years, the son got married. A child too born to him exactly after one year of his marriage. When the child turned to be of 5 years, then the average of the family became 36 years. What is the age of daughter-in-law at the time of marriage ? **(SSC)**

(*a*) 23 years (*b*) 24 years (*c*) 25 years (*d*) 26 years

Sol. (*c*) Average age of husband, wife and son at the time of birth of child = 42 + 1 = 43 years

∴ Their total age = 3 × 43 = 129 years

Average age of all the 5 members at the time of birth of child = 36 – 5 = 31 years

∴ Total age of all = 5 × 31 = 155 years

∴ Age of daughter-in-law at birth of child = 155 – 129 = 26 years

∴ Age of daughter-in-law at the time of marriage = 26 – 1 = 25 years

133. Average marks obtained by first 10 students out of 11 students in a class is 189. If the first student has obtained 163 marks and the 11th student 193 marks, find the average marks of the last 10 students.

(*a*) 190 (*b*) 192 (*c*) 199 (*d*) 219

Sol. (*b*) More marks obtained by 11th student than 1st student = 193 – 163 = 30 marks

∴ Average of last 10 students is more than that of first 10 students by $\frac{30}{10} = 3$ marks

∴ Average marks of last 10 students = 189 + 3 = 192 marks

134. The average temperature of Monday, Tuesday and Wednesday was 30°C and that of Tuesday, Wednesday and Thursday is 33°C. If the temperature on Monday was 32°C, then the temperature on Thursday was : **(SSC CGL Tier I 2012)**

(*a*) 33°C (*b*) 30°C (*c*) 31°C (*d*) 32°C

Sol. (*) Difference between temperature on Monday and Thursday = 3 × (33°C – 30°C) = 9°C

Temperature for 3 days including Thursday is more than that of 3 days including Monday.

∴ Temperature on Thursday is more than the temperature on Monday.

∴ Temperature on Thursday = Temperature on Monday + Difference between temperatures = 32°C + 9°C = 41°C

135. Out of four numbers, the average of the first three is 16 and that of the last three is 15. If the last number is 20 then the first number is **(SSC CGL Tier I 2015)**

(*a*) 28 (*b*) 23 (*c*) 25 (*d*) 21

Sol. (*b*) Difference between value of first and last number = 3 × (16 – 15) = 3

Average of 3 number including first is more than that of 3 numbers including the last one.

∴ First number is more than the last number ⇒ First number = 20 + 3 = 23

136. Average temperature from 9th to 16th of a month is 30°C and that from 10th to 17th is 31°C. What is the temperature on 17th, if temperature on 9th is 35°C?

(*a*) 29°C (*b*) 26°C (*c*) 32°C (*d*) 43°C

Sol. (*d*) Difference between temperature on 9th and 17th = 8 × (31°C – 30°C) = 8°C

Temperature for 8 days including 17th is more than that of 8 days including 9th.

∴ Temperature on 17th is more than the temperature on 9th.

∴ Temperature on 17th = Temperature on 9th + Difference = 35°C + 8°C = 43°C

137. Average rainfall on Monday, Tuesday, Wednesday and Thursday is 420.5 cm and average on Tuesday, Wednesday, Thursday and Friday is 440.5 cm. If the ratio of rainfall for Monday and Friday is 20 : 21, find the rainfall in cm on Monday and Friday. **(SSC CGL Tier II 2012)**

(*a*) 1700, 1470 (*b*) 1682, 1762 (*c*) 1800, 1890 (*d*) 1600, 1680

Sol. (*d*) Difference between the rainfall on Monday and Friday = 4 × (440.5 – 420.5) = 80

Difference between the two ratios = 21 – 20 = 1 ratio

∴ Rainfall on Monday and Friday = 20 × 80 and 21 × 80 = 1600 and 1680 respectively.

138. The average temperature from Monday to Friday is 38°C and that from Tuesday to Saturday is 35°C. Find temperature on Monday, if ratio between the temperatures on Monday and Saturday is 8 : 5?

(*a*) 25°C (*b*) 37°C (*c*) 40°C (*d*) 44°C

Sol. (*c*) Difference between the temperatures on Monday and Saturday = 5 × (38°C – 35°C) = 15°C
Difference between the two ratios = 8 – 5 = 3 ratio = 15°C ⇒ 1 ratio = 5°C
If difference between temperatures is 3°C, then temperature on Monday is 8°C.
∴ Temperature on Monday = 8 ratio = 8 × 5°C = 40°C

139. The mean temperature from Monday to Wednesday was 37°C and from Tuesday to Thursday was 34°C. If the temperature on Thursday was 4/5th that of Monday, the temperature on Thursday was : **(SSC FCI Asstt. 2012, SSC CGL Tier I 2013)**
(*a*) 34°C (*b*) 35°C (*c*) 36°C (*d*) 37°C

Sol. (*c*) Difference between the temperatures on Monday and Thursday = 3 × (37°C – 34°C) = 9°C
Difference between the two ratios = 5 – 4 = 1 ratio = 9°C
∴ Temperature on Thursday = 4 ratio = 4 × 9°C = 36°C.

140. The average of nine numbers is 50. The average of the first five numbers is 54 and that of the last three numbers is 52. Then the sixth number is **(SSC CGL Tier I 2013)**
(*a*) 44 (*b*) 34 (*c*) 14 (*d*) 24

Sol. (*d*) Sixth number = 50 + 5 × (50 – 54) + 3 × (50 – 52) = 50 – 20 – 6 = 24

141. The average of 20 numbers is 12. The average of the first 12 numbers is 11 and that of the next 7 numbers is 10. The last number is : **(SSC CGL Tier I 2004)**
(*a*) 40 (*b*) 38 (*c*) 48 (*d*) 50

Sol. (*b*) Last number = 12 + 12 × (12 – 11) + 7 × (12 – 10) = 12 + 12 + 14 = 38

142. The average age of 15 students of a class is 15 years. Out of these the average age of 5 students is 14 years and that of the other 9 students is 16 years. The age of the 15th student is : **(SSC CGL Tier I 2003)**
(*a*) 11 years (*b*) 15 years (*c*) $15\frac{2}{7}$ years (*d*) 14 years

Sol. (*a*) Required age = 15 + 5 × (15 – 14) + 9 × (15 – 16) = 15 + 5 – 9 = 11 years

143. The average of 9 numbers is 30. The average of first 5 numbers is 25 and that of the last 3 numbers is 35. What is the 6th number ? **(SSC CGL Tier I 2000)**
(*a*) 20 (*b*) 30 (*c*) 40 (*d*) 50

Sol. (*c*) 6th number = 30 + 5 × (30 – 25) + 3 × (30 – 35) = 30 + 25 – 15 = 40

144. The average age of 8 men, 7 women and 1 boy is 45 years. If the average age of 8 men is 48 years and that of 7 women is 46 years, then the age of the boy is **(SAS 2010)**
(*a*) 14 years (*b*) 15 years (*c*) 16 years (*d*) 18 years

Sol. (*a*) Age of boy = 45 + 8 × (45 – 48) + 7 × (45 – 46) = 45 – 24 – 7 = 14

145. The mean of 25 observations is 36. The mean of first 13 observations is 32 and that of last 13 observations is 39. What is the value of 13th observation ? **(CDS 2009)**
(*a*) 20 (*b*) 23 (*c*) 32 (*d*) 40

Sol. (*b*) 13th observation = 36 + 13 (32 – 36) + 13(39 – 36) = 36 – 4 × 13 + 3 × 13 = 36 – 13 = 23

146. The average of 11 results is 50. If the average of the first six results is 49 and that of the last six is 52, the sixth result is **(SSC CGL Tier II 2013)**
(*a*) 48 (*b*) 50 (*c*) 52 (*d*) 56

Sol. (*d*) Sixth result = 50 + (49 – 50) × 6 + (52 – 50) × 6 = 50 – 6 + 12 = 56

147. The average of 11 numbers is 10.8. If the average of the first six numbers be 10.4 and that of the last six is 11.5, the middle (6th) number is : **(SSC CGL Tier I 1999)**
(*a*) 10.3 (*b*) 12.6 (*c*) 13.5 (*d*) 15.5

Sol. (*b*) 6th number =10.8 + 6 × (10.4 – 10.8) + 6 × (11.5 – 10.8) = 10.8 – 2.4 + 4.2 = 12.6

148. The average of 11 numbers is 63. If the average of first six numbers is 60 and the last six numbers is 65, then the 6th number is: **(SSC CGL Tier I 2013)**

(*a*) 57 (*b*) 60 (*c*) 62 (*d*) 64

Sol. (*a*) 63 + 6 × (60 – 63) + 6 × (65 – 63) = 63 – 18 + 12 = 57

149. The average of marks of 13 papers is 40. The average of marks of first 7 papers is 42 and that of the last 7 papers is 35. What are the marks of the seventh paper ? **(Civil Pre. 2011)**

(*a*) 24 (*b*) 21 (*c*) 19 (*d*) 17

Sol. (*c*) 7 × 42 + 7 × 35 – 13 × 40 = 7 × 77 – 13 × 40 = 539 – 520 = 19

Alt : 40 + 7 × (42 – 40) + 7 × (35 – 40) = 40 + 14 – 35 = 19

150. The average of 15 numbers is 7. If the average of the first 8 numbers is 6.5 and the average of the last 8 numbers be 8.5, then the middle number is **(SSC CHSL 2013, CGL 2003)**

(*a*) 10 (*b*) 15 (*c*) 13 (*d*) 23

Sol. (*b*) Middle term = 7 + 8 × (6.5 – 7) + 8 × (8.5 – 7) = 7 – 4 + 12 = 15

151. In a certain year, the average monthly income of a person is ₹ 3400 and that for the first eight months is ₹ 3160 and for the last five months is ₹ 4120. The income in the eighth month of the year is : **(SSC CGL Tier I 2003 and 2008)**

(*a*) ₹ 5080 (*b*) ₹ 6080 (*c*) ₹ 5180 (*d*) ₹ 3880

Sol. (*a*) Income in eighth month = 3400 + 8 × (3160 – 3400) + 5 × (4120 – 3400)

= 3400 – (8 × 240) + (5 × 720) = 3400 – 1920 + 3600 = ₹ 5080

152. Nine numbers are written in ascending order. The middle number is the average of the nine numbers. The average of the first five large numbers is 68 and that of five smaller numbers is 44. What is the sum of all nine numbers ? **(CDS 2009)**

(*a*) 450 (*b*) 501 (*c*) 504 (*d*) 540

Sol. (*c*) Sum of 5 larger numbers = 5 × 68 = 340. Sum of 5 smaller numbers = 5 × 44 = 220

Sum of 9 numbers + 5th number (average) = 340 + 220 = 560

Average of all (10) numbers = 560 ÷ 10 = 56 ⇒ Sum of nine numbers = 560 – 56 = 504

153. On a test consisting 150 questions carrying 1 mark each, Meenal answered 80% of the first 75 questions correctly. What percent of the other 75 questions does she need to answer correctly to score 60% on the entire exam ? **(Maharashtra Specialist Officers 2008)**

(*a*) 60% (*b*) 50% (*c*) 20% (*d*) 40%

Sol. (*d*) Two parts have equal number of questions ⇒ Required % = $\frac{80+x}{2} = 60 \Rightarrow x = 40$

154. A shopkeeper purchased 25 kg. of sugar at ₹ 5 per kg. and 75 kg. of sugar at ₹ 9 per kg. He mixed the two varieties and sells the mixture at ₹ 10 per kg., find his profit percentage.

(*a*) 11.11% (*b*) 25% (*c*) 33.33% (*d*) 100%

Sol. (*b*) Let cost price of the mixture = ₹ 5 per kg. (assumed mean)

Extra price paid for 75 kg. = ₹ (9 – 5) = ₹ 4 per kg.

∴ Total extra amount paid = 75 × ₹ 4 = ₹ 300

This will increase the assumed average price by $\frac{300}{100}$ = ₹ 3 per kg.

∴ Average cost price per kg. of mixture = ₹ 5 + ₹ 3 = ₹ 8

Selling price per kg. of mixture = ₹ 10

∴ Profit = ₹ 10 – ₹ 8 = ₹ 2 per kg. ∴ Profit % = $\frac{2}{8} \times 100 = 25\%$

Alternative method : Average cost price = $\frac{25 \times 5 + 75 \times 9}{100} = \frac{125 + 675}{100} = \frac{800}{100} = 8$

155. A shopkeeper bought 80 kg. of sugar at the rate of ₹ 13.50 per kg. He mixed it with 120 kg of sugar costing ₹ 16 per kg. In order to make a profit of 20%, he must sell the mixture at :

(SSC CGL Tier I 2008)

(*a*) ₹ 18 per kg (*b*) ₹ 17 per kg (*c*) ₹ 16.40 per kg (*d*) ₹ 15 per kg

Sol. (*a*) Let assumed mean = ₹ 13 per kg ⇒ Excess CP = (80 × 0.50) + (120 × 3) = ₹ 400

∴ Average cost price = ₹ 13 + (400 ÷ 200) = ₹ 15 per kg

Selling price at 20% profit = $15 \times \frac{120}{100}$ = ₹18 per kg.

156. Sri Ganesh bought 40 kgs of wheat at ₹ 12.50 per kg and 25 kg of it at ₹ 15.10 per kg. He mixed them together. At what rate should he sell the mixture to earn 10% profit ?

(Bank PO 2004)

(*a*) ₹ 13.50 (*b*) ₹ 13.25 (*c*) ₹ 14.75 (*d*) ₹ 14.85

Sol. (*d*) We take assumed mean = ₹ 12.50

Then excess price over assumed mean = ₹ (15.10 – 12.50) × 25 = ₹ 2.60 × 25 = ₹ 65

Average cost price = ₹ 12.50 + (65 ÷ 65) = ₹ 13.50 per kg.

Desired profit = 10% of ₹ 13.50 = ₹ 1.35

∴ Required selling price per kg. = ₹ (13.50 + 1.35) = ₹ 14.85

157. Tea worth ₹ 126 per kg and ₹ 135 per kg are mixed with a third variety in the ratio 1 : 1 : 2. If the mixture is worth ₹ 153 per kg, the price of the third variety per kg will be

(NIFT 2002, SSC CPO SI 2003, SSC CHSL 2012)

(*a*) ₹ 169.50 (*b*) ₹ 170.00 (*c*) ₹ 175.50 (*d*) ₹ 180.00

Sol. (*c*) Let total quantity = 4 kg. Then cost of mixture = ₹ 153 × 4 = ₹ 612

CP of first two quantities = 126 × 1 + 135 × 1 = ₹ 261

CP of third quantity = ₹(612 – 261) = ₹ 351

⇒ Cost price (per kg) = ₹ 351 ÷ 2 = ₹ 175.50

158. The batting average of a batsman in 42 innings is 45 runs. He was out for a duck in 7 innings. His batting average for the remaining innings is:

(*a*) 50 (*b*) 54 (*c*) 55 (*d*) 60.25

Sol. (*b*) Total runs of 42 innings = 42 × 45

∴ Average of remaining (42 – 7) or 35 innings = $\frac{42 \times 45}{35} = 54$

Alt: Average runs of 7 innings = 45 Total runs of 7 innings = 45 × 7

∴ Average of remaining 35 innings is increased by $\frac{45 \times 7}{35}$ = 9 runs

∴ Average of remaining 35 innings = 45 + 9 = 54 runs

159. If average of *a* and *b* is $\frac{a^{n+1} + b^{n+1}}{a^n + b^n}$ and $a \neq b$, find the value of *n*.

(*a*) 0 (*b*) 1 (*c*) 2 (*d*) 5

Sol. (*a*) $\frac{a^{n+1} + b^{n+1}}{a^n + b^n} = \frac{a+b}{2} \Rightarrow 2a^{n+1} + 2b^{n+1} = a^{n+1} + b^{n+1} + ba^n + ab^n$

$\Rightarrow a^{n+1} + b^{n+1} = ba^n + ab^n \Rightarrow a^{n+1} - ba^n = ab^n - b^{n+1} \Rightarrow a^n(a - b) = b^n(a - b)$

$\Rightarrow a^n = b^n \Rightarrow n = 0$

160. In a factory, 600 men and 400 women workers are employed and their daily average wages is ₹2.55. If a woman gets daily 50 paise less than a man, what is the daily wages of a woman?

(*a*) ₹ 3.05 (*b*) ₹ 2.05 (*c*) ₹ 2.25 (*d*) ₹ 2.35

Sol. (*c*) Excess wages of 600 men = 600 × ₹ 0.50 = ₹ 300

This will increase average of all by $\frac{300}{600+400}$ = ₹ 0.30

∴ Average of a woman = ₹ 2.55 – ₹ 0.30 = ₹2.25.

Alt: Total wages = ₹ 2.55 (600 + 400) = ₹ 2550 ⇒ Balance = ₹ 2550 – ₹ 300 = ₹ 2250

∴ Daily wages of a woman = 2250 ÷ 1000 = ₹ 2.25.

161. 11 friends went to a hotel and decided to pay the bill amount equally. But 10 of them could pay ₹ 60 each as a result 11th has to pay ₹ 50 extra than his share. Find the amount paid by him.

(*a*) ₹ 105 (*b*) ₹ 110 (*c*) ₹ 115 (*d*) ₹ 120

Sol. (*c*) $11x = 600 + (x + 50) \Rightarrow x = 65 \Rightarrow$ Required amount = 65 + 50 = 115.

Alternative method : Average amount paid by 10 friends = ₹ 60

Increase in average due to ₹ 50 paid extra by the 11th friend = ₹ $\frac{50}{10}$ = ₹ 5

∴ Average expenditure of 11 friends = ₹ 60 + ₹ 5 = ₹ 65

∴ Amount paid by the 11th friends = ₹ 65 + ₹ 50 = ₹ 115

162. 9 men started a partnership business contributing ₹ 10,000 each. Later on 10th men joined the partnership contributing ₹ 3,600 more than the average contribution of all the 10 men. How much did the 10th men contribute?

(*a*) ₹ 3,600 (*b*) ₹ 10,400 (*c*) ₹ 14,000 (*d*) ₹ 15,000

Sol. (*c*) Average contribution of 9 partners = ₹ 10,000

Increase in average due to ₹ 3,600 paid extra by the 10th partner = ₹ $\frac{3600}{9}$ = ₹ 400

∴ Average contribution of 10 partners = ₹ 10,000 + ₹ 400 = ₹ 10,400

∴ Amount paid by the 10th partner = ₹ 10,400 + ₹ 3,600 = ₹ 14,000

Alternative method : $10x = 90{,}000 + (x + 3{,}600) \Rightarrow 9x = 93{,}600$

$\Rightarrow x = 10{,}400 \Rightarrow$ Required amount = 10,400 + 3,600 = 14,000

163. 5 men started a business contributing ₹ 20,000 each. Later on 6th men joined the partnership contributing ₹ 2,000 less than the average contribution of all the 6 men. Find the amount contributed by the 6th partner.

(*a*) ₹ 2,000 (*b*) ₹ 17,600 (*c*) ₹ 19,600 (*d*) ₹ 20,500

Sol. (*b*) Average contribution of 5 partners = ₹ 20,000

Decrease in average due to ₹ 2,000 less paid by the 6th partner = ₹ $\frac{2000}{5}$ = ₹ 400

∴ Average contribution of 6 partners = ₹ 20,000 – ₹ 400 = ₹ 19,600

∴ Amount paid by the 6th partner = ₹ 19,600 – ₹ 2,000 = ₹ 17,600

Alternative method : $6x = 100000 + (x - 2000) \Rightarrow 5x = 98{,}000 \Rightarrow x = 19{,}600$

⇒ Required amount = 19,600 – 2,000 = 17,600

164. A number of friends decided to go on a picnic and planned to spend ₹ 108 on eatables. Three of them however did not turn up. As a consequence each one of the remaining had to contribute ₹ 3 extra. The number of them who attended the picnic was : **(SSC CGL 2003)**

(*a*) 15 (*b*) 12 (*c*) 9 (*d*) 6

Sol. (*c*) **Trick:** $12 \times 9 = 108$, so we check $\frac{108}{9} - \frac{108}{12} = 12 - 9 = 3$

165. The average marks obtained by some students in an examination is 54. If 20% of the students got a mean score of 90 marks and the 30% of the students got a mean score of 20. Find the average marks of the remaining students.

(*a*) 42 (*b*) 45 (*c*) 50 (*d*) 60

Sol. (*d*) Remaining students = 100% – 20% – 30% = 50%

Let remaining students got a mean score of *x* marks.

Then 20% of 90 + 30% of 20 + 50% of *x* = 54

$\therefore$ 18 + 6 + 50% of *x* = 54 $\Rightarrow$ 50% of *x* = 54 – 18 – 6 = 30 $\Rightarrow$ $x = 30 \times 2 = 60$ marks

166. The arithmetic mean of the scores of a group of students in a test was 52. The brighest 20% of them secured a mean score of 80 and the dullest 25% a mean score of 31. The mean score of the remaining 55% is : **(SSC CGL Tier I 2000, 2013)**

(*a*) 45% (*b*) 50% (*c*) 51.4% approx (*d*) 54.6% approx

Sol. (*c*) Remaining students = 100% – 20% – 25% = 55%

Let remaining students got a mean score of *x* marks.

Then 20% of 80 + 25% of 31 + 55% of *x* = 52 $\Rightarrow$ 16 + 7.75 + 55% of *x* = 52

$\therefore$ 55% of *x* = 52 – 16 – 7.75 = 28.25 $\Rightarrow$ $x = \frac{2825}{100} \times \frac{100}{55} = \frac{565}{11} = 51.4\%$ (approx.)

167. A shopkeeper has sufficient money to buy 50 books. On reduction in price of one book by ₹ 4 each, he could buy 10 books more. How much money does he has? **(MBA)**

(*a*) ₹ 1000 (*b*) ₹ 1200 (*c*) ₹ 1500 (*d*) ₹ 2000

Sol. (*b*) Total savings on reduction in price = 50 × ₹ 4 = ₹ 200

With this amount, he could buy 10 extra books$\Rightarrow$Reduced price per book = ₹ $\frac{200}{10}$ = ₹ 20

$\therefore$ Original price per book = ₹ 20 + ₹ 4 = ₹ 24

He was buying 50 books originally. $\therefore$ Money with him = 50 × ₹ 24 = ₹ 1200

Alternative Method: Ratio of books = 50 : (50 + 10) = 5 : 6 $\Rightarrow$ Ratio of prices = 6 : 5

Difference = (6 – 5) ratio = 1 ratio ≡ ₹ 4 $\Rightarrow$ Original price = 6 ratio ≡ 6 × ₹ 4 = ₹ 24

He was buying 50 books originally. $\therefore$ Money with him = 50 × ₹ 24 = ₹ 1200

168. Ramesh can buy a certain number of pens for ₹ 450. If price of pens was reduced by ₹ 15 each, he could have purchased 5 more pens with the same amount. Find the original price of a pen. **(MBA)**

(*a*) ₹ 15 (*b*) ₹ 30 (*c*) ₹ 20 (*d*) ₹ 45

Sol. (*d*) Let cost price of a pen = *x*. Then $\frac{450}{x-15} - \frac{450}{x} = 5$

If we put *x* = 20 (one of the given options), in the above equation, we get:

$$\text{LHS} = \frac{450}{5} - \frac{450}{20} = 90 - 22.5 = 67.5$$

67.5 is much higher than 5, the desired difference. $\Rightarrow$ We need to take higher value of *x*.

If we take *x* = 45, we get: LHS = $\frac{450}{30} - \frac{450}{45} = 15 - 10 = 5$

$\therefore$ Original price of one pen = ₹ 45

169. A man buys a certain number of books for ₹ 300. He could have bought 5 more books for the same amount had the books been ₹ 2 cheaper. Find the number of books originally bought by him. **(MBA)**

(*a*) 20 (*b*) 25 (*c*) 30 (*d*) None of these

Sol. (*b*) Let number of books bought be x. Then $\frac{300}{x}-\frac{300}{x+5}=2$

If we put $x = 20$ (option *a*), we get : $\frac{300}{20}-\frac{300}{25}=15-12=3\neq \text{R.H.S.}$

If we put $x = 25$ (option *b*), we get : $\frac{300}{25}-\frac{300}{30}=12-10=2=\text{R.H.S.}$

$x = 25$ satisfies the equation $\Rightarrow$ Books originally bought = 25

170. Some students planned a trip and estimated their total expenses to be ₹ 500. However, 5 of them could not go for the trip and as a result average expenditure of the remaining students is increased by ₹ 5. How many students had gone for the trip? **(MBA)**

(*a*) 15 (*b*) 20 (*c*) 25 (*d*) 30

Sol. (*b*) Let total number of students be x. Then $\frac{500}{x}=\frac{500}{x-5}-5$

If we put $x-5 = 15$ (option *a*), $\frac{500}{x-5}$ is not a natural number. so option (*a*) is ruled out.

If we put $x-5 = 20$ (option *b*), we get : $\frac{500}{25}=\frac{500}{20}-5 \Rightarrow 20 = 25 - 5$, which is true.

CHAPTER

19

ALLIGATION

1. Alligation Rule: It is also called **Rule of Mixture**. Alligation Rule is used to find:
The ratio in which items from two different groups having different averages, should be mixed to get a certain combined average for all the items.

$$
\begin{array}{ccc}
\text{Avr I} & & \text{Avr. II} \\
(a) & & (b) \\
\searrow & & \swarrow \\
& (m)\text{combined average} & \\
\swarrow & & \searrow \\
|m-b| & & |m-a|
\end{array}
$$

⇒ If averages of two different groups are a and b. Then the items from these groups should be mixed in the ratio of $|(m-b)|$ and $|(m-a)|$ to get the combined average m.

$$\Rightarrow \frac{\text{Quantity of Higher Average}}{\text{Quantity of Lower Average}} = \frac{\text{Combined average} - \text{Lower Average}}{\text{Higher Average} - \text{Combined Average}}$$

Notes:

(i) $|(m-b)|$ and $|(m-a)|$ are absolute values of $(m-b)$ and $(m-a)$ respectively. Absolute value of a number doesn't carry negative sign. Absolute value of $|3-6|$ is 3 (not -3)

(ii) m always lies between a and b.

2. When three quantities having different averages a, b, c are mixed. Take any two quantities (say a and b) such that the combined average lies between them and then find the ratio using alligation rule. Similarly take third quantity and one of a, b such that combined average lies between the two averages. Now find combined ratio of two ratios calculated above.

Note:

(i) We will get infinite number of solutions by adding these ratio in different proportions.

(ii) In such cases, average of atleaset one group must be more than the combined average. Similarly, average of atleast one group must be less than the combined average.

EXAMPLES

Example 1. In what ratio two varieties of milk costing ₹ 8 and ₹ 9 per litre respectively mixed, so that the mixture costs ₹ 8.30 per litre?

(*a*) 9 : 8 (*b*) 7 : 3 (*c*) 3 : 7 (*d*) 8 : 91

Sol. (*b*)

$$
\begin{array}{ccc}
\text{I} & & \text{II} \\
₹\,8 & & ₹\,9 \\
\searrow & & \swarrow \\
& ₹\,8.30 & \\
\swarrow & & \searrow \\
0.70 & & 0.30
\end{array}
$$

∴ Ratio in which two mixtures are to be mixed = 0.70 : 0.30 = 7 : 3

Alternative method: Ratio = (9 – 8.30) : (8.30 – 8) = 0.70 : 0.30 = 7 : 3

Example 2. Two mixtures of milk and water contain 10% and 20% water. In what ratio two mixtures should be mixed so that the resulting mixture contains 16% water? **(SSC CGL Exam.)**

(*a*) 1 : 2 (*b*) 2 : 1 (*c*) 2 : 3 (*d*) 3 : 2

Sol. (*c*) Mix I Mix II

10 20

↘ ↙

16

↙ ↘

4 6

∴ Ratio in which two mixtures are to be mixed = 4 : 6 = 2 : 3

Alternative method: Ratio = (20 – 16) : (16 – 10) = 4 : 6 = 2 : 3.

Example. In what proprtion may three kinds of rice at ₹ 4, ₹ 6 and ₹ 8 per kg. may be mixed so that price of the resultant mixture is ₹ 7 per kg ? **(MBA Ent. Exam.)**

(*a*) 1 : 1 : 4 (*b*) 1 : 1 : 3 (*c*) 1 : 2 : 4 (*d*) 2 : 1 : 4

Sol. (*a*)

I III | II III

4 8 | 6 8

↘ ↙ | ↘ ↙

7 | 7

↙ ↘ | ↙ ↘

1 3 | 1 1

Whatever proportion of above two mixtures, we take, the price of the resulting mixture will always be same as both the mixture cost ₹ 7 per kg.

Let us take 4 kg of the first mixture and 2 kg of second mixture.

Then required ratio of the three quantities at ₹ 4, ₹ 6 and ₹ 8 is 1 : 1 : (3 + 1) = 1 : 1 : 4

We will get same average price, if we change the proportion of two mixtures, *e.g.*, let us take 8 kg of the first mixture and 6 kg of the second mixture.

Then the ratio is 2 : 3 : (6 + 3) = 2 : 3 : 9

SOLVED EXERCISE

1. Gold is 19 times as heavy as water and silver is 9 times as heavy as water. In what ratio, the two metals be mixed to get mixture which is 15 times as heavy as water ?

(*a*) 3 : 2 (*b*) 2 : 3 (*c*) 1 : 2 (*d*) 2 : 1

Sol. (*a*) Gold Silver

19 9

↘ ↙

15

↙ ↘

6 4

⇒ Required ratio = 6 : 4 = 3 : 2

2. A merchant mixes two varieties of wine containing 25% and 13% alcohol. The resultant mixture contains 17% alcohol. Find the quantity of second mixture, if 8 litre of first mixture is taken.

(*a*) 4 litres (*b*) 16 litres (*c*) 24 litres (*d*) 32 litres

Sol. (*b*)

Mix I Mix II

25% 13%

↘ ↙

17%

↙ ↘

4% 8%

∴ Ratio of two varieties = 4 : 8 = 1 : 2 ⇒ Quantity of first mixture = 1 ratio = 8 litres

∴ Quantity of second mixture = 2 ratio = 8 × 2 = 16 litres

3. The average of marks obtained by 120 candidates in a certain examination is 35. If the average marks obtained by passed candidates are 39 and those of the failed candidates are 15, what is the number of candidates who passed the examination ? **(FMS 2003, SSC CGL 2002)**

(*a*) 100 (*b*) 120 (*c*) 150 (*d*) 140

Sol. (*a*)

Passed Failed

39 15

↘ ↙

35

↙ ↘

20 4

∴ Ratio of passed and failed students = 20 : 4 = 5 : 1

∴ Passed candidates = $\frac{5}{6} \times 120 = 100$

4. The wheat sold by a grocer contained 10% low quality wheat. What quantity of good quality wheat should be added to 150 kgs of wheat so that the percentage of low quality wheat becomes 5% ? **(SSC CGL 2000, Bank PO Exam. 2002)**

(*a*) 150 kgs (*b*) 135 kgs (*c*) 50 kgs. (*d*) 85 kgs

Sol. (*a*) We consider quantity of low quality wheat in two varieties

Mixture Wheat added

10% 0%

↘ ↙

5%

↙ ↘

5 5

∴ Ratio of mixture and wheat added = 5 : 5 = 1 : 1 ⇒ Wheat added = Mixture = 150 kgs

5. In a town, the population was 8000. In one year, male population increased by 10% and female population increased by 8% but the total population increased by 9%, The number of males in the town was : **(SSC CGL Tier I 1999)**

(*a*) 4000 (*b*) 4500 (*c*) 5000 (*d*) 6000

Sol. (*a*)

Males Females

10% 8%

↘ ↙

9%

↙ ↘

1 1

∴ Number of males = $\frac{1}{2} \times 8000 = 4000$

6. The average of marks scored by the students of a class is 68. The average of marks of the girls in the class is 80 and that of boys is 60. What is the percentage of boys in the class ? **(SSC CGL Tier I 2005)**

(*a*) 40 (*b*) 60 (*c*) 65 (*d*) 70

Sol. (*b*)

$$\begin{array}{ccc} \text{Girls} & & \text{Boys} \\ 80 & & 60 \\ & \searrow \; \swarrow & \\ & 68 & \\ & \swarrow \; \searrow & \\ 8 & & 12 \end{array}$$

$\therefore$ Girls : Boys = 8 : 12 = 2 : 3 $\Rightarrow$ Percentage of boys $= \frac{3}{5} \times 100 = 60\%$

7. A class of 60 students contributed ₹ 4800 for a charity. If each boy has contributed ₹ 94 and each girl contributed ₹ 73, find the number of boys in the class.

(*a*) 20 (*b*) 30 (*c*) 40 (*d*) 45

Sol. (*a*) Average contribution per student $= \frac{4800}{60} =$ ₹ 80

$$\begin{array}{ccc} \text{Boys} & & \text{Girls} \\ 94 & & 73 \\ & \searrow \; \swarrow & \\ & 80 & \\ & \swarrow \; \searrow & \\ 7 & & 14 \end{array}$$

$\therefore$ Boys : Girls = 7 : 14 = 1 : 2 $\Rightarrow$ Boys $= \frac{1}{3} \times 60 = 20$

8. A man had 100 kgs of sugar, part of which he sold at 7% profit and rest at 17% profit. He gained 10% on the whole. How much did he sell at 7% profit : **(SSC CGL Tier I 2004)**

(*a*) 65 kg (*b*) 35 kg (*c*) 30 kg (*d*) 70 kg

Sol. (*d*)

$$\begin{array}{ccc} \text{Profit} & & \text{Profit} \\ 7\% & & 17\% \\ & \searrow \; \swarrow & \\ & 10\% & \\ & \swarrow \; \searrow & \\ 7 & & 3 \end{array}$$

$\therefore$ Sugar sold at 7% profit $= \frac{7}{10} \times 100 \text{ kg} = 70 \text{ kg}$

9. A man purchased a horse and a cow for ₹ 5000. He sells the horse at 20% profit and the cow at 10% loss. If he gains 2% on the whole transaction, the cost price of the horse is:

(*a*) ₹ 2000 (*b*) ₹ 2500 (*c*) ₹ 2800 (*d*) ₹ 3000

Sol. (*a*)

$$\begin{array}{ccc} \text{Horse} & & \text{Cow} \\ 20\% & & -10\% \\ & \searrow \; \swarrow & \\ & 2\% & \\ & \swarrow \; \searrow & \\ 12 & & 18 \end{array}$$

$\therefore$ Ratio between cost price of the horse and that of the cow = 12 : 18 = 2 : 3

$\therefore$ Cost price of the horse $= \frac{2}{5} \times 5000 =$ ₹ 2000

10. A merchant sold 80 dozen oranges. He sold some at 4% loss and rest at the cost price and thus losing 3% on the whole. What is the quantity sold at loss?

(*a*) 20 dozen (*b*) 40 dozen (*c*) 50 dozen (*d*) 60 dozen

Sol. (*d*) At loss At cost price

–4% 0%

↘ ↙

–3%

↙ ↘

3 1

∴ Ratio of oranges sold at loss and at cost price = 3 : 1

∴ Quantity sold at loss = $\frac{3}{4} \times 80 = 60$ dozen

11. A shopkeeper sold 40 kg. of goods – partly at 7% profit and rest at 5% loss, suffering 2% loss on the whole, what is the quantity sold at profit?

(*a*) 10 kg (*b*) 15 kg (*c*) 25 kg (*d*) 30 kg

Sol. (*a*) At profit Loss

7% –5%

↘ ↙

–2%

↙ ↘

3 9

Ratio of quantities sold at profit and at loss = 3 : 9 = 1 : 3

∴ Quantity sold at profit = $\frac{1}{4} \times 40 = 10$ kg.

12. A mixture of certain quantity of milk with 16 litre of water is worth 90 paise per litre. If pure milk costs 108 paise per litre, then the amount of milk (in litres) in the mixture is

(SSC CHSL 2012)

(*a*) 70 (*b*) 80 (*c*) 60 (*d*) 50

Sol. (*b*) Milk Water

108 0

↘ ↙

90

↙ ↘

90 18

Ratio of Milk and water = 90 : 18 = 5 : 1

Water = 1 ratio = 16 litres ⇒ Milk = 5 ratio = 80 litres

13. The average of the test scores of a class of 'm' students is 70 and that of 'n' students is 91. When the scores of both the classes are combined, the average is 80. What is n/m?

(SSC CGL Tier II 2012)

(*a*) 10/13 (*b*) 10/11 (*c*) 11/10 (*d*) 13/10

Sol. (*b*) '*m*' students '*n*' students

70 91

↘ ↙

80

↙ ↘

11 10

∴ $n/m = 10/11$

14. 40 litres of a mixture of milk and water contains 10% of water. The water to be added, to make the water content 20% in the new mixture is : **(SSC CGL 2003)**

(*a*) 6 litres (*b*) 6.5 litres (*c*) 5.5 litres (*d*) 5 litres

Sol. (*d*) Mixture Water added

10% 100%

↘ ↙

20%

↙ ↘

80% 10%

∴ Ratio of mixture and water added = 80 : 10 = 8 : 1

∴ Quantity of water to be added = $\frac{1}{8}$ × 40 litres = 5 litres

15. A sample of 50 litres of glycerine is found to be adulterated to the extent of 20%. How much pure glycerine should be added to it to so as to bring down the percentage of impurity to 5%? **(SSC CGL Tier I 2004)**

(*a*) 155 litres (*b*) 150 litres (*c*) 150.4 litres (*d*) 149 litres

Sol. (*b*) Sample Glycerine added

20% 0%

↘ ↙

5%

↙ ↘

5% 15%

∴ Ratio of original mixture and Glycerine added = 5 : 15 = 1 : 3

∴ Required quantity = 3 × 50 litres = 150 litres

Alternative mehod : Ratio of adulteration in two mixutres = 20 : 5 = 4 : 1

∴ Ratio of quantities of solution = 1 : 4

⇒ Glycerined added = (4 – 1) ratio = 3 ratio = 3 × 50 litres = 150 litres

16. How much pure alcohol has to be added to 400 ml of a solution containing 15% of alcohol to change the concentration of alcohol in the mixture to 32% ? **(SSC CGL Tier I 2003)**

(*a*) 60 ml (*b*) 100 ml (*c*) 128 ml (*d*) 68 ml

Sol. (*b*) Mixture Alcohol added

15% 100%

↘ ↙

32%

↙ ↘

68% 17%

∴ Mixture : Alcohol added = 68 : 17 = 4 : 1

Mixture = 4 ratio = 400 ml ⇒ 1 ratio = 100 ml ⇒ Alcohol added = 1 ratio = 100 ml

17. The average monthly salary of the workers in a workshop is ₹8500. If the average monthly salary of 7 technicians is ₹10000 and average monthly salary of the rest is ₹7800, then total number of workers in the workshop is : **(SSC CGL Tier I 2005)**

(*a*) 18 (*b*) 20 (*c*) 22 (*d*) 24

Sol. (*c*) Technicians Rest

10000 7800

↘ ↙

8500

↙ ↘

700 1500

∴ Ratio of number of technicians and rest workers = 700 : 1500 = 7 : 15

Number of technicians = 7 ratio = 7 ⇒ 1 ratio = 1

∴ Total number of workers = 7 + 15 = 22

18. The average salary per head of all workers of an institution is ₹ 60. The average salary per head of 12 officers is ₹ 400. The average salary per head of the rest is ₹ 56. Then the total number of workers in the institution is : **(SSC CGL Tier II 2012, SSC CGL Tier I 2014)**

(*a*) 1062 (*b*) 1060 (*c*) 1030 (*d*) 1032

Sol. (*d*) Officers Rest

400 56

↘ ↙

60

↙ ↘

4 340

∴ Ratio of number of officers and the rest workers = 4 : 340 = 1 : 85

Number of officers = 1 ratio = 12

∴ Total number of workers = (1 + 85) ratio = 12 × 86 = 1032

19. In a family of 8 adults and some minors, the average consumption of rice per head per month is 10.8 kg; while the average consumption for adults is 15 kg per head and for minors it is 6 kg per head. The number of minors in the family is : **(SSC CGL Tier I 2004)**

(*a*) 8 (*b*) 6 (*c*) 7 (*d*) 9

Sol. (*c*) Adults Minors

15 6

↘ ↙

10.8

↙ ↘

4.8 4.2

∴ Ratio of Adults and minors = 4.8 : 4.2 = 8 : 7

Number of adults = 8 ratio = 8 ⇒ 1 ratio = 1 ⇒ Number of minors = 7 ratio = 7

20. The mean weight of 150 students in a certain class is 60 kg. The mean weight of the boys from the class is 70 kg, while that of girls is 55 kg. What is the number of girls in the class ? **(CDS 2010)**

(*a*) 105 (*b*) 100 (*c*) 95 (*d*) 60

Sol. (*b*) Boys Girls

70 55

↘ ↙

60

↙ ↘

5 10

∴ Boys : Girls = 5 : 10 = 1 : 2 ⇒ Girls in class = $\frac{2}{3} \times 150 = 100$

21. How many litres of water must be evaporated from 20 litres of 3% solution to get a 5% solution?

(*a*) 6 (*b*) 8 (*c*) 10 (*d*) 12

Sol. (*b*) This problem can be manipulated as follows :

How much water must be added to 5% solution to get 3% solution :

5% 0%

3%

3 2

$\therefore$ Ratio of new solution and water added = 3 : 2 $\Rightarrow$Water added = $\frac{2}{5} \times 20 = 8$ litres.

Alternative method : Ratio of sugar in two solutions = 3 : 5

$\therefore$ Ratio of solutions (with same quantity of sugar) = 5 : 3

$\therefore$ Quantity of new solution = $\frac{3}{5} \times 20$ litres = 12 litres

$\therefore$ Water evaporated = (20 – 12) litres = 8 litres

Alternative method : 3% of 20 litres = 5% of new solution

$\therefore$ New solution = $\frac{3}{5} \times 20$ litres = 12 litres $\Rightarrow$ Water evaporated = (20 – 12) litres = 8 litres

22. In an examination, a student gets 3 marks for every correct answer and loses 1 mark for every wrong answer. If he scores 0 marks in a paper of 100 questions, how many of his answers were correct assuming that he attempts all the questions?

(*a*) 25 (*b*) 50 (*c*) 60 (*d*) 75

Sol. (*a*) Correct Wrong

3 –1

0

1 3

$\therefore$ Ratio of correct and wrong answers = 1 : 3 $\Rightarrow$ Correct answers = $\frac{1}{4} \times 100 = 25$

Alt: Suppose all the answers are correct. Then total marks = 100 × 3 = 300

For every wrong answer, he loses 4 marks (*i.e.* 3 + 1). Actual marks are 0.

$\therefore$ Wrong answers = 300 ÷ 4 = 75 $\Rightarrow$ Correct answers = 100 – 75 = 25

23. In an examination, a student gets 3 marks for every correct answer and loses 1 mark for every wrong answer. If he scores 120 marks in a paper of 100 questions, how many of his answers were correct assuming that he attempts all the questions?

(*a*) 15 (*b*) 40 (*c*) 55 (*d*) 60

Sol. (*c*) Let he scores 120 marks in the first *x* questions.

$\therefore$ Correct answer are $\frac{120}{3} = 40$ *i.e.* first 40 questions.

In the remaining 60 (*i.e.* 100 – 40) questions, he scores 0 marks.

Correct Wrong

3 ↘ ↙ −1

0

1 ↙ ↘ 3

$\therefore$ Ratio of correct and wrong answers = 1 : 3 $\Rightarrow$ Correct answers = $\frac{1}{4} \times 60 = 15$

$\therefore$ Total correct answers = 40 + 15 = 55.

Alt: Suppose all the answers are correct. Then total marks = 100 × 3 = 300

For every wrong answer he loses 4 marks (i.e. 3 + 1)

Actual marks short by 180 (*i.e.* 300 − 120).

Wrong answers = 180 ÷ 4 = 45 $\Rightarrow$ Correct answers = 100 − 45 = 55

24. Pure milk costs ₹ 10 per litre. A milkman adds some water to 20 litres of pure milk and sells the mixture at ₹ 8 per litre and hence making no profit or loss. How many litres of water does he add?

(*a*) 2 litres (*b*) 5 litres (*c*) 8 litres (*d*) 10 litres

Sol. (*b*) Cost of milk Cost of water

10 ↘ ↙ 0

8

8 ↙ ↘ 2

$\therefore$ Milk : Water = 8 : 2 = 4 : 1 $\Rightarrow$ For every 4 litres of pure milk, he adds 1 litre of water.

$\therefore$ For 20 litres of pure milk, he adds 5 litres of water.

Alternative Method: Cost price of 20 litres = ₹ 200

$\therefore$ Quantity sold = $\frac{200}{8}$ = 25 litres $\Rightarrow$ Water added = 25 − 20 = 5 litres

25. In what ratio a shopkeeper must add water to milk so that he gains 25% on selling it at cost price ?

(*a*) 2 : 3 (*b*) 3 : 2 (*c*) 3 : 1 (*d*) 4 : 1

Sol. (*d*) Let C.P. of pure milk is ₹ 1 per litre. Then S.P. of mixture = ₹ 1 per litre.

Cost price of 1 litre of mixture = $\frac{100}{125}$ = ₹ $\frac{4}{5}$

Milk Water

₹ 1 ↘ ↙ ₹ 0

₹ 4/5

4/5 ↙ ↘ 1/5

$\therefore$ Ratio of milk and water = $\frac{4}{5} : \frac{1}{5} = 4 : 1$

Alt: CP : SP = 4 : 5. Water added = (5 − 4) ratio = 1 ratio $\Rightarrow$ Required ratio = 4 : 1.

Alt: Profit = 25% = $\frac{1}{4} = \frac{\text{Water added}}{\text{Quantity of milk}}$

26. Pure milk costs ₹ 12 per litre. A milkman adds some water to 16 litres of pure milk and sells the mixture at ₹ 10 per litre making 25% profit. How many litres of water does he add?

(*a*) 5 litres (*b*) 6 litres (*c*) 8 litres (*d*) 10 litres

Sol. (*c*) Profit $= \frac{1}{4}$ of CP $= \frac{1}{5}$ of SP $= \frac{1}{5} \times$ ₹ 10 = ₹2/litre ⇒CP of mixture = ₹(10–2)=₹8/litre

By Alligation Method:

Milk Cost of Water

12 0

↘ ↙

8

↙ ↘

8 4

∴ Milk : Water = 8 : 4 = 2 : 1 ⇒ For every 2 litres of pure milk, he adds 1 litre of water.

∴ For 16 litres of pure milk, he adds 8 litres of water.

27. In what ratio must a grocer mix teas at ₹ 60 a kg and ₹ 65 a kg, so that by selling the mixture at ₹ 68.20 a kg, he may gain 10% ? **(SSC CGL Tier I 2004)**

(*a*) 3 : 2 (*b*) 3 : 4 (*c*) 3 : 5 (*d*) 4 : 5

Sol. (*a*) Cost price of mixture = ₹ 68.20 $\times \frac{10}{11}$ = ₹ 62

Mix. I Mix. II

60 65

↘ ↙

62

↙ ↘

3 2

∴ Required ratio = 3 : 2

28. A shopkeeper mixed 10 kg of rice costing ₹10 per kg with a certain quantity of another variety of rice costing ₹15 per kg. He sold the mixture at the rate of ₹14 per kg. If he gained by 5%, the quantity of the second variety of rice was: **(SSC CGL Tier I 2006)**

(*a*) 5 kg (*b*) 10 kg (*c*) 20 kg (*d*) 30 kg

Sol. (*c*) Cost price of rice = ₹ $14 \times \frac{100}{105} = \frac{40}{3}$

Rice I Rice II

10 15

↘ ↙

$40/3$

↙ ↘

$5/3$ $10/3$

∴ Ratio of two varieties $= \frac{5}{3} : \frac{10}{3}$ = 1 : 2 ⇒ Quantity of first variety of rice = 1 ratio = 10 kg

∴ Quantity of second variety of rice = 2 ratio = 2 × 10 kg = 20 kg

29. Two vessels A and B contain milk and water in the ratio 7 : 3 and 3 : 2 respectively. Find the ratio in which the quantities be taken from the two vessels so that the ratio of milk and water in the new mixture is 2 : 1.

(*a*) 1 : 1 (*b*) 1 : 2 (*c*) 2 : 1 (*d*) 2 : 3

Sol. (*c*) Milk in vessel $A = \frac{7}{10}$, milk in vessel $B = \frac{3}{5}$, milk in resultant Mixture $= \frac{2}{3}$

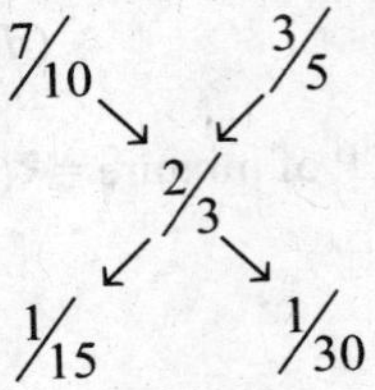

$\therefore$ Ratio of the mixture from vessels A and B $= \frac{1}{15} : \frac{1}{30} = 2 : 1$

Alternative Method : Let vessels A and B contain solutions $10x$ and $5y$ respectively.
Then milk in resultant mixture $= 7x + 3y$ And water in resultant mixture $= 3x + 2y$
According to question, $7x + 3y = 2(3x + 2y) \Rightarrow x = y \Rightarrow x : y = 1 : 1$
$\therefore$ Ratio of two mixtures $= 10x : 5y = 10 \times 1 : 5 \times 1 = 2 : 1$

Alternative Method :
Vessel A contains 1 litre more of milk than the required ratio, for every 10 litres of mixture.
Vessel B contains 1 litre less of mik than the required ratio, for every 5 litres of mixture.

$\therefore$ More milk in vessel $A = \frac{1}{10}$ and Less milk in vessel $B = \frac{1}{5}$

$\therefore$ Ratio in which mixtures are required to be taken from vessels A and B is $\frac{1}{5} : \frac{1}{10} = 2 : 1$

30. Two vessels A and B contain milk and water in the ratio 7 : 3 and 4 : 3 respectively. Find the ratio in which the mixtures must be taken from the two vessels to make the resultant mixture in the ratio 2 : 1.

(*a*) 10 : 7 (*b*) 1 : 2 (*c*) 2 : 1 (*d*) 20 : 7

Sol. (*d*) Milk in vessel $A = \frac{7}{10}$, milk in vessel $B = \frac{4}{7}$, milk in resultant Mixture $= \frac{2}{3}$

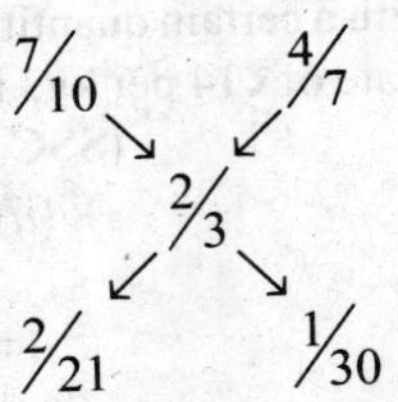

$\therefore$ Ratio of the mixture from vessels A and B $= \frac{2}{21} : \frac{1}{30} = \frac{2}{7} : \frac{1}{10} = 20 : 7$

Alternative Method : Let vessels A and B contain solutions $10x$ and $7y$ respectively.
Then milk in resultant mixture $= 7x + 4y$ And water in resultant mixture $= 3x + 3y$
According to question, $7x + 4y = 2(3x + 3y) \Rightarrow x = 2y \Rightarrow x : y = 2 : 1$
$\therefore$ Ratio of two mixtures $= 10x : 7y = 10 \times 2 : 7 \times 1 = 20 : 7$

Alternative Method : In vessel A, for every 7 litres of milk, there is 3 litre of water *i.e.* 1 litre extra milk than the required ratio.
In vessel B, for every 4 litres of milk, there is 3 litre of water *i.e.* 2 litres less milk than the required ratio.

More milk in vessel $A = \frac{1}{10}$. Less milk in vessel $B = \frac{2}{7}$

$\therefore$ Required ratio $= \frac{2}{7} : \frac{1}{10} = 20 : 7$

31. In an alloy, zinc and copper are in the ratio 1 : 2. In the second alloy, the same elements are in the ratio 2 : 3. If these two alloy be mixed to form a new alloy in which the two elements are in ratio 5 : 8, the ratio of these two alloys in the new alloy is : **(SSC CGL Tier I 2008)**

(*a*) 3 : 10 (*b*) 3 : 7 (*c*) 10 : 3 (*d*) 7 : 3

Sol. (*a*)

Alloy I Alloy II

$\frac{1}{3}$ $\frac{2}{5}$

↘ ↙

$\frac{5}{13}$

↙ ↘

$\frac{1}{65}$ $\frac{2}{39}$

$\therefore$ Required ratio $= \frac{1}{65} : \frac{2}{39} = \frac{1}{5} : \frac{2}{3} = 3 : 10$

Trick :	Zinc	:	Copper		Zinc	:	Copper
Alloy 1	1	:	2	=	1	:	2
Alloy 2	2	:	3	=	4	:	6 (i.e. twice the quantities)
Total					5	:	8

$\therefore$ Required ratio = (1 + 2) : (4 + 6) = 3 : 10

32. The batting average of a cricket player is 72 runs per inning. In the next 4 innings, he could score only 80 runs and thereby decreases his batting average by 2 runs. What is total number of innings played by him till last match?

(*a*) 72 (*b*) 88 (*c*) 92 (*d*) 104

Sol. (*d*) Average of last 4 innings = 80 ÷ 4 = 20 runs ⇒ New average = 72 – 2 = 70 runs

Earlier Last 4 innings

72 20

↘ ↙

70

↙ ↘

50 2

Ratio of innings played (before and after) = 50 : 2 = 25 : 1

Innings played (after) = 1 ratio = 4 ⇒ Innings played (before) = 25 ratio = 25 × 4 = 100

$\therefore$ Total innings played = 100 + 4 = 104

33. The batting average of a cricketer is 50 runs per inning. In the next 5 innings, he scores 400 runs and thereby increases his batting average by 5 runs. What is total number of innings played by him till last match?

(*a*) 25 (*b*) 30 (*c*) 40 (*d*) 50

Sol. (*b*) Average runs of last 5 innings $= \frac{400}{5} = 80$ runs. New average = 50 + 5 = 55 runs

Earlier Last 5 innings

50 80

↘ ↙

55

↙ ↘

25 5

Ratio of innings played (before and after) = 25 : 5 = 5 : 1

Innings played (after) = 1 ratio = 5 ⇒ Innings played (before) = 5 ratio = 5 × 5 = 25

$\therefore$ Total innings played = 25 + 5 = 30

34. In a school, the average age of students is 6 years, and the average age of 12 teachers is 40 years. If the average age of the combined group of all the teachers and students is 7 years, then the number of students is : **(SSC CGL Tier I 2005)**

(*a*) 396 (*b*) 400 (*c*) 408 (*d*) 416

Sol. (*a*)

Students Teachers

6 40

↘ ↙

7

↙ ↘

33 1

Students : Teacher = 33 : 1. Teacher = 1 ratio = 12

Students = 33 ratio = 33 × 12 = 396

Alt: $6x + (12 \times 40) = 7(x + 12) \Rightarrow 6x + 480 = 7x + 84 \Rightarrow 7x - 6x = 480 - 84 \Rightarrow x = 396$

Alternative method : Excess average age of 12 teachers = 40 – 7 = 33 years

Total excess age of 12 teachers = 12 × 33 = 396 years

Difference in two averages = 7 – 6 = 1 year

⇒ Number of students = 396 ÷ 1 = 396

35. The bowling average of a cricketer was 12.4. He improves his bowling average by 0.2 points when he takes 5 wickets for 26 runs in his last match. The number of wickets taken by him before the last match was : **(CBI 1991 M, SSC CGL Tier I 2008)**

(*a*) 125 (*b*) 150 (*c*) 175 (*d*) 200

Sol. (*c*) Average runs of last 5 wickets = 26 ÷ 5 = 5.2

New average for all the wickets = 12.4 – 0.2 = 12.2

Earlier Last 5 innings

12.4 5.2

↘ ↙

12.2

↙ ↘

7 0.2

Ratio of wickets taken (before and after) = 7 : 0.2 = 35 : 1

Wickets taken earlier = 35 × 5 = 175

36. A cricketer whose bowling average is 24.85 runs per wicket, takes 5 wickets for 52 runs and thereby decreases his average by 0.85. The number of wickets taken by him before the last match was : **(SSC CGL Tier I 2000)**

(*a*) 64 (*b*) 72 (*c*) 80 (*d*) 96

Sol. (*c*) Average for last 5 wickets = 52 ÷ 5 = 10.4 ⇒ New average = 24.85 – 0.85 = 24

Earlier wickets Last 5 wickets

24.85 10.4

↘ ↙

24

↙ ↘

13.6 0.85

∴ Ratio of wickets taken (before and after) = 13.6 : 0.85 = 16 : 1

Wickets taken in the last match = 1 ratio = 5 wickets

∴ Wickets taken before the last match = 16 ratio = 16 × 5 = 80 wickets

37. A vessel contains mixture of spirit and water. The percentage of spirit in the mixture is 18%. 6 litres of the mixture are replaced with pure water. If the vessel contains 15% spirit now, the capacity of the vessel is:

(*a*) 25 litres (*b*) 30 litres (*c*) 36 litres (*d*) None of these

Sol. (*c*)

Mixture Water added
18% 0%
↘ ↙
15%
↙ ↘
15% 3%

∴ Ratio of mixture left and water added = 15 : 3 = 5 : 1 ⇒ Water added = 1 ratio = 6 litres

∴ Total mixture = (5 + 1) ratio = 6 ratio = 6 × 6 = 36 litres

38. Ratio of syrup and water in a sweet drink contained in bottle is 7 : 5. What part of the drink is to be taken out and replaced by water so that in the new mixture, syrup and water are in equal proportion? **(SSC CHSL 2012)**

(*a*) $\frac{1}{3}$ (*b*) $\frac{1}{5}$ (*c*) $\frac{1}{6}$ (*d*) $\frac{1}{7}$

Sol. (*d*)

Mixture Water added
7/12 0
↘ ↙
1/2
↙ ↘
1/2 1/12

∴ Ratio of two quantities = $\frac{1}{2}:\frac{1}{12} = 6:1$ ⇒ Required quantity = $\frac{1}{1+6} = \frac{1}{7}$

39. A mixture of 40 litres contains milk and water in ratio 4 : 1. How many litres of mixture must be replaced with water so that the new ratio is 1 : 1 ? **(SSC CGL Tier I 2004)**

(*a*) 12 litres (*b*) 15 litres (*c*) 16 liters (*d*) 18 litres

Sol. (*b*) Water in the original mixture = $\frac{1}{5} = 20\%$. Water in the resultant mixture = $\frac{1}{2} = 50\%$

Mixture Water added
20% 100%
↘ ↙
50%
↙ ↘
50 30

∴ Ratio of mixture left and water added = 50 : 30 = 5 : 3

∴ Mixture replaced = Water added = $\frac{3}{8} \times 40 = 15$ litres

40. A barrel contains a mixture of wine and water in the ratio 3 : 1. How much fraction of the mixture must be drawn off and substituted by water so that the ratio of wine and water in the resultant mixture in the barrel becomes 1 : 1`? **(SSC CGL Tier I 2004)**

(*a*) $\frac{1}{4}$ (*b*) $\frac{1}{3}$ (*c*) $\frac{3}{4}$ (*d*) $\frac{2}{3}$

Sol. (*b*) Water in the original mixture = $\frac{1}{4}$ = 25%. Water in the resultant mixture = $\frac{1}{2}$ = 50%

Mixture Water added
25% 100%
↘ ↙
50%
↙ ↘
50% 25%

∴ Mixture : water added = 50 : 25 = 2 : 1 ⇒ Mixture replaced = water added = $\frac{1}{3}$

41. A mixture contains two liquids A and B in the ratio 4 : 1. If 10 litres of the mixture is replaced with 10 litres of liquid B, the ratio becomes 2 : 3. How many litres of liquid A was in the mixture? **(SSC CGL Tier I 2008, 2012)**

(*a*) 10 litres (*b*) 12 litres (*c*) 16 liters (*d*) 18 litres

Sol. (*c*) Percentage of liquid B in the original mixture = $\frac{1}{4+1} \times 100 = 20\%$

Percentage of liquid B in the new mixture = $\frac{3}{2+3} \times 100 = 60\%$

Mixture Liquid B
20% 100%
↘ ↙
60%
↙ ↘
40 40

∴ Ratio = 40 : 40 = 1 : 1

∴ Remaining quantity of original mixture = Quantity of liquid B added = 1 ratio = 10 litres

∴ Total mixture = (1 + 1) ratio = 2 ratio = 10 × 2 = 20 litres

∴ Liquid A in original mixture = $\frac{4}{5} \times 20$ litres = 16 litres

Alt : 10 lires of mixture replaced contains 8 litres of liquid A and 2 litres of liquid B.

Let the quantaties of liquids A and B in original mixture be $4x$ and x respectively.

Quantities of A and B in new mixture = $(4x - 8) : (x - 2 + 10) = (4x - 8) : (x + 8)$

Then $(4x - 8) : (x + 8) = 2 : 3 \Rightarrow 12x - 24 = 2x + 16 \Rightarrow 10x = 40 \Rightarrow x = 4$

∴ Quantity of liquid A in original mixture = $4x = 4 \times 4 = 16$ litres

42. The ratio of alcohol and water in 40 litre mixture is 5 : 3. 8 litres of the mixture is removed and replaced with water. Now, the ratio of the alcohol and water in the resultant mixture is **(SSC FCI Asst. 2012)**

(*a*) 1 : 2 (*b*) 1 : 1 (*c*) 2 : 1 (*d*) 1 : 3

Sol. (*b*) Alcohol in original mixture = $\frac{5}{8} \times 40$ = 25 litre ⇒ water = 40 – 25 = 15 litres

Alcohol in 8 litres removed = $\frac{5}{8} \times 8$ = 5 litres ⇒ water = 8 – 5 = 3 litres

∴ Required ratio = (25 – 5) : (15 – 3 + 8) = 20 : 20 = 1 : 1

43. A mixture contains two liquids A and B in the ratio 7 : 5. If 9 litres of the mixture is replaced with 9 litres of liquid B, then the new ratio becomes 7 : 9. How much of the liquid A was there originally ?

(*a*) 21 litres (*b*) 15 litres (*c*) 27 liters (*d*) None of these

Sol. (*a*) We can solve this question in following two ways :

(*i*) By comparing the quantities of Liquid B in the two solutions :

Mixture Liquid B

$5/12$ 1

↘ ↙

$9/16$

↙ ↘

$7/16$ $7/48$

∴ Ratio = $7/16 : 7/48 = 3 : 1$. Liquid B added = 1 ratio = 9 litres

∴ Total mixture = 4 ratio = 4 × 9 litres = 36 litres

∴ Quantity of liquid A in the original mixture = $\frac{7}{12} \times 36$ litres = 21 litres

(*ii*) By comparing the quantities of Liquid A in the two solutions :

Mixture Liquid A

$7/12$ 0

↘ ↙

$7/16$

↙ ↘

$7/16$ $7/48$

∴ Ratio = $7/16 : 7/48 = 3 : 1$. Liquid B added = 1 ratio = 9 litres

∴ Total mixture = (3 + 1) ratio = 4 ratio = 4 × 9 litres = 36 litres

∴ Quantity of liquid A in the original mixture = $\frac{7}{12} \times 36$ litres = 21 litres

44. Two products A and B costing ₹ 24 and ₹ 18 per kg. respectively are mixed. Find the ratio of two quantities, if 1% of the quantity is lost in the process of mixing and cost of mixture (after wastage) is ₹ 20 per kg. **(MBA)**

(*a*) 1 : 2 (*b*) 2 : 1 (*c*) 3 : 7 (*d*) 2 : 9

Sol. (*c*) Cost of one kg of mixture (before wastage) = $20 \times \frac{100-1}{100} = \frac{₹\,99}{5} = ₹\,19.80$

Prod. A Prod. B

24 18

↘ ↙

19.8

↙ ↘

1.8 4.2

∴ Product A : Product B = 1.8 : 4.2 = 3 : 7

45. **By mixing two qualities of pulses in the ratio 2 : 3 and the selling the mixture at the rate of ₹ 22 per kg, a shopkeeper makes a profit of 10%. If the cost of the smaller quantity is ₹ 14 per kg, the cost per kg of the larger quantity is:**

(*a*) ₹ 23 (*b*) ₹ 24 (*c*) ₹ 25 (*d*) ₹ 26

Sol. (*b*) CP of mixture (per kg.) = ₹ 22 × $\frac{100}{110}$ = ₹ 20 ⇒ Total CP of 5 kg. = ₹ 20 × 5 = ₹ 100

Total CP of smaller quantity = ₹14 × 2 = ₹28 ⇒ Total CP of larger quantity = 100 – 28 = ₹72

∴ Per kg cost price of larger quantity = ₹ 72 ÷ 3 = ₹ 24

Alternative Method: Let cost price of the larger quantity = ₹ *x* per kg.

14 x

20

x – 20 6

Actual ratio = 2 : 3 ⇒ (*x* – 20) : 6 = 2 : 3 ⇒ 3(*x* – 20) = 12 ⇒ (*x* – 20) = 4 ⇒ *x* = 24

46. **A vessel of 64 litres is filled with milk and water. If 70% of milk and 30% of water is drawn, the vessel is vacated by 55%. Find the initial quantity of the milk.**

(*a*) 24 litres (*b*) 40 litres (*c*) 45 litres (*d*) 50 litres

Sol. (*b*) Milk Water

70% 30%

55%

25 15

Milk : Water = 25 : 15 = 5 : 3 ⇒ Quantity of milk = $64 \times \frac{5}{8}$ = 40 litres

47. **Ramesh earns ₹ 4000 in a month. In the next month his income increases by 5%. Due to rise in prices his monthly expenses increases by 7% and his savings decreases by 1%. Find his increased expenditure.**

(*a*) ₹ 2800 (*b*) ₹ 3000 (*c*) ₹ 3210 (*d*) ₹ 3600

Sol. (*c*) Income = Expenditure + Saving ⇒ Income is sum of expenditure and saving and hence we can solve this problem using Alligation Method.

Expenditure Saving

7 –1

5

6 2

∴ Ratio of expenditure and saving = 6 : 2 = 3 : 1

Initial income = (3 + 1) ratio = 4 ratio = ₹ 4000 ⇒ 1 ratio = ₹ 1000

∴ Initial expenditure = 3 ratio = 3 × ₹ 1000 = ₹ 3000

Increased expenditure = 3000 + 7% of 3000 = ₹ 3,210

48. A solution contains 10% salt. One-fourth of this solution is replaced by a second solution resulting in a solution that contains 16% salt. The percentage of salt in the second solution is

(*a*) 16% (*b*) 22% (*c*) 28% (*d*) 34%

Sol. (*d*) 10 ↘ ↗ 34

16

18 ← ↗ ↘ 6

(**Note:** $10 < 16$ and $34 > 18$)

Alt: $16 \times 4 - 10 \times 3 = 64 - 30 = 34$

49. The ratio of land to water on the whole of the earth is 1 : 2 and it is 2 : 3 on the northern hemisphere. The ratio of land to water on the southern hemisphere is: **(SSC CGL 2004)**

(*a*) 11 : 4 (*b*) 4 : 11 (*c*) 15 : 4 (*d*) 4 : 15

Sol. (*b*) Ratio of land on the earth = 1/3. Ratio of land on the northern hemisphere = 2/5

Northern Southern hemisphere

$2/5$ ↘ ↗ $4/15$

$1/3$

$1/15$ ← ↗ ↘ $1/15$

(**Note:** $\frac{2}{5} > \frac{1}{3}$ and $\frac{4}{15} < \frac{1}{3}$)

(Note: See the direction of arrows to know the steps)

$\therefore$ Required ratio = 4 : (15 – 4) = 4 : 11

(Note: Southern hemisphere may be $\frac{1}{3} \pm \frac{1}{15} = \frac{6}{15}$ or $\frac{4}{15}$. But $\frac{6}{15} = \frac{2}{5}$ is not possible as the both Northern and Southern hemisphere become same and greater than $\frac{1}{3}$.)

Alt: Land on whole earth = $\frac{1}{3}$ and land on nortern hemisphere = $\frac{2}{5}$

Land on southern hemisphere = $2 \times \frac{1}{3} - \frac{2}{5} = \frac{4}{15}$ ⇒ Required ratio = 4 : (15 – 4) = 4 : 11

CHAPTER

20

SIMPLE INTEREST

BASIC TERMS

Principal: The money borrowed by the borrower from the lender. Principal remains the same throughout the loan period.

Rate of Interest: Rate at which the principal is charged.

Time: The period for which the money is borrowed.

Interest: The extra money paid by the borrower to the lender for the use of money.

Amount: The money returned by the borrower at the end of the certain period.

Amount = Principal + Interest. Interest = Amount – Principal

FORMULAE

1. Simple interest (S.I.) $= \dfrac{\text{Principal} \times \text{Rate} \times \text{Time}}{100}$

Simple interest is calculated on the principal uniformly for the entire period. In other words, principal remains same throughout the period.

Example: Simple interest on ₹ 5000 at 10% per annum for 4 years is: **(Bank Clerical)**

(*a*) ₹ 1500 (*b*) ₹ 2000 (*c*) ₹ 2500 (*d*) ₹ 5000

Sol. (*b*) Simple interest $= \dfrac{5000 \times 10 \times 4}{100} =$ ₹ 2000

Example 2: Find simple interest on ₹ 2000 at 5% p.a. for $2\frac{1}{2}$ years.

(*a*) ₹ 200 (*b*) ₹ 250 (*c*) ₹ 300 (*d*) ₹ 350

Sol. (*b*) Time $= 2\frac{1}{2}$ years $= \frac{5}{2}$ years $\Rightarrow$ Simple interest = ₹ $2000 \times \dfrac{5}{100} \times \dfrac{5}{2} =$ ₹ 250

2. Principal $= \dfrac{\text{Simple Interest} \times 100}{\text{Rate} \times \text{Time}}$

Example. If simple interest on a certain sum at 4% per annum for 4 years is ₹ 80, the sum is: **(Bank Clerical)**

(*a*) ₹ 320 (*b*) ₹ 400 (*c*) ₹ 500 (*d*) ₹ 600

Sol. (*c*) Principal $= \dfrac{80 \times 100}{4 \times 4} =$ ₹ 500

3. Rate of Interest $= \dfrac{\text{Simple Interest} \times 100}{\text{Principal} \times \text{Time}}$

Example. At what rate per cent per annum will ₹ 2000 earn ₹ 1000 simple interest in 5 years?

(*a*) 4% (*b*) 5% (*c*) 10% (*d*) 12.5%

Sol. (*c*) Rate of interest $= \frac{\text{Simple Interest} \times 100}{\text{Principal} \times \text{Time}} = \frac{1000 \times 100}{2000 \times 5} = 10\%$

4. Time $= \frac{\text{Simple Interest} \times 100}{\text{Principal} \times \text{Rate}}$

Example. In how many years will ₹ 5000 amount to ₹ 6500 at 10% p.a. simple interest?

(*a*) 2 years (*b*) $2\frac{1}{2}$ years (*c*) 3 years (*d*) $3\frac{1}{2}$ years

Sol. (*c*) Simple interest = ₹ 6500 – ₹ 5000 = ₹ 1500 ⇒ Time $= \frac{1500 \times 100}{5000 \times 10} = 3$ years

5. If rate of Simple interest differs from year to year, then

Simple interest = Principal $\times \frac{(R_1 + R_2 + R_3 + \text{.........})}{100}$

Example: Find simple interest on ₹ 500 for 3 years, if rate of interest for 3 years is 4%, 6% and 7% respectively.

(*a*) ₹ 60 (*b*) ₹ 85 (*c*) ₹ 90 (*d*) ₹ 170

Sol. (*b*) Equivalent rate of interest for 1 year = (4 + 6 + 7)% = 17%

∴ Interest $= \frac{500 \times 17 \times 1}{100} = ₹85$

6. Amount = Principal $\times \frac{100 + (\text{Rate} \times \text{Time})}{100}$

7. Principal = Amount $\times \frac{100}{100 + (\text{Rate} \times \text{Time})}$

Example. What sum will amount to ₹ 660 in 4 years at 8% p.a. of simple interest?

(*a*) ₹ 500 (*b*) ₹ 550 (*c*) ₹ 600 (*d*) ₹ 650

Sol. (*a*) Principal = ₹ 660 $\times \frac{100}{100 + (8 \times 4)} = 660 \times \frac{100}{132} =$ ₹ 500

Alt: ₹ 100 becomes ₹ 132 in 4 years. ⇒ Required Principal = ₹ 660 $\times \frac{100}{132}$ = ₹ 500

8. If a sum of money doubles in n years. Then it will become x times in $(x - 1) \times n$ years.

Example: A certain sum of money at a certain rate of simple interest becomes double in 10 years. It will become three times in: **(SSC CGL Tier I 2006)**

(*a*) 20 years (*b*) 15 years (*c*) 10 years (*d*) $7\frac{1}{2}$ years

Sol. (*a*) Principal will become 3 times in (3 – 1) × 10 = 20 years.

9. If ratio of simple interest on two equal sums = $x : y$.

Then ratio of sums to earn equal interest = $y : x$.

Example: A sum of ₹ 29000 is divided into two parts and invested at simple interest in such a way that first part invested at 5% p.a. for 6 years earns the same interest as the second part for 7 years @ 4% p.a. Find the amount invested in the first part.

(*a*) ₹ 14000 (*b*) ₹ 14500 (*c*) ₹ 15000 (*d*) ₹ 16500

Sol. (*a*) Let the amount invested in each part = ₹ 100

Then simple interest on the first part = 100 × 5% × 6 = ₹ 30

And simple interest on the second part = 100 × 4% × 7 = ₹ 28

∴ The ratio of interests earned on equal amount = 30 : 28 = 15 : 14

∴ Ratio of sums invested to get equal interest = 14 : 15

∴ Amount invested in first part = $\frac{14}{29}$ × ₹ 29000 = ₹ 14000

10. A sum of money is divided into n parts in such a way that the interest on the first part at r_1% for t_1 years, on second part at r_2% for t_2 years, on third part at r_3% for t_3 years and so on, are equal. Then the ratio in which the sum is divided in n parts is:

$$\frac{1}{r_1 \times t_1} : \frac{1}{r_2 \times t_2} : \frac{1}{r_3 \times t_3} \ldots\ldots\ldots : \frac{1}{r_n \times t_n}$$

Example. A man invested ₹ 600 into three parts at simple interest – first part at 3% for 4 years, second part at 4% for 5 years and the third part at 5% for 3 years. If interest received on each part is equal, find amount invested at 5%.

(*a*) ₹ 150 (*b*) ₹ 200 (*c*) ₹ 250 (*d*) ₹ 300

Sol. (*b*) Ratio of principals of three parts = $\frac{1}{3 \times 4} : \frac{1}{4 \times 5} : \frac{1}{5 \times 3} = 5 : 3 : 4$

Sum of ratios = 5 + 3 + 4 = 12 ⇒ Amount invested at 5% = $\frac{4}{12}$ × ₹ 600 = ₹ 200

11. If annual payment of ₹ x discharges a debt of ₹ A due in n years. Find the amount of each instalment, if rate of simple interest is r % p.a. Then

$$A = \frac{x\left(100n + \frac{n(n-1)r}{2}\right)}{100} \Rightarrow x = \frac{100A}{100n + \frac{n(n-1)r}{2}}$$

Example 1. Find the amount of equal instalment, annual payment of which will discharge a debt of ₹ 848 due in 4 years at 4% p.a. of simple interest.

(*a*) ₹ 180 (*b*) ₹ 200 (*c*) ₹ 225 (*d*) ₹ 250

Sol. (*b*) Amount of each instalment = $\frac{848 \times 100}{100 \times 4 + \frac{4 \times 3 \times 4}{2}} = \frac{848 \times 100}{424}$ = ₹ 200

Example 2. Find the amount of debt that will be discharged by 5 equal instalments of ₹ 200 each, if the debt is due in 5 year at 5% p.a.

(*a*) ₹ 900 (*b*) ₹ 1000 (*c*) ₹ 1100 (*d*) ₹ 1200

Sol. (*c*) Amount of loan = 200 × $\frac{100 \times 5 + \frac{5 \times 4 \times 5}{2}}{100}$ = 200 × $\frac{550}{100}$ = ₹ 1100

SOLVED EXERCISE

1. The simple interest on ₹7300 from 11 May, 1987 to 10 September, 1987 (both days included) at 5% p.a. is : **(SSC CGL Tier I 2005)**

(*a*) ₹ 123 (*b*) ₹ 103 (*c*) ₹ 200 (*d*) ₹ 223

Sol. (*a*) Number of days = 21 + 30 + 31 + 31 + 10 = 123 days

$\therefore$ Simple interest = ₹ 7300 × $\frac{5}{100} \times \frac{123}{365}$ = ₹ 123

2. Find Simple interest on ₹ 1500 for 2 years at the rate of 1 paise per rupee per month.

(*a*) ₹ 30 (*b*) ₹ 36 (c) ₹ 300 (*d*) ₹ 360

Sol. (*d*) Rate = 1% p.m. = 12% p.a. ⇒ Simple interest = $\frac{1500 \times 12 \times 2}{100}$ = ₹ 360

3. If simple interest on ₹ 8960 for 3 years is ₹ 1344, find the rate of interest in percent p.a. **(SBI Clerical)**

(*a*) 2 (*b*) 3 (*c*) 4 (*d*) 5

Sol. (*d*) Interest Rate = $\frac{1344 \times 100}{8960 \times 3}$ = 5% p.a.

4. Veena obtained an amount of ₹ 8,376 as simple interest on a certain amount as 8% p.a. after 6 years. What is the amount invested by Veena ? **(SBI Bank PO 2008)**

(*a*) ₹ 17,180 (*b*) ₹ 18,110 (*c*) ₹ 16,660 (*d*) ₹ 17,450

Sol. (*d*) Principal = $\frac{8376 \times 100}{8 \times 6}$ = ₹ 17,450

5. A lent ₹ 450 to B for 2 years, and ₹ 500 to C for 3 years, at a certain rate of simple interest. If he received altogether from both ₹ 120 as interest, find the rate of interest.

(*a*) 3% (*b*) 4% (*c*) 5% (*d*) 6%

Sol. (*c*) Equivalent principal for one year = ₹ 450 × 2 + ₹ 500 × 3 = ₹ 2400

Rate of interest = $\frac{120 \times 100}{2400}$ = 5% p.a.

6. Ravi borrowed some money at the rate of 4% p.a. for the first three years, at the rate of 8%. p.a. for the next two years and at the rate of 9% p.a. for the period beyond 5 years. If he pays a total simple interest of ₹ 19550 at the end of 7 years, how much money did he borrow ? **(Bank PO 2005)**

(*a*) ₹ 39500 (*b*) ₹ 42500 (*c*) ₹ 41900 (*d*) ₹ 43000

Sol. (*b*) Equivalent rate of interest for one year = 3 × 4% + 2 × 8% + (7 – 5) × 9%

= 12% + 16% + 18% = 46% ⇒ Total interest = 46% of sum borrowed = ₹ 19550

$\therefore$ Sum borrowed = ₹ 19550 × $\frac{100}{46}$ = ₹ 42500

7. Two equal sums of money are deposited in different banks – first part at 5% p.a. simple interest for 4 years and second part at 4% p.a. simple interest for 6 years. If difference between the two interests is ₹80, find the sum deposited in each bank.

(*a*) ₹ 500 (*b*) ₹ 1000 (*c*) ₹ 2000 (*d*) ₹ 2500

Sol. (*c*) Difference of interests on ₹ 100 is (4 × 6) – (5 × 4) = ₹ 4

But actual difference = ₹ 80 (*i.e.* 20 times 4) $\therefore$ Each part = 20 × 100 = ₹ 2000

8. Mohan lent some amount of money at 9% simple interest and an equal amount of money at 10% simple interest each for two years. If his total interest was ₹ 760, what amount was lent in each case ? **(SSC CGL Tier I 2004)**

(*a*) ₹ 1700 (*b*) ₹ 1800 (*c*) ₹ 1900 (*d*) ₹ 2000

Sol. (*d*) Let money lent in each case be ₹ 100.

Then total interest = $\frac{100\times9\times2}{100}+\frac{100\times10\times2}{100}$ = ₹ 18 + ₹ 20 = ₹ 38

If total interest is ₹ 38, sum lent in each case = ₹ 100

If total interest is ₹ 760, sum lent in each case = $\frac{100}{38}\times760$ = ₹ 2000

Alt: Simple interest for one year = 9% + 10% = 19% = ₹ 760 ÷ 2 = ₹ 380 (i.e. 20 times 19)

∴ Sum lent in each case = ₹ 200 × 10 = ₹ 2000

9. In what time will the simple interest be 2/5 of the principal at 8 per cent per annum ? **(SSC CGL Tier I 2001)**

(*a*) 8 years (*b*) 7 years (*c*) 5 years (*d*) 6 years

Sol. (*c*) Let principal = ₹5 ⇒ Interest = $\frac{2}{5}$ × ₹ 5 = ₹ 2 ⇒ Required time = $\frac{2\times100}{5\times8}$ = 5 years

10. At what rate of simple interest per annum will a sum become $\frac{7}{4}$ of itself in 4 years ? **(SSC CGL Tier I 2008)**

(*a*) 18% (*b*) $18\frac{1}{4}\%$ (*c*) $18\frac{3}{4}\%$ (*d*) $18\frac{1}{2}\%$

Sol. (*c*) Let principal = ₹ 4. Then amount = ₹ 4 × $\frac{7}{4}$ = ₹ 7 ⇒ Interest = ₹ 7 – ₹ 4 = ₹ 3

⇒ Rate of interest = $\frac{3\times100}{4\times4}=18\frac{3}{4}\%$

11. A sum of ₹ 1600 gives a simple interest of ₹ 252 in 2 years and 3 months. The rate of interest per annum is : **(SSC CGL Tier I 2000)**

(*a*) $5\frac{1}{2}\%$ (*b*) 8% (*c*) 7% (*d*) 6%

Sol. (*c*) Time = $2\frac{3}{12}=2\frac{1}{4}=\frac{9}{4}$ years ⇒ Rate of interest = $\frac{252\times100}{1600\times9/4}=\frac{252\times100\times4}{1600\times9}$ = 7%

12. John invested a sum of money at an annual simple interest rate of 10%. At the end of four years, the amount invested plus interest earned was 770. The amount invested was : **(CISF Head Constable 2011)**

(*a*) ₹ 650 (*b*) ₹ 350 (*c*) ₹ 550 (*d*) ₹ 500

Sol. (*c*) Amount invested = $770\times\frac{100}{100+(10\times4)}=770\times\frac{100}{140}$ = ₹ 550

Alternative method : ₹ 100 becomes ₹ 140 in 4 years at 10% rate of simple interest.

∴ Amount invested = ₹ $770\times\frac{100}{140}$ = ₹ 550

13. The rate per annum of simple interest, for which a sum of money will be double of itself in 10 years, is : **(SSC CGL Tier I 2006)**

(*a*) 1% (*b*) 5% (*c*) 10% (*d*) 20%

Sol. (*c*) Let principal = ₹ 1. ⇒ Amount after 10 years = 2 × ₹ 1 = ₹ 2

∴ Simple interest = ₹ 2 – ₹ 1 = ₹ 1 ∴ Rate of interest = $\frac{1 \times 100}{1 \times 10} = 10\%$

14. An amount doubles itself in 5 years with simple interest. What is the rate of interest per cent per annum ? **(Dena Bank PO 2006)**

(*a*) 20% (*b*) 35% (*c*) 25% (*d*) None of these

Sol. (*a*) Let principal ⇒ ₹ 1 ⇒ Amount after 5 years = 2 × ₹ 1 = ₹ 2

∴ Simple interest = ₹ 2 – ₹ 1 = ₹ 1 ∴ Rate of interest = $\frac{1 \times 100}{1 \times 5} = 20\%$

15. If a sum of money at simple interest doubles in 12 years, the rate of interest p.a. is : **(SSC CGL Tier I 2008)**

(*a*) $16\frac{2}{3}\%$ (*b*) 10% (*c*) $8\frac{1}{3}\%$ (*d*) 7.5%

Sol. (*c*) Let principal = ₹ 1. ⇒ Amount after 12 years = 2 × ₹ 1 = ₹ 2

∴ Simple interest = ₹ 2 – ₹ 1 = ₹ 1 ⇒ Rate of interest = $\frac{1 \times 100}{1 \times 12} = \frac{25}{3} = 8\frac{1}{3}\%$

16. At what rate per cent of simple interest per annum will a sum of money become three times in 10 years? **(SSC Data Entry Operator)**

(*a*) 5% (*b*) 10% (*c*) 20% (*d*) 25%

Sol. (*c*) Let principal = ₹ 1. ⇒ Amount after 10 years = 3 × ₹ 1 = ₹ 3

∴ Simple interest = ₹ 3 – ₹ 1 = ₹ 2 ∴ Rate of interest = $\frac{2 \times 100}{1 \times 10} = 20\%$

17. ₹ 600 amounts to ₹ 735 in 5 years at a certain rate of simple interest. If the rate of interest is increased by 2%, what will be the amount then ? **(Bank Clerical 2007)**

(*a*) ₹ 660 (*b*) ₹ 700 (*c*) ₹ 750 (*d*) ₹ 795

Sol. (*d*) Increase in simple interest = 2% of ₹ 600 for 5 years = $\frac{600 \times 2 \times 5}{100}$ = ₹ 60

∴ New amount = Original amount + Extra interest = ₹ 735 + ₹ 60 = ₹ 795.

18. A certain amount earns simple interest of ₹ 1750 after 7 years. Had the interest rate be 2% more, how much more interest would it have earned ? **(SBI Associate Bank PO 2007)**

(*a*) ₹ 35 (*b*) ₹ 350 (*c*) ₹ 245 (*d*) Data insufficient

Sol. (*d*) Principal and rate of interest are not given. Therefore, data is insufficient.

19. ₹ 800 becomes ₹ 956 in 3 years at certain simple rate of interest. If the rate of interest is increased by 4%, what amount will ₹ 800 become in 3 years ?

(SBI Bank PO 2006, SSC CGL Tier I 2011, BSRB PO 1999M)

(*a*) ₹ 1020.80 (*b*) ₹ 1025 (*c*) ₹ 1052 (*d*) None of these

Sol. (*c*) Increase in simple interest = 800 × 4% × 3 = ₹ 96

∴ New amount = Original amount + Extra interest = ₹ 956 + ₹ 96 = ₹ 1052.

20. ₹ 2000 amounts to ₹ 2600 in five years at a certain rate of simple interest. If interest is 3% more than the previous rate, then in the same time period, the same sum will amount to :

(SSC CGL Tier I 2006, 2000 M)

(*a*) ₹ 2700 (*b*) ₹ 2800 (*c*) ₹ 2900 (*d*) ₹ 3000

Sol. (*c*) Increase in simple interest = 2000 × 3% × 5 = ₹ 300

∴ New amount = Original amount + Extra interest = ₹ 2600 + ₹ 300 = ₹ 2900

21. A sum of money is invested for 3 years at a certain rate of simple interest. Had the rate of interest been 1% more, it would have fetched ₹ 18 more than before. Find the sum invested.

(*a*) ₹ 540 (*b*) ₹ 600 (*c*) ₹ 900 (*d*) ₹ 1800

Sol. (*b*) Amount invested $= \frac{\text{Increase in Simple Interest} \times 100}{\text{Increase in Rate} \times \text{Time}} = \frac{18 \times 100}{1 \times 3} = ₹\ 600$

22. If the annual rate of simple interest increases from 10% to $12\frac{1}{2}$%, a man's yearly income increases by ₹ 1250. His principal (in rupees) is : **(SSC CGL Tier I 2004)**

(*a*) 50000 (*b*) 45000 (*c*) 60000 (*d*) 65000

Sol. (*a*) Difference in interest rates $= 12\frac{1}{2}\% - 10\% = 2\frac{1}{2}\% = \frac{5}{2}\%$

Difference in interest = ₹ 1250

$\therefore$ Principal $= \frac{\text{Simple Interest} \times 100}{\text{Rate} \times \text{Time}} = \frac{1250 \times 100}{\frac{5}{2} \times 1} = 1250 \times 100 \times \frac{2}{5} = ₹\ 50000$

23. ₹1,000 becomes ₹1,200 in 2 years at a certain rate of simple interest. If the rate of interest is increased by 4%, what amount will ₹1,000 become in 3 years ? **(SSC MTS 2014)**

(*a*) ₹ 1300 (*b*) ₹ 1400 (*c*) ₹ 1100 (*d*) ₹ 1420

Sol. (*d*) Annual interest = (1200 – 1000) ÷ 2 = ₹100

Extra interest = 100 + 4% of ₹1000 for 3 years = ₹ 220

New amount = 1200 + 220 = ₹1420

24. The simple interest on a certain sum at 6.5% per annum for 3 years is ₹ 123.50. What will be additional interest if the amount was lent at 7% p.a. ?

(*a*) ₹ 4.25 (*b*) ₹ 6.75 (*c*) ₹ 9.50 (*d*) ₹ 19

Sol. (*c*) Additional interest rate = (7 – 6.5)% = 0.5% p.a.

$\therefore$ Additional interest $= 123.5 \times \frac{0.5}{6.5} = \frac{123.5}{13} = 9.50$

25. The simple interest accrued on an amount of ₹ 2500 at the end of six years is ₹ 1875. What would be the simple interest accrued on an amount of ₹ 6875 at the same rate and same period? **(PNB Officers Exam. 2009)**

(*a*) ₹ 4556.50 (*b*) ₹ 5022.25 (*c*) ₹ 4895.25 (*d*) None of these

Sol. (*d*) Required simple interest $= 1875 \times \frac{6875}{2500} = ₹\ 5156.25$

26. A person borrows ₹5000 for 2 years at 4% per annum simple interest. He immediately lends it to another person at $6\frac{1}{4}$ % per annum simple interest for 2 years. His gain in the transaction is : **(SSC CGL Tier I 2005)**

(*a*) ₹ 112.50 (*b*) ₹ 450 (*c*) ₹ 225 (*d*) ₹ 150

Sol. (*c*) Difference in interest rates $= 6\frac{1}{4}\% - 4\% = 2\frac{1}{4}\% = \frac{9}{4}\%$

$\therefore$ Gain $= 5000 \times \frac{9}{400} \times 2 = ₹\ 225$

27. A man loses ₹ 55.50 yearly when the annual rate of interest falls from 11.5% to 10%. His capital (in rupees) is : **(SSC CGL Tier I 2003)**

(*a*) 3700 (*b*) 7400 (*c*) 8325 (*d*) 11100

Sol. (*a*) Difference in interest rates = 11.5% – 10% = 1.5%

$\Rightarrow$ 1.5% of capital = ₹ 55.50 $\Rightarrow$ Capital = ₹ $\frac{555}{15} \times 100$ = ₹ 3700

28. The difference between the interests received from two different banks on ₹ 500 for two years is ₹2.50. The difference between their rates is **(SSC CGL Tier I 2014)**

(*a*) 1% (*b*) 0.5% (*c*) 2.5% (*d*) 0.25%

Sol. (*d*) Difference in interest rates = $\frac{2.5 \times 100}{500 \times 2} = 0.25\%$

29. The simple interest on a certain sum for 8 months at 4% per annum is ₹ 129 less than the simple interest on the same sum for 15 months at 5% per annum. The sum is : **(SSC CGL 2003)**

(*a*) ₹ 2580 (*b*) ₹ 2400 (*c*) ₹ 2529 (*d*) ₹ 3600

Sol. (*d*) Let the sum = ₹ 100*x*. Then $\frac{75}{12}x - \frac{32}{12}x$ = ₹ 129

$\Rightarrow \frac{43}{12}x = 129 \Rightarrow x = 129 \times \frac{12}{43} = 36 \Rightarrow$ Sum = 100*x* = 100 × 36 = ₹ 3600

30. A person lent two equal amounts of money at 12% p.a. for 3.5 years and 5.5 years respectively. If difference between two interests is ₹1800, find total sum lent.

(*a*) ₹ 6000 (*b*) ₹ 7500 (*c*) ₹ 12000 (*d*) ₹ 15000

Sol. (*d*) Difference in intersts = ₹ 1800, Rate = 12% p.a.

Time difference = 5.5 years – 3.5 years = 2 years

$\therefore$ Principal = $\frac{1800 \times 100}{12 \times 2}$ = ₹7500 $\Rightarrow$ Total sum lent = 2 × ₹ 7500 = ₹ 15000

31. The simple interest on a sum of money is $\frac{1}{4}$th of the principal and the number of years is equal to rate per cent per annum. The rate per cent is : **(SSC CPO SI 1999)**

(*a*) 2.5 (*b*) 5 (*c*) 7.5 (*d*) 10

Sol. (*b*) Let principal = ₹ 4, then interest = ₹ 1, and rate of interest and time be *x* each.

$x^2 = \frac{1 \times 100}{4} = 25 \Rightarrow x = 5\%$ per annum

32. The simple interest on a sum of money is $\frac{4}{9}$ of the principal and the number of years is equal to the rate per cent per annum. The rate per annum is : **(SSC CGL Tier I 2000)**

(*a*) 5% (*b*) $6\frac{2}{3}\%$ (*c*) 6% (*d*) $7\frac{1}{5}\%$

Sol. (*b*) Let principal = ₹ 9. Then interset = ₹ 4. Let rate of interest and time be *x* each.

Then $\frac{4 \times 100}{9 \times x \times x} = 1 \Rightarrow x^2 = \frac{400}{9} \Rightarrow x = \frac{20}{3} = 6\frac{2}{3}\%$

33. Simple interest on a certain sum at a certain annual rate of interest is 25/16 of the sum. If the numbers representing rate per cent and time in years be equal, then the rate of interest is : **(SSC CGL Tier I 2004)**

(*a*) 8% (*b*) $11\frac{1}{2}\%$ (*c*) $12\frac{1}{2}\%$ (*d*) $12\frac{1}{4}\%$

Sol. (*c*) Let principal be ₹ 1, and rate of interest and time be x each.

Then interest = ₹ $\frac{25}{16} = \frac{1 \times x \times x}{100} \Rightarrow x^2 = \frac{2500}{16} \Rightarrow x = \frac{50}{4} = 12\frac{1}{2}\%$

34. A sum of money invested at simple interest becomes ₹306 at the end of 5 years. If the interest is $\frac{9}{25}$ part of the principal, the rate of interest per annum is : **(SSC CGL Tier I 2006)**

(*a*) 6% (*b*) 7% (*c*) 7.2% (*d*) 7.5%

Sol. (*c*) Let principal = ₹ 25. Then interest = ₹ $\frac{9}{25} \times 25$ = ₹ 9

$\therefore$ Rate of interest = $\frac{\text{Simple Interest} \times 100}{\text{Principal} \times \text{Time}} = \frac{9 \times 100}{25 \times 5} = 7.2\%$

35. A sum of money becomes 41/40 of itself in 1/4 years at a certain rate of simple interest. The rate of interest per annum is : **(SSC CGL Tier I 2003)**

(*a*) 10% (*b*) 1% (*c*) 2.5% (*d*) 5%

Sol. (*a*) Let principal = ₹ 40. Then amount = ₹ 41. Interest = ₹ 41 – ₹ 40 = ₹ 1

$\therefore$ Rate of interest = $\frac{\text{Simple Interest} \times 100}{\text{Principal} \times \text{Time}} = \frac{1 \times 100}{40 \times \frac{1}{4}} = 10\%$

36. A sum of money doubles itself at the end of 8 years at certain rate of simple interest. What will be simple interest on ₹ 10,000 at the end of 4 years? **(Bank Clerical, 2005)**

(*a*) ₹ 2,500 (*b*) ₹ 5,000 (*c*) ₹ 10,000 (*d*) None of these

Sol. (*b*) Sum doubles in 8 years. $\Rightarrow$ It will earn 50% interest in 4 years

$\therefore$ Interest in 4 years = $\frac{1}{2} \times 10{,}000$ = ₹ 5,000

37. ₹ 6000 becomes ₹ 7200 in 4 years at a certain rate of simple interest. If the rate becomes 1.5 times of itself, the amount of same principal in 5 years will be : **(SSC CGL Tier I 2007)**

(*a*) ₹ 8000 (*b*) ₹ 8250 (*c*) ₹ 9250 (*d*) ₹9000

Sol. (*b*) Total interest = ₹ 7200 – ₹ 6000 = ₹1200 $\Rightarrow$ Interest for one year = ₹ 1200 ÷ 4 = ₹ 300

Interest at 1.5 times interest rate = ₹ 300 × 1.5 = ₹450

Interest for 5 years = 5 × ₹ 450 = ₹ 2250

$\therefore$ Amount after 5 years = ₹ 6000 + ₹ 2250 = ₹ 8250

Alternative method : Total interest in first case = ₹ 7200 – ₹ 6000 = ₹ 1200

New interest = 1200 × 1.5 × $\frac{5}{4}$ = ₹ 2250 $\Rightarrow$ Required amount = 6000 + 2250 = ₹ 8250

38. A certain sum becomes 3 fold at 4% rate of simple interest. At what rate, it will become 6 fold? **(CBI 2008)**

(*a*) 8% (*b*) 10% (*c*) 9% (*d*) 12%

Sol. (*b*) Let Prinicipal = ₹ 1 $\Rightarrow$ SI = 3 – 1 = 2

In second case, SI = 6 – 1 = 5 $\Rightarrow$ Rate of interest = $\frac{5}{2} \times 4\% = 10\%$

39. A certain sum of money at a certain rate of simple interest becomes double in 5 years. It will become four times in:

(*a*) $7\frac{1}{2}$ years (*b*) 10 years (*c*) 15 years (*d*) 20 years

Sol. (*c*) Principal will become 4 times in (4 – 1) × 5 = 15 years

40. A sum of money, at simple interest, trebles of itself in 15 years. It will become 5 times of itself in : **(SSC CGL Tier I 2006)**

(*a*) 40 years (*b*) 36 years (*c*) 30 years (*d*) 25 years

Sol. (*c*) Let principal = ₹ 1 ⇒ Simple interest in the first case = ₹ 3 – ₹ 1 = ₹ 2

SI in second case = ₹ 5 – ₹ 1 = ₹4 ⇒ Required time = $\frac{4}{2} \times 15$ years = 30 years

41. A sum of money becomes 3 times in 5 years. In how many years will the same sum become 6 times at the same rate of simple interest? **(CDS 2012)**

(*a*) 15 years (*b*) 12.5 years (*c*) 10 years (*d*) 7.5 years

Sol. (*b*) 2 times interest in 5 years ⇒ 5 times interest in $\frac{5}{2} \times 5$ = 12.5 years

42. A sum of money lent on simple interest trebles itself in 15 years and 6 months. In how many years will it be doubled ? **(CDS 2010)**

(*a*) 6 years, 3 months (*b*) 7 years, 9 months (*c*) 8 years, 3 months (*d*) 9 years, 6 months

Sol. (*b*) Required time = $\frac{1}{2}$ × (15 years and 6 months) = 7 years and 9 months

43. A person lent a sum of money @ 6% p.a. In 8 years, the simple interest was ₹ 260 less than the sum lent. Find out the sum lent.

(*a*) ₹ 400 (*b*) ₹ 450 (*c*) ₹ 500 (*d*) ₹ 600

Sol. (*c*) Let the sum lent be ₹ 100.

Then S.I. = $\frac{100 \times 6 \times 8}{100}$ = ₹ 48 which is ₹ 52 (i.e. ₹ 100 – ₹48) less than the principal.

If difference is ₹ 52, then Principal is ₹ 100.

∴ If difference is ₹ 260, then Principal is ₹ 260 × $\frac{100}{52}$ = ₹ 500

44. A man has ₹ 10000 to invest. He invests ₹ 4000 at 5% per annum and ₹ 3500 at 4% per annum simple interest. In order to have a yearly income of ₹ 500, he must invest the remaining amount of money at : **(SSC CGL Tier I 2006)**

(*a*) 6% p.a (*b*) 6.1% p.a. (*c*) 6.3% p.a. (*d*) 6.4% p.a.

Sol. (*d*) Interest on ₹ 4000 at 5% = ₹ 200. Interest on ₹ 3500 at 4% = ₹ 140

⇒ Remaining interest = 500 – (200 + 140) = ₹ 160

⇒ Remaining principal = 10000 – (4000 + 3500) = ₹ 2500

∴ Required rate of interest = $\frac{160 \times 100}{2500}$ = 6.4% p.a.

45. A lends ₹ 2500 to B and a certain sum to C at the same time at 7% annual simple interest. If after 4 years, A altogether receives ₹ 1120 as interest from B and C, the sum lent to C is : **(SSC CGL Tier I 2003)**

(*a*) ₹ 7000 (*b*) ₹ 6500 (*c*) ₹ 4000 (*d*) ₹ 1500

Sol. (*d*) Interest received from B = $\frac{2500 \times 7 \times 4}{100}$ = ₹ 700

⇒ Interest received from C = 1120 – 700 = ₹420⇒Sum lent to C = $\frac{420 \times 100}{7 \times 4}$ = ₹1500

Alternative method : Total amount lent $= \frac{1120 \times 100}{7 \times 4} =$ ₹ 4000

Amount lent to B = ₹ 2500 $\Rightarrow$ Amount lent to C = ₹ 4000 – ₹ 2500 = ₹ 1500

46. A certain sum of money amounts to ₹ 5182 in 2 years and ₹ 5832 in 3 years. If the simple interest is calculated annually, then the principal is **(SSC CHSL 2013)**

(*a*) ₹ 5000 (*b*) ₹ 3882 (*c*) ₹ 4000 (*d*) ₹ 2000

Sol. (*b*) Increase in amount is due to extra interest for 3 – 2 = 1 year

∴ SI for 1 year = 5832 – 5182 = ₹ 650

∴ Principal = Amount after 2 years – S.I. for 2 years = ₹ 5182 – 2 × ₹ 650 = ₹ 3882

47. A sum of money lent out at simple interest amounts to ₹ 720 after 2 years and ₹ 1020 after a further period of 5 years. Find the principal **(SSC CGL Tier I 2015)**

(*a*) ₹ 6000 (*b*) ₹ 1740 (*c*) ₹ 120 (*d*) ₹ 600

Sol. (*d*) Increase in amount is due to extra interest for 5 years

∴ SI for 1 year = (1020 – 720) ÷ 5 = 300 ÷ 5 = ₹ 60

∴ Principal = Amount after 2 years – S.I. for 2 years = ₹ 720 – 2 × ₹ 60 = ₹ 600

Alt: Choices (a) and (b) are higher than the amount. Choice (c) is too lower than the amount.

48. A certain sum of money amounts to ₹ 600 in 2 years and ₹ 700 in 4 years at a certain rate of simple interest. Find the rate of interest. **(Bank Clerical 2005)**

(*a*) 5% (*b*) 8.16% (*c*) 10% (*d*) 16.33%

Sol. (*c*) S.I. for 2 (*i.e.* 4 – 2) years = ₹ 700 – ₹ 600 = ₹ 100 $\Rightarrow$ S.I. for 1 year = 100 ÷ 2 = ₹ 50

∴ Principal = Amount after 2 years – Simple interest for 2 years = ₹ 600 – 2 × ₹ 50 = ₹ 500

Hence simple interest for 1 year is ₹ 50 and Principal is ₹ 500.

∴ Rate of interest $= \frac{50 \times 100}{500 \times 1} = 10\%$ p.a

49. A sum amounts to ₹ 5750 in 3 years at 5% p.a. simple interest. In how many years will the same sum amount to ₹ 7000 at the same rate of interest?

(*a*) 5 years (*b*) 7 years (*c*) 8 years (*d*) 10 years

Sol. (*c*) Principal $= 5750 \times \frac{100}{100 + (5 \times 3)} = 5750 \times \frac{100}{115} =$ ₹ 5000

∴ SI in second case = ₹ 7000 – ₹ 5000 = ₹ 2000 $\Rightarrow$ Required time $= \frac{2000 \times 100}{5000 \times 5} = 8$ years

50. Simple interest on ₹ 3500 at 10% p.a. for a certain period is ₹ 400 more than the interest on ₹ 2500 at 10% p.a. for the same period. Find the time.

(*a*) 2 years (*b*) 3 years (*c*) 4 years (*d*) 5 years

Sol. (*c*) Difference in two simple interests = ₹ 400

Difference in principals = ₹ 3500 – ₹ 2500 = ₹ 1000

∴ Time $= \frac{\text{Difference in SI} \times 100}{\text{Difference in Principals} \times \text{Rate}} = \frac{400 \times 100}{1000 \times 10} = 4$ years

51. If difference between simple interest on a certain sum at 4% for 6 years and at 5% for 4 years is ₹ 28, find the sum.

(*a*) ₹ 200 (*b*) ₹ 400 (*c*) ₹ 500 (*d*) ₹ 700

Sol. (*d*) Let the sum be ₹ 100.

Then simple interest in the first case = 100 × 4% × 6 = ₹ 24

And simple interest in the second case = 100 × 5% × 4 = ₹ 20

Difference in two simple interests = ₹ 24 – ₹ 20 = ₹ 4

Actual difference = ₹ 28 (*i.e.* 7 times ₹ 4) $\quad\therefore$ Principal = 7 × ₹ 100 = ₹ 700

52. Ram deposited a certain sum of money in a company at 12% per annum simple interest for 4 years and deposited equal amount in fixed deposit in a bank for 5 years at 15% per annum simple interest. If the difference in the interest from the two sources is ₹1350, then the sum deposited in each case is **(SSC CGL Tier I 2015)**

(*a*) ₹ 5000 (*b*) ₹ 4000 (*c*) ₹ 3000 (*d*) ₹ 6500

Sol. (*a*) Let the sum in each case be ₹ 100.

Then difference in interest = 4 × 12% – 5 × 15% = 27% = ₹1350 (*i.e.* 50 times)

Sum invested in each case = 50 × ₹100 = ₹5000

53. The difference of simple interest of ₹1200 for 3 years and that of ₹800 for 4 years at same rate is ₹20. The rate per cent per annum is : **(SSC CGL Tier I 2006)**

(*a*) 2.5 (*b*) 5 (*c*) 8 (*d*) 10

Sol. (*b*) Let rate of interest be 1% p.a.

Then difference of interests = 1200 × 1% × 3 – 800 × 1% × 4 = 36 – 32 = ₹ 4

But actual difference = ₹ 20 (i.e. 5 times 4) ⇒ Rate of interest = 5 × 1% = 5% p.a.

54. A person deposited ₹400 for 2 years, ₹550 for 4 years and ₹1200 for 6 years. He received the total simple interest of ₹1020. The rate of interest per annum is : **(SSC CGL Tier I 2007)**

(*a*) 10% (*b*) 5% (*c*) 15% (*d*) 20%

Sol. (*a*) Let the sum be lent at 1% p.a.

Then total interest = 400 × 1% × 2 + 550 × 1% × 4 + 1200 × 1% × 6 = ₹(8 + 22 + 72) = ₹102

But actual interest = ₹1020 (i.e. 10 × ₹102) ⇒ Rate of interest = 10 × 1% = 10%

55. ₹ 500 was invested at 12% per annum simple interest and a certain sum of money invested at 10% per annum simple interest. If the sum of the interests on both the sums after 4 years is ₹ 480, the latter sum of money is : **(SSC CGL Tier I 2003)**

(*a*) ₹ 450 (*b*) ₹ 750 (*c*) ₹ 600 (*d*) ₹ 550

Sol. (*c*) Total interest per annum = ₹ 480 ÷ 4 = ₹ 120

Annual interest on ₹ 500 at 12% p.a. = 500 × 12% = ₹ 60

Remaining interest = ₹ 120 – ₹ 60 = ₹ 60 (on investment at 10% p.a.)

Sum invested at 10% p.a. = $\frac{60 \times 100}{10}$ = ₹ 600

56. A man borrowed ₹ 2500 at 4% p.a. and ₹ 1800 at 5% simple interest for the same period. If he pays ₹ 570 as total interest, find the time for which the sums were borrowed.

(*a*) 2 years (*b*) 3 years (*c*) 4 years (*d*) 5 years

Sol. (*b*) Let the both sums are borrowed for one year each.

Then total interest = (2500 × 4% × 1) + (1800 × 5% × 1) = ₹ (100 + 90) = ₹ 190

Actual simple interest = ₹ 570 (*i.e.* 3 times ₹ 190) ⇒ Required time = 3 × 1 = 3 years

57. Ramesh withdraws ₹ 2,000 from his saving account when the rate of interest is reduced from 5% to 3%. If his annual interest is reduced by ₹ 150, how much money was initially deposited in his account?

(*a*) ₹ 7500 (*b*) ₹ 6000 (*c*) ₹ 4500 (*d*) ₹ 2500

Sol. (*c*) Reduction in interest due to withdrawl of ₹ 2,000 = 5% of ₹ 2,000 = ₹ 100

Rate reduced on the balance amount = 5% – 3% = 2% ⇒ 2% of balance = ₹ 150 – ₹ 100 = ₹50

∴ Balance amount = ₹ 50 × $\frac{100}{2}$ = ₹ 2500 ⇒ Initial deposit = ₹ 2000 + ₹ 2500 = ₹ 4500

58. A borrows ₹ 800 at rate of 12% per annum simple interest and B borrows ₹ 910 at the rate of 10% per annum, simple interest. In how many years will their amounts of debt be equal ? **(SSC CGL Tier I 2007)**

(*a*) 18 years (*b*) 20 years (*c*) 22 years (*d*) 24 years

Sol. (*c*) Difference between two principals = ₹ 910 – ₹ 800 = ₹ 110

Difference between annual interest = (800 × 12%) – (910 × 10%) = ₹ (96 – 91) = ₹ 5

∴ Required time = 110 ÷ 5 = 22 years

59. A person invested ₹ 700 at a certain rate of interest and ₹ 900 at 2% higher rate of interest. If total interest earned on both the investments for 3 years is ₹ 294, find the rate of interest on first investment.

(*a*) 3% (*b*) 4% (*c*) 5% (*d*) None of these

Sol. (*c*) Extra interest on second investment = $\frac{900 \times 2 \times 3}{100}$ = ₹ 54

Remaning interest earned on ₹ (700 + 900 = 1,600) is ₹ (294 – 54) = ₹ 240

∴ Interest for one year = ₹ 240 ÷ 3 = ₹ 80 ⇒ Rate of interest = $\frac{80 \times 100}{1600}$ = 5% p.a.

60. A person lent ₹ 5000 partly at the rate of 4 per cent and partly at the rate of 5 per cent per annum simple interest. The total interest after 2 years is ₹ 440. To find the sum of money lent at each of the above rates, ₹ 5000 is to be divided in the ratio : **(SSC CGL 2002)**

(*a*) 4 : 5 (*b*) 3 : 2 (*c*) 5 : 4 (*d*) 2 : 3

Sol. (*b*) Annual interest = ₹ 440 ÷ 2 = ₹ 220. Let the whole sum borrowed at 4% p.a.

Then total interest for one year = 5000 × 4% = ₹ 200

This is short from actual interest by ₹ 220 – ₹ 200 = ₹ 20

Which is equal to 1% (i.e. 5% – 4%) of amount lent at 5%.

∴ 1% of the amount lent at 5% = ₹ 20 ∴ Amount lent at 5% = ₹ 20 × 100 = ₹ 2000

∴ Amount lent at 4% = ₹5000 – ₹2000 = ₹3000 ⇒ Required ratio = 3000 : 2000 = 3 : 2

61. A person takes a loan of ₹10000 partly from a bank at 8% per annum and remaining from another bank at 10% per annum. He pays a total interest of ₹ 950 per annum. Amount of loan taken from the first bank (in ₹) is **(SSC CGL Tier I 2012)**

(*a*) 2500 (*b*) 5200 (*c*) 2050 (*d*) 5020

Sol. (*a*) Let the whole loan was taken from the second bank.

Then annual interest = 10% of 10000 = ₹ 1000

Difference from actual interest = ₹ (1000 – 950) = ₹ 50

Difference in interest rates = 10% – 8% = 2%

∴ 2% of amount invested at 8% = ₹ 50 ⇒ Amount invested at 8% = $\frac{50}{2} \times 100$ = ₹ 2500

62. Pratap borrowed a sum of money from Arun at simple interest, at the rate of 12% per annum for the first three years, 16% per annum for the next five years and 20% per annum for the period beyond eight years. If at the end of 11 years, the total interest is ₹6080 more than the sum, the sum borrowed was : **(SSC CGL Tier I 2006)**

(*a*) ₹ 6000 (*b*) ₹ 8000 (*c*) ₹10000 (*d*) ₹12000

Sol. (*b*) Let the sum be borrowed = ₹100.

Then total interest = 12 × 3 + 16 × 5 + 20 × (11 – 8) = ₹(36 + 80 + 60) = ₹176

Difference = 176 – 100 = ₹76 ∴ Principal = $\frac{100}{76}$ × 6080 = ₹8000

63. Ramesh deposited ₹ 15600 in a fixed deposit at the rate of 10% per annum simple interest. After every second year, he adds his interest to the principal. The interest for the fourth year is : **(SSC CGL Tier I 2006)**

(*a*) ₹1716 (*b*) ₹1560 (*c*) ₹3432 (*d*) ₹1872

Sol. (*d*) Principal for the first year = ₹15600

Simple interest for first 2 years = 15600 × 10% × 2 = ₹ 3120

Principal for the third and the fourth year = ₹(15600 + 3120) = ₹18720

Simple interest for fourth year = 187200 × 10% × 1 = ₹ 1872

65. At what rate per cent of simple interest per annum, ₹ 500 lent for 7 years will earn same interest as ₹ 700 earns in 6 years at 10% p.a.?

(*a*) 15% (*b*) 20% (*c*) 25% (*d*) 12%

Sol. (*d*)

Principal	Time	Rate
700	6	10
500	7	x

Less principal, more rate (to earn same interest) *i.e.* inverse proportion

More Time, less rate (to earn same interest) i.e. inverse proportion

$\therefore$ Required rate of interest $= 10 \times \frac{700}{500} \times \frac{6}{7} = 12\%$

Hint: If the proportion is inverse proportion, two values are respectively multiplied and divided. If the proportion is direct proportion, two values are respectively divided and multiplied.

66. What sum of money lent at 4% p.a. for 3 years will earn same interest as ₹ 1200 earns in 4 years at 5% p.a. of simple interest?

(*a*) ₹ 1000 (*b*) ₹ 1500 (*c*) ₹ 2000 (*d*) ₹ 2500

Sol. (*c*)

Principal	Time	Rate
1200	4	5
x	3	4

Less time, more principal, *i.e.* inverse proportion

Less rate, more principal, *i.e.* inverse proportion

$\therefore$ Required principal $= 1200 \times \frac{4}{3} \times \frac{5}{4} =$ ₹ 2000

Alternative method : Interest on ₹ 100 in two cases are ₹12 and ₹20 respectively.

$\therefore$ Amount must be invested in the ratio of 20 : 12 or 5 : 3 to earn the same interest in both the cases. $\Rightarrow$ Amount invested at 4% $= \frac{5}{3} \times 1200 =$ ₹ 2000

67. In what time will ₹8,000 at 3% per annum, produce the same interest as ₹ 6,000 does in 5 years at 4% simple interest ? **(SSC CGL Tier I 2014)**

(*a*) 5 years (*b*) 6 years (*c*) 3 years (*d*) 4 years

Sol. (*a*) Required time $= \frac{6000 \times 5 \times 4}{8000 \times 3} = 5$ years

68. If a sum of ₹ 500 amounts to ₹ 575 in 3 years, how much will ₹ 600 amount to in 4 years at the same rate of simple interest.

(*a*) ₹ 120 (*b*) ₹ 620 (*c*) ₹ 650 (*d*) ₹ 720

Sol. (*d*) Simple interest on ₹ 500 for 3 years = ₹ 575 – ₹ 500 = ₹ 75

Rate of interest = $\frac{75 \times 100}{500 \times 3}$ = 5% p.a.

Interest on ₹ 600 for 4 years = ₹ 600 × 5% × 4 = ₹ 120 ⇒ Amount = ₹ 600 + ₹ 120 = ₹ 720

Alternative method : Simple interest on ₹ 500 for 3 years = ₹ 575 – ₹ 500 = ₹ 75

∴ SI on ₹ 600 for 4 years = 75 × $\frac{600}{500} \times \frac{4}{3}$ = ₹ 120 ⇒ Amount = ₹ 600 + ₹ 120 = ₹ 720

69. A person invested part of ₹ 45,000 at 4% and the rest at 6%. If his annual income from both are equal, then what is the average rate of interest ? **(CDS 2009)**

(*a*) 4.6% (*b*) 4.8% (*c*) 5.0% (*d*) 5.2%

Sol. (*b*) Ratio of principals of two parts = 6 : 4 = 3 : 2

Average rate of interest = $\frac{3 \times 4\% + 2 \times 6\%}{3+2} = \frac{24\%}{5}$ = 4.8%

70. If ₹ 12000 is divided into two parts such that the simple interest on the first part for 3 years at 12% per annum is equal to the simple interest on the second part for $4\frac{1}{2}$ at 16% per annum, the greater part is : **(SSC CGL Tier I 2008)**

(*a*) ₹ 8000 (*b*) ₹ 6000 (*c*) ₹ 7000 (*d*) ₹ 7500

Sol. (*a*) Ratio of interest on equal amount = (12% × 3) : 16% × $4\frac{1}{2}$ = 36 : 72 = 1 : 2

∴ Ratio of principals for equal interest = 2 : 1 ⇒ Greater part = $\frac{2}{3}$ × ₹ 12000 = ₹ 8000

71. A sum of ₹ 1750 is divided into two parts such that the interests on the first part at 8% simple interest per annum and that on the other part at 6% simple interest per annum are equal. The interest on each part (in rupees) is : **(SSC CGL Tier I 2002)**

(*a*) 60 (*b*) 65 (*c*) 70 (*d*) 40

Sol. (*a*) Ratio of principals of two parts = 6 : 8 = 3 : 4

First part = $\frac{3}{7}$ × ₹ 1750 = ₹ 750 ⇒ Interest on first part = ₹ 750 × $\frac{8}{100}$ = ₹ 60

72. A sum of ₹ 5500 is divided into two parts and invested at 4% p.a. for 3 years and at 5% p.a. for 6 years respectively. If interest earned on the second part is three times of that on the first part, find the amount invested in the second part.

(*a*) ₹ 2000 (*b*) ₹ 2250 (*c*) ₹ 2500 (*d*) ₹ 3000

Sol. (*d*) Ratio of interest earned on equal amount = 4% × 3 : 5% × 6 = 12 : 30 = 2 : 5

∴ Ratio of amount to be invested to get equal interest = 5 : 2

∴ To get 3 times interest on the second part, ratio of principals = (5 × 1) : (2 × 3) = 5 : 6

∴ Amount invested in the second part = $\frac{6}{11}$ × ₹ 5500 = ₹ 3000

73. A man divided ₹ 6200 in two parts and lent the first part @ 6% for 5 years, and the second part @ 10% for 4 years. If interest earned on two parts is in the ratio of 4 : 5, find the first part.

(*a*) ₹ 2500 (*b*) ₹ 3000 (*c*) ₹ 3100 (*d*) ₹ 3200

Sol. (*d*) Ratio of interests on two parts (on equal amounts invested) = 6 × 5 : 10 × 4 = 3 : 4

Ratio of principals for equal interest = 4 : 3. But actual ratio = 4 : 5.

$\therefore$ Required ratio = (4 × 4) : (5 × 3) = 16 : 15 ⇒ First part = $\frac{16}{31}$ × ₹ 6200 = ₹ 3200

Alt: Ratio of interests (on equal amounts) = 6 × 5 : 10 × 4 = 3 : 4. But actual ratio = 4 : 5

$\therefore$ Ratio of two parts = $\frac{4}{3}:\frac{5}{4}$ = 16 : 15 ⇒ First part = $\frac{16}{31}$ × ₹ 6200 = ₹ 3200

74. Two equal sums were lent at simple interest for 5 years and 4 years respectively. If interests on two sums are equal and rate of interest on the latter sum is 2% higher than the former, find the lower rate of interest.

(*a*) 5% (*b*) 6% (*c*) 8% (*d*) 10%

Sol. (*c*) Let each principal = ₹ 100 and lower rate of interest = x% ⇒ $5x = 4 \times (x + 2) \Rightarrow x = 8\%$

75. A man invested a total sum of ₹ 8,500 in names of his sons at 4% and 5% respectively in such a way that they get equal amount on attaining the age of 20 years. If their present ages are 15 years and 13 years respectively, what is the sum invested in the name of elder son ?

(*a*) ₹ 3750 (*b*) ₹ 4000 (*c*) ₹ 4250 (*d*) ₹ 4500

Sol. (*d*) Let ₹ 100 is invested in each case.

Then amount received by elder son = 100 + (4 × 5) = ₹ 120

And amount received by younger son = 100 + (5 × 7) = ₹ 135

$\therefore$ Ratio of sum invested = $\frac{1}{120}:\frac{1}{135} = 135:120 = 9:8$

$\therefore$ Sum invested in name of elder son = $\frac{9}{17}$ × ₹ 8500 = ₹ 4500

76. ₹ 4310 is invested into three parts at simple interest so that the amounts received after 1, 2 and 3 years respectively in each part are equal. Find the amount invested for 3 years, if rate of interest is 10% p.a.

(*a*) ₹ 1320 (*b*) ₹ 1430 (*c*) ₹ 1560 (*d*) ₹ 1650

Sol. (*a*) Let ₹ 100 is invested in each part.

Then amount received in three parts on maturity are ₹ 110, ₹ 120 and ₹ 130 respectively.

$\therefore$ Ratio of amount invested in each part (to get equal amounts)

$= \frac{1}{110}:\frac{1}{120}:\frac{1}{130} = \frac{1}{11}:\frac{1}{12}:\frac{1}{13}$ = 156 : 143 : 132 (on multiplying each ratio by 11 × 12 × 13)

Sum of ratios = 156 + 143 + 132 = 431. But actual amount = ₹ 4310 (*i.e.* 10 times 431)

$\therefore$ Amount invested for 3 years = 10 × ₹ 132 = ₹ 1320

77. A man invested $\frac{1}{3}$ of his money at 5% p.a., $\frac{1}{4}$ of the money at 4% and the remaining $\frac{5}{12}$ money at 3%. If his annual income is ₹ 470, find total money invested by him.

(*a*) ₹ 3000 (*b*) ₹ 6000 (*c*) ₹ 12000 (*d*) ₹ 15000

Sol. (*c*) LCM of 3, 4 and 12 is 12. We assume that the total money invested = ₹ 1200

Money invested in three parts be ₹ 400, ₹ 300 and ₹ 500 respectively.

$\therefore$ Simple interest on the first part = 400 × 5% = ₹ 20

Simple interest on the second part = 300 × 4% = ₹ 12

$\therefore$ Simple interest on the third part = 500 × 3% = ₹ 15

Total SI = ₹ (20 + 12 + 15) = ₹ 47. Actual SI = ₹ 470 (*i.e.* 10 times 47)

∴ Total sum invested = 10 × ₹ 1200 = ₹ 12000

Alternative Method : Single equivalent rate = $\frac{1}{3}\times 5\% + \frac{1}{4}\times 4\% + \frac{5}{12}\times 3\% = \frac{47}{12}\%$

∴ Total sum invested × $\frac{47}{12}\%$ = ₹470 ∴ Total sum invested = $470\times\frac{12}{47}\times 100$ = ₹12,000

78. A man invested 1/5 of his money at 5%, 1/6 of the money at 8% and remaining money at 10% p.a. If his annual income from interest is ₹ 260, what is total money he had?

(*a*) ₹ 1000 (*b*) ₹ 1500 (*c*) ₹3000 (*d*) None of these

Sol. (*c*) Money invested at 10% p.a. = $1-\frac{1}{5}-\frac{1}{6}=\frac{19}{30}$

Single equivalent rate = $\frac{1}{5}\times 5\%+\frac{1}{6}\times 8\%+\frac{19}{30}\times 10\% = 1\%+\frac{4}{3}\%+\frac{19}{3}\% = \frac{26}{3}\%$

$\frac{26}{3}\%$ of the sum = ₹ 260 ∴ Total sum invested = $260\times\frac{3}{26}\times 100$ = ₹ 3000

79. A man invested ₹ 50000 at simple interest in three parts – one-third at 6% p.a., one-third at 7%. At what rate must he invest the remaining amount so that he earns 8% interest on the total amount.

(*a*) 9% (*b*) 10% (*c*) 11% (*d*) 12%

Sol. (*c*) Let total amount = ₹ 300. Then amount invested in each part = ₹ 300 ÷ 3 = ₹ 100

Total interest earned = ₹ 3 × 8 = ₹ 24

Interest earned from first and second part = ₹ 6 + ₹ 7 = ₹ 13

Interest required to be earned from the third part = ₹ 24 – ₹ 13 = ₹ 11

∴ Third part was invested at 11%.

Trick: 8% + (8% – 6%) + (8% – 7%) = 8% + 2% + 1% = 11%

80. A person lends 40% of his sum of money at 15% per annum, 50% of the rest at 10% per annum and the rest at 18% per annum rate of interest. What would be the annual rate of interest, if the interest is calculated on the whole sum ? **(SSC CGL Tier I 2007)**

(*a*) 13.4% (*b*) 14.33% (*c*) 14.4% (*d*) 13.33%

Sol. (*c*) Total interest on ₹ 100 = 40 × 15% + 30 × 10% + 30 × 18% = 6 + 3 + 5.4 = 14.4 = 14.4%

81. A person invested some amount at the rate of 12% simple interest and the remaining at 10%. He received yearly interest of ₹130. Had he interchanged the amounts invested, he would have received an interest of ₹134. How much money did he invest at different rates? **(CDS 2010)**

(*a*) ₹ 500 @ 10%, ₹ 800 @ 12% (*b*) ₹ 700 @ 10%, ₹ 600 @ 12%

(*c*) ₹ 800 @ 10%, ₹ 400 @ 12% (*d*) ₹ 700 @ 10%, ₹ 500 @ 12%

Sol. (*d*) Difference in two interest rates = 12% – 10% = 2%

Difference in interests = ₹ 134 – ₹ 130 = ₹ 4. ⇒ Difference in amounts = $\frac{4}{2}\times 100$ = ₹ 200

Difference is two amounts is ₹ 200 in option (*d*) only.

82. A man lent ₹ 2000 – partly at 5% and the balance at 4%. If he receives ₹ 92 towards annual interest, find the amount lent at 5%. **(SSC CGL Tier I Exam.)**

(*a*) ₹ 800 (*b*) ₹ 900 (*c*) ₹ 1000 (*d*) ₹ 1200

Sol. (*d*) Let the whole amount is invested at 4% p.a. Then simple interest = $\frac{2000 \times 4 \times 1}{100}$ = ₹ 80

This interest is short from actual interest by ₹ 92 – ₹ 80 = ₹ 12
The difference is because a part of the amount is also invested at 5% p.a.
Difference in two rates of interest = 5% – 4% = 1% p.a.
Difference in rate is 1%, and difference in interest = ₹ 12

$\therefore$ Amount invested at 5% = $12 \times \frac{100}{1}$ = ₹ 1200

Alternative Method: Simple interest on ₹ 2000 at 5% p.a. = $\frac{2000 \times 5 \times 1}{100}$ = ₹ 100

Simple interest on ₹ 2000 at 4% p.a. = $\frac{2000 \times 4 \times 1}{100}$ = ₹80
By Alligation Method:

At 5% At 4%
₹ 100 ₹ 80
↘ ↙
₹ 92
↙ ↘
12 8

$\therefore$ Ratio of amounts lent at 5% and 4% = 12 : 8 = 3 : 2 ⇒ Sum lent at 5% = $\frac{3}{5} \times 2000$ = ₹1200

Alt: Average rate of interest = $\frac{92 \times 100}{2000 \times 1} = 4.6\%$

5 4
↘ ↙
4.6
↙ ↘
0.6 0.4

$\therefore$ Required ratio = 0.6 = 0.4 = 3 : 2

83. A man had ₹16000 part of which he lent at 4% and the rest at 5% per annum simple interst. If the total interest received was ₹ 700 in one year, the money lent at 4% per annum was :

(SSC CGL Tier I 2005)

(*a*) ₹ 12000 (*b*) ₹ 8000 (*c*) ₹ 10000 (*d*) ₹ 6000

Sol. (*c*) Let money was invested at 4% p.a.
Then annual interest = 16000 × 4% = ₹ 640. Actual interest = ₹ 700
This interest is short from actual interest by ₹ 700 – ₹ 640 = ₹ 60
The difference is because the amount is also invested at 5% p.a.
Difference in two rates of interest = 5% – 4% = 1% p.a.
Difference in rate is 1%, and difference in interest = ₹ 60

$\therefore$ Amount invested at 5% = $\frac{60 \times 100}{1}$ = ₹ 6000

Money lent at 4% p.a. = ₹16000 – ₹6000 = ₹10000

84. A sum of ₹ 5000 was lent partly at 6% and partly at 9% simple interest. If the total interest received after one year was ₹ 390, the ratio in which the money was lent at 6% and 9% is **(SI Police 1997)**

(*a*) 1 : 1 (*b*) 1 : 2 (*c*) 2 : 3 (*d*) 3 : 2

Sol. (*c*) Interest on ₹5000 @ 6% = ₹300 Interest on ₹5000 @ 9% = ₹450

300 ↘ ↙ 450

390

60 ↙ ↘ 90

Required ratio = 60 : 90 = 2 : 3

85. Biswas borrowed a sum of ₹ 30000. He took a part of it at 12% per annum simple rate of interest and the remaining at 10% per annum. At the end of two years, he returned ₹ 36480 and discharged loan. What was the sum borrowed at 12% per annum rate of interest?

(*a*) ₹ 16000 (*b*) ₹ 18000 (*c*) ₹ 17500 (*d*) ₹ 12000

Sol. (*d*) Interest for 2 years = ₹ 36480 – ₹ 30000 = ₹ 6480

Annual interest = ₹ 6480 ÷ 2 = ₹ 3240. ⇒ Let the whole sum be borrowed at 10% p.a.

Then total interest for one year = 30000 × 10% = ₹ 3000

This is short from actual interest by ₹ 3240 – ₹ 3000 = ₹ 240

which is equal to 2% (i.e. 12% – 10%) of sum borrowed at 12% p.a.

∴ 2% of sum borrowed at 12% = ₹ 240 ⇒ Required sum = ₹ 240 × $\frac{100}{2}$ = ₹ 12000

86. A sum of ₹ 1550 was lent at simple interest partly at the rate of 5% and partly at 8% per annum. The total interest received after 3 years was ₹300. The ratio of the money lent at 5% to that lent at 8% is : **(SSC CGL Tier I 2006)**

(*a*) 5 : 8 (*b*) 15 : 16 (*c*) 8 : 5 (*d*) 16 : 15

Sol. (*d*) Annual interest = ₹300 ÷ 3 = ₹100.

Let the sum be lent at 5% p.a. Then total interest for one year = 1550 × 5% = ₹ 77.50

Short from actual interest by ₹100 – ₹77.50 = ₹22.50 = 3% (*i.e.* 8% – 5%) of second part.

∴ 3% of second part = ₹22.50 ⇒ Second part = $\frac{22.50\times100}{3}$ = ₹750

First part = 1550 – 750 = ₹ 800 ⇒ Required ratio = 800 : 750 = 16 : 15

87. Find the amount of equal instalment, annual payment of which will discharge a debt of ₹ 8600 due in 4 years at 5% p.a. simple interest.

(*a*) ₹ 1000 (*b*) ₹ 1500 (*c*) ₹ 2000 (*d*) ₹ 2500

Sol. (*c*) Instalment = $\frac{8600\times100}{100\times4+\frac{4\times3\times5}{2}}=\frac{8600\times100}{430}$ = ₹ 2,000

88. What annual instalment will discharge a debt of ₹ 6450 due in 4 years at 5% simple interest? **(SSC CGL Tier I 2010)**

(*a*) ₹ 1500 (*b*) ₹ 1835 (*c*) ₹ 1935 (*d*) ₹ 1950

Sol. (*a*) Instalment = $\frac{6450\times100}{100\times4+\frac{4\times3\times5}{2}}=\frac{6450\times100}{430}$ = ₹ 1500

89. With a given rate of simple interest, the ratio of principal and amount for a certain period of time is 4 : 5. After 3 years, with the same rate of interest, the ratio of principal and amount becomes 5 : 7. The rate of interest is : **(SSC CGL Tier I 2007)**

(*a*) 4% (*b*) 5% (*c*) 6% (*d*) 7%

Sol. (*b*)

Principal	Amount	Interest	Time
4	5	$5 - 4 = 1$	t
5	7	$7 - 5 = 2$	$t + 3$

We make principals equal in both cases by taking their LCM i.e. LCM of 4 and 5 = 20.

Principal	Amount	Interest	Time
$5 \times 4 = 20$	$5 \times 5 = 25$	$25 - 20 = 5$	t
$4 \times 5 = 20$	$4 \times 7 = 28$	$28 - 20 = 8$	$t + 3$

⇒ Interest on ₹ 20 for 3 years (i.e. $t + 3 - t$) is ₹ 3 (i.e. ₹ 8 – ₹ 5)

$$\therefore \text{ Rate of interest} = \frac{3 \times 100}{20 \times 3} = 5\% \text{ p.a.}$$

Alt: Interest in t years $= \frac{1}{4} = 25\% \Rightarrow$ Interest in $(t + 3)$ years $= \frac{2}{5} = 40\%$

Interest in 3 years = 40% – 25% = 15% ⇒ Interest in 1 year = 15% ÷ 3 = 5%

CHAPTER

21

COMPOUND INTEREST

In case of compound interest, interest due at the end of the first period is added to the principal and this amount becomes principal for the next period. Similarly, amount due at the end of each period becomes the principal for the next period. The difference between the final amount and the original principal is compound interest.

Compound interest and simple interest are same for the first period. For any period greater than one, compound interest is always greater than the simple interest for the same period.

Compound interest (C.I.) = Amount – Principal

Compound interest is compounded annually unless stated otherwise.

Compound interest for a particular year is $(100 + r)\%$ of compound interest of the previous year when interest is compounded annually.

Example. If compound interest on a certain sum of money for a particular year at 10% p.a. is ₹ 990, find compound interest for the succeeding year, assuming interest is compounded annually.

(*a*) ₹ 990 (*b*) ₹ 1000 (*c*) ₹ 1089 (*d*) ₹ 1200

Sol. (*c*) Interest for succeeding year = ₹ 990 × (100 + 10)% = ₹ 990 + ₹ 99 = ₹ 1089

Notes: If interest is compounded half-yearly, rate of interest is halved *i.e.* rate becomes $\frac{r}{2}$ and time is doubled *i.e.* time becomes $2t$, and

If interest is compounded quarterly, rate of interest is divided by 4 *i.e.* rate becomes $\frac{r}{4}$ and time is multiplied by 4 *i.e.* time becomes $4t$ and so on.

In solving problems based on compound interest, we are frequently required to solve $\left(1+\frac{r}{100}\right)$.

∴ It is better to remember the following values instead of calculating each time.

Rate = 2.5% p.a. $\Rightarrow (1 + r\%) = 1 + \frac{1}{40} = \frac{41}{40}$; Rate = 4% p.a. $\Rightarrow (1 + r\%) = 1 + \frac{1}{25} = \frac{26}{25}$

Rate = 5% p.a. $\Rightarrow (1 + r\%) = 1 + \frac{1}{20} = \frac{21}{20}$; Rate = 10% p.a. $\Rightarrow (1 + r\%) = 1 + \frac{1}{10} = \frac{11}{10}$

Rate = 20% p.a. $\Rightarrow (1 + r\%) = 1 + \frac{1}{5} = \frac{6}{5}$; Rate = 25% p.a. $\Rightarrow (1 + r\%) = 1 + \frac{1}{4} = \frac{5}{4}$

Rate = 50% p.a. $\Rightarrow (1 + r\%) = 1 + \frac{1}{2} = \frac{3}{2}$

FORMULAE

1. Amount = Principal $\left(1+\frac{\text{Rate}}{100}\right)^{\text{Time}}$

Example 1. Find the compound interest on ₹ 12,000 at 10% p.a. for 3 years, interest compounded annually.

(*a*) ₹ 1200 (*b*) ₹ 3600 (*c*) ₹ 3972 (*d*) None of these

Sol. (*c*) Amount $= 12,000\times\left(1+\frac{10}{100}\right)^3 = 12\times 11\times 11\times 11 = 132\times 11\times 11 = 1452\times 11 = 15,972$

$\therefore$ Compound interest = 15,972 – 12,000 = ₹ 3,972

Example 2: Find compound interest on ₹ 8000 at 20% p.a. for 9 months, interest being compounded quarterly.

(*a*) ₹ 1000 (*b*) ₹ 1200 (*c*) ₹ 1250 (*d*) ₹ 1261

Sol. (*d*) Principal = ₹ 8000. Time = 9 months = 3 quarters. Rate = 20% p.a. = 5% per quarter

$\therefore$ Amount $= 8000\times\frac{21}{20}\times\frac{21}{20}\times\frac{21}{20} = 9261$ $\therefore$ C.I. = 9261 – 8000 = ₹ 1261

Example 3: A certain sum will amount to ₹ 12,100 in 2 years at 10% per annum of compound interest, interest being compounded annually. The sum is **(SSC CGL Tier I 2015)**

(*a*) ₹ 12000 (*b*) ₹ 6000 (*c*) ₹ 8000 (*d*) ₹ 10000

Sol. (*d*) Principal $= 12100\div\left(\frac{11}{10}\right)^2 = 12100\times\frac{10}{11}\times\frac{10}{11}$ = ₹ 10000

2. Principal $= \dfrac{\text{Amount}}{\left(1+\frac{\text{Rate}}{100}\right)^{\text{Time}}}$

Example. A person returns ₹ 1331 after 3 years from the date of borrowing. If the rate of interest is 10% p.a., interest compounded annually, find the sum borrowed.

(*a*) ₹ 900 (*b*) ₹ 1000 (*c*) ₹ 1100 (*d*) ₹ 1200

Sol. (*b*) Principal $= \dfrac{\text{Amount}}{\left(\frac{1+\text{Rate}}{100}\right)^{\text{Time}}} = \dfrac{1331}{\left(\frac{11}{10}\right)^3} = 1331\times\left(\frac{10}{11}\right)^3$ = ₹ 1000

3. $\dfrac{\text{Amount}}{\text{Principal}} = \left(1+\frac{\text{Rate}}{100}\right)^{\text{Time}}$

Example. In how many years will ₹ 400 amount to ₹ 441 at 5% compound interest? **(SSC CGL Tier I Exam.)**

(*a*) 1 year (*b*) $1\frac{1}{2}$ years (*c*) 2 years (*d*) $2\frac{1}{2}$ years

Sol. (*c*) $\dfrac{\text{Amount}}{\text{Principal}} = \left(1+\frac{\text{Rate}}{100}\right)^{\text{Time}} \Rightarrow \frac{441}{400} = \left(\frac{21}{20}\right)^x \Rightarrow \left(\frac{21}{20}\right)^2 = \left(\frac{21}{20}\right)^x \Rightarrow x = 2$

4. If a sum of money becomes x times in n years, then it will become x^a times in an years.

Example 1. A sum borrowed under compound interest doubles itself in 10 years. When will it become fourfold of itself at the same rate of interest ? **(SSC CGL Tier I 2002)**

(*a*) 15 years (*b*) 20 years (*c*) 24 years (*d*) 40 years

Sol. (*b*) $2^2 = 4 \Rightarrow$ Required time $= 10 \times 2 = 20$ years

Example 2. A sum of money doubles itself in 4 years at compound interest. It will amount to 8 times itself at the same rate of interest in : **(SSC CGL Tier I 2006, 2002)**

(*a*) 18 years (*b*) 12 years (*c*) 16 years (*d*) 24 years

Sol. (*b*) $2^3 = 8 \Rightarrow$ The amount will become 8 times in $3 \times 4 = 12$ years

5. If rate of compound interest differs from year to year, then

$$\text{Amount} = \text{Principal} \times \frac{100 + R_1}{100} \times \frac{100 + R_2}{100} \times \frac{100 + R_3}{100} \ldots\ldots\ldots\ldots$$

Example. Find compound interest on ₹ 10000 for 3 years if rate of interest is 5%, 10% and 20% for first, second and third years respectively.

(*a*) ₹ 3310 (*b*) ₹ 3500 (*c*) ₹ 3860 (*d*) ₹ 3980

Sol. (*c*) Amount after 3 years $=$ ₹$10000 \times \frac{21}{20} \times \frac{11}{10} \times \frac{6}{5} =$ ₹ 13860

$\therefore$ Compound interest = ₹ 13860 – ₹ 10000 = ₹ 3860

6. If present value of a machine is ₹ P and rate of **annual depreciation** is r%, then

$$\text{Value of machine after } n \text{ years} = P \times \left(1 - \frac{r}{100}\right)^n$$

$$\text{Value of machine } n \text{ years before} = \frac{P}{\left(1 - \frac{r}{100}\right)^n}$$

Example. The value of a machine depreciates @ 25% p.a. If its present value is ₹ 14400, what will be its worth after 2 years.

(*a*) ₹ 8100 (*b*) ₹ 9216 (*c*) ₹ 22500 (*d*) ₹ 25600

Sol. (*a*) Value of machine after 2 years $= 14400 \times \left(1 - \frac{25}{100}\right)^2 = 14400 \times \frac{3}{4} \times \frac{3}{4} = 8100$

7. A man borrows ₹ A, which is to be discharged with compound interest @ r% p.a. in n equal yearly instalments, payable at the end of each year. Find the amount of each instalment.

Sol. Let amount of each instalment = ₹ x

Then principal of first instalment $= x \div \left(1 + \frac{r}{100}\right) = x \div \frac{a}{b} = x\left(\frac{b}{a}\right)$

Note : We substituted $\left(1 + \frac{r}{100}\right)$ with $\frac{a}{b}$

Similarly principal of second instalment $= x\left(\frac{b}{a}\right)^2$

Principal of third instalment $= x\left(\frac{b}{a}\right)^3 \Rightarrow$ Principal of nth instalment $= x\left(\frac{b}{a}\right)^n$

∴ Ratio between the principals

$$= x\left(\frac{b}{a}\right) : x\left(\frac{b}{a}\right)^2 : x\left(\frac{b}{a}\right)^3 : \ldots\ldots\ldots\ldots : x\left(\frac{b}{a}\right)^n = 1 : \frac{b}{a} : \left(\frac{b}{a}\right)^2 : \left(\frac{b}{a}\right)^3 : \ldots\ldots\ldots : \left(\frac{b}{a}\right)^{n-1}$$

Note : We have divided each ratio by $x\left(\frac{b}{a}\right)$

Example. A man borrows ₹ 2100 and undertakes to pay back with compound interest @ 10% p.a. in 2 equal yearly instalments at the end of first and second year. What is the amount of each instalment?

(*a*) ₹ 1000 (*b*) ₹ 1100 (*c*) ₹ 1200 (*d*) ₹ 1210

Sol. (*d*) $(1 + r\%) = 1 + \frac{1}{10} = \frac{11}{10}$. Ratio of principals of two instalments $= 1 : \frac{10}{11} = 11 : 10$

Sum of ratios = 11 + 10 = 21. ⇒ Principal of the first instalment $= 2100 \times \frac{11}{21} =$ ₹ 1100

∴ Instalment = Principal of the first instalment × $(1 + r\%) = 1100 \times \frac{11}{10} =$ ₹ 1210

8. Difference between compound interest on a sum for $1\frac{1}{2}$ years at $r\%$ p.a. when interest is compounded yearly and half-yearly is $P\left(1 + \frac{r}{2}\right)\left(\frac{r}{2}\right)^2$

Example. Find the difference between compound interest on ₹ 1000 for $1\frac{1}{2}$ years at 20% p.a. when interest is compounded annually and half-yearly respectively.

(*a*) ₹ 11 (*b*) ₹ 10 (*c*) ₹ 21 (*d*) ₹ 20

Sol. (*a*) Required difference $= P\left(1 + \frac{r}{2}\right)\left(\frac{r}{2}\right)^2 = 1000 \times \frac{11}{10} \times \frac{1}{10} \times \frac{1}{10} =$ ₹ 11

SOLVED EXERCISE

1. Find compound interest on ₹ 1600 at 2.5% p.a. for 2 years.

(*a*) ₹ 80 (*b*) ₹ 81 (*c*) ₹ 82 (*d*) ₹ 1681

Sol. (*b*) $(1 + r\%) = 1 + \frac{1}{40} = \frac{41}{40}$ ⇒ Amount $= 1600 \times \frac{41}{40} \times \frac{41}{40} = 1681$

∴ Compound interest = ₹ 1681 – ₹ 1600 = ₹ 81

Trick : $(1 + r\%) = 1 + \frac{1}{40} = \frac{41}{40}$ ⇒ Amount $= \left(\frac{41}{40}\right)^2$

⇒ Compound interest $= (41^2 - 40^2) =$ ₹ (41 + 40) = ₹ 81 on ₹ 40^2 = ₹ 1600

2. Find compound interest on ₹ 8000 at 5% p.a. for 3 years.

(*a*) ₹ 1200 (*b*) ₹ 1250 (*c*) ₹ 1261 (*d*) ₹ 9261

Sol. (*c*) $(1 + r\%) = 1 + \frac{1}{20} = \frac{21}{20}$ ⇒ Amount $= 8000 \times \left(\frac{21}{20}\right)^3 = 9261$

Compound interest = Amount – Principal = ₹ 9261 – ₹ 8000 = ₹ 1261

3. The present population of a city is 180000. If it increases at the rate of 10% per annum, its population after 2 years will be : **(SSC CGL Tier I 2003)**

(*a*) 207800 (*b*) 227800 (*c*) 217800 (*d*) 237800

Sol. (*c*) $(1 + r\%) = 1+\frac{1}{10}=\frac{11}{10}$ ⇒ Population after 2 years = $180000 \times \times\frac{11}{10}\times\frac{11}{10} = 217800$

4. The compound interest on ₹ 5000 for 3 years at 10% p.a will amount to **(SSC CGL Tier II 2013)**

(*a*) ₹ 1654 (*b*) ₹ 1655 (*c*) ₹ 1600 (*d*) ₹ 1565

Sol. (*b*) $1+\frac{10}{100}=\frac{11}{10}\Rightarrow\left(\frac{11}{10}\right)^3=\frac{1331}{1000}$

⇒ CI on ₹ 1000 = ₹ 1331 – ₹ 1000 = ₹ 331 ⇒ C.I. on ₹ 5000 = 5 × ₹ 331 = 1655

5. Find compound interest on ₹ 25000 at 20% p.a. for $2\frac{1}{2}$ years, if interest is compounded annually.

(*a*) ₹ 39600 (*b*) ₹ 14600 (*c*) ₹ 37500 (*d*) ₹ 12500

Sol. (*b*) Amount = $25000 \times \left(1+\frac{20}{100}\right)^2 \times \left(1+\frac{10}{100}\right)^1 = 25000 \times \left(\frac{6}{5}\right)^2 \times \frac{11}{10} = 39600$

∴ Compound interest = 39600 – 25000 = ₹ 14600

6. ₹ 10000 is borrowed at 20% p.a., interest compounded half-yearly. Find the amount repayable after one year.

(*a*) ₹ 11000 (*b*) ₹ 12000 (*c*) ₹ 12100 (*d*) ₹ 14400

Sol. (*c*) Principal = ₹ 10000. Time = 1 year = 2 half years. Rate = 20% p.a. = 10% per half-year

∴ Amount = $10000 \times\frac{11}{10}\times\frac{11}{10}=12100$

7. Sudharshan invested ₹15000 at interest @ 10% p.a. for one year. If the interest is compounded every six months, what amount will Sudharshan get at the end of the year ? **(SBI Bank PO 2002)**

(*a*) ₹ 16537.50 (*b*) ₹ 16500 (*c*) ₹ 16525.50 (*d*) ₹ 18150

Sol. (*a*) Rate of interest = 10% ÷ 2 = 5% per half-yearly. Time = 1 year = 2 half-year

∴ Amount $=15000\times\left(1+\frac{5}{100}\right)^2=15000\times\frac{21}{20}\times\frac{21}{20}=$ ₹ 16537.50

8. In what time will ₹ 1000 becomes ₹ 1331 at 10% per annum, compounded annually ? **(SSC CGL Tier I 2004, 2000 M)**

(*a*) 3 years (*b*) $2\frac{1}{2}$ years (*c*) 2 years (*d*) $3\frac{1}{2}$ years

Sol. (*a*) $\frac{\text{Amount}}{\text{Principal}}=\left(1+\frac{\text{Rate}}{100}\right)^{\text{Time}} \Rightarrow \frac{1331}{1000}=\left(1+\frac{10}{100}\right)^t \Rightarrow \left(\frac{11}{10}\right)^3=\left(\frac{11}{10}\right)^t \Rightarrow t = 3$ years

9. The population of a village increases @ 5% p.a. If present population is 8000, after how many years the population will be 9261?

(*a*) 1 years (*b*) 3 years (*c*) $1\frac{1}{2}$ years (*d*) $2\frac{1}{2}$ years

Sol. (*b*) $\frac{\text{Amount}}{\text{Principal}} = \left(1+\frac{\text{Rate}}{100}\right)^{\text{Time}} \Rightarrow \frac{9261}{8000} = \left(\frac{21}{20}\right)^3 \Rightarrow \text{Time} = 3 \text{ years}$

10. ₹ 8000, invested at compound interest, gives ₹ 1261 as interest after 3 years. The rate of interest per annum is : **(SSC CGL Tier I 2006)**

(*a*) 5% (*b*) 10% (*c*) 20% (*d*) 25%

Sol. (*a*) Amount after 3 years = 8000 + 1261 = ₹ 9261

$$\frac{\text{Amount}}{\text{Principal}} = \left(1+\frac{\text{Rate}}{100}\right)^{\text{Time}} \Rightarrow \frac{9261}{8000} = \left(\frac{21}{20}\right)^3 = \left(1+\frac{1}{20}\right)^3 \Rightarrow \text{Rate of interest} = \frac{1}{20} = 5\%$$

11. At what rate will a sum of ₹1,000 amount to ₹1,102.50 in 2 years at Compound interest ? **(SSC CGL Tier I 2014)**

(*a*) 5% (*b*) 5.5% (*c*) 6% (*d*) 6.5%

Sol. (*a*) $\frac{\text{Amount}}{\text{Principal}} = \frac{1102.5}{1000} = \frac{11025}{10000} = \frac{441}{400} = \left(\frac{21}{20}\right)^2 \Rightarrow \text{Rate} = \frac{21}{20} - 1 = \frac{1}{20} = 5\%$

12. The compound interest on ₹ 30000 at 7% per annum for a certain time is ₹ 4347. The time is **(SSC CHSL 2012)**

(*a*) 2 years (*b*) 2.5 years (*c*) 3 years (*d*) 4 years

Sol. (*a*) Amount = 30000 + 4347 = 34347

$$\frac{\text{Amount}}{\text{Principal}} = \frac{34347}{30000} = \frac{11449}{10000} = \left(\frac{107}{100}\right)^2 = \left(1+\frac{7}{100}\right)^2 \Rightarrow \text{Time} = 2 \text{ years}$$

13. A man invested ₹ 16000 at compound interest for 3 years, interest compounded annually. If he got ₹ 18522 at the end of 3 years, what is the rate of interest?

(*a*) 4% (*b*) 5% (*c*) 6% (*d*) 7%

Sol. (*b*) $(1 + r\%)^3 = \frac{18522}{16000} = \frac{9261}{8000} = \left(\frac{21}{20}\right)^3 = \left(1+\frac{1}{20}\right)^3 \Rightarrow \text{Rate of interest} = \frac{1}{20} = 5\%$

Trick: Compound interest = ₹ 18522 – ₹ 16000 = ₹ 2522

Let the amount is invested at 1% p.a. SI. Then SI of 3 years = 16000 × 1% × 3 = ₹ 480

∴ Rate of interest = $\frac{2522}{480}$ = 5 (Remainder is ₹ 122). ∴ Rate of interest is 5% p.a.

Note: If rate is 6%, then simple interest = 480 × 6 = 2880, which is more than the given compound interest and therefore is not possible.

14. The time in which ₹ 80,000 amounts to ₹ 92,610 at 10% p.a. at compound interest, interest being compounded semi annually is : **(SSC CGL Tier I 2013)**

(*a*) 3 years (*b*) $1\frac{1}{2}$ years (*c*) 2 years (*d*) $2\frac{1}{2}$ years

Sol. (*b*) $\left(1+\frac{\text{Rate}}{100}\right)^{\text{Time}} = \frac{\text{Amount}}{\text{Principal}}$. Rate of interest = 5% = $\frac{1}{20}$

$$\therefore (1 + r\%)^3 = \frac{92610}{80000} = \frac{9261}{8000} = \left(\frac{21}{20}\right)^3 \Rightarrow \text{Time} = 3 \text{ half years} = 1\frac{1}{2} \text{ years}$$

15. In how many years will ₹ 2,000 amount to ₹ 2,420 at 10% per annum compound interest ? **(SAS 2010, SSC CGL Tier I 2000)**

(*a*) 3 (*b*) $2\frac{1}{2}$ (*c*) 2 (*d*) $1\frac{1}{2}$

Sol. (*c*) $(1 + r\%)^2 = \frac{2420}{2000} = \frac{121}{100} = \left(\frac{11}{10}\right)^2 \Rightarrow$ Time = 2 years

16. In what time will ₹ 1000 amount to ₹ 1331 at 20% per annum, compounded half yearly ? **(SSC CGL Tier I 2003)**

(*a*) $1\frac{1}{2}$ years (*b*) 2 years (*c*) 1 year (*d*) $2\frac{1}{2}$ years

Sol. (*a*) Rate of interest = 20% p.a. = 10% per half-year

$\therefore \frac{1331}{1000} = \left(1+\frac{10}{100}\right)^{time} = \left(\frac{11}{10}\right)^{time} \Rightarrow \left(\frac{11}{10}\right)^3 = \left(\frac{11}{10}\right)^{time} \Rightarrow$ Time = 3 half-years = $1\frac{1}{2}$ years

17. A sum of ₹ 3200 invested at 10% p.a. compounded quarterly amount to ₹ 3,362. Compute the time period. **(SSC CGL Tier II 2013)**

(*a*) $\frac{1}{2}$ year (*b*) 1 year (*c*) 2 years (*d*) $\frac{3}{4}$ year

Sol. (*a*) $\frac{\text{Amount}}{\text{Principal}} = \frac{3362}{3200} = \frac{1681}{1600} = \left(\frac{41}{40}\right)^2 \Rightarrow$ Time = 2 quarters = $\frac{1}{2}$ year

18. A sum of money on compound interest amounts to ₹ 9,680 in 2 years and to ₹ 10,648 in 3 years. What is the rate of interest per annum ? **(CDS 2011)**

(*a*) 5% (*b*) 10% (*c*) 15% (*d*) 20%

Sol. (*b*) Interest for third year = 10648 – 9680 = ₹ 968 ⇒ Rate of interest = $\frac{968}{9680} \times 100 = 10\%$

19. A sum of money invested at compound interest amounts to ₹ 650 at the end of first year and ₹ 676 at the end of second year. The sum of money is : **(SSC Cons. (GD) 2012, SSC CGL 2002)**

(*a*) ₹ 600 (*b*) ₹ 540 (*c*) ₹ 625 (*d*) ₹ 560

Sol. (*c*) $(1 + r\%) = \frac{676}{650} = \frac{26}{25}$ $\therefore$ Principal = $650 \div \frac{26}{25} = 650 \times \frac{25}{26} =$ ₹ 625

20. A sum becomes ₹ 1352 in 2 years at 4% per annum compound interest. The sum is : **(SSC CGL Tier I 2005)**

(*a*) ₹ 1225 (*b*) ₹ 1270 (*c*) ₹ 1245 (*d*) ₹ 1250

Sol. (*d*) $(1 + r\%) = 1+\frac{4}{100} = \frac{26}{25} \Rightarrow$ Principal $= 1352 \div \left(\frac{26}{25}\right)^2 = 1352 \times \frac{25}{26} \times \frac{25}{26} =$ ₹ 1250

21. What sum lent at 5% per annum compound interest will amount to ₹ 441 in 2 years ? **(SSC CGL Tier I 2000)**

(*a*) ₹ 390 (*b*) ₹ 400 (*c*) ₹ 395 (*d*) ₹ 405

Sol. (*b*) $P = 441 \div \left(1+\frac{r}{100}\right)^t = 441 \div \left(1+\frac{5}{100}\right)^2 = 441 \times \frac{20}{21} \times \frac{20}{21} =$ ₹ 400

22. A certain sum, invested at 4% p.a. compound interest, compounded half-yearly, amounts to ₹ 7803 at the end of one year. The sum is : **(SSC CGL Tier I 2008)**

(*a*) ₹ 7000 (*b*) ₹ 7200 (*c*) ₹ 7500 (*d*) ₹ 7700

Sol. (*c*) Rate of interest = 2% per half-yearly

$(1 + r\%) = \left(1+\frac{2}{100}\right) = \frac{51}{50}$ $\therefore$ Principal $= 7803 \div \left(\frac{51}{50}\right)^2 = 7803 \times \left(\frac{50}{51}\right)^2 =$ ₹ 7500

23. At what rate per cent per annum will ₹ 2304 amount to ₹ 2500 in 2 years at compound interest? **(SSC CGL Tier I 2005)**

(*a*) $4\frac{1}{2}\%$ (*b*) $4\frac{1}{5}\%$ (*c*) $4\frac{1}{6}\%$ (*d*) $4\frac{1}{3}\%$

Sol. (*c*) $\frac{2500}{2304} = \left(1+\frac{r}{100}\right)^2 \Rightarrow \left(\frac{50}{48}\right)^2 = \left(1+\frac{r}{100}\right)^2 \Rightarrow \frac{r}{100} = \frac{50}{48} - 1 = \frac{1}{24}$

$\therefore$ Rate of interest $= \frac{1}{24} \times 100 = \frac{25}{6} = 4\frac{1}{6}\%$

24. At what rate percent per annum of compound interest, compounded annually, will a sum be 16 times of itself in 4 years ?

(*a*) 400% (*b*) 200% (*c*) 125% (*d*) 100%

Sol. (*d*) Let principal = ₹ 1, then amount = ₹ 16

$\therefore (1+r\%)^4 = \frac{16}{1} = 2^4 \Rightarrow 1 + r\% = 2 \Rightarrow r\% = 1 = 100\%$

25. A sum of money becomes eight times of itself in 3 years at compound interest. The rate of interest per annum is : **(SSC CGL Tier I 2004)**

(*a*) 100% (*b*) 80% (*c*) 20% (*d*) 10%

Sol. (*a*) Let principal be ₹ 1 and rate of interest be $r\%$ p.a.

$(1 + r\%)^3 = 8 = 2^3 \Rightarrow 1 + r\% = 2 \Rightarrow r\% = 1 = 100\%$

26. Find the rate percent of compound interest at which a sum of money will amount to $\frac{9}{4}$ times of itself in 2 years?

(*a*) 10% (*b*) 25% (*c*) 50% (*d*) 75%

Sol. (*c*) $(1 + r\%)^2 = \frac{9}{4} = \left(\frac{3}{2}\right)^2 \Rightarrow (1 + r\%) = \frac{3}{2} \Rightarrow r\% = \frac{3}{2} - 1 = \frac{1}{2} = 50\%$

27. If a sum of money compounded annually becomes 1.44 times of itself in 2 years, then the rate of interest per annum is **(SSC CGL Tier II 2013)**

(*a*) 25% (*b*) 22% (*c*) 21% (*d*) 20%

Sol. (*d*) $(1 + r\%)^2 = 1.44 \Rightarrow (1 + r\%) = \sqrt{1.44} = 1{\cdot}2 \Rightarrow$ Rate $= 1.2 - 1 = 0.2 = 20\%$

28. If the amount is 2.25 times of the sum after 2 years at compound interest (compounded annually), the rate of interest per annum is : **(SSC CGL Tier I 1999)**

(*a*) 25% (*b*) 30% (*c*) 45% (*d*) 50%

Sol. (*d*) $(1 + r\%)^2 = 2.25 \Rightarrow (1 + r\%) = \sqrt{2.25} = 1.5 \Rightarrow r\% = 1.5 - 1 = 0.5 = 50\%$

29. A sum of money invested at a certain rate of interest, compounded annually, doubles in 2 years. Find the rate of interest p.a.

(*a*) $100(\sqrt{2} - 1)\%$ (*b*) 50% (*c*) $100(\sqrt{2} + 1)\%$ (*d*) $50(\sqrt{2} + 1)\%$

Sol. (*a*) $\left(1+\frac{r}{100}\right)^2 = 2 \Rightarrow 1+\frac{r}{100} = \sqrt{2} \Rightarrow \frac{r}{100} = \sqrt{2}-1 \Rightarrow r = 100(\sqrt{2}-1)\%$

30. A man borrowed ₹ 5000 at 10% per annum compound interest. At the end of each year he has repaid ₹ 1500. The amount of money, he still owes after the third year, is : **(SSC CGL Tier I 2006)**

(*a*) ₹ 1600 (*b*) ₹ 1690 (*c*) ₹ 1700 (*d*) ₹ 1790

Sol. (*b*) Amount due after 1 year = ₹ 5000 + 10% of ₹ 5000 – 1500 = 5500 – 1500 = ₹ 4000

Amount due after 2 years = ₹ 4000 + 10% of ₹ 4000 – 1500 = 4400 – 1500 = ₹ 2900

Amount due after 3 years = ₹ 2900 + 10% of ₹ 2900 – 1500 = 3190 – 1500 = ₹ 1690

31. What sum of money will earn compound interest ₹ 1261 at 5% p.a. for 3 years ?

(*a*) ₹ 4000 (*b*) ₹ 6000 (*c*) ₹ 8000 (*d*) ₹ 9000

Sol. (*c*) $1+\frac{r}{100} = \frac{21}{20} \Rightarrow \left(\frac{21}{20}\right)^3 = \frac{9261}{8000}$ ⇒ C.I. on ₹ 8000 = ₹ 9261 – ₹ 8000 = ₹ 1261

32. What sum will earn ₹ 820 compound interest lent at 5% p.a. for 2 years?

(*a*) ₹ 4000 (*b*) ₹ 8000 (*c*) ₹ 8200 (*d*) None of these

Sol. (*b*) ₹ 400 amount to ₹ 441 in 2 years at 5% p.a.

∴ C.I. on ₹ 400 = ₹ 41 ⇒ Required principal = ₹ $820 \times \frac{400}{41}$ = ₹ 8000

33. The principal on which compound interest for 3 years at 5% amounts to ₹ 630.50, is

(*a*) ₹ 4000 (*b*) ₹ 4500 (*c*) ₹ 6000 (*d*) ₹ 8000

Sol. (*a*) Rate = 5% ⇒ $1 + r\% = \frac{21}{20} \Rightarrow \left(\frac{21}{20}\right)^3 = \frac{9261}{8000}$

⇒ Principal ₹ 8000 earns CI = 9261 – 8000 = ₹ 1261 (*i.e.* twice of 630.50)

∴ Principal to earn interest ₹ 630.50 = 8000 ÷ 2 = ₹ 4000

34. A sum of money invested at compound interest doubles itself in 6 years. At the same rate of interest it will amount to eight times of itself in : **(SSC CGL Tier I 2004, 2010M)**

(*a*) 15 years (*b*) 12 years (*c*) 18 years (*d*) 10 years

Sol. (*c*) $2^3 = 8$ ⇒ Required time = 3 × 6 = 18 years

35. A certain sum of money invested at a certain rate of compound interest triples in 7 years. In how many years will it become 81 times?

(*a*) 28 years (*b*) 36 years (*c*) 72 years (*d*) 77 years

Sol. (*a*) $3^4 = 81$ ⇒ The amount will become 81 times in 4 × 7 = 28 years

36. Mr. Rao invests a sum of ₹ 41,250 at the rate of 6% p.a. What approximate amount of compound interest will he obtain at the end of 3 years ? **(SBI Bank PO 2008)**

(*a*) ₹ 8100 (*b*) ₹ 7425 (*c*) ₹ 8210 (*d*) ₹ 7879

Sol. (*d*) Amount = $41250 \times \frac{106}{100} \times \frac{106}{100} \times \frac{106}{100}$ = ₹ 49129 (approx.)

∴ Compound interest = ₹ 49129 – ₹ 41250 = ₹ 7879

Approximate Method: Compound interest = simple interest + interest on S.I.

S.I. for one year = ₹ 41250 × 6% = ₹ 2475 ⇒ S.I. for 3 years = ₹ 2475 × 3 = ₹ 7425

Interest on interest = 6% of ₹ 2475 + 6% of ₹ (2475 × 2) + interest on interest
= 150 + 300 + interest on ₹ 150 (approx.) = 450 + interest on ₹ 150

∴ C.I. = ₹ 7425 + ₹ 450 + ₹ 9 = ₹ 7884

Out of given options, amount closest and little less to it is ₹ 7879.

37. What approximate compound interest can be obtained on an amount of ₹ 5850 at the rate of 8% p.a. at the end of 3 years ? **(RBI Note Printing Asstt. Manager 2006)**

(*a*) ₹ 1519 (*b*) ₹ 1429 (*c*) ₹ 1442 (*d*) ₹ 1500

Sol. (*a*) Compound interest = simple interest + interest on simple interest

S.I. for one year = ₹ 5850 × 8% = ₹ 468 ⇒ S.I. for 3 years = ₹ 468 × 3 = ₹ 1404

Interest on interest = 8% of ₹ 468 + 8% of ₹ (468 × 2) + interest on interest
= 37 + 75 + interest on ₹37 = ₹112 + ₹3 = ₹115 ⇒ C.I. = ₹ 1404 + ₹ 115 ₹75 ≈ ₹1519

38. Find the compound interest on ₹ 12,000 @ 5% p.a. for three years. **(Bank Clerical)**

(*a*) ₹ 1261 (*b*) ₹ 1,891.50 (*c*) ₹ 9261 (*d*) ₹ 13,891.50

Sol. (*b*) At 5% p.a., ₹ 8000 amounts to ₹ 9261 in 3 years.

∴ Compound interest on ₹ 8000 is 9261 – 8000 = ₹ 1261

∴ Compound interest on ₹ 12000 = $1261 \times \frac{12000}{8000}$ = ₹ 1891.50

39. The effective annual rate of interest equivalent to 8% p.a. interest compounded half yearly is :

(*a*) 8.4% (*b*) 8.8% (*c*) 8.16% (*d*) None of these

Sol. (*c*) Half yearly rate of interest = 8 ÷ 2 = 4%. Interest on interest = $\frac{4 \times 4}{100}$ = 0.16 %

∴ Equivalent interest rate = 8 + 0.16 = 8.16%

40. The effective annual rate of interest, corresponding to a nominal rate of 6% per annum payable half yearly, is : **(SSC CGL Tier I 2005)**

(*a*) 6.06% (*b*) 6.07% (*c*) 6.08% (*d*) 6.09%

Sol. (*d*) Half yearly rate of interest = 6 ÷ 2 = 3%. Interest on interest = $\frac{3 \times 3}{100}$ = 0.09%

∴ Equivalent interest rate = 6 + 0.09 = 6.09%

41. An automobile financer claims to be lending money at simple interest, but he includes the interest every six months for calculating the principal. If he is charging an interest at the rate of 10%, the effective rate of interest becomes **(CDS 2015)**

(*a*) 10.25% (*b*) 10.5% (*c*) 10.75% (*d*) 11%

Sol. (*a*) Half yearly rate of interest = 10 ÷ 2 = 5%. Interest on interest = $\frac{5 \times 5}{100}$ = 0.25%

∴ Equivalent interest rate = 10 + 0.25 = 10.25%

42. If compound interest on a certain sum at 10% p.a. for the first year is ₹ 95. Find the compound interest for the second year.

(*a*) ₹ 96 (*b*) ₹ 100 (*c*) ₹ 104.50 (*d*) ₹ 105

Sol. (*c*) Compound interest for second year = ₹ 95 + 10% of ₹ 95 = ₹ 95 + ₹ 9.50 = ₹ 104.50

43. An amount is lent at 15% p.a. compound interest for 2 years. Find the percent increase in the amount at the end of 2 years.

(*a*) 22.5% (*b*) 30% (*c*) 32.25% (*d*) 35.5%

Sol. (*c*) Increase of 15% and 15% again = 15 + 15 + $\frac{15 \times 15}{100}$ = 32.25%

44. If compound interest for the second year on a certain sum at 10% p.a. is ₹ 132, find the principal. **(SSC CGL Tier II 2014)**

(*a*) ₹ 600 (*b*) ₹ 1000 (*c*) ₹ 1100 (*d*) ₹ 1200

Sol. (*d*) Let principal = ₹ 100. Then amount after the first year = 110

Amount after the second year = $100 \times \left(\frac{11}{10}\right)^2$ = ₹ 121

∴ Compound interest for second year = ₹ 121 – ₹ 110 = ₹ 11

But actual compound interest for second year = ₹ 132 (*i.e.* 12 times ₹ 11)

∴ Principal = 12 × ₹ 100 = ₹ 1200

Alt: Let principal be ₹ 100. Then interest for the first year = 10% of ₹ 100 = ₹ 10

Compound interest for the second year = 10 × (100 + 10)% = ₹ 11

Trick: C.I. for second year = $10+\frac{10\times10}{100}=11\%$ = ₹ 132

45. If compound interest on a certain sum for 2 years at 10% p.a. is ₹ 315, what is the sum lent?

(*a*) ₹ 1500 (*b*) ₹ 1800 (*c*) ₹ 2000 (*d*) ₹ 2250

Sol. (*a*) Let principal = ₹ 100. Then amount after 2 years = ₹ 100 × $(1.1)^2$ = ₹ 121

C.I. = ₹ 121 – ₹ 100 = ₹ 21 ⇒ Required principal = $315\times\frac{100}{21}$ = ₹ 1500

46. The sum of money that yields a compound interest of ₹ 420 during the second year at 5% p.a. is **(SSC CGL Tier I 2012)**

(*a*) ₹ 4,000 (*b*) ₹ 42,000 (*c*) ₹ 8,000 (*d*) ₹ 21,000

Sol. (*c*) Let principal = ₹ 100. Then amount after the first year = 105

Amount after the second year = $100 \times \left(\frac{105}{100}\right)^2$ = ₹ 110.25

∴ CI for second year = ₹110.25 – ₹105 = ₹5.25 ⇒ Actual principal = $\frac{420}{5.25}$ × ₹100 = ₹8000

Trick: C.I. for second year = $5+\frac{5\times5}{100}=5.25\%$

47. The principal amount which yields a compound interest ₹ 208 in the second year at 4% is **(SSC CGL Tier II 2012)**

(*a*) ₹ 13,000 (*b*) ₹ 6,500 (*c*) ₹ 5,000 (*d*) ₹ 10,000

Sol. (*c*) Let principal be ₹ 100. The interest for the first year = 4% of ₹ 100 = ₹ 4

Compound interest for the second year = 4 + 4 × 4% = ₹ 4.16

But actual compound interest for second year = ₹ 208 (*i.e.* 50 times ₹ 4.16)

∴ Principal = 50 × ₹ 100 = ₹ 5,000

48. What is the least number of completed years in which a sum of money at 20% compound interest will be more than doubled. **(NIFT 2003, CDS 2011)**

(*a*) 7 (*b*) 6 (*c*) 5 (*d*) 4

Sol. (*d*) $(1 + r\%) = \frac{6}{5}$. We consider $\left(\frac{6}{5}\right)^n$. $\left(\frac{6}{5}\right)^4=\frac{1296}{625}>2$ ⇒ Required time = 4 years

49. Find the least number of completed years in which a sum of money lent out at 50% p.a. compound interest, will be more than 5 times of the principal. **(MBA Entrance)**

(*a*) 3 years (*b*) 4 years (*c*) 5 years (*d*) 6 years

Sol. (*c*) $(1 + r\%) = \frac{3}{2}$. We consider $\left(\frac{3}{2}\right)^n \cdot \left(\frac{3}{2}\right)^4 = \frac{81}{16} > 5 \Rightarrow$ Required time = 5 years

50. The value of a machine depreciates every year by 10%. If its present value is ₹ 50,000 then the value of the machine after 2 years is **(SSC CGL Tier I 2013)**

(*a*) ₹ 40,500 (*b*) ₹ 40,050 (*c*) ₹ 45,000 (*d*) ₹ 40,005

Sol. (*a*) Value of machine after 2 years $= 50000 \times \left(1 - \frac{10}{100}\right)^2 = 50000 \times \frac{9}{10} \times \frac{9}{10} =$ ₹ 40,500

51. Height of a tree is 1210 cm. What was its height two years ago, if it increases at the rate of 10% p.a.?

(*a*) 1331 cm (*b*) 1000 cm (*c*) 900 cm (*d*) None of these

Sol. (*b*) Two years ago, length of the tree $= \frac{1210}{\left(1 + \frac{10}{100}\right)^2} = 1210 \times \frac{10}{11} \times \frac{10}{11} = 1000$ cm

52. The value of a machine depreciates every year at the rate of 10% on its value at the beginning of that year. If the current value of the machine is ₹ 729, its worth 3 years ago was : **(SSC CGL Tier I 2013)**

(*a*) ₹ 800 (*b*) ₹ 1000 (*c*) ₹ 750.87 (*d*) ₹ 947.10

Sol. (*b*) $(1 - r\%) = 1 - \frac{1}{10} = \frac{9}{10}$ $729 = 9^3$ $10^3 = 1000$

53. The value of a machine depreciates @ 25% p.a. If its present value is ₹ 14400. Find the value of machine 2 years ago.

(*a*) ₹ 8100 (*b*) ₹ 9216 (*c*) ₹ 22500 (*d*) ₹ 25600

Sol. (*d*) Value of machine 2 years ago $= \frac{14400}{\left(1 - \frac{25}{100}\right)^2} = \frac{14400}{\left(\frac{3}{4}\right)^2} = 14400 \times \frac{4}{3} \times \frac{4}{3} = 25600$

54. A man bought a land for ₹ 72900 and spent ₹ 133100 on constructing a building on it. If value of land appreciates 10% p.a. and that of building depreciates 10% p.a., after how many years will the values of both be the same.

(*a*) 2 years (*b*) 3 years (*c*) 4 years (*d*) 5 years

Sol. (*b*) Let the required time = *n* years. Then $133100 \times \left(\frac{90}{100}\right)^n = 72900 \times \left(\frac{110}{100}\right)^n$

$$\Rightarrow \frac{133100}{72900} = \left(\frac{110}{100} \times \frac{100}{90}\right)^n \Rightarrow \frac{1331}{729} = \left(\frac{11}{9}\right)^n \Rightarrow \left(\frac{11}{9}\right)^3 = \left(\frac{11}{9}\right)^n \Rightarrow \text{Time} = 3 \text{ years}$$

55. A certain sum, invested at compound interest becomes ₹ 3000 in 3 years and ₹ 6000 in 6 years. Find the amount after 9 years.

(*a*) ₹7500 (*b*) ₹9000 (*c*) ₹10000 (*d*) ₹12000

Sol. (*d*) In 3 years (*i.e.* 6 years – 3 years), principal doubles itself *i.e.* from ₹ 3000 to ₹ 6000. In the next 3 years, amount will become double of ₹6000.

∴ Amount after 9 years = 2 × ₹ 6000 = ₹12000

56. If a certain sum on compound interest becomes ₹ 2000 in 2 years and ₹ 2200 in 3 years, find rate of interest p.a. **(Bank Clerical 2006)**

(*a*) 6.25% (*b*) 10% (*c*) 12.5% (*d*) 25%

Sol. (*b*) In case of C.I., amount after 2 years will become principal for the third year.

∴ Principal = ₹ 2000. Time = 3 years – 2 years = 1 year

We know that C.I. for 1 year (compounded annually) is equal to S.I. for 1 year.

∴ S.I. = ₹ 2200 – ₹ 2000 = ₹ 200 ⇒ Rate of interest $= \frac{200 \times 100}{2000} = 10\%$ p.a.

57. Compound interest on a certain sum for two successive years is ₹ 150 and ₹ 153 respectively. Find rate of interest.

(*a*) 2% (*b*) 3% (*c*) 5% (*d*) 6%

Sol. (*a*) Difference in compound interest of two successive years = ₹ 153 – ₹ 150 = ₹ 3

The difference is equal to simple interest on ₹ 150 for one year.

∴ Rate of interest $= \frac{3 \times 100}{150} = 2\%$ p.a.

58. A sum of money invested at compound interest amount in 3 years to ₹ 2400 and in 4 years to ₹ 2520. The interest rate per annum is : **(SSC CGL Tier I 2002)**

(*a*) 5% (*b*) 6% (*c*) 10% (*d*) 12%

Sol. (*a*) Difference in compound interest of two successive years = ₹ 2520 – ₹ 2400 = ₹ 120

The difference is equal to simple interest on ₹ 2400 for one year.

∴ Rate of interest $= \frac{120 \times 100}{2400} = 5\%$ p.a.

59. If a certain sum on compound interest becomes ₹ 2000 in 2 years and ₹ 2420 in 4 years. Rate of interest p.a. is:

(*a*) 6.25% (*b*) 10% (*c*) 12.5% (*d*) 25%

Sol. (*b*) In 2 years (i.e. 4 – 2), ₹ 2000 becomes ₹ 2420.

$$\left(1 + \frac{\text{Rate}}{100}\right)^{\text{Time}} = \frac{\text{Amount}}{\text{Principal}} \Rightarrow (1 + r\%)^2 = \frac{2420}{2000} = \frac{121}{100} = \left(\frac{11}{10}\right)^2$$

$$\Rightarrow 1 + r\% = \frac{11}{10} \Rightarrow r\% = \frac{11}{10} - 1 = \frac{1}{10} = 10\%$$

60. A certain sum of money amounts to ₹ 640 after 2 years and ₹ 720 after 3 years. How much will it amount to after 4 years ?

(*a*) ₹ 760 (*b*) ₹ 800 (*c*) ₹ 810 (*d*) ₹ 825

Sol. (*c*) In 1 year (*i.e.* 3 – 2), the amount becomes $\frac{720}{640}$ times of the principal.

∴ In next 1 year ₹ 720 will amount to $720 \times \frac{720}{640} =$ ₹ 810

61. Population of a town increases at a certain rate per cent per annum. Present population of the town is 3600 and in 5 years time it becomes 4800. How much will it be in 10 years time?

(*a*) 5000 (*b*) 6000 (*c*) 6400 (*d*) 7000

Sol. (*c*) Let rate of increase in population = r % p.a.

Then $4800 = 3600 \times \left(1+\frac{r}{100}\right)^5 \Rightarrow \left(1+\frac{r}{100}\right)^5 = \frac{4800}{3600} = \frac{4}{3}$

Population after next 5 years will become $4800 \times \frac{4}{3} = 6400$

62. ₹ 6000 becomes ₹ 9000 in 5 years at a certain rate of compound interest. In 10 years at the same rate of interest, a sum of ₹ 8000 will become :

(*a*) ₹ 15,000 (*b*) ₹ 12,000 (*c*) ₹ 13,500 (*d*) ₹ 18,000

Sol. (*d*) In 5 years, ₹ 6000 becomes ₹ 9000, i.e. $\frac{9000}{6000} = \frac{3}{2}$ times of itself.

In 10 years, ₹ 8000 will become ₹ $8000 \times \left(\frac{3}{2}\right)^2$ = ₹ 18000

63. A sum of money becomes ₹ 4500 after two years and ₹ 6750 after 4 years on compound interest. The sum is : **(SSC CGL Tier I 2002)**

(*a*) ₹ 4000 (*b*) ₹ 2500 (*c*) ₹ 3000 (*d*) ₹ 3050

Sol. (*c*) In 2 years (*i.e.* 4 – 2), ₹ 4500 becomes ₹ 6750 *i.e.* $\frac{6750}{4500} = \frac{3}{2}$ of itself

∴ Principal = $4500 \div \frac{3}{2} = 4500 \times \frac{2}{3}$ = ₹ 3000

64. A sum of money amounts to ₹ 2880 in 2 years and 3456 in 3 years at compound interest. Find the sum.

(*a*) ₹ 1800 (*b*) ₹ 2000 (*c*) ₹ 2400 (*d*) ₹ 2500

Sol. (*b*) ₹ 2880 amounts to ₹ 3456 in one year.

⇒ The sum amounts to $\frac{3456}{2880} = \frac{6}{5}$ times of itself in one year

∴ Principal = $2880 \div \left(\frac{6}{5}\right)^2 = 2880 \times \frac{5}{6} \times \frac{5}{6}$ = ₹ 2000

65. An amount is invested in a bank at compound rate of interest. The total amount, including interest, after first and third year is ₹ 1200 and ₹ 1587 respectively. What is the rate of interest? **(SSC CGL Tier II 2012)**

(*a*) 12% (*b*) 15% (*c*) 10% (*d*) 3.9%

Sol. (*b*) The sum amounts to $\frac{1587}{1200} = \frac{529}{400} = \left(\frac{23}{20}\right)^2$ times of itself in two years

∴ Rate of interest = $\frac{23}{20} - 1 = \frac{3}{20}$ = 15% per annum

66. Present population of a town is 12000. Increasing at a certain rate, the population becomes 27000 after six years. What will the size of population after 3 years from now ?

(*a*) 15000 (*b*) 18000 (*c*) 19500 (*d*) 20000

Sol. (*b*) Population becomes $\frac{27000}{12000} = \frac{9}{4}$ times of itself in 6 years.

$\therefore$ In 3 years it will become $\sqrt{\frac{9}{4}} = \frac{3}{2}$ times $\Rightarrow$ Population after 3 years = $12000 \times \frac{3}{2} = 18000$

67. Last year, the population of a town was x and if it increases at the same rate, next year it will be y. The present population of the town is **(SAS 2010)**

(*a*) $\frac{x+y}{2}$ (*b*) $\frac{y-x}{2}$ (*c*) $\frac{2xy}{x+y}$ (*d*) $\sqrt{xy}$

Sol. (*d*) Let present population be p. Then $x : p = p : y \Rightarrow p^2 = xy \Rightarrow p = \sqrt{xy}$

68. A man borrows a certain sum and pays it back in two equal annual instalments. If he pays back ₹ 676 annually and rate of interest is 4%, what is the sum borrowed ?

(*a*) ₹ 1352 (*b*) ₹ 1300 (*c*) ₹ 1275 (*d*) ₹ 1250

Sol. (*c*) $(1 + r\%) = 1 + \frac{1}{25} = \frac{26}{25}$ $\Rightarrow$ Principal of the first instalment $= 676 \times \frac{25}{26} =$ ₹ 650

Principal of the second instalment $= 676 \times \left(\frac{25}{26}\right)^2 =$ ₹ 625

$\therefore$ Sum borrowed = ₹ 650 + ₹ 625 = ₹ 1275

69. The difference between the compound interest on a certain sum for the second year and the third year at 5% p.a. is ₹ 10.50. Find the sum.

(*a*) ₹ 200 (*b*) ₹ 210 (*c*) ₹ 2000 (*d*) ₹ 4000

Sol. (*d*) The difference between interest for consecutive years is due to interest on interest.

$\therefore$ Interest for the second year = ₹ 10.50 ÷ 5% = ₹ 210

Let C.I. for the first year = ₹ 100. Then C.I. for the second year = ₹ 105

If C.I. for the second year is ₹210, then C.I. for the first year = ₹ $210 \times \frac{100}{105}$ = ₹200

Principal to earn interest of ₹ 200 @ 5% p.a. = ₹ $200 \times \frac{100}{5}$ = ₹ 4000

70. A man buys a scooter on making cash down payment of ₹ 16,224 and promises to pay two more yearly instalments of equivalent amounts at the end of first year and second year. If the rate of interest is 4% per annum, compunded annually, the cash value of the scooter, is : **(SSC CGL Tier I 2006)**

(*a*) ₹ 40,000 (*b*) ₹ 46,000 (*c*) ₹ 46,824 (*d*) ₹ 50,000

Sol. (*c*) Cash down payment = ₹ 16,224. Here $\left(1 + \frac{r}{100}\right) = \left(1 + \frac{4}{100}\right) = \frac{26}{25}$

$\therefore$ Principal of the first instalment $16224 \div \left(\frac{26}{25}\right)^1 = 16224 \times \left(\frac{25}{26}\right)^1 =$ ₹ 15600

Principal of the second instalment $= 16224 \div \left(\frac{26}{25}\right)^2 = 15600 \times \frac{25}{26} =$ ₹ 15000

$\therefore$ Cash value of the scooter = ₹ (16224 + 15600 + 15000) = ₹ 46,824

71: A builder borrows ₹ 2550 to be paid back with compound interest at the rate of 4% per annum by the end of 2 years in two equal yearly instalments. How much will each instalment be ? **(SSC CGL Tier I 2000)**

(*a*) ₹ 1352 (*b*) ₹ 1377 (*c*) ₹ 1275 (*d*) ₹ 1283

Sol. (*a*) $1 + r\% = 1 + \frac{4}{100} = \frac{26}{25}$

Ratio of principals of two instalments = $1 : \frac{25}{26} = 26 : 25 \Rightarrow$ Sum of ratios = 26 + 25 = 51

Principal of the first instalment = $\frac{26}{51} \times 2550 =$ ₹ 1300

Amount of each instalment = $\frac{26}{25} \times 1300 =$ ₹ 1352

Alternative method : Let amount of each instalment be ₹ x.

Then principal = $\left[x \div \frac{26}{25}\right] + \left[x \div \left(\frac{26}{25}\right)^2\right] = x \times \frac{25}{26} \times \left(1 + \frac{25}{26}\right) = \frac{25}{26} \times \frac{51}{26} \times x = 2550$

$\Rightarrow x = 2550 \times \frac{26}{25} \times \frac{26}{51} =$ ₹ 1352

72. Kamal took ₹ 6800 as a loan which along with interest is to be repaid in two equal annual instalments. If the rate of interest is $12\frac{1}{2}\%$, compounded annually, then the value of each instalment is : **(SSC CGL Tier I 2008)**

(*a*) ₹ 8100 (*b*) ₹ 4150 (*c*) ₹ 4050 (*d*) ₹ 4000

Sol. (*c*) Each instalment = $\frac{Amount}{\left(\frac{100}{100+r}\right) + \left(\frac{100}{100+r}\right)^2} = \frac{6800}{\left(\frac{8}{9}\right) + \left(\frac{8}{9}\right)^2} = 6800 \times \frac{81}{136} =$ ₹ 4050

Alternative method : $(1 + r\%) = 1 + \frac{1}{8} = \frac{9}{8}$

Ratio of principals of two instalments = $1 : \frac{8}{9} = 9 : 8 \Rightarrow$ Sum of ratios = 9 + 8 = 17

∴ Principal of the first instalment = $\frac{9}{17} \times 6800 =$ ₹ 3600

∴ Each instalment = Principal of the first instalment × $(1 + r\%) = 3600 \times \frac{9}{8} =$ ₹ 4050

73. A man borrows ₹ 1820 and undertakes to pay back with compound interest @ 20% p.a. in 3 equal yearly instalments at the end of first, second and third years. What is the amount of each instalment?

(*a*) ₹ 500 (*b*) ₹ 560 (*c*) ₹ 750 (*d*) ₹ 864

Sol. (*d*) $(1 + r\%) = 1 + \frac{1}{5} = \frac{6}{5}$

Ratio of principals for three years = $1 : \frac{5}{6} : \left(\frac{5}{6}\right)^2 = 6^2 : 6 \times 5 : 5^2 = 36 : 30 : 25$

Sum of ratios = 36 + 30 + 25 = 91 $\Rightarrow$ Principal of first instalment = $\frac{36}{91} \times 1820 =$ ₹ 720

∴ Instalment = Principal of first instalment × $(1 + r\%) = 720 \times \frac{6}{5} =$ ₹ 864

74. A father divides ₹ 5100 between his two sons, Mohan and Sohan who are 23 and 24 at present in such a way that if their shares are invested at compound interest @ 4% p.a., they will receive equal amount on attaining the age of 26 years. Find Mohan's share.

(*a*) ₹ 2400 (*b*) ₹ 2500 (*c*) ₹ 2550 (*d*) ₹ 2600

Sol. (*b*) $(1 + r\%) = 1 + \frac{1}{25} = \frac{26}{25}$. Let Mohan and Sohan receive ₹ x and ₹ y respectively.

$$x\left(\frac{26}{25}\right)^3 = y\left(\frac{26}{25}\right)^2 \Rightarrow \frac{x}{y} = \left(\frac{26}{25}\right)^{2-3} = \left(\frac{26}{25}\right)^{-1} = \frac{25}{26} \Rightarrow \text{Mohan's share} = \frac{25}{51} \times ₹5100 = ₹2500$$

75. Divide ₹ 15,494 between A and B so that A's share at the end of 9 years may be equal to B's share at the end of 11 years, compound interest being 20% per annum. Then A's share is

(SSC CGL Tier II 2012)

(*a*) ₹ 9,144 (*b*) ₹ 9,414 (*c*) ₹ 8,000 (*d*) ₹ 9,140

Sol. (*a*) $(1 + r\%) = 1 + \frac{1}{5} = \frac{6}{5}$. Let A and B received ₹ x and ₹ y respectively at present.

$$\text{Then } x\left(\frac{6}{5}\right)^9 = y\left(\frac{6}{5}\right)^{11} \Rightarrow \frac{y}{x} = \left(\frac{6}{5}\right)^{11-9} = \left(\frac{6}{5}\right)^2 = \frac{36}{25} \Rightarrow \text{A's share} = \frac{36}{61} \times ₹15494 = ₹9144$$

CHAPTER

22

RELATION BETWEEN C.I. AND S.I.

FORMULAE

1. Difference between compound interest and simple interest for two years :

Difference = Principal × $\left(\frac{\text{Rate}}{100}\right)^2 = Pr^2$ $\Rightarrow$ Principal = Difference × $\left(\frac{100}{\text{Rate}}\right)^2$

Proof : Compound interest for first year = simple interest for first year

CI for second year = SI for first year + Interest on SI for the first year

$\therefore$ Difference = Interest on simple interest first year

Proof : Compound interest for two years = $P(1 + r)^2 - P$

Simple interest for two years = $P \times r \times 2 = 2Pr$

Difference between compound interest and simple interest for two years

$= [P(1 + r)^2 - P] - (2Pr) = P(1 + 2r + r^2) - P - (2Pr) = P + 2Pr + Pr^2 - P - 2Pr = Pr^2$

$\Rightarrow$ Difference between compound interest and simple interest @ r% p.a. for two years = $P(r^2)$

Example 1. The difference between compound interest and simple interest on a sum of ₹4000 for 2 years at 5% payable yearly is: **(SSCCGLTierI2012)**

(*a*) ₹ 20 (*b*) ₹ 10 (*c*) ₹ 50 (*d*) ₹ 60

Sol. (*b*) Required difference = Pr^2 = ₹ 4000 × $\frac{1}{20} \times \frac{1}{20}$ = ₹ 10

Example 2. The difference between simple and compound interest on a certain sum of money at 5% p.a. for two years is ₹ 160. Find the principal. **(SSC CHSL 2011)**

(*a*) ₹ 60000 (*b*) ₹ 40000 (*c*) ₹ 64000 (*d*) ₹ 48000

Sol. (*c*) $r = 5\% = \frac{1}{20} \Rightarrow \frac{1}{r} \Rightarrow$ Principal = Difference × $\left(\frac{1}{r}\right)^2$ = ₹ 160 × 20 × 20 = ₹ 64000

2. Difference between CI and SI for 3 years = $Pr^2(3 + r)$

Example. Find the difference between compound interest and simple interest on ₹ 1000 for 3 years at 10% p.a., if interest is compounded annually.

(*a*) ₹ 30 (*b*) ₹ 31 (*c*) ₹ 50 (*d*) ₹ 331

Sol. (*b*) Difference = $Pr^2(3 + r)$ = ₹ 1000 × $\frac{1}{10} \times \frac{1}{10} \times \left(3 + \frac{1}{10}\right)$ = ₹ 31

Alternative Method: $r = \frac{1}{10} \Rightarrow 1 + r = \frac{11}{10}$

Let principal = 10^3 = ₹ 1000. Then compound interest = $11^3 - 10^3 = 1331 - 1000$ = ₹ 331

Simple interest for 3 years @ 10% on ₹ 1000 = ₹ 300 $\Rightarrow$ Difference = ₹ 31

3. Difference between CI and SI for 4 years = $Pr^2(6 + 4r + r^2) = Pr^2[(r + 2)^2 + 2]$

Example. Find the difference between compound interest and simple interest on ₹ 10000 for 4 years at 10% p.a., if interest is compounded annually.

(*a*) ₹ 4641 (*b*) ₹ 4000 (*c*) ₹ 641 (*d*) ₹ 600

Sol. (*c*) Difference between compound interest and simple interest for 4 years

$$= Pr^2(6 + 4r + r^2) = 10000 \times \frac{1}{10} \times \frac{1}{10} \times \left(6 + \frac{4}{10} + \frac{1}{100}\right) = 10000 \times \frac{1}{100} \times \frac{641}{100} = ₹\ 641$$

Alternative Method: $r = \frac{1}{10} \Rightarrow 1 + r = \frac{11}{10} \Rightarrow (1 + r)^4 = \left(\frac{11}{10}\right)^4$

Let principal = 10^4 = ₹ 10,000. Then CI = $11^4 - 10^4$ = 14,641 – 10,000 = ₹ 4,641

SI for 4 years @ 10% on ₹ 10,000 = ₹ 4,000 ⇒ Required difference = ₹ 641

4. How to **find compound interest** for two years, when **simple interest is given**:

Method: C.I. = Total simple interest + (Simple interest for one year × Rate of Interest)

Example. If simple interest on a certain sum at 10% p.a. for 2 years is ₹ 80. The compound interest on the same sum for the same period is: **(SSC CGL Tier I)**

(*a*) ₹ 168 (*b*) ₹ 164 (*c*) ₹ 84 (*d*) ₹ 88

Sol. (*c*) Simple interest for one year = ₹ 80 ÷ 2 = ₹ 40

C.I. = Simple interest × Interest on simple interest for one year = 80 + 40 × 10% = ₹ 84

5. Difference between Compound Interest and Simple interest, when Simple interest is given:

(*a*) For 2 years: $\frac{1}{2}r$ of SI (*b*) For 3 years: $\frac{1}{3}r(3 + r)$ of SI

Proof: (*a*) Difference between CI and SI for 2 years = $P(r^2) = Pr(r)$

But Pr = Principal × $\frac{\text{Rate}}{100}$ = Simple interest for one year = $\frac{1}{2}$ × Total simple interest

∴ Difference between CI and SI for 2 years = $Pr(r) = \frac{1}{2} \times \text{SI} \times (r) = \frac{1}{2}r$ of SI

(*b*) Difference between CI and SI for 3 years = $Pr^2(3 + r) = Pr\,(r)\,(3 + r)$

Now Pr = Principal × $\frac{\text{Rate}}{100}$ = SI for one year = $\frac{1}{3}$ × Total Simple Interest

∴ Difference between compound interest and simple interest for 3 years

$$= Pr(r)\,(3 + r) = \frac{1}{3} \times \text{Simple Interest} \times (r)\,(3 + r) = \frac{1}{3}r\,(3 + r) \text{ of Simple interest}$$

6. How to find simple interest when compound interest is given?

Method: If CI – SI = $\frac{1}{x}$ of simple interest, then Difference = $\frac{1}{x+1}$ of compound interest.

Hint: Compound Interest = Simple Interest + Difference

⇒ If we shift base from SI to CI, smaller fraction is required and vice versa.

SOLVED EXERCISE

1. Find the difference between compound interest and simple interest on ₹ 4000 for 1 year at 10% p.a., if the interest is compounded half-yearly.

(*a*) Nil (*b*) ₹ 5 (*c*) ₹ 10 (*d*) ₹ 40

Sol. (*c*) Since interest is compounded half-yearly, rate of interest is halved and time is doubled.

$\therefore$ Rate $= \frac{10}{2}\% = 5\% = \frac{1}{20}$ and time $= 2 \times 1 = 2$ half-years.

$\therefore$ Required difference $=$ ₹ $4000 \times \frac{1}{20} \times \frac{1}{20} =$ ₹ 10

2. The difference between simple and compound interest on a certain sum of money for 2 years at 4% p.a. is ₹ 1. The sum of money is : **(SSC CGL Tier I 2005, SSC CHSL 2011)**

(*a*) ₹ 600 (*b*) ₹ 625 (*c*) ₹ 560 (*d*) ₹ 650

Sol. (*b*) $r = 4\% = \frac{4}{100} = \frac{1}{25} \Rightarrow$ Principal = Difference $\times \left(\frac{1}{r}\right)^2 =$ ₹ $1 \times 25 \times 25 =$ ₹ 625

3. The difference between simple and compound interest on a certain sum of money at 5% rate of interest per annum for 2 years is ₹ 15, then the sum is : **(SSC CGL Tier I 2002)**

(*a*) ₹ 6500 (*b*) ₹ 5500 (*c*) ₹ 6000 (*d*) ₹ 7000

Sol. (*c*) $r = \frac{5}{100} = \frac{1}{20} \Rightarrow \frac{1}{r} = 20 \Rightarrow$ Principal = Difference $\times \left(\frac{1}{r}\right)^2 = 15 \times 20 \times 20 =$ ₹ 6000

4. If the difference between the compound interest, compounded every six months, and the simple interest on a certain sum of money at the rate of 12% per annum for one year is ₹ 36, the sum is : **(SSC CGL Tier I 2000, 2002 M)**

(*a*) ₹ 10,000 (*b*) ₹ 12,000 (*c*) ₹ 15,000 (*d*) ₹ 9,000

Sol. (*a*) $r\% = 12\% \div 2 = 6\% \Rightarrow$ Sum lent $= 36 \times \frac{100}{6} \times \frac{100}{6} =$ ₹ 10,000

5. If the difference between the compound interest and the simple interest on a certain sum for 2 years at 8% per annum is ₹ 32, then the sum is : **(SSC CHSL 2012)**

(*a*) ₹ 5,000 (*b*) ₹ 5,500 (*c*) ₹ 6,000 (*d*) ₹ 5,250

Sol. (*a*) Principal = ₹ $32 \times \frac{100}{8} \times \frac{100}{8} =$ ₹ 5000

6. If the difference between the simple interest and compound interest on a certain sum of money at 4% per annum in 2 years is ₹ 10, then the sum is equal to : **(CDS 2003)**

(*a*) ₹ 4000 (*b*) ₹ 5000 (*c*) ₹ 6250 (*d*) ₹ 8000

Sol. (*c*) Principal $= 10 \times \frac{100}{4} \times \frac{100}{4} =$ ₹ 6250

7. The difference between compound interest (compounded annually) and simple interest on a sum of ₹ 1000 at a certain rate of interest for 2 years is ₹ 10. The rate of interest per annum is: **(SSC CGL Tier I 2004)**

(*a*) 5% (*b*) 6% (*c*) 10% (*d*) 12%

Sol. (*c*) Difference $= Pr^2 \Rightarrow 10 = 1000 \times r^2 \Rightarrow r^2 = \frac{10}{1000} = \frac{1}{100} = \left(\frac{1}{10}\right)^2 \Rightarrow r = \frac{1}{10} = 10\%$

8. The difference between simple interest and compound interest at the same rate for ₹ 4000 for 2 years is ₹ 160. The rate of interest is :

(*a*) 15% (*b*) 20% (*c*) 25% (*d*) Data insufficient

Sol. (*b*) Difference = $Pr^2 \Rightarrow r^2 = \frac{160}{4000} = \frac{1}{25} \Rightarrow$ Rate $= \sqrt{\frac{1}{25}} = \frac{1}{5} = 20\%$

9. The difference between the compound interest and simple interest for the amount ₹ 5000 in 2 years is ₹ 32. The rate of interest is **(SSC CGL Tier I 2011)**

(*a*) 5% (*b*) 8% (*c*) 10% (*d*) 12%

Sol. (*b*) Difference = $Pr^2 \Rightarrow r^2 = \frac{32}{5000} = \frac{4}{625} \Rightarrow$ Rate $= \sqrt{\frac{4}{625}} = \frac{2}{25} = 8\%$

10. What will be the difference in simple and compound interest on ₹ 2000 after three years at the rate of 10% p.a. ? **(Bank PO 2002)**

(*a*) ₹ 160 (*b*) ₹ 42 (*c*) ₹ 62 (*d*) ₹ 20

Sol. (*c*) Difference = $Pr^2(3 + r)$ = ₹ 2000 × $\frac{1}{10} \times \frac{1}{10} \times \left(3 + \frac{1}{10}\right)$ = ₹ 62

Alternative Method: $r = \frac{1}{10} \Rightarrow 1 + r = \frac{11}{10} \Rightarrow (1 + r\%)^3 = \left(\frac{11}{10}\right)^3 = \frac{1331}{1000}$

Compound interest for 3 years = 2000 × $\frac{1331}{1000}$ – 2000 = 2662 – 2000 = ₹662

SI for three years = 2000 × 10% × 3 = ₹ 600 ⇒ Required difference = ₹ 662 – ₹ 600 = ₹ 62

11. B borrowed ₹5,000 from A at 6% p.a. simple interest and lends it to C at compound interest of 10% p.a. If B collects the money back from C after 2 years and repays A, the profit made by B in the transaction is **(SSC CGL Tier I 2012)**

(*a*) ₹ 500 (*b*) ₹ 450 (*c*) ₹ 1050 (*d*) ₹ 600

Sol. (b) S.I. = 5000 × 6% × 2 = ₹ 600; C.I. = 5000 × $\left(\frac{11}{10} \times \frac{11}{10} - 1\right)$ = 5000 × $\frac{21}{100}$ = ₹ 1050

⇒ Profit made by B = ₹(1050 – 600) = ₹450

12. What sum will give ₹ 244 as the difference between simple interest and compound interest at 10% in $1\frac{1}{2}$ years compounded half yearly? **(SSC CGL Tier II 2013)**

(*a*) ₹ 40000 (*b*) ₹ 36000 (*c*) ₹ 32000 (*d*) ₹ 28000

Sol. (*c*) Interest-rate = 5% ⇒ $\left(\frac{21}{20}\right)^3 = \frac{9261}{8000} \Rightarrow$ C.I = 9261 – 8000 = 1261

S.I. on 8000 = $\frac{8000 \times 5 \times 3}{100}$ = 1200 ⇒ C.I. – S.I. = 1261 – 1200 = 61

If difference = ₹ 244, Principal = 8000 × $\frac{244}{61}$ = ₹ 32,000

13. There is 100% increase to an amount in 8 years, at simple interest. Find the compound interest of ₹ 8000 after 2 years at the same rate of interest. **(SSC CGL Tier I 2013)**

(*a*) ₹ 2500 (*b*) ₹ 2000 (*c*) ₹ 2250 (*d*) ₹ 2125

Sol. (*d*) Rate of interest (simple) = $\frac{100}{8}\%$ = 12.5% p.a.

Amount in one year at Compound interest : $(1 + r\%) = 1 + \frac{1}{8} = \frac{9}{8}$

Amount after 2 years = ₹ $8000 \times \frac{9}{8} \times \frac{9}{8}$ = ₹ 10125 ⇒ CI = ₹ 10125 – ₹ 8000 = ₹ 2125

14. If simple interest on a certain sum for 2 years @ 10% p.a. is ₹ 200, the compound interest will be :

(*a*) ₹ 200 (*b*) ₹ 210 (*c*) ₹ 220 (*d*) ₹ 225

Sol. (*b*) C.I. = S.I. + $\frac{\text{S.I.}}{2}$ × Rate of interest = ₹ 200 + $\frac{200}{2} \times 10\%$ = ₹ 210

15. The simple interest accrued on an amount of ₹ 20,000 at the end of three years is ₹ 6000. What would be the compound interest accrued on the same amount at the same rate in the same period ? **(Bank PO Exam.)**

(*a*) ₹ 3300 (*b*) ₹ 3310 (*c*) ₹ 6620 (*d*) None of these

Sol. (*c*) Annual interest = ₹ 6000 ÷ 3 = ₹ 2000 ⇒ Rate of interest = $\frac{2000 \times 100}{20000 \times 1}$ = 10% p.a.

Compound interest = $20000 \times \frac{11}{10} \times \frac{11}{10} \times \frac{11}{10} - 20000 = 26620 - 20000$ = ₹ 6620

16. The compound interest on a sum for 2 years is ₹ 832 and the simple interest on the same sum at the same rate for the same period is ₹ 800. what is the rate of interest ? **(CDS 2009)**

(*a*) 6% (*b*) 8% (*c*) 10% (*d*) 12%

Sol. (*b*) Simple interest for one year = ₹ 800 ÷ 2 = ₹ 400

Difference between compound interest and simple interest = ₹ 832 – ₹ 800 = ₹ 32

∴ ₹ 32 is interest on simple interest for one year *i.e.* interest on ₹ 400.

∴ Rate of interest = $\frac{32 \times 100}{400 \times 1}$ = 8% p.a.

17. The compound interest on a certain sum for two years is ₹ 618 whereas the simple interest on the same sum at the same rate for two years is ₹ 600. The rate of interest per annum is : **(SSC CGL 2000)**

(*a*) 3% (*b*) 6% (*c*) 9% (*d*) 18%

Sol. (*b*) S.I. for one year = 600 ÷ 2 = ₹ 300 ⇒ Extra CI for second year = 618 – 600 = ₹ 18

⇒ Rate of interest = $\frac{18 \times 100}{300}$ = 6%

18. At a certain rate per annum, the simple interest on a sum of money for one year is ₹260 and compound interest for 2 years for the same sum at same rate of interest is ₹540.80. The rate of interest is **(SSC CGL Tier I 2008)**

(*a*) 4% (*b*) 6% (*c*) 8% (*d*) 10%

Sol. (*c*) S.I. for 2 years = 2 × ₹ 260 = ₹520. C.I. for 2 years = ₹ 540.80.

Difference = ₹540.80 – ₹520 = ₹20.80 which is one year simple interest on ₹260.

∴ Rate of interest = $\frac{20.80 \times 100}{260 \times 1}$ = 8% p.a.

19. On a certain sum, compound interest for 2 years is ₹ 4200 and simple interest on same sum for 3 years at same rate of interest is ₹ 6000. Find the rate of interest.

(*a*) 5% (*b*) 7.5% (*c*) 10% (*d*) 12.5%

Sol. (*c*) Simple interest for one year = ₹ 6000 ÷ 3 = ₹ 2000

Compound interest for second year = ₹ 4200 – ₹ 2000 = ₹ 2200

∴ ₹ 200 is interest on simple interest on ₹ 2000 for one year.

∴ Rate of interest = $\frac{200 \times 100}{2000}$ = 10% p.a.

20. If simple interest and compound interest on a certain sum for 2 years is ₹ 160 and ₹ 168 respectively. What is the principal?

(*a*) ₹ 800 (*b*) ₹ 1600 (*c*) ₹ 2000 (*d*) Data Insufficient

Sol. (*a*) Simple interest for one year = ₹ 160 ÷ 2 = ₹ 80

∴ ₹ 8 (*i.e.* ₹ 168 – ₹ 160) is interest on simple interest for one year *i.e.* interest on ₹ 80.

∴ Rate of interest = $\frac{8 \times 100}{80}$ = 10% p.a. ⇒ Principal = $\frac{80 \times 100}{10 \times 1}$ = ₹ 800

Alternative method : ₹ 8 is S.I. on ₹ 80 for one year. (Note : 8 × 10 = 80)

⇒ Principal to earn ₹ 80 as interest = ₹ 80 × 10 = ₹ 800

21. On a certain sum, compound interest and simple interest for two years are ₹ 500 and ₹ 520 respectively. Find the principal. **(SSC CGL Tier I 2006 M)**

(*a*) ₹ 2500 (*b*) ₹ 3125 (*c*) ₹ 5000 (*d*) ₹ 12500

Sol. (*b*) Difference between compound interest and simple interest = ₹ 520 – ₹ 500 = ₹ 20

∴ ₹ 20 is simple interest on simple interest for one year *i.e.* interest on ₹ 250 for one year.

∴ Rate of interest = $\frac{20 \times 100}{250 \times 1}$ = 8% p.a. ⇒ Principal = $\frac{250 \times 100}{8 \times 1}$ = ₹ 3125

Alternative method : ₹ 20 is S.I. on ₹ 250 for one year.

⇒ Principal to earn ₹ 250 as interest = $\frac{250}{20} \times 250$ = ₹ 3125

22. If the simple interest on a certain sum of money for 3 years is ₹ 225 and the compound interest on the same sum at the same rate for 2 years is ₹ 153, then the principal invested (in ₹) is **(SSC CPO SI 2012)**

(*a*) 3000 (*b*) 1875 (*c*) 1500 (*d*) 2250

Sol. (*b*) SI for one year = ₹ 225 ÷ 3 = ₹ 75 ⇒ Extra CI for second year = 153 – 150 = ₹ 3

If interest is ₹ 3, then principal = ₹ 75 (i.e. 25 times)

⇒ If interest is ₹ 75, then principal = 75 × 25 = ₹ 1875

23. A money-lender borrows money at 4% per annum and pays the interest at the end of the year. He lends it at 6% per annum compound interest, compounded half yearly and receives the interest at the end of the year. In this way, he gains ₹ 104.50 a year. The amount of money he borrows, is **(SSC CGL Tier I 2007)**

(*a*) ₹ 6000 (*b*) ₹ 5500 (*c*) ₹ 5000 (*d*) ₹ 4500

Sol. (*c*) Let the principal be ₹ 100. Then S.I. for 1 year = ₹ 100 × 4% = ₹ 4

Compound interest for 1 year = 3 + 3 + $\frac{3 \times 3}{100}$ = ₹ 6.09

Difference between compound interest and simple interest = ₹ 6.09 – ₹ 4 = ₹ 2.09

But actual difference = ₹ 104.50 ⇒ Actual principal = $\frac{104.50}{2.09}$ × 100 = ₹ 5000

24. On what sum does the difference between the compound interest and the simple interest for 3 years at 10% is ₹ 31 ? **(SSC CHSL 2011, SSC CGL Tier I 2000)**

(*a*) ₹ 1500 (*b*) ₹ 1200 (*c*) ₹ 1100 (*d*) ₹ 1000

Sol. (*d*) $r = \frac{1}{10} \Rightarrow 1 + r = \frac{11}{10} \Rightarrow (1+r)^3 = \left(\frac{11}{10}\right)^3 = \frac{1331}{1000}$

Let principal = 10^3 = ₹ 1,000. Then C.I. = $11^3 - 10^3 = 1331 - 1000$ = ₹ 331

Simple interest on ₹ 1000 = 1000 × 10% × 3 = ₹ 300

Difference = ₹ 331 – ₹ 300 = ₹ 31 which is equal to the required difference.

∴ Required Principal = ₹ 1000

25. The difference between compound and simple rates of interest on ₹ 10000 for 3 years at 5% p.a. is **(SSC CGL Tier II 2012)**

(*a*) ₹ 76.50 (*b*) ₹ 76 (*c*) ₹ 76.25 (*d*) ₹ 76.75

Sol. (*c*) Difference = $Pr^2(3+r)$ = ₹ $10{,}000 \times \frac{1}{20} \times \frac{1}{20} \times \left(3 + \frac{1}{20}\right)$ = ₹ 76.25

Alternative Method: $r = \frac{1}{20} \Rightarrow 1 + r = \frac{21}{20}$

Let principal = 20^3 = ₹ 8000. Then C.I. = $21^3 - 20^3 = 9261 - 8000$ = ₹ 1261

Simple interest on ₹ 8000 = ₹ 1200. Difference of interests on ₹ 8000 = ₹ 61

Difference of interests on ₹ 10,000 = ₹ 61 × $\frac{10000}{8000}$ = ₹ 76.25

26. The simple interest on a certain sum for 2 years is ₹ 800 and the compound interest on the same sum for the same period is ₹ 840. Then the difference between the compound interest and simple interest for 3 years will be :

(a) ₹ 60 (b) ₹ 80 (c) ₹ 124 (d) ₹ 136

Sol. (c) Simple interest for one year = ₹ 800 ÷ 2 = ₹ 400

⇒ S.I. on ₹ 400 for one year = 840 – 800 = 40 ⇒ Rate of interest = $\frac{40 \times 100}{400}$ = 10% p.a.

Difference of C.I. and S.I. for third year is equal to S.I. on ₹ 840 for 1 year = 10% of 840 = 84

Total difference = 40 + 84 = ₹ 124

27. The difference between compound interest and simple interest on a certain sum of money for 3 years at 5% p.a. is ₹122. Find the principal. **(SSC CGL Tier I 2008 & 2002 modified)**

(*a*) ₹ 8,000 (*b*) ₹ 10,000 (*c*) ₹ 12,000 (*d*) ₹ 16,000

Sol. (*d*) $r\% = \frac{1}{20} \Rightarrow 1 + r\% = \frac{21}{20} \Rightarrow (1 + r\%)^3 = \left(\frac{21}{20}\right)^3$

Let principal = 20^3 = ₹ 8000. Then compound interest = $21^3 - 20^3$ = 9,261 – 8,000 = ₹ 1261

S.I. on ₹ 8000 = 8000 × 5% × 3 = ₹1,200 ⇒ Difference = ₹ 1261 – ₹ 1200 = ₹ 61

∴ Required principal = ₹ 122 × $\frac{8000}{61}$ = ₹ 16,000

28. If simple interest for 3 years @ 10% is ₹ 900, then compound interest is:

(*a*) ₹ 933 (*b*) ₹ 966 (*c*) ₹ 993 (*d*) ₹ 1000

Sol. (*c*) Here $r\% = \frac{1}{10} \Rightarrow 1 + r\% = \frac{11}{10} \Rightarrow (1 + r\%)^3 = \left(\frac{11}{10}\right)^3$

Let principal = 10^3 = ₹ 1,000. Then C.I. = $11^3 - 10^3$ = 1,331 – 1,000 = ₹ 331
Simple interest on ₹ 1000 = ₹ 100 × 10% × 3 = ₹ 300
If simple interest is ₹ 300, then compound interest is ₹ 331.

If simple interest is ₹ 900, then compound interest = $\frac{331}{300} \times 900$ = ₹ 993

Alt : Principal $= \frac{900 \times 100}{3 \times 10}$ = ₹ 3000 ⇒ C.I. = $3000 \times \left(\frac{11}{10}\right)^3 - 3000 = 3993 - 3000$ = ₹993

29. Compound interest on a certain sum for 2 years at 3% per annum, compounded annually, is ₹ 101.50. The simple interest on the same sum at the same rate and for the same time duration will be : **(SSC CGL Tier I 2006)**

(*a*) ₹ 96 (*b*) ₹ 80 (*c*) ₹ 90 (*d*) ₹ 100

Sol. (*d*) Let SI for first year = ₹ 100 ⇒ SI for second year = ₹ 100 ⇒ SI for 2 years = ₹ 200
CI for first year = ₹ 100 ⇒ CI for second year = ₹ 100 + 3% of ₹ 100 = ₹ 103
Total compound interest for two years = ₹ (100 + 103) = ₹ 203
If compound interest is ₹ 203, simple interest = ₹ 200

If compound interest is ₹ 101.50, simple interest = $\frac{200}{203} \times 101.50$ = ₹100

30. The compound interest on a certain sum of money for 2 years at 5% per annum is ₹ 410. The simple interest on the same sum at the same rate and for the same time is : **(SSC CGL Tier I 2014)**

(*a*) ₹ 400 (*b*) ₹ 300 (*c*) ₹ 350 (*d*) ₹ 405

Sol. (*d*) Let SI for first year = ₹ 100 ⇒ SI for second year = ₹ 100 ⇒ SI for 2 years = ₹ 200
⇒ CI for first year = ₹ 100 ⇒ CI for second year = ₹ 100 + 5% of ₹ 100 = ₹ 105
Total compound interest for two years = ₹ (100 + 105) = ₹ 205
If compound interest is ₹ 205, simple interest = ₹ 200

If compound interest is ₹ 410, simple interest = $\frac{200}{205} \times 410$ = ₹400

31. If compound interest on a certain sum for 2 years @ 5% p.a. is ₹ 328, the simple interest for the same sum at the same rate for the same period will be?

(*a*) ₹ 300 (*b*) ₹ 310 (*c*) ₹ 320 (*d*) ₹ 325

Sol. (*c*) Suppose C.I. for the first year = ₹ 100. Then C.I. for the second year = ₹ 105
Total compound interest for two years = (₹ 100 + ₹ 105) = ₹ 205
And simple interest for two years = 2 × ₹ 100 = ₹ 200
If compound interest is ₹ 205, simple interest = ₹ 200

If compound interest is ₹ 328, simple interest = ₹ 328 × $\frac{200}{205}$ = ₹ 320

32. If the compound interest on a certain sum for 2 years at $12\frac{1}{2}$% per annum is ₹ 510, the simple interest on the same sum at the same rate for the same period of time is : **(SSC CGL 2004)**

(*a*) ₹ 400 (*b*) ₹ 480 (*c*) ₹ 450 (*d*) ₹ 460

Sol. (*b*) Let C.I. for the first year = ₹ 200. Then C.I. for the second year = ₹ 225

Total CI for two years = (₹ 200 + ₹ 225) = ₹ 425 and SI for two years = 2 × ₹ 200 = ₹ 400

If CI is ₹ 425, then SI = ₹ 400 ⇒ If CI is ₹ 510, then SI = ₹ 510 × $\frac{400}{425}$ = ₹ 480

33. If compound interest on a certain sum for 3 years at 10% is ₹ 662, find simple interest on that sum for 3 years.

(*a*) ₹ 570 (*b*) ₹ 600 (*c*) ₹ 625 (*d*) ₹ 630

Sol. (*b*) ₹ 1 will become ₹ $\left(\frac{11}{10}\right)^3$ or $\frac{1331}{1000}$ in 3 years ⇒ ₹ 1000 will become ₹ 1,331.

∴ Compound interest on ₹ 1,000 = 1,331 – 1,000 = ₹ 331

And simple interest on ₹ 1000 for 3 years at 10% = ₹ 100 × 10% × 3 = ₹ 300

If compound interest is ₹ 331, simple interest = ₹ 300

If compound interest is ₹ 662, simple interest = ₹ $\frac{300}{331}$ × 662 = ₹ 600

34. If compound interest on a certain sum for 3 years at 20% is ₹ 728, find simple interest for the same sum for 3 years.

(*a*) ₹ 600 (*b*) ₹ 650 (*c*) ₹ 700 (*d*) ₹ 750

Sol. (*a*) $r = \frac{1}{5} \Rightarrow 1 + r = \frac{6}{5} \Rightarrow (1 + r)^3 = \left(\frac{6}{5}\right)^3$

Let principal = 5^3 = ₹ 125. Then compound interest = $6^3 - 5^3$ = 216 – 125 = ₹ 91

Simple interest on ₹ 125 for 3 years at 20% p.a. = ₹ 125 × 20% × 3 = ₹ 75

If compound interest is ₹ 91, then simple interest is ₹ 75.

If compound interest is ₹ 728, then simple interest = $\frac{75}{91}$ × 728 = ₹ 600

35. The compound interest on a certain sum of money at 5% per annum for 2 years is ₹ 246. The simple interest on the same sum for 3 years at 6% p.a. is **(SSC CGL Tier I 2008)**

(*a*) ₹ 435 (*b*) ₹ 450 (*c*) ₹ 430 (*d*) ₹ 432

Sol. (*d*) $1 + r\% = 1 + \frac{1}{20} = \frac{21}{20} \Rightarrow (1 + r\%)^2 = \left(\frac{21}{20}\right)^2 = \frac{441}{400}$

⇒ ₹ 400 earns ₹ 41 as compound interest in 2 years.

∴ Principal to earn ₹ 246 as interest = ₹ 246 × $\frac{400}{41}$ = ₹ 2400

Simple interest on ₹ 2400 for 3 years at 6% p.a. = ₹ 2400 × 6% × 3 = ₹ 432.

CHAPTER 23

TRUE DISCOUNT AND BANKER DISCOUNT

Sum Due (S.D.) or Amount : The amount payable on the due date is Sum Due or Amount.
Present Worth (P.W.) : Suppose an amount is due after some time. If the amount is paid now, less money is required to be paid now. The money paid now is called Present Worth.

True Discount (T.D.) : True Discount is the interest on the Present Worth.

True Discount = Sum Due – Present Worth ⇒ Sum Due = Present Worth + True Discount

$$\text{True Discount} = \frac{\text{Present worth} \times \text{Rate} \times \text{Time}}{100} = \frac{\text{Sum Due} \times \text{Rate} \times \text{Time}}{100 + (\text{Rate} \times \text{Time})}$$

Example 1: Find the True Discount for the Present Worth of ₹ 5000 for a period of 6 months at 12% p.a.

Sol. True Discount $= \frac{5000 \times 12 \times 6}{100 \times 12} = ₹\ 300$

Example 2: The true discount on a sum of money due 2 years hence at 5% is ₹ 15. Find the sum. **(SSC FCI Asstt. 2013)**

(*a*) ₹ 150 (*b*) ₹ 165 (*c*) ₹ 170 (*d*) ₹ 160

Sol. (*a*) $15 = \frac{\text{P.W.} \times 5 \times 2}{100} \Rightarrow \text{P.W.} = \frac{15 \times 100}{5 \times 2} = ₹\ 150$

If interest is simple, $\text{Present Worth} = \frac{\text{Sum Due} \times 100}{100 + (\text{Rate} \times \text{Time})}$

Example. What is the present worth of ₹ 1200 due 4 years hence at 5% p.a. simple interest?

(*a*) ₹ 1000 (*b*) ₹ 900 (*c*) ₹ 800 (*d*) ₹ 1500

Sol. (*a*) $\text{Present Worth} = \frac{\text{Sum Due} \times 100}{100 + (\text{Time} \times \text{Rate})} = \frac{1200 \times 100}{100 + (5 \times 4)} = ₹\ 1000$

If interest is compounded, then : $\text{Sum Due} = \text{Present Worth} \times \left(1 + \frac{r}{100}\right)^{time}$

$$\text{Present Worth} = \text{Sum Due} \div \left(1 + \frac{r}{100}\right)^{n}$$

Bankers Discount (B.D) : Banker Discount is simple interest on the Sum Due for the due period. $\text{Banker Discount} = \frac{\text{Sum Due} \times \text{Rate} \times \text{Time}}{100}$

Banker Gain (B.G.) : Banker Gain is the difference between the simple interest on the Sum Due and that on the Present Worth. Banker Gain is the simple interest on the True Discount.
Bankers Gain = Baker Discount – True Discount

$$\text{Sum Due} = \frac{\text{Banker Discount} \times \text{True Discount}}{\text{Banker Discount} - \text{True Discount}} = \frac{\text{Simple Interest} \times \text{True Discount}}{\text{Simple Interest} - \text{True Discount}}$$

Simple interest on True Discount = Simple Interest – True Discount

$$\text{Sum Due} = \frac{\text{Banker Discount} \times 100}{\text{Rate} \times \text{Time}}$$

Example. The interest on a certain sum of money is ₹ 22 and the discount on the same sum for the same time and at the same rate is ₹ 20, find the sum. **(SSC CGL Tier I 2013)**

(*a*) ₹ 220 (*b*) ₹ 200 (*c*) ₹ 210 (*d*) ₹ 212

Sol. (*a*) Sum Due = $\frac{22 \times 20}{22 - 20}$ = ₹ 220

Example. The true discount on ₹ 5250 due 5 years hence, is ₹ 250. Find the rate of interest per annum.

(*a*) 25% (*b*) 5% (*c*) 2% (*d*) 1%

Sol. (*d*) P.W. = Amount – T.D. = 5250 – 250 = 5000 ⇒ Rate = $\frac{250 \times 100}{5 \times 5000}$ = 1%

SOLVED EXERCISE

1. The true discount on a certain sum of money due after 5 years is ₹ 80 and the simple interest on the same sum for the same period at the same rate of interest is ₹ 100. Find the sum due.

(*a*) ₹ 440 (*b*) ₹ 400 (*c*) ₹ 500 (*d*) ₹ 450

Sol. (*b*) Sum = $\frac{100 \times 80}{100 - 80}$ = ₹ 400

2. If the present worth of a certain sum due 2 years hence, at 10% per annum compound interest is ₹ 5000, find the sum.

(*a*) ₹ 4500 (*b*) ₹ 5000 (*c*) ₹ 6050 (*d*) ₹ 6000

Sol. (*c*) Sum Due = Present Worth × $\left(1 + \frac{r}{100}\right)^{time}$ = $5000 \times \frac{11}{10} \times \frac{11}{10}$ = ₹ 6050

3. Find the true discount on ₹ 575 due 3 years hence at 5% Simple Interest.

(*a*) ₹ 25 (*b*) ₹ 50 (*c*) ₹ 75 (*d*) ₹ 87.50

Sol. (*c*) Present Worth = $\frac{\text{Sum Due} \times 100}{100 + (\text{Rate} \times \text{Time})} = 575 \times \frac{100}{115}$ = ₹ 500

True Discount = ₹ 575 – ₹ 500 = ₹ 75

4. Find the present worth of ₹ 4410 due 2 years hence at 5% p.a. compound interest?

(*a*) ₹ 4000 (*b*) ₹ 3600 (*c*) ₹ 4400 (*d*) ₹ 4900

Sol. (*a*) Present Worth = Sum Due ÷ $\left(1 + \frac{r}{100}\right)^n$ = ₹ 4410 × $\frac{20}{21} \times \frac{20}{21}$ = ₹ 4000

5. Find the true discount on ₹ 6760 due in 2 years hence at 4% p.a. compound interest?

(*a*) ₹ 450 (*b*) ₹ 500 (*c*) ₹ 550 (*d*) ₹ 510

Sol. (*d*) Present Worth = Sum Due ÷ $\left(1 + \frac{r}{100}\right)^n$ = ₹ 6760 × $\frac{25}{26} \times \frac{25}{26}$ = ₹ 6250

True Discount = ₹ 6760 – ₹ 6250 = ₹ 510

6. If the true discount on ₹ 4600 due 3 years hence is ₹ 600, what is the rate percent?

(*a*) 2.5% (*b*) 5% (*c*) 10% (*d*) 12%

Sol. (*b*) P.W. = S.D. – T.D. = 4600 – 600 = ₹ 4,000

$$\text{Rate of interest} = \frac{\text{T.D} \times 100}{\text{P.W} \times \text{Time}} = \frac{600 \times 100}{4000 \times 3} = 5\%$$

7. The true discount of a bill due 3 years hence at 5% simple interest is ₹ 30. Find the amount of the bill.

(*a*) ₹ 200 (*b*) ₹ 170 (*c*) ₹ 230 (*d*) ₹ 250

Sol. (*c*) Let Present Worth = ₹ 100 ⇒ Simple interest on ₹ 100 for 3 years at 5% = ₹ 15

Sum Due = Present Worth + Interest = ₹ 100 + ₹ 15 = ₹ 115

If True Discount is ₹ 15, Sum Due = ₹ 115.

If True Discount is ₹ 30, Sum Due = $\frac{115}{15} \times 30$ = ₹ 230

8. The true discount on a certain sum due 2 years hence, is ₹ 200. If S.I is ₹ 240, find the rate of interest per annum.

(*a*) 2.5% (*b*) 5% (*c*) 10% (*d*) 12%

Sol. (*d*) Sum Due = $\frac{\text{SI} \times \text{TD}}{\text{SI} - \text{TD}} = \frac{240 \times 200}{240 - 200}$ = ₹ 1200 ⇒ P.W. = 1200 – 200 = 1000

T.D. is S.I. on the P.W. ⇒ ₹ 240 is S.I. on ₹ 1000 for 2 years.

$$\text{Rate of interest} = \frac{240 \times 100}{1000 \times 2} = 12\%$$

9. The banker discount and the true discount on a certain sum due at 5% simple interest are ₹ 110 and ₹ 100 respectively. The time is :

(*a*) 1 year (*b*) 2 years (*c*) 2.5 years (*d*) 5 years

Sol. (*b*) Sum Due = $\frac{110 \times 100}{110 - 100}$ = ₹ 1100 ⇒ B.D. on ₹ 1100 = ₹ 110 ⇒ Time = $\frac{110 \times 100}{1100 \times 5}$ = 2 years

Alt: Interest on ₹ 100 = ₹ 110 – ₹ 100 = ₹ 10 ⇒ Time = $\frac{10 \times 100}{100 \times 5}$ = 2 years

10. If the difference between banker discount and true discount on a sum due six months hence at 4% is ₹ 4, find the sum.

(*a*) ₹ 9000 (*b*) ₹ 10000 (*c*) ₹ 10200 (*d*) ₹ 12500

Sol. (*c*) Let the Sum Due = ₹ 100

Then Banker Discount = $\frac{\text{Sum Due} \times \text{Rate} \times \text{Time}}{100} = \frac{100 \times 4 \times 1}{100 \times 2}$ = ₹ 2

$$\text{True Discount} = \frac{\text{Sum Due} \times \text{Rate} \times \text{Time}}{100 + (\text{Rate} \times \text{Time})} = \frac{100 \times 4 \times ½}{100 + (4 \times ½)} = \frac{200}{102} = \frac{100}{51}$$

Banker Gain = Banker Discount – True Discount = $2 - \frac{100}{51} = \frac{2}{51}$

If B.G. is ₹ $\frac{2}{51}$, S.D. = ₹ 100 ⇒ If B.G. is ₹ 4, S.D. = ₹ 100 × $\frac{51}{2}$ × 4 = ₹ 10200

CHAPTER

24

TIME AND WORK

FORMULAE

1. If a man can do a piece of work in x days, then his work for one day $= \frac{1}{x}$.

2. If the ratio of time taken by A and B in doing a work $= x : y$.

   Then ratio of work done by A and B $= \frac{1}{x} : \frac{1}{y} = y : x$.

   And ratio in which the wages is to be distributed $= y : x$.

Example. The efficiencies of two workers A and B are in the ratio 5 : 4. If A can do a piece of work in 12 hours, then B can do it in : **(SSC CGL Tier I 2006)**

(*a*) 15 hours (*b*) 16 hours (*c*) 18 hours (*d*) 9.6 hours

Sol. (*a*) Ratio of work of A and B = 5 : 4

Ratio of time taken by A and B for the same work = 4 : 5

Time taken by A = 4 ratio = 12 hours ⇒ 1 ratio = 3 hours

⇒ Time taken by B = 5 ratio = 15 hours

Alt: $12 \times \frac{5}{4} = 15$ hours

Example. A and B can complete a piece of work in 15 days and 10 days respectively. They contracted to complete the work for ₹ 30000. The share of A in the contracted money will be: **(SSC CGL Tier I 2004)**

(*a*) ₹ 18000 (*b*) ₹ 16500 (*c*) ₹ 12500 (*d*) ₹ 12000

Sol. (*d*) Ratio of wages = 10 : 15 = 2 : 3 ⇒ A's share $= \frac{2}{5} \times 30000 =$ ₹ 12000

3. If three men can do a work in x, y and z days respectively.

   Then ratio in which the wages is to be distributed $= \frac{1}{x} : \frac{1}{y} : \frac{1}{z}$

Example. A, B and C can do a piece of work in 12, 18, and 30 days respectively. If they do a work together, find the ratio in which the amount is to be distributed.

(*a*) 2 : 3 : 5 (*b*) 5 : 3 : 2 (*c*) 6 : 10 : 15 (*d*) 15 : 10 : 6

Sol. (*d*) Ratio of wages $= \frac{1}{12} : \frac{1}{18} : \frac{1}{30} = 15 : 10 : 6$ or $\frac{1}{12} : \frac{1}{18} : \frac{1}{30} = \frac{1}{2} : \frac{1}{3} : \frac{1}{5} = 15 : 10 : 6$

4. If A is $\frac{x}{y}$ times as good a workman as B, then A will take $\frac{y}{x}$ of the time that B takes in doing the work.

5. If number of persons engaged in doing a certain job is changed in the ratio of $x : y$, the days required in completion of work will change in the ratio of $y : x$.

6. If m_1 men can do a work in d_1 days and m_2 men can do the same work in d_2 days, then man days $= m_1 \times d_1 = m_2 \times d_2$

Example 1: 12 men can do a work in 25 days. How long will 10 men take to complete the work? **(Bank Clerical)**

(*a*) 27 days (*b*) 28 days (*c*) 30 days (*d*) 50 days

Sol. (*c*) Time taken by 10 men $= \dfrac{12 \times 25}{10} = 30$ days

2. If a job is to be completed in 10 days, it requires 270 persons. If 180 persons take up the same job, they will finish it in **(SSC CGL Tier I 2014)**

(*a*) 15 days (*b*) 18 days (*c*) 27 days (*d*) 12 days

Sol. (*a*) Required number of days $= \dfrac{10 \times 270}{180} = 15$ days

7. If A can do a piece of work in x days and B in y days. In how many days, both working together will finish the work? **(SSC CHSL 2013, SSC CGL Tier I 2014)**

Sol. A's one day's work $= \dfrac{1}{x}$ and B's one day's work $= \dfrac{1}{y}$

A's and B's one day's work $= \dfrac{1}{x} + \dfrac{1}{y} = \dfrac{x+y}{xy}$ ⇒ A and B will do the work in $\dfrac{xy}{x+y}$ days

Example 1: If A can do a piece of work in 20 days and B in 30 days. In how many days, both working together will finish the job? **(Bank Clerical 2008)**

(*a*) 25 (*b*) 18 (*c*) 12 (*d*) None of these

Sol. (*c*) Time taken by A and B $= \dfrac{20 \times 30}{20 + 30} = 12$ days

2. A can do a piece of work in 4 days and B can do it in 12 days. In how many days will they finish the work, both working together? **(SSC CGL Tier I 2014)**

(*a*) 4 days (*b*) 6 days (*c*) 2 days (*d*) 3 days

Sol. (*d*) Time taken by A and B $= \dfrac{4 \times 12}{16} = 3$ days

3. Two pipes can separately fill a tank in 3 hours and 2 hours respectively. If both pipes are in operation, the tank will be filled in **(SSC CGL Tier I 2012)**

(*a*) 1.8 hours (*b*) 1 hour (*c*) 1.5 hours (*d*) 1.2 hours

Sol. (*d*) Required time $= \dfrac{3 \times 2}{3+2} = \dfrac{6}{5} = 1\dfrac{1}{5} = 1.2$ hours

8. A and B working together can do a piece of work in x days whereas B working alone can do the same work in y days. How many days will A alone take to do the work?

Sol. (A + B)'s one day's work $= \dfrac{1}{x}$; B's one day's work $= \dfrac{1}{y}$

∴ A's 1 day's work $= \dfrac{1}{x} - \dfrac{1}{y} = \dfrac{y-x}{xy}$ ⇒ **A alone will complete the work in** $\dfrac{xy}{y-x}$ **days.**

Example 1: A and B together can do a piece of work in 12 days, while B alone can finish it in 30 days. A alone can finish the work in : **(SSC CHSL 2012)**

(*a*) 15 days (*b*) 20 days (*c*) 25 days (*d*) 18 days

Sol. (*b*) Time taken by A = $\frac{30 \times 12}{30 - 12}$ = 20 days

2. A and B can together do a piece of work in 6 days. If B alone can do the work in 8 days, how many days will A take to do the work independently ? **(SSC FCI 2012)**

(a) 24 days (b) 14 days (c) 27 days (d) 22 days

Sol. (*a*) Time taken by A = $\frac{8 \times 6}{8 - 6}$ = 24 days

9. Pipe A can fill a tank in *x* hours and pipe B can empty it in *y* hours.

If both the pipes opened together, the tank will be filled in $\frac{xy}{y - x}$ hours.

And 1 hour's work of both the pipes = $\frac{1}{x} - \frac{1}{y} = \frac{y - x}{xy}$

Hint: Pipe B is doing negative work here.

Example: A tap can fill a tank in 15 minutes while another tap can empty it in 20 minutes. If both the taps are opened simultaneously, when the tank be full?

(*a*) 5 minutes (*b*) 35 minutes (*c*) 60 minutes (*d*) 70 minutes

Sol. (*c*) Time taken = $\frac{15 \times 20}{20 - 15} = \frac{15 \times 20}{5}$ = 60 minutes

10. A pipe can fill a cistern in *x* hours but due to leakage in the bottom, it is filled in *y* hours.

Then time taken by the leak to empty the cistern = $\frac{xy}{y - x}$

Example. A pipe can fill a tank in 20 minutes. Due to leakage in the bottom of the tank, it takes 25 minutes to fill the tank. If the tank is full, how long will the leak take to empty the tank?

(*a*) 50 minutes (*b*) 60 minutes (*c*) 80 minutes (*d*) 100 minutes

Sol. (*d*) Time taken = $\frac{20 \times 25}{25 - 20} = \frac{20 \times 25}{5}$ = 100 minutes

Alt: Tank filled in 5 extra minutes (i.e. 25 – 20) is emptied in 25 minutes (i.e. 5 × 5)

∴ Tank filled in 20 minutes is emptied in 20 × 5 = 100 minutes.

11. If *A* takes *a* days more than time taken by *A* and *B* together to complete a work and *B* takes *b* days more than the time taken by *A* and *B* to do the work. Then time taken by *A* and *B* together to do the work is $\sqrt{ab}$ days.

Example: A alone takes 9 days more than the time taken by *A* and *B* together. *B* alone takes 16 days more than the time taken by *A* and *B* to do the work. Find the time *A* and *B* together take to finish the work.

(*a*) 25 days (*b*) 15 days (*c*) 7 days (*d*) 12 days

Sol. (*d*) Required time = $\sqrt{9 \times 16}$ = 12 days.

12. If a men or b women can do a piece of work in x days.

Then m men and n women together can finish the work in $\frac{abx}{an+bm}$ or $\frac{x}{\frac{m}{a}+\frac{n}{b}}$ days

Example: 2 men or 4 women can do a work in 28 days. How long will 4 men and 8 women take to complete the work?

(*a*) 7 days (*b*) 14 days (*c*) 28 days (*d*) 56 days

Sol. (*a*) Time taken $= \frac{2 \times 4 \times 28}{2 \times 8 + 4 \times 4} = \frac{2 \times 4 \times 28}{32} = 7$ days

Alternative Method : Time taken $= \frac{28}{\frac{4}{2}+\frac{8}{4}} = \frac{28}{4} = 7$ days

13. If A and B can finish a work in x and ax days respectively. In other words, A is a times efficient as B or A can do a times the work as much as B can do.

Then working together they can finish the work in $\frac{ax}{a+1}$ days.

Example 1: A alone can do a piece of work in 8 days and B alone can do the same work in 24 days. Working together they will finish the work in:

(*a*) 4 days (*b*) 6 days (*c*) 12 days (*d*) 16 days

Sol. (*b*) Time taken by B is three times of that taken by A.

$\therefore$ Together they will finish the work in $\frac{24}{3+1} = 6$ days

Example 2: A and B can do a piece of work in 15 days and 30 days respectively. Working together, they will finish the work in:

(*a*) 10 days (*b*) 15 days (*c*) 22.5 days (*d*) 45 days

Sol. (*a*) A is twice as efficient as B. $\Rightarrow$ Time taken by A and B $= \frac{30}{1+2} = 10$ days

14. If A is x times efficient than B, and working together they finish a work in y days. Then

Time taken by A $= \frac{y(x+1)}{x}$ and Time taken by B $= y(x+1)$

Example: A and B can do a job together in 12 days. A is 2 times as efficient as B. In how many days can B alone complete the work **(SSC CHSL 2012)**

(*a*) 18 days (*b*) 12 days (*c*) 36 days (*d*) 9 days

Sol. (*c*) Time taken by B = 12 × (2 + 1) = 12 × 3 = 36 days

Alt: Ratio of A's and B's work = 2 : 1 $\Rightarrow$ 1 day work of B $= \frac{1}{12} \times \frac{1}{3} = \frac{1}{36}$

SOLVED EXERCISE

1. A work can be done by a certain number of workers in 8 days. How many days will they take to finish a work four times the original work?

(*a*) 2 days (*b*) 12 days (*c*) 24 days (*d*) 32 days

Sol. (*d*) More work, more days *i.e.* Direct Proportion $\Rightarrow$ Days required = 8 × 4 = 32 days

2. A single machine can produce 1 unit in 6 minutes. How many such machines work together if 1450 units are produced from 10 a.m. to 12.30 p.m.? **(SBI Clerical)**

(*a*) 2 (*b*) 29 (*c*) 58 (*d*) None of these

Sol. (*c*) Time taken = 10 a.m. to 12.30 p.m. = 150 minutes

Units produced by 1 machine in 150 minutes = 150 ÷ 6 = 25 units

∴ Machines required to produce 1450 units = 1450 ÷ 25 = 58

3. 18 children can do a piece of work in 12 days. How many children would be required to do the same work in 8 days ? **(Canara Bank PO 2009)**

(*a*) 12 (*b*) 18 (*c*) 24 (*d*) 27

Sol. (*d*) 18 children can do the work in 12 days.

1 child can do the work in 12 × 18 days (less children, more days)

8 children will take = $\frac{12 \times 18}{8}$ = 27 days. (more children, less days)

4. 56 men can complete a piece of work in 24 days. In how many days can 42 men complete the same piece of work ? **(Bank of Maharashtra PO 2009)**

(*a*) 18 (*b*) 32 (*c*) 98 (*d*) 48

Sol. (*b*) 56 men can complete the work in 24 days.

1 man can complete the work in 24 × 56 days. (less man, more days)

42 men will take = $\frac{24 \times 56}{42}$ = 32 days. (more men, less days)

5. If *m* persons can paint a house in *d* days, how many days will it take for (*m* + 2) persons to paint the same house ? **(CDS 2015)**

(*a*) $md + 2$ (*b*) $md - 2$ days (*c*) $\frac{m+2}{md}$ (*d*) $\frac{md}{m+2}$

Sol. (*d*) Man-days = $m \times d = md \Rightarrow$ Required days = $\frac{md}{m+2}$

6. If *x* men can do a piece of work in *x* days, then the number of days in which *y* men can do the same work is **(SSC CGL Tier II 2013)**

(*a*) xy days (*b*) $\frac{y^2}{x}$ days (*c*) $\frac{x^2}{y}$ days (*d*) x^2y days

Sol. (*c*) Man-days = $x \times x = x^2 \Rightarrow$ Required days = $\frac{x^2}{y}$

7. A stock of food is enough for 240 men for 48 days. How long will the same stock last for 160 men ? **(CDS 2012)**

(*a*) 54 days (*b*) 60 days (*c*) 64 days (*d*) 72 days

Sol. (*d*) $\frac{240 \times 48}{160}$ = 72 days

8. A farmer can plough a field working 6 hours per day in 18 days. The farmere has to work how many hours per day to finish the same work in 12 days ? **(SSC CGL Tier II 2013)**

(*a*) 7 (*b*) 9 (*c*) 11 (*d*) 13

Sol. (*b*) $\frac{6 \times 18}{12}$ = 9 hours

9. 40 men can finish a work in 15 days. If the work is to be completed 3 days earlier, how many more people should be employed? **(SSC CGL Tier I 2012)**

(*a*) 15 (*b*) 10 (*c*) 8 (*d*) 50

Sol. (*b*) Total men required = $\frac{40 \times 15}{12} = 50 \Rightarrow$ More men required = 50 – 40 = 10

10. A's 2 days work is equal to B's 3 days work. If A can complete the work in 8 days then to complete the work B will take : **(SSC CGL Tier I 2015)**

(*a*) 14 days (*b*) 15 days (*c*) 16 days (*d*) 12 days

Sol. (*d*) A's time : B's time = 2 : 3 $\Rightarrow$ Time taken by B = $8 \times \frac{3}{2} = 12$ days

11. If 5 men can do a work in 10 days and 12 women can do the same work in 15 days, then 5 men and 6 women working together, can complete the work in : **(CDS 2003)**

(*a*) 15 days (*b*) $7\frac{1}{2}$ days (*c*) 9 days (*d*) 6 days

Sol. (*b*) 12 women can do the work in 15 days $\Rightarrow$ 6 women can do the work in 2 × 15 = 30 days

5 men and 6 women can do the work in $\frac{10 \times 30}{40} = 7\frac{1}{2}$ days

12. A can do a work in 10 days and B in 20 days. If they together work on it for 5 days, then the fraction of the work that is left is : **(SSC CGL Tier I 2015)**

(*a*) 1/4 (*b*) 3/20 (*c*) 3/4 (*d*) 4/3

Sol. (*a*) Work done in 5 days = $5 \times \left(\frac{1}{10} + \frac{1}{20}\right) = 5 \times \frac{3}{20} = \frac{3}{4} \Rightarrow$ Remaining work = $1 - \frac{3}{4} = \frac{1}{4}$

13. Subhash can copy 50 pages in 10 hours; Subhash and Prakash can copy 300 pages in 40 hours. In how much time can Prakash copy 30 pages ? **(SSC CGL Tier I 2014)**

(*a*) 13 hours (*b*) 10 hours (*c*) 12 hours (*d*) 9 hours

Sol. (c) 1 hour work of Subhash = 50 ÷ 10 = 5 pages

$\Rightarrow$ Work done by Subhash in 40 hours = 40 × 5 = 200 pages

1 hour work of Prakash = (300 – 200) ÷ 40 = 2.5 $\Rightarrow$ Required time = 30 ÷ 2.5 = 12 hours

14. *A* does 20% less work than *B*. If *A* can complete a piece of work in $7\frac{1}{2}$ hours, then *B* can do it in **(SSC CGL Tier I 2013)**

(*a*) 6 hours (*b*) 8 hours (*c*) 10 hours (*d*) 4 hours

Sol. (*a*) Ratio of work = 80 : 100 = 4 : 5 $\Rightarrow$ Ratio of time = 5 : 4

$\Rightarrow$ Hours taken by *A* = $7\frac{1}{2} = \frac{15}{2} \Rightarrow$ Hours taken by B = $\frac{15}{2} \times \frac{4}{5} = 6$

15. A can do a piece of work in 24 days. If B is 60% more efficient than A, then how many days does B require to do the same work ? **(CDS 2010)**

(*a*) 12 (*b*) 15 (*c*) 16 (*d*) 18

Sol. (*b*) Ratio of work of A and B = 100 : 160 = 5 : 8 $\Rightarrow$ Ratio of time taken by A and B = 8 : 5

$\therefore$ Time taken by B = $\frac{5}{8} \times 24 = 15$ days

Alt: Time taken by B = 24 ÷ 160% = 24 × $\frac{100}{160}$ = 15 days

16. A can do a piece of work in 12 days. B is 50% more efficient than A. In how many days B will finish the same work ? **(SSC FCI 2012)**

(*a*) 6 days (*b*) 8 days (*c*) 12 days (*d*) 24 days

Sol. (*b*) $B = 12 \times \frac{100}{150} = 8$ days

17. Kamal can do a work in 15 days. Bimal is 50% more efficient than Kamal. The number of days, Bimal will take to do the same work, is **(SSC CPO SI 2006, MAT)**

(*a*) 10.5 days (*b*) 15 days (*c*) 12 days (*d*) 10 days

Sol. (*d*) Time taken by Bimal $= 15 \times \frac{100}{150} = 10$ days

18. Two pipes A and B can fill a tank in 20 minutes and 30 minutes respectively. If both pipes are opened together, the time taken to fill the tank is : **(SSC CGL Tier I 1999)**

(*a*) 50 minutes (*b*) 12 minutes (*c*) 25 minutes (*d*) 15 minutes

Sol. (*b*) Required time $= \frac{20 \times 30}{50} = 12$ minutes

19 A particular job can be completed by a team of 10 men in 12 days. The same job can be completed by a team of 10 women in 6 days. How many days are needed to complete the job if the two teams work together ? **(SSC CGL Tier I 2000)**

(*a*) 4 (*b*) 6 (*c*) 9 (*d*) 18

Sol. (*a*) Required time $= \frac{12 \times 6}{18} = 4$ days

20. A can do 1/4 of a work in 10. B can do 1/3 of the work in 20 days. In how many days can both A and B together do the work? **(SSC CGL Tier I 2014)**

(*a*) 30 days (*b*) 32 days (*c*) 24 days (*d*) 25 days

Sol. (*c*) Days taken by A = 10 × 4 = 40. Days taken by B = 20 × 3 = 60 days

$\therefore$ Required time $= \frac{40 \times 60}{100} = 24$ days

21. A alone can complete a work in 12 days. A and B together can complete it in 8 days. How long will B alone take to complete the work ? **(SSC CGL Tier I 2007 & SBI Bank PO 2008)**

(*a*) 24 days (*b*) 20 days (*c*) 18 days (*d*) 16 days

Sol. (*a*) Time taken by B $= \frac{12 \times 8}{12 - 8} = \frac{12 \times 8}{4} = 24$ days

22. A, B and C together can do a piece of work in 12 days. A and B together can do the same work in 20 days. In how many days C working alone can finish the work?

(*a*) 8 days (*b*) 16 days (*c*) 25 days (*d*) 30 days

Sol. (*d*) C alone can do the work in $\frac{20 \times 12}{20 - 12} = \frac{20 \times 12}{8} = 30$ days

23. Pipes A and B together can fill a tank in 18 hours. Pipe A alone can fill the tank in 24 hours. In how many hours, can pipe B alone fill the tank? **(EPFO SSA)**

(*a*) 6 hours (*b*) 36 hours (*c*) 48 hours (*d*) 72 hours

Sol. (*d*) Time taken by pipe B $= \frac{18 \times 24}{24 - 18} = \frac{18 \times 24}{6} = 72$ hours

24. A man can do a piece of work in 5 days, but with the help of his son, he can do it in 3 days. In what time can the son do it alone ? **(SSC CGL Tier I 2004)**

(*a*) 7 days (*b*) 8 days (*c*) $7\frac{1}{2}$ days (*d*) $6\frac{1}{2}$ days

Sol. (*c*) Son's one day's work $= \frac{1}{3} - \frac{1}{5} = \frac{2}{15} \Rightarrow$ Required time $= \frac{15}{2} = 7\frac{1}{2}$ days

25. A tap can empty a tank in 30 minutes while second tap empty it in 20 minutes. If both the taps are kept open, how much time is required to empty the cistern?

(*a*) 12 minutes (*b*) 25 minutes (*c*) 50 minutes (*d*) 1 hour

Sol. (*a*) Time taken $= \frac{30 \times 20}{50} = 12$ minute

Hint : Both taps are doing negative work. ∴ Work done by them is added.

26. A tap can empty a tank in one hour. A second tap can empty it in 30 minutes. If both the taps operate simultaneously, how much time is needed to empty the tank ? **(SSC CGL 2000)**

(*a*) 20 minutes (*b*) 30 minutes (*c*) 40 minutes (*d*) 45 minutes

Sol. (*a*) Required time $= \frac{60 \times 30}{60 + 30} = \frac{60 \times 30}{90} = 20$ minutes

Alt : 1 minute work of both pipes $= \frac{1}{60} + \frac{1}{30} = \frac{3}{60} = \frac{1}{20} \Rightarrow$ Required time = 20 minutes

27. Three pipes A, B and C can fill a tank in 30 min, 20 min and 10 min respectively. When the tank is empty, all the three pipes are opened. If A, B and C discharge chemical solutions P, Q and R respectively, then the part of solution R in the liquid in the tank after 3 min is **(SSC MTS 2013)**

(*a*) $\frac{8}{11}$ (*b*) $\frac{7}{11}$ (*c*) $\frac{6}{11}$ (*d*) $\frac{5}{11}$

Sol. (*c*) Ratio of work of three pipes $= \frac{1}{30} : \frac{1}{20} : \frac{1}{10} = \frac{60}{30} : \frac{60}{20} : \frac{60}{10} = 2 : 3 : 6$

∴ Solution R = ratio of work done by pipe C $= \frac{6}{11}$

(Hint: Ratio of solution R will remain constant at any given point of time)

28. Three pipes A, B and C can fill a cistern in 10, 12 and 15 hours respectively, while working alone. If all the three pipes are opened together, then find the time taken to fill the cistern: **(SSC CHSL 2012, SSC Constable GD2012 M)**

(*a*) 3 hours (*b*) 4 hours (*c*) 5 hours (*d*) 6 hours

Sol. (*b*) 1 hour's work $= \frac{1}{10} + \frac{1}{12} + \frac{1}{15} = \frac{6+5+4}{60} = \frac{15}{60} = \frac{1}{4}$

29. Four taps can individually fill a cistern of water in 1 hour, 2 hours, 3 hours and 6 hours respectively. If all the four taps are opened simultaneously, the cistern can be filled in how many minutes ? **(CDS 2011)**

(*a*) 20 (*b*) 30 (*c*) 35 (*d*) 40

Sol. (*b*) Required time $= \frac{1}{60} + \frac{1}{120} + \frac{1}{180} + \frac{1}{360} = \frac{6+3+2+1}{360} = \frac{12}{360} = \frac{1}{30}$

Alt: 1 hour work $= 1 + \frac{1}{2} + \frac{1}{3} + \frac{1}{6} = 2 \Rightarrow$ Required time = 60 ÷ 2 = 30 minutes.

30. A, B and C can do a work in 2 days. A alone can do the same work in 10 days, B alone in 5 days. In how many days C alone can do the work ?

(*a*) 5 days (*b*) 2 days (*c*) 2.5 days (*d*) 10 days

Sol. (*a*) (A + B + C)'s 1 day work = 100% ÷ 2 = 50%

A's 1 day work = 100% ÷ 10 = 10% B's 1 day work = 100% ÷ 5 = 20%

C's 1 day work = 50% – (10% + 20%) = 20% ⇒ Days = 100% ÷ 20% = 5 days

31. A and B can do a work in 12 days, B and C in 15 days, C and A in 20 days. If A, B and C work together, they will complete the work in **(SSC CGL Tier I 2003, 2013)**

(*a*) 5 days (*b*) 10 days (*c*) 12 days (*d*) 18 days

Sol. (*b*) $(A+B)+(B+C)+(A+C) = \frac{1}{12}+\frac{1}{15}+\frac{1}{20} = \frac{5+4+3}{60} = \frac{12}{60} = \frac{1}{5}$

$\Rightarrow 2(A+B+C) = \frac{1}{5} \Rightarrow (A+B+C) = \frac{1}{5}\times\frac{1}{2} = \frac{1}{10}$ ⇒ Required time = 10 days

32. A and B can complete a piece of work in 30 days, B and C in 20 days, and C and A in 15 days. If all of them work together, the time taken in completing the work will be : **(SSC CGL Tier I 2005)**

(*a*) 10 days (*b*) 12 days (*c*) $12\frac{2}{3}$ days (*d*) $13\frac{1}{3}$ days

Sol. (*d*) $(A+B)+(B+C)+(A+C) = \frac{1}{30}+\frac{1}{20}+\frac{1}{15} = \frac{2+3+4}{60} = \frac{9}{60} = \frac{3}{20}$

$\therefore 2(A+B+C) = \frac{3}{20} \Rightarrow (A+B+C) = \frac{3}{20}\times\frac{1}{2} = \frac{3}{40}$ ⇒ Required time = $\frac{40}{3} = 13\frac{1}{3}$ days.

33. A can complete a piece of work in 2 and B in 3 days. With the help of C, they completed the work in 1 day. C alone could complete the work in **(SAS 2010)**

(*a*) 4 days (*b*) 5 days (*c*) 6 days (*d*) 7 days

Sol. (*c*) $C = (A+B+C) - A - B = 1 - \frac{1}{2} - \frac{1}{3} = \frac{1}{6}$ ⇒ Required time = 6 days

34. Two pipes can fill a cisterm in 3 hours and 4 hours respectively and a waste pipe can empty it in 2 hours. If all the three pipes are kept open, then the cistern will be filled in : **(SSC CGL Tier I 2004)**

(*a*) 5 hours (*b*) 8 hours (*c*) 10 hours (*d*) 12 hours

Sol. (*d*) One hour work of three pipes $\frac{1}{3}+\frac{1}{4}-\frac{1}{2} = \frac{4+3-6}{12} = \frac{1}{12}$ ⇒ Required time = 12 hours

35. A man, a woman and a boy together finish a piece of work in 6 days. If a man and a woman can do the work in 10 and 24 days respectively. The days it will take by a boy to finish the work is **(SSC CGL Tier I 2013)**

(*a*) 30 days (*b*) 35 days (*c*) 40 days (*d*) 45 days

Sol. (*c*) Boy's 1 day work = $\frac{1}{6}-\frac{1}{10}-\frac{1}{24} = \frac{20-12-5}{120} = \frac{3}{120} = \frac{1}{40}$ ⇒ Required time = 40 days

36. Two taps can fill a tub in 5 minutes and 7 minutes respectively. A pipe can empty it in 3 minutes. If all the three taps are kept open simultaneously, when will the tub be full ? **(CDS 2009)**

(*a*) 60 min (*b*) 185 min (*c*) 90 min (*d*) 105 min

Sol. (*d*) 1 minute work of the three pipes $= \frac{1}{5}+\frac{1}{7}-\frac{1}{3}=\frac{21+15-35}{105}=\frac{1}{105}$

37. A leak can empty a full tank in 12 minutes. If the leak is absent, two pipes A and B can fill the tank separately in 15 minutes and 20 minutes respectively. If the leak is active and both the pipes are open, then the tank will be filled in **(SSC CGL Tier I 2013)**

(*a*) 30 minutes (*b*) 60 minutes (*c*) 23 minutes (*d*) 35 minutes

Sol. (*a*) 1 minute work of the three pipes $= \frac{1}{15}+\frac{1}{20}-\frac{1}{12}=\frac{4+3-5}{60}=\frac{1}{30}$

38. A cistern can be filled with water by a pipe in 5 hours and it can be emptied by a second pipe in 4 hours. If both the pipes are opened when the cistern is full, the time in which it will be emptied is : **(SSC CGL Tier I 2001)**

(*a*) 9 hours (*b*) 18 hours (*c*) 20 hours (*d*) $20\frac{1}{2}$ hours

Sol. (*c*) One hour work of both pipes $= \frac{1}{4}-\frac{1}{5}=\frac{1}{20}$ ⇒ Required time = 20 hours

39. A cistern is normally filled in 9 hours but due to leak in the bottom, it takes one hour more to be filled. If the cistern is full, in what time will the leak empty it?

(*a*) 1 hour (*b*) 19 hours (*c*) 90 hours (*d*) 100 hours

Sol. (*c*) When both pipes are opened, tank is filled in 9 + 1 = 10 hours.

∴ Required time $= \frac{9 \times 10}{10-9} = 90$ hours

Alt : Water filled in 1 extra hour = Water leaked in 10 hours (i.e., 9 + 1)

∴ Water filled in 9 hours = Water leaked in 9 × 10 hours = 90 hours

40. If a cistern generally takes 20 minutes to be filled by a pipe but due to a leak, it takes 10 extra minutes to be filled, then find the time in which the leak can empty the full cistern. **(SSC CHSL 2012)**

(*a*) 60 minutes (*b*) 50 minutes (*c*) 30 minutes (*d*) 40 minutes

Sol. (*a*) When both pipes are opened, cistern is filled in 20 + 10 = 30 minutes

∴ Required time $= \frac{30 \times 20}{30-20} = 60$ minutes

Alt: Water filled in 10 minutes = Water leaked in 30 minutes (i.e., 20 + 10)

∴ Water filled in 20 minutes = Water leaked in 2 × 30 minutes = 60 minutes

41. A pump can fill a tank with water in 2 hours. Because of a leak in the tank it was taking $2\frac{1}{3}$ hours to fill the tank. The leak can drain all the water off the tank in : **(SSC CGL 2002)**

(*a*) 8 hours (*b*) 7 hours (*c*) $4\frac{1}{3}$ hours (*d*) 14 hours

Sol. (*d*) One hour work of both pipes $= 1 \div 2\frac{1}{3} = 1 \div \frac{7}{3} = \frac{3}{7}$

One hour work of leak pipe $= \frac{1}{2}-\frac{3}{7}=\frac{7-6}{14}=\frac{1}{14}$ ⇒ Required time = 14 hours

42. A pipe can fill a tank with water in 3 hours. Due to leakage in bottom, it takes $3\frac{1}{2}$ hours to fill it. In what time the leak will empty the fully filled tank ? **(SSC CGL Tier I 2001)**

(*a*) 12 hours (*b*) 21 hours (*c*) $6\frac{1}{2}$ hours (*d*) $10\frac{1}{2}$ hours

Sol. (*b*) Water filled in $\frac{1}{2}$ hour = Water leaked in $3\frac{1}{2}$ hours

$\therefore$ Water filled in 3 hours = Water leaked in $\frac{3\frac{1}{2}}{\frac{1}{2}} \times 3$ hours = 21 hours

43. A can cultivate $\frac{2}{5}$th of a land in 6 days and B can cultivate $\frac{1}{3}$rd of the same land in 10 days. Working together A and B cultivate $\frac{4}{5}$th of the land in : **(SSC CGL Tier I 2001)**

(*a*) 4 days (*b*) 5 days (*c*) 8 days (*d*) 10 days

Sol. (*c*) One day work of A = $\frac{2}{5} \times \frac{1}{6} = \frac{1}{15}$. One day work of B = $\frac{1}{3} \times \frac{1}{10} = \frac{1}{30}$

One day work of A and B = $\frac{1}{15} + \frac{1}{30} = \frac{3}{30} = \frac{1}{10}$

$\therefore$ Days taken to do $\frac{4}{5}$th work = $\frac{4}{5} \times 10 = 8$ days

44. Two pipes A and B can separately fill a cistern in 60 minutes and 75 minutes respectively. There is a third pipe in the bottom of the cistern to empty it. If all the three pipes are simultaneously opened, then the cistern is full in 50 minutes. In how much time the third pipe alone can empty the cistern ? **(SSC CGL Tier I 2003)**

(*a*) 110 minutes (*b*) 100 minutes (*c*) 120 minutes (*d*) 90 minutes

Sol. (*b*) 1 minute work of third pipe = $\frac{1}{60} + \frac{1}{75} - \frac{1}{50} = \frac{10+8-12}{600} = \frac{6}{600} = \frac{1}{100}$

45. A tank can be filled by two pipes A and B separately in 3 hours, and 3 hours and 45 minutes respectively. A third pipe C can empty the full tank in 1 hour. When the tank was exact half-filled, all the three pipes are opened. The tank will become empty in :

(SSC CGL Tier I 2006)

(*a*) 1 hour 15 min (*b*) 2 hours 30 min (*c*) 3 hours 15 min (*d*) 4 hours 10 min

Sol. (*a*) One hour work of three pipes = $\frac{1}{3} + \frac{4}{15} - 1 = \frac{-6}{15} = \frac{-2}{5}$

$\therefore$ Time taken to empty full filled tank = $\frac{5}{2}$ hours

$\therefore$ Time taken to empty half-filled tank = $\frac{1}{2} \times \frac{5}{2} = \frac{5}{4} = 1\frac{1}{4}$ = 1 hour 15 minutes

46. A and B together can do $\frac{11}{19}$ of a work. In the same time B and C together can do $\frac{14}{19}$ of the same work. The ratio of work done by A, B and C is : **(SSC CPO SI 2011)**

(*a*) 3 : 4 : 5 (*b*) 4 : 5 : 7 (*c*) 5 : 6 : 8 (*d*) 5 : 7 : 8

Sol. (*c*) $(A+B)+(B+C)=\frac{11}{19}+\frac{14}{19}=\frac{25}{19} \Rightarrow B=\frac{25}{19}-1=\frac{6}{19}$

$A=(A+B)-B=\frac{11}{19}-\frac{6}{19}=\frac{5}{19}$. $C=(B+C)-B=\frac{14}{19}-\frac{6}{19}=\frac{8}{19}$

$\therefore$ Required ratio $=\frac{5}{19}:\frac{6}{19}:\frac{8}{19}=5:6:8$

47. A sum of money is sufficient to pay A's wages for 21 days and B's wages for 28 days. The same money is sufficient to pay the wages of both for :

(SSC constable GD 2011, SSC CGL Tier I 2013)

(*a*) 12 days (*b*) $12\frac{1}{4}$ days (*c*) 14 days (*d*) $24\frac{1}{2}$ days

Sol. (*a*) Number of days $=\frac{21\times 28}{21+28}=12$ days

48. A and B together can do a piece of work in T days. If A alone can do the same work in (T + 3) days and B alone cane do the work in (T + 12) days, find the value of T.

(Bank PO 2008)

(*a*) 3 (*b*) 6 (*c*) 9 (*d*) 12

Sol. (*b*) Required time $=\sqrt{3\times 12}=6$ days.

49. A boy and a girl together fill a cistern with water. The boy pours 4 litres of water every 3 minutes and the girl pours 3 litres every 4 minutes. How much time will it take to fill 100 litres of water in the cistern ? **(SSC CGL Tier I 2008)**

(*a*) 36 minutes (*b*) 42 minutes (*c*) 48 minutes (*d*) 44 minutes

Sol. (*c*) Boy's work for 12 minutes $=4\times\frac{12}{3}=16$ litres

Girl's work for 12 minutes $=3\times\frac{12}{4}=9$ litres

$\therefore$ Their work for 12 minutes each $=16+9=25$ litres

$\therefore$ Time taken to fill 100 litres $=\frac{100}{25}\times 12$ minutes $=48$ minutes

50. A can do a piece of work in 12 days while B alone can do it in 15 days. With the help of C they can finish it in 5 days. If they are paid ₹ 960 for the whole work how much money A gets? **(SSC CGL Tier I 2013)**

(*a*) ₹ 480 (*b*) ₹ 240 (*c*) ₹ 320 (*d*) ₹ 400

Sol. (*d*) A's share $=\frac{5}{12}\times$ ₹ 960 = ₹ 400

51. A, B and C did a work together and earned ₹ 729 for the work. If the ratio of work done by A, B and C is 2 : 3 : 4, find C's share.

(*a*) ₹ 81 (*b*) ₹ 162 (*c*) ₹ 243 (*d*) ₹ 324

Sol. (*d*) Wages is distributed in ratio of work done *i.e.* in the ratio of 2 : 3 : 4.

Sum of ratios = 2 + 3 + 4 = 9 ⇒ C's share = $\frac{4}{9} \times 729$ = ₹ 324

52. Suman can do a work in 3 days. Sumati can do the same work in 2 days. Both of them finish the work together and get ₹ 150. What is the share of Suman ? **(SSC CGL Tier I 2003)**

(*a*) ₹ 30 (*b*) ₹ 60 (*c*) ₹ 70 (*d*) ₹ 75

Sol. (*b*) Ratio of wages = $\frac{1}{3} : \frac{1}{2} = 2 : 3$ ⇒ Suman's wages = $\frac{2}{5}$ × ₹ 150 = ₹ 60

53. 'X' completes a job in 2 days and 'Y' completes it in 3 days and 'Z' takes 4 days to complete it. If they work together and get ₹ 3,900 for the job, then how much amount does 'Y' get? **(CDS 2011)**

(*a*) ₹ 1,800 (*b*) ₹ 1,200 (*c*) ₹ 900 (*d*) ₹ 800

Sol. (*b*) Ratio of wages = ratio of work done by X : Y : Z = $\frac{1}{2} : \frac{1}{3} : \frac{1}{4}$ = 6: 4 : 3

⇒ Y's share = $\frac{4}{13}$ × ₹ 3,900 = ₹ 1,200

54. A and B undertake to do a piece of work for ₹ 1,200. A alone can do it in 8 days, while B can do it in 6 days. With the help of C, they complete it in 3 days. Find C's share. **(SSC CGL Tier II 2012, CGL Tier I 2004 M)**

(*a*) ₹ 150 (*b*) ₹ 100 (*c*) ₹ 450 (*d*) ₹ 300

Sol. (*a*) A's work for 3 days = $\frac{3}{8}$. B's work for 3 days = $\frac{3}{6} = \frac{1}{2}$. C's work = $1 - \frac{3}{8} - \frac{1}{2} = \frac{1}{8}$

C's share = $\frac{1}{8}$ × ₹ 1200 = ₹ 150

Alt: A's wages = $\frac{3}{8}$ × ₹ 1200 = ₹ 450. B's wages = $\frac{3}{6}$ × ₹ 1200 = ₹ 600.

C's wages = 1200 – (450 + 600) = ₹ 150

55. Two men undertook to do a job for ₹ 1400. One of them alone can do it in 7 days, and the other in 8 days. With the assistance of a boy, they together completed the work in 3 days. How much money will the boy get ? **(SSC CGL Tier I 2006)**

(*a*) ₹ 200 (*b*) ₹ 250 (*c*) ₹ 275 (*d*) ₹ 325

Sol. (*c*) First man work for 3 days = 3/7. Second man work for 3 days = 3/8

Boy's work = $1 - \frac{3}{7} - \frac{3}{8} = \frac{56 - 24 - 21}{56} = \frac{11}{56}$ ⇒ Boy's share = $\frac{11}{56}$ × ₹ 1400 = ₹ 275

56. Three persons undertake to complete a piece of work for ₹ 1200. The first person can complete the work in 8 days, second person in 12 days and third person in 16 days. They complete the work with the help of a fourth person in 3 days. What does the fourth person get? **(SSC CGL Tier II 2013)**

(*a*) ₹ 180 (*b*) ₹ 200 (*c*) ₹ 225 (*d*) ₹ 250

Sol. (*c*) Fourth person's work = $1-\frac{3}{8}-\frac{3}{12}-\frac{3}{16}=1-\frac{3}{8}-\frac{1}{4}-\frac{3}{16}=\frac{16-6-4-3}{16}=\frac{3}{16}$

$\therefore$ His wages = $\frac{3}{16}\times$ ₹ 1200 = ₹ 225

57. Sunil completes a work in 4 days, whereas Dinesh completes the work in 6 days. Ramesh work $1\frac{1}{2}$ times as fast as Sunil. The three together can complete the work in

(SSC CGL Tier II 2013)

(*a*) $1\frac{5}{12}$ days (*b*) $1\frac{5}{7}$ days (*c*) $1\frac{3}{8}$ days (*d*) $1\frac{5}{19}$ days

Sol. (*d*) Ramesh = $\frac{1}{4}\times\frac{3}{2}=\frac{3}{8}$ One day work of all = $\frac{1}{4}+\frac{1}{6}+\frac{3}{8}=\frac{6+4+9}{24}=\frac{19}{24}$

$\Rightarrow$ Days taken = $\frac{24}{19}=1\frac{5}{19}$ days

58. A tank has three pipes. The first pipe can fill 1/2 part of the tank in 1 hour and the second pipe can fill 1/3 part in hour. The third pipe is for making the tank empty. When all the three pipes are open, 7/12 part of the tank is filled in 1 hour. How much time will the third pipe take to empty the completely filled tank ? **(SSC CGL Tier I 2006)**

(*a*) 3 hours (*b*) 4 hours (*c*) 5 hours (*d*) 6 hours

Sol. (*b*) One hour work of third pipe = $\frac{1}{2}+\frac{1}{3}-\frac{7}{12}=\frac{3}{12}=\frac{1}{4}$ $\Rightarrow$ Required time = 4 hours

59. Two pipes together can fill a cistern in 6 minutes. If one pipe takes 5 minutes longer than the other to fill the cistern, find the time in which the faster pipe would fill the cistern.

(*a*) 5 minutes (*b*) 6 minutes (*c*) 9 minutes (*d*) 10 minutes

Sol. (*d*) $\frac{x(x+5)}{x+(x+5)}=6\Rightarrow x^2+5x=12x+30\Rightarrow x^2-7x+30=0$

$\Rightarrow (x-10)(x+3)=0\Rightarrow x=10$ minutes

60. A tap takes 36 hours extra to fill a tank due to a leakage equivalent to half of its inflow. The inflow can fill the tank in how many hours ?

(*a*) 36 hour (*b*) 30 hours (*c*) 24 hours (*d*) 18 hours

Sol. (*d*) Let inflow can fill tank in x hours. Then $\frac{1}{x}-\frac{1}{2x}=\frac{1}{36}\Rightarrow\frac{2-1}{2x}=\frac{1}{36}\Rightarrow x=18$ hours

61. Two workers A and B working together completed a job in 5 days. If A worked twice as efficiently as he actually did and B worked $\frac{1}{3}$ as efficiently as he actually did, the work would have completed in 3 days. To complete the job alone, A would require

(SSC CGL Tier II 2013)

(*a*) $5\frac{1}{5}$ days (*b*) $6\frac{1}{4}$ days (*c*) $7\frac{1}{2}$ days (*d*) $8\frac{3}{4}$ days

Sol. (*b*) $\frac{1}{x}+\frac{1}{y}=\frac{1}{5} \Rightarrow \frac{x+y}{xy}=\frac{1}{5} \Rightarrow 5x+5y=xy$ ... (1)

$\frac{2}{x}+\frac{1}{3y}=\frac{1}{3} \Rightarrow \frac{6y+x}{3xy}=\frac{1}{3} \Rightarrow x+6y=xy \Rightarrow 5x+30y=5xy$ ... (2)

(2) – (1) gives, $25y=4xy \Rightarrow 4x=25 \Rightarrow x=6\frac{1}{4}$ days

62. A, B and C together can do a work in 8 days. If B takes double time of A, C takes double time of B, find the time taken by A alone to do the work.

(*a*) 14 days (*b*) 20 days (*c*) 28 days (*d*) 35 days

Sol. (*a*) Let A takes x days. Then B and C takes $2x$ and $4x$ days respectively.

(A + B + C)'s 1 day's work $=\frac{1}{x}+\frac{1}{2x}+\frac{1}{4x}=\frac{1}{8} \Rightarrow \frac{7}{4x}=\frac{1}{8} \Rightarrow x=\frac{8\times 7}{4}=14$ days.

63. Amit takes twice as much as Aryan and thrice as much as Raman to finish a piece of work. They all finish the work in one day. What is the time taken by Amit to finish the work ? **(UPSC Civil Services (Pre.) 2011)**

(*a*) 2 days (*b*) 3 days (*c*) 4 days (*d*) 6 days

Sol. (*d*) Ratio of time taken by Amit, Aryan and Raman = 6 : 3 : 2

According to question, $\frac{x}{6}+\frac{x}{3}+\frac{x}{2}=1 \Rightarrow \frac{(1+2+3)x}{6}=1 \Rightarrow x=1$

One day work of Amit = 1/6 ⇒ Days taken by Amit = 6

64. Tap A and B can fill a tank in 4 and 5 hours respectively. Tap C can empty the tank in 6 hours. All the three taps opened simultaneously and tap C is closed after 3 hours, how long would it take to fill the tank?

(*a*) 3 hours (*b*) 3 hours 20 min (*c*) 3 hours 40 min (*d*) 4 hours

Sol. (*b*) Let tap A and B were opened for x hours each.

Then $\frac{x}{4}+\frac{x}{5}-\frac{3}{6}=1 \Rightarrow \frac{9x}{20}=\frac{3}{2} \Rightarrow x=\frac{3}{2}\times\frac{20}{9}=\frac{10}{3}$ = 3 hours 20 minutes

65. A does half as much work as B in one-sixth of the time. If together they take 10 days to complete a work, how much time shall B take to do it alone ? **(SSC CGL Tier I 2001)**

(*a*) 70 days (*b*) 30 days (*c*) 40 days (*d*) 50 days

Sol. (*c*) Let B alone can do the work in x days. Then A alone can do it in $2\times\frac{x}{6}=\frac{x}{3}$ days

Their work for one day $=\frac{3}{x}+\frac{1}{x}=\frac{1}{10} \Rightarrow \frac{4}{x}=\frac{1}{10} \Rightarrow x=40$ days

66. A does half as much work as B in three-fourth of the time. If together they take 18 days to complete a work, how much time shall B take to do it alone? **(SSC CGL Tier I 2011)**

(*a*) 30 days (*b*) 35 days (*c*) 40 days (*d*) 45 days

Sol. (*a*) A : B $=\frac{1}{2}\times\frac{4}{3}:1=2:3 \Rightarrow$ (A + B)'s 1 day work $=\frac{1}{18} \Rightarrow$ B's 1 day work $=\frac{1}{18}\times\frac{3}{5}=\frac{1}{30}$

67. In a factory, there are equal number of women and children. Women work for 6 hours a day and children for 4 hours a day. During festival time, the work load goes up by 50%. The government rule does not allow children to work for more than 6 hours a day. If they are equally efficient and the extra work is done by women, then extra hours of effort put in by women every day are **(SSC CGL Tier II 2012)**

(*a*) 4 (*b*) 9 (*c*) 5 (*d*) 3

Sol. (*d*) Original hours of work = 6(women) + 4(children) = 10 hours

Increase in hours of work = 3(women) + 2(children) = 5 hours

Increased hours of work for children = 4 + 2 = 6

$\therefore$ Increase in working hours of women = 3 hours

68. 12 men can complete a work in 8 days. 3 days after they started the work, 3 more men joined. In how many days will all of them together complete the remaining work? **(LIC 1990)**

(*a*) 2 days (*b*) 3 days (*c*) 4 days (*d*) 7 days

Sol. (*c*) Let total work = 12 × 8 = 96 units ⇒ Work done in first 3 days = 12 × 3 = 36 units

Remaining work = 96 – 36 = 60 units ⇒ Required days = 60 ÷ (12 + 3) = 4 days

69. 12 men can complete a work in 18 days. 6 days after they started working, 4 more men joined them. How many days will all of them take to finish the remaining work?

(Central Excise & I. Tax 1993)

(*a*) 9 days (*b*) 12 days (*c*) 15 days (*d*) 18 days

Sol. (*a*) Work remaining after 6 days work = 12 × (18 – 6) = 12 × 12 man-days

$$\Rightarrow \text{Required days} = \frac{12\times12}{12+4} = \frac{144}{16} = 9 \text{ days}$$

70. A job can be completed by 12 men in 12 days. How many extra days will be needed to complete the job if 6 men leave after working for 6 days. **(SSC CGL Tier I 2005)**

(*a*) 3 (*b*) 6 (*c*) 12 (*d*) 24

Sol. (*b*) After 6 days remaining work is 12 men work for 6 days.

$$\text{Time taken by 6 men} = \frac{12\times6}{6} = 12 \text{ days.} \Rightarrow \text{Extra days} = 12 - 6 = 6 \text{ days}$$

71. If the work done by $(x - 1)$ men in $(x + 1)$ days is to the work done by $(x + 2)$ men in $(x - 1)$ days are in the ratio 9 : 10, then the value of x is equal to. **(SSC CHSL 2011)**

(*a*) 5 (*b*) 6 (*c*) 7 (*d*) 8

Sol. (*d*) $(x - 1) \times (x + 1) : (x + 2) \times (x - 1) = 9 : 10 \Rightarrow (x + 1) : (x + 2) = 9 : 10 \Rightarrow x = 8$

72. A certain number of men can do a work in 60 days. If there were eight more men, it could be completed in 10 days less. How many men were there in the beginning ? **(SSC CGL 2004)**

(*a*) 70 men (*b*) 55 men (*c*) 45 men (*d*) 40 men

Sol. (*d*) Let men employed originally = x. Then $60x = 50(x + 8) \Rightarrow 10x = 400 \Rightarrow x = 40$ men

Alt: Ratio of days = 60 : (60 – 10) = 6 : 5 ⇒ Ratio of men = Inverse ratio of 6 : 5 = 5 : 6

Difference = (6 – 5) ratio = 1 ratio = 8 men ⇒ Men (originally) = 5 ratio = 5 × 8 = 40 men

73. A certain number of men can complete a job in 30 days. If there were 5 men more, it could be completed in 10 days less. How many men were in the beginning ? **(SSC CGL Tier I 2000)**

(*a*) 10 (*b*) 15 (*c*) 20 (*d*) 25

Sol. (*a*) Let men employed originally = x. Then $30x = 20(x + 5) \Rightarrow 10x = 100 \Rightarrow x = 10$ men

Alt : Ratio of days = 30 : (30 – 10) = 3 : 2 ⇒ Ratio of men = Inverse ratio of 3 : 2 = 2 : 3

Difference = (3 – 2) ratio = 1 ratio = 5 men ⇒ Men (originally) = 2 ratio = 2 × 5 = 10 men

74. A certain number of persons can complete a piece of work in 55 days. If there were 6 persons more, the would could be finished in 11 days less. How many persons were originally there ? **(SSC CGL Tier I 2003)**

(*a*) 17 (*b*) 24 (*c*) 30 (*d*) 22

Sol. (*b*) Ratio of days = 55 : (55 – 11) = 5 : 4 ⇒ Ratio of workers = 4 : 5

Difference = (5 – 4) ratio = 1 ratio ≡ 6 persons

∴ Workers initially = 4 ratio = 4 × 6 = 24 persons

75. A certain number of men can finish a piece of work in 100 days. However, if there were 10 men less, then it would take 10 days more for the work to be finished. How many men were there originally. **(SSC CGL Tier I 2005 M, SSC CGL Tier I 2014)**

(*a*) 100 (*b*) 110 (*c*) 55 (*d*) 50

Sol. (*b*) Let original workers = x. Then $100x = 110(x - 10) \Rightarrow 10x = 1100 \Rightarrow x = 110$ workers

Alternative method : Ratio of days = 100 : 110 = 10 : 11 ⇒ Ratio of workers = 11 : 10

Difference of workers = (11 – 10) ratio = 1 ratio ≡ 10 workers

∴ Original number of workers = 11 ratio = 11 × 10 = 110 workers

76. 76 ladies complete a job in 33 days. Due to some reason some ladies did not join the work and therefore it was completed in 44 days. The number of ladies who did not report for the work is **(CDS 2011)**

(*a*) 17 (*b*) 18 (*c*) 19 (*d*) 20

Sol. (*c*) Ratio of days = 33 : 44 = 3 : 4 ⇒ Ratio of ladies = 4 : 3

⇒ 4 ratio = 76 ⇒ 1 ratio = 19 ladies ⇒ Ladies absent = (4 – 3) ratio = 1 ratio = 19

77. 10 persons started a work together. After a few days 4 persons left the work. Due to this, the work which was expected to finish in 40 days was finished in 50 days. After how many days from the starting of the work, 4 persons had left ? **(SSC CGL Tier I 2004)**

(*a*) 10 (*b*) 15 (*c*) 20 (*d*) 25

Sol. (*d*) Total man-days = 10 × 40 = 400. Let 4 men left after x days.

Then $400 = 10x + 6 \times (50 - x) \Rightarrow 400 = 10x + 300 - 6x \Rightarrow 4x = 100 \Rightarrow x = 25$

Alternative method : Total man-days (in first case) = 10 × 40 = 400

Total man-days (had 10 men worked for 50 days) = 10 × 50 = 500

Extra man-days = 500 – 400 = 100 ⇒ 4 men had not worked for last 100 ÷ 4 = 25 days

4 men had left the work after = 50 – 25 = 25 days

78. 40 men can do a piece of work in a given time. If only 30 men be engaged, 6 more days are needed to complete the work. In how many days can 60 men do the work ? **(SSC CGL Tier I 2006)**

(*a*) 9 (*b*) 10 (*c*) 12 (*d*) 15

Sol. (*c*) Let 40 men can do the work in x days. Then $40x = 30(x + 6) \Rightarrow 10x = 180 \Rightarrow x = 18$

⇒ 40 men can do the work in 18 days ⇒ 60 men can do the work in $18 \times \frac{40}{60} = 12$ days

Alt : Ratio of men = 40 : 30 = 4 : 3 ⇒ Ratio of days = Inverse ratio of 4 : 3 = 3 : 4

Difference of days = (4 – 3) ratio = 1 ratio ≡ 6 days

∴ Days taken by 40 men = 3 ratio = 3 × 6 = 18 days

∴ Days taken by 60 men = $18 \times \frac{40}{60} = 12$ days

79. 10 men and 15 women can complete a work in 6 days. 1 man can do the same job in 100 days. In how many days 1 women can do the job. **(SSC CGL Tier I 2006)**

(*a*) 80 days (*b*) 125 days (*c*) 180 days (*d*) 225 days

Sol. (*d*) 10 men + 15 women can do in 6 days ⇒ 60 men + 90 woman can do in 1 day

⇒ 60 men + 90 women = 100 men ⇒ 40 men = 90 women

$\Rightarrow 1 \text{ Man} = \frac{90}{40} \text{ women} = \frac{9}{4} \text{ women} \Rightarrow \text{Days taken by 1 women} = \frac{9}{4} \times 100 = 225 \text{ days}$

Alt: $10 \text{ men's work for 6 days} = \frac{10 \times 6}{100} = \frac{3}{5}$

$\text{Remaining work} = 1 - \frac{3}{5} = \frac{2}{5}$ done by 15 women in 6 days

$\therefore \text{Required time} = \frac{5}{2} \times (15 \times 6) = 225 \text{ days}$

80. 10 men working 6 hours a day can complete a work in 18 days. How many hours a day must 15 men work to complete the same work in 12 days ? **(SSC CGL Tier I 2004)**

(*a*) 6 (*b*) 10 (*c*) 12 (*d*) 15

Sol. (*a*)

Men	Hours	Days
10	6	18
15	x	12

More men, less hours, *i.e.* inverse proportion. Less days, more hours, *i.e.* inverse proportion

$\therefore \text{Required time} = 6 \times \frac{10}{15} \times \frac{18}{12} = 6 \text{ hours}$

81. 50 people consume 350 kgs of rice in 30 days. In how many approximate days will 35 people consume 50 kgs of rice ? **(Nabard Officers Exam. 2008)**

(*a*) 3 days (*b*) 5 days (*c*) 7 days (*d*) None of these

Sol. (*d*)

People	Rice	Days
50	350	30
35	50	x

Less people, more days, *i.e.* inverse proportion. Less rice, less days, *i.e.* direct proportion

$\therefore x = 30 \times \frac{50}{35} \times \frac{50}{350} = \frac{300}{49} \approx 6 \text{ days}$

82. If 100 men can do 100 work in 100 days. How long will it take 1 man to do 1 work? **(SSC)**

(*a*) 1 day (*b*) 10 days (*c*) 100 days (*d*) 200 days

Sol. (*c*)

Men	Work	Days
100	100	100
1	1	x

Less men, more days, *i.e.* inverse proportion. Less work, less days, *i.e.* direct proportion

$\therefore \text{Time taken} = 100 \times \frac{100}{1} \times \frac{1}{100} = 100 \text{ days}$

83. 12 men can do $\frac{1}{2}$ of a work in 15 days. In how many days will 10 men finish the work?

(*a*) 9 days (*b*) 18 days (*c*) 27 days (*d*) 36 days

Sol. (*d*)

Men	Work	Days
12	$\frac{1}{2}$	15
10	1	x

Less men, more days, *i.e.* inverse proportion. More work, more days, *i.e.* direct proportion

$\therefore$ Time taken $= 15\times\frac{12}{10}\times\frac{2}{1} = 36$ days

84. 5 persons can prepare an admission list in 8 days working 7 hours a day. If 2 persons join them so as to complete the work in 4 days, they need to work per day for : **(SSC CGL 2004)**

(*a*) 10 hours (*b*) 9 hours (*c*) 12 hours (*d*) 8 hours

Sol. (*a*)

Men	Days	Hours
5	8	7
7	4	x

More men, less hours, *i.e.* inverse proportion. Less days, more hours, *i.e.* inverse proportion

$\therefore$ Days taken $= 7\times\frac{5}{7}\times\frac{8}{4} = 10$ hours

85. If 72 men can build a wall of 280 m long in 21 days, how many men could take 18 days to build a similar type of wall of length 100 m ? **(SSC CGL Tier I 2003)**

(*a*) 30 (*b*) 10 (*c*) 18 (*d*) 28

Sol. (*a*)

Men	Length	Days
72	280	21
x	100	18

Less length, less men, *i.e.* direct proportion. Less days, more men, *i.e.* indirect proportion

$\therefore\ x = 72\times\frac{100}{280}\times\frac{21}{18} = 30$ men

86. If 30 labours working 5 hours a day can make a 300 metres long road in 32 days, how long will it take 20 labours to make the 100 metres long road working 8 hours a day?

(*a*) 10 days (*b*) 15 days (*c*) 20 days (*d*) 25 days

Sol. (*a*)

Men	Hours	Length	Days
30	5	300	32
20	8	100	x

Days taken $= 32\times\frac{30}{20}\times\frac{5}{8}\times\frac{100}{300} = 10$ days

87. If 15 workers can earn ₹ 1800 in 10 days. Find the earning of 5 workers in 8 days.

(*a*) ₹ 360 (*b*) ₹ 400 (*c*) ₹ 480 (*d*) ₹ 540

Sol. (*c*)

Worker	Income	Days
15	1800	10
5	x	8

Total earnings = ₹ $1800\times\frac{5}{15}\times\frac{8}{10} =$ ₹ 480

88. A wall of 100 metres can be built by 7 men or 10 women in 10 days. How many days will 14 men and 20 women take to build a wall of 600 metres ?

(*a*) 15 (*b*) 20 (*c*) 25 (*d*) 30

Sol. (*a*) 7 men ≡ 10 women ⇒ 14 men ≡ 20 women ⇒ 14 men + 20 women = 40 women

Length	Women	Days
100	10	10
600	40	x

$\therefore$ Required days $= 10\times\frac{600}{100}\times\frac{10}{40} = 15$ days

89. If 40 men working 9 hours a day can do a piece of work in 24 days. How many men working 8 hours a day can do double that work in 60 days?

(*a*) 30 men (*b*) 36 men (*c*) 45 men (*d*) 54 men

Sol. (*b*)

Men	Hours	Work	Days
40	9	1	24
x	8	2	60

$\therefore$ Men required $= 40\times\frac{9}{8}\times\frac{2}{1}\times\frac{24}{60} = 36$ men

90. If p men working p hours per day for p days produce p units of work, then the units work produced by n men working n hours a day for n days is : **(SSC CGL Tier I 2008)**

(*a*) $\frac{p^2}{n^2}$ (*b*) $\frac{p^3}{n^2}$ (*c*) $\frac{n^2}{p^2}$ (*d*) $\frac{n^3}{p^2}$

Sol. (*d*)

Men	Hours	Days	Works
p	p	p	p
n	n	n	?

Work $= p\times\frac{n\times n\times n}{p\times p\times p} = \frac{n^3}{p^2}$ units

91. A contractor undertakes to complete a road 360 m long in 120 days and employs 30 men for the work. After 60 days, he finds that only 120 m length of road has been made. How many more men should he employ so that the work may be completed in time? **(SSC CGL 2006)**

(*a*) 15 (*b*) 20 (*c*) 30 (*d*) 45

Sol. (*c*) Days remaining after 60 days work = 120 – 60 = 60 days

Work remaining = 360 m – 120 m = 240 m

Men	Days	Length
30	60	120
x	60	240

Men required $= 30\times\frac{60}{60}\times\frac{240}{120} = 60$ men ⇒ Additional men required = 60 – 30 = 30 men.

92. A contractor undertakes to make a road in 40 days and employs 25 men. After 24 days, he finds that only one-third of the road is made. How many extra men should he employ so that he is able to complete the work 4 days earlier ? **(SSC CGL Tier I 2000)**

(*a*) 100 (*b*) 60 (*c*) 75 (*d*) None of these

Sol. (*c*) Days remaining after 24 days' work = 40 – 24 = 16 days

25 men has completed $\frac{1}{3}$ work in 24 days.

Remaining work = $1-\frac{1}{3}=\frac{2}{3}$ which is equal to double of work already completed.

Remaining work needs to be completed in 16 – 4 = 12 days.

Men	Days	Work
25	24	1
x	12	2

$\therefore$ Men required = $25\times\frac{24}{12}\times\frac{2}{1}$ = 100 men $\Rightarrow$ Additional men required = 100 – 25 = 75 men

93. A man undertakes to do a certain work in 150 days. He employs 200 men. He finds that only a quarter of the work is done in 50 days. The number of additional men that should be appointed so that the whole work will be finished in time is : **(SSC CGL Tier I 2013)**

(*a*) 50 (*b*) 75 (*c*) 100 (*d*) 125

Sol. (*c*) Days remaining after 50 days' work = 150 – 50 = 100 days

50 men has completed $\frac{1}{4}$ work in 50 days.

Remaining work = $1-\frac{1}{4}=\frac{3}{4}$ which is equal to thrice of work already completed.

Men	Days	Work
200	50	1
x	100	3

$\therefore$ Men required = $200\times\frac{50}{100}\times\frac{3}{1}$ = 300 men $\Rightarrow$ Additional men required = 300 – 200 = 100

94. A contractor undertook to finish a certain work in 124 days and employed 120 men. After 64 days, he found that he had already done $\frac{2}{3}$ of the work. How many men can be discharged now so that the work may finish in time ? **(SSC CGL Tier I 2013)**

(*a*) 50 (*b*) 48 (*c*) 56 (*d*) 40

Sol. (*c*) Days remaining after 64 days' work = 124 – 64 = 60 days

120 men has completed $\frac{2}{3}$ work in 64 days.

Remaining work = $1-\frac{2}{3}=\frac{1}{3}$ which is equal to half of work already completed.

Men	Days	Work
120	64	2
x	60	1

$\therefore$ Men required = $120\times\frac{64}{60}\times\frac{1}{2}$ = 64 men $\Rightarrow$ Men discharged = 120 – 64 = 56 men

95. In a fort, there was sufficient food for 200 soliders for 31 days. After 27 days, 120 soliders left the fort. For how many extra days will the food last for the remaining soliders ?

(SSC CPO SI Exam 2012)

(*a*) 10 days (*b*) 6 days (*c*) 4 days (*d*) 12 days

Sol. (*b*) After 27 days, the remaining food is sufficient for 200 soliders for 31 – 27 = 4 days.

Remaining soliders = 200 – 120 = 80 soliders

Soliders	Days
200	4
80	x

Less soliders, provision will last for more days, i.e. inverse proportion

$\therefore$ The food will be sufficient for $4 \times \frac{200}{80} = 10$ days $\Rightarrow$ Extra days = 10 – 4 = 6 days

96. A fort has provision for 50 days. If after 10 days, they are strenghtened by 500 men and the remaining food lasts 35 days, how many men were there in the fort initially ? **(SSC CGL Tier I 2002)**

(*a*) 3500 men (*b*) 3000 men (*c*) 2500 men (*d*) 4000 men

Sol. (*a*) Ratio of days = (50 – 10) : 35 = 8 : 7 $\Rightarrow$ Ratio of men = Inverse of 8 : 7 = 7 : 8

Difference = (8 – 7) ratio = 1 ratio ≡ 500 men $\Rightarrow$ Men (initially) = 7 ratio = 7 × 500 = 3500 men

Alternative method: $40x = 35(x + 500) \Rightarrow 5x = 35 \times 500 \Rightarrow x = 3500$

97. A garrison of '*n*' men had enough food to last for 30 days. After 10 days, 50 more men joined them. If the food now lasted for 16 days, what is the value of *n* ? **(CDS 2011)**

(*a*) 200 (*b*) 240 (*c*) 280 (*d*) 320

Sol. (*a*) Ratio of days = (30 – 10) : 16 = 20 : 16 = 5 : 4 $\Rightarrow$ Ratio of men = 4 : 5

Difference = (5 – 4) ratio = 1 ratio ≡ 50 men

$\therefore$ Men (initially) = n = 4 ratio = 4 × 50 = 200 men

Alternative method : $20x = 16(x + 50) \Rightarrow 4x = 16 \times 50 \Rightarrow x = 200$

98. A garrison of 1500 men has provision for 55 days. After 15 days more men arrived and the remaining provision lasted for 30 days. The number of men who arrived were:

(*a*) 500 (*b*) 750 (*c*) 1000 (*d*) 3000

Sol. (*a*) After 15 days, the provision should have lasted for 55 – 15 = 40 days

$\therefore$ Ratio of days = 40 : 30 = 4 : 3 $\Rightarrow$ Ratio of men = 3 : 4

$\Rightarrow$ 3 ratio = 1500 $\Rightarrow$ 1 ratio = 500 $\Rightarrow$ Difference = (4 – 3) ratio = 1 ratio = 500

99. A garison has provision for a certain number of days. After 15 days, 1/4 of the men left and the remaining provision lasted for as long as before. How many days the remaining provision would have lasted before?

(*a*) 30 days (*b*) 45 days (*c*) 60 days (*d*) 75 days

Sol. (*a*) After 15 days, number of men reduced in the ratio 4 : 3

$\therefore$ Ratio of days (before and after) = 3 : 4 $\Rightarrow$ Difference = (4 – 3) ratio = 1 ratio = 15 days

$\therefore$ Original number of days = 3 ratio = 3 × 15 days = 45 days

Required number of days = 45 – 15 = 30 days

100. In a garrison, there was food for 1000 soldiers for one month. After 10 days, 1000 more soldiers joined the garrison. How long would the soldiers be able to carry on with the remaining food ? **(UPSC Civil Services (Pre.) 2013)**

(*a*) 25 days (*b*) 20 days (*c*) 15 days (*d*) 10 days

Sol. (*d*) After 10 days, the remaining provision is sufficient for 1000 men for 30 – 10 = 20 days.

Now total soliders = 1000 + 1000 = 2000 $\Rightarrow$ Soldiers are doubled, therefore days halved.

$\therefore$ The provision will last for 20 ÷ 2 = 10 days

101. A completes a work in 20 days. B completes the same work in 30 days. A starts the work and B joins him after 10 days. If the remuneration for the work is ₹ 480, what is A's share?

(RRB Allahabad TC 2006)

(*a*) ₹ 384 (*b*) ₹ 372 (*c*) ₹ 376 (*d*) ₹ 380

Sol. (*a*) Work done by A in first 10 days = $\frac{10}{20} = \frac{1}{2}$

A's share in wages for first 10 days = $\frac{1}{2} \times$ ₹ 480 = ₹ 240

Remaining wages is to distributed between A and B in $\frac{1}{20} : \frac{1}{30} = 3 : 2$

A's share in remaining wages = $\frac{3}{5} \times$ ₹ 240 = ₹ 144 ⇒ A's total share = 240 + 144 = ₹384

102. 20 men can do a work in 35 days. When should 5 men leave so as to finish in 45 days.

(*a*) 3 days (*b*) 4 days (*c*) 5 days (*d*) 6 days

Sol. (*c*) Total man days = 20 × 35 = 700

Let (20 – 5) men = 15 men did the work, then man days = 15 × 45 = 675

Remaining man days = 700 – 675 = 25 ∴ 5 men should leave after 25 ÷ 5 = 5 days

103. 25 men can do a piece of work in 17 days. They worked at it for 5 days and then some more men joined the group and all of them together finished the remaining work in 10 days. How many more men joined the group?

(*a*) 5 men (*b*) 6 men (*c*) 7 men (*d*) 8 men

Sol. (*a*) Days remaining after 5 days' work = 17 – 5 = 12 days

Work remaining = 25 men's work for 12 days

∴ In 10 days, remaining work will be finished by $\frac{25 \times 12}{10}$ = 30 men

∴ Additional men joined = 30 – 25 = 5 men

104. A contractor engaged 5 men to complete a work in 20 days. After 15 days, he engaged 3 more men in order to complete the work on time. If additional men were not engaged, how many total days it would have taken to complete the work ?

(*a*) 21 days (*b*) 22 days (*c*) 23 days (*d*) 24 days

Sol. (*c*) Total man-days = 5 × 15 + (5 + 3) × 5 = 75 + 40 = 115

Total days required = 115 ÷ 5 = 23 days

Alternative Method : Total extra man-days = 3 × 5 = 15

Total extra days = 15 ÷ 5 = 3 ⇒ Total days = 20 + 3 = 23

105. 25 men can reap a field in 20 days. After how many days 15 men should leave the work, so that the remaining field is reaped in 25 days after they leave the work?

(*a*) 5 days (*b*) 8 days (*c*) 10 days (*d*) 12 days

Sol. (*c*) Total man days = 25 × 20 = 500 man days

Man days after 15 men leave = (25 – 15) × 25 = 250

Remaining man days = 500 – 250 = 250 ⇒ Days required = 250 ÷ 25 = 10 days

106. 30 men can do a work in 20 days. After how many days should 10 men leave the work so that the whole work is finished in 25 days?

(*a*) 5 days (*b*) 6 days (*c*) 8 days (*d*) 10 days

Sol. (*d*) Total man days = 30 × 20 = 600

Man days worked by (30 – 10) men or 20 men = 20 × 25 = 500

Remaining man days worked by 10 men = 600 – 500 = 100

∴ 10 men should left after 100 ÷ 10 = 10 days

107. A pipe can fill a cistern in 12 minutes and another pipe can fill it in 15 minutes, but a third pipe can empty it in 6 minutes. The first two pipes are kept open for 5 minutes in the beginning and then the third pipe is also opened. Number of minutes taken to empty the cistern is **(SSC CGL Tier II 2012, MBAs)**

(*a*) 42 (*b*) 45 (*c*) 38 (*d*) 22

Sol. (*b*) 5 minute work of first two pipes = $5\times\left(\frac{1}{12}+\frac{1}{15}\right)=5\times\frac{9}{60}=\frac{3}{4}$

1 minute work of three pipes = $\frac{1}{12}+\frac{1}{15}-\frac{1}{6}=\frac{-1}{60}$ ⇒ Required time = $\frac{3}{4}\times 60$ = 45 minutes

108. Two pipes A anc B can fill a tank with water in 30 minutes and 45 minutes respectively. The water pipe C can empty the tank in 36 minutes. First A and B are opened. After 12 minutes C is opened. Total time (in minutes) in which the tank will be filled up is : **(SSC CGL Tier I 2015)**

(*a*) 12 (*b*) 24 (*c*) 36 (*d*) 30

Sol. (*b*) (A + B)'s work = $\frac{1}{30}+\frac{1}{45}=\frac{5}{90}=\frac{1}{18}$

⇒ Work finished in 12 minutes = $\frac{1}{18}\times 12=\frac{2}{3}$ ⇒ Remaining work = $1-\frac{2}{3}=\frac{1}{3}$

⇒ (A + B) – C = $\frac{1}{18}-\frac{1}{36}=\frac{1}{36}$ ⇒ Time taken to fill this = $\frac{1}{3}\times 36$ = 12 minutes

⇒ Total time taken = 12 + 12 = 24 minutes

109. A cistern has 3 pipes A, B and C. A and B can fill it in 3 hours and 4 hours respectively while C can empty the completely filled cistern in 1 hour. If the three pipes are opened at 3, 4 and 5 p.m. respectively, at what time will the cistern be empty ? **(SSC CGL Tier I 2007)**

(*a*) 6:15 p.m. (*b*) 7:12 p.m. (*c*) 8:12 p.m. (*d*) 8:35 p.m.

Sol. (*b*) Work of pipe A for 2 hours (*i.e.* from 3 to 5 p.m.) = $\frac{1}{3}\times 2=\frac{2}{3}$

Work of pipe B for 1 hour (*i.e.* from 4 to 5 p.m.) = $\frac{1}{4}\times 1=\frac{1}{4}$

Tank filled till 5 p.m. = $\frac{2}{3}+\frac{1}{4}=\frac{11}{12}$

Work of the three pipes for one hour each = $\frac{1}{3}+\frac{1}{4}-1=\frac{-5}{12}$

∴ $\frac{11}{12}$ part of the tank is emptited in $\frac{11}{12}\div\frac{5}{12}=\frac{11}{12}\times\frac{12}{5}=\frac{11}{5}=2\frac{1}{5}$ = 2 hours 12 minutes

∴ Required time = 5 p.m. + 2 hours 12 minutes = 7:12 p.m.

110. Two pipes can fill a tank with water in 15 and 12 hours respectively and a third pipe can empty it in 4 hours. if the pipes be opened in order at 8, 9 and 11 a.m. respectively, the tank will be emptied at : **(SSC CGL Tier I 2005)**

(*a*) 11:40 a.m. (*b*) 12:40 p.m. (*c*) 1:40 p.m. (*d*) 2: 40 p.m.

Sol. (*d*) Work of first pipe for 3 hours (*i.e.* from 8 to 11 a.m.) = $\frac{1}{15}\times 3=\frac{1}{5}$

Work of second pipe for 2 hours (*i.e.* from 9 to 11 a.m.) = $\frac{1}{12}\times 2=\frac{1}{6}$

Tank filled till 11 a.m. = $\frac{1}{5}+\frac{1}{6}=\frac{11}{30}$

Work of the three pipes for one hour each = $\frac{1}{15}+\frac{1}{12}-\frac{1}{4}=\frac{4+5-15}{60}=\frac{-6}{60}=\frac{-1}{10}$

$\therefore$ $\frac{11}{30}$ part of the tank is emptited in $\frac{11}{30}\div\frac{1}{10}=\frac{11}{30}\times\frac{10}{1}=\frac{11}{3}=3\frac{2}{3}$ = 3 hours 40 minutes

$\therefore$ Required time = 11 a.m. + 3 hours 40 minutes = 2:40 p.m.

111. A can do a piece of work in 4 hours, B and C can do it in 3 hours, A and C can do it in 2 hours. How long will B alone take to do it ? **(SSC CGL Tier I 2002)**

(*a*) 10 hours (*b*) 12 hours (*c*) 8 hours (*d*) 24 hours

Sol. (*b*) B = A + (B + C) – (A + C) = $\frac{1}{4}+\frac{1}{3}-\frac{1}{2}=\frac{3+4-6}{12}=\frac{1}{12}$

112. A and B can do a piece of work in 10 days, B and C in 15 days and C and A in 20 days. C alone can do the work in : **(SSC CGL Tier I 2001)**

(*a*) 60 days (*b*) 120 days (*c*) 80 days (*d*) 30 days

Sol. (*b*) (B + C) + (A + C) – (A + B) = 2C ⇒ 2C = $\frac{1}{15}+\frac{1}{20}-\frac{1}{10}=\frac{1}{60}\Rightarrow C=\frac{1}{60}\times\frac{1}{2}=\frac{1}{120}$

113. A and B can do a piece of work in 12 days, B and C in 8 days, C and A in 6 days. How long would B take to do the same work alone ? **(SSC CGL Tier I 2005)**

(*a*) 24 days (*b*) 32 days (*c*) 40 days (*d*) 48 days

Sol. (*d*) (A + B) + (B + C) – (A + C) = 2B ⇒ 2B = $\frac{1}{12}+\frac{1}{8}-\frac{1}{6}=\frac{1}{24}\Rightarrow B=\frac{1}{2}\times\frac{1}{24}=\frac{1}{48}$

114. A and B can do a piece of work in 8 days, B and C can do the same work in 12 days. If A, B and C can complete the same work in 6 days, in how many days can A and C complete the same work ? **(CDS 2009)**

(*a*) 8 (*b*) 10 (*c*) 12 (*d*) 16

Sol. (*a*) (A + B) + (B + C) + (A + C) = 2 (A + B + C) = $2\times\frac{1}{6}=\frac{1}{3}$

$\Rightarrow\frac{1}{8}+\frac{1}{12}+(A+C)=\frac{1}{3}\Rightarrow(A+C)=\frac{1}{3}-\frac{1}{8}-\frac{1}{12}=\frac{8-3-2}{24}=\frac{3}{24}=\frac{1}{8}$

115. A and B can do a piece of work in 12 days, B and C in 15 days. C and A in 20 days. A alone can finish the work in **(SSC FCI 2012)**

(*a*) 10 days (*b*) 5 days (*c*) 30 days (*d*) 12 days

Sol. (*c*) (A + B) + (B + C) + (C + A) = $\frac{1}{12}+\frac{1}{15}+\frac{1}{20}=\frac{5+4+3}{60}=\frac{12}{60}=\frac{1}{5}$

(A + B) + (B + C) + (C + A) – 2 (B + C) = $\frac{1}{5}-2\times\frac{1}{15}\Rightarrow 2A=\frac{1}{15}\Rightarrow A=\frac{1}{30}$

116. A, B and C are three taps connected to a tank. A and B together can fill the tank in 6 hours, B and C together can fill it in 10 hours and A and C together can fill it in 7½ hours. In how much time will C alone can fill the tank ? **(SSC CHSL 2013)**

(*a*) 10 hours (*b*) 12 hours (*c*) 30 hours (*d*) 20 hours

Sol. (*c*) 1 minute work of (B + C) + (A + C) – (A + B) = $\frac{1}{10}+\frac{2}{15}-\frac{1}{6}=\frac{3+4-5}{30}=\frac{2}{30}=\frac{1}{15}$

⇒ Required time = 2 × 15 days = 30 days

117. A completes 7/10 of a work in 15 days, then he completes the remaining work with the help of B in 4 days. The time required for A and B together to complete the entire work is : **(SSC CGL Tier I 2005)**

(*a*) $10\frac{1}{2}$ days (*b*) $12\frac{2}{2}$ days (*c*) $13\frac{1}{3}$ days (*d*) $8\frac{1}{4}$ days

Sol. (*c*) Work done by A = $\frac{7}{10}$ ⇒ Remaining work = $1-\frac{7}{10}=\frac{3}{10}$ is done by A and B in 4 days.

∴ Entire work can be done by A and B in $4\times\frac{10}{3}=\frac{40}{3}=13\frac{1}{3}$ days

118. A does 4/5 of a piece of work in 20 days. He then calls in B and they finish the remaining work in 3 days. How long B alone will take to do the whole work ? **(SSC CGL Tier I 2002)**

(*a*) $37\frac{1}{2}$ days (*b*) 37 days (*c*) 40 days (*d*) 23 days

Sol. (*a*) Work done by A in 20 days = $\frac{4}{5}$ ⇒ A work for 1 day = $\frac{4}{5}\times\frac{1}{20}=\frac{1}{25}$

Remaining work = $1-\frac{4}{5}=\frac{1}{5}$ is done by A and B in 3 days.

A and B work for 1 day = $\frac{1}{5}\times\frac{1}{3}=\frac{1}{15}$ ⇒ B work for 1 day = (A + B) – A = $\frac{1}{15}-\frac{1}{25}=\frac{2}{75}$

∴ Time taken by B = $\frac{75}{2}=37\frac{1}{2}$ days

119. A and B can do a piece of work in 12 days, B and C can do it in 15 days. If A is twice as good worker as C, find the time that C alone will take to do the work.

(*a*) 27 days (*b*) 33 days (*c*) 45 days (*d*) 60 days

Sol. (*d*) (A + B) = $\frac{1}{12}$ ⇒ (2C + B) = $\frac{1}{12}$ But (B + C) = $\frac{1}{15}$

∴ C's one day's work = (2C + B) – (B + C) = $\frac{1}{12}-\frac{1}{15}=\frac{1}{60}$

120. A is twice as good worker as B. B is thrice as good worker as C. A alone started a work and did the work for 5 minutes. Afterwards B alone did the work for 30 minutes till half of the total work is completed. How long will C alone take to finish the remaining work. **(SSC)**

(*a*) 1 hour (*b*) 90 minute (*c*) 2 hours (*d*) 150 minutes

Sol. (*c*) Ratio of work of A, B and C = 6 : 3 : 1

Time taken by C (to do work done by A in 5 minutes) = 6 × 5 minutes = 30 minutes

Time taken by C (to do work done by B in 30 minutes) = 3 × 30 minutes = 90 minutes

Total time taken by C = (30 + 90) minutes = 120 minutes = 2 hours

121. A tap can fill a tank in 6 hours. After half the tank is filled, three more similar taps are opened. What is the total time taken to fill the tank completely ? **(SSC CGL Tier I 2003)**

(*a*) 4 hours (*b*) 4 hours 15 min (*c*) 3 hours 15 min (*d*) 3 hours 45 min

Sol. (*d*) Half of the tank is filled in $\frac{1}{2}\times 6 = 3$ hours

$\therefore$ Time taken by 4 taps to fill the remaining half $= 3\times\frac{1}{4}$ hour $= 45$ minutes

$\therefore$ Total time taken = 3 hours and 45 minutes.

122. Two taps A and B can fill a cistern is 18 minutes and 24 minutes respectively. If both taps are opened together, when the tap B must be turned off so that the cistern is filled in 12 minutes?

(*a*) 2 minutes (*b*) 4 minutes (*c*) 6 minutes (*d*) 8 minutes

Sol. (*d*) Cistern filled in 12 minutes by tap A $= \frac{12}{18} = \frac{2}{3} \Rightarrow$ Remaining work $= 1 - \frac{2}{3} = \frac{1}{3}$

Time taken by tap B to fill $\frac{1}{3}$ of the cistern $= \frac{1}{3} \times 24 = 8$ minutes

123. Pipe A can fill a tank in 15 minutes and pipe B can drain 40 litre per minute. If both the pipes are opened together, the cistern is full in 45 minutes, find the capacity of the cistern.

(*a*) 600 litres (*b*) 750 litres (*c*) 900 litres (*d*) 1800 litres

Sol. (*c*) 1 minute's work of pipe A $= \frac{1}{15} \Rightarrow$ Water filled in 45 minutes $= \frac{1}{15} \times 45 = 3$ cisterns

Work of two pipes in 45 minutes = 1 cistern

$\Rightarrow$ Work of waste-pipe in 45 minutes = 3 – 1 = 2 cisterns

But waste-pipe drains (45 × 40) litres in 45 minutes.

$\therefore$ 2 cisterns = 45 × 40 litres $\qquad\therefore$ 1 cistern $= \frac{45 \times 40}{2} = 900$ litres

Alt : Leak in 1 minute $= \frac{1}{15} - \frac{1}{45} = \frac{2}{45}$ of cistern = 40 litre

$\therefore$ Capacity of cistern $= 40 \times \frac{45}{2} = 900$ litres

124. A tank has a leak which would empty the completely filled tank in 10 hours. If the tank is full and a tap which admits 4 litres of water per minute in the tank, the leak takes 15 hours to empty the tank. What is the capacity of the tank ? **(SSC CGL Tier I 2008)**

(*a*) 1200 litres (*b*) 2400 litres (*c*) 3600 litres (*d*) 7200 litres

Sol. (*d*) Water leaked in 15 hours $= \frac{15}{10} = \frac{3}{2}$ of tank

Water admitted in 15 hours = 15 × 60 × 4 litres = 3600 litres $= \left(\frac{3}{2}-1\right) = \frac{1}{2}$ of tank

$\therefore$ Capacity of tank = 2 × 3600 litres = 7200 litres

Alternative Method : Water leaked = 4 litre/min = 240 litre/hour

1 hour work of leakage pipe $= \frac{1}{10} - \frac{1}{15} = \frac{1}{30}$

$\therefore$ Capacity of tank = 30 × 240 litres = 7200 litres

125. **Ronald and Elan are working on an Assignment, Ronald takes 6 hours to type 32 pages on a computer, while Elan takes 5 hours to type 40 pages. How much time will they take, working together on two different computers to type an assignment of 110 pages?** **(SSC CGL Tier I 2013)**

(*a*) 7 hrs 30 minutes (*b*) 8 hours (*c*) 8 hours 15 min (*d*) 8 hours 25 min

Sol. (*c*) Ratio of work done by Ronal and Elan $= \frac{32}{6} : \frac{40}{5} = 160 : 240 = 2 : 3$

$\therefore$ Work done by Ronald $= \frac{2}{5} \times 110 = 44$ pages

$\Rightarrow$ Time taken $= 44 \times \frac{6}{32} = \frac{33}{4} = 8\frac{1}{4}$ hours $= 8$ hours 15 minutes.

126. **If 3 men or 4 women can plough a field in 43 days, how long will 7 men and 5 women take to plough it ?** **(SSC CGL Tier I 2003)**

(*a*) 10 days (*b*) 11 days (*c*) 9 days (*d*) 12 days

Sol. (*d*) Required time $= \frac{3 \times 4 \times 43}{3 \times 5 + 4 \times 7} = \frac{3 \times 4 \times 43}{43} = 12$ days

127. **If 3 men or 6 women can do a piece of work in 16 days, in how many days can 12 men and 8 women do the same piece of work ?** **(SSC CGL Tier I 2000)**

(*a*) 4 days (*b*) 5 days (*c*) 3 days (*d*) 2 days

Sol. (*c*) Required time $= \frac{3 \times 6 \times 16}{3 \times 8 + 6 \times 12} = \frac{3 \times 6 \times 16}{96} = 3$ days

Alt: 3 men's work $\equiv$ 6 women's work $\Rightarrow$ 1 man's work $\equiv$ 2 women's work

$\therefore$ 12 men + 8 women $\equiv$ 24 women + 8 women = 32 women

6 women can do the work in 16 days. $\Rightarrow$ 32 women can do the work in $\frac{6 \times 16}{32} = 3$ days

128. **If 2 men or 3 women can do a piece of work in 30 days, in how many days can the work be done by 2 men and 6 women?**

(*a*) 4 days (*b*) 5 days (*c*) 10 days (*d*) 15 days

Sol. (*c*) Time taken $= \frac{2 \times 3 \times 30}{2 \times 6 + 3 \times 2} = \frac{2 \times 3 \times 30}{18} = 10$ days

Alternative Method : Time taken $= \frac{30}{\frac{2}{2} + \frac{6}{3}} = \frac{30}{3} = 10$ days

Alt: 2 men can do the work in 30 days; 6 women can do the work in 15 days.

Days taken by 2 men and 6 women $= \frac{30 \times 15}{45} = 10$ days

129. **If 5 men or 8 women can do a piece of work in 12 days, how many days will be taken by 2 men and 4 women to do the same work ?** **(SSC CGL Tier I 2001 & 2004 modified)**

(*a*) 15 days (*b*) $13\frac{1}{2}$ days (*c*) $13\frac{1}{3}$ days (*d*) 10 days

Sol. (*c*) Time taken $= \frac{5 \times 8 \times 12}{5 \times 4 + 8 \times 2} = \frac{5 \times 8 \times 12}{36} = \frac{40}{3} = 13\frac{1}{3}$ days

130. 6 men or 12 women can do a piece of work in 20 days. In how many days can 8 men and 16 women do twice as big as this work ? **(SSC CGL Tier I 2004)**

(*a*) 2 days (*b*) 5 days (*c*) 15 days (*d*) 10 days

Sol. (*c*) 6 men = 12 women ⇒ 1 man = 2 women

8 men + 16 women = 16 women + 16 women = 32 women

Required time = $\frac{12\times20\times2}{32}$ = 15 days

131. If 4 men or 8 women can do a piece of work in 15 days, in how many days can 6 men and 12 women do the same piece of work ? **(SSC CGL Tier I 2015)**

(*a*) 45 days (*b*) 20 days (*c*) 15 days (*d*) 30 days

Sol. (*) $\frac{6}{4}=\frac{3}{2}$ and $\frac{12}{8}=\frac{3}{2}$ ⇒ Work can be done in $\frac{1}{2}\times\frac{2}{3}\times15=5$ days

132. 5 men can do a piece of work in 6 days while 10 women can do it in 5 days. In how many days can 5 women and 3 men do it ? **(SSC CGL Tier I 1999)**

(*a*) 4 days (*b*) 5 days (*c*) 6 days (*d*) 8 days

Sol. (*b*) (5 × 6) men ≡ (10 × 5) women ⇒ 30 men ≡ 50 women ⇒ 3 men ≡ 5 women

10 women can do the work in 5 days.

⇒ (3 men + 5 women) or (5 + 5) women can do the work in $\frac{10\times5}{10}$ = 5 days

133. If 10 men or 20 women or 40 children can do a piece of work in 7 months, then 5 men, 5 women and 5 children together can do half of the work in : **(SSC CGL Tier I 2013)**

(*a*) 8 months (*b*) 6 months (*c*) 4 months (*d*) 5 months

Sol. (*c*) 10 men = 20 women = 40 children ⇒ 1 man = 2 women = 4 children

⇒ 5 men + 5 women + 5 children = (20 + 10 + 5) children = 35 children

⇒ 35 children can do the work in = $\frac{40\times7}{35}$ = 8 months.

⇒ Required time = $\frac{1}{2}$ × 8 months = 4 months.

134. Seven men and four boys can complete a work in 6 days. A man completes double the work as that of a boy. In how many days will 5 men and 4 boys complete the work ? **(Bank PO 2002)**

(*a*) 5 (*b*) 4 (*c*) 6 (*d*) None of these

Sol. (*d*) 7 men + 4 boys = 7 × 2 boys + 4 boys = 18 boys

5 men + 4 boys = 5 × 2 boys + 4 boys = 14 boys

18 boys can do the work in 6 days. ⇒ 14 boys will take = $\frac{6\times18}{14}=\frac{54}{7}=7\frac{5}{7}$ days

135. 8 men can do a piece of work in 5 days whereas 6 women can do the same work in 10 days. In how many days can 4 men and 9 women finish the work?

(*a*) 2 days (*b*) 3 days (*c*) 4 days (*d*) 6 days

Sol. (*c*) (8 × 5) men's work ≡ (6 × 10) women's work ⇒ 40 men's work ≡ 60 women's work

∴ 1 woman's work ≡ $\frac{40}{60}=\frac{2}{3}$ man's work ⇒ 9 women's work ≡ $\frac{2}{3}\times9=6$ men's work

∴ 4 men + 9 women ≡ 4 men + 6 men ≡ 10 men

8 men can do the work in 5 days. ⇒ 10 men can do the work in $\frac{8 \times 5}{10} = 4$ days

136. 12 men take 36 days to do a work while 12 women complete $\frac{3}{4}$th of the same work in 36 days. In how many days 10 men and 8 women together will complete the same work ? **(SBI Bank PO 2006)**

(*a*) 6 days (*b*) 27 days (*c*) 12 days (*d*) Data inadequate

Sol. (*b*) The number of persons and number of days are same in both the cases.

⇒ 1 woman's work ≡ $\frac{3}{4}$ man's work ⇒ 8 women's work ≡ $8 \times \frac{3}{4} = 6$ men's work

∴ 10 men + 8 women ≡ 10 men + 6 men = 16 men

12 men can do the work in 36 days. ⇒ 16 men can do the work in $36 \times \frac{12}{16} = 27$ days

137. 4 men and 6 women can complete a work in 8 days, while 3 men and 7 women can complete it in 10 days. In how many days will 10 women complete it ? **(SSC CGL Tier I 2004)**

(*a*) 50 (*b*) 45 (*c*) 40 (*d*) 35

Sol. (*c*) (4 men + 6 women) × 8 ≡ (3 men + 7 women) × 10

⇒ 32 men + 48 women ≡ 30 men + 70 women

⇒ (32 – 30) men ≡ (70 – 48) women ⇒ 2 men ≡ 22 women ⇒ 1 man ≡ 11 women

∴ 4 men + 6 women ≡ 44 women + 6 women ≡ 50 women

50 women can do the work in 8 days ⇒ 10 women can do the work in $8 \times \frac{50}{10} = 40$ days

138. 4 men and 6 women complete a work in 8 days, 2 men and 9 women also complete in 8 days. The number of days 18 women complete the work is **(SSC CGL Tier I 2015)**

(*a*) $4\frac{2}{3}$ days (*b*) $5\frac{2}{3}$ days (*c*) $4\frac{1}{3}$ days (*d*) $5\frac{1}{3}$ days

Sol. (*d*) (4 men + 6 women) × 8 = (2 men + 9 women) × 8

⇒ (4 men + 6 women) = (2 men + 9 women)

⇒ (4 – 2) men ≡ (9 – 6) women ⇒ 2 men ≡ 3 women

∴ 2 men + 9 women ≡ 3 women + 9 women ≡ 12 women can do the work in 8 days.

∴ 18 women can do the work in $\frac{12 \times 8}{18} = \frac{16}{3} = 5\frac{1}{3}$ days.

139. If 6 men and 8 boys can do a piece of work in 10 days while 26 men and 48 boys can do the same in 2 days, what is the time taken by 15 men and 20 boys in doing the same type of work? **(CDS 2010)**

(*a*) 4 days (*b*) 5 days (*c*) 6 days (*d*) 7 days

Sol. (*a*) (6 men + 8 boys) × 10 ≡ (26 men + 48 boys) × 2

⇒ 60 men + 80 boys ≡ 52 men + 96 boys

⇒ (60 – 52) men ≡ (96 – 80) boys ⇒ 8 men ≡ 16 boys ⇒ 1 man ≡ 2 boys

∴ 6 men + 8 boys ≡ 12 boys + 8 boys ≡ 20 boys can do the work in 10 days.

∴ 15 men + 20 boys = 30 boys + 20 boys = 50 boys can do the work in = $\frac{20 \times 10}{50} = 4$ days

140. 8 men and 4 women together can complete a piece of work in 6 days. Work done by a man in one day is double the work done by a woman in one day. If 8 men and 4 women started working and after 2 days, 4 men left and 4 new women joined, in how many more days will the work be completed ? **(Bank P.O. 2009)**

(*a*) 5 days (*b*) 8 days (*c*) 6 days (*d*) 4 days

Sol. (*a*) 8 men + 4 women ≡ (8 × 2) women + 4 women = 20 women

8 men + 4 women ≡ 20 women worked for 2 days

Remaining work = 20 women's work for (6 – 2) = 4 days

(8 men + 4 women) – 4 men + 4 women = 4 men + 8 women ≡ 16 women

$\therefore$ Required time = $\frac{4 \times 20}{16}$ = 5 days

141. 4 men and 3 boys can do a work in 5 days whereas 3 men and 4 boys can do the same work in 6 days. If daily wages of a man is ₹ 180, find the daily wages of a boy.

(*a*) ₹ 40 (*b*) ₹ 60 (*c*) ₹ 90 (*d*) ₹ 180

Sol. (*a*) (4 men + 3 boys) × 5 ≡ (3 men + 4 boys) × 6

⇒ (20 – 18) men ≡ (24 – 15) boys ⇒ 2 men ≡ 9 boys ⇒ Wages of a boy = $\frac{2}{9} \times 180$ = ₹ 40

142. 18 men and 12 boys can do a piece of work in 5 days. 9 men and 12 boys can do the same work in 8 days. How many boys should assist 12 men so that the work is finished in 6 days?

(*a*) 4 boys (*b*) 8 boys (*c*) 16 boys (*d*) 20 boys

Sol. (*c*) (18 × 5 – 9 × 8) men's work ≡ (12 × 8 – 12 × 5) boys' work

$\therefore$ 18 men's work ≡ 36 boys' work ⇒ 1 man's work ≡ 2 boys' work

18 men + 12 boys ≡ 18 men + 6 men = 24 men

Men required to finish the work in 5 days = 24 men

$\therefore$ Men required to finish the work in 6 days = 24 × $\frac{5}{6}$ = 20 men

12 men are already working on it.

$\therefore$ Additional men required = 20 men – 12 men = 8 men ≡ 16 boys

143. 9 children can complete a piece of work in 360 days. 18 men can complete the same piece of work in 72 days and 12 women can complete the piece of work in 162 days. In how many days can 4 men, 12 women and 10 children together complete the piece of work ? **(Corporation Bank PO 2006)**

(*a*) 124 (*b*) 81 (c) 68 (*d*) 96

Sol. (*b*) (18 × 72) men's work = (12 × 162) women's work = (9 × 360) children's work

$\therefore$ 2 men ≡ 3 women ≡ 5 children

$\therefore$ 4 men + 12 women + 10 children = 4 men + 12 × $\frac{2}{3}$ men + 10 × $\frac{2}{5}$ men = 16 men

18 men can do the work in 72 days ⇒ 16 men can do the work in $\frac{18 \times 72}{16}$ = 81 days

144. 8 men can complete a piece of work in 4 days. 12 women can complete the same piece of work in 4 days whereas 8 children can complete the same piece of work in 8 days. 2 men, 8 children and 3 women work together for 2 days. If only women were to finish the remaining work in 2 days, how many total women would be required ? **(PNB Officers Exam. 2009)**

(*a*) 12 (*b*) 18 (c) 24 (*d*) 20

Sol. (*a*) (8×4) men's work = (12×4) women's work = (8×8) children's work

$\Rightarrow$ 2 men $\equiv$ 3 women $\equiv$ 4 children

2 men + 3 women + 8 children = 3 women + 3 women + 6 women = 12 women

12 women can do the work in 4 days. 12 women has worked for 2 days.

Remaining work = 12 women's work for 2 days. $\Rightarrow$ Required number of women = 12

145. 12 men can complete a piece of work in 36 days. 18 women can complete the same piece of work in 60 days. 8 men and 20 women work together for 20 days. If only women were to finish the remaining work in 4 days, how many total women would be required ?

(Union Bank of India PO 2005)

(*a*) 70 (*b*) 28 (c) 66 (*d*) 40

Sol. (*a*) (12×36) men's work = (18×60) women's work $\Rightarrow$ 1 man $\equiv \frac{18 \times 60}{12 \times 36} = \frac{5}{2}$ women

8 men + 20 women = $(8 \times \frac{5}{2} + 20)$ women = 40 women

Total work days = 18×60 woman days = 1080 woman days

Work already done = 40×20 = 800 woman days

Remaining work = 1080 – 800 = 280 woman days

Required number of women = $280 \div 4$ = 70 women

146. A man, a woman and a boy can do a work in 3, 4 and 12 days respectively. How many boys must assist 1 man and 1 woman so as to finish the work in 1 day? **(SSC Asstt. Grade 1996)**

(*a*) 3 boys (*b*) 5 boys (*c*) 10 boys (*d*) 17 boys

Sol. (*b*) 3 men $\equiv$ 12 boys $\Rightarrow$ 1 man $\equiv$ 4 boys. 4 women $\equiv$ 12 boys $\Rightarrow$ 1 woman $\equiv$ 3 boys

$\therefore$ 1 man + 1 woman $\equiv$ (4 + 3) boys = 7 boys

1 boy can do the work in 12 days. $\Rightarrow$ 12 boys can do the work in 1 day.

$\therefore$ 12 – 7 = 5 boys must assist 1 man and 1 woman to complete the work in 1 day.

Alt : 1 day work of a man and a woman = $\frac{1}{3} + \frac{1}{4} = \frac{7}{12} \Rightarrow$ Remaining work = $1 - \frac{7}{12} = \frac{5}{12}$

$\therefore$ Required number of boys = $\frac{5}{12} \div \frac{1}{12} = \frac{5}{12} \times \frac{12}{1} = 5$

147. A can do a certain work in the same time in which B and C together can do it. If A and B together could do it in 10 days and C alone in 50 days, then B alone could do the work in :

(SSC CGL Tier I 2002, 2013)

(*a*) 15 days (*b*) 20 days (*c*) 25 days (*d*) 30 days

Sol. (*c*) $(A + B + C) = (A + B) + C = \frac{1}{10} + \frac{1}{50} = \frac{6}{50} = \frac{3}{25}$

$A = B + C \Rightarrow A = \frac{1}{2}(A + B + C) = \frac{1}{2} \times \frac{3}{25} = \frac{3}{50} \Rightarrow B = (A + B) - A = \frac{1}{10} - \frac{3}{50} = \frac{2}{50} = \frac{1}{25}$

148. A, B and C working together can do a piece of work in 10 days. If A takes twice as long as B and C together, B takes thrice as long as A and C together, how many days will C alone take to finish the work?

(*a*) 20 days (*b*) 24 days (*c*) 30 days (*d*) 40 days

Sol. (*b*) Time taken by A is twice the time taken by B and C together.

$\therefore$ Days taken by A = 3 times of days taken by $(A + B + C) = 3 \times 10 = 30$ days

$\therefore$ One day's work of A = $\frac{1}{30}$

Time taken by B is thrice the time taken by A and C together.

$\therefore$ Days taken by B = 4 times of days taken by (A + B + C) = 4 × 10 = 40 days

$\therefore$ One day's work of B = $\frac{1}{40}$ But (A + B + C)'s one day's work = $\frac{1}{10}$

$\therefore$ One day's work of C = $\frac{1}{10} - \frac{1}{30} - \frac{1}{40} = \frac{12-4-3}{120} = \frac{5}{120} = \frac{1}{24}$

149. A can do a work in 40 days. He worked at it for 5 days and then with the help of B, he finishes the remaining work in 28 days. They can finish the work together in :

(*a*) 32 (*b*) 34 (*c*) 35 (*d*) 42

Sol. (*a*) A's work for 5 days = $\frac{1}{40} \times 5 = \frac{1}{8} \Rightarrow$ Remaining work = $1 - \frac{1}{8} = \frac{7}{8}$ done in 28 days

$\Rightarrow$ Required days = $28 \times \frac{8}{7}$ = 32 days

150. A can finish a work in 18 days and B can do the same work in 15 days. B worked for 10 days and left the job. In how many days, A alone can finish the remaining work ?

(SSC CGL Tier I 2013)

(*a*) 8 (*b*) 6 (*c*) 5½ (*d*) 5

Sol. (*b*) Work done by B in 10 days = $\frac{1}{15} \times 10 = \frac{2}{3} \Rightarrow$ Remaining work = $1 - \frac{2}{3} = \frac{1}{3}$

$\therefore$ Time taken by A = $\frac{1}{3} \times 18$ = 6 days

Alt : Let total work = LCM of 15 and 18 = 90. Work done by B in one day = 90 ÷ 15 = 6

Work of B for 10 days = 6 × 10 = 60 $\Rightarrow$ Remaining work = 90 – 60 = 30

Work done by A in one day = 90 ÷ 18 = 5 $\Rightarrow$ Time taken by A = 30 ÷ 5 = 6 days

151. A and B can together finish a work in 30 days. They worked at it for 20 days and then B left. The remaining work was done by A alone in 20 more days. A alone can finish the work in :

(SSC CGL Tier I 2003, 2014)

(*a*) 50 days (*b*) 60 days (*c*) 48 days (*d*) 54 days

Sol. (*b*) Work of A and B for 20 days = $\frac{1}{30} \times 20 = \frac{2}{3}$

Remaining work = $1 - \frac{2}{3} = \frac{1}{3}$ which is completed by A in 20 days.

$\therefore$ Required time = 3 × 20 days = 60 days

152. A can do a piece of work in 30 days. He worked at it for 7 days and then B finished the remaining work in 23 days. A and B working together can finish the work in

(*a*) 30 days (*b*) 24 days (*c*) 12 days (*d*) 15 days

Sol. (*d*) A could finish the remaining work in 30 – 7 = 23 days = Time taken by B.

Working capabilities of A and B are same $\Rightarrow$ Required time = 30 ÷ 2 = 15 days

153. A can do a piece of work in 25 days and B in 20 days. They worked together for 5 days and then A goes away. In how many days will B finish the remaining work ? **(MAT)**

(*a*) 10 days (*b*) 11 days (*c*) 16 days (*d*) 21 days

Sol. (b) A's work for 5 days = $\frac{1}{25} \times 5 = \frac{1}{5} \Rightarrow$ Remaining work = $1 - \frac{1}{5} = \frac{4}{5}$

Total time taken by B = 4/5 × 20 = 16 days ⇒ B alone worked for 16 – 5 = 11 days

154. A can do a piece of work in 60 days. He works at it for 15 days and then B alone finishes the remaining work in 30 days. The two together can finish the work in : (SSC CGL Tier I 2003)

(*a*) 24 days (*b*) 25 days (*c*) 30 days (*d*) 32 days

Sol. (*a*) A's 15 days work = $\frac{15}{60} = \frac{1}{4}$

Remaining work = $1 - \frac{1}{4} = \frac{3}{4}$ which is completed by B in 30 days.

∴ B alone can do the work in $30 \times \frac{4}{3} = 40$ days

∴ A and B together can do the work in $\frac{60 \times 40}{100} = 24$ days

155. A can finish a work in 24 days, B in 9 days and C in 12 days. B and C start the work but are forced to leave after 3 days. The remaining work was done by A in : (SSC CGL Tier I 2003)

(*a*) 5 days (*b*) 6 days (*c*) 10 days (*d*) $10\frac{1}{2}$ days

Sol. (*c*) B work for 3 days = $\frac{1}{9} \times 3 = \frac{1}{3}$ and C work for 3 days = $\frac{1}{12} \times 3 = \frac{1}{4}$

Their work for 3 days = $\frac{1}{3} + \frac{1}{4} = \frac{7}{12} \Rightarrow$ Remaining work = $1 - \frac{7}{12} = \frac{5}{12}$

Required time = $\frac{5}{12} \times 24$ days = 10 days

156. A and B can do a work in 15 days and 10 days respectively. They started doing the work together but after 2 days, B had to leave and A alone completed the remaining work. The whole work was completed in : (SSC CGL Tier I 2004)

(*a*) 10 days (*b*) 8 days (*c*) 12 days (*d*) 15 days

Sol. (*c*) B's work for 2 days = $\frac{1}{10} \times 2 = \frac{1}{5} \Rightarrow$ Remaining work done by A = $1 - \frac{1}{5} = \frac{4}{5}$

∴ Time taken by A = $\frac{4}{5} \times 15 = 12$ days

157. A and B can do a job in 6 and 12 days respectively. They began the work together but A leaves after 3 days. Then the total number of days needed for the completion of the work is: (SSC CGL Tier I 2000)

(*a*) 4 (*b*) 5 (*c*) 6 (*d*) 9

Sol. (*c*) A's work for 3 days = $\frac{3}{6} = \frac{1}{2} \Rightarrow$ Remaining work = $1 - \frac{1}{2} = \frac{1}{2}$

Days taken by B = $\frac{1}{2} \times 12 = 6$ days

158. A can do a work in 10 days and B can do the same work in 20 days. They work together for 5 days and then A leaves. How many days required by B to finish the remaining work?

(*a*) 8 days (*b*) 5 days (*c*) 12 days (*d*) 6 days

Sol. (*b*) A work for 1 day = 100% ÷ 10 = 10% B work for 1 day = 100% ÷ 20 = 5%

Their work for 5 days = 5 × (10% + 5%) = 75%

⇒ Remaining work = 100% – 75% = 25% ⇒ Time taken by B = 25% ÷ 5 = 5 days.

159. A can do a piece of work in 20 days and B in 30 days. They work together for 7 days and then both leave the work. Then C alone finishes the remaining work in 10 days. In how many days will C finish the full work? **(SSC CGL Tier II 2013)**

(*a*) 25 days (*b*) 30 days (*c*) 24 days (*d*) 20 days

Sol. (*c*) A and B's work for 7 days = $\left(\frac{1}{20}+\frac{1}{30}\right)\times 7=\frac{5}{60}\times 7=\frac{7}{12}$

Remaining work = $1-\frac{7}{12}=\frac{5}{12}$ done by C in 10 days

Time taken by C to do the full work = $10\times\frac{12}{5}=24$ days

160. A can complete a work in 9 days, B in 10 days and C in 15 days. B and C together started the work, but left the work after 2 days. The time taken to complete the remaining work by A will be : **(SSC CGL Tier I 2006)**

(*a*) 4 days (*b*) 6 days (*c*) 8 days (*d*) 9 days

Sol. (*b*) Work of B and C for 1 day each = $\frac{1}{10}+\frac{1}{15}=\frac{5}{30}=\frac{1}{6}$

Their work for 2 days = $\frac{1}{6}\times 2=\frac{1}{3}$ ⇒ Remaining work = $1-\frac{1}{3}=\frac{2}{3}$

∴ Time taken by A = $\frac{2}{3}\times 9=6$ days

161. A and B together can complete a work in 3 days. They start together. But, after 2 days, B left the work. If the work is completed after 2 more days, B alone could do the work in **(SSC CGL Tier I 2013)**

(*a*) 8 days (*b*) 10 days (*c*) 4 days (*d*) 6 days

Sol. (*d*) (A + B)'s work in 2 days = $\frac{1}{3}\times 2=\frac{2}{3}$ ⇒ Remaining work = $1-\frac{2}{3}=\frac{1}{3}$

A's 1 day work = $\frac{1}{2}\times\frac{1}{3}=\frac{1}{6}$ ⇒ B's 1 day work = $\frac{1}{3}-\frac{1}{6}=\frac{1}{6}$

162. First person can complete a work in 32 days and second person can complete the work in 40 days. Both the persons worked together for 10 days and then the second person leaves. The number of days taken by the first person to finish the balance work is **(SSC FCI 2012)**

(*a*) 16 (*b*) 14 (*c*) 21 (*d*) 7

Sol. (*b*) Work done by Second person (10 days) = $\frac{1}{40}\times 10=\frac{1}{4}$ ⇒ Remaining work = $1-\frac{1}{4}=\frac{3}{4}$

Days taken by first person = $\frac{3}{4}\times 32=24$ days

But the first person has already worked for 10 days with the second person

∴ Required number of days = 24 – 10 = 14 days

163. X and Y can do a piece of work in 30 days. They worked together for 6 days and then X quits and Y finishes the work in 32 more days. In how many days can Y do the piece of work alone ? **(SSC CGL Tier I 2012)**

(*a*) 30 days (*b*) 32 days (*c*) 34 days (*d*) 40 days

Sol. (*d*) Their work for 6 days = $\frac{6}{30} = \frac{1}{5} \Rightarrow$ Remaining work = $1 - \frac{1}{5} = \frac{4}{5}$

Y took 32 days to finish the remaining work.

$\therefore$ Y can finish the whole work in $32 \times \frac{5}{4} = 40$ days

164. Ram and Shyam can do a work in 8 days. Ram worked for 6 days and Shyam completed the remaining work in 10 days. In how many days can Shyam alone finish the work ?

(*a*) 12 days (*b*) 15 days (*c*) 16 days (*d*) 18 days

Sol. (*c*) Their work for 6 days $= \frac{6}{8} = \frac{3}{4}$

Work done by Shyam in (10 – 6) or 4 days $= 1 - \frac{3}{4} = \frac{1}{4}$

$\therefore$ Shyam can do the work in 4 × 4 days = 16 days

165. A and B can separately do a piece of work in 20 and 15 days respectively. They worked together for 6 days after which B was replaced by C. The work was finished in next 4 days. The number of days in which C alone could do the work is **(SSC CGL Tier I 2012)**

(*a*) 30 days (*b*) 35 days (*c*) 40 days (*d*) 45 days

Sol. (*c*) $A = \frac{6+4}{20} = \frac{1}{2}, B = \frac{6}{15} = \frac{2}{5} \Rightarrow$ Remaining work $= 1 - \frac{1}{2} - \frac{2}{5} = \frac{1}{10}$

$\Rightarrow$ Time taken by C = 10 × 4 = 40 days

166. A and B together can do a work in 25 days, C alone can do it in 15 days. They worked together for 6 days and then C leaves off. In how many days will the remaining work be finished by A and B?

(*a*) 9 days (*b*) 10 days (*c*) 12 days (*d*) 15 days

Sol. (*a*) C's 6 days work $= \frac{1}{15} \times 6 = \frac{2}{5} \Rightarrow$ Remaining Work $= 1 - \frac{2}{5} = \frac{3}{5}$

A and B will finish the remaining work in $\frac{3}{5} \times 25 = 15$ days

But A and B had already worked for 6 days with C.

$\therefore$ A and B alone will work at it for 15 – 6 = 9 days.

167. Two pipes, P and Q can fill a cisten in 12 and 15 minutes respectively. If both are opened together and at the end of 3 minutes, the first is closed, how much longer will the cistern take to fill ? **(SSC CGL Tier I 2013)**

(*a*) $8\frac{3}{4}$ minutes (*b*) 5 minutes (*c*) $8\frac{1}{2}$ minutes (*d*) $8\frac{1}{4}$ minutes

Sol. (*d*) Work of pipe P in 3 minutes $= \frac{3}{12} = \frac{1}{4} \Rightarrow$ Remaining work $= 1 - \frac{1}{4} = \frac{3}{4}$

$\Rightarrow$ Time taken by pipe Q $= \frac{3}{4} \times 15$ minutes $= \frac{45}{4} = 11\frac{1}{4}$ minutes

Required time $= 11\frac{1}{4} - 3 = 8\frac{1}{4}$ minutes

168. A and B can do a piece of work in 20 days and 12 days respectively. A started the work alone and then after 4 days B joined him till the completion of the work. How long did the work last ? **(SSC CGL Tier I 2004)**

(*a*) 10 days (*b*) 20 days (*c*) 15 days (*d*) 6 days

Sol. (*a*) Work of A for first 4 days $= \frac{1}{20} \times 4 = \frac{1}{5} \Rightarrow$ Remaining work $= 1 - \frac{1}{5} = \frac{4}{5}$

Time taken to finish the remaining work $= \frac{4}{5} \times \frac{20 \times 12}{32} = 6$ days

$\therefore$ Total time taken = 4 + 6 = 10 days

169. A and B can separately complete a piece of work in 20 days and 30 days respectively. They worked together for some time, then B left the work. If A completed the rest of the work in 10 days, then B worked for : **(SSC CGL Tier I 2008)**

(*a*) 6 days (*b*) 8 days (*c*) 12 days (*d*) 16 days

Sol. (*a*) A's work for last 10 days $= \frac{1}{20} \times 10 = \frac{1}{2}$

Remaining work $= 1 - \frac{1}{2} = \frac{1}{2}$ is done by A and B together.

$\therefore$ Time taken by them $= \frac{1}{2} \times \frac{30 \times 20}{50} = 6$ days

170. A and B each working alone can do a work in 15 days and 25 days respectively. They started the work together, but B left after some time and A finished the remaining work in 7 days. After how many days from the start did B leave ? **(SSC CGL Tier II 2012)**

(*a*) 7 (*b*) 9 (*c*) 3 (*d*) 5

Sol. (*d*) A's work in last 7 days $= \frac{7}{15}$. Remaining work $= 1 - \frac{7}{15} = \frac{8}{15}$

Time taken by A and B to finish $\frac{8}{15}$ of the work $= \frac{8}{15} \times \frac{15 \times 25}{40} = 5$ days

171. A, B, C can do a piece of work individually in 8, 10 and 15 days respectively. A and B start working but A quits after working for 2 days. After this, C joins B till the completion of the work. In how many days will the work be completed ? **(CDS 2010)**

(*a*) 53/9 (*b*) 34/7 (*c*) 85/3 (*d*) 53/10

Sol. (*d*) A's work for first 2 days $= \frac{2}{8} = \frac{1}{4}$. B's work for first 2 days $= \frac{2}{10} = \frac{1}{5}$

Work in first two days $= \frac{1}{4} + \frac{1}{5} = \frac{9}{20} \Rightarrow$ Remaining work $= 1 - \frac{9}{20} = \frac{11}{20}$

Time taken by B and C $= \frac{11}{20} \times \frac{10 \times 15}{25} = \frac{33}{10} \Rightarrow$ Total time taken $= \frac{33}{10} + 2 = \frac{53}{10}$

172. A alone can do a piece of work in 12 days, B alone in 15 days. Find the time taken to finish the work if A leave two days before the scheduled time of completion of the work.

(*a*) $6\frac{2}{3}$ days (*b*) $9\frac{2}{3}$ days (*c*) $8\frac{1}{6}$ days (*d*) $9\frac{1}{6}$ days

Sol. (*d*) 1 day work of (A + B) = $\frac{1}{12}+\frac{1}{15}=\frac{3}{20}$ ⇒ Scheduled time = $\frac{20}{3}=6\frac{2}{3}$ days

⇒ They worked together for $6\frac{2}{3}-2=4\frac{2}{3}$ days

Work to be completed in last 2 days = $\frac{3}{20}\times 2=\frac{3}{10}$

Time taken by B to finish this work = $\frac{3}{10}\times 15=\frac{9}{2}$ days = $4\frac{1}{2}$ days

∴ Total time taken = $4\frac{2}{3}+4\frac{1}{2}=9\frac{1}{6}$ days

173. A, B and C can do a piece of work in 20, 24 and 30 days respectively. They undertook to do the piece of work for ₹ 5400. They began the work together but B left 2 days before the completion of work and C left 5 days before the completion of work. The share of A from the assured money is **(SSC MTS 2013)**

(*a*) ₹ 2700 (*b*) ₹ 5400 (*c*) ₹ 600 (*d*) ₹ 1800

Sol. (*a*) Their work for 1 day each = $\frac{1}{20}+\frac{1}{24}+\frac{1}{30}=\frac{6+5+4}{120}=\frac{15}{120}=\frac{1}{8}$

Their work in last 5 days = $\frac{5}{20}+\frac{3}{24}+\frac{0}{30}=\frac{1}{4}+\frac{1}{8}=\frac{3}{8}$

Remaining work = $1-\frac{3}{8}=\frac{5}{8}$ Days taken to finish = $\frac{5}{8}\div\frac{1}{8}=\frac{5}{8}\times 8=5$ days

Total days taken = 5 + 5 = 10 days ⇒ A's share = $\frac{10}{20}\times$ ₹ 5400 = ₹ 2700

174. A and B can complete a job in 24 days working together. A alone can complete it in 32 days. Both of them worked together for 8 days and then A left. The number of days B will take to complete the remaining job is **(SSC CGL Tier II 2012)**

(*a*) 64 (*b*) 128 (*c*) 16 (*d*) 32

Sol. (*a*) Work finished by A and B in first 8 days = $8\times\frac{1}{24}=\frac{1}{3}$

Remaining work = $1-\frac{1}{3}=\frac{2}{3}$ B alone can complete the work in $\frac{24\times 32}{8}=96$ days

Time taken by B to complete $\frac{2}{3}$ work = $\frac{2}{3}\times 96=64$ days

175. A and B together can do a piece of work in 36 days. If A work alone for the last 10 days, it is completed in 40 days. B alone can do the work in : **(SSC CGL Tier I 2006)**

(*a*) 45 days (*b*) 60 days (*c*) 75 days (*d*) 90 days

Sol. (*d*) Work of A and B for 1 day = $\frac{1}{36}$

Work of A and B for (40 – 10=) 30 days = $\frac{1}{36}\times 30=\frac{5}{6}$

Remaining work = $1-\frac{5}{6}=\frac{1}{6}$ which is done by A in last 10 days

$\therefore$ A's work for 1 day = $\frac{1}{6}\times\frac{1}{10}=\frac{1}{60}$ $\Rightarrow$ B's work for 1 day = $\frac{1}{36}-\frac{1}{60}=\frac{2}{180}=\frac{1}{90}$

176. **A did three-fourth of a work in 15 days. With the help of B, he finished the remaining work in 3 days. Find the ratio in which the wages earned by them, is to be distributed?**

(*a*) 3 : 4 (*b*) 3 : 2 (*c*) 9 : 1 (*d*) 9 : 2

Sol. (*c*) A alone can finish the work in $\frac{4}{3}\times 15$ = 20 days

Work done by A in 18 (*i.e.* 15 + 3) days = $\frac{18}{20}=\frac{9}{10}$

$\therefore$ Work done by B = $1-\frac{9}{10}=\frac{1}{10}$ $\Rightarrow$ Ratio of wages = $\frac{9}{10}:\frac{1}{10}$ = 9 : 1

177. **Two pipes A and B can fill a cistern in 60 minutes and 48 minutes respectively. Both pipes are opened together. This cistern will be filled in half an hour, if the pipe B is turned off after:**

(*a*) 6 minutes (*b*) 18 minutes (*c*) 20 minutes (*d*) 24 minutes

Sol. (*d*) Work of pipe A in 30 minutes = $\frac{1}{60}\times 30=\frac{1}{2}$

Remaining work = $1-\frac{1}{2}=\frac{1}{2}$ $\Rightarrow$ Time taken by pipe B = $\frac{1}{2}\times 48$ = 24 minutes

178. **Two pipes A and B can fill a cistern in $37\frac{1}{2}$ minutes and 45 minutes respectively. Both pipes are opened. This cistern will be filled just in half an hour, if the pipe B is turned off after :** **(SSC CGL Tier I 2004)**

(*a*) 15 minutes (*b*) 10 minutes (*c*) 5 minutes (*d*) 9 minutes

Sol. (*d*) Work of pipe A in 30 minutes = $30\div 37\frac{1}{2}=30\times\frac{2}{75}=\frac{4}{5}$

Remaining work = $1-\frac{4}{5}=\frac{1}{5}$ $\Rightarrow$ Time taken by pipe B = $\frac{1}{5}$ × 45 minutes = 9 minutes

179. **A cistern can be filled by two supply pipes in 20 mins and 30 mins respectively. But a waste pipe can empty it in an hour. Both the supply pipes were opened together to fill the empty cistern, but by mistake the waste pipe was also open. How much later was the waste pipe closed, if the cistern is filled in 14 mins ?** **(SSC CGL Tier I 2013)**

(*a*) 6 mins (*b*) 8 mins (*c*) 10 mins (*d*) 12 mins

Sol. (*c*) One minute work of two pipes = $\frac{1}{20}+\frac{1}{30}=\frac{5}{60}=\frac{1}{12}$

Work done by two pipes in 2 extra (i.e. 14 – 12) minutes = $\frac{1}{12}\times 2=\frac{1}{6}$

Time taken by waste pipe to drain $\frac{1}{6}$ th tank $= \frac{1}{6} \times 1$ hour $= 10$ minutes

180. A and B can do a piece of work in 30 days. They worked together for 6 days and then A leaves off. If B finishes the remaining work in 36 days, how many days will A alone take to do the work?

(*a*) 45 days (*b*) 60 days (*c*) 80 days (*d*) 90 days

Sol. (*d*) 6 days' work of A and B $= \frac{6}{30} = \frac{1}{5}$

Remaining work $= 1 - \frac{1}{5} = \frac{4}{5}$ which is completed by B in 36 days.

$\therefore$ B alone can do the work in $36 \times \frac{5}{4} = 45$ days

$\therefore$ A alone can do the work in $\frac{30 \times 45}{15} = 90$ days

181. A can do a piece of work in 18 days and B in 12 days. They began the work together, but B left the work 3 days before its completion. In how many days, in all, was the work completed? **(SSC CGL Tier I 2008)**

(*a*) 12 days (*b*) 10 days (*c*) 9.6 days (*d*) 9 days

Sol. (*d*) For last 3 days, A alone has done the work.

$\therefore$ A's work for last 3 days $= \frac{1}{18} \times 3 = \frac{1}{6} \Rightarrow$ Remaining work $= 1 - \frac{1}{6} = \frac{5}{6}$

Time taken by A and B to do the remaining work $= \frac{5}{6} \times \frac{12 \times 18}{30} = 6$ days

$\therefore$ Total time taken $= 3 + 6 = 9$ days

182. A, B and C can do a piece of work in 10, 12 and 15 days respectively. They started the work together, but B leaves after 3 days. Find the days in which A and C will finish the remaining work.

(*a*) 1 day (*b*) 1.5 days (*c*) 3 days (*d*) 4.5 days

Sol. (*b*) B's work for first 3 days $= \frac{1}{12} \times 3 = \frac{1}{4} \Rightarrow$ Remaining work $= 1 - \frac{1}{4} = \frac{3}{4}$

Time taken by A and C $= \frac{3}{4} \times \frac{10 \times 15}{25} = \frac{9}{2} = 4.5$ days

But out of 4.5 days, A and C had already worked with B for 3 days.

$\therefore$ Time taken for the remaining work $= 4.5 - 3 = 1.5$ days

183. A, B and C together can finish a work in 12 days. They started the work together and then A leaves after 4 days. If B and C took 10 days to finish the remaining work, in how many days can A alone do the work?

(*a*) 60 days (*b*) 80 days (*c*) 100 days (*d*) 120 days

Sol. (*a*) 4 days' work of (A + B + C) $= \frac{1}{12} \times 4 = \frac{1}{3} \Rightarrow$ Remaining work $= 1 - \frac{1}{3} = \frac{2}{3}$

B and C finished $\frac{2}{3}$rd of the work in 10 days.

$\therefore$ One day's work of B and C $= \frac{2}{3} \times \frac{1}{10} = \frac{1}{15}$

A's 1 day's work = (A + B + C) – (B + C) $= \frac{1}{12} - \frac{1}{15} = \frac{5-4}{60} = \frac{1}{60}$

184. A can do a piece of work in 10 days and B in 20 days. They begin together but A leaves 2 days before the completion of the work. The whole work will be done in **(SSC CHSL 2012)**

(*a*) 8 days (*b*) $7\frac{2}{3}$ days (*c*) 7 days (*d*) 6 days

Sol. (*a*) Work done in last two days (done by B only) $= \frac{2}{20} = \frac{1}{10}$

Remaining work (done by A and B together) $= 1 - \frac{1}{10} = \frac{9}{10}$

Time taken by A and B together $= \frac{10 \times 20}{30} \times \frac{9}{10} = 6$ days

$\therefore$ Total days taken = 2 + 6 = 8 days

185. A can complete a work in 20 days and B in 30 days. A worked alone for 4 days and then B completed the remaining work along with C in 18 days. In how many days can C working alone complete the work ? **(SSC CGL Tier II 2012)**

(*a*) 72 (*b*) 90 (*c*) 12 (*d*) 68

Sol. (*b*) A's work for 4 days $= \frac{4}{20} = \frac{1}{5}$ B's work for 18 days $= \frac{18}{20} = \frac{3}{5}$

Remaining work $= 1 - \frac{1}{5} - \frac{3}{5} = \frac{1}{5}$ is done by C in 18 days.

$\Rightarrow$ C's work for 1 day $= \frac{1}{5} \times \frac{1}{18} = \frac{1}{90}$

186. A, B and C can do a work in 12, 16 and 24 days respectively. They started the work together but A left after working for 3 days and B left 3 days before the completion of the work. How many days would it take to finish the work ?

(*a*) 3 days (*b*) 6 days (*c*) 12 days (*d*) 9 days

Sol. (*d*) A's work for first 3 days $= \frac{1}{12} \times 3 = \frac{1}{4}$ C's work for last 3 days $= \frac{1}{24} \times 3 = \frac{1}{8}$

$\therefore$ Remaining work $= 1 - \frac{1}{4} - \frac{1}{8} = \frac{5}{8}$. This work was done by B and C together.

$\therefore$ Days taken by B and C to do $\frac{5}{8}$ of the work $= \frac{5}{8} \times \frac{16 \times 24}{40} = 6$ days

$\therefore$ Total time taken = 6 + 3 = 9 days.

Hint: For the first 3 days A has worked with B and C.

$\therefore$ Those 3 days are already included in 6 days and hence not added again.

187. A, B and C can do a piece of work in 24, 30 and 40 days respectively. They began the work together but C left 4 days before completion of the work. In how many days was the work done ? **(SSC CGL Tier I 2015)**

(*a*) 11 days (*b*) 12 days (*c*) 13 days (*d*) 14 days

Sol. (*a*) Work done by A and B in last 4 days $= 4\left(\frac{1}{24}+\frac{1}{30}\right) = 4\times\frac{9}{120} = \frac{3}{10}$

Remaining work $= 1-\frac{3}{10} = \frac{7}{10}$

(A + B + C)'s one day work $= \frac{1}{24}+\frac{1}{30}+\frac{1}{40} = \frac{5+4+3}{120} = \frac{12}{120} = \frac{1}{10}$

Time required to finish $\frac{7}{10}$ work $= \frac{7}{10}\times 10 = 7$ days $\Rightarrow$ Total time taken $= 7 + 4 = 11$ days

Trick: Loss of work due to absence of C in last 4 days = 4/40 = 1/10
Extra time required by A, B and C to finish this work = 1/10 × 10 = 1 days
(A + B + C) can finish the work in 10 days ⇒ Total days = 10 + 1 = 11 days

188. A, B and C can complete a work in 10, 12 and 15 days respectively. They started the work together. But A left the work before 5 days of its completion. B also left the work 2 days after A left. In how many days was the work completed ? **(SSC CPO SI 2007)**

(*a*) 4 days (*b*) 5 days (*c*) 7 days (*d*) 8 days

Sol. (*c*) Let total days = *x*. Then $\frac{x-5}{10}+\frac{x-3}{12}+\frac{x}{15} = 1 \Rightarrow \frac{6x-30+5x-15+4x}{60} = 1$

$\Rightarrow 15x - 45 = 60 \Rightarrow 15x = 105 \Rightarrow x = 7$ days

189. 30 men can do a work in 20 days. After how many days should 6 men leave so that the work is completed in 23 days.

(*a*) 5 days (*b*) 9 days (*c*) 7.5 days (*d*) 8 days

Sol. (*d*) $30 \times x + 24(23 - x) = 30 \times 20 \Rightarrow 6x = 600 - 552 = 48 \Rightarrow x = 8$ days

Alt : Total man-days = 30 × 20 = 600
Man-days by 30 men in 23 days = 30 × 23 = 690
Extra man-days = 690 – 600 = 90
6 men did not work for 90 ÷ 6 = 15 days ⇒ 6 men left after (23 – 15) = 8 days

190. A man and a woman working together can do a piece of work in 33 days. If they do the work in the ratio 14 : 11, how many days will a woman working alone take to do the work. **(SSC CGL Tier I 2009)**

(*a*) 11 (*b*) 14 (*c*) 75 (*d*) 90

Sol. (*c*) One day work of a man and a woman $= \frac{1}{33}$

Ratio of work done by a man and a woman = 14 : 11

$\therefore$ 1 day's work of a woman $= \frac{1}{33}\times\frac{11}{25} = \frac{1}{75}$

191. A man and a woman working together can do a piece of work in 30 days. If they do the work in the ratio 6 : 7, how many days will a man working alone take to do the work.

(*a*) 180 days (*b*) 5 days (*c*) 42 days (*d*) 65 days

Sol. (*d*) One day work of a man and a woman $= \frac{1}{30}$

Ratio of work done by a man and a woman = 6 : 7

$\therefore$ The man does $\frac{6}{13}$th of the work in the given period.

$\therefore$ 1 day's work of a man $= \frac{1}{30} \times \frac{6}{13} = \frac{1}{65}$

192. A works twice as fast as B. If both of them can together finish a work in 12 days, B alone can do it in : **(SSC CGL Tier I 2000, SSC CHSL 2012, SSC CGL 1994 M)**

(*a*) 48 days (*b*) 36 days (*c*) 27 days (*d*) 24 days

Sol. (*b*) Ratio of work done by A and B = 2 : 1 $\Rightarrow$ B's one day work $= \frac{1}{12} \times \frac{1}{1+2} = \frac{1}{36}$

193. A, B and C can complete a work in 6 days. If A can work twice as faster than B and thrice faster than C, then the number of days C alone can complete the work is : **(SSC CGL Tier I 2015)**

(*a*) 11 days (*b*) 44 days (*c*) 22 days (*d*) 33 days

Sol. (*d*) Ratio of work of A, B and C = 6 : 3 : 2 $\Rightarrow$ C's 1 day work $= \frac{2}{11} \times \frac{1}{6} = \frac{1}{33}$

194. Babu and Asha can do a job together in 7 days. Asha is $1\frac{3}{4}$ times as efficeint as Babu. The same job can be done by Asha alone in : **(SSC CGL Tier I 2003)**

(*a*) $\frac{49}{4}$ days (*b*) $\frac{49}{3}$ days (*c*) 11 days (*d*) $\frac{28}{3}$ days

Sol. (*c*) Babu : Asha $= 1 : 1\frac{3}{4} = 1 : \frac{7}{4} = 4 : 7$ $\Rightarrow$ One day work of Asha $= \frac{1}{7} \times \frac{7}{7+4} = \frac{1}{11}$

195. To complete a work, A takes 50% more time than B. If together they take 18 days to complete the work, how much time shall B take to do it ? **(SSC CGL Tier I 2007)**

(*a*) 30 days (*b*) 35 days (*c*) 40 days (*d*) 45 days

Sol. (*a*) A : B = 2 : 3 $\Rightarrow$ One day work of B $= \frac{3}{5} \times \frac{1}{18} = \frac{1}{30}$

196. A can finish a work in 18 days and B can do the same work in half the time taken by A. Then working together what part of the same work they can finish in a day ? **(SSC CGL 2002)**

(*a*) $\frac{1}{6}$ (*b*) $\frac{2}{5}$ (*c*) $\frac{1}{9}$ (*d*) $\frac{2}{7}$

Sol. (*a*) B is twice as efficient as A.

Time taken by A and B together $= \frac{18}{2+1} = 6$ days $\Rightarrow$ One day work of A and B $= \frac{1}{6}$

197. A is 4 times as good a workman as B. If A could finish a work in 15 days less than B, find the time taken by them to finish the work while working together.

(*a*) 4 days (*b*) 5 days (*c*) 6 days (*d*) $7\frac{1}{2}$ days

Sol. (*a*) Ratio of time taken by A and B = 1 : 4

$\Rightarrow$ Difference = (4 – 1) ratio = 3 ratio ≡ 15 $\Rightarrow$ 1 ratio ≡ 5 days

$\therefore$ Days taken by A and B are 5 × 1 and 5 × 4 or 5 and 20 respectively.

$\therefore$ Working together they will finish the work in $\frac{20}{4+1} = 4$ days

198. Two men A and B started a job in which A was thrice as good as B and therefore took 60 days less than B to finish the job. How many days will they take to finish the job if they start working together ? **(SSC CGL Tier I 2012, 1999)**

(*a*) 15 days (*b*) 20 days (*c*) $22\frac{1}{2}$ days (*d*) 25 days

Sol. (*c*) Let A and B take x and $3x$ days respectively. Then $3x - x = 60 \Rightarrow 2x = 60 \Rightarrow x = 30$

Days taken by B = $3x = 3 \times 30 = 90$ days

Required number of days = $\frac{90}{3+1} = \frac{90}{4} = 22\frac{1}{2}$ days

199. Harpal is thrice good as a workman as Kewal and takes 10 days less than Kewal to complete a piece of work. How many days will Kewal take to complete the same piece of work ? **(SSC Tax Assistant 2008)**

(*a*) 5 days (*b*) 10 days (*c*) 15 days (*d*) 20 days

Sol. (*c*) Let Harpal and Kewal working alone will take x and $3x$ days respectively.

Then $3x - x = 10 \Rightarrow x = 5 \Rightarrow 3x = 15$ days

Alternative Method : Ratio of work done by Harpal and Kewal = 3 : 1

Difference in time taken by them = (3 – 1) ratio = 2 ratio = 10 days $\Rightarrow$ 1 ratio = 5 days

Time taken by Kewal = 3 ratio = 3 × 5 days = 15 days

200. A is twice as good a workman as B. Working together they can finish a work in 16 days. How many days will A take to finish the work?

(*a*) 4 days (*b*) 8 days (*c*) 24 days (*d*) 48 days

Sol. (*c*) Time taken by A = $\frac{16 \times (2+1)}{2} = 24$ days

Alt: One day work of A and B = $\frac{1}{16}$. Ratio of work done by A and B = 2 : 1

A's one day work = $\frac{1}{16} \times \frac{2}{3} = \frac{1}{24}$ $\Rightarrow$ A alone will take 24 days.

Alternative Method: A = 2B $\Rightarrow$ A + B = 2B + B = 3B

$\therefore$ (A + B) or 3B can do the work in 16 days.

$\therefore$ B alone can do the work in 16 × 3 = 48 days.

$\therefore$ A alone can do the work in 48 ÷ 2 = 24 days.

201. P takes 5 hours more than Q to complete a work. Working together they completed the work in 6 hours. Q alone could complete the work in **(SAS 2010)**

(*a*) 8 hours (*b*) 10 hours (*c*) 12 hours (*d*) 15 hours

Sol. (*b*) We can check the given options Option (*a*) : $\frac{13 \times 8}{21} \neq 6$; Option (*b*) : $\frac{15 \times 10}{25} = 6$

202. Two pipes A and B can fill a water tank in 10 minutes and 15 minutes respectively. Pipes A, B and a waste pipe C are opened simultaneously. If the waste pipe is closed after 6 minutes, it took 4 minutes more to fill the tank. Find the time in which waste-pipe would empty the tank.

(*a*) 4 minutes (*b*) 6 minutes (*c*) 9 minutes (*d*) 12 minutes

Sol. (*c*) One minute work of pipe A and B = $\frac{1}{10} + \frac{1}{15} = \frac{1}{6}$

(6 + 4) minute or 10 minutes work of pipe A and B = $\frac{1}{6} \times 10 = \frac{5}{3} = 1 + \frac{2}{3}$

$\therefore$ In 6 minutes, pipe C has emptied $\frac{2}{3}$ of the tank

Pipe C can empty the tank in $6 \times \frac{3}{2} = 9$ minutes

Alt: $4 \times \left(\frac{1}{10} + \frac{1}{15}\right) = 6 \times \frac{1}{x} \Rightarrow \frac{2}{3} = \frac{6}{x} \Rightarrow x = \frac{6 \times 3}{2} = 9$ minutes

203. A and B can do a work in 10 days and 15 days respectively. If they work at it on an alternate day, A starting first, in how many days the work will be completed? **(SSC CGL Tier I)**

(*a*) 6 days (*b*) 10 days (*c*) 12 days (*d*) 15 days

Sol. (*c*) Work done by A on the first day = $\frac{1}{10}$ and work done by B on the second day = $\frac{1}{15}$

$\therefore$ Two days' work (one day each of A and B) = $\frac{1}{10} + \frac{1}{15} = \frac{25}{10 \times 15} = \frac{1}{6}$

$\therefore$ Time taken by A and B to do the work = 2 × 6 = 12 days

Alternative Method : Let total work = LCM of 10 and 15 = 60

Work done by A in one day = 60 ÷ 10 = 6. Work done by B in one day = 60 ÷ 15 = 4

Work for 2 days (1 day of each) = 6 + 4 = 10 $\Rightarrow$ Total days = $2 \times \frac{60}{10} = 12$ days

204. Pipe A can fill a tank in 4 hours and pipe B can fill it in 6 hours. If they are opened on alternate hours and if pipe A is opened first, in how many hours, the tank shall be full ? **(SSC CGL Tier I 2015)**

(*a*) 4 (2/3) (*b*) 3 (1/4) (*c*) 4 (1/2) (*d*) 3 (1/2)

Sol. (*a*) Work done by A in the first hour = 1/4 and work done by B in the second hour = 1/6

$\therefore$ Two hours' work (one hour each of A and B) = $\frac{1}{4} + \frac{1}{6} = \frac{10}{24} = \frac{5}{12}$

$\therefore$ Work done by A and B in 2 hours each = $\frac{5}{12} \times 2 = \frac{5}{6}$

Time taken to finish the remaining work by A = $\frac{1}{6} \times 4 = \frac{2}{3}$

$\Rightarrow$ Total time = 4 + (2/3) = 4 (2/3) hours

Alternative Method : Let total work = LCM of 4 and 6 = 12

Work done by A in 1 hour = 12 ÷ 4 = 3. Work done by B in 1 hour = 12 ÷ 6 = 2

Work in 2 hours (1 day of each) = 3 + 2 = 5 $\Rightarrow$ Work in 4 hours (2 hours each) = 5 × 2 = 10

$\Rightarrow$ Total hours = 2 × 2 + (2/3) = = 4 (2/3) hours

205. A, B and C can do a piece of work in 10, 30 and 60 days respectively. If A is assisted by B and C on an alternate day, find number of days taken to finish the work.

(*a*) 4 days (*b*) 6 days (*c*) 8 days (*d*) 12 days

Sol. (*c*) Work done by A and B on the first day = $\frac{1}{10} + \frac{1}{30} = \frac{4}{30}$

Work done by A and C on the second day = $\frac{1}{10} + \frac{1}{60} = \frac{7}{60}$

$\therefore$ Work done in first two days $= \frac{4}{30} + \frac{7}{60} = \frac{15}{60} = \frac{1}{4} \Rightarrow$ Total time $= 4 \times 2 = 8$ days

Alternative Method : Let total work = LCM of 10, 30 and 60 = 60

A = 60 ÷ 10 = 6. B = 60 ÷ 30 = 2. C = 60 ÷ 60 = 1

A + B = 6 + 2 = 8; A + C = 6 + 1 = 7 ⇒ Work in 2 days = 8 + 7 = 15

Days taken = (60 ÷ 15) × 2 = 8

206. A can do a piece of work in 20 days and B in 40 days. A begins the work and thereafter they work alternatively one on each day. On which day will the work be completed ? **(SSC CGL Tier I 2006)**

(*a*) 24th day (*b*) 25th day (*c*) 26th day (*d*) 27th day

Sol. (*d*) Work done by A on the first day $= \frac{1}{20}$ and work done by B on the second day $= \frac{1}{40}$

$\therefore$ Two days' work (one day each of A and B) $= \frac{1}{20} + \frac{1}{40} = \frac{3}{40}$

Their work for first 26 days $= \frac{3}{40} \times 13 = \frac{39}{40} \Rightarrow$ Remaining work $= 1 - \frac{39}{40} = \frac{1}{40}$

On 27th day, A will work and his work for one day is $\frac{1}{20}$ which is more than $\frac{1}{40}$.

$\therefore$ Work will be completed on 27th day.

207. Two men A and B can finish a work in 8 and 16 hours respectively. If they work on it alternatively for one hour each, A beginning first, in how many hours will the work be finished?

(*a*) 5 hours (*b*) $5\frac{1}{2}$ hours (*c*) 10 hours (*d*) $10\frac{1}{2}$ hours

Sol. (*d*) Work of A and B for 1 hour each $= \frac{1}{8} + \frac{1}{16} = \frac{3}{16}$

Work of A and B for 5 hours each $= \frac{3}{16} \times 5 = \frac{15}{16} \Rightarrow$ Remaining work $= 1 - \frac{15}{16} = \frac{1}{16}$

(**Hint:** Work of A and B for 6 hours each exceeds 1)

Time taken by A to finish $\frac{1}{16}$th of the work $= 8 \times \frac{1}{16} = \frac{1}{2}$

$\therefore$ Required time $= (2 \times 5) + \frac{1}{2} = 10\frac{1}{2}$ hours

Alternative Method : Let total work = LCM or 8 and 16 = 16 units

A's 1 hour work = 2 units, and B's 1 hour work = 1 unit

10 hours (5 hours each) work = 5 × 3 units = 15 units ⇒ Remaining unit = 1

Time taken by A to do 1 unit $= \frac{1}{2}$ hours $\Rightarrow$ Required time $= (2 \times 5) + \frac{1}{2} = 10\frac{1}{2}$ hours

208. A can type 30 words in 1 minute and B 25 words in 1 minute. First A typed for 1 minute then B typed for 1 minute and so on. In this way, typing alternatively they typed 600 words. The time taken by them in typing these 600 words was : **(SSC CGL 2006)**

(*a*) 22 minutes (*b*) 21 minutes (*c*) 21 min 48 sec (*d*) 21 min 25 sec

Sol. (*c*) Words typed by both in first 2 minutes = 30 + 25 = 55 words

∴ Words typed in first 20 minutes (10 minutes by each) = 55 × 10 = 550 words

Remaining words = 600 – 550 = 50 words

Word typed by A in the next minute = 30 words ⇒ Remaining words = 50 – 30 = 20 words

Time taken by B = $\frac{20}{25} \times 60$ = 48 seconds ⇒ Total time taken = 21 minutes 48 seconds

209. A and B can complete a piece of work in 12 and 18 days respectively. A begins to do the work and they work alternatively one at a time for one day each. The whole work will be completed in : **(SSC CGL Tier I 2007)**

(*a*) $14\frac{1}{3}$ days (*b*) $15\frac{2}{3}$ days (*c*) $16\frac{1}{3}$ days (*d*) $18\frac{2}{3}$ days

Sol. (*a*) Work of A and B for 1 day each = $\frac{1}{12}+\frac{1}{18}=\frac{5}{36}$

Their work for 7 days each = $\frac{5}{36}\times 7=\frac{35}{36}$ ⇒ Remaining work = $1-\frac{35}{36}=\frac{1}{36}$

A can do the remaining work in $\frac{1}{36}\times 12=\frac{1}{3}$ day ⇒ Total time = $2 \times 7+\frac{1}{3}=14\frac{1}{3}$ days

210. Ganga and Saraswati, working separately can mow a field in 8 and 12 hours respectively. If they work in stretches of one hour alternatively, Ganga beginning at 9 a.m. when will the mowing be completed ? **(SSC CGL Tier I 2008)**

(*a*) 6 a.m. (*b*) 6.30 p.m. (*c*) 5 p.m. (*d*) 5.30 p.m.

Sol. (*b*) One hour work of each = $\frac{1}{8}+\frac{1}{12}=\frac{5}{24}$ ⇒ 4 hours work of each = $\frac{5}{24}\times 4=\frac{5}{6}$

Remaining work = $1-\frac{5}{6}=\frac{1}{6}$

Work done by Ganga in 9th hour = $\frac{1}{8}$ ⇒ Remaining work = $\frac{1}{6}-\frac{1}{8}=\frac{1}{24}$

$\frac{1}{24}$ th of the work finished by Saraswati in $\frac{1}{24}\times 12=\frac{1}{2}$ hours

Total time taken = $8+1+\frac{1}{2}=9\frac{1}{2}$ hours

Time at completion of work = 9 a.m. + $9\frac{1}{2}$ hours = 6.30 p.m.

Alt: Let total work = LCM of 8 and 12 = 24 units

Work done by A in first hour = 24 ÷ 8 = 3 units

Work done by B in second hour = 24 ÷ 12 = 2 units

Total work done in 2 hours = 3 + 2 = 5 units ⇒ Total work done in 8 hours = 5 × 4 = 20 units

Work done by A in next hour = 3 units

Remaining work = 24 – (20 + 3) = 1 unit ⇒ Time taken by B = 1 ÷ 2 = 30 minutes

Required time = 9 + 9 hours 30 minutes = 6.30 p.m.

211. Two pipes A and B can fill a cistern in 15 and 20 minutes respectively, while third pipe C can empty the cistern in 30 minutes. If the three taps are opened successively for one minute each, how long will it take to fill the cistern?

(*a*) 34 minutes (*b*) $34\frac{1}{3}$ minutes (*c*) 35 minutes (*d*) 36 minutes

Sol. (*b*) Work of all pipes for 3 minutes (1 minute's of each pipe) $= \frac{1}{15} + \frac{1}{20} - \frac{1}{30} = \frac{4+3-2}{60} = \frac{1}{12}$

$\therefore$ Work of 3 pipes for 11 minutes each $= \frac{1}{12} \times 11 = \frac{11}{12} \Rightarrow$ Remaining work $= 1 - \frac{11}{12} = \frac{1}{12}$

Tap A will fill $\frac{1}{15}$ of the cistern in next 1 minute $\Rightarrow$ Remaining work $= \frac{1}{12} - \frac{1}{15} = \frac{1}{60}$

Time taken by tap B to fill $\frac{1}{60}$ of the cistern $= \frac{1}{60} \times 20 = \frac{1}{3}$ minute

$\therefore$ Total time taken $= 33 + 1 + \frac{1}{3} = 34\frac{1}{3}$ minutes

212. A and B together can do a piece of work in 30 days. B and C can together do it in 20 days. A starts the work and works on it for 5 days, then B takes up and works for 15 days. Finally C finishes the work in 18 days. The number of days in which C alone can do the work when doing it separately is : **(SSC CGL Tier I 2015)**

(*a*) 24 days (*b*) 40 days (*c*) 120 days (*d*) 60 days

Sol. (a) Total work done = 5A + 15B + 18C = 1 $\Rightarrow$ 5(A + B) + 10(B + C) + 8C

$\Rightarrow \frac{5}{30} + \frac{10}{20} + 8C = 1 \Rightarrow 8C = 1 - \frac{1}{6} - \frac{1}{2} = \frac{1}{3} \Rightarrow C = \frac{1}{3} \times \frac{1}{8} = \frac{1}{24}$

213. 20 men can finish a work in 30 days. They started the work and at the end of every 10 days, 20 additional men are employed. In how many days will the work be finished?

(*a*) 15 days (*b*) 20 days (*c*) 24 days (*d*) 25 days

Sol. (*b*) Work done by 20 men in the first 10 days $= \frac{1}{3}$

Work done by 40 men in next 10 days $= \frac{2}{3} \Rightarrow$ Work done in 20 days $= \frac{1}{3} + \frac{2}{3} = 1$

Alt : 20 men can do the work in 30 days. Total man days = 20 × 30 = 600 days
Work in the first 10 days by 20 men = 10 × 20 = 200 man days
Work in the next 10 days by 40 men = 10 × 40 = 400 man days

214. 25 men can finish a work in 150 days. They started the work and at the end of every 10 days, 25 additional men are employed. In how many days will the work be finished?

(*a*) 30 days (*b*) 50 days (*c*) 60 days (*d*) 75 days

Sol. (*b*) Work done by 25 men in first 10 days $= \frac{10}{150} = \frac{1}{15}$

Men engaged in next 10 days = 25 + 25 = 50 *i.e.* 2 times 25

$\therefore$ Work done by 50 men in next 10 days $= \frac{1}{15} \times 2 = \frac{2}{15}$

Men engaged in next 10 days = 50 + 25 = 75 *i.e.* 3 times 25

∴ Work done by 75 men in next 10 days = $\frac{1}{15} \times 3 = \frac{3}{15}$

Thus, work done in successive 10 days' period = $\frac{1}{15}, \frac{2}{15}, \frac{3}{15}, \frac{4}{15}, \frac{5}{15}$

$\frac{1}{15} + \frac{2}{15} + \frac{3}{15} + \frac{4}{15} + \frac{5}{15} = \frac{15}{15} = 1$, *i.e.* complete work

∴ Work will be done in 5 × 10 days = 50 days

215. 40 men can complete a work in 40 days. They started a work together. But at the end of each 10th day, 5 men left the job. The work would have been completed in :

(SSC CGL Tier I 2008)

(*a*) $56\frac{2}{3}$ days (*b*) $53\frac{1}{3}$ days (*c*) 52 days (*d*) 50 days

Sol. (*a*) Work done by 40 men in first 10 days = $\frac{10}{40} = \frac{1}{4} = \frac{8}{32}$

Men engaged in next 10 days = 40 – 5 = 35

∴ Work done by 35 men in next 10 days = $\frac{1}{4} \times \frac{35}{40} = \frac{7}{32}$

∴ Work done by 30 men in next 10 days = $\frac{1}{4} \times \frac{30}{40} = \frac{6}{32}$

Similarly work done in successive 10 days : $\frac{8}{32}, \frac{7}{32}, \frac{6}{32}, \frac{5}{32}, \frac{4}{32}$ ....

Work done in first 50 days = $\frac{8}{32} + \frac{7}{32} + \frac{6}{32} + \frac{5}{32} + \frac{4}{32} = \frac{30}{32}$

Remaining work = $1 - \frac{30}{32} = \frac{2}{32} = \frac{1}{16}$. Work done by 15 men in next 10 days = $\frac{3}{32}$

Work of 15 men in one day = $\frac{3}{32} \times \frac{1}{10} = \frac{3}{320}$

Time take to finish $\frac{1}{16}$ th work = $\frac{1}{16} \times \frac{320}{3} = \frac{20}{3} = 6\frac{2}{3}$

∴ Total time taken = $50 + 6\frac{2}{3} = 56\frac{2}{3}$ days

216. A started a work and left after working for 2 days. Then B finished the remaining work in 9 days. Had A left the work after working for 3 days, B would have finished the remaining work in 6 days. In how many days B alone can finish the work? **(MBA Ent. Exam.)**

(*a*) 5 days (*b*) 6 days (*c*) 10 days (*d*) 15 days

Sol. (*d*) Let 1 day work of A is x and 1 day work of B is y. Then $2x + 9y = 3x + 6y \Rightarrow x = 3y$

Now $2x + 9y = (2 \times 3y) + 9y = 6y + 9y = 15y \Rightarrow y = \frac{1}{15}$ ⇒ Time taken by B = 15 days

Alternative Method : 1 (*i.e.* 3 – 2) day's work of A = 3 (*i.e.* 9 – 6) days' work of B

∴ 2 days work of A + 9 days work of B = (6 + 9) days work of B = 15 days of B's work

⇒ One day work of B = $\frac{1}{15}$

217. A water tank has sufficient water for 60 days. If 50 litres of water is wasted everyday, then the water last for 50 days only. How many days will the supply last if 350 litres of water is wasted everyday ? **(MBA Ent. Exam.)**

(*a*) 8 days (*b*) 10 days (*c*) 15 days (*d*) 25 days

Sol. (*d*) Water wasted in 50 days at 50 litres per day = 50 × 50 = 2500 litres

This 2500 litres was sufficient for consumption for 60 – 50 = 10 days

∴ Daily consumption of water = 2500 ÷ 10 = 250 litres

Capacity of the tank = 60 × 250 litres = 15000 litres

In second situation, consumption + wastage = 250 + 350 = 600 litres per day

∴ Water supply will last for = 15000 ÷ 600 = 25 days

CHAPTER

25

TIME AND DISTANCE

1. **Distance covered** = Time taken × Average speed
2. **Time taken** = $\frac{\text{Distance covered}}{\text{Average speed}}$

Example: A train runs at an average speed of 75 km/hr. If the distance to be covered is 1050 kms. How long will the train take to cover it ? **(SSC CGL Tier I 2015)**

(*a*) 12 hours (*b*) 13 hours (*c*) 14 hours (*d*) 15 hours

Sol. (*c*) Required time = 1050 ÷ 75 = 14 hours

3. **Average speed** = $\frac{\text{Distance covered}}{\text{Time taken}}$

Example. A car covers a distance of 448 km in 28 hours. What is the average speed (in km/hour) of the car? **(SBI Clerical 2008)**

(*a*) 12 (*b*) 16 (*c*) 26 (*d*) None of these

Sol. (*b*) Average Speed = $\frac{\text{Distance covered}}{\text{Time taken}} = \frac{448}{28}$ = 16 km/hour

4. If the speed of two trains are in the ratio of $x : y$, then ratio of time taken by them to cover a certain distance = $\frac{1}{x} : \frac{1}{y} = y : x$.

Example 1. A train running at the average speed of 30 km/h covers a distance in 6 hours. Find the speed of another train which covers the same distance in 5 hours.

(*a*) 24 km/h (*b*) 25 km/h (*c*) 36 km/h (*d*) 42 km/h

Sol. (*c*) Ratio of time taken by two trains = 6 : 5 ⇒ Ratio of speeds = 5 : 6

Speed of first train = 30 km/h ⇒ Speed of second train = $30 \times \frac{6}{5}$ = 36 km/h

Example 2. The speeds of A and B are in the ratio 3 : 4. If A takes 20 minutes more than B to reach destination. In what time does A reach the destination? **(SSC CGL Tier I 2007)**

(*a*) $1\frac{1}{3}$ hours (*b*) 2 hours (*c*) $2\frac{2}{3}$ hours (*d*) $1\frac{2}{3}$ hours

Sol. (*a*) Ratio of speeds = 3 : 4 ⇒ Ratio of time taken = 4 : 3

Difference of time = (4 – 3) ratio = 1 ratio = 20 minutes

Time taken by A = 4 ratio = 80 minutes = $1\frac{1}{3}$ hours

5. To cross a pole or a standing man, a train is required to cover distance equal to its own length.
6. To cross a platform or any other stationary object of some length, a train is required to cover distance equal to sum of its own length and the length of the platform or the object.

7. 1 km per hour = $\frac{5}{18}$ metre per second

∴ To convert speed of an object from km/h to m/s, multiply the given speed by $\frac{5}{18}$.

Example 1: Convert speed of 36 km/hr into speed m/s.

Sol. 36 km/hr = 36 × $\frac{5}{18}$ m/s = 10 m/s

Example 2: A speed of 45 km per hour is the same as **(SSC CGL Tier I 2014)**

(a) 12.5 m/s (b) 13 m/s (c) 15 m/s (d) 12 m/s

Sol. (a) 45 km/hr = 45 × 5/18 m/s = 12.5 m/s

8. 1 metre per second = $\frac{18}{5}$ km per hour

∴ To convert speed of an object from m/s to km/h, multiply the given speed by $\frac{18}{5}$.

Example 1: Convert speed of 20 m/s into speed km/hr.

Sol. 20 m/s = 20 × $\frac{18}{5}$ km/hr = 72 km/hr

Ex.2. A car goes 10 metres in a second. Find its speed in km/hour. **(SSC CGL Tier I 2012)**

(*a*) 40 (*b*) 32 (*c*) 48 (*d*) 36

Sol. (*d*) Speed = 10 m/s = 10 × $\frac{18}{5}$ km/h = 36 km/hr

9. Relative Speed:

If two objects are moving in **the opposite directions** (either towards each other or away from each other), their relative speed is equal to **'sum of their speeds'**.

Example: Two trains are running in opposite directions at x km/h and y km/h, their relative speed is $x + y$.

If two objects are moving in the **same direction**, their relative speed is equal to **'difference of their speeds'**.

Example: Two trains are running in the same direction at x km/h and y km/h, their relative speed is $x - y$.

Example 1. Two persons start walking from a place in opposite detections at 3 km/h and 4 km/h respectively. How many kms will they be apart after 8 hours? **(SSC CHSL 2006)**

(*a*) 8 km (*b*) 24 km (*c*) 32 km (*d*) 56 km

Sol. (*d*) Their relative speed = 3 + 4 = 7 km/h. Required distance = 8 × 7 = 56 km

Example 2. Two persons start walking from a place in opposite directions at 5 km/h and 4 km/h respectively. After how many hours will they be 45 km apart? **(SSC CHSL)**

(*a*) 5 hours (*b*) 7 hours (*c*) 9 hours (*d*) 11 hours

(*a*) Relative speed = 5 + 4 = 9 km/h ⇒ Required time = 45 ÷ 9 = 5 hours

Example 3. Two persons start walking in the same direction at 5 km/h and 4 km/h respectively. In how many hours will they be 45 km apart?

(*a*) 5 hours (*b*) 7 hours (*c*) 9 hours (*d*) 45 hours

(*d*) Relative speed = 5 – 4 = 1 km/h ⇒ Required time = 45 ÷ 1 = 45 hours

MORE FORMULAE

1. Two persons left from place A for B at the speed of x km/h and y km/h respectively. The first person (faster one) reaches at place B and immediately returns for place A. How far from place B will both the persons meet?

Two methods of solving such problems :

First Method :

Let total distance be d km and distance between B and the point of meeting = m km

A $\quad d-m \quad$ C $\quad m \quad$ B

$$\textbf{Distance between B and meeting point} = \textbf{Distance (i.e. AB)} \times \frac{\textbf{Difference between speeds}}{\textbf{Sum of speeds}}$$

Sol : Let the total distance = d km. Distance between B and the point of meeting = m km

∴ Distance covered by first person = $d + m$ and distance covered by second person = $d - m$

Since time taken by both are equal. $\Rightarrow \frac{d+m}{x} = \frac{d-m}{y}$

$$\Rightarrow dy + my = dx - mx \Rightarrow m \times (x+y) = d \times (x-y) \Rightarrow m = d \times \frac{x-y}{x+y}$$

Second Method :

Let ratio of speeds of two persons = $x : y$. Then ratio of distance travelled by them = $x : y$

$$\textbf{Distance covered by first person} = \frac{x}{x+y} \times \textbf{Total distance covered}$$

$$\textbf{Distance covered by second person} = \frac{y}{x+y} \times \textbf{Total distance covered}$$

Note: Total distance covered = AB + BA = 2AB.

Example : Two men A and B walk from P to Q, a distance of 36 km at 7 km/h and 5 km/h respectively. A reaches at Q and immediately returns and meets B at a place R. Find the distance between P and R.

(*a*) 15 km (*b*) 21 km (*c*) 30 km (*d*) 42 km

Sol. (*c*)

P $\quad d-m \quad$ R $\quad m \quad$ Q

$$RQ = 36 \times \frac{7-5}{7+5} = 6 \text{ km} \Rightarrow PR = PQ - RQ = 36 - 6 = 30 \text{ km}$$

Alt : Total distance travelled = 2 × 36 = 72 km. Ratio of speeds of A and B = 7 : 5

$$\therefore PR = \text{Distance travelled by B} = \frac{5}{12} \times 72 = 30 \text{ km.}$$

2. A train leaves a station at x km/h and another train leaves the same station after T hours at y km/h. They will meet after $\frac{xT}{y-x}$ hours, and distance = $\frac{xyT}{y-x}$

Example. A train leaves station A towards B at an average speed of 80 km/h. After 3 hours another train leaves station A at an average speed of 110 km/h. Find the distance from station A where the two trains meet.

(*a*) 660 km (*b*) 770 km (*c*) 800 km (*d*) 880 km

Sol. (*d*) Distance covered till meeting place = $\frac{110 \times 80}{110 - 80} \times 3 = 880$ km

Alternative Method : Distance covered by the first train in 3 hours = 3 × 80 = 240 km.
Second train gains = 110 – 80 = 30 km/h

∴ Second train will meet the first train after $\frac{240}{30} = 8$ hours.

Distance covered by the faster train in 8 hours = 8 × 110 = 880 km.

3. A train leaves station A for B and another train leaves station B for station A at the same time. The trains cross each other en route and reach their destinations in *m* and *n* hours respectively after crossing each other. Then **Ratio of speeds of two trains =** $\sqrt{n} : \sqrt{m}$
Rule: Speeds of the first and second trains are in the ratio of square roots of time taken by the second and the first train in reaching their destination after crossing each other.

Example. A starts from a place P to go to a place Q. At the same time B starts from Q to P. If after meeting each other, A and B took 4 and 9 hours respectively to reach their destinations, the ratio of their speeds is : **(SSC CGL Tier I 2004)**

(*a*) 3 : 2 (*b*) 5 : 2 (*c*) 9 : 4 (*d*) 9 : 13

Sol. (*a*) A's speed : B's speed = $\sqrt{9} : \sqrt{4} = 3 : 2$

4. If two trains travels a certain distance at the speed of *a* km/h, and *b* km/h respectively and difference between time taken by two trains is *t* hours.

Then distance between the starting point and the destination = $\frac{a \times b}{a - b} \times t$

∴ **Distance =** $\frac{\textbf{Multiplication of Speeds} \times \textbf{Difference of Time taken}}{\textbf{Difference of speeds}}$

Proof: Let the distance covered = '*x*' km. Then $\frac{x}{a} - \frac{x}{b} = T \Rightarrow \frac{(a-b)x}{a \times b} = T \Rightarrow x = \frac{a \times b}{a - b} \times T$

Example: If I walk @ 4 km/h, I reach the office 5 minutes late and if I walk @ 5 km/h, I reach 10 minutes early. How far my office is?

(*a*) 2 km (*b*) 4 km (*c*) 5 km (*d*) 9 km

Sol. (*c*) Time difference = 5 + 10 = 15 minutes = $\frac{1}{4}$ hour

∴ Required distance = $\frac{5 \times 4}{5 - 4} \times \frac{1}{4} = 5$ km.

Alternative method : Ratio of speeds = 4 : 5 ⇒ Ratio of time taken = 5 : 4
Difference in time = (5 – 4) ratio = 1 ratio ≡ (5 + 10) = 15 minutes

∴ First time = 15 × 5 = 75 minutes ⇒ Required Distance = $4 \times \frac{75}{60} = 5$ km

5. If a person goes from A to B @ *x* km/h and returns @ *y* km/h, then his average speed during the whole journey is: $\frac{2xy}{x + y}$ **(CDS 2003)**

Total distance covered = $\frac{2xy}{x + y} \times$ Time taken

Distance from A to B = $\frac{2xy}{x + y} \times$ Time Taken $\times \frac{1}{2}$

Proof: Let distance between two places is 1 km. Then the time taken to travel from A to B $= \frac{1}{x}$ hours. And time taken to travel from B to A $= \frac{1}{y}$ hours.

Total distance covered = 1 + 1 = 2 km.

Total time taken $= \frac{1}{x} + \frac{1}{y}$. Average speed $= \frac{1+1}{\frac{1}{x}+\frac{1}{y}} = \frac{2}{\frac{x+y}{xy}} = \frac{2xy}{x+y}$

Alternative method: 1. Find ratio of two speeds.

2. Find difference of two speeds.
3. Divide the difference into the ratio calculated above.
4. Add/subtract the above values to/from original speeds.

Example 1: If a man goes to a place at an average speed of 10 km/hr and then returns at the average speed of 15 km/hr. Find his average speed during the whole journey?

(SSC CGL Tier I 2007)

(*a*) 12 km/hr (*b*) 12.5 km/hr (*c*) 13 km/hr (*d*) 15 km/hr

Sol. (*a*) Average speed $= \frac{2 \times 10 \times 15}{10+15} = \frac{2 \times 10 \times 15}{25} = 12$ km/h

Alt: 10 : 15 = 2 : 3 15 – 10 = 5 km/hr

5 is divided into 2 : 3 i.e. 2 and 3 Average speed = 10 + 2 = 15 – 3 = 12 km/hr

Example 2: A man goes to his office at the speed of 8 km/h and returns at the speed of 12 km/h. If he takes 10 hours in all. Find total distance covered by him.

(*a*) 24 km (*b*) 48 km (*c*) 50 km (*d*) 96 km

Sol. (*d*) Average speed $= \frac{2 \times 8 \times 12}{20}$ km/h

Distance covered in 10 hours $= \frac{2 \times 8 \times 12}{20} \times 10 = 96$ km

Example 3 : A man goes to his office at the speed of 5 km/h and returns at the speed of 20 km/hr. If he takes 4 hours in all, the distance between the residence and office is.

(*a*) 16 km (*b*) 25 km (*c*) 32 km (*d*) 50 km

Sol. (*a*) Total distance covered $= \frac{2 \times 5 \times 20}{25} \times 4$ km = 32 km

∴ Distance between office and residence = 32 ÷ 2 = 16 km

Note. If instead of distances covered, time taken in two cases is same, then average speed during the journey is $\frac{x+y}{2}$.

Note. If a km is covered at the speed of x km/hr and b km at the speed of y km/hr, then average speed $\frac{a+b}{\frac{a}{x}+\frac{b}{y}} = \frac{(a+b)\,xy}{ay+bx}$.

6. A man covers $\frac{1}{3}$rd of his journey at x km/h, next $\frac{1}{3}$rd of his journey at y km/h and remaining $\frac{1}{3}$rd journey at z km/h, then his average speed during the journey is $\frac{3xyz}{xy+yz+xz}$

(SSC Asstt. FCI 2012)

Example: A car travels the first one-third of a certain distance with a speed of 10km/hr, the next one-third distance with a speed of 20 km/hr and the last one-third distance with a speed of 60 km/hr. The average speed of the car for the whole journey is **(CDS 2015)**

(*a*) 18 km/hr (*b*) 30 km/hr (*c*) 24 km/hr (*d*) 36 km/hr

Sol. (*a*) Average speed $= \frac{3 \times 10 \times 20 \times 60}{10 \times 20 + 20 \times 60 + 60 \times 10} = \frac{3 \times 10 \times 2 \times 6}{2+12+6} = 18$ km/h

Alt. Let each stretch = LCM of 10, 20, 60 = 60 km

Then total time taken $= \frac{60}{10} + \frac{60}{20} + \frac{60}{60} = 6 + 3 + 1 = 10$ hours

Total distance covered = 60 × 3 = 180 km ⇒ Average speed = 180 ÷ 10 = 18 km/hr

7. If speeds of two trains are in the ratio $x : y$. The first train crosses a pole in a seconds and the second train in b seconds. The time taken by the two trains to cross each other completely is $\frac{ax + by}{x + y}$ i.e. $\frac{\text{Sum of lengths of the trains}}{\text{Relative speed of the trains}}$.

SOLVED EXERCISE

1. Raj and Prem walk in opposite direction at the rate of 3 km and 2 km per hour respectively. How far will they be from each other after 2 hours ? **(SSC CGL Tier I 2015)**

(*a*) 10 km (*b*) 2 km (*c*) 8 km (*d*) 6 km

Sol. (*a*) Relative speed = (3 + 2) = 5 km/hr ⇒ Required distance = 2 × 5 = 10 km

2. The speed $3\frac{1}{3}$ m/ sec when expressed in km/hour becomes **(SSC CGL Tier I 2012)**

(*a*) 8 (*b*) 9 (*c*) 10 (*d*) 12

Sol. (*d*) Speed $= \frac{10}{3}$ m/s $= \frac{10}{3} \times \frac{18}{5}$ km/h = 12 km/hr

3. A person can cross a standing bus in 18 seconds. The same bus can cross a pole in 4 seconds. The ratio of speeds of the bus and the person is : **(Bank PO 2010)**

(*a*) 9 : 2 (*b*) 9 : 5 (*c*) 9 : 4 (*d*) None of these

Sol. (*a*) Speed of bus : person $= \frac{1}{4} : \frac{1}{18} = 18 : 4 = 9 : 2$

4. If a bus travels 160 km in 4 hours and a train travels 320 km in 5 hours at uniform speeds, then what is the ratio of the distances travelled by them in one hour ? **(UPSC Civil Services Pre. 2011)**

(*a*) 8 : 5 (*b*) 5 : 8 (*c*) 4 : 5 (*d*) 1 : 2

Sol. (*b*) Required ratio = (160 ÷ 4) : (320 ÷ 5) = 40 : 64 = 5 : 8

5. A train travelling at a speed of 55 km/hr, travels from place X to place Y in 4 hours. If its speed is increased by 5 km/hr, then the time of journey is reduced by **(SSC CGL Tier I 2014)**

(*a*) 25 minutes (*b*) 35 minutes (*c*) 20 minutes (*d*) 30 minutes

Sol. (c) Total time taken $= \frac{55 \times 4}{60} = \frac{11}{3} = 3\frac{2}{3}$ hours = 3 hours 40 minutes

Time saved = 4 hours – 3 hours 40 minutes = 20 minutes

6. A train 120 m long runs through a station at the speed of 54 km/hr. The time taken by the train to pass a man standing on the platform is **(SSC CGL Tier I 2014)**

(*a*) 5 sec (*b*) 6 sec (*c*) 7 sec (*d*) 8 sec

Sol. (*d*) Speed of train = 54 km/hr = $54 \times \frac{5}{18}$ m/s = 15 m/s

∴ Required time to cover 120 m = 120 ÷ 15 = 8 sec

7. A train 180 m long is running at a speed of 90 km/hr. How long will it take to pass a post ? **(SSC CGL Tier I 2015)**

(*a*) 8.2 sec (*b*) 7.2 sec (*c*) 7.8 sec (*d*) 8 sec

Sol. (*b*) Speed of train = 90 km/hr = $90 \times \frac{5}{18}$ m/s = 25 m/s

∴ Required time to cover 180 m = 180 ÷ 25 = 7.2 sec

8. A train travelling with uniform speed crosses two bridges of lengths 300 m and 240 m in 21 seconds and 18 seconds respectively. The speed of the train is **(SSC CHSL 2010)**

(*a*) 72 km/hr (*b*) 68 km/hr (*c*) 65 km/hr (*d*) 60 km/hr

Sol. (*a*) (300 – 240) m = 60 m covered in (21 – 18) seconds = 3 seconds

∴ Speed = $\frac{60}{3}$ m/s = 20 m/s = $20 \times \frac{18}{5}$ km/h = 72 km/h

9. The speeds of three cars are in the ratio 4 : 3 : 2. What is the ratio between the times taken by the cars to cover the same distance ? **(CDS 2011, 2015)**

(*a*) 2 : 3 : 4 (*b*) 3 : 4 : 6 (*c*) 1 : 2 : 3 (*d*) 4 : 3 : 2

Sol. (*b*) Required ratio = $\frac{1}{4}:\frac{1}{3}:\frac{1}{2} = 3:4:6$

10. A certain distance is covered by a cyclist at a certain speed. If a jogger covers half the distance in double the time, the ratio of the speed of the jogger to that of the cyclist is **(SSC CGL Tier II 2013, CGL Tier I 2013)**

(*a*) 1 : 4 (*b*) 4 : 1 (*c*) 1 : 2 (*d*) 2 : 1

Sol. (*a*) Cyclist : Jagger = $\frac{x}{y}:\frac{x}{2\times 2y} = 1:\frac{1}{4} = 4:1$

11. A train running at 36 km/h, covers a certain distance in 50 minutes. At what speed train must run to cover the same distance in 30 minutes?

(*a*) 45 km/h (*b*) 50 km/h (*c*) 54 km/h (*d*) 60 km/h

Sol. (*d*) Required speed = $36 \times \frac{50}{30}$ minutes = 60 km/h

12. A car travelling at a speed of 40 km/hour can complete a journey in 9 hours. How long will it take to travel the same distance at 60 km/hour? **(SSC CGL Tier I 2003)**

(*a*) 6 hours (*b*) 3 hours (*c*) 4 hours (*d*) $4\frac{1}{2}$ hours

Sol. (*a*) Required time = $\frac{40 \times 9}{60}$ = 6 hours

13. A car covers its journey at the speed of 80 km/hr in 10 hours. If the same distance is to be covered in 4 hours, by how much the speed of the car will have to increase ? **(SBI Bank PO 2006)**

(*a*) 80 km/hr (*b*) 100 km/hr (*c*) 120 km/hr (*d*) 160 km/hr

Sol. (*c*) Total distance covered = 80 × 10 = 800 km

New speed = 800 ÷ 4 = 200 km/hr ⇒ Increase in speed = (200 – 80)km/hr = 120 km/hr

14. Two stations A and B are 60 km apart. One train starts from A and another train starts from B at the same time. If they go in the same direction, they meet in 6 hours and if they go in the opposite directions, they meet in 2 hours. The speed of the faster train is : **(SBI Bank PO 2006)**

(*a*) 10 km/hr (*b*) 20 km/hr (*c*) 30 km/hr (*d*) 40 km/hr

Sol. (*b*) Relative speed (when going in same direction) = 60 ÷ 6 = 10

Relative speed (when going in opposite directions) = 60 ÷ 2 = 30

Speed of faster train = (30 + 10) ÷ 2 = 20

15. Points A and B are 90 km apart from each other on a highway. A car starts from A and another from B at the same time. If they go in the same direction, they meet after 9 hours and if they go in opposite directions, they meet after $\frac{9}{7}$ hours. The speeds (in km/hr) of the two cars are **(SSC CGL Tier I 2012)**

(*a*) 25, 45 (*b*) 20, 50 (*c*) 30, 40 (*d*) 35, 65

Sol. (*c*) Relative speed (when going in same direction) = 90 ÷ 9 = 10 km/hr

Relative speed (when going in opposite direction) = $90 \div \frac{9}{7} = 70$ km/hr

Their speeds are (70 – 10) ÷ 2 and (70 + 10) ÷ 2 = 30 and 40 km/hr

16. Two trains measuring 120 metre and 180 metre in length, cross each other completely in 10 seconds when running in opposite directions. The faster train takes 30 seconds to cross the slower train when the two trains are running in the same direction. Find the speed of the faster train.

(*a*) 30 km/h (*b*) 36 km/h (*c*) 54 km/h (*d*) 72 km/h

Sol. (*d*) Total distance to be covered = 120 + 180 = 300 metre

Time taken to cross each other, while running in the opposite directions = 10 seconds

∴ Relative speed (*i.e.* sum of speeds) of the trains = $\frac{300}{10} = 30$ m/s

Time taken to cross each other, while running in the same direction = 30 seconds

∴ Relative speed (*i.e.* difference of the speeds) of the trains = $\frac{300}{30} = 10$ m/s

∴ Speed of the faster train = $\frac{1}{2}(30 + 10) = 20 \text{ m/s} = 20 \times \frac{18}{5} \text{ km/h} = 72$ km/h

17. A train 800 metres long is running at the speed of 78 km/hr. If it crosses a tunnel in 1 minute, then the length of the tunnel (in metres) is : **(SSC CGL Tier I 2003)**

(*a*) 500 (*b*) 800 (*c*) 1300 (*d*) 13

Sol. (*a*) Distance covered in one minute = 78000 ÷ 60 = 1300 m

∴ Length of tunnel = 1300 – 800 = 500 m

18. What is the ratio between times taken by a train 240 m long to cross an electric pole and a bridge of 80 m length ? **(CDS 2011)**

(*a*) 2 : 3 (*b*) 3 : 4 (*c*) 4 : 5 (*d*) 5 : 6

Sol. (*b*) Ratio of time taken = Ratio of distances covered = 240 : (240 + 80) = 240 : 320 = 3 : 4

19. In covering a certain distance, the speeds of A and B are in the ratio of 3 : 4. A taken 30 minutes more than B to reach the destination. The time taken by A to reach the destination is : **(SSC CGL Tier I 1999)**

(*a*) 1 hour (*b*) $1\frac{1}{2}$ hour (*c*) 2 hours (*d*) $2\frac{1}{2}$ hours

Sol. (*c*) Ratio of speeds = 3 : 4 ⇒ Ratio of time = 4 : 3

Difference of time = (4 – 3) ratio = 1 ratio = 30 minutes

⇒ Time taken by A = 4 ratio = 4 × 30 minutes = 120 minutes = 2 hours

20. The ratio between the rates of travelling of A and B is 2 : 3 and therefore A takes 10 minutes more than the time taken by B to reach a destination. If A had walked at double the speed, the time taken by him to reach the destination would have been **(SSC CGL Tier II 2012)**

(*a*) 25 minutes (*b*) 15 minutes (*c*) 30 minutes (*d*) 20 minutes

Sol. (*b*) Ratio of speeds = 2 : 3 ⇒ Ratio of time taken = 3 : 2.

Difference of time = (3 – 2) ratio = 1 ratio = 10 minutes

∴ A's normal time = 3 ratio = 30 minutes.

∴ Time taken by A at double the speed = 30 ÷ 2 = 15 minutes

21. Train A leaves Bangalore for Mysore at 9 A.M. @ 40 km/hr and another train leaves at 12 Noon from Bangalore for Mysore at a higher speed at 80 km/hr. At what time train B will overtake the train A ? **(SSC CGL Tier I 2012)**

(*a*) 4 P.M. (*b*) 6 P.M. (*c*) 5 P.M. (*d*) 3 P.M.

Sol. (*d*) Distance covered by slower train from 9 A.M. to Noon = 40 × 3 = 120 km

Relative speed of two trains = 80 – 40 = 40 km/hr

Time required to gain 120 km = 120 ÷ 40 = 3 hours

⇒ Required time = 12 Noon + 3 hours = 3 P.M.

22. A car covers the first 35 km of its journey in 45 minutes and covers the remaining 69 km in 75 minutes. What is the average speed of the car ? **(Bank of Baroda PO 2007)**

(*a*) 42 km/hr (*b*) 50 km/hr (*c*) 52 km/hr (*d*) 60 km/hr

Sol. (*c*) Total distance covered = 35 + 69 = 104 km

Total time = 45 + 75 = 120 minutes = 2 hours ⇒ Average speed = 104 ÷ 2 = 52 km/hr

23. A person travelled by train for 1 hour at a speed of 50 kmph. He then travelled by a taxi for 30 minutes at a speed of 32 kmph to complete his journey. What is the average speed at which he travelled during the journey ? **(CDS 2010)**

(*a*) 44 kmph (*b*) 42 kmph (*c*) 41 kmph (*d*) 33 kmph

Sol. (*a*) Total distance covered = $1 \times 50 + \frac{1}{2} \times 32 = 50 + 16 = 66$ km

Total time taken = 1 hour 30 minutes = $\frac{3}{2}$ hour

∴ Average speed during the journey = 66 ÷ 3/2 = 66 × 2/3 = 44 kmph

24. The distance between places A and B is 999 km. An express train leaves place A at 6 am and runs at a speed of 55.5 km/hr. The train stops on the way for 1 hour 20 minutes. It reaches B at **(SSC CGL Tier II 2013)**

(*a*) 1.20 am (*b*) 12 pm (*c*) 6 pm (*d*) 11 pm

Sol. (*a*) 999 ÷ 55.5 = 18 hours ⇒ Required time = 6 am + 18 hours + 1 hr 20 min = 1.20 am

25. A man can reach a certain place in 30 hours. If he reduces his speed by $\frac{1}{15}$th, he goes 10 km less in that time. Find his speed (in km/hour). **(SSC CGL Tier I 2002)**

(*a*) 6 (*b*) $5\frac{1}{2}$ (*c*) 4 (*d*) 5

Sol. (*d*) Ratio of speeds = 15 : 14 ⇒ Ratio of distance covered = 15 : 14

Difference in distance covered = (15 – 14) ratio = 1 ratio = 10 km

Original distance = 15 ratio = 15 × 10 km = 150 km ⇒ Original speed = 150 ÷ 30 = 5 km/hr

26. Two cars are moving with speeds V_1, V_2 towards a crossing along two roads. If their distances from the crossing be 40 metres and 50 metres at an instant of time then they do not collide if their speeds are such that: **(SSC CGL Tier I 2013)**

(*a*) $V_1 : V_2 = 25 : 16$ (*b*) $V_1 : V_2 = 16 : 25$ (*c*) $V_1 : V_2 \neq 4 : 5$ (*d*) $V_1 : V_2 \neq 5 : 4$

Sol. (*c*) Time taken by two cars to reach the crossing = $\frac{40}{V_1}$ and $\frac{50}{V_2}$

The cars do not collide if $\frac{40}{V_1} \neq \frac{50}{V_2} \Rightarrow \frac{V_1}{V_2} \neq \frac{40}{50} \Rightarrow \frac{V_1}{V_2} \neq \frac{4}{5}$

27. A car can cover a certain distance in $4\frac{1}{2}$ hours. If the speed of the car is increased by 5 km/hr, it will take half an hour less than before. Find the original speed of the car. **(SSC CPO SI 2009)**

(*a*) 50 km/hr (*b*) 40 km/hr (*c*) 60 km/hr (*d*) 45 km/hr

Sol. (*b*) Let original speed of car = x km/hr. Original time = $4\frac{1}{2} = \frac{9}{2}$ hours

Then $\frac{9}{2}x = 4(x + 5) \Rightarrow 9x = 8x + 40 \Rightarrow x = 40$

Alternative Method : Ratio of time = $\frac{9}{2} : 4 = 9 : 8 \Rightarrow$ Ratio of speeds = 8 : 9

Difference of speeds = (9 – 8) ratio = 1 ratio = 5 km/hr

Original speed = 8 ratio = 8 × 5 km/hr = 40 km/hr

28. A 240 metres long train crosses a platform twice its length in 40 seconds. What is the speed of the train ? **(PNB Officers Exam. 2009)**

(*a*) 6 m/s (*b*) 28 m/s (*c*) 18 m/s (*d*) 16 m/s

Sol. (*c*) Length of platform = 2 × 240 = 480 m

⇒ Total distance covered = 240 + 480 = 720 m ⇒ Speed of the train = 720 ÷ 40 = 18 m/s

29. A train 50 metres long passes a platform of length 100 metres in 10 seconds. The speed of the train in metres / second is : **(SSC CGL Tier I 2013)**

(*a*) 50 (*b*) 10 (*c*) 15 (*d*) 20

Sol. (*c*) Total distance covered = 100 + 50 = 150 m ⇒ Speed of train = 150 ÷ 10 = 15 m/s.

30. Train A took 35 minutes to cover a distance of 50 km. If the speed of train B is 25% faster than train A, it will cover the same distance in:

(*a*) 25 minutes (*b*) 28 minutes (*c*) 30 minutes (*d*) 36 minutes

Sol. (*b*) Let speed of train A = 100 km/h. Then speed of train B = 125 km/h

Ratio of speeds of trains A and B = 100 : 125 = 4 : 5

∴ Ratio of time taken by them to cover equal distance = 5 : 4

Time taken by train A = 35 minutes ⇒ Time take by train B = $\frac{4}{5}$ × 35 minutes = 28 minutes.

Alt: Required time = $35 \times \frac{100}{125}$ minutes = 28 minutes.

31. The lengths of two trains is 5 : 3 and the ratio of their speeds is 6 : 5. The ratio of times taken by them to cross a pole is : **(SSC CGL Tier I 2006)**

(*a*) 5 : 6 (*b*) 11 : 8 (*c*) 25 : 18 (*d*) 27 : 16

Sol. (*c*) Ratio of time taken = $\frac{\text{Ratio of distances}}{\text{Ratio of speeds}} = \frac{5}{6}:\frac{3}{5} = 25:18$

32. A car starts from A to B and another car starts from B to A. They meet at the exact middle of the places A and B. After meeting, they complete their journeys in 2 hours and 1 hour 20 minutes respectively. The ratio of the time taken by them to complete the whole journey is : **(SSC CGL Tier I 2006)**

(*a*) 3 : 2 (*b*) 16 : 24 (*c*) 13 : 16 (*d*) 16 : 13

Sol. (*a*) Time taken by two cars to cover equal distances = 2 hours and 1 hour 20 minutes = 120 minutes and 80 minutes respectively ∴ Required ratio = 120 : 80 = 3 : 2

33. A truck covers a distance of 550 metre in a minute whereas a bus covers a distance of 33 km in 45 minutes. The ratio of their speeds is : **(SSC CGL Tier I 2004)**

(*a*) 4 : 3 (*b*) 3 : 5 (*c*) 3 : 4 (*d*) 50 : 3

Sol. (*c*) Required ratio = $550 : \frac{33000}{45} = 550 \times 45 : 33000 = 3 : 4$

34. A person travels 285 km in 6 hours. In the first part of the journey, he travels at 40 km/hr by bus. In the second part, he travels at 55 km/hr by train. The distance travelled by train is **(SSC CGL Tier I 2012)**

(*a*) 165 km (*b*) 615 km (*c*) 561 km (*d*) 156 km

Sol. (*a*) Total distance covered @ 40 km/hr in 6 hours = 40 × 6 = 240 km

Remaining distance = 285 – 240 = 45 km

Difference in speeds = (55 – 40) km/hr = 15 km/hr

Time taken to Distance cover at 55 km/hr = 45 ÷ 15 = 3 hours

Distance travelled at 55 km/hr = 55 × 3 = 165 km

Trick: Only 165 is divisible by 55, therefore, we may try this value.

35. P and Q are 27 km away. Two trains with speeds of 24 km/hr and 18 km/hr respectively start simultaneously from P and Q and travel in the same direction. They meet at a point R beyond Q. Distance QR is **(SSC CGL Tier II 2012)**

(*a*) 48 km (*b*) 36 km (*c*) 126 km (*d*) 81 km

Sol. (*d*) Ratio of speeds of two trains = 24 : 18 = 4 : 3

Difference of distances = (4 – 3) ratio = 1 ratio = 27 km

Distance QR = 3 ratio = 3 × 27 km = 81 km

36. A and B, standing at places having distance 12 km, start walking towards each other and meet in 1 hour 15 minutes. If the speed of A is 4 km/hr, then the speed of B is : **(SSC CGL Tier I 2006)**

(*a*) $3\frac{3}{5}$ km/hr (*b*) $4\frac{3}{5}$ km/hr (*c*) $5\frac{3}{5}$ km/hr (*d*) $6\frac{3}{5}$ km/hr

Sol. (*c*) Time taken = 1 hour 15 minutes = $1\frac{1}{4}$ hour = $\frac{5}{4}$ hour

∴ Their relative speed = $12 \div \frac{5}{4} = 12 \times \frac{4}{5} = \frac{48}{5}$ km ⇒ B's Speed = $\frac{48}{5} - 4 = \frac{28}{5} = 5\frac{3}{5}$ km/hr

37. A train 100 metres long completely passes a man walking in the same direction at 6 km/hr in 5 seconds and a car travelling in the same direction in 6 seconds. At what speed was the car travelling? **(SSC CGL Tier I 2006)**

(*a*) 12 km/hr (*b*) 18 km/hr (*c*) 24 km/hr (*d*) 30 km/hr

Sol. (*b*) Relative speed of train and man = $\frac{100}{5} \times \frac{18}{5} = 72$ km/hr

$\Rightarrow$ Speed of train = 72 + 6 = 78 km/hr

Relative speed of train and car = $\frac{100}{6} \times \frac{18}{5} = 60$ km/hr $\Rightarrow$ Speed of car = 78 – 60 = 18 km/hr

38. A motor starts with the speed of 70 kmph with its speed increased every two hours by 10 kmph. In how many hours will it cover 345 kms ? **(Canara Bank PO 2003)**

(*a*) $2\frac{1}{4}$ hours (*b*) $4\frac{1}{2}$ hours (*c*) 4 hours 5 minutes (*d*) None of these

Sol. (*b*) Speed of motor in the 1, 2, 3, 4, 5, 6th hours = 70, 70, 80, 80, 90, 90 kmph respectively.

Distance covered in first four hours = 70 + 70 + 80 + 80 = 300 km

Remaining distance = 345 – 300 = 45 km

Time taken to cover 45 km @ 90 kmph = $\frac{45}{90} = \frac{1}{2}$ hour $\Rightarrow$ Total time = $4 + \frac{1}{2} = 4\frac{1}{2}$ hour

39. I walk a certain distance and ride back taking a total time of 37 minutes. I could walk both ways in 55 minutes. How long would it take me to ride both ways ?

(SSC CGL Tier I 2000, CPO SI Exam. 2012)

(*a*) 9.5 minutes (*b*) 19 minutes (*c*) 18 minutes (*d*) 20 minutes

Sol. (*b*) One way walking + One way riding = 37 minutes

2 × (One way walking + One way riding) = 74 minutes

Two ways walking time = 55 minutes

Two ways riding time = 74 minutes – 55 minutes = 19 minutes

40. A person can walk a certain distance and drive back in six hours. He can also walk both ways in 10 hours. How much time will he take to drive both ways ? **[Civil Services (Pre.) 2013]**

(*a*) Two hours (*b*) Two and a half hours (*c*) Five and a half hours (*d*) Four hours

Sol. (*a*) One way walking + One way driving = 6 hours

2 × (One way walking + One way driving) = 12 hours

Two ways walking time = 10 hours

Two ways driving time = 12 hours – 10 hours = 2 hours

41. A man takes 6 hours 15 minutes in walking a distance and riding back to the starting place. He could walk both ways in 7 hours 45 minutes. The time taken by him to ride both ways is: **(SSC CGL Tier I 2008)**

(*a*) 4 hours (*b*) 4 hours 30 minutes

(*c*) 4 hours 45 minutes (*d*) 5 hours

Sol. (*c*) One way walking + One way riding = 6 hours 15 minutes

2 × (One way walking + One way riding) = 12 hours 30 minutes

Two ways walking time = 7 hours 45 minutes

Two ways riding time = 12 hours 30 minutes – 7 hours 45 minutes = 4 hours 45 minutes

42. A man takes 6 hours 35 minutes in walking to a certain place and riding back. He would have taken 2 hours less by riding both ways. What time would he take to walk both ways ?

(Bank of Maharashtra PO 2008)

(*a*) 4 hours 35 min (*b*) 8 hours 35 min (*c*) 10 hours (*d*) 8 hours 25 min

Sol. (*b*) One way walking + One way riding = 6 hours 35 minutes

Two ways walking time = 6 hours 35 minutes + 2 hours = 8 hours 35 minutes

43. A train running at 7/11 of its own speed reached a place in 22 hours. How much time could be saved if the train would have run at its normal speed ? **(SSC CGL Tier I 2001)**

(*a*) 14 hours (*b*) 7 hours (*c*) 8 hours (*d*) 16 hours

Sol. (*c*) Running at 7/11 of speed, the train will take 11/7 of the normal time.

$\therefore \ \frac{11}{7}$ of normal time = 22 hours ⇒ Normal time = $22 \times \frac{7}{11}$ = 14 hours

∴ Time saved at normal speed = 22 – 14 = 8 hours

44. If a man reduces his speed to 2/3, he takes 1 hour more in walking a certain distance. The time (in hours) to cover the distance with his normal speed is : **(SSC CGL Tier I 2000)**

(*a*) 2 (*b*) 1 (*c*) 3 (*d*) 1.5

Sol. (*a*) Walking at 2/3 of speed, he will take 3/2 of the normal time.

∴ 3/2 of Usual time = Usual time + 1 hour ⇒ 1/2 Usual time = 1 hour

∴ Usual time = 2 × 1 hour = 2 hours

45. Walking at three-fourth of his usual speed, a man covers a certain distance in 2 hours more than the time he takes to cover the distance at his usual speed. The time taken by him to cover the same distance with his usual speed is : **(SSC CGL 2005, 2010M)**

(*a*) 4.5 hours (*b*) 5.5 hours (*c*) 6 hours (*d*) 5 hours

Sol. (*c*) Walking at $\frac{3}{4}$th of his usual speed, he will take $\frac{4}{3}$ of usual time.

$\therefore \ \frac{4}{3}$ of the usual time = Usual time + 2 hours ⇒ $\frac{1}{3}$ of the usual time = 2 hours

∴ Usual time = 3 × 2 hours = 6 hours

Alternative method : Ratio of speeds = 4 : 3 ⇒ Ratio of time taken = 3 : 4

Difference = (4 – 3) ratio = 1 ratio = 2 hours ⇒ Usual time = 3 ratio = 3 × 2 hours = 6 hours

46. Walking $\frac{4}{5}$ of his usual speed, a man is late by 20 minutes in reaching his destination. His usual time to reach his destination is : **(SSC CGL Tier I 2006)**

(*a*) $2\frac{1}{2}$ hours (*b*) 2 hours (*c*) $1\frac{1}{2}$ hours (*d*) $1\frac{1}{3}$ hours

Sol. (*d*) Time taken at $\frac{4}{5}$ of usual speed = $\frac{5}{4}$ of usual time

$\therefore \ \frac{5}{4}$ of the usual time = Usual time + $\frac{1}{3}$ hours ⇒ $\frac{1}{4}$ of the usual time = $\frac{1}{3}$ hours

∴ Usual time = $\frac{1}{3} \times \frac{4}{1} = \frac{4}{3} = 1\frac{1}{3}$ hours

47. A car travelling with 5/7 of its usual speed covers 42 km in 1 hour 40 minutes 48 seconds. What is the usual speed of the car ? **(SSC CGL Tier I 2005)**

(*a*) $17\frac{6}{7}$ km/hr (*b*) 35 km/hr (*c*) 25 km/hr (*d*) 30 km/hr

Sol. (*b*) 40 minutes = $40 \times \frac{1}{60}$ hour = $\frac{2}{3}$ hour; 48 seconds = $48 \times \frac{1}{60} \times \frac{1}{60}$ hour = $\frac{1}{75}$ hour

Total time = $1 + \frac{2}{3} + \frac{1}{75} = \frac{126}{75} = \frac{42}{25}$ hours

Speed of car = $42 \div \frac{42}{25} = 42 \times \frac{25}{42} = 25$ km/hr

Usual speed of the car = $\frac{7}{5} \times 25$ km/hr = 35 km/hr

48. Walking at $\frac{5}{4}$th of the usual speed, a person reaches his office 12 minutes too early. What is the usual time?

(*a*) 48 minutes (*b*) 60 minutes (*c*) 70 minutes (*d*) 80 minutes

Sol. (*b*) Walking at $\frac{5}{4}$th his usual speed, he will take $\frac{4}{5}$ of usual time.

$\therefore$ Usual time = $\frac{4}{5}$ of usual time + 12 minutes $\Rightarrow \frac{1}{5}$ of usual time = 12 minutes

$\therefore$ Usual time = 5 × 12 minutes = 60 minutes

49. A man started 20 minutes late and travelling at a speed of $1\frac{1}{2}$ times of his usual speed reaches his office in time. The time taken by the man to reach his office at his usual speed is **(SSC CGL Tier II 2012)**

(*a*) 1 hour (*b*) 30 minutes (*c*) 40 minutes (*d*) 1 hr 20 minutes

Sol. (*a*) Walking at $\frac{3}{2}$th his usual speed, he will take $\frac{2}{3}$ of usual time.

$\therefore$ Usual time = $\frac{2}{3}$ of usual time + 20 minutes $\Rightarrow \frac{1}{3}$ of usual time = 20 minutes

$\therefore$ Usual time = 3 × 20 minutes = 60 minutes = 1 hour

Alt: Ratio of speeds = 2 : 3 $\Rightarrow$ Ratio of time taken = 3 : 2 = $3x$, $2x$ (say)

$\Rightarrow 3x - 2x = 20$ minute $\Rightarrow x = 20$ minute $\Rightarrow 3x = 3 \times 20$ minutes = 60 minutes = 1 hour

50. A and B started at the same time from same place for a certain distance. B walking at 5/6 of A's speed reached destination 1 hour 15 minutes after A. Time taken by B is : **(SSC CGL Tier I 2008)**

(*a*) 6 hours 45 min (*b*) 7 hours (*c*) 7 hours 15 min (*d*) 7 hours 30 min

Sol. (*d*) Ratio of speeds of A and B = 6 : 5 $\Rightarrow$ Ratio of time taken by A and B = 5 : 6

Difference in time = (6 – 5) ratio = 1 ratio = 1 hour 15 minutes = 75 minutes

$\therefore$ Time taken by B = 6 ratio = 6 × 75 minutes = 450 minutes = 7 hours 30 minutes

51. A 250 metre train is running at the average speed of 90 km/h. How long will it take to pass a standing man.

(*a*) 8 seconds (*b*) 10 seconds (*c*) 12.5 seconds (*d*) 20 seconds

Sol. (*b*) Speed of the train = $\frac{5}{18} \times 90$ km/h = 25 m/s

∴ Time taken to cover 250 metre = $\frac{250}{25}$ = 10 seconds

Alternative method : Required time = $\frac{250}{90000} \times 3600$ = 10 seconds

52. A train 150 metre long crosses a 150 metre long platform, in 15 seconds. What is the speed of the train?

(*a*) 20 km/h (*b*) 36 km/h (*c*) 72 km/h (*d*) 144 km/h

Sol. (*c*) Total distance covered = (150 + 150) metres = 300 metres

Speed of train = $\frac{300}{15}$ m/s = $\frac{300}{15} \times \frac{18}{5}$ km/h = 72 km/h

53. If a train, with a speed of 60 km/hr, crosses a pole in 30 seconds, the length of the train (in metres) is : **(SSC CGL Tier I 2005)**

(*a*) 1000 (*b*) 900 (*c*) 750 (*d*) 500

Sol. (*d*) Speed of train = 60 km/hr = 60 × $\frac{5}{18}$ m/s

Length of train = Distance covered in 30 seconds = 60 × $\frac{5}{18}$ × 30 m = 500 metres

54. A train running at 36 km/ hr crosses a pole in 25 seconds. Length of the train is : **(SSC CHSL 2012)**

(*a*) 250 m (*b*) 300 m (*c*) 225 m (*d*) 275 m

Sol. (*a*) Speed = 36 km/hr = 36 × $\frac{5}{18}$ m/s = 10 m/s ⇒ Length of train = 10 × 25 = 250 m

55. A train passes a man standing on a platform in 8 seconds and also crosses the platform which is 264 metres long in 20 seconds. The length (in metres) of the train is : **(SSC CGL Tier I 2002)**

(*a*) 188 (*b*) 176 (*c*) 175 (*d*) 96

Sol. (*b*) Time taken to cover 264 metres = 20 – 8 = 12 seconds

⇒ Speed of train = 264 ÷ 12 = 22 m/s

Length of train = Distance covered in 8 seconds = 8 × 22 = 176 metres

56. A train running at the speed of 15 m/s takes 20 seconds to pass a man running at the speed of 5 m/s in the same direction. Find the length of the train.

(*a*) 50 metre (*b*) 100 metre (*c*) 200 metre (*d*) 250 metre

Sol. (*c*) Relative speed = (15 – 5) m/s = 10 m/s

∴ Length of the train = Distance covered in 20 seconds = 10 × 20 = 200 metre

57. A train running at the speed of 25 m/s takes 10 seconds to pass a man running at the speed of 10 m/s in the opposite direction. Find the length of the train.

(*a*) 150 metre (*b*) 200 metre (*c*) 250 metre (*d*) 350 metre

Sol. (*d*) Relative speed = (25 + 10) m/s = 35 m/s

⇒ Length of the train = Distance covered in 10 seconds = 35 × 10 = 350 metre

58. A train crosses 80 metre long platform in 30 seconds. If the speed of the train is 4 m/s, what is the length of the train ? **(Bank Specialist Officers Exam. 2004)**

(*a*) 40 metres (*b*) 55 metres (*c*) 80 metres (*d*) 60 metres

Sol. (*a*) Distance covered in 30 seconds = 4 × 30 = 120 m

∴ Length of train + platform = 120 m ⇒ Length of train + 80 m = 120 m

⇒ Length of train = 120 – 80 = 40 m

59. A 220 metre long train crosses a 260 metre long platform in 24 seconds. What is the speed of train in km/hr ? **(Bank PO 2003)**

(*a*) 60 (*b*) 76 (*c*) 72 (*d*) 64

Sol. (*c*) Total distance covered in 24 seconds = 220 + 260 = 480 m

∴ Speed of train = 480 ÷ 24 = 20 m/s = $20 \times \frac{18}{5}$ km/hr = 72 km/hr

60. A train 150 metres long passes a tree in 12 seconds. It will pass a tunnel 250 m in length in: **(SSC CHSL 2012)**

(*a*) 20 seconds (*b*) 25 seconds (*c*) 26 seconds (*d*) 32 seconds

Sol. (*d*) Total distance to be covered = 150 + 250 = 400 m

∴ Requird time = $12 \times \frac{400}{150} = 12 \times \frac{8}{3}$ = 32 seconds

61. A train moving at a rate of 36 km/hr crosses a standing man in 10 seconds. It will cross a platform 55 metres long, in : **(SSC CGL Tier I 2013)**

(*a*) $5\frac{1}{2}$ seconds (*b*) 6 seconds (*c*) 7 seconds (*d*) $15\frac{1}{2}$ seconds

Sol. (*d*) Speed of train = 36 km/hr = $36 \times \frac{5}{18}$ = 10 m/s

∴ Time taken to cross platform = 10 seconds + (55 ÷ 10) seconds = $15\frac{1}{2}$ seconds

Trick : Time should be more than time taken to cross standing man.

62. A 120 metre long train is running at a speed of 90 km per hour. It will cross a railway platform 230 m long in : **(SSC CGL Tier I 2005)**

(*a*) $4\frac{4}{5}$ seconds (*b*) $9\frac{1}{5}$ seconds (*c*) 7 seconds (*d*) 14 seconds

Sol. (*d*) Total distance = 120 + 230 = 350 metres

Speed of train = 90 km/hr = $90 \times \frac{5}{18}$ = 25 m/s ⇒ Required time = 350 ÷ 25 = 14 seconds

63. A train 280 m long is moving at a speed of 60 kmph. What is the time taken by the train to cross a platform 220 m long ? **(CDS 2011)**

(*a*) 45 s (*b*) 40 s (*c*) 35 s (*d*) 30 s

Sol. (*d*) Total distance = 280 + 220 = 500 m. Speed of train = 60 kmph = $60 \times \frac{5}{18}$ m/s

∴ Required time = $\frac{500 \times 18}{60 \times 5}$ = 30 seconds

64. A train moves past a telegraph post and a bridge 264 m long in 8 seconds and 20 seconds respectively. What is the speed of the train ? **(SSC CGL Tier I 2004)**

(*a*) 69.5 km/hr (*b*) 70 km/hr (*c*) 79 km/hr (*d*) 79.2 km/hr

Sol. (*d*) Time take to cover 264 m = (20 – 8) seconds = 12 seconds

∴ Speed of train = 264 ÷ 12 = 22 m/s = $22 \times \frac{18}{5}$ km/hr = 79.2 km/hr

65. A train passes a 50 metres long platform in 14 seconds and a man standing on the platform in 10 seconds. The speed of the train is : **(SSC CGL Tier I 2000)**

(*a*) 24 km/hr (*b*) 36 km/hr (*c*) 40 km/hr (*d*) 45 km/hr

Sol. (*d*) Time take to cover 50 m = (14 – 10) seconds = 4 seconds

$\therefore$ Speed of train = $\frac{50}{4} \times \frac{18}{5}$ km/hr = 45 km/hr

66. A train running at the speed of 20 m/s crosses a pole in 24 seconds less than the time it requires to cross a platform thrice its length at the same speed. What is the length of the train? **(Bank PO 2005)**

(*a*) 270 m (*b*) 340 m (*c*) 180 m (*d*) None of these

Sol. (*d*) Let length of train be x metres. Then length of the platform = $3x$

Time taken to cover $[(x + 3x) - x]$ metres = $3x$ metres = 24 seconds

Time taken to cover x metres = 24 ÷ 3 = 8 seconds

Length of train = 8 × 20 = 160 metres

67. Two trains are running in opposite directions with the same speed. If the length of each train is 120 metres and they cross each other in 12 seconds, the speed of each train (in km/hr) is : **(SSC CGL Tier I 2003)**

(*a*) 72 (*b*) 10 (*c*) 36 (*d*) 18

Sol. (*c*) Total distance covered = 120 + 120 = 240 metres

Relative speed = 240 ÷ 12 = 20 m/s

Speed of each train = 20 ÷ 2 = 10 m/s = $10 \times \frac{18}{5}$ km/hr = 36 km/hr

68. Two trains, of same length, are running in parallel tracks in opposite directions with speeds 65 km/hr and 85 km/hr respectively. They cross each other in 6 seconds. The length of each train is : **(SSC CHSL 2013)**

(*a*) 100 m (*b*) 125 m (*c*) 115 m (*d*) 150 m

Sol. (*b*) Relative speed = (65 + 85) km/hr = 150 km/hr = $150 \times \frac{5}{18}$ m/s

Distance covered in 6 seconds = $150 \times \frac{5}{18} \times 6$ = 250 m

Length of each train = 250 ÷ 2 = 125 m

69. A train crosses a pole and a bridge 300 metres long in 10 seconds and 25 seconds respectively. How long will the train take to cross a platform 200 metres long ? **(SSC CGL Tier I 2008)**

(*a*) 10 seconds (*b*) 18 seconds (*c*) 20 seconds (*d*) 25 seconds

Sol. (*c*) Time taken to cover 300 metres = (25 – 10) seconds = 15 seconds

$\therefore$ Speed of the train = 300 ÷ 15 = 20 m/s

$\Rightarrow$ Length of the train = 10 × 20 m = 200 metres

Time taken by train to cross a platform of 200 metres = (200 + 200) ÷ 20 = 20 seconds

Alt: 10 sec + 2/3 × 15 sec = 20 seconds

70. A train passes two bridges of lengths 800 m and 400 m in 100 seconds and 60 seconds respectively. The length of the train is : **(SSC CGL Tier I 2005)**

(*a*) 80 m (*b*) 90 m (*c*) 200 m (*d*) 150 m

Sol. (*c*) (800 m – 400 m) = 400 m is covered in (100 – 60) = 40 seconds

$\therefore$ Speed of the train = 400 ÷ 40 = 10 m/s

$\therefore$ Distance covered in 100 seconds = 100×10 m = 1000 metres

Length of the train = 1000 – 800 = 200 metres

Alt: $(800 + x)$ metres covered in 100 seconds. $(400 + x)$ metres covered in 60 seconds.

$(800 + x) - (400 + x) = 400$ metres covered in 100 – 60 = 40 seconds.

800 metres covered in $2 \times 40 = 80$ seconds.

Total distance covered in 100 seconds = $400/40 \times 100 = 1000$ metres

$\therefore$ Length of the train = 1000 – 800 = 200 metres

71. A train, 150 m long, passes a pole in 15 seconds and another train of the same length travelling in the opposite direction in 12 seconds. The speed of the second train is

(SSC CGL Tier I 2008)

(*a*) 45 km/hr (*b*) 48 km/hr (*c*) 52 km/hr (*d*) 54 km/hr

Sol. (*d*) Speed of the first train = $150 \div 15 = 10$ m/s

Relative speed of two trains = $\dfrac{150+150}{12} = 25$ m/s

$\therefore$ Speed of the second train = $(25 - 10)$ m/s = 15 m/s = $15 \times \dfrac{18}{5} = 54$ km/hr

72. A person started his journey in the morning. At 11 a.m., he covered 3/8 of the journey and on the same day at 4.30 p.m, he covered 5/6 of the journey. He started his journey at :

(SSC CGL Tier I 2008)

(*a*) 6 a.m. (*b*) 5.30 a.m. (*c*) 7 a.m. (*d*) 6.30 a.m.

Sol. (*d*) Journey covered between 11 a.m. and 4.30 p.m. = $\dfrac{5}{6} - \dfrac{3}{8} = \dfrac{11}{24}$

Time taken = 4.30 p.m. – 11 a.m. = $5\dfrac{1}{2}$ hours = $\dfrac{11}{2}$ hours

$\Rightarrow \dfrac{11}{24}$ of journey is covered in $\dfrac{11}{2}$ hours

$\therefore \dfrac{3}{8}$ of journey is covered in $\dfrac{11}{2} \times \dfrac{24}{11} \times \dfrac{3}{8} = \dfrac{9}{2}$ hours = $4\dfrac{1}{2}$ hours

Journey was started at 11 a.m. – $4\dfrac{1}{2}$ hours = 6.30 a.m.

73. A 300 metre long train crosses a platform in 42 seconds while it crosses a pole in 18 seconds. What is the length of the platform ? **(Bank PO Exam.)**

(*a*) 400 metre (*b*) 200 metre (*c*) 300 metre (*d*) None of these

Sol. (*a*) Time taken to cover distance equal to length of platform = 42 – 18 = 24 seconds

Time taken to cover distance equal to length (300 m) of train = 18 seconds

$\therefore$ Length of the platform = $\dfrac{300}{18} \times 24 = 400$ m

74. A train 150 metres long running at an average speed of 52 km/h crosses a man walking in the same direction at an average speed of 7 km/h in:

(*a*) 12 seconds (*b*) 13 seconds (*c*) 15 seconds (*d*) None of these

Sol. (*a*) Relative speed = $(52 - 7)$ km/h = 45 km/h = $45 \times \dfrac{5}{18}$ m/s = $\dfrac{25}{2}$ m/s

$\therefore$ Time taken to cover 150 metre $= 150 \div \frac{25}{2} = 150 \times \frac{2}{25} = 12$ seconds

75. Two trains of lengths 100 m and 150 m are travelling in opposite directions at speeds of 75 kmph and 50 kmph respectively. What is the time taken by them to cross each other ? **(CDS 2010, SSC CHSL 2013 M)**

(*a*) 7.4 s (*b*) 7.2 s (*c*) 7 s (*d*) 6.8 s

Sol. (*b*) Relative speed = (75 + 50) kmph $= 125 \times \frac{5}{18}$ m/s

Total distance = (100 + 150) m = 250 m $\Rightarrow$ Required time $= 250 \times \frac{1}{125} \times \frac{18}{5} = 7.2$ s

76. Two trains, one is of 121 m in length at the speed of 40 km/hour and other is of 99 m in length at the speed of 32 km/hour are running in opposite directions. In how much time will they be completely clear from each other from the moment they meet ? **(CDS 2015)**

(*a*) 10 s (*b*) 11 s (*c*) 16 s (*d*) 21 s

Sol. (*a*) Relative speed = (40 + 32) kmph $= 72 \times \frac{5}{18}$ m/s = 20 m/s

Total distance = (121 + 99) m = 220 m $\Rightarrow$ Required time = 220 ÷ 20 = 10 s

77. Two trains each 200 m long move towards each other on parallel lines with velocities 20 kmph and 30 kmph respectively. What is the time that elapses when they first meet until they have cleared each other ? **(CDS 2011)**

(*a*) 20 seconds (*b*) 24.8 seconds (*c*) 28.8 seconds (*d*) 30 seconds

Sol. (*c*) Total distance = 200 + 200 = 400 m

Relative speed = (20 + 30) km/hr = 50 km/hr $= 50 \times \frac{5}{18}$ m/s $= \frac{125}{9}$ m/s

$\therefore$ Required time $= 400 \div \frac{125}{9} = 400 \times \frac{9}{125} = 28.8$ seconds

78. Two trains one 160 m and the other 140 m long are running in opposite directions on parallel rails, the first at 77 km an hour and the other at 67 km an hour. How long (in seconds) will they take to cross each other ? **(SSC CGL Tier I 2003)**

(*a*) 7 (*b*) $7\frac{1}{2}$ (*c*) 6 (*d*) 10

Sol. (*b*) Total distance = (160 + 140) m = 300 m

Relative speed = (77 + 67) km/hr = 144 km/hr $= 144 \times \frac{5}{18}$ m/s = 40 m/s

$\therefore$ Required time $= 300 \div 40 = 7\frac{1}{2}$ seconds

79. Two trains of the length of 200 metre and 250 metre are running in the opposite directions at the speed of 35 km/h and 55 km/h respectively. Find the time in which the faster train will pass a man sitting in the slower train.

(*a*) 2 seconds (*b*) 8 seconds (*c*) 10 seconds (*d*) 18 seconds

Sol. (*c*) Relative speed = (35 + 55) km/h = 90 km/h $= \frac{5}{18} \times 90$ m/s = 25 m/s

The faster train will completely cross a man sitting in the slower train when a distance equal to length of the faster train is covered.

Distance to be covered = 250 metre ⇒ Required time = $\frac{250}{25}$ = 10 seconds

80. Two trains are running in the same direction at 74 km/h and 38 km/h. If the faster train crosses a man sitting in the slower train in 15 seconds, find the length of the faster train.

(SSC CGL Tier I 2008)

(*a*) 100 metre (*b*) 125 metre (*c*) 150 metre (*d*) 165 metre

Sol. (*c*) Relative speeds of the two trains = (74 – 38) km/h = 36 km/h

Distance covered in 15 seconds = $36 \times \frac{5}{18} \times 15 = 150$ m

∴ Length of the faster train = 150 metre

81. Two trains are running 40 km/hr and 20 km/hr respectively in the same direction. The fast train completely passes a man sitting in the slow train in 5 seconds. The length of the fast train is **(SSC CGL Tier II 2013)**

(*a*) $23\frac{2}{9}$ m (*b*) 27 m (*c*) $27\frac{7}{9}$ m (*d*) 23 m

Sol. (*c*) Relative speed = (40 – 20)km/hr = 20 km/hr = $20 \times \frac{5}{18}$ m/s

Distance covered in 5 seconds = $20 \times \frac{5}{18} \times 5 = \frac{250}{9} = 27\frac{7}{9}$ m

82. Two trains running in the same direction at the speeds of 68 km/h and 48 km/h take 36 seconds to cross each other. If the length of one train is 85 metre, what is the length of the other train?

(*a*) 115 metre (*b*) 125 metre (*c*) 150 metre (*d*) 160 metre

Sol. (*a*) Their relative speed = (68 – 48) km/h = 20 km/h

Distance covered in 36 seconds = $\frac{5}{18} \times 20 \times 36 = 200$ metre

∴ Length of the second train = (200 – 85) metre = 115 metre

83. Two trains of length of 250 metre and 200 metre are running in the same direction. If the trains cross each other in 45 seconds and the speed of the first train is 27 km/h, find speed of the second train.

(*a*) 36 km/h (*b*) 54 km/h (*c*) 63 km/h (*d*) 72 km/h

Sol. (*c*) Total distance covered = (250 + 200) metres = 450 metres

Relative speed of two trains = $\frac{450}{45}$ m/s = $\frac{450}{45} \times \frac{18}{5}$ km/h = 36 km/h

∴ Speed of the second train = (27 + 36) km/h = 63 km/h

84. A train of length 150 m takes 10s to cross another train 100 m long coming from the opposite direction. If the speed of first train is 30 kmph, what is the speed of second train?

(CDS 2009)

(*a*) 72 kmph (*b*) 60 kmph (*c*) 54 kmph (*d*) 48 kmph

Sol. (*b*) Total distance covered = (150 + 100) metres = 250 metres

Relative speed of two trains = $\frac{250}{10}$ m/s = $\frac{250}{10} \times \frac{18}{5}$ kmph = 90 kmph

∴ Speed of the second train = 90 – 30 = 60 kmph

85. A train travelling at 48 km/hr crosses another train, having half its length and travelling in opposite direction at 42 km/hr, in 12 seconds. It also passes a railway platform in 45 seconds. The length of the railway platform is : **(SSC CGL Tier I 2008)**

(*a*) 200 metres (*b*) 300 metres (*c*) 350 metres (*d*) 400 metres

Sol. (*d*) Relative speeds of two trains = (48 + 42) = 90 km/hr = $90 \times \frac{5}{18}$ m/s = 25 m/s

Total distance covered in 12 seconds = 12 × 25 = 300 metres

∴ Length of the first train = $300 \times \frac{2}{3} = 200$ metres

Distance covered by the first train in 45 seconds = $48 \times \frac{5}{18} \times 45 = 600$ metres

∴ Length of the platform = 600 – 200 = 400 metres

86. The driver of a car driving at the speed of 36 km/hr locates a bus 40 metres ahead of him. After 20 seconds the bus is 60 metres behind. The speed of the bus is **(SSC CHSL 2013)**

(*a*) 18 km/hr (*b*) 36 km/hr (*c*) 54 km/hr (*d*) 72 km/hr

Sol. (d) In 20 seconds, the car has covered (40 + 60) m = 100 m more than the bus.
Relative speed (extra) of car = 100 ÷ 20 = 5 m/s = 18 kmph
Speed of bus = (36 – 18) = 18 kmph

87. A starts from a place at 7 AM with a speed of 10 km per hour B starts from the same place at 9 AM with a speed of 15 km per hour. When will B meet A ?

(UPSC Civil Services Pre. 2011)

(*a*) 1 PM (*b*) 3 PM (*c*) 2 PM (*d*) 12 Noon

Sol. (*a*) Distance covered by A from 7 AM to 9 AM = 2 × 10 = 20 km
B gains (15 – 10) = 5 km/hr over A.
Time taken by B to cover 20 km extra = 20 ÷ 5 = 4 hours
Required time = 9 AM + 4 hours = 1 PM

88. A theft is reported to a policeman. The thief starts running and the policeman chases him. When the policeman starts chasing, the thief was at a distance of 250 metre. The thief and the policeman run at the speed of 8 km/hr and 9 km/hr respectively. Find the time the policeman will take to catch the thief. **(SSC CGL Tier I 2006)**

(*a*) 10 minutes (*b*) 15 minutes (*c*) 30 minutes (*d*) 60 minutes

Sol. (*b*) Policeman gains = (9 – 8) km/hr = 1 km/hr ⇒ He will gain 250 metre in 15 minutes.

89. A thief running at 8 km/hr is chased by a policeman whose speed is 10 km/hr. If the thief is 100m ahead of the policeman, then the time required for the policeman to catch the thief will be **[UPSC Civil Services (Pre.) 2013]**

(*a*) 2 min (*b*) 3 min (*c*) 4 min (*d*) 6 min

Sol. (*b*) Policeman gains = (10 – 8) km/hr = 2 km/hr ⇒ He will gain 100 metre in 3 minutes.

90. A theft is reported to a policeman. The thief starts running and the policeman chases him. When the policeman starts chasing, the thief was at a distance of 150 metre. The thief and the policeman run at the speed of 9 km/hr and 10 km/hr respectively. How far the policeman had to run to catch the thief ?

(*a*) 1000 m (*b*) 1250 m (*c*) 1500 m (*d*) None of these

Sol. (*c*) Ratio of speeds = 9 : 10
⇒ Policeman gains = (10 – 9) km = 1 km for every 10 km covered by him.
∴ Required distance = 10 × 150 m = 1500 m

91. A thief is noticed by a policeman from a distance of 200 m. The thief starts running and the policeman chases him. The thief and the policeman run at the rate of 10 km and 11 km per hour respectively. The distance (in metres) between them after 6 minutes ? **(SSC CGL Tier I 2000, 2012, CDS 2015)**

(*a*) 100 (*b*) 150 (*c*) 190 (*d*) 200

Sol. (*a*) Policeman gains = (11 – 10) km/hr = 1 km/hr = 1000 m/hr

In 6 minutes, he gains = $\frac{6}{60} \times 1000 = 100$ m ⇒ Required distance = 200 – 100 = 100 m

92. A, B, C walk 1 km in 5 minutes, 8 minutes and 10 minutes respectively. C starts walking from a point, at a certain time, B starts from the same point 1 minute later and A starts from the same point 2 minutes later than C. Then A meets B and C at times. **(SSC CGL Tier I 2013)**

(*a*) $\frac{4}{3}$ min, 3 min (*b*) $\frac{5}{3}$ min, 2 min (*c*) 1 min, 2 min (*d*) 2 min, 3 min

Sol. (*b*) Speeds of A, B, C (in m/min) = $\frac{1000}{5}, \frac{1000}{8}, \frac{1000}{10}$ = 200, 125, 100

Distance covered by B in 1 minute = 125 m. Distance covered by C in 2 minutes = 200 m.

A gains 75 m/min over B and 100 m/min over C.

Time taken by A to gain 125 m over B = $\frac{125}{75} = \frac{5}{3}$

Time taken by A to gain 200 m over C = $\frac{200}{100} = 2$

93. A and B can walk around a circular path in 4 hours and 6 hours respectively. If they start from the same point in the same direction, when will they be together at the starting point again?

(*a*) 2 hours (*b*) 6 hours (*c*) 8 hours (*d*) 12 hours

Sol. (*d*) LCM of time taken by A and B = LCM of 4 and 6 = 12 hours

∴ They will be together again at the starting point after 12 hours.

Alternative Method : In 12 hours A runs 3 rounds and B runs 2 rounds. Thus A takes 12 hours to gain one round over B.

94. A and B walk around a circular path of 900 metre in circumference, starting together from the same point in the same direction. If their speeds are 150 metre per minute and 200 metre per minute respectively, after how many minutes will they be again at the starting point?

(*a*) 4.5 minutes (*b*) 6 minutes (*c*) 15 minutes (*d*) 18 minutes

Sol. (*d*) They will be together at the starting point when B (the faster one) has gained 900 metre (*i.e.* one round) over A.

In 1 minute B gains (200 – 150) metre = 50 metre over A.

∴ B will gain 900 metres in 900 ÷ 50 = 18 minutes

Alternative method : LCM of $\frac{900}{150}$ and $\frac{900}{200} = \frac{900}{50} = 18$ minutes

95. A and B walk around a circular path of 5 km in circumference, starting together from the same point in the same direction. If their speeds are 4 km/h and 3 km/h respectively, after how many hours will they be again at the starting point?

(*a*) 3 hours (*b*) 4 hours (*c*) 5 hours (*d*) 6 hours

Sol. (*c*) They will be together at the starting point when A (the faster one) has gained 5 km (*i.e.*

one round) over B.

In 1 hour A gains (4 km – 3 km) *i.e.* 1 km over B $\Rightarrow$ A will gain 5 km in 5 hours.

Alt: Time taken by A to cover 5 km = $\frac{5}{4}$ hours. Time taken by B to cover 5 km = $\frac{5}{3}$ hours

LCM of $\frac{5}{4}$ and $\frac{5}{3}$ is 5 hours $\Rightarrow$ They will be together again at the starting point after 5 hours.

96. A, B and C walk round a circle of 900 metres in circumference at the rates of 60 m, 50 m and 45 m per minute respectively. If all of them start together in the same direction after how long will they all meet for the first time?

(*a*) 60 min (*b*) 90 min (*c*) 120 min (*d*) 180 min

Sol. (*d*) A gains 10 m over B in 1 minute. A gains 15 m over C in 1 minute.

A will gain one round round over B in 90 min. A will gain one round over C in 60 min.

Hence A and B will meet after 90 min while A and C will meet after 60 min.

All of them be together again after 180 minutes (*i.e.* LCM of 90 & 60 min).

Alt: Time taken by A, B and C to complete one round = $\frac{900}{60}, \frac{900}{50}, \frac{900}{45} = \frac{900}{5} = 180$ minutes

97. From a point on a circular track 5 km long A, B and C started running in the same direction at the same time with speeds of $2\frac{1}{2}$ km per hour, 3 km per hour and 2 km per hour respectively. Then on the starting point all three will meet again after : **(SSC CGL Tier I 2003)**

(*a*) 30 hours (*b*) 6 hours (*c*) 10 hours (*d*) 15 hours

Sol. (*c*) Time taken by A, B and C to complete one round = $\frac{5}{2\frac{1}{2}}, \frac{5}{3}, \frac{5}{2} = 2, \frac{5}{3}, \frac{5}{2}$ hours

Then will be at starting point again after LCM of $2, \frac{5}{3}, \frac{5}{2} = 10$ hours

98. A, B and C start together from the same place to walk round a circular path of length 12 km. A walks at the rate of 4 km/hr, B 3 km/hr and C 3/2 km/hr. They will meet together at the starting place at the end of : **(SSC CGL Tier I 2008)**

(*a*) 10 hours (*b*) 12 hours (*c*) 15 hours (*d*) 24 hours

Sol. (*d*) Time taken by A to complete a round = 12 ÷ 4 = 3 hours

Time taken by B to complete a round = 12 ÷ 3 = 4 hours

Time taken by C to complete a round = $12 \div \frac{3}{2} = 8$ hours

$\therefore$ Required time = LCM of 3, 4 and 8 hours = 24 hours

99. A and B can walk around a circular path in 5 minutes and 8 minutes respectively. If they start from the same point in the same direction, after how much time will they meet again for the first time?

(*a*) 13 minutes (*b*) 12 minutes (*c*) 40 minutes (*d*) 13 minutes 20 seconds

Sol. (*d*) Let length of the circular path = LCM of 5 and 8 = 40 metres.

Then speeds of A and B are 8 and 5 metres/minute.

They will meet for the first time when A (*i.e.* the faster one) has gained one round *i.e.* 40 metres over B.

Time taken by A = $\frac{40}{8-5} = 13\frac{1}{3}$ minutes = 13 minutes 20 seconds

Alt: A and B would again be together at starting poing after 40 minutes (= LCM of 5 & 8)

In 40 minutes A completes 8 rounds and B completes 5 rounds.

∴ A gains 3 rounds over B in 40 minutes.

A will gain 1 round in $\frac{40}{3}$ minutes = $13\frac{1}{3}$ minutes = 13 minutes 20 seconds

100. A man is walking at a speed of 10 km/h. After every km, he takes rest for 4 minutes. How much time will he take to cover a distance of 10 km? **(SSC CHSL)**

(*a*) 1 hour (*b*) 90 minutes (*c*) 96 minutes (*d*) 100 minutes

Sol. (*c*) He covers 10 km in 1 hour (*i.e.* in 60 minutes) and rests for 4 minutes after every km. He will take rest 9 times. ⇒ Total time taken = 60 + 9 × 4 = 96 minutes

101. A man rides at the rate of 18 km/hr, but stops for 6 mins. to change horses at the end of every 7th km. The time that he will take to cover a distance of 90 km is **(SSC CGL Tier I 2013)**

(*a*) 6 hrs. (*b*) 6 hrs. 12 min. (*c*) 6 hrs. 18 min. (*d*) 6 hrs. 24 min.

Sol. (*b*) He covers 90 km in 5 hours (@ 18 km/hr) and stops for 6 minutes at the end of every 7th km. He will stop = 90 ÷ 7 = 12 times ⇒ Total time = 5 hours + 6 × 12 min = 6 hours 12 min

102. Nitin rides a bicycle at the speed of 15 km/hr, but stops for 10 minutes to take rest every 20 km. How much time will he take to cover a distance of 150 kms ? **(RBI Note Printing Asstt. Manager 2006)**

(*a*) 12 hours 30 min (*b*) 11 hours 10 min (*c*) 10 hours 20 min (*d*) 12 hours 10 min

Sol. (*b*) Time taken to cover 150 km @ 15 km/hr = 10 hours

He will take rest 150 ÷ 20 = 7 times ⇒ Total rest time = 10 × 7 = 70 min = 1 hour 10 min

Total time taken = 10 hours + 1 hour 10 minutes = 11 hours 10 minutes

103. Excluding stoppage time, average speed of a bus is 64 km/hr and average speed of the bus including stoppage time is 48 km/hr. How much time the bus stops on an average (per hour)? **(Bank PO 2009)**

(*a*) 12.5 minutes (*b*) 10 minutes (*c*) 18 minutes (*d*) 15 minutes

Sol. (*d*) Time spent on stoppages (per hour)

= Time in which train could have travelled (64 – 48) km i.e. 16 km = $\frac{16}{64} \times 60 = 15$ min/hr

104. Excluding stoppages, the speed of a bus is 54 km/hr and including stoppages, the speed is 45 km/hr. For how many minutes does the bus stop per hour ? **(SAS 2010)**

(*a*) 8 (*b*) 10 (*c*) 12 (*d*) 14

Sol. (*b*) Stoppages = (54 – 45) = 9 km/hr ⇒ Time = $\frac{9}{54} \times 60 = 10$ min/hr

105. A policeman starts to chase a thief. When the thief goes 10 steps the policeman moves 8 steps. 5 steps of the policeman is equal to 7 steps of the thief. The ratio of the speeds of the policeman and the thief is **(SSC CGL Tier I 2014)**

(*a*) 25 : 28 (*b*) 25 : 56 (*c*) 28 : 25 (*d*) 56 : 25

Sol. (*c*) 5 steps of Policeman = 7 steps of thief ⇒ 8 steps of policeman = $\frac{7}{5} \times 8 = \frac{56}{5}$ steps of thief

∴ Speed of policeman : Speed of thief = $\frac{56}{5}$: 10 = 28 : 25

Alternative Method :

	Policeman	Thief
Frequency of steps	8	10
Ratio of length of leaps	7	5

$\therefore$ Speed of Policeman : Speed of Thief = $8 \times 7 : 10 \times 5 = 28 : 25$

106. A rat takes 5 leaps for every 4 leaps of cat, but 3 leaps of the cat are equal to 4 leaps of the rat. What is the ratio of the speed of the rat to that of the cat ? **(SSC CGL Tier I 2013)**

(*a*) 11 : 15 (*b*) 15 : 11 (*c*) 16 : 15 (*d*) 15 : 16

Sol. (*d*) 4 leaps of rat = 3 leaps of cat

5 leaps of rat = $\frac{3}{4} \times 5 = \frac{15}{4}$ leaps of cat $\Rightarrow$ Speed of rat : Speed of cat = $\frac{15}{4} : 4 = 15 : 16$

Alternative Method :

	Rat	Cat
Frequency of leaps	5	4
Ratio of length of leaps	3	4

$\therefore$ Speed of rat : Speed of cat = $5 \times 3 : 4 \times 4 = 15 : 16$

107. Two men A and B walk from P to Q, a distance of 21 km at 3 km/h and 4 km/h respectively. B reaches at point Q and returns immediately and he meets A at point R. Find the distance from P to R.

(*a*) 9 km (*b*) 18 km (*c*) 24 km (*d*) 27 km

Sol. (*b*) Both of them together have walked twice the distance from P to Q *i.e.* 42 km.

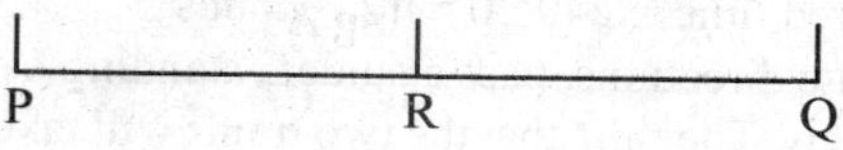

Ratio of speeds of A and B = 3 : 4 $\Rightarrow$ Distance travelled by A = PR = $\frac{3}{7} \times 42 = 18$ km.

108. Two men A and B start walking at the same time from place P to Q, at a distance of 20 km. A walks 2 km/h slower than B. B reaches at place Q and immediately starts walking towards P and meets A 4 km away from Q. Find the speed of A.

(*a*) 24 km/h (*b*) 4 km/h (*c*) 6 km/h (*d*) 8 km/h

Sol. (*b*) Distance covered by A = 20 – 4 = 16 km.

Distance covered by B = 20 + 4 = 24 km $\therefore$ Ratio of speeds of A and B = 16 : 24 = 2 : 3

Difference in ratio of speeds = (3 – 2) ratio = 1 ratio $\equiv$ 2 km/h

$\therefore$ Speed of A = 2×2 km/h = 4 km/h

109. Two persons P and Q start at the same time from city A for city B, 60 km away. P travels 4 km/hr slower than Q. Q reaches city B and at once turns back meeting P, 12 km from city B. What is the speed of P ? **(SSC CGL Tier I 2008 M, CDS 2009)**

(*a*) 8 km/hr (*b*) 12 km/hr (*c*) 16 km/hr (*d*) 20 km/hr

Sol. (*a*)

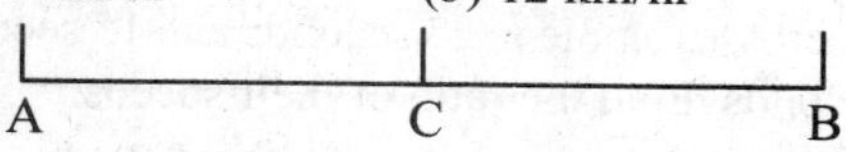

AB = 60 km, BC = 12 km. Total distance = 2 AB = 120 km

Ratio of distance covered by P and Q = (60 – 12) : (60 + 12) = 48 : 72 = 2 : 3

Difference of speeds = (3 – 2) ratio = 1 ratio = 4 km/hr

$\therefore$ Speed of P = 2 ratio = 2×4 km/hr = 8 km/hr

110. A train covers a distance of 3584 km in 2 days 8 hours. If it covers 1440 km on the first day and 1608 km on the second day, by how much does the average speed of the train for the remaining part of the journey differ from that for the entire journey ? **(SSC CGL Tier I 2008)**

(*a*) 3 km/h more (*b*) 3 km/h less (*c*) 4 km/h more (*d*) 4 km/h less

Sol. (*a*) Average speed for the whole journey = 3584 ÷ 56 = 64 km/hr

Remaining distance = 3584 – (1440 + 1608) = 536 km. Remaining time = 8 hours

Average speed for the remaining journey = 536 ÷ 8 = 67 km/h

∴ Required difference = 67 – 64 = 3 km/hr (more)

111. Two trains of equal length take 10 seconds and 15 seconds respectively to cross a telegraph post. If the length of each train be 120 metres, in what time (in seconds) will they cross each other travelling in opposite directions. **(SSC CGL Tier I 2004)**

(*a*) 16 (*b*) 15 (*c*) 12 (*d*) 10

Sol. (*c*) Required time $= \dfrac{2ab}{a+b} = \dfrac{2 \times 10 \times 15}{25} = 12$ seconds

Alternative method : One train covers in 10 seconds and other train in 15 seconds.

Total distance = 1 + 1 = 2 ⇒ Required time $= 2 \times \dfrac{10 \times 15}{25} = 12$ seconds

Alt : Speeds = 120/10 = 12 m/s, and 120/15 = 8 m/s.

Relative speed = 12 + 8 = 20 m/s ⇒ Required time = 240/20 = 12 seconds

112. Two trains of equal length running in opposite directions, pass a student standing by the side of railway line in 18 sec and 12 sec respectively. The time that the two trains will take to cross each other, is **(SSC CGL Tier I 2013)**

(*a*) 7.2 sec (*b*) 9.6 sec (*c*) 10.8 sec (*d*) 14.4 sec

Sol. (*d*) Required time $= \dfrac{2ab}{a+b} = \dfrac{2 \times 18 \times 12}{18+12} = 14.4$ seconds

Alternative method : One train covers in 18 seconds and other train in 12 seconds.

Total distance = 1 + 1 = 2 ⇒ Required time $= 2 \times \dfrac{18 \times 12}{30} = 14.4$ seconds

113. A man standing on a railway platform notices that a train going in one direction takes 9 seconds to pass him and other train of the same length takes 6 seconds to pass him. Find the time taken by the two trains to cross each other when they are running in the same direction.

(*a*) 3 seconds (*b*) 27 seconds (*c*) 30 seconds (*d*) 36 seconds

Sol. (*d*) Required time $= \dfrac{2ab}{a-b} = \dfrac{2 \times 9 \times 6}{3} = 36$ seconds

114. Two trains running in opposite directions crosses a pole in 23 seconds and 15 seconds respectively and they cross each other in 17 seconds. Find the ratio of their speeds.

(*a*) 1 : 2 (*b*) 2 : 1 (*c*) 1 : 3 (*d*) 3 : 1

Sol. (*c*) Let the speeds of the trains be x m/s and y m/s ⇒ Total lengths of two trains = $23x + 15y$

$\therefore \dfrac{23x+15y}{x+y} = 17 \Rightarrow 23x + 15y = 17x + 17y \Rightarrow 6x = 2y \Rightarrow x : y = 1 : 3$

115. Two trains start at the same time from stations A and B towards each other. Distance between the two stations is 420 km and the trains meet each other after 6 hours. If the difference between speeds of two trains is 10 km/h, find the speed of the slower train.

(*a*) 30 km/h (*b*) 40 km/h (*c*) 60 km/h (*d*) 70 km/h

Sol. (*a*) Relative speed (sum) of the trains $= \frac{420}{6} = 70$ km/h

$\therefore$ Speed of the slower train $= \frac{70-10}{2} = 30$ km/h

116. Two trains start at the same time from two stations 360 km apart. If the trains meet after 5 hours and difference between their speeds is 12 km/h, find the speed of the faster train.

(*a*) 30 km/h (*b*) 36 km/h (*c*) 42 km/h (*d*) 60 km/h

Sol. (*c*) Relative speed (sum) of the trains $= \frac{360}{5} = 72$ km/h

$\Rightarrow$ Speed of faster train $= \frac{72+12}{2} = 42$ km/h

117. A train leaves P for Q; at the same time another train leaves Q for P. The trains meet after 6 hours, the train from P to Q having travelled 8 km per hour more than the other. If the distance from P to Q is 528 km, the speed of train from P to Q in km/hour is **(SSC MTS 2014)**

(*a*) 40 (*b*) 48 (*c*) 45 (*d*) 50

Sol. (*b*) Sum of speeds = 528 ÷ 6 = 88 km/hr. Difference of speeds = 8 km/hr

Speeds from P to Q (faster train) $= \frac{88+8}{2} = 48$ km/hr

118. Two trains of length 150 m and 250 m run on parallel lines. When they run in the same direction it will take 20 seconds to cross each other and when they travel in opposite direction it will take 5 seconds. Then speeds (in kmph) of the two trains are **(SSC MTS 2014)**

(*a*) 108 & 180 (*b*) 30 & 50 (*c*) 120 & 100 (*d*) 72 & 18

Sol. (*a*) Sum of speeds $= \frac{150+250}{5} = 80$m/s. Difference of speeds $= \frac{150+250}{20} = 20$m/s

Speeds $= \frac{80+20}{2}$ m/s $= 50 \times \frac{18}{5}$ km/h $= 180$ km/h; $\frac{80-20}{2}$ m/s $= 30 \times \frac{18}{5}$ km/h $= 108$ km/h

119. A train 150 metre long passes a standing man in 15 seconds. While it crosses another train of the same length coming from the opposite direction in 12 seconds. Find the speed of the second train.

(*a*) 40 km/h (*b*) 45 km/h (*c*) 50 km/h (*d*) 54 km/h

Sol. (*d*) Distance covered = (150 + 150) metres = 300 metres

Relative speed of two trains $= \frac{300}{12} = 25$ m/s. Speed of the first train $= \frac{150}{15}$ m/s $= 10$ m/s

$\therefore$ Speed of the second train = (25 – 10) m/s = 15 m/s $= 15 \times \frac{18}{5}$ km/h = 54 km/h

120. The distance between two stations A and B is 300 km. One train leaves station A towards station B at the average speed of 40 km/h. At the same time, another train leaves station B towards A at the average speed of 80 km/h. The distance from station A where the two trains meet is:

(*a*) 80 km (*b*) 100 km (*c*) 200 km (*d*) 160 km

Sol. (*b*) Ratio of distance covered = Ratio of speeds = 40 : 80 = 1 : 2

∴ Distance covered by them are 100 km and 200 km respectively.

∴ Distance from station A = 100 km.

121. The distance between two stations A and B is 550 km. One train leaves station A towards station B at the average speed of 100 km/h. After half an hour, another train leaves station B towards A at the average speed of 150 km/h. The distance from station A where the two trains meet is: **(GMAT)**

(*a*) 100 km (*b*) 200 km (*c*) 250 km (*d*) 300 km

Sol. (*c*) Distance covered by the first train in half an hour = $\frac{1}{2} \times 100 = 50$ km

Remaining distance = 550 km – 50 km = 500 km

Ratio of speeds of two trains = 100 : 150 = 2 : 3 ⇒ First train covers = $\frac{2}{5} \times 500 = 200$ km

∴ Total distance covered by the first train = 50 + 200 = 250 km.

122. The distance between two cities A and B is 330 km. A train starts from A at 8 a.m. and travels towards B at 60 km/hr. Another tranin starts from B at 9 a.m. and travels towards A at 75 km/hr. At what time do they meet ? **(SSC CGL Tier I 2003)**

(*a*) 10 a.m. (*b*) 10:30 a.m. (*c*) 11 a.m. (*d*) 11:30 a.m.

Sol. (*c*) Distance travelled by first train in 1 hour = 60 km

⇒ Remaining distance = 330 – 60 = 270 km. Relative speed = (60 + 75) = 135 km/hr

Time taken to cover 270 km = 270 ÷ 135 = 2 hours

Required time = 9 a.m. + 2 hours = 11 a.m.

123. A and B are two stations 285 km apart. A train starts from A at 10 a.m. and travels towards B at 60 km/hr. Another train starts from B at 11 a.m. and travels towards A at 40 km/hr. At what time do they meet ?

(*a*) 1 p.m. (*b*) 2:15 p.m. (*c*) 1 : 15 p.m. (*d*) 1 : 25 p.m.

Sol. (*c*) Distance travelled by train A in 1 hour = 60 km

⇒ Remaining distance = 285 – 60 = 225 km. Relative speed = (60 + 40) = 100 km/hr

Time taken to cover 225 km = 225 ÷ 100 = 2.25 hours = 2 hours 15 minutes

Required time = 11 a.m. + 2 hours 15 minutes = 1 : 15 p.m.

124. Two trains start at the same time from two stations and proceed towards each other at the speeds of 15 km and 20 km per hour respectively. When the trains meet, it is found that one train has travelled 50 km more than the other. Find the distance between the two stations.

(*a*) 10 km (*b*) 35 km (*c*) 350 km (*d*) 500 km

Sol. (*c*) In one hour, faster train travels (20 – 15) = 5 km more than the slower train.

In 10 hours, faster train travels 50 km more than the slower train.

In one hour, total distance covered by the two trains = (15 + 20) = 35 km

∴ Total distance covered in 10 hours = 10 × 35 = 350 km.

125. Two trains start at the same time from station A and B towards station B and A at the speed of 50 km/h and 55 km/h respectively. When the two trains meet it is found that the faster train has travelled 30 km more than the slower one. What is the distance between the two stations.

(*a*) 300 km (*b*) 330 km (*c*) 550 km (*d*) 630 km

Sol. (*d*) Faster train covers 5 km/h (*i.e.* 55 – 50) extra than the slower train.

∴ To cover 30 km extra it will take 30 ÷ 5 = 6 hours

Total distance covered by the two trains in 1 hour = (55 + 50) km = 105 km

∴ Distance covered in 6 hours = 6 × 105 km = 630 km

126. A train travelling at 45 km per hour leaves Delhi at 5 a.m. and another train travelling at 55 km per hour leaves Delhi at 9 a.m. in the same direction. How many km from Delhi will the two trains be together? **(SSC CHSL 2006)**

(*a*) 180 km (*b*) 810 km (*c*) 900 km (*d*) 990 km

Sol. (*d*) Distance covered by first train in 4 hours (i.e. from 5 a.m. to 9 a.m.) = 4 × 45 = 180 km

Relative speed of two trains = 55 – 45 = 10 km/hour

Two trains will meet after 180 ÷ 10 = 18 hours

Distance travelled by the faster train in 18 hours = 18 × 55 = 990 km

127. A train leaves station A towards B at an average speed of 60 km/h at 9 a.m. After 2 hours another train leaves station A at an average speed of 100 km/h. When the two trains will meet?

(*a*) 11 a.m. (*b*) 1 p.m. (*c*) 2 p.m. (d) 3 p.m.

Sol. (*c*) Distance covered by the first train in 2 hours = 2 × 60 = 120 km.

Second train gains = (100 – 60) km/h = 40 km/h

∴ Second train will gain 120 km in 120 ÷ 40 = 3 hours.

∴ They will meet at 9 a.m. + 2 hours + 3 hours = 2 p.m.

128. Two trains start at the same time from two stations 117 km apart and proceed towards each other. If the ratio of speeds of the two trains is 4 : 5 and the speed of the first train is 60 km/h, find the time after which the two trains will meet.

(*a*) 47 minutes (*b*) 52 minutes (*c*) 57 minutes (*d*) 59 minutes

Sol. (*b*) Sum of ratio of speeds of two trains = 4 + 5 = 9

∴ Distance covered by the first train till meeting point = $\frac{4}{9} \times 117 = 52$ km

Time taken by the first train = $\frac{52}{60} \times 60 = 52$ minutes

Alternative Method : Speed of the faster train = $\frac{5}{4} \times 60$ km/hr = 75 km/hr

Relative speed = (60 + 75) km/hr = 135 km/hr

∴ Time taken = $\frac{117}{135} \times 60 = 52$ minutes

129. A train passes two persons who are walking in the opposite direction of the train at the rates of 5 m/s and 9 m/s in 7 seconds and 6 seconds respectively. Find the speed of the train?

(*a*) 4 m/s (*b*) 7 m/s (*c*) 14 m/s (*d*) 19 m/s

Sol. (*d*) Let speed of the train be x m/s.

Then $7(x + 5) = 6(x + 9) \Rightarrow 7x + 35 = 6x + 54 \Rightarrow x = 54 - 35 = 19$ m/s

Alternative method : Ratio of time = 7 : 6 ⇒ Ratio of relative speeds = 6 : 7

Difference of speeds = (7 – 6) ratio = 1 ratio = (9 – 5) m/s = 4 m/s

First relative speed = 6 ratio = 6 × 4 m/s = 24 m/s

Speed of the train = (24 – 5) m/s = 19 m/s

130. A train overtakes two persons who are walking in the same direction as the train is moving, at the rates of 5 km/hr and 9 km/hr in 7 seconds and 8 seconds respectively. Find the speed of train (in km/hr).

(*a*) 35 (*b*) 37 (*c*) 45 (*d*) 63

Sol. (*b*) Let speed of the train be x km/hr.

Then $7(x - 5) = 8(x - 9) \Rightarrow 7x - 35 = 8x - 72 \Rightarrow x = 72 - 35 = 37$ km/hr

131. A train overtakes two persons who are walking in the same direction as the train is moving, at the rates of 3 km/hr and 5 km/hr in 8 seconds and 9 seconds respectively. Find the length of the train.

(*a*) 36 m (*b*) 40 m (*c*) 45 m (*d*) 54 m

Sol. (*b*) Let speed of the train be x km/hr.

Then $8(x-3) = 9(x-5) \Rightarrow 8x - 24 = 9x - 45 \Rightarrow x = 45 - 24 = 21$ km/hr

$\therefore$ Length of the train $= (21-3) \times \frac{5}{18} \times 8 = 40$ m or $= (21-5) \times \frac{5}{18} \times 9 = 40$ m

Alternative method : Ratio of time = 8 : 9 ⇒ Ratio of relative speeds = 9 : 8

Difference of speeds = (9 – 8) ratio = 1 ratio = (5 – 3) km/hr = 2 km/hr

First relative speed = 9 ratio = 9 × 2 km/hr = 18 km/hr

$\therefore$ Length of the train $= 18 \times \frac{5}{18} \times 8 = 40$ m

132. A train overtakes two persons who are walking in the same direction in which the train is running, at the rate of 2 kmph and 4 kmph and passes them completely in 9 and 10 seconds respectively. The length of the train (in metres) is : **(SSC CGL Tier I 2012)**

(*a*) 72 (*b*) 45 (*c*) 54 (*d*) 50

Sol. (*d*) Let speed of the train be x km/hr.

Then $9(x-2) = 10(x-4) \Rightarrow 9x - 18 = 10x - 40 \Rightarrow x = 40 - 18 = 22$ km/hr

Relative speed in second case = (22 – 4) km/hr = 18 km/hr = 5 m/s

Length of the train = 10 × 5 = 50 m

133. A train running at 30 km/h passes a cyclist moving in the opposite direction in 5 seconds. If the cyclist goes in the same direction, the train takes 7 seconds to pass him. Find the speed of the cyclist.

(*a*) 2.5 km/h (*b*) 3 km/h (*c*) 4.5 km/h (*d*) 5 km/h

Sol. (*d*) Let speed of the cyclist = x km/h

Then relative speed when moving in the opposite directions = $30 + x$

And relative speed when moving in the same direction = $30 - x$

Ratio of time taken to travel at the two speeds = 5 : 7

$\therefore 5(30 + x) = 7(30 - x) \Rightarrow 5x + 7x = 30(7 - 5) \Rightarrow x = \frac{30 \times 2}{12} = 5$ km/h

134. Train A takes 6 hours more than train B to cover a distance of 800 km. If the speed of train A is doubled it takes 2 hours less than train B. What is the speed of train B?

(*a*) 50 km/h (*b*) 60 km/h (*c*) 75 km/h (*d*) 80 km/h

Sol. (*d*) Let time taken by train B = x hours. Then time taken by train A = $(x + 6)$ hours

Time taken by train A on doubling the speed = $(x - 2)$ hours

If speed is doubled, time taken is halved $\Rightarrow 2(x-2) = x + 6 \Rightarrow 2x - 4 = x + 6 \Rightarrow x = 10$ hours

$\therefore$ Speed of train B = 800 ÷ 10 = 80 km/h

Alt: Difference in time taken on doubling the speed by train B = 6 + 2 = 8 hours

$\therefore$ Time taken originally by train A = 8 × 2 = 16 hours

Time taken by train B = 16 – 6 = 10 hours ⇒ Speed of train B = 800 ÷ 10 = 80 km/h

135. In covering a distance of 30 km, Abhay takes 2 hours more than Sameer. If Abhay doubles his speed, then he would take 1 hour less than Sameer. Abhay's speed (in km/hr) is **(SSC Constable GD 2012)**

(*a*) 5 (*b*) 6 (*c*) 6.25 (*d*) 7.5

Sol. (*a*) Let Abhay's speed = x km/hr

Then time difference = $\frac{30}{x} - \frac{30}{2x} = (2 + 1) \Rightarrow \frac{30}{2x} = 3 \Rightarrow 6x = 30 \Rightarrow x = 5$ km/hr

136. A motor cyclist completes a certain journey of 360 km in 5 hours. He covers one part at 60 km/hr and the rest at 80 km/hr. Find the distance covered at 60 km/hr.

(*a*) 60 km (*b*) 120 km (*c*) 150 km (*d*) 240 km

Sol. (*b*) Distance covered in 5 hours @ 60 km/hr = 300 km

Remaining distance = 360 – 300 = 60 km, Difference of speeds = 80 – 60 = 20 km/hr

$\therefore$ Time taken to cover @ 80 km/hr = 60 ÷ 20 = 3 hours

$\Rightarrow$ Time taken to cover @ 60 km/hr = 5 – 3 = 2 hours

$\therefore$ Distance covered @ 60 km/hr = 2 × 60 = 120 km

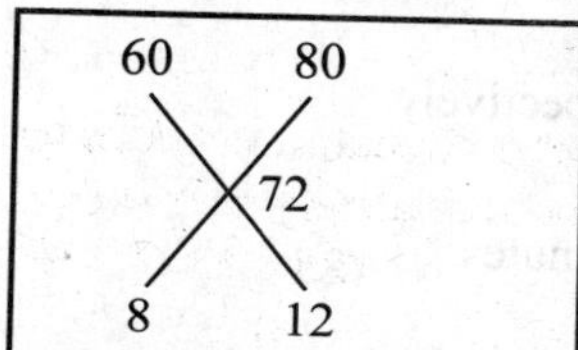

Alt: Average speed = 360 ÷ 5 = 72 km/hr

$\frac{\text{Time taken to cover @ 60 km/hr}}{\text{Time taken to cover @ 80 km/hr}} = \frac{8}{12} = \frac{2}{3}$

$\therefore$ Distance covered at 60 km/hr = 2 × 60 = 120 km

137. A man perform a journey of 750 km partly by bus partly by train. He takes 11 hours if he travels 150 km by bus and rest by the car. He saves 1 hour, if he travels all the distance by car. Find the speed of the car.

(*a*) 50 km/h (*b*) 60 km/h (*c*) 75 km/h (*d*) 90 km/h

Sol. (*c*) Time taken to cover the whole journey by the car = (11 – 1) hours = 10 hours

$\therefore$ Speed of the car = 750 ÷ 10 = 75 km/hr

138. A man travels 600 km partly by car and partly by bus. He takes 11 hours if he travels 360 km by car and rest by bus. If he travels all the distance by car, he would have saved 1 hour. Find the average speed of the bus ?

(*a*) 48 km/hr (*b*) 54 km/hr (*c*) 60 km/hr (*d*) 66 km/hr

Sol. (*a*) Average speed of car = 600 ÷ (11 – 1) = 60 km/hr

Time taken by car to cover 360 km = 360 ÷ 60 = 6 hours

Distance covered by bus (in first case) = 600 – 360 = 240 km

Time taken = 11 – 6 = 5 hours $\Rightarrow$ Speed of the bus = 240 ÷ 5 = 48 km/hr

139. Ravi travels 300 km partly by train and partly by car. He takes 4 hours to reach, if he travels 60 km by train and rest by car. He will take 10 minutes more if he were to travel 100 km by train and rest by car. The speed of the train is : **(SSC CGL Tier I 2012)**

(*a*) 50 km/hr (*b*) 60 km/hr (*c*) 100 km/hr (*d*) 120 km/hr

Sol. (*b*) If he travels 40 km (*i.e.* 100 km – 60 km) more by train, he takes 10 minutes = 1/6 hour extra.

$\therefore$ If he travels 240 km more by train, he would take $\frac{240}{40} \times \frac{1}{6} = 1$ hour extra.

$\therefore$ Time taken to cover the whole journey by the train = 4 + 1 = 5 hours.

$\therefore$ Speed of the train = 300 ÷ 5 = 60 km/hr

140. A man perform a journey of 630 km. He takes 8 hours if he travels 180 km by bus and rest by the car. He takes 8 hours 10 minutes, if he travels 210 km by the bus and rest by the car. Find the speed of the bus.

(*a*) 60 km/h (*b*) 75 km/h (*c*) 80 km/h (*d*) 90 km/h

Sol. (*a*) If he travels 30 km (*i.e.* 210 km – 180 km) more by bus, he takes 10 minutes = 1/6 hour extra.

$\therefore$ If he travels 450 km more by bus, he would take $\frac{450}{30} \times \frac{1}{6} = \frac{5}{2}$ hours extra.

$\therefore$ Time taken to cover the whole journey by the bus $= 8 + \frac{5}{2} = \frac{21}{2}$ hours.

$\therefore$ Speed of the bus $= 630 \div \frac{21}{2} = 630 \times \frac{2}{21} = 60$ km/h

141. A and B start from X and Y towards Y and X respectively. If they cover the distance in 4 and 6 hours respectively, when will they meet ?

(*a*) 2 hours 24 minutes (*b*) 2 hours 40 minutes (c) 5 hours (*d*) 10 hours

Sol. (*a*) Let the distance is 4 × 6 = 24 km

$\therefore$ Speeds of A and B are $\frac{24}{4}$ and $\frac{24}{6}$ = 6 and 4 km/h respectively

$\therefore$ They will meet after $\frac{24}{6+4}$ = 2.4 hour = 2 hours 24 minutes

Alternative method : Required time $= \frac{4 \times 6}{4+6}$ = 2.4 hour = 2 hours 24 minutes

142. Two trains left stations A and B in opposite directions towards station B and A respectively. If they complete the journey in 42 minutes and 56 minutes respectively. They will meet after:

(*a*) 24 minutes (*b*) 48 minutes (*c*) 72 minutes (*d*) 168 minutes

Sol. (*a*) Required time $= \frac{42 \times 56}{42+56} = \frac{42 \times 56}{98}$ = 24 minutes

143. A starts from X towards Y at 10 a.m. B starts from Y towards X at 11 a.m. If they reach their destinations at 2 p.m. and 5 p.m. respectively, at what time will they meet ?

(*a*) 12.24 p.m. (*b*) 12.48 p.m. (*c*) 1.48 p.m. (*d*) 2.24 p.m.

Sol. (*b*) Time taken by A = 4 hours. Time taken by B = 6 hours

Let the distance = 4 × 6 = 24 km. Then speeds of A and B are 6 and 4 km/h respectively

Distance covered by A in first hour = 6 km ⇒ Remaining distance = 24 – 6 = 18 km

$\therefore$ They will meet after $\frac{18}{6+4}$ = 1.8 hour = 1 hour 48 minutes

$\therefore$ They will meet at 11 a.m. + 1 hour 48 minutes = 12.48 p.m.

Alternative method : Time taken by A = 4 hours. Time taken by B = 6 hours

Distance covered by A in one hour (10 a.m. to 11 a.m.) $= \frac{1}{4}$

Time taken by A and B to cover remaining $\frac{3}{4}$ of distance $= \frac{3}{4} \times \frac{4 \times 6}{10}$

= 1.8 hour = 1 hour 48 minutes

$\therefore$ They will meet at 11 a.m. + 1 hour 48 minutes = 12.48 p.m.

144. A train leaves station A at 5 AM and reaches station B at 9 AM on the same day. Another train leaves station B at 7 AM and reaches station A at 10.30 AM on the same day. The time at which the two trains cross one another is : **(SSC CGL Tier I 2015)**

(*a*) 8.26 AM (*b*) 7.56 AM (*c*) 8 AM (*d*) 8.36 AM

Sol. (*b*) First train covers full journey in 9 – 5 = 4 hours

Journey covered in 2 hours (i.e. from 5 to 7) = $\frac{2}{4}=\frac{1}{2}$

Second train covers the distance in (10.30 – 7) = $3\frac{1}{2}$ hours

Time taken to cover the remaining distance = $\frac{1}{2}\times\frac{4\times\frac{7}{2}}{4+\frac{7}{2}}=\frac{1}{2}\times\frac{4\times7}{15}\times 60 = 56$ minutes

Required time = 7 a.m. + 56 min = 7.56 a.m.

Alt: LCM of times taken = LCM of 4 and 7/2 = 28. Let total distance = 28 km

First train covers (in 2 hours) = (28 ÷ 4) × 2 = 14 km

Remaining distance = 28 – 14 = 14 km

Relative speed of two trains for 1 hour = (28 ÷ 4) + 28 ÷ 7/2 = 7 + 8 = 15 km

Time to cover 15 km = 14/15 × 60 = 56 min ⇒ Required time = 7 a.m. + 56 min = 7.56 a.m.

145. Two trains X and Y starts from station A to station B at the same time towards each other. After passing each other, they complete their journey in 4 hours 12 minutes and $2\frac{1}{7}$ hours respectively. What is the speed of train Y, if speed train X is 70 km/hr. **(SSC CGL Tier I)**

(*a*) 50 km/h (*b*) 49 km/h (*c*) 98 km/h (*d*) None of these

Sol. (*c*) Time taken by train X = 4 hours 12 minutes = $4\frac{1}{5}$ hours = $\frac{21}{5}$ hours

Time taken by train Y = $2\frac{1}{7}$ hours = $\frac{15}{7}$ hours

X's speed : Y's speed = $\sqrt{\frac{15}{7}\div\frac{21}{5}}=\sqrt{\frac{15}{7}\times\frac{5}{21}}=\frac{5}{7}=5:7$

But speed of train X = 70 km/h ⇒ Speed of train Y = $\frac{7}{5}\times 70 = 98$ km/h.

146. Pratibha covers a distance of 24 kms at the speed of 8 km/hr and a distance of 18 kms at the speed of 9 km/hr. Further she covers a distance of 12 km at the speed of 3 km/hr. What is her average speed in covering the whole distance ? **(Nabard Officer Exam. 2008)**

(*a*) 8 km/h (*b*) 5.5 km/h (*c*) 3 km/h (*d*) 6 km/h

Sol. (*d*) Total distance covered = 24 + 18 +12 = 54 km

Total time taken = $\frac{24}{8}+\frac{18}{9}+\frac{12}{3}=3+2+4=9$ hours

∴ Average speed = 54 ÷ 9 = 6 km/hr

147. A man completed a certain journey by a car. If he covered 30% of the distance at the speed of 20 km/hr, 60% of the distance at 40 km/hr, and the remaining distance at 10 km/hr, his average speed for the whole journey was : **(SSC CGL Tier I 2008)**

(*a*) 25 km/h (*b*) 28 km/h (*c*) 30 km/h (*d*) 33 km/h

Sol. (*a*) Let total distance = 100 km. ⇒ Total time = $\frac{30}{20}+\frac{60}{40}+\frac{10}{10}=1.5+1.5+1=4$ hours

∴ Average speed = 100 ÷ 4 = 25 km/hr

148. A man covers three equal distances with speeds of 3 km/hr, 4 km/hr and 5 km/hr. If total time taken by him is 47 minutes, find the total distance covered by him.

(*a*) 3 km (*b*) 4 km (*c*) 5 km (*d*) 6 km

Sol. (*a*) Let each distance = LCM of 3, 4, 5 = 60 km ⇒ Total time = $\frac{60}{3}+\frac{60}{4}+\frac{60}{5}$ = 47 hours

⇒ Average speed = $\frac{60 \times 3}{47}$ ⇒ Distance covered in 47 minute = $\frac{60\times 3}{47}\times\frac{47}{60}$ = 3 km

149. A car covers four successive 6 km stretches at speeds of 25 kmph, 50 kmph, 75 kmph and 150 kmph respectively. Its average speed over this distance is **(SSC CGL Tier I 2012)**

(*a*) 25 kmph (*b*) 50 kmph (*c*) 75 kmph (*d*) 150 kmph

Sol. (*b*) Let each stretch = LCM of 25, 50, 75, 150 = 150 km

Then total time taken = $\frac{150}{25}+\frac{150}{50}+\frac{150}{75}+\frac{150}{150}$ = 6 + 3 + 2 + 1 = 12 hours

∴ Average speed for whole journey = (4 × 150) ÷ 12 = 50 kmph

150. A train travels from Dehradun to Delhi at a speed of 40 km/hr and returns at 60 km/hr. Find the average speed for the whole journey. **(SSC CGL Tier I 2014, SSC CHSL 2012, 2013)**

(*a*) 45 km/h (*b*) 48 km/h (*c*) 50 km/h (*d*) 54 km/h

Sol. (*b*) Average speed = $\frac{2 \times 60 \times 40}{60 + 40} = \frac{2 \times 60 \times 40}{100}$ = 48 km/h

Alt: Ratio of speeds = 40 : 60 = 2 : 3. Difference of speeds = 60 – 40 = 20 km = (8 + 12) km
Average speed = 40 + 8 = 60 – 12 = 48 km/hr

151. A train goes from Ballygunge to Sealdah at an average speed of 20 km/hour and comes back at an average speed of 30 km/hour. The average speed of the train for the whole journey is **(SSC CGL Tier II 2013, SSCCHSL 2006)**

(*a*) 27 km/hr (*b*) 26 km/hr (*c*) 25 km/hr (*d*) 24 km/hr

Sol. (*d*) Average speed = $\frac{2\times 20\times 30}{50}$ = 24 km/hr

152. A man covers a journey from place A to place B at the speed of 51 km/h and returns at the speed of 57 km/h. What is his average speed during the whole journey ? **(Bank Clerical 2007)**

(*a*) 26.92 km/h (*b*) 54 km/h (*c*) 55 km/h (*d*) None of these

Sol. (*d*) Average speed = $\frac{2xy}{x + y} = \frac{2 \times 51 \times 57}{108}$ = 53.83 (approx.)

Trick: The average speed in such cases is slightly less than the simple average of two speeds. It lies between the value of two given speeds.

Simple average = $\frac{51 + 57}{2}$ = 54 km/h ⇒ (*a*), (*b*) and (*c*) are not possible.

153. A & B are two towns, Mr. Chandra covers the distance from A to B on cycle at 16 km/hr. However, he covers the distance from B to A on foot at 9 km/hr. His average speed during the whole journey in km/hr is **(SSC CGL Tier I 2012)**

(*a*) 11.52 (*b*) 12.5 (*c*) 10.25 (*d*) 12.32

Sol. (*a*) Average speed = $\frac{2xy}{x + y} = \frac{2\times 16\times 9}{25}$ = 11.52

154. Ravi went to his college @ 60 km/hr. While returning home, he covered the half of the distance @ 10 km/hr and rest half of the distance @ 30 km/hr. Find his average speed during the whole journey. **(MBA)**

(*a*) 20 km/hr (*b*) 40 km/hr (*c*) 24 km/hr (*d*) 33.33 km/hr

Sol. (*c*) Average speed (for returning) = $\frac{2 \times 10 \times 30}{40} = 15$ km/ hr

Average speed (for whole journey) = $\frac{2 \times 60 \times 15}{75} = 24$ km/ hr

155. A man travelled a certain distance by train at the rate of 25 kmph and walked back at the rate of 4 kmph. If the whole journey took 5 hours 48 minutes, the distance was: **(SSC CGL Tier I 2004)**

(*a*) 25 km (*b*) 30 km (*c*) 20 km (*d*) 15 km

Sol. (*c*) Time = 5 hours + 48 minutes = $5\frac{48}{60}$ hours = $5\frac{4}{5}$ hours = $\frac{29}{5}$ hours

Average speed = $\frac{2 \times 25 \times 4}{29}$ ⇒ Required distance = $\frac{2 \times 25 \times 4}{29} \times \frac{29}{5} \times \frac{1}{2} = 20$ km

156. Ram goes from Delhi to Agara in his car at a speed of 90 km/hr and returns at a speed of x km/hr. If his average speed for the whole journey is 72 km/hr, find the value of x. **(SSC CGL Tier I 2012)**

(*a*) 60 (*b*) 70 (*c*) 80 (*d*) 81

Sol. (*a*) **Trick :** $\frac{90 + x}{2} > 72 \Rightarrow 90 + x > 144 \Rightarrow x > 54$

∴ We try a value of x slightly greater than 54 (say 60)

When $x = 60 \Rightarrow$ Average speed = $\frac{2 \times 90 \times 60}{150} = 72$ km/hr

157. A motor cyclist completes a certain journey in 5 hours. He covers one-third distance at 60 km/hr and the rest at 80 km/hr. The length of journey is : **(CDS 2004)**

(*a*) 180 km (*b*) 240 km (*c*) 300 km (*d*) 360 km

Sol. (*d*) Let the total distance be x km. Then time taken = $x \times \frac{1}{3} \times \frac{1}{60} + x \times \frac{2}{3} \times \frac{1}{80} = 5$

$\Rightarrow \frac{x}{180} + \frac{x}{120} = 5 \Rightarrow \frac{5x}{360} = 5 \Rightarrow x = 360$ km

Trick: Distance covered in 5 hours @ 60 km/hr = 5 × 60 = 300 km
Total distance covered > 300. Only option (*d*) > 300

158. A man starts from a place P and reaches the place Q in 7 hours. He travels one fourth of the distance at 10 km/hr and remaining distance at 12 km/hr. What is the distance between P and Q : **(SSC CGL Tier I 2013)**

(*a*) 70 (*b*) 80 (*c*) 72 (*d*) 90

Sol. (*b*) $\frac{x/4}{10} + \frac{3x/4}{12} = 7 \Rightarrow \frac{x}{40} + \frac{x}{16} = 7 \Rightarrow \frac{7x}{80} = 7 \Rightarrow x = 80$

159. The average speed of a train is 20% less on the return journey than on the onward journey. The train halt for half an hour at the destination station before starting on the return journey. If the total time taken for the to and fro journey is 23 hours, covering a distance of 1000 km, the speed of the train on the return journey is **(SSC CGL Tier I 2004)**

(*a*) 40 km/hr (*b*) 45 km/hr (*c*) 50 km/hr (*d*) 60 km/hr

Sol. (*a*) Let onward speed = 5*x* km/hr. Then speed during return journey = 4*x* km/hr.

Time taken for journey $= 23 - \frac{1}{2} = 22\frac{1}{2} = \frac{45}{2}$ hours

$\Rightarrow$ Total distance $= \frac{2 \times 5x \times 4x}{9x} \times \frac{45}{2} = 1000 \text{ km} \Rightarrow 100x = 1000$ km

$\Rightarrow x = 10$ km $\Rightarrow$ Required speed $= 4x = 4 \times 10 = 40$ km/hr

160. Ram arrives at a Bank 15 minutes earlier than scheduled time if drives his car at 42 km/hr. He arrives 5 minutes late if he drives his car at 35 km/hr. The distance of the Bank, from his starting point is : **(SSC CGL Tier I 2006)**

(*a*) 60 km (*b*) 70 km (*c*) 90 km (*d*) 140 km

Sol. (*b*) Difference in time = (15 + 5) minutes = 1/3 hour

$\therefore$ Required distance $= \frac{42 \times 35}{7} \times \frac{1}{3} = 70$ km

161. If a boy walks from his house to school at the rate of 4 km per hour, he reaches the school 10 minutes earlier than the scheduled time. However, if he walks at the rate of 3 km per hour, he reaches 10 minutes late. Find the distance of his school from his house. **(SSC CGL Tier II 2013)**

(*a*) 5 km (*b*) 4 km (*c*) 6 km (*d*) 4.5 km

Sol. (*b*) Time difference = (10 + 10) minutes $= \frac{1}{3}$ hour $\Rightarrow$ Distance $= \frac{4 \times 3}{4-3} \times \frac{1}{3} = 4$ km

162. A man has to be a certain place at a certain time. He finds that he shall be 20 minutes late if he walks at 3 km/h speed and 10 minutes earlier if he walks at a speed of 4 km/h. The distance he has to walk is **(SSC CPO SI 2011)**

(*a*) 24 km (*b*) 12.5 km (*c*) 10 km (*d*) 6 km

Sol. (*d*) Difference in time = (20 + 10) minutes = 1/2 hour

$\therefore$ Required distance $= \frac{4 \times 3}{1} \times \frac{1}{2} = 6$ km

Alt: Let total distance = 1 km. Then time taken to cover @ 3 km/hr = 20 min

And time taken to cover @ 4 km/hr = 15 min $\Rightarrow$ Difference in time = 20 – 15 = 5 min.

But actual difference = 30 min $\Rightarrow$ Actual distance = 6 × 1 km = 6 km

163. A boy walks from his house to school at 2.5 km/hr and arrives 12 minutes late. The next day he walks at 4 km/hr and reaches the school 15 minutes earlier. What is the distance from his house to school ? **(CDS 2009)**

(*a*) 2 km (*b*) 2.5 km (*c*) 3 km (*d*) 3.5 km

Sol. (*c*) Difference in time = (12 + 15) minutes $= \frac{27}{60}$ hour

$\therefore$ Required distance $= \frac{2.5 \times 4}{4 - 2.5} \times \frac{27}{60} = 3$ km

164. Raman drove from home to a neighboring town at the speed of 50 km/hr and on his returning journey, he drove at the speed of 45 km/hr and also took an hour longer to reach home. What distance did he cover each way ? **(SBI Bank PO 2008)**

(*a*) 450 km (*b*) 225 km (*c*) 900 km (*d*) 500 km

Sol. (*a*) Required distance $= \frac{50 \times 45}{50 - 45} \times 1 = 450$ km

165. With an average speed of 40 km/hr, a train reaches its destination in time. If it goes with an average speed of 35 km/hr, it is late by 15 minutes. The total journey is **(SSC CGL Tier I 2012)**

(*a*) 70 km (*b*) 50 km (*c*) 60 km (*d*) 35 km

Sol. (*a*) Total journey $= \frac{40 \times 35}{40 - 35} \times \frac{15}{60} = 70$ km

166. A train running with a speed of 40 km/hr takes $1\frac{1}{2}$ hours more to cover a certain distance than another train running at 60 km/hr. The distance is : **(SSC CGL Tier I 2006)**

(*a*) 180 km (*b*) 160 km (*c*) 200 km (*d*) 120 km

Sol. (*a*) Time difference $= 1\frac{1}{2} = \frac{3}{2}$ hours $\Rightarrow$ Required distance $= \frac{60 \times 40}{60 - 40} \times \frac{3}{2} = 180$ km

167. Two cars A and B start simultaneously from a certain place at the speed of 30 km/hr and 45 km/hr respectively. The car B reaches the destination 2 hours earlier than A. What is the distance between the starting point and destination ? **(CDS 2013)**

(*a*) 90 km (*b*) 180 km (*c*) 270 km (*d*) 360 km

Sol. (*b*) Required distance $= \frac{45 \times 30}{15} \times 2 = 180$ km

168. A and B starts at the same time with speeds of 40 km/hr and 50 km/hr respectively. If in covering the journey A takes 15 minutes longer than B, the total distance of the journey is : **(SSC CGL Tier I 2000)**

(*a*) 46 km (*b*) 48 km (*c*) 50 km (*d*) 52 km

Sol. (*c*) Distance $= \frac{50 \times 40}{50 - 40} \times \frac{1}{4} = 50$ km

169. A father and his son start at a point A with speeds of 12 kmph and 18 kmph respectively, and reach another point B. If his son starts 60 minutes after his father at A and reaches B 60 minutes before his father, what is the distance between A and B ? **(CDS 2010)**

(*a*) 90 km (*b*) 72 km (*c*) 36 km (*d*) None of these

Sol. (*b*) Time difference = (60 + 60) minutes = 2 hours

$\therefore$ Required distance $= \frac{18 \times 12}{6} \times 2 = 72$ km

170. If a train runs at 40 km/hr, it reaches its destination late by 11 min, but if it runs at 50 km/h, it is late by 5 min only. Find the correct time for the train to complete its journey **(SSC FCI Asstt 2012)**

(*a*) 19 min (*b*) 20 min (*c*) 21 min (*d*) 18 min

Sol. (*a*) Distance $= \frac{50 \times 40}{10} \times \frac{11 - 5}{60} = 20$ km

Time taken @ 40 km/hr = $20 \times \frac{60}{40}$ = 30 minutes ⇒ Normal time = (30 – 11)min = 19 min

or, Time taken @ 50 km/hr = $20 \times \frac{60}{50}$ = 24 min ⇒ Normal time = (24 – 5) min = 19 min

171. **A man cycles with a speed of 10 kmph and reaches his office at 1 p.m. However, when he cycles with a speed of 15 kmph, he reaches his office at 11 a.m. At what speed should he cycle so that he reaches his office at 12 noon ?** **(CDS 2004, 2013)**

(*a*) 12.5 kmph (*b*) 12 kmph (*c*) 13 kmph (*d*) 13.5 kmph

Sol. (*b*) Ratio of speeds = 10 : 15 = 2 : 3 ⇒ Ratio of time = 3 : 2

⇒ Time difference = (3 – 2)ratio = 1 ratio = 2 hours

⇒ Original time = 3 ratio = 3 × 2 hours = 6 hours

Required speed (to cover in 5 hours) = $\frac{10 \times 6}{5}$ = 12 kmph

172. **If a train runs at 30 km/hour, it reaches its destination late by 10 minute but if it runs at 45 km/hour, it is late by 5 minute only. Then the correct time for the train to complete its journey is** **(SSC FCI 2012)**

(*a*) 5 min (*b*) 7 min (*c*) 10 min (*d*) 15 min

Sol. (*d*) Distance = $\frac{45 \times 30}{15} \times \frac{5}{60} = \frac{15}{2}$ ⇒Time @ 45 km/hr = $\frac{15}{2} \times \frac{60}{45}$ = 10 minutes

⇒ Normal time = (10 + 5) minute = 15 minute

173. **A student walks from his house at a speed of $2\frac{1}{2}$ km per hour and reaches his school 6 minutes late. The next day he increase his speed by 1 km per hour and reaches 6 minutes before school time. How far is the school from his house ?** **(SSC CGL Tier 2006 & 2004)**

(*a*) $\frac{5}{4}$ km (*b*) $\frac{7}{4}$ km (*c*) $\frac{9}{4}$ km (*d*) $\frac{11}{4}$ km

Sol. (*b*) Difference in time = (6 + 6) minutes = 1/5 hour

$$\therefore \text{Required distance} = \frac{3\frac{1}{2} \times 2\frac{1}{2}}{3\frac{1}{2} - 2\frac{1}{2}} \times \frac{1}{5} = \frac{\frac{7}{2} \times \frac{5}{2}}{1} \times \frac{1}{5} = \frac{7}{2} \times \frac{5}{2} \times \frac{1}{5} = \frac{7}{4} \text{ km}$$

174. **Ramesh leaves his house daily at a particular time to go to his office. If he travels at 50 km/h, he reaches 5 minutes late and if travel at 60 km/h, he reaches 5 minutes early. At what speed should he travel to reach office on time?**

(*a*) 54.54 km/h (*b*) 55 km/h (*c*) 56.67 km/h (*d*) 57.5 km/h

Sol. (*a*) Ratio of speeds = 50 : 60 = 5 : 6 ⇒ Ratio of time = 6 : 5

Difference of time = (6 – 5) ratio = 1 ratio = 10 minutes

Time taken at 50 km/h = 6 ratio = 6 × 10 = 60 minutes

Time taken to reach in time = 60 – 5 = 55 minutes ⇒ Required speed = $50 \times \frac{60}{55}$ = 54.54 km

175. **A train covers a certain distance in 45 minutes. Had the train running 4 km/hr faster, it could cover the same distance in 42 minutes. Find the distance covered.**

(*a*) 36 km (*b*) 42 km (*c*) 45 km (*d*) 50 km

Sol. (*b*) Ratio of time taken = 45 : 42 = 15 : 14 ⇒ Ratio of speeds = 14 : 15

Difference of speeds = (15 – 14) ratio = 1 ratio = 4 km/hr

⇒ Original speed = 14 ratio = 14 × 4 km/hr = 56 km/hr ⇒ Distance = $56 \times \frac{45}{60} = 42$ km

176. With a uniform speed, a car covers a distance in 8 hours. Had the speed been increased by 4 km/hr, the same distance could have been covered in 7 hours 30 minutes. What is the distance covered ? **(CDS 2015)**

(*a*) 420 km (*b*) 480 km (*c*) 640 km (*d*) 520 km

Sol. (*b*) Ratio of time taken = 8 : 15/2 = 16 : 15 ⇒ Ratio of speeds = 15 : 16

Difference of speeds = (16 – 15) ratio = 1 ratio = 4 km/hr

⇒ Original speed = 15 ratio = 15 × 4 km/hr = 60 km/hr ⇒ Distance = 60 × 8 = 480 km

177. If a train runs at 40 km/hr, it reaches it destination late by 12 minutes. If it runs at 50 km/hr, its is late by 5 minutes. The correct time for the train to reach its destination is:

(*a*) 13 minutes (*b*) 19 minutes (*c*) 23 minutes (*d*) 17 minutes

Sol. (*c*) Let correct time = x min. Then $40(x + 12) = 50(x + 5)$

⇒ $40x + 480 = 50x + 250$ ⇒ $10x = 230$ ⇒ $x = 23$ minutes

178. If a man walks 20 km at 5 km/h, he will be late by 40 minutes. If he walks at 8 km/h, how early from the normal time will he reach?

(*a*) 15 minutes (*b*) 20 minutes (*c*) 50 minutes (*d*) 45 minutes

Sol. (*c*) Time taken at 5 km/h = 20 ÷ 5 = 4 hours

∴ Normal time = 4 hours – 40 minutes = 3 hours 20 minutes

Time taken at 8 km/h = 20 ÷ 8 = 2 hours 30 minutes

Hence time saved = 3 hours 20 minutes – 2 hours 30 minutes = 50 minutes

179. The time taken by a train to cover a distance of 150 km was 2.5 hours more than the time taken for the return journey. If the speed of the train during the return journey was 10 km/hr more than the former speed. What was the speed of the train during return journey ?

(*a*) 20 km/hr (*b*) 25 km/hr (*c*) 30 km/hr (*d*) 40 km/hr

Sol. (*c*) Option (*a*): $\frac{150}{10} - \frac{150}{20} = 15 - 7.5 = 7.5$ hours

Option (*b*) : $\frac{150}{15} - \frac{150}{25} = 10 - 6 = 4$ hours

Option (*c*) : $\frac{150}{20} - \frac{150}{30} = 7.5 - 5 = 2.5$ hours = Required time

180. From two places, 60 km apart, A and B start towards each other at the same time and meet each other after 6 hours. Had A travelled with 2/3 of his speed and B travelled with double his speed, they would have met after 5 hours. The speed of A is: **(SSC CGL Tier I 2008)**

(*a*) 4 km/hr (*b*) 6 km/hr (*c*) 10 km/hr (*d*) 12 km/hr

Sol. (*b*) Let speeds of A and B be x km/hr and y km/hr respectively.

Then $x + y = 60 \div 6 \Rightarrow x + y = 10$ km/hr ...(1)

And $\frac{2}{3}x + 2y = 60 \div 5 = 12 \Rightarrow 2x + 6y = 36$ ...(2)

(1) × 6 ⇒ $6x + 6y = 60$ ...(3)

Equation (3) – Equation (2) ⇒ $4x = 24 \Rightarrow x = 6$ km/hr

181. A plane left 30 minutes later than its scheduled time and in order to reach its destination 1500 km in time, it has to increase its speed by 250 km/h from its usual speed. What is its usual speed in km/h? **(MBA Entrance 2006)**

(*a*) 1000 km/h (*b*) 750 km/h (*c*) 600 km/h (*d*) 500 km/h

Sol. (*b*) We prepare a chart of time and speed, as given below

Hours	Speed
1	1500
$1\frac{1}{2}$	1000
2	750

We observe that difference between time $1\frac{1}{2}$ and 2 hours is $\frac{1}{2}$ hours and difference between two speeds is 250 km $\therefore$ Usual speed = 750 km/h

182. A car is required to cover a distance of 495 km at a certain speed. Due to some technical fault, car could run slower by 10 km/h and hence took 2 hour more than its usual time. What is the usual speed of the car? **(MBA Entrance)**

(*a*) 99 km/h (*b*) 55 km/h (*c*) 45 km/h (*d*) Can't say

Sol. (*b*)

Hours	Speed
1	495
3	165
5	99
9	55
11	45

Time difference between 9 and 11 is 2 hours and difference in two speeds is 10 km/hr.

$\therefore$ Usual speed = 55 km/h

183. A car covered a distance of 240 km at a certain speed. The journey would have taken 1 hour less, if the car had travelled 20 km/hr faster. What is the original speed of the car ? **(GMAT)**

(*a*) 40 km/hr (*b*) 50 km/hr (*c*) 60 km/hr (*d*) 80 km/hr

Sol. (*c*)

Hours	Speed
6	40
5	48
4	60
3	80

Time difference between 4 hours and 3 hours is 1 hour and difference in two speeds is 20 km/hr. $\therefore$ Original speed of the car = 60 km/hr

184. By increasing the speed of his car by 15 km/hour, a person covers 300 km distance by taking an hour less than before. The original speed of the car was **(CDS 2015)**

(*a*) 45 km/hr (*b*) 50 km/hr (*c*) 60 km/hr (*d*) 75 km/hr

Sol. (*c*)

Hours	Speed
6	50
5	60
4	75

Time difference between 5 hours and 4 hours is of 1 hour and difference in two speeds is 15 km/hr. $\therefore$ Original speed of the car = 60 km/hr

Alt: From choices : $\frac{300}{60}-\frac{300}{75}=5-4=1$ hour

185. In a flight of 600 km, an aircraft was slowed down due to bad weather. Its average speed for the trip was reduced by 200 km/hr and the time of flight increased by 30 minutes. The duration of the flight is **(CDA 2015)**

(*a*) 1 hour (*b*) 2 hours (*c*) 3 hours (*d*) 4 hours

Sol. (*a*) We observe, $\frac{600}{400}-\frac{600}{600}=\frac{3}{2}-1=\frac{1}{2}$ $\therefore$ Usual time = 1 hour

186. A train travels a distance of 300 km at a constant speed. If the speed of the train is increased by 5 km an hour, the journey would have taken 2 hours less. What is the original speed of the train in kmph ? **(NIFT)**

(a) 20 km/hr (b) 25 km/hr (c) 24 km/hr (d) 30 km/hr

Sol. (b) Option (a) : $\frac{300}{20}-\frac{300}{25}=15-12=3$ hours

Option (b) : $\frac{300}{25}-\frac{300}{30}=12-10=2$ hours

187. A train travels at a certain average speed for a distance of 63 km and then travels a distance of 72 km at an average speed of 6 km/hr more than its original speed. If it takes 3 hours to complete the total journey, what is the original speed of the train in km/hr ? **(UPSC Civil Services Pre. 2013)**

(a) 24 (b) 33 (c) 42 (d) 66

Sol. (c) $\frac{63}{42}+\frac{72}{48}=1.5+1.5=3$

188. Buses start from a bus terminal with a speed of 20 km/hr at intervals of 10 minutes. What is the speed of a man coming from the opposite direction towards the bus terminal if he meets the buses at intervals of 8 minutes ? **(SSC CGL Tier I 2010)**

(*a*) 3 km/hr (*b*) 4 km/hr (*c*) 5 km/hr (*d*) 7 km/hr

Sol. (*c*) Time interval between two buses = 10 minutes

Time interval between seeing two consecutive buses by the man = 8 minutes

Distance covered by man in 8 minutes = Distance covered by bus in (10 – 8) min = 2 minutes

Speed of man = $20\times\frac{2}{8}=5$ km/hr

189. Two guns were fired from the same place at an interval of 12 minutes. A person approaching the firing place in his car, hears the second shot 11 minutes after the first. If the speed of sound is 330 m/s, find the speed of the car.

(*a*) 50 km/h (*b*) 72 km/h (*c*) 90 km/h (*d*) 108 km/h

Sol. (*d*) The person hears the second fire in 11 minutes instead of 12 minutes since the car is moving towards the firing place.

$\therefore$ The distance travelled by car in 11 minutes is equal to distance covered by sound in 12 – 11 = 1 minute.

$\therefore$ Distance covered by car in 1 second = $\frac{330}{11}=30$ m

$\therefore$ Distance covered by car in 1 hour = $30\times\frac{18}{5}=108$ km/h

190. Two guns are fired from the same place at an interval of 6 minutes. A person approaching the firing place observes that 5 minutes 52 seconds have elapsed between the hearing of the sound of the two guns. If the velocity of the sound is 330 m/s, the man was approaching that place at what speed ? **(SSC CGL Tier I 2007)**

(*a*) 24 km/h (*b*) 27 km/h (*c*) 30 km/h (*d*) 36 km/h

Sol. (*b*) Difference in time of firing = 6 minutes

Difference in time of hearing sound = 5 min 52 seconds

Difference in time of two firings and hearing their sounds = 6 min – 5 min 52 s = 8 s

∴ Distance covered by person in 5 minutes 52 seconds or 352 seconds

= Distance covered by sound in 8 seconds = 8 × 330 m

∴ Speed of the person = $\frac{8 \times 330}{352} \times \frac{18}{5}$ km/hr = 27 km/hr.

191. In a 500 m race, the ratio of speeds of two runners A and B is 3 : 4. If A has a start of 170 m, then A wins by

(*a*) 30 m (*b*) 50 m (*c*) 60 m (*d*) 90 m

Sol. (*c*) A has to cover = 500 – 170 = 330 m

When A runs 330 m, B runs = $\frac{4}{3} \times 330 = 440$ m ⇒ A wins by 500 – 440 = 60 m

192. A runs $1\frac{2}{3}$ times as fast as B. If A gives B a start of 80 m, how far must the winning post from the starting point be so that A and B might reach it at the same time? **(CDS 2015)**

(*a*) 200 m (*b*) 300 m (*c*) 160 m (*d*) 270 m

Sol. (*a*) Ratio of speeds = $1\frac{2}{3} : 1 = 5 : 3$

Difference of speeds = (5 – 3) ratio = 2 ratio = 80 m ⇒ 1 ratio = 40 m

∴ Required distance = 5 ratio = 5 × 40 = 200 m

193. Three athletes run a 4 km race. Their speeds are in the ratio 16 : 15 : 11. When the winner wins the race, then the distance between the athlete in the second position to the athlete in the third position is **(CDS 2015)**

(*a*) 1000 m (*b*) 800 m (*c*) 750 m (*d*) 600 m

Sol. (*a*) First athelete covers = 16 ratio = 4000 m ⇒ 1 ratio = 250 m

∴ Difference = (15 – 11) ratio = 4 ratio = 4 × 250 = 1000 m

194. Two cyclists A and B cover a distance of 110 km, A reaches to the destination one hour before B. Even if B had given a start of 8 km, he would have reached 12 minutes late. Find the speed of A.

(*a*) 5 km/h (*b*) 7.5 km/h (*c*) 10 km/h (*d*) 11 km/h

Sol. (*d*) In the first case, B is late by 60 minutes.

In the second case, B after getting a start of 8 km, reaches 12 minutes late.

∴ Time taken by B to travel 8 km = 60 – 12 = 48 minutes

∴ B's speed = $\frac{8}{48} \times 60 = 10$ km/h ⇒ Time taken by B = $\frac{110}{10} = 11$ hours

∴ Time taken by A = 11 – 1 = 10 hours ⇒ A's speed = $\frac{110}{10} = 11$ km/h

195. In a 1000 metre race, A beats B by 100 metre and B beats C by 50 metre. Find out that by how many metre will A beat C in a 1000 metre race?

(*a*) 145 m (*b*) 150 m (*c*) 175 m (*d*) 200 m

Sol. (*a*) If A covers 1000 m, B covers 900 m. If B covers 1000 m, C covers 950 m.

If A covers 1000 metre, C will cover $= 1000 \times \frac{900}{1000} \times \frac{950}{1000} = 855$ metres

A beats C by 1000 – 855 = 145 metre

196. In a race of 200 metres, B can give a start of 10 metres to A, and C give a start of 20 metres to B. The start that C can give to A, in the same race, is : **(SSC CPO SI 2007)**

(*a*) 36 m (*b*) 30 m (*c*) 33 m (*d*) 29 m

Sol. (*b*) If B covers 200 m, then A covers = 200 – 10 = 190 m

If C covers 200 m, then B covers = 200 – 20 = 180 m

$\Rightarrow$ If C covers 200 m, then A covers $= \frac{180}{200} \times \frac{190}{200} \times 200 = 171$ m

$\Rightarrow$ C can give a start of 200 – 171 = 29 m

197. In a kilometre race, A beats B by 60 metres or 12 seconds. What is time taken by A to complete the race?

(*a*) 150 seconds (*b*) 160 seconds (*c*) 188 seconds (*d*) 200 seconds

Sol. (*c*) B covers 60 metres in 12 seconds.

$\therefore$ B covers 1000 metres in $\frac{12}{60} \times 1000 = 200$ seconds.

$\therefore$ Time taken by A = 200 – 12 = 188 seconds

198. In a kilometre race, A can give B 100 metres start and to C 154 metres start. How many metres start can B give C in a 2 km race?

(*a*) 120 m (*b*) 150 m (*c*) 175 m (*d*) 200 m

Sol. (*a*) In 1 km race, B runs 900 m (= 1000 – 100) and C runs 846 m (= 1000 – 154).

If B runs 900 m, then C runs = 846 m

If B runs 2000 m, then C runs $= \frac{846}{900} \times 2000$ m = 1880 m

$\therefore$ B can give C (2000 – 1880) = 120 m start.

199. A can run 200 m in 35 seconds and B in 40 seconds. By what distance can A beat B?

(*a*) 20 m (*b*) 25 m (*c*) 50 m (*d*) 75 m

Sol. (*b*) A beats B by 40 – 35 = 5 seconds

Distance covered by B in 5 seconds $= \frac{200}{40} \times 5 = 25$ metres

200. A can run 330 m in 41 seconds whereas B takes 44 seconds to cover the same distance. By how much seconds will B win if he has a 30 m start ?

(*a*) 1 second (*b*) 2 seconds (*c*) 3 seconds (*d*) 4 seconds

Sol. (*a*) B covers 330 – 30 i.e. 300 m in $\frac{44}{330} \times 300 = 40$ seconds

$\therefore$ Required time = (41 – 40) seconds = 1 second

201. A runs 100 metres in 11 seconds and B runs 100 metres in 12 seconds. The start, which must be given to B to make the race to be completed in 11 seconds a draw, will be : **(SSC CGL Tier I 2006)**

(*a*) 8 m (*b*) $8\frac{1}{2}$ m (*c*) $8\frac{1}{3}$ m (*d*) $8\frac{1}{2}$ m

Sol. (*c*) B took (12 seconds – 11 seconds) = 1 second more than A.

A can give B start of distance covered by B in 1 second.

Distance covered by B in 1 second = $\frac{100}{12} = \frac{25}{3} = 8\frac{1}{3}$ metres

202. A can run 1 km in 9 minute 30 seconds and B in 10 minutes. How many metres start can A give to B so that they complete the race at the same time?

(*a*) 20 m (*b*) 25 m (*c*) 50 m (*d*) 75 m

Sol. (*c*) B took (10 minutes – 9 minutes 30 seconds) = 30 seconds more than A.

A can give B start of distance covered by B in 30 seconds.

Distance covered by B in 30 seconds or $\frac{1}{2}$ minute = $\frac{1000}{10} \times \frac{1}{2} = 50$ metres

203. A covers a distance of 800 metres at 36 km/h. A gives B a start of 50 metres and still beats him by 10 seconds. Find the speed (in km/h) of B.

(*a*) 24 (*b*) 30 (*c*) 35 (*d*) 40

Sol. (*b*) Speed of A (in m/s) = $36 \times \frac{5}{18} = 10$ m/s

Time taken by A to cover 800 metres = $\frac{800}{10} = 80$ seconds

B covers (800 – 50) = 750 m in (80 + 10) = 90 seconds

Speed of B = $\frac{750}{90} \times \frac{18}{5} = 30$ km/h

204. In a kilometre race, A beats B by 30 seconds and B beats C by 15 seconds. If A beats C by 180 metres, the time taken by A to run 1 km race is **(SSC CPO SI 2006)**

(*a*) 200 seconds (*b*) 250 seconds (*c*) 205 seconds (*d*) None of these

Sol. (*c*) A beats B by 30 seconds and B beats C by 15 seconds

⇒ A beats C by (30 + 15) = 45 seconds or by 180 metres

⇒ Speed of C = 180 ÷ 45 = 4 m/s

⇒ Time taken by C to cover 1 km = 1000 ÷ 4 = 250 seconds

⇒ Time taken by A to cover 1 km = (250 – 45) = 205 seconds

205. A and B run a kilometre and A wins by 25 sec. A and C run a kilometre and A wins by 275 m. When B and C runs the same distance, B wins by 30 sec. The time taken by A to run a kilometre is : **(SSC CGL Tier I 2005)**

(*a*) 2 min 25 sec (*b*) 2 min 50sec (*c*) 3 min 20 sec (*d*) 3 min 30 sec

Sol. (*a*) Let the time taken by A to run 1 km = x seconds

Then time taken by B to run 1 km = $(x + 25)$ seconds

Then time taken by C to run 1 km = $(x + 25) + 30 = (x + 55)$ seconds

In the time A runs 1 km, C runs = 1000 – 275 = 725 m

Ratio of speeds of A and C = 1000 : 725 = 40 : 29

$40 \times x = 29 \times (x + 55) \Rightarrow 11x = 29 \times 55 \Rightarrow x = 145$ seconds = 2 minutes 25 seconds

206. A train can move at 45 km/hr without any wagon and speed is diminished by a constant that varies as the square root of the number of wagon attached. If with 9 wagons, the train can move at 30 km/hr, what is the greatest number of wagons with which the train can just move? **(IIFT 2012)**

(*a*) 63 (*b*) 64 (*c*) 80 (*d*) 81

Sol. (*c*) Speed = $45 - x\sqrt{\text{wagons}} \Rightarrow 30 = 45 - x\sqrt{9} \Rightarrow 3x = 15 \Rightarrow x = 5$

Minimum speed = $0 = 45 - 5\sqrt{\text{Wagons}} \Rightarrow$ Wagons = 81

207. If a motorist had driven 1 hour longer and at an average speed of 5 km/hr faster, he would have covered 50 km more than he actually did. How many more km would he have covered than he actually did if he had driven 2 hour longer and at an average rate of 10 km/hr faster?

(*a*) 90 km (*b*) 100 km (*c*) 110 km (*d*) None of these

Sol. (*c*) Let actual time = t and normal speed = s.

Then $(t + 1)(s + 5) = ts + 50 \Rightarrow 5t + s + 5 = 50$

In second case, Distance = $(t + 2)(s + 10) = ts + 10t + 2s + 20$

$\therefore$ Extra km = $10t + 2s + 20 = 2 \times (5t + s + 5) + 10 = 2 \times 50 + 10 = 110$ km

208. A man covered a certain distance at some speed. Had he moved 3 km per hour faster, he would have taken 40 minutes less. If he moved 2 km per hour slower, he would have taken 40 minutes more. The distance (in km) is : **(SSC CGL Tier I 2003)**

(*a*) 20 (*b*) 35 (*c*) $36\frac{2}{3}$ (*d*) 40

Sol. (*d*) Let total distance be D km and the normal speed be x km/h.

Then $\frac{D}{x} - \frac{D}{x+3} = \frac{2}{3}$ and $\frac{D}{x-2} - \frac{D}{x} = \frac{2}{3}$

$$\Rightarrow \frac{D}{x} - \frac{D}{x+3} = \frac{D}{x-2} - \frac{D}{x} \Rightarrow \frac{1}{x} - \frac{1}{x+3} = \frac{1}{x-2} - \frac{1}{x}$$

$$\Rightarrow \frac{(x+3)-x}{x(x+3)} = \frac{x-(x-2)}{x(x-2)} \Rightarrow \frac{3}{x(x+3)} = \frac{2}{x(x-2)}$$

$\Rightarrow 3(x-2) = 2(x+3) \Rightarrow 3x - 6 = 2x + 6 \Rightarrow x = 12$ (km/hr)

Now $\frac{D}{12} - \frac{D}{15} = \frac{2}{3} \Rightarrow \frac{D}{60} = \frac{2}{3} \Rightarrow D = \frac{2}{3} \times 60 = 40$ km

Trick: Normal speed = $\frac{2 \times \text{Increase in speeds} \times \text{Decrease in speed}}{\text{Difference in Increase and Decrease in speeds}} = \frac{2 \times 3 \times 2}{3-2} = 12$km/hr

$\Rightarrow$ Distance = $\frac{15 \times 12}{15-12} \times \frac{2}{3} = 40$ km or $\frac{12 \times 10}{12-10} \times \frac{2}{3} = 40$ km

209. A train covered a certain distance at a uniform speed. If the train had been 6 km/hr faster, it would have taken 4 hours less than the scheduled time. And, if the train were slower by 6 km/hr, the train would have taken 6 hours more than the scheduled time. The length of the journey is : **(MBA 2006)**

(*a*) 700 km (*b*) 720 km (*c*) 740 km (*d*) 760 km

Sol. (*b*) Let total distance be d km and the normal speed be x km/h

Then $\frac{d}{x} - \frac{d}{x+6} = 4$ and $\frac{d}{x-6} - \frac{d}{x} = 6 \Rightarrow \frac{d\{(x+6)-x\}}{x(x+6)} = 4$ and $\frac{d\{(x)-(x-6)\}}{x(x-6)} = 6$

$6d = 4x(x+6)$ and $6d = 6x(x-6) \Rightarrow 4x(x+6) = 6x(x-6) \Rightarrow 4(x+6) = 6(x-6)$

$\Rightarrow 4x + 24 = 6x - 36 \Rightarrow 2x = 60 \Rightarrow x = 30$

$\frac{d}{x} - \frac{d}{x+6} = 4 \Rightarrow \frac{d}{30} - \frac{d}{36} = 4 \Rightarrow 6d - 5d = 4 \times 180 \Rightarrow d = 720$ km

210. A train left station A for station B at a certain speed. After travelling for 100 km, the train meets with an accident and could travel at $\frac{4}{5}$th of the original speed and reaches 45 minutes late at station B. Had the accident taken place 50 km further on, it would have reached 30 minutes late at station B. What is the distance between station A and B?

(*a*) 125 km (*b*) 150 km (*c*) 200 km (*d*) 250 km

Sol. (*d*) First 100 km are covered at the same speed in both the cases. Speed after (100 + 50) km is also the same in both the cases. Hence the difference in the time (45 – 30 = 15 minutes) is due to difference in speeds during the journey of 50 km after first 100 km.

Let initial speed of the train = 5 km/h

Then speed after the accident $= \frac{4}{5} \times 5 = 4$ km/h

Time difference = (45 – 30) minutes = 15 minutes $= \frac{1}{4}$ hour

Then distance covered $= \frac{5 \times 4}{5 - 4} \times \frac{1}{4} = 5$ km

But actual distance is 50 km. ⇒ Speed is 10 times of assumed speed.

∴ Speed before accident = 10 × 5 = 50 km/h and Speed after accident = 10 × 4 = 40 km/h

Time difference = 45 minutes $= \frac{3}{4}$ hour

∴ Distance between place of accident and B $= \frac{50 \times 40}{50 - 40} \times \frac{3}{4} = 150$ km

∴ Distance between A and B = 100 + 150 = 250 km

Alt : Ratio of speeds (before and after the accident) = 5 : 4 ⇒ Ratio of time taken = 4 : 5

Difference in time taken = (5 – 4) ratio = 1 ratio = 15 minutes

First time = 4 ratio = 15 × 4 = 60 minutes

Distance = 50 km (given) ⇒ First speed = 50 km/hr and second speed = 40 km/hr

Time difference = 45 minutes $= \frac{3}{4}$ hour

∴ Distance between place of accident and B $= \frac{50 \times 40}{50 - 40} \times \frac{3}{4} = 150$ km

∴ Distance between A and B = 100 + 150 = 250 km.

CHAPTER

26

BOATS AND STREAMS

1. A boat is said to be moving downstream when it is going with the stream and it is said to be going upstream when it is going against the stream.

2. If speed of a boat in still water is x km/hr and speed of the stream (or current) is y km/hr,

Then speed of the boat downstream (i.e., in the direction of stream) = $(x + y)$ km/hr

And speed of the boat upstream (i.e., in the opposite direction of stream) = $(x - y)$ km/hr

3. If we are given speed of boat upstream and downstream then we can find speed of boat in still water and speed of stream as follows:

Speed of boat in still water = $\frac{1}{2}$ (Downstream + Upstream)

Speed of stream = $\frac{1}{2}$ (Downstream – Upstream)

Example 1. A man can row a boat 12 km/hr with the stream and 8 km/hr against the stream. Find his speed in still water.

(*a*) 2 km/hr (*b*) 4 km/hr (*c*) 8 km/hr (*d*) 10 km/hr

Sol. (*d*) Speed of boat in still water = $\frac{12+8}{2}$ = 10 km/hr

Example 2. A man can row a boat 27 km/hr with the stream and 11 km/hr against the stream. Find speed of stream.

(*a*) 2 km/hr (*b*) 4 km/hr (*c*) 8 km/hr (*d*) 10 km/hr

Sol. (*c*) Speed of stream = $\frac{27-11}{2}$ = 8 km/hr

FORMULAE

1. If ratio of downstream and upstream speeds of a boat is $a : b$. Then ratio of time taken = $b : a$.

Speed of stream $= \frac{a-b}{a+b} \times$ Speed in still water.

Speed in still water $= \frac{a+b}{a-b} \times$ Speed of stream

Example 1. A man can row a boat to a certain distance upstream in 4 hours and takes 3 hours to row downstream the same distance. What is the speed of boat in still water, if the speed of the stream is 2 km/hr ?

(*a*) 9 km/hr (*b*) 10 km/hr (*c*) 12 km/hr (*d*) 14 km/hr

Sol. (*d*) Ratio of speed downstream and upstream = 3 : 4

∴ Speed of boat in still water = $\frac{4+3}{4-3} \times 2$ = 14 km/hr

Example 2. A man can row a boat 120 km with stream in 5 hours. If speed of the boat is double the speed of the stream, find the speed of stream.

(*a*) 6 km/hr (*b*) 8 km/hr (*c*) 9 km/hr (*d*) 12 km/hr

Sol. (*b*) Speed of the boat downstream $= \frac{120}{5} = 24$ km/h

Ratio of speeds of boat and stream = 2 : 1 ⇒ Speed of the stream $= \frac{1}{3} \times 24 = 8$ km/h

2. A person rows a boat to a certain place and then comes back to the starting point. If he rows the boat at the speed of *x* km/h in still water and the rate of the current is *y* km/h, then Average speed $= \frac{2(x+y)(x-y)}{(x+y)+(x-y)} = \frac{x^2-y^2}{x} = \frac{(x+y)(x-y)}{x}$

Hint: $(x + y)$ and $(x - y)$ are speeds of boat downstream and upstream respectively.

Example. A man can row a boat at 5 kmph in still water. If the velocity of the current is 1 kmph and it takes him 1 hour to row to a place and come back, how far is the place ?

(SSC CGL Tier I 2004)

(*a*) 2.5 km (*b*) 3 km (*c*) 2.4 km (*d*) 3.6 km

Sol. (*c*) Required distance $= \frac{5^2 - 1^2}{5} \times \frac{1}{2} = 2.4$ km

Alternative method : Speed of boat downstream = 5 + 1 = 6 km/hr
Speed of boat upstream = 5 – 1 = 4 km/hr

∴ Distance between two places = Average Speed × Time $\times \frac{1}{2} = \frac{2 \times 6 \times 4}{10} \times \frac{1}{2} = 2.4$ km

SOLVED EXERCISE

1. A man can row a boat 40 km upstream in 8 hours and a distance of 36 km downstream in 6 hours. The speed of stream is: **(SSC CHSL 2011)**

(*a*) 0.5 km/hr (*b*) 1.5 km/hr (*c*) 3 km/hr (*d*) 2 km/hr

Sol. (*a*) Speed downstream = 36 ÷ 6 = 6 km/hr. Speed upstream = 40 ÷ 8 = 5 km/hr

∴ Speed of the current $= \frac{1}{2} \times (6 - 5) = 0.5$ km/hr

2. A boat running downstream covers a distance of 24 kms in 4 hours, while for covering the distance upstream it takes 6 hours. What is the speed of the boat in still water ?

(Dena Bank PO 2006)

(*a*) 5.5 km/hr (*b*) 6 km/hr (*c*) 3.5 km/hr (*d*) None of these

Sol. (*d*) Speed downstream = 24 ÷ 4 = 6 km/hr. Speed upstream = 24 ÷ 6 = 4 km/hr

∴ Speed of the boat in still water $= \frac{1}{2} \times (6 + 4) = 5$ km/hr

3. A boat can travel with a speed of 13 km/hr in still water. If the speed of stream is 4 km/hr in the same direction, time taken by boat to go 63 km in opposite direction is

(SSC CGL Tier I 2015)

(*a*) 3 (9/17) hours (*b*) 4 hours (*c*) 7 hours (*d*) 9 hours

Sol. (*c*) Speed upstream = 13 – 4 = 9 km/hr. ⇒ Required time = 63 ÷ 9 = 7 hours

4. The speed of a boat in still water is 6 km/hr and the speed of the stream is 1.5 km/hr. A man rows to a place 22.5 km away and come back to the starting point. The total time taken by him is : **(SSC CGL Tier I 2015)**

(*a*) 4 hours 10 min (*b*) 8 hours (*c*) 10 hours (*d*) 6 hours 10 minutes

Sol. (*b*) Time $= \frac{22.5}{4.5} + \frac{22.5}{7.5} = \frac{45}{9} + \frac{45}{15} = 5 + 3 = 8$

5. A man rows down a river 15 km in 3 hrs. with the stream and returns in $7\frac{1}{2}$ hrs. The rate at which he rows in still water is **(SSC CGL Tier I 2013)**

(*a*) 2.5 km/hr (*b*) 1.5 km/hr (*c*) 3.5 km/hr (*d*) 4.5 km/hr

Sol. (*c*) Speed downstream = 15 ÷ 3 = 5 km/hr. Speed upstream = 15 ÷ 7.5 = 2 km/hr

$\therefore$ Speed of the boat in still water $= \frac{1}{2} \times (5 + 2) = 3.5$ km/hr

6. A motorboat takes 2 hours to travel a distance of 9 km with the current and it takes 6 hours to travel the same distance against the current. What is the speed of the boat in still water in kmph?

(*a*) 3 (*b*) 2 (*c*) 1.5 (*d*) 1

Sol. (*a*) Downstream speed = 9 ÷ 2 = 4.5 kmph. Upstream speed = 9 ÷ 6 = 1.5 kmph

Speed in still water $= \frac{1}{2} \times$ (downstream speed + upstream speed) $= \frac{1}{2} \times (4.5 + 1.5) = 3$ kmph

7. A man can swim 3 km/hr in still water. If the velocity of the stream is 2 km/hr, the time taken by him to swim to a place 10 km upstream and back is : **(SSC CGL Tier I 2013)**

(*a*) $8\frac{1}{3}$ hr. (*b*) $9\frac{1}{3}$ hr. (*c*) 10 hr. (*d*) 12 hr.

Sol. (*d*) Speed upstream = (3 – 2) = 1 km/hr. Speed downstream = (3 + 2) = 5 km/hr

Required time = (10 ÷ 1) + (10 ÷ 5) = 10 + 2 = 12 hours

8. A sailor goes 12 km downstream in 48 minutes and returns in 1 hour and 20 minutes. The speed of the sailor in still water is : **(SSC CHSL 2010)**

(*a*) 12 km/hr (*b*) 12.5 km/hr (*c*) 13 km/hr (*d*) 15 km/hr

Sol. (*a*) Speed downstream $= 12 \times \frac{60}{48} = 15$ km/hr. Speed upstream $= 12 \times \frac{60}{80} = 9$ km/hr

$\therefore$ Speed in still water $= \frac{1}{2} \times (15 + 9) = 12$ km/hr

9. A man can row a boat 15 km/hr with the current and speed of current is 3 km/hr. Find his speed (in km/hr) against the current.

(*a*) 9 (*b*) 12 (*c*) 18 (*d*) 21

Sol. (*a*) Speed in still water = 15 – 3 = 12 km/hr. ⇒ Speed against current = 12 – 3 = 9 km/hr

10. A man rows a boat 18 kilometres in 4 hours downstream and returns upstream in 12 hours. The speed of the stream (in km per hour) is : **(SSC CGL Tier I 2003)**

(*a*) 1 km/h (*b*) 1.5 km/h (*c*) 2 km/h (*d*) 1.75 km/h

Sol. (*b*) Downstream speed $= \frac{18}{4} = \frac{9}{2}$ and upstream speed $= \frac{18}{12} = \frac{3}{2}$

Speed of stream $= \frac{1}{2} \times$ (Downstream speed – Upstream) $= \frac{1}{2} \times \left(\frac{9}{2} - \frac{3}{2}\right) = 1.5$ km/hr

11. A boat goes 8 km in one hour along the stream and 2 km in one hour against the stream. The speed (in km/hr) of the stream is : **(SSC CGL Tier I 2005)**

(*a*) 2 (*b*) 3 (*c*) 4 (*d*) 5

Sol. (*b*) Speed of stream = $\frac{8-2}{2}$ = 3 km/hr

12. The speed of a boat in still water is 12 km/hr and its speed against the current is 8.5 km/hr. Find the speed of boat with the current.

(*a*) 3.5 km/hr (*b*) 7 km/hr (*c*) 15.5 km/hr (*d*) 20.5 km/hr

Sol. (*c*) Speed of current = 12 – 8.5 = 3.5 km/hr
Downstream speed = 12 + 3.5 = 15.5 km/hr

13. On a river, B is equidistant from both ends A and C. If a boat can go from A to B and back in 6 hours and from A to C in 8 hours, how long would it take to go from C to A?

(*a*) 2 hours (*b*) 4 hours (*c*) 5 hours (*d*) 6 hours

Sol. (*b*) Time taken from A and C and back = Double the time from A to B and back
= 2 × 6 hours = 12 hours. But time taken from A to C = 8 hours
∴ Time taken from C to A = 12 hours – 8 hours = 4 hours.

14. A man rows a boat from point A against current for 10 minutes and then comes back with the current for next 10 minutes and reaches to a point B. If distance between A and B is 1 km, find the speed of the current.

(*a*) 1 km/h (*b*) 2 km/h (*c*) 2.5 km/h (*d*) 3 km/h

Sol. (*d*) Going with current and against current for the same time means speed of the boat in still water is neutralized. Therefore the boat is moved by the current only.

⇒ Boat is moved 1 km by the current in (10 + 10) minutes = 20 minutes = $\frac{1}{3}$ hour.

∴ Speed of current = 1 × 3 = 3 km/h

15. A swimmer swims from a point A against a current for 5 minutes and then swims backwards in favour of the current for the next 5 minutes and comes to the point B. If AB = 100 metres, the speed of the current (in km per hour) is : **(SSC CGL Tier I 2013)**

(*a*) 0.6 (*b*) 0.4 (*c*) 0.2 (*d*) 1

Sol. (*a*) Going with current and against current for the same time means work of the swimmer in is neutralized. Therefore he is moved by the current only.

⇒ Swimmer is moved 100 m by the current in (5 + 5) minutes = 10 minutes = $\frac{1}{6}$ hour.

∴ Speed of current = 6 × 0.1 = 0.6 km/h

16. A man rows a boat 3 hours downstream and 3 hours upstream. Find speed of boat in still water, if he covers a total distance of 12 km.

(*a*) 2 km/h (*b*) 3 km/h (*c*) 4 km/h (*d*) Data insufficient

Sol. (*a*) Let speed of boat and current are x km/h and y km/h respectively.
Then $3 \times (x + y) + 3 \times (x - y) = 12 \Rightarrow 6x = 12 \Rightarrow x = 2$ km/h

17. Two boats A and B start towards each other from two places, 108 km apart. Speeds of the boats A and B in still water are 12 km/hr and 15 km/hr respectively. If A proceeds down and B up the stream, they will meet after : **(SSC CGL Tier I 2008)**

(*a*) 4.5 hours (*b*) 4 hours (*c*) 5.4 hours (*d*) 6 hours

Sol. (*b*) Let speed of stream be x km/hr. Then relative speed = $(12 + x) + (15 - x)$ = 27 km/hr
∴ Required time = 108 ÷ 27 = 4 hours

18. A boat goes 6 km an hour in still water. It takes thrice as much time going upstream as much time going downstream. The speed of the current is : **(SSC CGL Tier I 2003)**

(*a*) 4 km/hr (*b*) 5 km/hr (*c*) 3 km/hr (*d*) 2 km/hr

Sol. (*c*) Speed downstream : Speed upstream = 3 : 1

$\therefore$ Speed of current $= \dfrac{3-1}{3+1} \times 6 = 3$ km/hr

19. Speed of a boat in still water is 16 km/h. If it can travel 20 km downstream in the same time as it can travel 12 km upstream, the rate of stream is.

(*a*) 1 km/h (*b*) 2 km/h (*c*) 4 km/h (*d*) 5 km/h

Sol. (*c*) Speed downstream : Speed upstream = 20 : 12 = 5 : 3

$\therefore$ Speed of current $= \dfrac{5-3}{5+3} \times 16 = 4$ km/h

20. A steamer goes downstream from one port to another in 4 hours. It covers the same distance upstream in 5 hours. If the speed of stream is 2 km/hr, the distance between the two ports is: **(SSC CGL Tier I 2007)**

(*a*) 50 km (*b*) 60 km (*c*) 70 km (*d*) 80 km

Sol. (*d*) Ratio of speeds = 5 : 4 $\Rightarrow$ Speed in still water $= \dfrac{5+4}{5-4} \times 2 = 18$ km/hr

$\therefore$ Speed downstream = 18 + 2 = 20 km/hr $\Rightarrow$ Distance = 4 × 20 km = 80 km

21. A boat covers a certain distance down-stream in 8 hours and comes back upstream in 10 hours. If the speed of the current be 1 km/hr, the distance (in km) of the one-way journey is: **(SSC CGL Tier I 2006)**

(*a*) 60 (*b*) 70 (*c*) 80 (*d*) 90

Sol. (*c*) Ratio of time = 8 : 10 = 4 : 5 $\Rightarrow$ Ratio of speeds = 5 : 4

Speed in still water $= \dfrac{5+4}{5-4} \times 1 = 9$ km/hr

$\therefore$ Speed down stream = 9 + 1 = 10 km/hr $\Rightarrow$ Distance = 8 × 10 km = 80 km

22. A man can row a boat at 12 km/h in still water. If he takes twice the time to go upstream as to row downstream, find the speed of the boat upstream. **(SSC CGL Tier I Exam)**

(*a*) 4 km/hr (*b*) 5 km/hr (*c*) 6 km/hr (*d*) 8 km/hr

Sol. (*d*) Ratio of time taken (downstream and upstream) = 1 : 2

$\therefore$ Ratio of speeds (downstream and upstream) = 2 : 1

$\therefore$ Speed of current $= \dfrac{2-1}{2+1} \times 12 = \dfrac{1}{3} \times 12 = 4$ km/hr

Speed of boat upstream = 12 – 4 = 8 km/hr

Alternative Method: Speed in still water $= \dfrac{1}{2}$ (speed upstream + speed downstream)

$\therefore$ Speed upstream + speed downstream = 2 × 12 = 24 km

But speed upstream : speed downstream = 1 : 2 $\Rightarrow$ Speed upstream $= \dfrac{1}{3} \times 24 = 8$ km/hr

23. A man goes downstream with a boat to some destination and returns upstream to his original place in 5 hours. If the speed of the boat in still water and the stream are 10 km/hr and 4 km/hr respectively, the distance of the destination from the starting place is : **(SSC CGL Tier I 2008)**

(*a*) 16 km (*b*) 18 km (*c*) 21 km (*d*) 25 km

Sol. (*c*) Speed downstream = 10 + 4 = 14 km/h. Speed upstream = 10 – 4 = 6 km/h

$\therefore$ Distance between two places = Average Speed × Time × $\frac{1}{2} = \frac{2 \times 14 \times 6}{20} \times \frac{5}{2} = 21$ km

24. Speed of a boat is 5 km/hr in still water and the speed of the stream is 3 km/hr. If the boat takes 3 hours to go to a place and come back, the distance of the place is **(SSC DEO 2011)**

(*a*) 3.75 km (*b*) 4 km (*c*) 4.8 km (*d*) 4.25 km

Sol. (*c*) Average speed $= \frac{2 \times (5+3) \times (5-3)}{(5+3)+(5-3)} = \frac{2 \times 8 \times 2}{10}$

Distance $= \frac{2 \times 8 \times 2}{10} \times 3 \times \frac{1}{2} = 4.8$ km

25. A man can row at 5 km/h in still water. If the velocity of current is 1 km/hr and it takes him 1 hour to row to a place and come back, how far is the place? **(CDS 2004)**

(*a*) 3 km (*b*) 3.6 km (*c*) 2.4 km (*d*) 1.8 km

Sol. (*c*) Speed downstream : Speed upstream = (5 + 1) : (5 – 1) = 6 : 4 = 3 : 2

$\Rightarrow$ Time downstream : time upstream = 2 : 3 = 24 min and 36 min respectively.

$\therefore$ Total distance $= \frac{24}{60} \times 6$ km = 2.4 km

26. A man rows boat to a place covering 72 km distance and back in 15 hours. He finds that he can row 3 km with the stream in the same time as 2 km against the stream. Find the speed of the stream.

(*a*) 1 km/hr (*b*) 2 km/hr (*c*) 3 km/hr (*d*) 4 km/hr

Sol. (*b*) Speed downstream : Speed upstream = 3 : 2

$\therefore$ Time taken to row downstream : Time taken to row upstream = 2 : 3

But total time taken = 15 hours

$\therefore$ Time taken to row downstream $= \frac{2}{5} \times 15$ hours = 6 hours

Time taken to row upstream = 15 – 6 = 9 hours

$\therefore$ Speed downstream $= \frac{72}{6} = 12$ km/hr. $\Rightarrow$ Speed upstream $= \frac{72}{9} = 8$ km/hr

$\therefore$ Speed of the stream $= \frac{12-8}{2} = 2$ km/hr

Alt: Let downstream and upstream speeds are 3 km/hr and 2 km/hr respectively.

Then distance covered in 15 hours $= \frac{2 \times 3 \times 2}{5} \times 15 = 36$ km

But actual distance covered = 2 × 72 km = 144 km (*i.e.* 4 times 36 km)

$\therefore$ Speeds are 4 × 3 and 4 × 2, *i.e.* 12 and 8 km/hr

$\Rightarrow$ Speed of the stream $= \frac{12-8}{2} = 2$ km/hr

27. A boat covers 20 km downstream and 6 km upstream taking 3 hours in all. It covers 30 km downstream and 12 km upstream taking 5 hours in all. Find the speed of the boat in still water.

(*a*) 2 km/hr (*b*) 4 km/hr (*c*) 6 km/hr (*d*) 8 km/hr

Sol. (*d*) Let speed downstream = u and speed upstream = v. Let $u = \frac{1}{x}$ and $v = \frac{1}{y}$

$20x + 6y = 3$ .... (1) $30x + 12y = 5$ .... (2)

(1) × 2 gives $40x + 12y = 6$ .... (3)

(3) – (2) gives $10x = 1 \Rightarrow x = \frac{1}{10} \Rightarrow u = 10$

In (1) : $20 \times \frac{1}{10} + 6y = 3 \Rightarrow 6y = 1 \Rightarrow y = \frac{1}{6} \Rightarrow v = 6$

Speed of the boat in still water = $\frac{u+v}{2} = \frac{10+6}{2} = 8$ km/hr

Trick : HCF of 20 and 30 is 10. HCF of 6 and 12 is 6.

We try these values in the equations (1) and (2) and find that these values satisfies both the equations.

28. A boat goes 24 km upstream and 28 km downstream in 6 hours. It goes 30 km upsteam and 21 km downstream in 6 hours and 30 minutes. Find the speed of the boat in still water.

(SSC CGL Tier I 2014)

(*a*) 8 km/hr (*b*) 9 km/hr (*c*) 10 km/hr (*d*) 12 km/hr

Sol. (*c*) Trial and Error Method: HCF of 24 and 30 is 6. HCF of 28 and 21 is 7.

Putting these values, we get : $\frac{24}{6} + \frac{28}{7} = 8 \neq 6$ and $\frac{30}{6} + \frac{21}{7} = 8 \neq 6\frac{1}{2}$

We need to take higher value for down-stream speed. (Down-stream > Up-stream speed)

Let us take down-stream speed = 7 × 2 = 14 km/hr

Then $\frac{24}{6} + \frac{28}{14} = 6 = 6$ and $\frac{30}{6} + \frac{21}{14} = 6\frac{1}{2} = 6\frac{1}{2}$ (Hence, satisfies the equations)

$\Rightarrow$ Speed in still water = $\frac{1}{2} \times (14 + 6) = 10$ km/hr.

29. A boat goes 30 km upstream and 44 km downstream in 10 hrs. In 13 hrs, it can go 40 km upstream and 55 km downstream, then the speed of the boat in still water is

(SSC CHSL 2013)

(*a*) 5 km/hr (*b*) 8 km/hr (*c*) 15 km/hr (*d*) 13 km/hr

Sol. (*b*) **Trick :** HCF of 30, 40 = 10 and HCF of 44, 55 = 11

If upstream speed = 10 km/hr and downstream speed = 11 km/hr,

then total time (in first case) = $\frac{30}{10} + \frac{44}{11} = 3 + 4 = 7 < 10$

$\therefore$ We try lower value of upstream speed = 5 km/hr (i.e., 10 km/hr ÷ 2),

then time = $\frac{30}{5} + \frac{44}{11} = 6 + 4 = 10$ hours = given time

$\therefore$ Required speed = $\frac{\text{Upstream} + \text{Downstream speed}}{2} = \frac{5+11}{2} = 8$ km/hr

30. A boat covers 12 km upstream and 18 km downstream in 3 hours while it covers 36 km upstream and 24 km downstream in $6\frac{1}{2}$ hours. What is the speed of the current?

(SSC CGL Tier II 2012)

(*a*) 2 km/hr (*b*) 2.5 km/hr (*c*) 1.5 km/hr (*d*) 1 km/hr

Sol. (*a*) $12x + 18y = 3$ .... (1) and $36x + 24y = 6\frac{1}{2}$ .... (2)

$3 \times (1) \Rightarrow 36x + 54y = 9$ .... (3)

$(3) - (2) \Rightarrow 30y = \frac{5}{2} \Rightarrow y = \frac{5}{2} \times \frac{1}{30} = \frac{1}{12}$

Substituting this value in (2), $36x + 2 = 6\frac{1}{2} \Rightarrow 36x = 6\frac{1}{2} - 2 \Rightarrow x = \frac{9}{2} \times \frac{1}{36} = \frac{1}{8}$

$\therefore$ Speed of current $= \frac{12-8}{2} = 2$ km/hr

31. A boat whose speed in the still water is 16 km/hr, covers a distance of 30 km upstream and then returns taking 4 hours in total. What is the speed of the current ?

(*a*) 2 km/hr (*b*) 3 km/hr (*c*) 4 km/hr (*d*) 5 km/hr

Sol. (*c*) Total distance covered = 2 × 30 km = 60 km. Time taken = 4 hours

$\therefore$ Average speed = 60 ÷ 4 = 15 km/hr. Let the speed of current = x km/hr

Then $\frac{16^2 - x^2}{16} = 15 \Rightarrow 16^2 - x^2 = 16 \times 15 \Rightarrow x^2 = 16^2 - 16 \times 15 = 16 \times (16 - 15) = 16$

$\therefore x = \sqrt{16} = 4 \Rightarrow$ Speed of current = 4 km/hr

Trick : Total distance covered = 2 × 30 = 60 km

Distance covered in 4 hours @ 16 km/h = 4 × 16 = 64 km⇒Difference = 64 – 60 = 4 km/h

32. A boat whose speed in still water is 8 km/hr, takes 5 hours to go 15 km upstream and 22 km downstream. The speed of the stream is

(*a*) 2 km/hr (*b*) 3 km/hr (*c*) 4 km/hr (*d*) 2.5 km/hr

Sol. (*b*) Let speed of stream = x km/hr. Then $\frac{15}{8-x} + \frac{22}{8+x} = 5$

Out of given options, when we take $x = 3$, then $8 - 3 = 5$ divides 15 and $8 + 3 = 11$ divides 22 exactly.

Now, $\frac{15}{8-x} + \frac{22}{8+x} = \frac{15}{5} + \frac{22}{11} = 3 + 2 = 5$ satisfies the equation.

Alternative Method : Total distance covered = 15 + 22 = 37 km

Distance covered in 5 hours @ 8 km/hr = 40 km ⇒ Difference = 40 – 37 = 3 km

Possible value of speed of the stream would be 3 km/hr or factor thereof.

We check, $x = 3$, $\frac{15}{8-x} + \frac{22}{8+x} = \frac{15}{5} + \frac{22}{11} = 3 + 2 = 5$ hours

33. A man can row at 10 kmph in still water. If it takes a total of 5 hours for him to go to a place 24 km away and return, then the speed of the water current is **(SSC CGL Tier II 2012)**

(*a*) $\frac{1}{2}$ kmph (*b*) 1 kmph (*c*) 2 kmph (*d*) 3 kmph

Sol. (*c*) Let speed of the current is x kmph. Then $\frac{24}{10+x}+\frac{24}{10-x}=5$

Out of given options, $x = 2$ satisfies this equation.

Trick : Total distance covered = 2 × 24 = 48 km

Distance covered in 5 hours @ 10 km/hr = 50 km

Difference = 50 – 48 = 2 km/hr

34. The speed of a boat in still water is 11 km/h. It can go 12 km upstream and return downstream to the starting point in 2 hours 45 minutes. What is the speed of the stream ? **(CDS 2010)**

(*a*) 5 km/h (*b*) 4 km/h (*c*) 3 km/h (*d*) 2 km/h

Sol. (*a*) Let speed of stream be x km/h. Time = $2\frac{3}{4}$ hours = $\frac{11}{4}$ hours

Then $\frac{11^2-x^2}{11}\times\frac{11}{4}=2\times 12\text{ km}\Rightarrow 121-x^2=96\Rightarrow x^2=121-96=25\Rightarrow x=5.$

35. A motor boat, whose speed is 15 km/hr in still water goes 30 km downstream and comes back in a total of 4 hours and 30 minutes. The speed of the stream is **(CDS 2015)**

(*a*) 4 km/hour (*b*) 5 km/hour (*c*) 6 km/hour (*d*) 10 km/hour

Sol. (*b*) From options, we try : $\frac{30}{10}+\frac{30}{20}=3+\frac{3}{2}=4\frac{1}{2}$ ⇒ Speed of stream = $\frac{20-10}{2}=5$ km/hr

36. The speed of current is 5 km/h. A boat goes 10 km upstream and then returns to the starting point in 50 minutes. What is the speed of boat in still water ? **(SSC CPO SI 2011)**

(*a*) 20 km/hr (*b*) 25 km/hr (*c*) 28 km/hr (*d*) 30 km/hr

Sol. (*b*) Total distance covered = 2 × 10 km = 20 km ⇒ Average speed $=20\times\frac{60}{50}=24$ km/hr

Let speed of boat in still water = x km/hr

Then $\frac{x^2-5^2}{x}=24\Rightarrow x^2-24x-25=0\Rightarrow x^2+x-25x-25=0$

$\Rightarrow x(x+1)-25(x+1)=0\Rightarrow x=25$ or -1

CHAPTER

27 CLOCK AND CALENDAR

CLOCK

The clock has two hands – smaller hand and bigger hand. The smaller hand is called hour hand and the bigger hand is called minute hand. In one hour, the minute hand covers 60 minute space whereas the hour hand covers 5 minute space. It means that in 60 minutes, the minute hand gains 55 minutes over the hour hand.

∴ To gain 55 minute space over the hour hand, the minute hand takes 60 minutes.

∴ To gain 1 minute space over the hour hand, the minute hand takes $\frac{60}{55}$ or $\frac{12}{11}$ minutes.

∴ We multiply $\frac{12}{11}$ with total minutes to be gained over the hand hour, or we can simply multiply the number by $\left(1+\frac{1}{11}\right)$ to make calculation easy.

In one hour minute hand moves 360° and hour hand moves 30°. The minute hand moves 6° per minute and the hour hand moves $\left(\frac{1}{2}\right)^{\circ}$ per minute. Hence minute hand gains $\left(5\frac{1}{2}\right)^{\circ}$ per minute.

Example 1. The ratio of the velocities of the hour-hand and the minute-hand of a clock is: **(SSC CGL Tier I 2006)**

(*a*) 1 : 1 (*b*) 5 : 1 (*c*) 1 : 12 (*d*) 12 : 1

Sol. (*c*) In one hour, hour-hand covers 5 minute space and minute-hand covers 60 minute space.

∴ Required ratio = 5 : 60 = 1 : 12

Example 2. At what time between 5'o clock and 6'o clock will the hands of a clock coincide?

Sol. At 5'o clock, the minute hand is 25 minute space behind the hour hand.

∴ In order to coincide, the minute hand must gain 25 minute space over hour hand.

Time taken by minute hand to gain 25 minutes

$= 25 \times \left(1+\frac{1}{11}\right) = 25 + \frac{25}{11} = 25 + 2\frac{3}{11} = 27\frac{3}{11}$ minutes

Hence the hands of the clock will coincide at $27\frac{3}{11}$ minutes past 5.

CALENDAR

An ordinary year has 365 days. In an ordinary year, February has 28 days whereas in a leap year, February has 29 days. Every year which is divisible by 4, is a leap year and has 366 days. But century year has 365 days except for the year divisible by 400 which has 366 days.

An ordinary year contains 365 days *i.e.* 52 weeks + 1 day *i.e.* 1 odd day.

A leap year contains 366 days *i.e.* 52 weeks + 2 days *i.e.* 2 odd days.

A century (100 years) contains = 24 leap years + 76 ordinary years
$= 24 \times 2 + 76 = 124$ odd days = 17 weeks + 5 odd days

Similarly,

200 years contains $2 \times 5 - 7 = 3$ odd days

300 years contains $3 \times 5 - 14 = 1$ odd day

400 years contains $4 \times 5 + 1 - 21 = 0$ odd days [1 more day has been added as 400th year is a leap year]

First January, 1 A.D. was Monday.

A solar year contains 365 days 5 hours 48 minutes 48 seconds.

The first day of a century must either be Monday, Tuesday, Thursday or Saturday.

SOLVED EXERCISE

1. At what time between 9'o clock and 10'o clock will the hands of a clock point in the opposite directions ?

Sol. At 9'o clock, the minute hand is ahead of hour hand by 15 minutes.
The hands will be opposite to each other when there is a space of 30 minutes between them.
This will happen when the minute hand gains 15 minute space over hour hand.
Time taken by minute hand to gain 15 minutes

$$= 15 \times \left(1 + \frac{1}{11}\right) = 15 + \frac{15}{11} = 15 + 1\frac{4}{11} = 16\frac{4}{11} \text{ minutes.}$$

Hence the hands are opposite to each other at $16\frac{4}{11}$ minutes past 9.

2. At what time between 4'o clock and 5'o clock will the hands of a clock be at right angle?

Sol. At 4'o clock, the minute hand is 20 minutes space behind the hour hand.

The hands of the clock will be at right angle twice between 4 and 5'o clock.

(*i*) When the minute hand is 15 minute space behind the hour hand, *i.e.*, when minute hand gains $(20 - 15) = 5$ minute space over hour hand.

Time taken by minute hand to gain 5 minutes $= 5 \times \left(1 + \frac{1}{11}\right) = 5 + \frac{5}{11} = 5\frac{5}{11}$ minutes

Hence the hands are at right angle at $5\frac{5}{11}$ minutes past 4.

(*ii*) The hands will be at right angle again when minute hand is 15 minute space ahead of hour hand *i.e.* when minute hand gains $(20 + 15) = 35$ minute space over hour hand.
Time taken by minute hand to gain 35 minutes

$$= 35 \times \left(1 + \frac{1}{11}\right) = 35 + \frac{35}{11} = 35 + 3\frac{2}{11} = 38\frac{2}{11} \text{ minutes}$$

$\therefore$ They will again be at right angle at $38\frac{2}{11}$ minutes past 4.

3. The angle formed by the hour-hand and the minute-hand of a clock at 2:15 p.m. is

(SSC CGL Tier I 2012)

(*a*) $22\frac{1}{2}^\circ$ (*b*) 30° (*c*) $27\frac{1}{2}^\circ$ (*d*) 45°

Sol. (*a*) At 2'O Clock, minute hand is $10 \times 6° = 60°$ behind the hour hand.

In 15 minutes, minute hand will gain $\left(5\frac{1}{2}\right)^{\circ} \times 15 = 75 + 7\frac{1}{2} = 82\frac{1°}{2}$

$\therefore$ Required angle $= 82\frac{1°}{2} - 60 = 22\frac{1°}{2}$

Alternative Method: Angle between 2 and 3 = $1 \times 30° = 30°$

Angle reduced in 15 minutes = $\left(\frac{1}{2}\right)^{\circ} \times 15 = 7\frac{1°}{2}$ $\Rightarrow$ Required angle $= 30° - 7\frac{1°}{2} = 22\frac{1°}{2}$.

4. What will be the acute angle between hands of a clock at 2:30?

Sol. At 2'O Clock, minute hand will be $10 \times 6° = 60°$ behind the hour hand.

In 30 minutes, minute hand will gain $\left(5\frac{1}{2}\right)^{\circ} \times 30 = 150° + 15° = 165°$

$\therefore$ Required angle = $165° - 60° = 105°$

Alternative Method: Angle between 2 and 6 = $4 \times 30° = 120°$

Angle reduced in 30 minutes = $\left(\frac{1}{2}\right)^{\circ} \times 30 = 15°$ $\Rightarrow$ Required angle = $120° - 15° = 105°$.

5. What will be the acute angle between hands of a clock at 5:25?

(*a*) $5\frac{1°}{2}$ (*b*) $7\frac{1°}{2}$ (*c*) $12\frac{1°}{2}$ (*d*) $37\frac{1°}{2}$

Sol. (*c*) The required angle is the angle made by hour hand in 25 minutes.

$\therefore$ Required angle = $\left(\frac{1}{2}\right)^{\circ} \times 25 = 12\frac{1°}{2}$

6. Find the angle between hands of a clock at 8:15?

(*a*) 150° (*b*) $157\frac{1°}{2}$ (*c*) $142\frac{1°}{2}$ (*d*) 165°

Sol. (*b*) At 8.15, minute hand is at 3. Angle between 3 and 8 is $5 \times 30° = 150°$

Angle made by hour hand in 15 minutes = $15 \times \frac{1°}{2}$

$\Rightarrow$ Required angle = $150 + \frac{15}{2} = 157\frac{1°}{2}$

7. At what time between 7 and 8 o'clock, corrected to the nearest minute, the hands of the clock will from an angle of 32°? **(MBA Ent. Exam.)**

(*a*) 7.33 & 7.44 (*b*) 7.06 & 7.54 (*c*) 7.32 & 7.44 (*d*) 7.33 & 7.43

Sol. (*c*) At 7'o clock, angle between the two hands is $30° \times 7 = 210°$

The minutes hand gains $\frac{11°}{2}$ per minute over hour hand.

The hands of the clock will form an angle of 32° twice between 7 and 8'o clock.

(*i*) When the Minute Hand is 32° behind the hour hand, *i.e.*, when minute hand gains $(210° - 32°) = 178°$ over hour hand.

Time taken by minute hand to gain 178° is $178 \times \frac{2}{11} = \frac{356}{11}$ = 32 minutes (approx)

Hence the two hands will make angle of 32° at 32 minutes past 7.

(*ii*) When minute hand is 32°ahead of hour hand i.e. when minute hand has gained (210° + 32°) = 242° over hour hand.

Time taken by minute hand to gain 242° is $242 \times \frac{2}{11} = \frac{484}{11} = 44$ minutes

Hence the two hands will again make angle of 32° at 44 minutes past 7.

8. A wall-clock takes 6 seconds to strike the number of hours at 6 o'clock. The time it will take to strike the number of hours at 11 o'clock is: **(SSC CGL Tier I 2007)**

(*a*) 10 seconds (*b*) 11 seconds (*c*) 12 seconds (*d*) 13 seconds

Sol. (*c*) Number of intervals in 6 strikes = 6 – 1 = 5 ⇒ 5 intervals = 6 seconds

Number of intervals in 11 strikes = 11 – 1 = 10

$\therefore$ 10 Intervals = $\frac{6}{5} \times 10 = 12$ seconds

9. The first day of the year 1998 was Thursday. On what day of the week would the last day of the year 2001 fall ? **(SSC CGL Tier I 2008)**

(*a*) Saturday (*b*) Sunday (*c*) Monday (*d*) Tuesday

Sol. (*c*) Number of odd days from 1.1.98 to 1.1.2002 = 1 + 1 + 2 + 1 = 5

$\therefore$ Number of odd days from 1.1.98 to 31.12.2001 = 5 – 1 = 4 days

$\therefore$ Day on 31.12.2001 is Thursday + 4 days *i.e.* Monday.

10. What day of the week was it on 26 January 1950?

(*a*) Monday (*b*) Tuesday (*c*) Thursday (*d*) Saturday

Sol. (*c*) Odd days upto 26.1.1950

= Odd days for 1600 years + odd days for 300 years + odd days for 49 years + odd days for 26 days

= (0 × 4) + (1) + (12 × 2 + 37) + 5 = 1 + 5 + 5 = 4 odd days

⇒ 26.1.1950 = Sunday + 4 days = Thursday

CHAPTER

28

LOG AND ANTILOG

Logarithm (log): The log of a given number to a given base is the power or exponent to which the base must be raised in order to get the given number.

If $b^x = y$, then $\log_b y = x$ where b is a positive number not equal to 0 or 1.

Example: $2^5 = 32 \Rightarrow \log_2 32 = 5$; $10^3 = 1000 \Rightarrow \log_{10} 1000 = 3$

$2^{-3} = \frac{1}{8} \Rightarrow \log_2 \frac{1}{8} = -3$ $\quad$ $8^{\frac{1}{3}} = 2 \Rightarrow \log_8 2 = \frac{1}{3}$

Note: When no base is mentioned with the log, it is assumed that the base is 10.

Opposite of logarithm is antilogarithms. If $\log x = y$, then antilog $y = x$.

FORMULAE

1. $\log (x \times y) = \log x + \log y$
2. $\log \left(\frac{x}{y}\right) = \log x - \log y$
3. $\log x^y = y \log x$
4. $\log \sqrt[y]{x} = \log x^{1/y} = \frac{1}{y} \log x$
5. $\log x + \log \left(\frac{1}{x}\right) = \log x \times \frac{1}{x} = \log 1 = 0$
6. $\log \frac{x}{y} = -\log \frac{y}{x}$
7. $\log_a a = 1$, since $a^1 = a$
8. $\log_a 1 = 0$, since $a^0 = 1$
9. $\log x^a \div \log x = a \log x \div \log x = a$
10. $\log_a b = \frac{1}{\log_b a} \Rightarrow \log_a b \times \log_b a = 1$
11. $\log_a x = \log_a b \times \log_b x$
12. $\log_a x = \frac{\log_b x}{\log_b a}$
13. $a^{\log_a x} = x$
14. $\log_a (a^x) = x$
15. $a^n = x \Rightarrow \log_x a = \frac{1}{n}$

Example : $3^2 = 9 \Rightarrow \log_9 3 = \frac{1}{2}$

SOLVED EXERCISE

1. Evaluate $\log_3 243$.

Sol. $\log_3 243 = \log_3 (3)^5 = 5 \log_3 3 = 5$

2. If $\log_5 x = 3$, find the value of x.

Sol. $x = 5^3 = 125$

3. $\log_3 \log_2 \log_{\sqrt{3}} 81 = ?$

Sol. G.E. = $\log_3 \log_2 \log_{\sqrt{3}} \left(\sqrt{3}\right)^8 = \log_3 \log_2 8 = \log_3 \log_2 2^3 = \log_3 3 = 1$

4. $\log_2 \log_3 \log_4 64 = ?$

Sol. G.E. = $\log_2 \log_3 \log_4 4^3 = \log_2 \log_3 3 = \log_2 1 = \log_2 2^0 = 0$

5. $\log_{81} \log_2 8 = ?$

Sol. G.E. = $\log_{81} \log_2 2^3 = \log_{81} 3 = \log_{81} (81)^{1/4} = 1/4$

Alt: G.E. = $\log_{81} \log_2 2^3 = \log_{81} 3 = \log_{3^4} 4 = \frac{1}{4}$

6. $\log_{32} 8 = ?$

(*a*) $1/4$ (*b*) $2/3$ (*c*) $3/5$ (*d*) $2/5$

Sol. (*c*) $\log_{32} 8 = \frac{\log 8}{\log 32} = \frac{\log 2^3}{\log 2^5} = \frac{3 \log 2}{5 \log 2} = \frac{3}{5}$

7. $\log 5 + \log 6 - \log 3 = ?$

Sol. $\log 5 + \log 6 - \log 3 = \log\left(\frac{5 \times 6}{3}\right) = \log 10 = 1$

8. Prove that $\log (1 + 2 + 3) = \log 1 + \log 2 + \log 3$

Sol. $\log (1 + 2 + 3) = \log 6 = \log (1 \times 2 \times 3) = \log 1 + \log 2 + \log 3$

9. $\log_{10}\left(\frac{26}{51}\right) + \log_{10}\left(\frac{119}{91}\right) - \log_{10}\left(\frac{13}{32}\right) - \log_{10}\left(\frac{64}{39}\right) = ?$ **(CDS 2001)**

Sol. G.E. = $\log_{10}\left(\frac{26}{51} \times \frac{119}{91} \times \frac{32}{13} \times \frac{39}{64}\right) = \log_{10} 1 = 0$

10. $\log\left(\frac{16}{15}\right) + \log\left(\frac{5}{8}\right) - \log\left(\frac{2}{3}\right) = ?$

Sol. $\log\left(\frac{16}{15}\right) + \log\left(\frac{5}{8}\right) - \log\left(\frac{2}{3}\right) = \log\left(\frac{16}{15} \times \frac{5}{8} \times \frac{3}{2}\right) = \log 1 = 0$

11. What is the value of 2 log (5/8) + log (128/125) + log (5/2) ? **(CDS 2009)**

(*a*) 0 (*b*) 1 (*c*) 2 (*d*) 5

Sol. (*a*) $\log\left(\frac{5}{8}\right)^2 + \log\frac{128}{125} + \log\frac{5}{2} = \log\left(\frac{5}{8} \times \frac{5}{8} \times \frac{128}{125} \times \frac{5}{2}\right) = \log 1 = 0$

12. What is the value of $\frac{1}{2} \log_{10} 25 - 2 \log_{10} 3 + \log_{10} 18$? **(CDS 2010)**

(*a*) 2 (*b*) 3 (*c*) 1 (*d*) 0

Sol. (*c*) G.E. = $\log_{10} 25^{1/2} - \log_{10} 3^2 + \log_{10} 18 = \log_{10} \frac{5 \times 18}{3 \times 3} = \log_{10} 10 = 1$

13. What is the value of $\frac{1}{3} \log_{10} 125 - 2 \log_{10} 4 + \log_{10} 32 + \log_{10} 1$? **(CDS 2010)**

(*a*) 0 (*b*) $\frac{1}{5}$ (*c*) 1 (*d*) $\frac{2}{5}$

Sol. (*c*) G.E. $= \log_{10} 5 - \log_{10} 16 + \log_{10} 32 + \log_{10} 1 = \log_{10} \dfrac{5 \times 32 \times 1}{16} = \log_{10} 10 = 1$

14. If $\log x = \log 5 + 2 \log 3 - \dfrac{1}{2} \log 25$, find the value of x.

Sol. $\log x = \log 5 + 2 \log 3 - \dfrac{1}{2} \log 25 \Rightarrow \log x = \log 5 + \log 3^2 - \log 25^{\frac{1}{2}}$

$\Rightarrow \log x = \log 5 + \log 9 - \log 5 \Rightarrow \log x = \log 9 \Rightarrow x = 9$

15. $\log 25 \div \log 5 = ?$

Sol. $\log 25 \div \log 5 = \log 5^2 \div \log 5 = 2 \log 5 \div \log 5 = 2$

16. $\dfrac{\log \sqrt{5}}{\log 5} = ?$

Sol. $\dfrac{\log \sqrt{5}}{\log 5} = \log 5^{\frac{1}{2}} \div \log 5 = \dfrac{1}{2} \log 5 \div \log 5 = \dfrac{1}{2}$

17. $\log 72 = ?$

Sol. $\log 72 = \log (2 \times 2 \times 2 \times 3 \times 3) = \log (2^3 \times 3^2) = \log (2^3) + \log (3^2) = 3 \log 2 + 2 \log 3$

18. $\log 20 = ?$

Sol. $\log 20 = \log \left(\dfrac{100}{5}\right) = \log 100 - \log 5 = \log (10^2) - \log 5 = 2 \log 10 - \log 5 = 2 - \log 5$

Alternative Solution : $\log 20 = \log (10 \times 2) = \log 10 + \log 2 = 1 + \log 2$

19. If $\log 2 = 0.3010$, then $\log 5 = ?$

Sol. $\log 5 = \log \dfrac{10}{2} = \log 10 - \log 2 = 1 - 0.3010 = 0.6990$

20. If $\log_{10} 2 = 0.301$, then $\log_{10} 125 = ?$ **(CDS 2002)**

Sol. $\log_{10} 125 = \log_{10} \dfrac{1000}{8} = \log_{10} 1000 - \log_{10} 8 = \log_{10} (10^3) - \log_{10} 2^3 = 3 \log_{10} 10 - 3 \log 2$

$= 3(1) - 3(0.301) = 3 - 0.903 = 2.097$

21. If $\log 2 = 0.3010$, $\log 3 = 0.4771$, then $\log 15 = ?$

Sol. $\log 15 = \log \left(\dfrac{10 \times 3}{2}\right) = \log 10 + \log 3 - \log 2 = 1 + 0.4771 - 0.3010 = 1.1761$

22. If $\log 3 = 0.4771$, then $\log (30)^4$

Sol. $\log (30)^4 = 4 \log 30 = 4 \log (10 \times 3) = 4 (\log 10 + \log 3)$

$= 4 (1 + \log 3) = 4 + 4 \log 3 = 4 + 4 (0.4771) = 4 + 1.9084 = 5.9084$

23. If $\log_{10}2 = 0.3010$ and $\log_{10}7 = 0.8451$, then the value of $\log_{10}2.8$ is **(SSC CGL 1999)**

(*a*) 0.4471 (*b*) 1.4471 (*c*) 2.4471 (*d*) 4.4471

Sol. (*a*) $\log_{10}2.8 = \log_{10}(2 \times 2 \times 7 \div 10) = \log_{10}2 + \log_{10}2 + \log_{10}7 - \log_{10}10$

$= 0.3010 + 0.3010 + 0.8451 - 1 = 1.4471 - 1 = 0.4471$

24. If log 27 = 1.431, log 9 = ?

Sol. log 27 = log 3^3 = 3 log 3 $\therefore \log 3 = \frac{\log 27}{3} = \frac{1.431}{3} = 0.477$

Now log 9 = log 3^2 = 2 log 3 = 2 × 0.477 = 0.954

25. If log 8 = 0.9031, and log 9 = 0.9542, then log 6 = ?

Sol. $\log 8 = \log 2^3 = 3 \log 2 \Rightarrow \log 2 = \frac{\log 8}{3} = \frac{0.9031}{3} = 0.3010$

$\log 9 = \log 3^2 = 2 \log 3 \Rightarrow \log 3 = \frac{\log 9}{2} = \frac{0.9542}{2} = 0.4771$

log 6 = log (2 × 3) = log 2 + log 3 = 0.3010 + 0.4771 = 0.7781

26. If log 72 = 1.8573 and log 288 = 2.4594, then log 2 = ?

Sol. $\log\left(\frac{288}{72}\right) = \log 288 - \log 72 = 2.4594 - 1.8573 = 0.6021$

$\Rightarrow \log 4 = 0.6021 \Rightarrow 2 \log 2 = 0.6021 \Rightarrow \log 2 = \frac{\log 4}{2} = \frac{0.6021}{2} = 0.30105$

27. $\log_{10} 2 = 0.3010$, then $\log_2 10$ = ? **(SSC CGL 2000, RRB 2002)**

(*a*) 0.3010 (*b*) 0.6990 (*c*) $\frac{1000}{301}$ (*d*) $\frac{699}{301}$

Sol. (*c*) $\log_2 10 = \frac{1}{\log_{10} 2} = \frac{1}{0.3010} = \frac{1000}{301}$

28. Find the value of $\log_8 16$.

(*a*) 2 (*b*) $\frac{1}{2}$ (*c*) $\frac{4}{3}$ (*d*) $\frac{3}{4}$

Sol. (*c*) Let $\log_8 16 = x \Rightarrow 16 = 8^x \Rightarrow 2^4 = 2^{3x} = 3x = 4 \Rightarrow x = \frac{4}{3}$

Alternative Method : $\log_8 16 = \frac{\log 16}{\log 8} = \frac{\log 2^4}{\log 2^3} = \frac{4 \log 2}{3 \log 2} = \frac{4}{3}$

29. What is the value of $2 \log_8 2 - \frac{1}{3} \log_3 9$? **(NDA 2012)**

(*a*) 0 (*b*) 1 (*c*) 2 (*d*) 1/3

Sol. (*a*) G.E. $= 2 \times \frac{1}{3} - \frac{1}{3} \times 2 = \frac{2}{3} - \frac{2}{3} = 0$

30. Find the value of $\log_2 3 \times \log_9 4$?

(*a*) 0 (*b*) 1 (*c*) 2 (*d*) 1/3

Sol. (*b*) $\log_2 3 \times \log_9 4 = \frac{\log 3}{\log 2} \times \frac{\log 4}{\log 9} = \frac{\log 3}{\log 2} \times \frac{\log 2^2}{\log 3^2} = \frac{\log 3}{\log 2} \times \frac{2 \log 2}{2 \log 3} = 1$

31. If log $(a + b)^2$ = log 9 + log a + log b, then $a^2 + b^2$ = ?

Sol. $\log (a + b)^2 = \log 9 + \log a + \log b \Rightarrow \log (a + b)^2 = \log 9ab$

$\Rightarrow (a + b)^2 = 9ab \Rightarrow a^2 + b^2 + 2ab = 9ab \Rightarrow a^2 + b^2 = 9ab - 2ab = 7ab$

32. If $\log x + \log(x-9) = 1$; find the value of x.

Sol. $\log x + \log(x-9) = 1 \Rightarrow \log[(x)(x-9)] = \log 10 \Rightarrow x^2 - 9x - 10 = 0 \Rightarrow x = 10, -1$

$\Rightarrow x = 10, -1$ ($\because x = -1$ is not valid number)

33. What are the roots of the equation $\log_{10}(x^2 - 6x + 45) = 2$? **(CDS 2010)**

(*a*) 9, –5 (*b*) –9, 5 (*c*) 11, –5 (*d*) –11, 5

Sol. (*c*) $\log_{10}(x^2 - 6x + 45) = 2 = \log_{10} 100 \Rightarrow x^2 - 6x + 45 = 100 \Rightarrow x^2 - 6x - 55 = 0$

$\Rightarrow x^2 + 5x - 11x - 55 = 0 \Rightarrow x(x+5) - 11(x+5) = 0 \Rightarrow (x-11)(x+5) = 0 \Rightarrow x = 11, -5$

34. If $\log_2(x^2 - 4x + 7) = 2$, find the value of x.

Sol. $\log_2(x^2 - 4x + 7) = 2 = \log_2 4 \Rightarrow x^2 - 4x + 7 = 4 \Rightarrow x^2 - 4x + 3 = 0$

$\Rightarrow x^2 - x - 3x + 3 = 0 \Rightarrow x(x-1) - 3(x-1) = 0 \Rightarrow (x-1)(x-3) = 0 \Rightarrow x = 1, 3$

35. If $\log 9 = 0.9542$, how many digits are there in 9^{20}?

Sol. 9^{20} = Antilog ($\log 9^{20}$) = Antilog (20 log 9) = Antilog (20 × 0.9542) = Antilog 19.084

In Antilog of 19.084, decimal will be after 19 + 1 = 20 digits from the left side.

$\therefore$ 9^{20} has 20 digits.

36. If log 2 = 0.3010, then number of digits in 2^{100} is **(CDS 2001)**

(*a*) 29 (*b*) 30 (*c*) 31 (*d*) 32

Sol. (*c*) 2^{100} = Antilog ($\log 2^{100}$) = Antilog (100 log 2) = Antilog (100 × 0.3010) = Antilog 30.10

In Antilog of 30.10, decimal will be after 30 + 1 = 31 digits from the left side.

$\therefore$ 2^{100} has 31 digits.

37. If $\log_{10} 3 = 0.4771$, the number of digits in 3^{30} is

(*a*) 13 (*b*) 16 (*c*) 14 (*d*) 15

Sol. (*d*) 3^{30} = Antilog ($\log 3^{30}$) = Antilog (30 log 3) = Antilog (30 × 0.4771) = Antilog 14.313

In Antilog of 14.313, decimal will be after 14 + 1 = 15 digits from the left side.

$\therefore$ 3^{30} has 15 digits.

38. If $\log_2[\log_3(\log_4 x) = 0$, find the value of x.

(*a*) 9 (*b*) 24 (*c*) 32 (*d*) 64

Sol. (*d*) $\log_2[\log_3(\log_4 x) = 0 = \log_2 1 \Rightarrow [\log_3(\log_4 x) = 1 = \log_3 3$

$\Rightarrow (\log_4 x) = 3 \Rightarrow x = 4^3 = 64$

CHAPTER

29

MENSURATION – TWO DIMENSIONAL

Two Dimensional figure: A figure which has two dimensions viz. length and breadth.

Area: Area of a two-dimensional figure is total surface enclosed within the boundaries of the figure. Area of a figure is always measured in square units.

Perimeter: Perimeter of a two-dimensional figure is the total length of the sides enclosing the figure.

Unit of measurement :

10 millimetres (mm) = 1 centrimetre (cm)   10 centrimetre (cm) = 1 decimetre (dm)

10 decimetre (dm) = 1 metre (m)   1 kilometre (km) = 1000 metre (m)

1 Acre = 100 m^2 1 hectare = 100 acres = 10,000 m^2

1. **Quadrilateral**: A two-dimensional figure enclosed by four straight lines. Sum of internal angles is 360°.

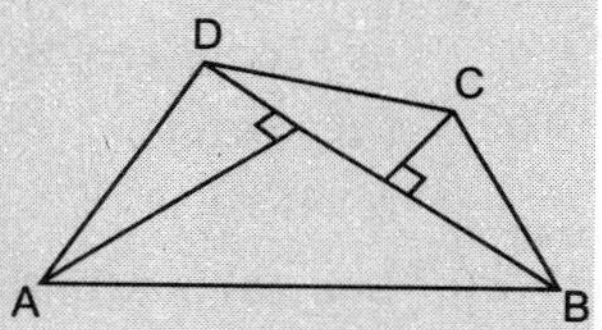

Area $= \frac{1}{2} \times$ One of the diagonal × Sum of lengths of perpendiculars drawn from the opposite vertices to that diagonal.

Example. In a quadrilateral, length of one diagonal is 15 cm and the lengths of the perpendiculars drawn from the opposite vertices are 7 cm and 9 cm. Find the area of the quadrilateral.

(*a*) 30 cm^2 (*b*) 60 cm^2 (*c*) 90 cm^2 (*d*) 120 cm^2

Sol. (*d*) Area of quadrilateral $= \frac{1}{2} \times 15 \times (7 + 9) = 120$ cm^2

2. **Trapezium**: A quadrilateral whose one pair of opposite sides are parallel.

Area $= \frac{1}{2} \times$ Sum of parallel sides × Perpendicular distance between them $= \frac{1}{2}(a + b) \times h$

Where a and b are the length of parallel sides, and h is the perpendicular distance between the parallel sides.

Example. Two parallel sides of a trapezium are 4 cm and 5 cm respectively. The perpendicular distance between the parallel sides is 6 cm. Find the area of the trapezium.

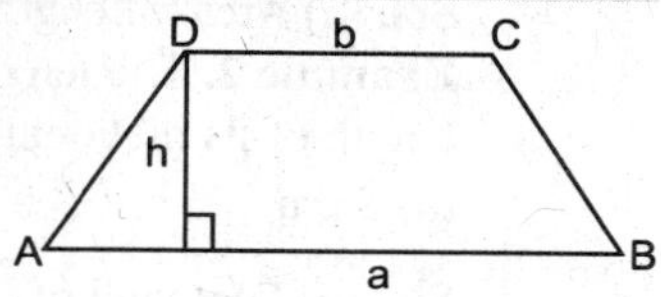

(*a*) 27 cm^2 (*b*) 30 cm^2 (*c*) 40 cm^2
(*d*) 54 cm^2

Sol. (*a*) Area of the trapezium $= \frac{1}{2} \times 6 \times (4 + 5) = 27$ cm^2

Isosceles Trapezium: If the non-parallel sides are equal in length. Then the length of its diagonals will also be equal.

3. **Parallelogram**: A quadrilateral whose opposite sides are equal and parallel. Its diagonals bisect each other. AO = OC and BO = OD Each diagonal divides it into two congruent triangles. Diagonals are of unequal lengths i.e. AC ≠ BD. Diagonals bisect the vertex angles. Sum of any two adjacent angles is 180°. Opposite angles are equal.

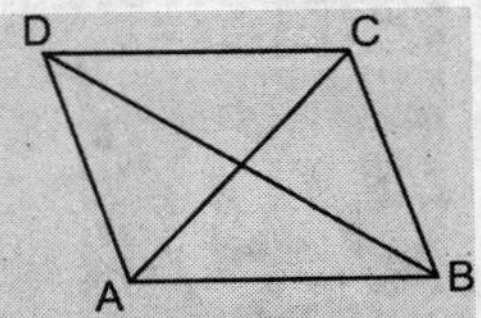

Area = Base × Height = 2 ΔABC = 2 ΔADC = 2ΔABD = 2ΔBCD

Sum of squares of the diagonals of a parallelogram is equal to the sum of the squares of its sides.

$AB^2 + BC^2 + CD^2 + DA^2 = AC^2 + BD^2 \Rightarrow AC^2 + BD^2 = 2(AB^2 + BC^2)$

4. **Rhombus**: A quadrilateral whose all sides are equal and opposite sides are parallel. Its diagonals bisect each other at 90° but are unequal in lengths. AC ≠ BD. A parallelogram is rhombus if its diagonals are perpendicular to each other.

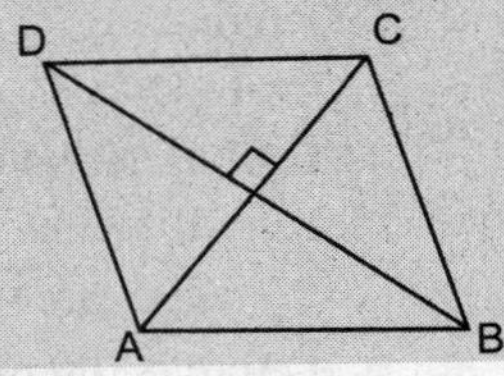

Area = Base × Height

Area = $\frac{1}{2}$ × Product of two diagonals = $\frac{1}{2}$ AC× BD

$4AB^2 = AC^2 + BD^2 \Rightarrow 4(\text{side})^2$ = sum of square of diagonals

Example. The perimeter of a rhombus is 40 m and its height is 5 m. Its area is : **(SSC CGL Tier I 2003)**

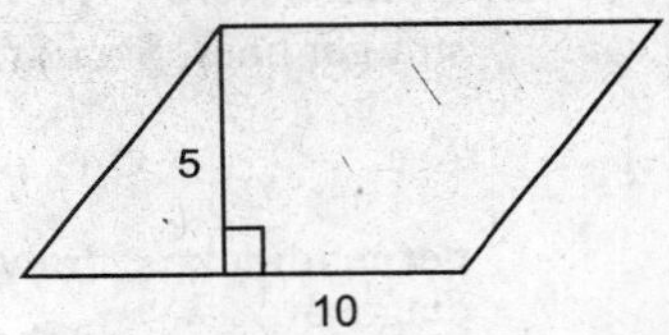

(*a*) 60 sq. m. (*b*) 50 sq. m.
(*c*) 45 sq. m. (*d*) 55 sq. m.

Sol. (*b*) Side of rhombus = 40 ÷ 4 = 10 m, Height = 5m.

Area = 10 × 5 = 50 m^2

5. **Rectangle**: A quadrilateral whose opposite sides are equal and parallel. All the four angles are of 90°. Its diagonals are equal in lengths and bisect each other (but not at right angle).

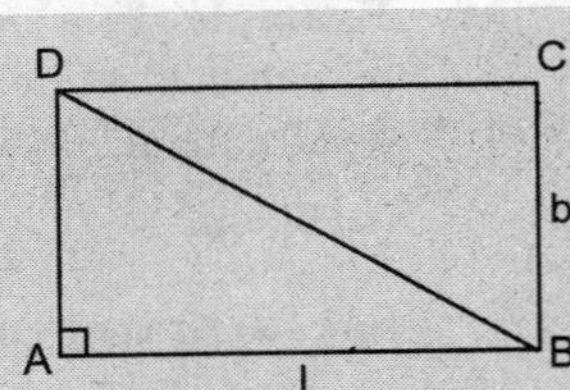

Perimeter = 2 (Length + Breadth)

Area = Length × Breadth

Diagonal = $\sqrt{\text{Length}^2 + \text{Breadth}^2}$

Example 1. The length and breadth of a room are 6 metres and 4 metres respectively. Find floor area of the room.

(*a*) 24 m^2 (*b*) 24 metre (*c*) 27 metre (*d*) 27 m^2

Sol. (*d*) Area = Length × Breadth = 6 × 4 = 24 m^2

Example 2. The length and breadth of a rectangle are 8 cm and 6 cm respectively. Find the length of its diagonal.

(*a*) 5 cm (*b*) 7 cm (*c*) 10 cm (*d*) 48 cm^2

Sol. (*c*) Diagonal of rectangle = $\sqrt{l^2 + b^2} = \sqrt{8^2 + 6^2} = \sqrt{100} = 10$ cm

6. **Square**: A quadrilateral whose all sides are equal and all the angles are of 90°. Its diagonals are equal in length and bisect each other at 90°. Converse also true.

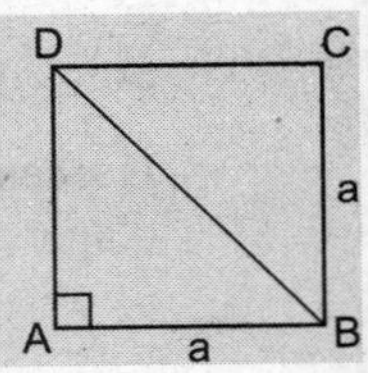

Perimeter = 4 × Side

Area = $(\text{Side})^2 = \frac{1}{2}$ $(\text{Diagonal})^2$

Diagonal = $\sqrt{2} \times$ Side = $\sqrt{2 \times \text{Area}}$

Example 1. If side of a square is 4 cm. Find the length of its diagonal.

(*a*) 4 cm (*b*) $4\sqrt{2}$ cm (*c*) 8 cm (*d*) 16 cm

Sol. (*b*) Diagonal of a square = $\sqrt{2}$ × side = $4\sqrt{2}$ cm

Example 2. If diagonal of a square is 12 cm, find its area.

(*a*) 36 cm^2 (*b*) 72 cm^2 (*c*) 144 cm^2 (*d*) 288 cm^2

Sol. (*b*) Area of square = $\frac{1}{2} \times 12 \times 12 = 72$ cm^2

7. Triangle: A closed figure formed by 3 line-segments

Sum of internal angles of a triangle is 180°. $\angle A + \angle B + \angle C = 180°$

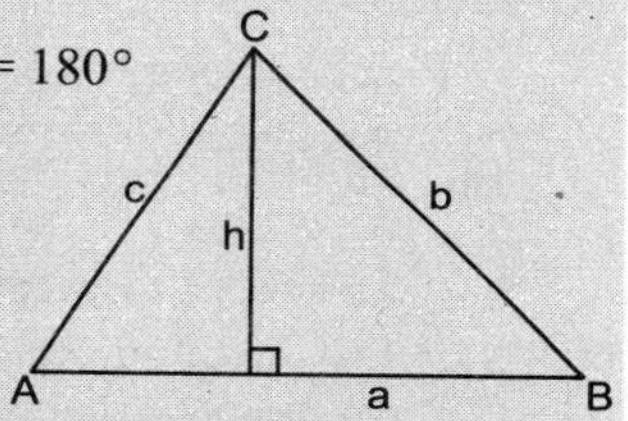

Area = $\frac{1}{2}$ × Base × Height

Area = $\sqrt{s\,(s-a)\,(s-b)\,(s-c)}$ (Hero's formula)

Where $s = \frac{1}{2}$ $(a + b + c)$ and a, b, c are the lengths of the three sides of the triangle.

Example 1. Find the area of a triangle whose base is 8 cm and height is 5 cm.

(*a*) 13 cm^2 (*b*) 20 cm^2 (*c*) 40 cm^2 (*d*) 80 cm^2

Sol. (*b*) Area of triangle = $\frac{1}{2} \times 8 \times 5 = 20$ cm^2

Example 2. The area of a triangular field, the length of whose sides are 50 m, 78 m and 112 m, is: **(SSC CGL Tier I 2006)**

(*a*) 1480 m^2 (*b*) 1600 m^2 (*c*) 1680 m^2 (*d*) 3360 m^2

Sol. (*c*) $s = \frac{50+78+112}{2} = \frac{240}{2} = 120$ cm

$\therefore$ Area = $\sqrt{120\times(120-50)\times(120-78)\times(120-112)} = \sqrt{120\times70\times42\times8}$

$= \sqrt{4\times3\times10\times7\times10\times7\times2\times3\times2\times4} = 2 \times 3 \times 10 \times 7 \times 2 \times 2 = 1680$ m^2

8. Isosceles Triangle: A triangle whose two sides are equal.

In isosceles triangles two angles are opposite to equal sides, also equal.

AB = AC and $\angle$ B = $\angle$ C.

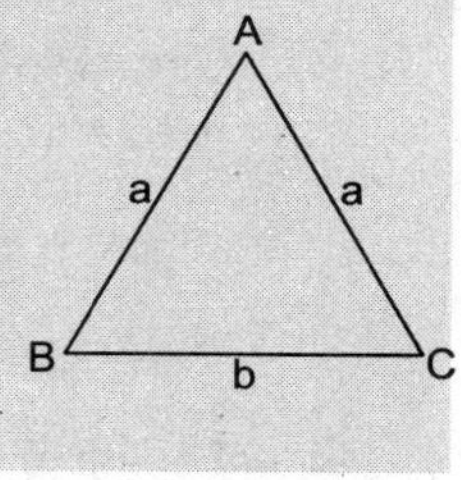

Area = $\frac{b}{4} \times \sqrt{4a^2 - b^2}$

where b is the base of the triangle and a is length of one of the two equal sides.

9. Equilateral Triangle: A triangle whose all sides are equal. Each angle is of 60° in an equilateral triangle.

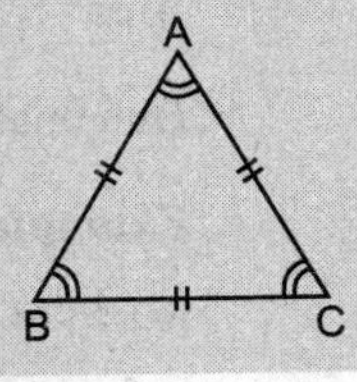

Area of the equilateral triangle = $\frac{\sqrt{3}}{4} \times (\text{Side})^2$

Altitude/Height of the equilateral triangle = $\frac{\sqrt{3}}{2} \times \text{side}$

Example. The area of an equilateral triangle is $400\sqrt{3}$ sq.m. Its perimeter is :

(SSC CGL Tier I 2003)

(*a*) 120 m (*b*) 150 m (*c*) 90 m (*d*) 135 m

Sol. (*a*) Area of the equilateral triangle = $\frac{\sqrt{3}}{4} \times (\text{Side})^2 = 400\sqrt{3}$

$\Rightarrow (\text{Side})^2 = 400\sqrt{3} \times \frac{4}{\sqrt{3}} = 1600 \Rightarrow \text{Side} = \sqrt{1600} = 40$ m

∴ Perimeter = 3 × 40 m = 120 m

10. Right angle triangle: A triangle whose one angle is of 90°.

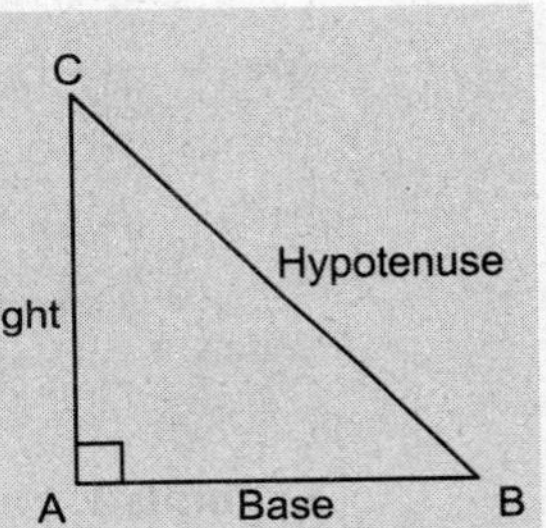

Area = $\frac{1}{2}$ × Base × Perpendicular Height

By Pythagorean Theorem : In a right triangle, the square of length of its hypotenuse is equal to the sum of squares of lengths of two other sides.

$(\text{Hypotenuse})^2 = (\text{Base})^2 + (\text{Height})^2$

If ABC is an obtuse triangle,

then $AC^2 = AB^2 + BC^2 + 2BC.BD$

Note: $AC^2 > AB^2 + BC^2$

If ABC is an acute angled triangle, then

$AC^2 = AB^2 + BC^2 - 2BC.BD$

Note: $AC^2 < AB^2 + BC^2$

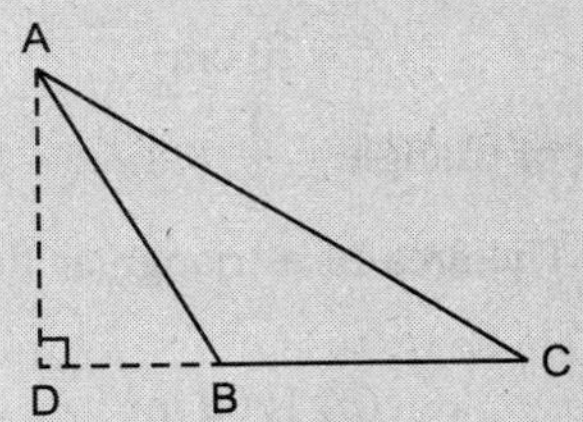

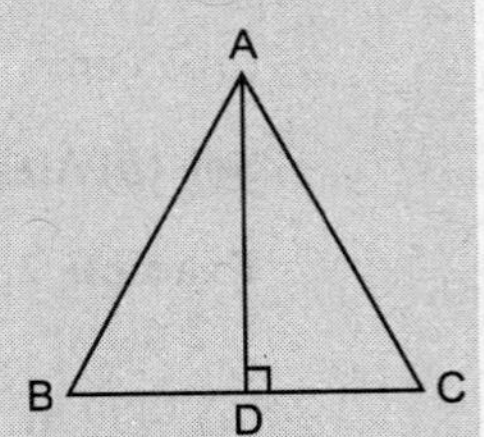

Triangle ABC is

Right triangle, if $a^2 + b^2 = c^2$

Obtuse triangle, if $a^2 + b^2 < c^2$

Acute triangle, if $a^2 + b^2 > c^2$

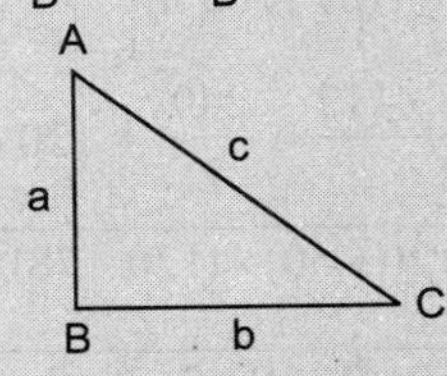

11. Circle: Circle is a set of points in plane that are located in the same distance from a fixed point *i.e.* center of the circle.

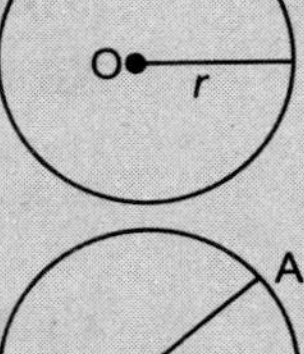

Diameter = $2r$

Circumference = $2\pi r$ = Diameter × π

Area = πr^2

Where $\pi = \frac{22}{7}$ or 3.1416 (approx)

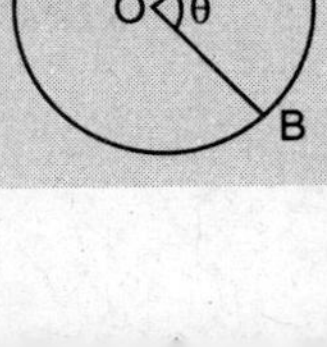

An Arc is a portion of circumference of the circle.

Length of Arc AB = $\frac{\theta}{360} \times 2\pi r$

Area of a sector AOB $= \frac{\theta}{360} \times \pi r^2$

Example. If diameter of a circle is 28 cm, then area of the circle is:

(*a*) 77 cm^2 (*b*) 154 cm^2 (*c*) 308 cm^2 (*d*) 616 cm^2

Sol. (*d*) Radius of the circle $= \frac{1}{2} \times 28$ cm $= 14$ cm

$\therefore$ Area of the circle $= \frac{22}{7} \times 14 \times 14 = 616$ cm^2

12. If **ratio of sides** of two two-dimensional figures (such as circle, square) $= x : y$. Then **ratio of their areas** $= x^2 : y^2$.

Example. If all the sides of a square are doubled. Ratio of the area of new square and that of the original square is:

(*a*) 1 : 1 (*b*) 2 : 1 (*c*) 4 : 1 (*d*) Data Insufficient

Sol. (*c*) Ratio of sides of two squares = 2 : 1 $\Rightarrow$ Ratio between their areas $= 2^2 : 1^2 = 4 : 1$

13. If **length** of rectangle **is increased by** x**%** and **breadth by** y**%**, then percent **increase in its area** $= x + y + \frac{xy}{100}$

Alternative Method: $\frac{(100+x)\times(100+y)}{100} - 100$

Example. If the length of a rectangle is increased by 30% and width by 20%, its area will increase by:

(*a*) 50% (*b*) 25% (*c*) 56% (*d*) 60%

Sol. (*c*) Increase in Area $= 30 + 20 + \frac{30 + 20}{100} = 56\%$

Alternative Method :

Let each side of the rectangle be 10 units. Then its original area = 10 × 10 = 100 sq. units.

New area = 13 × 12 = 156 sq. units $\Rightarrow$ Percent increase in area = 156 – 100 = 56%

14. If **all the sides** of a two-dimensional figure are **increased by** x**%**, then percent **increase in its area** $= x + x + \frac{x^2}{100} = 2x + \frac{x^2}{100}$

Alternative Method: $\frac{(100+x)^2}{100} - 100$

Example. The percentage increase in the area of a rectangle, if each of its sides is increased by 20%, is: **(SSC CGL Tier I 2004)**

(*a*) 40% (*b*) 42% (*c*) 44% (*d*) 46%

Sol. (*c*) Required percentage $= 20 + 20 + \frac{20 \times 20}{100} = 44\%$

Alternative method :

Let each side of the rectangle be 10 units. Then its original area = 10 × 10 = 100 sq. units

New area = 12 × 12 = 144 sq. units. $\Rightarrow$ Percent increase in area = 144 – 100 = 44%

15. If all sides of a quadrilateral are increased by $x\%$, then its corresponding diagonals will also increase by $x\%$.

16. Ratio between area of a square and area of the square drawn on its diagonal is 1 : 2. **(SSC CGL Tier I 2005)**

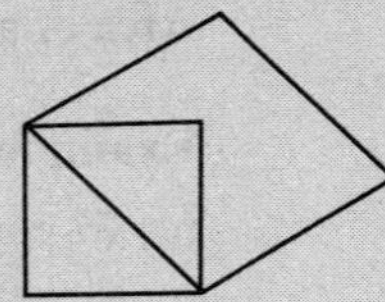

Proof : Let side of the small square = 1 cm

Then side of the bigger square = Diagonal of the smaller square = $\sqrt{2}$ cm

$\therefore$ Ratio between the areas = $(1)^2 : (\sqrt{2})^2 = 1 : 2$

17. Area of a square inscribed in a circle of radius r is $2r^2$.

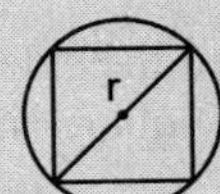

Proof : Diagonal of square = Diameter of circle = $2r$

Area of square = $\frac{1}{2}(\text{Diagonal})^2 = \frac{1}{2}(2r)^2 = 2r^2$

18. Area of a square circumscribing a circle is double to that of the one inscribed in the circle.

Proof : Let the side of square circumscribing the circle = a. Then diameter of the circle = a.
Diagonal of inner square = Diameter of circle = a

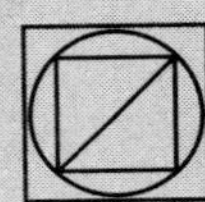

$\therefore$ Area of inner square = $\frac{1}{2}(a^2)$

$\therefore$ Ratio between areas of two squares = $a^2 : \frac{1}{2}a^2 = 2 : 1$

19. Area of a circle circumscribing a square is double to that of one inscribed in the square. **(SSC CGL Tier II 2012)**

Proof : Let radius of inner circle = r. Then side of square = $2r$.
Diagonal of square = $2\sqrt{2}r$

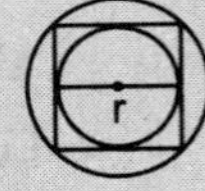

$\therefore$ Radius of outer circle = $\frac{1}{2} \times 2\sqrt{2}r = \sqrt{2}r$. Area of inner circle = πr^2

Area of outer circle = $\pi(\sqrt{2}r)^2 = 2\pi r^2$

$\therefore$ Ratio between area of two circles = $2\pi r^2 : \pi r^2 = 2 : 1$

20. Area of the largest circle inscribed in a square of side a is $\pi\left(\frac{a}{2}\right)^2$.

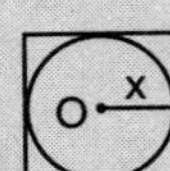

Ratio of areas of square and circle is $a^2 : \pi\left(\frac{a}{2}\right)^2 = 4 : \pi = 14 : 11$

Difference in areas = $\frac{3}{14}$ of area of square = $\frac{3}{11}$ of area of circle = $\frac{6}{7}r^2 = \frac{3}{14}a^2$

Proof : Let radius of circle = x. Then side of square = $2x$.
Area of square = $(2x^2) = 4x^2$ Area of circle = πx^2
$\therefore$ Area of square which is not covered by the circle
$= 4x^2 - \pi x^2 = x^2(4 - \pi) = x^2\left(4 - \frac{22}{7}\right) = \frac{6}{7}x^2$

Example. What is the area of the shaded portion if the side of the square measures 21 cms ? **(Dena Bank PO 2006)**

(*a*) 86.5 sq. cms (*b*) 102 sq. cms (*c*) 94.5 sq. cms (*d*) 81.5 sq. cms

Sol. (*c*) Area of square = $(\text{Side})^2 = (21)^2 = 441$ sq. cms

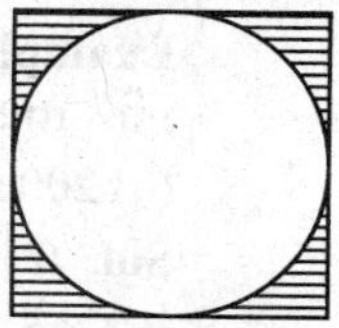

Radius of circle = $\frac{21}{2}$ cm

Area of circle = $\pi r^2 = \frac{22}{7}\times\frac{21}{2}\times\frac{21}{2} = \frac{693}{2} = 346.5$ sq. cm.

Area of the shaded portion = 441 – 346.5 = 94.5 sq. cm.

Alternative method :

Ratio of areas of square and circle = 14 : 11

Difference in areas = $\frac{3}{14}$ of area of square = $\frac{3}{14} \times 21 \times 21 = \frac{189}{2} = 94.5 \text{ cm}^2$

21. If four circles of equal radius are drawn at the each corner of a square, then the area of the square which is not covered by any of the circles is, $\frac{6}{7}x^2$, where x is radius of the circle.

Proof: Let side of square = $2x$. Then radius of one circle = x

Area of square = $(2x^2) = 4x^2$. Area of one circle = πx^2

Area of one circle which is covered by square = $\frac{1}{4}\pi x^2$

$\therefore$ Area covered by 4 circles = $4 \times \frac{1}{4}\pi x^2 = \pi x^2$

$\therefore$ Area of square which is not covered by any of the circle

$= 4x^2 - \pi x^2 = x^2 (4 - \pi) = x^2\left(4 - \frac{22}{7}\right) = \frac{6}{7}x^2$

22. Area of **uniform path outside the rectangular field** surrounding it:

= 2 × Width of path × [Length of field + Breadth of field + 2 (Width of path)]

or = Width × Perimeter of the field + 4 × Width^2

Example 1. A lawn 30 m long and 16 m wide is surronded by a path 2 m wide. What is the area of the path ? **(CDS 2011)**

(*a*) 200 m^2 (*b*) 280 m^2 (*c*) 300 m^2 (*d*) 320 m^2

Sol. (*a*) Required area = $2 \times 2 \times (30 + 16) + 4 \times 2^2 = 2 \times 2 \times 46 + 16 = 200 \text{ m}^2$

Alt: Area of lawn = $30 \times 16 = 480 \text{ m}^2$. Area of outer rectangle = $34 \times 20 = 680 \text{ m}^2$

Area of surrounding path = $680 - 480 = 200 \text{ m}^2$

Example 2. A field is 40 metre long and 35 metre wide. The field is surrounded by a path of uniform width of 2.5 metre runs round it on the outside. Find the area of the path.

(*a*) 250 m^2 (*b*) 400 m^2 (*c*) 500 m^2 (*d*) 600 m^2

Sol. (*b*) Area of path = 2 × Width × [Length + Breadth + (2 × Width)]

$= 2.5 \times 2 \times (40 + 35 + 2 \times 2.5) = 5 \times (75 + 5) = 400 \text{ m}^2$

Alternative Method : Required area = Width × Perimeter + 4 × Width^2

$= 2.5 \times 2 \times (40 + 35) + 4 \times (2.5)^2 = 5 \times 75 + 10 \times 2.5 = 375 + 25 = 400 \text{ m}^2$

Alternative Method : $45 \times 40 - 40 \times 35 = 1800 - 1400 = 400$
or $45 \times 40 - 40 \times 35 = 40 \times (45 - 35) = 400$

23. Area of **uniform path inside a rectangular field**, surrounded by the field:
$= 2 \times$ Width of path $\times$ [Length of field + Breadth of field – 2 (Width of path)]
$=$ Width $\times$ Perimeter $- 4 \times$ Width^2

Example : A field is 100 metre long and 60 metre wide. A path of uniform width of 5 metre runs round it on the inside. Find the area of the path.
(*a*) 200 m^2 (*b*) 800 m^2 (*c*) 1500 m^2 (*d*) 1600 m^2

Sol. (*c*) Area of path $= 2 \times$ Width $\times$ [Length + Breadth – (2 $\times$ Width)]
$= 2 \times 5 \times (100 + 60 - 10) = 10 \times 150 = 1500$ m^2

Alternative Method : Required area = Width $\times$ Perimeter $- 4 \times$ Width^2
$5 \times 2 \times (100 + 60) - 4 \times 5^2 = 1600 - 100 = 1500$ m^2

Alternative Method : $100 \times 60 - 90 \times 50 = 6000 - 4500 = 1500$

24. Area of **uniform path** of x metre width **running from the centre of each side of a rectangular field to the centre of opposite side**:
$=$ Width of path $\times$ (Length of field + Breadth of field) – (Width of path)^2

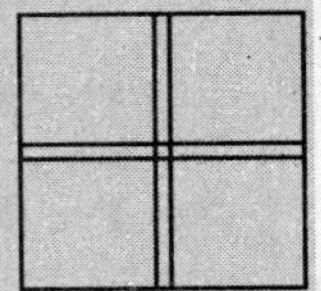

25. If a **square field** with x cm a side is **surrounded by a path of uniform width** of y cm. Then area of path $= 4y(x + y)$

Proof : Area of inner square $= x^2$, Side of outer square $= (x + 2y)$
Area of outer square $= (x + 2y)^2 = x^2 + 4xy + 4y^2$
$\therefore$ Area of path $= x^2 + 4xy + 4y^2 - x^2 = 4xy + 4y^2 = 4y(x + y)$

26. If a circle of radius r is surronded by a circle of radius R . Then area of path $= \pi(R^2 - r^2)$

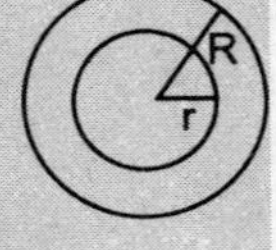

27. Area of regular polygon $= \dfrac{n}{4}(\text{side})^2 \times \cot\dfrac{\pi}{n}$

Area of regular hexagon $= \dfrac{1}{4} \times 6(\text{side})^2 \times (\sqrt{3}) = \dfrac{3\sqrt{3}}{2}(\text{side})^2$

Example: The area of a regular hexagon with side 10 cm is : **(SSC CGL Tier I 2006)**
(*a*) $50\sqrt{3}$ cm^2 (*b*) $150\sqrt{3}$ cm^2 (*c*) 50 cm^2 (*d*) 300 cm^2

Sol. (*b*) Area $= \dfrac{3\sqrt{3}}{2} \times 10^2 = 150\sqrt{3}$ cm^2.

28. Difference between the areas of two circles whose circumferences are x and y cm respectively
$= \dfrac{1}{4x}(x^2 - y^2) = \dfrac{1}{4x}(x - y)(x + y)$

Example: The area of ring between two concentric circles, whose circumferences are 88 cm and 132 cm, is : **(SSC CGL Tier I 2000)**
(*a*) 780 cm^2 (*b*) 770 cm^2 (*c*) 715 cm^2 (*d*) 660 cm^2

Sol. (*b*) Area of ring $= \dfrac{1}{4\pi}(x + y)(x - y) = \dfrac{1}{4} \times \dfrac{7}{22} \times 220 \times 44 = 770$ cm^2

Alternative method : Let radii of the circles are R and r respectively.

Then $2\pi r = 88 \Rightarrow r = 88 \times \dfrac{1}{2} \times \dfrac{7}{22} = 14$ cm and $2\pi R = 132 \Rightarrow R = 132 \times \dfrac{1}{2} \times \dfrac{7}{22} = 21$ cm

$\Rightarrow$ Area of ring $= \pi (R^2 - r^2) = \pi (21^2 - 14^2) = \frac{22}{7} \times 35 \times 7 = 770 \text{ cm}^2$

29 (*i*) Parallelograms on the same base and between the same parallels are equal in area.

(*ii*) Triangles on the same base and between the same parallels are equal in area.

(*iii*) If a triangle and a parallelogram are on the same base and between the same parallels, then the area of the triangle is equal to the half of area of the parallelogram.

Diagonals of a parallelogram divide it into four triangles of equal area.

30. Find **difference between area of a square and that of a rectangle, when their perimeters are same.**

Sol. Let side of square = x. Then perimeter of square and of rectangle = $4x$

Area of square = x^2

Let length of square is increased by a and breadth is decreased by a.

Then dimensions of the rectangle are $(x + a)$ and $(x - a)$

Area of the rectangle = $(x + a)(x - a) = x^2 - a^2$

$\therefore$ Difference between areas of square and rectangle = $x^2 - (x^2 - a^2) = a^2$

We observe that:

1. Area of square is always more than that of a rectangle, when both have same perimeter.

2. Difference between the areas is equal to the square of increase or decrease in the dimensions.

Example. A square and a rectangle each has permiter 48 m. Difference between the areas of two figures is 4 m^2. The dimensions of the rectangle (in meter) are;

(*a*) 20, 4 (*b*) 18, 6 (*c*) 16, 8 (*d*) 14, 10

Sol. (*d*) Side of square = $48 \div 4 = 12$ m. Difference in areas = $4\text{m}^2 = (2\text{m})^2$

$\therefore$ Dimensions of rectangle = 12 ± 2 or 14 m and 10 m

31. Difference between area of a square and that of a circle, when their perimeter and circumference numerically are same, is $\frac{3}{11}x^2$, where x is the side of the square.

Note 1 : Ratio of Area of circle and square = $\frac{4}{\pi}x^2 : x^2 = 14 : 11$

Note 2: Circle has more area than that of square with same perimeter.

Proof : Let side of square = x. Then perimeter of square = $4x$ = circumference of circle

$\therefore$ Radius of circle = $\frac{4x}{2\pi} = \frac{2x}{\pi}$. Area of square = x^2 Area of circle = $\pi\left(\frac{2x}{\pi}\right)^2 = \frac{4}{\pi}x^2$

$\therefore$ Difference between areas of circle and square = $\frac{4}{\pi}x^2 - x^2$

$$= x^2\left(\frac{4}{\pi} - 1\right) = x^2\left(4 \times \frac{7}{22} - 1\right) = x^2\left(\frac{14}{11} - 1\right) = \frac{3}{11}x^2$$

Example. The perimeters of a square and a circular field are the same. If the area of the circular field is 3850 square metres, what is the area (in m^2) of the square ?

(SSC CGL Tier I 2000)

(*a*) 4225 (*b*) 3025 (*c*) 2500 (*d*) 2025

Sol. (*b*) Area of square = $3850 \times \frac{11}{14} = 3025 \text{ m}^2$

SOLVED EXERCISE

1. ABCD is a rhombus with diagonals AC and BD. Then which one among the following is correct? **(CDS 2012)**

(*a*) AC and BD bisect each other but not necessarily perpendicular to each other

(*b*) AC and BD are perpendicular to each other but not necessarily bisect each other

(*c*) AC and BD bisect each other and perpendicular to each other

(*d*) AC and BD neither bisect each other nor perpendicular to each other

Sol. (*c*) Statement (c) is correct.

2. Area of a square field is 729 square metre, find its perimeter.

(*a*) 27 metre (*b*) 54 metre (*c*) 81 metre (*d*) 108 metre

Sol. (*d*) Side of the square field = $\sqrt{729} = 27$ metre

$\therefore$ Perimeter of the field = 4×27 metre = 108 metre

3. What will be the cost of fencing a square field, whose side is 20 metre, if the cost of fencing is ₹ 10 per metre ?

(*a*) ₹ 200 (*b*) ₹ 400 (*c*) ₹ 800 (*d*) ₹ 4000

Sol. (*c*) Cost of fencing = Perimeter of the field × Cost per metre = (4 × 20) × ₹ 10 = ₹ 800

4. On incrreasing the diameter of a circle by 75%, the percentage increase in the perimeter is **(SSC CGL Tier I 2014)**

(*a*) 65% (*b*) 70% (*c*) 75% (*d*) 80%

Sol. (*c*) Same percentage increase in perimeter = 75%

5. If area of a circle is 154 cm², find its radius. **(RRB ASM 2001)**

(*a*) 7 cm (*b*) 14 cm (*c*) 21 cm (*d*) 28 cm

Sol. (*a*) Area of circle = $\pi r^2 = 154 \Rightarrow r^2 = 154 \times \frac{7}{22} = 49 \Rightarrow r = 7$ cm

6. A wall is of the form of a trapezium with height 4 m and parallel sides being 3 m and 5 m. What is the cost of painting the wall, if the rate of painting is ₹ 25 per square metre ? **(CDS 2011)**

(*a*) ₹ 240 (*b*) ₹ 400 (*c*) ₹ 480 (*d*) ₹ 800

Sol. (*b*) Area = $\frac{1}{2} \times 4 \times (3 + 5) \Rightarrow$ Cost = $\frac{1}{2} \times 4 \times (3 + 5) \times 25 =$ ₹ 400

7. If the longer side of a rectangle is doubled and the other reduced to half, the area of the new rectangle goes up by **(SSC CGL Tier I 2013)**

(*a*) 50% (*b*) 100% (*c*) 150% (*d*) 0%

Sol. (d) Let the sides be x and y. Then $2x \times \frac{y}{2} = xy \Rightarrow$ No change in area.

8. If the breadth of a rectangle is decreased by 50%, then to double its area, its length must be increased by : **(CDS 2003)**

(*a*) 150% (*b*) 300% (*c*) 200% (*d*) 400%

Sol. (b) $A_1 = l_1 b_1$ $\quad A_2 = l_2 \times \frac{b_1}{2} = 2\, l_1 b_1$

$\Rightarrow l_2 = 4\, l_1 \Rightarrow l_2 : l_1 = 4 : 1 \Rightarrow$ Required % = $\frac{4-1}{1} \times 100 = 300\%$

9. Length of a rectangle is 1 cm more than breadth. If perimeter of the rectangle is 14 cm, find its area.

(*a*) 14 cm^2 (*b*) 15 cm^2 (*c*) 16 cm^2 (*d*) 12 m^2

Sol. (*d*) Perimeter = $2 \times (l + b) = 14 \Rightarrow (l + b) = 7, \quad l - b = 1$

$\therefore l = \frac{7+1}{2} = 4$ and $b = \frac{7-1}{2} = 3.$ Area = $l \times b = 4 \times 3 = 12$

10. If the perimeter of an equilateral triangle is 42 cm, its area, in sq. cm, will be **(SSC CGL Tier I 2014)**

(*a*) $7\sqrt{3}$ (*b*) $14\sqrt{3}$ (*c*) $24\sqrt{3}$ (*d*) $49\sqrt{3}$

Sol. (*d*) Side of equilateral triangle = 42 ÷ 3 = 14 cm ⇒ Area = $\frac{\sqrt{3}}{4} \times 14 \times 14 = 49\sqrt{3}$

11. The area of a regular hexagon of side $2\sqrt{3}$ cm is **(SSC CGL Tier I 2004)**

(*a*) $18\sqrt{3}$ cm^2 (*b*) $12\sqrt{3}$ cm^2 (*c*) $36\sqrt{3}$ cm^2 (*d*) $27\sqrt{3}$ cm^2

Sol. (*a*) Area = $\frac{3\sqrt{3}}{2} \times (2\sqrt{3})^2 = \frac{3\sqrt{3}}{2} \times 12 = 18\sqrt{3}$ cm^2

12. If perimeter and area of a square are equal, side of the square (in cm) is:

(*a*) 1 cm (*b*) 2 cm (*c*) 4 cm (*d*) 8 cm

Sol. (*c*) Here (Side)^2 = 4 × (Side) ⇒ Side = 4 cm

13. From a rectangular sheet of sides 18 cm and 14, a semicircular portion with smaller side as diameter is taken out. Then the area of the remaining sheet will be **(CDS 2015)**

(*a*) 98 cm^2 (*b*) 100 cm^2 (*c*) 108 cm^2 (*d*) 175 cm^2

Sol. (*d*) Area of rectangular sheet = 18 × 14 = 252 cm^2

Area of semi-circular sheet = $\frac{1}{2}\pi r^2 = \frac{1}{2} \times \frac{22}{7} \times 7^2 = 77$

⇒ Required area = 252 – 77 = 175 cm^2

14. Area of a square plot is 200 m^2. Find the length of its diagonal.

(*a*) 20 metre (*b*) $40\sqrt{2}$ metre (*c*) 10 metre (*d*) 15 metre

Sol. (*a*) Diagonal = $\sqrt{2 \times \text{Area of square}} = \sqrt{2 \times 200} = 20$ metre

15. Find the ratio of areas of a square and square drawn on its diagonal, with a side equal to diagonal of the first square.

(*a*) 1 : 1 (*b*) 1 : $\sqrt{2}$ (*c*) 1 : 2 (*d*) 1 : 4

Sol. (*c*) Ratio of sides of squares = 1 : $\sqrt{2}$ ⇒ Ratio of areas = $1^2 : (\sqrt{2})^2 = 1 : 2$

16. A rectangular field has 6 km perimeter and 2 sq. km area. What will be the difference between length and breadth of the rectangle?

(*a*) 0.5 km (*b*) 1 km (*c*) 1.5 km (*d*) 2 km

Sol. (*b*) Perimeter = $2 \times (l + b) = 6 \Rightarrow (l + b) = 3$. Area = $l \times b = 2$

Easily we can tell that the numbers whose sum is 3 and product is 2 are 2 and 1.

Required difference = $l - b = 2 - 1 = 1$ km

17. The length of diagonal of a rectangle is 25 cm and its area is 168 cm^2. Find the length (in cm) of the rectangle.

(*a*) 31 (*b*) 24 (*c*) 17 (*d*) 7

Sol. (*b*) Diagonal = $\sqrt{l^2+b^2} = 25 \Rightarrow l^2 + b^2 = 625$. Area = $l \times b = 168$

168 is divisible by 24 $\Rightarrow$ length = 24, breadth = 168 ÷ 24 = 7. $24^2 + 7^2 = 25^2$

18. If the length of a rectangle is 8 cm and its diagonal is 10 cm, find its area.

(*a*) 64 cm^2 (*b*) 40 cm^2 (*c*) 48 cm^2 (*d*) 80 cm^2

Sol. (*c*) Breadth of the rectangle = $\sqrt{10^2 - 8^2} = 6$ cm

$\therefore$ Area of the rectangle = 6 × 8 = 48 cm^2

19. Radius of a circle is equal to diagonal of a square. If the area of the square is 18 sq. units, then the area of the circle is **(SSC FCI 2012)**

(*a*) 9π sq. units (*b*) 16 π sq. units (*c*) 36 π sq. units (*d*) 64 π sq. units

Sol. (*c*) Diagonal of square = $\sqrt{2 \times \text{Area}} = \sqrt{2 \times 18} = 6 \Rightarrow$ Area of circle = $\pi r^2 = \pi(6)^2 = 36\pi$

20. If the length of the diagonal of a square is 20 cm, its perimeter is : **(SSC CGL Tier I 2006)**

(*a*) $10\sqrt{2}$ cm (*b*) $20\sqrt{2}$ cm (*c*) $40\sqrt{2}$ (*d*) 80 cm

Sol. (*c*) Side of square = $20 \div \sqrt{2} = 10\sqrt{2}$. Perimeter of square = $4 \times 10\sqrt{2} = 40\sqrt{2}$

21. What will be the area of a circle with circumference equal to 22 cms ? **(Vijaya Bank PO 2008)**

(*a*) 41.2 sq. cms (*b*) 26.4 sq. cms (*c*) 38.5 sq. cms (*d*) None of these

Sol. (*c*) Circumference = $2\pi r$ = 22 cms $\Rightarrow r = 22 \times \frac{1}{2} \times \frac{7}{22} = \frac{7}{2}$

Area = $\pi r^2 = \frac{22}{7} \times \frac{7}{2} \times \frac{7}{2} = \frac{77}{2} = 38.5$ sq cms

22. The radius of a circle is 10 cm. What percentage the area of circle is of its circumference?

(*a*) 100% (*b*) 200% (*c*) 300% (*d*) 500%

Sol. (*d*) Ratio between area and circumference = $\pi r^2 : 2\pi r = r : 2 = 10 : 2 = 5 : 1$

$\therefore$ Area of the given circle is 5 times i.e. 500% of its circumference.

23. The area of a circle and its circumference are equal, find its diameter.

(*a*) 2 cm (*b*) 4 cm (*c*) 6 cm (*d*) 10 cm

Sol. (*b*) Ratio between area and circumference = $\pi r^2 : 2\pi r = r : 2$

Given, area and circumference are equal. $\Rightarrow r = 2$ cm $\Rightarrow$ Diameter = $2r$ = 4 cm

24. What is the circumference of a circle whose area is 36π cm^2 ? **(SSC CGL Tier I 2006)**

(*a*) 9π cm (*b*) 12π cm (*c*) 16π cm (*d*) 18π cm

Sol. (*b*) Area of circle = $\pi r^2 = 36\pi\ cm^2 \Rightarrow r^2 = 36 \Rightarrow r = 6$ cm

Circumference = $2\pi r = 2\pi(6) = 12\pi$ cm

25. The sides of a triangle are 25 cm, 39 cm and 56 cm. The perpendicular from the opposite vertex on the side of 56 cm is **(CDS 2015)**

(*a*) 10 cm (*b*) 12 cm (*c*) 15 cm (*d*) 16 cm

Sol. (*c*) $s = \frac{25+39+56}{2} = \frac{120}{2} = 60$ ∴ Area = $\sqrt{60 \times 35 \times 21 \times 4} = 420$

Required altitude = $\frac{2 \times 420}{56} = 15$ cm

26. A circle and a square have same area. The ratio of the side of the square and the radius of the circle is :

(*a*) $1 : \pi$ (*b*) $\pi : 1$ (*c*) $\sqrt{\pi} : 1$ (*d*) $1 : \sqrt{\pi}$

Sol. (*c*) $a^2 = \pi r^2 \Rightarrow \frac{a^2}{r^2} = \pi \Rightarrow a : r = \sqrt{\pi} : 1$

27. The length and perimeter of a rectangle are in the ratio 5 : 18. Then length and breadth will be in the ratio **(SSC CGL Tier I 2012)**

(*a*) 4 : 3 (*b*) 3 : 5 (*c*) 5 : 4 (*d*) 4 : 7

Sol. (*c*) Perimeter = 18 ⇒ (length + breadth) = 18 ÷ 2 = 9 ⇒ breadth = 9 – 5 = 4

28. The length and the breadth of a rectangle are in the ratio 3 : 2. The perimeter and the area of the rectangle are 5 : 6. What is the length of the rectangle? **(Bank Clerical 2006)**

(*a*) 3 units (*b*) 4 units (*c*) 6 units (*d*) None of these

Sol. (*c*) Let the length and the breadth of the rectangle be $3x$ and $2x$.

Perimeter = $2(3x + 2x) = 10x$ ⇒ Area = $3x \times 2x = 6x^2$

∴ $10x : 6x^2 = 5 : 6 \Rightarrow 10x \times 6 = 6x^2 \times 5 \Rightarrow 60x = 30x^2 \Rightarrow x = 2$

∴ Length of the rectangle = $3x = 3 \times 2 = 6$ units

29. If the perimeters of a rectangle and a square are equal and the ratio of two adjacent sides of the rectangle is 1 : 2 then the ratio of area of the rectangle and that of the square is **(SSC CGL Tier I 2013)**

(*a*) 1 : 1 (*b*) 1 : 2 (*c*) 2 : 3 (*d*) 8 : 9

Sol. (*d*) Let adjacent sides of rectangle be 4 and 2.

Then area = 4 × 2 = 8. Perimeter = 2 × (4 + 2) = 12

Side of square = 12 ÷ 4 = 3. Area of square = $3^2 = 9$

Required ratio of areas = 8 : 9

30. The length of a rectangular plot is thrice its breadth. If the area of the rectangular plot is 6075 sq. metres, what is its length ? **(RBI Note Printing Asstt. Manager 2006)**

(*a*) 145 metres (*b*) 130 metres (*c*) 75 metres (*d*) None of these

Sol. (*d*) Let breadth of plot be x cm. Then length of the plot = $3x$ metres

Area of the field = $x \times 3x = 6075 \Rightarrow x^2 = 6075 \div 3 = 2025 \Rightarrow x = 45$ metres

Length of the plot = $3x = 3 \times 45 = 135$ metres

31. The area of rectangular field is 460 square metres. If the length is 15 per cent more than the breadth, what is breadth of the rectangular field ? **(SBI Associate Bank PO 2007)**

(*a*) 15 metres (*b*) 26 metres (*c*) 34.5 metres (*d*) None of these

Sol. (*d*) Let breadth of diagonal be x cm. Then length of the field = $1.15x$ metres

Area of the field = $x \times 1.15x = 460 \Rightarrow x^2 = 460 \times \frac{100}{115} = 400 \Rightarrow x = 20$

32. The area of a rectangle is 3 times the area of a square. If length of the rectangle is twice the side of the square and breadth of the rectangle is 30 cm, find the side of the square. **(Bank Clerical)**

(*a*) 20 cm (*b*) 24 cm (*c*) 30 cm (*d*) None of these

Sol. (*a*) Let side of the square be x cm. Then $3x^2 = 2x \times 30 \Rightarrow x = \frac{2 \times 30}{3} = 20$ cm

33. The area of a rectangle is thrice that of a square. The length of the rectangle is 20 cm and the breadth of the rectangle is 3/2 times that of the side of the square. The side of the square, in cm, is **(SSC CGL Tier II 2013)**

(*a*) 10 (*b*) 20 (*c*) 30 (*d*) 60

Sol. (*a*) Let side of square = x. Then $3x^2 = 20 \times \frac{3}{2}x \Rightarrow x = 20 \times \frac{3}{2} \times \frac{1}{3} = 10$ cm

34. The area (in m^2) of the square which has the same perimeter as a rectangle whose length is 48 m and is 3 times its breadth is : **(SSC CGL Tier I 2003)**

(*a*) 1000 (*b*) 1024 (*c*) 1600 (*d*) 1042

Sol. (*b*) Breadth of rectangle = 48 ÷ 3 = 16 m

Perimeter of square = Perimeter of rectangle = 2 × (48 + 16) = 128 m

Side of square = 128 ÷ 4 = 32 m ⇒ Area of square = 32^2 = 1024 m^2

35. The diameters of two circles are the side of a square and the diagonal of the square. The ratio of the areas of the smaller circle and the larger circle is **(SSC CGL Tier I 2013)**

(*a*) $\sqrt{2} : \sqrt{3}$ (*b*) $1 : \sqrt{2}$ (*c*) 1 : 2 (*d*) 1 : 4

Sol. (*c*) Ratio of radii of two circle = Side of square : Diagonal = $1 : \sqrt{2}$

⇒ Ratio of areas = $1^2 . (\sqrt{2})^2 = 1 : 2$.

36. A circle circumscribes a rectangle with sides 16 cm and 12 cm. What is the area of the circle ? **(CDS 2011)**

(*a*) 48 π cm^2 (*b*) 50 π cm^2 (*c*) 100 π cm^2 (*d*) 200 π cm^2

Sol. (*c*) Diameter of circle = Diagonal of rectangle

$= \sqrt{12^2 + 16^2} = \sqrt{20^2} = 20$ cm

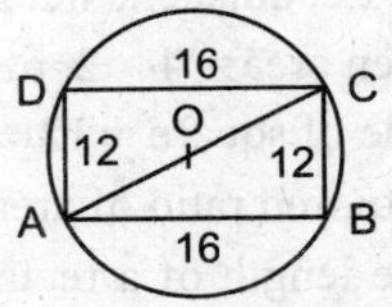

Area of circle = $\pi r^2 = \pi (10)^2 = 100\pi$ cm^2

37. The diameter of a circle is 8 cm less than its semi-circumference. Its radius is **(SAS 2010)**

(*a*) 3.5 cm (*b*) 7 cm (*c*) 10.5 cm (*d*) 14 cm

Sol. (*b*) $\pi r - 2r = 8 \Rightarrow r(\pi - 2) = 8 \Rightarrow r\left(\frac{22}{7} - 2\right) = 8 \Rightarrow \frac{8}{7}r = 8 \Rightarrow r = 8 \times \frac{7}{8} = 7$ cm

38. The length and breadth of a room are 15 metre and 8 metre respectively. Find the length of 3 metre wide carpet required to cover the floor of the room.

(*a*) 40 metre (*b*) 40 m^2 (*c*) 360 metre (*d*) 360 m^2

Sol. (*a*) Carpet required = $\frac{15 \times 8}{3}$ = 40 metre

39. How many tiles, each 4 decimeter square, will be required to cover the floor of a room 8m long and 6 m broad ? **(SSC CGL Tier II 2013)**

(*a*) 200 (*b*) 260 (*c*) 280 (*d*) 300

Sol. (*d*) Tiles = $\frac{80 \times 60}{4 \times 4}$ = 300 [1 metre = 10 decimeter]

40. The least number of square tiles required to cover the floor of a hall of length 68 metres and breadth is 51 metres is : **(SSC CGL Tier I 2007)**

(*a*) 3 (*b*) 4 (*c*) 12 (*d*) 17

Sol. (*c*) HCF of 68 and 51 is 17. ⇒ Size of square tile is 17 m × 17 m

$\therefore$ Number of tiles required = $\frac{68 \times 51}{17 \times 17} = 12$ tiles

41. What is the least number of square tiles required to pave the floor of a room 15 m 17 cm long and 9 m 2 cm broad ? **(SSC CGL Tier I 2003)**

(*a*) 840 (*b*) 841 (*c*) 820 (*d*) 814

Sol. (*d*) LCM of 1517 and 902 is 41.

$\therefore$ Required number of tiles = $\frac{1517 \times 902}{41 \times 41} = 37 \times 22 = 814$

42. The cost of fencing a circular plot at the rate of ₹ 15 per metre is ₹ 3300. What will be the cost of flooring the plot at the rate of ₹ 100 per square metre ? **(Bank P.O. 2009)**

(*a*) ₹ 3,85,000 (*b*) ₹ 2,20,000 (*c*) ₹ 3,50,000 (*d*) None of these

Sol. (*a*) Perimeter = $2\pi r = 3300 \div 15 = 220 \Rightarrow r = 220 \times \frac{1}{2} \times \frac{7}{22} = 35$ m

$\therefore$ Area = $\pi r^2 = \frac{22}{7} \times 35 \times 35 = 3850$ m² ⇒ Cost of flooring = 3850 × ₹ 100 = ₹ 3,85,000

43. In a circle of radius 21 cm, an arc subtends an angle of 72° at the centre. The length of the arc is **(SSC CHSL 2012)**

(*a*) 13.2 cm (*b*) 19.8 cm (*c*) 21.6 cm (*d*) 26.4 cm

Sol. (*d*) Length of the arc = $\frac{\theta}{360°} \times 2\pi r = \frac{72°}{360°} \times 2 \times \frac{22}{7} \times 21 = 26.4$ cm

44. The difference between two parallel sides of a trapezium is 10 cm. The length of the perpendicular distance between them is 18 cm. If the area of the trapezium is 252 cm², the longer side is:

(*a*) 7 cm (*b*) 16 cm (*c*) 19 cm (*d*) 38 cm

Sol. (*c*) Let the parallel sides be x and y cm. Then $252 = \frac{1}{2} \times (x + y) \times 18 \Rightarrow x + y = 28$ cm

Given $x - y = 10$ cm $\Rightarrow x = \frac{1}{2} \times (28 + 10) = 19$ cm

45. The area of a field in the shape of a trapezium measures 1440 m². The perpendicular distance between its parallel sides is 24 m. If the ratio of the parallel sides is 5 : 3, the length of the longer parallel side is : **(SSC CGL Tier I 2004)**

(*a*) 75 m (*b*) 45 m (*c*) 120 m (*d*) 60 m

Sol. (*a*) Let the parallel sides be $5x$ and $3x$. Then $\frac{1}{2} \times 24 \times (5x + 3x) = 1440 \Rightarrow x = \frac{1440 \times 2}{24 \times 8} = 15$ m

$\therefore$ Longer parallel side = $5x = 5 \times 15 = 75$ m

46. A man walking at the rate 3 km/hr crosses a square field diagonally in 1 minute. What is the area of the field ? **(CDS 2009)**

(*a*) 1000 m² (*b*) 1250 m² (*c*) 2500 m² (*d*) 5000 m²

Sol. (*b*) Diagonal of field = Distance covered in 1 minute = 3000 ÷ 60 = 50 m

Area of the field = $\frac{1}{2} \times 50 \times 50 = 1250\text{ m}^2$

47. A man walking at the rate 3 km/hr crosses a square field diagonally in 2 minutes. The area of the field is : **(SSC CGL Tier I 2013)**

(*a*) 5000 m^2 (*b*) 6000 m^2 (*c*) 2500 m^2 (*d*) 3000 m^2

Sol. (*a*) Diagonal of field = Distance covered in 2 minutes = 3000 ÷ 60 × 2 = 100 m

Area of the field = $\frac{1}{2} \times 100 \times 100 = 5000\text{ m}^2$

48. A and B walk from one corner of a rectangular field to the corner just diagonally opposite. A walks along the diagonal, and B walks along the sides, reaching the finishing point together. If the ratio of the length and breadth of the field is 3 : 4, the ratio of their speeds is **(SSC CHSL 2013)**

(*a*) 3 : 4 (*b*) 4 : 3 (*c*) 4 : 5 (*d*) 5 : 7

Sol. (*d*) Let sides be 3 cm and 4 cm. Then diagonal = $\sqrt{3^2 + 4^2} = 5$ cm

Ratio of speeds = Ratio of distances covered = 5 : (3 + 4) = 5 : 7

49. If P, R, T are the area of a parallelogram, a rhombus and a triangle standing on the same base and between the same parallels, which is the following is true ? **(SSC CGL Tier I 2012)**

(*a*) R < P < T (*b*) P > R > T (*c*) R = P = T (*d*) R = P = 2T

Sol. (*d*) Area of two parallelograms on same base and between the same parallels, is equal.

Area of a triangle is half of the parallelogram on the same base and between same paralles.

⇒ R = P and P = 2T ⇒ R = P = 2T

50. If the circumference of a circle is decreased by 50%, then the percentage of decrease in its area is **(SSC CGL Tier I 2010)**

(*a*) 25 (*b*) 50 (*c*) 60 (*d*) 75

Sol. (*d*) Decrease in radius = Decrease in circumference = 50%

⇒ Required % = $50 + 50 - \frac{50 \times 50}{100} = 75\%$

Alternative Method : $\frac{1}{2} \times \frac{1}{2} = \frac{1}{4} \Rightarrow \text{Decrease} = 1 - \frac{1}{4} = \frac{3}{4} = 75\%$

51. The ratio of the length and the breadth of a rectangular plot is 6 : 5 respectively. If the breadth of the plot is 34 metres less than the length, what is the perimeter of the rectangular plot ? **(SBI Bank PO 2008)**

(*a*) 374 metres (*b*) 408 metres (*c*) 814 metres (*d*) 748 metres

Sol. (*d*) Length – Breadth = (6 – 5) ratio = 1 ratio = 34 metres

Perimeter = 2 × (length + breadth) = 2 × (6 + 5) ratio = 22 ratio = 22 × 34 = 748 metres

52. The ratio of length and breadth of a rectangular plot is 8 : 5 respectively. If the breadth is 60 metre less than the length, what is the perimeter of the rectangular plot ? **(Bank of Maharashtra PO 2008)**

(*a*) 260 metres (*b*) 160 metres (*c*) 500 metres (*d*) None of these

Sol. (*d*) Length – Breadth = (8 – 5) ratio = 3 ratio = 60 metres ⇒ 1 ratio = 60 ÷ 3 = 20 metres

Perimeter = 2 × (length + breadth) = 2 × (8 + 5) ratio = 26 ratio = 26 × 20 = 520 metres

53. A rectangle has 30 cm as its length and 720 sq cm as its area. Its area is increased to $1\frac{1}{4}$ times its original area by increasing only its length. Its new perimeter is **(SSC FCI Asst. 2012)**

(*a*) 123 cm (*b*) 125 cm (*c*) 119 cm (*d*) 121 cm

Sol. (*a*) Breadth of rectangle = 720 ÷ 30 = 24 cm ⇒ Increased length = $30 \times \frac{5}{4} = \frac{75}{2}$ cm

New Perimeter = $2 \times 24 + 2 \times \frac{75}{2} = 48 + 75 = 123$ cm

54. If a circle and a semi-circle have the same radius of 14 cm, then the ratio of their perimeters is **(SAS 2010)**

(*a*) 3 : 1 (*b*) 2 : 1 (*c*) 10 : 9 (*d*) 11 : 9

Sol. (*d*) Ratio = $2\pi r : (\pi r + 2r) = 2\pi : (\pi + 2) = 2\times\frac{22}{7} : \left(\frac{22}{7}+2\right) = \frac{44}{7} : \frac{36}{7} = 11 : 9$

Note : The ratio is independent of r ⇒ The ratio will remain the same 11 : 9 irrespective of the length of radius.

55. If the area of a circle and a square are equal, then the ratio of their perimeters is

(*a*) 1 : 1 (*b*) 2 : π (*c*) π : 2 (*d*) $\sqrt{\pi}$: 2

Sol. (*d*) Let radius of circle = r and side of square = a. Then $a^2 = \pi r^2 \Rightarrow a = \sqrt{\pi}\, r$

∴ Required ratio = $2\pi r : 4a = 2\pi r : 4\sqrt{\pi}\, r = \sqrt{\pi} : 2$

56. The perimeter of a rectangle is 82 m and its area is 400 m². What is the breadth of the rectangle ? **(CDS 2012)**

(*a*) 18 m (*b*) 16 m (*c*) 14 m (*d*) 12 m

Sol. (*b*) Let the length and the breadth be x and y. Then $x + y = 82 \div 2 = 41$ and $xy = 400$

∴ $(x - y)^2 = (x + y)^2 - 4xy = 41^2 - 1600 = 1681 - 1600 = 81$

⇒ $(x - y) = \sqrt{81} = 9$ m ∴ $y = \frac{41-9}{2} = 16$ m

Trick : 400 is divisible by 16. 400 ÷ 16 = 25. Perimeter = 2 (25 + 16) = 82 m

57. The perimeters of two squares are 40 cm and 32 cm. The perimeter of a third square whose area is the difference of the areas of the two squares, is : **(SSC CGL 2003)**

(*a*) 24 cm (*b*) 42 cm (*c*) 40 cm (*d*) 20 cm

Sol. (*a*) Sides of the squares are 40 ÷ 4 and 32 ÷ 4 = 10 and 8 cm.

Difference in their areas = $10^2 - 8^2 = 100 - 64 = 36$ cm²

Side of new square = $\sqrt{36}$ = 6 cm ⇒ Perimeter of new square = 4 × 6 cm = 24 cm

58. The perimeters of two squares are 24 cm and 32 cm. The perimeter of a third square equal in area to the sum of the areas of these squares is : **(SSC CGL 2002)**

(*a*) 45 cm (*b*) 40 cm (*c*) 32 cm (*d*) 48 cm

Sol. (*b*) Sides of the squares are 24 ÷ 4 and 32 ÷ 4 = 6 and 8 cm.

Sum of areas of squares = $6^2 + 8^2 = 36 + 64 = 100$ cm^2

Side of new square = $\sqrt{100} = 10$ cm $\Rightarrow$ Perimeter of new square = 4×10 cm = 40 cm

59. The perimeters of five squares are 24 cm, 32 cm, 40 cm, 76 cam and 80 cm respectively. The perimeter of another square equal in area to sum of the areas of these squares is : **(SSC CGL Tier I 2004)**

(*a*) 31 cm (*b*) 62 cm (*c*) 124 cm (*d*) 961 cm

Sol. (*c*) Side of smaller squares = Perimeter ÷ 4 = 6 cm, 8 cm, 10 cm, 19 cm and 20 cm.

Sum of areas of smaller squares = 36 + 64 + 100 + 361 + 400 = 961

$\therefore$ Side of the new square = $\sqrt{961} = 31 \Rightarrow$ Perimeter of new square = 4×31 cm = 124 cm

60. The radii of two circles are 5 cm and 12 cm. The area of a third circle is equal to the sum of the areas of the two circles. The radius of the third circle is **(SSC CHSL 2012)**

(*a*) 17 cm (*b*) 13 cm (*c*) 21 cm (*d*) 30 cm

Sol. (*b*) $5^2 + 12^2 = 13^2 \Rightarrow \pi(5)^2 + \pi\,(12)^2 = \pi\,(13)^2$

61. Find the radius of a circle whose area is equal to the sum of areas of three circles with radii 8 cm, 9 cm, 12 cm respectively.

(*a*) 12 cm (*b*) 14 cm (*c*) 17 cm (*d*) 20 cm

Sol. (*c*) Let radius of new circle is *R* cm. Then $\pi R^2 = \pi 8^2 + \pi 9^2 + \pi 12^2 = 64\pi + 81\pi + 144\pi = 289\pi$

$\therefore R^2 = 289 \Rightarrow R = 17$ cm

62. The sides of two squares are in the ratio 3 : 4. If their combined area is 12.25 sq. m, the side of larger square in meters is **(SSC CHSL 2012)**

(*a*) 0.7 (*b*) 2.0 (*c*) 3.5 (*d*) 2.8

Sol. (*d*) Let sides of two squares be $3x$ and $4x$ respectively.

$\Rightarrow$ Total area of two squares = $(3x)^2 + (4x)^2 = 25x^2 = 12.25$ m$^2 \Rightarrow x^2 = 0.49 \Rightarrow x = 0.7$

$\Rightarrow$ Side of larger square = $4x = 4 \times 0.7 = 2.8$ cm

63. If the perimeters of an equilateral triangle and that of a square are equal, then the ratio of their areas will be **(SSC MTS 2014)**

(*a*) 4 : 3 (*b*) 4 : $\sqrt{3}$ (*c*) 4 : 3$\sqrt{3}$ (*d*) 2 : 3$\sqrt{3}$

Sol. (*c*) Let perimeter of each figure = 12. Then side of square = 3 and side of triangle = 4.

$\Rightarrow$ Ratio of areas = $\frac{\sqrt{3}}{4} \times 4^2 : 3^2 = 4\sqrt{3} : 9 = 4 : 3\sqrt{3}$

64. The area of the square inscribed in a circle of radius 8 cm is **(SSC CGL Tier II 2012)**

(*a*) 128 sq. cm (*b*) 125 sq. cm (*c*) 256 sq. cm (*d*) 250 sq. cm

Sol. (*a*) Diagonal of square = Diameter of circle = 2×8 cm = 16 cm

$\therefore$ Area of square = $\frac{16 \times 16}{2} = 128$ sq. cm

65. The area of the greatest circle, which can be inscribed in a square whose perimeter is 120 cm, is **(SSC CGL Tier I 2004)**

(a) $\pi(15)^2$ cm^2 (b) $\pi(7.5)^2$ cm^2 (c) $\pi(3.5)^2$ cm^2 (d) $\pi(4.5)^2$ cm^2

Sol. (a) Side of square = 120 ÷ 4 = 30 cm = Diameter of the greatest circle

Radius of the circle = 30 ÷ 2 = 15 cm Area of the circle = $\pi r^2 = \pi(15)^2$ cm^2

66. A paper is in the form of a rectangle ABCD where AB = 22 cm and BC = 14 cm. A semi-circular portion with BC as diameter is removed. Find the remaining area of the paper, (in cm^2) : **(SSC CHSL 2012)**

(*a*) 221 (*b*) 231 (*c*) 241 (*d*) 211

Sol. (*b*) Area of rectangle ABCD = 22 × 14 = 308

Area of semi-circle portion = $\frac{1}{2}\pi r^2 = \frac{1}{2} \times \frac{22}{7} \times 7^2 = 77 \Rightarrow$Required area = 308 – 77 = 231

67. The diagonal of a square is $4\sqrt{2}$ cm. The diagonal of another square, whose area is double that of the first square, is : **(SSC CGL 2002)**

(*a*) $8\sqrt{2}$ cm (*b*) 16 cm (*c*) $\sqrt{32}$ cm (*d*) 8 cm

Sol. (*d*) Area of first square = $\frac{1}{2} \times (4\sqrt{2})^2 = 16 \text{ cm}^2$

Area of second square = 2 × 16 cm^2 = 32 cm^2 ⇒ Diagonal of second square = $\sqrt{2 \times 32} = 8$ cm

Alt: Ratio of areas = 1 : 2 ⇒ Ratio of sides / diagonals = 1 : $\sqrt{2}$

New diagonal = $4\sqrt{2} \times \sqrt{2} = 8$ cm

68. The base of a triangle is $\frac{4}{3}$ times its height. If area of the triangle is 600 cm^2, find its base.

(*a*) 30 cm (*b*) 40 cm (*c*) 50 cm (*d*) 60 cm

Sol. (*b*) Let height of the traingle = $3x$ cm. Then base of the traingle = $4x$ cm

∴ Area = $\frac{1}{2} \times 3x \times 4x = 6x^2 = 600 \text{ cm}^2 \Rightarrow x = \sqrt{\frac{600}{6}} = 10$ cm

∴ Base of the triangle = 4 × 10 = 40 cm

69. A lawn is in the form of a rectangle having its breadth and length in the ratio 3 : 4. The area of the lawn is 1/12 hectare . The breadth of the lawn is **(SSC CGL Tier II 2013)**

(*a*) 25 metres (*b*) 50 metres (*c*) 75 metres (*d*) 100 metres

Sol. (*a*) Area = $3x \times 4x = \frac{10000}{12} \text{ m}^2 \Rightarrow x^2 = \frac{10000}{144} \Rightarrow x = \frac{100}{12} = \frac{25}{3}$ m

∴ Breadth = $3x = 3 \times \frac{25}{3} = 25$ m

70. Two equal sides of a right isosceles triangle are 8 cm each. What is its area ?

(*a*) 8 cm^2 (*b*) 16 cm^2 (*c*) 32 cm^2 (*d*) 64 cm^2

Sol. (*c*) Two equal sides are base and height of the triangle.

∴ Area of the triangle = $\frac{1}{2} \times$ Base × Height $\frac{1}{2} \times 8 \times 8 = 32 \text{ cm}^2$

71. The base of a right angle triangle is 8 cm and hypotenuse is 10 cm. Find its area.

(*a*) 12 cm^2 (*b*) 24 cm^2 (*c*) 40 cm^2 (*d*) 80 cm^2

Sol. (*b*) Height of the right-angle triangle = $\sqrt{10^2 - 8^2} = \sqrt{36} = 6$

∴ Area of triangle = $\frac{1}{2} \times 8 \times 6 = 24 \text{ cm}^2$

72. The lengths of the sides of a triangle are 5 cm, 12 cm and 13 cm. Its area is: **(SSC CGL Tier I 2006)**

(*a*) 30 cm^2 (*b*) 37.5 cm^2 (*c*) 60 cm^2 (*d*) 78 cm^2

Sol. (*a*) $5^2 + 12^2 = 13^2 \Rightarrow$ The triangle is a right angle triangle.

$\therefore$ Area of triangle $= \frac{1}{2} \times 5 \times 12 = 30$ cm^2

73. If sides of a triangle are 8 cm, 15 cm and 17 cm respectively. Find its area.

(*a*) 60 cm^2 (*b*) 120 cm^2 (*c*) 240 cm^2 (*d*) 360 cm^2

Sol. (*a*) The triangle is right angle triangle as $17^2 = 15^2 + 8^2$

$\therefore$ Area of right angle triangle $= \frac{1}{2} \times 8 \times 15 = 60$ cm^2

74. The area of a triangle is 216 cm^2 and its sides are in the ratio 3 : 4 : 5. The perimeter of the triangle is : **(SSC CGL Tier I 2004)**

(*a*) 6 cm (*b*) 12 cm (*c*) 36 cm (*d*) 72 cm

Sol. (*d*) $3^2 + 4^2 = 5^2 \Rightarrow$ The triangle is a right angle triangle.

Area of the triangle $= \frac{1}{2} \times 3x \times 4x = 216 \Rightarrow x^2 = \frac{216 \times 2}{3 \times 4} = 36 \Rightarrow x = 6$

Perimeter of triangle $= 3x + 4x + 5x = 12x = 12 \times 6$ cm $= 72$ cm

75. The ratio of sides of a triangle is 3 : 4 : 5. If area of the triangle is 72 sq units, then the length of the smallest side is **(SSC CHSL 2012)**

(*a*) $3\sqrt{3}$ units (*b*) $4\sqrt{3}$ units (*c*) $5\sqrt{3}$ units (*d*) $6\sqrt{3}$ units

Sol. (*d*) $3^2 + 4^2 = 5^2 \Rightarrow$ The triangle is right angle triangle.

Area $= \frac{1}{2} \times 3x \times 4x = 72 \Rightarrow 6x^2 = 72 \Rightarrow x^2 = 12 \Rightarrow x = 2\sqrt{3}$

Smallest side $= 3x = 3 \times 2\sqrt{3} = 6\sqrt{3}$ units

76. Three sides of a triangular field are of length 15 m, 20 m and 25 m long respectivly. Find the cost of sowing seeds in the field at the rate of 5 rupees per sq.m. **(SSC CGL Tier I 2013)**

(*a*) ₹ 150 (*b*) ₹ 300 (*c*) ₹ 600 (*d*) ₹ 750

Sol. (*d*) $15^2 + 20^2 = 25^2 \Rightarrow$ The triangle is a right angle triangle.

Area of the triangle $= \frac{1}{2} \times 15 \times 20 = 150$ $m^2 \Rightarrow$ Cost of sowing seeds = 150 × ₹ 5 = ₹ 750

77. If the perimeter of a right angled triangle is 56 cm and area of the triangle is 84 sq cm, then the length of the hypotenuse is **(SSC CHSL 2012)**

(*a*) 7 cm (*b*) 24 cm (*c*) 25 cm (*d*) 50 cm

Sol. (*c*) We consider option (c), $25^2 = 24^2 + 7^2$

Perimeter = 25 + 24 + 7 = 56 cm. Area of $\Delta = \frac{1}{2} \times 24 \times 7 = 84$ cm^2

78. The measure of the sides of a triangle are $(x^2 - 1)$, $(x^2 + 1)$ and $2x$ cm, then the triangle would be : **(SSC CGL Tier I 2014)**

(*a*) equilateral (*b*) acute-angled (*c*) right-angled (*d*) isosceles

Sol. (*c*) $(x^2 - 1)^2 + (2x)^2 = (x^2 + 1)^2 \Rightarrow$ The triangle is right-angled

79. The sides of a triangular field are 41 m, 40 m and 9 m. The number of rose beds that can be prepared in the field if each rose bed, on an average, needs 900 square cm space, is **(CDS 2015)**

(*a*) 2000 (*b*) 1800 (*c*) 900 (*d*) 800

Sol. (*a*) $40^2 + 9^2 = 41^2 \Rightarrow$ The triangle is a right angle triangle.

Area of the triangle = $\frac{1}{2} \times 40 \times 9 = 180 \text{ m}^2 = 180 \times 10000 \text{ cm}^2$

Number of rose bed = $(180 \times 10000) \div 900 = 2000$

80. What is the area of a right angled isosceles triangle whose hypotenuse is $6\sqrt{2}$ cm ? **(CDS 2009)**

(*a*) 12 cm² (*b*) 18 cm² (*c*) 24 cm² (*d*) 36 cm²

Sol. (*b*) Let equal sides be *a* cm each. Then $\sqrt{a^2 + a^2} = 6\sqrt{2} \Rightarrow 2a^2 = 72 \Rightarrow a^2 = 36 \Rightarrow a = 6$ cm

$\therefore$ Area of triange = $\frac{1}{2} \times 6 \times 6 = 18 \text{ cm}^2$

81. A cost of cultivating a square field at a rate of ₹ 135 per hectare is ₹ 1,215. The cost of putting a fence around it at the rate of 75 paise per metre would be **(SSC CGL Tier II 2012)**

(*a*) ₹ 900 (*b*) ₹ 1,800 (*c*) ₹ 360 (*d*) ₹ 810

Sol. (*a*) Area of field = $1215 \div 135 = 9$ hectares = $90000 \text{ m}^2 \Rightarrow$ Side of field = $\sqrt{90000} = 300$ m

$\Rightarrow$ Perimeter of the field = 4×300 m = 1200 m. Cost of fencing = 1200×0.75 = ₹ 900

82. The area of a square is $\frac{1}{2}$ hectare. The length of its diagonal is :

(*a*) 100 m (*b*) 50 m (*c*) 25 m (*d*) $50\sqrt{2}$ m

Sol. (*a*) Area of square = $\frac{1}{2} \times (\text{Diagonal})^2 = \frac{1}{2} \times 10000 \Rightarrow (\text{Diagonal})^2 = 10000 \Rightarrow \text{Diagonal} = 100$

83. If the numerical value of the perimeter of an equilateral triangle is $\sqrt{3}$ times the area of it, then the length of each side of the triangle is **(SSC FCI Asstt. 2012)**

(*a*) 2 unit (*b*) 3 unit (*c*) 4 unit (*d*) 6 unit

Sol. (*c*) Perimeter = $\frac{\sqrt{3}}{4} a^2 \times \sqrt{3} = 3a \Rightarrow \frac{3}{4} a^2 = 3a \Rightarrow a = 4$ units

84. A square and an equilateral triangle are drawn on the same base. The ratio of their areas is : **(SSC CGL Tier I 2005)**

(*a*) 2 : 1 (*b*) 1 : 1 (*c*) $\sqrt{3}$: 4 (*d*) 4 : $\sqrt{3}$

Sol. (*d*) Required ratio = $a^2 : \frac{\sqrt{3}}{4} a^2 = 4 : \sqrt{3}$

85. If diagonal of a square is 3 cm, then the area of an equilateral triangle drawn on a side of the square will be **(SSC FCI 2012)**

(*a*) $\frac{11\sqrt{3}}{8}$ Sq. cm (*b*) $\frac{9\sqrt{3}}{8}$ Sq. cm (*c*) $\frac{8\sqrt{3}}{9}$ Sq. cm (*d*) $\frac{5\sqrt{3}}{8}$ Sq. cm

Sol. (*b*) Side of square $= \dfrac{\text{Diagonal}}{\sqrt{2}} = \dfrac{3}{\sqrt{2}}$

Area of equilateral triangle $= \dfrac{\sqrt{3}}{4}(\text{side})^2 = \dfrac{\sqrt{3}}{4} \times \left(\dfrac{3}{\sqrt{2}}\right)^2 = \dfrac{9\sqrt{3}}{8}$

86. An equilateral triangle is drawn on the diagonal of a square. The ratio of the area of the triangle to that of the square is **(SSC FCI Asstt. 2012)**

(*a*) $\sqrt{3} : 2$ (*b*) $2 : \sqrt{3}$ (*c*) $\sqrt{2} : \sqrt{3}$ (*d*) $1 : \sqrt{2}$

Sol. (*a*) Let side of square = *a*. Then diagonal = $\sqrt{2}\, a$

$\therefore$ Required ratio $= \dfrac{\sqrt{3}}{4}(\sqrt{2}a)^2 : a^2 = \dfrac{\sqrt{3}}{2}a^2 : a^2 = \sqrt{3} : 2$

87. The area of an equilateral triangle is $4\sqrt{3}$ cm. The length of each side of the triangle is: **(SSC CHSL 2010)**

(*a*) 3 cm (*b*) $2\sqrt{2}$ cm (*c*) $2\sqrt{3}$ cm (*d*) 4 cm

Sol. (*d*) Area $= \dfrac{\sqrt{3}}{4}a^2 = 4\sqrt{3} \Rightarrow a^2 = 4\sqrt{3} \times \dfrac{4}{\sqrt{3}} = 16 \Rightarrow a = \sqrt{16} = 4$ cm

88. If the length of each median of an equilateral triangle is $6\sqrt{3}$ cm, the perimeter of the triangle is **(SSC CGL Tier I 2012)**

(*a*) 24 cm (*b*) 32 cm (*c*) 36 cm (*d*) 42 cm

Sol. (*c*) Median of equilateral triangle = Altiture of equilateral triangle $= \dfrac{\sqrt{3}}{2}$ side $= 6\sqrt{3}$

Side $= 6\sqrt{3} \times \dfrac{2}{\sqrt{3}} = 12$ cm. Perimter = 3 × 12 cm = 36 cm

89. What is the area of an equilateral triangle having altitude equal to $2\sqrt{3}$ cm ? **(CDS 2011)**

(*a*) $\sqrt{3}$ cm² (*b*) $2\sqrt{3}$ cm² (*c*) $3\sqrt{3}$ cm² (*d*) $4\sqrt{3}$ cm²

Sol. (*d*) Altitude $= \dfrac{\sqrt{3}}{2}a = 2\sqrt{3} \Rightarrow a = 2\sqrt{3} \times \dfrac{2}{\sqrt{3}} = 4$ cm

Area $= \dfrac{\sqrt{3}}{4}a^2 = \dfrac{\sqrt{3}}{4} \times 4^2 = 4\sqrt{3}$ cm²

90. If the altitude of an equilateral triangle is $12\sqrt{3}$ cm, then its area would be **(SSC CGL I 2015)**

(*a*) 12 cm² (*b*) 72 cm² (*c*) $56\sqrt{3}$ cm² (*d*) $144\sqrt{3}$ cm²

Sol. (*d*) Altitude $= \dfrac{\sqrt{3}}{2}a = 12\sqrt{3} \Rightarrow a = 12\sqrt{3} \times \dfrac{2}{\sqrt{3}} = 24$ cm

Area $= \dfrac{\sqrt{3}}{4}a^2 = \dfrac{\sqrt{3}}{4} \times 24^2 = 144\sqrt{3}$ cm²

91. The length of the median of an equilateral triangle be 6 cm, then its area (in sq. cm) is **(SSC FCI Asstt. 2012)**

(*a*) $6\sqrt{3}$ (*b*) $8\sqrt{3}$ (*c*) $4\sqrt{3}$ (*d*) $12\sqrt{3}$

Sol. (*d*) Median = Altitude = $\frac{\sqrt{3}}{2}$ side = 6 ⇒ side = 6 × $\frac{2}{\sqrt{3}}$ = $4\sqrt{3}$

$\Rightarrow$ Area = $\frac{\sqrt{3}}{4}a^2 = \frac{\sqrt{3}}{4}\times(4\sqrt{3}^2) = 12\sqrt{3}$ sq. cm.

92. If median of an equilateral triangle is x cm, find area of the triangle.

(*a*) x^2 (*b*) $\frac{1}{2}x^2$ (*c*) $\frac{\sqrt{3}}{2}x^2$ (*d*) $\frac{\sqrt{3}}{3}x^2$

Sol. (*d*) Let side of the equilateral triangle be a cm and the median be x.

Median of equilateral triangle = height = $\frac{\sqrt{3}}{2}a = x \Rightarrow a = \frac{2x}{\sqrt{3}} \Rightarrow \frac{a}{2} = \frac{x}{\sqrt{3}}$

Area of the triangle = $\frac{\sqrt{3}}{4}a^2 = \frac{\sqrt{3}}{2}a\times\frac{a}{2} = x\times\frac{x}{\sqrt{3}} = \frac{\sqrt{3}x^2}{3}$

93. The area of an equilateral triangle is $9\sqrt{3}$ m². The length (in m) of the median is **(SSC CGL Tier II 2013)**

(*a*) $2\sqrt{3}$ (*b*) $3\sqrt{3}$ (*c*) $3\sqrt{2}$ (*d*) $2\sqrt{2}$

Sol. (*b*) $\frac{\sqrt{3}}{4}a^2 = 9\sqrt{3} \Rightarrow a^2 = 9\sqrt{3}\times\frac{4}{\sqrt{3}} = 36 \Rightarrow a = 6$ m

Median = Altitude = $\frac{\sqrt{3}}{2}a = \frac{\sqrt{3}}{2}\times 6 = 3\sqrt{3}$ m

94. Difference between the area of a square and that of an equilateral trianagle on the same base is $\frac{1}{4}$ cm². What is the length of side of triangle ? **(CDS 2009)**

(*a*) $(4-\sqrt{3})^{1/2}$ cm (*b*) $(4+\sqrt{3})^{1/2}$ (*c*) $(4-\sqrt{3})^{-1/2}$ (*d*) $(4+\sqrt{3})^{-1/2}$

Sol. (*c*) Let side of triangle = a = Side of square

Then $a^2 - \frac{\sqrt{3}}{4}a^2 = \frac{1}{4} \Rightarrow 4a^2 - \sqrt{3}a^2 = 1 \Rightarrow (4-\sqrt{3})a^2 = 1 \Rightarrow a^2 = 1 \div (4-\sqrt{3})$

$\Rightarrow a^2 = (4-\sqrt{3})^{-1} \Rightarrow a = (4-\sqrt{3})^{-1/2}$

95. The area of a triangle ABC is equal to area of square of side length 6 cm. What is the length of the altitude to AB where AB = 9 cm? **(CDS 2011)**

(*a*) 18 cm (*b*) 14 cm (*c*) 12 cm (*d*) 8 cm

Sol. (*d*) Length of altitude = $\frac{6\times6\times2}{9}$ = 8 cm

96. The base of triangle is 15 cm and height is 12 cm. The height of another triangle of double the area having the base 20 cm is : **(SSC CGL Tier I 2002)**

(*a*) 9 cm (*b*) 18 cm (*c*) 8 cm (*d*) 12.5 cm

Sol. (*b*) Let height of another triangle = h cm

Then $\frac{1}{2} \times 20 \times h = 2 \times \left(\frac{1}{2} \times 15 \times 12\right) \Rightarrow h = 180 \div 10 = 18$ cm

Alternative method : Required height $= \frac{15 \times 12 \times 2}{20} = 18$ cm

97. If the area of triangle with base 12 cm is equal to the area of a square with side 12 cm, then the altitude of the triangle is: **(SSC CGL Tier I 2005)**

(*a*) 12 cm (*b*) 24 cm (*c*) 18 cm (*d*) 36 cm

Sol. (*b*) Area of triangle = Area of square (Given) $\Rightarrow \frac{1}{2} \times$ Base $\times$ Altitude = Side $\times$ Side

$\Rightarrow \frac{1}{2} \times 12 \times \text{Altitude} = 12 \times 12 \Rightarrow \text{Altitude} = \frac{12 \times 12}{12} \times 2 = 24$ cm

98. The area of an isosceles triangle ABC with AB = AC and altitude AD = 3 cm is 12 square cm. What is its perimeter? **(CDS 2013)**

(*a*) 18 cm (*b*) 16 cm (*c*) 14 cm (*d*) 12 cm

Sol. (*a*) Area $= \frac{1}{2}$BC $\times$ AD $= 12 \Rightarrow$ BC $= \frac{12 \times 2}{3} = 8$

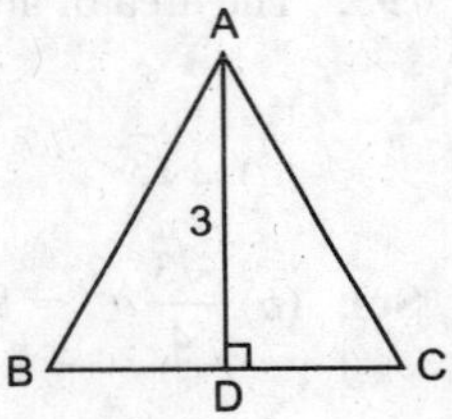

$\therefore$ BD $= \frac{1}{2}$ BC $= \frac{1}{2} \times 8 = 4$ cm

$AB^2 = AD^2 + BD^2 = 3^2 + 4^2 = 5^2$

$\therefore$ Perimeter = AB + AC + BC = 5 + 5 + 8 = 18 cm

99. The length of a room floor exceeds it breadth by 20 m. The area of the floor remains unaltered when the length is decreased by 10 m but the breadth is increased by 5 m. The area of the floor (in square metres) is : **(SSC CHSL 2011)**

(*a*) 280 (*b*) 325 (*c*) 300 (*d*) 420

Sol. (*c*) Let the original length and breadth be x and $x - 20$.

Then $x \times (x - 20) = (x - 10) \times (x - 15) \Rightarrow x^2 - 20x = x^2 - 25x + 150 \Rightarrow 5x = 150 \Rightarrow x = 30$

Area of floor = 30 × (30 – 20) = 30 × 10 = 300

100. The length of a rectangular hall is 5 m more than its breadth. The area of the hall is 750 m^2. The length of the hall is : **(SSC CGL Tier I 2004)**

(*a*) 15 m (*b*) 22.5 m (*c*) 25 m (*d*) 30 m

Sol. (*d*) Let breadth of the hall = x cm. Then length of the hall = $x + 5$

$\therefore x(x + 5) = 750 \Rightarrow x^2 + 5x - 750 = 0 \Rightarrow x^2 + 30x - 25x - 750 = 0$

$\Rightarrow x(x + 30) - 25(x + 30) = 0 \Rightarrow x = 25 \Rightarrow$ Length of the hall $= x + 5 = 25 + 5 = 30$ metres

Trick : $x(x + 5) = 750 \Rightarrow 25 \times 30 = 750$

101. If area of a rectangle with two adjacent sides measuring $(x + 3)$ and $(x + 4)$ is 50 sq. units. Find the area (in sq. units) of a square with a side $(2x + 7)$ units **(GMAT)**

(*a*) 50 (*b*) 100 (*c*) 200 (*d*) 201

Sol. (*d*) Area of rectangle $= (x + 3) \times (x + 4) = 50 \Rightarrow x^2 + 7x + 12 = 50 \Rightarrow x^2 + 7x = 38$

Area of Square $= (2x + 7)^2 = 4x^2 + 28x + 49 = 4(x^2 + 7x) + 49 = 4 \times 38 + 49 = 152 + 49 = 201$

102. The length of a plot is five times its breadth. A playground measuring 245 square metres occupies half of the total area of the plot. What is the length of the plot ? **(SSC CGL 2000)**

(*a*) $35\sqrt{2}$ metres (*b*) $175\sqrt{2}$ metres (*c*) 490 metres (*d*) $5\sqrt{2}$ metres

Sol. (*a*) Let the breadth and the length of the plot are x and $5x$ respectively.
Then area of plot $= x \times 5x = 2 \times 245 \Rightarrow x^2 = 2 \times 245 \div 5 = 98 \Rightarrow x = \sqrt{98} = 7\sqrt{2}$
Length of the plot $= 5x = 5 \times 7\sqrt{2} = 35\sqrt{2}$ metres

103. The length of the three sides of a right angled triangle are $(x - 2)$ cm, x cm and $(x + 2)$ cm respectively. The value of x is **(SSC Constable GD 2012)**

(*a*) 10 (*b*) 8 (*c*) 4 (*d*) 0

Sol. (*b*) Here, Difference between sides = 2 cm. $3^2 + 4^2 = 5^2$ and $6^2 + 8^2 = 10^2 \Rightarrow x = 8$

104. In the figure given area of square ABCD is 81 cm². If the length of MN is 5 cm, find the area of triangle ANB.

(*a*) 18 cm² (*b*) 27 cm² (*c*) 36 cm² (*d*) 72 cm²

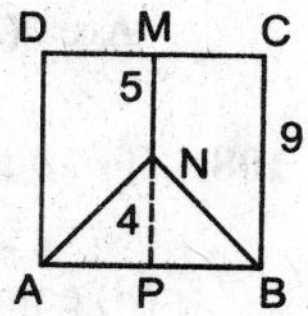

Sol. (*a*) NP = (9 – 5) cm = 4 cm ⇒ Area of triangle ANB $= \frac{1}{2} \times 9 \times 4 = 18$ cm²

105. If the sides of an equilateral triangle are increased by 20%, 30% and 50% respectively to form a new triangle, the increase in the perimeter of the equilateral triangle is : **(SSC CGL Tier I 2007)**

(*a*) 25% (*b*) $33\frac{1}{3}\%$ (*c*) 75% (*d*) 100%

Sol. (*b*) Let each side of the equilateral triangle be 10 cm. Then its perimeter = 3 × 10 cm = 30 cm
New dimensions = 12, 13 and 15 cm. ⇒ New perimeter = 12 + 13 + 15 = 40 cm
⇒ Increase = 40 – 30 = 10 cm ⇒ Percentage increase $= \frac{10}{30} \times 100 = 33\frac{1}{3}\%$

Alternative method : Let each side of the equilateral triangle be 10 cm.
∴ Total increase = 10 × 20% + 10 × 30% + 10 × 50% = (2 + 3 + 5) = 10 cm

∴ Percentage increase $= \frac{10}{30} \times 100 = 33\frac{1}{3}\%$

Alt: $\frac{20\% + 30\% + 50\%}{3} = \frac{100\%}{3} = 33\frac{1}{3}\%$

106. ABC is a triangle with base AB. D is a point on AB such that AB = 5 and DB = 3. What is the ratio of the areas of ΔADC to the area of ΔABC ? **(SSC CGL Tier I 2000)**

(*a*) 3 : 2 (*b*) 2 : 3 (*c*) 3 : 5 (*d*) 2 : 5

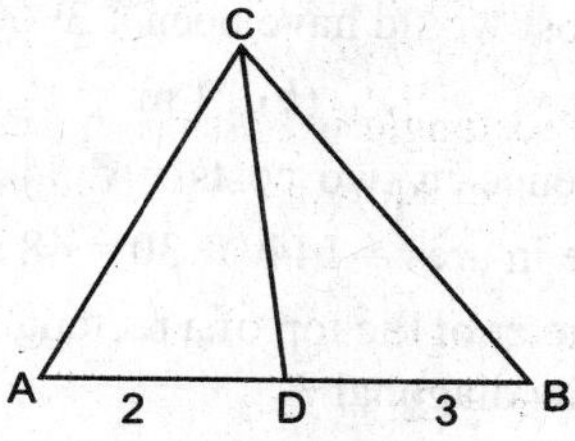

Sol. (*d*) Area of Δ ABC $= \frac{1}{2} \times AB \times h = \frac{5}{2}h$

Area of Δ ADC $= \frac{1}{2} \times AD \times h = h$

∴ Required ratio $= h : \frac{5}{2}h = 2 : 5$

Trick : Heights are equal. ⇒ Ratio of Areas = Ratio of bases = 2 : 5

107. Side AB of rectangle ABCD is divided into four equal parts by points X, Y, Z. The ratio of the $\frac{\text{Area}(\Delta XYC)}{\text{Area (Reactangle ABCD)}}$ is **(SSC CGL Tier I 2012)**

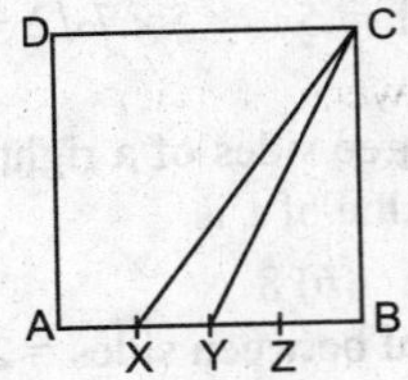

(a) $\frac{1}{6}$ (b) $\frac{1}{9}$

(c) $\frac{1}{8}$ (d) $\frac{1}{7}$

Sol. (c) Area (reactangle ABCD) = AB × BC XY = $\frac{1}{4}$AB

Area (Δ XYC) = $\frac{1}{2}$ × XY × BC = $\frac{1}{2}\times\frac{1}{4}$ AB × BC = $\frac{1}{8}$ AB × BC

108. If a, b and c are the sides of a triangle and $a^2 + b^2 + c^2 = ab + bc + ca$, then the triangle is **(SSC CHSL 2012)**

(a) equilateral (b) isosceles (c) right angled (d) obtuse angled

Sol. (a) $a^2 + b^2 + c^2 = ab + bc + ca \Rightarrow a = b = c \Rightarrow$ The triangle is equilateral.

109. If the perimeter of a semicircular field is 144 m, then the diameter of the field is : **(SSC CGL Tier I 2008)**

(a) 55 metres (b) 30 metres (c) 28 metres (d) 56 metres

Sol. (d) Perimeter of the semicircle = $\pi r + 2r = 144 \Rightarrow \left(\frac{22}{7}+2\right)r = 144 \Rightarrow r = 144\times\frac{7}{36} = 28$ m

$\Rightarrow$ Diameter = $2r = 2 \times 28$ m = 56 m

110. Perimeter of rectangular field is 160 metres and the difference between its two adjacent sides is 48 metres. The side of a square field, having the same area as that of the rectangle is : **(SSC CGL Tier I 2005, 2001)**

(a) 32 m (b) 8 m (c) 4 m (d) 16 m

Sol. (a) Let length and breadth of the rectangle be l and b. Then $2(l + b) = 160 \Rightarrow l + b = 80$

Difference between two adjacent sides = $l - b = 48$

$\Rightarrow l = \frac{80+48}{2} = 64$ and $b = \frac{80-48}{2} = 16$

Area of square = Area of rectangle = 64 × 16

∴ Side of square = $\sqrt{64\times16} = 8 \times 4 = 32$ metres

111. The cost of carpeting a hall at ₹ 30 per square metre is ₹ 5400. Had the length been 4 metre less, the cost would have been ₹ 3960. Find the breadth of the hall.

(a) 12 m (b) 14 m (c) 15 m (d) 16 m

Sol. (a) Difference in two costs = ₹ 5400 – ₹ 3960 = ₹ 1440

Difference in area = 1440 ÷ 30 = 48 m² ⇒ Breadth of hall = 48 ÷ 4 = 12 m

112. The perimeter of the top of a rectangular table is 28 m, whereas its area is 48 m². What is the length of its diagonal ? **(SSC CGL Tier I 2000)**

(a) 5 m (b) 10 m (c) 12 m (d) 12.5 m

Sol. (b) Perimeter = $2(l + b) = 28$ cm $\Rightarrow (l + b) = 14$ cm. Area = $(l \times b) = 48$

$l^2 + b^2 = (l + b)^2 - 2lb = (14)^2 - 2 \times 48 = 196 - 96 = 100$

$\therefore$ Diagonal $= \sqrt{l^2 + b^2} = \sqrt{100} = 10$ cm

Trick: $l + b = 14$ and $l \times b = 48 \Rightarrow l = 8$ and $b = 6 \Rightarrow$ Diagonal $= \sqrt{8^2 + 6^2} = 10$ m

113. The difference between the areas of two circles, one circumscribing and another inscribed in a square is 77 cm^2. The area of the square is **(SSC MTS 2014)**

(a) 14 cm^2 (b) 49 cm^2 (c) 98 cm^2 (d) 28 cm^2

Sol. (c) Ratio of areas of outer and inner circle = 2 : 1 $\Rightarrow$ Difference in areas = 1 ratio = 77 cm^2

Area of outer circle = 2 ratio = 154 cm^2 $\Rightarrow \pi r^2 = 154 \Rightarrow r^2 = 154 \times \frac{7}{22} \Rightarrow r = 7$ cm

Diagonal $= 2r = 14$ cm $\Rightarrow$ Area $= \frac{d^2}{2} = \frac{14 \times 14}{2} = 98$ cm^2

114. A circular swimmimg pool is surrounded by a concrete wall 4 m wide. If the area of the concrete wall surronding the pool is 11/25 that of the pool, then the radius (in m) of the pool is : **(SSC CGL Tier I 2015)**

(*a*) 16 (*b*) 20 (*c*) 30 (*d*) 8

Sol. (*b*) Ratio of areas of inner and outer circles = 25 : (25 + 11) = 25 : 36 = $5^2 : 6^2$

$\Rightarrow$ Ratio of radii of inner and outer circles = 5 : 6

$\Rightarrow$ Difference of radii = (6 – 5) ratio = 1 ratio = 4 m

$\Rightarrow$ Radius of the smaller circle (swimming pool) = 5 ratio = 20 m

115. The perimeter (in metres) of a semicircle is numerically equal to its area (in square metres). The length of its diameter is (Take $\pi = 22/7$) : **(SSC CGL Tier I 2008)**

(*a*) $3\frac{6}{11}$ m (*b*) $5\frac{6}{11}$ m (*c*) $6\frac{6}{11}$ m (*d*) $6\frac{1}{2}$ m

Sol. (*c*) Here $\pi r + 2r = \frac{1}{2}\pi r^2 \Rightarrow 2(\pi + 2) = \pi r \Rightarrow 2\left(\frac{22}{7} + 2\right) = \frac{22}{7} r$

$\Rightarrow r = 2 \times \frac{36}{7} \times \frac{7}{22} = \frac{36}{11}$ metres $\Rightarrow$ Diameter $= 2 \times \frac{36}{11} = \frac{72}{11} = 6\frac{6}{11}$ metres

116. If each side of an equilateral triangle is increased by 2 cm, its area is increased by $3\sqrt{3}$ cm^2. The length of each side of the original triangle is :

(*a*) 1 cm (*b*) 2 cm (*c*) 2.5 cm (*d*) 3 cm

Sol. (*b*) Increase in area $= \frac{\sqrt{3}}{4}\left[(a+2)^2 - a^2\right] = 3\sqrt{3} \Rightarrow \frac{1}{4}(4a + 4) = 3 \Rightarrow a + 1 = 3 \Rightarrow a = 2$

Alt: From options : $\frac{\sqrt{3}}{4}(3^2 - 1^2) = 2\sqrt{3} \neq 3\sqrt{3}$; $\frac{\sqrt{3}}{4}(4^2 - 2^2) = 3\sqrt{3}$

117. In two trianges, the ratio of the areas is 4 : 3 and the ratio of their heights is 3 : 4. Find the ratio of their bases. **(SSC CGL Tier I 2008, 2002)**

(*a*) 16 : 9 (*b*) 9 : 16 (*c*) 9 : 12 (*d*) 16 : 12

Sol. (*a*) Let the bases of the triangles be *x* and *y*.

Then $\frac{1}{2}(x)(3) : \frac{1}{2}(y)(4) = 4 : 3 \Rightarrow \frac{3x}{4y} = \frac{4}{3} \Rightarrow \frac{x}{y} = \frac{4}{3} \times \frac{4}{3} = \frac{16}{9} = 16 : 9$

Direct method : Required ratio $= \frac{\text{Area}}{\text{Height}} = \frac{4}{3} : \frac{3}{4} = 16 : 9$

118. The ratio of bases of two triangles is $x : y$ and that of their areas is $a : b$. Then the ratio of their corresponding altitudes will be : **(SSC CGL Tier I 2004)**

(*a*) $\frac{a}{x} : \frac{b}{y}$ (*b*) $ax : by$ (*c*) $ay : bx$ (*d*) $\frac{x}{a} : \frac{y}{b}$

Sol. (*a*) Area of triangle $= \frac{1}{2} \times \text{Base} \times \text{Altitude} \Rightarrow \text{Altitude} = \frac{2\ \text{Area}}{\text{Base}}$

$\therefore$ Required ratio $= \frac{2a}{x} : \frac{2b}{y} = \frac{a}{x} : \frac{b}{y}$

Direct method : Required ratio $= \frac{\text{Area}}{\text{Base}} = \frac{a}{x} : \frac{b}{y}$

119. There is a rectangular tank of length 180 m and breadth 120 m in a circular field. If the area of the land portion of the field is 40,000 m^2. What is the radius of the field ? (Take $\pi = 22/7$) **(SSC CGL Tier I 2004)**

(*a*) 130 m (*b*) 135 m (*c*) 140 m (*d*) 145 m

Sol. (*c*) Total area of circular field = 180 × 120 + 40,000 = 21,600 + 40,000 = 61,600

$\therefore \pi r^2 = 61600 \Rightarrow r^2 = 61,600 \times \frac{7}{22} = 19600 \Rightarrow r = \sqrt{19,600} = 140$ m

120. A horse is tied to a pole fixed at one corner of a 50 m × 50 m square field of grass by means of a 20 m long rope. what is the area of that part of the field which the horse can graze? (take $\pi = 3.14$) **(CDS 2009)**

(*a*) 1256 m^2 (*b*) 942 m^2 (*c*) 628 m^2 (*d*) 314 m^2

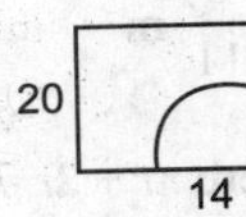

Sol. (*d*) Required area $= \frac{1}{4}\pi r^2 = \frac{1}{4} \times 3.14 \times 20 \times 20 = 314$ m^2

121. A horse is tied with a rope of length 7 m at one corner of a square field having side equal to 10 m. The minimum possible area of the square filed tha is left ungrazed is **(SSC CHSL 12)**

(*a*) 56.1 m^2 (*b*) 65.1 m^2 (*c*) 51.6 m^2 (*d*) 61.5 m^2

Sol. (*d*) Area of square field = (10 m)^2 = 100 m^2

Area of the field grazed by the horse $= \frac{1}{4}\pi r^2 = \frac{1}{4} \times \frac{22}{7} \times 7 \times 7 = \frac{77}{2}$ m^2

$\therefore$ Field ungrazed = (100 – 38.5) = 61.5 m^2

122. The area of the region swept by the minute-hand, 7 cm long, of a wall clock in an interval of 6 minutes is **(SAS 2010)**

(*a*) 154 cm^2 (*b*) 92.4 cm^2 (*c*) 15.4 cm^2 (*d*) 4.4 cm^2

Sol. (*c*) In 60 minutes, minute-hand moves one full round. In 6 minutes, the minute hand will move 1/10 of one round.

$\therefore$ Required area $= \frac{1}{10} \times \frac{22}{7} \times 7 \times 7 = 15.4$ cm^2

123. 4 horses are tied – one to each corner of a square with 14 m a side, and length of the rope is 7m. Find the area of ungrazed field.

(*a*) 21 m^2 (*b*) 42 m^2 (*c*) 49 m^2 (*d*) 154 m^2

Sol. (*b*) Ungrazed area = Area of square – Area of circle

$= 14^2 - \frac{22}{7} \times 7 \times 7 = 196 - 154 = 42 \text{ m}^2$

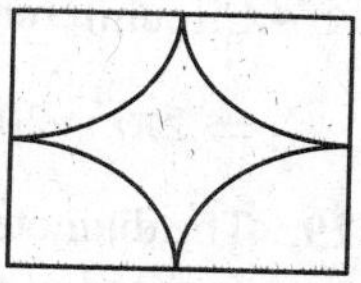

Short-Cut Method: Side of the square = 14 m

$\therefore$ Ungraged field = $\frac{3}{14} \times 14 \times 14 = 42 \text{ m}^2$

124. P, Q, R and S are the middle points of the sides AB, BC, CD and DA respectively of a square ABCD, with a side of 2 cm. With radius equal to 1 cm, arcs are drawn with centres at A, B, C and D . Find the area (in square cm) of the portion of the square bounded by the four arcs.

(SSC CGL Tier I 2006)

(*a*) $4 - \frac{\pi}{3}$ (*b*) $4 - \pi$

(*c*) $2 - \frac{\pi}{2}$ (*d*) $\pi - 4$

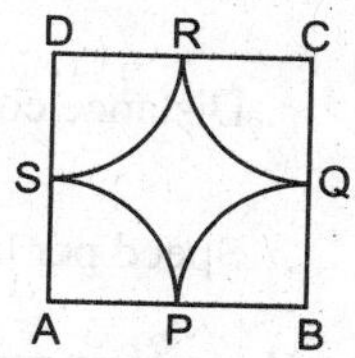

Sol. (*b*) Required area = $2^2 - 4 \times \left(\frac{1}{4} \times \pi \times 1^2\right) = 4 - \pi = 4 - \pi$

125. The diameter of the front wheel of an engine is $2x$ cm and that of rear wheel is $2y$ cm. To cover the same distance, find the number of times the rear wheel will revolve when the front wheel revolves n times

(SSC CHSL 2014)

(*a*) n/xy (*b*) ny/x (*c*) nx/y (*d*) xy/n

Sol. (*b*) Ratio of two radii = $x : y$

$\Rightarrow$ Ratio of the revolutions to cover equal distances = $y : x$

$\Rightarrow$ Required number of revolutions = ny/x

126. The number of revolutions, a wheel of diameter 40 cm makes in travelling a distance of 176 m, is (use $\pi = 22/7$):

(SSC CGL Tier I 2004)

(*a*) 140 (*b*) 150 (*c*) 160 (*d*) 166

Sol. (*a*) Radius of the wheel = $\frac{1}{2} \times 40 = 20$ cm

Distance covered in 1 revolution = Circumference of the wheel = $2\pi r = 2 \times \frac{22}{7} \times 20 = \frac{880}{7}$ cm

Distance to cover = 176 metres = 17600 cm

$\therefore$ Required number of revolutions = $17600 \div \frac{880}{7} = 17600 \times \frac{7}{880} = 140$

127. The diameter of a circular wheel is 7 m. How many revolutions will it make in travelling 22 km ?

(SSC CGL Tier II 2013)

(*a*) 100 (*b*) 400 (*c*) 500 (*d*) 1000

Sol. (*d*) Circumference of wheel = $2\pi r = 2 \times \frac{22}{7} \times \frac{7}{2} = 22$ m

Revolutions required = 22000 ÷ 22 = 1000

128. Find the diameter of a wheel that makes 113 revolutions to go 2 km 26 decametres : (Take $\pi = 22/7$)

(SSC CGL Tier I 2003, CDS 2015 M)

(*a*) $4\frac{4}{13}$ m (*b*) $6\frac{4}{11}$ m (*c*) $12\frac{4}{11}$ m (*d*) $12\frac{8}{11}$ m

Sol. (*b*) Total distance covered = 2 km 26 decametres = 2000 m + 260 m = 2260 m

Circumference of wheel = 2260 ÷ 113 = 20 m

$\Rightarrow 2\pi r = 20 \text{ m} \Rightarrow r = 20 \times \frac{1}{2} \times \frac{7}{22} = \frac{35}{11}$ m $\Rightarrow$ Diameter $= 2r = 2 \times \frac{35}{11} = \frac{70}{11} = 6\frac{4}{11}$ metres

129. The diameter of the wheel of a vehicle is 5 metre. It makes 7 revolutions per 9 seconds. What is speed of the vehicle in km/h?

(*a*) 30 km/h (*b*) 36 km/h (*c*) 40 km/h (*d*) 44 km/h

Sol. (*d*) Radius of the wheel = $\frac{5}{2}$ metre

Distance covered in 1 revolution = Circumference of the wheel = $2\pi r = 2 \times \frac{22}{7} \times \frac{5}{2}$ metre

$\therefore$ Distance covered in one second = $2 \times \frac{22}{7} \times \frac{5}{2} \times \frac{7}{9}$ metre

$\therefore$ Speed per hour = $2 \times \frac{22}{7} \times \frac{5}{2} \times \frac{7}{9} \times \frac{18}{5} = 44$ km/h

130. How many revolutions are required to be made by a wheel of 21 cm in radius, to cover the same distance as covered by a wheel of 49 cm in radius in 18 revolutions ?

(*a*) 21 (*b*) 42 (*c*) 49 (*d*) 84

Sol. (*b*) Ratio of two radii = 21 : 49 = 3 : 7

Ratio of the revolutions to cover equal distances = 7 : 3

$\therefore$ Revolutions required to be made by smaller wheel $\frac{7}{3} \times 18 = 42$

131. The difference of the areas of two squares drawn on two line segments of different lengths is 32 sq. cm. Find the length of the greater line segment if one is longer than the other by 2 cm? **(SSC CGL Tier I 2003)**

(*a*) 7 cm (*b*) 9 cm (*c*) 11 cm (*d*) 16 cm

Sol. (*b*) Let the length of smaller segment be *x* cm.

Then $(x + 2)^2 - x^2 = 32 \Rightarrow x^2 + 4x + 4 - x^2 = 32 \Rightarrow 4x = 28 \Rightarrow x = 7$

Length of the greater line segment = $x + 2 = 7 + 2 = 9$ cm

Trick: Option (*a*): $7^2 - 5^2 = 12 \times 2 = 24$. Option (*b*): $9^2 - 7^2 = 16 \times 2 = 32$

132. The diagonals of a rhombus are 32 cm and 24 cm respectively. The perimeter of the rhombus is : **(SSC CGL Tier I 2006, 2001)**

(*a*) 80 cm (*b*) 72 cm (*c*) 68 cm (*d*) 64 cm

Sol. (*a*) Let ABCD is the rhombus whose diagonals BD and AC intersecting at point 'O'.

$AO = \frac{1}{2} \times AC = \frac{1}{2} \times 32 = 16$ cm

$BO = \frac{1}{2} \times BD = \frac{1}{2} \times 24 = 12$ cm

$AB^2 = AO^2 + BO^2 = 16^2 + 12^2 = 20^2$

$\therefore$ Perimeter = 4 × AB = 4 × 20 = 80 cm

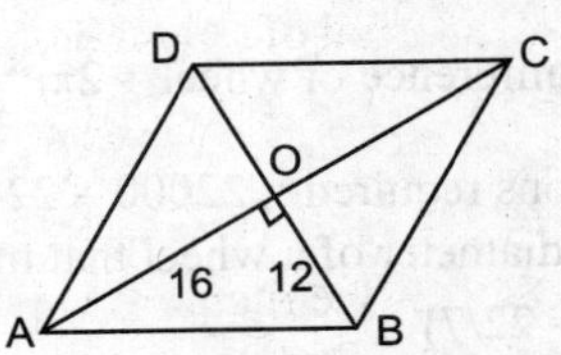

133. The diagonals of a rhombus are 24 cm and 10 cm. Ther perimeter of the rhombus (in cm is: **(CDS 2013M, SSC CGL Tier I 2002**

(*a*) 68 (*b*) 65 (*c*) 54 (*d*) 52

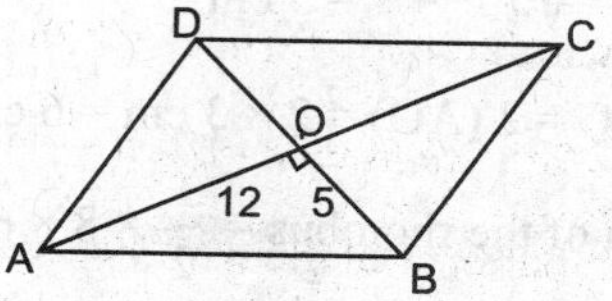

Sol. (*d*) $AO = \frac{1}{2} \times AC = \frac{1}{2} \times 24 = 12$ cm

$BO = \frac{1}{2} \times BD = \frac{1}{2} \times 10 = 5$ cm

$AB = \sqrt{AO^2 + BO^2} = \sqrt{12^2 + 5^2} = 13$ cm

Perimeter = 4 × 13 = 52 cm

134. The diagonals of a rhombus are 12 cm and 16 cm respectively. The length of one side is **(SSC CGL Tier II 2013)**

(*a*) 8 cm (*b*) 6 cm (*c*) 10 cm (*d*) 12 cm

Sol. (*c*) $Side^2 = 6^2 + 8^2 = 10^2 \Rightarrow$ side = 10 cm

135. If the sum of the diagonals of rhombus is 10 cm and its perimeter is $4\sqrt{13}$ cm, find the length of its diagonals?

(*a*) 5, 5 (*b*) 6, 4 (*c*) 7 , 3 (*d*) 8, 2

Sol. (*b*) Side of rhombus = $4\sqrt{13} \div 4 = \sqrt{13}$ cm

From options: $\left(\frac{d_1}{2}\right)^2 + \left(\frac{d_2}{2}\right)^2 = 13 \Rightarrow \left(\frac{6}{2}\right)^2 + \left(\frac{4}{2}\right)^2 = 9 + 4 = 13$

136. The perimeter of a rhombus is 40 cm. If the length of one of its diagonals be 12 cm, the length of the other diagonal is : **(SSC CGL Tier I 2003, 2001)**

(*a*) 14 cm (*b*) 15 cm (*c*) 16 cm (*d*) 12 cm

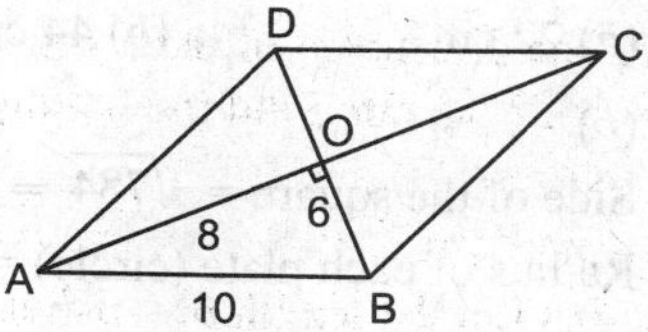

Sol. (*c*) Side of rhombus = 40 ÷ 4 = 10 cm

In right-angle triangle AOB,

$AO^2 = AB^2 - BO^2 = 10^2 - 6^2 = 8^2$

Other diagonal = AC = 2 × AO = 2 × 8 = 16 cm

137. The area of a rhombus with side 13 cm and one diagonal 10 cm will be **(CDS 2015)**

(*a*) 140 cm^2 (*b*) 130 cm^2 (*c*) 120 cm^2 (*d*) 110 cm^2

Sol. (*c*) $AO = \frac{1}{2} \times AC = \frac{1}{2} \times 10 = 5$ cm

$BO = \sqrt{AB^2 - AO^2} = \sqrt{13^2 - 5^2} = 12$ cm $\Rightarrow BD = 2 \times 12 = 24$ cm

Area of rhombus = $\frac{1}{2} \times AC \times BD = \frac{1}{2} \times 10 \times 24 = 120$

138. If side and diagonal of a rhombus are 5 cm and 8 cm respectively. Find area of the rhombus. **(LIC AAO 2009)**

(*a*) 20 cm^2 (*b*) 24 cm^2 (*c*) 40 cm^2 (*d*) 48 cm^2

Sol. (*b*) Let ABCD is the rhombus whose diagonals BD and AC intersecting at point 'O'.
Let AB = 5 cm, BD = 8 cm

$\therefore BO = \frac{1}{2} BD = 4$ cm

$AO = \sqrt{5^2 - 4^2} = 3$ cm

$\therefore AC = 2 (AO) = 2 \times 3$ cm $= 6$ cm

Area of the rhombus $= \frac{1}{2} \times 8 \times 6 = 24$ cm^2

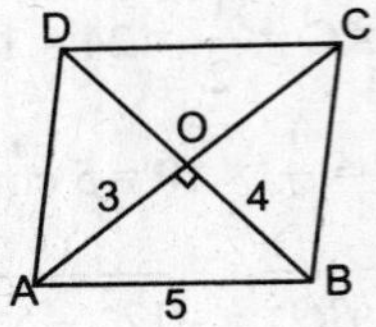

139. From a rectangular metal sheet of sides 25 cm and 20 cm, a circular sheet as large as possible is cut-off. What is the area of the remaining sheet ? **(CDS 2009)**

(*a*) 186 cm^2 (*b*) 144 cm^2 (*c*) 93 cm^2 (*d*) 72 cm^2

Sol. (*a*) Area of rectangular sheet = 25 × 20 = 500 cm^2

Diameter of circular sheet = 20 cm. Radius of circular sheet = 20 ÷ 2 = 10 cm

Area of circular sheet $= \pi r^2 = 3.14 \times (10)^2 = 314$ cm^2

Required area = 500 – 314 = 186 cm^2

140. Four coins of same size are placed on a table so that each coin touches the other two. Find the area of unoccupied space between them, if radius of each coin is 1 cm.

(*a*) 4 cm^2 (*b*) π cm^2

(*c*) $(4 - \pi)$ cm^2 (*d*) $(\pi - 4)$ cm^2

Sol. (*c*) Area unoccupied = Area of square – Area of circle

$= 2^2 - \pi(1)^2 = 4 - \pi$

141. Four equal sized maximum circular plates are cut off from a square paper sheet of area 784 sq. cm. The circumference of each plate is (Take $\pi = 22/7$):**(SSC CGL 2003, CDS 2015)**

(*a*) 22 cm (*b*) 44 cm (*c*) 66 cm (*d*) 88 cm

Sol. (*b*)

Side of the square $= \sqrt{784} = 28$ cm

Radius of each plate (circle) = 28 ÷ 4 = 7 cm

Circumference of each plate $= 2\pi r = 2 \times \frac{22}{7} \times 7 = 44$ cm

142. The area of two circle are in the ratio 1 : 2. If the two circles are bent in the form of squares, what is the ratio of their areas ? **(CDS 2011)**

(*a*) 1 : 2 (*b*) 1 : 3 (*c*) $1 : \sqrt{2}$ (*d*) 1 : 4

Sol. (*a*) The ratio of areas will remain the same in such cases.

143. If the circumference of a circle increases from 4π to 8π, what change occurs in its area ? **(SSC CGL Tier I 2000)**

(*a*) It doubles (*b*) It triples (*c*) It quadruples (*d*) It is halved

Sol. (*c*) Ratio of radii = Ratio of circumferences $= 4\pi : 8\pi = 1 : 2$

Ratio of areas $= 1^2 : 2^2 = 1 : 4$

144. If the ratio of areas of two squares is 225 : 256, then the ratio of their perimeters is : **(SSC CGL Tier I 2004)**

(*a*) 225 : 256 (*b*) 256 : 225 (*c*) 15 : 16 (*d*) 16 : 15

Sol. (*c*) Ratio between areas of two squares = 225 : 256

Ratio between sides of two squares = $\sqrt{225} : \sqrt{256} = 15 : 16$

Radius of perimeters = Ratio of sides = 15 : 16

145. The cost of levelling a square plot of a side 2 metre is ₹1000. What will be the cost of levelling a square plot of a side 10 metre?

(*a*) ₹5000 (*b*) ₹10000 (*c*) ₹20000 (*d*) ₹25000

Sol. (*d*) Ratio of sides = 2 : 10 = 1 : 5 ⇒ Ratio of areas = $1^2 : 5^2 = 1 : 25$

∴ Cost of levelling the bigger plot = 25 × ₹1000 = ₹ 25000

146. The radius of a circle is 7 cm. Find the radius of another circle whose area is 25 times of area of the first circle.

(*a*) 5 cm (*b*) 32 cm (*c*) 35 cm (*d*) 175 cm

Sol. (*c*) Ratio of areas of two circles = 1 : 25

Ratio of radii of two circles = $\sqrt{1} : \sqrt{25} = 1 : 5$

∴ Radius of second circle = 5 × 7 cm = 35 cm

147. The area of circle whose radius is 5 cm is trisected by two concentric circles. The ratio of the radii of the concentric circles in ascending order is

(*a*) 1 : 2 : 3 (*b*) 1 : 2 : 4 (*c*) $1 : \sqrt{2} : \sqrt{3}$ (*d*) $\sqrt{2} : \sqrt{3} : \sqrt{5}$

Sol. (*c*) Ratio of areas of three circles = 1 : 2 : 3

Ratio of radii = $\sqrt{1} : \sqrt{2} : \sqrt{3} = 1 : \sqrt{2} : \sqrt{3}$

148. If the radius of a circle is 9 times greater than the radius of the other circle, the area of the bigger circle is greater by how much times than the area of the smaller circle.

(*a*) 9 times (*b*) 81 times (*c*) 99 times (*d*) 100 times

Sol. (*c*) Let radius of smaller circle = 1 cm. Then radius of bigger circle = 1 + 9 = 10 cm

Area of smaller circle = 1 cm^2. Area of the bigger circle = 100 cm^2

∴ Difference in areas = 100 – 1 = 99 cm^2 *i.e.*, 99 times

149. Find area of uniform path of width 2 metre running from centre of each side to the opposite side of a rectangle field measuring 17 metre by 12 metre.

(*a*) 24 m^2 (*b*) 34 m^2 (*c*) 48 m^2 (*d*) 54 m^2

Sol. (*d*) Area of path = $2 \times (17 + 12) - (2)^2 = 58 - 4 = 54$ m^2

150. Two cross roads, each of width 5 cm, perpendicular to each other, run through the middle of a rectangular field of size 80 m × 60 m. The cost of gravelling the roads at the rate of ₹10 per square metre will be : **(SSC CGL Tier I 2006)**

(*a*) ₹ 6500 (*b*) ₹ 6750 (*c*) ₹ 7000 (*d*) ₹ 7250

Sol. (*b*) Area of cross roads = $5 \times (80 + 60) - (5)^2 = 700 - 25 = 675$ m^2

∴ Cost of gravelling = ₹ 10 × 675 = ₹ 6750

151. A path of uniform width runs outside a square field. Area of the path is 72 m^2. If the width of the path is 2 m, find the area of the field. **(BCA 2008)**

(*a*) 64 m^2 (*b*) 70 m^2 (*c*) 80 m^2 (*d*) 49 m^2

Sol. (*d*) Let side of the square field be *x* m.

Then $(x + 4)^2 - x^2 = 72 \Rightarrow (2x + 4) \times (4) = 72 \Rightarrow (2x + 4) = 72 \div 4 = 18 \Rightarrow 2x = 14 \Rightarrow x = 7$

Area of square field = $7^2 = 49$ m^2

Alt: Area of path = Width × Perimeter + 4 × Width^2

$\Rightarrow 72 - 4 \times 2^2 = 56 \Rightarrow$ Permiter = 56 ÷ Width = 28 ⇒ Side = Perimeter ÷ 4 = 7

153. A square lawn has a path 4 m wide around it. If the area of the path is 196 m^2, then each side of the lawn is **(SSC CGL Tier I 2014)**

(a) 8.25 m (b) 8 m (c) 7.25 m (d) 9.25 m

Sol. (a) $(x + 8)^2 - x^2 = 196 \Rightarrow 16x + 64 = 196 \Rightarrow 16x = 132 \Rightarrow x = 132 \div 16 = 8.25$ m

Alt: $196 - 4 \times 4^2 = 132 \Rightarrow$ Permiter $= 132 \div 4 = 33 \Rightarrow$ Side $= 33 \div 4 = 8.25$

153. A room of dimensions 6 m × 5 m is surrounded by a verandah. If the area of the verandah is 26 m^2, find the width of the verandah.

(*a*) 1 m (*b*) 2 m (*c*) 3 m (*d*) 4 m

Sol. (*a*) Total area $= 6 \times 5 + 26 = 56 = 8 \times 7 \Rightarrow$ Width of verandah $= \frac{8-6}{2}$ or $\frac{7-5}{2} = 1$ m

Alt: Perimeter $= 2 \times (6 + 5) = 22$, if width $= 1$, Area of path $= 22 + 4 \times 1^2 = 26$

154. A path of uniform width runs round the inside of a rectangular field 38 m long and 32 m wide. If the path occupies 600 m^2, the width of the path is: **(SSC CGL Tier I 2007)**

(*a*) 30 m (*b*) 5 m (*c*) 18.75 m (*d*) 10 m

Sol. (*b*) Perimeter of the field $= 2 \times (38 + 32) = 140$ m

Let the width of the path be x m. Then area of the path $= x \times 140 - 4x^2 = 600$ m^2

$\Rightarrow x^2 - 35x + 150 = 0 \Rightarrow x^2 - 5x - 30x + 150 = 0 \Rightarrow (x - 5)(x - 30) = 0 \Rightarrow x = 5$ or $x = 30$

∴ Width of the path = 5 m (as width 30 m is not possible)

Alt: Perimeter $= 2 \times (38 + 32) = 140$. Let width $= 5$, Area of path $= 140 \times 5 - 4 \times 5^2 = 600$

155. What will be the cost of gardening 1 metre broad boundary around a rectangular plot having perimeter of 340 metres at the rate of ₹ 10 per square metre ? **(Canara Bank PO 2003)**

(*a*) ₹ 3400 (*b*) ₹ 1700 (*c*) ₹ 3440 (*d*) None of these

Sol. (*c*) Area of boundary = Width × Perimeter + 4 × width^2 $= 1 \times 340 + 4 \times 1 = 344$.

∴ Required cost = ₹ 10 × 344 = ₹ 3440

156. A square hall is surrounded by a verandah of width 5 m on its outside. If the area of the verandah is 500 m^2, find the side of the hall.

(*a*) 15 m (*b*) 20 m (*c*) 25 m (*d*) 30 m

Sol. (*b*) Let side of hall = x metre

Then area of verandah $= 5 \times (4x) + 4 \times 5^2 = 500 \Rightarrow 20x + 100 = 500$

$\Rightarrow 20x = 500 - 100 = 400 \Rightarrow x = 400 \div 20 = 20$ metre

157. The difference between circumference and the radius of a circle is 74 cm. The radius of the circle is:

(*a*) 7 cm (*b*) 14 cm (*c*) 21 cm (*d*) 28 cm

Sol. (*b*) $2\pi r - r = r(2\pi - 1) = r\left(\frac{44}{7} - 1\right) = r\left(\frac{37}{7}\right) = 74 \text{ cm} \Rightarrow r = 74 \times \frac{7}{37} = 14$ cm

158. A person observed that he required 30 seconds less time to cross a circular ground along its diameter than to cover it once along the boundary. If his speed was 30m/minute, then the radius of the circular ground is (Take $\pi = 22/7$) **(SSC CGL Tier I 2013)**

(*a*) 3.5 m (*b*) 5.5 m (*c*) 7.5 m (*d*) 10.5 m

Sol. (*a*) Distance covered in 30 seconds $= 30 \div 2 = 15$ m

Difference between circumference and diameter $= 2\pi r - 2r = 15$ m

$\therefore 2r(\pi - 1) = 15 \Rightarrow 2r\left(\frac{22}{7} - 1\right) = 15 \Rightarrow 2r = 15 \times \frac{7}{15} = 7 \text{ m} \Rightarrow r = \frac{1}{2} \times 7 = 3.5$ m

159. If length of a rectangle is reduced by 20% and width remaining the same. Area of the resulting rectangle will be decreased by:

(*a*) 10% (*b*) 20% (*c*) 25% (*d*) 40%

Sol. (*b*) Decrease in area = 20%

160. If length of a rectangle is increased by 50%, by what percent should its width be decreased to get the same area as before?

(*a*) 25% (*b*) 33.33% (*c*) 50% (*d*) 100%

Sol. (*b*) Increase in length = 50% = $\frac{1}{2}$ ⇒ Decrease in width = $\frac{1}{2+1} = \frac{1}{3}$ = 33.33%

Alt: $\frac{50}{150} \times 100 = 33.33\%$

161. If length of a rectangle is decreased by 20%, by what percent should its width be increased to get the same area as before?

(*a*) 16.67% (*b*) 20% (*c*) 25% (*d*) 30%

Sol. (*c*) Decrease in length = 20% = $\frac{1}{5}$ ⇒ Increase in width = $\frac{1}{5-1} = \frac{1}{4}$ = 25%

Alt: $\frac{20}{80} \times 100 = 25\%$

162. In measuring the sides of a rectangle, there is an excess of 5% on one side and 2% deficit on the other. Then the error percent in the area is **(SSC MTS 2011)**

(*a*) 3.3 (*b*) 3 (*c*) 2.9 (*d*) 2.7

Sol. (*c*) Error = $5 - 2 - \frac{5 \times 2}{100} = 3 - 0.1 = 2.9\%$

163. In measuring the sides of a rectangle, its length is taken 5% in excess and the breadth 4% less. The percentage of error in the calculation of its area will be : **(SSC CGL Tier I 2006)**

(*a*) $\frac{4}{5}$ (*b*) $\frac{5}{4}$ (*c*) $\frac{3}{4}$ (*d*) $\frac{4}{3}$

Sol. (*a*) Percentage error = $5 - 4 - \frac{5 \times 4}{100} = 1 - \frac{1}{5} = \frac{4}{5}$

164. If each side of a rectangle is increased by 50%, its area will be increased by : **(SSC CGL Tier I 2011M, SSC CGL Tier I 2003)**

(*a*) 50% (*b*) 100% (*c*) 125% (*d*) 250%

Sol. (*c*) Required percentage = $50 + 50 + \frac{50 \times 50}{100} = 125\%$

Alternative method :

Let each side of the rectangle be 10 units. Then its original area = 10 × 10 = 100 sq. units

New area = 15 × 15 = 225 sq. units ⇒ Percent increase in area = 225 – 100 = 125%

165. If all the sides of a square are increased by 200%, its area is increased by :

(*a*) 200% (*b*) 400% (*c*) 600% (*d*) 800%

Sol. (*d*) Increase in Area = $2 \times 200 + \frac{(200)^2}{100} = 400 + 400 = 800\%$

Alternative Method: Ratio of sides = 1 : 3 ⇒ Ratio of areas = $1^2 : 3^2$ = 1 : 9

∴ Increase in area = 9 – 1 = 8 = 800%

166. If radius of a circle is increased by 10%. Area of the resulting circle will be increased by:

(*a*) 11% (*b*) 20% (*c*) 21% (*d*) 25%

Sol. (*c*) Increase in Area $= 2 \times 10 + \frac{(10)^2}{100} = 20 + 1 = 21\%$

Alternative Method : 11 × 11 = 121, 121 – 100% = 21%

167. If all the sides of a square are decreased by 10%, its area will be decreased by:

(*a*) 19% (*b*) 20% (*c*) 21% (*d*) 22%

Sol. (*a*) Increase in Area $= 2 \times (-10) + \frac{(-10)^2}{100} = -20 + 1 = (-)\ 19\%$

Alternative Method : 9 × 9 = 81, 81 – 100 = – 19%

168. If length of a rectangle is increased by 10% and width is decreased by 10%, its area will:

(*a*) 20% increase (*b*) 1% decrease (*c*) 20% decrease (*d*) Remains unchanged

Sol. (*b*) Decrease in Area $= \frac{10^2}{100} = 1\%$

Alternative Method : 11 × 9 = 99 99 – 100 = – 1%

169. If the radius of a circle is reduced by 50%. Its area is reduced by :

(*a*) 25% (*b*) 50% (*c*) 62.5% (*d*) 75%

Sol. (*d*) Required % $= -50 - 50 + \frac{50 \times 50}{100} = -75\%$

Alt: Ratio of radii = 2 : 1 Ratio of areas = $2^2 : 1^2 = 4 : 1$

Decrease in area $= \frac{3}{4} \times 100 = 75\%$

170. The length and breadth of a rectangle are increased by 20% and 25% respectively. The increase in area of the resulting rectangle will be : **(SSC CHSL 2010)**

(*a*) 60 (*b*) 50% (*c*) 40% (*d*) 30%

Sol. (*b*) Increase in Area $= 20 + 25 + \frac{20 \times 25}{100} = 45 + 5 = 50\%$

171. ABCD is a square. If the sides AB and CD are increased by 30%, sides BC and AD are increased by 20%, then the area of the resulting rectangle exceeds the area of the square by **(CDS 2015)**

(*a*) 50% (*b*) 52% (*c*) 54% (*d*) 56%

Sol. (*d*) Increase in area $= 30 + 20 + \frac{30 \times 20}{100} = 56\%$

Alt: 13 × 12 = 156 ⇒ 156 – 100 = 56%

172. A towel was 40 cm broad and 90 cm long. When bleached, it was found to have lost 20% of its length and 10% of its width. The percentage decrease in area is:

(*a*) 28% (*b*) 30% (*c*) 32% (*d*) 33%

Sol. (*a*) Increase in Area $= -20 - 10 + \frac{(-20) \times (-10)}{100} = -30 + 2 = (-)\ 28\%$

Alternative Method : 8 × 9 = 72 72 – 100 = – 28%

173. A gardener increased the area of his rectangular garden by increasing its length by 40% and decreasing its width by 20%. The area of the new garden **(UPSC Civil Services Pre. 2014)**

(*a*) has increased by 20% (b) has increased by 12%

(c) has increased by 8% (d) is exactly the same as the old area

Sol. (*b*) $+40 - 20 - \frac{40 \times 20}{100} = 20 - 8 = (+)\ 12\% = 12\%$ increase

174. If the length of a rectangular plot of land is increased by 5% and its breadth is decreased by 10%, by how much will its area change? **(SSC CGL Tier I 2008)**

(*a*) Increase by 5% (*b*) Decrease by 5% (*c*) Increase by 5.5% (*d*) Decrease by 5.5%

Sol. (*d*) Change in area $= 5 - 10 - \frac{5 \times 10}{100} = (-)\ 5.5\%$

175. If the length of a rectangle is increased by one-third and its width is decreased by one-third, the percentage of decrease in its area will be : **(SSC CGL Tier I 2006)**

(*a*) $66\frac{2}{3}$ (*b*) $33\frac{1}{3}$ (*c*) $16\frac{2}{3}$ (*d*) $11\frac{1}{9}$

Sol. (*d*) Let length and breadth of rectangle be 1 unit each.

Then new dimensions $= \frac{4}{3}$ and $\frac{2}{3}$ $\Rightarrow$ Change in area $= 1 \times 1 - \frac{4}{3} \times \frac{2}{3} = \frac{1}{9} = 11\frac{1}{9}\%$

176. A man walked diagonally across a square plot. Approximately what was the percent time saved by not walking along the sides ?

(*a*) 30% (*b*) 40% (*c*) 50% (*d*) 33.33%

Sol. (*a*) Let side of the square = 1 unit

Then diagonal of the square $= \sqrt{2}$ unit

Difference of distances covered $= (1 + 1) - \sqrt{2} = 2 - \sqrt{2}$

Percent time saved $= \frac{2 - \sqrt{2}}{2} \times 100\% = \frac{2 - 1.414}{2} \times 100\ \% = 29.3\% \approx 30\%$

177. The medians of a triangle are 9 cm, 12 cm, 15 cm, then the area of the triangle is

(*a*) 48 cm^2 (*b*) 24 cm^2 (*c*) 144 cm^2 (*d*) 72 cm^2

Sol. (*d*) Area of triangle $= \frac{4}{3}$ area of triangle formed by its medians

$9^2 + 12^2 = 15^2 \Rightarrow$ They form a right triangle $\Rightarrow$ Required area $= \frac{4}{3} \times \left(\frac{1}{2} \times 9 \times 12\right) = 72\ cm^2$

178. The medians AD, BE and CF of a triangle ABC are of lengths 18 cm, 24 cm and 30 cm respectively. The area in cm^2 of the triangle is **(SSC CHSL 2013)**

(*a*) 96 (*b*) 192 (*c*) 288 (*d*) 374

Sol. (c) Area of triangle $= \frac{4}{3} \times$ triangle formed by medians

$18^2 + 24^2 = 30^2 \Rightarrow$ Area $= \frac{1}{2} \times 18 \times 24 = 216\ cm^2$

$\Rightarrow$ Required area $= \frac{4}{3} \times 216 = 288\ cm^2$

179. An order was placed for supply of carpet of breadth 3 metres, the length of the carpet was 1.44 times of breadth. Subsequently the breadth and length were increased by 25 and 40 per cent respectively. At the rate of ₹ 45 per square metre, what would be the increase in the cost of the carpet ? **(Indian Overseas Bank P.O. 2009)**

(*a*) ₹ 1020.6 (*b*) ₹ 398.8 (*c*) ₹ 437.4 (*d*) ₹ 583.2

Sol. (*c*) Original area of the carpet = $3 \times (3 \times 1.44) = 12.96 \text{ m}^2$

Increased area of carpet = $\left(3 \times \frac{125}{100}\right) \times \left(3 \times 1.44 \times \frac{140}{100}\right) = 22.68 \text{ m}^2$

$\therefore$ Increase in area = $22.68 - 12.96 = 9.72 \text{ m}^2$

Increase in cost of carpet = ₹ 45×9.72 = ₹ 437.4

Alternative method : Original cost = $3 \times (3 \times 1.44) \times$ ₹ 45

Percentage increase in the cost = Percentage increase in area = $25 + 40 + \frac{25 \times 40}{100} = 75\%$

Total increase in cost = 75% of $3 \times (3 \times 1.44) \times 45 = \frac{3}{4} \times 3 \times 3 \times 1.44 \times 45$ = ₹ 437.4

180. The hypotenuse of an isosceles triangle is 5 cm. Find the area of the triangle.

(*a*) 5 cm^2 (*b*) 6.25 cm^2 (*c*) 6.5 cm^2 (*d*) 12.5 cm^2

Sol. (*b*) Let equal sides of the triangle be x cm each.

Then $x^2 + x^2 = 25 \Rightarrow 2x^2 = 25 \Rightarrow x^2 = \frac{25}{2}$

Area of triangle = $\frac{1}{2}x^2 = \frac{1}{2} \times \frac{25}{2} = \frac{25}{4} = 6.25 \text{ cm}^2$

181. PQR is an equilateral triangle. O is the point of intersection of altitudes PL, QM and RN. If OP = 8 cm, then what is the perimeter of the triangle PQR? **(CDS 2013)**

(*a*) $8\sqrt{3}$ cm (*b*) $12\sqrt{3}$ cm (*c*) $16\sqrt{3}$ cm (*d*) $24\sqrt{3}$ cm

Sol. (*d*) PL, QM, RN are altitudes and also medians.

$\Rightarrow$ O divides PL in 2 : 1

$\Rightarrow$ OL = OP ÷ 2 = 8 ÷ 2 = 4 cm

$\Rightarrow$ PL = (8 + 4) cm = 12 cm

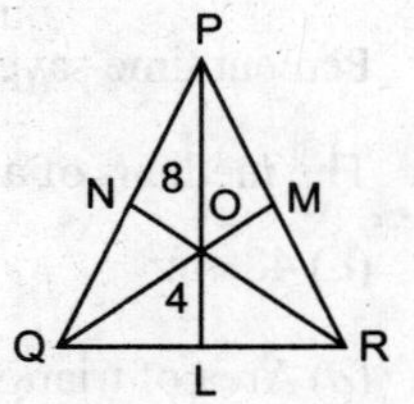

Altitude = $\frac{\sqrt{3}}{2}a = 12 \Rightarrow a = \frac{12 \times 2}{\sqrt{3}} = 8\sqrt{3}$

Preimeter of equilateral triangle = $3 \times 8\sqrt{3}$ cm = $24\sqrt{3}$ cm

182. What is the maximum area of a rectangle, the perimeter of which is 18 cm ? **(CDS 2010)**

(*a*) 20.25 cm^2 (*b*) 20.00 cm^2 (*c*) 19.75 cm^2 (*d*) 19.60 cm^2

Sol. (*a*) $2 \times (l + b) = 18 \Rightarrow (l + b) = 18 \div 2 = 9$

$(l \times b)$ is maximum when $l = b \Rightarrow l = b = 9 \div 2 = 4.5$

$\therefore$ Maximum area = $(l \times b) = 4.5 \times 4.5 = 20.25$

183. The perimeter of a square is 48 cm. The area of a rectangle is 4 cm less than the area of the square. If the length of the rectangle is 14 cm, then its perimeter is: **(SSC CGL Tier I 2012)**

(*a*) 44 cm (*b*) 40 cm (*c*) 52 cm (*d*) 48 cm

Sol. (*d*) Side of Square = 48 ÷ 4 = 12 cm $\Rightarrow$ Area of the rectangle = $12^2 - 4 = 140$ cm^2

$\Rightarrow$ Breadth of the rectangle = 140 ÷ 14 = 10 cm

$\Rightarrow$ Perimeter of the rectangle = $2 \times (14 + 10) = 48$ cm

184. The perimetres of a rectangle and a square are 160 m each. The area of the rectangle is less than that of the square by 100 sq. m. The length of the rectangle is : **(SSC CGL Tier I 2005)**

(*a*) 30 m (*b*) 60 m (*c*) 40 m (*d*) 50 m

Sol. (*d*) Side of the square = 160 ÷ 4 = 40 metre

Let the sides of rectangle be $(40 + x)$ and $(40 - x)$

Difference in areas of square and rectangle = $(40 \times 40) - (40 + x)(40 - x) = 100$

$\Rightarrow 1600 - 1600 + x^2 = 100 \Rightarrow x^2 = 100 \Rightarrow x = 10$

$\therefore$ Length of rectangle = $40 + x = 40 + 10 = 50$ metre

Alternative method : Side of the square = 160 ÷ 4 = 40 metre

Difference in areas of square and rectangle = 100 m^2

$\therefore$ Increase and decrease in dimensions = $\sqrt{100}$ = 10 metre

$\therefore$ Sides of rectangle are (40 + 10) and (40 – 10) metre

185. A square and a rectangle each has perimeter 60 metre. Difference between the areas of the two figures is 16 square metre. Find length of the rectangle.

(*a*) 4 m (*b*) 11 m (*c*) 16 m (*d*) 19 m

Sol. (*d*) Side of the square = 60 ÷ 4 = 15 metre

Difference in areas of square and rectangle = 16 m^2

$\therefore$ Increase and decrease in dimensions = $\sqrt{16}$ = 4 metre

$\therefore$ Sides of rectangle are (15 + 4) and (15 – 4) metre

186. If each of two opposite sides of a square is increased by 4 cm and each of other two sides is decreased by 4 cm, we obtain a rectangle of area 84 cm^2. What is the perimeter (in cm) of the original square ? **(SAS 2010)**

(*a*) 40 (*b*) 44 (*c*) 48 (*d*) 52

Sol. (*a*) Difference of areas = $4 \times 4 = 16$ cm^2 $\Rightarrow$ Area of square = $84 + 4^2 = 84 + 16 = 100$

Side of square = $\sqrt{100} = 10 \Rightarrow$ Perimeter = $4 \times 10 = 40$ cm

187. The areas of a square and a rectangle are equal. The length of the rectangle is greater than the length of a side of the square by 5 cm and the breadth is less than the length of the side of the square by 3 cm. The perimeter of the rectangle is : **(SSC CGL Tier I 2002)**

(*a*) 17 cm (*b*) 26 cm (*c*) 30 cm (*d*) 34 cm

Sol. (*d*) Let the side of the square be x cm.

Then length and breadth of the rectangle = $(x + 5)$ and $(x - 3)$

According to question, $(x + 5) \times (x - 3) = x^2 \Rightarrow 2x - 15 = 0 \Rightarrow x = 7.5$

Length = 7.5 + 5 = 12.5 cm, Breadth = 7.5 – 3 = 4.5 cm

Perimeter = $2 \times (12.5 + 4.5) = 34$ cm

188. The length and the breadth of a rectangular plot are in the ratio 5 : 4. If the length is reduced by 5 meters and breadth is increased by 3 meters, then the area is reduced by 40 m^2. The length and breadth of the plot are:

(*a*) 23, 20 (*b*) 25, 20 (*c*) 50, 40 (*d*) 25, 16

Sol. (*b*) Let the length and the breadth are $5x$ and $4x$ respectively.

Then $(5x)(4x) = (5x - 5)(4x + 3) + 40$

$\Rightarrow 20x^2 = 20x^2 + 15x - 20x - 15 + 40 \Rightarrow 5x = 25 \Rightarrow x = 5$ meters

Length of plot = $5x = 25$ m. Breadth of plot = $4x = 20$ m

189. The length and the breadth of a rectangle are in the ratio 2 : 1. If its length is decreased by 5 cm and breadth is decreased by 5 cm, area is increased by 75 cm^2. The original length of the rectangle is

(*c*) 20 cm (*b*) 30 cm (*c*) 40 cm (*d*) 50 cm

Sol. (*c*) Option (a) $20 \times 10 = 200$; $15 \times 15 = 225 \Rightarrow$ Difference = 25

Option (b) $30 \times 15 = 450$; $25 \times 20 = 500 \Rightarrow$ Difference = 50

Option (c) $40 \times 20 = 800$; $35 \times 25 = 875 \Rightarrow$ Difference = 75

Alt: Let breadth = x, length = $2x$. Then $(2x - 5)(x + 5) = (2x)(x) + 75$

$\Rightarrow 2x^2 + 5x - 25 = 2x^2 + 75 \Rightarrow 5x = 100 \Rightarrow x = 20 \Rightarrow$ length = $2x$ = 40 cm

190. The ratio of the outer and the inner perimeters of a circular path is 23 : 22. If the path is 5 metres wide, the diameter of the inner circle is : **(SSC CGL Tier I 2004)**

(*a*) 110 m (*b*) 55 m (*c*) 220 m (*d*) 230 m

Sol. (*c*) Let the inner and outer radii be r and R respectively.

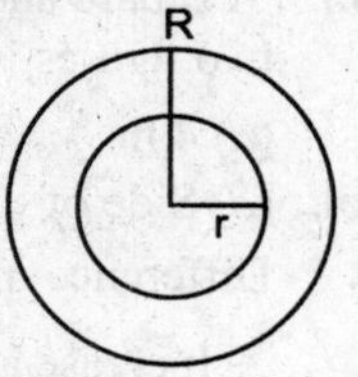

Then $2\pi R : 2\pi r = 23 : 22 \Rightarrow R : r = 23 : 22$

Difference in radii = (23 – 22) ratio = 1 ratio = 5 metres

$\therefore$ Inner radius = r = 22 ratio = 110 metres

$\therefore$ Inner diameter = $2r$ = 220 meters

191. A uniform path of 7 metre surrounds a circular field whose diameter is 42 metre. Find the area of the path.

(*a*) 154 m^2 (*b*) 1001 m^2 (*c*) 1078 m^2 (*d*) 2002 m^2

Sol. (*c*) Radius of field = $\frac{1}{2} \times 42 = 21$ m $\Rightarrow$ Radius of outer path = 21 + 7 = 28 m

$\therefore$ Area of path $= \frac{22}{7} \times (28^2 - 21^2) = \frac{22}{7} \times 49 \times 7 = 1078\ m^2$

192. The circumference of two circles are 176 metre and 132 metre respectively. What is the difference between their radii?

(*a*) 5 metre (*b*) 7 metre (*c*) 11 metre (*d*) 16 metre

Sol. (*b*) Let radii of the circles are R and r respectively.

Then $2\pi R = 176$ and $2\pi r = 132 \Rightarrow 2\pi R - 2\pi r = 176 - 132 \Rightarrow 2\pi(R - r) = 44$

$\Rightarrow (R - r) = \frac{44}{2\pi} = 44 \times \frac{1}{2} \times \frac{7}{22} = 7$ metre

193. A circular park has a path of uniform width around it. The difference between outer and inner circumferences of the circular path is 132 m. Its width is (Take $\pi = 22/7$) : **(SSC CGL Tier I 2003)**

(*a*) 22 m (*b*) 20 m (*c*) 21 m (*d*) 24 m

Sol. (*c*) Let external and internal radii be R and r respectively.

Then $2\pi R - 2\pi r = 2\pi(R - r) = 132 \Rightarrow (R - r) = 132 \times \frac{1}{2} \times \frac{7}{22} = 21$ m

194. A circular road runs around a circular ground. If the difference between the circumferences of the outer circle and the inner circle is 66 metres, the width of the road is : (Take $\pi = 22/7$): **(SSC CGL Tier I 2013)**

(*a*) 21 metres (*b*) 10.5 metres (*c*) 7 metres (*d*) 5.25 metres

Sol. (*b*) $2\pi R - 2\pi r = 2\pi(R - r) = 66 \Rightarrow (R - r) = 66 \times \frac{1}{2} \times \frac{7}{22} = 10.5$ metres

195. The area of a circle is increased by 22 cm^2 when its radius is increased by 1 cm. The original radius of the circle is: **(SSC CGL Tier I 2007)**

(*a*) 6 cm (*b*) 3.2 cm (*c*) 3 cm (*d*) 3.5 cm

Sol. (*c*) Let radius = r cm. Then $\pi(r + 1)^2 - \pi r^2 = 22$ cm $\Rightarrow \pi(2r + 1) = 22$ cm

$\Rightarrow 2r + 1 = 22 \times \frac{7}{22} = 7$ cm $\Rightarrow 2r = 7 - 1 = 6$ cm $\Rightarrow r = 3$ cm

196. Two circles touch externally. The sum of their areas is 130 π sq. cm and the distance between their centres is 14 cm. The radius of the smaller circle is **(SSC CGL Tier I 2015)**

(*a*) 5 cm (*b*) 2 cm (*c*) 3 cm (*d*) 4 cm

Sol. (*c*) $\pi(R^2 + r^2) = 130\,\pi \Rightarrow (R^2 + r^2) = 130$ $(R + r) = 14$ (given)

$(R + r)^2 - (R^2 + r^2) = 14^2 - 130 \Rightarrow 2Rr = 66 \Rightarrow Rr = 33 = 11 \times 3$

$\Rightarrow$ Radius of the smaller circle = 3 cm

197. A ladder 8.2 m long is placed at 8 m high wall. At what distance from the base should the ladder be placed so that it just reaches the wall?

(*a*) 0.2 m (*b*) 0.4 m (*c*) 0.9 m (*d*) 1.8 m

Sol. (*d*) Let AB is the wall and AC is the ladder.

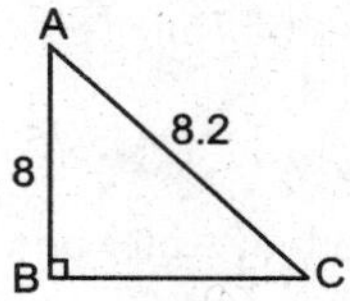

Then $BC^2 = AC^2 - AB^2 = 8.2^2 - 8^2$
$= (8.2 + 8) \times (8.2 - 8) = 16.2 \times 0.2 = 3.24$

$BC = \sqrt{3.24} = 1.8$ m

198. A ladder 25 m long is leaning against a wall which is perpendicular to the level ground. The bottom of the ladder is 7 m from the base of the wall. If the top of the ladder slips down 4 m, how much will the bottom of the ladder slip ? **(CDS 2009)**

(*a*) 7 m (*b*) 8 m (*c*) 10 m (*d*) 15 m

Sol. (*b*) Let height of wall = AC = x

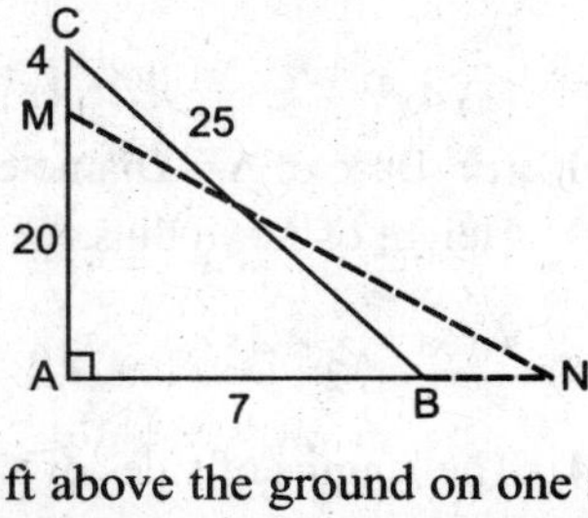

Then $AC^2 + AB^2 = BC^2 \Rightarrow x^2 + 7^2 = 25^2$

$\Rightarrow x^2 = 25^2 - 7^2 = 625 - 49 = 576 = 24^2 \Rightarrow x = 24$

AM = 24 – 4 = 20 m

$AM^2 + AN^2 = MN^2 \Rightarrow 20^2 + AN^2 = 25^2$

$\Rightarrow AN^2 = 25^2 - 20^2 = 625 - 400 = 225 = 15^2 \Rightarrow AN = 15$

$\therefore$ The bottom of the ladder slips = 15 – 7 = 8 m

199. A ladder of 17 feet length reaches a window which is 15 ft above the ground on one side of the street. Keeping its foot at the same point the ladder is turned to the other side of the street and now it reaches a window 8 ft high. What is the width of the street? **(CDS 2010)**

(*a*) 23ft (*b*) 15 ft (*c*) 25 ft (*d*) 30 ft

Sol. (*a*) In right-angle triangle ABC,

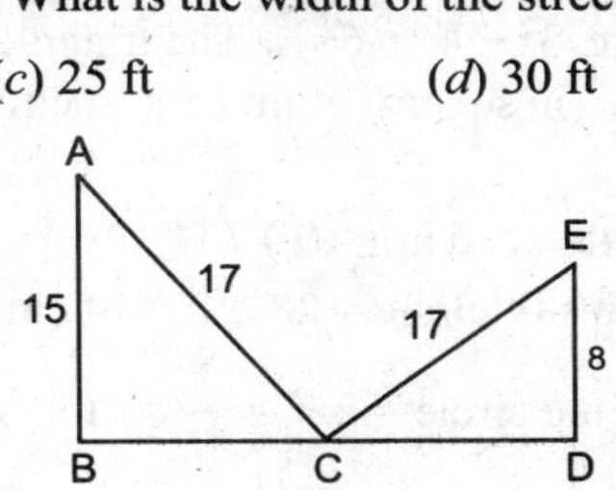

$BC^2 = AC^2 - AB^2 = 17^2 - 15^2$

$= 289 - 225 = 64 = 8^2 \Rightarrow BC = 8$

$CD^2 = 17^2 - 8^2 = 15^2 \Rightarrow CD = 15$

Width of street = BC + CD = 8 + 15 = 23 ft

200. Two poles whose heights are 11 metre and 5 metre stand upright in a field. If the distance between their feet is 8 metre, what is the distance between their tops ? **(CDS 2012)**

(*a*) 6 metre (*b*) 8 metre (*c*) 10 metre (*d*) 12 metre

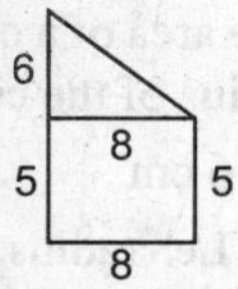

Sol. (*c*) Distance between feet of two poles = 8 metre

Difference between heights of two poles = 11 – 5 = 6 metre

Distance between the top of poles = $\sqrt{6^2+8^2}$ = 10 metre

201. Two poles of heights are 6 m and 11 m stand vertically upright on a plane ground. If the distance between their feet is 12 m, what is the distance between their tops ? **(CDS 2010)**

(*a*) 11 m (*b*) 12 m (*c*) 13 m (*d*) 14 m

Sol. (*c*)

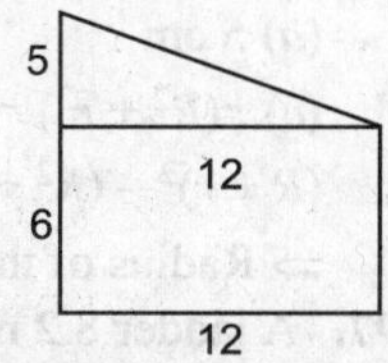

Distance between feet of two poles = 12 metre

Difference between heights of two poles = 11 – 6 = 5 metre

Distance between the top of poles

$= \sqrt{5^2+12^2} = \sqrt{25+144} = \sqrt{169} = 13$ m

202. A lotus is seen 5 cm above the water level of a lake. With the onset of the wind, it sinks in the water 10 cm away from its place. How deep is the water in that place?

(SSC CGL Tier I 2004)

(*a*) 5 cm (*b*) $5\sqrt{5}$ cm (*c*) 7.5 cm (*d*) 10 cm

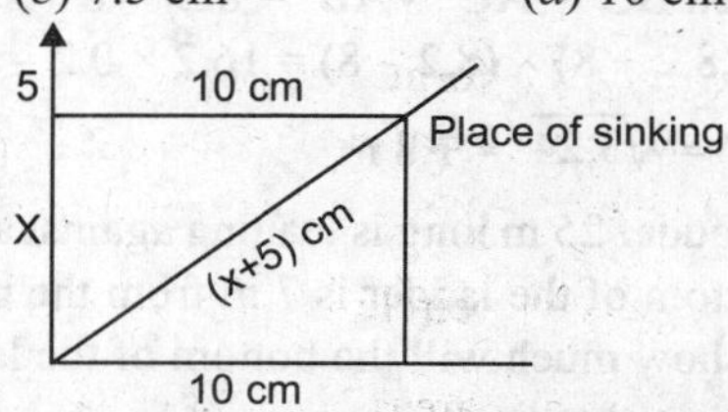

Sol. (*c*) Let the depth of water be *x* cm.

Then $(x+5)^2 = x^2 + 10^2$

$\Rightarrow x^2 + 10x + 25 = x^2 + 100$

$\Rightarrow 10x = 100 - 25 = 75$

$\Rightarrow x = 7.5$ cm.

203. The area of the largest triangle that can be inscribed in a semi circle of radius *x* in square unit is : **(SSC CGL Tier II 2010 SSC CGL Tier I 2012, CDS 2015)**

(*a*) $4x^2$ (*b*) $2x^2$ (*c*) x^2 (*d*) $3x^2$

Sol. (*c*) Base of Δ = Diameter = 2*x*

Height of Δ = radius = *x*

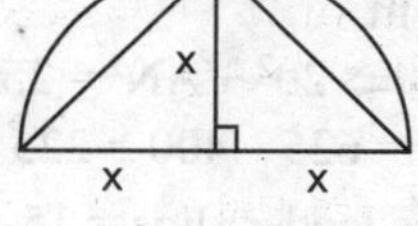

Area $= \frac{1}{2} \times 2x \times x = x^2$

204. The lengths of sides of a triangle are in the ratio 3 : 4 : 6, then the triangle is **(SSC CGL Tier I 2004)**

(*a*) right angled (*b*) Acute angled (*c*) obtuse angled (*d*) Can't say

Sol. (*c*) Here, $3^2 + 4^2 < 6^2 \Rightarrow$ The triangle is obtuse angled.

205. The area (in sq. cm) of the largest circle that can be drawn inside a square of side 28 cm, is: **(SSC CGL Tier I 2000)**

(*a*) 17248 (*b*) 784 (*c*) 8624 (*d*) 616

Sol. (*d*) Radius of circle = 28 ÷ 2 = 14 cm

Area of the circle $= \pi r^2 = \frac{22}{7} \times 14 \times 14 = 616$ cm^2

Alt: Area of square = 28 × 28. Ratio of areas of square and circle = 14 : 11

Area of circle = $28 \times 28 \times \frac{11}{14} = 616 \text{ cm}^2$

206. If the area of a circle inscribed in a square is 9π cm^2, then the area of the square is : **(SSC CGL Tier I 2008)**

(*a*) 24 cm^2 (*b*) 30 cm^2 (*c*) 36 cm^2 (*d*) 81 cm^2

Sol. (*c*) Ratio of areas of square and circle = $4 : \pi \Rightarrow$ Area of square = $\frac{4}{\pi} \times 9\pi = 36 \text{ cm}^2$

207. A wire when bent in the form of a square encloses an area of 484 sq. cm. What will be the enclosed area when the same wire is bent into the form of a circle ? (Take $\pi = 22/7$) **(SSC CGL Tier I 2005, 2001)**

(*a*) 462 cm^2 (*b*) 539 cm^2 (*c*) 616 cm^2 (*d*) 693 cm^2

Sol. (*c*) Side of square = $\sqrt{484}$ = 22 cm $\Rightarrow$ Perimeter of square = 4 × 22 = 88 cm

$\Rightarrow$ Perimeter of circle = $2\pi r$ = 88 cm $\Rightarrow r = 88 \times \frac{1}{2} \times \frac{7}{22} = 14$ cm

Area of circle = $\pi r^2 = \frac{22}{7} \times 14 \times 14 = 616 \text{ cm}^2$

Alternative method: Area of the square = 484 sq. cm.

Ratio of areas of circle and square = 14 : 11 $\Rightarrow$ Area of circle = $\frac{14}{11} \times 484 = 616$ sq. cm.

208. A wire in the form of a circle of radius 7 cm is bent in the form of square. What is the difference between area of the circle and that of the square?

(*a*) 33 cm^2 (*b*) 66 cm^2 (*c*) 77 cm^2 (*d*) 154 cm^2

Sol. (*a*) Area of the circle = $\pi r^2 = \frac{22}{7} \times 7 \times 7 = 154 \text{ cm}^2$

Ratio between areas of circle and square = 14 : 11

Difference between areas of two = (14 – 11) ratio = 3 ratio

Actual difference between areas of two figures = $\frac{3}{14} \times 154 = 33 \text{ cm}^2$

209. A wire, when bent in the form of a square, encloses a region having area of 121 cm^2. If the same wire is bent into the form of a circle, then area of the circle is : (take $\pi = 22/7$) **(SSC CGL Tier I 2008)**

(*a*) 144 cm^2 (*b*) 154 cm^2 (*c*) 176 cm^2 (*d*) 180 cm^2

Sol. (*b*) Ratio of area of square and circle with same perimeter = 11 : 14

$\therefore$ Area of circle = $\frac{14}{11} \times 121 = 154 \text{ cm}^2$

210. A circular wire of diameter 112 cm is cut and bent in the form of a rectangle whose sides are in the ratio of 9 : 7. The small side of the rectangle is **(SSC CGL Tier I 2014)**

(*a*) 77 cm (*b*) 97 cm (*c*) 67 cm (*d*) 87 cm

Sol. (*a*) $r = 112 \div 2 = 56$ cm $\Rightarrow 2\pi r = 2 \times \frac{22}{7} \times 56 = 352$

$\Rightarrow 2(9x + 7x) = 352 \Rightarrow x = 11 \Rightarrow$ Small side = $7x = 77$

211. A circular wire of diameter 42 cm is bent in the form of a rectangle whose sides are in the ratio 6 : 5. The area of the rectangle is : (use $\pi = 22/7$) **(SSC CGL Tier I 2005)**

(*a*) 540 cm^2 (*b*) 1080 cm^2 (*c*) 2160 cm^2 (*d*) 4320 cm^2

Sol. (*b*) Length of wire = circumference of circle = $2\pi r = 2 \times \frac{22}{7} \times \frac{42}{2} = 132$ cm

Then $2(6x + 5x) = 132$ cm $\Rightarrow x = 6$ cm

$\therefore$ Sides of the rectangle are $6x$ and $5x = 6 \times 6$ and $5 \times 6 = 36$ and 30 cm

$\therefore$ Area of the rectangle = $36 \times 30 = 1080$ cm^2

212. A wire when bent in the form of an equilateral triangle encloses a region having area of $121\sqrt{3}$ cm^2. If the same wire is rebent in the form of a circle, it radius will be: (take $\pi = 22/7$) **(SSC CGL Tier I 2008)**

(*a*) 21 cm (*b*) 15.75 cm (*c*) 10.5 cm (*d*) 9 cm

Sol. (*c*) Area of the equilateral triangle = $\frac{\sqrt{3}}{4} \times (\text{Side})^2 = 121\sqrt{3}$ cm^3

$\Rightarrow (\text{Side})^2 = 121\sqrt{3} \times \frac{4}{\sqrt{3}} = 121 \times 4 = 11^2 \times 2^2 \Rightarrow (\text{Side}) = 11 \times 2 = 22$

Perimeter of the triangle = $3 \times 22 = 66$ cm $\Rightarrow$ Perimeter of the circle = $2\pi r = 66$ cm

Radius of the circle = $r = 66 \times \frac{1}{2} \times \frac{7}{22} = 10.5$ cm

213. If a wire is bent into the shape of a square, the area of the square is 81cm^2. When the wire is bent into a semicircular shape, the area of the semicircle is : **(SSC CGL Tier I 2001, 2002, 2011)**

(*a*) 154 cm^2 (*b*) 77 cm^2 (*c*) 44 cm^2 (*d*) 22 cm^2

Sol. (*b*) Side of the square = $\sqrt{81} = 9$ cm $\Rightarrow$ Perimeter of the square = $4 \times 9 = 36$ cm

Perimeter of the semicircle = $\pi r + 2r = (\pi + 2)\, r = 36$ cm $\Rightarrow r = 36 \times \frac{7}{36} = 7$ cm

Area of semicircle = $\frac{1}{2}\pi r^2 = \frac{1}{2} \times \frac{22}{7} \times 7 \times 7 = 77$ cm^2

214. If the areas of a circle and a square are equal, then the ratio of their perimetres is : **(SSC CGL Tier I 2005)**

(*a*) 1 : 1 (*b*) 2 : π (*c*) π : 2 (*d*) $\sqrt{\pi}$: 2

Sol. (*d*) Let radius of the circle be r and side of square be x. Then $x^2 = \pi r^2 \Rightarrow x = \sqrt{\pi} r$

$\therefore$ Required ratio = $2\pi r : 4x = 2\pi r : 4(\sqrt{\pi}\, r) = \sqrt{\pi} : 2$

215. The perimeter of a triangle is 30 cm and its area is 30 cm^2. If the largest side measures 13 cm, what is the length of the smallest side of the triangle ? **(SSC CGL Tier I 2003)**

(*a*) 3 cm (*b*) 4 cm (*c*) 5 cm (*d*) 6 cm

Sol. (*c*) Let the smallest side of the triangle be x cm.

Then third side = $30 - (13 + x) = (17 - x)$ cm

$s = \frac{a+b+c}{2} = \frac{30}{2} = 15$ cm

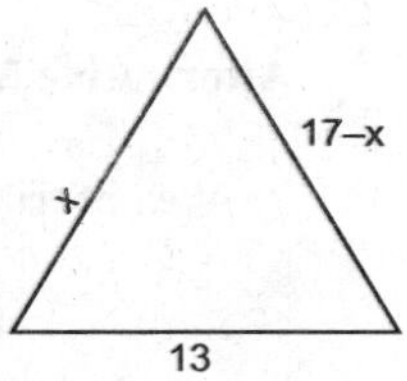

Area = $\sqrt{s(s-a)(s-b)(s-c)}$

$\Rightarrow 30 = \sqrt{15(15-13)(15-x)(15-17+x)}$

$\Rightarrow 900 = 15 \times 2 \times (15-x) \times (x-2) \Rightarrow 30 = 15x - x^2 - 30 + 2x$

$\Rightarrow x^2 - 17x + 60 = 0 \Rightarrow x^2 - 12x - 5x + 60 = 0 \Rightarrow x(x-12) - 5(x-12) = 0$

$\Rightarrow (x-5)(x-12) = 0 \Rightarrow x = 5$ or $x = 12 \Rightarrow$ Smallest side is 5 cm.

Trick : We know that $13^2 = 5^2 + 12^2$ and $5 + 12 + 13 = 30 \Rightarrow$ The smallest side = 5 cm

216. Find the length (in cm) of the base of an isosceles triangle whose area is 60 cm^2 and length of one of its equal sides is 13 cm.

(*a*) 12 or 5 (*b*) 24 or 5 (*c*) 24 or 10 (*d*) 12 or 10

Sol. (*c*) Draw AD ⊥ BC ⇒ Δ ABD is a right Δ.

AD is also median (isosceles Δ)

$\therefore$ Area (Δ ABD) = $\frac{1}{2} \times \Delta ABC = \frac{1}{2} \times 60 = 30$ cm^2

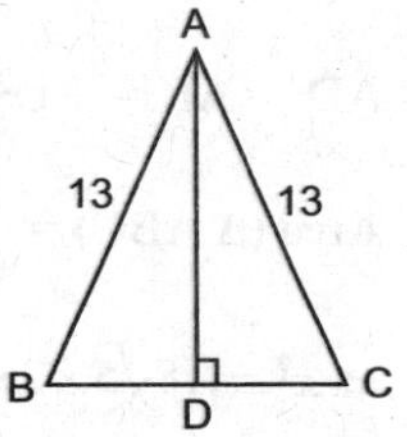

We know, $13^2 = 5^2 + 12^2$

If AD, BD are 5 or 12 cm, then

area (Δ ABD) = $\frac{1}{2} \times 5 \times 12 = 30$ cm^2

$\Rightarrow$ BC = 2 BD = 2 (5 or 12) = 10 or 24 cm

217. Between a square of perimeter 44 cm and a circle of circumference 44 cm, which figure has larger area and by how much? **(SSC CGL Tier I 2000)**

(*a*) Square by 33 cm^2 (*b*) Circle by 33 cm^2 (*c*) Both have equal area (*d*) Square by 495 cm^2

Sol. (*b*) Side of the square = 44 ÷ 4 = 11 cm. Area of square = 11^2 = 121 cm^2

Circumference of circle = $2\pi r$ = 44 cm ⇒ Radius of circle = $r = 44 \times \frac{1}{2} \times \frac{7}{22} = 7$ cm

Area of circle = $\pi r^2 = \frac{22}{7} \times 7 \times 7 = 154$ cm^2

$\therefore$ Circle has larger area by 154 – 121 = 33 cm^2

Alternative method : x = Side of square = 44 ÷ 4 = 11 cm

We know that between square and circle with same perimeter, circle always has more area.

Difference between areas of circle and square = $\frac{3}{11} \times x^2 = \frac{3}{11} \times 11 \times 11 = 33$ cm^2

218. An equilateral triangle and a regular hexagon have the same perimeter. The ratio of the area of the triangle to that of the hexagon is **(SSC CHSL 2011, SSC MTS 2011)**

(*a*) 3 : 2 (*b*) 2 : 3 (*c*) 1 : 2 (*d*) 1 : 4

Sol. (*b*) $3a = 6b \Rightarrow a : b = 2 : 1$

$\Rightarrow$ Required ratio = $\frac{\sqrt{3}}{4}a^2 : 6\times\frac{\sqrt{3}}{4}b^2 = a^2 : 6b^2 = a^2 : 6 \times \frac{a^2}{4} = 4 : 6 = 2 : 3$

219. Find the area of the isosceles triangle whose base is 6 cm and each of the equal side is 5 cm.

(*a*) 25 cm^2 (*b*) 12 cm^2 (*c*) 15 cm^2 (*d*) 24 cm^2

Sol. (*b*) Area of isosceles triangle (Here $a = 6$, $b = 5$)

$= \frac{a}{4}\times\sqrt{4b^2 - a^2} = \frac{6}{4}\times\sqrt{4\times 25 - 36} = \frac{6}{4}\times 8 = 12$ cm^2

Alternative Method : AD = $\sqrt{5^2 - 3^2} = 4$

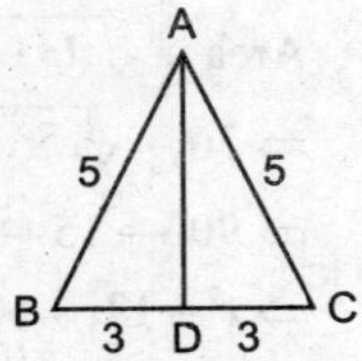

$\therefore$ Area of triangle = $\frac{1}{2} \times 6 \times 4 = 12$ cm²

220. The ratio of lenght of each equal side and the third side of an isosceles triangle is 3 : 4. If the area of the triangle is $18\sqrt{5}$ sq units, the third side is **(SSC CHSL 2012)**

(*a*) $8\sqrt{2}$ units (*b*) 12 units (*c*) 16 units (*d*) $5\sqrt{10}$ units

Sol. (*b*) Draw AD ⊥ BC

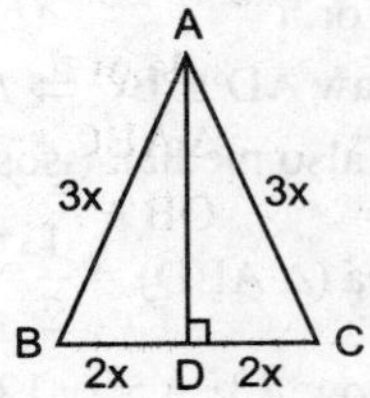

Then BD = $\frac{1}{2}$ BC = $\frac{1}{2} \times 4x = 2x$

AD² = AB² – BD² = $(3x)^2 - (2x)^2 = 5x^2 \Rightarrow$ AD = $\sqrt{5}\,x$

Area (Δ ABC) = $\frac{1}{2}$ BC × AD = $\frac{1}{2} \times 4x \times \sqrt{5}\,x = 18\sqrt{5}$

$\Rightarrow x^2 = 18\sqrt{5} \times 2 \times \frac{1}{4} \times \frac{1}{\sqrt{5}} = 9 \Rightarrow x = 3$ ⇒ The third side = $4x = 4 \times 3 = 12$ units

221. The lengths of the perpendiculars drawn from any point in the interior of an equilateral triangle to the respective sides are p_1, p_2 and p_3. The length of each side of the triangle is : **(SSC CGL Tier I 2004)**

(*a*) $\frac{2}{\sqrt{3}} \times (p_1 + p_2 + p_3)$ (*b*) $\frac{1}{3} \times (p_1 + p_2 + p_3)$

(*c*) $\frac{1}{\sqrt{3}} \times (p_1 + p_2 + p_3)$ (*d*) $\frac{4}{\sqrt{3}} \times (p_1 + p_2 + p_3)$

Sol. (*a*) Let each side of the triangle be *a* cm.

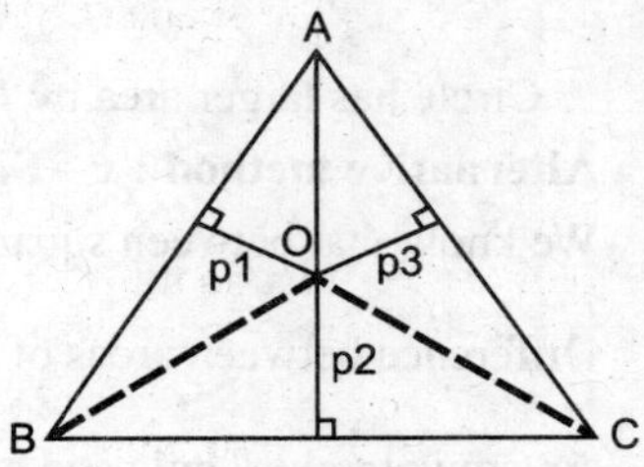

Then area of ΔABC

= Area of ΔAOB + Area of ΔBOC + Area of ΔAOC

$\Rightarrow \frac{\sqrt{3}}{4}a^2 = \frac{1}{2} \times a \times p_1 + \frac{1}{2} \times a \times p_2 + \frac{1}{2} \times a \times p_3$

$\Rightarrow \frac{\sqrt{3}}{4}a^2 = \frac{1}{2} \times a \times (p_1 + p_2 + p_3)$

$\Rightarrow a = \frac{1}{2} \times \frac{4}{\sqrt{3}} \times (p_1 + p_2 + p_3) = \frac{2}{\sqrt{3}} \times (p_1 + p_2 + p_3)$

222. From a point in the interior of an equilateral triangle the perpendicular distances of the sides are $\sqrt{3}$ cm, $2\sqrt{3}$ cm and $5\sqrt{3}$ cm. The perimeter (in cm) of the triangle is : **(SSC CGL Tier I 2003)**

(*a*) 64 (*b*) 32 (*c*) 48 (*d*) 24

Sol. (*c*) Let each side of the triangle be *a* cm. Then area of ΔABC

= Area of ΔAOB + Area of ΔAOC + Area of ΔBOC

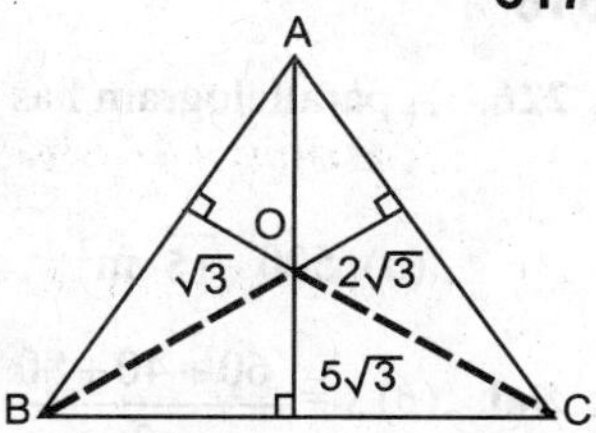

$$\Rightarrow \frac{\sqrt{3}}{4}a^2 = \frac{1}{2}\times a\times\sqrt{3} + \frac{1}{2}\times a\times 2\sqrt{3} + \frac{1}{2}\times a\times 5\sqrt{3} = 4\sqrt{3}\,a$$

$$\Rightarrow a = 4\sqrt{3}\times\frac{4}{\sqrt{3}} = 16$$

Perimeter of triangle = 3 × 16 cm = 48 cm

223. The three perpendicular distances of three sides of an equilateral triangle from a point which lies inside the triangle are 6 cm, 9 cm and 12 cm respectively. The perimeter of the triangle is **(SSC CGL Tier II 2012)**

(*a*) $52\sqrt{2}$ cm (*b*) $54\sqrt{3}$ cm (*c*) $42\sqrt{2}$ cm (*d*) $45\sqrt{3}$ cm

Sol. (*b*) Let each side of the triangle be *a* cm.

Then area of Δ ABC

= Area of ΔAOB + Area of ΔAOC + Area of ΔBOC

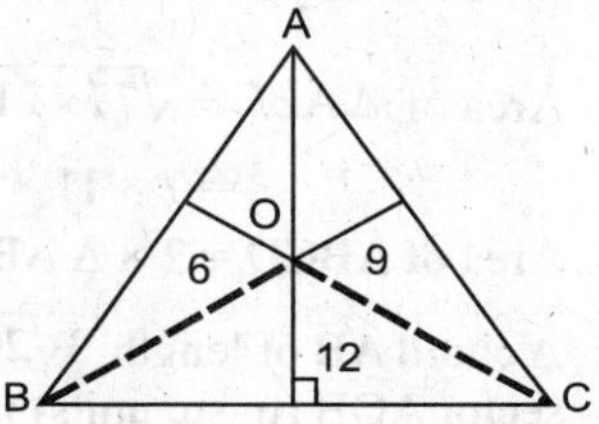

$$\Rightarrow \frac{\sqrt{3}}{4}a^2 = \frac{1}{2}\times a\times 6 + \frac{1}{2}\times a\times 9 + \frac{1}{2}\times a\times 12 = \frac{27}{2}a$$

$$\Rightarrow a = \frac{27}{2}\times\frac{4}{\sqrt{3}} = 18\sqrt{3}$$

Perimeter of triangle = 3 × $18\sqrt{3}$ cm = $54\sqrt{3}$ cm

224. If from a point inside an equilateral triangle, the lengths of perpendicular on the sides are 5 cm, 10 cm and 15 cm respectively, then the height of the tirangle is **(SSC CPO SI 2012)**

(*a*) 40 cm (*b*) 20 cm (*c*) 10 cm (*d*) 30 cm

Sol. (*d*) Area = $\frac{\sqrt{3}}{4}a^2 = \frac{1}{2}\times a\times(5+10+15) \Rightarrow a = \frac{30}{2}\times\frac{4}{\sqrt{3}} = 20\sqrt{3}$

Height of triangle = $\frac{\sqrt{3}}{2}a = \frac{\sqrt{3}}{2}\times 20\sqrt{3} = 30$ cm

225. The area of the shaded region in the figure given below is : **(SSC CGL Tier I 2007)**

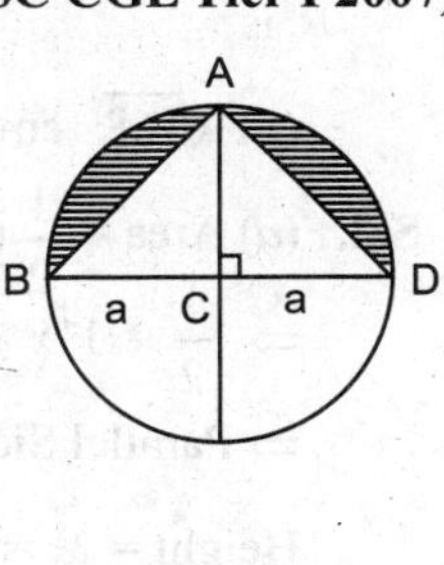

(*a*) $\frac{a^2}{2}\left(\frac{\pi}{2}-1\right)$ sq units (*b*) $a^2(\pi - 1)$ sq units

(*c*) $a^2\left(\frac{\pi}{2}-1\right)$ sq. units (*d*) $\frac{a^2}{2}(\pi-1)$ sq units

Sol. (*c*) In Δ ABD, AC ⊥ BD, Area of Δ ABD = $\frac{1}{2}$ × BD × AC = $\frac{1}{2}$ × $2a \times a = a^2$ sq. units

Area of semi-circle = $\frac{1}{2}\pi a^2$ sq. units

Area of the shaded region = $\frac{1}{2}\pi a^2 - a^2 = a^2\left(\frac{\pi}{2}-1\right)$ sq. units

226. A parallelogram has sides 60 m and 40 m and one of its diagonals is 80 m long. Its area is **(SSC CGL Tier I 2014)**

(*a*) $500\sqrt{15}$ m^2 (*b*) $600\sqrt{15}$ m^2 (*c*) $400\sqrt{15}$ m^2 (*d*) $450\sqrt{15}$ m^2

Sol. (*b*) $s = \frac{60+40+80}{2} = 90$ cm $\Rightarrow$ Area of $\Delta = \sqrt{90\times30\times50\times10} = 300\sqrt{15}$

Area of parallelogram = $2 \times 300\sqrt{15} = 600\sqrt{15}$

227. In a parallelogram two adjacent sides are 33 cm and 56 cm and one diagonal is 65 cm. What is its area?

(*a*) 824 cm^2 (*b*) 924 cm^2 (*c*) 1648 cm^2 (*d*) 1848 cm^2

Sol. (*d*) In traingle ABC, $s = \frac{1}{2}\times(56+33+65) = 77$ cm

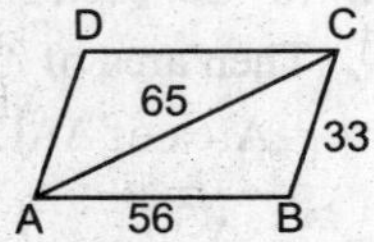

Area of ΔABC = $\sqrt{77\times21\times44\times12} = \sqrt{7\times11\times3\times7\times4\times11\times3\times4}$

$= 7 \times 11 \times 3 \times 2 \times 2 = 924$ cm^2

Area of ABCD = 2 × Δ ABC = 2 × 924 = 1848 cm^2

228. A chord AB of length $3\sqrt{2}$ unit subtends a right angle at the centre O of a circle. Area of the sector AOB (in sq. units) is **(SSC CGL Tier I 2013)**

(*a*) $\frac{9}{4}\pi$ (*b*) 5π (*c*) 9π (*d*) $\frac{9}{2}\pi$

Sol. (*a*) OA^2 + OB^2 = AB^2 $\Rightarrow r^2 + r^2 = (3\sqrt{2})^2$

$\Rightarrow 2r^2 = 18 \Rightarrow r^2 = 9 \Rightarrow r = 3$

Area of sector AOB = $\frac{90}{360}\pi r^2 = \frac{1}{4}\pi(3)^2 = \frac{9}{4}\pi$

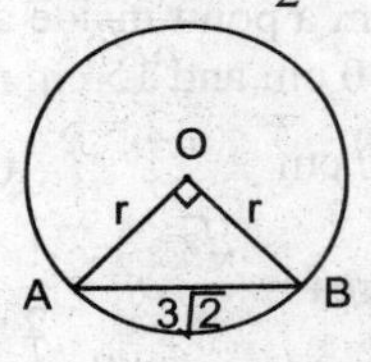

229. The area of an isosceles trapezium is 176 cm^2 and the height is 2/11th of the sum of its parallel sides. If the ratio of the length of the parallel sides be 4 : 7, then the length of a diagonal is **(SSC CHSL 2013)**

(*a*) $\sqrt{168}$ cm (*b*) $2\sqrt{147}$ cm (*c*) $\sqrt{137}$ cm (*d*) $2\sqrt{137}$ cm

Sol. (*d*) Area = $\frac{1}{2}(a + b)\,h = 1176$

$\Rightarrow \frac{1}{2} \times 11x \times 2x = 176 \Rightarrow x^2 = 16 \Rightarrow x = 4$

$\Rightarrow$ Parallel Sides = $4x$ and $7x$ = 16 cm and 28 cm

Height = $2x$ = 8 cm

In ΔAFC, AF = AE + EF = 6 + 16 = 22 cm

Diagonal AC = $\sqrt{AF^2 + CF^2} = \sqrt{22^2 + 8^2}$

$= \sqrt{484+64} = \sqrt{548} = 2\sqrt{137}$

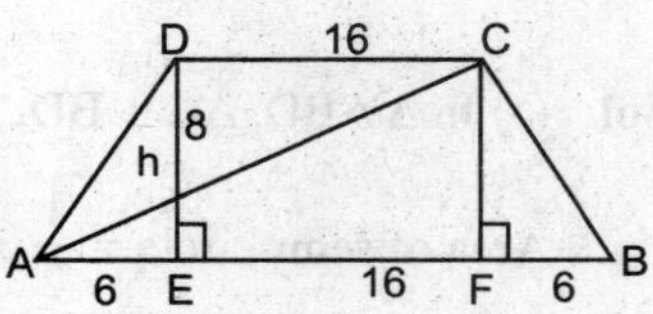

230. In an infinite set of circles, the radius of the first circle is 1 metre, that of the second $\frac{1}{2}$ metre; that of the third $\frac{1}{4}$ metre and so on indefinitely. The sum of the areas of the circles is

(SSC CGL Tier II 2012)

(*a*) $\frac{2\pi}{3}$ sq. m (*b*) $\frac{\pi}{4}$ sq. m (*c*) $\frac{3\pi}{4}$ sq. m (*d*) $\frac{4\pi}{3}$ sq. m

Sol. (*d*) Radius of circles = $1, \frac{1}{2}, \frac{1}{4}, \frac{1}{8}$ ....

Sum of areas of circles = $\pi\left(1+\frac{1}{4}+\frac{1}{16}+\frac{1}{64}\text{........}\right)$ (an Geometric Progresssion)

Sum of GP = $\frac{b}{1-r} = \pi\frac{1}{1-\frac{1}{4}} = \pi\frac{1}{\frac{3}{4}} = \frac{4\pi}{3}$

231. The parallel sides of a trapezium are 15 cm and 29 cm. If the non-parallel sides are 13 cm and 15 cm, what is the area of trapezium ?

(*a*) 225 cm^2 (*b*) 264 cm^2 (*c*) 308 cm^2 (*d*) 616 cm^2

Sol. (*b*) Let ABCD is a trapezium and draw a line CE is parallel to AD.

AECD is parallelogram. CE = AD = 13 cm

EB = AB – AE = 29 – 15 = 14 cm, CB = 15 cm

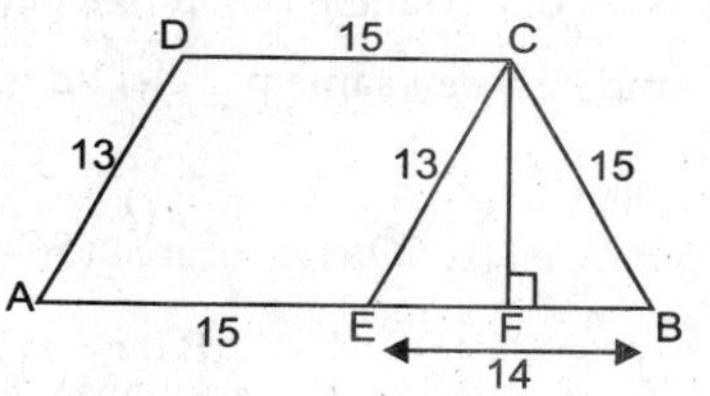

In triangle BCE : $s = \frac{15+13+14}{2} = \frac{42}{2} = 21$

Area = $\sqrt{21\times6\times8\times7} = \sqrt{7\times3\times2\times3\times2\times4\times7}$

$= 7 \times 3 \times 2 \times 2 = 84$

Area of ΔBCE = $\frac{1}{2}$ × Base × Altitude = $\frac{1}{2}$ × 14 × CF = 84 ⇒ CF = $\frac{84\times2}{14}$ = 12 cm

Area of trapezium ABCD = $\frac{1}{2}$ × Sum of parallel sides × Perpendicular distance between them $= \frac{1}{2}\times(15+29)\times12 = 264\ cm^2$

232. Three coins of the same size (radius 1 cm) are placed on a table such that each of them touches the other two. The area enclosed by the coins is : **(SSC CGL Tier I 2004, 2008)**

(*a*) $\left(\frac{\pi}{2}-\sqrt{3}\right)cm^2$ (*b*) $\left(\sqrt{3}-\frac{\pi}{2}\right)cm^2$ (*c*) $\left(2\sqrt{3}-\frac{\pi}{2}\right)cm^2$ (*d*) $\left(3\sqrt{3}-\frac{\pi}{2}\right)cm^2$

Sol. (*b*) AB = BC = AC = 2 cm ⇒ ∠A = ∠B = ∠C = 60°

$\Delta ABC = \frac{\sqrt{3}}{4}\times(2)^2 = \sqrt{3}\ cm^2$

Area of three sectors $= 3\times\pi\times(1)^2\times\frac{60}{360} = \frac{\pi}{2}\ cm^2$

∴ The area enclosed by the coins $= \left(\sqrt{3}-\frac{\pi}{2}\right)cm^2$

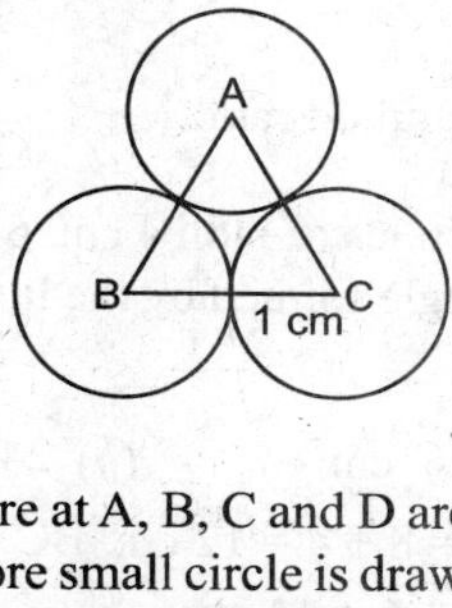

233. ABCD is a squre, whose side is 4 cm. 4 circle with centre at A, B, C and D are drawn in such a way that each circle touches two other circles. One more small circle is drawn in the middle of the square such that it touches all the four circles. The area of the smaller circle is :

(*a*) 4 π (*b*) 4π (3 – $2\sqrt{2}$) (*c*) 4π – (3 + $2\sqrt{2}$) (*d*) 4π (3 + $2\sqrt{2}$)

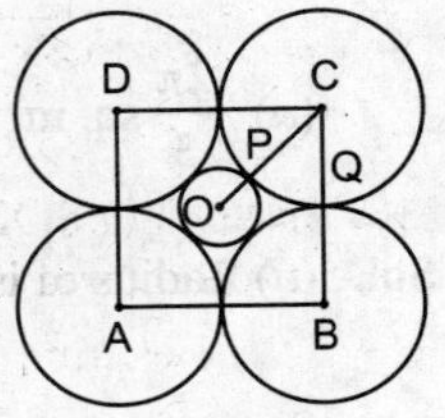

Sol. (*b*) PC = CQ = BC ÷ 2 = 4 ÷ 2 = 2 cm

OC = AC ÷ 2 = $4\sqrt{2} \div 2 = 2\sqrt{2}$ cm

OP = OC – PC = $2\sqrt{2} - 2 = 2(\sqrt{2} - 1)$

Area of smaller circle

$= \pi[2(\sqrt{2}-1)]^2 = 4\pi(2 + 1 - 2\sqrt{2}) = 4\pi(3 - 2\sqrt{2})$

234. If *ABCD* be a rectangle and *P, Q, R, S* be the mid-points of $\overline{AB}$, $\overline{BC}$, $\overline{CD}$ and $\overline{DA}$ respectively, then the area of the quadrilateral *PQRS* is equal to : **(SSC CGL Tier I 2013)**

(*a*) $\frac{1}{3}$ area (*ABCD*) (*b*) $\frac{3}{4}$ area (*ABCD*) (*c*) $\frac{1}{2}$ area (*ABCD*) (*d*) area (*ABCD*)

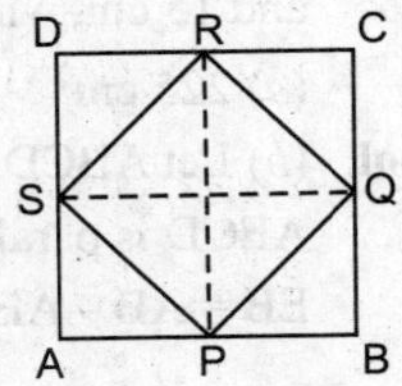

Sol. (*c*) Area (*PQRS*) = $\frac{RP \times QS}{2}$ $\left[\frac{1}{2}d_1d_2\right]$

Alternative Method :

Area of a triangle is half of a parallelogram on the same base and between same parallel lines.

$\Rightarrow$ Area (Δ *SQR*) = $\frac{1}{2}$ area (*SQCD*) and area (Δ *SPQ*) = $\frac{1}{2}$ (area of *ABQS*)

$\Rightarrow$ Area (Δ *SQR* + Δ *SPQ*) = $\frac{1}{2}$(Area *SQCD* + area *ABQS*) $\Rightarrow$ Area (*PQRS*) = $\frac{1}{2}$ (*ABCD*)

235. The ratio of inradius and circumradius of a square is : **(SSC CGL Tier I 2013, CHSL 2011)**

(*a*) $\sqrt{2}:\sqrt{3}$ (*b*) 1 : 3 (*c*) 1 : 2 (*d*) $1:\sqrt{2}$

Sol. (*d*) Let Side of square = *a* = Diameter of incircle

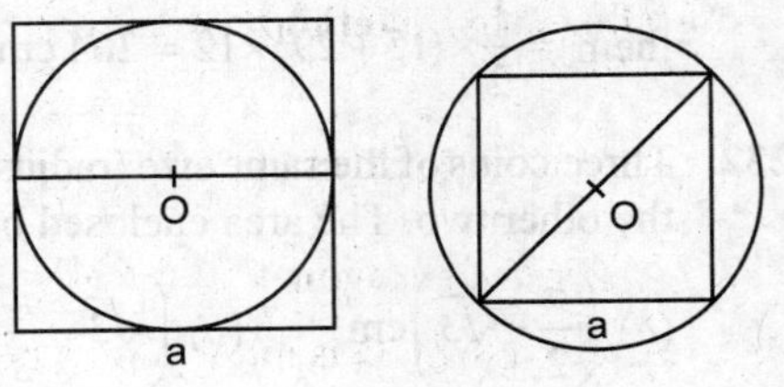

$\therefore$ Inradius = $\frac{a}{2}$

Diagonal of square = $\sqrt{2}a$

= Diameter of circumcircle

$\therefore$ Circumradius = $\frac{\sqrt{2}a}{2} = \frac{a}{\sqrt{2}}$

$\therefore$ Required ratio = $\frac{a}{2} : \frac{a}{\sqrt{2}} = 1 : \sqrt{2}$.

236. Three circles of radii 4 cm, 6 cm and 8 cm touch each other pairwise externally. The area of the triangle formed by the line segments joining the centres of the three circles is **(SSC CHSL 2012)**

(*a*) $6\sqrt{6}$ cm² (*b*) $24\sqrt{6}$ cm² (*c*) $144\sqrt{13}$ cm² (*d*) $12\sqrt{105}$ cm²

Sol. (*b*) AB = 8 + 4 = 12 cm, BC = 4 + 6 = 10 cm

AC = 8 + 6 = 14 cm

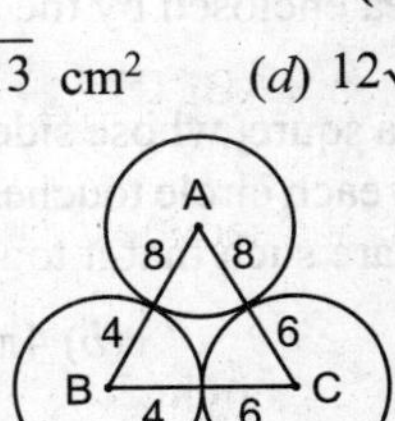

$s = \frac{12+10+14}{2} = \frac{36}{2} = 18$

$\Delta ABC = \sqrt{18(18-12)(18-10)(18-14)}$

$= \sqrt{18 \times 6 \times 8 \times 4} = \sqrt{6 \times 3 \times 6 \times 2 \times 4 \times 4}$

$= 6 \times 4 \times \sqrt{3 \times 2} = 24\sqrt{6}$ cm^2

237. In the figure, XYZ is a right isosceles triangle and $\angle X = 90°$. ABCD is a square inscribed in it whose area is 64 cm^2. Find the area of ΔXYZ.

(*a*) 128 cm^2 (*b*) 144 cm^2 (*c*) 156 cm^2 (*d*) 164 cm^2

Sol. (*b*) Side of square $= \sqrt{64} = 8$ cm

In Δ ADY, $\angle ADY = 180° - 45° - 90° = 45° = \angle AYD$

$\Rightarrow \angle ADY = \angle AYD \Rightarrow AD = AY \Rightarrow \Delta ADY$ is a right isosceles triangle.

$DY^2 = AD^2 + AY^2 = 8^2 + 8^2 = 2(8)^2 = (8\sqrt{2})^2$

In Δ XDC, $\angle XDC = \angle XCD = 45° \Rightarrow XD = XC$ and $\angle CXD = 90°$

$\Rightarrow XD^2 + XC^2 = 8^2 \Rightarrow 2 \times XD^2 = 8^2 \Rightarrow XD = 4\sqrt{2}$

$XY = XD + DY = 4\sqrt{2} + 8\sqrt{2} = 12\sqrt{2}$

Area $(\Delta XYZ) = \frac{1}{2} \times (12\sqrt{2})^2 = 144$ cm^2

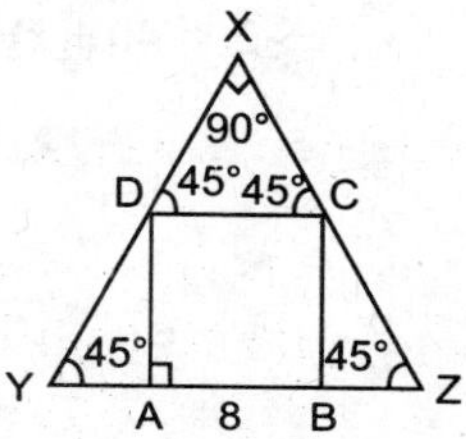

Trick : Area of triangle : Area of square = 9 : 4

$\Rightarrow$ Area of triangle $= \frac{9}{4} \times 64$ cm$^2 = 144$ cm^2

238. An equilateral triangle of side 6 cm has its corners cut off to form a regular hexagon. Area (in cm^2) of this regular hexagon will be **(SSC CGL Tier I 2011)**

(*a*) $3\sqrt{3}$ (*b*) $3\sqrt{6}$ (*c*) $6\sqrt{3}$ (*d*) $5\sqrt{3}$

Sol. (*c*) Area of hexagon $= \frac{6\sqrt{3}}{4}a^2 = \frac{6\sqrt{3}}{4}(2)^2 = 6\sqrt{3}$ cm^2

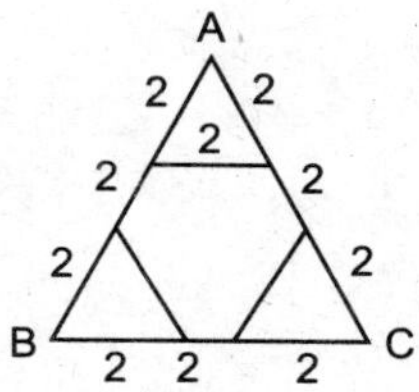

Alternative Method : Area of Δ ABC $= \frac{\sqrt{3}}{4}a^2 = \frac{\sqrt{3}}{4} \times 36 = 9\sqrt{3}$

Area of one corner cut off $= \frac{\sqrt{3}}{4} \times 2^2 = \sqrt{3}$

$\therefore$ Area of hexagon $= 9\sqrt{3} - 3\sqrt{3} = 6\sqrt{3}$ cm^2

239. A square ABCD is incribed in a circle of unit radius. Semicircles are described on each side as a diameter. The area of the region bounded by the four semicircles and the circle is **(SSC CGL Tier I 2013)**

(*a*) 1 sq. unit (*b*) 2 sq. unit (*c*) 1.5 sq. unit (*d*) 2.5 sq. unit

Sol. (*b*) Area of circle $= \pi(1)^2 = \pi$ sq. unit

Side of square ABCD $= \frac{\text{Diameter}}{\sqrt{2}} = \frac{2}{\sqrt{2}} = \sqrt{2}$ unit

Area of square ABCD $= (\sqrt{2})^2 = 2$ sq. unit

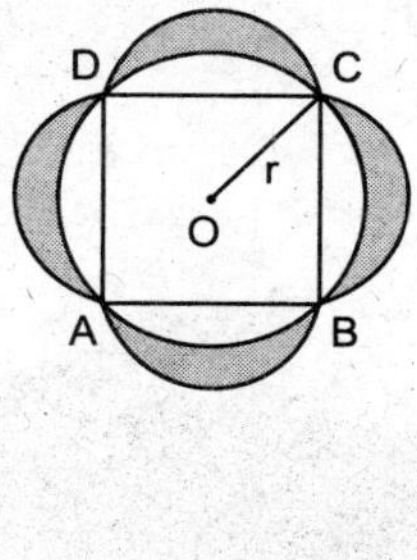

Radius of each semicircle $= \frac{AB}{2} = \frac{\sqrt{2}}{2} = \frac{1}{\sqrt{2}}$

Area of 4 Semicircles $= 4 \times \frac{1}{2}\pi\left(\frac{1}{\sqrt{2}}\right)^2 = \pi$

$\therefore$ Required area (shaded area) = Area of 4 semicircles – (Area of circle – area of square (ABCD)
$= \pi - (\pi - 2) = 2$ sq. unit

240. A circular grass plot, 40 m in radius is surrounded by a path. If the area of the grass plot is twice the area of the path, the width of the path, in metres, would be: **(SSC CGL Tier I 06)**

(*a*) $40\left(1+\sqrt{\frac{2}{3}}\right)$ (*b*) $40\left(1-\sqrt{\frac{2}{3}}\right)$ (*c*) $40\left(\sqrt{\frac{3}{2}}-1\right)$ (*d*) $40\left(\sqrt{\frac{3}{2}}+1\right)$

Sol. (*c*) Let width of the path be x metres. Area of plot = 2 × Area of path

$\Rightarrow \pi(40)^2 = 2\pi\,[(40+x)^2 - (40)^2] \Rightarrow (40)^2 = 2 \times (40+x)^2 - 2 \times (40)^2$

$\Rightarrow 2 \times (40+x)^2 = 3 \times (40)^2 \Rightarrow 40 + x = 40 \times \sqrt{\frac{3}{2}} \Rightarrow x = 40 \times \sqrt{\frac{3}{2}} - 40 = 40\left(\sqrt{\frac{3}{2}}-1\right)$

CHAPTER

30

GEOMETRY

1. Types of angles :

Right Angle :

(*i*) An angle whose measure is 90°, is called a right angle.

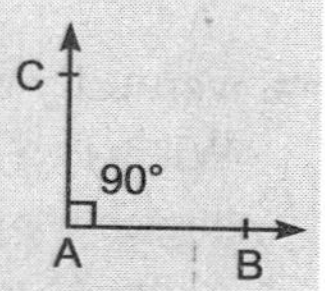

(*ii*) Acute Angle: An angle whose measure is less than 90° is called an acute angle.

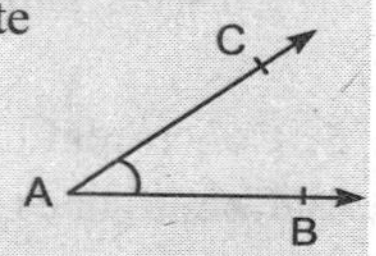

(*iii*) Obtuse Angle: An angle whose measure is more than 90° but less than 180° is called an obtuse angle.

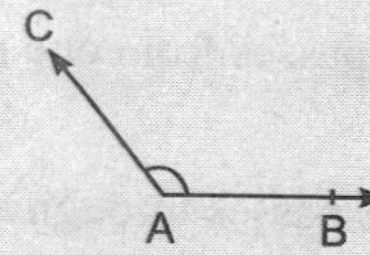

(*iv*) Reflex Angle: An angle whose measure is more than 180° but less than 360° is called reflex angle.

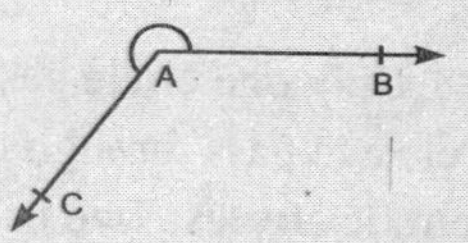

(*v*) Straight line angle: An angle whose measure is 180° is called a straight line.

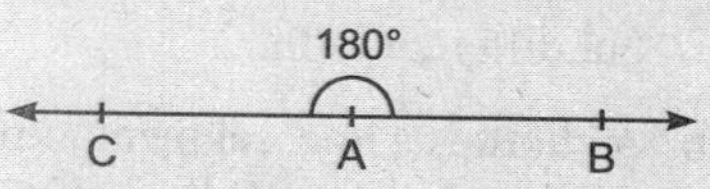

2. Supplementary angles :

Two angles, which add upto 180°

$\angle AOC + \angle BOC = 130° + 50° = 180°$.

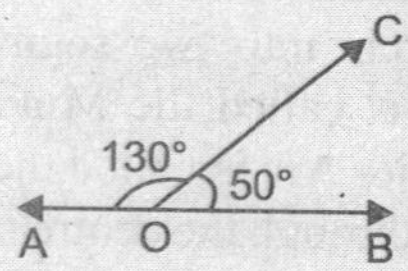

Example 1. If $5x°$ be the supplementary angle of $x°$, then the value of x is **(SSC FCI 2012)**

(*a*) 45° (*b*) 36° (*c*) 30° (*d*) 15°

Sol.(*c*) $5x + x = 180° \Rightarrow 6x = 180° \Rightarrow x = 30°$

2: If two supplementary angles differ by 44°, then one of the angle is **(SSC CGL Tier I 2015)**

(*a*) 68° (*b*) 72° (*c*) 102° (*d*) 65°

Sol.(*a*) Two angles are $\frac{180+44}{2}$ and $\frac{180-44}{2} = 112°$ and 68°

3. Complementary angles :

Two angles, which add upto 90°.

$\angle AOB + \angle BOC = 50° + 40° = 90°$.

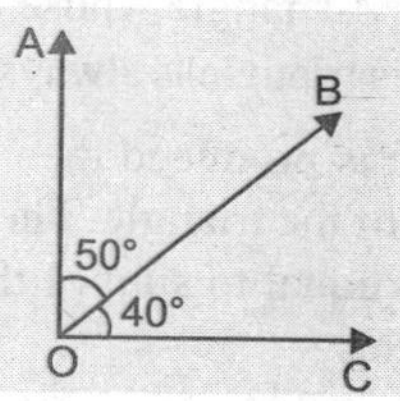

Example 1 : An angle is 20° less than its complement. Find the angle

(*a*) 70° (*b*) 50° (*c*) 45° (*d*) 35°

Sol.(*d*) Let the angle be x. Then its complement = $90° - x$.

$(90° - x) - x = 20° \Rightarrow 90° - 2x = 20° \Rightarrow 2x = 70° \Rightarrow x = 35°$

Ex. 2. How many degrees are there in an angle which equals two-third of its complement ? **(CDS 2011)**

(*a*) 36° (*b*) 45° (*c*) 48° (*d*) 60°

Sol.(*a*) $x = \frac{2}{3}(90° - x) \Rightarrow 3x = 180° - 2x \Rightarrow 5x = 180° \Rightarrow x = 36°$

4. Vertically opposite angles :
When two lines intersect each other, the vertically opposite angles are equal.
$\angle a = \angle b$ and $\angle c = \angle d$.

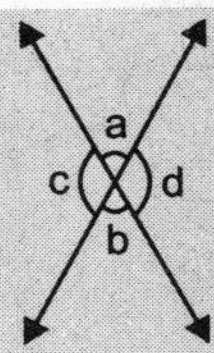

5. Concurrent Line: Three or more lines intersecting at the same point.
Infinite lines passes from one point on a plane.

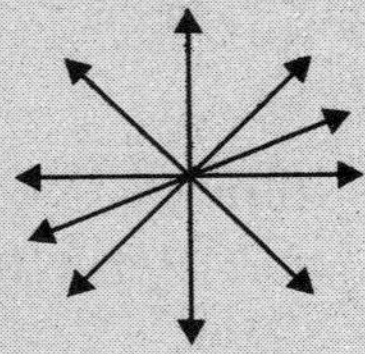

6. Only one line passes through two points on a plane.

7. Circle : Only one circle passes through three non-collinear points.
No circle can pass through three collinear points.
Concentric circles: Two or more circles having the same centre but different radii.

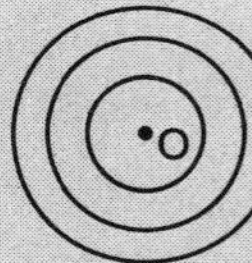

8. A chord is a line segment joining two points on the circumference of a circle. Let PQ be a chord of the circle. Then PQ divides the region enclosed by the circle into two segments. The segment containing the Minor Arc is called the Minor segment and the segment containing the Major Arc is called the Major Segment of the circle. A chord passing through the centre of the circle is called diameter. It divides the circle into two equal segments.

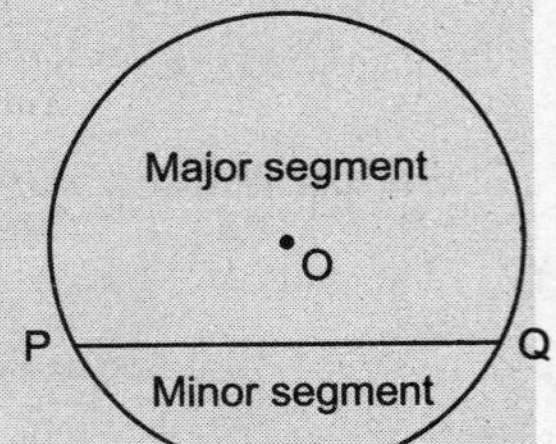

Angle in a major segment of a circle is less than 90°. Angle in a minor segment of a circle is more than 90°.

9. Triangle : A closed figure formed by three line segments.

(*i*) Scalene triangle is a triangle having all sides of different lengths.

(*ii*) Acute triangle : A triangle whose each angle is acute *i.e.*, less than 90°.

(*iii*) Right triangle: A triangle whose one angle is a right angle. The side opposite to right angle is the longest side of the triangle. It is called hypotenuse.

(*iv*) Obtuse triangle : A triangle whose one of angles is obtuse *i.e.*, between 90° and 180°.

(*vi*) The perimeter of a triangle is always greater than the sum of the lengths of its three medians.

10. If a side of a triangle is produced then the angle so formed is called exterior angle of the triangle. The measure of an exterior angle of a triangle is equal to sum of the measures of the two opposite interior angles.
$\angle 1 + \angle 2 = \angle 3$
Sum of three exterior angles of a triangle = 360°.

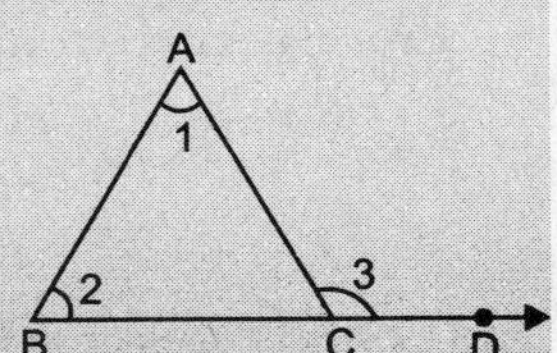

Example. ABC is a triangle and the sides AB, BC and CA are produced to E, F and G respectively. If $\angle CBE = \angle ACF = 130°$, then the value of $\angle GAB$ is **(SSC CGL Tier I 2015)**

(*a*) 100° (*b*) 90° (*c*) 30° (*d*) 80°

Sol.(*a*) $\angle CBE + \angle ACF + \angle GAB = 360° \Rightarrow \angle GAB = 360° - 130° - 130° = 100°$

11. Sum of the lengths of any two sides of a triangle is always greater than the third side of the triangle. i.e. $a + b > c, b + c > a, a + c > b$

Difference between the lengths of any two sides of a triangle is always less than the third side. i.e. $|a - b| < c, |b - c| < a, |c - a| < b$

If two sides of a triangle are unequal, then the longer side has greater angle opposite to it. Conversely, in a triangle, the greater angle has the longer side opposite to it.

If $\angle B > \angle C$, then AC > AB

If AB < AC, then $\angle C < \angle B$

The shortest side is always opposite to the smallest interior angle.

The largest side is always opposite the largest interior angle.

12. Median: A line segment, joining a vertex of a triangle to the mid-point of the opposite side is called median. Median of a triangle divides it into two triangles having equal areas.

Angle bisector: Angle bisector is a line segment which bisects an angle of a triangle and meets opposite side of the triangle.

Perpendicular bisector : A line segment, which bisects a side of triangle and is also perpendicular to it, is called perpendicular bisector of a side of the triangle.

Altitude: Perpendicular drawn from a vertex of a triangle to the opposite side is called an altitude.

Note: In an isosceles triangle, median drawn on unequal side, is also angle bisector and perpendicular to the base.

Note: Sum of any two sides a triangle is always greater than twice the median drawn on the third side of the triangle.

Example. If D is the mid-point of the side BC of ΔABC and the area of ΔABD is 16 cm^2, then the area of ΔABC is **(SSC CGL Tier I 2012)**

(*a*) 16 cm^2 (*b*) 24 cm^2

(*c*) 32 cm^2 (*d*) 48 cm^2

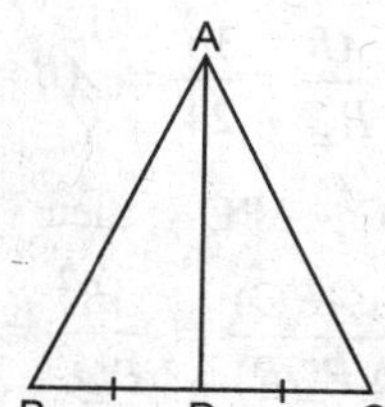

Sol.(*c*) $BD = DC$

$\therefore \Delta ABD = \Delta ADC$ (Equal base and height)

$\therefore \Delta ABC = 2 \times 16 \text{ cm}^2 = 32 \text{ cm}^2$.

13. Congruent triangles : $\Delta ABC \cong \Delta DEF$

Every corresponding side has same length and every corresponding angle has the same measure. i.e. AB = DE, BC = EF, AC = DF and $\angle A = \angle D, \angle B = \angle E, \angle C = \angle F$,

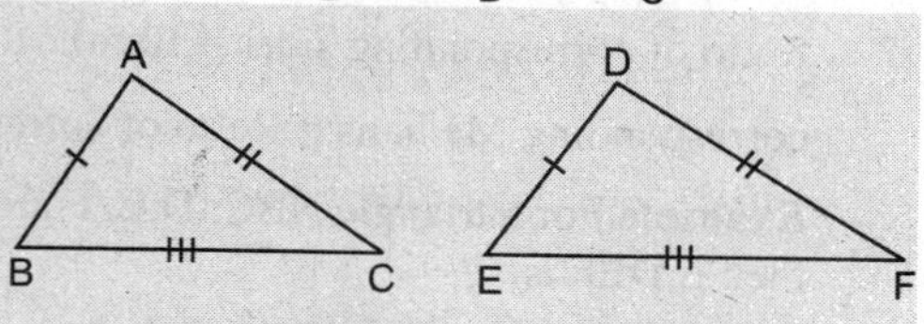

Criterion for congruence of triangles :

(1) **Side Angle Side (SAS):** If two sides and the included angle of one triangle are equal to corresponding two sides and the included angle of the other triangle.

(2) **Side Side Side (SSS):** If three sides of one triangle are equal to corresponding three sides of the other triangle.

(3) **Angle Side Angle (ASA):** If two angles and the included side of one triangle are equal to corresponding two angles and included side of the other triangle.

(4) **Angle Angle Side:** Two angles and a non-included side of one triangle are equal to corresponding angles and side of the other triangle.

(5) **Right Angle-Hypotenuse-Side (RHS):** If the hypotenuse and a side of one right angle triangle are respectively equal to the hypotenuse and a side of the other right angle triangle.

14. Similar triangles: $\Delta ABC \sim \Delta DEF$

– Pairs of corresponding sides of triangle, are in same proportion.

– Corresponding angles are equal

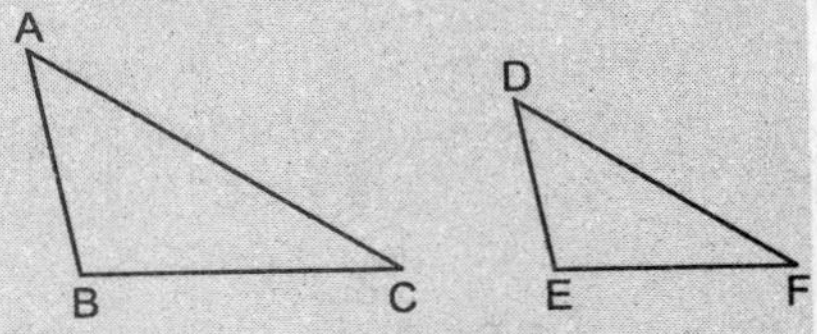

(*i*) Equiangular or AAA or AA Theorem

$\angle A = \angle D, \angle B = \angle E, \angle C = \angle F$

(*ii*) Corresponding three sides are proportional:

$$\frac{AB}{DE} = \frac{BC}{EF} = \frac{AC}{DF} = \frac{\text{Perimeter of } \Delta ABC}{\text{Perimeter of } \Delta DEF}$$

Note : Two triangles are similar if either of two conditions mentioned above is satisfied. Other condition will hold automatically.

(*ii*) Two sides are proportional and included angle is equal.

Example 1: If ΔABC and ΔDEF are similar such that $2AB = DE$, $BC = 8$ cm, then EF is **(SSC CHSL 2012)**

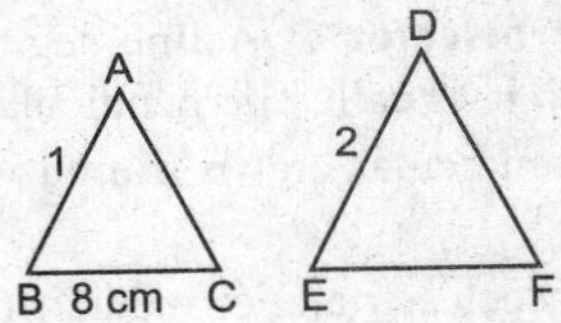

(*a*) 4 cm (*b*) 8 cm

(*c*) 16 cm (*d*) 12 cm

Sol.(*c*) $\frac{AB}{DE} = \frac{BC}{EF}$

$\Rightarrow \frac{1}{2} = \frac{8}{EF} \quad \Rightarrow EF = 8 \times 2 = 16$ cm

Example 2. The perimeters of two similar triangles ΔABC and ΔPQR are respectively 36 cm and 24 cm respectively. If $PQ = 10$ cm then AB is. **(SSC CGL Tier I 2012)**

(*a*) 7.5 cm (*b*) 10 cm (*c*) 12 cm (*d*) 15 cm

Sol.(*d*) $\frac{AB}{PQ} = \frac{36}{24} \Rightarrow AB = \frac{36}{24} \times 10 = 15$ cm

15. If $\Delta ABC \sim \Delta PQR$, then

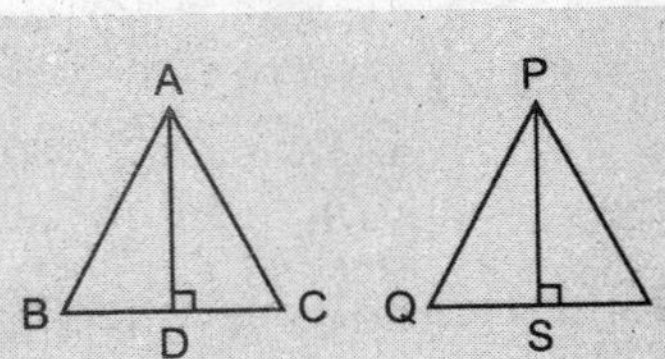

$$\frac{\text{Area}(\Delta ABC)}{\text{Area}(\Delta PQR)} = \frac{AB^2}{PQ^2} = \frac{BC^2}{QR^2} = \frac{AC^2}{PR^2} = \frac{AD^2}{PS^2}$$

Ratio of corresponding sides = Ratio of corresponding Medians = Ratio of corresponding Angle Bisectors

Example. For a triangle ABC, D,E, F are the mid-points of its sides. If Δ ABC = 24 sq. units, then Δ DEF is **(SSC CGL Tier I 2013)**

(*a*) 4 sq. units (*b*) 8 sq. units (*c*) 6 sq. units (*d*) 12 sq. units

Sol. (*c*) Δ DEF $= 24 \div 4 = 6$ sq. units

16. Basic Proportional Theorem. If a line parallel to one side of a triangle and intersects the other two sides, it divides those sides proportionally.

If DE || BC, then $\frac{AD}{BD} = \frac{AE}{CE}$ and $\frac{AD}{AB} = \frac{AE}{AC} = \frac{DE}{BC}$. Also $\Delta ABC \sim \Delta ADE$

Conversely, **Thales Theorem**: If a line divides any two sides of a triangle in the same ratio, then the line is parallel to the third side.

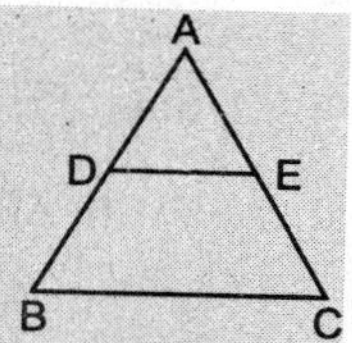

If $\frac{AD}{BD} = \frac{AE}{CE}$, then $DE \parallel BC$

Mid-Point Theorem: A line joining the mid-points of any two sides of a triangle is parallel to the third side and the length of the line segment is half the third side. A line drawn from the mid-point of one side of a triangle parallel to another side, bisects the third side.

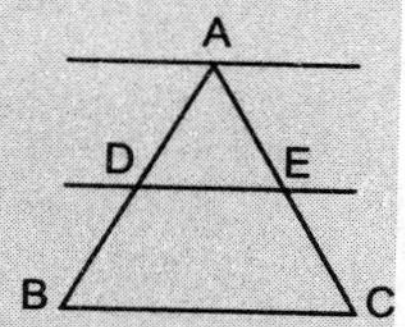

If $AD = DB$ and $DE \parallel BC$, then $AE = EC$

Converse : A line joining the mid-points of two sides of a triangle must be parallel to the third side.

If $AD = DB$ and $AE = EC$, then $DE \parallel BC$.

17. If there are three parallel lines and the intercepts made by them on one transversal are equal, then the intercepts on any other transversal are equal.

The line segment joining the mid-points of any two sides of a triangle is parallel to the third side and equal to half of it in length. Conversely, the line drawn through the mid-point of one side of a triangle, and parallel to other side, intersects the third side at its mid-point.

18. If three or more parallel lines are intersected by two or more transversals, the intercepts made by them on the transversals are porportional.

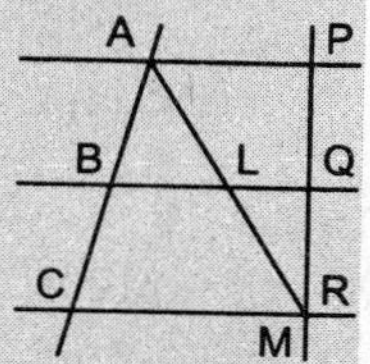

i.e., $\frac{AB}{BC} = \frac{AL}{LM} = \frac{PQ}{QR}$

19. The line segments joining the midpoints of three sides of a triangle divides the triangle into four congruent triangles, each of which is similar to the original triangle. Area of each smaller triangle is equal to one-fourth of the original triangle. Each smaller Δ has same area.

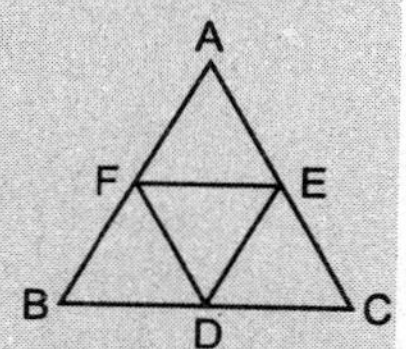

Δ ABC ~ Δ AFE, Δ CED ~ Δ CAB,

Δ BDF ~ Δ BCA and Δ DEF ~ Δ ABC

BDEF, DCEF, AFDE are parallelograms

Example. If D, E and F are the mid points of BC, CA and AB respectively of the ΔABC then the ratio of area of the parallelogram DEFB and area of the trapezium CAFD is :

(SSC CGL Tier I 2015)

(*a*) 1 : 3 (*b*) 1 : 2 (*c*) 3 : 4 (*d*) 2 : 3

Sol.(*d*) Refer to diagram above.

Required ratio = Area of 2 triangles : Area of 3 triangles = 2 : 3

20. If a line bisects an angle of a triangle, then it divides the opposite sides into segments that are proportional to the sides that formed the angle.

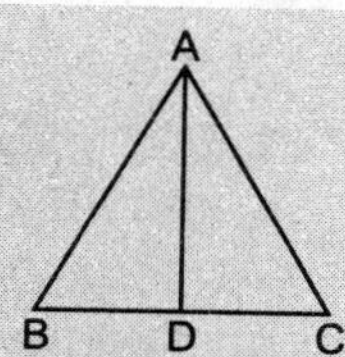

If ∠BAD = ∠CAD, then $\frac{BD}{DC} = \frac{AB}{AC}$

Example. In the figure, $AD : DC = 2 : 3$. Find $\angle ABC$

(*a*) 30° (*b*) 50° (*c*) 60° (*d*) 75°

Sol.(*c*) Here $\frac{AD}{DC} = \frac{AB}{BC} \Rightarrow BD$ is angle bisector of $\angle B$

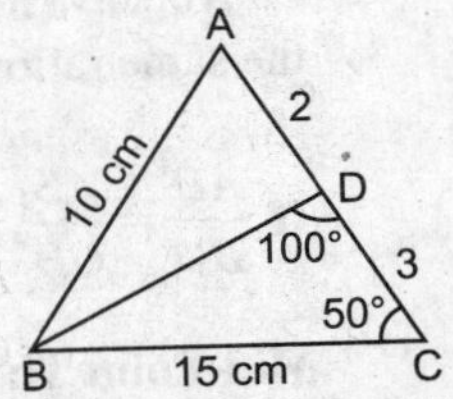

In ΔBDC, $\angle DBC = 180° - (100° + 50°) = 30°$

$\therefore \quad \angle ABC = 2 \times 30° = 60°$

21. Two parallel lines intersected by a transversal line

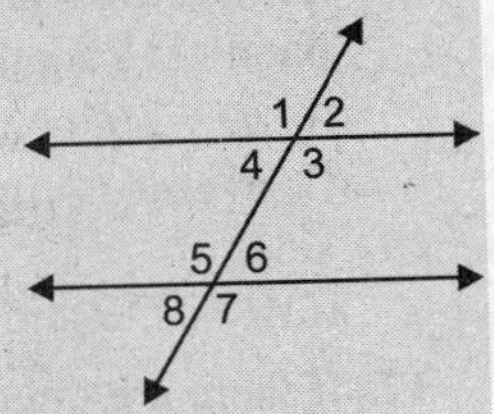

(1) Alternate interior angles are equal
$\angle 4 = \angle 6$ and $\angle 3 = \angle 5$

(2) Alternate exterior angles are equal
$\angle 1 = \angle 7$ and $\angle 2 = \angle 8$

(3) Corresponding angles are equal :
$\angle 1 = \angle 5$, $\angle 2 = \angle 6$, $\angle 3 = \angle 7$, and $\angle 4 = \angle 8$

(4) Interior angles on the same side of the transversal are supplementary.
$\angle 4 + \angle 5 = 180°$ and $\angle 3 + \angle 6 = 180°$

Trick to remember: All acute angles are equal. All obtuse angles are equal. Any acute angle is supplementary to any obtuse angle.

Example 1: In the given figure $AB \parallel CD$ and $CD \parallel EF$, then $\angle CEF = ?$

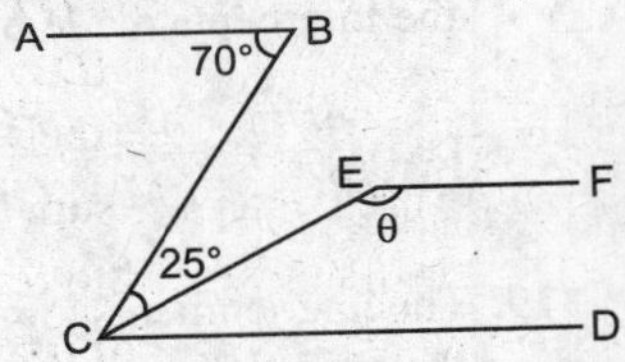

(*a*) 95° (*b*) 105°

(*c*) 120° (*d*) 135°

Sol.(*d*) $\angle BCD = \angle ABC = 70°$ (Alternate angles)

$\Rightarrow \angle ECD = 70° - 25° = 45°$

$\angle ECD + \angle CEF = 180°$

$\Rightarrow \angle CEF = 180° - 45° = 135°$

Example 2. In the given figure, $AB \parallel CD$, $\angle ABE = 30°$,

$\angle CDE = 50°$, $x = ?$

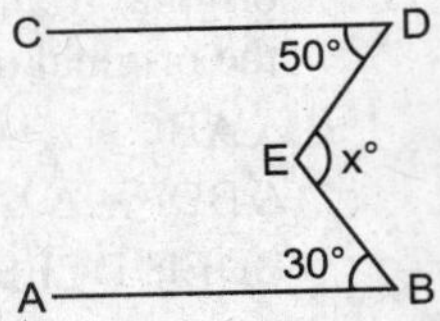

(*a*) 20° (*b*) 75°
(*c*) 80° (*d*) 85°

Sol. (*c*) Draw a line $PEQ \parallel AB$

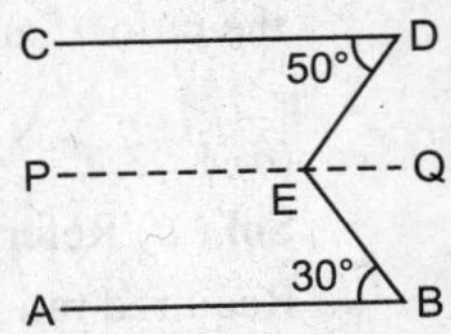

$\angle BEQ = \angle ABE = 30°$ (Alternate angles)

$\angle DEQ = \angle CDE = 50°$ (Alternate angles)

$\angle DEB = \angle DEQ + BEQ$

$= 50° + 30° = 80°$

22. If *ABCD* is a quadrilateral in which *P*, *Q*, *R* and *S* are the mid-points of sides *AB*, *BC*, *CD* and *DA* respectively, then *PQRS* is a parallelogram.

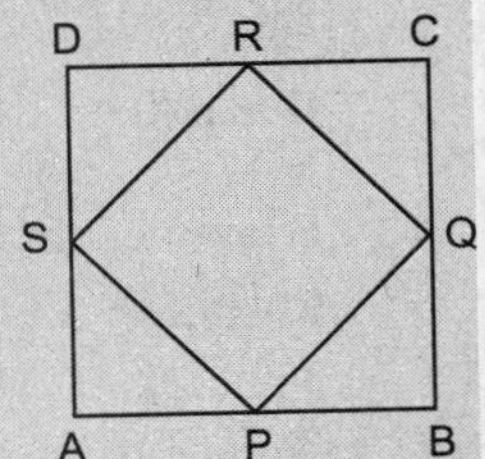

ABCD	*PQRS*
Quadrilateral	Parallelogram
Rectangle	Rhombus
Rhombus	Rectangle
Square	Square

(CDS 2003)

Example: $ABCD$ is a quadrilateral in which P, Q, R and S are the mid-points of AB, BC, CD and DA respectively. Then $PQRS$ is :

(*a*) trapezium (*b*) parallelogram (*c*) rectangle (*d*) square

Sol.(*b*) Join AC.

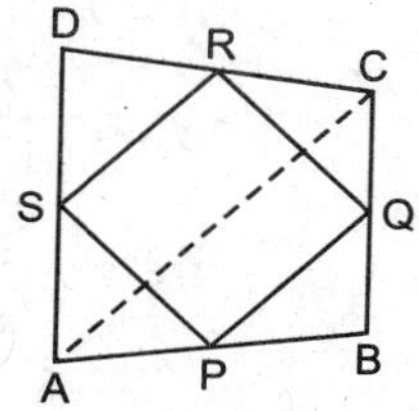

In ΔABC, P and Q are the mid-points of sides AB and BC respectively.

$\therefore PQ \parallel AC$

In ΔACD, R and S are the mid-points of CD and DA respectively

$\therefore SR \parallel AC$

$\therefore PQ \parallel AC$ and $SR \parallel AC \Rightarrow PQ \parallel SR$

Similarly, we can prove that $PS \parallel QR$ $\therefore PQRS$ is a parallelogram.

23. Isosceles triangle :

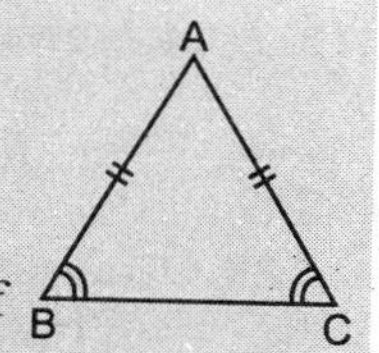

Two sides of a triangle are equal in length. $AB = AC$.

$\Rightarrow$ Opposite angles are also equal. $\angle B = \angle C$

Converse: If $\angle B = \angle C \Rightarrow AB = AC$

In isosceles triangle, the two medians drawn from the vertices at the base of the equal sides are of equal lengths.

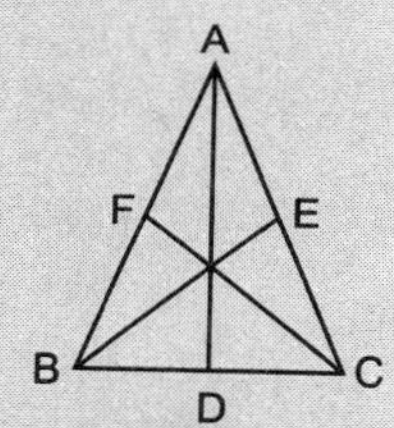

In an isosceles triangle, the median drawn on the unequal side of the triangle, is also the angle bisector and altitude to the base.

Proof : $AD = AD$ (common); $AB = AC$ (Given); $BD = DC$ ($\because$ AD is median)

$\therefore \Delta ABD \cong \Delta ACD$ (SSS-Criterian)

$\Rightarrow \angle BAD = \angle CAD \Rightarrow AD$ is angle bisector and $\angle ADB = \angle ADC = 90°$

$\Rightarrow AD$ is altitude on BC.

24. Three times the sum of the squares of the sides of a triangle is equal to four times the sum of the squares of the medians of triangle.

$3(AB^2 + BC^2 + AC^2) = 4(AD^2 + BE^2 + CF^2)$

In case of equilateral triangle, three times the square of a side is equal to four times the square of the median (or altitude) i.e., $3AB^2 = 4AD^2$

25. Trapezium :

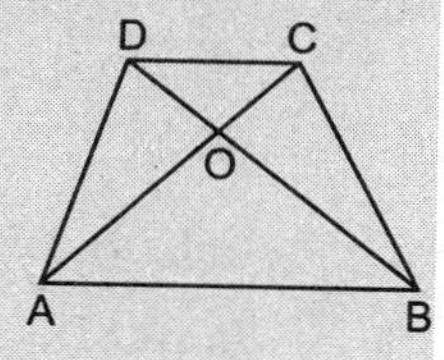

A pair of opposite sides are parallel.

$AB \parallel DC$. Δ OAB ~ Δ OCD

Diagonals of a trapezium divide each other proportionally.

$$\frac{OA}{OC} = \frac{OB}{OD}$$

Conversely if the diagonals of a quadrilateral divides each other proportionally, then it must be a trapezium.

Example 1: $ABCD$ is a trapezium in which $AB \parallel DC$ and $AB = 2\ DC$. Find the ratio of the areas of $\Delta\ AOB$ and $\Delta\ COD$.

(*a*) 1 : 4 (*b*) 4 : 1 (*c*) 2 : 1 (*d*) 1 : 1

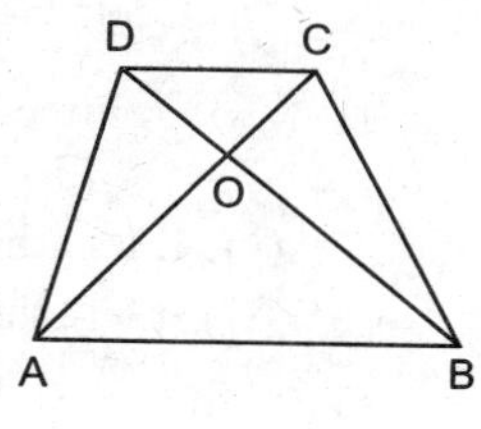

Sol.(*b*) $\angle AOB = \angle COD$ (Vertically opposite angles)

$\angle OAB = \angle OCD$ (Alternate angles)

$\therefore \Delta\ AOB \sim \Delta\ COD$ (AA criterion)

$$\frac{\Delta\ AOB}{\Delta COD} = \frac{AB^2}{CD^2} = \frac{2^2}{1^2} = \frac{4}{1}$$

Example 2. *ABCD* is a trapezium whose side *AD* is parallel to *BC*. Diagonals *AC* and *BD* intersect at *O*. If $AO = 3$, $CO = x - 3$, $BO = 3x - 19$ and $DO = x - 5$, the value of *x* will be **(SSC CHSL 2012)**

(*a*) 8, 9 (*b*) 7, 6 (*c*) 12, 6 (*d*) 7, 10

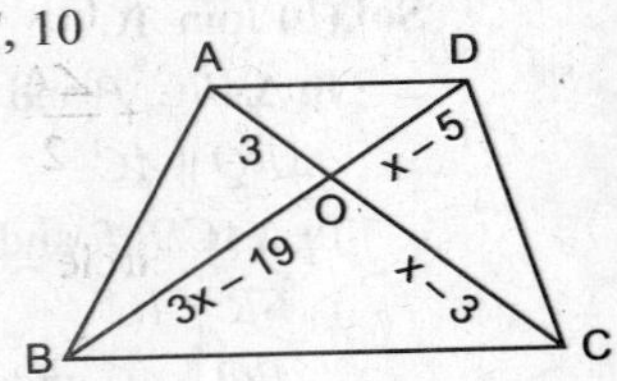

Sol.(*a*) $\frac{AO}{OC} = \frac{OD}{OB} \Rightarrow \frac{3}{x-3} = \frac{x-5}{3x-19}$

$\Rightarrow 3(3x - 19) = (x - 3)(x - 5) \Rightarrow 9x - 57 = x^2 - 8x + 15$

$\Rightarrow x^2 - 17x + 72 = 0$

$\Rightarrow (x - 8)(x - 9) = 0 \Rightarrow x = 8, 9$

26. A line drawn parallel to the parallel sides of a trapezium divides the non-parallel sides proportionally. If $EF \parallel DC$,

then $\frac{AE}{ED} = \frac{BF}{FC}$

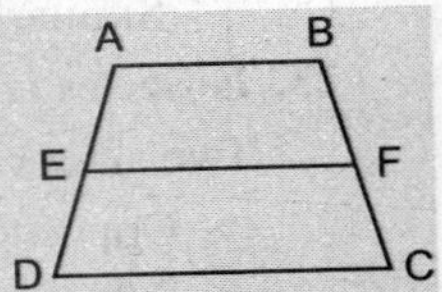

27. Parallelogram:

(*i*) Opposite sides are equal and parallel.

(*ii*) Opposite angles are equal.

$\angle A = \angle C$ and $\angle B = \angle D$

(*iii*) Diagonals bisect each other.

$OA = OC$ and $OB = OD$

(*iv*) Sum of any two adjacent angles is 180°.

$\angle A + \angle B = \angle B + \angle C = \angle C + \angle D = \angle D + \angle A = 180°$

(*v*) Diagonal divides it into two congruent triangles.

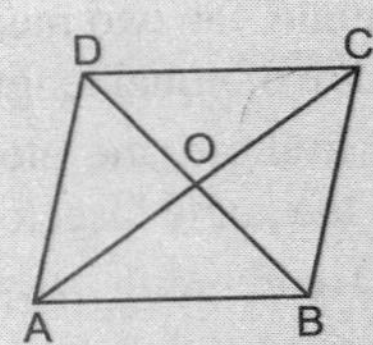

28. Rhombus : A quadrilateral whose all sides are equal. Opposite sides are parallel. Diagonals bisect vertex angles. Diagonals are perpendicular bisectors to each other. Opposite angles are congruent. Adjacent angles are supplementary. AC ≠ BD

Note: A rhombus is always a parallelogram.

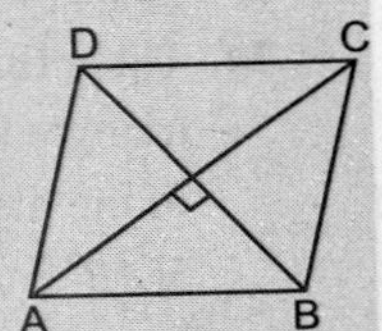

Example: If PQRS is a rhombus and $\angle SPQ = 50°$, then $\angle RSQ$ is **(SSC CGL Tier I 2015)**

(*a*) 65° (*b*) 75°

(*c*) 55° (*d*) 45°

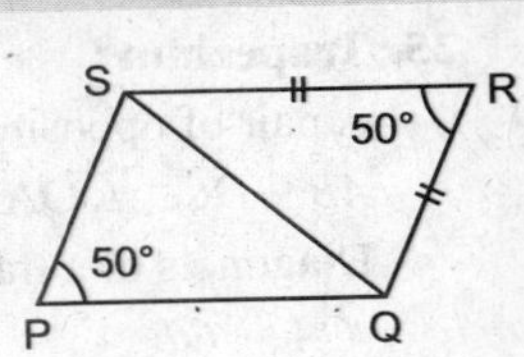

Sol. (*a*) $\angle RSQ = \angle QPS = 50°$ (Opposite angles)

In ΔRSQ, $\angle RSQ + \angle RQS = 180° - 50° = 130°$

$RS = QR \Rightarrow \angle RSQ = \angle RQS = 130° \div 2 = 65°$

29. Centroid : Centroid is a point where three medians of the triangle intersect. Three medians of a triangle are concurrent.

Properties :

(*i*) Centroid is always inside the triangle.

(*ii*) Centroid always divides the medians into two segments whose lengths are in the ratio 2 : 1 with longer side near the vertex. $\frac{AO}{OD} = \frac{BO}{OE} = \frac{CO}{FO} = \frac{2}{1}$

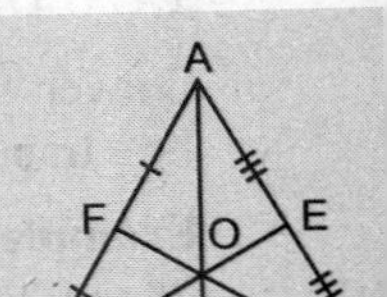

(*iii*) If lengths of two medians of a triangle are equal, the triangle is an isosceles triangle.

(*iv*) If lengths of all the medians of a triangle are equal, the triangle is an equilateral triangle

30. Incentre : The point of concurrence of three angle bisectors of a triangle. Incentre is also the centre of triangle incircle, the largest circle that will fit inside the triangle and touch all the three sides. Incenter alway lies inside the triangle and is equidistant from sides.

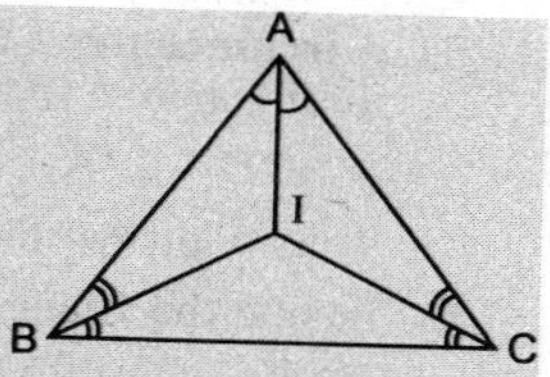

$\angle BIC = 90° + \frac{\angle A}{2}$.

Radius of incircle $= \frac{2 \times \text{Area}}{\text{Perimeter}} = \frac{\text{Area}}{\text{Semi-perimeter}}$

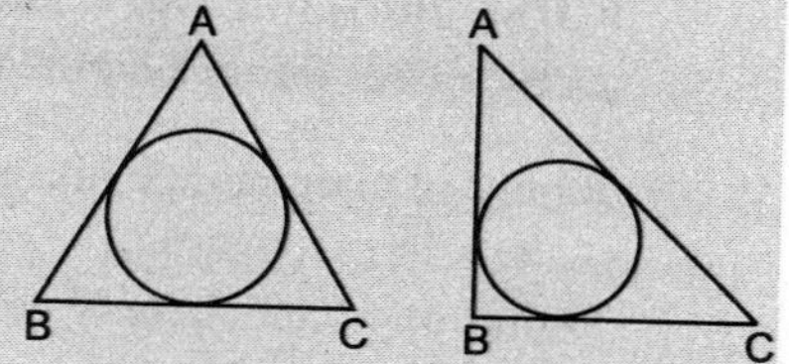

Radius of incircle of right triangle

$= \frac{AB \times BC}{AB + BC + AC} = \frac{AB + BC - AC}{2}$ **(CDS 2015)**

Radius of incircle of equilateral triangle $= \frac{\text{Side}}{2\sqrt{3}}$

Radius of circum-circle of equilateral triangle $= \frac{\text{Side}}{\sqrt{3}}$

Example 1. O is the incentre of ΔPQR and $\angle QPR = 50°$, then the measure of $\angle QOR$ is **(SSC CGL Tier I 2015)**

(*a*) 130° (*b*) 125° (*c*) 100° (*d*) 115°

Sol. (*d*) $\angle QOR = 90° + \frac{\angle P}{2} = 90° + 25° = 115°$

Example 2. Internal bisectors of ∠Q and ∠R of ΔPQR intersect at O. If ∠ROQ = 96°, then the value of $\angle RPQ$ is : **(SSC CGL Tier I 2015)**

(*a*) 12° (*b*) 6° (*c*) 36° (*d*) 24°

Sol. (*a*) $96° = 90° + \frac{\angle P}{2} \Rightarrow \angle RPQ = 12°$

Example 3. In a triangle, the area is numerically equal to the perimeter, then the radius of the inscribed circle of the triangle is: **(SSC CGL Tier I 2000)**

(*a*) 1 (*b*) 2 (*c*) 3 (*d*) 1.5

Sol. (*b*) Inradius $= \frac{2 \times \text{Area}}{\text{Perimeter}} = \frac{2 \times \text{Area}}{\text{Area}} = 2$

Example 4. If the inradius of a triangle with perimeter 32 cm is 6 cm, then the area of the triangle in sq. cm is **(SSC CGL Tier I 2013)**

(*a*) 48 (*b*) 100 (*c*) 64 (*d*) 96

Sol.(*d*) Area $= \frac{\text{Perimeter} \times \text{Inradius}}{2} = \frac{32 \times 6}{2} = 96$ sq. cm.

Example 5. *ABC* is a right-angled triangle. If the lengths of two sides containing the right angle are 6 cm and 8 cm. Find radius of its incircle.

(*a*) 1 cm (*b*) 1.25 cm (*c*) 1.5 cm (*d*) 2 cm

Sol.(*d*) Hypotenuse of $\Delta = \sqrt{6^2 + 8^2} = 10$ cm $\Rightarrow$ Radius of incircle $= \frac{6 \times 8}{6 + 8 + 10} = \frac{48}{24} = 2$ cm

Alt: Inradius $= \frac{6+8-10}{2} = 2$ cm

31. Circumcentre : The point of concurrence of three perpendicular bisectors of a triangle.

$\angle BOC = 2\angle A$ $\quad\quad$ $OA = OB = OC$

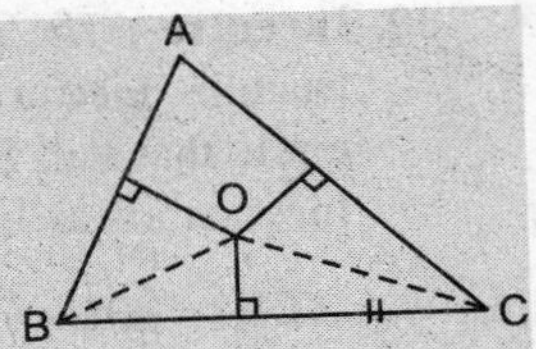

It lies outside the triangle if the triangle is obtuse-angled. It lies inside the triangle if the triangle is acute angled. It lies on the mid-point of the hypotenuse, if the triangle is right-angled. **(SSC CHSL 2013)** It is also centre of triangle's circumcircle – i.e. the circle passes through all the three vertices of the triangle.

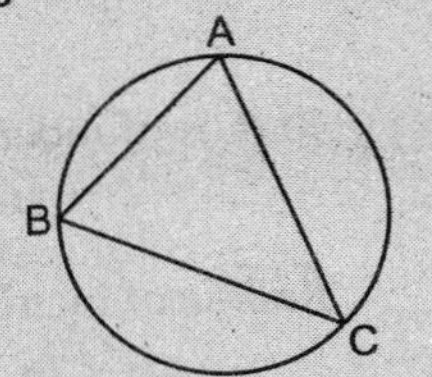

Radius of circumcircle of right-angled triangle = $\dfrac{\text{Hypotenuse}}{2}$

Radius of circumcircle of equilateral triangle = $\dfrac{\text{Side}}{\sqrt{3}}$

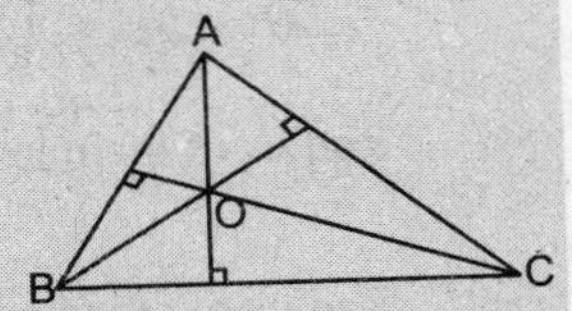

32. Orthocentre : The point where three altitudes of a triangle intersect.
Orthocentre is equidistant from three vertices.
It lies outside the triangle when the triangle is obtuse.
Angle made by any side at the orthocentre and vertical angles are supplementary.
$\angle A + \angle BOC = \angle B + \angle AOC = \angle C + \angle AOB = 180°$

33. Cyclic Quadrilateral :

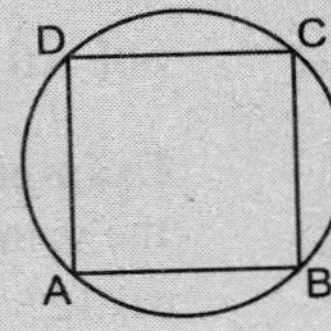

A quadrilateral whose all the four vertices lie on a single circle. Opposite angles of a cyclic quadrilateral are supplementary. The converse is also true.
$\angle A + \angle C = \angle B + \angle D = 180°$ **(SSC CGL I 2014)**

34. If one side of a cyclic quadrilateral is produced, then the exterior angle is equal to interior opposite angle

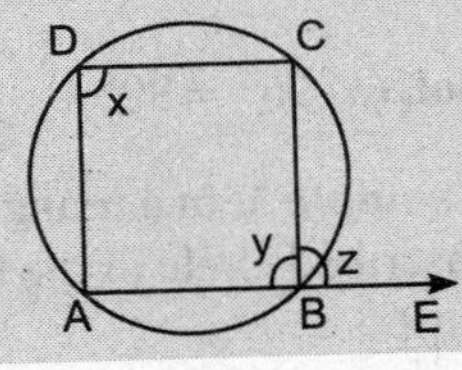

Proof : $x + y = 180°$ (cyclic quadrilateral)
$y + z = 180°$ (Linear Pair) $\Rightarrow x + y = y + z \Rightarrow x = z$

Example. If an exterior angle of a cyclic quadrilateral be 50°, then the interior opposite angle is : **(SSC CHSL 2011)**

(*a*) 130° $\quad$ (*b*) 40°
(*c*) 50° $\quad$ (*d*) 90°

Sol.(*c*) Interior $\angle D$ = Exterior $\angle B = 50°$

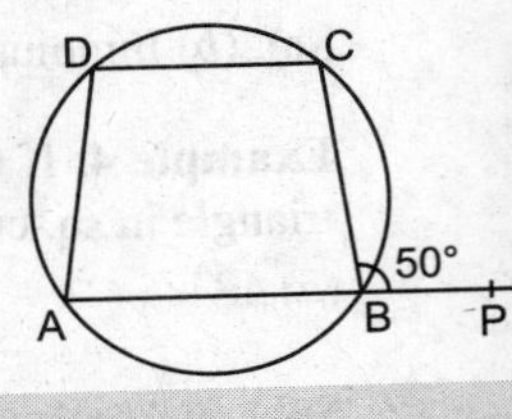

35. If ABCD is a cyclic quadrilateral. Then AB + CD = BC + AD

Note: $a + b + c + d = AB + CD = BC + AD$

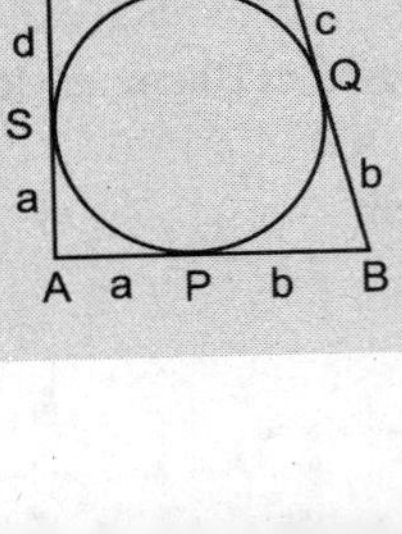

Example. *ABCD* is a quadrilateral. A circle is drawn in it touching its sides *AB*, *BC*, *CD* and *DA* at *P*, *Q*, *R*, and *S* respectively. If $AB = 6$ cm, $BC = 5$ cm, $CD = 4$ cm, find *AD*.

(*a*) 3 cm (*b*) 4 cm (*c*) 5 cm (*d*) 6 cm

Sol. (*c*) $AB + CD = BC + AD$

$\Rightarrow AD = AB + CD - BC = 6 + 4 - 5 = 5$ cm

36. Angle subtended by an arc at the centre is double the angle subtended by it at any point on the remaining part of the circle.

$\angle AOB = 2\angle ACB$

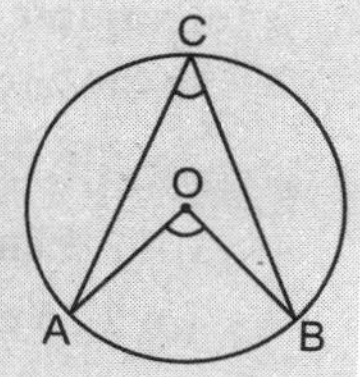

37. Angles subtended by an arc in the same segment of a circle, are equal. Equal chords of a circle subtend equal angles in the same segment. Conversely if the two chords subtend equal angles at the centre, they are equal in lengths.

$\angle ACB = \angle ADB$

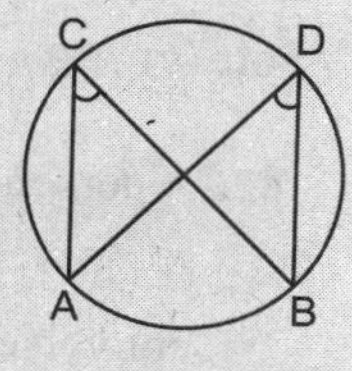

38. If *AB* is diameter of the circle and C is a point on the circle.

An angle inscribed in a semi-circle is a right angle. **(SSC CGL Tier I 2013)**

$\angle ACB = \angle ADB = 90°$

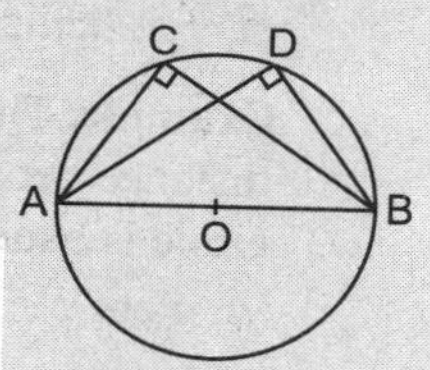

Example 1. *ABCD* is a cyclic trapezium with $AB \parallel DC$ and *AB* diameter of the circle. If $\angle CAB = 30°$, then $\angle ADC$ is **(SSC CGL Tier I 2013)**

(*a*) 60° (*b*) 120°

(*c*) 150° (*d*) 30°

Sol.(*b*) $\angle ACB = 90°$ (Angle in semi-circle)

$\angle B = 180° - 30° - 90° = 60°$

$\angle D = 180° - \angle B$

$= 180° - 60° = 120°$

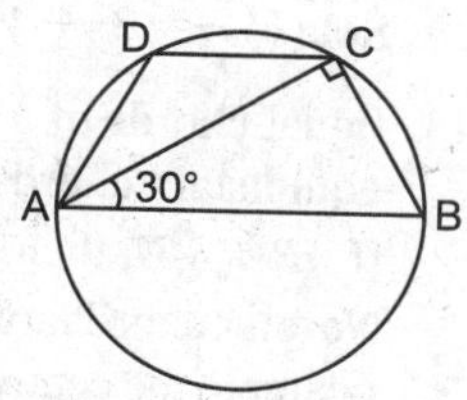

Example 2. In the given figure, *AB* is a diameter of a circle with centre *O*. Find the value of *x*.

(*a*) 90° (*b*) 60° (*c*) 30° (*d*) 15°

Sol.(*c*) $\angle BAC = \frac{1}{2} \angle BOC = \frac{1}{2} \times 60° = 30°$

In ΔAOC, $OA = OC$ (radii)

$\therefore \angle OCA = \angle OAC = 30°$

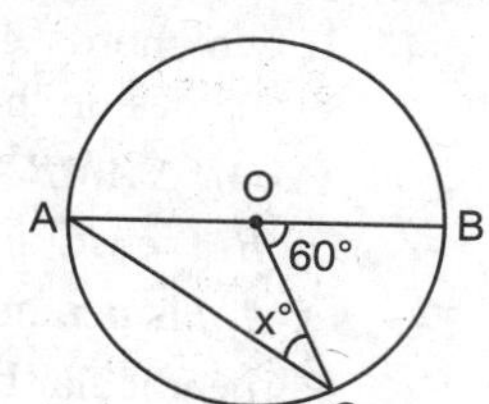

39. If AB and *CD* are two chords of a circle intersecting intenally at *P*.

$AP \times BP = CP \times DP$

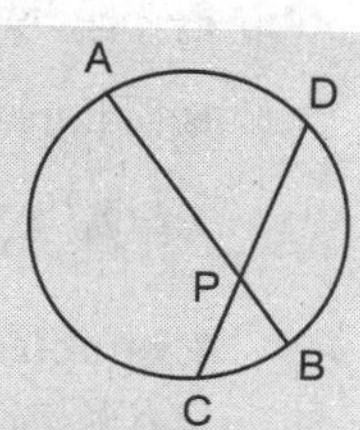

40. If two chords of a circle AB and CD intersect each other externally at P i.e. *PAB* and *PCD* are secants to a circle. Then, $PA \times PB = PC \times PD$

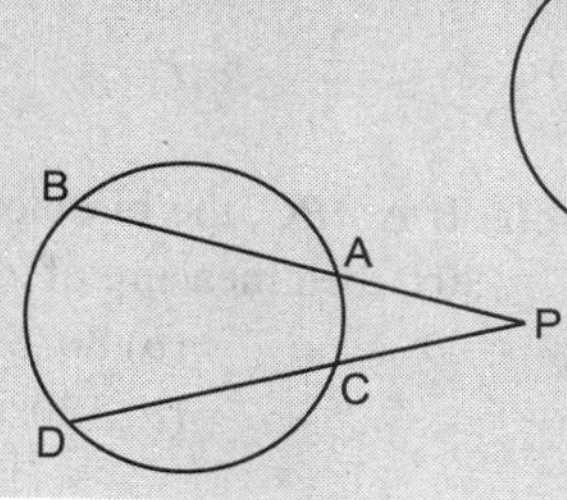

41. *PAB* is a secant to a circle at *A* and *B* and *PT* is tangent

Then, $PT^2 = PA \times PB$

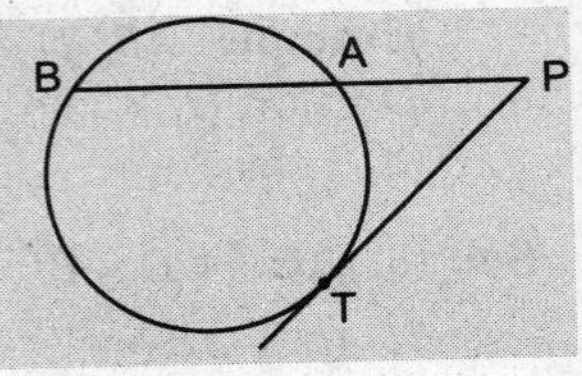

Example 1. In the given figure, *PAB* is a secant and *PT* is a tangent to the circle from a external point *P*. If *PT* = 5 cm, *PA* = 4 cm, *AB* = *x* cm, then *x* = ?

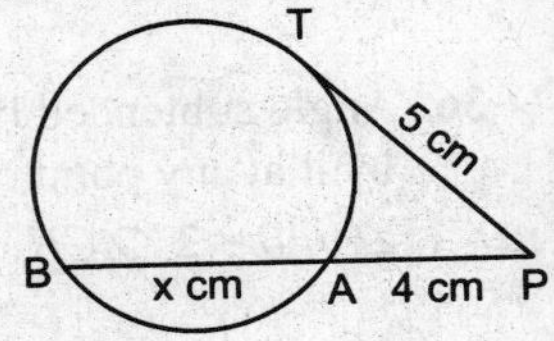

(*a*) 2.5 cm (*b*) 2.6 cm
(*c*) 2.25 cm (*d*) 2.75 cm

Sol. (*c*) $PB = \frac{PT^2}{PA} = \frac{5^2}{4} = 6.25 \Rightarrow x = 6.25 - 4 = 2.25$ cm

42. A perpendicular from the centre of a circle to a chord bisects the chord .

If $OP \perp MN$ then $MP = PN$.
conversely, the line joining the mid-point of a chord to the centre is perpendicular to the chord. If $MP = PN$ then $OP \perp MN$

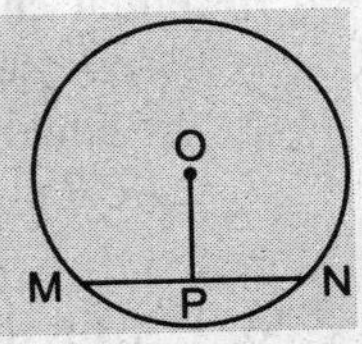

Example. The length of the chord of a circle is 8 cm and perpendicular distance between centre and the chord is 3 cm. Then the radius of the circle is equal to **(SSC FCI 2012, SSC LDC 2012)**

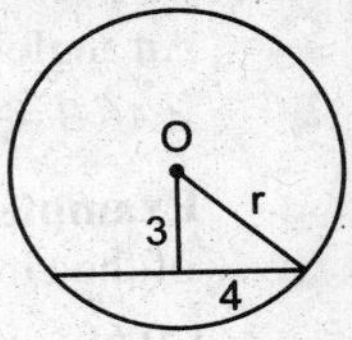

(*a*) 4 cm (*b*) 5 cm
(*c*) 6 cm (*d*) 8 cm

Sol.(*b*) $r^2 = 3^2 + 4^2 = 9 + 16 = 25 = 5^2$

43. Equal chords of a circle are equidistant from the centre. conversely, equidistant chords from the centre of a circle are equal in lengths.
If $AB = MN$, then $OC = OP$ and if $OC = OP$, then $AB = MN$
Note: A chord farther from the centre in a circle is shorter than the one nearer to the centre. Diameter, a chord through the centre, is the longest chord of a circle. Diameter divides the circle into two semicircles.

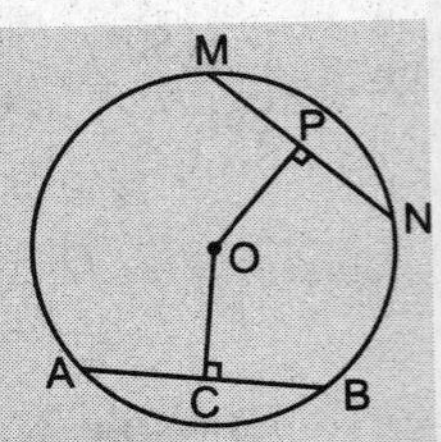

44. If two chords *AB* and *AC* of a circle are equal. Then the centre of the cirlce lies on the angle bisector of $\angle BAC$.
Proof. $\Delta AMB \cong \Delta AMC$ (SAS Criterion)
$\Rightarrow BM = MC$ and $\angle AMB = \angle AMC = 90°$
$\therefore$ *AM* is perendicular bisector of chord BC
Perpendicular bisector of a chord always passes through the centre.

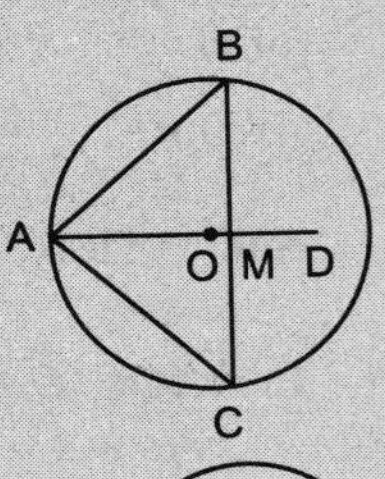

45. Perpendicular bisectors of two chords of a circle intersect at its centre.

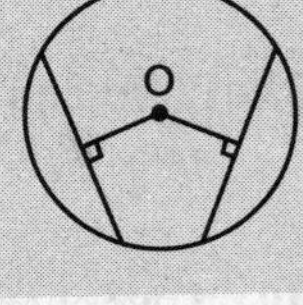

46. Bisectors of angles $\angle ABC$ and $\angle ACB$ meet at a point *O*, then:

$\angle BOC = 90° + \frac{1}{2} \angle BAC$

Example. If Δ *ABC*, the bisector of $\angle B$ and $\angle C$ meet at centre *O* and $\angle BAC = 80°$, then measure of $\angle BOC$ is: **(SSC CGI Tier II 2012)**

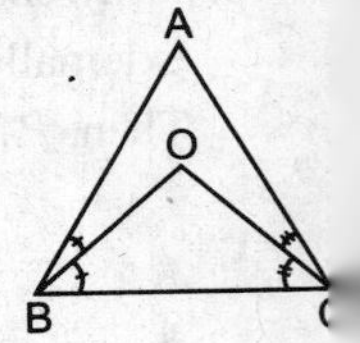

(*a*) 120° (*b*) 80°
(*c*) 90° (*d*) 130°

Sol. (*d*) $\angle BOC = 90° + \frac{1}{2}\angle BAC = 90° + \frac{1}{2} \times 80° = 130°$

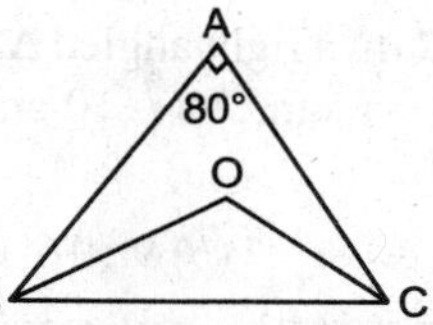

47. The sides AB and AC are produced to P and Q respectively. The bisectors of $\angle PBC$ *and* $\angle QCB$ intersect at D.

Then, $\angle BDC = 90° - \frac{1}{2}\angle BAC$

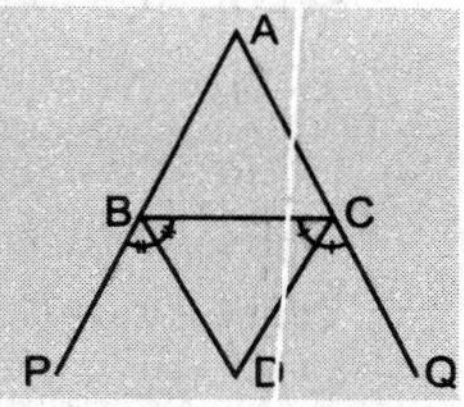

Example. The external bisector of $\angle B$ and $\angle C$ of ΔABC (where AB and AC extended to E and F respectively) meet at point P. If $\angle BAC = 100°$, then $\angle BPC$ is **(SSC CHSL 2011)**

(*a*) 50° (*b*) 80°

(*c*) 40° (*d*) 100°

Sol.(*c*) $\angle BPC = 90° - \frac{\angle A}{2} = 90° - \frac{100°}{2} = 90° - 50° = 40°$

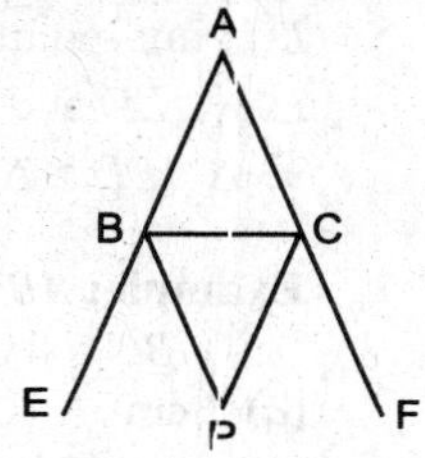

48. Angle made by a tangent of a circle with a chord from the point of contact is equal to the angle subtended by the chord in the alternate segment of the circle.

If PQ is tangent of a circle and touches it at A

Then $\angle CAP = \angle CBA$ and $\angle BAQ = \angle ACB$

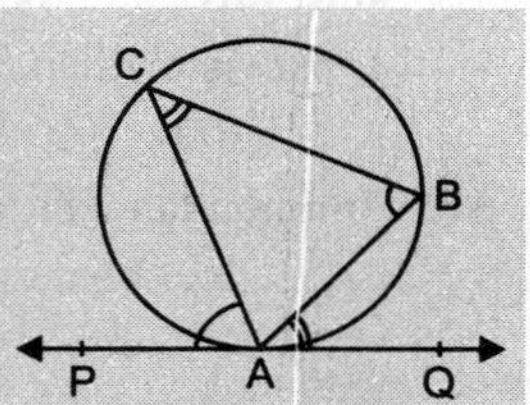

49. ABC is a right angled triangle with $\angle ABC = 90°$ and $BD \perp AC$. Then,

(*i*) $\Delta ABC \sim \Delta ADB$ (*ii*) $\Delta ABC \sim \Delta BDC$

(*iii*) $\Delta ADB \sim \Delta BDC$ (*iv*) $BD^2 = AD \times DC$

(*v*) $AB^2 = AC \times AD$ (*vi*) $BC^2 = AC \times CD$

(*vii*) $\frac{1}{BD^2} = \frac{1}{AB^2} + \frac{1}{BC^2}$ **(CDS 2001, SSC CGL Tier I 2013)**

(*viii*) Derived from Area Formula:

$$\Delta ABC = \frac{1}{2}(AC \times BD) = \frac{1}{2}(AB \times BC) \Rightarrow BD = \frac{AB \times BC}{AC}$$

(*viii*) Derived from (v) & (vii) : $\frac{AB^2}{BC^2} = \frac{AD}{CD}$

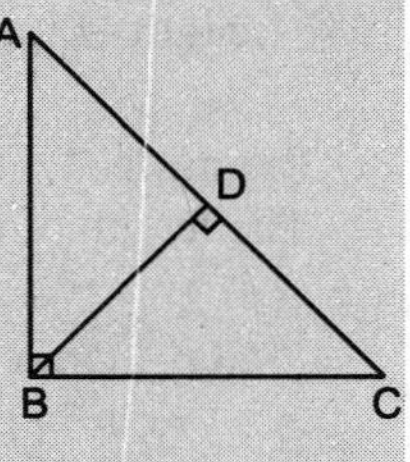

Example 1: The area of a right-angled triangle is 24 cm^2 and one of the sides containing the right angle is 6 cm. The altitude on the hypotenuse is **(SSC CGL Tier II 2012)**

(*a*) 5.2 cm (*b*) 12 cm (*c*) 3.6 cm (*d*) 4.8 cm

Sol.(*d*) Let $AB = x$, then $\frac{1}{2} \times 6 \times x = 24 \Rightarrow x = 8$

$AC^2 = 6^2 + 8^2 = 10^2$

Area $= \frac{1}{2} \times AC \times BD = 24 \Rightarrow BD = \frac{24 \times 2}{10} = 4.8$ cm

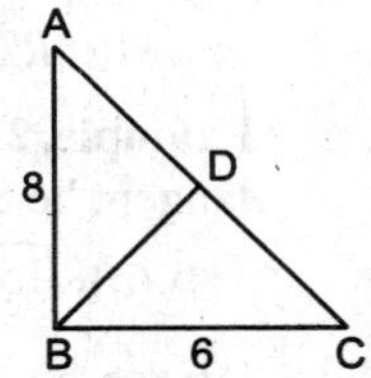

Example 2. In a right angled ΔABC, $\angle ABC = 90°$; BN is perpendicular to AC, $AB = 6$ cm, $AC = 10$ cm. Then $AN : NC$ is **(SSC CGL Tier I 2012)**

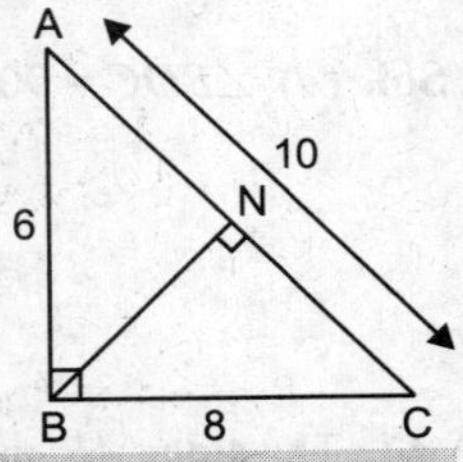

(*a*) 3 : 4 (*b*) 9 : 16 (*c*) 3 : 16 (*d*) 1 : 4

Sol.(*b*) $BC^2 = AC^2 - AB^2 = 10^2 - 6^2 = 8^2$

$\Rightarrow AN : NC = AB^2 : BC^2 = 6^2 : 8^2 = 3^2 : 4^2 = 9 : 16$

50. If ABC is a right triangle and D is the mid-point of hypotenuse AC. Then $AD = CD = BD$

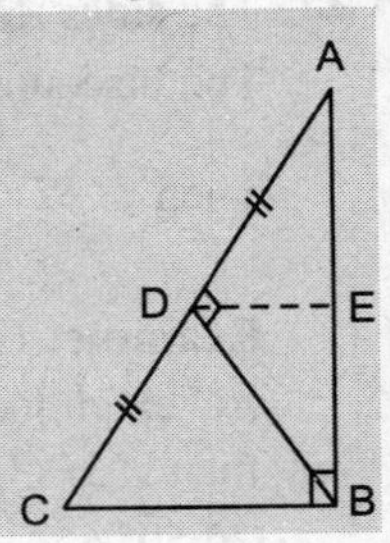

Proof. Draw $DE \parallel BC$

$\Rightarrow \angle AED = \angle ABC = 90°$

D is mid-point of $AC \Rightarrow E$ is mid-point of AB $\Rightarrow$ AE = BE

In ΔAED and ΔBED, $AE = BE$, $ED = ED$, $\angle AED = \angle BED = 90°$

$\Rightarrow \Delta AED \cong \Delta BED$ (SAS-criterion) $\Rightarrow DA = DB$

Example: ABC is a right angled triangle and $\angle B = 90°$. $AB = 6$ cm, $BC = 8$ cm and BD is median. Find length of BD

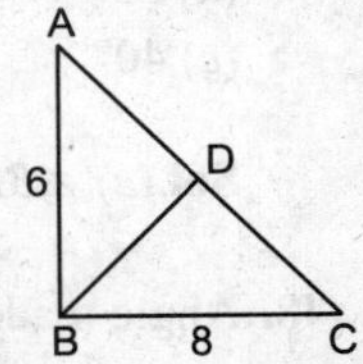

(*a*) 7 cm (*b*) 14 cm (*c*) 5 cm (*d*) 10 cm

Sol.(*c*) $AC^2 = 6^2 + 8^2 = 10^2$

BD is median $\Rightarrow AD = DC = 5$ cm

$\therefore BD = AD = DC = 5$ cm

51. Common tangent to two circles :

If r_1, r_2 are the radii of two circles and d be distance between their centres, then **(SSC CGL Tier I 2012)**

(*i*) Length of direct common tangent = $\sqrt{d^2 - (r_1 - r_2)^2}$

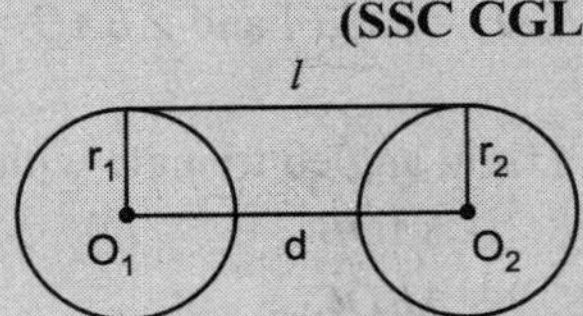

(*ii*) Length of transverse common tangent = $\sqrt{d^2 - (r_1 + r_2)^2}$

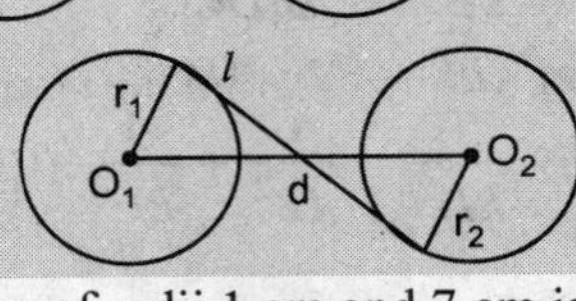

Example. 1. The distance between the centres of two circles of radii 1 cm and 7 cm is 10 cm. The length of direct common tangent (in cm) is : **(SSC CGL Tier I 2012)**

(*a*) 2 (*b*) 6 (*c*) 8 (*d*) 18

Sol.(*c*) Draw $RS \parallel PQ$

$RS = PQ = 10$ cm

$TS = TQ - SQ = 7 - 1 = 6$ cm

In ΔRST, $RS = 10$ cm, $TS = 6$ cm

$\therefore RT^2 = RS^2 - ST^2 = 10^2 - 6^2 = 8^2$ [$\angle RTS = 90°$]

Alt: $RT = \sqrt{10^2 - (7-1)^2} = \sqrt{10^2 - 6^2} = 8$ cm

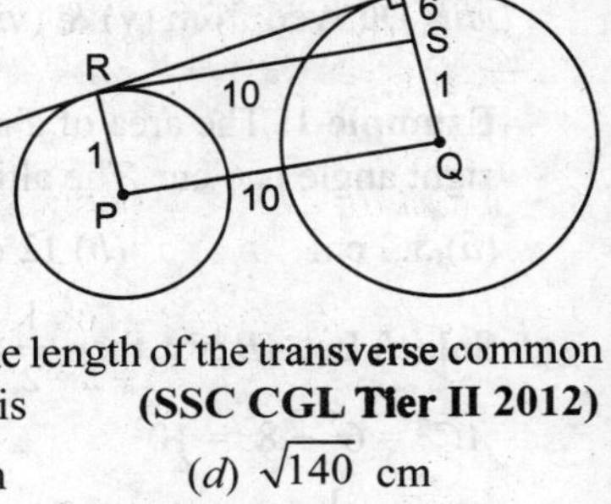

Example. 2. If the radii of two circles be 6 cm and 3 cm and the length of the transverse common tangent be 8 cm, then the distance between the two centres is **(SSC CGL Tier II 2012)**

(*a*) $\sqrt{150}$ cm (*b*) $\sqrt{135}$ cm (*c*) $\sqrt{145}$ cm (*d*) $\sqrt{140}$ cm

Sol. (*c*) $8 = \sqrt{d^2 - (6+3)^2} \Rightarrow 64 = d^2 - 81 \Rightarrow d^2 = 145 \Rightarrow d = \sqrt{145}$ cm

52. A tangent to a circle is a line which intersects the circle in only one point. A tangent to a circle is perpendicular to the radius through the point of contact. $OP \perp AB$.

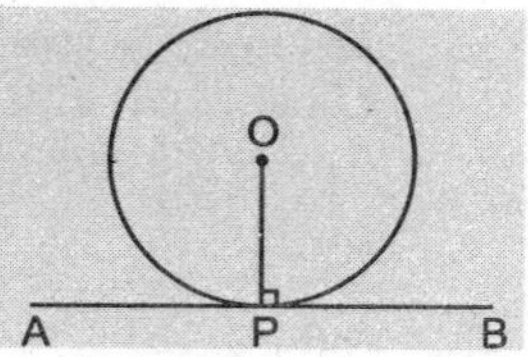

Example: The length of tangent (up to the point of contact) drawn from an external point P to a circle of radius 5 cm is 12 cm. The distance of *P* from the centre of the circle is

(SSC CGL Tier I 2013)

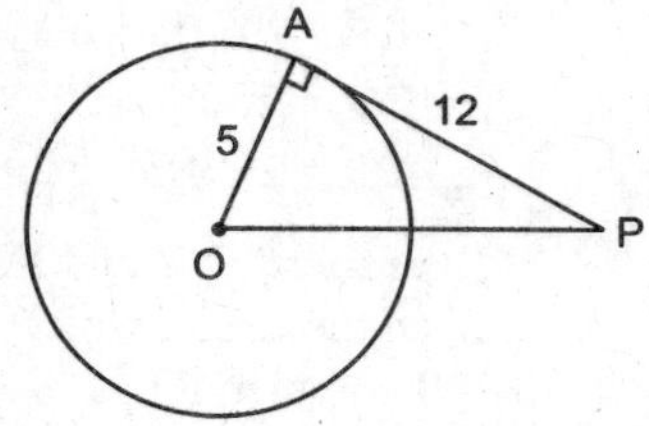

(*a*) 11 cm (*b*) 12 cm
(*c*) 13 cm (*d*) 14 cm

Sol.(*c*) $OP^2 = OA^2 + AP^2 = 5^2 + 12^2 = 13^2 \Rightarrow OP = 13$ cm

53. Secant : A line which intersects a circle at two distinct points.

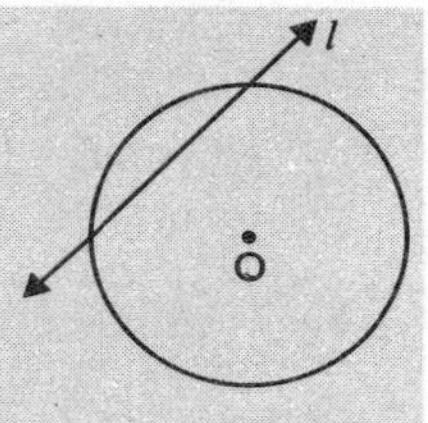

54. From an external point *A*, exactly two tangents say *AP* and *AQ* can be drawn to a circle with centre *O*.

Then $\Delta APB \cong \Delta AQB$

$\angle APB = \angle AQB$

$AP = AQ$ **(Tangents are equal in length)**

$\angle PAB = \angle BAQ$

$PB = BQ$

$\angle ABP = \angle ABQ = 90° \Rightarrow AB \perp PQ$

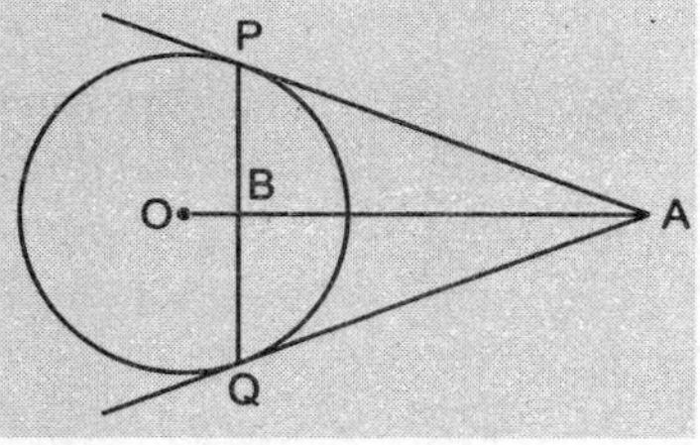

Example 1. Two circles touch externally at a point *C*. From a point *A* outside the circles, tangents at *B*, *C* and *D* are drawn. If $AB = 5$ cm, find the length of *AD*.

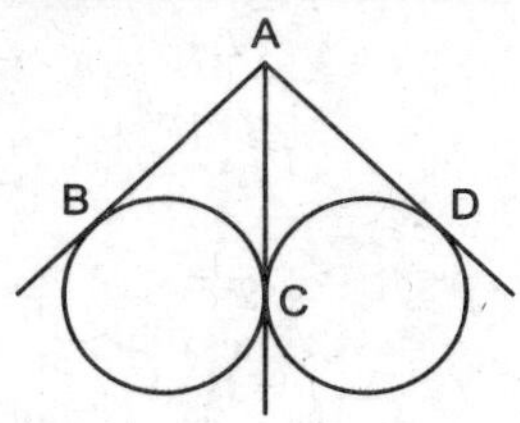

(*a*) 3 cm (*b*) 4 cm
(*c*) 6 cm (*d*) 5 cm

Sol.(*d*) $AB = AC$ (tangents to same circle)

$AC = AD$ (tangents to the same circle)

$\therefore AD = AC = AB = 5$ cm

Example 2: The tangents drawn at *P* and *Q* on the circumference of a circle intersect at *A*. If $\angle PAQ = 68°$, then the measure of $\angle APQ$ is **(SSC FCI Asstt 2012)**

(*a*) 56° (*b*) 68°
(*c*) 28° (*d*) 34°

Sol.(*a*) $AP = AQ$ (tangents)

$\therefore \angle P = \angle Q$

$\angle P + \angle Q = 180° - 68° = 112°$

$\therefore \angle APQ = 112° \div 2 = 56°$

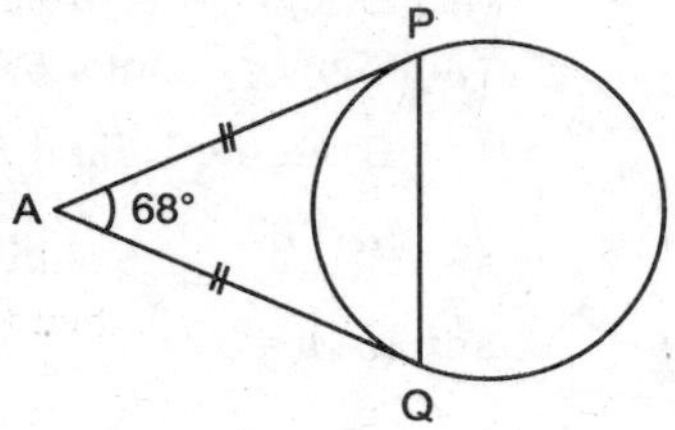

Example 3. Two tangents are drawn from a point P to a circle at A and B. O is the centre of the circle. If $\angle AOP = 60°$, then $\angle APB$ is **(SSC CGL Tier I 2012)**

(*a*) 60° (*b*) 30°

(*c*) 120° (*d*) 90°

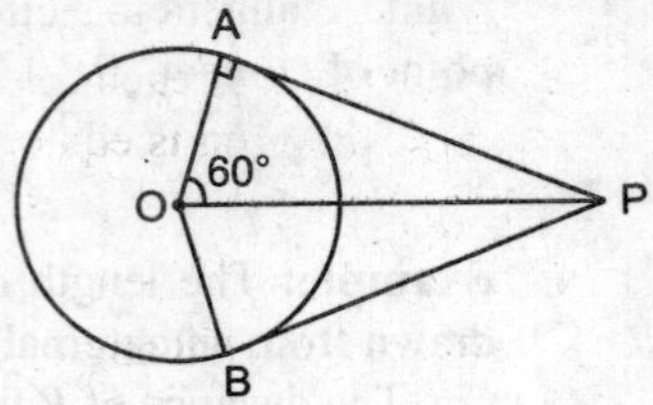

Sol.(*a*) $\angle AOP = 60°$

$\angle OAP = 90°$ (tangent)

$\angle APO = 180° - (90° + 60°) = 30°$

$\angle APB = 2\angle APO = 2 \times 30° = 60°$

Example 4. XY and XZ are tangents to a circle, ST is another tangent to the circle at the point R on the circle, which intersects XY and XZ at S and T respectively. If XY = 15 cm and TX = 9 cm, then RT is **(SSC CGL Tier I 2015)**

(*a*) 4.5 cm (*b*) 6 cm

(*c*) 7.5 cm (*d*) 3 cm

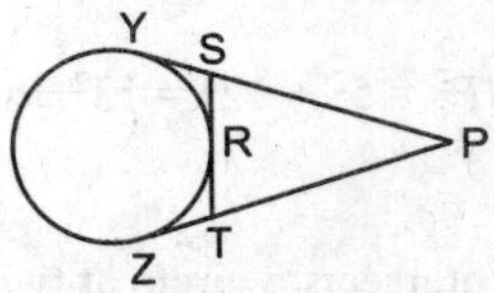

Sol.(*b*) XZ = XY = 15 cm

ZT = XZ – TX = 15 – 9 = 6

RT = TX = 6

55. If two circles intersect in two points, then the line through the centres of the circles, is the perpendicular bisector of the common chord.

$AM = BM$ and $\angle AMO = \angle BMO = 90°$

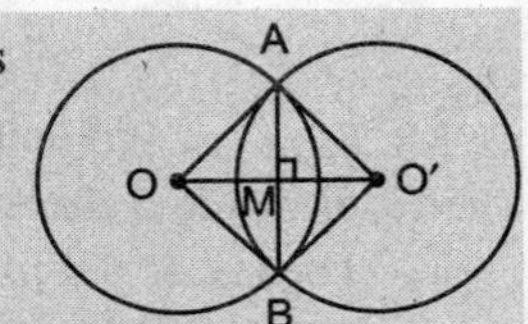

56. Exactly two tangents can be drawn through a point outside the circle. The lengths of two tangents drawn from an external point to a circle are equal.

$AP = AQ;\ AO = AO$ (common)

$OP = OQ$ (Radii)

$\Rightarrow \Delta OPA \cong \Delta OQA$

$\angle AOP = \angle AOQ$

$\angle OAP = \angle OAQ$

$\angle PAQ + \angle POQ = 180°$

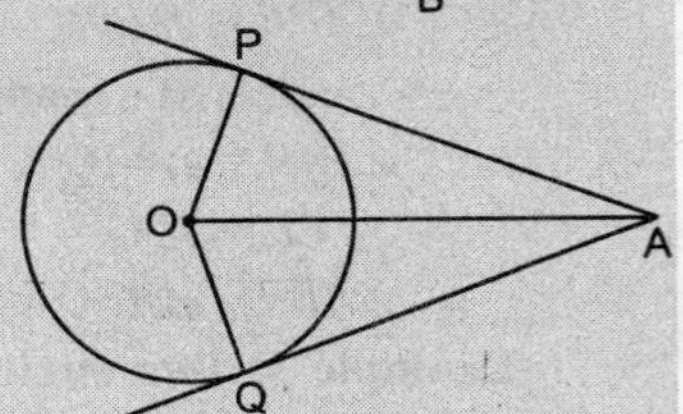

57. Polygon : Polygon is a closed plane figure bounded by three or more straight line segments e.g. triangle (3 lines), Quadrilateral (4 lines), Pentagon (5 lines), Hexagon (6 lines). Heptagon (7 sides), Octagon (8 sides), Nonagon (9 sides), Decagon (10 sides).

Regular polygon is a polygon having all sides equal and all interior angles equal.

(*i*) Each Exterior angle of regular polygon of '*n*' sides = $\dfrac{360°}{n}$

(*ii*) Each Interior angle of regular polygon of '*n*' sides = 180° – Exterior angle or $\dfrac{n-2}{n} \times 180°$

(*iii*) Sum of all exterior angles of a regular polygon = 360°

(*iv*) Sum of all interior angles of polygon of '*n*' sides = $(n - 2) \times 180°$

Example 1. Find each angle of a regular pentagon.

(*a*) 90° (*b*) 108° (*c*) 144° (*d*) 120°

Sol. (*b*) $n = 5$, $\therefore$ Each angle $= \dfrac{5-2}{5} \times 180° = 108°$

Alternative method : Each angle $= 180° - \dfrac{360°}{5} = 180° - 72° = 108°$

(*v*) Number of diagonals of a polygon of *n* sides = $\left[\frac{n(n-1)}{2} - n\right]$ or $\frac{n(n-3)}{2}$

Example 2. If each interior angle of a regular polygon is 135°, then the number of diagonals of the polygon is equal to **(CDS 2015)**

(*a*) 54 (*b*) 48 (*c*) 18 (*d*) 20

Sol.(*d*) External angle = 180° – 135° = 45° ⇒ Number of sides = 360° ÷ 45° = 8

Number of diagonals = $\frac{8(8-3)}{2} = \frac{8 \times 5}{2} = 20$

58. *ABCD* is a parallelogram and O is any point in its interior, then

$\Delta AOB + \Delta COD = \Delta BOC + \Delta AOD$

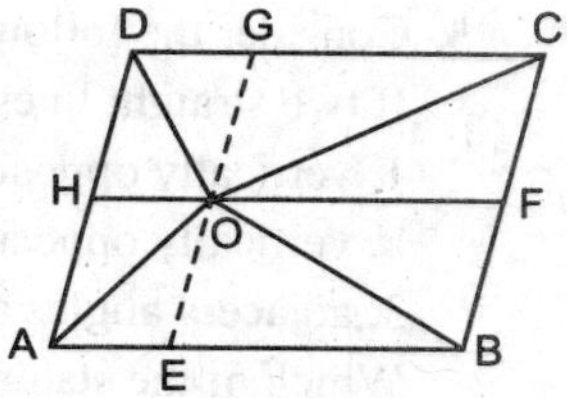

Proof: Draw $HOF \parallel AB$, $EOG \parallel AD$

⇒ *AEOH, EBFO, OFCG, HOGD* are parallelograms

OA is a diagonal of parallelogram *AEOH*

$\Rightarrow \Delta AEO = \Delta AOH$

Similarly, $\Delta EOB = \Delta OBF$, $\Delta OFC = \Delta OCG$,

$\Delta DOG = \Delta DOH$

$\Rightarrow \Delta AEO + \Delta OBE + \Delta GOC + \Delta DOG = \Delta AOH + \Delta OBF + \Delta OFC + \Delta DOH$

$\Rightarrow \Delta AOB + \Delta COD = \Delta BOC + \Delta AOD$

59. If the medians of a triangle *ABC* intersect at *G*, then Area (ΔAGB) = Area (*BGC*) = Area (ΔAGC)

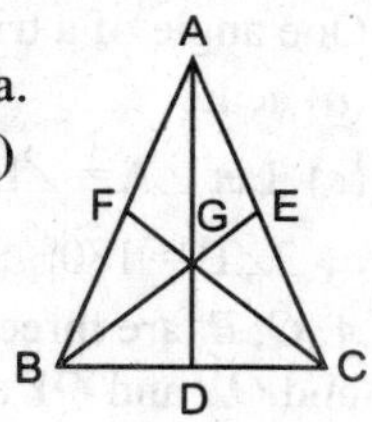

Proof: Median of a triangle divides it into two triangles of equal area.

⇒ Area (ΔABD) = Area (ΔACD) and area (ΔBDG) = area (ΔDCG)

$\Rightarrow \Delta ABD - \Delta BDG = \Delta ACD - \Delta DCG$

$\Rightarrow \Delta AGB = \Delta AGC$

Similarly, $\Delta AGB = \Delta BGC$

⇒ Area (ΔAGB) = area (ΔBGC) = area (ΔAGC)

SOLVED EXERCISE

1. The angles of a triangle are in the ratio of 3 : 4 : 5. What is degree of the smallest angle? **(SSCCGLTierI2015)**

(*a*) 15° (*b*) 30° (*c*) 45° (*d*) 60°

Sol. (*c*) Sum of ratios = 3 + 4 + 5 = 12 ⇒ Smallest angle = $\frac{3}{12} \times 180° = 45°$

2. If one angle of a triangle is equal to the sum of the other two angles, then the triangle is : **(SSC CHSL 2012)**

(*a*) Isosceles (*b*) Obtuse-angled (*c*) Acute-angled (*d*) Right-angled

Sol. (*d*) One angle = $\frac{1}{2} \times 180° = 90°$

3. If the measures of three angles of a triangle are in the ratio 2 : 3 : 5, then the triangle is : **(SSC CGL Tier I 2015)**

(*a*) equilateral (*b*) isosceles (*c*) obtuse angled (*d*) right angled

Sol. (*d*) Ratio of three angles = 2 : 3 : 5 = 36°, 54°, 90° ⇒ Triangle is right angled

4. Which one among the following is not correct ? **(CDS 2012)**

(*a*) Two congruent triangles are necessarily similar (b) All equiangular triangles are similar

(*c*) Two isosceles right triangles are similar (d) All isosceles triangles are similar

Sol. (*d*) Statement (d) is incorrect.

5. Consider the following statements : **(CDS 2012)**

If two straight lines intersect, then :

1. vertically opposite angles are equal.
2. vertically opposite angles are supplementary.
3. adjacent angles are supplementary

Which of the statements given above is/are correct ?

(*a*) 3 only (*b*) 1 only (*c*) 1 and 3 (*d*) 2 and 3

Sol. (*c*) Statement 2 is incorrect.

6. The measures of two angles of a triangle are in the ratio 4 : 5. If the sum of these two numbers is equal to the measure of the third angle. Find the smallest angle. **(SSC CGL Tier I 2015)**

(*a*) 40° (*b*) 50° (*c*) 10° (*d*) 90°

Sol. (*a*) Ratio of three angles = 4 : 5 : (4 + 5) = 4 : 5 : 9 ⇒ Smallest angle = $\frac{4}{18} \times 180° = 40°$

7. One angle of a triangle ABC is equal to difference of other two angles, then the triangle is

(*a*) acute (*b*) obtuse (*c*) right (*d*) equilateral

Sol. (*c*) Let ∠A = ∠B – ∠C, then (∠B – ∠C) + ∠B + ∠C = 180°

⇒ 2∠B = 180° ⇒ ∠B = 90° ⇒ triangle is right angled.

8. *A*, *O*, *B*, are three points on a line segment and *C* is a point not lying on *AOB*. If ∠*AOC* = 40° and *OX* and *OY* are the internal and external bisectors of ∠*AOC* respectively, then ∠*BOY* is: **(SSC CGL Tier I 2012)**

(*a*) 72° (*b*) 68°

(*c*) 70° (*d*) 80°

Sol. (*c*) ∠*BOC* = 180° – 40° = 140°

$$\angle BOY = \frac{1}{2} \times \angle BOC = \frac{1}{2} \times 140° = 70°$$

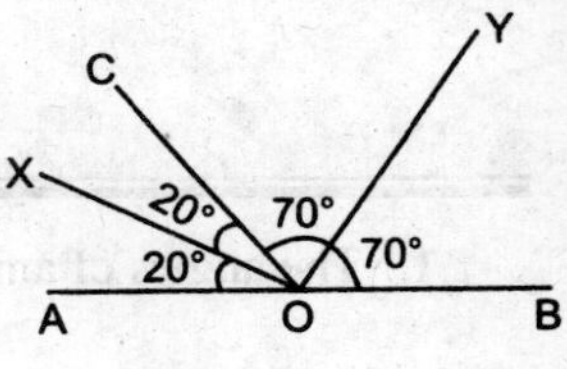

9. ∠*A*, ∠*B*, ∠*C* are three angles of a triangle. If ∠*A* – ∠*B* = 15°, ∠*B* – ∠*C* = 30°, then ∠*A*, ∠ *B* and ∠*C* are **(SSC CGL Tier II 2013)**

(*a*) 80°, 60°, 40° (*b*) 70°, 50°, 60° (*c*) 80°, 65°, 35° (*d*) 80°, 55°, 45°

Sol. (*c*) 80° – 65° = 15° and 65° – 35° = 30°

10. The angles of a triangle are in Arithmetic Progression. The ratio of the least angle in degrees to the number of radians in the greatest angle is 60 : π. The angles in degrees are **(SSC CGL Tier II 2012)**

(*a*) 40°, 50°, 90° (*b*) 40°, 55°, 85° (*c*) 30°, 60°, 90° (*d*) 35°, 55°, 90°

Sol. (*c*) Ratio between least and greatest angle = 60 : π = 60 : 180 = 1 : 3

∴ Ratio of three angles = 1 : 2 : 3 (in A.P.) ⇒ The angles are 30°, 60°, 90°

11. Two sides of a triangle are of length 4 cm and 10 cm. If the length of the third side is 'a' cm, then **(SSC CGL Tier I 2012)**

(*a*) $a < 6$ (*b*) $6 < a < 14$ (*c*) $a > 5$ (*d*) $6 \leq a \leq 12$

Sol. (*b*) Shortest length = 10 – 4 = 6 cm (excluding 6)
Largest length = 10 + 4 = 14 cm (excluding 14)

12. Taking any three of the line segments of length 2 cm, 3 cm, 5 cm and 6 cm, the number of triangles that can be formed is : **(SSC CGL Tier I 2013)**

(*a*) 4 (*b*) 3 (*c*) 2 (*d*) 1

Sol. (*c*) Total combinations : (2, 3, 5), (2, 3, 6), (2, 5, 6), (3, 5, 6)
Sum of lengths of two sides of the triangle is always more than the length of the third side.
∴ Combinations (2, 3, 5) and (2, 3, 6) are not possible.

13. Two unequal sides of an isosceles triangles are 7 cm and 15 cm. Find perimeter of the triangle. **(Railways, 2006)**

(*a*) 29 cm (*b*) 37 cm (*c*) 22 cm (*d*) None of these

Sol. (*b*) Sides can't be 7, 7, 15 as 7 + 7 < 15.
∴ Sides are 7, 15, 15 ⇒ Perimeter = 7 + 15 + 15 = 37 cm

14. In a triangle, if three altitudes are equal, then the triangle is **(SSC CGL Tier I 2013)**

(*a*) Right (*b*) Isosceles (*c*) Obtuse (*d*) Equilateral

Sol. (*d*) The triangle is an equilateral triangle.

15. The three equal circles touches each other externally. If the centres of these circles be A, B, C then ΔABC is: **(SSC CGL Tier I 2013)**

(*a*) a right angle triangle (*b*) an equilateral triangle
(*c*) an isosceles triangle (*d*) a scalene triangle

Sol. (*b*) Let radius of each circle = r cm
Then each side of the triangle = $2r$ cm ⇒ Equilateral triangle.

16. If ABC is an equilateral triangle and D is a point on BC such that $AD \perp BC$, then $AB : BD = ?$ **(SSC CGL Tier II 2013)**

(*a*) 1 : 1 (*b*) 1 : 2
(*c*) 2 : 1 (*d*) 3 : 2

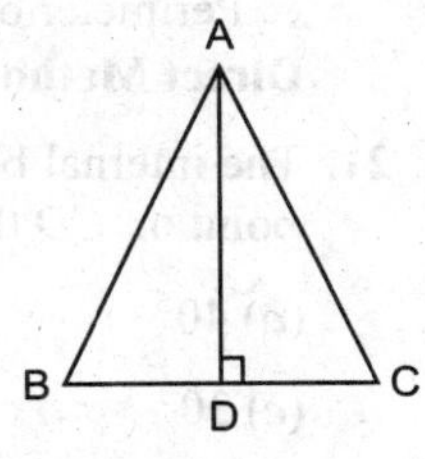

Sol. (*c*) $AB = BC = AC$

$BD = \frac{1}{2}BC$ ∴ $AB : BD = 2 : 1$

17. In an isosceles triangle ABC, $AB = AC = 25$ cm, and $BC = 14$ cm. Find the area of the triangle.

(*a*) 168 cm² (*b*) 336 cm² (*c*) 175 cm² (*d*) 350 cm²

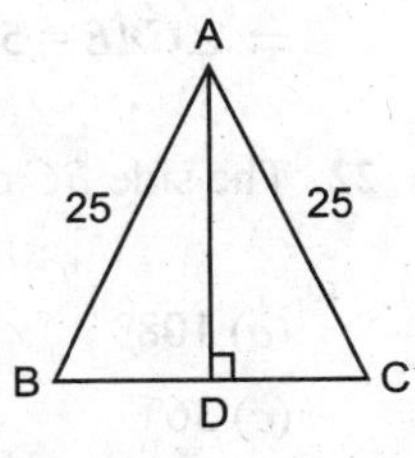

Sol. (*a*) In an isosceles triangle, AD is altitude and also median
∴ $BD = DC = 14 \div 2 = 7$ cm

$AD = \sqrt{25^2 - 7^2} = 24$ cm

Area of $\Delta ABC = \frac{1}{2} \times BC \times AD = \frac{1}{2} \times 14 \times 24 = 168$ cm²

18. $ABCD$ is a square, F is the mid-point of AB, BE is one-third of BC. If the area of ΔFBE is 108 cm², find the length of the diagonal of the square.

(*a*) 54 cm (*b*) 36 cm (*c*) $36\sqrt{2}$ cm (*d*) $27\sqrt{2}$ cm

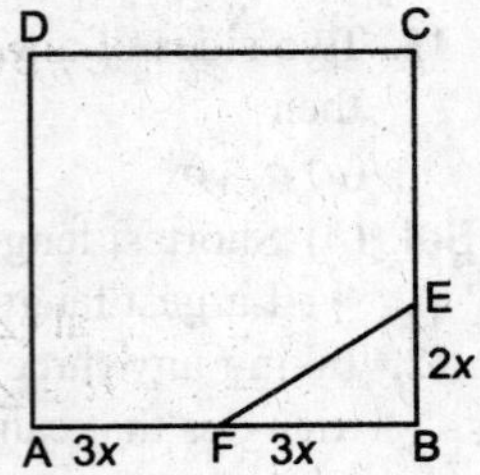

Sol. (*c*) Let side of square = $6x$

Then $FB = 6x \div 2 = 3x$

and $BE = 6x \div 3 = 2x$

$\Delta FBE = \frac{1}{2} \times FB \times BE = \frac{1}{2} \times 3x \times 2x = 108$

$\Rightarrow 3x^2 = 108 \Rightarrow x^2 = 36 \Rightarrow x = 6$ cm

$\therefore$ Side of square = $6x$ = 36 cm

$\therefore$ Diagonal of square = $36\sqrt{2}$

19. The mid-points of AB and AC of a ΔABC are respectively X and Y. If $BC + XY = 12$ units, then the value of $BC - XY$ is **(SSC FCI Asstt. 2012)**

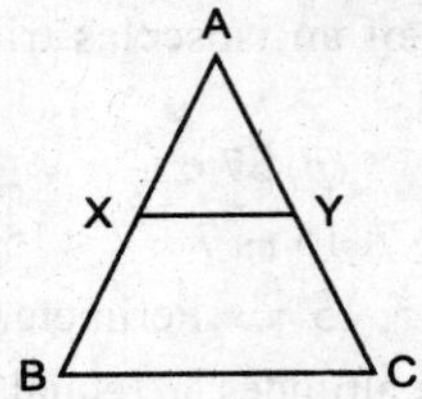

(*a*) 6 (*b*) 8

(*c*) 4 (*d*) 12

Sol. (*c*) If $XY = x$, then $BC = 2x$

$BC + XY = 12 \Rightarrow 2x + x = 12 \Rightarrow x = 4$

$BC - XY = 2x - x = x = 4$ units

20. In the given figure, radius of each circle is 7 cm and P, Q, R are centres. If $AB = 4$ cm, $CD = 2$ cm and $EF = 3$ cm, find the perimeter of ΔPQR.

(*a*) 21 cm (*b*) 24 cm

(*c*) 42 cm (*d*) 33 cm

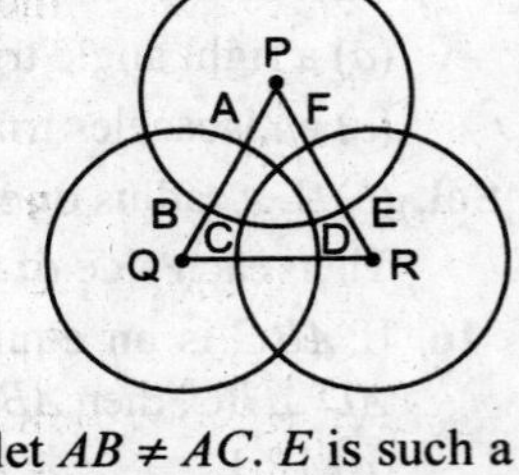

Sol. (*d*) $PQ = PB + AQ - AB = 7 + 7 - 4 = 10$ cm

$QR = QD + RC - CD = 7 + 7 - 2 = 12$ cm

$PR = PE + RF - EF = 7 + 7 - 3 = 11$ cm

$\therefore$ Perimeter of ΔPQR = (10 + 12 + 11)cm = 33 cm

Direct Method = $(7 + 7) \times 3 - (4 + 2 + 3) = 33$ cm

21. The internal bisector of $\angle C$ of ΔABC intersects AB at point D and let $AB \neq AC$. E is such a point on CD that $AE = AD$. If $\angle ABC = 50°$, then $\angle CAE$ is **(SSC FCI Asstt 2012)**

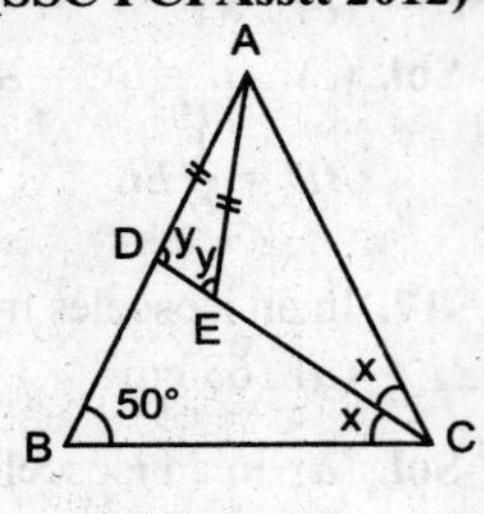

(*a*) 40° (*b*) 50°

(*c*) 30° (*d*) 25°

Sol. (*b*) In ΔBCD, $50° + x = y$ (external angle)

In ΔACE, $\angle CAE + x = y$ (external angle)

$\Rightarrow \angle CAE + x = 50° + x$

$\Rightarrow \angle CAE = 50°$

22. The side BC of ΔABC is produced to D. If $\angle ACD = 108°$ and $\angle B = \frac{1}{2}\angle A$, then $\angle A$ is **(SSC CGL Tier I 2011)**

(*a*) 108° (*b*) 59°

(*c*) 36° (*d*) 72°

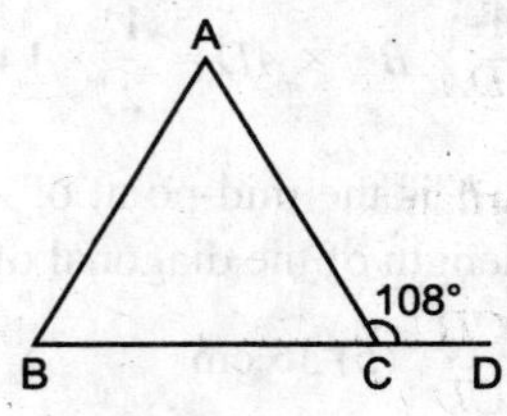

Sol. (*d*) $\angle ABC + \angle BAC = 108°$

$\Rightarrow \angle ABC + 2\angle ABC = 108°$

$\Rightarrow \angle ABC = 36°$

$\Rightarrow \angle A = 2 \times 36° = 72°$

23. In a triangle ABC, $AB = AC$, $\angle BAC = 40°$, Then the external angle at B is : **(SSC CGL Tier I 2013)**

(*a*) 80° (*b*) 90° (*c*) 70° (*d*) 110°

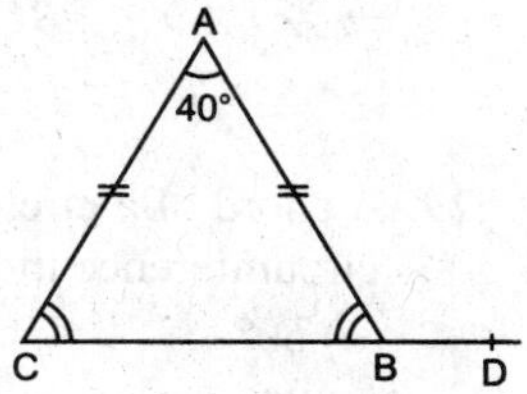

Sol. (*d*) Internal $(\angle A + \angle B + \angle C) = 180°$

$\Rightarrow \angle B + \angle C = 180° - 40° = 140°$

$\Rightarrow \angle B = \frac{1}{2} \times 140° = 70°$

$\therefore \angle ABD = 180° - 70° = 110°$

24. If in any ΔABC, the base BC is produced in both ways, the sum of the exterior angles at B and C is **(SSC CPO SI 2012, CDS 2015)**

(*a*) $\frac{\pi}{2} + A$ (*b*) $\pi - \frac{A}{2}$ (*c*) $\pi - A$ (*d*) $\pi + A$

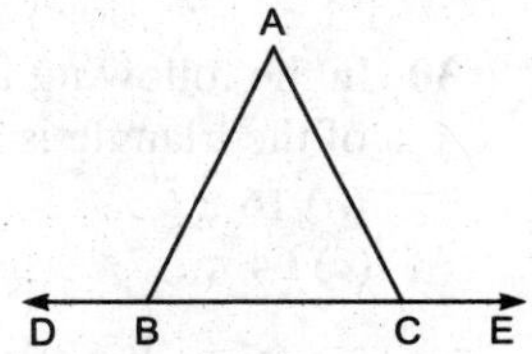

Sol. (*d*) $\angle ABD = \angle A + \angle C$

$\angle ACE = \angle A + \angle B$

$\therefore \angle ABD + \angle ACE = \angle A + \angle C + \angle A + \angle B$

$= (\angle A + \angle B + \angle C) + \angle A$

$180° + \angle A = \pi + A$

25. In the figure, find $\angle A$.

(*a*) 30° (*b*) 45°

(*c*) 70° (*d*) 80°

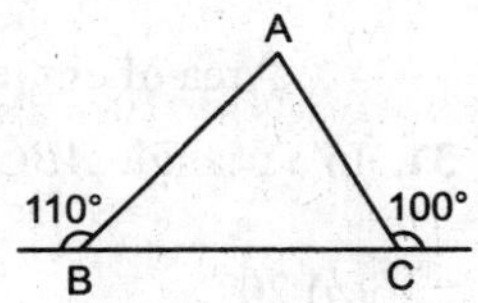

Sol. (*a*) Internal $\angle B = 180° - 110° = 70°$

Internal $\angle C = 180° - 100° = 80°$

$\angle A = 180° - \angle B - \angle C = 180° - 70° - 80° = 30°$

26. In ΔABC, D and E are two mid points of sides AB and AC respectively. If ∠BAC = 40° and ∠ABC = 65°, then ∠CED is **(SSC CGL Tier I 2015)**

(*a*) 75° (*b*) 125° (*c*) 130° (*d*) 105°

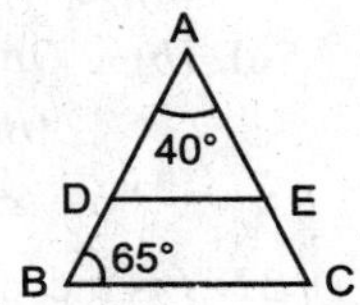

Sol. (*d*) In ΔABC, ∠C = 180° – ∠A – ∠B

180° – 40° – 65° = 75°

$DE \parallel BC \Rightarrow$ ∠BCE + ∠CED = 180°

$\Rightarrow$ ∠CED = 180° – ∠CED = 180° – 75° = 105°

27. All sides of a quadrilateral $ABCD$ touch a circle. If $AB = 6$ cm, $BC = 7.5$ cm, $CD = 3$ cm, then DA is **(SSC CGL Tier II 2013)**

(*a*) 3.5 cm (*b*) 4.5 cm

(*c*) 2.5 cm (*d*) 1.5 cm

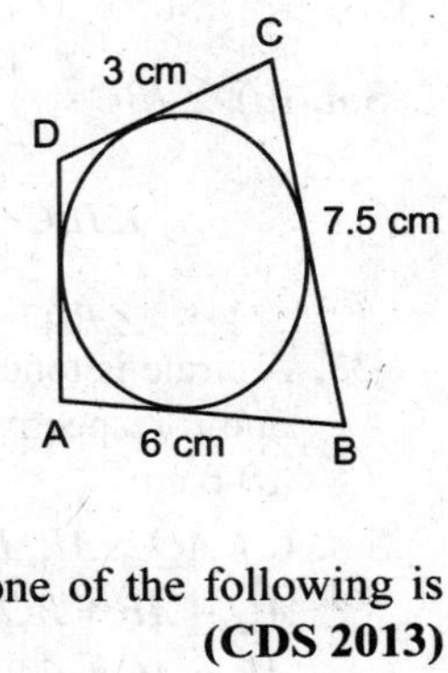

Sol. (*d*) $AB + CD = BC + DA$

$\Rightarrow 6 + 3 = 7.5 + DA \Rightarrow DA = 9 - 7.5 = 1.5$ cm

28. $ABCD$ is a quadrilateral such that $BC = BA$ and $CD > AD$. Which one of the following is correct? **(CDS 2013)**

(*a*) $\angle BAD = \angle BCD$ (*b*) $\angle BAD < \angle BCD$

(*c*) $\angle BAD > \angle BCD$ (*d*) $2\angle BAD = \angle BCD$

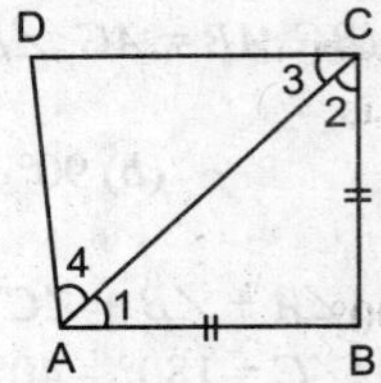

Sol. (*c*) In ΔABC, $AB = BC \Rightarrow \angle 1 = \angle 2$

In ΔACD, $CD > AD \Rightarrow \angle 4 > \angle 3$

$\Rightarrow \angle 1 + \angle 4 > \angle 2 + \angle 3$

$\Rightarrow \angle BAD > \angle BCD$

29. A chord of a circle is equal to its radius. The angle subtended by this chord at a point on the circumference in the major segment is **(SSC CGL Tier I 2011)**

(*a*) 60° (*b*) 120°
(*c*) 90° (*d*) 30°

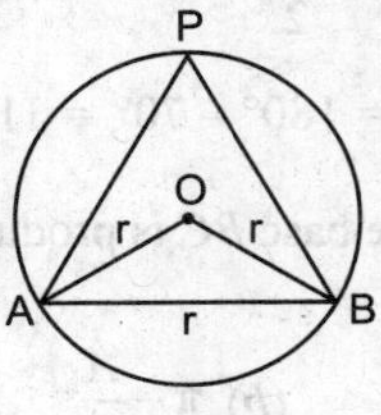

Sol. (*d*) ΔOAB is equilateral triangle.

$\therefore$ Each angle = 60°

$$\therefore \quad \angle APB = \frac{1}{2} \times 60° = 30°$$

30. In the following figure, *O* is the centre of the circle and *XO* is perpendicular to *OY*. If the area of the triangle is 32, then the area of the circle is **(SSC CGL Tier I 2011)**

(*a*) 16 π (*b*) 32 π
(*c*) 64 π (*d*) 256 π

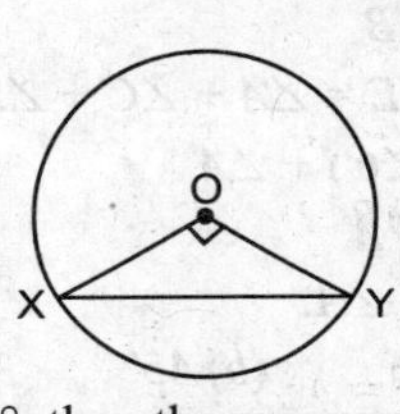

Sol. (*c*) Area of $\Delta OXY = \frac{1}{2} \times r \times r = 32 \Rightarrow r = 8$

Area of circle = $\pi r^2 = \pi (8)^2 = 64\pi$

31. In a triangle *ABC*, incentre is *O* and $\angle BOC = 110°$, then the measure of $\angle BAC$ is : **(SSC CHSL 2011)**

(*a*) 20° (*b*) 40°
(*c*) 55° (*d*) 110°

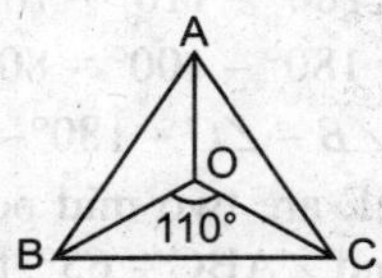

Sol. (*b*) $\angle OBC + \angle OCB = 180° - 110° = 70°$

$\angle ABC + \angle ACB = 2 \times 70° = 140°$

$\therefore \angle BAC = 180° - 140° = 40°$

32. *I* is incenter of a triangle *ABC*. If $\angle ABC = 65°$ and $\angle ACB = 55°$, then the value of $\angle BIC$ is **(SSC CGL Tier II 2012)**

(*a*) 140° (*b*) 110°
(*c*) 130° (*d*) 120°

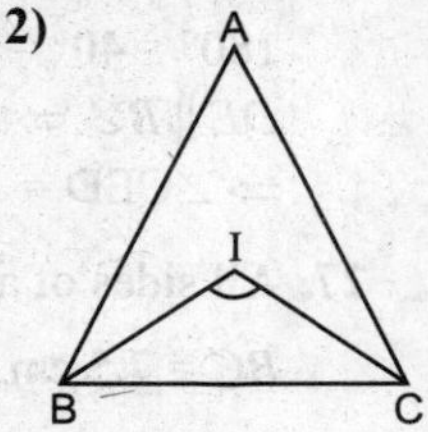

Sol. (*d*) $\angle IBC = \frac{1}{2}\angle B$ and $\angle ICB = \frac{1}{2}\angle C$ (Angle bisectors)

$\therefore \angle IBC + \angle ICB = \frac{1}{2}(65° + 55°) = 60°$

$\therefore \angle BIC = 180° - 60° = 120°$

33. A circle is touching the side *BC* of ΔABC at *P* and is also touching *AB* and *AC* produced at *Q* and *R* respectively. If *AQ* = 6 cm, then the perimeter of ΔABC is **(SSC CHSL 2013)**

(*a*) 6 cm (*b*) 10 cm (*c*) 12 cm (*d*) 18 cm

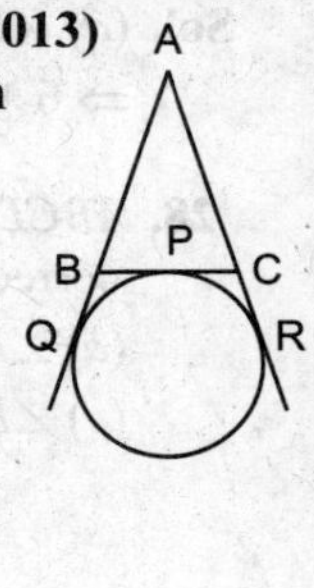

Sol. (*c*) $AQ = AR$, $BP = BQ$, $CP = CR$ (tangents)

$AQ = AB + BQ = AB + BP$

$AR = AC + CR = AC + CP$

$\therefore AB + BC + AC = (AB + BP) + (AC + PC) = AQ + AR = 2AQ = 2 \times 6 = 12$ cm

Note : Perimeter of triangle is double the length of the tangent.

34. The angle between the external bisectors of two angles of a triangle is 60°. Then the third angle of the triangle is **(SSC CGL Tier I 2012)**

(*a*) 40° (*b*) 50° (*c*) 60° (*d*) 80°

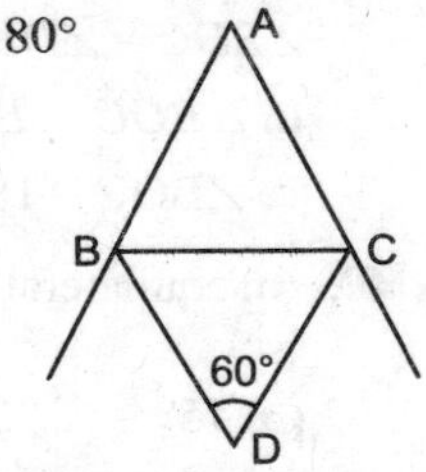

Sol. (*c*) $\angle BDC = 90° - \frac{1}{2}\angle BAC$

$\Rightarrow \frac{1}{2}\angle BAC = 90° - 60° = 30°$

$\Rightarrow \angle BAC = 2 \times 30° = 60°$

35. A rail road curve is to be laid out on a circle. What radius should be used if the track is to change direction by 25° in a distance of 40 metres? **(SSC CGL Tier I 2012)**

(*a*) 91.64 m (*b*) 93.64 cm (*c*) 90.64 m (*d*) 89.64 m

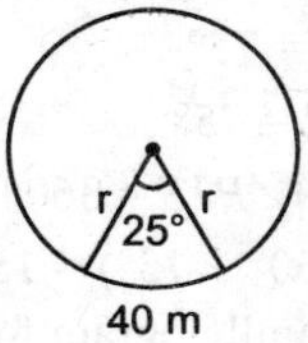

Sol. (*a*) Here, $2\pi r \times \frac{25}{360} = 40$

$\Rightarrow r = 40 \times \frac{360}{25} \times \frac{1}{2} \times \frac{7}{22} = 91.64$ (approx)

36. In a quadrilateral *ABCD*, $\angle B = 90°$ and $AD^2 = AB^2 + BC^2 + CD^2$, then $\angle ACD$ is equal to

(*a*) 30° (*b*) 60° (*c*) 90° (*d*) 45°

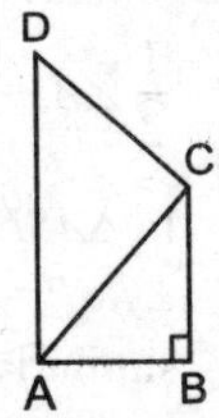

Sol. (*c*) In ΔABC, $AC^2 = AB^2 + BC^2$

Given, $AD^2 = AB^2 + BC^2 + CD^2$

$\Rightarrow AD^2 = AC^2 + CD^2$

$\Rightarrow \Delta ACD$ is right triangle and $\angle ACD = 90°$

37. In ΔABC, *O* is the point of intersection of external bisectors of $\angle B$ and $\angle C$ and $\angle BOC = 45°$, then $\angle A =$ **(SSC FCI 2012)**

(*a*) 45° (*b*) 60°

(*c*) 90° (*d*) 110°

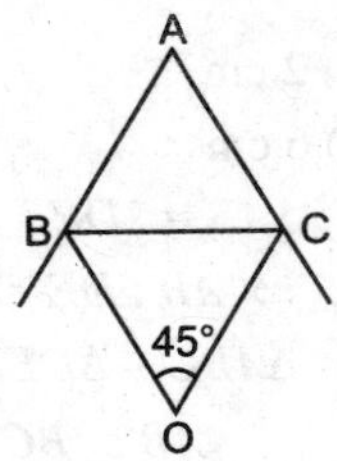

Sol. (*c*) $\angle BOC = 90° - \frac{1}{2}\angle A$

$\Rightarrow 45° = 90° - \frac{1}{2}\angle A \quad \Rightarrow \quad \frac{1}{2}\angle A = 45° \quad \Rightarrow \quad \angle A = 90°$

38. When a pendulum of length 50 cm oscillates, it produces an arc of 16 cm. The angle so formed in degree measure is (approx)? **(SSC CGL Tier I 2012)**

(*a*) 18° 25′ (*b*) 18° 20′ (*c*) 18° 8′ (*d*) 18° 35′

Sol. (*b*) $\frac{\theta}{360} = \frac{\text{Arc}}{2\pi r} \Rightarrow \theta = \frac{16 \times 360 \times 7}{2 \times 22 \times 50} = 18° 20'$ (approx)

39. Inside a square *ABCD*, ΔBCE is an equilateral triangle. If *CE* and *BD* intersect at *O*, then $\angle BOC$ is equal to **(SSC CGL Tier II 2013)**

(*a*) 60° (*b*) 75° (*c*) 90° (*d*) 120°

Sol. (*b*) $\Delta\, BCE$ is equilateral

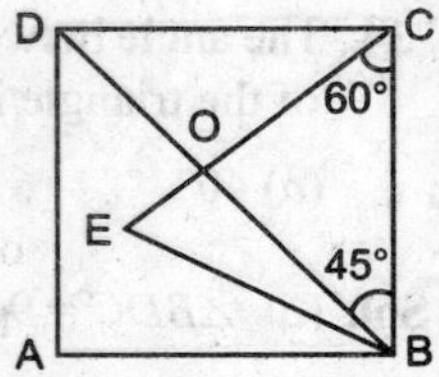

$\therefore$ In $\Delta\, BCE$, $\angle B = \angle C = \angle E = 60°$

$\angle BDC = \angle DBC = 45°$

In $\Delta\, BOC$, $\angle B = \angle 45°$, $\angle C = 60°$

$\therefore\ \angle BOC = 180° - (45° + 60°) = 75°$

40. An equilateral triangle *TQR* is drawn inside a square *PQRS*. The value of the $\angle PTS$ in degrees, is **(SSC CHSL 2012)**

(*a*) 75° (*b*) 90° (*c*) 120° (*d*) 150°

Sol. (*d*) $\Delta\, QRT$ is equilateral $\Rightarrow \angle Q = \angle R = \angle T = 60°$

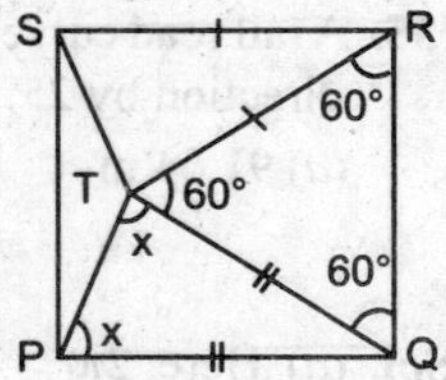

$\angle PQT = \angle PQR - \angle RQT = 90° - 60° = 30°$

In $\Delta\, PTQ$, $PQ = QT$ ($\because$ $QT = RQ$)

$\therefore\ x + x = 180° - 30° \Rightarrow x = 75°$

Similarly, in $\Delta\, STR$, $\angle STR = 75°$

$\angle PTQ + \angle QTR + \angle RTS + \angle PTS = 360°$

$\therefore\ \angle PTS = 360° - (75° + 60° + 75°) = 150°$

41. The angle bisectors of a parallelogram form a

(*a*) parallelogram (*b*) Rectangle (*c*) Rhombus (*d*) Square

Sol. (*b*) $\angle A + \angle D = 180°$

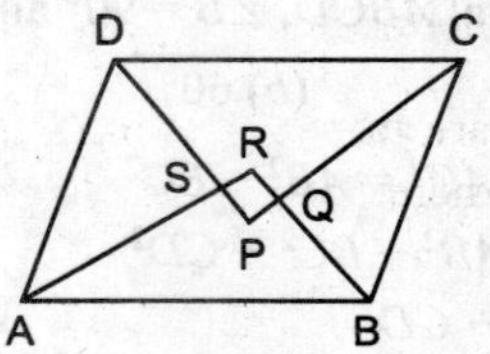

$\frac{1}{2}\angle A + \frac{1}{2}\angle D = 90°$

In ΔASD, $\angle ASD = 180° - 90° = 90°$

$\angle RSP = \angle ASD = 90°$ (Vertically opposite)

similarly, $\angle PQR = 90°$, $\angle QRS = 90°$, $\angle SPQ = 90° \Rightarrow$ *PQRS* is a rectangle.

42. ΔABC and ΔDEF are similar. If $5AB = 2DE$ and $EF = 10$ cm, then *BC* is equal to : **(SSC CGL Tier I 2013)**

(*a*) 2 cm (*b*) 4 cm

(*c*) 6 cm (*d*) 7 cm

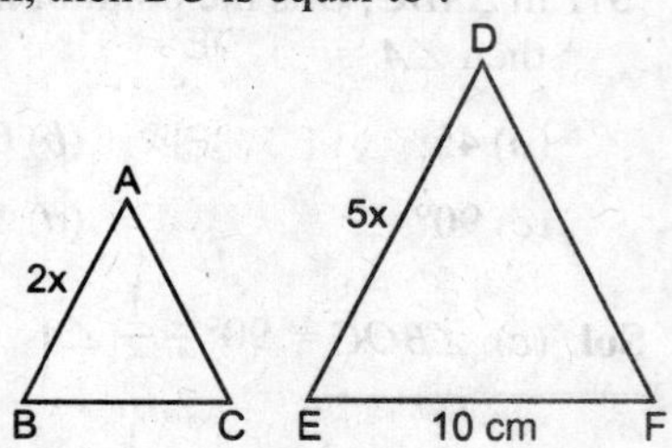

Sol. (*b*) $5AB = 2DE$

$\Rightarrow AB : DE = 2 : 5$

$\Delta ABC \sim \Delta DEF$

$$\therefore\ \frac{AB}{DE} = \frac{BC}{EF} \Rightarrow \frac{2}{5} = \frac{BC}{10} \Rightarrow BC = \frac{2}{5} \times 10 = 4 \text{ cm}$$

43. If ΔABC is similar to ΔDEF such that $BC = 3$ cm, $EF = 4$ cm and area of $\Delta ABC = 54\ cm^2$, then the area of ΔDEF is : **(SSC CGL Tier I 2013)**

(*a*) 54 cm^2 (*b*) 66 cm^2 (*c*) 78 cm^2 (*d*) 96 cm^2

Sol. (*d*) Ratio of sides = 3 : 4 $\Rightarrow$ Ratio of areas = $3^2 : 4^2 = 9 : 16$

$$\therefore\ \text{Area of } \Delta DEF = \frac{16}{9} \times 54 = 96 \text{ cm}^2$$

44. Given that $\Delta ABC \sim \Delta PQR$. If Area ($\Delta PQR$) : Area ($\Delta ABC$) = 256 : 441 and PR = 12 cm, then AC is equal to would be : **(SSC CGL Tier I 2015)**

(*a*) 16 cm (*b*) 15.75 cm (*c*) $12\sqrt{2}$ cm (*d*) 15.5 cm

Sol. (*b*) (ΔPQR) : Area (ΔABC) = $256 : 441 = 16^2 : 21^2 \Rightarrow PR : AC = 16 : 21$

$\Rightarrow AC = \frac{21}{16} \times 12 = 15.75$ cm

45. The sides of a triangle are 3 cm, 4 cm and 5 cm. The area (in cm^2) of the triangle formed by joining the mid-points of this triangle is : **(SSC CGL Tier I 2003)**

(*a*) 6 (*b*) 3 (*c*) $\frac{3}{2}$ (*d*) $\frac{3}{4}$

Sol. (*c*) The triangle is a right angled triangle as $3^2 + 4^2 = 5^2$

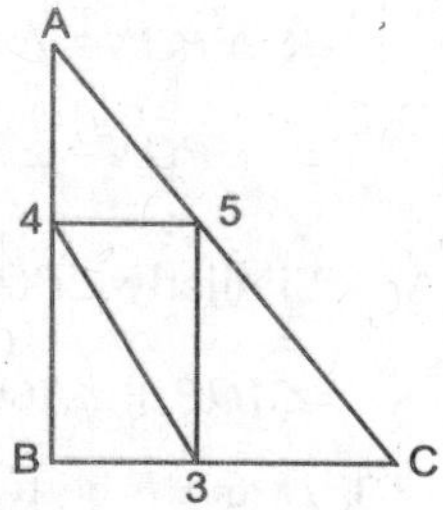

The sides of the new triangle are half of opposite parallel side.

$\therefore$ The sides of the new triangle are $\frac{3}{2}$, 2 and $\frac{5}{2}$

$\therefore$ Required area = $\frac{1}{2} \times \frac{3}{2} \times 2 = \frac{3}{2}$

Alternative method :

Area of right triangle ABC = $\frac{1}{2} \times 3 \times 4 = 6$

Each side of smaller triangle is half of corresponding side of triangle ABC.

Area of smaller triangle = $\left(\frac{1}{2}\right)^2 \times \Delta ABC = \frac{1}{4} \times \Delta ABC = \frac{1}{4} \times 6 = \frac{3}{2}$

46. If D and E are the mid-points of the sides AB and AC respectively of the ΔABC in the figure given here, the shaded region of the triangle is what per cent of the whole triangular region? **(SSC CGL Tier I 2004)**

(*a*) 50% (*b*) 25% (*c*) 75% (*d*) 60%

Sol. (*c*) AD = $\frac{1}{2}$ AB and AE = $\frac{1}{2}$ AC $\Rightarrow$ DE $\parallel$ BC and Δ ADE ~ ΔABC.

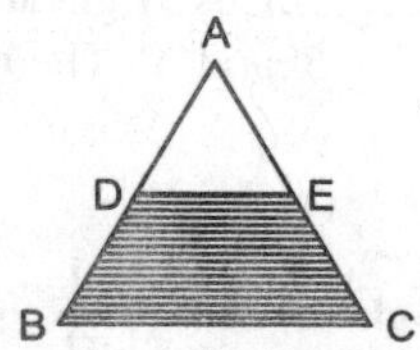

$\therefore$ Area of ADE = $\frac{1}{2} \times \frac{1}{2} \times$ ABC = $\frac{1}{4} \times \Delta$ABC = 25% of ΔABC

$\therefore$ Area of BCED = (100 – 25)% = 75% of ΔABC

47. A point *D* is taken on the side *BC* of a right-angled triangle *ABC* where *AB* is hypotenuse. Then **(SSC CGL Tier II 2013)**

(*a*) $AB^2 + CD^2 = BC^2 + AD^2$ (*b*) $CD^2 + BD^2 = 2AD^2$

(*c*) $AB^2 + AC^2 = 2AD^2$ (*d*) $AB^2 = AD^2 + BD^2$

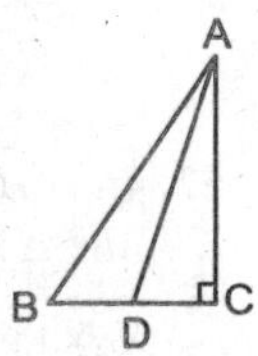

Sol. (*a*) In ΔABC, $AB^2 = AC^2 + BC^2$

In ΔADC, $AD^2 = AC^2 + DC^2$

$\Rightarrow AB^2 + (AC^2 + DC^2) = (AC^2 + BC^2) + AD^2$

$\Rightarrow AB^2 + DC^2 = AC^2 + BC^2 + AD^2 - AC^2 \Rightarrow AB^2 + CD^2 = BC^2 + AD^2$

48. In ΔABC, $AB = AC$, BC is produced to D in such a way that $AC = CD$. If $\angle BAD = 84°$, then $\angle BAC$ is **(SSC FCI Asstt. 2012)**

(*a*) 32° (*b*) 48° (*c*) 42° (*d*) 52°

Sol. (*d*) In ΔACD, $y + y = x$ (external angle)

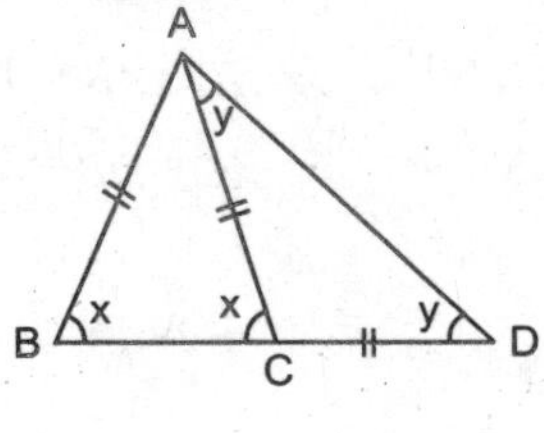

In ΔABD, $\angle A + \angle B + \angle D = 180°$

$\Rightarrow x + y = 180° - \angle A = 180° - 84° = 96°$

$\Rightarrow 2y + y = 96° \Rightarrow y = 96° \div 3 = 32°$

$\angle BAC = 84° - y = 84° - 32° = 52°$

49. The tangents are drawn from the points A and B of a diameter of a circle with centre O. If another tangent to the circle at the point P intersects the other two tangents at M and N, $\angle MON = ?$

(*a*) 45° (*b*) 60° (*c*) 90° (*d*) 120°

Sol. (*c*) In ΔPON and ΔBON,

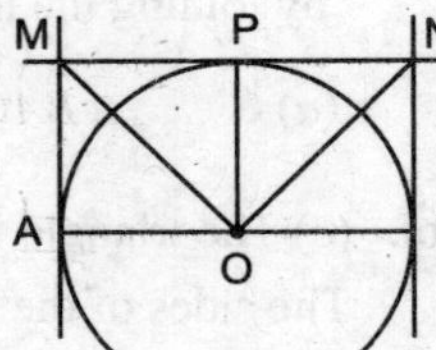

$PO = BO$ (radii)

$NP = NB$ (tangents from same point)

$ON = ON$ (common)

$\Rightarrow \Delta PON \cong \Delta BON$ (SSS-criterion)

$\Rightarrow \angle PON = \angle BON \Rightarrow \angle PON = \frac{1}{2} \angle BOP$

Similarly $\angle POM = \angle AOM \Rightarrow \angle POM = \frac{1}{2} \angle AOP$

$\angle AOP + \angle BOP = 180° \Rightarrow \angle MOP + \angle NOP = 180° \div 2 = 90°$

50. D and E be two points on the sides $\overline{AB}$ and $\overline{AC}$ of the ΔABC such that $\overline{DE} \parallel \overline{BC}$ and $\frac{\overline{AD}}{\overline{BD}} = \frac{2}{3}$. Find the $\frac{\text{area of trapezium } DECB}{\Delta ABC}$. **(SSC CHSL 2012)**

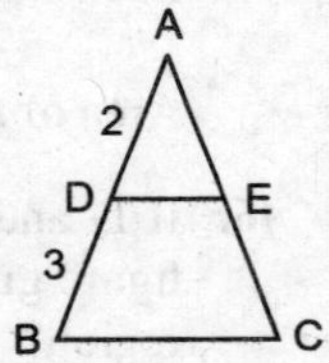

(*a*) 4 : 25 (*b*) 2 : 5
(*c*) 4 : 9 (*d*) 21 : 25

Sol. (*d*) $\Delta ADE : \Delta ABC = 2^2 : 5^2 = 4 : 25$

Area of $DECB = 25 - 4 = 21 \Rightarrow$ Required ratio = 21 : 25

51. ABC is a right angled triangle, B being the right angle. Mid-points of BC and AC are respectively B' and A'. The ratio of the area of the quadrilateral $AA'BB'$ to the area of the triangle ABC is **(SSC CGL Tier I 2013)**

(*a*) 1 : 2 (*b*) 2 : 3 (*c*) 3 : 4 (*d*) None of these

Sol. (*c*) $A'B' = \frac{1}{2} AB$

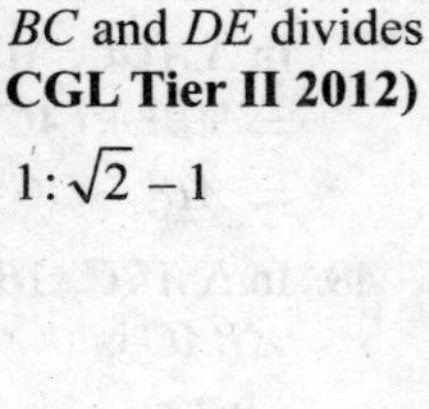

$\therefore AB : A'B' = 2 : 1$

$\therefore \Delta ABC : \Delta A'B'C' = 2^2 : 1^2 = 4 : 1$

$\therefore$ Area of $AA'B'B = 4 - 1 = 3$

$\therefore$ Required ratio = 3 : 4

52. In ΔABC, D and E are points on AB and AC respectively such that $DE \parallel BC$ and DE divides the ΔABC into two parts of equal areas. The ratio of AD and BD is **(SSC CGL Tier II 2012)**

(*a*) $1 : \sqrt{2}$ (*b*) $1 : \sqrt{2} + 1$ (*c*) 1 : 1 (*d*) $1 : \sqrt{2} - 1$

Sol. (*d*) $\Delta ADE : \Delta ABC = 1 : (1 + 1) = 1 : 2$

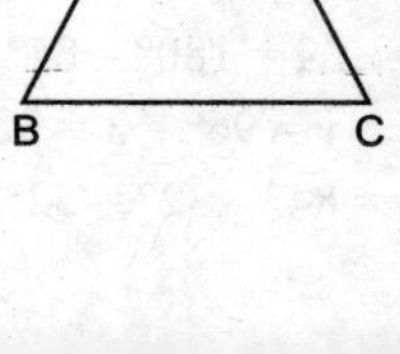

$\therefore$ Ratio of sides $= 1 : \sqrt{2}$

$\therefore AD : AB = 1 : \sqrt{2}$

$\therefore AD : DB = 1 : (\sqrt{2} - 1)$

53. In ΔABC, X and Y are points on sides AB and BC respectively such that $XY \parallel AC$ and XY divides triangular region ABC into two parts equal in area. Then $\frac{AX}{AB}$ is equal to **(SSC CHSL 2013)**

(*a*) $\frac{2+\sqrt{2}}{2}$ (*b*) $\frac{\sqrt{2}+3}{2}$ (*c*) $\frac{2-\sqrt{2}}{2}$ (*d*) $\frac{3-\sqrt{2}}{2}$

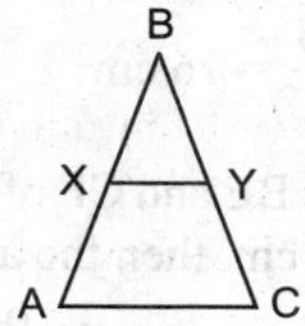

Sol. (*c*) $\Delta BXY : \Delta BAC = 1 : 2 \Rightarrow BX : BA = 1 : \sqrt{2}$

$\Rightarrow \frac{AX}{AB} = \frac{\sqrt{2}-1}{\sqrt{2}} = \frac{2-\sqrt{2}}{2}$

54. ΔABC is an isosceles triangle and $AB = AC = 2a$ unit, $BC = a$ unit. Draw $AD \perp BC$, and find the length of AD. **(SSC CGL Tier II 2013)**

(*a*) $\sqrt{15}\ a$ unit (*b*) $\frac{\sqrt{15}}{2}\ a$ unit (*c*) $\sqrt{17}\ a$ unit (*d*) $\frac{\sqrt{17}}{2}\ a$ unit

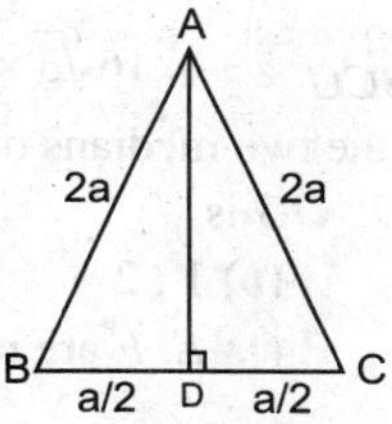

Sol. (*b*) In ΔABD, $AD^2 = AB^2 - BD^2$

$= (2a)^2 - \left(\frac{a}{2}\right)^2 = 4a^2 - \frac{a^2}{4} = \frac{15}{4}a^2$

$\Rightarrow AD = \frac{\sqrt{15}}{2}\ a$ unit

55. The medians of ΔABC meet at G. If the area of the triangle ΔABC is 120 cm^2, then the area of ΔGBC is **(SSC FCI Asstt. 2012)**

(*a*) 20 cm^2 (*b*) 40 cm^2 (*c*) 30 cm^2 (*d*) 50 cm^2

Sol. (*b*) $AG : GD = 2 : 1 \Rightarrow AD : GD = 3 : 1$

$\therefore$ Area ΔABC : Area $\Delta GBC = 3 : 1$

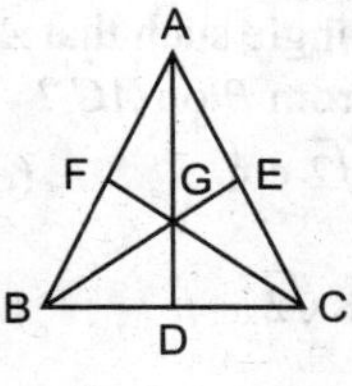

$\therefore$ Area $\Delta GBC = \frac{1}{3} \times 120 = 40\ cm^2$

56. Three medians $\overline{AD}$, $\overline{BE}$ and $\overline{CF}$ of Δ ABC intersect at G. Area of Δ ABC is 108 sq cm. The area of Δ AGB is : **(SSC CHSL 2012)**

(*a*) 54 cm^2 (*b*) 36 cm^2 (*c*) 48 cm^2 (*d*) 24 cm^2

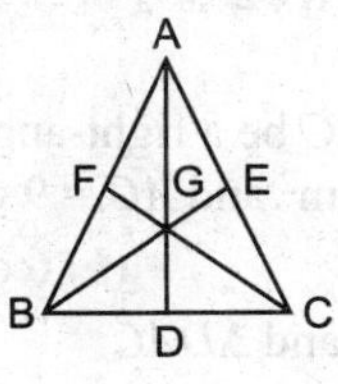

Sol. (*b*) FG : GC = 1 : 2 ⇒ FG : CF = 1 : 3

$\therefore$ Area (Δ ABG) : Area (Δ ABC) = 1 : 3

$\therefore$ Area (Δ ABG) = 108 ÷ 3 = 36 cm^2

57. If G is the centroid of ΔABC and $\Delta ABC = 48\ cm^2$, then the area of ΔBGC is **(SSC CGL Tier I 2013)**

(*a*) 16 cm^2 (*b*) 24 cm^2 (*c*) 32 cm^2 (*d*) 8 cm^2

Sol. (*a*) $DG = \frac{1}{3}AD$

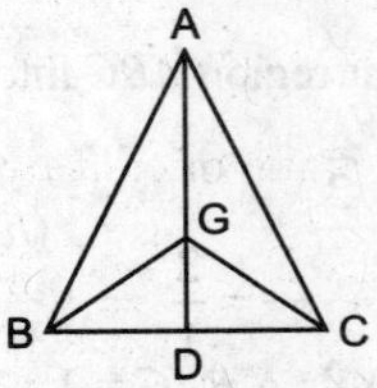

$$\Delta BGC = \frac{1}{2} \times BC \times DG = \frac{1}{2} \times BC \times \frac{AD}{3}$$

$$= \frac{\Delta ABC}{3} = \frac{48}{3} = 16 \text{ cm}^2$$

58. If two medians BE and CF of a triangle ABC, intersect each other at G and if BG = CG, $\angle BGC$ = 60°, BC = 8 cm, then the area of triangle ABC is **(SSC CGL Tier I 2015)**

(*a*) $48\sqrt{3}$ cm² (*b*) $96\sqrt{3}$ cm² (*c*) 48 cm² (*d*) $64\sqrt{3}$ cm²

Sol. (*a*) $\angle CBG + \angle BCG = 180° - 60° = 120°$

BG = CG $\Rightarrow \angle CBG = \angle BCG = 60° \Rightarrow \Delta BCG$ is equilateral.

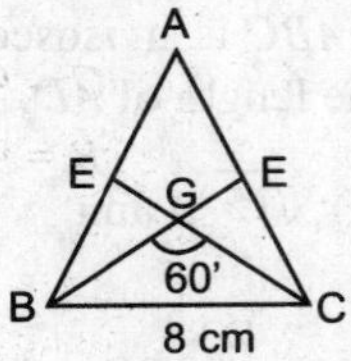

Area of $\Delta BCG = \frac{\sqrt{3}}{4}a^2 = \frac{\sqrt{3}}{4} \times 8^2 = 16\sqrt{3}$

Area $\Delta ABC = \Delta BCG \times 3 = 16\sqrt{3} \times 3 = 48\sqrt{3}$

59. Let *BE* and *CF* be the two medians of a ΔABC and *G* be point of intersection. Also let *EF* cut *AG* at *O*. Then, *AO* : *OG* is **(SSC CGL Tier I 2012)**

(*a*) 1 : 1 (*b*) 1 : 2 (*c*) 2 : 1 (*d*) 3 : 1

Sol. (*d*) EF || BC ($\because$ *E*, *F* are mid-points of *AC* and *AB* respectively)

$$\therefore \frac{AO}{OD} = \frac{AF}{FB} = \frac{1}{1} \Rightarrow AO = OD$$

$GD = \frac{1}{3}AD$ [*G* is centroid]

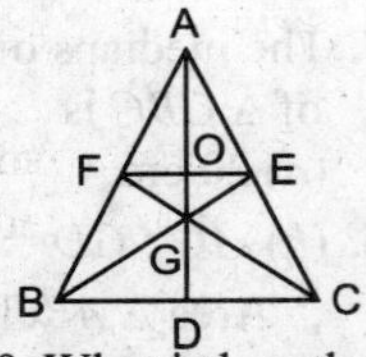

Let *AD* = 6, then *AO* = 3, *GD* = 2

$OG = AD - AO - GD = 6 - 3 - 2 = 1 \Rightarrow AO : OG = 3 : 1$

60. *ABC* is an isosceles triangle such that *AB* = *BC* = 8 cm and $\angle ABC = 90°$. What is length of the perpendicular drawn from *B* on *AC* ? **(CDS 2011)**

(*a*) 4 cm (*b*) $4\sqrt{2}$ cm (*c*) $2\sqrt{2}$ cm (*d*) 2 cm

Sol. (*b*) $AC = \sqrt{8^2 + 8^2} = 8\sqrt{2}$

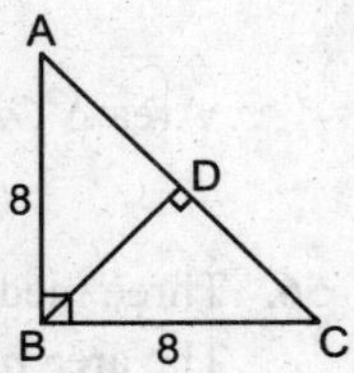

$BD \times AC = AB \times BC$

$$\Rightarrow BD = \frac{8 \times 8}{8\sqrt{2}} = 4\sqrt{2} \text{ cm}$$

61. Suppose ΔABC be a right-angled triangle where $\angle A = 90°$ and $AD \perp BC$. If $\Delta ABC = 40$ cm², $\Delta ACD = 10$ cm² and *AC* = 9 cm, then the length of *BC* is **(SSC CGL Tier II 2012)**

(*a*) 4 cm (*b*) 6 cm (*c*) 12 cm (*d*) 18 cm

Sol. (*d*) In ΔABC and ΔDAC

$\angle CAB = \angle ADC = 90°$

$\angle C = \angle C$ (Common)

$\therefore \ \Delta ABC \sim \Delta DAC$ (*AA* criterion)

$\Delta ABC : \Delta DAC = 40 : 10 = 4 : 1 = 2^2 : 1^2$

$\therefore$ Ratio of corresponding sides = 2 : 1

$$\Rightarrow \frac{BC}{AC} = \frac{2}{1} \Rightarrow \frac{BC}{9} = \frac{2}{1} \Rightarrow BC = 18 \text{ cm}$$

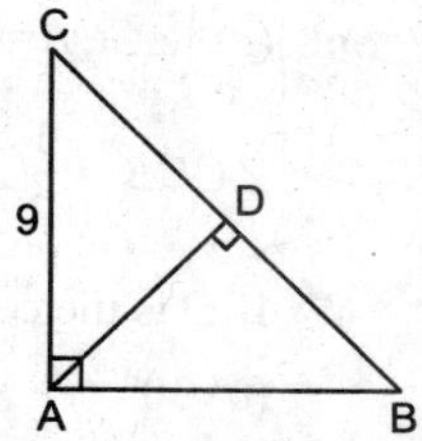

62. *AC* is the diameter of the circum-circle of the cyclic quadrilateral *ABCD*. If $\angle BDC = 42°$, then what is $\angle ACB$ equal to? **(CDS 2011)**

(*a*) 42° (*b*) 45°
(*c*) 48° (*d*) 58°

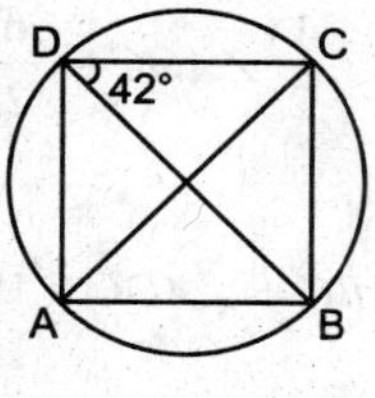

Sol. (*c*) $\angle ADC = 90°$ (Angle in semi-circle)

$\therefore \angle ADB = 90° - 42° = 48°$

$\angle ACB = \angle ADB = 48°$ (Angles of same chord)

63. *ABCD* is a quadrilateral inscribed in a circle with centre O. If $\angle COD = 120°$ and $\angle BAC = 30°$, then $\angle BCD$ is **(SSC CHSL 2012)**

(*a*) 60° (*b*) 75° (*c*) 90° (*d*) 120°

Sol. (*c*) $\angle CAD = \frac{1}{2}\angle COD = \frac{1}{2} \times 120° = 60°$ (Angle on circle)

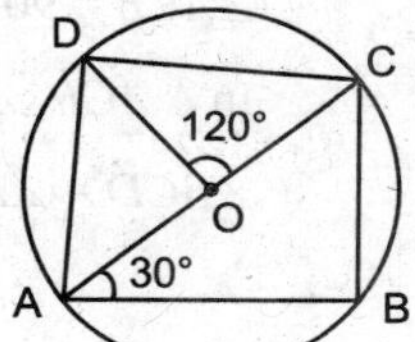

$\angle BAD = \angle BAC + \angle CAD = 30° + 60° = 90°$

$\angle A + \angle C = 180°$ (Cyclic quadrilateral)

$\therefore \angle C = 180° - 90° = 90°$

64. A horse is tied to a post by a rope. If the horse moves along a circular path always keeping the rope stretched and describes 88 metres when it has traced out 72° at the centre the length of the rope is $\left(\text{Take } \pi = \frac{22}{7}\right)$ **(SSC CGL Tier I 2013)**

(*a*) 70 m (*b*) 75 m (*c*) 80 m (*d*) 65 m

Sol. (*a*) $2\pi r \times \frac{72}{360} = 88$ metres $\Rightarrow r = 88 \times \frac{360}{72} \times \frac{1}{2} \times \frac{7}{22} = 70$ m

65. In the figure given above, *PQ* is parallel to *RS*. What is $\angle NMS$ equal to ? **(CDS 2011)**

(*a*) 20° (*b*) 23° (*c*) 27° (*d*) 47°

Sol. (*a*) $\angle MLN = 180° - 55° - 47° = 78°$

$\angle LMN = 180° - 78° - 67° = 35°$

$\angle LMS = \angle PLM = 55°$ (Alternate angles)

$\therefore \angle NMS = \angle LMS - \angle LMN = 55° - 35° = 20°$

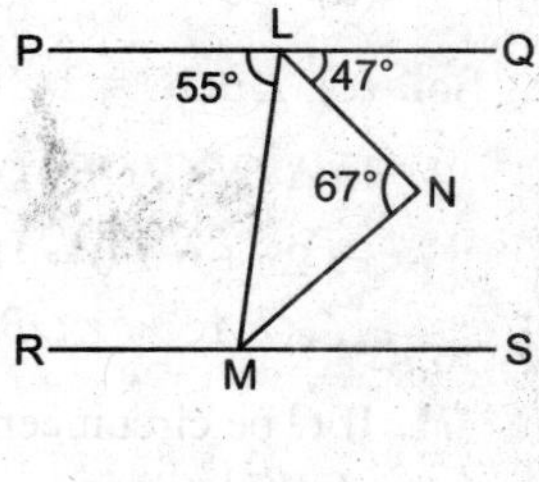

66. In the figure, *AOB* is the diameter of the circle. Find the value of *x*.

(*a*) 65° (*b*) 50° (*c*) 25° (*d*) 20°

Sol. (*c*) $\angle COB = 180° - \angle AOC = 180° - 130° = 50°$

$$\angle CDB = \frac{1}{2}\angle COB = \frac{1}{2} \times 50° = 25°$$

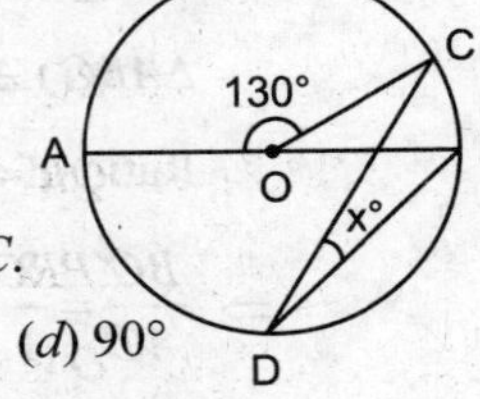

67. If O is the centre of a circle and $OABC$ is a parallelogram, find $\angle ABC$.

(*a*) 30° (*b*) 60° (*c*) 120° (*d*) 90°

Sol. (*c*) Let $\angle ABC = x°$. Then $\angle AOC = \angle ABC = x°$

$$\angle ADC = \frac{1}{2}\angle AOC = \frac{x°}{2}$$

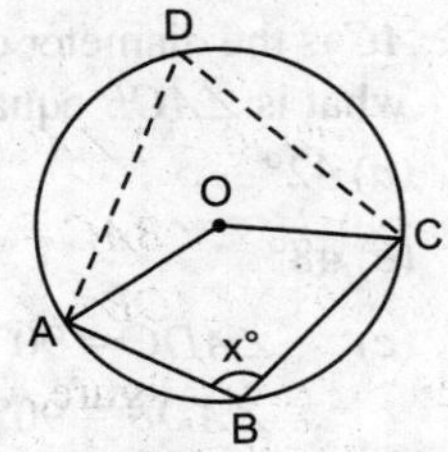

$$\Rightarrow \angle ABC + \angle ADC = 180° \Rightarrow x° + \frac{x°}{2} = 180° \Rightarrow x = 120°$$

68. In the given figure, AB is diameter and chord CD is parallel to it. If $\angle CEB = 60°$; find $\angle BCD$.

(*a*) 30° (*b*) 60° (*c*) 90° (*d*) 120°

Sol. (*a*) $\angle BAC = \angle BEC = 60°$ (angle in the same segment)

$\angle ACB = 90°$ (angle in semi-circle)

In ΔACB, $\angle ABC = 180° - \angle BAC - \angle ACB = 180° - 60° - 90° = 30°$

$\angle BCD = \angle ABC = 30°$ (alternate angles)

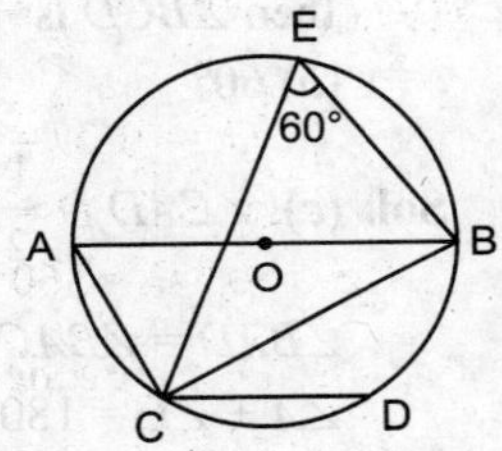

69. Consider ΔABD such that $\angle ADB = 20°$ and C is a point on BD such that $AB = AC$ and $CD = CA$. Then, the measure of $\angle ABC$ is **(SSC CGL Tier I 2012)**

(*a*) 45° (*b*) 60° (*c*) 30° (*d*) 40°

Sol. (*d*) In ΔACD, $CD = CA \Rightarrow \angle CAD = \angle CDA = 20°$

$\angle ACB = \angle CAD + \angle CDA = 20° + 20° = 40°$ (External angle)

In ΔABC, $AC = AB$

$\therefore \angle ABC = \angle ACB = 40°$

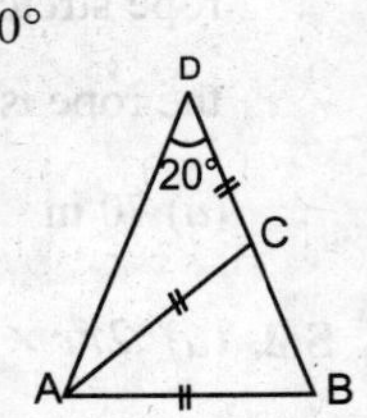

70. BC is the chord of a circle with centre O. A is a point on major arc BC as shown in the figure. What is the value of $\angle BAC + \angle OBC$? **(SSC CGL Tier I 2012)**

(*a*) 120° (*b*) 60° (*c*) 90° (*d*) 180°

Sol. (*c*) $\angle BOC = 2\angle BAC = 2x$ (angle at centre)

In ΔOBC, $OB = OC \Rightarrow \angle O + \angle B + \angle C = 180°$

$\Rightarrow 2x + \theta + \theta = 180° \Rightarrow x + \theta = 90°$

$\therefore \angle BAC + \angle OBC = x + \theta = 90°$

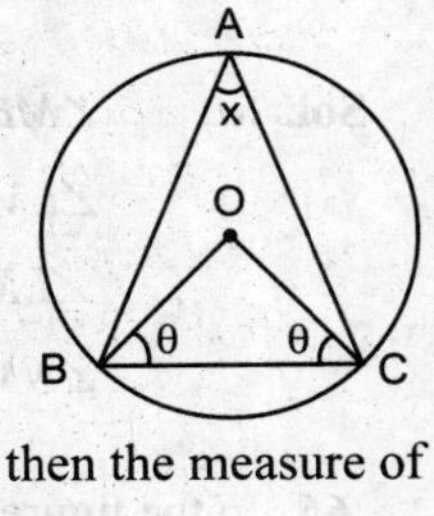

71. If O be circumcentre of a triangle PQR and $\angle QOR = 110°$, $\angle OPR = 25°$, then the measure of $\angle PRQ$ is

(*a*) 55° (*b*) 60° (*c*) 65° (*d*) 50°

Sol. (*b*) $\angle ORP = \angle OPR = 25°$

$\angle OQR + \angle ORQ = 180° - 110° = 70°$

$\angle ORQ = \angle OQR = 70° \div 2 = 35°$

$\therefore \angle PRQ = \angle PRO + \angle ORQ = 25° + 35° = 60°$

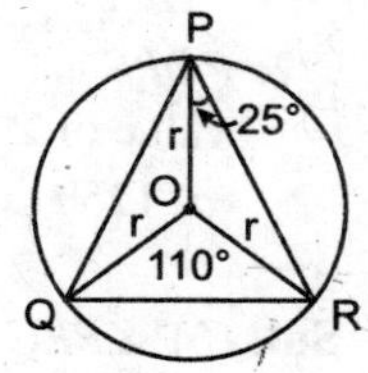

72. In the figure, *O* is the centre of the circle, find $\angle BOC$.

(*a*) 75° (*b*) 105° (*c*) 170° (*d*) 150°

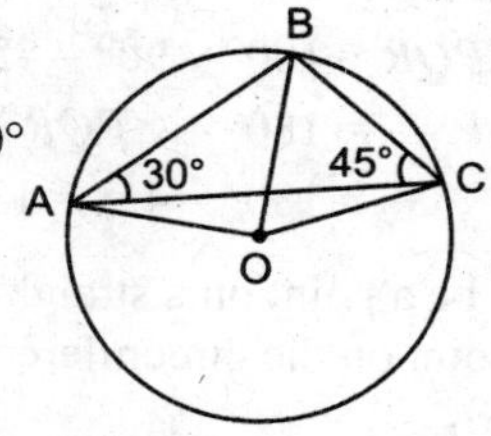

Sol. (*d*) $\angle AOB = 2\ \angle ACB = 2 \times 45° = 90°$

$\angle BOC = 2\ \angle BAC = 2 \times 30° = 60°$

$\angle AOC = \angle AOB + \angle BOC = 90° + 60° = 150°$

73. In the given figure, *O* and *O′* are centres of two circles, intersecting at *B* and *C*. *AOBD* is a straight line, find $\angle CO'D$.

(*a*) 130° (*b*) 150° (*c*) 200° (*d*) 230°

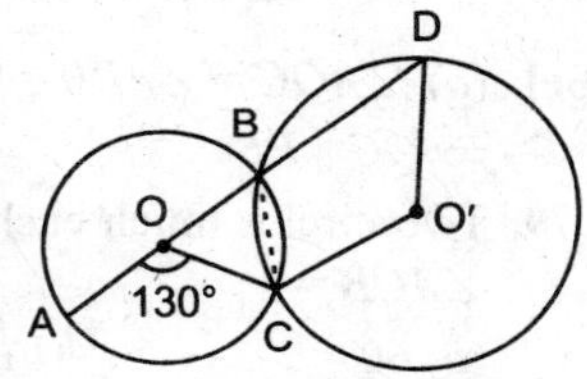

Sol. (*a*) $\angle ABC = \frac{1}{2}\angle AOC = \frac{1}{2}\times 130° = 65°$

$\angle CBD = 180° - \angle ABC = 180° - 65° = 115°$

Reflex $\angle CO'D = 2\angle CBD = 2 \times 115° = 230°$

$\Rightarrow \angle CO'D = 360° - 230° = 130°$

74. *ABCD* is a cyclic quadrilateral, *AB* is a diameter of the circle. If $\angle ACD = 50°$, the value of $\angle BAD$ is **(SSC CGL Tier I 2013)**

(*a*) 30° (*b*) 40°

(*c*) 50° (*d*) 60°

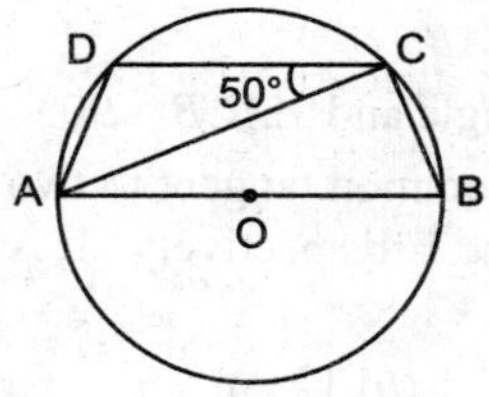

Sol. (*b*) $\angle ACB = 90°$ (Angle in semi-circle)

$\angle BCD = 90° + 50° = 140°$

$\angle BAD = 180° - 140° = 40°$

75. In a rhombus ABCD, $\angle A = 60°$ and AB = 12 cm. Then the diagonal BD is : **(SSC CPO SI 2007M, SSC CGL Tier I 2015)**

(*a*) 10 cm (*b*) 12 cm (*c*) 6 cm (*d*) $2\sqrt{3}$ cm

Sol. (*b*) In $\Delta\ ABD$, $AB = AD = 12$ cm

$\Rightarrow \angle ABD + \angle ADB = 180° - 60° = 120°$

$\Rightarrow \angle ABD = \angle ADB = 120° \div 2 = 60°$

$\Rightarrow \Delta\ ABD$ is an equilateral triangle

$\therefore BD = AB = 12$ cm

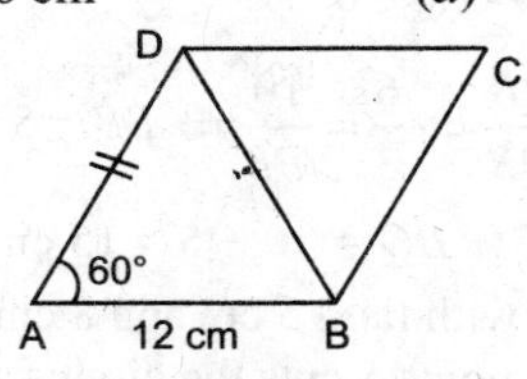

76. If the length of the side *PQ* of the rhombus *PQRS* is 6 cm and $\angle PQR = 120°$, then the length of *QS*, in cm, is **(SSC CGL Tier I 2012)**

(*a*) 3 (*b*) 5

(*c*) 4 (*d*) 6

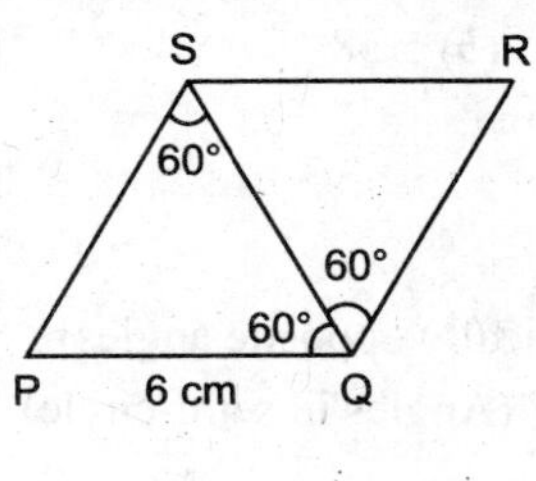

Sol. (*d*) $\angle PSQ = \angle PQS = 60°$

$\angle QPS = 180° - 60° - 60° = 60°$

$\Rightarrow \Delta\ PQS$ is an equilateral triangle

$\therefore SQ = PS = PQ = 6$ cm

77. $PQRS$ is a cyclic quadrilateral and PQ is the diameter of the circle. If $\angle QPR = 45°$, then the value of $\angle PSR$ is equal to **(SSC CHSL 2012)**

(*a*) 125° (*b*) 135° (*c*) 45° (*d*) 90°

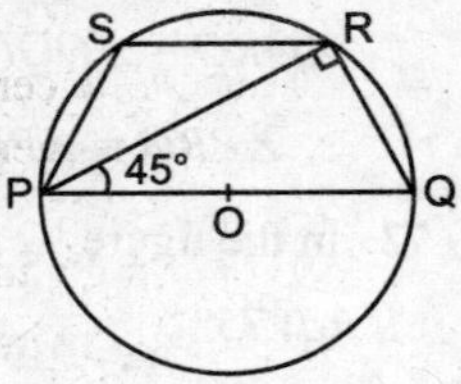

Sol. (*b*) $\angle PRQ = 90°$ (Angle in semi-circle)

$\angle PQR = 180° - 90° - 45° = 45°$

$\angle PSR = 180° - \angle PQR$

$= 180° - 45° = 135°$

78. Let C be a point on a straight line AB. Circles are drawn with diameters AC and AB. Let P be any point on the circumference of the circle with diameter AB. If AP meets the other circle at Q, then **(SSC CGL Tier II 2013)**

(*a*) $QC \parallel PB$ (*b*) QC is never parallel to PB

(c) $QC = \frac{1}{2}PB$ (*d*) $QC \parallel PB$ and $QC = \frac{1}{2}PB$

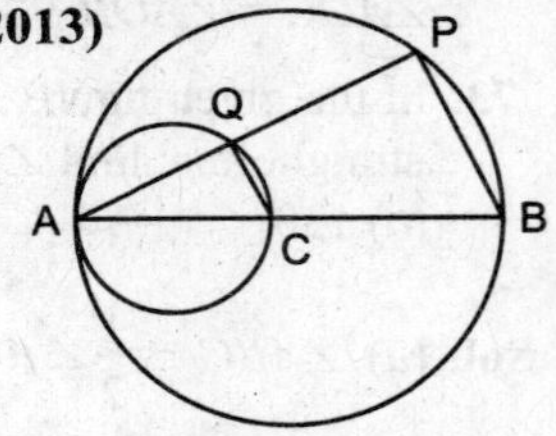

Sol. (*a*) $\angle AQC = \angle APB = 90°$ (Angles in semi-circles)

$\Rightarrow QC \parallel PB$

79. Two circles touch each other externally at C and AB is common tangent to the circle. Then $\angle ACB = ?$

(*a*) 60° (*b*) 30° (*c*) 45° (*d*) 90°

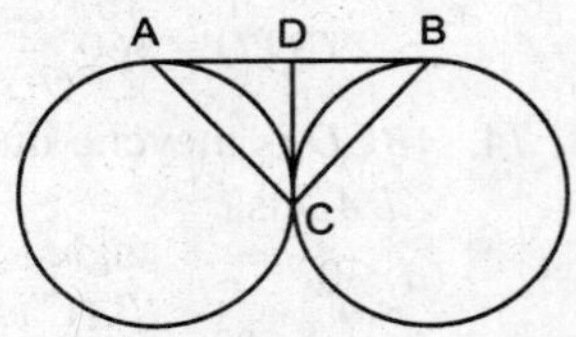

Sol. (*d*) $AD = CD$ and $BD = CD$ (tangents)

In ΔABC, $AD = BD = CD$

$\therefore$ CD is median on AB

$\therefore$ ABC is right triangle and $\angle ACB = 90°$

80. AC is a transverse common tangent to two circles with centres P and Q and radii 6 cm and 3 cm at the point A and C respectively. If AC cuts PQ at the point B and AB = 8 cm, then the length of PQ is : **(SSC CGL Tier I 2015)**

(*a*) 15 cm (*b*) 12 cm (*c*) 13 cm (*d*) 10 cm

Sol. (*a*) In ΔPAB, $PB^2 = PA^2 + AB^2$

$\Rightarrow PB = \sqrt{6^2 + 8^2} = 10$ cm

Now, $\Delta PAB \sim \Delta QCB$

$\Rightarrow \frac{PA}{QC} = \frac{PB}{QB} \Rightarrow \frac{6}{3} = \frac{10}{PB} \Rightarrow PB = 5$

$\Rightarrow PQ = PB + BQ = 10 + 5 = 15$ cm

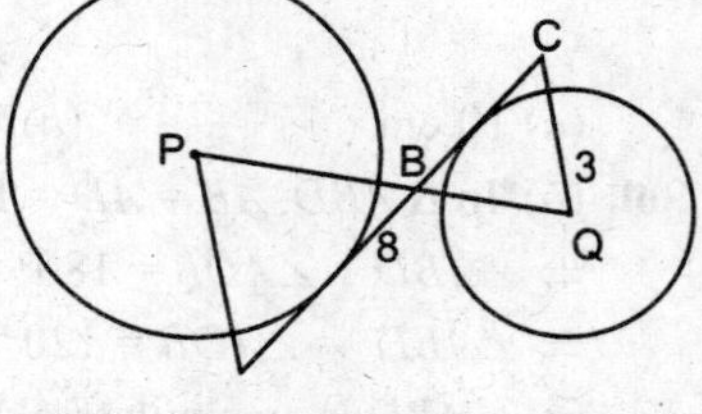

81. Two circles with radii 5 cm and 8 cm touch each other externally at a point A. If a straight line through the point A cuts the circles at points P and Q respectively, then $AP : AQ$ is **(SSC CGL Tier I 2012)**

(*a*) 8 : 5 (*b*) 5 : 8 (*c*) 3 : 4 (*d*) 4 : 5

Sol. (*b*) $AB = 2 \times 5 = 10$ cm

$AC = 2 \times 8 = 16$ cm

In Δ APB and Δ AQC,

$\angle BAP = \angle CAQ$ (Vertically opposite angles)

$\angle APB = \angle AQC = 90°$ (Angles in semi-circle)

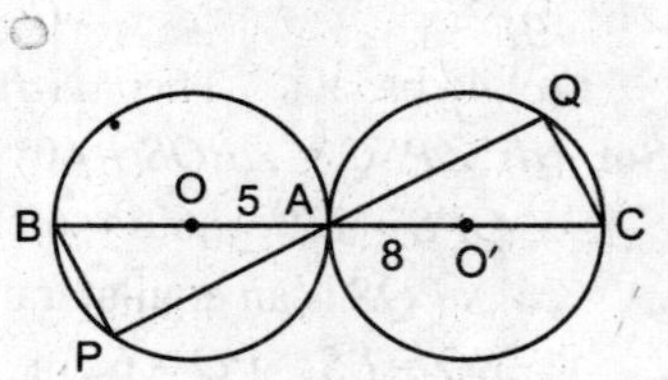

$\therefore \Delta APB \sim \Delta AQC$ (AA-criterion)

$\therefore \dfrac{AP}{AQ} = \dfrac{AB}{AC} = \dfrac{10}{16} = \dfrac{5}{8}$

82. A and B are centres of the two circles whose radii are 5 cm and 2 cm respectively. The direct common tangents to the circles meet AB extended at P. Then P divides AB

(SSC CGL Tier I 2012)

(*a*) externally in the ratio 5 : 2 (*b*) internally in the ratio 2 : 5
(*c*) internally in the ratio 5 : 2 (*d*) externally in the ratio 7 : 2

Sol. (*a*) In ΔBNP and ΔAMP,

$\angle BNP = \angle AMP = 90°$ (tangents)

$\angle BPN = \angle APM$ (common)

$\Rightarrow \Delta AMP \sim \Delta BNP \Rightarrow \dfrac{AP}{BP} = \dfrac{AM}{BN} = \dfrac{5}{2}$

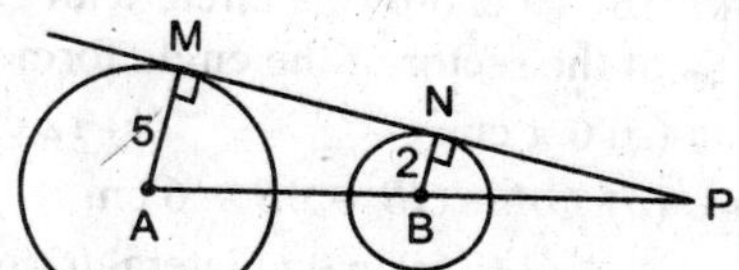

83. If O is the orthocentre of the ΔABC and $\angle BAC = 80°$, then measure of $\angle BOC$ is

(SSC CGL Tier II 2012)

(*a*) 120° (*b*) 90°
(*c*) 80° (*d*) 100°

Sol. (*d*) In quadrilateral $APOQ$,

$\angle POQ = 360° - (80° + 90° + 90°) = 100°$

$\angle BOC = \angle POQ = 100°$ (Vertically opposite angles)

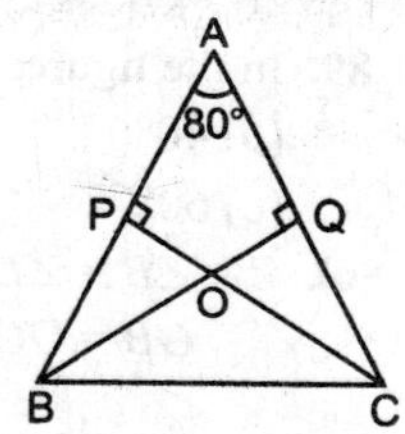

84. In an obtuse-angled triangle ABC, $\angle A$ is the obtuse angle and O is the orthocentre. If $\angle BOC = 54°$, then $\angle BAC$ is **(SSC CGL Tier I 2012)**

(*a*) 108° (*b*) 116°
(*c*) 126° (*d*) 136°

Sol. (*c*) In quadrilateral, OEAF

$\angle E = \angle F = 90°$

$\therefore \angle A = 180° - 54° = 126°$

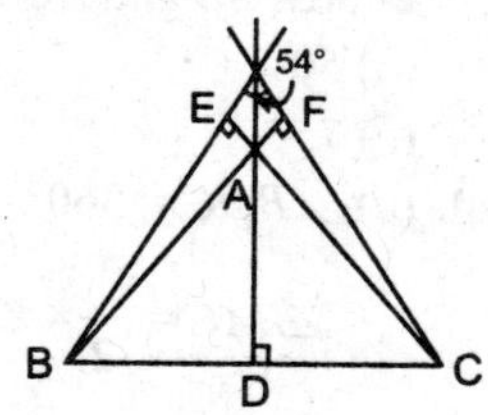

85. In a ΔABC, $\angle A : \angle B : \angle C = 2 : 3 : 4$. A line CD drawn || to AB, then the $\angle ACD$ is :

(SSC CGL Tier I 2013)

(*a*) 20° (*b*) 40° (*c*) 60° (*d*) 80°

Sol. (*b*) $\angle A = \dfrac{2}{9} \times 180° = 40°$

$AB \parallel CD$

$\therefore \angle ACD = \angle BAC = 40°$ (Alternate interior angles)

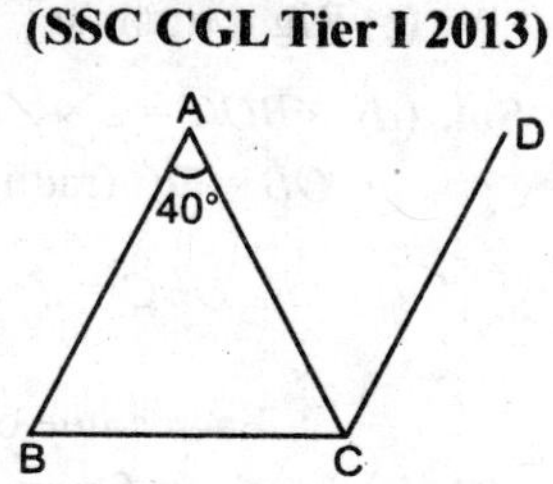

86. ABC is an isosceles triangle such that $AB = AC$ and $\angle B = 35°$. AD is median to the base BC. Then $\angle BAD$ is : **(SSC CGL Tier I 2013)**

(*a*) 35° (*b*) 110° (*c*) 55° (*d*) 70°

Sol. (*c*) $\angle BAD = 180° - 35° - 90° = 55°$

[In an isosceles triangle, altitude on unequal side is also angle bisector and median]

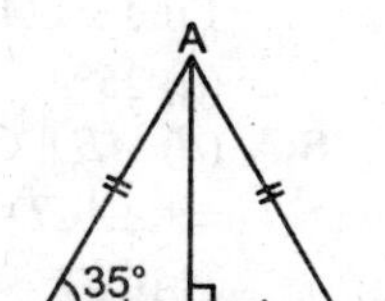

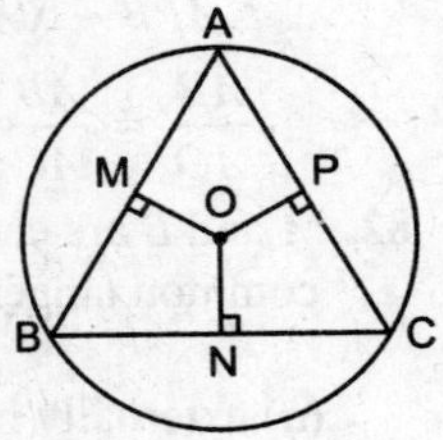

87. In the given figure, sides of a triangle ABC are equidistant from centre of the centre, What type of triangle is ΔABC ?

(*a*) equilateral (*b*) isosceles

(*c*) right (*d*) scalene

Sol. (*a*) $OM = ON = OP$ (Given)

$\therefore AB = BC = AC$ (Chords equidistant from centre)

$\therefore \Delta ABC$ is an equilateral triangle.

88. AB is a chord of a circle with centre O and radius 6 cm. If length of AB is 6 cm, find the area of the sector of the circle formed by chord AB.

(*a*) 6 π cm² (*b*) 12 cm² (*c*) 24 cm² (*d*) 36 cm²

Sol. (*a*) $OA = OB = AB = 6$ cm

$\therefore \Delta\, OAB$ is an equilateral triangle.

$\therefore \angle AOB = 60°$

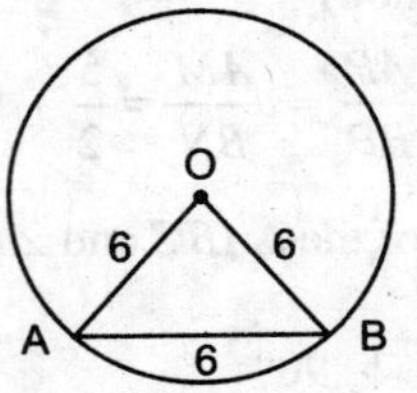

$$\text{Required Area} = \frac{60}{360} \times \pi(6)^2 = 6\pi$$

89. In the figure, find value of x.

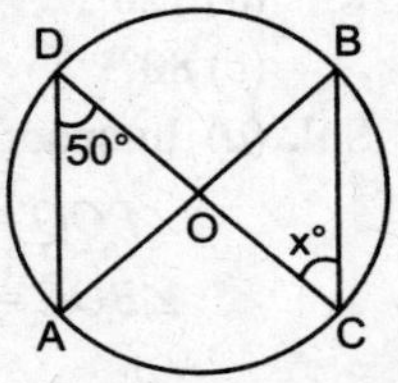

(*a*) 40° (*b*) 50°

(*c*) 60° (*d*) 70°

Sol. (*b*) $\angle B = \angle D = 50°$

$OB = OC$ (radii) $\Rightarrow \angle C = \angle B = 50°$

90. In the given figure, A, B, C are three points on a circle such that the angles subtended by the chords AB and AC at the centre O are 110° and 100° respectively. Find $\angle BAC$.**(NIFT 2007)**

(*a*) 50° (*b*) 55°

(*c*) 75° (*d*) 105°

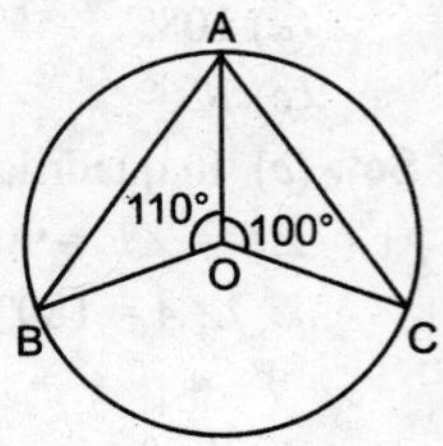

Sol. (*c*) $\angle BOC = 360° - (\angle BOA + \angle AOC) = 360° - (110° + 100°) = 150°$

$$\angle BAC = \frac{1}{2} \times \angle BOC = \frac{1}{2} \times 150° = 75°$$

91. In the given figure, $\angle BAC = 30°$, then ΔOBC is :

(*a*) Scalene triangle (*b*) isosceles triangle

(*c*) Right triangle (*d*) equilateral triangle

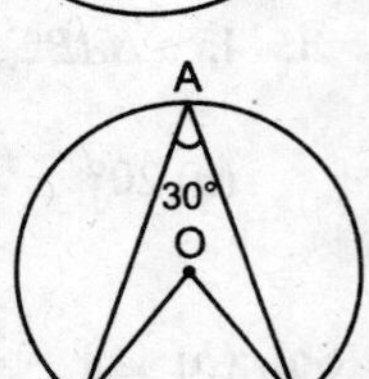

Sol. (*d*) $\angle BOC = 2 \times \angle BAC = 2 \times 30° = 60°$

$OB = OC$ (radii of same circle)

$$\therefore \angle OBC = \angle OCB = \frac{1}{2} \times (180° - 60°) = 60° \text{ each}$$

$\therefore$ Each angle of ΔOBC is equal to 60° $\Rightarrow \Delta OBC$ is an equilateral triangle.

92. In the given figure, AB is diameter, $ABCD$ is a cyclic quadrilateral, $AB \parallel CD$, $\angle CAB = 25°$. Find $\angle CAD$.

(*a*) 25° (*b*) 65° (*c*) 50° (*d*) 40°

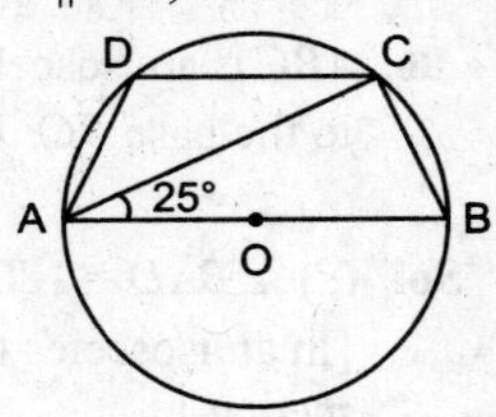

Sol. (*d*) $AB \parallel CD$ and AC is transversal

$\Rightarrow \angle ACD = \angle CAB = 25°$

$\angle ACB = 90°$ (Angle in semi-circle)

$\angle BAD + \angle BCD = 180° \Rightarrow \angle CAD = 180° - (90° + 25°) - 25° = 40°$

93. In a ΔABC, $AB = AC$ and D is a point on AB, such that $AD = DC = BC$. Then, the $\angle BAC$ is **(SSC FCI Asstt. 2012)**

(*a*) 40° (*b*) 45°
(*c*) 30° (*d*) 36°

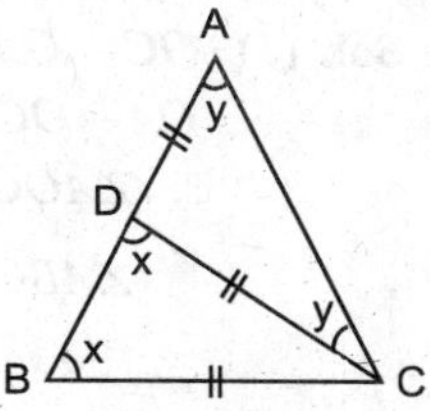

Sol. (*d*) In ΔADC, $AD = CD \Rightarrow \angle DAC = \angle ACD = y°$
In ΔBCD, $BC = CD \Rightarrow \angle CBD = \angle CDB = x°$
$\angle CAD + \angle ACD = \angle CDB \Rightarrow y + y = x \Rightarrow 2y = x$
In ΔABC, $AB = AC \Rightarrow \angle ABC = \angle ACB = x°$
In ΔABC, $\angle ABC + \angle ACB + \angle BAC = 180°$
$\Rightarrow x + x + y = 180° \Rightarrow 5y = 180° \Rightarrow y = 36°$

94. $ABCD$ is a cyclic quadrilateral. $\angle AOC = 150°$, find $\angle ABC$.

(*a*) 75° (*b*) 30° (*c*) 60° (*d*) 105°

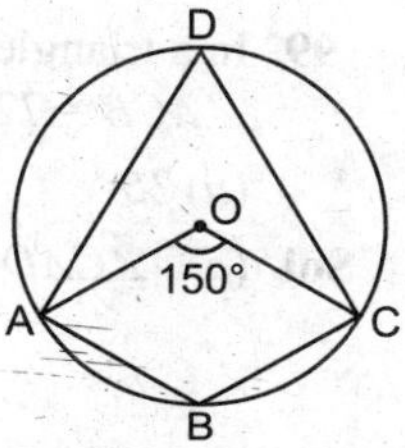

Sol. (*d*) $\angle ADC = \frac{1}{2} \times \angle AOC = \frac{1}{2} \times 150° = 75°$
$\angle ADC + \angle ABC = 180°$
$\Rightarrow \angle ABC = 180° - 75° = 105°$

95. O is the centre of the circle passing through the points A, B and C such that $\angle BAO = 30°$, $\angle BCO = 40°$ and $\angle AOC = x°$. What is the value of x? **(SSC CGL Tier I 2012)**

(*a*) 70° (*b*) 140° (*c*) 220° (*d*) 280°

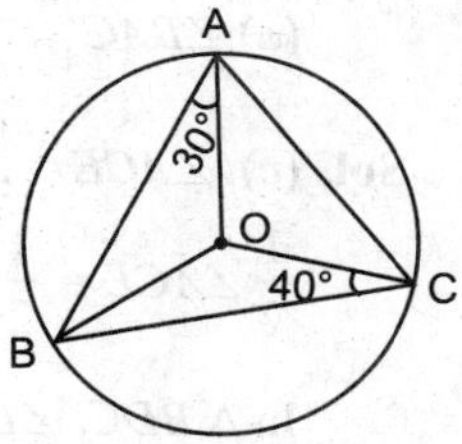

Sol. (b) In ΔAOC, $AO = OB \Rightarrow \angle ABO = 30°$
$\angle AOB = 180 - (30° + 30°) = 120°$
In ΔBOC, $\angle BOC = 180 - (40° + 40°) = 100°$
$\angle AOC = 360° - (\angle AOB + \angle BOC)$
$= 360° - (120° + 100°) = 140°$

96. In the figure, find $\angle AOC$.

(*a*) 50° (*b*) 100° (*c*) 60° (*d*) 90°

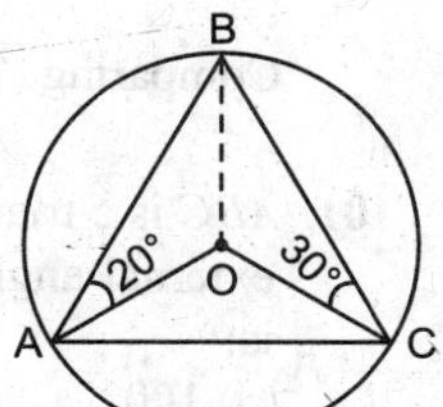

Sol. (*b*) Join OB

$OA = OB \Rightarrow \angle OBA = \angle OAB = 20°$
$OC = OB \Rightarrow \angle OBC = \angle OCB = 30°$
$\therefore \quad \angle B = \angle OBA + \angle OBC = 20° + 30° = 50°$
$\angle AOC = 2 \times \angle B = 2 \times 50° = 100°$

97. In the given figure, $AD = BD = AC$; $\angle CAE = 75°$, $\angle ABD = ?$ **(CAT 2003)**

(*a*) 15° (*b*) 25° (*c*) 30° (*d*) 50°

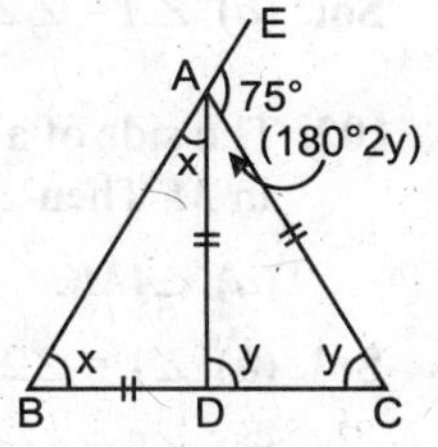

Sol. (*b*) In ΔABD, external $\angle D = \angle A + \angle B$
$\Rightarrow y = 2x$
$x + (180° - 2y) + 75° = 180°$ (Linear Angles)
$\Rightarrow x - 4x + 75 = 0 \Rightarrow 3x = 75° \Rightarrow x = 25°$

98. In the figure given here, O is the centre of the circle, $OA = 3$ cm, $AC = 3$ cm and OM is perpendicular to AC. What is $\angle ABC$ equal to **(CDS 2011)**

(*a*) 60° (*b*) 45° (*c*) 30° (*d*) None of these

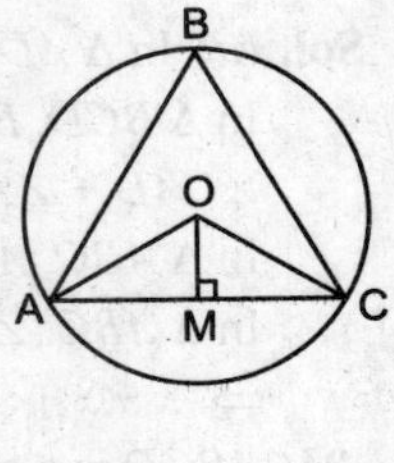

Sol. (*c*) $OC = OA = 3$ cm (radii)

$OA = OC = AC \Rightarrow \Delta AOC$ is equilateral triangle

$\therefore \angle AOC = 60°$

$$\angle ABC = \frac{1}{2}\angle AOC$$

$$= \frac{1}{2} \times 60° = 30° \quad \left[\begin{array}{l}\text{Angle at circle is half} \\ \text{of angle at centre}\end{array}\right]$$

99. In a triangle ABC, the side BC is extended up to D. Such that $CD = AC$, if $\angle BAD° = 109°$ and $\angle ACB = 72°$ then the value of $\angle ABC$ is **(SSC CGL Tier I 2013)**

(*a*) 35° (*b*) 60° (*c*) 40° (*d*) 45°

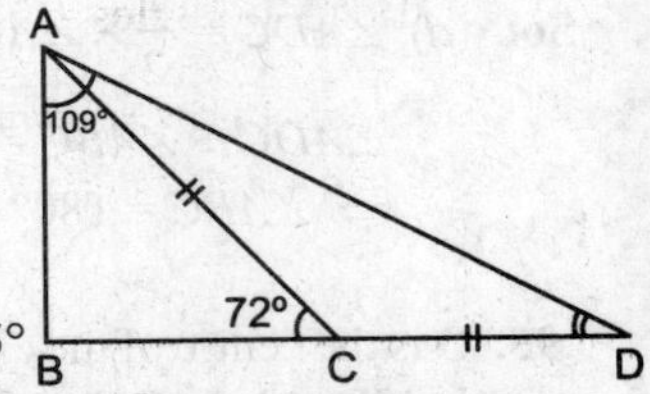

Sol. (*a*) $\angle CAD + \angle CDA =$ (Exterior angle) 72°

$$\therefore \angle CDA = \frac{1}{2} \times 72° = 36°$$

In ΔABC, $= \angle B = 180° - \angle B - \angle D = 180° - 190° - 36° = 35°$

100. In the given figure, BD is angle bisector of $\angle ABC$ and CD is angle bisector of $\angle ACE$. $\angle BDC = ?$

(*a*) $\angle BAC$ (*b*) 60° (*c*) $\frac{1}{2}\angle BAC$ (*d*) 30°

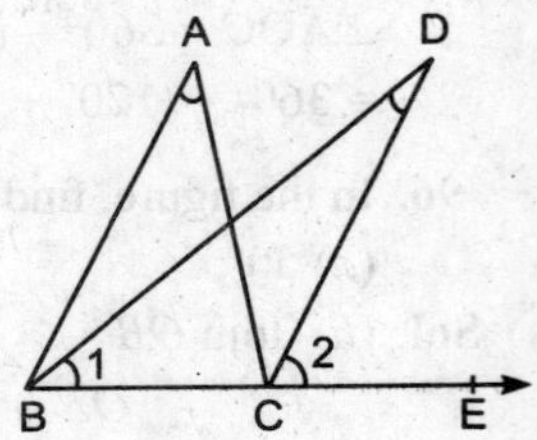

Sol. (*c*) $\angle ACE = \angle BAC + \angle \text{ABC}$

$$\frac{1}{2}\angle ACE = \frac{1}{2}\angle BAC + \frac{1}{2}\angle ABC \Rightarrow \angle 2 = \frac{1}{2}\angle BAC + \angle 1 \quad \text{... (1)}$$

In ΔBDC, $\angle DCE = \angle 1 + \angle BDC$

$\Rightarrow \angle 2 = \angle 1 + \angle BDC$ ... (2)

Comparing (1) and (2), $\frac{1}{2}\angle BAC = \angle BDC$

101. ABC is a triangle. The bisectors of the internal angle $\angle B$ and external angle $\angle C$ intersect at D. If $\angle BDC = 50°$, then $\angle A$ is: **(SSC CGL Tier I 2013)**

(*a*) 100° (*b*) 90°

(*c*) 120° (*d*) 60°

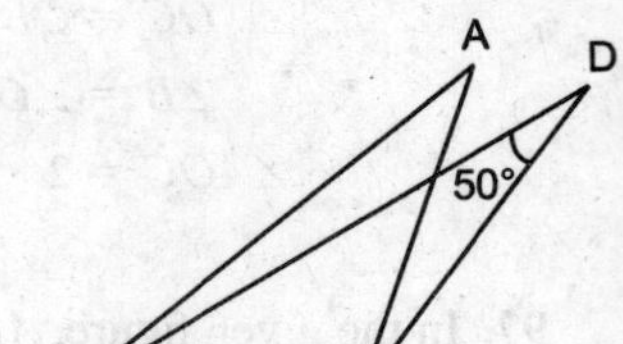

Sol. (*a*) $\angle A = 2\angle D = 2 \times 50° = 100°$

102. The side of a triangle ABC is produced such that D is on ray BC. The bisector of $\angle A$ meets BC in M. Then $\angle ABC + \angle ACD = ?$ **(SSC CHSL 2013)**

(*a*) $\angle AMC$ (*b*) $2\angle AMC$ (*c*) $2\angle ABC$ (*d*) $2\angle AZC$

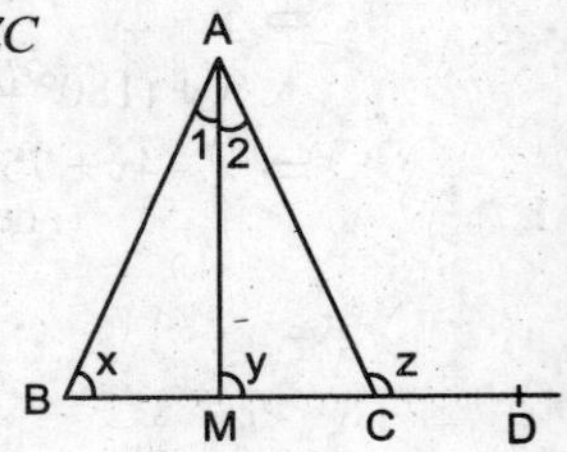

Sol. (*b*) $\angle 1 = \angle 2$ ($\because$ AM is the angle bisector)

$x + \angle 1 = y$ (external angle) $\Rightarrow x = y - \angle 1$

$z = y + \angle 2 = y + \angle 1$

$\therefore x + z = (y - \angle 1) + (y + \angle 1) = 2y = 2\angle AMC$

103. $ABCD$ is a cyclic quadrilateral. AB and DC are produced to meet at P. If $\angle ADC = 70°$ and $\angle DAB = 60°$, then the $\angle PBC + \angle PCB$ is : **(SSC CGL Tier I 2013)**

(*a*) 130° (*b*) 150° (*c*) 155° (*d*) 180°

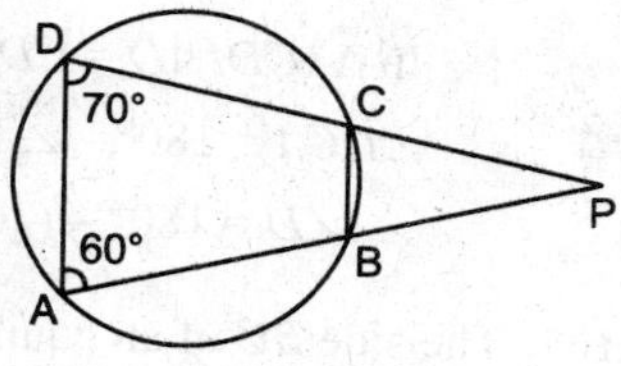

Sol. (*a*) $\angle PBC = \angle ADC = 70°$

$\angle PCB = \angle DAB = 60°$

$\therefore \quad \angle PBC + \angle PCB = 70° + 60° = 130°$

104. If two arcs of the same length in two circles subtend angles of 60° and 75° at their centres, the ratio of their radii is : **(SSC CGL Tier I 2013)**

(*a*) 3 : 4 (*b*) 4 : 5 (*c*) 5 : 4 (*d*) 3 : 5

Sol. (*c*) Ratio of angles = 60° : 75° = 4 : 5 $\Rightarrow$ Ratio of radii = $\frac{1}{4} : \frac{1}{5} = 5 : 4$

105. AB is a diameter of a circle with centre at O. DC is a chord such that $DC \| AB$. If $\angle BAC = 20°$, then the measure $\angle COD$ is : **(SSC FCI Astt. 2012)**

(*a*) 90° (*b*) 60° (*c*) 100° (*d*) 120°

Sol. (*c*) $\angle ACB = 90°$ (Angle in semi-circle)

$\angle ABC = 180° - 20° - 90° = 70°$

$\angle ADC = 180° - \angle ABC = 180° - 70° = 110°$

$\angle ACD = \angle CAB = 20°$ (Alternate angles)

$\angle CAD = 180° - 110° - 20° = 50°$

$\angle DOC = 2\ \angle CAD = 2 \times 50° = 100°$

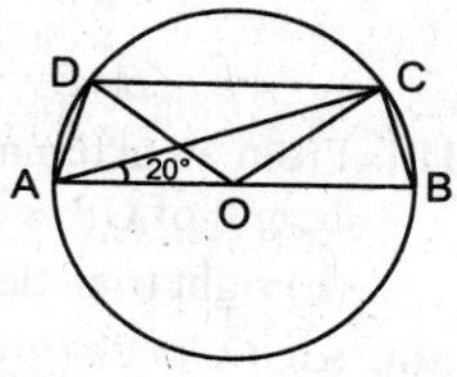

106. In the given figure $ABCD$ is a rhombus. If $\angle A = 70°$, then $\angle BDC = ?$

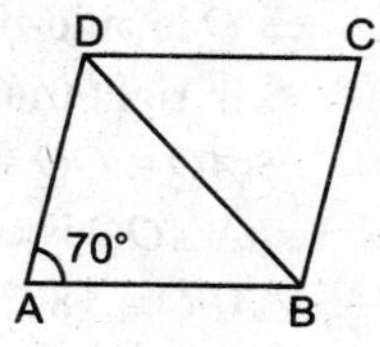

(*a*) 35° (*b*) 45°

(*c*) 55° (*d*) 70°

Sol. (*c*) $\angle D = 180° - 70° = 110°$ (Adjacent angles are supplementary)

$\angle BDC = \frac{1}{2} \times \angle D = \frac{1}{2} \times 110° = 55°$ ($\because BD$ is angle bisector)

Alt: $\Delta ABD, AB = AD \Rightarrow \angle B = \angle D \Rightarrow \angle B + \angle D = 180° - 70° = 110°$

$\Rightarrow \angle BDC = \frac{1}{2} \times 110° = 55°$

107. From an external point, a tangent to a circle is drawn. If length of this tangent be 8 cm and radius of the circle is 6 cm, the distance of the external point from the centre of the circle is **(SSC CHSL 2012)**

(*a*) $\frac{12}{\sqrt{3}}$ cm (*b*) 10 cm

(*c*) $8\sqrt{3}$ cm (*d*) $6\sqrt{3}$ cm

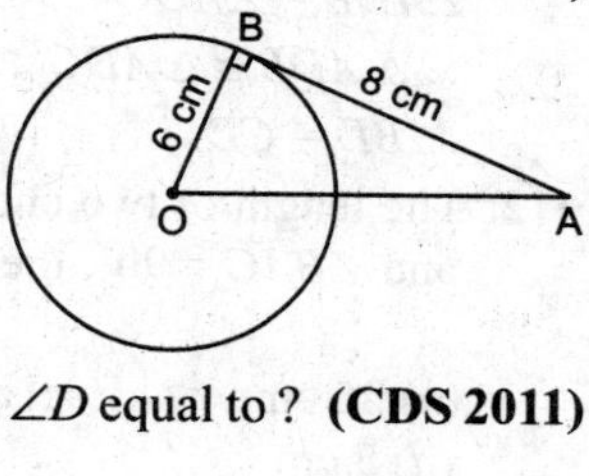

Sol. (*b*) $AO^2 = AB^2 + OB^2 = 8^2 + 6^2 = 10^2$

$\Rightarrow AO = 10$ cm

108. In the figure given here, $\angle B = 38°$, $AC = BC$ and $AD = CD$. What is $\angle D$ equal to? **(CDS 2011)**

(*a*) 26° (*b*) 28°

(*c*) 38° (*d*) 52°

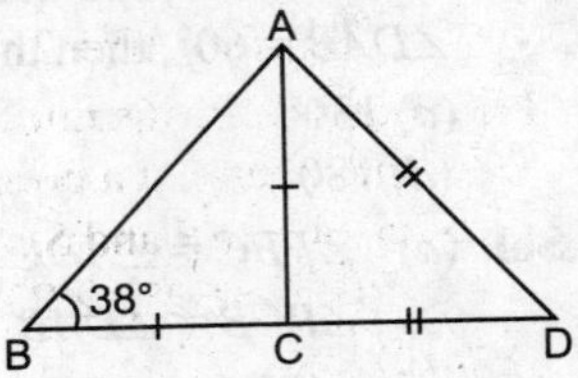

Sol. (*b*) In ΔABC, $AC = BC \Rightarrow \angle BAC = \angle CBA = 38°$

$\angle ACB = 180° - (38° + 38°) = 104°$

In ΔACD, $AD = CD \Rightarrow \angle DAC = \angle DCA$

$\angle DCA = 180° - \angle ACB = 180° - 104° = 76°$

$\therefore \angle D = 180° - (76° + 76°) = 28°$

109. The side BC of an equilateral triangle ABC is produced upto D in such a way that $BC = CD$. Then the value of $\angle BAD$ is : **(SSC CHSL 2012)**

(*a*) 120° (*b*) 135° (*c*) 90° (*d*) 145°

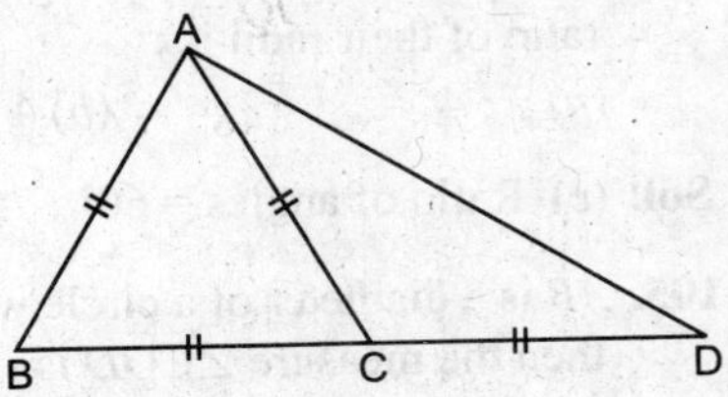

Sol. (*c*) In ΔABC, $AB = BC = AC$

$\therefore \angle A = \angle B = \angle C = 60°$

In ΔACD, $AC = CD$

$\therefore \angle CAD = \angle CDA$

$\angle ACD = 180° - \angle ACB = 180° - 60° = 120°$

$\therefore \angle CAD + \angle CDA = 180° - 120° = 60°$

$\therefore \angle CAD = \frac{1}{2} \times 60° = 30°$

$\therefore \angle BAD = \angle BAC + \angle CAD = 60° + 30° = 90°$

110. From an external point P, two tangents PA and PB are drawn to a circle with centre O. If the length of OP is equal to diameter of the circle, then ΔABP is

(*a*) right triangle (*b*) Isosceles (*c*) equilateral (*d*) None of these

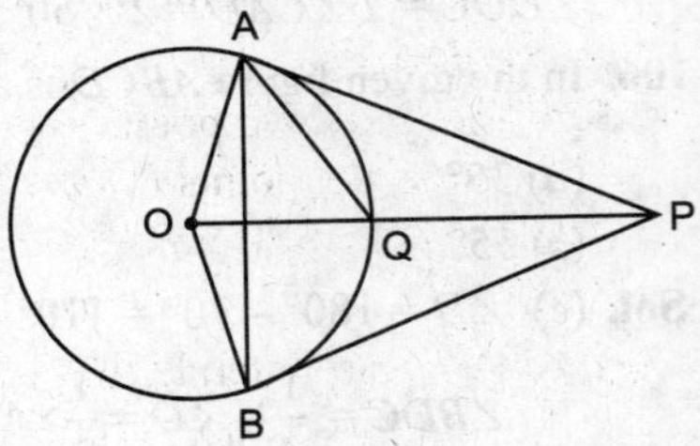

Sol. (*c*) $OP = 2r$ (given), $OQ = r$ (radius)

$PQ = OP - OQ = 2r - r = r$

$\Rightarrow Q$ is mid-point of OP.

$\therefore$ In right triangle OAP, AQ is median on OP.

$\Rightarrow AQ = OQ = r = OA$

$\therefore \Delta AOQ$ is equilateral triangle $\therefore \angle AOQ = 60°$

$\angle APO = 180° - 90° - 60° = 30°$

$\angle APB = 2\angle APO = 2 \times 30° = 60°$ $\therefore \Delta ABP$ is equilateral

111. A circle is drawn on one of the equal side of an isosceles triangle as diameter, as shown in the figure. Then

(*a*) $BD = CD$ (*b*) $BD > CD$ (*c*) $BD < CD$ (*d*) $BD \neq CD$

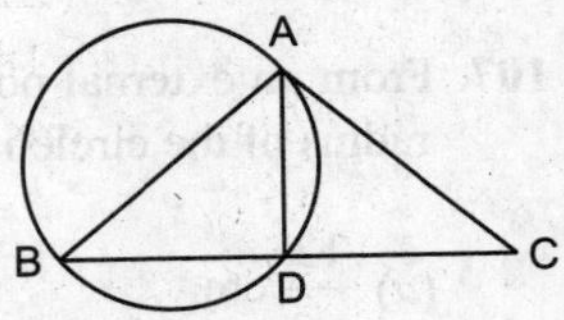

Sol. (*a*) $\angle ADB = 90°$ (Angle in semi-circle)

$\angle ADC = 180° - 90° = 90°$

$AB = AC$ and $AD = AD$

$\angle ADB = \angle ADC = 90°$

$\therefore \Delta ADB \cong \Delta ADC$

$\therefore BD = CD$

112. The length of two chords AB and AC of a circle are 8 cm and 6 cm and $\angle BAC = 90°$, then the radius of the circle is **(SSC CHSL 2011)**

(*a*) 25 cm (*b*) 20 cm (*c*) 4 cm
(*d*) 5 cm

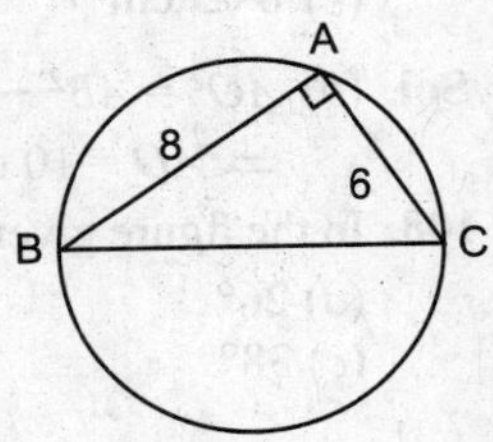

Sol. (*d*) $BC = \sqrt{8^2 + 6^2} = 10$ cm

$\angle BAC = 90° \Rightarrow BC$ is diameter of circle

$\therefore$ Radius = 10 ÷ 2 = 5 cm

113. *P* and *Q* are the middle points of two chords (not diameters) *AB* and *AC* respectively of a circle with centre at a point *O*. The lines *OP* and *OQ* are produced to meet the circle respectively at the points *R* and *S*. *T* is any point on the major arc between the points *R* and *S* on the circle. If $\angle BAC = 32°$, $\angle RTS$ = ? **(SSC CGL Tier I 2012)**

(*a*) 74° (*b*) 106°

(*c*) 64° (*d*) 32°

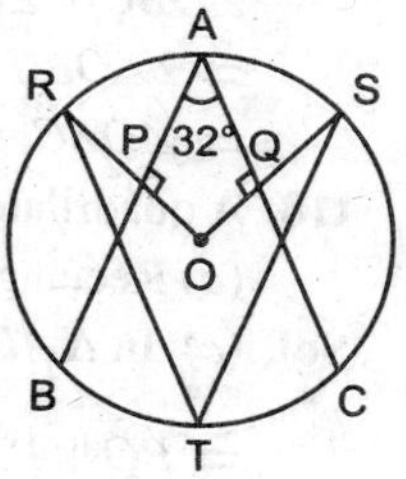

Sol. (*a*) $OP \perp AB$, $OQ \perp AC$ [Joining mid-points of chords]

$\angle POQ = 180° - \angle PAQ = 180° - 32° = 148°$

$\angle ROS = \angle POQ = 148°$

$\angle RTS = \frac{1}{2} \times 148° = 74°$ (Angle on circle)

114. *ABC* is an isosceles triangle with *AB* = *AC* and *D*, *E*, *F* are mid-points of *BC*, *CA* and *AB* respectively, then

(*a*) $AD \perp FE$ (*b*) $AD = FE$ (*c*) $AD = AB$ (*d*) $AD = AC$

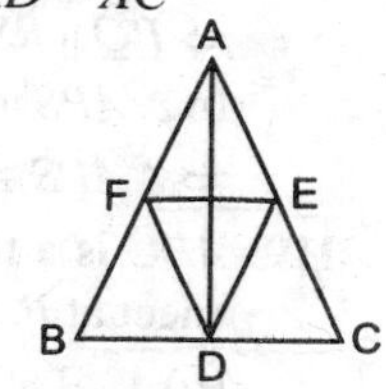

Sol. (*a*) $DE \parallel AB$ and $DE = \frac{1}{2}AB = AF$

$DF \parallel AC$ and $DF = \frac{1}{2} AC = AE$

$\Rightarrow DE = AF = DF = AE$ ($\because AB = AC$)

$\therefore$ AFDE is a rhombus and diagonals of a rhombus bisect each other at right angle.

$\Rightarrow AD \perp FE$

115. *P* and *Q* are two points on a circle with centre at *O*. *R* is a point on the minor arc of the circle, between the points *P* and *Q*. The tangents to the circle at the points *P* and *Q* meet each other at the point *S*. If $\angle PSQ = 20°$, $\angle PRQ$ = ? **(SSC CGL Tier I 2013)**

(*a*) 200° (*b*) 160° (*c*) 100° (*d*) 80°

Sol. (*c*) In quadrilateral *OPSQ*

$\angle SPO = \angle SQO = 90°$ (tangents)

$\therefore \angle POQ = 360° - 90° - 90° - 20° = 160°$

$\angle PAQ = \frac{1}{2} \times 160° = 80°$ (Angle at circle)

$\angle PAQ + \angle PRQ = 180°$ (cyclic quadrilateral)

$\therefore \angle PRQ = 180° - 80° = 100°$

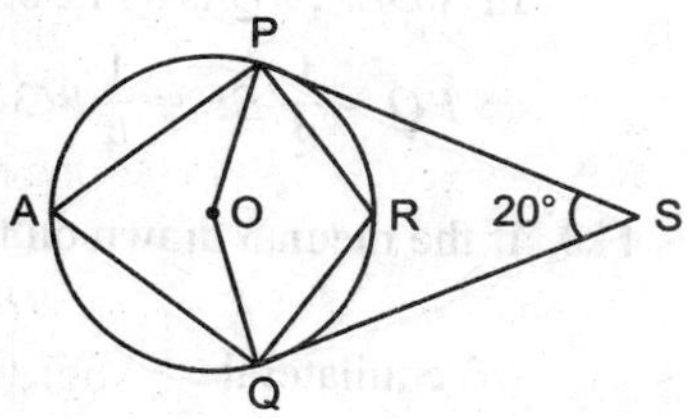

116. In the given figure, $\angle ADE = 100°$ and $\angle ECD = 40°$, find $\angle BFE$.

(*a*) 10° (*b*) 20°

(*c*) 30° (*d*) 40°

Sol. (*b*) In Δ *DCE*, $\angle DCE + \angle CED = \angle EDA$ (External angle)

$\Rightarrow \angle CED = 100° - 40° = 60°$

$\angle BEF = \angle CED = 60°$ (vertically opposite angles)

$\angle ADE + \angle ABE = 180°$ (cyclic quadrilateral)

$\Rightarrow \angle ABE = 180° - 100° = 80°$

$\angle BFE = 180° - \angle EBF - \angle BEF = 180° - 100° - 60° = 20°$

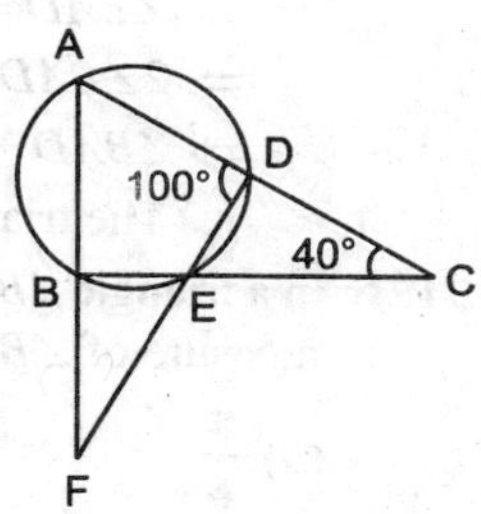

117. O is the circumcentre of the triangle ABC. If $\angle BAC = 50°$, then $\angle OBC$ is

(SSC CHSL 2012, SSC CGL Tier I 2012)

(a) 30° (b) 40° (c) 50° (d) 60°

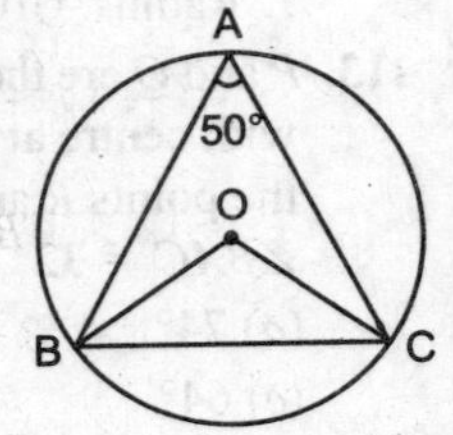

Sol. (b) $\angle BOC = 2\angle BAC = 2 \times 50° = 100°$ (Angle at centre)

$OB = OC$ (radii) $\Rightarrow \angle OBC = \angle OCB$

$\angle OBC + \angle OCB = 180° - 100°$

$\Rightarrow 2\angle OBC = 80°$

$\Rightarrow \angle OBC = 40°$

118. A quadrilateral formed by joining the mid-points of sides of a rectangle is

(a) Rectangle (b) Square (c) Rhombus (d) Parallelogram

Sol. (c) In ΔABC, P and Q are mid-points of sides AB and AC respectively.

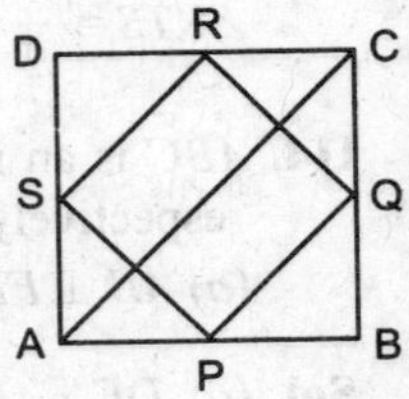

$\Rightarrow PQ \parallel AC$ and $PQ = \frac{1}{2} AC$

Similarly $SR \parallel AC$ and $SR = \frac{1}{2} AC$

$\Rightarrow PQ \parallel RS$ and $PQ = RS \Rightarrow PQRS$ is a parallelogram

In ΔAPS and ΔBPQ, $AP = BP$, $AS = BQ$, $\angle PAS = \angle PBQ = 90°$

$\Rightarrow \Delta APS \cong \Delta BPQ \Rightarrow PS = PQ \Rightarrow PQRS$ is a paralellogram and $PS = PQ \Rightarrow PQRS$ is a rhombus.

119. ABC is a triangle and D, E, F are mid-points of sides BC, CA, AB respectively. DF and BE meet at P, DE and CF meet at Q, then $PQ : BC = ?$

(a) 1 : 2 (b) 1 : 3 (c) 1 : 4 (d) 1 : 5

Sol. (c) $EF \parallel BC$ and $EF = \frac{1}{2} BC = BD$

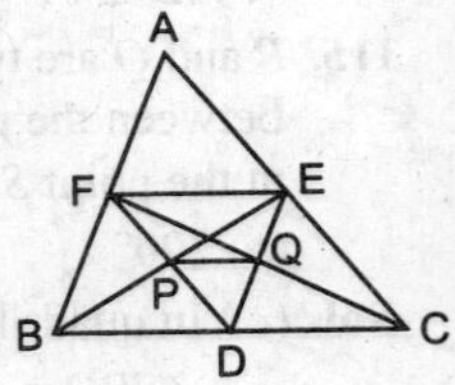

$\Rightarrow BDEF$ is a parallelogram. Diagonals of a parallelogram bisects each other

$\Rightarrow P$ is a mid-point of DF. Similarly, Q is mid-point of DE.

In ΔDEF, PQ is a line segment joining mid-points of sides DF and DE

$\Rightarrow PQ = \frac{1}{2} EF = \frac{1}{4} BC \Rightarrow PQ : BC = 1 : 4$

120. If the median drawn on the base of a triangle is half its base, the triangle will be :

(SSC CGL Tier I 2013)

(a) equilateral (b) right-angled (c) acute-angled (d) obtuse-angled

Sol. (b) In ΔABD, $AD = BD \Rightarrow \angle BAD = \angle ABD$

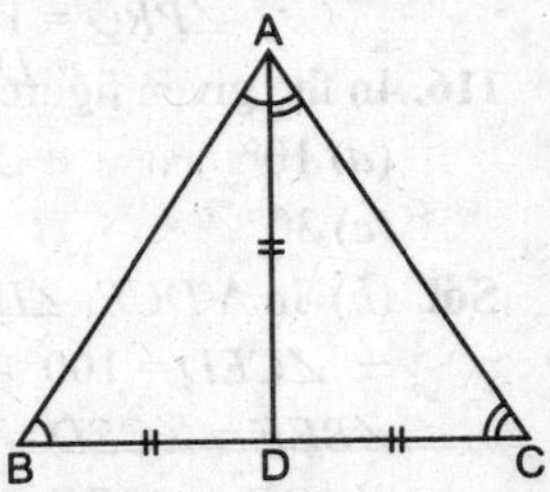

In ΔADC, $AD = DC \Rightarrow \angle DCA = \angle DAC$

$\angle BDA + \angle ADC = 180°$

$\therefore \angle BAD + \angle ABD + \angle DCA + \angle DAC = 180°$

$\Rightarrow 2\angle BAD + 2\angle DAC = 180°$

$\Rightarrow \angle BAD + \angle DAC = 90° \Rightarrow \angle BAC = 90°$

$\therefore$ The triangle is a right-angled triangle.

121. In a triangle ABC, $AB = AC$. BA is produced to D in such a manner that $AC = AD$. The circular measure of $\angle BCD$ is

(SSC CGL Tier I 2012)

(a) $\frac{\pi}{6}$ (b) $\frac{\pi}{3}$ (c) $\frac{2\pi}{3}$ (d) $\frac{\pi}{2}$

Sol. (*d*) In ΔABC, $\angle B = \angle C$

In ΔACD, $\angle C = \angle D$

$\angle CAB + \angle CAD = 180°$

$\Rightarrow \angle ABC + \angle ACB + \angle ADC + \angle ACD = 180°$

$\Rightarrow 2\angle ACB + 2\angle ACD = 180°$

$\Rightarrow \angle ACB + \angle ACD = 90° \Rightarrow \angle BCD = 90° = \frac{\pi}{2}$

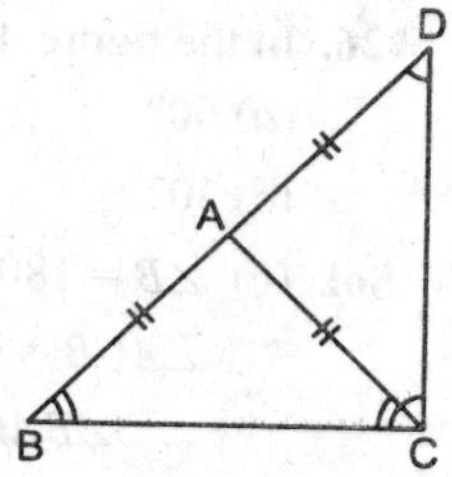

122. In the figure, PQ is a diameter of a circle with centre O. If $\angle ROS = 40°$, find $\angle RTS$.

(*a*) 50° (*b*) 60° (*c*) 70° (*d*) 40°

Sol. (*c*) $\angle RQS = \frac{1}{2}\angle ROS = \frac{1}{2} \times 40° = 20°$

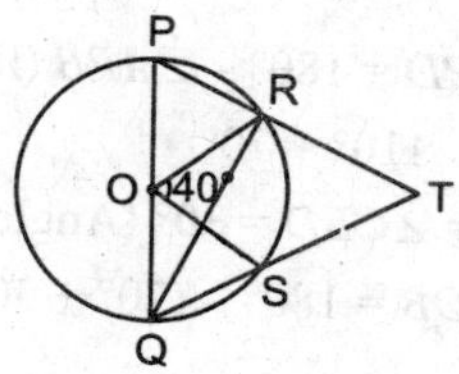

$\angle PRQ = 90°$ (Angle in semi-circle)

In $\Delta\ QRT$, $\angle RTQ + \angle RQT = \angle PRQ = 90°$

$\Rightarrow \angle RTQ + 20° = 90° \Rightarrow \angle RTQ = 70°$

123. In ΔABC, AE is angle bisector of $\angle A$ and $AD \perp BC$. If $\angle B = 40°$ and $\angle C = 60°$, find $\angle AED$.

(*a*) 40° (*b*) 60° (*c*) 80° (*d*) 75°

Sol. (*c*) In $\Delta ABC = \angle A = 180° - (40° + 60°) = 80°$

In ΔAEC, $\angle EAC = \frac{1}{2} \times 80° = 40°$

$\therefore \angle AED = 180° - (40° + 60°) = 80°$

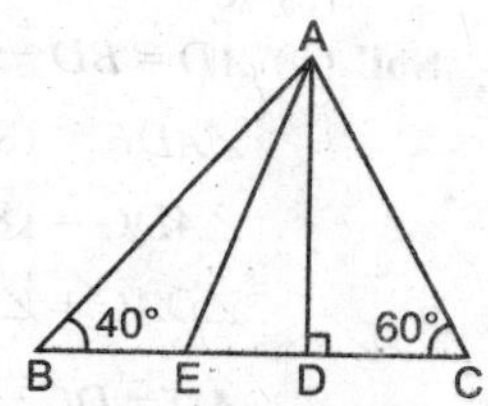

124. $ABCD$ is a parallelogram. E is mid-point of BC. DE is produced to meet extended AB at F. Then,

(*a*) $BF = AB$ (*b*) $BF = 2AB$ (*c*) $2AB = BF$ (*d*) None of these

Sol. (*a*) In ΔDCE and ΔFBE

$CE = EB$ (Given)

$\angle DEC = \angle BEF$ (Vertically opposite angles)

$\angle DCE = \angle FBE$ (Alternate angles)

$\therefore \Delta DCE \cong \Delta FBE$

$\therefore BF = DC = AB$

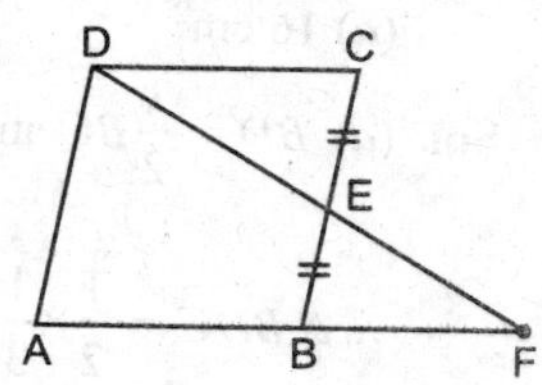

125. In the figure, find $\angle ACB$.

(*a*) 30° (*b*) 60° (*c*) 90° (*d*) 120°

Sol. (*b*) In ΔOAB, $OA = OB$ (radii)

$\therefore \angle OBA = \angle OAB = 30°$

$\therefore \angle AOB = 180° - (30° + 30°) = 120°$

$\angle ACB = \frac{1}{2} \times \angle AOB = \frac{1}{2} \times 120° = 60°$

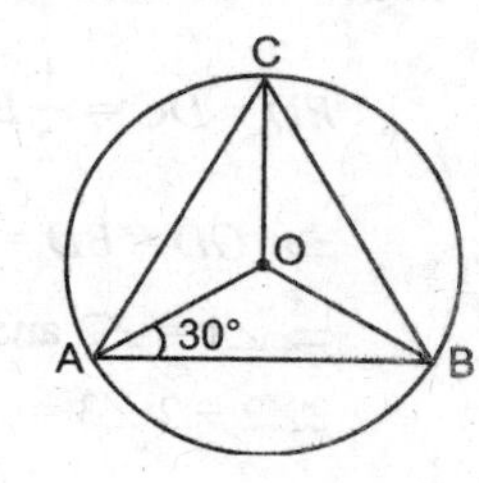

126. In the figure, find $\angle BAC$

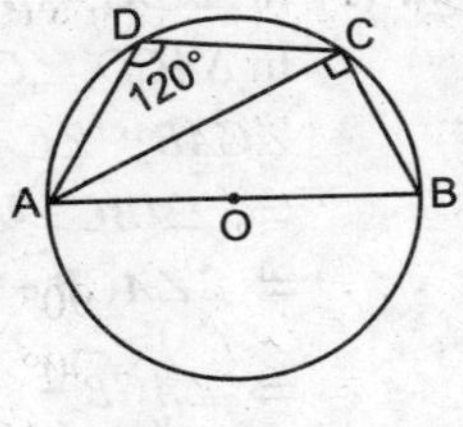

(*a*) 60° (*b*) 40°

(*c*) 30° (*d*) 50°

Sol. (*c*) $\angle B = 180° - \angle D = 180° - 120° = 60°$

$\angle ACB = 90°$ (Angle in semi-circle)

$\therefore \angle BAC = 180° - (90° + 60°) = 30°$

127. ABCD is a cyclic quadrilateral. Diagonals AC and BD meet at P. If $\angle APB = 110°$ and $\angle CBD = 30°$, then $\angle ADB$ measures **(SSC CGL Tier I 2015)**

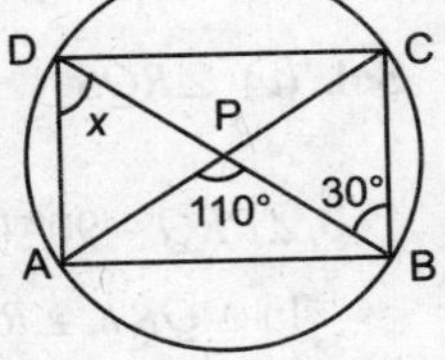

(*a*) 55° (*b*) 30° (*c*) 80° (*d*) 70°

Sol. (*c*) $\angle APD = 180° - \angle APB$ (linear pair)

$= 180° - 110° = 70°$

$\angle PAD = \angle CBD = 30°$ (Angle in same segment)

$\therefore \angle ADB = 180° - (70° + 30°) = 80°$

128. In ΔABC, AD is the median and $AD = \frac{1}{2}BC$. If $\angle BAD = 30°$, then measure of $\angle ACB$ is **(SSC CGL Tier I 2012)**

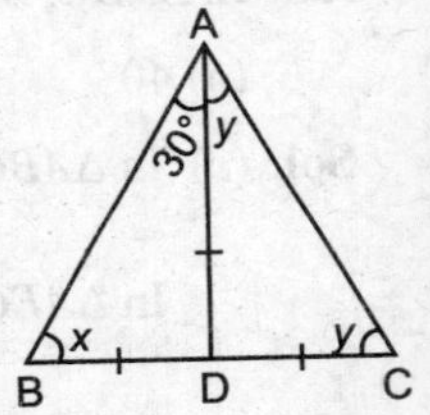

(*a*) 30° (*b*) 60°

(*c*) 90° (*d*) 45°

Sol. (*b*) $AD = BD \Rightarrow \angle ABD = \angle BAD = 30°$

$\angle ADB = 180° - 30° - 30° = 120°$

$\angle ADC = 180° - 120° = 60°$

$\angle DAC + \angle ACD = 180° - 60° = 120°$

$AD = DC \Rightarrow \angle DAC = \angle ACB = \frac{1}{2} \times 120° = 60°$

129. If G is the centroid and AD, BE, CF are three medians of ΔABC with area 72 cm², then the area of ΔBDG is **(SSC Constable GD 2012)**

(*a*) 16 cm² (*b*) 8 cm² (*c*) 24 cm² (*d*) 12 cm²

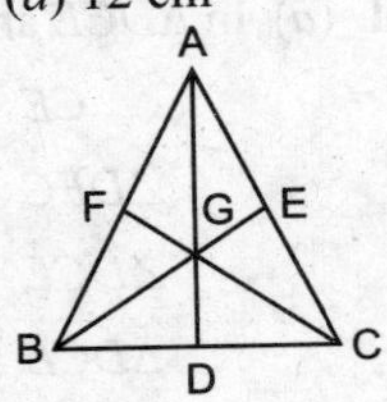

Sol. (*d*) $BD = \frac{1}{2}BC$ and $GD = \frac{1}{3}AD$

$\therefore \Delta BDG = \frac{1}{2} \times \frac{1}{3} \times \Delta ABC = \frac{1}{6} \times 72 = 12 \text{ cm}^2$

130. If G is the cenroid of ΔABC, and $AG = BC$, find the angle BGC. **(SSC CGL Tier I 2012)**

(*a*) 60° (*b*) 90° (*c*) 45° (*d*) 30°

Sol. (*b*) $AG : GD = 2 : 1$ (centroid)

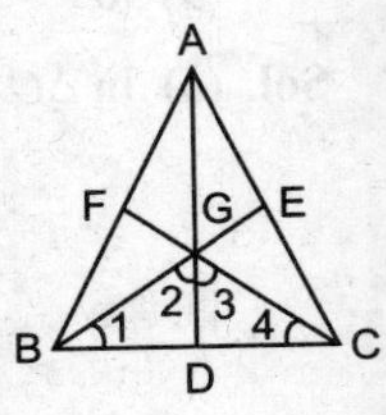

$BD = DC = \frac{1}{2}BC = \frac{1}{2}AG$

$\Rightarrow GD = BD = DC$

$\Rightarrow \angle 1 = \angle 2$ and $\angle 3 = \angle 4$. $\angle 1 + \angle 2 + \angle 3 + \angle 4 = 360° - 180° = 180°$

$2\angle 2 + 2\angle 3 = 180° \Rightarrow \angle 2 + \angle 3 = 90°$

131. Two circles intersect each other at P and Q. PA and PB are two diameters. Then $\angle AQB$ is **(SSC CGL Tier I 2012)**

(*a*) 120° (*b*) 135°
(*c*) 160° (*d*) 180°

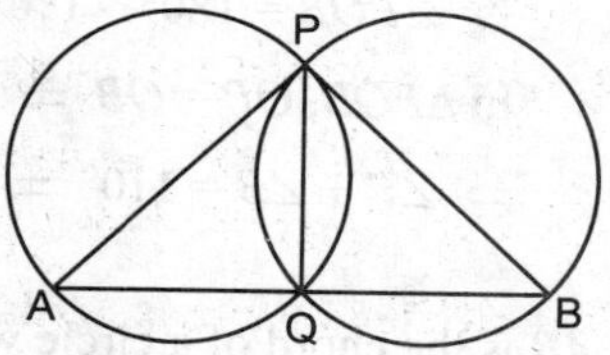

Sol. (*d*) $\angle AQP = 90°$ (Angle in semi-circle)
$\angle BQP = 90°$ (Angle in semi-circle)
$\angle AQB = \angle AQP + \angle BQP$
$= 90° + 90° = 180°$

132. ABCD is a square and DEC is an equilateral triangle on its top. Find $\angle BEC$.

(*a*) 15° (*b*) 30°
(*c*) 45° (*d*) 60°

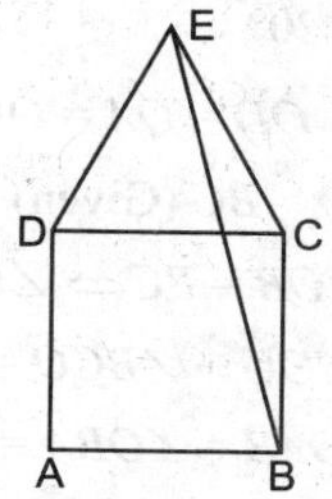

Sol. (*a*) In square $ABCD$, $BC = DC$ and $\angle C = 90°$.
In equilateral triangle DEC, $\angle C = 60°$, $EC = DC$
$\therefore$ In ΔCEB, $EC = BC$
$\therefore$ $\angle CEB = \angle CBE$
$\angle C = 90° + 60° = 150°$
$\therefore$ $\angle BEC = \frac{1}{2}(180° - 150°) = 15°$

133. In the figure, $ABCD$ is a square and DEC is an equilateral triangle on its top. Find $\angle AEB$.

(*a*) 15° (*b*) 30° (*c*) 45° (*d*) 60°

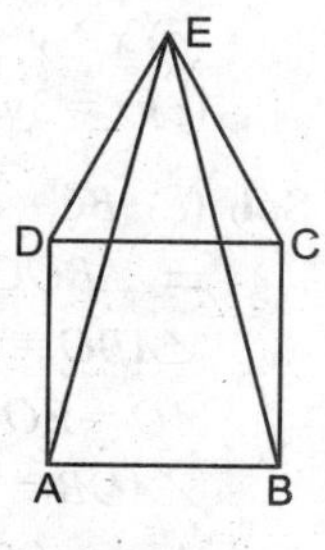

Sol. (*b*) In square $ABCD$, $\angle A = \angle B = \angle C = \angle D = 90°$
In equilateral ΔDEC, $\angle D = \angle E = \angle C = 60°$
In ΔDEA, $\angle ADC + \angle EDC = 90° + 60° = 150°$
$AD = DE \Rightarrow \angle DAE = \angle AED$
$\therefore$ $\angle AED = \frac{1}{2} \times (180° - 150°) = 15°$, similarly $\angle BEC = 15°$
$\therefore$ $\angle AEB = \angle DEC - \angle DEA - \angle CEB = 60° - 15° - 15° = 30°$

134. In the figure, O is any point inside ΔABC. OD, OE, OF are angle bisectors of $\angle AOB$, $\angle BOC$, $\angle AOC$ respectively. Then $AD \times BE \times CF = ?$

(*a*) $DB + EC + FA$ (*b*) $OD \times OE \times OF$ (*c*) $DB \times EC \times FA$ (*d*) $OD + OE + OF$

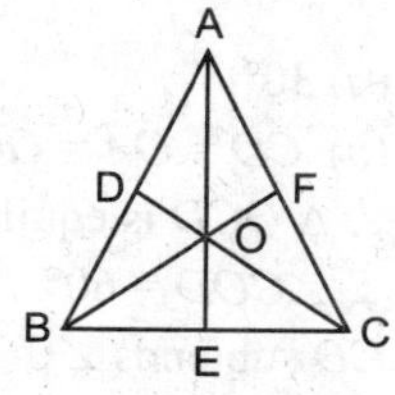

Sol. (*c*) $\frac{OA}{OB} = \frac{AD}{DB}, \frac{OB}{OC} = \frac{BE}{EC}, \frac{OC}{OA} = \frac{CF}{AF}$

$\Rightarrow \frac{OA}{OB} \times \frac{OB}{OC} \times \frac{OC}{OA} = \frac{AD}{DB} \times \frac{BE}{EC} \times \frac{CF}{AF}$

$\Rightarrow 1 = \frac{AD}{DB} \times \frac{BE}{EC} \times \frac{CF}{AF} \Rightarrow DB \times EC \times AF = AD \times BE \times CF$

135. Two circles touch each other externally at P. AB is a direct common tangent to the two circles, A and B are points of contact and $\angle PAB = 35°$. Then $\angle ABP$ is **(SSC CGL Tier II 2012)**

(*a*) 65° (*b*) 75° (*c*) 35° (*d*) 55°

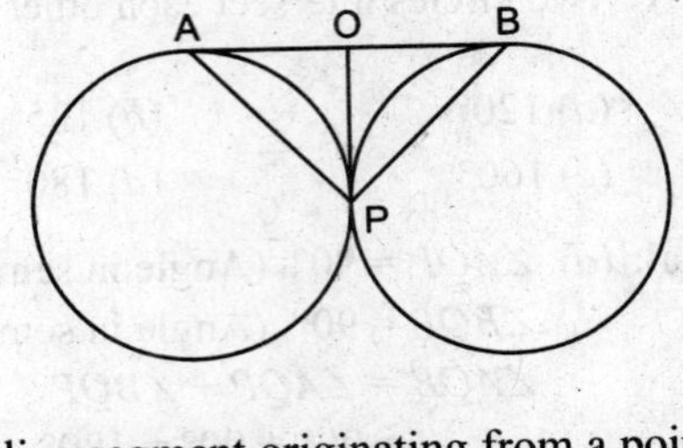

Sol. (*b*) In ΔAOP, $OA = OP \Rightarrow \angle A = \angle P = 35°$

$\therefore\ \angle AOP = 180° - (35° + 35°) = 110°$

$\therefore\ \angle POB = 180° - 110° = 70°$

In ΔPOB, $OP = OB \Rightarrow \angle P = \angle B$

$\Rightarrow \angle P + \angle B = 110° \Rightarrow \angle ABP = \frac{1}{2} \times 110° = 55°$

136. *AB* is the chord of a circle with centre *O* and *DOC* is a line segment originating from a point *D* on the circle and intersecting *AB* produced at *C* such that $BC = OD$. If $\angle BCD = 20°$, then $\angle AOD = ?$ **(CAT 2003, SSC CGL Tier I 2013)**

(*a*) 20° (*b*) 30° (*c*) 40° (*d*) 60°

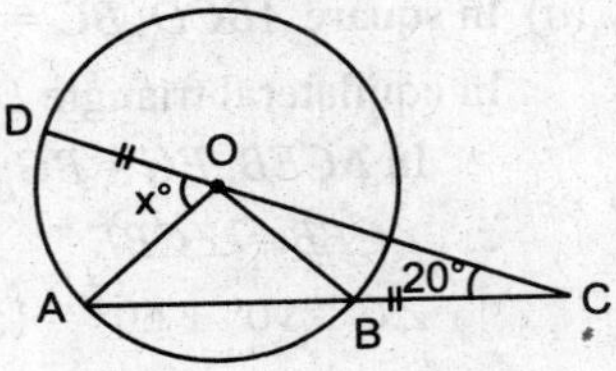

Sol. (*d*) $OD = OA = OB$ (radii)

$OD = BC$ (Given)

$\Rightarrow OB = BC \Rightarrow \angle BOC = \angle BCO = 20°$

$\angle BOC + \angle BCO = 20° + 20° = 40°$

$\angle OAB = \angle OBA = 40°$ ($\because OA = OB$)

$\angle AOB = 180° - 40° - 40° = 100°$

$\angle AOD = 180° - (\angle AOB + \angle BOC) = 180° - (100° + 20°) = 60°$

137. In the given figure, *O* is the centre and $BC = AO$, then **(SSC CHSL 2013)**

(*a*) $x = y$ (*b*) $x = 2y$

(*c*) $x = 3y$ (*d*) $2x = y$

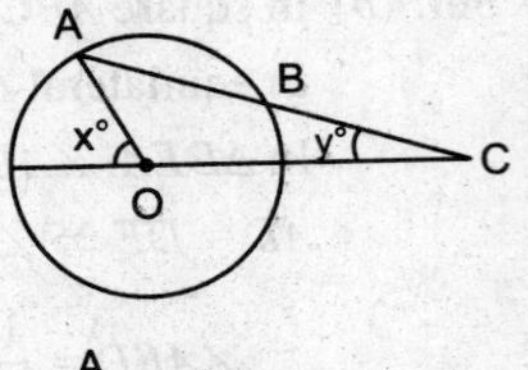

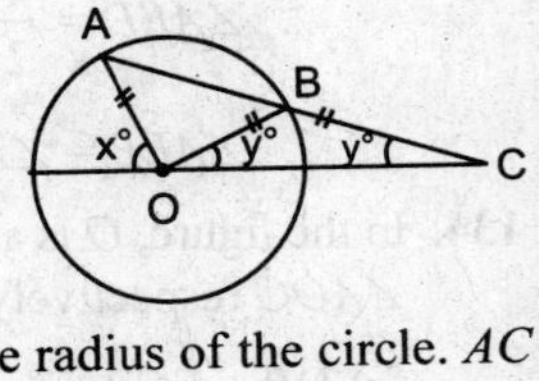

Sol. (*c*) $BC = AO = BO \Rightarrow BO = BC$

$\Rightarrow \angle BOC = \angle BCO = y°$

$\angle ABO = \angle BOC + \angle BCO = y° + y° = 2y°$

$AO = BO \Rightarrow \angle OAB = \angle OBA = 2y°$

$\angle AOB = 180° - \angle OAB - \angle OBA = 180° - 4y°$

Also, $\angle AOB = 180° - x° - y°$

$\Rightarrow 180° - 4y° = 180° - x° - y° \Rightarrow x = 3y$

138. *AB* is a diameter of a circle with centre *O*. *CD* is a chord equal to the radius of the circle. *AC* and *BD* are produced to meet at *P*. Then the measure of $\angle APB$ is. **(SSC CGL Tier I 2012, CDS 2012)**

(*a*) 30° (*b*) 60° (*c*) 90° (*d*) 120°

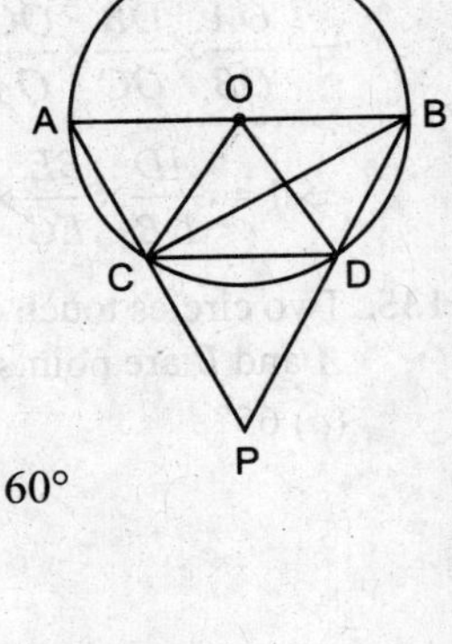

Sol. (*b*) $CD = OA = OC = OD$ = radii

$\therefore\ \Delta\ OCD$ is equilateral triangle

$\therefore\ \angle COD = 60°$

CD subtends $\angle COD$ at centre and $\angle CBD$ on circle

$\therefore\ CBD = \frac{1}{2}\ \angle COD = \frac{1}{2} \times 60° = 30°$

$\angle ACB = 90°$ (Angle in semi-circle)

$\angle BCP = 180° - 90° = 90°$

In $\Delta\ BCP$, $\angle CPB = 180° - (\angle CBP + \angle BCP) = 180° - (30° + 90°) = 60°$

139. PQ is a direct common tangent of two circles of radii r_1 and r_2 touching each other externally at A. Then, the value of PQ^2 is **(SSC CGL Tier I 2012)**

(*a*) $r_1 r_2$ (*b*) $2\, r_1 r_2$
(*c*) $3\, r_1 r_2$ (*d*) $4\, r_1 r_2$

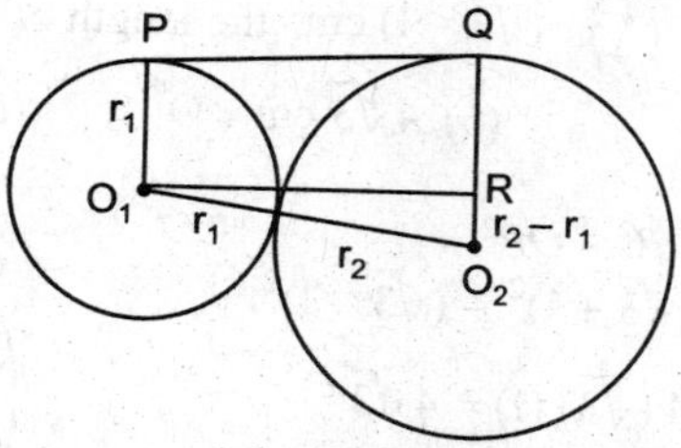

Sol. (*d*) $PQ^2 = OR^2$
$= (O_1O_2)^2 - (O_2R)^2$
$= (r_1 + r_2)^2 - (r_2 - r_1)^2 = 4r_1 r_2$

140. Two circles of radii 4 cm and 9 cm respectively touch each other externally at a point and a common tangent touches them at the points P and Q respectively. Then the area of a square with one side PQ, is **(SSC CGL Tier I 2011)**

(*a*) 72 sq. cm (*b*) 144 sq. cm
(*c*) 97 sq. cm (*d*) 194 sq. cm

Sol. (*b*) Draw a line $PR \parallel AB$

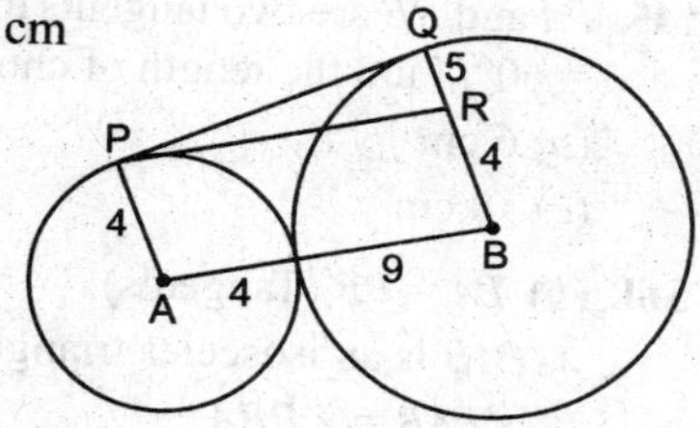

In right triangle PQR, $PR = 13$ cm, $QR = 5$ cm

$\therefore\; PQ^2 = 13^2 - 5^2 = 169 - 25 = 144$ sq. cm.

Alt: $PQ = \sqrt{13^2 - (9-4)^2} = \sqrt{13^2 - 5^2} = 12$ cm

141. A chord of length 16 cm is drawn in a circle of radius 10 cm. The distance of the chord from the centre of the circle is **(SSC CHSL 2012)**

(*a*) 9 cm (*b*) 12 cm
(*c*) 6 cm (*d*) 8 cm

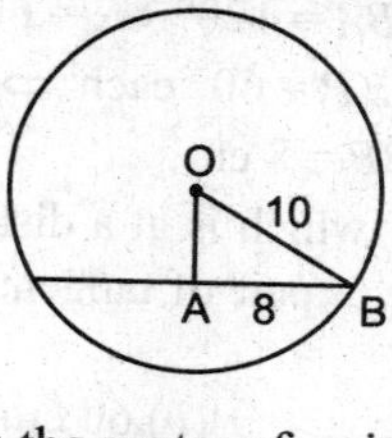

Sol. (*c*) $OA^2 = OB^2 - AB^2 = 10^2 - 8^2 = 6^2$
$\Rightarrow OA = 6$ cm

142. A chord of length 30 cm is at a distance of 8 cm from the centre of a circle. The radius of the circle is : **(SSC CGL Tier I 2013)**

(*a*) 19 cm (*b*) 17 cm
(*c*) 23 cm (*d*) 21 cm

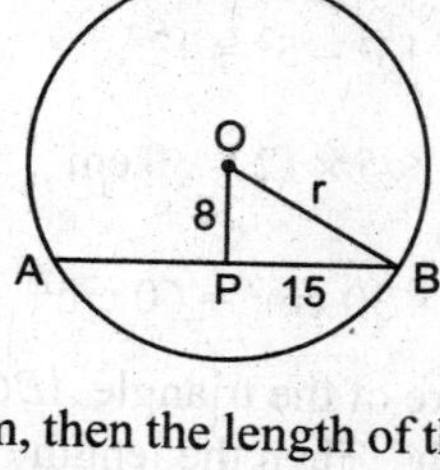

Sol. (*b*) $PB = \dfrac{1}{2} \times AB = \dfrac{1}{2} \times 30 \text{ cm} = 15$ cm

$r^2 = 8^2 + 15^2 = 64 + 225 = 289 = 17^2$

143. If two concentric circles of radii 5 cm and 3 cm, then the length of the chord of the larger circle which touches the smaller circle is **(SSC CGL Tier II 2013)**

(*a*) 6 cm (*b*) 7 cm
(*c*) 10 cm (*d*) 8 cm

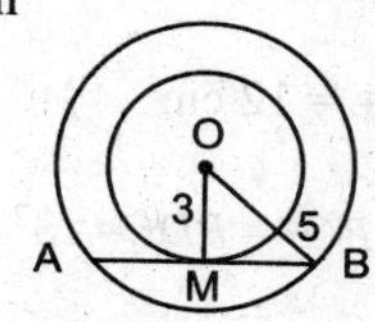

Sol. (*d*) $MB^2 = OB^2 - OM^2 = 5^2 - 3^2 = 4^2$
$AB = 2 \times MB = 2 \times 4 \text{ cm} = 8$ cm

144. A chord AB of a circle C_1 of radius $(\sqrt{3}+1)$ cm touches a circle C_2 which is concentric to C_1. If the radius of C_2 is $(\sqrt{3}-1)$ cm, the length of AB is : **(SSC CGL Tier I 2013)**

(*a*) $8\sqrt{3}$ cm (*b*) $4\sqrt[4]{3}$ cm (*c*) $4\sqrt{3}$ cm (*d*) $2\sqrt[4]{3}$ cm

Sol. (*b*) $PB^2 = OB^2 - OP^2$
$= (\sqrt{3}+1)^2 - (\sqrt{3}-1)^2$
$= 4(\sqrt{3})(1) = 4\sqrt{3}$
$\therefore\ PB = 2\sqrt[4]{3}$
$AB = 2 \times PB = 4\sqrt[4]{3}$

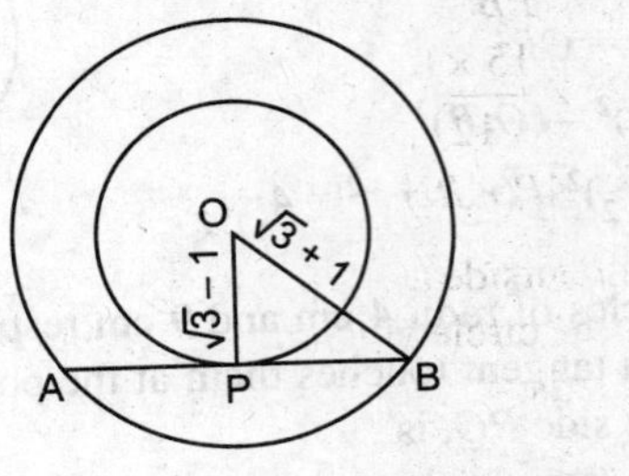

145. PA and PB are two tangents to a circle from an external point P such than $PA = 8$ cm and $\angle APB = 60°$. Find the length of chord AB.

(*a*) 6 cm (*b*) 8 cm
(*c*) 10 cm (*d*) 16 cm

Sol. (*b*) $PA = PB$ (Tangents)
$\therefore$ PAB is an isosceles triangle
$\therefore \angle PAB = \angle PBA$
$\angle PAB + \angle PBA = 180° - \angle APB = 180° - 60° = 120°$
$\therefore \angle PAB = \angle PBA = 120° \div 2 = 60°$
$\therefore \angle A = \angle B = \angle P = 60°$ each $\Rightarrow \Delta\ PAB$ is an equilateral triangle
$\therefore AB = AP = BP = 8$ cm

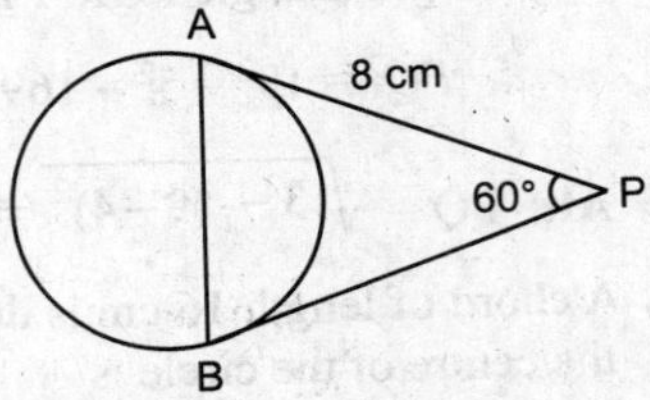

146. From a point P which is at a distance of 13 cm from centre O of a circle of radius 5 cm, in the same plane, a pair of tangents PQ and PR are drawn to the circle. Area of quadrilateral $PQOR$ is **(SSC CGL Tier I 2013)**

(*a*) 65 cm² (*b*) 60 cm²
(*c*) 30 cm² (*d*) 90 cm²

Sol. (*b*) In ΔPOQ, $PQ^2 = 13^2 - 5^2 = 12^2$

Area $\Delta POQ = \dfrac{1}{2} \times 5 \times 12 = 30$ cm²

Area $PQOR = 2 \times 30$ cm² $= 60$ cm²

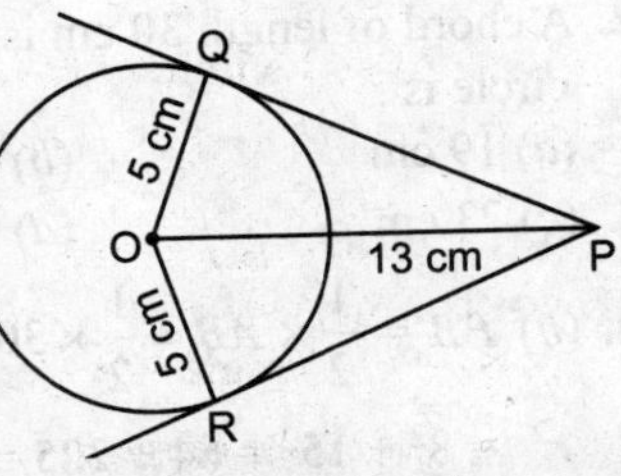

147. O is the circum-centre of the triangle ABC with circumradius 13 cm. Let $BC = 24$ cm and OD is perpendicular to BC. Then the length of OD is: **(SSC CGL Tier I 2012)**

(*a*) 3 cm (*b*) 7 cm
(*c*) 5 cm (*d*) 4 cm

Sol. (*c*) $BD = \dfrac{1}{2}BC = \dfrac{1}{2} \times 24 = 12$ cm

In right $\Delta\ BOD$, $OD^2 = BO^2 - BD^2 = 13^2 - 12^2 = 5^2$

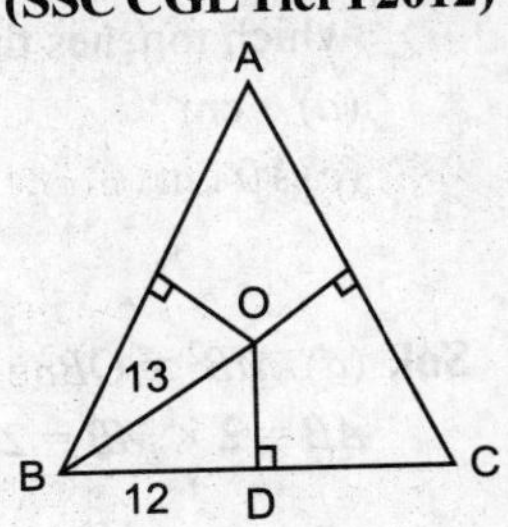

148. In the given figure, $AP = 12$ cm, $PC = 15$ cm and $CD = 7$ cm. Find AB.

(*a*) 2 cm (*b*) 5 cm (*c*) 8 cm (*d*) 10 cm

Sol. (*a*) $PD = CP - DP = 15 - 7 = 8$ cm

$PA \times PB = PC \times PD$

$\Rightarrow 12 \times PB = 15 \times 8$

$\Rightarrow PB = \dfrac{15 \times 8}{12} = 10$ cm

$\therefore AB = AP - BP = 12 - 10 = 2$ cm

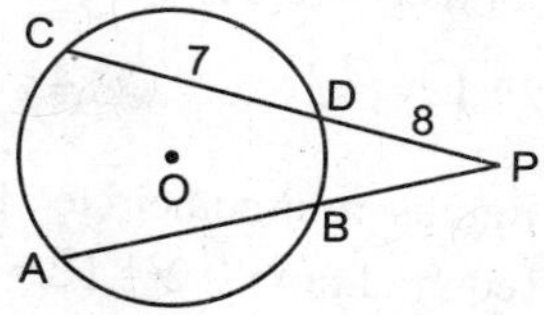

149. P is a point outside a circle and is 13 cm away from its centre. A secant drawn from the point P intersects the circle at points A and B in such a way that $PA = 9$ cm and $AB = 7$ cm. The radius of the circle is: **(SSC CGL Tier I 2012)**

(*a*) 4 cm (*b*) 4.5 cm
(*c*) 5.5 cm (*d*) 5 cm

Sol. (*d*) $PT^2 = PA \times PB = 9 \times 16 = 144 \Rightarrow PT = \sqrt{144} = 12$ cm

In right Δ OTP, $OT^2 = OP^2 - PT^2 = 13^2 - 12^2 = 5^2$

$\Rightarrow$ radius $= OT = 5$ cm

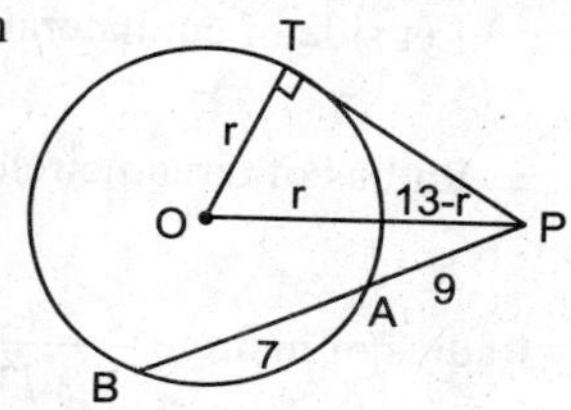

Alt: $9 \times 16 = (13 - r) \times (13 + r) \Rightarrow r^2 = 169 - 144 = 25 \Rightarrow r = 5$

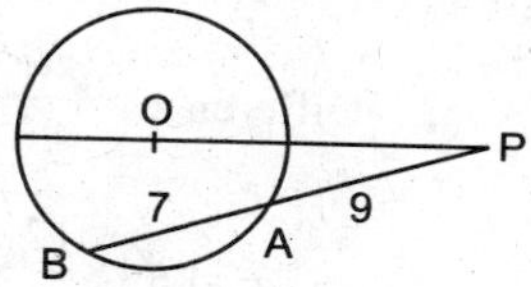

150. Find ratio of side of an equilateral triangle and diameter of its circumcircle.

(*a*) $2 : \sqrt{3}$ (*b*) $1 : 2$ (*c*) $\sqrt{3} : 1$ (*d*) $\sqrt{3} : 2$

Sol. (*d*) Radius of circumcircle $= \dfrac{\text{Side}}{\sqrt{3}} \Rightarrow$ Diameter $= \dfrac{2 \times \text{side}}{\sqrt{3}}$

Let side of equilateral triangle be x. Then required ratio $= x : \dfrac{2x}{\sqrt{3}} = \sqrt{3} : 2$

151. If area of an incircle of an equilateral triangle is 462 cm^2, find perimeter of the triangle.

(*a*) 126 cm (*b*) 168 cm (*c*) $21\sqrt{3}$ (*d*) $42\sqrt{3}$

Sol. (*a*) $\pi r^2 = 462 \Rightarrow r = 462 \times \dfrac{7}{22} = 21 \times 7 \Rightarrow r = 7\sqrt{3}$

Side of triangle = Radius of incircle $\times\ 2\sqrt{3} = 7\sqrt{3} \times 2\sqrt{3} = 42$ cm

$\therefore$ Perimeter of triangle $= 3 \times 42 = 126$ cm

152. If each side of an equilateral triangle is 24 cm, find area of its incircle.

(*a*) 24π cm^2 (*b*) 36π cm^2 (*c*) 48π cm^2 (*d*) 60π cm^2

Sol. (*c*) Radius of incircle $= \dfrac{\text{Side of triangle}}{2\sqrt{3}} = \dfrac{24}{2\sqrt{3}} = 4\sqrt{3}$ cm

Area of incircle $= \pi r^2 = \pi (4\sqrt{3})^2 = 48\,\pi$ cm^2

153. The area of the incircle of an equilateral triangle of side 42 cm is : **(SSC CGL Tier I 2004)**

(*a*) 231 cm^2 (*b*) 462 cm^2 (*c*) $22\sqrt{3}$ cm^2 (*d*) 924 cm^2

Sol. (*b*) Radius of incircle $= \dfrac{42}{2\sqrt{3}} = 7\sqrt{3}$

Area of incircle $= \pi r^2 = \dfrac{22}{7} \times 7\sqrt{3} \times 7\sqrt{3} = 462$ cm^2

154. The height of an equilateral triangle is $4\sqrt{3}$ cm. The ratio of the area of its circumcircle to that of its incircle is : **(SSC CGL Tier I 2008)**

(*a*) 2 : 1 (*b*) 4 : 1 (*c*) 4 : 3 (*d*) 3 : 2

Sol. (*b*) Radius of circumcircle : Radius of incircle = $\frac{\text{Side}}{\sqrt{3}} : \frac{\text{Side}}{2\sqrt{3}} = 2 : 1$

$\therefore$ Required ratio = $2^2 : 1^2 = 4 : 1$

155. If the difference between areas of circumcircle and the incircle of an equilateral triangle is 44 cm², then the area of the triangle is : **(SSC CGL Tier II 2012)**

(*a*) $7\sqrt{3}$ cm² (*b*) 14 cm² (*c*) $14\sqrt{3}$ cm² (*d*) 22 cm²

Sol. (*c*) Let side of equilateral triangle = *a*

$$\Rightarrow \text{Radius of circumcircle} = \frac{a}{\sqrt{3}} \Rightarrow \text{Area of circumcircle} = \pi\left(\frac{a}{\sqrt{3}}\right)^2 = \frac{\pi a^2}{3}$$

$$\text{Radius of incircle} = \frac{a}{2\sqrt{3}} \Rightarrow \text{Area of incircle} = \pi\left(\frac{a}{2\sqrt{3}}\right)^3 = \frac{\pi a^2}{12}$$

$$\text{Difference} = \frac{\pi a^2}{3} - \frac{\pi a^2}{12} = \pi a^2\left(\frac{1}{3} - \frac{1}{12}\right) = \pi a^2\left(\frac{3}{12}\right) = 44$$

$$\Rightarrow a^2 = 44 \times \frac{12}{3} \times \frac{7}{22} = 56 \Rightarrow \text{Area of triangle} = \frac{\sqrt{3}}{4}a^2 = \frac{\sqrt{3}}{4} \times 56 = 14\sqrt{3} \text{ cm}^2$$

156. The length of the circum-radius of a triangle having sides of lengths 12 cm, 16 cm and 20 cm is **(SSC CGL Tier I 2012)**

(*a*) 15 cm (*b*) 10 cm (*c*) 18 cm (*d*) 16 cm

Sol. (*b*) The given triangle is a right-angled triangle as $12^2 + 16^2 = 20^2$.

$$\therefore \text{Circumradius} = \frac{\text{Hypotenuse}}{2} = \frac{20}{2} = 10 \text{ cm}$$

157. Let C1 and C2 be the inscribed and circumscribed circles of a triangle with sides 3 cm, 4 cm and 5 cm, then Area of C1 : Area of C2 is **(SSC CGL Tier I 2015)**

(*a*) 9 : 16 (*b*) 9 : 25 (*c*) 4 : 25 (*d*) 16 : 25

Sol. (*c*) The given triangle is a right-angled triangle as $3^2 + 4^2 = 5^2$.

$$\text{Inradius} = \frac{3+4-5}{2} = 1 \qquad \text{Circumradius} = \frac{\text{Hypotenuse}}{2} = \frac{5}{2} = 2.5 \text{ cm}$$

Area of C1 : Area of C2 = $1^2 : 2.5^2 = 2^2 : 5^2 = 4 : 25$

158. The radius of the incircle of a triangle is 2 cm. If the area of the triangle is 6 cm², then its perimeter is **(SSC CPO SI 2011)**

(*a*) 2 cm (*b*) 3 cm (*c*) 6 cm (*d*) 9 cm

Sol. (*c*) Radius of incircle = $\frac{2 \times \text{Area}}{\text{Perimeter}} \Rightarrow 2 = \frac{2 \times \text{Area}}{\text{Perimeter}} \Rightarrow \text{Perimeter} = \frac{2 \times 6}{2} = 6$ cm

159. The radii of two concentric circles are 5 cm and 3 cm. *AB* is diameter of the bigger circle and *AC* is a a tangent to the smaller circle touches it at *P*. Find the length of *BP*.

(*a*) $\sqrt{52}$ cm (*b*) $\sqrt{45}$ cm (*c*) $\sqrt{42}$ cm (*d*) $\sqrt{57}$ cm

Sol. (*a*) $AO = 5$ cm, $OP = 3$ cm

In ΔAOP, $AP = \sqrt{5^3 - 3^2} = 4$ cm

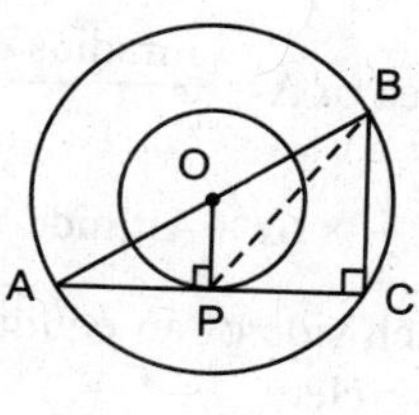

$\angle ACB = 90°$ (Angle in semi-circle)

$\angle APO = 90°$ ($\because OP \perp AC$)

In $\angle ABC$, O and P are mid points of AB and AC.

$\Rightarrow BC = 2OP = 2 \times 3$ cm $= 6$ cm

In ΔBCP, $BP = \sqrt{BC^2 + CP^2} = \sqrt{6^2 + 4^2} = \sqrt{36+16} = \sqrt{52}$

160. PA, QB and RC are three perpendiculars to AC, then $y = ?$

(*a*) $\dfrac{x+z}{xz}$ (*b*) $\dfrac{xz}{x+z}$ (*c*) $\dfrac{xz}{x-z}$ (*d*) $\dfrac{x-z}{xz}$

Sol. (*b*) $\Delta ABQ \sim \Delta ACR$

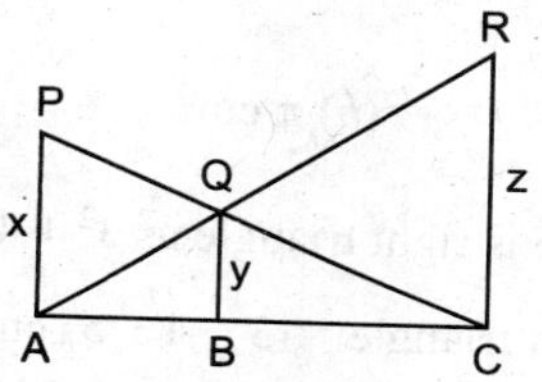

$\Rightarrow \dfrac{AB}{AC} = \dfrac{y}{z}$

$\Delta CBQ \sim \Delta CAP \Rightarrow \dfrac{BC}{AC} = \dfrac{y}{x}$

$$\frac{y}{z} + \frac{y}{x} = \frac{AB}{AC} + \frac{BC}{AC} = \frac{AB+BC}{AC} = \frac{AC}{AC} = 1 \Rightarrow y\left(\frac{1}{z} + \frac{1}{x}\right) = 1 \Rightarrow y\left(\frac{x+z}{xz}\right) = 1 \Rightarrow y = \frac{xz}{x+z}$$

161. ABC is a right-angled triangle. A circle is inscribed in it. If the lengths of two sides containing the right angle are 6 cm and 8 cm. Find the radius of the incircle?

(*a*) 1 cm (*b*) 2 cm (*c*) 3 cm (*d*) 4 cm

Sol. (*b*) $AC = \sqrt{6^2 + 8^2} = 10$ cm

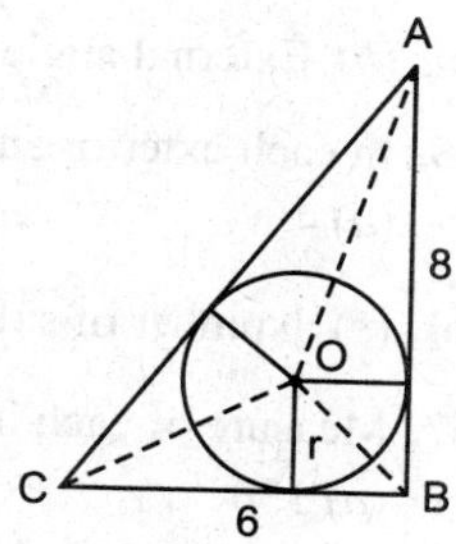

Area $\Delta ABC = \Delta AOB + \Delta BOC + \Delta COA$

$\Rightarrow \dfrac{1}{2} \times AB \times BC = \dfrac{1}{2} AB \times r + \dfrac{1}{2} BC \times r + \dfrac{1}{2} AC \times r$

$\Rightarrow \dfrac{1}{2} \times AB \times BC = \dfrac{1}{2}(AB + BC + AC) \times r$

$\Rightarrow \dfrac{1}{2} \times 6 \times 8 = \dfrac{1}{2} \times (6 + 8 + 10) \times r \Rightarrow 48 = 24\, r \Rightarrow r = 2$ cm

Alt: Area $\Delta ABC = \dfrac{1}{2} \times 6 \times 8 = 24$ cm^2

$AC^2 = 6^2 + 8^2 = 10^2$

Perimeter $= 6 + 8 + 10 = 24$ cm

Inradius $= \dfrac{2 \times \text{Area}}{\text{Perimeter}} = \dfrac{2 \times 24}{24} = 2$ cm

162. If the lengths of the sides of a triangle are in the ratio 4 : 5 : 6 and the inradius of the triangle is 3 cm, then the altitude of the triangle corresponding to the largest side as base is :

(SSC CGL Tier I 2013)

(*a*) 8 cm (*b*) 7.5 cm (*c*) 6 cm (*d*) 10 cm

Sol. (*b*) Let sides of a triangle be $4x$, $5x$, $6x$. Then perimeter $= 4x + 5x + 6x = 15x$

Area of $\Delta = \frac{\text{Inradius} \times \text{Perimeter}}{2} = \frac{3 \times 15x}{2} = \frac{45}{2}x$

$\therefore \frac{1}{2} \times 6x \times \text{Altitude} = \frac{45}{2}x \Rightarrow \text{Altitude} = \frac{45}{2}x \times 2 \times \frac{1}{6x} = 7.5 \text{ m}$

163. Each side of an equilateral triangle is 6 cm. Find difference of areas of its circumcircle and incircle.

(*a*) 8π cm^2 (*b*) 9π cm^2 (*c*) 10π cm^2 (*d*) 12π cm^2

Sol. (*b*) Radius of circumcirle $= \frac{6}{\sqrt{3}} = 2\sqrt{3}$ cm. Radius of incircle $= \frac{6}{2\sqrt{3}} = \sqrt{3}$ cm

$\pi R^2 - \pi r^2 = \pi[(2\sqrt{3})^2 - (\sqrt{3})^2] = \pi(12 - 3) = 9\pi \text{ cm}^2$

164. A triangle has sides of lengths 3 cm, 4 cm and 5 cm. What is the area of the circle inscribed in the triangle? **(CDS 2004)**

(*a*) 2π cm^2 (*b*) π cm^2 (*c*) $\frac{\pi}{2}$ cm^2 (*d*) $\frac{\pi}{4}$ cm^2

Sol. (*b*) The triangle is right triangle as $3^2 + 4^2 = 5^2$. Area of triangle $= \frac{1}{2} \times 3 \times 4 = 6 \text{ cm}^2$

Perimeter of triangle = (3 + 4 + 5) cm = 12 cm

Radius of incircle $= \frac{2 \times \text{Area}}{\text{Perimeter}} = \frac{2 \times 6}{12} = 1$ cm

Area of incircle $= \pi r^2 = \pi(1)^2 = \pi$

165. Each interior angle a regular octagon in radians is **(SSC CGL Tier II 2013)**

(*a*) $\frac{\pi}{4}$ (*b*) $\frac{3\pi}{4}$ (*c*) $\frac{2\pi}{3}$ (*d*) $\frac{\pi}{3}$

Sol. (*b*) External angle $= \frac{360}{8} = \frac{180}{4} = \frac{\pi}{4} \Rightarrow$ Internal angle $= \pi - \frac{\pi}{4} = \frac{3\pi}{4}$

166. In each exterior angle of a regular polygon is 60°, the number of sides of the polygon is :

(*a*) 4 (*b*) 5 (*c*) 6 (*d*) 3

Sol. (*c*) Number of sides $= \frac{360°}{60°} = 6$

167. Measure of each interior angle of a regular polygon can never be : **(SSC CHSL 2011)**

(*a*) 150° (*b*) 105° (*c*) 108° (*d*) 144°

Sol. (*b*) External angles = 180° – Internal angles = 30°, 75°, 72°, 36° respectively.

180 ÷ 75 is not a natural number

168. If each interior angle of a regular polygon is 144°, find the number of sides of the polygon is :

(*a*) 10 (*b*) 20 (*c*) 24 (*d*) 36

(SSC CHSL 2011)

Sol. (*a*) $\frac{n-2}{n} \times 180° = 144° \Rightarrow \frac{n-2}{n} = \frac{144°}{180°} = \frac{8}{10} \Rightarrow n = 10$

Alt: External angle = 180° – 144° = 36° ⇒ Number of sides = 360° ÷ 36° = 10

169. If each interior angle of a regular polygon is 135°, the number of sides of the polygon is :

(*a*) 5 (*b*) 6 (*c*) 8 (*d*) 9

Sol. (*c*) $\frac{n-2}{n} \times 180° = 135° \Rightarrow \frac{n-2}{n} = \frac{135°}{180°} = \frac{3}{4} = \frac{6}{8} \Rightarrow n = 8$

Alt: External angle = 180° – 135° = 45° ⇒ Number of sides = 360° ÷ 45° = 8

170. The ratio of each interior angle to each exterior angle of a regular polygon is 3 : 1. The number of sides of the polygon is : **(SSC CGL Tier I 2015)**

(*a*) 7 (*b*) 8 (*c*) 6 (*d*) 9

Sol. (*b*) Each external angle = $\frac{1}{4} \times 180° = 45 \Rightarrow$ Number of sides = 360° ÷ 45° = 8

171. An exterior angle of a regular polygon is 72°. The sum of all the interior angles is **(SSC CGL Tier I 2012)**

(*a*) 360° (*b*) 480° (*c*) 520° (*d*) 540°

Sol. (*d*) Number of sides = $\frac{360°}{72°} = 5$

Sum of interior angles = $(n-2) \times 180° = (5-2) \times 180° = 540°$

172. The difference between the interior and exterior angle of regular polygon is 36°. Find the number of sides of the polygon.

(*a*) 4 (*b*) 5 (*c*) 6 (*d*) 8

Sol. (*b*) External angle = $\frac{360°}{n}$, Internal angle = $180° - \frac{360°}{n}$

$\therefore\ 180° - \frac{360°}{n} - \frac{360°}{n} = 36 \Rightarrow 144 = \frac{720}{n} \Rightarrow n = 5$

Alt: Let external angle = x, then internal angle = $36 + x$

$\Rightarrow\ x + (36 + x) = 180° \ \Rightarrow\ 2x = 144 \ \Rightarrow\ x = 72°$

$\therefore$ Number of sides = 360° ÷ 72° = 5

173. The difference between the exterior and interior angles at a vertex of a regular polygon is 150°. The number of sides of the polygon is **(SSC CHSL 2011)**

(*a*) 10 (*b*) 15 (*c*) 24 (*d*) 30

Sol. (*c*) $x + (150° + x) = 180° \Rightarrow 2x = 30° \Rightarrow x = 15° \Rightarrow$ Number of sides = 360° ÷ 15° = 24.

174. If the difference between an exterior angle of a regular polygon on n sides and an exterior angle of another regular polygon of $(n + 1)$ sides is equal to 5°, then the value of n is **(SSC CHSL 2013)**

(*a*) 5 (*b*) 6 (*c*) 7 (*d*) 8

Sol. (*d*) From options, if $n = 5$, then difference = $\frac{360}{5} - \frac{360}{6} = 72° - 60° = 12°$

To get smaller difference, we should check larger option.

If $n = 8$, then difference = $\frac{360}{8} - \frac{360}{9} = 45° - 40° = 5°$

175. If each interior angle is double of each exterior angle of a regular polygon with n sides, then the value of n is **(SSC CGL Tier II 2012)**

(*a*) 5 (*b*) 6 (*c*) 8 (*d*) 10

Sol. (*b*) Sum of all exterior angles = 360°

Sum of all interior angles = $2 \times 360° = (n-2) \times 180° \Rightarrow n - 2 = 4 \Rightarrow n = 6$

Alternative Method: Exterior angle = $\frac{1}{3} \times 180° = 60°$ $\therefore$ Number of sides = $\frac{360}{60} = 6$.

176. Each interior angle of a regular polygon is three times its exterior angle, then the number of sides of the regular polygon is: **(SSC FCI 2012)**

(*a*) 7 (*b*) 8 (*c*) 9 (*d*) 10

Sol. (*b*) Exterior angle : Interior angle = 1 : 3 ⇒ Each exterior angle = $\frac{1}{4} \times 180 = 45°$

∴ Number of sides = 360° ÷ 45° = 8

177. The ratio of the number of sides of two regular polygons is 1 : 2. If each interior angle of first polygon is 120°, then the measure of each interior angle of the second polygon is : **(SSC CGL Tier II 2012)**

(*a*) 150° (*b*) 160° (*c*) 140° (*d*) 135°

Sol. (*a*) Exterior angle of first polygon = 180° – 120° = 60°

∴ Sides of first polygon = 360° ÷ 60° = 6

∴ Sides of second polygon = 2 × 6 = 12

Exterior angle of second polygon = 360° ÷ 12 = 30°

Interior angle of second polygon = 180° – 30° = 150°

178. Each of the five angles of a polygon is equal to 172° and the other angles are 160° each. The number of sides of the polygon is **(SSC CPO SI 2012)**

(*a*) 22 (*b*) 23 (*c*) 20 (*d*) 21

Sol. (*d*) Let sides of the polygon be *n*.

Then sum of angles = $(n-2) \times 180° = 5 \times 172° + (n-5) \times 160°$

$\Rightarrow 180n - 360° = 860° + 160n - 800° \Rightarrow 20n = 420° \Rightarrow n = 420 \div 20 = 21$ sides

179. The interior angles of a polygon are in A.P. The smallest angle is 120° and the common difference is 5°. Find the number of sides of the polygon.

(*a*) 5 (*b*) 7 (*c*) 9 (*d*) 11

Sol. (*c*) The angles of the polygon are 120°, 125°, 130°, 135°, 140°, ...

Option	*Sides*	*Sum of angles = (n – 2) × 180°*	*Sum = Middle term × No. of sides*
a	5	(5 – 2) × 180° = 540°	3rd term × 5 = 130° × 5 = 650°
b	7	(7 – 2) × 180° = 900°	4th term × 7 = 135° × 7 = 945°
c	9	(9 – 2) × 180° = 1260°	5th term × 9 = 140° × 9 = 1260°

When sides = 9, sum of angles = 1260° by both methods.

180. The ratio between the number of sides of two regular polygons is 1 : 2 and the ratio between their interior angles is 2 : 3. The number of sides of these polygons is respectively. **(SSC CGL Tier II 2012)**

(*a*) 4, 8 (*b*) 7, 14 (*c*) 6, 12 (*d*) 5, 10

Sol. (*a*) Let number of sides of polygons are *n*, 2*n*.

Then ratio of their angles = $\frac{n-2}{n} \times 180° : \frac{2n-2}{2n} \times 180° = \frac{n-2}{n} : \frac{n-1}{n} = (n-2) : (n-1)$

$\therefore \frac{n-2}{n-1} = \frac{2}{3} \Rightarrow 3n - 6 = 2n - 2 \Rightarrow n = 4 \Rightarrow 2n = 8$

181. The ratio between the number of sides of two polygons is 1 : 2 and the ratio between their interior angles is 3 : 4. The number of sides of these polygons are respectively.

(*a*) 4, 8 (*b*) 5, 10 (*c*) 6, 12 (*d*) 7, 14

Sol. (*b*) $\left(\frac{n-2}{n} \times 180°\right) : \left(\frac{2n-2}{2n} \times 180°\right) = 3 : 4$

$\Rightarrow (n-2) : (n-1) = 3 : 4 \Rightarrow 4(n-2) = 3(n-1) \Rightarrow 4n - 8 = 3n - 3 \Rightarrow n = 5 \Rightarrow 2n = 10$

182. Find the number of diagonals of a polygon of 100 sides.

(*a*) 50 (*b*) 100 (*c*) 4850 (*d*) 4950

Sol. (*c*) Diagonals = $\frac{100 \times 99}{2} - 100 = 4950 - 100 = 4850$

183. If the ratio of external angle and internal angle of a regular polygon is 1 : 4, the number of diagonals of the polygon is.

(*a*) 5 (*b*) 10 (*c*) 35 (*d*) 45

Sol. (*c*) External angle = $\frac{1}{5} \times 180° = 36°$ Number of sides = $\frac{360°}{36°} = 10$

Number of diagonals = $\frac{10 \times 9}{2} - 10 = 45 - 10 = 35$

184. A polygon has 27 diagonals. The number of sides of the polygon is :

(*a*) 8 (*b*) 9 (*c*) 10 (*d*) 12

Sol. (*b*) $\frac{n(n-3)}{2} = 27 \Rightarrow n(n-3) = 54 = 9 \times 6 \Rightarrow n = 9$

Alternative Method (from options):

It sides are 8, diagonals = $\frac{8 \times 5}{2} = 20$; If sides are 9, diagonals = $\frac{9 \times 6}{2} = 27$

185. *ABCDE* is a regular pentagon. Bisector of $\angle BAE$ meets *CD* at *M* and bisector of $\angle BCD$ meets *AM* at *P*, find $\angle CPM$.

(*a*) 36° (*b*) 54° (*c*) 126° (*d*) 108°

Sol. (*a*) Each interior angle = $\frac{3}{5} \times 180 = 108°$

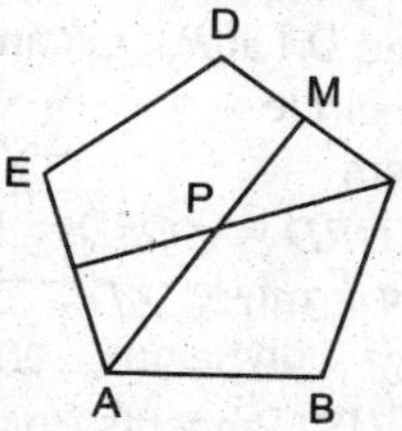

$\therefore \angle BAM = 108° \div 2 = 54°$

In quadrilateral *ABCM*,

$\angle CMA = 360° - \angle BAM - \angle ABC - \angle BCM$

$= 360° - 54° - 108° - 108° = 90°$

In Δ *CMP*, $\angle CPM = 180° - \angle MCP - \angle CMP = 180° - 54° - 90° = 36°$

186. In the figure, DEFG is a square and Δ ABC is a right angle triangle and $\angle ABC = 90°$. If CG = 9 cm and AF = 16 cm, find the length of FG.

(*a*) 8 cm (*b*) 12 cm

(*c*) 15 cm (*d*) 20 cm

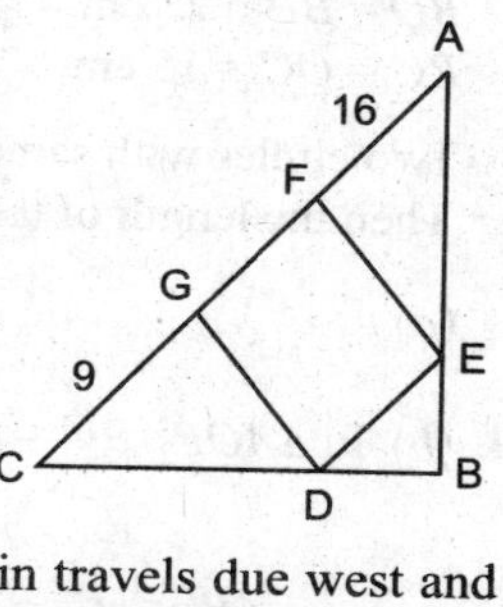

Sol. (*b*) Δ CGD ~ Δ EFA

$\Rightarrow \frac{CG}{EF} = \frac{GD}{FA} \Rightarrow \frac{9}{EF} = \frac{GD}{16} = EF^2 = 9 \times 16$ ($\because$ GD = EF)

$\Rightarrow EF = \sqrt{9 \times 16} = 3 \times 4 = 12$ cm

187. Two trains leaves a railway station at the same time. The first train travels due west and the second train due north. The first trains 5 km/hr faster than the second train. If after two hours, they are 50 km apart. The average speeds (in km/hr) of the two trains are :

(*a*) 10, 15 (*b*) 15, 20

(*c*) 15, 10 (*d*) 20, 15

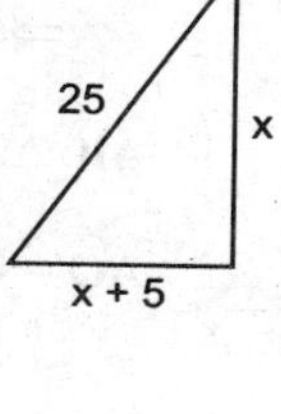

Sol. (*d*) Distance between trains in 1 hour = 50 ÷ 2 = 25 km

$25^2 = 15^2 + 20^2$

188. $ABCD$ is a rhombus. AB is produced to F and BA is produced to E such that $AB = AE = BF$. Then **(SSC CGL Tier I 2013)**

(a) $ED \parallel CF$ (b) $ED > CF$ (c) $ED \perp CF$ (d) $ED^2 + CF^2 = EF^2$

Sol. (c) $AE = AB, AB = AD \Rightarrow AE = AD \Rightarrow \angle 1 = \angle 2$

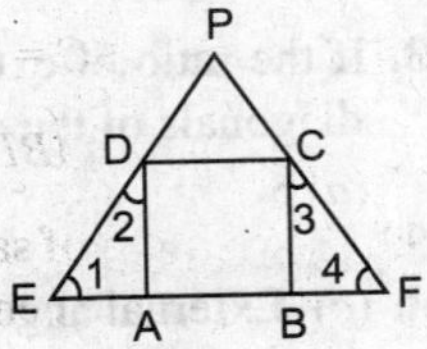

$BF = AB, AB = BC \Rightarrow BF = BC \Rightarrow \angle 3 = \angle 4$

$\angle 1 + \angle 2 = \angle DAB$ (external angle)

$\angle 3 + \angle 4 = \angle ABC$ (external angle)

$\angle DAB + \angle ABC = 180° \Rightarrow \angle 1 + \angle 4 = 90°$

In ΔEFP, $\angle P = 180° - (\angle 1 + \angle 4) = 180° - 90° = 90°$

$\Rightarrow EP \perp FP \Rightarrow ED \perp CF$ (Produced)

189. What is the length of the perpendicular drawn from the centre of circle of radius r on the chord of length $\sqrt{3}r$? **(CDS 2011)**

(a) r (b) $\sqrt{2}r$

(c) $r/2$ (d) $r/4$

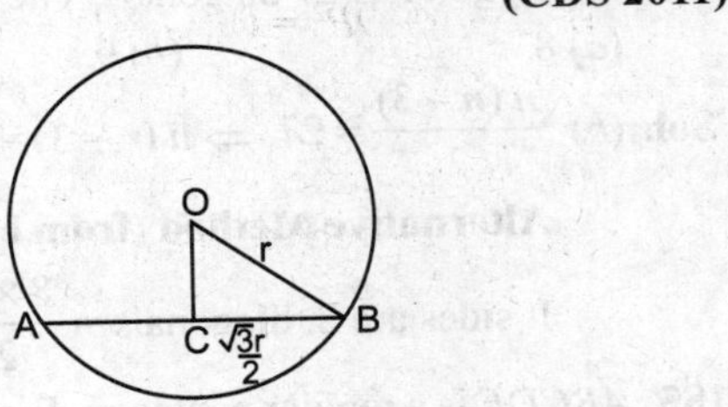

Sol. (c) $OC^2 = r^2 - \left(\frac{\sqrt{3}r}{2}\right)^2$

$= r^2 - \frac{3}{4}r^2 = \frac{1}{4}r^2 \Rightarrow OC = \frac{r}{2}$

190. $ABCD$ is a quadrilateral in which $\angle D = 90°$. A circle is drawn in it touching the sides AB, BC, CD and DA at P, Q, R and S respectively. If $BP = 23$ cm, $BC = 38$ cm, $CD = 23$ cm, find radius of the circle.

(a) 8 cm (b) 12 cm (c) 15 cm (d) 18 cm

Sol. (a) $\angle ORD = \angle OSD = 90°$ (Tangents)

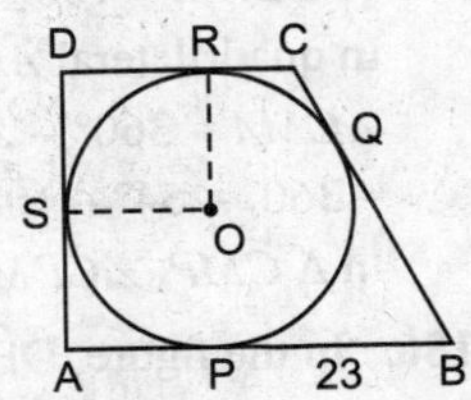

$\angle ROS = 360° - \angle D - \angle ORD - \angle OSD$

$= 360° - 90° - 90° - 90° = 90°$

$DR = DS$ (Tangents from same point D)

$OR = OS$ (Radii)

$\therefore SORD$ is a square

$BQ = BP = 23$ cm $QC = BC - BQ = 38 - 23 = 15$ cm

$RC = QC = 15$ cm $DR = CD - RC = 23 - 15 = 8$ cm $\therefore$ OR = DR = 8 cm

191. Two circles with same radius r intersect each other and passes through the centre of the other. Then the length of the common chord is **(SSC CGL Tier I 2012, CDS 2015)**

(a) r (b) $\sqrt{3}r$ (c) $\frac{\sqrt{3}}{2}r$ (d) $\sqrt{5}r$

Sol. (b) In ΔAOP, $AO = r, OP = \frac{r}{2}$

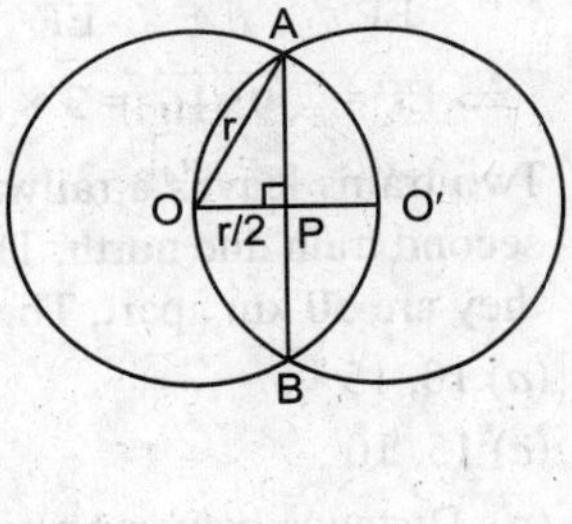

$$AP^2 = r^2 - \frac{r^2}{4} = \frac{3r^2}{4}$$

$$\Rightarrow \quad AP = \frac{\sqrt{3}r}{2} \qquad \therefore AB = 2 \times \frac{\sqrt{3}}{2}r = \sqrt{3}r$$

Alt: AOO′ is an equilateral triangle

$$AO = OO' = AO' = r \Rightarrow AP = \frac{\sqrt{3}}{2}r \Rightarrow AB = \sqrt{3}r$$

192. Each of two circles of equal radii with centres at A and B pass through the centre of one another. If they cut at C and D then $\angle DBC$ is equal to: **(SSC CGL Tier I 2013)**

(*a*) 60° (*b*) 100°
(*c*) 120° (*d*) 140°

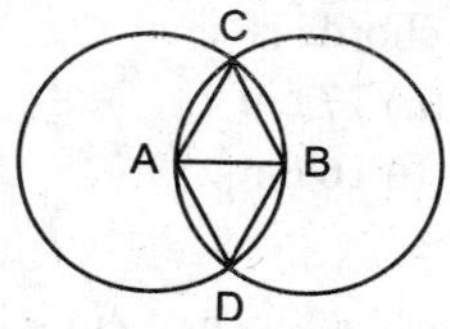

Sol. (*c*) $AC = BC = AC$ (radii) $\Rightarrow \angle ABC = 60°$.

Similarly $\angle ABD = 60° \Rightarrow \angle DBC = 60° + 60° = 120°$

193. Two circles of same radius 5 cm, intersect each other at *A* and *B*. If AB = 8 cm, then the distance between the centres is **(SSC CGL Tier I 2013)**

(*a*) 10 cm (*b*) 4 cm
(*c*) 6 cm (*d*) 8 cm

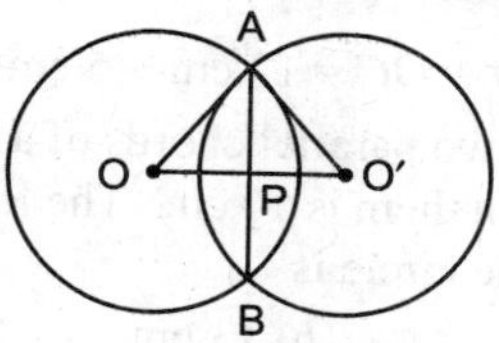

Sol. (*c*) $AB = 8$ cm $\Rightarrow AP = 8 \div 2 = 4$ cm

In ΔAOP, $OP^2 = OA^2 - AP^2 = 5^2 - 4^2 = 3^2$

$\therefore OO' = 2 \times OP = 2 \times 3$ cm $= 6$ cm

194. Two equal circles pass through each other's centre. If the radius of each circle is 5 cm, what is the length of the common chord? **(SSC CGL Tier I 2012)**

(*a*) $5\sqrt{3}$ (*b*) $10\sqrt{3}$

(*c*) $\frac{5\sqrt{3}}{2}$ (*d*) 5

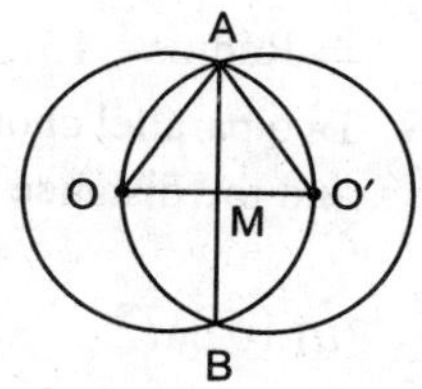

Sol. (*a*) $AO = OO' = AO'$ = radius = 5 cm

$\therefore AOO'$ is an equilateral Δ

AM = Altitude = $\frac{\sqrt{3}}{2} \times 5$ cm. $\Rightarrow AB = 2 \times \frac{\sqrt{3}}{2} \times 5$ cm $= 5\sqrt{3}$ cm

195. Each of the two circles of same radius *a* passes through the centre of the other. If the circles cut each other at the points *A* and *B* and *O* and *O′* be their centres, area of the quadrilateral *AOBO′* is : **(SSC CGL Tier I 2013)**

(*a*) $\frac{1}{2}a^2$ (*b*) $\frac{\sqrt{3}}{2}a^2$ (*c*) a^2 (*d*) $\frac{1}{4}a^2$

Sol. (*b*) In $\Delta AOO'$

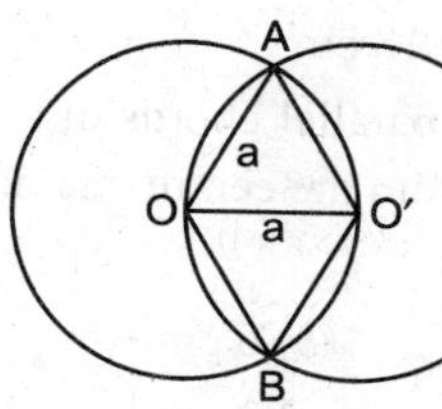

$AO = OO' = AO' = a$

Area of equilateral $\Delta AOO' = \frac{\sqrt{3}}{4}a^2$

Area of $AOBO' = 2 \times \frac{\sqrt{3}}{4}a^2 = \frac{\sqrt{3}}{2}a^2$

196. If a chord of length 16 cm is at a distance of 15 cm from the centre of the circle, then the length of the chord of the same circle which is a distance of 8 cm from the centre is equal to **(SSC CPO SI 2011)**

(*a*) 10 cm (*b*) 20 cm
(*c*) 30 cm (*d*) 40 cm

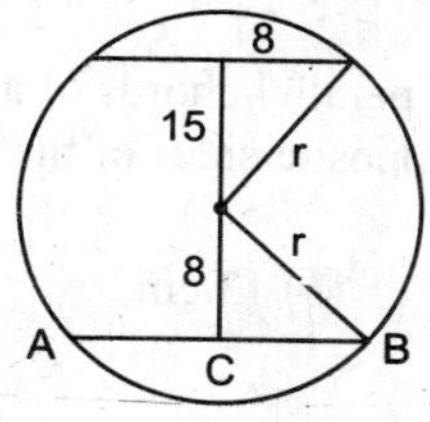

Sol. (*c*) $r^2 = 8^2 + 15^2 = 64 + 225 = 17^2$

$BC^2 = 17^2 - 8^2 = 15^2$

$AB = 2 \times BC = 2 \times 15$ cm $= 30$ cm

197. *AB* and *CD* are two parallel chords on the opposite sides of the centre of the circle. If $\overline{AB}$ = 10 cm, $\overline{CD}$ = 24 cm and the radius of the circle is 13 cm, the distance between the chords is **(SSC CHSL 2013, SSC CGL Tier I 2012)**

(*a*) 17 cm (*b*) 15 cm
(*c*) 16 cm (*d*) 18 cm

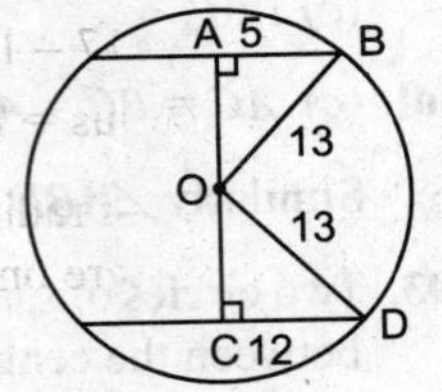

Sol. (*a*) In ΔABO, $OA = \sqrt{13^2 - 5^2} = 12$ cm

In ΔOCD, $OC = \sqrt{13^2 - 12^2} = 5$ cm

$\therefore\ AC = OA + OC = 12 \text{ cm} + 5 \text{ cm} = 17$ cm

198. *AB* and *CD* are two parallel chords of a circle lying on the opposite sides of the centre and the distance between them is 17 cm. The length of AB and CD are 10 cm and 24 cm respectively. The radius of the circle is **(SSC CGL Tier I 2015)**

(*a*) 18 cm (*b*) 13 cm (*c*) 15 cm (*d*) 9 cm

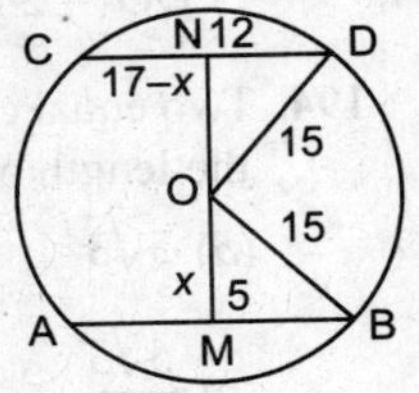

Sol. (*b*) In ΔMBO, $MB = 10 \div 2 = 5$ cm

In ΔOND, $ND = 24 \div 2 = 12$ cm

$MN = OM + ON = 17$ cm (given)

We know that, $5^2 + 12^2 = 13^2$

$\Rightarrow$ Radius = 13

199. Two parallel chords are drawn in a circle of diameter 30 cm. The length of one chord is 24 cm and the distance between the two chords is 21 cm. The length of the other chord is **(SSC CGL Tier I 2012)**

(*a*) 10 cm (*b*) 18 cm (*c*) 12 cm (*d*) 16 cm

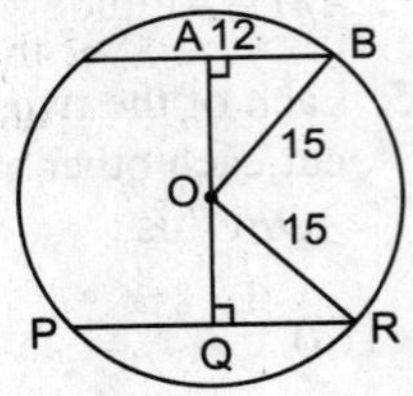

Sol. (*b*) $OA = \sqrt{15^2 - 12^2} = \sqrt{225 - 144} = 9$ cm

$OQ = (21 - 9)$ cm = 12 cm

$QR = \sqrt{15^2 - 12^2} = 9$ cm

$PR = 2 \times QR = 18$ cm

200. The length of two parallel chords of a circle are 6 cm and 8 cm. If the smaller chord is at a distance of 4 cm from the centre, the distance of the other chord from the centre is **(SSC CHSL 2013)**

(*a*) 5 cm (*b*) 6 cm
(*c*) 4 cm (*d*) 3 cm

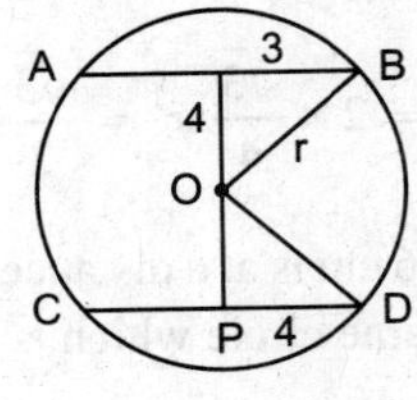

Sol. (*d*) $r^2 = 3^2 + 4^2 = 5^2$

$OP^2 = r^2 - 4^2 = 5^2 - 4^2 = 3^2$

201. *AB* and *CD* are two parallel chords of a circle such than *AB* = 10 cm and *CD* = 24 cm. If the chords are on the opposite sides of the centre and distance between them in 17 cm, then the radius of the circle : **(SSC CGL Tier I 2013)**

(*a*) 12 cm (*b*) 13 cm (*c*) 10 cm (*d*) 11 cm

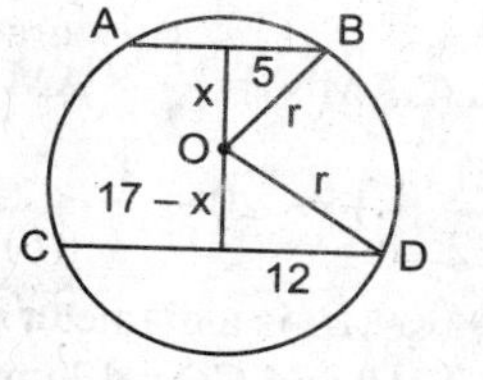

Sol. (*b*) Trick :

$$5^2 + 12^2 = 13^2$$
$$12^2 + (17 - 12)^2 = 13^2$$
$$\Rightarrow \text{radius} = 13 \text{ cm}$$

202. In a circle of radius 17 cm, two parallel chords of lengths 30 cm and 16 cm are drawn. If both the chords are on the same side of the centre, then the distance between the chords is **(SSC CGL Tier I 2013)**

(*a*) 9 cm (*b*) 7 cm
(*c*) 23 cm (*d*) 11 cm

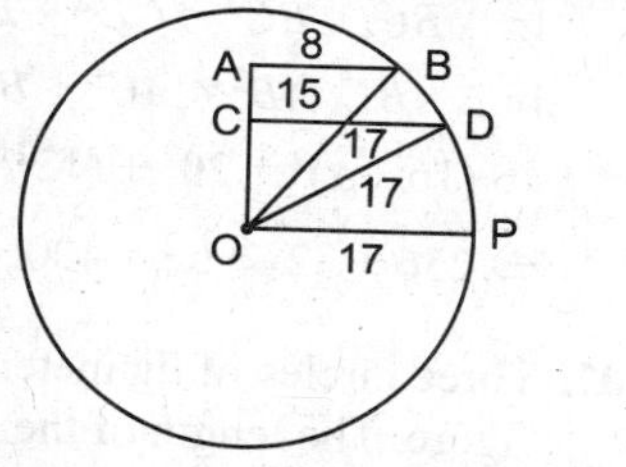

Sol. (*b*) $AO^2 = OB^2 - BA^2 = 17^2 - 8^2 = 15^2$

$CO^2 = OD^2 - CD^2 = 17^2 - 15^2 = 8^2$

$\therefore AC = (15 - 8) \text{ cm} = 7 \text{ cm}$

203. *AB* and *CD* are two chords of a circle such that *AB* = 6 cm, *CD* = 12 cm and *AB* || *CD*. If distance between *AB* and *CD* is 3 cm, find the radius of the circle. **(SSC CGL Tier I 2013)**

(*a*) 6 cm (*b*) 9 cm (*c*) 4.5 cm (*d*) $3\sqrt{5}$ cm

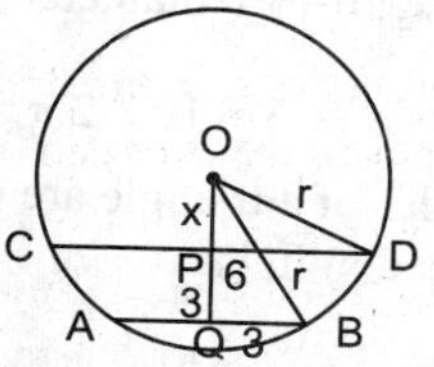

Sol. (*d*) $r^2 = OD^2 = OP^2 + PD^2 = x^2 + 6^2$

$r^2 = OB^2 = OQ^2 + QB^2 = (x + 3)^2 + 3^2$

$\Rightarrow r^2 = x^2 + 6^2 = (x + 3)^2 + 3^2 \Rightarrow x = 3$

$\therefore r^2 = 3^2 + 6^2 = 9 + 36 = 45 \Rightarrow r = \sqrt{45} = 3\sqrt{5}$ cm

204. The length of the common chord of two intersecting circles is 24 cm. If the diameters of the circles are 30 and 26 cm, then the distance between the centres in cm is **(SSC CGL Tier II 2012)**

(*a*) 15 (*b*) 16
(*c*) 13 (*d*) 14

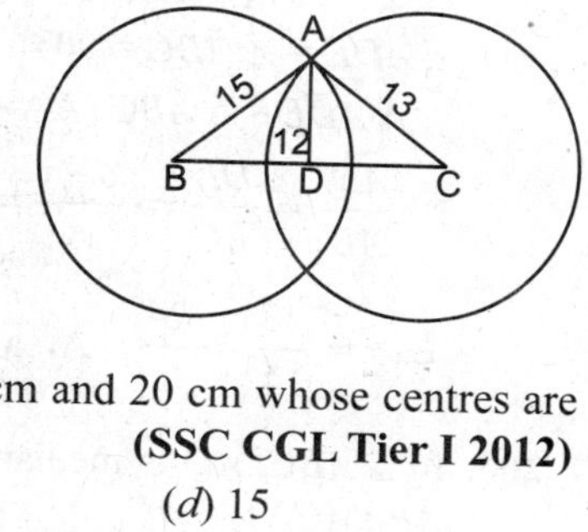

Sol. (*d*) $BD^2 = AB^2 - AD^2 = 15^2 - 12^2 = 9^2$

$DC^2 = AC^2 - AD^2 = 13^2 - 12^2 = 5^2$

$\therefore BC = BD + DC = 9 + 5 = 14$ cm

205. The length of the common chord of two circles of radii 15 cm and 20 cm whose centres are 25 cm apart is (in cm) : **(SSC CGL Tier I 2012)**

(*a*) 20 (*b*) 24 (*c*) 25 (*d*) 15

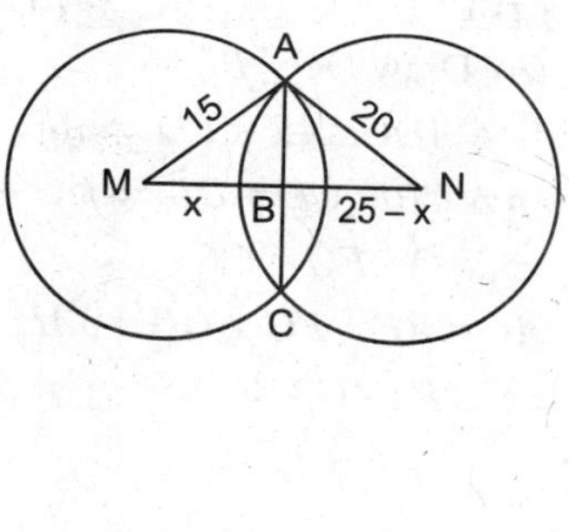

Sol. (*b*) In ΔAMB, $AB^2 = 15^2 - x^2$

In ΔANB, $AB^2 = 20^2 - (25 - x)^2$

$\Rightarrow 15^2 - x^2 = 20^2 - (25 - x)^2$

$\Rightarrow 225 - x^2 = 400 - 625 - x^2 + 50x$

$\Rightarrow 50x = 450 \Rightarrow x = 9$

$\therefore AB^2 = 15^2 - x^2 = 15^2 - 9^2 = 225 - 81 = 144 = 12^2$

$\therefore AC = 2AB = 2 \times 12 \text{ cm} = 24 \text{ cm}$

Trick: $15^2 + 20^2 = 25^2 \Rightarrow$ AMN is right triangle.

Area (ΔAMN) $= \frac{1}{2} \times AM \times AN = \frac{1}{2} \times MN \times AB$

$\Rightarrow \frac{1}{2} \times 15 \times 20 = \frac{1}{2} \times 25 \times x \Rightarrow x = \frac{15 \times 20}{25} = 12 \Rightarrow AC = 2x = 24$ cm

206. Suppose AB is a diameter of a circle, whose centre is at O and C be any point on the circle. If $CD \perp AB$ and $CD = 12$ cm, $AD = 16$ cm, then BD is **(SSC FCI Asstt. 2012)**

(*a*) 10 cm (*b*) 12 cm (*c*) 8 cm (*d*) 9 cm

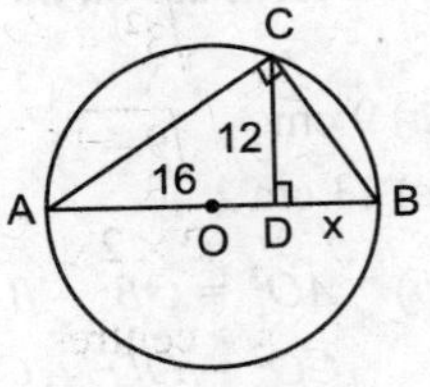

Sol. (*d*) In $\Delta ADC, AC^2 = AD^2 + DC^2 = 16^2 + 12^2 = 20^2$

In $\Delta BCD, BC^2 = CD^2 + BD^2 = 12^2 + x^2$

In $\Delta ABC, AB^2 = AC^2 + BC^2$

$\Rightarrow (16 + x)^2 = 20^2 + (12^2 + x^2)$

$\Rightarrow 256 + 32x + x^2 = 400 + 144 + x^2 \Rightarrow 32x = 288 \Rightarrow x = 9$ cm

207. Three circles of diameter 10 cm each, are bound together by a rubber band, as shown in the figure. The length of the rubber band, in cm, if it is stretched as shown, is

(*a*) 30 (*b*) $30 + 10\pi$

(*c*) 10π (*d*) $60 + 20\pi$

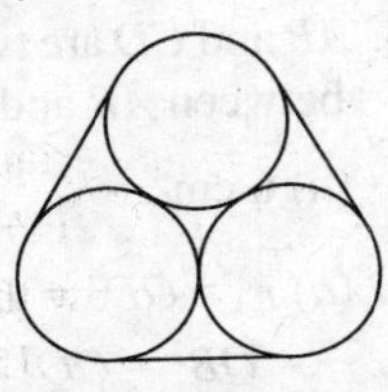

Sol. (*b*) Length $= 3 \times$ diameter $+ 3 \times \frac{1}{3} 2\pi r$

$= 3 \times 10 + 2.\pi.5 = 30 + 10\pi$

208. The sides of a triangle are 6 cm, 8 cm and 10 cm. The area of the greatest square that can be inscribed in it, is **(SSC CGL Tier I 2008)**

(*a*) 18 cm^2 (*b*) 15 cm^2 (*c*) $\frac{2304}{49}$ cm^2 (*d*) $\frac{576}{49}$ cm^2

Sol. (*d*) $6^2 + 8^2 = 10^2 \Rightarrow \Delta ABC$ is right triangle

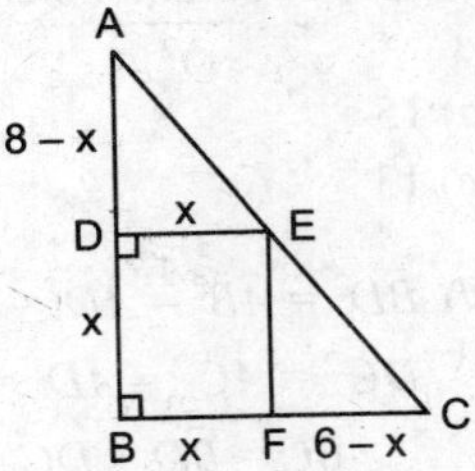

In ΔADE and ΔABC

$\angle A = \angle A$ (Common)

$\angle ADE = \angle ABC = 90°$

$\therefore \Delta ADE \sim \Delta ABC$ (AA-criterion)

$\therefore \frac{AD}{AB} = \frac{DE}{BC} \Rightarrow \frac{8-x}{8} = \frac{x}{6} \Rightarrow 48 - 6x = 8x \Rightarrow 14x = 48$

$\Rightarrow x = \frac{48}{14} = \frac{24}{7}$. Area of square $= \left(\frac{24}{7}\right)^2 = \frac{576}{49}$ cm^2

209. In ΔABC, AD is median and E is the mid-point of AD . BE produced meets AC in F, then $AF : AC$

(*a*) 1 : 2 (*b*) 1 : 3 (*c*) 1 : 4 (*d*) None of these

Sol. (*b*) Draw $DG \parallel EF$

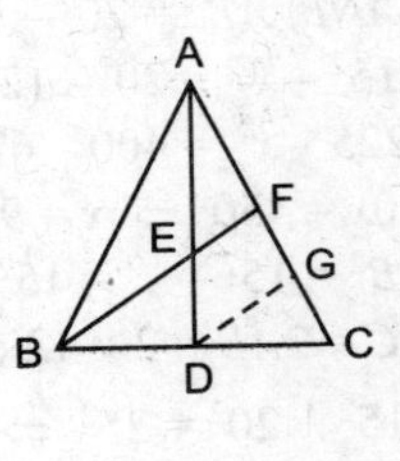

In ΔADG, $AE = ED \Rightarrow AF = FG$

In ΔCBF, $CD = DB \Rightarrow CG = GF$

$\Rightarrow AF = FG = GC$

$AC = AF + FG + GC = 3AF$

$\Rightarrow AF : AC = 1 : 3$

210. Two circles touch each other internally. Their radii are 2 cm and 3 cm. The biggest chord of the greater circle which is outside the inner circle is of length **(SSC CGL Tier I 2013)**

(*a*) $2\sqrt{2}$ cm (*b*) $3\sqrt{2}$ cm (*c*) $2\sqrt{3}$ cm (*d*) $4\sqrt{2}$ cm

Sol. (*d*) PQ = Diameter of smaller circle = 4 cm

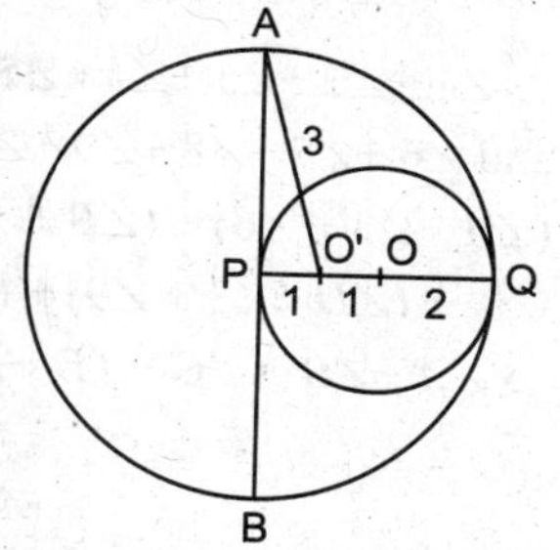

$PO' = (4 - 2 - 1)$ cm = 1 cm

$AO' = 3$ cm (radius of bigger circle)

$$AP = \sqrt{3^2 - 1^2}$$
$$= \sqrt{9-1} = \sqrt{8} = 2\sqrt{2} \text{ cm}$$
$$AB = 2 \times 2\sqrt{2} \text{ cm} = 4\sqrt{2} \text{ cm}$$

211. X and Y are centres of circles of radii 9 cm and 2 cm respectively $XY = 17$ cm. Z is centre of a circle of radius r cm which touches the above circles externally. Given that $\angle XZY = 90°$, the value of r is **(SSC FCI 2012, SSC CGL Tier II 2012)**

(*a*) 9 cm (*b*) 8 cm (*c*) 13 cm (*d*) 6 cm

Sol. (*d*) In right-triangle XYZ, $XZ^2 + YZ^2 = XY^2$

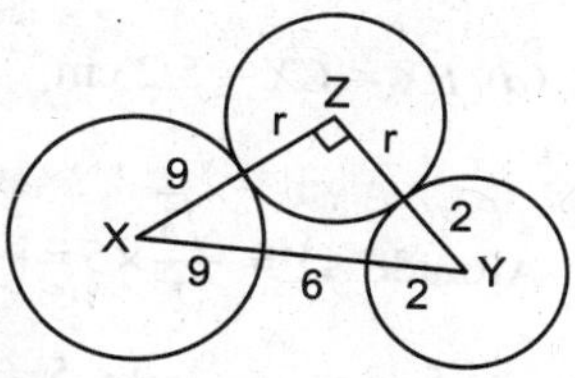

$\Rightarrow (r+9)^2 + (r+2)^2 = 17^2$

[We know that $8^2 + 15^2 = 17^2$]

$\Rightarrow r + 9 = 15$ and $r + 2 = 8 \Rightarrow r = 6$ cm

212. AB is a chord of length 8 cm of a circle of radius 5 cm. Tangents PA and PB are drawn to the circle. Find the length of PA.

(*a*) 9 cm (*b*) 20/3 cm (*c*) 8 cm (*d*) None of these

Sol. (*b*) In ΔARO, $\angle ARO = 90°$

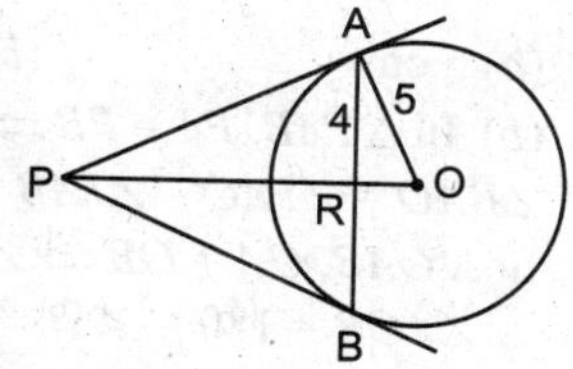

$$\Rightarrow OR = \sqrt{AO^2 - AR^2} = \sqrt{5^2 - 4^2} = 3 \text{cm}$$

$$\Rightarrow \Delta PAO \sim \Delta ARO \Rightarrow \frac{AP}{AR} = \frac{AO}{RO} \Rightarrow \frac{AP}{4} = \frac{5}{3} \Rightarrow AP = \frac{20}{3}$$

213. In the given figure, from a point P, 13 cm away from the centre O of a circle of radius 5 cm, tangents PM and PN are drawn. Find the length of tangent AB.

(*a*) $\frac{10}{3}$ cm (*b*) $\frac{20}{3}$ cm (*c*) $\frac{40}{3}$ cm (*d*) 6 cm

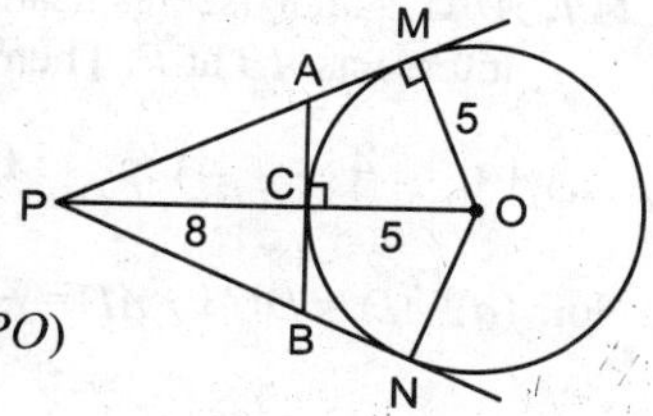

Sol. (*b*) $PM^2 = OP^2 - OM^2 = 13^2 - 5^2 = 12^2 \Rightarrow PM = 12$ cm

$PM = PN$, $AM = AC$, $BC = BN$

$\Delta PCA \sim \Delta PMO$ ($\because \angle ACP = \angle OMP = 90°$ and $\angle CPA = \angle MPO$)

$$\Rightarrow \frac{AC}{OM} = \frac{PC}{PM} \Rightarrow \frac{AC}{5} = \frac{8}{12} \Rightarrow AC = \frac{8}{12} \times 5 = \frac{10}{3}$$

$$\Rightarrow AB = 2AC = 2 \times \frac{10}{5} = \frac{20}{3} \text{ cm}$$

214. In the given figure, find $\angle A + \angle B + \angle C + \angle D + \angle E$ **(SSC CHSL I 2013)**

(*a*) 90° (*b*) 180° (*c*) 270° (*d*) 360°

Sol. (*b*) PQRST is a polygon and therefore, sum of exterior angles is 360°.

$\therefore \angle 1 + \angle 2 + \angle 3 + \angle 4 + \angle 5 = 360°$.

and $\angle 6 + \angle 7 + \angle 8 + \angle 9 + \angle 10 = 360°$

$(\angle A + \angle 5 + \angle 6) + (\angle B + \angle 1 + \angle 7) + (\angle C + \angle 2 + \angle 8) + (\angle D + \angle 3 + \angle 9) + (\angle E + \angle 4 + \angle 10) = 180° \times 5$

$\therefore \angle A + \angle B + \angle C + \angle D + \angle E = 900° - 360° - 360° = 180°$

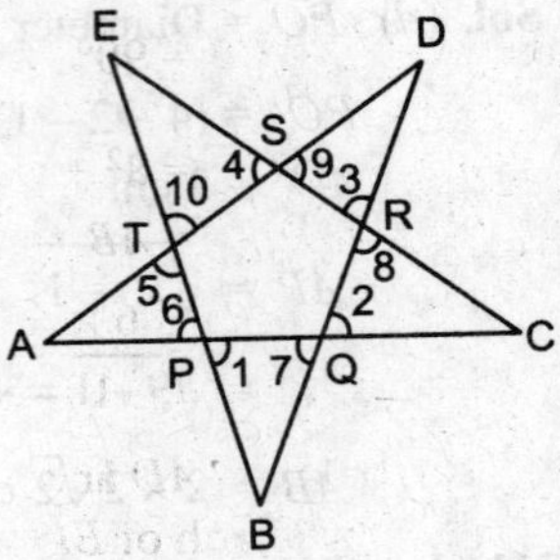

215. *ABC* is an equilateral triangle, with a side 5 cm, inscribed in a circle. The bisector of $\angle A$ meets *BC* in *X* and the circle in *Y*. Find the length of *XY*.

(*a*) $5\sqrt{3}$ cm (*b*) $2\sqrt{3}$ cm (*c*) $\frac{5}{\sqrt{3}}$ cm (*d*) $\frac{5}{2\sqrt{3}}$ cm

Sol. (*d*) $BX = CX = 5/2$ cm

Altitude $AX = \frac{\sqrt{3}}{2} \times 5 = \frac{5\sqrt{3}}{2}$ cm $\quad AX \times XY = BX \times XC$

$\Rightarrow \frac{5\sqrt{3}}{2}$ cm $\times XY = \frac{5}{2} \times \frac{5}{2} \Rightarrow XY = \frac{25}{4} \times \frac{2}{5\sqrt{3}} = \frac{5}{2\sqrt{3}}$ cm

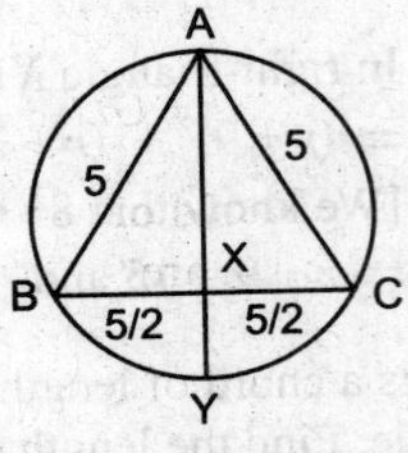

216. If the length of a chord of a circle, which makes an angle 45° with the tangent drawn at one end point of the chord, is 6 cm, then the radius of the circle is **(SSC CGL Tier I 2012)**

(*a*) 5 cm (*b*) $3\sqrt{2}$ cm (*c*) 6 cm (*d*) $6\sqrt{2}$ cm

Sol. (*b*) In Δ *PAB*, $PA = PB \Rightarrow \angle PBA = \angle PAB = 45°$

$\angle BAO = \angle PAO - \angle PAB = 90° - 45° = 45°$

In Δ *OAB*, $OA = OB \Rightarrow \angle OBA = \angle OAB = 45°$

$\therefore \angle AOB = 180° - 45° - 45° = 90°$

$\Rightarrow OA^2 + OB^2 = AB^2 \Rightarrow OA^2 = AB^2 \div 2$

$= 6^2 \div 2 = 18 \Rightarrow OA = \sqrt{18} = 3\sqrt{2}$ cm

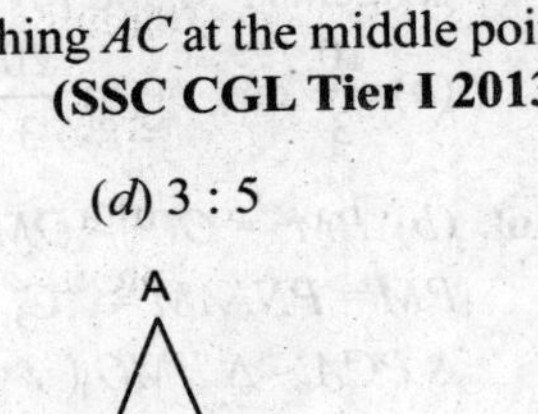

217. *ABC* is an isosceles triangle with *AB* = *AC*. A circle through *B* touching *AC* at the middle point intersects *AB* at *P*. Then *AP* : *AB* is : **(SSC CGL Tier I 2013)**

(*a*) 1 : 4 (*b*) 4 : 1 (*c*) 2 : 3 (*d*) 3 : 5

Sol. (*a*) $AD = DC \Rightarrow AD = \frac{1}{2}AC = \frac{1}{2}AB$

$AD^2 = AP \times AB \Rightarrow \left(\frac{AB}{2}\right)^2 = AP \times AB$

$AP = \frac{AB^2}{4} \times \frac{1}{AB} = \frac{AB}{4} \Rightarrow AP : AB = 1 : 4$

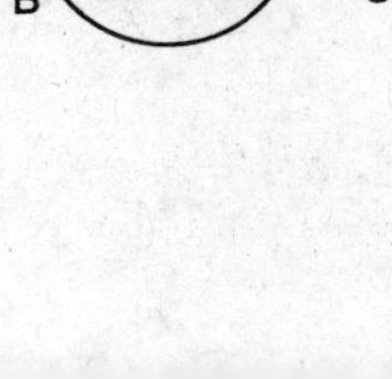

218. AB is a diameter of the circumcircle of ΔAPB; N is the foot of the perpendicular drawn from the point P on AB. If $AP = 8$ cm and $BP = 6$ cm, then the length of BN is

(SSC CGL Tier I 2012)

(*a*) 3.6 cm (*b*) 3 cm (*c*) 3.4 cm (*d*) 3.5 cm

Sol. (*a*) $\angle APB = 90°$ (Angle in semi-circle)

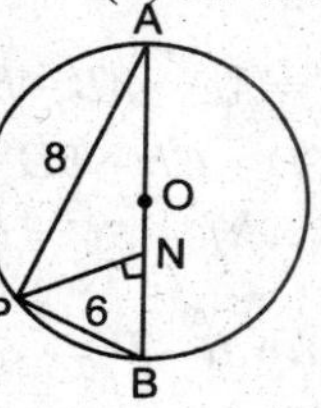

$$AB^2 = 8^2 + 6^2 = 10^2$$
$$PB^2 = AB \times NB$$
$$\Rightarrow \quad NB = \frac{6 \times 6}{10} = 3.6 \text{ cm}$$

219. Two medians AD and BE of ΔABC intersect at G at right angles. If $AD = 9$ cm and $BE = 6$ cm, then the length of BD, in cm, is

(SSC CPO SI 2011)

(*a*) 10 (*b*) 6 (*c*) 5 (*d*) 3

Sol. (*c*) $GD = \frac{1}{3} \times AD = \frac{1}{3} \times 9 = 3$ cm

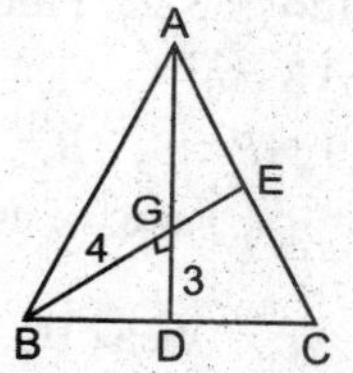

$$BG = \frac{2}{3} \times BE = \frac{2}{3} \times 6 = 4 \text{ cm}$$
$$BD^2 = BG^2 + GD^2 = 4^2 + 3^2 = 5^2$$

220. G is the centroid of ΔABC. The medians AD and BE intersect at right angles. If the lengths of AD and BE are 9 cm and 12 cm respectively; then the length of AB (in cm) is :

(SSC CGL Tier I 2015)

(*a*) 10 (*b*) 6 (*c*) 5 (*d*) 3

Sol. (*a*) $AG = \frac{2}{3} \times AD = \frac{2}{3} \times 9 = 6$ cm

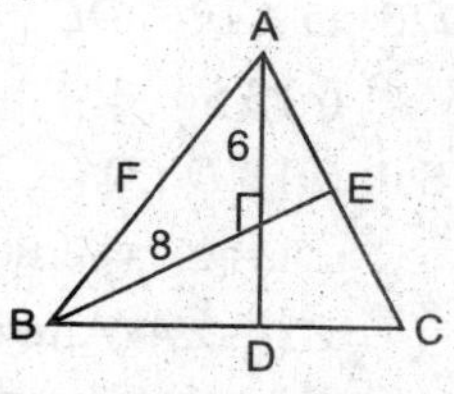

$$BG = \frac{2}{3} \times BE = \frac{2}{3} \times 12 = 8 \text{ cm}$$
$$AB^2 = AM^2 + BM^2 = 6^2 + 8^2 = 10^2$$

221. The radius of the circumcircle of a right angled triangle is 15 cm and the radius of its inscribed circle is 6 cm. Find the sides of the triangle.

(SSC CGL Tier I 2013)

(*a*) 24, 36, 30 (*b*) 30, 40, 41 (*c*) 18, 24, 30 (*d*) 30, 24, 25

Sol. (*c*) Radius of circumcircle $= \frac{\text{Hypotenuse}}{2} = 15$ cm $\Rightarrow$ Hypotenuse = 30 cm

Hypotenuse is largest side of a right triangle. In option (*c*), $18^2 + 24^2 = 30^2$. and $\frac{18 + 24 - 30}{2} = 6$

222. Find the area of a right angled triangle, if the radius of its circumcircle is 10 cm and altitude drawn to hypotenuse is 8 cm.

(*a*) 32 cm^2 (*b*) 40 cm^2 (*c*) 80 cm^2 (*d*) 160 cm^2

Sol. (*c*) Diameter AC = 2 AO = 2 × 10 = 20 cm

Area Δ ABC $= \frac{1}{2} \times AC \times BD = \frac{1}{2} \times 20 \times 8 = 80$ cm^2

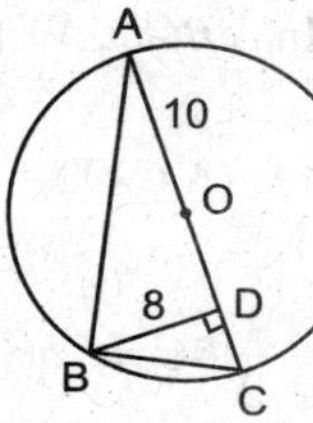

223. In the figure a, b, c are the sides of a right triangle and r is the radius of its incircle, then the radius 'r' in terms of a, b, c is :

(*a*) $\frac{a+b+c}{2}$ (*b*) $\frac{a+b-c}{2}$ (*c*) $\frac{a+c-b}{2}$ (*d*) $\frac{b+c-a}{2}$

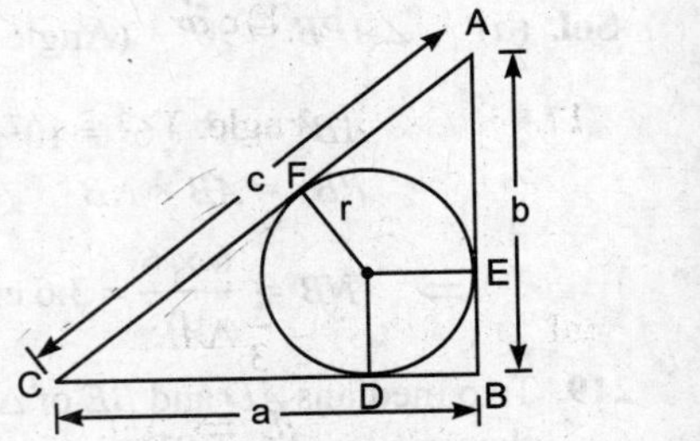

Sol. (*b*) $AF = AE = AB - BE = b - r$

$CF = CD = CB - BD = a - r$

$c = AC = AF + FC = (b - r) + (a - r)$

$\Rightarrow c = a + b - 2r \Rightarrow 2r = a + b - c \Rightarrow r = \frac{a+b-c}{2}$

224. If area of a triangle ABC is 32 cm². PQ is drawn to parallel of BC such that it divides AB in the ratio of 3 : 5. Find the area of ΔPQB.

(*a*) 8 cm² (*b*) 16 cm² (*c*) 7.5 cm² (*d*) 3.75 cm²

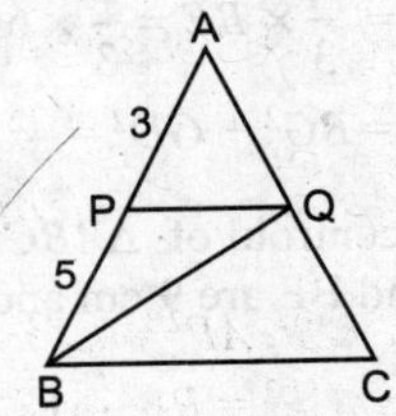

Sol. (*c*) $AP = 3, AB = 3 + 5 = 8$

$\therefore$ Ratio of areas of ΔAPQ and $\Delta ABC = 3^2 : 8^2 = 9 : 64$

$\therefore$ Area of $\Delta APQ = \frac{9}{64} \times 32 = 4.5 \text{ cm}^2$

Ratio of areas of ΔAPQ and $\Delta PQB = 3 : 5$

$\therefore$ Area of $\Delta PQB = \frac{5}{3} \times 4.5 \text{ cm}^2 = 7.5 \text{ cm}^2$

225. In ΔABC, $DE \parallel BC$ and $AD : DB = 3 : 4$. Find the ratio of areas of ΔDEF and ΔBFC.

(*a*) 3 : 4 (*b*) 9 : 16 (*c*) 9 : 25 (*d*) 9 : 49

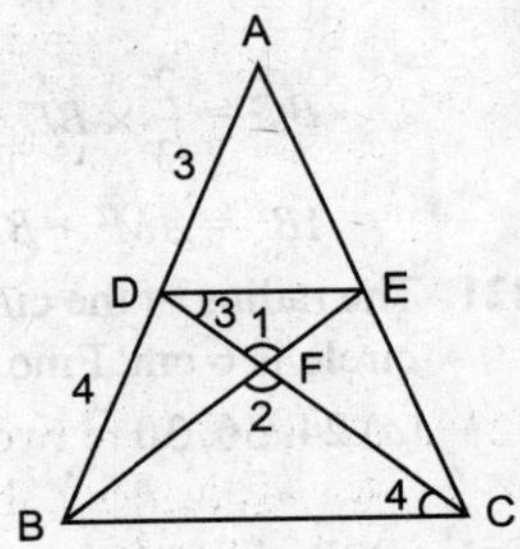

Sol. (*d*) $AD : DB = 3 : 4 \Rightarrow AD : AB = 3 : (3 + 4) = 3 : 7$

$\angle 1 = \angle 2$ (Vertically opposite angles)

$\angle 3 = \angle 4$ (Alternate angles)

$\therefore \angle DEF \sim \Delta CBF$ (AA-criterion)

Also $\Delta ADE \sim \Delta ABC$

$\therefore \frac{DE}{BC} = \frac{AD}{AB} = \frac{3}{7}$

$\therefore \frac{\Delta DEF}{\Delta BFC} = \frac{DE^2}{BC^2} = \frac{3^2}{7^2} = \frac{9}{49}$

226. $ABCD$ is a parallelogram and X and Y are mid-points of sides BC and CD respectively. Then $ar\ (\Delta\ AXY) : ar$ (Parallelogram $ABCD$) = ?

(*a*) 1 : 3 (*b*) 1 : 4 (*c*) 3 : 5 (*d*) 3 : 8

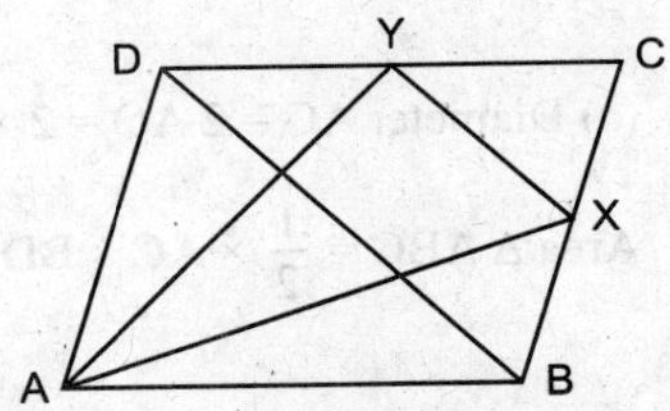

Sol. (*d*) In $\Delta\ BCD$, $XY \parallel BD$ and $XY = \frac{1}{2}$BD

$\Rightarrow$ area $(\Delta\ CXY) = \frac{1}{4} ar\ (\Delta\ BCD) = \frac{1}{8} ar\ (ABCD)$

$BX = \frac{1}{2} BC \Rightarrow ar(\Delta\ ABX) = \frac{1}{4} ar(ABCD)$

Similarly, $ar(\Delta AYD) = \frac{1}{4}\, ar(ABCD)$

$ar(\Delta AXY) = ar(ABCD) - ar(\Delta ABX) - ar(\Delta CXY) - ar(\Delta AYD) = 1 - \frac{1}{4} - \frac{1}{8} - \frac{1}{4} = \frac{3}{8}.$

$\Rightarrow ar(\Delta AXY) : ar(ABCD) = 3 : 8$

227. *ABC* is a triangle. The medians *CD* and *BE* intersect each other at *O*. Then $\Delta ODE : \Delta ABC$ is **(SSC CGL Tier I 2012)**

(*a*) 1 : 3 (*b*) 1 : 4 (*c*) 1 : 6 (*d*) 1 : 12

Sol. (*d*) $\Delta BOC = \frac{1}{3} \Delta ABC$

$BC = 2DE \Rightarrow \Delta DOE = \frac{1}{4} \Delta BOC$

$\therefore \Delta DOE = \frac{1}{4} \times \frac{1}{3} \times \Delta ABC$

$\therefore \Delta DOE : \Delta ABC = 1 : 12$

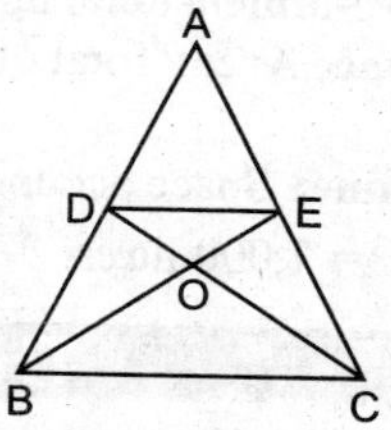

228. *N* is the foot of the perpendicular from a point *P* of a circle with radius 7 cm, on a diameter *AB* of the circle. If the length of the chord *PB* is 12 cm, the distance of the point *N* from the point *B* is **(SSC CGL Tier I 2013)**

(*a*) $3\frac{5}{7}$ cm (*b*) $10\frac{2}{7}$ cm (*c*) $6\frac{5}{7}$ cm (*d*) $12\frac{2}{7}$ cm

Sol. (*b*) In ΔAPB, $AP^2 = AB^2 - BP^2 = 14^2 - 12^2 = 196 - 144 = 52$

In ΔPNB, $NP^2 = PB^2 - NB^2 = 12^2 - (7 + x)^2$

$= 144 - 49 - 14x - x^2 = 95 - 14x - x^2$

In ΔANP, $NP^2 = AP^2 - AN^2$

$= 52 - (7 - x)^2 = 52 - (49 - 14x + x^2) = 3 + 14x - x^2$

$\Rightarrow 95 - 14x - x^2 = 3 + 14x - x^2 \Rightarrow 28x = 92 \Rightarrow x = \frac{92}{28} = \frac{23}{7}$

$\therefore\ NB = 7 + x = 7 + \frac{23}{7} = 7 + 3\frac{2}{7} = 10\frac{2}{7}$ cm

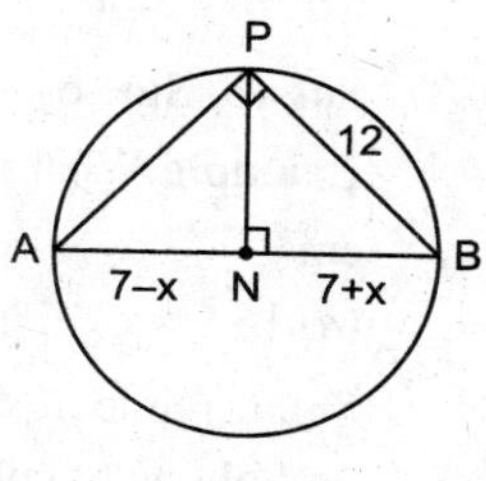

230. *AB* and *BC* are two chords of a circle with centre *O*. If *P* and *Q* are the mid-points of *AB* and *BC* respectively, then the quadrilateral *OQBP* must be **(SSC CGL Tier I 2012)**

(*a*) a rectangle (*b*) a square (*c*) a rhombus (*d*) concylic

Sol. (*d*) $OP \perp AB$ and $OQ \perp BC$

$\therefore\ \angle OPB + \angle OQB = 90° + 90° = 180°$

$\therefore\ \angle POQ + \angle PBQ = 360° - 180° = 180°$

$\therefore$ *OQBP* is concylic

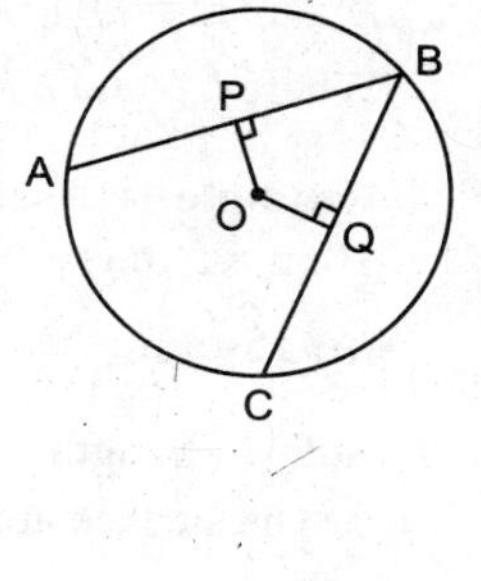

CHAPTER

31

MENSURATION – THREE DIMENSIONAL

Three-dimensional figures: These figures have three dimensions viz. length, breadth and height.

Surface Area: Total of surface areas of each side of the figure. It is always measured in square units.

Volume: Space occupied by the object is called its volume. It is always measured in cubic units.

$1\ m^3 = 1,000$ litres $\quad 1\ cm^3 = 1$ ml $\quad 1,000\ cm^3 = 1$ litre $= 1\ dm^3$

THREE - DIMENSIONAL FIGURES

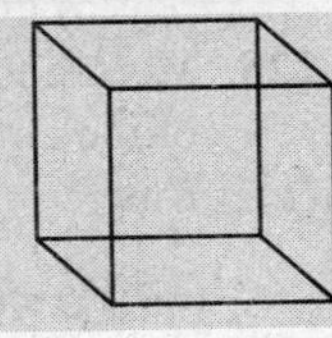

1. **Cube**: A figure with six square sides.

   Surface Area $= 6 \times (\text{Side})^2$

   Volume $= (\text{Side})^3$

   Diagonal $= \sqrt{3} \times$ Side

Example : The length of a diagonal of a cube with volume 729 cm^3 is **(SSC FCI 2012)**

(*a*) $4\sqrt{3}$ cm (*b*) $9\sqrt{3}$ cm (*c*) $\sqrt{3}$ cm (*d*) 1 cm

Sol. (*b*) Side of cube $= \sqrt[3]{729} = 9$ cm. Diagonal of cube $= 9\sqrt{3}$ cm

Example: What is the volume of a cube (in cubic cm) whose diagonal measures $4\sqrt{3}$ cm ? **(SSC CGL Tier I 2001, CDS 2010)**

(*a*) 16 (*b*) 27 (*c*) 64 (*d*) 8

Sol. (*c*) Side of cube = Diagonal $\div \sqrt{3} = 4\sqrt{3} \div \sqrt{3} = 4$ cm

$\Rightarrow$ Volume of cube $= (\text{Side})^3 = 4^3 = 64\ cm^3$

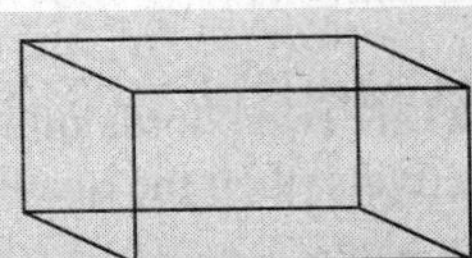

2. **Cuboid**: A figure with six rectangular sides.

   Surface Area $= 2\,(lb + bh + lh)$

   Volume = Area of Base × Height $= l \times b \times h$

   Diagonal $= \sqrt{l^2 + b^2 + h^2}$

   Where *l*, *b* and *h* respectively are length, breadth and height of the cuboid.

Example 1: The length of the longest rod that can be put in a room of dimensions 10 m × 10 m × 5 m is **(SSC CGL Tier I 2011)**

(*a*) $15\sqrt{3}$ m (*b*) 15 m (*c*) $10\sqrt{2}$ m (*d*) $5\sqrt{3}$ m

Sol. (*b*) Length of rod $= \sqrt{10^2 + 10^2 + 5^2} = \sqrt{225} = 15$ m

2. The surface area of a cube is 486 cm^2. Find its volume.

(*a*) 81 cm^3 (*b*) 243 cm^3 (*c*) 529 cm^3 (*d*) 729 cm^3

Sol. (*d*) Surface Area of cube $= 6\,(\text{Side})^2 = 486 \Rightarrow (\text{Side})^2 = 486 \div 6 = 81\ cm^3$

Side of the cube = $\sqrt{81}$ = 9 cm ⇒ Volume of the cube = $(9 \text{ cm})^3 = 729 \text{ cm}^3$

3. Room:

Area of four walls = Perimeter × Height = 2 × (Length + Breadth) × Height

Floor area = Length × Breadth

Volume of Room = Floor Area × Height = Length × Breadth × Height

Diagonal of the room = $\sqrt{l^2 + b^2 + h^2}$.

Where l, b and h respectively are length, breadth and height of the room.

Example 1: The length, breadth and height of a room are 7 metre, 6 metre and 5 metre respectively. Find the area of four walls.

(*a*) 130 m² (*b*) 144 m² (*c*) 154 m² (*d*) 210 m²

Sol. (*a*) Area of the walls = 2 × (Length + Breadth) × Height = 2 × (7 + 6) × 5 = 130 m²

Example 2: Find the volume of air in a room whose floor area is 40 m² and the height is 5 metre.

(*a*) 8 m³ (*b*) 100 m³ (*c*) 200 m³ (*d*) 400 m³

Sol. (*c*) Volume of air in the room = Floor area × Height = 40 × 5 = 200 m³

Example 3. The volume of air in a room is 204 m³. The height of the room is 6 m. What is the floor area of the room ? **(SSC CHSL 2013)**

(*a*) 34 m² (*b*) 32 m² (*c*) 46 m² (*d*) 44 m²

Sol. (*a*) Floor area = Volume ÷ height = 204 ÷ 6 = 34 m²

4. Sphere:

Surface Area = $4\pi r^2$

Volume = $\frac{4}{3}\pi r^3$

Example : If the volume and surface area of a sphere are numerically equal, then the numerical value of the radius of the sphere is:**(SSC CPO SI 2012, FCI 2012, SSC CGL Tier I 2015)**

(*a*) 1 (*b*) 2 (*c*) 3 (*d*) 4

Sol. (*c*) $\frac{4}{3}\pi r^3 = 4\pi r^2 \Rightarrow r = 3$

5. Hemisphere:

Curved Surface Area = $2\pi r^2$

Total Surface Area = $3\pi r^2$

Volume = $\frac{2}{3}\pi r^3$

Example. Find the total surface area of a hemisphere of radius of 7 cm.

(*a*) 154 cm² (*b*) 308 cm² (*c*) 462 cm² (*d*) 616 cm²

Sol. (*c*) Total surface area = $3\pi r^2 = 3 \times \frac{22}{7} \times 7 \times 7 = 462 \text{ cm}^2$

6. Cylinder:

Curved Surface Area = Circumference of Base × Height = $2\pi rh$

Total Surface Area = $2\pi rh + 2\pi r^2 = 2\pi r(h + r)$

Volume = Area of Base × Height = $\pi r^2 h$

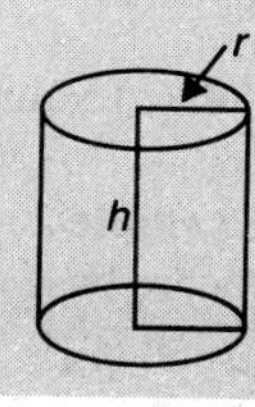

7. Cone:

Curved Surface Area = πrl

Where, l is slant height of the Cone and $l = \sqrt{h^2 + r^2}$

Total Surface Area $= \pi rl + \pi r^2 = \pi r(l + r)$

Volume $= \frac{1}{3} \times$ Area of Base $\times$ Height $= \frac{1}{3}\pi r^2 h$

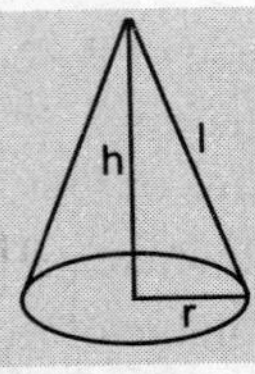

Example: The diameter and height of a cone are 14 cm and 24 cm respectively. Find curved surface area of the cone.

(*a*) 168 cm^2 (*b*) 528 cm^2 (*c*) 550 cm^2 (*d*) 1056 cm^2

Sol. (*c*) Radius $= \frac{1}{2} \times 14$ cm $= 7$ cm

Slant height $= l = \sqrt{h^2 + r^2} = \sqrt{24^2 + 7^2} = \sqrt{625} = 25$ cm

Surface area $= \pi rl = \frac{22}{7} \times 7 \times 25 = 550$ cm^2

8. Frustum : A frustum (e.g. bucket) is obtained by cutting a cone.

R = Larger radius, r = Smaller radius

h = Height of the frustum

l = Slant height of the frustum $= \sqrt{h^2 + (R - r)^2}$

$V = \frac{1}{3}\pi(R^2 + r^2 + Rr)h$ **(CDS 2010)**

Curved surface area $= \pi(R + r)l$

Total Surface Area $= \pi[(R + r)l + R^2 + r^2] = \pi(R^2 + r^2 + Rl + rl)$

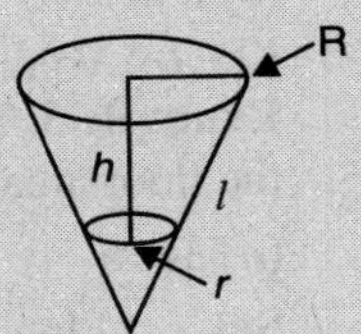

Example: The radii of two ends of a bucket are 8 cm and 5 cm and its height is 7 cm. Find the capacity of the bucket. **(MBA Ent. Exam.)**

(*a*) 280 cm^3 (*b*) 560 cm^3 (*c*) 616 cm^3 (*d*) 946 cm^3

Sol. (*d*) Capacity of the bucket $= \frac{1}{3}\pi h(R^2 + r^2 + Rr)$

$= \frac{1}{3} \times \frac{22}{7} \times 7 \times (8^2 + 5^2 + 8 \times 5) = \frac{1}{3} \times \frac{22}{7} \times 7 \times 129 = 946$ cm^3

9. Pyramid : Lateral Surface area $= \frac{1}{2} \times$ Perimeter of Base $\times$ Slant Height

Total surface area $= \frac{1}{2} \times$ Perimeter of Base $\times$ Slant Height + Area of Base

Volume $= \frac{1}{3} \times$ Area of Base $\times$ Height

Example 1 : What is the volume of a pyramid that has a rectangular base 4 m by 3 m and a height of 5 m?

(*a*) 20 m^3 (*b*) 40 m^3 (*c*) 60 m^3 (*d*) 75 m^3

Sol. (*a*) Volume $= \frac{1}{3} \times$ Area of base $\times$ Height $= \frac{1}{3} \times (4 \times 3) \times 5 = 20$ m^3

Example 2. Find the slant surface area of a pyramid having height 12 cm and standing on a square base of 10 cm side is

(*a*) 60 cm^3 (*b*) 120 cm^3 (*c*) 260 cm^3 (*d*) 520 cm^2

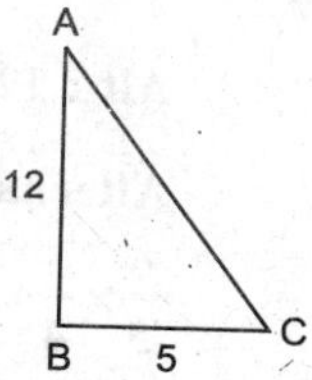

Sol. (*c*) Slant height (AC) = $\sqrt{12^2+5^2} = 13$ cm

$\therefore$ Slant surface area = $\frac{1}{2} \times p \times l$

$= \frac{1}{2} \times (4 \times 10) \times 13 = 260 \text{ cm}^3$

10. Prism : Volume of Prism = Area of Base × Height
Total surface area of Prism = (2 × Area of Base) + (Perimeter of Base × Height)
Lateral surface area = Perimeter of Base × Height

Example 1. The base of a solid right prism is a triangle whose sides are 9 cm, 12 cm and 15 cm. The height of the prism is 5 cm. Then the total surface area of the prism is

(SSC CPO SI 2011)

(*a*) 180 cm^2 (*b*) 234 cm^2 (*c*) 288 cm^2 (*d*) 270 cm^2

Sol. (*c*) $9^2 + 12^2 = 15^2 \Rightarrow$ The triangle is a right angle triangle.

Area of the triangle = $\frac{1}{2} \times 9 \times 12 = 54$ cm^2

Perimeter of triangle = 9 + 12 + 15 = 36 cm

Total surface area of Prism = (2 × 54) + (36 × 5) = 108 + 180 = 288 cm^2

Example 2. The perimeter of the triangular base of a right prism is 60 cm and the sides of the base are in the ratio 5 : 12 : 13. Then its volume will be (height of the prism being 50 cm)

(SSC CGL Tier II 2012)

(*a*) 5400 cm^3 (*b*) 9600 cm^3 (*c*) 6000 cm^3 (*d*) 6600 cm^3

Sol. (*c*) Sides of triangular base are 10 cm, 24 cm, 26 cm (ratio = 5 : 12 : 13 and sum = 60)

Area of triangle = $\frac{1}{2} \times 10 \times 24 = 120$ ($\because$ Triangle is a right angle triangle)

$\therefore$ Volume of prism = Area of Base × Height = 120 × 50 = 6000 cm^3

11. If **ratio of sides** of two three-dimensional figures such as spheres, cubes etc. = $x : y$. Then **ratio of their surface areas** = $x^2 : y^2$. And **ratio of their volumes** = $x^3 : y^3$.

Example. If the radius of a sphere is doubled, its volume becomes:

(SSC CGL Tier I 2006)

(*a*) Double (*b*) Four times (*c*) Six times (*d*) Eight times

Sol. (*d*) Ratio of radii = 1 : 2 $\Rightarrow$ Ratio of volumes = $1^3 : 2^3 = 1 : 8$

12. If length, breadth and height of a cuboid are increased by x%, y% and z% respectively. Then percentage increase in its volume = $(x+y+z) + \frac{xy+yz+xz}{100} + \frac{xyz}{100^2}$

Example. Dimensions of a cuboid are 29 cm, 18 cm and 15 cm. If its sides are increased by 20%, 25% and 40% respectively, find percent increase in its volume.

(*a*) 85% (*b*) 104% (*c*) 110% (*d*) 121%

Sol. (*c*) Percent increase in Volume

$= (20+25+40) + \frac{20 \times 25 + 25 \times 40 + 20 \times 40}{100} + \frac{20 \times 25 \times 40}{100 \times 100}$

$= (85) + (5 + 10 + 8) + (2) = 85 + 23 + 2 = 110\%$

Alt : $12 \times 12.5 \times 14 - 1000 = 2100 - 1000 = 1100 \Rightarrow$ % increase $= \frac{1100}{1000} \times 100 = 110\%$

Alternative Method (by converting into fractions) :

$\frac{6}{5} \times \frac{5}{4} \times \frac{7}{5} = \frac{21}{10} \Rightarrow$ Increase $= \frac{21}{10} - 1 = \frac{11}{10} = \frac{11}{10} \times 100\% = 110\%$

13. If all the sides of a three-dimensional figure are increased by x%.

Then increase in surface area $= x + x + \frac{x^2}{100} = 2x + \frac{x^2}{100}$

And increase in volume $= (x + x + x) + \frac{x^2 + x^2 + x^2}{100} + \frac{x^3}{100^2} = 3x + \frac{3x^2}{100} + \frac{x^3}{100^2}$

Example 1. If each side of a cube is increased by 30%, then its surface area will be increased by:

(*a*) 50% (*b*) 51% (*c*) 60% (*d*) 69%

Sol. (*d*) Percent increase in surface area $= 2 \times 30 + \frac{30^2}{100} = 60 + 9 = 69\%$

Alternative Method : $13 \times 13 = 169$; $169 - 100 = 69\%$

Example 2. If both the radius and height of a right circular cylinder are increased by 20%, its volume will increase by : **(SSC CGL Tier I 2004)**

(*a*) 40% (*b*) 44% (*c*) 66% (*d*) 72.8%

Sol. (*d*) Percent increase in Volume $= (3 \times 20) + \frac{3 \times 20^2}{100} + \frac{20^3}{100^2} = 60 + 12 + 0.8 = 72.8\%$

Alternative Method : $12^3 - 10^3 = 728 \Rightarrow$ Percent increase $= \frac{728}{1000} \times 100 = 72.8\%$

Hint: We assumed that each side = 10 cm, then increased side = 10 + 20% of 10 = 12 cm

14. If outer and inner radii of a cylinder are R and r respectively.

Then metal required to make the cylinder $= \pi (R^2 - r^2) h$

Proof: Volume (outer) of the cylinder $= \pi R^2 h$. Volume (inner) of the cylinder $= \pi r^2 h$

$\therefore$ Volume of metal $= \pi R^2 h - \pi r^2 h = \pi(R^2 - r^2)h$

15. A spherical ball is fitted into a cylinder. Then ratio between volumes of cylinder and sphere is 3 : 2.

Difference between volumes of cylinder and sphere = 3 – 2 = 1

$\therefore$ Difference $= \frac{1}{3}$ of volume of cylinder Or Difference $= \frac{1}{2}$ of volume of sphere

Proof: Let radius of sphere $= x$. Then radius of cylinder $= x$, and height of cylinder $= 2x$

$\therefore$ Volume of cylinder $= \pi(x^2)(2x) = 2\pi x^3$ and, volume of sphere $= \frac{4}{3}\pi x^3$

$\therefore$ Ratio between volumes of cylinder and sphere $= 2\pi x^3 : \frac{4}{3}\pi x^3 = 1 : \frac{2}{3} = 3 : 2$

16. A **solid cube** with a side of '*a*' cm has been **painted on all sides.** Now the cube has been **cut into cubes of 1 cm each.** We get a^3 cubes of 1 cm each side (if $a \geq 2$), out of which :

$(a-2)^3$ Cubes have **no side painted**

$6(a-2)^2$ Cubes have **one side painted**

$12(a-2)$ Cubes have **two sides painted**

8 cubes have three sides painted

Note : $(a-2)^3 + 6(a-2)^2 + 12(a-2) + 8 = a^3$

SOLVED EXERCISE

1. The length, breadth and height of a room are respectively 5 metres, 4 metres and 3 metres. The length of the longest bamboo stick that can be kept entirely into the room, is: **(SSC CGL Tier I 2006 and 2001)**

(*a*) 5 m (*b*) 7 m (*c*) $5\sqrt{2}$ m (*d*) 60 m

Sol. (*c*) Length of the stick = $\sqrt{5^2+4^2+3^2} = \sqrt{25+16+9} = \sqrt{50} = 5\sqrt{2}$ m

2. Find the length of the longest rod that can be placed in a hall of 10 metres length, 6 metres breadth and 4 metres height. **(SSC CGL Tier I 2002)**

(*a*) $2\sqrt{38}$ metres (*b*) $4\sqrt{38}$ metres (*c*) $2\sqrt{19}$ metres (*d*) 19 metres

Sol. (*a*) Length of rod = $\sqrt{10^2+6^2+4^2} = \sqrt{100+36+16} = \sqrt{152} = 2\sqrt{38}$ metres

3. A soap cake is of size 8 cm × 5 cm × 4 cm. The number of such soap cakes that can be packed in a box measuring 56 cm × 35 cm × 28 cm is : **(SSC CGL Tier I 2004)**

(*a*) 49 (*b*) 196 (*c*) 243 (*d*) 343

Sol. (*d*) Required number of soap cakes = $\dfrac{56 \times 35 \times 28}{8 \times 5 \times 4} = 343$

4. Some bricks are arranged in an area measuring 20 cu. m. If the length, breadth and height of each brick is 25 cm, 12.5 cm and 8 cm respectively, then in that pile the number of bricks are (suppose there is no gap in between two bricks) **(SSC CGL Tier I 2014)**

(*a*) 6,000 (*b*) 8,000 (*c*) 4,000 (*d*) 10,000

Sol. (*b*) Number of bricks = $\dfrac{20 \times 100 \times 100 \times 100}{25 \times 12.5 \times 8} = 8000$

5. A village having a population of 4000 requires 150 litres of water per head per day. It has a tank measuring 20 m × 15 m × 6 m. The water of this tank will last for **(UPSC Civil Services Pre. 2011)**

(*a*) 2 days (*b*) 3 days (*c*) 4 days (*d*) 5 days

Sol. (*b*) Capacity of tank = 20 m × 15 m × 6 m = 1800 m = 1800 × 1000 litres

Required days = $\dfrac{1800 \times 1000}{4000 \times 150} = 3$ days

6. A metal pipe of negligible thickness has radius 21 cm and length 90 cm. The outer curved surface area of the pipe in square cm is **(SSC CGL Tier I 2012)**

(*a*) 11880 (*b*) 11680 (*c*) 11480 (*d*) 10080

Sol. (*a*) Curved surface area $= 2\pi rh = 2 \times \frac{22}{7} \times 21 \times 90 = 11880$

7. Find the height of a cuboid whose volume and base area are 144 m^3 and 18 m^2 respectively.

(*a*) 6 metre (*b*) 7 metre (*c*) $7\frac{1}{2}$ metre (*d*) 8 metre

Sol. (*d*) Height of the cuboid $= \frac{\text{Volume}}{\text{Base Area}} = \frac{144}{18} = 8$ metre

8. The volume of a cuboid is twice the volume of a cube. If the dimensions of the cuboid are 9 cm, 8 cm and 6 cm, the total surface area of the cube is : **(SSC CGL Tier I 2002)**

(*a*) 72 cm^2 (*b*) 216 cm^2 (*c*) 432 cm^2 (*d*) 108 cm^2

Sol. (*b*) Volume of the cuboid $= 9 \times 8 \times 6 = 432$ $cm^3 \Rightarrow$ Volume of the cube $= 432 \div 2 = 216$ cm^3

Side of cube $= \sqrt[3]{216} = 6$ cm $\Rightarrow$ Surface area of cube $= 6 \times (\text{side})^2 = 6 \times (6)^2 = 216$ cm^2

9. A large water tank has the shape of a cube. If 128 m^3 of water is pumped out, the water level goes down by 2 m. Then the maximu capacity of the tank is **(CDS 2015)**

(*a*) 512 m^3 (*b*) 480 m^3 (*c*) 324 m^3 (*d*) 256 m^3

Sol. (*a*) Base area of cubic tank $= 128 \div 2 = 64$ $cm^2 \Rightarrow$ Side of cube $= 8$ cm

$\Rightarrow$ Volume of cubic tank $= 8^3 = 512$ m^3

10. If the surface area of a sphere is 616 square cm, what is its volume ? **(CDS 2012)**

(*a*) 4312/3 cm^3 (*b*) 4102 cm^3 (*c*) 1257 cm^3 (*d*) 1023 cm^3

Sol. (*a*) Surface Area $= 4\pi r^2 = 616 \Rightarrow r^2 = 616 \times \frac{1}{4} \times \frac{7}{22} = 49 = 7^2 \Rightarrow r = 7$

Volume $= \frac{4}{3}\pi r^3 = \frac{4}{3} \times \frac{22}{7} \times 7 \times 7 \times 7 = \frac{4312}{3}$

11. The length, breadth and height of a cuboid are in the ratio 3 : 4 : 6 and its volume is 576 cm^3. the whole surface area of the cuboid is **(SSC CHSL 2014)**

(*a*) 216 cm^2 (*b*) 324 cm^2 (*c*) 432 cm^2 (*d*) 460 cm^2

Sol. (*c*) Volume of cuboid $= (3x) \times (4x) \times (6x) = 576 \Rightarrow 72x^3 = 576 \Rightarrow x^3 = 8 \Rightarrow x = 2$

Dimensions of cuboid $= (3x), (4x), (6x) = 6, 8, 12$ cm

Total Surface Area $= 2(lb + bh + lh) = 2 \times (48 + 96 + 72) = 2 \times 216 = 432$ cm^2

12. The areas of three adjacent sides of a cuboid are 15 m^2, 12 m^2 and 5 m^2. What is the volume of the cuboid ?

(*a*) 30 m^3 (*b*) 60 m^3 (*c*) 180 m^3 (*d*) 75 m^3

Sol. (*a*) Area of the sides are $l \times b$, $b \times h$ and $l \times h$

Volume of cuboid $= l \times b \times h = \sqrt{15 \times 12 \times 5} = 30$ m^3

13. If the areas of three adjacent faces of a cuboid are *x*, *y*, *z* respectively, then the volume of the cuboid is : **(CDS 2003, MBA 2007, SSC CHSL 2010)**

(*a*) xyz (*b*) $2xyz$ (*c*) $\sqrt{xyz}$ (*d*) $3\sqrt{xyz}$

Sol. (*c*) Volume $= l \times b \times h = \sqrt{l^2 \times b^2 \times h^2} = \sqrt{lb \times bh \times hl} = \sqrt{xyz}$

14. A solid hemisphere is of radius 11 cm. The curved surface area in sq. cm is **(SSC CGL Tier I 2012, CDS 2015 M)**

(*a*) 1140.85 (*b*) 1386.00 (*c*) 760.57 (*d*) 860.57

Sol. (*c*) Curved surface Area = $2\pi r^2 = 2 \times \frac{22}{7} \times 11 \times 11 = 760.57$ cm^2

15. The height of a room is *a* and the areas of two adjacent walls of the room are *b* and *c*. The area of the roof will be: **(SSC CGL Tier I 2006)**

(*a*) $\frac{bc}{a}$ (*b*) $\frac{bc}{a^2}$ (*c*) $\frac{ac}{b^2}$ (*d*) $\sqrt{bc}$

Sol. (*b*) Let the length and breadth of the room are *x* and *y* respectively.

Then area of two adjacent walls = ax and $ay = b$ and c respectively.

Area of roof = $x \times y = \frac{ax \times ay}{a \times a} = \frac{b \times c}{a \times a} = \frac{bc}{a^2}$

16. The length, breadth and height of a room are 5 metre, 4 metre and 4 metre respectively. If all the four walls of the room are to be covered with 50 cm wide wall paper, find the length of the paper required.

(*a*) 36 metre (*b*) 72 metre (*c*) 144 metre (*d*) 288 metre

Sol. (*c*) Area of four walls = 2 × (Length + Breadth) × Height = 2 × (5 + 4) × 4 = 72 m^2

Width of the wall paper = 50 cm = 0.5 metre

∴ Length of the paper required = $72 \div 0.5 = 72 \div \frac{1}{2} = 72 \times 2 = 144$ metre

17. The area of the base of a cuboidal box is 21 sq. cm. and the area of one of the faces is 30 sq. cm. The numerical value of each of the dimensions of this box is an integer greater than 1. Then the volume of the cuboidal box, in cu. cm, is **(SSC CHSL 2011)**

(*a*) 210 (*b*) 630 (*c*) 105 (*d*) 120

Sol. (*a*) Area of base = length × breadth = 21 = 7 × 3

Area of face = 30 = 3 × 10 = Breadth × height ⇒ Volume = $l \times b \times h = 7 \times 3 \times 10 = 210$

18. What are the dimensions (length, breadth and height respectively) of a cuboid with volume 720 cubic cm, surface area 484 cm^2 and the area of the base 72 cm^2 ? **(CDS 2012)**

(*a*) 9, 8 and 10 (*b*) 12, 6 and 10 (*c*) 18, 4 and 10 cm (*d*) 30, 2 and 12

Sol. (*a*) Area of base in each of the option (a), (b), (c) is 72.

Volume of cuboid in each of the option (a), (b), (c) is 720.

Surface area for option (a) = 2 × (9 × 8 + 8 × 10 + 10 × 9) = 2 × (72 + 80 + 90) = 484

Option (a) satisfies all the three conditions, hence the answer.

19. If the sum of three dimensions and the total surface area of a rectangular box are 12 cm and 94 cm^2 respectively, then the maximum length of a stick that can be placed inside the box is **(SSC CPO SI 2011)**

(*a*) $5\sqrt{2}$ cm (*b*) 5 cm (*c*) 6 cm (*d*) $2\sqrt{5}$ cm

Sol. (*a*) Let the three dimensions be *l*, *b* and *h*.

Then $l + b + h = 12$ and $2 \times (lb + bh + hl) = 94$

$(l^2 + b^2 + h^2) = (l + b + h)^2 - 2 \times (lb + bh + hl) = 12^2 - 94 = 144 - 94 = 50$

∴ $\sqrt{l^2 + b^2 + h^2} = \sqrt{50} = 5\sqrt{2}$ cm

20. The sum of length, breadth and height of a cuboid is 19 cm and its diagonal is 11 cm. Find the surface area of the cuboid, in cm^2. **(SSC FCI Asstt. 2012)**

(*a*) 120 (*b*) 140 (*c*) 160 (*d*) 240

Sol. (*d*) $l + b + h = 19$ and $\sqrt{l^2 + b^2 + h^2} = 11 \Rightarrow l^2 + b^2 + h^2 = 121$

$(l + b + h)^2 = l^2 + b^2 + h^2 + 2(lb + bh + lh)$

$\Rightarrow 19^2 = 121 + 2(lh + bh + lb) \Rightarrow 2(lh + bh + lb) = 361 - 121 = 240$

21. A cylinder has a diameter of 14 cm and the area of its curved surface is 220 sq. cm. The volume of the cylinder is **(SSC CGL Tier I 2013)**

(*a*) 620 cm^3 (*b*) 670 cm^3 (*c*) 707 cm^3 (*d*) 770 cm^3

Sol. (*d*) Radius = 14 ÷ 2 = 7 cm

Curved Surface Area = $2\pi rh = 220 \Rightarrow h = 220 \times \frac{1}{2} \times \frac{7}{22} \times \frac{1}{7} = 5$ cm

Volume = $\pi r^2 h = \frac{22}{7} \times 7 \times 7 \times 5 = 770$ cm^3

22. The base of a conical tent is 19.2 metres in diameter and its height is 2.8 metres. The area of the canvas required to put up such a tent (in square metres) (Taking $\pi = 22/7$) is nearly : **(SSC CGL Tier I 2008)**

(*a*) 3017.1 (*b*) 3170 (*c*) 301.7 (*d*) 30.17

Sol. (*c*) Radius = $\frac{1}{2} \times 19.2$ m = 9.6 m and height = 2.8 m

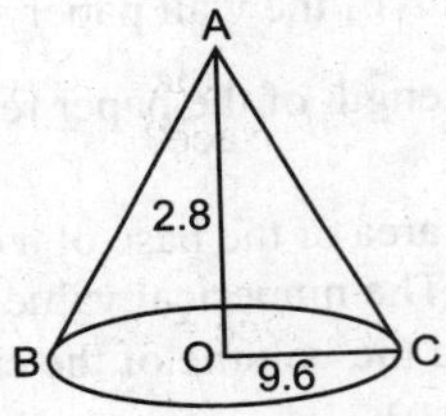

Slant height = $l = \sqrt{h^2 + r^2}$

$= \sqrt{(2.8)^2 + (9.6)^2} = \sqrt{(4 \times 0.7)^2 + (4 \times 2.4)^2}$

$= \sqrt{4^2 \times (0.7^2 + 2.4^2)} = \sqrt{4^2 \times (2.5)^2} = 4 \times 2.5 = 10$ m

Area of canvas = $\pi rl = \frac{22}{7} \times 9.6 \times 10 = 301.7$ m^2

23. The difference in radii of the circular ends of a bucket is 20 cm. If the height of the bucket is 15 cm, its slant height will be **(SAS 2010)**

(*a*) 25 cm (*b*) 28 cm (*c*) 30 cm (*d*) 36 cm

Sol. (*a*) Slant height = $\sqrt{h^2 + (R^2 - r^2)} = \sqrt{15^2 + 20^2} = 25$ cm

24. A bucket has top and bottom radii of 20 cm and 10 cm respectively. Find the volume of bucket if its depth is 12 cm.

(*a*) 25400 cm^3 (*b*) 8800 cm^3 (*c*) 4400 cm^3 (*d*) 17600 cm^3

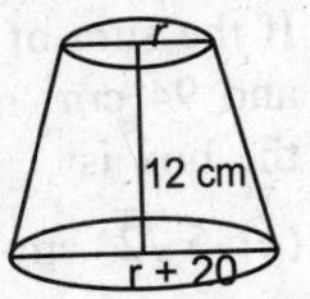

Sol. (*b*) Volume = $\frac{\pi h}{3}(R^2 + r^2 + rR) = \frac{22}{7} \times \frac{12}{3} \times (400 + 100 + 200)$

$= \frac{22}{7} \times 4 \times 700 = 8800$ cm^3

25. If the length of a rectangular parallelopiped is 3 times of its breadth and 5 times of its height and its volume is 14400 cu. cm, then area of the total surface will be: **(SSC CGL Tier II 2012)**

(*a*) 4320 cm^2 (*b*) 5320 cm^2 (*c*) 3320 cm^2 (*d*) 2420 cm^2

Sol. (*a*) Let dimensions of parallelopiped be $15x$, $5x$, $3x$.

Then $(15x)(5x)(3x) = 14400 \Rightarrow 225x^3 = 14400 \Rightarrow x^3 = 64 \Rightarrow x = 4$

$\Rightarrow$ The dimensions are 60, 20, 12.

$\therefore$ Surface area = $2(60 \times 20 + 20 \times 12 + 12 \times 60) = 4320$ cm^2

26. If the base of a right pyramid is a triangle of sides 21 cm, 20 cm, 13 cm and its altitude is 30 cm, then its volume (in cm^3) will be **(SSC CHSL 2013)**

(*a*) 430 (*b*) 630 (*c*) 1068 (*d*) 1260

Sol. (*d*) $s = \frac{21+20+13}{2} = \frac{54}{2} = 27$

$\Rightarrow$ Area of triangular base $= \sqrt{27 \times 6 \times 7 \times 14} = 3 \times 3 \times 2 \times 7 = 126 \text{ cm}^2$

Area of pyramid $= \frac{1}{3} \times$ Area of base $\times$ Height $= \frac{1}{3} \times 126 \times 30 = 1260 \text{ cm}^3$

27. The volume of a right circular cylinder whose height is 40 cm, and circumference of its base is 66 cm, is : **(SSC CGL Tier I 2000)**

(*a*) 55440 cm^3 (*b*) 3465 cm^3 (*c*) 7720 cm^3 (*d*) 13860 cm^3

Sol. (*d*) Circumference of base $= 2\pi r = 66 \Rightarrow r = 66 \times \frac{1}{2} \times \frac{7}{22} = \frac{21}{2}$ cm

Volume of cylinder $= \pi r^2 h = \frac{22}{7} \times \frac{21}{2} \times \frac{21}{2} \times 40 = 13860 \text{ cm}^3$

28. A cylindrical tank 7 m in diameter, contains water to a depth of 4 m. What is the total area of wetted surface? **(CDS 2011)**

(*a*) 110 m^2 (*b*) 126.5 m^2 (*c*) 131.5 m^2 (*d*) 136.5 m^2

Sol. (*b*) Wet surface $= 2\pi rh + \pi r^2 = \pi r(2h + r) = \frac{22}{7} \times \frac{7}{2} \times \left(8 + \frac{7}{2}\right) = \frac{22}{7} \times \frac{7}{2} \times \frac{23}{2} = 126.5 \text{ m}^2$

29. The total outer surface area of a right circular cone of height 24 cm with a hemisphere of radius 7 cm upon its base is **(CDS 2015)**

(*a*) 327π cm^2 (*b*) 307π cm^2 (*c*) 293π cm^2 (*d*) 273π cm^2

Sol. (*d*) Slant height of cone $= \sqrt{24^2 + 7^2} = 25$ cm

Total surface area $= \pi rl + 2\pi r^2 = 175\pi + 98\pi = 273\pi$

30. A cistern, 6 m long and 4 m wide, contains water upto a depth of 1 m 25 cm. The total area of the wet surface is : **(SSC CGL Tier I 2004)**

(*a*) 55 m^2 (*b*) 53.5 m^2 (*c*) 50 m^2 (*d*) 49 m^2

Sol. (*d*) Wet surface $= (6 \times 4) + 2 \times 1.25 \times (6 + 4) = 24 + 25 = 49 \text{ m}^2$

31. The volume of a solid hemisphere is numerically equal to its total surface area. Its radius is **(SSC CPO SI 2012)**

(*a*) $4\frac{1}{2}$ units (*b*) 9 units (*c*) 3 units (*d*) $1\frac{1}{2}$ units

Sol. (*a*) $\frac{2}{3}\pi r^3 = 3\pi r^2 \Rightarrow 2r = 9 \Rightarrow r = 4\frac{1}{2}$ units

32. A cone and a hemisphere have equal bases and equal volumes. Find the ratio of their heights.

(*a*) 1 : 1 (*b*) 1 : $\sqrt{2}$ (*c*) 2 : 1 (*d*) $\sqrt{2}$: 1

Sol. (*c*) $\frac{1}{3}\pi r^2 h = \frac{2}{3}\pi r^3 \Rightarrow h = 2r \Rightarrow h : r = 2 : 1$

33. The radii of a sphere and a right circular cylinder are equal and their curved surface areas are also equal. The ratio of their volumes is **(SSC MTS 2013)**

(*a*) 3 : 4 (*b*) 2 : 3 (*c*) 3 : 2 (*d*) 4 : 3

Sol. (*b*) Given, $4\pi r^2 = 2\pi rh \Rightarrow h = 2r$

Required ratio = $\frac{4}{3}\pi r^3 : \pi r^2 h = 4r : 3h = 4r : 6r = 2 : 3$

34. A sphere, a cylinder and a cone are of the same radius and same height. Find the ratio of their curved surface areas.

(*a*) $2 : 2 : \sqrt{3}$ (*b*) 3 : 3 : 2 (*c*) 2 : 2 : 3 (*d*) $4 : 4 : \sqrt{5}$

Sol. (*d*) Height of cylinder = Height of cone = $2r$

$\Rightarrow$ Slant height of cone = $l = \sqrt{r^2 + (2r)^2} = \sqrt{5}\, r$

Required ratio = $4\pi r^2 : 2\pi r(2r) : \pi r(\sqrt{5}\, r) = 4 : 4 : \sqrt{5}$

35. If S_1 and S_2 be the surface area of a sphere and the curved surface area of the circumscribed cylinder respectively, then S_1 is equal to : **(SSC CGL Tier I 2008)**

(*a*) $\frac{3}{4}S_2$ (*b*) $\frac{1}{2}S_2$ (*c*) $\frac{2}{3}S_2$ (*d*) S_2

Sol. (*d*) Let r be the radius of sphere and cylinder. Then height of the cylinder = $2r$

Surface area of the sphere = $S_1 = 4\pi r^2$

Curved surface area of a cylinder = $S_2 = 2\pi rh = 2\pi r(2r) = 4\pi r^2 \Rightarrow S_1 = S_2$

36. A cube of maximum volume is cut from a sphere. Find the ratio of their volumes.

(*a*) $2 : \sqrt{3}\pi$ (*b*) $2 : \sqrt{3}$ (*c*) 1 : 2 (*d*) $1 : \sqrt{2}$

Sol. (*a*) Let side of cube = x, radius of sphere = r

Diagonal of cube = Diameter of sphere $\Rightarrow \sqrt{3}\,x = 2r \Rightarrow x = \frac{2r}{\sqrt{3}}$

$\therefore$ Required ratio = $x^3 : \frac{4}{3}\pi r^3 = \left(\frac{2r}{\sqrt{3}}\right)^3 : \frac{4}{3}\pi r^3 = \frac{8r^3}{3\sqrt{3}} : \frac{4}{3}\pi r^3 = 2 : \sqrt{3}\pi$

37. The total surface area of a solid hemisphere is 108 π cm². The volume of the hemisphere is: **(SSC CGL Tier I 2008)**

(*a*) 72π cm³ (*b*) 144π cm³ (*c*) $108\sqrt{6}\,\pi$ cm³ (*d*) 288π cm³

Sol. (*b*) Total surface area of hemisphere = $3\pi r^2 = 108\,\pi$ cm² $\Rightarrow r^2 = 108\pi \div 3\pi = 36 \Rightarrow r = 6$ cm

Volume of hemisphere = $\frac{2}{3}\pi r^3 = \frac{2}{3}\pi(6)^3 = 144\pi$ cm³

38. Find the cost of canvas cloth required for conical tent whose height is 3 m and radius is 4 m and cost of cloth is ₹ 140 per square metre.

(*a*) ₹ 7040 (*b*) ₹ 8800 (*c*) ₹ 5280 (*d*) ₹ 15840

Sol. (*b*) Slant height = $l = \sqrt{3^2 + 4^2} = 5$ m $\Rightarrow$ Curved Surface Area = $\pi rl = \frac{22}{7} \times 4 \times 5$

Cost of canvas cloth = $\frac{22}{7} \times 4 \times 5 \times 140 =$ ₹ 8800

39. A cube of edge 5 cm is cut into cubes each of edge 1 cm. The ratio of the total surface area of one of the small cube to that of the large cube is equal to : **(SSC CGL Tier I 2004)**

(*a*) 1 : 125 (*b*) 1 : 5 (*c*) 1 : 625 (*d*) 1 : 25

Sol. (*d*) Total surface area of the large cube = $6 \times (5)^2 = 150$ cm^3

Total surface area of the small cube = $6 \times (1)^2 = 6$ cm^3

$\therefore$ Required ratio = 6 : 150 = 1 : 25

Alternative method : Ratio of sides = 1 : 5 $\Rightarrow$ Ratio of surface areas = $1^2 : 5^2 = 1 : 25$

40. The percentage increase in the surface area of a cube when each side is doubled is : **(SSC CGL Tier I 2015)**

(*a*) 200% (*b*) 150% (*c*) 50% (*d*) 300%

Sol. (*d*) Ratio of sides = 1 : 2 $\Rightarrow$ Ratio of surface areas = $1^2 : 2^2 = 1 : 4$

$\Rightarrow$ Percentage increase = (4 – 1) = 3 = 300%

41. Spheres A and B have their radii 40 cm and 10 cm respectively. Ratio of surface area of A to surface area of B is : **(SSC CGL Tier I 2003)**

(*a*) 1 : 16 (*b*) 4 : 1 (*c*) 1 : 4 (*d*) 16 : 1

Sol. (*d*) Ratio of radii of spheres = 40 : 10 = 4 : 1 $\Rightarrow$ Ratio of surface areas = $4^2 : 1^2 = 16 : 1$

42. If two spheres have their radii in the ratio 2 : 3, then the ratio of their volumes is **(SSC CGL)**

(*a*) 8 : 27 (*b*) 4 : 9 (*c*) 2 : 3 (*d*) None of these

Sol. (*a*) Ratio of volumes = $2^3 : 3^3 = 8 : 27$

43. The diameter of the moon is approximately one-fourth of that of the earth. What is the (approximate) ratio of the volume of the moon to that of the earth ? **(CDS 2009, 2013)**

(*a*) 1/16 (*b*) 1/32 (*c*) 1/48 (*d*) 1/64

Sol. (*d*) Ratio of radii of the moon and the earth = 1 : 4

$\therefore$ Ratio of volumes of the moon and the earth = $1^3 : 4^3 = 1 : 64$

44. If the surface areas of two spheres are in the ratio 4 : 9, then the ratio of their volumes will be **(SSC CGL Tier II 2013; SSC CGL Tier I 2014)**

(*a*) 4 : 9 (*b*) 16 : 27 (*c*) 8 : 27 (*d*) 16 : 81

Sol. (*c*) Ratio of radii = $\sqrt{4} : \sqrt{9} = 2 : 3 \Rightarrow$ Ratio of volumes = $2^3 : 3^3 = 8 : 27$

45. What is the number of wax balls, each of radius 1 cm, that can be molded out of a sphere of radius 8 cm? **(CDS 2011)**

(*a*) 256 (*b*) 512 (*c*) 768 (*d*) 1024

Sol. (*b*) Ratio of radii = 1 : 8 $\Rightarrow$ Ratio of volumes = $1^3 : 8^3 = 1 : 512$

46. Volume of two spheres are in the ratio of 27 : 125. Find radius of the bigger sphere, if radius of smaller sphere is 27 cm. **(SSC CGL 1999 M)**

(*a*) 45 cm (*b*) 125 cm (*c*) 135 cm (*d*) 625 cm

Sol. (*a*) Let radius of bigger and smaller sphere are *R* and *r* respectively.

Then $R : r = \sqrt[3]{125} : \sqrt[3]{27} = 5 : 3 \Rightarrow R = \frac{5}{3} \times 27 = 45$ cm

47. The volumes of two spheres are in the ratio 8 : 27. The ratio of their surface areas is : **(SSC CGL Tier I 2000)**

(*a*) 4 : 9 (*b*) 2 : 3 (*c*) 4 : 5 (*d*) 5 : 6

Sol. (*a*) Ratio of radii = $\sqrt[3]{8} : \sqrt[3]{27} = 2 : 3 \Rightarrow$ Ratio of surface areas = $2^2 : 3^2 = 4 : 9$

48. If the volumes of two cubes are in the ratio 27 : 64, then the ratio of their total surface areas is: **(SSC CGL Tier I 2002)**

(*a*) 27 : 64 (*b*) 3 : 4 (*c*) 9 : 16 (*d*) 3 : 8

Sol. (*c*) Ratio of sides = $\sqrt[3]{27} : \sqrt[3]{64}$ = 3 : 4 ⇒ Ratio of surface areas = $3^2 : 4^2$ = 9 : 16

49. Volume of two spheres are in the ratio of 343 : 8. Find the ratio between their surface areas.

(*a*) 7 : 2 (*b*) 49 : 4 (*c*) 81 : 45 (*d*) 343 : 8

Sol. (*b*) Ratio of sides = $\sqrt[3]{343} : \sqrt[3]{8}$ = 7 : 2 ⇒ Ratio of their surface areas = $7^2 : 2^2$ = 49 : 4

50. The capacities of two hemispherical bowls are 6.4 litres and 21.6 litres respectively. Then the ratio of their internal curved surface areas will be **(SSC CGL Tier II 2012, CGL I 2012)**

(*a*) $\sqrt{2} : \sqrt{3}$ (*b*) 16 : 81 (*c*) 4 : 9 (*d*) 2 : 3

Sol. (*c*) Ratio of volumes = Ratio of capacity = 6.4 : 21.6 = 8 : 27

Ratio of radii = $\sqrt[3]{8} : \sqrt[3]{27}$ = 2 : 3 ⇒ Ratio of surface areas = $2^2 : 3^2$ = 4 : 9

51. The volume of a sphere is 8 times that of another sphere. What is the ratio of their surface areas ? **(CDS 2010)**

(*a*) 8 : 1 (*b*) 4 : 1 (*c*) 2 : 3 (*d*) 4 : 3

Sol. (*b*) Ratio of sides = $\sqrt[3]{8} : \sqrt[3]{1}$ = 2 : 1 ⇒ Ratio of their surface areas = $2^2 : 1^2$ = 4 : 1

52. A solid metallic sphere of radius 8 cm is melted to form 64 equal small solid spheres. The ratio of the surface area of bigger sphere to that of small sphere is : **(SSC CGL Tier 2008)**

(*a*) 1 : 4 (*b*) 4 : 1 (*c*) 16 : 1 (*d*) 1 : 16

Sol. (*c*) Ratio of volumes of two spheres = 64 : 1

⇒ Ratio of radii of two spheres = $\sqrt[3]{64} : \sqrt[3]{1}$ = 4 : 1

⇒ Ratio of surface area of two spheres = $4^2 : 1^2$ = 16 : 1

53. A solid cube of 1.8 cm a side, weights 135 g. What will be weight of another solid cube of the same metal, whose side is 2.4 cm

(*a*) 320 g (*b*) 240 g (*c*) 180 g (*d*) None of these

Sol. (*a*) Ratio of sides = 1.8 : 2.4 = 3 : 4

Ratio of weights = Ratio of volumes = $3^3 : 4^3$ = 27 : 64

Weight of second cube = $\frac{64}{27} \times 135$ = 320 g

54. Let A and B be two solid spheres such that the surface area of B is 300% higher than the surface area of A. The volume of A is found to be *k*% lower than the volume of B. The value of *k* must be **(SSC CGL Tier II 2012)**

(*a*) 90.5 (*b*) 87.5 (*c*) 85.5 (*d*) 92.5

Sol. (*b*) Ratio of surface areas = 100 : 400 = 1 : 4 = $1^2 : 2^2$ ⇒ Ratio of sides = 1 : 2

Ratio of volumes = $1^3 : 2^3$ = 1 : 8 ⇒ $k\% = \frac{7}{8} \times 100 = 87.5\%$

55. If the radius of a sphere is increased by 2 cm, its surface area increases by 352 cm². Find the length of radius of original sphere. **(SSC CPO SI 2002)**

(*a*) 3 cm (*b*) 6 cm (*c*) 5 cm (*d*) 4 cm

Sol. (*b*) Let original radius = *r*. Then $4\pi(r+2)^2 - 4\pi r^2 = 352$

$\Rightarrow 4r + 4 = \frac{352}{4\pi} = \frac{352}{4} \times \frac{7}{22} \Rightarrow 4r + 4 = 28 \Rightarrow r = 6$ cm

56. The ratio of the volume of a cube to that of a sphere, which will exactly fit inside the cube is : **(SAS 2010, SSC CGL Tier I 2008, 2005)**

(*a*) 6 : π (*b*) π : 6 (*c*) 4 : 3 (*d*) 2 : 1

Sol. (*a*) Let radius of sphere be *r* cm. Then side of the cube is 2*r*.

Volume of cube = $(2r)^3 = 8r^3$ and volume of sphere = $\frac{4}{3}\pi r^3$

$\therefore$ Ratio of volumes = $8r^3 : \frac{4}{3}\pi r^3 = 8 : \frac{4}{3}\pi = 2 : \frac{1}{3}\pi = 6 : \pi$

57. A sphere and a hemisphere have the same surface area. The ratio of their volumes is **(SSC CHSL 2012)**

(*a*) $\frac{\sqrt{3}}{4} : 1$ (*b*) $\frac{3\sqrt{3}}{4} : 1$ (*c*) $\frac{\sqrt{3}}{8} : 1$ (*d*) $\frac{3\sqrt{3}}{8} : 1$

Sol. (*d*) Let the radii of sphere and hemisphere be *r* and R respectively.

Then $4\pi r^2 = 3\pi R^2 \Rightarrow r : R = \sqrt{3} : 2$

$\therefore$ Ratio of volumes = $(\sqrt{3})^3 : 2^3 = 3\sqrt{3} : 8 = \frac{3\sqrt{3}}{8} : 1$

58. A hemispherical cup of radius 4 cm is filled to the brim with coffee. The coffee is then poured into a vertical cone of radius 8 cm and height 16 cm. The percentage of the volume of the cone that remains empty is **(SSC CHSL 2012)**

(*a*) 88.2% (*b*) 87.5% (*c*) 80.5% (*d*) 81.6%

Sol. (*b*) $\frac{\text{Volume of cup}}{\text{Volume of cone}} = \frac{\frac{2}{3}\pi r^3}{\frac{1}{3}\pi R^2 H} = \frac{2r^3}{R^2 H} = \frac{2(4)^3}{(8)^2 16} = \frac{1}{8}$

$\therefore$ Cone remained empty = $1 - \frac{1}{8} = \frac{7}{8} = \frac{7}{8} \times 100\% = 87.5\%$

59. Surface area of a cube is equal to surface area of a sphere. Find ratio between their volumes.

(*a*) $\sqrt{\pi} : \sqrt{6}$ (*b*) $\sqrt{\pi} : \sqrt{3}$ (*c*) $\sqrt{3} : \sqrt{\pi}$ (*d*) $\sqrt{6} : \sqrt{\pi}$

Sol. (*a*) Let side of cube = *a* and radius of sphere = *r*.

Given, $6a^2 = 4\pi r^2 \Rightarrow \frac{a^2}{r^2} = \frac{4\pi}{6} \Rightarrow \frac{a}{r} = \sqrt{\frac{4\pi}{6}}$

Ratio of volumes = $\frac{a^3}{\frac{4}{3}\pi r^3} = \frac{3a^3}{4\pi r^3} = \frac{3}{4\pi} \times \frac{a^2}{r^2} \times \frac{a}{r} = \frac{3}{4\pi} \times \frac{4\pi}{6} \times \sqrt{\frac{4\pi}{6}} = \sqrt{\frac{\pi}{6}}$

60. A sphere and a hemisphere have the same volume. The ratio of their curved surface area is **(SSC CHSL 2012)**

(*a*) $2^{1/3} : 1$ (*b*) $2^{-3/2} : 1$ (*c*) $2^{2/3} : 1$ (*d*) $4^{-2/3} : 1$

Sol. (*a*) $\frac{4}{3}\pi r_1^3 = \frac{2}{3}\pi r_2^3 \Rightarrow 2r_1^3 = r_2^3 \Rightarrow 2^{1/3} r_1 = r_2 \Rightarrow r_1 : r_2 = 1 : 2^{1/3}$

Required ratio = $4\pi r_1^2 : 2\pi r_2^2 = 2r_1^2 : r_2^2 = 2 : (2^{1/3})^2 = 2^{1/3} : 1$

61. The base of a right pyramid is a square of side 16 cm long. If its height be 15 cm, then the area of the lateral surface in square cm is **(SSC CHSL 2011)**

(*a*) 136 (*b*) 544 (*c*) 800 (*d*) 1280

Sol. (*b*) Slant height of pyramid = $\sqrt{8^2+15^2}$ = 17 cm

Lateral surface area = $\frac{1}{2}\times(16\times4)\times17$ = 544 cm²

62. The circumference of the base of a circular cylinder is 6π cm². The height of the cylinder is equal to the diameter of the base. How many litres of water can it hold ? **(SSC CGL Tier I 2000)**

(*a*) 54π cc (*b*) 36π cc (*c*) 0.054π cc (*d*) 0.54π cc

Sol. (*a*) Circumference of Base = $2\pi r = 6\pi \Rightarrow r = 3$ cm $\Rightarrow d = 6$ cm $\Rightarrow h = d = 6$ cm

∴ Volume of cylinder = $\pi r^2 h = \pi (3)^2 6 = 54\pi$ cc

63. From a solid cylinder of height 10 cm and radius of the base 6 cm, a cone of same height and same base is removed. The volume of the remaining solid is:. **(SSC CHSL 2011)**

(*a*) 240 π cm³ (*b*) 528 π cm³ (*c*) 620 π cm³ (*d*) 360 π cm³

Sol. (*a*) Remaining volume = $\pi r^2 h - \frac{1}{3}\pi r^2 h = \frac{2}{3}\pi r^2 h = \frac{2}{3}\pi \times (6)^2 \times 10 = 240\pi$ cm³

64. The base a right prism is an equilateral triangle of area 173 cm² and the volume of the prism is 10380 cm³. The area of the lateral surface of the prism is (use $\sqrt{3}$ = 1.73) **(SSC CHSL 2011)**

(*a*) 1200 cm² (*b*) 2400 cm² (*c*) 3600 cm² (*d*) 4800 cm²

Sol. (*c*) Height of prism = $\frac{\text{Volume}}{\text{Area of base}} = \frac{10380}{173} = 60$ cm

Area of base (equilateral triangle) = $\frac{\sqrt{3}}{4}a^2 = 173 \Rightarrow a^2 = \frac{173\times4}{\sqrt{3}}$

$= \frac{173\times4}{1.73} = 400 \Rightarrow a = 20$ cm

Perimeter of base (equilateral triangle) = 3 × 20 = 60 cm

Required area = Perimeter of base × Height = 60 × 60 = 3600 cm²

65. If a right circular cone of height 24 cm has a volume of 1232 cm³, then the area of its curved surface is : (Take π = 22/7) **(SSC CGL Tier I 2003, SSC CGL Tier II 2013)**

(*a*) 1254 cm² (*b*) 704 cm² (*c*) 550 cm² (*d*) 154 cm²

Sol. (*c*) Volume of cone = $\frac{1}{3}\pi r^2 h = 1232$ cm³ $\Rightarrow r^2 = 1232 \times 3 \times \frac{7}{22}\times\frac{1}{24} = 49 \Rightarrow r = 7$ cm

$l = \sqrt{24^2+7^2} = 25 \Rightarrow$ Curved surface area = $\pi rl = \frac{22}{7}\times 7 \times 25 = 550$ cm²

66. If height of a cone is 24 cm and its surface area is 550 cm². Find its volume.

(*a*) 13200 cm² (*b*) 4400 cm² (*c*) 616 cm² (*d*) 1232 cm²

Sol. (*d*) $\pi rl = 550 \Rightarrow rl = 550\times\frac{7}{22} = 175 = 25 \times 7$

$l^2 = r^2 + h^2$ and $25^2 = 7^2 + 24^2 \Rightarrow r = 7$ and $l = 25$

Volume $= \frac{1}{3}\pi r^2 h = \frac{1}{3}\times\frac{22}{7}\times 7^2 \times 24 = 1232\text{ cm}^2$

67. If each edge of a cube is increased by 50%, the percentage increase in surface area is **(SSC CGL Tier II 2013, SSC CGL Tier I 2006)**

(*a*) 125% (*b*) 50% (*c*) 100% (*d*) 75%

Sol. (*a*) Required increase $= 50 + 50 + \frac{50\times 50}{100} = 125\%$

Alternative Method : $15 \times 15 = 225$; $225 - 100 = 125\%$

68. The diameter of a sphere is decreased by 25%. The percentage decrease in its surface area is

(*a*) 50% (*b*) 43.75% (*c*) 56.25% (*d*) 60%

Sol. (*b*) % Decrease $= -25 - 25 + \frac{25\times 25}{100} = -50 + 6.25 = 43.75\%$

Alternative Method : New area $= \frac{3}{4}\times\frac{3}{4} = \frac{9}{16}$; Decrease $= 1 - \frac{9}{16} = \frac{7}{16}\times 100 = 43.75\%$

69. The length of a cuboid is decreased by 10%, breadth is increased by 25% and the height is decreased by 20%. What is the increase or decrease in its volume?

(*a*) 15 % decrease (*b*) 10% decrease (*c*) 10% increase (*d*) 15% increase

Sol. (*b*) $\frac{100-10}{100}\times\frac{100+25}{100}\times\frac{100-20}{100} = \frac{9}{10}\times\frac{5}{4}\times\frac{4}{5} = \frac{9}{10} \Rightarrow$ Net decrease $= 1 - \frac{9}{10} = \frac{1}{10} = 10\%$

70. If each side of a cube is decreased by 10%, find percentage decrease in its volume.

(*a*) 30% (*b*) 27.1% (*c*) 33% (*d*) 39%

Sol. (*b*) Decrease in Volume $= 3 \times (-10) + \frac{3\times(-10)^2}{100} + \frac{(-10)^3}{100^2} = -30 + 3 - 0.1 = -27.1\%$

Alt: $9^3 - 10^3 = 729 - 1000 = -271 \Rightarrow$ Percent decrease $= \frac{271}{1000}\times 100 = 27.1\%$

71. If the radius of a right circular cylinder is decreased by 50% and its height is increased by 60%, its volume will be decreased by: **(SSC CGL Tier I 2006)**

(*a*) 10% (*b*) 20% (*c*) 40% (*d*) 60%

Sol. (*d*) $(10 - 50\%) \times (10 - 50\%) \times (10 + 60\%) = 5 \times 5 \times 16 = 400$

$\therefore$ Required percentage $= \frac{1000-400}{1000}\times 100 = \frac{6}{10}\times 100 = 60\%$

72. The surface area of a sphere is 616 square cm. If its radius is changed so that the area gets reduced by 75%, then the radius becomes **(CDS 2011)**

(*a*) 1.6 cm (*b*) 2.3 cm (*c*) 2.5 cm (*d*) 3.5 cm

Sol. (*d*) Surface Area of sphere $= 4\pi r^2 = 616 \Rightarrow r^2 = 616\times\frac{1}{4}\times\frac{7}{22} = 7^2$

Ratio of surface areas of two spheres $= 100\% : (100 - 75)\% = 4 : 1 = 2^2 : 1^2$

$\therefore$ New radius $= 7 \div 2 = 3.5$ cm

73. Two cubes of sides 6 cm each are kept side by side to form a rectangular parallelepiped. The area (in sq. cm) of the whole surface of the rectangular parallelepiped is **(SSC CGL Tier I 2012)**

(*a*) 432 (*b*) 360 (*c*) 396 (*d*) 340

Sol. (*b*) Dimensions of resulting cuboid, $l = 6 + 6 = 12$ cm, $b = 6$ cm, $h = 6$ cm

Surface Area $= 2 \times (12 \times 6 + 6 \times 6 + 12 \times 6) = 2\,(72 + 36 + 72) = 2 \times 180 = 360$ cm^2

Alternative method : Two cubes have total 12 sides. When two cubes are joined together, two of the sides are joined and 10 sides are exposed.

Area of one side $= 6 \times 6 = 36$ cm^2 $\quad\therefore$ Area of 10 sides $= 10 \times 36$ cm^2 $= 360$ cm^2

74. Three cubes each of side 5 cm are joined end to end. What is the surface area of the resulting cuboid ? **(CDS)**

(*a*) 300 cm^2 (*b*) 350 cm^2 (*c*) 375 cm^2 (*d*) 400 cm^2

Sol. (*b*) Dimensions of resulting cuboid, $l = 5 + 5 + 5 = 15$ cm, $b = 5$ cm, $h = 5$ cm

Surface Area $= 2 \times (15 \times 5 + 5 \times 5 + 5 \times 15) = 2\,(75 + 25 + 75) = 2 \times 175 = 350$ cm^2

Alternative method : Surface area of one cube $= 6 \times (5)^2 = 6 \times 25 = 150$ cm^2

Total surface area of three cubes $= 3 \times 150 = 450$ cm^2

Surface area of one side of single cube $= (5)^2 = 25$ cm^2

When three cubes are joined, four sides are hidden.

$\therefore$ Required area $= 450 - 4 \times 25 = 350$ cm^2

75. The material of a solid cone is converted into the shape of a solid cylinder of equal radius. If the height of the cylinder is 5 cm, what is the height of the cone ? **(CDS 2011)**

(*a*) 15 cm (*b*) 20 cm (*c*) 25 cm (*d*) 30 cm

Sol. (*a*) Volume of cone $= \frac{1}{3}\pi r^2 h$. Volume of cylinder $= \pi r^2 h$. $\Rightarrow \frac{1}{3}\pi r^2 h : \pi r^2 h = 1 : 3$

$\therefore$ Height of cone $= 3 \times 5 = 15$ cm

76. The radii and total surface areas of a sphere, a cylinder and a cone are equal. The ratio of their heights will be

(*a*) 1 : 1 : 1 (*b*) $2:1:2\sqrt{2}$ (*c*) $1:1:2\sqrt{2}$ (*d*) $1:1:\sqrt{2}$

Sol. (*b*) $4\pi r^2 = 2\pi r\,(r + h) = \pi r(l + r) \;\Rightarrow\; 4r = 2(r + h) = l + r$

Now, $4r = 2r + 2h \Rightarrow 2r = 2h \Rightarrow r = h$

Again, $4r = l + r \;\Rightarrow\; 3r = l \Rightarrow 3r = \sqrt{r^2 + h^2} \;\Rightarrow 9r^2 = r^2 + h^2$

$\Rightarrow h^2 = 8r^2 \;\Rightarrow\; h = 2\sqrt{2} \Rightarrow$ Required ratio $= 2:1:2\sqrt{2}$

77. The base radii of two cylinders are in the ratio 2 : 3 and their heights are in the ratio 5 : 3. The ratio of their volumes is : **(SSC CGL Tier I 2002)**

(*a*) 27 : 20 (*b*) 20 : 27 (*c*) 9 : 4 (*d*) 4 : 9

Sol. (*b*) Volume of cylinder $= \pi r^2 h \Rightarrow$ Required ratio $= \pi\,(2)^2 5 : \pi\,(3)^2 3 = 20 : 27$

78. If the height of a cone is doubled and its base diameter is trebled, then the ratio of the volume of the resultant cone to that of the original cone is : (CDS 2003)

(a) 6 : 1 (b) 9 : 1 (c) 18 : 1 (d) 9 : 2

Sol. (c) Volume of cone $= \pi r^2 h$

Required ratio $= \pi R^2 H : \pi r^2 h = (3)^2(2) : 1 = 18 : 1$

80. If the radius of a right circular cylinder is doubled and the height is halved, then the ratio between the new volume and the previous volume of the cylinder is **(SSC FCI 2012)**

(*a*) 2 : 1 (*b*) 1 : 2 (*c*) π : 2 (*d*) 2 : π

Sol. (*b*) Ratio of radii = 1 : 2, ratio of heights = 2 : 1

Volume $= \pi r^2 h \Rightarrow$ Required ratio $= (1)^2\,2 : (2)^2 1 = 1 : 2$

81. The radii of the base of a cylinder and a cone are in the ratio $\sqrt{3}:\sqrt{2}$ and their heights are in the ratio $\sqrt{2}:\sqrt{3}$. Their volumes are in the ratio of **(SSC CGL Tier I 2012)**

(*a*) $\sqrt{3}:\sqrt{2}$ (*b*) $3\sqrt{3}:\sqrt{2}$ (*c*) $\sqrt{3}:2\sqrt{2}$ (*d*) $\sqrt{2}:\sqrt{6}$

Sol. (*b*) Volume of cylinder $= \pi r^2 h$ and volume of cone $= \frac{1}{3}\pi r^2 h$

$\Rightarrow$ Required ratio $= 3(\sqrt{3})^2(\sqrt{2}):(\sqrt{2})^2(\sqrt{3}) = 9\sqrt{2}:2\sqrt{3} = 3\sqrt{3}:\sqrt{2}$

82. If two cylinders of equal volume have their heights in the ratio 2 : 3, then the ratio of their radii is **(CDS 2003)**

(a) 2 : 3 (b) 3 : 2 (c) $\sqrt{3}:\sqrt{2}$ (d) $\sqrt{2}:\sqrt{3}$

Sol. (c) $V = \pi r^2 h \Rightarrow \pi r_1^2 h = \pi r_2^2 h$

$\Rightarrow 2r_1^2 = 3r_2^2 \Rightarrow r_1^2 : r_2^2 = 3:2 \Rightarrow r_1 : r_2 = \sqrt{3}:\sqrt{2}$

83. A sphere and a cylinder have equal volume and equal radius. The ratio of the curved surface area of the cylinder to that of the sphere is **(SSC MTS 2011)**

(*a*) 4 : 3 (*b*) 2 : 3 (*c*) 3 : 2 (*d*) 3 : 4

Sol. (*b*) $\frac{4}{3}\pi r^3 = \pi r^2 h \Rightarrow 4r = 3h \Rightarrow r : h = 3 : 4$

Required ratio $= 2\pi rh : 4\pi r^2 = h : 2r = 4 : 6 = 2 : 3$

84. The volumes of a sphere and a right circular cylinder having the same radius are equal. The ratio of the diameter of the sphere to the height of the cylinder is : **(SSC CGL Tier I 2007)**

(*a*) 3 : 2 (*b*) 2 : 3 (*c*) 1 : 2 (*d*) 2 : 1

Sol. (*a*) Volume of sphere = Volume of cylinder $\Rightarrow \frac{4}{3}\pi r^3 = \pi r^2 h \Rightarrow r : h = 3 : 4$

$\therefore d : h = (3 \times 2) : 4 = 3 : 2$

85. The volume of a cone is equal to that of a sphere. If the diameter of base of cone is equal to the diameter of the sphere, what is the ratio of height of cone to the diameter of the sphere sphere ? **(CDS 2010)**

(*a*) 2 : 1 (*b*) 1 : 2 (*c*) 3 : 1 (*d*) 4 : 1

Sol. (*a*) $\frac{1}{3}\pi r^2 h = \frac{4}{3}\pi r^3 \Rightarrow h = 4r \Rightarrow h : r = 4 : 1 \Rightarrow h : d = 4 : 2 = 2 : 1$

86. The total surface area of a solid right circular cylinder is twice that of a solid sphere. If they have the same radii, the ratio of the volume of the cylinder to that of the sphere is given by **(SSC CPO SI 2011)**

(*a*) 9 : 4 (*b*) 2 : 1 (*c*) 3 : 1 (*d*) 4 : 9

Sol. (*a*) $2\pi r(h + r) = 2 \times 4\pi r^2 \Rightarrow h + r = 4r \Rightarrow h = 3r$

Ratio of volumes $= \pi r^2 h : \frac{4}{3}\pi r^3 = \pi r^2(3r) : \frac{4}{3}\pi r^3 = 9 : 4$

87. If the height of a given cone be doubled and radius of the base remains same, the ratio of the volumes of the given cone to that of the second will be : **(SSC CGL Tier I 2003)**

(*a*) 2 : 1 (*b*) 1 : 8 (*c*) 1 : 2 (*d*) 8 : 1

Sol. (*c*) Volume of cone $= \frac{1}{3}\pi r^2 h \Rightarrow$ Required ratio $= 1h : 2h = 1 : 2$

88. If the radius of the base of a cone be doubled and height is left unchanged, then the ratio of the volume of new cone to that of the original cone will be : **(SSC CGL Tier I 2004)**

(*a*) 1 : 4 (*b*) 2 : 1 (*c*) 1 : 2 (*d*) 4 : 1

Sol. (*d*) Volume of cone $= \frac{1}{3}\pi r^2 h$. Radius of new cone $= 2r$

Required ratio $= \frac{1}{3}\pi(2r)^2 h : \frac{1}{3}\pi r^2 h = 4 : 1$

89. The ratio of the volumes of two cones is 2 : 3 and the ratio of radii of their bases is 1 : 2. The ratio of their heights is : **(SSC CGL Tier I 2002, SSC CGL Tier I 2013)**

(*a*) 3 : 8 (*b*) 8 : 3 (*c*) 4 : 3 (*d*) 3 : 4

Sol. (*b*) Volume of cone $= \frac{1}{3}\pi r^2 h$. Let heights of two cones be h_1 and h_2.

Then $\frac{(1)^2 h_1}{(2)^2 h_2} = \frac{2}{3} \Rightarrow \frac{h_1}{h_2} = \frac{2}{3} \times \frac{4}{1} = \frac{8}{3}$ **Trick:** Ratio $= \frac{2}{1^2} : \frac{3}{2^2} = \frac{2}{1} : \frac{3}{4} = 8 : 3$

90. The diameters of two cylinders, whose volumes are equal, are in the ratio 3 : 2. Their height will be in the ratio **(SSC CHSL 2010)**

(*a*) 4 : 9 (*b*) 5 : 6 (*c*) 5 : 8 (*d*) 8 : 9

Sol. (*a*) Volume of cylinder $= \pi r^2 h$

Let heights of two cylinders be h_1 and h_2. Then $\pi(3)^2 h_1 = \pi(2)^2 h_2 \Rightarrow h_1 : h_2 = 4 : 9$

91. The radii of two cylinders are in the ratio 2 : 3 and their curved surface areas are in the ratio 5 : 3. What is the ratio of their volumes? **(CDS 2011)**

(*a*) 20 : 27 (*b*) 10 : 9 (*c*) 9 : 10 (*d*) 27 : 20

Sol. (*b*) Curved surface area $= 2\pi rh \Rightarrow$ Ratio of $rh = 5 : 3$, Ratio of radii $= 2 : 3$

$\therefore$ Ratio of heights $= \frac{5}{2} : \frac{3}{3} = 5 : 2$

Ratio of volumes $= \pi r^2 h$ or $r^2 h = 2^2 \times 5 : 3^2 \times 2 = 10 : 9$

92. The height of two cylinders, whose volumes are equal, are in the ratio 1 : 3. Their radii will be in the ratio

(*a*) 3 : 1 (*b*) 1 : 3 (*c*) 1 : $\sqrt{3}$ (*d*) 3 : $\sqrt{3}$

Sol. (*d*) Volume of cylinder $= \pi r^2 h$. Let radii of two cylinders be r_1 and r_2.

Then $\pi(r_1)^2 1 = \pi(r_2)^2 3 \Rightarrow r_1 : r_2 = \sqrt{3} : 1 = 3 : \sqrt{3}$

Short-cut : $\frac{1}{\sqrt{1}} : \frac{1}{\sqrt{3}} = \sqrt{3} : 1 = 3 : \sqrt{3}$

93. The base of a pyramid of volume $48\sqrt{3}$ cm³ is an equilateral triangle. If the height of the pyramid is 4 cm, then each side of the equilateral triangle at the base is **(SSC CHSL 2013)**

(*a*) 3 cm (*b*) 4 cm (*c*) 6 cm (*d*) 12 cm

Sol. (*d*) Volume of pyramid $= \frac{1}{3} \times$ Base Area $\times$ Height $= \frac{1}{3} \times$ Base Area $\times 4 = 48\sqrt{3}$ cm³

$\Rightarrow$ Area of base $= \frac{\sqrt{3}}{4} a^2 = 48\sqrt{3} \times \frac{3}{4} = 36\sqrt{3} \Rightarrow a^2 = 36\sqrt{3} \times \frac{4}{\sqrt{3}} = 144 \Rightarrow a = 12$

94. A pipe of diameter d can drain a certain water tank in 40 minutes. The time taken by a pipe of diameter $2d$ for doing the same job is : **(SSC CGL Tier I 2000)**

(*a*) 5 minutes (*b*) 10 minutes (*c*) 20 minutes (*d*) 80 minutes

Sol. (*b*) Ratio of diameter/radii of two pipes $= d : 2d = 1 : 2$

Ratio of work $= 1^2 : 2^2 = 1 : 4 \Rightarrow$ Ratio of time $= 4 : 1$

$\therefore$ Time taken by second pipe $= \frac{1}{4} \times 40 = 10$ minutes

95. A square card board has a side 10 cm. From each corner a square of 3 cm is cut down and then the corners are folded to make an open box. What is the volume of open box ? **(SSC CGL)**

(*a*) 27 cm^3 (*b*) 48 cm^3

(*c*) 64 cm^3 (*d*) 90 cm^3

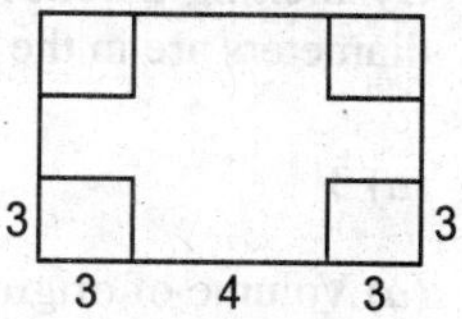

Sol. (*b*) Volume of Box $= l \times b \times h = 4 \times 4 \times 3 = 48$ cm^3

96. A square of side 3 cm is cut off from each corner of a rectangular sheet of length 24 cm and breadth 18 cm and the remaining sheet is folded to form an open rectangular box. The surface area of the box is **(SSC CHSL 2013)**

(*a*) 423 cm^2 (*b*) 468 cm^2 (*c*) 396 cm^2 (*d*) 612 cm^2

Sol. (*c*) Area of rectangular sheet $= 24 \times 18 = 432$ cm^2

Area of one square cut off $= 3^2 = 9$ cm^2

$\therefore$ Required area $= 432 - 4 \times 9 = 432 - 36 = 396$ cm^2

97. A rectangular sheet of paper 44 cm × 20 cm is rolled along its length to form a cylinder. The volume of the cylinder thus formed is : **(CDS 2003, CDS 2015 M)**

(*a*) 1540 cm^3 (*b*) 1450 cm^3 (*c*) 2900 cm^3 (*d*) 3080 cm^3

Sol. (*d*) Circumference of base of cylinder $= 2\pi r = 44$

$\Rightarrow r = 44 \times \frac{1}{2} \times \frac{7}{22} = 7$ cm $\Rightarrow V = \pi r^2 h = \frac{22}{7} \times 7^2 \times 20 = 3080$ cm^3

98. Three solid spheres of metal whose radii are 1 cm, 6 cm and 8 cm are melted to form an other solid sphere. The radius of new sphere is : **(SSC CGL Tier I 2001)**

(*a*) 10.5 cm (*b*) 9.5 cm (*c*) 10 cm (*d*) 9 cm

Sol. (*d*) Total volume $= \frac{4}{3}\pi\,(1^3 + 6^3 + 8^3) = \frac{4}{3}\pi\,(1 + 216 + 512) = \frac{4}{3}\pi\,(729) = \frac{4}{3}\pi\,(9)^3$

$\therefore$ Radius of new sphere = 9 cm

99. Three solid metallic spheres of diameters 6 cm, 8 cm and 10 cm are melted and recast into a new solid sphere. The diameter of the new sphere is : **(SSC CGL Tier I 2002, CDS 2015)**

(*a*) 4 cm (*b*) 6 cm (*c*) 8 cm (*d*) 12 cm

Sol. (*d*) Radii of three spheres = 3 cm, 4 cm, 5 cm

Total volume $= \frac{4}{3}\pi\,(3^3 + 4^3 + 5^3) = \frac{4}{3}\pi\,(27 + 64 + 125) = \frac{4}{3}\pi\,(216) = \frac{4}{3}\pi\,(6)^3$

$\therefore$ Radius of new sphere = 6 cm $\Rightarrow$ Diameter of the new sphere $= 2 \times 6$ cm $= 12$ cm

100. A metal sphere of radius 12 cm is melted into 3 smaller spheres. If the radii of two smaller spheres are 6 cm and 8 cm, the radius of the third is **(SSC FCI Asstt. 2012)**

(*a*) 10 cm (*b*) 12 cm (*c*) 14 cm (*d*) 16 cm

Sol. (*a*) $\frac{4}{3}\pi\ (12)^3 = \frac{4}{3}\pi(6)^3 + \frac{4}{3}\pi(8)^3 + \frac{4}{3}\pi r^3 \Rightarrow 12^3 = 6^3 + 8^3 + r^3$

$\Rightarrow 1728 = 216 + 512 + r^3 \Rightarrow r^3 = 1728 - 216 - 512 = 1000 = 10^3$

101. 3 cubes whose edges are 3 cm, 4 cm and 5 cm respectively are melted to form a single cube. The surface area of the new cube will be **(SSC CHSL 2012)**

(*a*) 215 sq. m. (*b*) 216 sq. m. (*c*) 115 sq. m. (*d*) 150 sq. m.

Sol. (*b*) Total volume of small cubes = $3^3 + 4^3 + 5^3 = 27 + 64 + 125 = 216 = 6^3$

Surface area of new cube = $6 \times (\text{side})^2 = 6 \times (6)^2 = 216$ sq. m.

102. By melting a solid lead sphere of diameter 12 cm, three small spheres are made whose diameters are in the ratio 3 : 4 : 5. The radius (in cm) of the smallest sphere is **(SSC CGL Tier I 2005)**

(*a*) 3 (*b*) 6 (*c*) 1.5 (*d*) 4

Sol. (*a*) Volume of original sphere = $\frac{4}{3}\pi r^3 = \frac{4}{3}\pi(6)^3 = \frac{4}{3}\pi(216)$

Let the radii of small spheres be $3x$, $4x$ and $5x$ cm respectively.

Total volumes of small spheres = $\frac{4}{3}\pi(3x)^3 + \frac{4}{3}\pi(4x)^3 + \frac{4}{3}\pi(5x)^3$

$= \frac{4}{3}\pi\ (27x^3 + 64x^3 + 125x^3) = \frac{4}{3}\pi\ (216x^3) \Rightarrow \frac{4}{3}\pi\ (216x^3) = \frac{4}{3}\pi(216) \Rightarrow x^3 = 1 \Rightarrow x = 1$

Radius of smallest sphere = $3x = 3 \times 1$ cm = 3cm

103. How many cubes of 2 cm, a side can be made by melting a solid cube of 6 cm, a side?

(*a*) 8 (*b*) 9 (*c*) 27 (*d*) 216

Sol. (*c*) Ratio of sides of two cubes = 6 : 2 = 3 : 1

Ratio between volumes of two cubes = $3^3 : 1^3 = 27 : 1$

$\therefore$ 27 small cubes can be made from the bigger cube.

104. How many spheres of radius 2 cm can be made by melting two spheres of radius 12 cm each?

(*a*) 18 (*b*) 108 (*c*) 216 (*d*) 432

Sol. (*d*) Radius of big sphere = 12 cm. Radius of smaller sphere = 2 cm

Ratio between radii of two spheres = 12 : 2 = 6 : 1

Ratio between volumes of two spheres = $6^3 : 1^3 = 216 : 1$

$\therefore$ 216 spheres can be made from one big sphere.

$\therefore$ From two big spheres, smaller spheres made = 216 × 2 = 432

105. How many cubes, each of edge 3 cm, can be cut from a cube of edge 15 cm? **(SSC CGL 2000)**

(*a*) 25 (*b*) 27 (*c*) 125 (*d*) 144

Sol. (*c*) Ratio of radii = 3 : 15 = 1 : 5 $\Rightarrow$ Ratio of volumes = $1^3 : 5^3 = 1 : 125$

106. A solid cube of side 3 cm of metal weights 15 g. Find the weight of similar cube of side 12 cm.

(*a*) 60 g (*b*) 300 g (*c*) 240 g (*d*) None of these

Sol. (*d*) Ratio of sides of cubes = 3 : 12 = 1 : 4 $\Rightarrow$ Ratio of their volumes = $1^3 : 4^3 = 1 : 64$

$\therefore$ Weight of bigger cube = 64 × 15g = 960 g

107. A cube of 9 cm a side is melted and 27 small cubes are made from that. Find side of the smaller cubes.

(*a*) 1.5 cm (*b*) 2 cm (*c*) 2.5 cm (*d*) 3 cm

Sol. (*d*) Volume of cube = Side^3

$\therefore$ Volume of the small cube = $\frac{9^3}{27} = 27 \Rightarrow$ Side of the small cube = $\sqrt[3]{27} = 3$ cm

108. A cylindrical iron rod whose height is four times its radius is melted and cast into spherical balls, each having the same radius as that of the rod. The number of spherical balls will be : **(SSC CGL Tier I 2006)**

(*a*) 1 (*b*) 2 (*c*) 3 (*d*) 4

Sol. (*c*) Volume of rod = $\pi r^2 h = \pi r^2(4r) = 4\pi r^3$

Volume of one ball = $\frac{4}{3}\pi r^3 \Rightarrow$ 3 balls can be made.

109. A large solid metallic cylinder whose radius and height are equal to each other is to be melted and 48 identical solid balls are to be recast from the liquid metal so formed. What is the ratio of the radius of a ball to the radius of the cylinder ? **(CDS 2012)**

(*a*) 1 : 16 (*b*) 1 : 12 (*c*) 1 : 8 (*d*) 1 : 4

Sol. (*d*) Volume of cylinder = $\pi R^2 h = \pi R^3$ ($\because R = h$)

Volume of each ball = $\frac{4}{3}\pi r^3 = \pi R^3 \div 48 \Rightarrow \pi r^3 = \pi R^3 \times \frac{1}{48} \times \frac{3}{4} \Rightarrow = \pi R^3 \times \frac{1}{64}$

$\Rightarrow r^3 = R^3 \times \frac{1}{64} \Rightarrow r = R \times \frac{1}{4} \Rightarrow r : R = 1 : 4$

110. The volume of a conical tent is 1232 cu.m and the area of its base is 154 sq.m. Find the length of the canvas required to build the tent, if the canvas is 2 m in width. **(SSC CGL Tier II 2013)**

(*a*) 270 m (*b*) 272 m (*c*) 276 m (*d*) 275 m

Sol. (*d*) $\frac{1}{3}\pi r^2 h = 1232$ and $\pi r^2 = 154 \Rightarrow h = \frac{1232}{154} \times 3 = 24$

$\pi r^2 = 154 \Rightarrow r^2 = 154 \times \frac{7}{22} = 7^2 \Rightarrow r = 7 \Rightarrow l = \sqrt{r^2 + h^2} = \sqrt{7^2 + 24^2} = 25$

$\pi r l = \frac{22}{7} \times 7 \times 25 = 550 \Rightarrow$ Cloth required = $550 \div 2 = 275$ m

111. The length of canvas, 75 cm wide required to build a conical tent of height 14 m and the floor area 346.5 m^2 is **(SSC CGL Tier I 2015)**

(*a*) 490 m (*b*) 770 m (*c*) 860 m (*d*) 665

Sol. (*b*) $\pi r^2 = 346.5 \Rightarrow r^2 = 110.25 \Rightarrow r = 10.5$ m $h = 14$ m (given)

$l = \sqrt{10.5^2 + 14^2} = \sqrt{(3.5 \times 3)^2 + (3.5 \times 4)^2} = 3.5\sqrt{3^2 + 4^2} = 3.5 \times 5 = 17.5$

Slant surface area = $\pi r l = \frac{22}{7} \times 10.5 \times 17.5$

Length of cloth = $\frac{22}{7} \times 10.5 \times 17.5 \times \frac{4}{3} = 770$ m

112. A hollow sphere of internal and external diameters 4 cm and 8 cm respectively is melted into a cone of base diameter 8 cm. What is the height of the cone ? **(CDS 2011)**

(*a*) 12 cm (*b*) 14 cm (*c*) 17 cm (*d*) 18 cm

Sol. (*b*) Volume of metal of sphere = $\frac{4}{3}\pi(4^3 - 2^3) = \frac{4}{3}\pi(56)$

Volume of cone = $\frac{1}{3}\pi(4)^2 h = \frac{4}{3}\pi(56) \Rightarrow h = 14$ cm

112. 5 persons will live in a tent. If each person requires 16 m^2 of floor area and 100 m^3 space for air then the height of cone of smallest size to accommodate these persons would be ? **(SSC CGL Tier I 2015)**

(*a*) 18.75 m (*b*) 16 m (*c*) 10.25 m (*d*) 20 m

Sol. (*a*) Volume of cone = $\frac{1}{3}\pi r^2 h = \frac{1}{3} h \times \pi r^2 = \frac{1}{3} \times$ height × Floor area

Floor Area : Volume = 16 : 100 ⇒ Height = $\frac{100 \times 3}{16} = \frac{75}{4} = 18.75$

113. A solid metal sphere, 6 cm in diameter, is formed into a cylindrical tube 10 cm, in external diameter and 4 cm in length. Find the thickness of the tube. **(NDA)**

(*a*) 0.5 cm (*b*) 1 cm (*c*) 2 cm (*d*) 2.5 cm

Sol. (*b*) Volume of original sphere = $\frac{4}{3}\pi r^3 = \frac{4}{3}\pi(3)^3 = 36\pi$

Volume of metal used in tube = $\pi(R^2 - r^2)h = \pi(5^2 - r^2) \times 4 = 36\pi$

$\Rightarrow 25 - r^2 = 9 \Rightarrow r^2 = 16 \Rightarrow r = 4 \Rightarrow$ Thickness of tube = 5 – 4 = 1 cm

114. If the area of the circular shell having inner and outer radii of 8 cm and 12 cm respectively is equal to the surface area of a cylinder of radius R_1 and height h, then h, in terms of R_1 will be: **(SSC CGL Tier II 2012)**

(*a*) $\frac{30 - R_1}{R_1^2}$ (*b*) $\frac{40 - R_1^2}{R_1}$ (*c*) $\frac{3R_1^2 - 30}{7R_1}$ (*d*) $\frac{R_1^2 - 40}{R_1^2}$

Sol. (*b*) $2\pi R_1(R_1 + h) = \pi(r_2^2 - r_1^2)$

$\Rightarrow 2R_1(R_1 + h) = 12^2 - 8^2 = 144 - 64 = 80$

$\Rightarrow R_1^2 + R_1 h = 40 \Rightarrow R_1 h = 40 - R_1^2 \Rightarrow h = \frac{40 - R_1^2}{R_1}$

115. The sum of radius of the base and height of a cylinder is 37 cm. If the total surface area of cylinder is 1628 cm^2. Find volume of the cylinder. **(CDS 2011)**

(*a*) 3620 cm^2 (*b*) 4800 cm^2 (*c*) 4620 cm^2 (*d*) 5400 cm^2

Sol. (c) Total surface Area = $2\pi r(h + r) = 1628 \Rightarrow 2\pi r = 1628 \div 37 = 44 \Rightarrow r = 44 \times \frac{1}{2} \times \frac{7}{22} = 7$ cm

$h + r = 37 \Rightarrow h = 37 - 7 = 30 \Rightarrow$ Volume = $\pi r^2 h = \frac{22}{7} \times 7 \times 7 \times 30 = 4620$ cm^2

116. A copper rod of 1 cm diameter and 8 cm length is drawn into a wire of uniform diameter and 18 m length. The radius (in cm) of the wire is : **(SSC CGL Tier I 2001)**

(*a*) $\frac{1}{15}$ (*b*) $\frac{1}{30}$ (*c*) $\frac{2}{15}$ (*d*) 15

Sol. (*b*) $\pi\left(\frac{1}{2}\right)^2 8 = \pi (x)^2 (1800) \Rightarrow 2 = 1800 x^2 \Rightarrow x^2 = \frac{2}{1800} = \frac{1}{900} \Rightarrow x = \sqrt{\frac{1}{900}} = \frac{1}{30}$

117. A solid cylinder of metal, whose radius is 6 cm and height is 125 cm is melted into a sphere. Find the radius of the sphere.

(*a*) 3 cm (*b*) 5 cm (*c*) 12 cm (*d*) 15 cm

Sol. (*d*) Volume of cylinder = $\pi r^2 h = \pi (6)^2 125$

Volume of sphere = $\frac{4}{3}\pi r^3 = \pi (6)^2 125 \Rightarrow r^3 = \pi \times 6 \times 6 \times 125 \times \frac{3}{4} \times \frac{1}{\pi} \Rightarrow r = 3 \times 5 = 15$ cm

118. A wire is drawn from a solid iron sphere of radius 6 cm. Find the length of the wire, if its diameter is 6 cm.

(*a*) 16 cm (*b*) 32 cm (*c*) 50 cm (*d*) 64 cm

Sol. (*b*) An iron wire is a form of cylinder. Here, Volume of cylinder = Volume of Sphere

$\therefore \pi$ (Radius of wire)^2 (length of wire) = $\frac{4}{3}\pi$ (Radius of sphere)^3

$\therefore$ Length of wire = $\frac{4}{3} \times \frac{\pi \text{ (Radius of sphere)}^3}{\pi \text{ (Radius of cylinder)}^2} = \frac{4}{3} \times \frac{6 \times 6 \times 6}{3 \times 3} = 32$ cm

119. A rectangular block of metal has dimensions 21 cm, 77 cm and 24 cm. The block has been melted into a sphere. The radius of the sphere is **(SSC CGL Tier II 2013)**

(*a*) 21 cm (*b*) 7 cm (*c*) 14 cm (*d*) 28 cm

Sol. (*a*) $\frac{4}{3}\pi r^3 = 21 \times 77 \times 24 \Rightarrow r^3 = 21 \times 77 \times 24 \times \frac{3}{4} \times \frac{7}{22} = 21^3 \Rightarrow r = 21$ cm

120. A right circular cone is 3.6 cm high and radius of its base is 1.6 cm. It is melted and recast into a right circular cone with radius of its base as 1.2 cm. Then the height of the cone (in cm) is **(SSC CGL Tier II 2013)**

(*a*) 3.6 (*b*) 4.8 (*c*) 6.4 (*d*) 7.2

Sol. (*c*) Required height $= \frac{1.6 \times 1.6 \times 3.6}{1.2 \times 1.2} = 6.4$

121. A solid consists of a circular cylinder with exact fitting right circular cone placed on the top. The height of the cone is *h*. If total volume of the solid is three times the volume of the cone, then the height of the circular cylinder is **(SSC CGL Tier II 2012)**

(*a*) $4h$ (*b*) $\frac{3h}{2}$ (*c*) $2h$ (*d*) $\frac{2h}{3}$

Sol. (*d*) Volume of cone : Volume of cylinder (with equal radius and height) = 1 : 3

$\Rightarrow$ Ratio of height of cone and cylinder (with equal volumes) = 3 : 1

Here, volume of cylinder = 3 – 1 = 2 times of volume of cone.

$\therefore$ Height of cylinder $= \frac{2h}{3}$

122. Water flows at the rate of 10 metres per minute from a cylindrical pipe 5 mm in diameter. How long will it take to fill up a conical vessel whose diameter at the base is 40 cm and depth 24 cm? **(SSC CGL Tier II 2012)**

(*a*) 48 mins 15 secs (*b*) 55 mins (*c*) 51 mins 12 secs (*d*) 52 mins 1 sec

Sol. (*c*) Volume of conical vessel $= \frac{1}{3}\pi r^2 h = \frac{1}{3}\pi(20)^2(24)$

Water flow per minute $= \pi r^2 h = \pi\left(\frac{1}{4}\right)^2(1000)$

Required time $= \frac{1}{3}\pi(20)^2(24) \div \pi\left(\frac{1}{4}\right)^2(1000) = 51.2$ mins = 51 mins 12 secs

123. If V be the volume and S the surface area of a cuboid of dimensions *a*, *b*, *c* then : $\frac{2}{S}\left(\frac{1}{a}+\frac{1}{b}+\frac{1}{c}\right)$ = ? **(NIFT 2004)**

(*a*) V (*b*) $\frac{1}{V}$ (*c*) S (*d*) $\frac{1}{S}$

Sol. (*b*) Here, V = *abc* and S = 2(*ab* + *bc* + *ca*)

$$\Rightarrow \frac{2}{S}\left(\frac{1}{a}+\frac{1}{b}+\frac{1}{c}\right) = \frac{2}{S}\left(\frac{ab+bc+ca}{abc}\right) = \frac{S}{S \times V} = \frac{1}{V}$$

124. The volume of a right circular cylinder is equal to the volume of that right circular cone whose height is 108 cm and diameter of base is 30 cm. If the height of the cylinder is 9 cm, the diameter of its base is : **(SSC CGL Tier I 2002)**

(*a*) 30 cm (*b*) 60 cm (*c*) 50 cm (*d*) 40 cm

Sol. (*b*) Radius of cone = 30 ÷ 2 = 15 cm. Let radius of cylinder be *r* cm.

Then $\frac{1}{3}\pi(15)^2 108 = \pi r^2(9) \Rightarrow r^2 = \frac{15 \times 15 \times 108}{3 \times 9} = 900 \Rightarrow r = 30$ cm ⇒ Diameter = 60 cm

125. 12 spheres of same size are made by melting a solid cylinder of 16 cm diameter and 2 cm height. The diameter of each sphere is: **(SSC CGL Tier I 2005)**

(*a*) 2 cm (*b*) 4 cm (*c*) 3 cm (*d*) $\sqrt{3}$ cm

Sol. (*b*) Volume of cylinder = $\pi r^2 h = \pi (8)^2 2 = 128\pi$

Volume of 1 sphere = $\frac{4}{3}\pi r^3 = \frac{128}{12}\pi \Rightarrow r^3 = \frac{128}{12} \times \frac{3}{4} = 8 \Rightarrow r = 2$ cm ⇒ Diameter = 4 cm

126. A right circular cylinder is formed by rolling a rectangular sheet of length 24 cm and breadth 22 cm along its breadth. The volume of the cylinder is : (use π = 22/7) **(SSC CGL Tier I 2006)**

(*a*) 924 cm³ (*b*) 462 cm³ (*c*) 264 cm³ (*d*) 528 cm³

Sol. (*a*) Let radius of the cylinder is *r* cm.

$2\pi r$ = Breadth $\Rightarrow 2\pi r = 22 \Rightarrow r = 22 \div 2\pi = 22 \times \frac{1}{2} \times \frac{7}{22} = \frac{7}{2}$ cm

Volume of cylinder = $\pi r^2 h = \frac{22}{7} \times \frac{7}{2} \times \frac{7}{2} \times 24 = 924$ cm³

127. The size of a rectangular piece of paper is 100 cm × 44 cm. A cylinder is formed by rolling the paper along its length. The volume of the cylinder is : **(SSC CGL Tier I 2007)**

(*a*) 4400 cm³ (*b*) 15400 cm³ (*c*) 35000 cm³ (*d*) 144 cm³

Sol. (*c*) Circumference of base of cylinder = length of sheet = 100 cm

$\Rightarrow 2\pi r = 100$ cm $\Rightarrow r = \frac{50}{\pi}$

Volume of cylinder = $\pi r^2 h = \pi\left(\frac{50}{\pi}\right)^2 h = \pi \times \frac{2500}{\pi} \times \frac{1}{\pi} \times 44 = 2500 \times \frac{7}{22} \times 44 = 35000$ cm³

128. A rectangular sheet of paper 30 cm × 18 cm can be transformed into right circular cylinder in two ways i.e. by rolling the paper along its length or by rolling it along its breadth. Find the ratio of volumes of the two cylinders thus formed.

(*a*) 5 : 3 (*b*) 3 : 5 (*c*) 9 : 25 (*d*) 25 : 9

Sol. (*a*) 2πR = 30 and 2π*r* = 18 ⇒ R : *r* = 30 : 18 = 5 : 3; Ratio of heights = 18 : 30 = 3 : 5

Required ratio ($\pi r^2 h$) = $\pi(5)^2 3 : \pi(3)^2 5 = 5 : 3$

129. A right triangle with sides 3 cm, 4 cm and 5 cm is rotated about the side 3 cm to form a cone. The volume of the cone so formed is : **(SSC CGL Tier I 2007)**

(*a*) 16π cm³ (*b*) 12π cm³ (*c*) 15π cm³ (*d*) 20π cm³

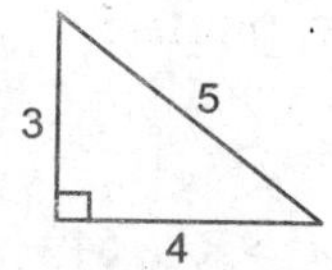

Sol. (*a*) Radius of cone = 4 cm, and height of the cone = 3 cm

Volume of cone = $\frac{1}{3}\pi r^2 h = \frac{1}{3} \times \pi \times 4^2 \times 3 = 16\pi \text{ cm}^3$

130. A wooden box measures 20 cm by 12 cm by 10 cm. Thickness of wood is 1 cm. Volume of wood to make the box (in cubic cm) is : **(SSC CGL Tier I 2003)**

(*a*) 960 (*b*) 519 (*c*) 2400 (*d*) 1120

Sol. (*a*) Outer dimensions of box = 20 cm, 12 cm, 10 cm

Inner dimensions of box = 18 cm, 10 cm, 8 cm

$\therefore$ Volume of wood = $(20 \times 12 \times 10) - (18 \times 10 \times 8) = (2400 - 1440) \text{ cm}^3 = 960 \text{ cm}^3$

131. A wooden box without lid has external length, breadth and height 11 cm, 9 cm and 7 cm respectively. If wood is 5 mm thick, find the volume of wood used in making the box.

(*a*) 173 cm³ (*b*) 213 cm³ (*c*) 273 cm³ (*d*) 693 cm³

Sol. (*a*) External volume of box = $11 \times 9 \times 7 = 693$ cm

Internal volume of box = $(11 - 1) \times (9 - 1) \times (7 - 0.5) = 10 \times 8 \times 6.5 = 520 \text{ cm}^3$

$\therefore$ Volume of wood = $693 - 520 = 173 \text{ cm}^3$

132. The curved surface area and the total surface area of a cylinder are in the ratio 1 : 2. If the total surface area of the right cylinder is 616 cm², then its volume is : **(SSC CGL Tier I 2013)**

(*a*) 1078 cm³ (*b*) 1232 cm³ (*c*) 1848 cm³ (*d*) 1632 cm³

Sol. (*a*) $2\pi rh : (2\pi rh + 2\pi r^2) = 1 : 2 \Rightarrow 2\pi rh = 2\pi r^2 \Rightarrow h = r$

Now, $(2\pi rh + 2\pi r^2) = 616 \Rightarrow 4\pi r^2 = 616 \Rightarrow r^2 = 616 \times \frac{1}{4} \times \frac{7}{22} = 49 \Rightarrow r = 7$

Volume of cylinder = $\pi r^2 h = \frac{22}{7} \times 7 \times 7 \times 7 = 1078 \text{ cm}^3$

133. A hollow cylindrical tube 20 cm long, is made of iron and its external and internal diameters are 8 cm and 6 cm respectively. The volume of iron used in making the tube is : (Take $\pi = 22/7$) **(SSC CGL Tier I 2003)**

(*a*) 1760 cu. cm. (*b*) 880 cu. cm. (*c*) 440 cu. cm. (*d*) 220 cu. cm

Sol. (*c*) External radius = 4 cm and internal radius = 3 cm

Volume of iron used = $\pi(R^2 - r^2)h = \pi(4^2 - 3^2) \times 20 = \frac{22}{7} \times 7 \times 20 = 440 \text{ cm}^3$

134. Weight of a solid metal sphere of radius 4 cm is 4 kg. The weight of a hollow sphere made with same metal, whose outer diameter is 16 cm and inner diameter is 12 cm, is **(SSC MTS 2013)**

(*a*) 20.5 kg (*b*) 15.5 kg (*c*) 16.5 kg (*d*) 18.5 kg

Sol. (*d*) Ratio of volumes = $\frac{4}{3}\pi(4)^3 : \frac{4}{3}\pi(8^3 - 6^3) = 2^3 : (4^3 - 3^3) = 8 : 37$

$\therefore$ Weight of hollow sphere = $\frac{37}{8} \times 4 \text{ kg} = \frac{37}{2} = 18.5 \text{ kg}$

135. 1496 cm³ of metal is used to cast a pipe of length 28 cm. If the internal radius of the pipe is 8 cm, the outer radius of the pipe is

(*a*) 7 cm (*b*) 8 cm (*c*) 9 cm (*d*) 12 cm

Sol. (*c*) $\pi(R^2 - r^2)h = 1496 \Rightarrow \frac{22}{7}(R^2 - 8^2)\,28 = 1496$

$\Rightarrow R^2 - 64 = 1496 \times \frac{7}{22} \times \frac{1}{28} = 17 \Rightarrow R^2 = 17 + 64 = 81 \Rightarrow R = \sqrt{81} = 9$ cm

136. A hollow iron pipe is 21 cm long and its exterior diameter is 8 cm. If the thickness of the pipe is 1 cm and iron weighs 8 g/cm^3, then the weight of the pipe (Take $\pi = 22/7$) is : **(SSC CGL Tier I 2004)**

(*a*) 3.696 kg (*b*) 3.6 kg (*c*) 36 kg (*d*) 36.9 kg

Sol. (*a*) Volume of metal used $= \pi(R^2 - r^2)h = \pi(4^2 - 3^2) \times 21 = \frac{22}{7} \times 7 \times 21 = 462$ cm^3

$\therefore$ Weight of pipe $= 462 \times 8$ g $= 3696$ g $= 3.696$ kg

137. The inner and outer radii of a 7 m long hollow iron right circular cylindrical pipe are 2 cm and 4 cm respectively. If 1000 cm^3 of iron weighs 5 kg, what is the weight of the pipe ? **(CDS 2010)**

(*a*) 264 kg (*b*) 132 kg (*c*) 396 kg (*d*) None of these

Sol. (*b*) Volume of metal $= \pi(R^2 - r^2)h = \pi(4^2 - 2^2)700 = \frac{22}{7} \times 6 \times 2 \times 700 = 26400$ cm^3

Weight of the pipe $= 26.4 \times 5$ kg. $= 132$ kg.

138. The ratio of the volume of a cube and of a solid sphere is 363 : 49. The ratio of an edge of the cube and the radius of the sphere is **(SSC CHSL 2011)**

(*a*) 7 : 11 (*b*) 22 : 7 (*c*) 11 : 7 (*d*) 7 : 22

Sol. (*b*) Here $\frac{a^3}{\frac{4}{3}\pi r^3} = \frac{363}{49} \Rightarrow \frac{a^3}{r^3} = \frac{363}{49} \times \frac{4}{3} \times \frac{22}{7} = \frac{121 \times 8 \times 11}{49 \times 7} = \frac{11^3 \times 2^3}{7^3} \Rightarrow \frac{a}{r} = \frac{11 \times 2}{7} = \frac{22}{7}$

139. A cistern of capacity 8000 litres measures externally 3.3 m by 2.6 m by 1.1. m and its walls are 5 cm thick. The thickness of the bottom is : **(SSC CGL Tier I 2003)**

(*a*) 1 m (*b*) 1.1 m (*c*) 1 dm (*d*) 90 cm

Sol. (*c*) 1 m^3 = 1000 litres $\Rightarrow$ 8000 litres = 8 m^3

Let internal height of cistern be x cm.

Then internal volume of cistern $= (3.3 - 2 \times 0.05) \times (2.6 - 2 \times 0.05) \times x = 8$

$\Rightarrow 3.2 \times 2.5 \times x = 8 \Rightarrow x = 1$ m

$\therefore$ Thickness of bottom $= (1.1 - 1)$ m $= 0.1$ m $= 1$ dm

140. A cylindrical vessel of base radius 14 cm is filled with water to some height. if a rectangular solid of dimensions 22 cm × 7 cm × 5 cm is immersed in it, what is the rise in water lever? **(CDS 2009)**

(*a*) 0.5 cm (*b*) 1.0 cm (*c*) 1.25 cm (*d*) 1.5 cm

Sol. (*c*) Let rise in water level $= h$

Then $\pi r^2 h = 22 \times 7 \times 5 \Rightarrow h = 22 \times 7 \times 5 \times \frac{7}{22} \times \frac{1}{14} \times \frac{1}{14} = \frac{5}{4} = 1.25$ cm

141. A hemispherical bowl of internal radius 20 cm contains sauce. The sauce is to be filled in conical shaped bottles of radius 5 cm and height 8 cm. What is the number of bottles required ? **(CDS 2010)**

(*a*) 100 (*b*) 90 (*c*) 80 (*d*) 75

Sol. (*c*) Required number of bottles = $\dfrac{\frac{2}{3}\pi r^3}{\frac{1}{3}\pi r^2 h} = \dfrac{\frac{2}{3}\pi(20)^3}{\frac{1}{3}\pi(5)^2(8)} = 80$

142. A sphere of radius 3 cm is put into water contained in a cylinder of 6 cm radius. If the sphere is completely immersed in the water, the water level in the cylinder rises by : **(SSC CGL Tier I 1996)**

(*a*) 1 cm (*b*) $\frac{1}{3}$ cm (*c*) $\frac{1}{2}$ cm (*d*) $\frac{2}{3}$ cm

Sol. (*a*) Volume of sphere = $\frac{4}{3}\pi r^3 = \frac{4}{3}\pi(3)^3 = 36\pi$

∴ On immersing the ball in the cylinder, volume of water will rise by volume of ball.

∴ π (Radius of cylinder)^2 (Rise in water level) = 36π

π $(6)^2$ (Rise in water level) = 36π ⇒ Rise in water level = $\frac{36}{(6)^2} = 1$

143. Two iron shots each of diameter 6 cm are immersed in the water contained in a cylindrical vessel of radius 6 cm. The level of the water in the vessel will be raised by : **(SSC CGL Tier I 2009)**

(*a*) 1 cm (*b*) 2 cm (*c*) 3 cm (*d*) 6 cm

Sol. (*b*) Volume of one shot (sphere) = $\frac{4}{3}\pi r^3 = \frac{4}{3}\pi(3)^3 = 36\pi$ cm^3

Volume of two shots = 2 × 36π cm^3 = 72π cm^3

Let the level of water is raised by *h* cm. Then $\pi r^2 h = 72\pi \Rightarrow \pi(6)^2 h = 72\pi \Rightarrow h = 2$ cm

144. A cylindrical rod of iron whose radius is one-fourth of its height is melted and cast into spherical balls of same radius as that of the cylinder. What is the number of spherical balls? **(CDS 2011)**

(*a*) 2 (*b*) 3 (*c*) 4 (*d*) 5

Sol. (*b*) Let radius of cylindrical rod = *r*. Then its height = 4*r*

Volume of cylindrical rod = $\pi r^2 h = \pi(r)^2(4r) = 4\pi r^3$

Volume of sphere = $\frac{4}{3}\pi r^3$ ⇒ Required number of balls = 3

145. Some oil is filled in a right circular cylinder vessel. The radius of the base of the vessel is 6 cm. Some iron balls, each of diameter 3 cm, are completely dipped into the oil. If the level of oil in the vessel rises by 2 cm, the number of iron balls dipped in the oil is: **(SSC CGL Tier I 06)**

(*a*) 4 (*b*) 8 (*c*) 16 (*d*) 32

Sol. (*c*) Radius of each ball = $\frac{3}{2}$ cm ⇒ Volume of each ball = $\frac{4}{3}\pi r^3 = \frac{4}{3}\pi\left(\frac{3}{2}\right)^3 = \frac{9}{2}\pi$ cm^3

Volume of oil level raised in cylindrical vessel = $\pi r^2 h = \pi(6)^2 2 = 72\pi$

Number of balls required = $72\pi \div \frac{9}{2}\pi = 72 \times \frac{2}{9} = 16$ balls

146. A rectangular water tank is 2.1 m long and 1.5 m broad. If 630 litres of water are poured into the tank, how much will the water level rise? **(SSC CGL Tier I 2006)**

(*a*) 0.2 m (*b*) 0.6 m (*c*) 1.5 m (*d*) 2 m

Sol. (*a*) Le x be the rise in water level. Then $2.1 \times 1.5 \times x\ m^3 = 630$ litres

$\Rightarrow 3.15 \times x \times 1000$ litres $= 630$ litres $\Rightarrow x = 630 \times \frac{100}{315} \times \frac{1}{1000} = 0.2$ m

147. Assume that a drop of water is spherical and its diameter is one-tenth of a cm. A conical glass has a height equal to the diameter of its rim. If 32,000 drops of water fill the glass completely, then the height of the glass, in cm, is **(SSC CGL Tier II 2013)**

(*a*) 1 (*b*) 2 (*c*) 3 (*d*) 4

Sol. (*d*) Volume of water $= 32000 \times \frac{4}{3}\pi r^3 = 32000 \times \frac{4}{3}\pi\left(\frac{1}{20}\right)^3 = 32000 \times \frac{4}{3}\pi \times \frac{1}{8000} = \frac{16}{3}\pi$

$\Rightarrow \frac{1}{3}\pi r^2 h = \frac{16}{3}\pi \Rightarrow \frac{1}{3}\pi r^2(2r) = \frac{16}{3}\pi \Rightarrow 2r^3 = 16 \Rightarrow r = 2 \Rightarrow h = 4$ cm

148. The total number of spherical bullets, each of diameter 5 decimeter, that can be made by utilizing the maximum of a rectangular block of lead with 11 metre length, 10 metre breadth and 5 metre width is (assume that $\pi > 3$) **(SSC CGL Tier II 2013)**

(*a*) equal to 8800 (*b*) less than 8800 (*c*) equal to 8400 (*d*) greater than 9000

Sol. (*b*) Volume of block $= 11 \times 10 \times 5 = 550\ m^3$

Volume of 1 bullet $= \frac{4}{3}\pi\left(\frac{1}{4}\right)^3 > \frac{4}{3} \times 3 \times \frac{1}{64} = \frac{1}{16}$ (more than)

$\therefore$ Total number of bullets $< 550 \times 16 =$ less than 8800

149. A solid iron sphere of diameter 6 cm is melted and made to a hollow cylindrical pipe of external diameter 10 cm and of length 4 cm. The width of the sheet of pipe is **(SSC CHSL 2013)**

(*a*) 1 cm (*b*) 2 cm (*c*) 3 cm (*d*) 4 cm

Sol. (*a*) Volume of sphere $= \frac{4}{3}\pi r^3 = \frac{4}{3}\pi(3)^3 = 36\pi$

Let internal radius of cylindrical pipe $= r$ cm.

Then $\pi(5^2 - r^2)4 = 36\pi \Rightarrow 25 - r^2 = 9 \Rightarrow r^2 = 16 \Rightarrow r = 4$

$\Rightarrow$ Width of sheet $= (5 - 4)$cm $= 1$ cm

150. A well of radius 3.5 m is dug 16 m deep. The earth removed is spread over an area of 400 m^2 to form a platform. Height of the platform is **(SSC CGL Tier II 2012)**

(*a*) 7.7 m (*b*) 77 m (*c*) 1.54 m (*d*) 154 m

Sol. (*c*) Volume of earth dug out $= \pi r^2 h = \frac{22}{7} \times \frac{7}{2} \times \frac{7}{2} \times 16 = 616\ m^3$

$\Rightarrow$ Height of platform $= 616 \div 400 = 1.54$ m

151. A well 20 m in diameter is dug 14 m deep and the earth taken out is spread all round it to a width of 5 m to form an embankment. The height of the embankment is : **(SSC CGL Tier I 2004)**

(*a*) 10 metre (*b*) 11 metre (*c*) 11.2 metre (*d*) 11.5 metre

Sol. (*c*) Volume of earth taken out $= \pi r^2 h = \pi(10)^2 \times 14 = 1400\pi$

Area of embankment $= \pi R^2 - \pi r^2 = \pi(15^2 - 10^2) = 125\pi$

$\therefore$ Height of embankment $= \frac{1400\pi}{125\pi} = 11.2$ metre

152. A field is 90 metre long and 50 metre broad. A 25 metre long, 20 metre broad and 4 metre deep tank dug in the field and the earth taken out is spread evenly over the remaining field. How much the level of field will rise? **(SSC CGL Tier I Exam.)**

(*a*) 0.5 metre (*b*) 1 metre (*c*) 1.5 metre (*d*) 2 metre

Sol. (*a*) Area of field = 90 × 50 = 4500 m^2. Area of field dug out = 25 × 20 = 500 m^2

∴ Area of remaining field = 4500 m^2 – 500 m^2 = 4000 m^2

Volume of the earth dug out = 25 × 20 × 4 = 2000 m^3

∴ Field will rise by $\frac{2000}{4000} = 0.5$ metre

153. The height and the radius of the base of a right circular cone are 12 cm and 6 cm respectively. The radius of the circular cross-section of the cone cut by a plane parallel to its base at a distance of 3 cm from the base is : **(SSC CGL Tier I 2008)**

(*a*) 3.5 cm (*b*) 4 cm

(*c*) 4.5 cm (*d*) 5.5 cm

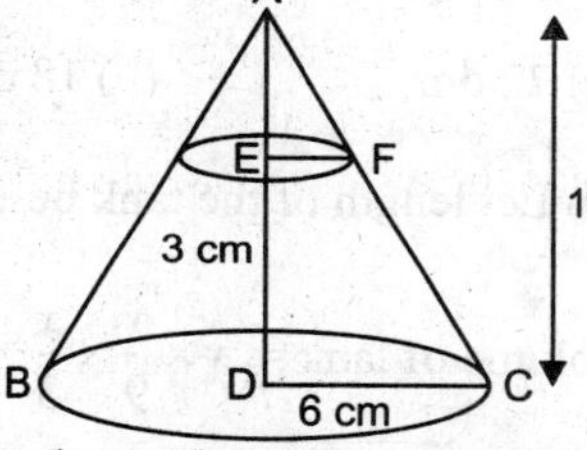

Sol. (*c*) ΔADC and ΔAEF are similar triangles.

$$\therefore \frac{EF}{DC} = \frac{AE}{AD} \Rightarrow \frac{EF}{6} = \frac{9}{12}$$

$$\Rightarrow EF = \frac{9}{12} \times 6 = 4.5 \text{ cm}$$

154. The height of a cone is 30 cm. A small cone is cut off at the top by a plane parallel to its base. If its volume is 1/27 of the volume of the cone, at what height, above the base, is section made? **(CDS 2010)**

(*a*) 6 cm (*b*) 8 cm (*c*) 10 cm (*d*) 20 cm

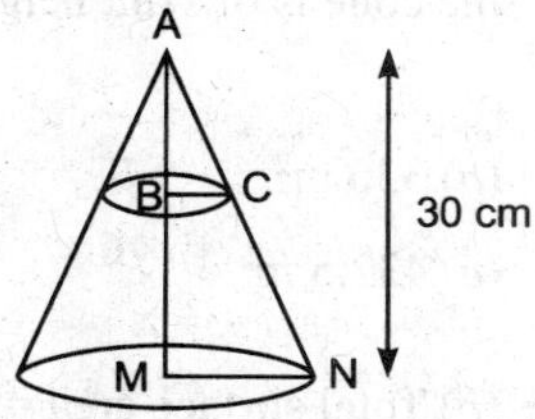

Sol. (*d*) ΔABC ~ ΔAMN $\Rightarrow \frac{AB}{AM} = \frac{BC}{MN}$

Ratio of volumes = 1 : 27

$\Rightarrow$ Ratio of heights/radii = $\sqrt[3]{1} : \sqrt[3]{27}$ = 1 : 3

∴ Height of upper cone = $\frac{1}{3} \times 30 = 10$ cm

$\Rightarrow$ Required Height = (30 – 10) cm = 20 cm

155. A solid cone of height 9 cm with diameter of its base 18 cm is cut out from a wooden solid sphere of radius 9 cm. The per centage of wood wasted is **(SSC FCI 2012)**

(*a*) 25 (*b*) 30 (*c*) 50 (*d*) 75

Sol. (*d*) Ratio of volumes of sphere and cone

$$= \frac{4}{3}\pi r^3 : \frac{1}{3}\pi r^2 h = 4(9)^3 : (9)^2 9 = 4 : 1 \Rightarrow \text{Wood wasted} = \frac{4-1}{4} \times 100 = 75\%$$

156. A cone of height 7 cm and base-radius 1 cm is carved out from a cuboidal block of wood of size 10 cm × 5 cm × 2 cm. In the process, the wood wasted will be : **(SSC CGL Tier I 2006)**

(*a*) $92\frac{2}{3}\%$ (*b*) $46\frac{1}{3}\%$ (*c*) $53\frac{2}{3}\%$ (*d*) $7\frac{1}{3}\%$

Sol. (*a*) Volume of the cuboidal block = 10 × 5 × 2 = 100 cm^3

$$\text{Volume of cone} = \frac{1}{3}\pi r^2 h = \frac{1}{3} \times \frac{22}{7} \times 1^2 \times 7 = \frac{22}{3} = 7\frac{1}{3} \text{ cm}^3$$

Wood wasted = $100 - 7\frac{1}{3} = 92\frac{2}{3}\%$

157. If h, c, v are respectively the height, curved surface area and volume of a right circular cone then the value of $3\pi v h^3 - c^2h^2 + 9v^2$ is **(SSC CGL Tier II 2013)**

(*a*) 2 (*b*) –1 (*c*) 1 (*d*) 0

Sol. (*d*) $3\pi\left(\frac{1}{3}\pi r^2 h\right)h^3 - (\pi rl)^2 h^2 + 9\left(\frac{1}{3}\pi r^2 h\right)^2$

$= \pi^2r^2h^4 - \pi^2r^2l^2h^2 + \pi^2r^4h^2 = \pi^2r^2h^2(h^2 - l^2 + r^2) = 0$

158. A cuboidal water tank contains 216 litres of water. Its depth is $\frac{1}{3}$ of its length and breadth is $\frac{1}{2}$ of $\frac{1}{3}$ of their difference between length and depth. The length of the tank is : **(SSC CGL Tier I 2005)**

(*a*) 72 dm (*b*) 18 dm (*c*) 6 dm (*d*) 2 dm

Sol. (*b*) Let length of the tank be x. Then depth $= \frac{x}{3}$. Breadth $= \frac{1}{2}\times\frac{1}{3}\left(x - \frac{x}{3}\right) = \frac{x}{9}$

Volume of tank $= x \times \frac{x}{9} \times \frac{x}{3} = 216$ litres $\Rightarrow \frac{x^3}{27} = 216\ \text{dm}^3$

$\Rightarrow x = \sqrt[3]{216 \times 27}$ dm $= 6 \times 3$ dm $= 18$ dm

159. A rocket is in the form a right circular cylinder, closed at the lower end, with a cone of the same base-radius attached to the top. The cylinder is of radius $2\frac{1}{2}$ m and height 21 m, and the cone is of slant height 8 m. The total surface area of the rocket is : (use $\pi = 22/7$) : **(SSC CGL Tier I 2006)**

(*a*) 330 m² (*b*) 412.5 m²

(*c*) 432.5 m² (*d*) 445 m²

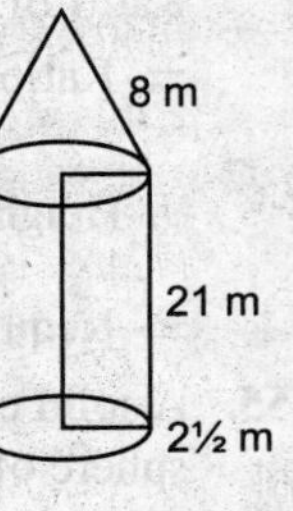

Sol. (*b*) Total surface area $= \pi rl + 2\pi rh + \pi r^2 = \pi r(l + 2h + r)$

$= \frac{22}{7}\times\frac{5}{2}\times\left(8 + 42 + \frac{5}{2}\right) = \frac{22}{7}\times\frac{5}{2}\times\frac{105}{2} = 412.5\ \text{m}^2$

160. The curved surface of a cylindrical pillar is 264 m² and its volume is 924 m³. The ratio of its diameter to its height is (use $\pi = 22/7$): **(SSC CGL Tier I 2005, SSC CHSL 2011)**

(*a*) 7 : 6 (*b*) 6 : 7 (*c*) 3 : 7 (*d*) 7 : 3

Sol. (*d*) Curved surface area of cylinder $= 2\pi rh = 264$. Volume of cylinder $= \pi r^2 h = 924$

$\frac{\text{Volume}}{\text{Area}} = \frac{\pi r^2 h}{2\pi rh} = \frac{924}{264} \Rightarrow \frac{r}{2} = \frac{7}{2} \Rightarrow r = 7$ metres

$2\pi rh = 264 \Rightarrow h = 264 \times \frac{1}{2}\times\frac{7}{22}\times\frac{1}{7} = 6$ metres $\Rightarrow$ Required ratio $= \frac{2r}{h} = \frac{2\times 7}{6} = \frac{7}{3} = 7:3$

161. A hemisphere and a cone have equal bases. If their heights are also equal, the ratio of their curved surfaces will be : **(SSC CGL Tier I 2002, 2011)**

(*a*) $1 : \sqrt{2}$ (*b*) $\sqrt{2} : 1$ (*c*) 1 : 2 (*d*) 2 : 1

Sol. (*b*) Let radius of hemisphere and cone be *r*. Curved surface area of hemisphere = $2\pi r^2$

Height of cone = *r* $\Rightarrow$ Slant height of cone = $l = \sqrt{r^2 + r^2} = \sqrt{2}r$

Curved surface of cone = $\pi rl = \pi r \times \sqrt{2}r = \sqrt{2}\,\pi r^2$

Required ratio = $2\pi r^2 : \sqrt{2}\,\pi r^2 = \sqrt{2} : 1$

162. A toy is made in the form of a hemisphere surmounted by a right circular cone whose base coincides with the plane surface of the hemisphere. The radius of the base of the cone is 1.5 m and its height is 4 m. The volume of the toy is (π = 22/7): **(SSC CGL Tier I 2005)**

(*a*) 14.75 m^3 (*b*) 16.75 m^3 (*c*) 16.5 m^3 (*d*) 17.5 m^3

Sol. (*c*) Volume of toy = $\frac{1}{3}\pi r^2 h + \frac{2}{3}\pi r^3 = \frac{1}{3}\pi r^2 (h + 2r)$

$= \frac{1}{3} \times \frac{22}{7} \times \frac{3}{2} \times \frac{3}{2} \times 7 = 16.5 \text{ m}^3$

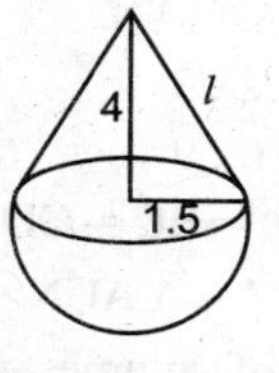

163. A toy is in the form of a cone mounted on a hemisphere such that the diameter of the base of the cone is equal to that of the hemisphere. If the diameter of the base of the cone is 6 cm and its height is 4 cm, what is the surface area of the toy in cm^2 ? (take π = 3.14)

(CDS 2011, SSC CHSL 2011 M)

(*a*) 93.62 (*b*) 103.62

(*c*) 113.62 (*d*) 115.50

Sol. (*b*) $l = \sqrt{3^2 + 4^2} = 5$

Total surface area of toy = $\pi rl + 2\pi r^2 = \pi r (l + 2r)$

$= 3.14 \times 3 \times 11 = 103.62 \text{ cm}^2$

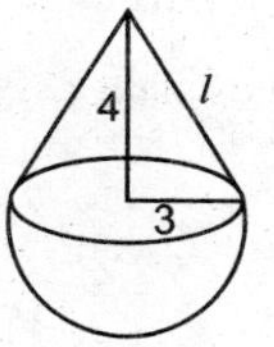

164. Semi-vertical angle of a right circular cone is 30°. If the ratio of the numerical value of the volume and slant surface area be 1 : 3, the radius of the base is **(SSC CPO SI 2012)**

(*a*) $\sqrt{3}$ units (*b*) 3 units (*c*) 1 unit (*d*) 2 units

Sol. (*d*) $\tan 30° = \frac{h}{r} \Rightarrow \frac{1}{\sqrt{3}} = \frac{h}{r} \Rightarrow h = \frac{r}{\sqrt{3}}$

Given, $\frac{\text{Volume}}{\text{Slant surface area}} = \frac{1}{3}$

$\Rightarrow \frac{\frac{1}{3}\pi r^2 h}{\pi r l} = \frac{1}{3} \Rightarrow rh = l \Rightarrow rh = \sqrt{r^2 + h^2}$

$\Rightarrow r^2h^2 = r^2 + h^2 \Rightarrow r^2\left(\frac{r}{\sqrt{3}}\right)^2 = r^2 + \left(\frac{r}{\sqrt{3}}\right)^2$

$\Rightarrow \frac{r^4}{3} = \frac{4r^2}{3} \Rightarrow r^2 = 4 \Rightarrow r = 2 \text{ units}$

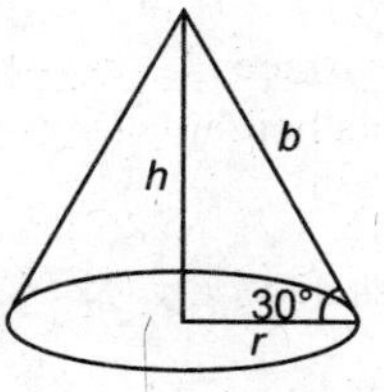

165. From a circular sheet of paper of radius 10 cm, a sector of area 40% is removed. If the remaining part is used to make a conical surface, then the ratio of radius and the height of the cone is : **(SSC CGL Tier II 2012)**

(*a*) 3 : 4 (*b*) 4 : 3 (*c*) 1 : 2 (*d*) 1 : 1

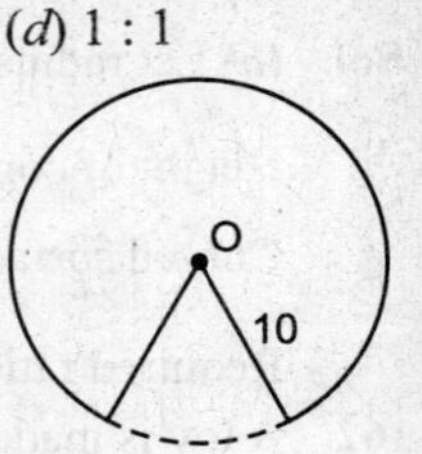

Sol. (*a*) Area of circle $= \pi(10)^2 = 100\pi$

Area of remaining sheet = 60% of $100\pi = 60\pi$

Slant height of cone = Radius of circle = 10 cm

S.A. of cone $= \pi rl = 60\pi \Rightarrow \pi r(10) = 60\pi \Rightarrow r = 6$

$l^2 = r^2 + h^2 \Rightarrow 10^2 = 6^2 + h^2 \Rightarrow h = 8$ cm

$\therefore$ Required ratio $= r : h = 6 : 8 = 3 : 4$

166. If a right circular cone is separated into solids of volumes V_1, V_2, V_3 by two planes parallel to the base, which also trisect the altitude, then $V_1 : V_2 : V_3$ is **(SSC CHSL 2011)**

(*a*) 1 : 2 : 3 (*b*) 1 : 4 : 6

(*c*) 1 : 6 : 9 (*d*) 1 : 7 : 19

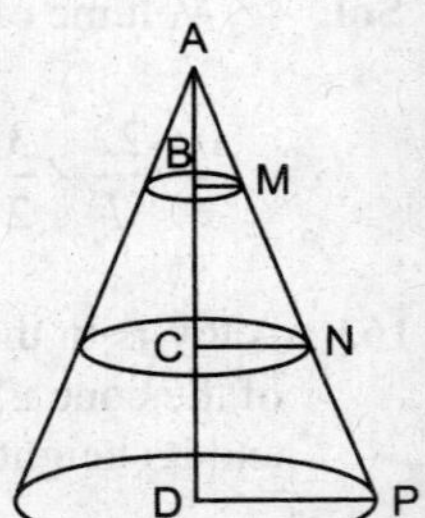

Sol. (*d*) AB = BC = CD ⇒ AB : AC : AD = 1 : 2 : 3

$\Delta ABM \sim \Delta ACN \sim \Delta ADP \Rightarrow BM : CN : DP = 1 : 2 : 3$

Ratio of volumes $= 1^3 : 2^3 : 3^3 = 1 : 8 : 27$

$\therefore V_1 : V_2 : V_3 = 1 : (8 - 1) : (27 - 8) = 1 : 7 : 19$

CHAPTER

32

TRIGONOMETRY

1. (*i*) $\sin\theta = \dfrac{\text{Perpendicular}}{\text{Hypotenuse}} = \dfrac{\text{Opposite Side}}{\text{Hypotenuse}}$

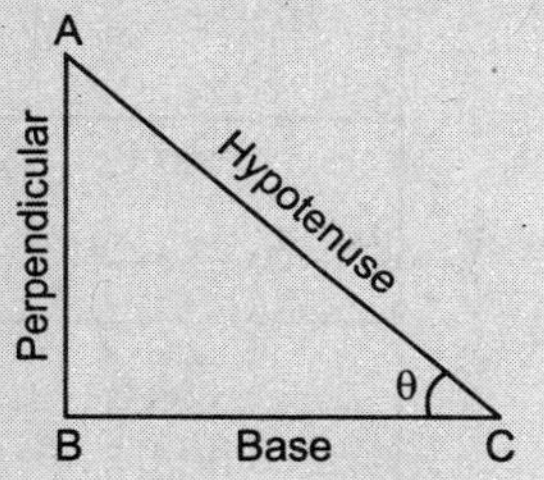

(*ii*) $\cos\theta = \dfrac{\text{Base}}{\text{Hypotenuse}} = \dfrac{\text{Adjacent Side}}{\text{Hypotenuse}}$

(*iii*) $\tan\theta = \dfrac{\text{Perpendicular}}{\text{Base}} = \dfrac{\text{Opposite Side}}{\text{Adjacent Side}} = \dfrac{\sin\theta}{\cos\theta}$

(*iv*) $\text{cosec}\,\theta = \dfrac{\text{Hypotenuse}}{\text{Perpendicular}} = \dfrac{1}{\sin\theta}$

(*v*) $\sec\theta = \dfrac{\text{Hypotenuse}}{\text{Base}} = \dfrac{1}{\cos\theta}$

(*vi*) $\cot\theta = \dfrac{\text{Base}}{\text{Perpendicular}} = \dfrac{1}{\tan\theta} = \dfrac{\cos\theta}{\sin\theta}$

Note: Base is side adjacent to the angle and perpendicular is side opposite to the angle. The above trigonometric functions are defined for an acute angle θ.

2. Pythagorean Identities :

(*i*) $\sin^2\theta + \cos^2\theta = 1$ (*ii*) $\tan^2\theta + 1 = \sec^2\theta$

(*iii*) $\cot^2\theta + 1 = \text{cosec}^2\theta$ (*iv*) $\sin 2\theta = 2\sin\theta\cos\theta$

(*v*) $\cos 2\theta = \cos^2\theta - \sin^2\theta$

(*vi*) $\sin(\alpha+\beta) = \sin\alpha\cos\beta + \cos\alpha\sin\beta$ (*vii*) $\sin(\alpha-\beta) = \sin\alpha\cos\beta - \cos\alpha\sin\beta$

(*vii*) $\cos(\alpha+\beta) = \cos\alpha\cos\beta - \sin\alpha\sin\beta$ (*ix*) $\cos(\alpha-\beta) = \cos\alpha\cos\beta + \sin\alpha\sin\beta$

3. **Measurement of angle in radian**:

Radian is the angle subtended at the centre of the circle, by an arc equal to length of the radius of the circle.

$\pi \text{ radian} = 180° \Rightarrow 1 \text{ radian} = 180° \times \dfrac{7}{22} = 90° \times \dfrac{7}{11} = 57° + \dfrac{3°}{11}$

$= 57° + \dfrac{3}{11} \times 60' = 57° + 16' + \dfrac{4}{11} \times 60'' = 57°\,16'\,22''$ **(SSC CGL Tier I 2013)**

Example. In a triangle *ABC*, $\angle ABC = 75°$ and $\angle ACB = \dfrac{\pi^C}{4}$. The circular measure of $\angle BAC$ is **(SSC CGL Tier I 2012)**

(*a*) $\dfrac{5\pi}{12}$ radian (*b*) $\dfrac{\pi}{3}$ radian (*c*) $\dfrac{\pi}{6}$ radian (*d*) $\dfrac{\pi}{2}$ radian

Sol.(*b*) $\angle ACB = \frac{\pi}{4} = 45° \Rightarrow \angle BAC = 180° - (45° + 75°) = 60° = \frac{\pi}{3}$ radian

4. Some common angles in degree and equivalent radian

Degree	30°	45°	60°	90°	180°	270°	360°
Radian	$\frac{\pi}{6}$	$\frac{\pi}{4}$	$\frac{\pi}{3}$	$\frac{\pi}{2}$	π	$\frac{3\pi}{2}$	2π

Note: $0 \leq |\sin\theta| \leq 1$ and $0 \leq |\cos\theta| \leq 1$

$\Rightarrow$The value of Sin θ, Cos θ can never be greater than 1.

5. Value of trigonometric functions :

	0°	30°	45°	60°	90°
$\sin\theta$	$\sqrt{\frac{0}{4}} = 0$	$\sqrt{\frac{1}{4}} = \frac{1}{2}$	$\sqrt{\frac{2}{4}} = \frac{1}{\sqrt{2}}$	$\frac{\sqrt{3}}{4} = \frac{\sqrt{3}}{2}$	$\sqrt{\frac{4}{4}} = 1$
$\cos\theta = \sin(90° - \theta)$	1	$\frac{\sqrt{3}}{2}$	$\frac{1}{\sqrt{2}}$	$\frac{1}{2}$	0
$\tan\theta = \frac{\sin\theta}{\cos\theta}$	$\frac{0}{1} = 1$	$\frac{\frac{1}{2}}{\frac{\sqrt{3}}{2}} = \frac{1}{\sqrt{3}}$	$\frac{\frac{1}{\sqrt{2}}}{\frac{1}{\sqrt{2}}} = 1$	$\frac{\frac{\sqrt{3}}{2}}{\frac{1}{2}} = \sqrt{3}$	$\frac{1}{0}$ = not defined

6. Trigonometric ratios of complementary angles: (θ and 90 – θ) are complimentary angles.

$\sin(90° - \theta) = \cos\theta$ $\cos(90° - \theta) = \sin\theta$

$\tan(90° - \theta) = \cot\theta$ $\cot(90° - \theta) = \tan\theta$

$\sec(90° - \theta) = \text{cosec}\,\theta$ $\text{cosec}(90° - \theta) = \sec\theta$

Hint to remember :

sine ↔ cosine, tangent ↔ cotangent, sectant ↔ cosectant *i.e.*, prefix 'Co' or remove 'Co'.

7. If $\sin\alpha = \cos\beta$, then $\alpha + \beta = 90°$.

If $\tan\alpha.\tan\beta = 1$, or If $\tan\alpha = \cot\beta$ then $\alpha + \beta = 90°$.

$\sin^2\theta + \sin^2(90° - \theta) = \sin^2\theta + \cos^2\theta = 1$

Example 1. If $\sin\theta = \cos 4\theta$, $0 < \theta < 90°$, the value of θ is **(SSC CGL Tier I 2012**

(*a*) 15° (*b*) 18° (*c*) 30° (*d*) 36°

Sol.(*b*) $\theta + 4\theta = 90° \Rightarrow 5\theta = 90° \Rightarrow \theta = 18°$

Alt: $\sin\theta = \cos 4\theta \Rightarrow \sin\theta = \sin(90° - 4\theta) \Rightarrow \theta = 90° - 4\theta \Rightarrow 5\theta = 90° \Rightarrow \theta = 18°$

Example 2. If $\tan 7\theta \tan 2\theta = 1$, then the value of $\tan 3\theta$ is **(SSC CGL Tier I 2012**

(*a*) $\sqrt{3}$ (*b*) $-\frac{1}{\sqrt{3}}$ (*c*) $\frac{1}{\sqrt{3}}$ (*d*) $-\sqrt{3}$

Sol.(*c*) $7\theta + 2\theta = 90° \Rightarrow 9\theta = 90° \Rightarrow \theta = 10° \Rightarrow 3\theta = 30° \Rightarrow \tan 3\theta = \tan 30° = \frac{1}{\sqrt{3}}$

Alt: $\tan 7\theta \tan 2\theta = 1 \Rightarrow \tan 7\theta = \frac{1}{\tan 2\theta} = \cot 2\theta = \tan(90° - 2\theta)$

$\Rightarrow 7\theta = 90° - 2\theta \Rightarrow 9\theta = 90° \Rightarrow \theta = 10° \Rightarrow \tan 3\theta = \tan 30° = \frac{1}{\sqrt{3}}$

Example 3. The value of $\text{Sin}^2 22° + \text{Sin}^2 68° + \text{Cot}^2 30°$ is **(SSC CGL Tier I 2015)**

(*a*) 3/4 (*b*) 4 (*c*) 3 (*d*) 5/4

Sol.(*b*) $\text{Sin}^2 22° + \text{Sin}^2 68° + \text{Cot}^2 30° = \text{Sin}^2 22° + \text{Cos}^2 22° + \text{Cot}^2 30° = 1 + (\sqrt{3})^2 = 1 + 3 = 4$

8. $\sin 2\theta = 2 \sin\theta \cos\theta = \frac{2\tan\theta}{1+\tan^2\theta}$

$\cos 2\theta = \cos^2\theta - \sin^2\theta = 2\cos^2\theta - 1 = 1 - 2\sin^2\theta$

$\tan 2\theta = \frac{2\tan\theta}{1-\tan^2\theta}$

9. 1° = 60 minutes (= 60′) 1′ = 60 seconds (= 60″)

10. Law of sines:

$$\frac{a}{\sin A} = \frac{b}{\sin B} = \frac{c}{\sin c}$$

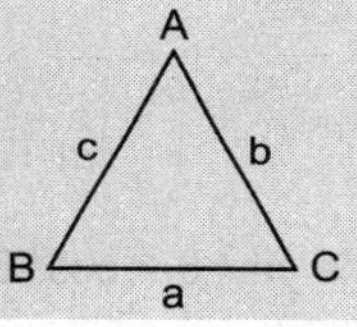

SOLVED EXERCISE

1. If $\tan(\theta - 30°) = \frac{1}{\sqrt{3}}$, then the value of θ is **(SSC CGL Tier I 2013)**

(*a*) 0° (*b*) 30° (*c*) 60° (*d*) 90°

Sol. (*c*) $\tan(\theta - 30°) = \frac{1}{\sqrt{3}} = \tan 30° \Rightarrow (\theta - 30°) = 30° \Rightarrow \theta = 60°$

2. If $3\tan^2\theta - 1 = 0$, find the value of θ.

(*a*) 0° (*b*) 15° (*c*) 30° (*d*) 45°

Sol. (*c*) $3\tan^2\theta - 1 = 0 \Rightarrow 3\tan^2\theta = 1 \Rightarrow \tan\theta = \frac{1}{\sqrt{3}} = \tan 30°$

3. The value of cos 1° cos 2° cos 3° ..... cos 177° cos 178° cos 179° is **(SSC CGL Tier I 2013)**

(*a*) $\frac{1}{\sqrt{2}}$ (*b*) 0 (*c*) $\frac{1}{2}$ (*d*) 1

Sol. (*b*) cos 1° cos 2° cos 3° ..... × 0 × ..... cos 177° cos 178° cos 179° = 0 [∵ cos90° = 0]

4. Value of tan 1° . tan 2° . tan 3° ..... tan 89° is : **(SSC CGL Tier I 2013 & 2015, SSC FCI 2012)**

(*a*) 0 (*b*) 1 (*c*) –1 (*d*) 2

Sol. (*b*) tan 1° . tan 2° . tan 3° ..... tan 89°

= (tan 1° . tan 89°) (tan 2° . tan 88°) (tan 3° . tan 87°) ..... (tan 45°)

= (tan 1° . cot 1°) (tan 2° . cot 2°) (tan 3° . cot 3°) ..... (tan 45°)

= 1 × 1 × 1 × 1 × ..... 1 = 1

5. The value of $(\sin^2 25° + \sin^2 65)$ is : **(SSC CGL Tier I 2013)**

(*a*) $\frac{2}{\sqrt{3}}$ (*b*) $\frac{\sqrt{3}}{2}$ (*c*) 1 (*d*) 0

Sol. (*c*) $\sin^2 25° + \sin^2 65° = \sin^2 25° + \sin^2 (90° - 65°) = \sin^2 25° + \cos^2 25° = 1$

6. If $\cos\alpha = \dfrac{5}{13}$, and $0 < \alpha < 90°$, then the value of $\cot\alpha + \operatorname{cosec}\alpha$ is **(SSC CGL Tier I 2013)**

(*a*) 2 (*b*) 3.5 (*c*) 4 (*d*) 1.5

Sol. (*d*) $\cot\alpha + \operatorname{cosec}\alpha = \dfrac{\cos\alpha}{\sin\alpha} + \dfrac{1}{\sin\alpha} = \dfrac{1+\cos\alpha}{\sin\alpha} = \dfrac{1+\frac{5}{13}}{\frac{12}{13}} = \dfrac{18}{12} = 1.5$

7. If $\sin\theta + \cos\theta = 1$, then $\sin\theta\cos\theta$ is equal to :

(*a*) 0 (*b*) 1 (*c*) 2 (*d*) –1

Sol. (*a*) $\sin\theta + \cos\theta = 1 \Rightarrow (\sin\theta + \cos\theta)^2 = 1 \Rightarrow \sin^2\theta + \cos^2\theta + 2\sin\theta\cos\theta = 1$

$\Rightarrow 1 + 2\sin\theta\cos\theta = 1 \Rightarrow 2\sin\theta\cos\theta = 0 \Rightarrow \sin\theta\cos\theta = 0$

8. $\dfrac{\sin^3\theta + \cos^3\theta}{\sin\theta + \cos\theta} = ?$

(*a*) $\sin\theta\cos\theta$ (*b*) $\tan\theta$ (*c*) $1 - \sin\theta\cos\theta$ (*d*) $\cot\theta$

Sol. (*c*) G.E. $= \dfrac{a^3+b^3}{a+b} = a^2 + b^2 - ab = \sin^2\theta + \cos^2\theta - \sin\theta\cos\theta = 1 - \sin\theta\cos\theta$

9. $\dfrac{1+\tan^2\theta}{1+\cot^2\theta} = ?$

(*a*) $\tan^2\theta$ (*b*) $\cot^2\theta$ (*c*) $\left(\dfrac{1+\tan\theta}{1-\tan\theta}\right)^2$ (*d*) $\left(\dfrac{1-\tan\theta}{1+\tan\theta}\right)^2$

Sol. (*a*) $\dfrac{1+\tan^2\theta}{1+\cot^2\theta} = \dfrac{\sec^2\theta}{\operatorname{cosec}^2\theta} = \dfrac{\sin^2\theta}{\cos^2\theta} = \tan^2\theta$

10. The value of $\dfrac{1}{(1+\tan^2\theta)} + \dfrac{1}{(1+\cot^2\theta)}$ is **(SSC CGL Tier I 2013, SSC LDC 201**

(*a*) 2 (*b*) $\dfrac{1}{2}$ (*c*) $\dfrac{1}{4}$ (*d*) 1

Sol. (*d*) G.E. $= \dfrac{1}{\sec^2\theta} + \dfrac{1}{\operatorname{cosec}^2\theta} = \cos^2\theta + \sin^2\theta = 1$

11. The numerical value of $\dfrac{9}{\operatorname{cosec}^2\theta} + 4\cos^2\theta + \dfrac{5}{1+\tan^2\theta}$ is : **(SSC CHSL 201**

(*a*) 4 (*b*) 5 (*c*) 9 (*d*) 1

Sol. (*c*) $\dfrac{9}{\operatorname{cosec}^2\theta} + 4\cos^2\theta + \dfrac{5}{1+\tan^2\theta} = 9\sin^2\theta + 4\cos^2\theta + \dfrac{5}{\sec^2\theta}$

$= 9\sin^2\theta + 4\cos^2\theta + 5\cos^2\theta = 9\sin^2\theta + 9\cos^2\theta = 9(\sin^2\theta + \cos^2\theta) = 9$

12. Find the value of $1 - 2\sin^2\theta + \sin^4\theta$ **(SSC CGL Tier I 201**

(*a*) $\sin^4\theta$ (*b*) $\cos^4\theta$ (*c*) $\operatorname{cosec}^4\theta$ (*d*) $\sec^4\theta$

Sol. (*b*) $1 - 2\sin^2\theta + \sin^4\theta = (1 - \sin^2\theta)^2 = \cos^4\theta$

13. $(1 + \tan^2\theta)(1 - \sin\theta)(1 + \sin\theta) = ?$

(*a*) 0 (*b*) 1 (*c*) –1 (*d*) 2

Sol. (*b*) G.E. $= \sec^2\theta(1 - \sin^2\theta) = \sec^2\theta \times \cos^2\theta = 1$

14. If $\cos x + \cos y = 2$, the value of $\sin x + \sin y$ is **(SSC FCI Asstt. 2012)**

(*a*) 0 (*b*) 1 (*c*) 2 (*d*) – 1

Sol. (*a*) $\cos x + \cos y = 2 \Rightarrow \cos x = \cos y = 1 = \cos 0°$ $[\because -1 \le \cos\theta \le 1]$

$\Rightarrow \sin x + \sin y = \sin 0° + \sin 0° = 0 + 0 = 0$

15. If $\cos^2\alpha + \cos^2\beta = 2$, then the value of $\tan^3\alpha + \sin^5\beta$ is : **(SSC CHSL 2011)**

(*a*) –1 (*b*) 0 (*c*) 1 (*d*) $1/\sqrt{3}$

Sol. (*b*) $-1 \le \cos\theta \le 1 \Rightarrow 0 \le \cos^2\theta \le 1$

$\cos^2\alpha + \cos^2\beta = 2 \Rightarrow \cos\alpha = \cos\beta = 1 \Rightarrow \alpha = \beta = 0° \Rightarrow \tan^3\alpha + \sin^5\beta = 0 + 0 = 0$

16. If $\sin\theta = \frac{4}{5}$, then $\tan\theta = ?$

(*a*) $\frac{3}{5}$ (*b*) $\frac{3}{4}$ (*c*) $\frac{4}{3}$ (*d*) $\frac{5}{4}$

Sol. (*c*) $\cos^2\theta = 1 - \sin^2\theta = 1 - \left(\frac{4}{5}\right)^2 = \frac{9}{25} \Rightarrow \cos\theta = \frac{3}{5}$

$\therefore \tan\theta = \sin\theta \div \cos\theta = \frac{4}{5} \div \frac{3}{5} = \frac{4}{5} \times \frac{5}{3} = \frac{4}{3}$

17. If $\tan\theta = 3\cot\theta$, then $\theta = ?$

(*a*) 30° (*b*) 45° (*c*) 60° (*d*) 90°

Sol. (*c*) $\tan\theta = 3\cot\theta = \frac{3}{\tan\theta} \Rightarrow \tan^2\theta = 3 \Rightarrow \tan\theta = \sqrt{3} = \tan 60° \Rightarrow \theta = 60°$

18. If $\tan\theta \,.\, \cos 60° = \frac{\sqrt{3}}{2}$, then the value of $\sin(\theta - 15°)$ is **(SSC FCI Asstt. 2012)**

(*a*) $\frac{\sqrt{3}}{2}$ (*b*) $\frac{1}{2}$ (*c*) 1 (*d*) $\frac{1}{\sqrt{2}}$

Sol. (*d*) $\tan\theta.\cos 60° = \frac{\sqrt{3}}{2} \Rightarrow \tan\theta\cdot\frac{1}{2} = \frac{\sqrt{3}}{2} \Rightarrow \tan\theta = \sqrt{3} = \tan 60°$

$\Rightarrow \sin(\theta - 15°) = \sin(60° - 15°) = \sin 45° = \frac{1}{\sqrt{2}}$

19. If $\sin(A - B) = \frac{1}{2}$ and $\cos(A + B) = \frac{1}{2}$ where $A > B > 0$ and $A + B$ is an acute angle, then the value of B is **(SSC CGL Tier I 2013)**

(*a*) $\frac{\pi}{6}$ (*b*) $\frac{\pi}{12}$ (*c*) $\frac{\pi}{4}$ (*d*) $\frac{\pi}{2}$

Sol. (*b*) $\sin(A - B) = \frac{1}{2} = \sin 30° \Rightarrow A - B = 30°$

$\cos(A + B) = \frac{1}{2} = \cos 60° \Rightarrow A + B = 60°$ $\therefore B = \frac{60° - 30°}{2} = 15° = \frac{\pi}{12}$

20. If $\tan (A + B) = \sqrt{3}$ and $\tan A = 1$, then $\tan (A - B)$ is equal to **(CDS 2015)**

(*a*) 0 (*b*) $\frac{1}{\sqrt{3}}$ (*c*) 1 (*d*) $\sqrt{2}$

Sol. (*b*) $\tan (A + B) = \sqrt{3} = \tan 60° \Rightarrow (A + B) = 60°$ $\tan A = 1 = \tan 45° \Rightarrow A = 45°$

$\Rightarrow B = 15°$ and $\Rightarrow A - B = 30° \Rightarrow \tan (A - B) = \tan 30° = \frac{1}{\sqrt{3}}$

21. If $\sin (A + B) = 1$ and $\cos (A - B) = \frac{\sqrt{3}}{2}$, then A and B are **(SSC CPO SI 2012)**

(*a*) $A = 60°, B = 30°$ (*b*) $A = 45°, B = 45°$ (*c*) $A = 75°, B = 15°$ (*d*) None of these

Sol. (*a*) In each option, $A + B = 90°$ and $\sin 90° = 1$

$\cos (A - B) = \frac{\sqrt{3}}{2} = \cos 30° \Rightarrow A - B = 30°$ which is in case of option (*a*).

22. $\frac{\sin 30° \cos 45°}{\tan 60°} = ?$

(*a*) $\frac{2}{\sqrt{3}}$ (*b*) $2\sqrt{3}$ (*c*) $\sqrt{\frac{2}{3}}$ (*d*) $\frac{\sqrt{6}}{12}$

Sol. (*d*) $\frac{\sin 30° \cos 45°}{\tan 60°} = \frac{\frac{1}{2} \times \frac{1}{\sqrt{2}}}{\sqrt{3}} = \frac{1}{2\sqrt{6}} = \frac{\sqrt{6}}{12}$

23. What is $\sin A \cos A \tan A + \cos A \sin A \cot A$ equal to ? **(NDA 2012)**

(*a*) $\sin A$ (*b*) $\cos A$ (*c*) $\tan A$ (*d*) 1

Sol. (*d*) $\sin A \cos A \times \frac{\sin A}{\cos A} + \cos A \sin A \times \frac{\cos A}{\sin A} = \sin^2 A + \cos^2 A = 1$

24. $(\text{cosec } \theta + \cot \theta)(1 - \cos \theta) = ?$

(*a*) $\sin \theta$ (*b*) $\cos \theta$ (*c*) $\tan \theta$ (*d*) $\cot \theta$

Sol. (*a*) G.E. $= \left(\frac{1}{\sin \theta} + \frac{\cos \theta}{\sin \theta}\right)(1 - \cos \theta) = \frac{(1 + \cos \theta)(1 - \cos \theta)}{\sin \theta} = \frac{1 - \cos^2 \theta}{\sin \theta} = \frac{\sin^2 \theta}{\sin \theta} = \sin \theta$

25. If $0° \leq A \leq 90°$, the simplest form of the expression Sin A Cos A (tan A – cot A) is **(SSC CGL Tier I 2015)**

(*a*) $1 - \text{Cos}^2 A$ (*b*) $2 \text{Sin}^2 A - 1$ (*c*) $1 - 2 \text{Sin}^2 A$ (*d*) 1

Sol. (*b*) Sin A Cos A (tan A – cot A) = Sin ACos A $\left(\frac{\sin A}{\cos A} - \frac{\cos A}{\sin A}\right) = \text{Sin}^2 A - \text{Cos}^2 A = 2 \text{Sin}^2 A - 1$

26. If $\tan 3x = \sin 45° \cos 45° + \sin 30°$, find the value of x.

(*a*) 15° (*b*) 30° (*c*) 45° (*d*) 90°

Sol. (*a*) $\tan 3x = \frac{1}{\sqrt{2}} \times \frac{1}{\sqrt{2}} + \frac{1}{2} = \frac{1}{2} + \frac{1}{2} = 1 = \tan 45° \Rightarrow 3x = 45° \Rightarrow x = 15°$

27. If $\sec^2 \theta + \tan^2 \theta = 7$, then the value of θ when $0° \leq \theta \leq 90°$ is **(SSC CHSL 2011)**

(*a*) 0° (*b*) 30° (*c*) 60° (*d*) 90°

Sol. (*c*) $1 + \tan^2 \theta + \tan^2 \theta = 7 \Rightarrow 2 \tan^2 \theta = 7 - 1 = 6 \Rightarrow \tan^2 \theta = 3 \Rightarrow \tan \theta = \sqrt{3} = \tan 60°$

28. If $P \sin \theta = \sqrt{3}$ and $P \cos \theta = 1$, then the value of P is **(SSC CGL Tier I 2012)**

(*a*) $\frac{2}{\sqrt{3}}$ (*b*) $\frac{1}{\sqrt{3}}$ (*c*) 2 (*d*) $\frac{1}{2}$

Sol. (*c*) $\frac{P\sin\theta}{P\cos\theta} = \frac{\sqrt{3}}{1} \Rightarrow \tan\theta = \sqrt{3} = \tan 60° \Rightarrow \theta = 60°$

Now, $P \sin \theta = \sqrt{3} \Rightarrow P \sin 60° = \sqrt{3} \Rightarrow P \times \frac{\sqrt{3}}{2} = \sqrt{3} \Rightarrow P = 2$

Alt: $P^2 \sin^2 \theta = 3$ and $P^2 \cos^2 \theta = 1$

$\therefore P^2 \sin^2 \theta + P^2 \cos^2 \theta = 3 + 1 \Rightarrow P^2(\sin^2 \theta + \cos^2 \theta) = 4 \Rightarrow P^2 = 4 \Rightarrow P = 2$

29. If $2 \text{ Sin } \theta + \cos \theta = 7/3$, then the value of $(\tan^2\theta - \sec^2\theta)$ is **(SSC CGL Tier I 2014)**

(*a*) 0 (*b*) – 1 (*c*) 3/7 (*d*) 7/3

Sol. (*b*) $(\tan^2\theta - \sec^2\theta) = -(\sec^2\theta - \tan^2\theta) = -1$

30. If $\tan \theta + \cot \theta = 5$, then $\tan^2 \theta + \cot^2 \theta$ is **(SSC CGL Tier I 2015, CDS 2015 M)**

(*a*) 23 (*b*) 25 (*c*) 26 (*d*) 24

Sol. (*a*) $\tan^2 \theta + \cot^2 \theta = (\tan \theta + \cot \theta)^2 - 2 = 5^2 - 2 = 23$

31. If $\tan \theta + \cot \theta = 2$, find value of $\tan^{100} \theta + \cot^{100} \theta$. **(SSC FCI Asstt. 2013)**

(*a*) 1 (*b*) 2 (*c*) 4 (*d*) 8

Sol. (*b*) $\cot \theta = \frac{1}{\tan \theta}$ $\left[x + \frac{1}{x} = 2 \Rightarrow x = 1\right]$

$\Rightarrow \tan \theta = \cot \theta = 1$ $\therefore$ G.E. $= 1^{100} + 1^{100} = 2$

32. If $\sin\theta + \text{cosec}\theta = 2$, then the value of $\sin^9\theta + \text{cosec}^9\theta$ is:**(SSC CGL Tier I 2013, CHSL 2011)**

(*a*) 1 (*b*) 3 (*c*) 2 (*d*) 4

Sol. (*c*) $\text{cosec } \theta = \frac{1}{\sin \theta}$ $\left[x + \frac{1}{x} = 2 \Rightarrow x = 1\right]$ $\therefore$ Given expression $= 1^9 + 1^9 = 2$

33. If $\cos \theta + \sec \theta = 2(0° \le \theta \le 90°)$ then the value of $\cos 10\theta + \sec 11\theta$ is **(SSC CHSL 2013)**

(*a*) 0 (*b*) 1 (*c*) 2 (*d*) – 1

Sol. (*c*) $\cos\theta + \frac{1}{\cos\theta} = 2 \Rightarrow \cos \theta = 1 = \cos 0° \Rightarrow \theta = 0°$

$\therefore \cos 10\theta + \sec 11\theta = \cos 0° + \sec 0° = 1 + 1 = 2$

34. If $\frac{\sin \theta}{\cos \theta} + \frac{\cos \theta}{\sin \theta} = 2$ with $0 < \theta < 90°$ then what is θ equal to ? **(CDS 2011)**

(*a*) 30° (*b*) 45° (*c*) 60° (*d*) 75°

Sol. (*b*) $\tan \theta + \cot \theta = 2 \Rightarrow \tan \theta = 1 = \tan 45° \Rightarrow \theta = 45°$

35. If $\tan^2 \theta + 1/\tan^2 \theta = 2$, then the value of θ is **(SSC CGL Tier I 2015)**

(*a*) 30° (*b*) 45° (*c*) 60° (*d*) 15°

Sol. (*b*) Here, $\left[x + \frac{1}{x} = 2 \Rightarrow x = 1\right]$ $\Rightarrow \tan^2 \theta = 1 \Rightarrow \tan \theta = 1 = \tan 45°$

36. If $r \sin \theta = 1$, $r \cos \theta = \sqrt{3}$, then the value of $(\sqrt{3} \tan \theta + 1)$ is **(SSC CGL Tier II 2013)**

(*a*) $\sqrt{3}$ (*b*) $\frac{1}{\sqrt{3}}$ (*c*) 1 (*d*) 2

Sol. (*d*) $\tan\theta = \frac{\sin\theta}{\cos\theta} = \frac{r\sin\theta}{r\cos\theta} = \frac{1}{\sqrt{3}}$ $\Rightarrow$ $(\sqrt{3}\tan\theta + 1) = \sqrt{3}\times\frac{1}{\sqrt{3}} + 1 = 1 + 1 = 2$

37. If $x \sin 45° = y \operatorname{cosec} 30°$, then $\frac{x^4}{y^4}$ is equal to **(SSC CGL Tier II 2013)**

(*a*) 4^3 (*b*) 6^3 (*c*) 2^3 (*d*) 8^3

Sol. (*a*) $x\times\frac{1}{\sqrt{2}} = y(2) \Rightarrow \frac{x}{y} = 2\sqrt{2} \Rightarrow \left(\frac{x}{y}\right)^4 = \left(2\sqrt{2}\right)^4 = 64 = 4^3$

38. If $\sec\theta - \operatorname{cosec}\theta = 0$, then the value of $\tan\theta + \cot\theta$ is **(SSC CHSL 2013)**

(*a*) 0 (*b*) 1 (*c*) – 1 (*d*) 2

Sol. (*d*) $\sec\theta - \operatorname{cosec}\theta = 0 \Rightarrow \sec\theta = \operatorname{cosec}\theta \Rightarrow \frac{1}{\cos\theta} = \frac{1}{\sin\theta}$

$\Rightarrow \tan\theta = 1 \Rightarrow \cot\theta = \frac{1}{\tan\theta} = 1 \quad \therefore \quad \tan\theta + \cot\theta = 1 + 1 = 2$

39. If $\sec\theta - \operatorname{cosec}\theta = 0$, then the value of $(\sec\theta + \operatorname{cosec}\theta)$ is **(SSC CGL Tier I 2012)**

(*a*) $\frac{2}{\sqrt{3}}$ (*b*) 0 (*c*) $2\sqrt{2}$ (*d*) $\frac{\sqrt{3}}{2}$

Sol. (*c*) $\sec\theta - \operatorname{cosec}\theta = 0 \Rightarrow \sec\theta = \operatorname{cosec}\theta \Rightarrow \frac{1}{\cos\theta} = \frac{1}{\sin\theta} \Rightarrow \tan\theta = 1 = \tan 45°$

$\therefore \sec\theta + \operatorname{cosec}\theta = \sec 45° + \operatorname{cosec} 45° = \sqrt{2} + \sqrt{2} = 2\sqrt{2}$

40. If $\sec\theta - \cos\theta = 3/2$, ($\theta$ is positive acute angle) then the value of $\sec\theta$ is **(SSC CHSL 2013)**

(*a*) 0 (*b*) 1/2 (*c*) – 1/2 (*d*) 2

Sol. (*d*) $(\sec\theta + \cos\theta)^2 = \left(\frac{3}{2}\right)^2 + 4 = \frac{25}{4} \Rightarrow \sec\theta + \cos\theta = \frac{5}{2} = 2 + \frac{1}{2} \Rightarrow \sec\theta = 2$

Trick : $x - \frac{1}{x} = \frac{3}{2} = 2 - \frac{1}{2} \Rightarrow \sec\theta = 2$

41. The value of $\sin^2 1° + \sin^2 5° + \sin^2 9° + ... + \sin^2 89°$ is **(SSC CHSL 2011)**

(*a*) $11\frac{1}{2}$ (*b*) $11\sqrt{2}$ (*c*) 11 (*d*) $\frac{11}{\sqrt{2}}$

Sol. (*a*) G.E. $= (\sin^2 1° + \sin^2 89°) + (\sin^2 5° + \sin^2 85°) + (\sin^2 9° + \sin^2 81°) + ... + (\sin^2 41° + \sin^2 49°) + \sin^2 45°$

$= (\sin^2 1° + \cos^2 1°) + (\sin^2 5° + \cos^2 5°) + ... + (\sin^2 41° + \cos^2 41°) + \left(\frac{1}{\sqrt{2}}\right)^2$

$= \underbrace{1+1+1+...+1}_{\text{11 times}} + \frac{1}{2} = 11\frac{1}{2}$

Hint: Total number of pairs $= \frac{41-1}{4} + 1 = 10 + 1 = 11$ pairs

42. The value of $\sin^2 1° + \sin^2 3° + \sin^2 5° + ... + \sin^2 87° + \sin^2 89°$ is **(SSC CHSL 201**

(*a*) 22 (*b*) $22\frac{1}{2}$ (*c*) $22\frac{1}{4}$ (*d*) 23

Sol.(*b*) G.E. = $(\sin^2 1° + \sin^2 89°) + (\sin^2 3° + \sin^2 87°) + (\sin^2 5° + \sin^2 85°) + ... + (\sin^2 43° + \sin^2 47°) + \sin^2 45°$

$$= (\sin^2 1° + \cos^2 1°) + (\sin^2 3° + \cos^2 3°) + ... + (\sin^2 43° + \cos^2 43°) + \left(\frac{1}{\sqrt{2}}\right)^2$$

$$= \underbrace{1+1+1+...+1}_{22 \text{ times}} + \frac{1}{2} = 22\frac{1}{2}$$

Hint: Total number of pairs = $\frac{43-1}{2} + 1 = 21 + 1 = 22$ pairs

43. If $\sec x + \cos x = 3$, then $\tan^2 x - \sin^2 x$ is **(SSC CGL Tier I 2012)**

(*a*) 9 (*b*) 4 (*c*) 5 (*d*) 13

Sol. (*c*) $\tan^2 x - \sin^2 x = (\sec^2 x - 1) - (1 - \cos^2 x) = \sec^2 x + \cos^2 x - 2$

$= (\sec x + \cos x)^2 - 2 \sec x \cos x - 2 = (3)^2 - 2 - 2 = 9 - 4 = 5$

44. Find the maximum value of $\sin\theta \,.\, \cos\theta$.

(*a*) 1 (*b*) $\frac{1}{2}$ (*c*) $\frac{1}{4}$ (*d*) 0

Sol. (*b*) $\sin\theta.\cos\theta$ is maximum, when $\sin\theta = \cos\theta$

$\sin\theta = \cos\theta = \frac{1}{\sqrt{2}}$, when $\theta = 45° \Rightarrow$ Maximum value $= \frac{1}{\sqrt{2}} \times \frac{1}{\sqrt{2}} = \frac{1}{2}$

Alt: $\sin\theta \,.\, \cos\theta = \frac{1}{2}(2\sin\theta \,.\, \cos\theta) = \frac{1}{2}(\sin 2\theta) = \frac{1}{2} \times 1 = \frac{1}{2}$

45. Maximum value of $(2\sin\theta + 3\cos\theta)$ is **(SSC CGL Tier I 2013)**

(*a*) 2 (*b*) $\sqrt{13}$ (*c*) $\sqrt{15}$ (*d*) 1

Sol. (*b*) Maximum value of $a\sin\theta + b\cos\theta$ is $\sqrt{a^2 + b^2}$

$\therefore$ Maximum value of $2\sin\theta + 3\cos\theta = \sqrt{2^2 + 3^2} = \sqrt{13}$

46. Find minimum value of $5\sin\theta + 12\cos\theta$

(*a*) 0 (*b*) –1 (*c*) $-\sqrt{3}$ (*d*) –13

Sol. (*d*) Minimum value $= -\sqrt{5^2 + 12^2} = -13 \Rightarrow$ Minimum value = –13

47. If $0° < \theta < 90°$, the value of $\sin\theta + \cos\theta$ is **(SSC CGL Tier II 2012)**

(*a*) less than 1 (*b*) equal to 2 (*c*) equal to 1 (*d*) greater than 1

Sol. (*d*) Greater than 1.

48. The least value of $2\sin^2\theta + 3\cos^2\theta$ is **(SSC CPOSI 2011, SSC CGL Tier I 2012 and 2014)**

(*a*) 1 (*b*) 2 (*c*) 3 (*d*) 5

Sol. (*b*) $2\sin^2\theta + 3\cos^2\theta = 2(\sin^2\theta + \cos^2\theta) + \cos^2\theta = 2 + \cos^2\theta$

$= 2 + 0 = 2$ [when $\theta = 90°$, $\cos\theta = 0$]

49. The least value of $4\,\text{cosec}^2\alpha + 9\sin^2\alpha$ is **(SSC CGL Tier I 2012, SSC CHSL 2013 M)**

(*a*) 10 (*b*) 12 (*c*) 14 (*d*) 11

Sol. (*b*) G.E = $(2\,\text{cosec}\,\alpha - 3\sin\alpha)^2 + 12 \Rightarrow$ Least value = 12

Alt: Geometric Mean = $\sqrt{4\text{cosec}^2\alpha \times 9\sin^2\alpha} = \sqrt{36} = 6 \Rightarrow$ Least value = 2 G.M. = $2 \times 6 = 12$

50. The least value of $(4\sec^2\theta + 9\,\text{cosec}^2\theta)$ is: **(SSC CGL Tier I 2012)**

(*a*) 1 (*b*) 25 (*c*) 7 (*d*) 19

Sol. (*b*) G.E. $= 4(1 + \tan^2\theta) + 9(1 + \cot^2\theta) = 4 + 4\tan^2\theta + 9 + 9\cot^2\theta = 13 + 4\tan^2\theta + 9\cot^2\theta$

G.M. of $4\tan^2\theta + 9\cot^2\theta = \sqrt{4\tan^2\theta . 9\cot^2\theta} = \sqrt{36} = 6$

$\because\ AM \geq GM \Rightarrow 4\tan^2\theta + 9\cot^2\theta \geq 2 \times 6$

$\therefore$ The least value of G.E. $= 13 + 2 \times 6 = 25$

51. The minimum value of $\sin^2\theta + \cos^2\theta + \sec^2\theta + \text{cosec}^2\theta + \tan^2\theta + \cot^2\theta$ is **(SSC CGL Tier II 2012**

(*a*) 5 (*b*) 7 (*c*) 1 (*d*) 3

Sol. (*b*) G.E $= \sin^2\theta + \cos^2\theta + 1 + \tan^2\theta + 1 + \cot^2\theta + \tan^2\theta + \cot^2\theta$

$= 3 + 2(\tan^2\theta + \cot^2\theta)$ G.M. $= \sqrt{\tan^2\theta \times \cot^2\theta} = 1$

$\therefore$ Minimum value of $(\tan^2\theta + \cot^2\theta)$ is $2 \times 1 = 2$

$\therefore$ Minimum value of G.E. $= 3 + (2 \times 2) = 7$

Alt: G.E. $= (\sin^2\theta + \cos^2\theta) + (1 + \tan^2\theta) + (1 + \cot^2\theta) + \tan^2\theta + \cot^2\theta$

$= 3 + 2(\tan^2\theta + \cot^2\theta) = 3 + 2[(\tan\theta - \cot\theta)^2 + 2\tan\theta\cot\theta]$

$= 3 + 2[0 + 2] = 7$ [When $\tan\theta = \cot\theta$]

52. If $\tan(x + y)\tan(x - y) = 1$, then the value of $\tan x$ is : **(SSC CGL Tier I 201.**

(*a*) $\sqrt{3}$ (*b*) 1 (*c*) $\frac{1}{2}$ (*d*) $\frac{1}{\sqrt{3}}$

Sol. (*b*) $(x + y) + (x - y) = 90° \Rightarrow 2x = 90° \Rightarrow x = 45° \Rightarrow \tan x = \tan 45° = 1$

Alt: $\tan(x + y) = \dfrac{1}{\tan(x - y)} = \cot(x - y) = \tan[90° - (x - y)]$

$\Rightarrow x + y = 90° - (x - y) \Rightarrow x + y = 90° - x + y \Rightarrow 2x = 90° \Rightarrow x = 45°$

$\therefore\ \tan x = \tan 45° = 1$

53. If θ be acute angle, $\tan(4\theta - 50°) = \cot(50° - \theta)$, then the value of θ (in degrees) is **(SSC FCI 2012, SSC CGL Tier I 201**

(*a*) 20 (*b*) 30 (*c*) 40 (*d*) 50

Sol. (*b*) $(4\theta - 50°) + (50° - \theta) = 90° \Rightarrow 3\theta = 90° \Rightarrow \theta = 30°$

Alt: $\tan(4\theta - 50°) = \cot(50° - \theta) \Rightarrow \tan(4\theta - 50°) = \tan[90° - (50° - \theta)]$

$\Rightarrow \tan(4\theta - 50°) = \tan(40° + \theta) \Rightarrow 4\theta - 50° = 40° + \theta \Rightarrow 3\theta = 90° \Rightarrow \theta = 30°$

54. The value of $\sin(45° + \theta) - \cos(45° - \theta)$ is equal to **(SSC CHSL 201**

(*a*) 1 (*b*) 0 (*c*) $2\cos\theta$ (*d*) $2\sin\theta$

Sol. (*b*) $\sin(45° + \theta) - \cos(45° - \theta) = \sin(45° + \theta) - \sin[90 - (45° - \theta)]$

$= \sin(45° + \theta) - \sin(45° + \theta) = 0$

55. If x, y are positive acute angles, $x + y < 90°$ and $\sin(2x - 20°) = \cos(2y + 20°)$, then the va of $\sec(x + y)$ is **(SSC CGL Tier I 20**

(*a*) 0 (*b*) 1 (*c*) $\sqrt{2}$ (*d*) $\frac{1}{\sqrt{2}}$

Sol. (*c*) $(2x - 20°) + (2y + 20°) = 2x + 2y = 90° \Rightarrow x + y = 45° \Rightarrow \sec(x + y) = \sec 45° = \sqrt{2}$

Alt: $\sin(2x - 20°) = \sin[90° - (2y + 20°)] \Rightarrow 2x - 20° = 90° - 2y - 20°$

$\Rightarrow 2x + 2y = 90° \Rightarrow x + y = 45° \Rightarrow \sec(x + y) = \sec 45° = \sqrt{2}$

56. It θ is a positive acute angle and $\tan 2\theta \tan 3\theta = 1$, then the value of $\left(2\cos^2\frac{5\theta}{2} - 1\right)$ is **(SSC CGL Tier II 2012)**

(*a*) 0 (*b*) $\frac{1}{2}$ (*c*) $-\frac{1}{2}$ (*d*) 1

Sol. (*a*) $2\theta + 3\theta = 90° \Rightarrow \theta = 18° \Rightarrow 2\cos^2\frac{5\theta}{2} - 1 = 2\cos^2 45° - 1 = 2\left(\frac{1}{\sqrt{2}}\right)^2 - 1 = 1 - 1 = 0$

57. If $\sin(x - 2y) = \cos(4y - x)$, the value of $\cot 2y$ is **(SSC FCI Asstt. 2012)**

(*a*) 0 (*b*) 1 (*c*) $\sqrt{3}$ (*d*) Undefined

Sol. (*a*) $(x - 2y) + (4y - x) = 90° \Rightarrow 2y = 90° \Rightarrow \cot 2y = \cot 90° = 0$

58. If $\sin(x + y) = \cos[3(x + y)]$, then the value of $\tan[2(x + y)]$ is **(SSC CGL Tier I 2012)**

(*a*) 0 (*b*) 1 (*c*) $\frac{1}{\sqrt{3}}$ (*d*) $\sqrt{3}$

Sol. (*b*) $(x + y) + 3(x + y) = 90° \Rightarrow 4(x + y) = 90° \Rightarrow 2(x + y) = 45° \Rightarrow \tan 45° = 1$

59. If $\sin\theta + \cos\theta = \sqrt{2}\cos(90 - \theta)$, then $\cot\theta$ is **(SSC CGL Tier I 2012)**

(*a*) $\sqrt{2}$ (*b*) $\sqrt{2} - 1$ (*c*) $\sqrt{2} + 1$ (*d*) 0

Sol. (*b*) $\sin\theta + \cos\theta = \sqrt{2}\sin\theta \Rightarrow \cos\theta = (\sqrt{2} - 1)\sin\theta \Rightarrow \cot\theta = \sqrt{2} - 1$

Alt: Dividing both sides by $\sin\theta$: $1 + \cot = \sqrt{2} \Rightarrow \cot\theta = \sqrt{2} - 1$

60. If $\sin\theta + \cos\theta = \sqrt{2}\sin(90° - \theta)$, then $\cot\theta$ = ? **(SSC CGL Tier I 2015)**

(*a*) $\sqrt{2} + 1$ (*b*) $\sqrt{2} - 1$ (*c*) $1 - \sqrt{2}$ (*d*) $\sqrt{2}$

Sol. (*a*) $\sin\theta + \cos\theta = \sqrt{2}\cos\theta \Rightarrow \sin\theta = (\sqrt{2} - 1)\cos\theta \Rightarrow \cot\theta = \frac{1}{\sqrt{2} - 1} = \sqrt{2} + 1$

61. If $\cos^4\theta - \sin^4\theta = \frac{2}{3}$, then the value of $1 - 2\sin^2\theta$ is **(SSC CGL Tier I 2013)**

(*a*) $\frac{2}{3}$ (*b*) $\frac{1}{3}$ (*c*) $\frac{4}{3}$ (*d*) 0

Sol. (*a*) $\cos^4\theta - \sin^4\theta = (\cos^2\theta + \sin^2\theta)(\cos^2\theta - \sin^2\theta)$

$= (\cos^2\theta - \sin^2\theta) = 1 - \sin^2\theta - \sin^2\theta = 1 - 2\sin^2\theta = \frac{2}{3}$

62. If $\cos^2\theta - \sin^2\theta = 1/3$ then the value of $\cos^4\theta - \sin^4\theta$ is equal to **(SSC FCI 2012, SSC CGL Tier I 2014)**

(*a*) $\frac{1}{3}$ (*b*) $\frac{2}{\sqrt{3}}$ (*c*) $\frac{1}{2}$ (*d*) $\sqrt{3}$

Sol. (*a*) $\cos^4\theta - \sin^4\theta = (\cos^2\theta + \sin^2\theta)(\cos^2\theta - \sin^2\theta) = 1 \times \frac{1}{3} = \frac{1}{3}$

63. If $\sec^2\theta + \tan^2\theta = \frac{5}{3}$, then the value of θ is:

(*a*) 30° (*b*) 45° (*c*) 60° (*d*) 90°

Sol. (*a*) $1 + \tan^2\theta + \tan^2\theta = \frac{5}{3} \Rightarrow 2\tan^2\theta = \frac{5}{3} - 1 = \frac{2}{3} \Rightarrow \tan^2\theta = \frac{1}{3} \Rightarrow \tan\theta = \frac{1}{\sqrt{3}} = \tan 30°$

64. $\sec^4\theta - \sec^2\theta$ is equal to **(SSC CGL Tier I 2011)**

(*a*) $\cos^4\theta - \cos^2\theta$ (*b*) $\cos^2\theta - \cos^4\theta$ (*c*) $\tan^2\theta - \tan^4\theta$ (*d*) $\tan^2\theta + \tan^4\theta$

Sol. (*d*) $\sec^4\theta - \sec^2\theta = \sec^2\theta(\sec^2\theta - 1) = (1 + \tan^2\theta)(\tan^2\theta) = \tan^2\theta + \tan^4\theta$

65. $\sec^4\theta(1 - \sin^4\theta) - 2\tan^2\theta = ?$

(*a*) 0 (*b*) 1 (*c*) –1 (*d*) $\sec\theta$

Sol. (*b*) $\sec^4\theta - \tan^4\theta - 2\tan^2\theta = (1 + \tan^2\theta)^2 - \tan^4\theta - 2\tan^2\theta$
$= 1 + \tan^4\theta + 2\tan^2\theta - \tan^4\theta - 2\tan^2\theta = 1$

66. $\sec^2\theta\,\text{cosec}^2\theta - (\tan^2\theta + \cot^2\theta) = ?$

(*a*) 1 (*b*) 0 (*c*) – 1 (*d*) 2

Sol. (*d*) G.E. $= (1 + \tan^2\theta)(1 + \cot^2\theta) - (\tan^2\theta + \cot^2\theta)$
$= 1 + \tan^2\theta + \cot^2\theta + 1 - \tan^2\theta - \cot^2\theta = 2$

67. If $\sin\theta + \cos\theta = \sqrt{3}$, then what is $\tan\theta + \cot\theta$ equal to? **(CDS 2013**

(*a*) 1 (*b*) $\sqrt{2}$ (*c*) 2 (*d*) $\sqrt{3}$

Sol. (*a*) $\sin\theta + \cos\theta = \sqrt{3} \Rightarrow (\sin\theta + \cos\theta)^2 = (\sqrt{3})^2$

$\Rightarrow \sin^2\theta + \cos^2\theta + 2\sin\theta\cos\theta = 3 \Rightarrow \sin\theta\cos\theta = 1$

$\therefore \tan\theta + \cot\theta = \frac{\sin\theta}{\cos\theta} + \frac{\cos\theta}{\sin\theta} = \frac{\sin^2\theta + \cos^2\theta}{\sin\theta\cos\theta} = \frac{1}{1} = 1$

68. If $x = a\sec\theta\cos\phi$, $y = b\sec\theta\sin\phi$, $z = c\tan\theta$, then the value of $\frac{x^2}{a^2} + \frac{y^2}{b^2} - \frac{z^2}{c^2}$ is : **(SSC CGL Tier I 2013**

(*a*) 0 (*b*) 1 (*c*) 4 (*d*) 9

Sol. (*b*) $\frac{x^2}{a^2} + \frac{y^2}{b^2} - \frac{z^2}{c^2} = \frac{a^2\sec^2\theta\cos^2\phi}{a^2} + \frac{b^2\sec^2\theta\sin^2\phi}{b^2} - \frac{c^2\tan^2\theta}{c^2}$

$= \sec^2\theta(\cos^2\phi + \sin^2\phi) - \tan^2\theta = \sec^2\theta - \tan^2\theta = 1$

69. If $\sin\theta - \cos\theta = 0$, find the value of $\sin\left(\frac{\pi}{2} - \theta\right) + \cos\left(\frac{\pi}{2} - \theta\right)$. **(SSC CHSL 201**

(*a*) 0 (*b*) 1 (*c*) $\sqrt{2}$ (*d*) $2\sqrt{2}$

Sol. (*c*) $\sin\theta - \cos\theta = 0 \Rightarrow \sin\theta = \cos\theta \Rightarrow \theta = 45°$

$\sin\left(\frac{\pi}{2} - \theta\right) + \cos\left(\frac{\pi}{2} - \theta\right) = \cos\theta + \sin\theta = \cos 45° + \sin 45° = \frac{1}{\sqrt{2}} + \frac{1}{\sqrt{2}} = \frac{2}{\sqrt{2}} = \sqrt{2}$

70. If $\sin\theta - \cos\theta = 0$, then what is $\sin^4\theta + \cos^4\theta$ equal to ? **(CDS 201**

(*a*) 1 (*b*) $\frac{3}{4}$ (*c*) $\frac{1}{2}$ (*d*) $\frac{1}{4}$

Sol. (*c*) $\sin\theta - \cos\theta = 0 \Rightarrow \sin\theta = \cos\theta \Rightarrow \theta = 45°$

$\Rightarrow \sin^4\theta + \cos^4\theta = \left(\frac{1}{\sqrt{2}}\right)^4 + \left(\frac{1}{\sqrt{2}}\right)^4 = \frac{1}{4} + \frac{1}{4} = \frac{1}{2}$

71. If $7\cos^2\theta + 3\sin^2\theta = 4$ and $0 < \theta < \frac{\pi}{2}$, what is the value of $\tan\theta$?

(SSC CPO SI 2012, SSC CHSL 2012, CDS 2012, SSC CGL Tier I 2012)

(*a*) $\sqrt{7}$ (*b*) $\sqrt{3}$ (*c*) 3 (*d*) $\frac{7}{3}$

Sol. (*b*) $7\cos^2\theta + 3\sin^2\theta = 4 \Rightarrow 7 + 3\tan^2\theta = 4\sec^2\theta$ (on dividing both sides by $\cos^2\theta$)

$\Rightarrow 7 + 3\tan^2\theta = 4(1 + \tan^2\theta) \Rightarrow \tan^2\theta = 3 \Rightarrow \tan\theta = \sqrt{3}$

Alt: $7\cos^2\theta + 3\sin^2\theta = 4(\cos^2\theta + \sin^2\theta) \Rightarrow 3\cos^2\theta = \sin^2\theta \Rightarrow \tan^2\theta = 3 \Rightarrow \tan\theta = \sqrt{3}$

72. If $7\sin^2\theta + 3\cos^2\theta = 4$ $(0° \le \theta \le 90°)$, then the value of θ is

(SSC CGL Tier I 2012, CDS 2003, SSC CHSL 2011)

(*a*) $\frac{\pi}{2}$ (*b*) $\frac{\pi}{6}$ (*c*) $\frac{\pi}{6}$ (*d*) $\frac{\pi}{4}$

Sol. (*b*) $7\sin^2\theta + 3\cos^2\theta = 4 \Rightarrow 4\sin^2\theta + (3\sin^2\theta + 3\cos^2\theta) = 4 \Rightarrow 4\sin^2\theta + 3 = 4$

$\Rightarrow \sin^2\theta = \frac{1}{4} \Rightarrow \sin\theta = \frac{1}{2} = \sin 30° \Rightarrow \theta = \frac{\pi}{6}$

73. If $2(\cos^2\theta - \sin^2\theta) = 1$ (θ is a positive acute angle), then $\cot\theta$ is equal to **(SSC CHSL 2013)**

(*a*) $\sqrt{3}$ (*b*) $-\sqrt{3}$ (*c*) $\frac{1}{\sqrt{3}}$ (*d*) 1

Sol. (*a*) $2(\cos^2\theta - \sin^2\theta) = \sin^2\theta + \cos^2\theta \Rightarrow 2\cos^2\theta - 2\sin^2\theta = \sin^2\theta + \cos^2\theta$

$\Rightarrow \cos^2\theta = 3\sin^2\theta \Rightarrow \frac{\cos^2\theta}{\sin^2\theta} = 3 \Rightarrow \cot^2\theta = 3 \Rightarrow \cot\theta = \sqrt{3}$

Alt: $2(\cos 2\theta) = 1 \Rightarrow (\cos 2\theta) = 1/2 = \cos 60° \Rightarrow \theta = 30° \Rightarrow \cot\theta = \sqrt{3}$

74. If $\text{cosec}^4\theta = 17 + \cot^4\theta$, find the value of $\sin\theta$.

(*a*) 3 (*b*) 1 (*c*) $\frac{1}{3}$ (*d*) $\frac{1}{9}$

Sol. (*c*) $(1 + \cot^2\theta)^2 = 17 + \cot^4\theta \Rightarrow 1 + \cot^4\theta + 2\cot^2\theta = 17 + \cot^4\theta$

$\Rightarrow 2\cot^2\theta = 17 - 1 = 16 \Rightarrow \cot^2\theta = 8 \Rightarrow \text{cosec}^2\theta - 1 = 8 \Rightarrow \text{cosec}^2\theta = 9$

$\Rightarrow \text{cosec}\,\theta = 3 \Rightarrow \sin\theta = 1/3$

75. A wheel rotates 3.5 times in one second. What time (in seconds) does the wheel take to rotate 55 radian of angle? **(SSC CGL Tier I 2012)**

(*a*) 1.5 (*b*) 2.5 (*c*) 3.5 (*d*) 4.5

Sol. (*b*) 1 complete revolution = 2π radian

$\Rightarrow$ 3.5 revolutions = 7π radian = $7 \times \frac{22}{7}$ = 22 radian $\Rightarrow$ Required time = $\frac{55}{22} = \frac{5}{2}$ = 2.5 seconds

76. From a point P, two tangents PA and PB are drawn to a circle with centre O. If OP is equal to diameter of the circle, then $\angle APB$ is. **(SSC CHSL 2013)**

(*a*) 60° (*b*) 45° (*c*) 90° (*d*) 30°

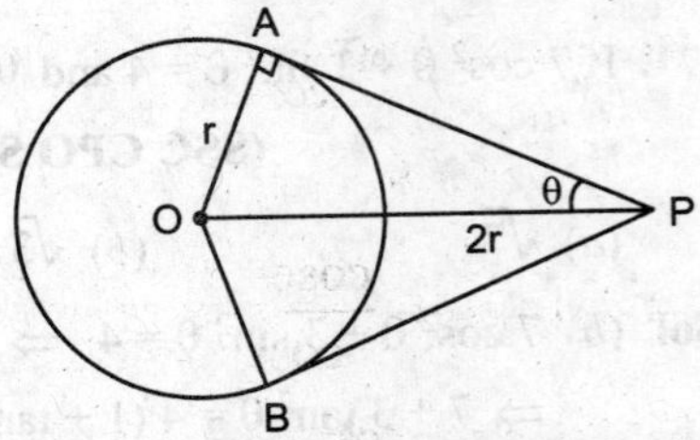

Sol. (*a*) If $OA = r$, then $OP = 2r$

In ΔAOP, $\sin \angle OPA = \frac{AO}{OP} = \frac{r}{2r} = \frac{1}{2} = \sin 30°$

$\Rightarrow \angle OPA = 30° \Rightarrow \angle APB = 2\ \angle OPA = 2 \times 30° = 60°$

77. If $\sec \theta + \text{cosec}\,(90° - \theta) = 4$, $(0 < \theta < 90°)$, then the value of $\tan \theta$ is : **(SSC CHSL 2012)**

(*a*) $\sqrt{3}$ (*b*) $\frac{1}{\sqrt{2}}$ (*c*) $\frac{1}{\sqrt{3}}$ (*d*) 1

Sol. (*a*) $\sec \theta + \text{cosec}\,(90° - \theta) = \sec \theta + \sec \theta = 2 \sec \theta = 4$

$\Rightarrow \sec \theta = 2 = \sec 60° \Rightarrow \theta = 60°$ $\therefore \tan \theta = \tan 60° = \sqrt{3}$

Alternative method: $\tan \theta = \sqrt{\sec^2 \theta - 1} = \sqrt{2^2 - 1} = \sqrt{3}$

78. If $\frac{\cos^2 \theta}{\cot^2 \theta - \cos^2 \theta} = 3$ and $0° < \theta < 90°$, then the value of θ is **(SSC CHSL 2011)**

(*a*) 30° (*b*) 45° (*c*) 60° (*d*) None of these

Sol. (*c*) $\frac{\cos^2 \theta}{\frac{\cos^2 \theta}{\sin^2 \theta} - \cos^2 \theta} = \frac{1}{\frac{1}{\sin^2 \theta} - 1} = \frac{\sin^2 \theta}{1 - \sin^2 \theta} = \frac{\sin^2 \theta}{\cos^2 \theta} = \tan^2 \theta = 3 \Rightarrow \tan \theta = \sqrt{3} = \tan 60°$

79. Eliminating t from the equations, $x = \cos t$, $y = 2 \sin t \cos t$, we get **(SSC CHSL 2012)**

(*a*) $y^2 = 2x^2 + 4x^4$ (*b*) $y^2 = 2x^2 - 4x^4$ (*c*) $y^2 = 4x^2 + 4x^4$ (*d*) $y^2 = 4x^2 - 4x^4$

Sol. (*d*) $y^2 = 4 \sin^2 t \cos^2 t = 4\,(1 - \cos^2 t) \cos^2 t = 4\,(1 - x^2)\,x^2 = 4x^2 - 4x^4$

103. $\frac{\tan \theta + \sin \theta}{\tan \theta - \sin \theta} = ?$

(*a*) $\tan^2 \theta + \sin^2 \theta$ (*b*) $\frac{\sec \theta + 1}{\sec \theta - 1}$ (*c*) $\frac{\sec \theta - 1}{\sec \theta + 1}$ (*d*) 1

Sol. (*b*) $\frac{\tan \theta + \sin \theta}{\tan \theta - \sin \theta} = \frac{\frac{\sin \theta}{\cos \theta} + \sin \theta}{\frac{\sin \theta}{\cos \theta} - \sin \theta} = \frac{\frac{1}{\cos \theta} + 1}{\frac{1}{\cos \theta} - 1} = \frac{\sec \theta + 1}{\sec \theta - 1}$.

80. If $x = a \cos \theta - b \sin \theta$, $y = b \cos \theta + a \sin \theta$, then find the value of $x^2 + y^2$. **(SSC CGL Tier I 2013, 2014 M)**

(*a*) a^2 (*b*) b^2 (*c*) $a^2 - b^2$ (*d*) $a^2 + b^2$

Sol. (*d*) $x^2 + y^2 = a^2 \cos^2 \theta + b^2 \sin^2 \theta - 2ab \sin \theta \cos \theta + b^2 \cos^2 \theta + a^2 \sin^2 \theta + 2ab \sin \theta \cos \theta$

$= (a^2 + b^2)\cos^2 \theta + (b^2 + a^2)\sin^2 \theta = (a^2 + b^2)(\sin^2 \theta + \cos^2 \theta) = a^2 + b^2$

81. If $a \sec \theta = x$ and $b \tan \theta = y$, then how are x and y connected with a and b **(SSC CGL Tier I 2014)**

(*a*) $a^2x^2 + b^2y^2 = a^2b^2$ (*b*) $b^2x^2 + a^2y^2 = a^2b^2$ (*c*) $a^2x^2 - b^2y^2 = a^2b^2$ (*d*) $b^2x^2 - a^2y^2 = a^2b^2$

Sol. (*d*) $b^2x^2 - a^2y^2 = b^2\,(a \sec \theta)^2 - a^2\,(b \tan \theta)^2 = a^2b^2 \sec^2 \theta - a^2b^2 \tan^2 \theta$

$= a^2b^2\,(\sec^2 \theta - \tan^2 \theta) = a^2b^2 \times 1 = a^2b^2$

82. The value of $\dfrac{1}{\text{cosec}\,\theta - \cot\theta} - \dfrac{1}{\sin\theta}$ is **(SSC CGL Tier I 2013)**

(*a*) cosec θ (*b*) tan θ (*c*) 1 (*d*) cot θ

Sol. (*d*) G.E. = $\dfrac{\text{cosec}^2\theta - \cot^2\theta}{\text{cosec}\,\theta - \cot\theta}$ – Cosec θ = (Cosec θ + cot θ) – Cosec θ = cot θ

83. If $\cos A + \cos^2 A = 1$, then $\sin^2 A + \sin^4 A$ is equal to **(SSC CGL Tier I 2012, 2014 M)**

(*a*) 0 (*b*) –1 (*c*) 1 (*d*) $\frac{1}{2}$

Sol. (*c*) $\cos A + \cos^2 A = 1 \Rightarrow \cos A = 1 - \cos^2 A = \sin^2 A$

$\therefore\ \sin^2 A + \sin^4 A = \sin^2 A + \cos^2 A = 1$

84. If $1 + \cos^2\theta = 3\sin\theta\cos\theta$, then the integral value of cot θ is **(SSC CGL Tier I 2015)**

(*a*) 3 (*b*) 1 (*c*) 0 (*d*) 2

Sol. (*c*) If we assume cot θ = 1 = cot 45°, we put these value in equation $1 + \cos^2\theta = 3\sin\theta\cos\theta$

$1+\frac{1}{2} = 3\times\frac{1}{\sqrt{2}}\times\frac{1}{\sqrt{2}} \Rightarrow \frac{3}{2} = \frac{3}{2}$, which is true

85. $\sqrt{\dfrac{1+\sin\theta}{1-\sin\theta}} = ?$ **(SSC CGL Tier I 2014)**

(*a*) sec θ + tan θ (*b*) sec θ – tan θ (*c*) cosec θ + cot θ (*d*) cosec θ – cot θ

Sol. (*a*) $\sqrt{\dfrac{1+\sin\theta}{1-\sin\theta}} = \sqrt{\dfrac{(1+\sin\theta)^2}{1-\sin^2\theta}} = \dfrac{1+\sin\theta}{\cos\theta} = \sec\theta + \tan\theta$

86. $\sqrt{\sec^2\theta + \text{cosec}^2\theta} = ?$

(*a*) tan θ (*b*) cot θ (*c*) sec θ + cosec θ (*d*) tan θ + cot θ

Sol. (*d*) $\sqrt{\sec^2\theta+\text{cosec}^2\theta} = \sqrt{\tan^2\theta+1+\cot^2\theta+1}$

$= \sqrt{\tan^2\theta+\cot^2\theta+2} = \sqrt{\tan^2\theta+\cot^2\theta+2\tan\theta\cot\theta} = \sqrt{(\tan\theta+\cot\theta)^2} = \tan\theta + \cot\theta$

87. If $\tan^4\theta + \tan^2\theta = 1$, then value of $\cos^4\theta + \cos^2\theta$ is equal to **(SSC CHSL 2012)**

(*a*) 3 (*b*) 10 (*c*) 1 (*d*) 2

Sol. (*c*) $\tan^4\theta + \tan^2\theta = 1 \Rightarrow \tan^2\theta(\tan^2\theta + 1) = 1$

$\Rightarrow \tan^2\theta\sec^2\theta = 1 \Rightarrow \dfrac{\sin^2\theta}{\cos^2\theta}\times\dfrac{1}{\cos^2\theta} = 1 \Rightarrow \sin^2\theta = \cos^4\theta$

$\therefore\ \cos^4\theta + \cos^2\theta = \sin^2\theta + \cos^2\theta = 1$

88. If $\cos^3\theta + 3\cos\theta\sin^2\theta = m$, $\sin^3\theta + 3\cos^2\theta\sin\theta = n$, $(m+n)^{2/3} + (m-n)^{2/3} = ?$

(*a*) *mn* (*b*) 0 (*c*) 1 (*d*) 2

Sol. (*d*) $(m+n) = \cos^3\theta + \sin^3\theta + 3\sin\theta\cos\theta(\sin\theta+\cos\theta) = (\cos\theta+\sin\theta)^3$

$(m-n) = \cos^3\theta - \sin^3\theta + 3\cos\theta\sin^2\theta - 3\cos^2\theta\sin\theta = (\cos\theta - \sin\theta)^3$

$\therefore\ (m+n)^{2/3} + (m-n)^{2/3} = (\cos\theta+\sin\theta)^2 + (\cos\theta-\sin\theta)^2 = 2(\cos^2\theta+\sin^2\theta) = 2$

89. If $\sin^2\alpha = \cos^3\alpha$, then the value of $(\cot^6\alpha - \cot^2\alpha)$ is **(SSC CGL Tier I 2012)**

(*a*) 1 (*b*) 0 (*c*) –1 (*d*) 2

Sol. (*a*) $\sin^2 \alpha = \cos^3 \alpha \Rightarrow \frac{1}{\sin \alpha} \times \sin^3 \alpha = \cos^3 \alpha \Rightarrow \text{cosec } \alpha = \frac{\cos^3 \alpha}{\sin^3 \alpha} = \cot^3 \alpha$

$\therefore \cot^6 \alpha - \cot^2 \alpha = (\cot^3 \alpha)^2 - \cot^2 \alpha = \text{cosec}^2 \alpha - \cot^2 \alpha = 1$

90. $(\text{cosec } A - \sin A)(\sec A - \cos A)(\tan A + \cot A) = ?$ **(SSC CPO 2012)**

(*a*) –1 (*b*) 0 (*c*) 1 (*d*) ±1

Sol. (*c*) G.E. $= \left(\frac{1}{\sin A} - \sin A\right)\left(\frac{1}{\cos A} - \cos A\right)\left(\frac{\sin A}{\cos A} + \frac{\cos A}{\sin A}\right)$

$= \left(\frac{1 - \sin^2 A}{\sin A}\right)\left(\frac{1 - \cos^2 A}{\cos A}\right)\left(\frac{\sin^2 A + \cos^2 A}{\sin A \cos A}\right) = \frac{\cos^2 A}{\sin A} \times \frac{\sin^2 A}{\cos A} \times \frac{1}{\sin A \cos A} = 1$

91. $\cot^4 \theta - \text{cosec}^4 \theta + \cot^2 \theta + \text{cosec}^2 \theta = ?$

(*a*) 0 (*b*) 1 (*c*) $\cot \theta$ (*d*) $\text{cosec } \theta$

Sol. (*a*) G.E. $= \cot^4 \theta - (\cot^2 + 1)^2 + \cot^2 \theta + \text{cosec}^2 \theta$

$= \cot^4 \theta - \cot^4 \theta - 1 - 2 \cot^2 \theta + \cot^2 \theta + \text{cosec}^2 \theta = -1 + (\text{cosec}^2 \theta - \cot^2 \theta) = -1 + 1 = 0$

92. If $A = \tan 11° \tan 29°$, $B = 2 \cot 61° \cot 79°$, then : **(SSC CHSL 2011)**

(*a*) $A = 2B$ (*b*) $A = -2B$ (*c*) $2A = B$ (*d*) $2A = -B$

Sol. (*c*) $B = 2 \cot 61° \cot 79° = 2 \tan (90° - 61°) \tan (90° - 79°) = 2 \tan 29° \tan 11° = 2A$

93. If $\sin \theta + \cos \theta = \sqrt{2}$, then the value of θ $(0° < \theta < 90°)$ is : **(SSC CGL Tier I 2013)**

(*a*) 60° (*b*) 90° (*c*) 45° (*d*) 30°

Sol. (*c*) $\sin \theta + \cos \theta = \sqrt{2} \Rightarrow \sin^2 \theta + \cos^2 \theta + 2 \sin \theta \cos \theta = 2$

$\Rightarrow 2 \sin \theta \cos \theta = 2 - 1 = 1 \Rightarrow \sin 2\theta = 1 = \sin 90° \Rightarrow 2\theta = 90° \Rightarrow \theta = 45°$

94. If $\tan \theta - \cot \theta = a$ and $\cos \theta - \sin \theta = b$, then the value of $(a^2 + 4)(b^2 - 1)^2$ is : **(SSC CGL Tier I 2012)**

(*a*) 4 (*b*) 1 (*c*) 2 (*d*) 3

Sol. (*a*) $(a^2 + 4)(b^2 - 1)^2 = [(\tan \theta - \cot \theta)^2 + 4][(\cos \theta - \sin \theta)^2 - 1]^2$

$= (\tan \theta + \cot \theta)^2 (2 \sin \theta \cos \theta)^2 = 4 [(\tan \theta + \cot \theta)(\sin \theta \cos \theta)]^2$

$= 4 (\sin^2 \theta + \cos^2 \theta)^2 = 4 \times 1^2 = 4$

95. If $x \sin^3 \theta + y \cos^3 \theta = \sin \theta \cos \theta \neq 0$ and $x \sin \theta - y \cos \theta = 0$, then value of $(x^2 + y^2)$ is **(SSC CGL Tier I 2012)**

(*a*) $\sin \theta - \cos \theta$ (*b*) $\sin \theta + \cos \theta$ (*c*) 0 (*d*) 1

Sol. (*d*) $x \sin \theta - y \cos \theta = 0 \Rightarrow x \sin \theta = y \cos \theta$

Now, $x \sin^3 \theta + y \cos^3 \theta = \sin \theta \cos \theta \Rightarrow y \cos \theta . \sin^2 \theta + y \cos^3 \theta = \sin \theta \cos \theta$

$\Rightarrow y \cos \theta (\sin^2 \theta + \cos^2 \theta) = \sin \theta \cos \theta \Rightarrow y \cos \theta = \sin \theta \cos \theta \Rightarrow y = \sin \theta$

Now, $x \sin \theta = y \cos \theta \Rightarrow x \sin \theta = \sin \theta \cos \theta \Rightarrow x = \cos \theta$

$\therefore x^2 + y^2 = \cos^2\theta + \sin^2\theta = 1$

Alt : $x \sin \theta - y \cos \theta = 0 \Rightarrow x \sin \theta = y \cos \theta$

$x \sin^3 \theta + y \cos^3 \theta = \sin \theta . \cos \theta$

$\Rightarrow x \sin^3 \theta + y \cos^3 \theta \times \frac{x \sin \theta}{y \cos \theta} = \sin \theta . \cos \theta$

$\Rightarrow x \sin^3 \theta + x \sin \theta . \cos^2 \theta = \sin \theta . \cos \theta$

$\Rightarrow x \sin^2 \theta + x \cos^2 \theta = \cos \theta \Rightarrow x(\sin^2 \theta + \cos^2 \theta) = \cos \theta$

$\Rightarrow x = \cos \theta \Rightarrow x \sin \theta = y \cos \theta \Rightarrow \cos \theta. \sin \theta = y \cos \theta \Rightarrow y = \sin \theta$

$\therefore x^2 + y^2 = \cos^2\theta + \sin^2\theta = 1$

96. If $\tan \theta = \frac{2}{3}$, find the value of $\frac{2 \sin \theta \cos \theta}{\cos^2 \theta - \sin^2 \theta}$.

(*a*) $\frac{7}{12}$ (*b*) $\frac{5}{12}$ (*c*) $\frac{12}{5}$ (*d*) $\frac{12}{7}$

Sol. (*c*) $\frac{2 \sin \theta \cos \theta}{\cos^2 \theta - \sin^2 \theta} = \frac{2 \tan \theta}{1 - \tan^2 \theta}$ (on dividing N and D by $\cos^2 \theta$) $= \frac{2 \times \frac{2}{3}}{1 - \left(\frac{2}{3}\right)^2} = \frac{\frac{4}{3}}{\frac{5}{9}} = \frac{4}{3} \times \frac{9}{5} = \frac{12}{5}$.

97. ABC is a triangle right angled at B and $AB : BC = 3 : 4$. What is $\sin A + \sin B + \sin C$ equal to? **(CDS 2015)**

(*a*) 2 (*b*) $\frac{11}{5}$ (*c*) $\frac{12}{5}$ (*d*) 3

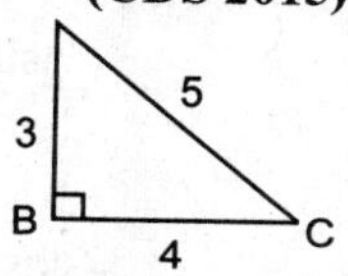

Sol. (*c*) $\sin A + \sin B + \sin C = \frac{BC}{AC} + \sin 90° + \frac{AB}{AC} = \frac{4}{5} + 1 + \frac{3}{5} = \frac{12}{5}$

98. If Sin (A + B) = sin A cos B + cos A sin B, then the value of sin 75° is **(SSC CHSL 2014)**

(*a*) $\frac{\sqrt{3}+1}{\sqrt{2}}$ (*b*) $\frac{\sqrt{2}+1}{2\sqrt{2}}$ (*c*) $\frac{\sqrt{3}+1}{2\sqrt{2}}$ (*d*) $\frac{\sqrt{3}+1}{2}$

Sol. (*c*) $\sin (75°) = \sin (30° + 45°) = \sin 30° \cos 45° + \cos 30° \sin 45°$

$= \frac{1}{2} \times \frac{1}{\sqrt{2}} + \frac{\sqrt{3}}{2} \times \frac{1}{\sqrt{2}} = \frac{1}{2\sqrt{2}} + \frac{\sqrt{3}}{2\sqrt{2}} = \frac{1+\sqrt{3}}{2\sqrt{2}}$

99. If A and B are complementary angles, then the value of $\sin A \cos B + \cos A \sin B - \tan A \tan B + \sec^2 A - \cot^2 B$ is **(SSC CGL Tier I 2012)**

(*a*) 1 (*b*) –1 (*c*) 2 (*d*) 0

Sol. (*a*) $\sin A \cos B + \cos A \sin B - \tan A \tan B + \sec^2 A - \cot^2 B$

$= \sin A \cos (90° - A) + \cos A \sin (90° - A) - \tan A \tan (90° - A) + \sec^2 A - \cot^2 (90° - A)$

$= \sin A \sin A + \cos A \cos A - \tan A \cot A + \sec^2 A - \tan^2 A$

$= (\sin^2 A + \cos^2 A) - (\tan A \cot A) + (\sec^2 A - \tan^2) = 1 - 1 + 1 = 1$

100. Find the value of $\frac{\sin\theta}{\cot\theta + \text{cosec}} - \frac{\sin\theta}{\cot\theta - \text{cosec}\,\theta}$ **(SSC CHSL 2013)**

(*a*) 1 (*b*) 2 (*c*) $\sin \theta$ (*d*) 0

Sol. (*b*) G.E. $= \frac{\sin\theta(\cot\theta - \text{cosec}\,\theta - \cot\theta - \text{cosec}\,\theta)}{\cot^2\theta - \text{cosec}^2\theta} = \frac{\sin\theta(-2\,\text{cosec}\,\theta)}{-1} = 2 \sin \theta. \text{cosec}\, \theta = 2$

101. Find the value of $\cot^2 A \left(\frac{\sec A - 1}{1 + \sin A}\right) + \sec^2 A \left(\frac{\sin A - 1}{1 + \sec A}\right)$

(*a*) 0 (*b*) – 1 (*c*) 1 (*d*) 2

Sol. (*a*) G.E. $= \frac{\cot^2 A(\sec^2 A - 1) + \sec^2 A(\sin^2 A - 1)}{(1 + \sin A)(1 + \sec A)}$

$= \frac{\cot^2 A(\tan^2 A) + \sec^2 A(-\cos^2 A)}{(1 + \sin A)(1 + \sec A)} = \frac{1 - 1}{(1 + \sin A)(1 + \sec A)} = 0$

103. $\dfrac{\sin A}{1+\cos A}+\dfrac{\sin A}{1-\cos A}$ is $(0° < A < 90°)$ **(SSC CGL Tier II 2013)**

(*a*) 2 cosec *A* (*b*) 2 sec *A* (*c*) 2 sin *A* (*d*) 2 cos *A*

Sol. (*a*) $= \dfrac{\sin A - \sin A\cos A + \sin A + \sin A\cos A}{1-\cos^2 A} = \dfrac{2\sin A}{\sin^2 A} = \dfrac{2}{\sin A} = 2\text{ cosec } A$

104. $\dfrac{\sin}{1+\cos\theta}+\dfrac{1+\cos\theta}{\sin\theta}=?$ **(CDS 2012)**

(*a*) 2 (*b*) 2 sin θ (*c*) 2 cosec θ (*d*) tan θ

Sol. (*c*) $\dfrac{\sin}{1+\cos\theta}+\dfrac{1+\cos\theta}{\sin\theta}=\dfrac{\sin^2\theta+(1+\cos\theta)^2}{(1+\cos\theta)\sin\theta}$

$$=\frac{\sin^2\theta+1+\cos^2\theta+2\cos\theta}{\sin\theta(1+\cos\theta)}=\frac{2+2\cos\theta}{\sin\theta(1+\cos\theta)}=\frac{2(1+\cos\theta)}{\sin\theta(1+\cos\theta)}=\frac{2}{\sin\theta}=2\text{ cosec }\theta.$$

105. If $\dfrac{\cos x}{1-\sin x}+\dfrac{\cos x}{1+\sin x}=4$, find value of *x*. **(SSC)**

(*a*) 60° (*b*) 0° (*c*) 30° (*d*) 45°

Sol. (*a*) $\dfrac{\cos x}{1-\sin x}+\dfrac{\cos x}{1+\sin x}=\cos x\left(\dfrac{1+\sin x+1-\sin x}{1-\sin^2 x}\right)=\cos x\left(\dfrac{2}{\cos^2 x}\right)=\dfrac{2}{\cos x}=4$

$\Rightarrow \cos x=\dfrac{1}{2}=\cos 60° \Rightarrow x=60°$

106. If $\dfrac{\cos\alpha}{\sin\beta}=n$ and $\dfrac{\cos\alpha}{\cos\beta}=m$, then the value of $\cos^2\beta$ is **(SSC CGL Tier I 2015)**

(*a*) $\dfrac{1}{m^2+n^2}$ (*b*) 0 (*c*) $\dfrac{n^2}{m^2+n^2}$ (*d*) $\dfrac{m^2}{m^2+n^2}$

Sol. (*c*) $\Rightarrow \tan\beta=\dfrac{m}{n}\Rightarrow \tan^2\beta=\dfrac{m^2}{n^2}\Rightarrow \sec^2\beta=\dfrac{m^2}{n^2}+1=\dfrac{m^2+n^2}{n^2}\Rightarrow \cos^2\beta=\dfrac{n^2}{m^2+n^2}$

107. If θ is acute and $\sin^2\theta - 2\cos\theta + \dfrac{1}{4}=0$, find the value of θ.

(*a*) 30° (*b*) 45° (*c*) 60° (*d*) 90°

Sol.(*c*) $(1-\cos^2\theta)-2\cos\theta+\dfrac{1}{4}=0 \Rightarrow 4-4\cos^2\theta-8\cos\theta+1=0$

$\Rightarrow 4\cos^2\theta+8\cos\theta-5=0 \Rightarrow 4\cos^2\theta+10\cos\theta-2\cos\theta-5=0$

$\Rightarrow (2\cos\theta+5)(2\cos\theta-1)=0 \Rightarrow 2\cos\theta=1 \Rightarrow \cos\theta=\dfrac{1}{2}=\cos 60°$

108. Value of $2(\sin^6\theta+\cos^6\theta)-3(\sin^4\theta+\cos^4\theta)+1$ is **(SSC CGL Tier I 2012)**

(*a*) 0 (*b*) 4 (*c*) 1 (*d*) 2

Sol. (*a*) G.E. $= 2[(\sin^2\theta + \cos^2\theta)(\sin^4\theta + \cos^4\theta - \sin^2\theta\cos^2\theta)] - 3(\sin^4\theta + \cos^4\theta) + 1$
$= 2\sin^4\theta + 2\cos^4\theta - 2\sin^2\theta\cos^2\theta - 3\sin^4\theta - 3\cos^4\theta + 1$
$= -(\sin^4\theta + \cos^4\theta + 2\sin^2\theta\cos^2\theta) + 1 = -(\sin^2\theta + \cos^2\theta)^2 + 1 = -1 + 1 = 0$

109. $\dfrac{\tan\theta + \sec\theta - 1}{\tan\theta - \sec\theta + 1} = ?$

(*a*) $\sec\theta + 1$ (*b*) $\sec\theta - 1$ (*c*) $\sec\theta - \tan\theta$ (*d*) $\sec\theta + \tan\theta$

Sol. (*d*) G.E. $= \dfrac{\sec\theta + \tan\theta - (\sec^2\theta - \tan^2\theta)}{\tan\theta - \sec\theta + 1}$

$= \dfrac{(\sec\theta + \tan\theta) - (\sec\theta + \tan\theta)(\sec\theta - \tan\theta)}{\tan\theta - \sec\theta + 1}$

$= \dfrac{(\sec\theta + \tan\theta)(1 - \sec\theta + \tan\theta)}{1 - \sec\theta + \tan\theta} = \sec\theta + \tan\theta$

110. $\dfrac{\sin\theta - \cos\theta + 1}{\sin\theta + \cos\theta - 1}$ (where $\theta \neq \dfrac{\pi}{2}$) is equal to **(SSC CGL Tier I 2012)**

(*a*) $\dfrac{1+\sin\theta}{\cos\theta}$ (*b*) $\dfrac{1-\sin\theta}{\cos\theta}$ (*c*) $\dfrac{1-\cos\theta}{\sin\theta}$ (*d*) $\dfrac{1+\cos\theta}{\sin\theta}$

Sol. (*a*) G.E. $= \dfrac{\sin\theta - (\cos\theta - 1)}{\sin\theta + (\cos\theta - 1)} \times \dfrac{\sin\theta - (\cos\theta - 1)}{\sin\theta - (\cos\theta - 1)}$

$= \dfrac{\sin^2\theta + \cos^2\theta + 1 - 2\cos\theta - 2\sin\theta(\cos\theta - 1)}{\sin^2\theta - \cos^2\theta - 1 + 2\cos\theta} = \dfrac{2 - 2\cos\theta + 2\sin\theta(1 - \cos\theta)}{-2\cos^2\theta + 2\cos\theta}$

$= \dfrac{2(1-\cos\theta) + 2\sin\theta(1-\cos\theta)}{2\cos\theta(1-\cos\theta)} = \dfrac{2(1-\cos\theta)(1+\sin\theta)}{2\cos\theta(1-\cos\theta)} = \dfrac{1+\sin\theta}{\cos\theta}$

Alt: Put $\theta = 45°$, then G.E. $= \dfrac{\frac{1}{\sqrt{2}} - \frac{1}{\sqrt{2}} + 1}{\frac{1}{\sqrt{2}} + \frac{1}{\sqrt{2}} - 1} = \dfrac{1}{\sqrt{2} - 1} = \sqrt{2} + 1$ Option (*a*) : $\dfrac{1 + \frac{1}{\sqrt{2}}}{\frac{1}{\sqrt{2}}} = \sqrt{2} + 1$

111. If $x\cos\theta + y\sin\theta = 2$ and $x\cos\theta - y\sin\theta = 0$, then which of the following in correct?

(*a*) $x^2 + y^2 = 1$ (*b*) $\dfrac{1}{x^2} + \dfrac{1}{y^2} = 1$ (*c*) $x + y = 1$ (*d*) $x - y = 0$

Sol. (*b*) Adding two equations, $2x\cos\theta = 2 \Rightarrow \cos\theta = 1/x$
$(x\cos\theta + y\sin\theta) - (x\cos\theta - y\sin\theta) = 2 - 0 \Rightarrow 2y\sin\theta = 2 \Rightarrow \sin\theta = 1/y$

$\sin^2\theta + \cos^2\theta = 1 \Rightarrow \left(\dfrac{1}{y}\right)^2 + \left(\dfrac{1}{x}\right)^2 = 1 \Rightarrow \dfrac{1}{x^2} + \dfrac{1}{y^2} = 1$

112. If $3\sin\theta + 4\cos\theta = 5$, then find the value of $3\cos\theta - 4\sin\theta$.
(*a*) 0 (*b*) 1 (*c*) – 1 (*d*) 2

Sol. (*a*) $(3\sin\theta + 4\cos\theta)^2 + (3\cos\theta - 4\sin\theta)^2 = 5^2 + x^2$
$\Rightarrow 9\sin^2\theta + 16\cos^2\theta + 9\cos^2\theta + 16\sin^2\theta = 25 + x^2$
$\Rightarrow 9(\sin^2\theta + \cos^2\theta) + 16(\sin^2\theta + \cos^2\theta) = 25 + x^2 \Rightarrow 25 = 25 + x^2 \Rightarrow x = 0$

113. If $2\cos\theta - \sin\theta = \frac{1}{\sqrt{2}}$, $(0° < \theta < 90°)$, the value of $2\sin\theta + \cos\theta$ is: **(SSC CHSL 2011)**

(*a*) $\frac{1}{\sqrt{2}}$ (*b*) $\sqrt{2}$ (*c*) $\frac{\sqrt{3}}{2}$ (*d*) $\frac{3}{\sqrt{2}}$

Sol. (*d*) $(2\cos\theta - \sin\theta)^2 + (2\sin\theta + \cos\theta)^2 = \left(\frac{1}{\sqrt{2}}\right)^2 + x^2$

$\Rightarrow 4\cos^2\theta + \sin^2\theta - 4\sin\theta.\cos\theta + 4\sin^2\theta + \cos^2\theta + 4\sin\theta\cos\theta = 1/2 + x^2$

$\Rightarrow 4(\cos^2\theta + \sin^2) + (\sin^2\theta + \cos^2\theta) = 1/2 + x^2 \Rightarrow x^2 = 5 - \frac{1}{2} = \frac{9}{2} \Rightarrow x = \frac{3}{\sqrt{2}}$

114. If $\sin\theta + \cos\theta = \sqrt{2}\cos\theta$, then the value of $(\cos\theta - \sin\theta)$ is **(SSC CGL Tier II 2013, SSC CGL Tier I 2013)**

(*a*) $\sqrt{3}\cos\theta$ (*b*) $\sqrt{3}\sin\theta$ (*c*) $\sqrt{2}\cos\theta$ (*d*) $\sqrt{2}\sin\theta$

Sol. (*d*) Let $\cos\theta - \sin\theta = x$

Then $(\sin\theta + \cos\theta)^2 + (\cos\theta - \sin\theta)^2 = (\sqrt{2}\cos\theta)^2 + x^2 \Rightarrow 2(\sin^2\theta + \cos^2\theta) = 2\cos^2\theta + x^2$

$\Rightarrow 2 = 2\cos^2\theta + x^2 \Rightarrow x^2 = 2(1 - \cos^2\theta) = 2\sin^2\theta \Rightarrow x = \sqrt{2}\sin\theta$

Alt: $(\cos\theta + \sin\theta)^2 = (\sqrt{2}\cos\theta)^2 \Rightarrow \cos^2\theta + \sin^2\theta + 2\sin\theta\cos\theta = 2\cos^2\theta$

$\Rightarrow \cos^2\theta - 2\sin\theta\cos\theta = \sin^2\theta \Rightarrow \cos^2\theta - 2\sin\theta\cos\theta + \sin^2\theta = 2\sin^2\theta$

$\Rightarrow (\cos\theta - \sin\theta)^2 = 2\sin^2\theta \Rightarrow \cos\theta - \sin\theta = \sqrt{2}\sin\theta$

Alt: $\sin\theta + \cos\theta = \sqrt{2}\cos\theta \Rightarrow \sin\theta = (\sqrt{2} - 1)\cos\theta$

$\Rightarrow \cos\theta = \frac{1}{\sqrt{2-1}}\sin\theta = (\sqrt{2} + 1)\sin\theta = \sqrt{2}\sin\theta + \sin\theta \Rightarrow \cos\theta - \sin\theta = \sqrt{2}\sin\theta$

115. $\frac{\sin A + \sin B}{\cos A - \cos B} + \frac{\cos A + \cos B}{\sin A - \sin B} = ?$ **(Railways 2003)**

(*a*) 0 (*b*) 1 (*c*) $\sin A\cos B$ (*d*) $\tan A\tan B$

Sol. (*a*) G.E. $= \frac{(\sin^2 A - \sin^2 B) + (\cos^2 A - \cos^2 B)}{(\sin A - \sin B)(\cos A - \cos B)} = \frac{(\sin^2 A + \cos^2 A) - (\sin^2 B + \cos^2 B)}{(\sin A - \sin B)(\cos A - \cos B)}$

$= \frac{1-1}{(\sin A - \sin B)(\cos A - \cos B)} = 0$

116. If $\frac{\sec\theta + \tan\theta}{\sec\theta - \tan\theta} = \frac{5}{3}$, then $\sin\theta$ is equal to : **(SSC CGL Tier I 2013)**

(*a*) $\frac{3}{4}$ (*b*) $\frac{1}{4}$ (*c*) $\frac{1}{3}$ (*d*) $\frac{2}{3}$

Sol. (*b*) $\frac{\sec\theta + \tan\theta}{\sec\theta - \tan\theta} = \frac{5}{3} \Rightarrow \frac{\frac{1}{\cos\theta} + \frac{\sin\theta}{\cos\theta}}{\frac{1}{\cos\theta} - \frac{\sin\theta}{\cos\theta}} = \frac{5}{3} \Rightarrow \frac{1 + \sin\theta}{1 - \sin\theta} = \frac{5}{3}$

$\Rightarrow 3 + 3\sin\theta = 5 - 5\sin\theta \Rightarrow 8\sin\theta = 2 \Rightarrow \sin\theta = \frac{1}{4}$

Alt: $3(\sec\theta + \tan\theta) = 5(\sec\theta - \tan\theta) \Rightarrow 8\tan\theta = 2\sec\theta$

$$8\frac{\sin\theta}{\cos\theta} = \frac{2}{\cos\theta} \Rightarrow 8\sin\theta = 2 \Rightarrow \sin\theta = \frac{1}{4}$$

117. If $\dfrac{\sin\theta+\cos\theta}{\sin\theta-\cos\theta} = 3$, then the value of $\sin^4\theta - \cos^4\theta$ is **(SSC)**

(*a*) 1/5 (*b*) 2/5 (*c*) 3/5 (*d*) 4/5

Sol. (*c*) $\dfrac{\sin\theta+\cos\theta}{\sin\theta-\cos\theta} = 3 \Rightarrow \sin\theta + \cos\theta = 3\sin\theta - \cos\theta \Rightarrow 2\sin\theta = 4\cos\theta \Rightarrow \tan\theta = 2$

$\Rightarrow \sin^4\theta - \cos^4\theta = (\sin^2\theta - \cos^2\theta)(\sin^2\theta + \cos^2\theta) = (\sin^2\theta - \cos^2\theta)$

$= (\cos^2\theta)(\tan^2\theta - 1) = \dfrac{\tan^2\theta-1}{\tan^2\theta+1} = \dfrac{2^2-1}{2^2+1} = \dfrac{3}{5}$

118. If $\dfrac{\sec\theta+\tan\theta}{\sec\theta-\tan\theta} = 2\dfrac{51}{79}$, then the value of $\sin\theta$ is **(SSC CGL Tier I 2015)**

(*a*) $\dfrac{65}{144}$ (*b*) $\dfrac{39}{72}$ (*c*) $\dfrac{35}{72}$ (*d*) $\dfrac{91}{144}$

Sol. (*a*) $\dfrac{\sec\theta+\tan\theta}{\sec\theta-\tan\theta} = 2\dfrac{51}{79} = \dfrac{209}{79} \Rightarrow 79(\sec\theta + \tan\theta) = 209(\sec\theta - \tan\theta)$

$\Rightarrow 79\sec\theta + 79\tan\theta = 209\sec\theta - 209\tan\theta \Rightarrow 130\sec\theta = 288\tan\theta \Rightarrow \dfrac{\tan\theta}{\sec\theta} = \dfrac{130}{288} = \dfrac{65}{144}$

119. If $\tan\theta + \sin\theta = m$ and $\tan\theta - \sin\theta = n$, then $m^2 - n^2 = ?$

(*a*) 0 (*b*) 1 (*c*) –1 (*d*) $4\sqrt{mn}$

Sol. (*d*) $m^2 - n^2 = (\tan\theta + \sin\theta)^2 - (\tan\theta - \sin\theta)^2 = 4\tan\theta\sin\theta$

option (*d*) $= 4\sqrt{mn} = 4\sqrt{\tan^2\theta - \sin^2\theta} = 4\sqrt{\dfrac{\sin^2\theta}{\cos^2\theta} - \sin^2\theta} = 4\sin\theta\sqrt{\dfrac{1-\cos^2\theta}{\cos^2\theta}}$

$= 4\sin\theta\dfrac{\sin\theta}{\cos\theta} = 4\sin\theta\tan\theta$

120. If $x\cos\theta = 3$ and $4\tan\theta = y$, then the relation between x and y, which is independent of θ, is given by **(SSC CGL Tier I 2013)**

(*a*) $\dfrac{x^2}{9}+\dfrac{y^2}{16}=1$ (*b*) $\dfrac{9}{x^2}-\dfrac{16}{y^2}=1$ (*c*) $\dfrac{x^2}{9}-\dfrac{y^2}{16}=1$ (*d*) $\dfrac{9}{x^2}+\dfrac{16}{y^2}=1$

Sol. (*c*) $x\cos\theta = 3 \Rightarrow x = 3\sec\theta \Rightarrow x^2 = 9\sec^2\theta;\ \ y^2 = 16\tan^2\theta$

$\Rightarrow \dfrac{x^2}{9} - \dfrac{y^2}{16} = \dfrac{9\sec^2\theta}{9} - \dfrac{16\tan^2\theta}{16} = \sec^2\theta - \tan^2\theta = 1$

121. If $\sin 17° = \dfrac{x}{y}$, then the value of $(\sec 17° - \sin 73°)$ is **(SSC CGL Tier II 2012)**

(*a*) $\dfrac{x^2}{y\sqrt{x^2-y^2}}$ (*b*) $\dfrac{y^2}{x\sqrt{x^2-y^2}}$ (*c*) $\dfrac{y^2}{x\sqrt{y^2-x^2}}$ (*d*) $\dfrac{x^2}{y\sqrt{y^2-x^2}}$

Sol. (*d*) $\sin 17° = \frac{x}{y} \Rightarrow \cos 17° = \sqrt{1 - \frac{x^2}{y^2}} = \frac{\sqrt{y^2 - x^2}}{y}$

Now $\sec 17° - \sin 73° = \frac{1}{\cos 17°} - \cos 17° = \frac{1 - \cos^2 17°}{\cos 17°}$

$$= \frac{\sin^2 17°}{\cos 17°} = \frac{\left(\frac{x}{y}\right)^2}{\frac{\sqrt{y^2 - x^2}}{y}} = \frac{x^2}{y\sqrt{y^2 - x^2}}$$

122. If $4x = \sec\theta$ and $\frac{4}{x} = \tan\theta$, then $8\left(x^2 - \frac{1}{x^2}\right)$ is **(SSC CGL Tier I 2012)**

(*a*) $\frac{1}{2}$ (*b*) $\frac{1}{4}$ (*c*) $\frac{1}{16}$ (*d*) $\frac{1}{8}$

Sol. (*a*) $\sec^2\theta - \tan^2\theta = 1$

$$\Rightarrow (4x)^2 - \left(\frac{4}{x}\right)^2 = 1 \Rightarrow 16x^2 - \frac{16}{x^2} = 1 \Rightarrow 16\left(x^2 - \frac{1}{x^2}\right) = 1 \Rightarrow 8\left(x^2 - \frac{1}{x^2}\right) = \frac{1}{2}$$

123. If $2\sin\left(\frac{\pi x}{2}\right) = x^2 + \frac{1}{x^2}$, then the value of $\left(x - \frac{1}{x}\right)$ is : **(SSC CGL Tier I 2012)**

(*a*) 1 (*b*) 0 (*c*) –1 (*d*) 2

Sol. (*b*) Maximum value of $2\sin\theta = 2 \times 1 = 2$. Minimum value of $x^2 + \frac{1}{x^2} = 2$

If $x = 1$, then $2\sin\frac{\pi}{2} = 1 + 1 \Rightarrow 2 = 2 \Rightarrow$ LHS = RHS $\therefore x - \frac{1}{x} = 1 - 1 = 0$

124. If $\cos x + \cos^2 x = 1$, the numerical value of $(\sin^{12} x + 3\sin^{10} x + 3\sin^8 x + \sin^6 x - 1)$ is : **(SSC CGL Tier I 2012 M, SSC CGL Tier I 2013)**

(*a*) 1 (*b*) –1 (*c*) 2 (*d*) 0

Sol. (*d*) $\cos x + \cos^2 x = 1 \Rightarrow \cos x = 1 - \cos^2 x = \sin^2 x$

$\Rightarrow (\sin^{12} x + 3\sin^{10} x + 3\sin^8 x + \sin^6 x) - 1$

$= (\sin^4 x + \sin^2 x)^3 - 1 = (\cos^2 x + \sin^2 x)^3 - 1 = 1^3 - 1 = 0$

125. If $\sin\theta + \sin^2\theta = 1$, then $\cos^{12}\theta + 3\cos^{10}\theta + 3\cos^8\theta + \cos^6\theta + 2\cos^4\theta + 2\cos^2\theta = ?$

(*a*) 4 (*b*) 1 (*c*) 2 (*d*) 3

Sol. (*d*) $\sin\theta + \sin^2\theta = 1 \Rightarrow \sin\theta = 1 - \sin^2\theta = \cos^2\theta$

$(\cos^{12}\theta + 3\cos^{10}\theta + 3\cos^8\theta + \cos^6\theta) + 2(\cos^4\theta + \cos^2\theta)$

$= (\cos^4\theta + \cos^2\theta)^3 + 2(\cos^4\theta + \cos^2\theta) = (\sin^2\theta + \cos^2\theta)^3 + 2(\sin^2\theta + \cos^2\theta) = 1^3 + 2 \times 1 = 3$

126. If $\sin\theta + \sin^2\theta + \sin^3\theta = 1$, $\cos^6\theta - 4\cos^4\theta + 8\cos^2\theta = ?$ **(SSC CGL Tier I 2012)**

(*a*) 0 (*b*) 1 (*c*) 2 (*d*) 4

Sol. (*d*) $\sin\theta + \sin^3\theta = 1 - \sin^2\theta \Rightarrow \sin\theta(1 + \sin^2\theta) = \cos^2\theta$

$\Rightarrow \sin^2\theta(1 + \sin^2\theta)^2 = \cos^4\theta \Rightarrow (1 - \cos^2\theta)(2 - \cos^2\theta)^2 = \cos^4\theta$

$\Rightarrow (1 - \cos^2\theta)(4 + \cos^4\theta - 4\cos^2\theta) = \cos^4\theta$

$\Rightarrow 4 + \cos^4\theta - 4\cos^2\theta - 4\cos^2\theta - \cos^6\theta + 4\cos^4\theta = \cos^4\theta$

$\Rightarrow -\cos^6\theta + 4\cos^4\theta - 8\cos^2\theta + 4 = 0 \Rightarrow \cos^6\theta - 4\cos^4\theta + 8\cos^2\theta = 4$

127. If $u_n = \cos^n\alpha + \sin^n\alpha$, then the value of $2u_6 - 3u_4 + 1$ is **(SSC CGL Tier I 2012)**

(*a*) 4 (*b*) 6 (*c*) 0 (*d*) 1

Sol. (*c*) $2u_6 - 3u_4 + 1 = 2(\cos^6\alpha + \sin^6\alpha) - 3(\cos^4\alpha + \sin^4\alpha) + 1$

$= 2(\cos^2\alpha + \sin^2\alpha)(\cos^4\alpha + \sin^4\alpha - \cos^2\alpha\sin^2\alpha) - 3(\cos^4\alpha + \sin^4\alpha) + 1$

$= 2\cos^4\alpha + 2\sin^4\alpha - 2\cos^2\alpha\sin^2\alpha - 3\cos^4\alpha - 3\sin^4\alpha + 1$

$= -(\cos^4\alpha + \sin^4\alpha + 2\cos^2\alpha\sin^2\alpha) + 1 = -(\cos^2\alpha + \sin^2\alpha)^2 + 1 = -1 + 1 = 0$

128. The value of $3(\sin^4\theta + \cos^4\theta) + 2(\sin^6\theta + \cos^6\theta + 12\sin^2\theta\cos^2\theta)$ is **(SSC CGL Tier I 2015)**

(*a*) 2 (*b*) 3 (*c*) 0 (*d*) 5

Sol. (*d*) $= 3(\sin^4\theta + \cos^4\theta) + 2\{(\sin^2\theta + \cos^2\theta)(\sin^4\theta + \cos^4\theta - 2\sin^2\theta\cos^2\theta)\} + 12\sin^2\theta\cos^2\theta$

$= 3(\sin^4\theta + \cos^4\theta) + 2(\sin^4\theta + \cos^4\theta - 2\sin^2\theta\cos^2\theta) + 12\sin^2\theta\cos^2\theta$

$= 5(\sin^4\theta + \cos^4\theta + 2\sin^2\theta\cos^2\theta) = 5(\sin^2\theta + \cos^2\theta)^2 = 5$

129. The simplified value of $1 - \frac{\sin^2 A}{1+\cos A} + \frac{1+\cos A}{\sin A} - \frac{\sin A}{1-\cos A}$ is : **(SSC CGL Tier I 2012)**

(*a*) $\cos A$ (*b*) 0 (*c*) 1 (*d*) $\sin A$

Sol. (*a*) $1 - \frac{1-\cos^2 A}{1+\cos A} + \frac{1+\cos A}{\sin A} - \frac{\sin A}{1-\cos A} = 1 - (1-\cos A) + \frac{1-\cos^2 A - \sin^2 A}{\sin A(1-\cos A)}$

$$= 1 - 1 + \cos A + \frac{1-(\sin^2 A + \cos^2 A)}{\sin A(1-\cos A)} = \cos A$$

130. The simplified value of $(1 + \tan\theta + \sec\theta)(1 + \cot\theta - \operatorname{cosec}\theta)$ is **(SSC CGL Tier I 2012)**

(*a*) –2 (*b*) 2 (*c*) 1 (*d*) –1

Sol. (*b*) $(1 + \tan\theta + \sec\theta)(1 + \cot\theta - \operatorname{cosec}\theta)$

$$= \left(1+\frac{\sin\theta}{\cos\theta}+\frac{1}{\cos\theta}\right)\left(1+\frac{\cos\theta}{\sin\theta}-\frac{1}{\sin\theta}\right) = \left[\frac{(\cos\theta+\sin\theta)+1}{\cos\theta}\right]\left[\frac{(\cos\theta+\sin\theta)-1}{\sin\theta}\right]$$

$$= \left[\frac{(\cos\theta+\sin\theta)^2-1}{\sin\theta\cos\theta}\right] = \left[\frac{\cos^2\theta+\sin^2\theta+2\sin\theta\cos\theta-1}{\sin\theta\cos\theta}\right] = \frac{2\sin\theta\cos\theta}{\sin\theta\cos\theta} = 2$$

131. If $\frac{\sec\theta+\tan\theta}{\sec\theta-\tan\theta} = \frac{2+\sqrt{3}}{2-\sqrt{3}}$, then the value of θ in circular measure will be: **(SSC CHSL 2013)**

(*a*) $\frac{\pi}{2}$ (*b*) $\frac{\pi}{3}$ (*c*) $\frac{\pi}{4}$ (*d*) $\sqrt{3}$

Sol. (*b*) Comparing two sides, $\sec\theta = 2 = \sec 60°$ and $\tan\theta = \sqrt{3} = \tan 60° \Rightarrow \theta = 60° = \frac{\pi}{3}$

Alt: Applying Componendo and Dividendo : $\frac{(\sec\theta+\tan\theta)+(\sec\theta-\tan\theta)}{(\sec\theta+\tan\theta)-(\sec\theta-\tan\theta)} = \frac{(2+\sqrt{3})+(2-\sqrt{3})}{(2+\sqrt{3})-(2-\sqrt{3})}$

$$\frac{2\sec\theta}{2\tan\theta} = \frac{4}{2\sqrt{3}} \Rightarrow \frac{1}{\sin\theta} = \frac{2}{\sqrt{3}} \Rightarrow \sin\theta = \frac{\sqrt{3}}{2} = \sin\frac{\pi}{3}$$

132. Find the acute angle θ when $\frac{\cos\theta-\sin\theta}{\cos\theta+\sin\theta} = \frac{1-\sqrt{3}}{1+\sqrt{3}}$

(*a*) 30° (*b*) 45° (*c*) 60° (*d*) 75°

Sol. (*c*) $\frac{\cos\theta-\sin\theta}{\cos\theta+\sin\theta} = \frac{1-\sqrt{3}}{1+\sqrt{3}} \Rightarrow \frac{1-\tan\theta}{1+\tan\theta} = \frac{1-\sqrt{3}}{1+\sqrt{3}}$ (on dividing *N* and *D* each by $\cos\theta$)

Comparing two sides, $\tan\theta = \sqrt{3} = \tan 60° \Rightarrow \theta = 60°$

Alt: Applying Componendo and Dividendo: $\dfrac{(\cos\theta-\sin\theta)+(\cos\theta+\sin\theta)}{(\cos\theta-\sin\theta)-(\cos\theta+\sin\theta)}=\dfrac{(1-\sqrt{3})+(1+\sqrt{3})}{(1-\sqrt{3})-(1+\sqrt{3})}$

$$\Rightarrow \frac{2\cos\theta}{-2\sin\theta}=\frac{2}{-2\sqrt{3}}\Rightarrow \cot\theta=\frac{1}{\sqrt{3}}\Rightarrow \tan\theta=\sqrt{3}=\tan 60^\circ$$

Alt: When $\theta = 60^\circ$, $\cos\theta=\dfrac{1}{2}$ and $\sin\theta=\dfrac{\sqrt{3}}{2}$ $\Rightarrow$ G.E. $=\dfrac{\frac{1}{2}-\frac{\sqrt{3}}{2}}{\frac{1}{2}+\frac{\sqrt{3}}{2}}=\dfrac{1-\sqrt{3}}{1+\sqrt{3}}$

133. If $1+\sin x+\sin^2 x+\sin^3 x+\ldots=4+2\sqrt{3}$ and $0<x<90^\circ$, then the value of x is equal to

(*a*) 0° (*b*) 30° (*c*) 60° (*d*) 90°

Sol. (*c*) $1+\sin x+\sin^2 x+\sin^3+\ldots=\dfrac{1}{1-\sin x}=4+2\sqrt{3}$

$$\Rightarrow 1-\sin x=\frac{1}{4+2\sqrt{3}}=\frac{4-2\sqrt{3}}{4}=1-\frac{\sqrt{3}}{2}\Rightarrow \sin x=\frac{\sqrt{3}}{2}=\sin 60^\circ\Rightarrow x=60^\circ$$

134. If x lies in the first quadrant and $\cos x=\dfrac{5}{13}$, what is $\tan x-\cot x$ equal to? **(CDS 2012)**

(*a*) $-\dfrac{139}{60}$ (*b*) $\dfrac{139}{60}$ (*c*) $\dfrac{119}{60}$ (*d*) None of these

Sol. (*c*) $\sin^2 x=1-\cos^2 x=1-\left(\dfrac{5}{13}\right)^2=\left(\dfrac{12}{13}\right)^2\Rightarrow \sin x=\dfrac{12}{13}$

$$\tan x=\frac{\sin x}{\cos x}=\frac{12}{13}\times\frac{13}{5}=\frac{12}{5}\Rightarrow \cot x=\frac{5}{12}\Rightarrow \tan x-\cot x=\frac{12}{5}-\frac{5}{12}=\frac{144-25}{60}=\frac{119}{60}$$

135. If $3\cot\theta=4$, then $\left(\dfrac{5\sin\theta+3\cos\theta}{5\sin\theta-3\cos\theta}\right)=?$ **(Railways, 2002)**

(*a*) $\dfrac{1}{9}$ (*b*) $\dfrac{1}{3}$ (*c*) 3 (*d*) 9

Sol. (*d*) $\dfrac{5\sin\theta+3\cos\theta}{5\sin\theta-3\cos\theta}=\dfrac{5+3\cot\theta}{5-3\cot\theta}$ [Dividing N and D by $\sin\theta$] $=\dfrac{5+4}{5-4}=\dfrac{9}{1}=9$

136. If $5\tan\theta=4$, then $\left(\dfrac{5\sin\theta-3\cos\theta}{5\sin\theta+3\cos\theta}\right)=?$ **(SSC CGL Tier I 2012)**

(*a*) $\dfrac{1}{7}$ (*b*) $\dfrac{2}{7}$ (*c*) $\dfrac{2}{5}$ (*d*) $\dfrac{5}{7}$

Sol. (*a*) $\dfrac{5\sin\theta-3\cos\theta}{5\sin\theta+3\cos\theta}=\dfrac{5\tan\theta-3}{5\tan\theta+3}$ [On dividing N and D by $\cos\theta$] $=\dfrac{4-3}{4+3}=\dfrac{1}{7}$

137. If $5\sin\theta=3$, the numerical value of $\dfrac{\sec\theta-\tan\theta}{\sec\theta+\tan\theta}$ is **(SSC CGL Tier I 2015)**

(*a*) 1/4 (*b*) 1/5 (*c*) 1/3 (*d*) 1/2

Sol. (*a*) $\dfrac{\sec\theta-\tan\theta}{\sec\theta+\tan\theta}=\dfrac{1-\sin\theta}{1+\sin\theta}=\dfrac{5-5\sin\theta}{5+5\sin\theta}=\dfrac{5-3}{5+3}=\dfrac{2}{8}=\dfrac{1}{4}$

138. If $\cot\theta = \dfrac{x}{y}$, then $\dfrac{y\cos\theta - x\sin\theta}{y\cos\theta + x\sin\theta}$ is : **(SSC CHSL 2012)**

(*a*) 0 (*b*) $\dfrac{x^2 - y^2}{x^2 + y^2}$ (*c*) $\dfrac{x}{x^2 + y^2}$ (*d*) $\dfrac{y}{x^2 + y^2}$

Sol. (*a*) $\cot\theta = \dfrac{x}{y} \Rightarrow y\cot\theta = x \Rightarrow \dfrac{y\cos\theta - x\sin\theta}{y\cos\theta + x\sin\theta} = \dfrac{y\cot\theta - x}{y\cot\theta + x} = \dfrac{x-x}{x+x} = 0$

139. If $\text{cosec}\,\theta - \sin\theta = x$ and $\sec\theta - \cos\theta = y$, then $(x^2y)^{2/3} + (xy^2)^{2/3} = ?$

(*a*) 0 (*b*) 1 (*c*) $\tan\theta$ (*d*) $\sec\theta$

Sol. (*b*) $x = \text{cosec}\,\theta - \sin\theta = \dfrac{1}{\sin\theta} - \sin\theta = \dfrac{1-\sin^2\theta}{\sin\theta} = \dfrac{\cos^2\theta}{\sin\theta}$

$y = \sec\theta - \cos\theta = \dfrac{1}{\cos\theta} - \cos\theta = \dfrac{1-\cos^2\theta}{\sin\theta} = \dfrac{\sin^2\theta}{\cos\theta}$

$\Rightarrow (x^2y)^{2/3} + (xy^2)^{2/3} = \left(\dfrac{\cos^4\theta}{\sin^2\theta} \times \dfrac{\sin^2\theta}{\cos\theta}\right)^{2/3} + \left(\dfrac{\cos^2\theta}{\sin\theta} \times \dfrac{\sin^4\theta}{\cos^2\theta}\right)^{2/3}$

$= (\cos^3\theta)^{2/3} + (\sin^3\theta)^{2/3} = \cos^2\theta + \sin^2\theta = 1$

Alt: Take $\theta = 45°$. Then $x = \sqrt{2} - \dfrac{1}{\sqrt{2}} = \dfrac{2-1}{\sqrt{2}} = \dfrac{1}{\sqrt{2}}$, $y = \sqrt{2} - \dfrac{1}{\sqrt{2}} = \dfrac{2-1}{\sqrt{2}} = \dfrac{1}{\sqrt{2}}$

$\Rightarrow (x^2y)^{2/3} + (xy^2)^{2/3} = \left(\dfrac{1}{\sqrt{2}}\right)^2 + \left(\dfrac{1}{\sqrt{2}}\right)^2 = \dfrac{1}{2} + \dfrac{1}{2} = 1$

140. If $2y\cos\theta = x\sin\theta$ and $2x\sec\theta - y\,\text{cosec}\,\theta = 3$, than the relation between x and y is **(SSC CGL Tier I 2012)**

(*a*) $2x^2 + y^2 = 2$ (*b*) $x^2 + 4y^2 = 4$ (*c*) $x^2 + 4y^2 = 1$ (*d*) $4x^2 + y^2 = 4$

Sol. (*b*) $2y\cos\theta = x\sin\theta \Rightarrow \dfrac{2y}{\sin\theta} = \dfrac{x}{\cos\theta} \Rightarrow 2y\,\text{cosec}\,\theta = x\sec\theta$

Now, $2x\sec\theta - y\,\text{cosec}\,\theta = 3 \Rightarrow 2(2y\,\text{cosec}\,\theta) - y\,\text{cosec}\,\theta = 3$

$\Rightarrow 3y\,\text{cosec}\,\theta = 3 \Rightarrow y\,\text{cosec}\,\theta = 1 \Rightarrow y = \sin\theta$

$2y\cos\theta = x\sin\theta \Rightarrow 2(\sin\theta)\cos\theta = x\sin\theta \Rightarrow x = 2\cos\theta$

$\therefore\ x^2 + 4y^2 = (2\cos\theta)^2 + 4\sin^2\theta = 4(\sin^2\theta + \cos^2\theta) = 4$

141. The value of 152 ($\sin 30° + 2\cos^2 45° + 3\sin 30° + 4\cos^2 45° + \ldots + 17\sin 30° + 18\cos^2 45°$) is **(SSC CGL Tier I 2013)**

(*a*) an integer but not a perfect square (*b*) a rational number but not an integer
(*c*) a perfect square of an integer (*d*) irrational

Sol. (*c*) $= 152\left[\dfrac{1}{2} + 2\times\left(\dfrac{1}{\sqrt{2}}\right)^2 + 3\times\dfrac{1}{2} + 4\times\left(\dfrac{1}{\sqrt{2}}\right)^2 + \ldots 17\times\dfrac{1}{2} + 18\times\left(\dfrac{1}{\sqrt{2}}\right)^2\right]$

$= 152 \times \dfrac{1}{2} \times (1 + 2 + 3 + 4 + \ldots + 17 + 18)$

$= 76 \times \left(\frac{18 \times 19}{2}\right) = 76 \times 9 \times 19 = 2^2 \times 3^2 \times 19^2$

142. If $5 \sin\theta + 12\cos\theta = 13$, then what is $5\cos\theta - 12\sin\theta$ equal to ? **(CDS 2013)**

(*a*) –2 (*b*) –1 (*c*) 0 (*d*) 1

Sol. (*c*) $(5\sin\theta + 12\cos\theta)^2 + (5\cos\theta - 12\sin\theta)^2 = 13^2 + x^2$

$\Rightarrow 25\sin^2\theta + 144\cos^2\theta + 25\cos^2\theta + 144\sin^2\theta = 169 + x^2$

$\Rightarrow 25 + 144 = 169 + x^2 \Rightarrow x^2 = 0 \Rightarrow x = 0$

143. If $(1 + \sin\alpha)(1 + \sin\beta)(1 + \sin\gamma) = (1 - \sin\alpha)(1 - \sin\beta)(1 - \sin\gamma)$, then each side is equal to : **(SSC CGL Tier I 2013)**

(*a*) $\pm \cos\alpha \cos\beta \cos\gamma$ (*b*) $\pm \sin\alpha \sin\beta \sin\gamma$ (*c*) $\pm \sin\alpha \cos\beta \cos\gamma$ (*d*) $\pm \sin\alpha \sin\beta \cos\gamma$

Sol. (*a*) Multiplying both sides by $(1 + \sin\alpha)(1 + \sin\beta)(1 + \sin\gamma)$, we get :

$(1 + \sin\alpha)^2 (1 + \sin\beta)^2 (1 + \sin\gamma)^2 = (1 - \sin^2\alpha)(1 - \sin^2\beta)(1 - \sin^2\gamma)$

$= \cos^2\alpha \cos^2\beta \cos^2\gamma$

$\Rightarrow (1 + \sin\alpha)(1 + \sin\beta)(1 + \sin\gamma) = \pm \cos\alpha \cos\beta \cos\gamma$

144. If $a\sin\theta + b\cos\theta = c$, then the value of $a\cos\theta - b\sin\theta$ is : **(SSC CGL Tier I 2013, 2015)**

(*a*) $\pm\sqrt{a^2 - b^2 + c^2}$ (*b*) $\pm\sqrt{-a^2 + b^2 + c^2}$ (*c*) $\pm\sqrt{a^2 + b^2 - c^2}$ (*d*) $\pm\sqrt{a^2 - b^2 - c^2}$

Sol. (*c*) $(a\sin\theta + b\cos\theta)^2 + (a\cos\theta - b\sin\theta)^2$

$= a^2\sin^2\theta + b^2\cos^2\theta + 2ab\sin\theta\cos\theta + a^2\cos^2\theta + b^2\sin^2\theta - 2ab\sin\theta\cos\theta$

$= a^2(\sin^2\theta + \cos^2\theta) + b^2(\sin^2\theta + \cos^2\theta) = a^2 + b^2$

$\therefore c^2 + (a\cos\theta - b\sin\theta)^2 = a^2 + b^2 \Rightarrow (a\cos\theta - b\sin\theta) = \pm\sqrt{a^2 + b^2 - c^2}$

Alt : $(a\sin\theta + b\cos\theta)^2 = c^2 \Rightarrow a^2\sin^2\theta + b^2\cos^2\theta + 2ab\sin\theta\cos\theta = c^2$

$\Rightarrow a^2(1 - \cos^2\theta) + b^2(1 - \sin^2\theta) + 2ab\sin\theta\cos\theta = c^2$

$\Rightarrow a^2 + b^2 - c^2 = a^2\cos^2\theta + b^2\sin^2\theta - 2ab\sin\theta\cos\theta$

$\Rightarrow a^2 + b^2 - c^2 = (a\cos\theta - b\sin\theta)^2 \Rightarrow (a\cos\theta - b\sin\theta) = \pm\sqrt{a^2 + b^2 - c^2}$

145. The expression $\dfrac{\tan 57° + \cot 37°}{\tan 33° + \cot 53°}$ is equal to **(SSC CGL Tier II 2012)**

(*a*) tan 33° cot 53° (*b*) tan 53° cot 37° (*c*) tan 33° cot 57° (*d*) tan 57° cot 37°

Sol. (*d*) $\dfrac{\tan 57° + \cot 37°}{\tan 33° + \cot 53°} = \dfrac{\cot(90 - 57°) + \tan(90 - 37°)}{\tan 33° + \cot 53°}$

$= \dfrac{\cot 33° + \tan 53°}{\tan 33° + \cot 53°} = \dfrac{\dfrac{1}{\tan 33°} + \tan 53°}{\tan 33° + \dfrac{1}{\tan 53°}} = \dfrac{\dfrac{1 + \tan 33° \tan 53°}{\tan 33°}}{\dfrac{1 + \tan 33° \tan 53°}{\tan 53°}} = \dfrac{\tan 53°}{\tan 33°}$

$= \tan 53° \cot 33° = \cot(90° - 53°)\tan(90° - 33°) = \cot 37° \tan 57°$

146. The value of $(1 + \sec 20° + \cot 70°)(1 - \text{cosec } 20° + \tan 70°)$ is **(SSC CGL Tier I 2012)**

(*a*) 0 (*b*) – 1 (*c*) 2 (*d*) 1

Sol. (*c*) G.E $= (1 + \sec 20° + \tan 20°)(1 - \text{cosec } 20° + \cot 20°)$

$= \left(1 + \dfrac{1}{\cos 20°} + \dfrac{\sin 20°}{\cos 20°}\right)\left(1 - \dfrac{1}{\sin 20°} + \dfrac{\cos 20°}{\sin 20°}\right)$

$$= \left(\frac{\cos 20° + 1 + \sin 20°}{\cos 20°}\right)\left(\frac{\sin 20° - 1 + \cos 20°}{\sin 20°}\right)$$

$$= \frac{(\sin 20° + \cos 20°)^2 - 1}{\sin 20° \cos 20°} = \frac{2\sin 20° \cos 20°}{\sin 20° \cos 20°} = 2$$

147. In ΔABC, $\angle C = 90°$ and $AC : BC = 3 : 4$, then the value of cosec B is **(SSC FCI 2012)**

(*a*) $\frac{3}{5}$ (*b*) $\frac{5}{3}$ (*c*) $\frac{3}{4}$ (*d*) $\frac{4}{3}$

Sol. (*b*) $AB^2 = 3^2 + 4^2 = 5^2$

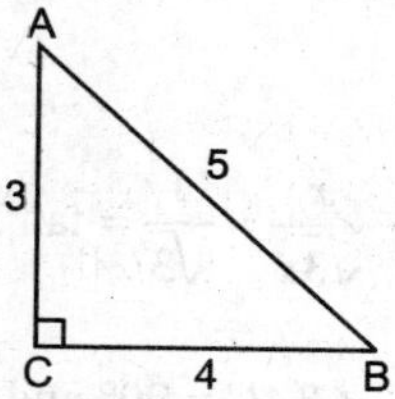

$$\text{cosec } B = \frac{\text{Hypotenuse}}{\text{Perpendicular}} = \frac{AB}{AC} = \frac{5}{3}$$

148. If cosec $\theta = \sqrt{10}$ and θ is acute, then sec θ = ?

(*a*) $\frac{1}{\sqrt{10}}$ (*b*) $\frac{3}{\sqrt{10}}$ (*c*) $3\sqrt{10}$ (*d*) $\frac{\sqrt{10}}{3}$

Sol. (*d*) $\text{cosec } \theta = \frac{\text{Hypotenuse}}{\text{Perpendicular}} = \sqrt{10} = \frac{\sqrt{10}}{1}$

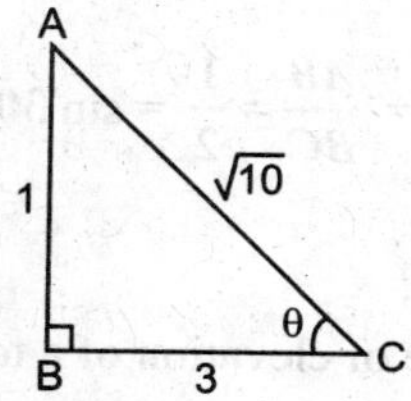

$\text{Base} = \sqrt{10 - 1} = 3$

$\sec \theta = \frac{\text{Hypotenuse}}{\text{Base}} = \frac{\sqrt{10}}{3}$

149. If $\tan \theta = \frac{3}{4}$ and θ is acute, sec θ = ? **(SSC CGL Tier I 2013)**

(*a*) $\frac{4}{5}$ (*b*) $\frac{5}{4}$ (*c*) $\frac{4}{3}$ (*d*) $\frac{5}{3}$

Sol. (*b*) $\tan \theta = \frac{\text{Perpendicular}}{\text{Base}} = \frac{3}{4} \Rightarrow \text{Hypotenuse} = \sqrt{3^2 + 4^2} = 5$

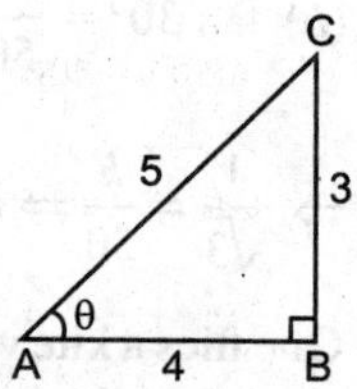

$\sec \theta = \frac{\text{Hypotenuse}}{\text{Base}} = \frac{5}{4}$

Alt: $\sec\theta = \sqrt{\tan^2 \theta + 1} = \sqrt{\left(\frac{3}{4}\right)^2 + 1} = \sqrt{\frac{9}{16} + 1} = \sqrt{\frac{25}{16}} = \frac{5}{4}$

150. If $\tan \theta = \frac{3}{4}$ and θ is acute, then cosec θ is **(SSC CGL Tier I 2013)**

(*a*) $\frac{5}{4}$ (*b*) $\frac{4}{3}$ (*c*) $\frac{4}{5}$ (*d*) $\frac{5}{3}$

Sol. (*d*) $\tan \theta = \frac{AB}{BC} = \frac{3}{4}$

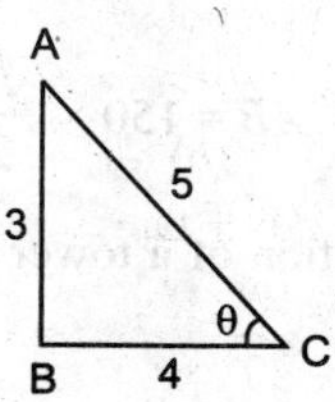

$AC^2 = 3^2 + 4^2 = 5^2$

$\text{cosec } \theta = \frac{AC}{AB} = \frac{5}{3}$

151. If the angle of elevation of the sun is 60°, then the ratio of the height of a wall and its shadow is **(SSC CHSL 2012)**

(*a*) $1:\sqrt{3}$ (*b*) $\sqrt{3}:1$

(*c*) $\sqrt{2}:3$ (*d*) $\sqrt{3}:\sqrt{2}$

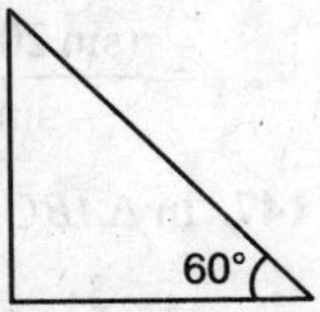

Sol. (*b*) $\frac{\text{Perpendicular}}{\text{Base}} = \tan 60° = \sqrt{3} \Rightarrow$ Height : Shadow $= \sqrt{3}:1$

152. The shadow of a tower is $\sqrt{3}$ times its height. Then the angle of elevation of the top of the tower is **(SSC CGL Tier I 2012)**

(*a*) 45° (*b*) 30°

(*c*) 60° (*d*) 90°

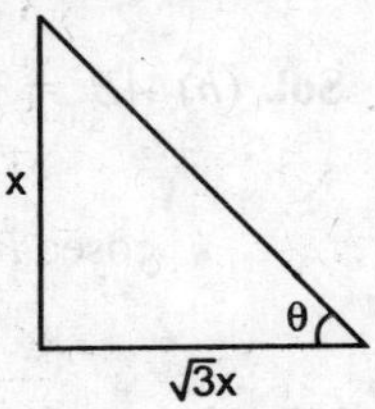

Sol. (*b*) $\tan\theta = \frac{x}{\sqrt{3}x} = \frac{1}{\sqrt{3}} = \tan 30°$

153. In ΔABC, $\angle BAC = 90°$ and $AB = \frac{1}{2}BC$. Then the measure of $\angle ACB$ is: **(SSC FCI 2012)**

(*a*) 60° (*b*) 30° (*c*) 45° (*d*) 15°

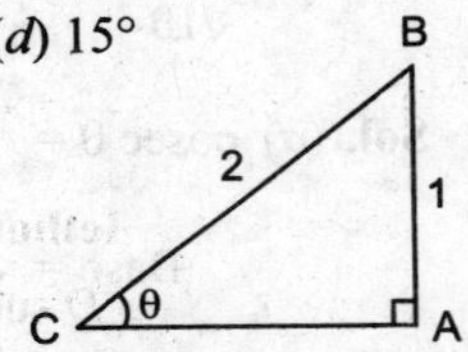

Sol. (*b*) $\sin\theta = \frac{AB}{BC} = \frac{1}{2} = \sin 30° \Rightarrow \angle ACB = 30°$

154. The angle of elevation of a tower from a distance 50 m from its foot is 30°. The height of the tower is **(SSC CGL Tier II 2013)**

(*a*) $50\sqrt{3}$ m (*b*) $\frac{50}{\sqrt{3}}$ m (*c*) $75\sqrt{3}$ m (*d*) $\frac{75}{\sqrt{3}}$ m

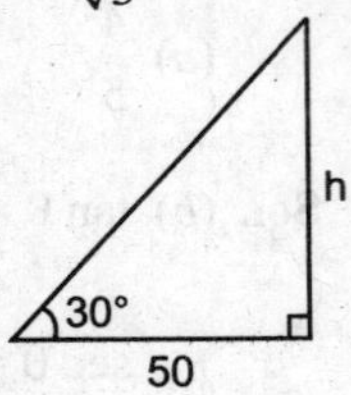

Sol. (*b*) $\tan 30° = \frac{h}{50}$

$$\Rightarrow \frac{1}{\sqrt{3}} = \frac{h}{50} \Rightarrow h = \frac{50}{\sqrt{3}}m$$

155. One flies a kite with a thread 150 m long. If the thread makes an angle of 60° with the horizontal line, then the height of the kite from the ground (assuming the thread to be in a straight line) is **(SSC FCI Asstt. 2012)**

(*a*) 50 m (*b*) $25\sqrt{3}$ m (*c*) $75\sqrt{3}$ m (*d*) 80 m

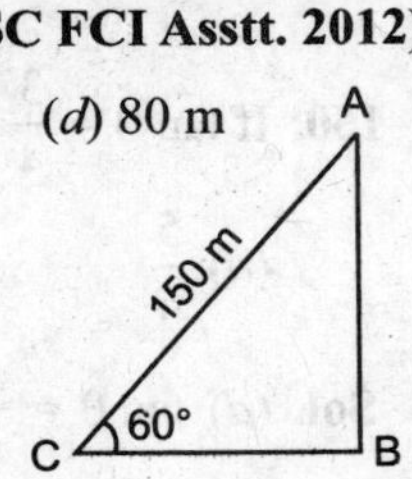

Sol. (*c*) $\sin 60° = \frac{AB}{AC}$

$$\Rightarrow \frac{\sqrt{3}}{2} = \frac{AB}{150} \Rightarrow AB = 150 \times \frac{\sqrt{3}}{2} = 75\sqrt{3}$$

156. The angle of elevation of a tower from a distance 100 m from its foot is 30°. Height of the tower is : **(SSC CGL Tier I 2013, 2015)**

(*a*) $50\sqrt{3}$ m (*b*) $\frac{200}{\sqrt{3}}$ m (*c*) $100\sqrt{3}$ m (*d*) $\frac{100}{\sqrt{3}}$ m

Sol. (*d*) $\tan 30° = \frac{h}{100}$

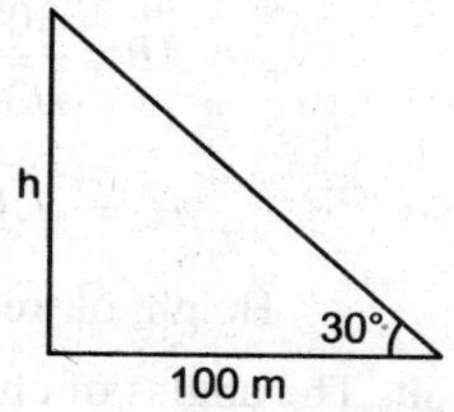

$\Rightarrow \frac{1}{\sqrt{3}} = \frac{h}{100}$

$\Rightarrow h = \frac{100}{\sqrt{3}}$ m

157. In a right angle triangle, one angle is of 30°. Find the ratio of lengths of hypotenuse and the smallest side of the triangle.

(*a*) 1 : 1 (*b*) 2 : 1 (*c*) $\sqrt{2}$: 1 (*d*) 3 : 1

Sol. (*b*) $\sin A = \frac{BC}{AC} \Rightarrow \sin 30° = \frac{BC}{AC}$

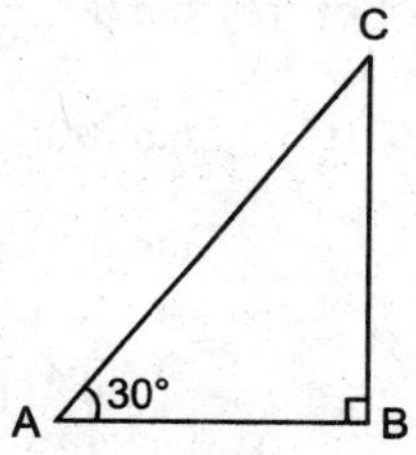

$\Rightarrow \frac{1}{2} = \frac{BC}{AC} \Rightarrow AC = 2BC \Rightarrow AC : BC = 2 : 1$

Note: $AB = \sqrt{2^2 - 1^2} = \sqrt{4-1} = \sqrt{3} > BC$

Alternative Method :

Produce AB to D such that $\angle BCD = 30°$

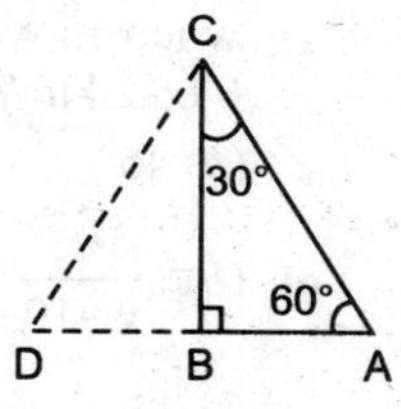

In ΔABC and ΔDBC

$\angle BCD = \angle BCA = 30°$ (By construction)

$BC = BC$ (common)

$\angle CBD = \angle CBA = 90°$

$\therefore \Delta ABC \cong \Delta DBC \quad \therefore \angle CDB = \angle CAB = 60°$

$\angle ACD = \angle ACB + \angle DCB = 30° + 30° = 60°$ $\therefore \Delta ACD$ is an equilateral triangle

$\Rightarrow AD = AC \Rightarrow 2\,BA = AC \Rightarrow AC : BA = 2 : 1$

158. *ABCD* is a rectangle where the ratio of the lengths of *AB* and *BC* is 3 : 2. If *P* is the mid-point of *AB*, then the value of sin ∠*CPB* is **(SSC CGL Tier II 2013)**

(*a*) $\frac{3}{5}$ (*b*) $\frac{2}{5}$ (*c*) $\frac{3}{4}$ (*d*) $\frac{4}{5}$

Sol. (*d*) Let $AB = 6$, then $BC = 4$

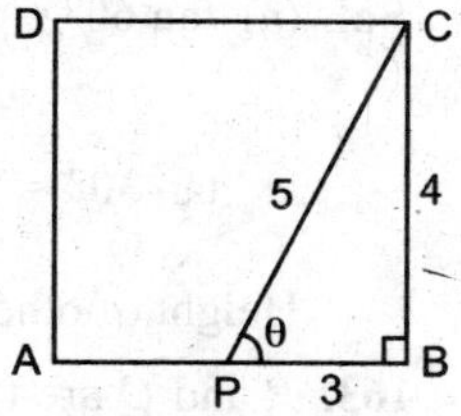

$PB = \frac{1}{2} \times 6 = 3$

ΔPCB is a right angled at $B \Rightarrow PC^2 = PB^2 + BC^2 = 3^2 + 4^2 = 5^2$

$\sin \angle CPB = \frac{BC}{PC} = \frac{4}{5}$

159. A tree is broken by the wind. If the top of the tree struck the ground at an angle of 30° and at a distance of 30 m from the root, then the height of the tree is **(SSC CGL Tier I 2012)**

(*a*) $15\sqrt{3}$ m (*b*) $20\sqrt{3}$ m (*c*) $25\sqrt{3}$ m (*d*) $30\sqrt{3}$ m

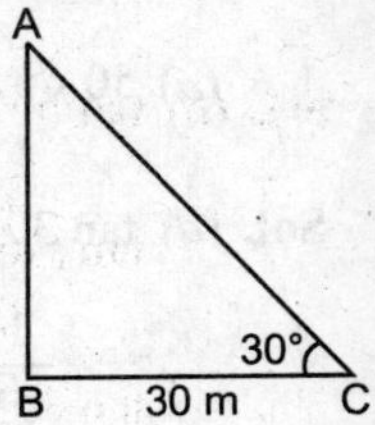

Sol. (*d*) $\tan 30° = \frac{AB}{30} \Rightarrow \frac{1}{\sqrt{3}} = \frac{AB}{30}$

$\Rightarrow AB = \frac{30}{\sqrt{3}} = 10\sqrt{3}$ m

$AC = \sqrt{AB^2 + BC^2} = \sqrt{1200} = 20\sqrt{3}$ m

Height of tree = $AB + AC = 10\sqrt{3} + 20\sqrt{3} = 30\sqrt{3}$ m

160. The angles of elevation of the top of a tower standing on a horizontal plane from two points on a line passing through the foot of the tower at a distance 9 ft and 16 ft respectively are complementary angles. Then the height of the tower is **(SSC CGL Tier I 2012)**

(*a*) 9 ft (*b*) 12 ft (*c*) 16 ft (*d*) 144 ft

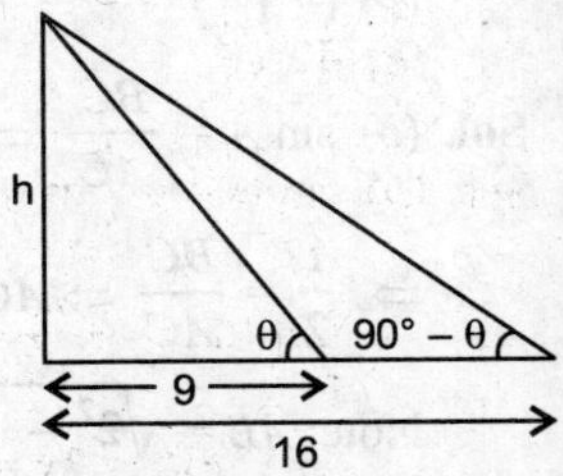

Sol. (*b*) $\tan\theta = \frac{h}{9}$

$\tan(90° - \theta) = \frac{h}{16} \Rightarrow \cot\theta = \frac{h}{16}$

$\tan\theta \times \cot\theta = 1$

$\therefore \frac{h}{9} \times \frac{h}{16} = 1 \Rightarrow h = \sqrt{9 \times 16} = 12$ ft.

161. The angle of elevation of a tower at a level ground is 30°. The angle of elevation become θ when moved 10 m towards the tower. If the height of tower is $5\sqrt{3}$ m, then what is θ equal to ? **(NDA 2012)**

(*a*) 45° (*b*) 60° (*c*) 75° (*d*) None of these

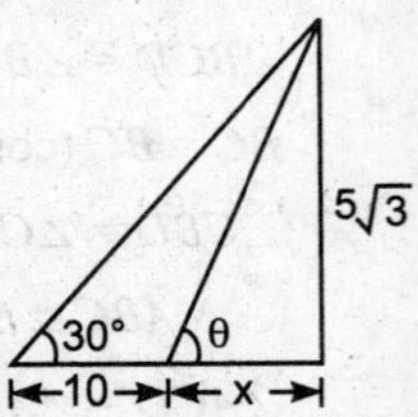

Sol. (*b*) $\frac{5\sqrt{3}}{x+10} = \tan 30° = \frac{1}{\sqrt{3}}$

$\Rightarrow 15 = x + 10 \Rightarrow x = 5$ m

$\tan\theta = \frac{5\sqrt{3}}{5} = \sqrt{3} = \tan 60°$

162. There are two temples, one on each bank of a river, just opposite to each other. One temple is 54 m high. From the top of this temple, the angles of depression of the top and the foot of the other temple are 30° and 60° respectively. The height of the other temple is **(SSC CHSL 2012)**

(*a*) $18\sqrt{3}$ (*b*) 18 m (*c*) 36 m (*d*) $36\sqrt{3}$ m

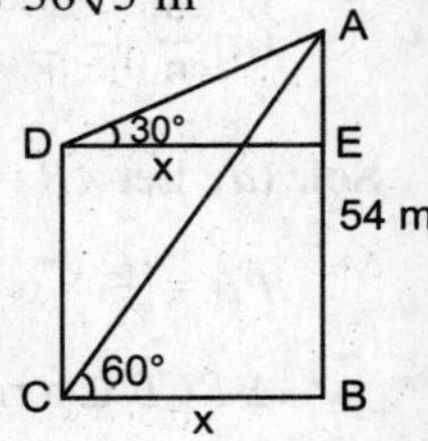

Sol. (*c*) $\tan 60° = \frac{AB}{BC} \Rightarrow \sqrt{3} = \frac{54}{x} \Rightarrow x = \frac{54}{\sqrt{3}} = 18\sqrt{3}$

$\tan 30° = \frac{AE}{DE} \Rightarrow \frac{1}{\sqrt{3}} = \frac{AE}{x} \Rightarrow AE = \frac{x}{\sqrt{3}} = \frac{18\sqrt{3}}{\sqrt{3}} = 18$

Height of other temple = $DC = EB = AB - BE = 54 - 18 = 36$ m

163. *P* and *Q* are two points observed from the top of a building $10\sqrt{3}$ m high. If the angles of depression of the points are complementary and *PQ* = 20 m, then the distance of *P* from the building is **(SSC CGL Tier I 2012)**

(*a*) 30 m (*b*) 40 m (*c*) 25 m (*d*) 45 m

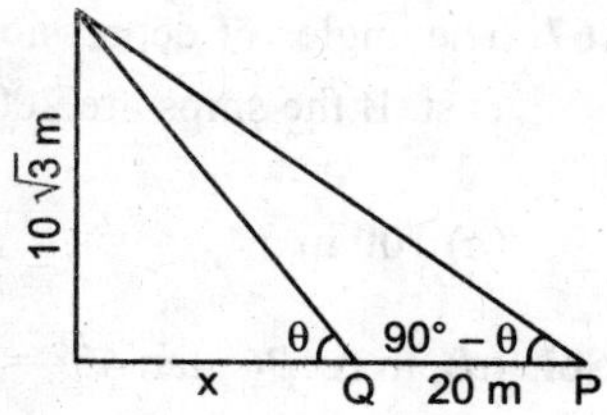

Sol. (*a*) $\tan\theta = \dfrac{10\sqrt{3}}{x}$

$\tan(90° - \theta) = \dfrac{10\sqrt{3}}{x+20} \Rightarrow \cot\theta = \dfrac{10\sqrt{3}}{x+20}$

$\tan\theta \times \cot\theta = 1 \Rightarrow \dfrac{10\sqrt{3}}{x} \times \dfrac{10\sqrt{3}}{x+20} = 1 \Rightarrow 300 = x(x+20)$

Solving, $x = 10$ m $\Rightarrow$ Reqd. Distance = (20 + 10) m = 30 m

164. A man standing in one corner of a square football field observes that the angle subtended by a pole in the corner just diagonally opposite to this corner is 60°. When he retires 80 m from the corner, along the same straight line, he finds the angle to be 30°. The length of the field, in m, is : **(SSC CGL Tier I 2013)**

(*a*) $40\sqrt{2}$ (*b*) 40 (*c*) $20\sqrt{2}$ (*d*) 20

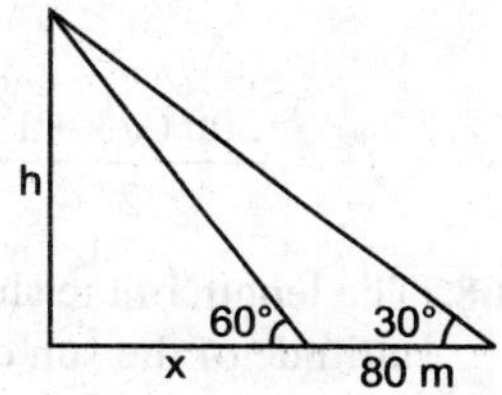

Sol. (*b*) $\tan 60° = \dfrac{h}{x} = \sqrt{3} \Rightarrow h = \sqrt{3}x$

$\tan 30° = \dfrac{h}{x+80} = \dfrac{1}{\sqrt{3}} \Rightarrow \sqrt{3}h = x + 80$

$\Rightarrow \sqrt{3}h = 3x = x + 80 \Rightarrow x = 40$ m

165. A man from the top of a 100 m high tower sees a car moving towards the tower at an angle of depression of 30°. After some time, the angle of depression becomes 60°. The distance (in metre) travelled by the car during this time is **(SSC FCI Asstt. 2012)**

(*a*) $100\sqrt{3}$ (*b*) $\dfrac{200\sqrt{3}}{3}$ (*c*) $\dfrac{100\sqrt{3}}{3}$ (*d*) $200\sqrt{3}$

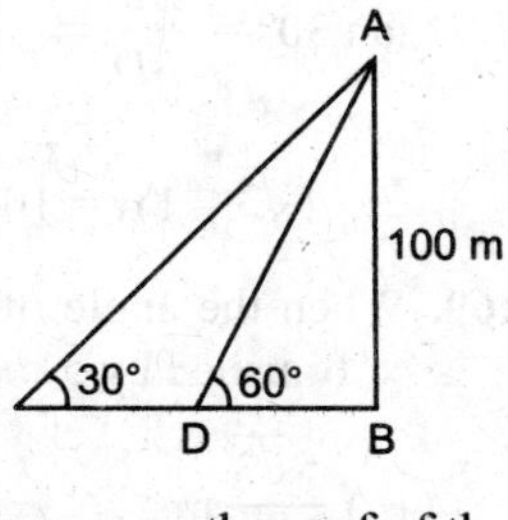

Sol. (*b*) $\tan 60° = \dfrac{AB}{BD} \Rightarrow \sqrt{3} = \dfrac{100}{BD} \Rightarrow BD = \dfrac{100}{\sqrt{3}} = \dfrac{100\sqrt{3}}{3}$

$\tan 30° = \dfrac{AB}{BC} \Rightarrow \dfrac{1}{\sqrt{3}} = \dfrac{100}{BC} \Rightarrow BC = 100\sqrt{3}$

$\therefore CD = BC - BD = 100\sqrt{3} - \dfrac{100\sqrt{3}}{3} = \dfrac{200\sqrt{3}}{3}$ m

166. The angles of elevation of the top of a building and the top of the chimney on the roof of the building from a point on the ground are x and 45° respectively. The height of building is h m. Then the height of the chimney in m, is : **(SSC CHSL 2011)**

(*a*) $h\cot x + h$ (*b*) $h\cot x - h$ (*c*) $h\tan x - h$ (*d*) $h\tan x + h$

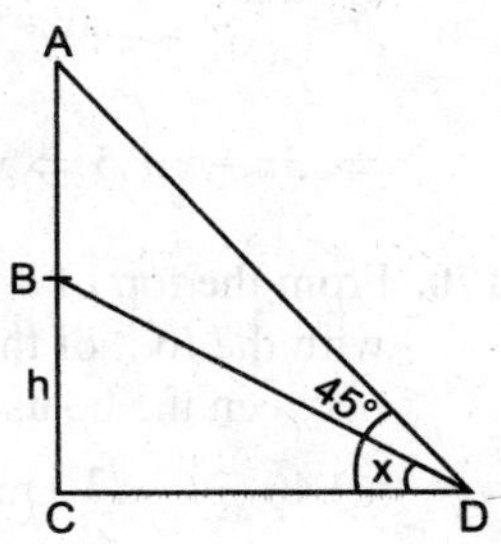

Sol. (*b*) $\tan x = \dfrac{h}{CD} \Rightarrow CD = \dfrac{h}{\tan x} = h\cot x$ ...(1)

$\tan 45° = \dfrac{h + AB}{CD}$

$\Rightarrow 1 = \dfrac{h+AB}{CD} \Rightarrow CD = h + AB$ ...(2)

From (1) and (2), $h\cot x = h + AB \Rightarrow AB = h\cot x - h$.

167. The angles of depression of two ships from the top of a light house are 45° and 30° towards east. If the ships are 200 m apart, the height of the light house is : (Take $\sqrt{3} = 1.73$)

(SSC LDC 2012, Railways 2006)

(*a*) 100 m (*b*) 173 m (*c*) 200 m (*d*) 273 m

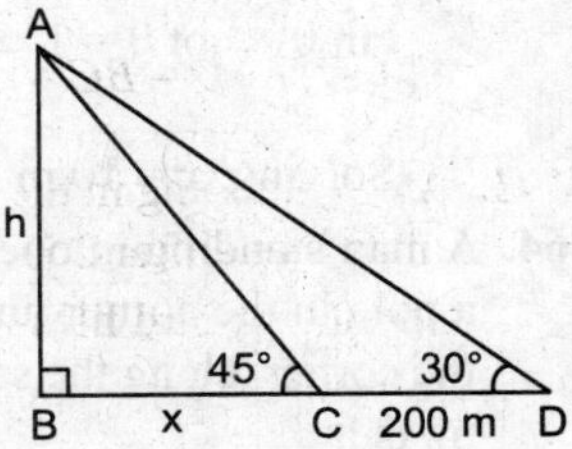

Sol. (*d*) In ΔABC, $\tan 45° = \frac{AB}{BC} \Rightarrow 1 = \frac{h}{x} \Rightarrow h = x$

In ΔABD, $\tan 30° = \frac{h}{x+200}$

$\Rightarrow \frac{1}{\sqrt{3}} = \frac{x}{x+200} \Rightarrow \sqrt{3}x = x + 200$

$\Rightarrow (\sqrt{3}-1)\,x = 200 \Rightarrow x = \frac{200}{\sqrt{3}-1} = \frac{200}{\sqrt{3}-1} \times \frac{\sqrt{3}+1}{\sqrt{3}+1}$

$= \frac{200\,(\sqrt{3}+1)}{2} = 100\,(\sqrt{3}+1) = 100\,(1.73+1) = 273$ m

168. The length of the shadow of a vertical tower on level ground increases by 10 metres when the altitude of the sun changes from 45° to 30°. Then the height of the tower is

(SSC CHSL 2013)

(*a*) $10\sqrt{3}$ m (*b*) $5\sqrt{3}$ m (*c*) $10\,(\sqrt{3}+1)$ m (*d*) $5(\sqrt{3}+1)$ m

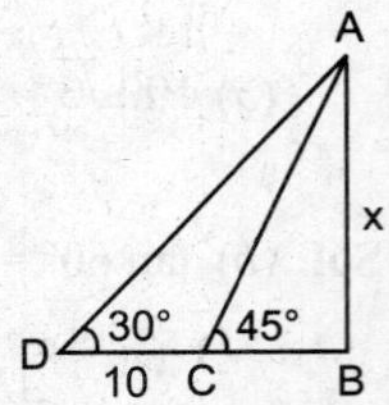

Sol. (*d*) $\tan 45° = \frac{AB}{BC} \Rightarrow 1 = \frac{x}{BC} \Rightarrow BC = x$

$\tan 30° = \frac{AB}{BD} \Rightarrow \frac{1}{\sqrt{3}} = \frac{x}{x+10} \Rightarrow x + 10 = \sqrt{3}\,x$

$\Rightarrow (\sqrt{3}-1)x = 10 \Rightarrow x = \frac{10}{\sqrt{3}-1} = \frac{10(\sqrt{3}+1)}{3-1} = 5\,(\sqrt{3}+1)$m

169. When the angle of elevation of the sun increases from 30° to 60°, the shadow of a post is minished by 5 metres. Then the height of the post is: **(SSC CGL Tier II 2012)**

(*a*) $\frac{5\sqrt{3}}{2}$ m (*b*) $\frac{2\sqrt{2}}{5}$ m (*c*) $\frac{2}{5\sqrt{3}}$ m (*d*) $\frac{4}{5\sqrt{3}}$ m

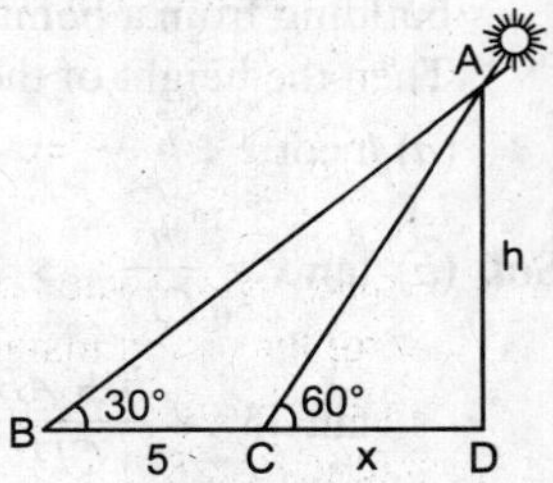

Sol. (*a*) $\tan 60° = \frac{h}{x} \Rightarrow \sqrt{3} = \frac{h}{x} \Rightarrow h = \sqrt{3}x$

$\tan 30° = \frac{h}{x+5} \Rightarrow \frac{1}{\sqrt{3}} = \frac{\sqrt{3}x}{x+5}$

$\Rightarrow 3x = x + 5 \Rightarrow x = \frac{5}{2} \Rightarrow h = \sqrt{3}x = \frac{5\sqrt{3}}{2}m$

170. From the top of a tree of height 120 m, the angle of depression of two boats in the same line with the foot of the tree and on the same side of it are 45° and 60° respectively. The distance between the boats is **(SSC CHSL 2013)**

(*a*) $40\,(3-\sqrt{3})$ m (*b*) $40\,(3\sqrt{3}-1)$m (*c*) $120\,(\sqrt{3}-1)$m (*d*) $120\,(3-\sqrt{3})$m

Sol. (*a*) $\tan 45^\circ = \dfrac{AB}{BD} \Rightarrow 1 = \dfrac{120}{BD} \Rightarrow BD = 120$ m

$$\tan 60^\circ = \frac{AB}{BC} \Rightarrow \sqrt{3} = \frac{120}{BC} \Rightarrow BC = \frac{120}{\sqrt{3}} = 40\sqrt{3}$$

$$\therefore CD = BD - BC = 120 - 40\sqrt{3} = 40\,(3 - \sqrt{3}) \text{ m}$$

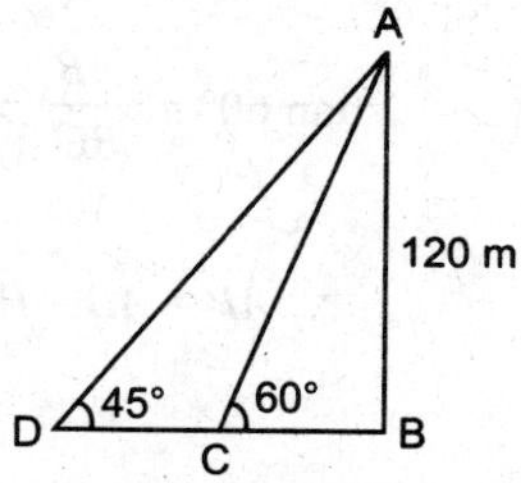

171. A boy standing in the middle of field, observes a flyind bird in the north at an angle of elevation of 30° and after 2 minutes, he observes the same bird in the south at an angle of elevation of 60°. If the bird files all along in a straight line at a height of 50 $\sqrt{3}$ m, then its speed in km/hr is : **(SSC CGL Tier I 2012)**

(*a*) 4.5 (*b*) 3 (*c*) 9 (*d*) 6

Sol. (*d*) $\tan 30^\circ = \dfrac{50\sqrt{3}}{AB} \Rightarrow \dfrac{1}{\sqrt{3}} = \dfrac{50\sqrt{3}}{AB}$

$$\Rightarrow AB = (50\sqrt{3})(\sqrt{3}) = 150$$

$$\tan 60^\circ = \frac{50\sqrt{3}}{AC} \Rightarrow \sqrt{3} = \frac{50\sqrt{3}}{AC} \Rightarrow AC = 50$$

BC = AB + AC = 150 + 50 = 200 $\therefore$ Speed $= \dfrac{200}{120}$ m/s $= \dfrac{200}{120} \times \dfrac{18}{5} = 6$ km/hr

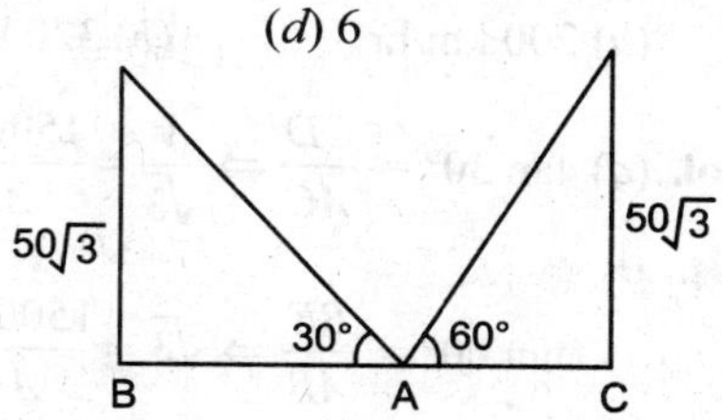

172. Two posts are *x* metres apart and the height of one is double that of the other. If from the mid-point of the line joining their feet, an observer finds the angular elevations of their tops to be complementary, then the height (in metres) of the shorter post is **(SSC CGL Tier II 2012)**

(*a*) $x\sqrt{2}$ (*b*) $\dfrac{x}{\sqrt{2}}$ (*c*) $\dfrac{x}{2\sqrt{2}}$ (*d*) $\dfrac{x}{4}$

Sol. (*c*) $\tan\theta = \dfrac{2h}{x/2} = \dfrac{4h}{x}$ ...(1)

$$\tan(90 - \theta) = \frac{h}{x/2} \Rightarrow \cot\theta = \frac{h}{x/2} = \frac{2h}{x} \quad ...(2)$$

On multiplying (1) and (2)

$$\tan\theta \,.\, \cot\theta = \frac{4h}{x} \times \frac{2h}{x} \Rightarrow \frac{8h^2}{x^2} = 1$$

$$\Rightarrow h^2 = \frac{x^2}{8} \Rightarrow h = \frac{x}{2\sqrt{2}}$$

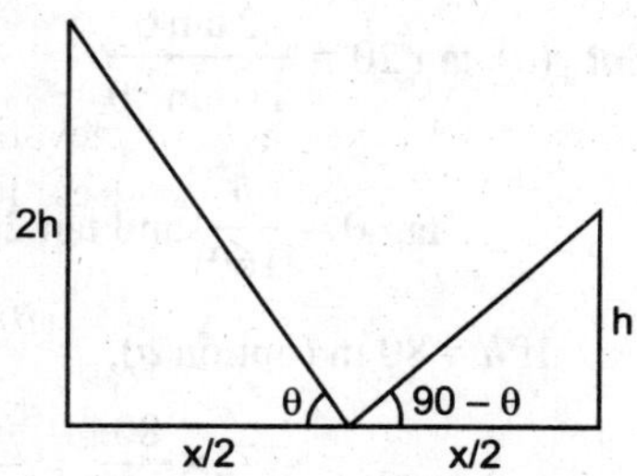

173. An aeroplane when flying at a height of 5000 m from the ground passes vertically above another aeroplane at an instant, when the angles of elevation of the two aeroplanes from the same point on the ground are 60° and 45° respectively. The vertical distance between the aeroplanes at that instant is **(SSC CGL Tier II 2012)**

(*a*) $5000\left(1 - \dfrac{1}{\sqrt{3}}\right)$ m (*b*) 4500 m (*c*) $5000\,(\sqrt{3} - 1)$ m (*d*) $5000\,(3 - \sqrt{3})$ m

Sol. (*a*) $\tan 45^\circ = \dfrac{PB}{BC} \Rightarrow 1 = \dfrac{PB}{BC} \Rightarrow PB = BC$

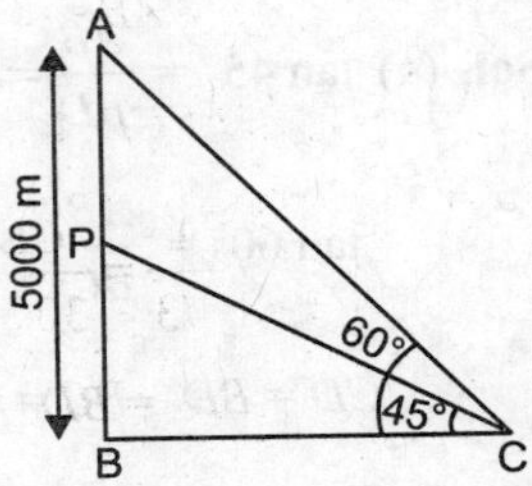

$$\tan 60° = \frac{AB}{BC} \Rightarrow \sqrt{3} = \frac{5000}{PB} \Rightarrow PB = \frac{5000}{\sqrt{3}}$$

$$\therefore\ AP = AB - PB = 5000 - \frac{5000}{\sqrt{3}} = 5000\left(1 - \frac{1}{\sqrt{3}}\right)$$

174. The angle of elevation of an aeroplane from a point on the ground is 60°. After a gap of 15 seconds, the angle of elevation changes to 30°. If the aeroplane is flying at constant height of $1500\sqrt{3}$, find the speed of the aeroplane.

(*a*) 200 km/hr (*b*) 320 km/hr (*c*) 450 km/hr (*d*) 720 km/hr

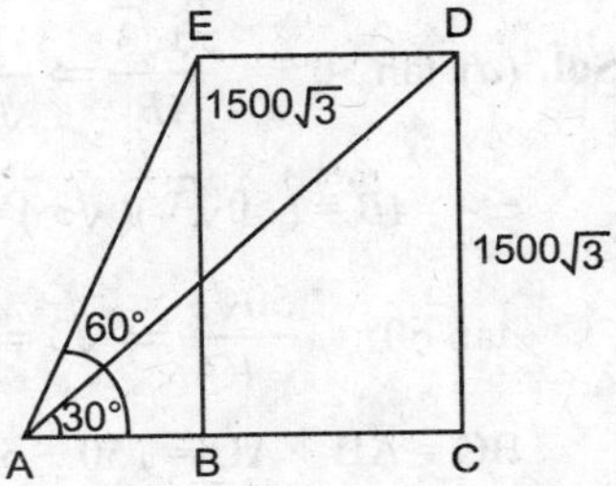

Sol. (*d*) $\tan 30° = \frac{CD}{AC} \Rightarrow \frac{1}{\sqrt{3}} = \frac{1500\sqrt{3}}{AC} \Rightarrow AC = 4500$

$$\tan 60° = \frac{BE}{AB} \Rightarrow \sqrt{3} = \frac{1500\sqrt{3}}{AB} \Rightarrow AB = \frac{1500\sqrt{3}}{\sqrt{3}} = 1500$$

$$\therefore\ BC = 4500 - 1500 = 3000 \text{ m}$$

$$\therefore\ \text{Speed of aeroplane} = \frac{3000}{15}\text{ m/s} = \frac{3000}{15} \times \frac{18}{5}\text{ km/hr} = 720 \text{ km/hr}$$

175. A tower standing on a horizontal plane subtends a certain angle at a point 160 m apart from the foot of the tower. On advancing 100 m towards it, the tower is found to subtend an angle twice as before. The height of the tower is **(SSC CGL Tier II 2013)**

(*a*) 80 m (*b*) 100 m (*c*) 160 m (*d*) 200 m

Sol. (*a*) $\tan 2\theta = \frac{2\tan\theta}{1-\tan^2\theta}$

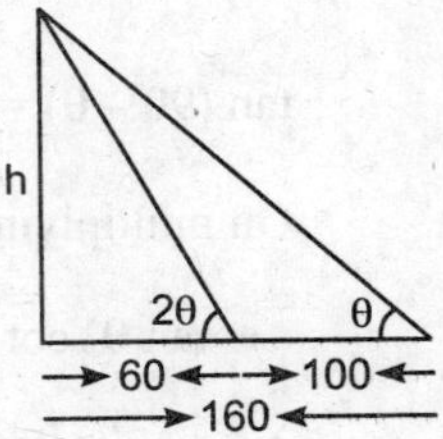

$$\tan\theta = \frac{h}{160} \text{ and } \tan 2\theta = \frac{h}{60}$$

If $h = 80$ m (option *a*),

$$\frac{2\tan\theta}{1-\tan^2\theta} = \frac{2\times\frac{80}{160}}{1-\left(\frac{80}{160}\right)^2} = \frac{1}{3/4} = \frac{4}{3} \Rightarrow \tan 2\theta = \frac{h}{60} = \frac{80}{60} = \frac{4}{3}$$

176. From the top of a cliff 90 m high, the angles of depression of the top and bottom of a tower are observed to be 30° and 60° respectively. What is the height of the tower? **(Railways 2005)**

(*a*) 30 m (*b*) 45 m (*c*) 60 m (*d*) 75 m

Sol. (*c*) In ΔABD, $\tan 60° = \frac{DB}{AB}$

$$\Rightarrow \sqrt{3} = \frac{90}{AB} \Rightarrow AB = \frac{90}{\sqrt{3}} = 30\sqrt{3} \text{ m}$$

$EC = AB = 30\sqrt{3}$

In ΔECD, $\tan 30° = \dfrac{CD}{EC}$

$\Rightarrow \dfrac{1}{\sqrt{3}} = \dfrac{CD}{30\sqrt{3}} \Rightarrow CD = \dfrac{30\sqrt{3}}{\sqrt{3}} = 30$ m

$\therefore BC = BD - CD = (90 - 30)$ m $= 60$ m

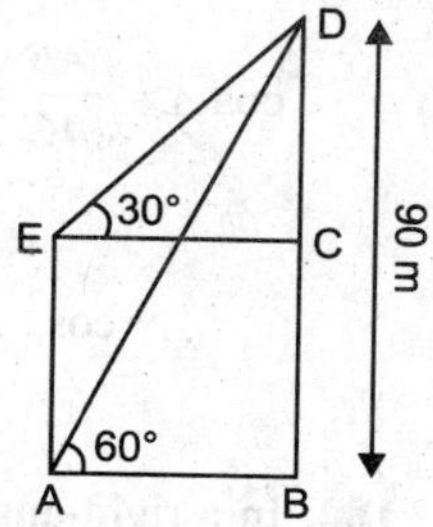

177. At a point on level ground, the angle of elevation of a tower is such that its tangent is 5/12. On moving 240 metres towards the tower, the tangent of the angle of elevation is 3/4. Find the height of the tower. **(SSC)**

(*a*) 220 m (*b*) 200 m (*c*) 225 m (*d*) 240 m

Sol. (*c*) $\tan\beta = \dfrac{h}{x} \Rightarrow \dfrac{3}{4} = \dfrac{h}{x} \Rightarrow 3x = 4h$

$\tan\alpha = \dfrac{h}{x+240} \Rightarrow \dfrac{5}{12} = \dfrac{h}{x+240}$

$\Rightarrow \quad 12h = 5(x + 240) \Rightarrow 9x = 5x + 5 \times 240$

$\Rightarrow \quad 4x = 5 \times 240 \Rightarrow x = 5 \times 60 = 300$ m

$\Rightarrow \quad h = \dfrac{3}{4}x = \dfrac{3}{4} \times 300 = 225$ m

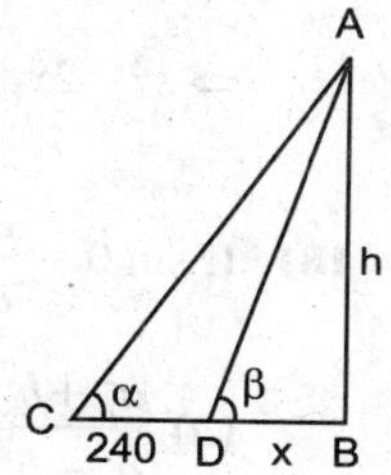

178. A man standing at the top of a tower observes a car at angle of depression of 30°. The car is approaching towards the tower with a uniform speed. 6 seconds later, the angle of depression of the car is found to be 60°. Find the further time taken by the car to reach the foot of the tower.

(*a*) 2 seconds (*b*) 3 seconds (*c*) 4 seconds (*d*) 6 seconds

Sol. (*b*) $\tan 60° = \dfrac{AB}{BC} \Rightarrow \sqrt{3} = \dfrac{AB}{BC} \Rightarrow BC = \dfrac{AB}{\sqrt{3}}$

$\tan 30° = \dfrac{AB}{BD} \Rightarrow \dfrac{1}{\sqrt{3}} = \dfrac{AB}{BD} \Rightarrow BD = \sqrt{3}AB$

$\Rightarrow BD : BC = \sqrt{3}\ AB : \dfrac{AB}{\sqrt{3}} = 3 : 1$

$\Rightarrow CD : BC = (BD - BC) : BC = 2 : 1$

$\therefore$ Required time $= \dfrac{1}{2} \times 6$ seconds $= 3$ seconds

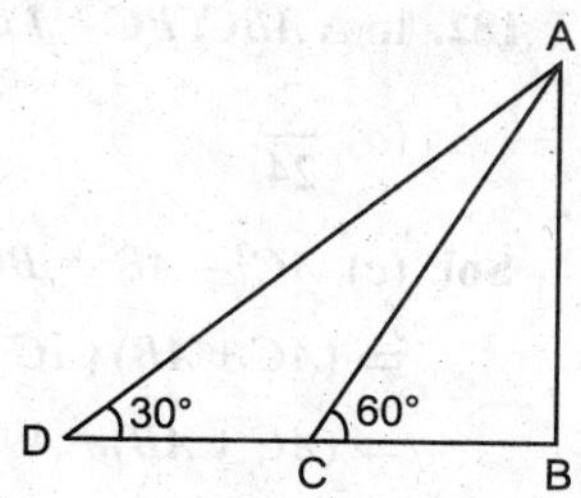

179. In a right-angled triangle ABC, $\angle B$ is the right angle and $AC = 2\sqrt{5}$ cm. If $AB - BC = 2$ cm, then the value of $(\cos^2 A - \cos^2 C)$ is : **(SSC CGL Tier I 2012)**

(*a*) $\dfrac{2}{5}$ (*b*) $\dfrac{3}{5}$ (*c*) $\dfrac{6}{5}$ (*d*) $\dfrac{3}{10}$

Sol. (*b*) $x^2 + (x + 2)^2 = 20 \Rightarrow 2x^2 + 4x + 4 = 20 \Rightarrow x^2 + 2x = 8 \Rightarrow x = 2$

or, $AC^2 = (2\sqrt{5})^2 = 20 = 2^2 + 4^2 \Rightarrow x = 2$ and $x + 2 = 4$

$$\cos A = \frac{AB}{AC} = \frac{4}{2\sqrt{5}} = \frac{2}{\sqrt{5}} \qquad \cos C = \frac{BC}{AC} = \frac{2}{2\sqrt{5}} = \frac{1}{\sqrt{5}}$$

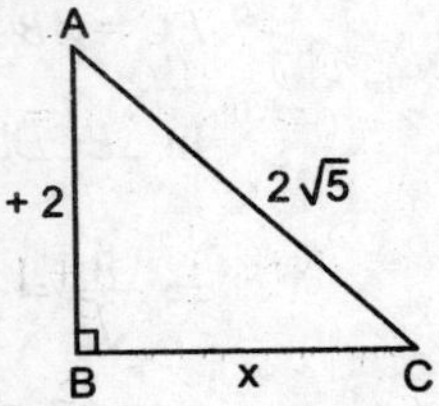

$$\therefore \cos^2 A - \cos^2 C = \left(\frac{2}{\sqrt{5}}\right)^2 - \left(\frac{1}{\sqrt{5}}\right)^2 = \frac{4}{5} - \frac{1}{5} = \frac{3}{5}$$

180. In a right-angled triangle, the product of two sides is equal to half of the square of the third side, i.e., hypotenuse. One of the acute angles must be **(SSC CGL Tier II 2013)**

(*a*) 60° (*b*) 30° (*c*) 45° (*d*) 15°

Sol. (*c*) Given, $xy = \frac{1}{2}z^2 \Rightarrow xy = \frac{1}{2}(x^2 + y^2) \Rightarrow 2xy = x^2 + y^2$

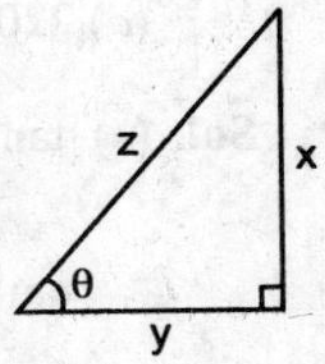

$$\Rightarrow x^2 - 2xy + y^2 = 0 \Rightarrow (x - y)^2 = 0 \Rightarrow x = y \Rightarrow \tan\theta = \frac{x}{y} = \frac{x}{x} = 1 = \tan 45°$$

181. If $\sin\theta = \frac{a^2 - b^2}{a^2 + b^2}$, then $\sec\theta = ?$

(*a*) $\frac{a^2+b^2}{a^2-b^2}$ (*b*) $\frac{2ab}{a^2+b^2}$ (*c*) $\frac{4a^2b^2}{a^2+b^2}$ (*d*) $\frac{a^2+b^2}{2ab}$

Sol. (*d*) $\sin\theta = \frac{\text{Perpendicular}}{\text{Hypotenuse}} = \frac{a^2 - b^2}{a^2 + b^2}$

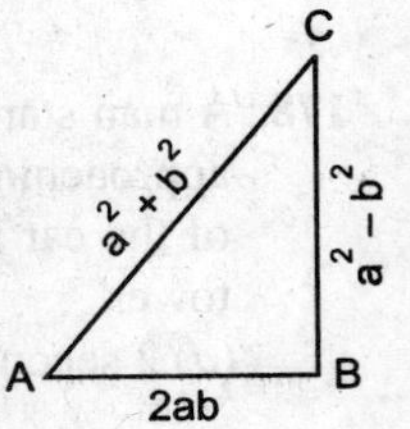

$$AB^2 = AC^2 - BC^2 = (a^2 + b^2)^2 - (a^2 - b^2)^2 = 4a^2b^2 = (2ab)^2$$

$$\sec\theta = \frac{1}{\cos\theta} = \frac{\text{Hypotenuse}}{\text{Base}} = \frac{a^2 + b^2}{2ab}$$

182. In ΔABC, $BC = 7$ cm, $AC - AB = 1$ cm. Find the value of $\sin\theta$.

(*a*) $\frac{7}{24}$ (*b*) $\frac{7}{25}$ (*c*) $\frac{24}{25}$ (*d*) $\frac{25}{24}$

Sol. (*c*) $AC^2 - AB^2 = BC^2$

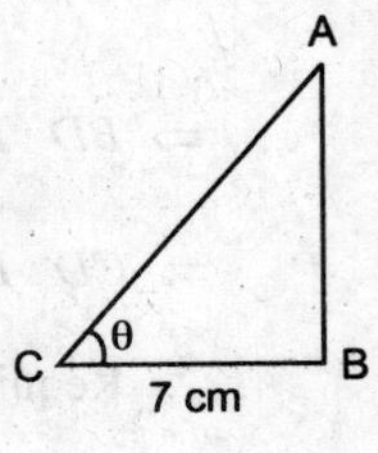

$\Rightarrow (AC + AB)(AC - AB) = 7^2 = 49$

$\Rightarrow (AC + AB) = 49$ and $(AC - AB) = 1$

$\Rightarrow AC = 25, AB = 24$

$$\sin\theta = \frac{\text{Perpendicular}}{\text{Hypotenuse}} = \frac{24}{25}$$

183. ΔABC is a right angled triangle. The length of its perpendicular and its hypotenuse are consecutive integers, then $\sin\theta = ?$

(*a*) $\frac{a^2-1}{a^2+1}$ (*b*) $\frac{a^2+1}{a^2-1}$ (*c*) $\frac{a-1}{a+1}$ (*d*) $\frac{a+1}{a-1}$

Sol. (*a*) $c = b + 1$ and $c^2 = a^2 + b^2$

$\Rightarrow (b+1)^2 = a^2 + b^2 \Rightarrow a^2 = 2b + 1 \Rightarrow b = \dfrac{a^2-1}{2}$

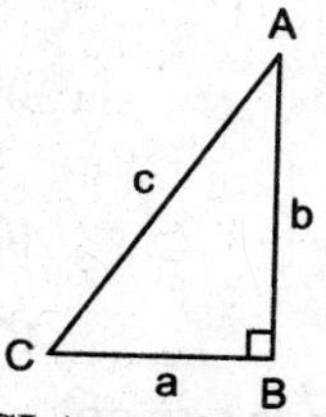

Now, $c = b + 1 = \dfrac{a^2-1}{2} + 1 = \dfrac{a^2+1}{2} \Rightarrow \sin\theta = \dfrac{b}{c} = \dfrac{a^2-1}{2} \times \dfrac{2}{a^2+1} = \dfrac{a^2-1}{a^2+1}$

184. If $\sin\theta = \dfrac{a^2-1}{a^2+1}$, then $\sec\theta + \tan\theta = ?$ **(SSC FCI Asstt. 2012)**

(*a*) $\dfrac{a}{\sqrt{2}}$ (*b*) $\dfrac{a}{a^2+1}$ (*c*) $\sqrt{2}a$ (*d*) a

Sol. (*d*) $\sin\theta = \dfrac{AB}{AC} = \dfrac{a^2-1}{a^2+1}$

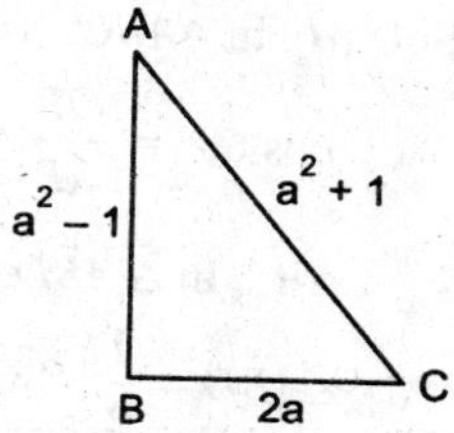

$BC^2 = AC^2 - AB^2 = (a^2+1)^2 - (a^2-1)^2 = 4a^2$

$BC = \sqrt{4a^2} = 2a$

$\therefore \sec\theta + \tan\theta = \dfrac{a^2+1}{2a} + \dfrac{a^2-1}{2a} = \dfrac{2a^2}{2a} = a$

185. In a right-angled triangle *XYZ*, right-angled at *Y*, if $XY = 2\sqrt{6}$ and $XZ - YZ = 2$, then $\sec X + \tan X$ is **(SSC CGL Tier II 2012)**

(*a*) $2\sqrt{6}$ (*b*) $\dfrac{\sqrt{6}}{2}$ (*c*) $\dfrac{1}{\sqrt{6}}$ (*d*) $\sqrt{6}$

Sol. (*d*) $XZ^2 = XY^2 + YZ^2$

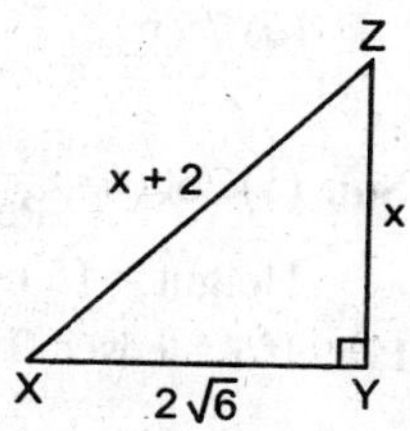

$\Rightarrow (x+2)^2 = x^2 + (2\sqrt{6})^2$

$\Rightarrow x^2 + 4x + 4 = x^2 + 24$

$\Rightarrow 4x = 20 \Rightarrow x = 5 \Rightarrow x + 2 = 7$

$\sec X + \tan X = \dfrac{7}{2\sqrt{6}} + \dfrac{5}{2\sqrt{6}} = \dfrac{12}{2\sqrt{6}} = \sqrt{6}$

186. A rhombus of side 20 cm has one angle of 60°. Find the length of longer diagonal.

(*a*) $10\sqrt{2}$ cm (*b*) $10\sqrt{3}$ cm (*c*) $20\sqrt{2}$ cm (*d*) $20\sqrt{3}$ cm

Sol. (*d*) $\sin 30° = \dfrac{BO}{AB} \Rightarrow \dfrac{1}{2} = \dfrac{BO}{20} \Rightarrow BO = \dfrac{20}{2} = 10$ cm

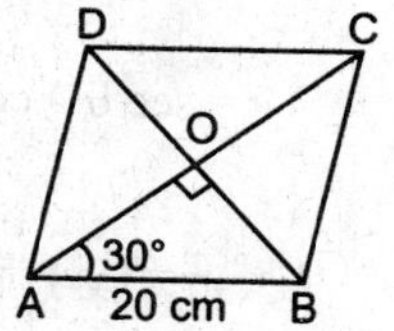

$\Rightarrow BD = 2 \times 10 \text{ cm} = 20$ cm

$\cos 30° = \dfrac{AO}{AB} \Rightarrow \dfrac{\sqrt{3}}{2} = \dfrac{AO}{20}$

$AO = 20 \times \dfrac{\sqrt{3}}{2} = 10\sqrt{3} \Rightarrow AC = 2 \times 10\sqrt{3} = 20\sqrt{3}$ cm

187. One of the four angles of a rhombus is 60°. If the length of each side of the rhombus is 8 cm, then the length of the longer diagonal is **(SSC CGL Tier I 2013)**

(*a*) $8\sqrt{3}$ cm (*b*) 8 cm (*c*) $4\sqrt{3}$ cm (*d*) $\dfrac{8}{\sqrt{3}}$ cm

Sol. (*a*) In ΔOAB, $AO \perp OB$

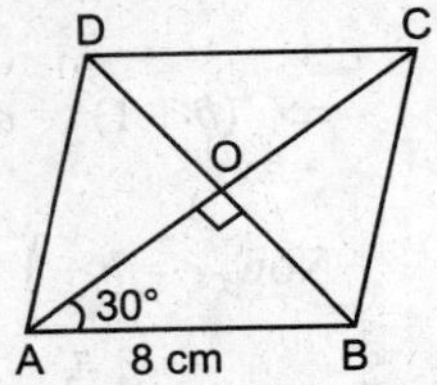

$$\sin 30° = \frac{OB}{AB} \Rightarrow \frac{1}{2} = \frac{OB}{8} \Rightarrow OB = 4 \text{ cm}$$

$$AO = \sqrt{8^2 - 4^2} = \sqrt{64 - 16} = \sqrt{48} = 4\sqrt{3}$$

$$\therefore \; AC = 2 \times 4\sqrt{3} = 8\sqrt{3} \text{ cm}$$

188. *ABCD* is a rhombus whose side *AB* = 4 cm and $\angle ABC = 120°$, then the length of diagonal *BD* is equal to : **(SSC CHSL 2011)**

(*a*) 1 cm (*b*) 2 cm

(*c*) 3 cm (*d*) 4 cm

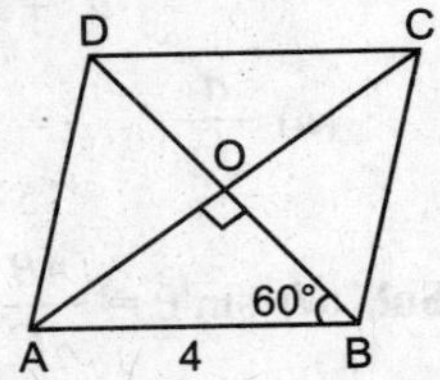

Sol. (*d*) In ΔABO, $AB = 4$ cm (rhombus)

$$\cos 60° = \frac{OB}{AB} \Rightarrow \frac{1}{2} = \frac{OB}{4} \Rightarrow OB = 2 \text{ cm} \Rightarrow BD = 2 \times 2 \text{ cm} = 4 \text{ cm}$$

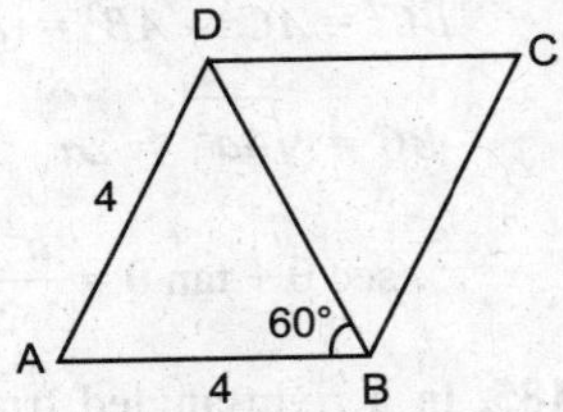

Alt: In $\Delta\, ABD$, $AB = AD = 4$ cm

$$\angle ABD = \frac{1}{2}\angle ABC = \frac{1}{2} \times 120° = 60°$$

$$\angle ADB = \angle ABD = 60° \Rightarrow \angle BAD = 180° - 60° - 60° = 60°$$

$\Rightarrow \Delta\, ABD$ is an equilateral triangle $\Rightarrow BD = AB = AD = 4$ cm

189. A kite is flying at the height of 75 m from the ground. The string makes an angle θ where ($\cot\theta = 8/15$) with the level ground. Assuming that there is no slack in the string, the length of the string is equal to **(SSC CGL Tier I 2015)**

(*a*) 75 m (*b*) 85 m (*c*) 40 m (*d*) 65 m

Sol. (*b*) $\cot\theta = \dfrac{\text{Base}}{\text{Perpendicular}} = \dfrac{8}{15} \Rightarrow \text{Hypotenuse} = \sqrt{8^2 + 15^2} = 17$ ratio

Height = 15 ratio = 75 m ⇒ 1 ratio = 5 m ⇒ Hypotenuse = 17 ratio = 17 × 5 m = 85 m

190. If $x = \text{cosec}\,\theta - \sin\theta$ and $y = \sec\theta - \cos\theta$, then the value of $x^2y^2(x^2 + y^2 + 3)$ is **(SSC CGL Tier I 2012)**

(*a*) 0 (*b*) 1 (*c*) 2 (*d*) 3

Sol. (*b*) $x = \text{cosec}\,\theta - \sin\theta = \dfrac{1}{\sin\theta} - \sin\theta = \dfrac{1-\sin^2\theta}{\sin\theta} = \dfrac{\cos^2\theta}{\sin\theta}$

$$y = \sec\theta - \cos\theta = \frac{1}{\cos\theta} - \cos\theta = \frac{1-\cos^2\theta}{\cos\theta} = \frac{\sin^2\theta}{\cos\theta}$$

$$x^2y^2 = (xy)^2 = \left(\frac{\cos^2\theta}{\sin\theta} \times \frac{\sin^2\theta}{\cos\theta}\right)^2 = (\cos\theta\,.\,\sin\theta)^2 = \sin^2\theta.\,\cos^2\theta$$

$$x^2 + y^2 + 3 = (\text{cosec}\,\theta - \sin\theta)^2 + (\sec\theta - \cos\theta)^2 + 3$$

$$= \text{cosec}^2\theta + \sin^2\theta - 2 + \sec^2\theta + \cos^2\theta - 2 + 3 = \text{cosec}^2 + \sec^2\theta - 4 + 4$$

$$= \text{cosec}^2\theta + \sec^2\theta = \frac{1}{\sin^2\theta} + \frac{1}{\cos^2\theta} = \frac{1}{\sin^2\theta\cos^2\theta}$$

$$\therefore\; x^2y^2(x^2 + y^2 + 3) = \sin^2\theta.\cos^2\theta\left(\frac{1}{\sin^2\theta.\cos^2\theta}\right) = 1$$

Alt: Put some value of θ, say θ = 45°, then $x=\sqrt{2}-\frac{1}{\sqrt{2}}=\frac{1}{\sqrt{2}}, y=\sqrt{2}-\frac{1}{\sqrt{2}}=\frac{1}{\sqrt{2}}$

$\therefore\ x^2y^2(x^2+y^2+3)=\frac{1}{2}\times\frac{1}{2}\times\left(\frac{1}{2}+\frac{1}{2}+3\right)=\frac{1}{4}\times 4=1$

191. If $0\le\alpha\le\frac{\pi}{2}$ and $2\sin\alpha+15\cos^2\alpha=7$, then the value of cot α is **(SSC CGL Tier I 2012)**

(*a*) 5/4 (*b*) 3/4 (*c*) 1/4 (*d*) 1/2

Sol. (*b*) $2\sin\alpha+15(1-\sin^2\alpha)=7$

$\Rightarrow 2\sin\alpha+15-15\sin^2\alpha=7 \Rightarrow 15\sin^2\alpha-2\sin\alpha-8=0$

$\Rightarrow 15\sin^2\alpha-12\sin\alpha+10\sin\alpha-8=0 \Rightarrow (5\sin\alpha-4)(3\sin\alpha+2)=0$

$\Rightarrow 5\sin\alpha=4\Rightarrow\sin\alpha=\frac{4}{5}\Rightarrow\cos\alpha=\sqrt{1-\left(\frac{4}{5}\right)^2}=\frac{3}{5}\Rightarrow\cot\alpha=\frac{\cos\alpha}{\sin\alpha}=\frac{3}{5}\times\frac{5}{4}=\frac{3}{4}$

192. If $2-\cos^2\theta=3\sin\theta\cos\theta$, $\sin\theta\ne\cos\theta$, then tan θ is **(SSC CGL Tier I 2012)**

(*a*) $\frac{2}{3}$ (*b*) $\frac{1}{3}$ (*c*) $\frac{1}{2}$ (*d*) 0

Sol. (*c*) $2-\cos^2\theta=3\sin\cos\theta\Rightarrow\frac{2}{\cos^2\theta}-1=\frac{3\sin\theta}{\cos\theta}$

$\Rightarrow 2\sec^2\theta-1=3\tan\theta\Rightarrow 2(\tan^2\theta+1)-1-3\tan\theta=0$

$\Rightarrow 2\tan^2\theta-3\tan\theta+1=0\Rightarrow 2\tan^2\theta-2\tan\theta-\tan\theta+1=0$

$\Rightarrow 2\tan\theta(\tan\theta-1)-1(\tan\theta-1)=0$

$\Rightarrow(2\tan\theta-1)(\tan\theta-1)=0\Rightarrow\tan\theta=\frac{1}{2}, 1.$ [But $\tan\theta\ne 1$ as $\sin\theta\ne\cos\theta$]

193. If tan θ + sec θ = 2, then tan θ is equal to **(CDS 2015)**

(*a*) $\frac{3}{4}$ (*b*) $\frac{5}{4}$ (*c*) $\frac{3}{2}$ (*d*) $\frac{5}{2}$

Sol. (*a*) $\sec^2\theta-\tan^2\theta=1$ ...(1) $\sec\theta+\tan\theta=2$ ..(2) (given)

$\therefore\ \sec\theta-\tan\theta=\frac{1}{2}$ ...(3) [Dividing (1) by (2)]

Subtracting (3) from (2) $\Rightarrow 2\tan\theta=2-\frac{1}{2}=\frac{3}{2}\Rightarrow\tan\theta=\frac{3}{4}$

194. If tan θ + sec θ = 4, then what is the value of sin θ ? **(NDA 2012)**

(*a*) $\frac{8}{17}$ (*b*) $\frac{8}{15}$ (*c*) $\frac{15}{17}$ (*d*) $\frac{23}{32}$

Sol. (*c*) $\sec^2\theta-\tan^2\theta=1$ ...(1) $\sec\theta+\tan\theta=4$ ...(2) (given)

$\therefore\ \sec\theta-\tan\theta=\frac{1}{4}$ ...(3) [Dividing (1) by (2)]

Adding (2) and (3) $\Rightarrow 2\sec\theta=4+\frac{1}{4}=\frac{17}{4}\Rightarrow\sec\theta=\frac{17}{8}\Rightarrow\cos\theta=\frac{8}{17}$

$\Rightarrow\ \sin\theta=\sqrt{1-\cos^2\theta}=\sqrt{1-\left(\frac{8}{17}\right)^2}=\sqrt{1-\frac{64}{289}}=\sqrt{\frac{225}{289}}=\frac{15}{17}$

195. If $\sec\theta + \tan\theta = 2 + \sqrt{5}$, then the value of $\sin\theta + \cos\theta$ is : **(SSC CGL Tier I 2013, 2015)**

(*a*) $\sqrt{5}$ (*b*) $\frac{7}{\sqrt{5}}$ (*c*) $\frac{1}{\sqrt{5}}$ (*d*) $\frac{3}{\sqrt{5}}$

Sol. (*d*) $\sec\theta + \tan\theta = \sqrt{5} + 2$, $\sec^2\theta - \tan^2\theta = 1 \Rightarrow (\sec\theta + \tan\theta)(\sec\theta + \tan\theta) = 1$

$$\therefore \sec\theta - \tan\theta = \frac{1}{\sqrt{5}+2} = \sqrt{5} - 2 \quad \Rightarrow \sec\theta = \sqrt{5} \text{ and } \tan\theta = 2$$

$$\sin\theta = \tan\theta \div \sec\theta = 2 \div \sqrt{5} = \frac{2}{\sqrt{5}} \qquad \cos\theta = \frac{1}{\sec\theta} = \frac{1}{\sqrt{5}}$$

$$\therefore \sin\theta + \cos\theta = \frac{2}{\sqrt{5}} + \frac{1}{\sqrt{5}} = \frac{3}{\sqrt{5}}$$

196. If $\sec\theta + \tan\theta = p$, then $\sin\theta$ is **(SSC CHSL 2012, SSC CGL Tier I 2015)**

(*a*) $\frac{1}{p^2+1}$ (*b*) $\frac{1}{p^2-1}$ (*c*) $\frac{p^2+1}{p^2-1}$ (*d*) $\frac{p^2-1}{p^2+1}$

Sol. (*d*) $\sec^2\theta - \tan^2\theta = 1$ $\sec\theta + \tan\theta = p$ $\Rightarrow \sec\theta - \tan\theta = \frac{1}{p}$

$$2\sec\theta = p + \frac{1}{p} = \frac{p^2+1}{p} \Rightarrow \sec\theta = \frac{p^2+1}{2p}$$

$$2\tan\theta = p - \frac{1}{p} = \frac{p^2-1}{p} \Rightarrow \tan\theta = \frac{p^2-1}{2p}$$

$$\sin\theta = \tan\theta \div \sec\theta = \frac{p^2-1}{2p} \times \frac{2p}{p^2+1} = \frac{p^2-1}{p^2+1}$$

197. If $\sec\theta + \tan\theta = p$, $\frac{p^2-1}{p^2+1} = ?$

(*a*) $\sin\theta$ (*b*) $\cos\theta$ (*c*) $\sec\theta$ (*d*) $\tan\theta$

Sol. (*a*) $\frac{p^2-1}{p^2+1} = \frac{(\sec\theta+\tan\theta)^2 - 1}{(\sec\theta+\tan\theta)^2+1}$

$$= \frac{\sec^2\theta + \tan^2\theta + 2\sec\theta\tan\theta - 1}{\sec^2\theta + \tan^2\theta + 2\sec\theta\tan\theta + 1} = \frac{(\sec^2\theta - 1) + \tan^2\theta + 2\sec\theta\tan\theta}{\sec^2\theta + (\tan^2\theta + 1) + 2\sec\tan\theta}$$

$$= \frac{2\tan^2\theta + 2\sec\theta\tan\theta}{2\sec^2\theta + 2\sec\theta\tan\theta} = \frac{2\tan\theta(\tan\theta + \sec\theta)}{2\sec\theta(\sec\theta + \tan\theta)} = \frac{\tan\theta}{\sec\theta} = \frac{\sin\theta}{\cos\theta} \times \frac{1}{\sec\theta} = \sin\theta$$

198. If $\cot A + \operatorname{cosec} A = 3$ and A is an acute angle, then the value of $\cos A$ is : **(SSC CGL Tier I 2012)**

(*a*) $\frac{4}{5}$ (*b*) 1 (*c*) $\frac{1}{2}$ (*d*) $\frac{3}{4}$

Sol. (*a*) $\operatorname{cosec} A + \cot A = 3$ ...(1) $\operatorname{cosec}^2 A - \cot^2 A = 1 \Rightarrow \operatorname{cosec} A - \cot A = \frac{1}{3}$ ...(2)

$(1) + (2) : 2 \text{ cosec } A = \frac{10}{3} \Rightarrow \text{cosec } A = \frac{10}{6}$

$(1) - (2) : 2 \cot A = 3 - \frac{1}{3} = \frac{8}{3} \Rightarrow \cot A = \frac{8}{6}$

$\cos A = \cot A \div \text{cosec } A = \frac{8}{6} \div \frac{10}{6} = \frac{8}{6} \times \frac{6}{10} = \frac{4}{5}$

Alt : $\frac{\cos A + 1}{\sin A} = 3 \Rightarrow (\cos A + 1)^2 = (3 \sin A)^2$

$\Rightarrow \cos^2 A + 1 + 2 \cos A = 9 \sin^2 A \Rightarrow \cos^2 A + 1 + 2 \cos A = 9(1 - \cos^2 A)$

$\Rightarrow 10 \cos^2 A + 2 \cos A - 8 = 0 \Rightarrow 5 \cos^2 A + \cos A - 4 = 0$

$\Rightarrow (\cos A + 1)(5 \cos A - 4) = 0 \Rightarrow \cos A = \frac{4}{5}$ ($\because$ A is acute angle)

199. If $\text{cosec } \theta - \cot \theta = \frac{1}{\sqrt{3}}$ where $\theta \neq 0$, then what is the value of $\cos \theta$?

(*a*) 0 (*b*) $\frac{\sqrt{3}}{2}$ (*c*) $\frac{1}{2}$ (*d*) $\frac{1}{\sqrt{2}}$

Sol. (*c*) $\text{cosec}^2 \theta - \cot^2 \theta = 1 \Rightarrow (\text{cosec } \theta + \cot \theta)(\text{cosec } \theta - \cot \theta) = 1$

$\Rightarrow \text{cosec } \theta + \cot \theta = \frac{1}{\text{cosec}\,\theta - \cot\theta} = 1 \div \frac{1}{\sqrt{3}} = \sqrt{3}$

$\Rightarrow (\text{cosec } \theta + \cot \theta) + (\text{cosec } \theta - \cot \theta) = \sqrt{3} + \frac{1}{\sqrt{3}}$

$\Rightarrow 2 \text{ cosec } \theta = \frac{3+1}{\sqrt{3}} \Rightarrow \text{cosec } \theta = \frac{1}{2} \times \frac{4}{\sqrt{3}} = \frac{2}{\sqrt{3}}$

Now, $(\text{cosec } \theta + \cot \theta) - (\text{cosec } \theta - \cot \theta) = \sqrt{3} - \frac{1}{\sqrt{3}}$

$\Rightarrow 2 \cot \theta = \frac{3-1}{\sqrt{3}} \Rightarrow \cot \theta = \frac{1}{2} \times \frac{2}{\sqrt{3}} = \frac{1}{\sqrt{3}}$

$\cos \theta = \frac{\cos\theta}{\sin\theta} \times \sin\theta = \frac{\cot\theta}{\text{cosec}\,\theta} = \frac{1}{\sqrt{3}} \div \frac{2}{\sqrt{3}} = \frac{1}{\sqrt{3}} \times \frac{\sqrt{3}}{2} = \frac{1}{2}$

200. If $\tan \theta = 1$, then $\frac{8\sin\theta + 5\sin\theta}{\sin^3\theta - 2\sin^3\theta + 7\sin\theta} = ?$ **(SSC CHSL 2011)**

(*a*) 2 (*b*) $\frac{1}{2}$ (*c*) 1 (*d*) 3

Sol. (*a*) $\tan \theta = 1 \Rightarrow \frac{\sin\theta}{\cos\theta} = 1 \Rightarrow \sin \theta = \cos \theta \Rightarrow \theta = 45° \Rightarrow \sin 45° = \frac{1}{\sqrt{2}}$

$$\text{G.E.} = \frac{8\sin\theta + 5\sin\theta}{\sin^3\theta - 2\sin^3\theta + 7\sin\theta}$$

$$= \frac{13\sin\theta}{\sin^3\theta + 7\sin\theta} = \frac{13\sin\theta}{-\sin^3\theta + 7\sin\theta} = \frac{13}{-\sin^2\theta + 7} = \frac{13}{\frac{-1}{2} + 7} = 13 \times \frac{2}{13} = 2$$

201. If $\tan\theta = 2$, then the value of $\dfrac{8\sin\theta + 5\cos\theta}{\sin^3\theta + 2\cos^3\theta + 3\cos\theta}$ is **(SSC CGL Tier I 2012)**

(*a*) $\dfrac{21}{5}$ (*b*) $\dfrac{8}{5}$ (*c*) $\dfrac{7}{5}$ (*d*) $\dfrac{16}{5}$

Sol. (*a*) $\text{Sec}^2\theta = \tan^2\theta + 1 = 2^2 + 1 = 5 \Rightarrow$ G.E. $= \dfrac{\cos\theta\left(\dfrac{8\sin\theta}{\cos\theta} + \dfrac{5\cos\theta}{\cos\theta}\right)}{\cos^3\theta\left(\dfrac{\sin^3\theta}{\cos^3\theta} + \dfrac{2\cos^3\theta}{\cos^3\theta} + \dfrac{3\cos\theta}{\cos^3\theta}\right)}$

$$= \frac{\sec^2\theta\,(8\tan\theta + 5)}{\tan^3\theta + 2 + 3\sec^2\theta} = \frac{5\times(8\times2+5)}{8+2+3\times5} = \frac{5\times21}{25} = \frac{21}{5}$$

202. If $\sin\theta$ and $\cos\theta$ are the roots of the equation $ax^2 - bx + c = 0$, then which one of the following is correct ? **(CDS 2011)**

(*a*) $a^2 + b^2 + 2ac = 0$ (*b*) $a^2 - b^2 + 2ac = 0$ (*c*) $a^2 + c^2 + 2ab = 0$ (*d*) $a^2 - b^2 - 2ac = 0$

Sol. (*b*) Product of roots $= \sin\theta \,.\, \cos\theta = \dfrac{c}{a}$; Sum of roots $= \sin\theta + \cos\theta = \dfrac{b}{a}$

$\Rightarrow \sin^2\theta + \cos^2 + 2\sin\theta\cos\theta = \dfrac{b^2}{a^2} \Rightarrow 1 + 2\times\dfrac{c}{a} = \dfrac{b^2}{a^2} \Rightarrow a^2 + 2ac = b^2 \Rightarrow a^2 - b^2 + 2ac = 0$

203. If $\dfrac{\cos^4\alpha}{\cos^2\beta} + \dfrac{\sin^4\alpha}{\sin^2\beta} = 1$, then the value of $\dfrac{\cos^4\beta}{\cos^2\alpha} + \dfrac{\sin^4\beta}{\sin^2\alpha}$ is **(SSC CGL Tier I 2012)**

(*a*) 4 (*b*) 0 (*c*) $\dfrac{1}{8}$ (*d*) 1

Sol. (*d*) If we take $\alpha = \beta$, then $\dfrac{\cos^4\alpha}{\cos^2\beta} + \dfrac{\sin^4\alpha}{\sin^2\beta} = \cos^2\alpha + \sin^2\alpha = 1$, which satisfies the equation.

$\therefore \dfrac{\cos^4\beta}{\cos^2\alpha} + \dfrac{\sin^4\beta}{\sin^2\alpha} = \cos^2\alpha + \sin^2\alpha = 1$

204. If $\tan\alpha = n\tan\beta$ and $\sin\alpha = m\sin\beta$, then $\cos^2\alpha$ is **(SSC CGL Tier I 2013)**

(*a*) $\dfrac{m^2-1}{n^2-1}$ (*b*) $\dfrac{m^2+1}{n^2+1}$ (*c*) $\dfrac{m^2}{n^2+1}$ (*d*) $\dfrac{m^2}{n^2}$

Sol. (*a*) $\tan\alpha = n\tan\beta \Rightarrow \dfrac{\sin\alpha}{\cos\alpha} = n\dfrac{\sin\beta}{\cos\beta} \Rightarrow \dfrac{m\sin\beta}{\cos\alpha} = \dfrac{n\sin\beta}{\cos\beta}$ $[\because \sin\alpha = m\sin\beta]$

$$\Rightarrow \frac{m}{n} = \frac{\cos\alpha}{\cos\beta} \Rightarrow \frac{m^2}{n^2} = \frac{\cos^2\alpha}{\cos^2\beta} = \frac{\cos^2\alpha}{1-\sin^2\beta} = \frac{\cos^2\alpha}{1-\dfrac{\sin^2\alpha}{m^2}}$$

$$\Rightarrow \frac{m^2}{n^2} = \frac{m^2\cos^2\alpha}{m^2-\sin^2\alpha} \Rightarrow m^2 - \sin^2\alpha = n^2\cos^2\alpha$$

$$\Rightarrow m^2 - (1-\cos^2\alpha) = n^2\cos^2\alpha \Rightarrow m^2 - 1 + \cos^2\alpha = n^2\cos^2\alpha$$

$$\Rightarrow m^2 - 1 = (n^2-1)\cos^2\alpha \Rightarrow \cos^2\alpha = \frac{m^2-1}{n^2-1}$$

Alt: Let $\alpha = 60°$ and $\beta = 30°$. Then $\tan 60° = n \tan 30° \Rightarrow \sqrt{3} = n \times \frac{1}{\sqrt{3}} \Rightarrow n = 3 \Rightarrow n^2 = 9$

$\sin 60° = m \sin 30° \Rightarrow \frac{\sqrt{3}}{2} = m \times \frac{1}{2} \Rightarrow m = \sqrt{3} \Rightarrow m^2 = 3$

$\cos^2 \alpha = \cos^2 60° = \frac{1}{4} = \frac{2}{8} = \frac{3-1}{9-1} = \frac{m^2-1}{n^2-1}$

205. If $(a^2 - b^2) \sin\theta + 2ab \cos\theta = a^2 + b^2$, then the value of $\tan\theta$ is : **(SSC CGL Tier I 2012)**

(*a*) $\frac{a^2-b^2}{2}$ (*b*) $\frac{a^2+b^2}{2}$ (*c*) $\frac{a^2-b^2}{2ab}$ (*d*) $\frac{a^2+b^2}{2ab}$

Sol. (*c*) $(a^2 - b^2)\tan\theta + 2ab = (a^2 + b^2)\sec\theta$ (on dividing both sides by $\cos\theta$)

$\Rightarrow (a^2 - b^2)^2 \tan^2\theta + 4ab(a^2 - b^2)\tan\theta + 4a^2b^2 = (a^2 + b^2)^2 (1 + \tan^2\theta)$

$\Rightarrow [(a^2 - b^2)^2 - (a^2 + b^2)^2]\tan^2\theta + 4ab(a^2 - b^2)\tan\theta = (a^2 + b^2)^2 - 4a^2b^2$

$\Rightarrow (-4a^2b^2)\tan^2\theta + 4ab(a^2 - b^2)\tan\theta = (a^2 - b^2)^2$

$\Rightarrow 4a^2b^2\tan^2\theta + (a^2 - b^2)^2 - 4ab(a^2 - b^2)\tan\theta = 0$

$\Rightarrow [2ab\tan\theta - (a^2 - b^2)]^2 = 0 \Rightarrow 2ab\tan\theta = a^2 - b^2 \Rightarrow \tan\theta = \frac{a^2-b^2}{2ab}$

206. If *A, B, C, D* are the successive angles of a cyclic quadrilateral, then what is $\cos A + \cos B + \cos C + \cos D$ equal to ? **(CDS 2011)**

(*a*) 4 (*b*) 2 (*c*) 1 (*d*) 0

Sol. (*d*) $\cos A + \cos C = \cos A + \cos(180° - A) = \cos A - \cos A = 0$ $[\because A + C = 180°]$

Similarly $\cos B + \cos D = 0$ [180° – *A* lies in second quadrant when *A* is an acute angle]

207. If $a = \sin\frac{\pi}{4}$, $b = \cos\frac{\pi}{4}$, and $c = -\text{cosec}\frac{\pi}{4}$, then the value of $a^3 + b^3 + c^3$ is:

(*a*) 0 (*b*) 1 (*c*) $\frac{3\sqrt{2}}{2}$ (*d*) $\frac{-3\sqrt{2}}{2}$

Sol. (*d*) $a + b + c = \sin\frac{\pi}{4} + \cos\frac{\pi}{4} - \text{cosec}\frac{\pi}{4} = \frac{1}{\sqrt{2}} + \frac{1}{\sqrt{2}} - \sqrt{2} = \frac{1+1-2}{\sqrt{2}} = 0$

$\because a + b + c = 0 \Rightarrow a^3 + b^3 + c^3 = 3abc = 3\left(\frac{1}{\sqrt{2}}\right)\left(\frac{1}{\sqrt{2}}\right)(-\sqrt{2}) = \frac{-3\sqrt{2}}{2}$

208. *D* and *E* are two points on the sides *AC* and *BC* respectively of ΔABC such that *DE* = 18 cm, *CE* = 5 cm and $\angle DEC = 90°$. If $\tan\angle ABC = 3.6$, then *AC* : *CD* = **(SSC CGL Tier II 2013)**

(*a*) *BC* : 2 *CE* (*b*) 2 *CE* : *BC* (*c*) 2 *BC* : *CE* (*d*) *CE* : 2 *BC*

Sol. (*a*) Draw $AF \perp BC$ then AF || DE

$\tan\angle DCE = \frac{AF}{FC} = \frac{DE}{EC} = \frac{18}{5} = 3.6$

$\tan\angle ABC = 3.6 \Rightarrow \frac{AF}{BF} = 3.6$

$\Rightarrow \frac{AF}{FC} = \frac{AF}{BF} \Rightarrow BF = FC \Rightarrow AB = AC$

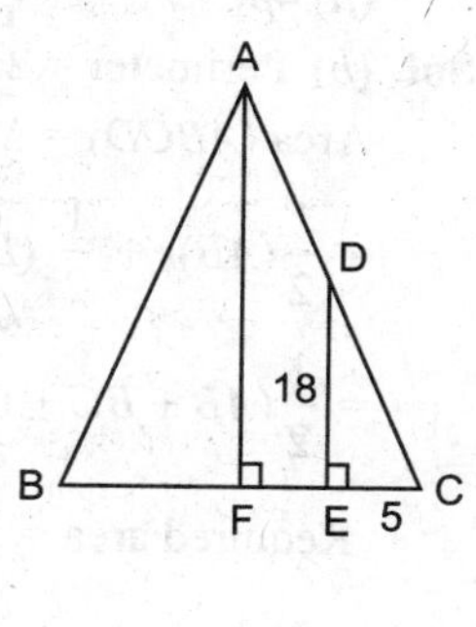

Δ AFC ~ Δ DEC $\quad \therefore \dfrac{AC}{CD} = \dfrac{CF}{CE} = \dfrac{2CF}{2CE} = \dfrac{BC}{2CE}$

209. *D* is a point on the side *BC* of a triangle *ABC* such that $AD \perp BC$. *E* is a point on *AD* for which $AE : ED = 5 : 1$. If $\angle BAD = 30°$ and $\tan(\angle ACB) = 6 \tan(\angle DBE)$, then $\angle ACB =$ **(SSC CGL Tier II 2013)**

(*a*) 30° (*b*) 45° (*c*) 60° (*d*) 15°

Sol. (*c*) $AE : ED = 5 : 1 \Rightarrow AD : ED = 6 : 1$

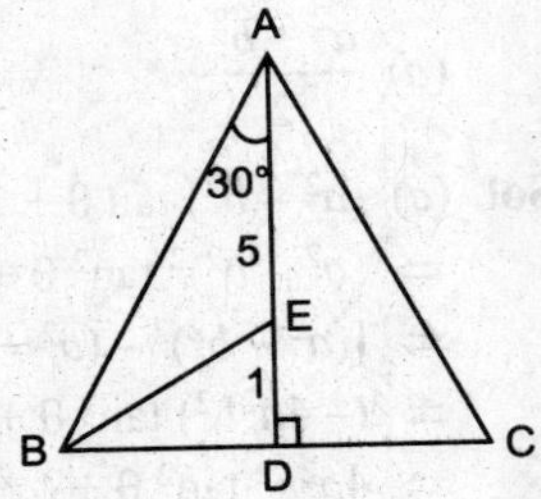

$\tan 30° = \dfrac{BD}{AD} \Rightarrow \dfrac{1}{\sqrt{3}} = \dfrac{BD}{AD} \Rightarrow BD = \dfrac{AD}{\sqrt{3}}$

$\tan(\angle ACB) = \tan(\angle ACD) = \dfrac{AD}{CD}$

$\tan(\angle DBE) = \dfrac{ED}{BD} = \dfrac{6ED}{6BD} = \dfrac{AD}{6BD}$

Given, $\tan(\angle ACB) = 6 \tan(\angle DBE)$

$\Rightarrow \dfrac{AD}{CD} = 6 \times \dfrac{AD}{6BD} \Rightarrow \dfrac{AD}{CD} = \dfrac{AD}{BD} \Rightarrow CD = BD = \dfrac{AD}{\sqrt{3}}$

$\tan(\angle ACB) = \dfrac{AD}{CD} = \dfrac{AD}{BD} = \dfrac{AD}{AD/\sqrt{3}} = \sqrt{3} = \tan 60°$

210. An isosceles triangle *ABC* is right angled at *B*. *D* is a point inside the triangle *ABC*. *P* and *Q* are the feet of the perpendicular drawn from *D* on the sides *AB* and *AC* respectively of ΔABC. If $AP = a$ cm, $AQ = b$ cm and $\angle BAD = 15°$, sin 75° = ? **(SSC CGL Tier II 2013)**

(*a*) $\dfrac{2b}{\sqrt{3}a}$ (*b*) $\dfrac{a}{2b}$ (*c*) $\dfrac{\sqrt{3}a}{2b}$ (*d*) $\dfrac{2a}{\sqrt{3}b}$

Sol. (*c*) In ΔABC, $AB = BC$ (isosceles triangle)

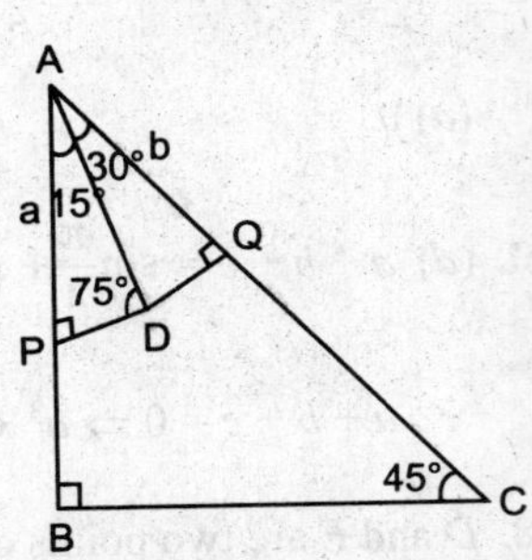

$\therefore \angle BAC = \angle BCA = 90° \div 2 = 45°$

$\angle DAQ = \angle BAC - \angle PAD = 45° - 15° = 30°$

In ΔAQD, $\cos 30° = \dfrac{AQ}{AD} \Rightarrow \dfrac{\sqrt{3}}{2} = \dfrac{b}{AD} \Rightarrow AD = \dfrac{2b}{\sqrt{3}}$

$\sin(\angle PDA) = \sin 75° = \dfrac{AP}{AD} = \dfrac{a}{2b/\sqrt{3}} = \dfrac{\sqrt{3}a}{2b}$

211. If a circular sheet of a perimeter $2\pi r$ touching each side of a given quadrilateral sheet of perimeter $2p$ is cut off from the quadrilateral, then the area of the remaining sheet is :

(*a*) $4p^2 - r^2$ (*b*) $r(p - \pi r)$ (*c*) r^2 (*d*) p^2

Sol. (*b*) Perimeter $= 2\pi r \Rightarrow$ radius $= r$

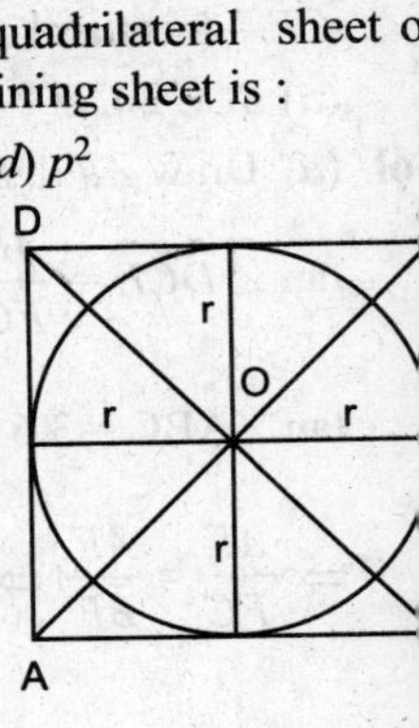

Area $(ABCD) = \Delta OAB + \Delta OBC + \Delta OCD + \Delta ODA$

$= \dfrac{1}{2}(AB)r + \dfrac{1}{2}(BC)r + \dfrac{1}{2}(CD)r + \dfrac{1}{2}(DA)r$

$= \dfrac{1}{2}(AB + BC + CD + DA)r = \dfrac{1}{2}(2p)r = pr$

Required area $= pr - \pi r^2 = r(p - \pi r)$

212. The height of a tower is h and the angle of elevation of the top of the tower is α. On moving a distance $h/2$ towards the tower, the angle of elevation becomes β. The value of $\cot\alpha - \cot\beta$ is **(SSC CGL Tier I 2012)**

(*a*) 1/2 (*b*) 2 (*c*) 1 (*d*) 2/3

Sol. (*a*) $\cot\alpha = \dfrac{x+h/2}{h}$

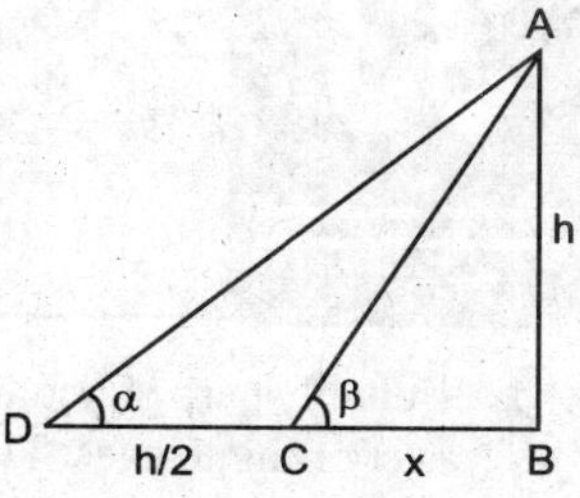

$$\cot\beta = \frac{x}{h}$$

$$\cot\alpha - \cot\beta = \frac{x+h/2}{h} - \frac{x}{h} = \frac{h/2}{h} = \frac{1}{2}.$$

CHAPTER

33

COORDINATE GEOMETRY

1. Mutually perpendicular lines in a plane are known as coordinate axes. The horizontal line ($X'OX$) is known as x-axis and the vertical line (YOY') is known as y-axis. The intersection point of the two axes is known as the Origin. The axes divide the plane into four parts called quadrants. $X'OX$ is positive to the right of the origin O and negative to the left of origin O. YOY' is positive upwards and negative downwards the origin O.

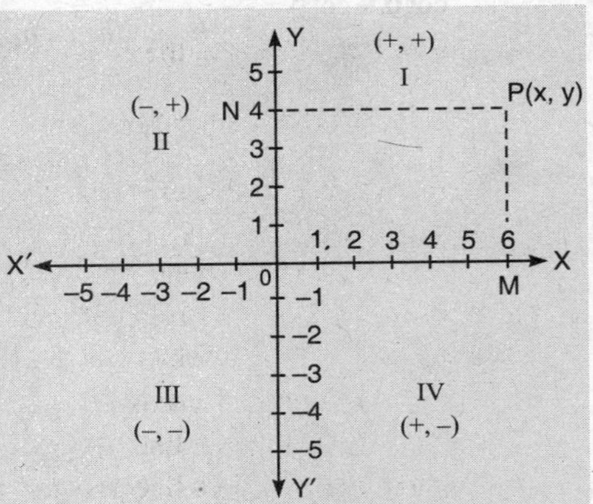

   Let P be any point in plane. Draw perpendiculars PM and PN on x-axis and y-axis respectively. Length of OM is called x-coordinate or **abscissa** and length of ON is called y-coordinate or **ordinate** of point P represented by (x, y).

   Ordered pair : $(2,3) \neq (3,2)$

2. The point of intersection of the coordinates axes is called origin. The coordinates of origin are $(0, 0)$. The coordinates of any point on x-axis are $(x, 0)$ and the coordinates of any point on y-axis are $(0, y)$.

3. XOY, $X'OY$, $X'OY'$, $Y'OX$ are known as the first, the second, the third, the fourth quadrant respectively.

   In the first quadrant, $x > 0$ and $y > 0$

   In the second quadrant, $x < 0$ and $y > 0$

   In the third quadrant, $x < 0$ and $y < 0$

   In the fourth quadrant, $x > 0$ and $y < 0$

4. Graphs of a pair of linear equations in two variables:

   (*i*) Two lines intersects only at one point, when the equations have one solution.

   (*ii*) Two lines are parallel, when the equations have no common solution, or

   (*iii*) The lines are coincident, when the equations have infinite solutions.

5. **Distance Formula:** If $P(x_1, y_1)$ and $Q(x_2, y_2)$ are two points in a plane, then

   $PQ = \sqrt{(x_2 - x_1)^2 + (y_2 - y_1)^2}$ or $\sqrt{(x_1 - x_2)^2 + (y_1 - y_2)^2}$

   Distance of a point $P(x, y)$ from the origin $O(0, 0) = OP = \sqrt{x^2 + y^2}$

Example. Find the distance between the points A (–2, 5) and B (3, –7)

(*a*) 13 units (*b*) 12 units (*c*) 20 units (*d*) 26 units

Sol. (*a*) $AB = \sqrt{(3+2)^2 + (-7-5)^2} = \sqrt{5^2 + 12^2} = \sqrt{25+144} = \sqrt{169} = 13$

6. **Section Formula:** If a point $P(x, y)$ divides a line segment joining the points $A\ (x_1, y_1)$ and $B\ (x_2, y_2)$ in the ratio $m : n$ internally, then

$$x = \frac{mx_2 + nx_1}{m+n},\ y = \frac{my_2 + ny_1}{m+n}$$

Note : If P is the mid-point of AB, then its coordinates are $\left(\frac{x_1 + x_2}{2}, \frac{y_1 + y_2}{2}\right)$

7. The coordinates of the centroid of a triangle whose vertices are $A\ (x_1, y_1)$, $B(x_2, y_2)$ and $C(x_3, y_3)$ are $\left(\frac{x_1 + x_2 + x_3}{3}, \frac{y_1 + y_2 + y_3}{3}\right)$

8. The area of a triangle, the coordinates of whose vertices are (x_1, y_1), (x_2, y_2), (x_3, y_3) is $\frac{1}{2}|x_1(y_2 - y_3) + x_2(y_3 - y_1) + x_3(y_1 - y_2)|$

Note : Modulus value or absolute value can't be negative.

9. Three points (x_1, y_1), (x_2, y_2) and (x_3, y_3) are collinear iff $x_1\ (y_2 - y_3) + x_2\ (y_3 - y_1) + x_3(y_1 - y_2) = 0$ i.e., area of the triangle is zero.

10. Equations of lines: A linear equation in two variables represents a straight line. Every point on the straight line is a solution of the linear equation.
    (*i*) The equation of x-axis is $y = 0$ (*ii*) The equation of y-axis is $x = 0$
    (*iii*) The equation of a line parallel to y-axis at a distance of a from it, is $x = a$
    (*iv*) The equation of a line parallel to x-axis at a distance of b from it, is $y = b$
    (*v*) The equation of a line passing through the points $A(x_1, y_1)$ and $B(x_2, y_2)$ is $\frac{y - y_1}{x - x_1} = \frac{y_2 - y_1}{x_2 - x_1}$
    (*vi*) Slope of line $AB = \frac{y_2 - y_1}{x_2 - x_1}$ *or* $\frac{y_1 - y_2}{x_1 - x_2}$
    (*vii*) **Slope-intercept form**: Equation of a line with slope m is $y = mx + c$ Example: $y = 3x + 5$

11. Two lines $a_1x + b_1y + c_1 = 0$ and $a_2x + b_2y + c_2 = 0$ are
    (i) Parallel: if $m_1 = m_2 \Rightarrow \frac{-a_1}{b_1} = \frac{-a_1}{b_2} \Rightarrow \frac{a_1}{a_2} = \frac{b_1}{b_2}$
    (ii) **Perpendicular**: Perpendicular lines have negative inverse slope.
    $m_1\ m_2 = -1 \Rightarrow m_1 = -\frac{1}{m_2} \Rightarrow \frac{-a_1}{b_1} = \frac{b_2}{a_2} \Rightarrow a_1a_2 + b_1b = 0$

12. The equation of a line parallel to $ax + by + c_1 = 0$ is $ax + by + c_2 = 0$
    The equation of a line perpendicular to $ax + by + c_1 = 0$ is $bx - ay + c_2 = 0$

13. **Point-slope form**: Equation of a line with a slope m and passing through the point (x_1, y_1) is $y - y_1 = m(x - x_1)$ or $y - y_1 = \left(\frac{y_2 - y_1}{x_2 - x_1}\right)(x - x_1)$

14. Three points A, B and C are collinear, if any of the following condition is satisfied.
    (*a*) $\Delta\ ABC = 0$ (*b*) $AB + BC = AC$ (*c*) Slope of AB = Slope of AC = Slope of BC

Example: $A(5, 2)$, $B(0, 0)$, $C(-5, -2)$

Sol.(*i*) Area $= |5(0 + 2) + 0(-2 - 2) - 5(2 - 0)| \div 2 = 0$

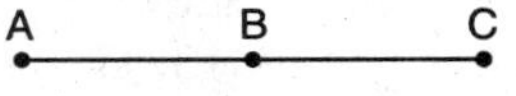

(*ii*) $AB = \sqrt{5^2+2^2} = \sqrt{29}$; $BC = \sqrt{5^2+2^2} = \sqrt{29}$;

$AC = \sqrt{(5+5)^2+(2+2)^2} = \sqrt{100+4} = \sqrt{104} = 2\sqrt{29} = AB + BC$

(*iii*) Slope of $AB = \frac{0-2}{0-5} = \frac{2}{5}$; Slope of $AC = \frac{-2-2}{-5-5} = \frac{2}{5}$

15. Intercept Form : Intercept is point where the graph crosses the axes.
A graph may have two intercepts.
x-intercept: To find *x*-intercept, put $y = 0$ in the equation.
y-intercept: To find *y*-intercept, put $x = 0$ in the equation.
The equation of a line which makes intercepts of lengths *a* and *b* on *x*-axis and *y*-axis respectively, is $\frac{x}{a}+\frac{y}{b}=1$.

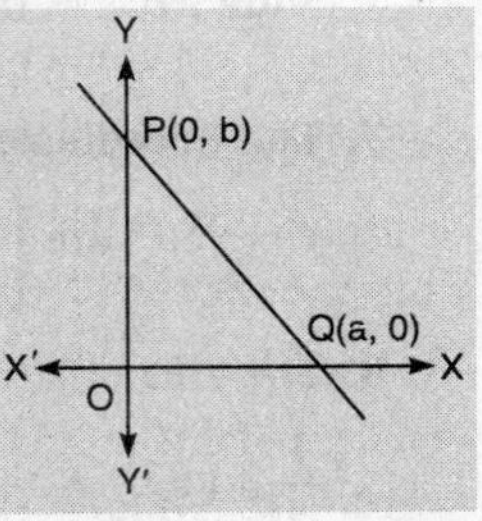

Example : The *x*-intercept of the graph of $5x - 4y = 20$ is **(SCC CGL Tier I 2012)**

(*a*) 4 units (*b*) 5 units (*c*) 9 units (*d*) 1 unit

Sol. (*a*) *x*-intercept: when the graph of the equation meets the *x*-axis *i.e.*, when $y = 0$

$\therefore \quad 5x - 4y = 20 \Rightarrow 5x - 4(0) = 20 \Rightarrow 5x = 20 \Rightarrow x = 4.$

SOLVED EXERCISE

1. The lines $2x + y = 5$ and $x + 2y = 4$ intersect at the point. **(SSC FCI Astt. 2012)**

(*a*) (1, 2) (*b*) (2, 1) (*c*) (0, 2) (*d*) (2, 0)

Sol. (*b*) $(x, y) = (2, 1)$ satisfies both the equations.

2. Find the value of *x*, if the distance between the points $(x, -1)$ and $(3, 2)$ is 5.

(*a*) – 7 or – 1 (*b*) –7 or 1 (*c*) 7 or 1 (*d*) 7 or – 1

Sol. (*d*) $\sqrt{(x-3)^2+(-1-2)^2} = 5 \Rightarrow (x-3)^2 + (-3)^2 = 25 \Rightarrow x^2 - 6x + 9 + 9 = 25$

$\Rightarrow x^2 - 6x - 7 = 0 \Rightarrow (x-7)(x+1) = 0 \Rightarrow x = 7$ or $x = -1$

3. If the distance between two points $(0, -5)$ and $(x, 0)$ is 13 unit, then $x = ?$

(*a*) ± 12 (*b*) ± 9 (*c*) ± 6 (*d*) ± 8

(SSC CGL Tier I 2012)

Sol. (*a*) $(x-0)^2 + (0+5)^2 = 13^2 \Rightarrow x^2 + 5^2 = 13^2 \Rightarrow x^2 = 169 - 25 = 144 \Rightarrow x = \sqrt{144} = \pm 12$

4. Find a point on *y*-axis which is equidistant from the points $A(6, 5)$ and $B(-4, 3)$.

(*a*) (1, 4) (*b*) (2, 8) (*c*) (0, 9) (*d*) (0, 5)

Sol. (*c*) Let the required point be $P(0, y)$.

Then $PA = PB \Rightarrow \sqrt{(0-6)^2+(y-5)^2} = \sqrt{(0+4)^2+(y-3)^2}$

$\Rightarrow 36 + y^2 - 10y + 25 = 16 + y^2 - 6y + 9 \Rightarrow 4y = 36 \Rightarrow y = 9 \Rightarrow (0, y) = (0, 9)$

5. The distance between the points (0, 0) and the intersecting point of the graphs of $x = 3$ and $y = 4$ is **(SSC CGL Tier I 2012)**

(*a*) 3 unit (*b*) 5 unit (*c*) 2 unit (*d*) 4 unit

Sol. (*b*) Coordinates of point of intersection are (3, 4)

$\therefore$ Required distance $= \sqrt{(3-0)^2+(4-0)^2} = \sqrt{9+16} = \sqrt{25} = 5$ unit

6. If the distance between the points $(x, 0)$ and $(-7, 0)$ be 10 units, then the possible values of *x* are **(SSC FCI Asstt. 2012)**

(*a*) 3, 17 (*b*) – 3, 17 (*c*) 3, – 17 (*d*) –3, –17

Sol. (*c*) Distance = $\sqrt{(x+7)^2+(0-0)^2} = 10 \Rightarrow \pm (x+7) = 10$

$\Rightarrow x + 7 = 10$ or $-(x + 7) = 10 \Rightarrow x = 3$ or $x = -17$

7. What is the distance between the lines $3x + 4y = 9$ and $6x + 8y = 18$? **(NDA 2012)**

(*a*) 0 (*b*) 3 units (*c*) 9 units (*d*) 18 units

Sol. (*a*) $\frac{a_1}{a_2} = \frac{b_1}{b_2} = \frac{c_1}{c_2} = \frac{3}{6} = \frac{4}{8} = \frac{9}{18} = \frac{1}{2} \Rightarrow$ The lines are coincident

8. Equation of the straight line parallel to x-axis and also 3 units below x-axis is **(SSC CGL Tier I 2013)**

(*a*) $y = 3$ (*b*) $y = -3$

(*c*) $x = 3$ (*d*) $x = -3$

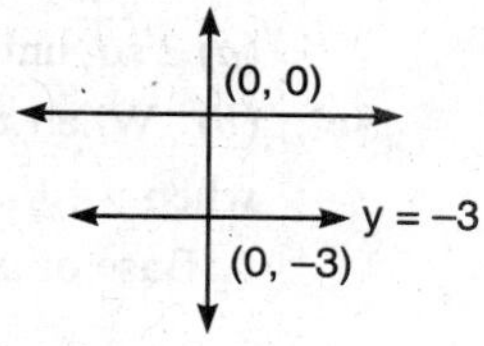

Sol. (*b*) Required equation is $y = -3$

9. An equation of the form $ax + by + c = 0$ where $a \neq 0, b \neq 0, c = 0$ represents a straight line which passes through **(SSC CGL Tier I 2013)**

(*a*) (0, 0) (*b*) (3, 2) (*c*) (2, 4) (*d*) None of these

Sol. (*a*) $ax + by + c = 0 \Rightarrow ax + by = 0$ ($\because c = 0$)

when $x = 0, y = 0$, then $a(0) + b(0) = 0 + 0 = 0$

10. The x-intercept of the graph of $7x - 3y = 2$ is **(SSC CHSL 2012)**

(*a*) $\frac{2}{5}$ (*b*) $\frac{2}{7}$ (*c*) $\frac{3}{4}$ (*d*) $\frac{3}{7}$

Sol. (*b*) x-intercept (when $y = 0$) : $7x - 3y = 2 \Rightarrow 7x = 2 \Rightarrow x = \frac{2}{7}$

11. If the graphs of $3x - 5y = -8$ and $3x + 5y = 32$ intersect at the point (p, q), then the value of $p - q$ is **(SSC CHSL 2013)**

(*a*) 3 (*b*) 2 (*c*) 1 (*d*) 0

Sol. (*d*) Solving equation, $x = 4$ and $y = 4 \Rightarrow (p, q) = (4, 4) \Rightarrow p - q = 4 - 4 = 0$

12. If $a = 0, b \neq 0, c \neq 0$, then the equation $ax + by + c = 0$ represents a line parallel to **(SSC CGL Tier I 2013)**

(*a*) $x + y = 0$ (*b*) x-axis (*c*) y-axis (*d*) None of these

Sol. (*b*) $ax + by + c = 0 \Rightarrow 0(x) + by + c = 0 \Rightarrow by + c = 0$

$\Rightarrow by = -c \Rightarrow y = \frac{-c}{b} \Rightarrow$ Parallel to x-axis

13. In xy-plane, P and Q are two points having co-ordinates (2, 0) and (5, 4) respectively. Then the numerical value of the area of the circle with radius PQ, is **(SSC CGL Tier I 2012)**

(*a*) 14π (*b*) 25π (*c*) 16π (*d*) 32π

Sol. (*b*) $PQ = \sqrt{(5-2)^2+(4-0)^2} = \sqrt{3^2+4^2} = \sqrt{9+16} = \sqrt{25} = 5$

Required area $= \pi r^2 = \pi(5)^2 = 25\pi$

14. The radius of the circumcircle of the triangle made by x-axis, y-axis and $4x + 3y = 12$ is **(SSC CGL Tier I 2012)**

(*a*) 2 unit (*b*) 2.5 unit (*c*) 3 unit (*d*) 4 unit

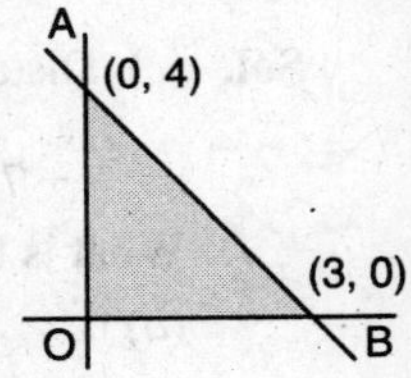

Sol. (*b*) *x*-intercept : when $y = 0$, $4x = 12 \Rightarrow x = 3$

y-intercept : when $x = 0$, $3y = 12 \Rightarrow y = 4$

$\therefore AB = \sqrt{(3-0)^2 + (0-4)^2} = \sqrt{25} = 5$ unit

Radius of circumcircle of right triangle $= \dfrac{\text{Hypotenuse}}{2} = \dfrac{5}{2} = 2.5$ unit

15. If the graph of the equation $2x + 3y = 6$ form a triangle with co-ordinates axes, then the area of triangle will be **(SSC CHSL 2012, SSC CGL 2012M)**

(*a*) 2 sq. unit (*b*) 3 sq. unit (*c*) 6 sq. unit (*d*) 1 sq. unit

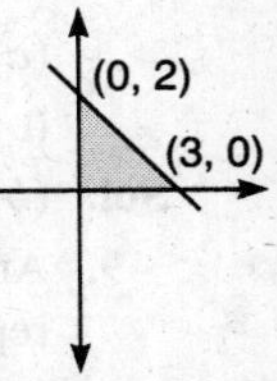

Sol. (*b*) When $x = 0$, then $3y = 6 \Rightarrow y = 2$

when $y = 0$, then $2x = 6 \Rightarrow x = 3$

$\therefore$ Base of $\Delta = 3$ units, Height = 2 units

Required area $= \dfrac{1}{2} \times$ Base $\times$ Height $= \dfrac{1}{2} \times 3 \times 2 = 3$ sq. units

16. The area (in sq. unit) of the triangle formed by the three graphs of the equations $x = 4$, $y = 3$ and $3x + 4y = 12$, is **(SSC CGL Tier I 2012, 2015)**

(*a*) 12 (*b*) 10 (*c*) 6 (*d*) 8

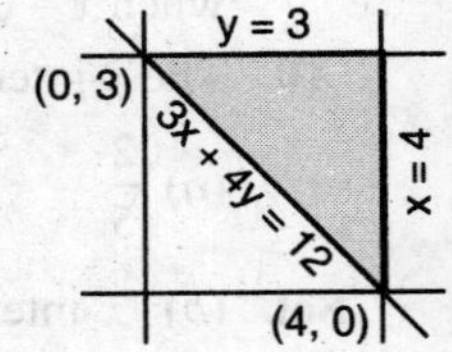

Sol. (*c*) Area of triangle

$= \dfrac{1}{2} \times 3 \times 4 = 6$

17. The area of the triangle formed by the graph of $ax + by = c$ (where *a* and *b* are two positive real numbers) and the coordinate axes, is **(SSC FCI Asstt. 2012)**

(*a*) $\dfrac{c^2}{ab}$ sq. unit (*b*) $\dfrac{a^2}{2bc}$ sq. unit (*c*) $\dfrac{c^2}{2ab}$ sq. unit (*d*) $\dfrac{a^2}{bc}$ sq. unit

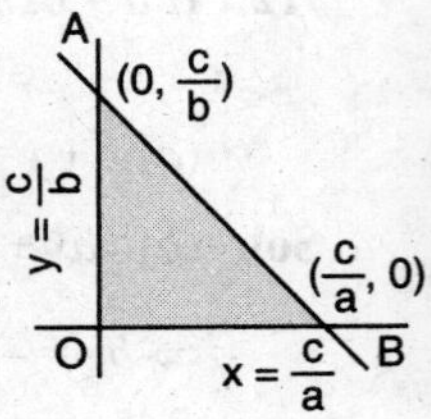

Sol. (*c*) *x*-intercept : when $y = 0$, $ax = c \Rightarrow x = \dfrac{c}{a}$

y-intercept : when $x = 0$, $by = c \Rightarrow y = \dfrac{c}{b}$

Required area $= \dfrac{1}{2} \times \dfrac{c}{a} \times \dfrac{c}{b} = \dfrac{c^2}{2ab}$

19. Area of the triangle formed by straight line $\dfrac{x}{\tan\theta} + \dfrac{y}{\cot\theta} = 2$ with *x*-axis and *y*-axis is

(*a*) 1 sq. units (*b*) 2 sq. units (*c*) 4 sq. units (*d*) 1/2 sq units.

Sol. (*b*) $\dfrac{x}{\tan\theta} + \dfrac{y}{\cot\theta} = 2 \Rightarrow \dfrac{x}{2\tan\theta} + \dfrac{y}{2\cot\theta} = 1$

$\Rightarrow$ *x*-intercept = $2\tan\theta$ $\Rightarrow$ *y*-intercept = $2\cot\theta$

$\Rightarrow$ Area $= \dfrac{1}{2} \times 2\tan\theta \times 2\cot\theta = 2$ sq. units.

19. The area bounded by the lines $x = 0, y = 0, x + y = 1, 2x + 3y = 6$ (in square units) is

(SSC CGL Tier I 2012)

(*a*) 2 (*b*) $2\frac{1}{3}$ (*c*) $2\frac{1}{2}$ (*d*) 3

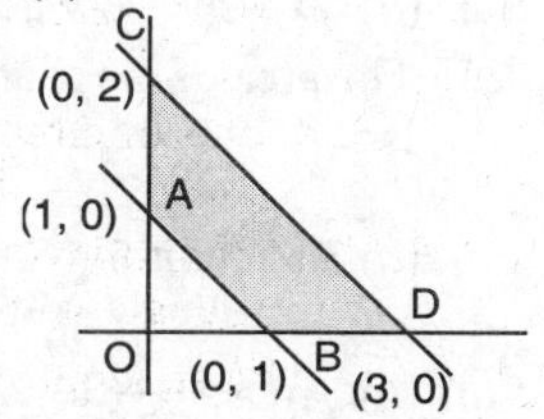

Sol. (*c*) Required area = $\Delta\, COD - \Delta\, AOB$

$$= \frac{1}{2}(2\times3-1\times1) = \frac{5}{2} = 2\frac{1}{2}$$

20. Area of the trapezium formed by *x*-axis, *y*-axis and the lines $3x + 4y = 12$ and $6x + 8y = 60$ is:

(*a*) 31.5 sq unit (*b*) 36.5 sq unit (*c*) 37.5 sq unit (*d*) 48 sq unit.

(SSC CGL Tier I 2012)

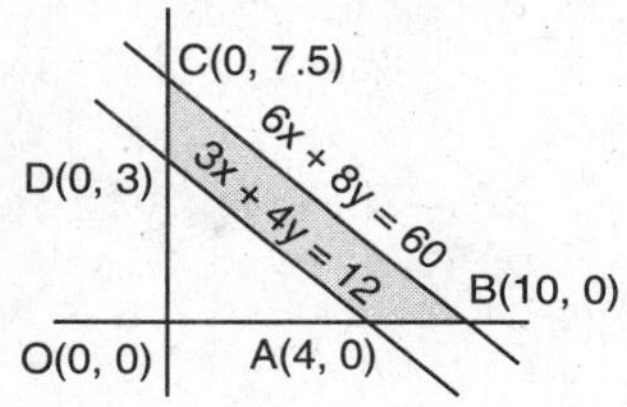

Sol. (*a*) Required area = Area of trapezium ABCD

$$= \Delta\, BOC - \Delta\, AOD = \frac{1}{2}(10 \times 7.5 - 4 \times 3)$$

$$= \frac{1}{2}(75 - 12) = \frac{63}{2} = 31.5 \text{ sq. units}$$

21. Area of the rectangle formed by lines $x = 2, y = 1, x = -3$, and $y = -2$ is

(*a*) 15 sq. units (*b*) 12 sq. units (*c*) 9 sq. units (*d*) 24 sq. units

Sol. (*a*) length = $|-3 - 2| = 5$ units;

Width = $|-2 - 1| = 3$ units

$\therefore$ Area of rectangle = $5 \times 3 = 15$ sq. units.

22. The area of the triangle formed by lines $5x + 7y = 35$, $4x + 3y = 12$ and *x*-axis is

(SSC CGL Tier I 2012)

(*a*) $\frac{160}{13}$ sq. units (*b*) $\frac{150}{13}$ sq. units (*c*) $\frac{140}{13}$ sq. units (*d*) 10 sq. units.

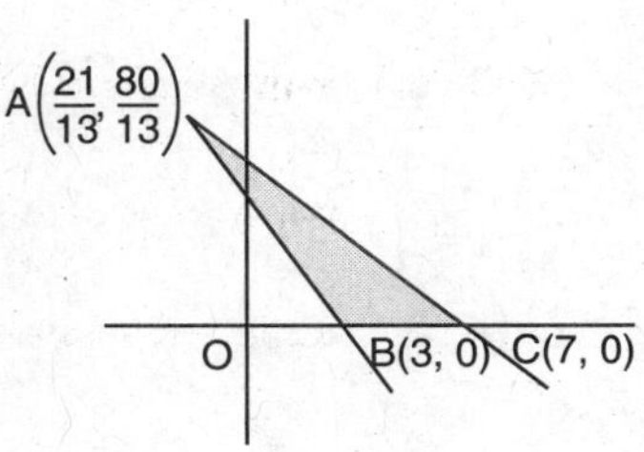

Sol. (*a*) Solving two equations, we get $y = \frac{80}{13}$ (Height)

x-intercept of $5x + 7y = 35$: $5x = 35,\ x = 7$

x-intercept of $4x + 3y = 12$: $4x = 12,\ x = 3$

Base = $7 - 3 = 4$ units

Area = $\frac{1}{2} \times$ Base $\times$ Height = $\frac{1}{2}\times4\times\frac{80}{13} = \frac{160}{13}$

23. The area of the region bounded by $y = |x| - 5$ with the co-ordinate axes is:

(SSC CGL Tier II 2012)

(*a*) 50 sq. units (*b*) 25 sq. units (*c*) 20 sq. units (*d*) 52 sq. units

Sol. (*b*) $y = x - 5$ when $x > 0$

$y = -x - 5$ when $x < 0$

$BC = (5 + 5)$ units = 10 units

$AO = 5$ units

Area of $\Delta\, ABC = \frac{1}{2} \times 10 \times 5 = 25$ sq. units

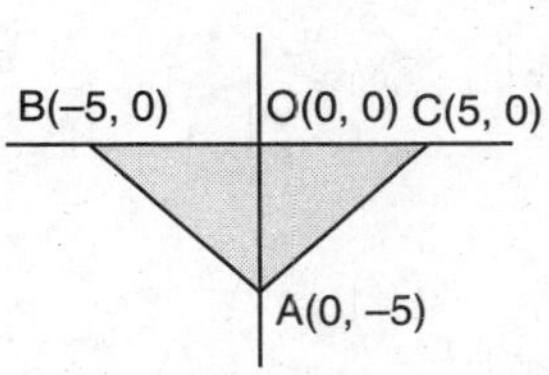

24. Among the following equations, the equation of the straight line passing through the origin is **(SSC CGL Tier I 2015)**

(*a*) $2x - 3y = 0$ (*b*) $5x - 4 = 0$ (*c*) $x + 2y + 9 = 0$ (*d*) $2y - 13 = 0$

Sol. (*a*) $2x - 3y = 0$ when $x = 0$ and $y = 0$ $\Rightarrow$ Line passes through (0, 0)

25. The area of the triangle formed by the graphs of the equations $x = 0$, $2x + 3y = 6$ and $x + y = 3$ is : **(SSC CGL Tier I 2015)**

(*a*) 3 sq. unit (*b*) $1\frac{1}{2}$ sq. unit (*c*) 1 sq. unit (*d*) $4\frac{1}{2}$ sq. unit

Sol. (*b*) Area $= \frac{1}{2}(3-2)\times 3 = \frac{3}{2} = 1\frac{1}{2}$ sq. unit

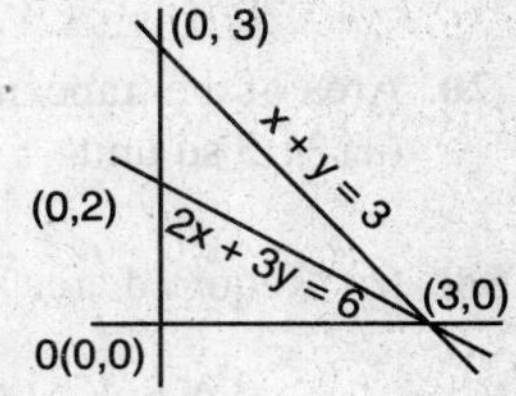

CHAPTER

34

PERMUTATIONS AND COMBINATIONS

NOT FOR SSC EXAMS

If a work can be done in m different ways, second work can be done in n different ways and the third work in p ways and so on, then total number of ways in which all the works can be done is $m \times n \times p$ ....

Example : There are four routes to travel from city A to city B and six routes from city B to city C. How many routes are possible to travel from the city A to city C ?

(UPSC Civil Service Pre. 2011)

(*a*) 24 (*b*) 12 (*c*) 10 (*d*) 8

Sol. (*a*) Total possible routes $4 \times 6 = 24$

If a work can be done in m different ways, second work (independent of the first work) can be done in n different ways and the third work in p ways and so on, then either of the works can be done in $m + n + p$ .... ways

Permutation : Permutation is different arrangements of n objects taking all or some objects (say r where $r \leq n$) at a time. Here order of objects is material. For example, *abc*, *bac*, *bca* .... are different.

If repetition is allowed, then number of permutations of n distinct objects, taking r objects at a time, is given by n^r *i.e.* $n \times n \times n$ .... (r times)

Example. How many three digit numbers can be formed with the digits 1, 2, 3 and 5 if repetition of digits is allowed?

(*a*) 64 (*b*) 60 (*c*) 24 (*d*) 125

Sol. (*a*) Required numbers $= n^r = 4^3 = 4 \times 4 \times 4 = 64$

If repetition is not allowed, then number of permutations (arrangements) of n different (distinct) objects, taking r objects at a time, is given by $P(n, r)$ or ^nP_r

where $^nP_r = \dfrac{n!}{(n-r)!}$ (and $0 \leq r \leq n$)

$n!$ is read as n factorial and $n! = n \times (n-1) \times (n-2) \times$ .... $\times 2 \times 1$ (where n is a natural number) **Note:** $0! = 1$ and $1! = 1$

Example. How many three digit numbers can be formed with the digits 1, 2, 3 and 5 if repetition of digits is not allowed?

(*a*) 64 (*b*) 60 (*c*) 24 (*d*) 125

Sol. (*c*) Total numbers $= {^nP_r} = {^4P_3} = \dfrac{4!}{(4-3)!} = 4 \times 3 \times 2 = 24$

Note: Number of permutations of n different objects, taken all at a time is $^nP_n = \dfrac{n!}{(n-n)!} = \dfrac{n!}{0!} = n!$

Example. In how many different ways can the letters of the word 'PRIDE' be arranged?

(Bank of Baroda PO 2007)

(*a*) 60 (*b*) 120 (*c*) 15 (*d*) 360

Sol. (*b*) Letters in the word are 5 which can be arranged in $5! = 5 \times 4 \times 3 \times 2 = 120$ ways

Permutation of n objects, taking all at a time, where all objects are not distinct, p objects are of one type, q objects are of second type, r objects of third type and so on (while rest all are distinct) = $\dfrac{n!}{p!q!r!\ldots\ldots\ldots}$

Example. How many words can be formed using all the letters of the word TOMORROW?

(Bank PO)

(*a*) 1680 (*b*) 3360 (*c*) 7620 (*d*) 5340

Sol. (*b*) There are 8 letters in the given word of which there are 3 O's, 2 R's and rest are all single *i.e.* distinct.

$$\therefore \text{ Total arrangements} = \frac{8!}{3!2!} = \frac{8\times7\times6\times5\times4\times3!}{3!\times2} = 3360$$

Circular Permutations : The number of circular permutations (such as arrangements of different objects around a table or on a circle) of n different objects is $(n-1)!$.

Example : Find the number of ways 5 persons can be seated around a circular table.

(Bank PO)

(*a*) 120 (*b*) 60 (*c*) 24 (*d*) 6

Sol. (*c*) Required number of ways $= (5-1)! = 4! = 4 \times 3 \times 2 \times 1 = 24$

Combinations: Different ways of selecting n objects chosing all or some (say r where $r \leq n$) at a time. Total number of combinations is given by $C(n, r)$ or ^nC_r where $^nC_r = \dfrac{n!}{r!(n-r)!}$ (and $0 \leq r \leq n$). Here order of arrangement is immaterial. Note: $^nP_r = {^nC_r} \times r!$

$$^nC_r = {^nC_{n-r}} = \frac{n!}{r!(n-r)!} \qquad [^nC_n = 1, {^nC_0} = 1]$$

Example: In how many ways can a committee of 4 people be chosen out of 8 people ?

(Bank Officers Exam 2007)

(*a*) 118 (*b*) 36 (*c*) 70 (*d*) 126

Sol. (*c*) Required number of ways $= {^8C_4} = \dfrac{8!}{4!(8-4)!} = \dfrac{8!}{4!4!} = \dfrac{8\times7\times6\times5}{4\times3\times2} = 70$

If $^nC_x = {^nC_y} \Rightarrow x + y = n$ or $x = y$ $\qquad$ $^nC_{r-1} + {^nC_r} = {^{n+1}C_y}$

Some or none objects out of n different objects are taken, then:

$^nC_0 + {^nC_1} + {^nC_2} + {^nC_3} + \ldots\ldots\ldots\ldots + {^nC_n} + = 2^n$

There are two ways for each object, include it or exclude it. Total ways $= 2 \times 2 \times 2 \ldots\ldots = 2^n$

Probability : Probability is measure of chances / likelihood of occurence of an event. Probability of two events that are independent of each other is P(A and B) = P(A) × P(B)

Example : What is the probability of obtaining two heads on tossing a coin twice?

(*a*) 1 (*b*) $\frac{1}{2}$ (*c*) $\frac{1}{3}$ (*d*) $\frac{1}{4}$

Sol. (*d*) Required probability $= \frac{1}{2} \times \frac{1}{2} = \frac{1}{4}$

If two events A and B are mutually exclusive, then probability of A or B is sum of probability of A and B i.e. P(A or B) = P(A) + P(B)

If two events A and B are not mutually exclusive, then probability of A or B is P(A or B) = P(A) + P(B) – P(A and B)

Example : A and B runs in the same race. The probability of their winning are 1/2 and 1/3 respectively. Find the probability that either of them wins the race.

(*a*) $\frac{5}{6}$ (*b*) $\frac{1}{2}$ (*c*) $\frac{1}{6}$ (*d*) $\frac{1}{3}$

Sol. (*a*) P(A or B) = P(A) + P(B) = $\frac{1}{2} + \frac{1}{3} = \frac{5}{6}$

Example : If the probability of A hitting a target is 1/2 and that of B hitting the target is 1/3, then

(*i*) The probability that both hit the target = P(A) × P(B) = $\frac{1}{2} \times \frac{1}{3} = \frac{1}{6}$

(*ii*) The probability that either or both hits the target = P(A) + P(B) = $\frac{1}{2} + \frac{1}{3} = \frac{5}{6}$

(*ii*) The probability that neither hits the target = $\left(1 - \frac{1}{2}\right) \times \left(1 - \frac{1}{3}\right) = \frac{1}{2} \times \frac{2}{3} = \frac{1}{3}$

Example. The probabilities of two events are 0.25 and 0.50. The probability of both happening together is 0.14. Find the probability of none of the events happening. **(IIT 1980)**

(*a*) 0.25 (*b*) 0.39 (*c*) 0.61 (*d*) 0.86

Sol. (*b*) $P(A \cup B) = P(A) + P(B) - P(A \cap B) = 0.25 + 0.50 - 0.14 = 0.61$

$P(A' \cap B') = 1 - P(A \cup B) = 1 - 0.61 = 0.39$

SOLVED EXERCISE

1. The largest integer that divides product of any four consecutive integer is **(CDS 2011)**

(*a*) 4 (*b*) 6 (*c*) 12 (*d*) 24

Sol. (*d*) 4! = 4 × 3 × 2 × 1 = 24

2. Product of any five consecutive numbers is always divisible by:

(*a*) 5 (*b*) 15 (*c*) 20 (*d*) 120

Sol. (*d*) Product of any five consecutive numbers is always divisible by 5!.

5! = 1 × 2 × 3 × 4 × 5 = 120

3. There are 10 points in a plane, out of which 4 points are collinear. The number of lines obtained from the pairs of these points is **(NIFT 2003)**

(*a*) 39 (*b*) 40 (*c*) 41 (*d*) 45

Sol. (*b*) Lines = ${}^{10}C_2 - {}^4C_2 + 1 = 45 - 6 + 1 = 40$

Note: A straight line can be formed by joining a pair of two points on a plane. We can form only one line by joining two or more collinear points.

4. Find the number of lines that can be formed by joining 15 points on a plane out of which 5 are collinerar.

(*a*) 105 (*b*) 10 (*c*) 95 (*d*) 96

Sol. (*d*) Required number of lines = ${}^{15}C_2 - {}^{5}C_2 + 1 = 105 - 10 + 1 = 96$

5. If $1 \times 2 \times 3 \times \ldots \times n$ is denoted by $n!$, then $(8! - 7! - 6!) = ?$ **(SSC CGL 2000)**

(*a*) $6 \times 8 \times 6!$ (*b*) $7 \times 8 \times 6!$ (*c*) $6 \times 7 \times 8!$ (*d*) $7 \times 8 \times 7!$

Sol. (*a*) $8! - 7! - 6! = 8 \times 7 \times 6! - 7 \times 6! - 6! = (8 \times 7 - 7 - 1) \times 6! = (8 \times 7 - 8) \times 6! = 8 \times 6 \times 6!$

6. If $(1 \times 2 \times 3 \times 4 \times \ldots \times n) = n!$, then $(14! - 13! - 12!)$ is equal to **(SSC CHSL 2012)**

(*a*) $14 \times 12 \times 12!$ (*b*) $14 \times 12 \times 13!$ (*c*) $14 \times 13 \times 13!$ (*d*) $13 \times 12 \times 12!$

Sol. (*a*) $14! - 13! - 12! = 14 \times 13 \times 12! - 13 \times 12! - 12!$

$= (14 \times 13 - 13 - 1)\ 12! = (14 \times 13 - 14)12! = 14(13 - 1)12\ ! = 14 \times 12 \times 12!$

7. Find the remainder on dividing $(1! + 2! + 3! + 4! \ldots + 100!)$ by 12.

(*a*) 1 (*b*) 6 (*c*) 9 (*d*) 11

Sol. (*c*) $4! = 4 \times 3 \times 2 \times 1 = 24$, divisible by 12 $5! = 5 \times 4!$, divisible by 12

Similarly 6!, 7!, 8! ... 100! are divisible by 12

$\therefore$ Required remainder $= 1! + 2! + 3! = 1 + 2 + 6 = 9$

8. There are four envelopes, each carrying a specific address for each of the four different letters. If the letters are randomly put into envelopes, what the probability that all the four letters will be put into their respective correctly addressed envelope? **(UPSC Civil Services Pre. 2011)**

(*a*) 1/30 (*b*) 1/24 (*c*) 1/18 (*d*) 1/12

Sol. (*b*) Four letters can be put in four envelopes in ${}^4P_4$ ways $= 4! = 4 \times 3 \times 2 \times 1 = 24$ ways

All ways except one unique way are incorrect. $\Rightarrow$ Required probability = 1/24

9. In how many ways 5 balls can be kept in 4 boxes, assuming that any number of balls can be kept in one box.

(*a*) 20 (*b*) 1024 (*c*) 625 (*d*) 120

Sol. (*b*) Each ball can be kept in any of the 4 boxes i.e. in 4 ways.

Total ways $= 4^5 = 4 \times 4 \times 4 \times 4 \times 4 = 1024$ ways

10. How many three digit numbers can be formed with the digits 0, 1, 2, 3, 5, 6, if repetition of digits is allowed?

(*a*) 120 (*b*) 180 (*c*) 100 (*d*) 216

Sol. (*b*) Options for hundred's place = 5 (all digits excluding 0)

Options for ten's place = 6 (all digits including 0)

Options for unit's place = 6 (all digits including 0)

$\therefore$ Total numbers possible $= 5 \times 6 \times 6 = 180$

11. How many three-digit numbers can be formed with the digits 1, 2, 3, 4 and 5?

(*a*) 125 (*b*) 15 (*c*) 120 (*d*) 60

Sol. (*d*) ${}^5P_3 = \dfrac{5!}{2!} = 5 \times 4 \times 3 = 60$

12. How many three digit numbers can be formed with the digits 0, 1, 2, 3, 5, 6, if repetition of digits is not allowed?

(*a*) 120 (*b*) 180 (*c*) 100 (*d*) 216

Sol. (*c*) Options for hundred's place = 5 (all digits excluding 0)

Options for ten's place = 5 (*i.e.* 6 – 1 digits)

Options for unit's place = 4 (*i.e.* 5 – 1 digits)

$\therefore$ Total numbers possible = $5 \times 5 \times 4 = 100$

13. How many three digit odd numbers can be formed with the digits 1, 2, 3, 5, 6, if digits can not be repeated?

(*a*) 30 (*b*) 36 (*c*) 48 (*d*) 50

Sol. (*b*) Odd numbers ends in odd number at unit's place.

Options for unit's place = 3 (*i.e.* 1, 3 or 5)

Options for hundred's place = 4 (*i.e.* 5 – 1 digits)

Options for unit's place = 3 (*i.e.* 4 – 1 digits)

$\therefore$ Total numbers possible = $3 \times 4 \times 3 = 36$

14. How many three digit even numbers can be formed with the digits 0, 1, 2, 3, 4, 5 and 6, if digits can not be repeated?

(*a*) 100 (*b*) 105 (*c*) 120 (*d*) 135

Sol. (*b*) Even numbers ends in even number at unit's place.

Options for unit's place = 4 (*i.e.* 0, 2, 4 or 6)

Options for hundred's place = 6 (*i.e.* 7 – 1 digits)

Options for ten's place = 5 (*i.e.* 6 – 1 digits)

$\therefore$ Total numbers possible = $4 \times 6 \times 5 = 120$

But these numbers include numbers starting with 0.

Options for hundred's place = 1 (*i.e.* 0 only)

Options for unit's place = 3 (*i.e.* 2, 4, or 6)

Options for ten's place = 5 (*i.e.* 7 – 2 digits)

$\therefore$ Total such numbers possible = $1 \times 3 \times 5 = 15$

$\therefore$ Required numbers = $120 - 15 = 105$

15. How many three digit numbers, multiple of 5, can be formed from the digits 0, 1, 3, 4, 5, 6, if digits can not be repeated?

(*a*) 30 (*b*) 36 (*c*) 48 (*d*) 50

Sol. (*b*) Multiples of 5, always end in 5 or 0.

Numbers ending in 5 : Options for unit's place = 1 (*i.e.* 5 only)

Options for hundred's place = 4 (excluding 0 and 5)

Options for ten's place = 4 (*i.e.* 6 – 2 digits) $\therefore$ Such numbers = $1 \times 4 \times 4 = 16$.

Numbers ending in 0: Options for unit's place = 1 (*i.e.* 0 only)

Options for hundred's place = 5 (*i.e.* 6 – 1 digits)

Options for ten's place = 4 (*i.e.* 6 – 2 digits) $\therefore$ Such numbers = $1 \times 4 \times 5 = 20$.

Total required numbers = $16 + 20 = 36$.

16. How many three digit numbers are there such that atleast one of their digits is 6?

(*a*) 900 (*b*) 252 (*c*) 648 (*d*) 300

Sol. (*b*) Total three digit numbers = $9 \times 10 \times 10 = 900$

Numbers that do not have 6 as one of their digits:

Unit's digit options = 9 (other than 6)

Ten's digit options = 9 (other than 6)

Hundred's digit options = 8 (other than 0 and 6)

Total such numbers = 9 × 9 × 8 = 648

∴ Required numbers = 900 – 648 = 252

17. In how many ways can 6 people be seated on a sofa if there are only 3 seats available ? **(Bank Officers Exam 2007)**

(*a*) 18 (*b*) 120 (*c*) 2 (*d*) 216

Sol. (*b*) Required number of ways = $^6P_3 = \frac{6!}{3!} = 6 \times 5 \times 4 = 120$

18. There are 25 persons in a room. Each of them shakes hand with other persons. How many hand shakes took place?

(*a*) 250 (*b*) 300 (*c*) 600 (*d*) 625

Sol. (*b*) Total number of hand shakes = $^nC_r = {}^{25}C_2 = \frac{25!}{2!(25-2)!} = \frac{25!}{2!23!} = \frac{25 \times 24 \times 23!}{2 \times 23!} = 300$

Alternative method : $\frac{25 \times 24}{2} = 300$

19. In a football tournament there are 50 teams. Each team plays with every other team exactly once. How many matches will have to be organised ? **(SSC CGL Tier I 2013)**

(*a*) 1025 (*b*) 1225 (*c*) 1026 (*d*) 1226

Sol. (*b*) Number of matches = $\frac{50 \times 49}{2} = 1225$ matches

20. There are some persons in a room. Each of them shakes hand with other persons. If 66 hand shakes took place, how many persons are in the party ? **[UPSC Civil Services (Pre.) 2008]**

(*a*) 11 (*b*) 12 (*c*) 22 (*d*) None of these

Sol. (*b*) Option (*a*) : $\frac{11 \times 10}{2} = 55$ Option (*b*) : $\frac{12 \times 11}{2} = 66$

21. In how many ways a selection can be made of at least one fruit out of 4 apples and 5 bananas?

(*a*) 30 (*b*) 29 (*c*) 20 (*d*) 19

Sol. (*b*) 4 apples can be selected as 0, 1, 2, 3, 4 = 5 ways

Similarly 5 bananas can be selected in 6 ways.

Total number of selection = 5 × 6 = 30 ways.

Required number of ways = 30 – 1 (when no fruit is selected) = 29

22. There are 5 bulbs in a room. Each of them can be switched on or off independently. The number of ways in which the room can be illuminated is

(*a*) 5! (*b*) 32 (*c*) 31 (*d*) 25

Sol. (*c*) Each bulb can either be switched on or off i.e. 2 ways for each bulb.

⇒ Required number of ways = $2^5 - 1 = 31$

23. From 10 members in a club, 5 executive members are to be chosen. In how may ways they can be chosen, if two particular members will always be included?

(*a*) 64 (*b*) 60 (*c*) 56 (*d*) 120

Sol. (*c*) Two particular members are always selected.

So we need to select 5 – 2 = 3 members out of 10 – 2 = 8 members

Their selection can be done in ways $= {}^8C_3 = \frac{8!}{3!(8-3)!} = \frac{8\times7\times6}{3\times2} = 56$ ways

24. From 10 members in a club, 5 executive members are to be chosen. In many how ways they can be chosen, if two particular members will always be excluded?

(*a*) 64 (*b*) 60 (*c*) 56 (*d*) 120

Sol. (*c*) Two particular members are always excluded.

So we need to select 5 members out of 10 – 2 = 8 members

Their selection ca n be done in ${}^8C_5$ ways $= \frac{8!}{5!(8-5)!} = \frac{8\times7\times6}{3\times2} = 56$ ways

25. In how many ways 5 boys and 2 girls can be seated in a row so that the two girls do not sit together ?

(*a*) 7! (*b*) 6! (*c*) 7! – 6! (*d*) 5 × 6!

Sol. (*d*) Total persons = 5 + 2 = 7 can be seated in 7 ways (without restriction).

When two girls sit together can be treated as one set (or one person).

6 persons can be seated in 6! ways and 2 girls (among themselves) can be seated in 2! ways.

Total number of ways, when the girls sit together = 6! × 2!

⇒ Required number of ways = 7! – 6! × 2! = 6! (7 – 2) = 5 × 6!

26. In how many different ways can the letters of the word ARISE can be arranged?

(SBI Clerical 2009)

(*a*) 24 (*b*) 120 (*c*) 720 (*d*) None of these

Sol. (*b*) 5 letters can be arranged in 5! = 5 × 4 × 3 × 2 × 1 = 120 ways

27. In how many different ways can the letters of the word 'SMART' be arranged ?

(Indian Overseas Bank P.O. 2009)

(*a*) 60 (*b*) 180 (*c*) 200 (*d*) None of these

Sol. (*d*) 5 letters can be arranged in 5! = 5 × 4 × 3 × 2 × 1 = 120 ways

28. In how many ways can the leters of the word WINDOW be arranged? **(Bank Clerical)**

(*a*) 120 (*b*) 360 (*c*) 720 (*d*) None of these

Sol. (*c*) 6 letters can be arranged in 6! = 6 × 5 × 4 × 3 × 2 = 720 ways

29. How many words can be formed from the letters of the word INFORMAL which begin with I and end with L ?

(*a*) 8! (*b*) 7! (*c*) 6! (*d*) 2 × 6!

Sol. (*c*) Total letters = 8 but I and L have fixed positions.

Remaining letters = 8 – 2 = 6 can be arranged in 6! ways.

25. In how many different ways can the letters of the word 'CREATE' be arranged ?

(PNB Officers Exam 2009)

(*a*) 25 (*b*) 36 (*c*) 710 (*d*) 360

Sol. (*d*) There are 6 letters in the given word of which there are 2 E, and rest are distinct.

∴ Total arrangements $= \frac{6!}{2!} = \frac{6\times5\times4\times3\times2!}{2!} = 360$

30. In how many different ways can be the letters of the word SOFTWARE be arranged in such a way that the vowels always come together ? **(Canara Bank PO 2007)**

(*a*) 13440 (*b*) 1440 (*c*) 360 (*d*) None of these

Sol. (*d*) The given word has 3 vowels (O, A and E) and 5 consonants (all distinct).

Since the vowels occur together, they can be treated as one letter.

$5 + 1 = 6$ letters can be arranged in $6! = 6 \times 5 \times 4 \times 3 \times 2 \times 1 = 720$ ways.

3 vowels can be arranged in $3! = 3 \times 2 \times 1 = 6$ ways

$\therefore$ Required number of ways $= 6! \times 3! = 720 \times 6 = 4320$

31. In how many ways can the letters of the word CAPITAL be arranged in such a way that all the vowels always come together ? **(Bank PO 2004)**

(*a*) 360 (*b*) 720 (*c*) 120 (*d*) 840

Sol. (*a*) The given word has 3 vowels (A, I and A) and 4 consonants (all distinct).

Since the vowels occur together, they can be treated as one letter.

$4 + 1 = 5$ letters can be arranged in $5! = 5 \times 4 \times 3 \times 2 \times 1 = 120$ ways

3 vowels (2 of which are same) can be arranged in $\frac{3!}{2!} = 3$ ways

$\therefore$ Required number of ways $= 5! \times \frac{3!}{2!} = 120 \times 3 = 360$

32. How many different letters arrangement can be made from the letters of the word ARRANGE so that all the vowels occur together?

(*a*) 160 (*b*) 180 (*c*) 1260 (*d*) 5040

Sol. (*b*) The given word has 3 vowels (including 2 A) and 4 consonants (incldUing 2 R) since the vowels occur together, they makes one set.

3 vowels (1 set) + 4 consonants = 5 letters

5 letters including 2 R can be arranged in $\frac{5!}{2!}$ ways

3 vowels including 2 A can be arranged in $\frac{3!}{2!}$ ways

$\therefore$ Total ways $= \frac{5!}{2!} \times \frac{3!}{2!} = (5 \times 4 \times 3) \times (3) = 180$ ways

33. How many words, beginning with vowels, can be formed using all the letters of the word DAUGHTER?

(*a*) 40320 (*b*) 15120 (*c*) 5040 (*d*) 20160

Sol. (*b*) There are 8 letters in the given word of which three are vowels.

Choices for the first letter $= {}^3P_1 = 3$

Remaining letter arrangement $= {}^7P_7 = 7! = 5040$

$\therefore$ Total arrangements $= 3 \times 5040 = 15120$

34. In how many different ways can the letters of the word SUBJECT be arranged so that the vowels are at the two ends?

(*a*) 120 (*b*) 240 (*c*) 720 (*d*) 5040

Sol. (*b*) Two vowels i.e. U and E can be arranged at two ends in 2! ways.

Remaining 5 letters can be arranged in 5! ways.

$\therefore$ Total arrangements $= 2! \times 5! = 2 \times 5 \times 4 \times 3 \times 2 = 240$

35. How many words can be formed from the letters of the word NATION so that vowels occupy the even places?

(*a*) 18 (*b*) 36 (*c*) 24 (*d*) 30

Sol. (*b*) The given word has 3 vowels and 3 consonants.

The 3 consonants can be arranged at 3 odd places in $3! = 6$ ways.

The 3 vowels can be arranged at 3 even places in $3! = 6$ ways.

$\therefore$ Total words $= 6 \times 6 = 36$

36. How many words, each of 3 consonants and 2 vowels can be formed from the letters of the word COMPUTER?

(*a*) 1680 (*b*) 2160 (*c*) 7620 (*d*) 3600

Sol. (*d*) The given word has 5 consonants and 3 vowels.

Ways of selecting 3 consonants $= {}^5C_3 = 10$

Ways of selecting 2 vowels $= {}^3C_2 = 3$

$\therefore$ Total number of combinations $= 10 \times 3 = 30$

But each combination of 5 letters can be done in $5! = 120$ ways

$\therefore$ Total number of words $= 30 \times 120 = 3600$

37. A team of 11 cricketers is to be selected from 15 players in which 7 are bowlers and rest are batsman. In how may ways, the team can be selected if team must have 5 bowlers and 6 batsmen? **(Bank P.O.)**

(*a*) 3003 (*b*) 1176 (*c*) 588 (*d*) 580

Sol. (*c*) 5 bowlers can be selected in ${}^7C_5$ ways. 6 batsman can be selected in ${}^8C_6$ ways.

$\Rightarrow$ Required number of ways $= {}^7C_5 \times {}^8C_6 = \dfrac{7!}{5!2!} \times \dfrac{8!}{6!2!} = \dfrac{7\times6}{2} \times \dfrac{8\times7}{2} = 588$ ways

38. Out of 5 men and 3 women, a committee of 3 members is to be formed so that it has 1 woman and 2 men. In how many different ways can it be done ? **(Bank P.O. 2009)**

(*a*) 20 (*b*) 10 (*c*) 23 (*d*) 30

Sol. (*d*) 1 woman can be selected in ${}^3C_1$ ways. 2 men can be selected in ${}^5C_2$ ways.

$\Rightarrow$ Required number of ways $= {}^3C_1 \times {}^5C_2 = \dfrac{3!}{1!2!} \times \dfrac{5!}{2!3!} = \dfrac{3}{1} \times \dfrac{5\times4}{2} = 3 \times 10 = 30$

39. In how many different ways a group of 4 men and 4 women be formed out of 7 men and 8 women? **(SBI Associate Bank P.O. 2007)**

(*a*) 2450 (*b*) 105 (*c*) 1170 (*d*) None of these

Sol. (*a*) Required number of ways $= {}^7C_4 \times {}^8C_4 = \dfrac{7\times6\times5}{3\times2\times1} \times \dfrac{8\times7\times6\times5}{4\times3\times2\times1} = 35\times70 = 2450$

40. Find the number of parallelogram that can be formed from a set of four parallel line-segments intersecting another sets of three line segments. **(SBI Associate Bank P.O. 2007)**

(*a*) 6 (*b*) 12 (*c*) 18 (*d*) 24

Sol. (*c*) 2 lines can be selected in ${}^4C_2$ ways and 2 lines can be selected in ${}^3C_2$ ways.

Total ways $= {}^4C_2 \times {}^3C_2 = \dfrac{4\times3}{2} \times \dfrac{3\times2}{2} = 18.$

41. A team of 5 children is to be selected out of 4 girls and 5 boys such that it contains at least 2 girls. In how many different ways the selection can be done ? **(Bank P.O. 2004)**

(*a*) 105 (*b*) 60 (*c*) 100 (*d*) 120

Sol. (*a*) The group can be formed in one of the following ways :

(*i*) 2 girls + 3 boys (*ii*) 3 girls + 2 boys (*iii*) 4 girls + 1 boy

Required number of ways = $^4C_2 \times {}^5C_3 + {}^4C_3 \times {}^5C_2 + {}^4C_4 \times {}^5C_1$

$= \frac{4!}{2! \times 2!} \times \frac{5!}{3!2!} + \frac{4!}{3! \times 1!} \times \frac{5!}{2! \times 3!} + \frac{4!}{4!} \times \frac{5!}{1! \times 4!} = \frac{4 \times 3}{2} \times \frac{5 \times 4}{2} + \frac{4}{1} \times \frac{5 \times 4}{2} + 1 \times 5$

$= 6 \times 10 + 4 \times 10 + 1 \times 5 = 60 + 40 + 5 = 105$

42. A select group of 4 is to be formed from 8 men and 6 women in such a way that the group must have atleast one woman. In how many different ways can it be done ? **(SBI Bank P.O. 2005)**

(*a*) 364 (*b*) 1001 (*c*) 728 (*d*) 931

Sol. (*d*) The group can be formed in one of the following ways :

(*i*) 1 woman + 3 men (*ii*) 2 women + 2 men (*iii*) 3 women + 1 man (*iv*) 4 women

Required number of ways = $^6C_1 \times {}^8C_3 + {}^6C_2 \times {}^8C_2 + {}^6C_3 \times {}^8C_1 + {}^6C_4$

$= \frac{6!}{1!5!} \times \frac{8!}{3!5!} + \frac{6!}{2!4!} \times \frac{8!}{2!6!} + \frac{6!}{3!3!} \times \frac{8!}{1!7!} + \frac{6!}{4!2!}$

$= 6 \times \frac{8 \times 7 \times 6}{3 \times 2} + \frac{6 \times 5}{2} \times \frac{8 \times 7}{2} + \frac{6 \times 5 \times 4}{3 \times 2} \times 8 + \frac{6 \times 5}{2}$

$= 6 \times 56 + 15 \times 28 + 20 \times 8 + 15 = 336 + 420 + 160 + 15 = 931$

Alt: Number of ways = $^{14}C_4 - {}^8C_4 = 1001 - 70 = 931$

43. In how many ways can 4 boys and 3 girls be seated in a row so that all the girls do not sit together?

(*a*) 5040 (*b*) 4320 (*c*) 720 (*d*) 5760

Sol. (*b*) Total persons = 4 + 3 = 7 can be seated in 7! ways (without any restriction)

3 girls when seated together, makes one set, then total number of persons = 4 + 1 = 5 can be seated in 5! and girls among themselves can be arranged in 3! ways.

∴ Required number of ways = 7! – 5! × 3! = 5040 – 120 × 6 = 5040 – 720 = 4320.

44. Three dice are rolled simultaneously. What is the number of possible outcomes such that atleast one show the number 2 ? **[UPSC Civil Services (Pre.)]**

(*a*) 36 (*b*) 81 (*c*) 91 (*d*) 116

Sol. (*c*) Total possible outcomes = $6^3 = 216$

The number of outcomes in which none of the dice show 2 = $5^3 = 125$

∴ Number of outcomes in which atleast one dice show 2 = 216 – 125 = 91

45. Uma has three children, what is the probability that none of the three children is a girl ? **(Dena Bank PO 2006)**

(*a*) $\frac{1}{2}$ (*b*) $\frac{1}{16}$ (*c*) $\frac{1}{3}$ (*d*) None of these

Sol. (*d*) Probability of a non-girl child = $\frac{1}{2}$ ⇒ Required probability = $\frac{1}{2} \times \frac{1}{2} \times \frac{1}{2} = \frac{1}{8}$

46. Three dice are thrown. What is the probability that the same number will appear on each of them? **(NDA 2012)**

(*a*) 1/6 (*b*) 1/8 (*c*) 1/24 (*d*) 1/36

Sol. (*d*) First dice may have any of six numbers.

Probability of having same number on second and third dice = $\frac{1}{6}\times\frac{1}{6} = \frac{1}{36}$

47. What is the probability that a leap year selected at random contains 53 Mondays? (NDA 2012)

(*a*) 1/7 (*b*) 2/7 (*c*) 7/366 (*d*) 26/183

Sol. (*b*) A leap year has 366 days i.e., 52 weeks + 2 odd days.

48. A man hit a target once in 3 shots. If he fires 4 shots in succession, what is the probability that he will hit the target? (MBA Ent. Exam.)

(*a*) $\frac{4}{3}$ (*b*) $\frac{3}{4}$ (*c*) 1 (*d*) $\frac{65}{81}$

Sol. (*d*) The man will hit the target if he hits it one or more time. He fails to hit it, if he could not hit the target even once.

Probability of not hitting the target in 1 shot = $1-\frac{1}{3}=\frac{2}{3}$

Probability of not hitting the target in 4 shots = $\left(\frac{2}{3}\right)^4 = \frac{16}{81}$

∴ The probability of hitting the target = $1-\frac{16}{81}=\frac{65}{81}$

49. Two balls are drawn one after another from a bag containing 3 white and 5 black balls, what is the probability that the first ball is black and the second is white ? (MBA 1990)

(*a*) 1/15 (*b*) 2/15 (*c*) 15/56 (*d*) 8/15

Sol. (*c*) P(First black ball) = $\frac{5}{8}$ and P (Second ball is white) = $\frac{3}{7}$

Required probability = $\frac{5}{8}\times\frac{3}{7}=\frac{15}{56}$

50. An urn contains 9 red, 7 white and 4 black balls. If two balls are drawn at random, find the probability that both balls are red. (MBA Ent. Exam.)

(*a*) $\frac{17}{95}$ (*b*) $\frac{18}{95}$ (*c*) $\frac{1}{12}$ (*d*) $\frac{91}{190}$

Sol. (*b*) Total number of balls = 9 + 7 + 4 = 20

Out of 20 balls, 2 balls can be selected in $^{20}C_2$ ways.

Out of 9 red balls, 2 balls can be selected in $^9C_2$ ways.

∴ Required probability = $\frac{^9C_2}{^{20}C_2}=\frac{9\times8}{2\times1}\times\frac{2\times1}{20\times19}=\frac{18}{95}$

Alt: Probability of first ball being red = $\frac{9}{20}$. Probability of second ball being red = $\frac{8}{19}$

∴ Required probability = $\frac{9}{20}\times\frac{8}{19}=\frac{18}{95}$

51. A bowl contains 4 red, 3 green, 2 blue and 5 black marbles. If two marbles are drawn at randon, what is the probaility that they are green ? **(Dena Bank PO 2006)**

(*a*) $\frac{4}{91}$ (*b*) $\frac{51}{91}$ (*c*) $\frac{4}{13}$ (*d*) None of these

Sol. (*d*) Total number of marbles = 4 + 3 + 2 + 5 = 14

Out of 14 marbles, 2 marbles can be drawn in $^{14}C_2 = \frac{14\times13}{1\times2} = 91$ ways

Out of 3 green marbles, 2 marbles can be drawn in $^{3}C_2 = \frac{3\times2}{1\times2} = 3$ ways

$\therefore$ Required probability = $\frac{3}{91}$

Alternative method : Required probability = $\frac{3}{14}\times\frac{2}{13}=\frac{3}{91}$

52. A basket contains 6 blue, 2 red, 4 green and 3 yellow balls. If 2 balls are picked at random, what is the probability that both are blue ? **(PNB Officers Exam. 2009)**

(*a*) $\frac{1}{5}$ (*b*) $\frac{8}{91}$ (*c*) $\frac{2}{15}$ (*d*) $\frac{1}{7}$

Sol. (*d*) Total number of balls = 6 + 2 + 4 + 3 = 15

Probability of first blue ball = $\frac{6}{15}$. Probability of second blue ball = $\frac{5}{14}$

Probability of both blue balls = $\frac{6}{15}\times\frac{5}{14}=\frac{1}{7}$

53. A basket contains 6 blue, 2 red, 4 green and 3 yellow balls. If 3 balls are picked at random, what is the probability that none is yellow ? **(PNB Officers Exam. 2009)**

(*a*) $\frac{3}{455}$ (*b*) $\frac{1}{5}$ (*c*) $\frac{44}{91}$ (*d*) $\frac{4}{5}$

Sol. (*c*) Total number of balls = 6 + 2 + 4 + 3 = 15. Non-yellow balls = 15 – 3 = 12

$\therefore$ Required probability = $\frac{12}{15}\times\frac{11}{14}\times\frac{10}{13}=\frac{44}{91}$

54. A bowl contains 4 red, 3 green, 2 blue and 5 black marbles. If four marbles are drawn at randon, what is the probaility that 2 are red and 2 are blue ? **(Dena Bank PO 2006)**

(*a*) $\frac{6}{1001}$ (*b*) $\frac{1}{143}$ (*c*) $\frac{1}{13}$ (*d*) $\frac{6}{91}$

Sol. (*a*) Total number of marbles = 4 + 3 + 2 + 5 = 14

Exhaustive number of cases = $^{14}C_4 = \frac{14\times13\times12\times11}{1\times2\times3\times4} = 1001$

Favourable number of cases = $^{4}C_2 \times {^{2}C_2} = \frac{4\times3}{1\times2}\times1 = 6$

$\therefore$ Required probability = $\frac{6}{1001}$

55. A basket contains 6 blue, 2 red, 4 green and 3 yellow balls. If 4 balls are picked at random, what is the probability that 2 are red and 2 are green ? **(PNB Officers Exam. 2009)**

(*a*) $\frac{4}{15}$ (*b*) $\frac{5}{27}$ (*c*) $\frac{1}{3}$ (*d*) $\frac{2}{455}$

Sol. (*d*) Total number of balls = 6 + 2 + 4 + 3 = 15

Total number of ways of selecting 4 balls = $^{15}C_4 = \frac{15\times14\times13\times12}{4\times3\times2\times1} = 1365$

Favourbale number of ways = $^2C_2 \times {}^4C_2 = 6$

$\therefore$ Required probability = $\frac{6}{1365} = \frac{2}{455}$

56. A bowl contains 4 red, 3 green, 2 blue and 5 black marbles. If three marbles are drawn at randon, what is the probaility that none of them are black? **(Dena Bank PO 2006)**

(*a*) $\frac{6}{13}$ (*b*) $\frac{3}{91}$ (*c*) $\frac{5}{91}$ (*d*) $\frac{3}{13}$

Sol. (*d*) Total number of marbles = 4 + 3 + 2 + 5 = 14. Non-black marbles = 14 – 5 = 9

Exhaustive number of cases = $^{14}C_3 = \frac{14\times13\times12}{3\times2\times1} = 364$

Favourable number of cases = $^9C_3 = \frac{9\times8\times7}{3\times2\times1} = 84$

$\therefore$ Required probability = $\frac{84}{364} = \frac{3}{13}$

57. A bowl contains 4 red, 3 green, 2 blue and 5 black marbles. If three marbles marbles are drawn at randon, what is the probability that at least one of them is red ? **(Dena Bank PO 2006)**

(*a*) $\frac{2}{7}$ (*b*) $\frac{4}{91}$ (*c*) $\frac{61}{91}$ (*d*) None of these

Sol. (*c*) Total number of marbles = 4 + 3 + 2 + 5 = 14. Non-red marbles = 14 – 4 = 10

Exhaustive number of cases = $^{14}C_3 = \frac{14\times13\times12}{3\times2\times1} = 364$

Number of cases, when all marbles are non-red = $^{10}C_3 = \frac{10\times9\times8}{3\times2\times1} = 120$

Probability of all non-red marbles = $\frac{120}{364} = \frac{30}{91}$ $\Rightarrow$ Required probability = $1 - \frac{30}{91} = \frac{61}{91}$

58. A basket contains 6 blue, 2 red, 4 green and 3 yellow balls. If 5 balls are picked at random, what is the probaility that at least one is blue ? **(PNB Officers Exam. 2009)**

(*a*) $\frac{137}{143}$ (*b*) $\frac{9}{91}$ (*c*) $\frac{18}{455}$ (*d*) $\frac{2}{5}$

Sol. (*a*) Total number of balls = 6 + 2 + 4 + 3 = 15. Number of non-blue balls = 15 – 6 = 9

Number of ways of selecting 5 non-blue balls = $^9C_5$

Number of ways of selecting 5 balls out of 15 balls = $^{15}C_5$

Probability of selection of all non-blue balls = $\frac{^9C_5}{^{15}C_5} = \frac{9!}{5!\times4!} \div \frac{15!}{5!\times10!}$

$= \frac{9!}{5!\times4!} \times \frac{5!\times10!}{15!} = \frac{9!}{4!} \times \frac{10!}{15!} = \frac{9\times8\times7\times6\times5}{15\times14\times13\times12\times11} = \frac{6}{143}$

$\therefore$ Required probability $= 1 - \frac{6}{143} = \frac{137}{143}$

Alternative method :

Total number of balls = 6 + 2 + 4 + 3 = 15. Number of non-blue balls = 15 – 6 = 9

Probability of first ball being non-blue $= \frac{9}{15}$

Probability of second ball being non-blue $= \frac{8}{14}$ and so on

Probability of first 5 balls being non-blue $= \frac{9}{15} \times \frac{8}{14} \times \frac{7}{13} \times \frac{6}{12} \times \frac{5}{11} = \frac{6}{143}$

$\therefore$ Required probability $= 1 - \frac{6}{143} = \frac{137}{143}$

59. A basket contains 6 blue, 2 red, 4 green and 3 yellow balls. If 2 balls are picked at random what is the probability that either both are green or both are yellow ?

(PNB Officers Exam. 2009)

(*a*) $\frac{2}{5}$ (*b*) $\frac{3}{35}$ (*c*) $\frac{1}{3}$ (*d*) $\frac{3}{91}$

Sol. (*b*) Total number of balls = 6 + 2 + 4 + 3 = 15

Probability of both green balls $= \frac{4}{15} \times \frac{3}{14} = \frac{2}{35}$.

Probability of both yellow balls $= \frac{3}{15} \times \frac{2}{14} = \frac{1}{35}$

$\therefore$ Required probability $= \frac{2}{35} + \frac{1}{35} = \frac{3}{35}$

60. A basket contains 6 red, 5 green and 8 blue balls. If four balls are picked at randon, what is the probability that all four of them are either red or any two out of the four are green ?

(Union Bank of India PO 2005)

(*a*) $\frac{5}{1292}$ (*b*) $\frac{925}{3876}$ (*c*) $\frac{359}{1938}$ (*d*) $\frac{11}{3876}$

Sol. (*b*) Total number of balls = 6 + 5 + 8 = 19

Exhaustive number of cases $= {}^{19}C_4 = \frac{19 \times 18 \times 17 \times 16}{4 \times 3 \times 2 \times 1} = 3876$

Number of ways of selecting 4 red balls out of 6 red balls $= {}^{6}C_4 = \frac{6 \times 5}{2 \times 1} = 15$

Number of ways of selecting 2 green balls and 2 other balls

$= {}^{5}C_2 \times {}^{14}C_2 = \frac{5 \times 4}{2 \times 1} \times \frac{14 \times 13}{2 \times 1} = 10 \times 91 = 910$

Total number of favourable ways = 15 + 910 = 925 $\Rightarrow$ Required probability $= \frac{925}{3876}$

61. A box contains 5 red balls, 3 blue balls and 4 green balls. If 2 balls are drawn at randon, what is the probability that both the balls are of the same colour ? **(Union Bank of India PO 2005)**

(*a*) $\frac{5}{33}$ (*b*) $\frac{1}{6}$ (*c*) $\frac{2}{11}$ (*d*) $\frac{19}{66}$

Sol. (*d*) Total number of balls = 5 + 3 + 4 = 12

Exhaustive number of cases = $^{12}C_2 = \frac{12\times11}{2\times1} = 66$

Favourable number of ways = $^5C_2 + {}^3C_2 + {}^4C_2 = 10 + 3 + 6 = 19$

$\Rightarrow$ Required probability = $\frac{19}{66}$

62. Ajay appeared in three exams. The probability of his passing the three exams are $\frac{1}{2}, \frac{2}{3}$ and $\frac{3}{4}$ respectively. What is the probability that he will pass at least one of these exams ?

(*a*) $\frac{1}{24}$ (*b*) $\frac{7}{24}$ (*c*) $\frac{13}{24}$ (*d*) $\frac{23}{24}$

Sol. (*d*) Probability of not passing all the three exams = $\left(1-\frac{1}{2}\right)\times\left(1-\frac{2}{3}\right)\times\left(1-\frac{3}{4}\right) = \frac{1}{2}\times\frac{1}{3}\times\frac{1}{4} = \frac{1}{24}$

$\Rightarrow$ Probability of passing at least one exam = $1-\frac{1}{24} = \frac{23}{24}$

63. A speaks truth in 75% of cases and B in 80% of cases. In what percent of cases are likely to contradict each other in narrating the same incident ? **(IIT 1975)**

(*a*) 35% (*b*) 15% (*c*) 45% (*d*) 55%

Sol. (*a*) A speaks truth = $P(A) = \frac{75}{100} = \frac{3}{4} \Rightarrow P(\bar{A}) = \frac{1}{4}$

B speaks truth = $P(B) = \frac{80}{100} = \frac{4}{5} \Rightarrow P(\bar{B}) = \frac{1}{5}$

A and B will contradict when A speaks truth and B lies or when A lies and B speaks truth.

Required probability = $P(A).P(\bar{B})+P(\bar{A}).P(B) = \frac{3}{4}\times\frac{1}{5}+\frac{1}{4}\times\frac{4}{5} = \frac{3}{20}+\frac{4}{20} = \frac{7}{20} = 35\%$

64. An elevator starts with 5 passengers and stops at 8 different floors of a building. Find the probability of the all the 5 passengers alignting at different floors. **(IIT Roorkee 1990)**

(a) 105/1024 (b) 105/512 (c) 5/8 (d) 25/64

Sol. (*b*) Favourable cases : 5 passengers can alight at different floors in $^8P_5$ ways

Exhaustive cases : 8 persons can alight in 8^5 ways.

$\Rightarrow$ Required probability = $\frac{8\times7\times6\times5\times4}{8\times8\times8\times8\times8} = \frac{105}{512}$

CHAPTER

35

DATA INTERPRETATION

SSC CGL TIER I 2012

The graph shows the production of wheat in different years of a particular State. Study the graph and answer Questions (1 to 4).

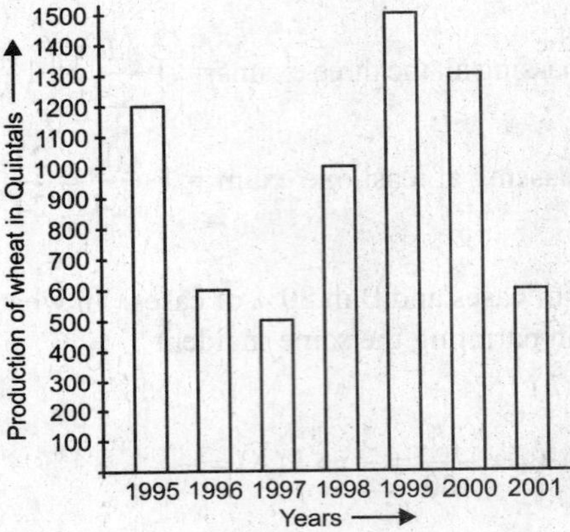

1. The year in which the production reached maximum is
(*a*) 1995 (*b*) 1997 (*c*) 1999 (*d*) 2000

2. The percentage increase in production of wheat from 1997-1998 is
(*a*) 100% (*b*) 150% (*c*) 90% (*d*) 120%

3. The year which had the percentage of decrease vis-a-vis its previous year in production as $13\frac{1}{3}\%$ is
(*a*) 96-97 (*b*) 95-96 (*c*) 99-2000 (*d*) 2000-01

4. The total production from the year 1995 to 1998 (in quintals) is
(*a*) 3,000 (*b*) 3,100 (*c*) 3,200 (*d*) 3,300

SSC CHSL 2012

Study the following bar-diagram carefully and answer question numbers (5 to 9).
Number of students passed (in thousands) from two Universities over the years.

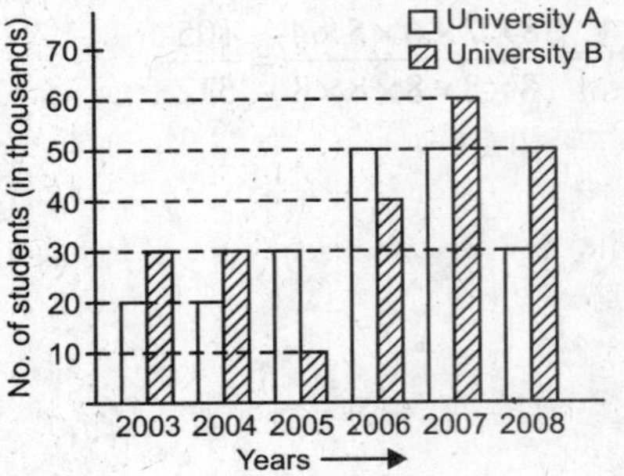

5. The sum of the students passed from University 'B' in years 2003, 2005 and 2006 combined together is:
(*a*) 80,000 (*b*) 60,000 (*c*) 50,000 (*d*) 75,000

6. The ratio between the number of students passed from University 'A' in the year 2007 and that from University 'B' in 2004 is:
(*a*) 5 : 4 (*b*) 5 : 3 (*c*) 5 : 6 (*d*) 3 : 5

7. Number of students passed from University 'B' in the year 2008 expressed as a percentage of the total number of students passed from University 'A' over the 6 years is:
(*a*) 30 (*b*) 20 (*c*) 25 (*d*) 35

8. The respective ratio between the number of students passed in year 2007, 2008 and 2009 from University 'B' is:
(*a*) 6 : 5 : 3 (*b*) 6 : 3 : 5 (*c*) 3 : 5 : 6 (*d*) 6 : 3 : 3

9. The difference between the total number of students passed from both the Universities in the year 2008 combined together and the total number of students passed in the year 2004 from both the Universities taken together is :
(*a*) 35,000 (*b*) 25,000 (*c*) 20,000 (*d*) 30,000

SSC CGL TIER II 2012

Directions : The bar chart given below shows the percentage distribution of the production of various models of a mobile manufacturing company in 2007 and 2008. The total production in 2007 was 35 lakh mobile phones and in 2008 the production was 44 lakh. Study the chart and answer questions (10 to 14).

Percentage of six different types of mobiles manufactured by a company over two years

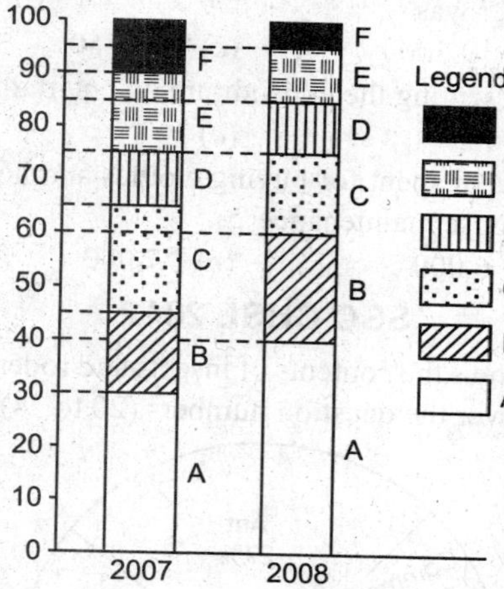

10. What was the difference in the number of B type mobiles produced in 2007 and 2008 ?
(*a*) 2,25,000 (*b*) 1,75,000 (*c*) 3,55,000 (*d*) 2,70,000

11. If the percentage production of A type mobiles in 2008 was same as that in 2007, then the number of A type mobiles produced in 2008 would have been
(*a*) 11,70,000 (*b*) 10,50,000 (*c*) 14,00,000 (*d*) 13,20,000

12. If 85% of the D type mobiles produced in each year were sold by the company, how many D type mobiles remained unsold ?
(*a*) 1,18,500 (*b*) 1,22,500 (*c*) 76,500 (*d*) 93,500

13. Total number of mobiles of models A, B and E manufactured in 2007 was
(*a*) 21,00,000 (*b*) 19,25,000 (*c*) 24,50,000 (*d*) 22,75,000

14. For which models was the percentage variation in production from 2007 to 2008 the maximum?

(*a*) D and E (*b*) A and B (*c*) B and C (*d*) C and D

SSC CGL TIER II 2012

Following is the pie-chart showing the spendings of a family on various items in particular year. Study the pie-chart and answer questions (15 to 19).

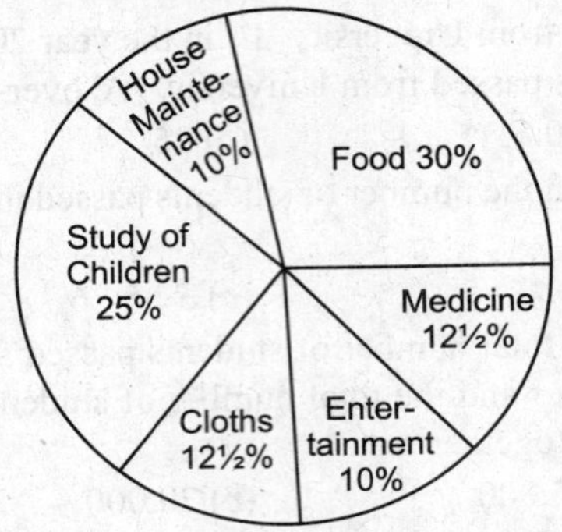

15. The ratio of the total amount spent for food and medicine is

(*a*) 1 : 2 (*b*) 3 : 1 (*c*) 12 : 5 (*d*) 11 : 5

16. If the total amount spent on the family during the year was ₹ 50,000 how much they spent for buying clothes ?

(*a*) ₹ 6,250 (*b*) ₹ 6,500 (*c*) ₹ 7,250 (*d*) ₹ 7,500

17. If the total amount spent on the family during the year was ₹ 35,000 the amount spent for study of children and food together was

(*a*) ₹ 19,250 (*b*) ₹ 19,500 (*c*) ₹ 19,750 (*d*) ₹ 19,850

18. Angle of the pie-chart representing the expenditure on entertainment is

(*a*) 15° (*b*) 10° (*c*) 36° (*d*) 12½°

19. If the difference in the amount spent for buying clothes and house maintenance was ₹ 1,500, how much they spent for house maintenance ?

(*a*) ₹ 5,000 (*b*) ₹ 6,000 (*c*) ₹ 7,000 (*d*) ₹ 8,000

SSC CHSL 2012

The following pie-chart shows the contents of insects and rodents in an average Indian Household. Examine the chart and answer the question numbers (20 to 24).

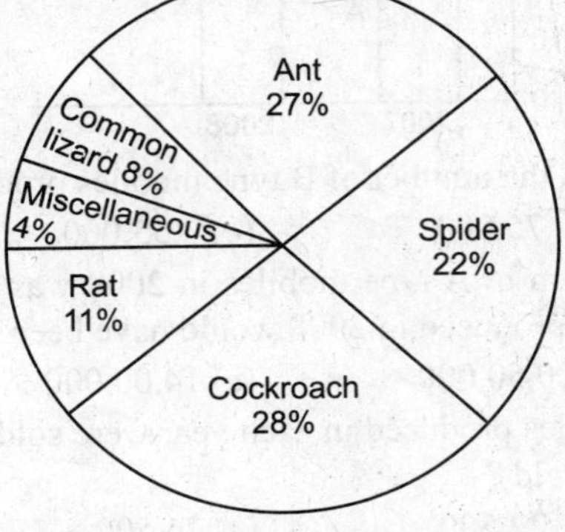

20. If the percentage of spider is x% of the percentage of cockroach, then x is equal to:

(*a*) $73\frac{4}{7}\%$ (*b*) $75\frac{2}{7}\%$ (*c*) $78\frac{4}{7}\%$ (*d*) $75\frac{4}{7}\%$

21. The total percentage of spider, rat and cockroach is greater than the percentage of ant by

(*a*) 34 (*b*) 32 (*c*) 28 (*d*) 24

22. If the percentage of rat is *y*% of the total percentage of ant and cockroach, then *y* is equal to:

(*a*) 30 (*b*) 35 (*c*) 25 (*d*) 20

23. The total percentage of common lizard, spider and cockroach is greater than the percentage of rat by:

(*a*) 37 (*b*) 47 (*c*) 35 (*d*) 25

24. If the difference of the percentage of rat and miscellaneous by *z*% of the percentage of cockroach, then *z* is equal to:

(*a*) 25% (*b*) 35% (*c*) 20% (*d*) 15%

SSC CGL TIER I 2013

Directions : Questions no. 25 to 27, the following pie-chart show the number of students admitted in different faculties of a college. Study the chart and answer.

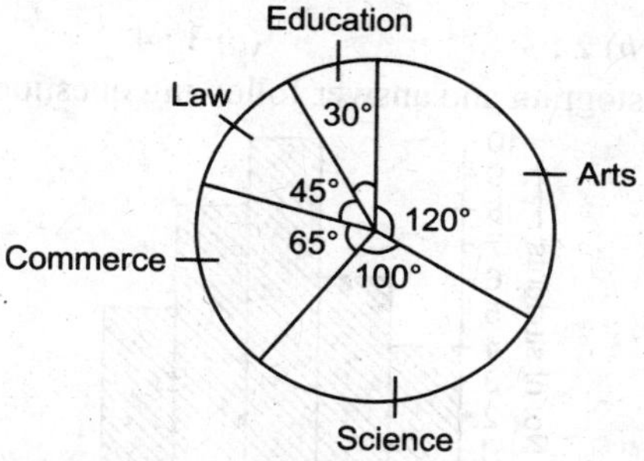

25. If 1000 students are admitted in science, what is the ratio of students in science and arts ?

(*a*) 7 : 5 (*b*) 7 : 6 (*c*) 5 : 6 (*d*) 6 : 5

26. If 1000 students are admitted in science, what is the total number of students ?

(*a*) 1800 (*b*) 3600 (*c*) 360 (*d*) 180

27. How many students are more in commerce than in law if 1000 students are in science ?

(*a*) 2000 (*b*) 500 (*c*) 20 (*d*) 200

Directions: Questions no. 28 to 31, study the two pie-charts and answer the questions.

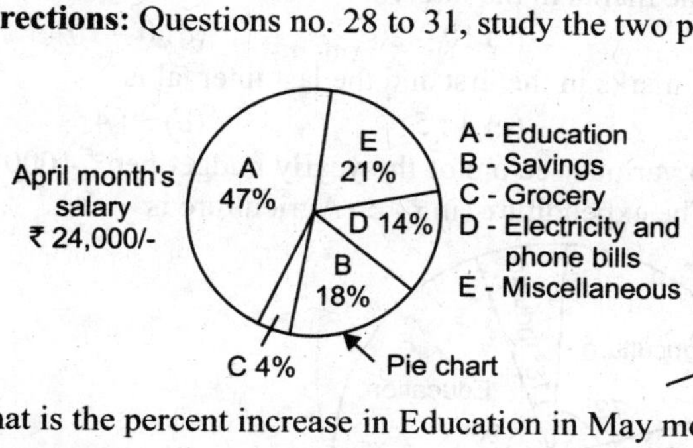

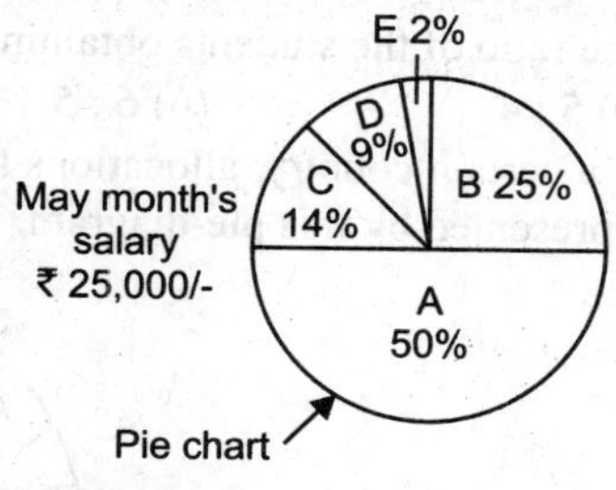

28. What is the percent increase in Education in May month than April month ?

(*a*) 12.35% (*b*) 20% (*c*) 10.82% (*d*) 9.56%

29. The ratio of amount spent for savings in April month's salary and miscellaneous in May month's salary is :

(*a*) 217 : 26 (*b*) 205 : 13 (*c*) 235 : 50 (*d*) 216 : 25

30. From the salary of May, the amount spent on Grocery and Electricity are:

(*a*) ₹ 960, ₹ 5040 (*b*) ₹ 3500, ₹2250 (*c*) ₹ 2160, ₹ 480 (*d*) ₹ 6250, ₹ 3360

31. The average amount spent on Education, Grocery and Saving from April month's salary is:
(*a*) ₹ 6000/- (*b*) ₹ 6325/- (*c*) ₹ 5520/- (*d*) ₹ 5800/-

SSC CGL TIER I 2013

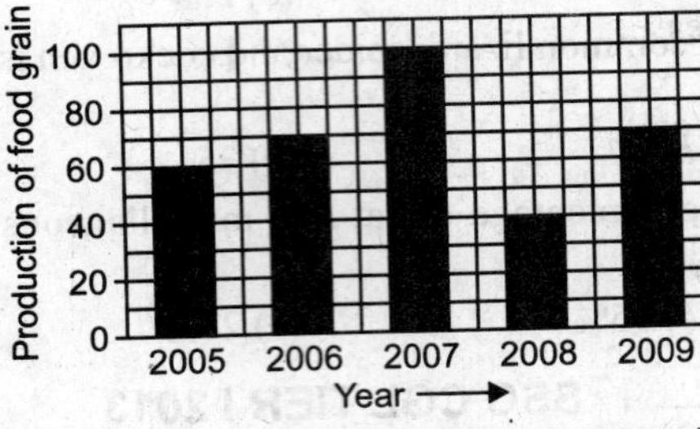

Study the above bar graph showing the production of food grains (in million tons).

32. What is the ratio between the maximum production and the minimum production during the given period ?
(*a*) 1 : 2 (*b*) 2 : 3 (*c*) 3 : 4 (*d*) 5 : 2

Study the following Histogram and answer following question nos. 33 to 36.

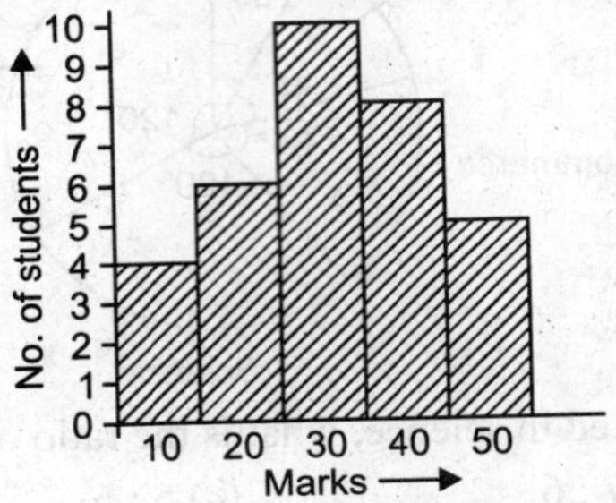

33. The total number of students involved in the data is
(*a*) 33 (*b*) 32 (*c*) 43 (*d*) 42

34. The maximum number of students got the marks in the interval of
(*a*) 10 – 20 (*b*) 20 – 30 (*c*) 30 – 40 (*d*) 40 – 50

35. The least number of students got the marks in the interval
(*a*) 40 – 50 (*b*) 20 – 30 (*c*) 10 – 20 (*d*) 0 – 10

36. The ratio of the students obtaining marks in the first and the last interval is
(*a*) 5 : 4 (*b*) 6 : 5 (*c*) 4 : 5 (*d*) 3 : 4

37. In a certain country, allocations to various sectors of the yearly budget per ₹ 1000 crores are represented by this pie-diagram. The expenditure (in ₹) on Agriculture is

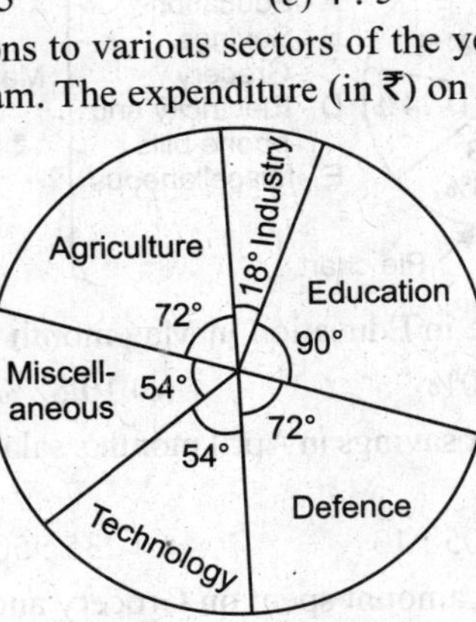

(*a*) 250 crores (*b*) 150 crores (*c*) 300 crores (*d*) 200 crores

SOLUTION

1. Sol. (*c*) Production was maximum in 1999 at 1500 quintals.

2. Sol. (*a*) Production in 1997 = 500, Production in 1998 = 1000.

Percentage increase = $\frac{500}{500} \times 100 = 100\%$

3. Sol. (*c*) % Decrease in 1999-2000 = $\frac{1500 - 1300}{1500} \times 100 = \frac{200}{1500} \times 100 = \frac{40}{3} = 13\frac{1}{3}\%$

4. Sol. (*d*) Total production from 1995 to 1998 = 1200 + 600 + 500 + 1000 = 3300.

5. Sol. (*a*) 30,000 + 10,000 + 40,000 = 80,000

6. Sol. (*b*) 50 : 30 = 5 : 3

7. Sol. (*c*) Reqd.% = $\frac{50}{20 + 20 + 30 + 50 + 50 + 30} \times 100 = \frac{50}{200} \times 100 = 25\%$

8. Sol. (*a*) 60,000 : 50,000 : 30,000 = 6 : 5 : 3

9. Sol. (*d*) Students passed in 2008 : 30,000 + 50,000 = 80,000

Students passed in 2004 : 20,000 + 30,000 = 50,000

Difference = 80,000 – 50,000 = 30,000

10. Sol. (*c*) In 2007 : (45 – 30)% of 35 lakhs = 5,25,000

In 2008 : (60 – 40)% of 44 lakhs = 8,80,000

Required difference = 8,80,000 – 5,25,000 = 3,55,000

11. Sol. (*d*) 30% of 44,00,000 = 13,20,000

12. Sol. (*a*) D type mobile produced in 2007 = (75 – 65)% of 35,00,000 = 3,50,000

D type mobile produced in 2008 = (85 – 75)% of 44,00,000 = 4,40,000

Total D type mobile = 3,50,000 + 4,40,000 = 7,90,000

Mobiles left unsold = 15% of 7,90,000 = 1,18,500

13. Sol. (*a*) A + B = 45%, E = (90 – 75)% = 15%

Required number of mobiles = (45 + 15)% of 35 lakhs = 21 lakhs

14. Sol. (*b*) In 2007, A = 30%, B = 15%, C = 20%, D = 10%, E = 15%, F = 10%

In 2008, A = 40%, B = 20%, C = 15%, D = 10%, E = 10%, F = 5%

% variation in A = $\frac{10}{30} \times 100 = 33\frac{1}{3}\%$

% variation in B = $\frac{5}{15} \times 100 = 33\frac{1}{3}\%$

15. Sol. (*c*) $30 : 12\frac{1}{2} = 30 \times 2 : 25 = 12 : 5$

16. Sol. (*a*) Expenditure on clothes = $12\frac{1}{2}\% = \frac{1}{8}$ of total expenditure = $\frac{1}{8} \times 50,000 =$ ₹ 6,250

17. Sol. (*a*) Expenditure on study and food = 25% + 30% = 55% = 55% of ₹ 35,000 = ₹ 19,250

18. Sol. (*c*) $\frac{10}{100} \times 360° = 36°$

19. Sol. (*b*) Difference = $\left(\frac{25}{2} - 10\right)\% = \frac{5}{2}\% \Rightarrow \frac{5}{2}\%$ of expenditure = ₹ 1500

$\Rightarrow$ Expenditure of house = $1500 \times \frac{2}{5} \times 100 \times \frac{10}{100} = 6,000$

20. Sol. (*c*) $x\% = \frac{22}{28} \times 100 = \frac{550}{7} = 78\frac{4}{7}\%$

21. Sol. (*a*) $(22 + 11 + 28)\% - 27\% = 34\%$

22. Sol. (*d*) $y\% = \frac{11}{27+28} \times 100 = 20\%$

23. Sol. (*b*) $(8 + 22 + 28)\% - 11\% = 47\%$

24. Sol. (*a*) $z\% = \frac{11-4}{28} \times 100 = 25\%$

25. Sol. (*c*) $100 : 120 = 5 : 6$

26. Sol. (*b*) $\frac{1000}{100} \times 360 = 3600$

27. Sol. (*d*) $100° = 1000 \Rightarrow 1° = 10$

Difference = $(65° - 45°) = 20° = 200$

28. Sol. (*c*) Education (April) = 47% of 24,000 = ₹ 11,280

Education (May) = 50% of 25,000 = ₹ 12,500

Required % = $\frac{1,220}{11,280} \times 100 = 10.82\%$

29. Sol. (*d*) $18 \times 24,000 : 2 \times 25,000 = 216 : 25$

30. Sol. (*b*) Grocery = 14% of ₹ 25,000 = ₹ 3,500

Electricity = 9% of ₹ 25,000 = ₹ 2,250

31. Sol. (*c*) Average = $\frac{47\% + 4\% + 18\%}{3}$ = 23% of ₹ 24,000 = ₹ 5,520

32. Sol. (*d*) $100 : 40 = 5 : 2$

33. Sol. (*a*) $4 + 6 + 10 + 8 + 5 = 33$

34. Sol. (*b*) 10 students

35. Sol. (*d*) 4 students

36. Sol. (*c*) 4 : 5

37. Sol. (*d*) $\frac{72}{360} \times 1000$ crores = 200 crores

CHAPTER

36 SOLVED SSC PREVIOUS PAPERS 2009–2015

SSC DATA ENTRY OPERATOR EXAM. 2009

1. Two numbers are in the ratio 3 : 4. The product of their H.C.F. and L.C.M. is 2028. The sum of the numbers is

(*a*) 68 (*b*) 72 (*c*) 86 (*d*) 91

2. In the solution of the multiplication question given below, '*a*' is a digit. The value of '*a*' is

$$\begin{array}{r} a\ \ 2 \\ \times\ \ 7\ \ a \\ \hline 6\ \ 3\ \ 9\ \ 6 \\ \hline \end{array}$$

(*a*) 3 (*b*) 8 (*c*) 6 (*d*) 4

3. $(256)^{0.16} \times (4)^{0.36}$ is equal to

(*a*) 64 (*b*) 16 (*c*) 256.25 (*d*) 4

4. $[1^3 + 2^3 + 3^3 + ... + 9^3 + 10^3]$ is equal to

(*a*) 3575 (*b*) 2525 (*c*) 5075 (*d*) 3025

5. The smallest number that must be subtracted from 1000 to make the resulting number a perfect square is

(*a*) 37 (*b*) 38 (*c*) 39 (*d*) 40

6. The sum and product of two numbers are 11 and 18 respectively. The sum of their reciprocals is

(*a*) $\frac{2}{11}$ (*b*) $\frac{11}{2}$ (*c*) $\frac{18}{11}$ (*d*) $\frac{11}{18}$

7. If $3^{x+y} = 81$ and $81^{x-y} = 3$, then the value of x is

(*a*) 42 (*b*) $\frac{15}{8}$ (*c*) $\frac{17}{8}$ (*d*) 39

8. Instead of multiplying a number by 0.72, a student multiplied it by 7.2. If his answer was 2592 more than the correct answer, then the original number was

(*a*) 400 (*b*) 420 (*c*) 500 (*d*) 560

9. 7,77,77,777 ÷ 77 equals

(*a*) 1111 (*b*) 101001 (*c*) 10101 (*d*) 1010101

10. A man ate 100 grapes in 5 days. Each day, he ate 6 more grapes than those he ate on the earlier day. How many grapes did he eat on the first day ?

(*a*) 8 (*b*) 12 (*c*) 54 (*d*) 76

11. The missing number in the sequence 5, 6, 15, __?__, 89, 170, 291 is

(*a*) 50 (*b*) 40 (*c*) 42 (*d*) 32

12. Numbers 1, 2, 3, 4, ..., 98, 99, 100 are multiplied together. The number of zeroes at the end of the product on the right will be equal to

(*a*) 24 (*b*) 22 (*c*) 21 (*d*) 11

13. A and B can do a piece of work in 12 days and 15 days respectively. They began to work together but A left after 4 days. In how many more days would B alone complete the remaining work?

(*a*) $\frac{20}{3}$ (*b*) $\frac{25}{3}$ (*c*) 6 (*d*) 5

14. Two pipes can fill a cistern separately in 10 hours and 15 hours. They can together fill the cistern in

(*a*) 6 hours (*b*) 7 hours (*c*) 8 hours (*d*) 9 hours

15. A man, a woman and a boy can complete a work in 20 days, 30 days and 60 days respectively. How many boys must assist 2 men and 8 women so as to complete the work in 2 days ?

(*a*) 8 (*b*) 12 (*c*) 4 (*d*) 6

16. The measures (in cm) of sides of a right angled triangle are given by consecutive integers. Its area (in cm^2) is

(*a*) 9 (*b*) 8 (*c*) 5 (*d*) 6

17. Lengths of the perpendiculars from a point in the interior of an equilateral triangle on its sides are 3 cm, 4 cm and 5 cm. Area of the triangle is

(*a*) $48\sqrt{3}$ cm^2 (*b*) $54\sqrt{3}$ cm^2 (*c*) $72\sqrt{3}$ cm^2 (*d*) $80\sqrt{3}$ cm^2

18. If a triangle with base 8 cm has the same area as a circle with radius 8 cm, then the corresponding altitude (in cm) of the triangle is

(*a*) 12π (*b*) 20π (*c*) 16π (*d*) 32π

19. The radius of the base and height of a right circular cone are in the ratio 5 : 12. If the volume of the cone is 314 cm^3, the slant height (in cm) of the cone will be

(*a*) 12 (*b*) 13 (*c*) 15 (*d*) 17

20. A solid metallic cone is melted and recast into a solid cylinder of the same base as that of the cone. If the height of the cylinder is 7 cm, the height of the cone was

(*a*) 20 cm (*b*) 21 cm (*c*) 28 cm (*d*) 24 cm

21. A single discount equivalent to three successive discounts of 20%, 25% and 10% is

(*a*) 55% (*b*) 50% (*c*) 48% (*d*) 46%

22. The marked price of an article is 20% more than its cost price. A discount of 20% is given on the marked price. In this kind of sale, the seller bears

(*a*) no gain, no loss (*b*) a loss of 4% (*c*) a gain of 4% (*d*) a gain of 8%

23. The marked price of a radio is ₹ 4,800. The shopkeeper allows a discount of 10% and gains 8%. If no discount is allowed, his gain per cent will be **(SSC CGL Tier I 2013)**

(*a*) 18 (*b*) 20 (*c*) 22 (*d*) 25

24. If A = $\frac{4}{5}$ of B and B = $\frac{5}{2}$ of C, then the ratio of A : C is

(*a*) 1 : 2 (*b*) 2 : 1 (*c*) 2 : 3 (*d*) 1 : 3

25. Which number when added to each of the numbers 6, 7, 15, 17 will make the resulting numbers proportional ?

(*a*) 6 (*b*) 5 (*c*) 4 (*d*) 3

26. The ratio of the first and second class fares between two railway stations is 4 : 1 and that of the number of passengers travelling by first and second classes is 1 : 40. If on a day ₹ 1,100 are collected as total fare, the amount collected from the first class passengers is

(SSC CGL Tier I 2014)

(*a*) ₹ 315 (*b*) ₹ 275 (*c*) ₹ 137.50 (*d*) ₹ 100

27. Annual incomes of A and B are in the ratio 4 : 3 and their annual expenses in the ratio 3 : 2. If each saves ₹ 60,000 at the end of the year, the annual income of A is

(*a*) ₹ 1,20,000 (*b*) ₹ 1,50,000 (*c*) ₹ 2,40,000 (*d*) ₹ 3,60,000

28. In one glass, milk and water are mixed in the ratio 3 : 5 and in another glass they are mixed in the ratio 6 : 1. In what ratio should the contents of the two glasses be mixed together so that the new mixture contains milk and water in the ratio 1 : 1?

(*a*) 20 : 7 (*b*) 8 : 3 (*c*) 27 : 4 (*d*) 25 : 9

29. The average of all odd numbers less than 100 is

(*a*) 49.5 (*b*) 50 (*c*) 50.5 (*d*) 51

30. The average marks of 32 boys of section A of class X is 60 whereas the average marks of 40 boys of section B of class X is 33. The average marks for both the sections combined together is

(*a*) 44 (*b*) 45 (*c*) $46\frac{1}{2}$ (*d*) $45\frac{1}{2}$

31. The average age of a husband and a wife was 27 years when they married 4 years ago. The average age of the husband, the wife and a new-born child is 21 years now. The present age of the child is

(*a*) 4 years (*b*) 3 years (*c*) 2 years (*d*) 1 year

32. The average age of eleven cricket players is 20 years. If the age of the coach is also included, the average age increases by 10%. The age of the coach is

(*a*) 48 years (*b*) 44 years (*c*) 40 years (*d*) 36 years

33. When an article is sold at a gain of 20%, it yields ₹ 60 more than when it is sold at a loss of 20%. The cost price of the article is

(*a*) ₹ 200 (*b*) ₹ 150 (*c*) ₹ 140 (*d*) ₹ 120

34. By selling 1 dozen ball pens, a shopkeeper earned the profit equal to the selling price of 4 ball pens. His profit percent is

(*a*) 50 (*b*) 40 (*c*) $33\frac{1}{3}$ (*d*) $31\frac{1}{4}$

35. A reduction of 25% in the price of sugar enables a man to buy $7\frac{1}{2}$ kg more sugar for ₹ 360. The original price per kg of sugar was

(*a*) ₹ 16 (*b*) ₹ 18 (*c*) ₹ 20 (*d*) ₹ 25

36. A man buys some articles at ₹ P per dozen and sells them at ₹ $\frac{P}{8}$ per piece. His profit percent is

(*a*) 30 (*b*) 40 (*c*) 50 (*d*) 60

37. A reduction of 10% in the price of a T.V. set brought down its price by ₹ 1,650. The original price of the set (in rupees) was

(*a*) 16,500 (*b*) 16,000 (*c*) 15,000 (*d*) 16,550

38. The salary of a person is decreased by 25% and then the decreased salary is increased by 25%. His new salary in comparison with his original salary is **(SSC CHSL 2013)**

(*a*) the same (*b*) 6.25% more (*c*) 6.25% less (*d*) 0.625% less

39. Two numbers x and y are respectively 20% and 50% more than a third number. x is how much per cent of y?

(*a*) 30 (*b*) 45 (*c*) 60 (*d*) 80

40. A man received ₹ 8,80,000 as his annual salary in the year 2007 which was 10% more than his annual salary in 2006. His annual salary in the year 2006 was

(*a*) ₹ 4,80,000 (*b*) ₹ 8,00,000 (*c*) ₹ 4,00,000 (*d*) ₹ 8,40,000

41. The numbers of boys and girls in a college are in the ratio of 3 : 2. If 20% of the boys and 25% of the girls are adults, the percentage of students, who are ***not*** adults, is

(A) 58 (*a*) $66\frac{2}{3}$ (*c*) 78 (*d*) $83\frac{1}{3}$

42. A jeep is chasing a car which is 5 km ahead. Their respective speeds are 90 km/hr and 75 km/hr. After how many minutes will the jeep catch the car ?

(*a*) 18 (*b*) 20 (*c*) 24 (*d*) 25

43. A train, with a uniform speed, crosses a platform, 162 metres long, in 18 seconds and another platform, 120 metres long, in 15 seconds. The speed of the train is

(*a*) 14 km/hr (*b*) 42 km/hr (*c*) 50.4 km/hr (*d*) 67.2 km/hr

44. In how many years will a sum of ₹ 3,000 yield a simple interest of ₹ 1,080 at 12% per annum?

(*a*) 3 (*b*) $2\frac{1}{2}$ (*c*) 2 (*d*) $3\frac{1}{2}$

45. The difference between the compound and simple interests on ₹ 1,000 for 2 years at the rate of 5% per annum is

(*a*) ₹ 2 (*b*) ₹ 2.50 (*c*) ₹ 3 (*d*) ₹ 3.50

46. What sum of money at compound interest will amount to ₹ 650 at the end of one year and ₹ 676 at the end of 2 years ?

(*a*) ₹ 520 (*b*) ₹ 572 (*c*) ₹ 600 (*d*) ₹ 625

Directions: For Q. No. *47 to 50.*

The pie-chart given here shows expenditures incurred by a family on various items and their savings, which amount to ₹ 8,000 in a month.

Study the chart and answer the questions number 47-50 based on the pie-chart.

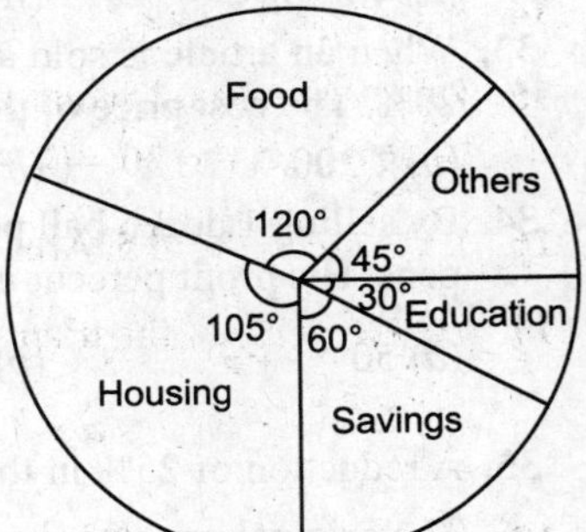

47. What is the total expenditure of the family for the month ?

(*a*) ₹ 40,000 (*b*) ₹ 48,000
(*c*) ₹ 45,000 (*d*) ₹ 50,000

48. How much more amount is spent on food than on housing ?

(*a*) ₹ 1,000 (*b*) ₹ 3,000 (*c*) ₹ 2,000 (*d*) ₹ 2,500

49. How much expenditure is incurred on education ?

(*a*) ₹ 3,000 (*b*) ₹ 5,000 (*c*) ₹ 4,000 (*d*) ₹ 7,000

50. The ratio of the expenditure on food to the savings is

(*a*) 3 : 2 (*b*) 2 : 1 (*c*) 4 : 3 (*d*) 3 : 4

ANSWER

1. (*d*)	**2.** (*b*)	**3.** (*d*)	**4.** (*d*)	**5.** (*c*)	**6.** (*d*)	**7.** (*c*)	**8.** (*a*)
9. (*d*)	**10.** (*a*)	**11.** (*b*)	**12.** (*a*)	**13.** (*c*)	**14.** (*a*)	**15.** (*a*)	**16.** (*d*)
17. (*a*)	**18.** (*c*)	**19.** (*b*)	**20.** (*b*)	**21.** (*d*)	**22.** (*b*)	**23.** (*b*)	**24.** (*b*)
25. (*d*)	**26.** (*d*)	**27.** (*c*)	**28.** (*a*)	**29.** (*b*)	**30.** (*b*)	**31.** (*d*)	**32.** (*b*)
33. (*b*)	**34.** (*a*)	**35.** (*a*)	**36.** (*c*)	**37.** (*a*)	**38.** (*c*)	**39.** (*d*)	**40.** (*b*)
41. (*c*)	**42.** (*b*)	**43.** (*c*)	**44.** (*a*)	**45.** (*b*)	**46.** (*d*)	**47.** (*a*)	**48.** (*c*)
49. (*c*)	**50.** (*b*)						

HINTS

1. $(3x)(4x) = 2028 \Rightarrow x = \sqrt{169} = 13 \Rightarrow$ Sum $= 13(3+4) = 91$

2. $2 \times a = 6 \Rightarrow a = 3$ or 8. $a \times 7$ + carry over $= 63 \Rightarrow a = 8$.

3. $= (4^4)^{0.16} \times (4)^{0.36} = 4^{0.64+0.36} = 4^1 = 4$

4. $= \left[\dfrac{n(n+1)}{2}\right]^2 = \left(\dfrac{10\times11}{2}\right)^2 = 55^2 = 3025$

5. $31^2 = 961$, $\quad 1000 - 961 = 39$

6. $\dfrac{\text{Sum}}{\text{Product}} = \dfrac{11}{18}$

7. $3^{x+y} = 3^4$ and $3^{4(x-y)} = 3^1 \Rightarrow x + y = 4$ and $x - y = 1/4 \Rightarrow x = \dfrac{1}{2}\left(4 + \dfrac{1}{4}\right) = 2 + \dfrac{1}{8} = \dfrac{17}{8}$

8. $(7.2 - 0.72)x = 2592 \Rightarrow 0.72 \times 9 \times x = 2592 \Rightarrow x = 400$

10. Average (on 3rd day) $100 \div 5 = 20$; $\quad$ First day $= 20 - 2 \times 6 = 8$

11. Difference = squares of 1, 3, 5, 7, 9, 11

12. Multiples of $5 = \dfrac{100}{5} + \dfrac{100}{5^2} = 20 + 4 = 24$

13. A (4 days) $= \dfrac{4}{12} = \dfrac{1}{3}$, Remaining $= \dfrac{2}{3}$, B $= \dfrac{2}{3} \times 15 = 10$ days; B alone $= 10 - 4 = 6$ days

14. $\dfrac{10\times15}{25} = 6$ hours

15. 1 man ≡ 3 boys, 1 woman ≡ 2 boys. $\quad$ 30 boys can do the work in 2 days.
Boys required $= 30 - (2 \times 3) - (8 \times 2) = 8$ boys

16. Sides are 3, 4 and 5. Area $= \dfrac{1}{2} \times 3 \times 4 = 6$

17. Let each side of the triangle be a cm. Then $\Delta ABC = \Delta AOB + \Delta AOC + \Delta BOC$

$\Rightarrow \dfrac{\sqrt{3}}{4}a^2 = \dfrac{1}{2} \times a \times 3 + \dfrac{1}{2} \times a \times 4 + \dfrac{1}{2} \times a \times 5 = 6a \Rightarrow a = 8\sqrt{3} \Rightarrow$ Area $= 6a = 48\sqrt{3}$

18. Area of circle $= \pi(8)^2 = 64\pi = \dfrac{1}{2} \times 8 \times h \Rightarrow h = 64\pi \times 2 \div 8 = 16\pi$

19. Volume of cone $= \dfrac{1}{3}\pi r^2 h = \dfrac{1}{3} \times 3.14 \times 5^2 \times 12 = 314 \Rightarrow l = \sqrt{5^2 + 12^2} = 13$

20. Height of cone $= 3 \times 7 = 21$ c.m.

21. $20 + 25 - \dfrac{20\times25}{100} = 40$; $\quad 40 + 10 - \dfrac{40\times10}{100} = 46$

22. $\dfrac{20\times20}{100} = 4\%$

23. 90% of MP = 108% of CP $\Rightarrow$ 100% of MP $= 108 \times \dfrac{100}{90} = 120\%$ of CP

24. $\dfrac{A}{C} = \dfrac{4}{5} \times \dfrac{5}{2} = \dfrac{2}{1}$

25. $\dfrac{7\times15 - 6\times17}{(6+17)-(7+15)} = 3$

26. Ratio of fares $= (4 \times 1) : (1 \times 40) = 1 : 10$; $\quad \dfrac{1}{11} \times 1100 = 100$

27. Both saves (4 – 3) ratio = (3 – 2) ratio = 1 ratio = ₹ 60,000 ⇒ A's income = 4 ratio = ₹ 2,40,000

28. $\left(\frac{6}{7}-\frac{1}{2}\right):\left(\frac{1}{2}-\frac{3}{8}\right)=\frac{5}{14}:\frac{1}{8}=20:7$

29. 99 = 50 × 2 – 1 ⇒ Number of terms = 50.

Average = Average of 25 and 26th terms = $\frac{49+51}{2}=50$ Alt : $\frac{1+99}{2}=50$

30. Assumed mean = 30. Excess = 32 × 30 + 40 × 3 = 1080, Excess average = 1080 ÷ 72 = 15 ⇒ Actual mean = 30 + 15 = 45

31. Husband and wife (total present age) = 2 × (27 + 4) = 62 years
Total present age of all = 3 × 21 = 63 years ⇒ Child age = 63 – 62 = 1 year

32. New average = 20 + 2 = 22 years ⇒ Age of coach = 22 + (2 × 11) = 44 years

33. 40% of CP = ₹ 60 ⇒ CP = $60\times\frac{100}{40}$ = ₹ 150

34. Profit = $\frac{4}{12-4}\times100=50\%$

35. Savings = 360 × 25% = ₹ 90. Reduced price = 90 ÷ $7\frac{1}{2}$ = ₹ 12. Original price = $12\times\frac{100}{75}=16$

36. CP = $\frac{P}{12}$, SP = $\frac{P}{8}$, $\frac{SP}{CP}=\frac{12}{8}=\frac{3}{2}$, Profit = $\frac{1}{2}=50\%$

37. 10% of price = ₹ 1650 ⇒ Price = ₹ 16,500

38. $\frac{25\times25}{100}=6.25\%$ less

39. $\frac{120}{150}\times100=80\%$

40. $8{,}80{,}000\times\frac{100}{110}=8{,}00{,}000$

41. Adult = $\frac{3}{5}\times20\%+\frac{2}{5}\times25\%=22\%$; Non-adult = 100% – 22% = 78%

42. $\frac{5}{15}\times60=20$ minutes

43. Difference in distance = 162 – 120 = 42 m; Difference in times = 18 – 15 = 3 seconds

Speed = $\frac{42}{3}\times\frac{18}{5}=50.4$ km/hr

44. $\frac{1080\times100}{3000\times12}=3$ years

45. $1000\times\frac{1}{20}\times\frac{1}{20}$ = ₹ 2.50

46. $650\times\frac{650}{676}$ = ₹ 625

47. 60° = ₹ 8,000 ⇒ 300° = ₹ 8,000 × 5 = ₹ 40,000

48. 120° – 105° = 15° = 8000 ÷ 4 = ₹ 2,000

49. 30° = 8000 ÷ 2 = ₹ 4,000

50. 120 : 60 = 2 : 1

SSC SECTION OFFICER (CMM. AUDIT), 2009
ARITHMETIC ABILITY

1. $\left(\frac{1}{1\times3}+\frac{1}{3\times5}+\frac{1}{5\times7}+\frac{1}{7\times9}+\ldots+\frac{1}{19\times21}\right)$ is equal to

(a) $\frac{20}{21}$ (b) $\frac{10}{21}$ (c) $\frac{21}{40}$ (d) $\frac{21}{20}$

2. $\left(\sqrt{7+4\sqrt{3}}-\sqrt{7-4\sqrt{3}}\right)$ is equal to

(a) 4 (b) 2 (c) $4\sqrt{3}$ (d) $2\sqrt{3}$

3. If the cube of a number equals the difference of the squares of 255 and 195, then the number is

(a) 20 (b) 25 (c) 30 (d) 40

4. One's digit of the number $(17)^{34}$ is

(a) 1 (b) 3 (c) 7 (d) 9

5. The square root of a natural number is equal to k times its cube root. The number must be

(a) k^{-12} (b) k^6 (c) k^{12} (d) k^{-6}

6. The missing number of the sequence 1, 2, 5, 10, 17, ?, 37 is

(a) 27 (b) 26 (c) 25 (d) 24

7. The sum $\left(\frac{1}{2}+\frac{1}{6}+\frac{1}{12}+\frac{1}{20}+\frac{1}{30}+\frac{1}{42}+\frac{1}{56}+\frac{1}{72}+\frac{1}{90}+\frac{1}{110}\right)$ is equal to

(a) $\frac{5}{6}$ (b) $\frac{10}{11}$ (c) $\frac{9}{10}$ (d) $1\frac{1}{11}$

8. The sum of all the numbers, formed by using all the four digits 1, 2, 3, 4 only once, is

(a) 66666 (b) 66662 (c) 66661 (d) 66660

9. The smallest number among 5^{20}, 4^{30}, 3^{40} and 2^{50} is

(a) 5^{20} (b) 4^{30} (c) 3^{40} (d) 2^{50}

10. The number $(10^{25}-7)$ is divisible by

(a) 3 (b) 7 (c) 11 (d) 13

11. If $a = 2.3$, $b = 3.2$ and $c = -5.5$, then the value of $a^3 + b^3 + c^3 - 3abc$ is

(a) 11 (b) 10 (c) 3 (d) 0

12. The greatest number among, $\sqrt[3]{2}$, $\sqrt[5]{3}$, $\sqrt[7]{4}$ and $\sqrt[8]{9}$ is

(a) $\sqrt[3]{2}$ (b) $\sqrt[5]{3}$ (c) $\sqrt[7]{4}$ (d) $\sqrt[8]{9}$

13. The sum '$0.\overline{25}+0.\overline{75}$' (where $0.\overline{25}$ denotes the recurring decimal 0.252525...) equals

(a) 1 (b) $1.\overline{01}$ (c) $1.0\overline{1}$ (d) $1.0\overline{01}$

14. A whole number N, when divided by 4 leaves remainder 3; and the quotient thus obtained leaves remainder 2 when divided by 3. The remainder, when N is divided by 6, will be

(a) 2 (b) 3 (c) 4 (d) 5

15. If B's salary is 30% less than that of A, then A's salary is more than that of B by

(a) 30% (b) $42\frac{6}{7}$% (c) $23\frac{1}{13}$% (d) 35%

16. When the price of rice increases by $12\frac{1}{2}$ % a man can get 250 g less rice for ₹ 18. The present cost of rice per kg is

(*a*) ₹ 8 (*b*) ₹ 9 (*c*) ₹ 16 (*d*) ₹ 18

17. The population of a city increases by 10% in every 5 years. If the population of the city in the year 1990 was 1,34,310, the population in the year 1980 was

(*a*) 1,00,000 (*b*) 1,10,000 (*c*) 1,11,000 (*d*) 1,21,000

18. The ratio of bus fare to train fare from Kolkata to a certain place is 3 : 4. If the train fare is increased by 20% and bus fare by 10%, then the new ratio of bus fare to train fare will be

(*a*) 11 : 16 (*b*) 11 : 8 (*c*) 11 : 32 (*d*) 11 : 4

19. A solution of salt and water contains 12% salt by weight. After evaporation of 25 kg of water from the solution, the concentration of salt is 22%. What is the original quantity of the solution?

(*a*) 55 kg (*b*) 47 kg (*c*) 37 kg (*d*) 34 kg

20. The average of all odd numbers upto 50 is

(*a*) 24 (*b*) 50 (*c*) 49 (*d*) 25

21. If $x^2 + y^2 = 41$, and $xy = 20$, $x > 0$, $y > 0$, then the average of x and y is

(*a*) 10 (*b*) 5 (*c*) 4.5 (*d*) 3.5

22. Four years ago, the average age of a family consisting of a father, a mother and a son was 28 years. A daughter now joins the family and the average age of the family is $24\frac{1}{2}$ years now. The present age of the daughter is

(*a*) 3 year (*b*) 2.5 years (*c*) 2 years (*d*) 1 year

23. In a family, average ages of father and son, father and mother, and mother and son are 25, 40 and 24 years respectively. The age of the son is

(*a*) 9 years (*b*) $9\frac{1}{2}$ years (*c*) 10 years (*d*) $10\frac{1}{2}$ years

24. There are 50 students in a class. One of them weighing 51 kg goes away and a new student joins. By this, the average weight of the class increases by $\frac{1}{2}$ kg. The weight of the new student is

(*a*) 73 kg (*b*) 76 kg (*c*) 74 kg (*d*) 75 kg

25. A man takes 4 hours 30 minutes to walk a certain distance and riding back. He could walk both ways in 8 hours 20 minutes. The time he will take to ride both ways is

(*a*) 20 minutes (*b*) 30 minutes (*c*) 40 minutes (*d*) 1 hour

26. 40 minutes more is required to cover a certain distance at a speed of 6 km/hr than covering the same distance at a speed of 8 km/hr. The distance covered in km is

(*a*) 8 (*b*) 12 (*c*) 14 (*d*) 16

27. A shopkeeper professes to sell his articles at discount of 10% but increases the cost price of each article by 20%. His gain on each article is

(*a*) 8% (*b*) 10% (*c*) 12% (*d*) 20%

28. A single discount equivalent to successive discounts of 10%, 10% and 10% is

(*a*) 30% (*b*) 10% (*c*) 72.9% (*d*) 27.1%

29. What should be the least number of years in which the simple interest on ₹ 2,600 at $6\frac{2}{3}$% per annum will be an exact number of rupees ?

(*a*) 6 (*b*) 5 (*c*) 3 (*d*) 2

30. The compound interest on ₹ 2,400 at the rate of 8% per annum for 6 months, when the interest is to be calculated half-yearly, will be

(*a*) ₹ 192 (*b*) ₹ 196 (*c*) ₹ 96 (*d*) ₹ 48

31. A and B start a business in partnership. A invests ₹ 500 for 9 months and B his capital for 6 months. If the total profit be ₹ 69 and B got ₹ 46 as his share, then the invested capital of B was

(*a*) ₹ 1,000 (*b*) ₹ 1,500 (*c*) ₹ 2,000 (*d*) ₹ 2,500

32. In a partnership business, A invests $\frac{1}{4}$ of the capital for $\frac{1}{3}$ of the time, B invests $\frac{1}{6}$ of the capital for $\frac{1}{4}$ of the time and C, the rest of the capital for the whole time. Out of a profit of ₹ 9,962, A's share is **(SSC CGL Tier I 2012 M, SSC CHSL 2011 M)**

(*a*) ₹ 1,172 (*b*) ₹ 1,586 (*c*) ₹ 2,804 (*d*) ₹ 4,102

33. If $2x = 3y = 4z$, then $x : y : z$ is

(*a*) 4 : 3 : 2 (*b*) 6 : 4 : 3 (*c*) 9 : 4 : 3 (*d*) 6 : 2 : 1

34. In what ratio, must alcohol of 30% strength be mixed with that of 50% strength to get a mixture of alcohol of 35% strength ?

(*a*) 2 : 1 (*b*) 3 : 1 (*c*) 3 : 3.5 (*d*) 3 : 5

35. The ratio of A to B is 2 : 3, that of B to C is 4 : 5, and of C to D is 6 : 7. The ratio A : B : C : D is

(*a*) 16 : 24 : 30 : 35 (*b*) 16 : 24 : 50 : 70 (*c*) 16 : 12 : 30 : 35 (*d*) 8 : 24 : 30 : 35

36. ₹ 6,800 have been divided among A, B, C such that A gets $\frac{2}{3}$ of what B gets and B gets $\frac{1}{4}$ of what C gets. B's share is

(*a*) ₹ 1,600 (*b*) ₹ 1,400 (*c*) ₹ 1,200 (*d*) ₹ 1,000

37. Three spheres of radii 6 cm, 8 cm, 10 cm are melted to form a single solid sphere. The radius of the sphere so formed is

(*a*) 10 cm (*b*) 11 cm (*c*) 12 cm (*d*) 13 cm

38. The area of a circle inscribed in a square is 154 cm². The area of the square is $\left(\text{Take } \pi = \frac{22}{7}\right)$

(*a*) 216 cm² (*b*) 205 cm² (*c*) 196 cm² (*d*) 184 cm²

39. Two parallel chords in a circle are 6 cm and 8 cm long and 1 cm apart. The length of the radius of the circle is

(*a*) 8 cm (*b*) 7 cm (*c*) 6 cm (*d*) 5 cm

40. The lengths of the sides of a triangle are 10 cm, 26 cm and 24 cm. The length of the shortest altitude of the triangle is

(*a*) $4\frac{8}{13}$ (*b*) $9\frac{3}{13}$ (*c*) 10 cm (*d*) 12 cm

41. If the length of the diameter of the base of a right circular cone is 42 cm and its volume is 12936 cm³, the length of its slant height (in cm) will be $\left(\text{Use } \pi = \frac{22}{7}\right)$

(*a*) 30.00 (*b*) 42.60 (*c*) 67.48 (*d*) 35.00

The table given below shows the number of students who appeared and qualified in a Selection Test for admission to the First year classes in Arts and Science streams of a college in the years 2000 to 2003.

Year	Arts Stream		Science Stream	
	Appeared	Percent qualified	Appeared	Percent qualified
2000	850	24	560	22
2001	1020	25	835	40
2002	960	40	720	35
2003	765	30	1085	20

Study the table and answer questions number 42 to 45 based on it.

42. What was the percent increase in the number of students appearing in Arts stream from the year 2000 to the year 2001 ?
(*a*) 1% (*b*) 12% (*c*) 17% (*d*) 20%

43. Total number of students declared qualified in both the streams for the year 2001 was
(*a*) 472 (*b*) 512 (*c*) 589 (*d*) 639

44. The number of students, who failed to qualify in Science stream in 2003, was
(*a*) 434 (*b*) 651 (*c*) 868 (*d*) 967

45. In the year 2002, how many more students qualified in Arts stream than those who qualified in Science stream ?
(*a*) 32 (*b*) 132 (*c*) 252 (*d*) 384

46. When the price of a toy was increased by 30%, the number of toys sold fell by 30%. The net sale of the shop
(*a*) increased by 9% (*b*) decreased by 9% (*c*) increased by 7% (*d*) decreased by 7%

47. A bicycle is sold at a gain of 16%. Had it been sold for ₹ 100 more, the gain would have been 20%. The cost price of the bicycle is
(*a*) ₹ 1,500 (*b*) ₹ 2,000 (*c*) ₹ 2,050 (*d*) ₹ 2,500

48. The difference between two selling prices of a shirt at profits of 5% and 4% is ₹ 6. The cost price of the shirt is
(*a*) ₹ 800 (*b*) ₹ 600 (*c*) ₹ 500 (*d*) ₹ 300

49. If even after giving a discount of 10% on marked price, a profit of 10% is to be made, then the marked price should exceed the cost price by
(*a*) $33\frac{1}{3}\%$ (*b*) $22\frac{2}{9}\%$ (*c*) 21% (d) 20%

50. A man sold two articles for ₹ 4,000 each, neither losing nor gaining in the deal. If he had sold one article at a gain of 25%, then the other article must have been sold at a loss of
(*a*) 20% (*b*) $14\frac{2}{7}\%$ (*c*) $16\frac{2}{3}\%$ (*d*) 25%

ANSWER

1. (*b*)	**2.** (*d*)	**3.** (*c*)	**4.** (*d*)	**5.** (*b*)	**6.** (*b*)	**7.** (*b*)	**8.** (*d*)
9. (*a*)	**10.** (*a*)	**11.** (*d*)	**12.** (*d*)	**13.** (*b*)	**14.** (*d*)	**15.** (*b*)	**16.** (*b*)
17. (*c*)	**18.** (*a*)	**19.** (*a*)	**20.** (*d*)	**21.** (*c*)	**22.** (*c*)	**23.** (*a*)	**24.** (*b*)
25. (*c*)	**26.** (*d*)	**27.** (*a*)	**28.** (*d*)	**29.** (*c*)	**30.** (*c*)	**31.** (*b*)	**32.** (*a*)
33. (*b*)	**34.** (*b*)	**35.** (*a*)	**36.** (*c*)	**37.** (*c*)	**38.** (*c*)	**39.** (*d*)	**40.** (*b*)
41. (*d*)	**42.** (*d*)	**43.** (*c*)	**44.** (*c*)	**45.** (*b*)	**46.** (*b*)	**47.** (*d*)	**48.** (*b*)
49. (*b*)	**50.** (*c*)						

HINTS

1. $\frac{1}{2}\left[\left(1-\frac{1}{3}\right)+\left(\frac{1}{3}-\frac{1}{5}\right)+\left(\frac{1}{5}-\frac{1}{7}\right)+\ldots\ldots\ldots\ldots+\left(\frac{1}{19}-\frac{1}{21}\right)\right]=\frac{10}{21}$
2. $\left(2+\sqrt{3}\right)-\left(2-\sqrt{3}\right)=2\sqrt{3}$
3. $255^2 - 195^2 = 450 \times 60 = 27000 = 30^3$
4. $7^2 = 49$
5. $x^{1/2} = kx^{1/3} \Rightarrow x^{1/2-1/3} = x^{1/6} = k \Rightarrow x = k^6$
6. Difference = 1, 3, 5, 7, 9, 11 $\Rightarrow$? = 17 + 9 = 26
7. $\left(1-\frac{1}{2}\right)+\left(\frac{1}{2}-\frac{1}{3}\right)\left(\frac{1}{3}-\frac{1}{4}\right)\ldots\ldots\ldots\left(\frac{1}{10}-\frac{1}{11}\right)=\frac{10}{11}$
8. Total numbers possible = 4! = 4 × 3 × 2 × 1 = 24.
   Each digit at each place = 24 ÷ 4 = 6 times
   Sum of digits = 1 + 2 + 3 + 4 = 10
   Sum of digits at each place = 6 × 10 = 60
   Sum of all the numbers = 60 | 60 | 60 | 60 = 66660
9. $(5^2)^{10}, (4^3)^{10}, (3^4)^{10}, (2^5)^{10} = 25^{10}, 64^{10}, 81^{10}, 32^{10}$,
10. Number is 999....3 which is divisible by 3.
11. $a + b + c = 0$
12. $\sqrt[3]{2} > \sqrt[5]{3}$ as $\sqrt[15]{2^5} > \sqrt[15]{3^3}$; $\sqrt[3]{2} > \sqrt[7]{4}$ as $\sqrt[21]{2^7} > \sqrt[21]{4^3}$; $\sqrt[3]{2} < \sqrt[8]{9}$ as $\sqrt[24]{2^8} < \sqrt[24]{9^3}$
13. $\frac{25}{99}+\frac{75}{99}=\frac{100}{99}=1+\frac{1}{99}=1.\overline{01}$
14. Remainder on dividing the number by 4 × 3 or 12 = $D_1 \times R_2 + R_1$ = 4 × 2 + 3 = 11
    Remainder on dividing the number by 6 = 11 – 6 = 5
15. $\frac{30}{100-30}=\frac{3}{7}=42\frac{6}{7}\%$ or $\frac{3}{10-3}=\frac{3}{7}=42\frac{6}{7}\%$
16. $12\frac{1}{2}\%=\frac{1}{8}$, 250 g = $\frac{1}{4}$ kg; Present price = $18\times\frac{1}{8}\times4$ = ₹ 9 per kg.
17. $1{,}34{,}310 \times \left(\frac{10}{11}\right)^2 = 1{,}11{,}000$
18. (3 + 10%) : (4 + 20%) = 33 : 48 = 11 : 16
19. Ratio of salt = 12 : 22 = 6 : 11 $\Rightarrow$ Ratio of solution = 11 : 6
    Difference = (11 – 6) ratio = 5 ratio = 25 kg $\Rightarrow$ Original solution = 11 ratio = 55 kg.
20. Number of terms = 25 $\Rightarrow$ Average = Mid-term = 13th term = 13 × 2 – 1 = 25
21. $(x + y)^2 = x^2 + y^2 + 2xy = 41 + 2 \times 20 = 81 = 9^2 \Rightarrow x + y = 9 \Rightarrow$ Average = 9 ÷ 2 = 4.5
22. Present total age of father, mother and son = 3 × 32 = 96 years
    Present total age of all = $4 \times 24\frac{1}{2}$ = 98 years $\Rightarrow$ Daughter's age = 98 – 96 = 2 years

Alternative method:

$$24\frac{1}{2} - 3 \times \left(32 - 24\frac{1}{2}\right) = 2 \text{ years}$$

23. Father + Son = 50....(1) Father + Mother = 80....(2) Mother + Son = 48......(3)

(1) + (3) – (2) ⇒ 2 × son = 50 + 48 – 80 = 18 ⇒ Son = 9 years

24. Increase = $\frac{1}{2} \times 50 = 25$ kg. New weight = 51 + 25 = 76 kg.

25. 2 × (Walk + Ride) = 2 × (4 hours 30 minutes) = 9 hours

2 × Ride = 2 × (Walk + Ride) – 2 × Walk = 9 hours – 8 hours 20 minutes = 40 minutes

26. $\frac{8\times6}{2}\times\frac{2}{3} = 16$ km

27. $20 - 10 - \frac{20\times10}{100} = 8\%$ or 120 – 10% = 108

28. 10 + 10 – 1 = 19 19 + 10 – 1.9 = 27.1

29. $2600 \times \frac{2}{3}$ (Denominator is 3)

30. 2400 × 4% = ₹ 96

31. Ratio of capitals / profits = 23 : 46 = 1 : 2 ⇒ B's capital = $\frac{500\times9\times2}{6}$ = ₹ 1500

32. C's capital = $1-\left(\frac{1}{4}+\frac{1}{6}\right)=\frac{7}{12}$ ⇒ Ratio of profits = $\frac{1}{4}\times\frac{1}{3}:\frac{1}{6}\times\frac{1}{4}:\frac{7}{12}\times1 = 2:1:14$

A's share = $\frac{2}{17} \times 9962$ = ₹ 1172

33. $2x = 3y = 4z = 12$ when $x = 6, y = 4, z = 3$

$\therefore\ x : y : z = 6 : 4 : 3$

34.
30 50
35
15 5

Required ratio = 15 : 5 = 3 : 1

35. A : B : C = 8 : 12 : 15 C : D = 6 : 7

A : B; C : D = 16: 24 : 30 : 35

36. A : B = 2 : 3 and B : C = 1 : 4 ⇒ A : B : C = 2 : 3 : 12

⇒ B's share = $\frac{3}{17} \times 6800$ = ₹ 1200

37. $6^3 + 8^3 + 10^3 = 216 + 512 + 1000 = 1728 = 12^3$

38. Ratio of areas of square and circle = 14 : 11

⇒ Area of square = $\frac{14}{11} \times 154 = 196$

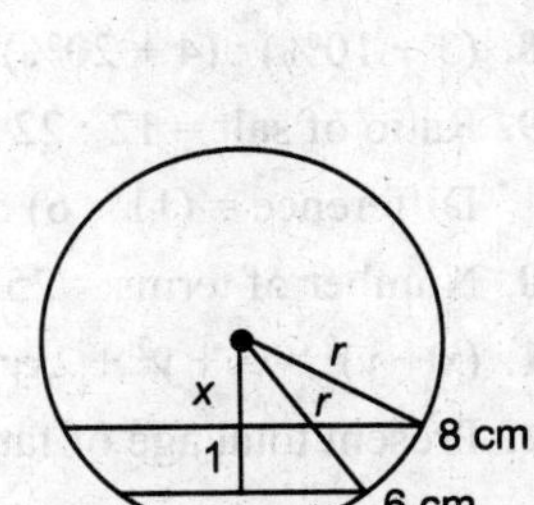

39. $r^2 = x^2 + 4^2$ and $r^2 = (x + 1)^2 + 3^2$

$\Rightarrow x^2 + 4^2 = (x + 1)^2 + 3^2 \Rightarrow x = 3$

$\Rightarrow r^2 = 3^2 + 4^2 = 5^2 \Rightarrow r = 5$

40. $S = \frac{1}{2} \times (10 + 26 + 24) = 30 \Rightarrow \text{Area} = \sqrt{30\times20\times4\times6} = 120$

Shortest altitude $= \frac{120\times2}{26} = 9\frac{3}{13}$

Alt: $5^2 + 12^2 = 13^2 \Rightarrow 10^2 + 24^2 = 26^2$

$\therefore$ the triangle is right angle.

Also area of traiangle $= \frac{1}{2}\times10\times24 = 120.$

41. Volume $= \frac{1}{3}\pi r^2 h = 12936 \Rightarrow h = 12936 \times 3 \times \frac{7}{22}\times\frac{1}{21}\times\frac{1}{21} = 28$

$l = \sqrt{21^2 + 28^2} = 35$ (Note: $21 = 3 \times 7$, $28 = 4 \times 7$, $35 = 5 \times 7$ and $3^2 + 4^2 = 5^2$)

42. $\frac{170}{850} \times 100 = 20\%$

43. 25% of 1020 + 40% of 835 = 255 + 334 = 589

44. 80% of 1085 = 868

45. 40% of 960 – 35% of 720 = 384 – 252 = 132

46. Decrease $= \frac{30\times30}{100} = 9\%$

47. 4% of cost price = ₹ 100 ⇒ Cost price = ₹ 2,500

48. 1% of cost price = ₹ 6 ⇒ Cost price = ₹ 600

49. Selling price = 110 ⇒ Marked price $= 110\times\frac{100}{90} = 122\frac{2}{9}$

50. Profit on first article $= \frac{1}{4}$ of CP $= \frac{1}{5}$ of SP

⇒ Loss on second article $= \frac{1}{5}$ of SP $= \frac{1}{6}$ of CP

SSC Combined Graduate Level (Tier-I) 2010
ARITHMETIC ABILITY (HELD ON 16.5.2010 : SECOND SHIFT)

1. $(1^2 + 2^2 + 3^2 + \ldots\ldots\ldots\ldots = 10^2)$ is equal to
   (*a*) 380 (*b*) 385 (*c*) 390 (*d*) 392
2. The sixth term of the sequence 2, 6, 11, 17, ... is
   (*a*) 24 (*b*) 30 (*c*) 32 (*d*) 36
3. Two numbers are in the ratio 7 : 11. If 7 is added to each of the numbers, the ratio becomes 2 : 3. The smaller number is
   (*a*) 39 (*b*) 49 (*c*) 66 (*d*) 77
4. $\left(1-\frac{1}{3}\right)\left(1-\frac{1}{4}\right)\left(1-\frac{1}{5}\right)\ldots\ldots\ldots\ldots\left(1-\frac{1}{25}\right)$ is equal to
   (*a*) $\frac{2}{25}$ (*b*) $\frac{1}{25}$ (*c*) $1\frac{19}{25}$ (*d*) $\frac{1}{325}$
5. A number, when divided by 136, leaves remainder 36. If the same number is divided by 17, the remainder will be
   (*a*) 9 (*b*) 7 (*c*) 3 (*d*) 2
6. Simplified form of $\left[\left(\sqrt[5]{x^{-3/5}}\right)^{-5/3}\right]^5$
   (*a*) x^5 (*b*) x^{-5} (*c*) x (*d*) $\frac{1}{x}$
7. A 4-digit number is formed by repeating a 2-digit number such as 1515, 3737 etc. Any number of this form is exactly divisible by
   (*a*) 7 (*b*) 11 (*c*) 13 (*d*) 101
8. $(0.1 \times 0.01 \times 0.001 \times 10^7)$ is equal to
   (*a*) 100 (*b*) $\frac{1}{10}$ (*c*) $\frac{1}{100}$ (*d*) 10
9. If $2p + \frac{1}{p} = 4$, the value of $p^3 + \frac{1}{8p^3}$ is
   (*a*) 4 (*b*) 5 (*c*) 8 (*d*) 15
10. If p and q represent digits, what is the possible maximum value of q in the statement $5p9 + 327 + 2q8 = 1114$?
   (*a*) 9 (*b*) 8 (*c*) 7 (*d*) 6
11. The least among the fractions $\frac{15}{16}, \frac{19}{20}, \frac{24}{25}, \frac{34}{35}$ is
   (*a*) $\frac{34}{35}$ (*b*) $\frac{15}{16}$ (*c*) $\frac{19}{20}$ (*d*) $\frac{24}{25}$
12. $1.\overline{27}$ in the form $\frac{p}{q}$ is equal to
   (*a*) $\frac{127}{100}$ (*b*) $\frac{73}{100}$ (*c*) $\frac{14}{11}$ (*d*) $\frac{11}{14}$

13. $\dfrac{3.25\times3.20-3.20\times3.05}{0.064}$ is equal to

(*a*) 1 (*b*) $\dfrac{1}{2}$ (*c*) $\dfrac{1}{10}$ (*d*) 10

14. Out of six consecutive natural numbers, if sum of first three is 27, what is the sum of the other three?

(*a*) 36 (*b*) 35 (*c*) 25 (*d*) 24

15. The H.C.F. and L.C.M. of two numbers are 12 and 336 respectively. If one of the numbers is 84, the other is

(*a*) 36 (*b*) 48 (*c*) 72 (*d*) 96

16. The sum of two numbers is 36 and their H.C.F. and L.C.M. are 3 and 105 respectively. The sum of the reciprocals of two numbers is **(SSC CHSL 2010)**

(*a*) $\dfrac{2}{35}$ (*b*) $\dfrac{3}{25}$ (*c*) $\dfrac{4}{35}$ (*d*) $\dfrac{2}{25}$

17. In '*n*' be any natural number, then by which largest number (n^3-n) is always divisible.

(*a*) 3 (*b*) 6 (*c*) 12 (*d*) 18

18. If 1.5 a = 0.04 b, then $\dfrac{b-a}{b+a}$ is equal to

(*a*) $\dfrac{73}{77}$ (*b*) $\dfrac{77}{33}$ (*c*) $\dfrac{2}{75}$ (*d*) $\dfrac{75}{2}$

19. How many perfect squares lie between 120 and 300 ?

(*a*) 5 (*b*) 6 (*c*) 7 (*d*) 8

20. $\left\{\dfrac{(0.1)^2-(0.01)^2}{0.0001}+1\right\}$ is equal to

(*a*) 1010 (*b*) 110 (*c*) 101 (*d*) 100

21. In how many years will a sum of ₹ 800 at 10% per annum compound interest, compounded semi-annually becomes ₹ 926.10 ?

(*a*) $1\frac{1}{2}$ (*b*) $1\frac{2}{3}$ (*c*) $2\frac{1}{3}$ (*d*) $2\frac{1}{2}$

22. In a 100 m race, Kamal defeats Bimal by 5 seconds. If the speed of Kamal is 18 km/hr, then the speed of Bimal is

(*a*) 15.4 km/hr (*b*) 14.5 km/hr (*c*) 14.4 km/hr (*d*) 14 km/hr

23. A train, 240 m long crosses a man walking along the line in opposite direction at the rate of 3 km/hr in 10 seconds. The speed of the train is

(*a*) 63 km/hr (*b*) 75 km/hr (*c*) 83.4 km/hr (*d*) 86.4 km/hr

24. A boatman rows 1 km in 5 minutes, along the stream and 6 km in 1 hour against the stream. The speed of the stream is

(*a*) 3 km/hr (*b*) 6 km/hr (*c*) 10 km/hr (*d*) 12 km/hr

25. A can complete $\dfrac{1}{3}$ of a work in 5 days and B $\dfrac{2}{5}$ of the work in 10 days. In how many days both A and B together can complete the work ?

(*a*) 10 (*b*) $9\frac{3}{8}$ (*c*) $8\frac{4}{5}$ (*d*) $7\frac{1}{2}$

26. 7 men can complete a piece of work in 12 days. How many additional men will be required to complete double the work in 8 days ?

(*a*) 28 (*b*) 21 (*c*) 14 (*d*) 7

27. One pipe fills a water tank three times faster than another pipe. If the two pipes together can fill the empty tank in 36 minutes, then how much time will the slower pipe alone take to fill the tank

(*a*) 1 hour 21 minutes (*b*) 1 hour 48 minutes (*c*) 1 hour (*d*) 2 hours 24 minutes

28. In an examination, a student scores 4 marks for every correct answer and loses 1 mark for every wrong answer. A student attempted all the 200 questions and scored, in all 200 marks. The number of questions, he answered correctly was

(*a*) 82 (*b*) 80 (*c*) 68 (*d*) 60

29. The average of odd numbers up to 100 is

(*a*) 50.5 (*b*) 50 (*c*) 49.5 (*d*) 49

30. If A's income is 25% less than B's income, by how much percent is B's income more then that of A ?

(*a*) 25 (*b*) 30 (*c*) $33\frac{1}{3}$ (*d*) $66\frac{2}{3}$

31. If there is a profit of 20% on the cost price of an article, the percentage of profit calculated on its selling price will be

(*a*) 24 (*b*) $16\frac{2}{3}$ (*c*) $8\frac{1}{3}$ (*d*) 20

32. If the cost price of 15 books is equal to selling price of 20 books, the loss percent is

(*a*) 16 (*b*) 20 (*c*) 24 (*d*) 25

33. If an article is sold at 200% profit, then the ratio of its cost price to its selling price will be

(*a*) 1 : 2 (*b*) 2 : 1 (*c*) 1 : 3 (*d*) 3 : 1

34. If on a marked price, the difference of selling prices with a discount of 30% and two successive discounts of 20% and 10% is ₹ 72, then the marked price (in rupees) is

(*a*) 3,600 (*b*) 3,000 (*c*) 2,500 (*d*) 2,400

35. If an electricity bill is paid before due date, one gets a reduction of 4% on the amount of the bill. By paying the bill before the due date, a person got a reduction of ₹ 13. The amount of his electricity bill was

(*a*) ₹ 125 (*b*) ₹ 225 (*c*) ₹ 325 (*d*) ₹ 425

36. Successive discounts of 10%, 20% and 30% is equivalent to a single discount of

(*a*) 60% (*b*) 49.6% (*c*) 40.5% (*d*) 36%

37. The price of an article was first increased by 10% and then again by 20%. If the last increased price be ₹ 33, the original price was

(*a*) ₹ 30 (*b*) 27.50 (*c*) ₹ 26.50 (*d*) ₹ 25

38. If each side of a square is increased by 10%, its area will be increased by

(*a*) 10% (*b*) 21% (*c*) 44% (*d*) 100%

39. The ratio of milk and water in mixtures of four containers are 5 : 3, 2 : 1, 3 : 2 and 7 : 4 respectively. In which container is the quantity of milk, relative to water, minimum ?

(*a*) First (*b*) Second (*c*) Third (*d*) Fourth

40. Two numbers are in the ratio 1 : 3. If their sum is 240, then their difference is

(*a*) 120 (*b*) 108 (*c*) 100 (*d*) 96

41. The ratio of income and expenditure of a person is 11 : 10. If he saves ₹ 9,000 per annum, his monthly income is

(*a*) ₹ 8,000 (*b*) ₹ 8,800 (*c*) ₹ 8,500 (*d*) ₹ 8,250

42. If $W_1 : W_2 = 2 : 3$ and $W_1 : W_3 = 1 : 2$, then $W_2 : W_3$ is

(*a*) 3 : 4 (*b*) 4 : 3 (*c*) 2 : 3 (*d*) 3 : 2

43. A copper wire of length 36 m and diameter 2 mm is melted to form a sphere. The radius of the sphere (in cm) is

(*a*) 2.5 (*b*) 3 (*c*) 3.5 (*d*) 4

44. The ratio of the radii of two wheels is 3 : 4. The ratio of their circumferences is

(*a*) 4 : 3 (*b*) 3 : 4 (*c*) 2 : 3 (*d*) 3 : 2

45. If the length of a rectangle is increased by 10% and its breadth is decreased by 10%, the change in its area will be

(*a*) 1% increase (*b*) 1% decrease (*c*) 10% increase (*d*) No change

46. In how many years will a sum of money double itself at $6\frac{1}{4}\%$ simple interest per annum ?

(*a*) 24 (*b*) 20 (*c*) 16 (*d*) 12

47. A sum of ₹ 12,000 deposited at compound interest becomes double after 5 years. How much will it be after 20 years

(*a*) ₹ 1,44,000 (*b*) ₹ 1,20,000 (*c*) ₹ 1,50,000 (*d*) ₹ 1,92,000

***Directions**: The pie chart, given here, shows the amount of money spent on various sports by a school administration in a particular year. Observe the pie chart and answer the questions (No. 48 to 50) based on this graph.*

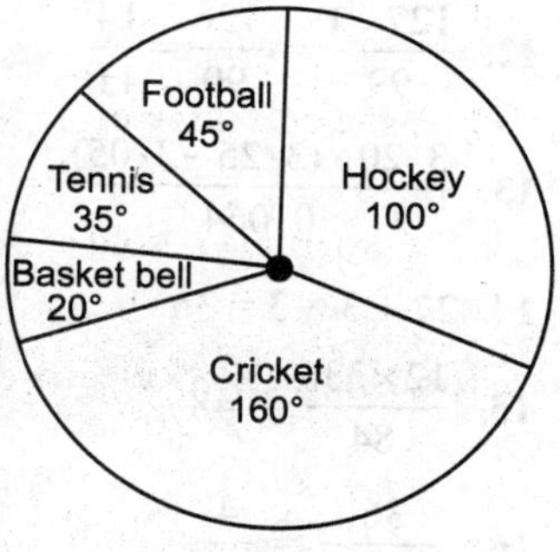

48. If the money spent on football was ₹ 9,000, how much more money was spent on hockey than on football ?

(*a*) ₹ 11,000 (*b*) ₹ 11,500
(*c*) ₹ 12,000 (*d*) ₹ 12,500

49. If the money spent on football was ₹ 9,000, what amount was spent on cricket ?

(*a*) ₹ 31,000 (*b*) ₹ 31,500 (*c*) ₹ 32,000 (*d*) ₹ 32,500

50. If the money spent on football is ₹ 9,000, then what was the total amount spent on all sports?

(*a*) ₹ 73,000 (*b*) ₹ 72,800 (*c*) ₹ 72,500 (*d*) ₹ 72,000

ANSWER

1. (*b*)	2. (*c*)	3. (*b*)	4. (*a*)	5. (*d*)	6. (*c*)	7. (*d*)	8. (*d*)
9. (*b*)	10. (*c*)	11. (*b*)	12. (*c*)	13. (*d*)	14. (*a*)	15. (*b*)	16. (*c*)
17. (*b*)	18. (*a*)	19. (*c*)	20. (*d*)	21. (*a*)	22. (*c*)	23. (*c*)	24. (*a*)
25. (*b*)	26. (*c*)	27. (*d*)	28. (*b*)	29. (*b*)	30. (*c*)	31. (*b*)	32. (*d*)
33. (*c*)	34. (*a*)	35. (*c*)	36. (*b*)	37. (*d*)	38. (*b*)	39. (*c*)	40. (*a*)
41. (*d*)	42. (*a*)	43. (*b*)	44. (*b*)	45. (*b*)	46. (*c*)	47. (*d*)	48. (*a*)
49. (*c*)	50. (*d*)						

HINTS

1. $\frac{n(n+1)(2n+1)}{6} = \frac{10\times11\times21}{6} = 385$

2. $2 + 4 = 6, 6 + 5 = 11, 11 + 6 = 17, 17 + 7 = 24, 24 + 8 = 32$

3. $(2 : 3)$ or $(8 : 12) - (7 : 11) = 1$ ratio $= 7 \Rightarrow$ smaller number $= 7$ ratio $= 7 \times 7 = 49$

4. $\frac{2}{3}\times\frac{3}{4}\times\frac{4}{5}\times........\times\frac{24}{25}=\frac{2}{25}$

5. $36 \div 17 \rightarrow$ Remainder $= 2$

6. $(x^{-3/25})^{-25/3} = x^1 = x$

7. 101

8. $10^{-6} \times 10^7 = 10^1 = 10$

9. $p+\frac{1}{2p} = 2 \Rightarrow p^3+\frac{1}{8p^3}=\left(p+\frac{1}{2p}\right)^3-\frac{3p}{2p}\left(p+\frac{1}{2p}\right) = 2^3 - 3 = 8 - 3 = 5$

10. $p + q = 7$; $q = 7$ when $p = 0$

11. $\frac{15}{16}$: $D - N = 1$ in each case.

12. $\frac{127-1}{99} = \frac{126}{99} = \frac{14}{11}$

13. $\frac{3\cdot20\times(3\cdot25-3\cdot05)}{0\cdot064} = \frac{3\cdot2\times0\cdot2}{0\cdot064} = \frac{0\cdot64}{0\cdot064} = 10$

14. $27 + 3 \times 3 = 36$

15. $\frac{12\times336}{84} = 48$

16. $\frac{36}{3\times105} = \frac{4}{35}$

17. $n^3 - n = (n-1)(n)(n+1)$ is divisible by $3! = 1 \times 2 \times 3 = 6$

18. $a : b = 0.4 : 1.5 = 4 : 150 = 2 : 75 \Rightarrow \frac{b-a}{b+a} = \frac{75-2}{75+2} = \frac{73}{77}$

19. $10^2 < 120 < 11^2$ and $17^2 < 300 < 18^2 \Rightarrow$ Required numbers $= 17 - 10 = 7$

20. $\frac{0\cdot11\times0\cdot009}{0\cdot0001}+1 = 99 + 1 = 100$

21. $\frac{9261}{8000}=\left(\frac{21}{20}\right)^3 \Rightarrow$ Time $= \frac{3}{2} = 1\frac{1}{2}$ years

22. Kamal: $18\times\frac{5}{18} = 5$ m/s $\Rightarrow$ Time $= 100 \div 5 = 20$ seconds

Bimal: Time $= 20 + 5 = 25$ seconds $\Rightarrow$ speed $= \frac{100}{25}$ m/s $= 4 \times \frac{18}{5}$ km/h $= 14.4$ km/h

23. Relative speed = $\frac{240}{10}$ m/s = $24\times\frac{18}{5}$ km/h = 86.4 km/hr

⇒ Speed of train = (86.4 – 3) = 83.4 km/hr.

24. Speed downstream = 12 km/hr and speed upstream = 6 km/hr

⇒ Speed of stream = $\frac{1}{2}\times(12-6)$ = 3 km/hr

25. $A = 5 \times 3 = 15$ days; $B = 10 \times \frac{5}{2} = 25$ days

$A + B = \frac{15\times25}{40} = \frac{75}{8} = 9\frac{3}{8}$ days

26. $\frac{7\times12\times2}{8} = 21$ ⇒ Additional men = 21 – 7 = 14 men

27. $\frac{1}{36}\times\frac{1}{4} = \frac{1}{144}$; 144 minutes = 2 hours 24 minutes

28. 200 ÷ 4 = 50 questions ⇒ Rest questions = 200 – 50 = 150

In rest questions : correct : wrong = 1 : 4

⇒ Correct answers = $\frac{1}{5}\times150 = 30$

Total correct answers = 50 + 30 = 80

29. Avr = $\frac{1+99}{2} = 50$

30. $B = 100$, then $A = 75$; $\frac{25}{75}\times100 = 33\frac{1}{3}\%$

31. $\frac{1}{5}$ of CP = $\frac{1}{6}$ of SP = $16\frac{2}{3}\%$

32. CP : SP = 20 : 15 = 4 : 3 ⇒ Loss = $\frac{1}{4}$ = 25%

33. 100 : 300 = 1 : 3

34. Difference = $\frac{20\times10}{100}$ = 2% = ₹ 72 ⇒ $MP = \frac{72}{2}\times100$ = ₹ 3600

35. 4% = ₹13 ⇒ Bill = $13 \times \frac{100}{4}$ = ₹ 325

36. $10 + 20 - \frac{10\times20}{100} = 28\%$; $28 + 30 - \frac{28\times30}{100} = 49.6\%$

37. $33 \times \frac{100}{110}\times\frac{100}{120}$ = ₹ 25

38. 11 × 11 – 100 = 21%

39. Third mixture

40. (1 + 3) ratio = 4 ratio = 240 ⇒ 1 ratio = 60

Difference = (3 – 1) ratio = 2 ratio = 2 × 60 = 120

41. Savings = (11 – 10) ratio = 1 ratio = ₹ 9,000 p.a.

Income = 11 ratio = ₹ 99,000 p.a. = 99,000 ÷ 12 = ₹ 8250 pm

42. $\frac{W_2}{W_3} = \frac{W_1}{W_3} \times \frac{W_2}{W_1} = \frac{1}{2} \times \frac{3}{2} = \frac{3}{4}$

43. $\frac{4}{3}\pi r^3 = \pi\,(0.01) \times 3600 \Rightarrow r^3 = 36 \times \frac{3}{4} = 27 \Rightarrow r = 3$ cm

44. Ratio of circumference = Ratio of radii = 3 : 4

45. $\frac{10 \times 10}{100}$ = 1% decrease

46. $\frac{1 \times 100 \times 4}{25}$ = 16 years

47. 12,000 × 2^4 = ₹ 1,92,000

48. 100° – 45° = 55° = 55 × ₹ 200 = ₹ 11,000

49. 45° = 9,000 ⇒ 1° = ₹ 200 ⇒ Cricket = 160° = 160 × ₹ 200 = ₹ 32,000

50. 1° = ₹ 200 ⇒ 360° = ₹ 72, 000

SSC Combined Graduate Level (Tier-I) 2011
ARITHMETIC ABILITY (HELD ON 26.6.2011 : FIRST SHIFT)

1. The value of $\frac{3\sqrt{2}}{\sqrt{3}+\sqrt{6}}-\frac{4\sqrt{3}}{\sqrt{6}+\sqrt{2}}+\frac{\sqrt{6}}{\sqrt{3}+\sqrt{2}}$ is

(a) 4 (b) 0 (c) $\sqrt{2}$ (d) $3\sqrt{6}$

2. $\sqrt{6+\sqrt{6+\sqrt{6+\ldots\ldots}}}=?$

(a) 2.3 (b) 3 (c) 6 (d) 6.3

3. The square root of $\left(\frac{\sqrt{3}+\sqrt{2}}{\sqrt{3}-\sqrt{2}}\right)$ is

(a) $\sqrt{3}+\sqrt{2}$ (b) $\sqrt{3}-\sqrt{2}$ (c) $\sqrt{2}+\sqrt{3}$ (d) $\sqrt{2}-\sqrt{3}$

4. The remainder when 3^{21} is divided by 5 is

(a) 1 (b) 2 (c) 3 (d) 4

5. The value of $\frac{2\frac{1}{3}-1\frac{2}{11}}{3+\frac{1}{3+\frac{1}{3+\frac{1}{3}}}}$

(a) $\frac{38}{109}$ (b) $\frac{109}{38}$ (c) 1 (d) $\frac{116}{109}$

6. The last digit of $(1001)^{2008}+1002$ is

(a) 0 (b) 3 (c) 4 (d) 6

7. $\sqrt[3]{1-\frac{127}{343}}$ is equal to

(a) $\frac{5}{9}$ (b) $1-\frac{1}{7}$ (c) $\frac{4}{7}$ (d) $1-\frac{2}{7}$

8. If the sum of two numbers be multiplied by each number separately, the products so obtained are 247 and 114. The sum of the numbers is

(a) 19 (b) 20 (c) 21 (d) 23

9. Find a number, one-seventh of which exceeds its eleventh part by 100. **(CDS 1994)**

(a) 1925 (b) 1825 (c) 1540 (d) 1340

10. If $x * y=(x+3)^2(y-1)$, then the value of $5 * 4$ is

(a) 192 (b) 182 (c) 180 (d) 172

11. $\frac{(0.05)^2+(0.41)^2+(0.073)^2}{(0.005)^2+(0.041)^2+(0.0073)^2}$

(a) 10 (b) 100 (c) 1000 (d) None of these

12. If $9\sqrt{x}=\sqrt{12}+\sqrt{147}$, then $x=?$

(a) 2 (b) 3 (c) 4 (d) 5

13. A and B can complete a piece of work in 8 days, B and C can do it in 12 days, C and A can do it in 8 days. A, B and C together can complete it in

(*a*) 4 days (*b*) 5 days (*c*) 6 days (*d*) 7 days

14. X is 3 times as fast as Y and is able to complete the work in 40 days less than Y. Then the time in which they can complete the work together is

(*a*) 15 days (*b*) 10 days (*c*) $7\frac{1}{2}$ days (*d*) 5 days

15. A copper wire is bent in the shape of a square of area 81 cm^2. If the same wire is bent in the form of a semicircle, the radius (in cm) of the semicircle is $\left(\text{Take } \pi = \frac{22}{7}\right)$

(*a*) 16 (*b*) 14 (*c*) 10 (*d*) 7

16. The volume (in m^3) of rain water that can be collected from 1.5 hectares of ground in a rainfall of 5 cm is

(*a*) 75 (*b*) 750 (*c*) 7500 (*d*) 75000

17. A river 3 m deep and 40 m wide is flowing at the rate of 2 km per hour. How much water (in litres) will fall into the sea in a minute ?

(*a*) 4,00,000 (*b*) 40,00,000 (*c*) 40,000 (*d*) 4,000

18. The L.C.M. of three different numbers is 120. Which of the the following ***cannot*** be their H.C.F.?

(*a*) 8 (*b*) 12 (*c*) 24 (*d*) 35

19. A number when divided by 49 leaves 32 as remainder. This number when divided by 7 will have the reaminder as

(*a*) 4 (*b*) 3 (*c*) 2 (*d*) 5

20. In an examination a student scores 4 marks for every correct answer and loses 1 mark for every wrong answer. If he attempts all 75 questions and secures 125 marks, the number of questions he attempts correctly is

(*a*) 35 (*b*) 40 (*c*) 42 (*d*) 46

21. The traffic lights at three different road crossing change after 24 seconds, 36 seconds and 54 seconds respectively. If they all change simultaneously at 10:15:00 AM, then at what time will they again change simultaneously ?

(*a*) 10:16:54 AM (*b*) 10:18:36 AM (*c*) 10:17:02 AM (*d*) 10:22:12 AM

22. A can do a work in 12 days. When he had worked for 3 days, B joined him. If they complete the work in 3 more days, in how many days can B alone finish the work ?

(*a*) 6 days (*b*) 12 days (*c*) 4 days (*d*) 8 days

23. Among three number, the first is twice the second and thrice the third. If the average of three numbers is 49.5, then the difference between the first and the third number is

(*a*) 54 (*b*) 28 (*c*) 39.5 (*d*) 41.5

24. The mean of 50 numbers is 30. Later is was discovered that two entries were wrongly entered as 82 and 13 instead of 28 and 31. Find the correct mean.

(*a*) 36.12 (*b*) 30.66 (*c*) 29.28 (*d*) 38.21

25. A bicycle wheel makes 5000 revolutions in moving 11 km. Then the radius of the wheel (in cm) is $\left(\text{Take } \pi = \frac{22}{7}\right)$ **(SSC CGL Tier I 2012)**

(*a*) 70 (*b*) 35 (*c*) 17.5 (*d*) 140

26. The perimeter of a triangle is 40 cm and its area is 60 cm^2. If the largest side measures 17 cm, then the length (in cm) of the smallest side of the triangle is

(*a*) 4 (*b*) 6 (*c*) 8 (*d*) 15

27. A shopkeeper allows a discount of 10% to his customers and still gains 20%. Find the marked price of the article which costs ₹ 450.

(*a*) ₹ 600 (*b*) ₹ 540 (*c*) ₹ 660 (*d*) ₹ 580

28. What single discount is equivalent to two successive discounts of 20% and 15% ?

(*a*) 35% (*b*) 32% (*c*) 34% (*d*) 30%

29. In a business partnership among A, B, C and D, the profit is shared as follows:

$\frac{\text{A's share}}{\text{B's share}} = \frac{\text{B's share}}{\text{C's share}} = \frac{\text{C's share}}{\text{D's share}} = \frac{1}{3}$. If the total profit is ₹ 4,00,000, the share of C is:

(*a*) ₹ 1,12,500 (*b*) ₹ 1,37,500 (*c*) ₹ 90,000 (*d*) ₹ 2,70,000

30. A can contains a mixture of two liquids A and B in the ratio 7 : 5. When 9 litres of mixutre are drawn off and the can is filled with B, the ratio of A and B becomes 7 : 9. Litres of liquid A contained by the can intially was

(*a*) 10 (*b*) 20 (*c*) 21 (*d*) 25

31. What number should be added to or subtracted from each term of the ratio 17 : 24 so that it becomes equal to 1 : 2 ?

(*a*) 5 is subtracted (*b*) 10 is added (*c*) 7 is added (*d*) 10 is subtracted

32. The ratio of weekly incomes of A and B is 9 : 7 and the ratio of their expenditures is 4 : 3. If each saves ₹ 200 per week, then the sum of their weekly incomes is

(*a*) ₹ 3,200 (*b*) ₹ 4,200 (*c*) ₹ 4,800 (*d*) ₹ 5,600

33. If 30% of A is added to 40% of B, the answer is 80% of B. What percentage of A is B ?

(*a*) 30% (*b*) 40% (*c*) 70% (*d*) 75%

34. A man can row 6 km/h in still water. If the speed of the current is 2 km/h, it takes 3 hours more in upstream than in the dwonstream for the same distance. The distance is

(*a*) 30 km (*b*) 24 km (*c*) 20 km (*d*) 32 km

35. If the selling price of 10 articles is equal to the cost price of 11 articles, then the gain percent is

(*a*) 10 (*b*) 11 (*c*) 15 (*d*) 25

36. While selling a watch, a shopkeeper gives a discount of 5%. If he gives a discount of 6%, he earns ₹ 15 less as profit. What is the marked price of the watch ?

(*a*) ₹ 1,250 (*b*) ₹ 1,400 (*c*) ₹ 1,500 (*d*) ₹ 750

37. Krishna purchased a number of articles at ₹ 10 for each and the same number for ₹ 14 each. He mixed them together and sold them for ₹ 13 each. Then his gain or loss percent is

(*a*) Loss $8\frac{1}{3}\%$ (*b*) Gain $8\frac{2}{3}\%$ (*c*) Loss $8\frac{2}{3}\%$ (*d*) Gain $8\frac{1}{3}\%$

38. A trader bought two horses for ₹ 19,500. He sold one at a loss of 20% and other at a profit of 15%. If the selling price of each horse is the same, then their cost prices are respectively.

(*a*) ₹ 10,00 and ₹ 9,500 (*b*) ₹ 11,500 and ₹ 8,000

(*c*) ₹ 12,000 and ₹ 7,500 (*d*) ₹ 10,500 and ₹ 9,000

39. The cost price of an article is 40% of the selling price. What percent of the cost price is the selling price ?

(*a*) 140% (*b*) 200% (*c*) 220% (*d*) 250%

40. If 90% of A = 30% of B and B = 2x% of A then the value of x is

(*a*) 450 (*b*) 400 (*c*) 300 (*d*) 150

41. When the price of sugar decreases by 10%, a man could buy 1 kg more for ₹ 270. Then the original price of sugar per kg is

(*a*) ₹ 25 (*b*) ₹ 30 (*c*) ₹ 27 (*d*) ₹ 32

42. If the price of sugar is raised by 25%, find by how much percent a householder must reduce his consumption of sugar so as not to increase his expenditure ?

(*a*) 10 (*b*) 20 (*c*) 18 (*d*) 25

43. The difference between the compound interest and simple interest on ₹ 10,000 for 2 years is ₹ 25. The rate of interest per annum is

(*a*) 5% (*b*) 7% (*c*) 10% (*d*) 12%

44. A student goes to school at the rate of $2\frac{1}{2}$ km/h and reaches 6 minutes late. If he travels at the speed of 3 km/h, he is 10 minutes early. The distance (in km) between the school and his house is

(*a*) 5 (*b*) 4 (*c*) 3 (*d*) 1

45. A sum of money placed at compound interest doubles itself in 4 years. In how many years will it amount to four times itself ?

(*a*) 12 years (*b*) 13 years (*c*) 8 years (*d*) 16 years

46. The simple interest on a sum for 5 years is one-fourth of the sum. The rate of interest per annum is

(*a*) 5% (*b*) 6% (*c*) 4% (*d*) 8%

***Directions**: (Questions No. 47 to 50). The following graph shows the demand and production of cotton by 5 companies A, B, C, D and E. Study the graph and answer questions 147 to 150.*

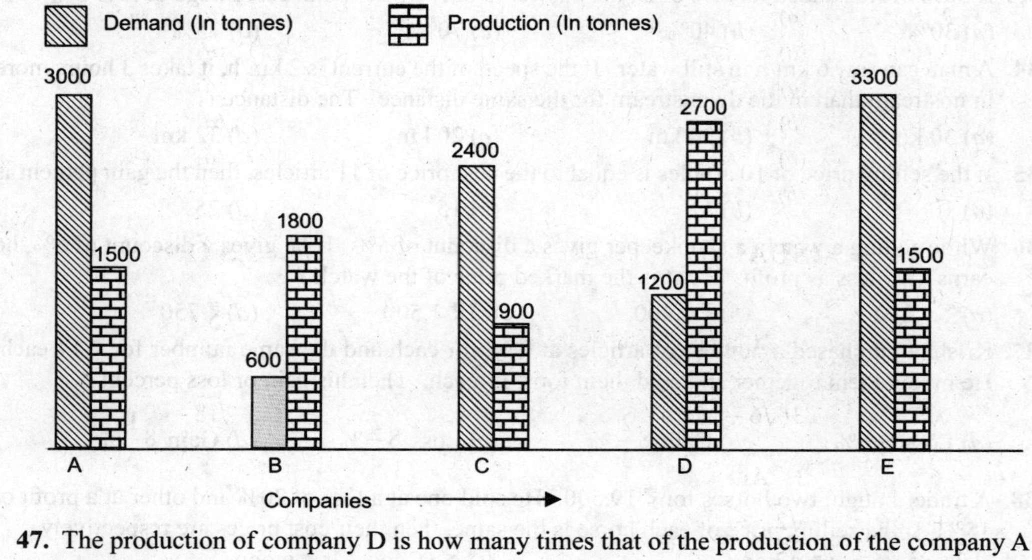

47. The production of company D is how many times that of the production of the company A ?

(*a*) 1.8 (*b*) 1.5 (*c*) 0.5 (*d*) 0.4

48. The demand for company B is what percent of the demand for company C ?

(*a*) 1.5 (*b*) 2.5 (*c*) 25 (*d*) 30

49. What is the ratio of companies having more demand than production to those having more production than demand ?

(*a*) 2 : 3 (*b*) 4 : 1 (*c*) 3 : 2 (*d*) 1 : 4

50. What is the difference (in tonnes) between average demand and average production of the five companies taken together ?

(*a*) 320 (*b*) 420 (*c*) 2100 (*d*) 1050

FOR VISUALLY HANDICAPPED CANDIDATES ONLY

47. A, B and C enter into a partnership with investments in the ratio $\frac{7}{2}:\frac{4}{3}:\frac{6}{5}$. After 4 months, A increases his share by 50%. If the total profit at the end of one year be ₹ 21,600, then B's share in the profit is **(SSC CHSL 2013)**

(*a*) ₹ 2,100 (*b*) ₹ 2,400 (*c*) ₹ 3,600 (*d*) ₹ 4,000

48. A person travels 48 km at 12 km/h and further 48 km at 16 km/h. His average speed for the whole journey, in km/h, is

(*a*) 14 (*b*) $13\frac{4}{7}$ (*c*) $12\frac{3}{7}$ (*d*) $13\frac{5}{7}$

49. The H.C.F. and L.C.M. of two numbers are 13 and 1989 respectively. If one of the numbers is 117, determine the other.

(*a*) 121 (*b*) 143 (*c*) 217 (*d*) 221

50. A sphere of radius *r* and a cylinder of the base radius *r* have equal volumes. If the height of the cylinder is *h*, then which of the following is true ?

(*a*) $4r = 3h$ (*b*) $4r = h$ (*c*) $h = 3r$ (*d*) $h = 2r$

ANSWER

1. (*b*)	2. (*b*)	3. (*a*)	4. (*c*)	5. (*a*)	6. (*b*)	7. (*b*)	8. (*a*)
9. (*a*)	10. (*a*)	11. (*b*)	12. (*b*)	13. (*c*)	14. (*a*)	15. (*d*)	16. (*b*)
17. (*b*)	18. (*d*)	19. (*a*)	20. (*b*)	21. (*b*)	22. (*a*)	23. (*a*)	24. (*c*)
25. (*b*)	26. (*c*)	27. (*a*)	28. (*b*)	29. (*c*)	30. (*c*)	31. (*d*)	32. (*a*)
33. (*d*)	34. (*b*)	35. (*a*)	36. (*c*)	37. (*d*)	38. (*b*)	39. (*d*)	40. (*d*)
41. (*b*)	42. (*b*)	43. (*a*)	44. (*b*)	45. (*c*)	46. (*a*)	47. (*a*)	48. (*c*)
49. (*c*)	50. (*b*)						

FOR VISUALLY HANDICAPPED CANDIDATES ONLY

47. (*d*)	48. (*d*)	49. (*d*)	50. (*a*)

HINTS

1. $\sqrt{2}(\sqrt{6}-\sqrt{3})-\sqrt{3}(\sqrt{6}-\sqrt{2})+\sqrt{6}(\sqrt{3}-\sqrt{2})=\sqrt{12}-\sqrt{6}-\sqrt{18}+\sqrt{6}+\sqrt{18}-\sqrt{12}=0$

2. $x^2 = 6 + x \Rightarrow x = 3$ Alt: $6 = 2 \times 3$

3. $\frac{\sqrt{3}+\sqrt{2}}{\sqrt{3}-\sqrt{2}}=\frac{\sqrt{3}+\sqrt{2}}{\sqrt{3}+\sqrt{2}}\times\frac{\sqrt{3}+\sqrt{2}}{\sqrt{3}+\sqrt{2}}=\left(\sqrt{3}+\sqrt{2}\right)^2$

4. $3^{21} = 3^{20} \times 3^1 = (3^2)^{10} \times 3^1 = (-1)10 \times 3^1 = 3$

Trick: Remainder on dividing by 5, depends on unit's digit.

Units's digit of $3^{21} = 3^1 = 3$

5. $\dfrac{1\frac{5}{33}}{\frac{109}{33}} = \dfrac{38}{109}$

6. Unit's digit = 1 + 2 = 3

7. $\sqrt[3]{1-\dfrac{127}{343}} = \sqrt[3]{\dfrac{216}{343}} = \dfrac{6}{7} = 1-\dfrac{1}{7}$

8. $(x + y)^2 = 247 + 114 = 361 = 19^2$

Trick: HCF of 114 and 247 is 19.

9. $\dfrac{1}{7}-\dfrac{1}{11}=\dfrac{4}{77}$ of number = 100 ⇒ Number = $100 \times \dfrac{77}{4} = 1925$

10. $(5 + 3)^2 (4 - 1) = 64 \times 3 = 192$

11. 100

12. $\sqrt{12}+\sqrt{147} = 2\sqrt{3}+7\sqrt{3} = 9\sqrt{3}$

13. $2(A + B + C) = \dfrac{1}{8}+\dfrac{1}{12}+\dfrac{1}{8}=\dfrac{1}{3} \Rightarrow (A + B + C) = 2 \times 3 = 6$ days

14. $X : Y = 1 : 3 \Rightarrow X - Y = (3 - 1)$ ratio = 40 days ⇒ 1 ratio = 20 days

⇒ $X = 20$ days and $Y = 60$ days ⇒ $X + Y = \dfrac{60}{3+1} = 15$ days

15. $a^2 = 81 \Rightarrow a = 9 \Rightarrow 4a = 36$ cm

$\pi r + 2r = 36 \Rightarrow r\left(\dfrac{22}{7}+2\right) = 36 \Rightarrow r = 36\times\dfrac{7}{36} = 7$ cm

16. $(1.5 \times 10000) \times \dfrac{5}{100} = 750\ \text{m}^3$

17. $3 \times 40 \times 2000 \times \dfrac{1}{60} = 4000\ \text{m}^3$ = 40,00,000 litres

18. 120 is not divisible by 35.

19. $32 \div 7$ gives remainder 4

20. If all correct = $75 \times 4 = 300$ marks; $300 - 125 = 175$

Wrong questions = $175 \div 5 = 35$ ⇒ correct = 40

21. LCM of 24, 36, 54 = $12 \times 2 \times 3 \times 3 = 216$ seconds = 3 minutes 36 seconds

⇒ Required time = 10:18:36 AM

22. $A(6) = \dfrac{6}{12} = \dfrac{1}{2}$; $B(3) = 1-\dfrac{1}{2} = \dfrac{1}{2}$; $B = 3 \times 2 = 6$ days

23. Numbers = $6x, 3x, 2x$; Average = $\dfrac{11x}{3} = 49.5 \Rightarrow x = \dfrac{49{\cdot}5\times3}{11} = 13.5$

First number – third number = $4x = 4 \times 13.5 = 54$

24. Difference = (82 + 13) – (28 + 31) = 34; $34 \div 50 = 0.68$;

30 – 0.72 = 29.28

25. Radius = $\frac{11000}{5000} \times \frac{1}{2} \times \frac{7}{22} \times 100 = 35$

26. Trick: $8^2 + 15^2 = 17^2$ and $8 + 15 + 17 = 40 \Rightarrow$ The triangle is right triangle.

Trick: (17, 4, 19) not possible as 17 < 19; (17, 6, 17) not posssible as 17 = 17; (17, 8, 15) possible; (17, 15, 8) not possible as 15 > 8

27. $450 \times \frac{120}{100} \times \frac{100}{90} =$ ₹ 600

28. $20 + 15 - \frac{20 \times 15}{100} = 32\%$

29. $A : B = B : C = C : D = 1 : 3 \Rightarrow A : B : C : D = 1 : 3 : 9 : 27$

$\Rightarrow C = \frac{9}{40} \times 4{,}00{,}000 =$ ₹ 90,000

30. $\left(\frac{7}{16} - 0\right) : \left(\frac{7}{12} - \frac{7}{16}\right) = 3 : 1$

1 ratio = 9 litres $\Rightarrow$ 4 ratio = 36 litres $\Rightarrow A = \frac{7}{12} \times 36 = 21$ litres

31. $\frac{17 - x}{24 - x} = \frac{1}{2} \Rightarrow 34 - 2x = 24 - x \Rightarrow x = 10$

Alt: $\frac{17-5}{24-5} \neq \frac{1}{2}; \frac{17+10}{24+10} \neq \frac{1}{2}; \frac{17+7}{24+7} \neq \frac{1}{2}; \frac{17-10}{24-10} = \frac{7}{14} = \frac{1}{2}$

32. Income = 9 : 7; Expenditure = 4 : 3 = 8 : 6 $\Rightarrow$ savings ratio = ₹ 200

sum of incomes = 16 ratio = ₹ 3,200

33. 30% of A = 40% of $B \Rightarrow A : B = 4 : 3 \Rightarrow B = \frac{3}{4} \times 100 = 75\%$ of A

34. Downstream = 8 km/hr, upstream = 4 km/hr $\Rightarrow$ Distance = $\frac{8 \times 4}{8 - 4} \times 3 = 24$ km

35. CP : SP = 10 : 11 $\Rightarrow$ Profit = $\frac{1}{10} = 10\%$

36. 1% of MP = ₹ 15 $\Rightarrow$ MP = ₹ 1500

37. CP = $\frac{10 + 14}{2}$ = ₹ 12; SP = ₹ 13 $\Rightarrow$ Profit = $\frac{1}{12} = 8\frac{1}{3}\%$

38. 80% of I = 115% of II $\Rightarrow$ I : II = 115 : 80 = 23 : 16

I = $\frac{23}{39} \times 19500$ = ₹ 11, 500; II = $\frac{16}{39} \times 19{,}500$ = ₹ 8000

39. CP : SP = 40 : 100 = 2 : 5 $\Rightarrow \frac{SP}{CP} = \frac{5}{2} = 250\%$

40. $A : B = 30 : 90 = 1 : 3 \Rightarrow \frac{B}{A} = \frac{3}{1} = 300\ \% = 2x\ \% \Rightarrow x = 150$

41. New price = $\frac{10\% \text{ of } 270}{1}$ = ₹ 27; original price = ₹ 27 × $\frac{10}{9}$ = ₹ 30

42. Increase = 25% = $\frac{1}{4}$ ⇒ Decrease = $\frac{1}{5}$ = 20%

Alt: $\frac{25}{100+25}\times 100 = 20\%$

43. $r^2 = \frac{25}{10000} = \frac{1}{400} \Rightarrow r = \frac{1}{20} = 5\%$

44. $\frac{\frac{5}{2}\times 3}{3-\frac{5}{2}}\times\frac{16}{60} = \frac{5}{2}\times 3\times 2\times\frac{16}{60} = 4$ km

45. $2^2 = 4 \Rightarrow$ Time = 2 × 4 years = 8 years

46. $\frac{1\times 100}{4\times 5} = 5\%$

47. $\frac{2700}{1500} = 1.8$

48. $\frac{600}{2400}\times 100 = 25\%$

49. $(A, C, E) : (B : D) = 3 : 2$

50. Demand = 3000 + 600 + 2400 + 1200 + 3300 = 10,500
Production = 1500 + 1800 + 900 + 2,700 + 1500 = 8,400
Difference = 10,500 – 8,400 = 2,100 ⇒ Average = 2100 ÷ 5 = 420

FOR VISUALLY HANDICAPPED CANDIDATES ONLY

47. $A : B : C = \left(\frac{7}{2}\times 12+\frac{7}{4}\times 8\right) : \frac{4}{3}\times 12 : \frac{6}{5}\times 12 = 56 : 16 : \frac{72}{5} = 280 : 80 : 72$

$\Rightarrow B = \frac{80}{432}\times 21600 =$ ₹ 4,000

48. Time = $\frac{48}{12}+\frac{48}{16} = 4 + 3 = 7$ hours; Average speed = $\frac{96}{7} = 13\frac{5}{7}$ km/hr

49. $\frac{13\times 1989}{117} = 221$

50. $\frac{4}{3}\pi r^3 = \pi r^2 h \Rightarrow 4r = 3h$

SSC CGL TIER II 2011

1. When $(67^{67} + 67)$ is divided by 68, the remainder is

(*a*) 1 (*b*) 63 (*c*) 66 (*d*) 67

2. In a division sum, the divisor is 4 times the quotient and twice the remainder. If a and b are respectively the divisor and the dividend, then

(*a*) $\frac{4b - a^2}{a} = 3$ (*b*) $\frac{4b - 2a}{a^2} = 2$ (*c*) $(a + 1)^2 = 4b$ (*d*) $\frac{a(a + 2)}{b} = 4$

3. If 738A6A is divisible by 11, then the value of A is

(*a*) 6 (*b*) 3 (*c*) 9 (*d*) 1

4. The product of 2 numbers is 1575 and their quotient is 9/7. Then the sum of the numbers is

(*a*) 74 (*b*) 78 (*c*) 80 (*d*) 90

5. The value of $\frac{(81)^{3.6} \times (9)^{2.7}}{(81)^{4.2} \times (3)}$ is

(*a*) 3 (*b*) 6 (*c*) 9 (*d*) 8.2

6. $\sqrt{6 + \sqrt{6 + \sqrt{6 + \ldots}}}$ is equal to

(*a*) 2 (*b*) 5 (*c*) 4 (*d*) 3

7. The sum of the squares of two natural consecutive odd numbers is 394. The sum of the numbers is

(*a*) 24 (*b*) 32 (*c*) 40 (*d*) 28

8. The square root of $\frac{(0.75)^3}{1 - 0.75} + [0.75 + (0.75)^2 + 1]$ is

(*a*) 1 (*b*) 2 (*c*) 3 (*d*) 4

9. Given that $\sqrt{4096} = 64$, the value of $\sqrt{4096} + \sqrt{40.96} + \sqrt{0.004096}$ is **(SSC CGL 2005)**

(*a*) 70.4 (*b*) 70.464 (*c*) 71.104 (*d*) 71.4

10. The least positive integer that should be subtracted from 3011×3012 so that the difference is a perfect square is

(*a*) 3009 (*b*) 3010 (*c*) 3011 (*d*) 3012

11. P, Q, R are employed to do a work for ₹ 5750. P and Q together finished $\frac{19}{23}$ of work and Q and R together finished $\frac{8}{23}$ of work. Wage of Q, in rupees, is

(*a*) 2850 (*b*) 3750 (*c*) 2750 (*d*) 1000

12. A can do a piece of work in 24 days, B in 32 days and C in 64 days. All begin to do it together, but A leaves after 6 days and B leaves 6 days before the completion of the work. How many days did the work last ?

(*a*) 15 (*b*) 20 (*c*) 18 (*d*) 30

13. A cistern has 3 pipes A, B and C. A and B can fill it in 3 and 4 hours respectively, and C can empty it in 1 hour. If the pipes are opened at 3 p.m., 4 p.m. and 5 p.m. respectively on the same day, the cistern will be empty at

(*a*) 7.12 p.m. (*b*) 7.15 p.m. (*c*) 7.10 p.m. (*d*) 7.18 p.m.

14. If A works alone, he would take 4 days more to complete the job than if both A and B worked together. If B worked alone, he would take 16 days more to complete the job than if A and B

work together. How many days would they take to complete the work if both of them worked together ?

(*a*) 10 days (*b*) 12 days (*c*) 6 days (*d*) 8 days

15. 250 men can finish a work in 20 days working 5 hours a day. To finish the work within 10 days working 8 hours a day, the minimum number of men required is

(*a*) 310 (*b*) 300 (*c*) 313 (*d*) 312

16. 2 men and 5 women can do a work in 12 days. 5 men and 2 women can do that work in 9 days. Only 3 women can finish the same work in

(*a*) 36 days (*b*) 21 days (*c*) 30 days (*d*) 42 days

17. While selling, a businessman allows 40% discount on the marked price and there is a loss of 30%. If it is sold at the marked price, profit percent will be

(*a*) 10% (*b*) 20% (*c*) $16\frac{2}{3}\%$ (*d*) $16\frac{1}{3}\%$

18. Successive discounts of 10%, 20% and 50% will be equivalent to a single discount of

(*a*) 36% (*b*) 64% (*c*) 80% (*d*) 56%

19. A retailer offers the following discount schemes for buyers on an article:

I. Two successive discounts of 10%.

II. A discount of 12% followed by a discount of 8%.

III. Successive discounts of 15% and 5%.

IV. A discount of 20%.

The selling price will be minimum under the scheme

(*a*) I (*b*) II (*c*) III (*d*) IV

20. By selling an article at $\frac{3}{4}$ th of the marked price, there is a gain of 25%. The ratio of the marked price and the cost price is

(*a*) 5 : 3 (*b*) 3 : 5 (*c*) 3 : 4 (*d*) 4 : 3

21. *A* and *B* earn in the ratio 2 : 1. They spend in the ratio 5 : 3 and save in the ratio 4 : 1. If the total monthly savings of both *A* and *B* are ₹ 5,000, the monthly income of *B* is

(*a*) ₹ 7,000 (*b*) ₹ 14,000 (*c*) ₹ 5,000 (*d*) ₹ 10,000

22. The ratio of the sum of two numbers and their difference is 5 : 1. The ratio of the greater number to the smaller number is

(*a*) 2 : 3 (*b*) 3 : 2 (*c*) 5 : 4 (*d*) 1 : 5

23. *A* and *B* are two alloys of gold and copper prepared by mixing metals in the ratio 7 : 2 and 7 : 11 respectively. If equal quantities of the alloys are melted to form a third alloy *C*, the ratio of gold and copper in *C* will be

(*a*) 5 : 7 (*b*) 5 : 9 (*c*) 7 : 5 (*d*) 9 : 5

24. In a laboratory, two bottles contain mixture of acid and water in the ratio 2 : 5 in the first bottle and 7 : 3 in the second. The ratio in which the contents of these two bottles be mixed such that the new mixture has acid and water in the ratio 2 : 3 is

(*a*) 4 : 15 (*b*) 9 : 8 (*c*) 21 : 8 (*d*) 1 : 2

25. A mixture contains 80% acid and rest water. Part of the mixture that should be removed and replaced by same amount of water to make the ratio of acid and water 4 : 3 is

(*a*) $\frac{1}{3}$rd (*b*) $\frac{3}{7}$th (*c*) $\frac{2}{3}$rd (*d*) $\frac{2}{7}$th

26. An employer reduces the number of his employees in the ratio 9 : 8 and increases their wages in the ratio 14 : 15. If the original wage bill was ₹ 18,900, find the ratio in which the wage bill is decreased.

(*a*) 20 : 21 (*b*) 21 : 20 (*c*) 20 : 19 (*d*) 19 : 21

27. The batting average for 40 innings of a cricketer is 50 runs. His highest score exceeds his lowest score by 172 runs. If these two innings are excluded, the average of the remaining 38 innings is 48 runs. The highest score of the player is **(SSC CHSL 2011)**

(*a*) 165 (*b*) 170 (*c*) 172 (*d*) 174

28. Of three numbers, the second is thrice the first and the third number is three-fourth the first. If the average of the three number is 114, the largest number is

(*a*) 72 (*b*) 216 (*c*) 354 (*d*) 726

29. A car covers $\frac{1}{5}$ of the distance from *A* to *B* at the speed of 8 km/hr, $\frac{1}{10}$ of the distance at 25 km per hour and the remaining at the speed of 20 km per hour. Find the average speed of the whole journey.

(*a*) 12.625 km/hr (*b*) 13.625 km/hr (*c*) 14.625 km/hr (*d*) 15.625 km/hr

30. The average of 3 numbers is 154. The first number is twice the second and the second number is twice the third. The first number is

(*a*) 264 (*b*) 132 (*c*) 88 (*d*) 66

31. The average salary of all the staff in a office of a corporate house is ₹ 5,000. The average salary of the officers is ₹ 14,000 and that of the rest is ₹ 4,000. If the total number of staff is 500, the number of officers is

(*a*) 10 (*b*) 15 (*c*) 25 (*d*) 50

32. The average marks of 40 students in a English exam is 72. Later it is found that three marks 64, 62 and 84 were wrongly entered as 68, 65 and 73. The average after mistakes were rectified is

(*a*) 70 (*b*) 72 (*c*) 71.9 (*d*) 72.1

33. Nitin bought some oranges at ₹ 40 a dozen and an equal number at ₹ 30 a dozen. He sold them at ₹ 45 a dozen and made a profit of ₹ 480. The number of oranges, he bought, was

(*a*) 48 dozen (*b*) 60 dozen (*c*) 72 dozen (*d*) 84 dozen

34. A man buys two chairs for a total cost of ₹ 900. By selling one for $\frac{4}{5}$ of its cost and the other for $\frac{5}{4}$ of its cost, he makes a profit of ₹ 90 on the whole transaction. The cost of the lower priced chair is

(*a*) ₹ 360 (*b*) ₹ 400 (*c*) ₹ 420 (*d*) ₹ 300

35. By selling 100 oranges, a vendor gains the selling price of 20 oranges. His gain percent is

(*a*) 20 (*b*) 25 (*c*) 30 (*d*) 32

36. 60% of the cost price of an article is equal to 50% of its selling price. Then the percentage of profit or loss on the cost price is

(*a*) 20% loss (*b*) $16\frac{2}{3}$% profit (*c*) 20% profit (*d*) 10% loss

37. Maninder bought two horses at ₹ 40,000 each. He sold one horse at 15% gain, but had to sell the second horse at a loss. If he had suffered a loss of ₹ 3,600 on the whole transaction, then the selling price of the second horse is

(*a*) ₹ 30,000 (*b*) ₹ 30,200 (*c*) ₹ 30,300 (*d*) ₹ 30,400

38. A fruit-seller buys x guavas for ₹ y and sell y guavas for ₹ x. If $x > y$, then he made

(a) $\frac{x^2 - y^2}{xy}$% loss (b) $\frac{x^2 - y^2}{xy}$% gain (c) $\frac{x^2 - y^2}{y^2}$% loss (d) $\frac{x^2 - y^2}{y^2}$% gain

39. A jar contain 10 red marbles and 30 green ones. How many red marbles must be added to the jar so that 60% of the marbles will be red ?

(a) 25 (b) 30 (c) 35 (d) 40

40. If a number multiplied by 25% of itself gives a number which is 200% more than the number, then the number is

(a) 12 (b) 16 (c) 20 (d) 24

41. The value of an article depreciates every year at the rate of 10% of its value. If the present value of the article is ₹ 729, then its worth 3 years ago was

(a) ₹ 1250 (b) ₹ 1000 (c) ₹ 1125 (d) ₹ 1200

42. The price of onions has been increased by 50%. In order to keep the expenditure on onions the same the percentage of reduction in consumption has to be

(a) 50% (b) $33\frac{1}{3}$% (c) 33% (d) 30%

43. Walking at 3 km per hour, Pintu reaches his school 5 minutes late. If he walks at 4 km per hour he will be 5 minutes early. The distance of Pintu's school from his house is

(a) $1\frac{1}{2}$ km (b) 2 km (c) $2\frac{1}{2}$ km (d) 5 km

44. A motor-boat can travel at 10 km/hour in still water. It travelled 91 km downstream in a river and then returned to the same place, taking altogether 20 hours. Find the rate of flow of river.

(a) 3 km/hour (b) 4 km/hour (c) 2 km/hour (d) 5 km/hour

45. A man driving at $\frac{3}{4}$th of his original speed reaches his destination 20 minutes later than the usual time. Then the usual time is

(a) 45 minutes (b) 60 minutes (c) 75 minutes (d) 120 minutes

46. A motor-boat, travelling at the same speed, can cover 25 km upstream and 39 km downstream in 8 hours. At the same speed, it can travel 35 km upstream and 52 km downstream in 11 hours. The speed of the stream is

(a) 2 km/hr (b) 3 km/hr (c) 4 km/hr (d) 5 km/hr

47. If a sum of money placed at compound interest, compounded annually, doubles itself in 5 years, then the same amount of money will be 8 times of itself in **(SSC CGL 2002)**

(a) 26 years (b) 20 years (c) 15 years (d) 10 years

48. A person has left an amount of ₹ 1,20,000 to be divided between his 2 sons aged 14 years and 12 years such that they get equal amounts when each attains 18 years of age. If the amount gets a simple interest of 5% per annum, the younger son's share at present is

(a) ₹ 48,800 (b) ₹ 57,600 (c) ₹ 62,400 (d) ₹ 84,400

49. If the simple interest on ₹ x at a rate of a% for m years is same as that on ₹ y at a rate of a^2 % for m^2 years, then $x : y$ is equal to

(a) $m : a$ (b) $am : 1$ (c) $\frac{1}{m} : \frac{1}{a}$ (d) $\frac{1}{am} : 1$

50. A took two loans altogether of ₹ 1200 from *B* and *C*. *B* claimed 14% simple interest per annum, while *C* claimed 15% per annum. The total interest paid by *A* in one year was ₹ 172. Then, *A* borrowed

(a) ₹ 800 from *C* (b) ₹ 625 from *C* (c) ₹ 400 from *B* (d) ₹ 800 from *B*

51. If a regular polygon has each of its angle equal to $\frac{3}{5}$ times of two right angles, the number of sides is

(*a*) 3 (*b*) 5 (*c*) 6 (*d*) 8

52. A square is of area 200 sq. m. A new square is formed in such a way that the length of its diagonal is $\sqrt{2}$ times of the diagonal of the given square. Then the area of the new square formed is

(*a*) $200\sqrt{2}$ sq. m (*b*) $400\sqrt{2}$ sq. m (*c*) 400 sq. m (*d*) 800 sq. m

53. The heights of a cone, cylinder and hemisphere are equal. If their radii are in the ratio 2 : 3 : 1, then the ratio of their volumes is

(*a*) 2 : 9 : 2 (*b*) 4 : 9 : 1 (*c*) 4 : 27 : 2 (*d*) 2 : 3 : 1

54. Base of a right pyramid is a square, length of diagonal of the base is $24\sqrt{2}$ m. If the volume of the pyramid is 1728 cu. m, its height is

(*a*) 7 m (*b*) 8 m (*c*) 9 m (*d*) 10 m

55. The height of a right circular cone and the radius of its circular base are respectively 9 cm and 3 cm. The cone is cut by a plane parallel to its base so as to divide it into two parts. The volume of the frustum (*i.e.*, the lower part) of the cone is 44 cubic cm. The radius of the upper circular surface of the frustum $\left(\text{taking } \pi = \frac{22}{7}\right)$ is

(*a*) $\sqrt[3]{12}$ cm (*b*) $\sqrt[3]{13}$ cm (*c*) $\sqrt[3]{6}$ cm (*d*) $\sqrt[3]{20}$ cm

56. The ratio of radii of two right circular cylinders is 2 : 3 and their heights are in the ratio 5 : 4. The ratio of their curved surface area is

(*a*) 5 : 6 (*b*) 3 : 4 (*c*) 4 : 5 (*d*) 2 : 3

57. A solid cylinder has total surface area of 462 sq. cm. Curved surface area is $\frac{1}{3}$rd of its total surface area. The volume of the cylinder is

(*a*) 530 cm^3 (*b*) 536 cm^3 (*c*) 539 cm^3 (*d*) 545 cm^3

58. A cylinder and a cone have equal radii of their bases and equal heights. If their curved surface areas are in the ratio 8 : 5, the ratio of their radius and height is

(*a*) 1 : 2 (*b*) 1 : 3 (*c*) 2 : 3 (*d*) 3 : 4

59. A solid is hemispherical at the bottom and conical above. If the surface areas of the two parts are equal, then the ratio of radius and height of its conical part is

(*a*) 1 : 3 (*b*) 1 : 1 (*c*) $\sqrt{3} : 1$ (*d*) $1 : \sqrt{3}$

60. Base of a right prism is an equilateral triangle of side 6 cm. If the volume of the prism is $108\sqrt{3}$ cm, its height is

(*a*) 9 cm (*b*) 10 cm (*c*) 11 cm (*d*) 12 cm

61. If $a + \frac{1}{a} + 2 = 0$, then the value of $\left(a^{37} - \frac{1}{a^{100}}\right)$ is

(*a*) 0 (*b*) –2 (*c*) 1 (*d*) 2

62. The value of *k* for which the graphs of $(k - 1)x + y - 2 = 0$ and $(2 - k)x - 3y + 1 = 0$ are parallel is

(*a*) $\frac{1}{2}$ (*b*) $-\frac{1}{2}$ (*c*) 2 (*d*) –2

63. If $a^2 + b^2 + c^2 = 2(a - b - c) - 3$, then the value of $(a - b + c)$ is

(*a*) –1 (*b*) 3 (*c*) 1 (*d*) –2

64. If $x^2 + 3x + 1 = 0$, then the value of $x^3 + \frac{1}{x^3}$ is

(*a*) –18 (*b*) 18 (*c*) 36 (*d*) –36

65. If $x^a . x^b . x^c = 1$, then the value of $a^3 + b^3 + c^3$ is

(*a*) 9 (*b*) abc (*c*) $a + b + c$ (*d*) $3abc$

66. The graphs of $x + 2y = 3$ and $3x - 2y = 1$ meet the *Y*-axis at two points having distance

(*a*) $\frac{8}{3}$ units (*b*) $\frac{4}{3}$ units (*c*) 1 unit (*d*) 2 units

67. If $x + \frac{1}{16x} = 1$, then the value of $64x^3 + \frac{1}{64x^3}$ is

(*a*) 4 (*b*) 52 (*c*) 64 (*d*) 76

68. If *a*, *b*, *c* are three non-zero real numbers such that $a + b + c = 0$, and $b^2 \neq ca$, then the value of $\frac{a^2 + b^2 + c^2}{b^2 - ca}$ is

(*a*) 3 (*b*) 2 (*c*) 0 (*d*) 1

69. If $a^4 + a^2b^2 + b^4 = 8$ and $a^2 + ab + b^2 = 4$, then the value of ab is

(*a*) – 1 (*b*) 0 (*c*) 2 (*d*) 1

70. If $a = 25$, $b = 15$, $c = -10$, then the value of $\frac{a^3 + b^3 + c^3 - 3abc}{(a-b)^2 + (b-c)^2 + (c-a)^2}$ is

(*a*) 30 (*b*) –15 (*c*) –30 (*d*) 15

71. *A*, *B*, *C* are three points on a circle. The tangent at *A* meets *BC* produced at *T*, $\angle BTA = 40°$, $\angle CAT = 44°$. The angle subtended by *BC* at the centre of the circle is

(*a*) 84° (*b*) 92° (*c*) 96° (*d*) 104°

72. If the length of a chord of a circle at a distance of 12 cm from the centre is 10 cm, then the diameter of the circle is

(*a*) 13 cm (*b*) 15 cm (*c*) 26 cm (*d*) 30 cm

73. In ΔABC, *P* and *Q* are the middle points of the sides *AB* and *AC* respectively. *R* is a point on the segment *PQ* such that $PR : RQ = 1 : 2$. If $PR = 2$ cm, then $BC = ?$

(*a*) 4 cm (*b*) 2 cm (*c*) 12 cm (*d*) 6 cm

74. If *O* is the circumcentre of ΔABC and $\angle OBC = 35°$, then the $\angle BAC$ is equal to

(*a*) 55° (*b*) 110° (*c*) 70° (*d*) 35°

75. If *I* is the incentre of ΔABC and $\angle BIC = 135°$, then ΔABC is

(*a*) acute angled (*b*) equilateral (*c*) right angled (*d*) obtuse angled

76. If $\sin^2 \alpha + \sin^2 \beta = 2$, then the value of $\cos\left(\frac{\alpha + \beta}{2}\right)$ is

(*a*) 1 (*b*) – 1 (*c*) 0 (*d*) 0.5

77. The length of a shadow of a vertical tower is $\frac{1}{\sqrt{3}}$ times its height. The angle of elevation of the Sun is

(*a*) 30° (*b*) 45° (*c*) 60° (*d*) 90°

78. If $\sin\theta + \cos\theta = \frac{17}{13}, 0 < \theta < 90°$, then the value of $\sin\theta - \cos\theta$ is

(a) $\frac{5}{17}$ (b) $\frac{3}{19}$ (c) $\frac{7}{10}$ (d) $\frac{7}{13}$

79. The value of $\cot\frac{\pi}{20}\cot\frac{3\pi}{20}\cot\frac{5\pi}{20}\cot\frac{7\pi}{20}\cot\frac{9\pi}{20}$ is

(a) –1 (b) $\frac{1}{2}$ (c) 0 (d) 1

80. If $\tan\theta \,.\, \tan 2\theta = 1$, then the value of $\sin^2 2\theta + \tan^2 2\theta$ is equal to

(a) $\frac{3}{4}$ (b) $\frac{10}{3}$ (c) $3\frac{3}{4}$ (d) 3

***Directions:** (Q. No. 81 to 85). The following line diagram represents the yearly sales figures of a company in the years 2001–2010. Examine the diagram and answer the questions 81 to 85.*

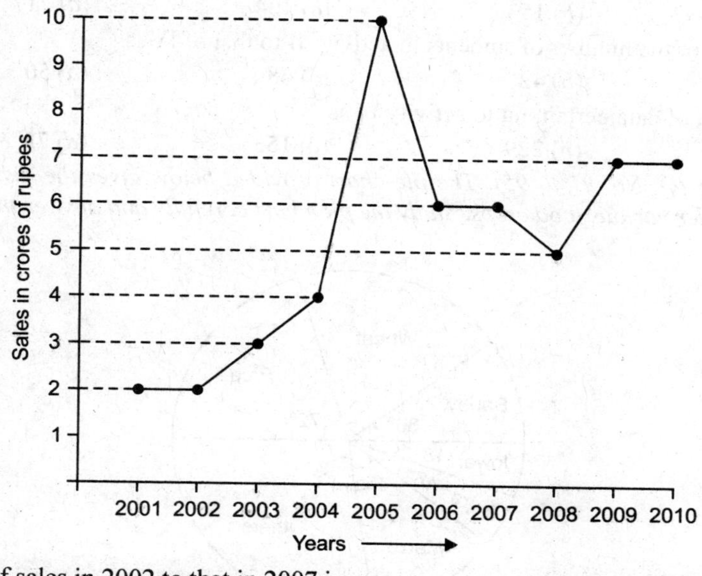

81. The ratio of sales in 2002 to that in 2007 is

(a) 2 : 3 (b) 1 : 3 (c) 1 : 1 (d) 3 : 2

82. Average sale (in crores of ₹) of the company during the period 2003 – 2007 is

(a) 5.8 (b) 5 (c) 6 (d) 5.5

83. The percentage increase in sales in the year 2005 with respect to the previous year is

(a) 80 (b) 100 (c) 120 (d) 150

84. Total sales (in crores of ₹) from 2005 to 2008 is

(a) 17 (b) 27 (c) 22 (d) 31

85. By what percent did the sales in 2008 decrease in comparison to the sales in 2006 ?

(a) 20 (b) 18 (c) $16\frac{2}{3}$ (d) $15\frac{2}{3}$

***Directions:** (Q. No. 86 to 90). The following table shows the number of students of 7 colleges participating in extra curricular activities:*

Extra Curricular Activities	Colleges						
	A	B	C	D	E	F	G
I	200	300	500	100	400	300	200
II	100	200	200	100	100	100	100
III	65	130	420	75	540	220	153
IV	317	155	438	105	385	280	120

Read the table and answer questions 86 to 90.

86. The college in which minimum number of students participate in extra curricular activities is
(*a*) D (*b*) G (*c*) F (*d*) A

87. The ratio of total number of students in II and I is
(*a*) 1 : 2 (*b*) 9 : 20 (*c*) 19 : 7 (*d*) 21 : 10

88. The difference of the range of number of students in activity IV and the average of number of students in activity III per college is
(*a*) 111 (*b*) 153 (*c*) 104 (*d*) 217

89. Percentage of the number of students in activity II to that of IV is
(*a*) 37 (*b*) 42 (*c*) 48 (*d*) 50

90. The median of data pertaining to activity III is
(*a*) 540 (*b*) 229 (*c*) 153 (*d*) 75

Directions: *(Q. No. 91 to 95). The pie-chart provided below gives the distribution of land (in a village) under various food crops. Study the pie-chart carefully and answer the questions from 91 to 95.*

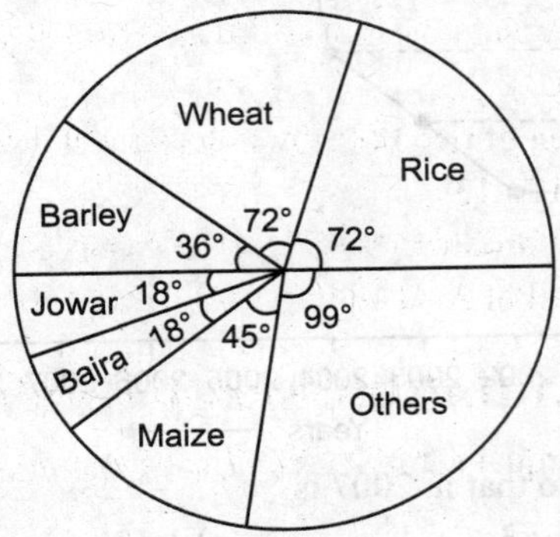

91. The combination of three crops which contribute to more than 50% of the total area under the food crops is
(*a*) wheat, rice and maize (*b*) wheat, rice and jowar
(*c*) wheat, rice and bajra (*d*) rice, barley and maize

92. The ratio of the land used for rice and barley is
(*a*) 3 : 1 (*b*) 1 : 2 (*c*) 2 : 1 (*d*) 3 : 2

93. If 10% of the land reserved for rice be distributed to wheat and barley in the ratio 2 : 1, then the angle corresponding to wheat in the new pie-chart will be
(*a*) 38.4° (*b*) 76.8° (*c*) 75.6° (*d*) 45.5°

94. If the production of rice is 5 times that of jowar and the production of jowar is 2 times that of bajra, then the ratio between the yield per acre of rice and bajra is
(*a*) 5 : 2 (*b*) 3 : 1 (*c*) 4 : 1 (*d*) 6 : 1

95. If the total area under bajra was three hundred acres, then the total area (in hundred acres) under rice and barley together is

(*a*) 18 (*b*) 12 (*c*) 15 (*d*) 20

Directions: *(Q. No. 96 to 100). The bar graph provided below represents the production of rice and wheat in different states of a country in a certain year. Answer questions no. 96 to 100 based on the bar graph.*

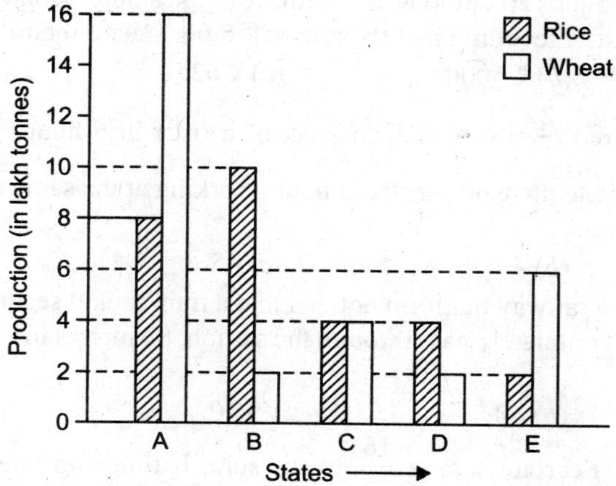

96. The state which is the largest producer of rice is

(*a*) A (*b*) B (*c*) C (*d*) D

97. The average of production of rice in the mentioned states (in lakh tonnes) is

(*a*) 5.5 (*b*) 5.6 (*c*) 5.7 (*d*) 5.8

98. The total production of rice and wheat in all the mentioned states is minimum in the state

(*a*) B (*b*) C (*c*) D (*d*) E

99. The ratio of total production of rice in the mentioned states to that of wheat in those states, is

(*a*) 15 : 16 (*b*) 12 : 13 (*c*) 13 : 14 (*d*) 14 : 15

100. The difference between the production in rice and wheat is maximum in

(*a*) A only (*b*) all of A, B and E (*c*) B and E both (*d*) A a nd B both

FOR VISUALLY HANDICAPPED CANDIDATES ONLY

81. A cube and a sphere have equal surface area. Then the ratio of the volume of the cube to that of the sphere is **(SSC MTS 2011)**

(*a*) $\pi : 6$ (*b*) $\sqrt{\pi} : \sqrt{6}$ (*c*) $\sqrt{6} : \sqrt{\pi}$ (*d*) $6 : \pi$

82. If $x = \sqrt{11} + \sqrt{5}$, $y = \sqrt{10} + \sqrt{6}$, $z = \sqrt{3} + \sqrt{13}$, then

(*a*) $x > y > z$ (*b*) $y > z > x$ (*c*) $x > z > y$ (*d*) $y > x > z$

83. The ratio of two numbers is 4 : 5. Their H.C.F. is 5. Their L.C.M. is

(*a*) 20 (*b*) 25 (*c*) 100 (*d*) 400

84. $\sqrt{\dfrac{1}{\sqrt{4}-\sqrt{3}} - \dfrac{1}{\sqrt{3}+\sqrt{2}} + \dfrac{1}{\sqrt{2}-1}}$ is equal to

(*a*) $\sqrt{\sqrt{3}+\sqrt{2}}$ (*b*) $1+\sqrt{2}$ (*c*) $\sqrt{3}$ (*d*) $\sqrt{2}\,(\sqrt{2}-\sqrt{3})$

85. The number of prime factors of the number $(60)^{15} \times (1.5)^8 \times (3.5)^7$ is

(*a*) 58 (*b*) 60 (*c*) 53 (*d*) 57

86. The value of $(\sqrt{2}+\sqrt{3}-\sqrt{5})^2 - 2\sqrt{6} + 2\sqrt{15} + 2\sqrt{10}$ is

(a) 0 (b) 4 (c) 10 (d) 3

87. The average of 8 numbers is a and the average of 2 numbers among them is b. If the average of remaining 6 numbers is c, then

(a) $4a - c = 3b$ (b) $2c - b = 2a$ (c) $3a + b = 4c$ (d) $4a - b = 3c$

88. The diameter of a roller 120 cm long is 84 cm. If it takes 500 complete revolutions to level a playground, the cost of levelling it at the rate of ₹ 5 per square metre is

(a) ₹ 4750 (b) ₹ 5000 (c) ₹ 6250 (d) ₹ 7920

89. 100 villagers repaired $\frac{2}{3}$rd of a broken bank of a river in 6 hours. More villagers join the repairing and complete the work in next 2 hours working at the same rate. The number of new villagers involved is

(a) 30 (b) 40 (c) 45 (d) 50

90. A man standing on a railway platform noticed that a train took 9 seconds in passing him and it took 21 seconds to completely pass through the station, 88 metres long. The length of the train, in metres, was

(a) 66 (b) 60 (c) 70 (d) 65

91. Simple interest on a certain sum is $\frac{16}{25}$ of the sum. If time and rate percent are equal, then time is

(a) 6 years (b) 6.5 years (c) 7 years (d) 8 years

92. The difference between the compound interest and simple interest on a certain sum of money at 10% per annum for 2 years is ₹ 500. The sum is

(a) ₹ 45,000 (b) ₹ 50,000 (c) ₹ 55,000 (d) ₹ 60,000

93. Given $a + b + c = 18$, $b + c + d = 22$, $c + d + a = 26$, $d + a + b = 30$. The average of $a + b + c + d$ is

(a) 32 (b) 24 (c) 16 (d) 8

94. In a test consisting of 80 questions, a student answered 75% of the first 60 questions correctly. The percentage of the other questions she needs to answer correctly so that her score on the entire examination be 80% is

(a) 20% (b) 25% (c) 50% (d) 95%

95. If 8% of x is y, then y% of 25 is

(a) 6% of x (b) 2% of x (c) 5% of x (d) 10% of x

96. A 30 litre solution has 20% salt content. The quantity of salt content to be added with the solution to make the salt content in it as 25% is

(a) 4 litres (b) 3 litres (c) 2 litres (d) 1 litre

97. If $\frac{2}{3}$ of the selling price of article is loss, loss percentage is

(a) 60% (b) 50% (c) 40% (d) 30%

98. P being a working partner gets 10% of the profit as salary, the remaining is shared between P, Q, R in the ratio 2 : 3 : 4. If P gets ₹ 3,00,000, the share of Q, in rupees, is

(a) 2,00,000 (b) 3,00,000 (c) 4,00,000 (d) 5,00,000

99. A invested ₹ 76,000 in a business. After a few months, B joined him with ₹ 57,000. At the end of the year, the total profit was divided between them in the ratio 2 : 1. If the profit was divided according to their equivalent capital share, B joined after

(a) 4 months (b) 6 months (c) 3 months (d) 5 months

100. *A* and *B* can do a work in 18 days, *B* and *C* in 24 days and *A* and *C* in 36 days. *C* can do the work alone in

(*a*) $\frac{144}{5}$ days (*b*) 48 days (*c*) 96 days (*d*) 144 days

ANSWER

1. (*c*) **2.** (*d*) **3.** (*c*) **4.** (*c*) **5.** (*c*) **6.** (*d*) **7.** (*d*) **8.** (*b*)
9. (*b*) **10.** (*c*) **11.** (*d*) **12.** (*b*) **13.** (*a*) **14.** (*d*) **15.** (*c*) **16.** (*a*)
17. (*c*) **18.** (*b*) **19.** (*d*) **20.** (*a*) **21.** (*a*) **22.** (*b*) **23.** (*c*) **24.** (*c*)
25. (*d*) **26.** (*b*) **27.** (*d*) **28.** (*b*) **29.** (*d*) **30.** (*a*) **31.** (*d*) **32.** (*d*)
33. (*a*) **34.** (*d*) **35.** (*b*) **36.** (*c*) **37.** (*d*) **38.** (*d*) **39.** (*c*) **40.** (*a*)
41. (*b*) **42.** (*b*) **43.** (*b*) **44.** (*a*) **45.** (*b*) **46.** (*c*) **47.** (*c*) **48.** (*b*)
49. (*b*) **50.** (*d*) **51.** (*b*) **52.** (*c*) **53.** (*c*) **54.** (*c*) **55.** (*b*) **56.** (*a*)
57. (*c*) **58.** (*d*) **59.** (*d*) **60.** (*d*) **61.** (*b*) **62.** (*a*) **63.** (*c*) **64.** (*a*)
65. (*d*) **66.** (*d*) **67.** (*b*) **68.** (*b*) **69.** (*d*) **70.** (*d*) **71.** (*d*) **72.** (*c*)
73. (*c*) **74.** (*a*) **75.** (*c*) **76.** (*c*) **77.** (*c*) **78.** (*d*) **79.** (*d*) **80.** (*c*)
81. (*b*) **82.** (*a*) **83.** (*d*) **84.** (*b*) **85.** (*c*) **86.** (*a*) **87.** (*b*) **88.** (*c*)
89. (*d*) **90.** (*c*) **91.** (*a*) **92.** (*c*) **93.** (*b*) **94.** (*a*) **95.** (*a*) **96.** (*b*)
97. (*b*) **98.** (*c*) **99.** (*d*) **100.** (*d*)

FOR VISUALLY HANDICAPPED CANDIDATES

81. (*b*) **82.** (*d*) **83.** (*c*) **84.** (*) **85.** (*b*) **86.** (*c*) **87.** (*d*) **88.** (*d*)
89. (*d*) **90.** (*a*) **91.** (*d*) **92.** (*b*) **93.** (*d*) **94.** (*d*) **95.** (*b*) **96.** (*c*)
97. (*c*) **98.** (*b*) **99.** (*a*) **100.** (*d*)

HINT

1. $(-1)^{67} + 67 = 66$

2. $a\,\overline{)\,b\,}$ quotient $\frac{a}{4}$, remainder $\frac{a}{2}$ $\quad b = a\times\frac{a}{4}+\frac{a}{2}=\frac{a^2}{4}+\frac{a}{2}=\frac{a(a+2)}{4} \Rightarrow \frac{a(a+2)}{b}=4$

3. 7 3 8 A 6 A (3 + 2A above; 21 below) $\quad 21 = 3 + 2A \Rightarrow A = 9$

4. $xy = 1575$ and $\frac{x}{y}=\frac{9}{7} \Rightarrow x^2 = xy\times\frac{x}{y}=1575\times\frac{9}{7} = 225 \times 9 \Rightarrow x = 15 \times 3 = 45$

$y = x\times\frac{7}{9}=45\times\frac{7}{9}=35 \quad x + y = 45 + 35 = 80$

5. $81^{3.6-4.2} = 81^{-0.6}$; $81^{-0.6} \times 9^{2.7} = 9^{2.7-1.2} = 9^{1.5} = 3^3$; $3^3 \div 3 = 3^2 = 9$

6. $x^2 = x + 6 \Rightarrow x = 3$ Alt: $6 = 2 \times 3 \Rightarrow$ Ans. = 3

7. $(x-1)^2 + (x+1)^2 = 394 \Rightarrow 2x^2 + 2 = 394 \Rightarrow x^2 = 197 - 1 = 196 \Rightarrow x = 14$

$\therefore$ The numbers are 13 and 15. $\Rightarrow$ sum of numbers = 13 + 15 = 28

8. $\dfrac{x^3}{1-x}+(x+x^2+1)=\dfrac{-x^3}{x-1}+(x^2+x+1)=\dfrac{-x^3+x^3-1}{x-1}=\dfrac{-1}{-0.25}=4, \sqrt{4}=2$

9. $64+6.4+0.064=70.464$

10. $3011\times 3012=3011\times(3011+1)=3011^2+3011$

11. $\dfrac{19}{23}+\dfrac{8}{23}=\dfrac{27}{23}$ $\qquad Q=\dfrac{4}{23}\times 5750=₹\,1000$

12. A's work for first 6 days $=\dfrac{1}{24}\times 6=\dfrac{1}{4}$

C's work for last 6 days $=\dfrac{1}{64}\times 6=\dfrac{3}{32}$

Total work done $=\dfrac{1}{4}+\dfrac{3}{32}=\dfrac{11}{32}\Rightarrow$ Remaining work $=1-\dfrac{11}{32}=\dfrac{21}{32}$

Days taken by B and C $=\dfrac{21}{32}\times\dfrac{32\times 94}{96}=14$ days

Total days taken = 6 + 14 = 20 days

13. $\dfrac{1}{3}+\dfrac{1}{4}-1=\dfrac{-5}{12}$; Work done till 5 p.m. $=\dfrac{2}{3}+\dfrac{1}{4}=\dfrac{11}{12}$

Time taken to empty the tank $=\dfrac{11}{12}\times\dfrac{12}{5}=\dfrac{11}{5}=2\dfrac{1}{5}$ hour = 2 hour 12 minutes.

14. $A=(A+B)+4$ and $B=(A+B)+16$

$\dfrac{(x+4)(x+16)}{(x+4)+(x+16)}=x$ $\qquad$ **Alt:** $\sqrt{16\times 4}=8$

$\Rightarrow x^2+20x+64=2x^2+20x \Rightarrow x^2=64\Rightarrow x=8$

15.

Men	Days	Hours
250	20	5
?	10	8

$?=250\times\dfrac{20}{10}\times\dfrac{5}{8}=\dfrac{625}{2}\sim 313$

16.

Men	Women	Days	$\Rightarrow$	Men	Women	Days
2	5	12		24	60	1
5	2	9		45	18	1

$\Rightarrow$ 21 men = 42 women $\Rightarrow$ 1 men = 2 women $\qquad \dfrac{9\times 12}{3}=36$ days

17. 60% of MP = 70% of CP $\Rightarrow$ MP $=70\times\dfrac{100}{60}=\dfrac{350}{3}$ $\qquad$ Profit $=\dfrac{350}{3}-1=\dfrac{50}{3}=16\dfrac{2}{3}\%$

18. 10% and 20% = (10 + 20 – 2)% = 28% $\qquad$ 28% and 50% = 28 + 50 – 14 = 64%

19. 20%

20. $\dfrac{4}{3}\times\dfrac{125}{100}=\dfrac{5}{3}$

21. Saving ratio = 4 : 1 $\Rightarrow$ saving = 4000 and 1000

Let their incomes be $2x$ and x. Their expenditures be $5y$ and $3y$.

Then $2x - 5y = 4000$ and $x - 3y = 1000 \Rightarrow 2x - 6y = 2000$

Solving $y = 2000 \Rightarrow x = 7000$

22. $(5 + 1) : (5 - 1) = 3 : 2$

Alt: $\frac{x+y}{x-y} = \frac{5}{1} \Rightarrow \frac{x}{y} = \frac{3}{2}$

23. $\frac{7}{9} + \frac{7}{18} = \frac{21}{18} \Rightarrow$ Required ratio $= 21 : (2 \times 18 - 21) = 21 : 15 = 7 : 5$

24.

2/7 7/10

2/5

3/10 4/35

Ratio $= \frac{3}{10} : \frac{4}{35} = \frac{3}{2} : \frac{4}{7} = 21 : 8$

25.

4/5 0

4/7

4/7 8/35

Ratio $= \frac{4}{7} : \frac{8}{35} = 1 : \frac{2}{5} = 5 : 2 \Rightarrow$ Part repplaced $= \frac{2}{7}$

26.

9 : 8

× 14 : 15

126 : 120

Ratio = 126 : 120 = 21 : 20

27. Score of 2 innings $= 2 \times 50 + 38 \times (50 - 48) = 176$

Highest score $= \frac{1}{2} \times (176 + 172) = 174$

28. Ratio of three numbers = 4 : 12 : 3 $\Rightarrow$ Largest number $= (114 \times 3) \times \frac{12}{19} = 216$

29. $\frac{1}{5}$ @ 8 km/hr ; $\frac{1}{10}$ @ 25 km/hr ; $\frac{7}{10}$ @ 20 km/hr

$\frac{200}{8} + \frac{100}{25} + \frac{700}{20} = 25 + 4 + 35 = 64$ hours $\Rightarrow$ Average speed $= \frac{1000}{64} = 15.625$ km/hr

30. Ratio of numbers = 4 : 2 : 1 $\Rightarrow$ First number $= (154 \times 3) \times \frac{4}{7} = 264$

31.

4 14

5

9 1

Officers $= \frac{1}{10} \times 500 = 50$

32. Difference $= (64 + 62 + 84) - (68 + 65 + 73) = -4 - 3 + 11 = 4$

Actual average $= 72 + \frac{4}{40} = 72.1$

33. Profit on selling 2 dozen oranges $= 2 \times 45 - (40 + 30) =$ ₹ 20

Required number of oranges $= \frac{2}{20} \times 480 = 48$ dozen

34. 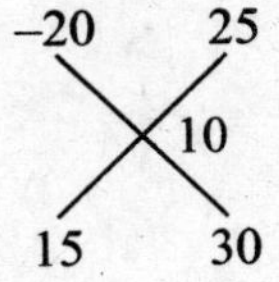

Ratio = 1 : 2

Cost price of lower priced chair = $\frac{1}{3} \times 900 =$ ₹ 300

35. $\frac{20}{100-20} = \frac{20}{80} = \frac{1}{4} = 25\%$

36. CP : SP = 5 : 6 ⇒ Profit = $\frac{1}{5} = 20\%$

37. Profit on first horse = 15% of 40,000 = ₹ 6,000. Total loss = ₹ 3,600
Loss on second horse = 6,000 + 3,600 = ₹ 9,600
∴ Selling price of second horse = 40,000 – 9,600 = ₹ 30,400

38. CP = $\frac{y}{x}$ and SP = $\frac{x}{y}$ ⇒ Profit = $\frac{x^2 - y^2}{xy}$ ⇒ profit % = $\frac{x^2 - y^2}{xy} \div \frac{y}{x} = \frac{x^2 - y^2}{y^2}\%$

39. Green marbles = 30 = 40% of total marbles
Red marbles = 60% of total marbles = $30 \times \frac{60}{40} = 45$ marbles
∴ Red marbles added = 45 – 10 = 35

40. $\frac{1}{4}x^2 = 3x \Rightarrow x = 4 \times 3 = 12$

41. $729 = 9^3$ ⇒ Required value = $10^3 = 1000$

42. Increase = 50% = $\frac{1}{2}$ ⇒ Decrease = $\frac{1}{3} = 33\frac{1}{3}\%$ **Alt:** $\frac{50}{100+50} = \frac{1}{3} = 33\frac{1}{3}\%$

43. $\frac{3 \times 4}{1} \times \frac{1}{6} = 2$ km

44. Trick: 10 ± 3 = 13 and 7 both divisible by 91
$\frac{91}{13} + \frac{91}{7} = 7 + 13 = 20$ hours ⇒ Speed of stream = $\frac{1}{2}(13-7) = 3$ km/hr

45. $\frac{1}{3}$ of usual time = 20 minutes ⇒ usual time = (3 × 20) minutes = 60 minutes

46. 25 up + 39 down = 8 hours
35 up + 52 down = 11 hours
Solving, 13 down-stream in 1 hours and 5 up-stream in 1 hour
Speed of stream = $\frac{13-5}{2} = 4$ km/hr
Trick: HCF of 25 and 35 is 5. HCF of 39 and 52 is 13.

47. $2^3 = 8$ ⇒ Required Time = 3 × 5 years = 15 years.

48. Ratio of capitals = $\frac{1}{120} : \frac{1}{130} = 13 : 12$
Share of younger son = $\frac{12}{25} \times 1{,}20{,}000 =$ ₹ 57,600

49. $\frac{1}{am}:\frac{1}{a^2m^2}=a^2m^2:am=am:1$

50. 14% of 1200 = ₹ 168. Extra interest = ₹ 172 – ₹ 168 = ₹ 4

Loan at 15% = $\frac{4\times100}{1}$ = ₹ 400 $\Rightarrow$ Loan from B = 1200 – 400 = ₹ 800

51. Each angle of regular polygon = $\frac{n-2}{n}\times180°=\frac{3}{5}\times180°$ $\Rightarrow$ No. of sides = 5

52. Ratio of sides = Ratio of diagonals = $1:\sqrt{2}$ $\Rightarrow$ Ratio of areas = $1^2:(\sqrt{2})^2=1:2$

Required area = 2 × 200 = 400

53. Ratio of volumes = $\frac{1}{3}\pi r_1^2h:\pi r_2^2h:\frac{2}{3}\pi r_3^3$

$=\frac{1}{3}r_1^2:r_2^2:\frac{2}{3}r_3^2=\frac{1}{3}(2)^2:(3)^2:\frac{2}{3}(1)^2=\frac{4}{3}:9:\frac{2}{3}=4:27:2$

54. Area of base = $\frac{(24\sqrt{2})^2}{2}=24\times24$

Volume of pyramid = $\frac{1}{3}$ × Area of base × height = 1728 m^3 $\Rightarrow$ Height = $\frac{1728\times3}{24\times24}=9$ m

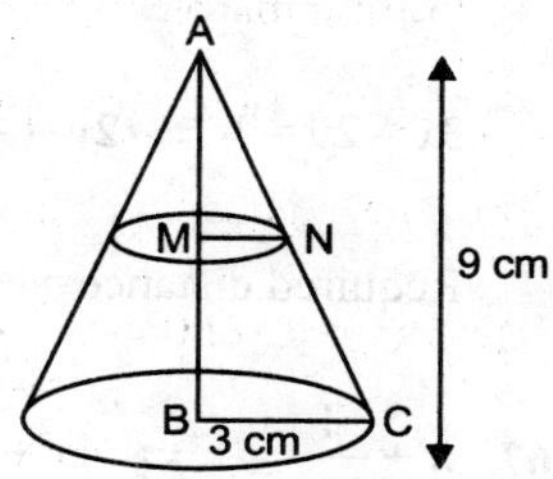

55. Volume of larger cone = $\frac{1}{3}\pi r^2h=\frac{1}{3}\pi(3)^2 9=27\pi$

Volume of frustum = 44 cm^3 = $44\times\frac{7}{22}\pi=14\pi$

Volume of smaller cone = $27\pi-14\pi=13\pi$

Ratio of volumes = $27\pi:13\pi=27:13$

Ratio of radii = $\sqrt[3]{27}:\sqrt[3]{13}=3:\sqrt[3]{13}$

[**Note:** Radius and Height of two cones are propotional as $\Delta ABC\sim\Delta AMN$]

56. Curved surface area of cylinder = $2\pi rh$ $\Rightarrow$ Reqd. ratio = 2 × 5 : 3 × 4 = 5 : 6

57. $2\pi rh+2\pi r^2=462$ $\Rightarrow$ $2\pi rh=\frac{1}{3}\times462=154$ $\Rightarrow$ $\pi rh=77$

$2\pi r^2=462-154=308$ $\Rightarrow$ $r^2=308\times\frac{1}{2}\times\frac{7}{22}=7^2$ $\Rightarrow$ $r=7$

Volume = $\pi r^2h=\pi rh\times r=77\times7=539$ cm^3

58. Ratio of curved surface areas = $2\pi rh:\pi rl=8:5$ $\Rightarrow$ $h:l=4:5$

$r=\sqrt{5^2-4^2}=3$ $\Rightarrow$ $r:h=3:4$

59. $2\pi r^2=\pi rl$ $\Rightarrow$ $l=2r$ $\Rightarrow$ $h=\sqrt{(2r)^2-r^2}=\sqrt{3}r$ $\Rightarrow$ $r:h=1:\sqrt{3}$

60. Area of base (equilateral triangle) = $\frac{\sqrt{3}}{4}\times6^2=9\sqrt{3}$

Volume of prism = Area of base × height = $108\sqrt{3}$ $\Rightarrow$ Height = $\frac{108\sqrt{3}}{9\sqrt{3}}=12$ cm

61. $a + \frac{1}{a} + 2 = 0 \Rightarrow a^2 + 2a + 1 = 0 \Rightarrow (a+1)^2 = 0 \Rightarrow a = -1$

$$\left(a^{37} - \frac{1}{a^{100}}\right) = (-1)^{37} - \frac{1}{1^{100}} = -1 - 1 = -2$$

62. Lines are parallel if $\frac{k-1}{2-k} = \frac{1}{-3} \Rightarrow -3k + 3 = 2 - k \Rightarrow 2k = 1 \Rightarrow k = 1/2$

63. $a^2 + b^2 + c^2 = 2(a - b - c) - 3 \Rightarrow a^2 + b^2 + c^2 - 2a + 2b + 2c + 3 = 0$

$\Rightarrow (a^2 - 2a + 1) + (b^2 + 2b + 1) + (c^2 + 2c + 1) = 0$

$\Rightarrow (a-1)^2 + (b+1)^2 + (c+1)^2 = 0 \Rightarrow a = 1, b = -1, c = -1$

$\therefore (a - b + c) = 1 + 1 - 1 = 1$

64. $x^2 + 3x + 1 = 0 \Rightarrow x^2 + 1 = -3x \Rightarrow x + \frac{1}{x} = -3$

$$\Rightarrow x^3 + \frac{1}{x^3} = \left(x + \frac{1}{x}\right)^3 - 3\left(x + \frac{1}{x}\right) = (-3)^3 - 3(-3) = -27 + 9 = -18$$

65. $x^a . x^b . x^c = 1 \Rightarrow x^{a+b+c} = x^0 \Rightarrow a + b + c = 0$

66. Lines meet at Y-axis when $x = 0$

$x + 2y = 3 \Rightarrow 2y = 3 \Rightarrow y = \frac{3}{2}$

$3x - 2y = 1 \Rightarrow 2y = -1 \Rightarrow y = -\frac{1}{2}$

Required distance $= \frac{3}{2} - \left(-\frac{1}{2}\right) = 2$

67. $x + \frac{1}{16x} = 1 \Rightarrow 4x + \frac{1}{4x} = 4$

$$\left(4x + \frac{1}{4x}\right)^3 = 64x^3 + \frac{1}{64x^3} + 3\left(4x + \frac{1}{4x}\right)$$

$$\Rightarrow 64x^3 + \frac{1}{64x^3} = 4^3 - 3 \times 4 = 64 - 12 = 52$$

68. $a + b + c = 0 \Rightarrow a + c = -b \Rightarrow a^2 + c^2 + 2ac = b^2 \Rightarrow a^2 + c^2 = b^2 - 2ac$

$$\therefore \frac{a^2 + b^2 + c^2}{b^2 - ca} = \frac{b^2 + b^2 - 2ac}{b^2 - ac} = 2$$

69. $(a^2 + ab + b^2)(a^2 - ab + b^2) = a^4 + a^2b^2 + b^4 \Rightarrow (a^2 - ab + b^2) = 8 \div 4 = 2$

$(a^2 + ab + b^2) - (a^2 - ab + b^2) = 4 - 2 \Rightarrow 2ab = 2 \Rightarrow ab = 1$

70. $$\frac{a^3 + b^3 + c^3 - 3abc}{(a-b)^2 + (b-c)^2 + (c-a)^2} = \frac{\frac{1}{2}(a+b+c)[(a-b)^2 + (b-c)^2 + (c-a)^2]}{(a-b)^2 + (b-c)^2 + (c-a)^2}$$

$$= \frac{1}{2}(a+b+c) = \frac{1}{2}(25 + 15 - 10) = \frac{1}{2} \times 30 = 15$$

71. $\angle ACT = 180° - 44° - 40° = 96°$
$\angle ACB = 180° - 96° = 84°$
$\angle ABC = \angle CAT = 44°$
$\angle BAC = \angle 180° - 84° - 44° = 52°$
Angle subtended by BC at centre $= 2 \times 52° = 104°$

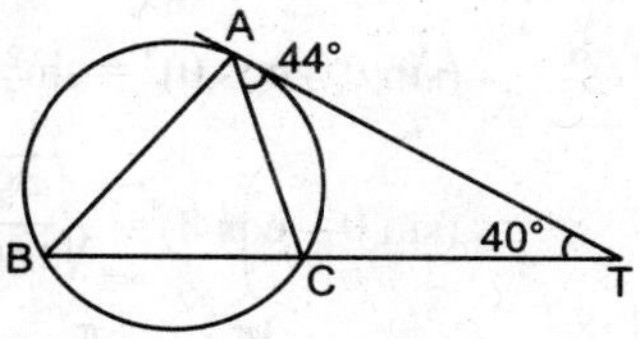

72. $r = \sqrt{12^2 + 5^2} = 13$ cm
Diameter = 26 cm

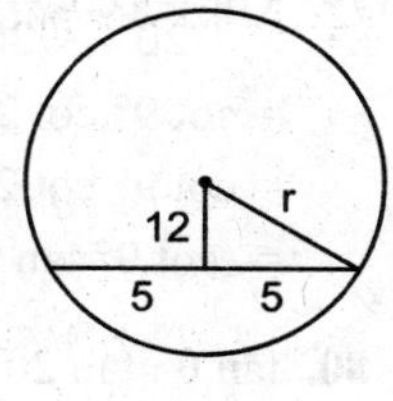

73. $RQ = 2 \times 2 = 4$ cm
$PQ = PR + RQ = 2 + 4 = 6$ cm
$\therefore BC = 2 \times 6$ cm $= 12$ cm

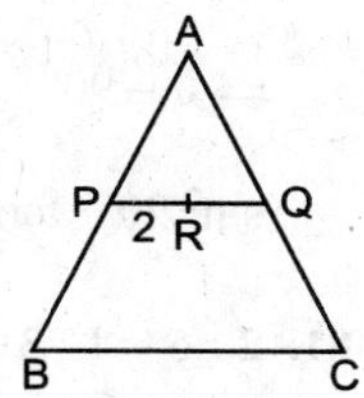

74. $\angle OCB = \angle OBC = 35°$
$\angle BOC = 180° - 35° - 35° = 110°$
$\angle BAC = \angle BOC \div 2 = 110° \div 2 = 55°$

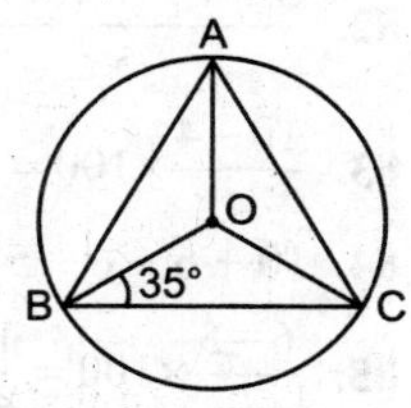

75. $\frac{1}{2}(\angle ABC + \angle BCA) = 180° - \angle BIC = 180° - 135° = 45°$
$\angle ABC + \angle BCA = 90°$
$\angle BAC = 180° - 90° = 90°$
$\therefore \Delta ABC$ is right angled

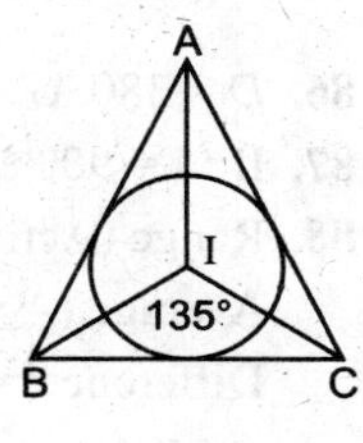

76. Given $\sin^2 \alpha + \sin^2 \beta = 2$
Maximum value of $\sin \theta$ is 1 when $\theta = 90°$
$\therefore \sin \alpha = \sin \beta = 1 \Rightarrow \alpha = \beta = 90°$
$\therefore \cos\left(\frac{\alpha + \beta}{2}\right) = \cos 90° = 0$

77. $\tan \theta = \frac{h}{h/\sqrt{3}} = \sqrt{3} = \tan 60° \Rightarrow \theta = 60°$

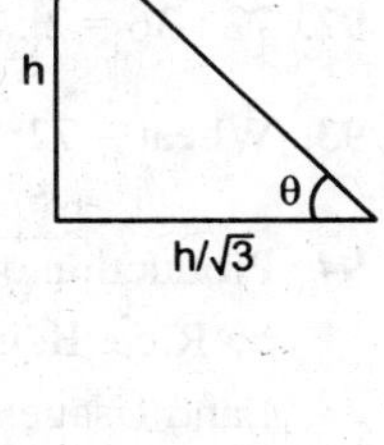

78. $\sin \theta + \cos \theta = \frac{17}{13} \Rightarrow \sin^2 \theta + \cos^2 \theta + 2 \sin \theta \cos \theta = \left(\frac{17}{13}\right)^2$

$\Rightarrow 2 \sin \theta \cos \theta = \frac{289}{169} - 1 = \frac{120}{169}$

$$(\sin\theta - \cos\theta)^2 = \sin^2\theta + \cos^2\theta - 2\sin\theta\cos\theta = 1 - \frac{120}{169} = \frac{49}{169}$$

$$\therefore\ (\sin\theta - \cos\theta) = \sqrt{\frac{49}{169}} = \frac{7}{13}$$

79. $\cot\frac{\pi}{20}\cot\frac{3\pi}{20}\cot\frac{5\pi}{20}\cos\frac{7\pi}{20}\cot\frac{9\pi}{20}$

$= \cot 9° \cot 27° \cot 45° \cot 63° \cot 81°$

$= \cot 9° \cot 27° \cot 45° \tan(90° - 63°) \tan(90° - 81°)$

$= (\cot 9° \tan 9°)(\cot 27° \tan 27°)\cot 45° = 1 \times 1 \times 1 = 1$

80. $\tan\theta \,.\, \tan 2\theta = 1 \Rightarrow \tan\theta = \frac{1}{\tan 2\theta} = \cot 2\theta = \tan(90° - 2\theta)$

$\Rightarrow \theta = 90° - 2\theta \Rightarrow 3\theta = 90° \Rightarrow \theta = 30°$

$\sin^2 2\theta + \tan^2 2\theta = \sin^2 60° + \tan^2 60° = \left(\frac{\sqrt{3}}{2}\right)^2 + (\sqrt{3})^2 = \frac{3}{4} + 3 = 3\frac{3}{4}$

81. $2 : 6 = 1 : 3$

82. $\frac{3+4+10+6+6}{5} = \frac{29}{5} = 5.8$

83. $\frac{10-4}{4}\times 100 = \frac{6}{4}\times 100 = 150\%$

84. $10 + 6 + 6 + 5 = 27$

85. $\frac{6-5}{6}\times 100 = \frac{1}{6}\times 100 = 16\frac{2}{3}\%$

86. $D = 380$, $G = 573$, $F = 900$, $A = 682 \Rightarrow D$ is minimum

87. II : I = 900 : 2000 = 9 : 20

88. Range (Activity IV) = 438 – 105 = 333

Average (Activity III) = 1603 ÷ 7 = 229

Difference = 333 – 229 = 104

89. $\frac{900}{1800}\times 100 = 50\%$

90. In ascending order: 65, 75, 130, 153, 220, 420, 540

91. Wheat, rice is common in (*a*), (*b*) and (*c*)

Maize > Jowar, Bajra;

In option (*a*) and (*d*): Rice > Maize

92. 72 : 36 = 2 : 1

93. Wheat $= 72° + 7.2° \times \frac{2}{3} = 72° + 4.8° = 76.8°$

94. Production of Rice: Jowar = 5 : 1, Jowar : Bajra = 2 : 1

$\Rightarrow$ Rice: Bajra = 10 : 1

Land Usage: Rice: Bajra = 72° : 18° = 4 : 1

$\therefore$ Required ratio $= \frac{10}{4} : \frac{1}{1} = 5 : 2$

95. Bajra = 18° = 3 ⇒ Rice + Barley = 72° + 36° = 108° = 6 × 18° = 6 × 3 = 18

96. B

97. $\frac{8+10+4+4+2}{5}=\frac{28}{5}=5\cdot6$

98. $B = 10 + 2 = 12$; $C = 4 + 4 = 8$; $D = 4 + 2 = 6$; $E = 2 + 6 = 8$

99. (8 + 10 + 4 + 4 + 2) : (16 + 2 + 4 + 2 + 6) = 28 : 30 = 14 : 15

100. $A = 16 - 8 = 8$; $B = 10 - 2 = 8$; $E = 6 - 2 = 4$

FOR VISUALLY HANDICAPPED CANDIDATES ONLY

81. $6a^2 = 4\pi r^2 \Rightarrow \frac{a^2}{r^2}=\frac{4\pi}{6}=\frac{2\pi}{3}\Rightarrow\frac{a}{r}=\sqrt{\frac{2\pi}{3}}$

$\Rightarrow a^3:\frac{4}{3}\pi r^3=\left(\frac{\sqrt{2\pi}}{3}\right)^3:\frac{4}{3}\pi=\frac{2\pi}{3}\left(\sqrt{\frac{2\pi}{3}}\right):\frac{4}{3}\pi=\sqrt{\frac{2\pi}{3}}:2\Rightarrow\sqrt{\pi}:\sqrt{6}$

82. $x^2 = 16 + 2\sqrt{55}$; $y^2 = 16 + 2\sqrt{60}$; $z^2 = 16 + 2\sqrt{39} \Rightarrow y > x > z$

83. LCM = 4 × 5 × 5 = 100

84. $=\sqrt{(\sqrt{4}+\sqrt{3})-(\sqrt{3}-\sqrt{2})+(\sqrt{2}+1)}$

$=\sqrt{\sqrt{4}+\sqrt{3}-\sqrt{3}+\sqrt{2}+\sqrt{2}+1}=\sqrt{3+2\sqrt{2}}=\sqrt{(\sqrt{2})^2+1^2+2\sqrt{2}}=\sqrt{(\sqrt{2}+1)^2}=\sqrt{2}+1$

85. $=(2\times2\times3\times5)^{15}\times\left(\frac{3}{2}\right)^8\times\left(\frac{7}{2}\right)^7=2^{15}\times3^{15}\times5^{15}\times3^8\times7^7$

⇒ Prime factors = 15 + 15 + 15 + 8 + 7 = 60

86. $(\sqrt{2}+\sqrt{3}-\sqrt{5})^2-2\sqrt{6}+2\sqrt{15}+2\sqrt{10}$

$=2+3+5+2\sqrt{6}-2\sqrt{15}-2\sqrt{10}-2\sqrt{6}+2\sqrt{15}+2\sqrt{10}=2+3+5=10$

87. $8a = 2b + 6c \Rightarrow 4a = b + 3c \Rightarrow 4a - b = 3c$

88. $2\pi rh\times500\times\frac{5}{10000}=2\times\frac{22}{7}\times60\times84\times500\times\frac{5}{10000}=$ ₹ 7920

89. $100\times\frac{1}{2}\times\frac{6}{2}=150$; $150-100=50$

90. speed of train = $\frac{88}{21-9}=\frac{88}{12}$ m/s ⇒ Length of train = $\frac{88}{12}\times9=66$ m

91. $x^2=\frac{16\times100}{25}=64\Rightarrow x=8$

92. 500 × 10 × 10 = ₹ 50,000

93. $(a+b+c)+(b+c+d)+(c+d+a)+(d+a+b)=18+22+26+30$

$\Rightarrow 3(a+b+c+d)=96\Rightarrow(a+b+c+d)=32\Rightarrow$ Average = 32 ÷ 4 = 8

94. Total score = 80% × 80 = 64

Score in first 60 questions = 75% × 60 = 45

Remaining score = 64 – 45 = 19

Required % = $\frac{19}{20} \times 100 = 95$

95. $y = 8\%$ of $x \Rightarrow y\%$ of $25 = 8\%$ of $x\%$ of $25 = 8\% \times \frac{25}{100}x = 2\%$ of x

96.

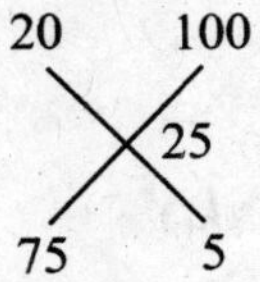

Ratio = 75 : 5 = 15 : 1

$\Rightarrow$ 15 ratio = 30 litres $\Rightarrow$ 1 ratio = 2 litres

97. CP = $1 + \frac{2}{3} = \frac{5}{3}$ of SP $\Rightarrow$ CP : SP = 5 : 3 $\Rightarrow$ Loss = $\frac{2}{5} \times 100 = 40\%$

98. If total profit = ₹ 10, then 10% of ₹ 10 = ₹ 1

Remaining profit = ₹ 9 distributed among A, B, C in 2 : 3 : 4

$\therefore P : Q : R = (2 + 1) : 3 : 4 = 3 : 3 : 4 \Rightarrow$ Q's share = P' share = ₹ 3,00,000

99. $\frac{76\times12}{57\times x} = \frac{2}{1} \Rightarrow x = \frac{76\times12}{57\times2} = 8$ months $\Rightarrow$ Required time = 12 – 8 = 4 months

100. $(B + C) + (A + C) - (A + B) = \frac{1}{24} + \frac{1}{36} - \frac{1}{18} \Rightarrow 2C = \frac{3+2-4}{72} = \frac{1}{72} \Rightarrow C = \frac{1}{144}$

SSC FCI, 2013 EXAM

PART - A
QUANTITATIVE METHODS

1. A sells an article to B at a gain of 20% and B sells it to C at a gain of 10% and C sells it to D at a gain of $12\frac{1}{2}\%$. If D pays ₹ 29.70, A purchased the article for,

(*a*) ₹ 20 (*b*) ₹ 30 (*c*) ₹ 40 (*d*) ₹ 10

2. By selling 80 ball pens for ₹ 140/-, a retailer loses 30%. How many ball pens should he sell for ₹ 104/- so as to make a profit of 30% ?

(*a*) 48 (*b*) 42 (*c*) 32 (*d*) 52

3. Nitin borrowed some money at the rate of 6% p.a. for the first three years, 9% p.a. for the next five years and 13% p.a. for the period beyond eight years. If the total interest paid by him at the end of eleven years is ₹ 8,160, the money borrowed by him (in ₹) was

(*a*) 8,000 (*b*) 10,000 (*c*) 12,000 (*d*) 6,000

4. A boy started from his house by bicycle at 10 a.m. at a speed of 12 km per hour. His elder brother started after 1 hr 15 mins by scooter along the same path and caught him at 1.30 p.m. The speed of the scooter will be (in km/hr)

(*a*) $18\frac{2}{3}$ (*b*) 9 (*c*) 4.5 (*d*) 36

5. Two pipes A and B can fill a cistern in 3 hours and 5 hours respectively. Pipe C can empty in 2 hours. If all the three pipes are open, in how many hours the cistern will be full ?

(*a*) 15 hours (*b*) 30 hours (*c*) can't be filled (*d*) 10 hours

6. Two men can do a piece of work in x days. But y women can do that in 3 days. Then the ratio of the work done by 1 man and 1 woman is

(*a*) $x : y$ (*b*) $2y : 3x$ (*c*) $3y : 2x$ (*d*) $2x : 3y$

7. A clerk received as annual salary of ₹ 3,660 in the year 1975. This was 20% more than his salary in 1974. What was his salary in 1974 ?

(*a*) 3,500 (*b*) 3,050 (*c*) 3,005 (*d*) 3,000

8. Out of his total income, Mr. Kapur spends 20% on house rent and 70% of the rest on house-hold expenses. If he saves ₹ 1,800, what is his total income (in rupees) ?

(*a*) 8,000 (*b*) 7,500 (*c*) 7,800 (*d*) 7,000

9. The sides of a triangle are in the ratio $\frac{1}{4}:\frac{1}{6}:\frac{1}{8}$ and its perimeter is 91 cm. The difference of the length of longest side and that of shortest side is

(*a*) 28 (*b*) 21 (*c*) 19 (*d*) 20

10. A box contains ₹ 56 in the form of coins of one rupee, 50 paise and 25 paise. The number of 50 paise coins is double the number of 25 paise coins and four times the number of one rupee coins. How many 50 paise coins are there in the box ?

(*a*) 32 (*b*) 16 (*c*) 52 (*d*) 64

11. The perimeter of an isosceles right angled triangle is $2p$ cm. Its area is **(SSC CGL Tier I 2012)**

(*a*) $(2-\sqrt{2})p$ sq cm (*b*) $(2+\sqrt{2})p^2$ sq cm

(*c*) $(3+2\sqrt{2})p$ sq cm (*d*) $(3-2\sqrt{2})p^2$ sq cm

12. The ratio between the areas of two circles are 4 : 7. What will be the ratio of their radii ?
(*a*) 16 : 49 (*b*) $4:\sqrt{7}$ (*c*) $2:\sqrt{7}$ (*d*) 4 : 7

13. The perimeter of a non-square rhombus is 20 cm. One of its diagonal is 8 cm. The area of the rhombus is
(*a*) 22 sq cm (*b*) 24 sq cm (*c*) 28 sq cm (*d*) 20 sq cm

14. If each edge of a cube is increased by 50%, the percentage increase in its surface area is
(*a*) 100% (*b*) 125% (*c*) 150% (*c*) 75%

15. The diameter of a copper sphere is 18 cm. The sphere is melted and is drawn into a long wire of uniform circular cross-section. If the length of the wire is 108 m, the diameter of the wire is
(*a*) 0.3 cm (*b*) 0.6 cm (*c*) 1 cm (*d*) 0.9 cm

16. A semicircular sheet of metal of diameter 28 cm is bent into an open conical cup. The capacity of the cup $\left(\text{taking } \pi = \frac{22}{7}\right)$ is
(*a*) 622.56 cm^3 (*b*) 623.20 cm^3 (*c*) 624.26 cm^3 (*d*) 622.26 cm^3

17. If surface area and volume of a sphere are S and V respectively, then value of $\frac{S^3}{V^2}$ is
(*a*) 18 unit (*b*) 27 unit (*c*) 32 unit (*d*) 9 unit

18. $\frac{\tan\theta}{1-\cot\theta} + \frac{\cot\theta}{1-\tan\theta}$ is equal to
(*a*) $1 - \tan\theta + \cot\theta$ (*b*) $1 + \tan\theta + \cot\theta$ (*c*) $1 - \tan\theta - \cot\theta$ (*d*) $1 + \tan\theta - \cot\theta$

19. If $\sec\theta = x + \frac{1}{4x}$ $(0° < \theta < 90°)$, then $\sec\theta + \tan\theta$ is equal to
(*a*) x (*b*) $\frac{1}{2x}$ (*c*) $\frac{x}{2}$ (*d*) $2x$

20. The circular measure of an angle of an isosceles triangle is $\frac{5\pi}{9}$. Circular measure of one of the other angles must be
(*a*) $\frac{2\pi}{9}$ (*b*) $\frac{4\pi}{9}$ (*c*) $\frac{5\pi}{18}$ (*d*) $\frac{5\pi}{9}$

21. If $\tan\theta + \cot\theta = 2$, then the value of $\tan^{100}\theta + \cot^{100}\theta$ is
(*a*) 1 (*b*) $\sqrt{3}$ (*c*) 2 (*d*) 0

22. The angles of elevation of the top of a tower from the points P and Q at distances of 'a' and 'b' respectively from the base of the tower and in the same straight line with it are complementary. The height of the tower is **(SSC CPO SI)**
(*a*) ab (*b*) a^2b^2 (*c*) $\sqrt{ab}$ (*d*) $\frac{a}{b}$

23. The value of $(3+\sqrt{8}) + \frac{1}{3-\sqrt{8}} - (6+4\sqrt{2})$ is
(*a*) $\sqrt{2}$ (*b*) 0 (*c*) 8 (*d*) 1

24. The area (in sq. unit) of the triangle formed in the first quadrant by the line $3x + 4y = 12$ is
(*a*) 6 (*b*) 4 (*c*) 8 (*d*) 12

25. If A and B are the H.C.F. and L.C.M. respectively of two algebraic expressions x and y, and $A + B = x + y$, then the value of $A^3 + B^3$ is
(*a*) y^3 (*b*) $x^3 + y^3$ (*c*) $x^3 - y^3$ (*d*) x^3

26. If $x > 1$ and $x^2 + \frac{1}{x^2} = 83$, then $x^3 - \frac{1}{x^3}$ is

(*a*) 756 (*b*) 760 (*c*) 764 (*d*) 750

27. If $\left(a + \frac{1}{a}\right)^2 = 3$, then $a^3 + \frac{1}{a^3} = ?$

(*a*) $3\sqrt{3}$ (*b*) 0 (*c*) $2\sqrt{3}$ (*d*) 2

28. If the internal bisectors of the $\angle ABC$ and $\angle ACB$ of ΔABC meet at O and also $\angle BAC = 80°$, then $\angle BOC$ is equal to **(SSC CHSL 2013)**

(*a*) 40° (*b*) 130° (*c*) 50° (*d*) 160°

29. O is the in-centre of the ΔABC, if $\angle BOC = 116°$, then $\angle BAC$ is

(*a*) 58° (*b*) 52° (*c*) 42° (*d*) 62°

30. The area of the circumcircle of an equilateral triangle is 3π sq. cm. The perimeter of the triangle is

(*a*) 18 cm (*b*) 3 cm (*c*) $3\sqrt{3}$ cm (*d*) 9 cm

31. In ΔABC, $\angle A = 90°$ and $AD \perp BC$ where D lies on BC. If $BC = 8$ cm, $AC = 6$ cm, then $\Delta ABC : \Delta ACD = ?$

(*a*) 16 : 9 (*b*) 25 : 9 (*c*) 4 : 3 (*d*) 25 : 16

32. Inside a triangle ABC, a straight line parallel to BC intersects AB and AC at the points P and Q respectively. If $AB = 3PB$, then $PQ : BC$ is

(*a*) 1 : 2 (*b*) 2 : 3 (*c*) 1 : 3 (*d*) 3 : 4

33. There are 50 boxes and 50 persons. Person 1 keeps 1 marble in every box. Person 2 keeps 2 marbles in every 2nd box, person 3 keeps 3 marbles in every third box. This process goes on till person 50 keeps 50 marbles in the 50th box. Find the total no. of marbles kept in the 50th box.

(*a*) 6 (*b*) 93 (*c*) 43 (*d*) 78

34. 252 m of pant cloth and 141 m of shirt cloth are available in a cloth store. To stitch one pant and one shirt, $2\frac{1}{2}$ m and $1\frac{3}{4}$ m of cloth are needed respectively. Then the approximate number of pants and shirts that can be made out of it are

(*a*) (100, 90) (*b*) (90, 80) (*c*) (80, 100) (*d*) (100, 80)

35. The greatest number that divides 411, 684, 821 and leaves 3, 4 and 5 as remainders, respectively, is

(*a*) 136 (*b*) 204 (*c*) 254 (*d*) 146

36. If $a = 7 - 4\sqrt{3}$, the value of $a^{\frac{1}{2}} + a^{-\frac{1}{2}}$ is

(*a*) 7 (*b*) $2\sqrt{3}$ (*c*) $3\sqrt{3}$ (*d*) 4

37. Terms a, 1, b are in Arithmetic Progression and terms 1, a, b are in Geometric Progression. Find 'a' and 'b' given $a \neq b$.

(*a*) 4, 1 (*b*) –2, 4 (*c*) 2, 4 (*d*) –2, 1

38. There are two groups A and B of a class, consisting of 42 and 28 students respectively. If the average weight of group A is 25 kg and that of group B is 40 kg, find the average weight of the whole class.

(*a*) 70 (*b*) 30 (*c*) 69 (*d*) 31

39. The average monthly salary of all the employees in an industry is ₹ 12,000. The average salary of male employees is ₹ 15,000 and that of female employees is ₹ 8,000. What is the ratio of male employees to female employees

(*a*) 4 : 3 (*b*) 2 : 5 (*c*) 5 : 2 (*d*) 3 : 4

40. Arun marks up the computer he is selling by 20% profit and sells them at a discount of 15%. Arun's net gain percent is

(*a*) 3.5 (*b*) 2.5 (*c*) 4 (*d*) 2

41. A dealer buys a table listed at ₹ 1,500 and gets successive discounts of 20% and 10%. He spends ₹ 20 on transportation and sells at a profit of 20%. Find the Selling Price of the table (in rupees).

(*a*) 1200 (*b*) 1230 (*c*) 1320 (*d*) 1080

Bar-chart showing the Sales of Books (in 1000) from six-branches B_1, B_2, B_3, B_4, B_5 and B_6 of a Publishing Company in 2000 and 2001 is given below. Study the chart and answer questions (42-45).

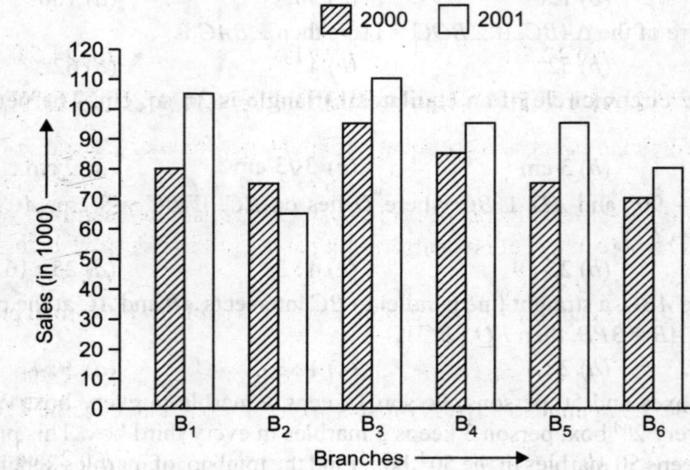

42. Total sales of branch B_6 for both the years is what percent of the total sales of branch B_3 for both the years ?

(*a*) 71.11% (*b*) 73.17% (*c*) 68.54% (*d*) 77.26%

43. What is the ratio of the total sales of branch B_2 for both the years to the total sales of branch B_4 for both years ?

(*a*) 2 : 3 (*b*) 3 : 5 (*c*) 5 : 7 (*d*) 7 : 9

44. What percent of the average sales of branches B_1, B_2 and B_3 in 2001 is the average sales of branches B_1, B_3 and B_6 in 2000 ?

(*a*) 87.5% (*b*) 75% (*c*) 77.5% (*d*) 85%

45. What is the average sale of books from all the branches for the year 2000 ?

(*a*) 70 (*b*) 80 (*c*) 70.5 (*d*) 80.5

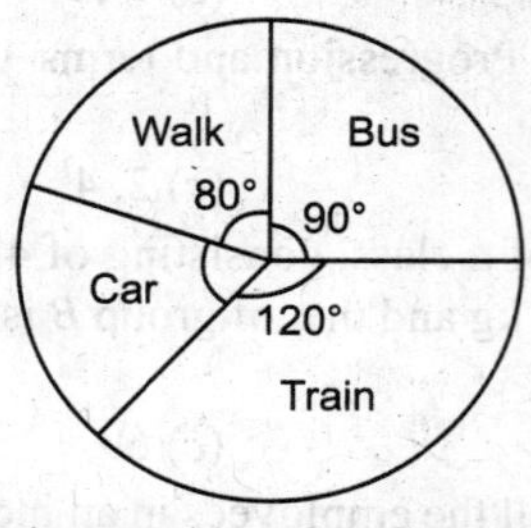

The above pie-chart represents the number of students using different transport to a school in which total no. of students is 2160.

Answer Qn. Nos. 46-50 based on the above diagram.

46. The no. of students who come to school by car is

(*a*) 70 (*b*) 290 (*c*) 420 (*d*) 480

47. The ratio of the number of students who came to school by car to the number of students who came to school by bus is

(*a*) 21 : 24 (*b*) 21 : 27 (*c*) 36 : 27 (*d*) 36 : 21

48. The total no. of students coming to school either by walking or by bus is

(*a*) 480 (*b*) 540 (*c*) 1020 (*d*) 170

49. The no. of students who don't come to school by train is

(*a*) 720 (*b*) 1020 (*c*) 2040 (*d*) 1440

50. The no. of students coming to school by bus exceeds the no. of students coming to school walking, by

(*a*) 10% (*b*) 12.5% (*c*) 11% (*d*) 11.5%

FOR VISUALLY HANDICAPPED CANDIDATES ONLY

42. Which of the following integers is the square of an integer for every integer n ?

(*a*) $n^2 + 1$ (*b*) $n^2 + n$ (*c*) $n^2 + 2n$ (*d*) $n^2 + 2n + 1$

43. In an examination, 65% of students pass in Physics, 55% pass in Chemistry and 25% fail in both subjects. The percentage of students having passed in Physics and Chemistry is

(*a*) 65% (*b*) 35% (*c*) 45% (*d*) 55%

44. In a partnership, 4 (A's capital) = 6 (B's capital) = 10 (C's capital). If the total profit is ₹ 4,650, the share of C is

(*a*) ₹ 1,200 (*b*) ₹ 1,100 (*c*) ₹ 800 (*d*) ₹ 900

45. A sum of ₹ 6,400, compounded quarterly at the rate of 5% p.a. will become ₹ 6,561 in

(*a*) Half year (*b*) One year (*c*) $1\frac{1}{2}$ years (*d*) 2 years

46. If each side of a square be increased by 10%, then area of the square is increased by

(*a*) 10% (*a*) 21% (*c*) 20% (*d*) 15%

47. If 25% of a number is 150, and x% of 150 is that number, then x is

(*a*) 200 (*b*) 40 (*c*) 120 (*d*) 400

48. In a right angled triangle ABC, $\sin(A + B) = 1$, $\cos A = \frac{1}{2}$, then value of $\cot(C - B)$ is

(*a*) $\sqrt{3}$ (*b*) 1 (*c*) 0 (*d*) $\frac{1}{\sqrt{3}}$

49. In ΔABC, I is the in-centre. If $\angle BAC = 50°$, then $\angle BIC =$

(*a*) 140° (*b*) 80° (*c*) 115° (*d*) 125°

50. In a class, there are 10 students at the age of 15, fifteen at the age of 16 and five at the age of 14. What is the average age of a student ?

(*a*) $15\frac{1}{3}$ (*b*) $15\frac{2}{3}$ (*c*) $14\frac{3}{4}$ (*d*) 15

ANSWER

1. (*a*)	**2.** (*c*)	**3.** (*a*)	**4.** (*a*)	**5.** (*b*)	**6.** (*c*)	**7.** (*b*)	**8.** (*b*)
9. (*b*)	**10.** (*d*)	**11.** (*d*)	**12.** (*c*)	**13.** (*b*)	**14.** (*b*)	**15.** (*b*)	**16.** (*d*)
17. (*)	**18.** (*b*)	**19.** (*d*)	**20.** (*a*)	**21.** (*c*)	**22.** (*c*)	**23.** (*b*)	**24.** (*a*)
25. (*b*)	**26.** (*a*)	**27.** (*b*)	**28.** (*b*)	**29.** (*b*)	**30.** (*d*)	**31.** (*a*)	**32.** (*b*)
33. (*b*)	**34.** (*d*)	**35.** (*a*)	**36.** (*d*)	**37.** (*b*)	**38.** (*d*)	**39.** (*a*)	**40.** (*d*)
41. (*c*)	**42.** (*b*)	**43.** (*d*)	**44.** (*a*)	**45.** (*b*)	**46.** (*c*)	**47.** (*b*)	**48.** (*c*)
49. (*d*)	**50.** (*b*)						

HINTS

1. $29.70\times\frac{5}{6}\times\frac{10}{11}\times\frac{8}{9}=₹\,20$

2. $\frac{140}{80}\times\frac{13}{7}=\frac{13}{4}; 104\times\frac{4}{13}=32$

3. Equivalent rate of interest = 3 × 6% + 5 × 9% + 3 × 13% = 102%

Amount borrowed = $\frac{8160\times100}{102}=₹\,8000$

4. 10 a.m. to 1.30 p.m. = $\frac{7}{2}$ hours

Distance travelled = $\frac{7}{2}\times12=42$ km

Speed of elder brother = $42\div\left(\frac{7}{2}-\frac{5}{4}\right)=42\times\frac{4}{9}=\frac{56}{3}=18\frac{2}{3}$ km/hr

5. $\frac{1}{3}+\frac{1}{5}-\frac{1}{2}=\frac{10+6-15}{30}=\frac{1}{30}$

6. Ratio = $\frac{1}{2x}:\frac{1}{3y}=3y:2x$

7. $3660\times\frac{100}{120}=₹\,3050$

8. $1800\times\frac{100}{80}\times\frac{100}{30}=₹\,7,500$

9. $\frac{1}{4}:\frac{1}{6}:\frac{1}{8}=6:4:3$

Difference = $\frac{6-3}{6+4+3}\times91$ cm = 21 cm

10.

Denomination	100	50	25
Ratio	1	4	2
Value	100	200	50

Ratio of value = 100 : 200 : 50 = 2 : 4 : 1

Value of 50 paise coins = $\frac{4}{7}\times56=₹\,32$

Number of 50 paise coins = $32\div\frac{1}{2}=64$ coins

11. Perimeter = $x+x+\sqrt{2}x=2p\ \Rightarrow\ x=\frac{2}{2+\sqrt{2}}p$

Area of triangle = $\frac{1}{2}x^2=\frac{1}{2}\times\left(\frac{2}{2+\sqrt{2}}p\right)^2$

$=\frac{1}{2}\times\frac{4}{4+2+4\sqrt{2}}p^2=\frac{1}{3+2\sqrt{2}}p^2=(3-2\sqrt{2})\,p^2$

12. Ratio of radii = $\sqrt{4} : \sqrt{7} = 2 : \sqrt{7}$

13. Side = $20 \div 4 = 5$ cm

$5^2 - 4^2 = 3^2$

$\therefore$ Diagonal = 2×3 cm = 6 cm

Area = $\frac{1}{2} \times 8 \times 6 = 24 \text{ cm}^2$

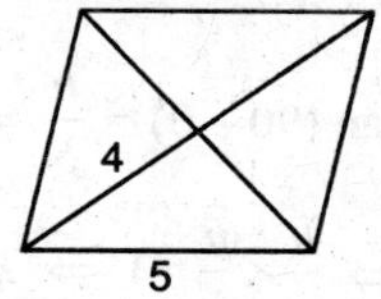

14. $50 + 50 + \frac{50 \times 50}{100} = 125\%$

15. $r^2 = \frac{\frac{4}{3}\pi (9)^3}{\pi (10800)} = \frac{4}{3} \times \frac{9 \times 9 \times 9}{10800} \Rightarrow r = 0.3 \Rightarrow d = 0.6$ cm

16. $h = \sqrt{14^2 - 7^2} = 7\sqrt{3} = 12.124$

$V = \frac{1}{3}\pi r^2 h = \frac{1}{3} \times \frac{22}{7} \times 7^2 \times 12.124 = 622.36 \text{ cm}^3$

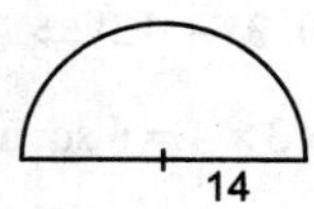

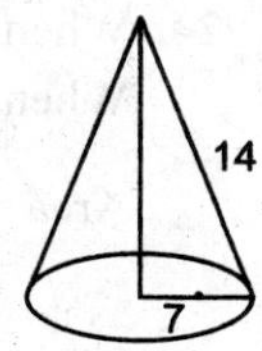

17. $S = 4\pi r^2$, $V = \frac{4}{3}\pi r^3$

$$\frac{S^3}{V^2} = \frac{(4\pi r^2)^3}{\left(\frac{4}{3}\pi r^3\right)^2} = 36\pi$$

18. $$\frac{\tan\theta}{1-\cot\theta} + \frac{\cot\theta}{1-\tan\theta} = \frac{\tan\theta}{1-\frac{1}{\tan\theta}} + \frac{\frac{1}{\tan\theta}}{1-\tan\theta} = \frac{\tan\theta}{\frac{\tan\theta-1}{\tan\theta}} + \frac{1}{\tan\theta(1-\tan\theta)}$$

$$= \frac{\tan^2\theta}{\tan\theta-1} - \frac{1}{\tan\theta(\tan\theta-1)} = \frac{\tan^3\theta-1}{\tan\theta(\tan\theta-1)}$$

$$= \frac{(\tan\theta-1)(\tan^2\theta+1+\tan\theta)}{\tan\theta(\tan\theta-1)} = \frac{\tan^2\theta+1+\tan\theta}{\tan\theta} = \tan\theta + \cot\theta + 1$$

19. $\tan^2 = \sec^2\theta - 1 = \left(x + \frac{1}{4x}\right)^2 - 1 = x^2 + \frac{1}{16x^2} + \frac{1}{2} - 1 = x^2 + \frac{1}{16x^2} - \frac{1}{2} = \left(x - \frac{1}{4x}\right)^2$

$\Rightarrow \tan\theta = x - \frac{1}{4x} \Rightarrow \sec\theta + \tan\theta = \left(x + \frac{1}{4x}\right) + \left(1 - \frac{1}{4x}\right) = 2x$

20. $\frac{\pi}{9} = 20° \Rightarrow \frac{5\pi}{9} = 100°$

Other angle = $\frac{1}{2}(180° - 100°) = 40° = \frac{2\pi}{9}$

21. $\tan\theta + \cot\theta = 2 \Rightarrow \tan\theta = \cot\theta = 1$

$\tan^{100}\theta + \cot^{100}\theta = 1 + 1 = 2$

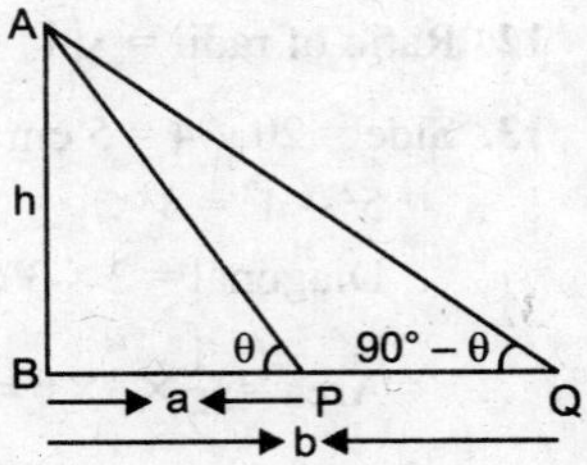

22. $\tan\theta = \dfrac{h}{a}$

$\tan(90-\theta) = \dfrac{h}{b} \Rightarrow \cot\theta = \dfrac{h}{b}$

$\Rightarrow \dfrac{h}{a}\times\dfrac{h}{b} = 1 \Rightarrow h^2 = ab \Rightarrow h = \sqrt{ab}$

23. $(3+\sqrt{8}) + \dfrac{1}{3-\sqrt{8}} - (6+4\sqrt{2}) = (3+\sqrt{8}) + (3+\sqrt{8}) - (6+4\sqrt{2})$

$= 3+2\sqrt{2}+3+2\sqrt{2}-6-4\sqrt{2} = 0$

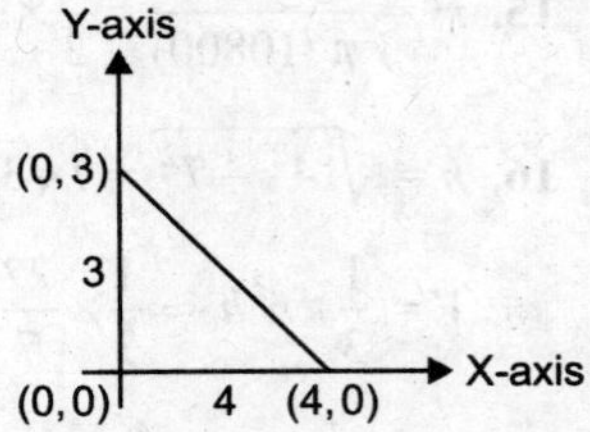

24. When $x = 0$, $4y = 12 \Rightarrow y = 3$
When $y = 0$, $3x = 12 \Rightarrow x = 4$

Area $= \dfrac{1}{2}\times 3\times 4 = 6$ sq. units

25. Product of numbers = H.C.F × L.C.M $= x\times y = A + B$
$(A^3 + B^3) = (A+B)^3 - 3AB\,(A+B) = (x+y)^3 - 3xy\,(x+y)$
$= x^3 + y^3 + 3xy\,(x+y) - 3xy\,(x+y) = x^3 + y^3$

26. $\left(x-\dfrac{1}{x}\right)^2 = x^2 + \dfrac{1}{x^2} - 2 = 83 - 2 = 81 \Rightarrow \left(x-\dfrac{1}{x}\right) = \sqrt{81} = 9$

$\therefore \left(x^3 - \dfrac{1}{x^3}\right) = \left(x-\dfrac{1}{x}\right)^3 + 3\left(x-\dfrac{1}{x}\right) = 9^3 + 3\times 9 = 729 + 27 = 756$

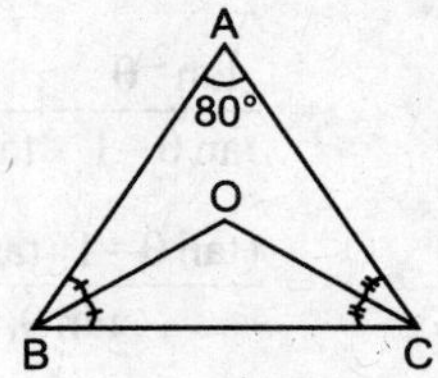

27. $a + \dfrac{1}{a} = \sqrt{3}$

$\left(a^3 + \dfrac{1}{a^3}\right) = \left(a+\dfrac{1}{a}\right)^3 - 3\left(a+\dfrac{1}{a}\right) = (\sqrt{3})^3 - 3\sqrt{3} = 0$

28. $\angle B + \angle C = 180° - \angle A = 180° - 80° = 100°$

$\angle BOC = 180° - \dfrac{1}{2}\times 100° = 130°$

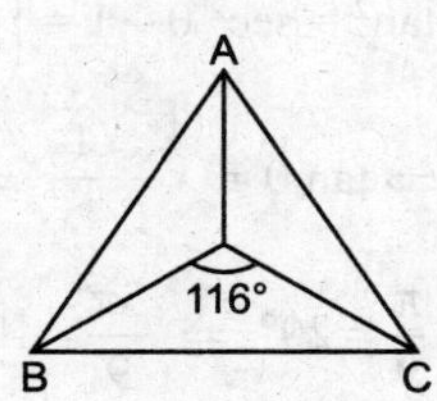

29. $\angle A = 2\times 116° - 180° = 232° - 180° = 52°$

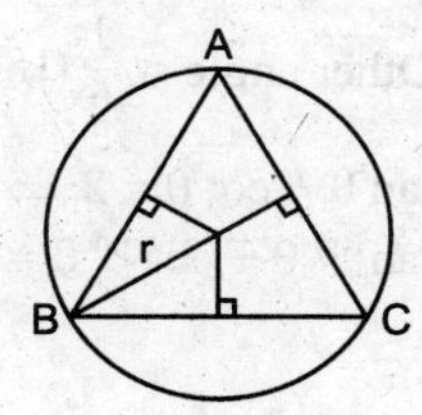

30. $\pi r^2 = 3\pi \Rightarrow r = \text{Side} \div \sqrt{3}$

Side of triangle $= r\times\sqrt{3} = \sqrt{3}\times\sqrt{3} = 3$ cm

Perimeter of triangle $= 3\times 3$ cm $= 9$ cm

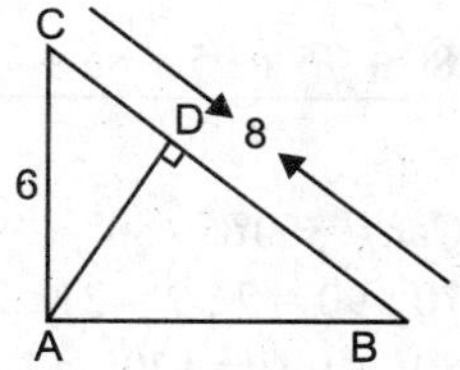

31. Ratio of hypotenuses of ΔABC and $\Delta DCA = 8 : 6 = 4 : 3$

$\therefore$ Ratio of areas $= 4^2 : 3^2 = 16 : 9$

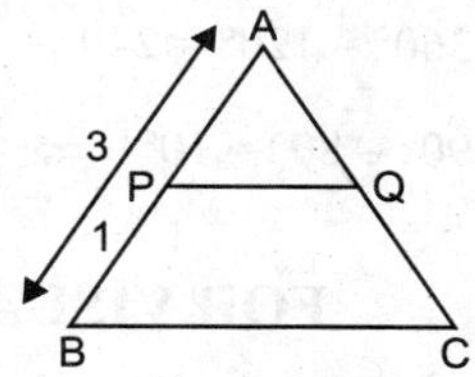

32. $AB = 3PB \Rightarrow \dfrac{AB}{PB} = 3 \Rightarrow \dfrac{AB}{AP} = \dfrac{3}{2}$

$\Rightarrow \dfrac{BC}{PQ} = \dfrac{3}{2} \Rightarrow \dfrac{PQ}{BC} = \dfrac{2}{3}$

33. $1 + 2 + 5 + 10 + 25 + 50 = 93$

34. $\left(252 \div \dfrac{5}{2}, 141 \div \dfrac{7}{4}\right) \sim (100, 80)$

35. $411 - 3 = 408,\ 684 - 4 = 680,\ 821 - 5 = 816$

HCF of 408, 680, 816 is 136.

36. $a = 7 - 4\sqrt{3} = (2 - \sqrt{3})^2$

$\therefore$ G..E. $= (2 - \sqrt{3}) + \dfrac{1}{2 - \sqrt{3}} = (2 - \sqrt{3}) + (2 + \sqrt{3}) = 4$

37. a, 1, b are in A.P. $= 1 \Rightarrow \dfrac{a+b}{2} = 1 \Rightarrow a + b = 2$

1, a, b are in G.P. $= 1 \Rightarrow \dfrac{a+b}{2} = 1 \Rightarrow 1 \times b = a^2$

Trick : –2, 1, 4 are in A.P. (common difference = 3)

1, –2, 4 are in G.P. (common ratio = –2)

38. $\dfrac{42 \times 25 + 28 \times 40}{42 + 28} = \dfrac{2170}{70} = 31$

39.

40. $20 - 15 - \dfrac{20 \times 15}{100} = 2\%$

41. CP $= 1500 \times \dfrac{80}{100} \times \dfrac{90}{100} + ₹\ 20 = ₹\ 1100$

Profit = 1100 × 20% = ₹ 220 $\Rightarrow$ S.P. = 1100 + 220 = ₹ 1320

42. $\dfrac{70 + 80}{95 + 110} \times 100 = \dfrac{3000}{41} = 73.17\%$

43. $(75 + 65) : (85 + 95) = 140 : 180 = 7 : 9$

44. $\% = \dfrac{80 + 95 + 70}{105 + 65 + 110} \times 100 = 87.5\%$

45. $\dfrac{80+75+95+85+75+70}{6}=\dfrac{480}{6}=80$

46. Car = 360° – (80° + 90° + 120°) = 70° = 420 students [1° = 2160 ÷ 360 = 6 students]

47. 70 : 90 = 7 : 9 = 21 : 27

48. 80° + 90° = 170° = 170 × 6 = 1020

49. 360° – 120° = 240° = 240 × 6 = 1440

50. 90° – 80° = 10° $\Rightarrow$ $\dfrac{10}{80}\times 100 = 12.5\%$

FOR VISUALLY HANDICAPPED CANDIDATES ONLY

42. $n^2 + 2n + 1 = (n + 1)^2$

43. (35 – 25)% + (45 – 25)% + 25% = 55%

$\therefore$ Passed = 100% – 55% = 45%

44. $4A = 6B = 10C \Rightarrow A : B : C = 15 : 10 : 6$

$\therefore C = \dfrac{6}{31}\times$ ₹ 4650 = ₹ 900

45. $r = \dfrac{5}{4}\% \Rightarrow (1 + r\%) = \left(1+\dfrac{5}{400}\right)=\dfrac{81}{80}$

$\dfrac{6561}{6400}=\left(\dfrac{81}{80}\right)^2 \Rightarrow$ Time = 2 quarters = Half-year

46. 11 × 11 = 121 ; 121 – 100 = 21%

47. Number = $150\times\dfrac{100}{25}=600$

$x\% \times 150 = 600 \Rightarrow x\% = 600 \div 150 = 4 \Rightarrow x = 400\%$

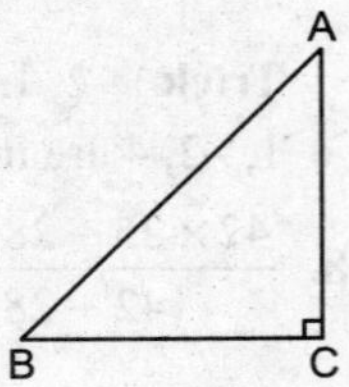

48. $\sin(A + B) = 1 = \sin 90° \Rightarrow A + B = 90°$

$\cos A = \dfrac{1}{2} = \cos 60° \Rightarrow A = 60°$

$\cot(C - B) = \cot(90° - 30°) = \cot 60° = \dfrac{1}{\sqrt{3}}$

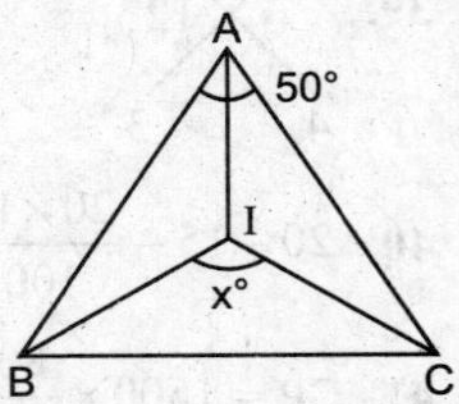

49. $\angle A = 2 \times x° - 180° \Rightarrow 2x° = 50° + 180° = 230° \Rightarrow x° = 115°$

50. $\dfrac{10\times15+15\times16+5\times14}{10+15+5}=\dfrac{460}{30}=15\dfrac{1}{3}.$

SSC COMBINED HIGHER SECONDARY LEVEL 2012

PART - III

QUANTITATIVE APTITUDE (012PG9)

1. The number of pairs of natural numbers, the difference of whose squares is 45 will be

(*a*) 2 (*b*) 3 (*c*) 6 (*d*) 5

2. The middle term(s) of the following series 2 + 4 + 6 + ... + 198 is

(*a*) 98 (*b*) 96 (*c*) 94 (*d*) 100

3. $\dfrac{0.3555 \times 0.5555 \times 2.025}{0.225 \times 1.7775 \times 0.2222}$ is equal to

(*a*) 5.4 (*b*) 4.58 (*c*) 4.5 (*d*) 5.45

4. A and B can do a piece of work in 36 days, B and C can do it in 60 days, A and C can do it in 45 days. C alone can do it in

(*a*) 90 days (*b*) 180 days (*c*) 120 days (*d*) 150 days

5. Two men undertake a job for ₹ 960. They can complete it in 16 days and 24 days respectively. They work along with a third man and take 8 days to complete it. Then the share of the third man should be

(*a*) ₹ 155 (*b*) ₹ 165 (*c*) ₹ 160 (*d*) ₹ 150

6. A toy train is marked at ₹ 400 and sold at a discount of 8% during Ganesh puja. A shopkeeper announces a discount of 8%. The amount he will loose if he announces a single discount of 16% is

(*a*) ₹ 2.56 (*b*) ₹ 3.84 (*c*) ₹ 4.16 (*d*) ₹ 5.78

7. Alex sold his goods after announcing two successive discounts of 30% each. The effective discount altogether is

(*a*) 52% (*b*) 49% (*c*) 50% (*d*) 51%

8. A publisher printed 2000 copies of a book at a cost of ₹ 70,000. He distributes 400 copies free as specimen copies. He gave 30% discount on printed price and the printed price of each book is ₹ 75. What is his gain or loss percentage ?

(*a*) 20% gain (*b*) 20% loss (*c*) 10% loss (*d*) 10% gain

9. If there is a reduction in the number of workers in a factory in the ratio 15: 11 and an increment in their wages in the ratio 22 : 25, then the ratio by which the total wages of the workers should be decreased is

(*a*) 6 : 5 (*b*) 5 : 6 (*c*) 3 : 7 (*d*) 3 : 5

10. A and B are two alloys of gold and copper in the ratio 7 : 2 and 7 : 11 respectively. If equal quantities of these two alloys are melted to form a new alloy C, then the ratio of gold and copper in C is

(*a*) 6 : 5 (*b*) 9 : 4 (*c*) 12 : 7 (*d*) 7 : 5

11. A Can contains a mixture of two liquids A and B in the ratio 7 : 5. When 9 litres of mixture are drain off and the Can is filled with B, the ratio of A and B becomes 7 : 9. How many litres of liquid A was contained by the Can initially ?

(*a*) 10 (*b*) 20 (*c*) 21 (*d*) 25

12. On mixing two classes A and B of students having average marks 25 and 40 respectively, the overall average obtained is 30. Find the ratio of the students in the classes A and B.

(*a*) 2 : 1 (*b*) 5 : 8 (*c*) 5 : 6 (*d*) 3 : 4

13. If a man receives on one-fourth of his capital 3% interest, on two third 5% and on the remainder 11%, the percentage he receives on the whole is

(*a*) 4.5 (*b*) 5 (*c*) 5.5 (*d*) 5.2

14. Average of first five odd multiple of 3 is

(*a*) 12 (*b*) 16 (*c*) 15 (*d*) 21

15. A man purchases two fans for ₹ 2,160. By selling one fan at a profit of 15% and the other at a loss of 9% he neither gains nor loses in the whole transaction. Find the cost price of each fan in ₹.

(*a*) 710,1450 (*b*) 1530,630 (*c*) 810,1350 (*d*) 1340,820

16. If a man were to sell his handcart for ₹ 720, he would lose 25%. To gain 25%, the selling price is

(*a*) ₹ 960 (*b*) ₹ 1,200 (*c*) ₹ 1,000 (*d*) ₹ 2,100

17. A sells a cycle to B at a profit of 10%, B sells to C at a profit of 20%. If C pays ₹ 264 for it, how much did A pay for it ?

(*a*) ₹ 200 (*b*) ₹ 220 (*c*) ₹ 225 (*d*) ₹ 234

18. A's salary is increased by 10% and then decreased by 10%. The change in salary is

(*a*) 0% (*b*) 1% decrease (*c*) 1% increase (*d*) 2% decrease

19. Kishan spends 30% of his salary on food and donates 3% in a Charitable Trust. He spends ₹ 2,310 on these two items, then total salary for that month is

(*a*) ₹ 6,000 (*b*) ₹ 8,000 (*c*) ₹ 9,000 (*c*) ₹ 7,000

20. A train is running at a speed of 90 km/hr. If it crosses a signal in 10 sec., the length of the train in metres is

(*a*) 150 (*b*) 324 (*c*) 900 (*d*) 250

21. A man can swim at the rate of 4 km/hr in still water. If the speed of the water is 2 km/hr, then the time taken by him to swim 10 km upstream is

(*a*) 2½ hrs (*b*) 3½ hrs (*c*) 5 hrs (*d*) 4 hrs

22. A person who pays income tax at the rate of 4 paise per rupee, find that a fall of interest rate from 4% to 3.75% diminishes his net yearly income by ₹ 48. What is his capital ?

(*a*) ₹ 24,000 (*b*) ₹ 25,000 (*c*) ₹ 20,000 (*d*) ₹ 18,000

23. The population of a village decreases at the rate of 20% per annum. If its population 2 years ago was 10,000, the present population is

(*a*) 4600 (*b*) 6400 (*c*) 7600 (*d*) 6000

24. If two adjacent sides of a rectangular parallelepiped are 1 cm and 2 cm and the total surface area of the parallelepiped is 22 square cm, then the diagonal of the parallelepiped is

(*a*) $\sqrt{10}$ cm (*b*) $2\sqrt{3}$ cm (*c*) $\sqrt{14}$ cm (*d*) 4 cm

25. Two circles with centres A and B and radius 2 units touch each other externally at 'C'. A third circle with centre 'C' and radius '2' units meets other two at D & E. Then the area of the quadrilateral ABDE is

(*a*) $2\sqrt{2}$ sq. unit (*b*) $3\sqrt{3}$ sq. unit (*c*) $3\sqrt{2}$ sq. unit (*d*) $2\sqrt{3}$ sq. unit

26. A sphere and a hemisphere have the same volume. The ratio of their radii is

(*a*) 1 : 2 (*b*) 1 : 8 (*c*) $1 : \sqrt{2}$ (*d*) $1 : \sqrt[3]{2}$

27. A right circular cylinder and a sphere have same radius and same volume. The ratio of their surface area is

(*a*) 8 : 3 (*b*) 9 : 4 (*c*) 4 : 3 (*d*) 7 : 6

28. The ratio of sides of a triangle is 3:4:5 and area of the triangle is 72 square unit. Then the area of an equilateral triangle whose perimeter is same as that of the previous triangle is

(*a*) $32\sqrt{3}$ square unit (*b*) $48\sqrt{3}$ square unit
(*c*) 96 square unit (*d*) $60\sqrt{3}$ square unit

29. The perimeter of a semicircular path is 36 m. Find the area of this semicircular path.

(*a*) 42 sq. m (*b*) 54 sq. m (*c*) 63 sq. m (*d*) 77 sq. m

30. The graphs of the equations $25x + 75y = 225$ and $x = 9$ meet at the point

(*a*) (0,9) (*b*) (9,0) (*c*) (3,0) (*d*) (0,3)

31. If $a = x + y$, $b = x - y$, $c = x + 2y$, then $a^2 + b^2 + c^2 - ab - bc - ca$ is

(*a*) $4y^2$ (*b*) $5y^2$ (*c*) $6y^2$ (*d*) $7y^2$

32. If $a + b = 1$ and $a^3 + b^3 + 3ab = k$, then the value of k is

(*a*) 1 (*b*) 3 (*c*) 5 (*d*) 7

33. If $ax + by = 3$, $bx - ay = 4$ and $x^2 + y^2 = 1$, then the value of $a^2 + b^2$ is

(*a*) 17 (*b*) 16 (*c*) 9 (*d*) 25

34. If the circumcentre of a triangle lies outside it, then the triangle is

(*a*) Equilateral (*b*) Acute angled (*c*) Right angled (*d*) Obtuse angled

35. In a triangle ABC, $\angle BAC = 90°$ and AD is perpendicular to BC. If $AD = 6$ cm and $BD = 4$ cm, then the length of BC is

(*a*) 8 cm (*b*) 10 cm (*c*) 9 cm (*d*) 13 cm

36. $ABCD$ is a cyclic quadrilateral and O is the centre of the circle. If $\angle COD = 140°$ and $\angle BAC = 40°$, then the value of $\angle BCD$ is equal to

(*a*) 70° (*b*) 90° (*c*) 60° (*d*) 80°

37. Two triangles ABC and DEF are similar to each other in which $AB = 10$ cm, $DE = 8$ cm. Then the ratio of the area of triangle ABC and DEF is

(*a*) 4 : 5 (*b*) 25 : 16 (*c*) 64 : 125 (*d*) 4 : 7

38. If $\cos\theta \operatorname{cosec} 23° = 1$, the value of θ is

(*a*) 23° (*b*) 37° (*c*) 63° (*d*) 67°

39. The value of $\sin^2 5° + \sin^2 10° + \sin^2 15° + \ldots \sin^2 85° + \sin^2 90°$ is

(*a*) $7\frac{1}{2}$ (*b*) $8\frac{1}{2}$ (*c*) $10\frac{1}{2}$ (*d*) $9\frac{1}{2}$

40. A vertical post 15 ft high is broken at a certain height and its upper part, not completely separated, meets the ground at an angle to 30°. Find the height at which the post is broken.

(*a*) 10 ft (*b*) 5 ft (*c*) $15\sqrt{3}(2 - \sqrt{3})$ ft (*d*) $5\sqrt{3}$ ft

Directions: The following Pie Chart shows the export of different foodgrains from India in 2010. Study the chart and answer the questions (41-45):

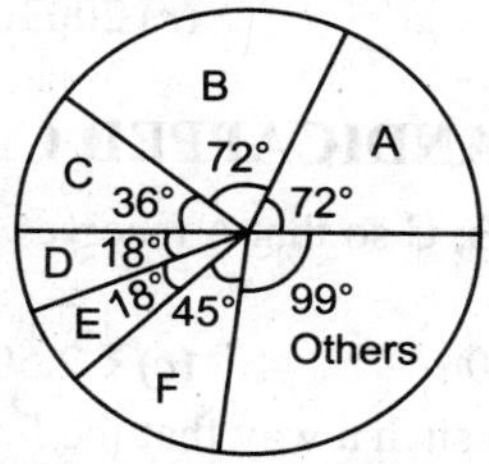

41. The three crops which combine to contribute to exactly 50% of the total export of foodgrains are

(*a*) A, F and others (*b*) B, C and F (*c*) A, B and C (*d*) C, F and others

42. If a total of 1.5 million quintals of crop F was exported, then the total quantity of D and E that was exported, in million quintals, was

(*a*) 1.2 (*b*) 1.5 (*c*) 4.5 (*d*) 6.5

43. If the revenue from 1 quintal of crop A is thrice that from 1 quintal of crop C, then the ratio of the total revenues of A and C is

(*a*) 1 : 6 (*b*) 2 : 3 (*c*) 3 : 2 (*d*) 6 : 1

44. Of the total export of foodgrains, the percentage of crop B exported is

(*a*) 15 (*b*) 20 (*c*) 18 (*d*) 10

45. The amount of total foodgrains exported in million quintals is

(*a*) 8.7 (*b*) 12 (*c*) 10.8 (*d*) 9.6

Directions: The following bar diagram analyses the sale of a company from 2000 to 2005. Examine the diagram and answer the questions (46 -50):

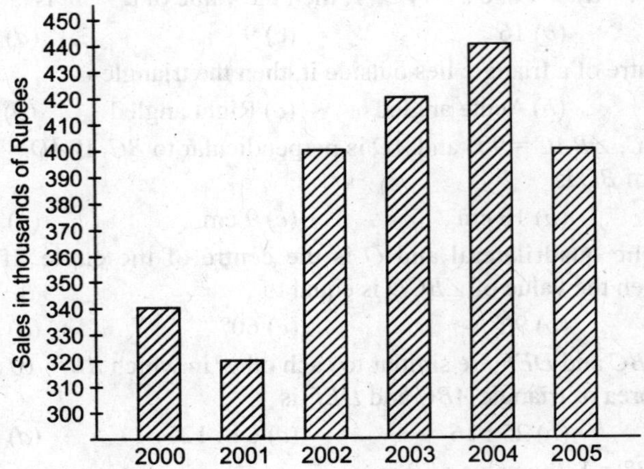

46. By what amount are the sales in 2003 more than those in 2001 ?

(*a*) ₹ one hundred (*b*) ₹ ten thousand (*c*) ₹ one lakh (*d*) ₹ ten lakhs

47. The sales in 2001 are how many times those of 2002 ?

(*a*) 8 times (*b*) 0.8 times (*c*) 2.5 times (*d*) 3 times

48. In which year did the sales show the least percent increase to that of the preceding year ?

(*a*) 2000 (*b*) 2002 (*c*) 2003 (*d*) 2004

49. The sales in 2004 are what percentage of those in 2002 ?

(*a*) 40 (*b*) 4 (*c*) 110 (*d*) 1.1

50. In which year did the sales show the least decrease to that of the preceding year ?

(*a*) 2004 (*b*) 2001 (*c*) 2003 (*d*) 2005

FOR VISUALLY HANDICAPPED CANDIDATES ONLY

41. If ₹ 6,900 be divided among A, B, C so that A receives 20% more than B and B receives 20% less than C, then C's share is

(*a*) ₹ 2,000 (*b*) ₹ 2,400 (*c*) ₹ 2,500 (*d*) ₹ 3,200

42. 170 is divided into three parts in such a way that the first part is 10 more than the second and its ratio with third part is 2 : 5. The three parts are

(*a*) 35,35,100 (*b*) 40,30,100 (*c*) 50,50,70 (*d*) 45,45,80

43. A man gets ₹ 18 more on selling an article at a gain of 10% than selling it at a loss of 10%. Cost price of the article is (in ₹)

(*a*) 90 (*b*) 100 (*c*) 101 (*d*) 105

44. In a stream running at 2 km/hr, a motorboat goes 10 km upstream and then back again to the starting point in 55 minutes. The speed of the motorboat in still water is

(*a*) 24 km/hr (*b*) 22 km/hr (*c*) 32 km/hr (*d*) 28 km/hr

45. If $x + \frac{1}{x} = 2$, then the value of $\left(x^5 + \frac{1}{x^5}\right)$ is

(*a*) 2 (*b*) 3 (*c*) 4 (*d*) 5

46. The average of first 10 multiples of 10 is 55. If each number is divided by 2 and multiplied by 3, then the new average is

(*a*) 82.5 (*b*) 83.5 (*c*) 84.5 (*d*) 85.5

47. If $\sin 61° = \frac{a}{\sqrt{a^2+b^2}}$, then the value of $\tan 61° + \tan 29°$ is

(*a*) $\frac{a}{b}$ (*b*) $\frac{(a+b)^2}{ab}$ (*c*) $\frac{a}{b}+1$ (*d*) $\frac{a}{b}+\frac{b}{a}$

48. Find the number between 1500 and 2000 which is divisible by 36 and 102.

(*a*) 1632 (*b*) 1734 (*c*) 1836 (*d*) 1944

49. *AB* and *CD* are two parallel chords of a circle with centre *O* such that *AB* = 6 cm and *CD* = 12 cm. The chords are on the same side of the centre and the distance between them is 3 cm. Then, the radius of the circle is

(*a*) 6 cm (*b*) $5\sqrt{2}$ cm (*c*) 7 cm (*d*) $3\sqrt{5}$ cm

50. The cost of painting on four walls of a room is ₹ 475. Each one of the length, breadth and height of another room is double that of the room. The cost of painting on four walls of this new room is

(*a*) ₹ 950 (*b*) ₹ 1,425 (*c*) ₹ 1,900 (*d*) ₹ 1,800

ANSWER

1. (*b*)	**2.** (*d*)	**3.** (*c*)	**4.** (*b*)	**5.** (*c*)	**6.** (*a*)	**7.** (*d*)	**8.** (*a*)
9. (*a*)	**10.** (*d*)	**11.** (*c*)	**12.** (*a*)	**13.** (*b*)	**14.** (*c*)	**15.** (*c*)	**16.** (*b*)
17. (*a*)	**18.** (*b*)	**19.** (*d*)	**20.** (*d*)	**21.** (*c*)	**22.** (*c*)	**23.** (*b*)	**24.** (*c*)
25. (*b*)	**26.** (*d*)	**27.** (*d*)	**28.** (*b*)	**29.** (*d*)	**30.** (*b*)	**31.** (*d*)	**32.** (*a*)
33. (*d*)	**34.** (*d*)	**35.** (*d*)	**36.** (*a*)	**37.** (*b*)	**38.** (*d*)	**39.** (*d*)	**40.** (*b*)
41. (*c*)	**42.** (*a*)	**43.** (*d*)	**44.** (*b*)	**45.** (*b*)	**46.** (*c*)	**47.** (*b*)	**48.** (*d*)
49. (*c*)	**50.** (*b*)						

SOLUTIONS

1. $x^2 - y^2 = (x - y)(x + y)$; $45 = 1 \times 45$; 3×15; 5×9 *i.e.*, 3 pairs

2. Middle term $= \frac{2+198}{2} = 100$

3. $\frac{3555 \times 5555 \times 2025}{225 \times 17775 \times 2222} = \frac{9}{2} = 4.5$

4. $(B+C)+(A+C)-(A+B) = \frac{1}{60}+\frac{1}{45}-\frac{1}{36} \Rightarrow 2C = \frac{2}{180} \Rightarrow C = \frac{1}{180}$

5. Third man $= 1-\frac{8}{16}-\frac{8}{24} = 1-\frac{1}{2}-\frac{1}{3} = \frac{1}{6} \Rightarrow \frac{1}{6} \times$ ₹ 960 = ₹ 160

6. Loss $= 400 \times \frac{8 \times 8}{100}\%$ = ₹ 2.56

7. Discount $= 30 + 30 - \frac{30 \times 30}{100} = 51\%$

8. Selling price $= (2000 - 400) \times 75 \times \frac{70}{100}$ = ₹ 84000

⇒ Profit = (₹ 84000 – ₹ 70, 000) = ₹ 14000

⇒ Profit% $= \frac{14000}{70000} \times 100 = 20\%$

9. $(15 \times 22) : (11 \times 25) = 330 : 275 = 6 : 5$

10. $\frac{7}{9}+\frac{7}{18} = \frac{7}{9}\left(1+\frac{1}{2}\right) = \frac{7}{9} \times \frac{3}{2} = \frac{7}{6}$ ⇒ Required ratio $= 7 : (2 \times 6 - 7) = 7 : 5$

11.

7/12   0

      7/16

7/16   7/48

Ratio $= \frac{7}{16} : \frac{7}{48} = 3 : 1$

⇒ Total mixture = 4 ratio = 4×9 lit = 36 litres

Quantity of liquid $A = \frac{7}{12} \times 36 = 21$ litres

12.

25   40

     30   Required ratio = 10 : 5 = 2 : 1

10   5

13. $\frac{1}{4} \times 3\% + \frac{2}{3} \times 5\% + \frac{1}{12} \times 11\% = \left(\frac{3}{4}+3\frac{1}{3}+\frac{11}{12}\right)\% = \left(3+\frac{9+4+11}{12}\right)\% = 5\%$

14. Average = 3rd term = $3 \times 5 = 15$

15. 15% of cost of Fan I = 9% of cost of Fan II ⇒ I : II = 9 : 15 = 3 : 5

⇒ $I = \frac{3}{8} \times 2160$ = ₹ 810; II $= \frac{5}{8} \times 2160$ = ₹ 1350

16. New selling price $= 720 \times \frac{125}{75}$ = ₹ 1200

17. $264 \times \frac{100}{120} \times \frac{100}{110}$ = ₹ 200

18. Change = $\frac{10\times10}{100}$ = 1% decrease

19. Salary = $2310 \times \frac{100}{33}$ = ₹ 7000

20. 90 km/hr = $90 \times \frac{5}{18}$ m/s = 25 m/s ⇒ length = 25 × 10 = 250 m

21. Speed upstream = 4 – 2 = 2 km/hr ⇒ Time = 10 ÷ 2 = 5 hours

22. (4 – 3.75)% of 96% of capital ₹ 48

⇒ Capital = $48 \times \frac{100}{96}\times\frac{100}{0\cdot25}$ = ₹ 20,000

23. $10000 \times \frac{8}{10}\times\frac{8}{10} = 6400$

24. $2(lb + bh + lh) = 22 \Rightarrow 2 \times 1 + 1 \times h + 2 \times h = 11 \Rightarrow 3h = 9 \Rightarrow h = 3$ cm

Diagonal = $\sqrt{l^2+b^2+h^2} = \sqrt{1+4+9} = \sqrt{14}$

25. In ΔACE, $AC = AE = CE = 2$ cm (radii)

⇒ ΔACE is an equilateral triangle

Similarly ΔCBD and ΔCDE are equilateral triangles with side 2 cm each.

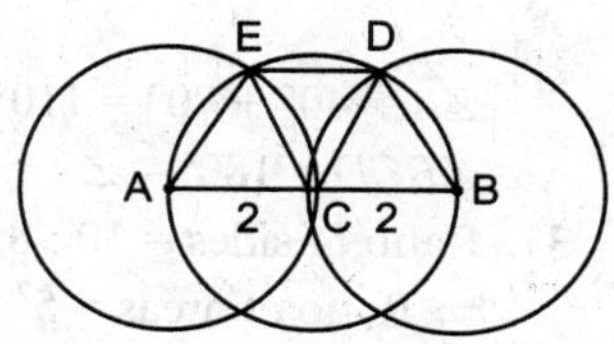

Area of each triangle = $\frac{\sqrt{3}}{4}(\text{side})^2 = \frac{\sqrt{3}}{4}\times2^2 = \sqrt{3}$

⇒ Area of three triangle = $3 \times \sqrt{3} = 3\sqrt{3}$ sq. unit

26. $\frac{4}{3}\pi r^3 = \frac{2}{3}\pi R^3 \Rightarrow 2r^3 = R^3 \Rightarrow r : R = 1 : \sqrt[3]{2}$

27. $\pi r^2 h = \frac{4}{3}\pi r^3 \Rightarrow h = \frac{4}{3}r \Rightarrow 2\pi r(r + h) : 4\pi r^3 = (r + h) : 2r^2 = \left(1+\frac{4}{3}\right) : 2 = 7 : 6$

28. $3^2 + 4^2 = 5^2$ ⇒ The triangle is right triangle.

Area = $\frac{1}{2} \times 3x \times 4x = 72 \Rightarrow x^2 = \frac{72\times2}{6} = 12 \Rightarrow x = 2\sqrt{3}$

Perimeter = $3x + 4x + 5x = 12x = 24\sqrt{3}$ ⇒ Side of equilateral triangle = $24\sqrt{3} \div 3 = 8\sqrt{3}$

⇒ Required area = $\frac{\sqrt{3}}{4}a^2 = \frac{\sqrt{3}}{4}(8\sqrt{3})^2 = 48\sqrt{3}$ sq. unit

29. $\pi r + 2r = 36 \Rightarrow r = 36 \times \frac{7}{36} = 7$ cm.

Required area = $\frac{1}{2}\pi r^2 = \frac{1}{2}\times\frac{22}{7}\times7\times7 = 77$ m^2

30. **Trick:** $x = 9 \Rightarrow y = 0 \Rightarrow (x, y) = (9, 0)$

31. G.E. $= \frac{1}{2}\left[2(a^2+b^2+c^2-ab-bc-ca)\right] = \frac{1}{2}\left[(a-b)^2+(b-c)^2+(c-a)^2\right]$

$\frac{1}{2}\left[(2y)^2+(-3y)^2+(y)^2\right] = \frac{1}{2}\left[4y^2+9y^2+y^2\right] = \frac{1}{2}\times 14y^2 = 7y^2$

32. $(a+b)^3 = 1^3 \Rightarrow a^3+b^3+3ab(a+b) = 1 \Rightarrow k = 1$

33. $a^2x^2+b^2y^2+2abxy = 9$ ... (1)

$b^2x^2+a^2y^2-2ab\,xy = 16$ ... (2)

$(1)+(2) \Rightarrow x^2(a^2+b^2)+y^2(a^2+b^2) = 25$

$\Rightarrow (x^2+y^2)(a^2+b^2) = 25 \Rightarrow (a^2+b^2) = 25 \div 1 = 25$

34. Circumcentre of a triangle lies inside the triangle for an acute triangle, outside for obtuse triangle and at the centre of the hypotenuse for the right triangle.

35. $AD^2 = BD \times DC$

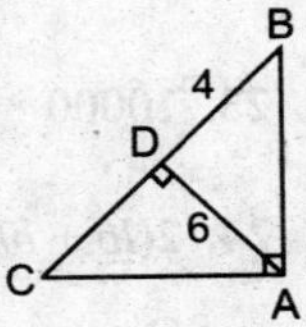

$\Rightarrow DC = \frac{6\times 6}{4} = 9$ cm $\Rightarrow BC = BD + DC = (4+9)$ cm $= 13$ cm

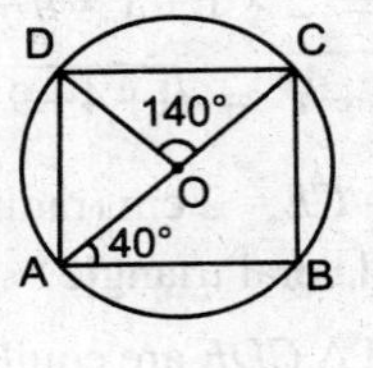

36. $\angle CAD = \frac{1}{2}\angle COD = \frac{1}{2}\times 140° = 70°$

$\angle A = 40° + 70° = 110°$

$\angle BCD = 180° - \angle A = 180° - 110° = 70°$

37. Ratio of sides = 10 : 8 = 5 : 4

$\Rightarrow$ Ratio of areas $= 5^2 : 4^2 = 25 : 16$

38. $\cos\theta = \frac{1}{\text{cosec } 23°} = \sin 23° = \cos(90° - 23°) = \cos 67°$

39. $(\sin^2 5° + \sin^2 85°) + (\sin^2 10° + \sin^2 80°) + ... + (\sin^2 40° + \sin^2 50°) + \sin^2 45° + \sin^2 90°$

$= (\sin^2 5° + \cos^2 5°) + (\sin^2 10° + \cos^2 10°) + ... + (\sin^2 40° + \cos^2 40°) + \left(\frac{1}{\sqrt{2}}\right)^2 + 1$

$= (1+1+1 \text{ .... 8 times}) + \frac{1}{2} + 1 = 8 + \frac{1}{2} + 1 = 9\frac{1}{2}$

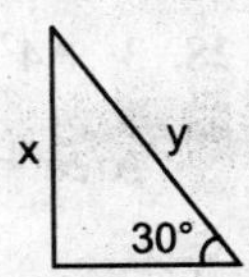

40. $\sin 30° = \frac{1}{2} = \frac{x}{y}$

$\Rightarrow x : y = 1 : 2 \Rightarrow x = \frac{1}{3}\times 15$ cm $= 5$ cm

41. $A + B + C = 72° + 72° + 36° = 180°$

42. $F = 45° = 1.5 \Rightarrow D + E = 18° + 180° = 36° = \frac{36}{45}\times 1{\cdot}5 = 1.2$

43. $72° \times 3 : 36° \times 1 = 6 : 1$

44. $\frac{72}{360}\times 100 = 20\%$

45. $36° = 1.2$ (see solution 42) $\Rightarrow 360° = 12$

46. $420 - 320 = 100$ thousand $= 1$ lakh

47. $\frac{320}{400} = 0 \cdot 8$ times **48.** 2003 : $\frac{20}{400} = \frac{1}{20}$; 2004 $= \frac{20}{420} = \frac{1}{21}$

49. $\frac{440}{400} \times 100 = 110\ \%$

50. 2001: $\frac{20}{340} = \frac{1}{17}$; 2005 : $\frac{40}{440} = \frac{1}{11}$

FOR VISUALLY HANDICAPPED CANDIDATES

41. $A : B = 120 : 100 = 6 : 5$; $B : C = 80 : 100 = 4 : 5$

$\Rightarrow A : B : C = 24 : 20 : 25 \Rightarrow C = \frac{25}{69} \times 6900 =$ ₹ 2500

42. $(2x + 10) + (2x) + (5x + 25) = 170 \Rightarrow 9x = 135 \Rightarrow x = 15$

$\Rightarrow$ Three parts = 40, 30, 100

Trick: In option (*b*), First : Third = 40 : 100 = 2 : 5,

First - second = 40 – 30 = 10

43. 20% of CP = ₹ 18 $\Rightarrow$ CP = ₹ 18 × 5 = ₹ 90

44. Total distance = 20 km; Average speed $= 20 \times \frac{60}{55} = \frac{240}{11}$

$\frac{10}{x-2} = \frac{55}{60} \Rightarrow \frac{x^2-4}{x} = \frac{240}{11}$

$\Rightarrow 11x^2 - 44 = 240x \Rightarrow 11x^2 = 240x + 44 \Rightarrow x = 22$ (from options)

45. $x + \frac{1}{x} = 2 \Rightarrow x = 1 \Rightarrow x^5 + \frac{1}{x^5} = 1 + 1 = 2$

46. $(55 \div 2) \times 3 = 82.5$

47. $\cos 61° = \sqrt{1 - \frac{a^2}{a^2+b^2}} = \sqrt{\frac{b^2}{a^2+b^2}} = \frac{b}{\sqrt{a^2+b^2}}$

$\tan 61° = \frac{\sin 61°}{\cos 61°} = \frac{a}{b} \Rightarrow \tan 29° = \cot(90° - 29°) = \cot 61° = \frac{b}{a}$

$\Rightarrow \tan 61° + \tan 29° = \frac{a}{b} + \frac{b}{a}$

48. LCM of 36 and 102 = 9 × 4 × 17 = 612; 612 × 3 = 1836

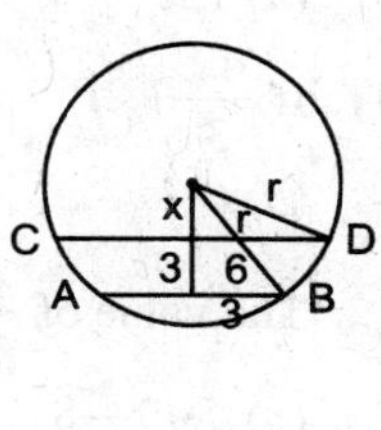

49. $r^2 = x^2 + 6^2$ and $r^2 = (x + 3)^3 + 3^2$

$\Rightarrow x^2 + 6^2 = (x + 3)^2 + 3^2$

Comparing we get, $x = 3$. $r = \sqrt{3^2 + 6^2} = \sqrt{45} = 3\sqrt{5}$

50. Ratio of areas = $1^2 : 2^2 = 1 : 4$

$\Rightarrow$ New cost = 4 × ₹ 475 = ₹ 1900

SSC CGL TIER I 2013

PART – C

QUANTITATIVE APTITUDE (123NM2)

1. If a, b, c, d, e are five consecutive odd numbers, their average is

(a) $5(a+4)$ (b) $\frac{abcde}{5}$ (c) $5(a+b+c+d+e)$ (d) $a+4$

2. The average of 20 numbers is 15 and the average of first 5 is 12. The average of the rest is

(a) 16 (b) 15 (c) 14 (d) 13

3. A tradesman sold an article at a loss of 20%. If the selling price had been increased by ₹ 100, there would have been a gain of 5%. The cost price of the article (in ₹) was

(a) 100 (b) 200 (c) 400 (d) 500

4. The price of an article is first decreased by 20% and then increased by 30%. If the resulting price is ₹ 416, the original price of the article is

(a) ₹ 350 (b) ₹ 405 (c) 400 (d) ₹ 450

5. A man performs $\frac{2}{15}$ of the total journey by train, $\frac{9}{20}$ by bus and the remaining 10 km on foot. His total journey in km is

(a) 15.6 (b) 24 (c) 16.4 (d) 12.8

6. By walking at $\frac{3}{4}$ of his usual speed, a man reaches his office 20 minutes later than usual. His usual time is

(a) 30 min. (b) 75 min. (c) 90 min. (d) 60 min

7. If the compound interest on a certain sum for two years at 12% per annum is ₹ 2,544, the simple interest on it at the same rate for 2 years will be

(a) ₹ 2,400 (b) ₹ 2,500 (c) ₹ 2,480 (d) ₹ 2,440

8. The total cost of 8 buckets and 5 mugs is ₹ 92 and the total cost of 5 buckets and 8 mugs is ₹ 77. Find the cost of 2 mugs and 3 buckets,

(a) ₹ 35 (b) ₹ 70 (c) ₹ 30 (d) ₹ 38

9. If $\frac{a}{1-a}+\frac{b}{1-b}+\frac{c}{1-c}=1$, then the value of $\frac{1}{1-a}+\frac{1}{1-b}+\frac{1}{1-c}$ is

(a) 1 (b) 3 (c) 4 (d) 0

10. If $(x-3)^2+(y-5)^2+(z-4)^2=0$, then the value of $\frac{x^2}{9}+\frac{y^2}{25}+\frac{z^2}{16}$ is

(a) 12 (b) 9 (c) 3 (d) 1

11. If $\frac{4x}{3}+2P=12$ for what value of P, $x=6$?

(a) 6 (b) 4 (c) 2 (d) 1

12. The value of $\frac{4+3\sqrt{3}}{7+4\sqrt{3}}$ is

(a) $5\sqrt{3}-8$ (b) $5\sqrt{3}+8$ (c) $8\sqrt{3}+5$ (d) $8\sqrt{3}-5$

13. If $x\left(3-\frac{2}{x}\right)=\frac{3}{x}$, then the value of $x^2+\frac{1}{x^2}$ is

(a) $2\frac{1}{9}$ (b) $2\frac{4}{9}$ (c) $3\frac{1}{9}$ (d) $3\frac{4}{9}$

14. What number must be added to the expression $16a^2 - 12a$ to make it a perfect square ?

(*a*) $\frac{9}{4}$ (*b*) $\frac{11}{2}$ (*c*) $\frac{13}{2}$ (*d*) 16

15. The straight line $4x + 3y = 12$ passes through

(*a*) 1st, 2nd and 3rd quadrant (*b*) 1st, 2nd and 4th quadrant
(*c*) 2nd, 3rd and 4th quadrant (*d*) 1st, 3rd and 4th quadrant

16. The sum of three altitudes of a triangle is

(*a*) equal to the sum of three sides (*b*) less than the sum of sides
(*c*) greater than the sum of sides (*d*) twice the sum of sides

17. In ΔABC, $\angle A + \angle B = 65°$, $\angle B + \angle C = 140°$, then find $\angle B$.

(*a*) 40° (*b*) 25° (*c*) 35° (*d*) 20°

18. The length of the tangent drawn to a circle of radius 4 cm from a point 5 cm away from the centre of the circle is

(*a*) 3 cm (*b*) $4\sqrt{2}$ cm (*c*) $5\sqrt{2}$ cm (*d*) $3\sqrt{2}$ cm

19. A cyclic quadrilateral *ABCD* is such that $AB = BC$, $AD = DC$, $AC \perp BD$, $\angle CAD = \theta$. Then the angle $\angle ABC =$

(*a*) 6 (*b*) $\frac{\theta}{2}$ (*c*) 2θ (*d*) 3θ

20. The height of an equilateral triangle is 15 cm. The area of the triangle is

(*a*) $50\sqrt{3}$ sq. cm. (*b*) $70\sqrt{3}$ sq. cm. (*c*) $75\sqrt{3}$ sq. cm. (*d*) $150\sqrt{3}$ sq. cm.

21. Two parallel chords of a circle, of diameter 20 cm lying on the opposite sides of the centre are of lengths 12 cm and 16 cm. The distance between the chords is

(*a*) 16 cm (*b*) 24 cm (*c*) 14 cm (*d*) 20 cm

22. In ΔABC, $DE \parallel AC$. *D* and *E* are two points on *AB* and *CB* respectively. If $AB = 10$ cm and $AD = 4$ cm, then $BE : CE$ is

(*a*) 2 : 3 (*b*) 2 : 5 (*c*) 5 : 2 (*d*) 3 : 2

23. *A*, *B* and *C* are the three points on a circle such that the angles subtended by the chords *AB* and *AC* at the centre *O* are 90° and 110° respectively. $\angle BAC$ is equal to

(*a*) 70° (*b*) 80° (*c*) 90° (*d*) 100°

24. If the angles of elevation of a balloon from two consecutive kilometre-stones along a road are 30° and 60° respectively, then the height of the balloon above the ground will be

(*a*) $\frac{\sqrt{3}}{2}$ km (*b*) $\frac{1}{2}$ km (*c*) $\frac{2}{\sqrt{3}}$ km (*d*) $3\sqrt{3}$ km

25. Evaluate: 3 cos 80° cosec 10° + 2 cos 59° cosec 31°

(*a*) 1 (*b*) 3 (*c*) 2 (*d*) 5

26. $\left(\frac{3\pi}{5}\right)$ radians is equal to

(*a*) 100° (*b*) 120° (*c*) 108° (*d*) 180°

27. If $\tan\theta + \cot\theta = 2$, then the value of $\tan^2\theta + \cot^2\theta$ is

(*a*) 2 (*b*) 1 (*c*) $\sqrt{2}$ (*d*) 0

28. The eliminant of θ from $x\cos\theta - y\sin\theta = 2$ and $x\sin\theta + y\cos\theta = 4$ will give

(*a*) $x^2 + y^2 = 20$ (*b*) $3x^2 + y^2 = 20$ (*c*) $x^2 - y^2 = 20$ (*d*) $3x^2 - y^2 = 10$

29. $\sin^2\theta - 3\sin\theta + 2 = 0$ will be true if

(*a*) $0 \le \theta < 90$ (*b*) $0 < \theta < 90$ (*c*) $\theta = 0°$ (*d*) $\theta = 90°$

30. The value of $\left[\frac{\cos^2 A(\sin A + \cos A)}{\text{cosec}^2 A(\sin A - \cos A)} + \frac{\sin^2 A(\sin A - \cos A)}{\sec^2 A(\sin A + \cos A)}\right](\sec^2 A - \text{cosec}^2 A)$ is

(*a*) 1 (*b*) 3 (*c*) 2 (*d*) 4

31. The fourth root of 24010000 is

(*a*) 7 (*b*) 49 (*c*) 490 (*d*) 70

32. The greatest 4 digit number which is a perfect square, is

(*a*) 9999 (*b*) 9909 (*c*) 9801 (*d*) 9081

33. A piece of work can be done by Ram and Shyam in 12 days, by Shyam and Hari in 15 days and by Hari and Ram in 20 days. Ram alone will complete the work in

(*a*) 30 days (*b*) 32 days (*c*) 36 days (*d*) 42 days

34. 3 men or 5 women can do a work in 12 days. How long will 6 men and 5 women take to finish the work ?

(*a*) 4 days (*b*) 5 days (*c*) 6 days (*d*) 7 days

35. *A* and *B* can do a job in 12 days, *B* and *C* in 15 days and *C* and *A* in 20 days. How long would *A* take to do that work ?

(*a*) 20 days (*b*) 60 days (*c*) 30 days (*d*) 40 days

36. The difference of perimeter and diameter of a circle is X unit. The diameter of the circle is

(*a*) $\frac{X}{\pi - 1}$ unit (*b*) $\frac{X}{\pi + 1}$ unit (*c*) $\frac{X}{\pi}$ unit (*d*) $\left(\frac{X}{\pi} - 1\right)$ unit

37. The perimeter of the base of a right circular cylinder is '*a*' unit. If the volume of the cylinder is *V* cubic unit, then the height of the cylinder is

(*a*) $\frac{4a^2 V}{\pi}$ unit (*b*) $\frac{4\pi a^2}{V}$ unit (*c*) $\frac{\pi a^2 V}{4}$ unit (*d*) $\frac{4\pi V}{a^2}$ unit

38. A sphere of diameter 6 cm is dropped in a right circular cylindrical vessel partly filled with water. The diameter of the cylindrical vessel is 12 cm. If the sphere is just completely submerged in water, then the rise of water level in the cylindrical vessel is

(*a*) 2 cm (*b*) 1 cm (*c*) 3 cm (*d*) 4 cm

39. A shopkeeper marks his goods 20% above his cost price and gives 15% discount on the marked price. His gain percent is

(*a*) 5% (*b*) 4% (*c*) 2% (*d*) 1%

40. A shopkeeper earns a profit of 12% on selling a book at 10% discount on printed price. The ratio of the cost price to printed price of the book is **(SSC CGL Tier II 2012)**

(*a*) 45 : 56 (*b*) 50 : 61 (*c*) 90 : 97 (*d*) 99 : 125

41. The list price of an article is ₹ 160 and a customer buys it for ₹ 122.40 after two successive discounts. If the first discount is 10%, then second discount is

(*a*) 12% (*b*) 10% (*c*) 14% (*d*) 15%

42. In a school, 10% of number of girls is equal to $\frac{1}{20}^{th}$ of number of boys. Ratio between the number of boys to number of girls is

(*a*) 1 : 2 (*b*) 2 : 1 (*c*) 1 : 4 (*d*) 4 : 1

43. The ratio of $25^{2.5} : 5^3$ is same as

(*a*) 5 : 3 (*b*) 5 : 6 (*c*) 1 : 25 (*d*) 25 : 1

Direction: The sub divided bar diagram given below depicts H.S. Students of a school for three years. Study the diagram and answer the Question Nos. 44 to 48.

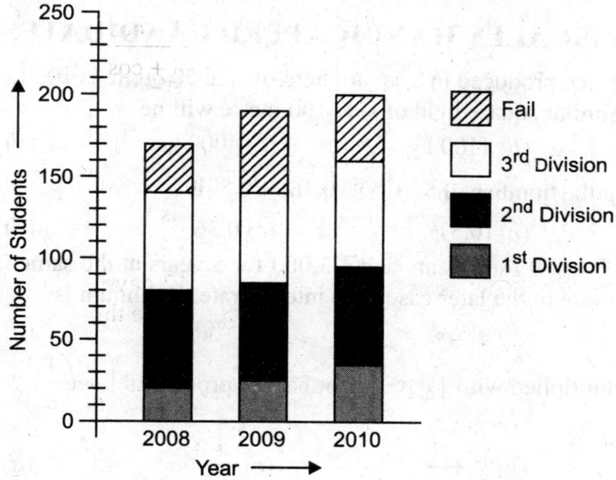

44. The percentage passed in 1st division in 2008 was

(*a*) 27% (*b*) 32% (*c*) $15\frac{3}{8}\%$ (*d*) $11\frac{13}{17}\%$

45. The pass percentage in 2008 was

(*a*) 67% (*b*) 73% (*c*) $79\frac{2}{3}\%$ (*d*) $82\frac{6}{17}\%$

46. In which year the school had the best result for H.S. in respect of percentage of pass candidates ?

(*a*) 2008 (*b*) 2009 (*c*) 2010

(*d*) The percentage of pass candidates are same for the three years.

47. The number of students passed in third division in the year 2008 was

(*a*) 50 (*b*) 60 (*c*) 70 (*d*) 80

48. The percentage of the students passed in 2nd division in the year 2010 was

(*a*) 30% (*b*) 40% (*c*) 50% (*d*) 60%

Following figure in Pie-chart representing itemwise cost of manufacturing certain product. Study the chart and answer Question Nos. 49 and 50.

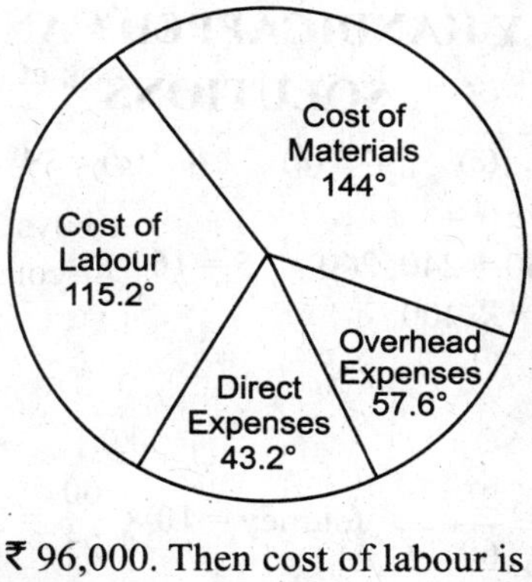

49. Total manufacturing cost is ₹ 96,000. Then cost of labour is

(*a*) ₹ 30,720 (*b*) ₹ 38,400 (*c*) ₹ 11,520 (*d*) ₹ 15,000

50. The difference of cost of material and direct expenses is
(*a*) ₹ 26,000 (*b*) ₹ 10,000 (*c*) ₹ 26,500 (*d*) ₹ 26,880

FOR VISUALLY HANDICAPPED CANDIDATES ONLY

44. The amount of rice produced in a square field of side 50 metre is 1000 kg. The amount of rice produced in a similar square field of side 100 metre will be
(*a*) 2000 kg (*b*) 1100 kg (*c*) 4000 kg (*d*) 1150 kg

45. Greatest among the numbers 0.5, $(0.5)^2$, 0.36, $\sqrt{0.5}$ is
(*a*) 0.5 (*b*) $(0.5)^2$ (*c*) 0.36 (*d*) $\sqrt{0.5}$

46. A man invests ₹ 4,000 for 3 years and ₹ 3,000 for 5 years at the same rate of simple interest. He gets ₹ 150 more in the later case. The interest rate per annum is
(*a*) 4% (*b*) 4½% (*c*) 5% (*d*) 6%

47. What is to be multiplied with $\left(b \times \frac{1}{a}\right)$ so that the product is $\left(b + \frac{1}{a}\right)$?
(*a*) $a + \frac{1}{b}$ (*b*) $b + \frac{1}{a}$ (*c*) $\frac{a}{b}$ (*d*) ab

48. If $a - b = 2$, $ab = 15$, the value of $(a^2 - b^2)(a^3 - b^3)$ is
(*a*) 1450 (*b*) 1500 (*c*) 1528 (*d*) 1568

49. A chord of a circle of radius 7.5 cm with centre O is of length 9 cm. Its distance from the centre is
(*a*) 12 cm (*b*) 6 cm (*c*) 3 cm (*d*) 2 cm

50. If cos 23° sec (90° – θ) = 1, then the value θ is
(*a*) 23° (*b*) 67° (*c*) 90° (*d*) 30°

ANSWER

1. (*d*)	**2.** (*a*)	**3.** (*c*)	**4.** (*c*)	**5.** (*b*)	**6.** (*d*)	**7.** (*a*)	**8.** (*a*)
9. (*c*)	**10.** (*c*)	**11.** (*c*)	**12.** (*a*)	**13.** (*b*)	**14.** (*a*)	**15.** (*b*)	**16.** (*b*)
17. (*b*)	**18.** (*a*)	**19.** (*c*)	**20.** (*c*)	**21.** (*c*)	**22.** (*d*)	**23.** (*b*)	**24.** (*a*)
25. (*d*)	**26.** (*c*)	**27.** (*a*)	**28.** (*a*)	**29.** (*d*)	**30.** (*c*)	**31.** (*d*)	**32.** (*c*)
33. (*a*)	**34.** (*a*)	**35.** (*c*)	**36.** (*a*)	**37.** (*d*)	**38.** (*b*)	**39.** (*c*)	**40.** (*a*)
41. (*d*)	**42.** (*b*)	**43.** (*d*)	**44.** (*d*)	**45.** (*d*)	**46.** (*a*)	**47.** (*b*)	**48.** (*a*)
49. (*a*)	**50.** (*d*)						

FOR VISUALLY HANDICAPPED CANDIDATES ONLY

SOLUTIONS

44. (*c*) **45.** (*d*) **46.** (*c*) **47.** (*a*) **48.** (*d*) **49.** (*b*) **50.** (*b*)

1. Average = mid-term = $c = a + 4$

2. $20 \times 15 - 5 \times 12 = 300 - 60 = 240$; $240 \div 15 = 16$

3. 25% of CP = ₹ 100 ⇒ CP = ₹ 400

4. $416 \times \frac{100}{80} \times \frac{100}{130} =$ ₹ 400

5. $1 - \frac{2}{15} - \frac{9}{20} = \frac{60-8-27}{60} = \frac{25}{60} \Rightarrow$ Journey $= 10 \times \frac{60}{25} = 24$ km

6. $\left(\frac{4}{3} - 1\right)$ usual time = 20 minutes ⇒ usual time = 3 × 20 min = 60 minutes

7. If S.I. = ₹ 200, then C.I. = ₹ 212 $\Rightarrow$ S.I. = $\frac{200}{212} \times 2544 =$ ₹ 2400

8. $8x + 5y = 92$ and $5x + 8y = 77 \Rightarrow 13x + 13y = 169 \Rightarrow x + y = 13$
$\Rightarrow 3x = 92 - 65 = 27 \Rightarrow x = 9 \Rightarrow y = 13 - 9 = 4 \quad \therefore 3x + 2y = 3(9) + 2(4) = 27 + 8 =$ ₹ 35

9. $\frac{a}{1-a}+1+\frac{b}{1-b}+1+\frac{c}{1-c}+1 = 1 + 3 = 4$

$\Rightarrow \frac{a+1-a}{1-a}+\frac{b+1-b}{1-b}+\frac{c-1-c}{1-c} = 4 \Rightarrow \frac{1}{1-a}+\frac{1}{1-b}+\frac{1}{1-c} = 4$

10. $(x-3) = 0, (y-5) = 0, (z-4) = 0 \Rightarrow x = 3, y = 5, z = 4$

$\Rightarrow \frac{x^2}{9}+\frac{y^2}{25}+\frac{z^2}{16} = \frac{9}{9}+\frac{25}{25}+\frac{16}{16} = 3$

11. $\frac{4\times 6}{3} + 2P = 12 \Rightarrow 2P = 12 - 8 = 4 \Rightarrow P = 2$

12. G.E. = $(4 + 3\sqrt{3})(7 - 4\sqrt{3}) = 28 - 36 + 21\sqrt{3} - 16\sqrt{3} = 5\sqrt{3} - 8$

13. $x\left(3-\frac{2}{x}\right) = \frac{3}{x} \Rightarrow 3x - 2 = \frac{3}{x} \Rightarrow 3x - \frac{3}{x} = 2 \Rightarrow x - \frac{1}{x} = \frac{2}{3}$

$\Rightarrow x^2 + \frac{1}{x^2} = \left(x - \frac{1}{x}\right)^2 + 2 = \left(\frac{2}{3}\right)^2 + 2 = \frac{4}{9} + 2 = 2\frac{4}{9}$

14. $16a^2 - 12a = (4a)^2 - 2(4a)\left(\frac{3}{2}\right) \Rightarrow$ Required number = $\left(\frac{3}{2}\right)^2 = \frac{9}{4}$

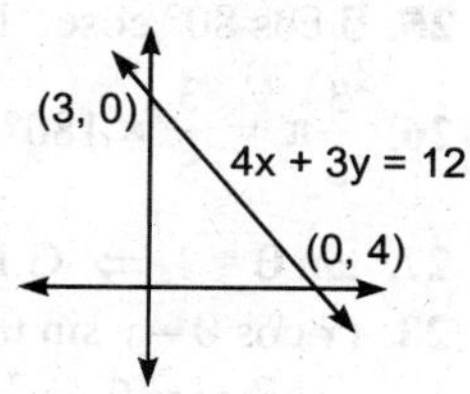

15. When $x = 0$, $3y = 12 \Rightarrow y = 4$
When $y = 0$, $4x = 12 \Rightarrow x = 3$
$\therefore$ Quadrants: I, II, IV

17. $\angle B = (\angle A + \angle B) + (\angle B + \angle C) - (\angle A + \angle B + \angle C) = 65° + 140° - 180° = 25°$

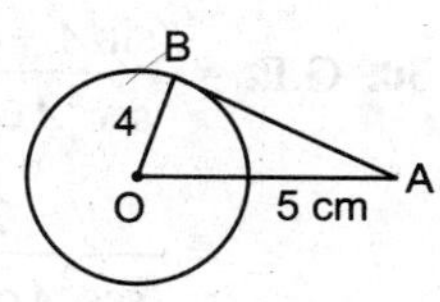

18. $AB^2 = AO^2 - BO^2 = 5^2 - 4^2 = 3^2$

19. In ΔACD, $AD = CD \Rightarrow \angle ACD = \angle CAD = \theta$
$\angle ABD = \angle ACD = \theta$ (Angle in same segment)
$\angle CBD = \angle CAD = \theta$ (Angle in same segment)
$\angle ABC = \angle ABD + \angle CBD = \theta + \theta = 2\theta$
Alt: $\angle ADC = 180° - 2\theta \Rightarrow \angle ABC = 180° - (180° - 2\theta) = 2\theta$

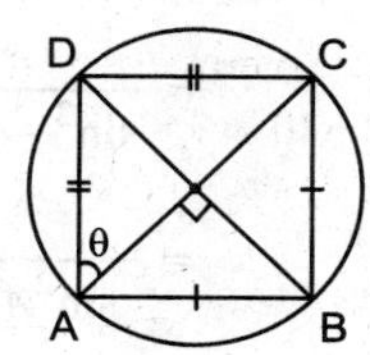

20. Height = $\frac{\sqrt{3}}{2}a = 15 \Rightarrow \frac{a}{2} = 15\times\frac{1}{\sqrt{3}} = 5\sqrt{3}$

$\Rightarrow$ Area = $\frac{\sqrt{3}}{4}a^2 = \frac{\sqrt{3}}{2}a\times\frac{a}{2} = 15\times 5\sqrt{3} = 75\sqrt{3}$

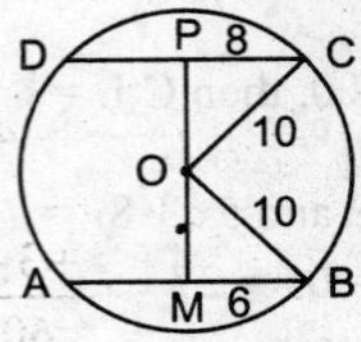

21. $OM = \sqrt{10^2 - 6^2} = \sqrt{8^2} = 8$ cm

$OP = \sqrt{10^2 - 8^2} = \sqrt{6^2} = 6$ cm

$\therefore PM = OP + OM = 6 + 8 = 14$ cm

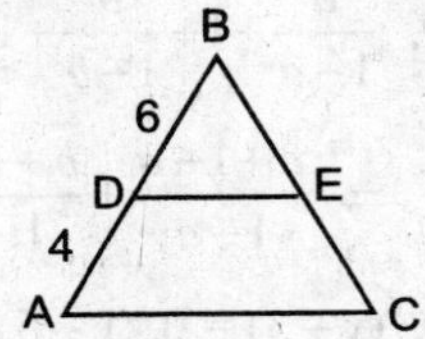

22. $BD = AB - AD = 10 - 4 = 6$ cm

$$\frac{BE}{CE} = \frac{BD}{AD} = \frac{6}{4} = \frac{3}{2}$$

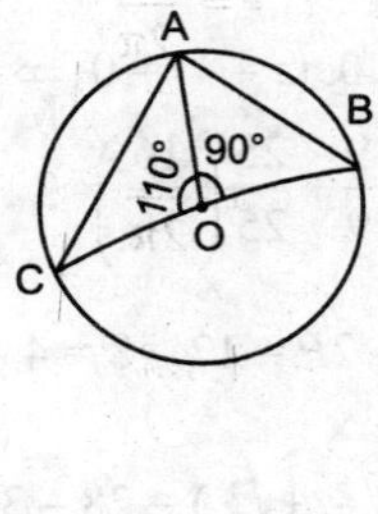

23. $\angle BOC = 90° + 110° = 200°$

$\angle BDC = \frac{1}{2}\angle BOC = \frac{1}{2} \times 200° = 100°$

$\angle BAC = 180° - \angle BDC = 180° - 100° = 80°$

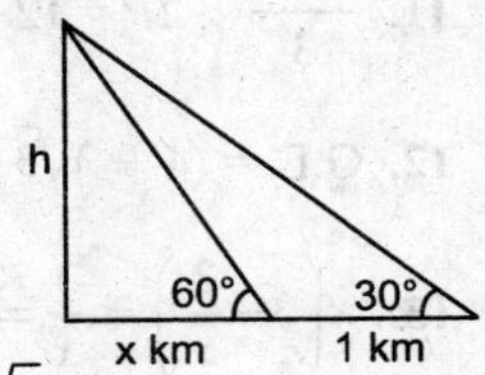

24. $\tan 60° = \frac{h}{x} \Rightarrow \sqrt{3} = \frac{h}{x} \Rightarrow h = \sqrt{3}x$

$\tan 30° = \frac{1}{\sqrt{3}} = \frac{h}{x+1} \Rightarrow \sqrt{3}h = x + 1$

$\Rightarrow \sqrt{3}(\sqrt{3}x) = x + 1 \Rightarrow 3x = x + 1 \Rightarrow x = \frac{1}{2} \Rightarrow h = \sqrt{3}x = \sqrt{3} \times \frac{1}{2} = \frac{\sqrt{3}}{2}$

25. $3 \cos 80° \text{ cosec } 10° + 2 \cos 59° \text{ cosec } 31° = 3 \cos 80° \sec 80° + 2 \cos 59° \sec 59° = 3 + 2 = 5$

26. $\frac{3}{5}\pi = \frac{3}{5} \times 180° = 108°$

27. $\tan\theta = 1 \Rightarrow$ G.E. $= 1^2 + 1^2 = 2$

28. $(x \cos\theta - y \sin\theta)^2 + (x \sin\theta - y \cos\theta)^2 = 2^2 + 4^2$

$\Rightarrow x^2 \cos^2\theta + y^2 \sin^2\theta + x^2 \sin^2\theta + y^2 \cos^2\theta = 20$

$\Rightarrow x^2(\sin^2\theta + \cos^2\theta) + y^2(\sin^2\theta + \cos^2\theta) = 20 \Rightarrow x^2 + y^2 = 20$

29. When $\theta = 90$, G.E. $= 1^2 - 3 + 2 = 0$

30. G.E. $= \dfrac{(\sin A + \cos A)^2 + (\sin A - \cos A)^2}{\sec^2 A \operatorname{cosec}^2 A (\sin^2 A - \cos^2 A)}(\sec^2 A - \operatorname{cosec}^2 A)$

$= \dfrac{2(\sin^2 A + \cos^2 A)}{\sec^2 A \operatorname{cosec}^2 A(\sin^2 A - \cos^2 A)}(\sec^2 A - \operatorname{cosec}^2 A)$

$= \dfrac{2}{\sin^2 - A\cos^2 A} \times \dfrac{\sec^2 A - \operatorname{cosec}^2 A}{\sec^2 A \operatorname{cosec}^2 A} = \dfrac{2}{\sin^2 A - \cos^2 A} \times \left(\dfrac{1}{\operatorname{cosec}^2 A} - \dfrac{1}{\sec^2 A}\right)$

$= \dfrac{2}{\sin^2 A - \cos^2 A}(\sin^2 A - \cos^2 A) = 2$

31. $\sqrt{2401} = 49; \sqrt{49} = 7$

32. $99^2 = 99 - 1 \mid 1^2 = 9801$

33. $\frac{1}{12} + \frac{1}{20} - \frac{1}{15} = \frac{5+3-4}{60} = \frac{4}{60} = \frac{1}{15} \Rightarrow$ Ram $= 15 \times 2 = 30$ days

34. $\frac{3\times5\times12}{3\times5+5\times6}=\frac{3\times5\times12}{45}=4$ days

35. $A(2)=\frac{1}{12}+\frac{1}{20}-\frac{1}{15}=\frac{5+3-4}{60}=\frac{4}{60}=\frac{1}{15}\Rightarrow A(1)=2\times15=30$ days

36. $2\pi r-2r=X\Rightarrow r=\frac{X}{2(\pi-1)}\Rightarrow d=\frac{X}{\pi-1}$

37. Perimeter $=2\pi r=a\Rightarrow r=\frac{a}{2\pi}$

$\Rightarrow$ Volume $=V=\pi r^2h=\pi\left(\frac{a}{2\pi}\right)^2h=\frac{a^2}{4\pi}h\Rightarrow h=\frac{4\pi V}{a^2}$

38. $\frac{4}{3}\pi r^3=\frac{4}{3}(\pi)(3)^2=36\pi=\pi R^2h=\pi(6)^2h\Rightarrow h=1$ cm

39. $20-15-\frac{20\times15}{100}=20-15-3=2\%$

40. 90% of MP = 112% of CP $\Rightarrow$ MP : CP = 112 : 90 = 56 : 45 $\Rightarrow$ CP : MP = 45 : 56

41. 160 – 10% = 144; 144 – 122.40 = 21.60; $\frac{21\cdot60}{144}\times100=15\%$

42. $\frac{\text{Boys}}{20}=\frac{\text{Girls}}{10}\Rightarrow$ Boys : Girls = 20 : 10 = 2 : 1

43. $25^{2.5}:5^3=(5^2)^{2.5}:5^3=5^5:5^3=5^2:5^0=25:1$

44. $\frac{20}{70}\times100=\frac{200}{17}=11\frac{13}{17}\%$

45. $\frac{140}{170}\times100=\frac{1400}{7}=82\frac{6}{17}\%$

46. 2008 : $\frac{140}{170}$; 2009 : $\frac{140}{190}<\frac{140}{170}$; 2010 : $\frac{160}{200}=80\%<82\frac{6}{17}\%$ or $\frac{140}{170}$

47. 140 – 80 = 60

48. $\frac{90-30}{2000}\times100=30\%$

49. $\frac{115\cdot2}{360}\times96000=$ ₹ 30,720

50. $\frac{144-43\cdot2}{360}\times96000=$ ₹ 26880

For VH only

44. 4 × 1000 = 4000 kg

45. $\sqrt{0.5}=0.7$ (Note if $0<x<1$, then $\sqrt{x}>x>x^2$

46. 150 – 120 = 30 $\Rightarrow$ Rate = 150 ÷ 30 = 5%

47. $b+\frac{1}{a}=\frac{ab+1}{a}\div\frac{b}{a}=\frac{ab+1}{b}=a+\frac{1}{b}$

48. $a=5$, $b=3$, G.E. = 16 × 98 = 1568

49. $\sqrt{7.5^2-4.5^2}=\sqrt{56.25-20.25}=\sqrt{36}=6$ cm

50. $\cos23°=\cos(90°-\theta)\Rightarrow23°=90°-\theta\Rightarrow\theta=90°-23°=67°$

SSC CGL TIER II (RE-EXAM) 2013

345 RN3 : HELD ON 21/9/2014

1. The compound interest on a certain sum of money for 2 years at 5% is ₹ 328, then the sum is

(*a*) ₹ 3000 (*b*) ₹ 3600 (*c*) ₹ 3200 (*d*) ₹ 3400

2. The height of a cone is 30 cm. A small cone is cut off at the top by a plane parallel to the base. If its volume be $\frac{1}{27}$th of the volume of the given cone, at what height above the base is the section made?

(*a*) 19 cm (*b*) 20 cm (*c*) 12 cm (*d*) 15 cm

3. *ABCD* is a trapezium with *AD* and *BC* parallel sides. *E* is a point on *BC*. The ratio of the area of *ABCD* to that of *AED* is

(*a*) $\frac{\overline{AD}}{\overline{BC}}$ (*b*) $\frac{\overline{BE}}{\overline{EC}}$ (*c*) $\frac{\overline{AD}+\overline{BE}}{\overline{AD}+\overline{CE}}$ (*d*) $\frac{\overline{AD}+\overline{BC}}{\overline{AD}}$

4. If the surface area of a sphere is 346.5 cm^2, then its radius $\left[\text{taking } \pi = \frac{22}{7}\right]$ is

(*a*) 7 cm (*b*) 3.25 cm (*c*) 5.25 cm (*d*) 9 cm

5. An interior angle of a regular polygon is 5 times its exterior angle. Then the number of sides of the polygon is

(*a*) 14 (*b*) 16 (*c*) 12 (*d*) 18

6. The height of the right pyramid whose area of the base is 30 m^2 and volume is 500 m^3, is

(*a*) 50 m (*b*) 60 m (*c*) 40 m (*d*) 20 m

7. The base of a prism is a right angled triangle with two sides 5 cm and 12 cm. The height of the prism is 10 cm. The total surface area of the prism is

(*a*) 360 sq cm (*b*) 300 sq cm (*c*) 330 sq cm (*d*) 325 sq cm

8. In an equilateral triangle of side 24 cm, a circle is inscribed touching its sides. The area of the remaining portion of the triangle is ($\sqrt{3}$ = 1.732)

(*a*) 98.55 sq cm (*b*) 100 sq cm (*c*) 101 sq cm (*d*) 95 sq cm

9. The base of a right prism is an equilateral triangle. If the lateral surface area and volume is 120 cm^2, $40\sqrt{3}$ cm^3 respectively then the side of base of the prism is

(*a*) 4 cm (*b*) 5 cm (*c*) 7 cm (*d*) 40 cm

10. Perimeter of a rhombus is 2*p* unit and sum of length of diagonals is *m* unit, then area of the rhombus is

(*a*) $\frac{1}{4}m^2p$ sq unit (*b*) $\frac{1}{4}mp^2$ sq unit (*c*) $\frac{1}{4}(m^2-p^2)$ sq unit (*d*) $\frac{1}{4}(p^2-m^2)$ sq unit

11. A ball of lead 4 cm in diameter is covered with gold. If the volume of the gold and lead are equal, then the thickness of gold [given $\sqrt[3]{2}$ = 1.259] is approximately.

(*a*) 5.038 cm (*b*) 5.190 cm (*c*) 1.038 cm (*d*) 0.518 cm

12. A large solid sphere is melted and moulded to form identical right circular cones with base radius and height same as the radius of the sphere. One of these cones is melted and moulded to form a smaller solid sphere. Then the ratio of the surface area of the smaller to the surface area of the larger sphere is

(*a*) $1 : 3^{4/3}$ (*b*) $1 : 2^{3/2}$ (*c*) $1 : 3^{2/3}$ (*d*) $1 : 2^{4/3}$

13. Two sides of a plot measuring 32 m and 24 m and the angle between them is a perfect right angle. The other two sides measure 25 m each and the other three angles are not right angles. The area of the plot in m^2 is

(*a*) 768 (*b*) 534 (*c*) 696.5 (*d*) 684

14. a and b are two sides adjacent to the right angle of a right-angled triangle and p is the perpendicular drawn to the hypotenuse from the opposite vertex. Then p^2 is equal to

(*a*) $a^2 + b^2$ (*b*) $\frac{1}{a^2}+\frac{1}{b^2}$ (*c*) $\frac{a^2b^2}{a^2+b^2}$ (*d*) $a^2 - b^2$

15. A conical cup is filled with ice-cream. The ice-cream forms a hemispherical shape on its open top. The height of the hemispherical part is 7 cm. The radius of the hemispherical part equals the height of the cone. Then the volume of the ice-cream is $\left[\pi=\frac{22}{7}\right]$.

(*a*) 1078 cubic cm (*b*) 1708 cubic cm (*c*) 7108 cubic cm (*d*) 7180 cubic cm

16. A is the centre of circle whose radius is 8 and B is the centre of a circle whose diameter is 8. If these two circles touch externally, then the area of the circle with diameter AB is

(*a*) 36π (*b*) 64π (*c*) 144π (*d*) 256π

17. If $a^2 + b^2 + c^2 = ab + bc + ac$ then the value of $\frac{a+c}{b}$ is

(*a*) 0 (*b*) 2 (*c*) 1 (*d*) -1

18. If $ab + bc + ca = 0$ then the value of $\left(\frac{1}{a^2-bc}+\frac{1}{b^2-ca}+\frac{1}{c^2-ab}\right)$ is

(*a*) 0 (*b*) 1 (*c*) 3 (*d*) $a + b + c$

19. If $(2+\sqrt{3})\,a = (2-\sqrt{3})\,b = 1$ then the value of $\frac{1}{a}+\frac{1}{b}$ is

(*a*) 1 (*b*) 2 (*c*) $2\sqrt{3}$ (*d*) 4

20. If $3x+\frac{3}{x}=1$ then $x^3+\frac{1}{x^3}+1$ is

(*a*) 0 (*b*) $\frac{1}{27}$ (*c*) $\frac{5}{27}$ (*d*) $\frac{28}{27}$

21. The factors of $(a^2 + 4b^2 + 4b - 4ab - 2a - 8)$ are

(*a*) $(a - 2b - 4)(a - 2b + 2)$ (*b*) $(a - b + 2)(a - 4b - 4)$
(*c*) $(a + 2b - 4)(a + 2b + 2)$ (*d*) $(a + 2b - 1)(a - 2b + 1)$

22. Area of the triangle formed by the graph of the straight lines $x - y = 0$, $x + y = 2$ and the x-axis is

(*a*) 1 sq unit (*b*) 2 sq units (*c*) 4 sq units (*d*) None of these

23. The value of $\frac{1}{a^2+ax+x^2}-\frac{1}{a^2-ax+x^2}+\frac{2ax}{a^4+a^2x^2+x^4}$ is

(*a*) 2 (*b*) 1 (*c*) -1 (*d*) 0

24. If $4x + 5y = 83$ and $3x : 2y = 21 : 22$ then $y - x$ equals

(*a*) 3 (*b*) 4 (*c*) 7 (*d*) 11

25. If $x = 11$, then the value of $x^5 - 12x^4 + 12x^3 - 12x^2 + 12x - 1$ is

(*a*) 5 (*b*) 10 (*c*) 15 (*d*) 20

26. If $p = 99$ then the value of $p(p^2 + 3p + 3)$ is **(SSC CGL Tier I 2015)**

(*a*) 10000000 (*b*) 999000 (*c*) 999999 (*d*) 990000

27. Two chords of lengths *a* metre and *b* metre subtend angles 60° and 90° at the centre of the circle respectively. Which of the following is true?

(*a*) $b = \sqrt{2}a$ (*b*) $a = \sqrt{2}b$ (*c*) $a = 2b$ (*d*) $b = 2a$

28. In a triangle *ABC*, $\angle A + \frac{1}{2}\angle B + \angle C = 140°$ then $\angle B$ is

(*a*) 50° (*b*) 80° (*c*) 40° (*d*) 60°

29. The radius of a circle is 6 cm. The distance of a point lying outside the circle from the centre is 10 cm. The length of the tangent drawn from the outside point to the circle is

(*a*) 5 cm (*b*) 6 cm (*c*) 7 cm (*d*) 8 cm

30. If *ABCD* be a cyclic quadrilateral in which $\angle A = 4x°$, $\angle B = 7x°$, $\angle C = 5y°$, $\angle D = y°$ then $x : y$ is

(*a*) 3 : 4 (*b*) 4 : 3 (*c*) 5 : 4 (*d*) 4 : 5

31. *G* is the centroid of the equilateral ΔABC, if $AB = 10$ cm then length of *AG* is

(*a*) $\frac{5\sqrt{3}}{3}$ cm (*b*) $\frac{10\sqrt{3}}{3}$ cm (*c*) $5\sqrt{3}$ cm (*d*) $10\sqrt{3}$ cm

32. Two chords *AB* and *CD* of a circle with centre *O*, intersect each other at *P*. If $\angle AOD = 100°$ and $\angle BOC = 70°$ then the value of $\angle APC$ is

(*a*) 80° (*b*) 75° (*c*) 85° (*d*) 95°

33. *ABCD* is a cyclic quadrilateral and AD is a diameter. If $\angle DAC = 55°$ then value of $\angle ABC$ is

(*a*) 55° (*b*) 35° (*c*) 145° (*d*) 125°

34. In triangle *ABC* is a straight line parallel to *BC* intersects *AB* and *AC* at *D* and *E* respectively. If $AB = 2AD$ then $DE : BC$ is

(*a*) 2 : 3 (*b*) 2 : 1 (*c*) 1 : 2 (*d*) 1 : 3

35. *ABC* is an isosceles triangle such that $AB = AC$ and *AD* is the median to the base *BC* with $\angle ABC = 35°$. Then $\angle BAD$ is

(*a*) 35° (*b*) 55° (*c*) 70° (*d*) 110°

36. A man goes 24 m due west and then 10 m due north. Then the distance of him from the starting point is

(*a*) 17 m (*b*) 26 m (*c*) 28 m (*d*) 34 m

37. From the top of a tower of height 180 m the angles of depression of two objects on either sides of the tower are 30° and 45°. Then the distance between the objects are

(*a*) $180(3 + \sqrt{3})$ (*b*) $180(3 - \sqrt{3})$ m (*c*) $180(\sqrt{3} - 1)$ m (*d*) $180(\sqrt{3} + 1)$ m

38. *ABCD* is a reactangle of which *AC* is a diagonal. The value of $(\tan^2 \angle CAD + 1) \sin^2 \angle BAC$ is

(*a*) 2 (*b*) $\frac{1}{4}$ (*c*) 1 (*d*) 0

39. If tan x = sin 45° . cos 45° + sin 30° then the value of x is.

(*a*) 30° (*b*) 45° (*c*) 60° (*d*) 90°

40. For any real values of θ, $\sqrt{\dfrac{\sec\theta-1}{\sec\theta+1}} = ?$

(*a*) cot θ – cosec θ (*b*) sec θ – tan θ (*c*) cosec θ – cot θ (*d*) tan θ – sec θ

41. If the sum and difference of two angles are 135° and $\dfrac{\pi}{12}$ respectively, then the value of the angles in degree measure are

(*a*) 70°, 65° (*b*) 75°, 60° (*c*) 45°, 90° (*d*) 80°, 55°

42. In a Δ ABC, $\angle B = \dfrac{\pi}{3}$, $\angle C = \dfrac{\pi}{4}$ and D divides BC internally in the ratio 1 : 3 then $\dfrac{\sin\angle BAD}{\sin\angle CAD}$ is equal to

(*a*) $\dfrac{1}{\sqrt{2}}$ (*b*) $\dfrac{1}{\sqrt{3}}$ (*c*) $\dfrac{1}{\sqrt{6}}$ (*d*) $\sqrt{6}$

43. If sin 3A = cos (A – 26°), where 3A is an acute angle then the vlaue of A is

(*a*) 29° (*b*) 26° (*c*) 23° (*d*) 28°

44. Value of $\sec^2\theta - \dfrac{\sin^2\theta - 2\sin^4\theta}{2\cos^4\theta - \cos^2\theta}$ is

(*a*) 1 (*b*) 2 (*c*) – 1 (*d*) 0

45. If $x = a(\sin\theta + \cos\theta)$, $y = b(\sin\theta - \cos\theta)$ then the value of $\dfrac{x^2}{a^2} + \dfrac{y^2}{b^2}$ is

(*a*) 0 (*b*) 1 (*c*) 2 (*d*) – 2

46. If sin 5θ = cos 20° (0° < θ < 90°) then the value of θ is

(*a*) 4° (*b*) 22° (*c*) 10° (*d*) 14°

47. Find the least number which when divided separately by 15, 20, 36 and 48 leaves 3 as remainder in each case.

(*a*) 183 (*b*) 243 (*c*) 483 (*d*) 723

48. Find the sum of all positive multiples of 3 less than 50

(*a*) 400 (*b*) 404 (*c*) 408 (*d*) 412

49. If a = 64 and b = 289, then the value of $\left(\sqrt{\sqrt{a}+\sqrt{b}} - \sqrt{\sqrt{b}-\sqrt{a}}\right)^{1/2}$ is

(*a*) $2^{1/2}$ (*b*) 2 (*c*) 4 (*d*) – 2

50. If the L.C.M. and H.C.F. of two expressions are $(x^2 + 6x + 8)(x + 1)$ and $(x + 1)$ respectively and one of the expression is $x^2 + 3x + 2$ find the other.

(*a*) $x^2 + 5x + 4$ (*b*) $x^2 - 5x + 4$ (*c*) $x^2 + 4x + 5$ (*d*) $x^2 - 4x + 5$

51. If the number of items of a set A $n(A) = 40$, $n(B) = 26$ and $n(A \cap B) = 16$ then $n(A \cup B)$ is equal to

(*a*) 30 (*b*) 40 (*c*) 50 (*d*) 60

52. $\sqrt{64009}$ is equal to

(*a*) 352 (*b*) 523 (*c*) 253 (*d*) 532

53. What is the smallest number by which 625 must be divided so that the quotient is a perfect cube?

(*a*) 25 (*b*) 5 (*c*) 2 (*d*) 3

54. If a distance of 50 m is covered in 1 min that 90 m in 2 min and 130 m in 3 min find the distance covered in 15 min.

(*a*) 610 m (*b*) 750 m (*c*) 1000 m (*d*) 650 m

55. Three men step off together from the same spot. Their steps measure 63 cm, 70 cm and 77 cm respectively. The minimum distance each should cover so that all can cover the distance in complete steps is

(*a*) 9630 cm (*b*) 9360 cm (*c*) 6930 cm (*d*) 6950 cm

56. Find the greatest number which will exactly divide 200 and 320.

(*a*) 10 (*b*) 20 (*c*) 16 (*d*) 40

57. *A*, *B* and *C* are employed to do a piece of work for ₹ 575. *A* and *C* are supposed to finish $\frac{19}{23}$ of the work together. Amount shall be paid to *B* is

(*a*) ₹ 210 (*b*) ₹ 100 (*c*) ₹ 200 (*d*) ₹ 475

58. A man is twice as fast as a woman and a woman is twice as fast as a boy in doing a work. If all of them, a man, a woman and a boy can finish the work in 7 days, in how many days a boy will do it alone?

(*a*) 49 (*b*) 7 (*c*) 6 (*d*) 42

59. *A*, *B* and *C* can do a job in 6 days, 12 days and 15 days respectively. After $\frac{1}{8}$ of the work is completed, *C* leaves the job. Rest of the work is done by *A* and *B* together. Time taken to finish the work is

(*a*) $5\frac{5}{6}$ days (*b*) $5\frac{1}{4}$ days (*c*) $3\frac{1}{2}$ days (*d*) $3\frac{3}{4}$ days

60. 15 men take 20 days to complete a job working 8 hours a day. The number of hours a day should 20 men take to complete the job in 12 days.

(*a*) 5 hours (*b*) 10 hours (*c*) 15 hours (*d*) 18 hours

61. Having the same capacity 9 taps fill up a water tank in 20 minutes. How many taps of the same capacity required to fill up the same water tank in 15 minutes?

(*a*) 10 (*b*) 12 (*c*) 15 (*d*) 18

62. Raj and Ram working together do a piece of work in 10 days. Raj alone can do it in 12 days. Ram alone will do the work in

(*a*) 20 days (*b*) 40 days (*c*) 50 days (*d*) 60 days

63. A shopkeeper sold an item at 10% loss after giving a discount equal to half the marked price. Then the cost price is

(*a*) $\frac{1}{9}$th of the market price (*b*) $\frac{4}{9}$th of the market price

(c) $\frac{5}{9}$ th of the market price (d) $\frac{7}{9}$ th of the market price

64. A person purchased a saree for ₹ 7710 after availing a net discount of ₹ 1285. The percentage of discount, the saree shop offers, is

(a) $14\frac{1}{7}\%$ (b) $14\frac{2}{7}\%$ (c) $14\frac{3}{7}\%$ (d) $14\frac{4}{7}\%$

65. A cycle dealer offers a discount of 10% and still makes a profit of 26%. What does he pay for a cycle whose marked price is ₹ 840?

(a) ₹ 600 (b) ₹ 650 (c) ₹ 700 (d) ₹ 750

66. If the cost price of an item is two fifth of its marked price and if it is sold at a discount of 10%, then there will be

(a) 25% profit (b) 40% profit (c) 50% profit (d) 125% profit

67. Prakash lends a part of ₹ 20,000 at 8% simple interest and remaining at $\frac{4}{3}\%$ simple interest. His total income after a year was ₹ 800. Find the sum lent at 8%.

(a) ₹ 8,000 (b) ₹ 12,000 (c) ₹ 6,000 (d) ₹ 10,000

68. 20 litres of a mixture contains 20% alcohol and the rest water. If 4 litres of water be mixed in it, the percentage of alcohol in the new mixture will be

(a) $33\frac{1}{3}\%$ (b) $16\frac{2}{3}\%$ (c) 25% (d) $12\frac{1}{2}\%$

69. A man divides his property so that his son's share to his wife's and wife's share to his daughter's are both as in the ratio 3 : 1. If the daughter gets ₹ 10,000 less than son, the value (in rupees) of the whole property is

(a) ₹ 16,250 (b) ₹ 16,000 (c) ₹ 18,250 (d) ₹ 17,000

70. There are two containers of equal capacity. The ratio of milk to water in the first container is 3 : 1, in the second container 5 : 2. If they are mixed up, the ratio of milk to water in the mixture will be

(a) 28 : 41 (b) 41 : 28 (c) 15 : 41 (d) 41 : 15

71. The sum of two numbers is equal to 200 and their difference is 25. The ratio of the two numbers is

(a) 9 : 7 (b) 7 : 9 (c) 3 : 5 (d) 2 : 7

72. A man travelled a distance of 80 km in 7 hrs partly on foot at the rate of 8 km per hour and partly on bicycle at 16 km per hour. The distance travelled on the foot is

(a) 32 km (b) 48 km (c) 36 km (d) 44 km

73. The frequency distribution data is given below. If the average age is 17 years, the value of m is

Age (in years) :	8	20	26	29
Number of people:	3	2	m	1

(a) 1 (b) 2 (c) 3 (d) 4

74. The average monthly expenditure of a family for the first four months is ₹ 2570, for the next three months ₹ 2490 and for the last five months ₹ 2490 and for the last five months ₹ 3030. If the family saves ₹ 5320 during the whole year, the average monthly income of the family during the year is?

(*a*) ₹ 3000 (*b*) ₹ 3185 (*c*) ₹ 3200 (*d*) ₹ 3580

75. After replacing an old member by a new member, it was found that the average age of five members of a club is the same as it was 3 years ago. The difference between the ages of the replaced and the new member is

(*a*) 2 years (*b*) 4 years (*c*) 8 years (*d*) 15 years

76. A man spends ₹ 1800 monthly on an average for the first four months and ₹ 2000 monthly for the next eight months and saves ₹ 5600 a year. His average monthly income is

(*a*) ₹ 2000 (*b*) ₹ 2200 (*c*) ₹ 2400 (*d*) ₹ 2600

77. The arithmetic mean of the following numbers

1, 2, 2, 3, 3, 3, 4, 4, 4, 4, 5, 5, 5, 5, 5, 6, 6, 6, 6, 6, 6, and 7, 7, 7, 7, 7, 7, 7 is

(*a*) 4 (*b*) 5 (*c*) 14 (*d*) 20

78. The average of 6 numbers is 20. If one number is removed, the average becomes 15. What is the number removed?

(*a*) 5 (*b*) 35 (*c*) 112 (*d*) 45

79. An item costing ₹ 200 is being sold at 10% loss. If the price is further reduced by 5% the selling price will be

(*a*) ₹ 170 (*b*) ₹ 171 (*c*) ₹ 175 (*d*) ₹ 179

80. A shopkeeper buys 144 items at 90 paise each. On the way 20 items are broken. He sells the remainder at ₹ 1.20 each. His gain per cent correct to one place of decimal is

(*a*) 13.8% (*b*) 14.6% (*c*) 14.8% (*d*) 15.8%

81. There is a profit of 20% on the cost price of an article. The % of profit, when calculated on selling price is

(*a*) $16\frac{2}{3}\%$ (*b*) 20% (*c*) $33\frac{1}{3}\%$ (*d*) None of these

82. By selling an article for ₹ 102, there is a loss of 15%, when the article is sold for ₹ 134.40, the net result in the transaction is

(*a*) 12% gain (*b*) 12% loss (*c*) 10% loss (*d*) 15% gain

83. Two toys are sold ₹ 504 each. One toy brings the dealer a gain of 12% and the other a loss of 4%. The gain or loss percent by selling both the toys is

(*a*) $3\frac{5}{13}\%$ profit (*b*) $4\frac{5}{13}\%$ profit (*c*) $5\frac{1}{13}\%$ profit (*d*) $2\frac{3}{13}\%$ loss

84. *A* sold a horse to *B* for ₹ 4800 by losing 20%. *B* sells it *C* at a price which would have given *A* a profit of 15%. B's gain is **(CDS 2015)**

(*a*) ₹ 1800 (*b*) ₹ 1900 (*c*) ₹ 2000 (*d*) ₹ 2100

85. If each side of a cube is increased by 10% the volume of the cube will increase by

(*a*) 30% (*b*) 10% (*c*) 33.1% (*d*) 25%

86. A reduction of 21% in the price of an item enables a person to buy 3 kg more for ₹ 100. The reduced price of item per kg is

(*a*) ₹ 5.50 (*b*) ₹ 7.50 (*c*) ₹ 10.50 (*d*) ₹ 7.00

87. The number that is to be added to 10% of of 320 to have the sum as 30% of 230 is

(*a*) 37 (*b*) 32 (*c*) 23 (*d*) 73

88. The strength of a school increases and decreases in every alternate year by 10%. It started with increase in 2000. Then the strength of the school in 2003 as compared to that in 2000 was.

(*a*) increased by 8.9% (*b*) decreased by 8.9% (*c*) increased by 9.8% (*d*) decreased by 9.8%

89. Two trains of equal length are running on parallel lines in the same direction at the rate of 46 km/hr and 36 km/hr. The faster train passes the slower train in 36 seconds. The length of each train is

(*a*) 50 m (*b*) 72 m (*c*) 80 m (*d*) 82 m

90. A car driver leaves Bangalore at 8 : 30 A.M. and expects to reach a place 300 km from Bangalore at 12:30 P.M. At 10:30 he finds that he has covered only 40% of the distance. By how much he has to increase the speed of the car in order to keep up his schedule?

(*a*) 45 km/hr (*b*) 40 km/hr (*c*) 35 km/hr (*d*) 30 km/hr

91. A train 300 m long is running with a speed of 54 km/hr. In what time will it cross a telephone pole?

(*a*) 20 sec (*b*) 15 sec (*c*) 17 sec (*d*) 18 sec

92. A man is walking at a speed of 10 kmph. After every km, he takes a rest for 5 min. How much time will he take to cover a distance of 5 km?

(*a*) 60 min (*b*) 50 min (*c*) 40 min (*d*) 70 min

93. A man borrows money at 3% per annum interest payable yearly and lend it immediately at 5% interest (compound) payable half-yearly and thereby gains ₹ 330 at the end of the year. The sum borrowed is

(*a*) ₹ 17000 (*b*) ₹ 16500 (*c*) ₹ 15000 (*d*) ₹ 16000

94. Two years ago, the value of my motorbike was ₹ 62500. If the value depreciates by 4% every year, now its value is

(*a*) ₹ 56700 (*b*) ₹ 57600 (*c*) ₹ 57500 (*d*) ₹ 55700

95. If the compound interest on a sum for 2 years at $12\frac{1}{2}\%$ p.a. is ₹ 510, the simple interest on the sample sum at the same rate for the same period of time is

(*a*) ₹ 400 (*b*) ₹ 450

(*c*) ₹ 460 (*d*) ₹ 480

A watch company produces four different products. The sale of these products in lakhs during 2005 and 2010 are shown in the following bar diagram. Study the graph and answer the question Nos. 96 to 100.

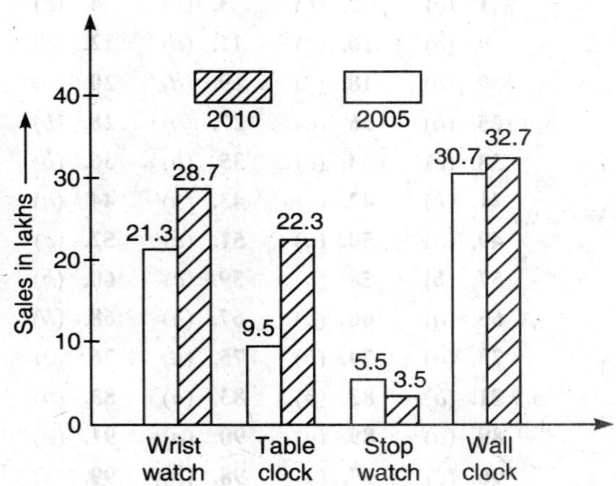

96. The sale in percentage of wrist watch in 2010 more than the sales of table clock in 2010 was nearly by

(*a*) 26.7% (*b*) 27.7%

(*c*) 28.7% (*d*) 21.7%

97. The ratio of sales of stop watch in 2010 to the sale of table clock in 2005 is

(*a*) 6 : 19 (*b*) 7 : 6 (*c*) 19 : 6 (*d*) 7 : 19

98. The sales of table clock in 2005 was less than the sales of wall clock in 2005 is nearly by

(*a*) 70.05% (*b*) 69.05% (*c*) 68.05% (*d*) 62.05%

99. During the period 2005-2010 the minimum rate of increase in sales is in the product of

(*a*) Wrist watch (*b*) Table clock (*c*) Stop watch (*d*) Wall clock

100. The sales have increased by nearly 135% from 2005 to 2010 in the product of

(*a*) Table clock (*b*) Wrist watch (*c*) Stop watch (*d*) Wall clock

FOR VISUALLY HANDICAPPED CANDIDATES ONLY

96. The area of the largest triangle inscribed in a semi-circular region of radius r cm is

(*a*) $2r^2$ square cm (*b*) r^2 square cm (*c*) $\frac{r^2}{2}$ square cm (*d*) $\frac{3r^2}{2}$ square cm

97. The length of the tangent from an external point which is at a distance of 13 cm from the centre of the circle of radius 5 cm is

(*a*) 8 cm (*b*) 12 cm (*c*) 10 cm (*d*) 14 cm

98. If $x = \frac{1}{2-\sqrt{3}}$ then $x^3 - 3(x-1)(x+2) =$

(*a*) 6 (*b*) 4 (*c*) 5 (*d*) 3

99. If $2\cos^2\theta = 3\sin\theta$, $0 < \theta < 90°$ then the value of $\sin\theta$ is

(*a*) $\frac{\sqrt{3}}{2}$ (*b*) $\frac{\sqrt{5}-1}{\sqrt{5}+1}$ (*c*) $\frac{1}{2}$ (*d*) $\frac{1}{\sqrt{3}}$

100. A sum of money was invested for 3 years at r% compound interest. An equal sum was invested for 1 year at r% simple interest. It was observed that the amount in the 1st investment is twice that in the 2nd. Taking $\sqrt{2} = 1.41$, r % is equal to

(*a*) 11% (*b*) 21% (*c*) 31% (*d*) 41%

ANSWER

1. (*c*)	**2.** (*b*)	**3.** (*d*)	**4.** (*c*)	**5.** (*c*)	**6.** (*a*)	**7.** (*a*)	**8.** (*a*)
9. (*a*)	**10.** (*c*)	**11.** (*d*)	**12.** (*d*)	**13.** (*d*)	**14.** (*c*)	**15.** (*a*)	**16.** (*a*)
17. (*b*)	**18.** (*a*)	**19.** (*d*)	**20.** (*b*)	**21.** (*a*)	**22.** (*a*)	**23.** (*d*)	**24.** (*b*)
25. (*b*)	**26.** (*c*)	**27.** (*a*)	**28.** (*b*)	**29.** (*d*)	**30.** (*b*)	**31.** (*b*)	**32.** (*d*)
33. (*c*)	**34.** (*c*)	**35.** (*b*)	**36.** (*b*)	**37.** (*d*)	**38.** (*c*)	**39.** (*b*)	**40.** (*c*)
41. (*b*)	**42.** (*c*)	**43.** (*a*)	**44.** (*a*)	**45.** (*c*)	**46.** (*d*)	**47.** (*d*)	**48.** (*c*)
49. (*a*)	**50.** (*a*)	**51.** (*c*)	**52.** (*c*)	**53.** (*b*)	**54.** (*a*)	**55.** (*c*)	**56.** (*d*)
57. (*b*)	**58.** (*a*)	**59.** (*c*)	**60.** (*b*)	**61.** (*b*)	**62.** (*d*)	**63.** (*c*)	**64.** (*b*)
65. (*a*)	**66.** (*d*)	**67.** (*a*)	**68.** (*b*)	**69.** (*a*)	**70.** (*d*)	**71.** (*a*)	**72.** (*a*)
73. (*a*)	**74.** (*b*)	**75.** (*d*)	**76.** (*c*)	**77.** (*b*)	**78.** (*d*)	**79.** (*b*)	**80.** (*c*)
81. (*a*)	**82.** (*a*)	**83.** (*a*)	**83.** (*a*)	**84.** (*d*)	**85.** (*c*)	**86.** (*d*)	**87.** (*a*)
88. (*a*)	**89.** (*a*)	**90.** (*d*)	**91.** (*a*)	**92.** (*b*)	**93.** (*d*)	**94.** (*b*)	**95.** (*d*)
96. (*c*)	**97.** (*d*)	**98.** (*b*)	**99.** (*d*)	**100.** (*a*)			

SOLUTIONS

1. $1 + r\,\% = 1 + \frac{1}{20} = \frac{21}{20} \Rightarrow (1+r\%)^2 = \left(\frac{21}{20}\right)^2 = \frac{441}{400}$

$\Rightarrow$ C.I. = 441 – 400 = 41 $\Rightarrow$ Principal = $\frac{328}{41} \times 400$ = ₹ 3200

2. $\Delta\, ABC\ \Delta\, ADE$

Ratio of volumes = 1 : 27

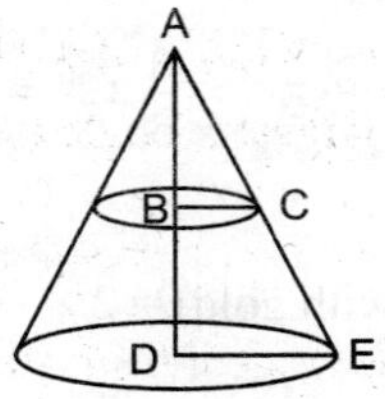

Ratio of sides = $\sqrt[3]{1} : \sqrt[3]{27} = 1 : 3$

Required height = $\frac{2}{3} \times 30 = 20$ cm

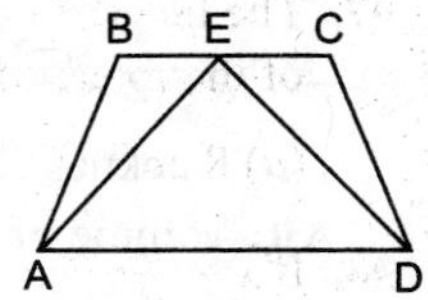

3. Area $(ADCB) = \frac{1}{2}h(AD + BC)$

Area $(\Delta\ ADE) = \frac{1}{2}h\ (AD)$

$\therefore$ Required ratio = $\frac{AD + BC}{AD}$

4. $4\pi r^2 = 346.5 \Rightarrow r^2 = 346.5 \times \frac{1}{4} \times \frac{7}{22} = 27.5625 \Rightarrow r = 5.25$ cm

5. Exterior angle = $\frac{1}{6} \times 180° = 30° \Rightarrow$ Number of sides = $\frac{360°}{30} = 12$

6. $\frac{1}{3} \times 30 \times$ Height = 500 $\Rightarrow$ Height = $\frac{500}{10} = 50$ m

7. Third side of triangle = $\sqrt{5^2 + 12^2} = 13$ cm

Total SA = (2 × Area of Base) + (Perimeter of Base × Height)

$= 2 \times \frac{1}{2} \times 5 \times 12 + (5 + 12 + 13) \times 10 = 60 + 300 = 360$ sq cm

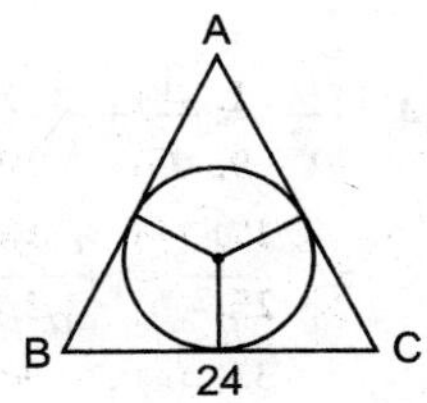

8. Area of $\Delta = \frac{\sqrt{3}}{4}(24)^2 = 144\sqrt{3} = 249.408$

Area of circle = $\pi\left(\frac{24}{2\sqrt{3}}\right)^2 = \pi(48) = 150.857$

Required Area = 249.408 – 150.857

= 98.55 cm^2

9. Lateral surface area = Perimeter of Base × Height = 120

Volume = Area of Base × Height = $40\sqrt{3}$ $\Rightarrow$ $\frac{\text{Perimeter of Base}}{\text{Area of Base}} = \frac{120}{40\sqrt{3}} = \sqrt{3}$

If side = 4 cm, then perimeter = 12, Area = $\frac{\sqrt{3}}{4}(4)^2 = 4\sqrt{3}$ $\Rightarrow$ $\frac{\text{Perimeter}}{\text{Area}} = \frac{12}{4\sqrt{3}} = \sqrt{3}$

10. Given, $4a = 2p \Rightarrow a = \frac{p}{2}$ and $d_1 + d_2 = m$

We know that, $\left(\frac{d_1}{2}\right)^2 + \left(\frac{d_2}{2}\right)^2 = a^2 \Rightarrow \frac{1}{4}\left(d_1^2 + d_2^2\right) = \left(\frac{p}{2}\right)^2 \Rightarrow {d_1}^2 + {d_2}^2 = p^2$

$\Rightarrow 2d_1 d_2 = (d_1 + d_2)^2 - ({d_1}^2 + {d_2}^2) = m^2 - p^2$

$\Rightarrow$ Area of rhombus = $\frac{1}{2}d_1 d_2 = \frac{1}{4}(m^2 - p^2)$

11. Volume of lead = $\frac{4}{3}\pi r^3 = \frac{4}{3}\pi 2^3 = \frac{32}{3}\pi$

Volume of ball (with gold) = $2 \times \frac{32}{3}\pi = \frac{64}{3}\pi \Rightarrow \frac{4}{3}\pi R^3 = \frac{64}{3}\pi \Rightarrow R^3 = \frac{64}{3}\times\frac{3}{4} = 16$

$\Rightarrow R = 2\sqrt[3]{2} = 2 \times 1.259 = 2.518$

$\therefore$ Thickness of gold = 2.518 – 2 = 0.518

Alt: Volume of lead : Total volume = 1 : 2

$\Rightarrow$ Ratio of radii = $\sqrt[3]{1} : \sqrt[3]{2} = 1 : 1.259$

Actual raddi = 2 : 2.518

$\Rightarrow$ Thickness of gold = 2.518 – 2 = 0.518

12. Ratio of volumes = $\frac{4}{3}\pi r^3 : \frac{1}{3}\pi r^2 . r = 4 : 1$

$\Rightarrow$ Ratio of radii = $\sqrt[3]{1} : \sqrt[3]{4} = 1 : 2^{2/3}$

$\Rightarrow$ Ratio of surface areas = $1^2 : 2^{4/3}$

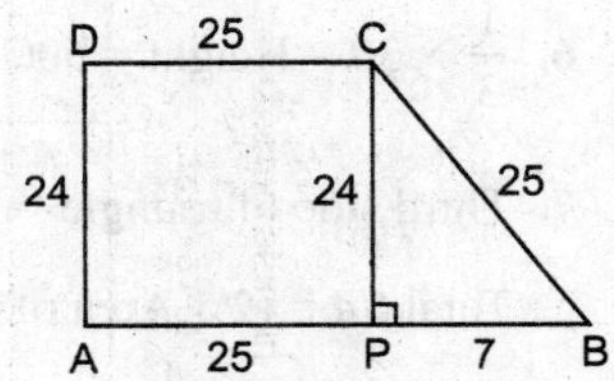

13. Draw $CP \perp AB$

In ΔCPB, $PB = 32 - 25 = 7$ m

$CP^2 = 25^2 - 7^2 = 24^2 \Rightarrow ABCD$ is a trapezium

$\therefore$ Area $(ABCD) = \frac{1}{2} \times 24 \times (25 + 32) = 684$ m^2

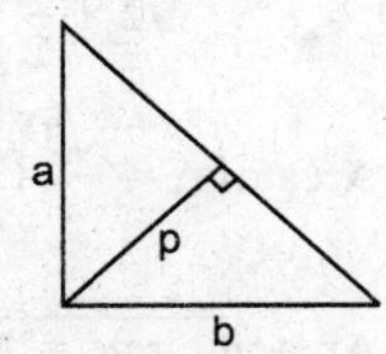

14. $\frac{1}{p^2} = \frac{1}{a^2} + \frac{1}{b^2}$ [By formula]

$\Rightarrow \frac{1}{p^2} = \frac{a^2+b^2}{a^2b^2} \Rightarrow p^2 = \frac{a^2b^2}{a^2+b^2}$

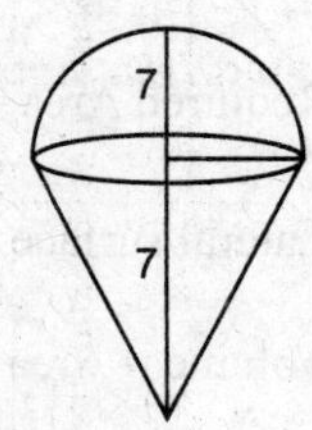

15. Volume = $\frac{1}{3}\pi r^2 h + \frac{2}{3}\pi r^3$

$= \pi r^3$ $[\because r = h]$

$= \frac{22}{7} \times 7^3 = 1078$ cm^3

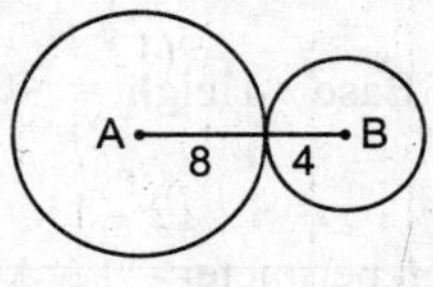

16. New radius = (8 + 4) ÷ 2 = 6

Area of new circle = $\pi(6)^2 = 36\pi$

17. $a^2 + b^2 + c^2 = ab + bc + ac \Rightarrow a = b = c = k$(say)

$\therefore \frac{a+c}{b} = \frac{k+k}{k} = 2$

18. $ab + bc + ca = 0 \Rightarrow a(b + c) = -bc$

$\Rightarrow a^2 - bc = a^2 + a(b + c) = a(a + b + c)$

Similarly, $b^2 - ca = b(a + b + c)$ and $c^2 - ab = c(a + b + c)$

$\therefore$ G.E = $\frac{1}{a(a+b+c)} + \frac{1}{b(a+b+c)} + \frac{1}{c(a+b+c)} = \frac{bc+ac+ab}{abc(a+b+c)} = 0$

19. $(2+\sqrt{3})a = (2-\sqrt{3})b = 1 \Rightarrow \frac{1}{a} = 2+\sqrt{3}$ and $\frac{1}{b} = 2-\sqrt{3}$

$\therefore \frac{1}{a}+\frac{1}{b} = (2+\sqrt{3})+(2-\sqrt{3}) = 4$

20. $3x + \frac{3}{x} = 1 \Rightarrow x + \frac{1}{x} = \frac{1}{3} \Rightarrow \left(x+\frac{1}{x}\right)^3 = \left(\frac{1}{3}\right)^3$

$\Rightarrow x^3 + \frac{1}{x^3} + 3(x)\left(\frac{1}{x}\right)\left(x+\frac{1}{x}\right) = \frac{1}{27}$

$\Rightarrow x^3 + \frac{1}{x^3} + 1 = \frac{1}{27}$

21. $+4b^2$ suggests $(-2b)(-2b)$ or $(2b)(+2b) \Rightarrow$ options (*a*) or (*c*)

$-4ab$ suggests $(-2b)(a) + (a)(-2b) \Rightarrow$ option (*a*)

22. Base = 2 unit, Height = 1 unit

Required area = $\frac{1}{2}$ × Base × Height

$= \frac{1}{2} \times 2 \times 1 = 1$ sq unit

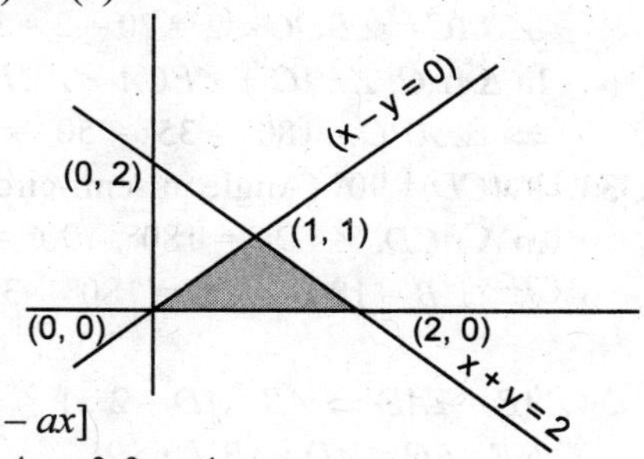

23. $(a^2+ax+x^2)(a^2-ax+x^2) = [(a^2+x^2)+ax][(a^2+x^2)-ax]$

$= (a^2+x^2)^2 - (ax)^2 = a^4 + 2a^2x^2 + x^4 - a^2x^2 = a^4 + a^2x^2 + x^4$

$\therefore$ G.E $= \frac{(a^2-ax+x^2)-(a^2+ax+x^2)+2ax}{a^4+a^2x^2+x^4} = 0$

24. $3x : 2y = 21 : 22 \Rightarrow x : y = 7 : 11$

$\Rightarrow 4x + 5y = 4 \times 7 + 5 \times 11 = 28 + 55 = 83 =$ RHS

$\therefore y - x = 11 - 7 = 4$

25. $x^5 - 12x^4 = x^4(x-12) = x^4(11-12) = -x^4$

$-x^4 + 12x^3 = x^3(-x+12) = x^3(12-11) = x^3$

$x^3 - 12x^2 = x^2(x-12) = x^2(11-12) = -x^2$

$-x^2 + 12x = x(-x+12) = 11(12-11) = 11$

$\therefore$ G.E. $= 11 - 1 = 10$

26. $p(p^2+3p+3) = p^3+3p^2+3p+1-1 = (p+1)^3 - 1$

$= (99+1)^3 - 1 = 100^3 - 1 = 999999$

27. Circle I : Each angle = 60°

$\Rightarrow$ Each side $= a = r$

Circle II : $b^2 = 2r^2 \Rightarrow b = \sqrt{2}r$

$\Rightarrow b = \sqrt{2}a$

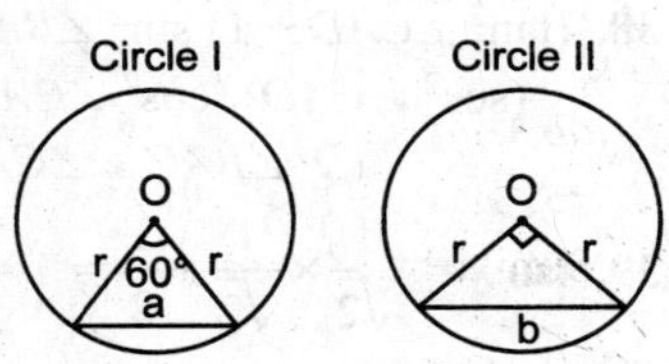

28. $\frac{1}{2}\ \angle B = 180° - 140° = 40° \Rightarrow \angle B = 80°$

29. Tangent = $\sqrt{10^2 - 6^2}$ = 8 cm

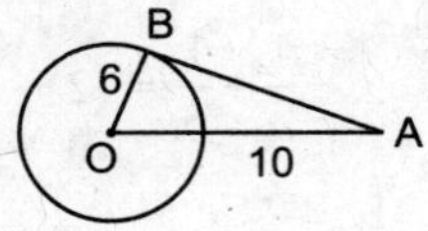

30. $4x + 5y = 7x + y = 180°$

$\Rightarrow\ \ 3x = 4y \Rightarrow x : y = 4 : 3$

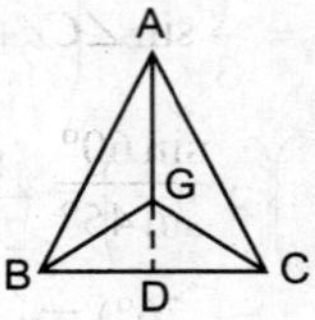

31. $AD = \frac{\sqrt{3}}{2}a = \frac{\sqrt{3}}{2} \times 10 = 5\sqrt{3}$

$AG = \frac{2}{3} \times 5\sqrt{3} = \frac{10\sqrt{3}}{3}$

32. $\angle DCA = \angle AOD \div 2 = 100 \div 2 = 50°$ (Angle at centre, circle)

$\angle BAC = \angle BOC \div 2 = 70 \div 2 = 35°$ (Angle at centre, circle)

In ΔAPC, $\angle PAC + \angle PCA + \angle APC = 180°$

$\Rightarrow\ \ \angle APC = 180° - 35° - 50° = 95°$

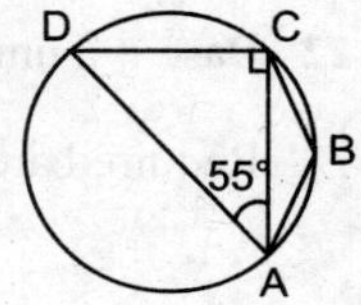

33. $\angle ACD = 90°$ (Angle in semi-circle)

In $\Delta\ ACD$, $\angle ADC = 180° - 90° - 55° = 35°$

$\therefore\ \ \angle B = 180° - \angle D = 180° - 35° = 145°$

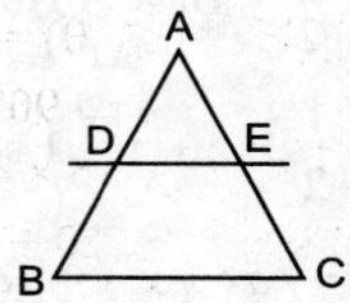

34. $AB = 2AD \Rightarrow AB : AD = 2 : 1$

$DE : BC = AD : AB = 1 : 2$

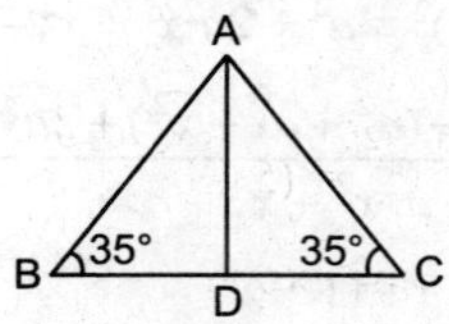

35. $\angle C = \angle B = 35°$

$\Rightarrow\ \ \angle A = 180° - (35° + 35°) = 110°$

$\therefore\ \ \angle BAD = 110° \div 2 = 55°$

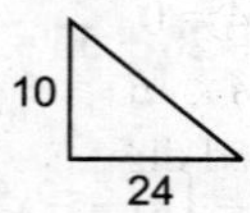

36. $24^2 + 10^2 = 26^2$ $\quad (\because 12^2 + 5^2 = 13^2)$

37. $\tan 45° = \frac{180}{a} \Rightarrow 1 = \frac{180}{a} \Rightarrow a = 180$

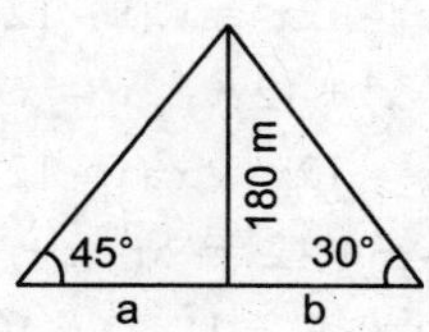

$\tan 30° = \frac{180}{b} \Rightarrow \frac{1}{\sqrt{3}} = \frac{180}{a} \Rightarrow a = 180\sqrt{3}$

Distance = $a + b = 180 + 180\sqrt{3} = 180(\sqrt{3} + 1)$m

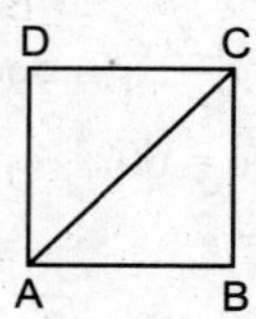

38. $(\tan^2 \angle CAD + 1) \sin^2 \angle BAC$

$= (\sec^2 \angle CAD)(\cos^2 \angle CAD) = 1$

$(\because \angle BAC + \angle CAD = 90°)$

39. $\tan x = \frac{1}{\sqrt{2}} \times \frac{1}{\sqrt{2}} + \frac{1}{2} = 1 = \tan 45°$

40. $\sqrt{\frac{\sec\theta - 1}{\sec\theta + 1} \times \frac{\sec\theta - 1}{\sec\theta - 1}} = \frac{\sec\theta - 1}{\tan\theta} = \text{cosec}\ \theta - \cot\theta$

41. Sum = 135°, Difference = $\frac{\pi}{12} = \frac{180}{2}$ = 15°

Option (*b*) satisfies both the conditions.

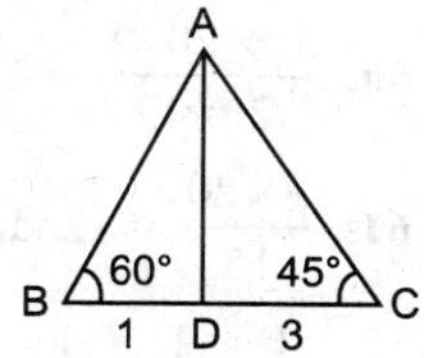

42. $\frac{\sin\angle BAD}{BD} = \frac{\sin\angle B}{AD} \Rightarrow \sin\angle BAD = \frac{BD\times\sin\angle B}{AD}$

$\frac{\sin\angle CAD}{CD} = \frac{\sin\angle C}{AD} \Rightarrow \sin\angle CAD = \frac{CD\times\sin\angle C}{AD}$

$\Rightarrow \frac{\sin\angle BAD}{\sin\angle CAD} = \frac{BD}{CD}\times\frac{\sin 60°}{\sin 45°} = \frac{1}{3}\times\frac{\sqrt{3}/2}{1/\sqrt{2}} = \frac{\sqrt{6}}{6} = \frac{1}{\sqrt{6}}$

43. cos (90° – 3*A*) = cos (*A* – 26°) ⇒ 90° – 3*A* = *A* – 26°

⇒ 4*A* = 116° ⇒ *A* = 29°

44. $\sec^2\theta - \frac{\sin^2\theta(1-2\sin^2\theta)}{\cos^2\theta(2\cos^2\theta-1)} = \sec^2\theta - \tan^2\theta = 1$ [$1 - 2\sin^2\theta = 2\cos^2\theta - 1$]

45. $\frac{x^2}{a^2}+\frac{y^2}{b^2} = (\sin\theta + \cos\theta)^2 + (\sin\theta - \cos\theta)^2$

$= 2(\sin^2\theta + \cos^2\theta) = 2$

46. cos (90° – 5θ) = cos 20° ⇒ 90° – 5θ = 20° ⇒ 5θ = 70° ⇒ θ = 14°

47. LCM of 15, 20, 36, 48 = 15 × 4 × 3 × 4 = 720

⇒ Required number = 720 + 3 = 723

48. $\frac{3+48}{2}\times 16 = 408$

49. G.E. = $\left(\sqrt{8+17}-\sqrt{17-8}\right)^{1/2} = (5-3)^{1/2} = 2^{1/2}$

50. (*x* + 1) must be a factor of the expression.

Option (*a*) : *f*(– 1) = 1 – 5 + 4 = 0

51. $n(A \cup B) = n(A) + n(B) - n(A \cap B) = 40 + 26 - 16 = 50$

52. Trick : $2^2 < 6 < 3^2$ ⇒ Hundred digit is 2.

53. $625 = 5^4 = 5^3 \times 5$ ⇒ To be divided by 5

54. 50, 90, 130, ... is an A.P. with *d* = 40

⇒ $a_n = a + (n-1)d \Rightarrow a_{15} = 50 + 14 \times 40 = 610$

55. LCM of 63, 70, 77 = 9 × 7 × 10 × 11 = 6930

56. HCF of 200 and 320 = 40

57. B's work = $1 - \frac{19}{23} = \frac{4}{23}$ ⇒ B's Wages = $\frac{4}{23}\times 575 = 100$

58. *M* = 2*W*, *W* = 2*B* ⇒ M = 4*B*

M + *W* + *B* = (4 + 2 + 1)*B* = 7 boys can do the workin 7 days

One boy will do it in 7 × 7 = 49 days

59. Remaining work = $1-\frac{1}{8} = \frac{7}{8}$

$(A + B) = \frac{1}{6}+\frac{1}{12} = \frac{3}{12} = \frac{1}{4}$

Required time = $\frac{7}{8}\times 4 = \frac{7}{2} = 3\frac{1}{2}$ days

60. $\frac{15\times 20\times 8}{20\times 12}$ = 10 hours

61. $\frac{9\times 20}{15}$ = 12 taps

62. $\frac{1}{10}-\frac{1}{12}=\frac{1}{60} \Rightarrow$ 60 days

63. 90% of CP = 50% of MP $\Rightarrow$ CP : MP = 5 : 9

64. $\frac{1285}{7710+1285}\times 100 = \frac{100}{7}\% = 14\frac{2}{7}\%$

65. CP = $840\times\frac{90}{100}\times\frac{100}{126}$ = ₹ 600

66. CP : MP = 2 : 5 $\Rightarrow$ SP = MP – 10% = 5 – 10% = 4.50

Profit = 4.50 – 2 = 2.5 on ₹ 2 = $\frac{2.5}{2}\times 100\%$ = 125%

67.

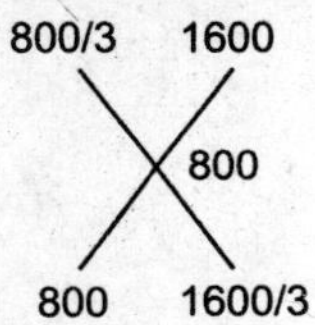

SI on 20000 @ $\frac{4}{3}\% = \frac{800}{3}$

S.I. on 20000 @ 8% = 1600

Ratio = 800 : $\frac{1600}{3}$ = 3 : 2

Sum lent at 8% = $\frac{2}{5}\times 20000$ = 8000

68. 20% × $\frac{20}{20+4} = \frac{50}{3}\% = 16\frac{2}{3}\%$

69. Son : Wife = Wife : Daughter = 3 : 1

$\Rightarrow$ Son : Wife : Daugher = 9 : 3 : 1

$\Rightarrow$ (Son – Daughter) = (9 – 1)ratio = 8 ratio = ₹ 10,000

$\therefore$ Total property = (9 + 3 + 1) ratio = 13 ratio = $\frac{13}{8}$ × 10000 = 16250

70. Milk = $\frac{3}{4}+\frac{5}{7}=\frac{21+20}{28}=\frac{41}{28}$;

$\Rightarrow$ Water = $2-\frac{41}{28}=\frac{15}{28}$

$\Rightarrow$ Required ratio = 41 : 15

71. Ratio = (200 + 25) : (200 – 25) = 225 : 175 = 9 : 7

72. From options : $\frac{32}{8}+\frac{48}{16}$ = 4 + 3 = 7 hours

73. 8 × 3 + 20 × 2 + 26 m + 29 × 1 = 17 (6 + m)

$\Rightarrow$ 93 + 26 m = 102 + 17 $m \Rightarrow 9m = 9 \Rightarrow m = 1$

74. $\frac{2570\times4+2490\times3+3030\times5+5320}{12}=\frac{38220}{12}=₹\,3185$

Alt : Assumed mean = 2490

Excess = $\frac{80\times4+0\times3+540\times5+5320}{12}=\frac{8340}{12}=₹\,695$

$\therefore$ Average = 2490 + 695 = ₹ 3185

75. 5 × 3 = 15 years

76. Assumed mean = 1800 ⇒ Average = $1800+\frac{0\times4+200\times8+5600}{12}$

$= 1800+\frac{7200}{12} = 1800+600 = 2400$

77. Number of terms = $1+2+3+4+5+6+7=\frac{7\times8}{2}=28$

Sum = $1\times1+2\times2+...+7\times7=\frac{7\times8\times15}{6}=140$ ⇒ Average = 140 ÷ 28 = 5

78. 6 × 20 – 5 × 15 = 120 – 75 = 45

79. SP = $200\times\frac{90}{100}\times\frac{95}{100}=171$

80. CP = 144 × 0.90 = 129.60; SP = 124 × 1.20 = 148.80

⇒ Profit % = $\frac{19.2}{129.6}\times100=14.8\%$

81. $\frac{20}{100+20}\times100=\frac{50}{6}=16\frac{2}{3}\%$

82. $\frac{85}{102}\times134.40=112\%$ ⇒ Gain = 12%

83. $\frac{100(12-4)+2(12)(-4)}{200+12-4}=\frac{800-96}{208}=\frac{44}{13}=3\frac{5}{13}\%$

Alt: Let SP in each case = LCM of 112 and 96 = 16 × 7 × 6 = 672

Total CP = $672\times\frac{100}{112}+672\times\frac{100}{96}=600+700=1300$

Total profit = 2 × 672 – 1300 = ₹ 44 ⇒ Profit% = $\frac{44}{1300}\times100=\frac{44}{13}=3\frac{5}{13}\%$

84. $4800\times\frac{100}{80}\times\frac{115}{100}=6900$ ⇒ B's gain = 6900 – 4800 = 2100

85. $\left(\frac{11}{10}\right)^3==\frac{1331}{100}\Rightarrow\%=\frac{1331}{1000}\times100=33.1\%$

86. Saving due to reduction = 21% of ₹ 100 = ₹ 21

Reduced price = 21÷ 3 = ₹ 7 per kg

87. 30% of 230 – 10% of 320 = 69 – 32 = 37

88. Net effect of 2000 and 2001 = $10-10-\frac{10\times10}{100}=1\%$ less

Net effect of this and 2002 $= -1 + 10 - \frac{1\times10}{100} = 8.9\%$ increase

89. Relative speed = 46 – 36 = 10 km/hr

Distance $= 10 \times \frac{5}{18} \times 36 = 100$m $\Rightarrow$ Length of each train $= 100 \div 2 = 50$m

90. Total time = 4 hours

Distance covered in first 2 hours = 40% = 120

Remaining distance = 300 – 120 = 180

Required speed $= \frac{180}{2} - \frac{120}{2} = 90 - 60 = 30$ km/hr

91. Speed = 54 km/hr $= 54 \times \frac{5}{18}$ m/s = 15 m/s

Required time = 300 ÷ 15 = 20 seconds

92. Required time $= \frac{5}{10} \times 60 + 4 \times 5 = 50$ min

93. Rate = 5% p.a. = 2.5% per half -yearly

$\Rightarrow 1 + r\% = 1 + \frac{1}{40} = \frac{41}{40}$

$\Rightarrow$ CI on $40^2 = 41^2 - 40^2 =$ ₹ 81

Interest on 40^2 @ 3% p.a. = 1600 × 3% = ₹ 48

Difference = 81 – 48 = 33

Actual difference = ₹ 330 (i.e., 10 times 33)

$\therefore$ Required principal = 1600 × 10 = ₹ 16000

94. $62500\times\left(1-\frac{4}{100}\right)^2 = 62500\times\frac{24}{25}\times\frac{24}{25} = 57600$

95. Let interest for first year = ₹ 8, then C.I. for second year = ₹ 8 + 1 = ₹ 9

$\therefore$ C.I. : S.I. = (8 + 9) : (8 + 8) = 17 : 16

$\therefore$ S.I. $= \frac{16}{17}\times510 =$ ₹480

96. $\frac{28.7-22.3}{22.3}\times100 = 28.7\%$

97. 3.5 : 9.5 = 7 : 19

98. $\frac{30.7-9.5}{30.7}\times100 = 69.05\%$

99. Wall clock : Minimum increase

100. Table clock

SSC MTS - 2014

1. If the quotient $\frac{a}{b}$ is positive, which of the following must be true?
(a) $a > 0$ (b) $b > 0$ (c) $ab > 0$ (d) $a + b > 0$

2. The smallest number which when divided by 5, 7, 11 and 13 leaves respective remainders of 2, 4, 8 and 10 is
(a) 5005 (b) 5002 (c) 5008 (d) 5029

3. *A*, *B* and *C* can complete a work in 2 hours. If *A* does the job alone in 6 hours and *B* in 5 hours, how long will it take for *C* to finish the job alone?
(a) $5\frac{1}{2}$ hours (b) $7\frac{1}{2}$ hours (c) 9 hours (d) $4\frac{1}{2}$ hours

4. 4 men or 6 women or 10 children can paint a house in 5 days. The painting is given to a couple and their 5 sons. They finish the job in.
(a) $\frac{11}{60}$ days (b) $5\frac{5}{11}$ days (c) $5\frac{6}{11}$ days (d) $11\frac{1}{5}$ days

5. A railing of 288 meteres is required for fencing a semicircular park. Find the area of the park. $\left(\pi = \frac{22}{7}\right)$
(a) 6589 m^2 (b) 9865 m^2 (c) 8956 m^2 (d) 9856 m^2

6. The radius of the driving wheel of a bus is 35 cm. How many revolutions per minute must the wheel make in order to keep a speed of 33 km/h?
(a) 350 (b) 250 (c) 200 (d) 150

7. A merchant advertises 10% off on the items bought from his store. Find the discount got by a customer who bought a suitcase worth ₹ 560, a bag worth worth ₹ 90 a towel worth ₹ 45.
(a) ₹ 70.00 (b) ₹ 70.50 (c) ₹ 69.50 (d) ₹ 72.00

8. Cost of a CD was reduced by 25%. A person paid ₹ 24 for the CD. What was the cash discount?
(a) ₹ 25 (b) ₹ 6 (c) ₹ 1 (d) ₹ 8

9. A dealer buys an old cooler listed at ₹ 950 and gets successive discounts of 20% and 10%. He spends ₹ 66 on painting. He sells the cooler at a profit of 25%. Find its selling price.
(a) ₹ 937.50 (b) ₹ 935.00 (c) ₹ 940.00 (d) ₹ 936.50

10. Ram earns ₹ 125 in 8 days and Shyam earns ₹ 140 in 10 days, the ratio of their earning is
(a) 125 : 110 (b) 112 : 125 (c) 125 : 112 (d) 100 : 112

11. If a quantity is divided in the ratio 3 : 5, the larger part is 75. Find the quantity.
(a) 120 (b) 130 (c) 240 (d) 150

12. In a money bag there are ₹ 1 coins, ₹ 2 coins and ₹ 5 coins. If the ratio of the number of coins be 4 : 3 : 2 and the total money in the bag is ₹ 500, the amount of money in ₹ 2 denomination is
(a) ₹ 250 (b) ₹ 200 (c) ₹ 150 (d) ₹ 100

13. The average of 5 consecutive odd positive integers is 9. The least one among them is.
(a) 5 (b) 3 (c) 1 (d) 7

14. The average age of two boys and their father is greater than the average age of those two boys and their mother by 3 years. The average age of the four is 19 years. If the average age of the two boys be $5\frac{1}{2}$ years, then find the age of the father and the mother.

(*a*) 37 years and 28 years (*b*) 47 years and 38 years
(*c*) 50 years and 41 years (*d*) 35 years and 32 years

15. A fruit seller buys 700 oranges at the rate of ₹ 500 for 100 oranges, and another variety of 500 oranges at the rate of ₹ 700 for 100 oranges and sells them at ₹ 84 per dozen. The profit percent is

(*a*) 20 (*b*) 40 (*c*) 30 (*d*) 10

16. A person sold his watch for ₹ 75 and got a percentage profit equal to the cost price. The cost price of the watch is

(*a*) ₹ 40 (*b*) ₹ 45 (*c*) ₹ 50 (*d*) ₹ 55

17. If the cost price of 16 pens is equal to the selling price of 12 pens, then the gain or loss percent is

(*a*) $33\frac{1}{3}$% gain (*b*) 25% gain (*c*) 25% loss (*d*) $33\frac{1}{3}$% loss

18. A number is increased by 10% and then the increased number is decreased by 10%. The net increase or decrease is

(*a*) 1% decrease (*b*) 2% increase (*c*) 0.1% increase (*d*) 0.2% decrease

19. During the first year the population of a village is increased by 5% and the second year it is diminished by 5%. At the end of the second year its population was 47,880. What was the population at the begininig of the first year?

(*a*) 45,500 (*b*) 48,000 (*c*) 43,500 (*d*) 53,000

20. A car starts running with the initial speed of 40 km/h with its speed increasing every hour by 5 km/h. How many hours will it take to cover a distance of 385 km?

(*a*) $8\frac{1}{2}$ hours (*b*) $9\frac{1}{2}$ hours (*c*) 9 hours (*d*) 7 hours

21. Two trains *A* and *B* start from Howrah and Patna towards Patna and Howrah respectively at the same time. After passing each other they take 4 hours 48 minutes and 3 hours 20 minutes to reach Patna & Howrah respectively. If the train from Howrah is moving at 45 km/hr, then the speed of the other train is:

(*a*) 60 km/hr (*b*) 45 km/hr (*c*) 35 km/hr (*d*) 54 km/hr

22. Shilpa deposited same amount of money in two different banks, which pay simple interest at the rate of 10% and 12% respectively. After 3 years total interest received by her is ₹ 3,300. The total amount deposited is

(*a*) ₹ 10,000 (*b*) ₹ 11,000 (*c*) ₹ 12,000 (*d*) ₹ 12,250

23. a sum invested for 3 years compounded at 5%, 10% and 20% respectively. In three years, if the sum amounts to ₹ 16,632, find the sum.

(*a*) ₹ 11,000 (*b*) ₹ 12,000 (*c*) ₹ 13,000 (*d*) ₹ 14,000

24. The bar graph of the tickets of different state lotteries sold by an agent on a day is given. Read the graph and answer the following:

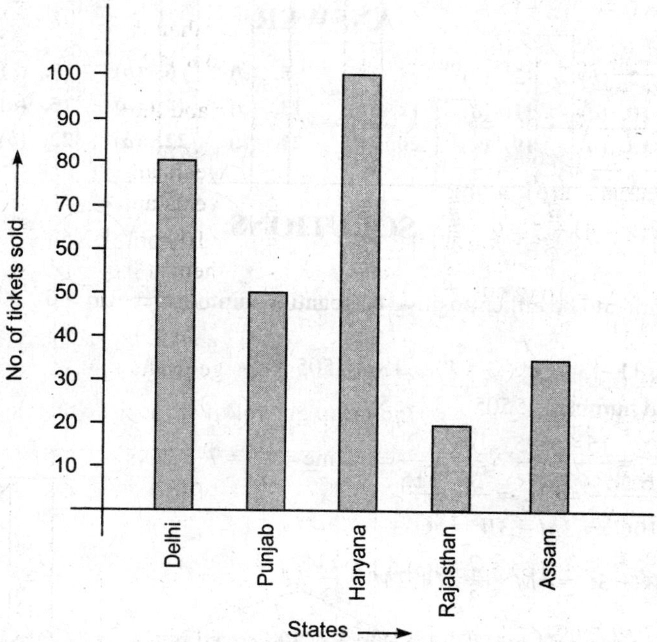

The numbe of tickets sold in a day is

(*a*) 230 (*b*) 250 (*c*) 285 (*d*) 280

25. Study the graph and answer the following question:

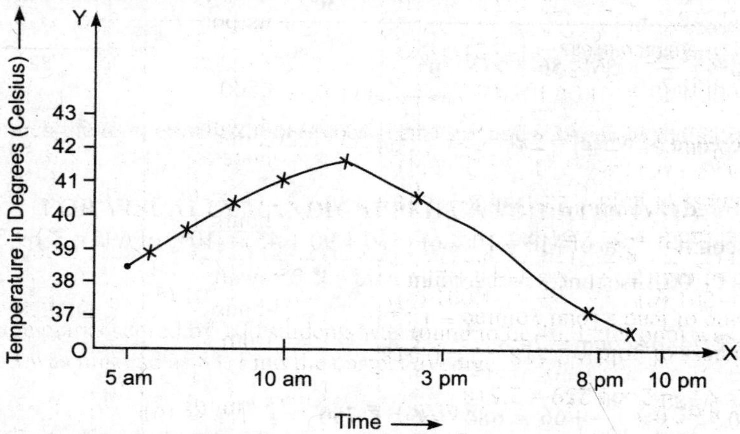

By how much the temperature shoot up between 8 am and 12 noon to the patient?

(*a*) 0.5° C (*b*) 2.5° C (*c*) 2° C (*d*) 1.5° C

FOR VISUALLY HANDICAPPED CANDIDATES ONLY

24. The cube root of $(72.8)^3 + (27.2)(27.2)(27.2) + 3(72.8)(27.2)(72.8 + 27.2)$ is

(*a*) 100 (*b*) 10 (*c*) 1000 (*d*) 102

25. The average marks scored by 100 students was found to be 40. Later on it was observed that a score of 53 was misread as 83. Find the correct average.

(*a*) 39.5 (*b*) 39.6 (*c*) 39.7 (*d*) 39.8

ANSWER

1. (*c*)	**2.** (*b*)	**3.** (*b*)	**4.** (*b*)	**5.** (*d*)	**6.** (*b*)	**7.** (*c*)	**8.** (*d*)
9. (*a*)	**10.** (*c*)	**11.** (*a*)	**12.** (*c*)	**13.** (*a*)	**14.** (*a*)	**15.** (*a*)	**16.** (*c*)
17. (*a*)	**18.** (*a*)	**19.** (*b*)	**20.** (*d*)	**21.** (*d*)	**22.** (*a*)	**23.** (*b*)	**24.** (*c*)
25. (*c*)							

SOLUTIONS

1. $\frac{a}{b} > 0 \Rightarrow$ Both and be either positive or negative numbers $\Rightarrow ab > 0$

2. LCM of 5, 7, 11, 13 $= 5 \times 7 \times 11 \times 13 = 5505$

$\Rightarrow$ Required number $= 5505 - 3 = 5502$

3. $C = \frac{1}{2} - \frac{1}{5} - \frac{1}{6} = \frac{15-6-5}{30} = \frac{4}{30} = \frac{2}{15} \Rightarrow$ Time $= \frac{15}{2} = 7\frac{1}{2}$ days

4. $4M = 6W = 10C \Rightarrow 2M = 3W = 5C$

$\Rightarrow \quad 1M + 1W + 5C = 1M + \frac{2}{3}M + 2M = \frac{11}{3}M$

$\therefore$ Required time $= (4 \times 5) \div \frac{11}{3} = \frac{20 \times 3}{11} = \frac{60}{11} = 5\frac{5}{11}$ days

5. $\pi r + 2r = 288 \Rightarrow (\pi + 2)r = 288 \Rightarrow r = 288 \times \frac{7}{36} = 56$

Area $= \pi r^2 = \frac{22}{7} \times 56 \times 56 = 9856 \text{ m}^2$

6. Circumference of wheel $= 2\pi r = 2 \times \frac{22}{7} \times 35 = 220$ cm

$\Rightarrow$ Required revolutions $= 33000 \div 220 = 150$

7. Total discount = 10% of MP = 10% of (560 + 90 + 45) = 10% of 695 = ₹ 69.50

8. If M.P. = ₹ 100, discount = 25, Amount paid = ₹ 75

If M.P. = ₹ 24, discount $= 24 \times \frac{25}{75} =$ ₹ 8

9. CP $= 950 \times \frac{80}{100} \times \frac{90}{100} + 66 = 684 + 66 =$ ₹ 750

$\Rightarrow$ SP $= 750 \times \frac{125}{100} =$ ₹ 937.50

10. Ratio $= \frac{125}{8} : \frac{140}{10} = \frac{125}{8} : 14 = 125 : 112$

11. $75 \times \frac{8}{5} = 120$

12. Ratio of value $= (1 \times 4) : (2 \times 3) : (5 \times 2) = 2 : 3 : 5$

$\Rightarrow$ Required amount $= \frac{3}{10} \times$ ₹ 500 = ₹ 150

13. Middle term = 3rd term = 9 $\Rightarrow$ First term = $9 - 2 \times 2 = 5$

14. $\frac{2B+F}{3} = \frac{2B+M}{3} + 3 \Rightarrow 2B + F = 2B + M + 9$

$\Rightarrow \quad F = M + 9 \Rightarrow F - M = 9$

$F + M = 4 \times 19 - 2 \times 5\frac{1}{2} = 76 - 11 = 65$

In option (*a*) : Sum = 65 and difference = 9

15. Cost of 12 oranges = $7 \times 5 + 5 \times 7$ = ₹ 70

SP = ₹ 84 per dozen $\Rightarrow$ Profit = $\frac{14}{70} \times 100 = 20\%$

16. From options : $\frac{150 \times 50}{100} = 75$

17. S.P : C.P = 16 : 12 = 4 : 3 $\Rightarrow$ Gain % = $\frac{1}{3} \times 100 = 33\frac{1}{3}\%$

18. Net decrease = $\frac{10 \times 10}{100} = 1\%$

19. $47880 \times \frac{100}{105} \times \frac{100}{95} = 48000$

20. $40 + 45 + 50 + 55 + 60 + 65 + 70 = 55 \times 7 = 385$ km

21. Ratio of time = 288 : 200 = 144 : 100

$\Rightarrow$ Ratio of speeds = $\sqrt{100} : \sqrt{144} = 10 : 12 = 5 : 6$

$\Rightarrow$ Required speed = $\frac{6}{5} \times 45 = 54$ km/hr

22. Interest rate = 10% + 12% = 22% p.a.

Annual interest = ₹ 3300 ÷ 3 = ₹ 1100

$\Rightarrow$ Deposit = $2 \times \frac{1100 \times 100}{22}$ = ₹ 10,000

23. $16632 \times \frac{100}{105} \times \frac{100}{110} \times \frac{100}{120} = 12000$

24. $80 + 50 + 100 + 20 + 35 = 285$

25. 42°C – 40°C = 2°C

FOR VISUALLY HANDICAPPED ONLY

24. G.E = $x^3 + y^3 + 3xy(x + y) = (x + y)^3 = (72.8 + 27.2)^3 = 100^3$

$\Rightarrow$ Cube root = $(100^3)^{1/3} = 100$

25. Difference = $(53 - 83) = -30 \Rightarrow$ Correct average = $40 - \frac{30}{100} = 39.7$

CGL TIER I 2014 (HELD ON 26/10/14 : 987RH1)

1. The least prime number is

(*a*) 0 (*b*) 1 (*c*) 3 (*d*) 2

2. Find the least number which when divided by 12, 18, 36 and 45 leaves the remainder 8, 14, 32 and 41 respectively.

(*a*) 176 (*b*) 180 (*c*) 178 (*d*) 186

3. An empty pool being filled with water at a constant rate takes 8 hours to fill $\frac{3^{th}}{5}$ of its capacity. How much more time will it take to finish filling the pool ?

(*a*) 5 hours 30 minutes (*b*) 5 hours 20 minutes

(*c*) 4 hours 48 minutes (*d*) 4 hours 50 minutes

4. *P* can be do a piece of work in 9 days. *Q* is 50% more efficient than *P*. The number of days it takes for *Q* to do the same piece of work is

(*a*) $13\frac{1}{2}$ (*b*) $4\frac{1}{2}$ (*c*) 6 (*d*) 3

5. Sixteen men can complete a work in fifteen days, twenty-four children can do the same work in twenty days. In how many days will eight men and eight children, complete the same work?

(*a*) 16 days (*b*) $13\frac{1}{3}$ days (*c*) 20 days (*d*) 18 days

6. The area of the iron sheet required to prepare a cone 24 cm high with base radius 7 cm is (Take $\pi = \frac{22}{7}$)

(*a*) 708 cm^2 (*b*) 804 cm^2 (*c*) 704 cm^2 (*d*) 408 cm^2

7. On decreasing each side of an equilateral triangle by 2 cm, there is a decrease of $4\sqrt{3}$ cm^2 in its area. The length of each side of the triangle is

(*a*) 3 cm (*b*) 5 cm (*c*) 6 cm (*d*) 8 cm

8. A rectangular tin sheet is 12 cm long and 5 cm broad. It is rolled along its length to form a cylinder by making the opposite edges just to touch each other. Then the volume of the cylinder is

(*a*) $\frac{180}{\pi}$cm^3 (*b*) $\frac{120}{\pi}$cm^3 (*c*) $\frac{100}{\pi}$cm^3 (*d*) $\frac{60}{\pi}$cm^3

9. A merchant advertises 10% off on the items bought from his store. The total discount got by a customer who bought a cooker worth ₹650, a heater worth ₹500 and a bag worth ₹65 is

(*a*) ₹120.50 (*b*) ₹123.50 (*c*) ₹128.50 (*d*) ₹121.50

10. What is better investment, 4% stock at ₹120 or 3% stock at ₹80?

(*a*) First (*b*) Second (*c*) Both (*d*) None of these

11. 20% raise of price followed by a discount of 25% of raised portion will

(*a*) increase the price by 10% (*b*) decrease the price by 5%

(*c*) increase the price by 15% (*d*) decrease the price by 15%

12. The sum of the ages of two brothers, having a difference of 8 years between them, will double after 10 years. What is the ratio of the age of the younger brother to that of the elder brother?

(*a*) 7 : 11 (*b*) 3 : 7 (*c*) 8 : 9 (*d*) 10 : 13

13. The ratio of two numbers is 3 : 4 and their LCM is 180. The second number is
(*a*) 60 (*b*) 45 (*c*) 90 (*d*) 30

14. Out of 30 teachers of a school, a teacher of age 60 years retired. In his place another teacher of age 30 years was appointed. As a result, the mean age of the teachers will
(*a*) decrease by 6 months (*b*) decrease by 1 year
(c) remain same. (*d*) decrease by 2 years

15. Average age of *A*, *B* and *C* is 84 years. When *D* joins them the average age becomes 80 years. A new person, *E*, whose age is 4 years more than *D*, replace *A* and the average of *B*, *C*, *D* and *E* becomes 78 years. What is the age of *A*?
(*a*) 60 years (*b*) 70 years (*c*) 80 years (*d*) 50 years

16. By selling an article for ₹450, a man loses 10%. The gain or loss percent if he sells it for ₹540 is
(*a*) loss 8% (*b*) gain 9% (*c*) loss 9% (*d*) gain 8%

17. A man loses $20\frac{1}{2}\%$ of his money and after spending 80% of the remainder, he is left with ₹159. How much did he have at first?
(*a*) ₹500 (*b*) ₹800 (*c*) ₹1000 (*d*) ₹1200

18. A gun is fired at a distance of 1.34 km from Geeta. She hears the sound after 4 secs. The speed at which sound travels is
(*a*) 335 m/sec (*b*) 330 m/sec (*c*) 300 m/sec (*d*) 325 m/sec

19. If I walk at 5 km/hr, I miss a train by 7 minutes. However, if I walk at 6 km/hr I reach the station 5 minutes before the departure of the train. The distance between my house and the station is
(*a*) 6.5 km (*b*) 7 km (*c*) 6 km (*d*) 5 km

20. The compound interest on ₹ 1,800 at 10% per annum for a certain period of time is ₹378. Find the time in years.
(*a*) 2.8 years (*b*) 3.0 years (*c*) 2.5 years (*d*) 2.0 years

21. The value of 204 × 197 is
(*a*) 40188 (*b*) 40212 (*c*) 39812 (*d*) 40218

22. The value of $\left\{\left(\sqrt[n]{x^2}\right)^{n/2}\right\}^2$ is
(*a*) x (*b*) $x^{n/2}$ (*c*) x^2 (*d*) $\frac{1}{x^2}$

23. If $(\sqrt{3})^5 \times 9^2 = 3^n \times 3\sqrt{3}$, then find the value of *n*.
(*a*) 2 (*b*) 3 (*c*) 4 (*d*) 5

24. If $p = 99$, then the value of $p(p^2 + 3p + 3)$ is
(*a*) 9999999 (*b*) 1000001 (*c*) 999999 (*d*) 10000001

25. If $2x + 3y = \frac{11}{2}$ and $xy = \frac{5}{6}$, then the value of $8x^3 + 27y^3$ is
(*a*) $\frac{583}{4}$ (*b*) 187 (*c*) $\frac{671}{8}$ (*d*) 583

26. If $x^4 + \frac{1}{x^4} = 119$ and $x > 1$, then the value of $x^3 - \frac{1}{x^3}$ is **(SSC CHSL 2011)**
(*a*) 36 (*b*) – 36 (*c*) ± 36 (*d*) ± 33

27. If $\{\frac{1}{2}(a-b)\}^2 + ab = p(a+b)^2$, then the value of p is (assume that $a \neq -b$)

(a) 1 (b) $\frac{1}{2}$ (c) $\frac{1}{4}$ (d) $\frac{1}{8}$

28. If $999x + 888y = 1332$ and $888x + 999y = 555$, then $x^2 - y^2$ is equal to

(a) 5 (b) 7 (c) 8 (d) 9

29. In two similar triangles ABC and MNP, if AB = 2.25 cm, MP = 4.5 cm and PN = 7.5 cm and $m\angle ACB = m\angle MNP$ and $m\angle ABC = m\angle MPN$, then the length of side $\overline{BC}$, in cm, is

(a) 3.75 (b) 4.75 (c) 3.5 (d) 4.5

30. Given an equilateral DABC, D, E, F are the mid-points of AB, BC and AC respectively. Then the quadrilateral $BEFD$ is exactly a

(a) square (b) rectangle (c) trapezium (d) rhombus

31. If $ABCD$ is a cyclic parallelogram, then the $\angle A$ is

(a) 60° (b) 80° (c) 90° (d) 100°

32. AC is a chord of circle whose centre is at O. If B is any point on the arc AC and $\angle OCA = 20°$, then the magnitude of $\angle ABC$ is

(a) 100° (b) 40° (c) 140° (d) 110°

33. The co-ordinates of the vertices of a right-angled triangle are $P(3, 4)$, $Q(7, 4)$ and $R(3, 8)$, the right-angle being at P. The co-ordinates of the orthocentre of $\Delta\, PQR$ are

(a) (3, 4) (b) (7, 4) (c) (3, 8) (d) (5, 6)

34. The value of $(\sec\theta + \text{cosec}\,\theta)$ when $\theta = 45°$, is

(a) $3\sqrt{2}$ (b) $4\sqrt{2}$ (c) $2\sqrt{2}$ (d) $5\sqrt{2}$

35. $\frac{\tan^2\theta}{\sec\theta+1} - \sec\theta$ is equal to

(a) -1 (b) 1 (c) 0 (d) None of these

36. $\sin^6\theta + \cos^6\theta$ is equal to

(a) $1 - 3\sin^2\theta\cos^2\theta$ (b) $1 - 3\sin\theta\cos\theta$

(c) $1 + 3\sin^2\theta\cos^2\theta$ (d) 1

37. If $\tan\theta + \cot\theta = 2$, $0 < \theta < 90°$, then the value of θ is

(a) 30° (b) 45° (c) 60° (d) 75°

38. The expression $1 + \frac{\cot^2 A}{1+\text{cosec}\,A}$ is equal to

(a) $\sin A$ (b) $\cos A$ (c) $\tan A$ (d) $\text{cosec}\,A$

39. The shadow of a tower standing on a level ground is found to be 40 m longer when the sun's altitude is 30° than when it is 60°. Find the length of the tower.

(a) 10 m (b) $10\sqrt{3}$ m (c) 20 m (d) $20\sqrt{3}$ m

40. The length of the shadow of a tower is 9 metres when the sun's altitude is 30°. What is the height of the tower?

(a) $4\frac{1}{2}$ m (b) $9\sqrt{3}$ m (c) $\frac{9\sqrt{3}}{2}$ m (d) $3\sqrt{3}$ m

41. Which of the following graphical representation of data represents cumulative frequencies?
(*a*) Histogram (*b*) Frequency polygon (*c*) Ogive (*d*) Pie-chart

42. A line graph.
(*a*) makes comparisons (*b*) show trend over time
(*c*) compares structures (*d*) None of the above

43. The heights of two similar right-angled triangles $\Delta\, LMN$ *and* $\Delta\, OPQ$ are 48 cm and 36 cm. If $OP = 12$ cm, then LM is
(*a*) 16 cm (*b*) 20 cm (*c*) 12 cm (*d*) $\frac{10\sqrt{6}}{3}$ cm

44. In the given figure, O is the centre. Then x is equal to

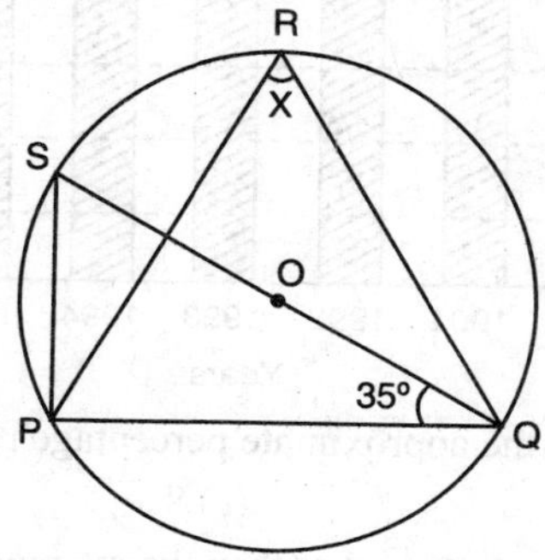

(*a*) 45° (*b*) 55° (*c*) 65° (*d*) 60°

45. In the given figures, the lengths of the sides of $\Delta\, ABC$ and $\Delta\, PQR$ are given and they are given in the same units. Also $\angle A$ and $\angle B$ are given. The value of $\angle P$ is

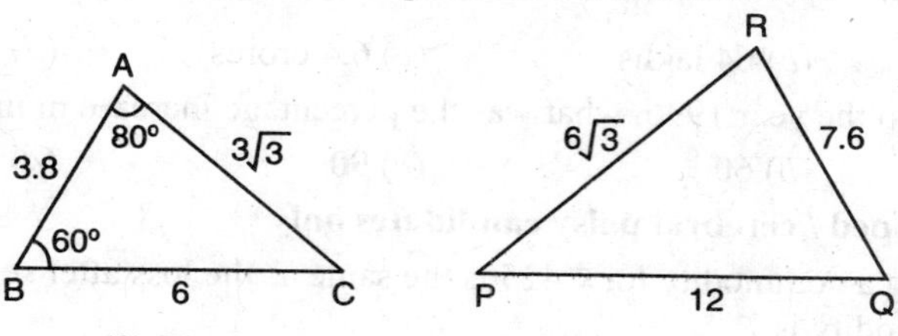

(*a*) 36° (*b*) 38° (*c*) 40° (*d*) 42°

46. In the given figure, POQ is a diameter and PQRS is a cyclic quadrilateral. If $\angle PSR = 130°$, then the value of $\angle RPQ$ is

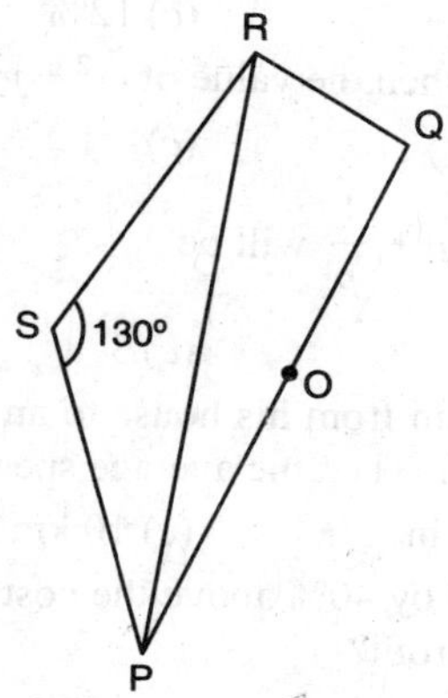

(*a*) 40° (*b*) 45° (*c*) 35° (*d*) 30°

Directions : Study the following graph and answer the questions no. 47 to 50.

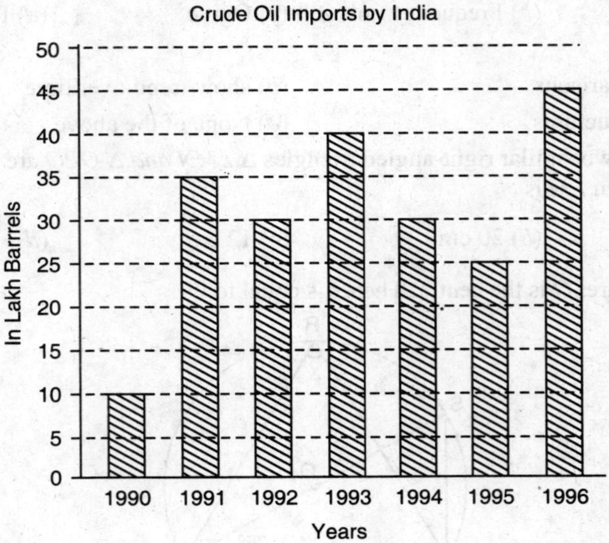

47. From 1991 to 1992, what was the approximate percentage reduction in import of crude oil?

(*a*) 22 (*b*) 18 (*c*) 8 (*d*) 14

48. In how many years were the imports higher than the average value of imports.

(*a*) 1 (*b*) 2 (*c*) 3 (*d*) 4

49. If the rate of one barrel of crude oil was ₹ 25.60 in 1995, then what was the total value of imports (in ₹) in that year?

(*a*) 64 crores (*b*) 64 lakhs (*c*) 6.4 crores (*d*) 8.2 crores

50. From year 1995 to the year 1996, what was the percentage increase in imports of crude oil?

(*a*) 180 (*b*) 60 (*c*) 90 (*d*) None of these

For visually handicapped / cerebral palsy candidates only

44. Profit after selling a commodity for ₹ 425 is the same as the loss after selling it for ₹ 355. The cost of the commodity is

(*a*) ₹ 385 (*b*) ₹ 390 (*c*) ₹ 395 (*d*) ₹ 400

45. A watch is purchased for ₹ 400 and sold for ₹ 460. The profit percent is

(*a*) 15% (*b*) 15.5% (*c*) 12% (*d*) 13%

46. If x = 997, y = 998 and z = 999, then the value of $(x^2 + y^2 + z^2 - xy - yz - zx)$ is

(*a*) 0 (*b*) 1 (*c*) – 1 (*d*) 3

47. If $p = 2 + \sqrt{3}$, then the value of $p + \dfrac{1}{p}$ will be

(*a*) 8 (*b*) 4 (*c*) 3 (*d*) 2

48. A man travels a distance of 18 km from his house to an exhibition by tonga at 15 km/hr and returns back on cycle at 10 km/hr. Then the average speed for the whole journey is

(*a*) 18 km/hr (*b*) 12 km/hr (*c*) 10 km/hr (*d*) 15 km/hr

49. The price of a radio set is raised by 40% above the cost price and sold at a discount of 10%. What will be the percentage of profit?

(*a*) 26% (*b*) 14% (*c*) 30% (*d*) 25%

50. If $2x = 3 + \frac{2}{x}$, then the value of $\frac{4x}{x^2 - x - 1}$ is

(a) 8 (b) 4 (c) 2 (d) $\frac{8}{5}$

ANSWER

1. (d)	2. (a)	3. (b)	4. (c)	5. (c)	6. (c)	7. (b)	8. (a)
9. (d)	10. (b)	11. (c)	12. (b)	13. (a)	14. (b)	15. (c)	16. (d)
17. (c)	18. (a)	19. (c)	20. (d)	21. (a)	22. (c)	23. (d)	24. (c)
25. (c)	26. (a)	27. (c)	28. (b)	29. (a)	30. (d)	31. (c)	32. (d)
33. (a)	34. (c)	35. (a)	36. (a)	37. (b)	38. (d)	39. (d)	40. (d)
41. (c)	42. (b)	43. (a)	44. (b)	45. (c)	46. (a)	47. (d)	48. (c)
49. (c)	50. (d)						

HINTS

1. 2 is the least prime number. 1 is neither a prime number nor a composite number.

2. LCM of 12, 18, 36, 45 = 12 × 3 × 5 = 180; Required number = 180 – 4 = 176

3. Time $= 8\times \frac{2}{3} = \frac{16}{3} = 5\frac{1}{3}$ hours = 5 hours 20 minutes.

4. $9 \times \frac{100}{150} = 6$ days

5. 16 × 15 men = 24 × 20 children ⇒ 1 man = 2 children ⇒ 8 men = 16 children;

Days $= \frac{24\times20}{24} = 20$ days

6. $l = \sqrt{24^2 + 7^2} = 25$ cm; $\pi rl = 550$, $\pi r^2 = 154 \Rightarrow$ Total = 550 + 154 = 704

7. From given options : $\frac{\sqrt{3}}{4}(5^2 - 3^2) = \frac{\sqrt{3}}{4}\times16 = 4\sqrt{3}.$

8. $2\pi r = 12 \Rightarrow r = \frac{6}{\pi}$; volume $= \pi r^2 h = \pi\left(\frac{36}{\pi^2}\right)5 = \frac{180}{\pi}$

9. Total MP = 650 + 500 + 65 = 1215 ⇒ Discount = 10% of ₹ 1215 = ₹ 121.50

10. $I: \frac{4}{120}\times100 = \frac{10}{3}$; $II: \frac{3}{80}\times100 = \frac{15}{4}$; $\frac{10}{3} < \frac{15}{4}$

11. 20% – 25% of 20 = 20 – 5 = 15%

12. Ages are 6 and 14.

13. LCM = $12x = 180 \Rightarrow x = 15 \Rightarrow 4x = 60$

14. Difference = 60 – 30 = 30 ⇒ Average is reduced by 30 ÷ 30 = 1

15. $D = 80 \times 4 - 84 \times 3 = 68$; $E = 68 + 4 = 72$; $B + C + D + E = 78 \times 4 = 312$

$\Rightarrow A = 320 + 72 - 312 = 80$

16. $\frac{90}{450} \times 540 = 108\% \Rightarrow$ Gain = 8%

17. $159 \times \frac{200}{159}\times\frac{100}{20} = 1000$

18. $1340 \div 4 = 335$ m/s

19. $\frac{6\times5}{6-5}\times\frac{1}{5} = 6$ km

20. $\frac{\text{Amount}}{\text{Principal}} = \frac{2178}{1800} = \frac{121}{100} = \left(\frac{11}{10}\right)^2 \Rightarrow$ Time = 2 years

21. $204 \times 197 = 40188$

22. G.E. $= \left[\left(x^{\frac{2}{n}}\right)^{\frac{n}{2}}\right]^2 = x^2$

23. $9\sqrt{3}\times9^2 = 3^n \times 3\sqrt{3} \Rightarrow 3^n = 3^5 \Rightarrow n = 5$

24. $p(p^2 + 3p + 3) = (p + 1)^3 - 1 = 999999$

25. $(2x + 3y)^3 = \left(\frac{11}{2}\right)^3 \Rightarrow 8x^3 + 27y^3 + 3(2x)(3y)(2x + 3y) = \frac{1331}{8}$

$\Rightarrow 8x^3 + 27y^3 = \frac{1331}{8} - 18\times\frac{5}{6}\times\frac{11}{2} = \frac{1331}{8} - \frac{165}{2} = \frac{671}{8}$

26. $x^2 + \frac{1}{x^2} = \sqrt{119+2} = 11 \Rightarrow x - \frac{1}{x} = \sqrt{11-2} = 3 \Rightarrow x^3 - \frac{1}{x^3} = \left(x-\frac{1}{x}\right)^3 + 3x\left(x-\frac{1}{x}\right) = 3^3 + 3\times3 = 36$

27. $\frac{1}{4}(a^2 + b^2 - 2ab) + ab = p(a^2 + b^2 + 2ab)$

$\Rightarrow a^2 + b^2 - 2ab + 4ab = 4p(a^2 + b^2 + 2ab) \Rightarrow 4p = 1 \Rightarrow p = \frac{1}{4}$

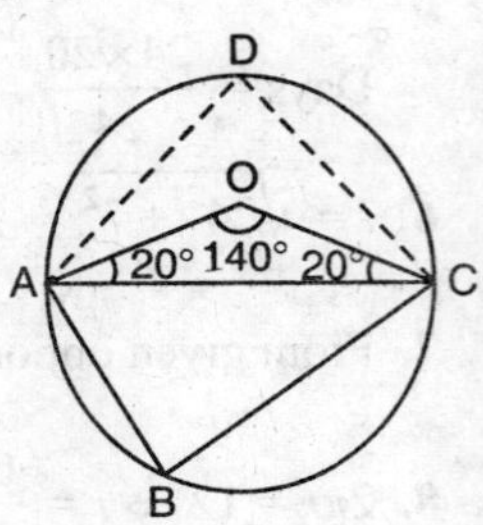

28. $111(x - y) = 777 \Rightarrow x - y = 7; x + y = 1 \Rightarrow x = 4, y = -3$

$\Rightarrow x^2 - y^2 = 4^2 - (-3)^2 = 16 - 9 = 7$

29. $AB : MP = 2.25 : 4.5 = 1 : 2 \Rightarrow BC = 7.5 \div 2 = 3.75$

30. All sides are equal $\Rightarrow$ Rhombus

31. $\angle A = C$ and $\angle A + \angle C = 180° \Rightarrow \angle A = 90°$

32. $\angle OAC = \angle OCA = 20°$

$\angle AOC = 180° - (20° + 20°) = 140°$

$\angle ADC = \frac{1}{2} \times \angle AOC = 70°$

$\angle ABC = 180° - \angle ADC = 180° - 70° = 110°$

33. In a right-angled triangle, the orthocentre lies at the vertex containing the right angle.

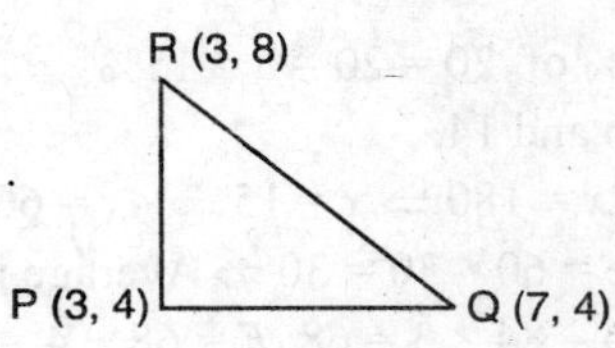

34. $\sqrt{2} + \sqrt{2} = 2\sqrt{2}$

35. $\frac{\tan^2\theta - \sec^2\theta - \sec\theta}{\sec\theta + 1} = \frac{-(\sec\theta + 1)}{(\sec\theta + 1)} = -1$

Alt: $\frac{\sec^2\theta - 1}{\sec\theta + 1} - \sec\theta = (\sec\theta - 1) - \sec\theta = -1$

36. $(\sin^2\theta + \cos^2\theta)^3 = \sin^6\theta + \cos^6\theta + 3\sin^2\theta\cos^2\theta(\sin^2\theta + \cos^2\theta)$

$\Rightarrow \sin^6\theta + \cos^6\theta = 1 - 3\sin^2\theta\cos^2\theta$ $\quad (\sin^2\theta + \cos^2\theta) = 1$

37. $\tan\theta = 1 = \tan 45°$

38. $1 + \dfrac{\text{cosec}^2 A - 1}{\text{cosec}\, A + 1} = 1 + \text{cosec}\, A - 1 = \text{cosec}\, A$

39. $\dfrac{h}{x} = \sqrt{3} \Rightarrow h = \sqrt{3}x$

$\dfrac{h}{x+40} = \dfrac{1}{\sqrt{3}} \Rightarrow \sqrt{3}h = x + 40$

$\Rightarrow 3x = x + 40$

$\Rightarrow x = 20$

$\Rightarrow h = 20\sqrt{3}\, m$

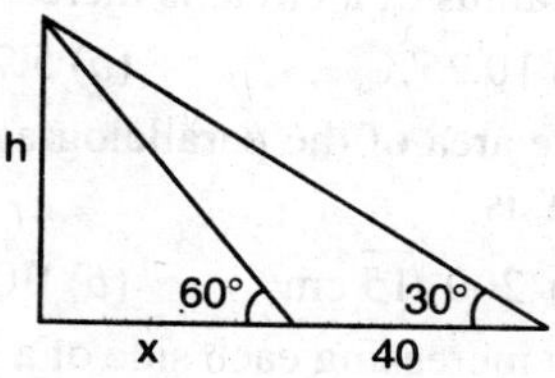

40. $\dfrac{h}{9} = \dfrac{1}{\sqrt{3}} \Rightarrow h = \dfrac{9}{\sqrt{3}} = 3\sqrt{3}$ m

43. 48 : 36 = 4 : 3 = 16 : 12

44. $\angle PSQ = 180° - 35° - 90° = 55° \Rightarrow x = \angle PSQ = 55°$

45. $\Delta ABC \sim \Delta RQP \Rightarrow \angle P = \angle C = 40°$

46. $\angle PRQ = 90°, \angle PQR = 180° - 130° = 50° \Rightarrow \angle RPQ = 180° - 50° - 90° = 40°$

47. $35 \to 30;\ \dfrac{5}{35} \times 100 = 14\%$ (approx)

48. Sum = 10 + 35 + 30 + 40 + 30 + 25 + 45 = 215 ⇒ Average = 215 ÷ 7 = 31 approx

⇒ 1991, 1993, 1996

49. 25.60 × 25 lakhs = 640 lakhs = 6.4 crores

50. $25 \to 45 \Rightarrow \dfrac{20}{25} \times 100 = 80\%$

For Visually Handicaped only

44. (425 + 355) ÷ 2 = 390

45. 60/400 × 100 = 15%

46. 3

47. $(2 + \sqrt{3}) + (2 - \sqrt{3}) = 4$

48. $\dfrac{2 \times 15 \times 10}{25} = 12$ km/hr

49. $40 - 10 - \dfrac{40 \times 10}{100} = 26$

50. $2x = 3 + \dfrac{2}{x} \Rightarrow 2x - \dfrac{2}{x} = 3 \Rightarrow x - \dfrac{1}{x} = \dfrac{3}{2} \Rightarrow \dfrac{4x}{x^2 - x - 1} = \dfrac{4}{\left(x - \dfrac{1}{x}\right) - 1} = \dfrac{4}{\dfrac{3}{2} - 1} = \dfrac{4}{\dfrac{1}{2}} = 8$

CGL TIER II 2014

1. The outer and inner diameter of a circular path be 728 m and 700 m respectively. The breadth of the path is
(a) 7 m (b) 14 m (c) 28 m (d) 20 m
2. A piece of wire when bent to form a circle will have a radius of 84 cm. If the wire is bent to form a square, the length of a side of the square is
(a) 152 cm (b) 168 cm (c) 132 cm (d) 225 cm
3. If radius of a circle is increased by 5%, then the increase in its area is
(a) 10.25% (b) 5.75% (c) 10% (d) 5%
4. The area of the parallelogram whose length is 30 cm, width is 20 cm and one diagonal is 40 cm, is
(a) $200\sqrt{15}$ cm^2 (b) $300\sqrt{15}$ cm^2 (c) $100\sqrt{15}$ cm^2 (d) $150\sqrt{15}$
5. On increasing each side of a square by 50%, the ratio of the area of new formed square and the given square will be
(a) 9 : 5 (b) 9 : 7 (c) 9 : 3.5 (d) 9 : 4
6. The area of a circle is 324 π sq. cm. The length of its longest chord (in cm) is
(a) 36 (b) 38 (c) 28 (d) 32
7. The area of a rhombus is 256 sq. cm. and one of its diagonal is twice the other in length. Then length of its larger diagonal is
(a) 32 cm (b) 48 cm (c) 16 cm (d) 24 cm
8. If $m=\sqrt{6+\sqrt{6+\sqrt{6+\ldots\ldots}}}$ and $n=\sqrt{6-\sqrt{6-\sqrt{6-\ldots\ldots}}}$, then among the following the relation between m and n holds is
(a) $m-n+1=0$ (b) $m+n+1=0$ (c) $m+n-1=0$ (d) $m-n-1=0$
9. If $\frac{3-5x}{2x}+\frac{3-5y}{2y}+\frac{3-5z}{2z}=0$, then the value of $\frac{2}{x}+\frac{2}{y}+\frac{2}{z}$ is
(a) 20 (b) 10 (c) 5 (d) 15
10. The distance between 2 places R and S is 42 km. Anita starts from R with a uniform speed of 4 km/h towards S and at the same time Romita starts from S towards R also with same uniform speed. They meet each other after 6 hours. The speed of Romita is
(a) 18 km/hr (b) 20 km/hr (c) 6 km/hr (d) 8 km/hr
11. If the side of a square is $\frac{1}{2}(x+1)$ units and its diagonal is $\frac{3-x}{\sqrt{2}}$ units, then the length of the side of the square would be
(a) 4/3 units (b) 1 unit (c) 1/2 units (d) 2 units
12. If $2s = a + b + c$, then the value of $s(s-c)+(s-a)(s-b)$ is
(a) ab (b) 0 (c) abc (d) $\frac{a+b+c}{2}$
13. If $p + m = 6$ and $p^3 + m^3 = 72$, then the value of pm is
(a) 6 (b) 9 (c) 12 (d) 8
14. When x^m is multiplied by x^n, product is 1. The relation between m and n is
(a) $mn = 1$ (b) $m + n = 1$ (c) $m = n$ (d) $m = -n$

15. The area (in sq. unit) of the triangle formed by the graphs of the equations $x = 4$, $y = 3$ and $3x + 4y = 12$; is

(*a*) 24 (*b*) 6 (*c*) 12 (*d*) 3

16. If $\frac{2p}{p^2 - 2p + 1} = \frac{1}{4}$, then the value of $\left(p + \frac{1}{p}\right)$ is

(*a*) 7 (*b*) 1 (*c*) 2/5 (*d*) 10

17. If the ordinate and abscissa of the point $(K^2, 2K - 1)$ be equal, then the value of K is

(*a*) 0 (*b*) 1 (*c*) – 1 (*d*) 1/2

18. ABC is a triangle in which $DE \parallel BC$ and $AD : DB = 5 : 4$. Then $DE : BC$ is

(*a*) 4 : 5 (*b*) 9 : 5 (*c*) 4 : 9 (*d*) 5 : 9

19. The radii of two concentric circles are 17 cm and 25 cm. A straight line PQRS intersects the larger circle at the points P and S and intersects the smaller circle at the points Q and R. If QR = 16 cm, then the length (in cm) of PS is

(*a*) 41 (*b*) 33 (*c*) 32 (*d*) 40

20. AB is a diameter of a circle with centre O. The tangents at C meets AB produced at Q. If $\angle CAB = 34°$, then measure of $\angle CBA$ is

(*a*) 56° (*b*) 68° (*c*) 34° (*d*) 124°

21. For an equilateral triangle, the ratio of in-radius and the outer-radius is

(*a*) 1 : 2 (*b*) 1 : 3 (*c*) $1 : \sqrt{2}$ (*d*) $1 : \sqrt{3}$

22. If a and b are the lengths of the sides of a right triangle whose hypotenuse is 10 and whose area is 20, then the value of $(a + b)^2$ is

(*a*) 140 (*b*) 120 (*c*) 180 (*d*) 160

23. The circumference of a triangle is 24 cm and the circumference of its in-circle is 44 cm. Then the area of the triangle is

(*a*) 56 cm^2 (*b*) 48 cm^2 (*c*) 84 cm^2 (*d*) 68 cm^2

24. If the length of each of two equal sides of an isosceles triangle is 10 cm and the adjacent angle is 45°, then the area of the triangle is

(*a*) $20\sqrt{2}$ cm^2 (*b*) $25\sqrt{2}$ cm^2 (*c*) $12\sqrt{2}$ cm^2 (*d*) $15\sqrt{2}$ cm^2

25. Let P and Q be two points on a circle with centre O. If two tangents of the circle through P and Q meet at A with $\angle PAQ = 48°$, then $\angle APQ$ is

(*a*) 96° (*b*) 66° (*c*) 48° (*d*) 60°

26. The length of the diagonal of a rectangle with sides 4 m and 3 m would be

(*a*) 12 m (*b*) 5 m (*c*) 7 m (*d*) 14 m

27. If the sides of a triangle are in the ratio $3 : 1\frac{1}{4} : 3\frac{1}{4}$, then the triangle is

(*a*) Right triangle (*b*) Equiangular (*c*) Obtuse triangle (*d*) Acute triangle

28. The value of θ $(0° \le \theta \le 90°)$ satisfying $2 \sin^2\theta = 3 \cos \theta$ is

(*a*) 60° (*b*) 90° (*c*) 30° (*d*) 45°

29. *a*, *b*, *c* are the lengths of three sides of a triangle ABC. If *a*, *b*, *c* are related by the relation $a^2 + b^2 + c^2 = ab + bc + ca$, then the value of $(\text{Sin}^2 A + \text{Sin}^2 B + \text{Sin}^2 C)$ is (*a*) $\frac{3}{4}$ (*b*) $\frac{3}{2}$ (*c*) $\frac{3\sqrt{3}}{2}$ (*d*) $\frac{9}{4}$

30. If $a \sin\theta + b\cos\theta = c$, then $a\cos\theta - b\sin\theta$ is equal to

(*a*) $\pm\sqrt{a+b-c}$ (*b*) $\pm\sqrt{a^2+b^2-c^2}$ (*c*) $\pm\sqrt{a^2+b^2+c^2}$ (*d*) $\pm\sqrt{c^2+a^2-b^2}$

31. If $\sin\theta + \cos\theta = \sqrt{2}\sin(90° - \theta)$, then the value of $\cot\theta$ is

(*a*) $-\sqrt{2}-1$ (*b*) $\sqrt{2}+1$ (*c*) $\sqrt{2}-1$ (*d*) $-\sqrt{2}+1$

32. If θ is positive acute angle and $3(\sec^2\theta + \tan^2\theta) = 5$, then the value of $\cos 2\theta$ is

(*a*) $\frac{1}{2}$ (*b*) $\frac{\sqrt{3}}{2}$ (*c*) $\frac{1}{\sqrt{2}}$ (*d*) 1

33. If $x\cos^2 30° \,.\, \sin 60° = \frac{\tan^2 45°.\sec 60°}{\text{cosec } 60°}$, then the value of x is

(*a*) $\frac{1}{\sqrt{3}}$ (*b*) $2\frac{2}{3}$ (*c*) $\frac{1}{\sqrt{2}}$ (*d*) $\frac{1}{2}$

34. If $\tan\alpha = 2$, then the value of $\frac{\text{cosec}^2\alpha - \sec^2\alpha}{\text{cosec}^2\alpha + \sec^2\alpha}$ is

(*a*) $-\frac{15}{9}$ (*b*) $\frac{3}{5}$ (*c*) $-\frac{3}{5}$ (*d*) $\frac{17}{5}$

35. If $\sin(\theta + 30°) = \frac{3}{\sqrt{12}}$, then the value of $\cos^2\theta$ is

(*a*) $\frac{1}{4}$ (*b*) $\frac{3}{4}$ (*c*) $\frac{\sqrt{3}}{2}$ (*d*) $\frac{1}{2}$

36. In a right angled triangle Δ PQR, PR is the hypotenuse of length 20 cm, $\angle PRQ = 30°$, the area of triangle is

(*a*) $50\sqrt{3}$ cm² (*b*) $25\sqrt{3}$ cm² (*c*) $100\sqrt{3}$ cm² (*d*) $100/\sqrt{3}$ cm²

37. If $(0° \le \theta \le 90°)$ and $4\cos^2\theta - 4\sqrt{3}\cos\theta + 3 = 0$, then the value of θ is

(*a*) 30° (*b*) 90° (*c*) 45° (*d*) 60°

38. If $x : y = 3 : 2$, then value of $\frac{x+y}{x-y}$ is

(*a*) 5 : 1 (*b*) 1 : 5 (*c*) 1 : 3 (*d*) 3 : 1

39. If 50% of x = 30% of y, then $x : y$ is

(*a*) 2 : 3 (*b*) 3 : 2 (*c*) 5 : 3 (*d*) 3 : 5

40. The value of $\frac{1}{1+\sqrt{2}} + \frac{1}{\sqrt{2}+\sqrt{3}} + \frac{1}{\sqrt{3}+\sqrt{4}} + \frac{1}{\sqrt{4}+\sqrt{5}} + ... + \frac{1}{\sqrt{8}+\sqrt{9}}$ is

(*a*) 2 (*b*) 4 (*c*) 0 (*d*) 1

41. While solving a problem, by mistake, Anita squared a number and then subtracted 25 from it rather than first subtracting 25 from the number and then squaring it. But she got the right answer. What was the number ?

(*a*) 48 (*b*) Cannot be determined (*c*) 13 (*d*) 38

42. A General of Army wants to form a square from 36562 armies. After arrangement, he found some armies left. How many armies were left ?

(*a*) 81 (*b*) 97 (*c*) 36 (*d*) 65

43. Square root of $\frac{2+\sqrt{3}}{2}$ is

(*a*) $\pm\frac{1}{\sqrt{2}}(\sqrt{3}+1)$ (*b*) None of these (*c*) $\pm\frac{1}{2}(\sqrt{3}-2)$ (*d*) $\pm\frac{1}{2}(\sqrt{3}-1)$

44. The value of $\sqrt{72+\sqrt{72+\sqrt{72+........}}}$ is

(*a*) 9 (*b*) 18 (*c*) 8 (*d*) 12

45. In a farm there are cows and hens. If heads are counted there are 180, if legs are counted there are 420. The number of cows in the farm is

(*a*) 130 (*b*) 50 (*c*) 150 (*d*) 30

46. The number which can be written in the form of $n(n+1)(n+2)$, where n is a natural number, is

(*a*) 7 (*b*) 5 (*c*) 3 (*d*) 6

47. The value of $(11111)^2$ is

(*a*) 12344321 (*b*) 123454321 (*c*) 121212121 (*d*) 11344311

48. Two workers A and B are engaged to do a piece of work. A working alone would take 8 hours more to complete the work than when work together. If B worked alone, would take $4\frac{1}{2}$ hours more than when work together. The time required to finish the work together is

(*a*) 5 hours (*b*) 8 hours (*c*) 4 hours (*d*) 6 hours

49. If a man earns ₹2000 for his first 50 hours of work in a week and is then paid one and one half times his regular hourly rate for additional hours, then the hours must he work to make ₹2300 in a week is

(*a*) 6 hours (*b*) 7 hours (*c*) 4 hours (*d*) 5 hours

50. If X can finish a job in 4 hours and Y can finish the same job in 8 hours independently, then they together will finish the job in

(*a*) 140 minutes (*b*) 120 minutes (*c*) 160 minutes (*d*) 150 minutes

51. A company employed 200 workers to complete a certain work in 150 days. If only 1/4th of the work has been done in 50 days, then in order to complete the whole work in time, the number of additional workers to be employed was

(*a*) 100 (*b*) 600 (*c*) 300 (*d*) 200

52. X can copy 80 pages in 20 hours, X and Y together can copy 135 pages in 27 hours. Then Y can copy 20 pages in

(*a*) 20 hours (*b*) 24 hours (*c*) 3 hours (*d*) 12 hours

53. X does 1/4 of a job in 6 days. Y completes rest of the job in 12 days. Then X and Y could complete the job together in

(*a*) 9 days (*b*) $8\frac{1}{8}$ days (*c*) $9\frac{3}{5}$ days (*d*) $7\frac{1}{3}$ days

54. Mr. X and Mr. Y each bought the same motorcycle using a 10% off coupon. Mr. X's cashier took 10% off the price and then added 8.5% sales tax whereas Mr. Y's cashier first added the sales tax and then took 10% off the total price. The amount Mr. X paid is

(*a*) less than ₹550 as the the amount Mr. Y paid

(*b*) greater than ₹ 85 as the the amount Mr. Y paid

(*c*) same as the amount Mr. Y paid (*d*) greater than ₹850 as the the amount Mr. Y paid

55. If a person marked a product 25% above the cost price but allows 10% discount, then the percentage of profit is

(*a*) 35% (*b*) 17.5% (*c*) 15% (*d*) 12.5%

56. The price of an antique is reduced by 20% and then this price is again reduced by 10%. The total reduction of the price is

(*a*) 25% (*b*) 23% (*c*) 30% (*d*) 28%

57. A builder purchases 25 windows at 25% off the total price of ₹1,20,000. If the builder receives an additional discount of ₹7,500 for the purchase, then the cost price of each window is

(*a*) ₹3300 (*b*) ₹3400 (*c*) ₹3100 (*d*) ₹3200

58. If A : B = $\frac{1}{2}:\frac{1}{3}$, B : C = $\frac{1}{5}:\frac{1}{3}$, then (A + B) : (B + C) is equal to

(*a*) 5 : 8 (*b*) 15 : 16 (*c*) 9 : 10 (*d*) 6 : 15

59. In a partnership business, B's capital was half of A's. If after 8 months, B withdrew half of his capital and after 2 month more A withdrew 1/4 th of his capital, then the profit ratio of A and B will be

(*a*) 5 : 2 (*b*) 2 : 5 (*c*) 10 : 23 (*d*) 23 : 10

60. A school group charters three identical buses and occupies 4/5 of the seats. After 1/4 of the passengers leave, the remaining passengers use only two of the buses. The fraction of the seats on the two buses that are now occupied is

(*a*) 8/9 (*b*) 7/9 (*c*) 7/10 (*d*) 9/10

61. A and B invest in the ratio 3 : 5. After 6 months, C joins the business investing an amount equal to B's. At the end of the year what will be the ratio of their profits ?

(*a*) 6 : 10 : 5 (*b*) 8 : 10 : 5 (*c*) 3 : 5 : 2 (*d*) 3 : 5 : 5

62. In a library the ratio of story books and other books is 7 : 2 and there are 1512 story books. Due to collection of some more story books the said ratio becomes 15 : 4. The number of story books collected is

(*a*) 108 (*b*) 205 (*c*) 100 (*d*) 97

63. If the product of two positive numbers be 1575 and their ratio is 7 : 9, then the greater number is

(*a*) 45 (*b*) 135 (*c*) 35 (*d*) 63

64. The average weight of first 11 persons among 12 persons is 95 kg. The weight of the 12th person is 33 kg more than the average weight of all the 12 persons. The weight of the 12th person is

(*a*) 128.75 kg (*b*) 131 kg (*c*) 128 kg (*d*) 97.45 kg

65. The average of the largest and smallest 3 digit numbers formed by 0, 2 and 4 would be

(*a*) 312 (*b*) 222 (*c*) 213 (*d*) 303

66. The average of six numbers is 3.95. The average of two of them is 3.4, while the average of the other two is 3.85. The average of the remaining two numbers is

(*a*) 4.6 (*b*) 4.8 (*c*) 4.5 (*d*) 4.7

67. If the average of eight consecutive even numbers be 93, then the greatest number among them is

(*a*) 100 (*b*) 102 (*c*) 86 (*d*) 98

68. If the arithmetic mean of 3*a* and 4*b* is greater than 50, and *a* is twice *b*, then the smallest possible integer value of *a* is

(*a*) 20 (*b*) 21 (*c*) 18 (*d*) 19

69. The average (aritmetic mean) of 3^{30}, 3^{60} and 3^{90} is

(*a*) $3^{27} + 3^{57} + 3^{87}$ (*b*) $3^{29} + 3^{59} + 3^{89}$ (*c*) 3^{60} (*d*) 3^{177}

70. Pooja wants to sell a watch at a profit of 20%. She bought it at 10% less and sold it at ₹30 less, but still she gained 20%. The cost price of watch is

(*a*) ₹240 (*b*) ₹250 (*c*) ₹220 (*d*) ₹225

71. A shopkeeper sold his goods at half the list price and thus lost 20%. If he had sold on the listed price, his gain percentage would be

(*a*) 60% (*b*) 72% (*c*) 20% (*d*) 35%

72. Two types of tea costing ₹180/kg and ₹280/kg should be mixed in the ratio so that the mixture obtained, sold at ₹320/kg to earn a profit of 20% is

(*a*) 3 : 13 (*b*) 4 : 13 (*c*) 1 : 13 (*d*) 2 : 13

73. A profit of 12% is made when a mobile phone is sold at ₹P and there is 4% loss when the phone is sold at ₹Q. Then Q : P is

(*a*) 1 : 1 (*b*) 6 :7 (*c*) 4 : 5 (*d*) 3 : 1

74. A sells a cycle to B at a profit of 20% and B sells it to C at a loss 25%. If C bought the cycle for ₹P, then the cost price of it for A was

(*a*) ₹$\frac{1}{20}$P (*b*) ₹$\frac{9}{20}$P (*c*) ₹$\frac{9}{10}$P (*d*) ₹$\frac{10}{9}$P

75. A man buys a toy for ₹25 and sells it for ₹30. His gain percent is

(*a*) 20% (*b*) 10% (*c*) 5% (*d*) 2.5%

76. From 1980-1990, the production of a country was increased by 20%. From 1990-2000, the population of the country was increased by 20%. From 2000-2010, the population of the country was increased by 20%. Then the overall increased population of the country from 1980-2010 was

(*a*) 72.2% (*b*) 72.8% (*c*) 60% (*d*) 62.8%

77. A supply of juice lasts for 35 days. If its use is increased by 40% the number of days would the same amount of juice lasts, is

(*a*) 25 days (*b*) 24 days (*c*) 30 days (*d*) 27 days

78. A store offers a variety of discounts that range between 20% and 25% inclusive. If a book is discounted to a price of ₹270, then its greatest possible original price was

(*a*) ₹345.5 (*b*) ₹360 (*c*) ₹324 (*d*) ₹337.5

79. If the ratio of cost price to selling price is 10 : 11, then the rate of percent of profit is

(*a*) 1.1% (*b*) 10% (*c*) 0.1% (*d*) 1%

80. A boat takes half times in moving a certain distance downstream than upstream. The ratio of the speed of the boat in still water and that of the current is

(*a*) 2 : 1 (*b*) 4 : 3 (*c*) 1 : 2 (*d*) 3 : 1

81. A tank has two pipes. The first pipe can fill it in 4 hours and the second can empty it in 16 hours. If two pipes be opened together at a time, then the tank will be filled in

(*a*) $5\frac{1}{2}$ hours (*b*) 6 hours (*c*) 10 hours (*d*) $5\frac{1}{3}$ hours

82. A farmer travelled a distance of 61 km in 9 hours. He travelled partly on foot at the rate of 4 km/hour and partly on bicycle at the rate of 9 km/hour. The distance travelled on foot is

(*a*) 16 km (*b*) 17 km (*c*) 14 km (*d*) 15 km

83. A man rows upstream 36 km and downstream 48 km taking 6 hours each time. The speed of the current is

(*a*) 0.5 km/hour (*b*) 1 km/hour (*c*) 2 km/hour (*d*) 1.5 km/hour

84. On a certain principal the compound interest compounded annually for the second year at 10% per annum is ₹132. The principal is

(*a*) ₹1250 (*b*) ₹1200 (*c*) ₹1000 (*d*) ₹1320

85. The principal which gives ₹1 interst per day at a rate of 5% simple interest per annum is

(*a*) ₹5000 (*b*) ₹36500 (*c*) ₹7300 (*d*) ₹3650

86. When principal = ₹S, rate of interest = $2r$% p.a., then a person will get after 3 years at compound interest

(*a*) ₹ $\frac{6Sr}{100}$ (*b*) ₹ $S\left(1+\frac{r}{50}\right)^3$ (*c*) ₹ $S\left(1+\frac{r}{100}\right)^3$ (*d*) ₹ $3S\left(1+\frac{r}{100}\right)^3$

87. The sum of money which becomes ₹2420 at 10% rate of compound interest after two years is

(*a*) ₹2000 (*b*) ₹2500 (*c*) ₹1000 (*d*) ₹1500

88. From a solid right circular cylinder of length 4 cm and diameter 6 cm, a conical cavity of the same height and base is hollowed out. The whole surface of the remaining solid (in cm^3) is

(*a*) 48π (*b*) 63π (*c*) 15π (*d*) 24π

89. A spherical ball of radius 1 cm is dropped into a conical vessel of radius 3 cm and slant height 6 cm. The volume of water (in cm^3), that can just immerse the ball, is

(*a*) $\frac{5\pi}{3}$ (*b*) 3π (*c*) $\frac{\pi}{3}$ (*d*) $\frac{4\pi}{3}$

90. The in-radius of a triangle is 6 cm, and the sum of the lengths of its sides is 50 cm. The area of the triangle (in sq. cm.) is

(*a*) 150 (*b*) 300 (*c*) 50 (*d*) 56

91. One of the angles of a right-angled triangle is 15°, and the hypotenuse is 1 m. The area of the triagnle (in sq. cm.) is

(*a*) 1220 (*b*) 1250 (*c*) 1200 (*d*) 1215

92. Assume that a drop of water is spherical and its diameter is one-tenth of a cm. A conical glass has a height equal to the diameter of its rim. If 32000 drops of water fill the glass completely, then the height of the glass (in cm.) is

(*a*) 3 (*b*) 4 (*c*) 1 (*d*) 2

93. If the height of a cylinder is 4 times its circumference, the volume of the cylinder in terms of its circumference *c*, is

(*a*) $\frac{2c^3}{\pi}$ (*b*) $\frac{c^3}{\pi}$ (*c*) $4\pi c^3$ (*d*) $2\pi c^3$

94. If the ratio of the angles of a quadrilateral is 2 : 7 : 2 : 7, then it is a

(*a*) trapezium (*b*) square (*c*) parallelogram (*d*) rhombus

95. If for an isosceles triangle the length of each equal side is *a* units and that of the third side is *b* units, then its area (in sq. units) will be

(*a*) $\frac{a}{4}\sqrt{4b^2-a^2}$ (*b*) $\frac{b}{4}\sqrt{4a^2-b^2}$ (*c*) $\frac{a}{2}\sqrt{2a^2-b^2}$ (*d*) $\frac{b}{2}\sqrt{a^2-2b^2}$

Study the bar diagram and answer Question numbers 96 and 97 :

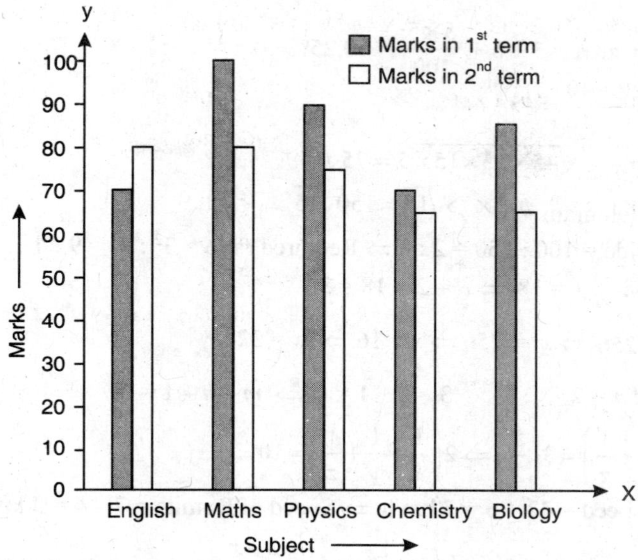

96. Ratio of highest and lowest marks obtained in first term among all the subjects is

(*a*) 7 : 9 (*b*) 9 : 7 (*c*) 10 : 7 (*d*) 7 : 10

97. Average marks obtained by the students for all subjects in second term is

(*a*) 65 (*b*) 73 (*c*) 62 (*d*) 72

In an institution there are 800 students. Students use different modes of transport for going to the institution and return. The given pie diagram represents the requisite data. Study the diagram carefully and answer the question numbers 98 to 100.

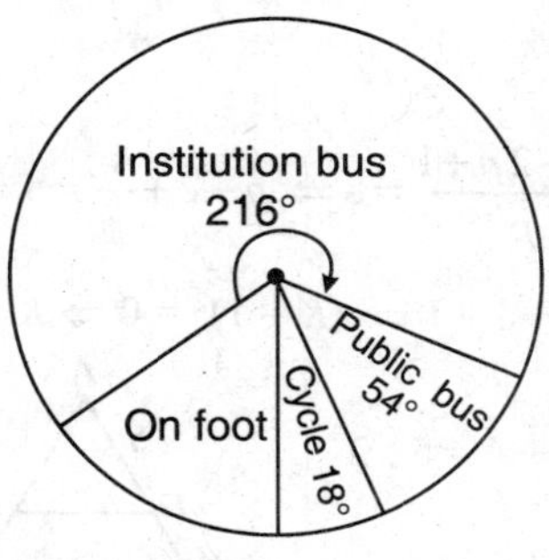

98. Number of students travel in public bus is

(*a*) 150 (*b*) 120 (*c*) 130 (*d*) 125

99. Number of students who do not use institute bus is

(*a*) 330 (*b*) 350 (*c*) 480 (*d*) 320

100. Number of students who go to institute on foot is

(*a*) 160 (*b*) 170 (*c*) 120 (*d*) 106

ANSWERS AND HINTS

1. **(*b*)** Breadth of path = (728 – 700) ÷ 2 = 14 m

2. **(*c*)** Length of wire = $2\pi r = 2 \times \frac{22}{7} \times 84 = 528 \Rightarrow$ Side = 132 cm

3. (*a*) Increase in area = $5 + 5 + \frac{5 \times 5}{100} = 10.25\%$

4. (*d*) $s = \frac{30+20+40}{2} = 45$

Area of triangle = $\sqrt{45 \times 25 \times 15 \times 5} = 75\sqrt{15}$

Area of parallelogram = $2 \times 75\sqrt{15} = 150\sqrt{15}$

5. (*d*) Ratio of side = 100 : 150 = 2 : 3 $\Rightarrow$ Required ratio = $3^2 : 2^2 = 9 : 4$

6. (*d*) $\pi r^2 = 324\pi \Rightarrow r = 18 \Rightarrow d = 2 \times 18 = 36$

7. (*a*) $\frac{x \times 2x}{2} = 256 \Rightarrow x^2 = 256 \Rightarrow x = 16 \Rightarrow 2x = 32$

8. (*d*) $m = 3$ and $n = 2$ $\quad 3 - 2 + 1 = 0 \Rightarrow m - n - 1 = 0$

9. (*b*) $\frac{3}{2}\left(\frac{1}{x}+\frac{1}{y}+\frac{1}{z}\right) = 3\left(\frac{5}{2}\right) \Rightarrow 2\left(\frac{1}{x}+\frac{1}{y}+\frac{1}{z}\right) = 10$

10. (*) Relative speed = 42 ÷ 6 = 7 km/h $\Rightarrow$ Speed of Romita = 7 – 4 = 3 km/hr

11. (*b*) Diagonal = $\sqrt{2}$ × side $\Rightarrow \sqrt{2} \times \frac{1}{2}(x+1) = \frac{3-x}{\sqrt{2}} \Rightarrow x + 1 = 3 - x \Rightarrow x = 1$

12. (*a*) Put some values and check: Le $a = 3$, $b = 4$, $c = 5$, then $s = (3 + 4 + 5) \div 2 = 6$

$s(s - c) + (s - a)(s - b) = 6 \times 1 + 3 \times 2 = 12 = 3 \times 4 = ab$

13. (*d*) $(p + m)^3 = 6^3 \Rightarrow p^3 + m^3 + 3pm(p + m) = 216 \Rightarrow 72 + 3pm(6) = 216 \Rightarrow 18pm = 144 \Rightarrow pm = 8$

14. (*d*) $x^m \times x^n = 1 \Rightarrow x^{m+n} = 1 = x^0 \Rightarrow m + n = 0 \Rightarrow m = -n$

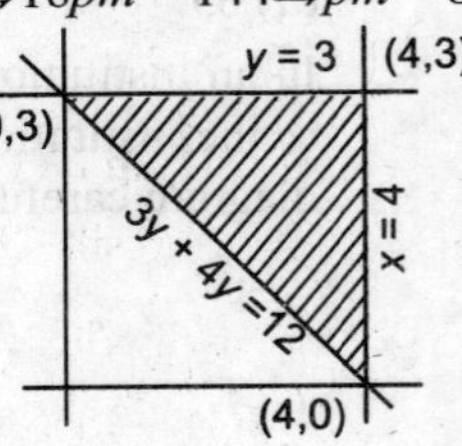

15. (*b*) Area = $\frac{1}{2} \times 4 \times 3 = 6$

16. (*d*) $\frac{p^2-2p+1}{2p} = 4 \Rightarrow \frac{p^2-2p+1}{p} = 8 \Rightarrow p - 2 + \frac{1}{p} = 8 \Rightarrow p + \frac{1}{p} = 10$

17. (*b*) $K^2 = 2K - 1 \Rightarrow K^2 - 2K + 1 = 0 \Rightarrow (K - 1)^2 = 0 \Rightarrow K = 1$

18. (*d*) $AD : DB = 5 : 4$

$\Rightarrow AD : (AD + DB) = 5 : 5 + 4 = 5 : 9$

$\Rightarrow DE : BC = 5 : 9$

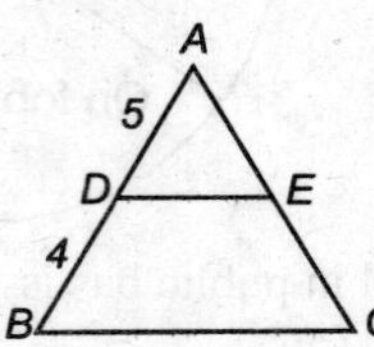

19. (*d*) $17^2 - 8^2 = 15^2$

$25^2 - 15^2 = 625 - 225 = 400 = 20^2$

$\Rightarrow$ PS = 2 × 20 = 40 cm

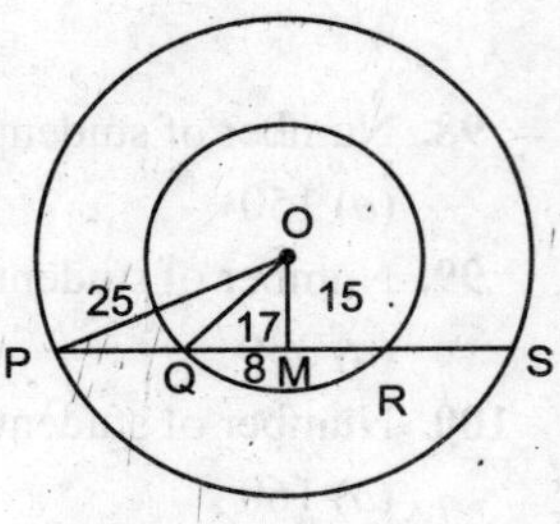

20. (*a*) $\angle CBA = 90° - 34° = 56°$

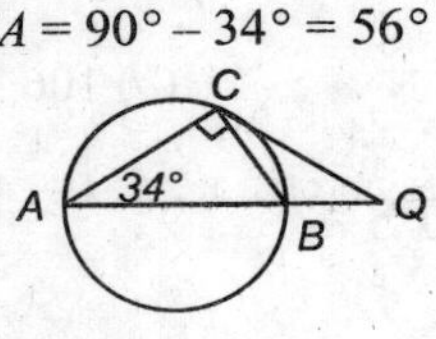

21. (*a*) $\frac{\text{Side}}{2\sqrt{3}} : \frac{\text{Side}}{\sqrt{3}} = 1 : 2$

22. **(*c*)** $\frac{1}{2}ab = 20 \Rightarrow ab = 40$ $\sqrt{a^2+b^2} = 10 \Rightarrow a^2 + b^2 = 100 \Rightarrow (a+b)^2 = 100 + 80 = 180$

23. **(*c*)** $2\pi r = 44 \Rightarrow r = 44 \times \frac{1}{2} \times \frac{7}{22} = 7 \Rightarrow$ **Area = Inradius × semi-perimeter** $= 7 \times 12 = 84$

24. **(*b*)** $\frac{1}{2} \times \sin\theta \times a^2 = \frac{1}{2} \times \frac{1}{\sqrt{2}} \times 10^2 = 25\sqrt{2}$

25. **(*b*)** $(180° - 48°) \div 2 = 66°$

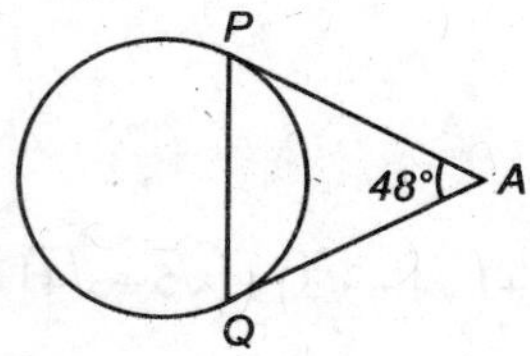

26. **(*b*)** Diagonal $= \sqrt{4^2+3^2} = 5$ m

27. **(*a*)** Ratio of sides $= 3 : 1\frac{1}{4} : 3\frac{1}{4} = 3 : \frac{5}{4} : \frac{13}{4} = 12 : 5 : 13$ $\because 12^2 + 5^2 = 13^2 \Rightarrow$ **It is right triangle.**

28. **(*a*)** If we put $\theta = 60°$, then $LHS = 2 \times \frac{3}{4} = \frac{3}{2}$, $RHS = 3 \times \frac{1}{2} = \frac{3}{2} \Rightarrow LHS = RHS$

29. **(*d*)** $a^2 + b^2 + c^2 = ab + bc + ca \Rightarrow a = b = c \Rightarrow$ **Equilateral triangle** $\Rightarrow \angle A = \angle B = \angle C = 60°$

$\Rightarrow (\text{Sin}^2 A + \text{Sin}^2 B + \text{Sin}^2 C) = 3\,\text{Sin}^2 60° = 3 \times \frac{3}{4} = \frac{9}{4}$

30. **(*b*)** $(a\sin\theta + b\cos\theta)^2 + (a\cos\theta - b\sin\theta)^2$

$= a^2\sin^2\theta + b^2\cos^2\theta + 2ab\sin\theta\cos\theta + a^2\cos^2\theta + b^2\sin^2\theta - 2ab\sin\theta\cos\theta$

$= a^2(\sin^2\theta + \cos^2\theta) + b^2(\sin^2\theta + \cos^2\theta) = a^2 + b^2$

$\therefore c^2 + (a\cos\theta - b\sin\theta)^2 = a^2 + b^2 \Rightarrow (a\cos\theta - b\sin\theta) = \pm\sqrt{a^2+b^2-c^2}$

31. **(*b*)** $\sin\theta + \cos\theta = \sqrt{2}\cos\theta \Rightarrow \cot\theta = \frac{1}{\sqrt{2}-1} = \sqrt{2}+1$

32. **(*a*)** $3(\sec^2\theta + \tan^2\theta) = 5 \Rightarrow 3(1 + \tan^2\theta + \tan^2\theta) = 5 \Rightarrow 3(1 + 2\tan^2\theta) = 5$

$\Rightarrow 3 + 6\tan^2\theta = 5 \Rightarrow 6\tan^2\theta = 2 \Rightarrow \tan^2\theta = \frac{1}{3} \Rightarrow \tan\theta = \frac{1}{\sqrt{3}} = \tan 30° \Rightarrow \cos 2\theta = \cos 60° = \frac{1}{2}$

33. **(*b*)** $x\left(\frac{3}{4}\right)\sin 60° = 1 \times 2 \times \sin 60° \Rightarrow x = 2 \times \frac{4}{3} = \frac{8}{3} = 2\frac{2}{3}$

34. **(*c*)** G.E. $= \frac{1+\cot^2\alpha - 1 - \tan^2\alpha}{1+\cot^2\alpha + 1 + \tan^2\alpha} = \frac{\cot^2\alpha - \tan^2\alpha}{2+\cot^2\alpha+\tan^2\alpha} = \frac{\frac{1}{4}-4}{2+\frac{1}{4}+4} = \frac{1-16}{8+1+16} = \frac{-15}{25} = \frac{-3}{5}$

35. **(*b*)** $\sin(\theta + 30°) = \frac{3}{\sqrt{12}} = \frac{3}{2\sqrt{3}} = \frac{\sqrt{3}}{2} = \sin 60° \Rightarrow \theta = 30° \Rightarrow \cos^2\theta = \cos^2 30° = \left(\frac{\sqrt{3}}{2}\right)^2 = \frac{3}{4}$

36. **(*a*)** $\frac{\text{Perpendicular}}{\text{Base}} = \tan 30° = \frac{1}{\sqrt{3}}$

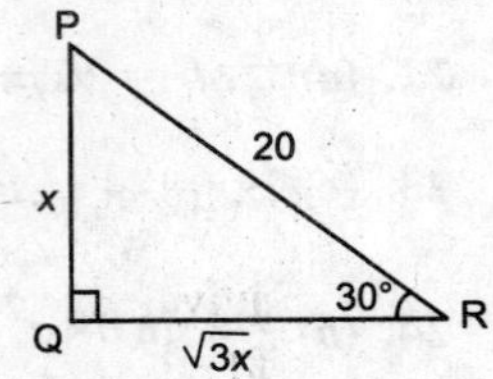

$(x)^2 + (\sqrt{3}\,x)^2 = 20^2$

$\Rightarrow 4x^2 = 400 \Rightarrow x = 10$

$\Rightarrow \text{Area} = \dfrac{10 \times 10\sqrt{3}}{2} = 50\sqrt{3}$

37. **(*a*)** G.E. $= (2\cos\theta - \sqrt{3})^2 = 0 \Rightarrow (2\cos\theta - \sqrt{3}) = 0 \Rightarrow \cos\theta = \dfrac{\sqrt{3}}{2} = \cos 30°$

38. **(*a*)** $\dfrac{x+y}{x-y} = \dfrac{3+2}{3-2} = \dfrac{5}{1}$

39. **(*d*)** $x : y = 30 : 50 = 3 : 5$

40. **(*a*)** G.E. $= (\sqrt{2}-1)+(\sqrt{3}-\sqrt{2})+(\sqrt{4}-\sqrt{3})+(\sqrt{5}-\sqrt{4})+\ldots\ldots+(\sqrt{9}-\sqrt{8}) = \sqrt{9}-1 = 3-1 = 2$

41. **(*c*)** $(x^2 - 25) = (x-25)^2 \Rightarrow x^2 - 25 = x^2 - 50x + 625 \Rightarrow 50x = 650 \Rightarrow x = 13$

42. **(*a*)** $190^2 = 36100,\ 191^2 = 36100 + (190 + 191) = 36481$

$\Rightarrow$ Required number = 36562 – 36481 = 81

43. **(*b*)** Square of $\pm\frac{1}{2}(\sqrt{3}+1) = \frac{1}{4}(3+1+2\sqrt{3}) = \dfrac{4+2\sqrt{3}}{4} = \dfrac{2+\sqrt{3}}{2}$

44. **(*a*)** $72 = 8 \times 9 \Rightarrow$ G.E. = 9

45. **(*d*)** $180 \times 2 = 360 \Rightarrow$ Short = 420 – 360 = 60 $\Rightarrow$ Cows = 60 ÷ 2 = 30

46. **(*d*)** $1 \times 2 \times 3 = 6$

47. **(*b*)** $(11111)^2 = 123454321$

48. **(*d*)** Required number of days $= \sqrt{8 \times \frac{9}{2}} = \sqrt{36} = 6$

49. **(*d*)** Regular wage rate = 2000 ÷ 50 = 40 $\Rightarrow$ Overtime rate = 40 × 1.5 = 60

Additional hours = (2300 – 2000) ÷ 60 = 5 hours

50. **(*c*)** $\dfrac{4 \times 8}{12}$ hours $= \dfrac{8}{3} \times 60$ minutes = 160 minutes

51. **(*a*)** $\dfrac{200 \times 50 \times 3}{100} = 300$ $\Rightarrow$ Additional men = 300 – 200 = 100

52. **(*a*)** Work per hour: X = 80/20 = 4; (X + Y) = 135/27 = 5 $\Rightarrow$ Y = 5 – 4 = 1 $\Rightarrow$ Time = 20/1 = 20

53. **(*c*)** Days taken by X = 6 × 4 = 24; Y = $12 \times \dfrac{4}{3} = 16 \Rightarrow$ (X + Y) $= \dfrac{24 \times 16}{40} = \dfrac{48}{5} = 9\dfrac{3}{5}$

54. **(*c*)** Note: $xy = yx \Rightarrow$ Change of order of multiplier will not affect the final result.

55. **(*d*)** 25 – 10% of 125 = 12.5%

56. **(*d*)** Net reduction $= -20 - 10 + \dfrac{20 \times 10}{100} = -28\%$

57. **(*a*)** CP = 120000 × 3/4 – 7500 = 82500 $\Rightarrow$ CP of each window 82500 ÷ 25 = ₹3300

58. **(*b*)** A : B = 3 : 2 and B : C = 3 : 5 $\Rightarrow$ A : B : C = 9 : 6 : 10

$\Rightarrow$ (A + B) : (B + C) = (9 + 6) : (6 + 10) = 15 : 16

59. **(*d*)** Let ratio of initial capitals of A : B = 4 : 2

Ratio of equivalent capitals of A and B = (4 × 10 + 3 × 2) : (2 × 8 + 1 × 4) = 46 : 20 = 23 : 10

60. **(*d*)** Total occupancy = $3\times\frac{4}{5}=\frac{12}{5}$ $\Rightarrow$ Total occupancy in two buses = $\frac{12}{5}\times\frac{3}{4}=\frac{9}{5}$

$\Rightarrow$ Occupancy per bus = $\frac{1}{2}\times\frac{9}{5}=\frac{9}{10}$

61. **(*a*)** Equivalent capitals = (3 × 2) : (5 × 2) : (5 × 1) = 6 : 10 : 5

62. **(*a*)** Old ratio = 7 : 2 = 14 : 4; New ratio = 15 : 4; Difference = 1 ratio = 1512 ÷ 14 = 108

63. **(*a*)** $(7x)\times(9x) = 1575 \Rightarrow x^2 = 1575 \div 63 = 25 \Rightarrow x = 5 \Rightarrow 9x = 45$

64. **(*b*)** Excess weight of 12th person = 33 kg.

This will increase average weight of other 11 persons by 33 ÷ 11 = 3 kg.

Average of all = 33 + 3 = 36 $\Rightarrow$ Weight of 12th person = 95 + 36 = 131 kg

65. **(*a*)** Sum of numbers = 420 + 204 = 624 $\Rightarrow$ Average = 624 ÷ 2 = 312

66. **(*a*)** 3.95 + 0.55 + 0.10 = 4.60

67. **(*a*)** Average of all = Average of 4 and 5 terms = 93 $\Rightarrow$ 5th term = 94 $\Rightarrow$ 8th term = 94 + 6 = 100

68. **(*b*)** $3a + 4b > 100$, and $a = 2b \Rightarrow 5a > 100 \Rightarrow a > 20$

69. **(*b*)** Average = $(3^{30} + 3^{60} + 3^{90}) \div 3 = 3^{29} + 3^{59} + 3^{89}$

70. **(*b*)** Let CP = 100, then SP = 120; New CP = 90 $\Rightarrow$ New SP = 120% of 90 = 108

$\Rightarrow$ Difference = 120 – 108 = 12 $\Rightarrow$ CP = $100 \times \frac{30}{12}$ = ₹250

71. **(*a*)** Half of SP = 80% of CP $\Rightarrow$ SP = 2 × 80% = 160% of CP $\Rightarrow$ New profit = 60% of CP

72. **(*d*)** Average CP = $320 \times \frac{5}{6}=\frac{800}{3}$

$\Rightarrow$ Required Ratio = $\left(280-\frac{800}{3}\right):\left(\frac{800}{3}-180\right)$ = 40 : 260 = 2 : 13

73. **(*b*)** P = 112% of CP; Q = 96% of CP $\Rightarrow$ Q : P = 96 : 112 = 6 : 7

74. **(*d*)** $P=\frac{6}{5}\times\frac{3}{4}\times x \Rightarrow x = P\times\frac{5}{6}\times\frac{4}{3}=\frac{10}{9}P$

75. **(*a*)** Gain = 30 – 25 = 5 $\Rightarrow$ Gain percent = $\frac{5}{25}\times100 = 20\%$

76. **(*b*)** $(1.2)^3 = 1.728 \Rightarrow$ Overall increase = 72.8%

77. **(*a*)** $\frac{100}{140}\times35 = 25$

78. **(*b*)** Greatest possible original price = $270\times\frac{100}{75}=360$

79. **(*b*)** Profit = 11 – 10 = 1 on 10 = 10%

80. **(*d*)** Downstream speed : Upstream speed = 2 : 1 $\Rightarrow$ Required ratio = (2 + 1) : (2 – 1) = 3 : 1

81. **(*d*)** $\frac{16\times4}{12}=\frac{16}{3}=5\frac{1}{3}$

82. **(*a*)** Option (*a*) : $\frac{16}{4}+\frac{61-16}{9}=4+5=9$

83. **(*b*)** Speed Upstream = $\frac{36}{6}$ = 6; Downstream = $\frac{48}{6}$ = 8 $\Rightarrow$ Speed of current = $\frac{8-6}{2}$ = 1 km/hr

84. (*b*) Let Principal = 100 $\Rightarrow$ Amount after 1 year = 110 $\Rightarrow$ Amount after 2 year = 121

Interest for second year = 121 – 110 = 11 $\Rightarrow$ Principal = $132 \times \frac{100}{11} = 1200$

85. (*c*) $\frac{1 \times 100}{5} \times 365 = ₹7300$

86. (*b*) $S\left(1+\frac{2r}{100}\right)^3 = S\left(1+\frac{r}{50}\right)^3$

87. (*a*) $₹2420 \times \frac{10}{11} \times \frac{10}{11} = ₹2000$

88. (*a*) $\pi r^2 + 2\pi rh + \pi rl = 9\pi + 24\pi + 15\pi = 48\pi$

89. (*a*) Given, AB = 3 cm, BH = 6 cm, DF = FG = 1 cm

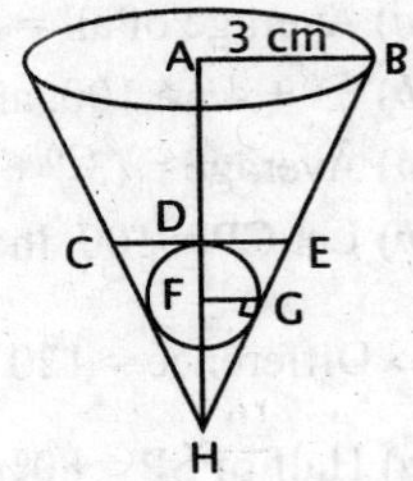

$AH^2 = BH^2 - AB^2 = 6^2 - 3^2 = 27 \Rightarrow AH = 3\sqrt{3}$

Δ ABH ~ Δ GFH

$\Rightarrow \frac{AB}{GF} = \frac{BH}{FH} \Rightarrow \frac{3}{1} = \frac{6}{FH} \Rightarrow FH = 2cm$

DH = DF + FH = 1 + 2 = 3 cm

Δ ABH ~ Δ DEH

$\frac{AB}{DE} = \frac{AH}{DH} \Rightarrow \frac{3}{DE} = \frac{3\sqrt{3}}{3} \Rightarrow DE = \sqrt{3}$

Volume of cone (HCE) = $\frac{1}{3}\pi\left(\sqrt{3}\right)^2(3) = 3\pi$

Volume of ball = $\frac{4}{3}\pi r^3 = \frac{4}{3}\pi(1)^3 = \frac{4}{3}\pi$

Required volume = $3\pi - \frac{4}{3}\pi = \frac{5}{3}\pi$

90. (*a*) Area = $\frac{50 \times 6}{2} = 150$

91. (*b*) $\sin 15° = \sin(45° - 30°)$

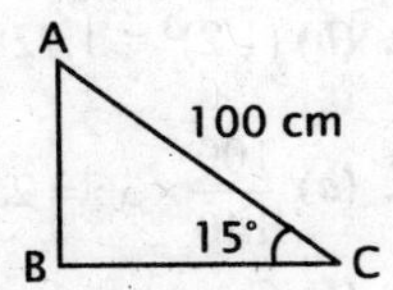

$= \sin 45° \cos 30° - \cos 45° \sin 30°$

$= \frac{1}{\sqrt{2}} \times \frac{\sqrt{3}}{2} - \frac{1}{\sqrt{2}} \times \frac{1}{2} = \frac{1}{\sqrt{2}}\left(\frac{\sqrt{3}}{2} - \frac{1}{2}\right) = \frac{\sqrt{3}-1}{2\sqrt{2}}$

$\cos 15° = \cos(45° - 30°)$

$= \cos 45° \cos 30° + \sin 45° \sin 30°$

$= \frac{1}{\sqrt{2}} \times \frac{\sqrt{3}}{2} + \frac{1}{\sqrt{2}} \times \frac{1}{2} = \frac{\sqrt{3}+1}{2\sqrt{2}}$

$\sin 15° = \frac{AB}{AC} \Rightarrow \frac{\sqrt{3}-1}{2\sqrt{2}} = \frac{AB}{100}$

$= AB = \frac{\sqrt{3}-1}{2\sqrt{2}} \times 100$

$$\cos 15° = \frac{BC}{AC} \Rightarrow \frac{\sqrt{3}+1}{2\sqrt{2}} = \frac{BC}{100}$$

$$\Rightarrow BC = \frac{\sqrt{3}+1}{2\sqrt{2}} \times 100$$

$$\text{Area } (\Delta ABC) = \frac{1}{2} \times AB \times BC$$

$$= \frac{1}{2} \times \frac{\sqrt{3}-1}{2\sqrt{2}} \times \frac{\sqrt{3}+1}{2\sqrt{2}} \times 100^2$$

$$= \frac{1}{2} \times \frac{3-1}{8} \times 100^2 = 1250$$

92. **(*b*)** Volume of water $= \frac{4}{3}\pi r^3 = \frac{4}{3}\pi\left(\frac{1}{20}\right)^3 \times 32000 = \frac{16}{3}\pi$

Volume of conical glass $= \frac{1}{3}\pi r^2 h = \frac{1}{3}\pi (r)^2 (2r) = \frac{2}{3}\pi r^3 = \frac{16}{3}\pi$

$$\Rightarrow r^3 = \frac{16}{3} \times \frac{3}{2} = 8 \Rightarrow r = 2 \Rightarrow h = 2r = 4$$

93. **(*b*)** $h = 4 \times 2\pi r = 8\pi r \Rightarrow \text{Volume} = \pi r^2 h = \pi(r)^2(8\pi r) = 8\pi^2 r^3 = \frac{(2\pi r)^3}{\pi} = \frac{c^3}{\pi}$

94. **(*c*)** Angle ratio = 2 : 7 : 2 : 7 ⇒ Opposite angles are equal ⇒ parallelogram

95. **(*b*)** Formula for area of isosceles triangle

96. **(*c*)** 100 : 70 = 10:7

97. **(*b*)** $(80 + 80 + 75 + 65 + 65) \div 5 = 365 \div 5 = 73$

98. **(*b*)** $\frac{54}{360} \times 800 = 120$

99. **(*d*)** $800 - \frac{216}{360} \times 800 = 800 - 480 = 320$

100. **(*a*)** $360° (216° + 54 + 18°) = 72$

$$\frac{72}{360} \times 800 = 160$$

CGL TIER I 2015 (HELD ON 9/8/2015

Study the following bar diagran carefully and answer the following four questions:

The number of the production of electronic items (TVs and LCDs) in a factory during the period from 2009 to 2013.

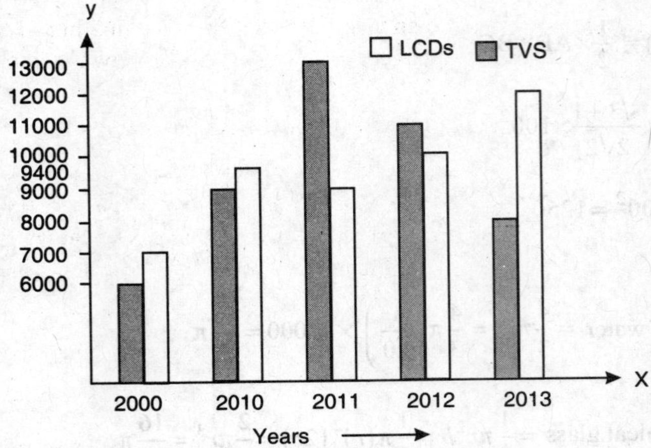

1. The ratio of production of TVs in the year 2009 and 2010 is
   (*a*) 6 : 7 (*b*) 2 : 3 (*c*) 7 : 6 (*d*) 3 : 2
2. The total number of production of electronic items is maximum in the year
   (*a*) 2013 (*b*) 2011 (*c*) 2010 (*d*) 2009
3. The difference between averages of production of TVs and LCDs from 2009 to 2012 is
   (*a*) 700 (*b*) 800 (*c*) 900 (*d*) 600
4. The ratio of production of LCDs in the year 2011 and 2013 is
   (*a*) 3 : 4 (*b*) 2 : 3 (*c*) 4 : 3 (*d*) 1 : 4

The following pie-chart shows the sources of funds to be collected by the National Highways Authority of India (NHAI) for its Phase II projects. Study the pie-chart and answer the following three questions:

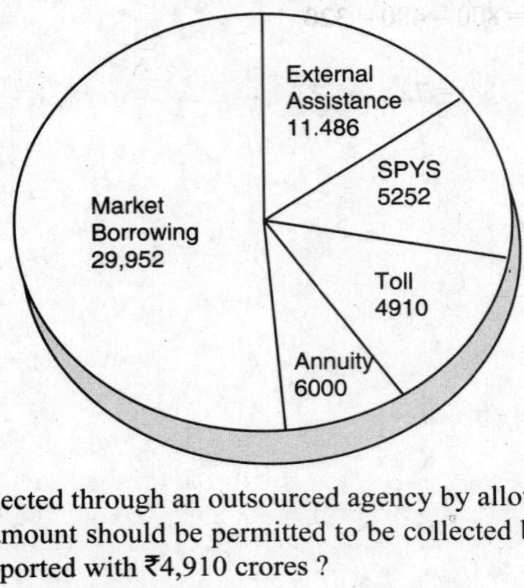

5. If the toll is to be collected through an outsourced agency by allowing a maximum 10% commission, how much amount should be permitted to be collected by the outsourced agency so that the project is supported with ₹4,910 crores ?
   (*a*) ₹ 5,401 crores (*b*) ₹ 5,827 crores (*c*) ₹ 5,361 crores (*d*) ₹ 6,213 crores

6. If NHAI could receive a total of ₹ 9,695 crores as External Assistance, by what percent (approximately) should it increase the Market Borrowing to arrange for the shortage of funds?
(a) 7.5% (b) 6% (c) 4.5% (d) 8%

7. The central angle corresponding to Market Borrowings is
(a) 52° (b) 137.8° (c) 192.4° (d) 187.2°

8. Walking at the rate of 4 kmph a man covers certain distance in 2 hrs 45 min. Running at the speed of 16.5 kmph the man will cover the same distance in how many minutes ?
(a) 45 min (b) 35 min (c) 40 min (d) 50 min

9. The sides of a triangle having area 7776 sq. cm are in the ratio 3 : 4 : 5. The perimeter of the triangle is :
(a) 412 cm (b) 432 cm (c) 400 cm (d) 424 cm

10. A and B together can do a piece of work in 6 days. If A can alone do the work in 18 days, then the number of days required for B to finish the work is :
(a) 9 (b) 12 (c) 15 (d) 20

11. $2x - ky + 7 = 0$ and $6x - 12y + 15 = 0$ has no solution for
(a) $k = 1$ (b) $k = 4$ (c) $k = -4$ (d) $k = -1$

12. If $x = 332$, $y = 333$ and $z = 335$, then the value of $x^3 + y^3 + z^3 - 3xyz$ is
(a) 8000 (b) 10000 (c) 9000 (d) 7000

13. 729 ml mixture of a mixture contains milk and water in the ratio 7 : 2. How much more water is to added to get a new mixture containing milk and water in the ratio 7 : 3 ?
(a) 60 ml (b) 52 ml (c) 71 ml (d) 81 ml

14. The sum of four numbers is 48. When 5 and 1 are added to the first two; and 3 and 7 subtracted from the third and the fourth, the numbers will be equal. The numbers are
(a) 6, 10, 14, 18 (b) 5, 11, 13, 19 (c) 9, 7, 15, 17 (d) 4, 12, 12, 20

15. 18 years ago, the ratio of A's age to B's age was 8 : 13. Their present ratio are 5 : 7. What is the present age of A ?
(a) 60 years (b) 40 years (c) 70 years (d) 50 years

16. A shopkeeper bought 30 kg of rice at the rate of ₹70 per kg and 20 kg of rice at the rate of ₹70.75 per kg. If he mixed the two brand of rice and sold the mixture at ₹80.50 per kg, his gain is
(a) ₹450 (b) ₹485 (c) ₹510 (d) ₹525

17. A ship after sailing 12 km towards south from a particular place covered 5 km more towards east. Then the straightway distance of the ship from that place is
(a) 11 km (b) 15 km (c) 13 km (d) 18 km

18. If $\sin\theta + \sin^2\theta = 1$, then $\cos^2\theta + \cos^4\theta$ is equal to **(SSC CGL Tier I 2014)**
(a) 1 (b) $\frac{\sin\theta}{\cos^2\theta}$ (c) $\frac{\cos^2\theta}{\sin\theta}$ (d) None of these

19. In ΔABC, a line through A cuts the side BC at D such that BD : DC = 4 : 5. If the area of $\Delta ABD = 60\ cm^2$, then the area of ΔADC is
(a) 60 cm^2 (b) 75 cm^2 (c) 90 cm^2 (d) 50 cm^2

20. If a clock started at noon, then the angle (in degrees) turned by the hour hand at 3 : 45 PM is
(a) $97\frac{1}{2}$ (b) $112\frac{1}{2}$ (c) $117\frac{1}{2}$ (d) $104\frac{1}{2}$

21. The length of the portion of the straight line $3x + 4y = 12$ incercepted between the axes is
(a) 5 (b) 7 (c) 3 (d) 4

22. Allowing 20% and 15% successive discounts, the selling price of an article beomes ₹3060, then the marked price will be

(*a*) ₹4500 (*b*) ₹4400 (*c*) ₹5000 (*d*) ₹4000

23. Choose the incorrect relation(s) from the following :

(*i*) $\sqrt{6}+\sqrt{2}=\sqrt{5}+\sqrt{3}$ (*ii*) $\sqrt{6}+\sqrt{2}<\sqrt{5}+\sqrt{3}$ (*iii*) $\sqrt{6}+\sqrt{2}>\sqrt{5}+\sqrt{3}$

(*a*) (*i*) (*b*) (*ii*) (*c*) (*i*) and (*iii*) (*d*) (*ii*) and (*iii*)

24. If $2+x\sqrt{3}=\dfrac{1}{2+\sqrt{3}}$, then the simplest value of x is

(*a*) 1 (*b*) 2 (*c*) – 1 (*d*) – 2

25. The average weight of 15 oarsmen in a boat is increased by 1.6 kg when one of the crew, who weighs 42 kg is replaced by a new man. Find the weight of the new man (in kg.)

(*a*) 65 (*b*) 67 (*c*) 43 (*d*) 66

26. A tangent is drawn to a circle of radius 6 cm from a point situated at a distance of 10 cm from the centre of the circle. The length of the tangent will be

(*a*) 5 cm (*b*) 7 cm (*c*) 8 cm (*d*) 4 cm

27. If $m=-4$, $n=-2$, then the value of $m^3-3m^2+3m+3n+3n^2+n^3$ is

(*a*) –126 (*b*) –124 (*c*) 124 (*d*) 126

28. A 10 m long ladder is placed against a wall. It is inclined at an angle of 30° to the ground. The distance (in m) of the foot of the ladder from the wall is (Given $\sqrt{3}=1.732$)

(*a*) 7.32 (*b*) 8.66 (*c*) 8.16 (*d*) 8.26

29. A tap can empty a tank in 30 minutes. A second tap can empty it in 45 minutes. If both the taps operate simultaneously, how much time is needed to empty the tank?

(*a*) 30 minutes (*b*) 18 minutes (*c*) 14 minutes (*d*) 15 minutes

30. If the area of a circle is A, radius of the circle is r and circumference of it is C, then

(*a*) $\dfrac{C}{A}=\dfrac{r}{2}$ (*b*) $AC=\dfrac{r^2}{2}$ (*c*) $\dfrac{A}{r}=C$ (*d*) $rC=2A$

31. Find the value of tan 4° tan 43° tan 47° tan 86°

(*a*) 1 (*b*) 2 (*c*) $\dfrac{1}{2}$ (*d*) $\dfrac{2}{3}$

32. If $x\cos\theta-\sin\theta=1$, then $x^2+(1+x^2)\sin\theta$ is equal to

(*a*) –1 (*b*) 2 (*c*) 1 (*d*) 0

33. If $\dfrac{m-a^2}{b^2+c^2}+\dfrac{m-b^2}{c^2+a^2}+\dfrac{m-c^2}{a^2+b^2}=3$, then the value of m is

(*a*) $a^2-b^2-c^2$ (*b*) $a^2+b^2+c^2$ (*c*) $a^2+b^2-c^2$ (*d*) $a^2-b^2+c^2$

34. The least number that should be added to 2055, so that the sum is exactly divisible by 27.

(*a*) 27 (*b*) 28 (*c*) 31 (*d*) 24

35. 10% discount and then 20% discount in succession is equivalent to total discount of

(*a*) 30% (*b*) 28% (*c*) 24% (*d*) 15%

36. Two chords of length a unit and b unit of a circle make angles 60° and 90° at the centre of a circle respectively, then the correct relation is

(*a*) $b=2a$ (*b*) $b=\dfrac{3}{2}a$ (*c*) $b=\sqrt{2}a$ (*d*) $b=\sqrt{3}a$

37. The measure of an angle whose supplement is three times as large as its complement, is

(*a*) 45° (*b*) 75° (*c*) 60° (*d*) 30°

38. A farmer travelled a distance of 61 km in 9 hours. He travelled partly on foot at the rate of 4 km/hr and partly at bicycle at the rate of 9 km/hr. The distance travelled on foot is

(*a*) 17 km (*b*) 16 km (*c*) 14 km (*d*) 15 km

39. In certain years a sum of money is doubled itself at $6\frac{1}{4}\%$ simple interest per annum, the required time will be

(*a*) $12\frac{1}{2}$ years (*b*) 16 years (*c*) 8 years (*d*) $10\frac{2}{3}$ years

40. The perimeter of one face of a cube is 20 cm. Its volume will be

(*a*) 125 cm^3 (*b*) 400 cm^3 (*c*) 625 cm^3 (*d*) 100 cm^3

41. What is the Arithmetic mean of the first n natural numbers ?

(*a*) $\frac{n^2(n+1)}{2}$ (*b*) $\frac{n+1}{2}$ (*c*) $\frac{n(n+1)}{2}$ (*d*) $2(n+1)$

42. A square is inscribed in a quarter-circle in such a manner that two of its adjacent vertices lie on the two radii at an equal distance from the centre, while the other two vertices lie on the circular arc. If the square has sides of length x, then the radius of the circle is

(*a*) $\sqrt{2}x$ (*b*) $\frac{2x}{\sqrt{\pi}}$ (*c*) $\frac{16x}{\pi+4}$ (*d*) $\frac{\sqrt{5}x}{\sqrt{2}}$

43. Two poles of height 7 m and 12 m stand on a plane ground. If the distance between their feet is 12 m, the distance between their top will be

(*a*) 13 m (*b*) 19 m (*c*) 17 m (*d*) 15 m

44. In a parallelogram PQRS, angle P is four times of angle Q, then the measure of ∠R is

(*a*) 72° (*b*) 144° (*c*) 36° (*d*) 130°

45. The population of a town increases by 5% every year. If the present population is 9261, the population 3 years ago was

(*a*) 7500 (*b*) 5700 (*c*) 6000 (*d*) 8000

46. The numerical value of $\frac{\cos^2 45^\circ}{\sin^2 60^\circ}+\frac{\cos^2 60^\circ}{\sin^2 45^\circ}-\frac{\tan^2 30^\circ}{\cot^2 45^\circ}-\frac{\sin^2 30^\circ}{\cot^2 30^\circ}$

(*a*) $\frac{3}{4}$ (*b*) $1\frac{1}{4}$ (*c*) $\frac{1}{4}$ (*d*) $\frac{1}{2}$

47. The value of $\frac{1}{\sqrt{7}-\sqrt{6}}-\frac{1}{\sqrt{6}-\sqrt{5}}+\frac{1}{\sqrt{5}-2}-\frac{1}{\sqrt{8}-\sqrt{7}}+\frac{1}{3-\sqrt{8}}$

(*a*) 5 (*b*) 1 (*c*) 7 (*d*) 0

48. The maximum value of $\sin^4\theta+\cos^4\theta$ is

(*a*) 1 (*b*) 3 (*c*) $\frac{1}{3}$ (*d*) 2

49. A pipe can fill a tank in x hours and another can empty it in y hours. They can together fill it in ($y > x$)

(*a*) $y-x$ (*b*) $\frac{xy}{x-y}$ (*c*) $\frac{xy}{y-x}$ (*d*) $x-y$

50. The marked price of a watch was ₹720. A man bought the same for ₹550.80 after getting two successive discounts, the first being 10%. The second discount rate is

(*a*) 12% (*b*) 18% (*c*) 15% (*d*) 14%

ANSWERS AND HINTS

1. (*b*) 600 : 9000 = 2 : 3
2. (*b*) Year 2011
3. (*c*) Total TVs = 6 + 9 + 13 + 11 = 39,000; Total LCDs = 7 + 9.4 + 9 + 10 = 35,400

   $\Rightarrow$ Difference = 39,000 – 35,400 = 3,600 $\Rightarrow$ Average = 3600 ÷ 4 = 900
4. (*a*) 9000 : 12000 = 3 : 4
5. (*a*) 4910 × 110% = 5401 crores
6. (*b*) Shortage = 11486 – 9695 = 1791 $\Rightarrow \% = \frac{1791}{29952} \times 100 = 6\%$ (approx)
7. (*d*) $\frac{29952}{57600} \times 360° = 187.2\%$
8. (*c*) Distance $= 4 \times 2\frac{3}{4} = 4 \times \frac{11}{4} = 11$ km $\Rightarrow$ Required time $= \frac{11}{16.5} \times 60 = \frac{2}{3} \times 60 = 40$ minutes
9. (*b*) $3^2 + 4^2 = 5^2 \Rightarrow$ The triangle is right triangle. $\Rightarrow \frac{1}{2} \times (3x) \times (4x) = 7776$

   $\Rightarrow 6x^2 = 7776 \Rightarrow x^2 = 1296 \Rightarrow x = 36$

   Perimeter $= 3x + 4x + 5x = (3 + 4 + 5) \times 36 = 432$
10. (*a*) $\frac{6 \times 18}{18 - 6} = 9$
11. (*b*) The equations have no solution, $\frac{2}{6} = \frac{-k}{-12} \neq \frac{7}{15} \Rightarrow \frac{1}{3} = \frac{k}{12} \Rightarrow 3k = 12 \Rightarrow k = 4$
12. (*d*) $x^3 + y^3 + z^3 - 3xyz = \frac{1}{2}(x + y + z)[(x - y)^2 + (y - z)^2 + (z - x)^2]$

    $= \frac{1}{2} \times (332 + 333 + 335) \times (1^2 + 2^2 + 3^2) = \frac{1}{2} \times 1000 \times 14 = 7000$
13. (*d*) Sum of ratios in the original mixture = 7 + 2 = 9 ratios

    $\therefore$ 9 ratio = 729 ml $\Rightarrow$ 1 ratio = 729 ÷ 9 = 81 ml

    Change in ratio of water = 3 – 2 = 1 ratio $\Rightarrow$ Water to be added = 1 ratio = 81 ml
14. (*a*) Choice (a) satisfies the given conditions.
15. (*d*) $\frac{(7-5) \times 18}{65 - 56} = 4 \Rightarrow$ A's present age $= 8x + 18 = 8 \times 4 + 18 = 50$ years
16. (*c*) Average cost price (per kg) $= 70 + \frac{0.75 \times 20}{50} = 70.30$

    SP (per kg) = 80.50 $\Rightarrow$ Gain (per kg) = 80.50 – 70.30 = 10.20

    Total gain = 50 × 10.20 = 510

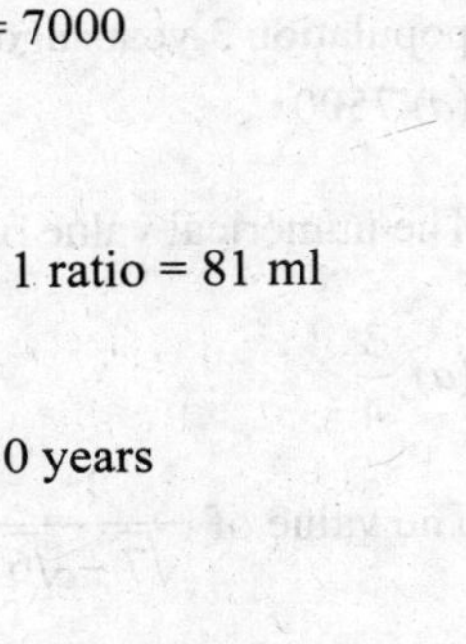

17. (*c*) Required distance $= \sqrt{12^2 + 5^2} = 13$ km
18. (*a*) $\sin\theta = 1 - \sin^2\theta = \cos^2\theta \Rightarrow \cos^2\theta + \cos^4\theta = \cos^2\theta + \sin^2\theta = 1$

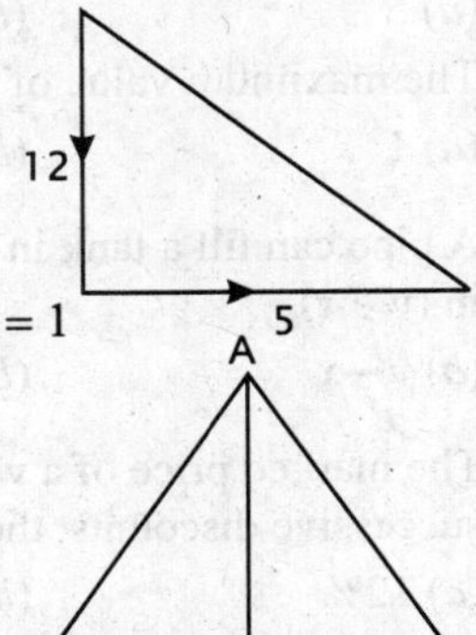

19. (*b*) Required area $= \frac{60}{4} \times 5 = 75$ cm^2

20. (*b*) Angle turned by hour hand in 1 hour = 360° ÷ 12 = 30°

$\Rightarrow$ Required angle = $3\frac{45}{60}\times 30 = 3\frac{3}{4}\times 30 = 90+\frac{45}{2} = 112\frac{1}{2}$

21. (*a*) *x*-incercept = 12 ÷ 3 = 4 and *y*-incercept = 12 ÷ 4 = 3

$\Rightarrow$ Required length = $\sqrt{4^2+3^2}$ = 5 units

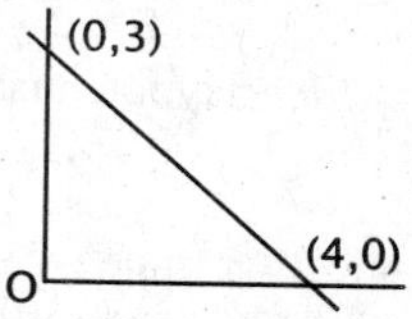

22. (*a*) Marked price = $3060\times\frac{100}{80}\times\frac{100}{85} = 4500$

23. (*c*) Squaring both sides, we get: $6+2+2\sqrt{12} < 5+3+2\sqrt{15}$ $\left(\because \sqrt{12} < \sqrt{15}\right)$

24. (*c*) $2+x\sqrt{3} = \frac{1}{2+\sqrt{3}} = 2-\sqrt{3} \Rightarrow x = -1$

25. (*d*) Increase in weight = 15 × 1.6 = 24 kg $\Rightarrow$ Weight of new man = 42 + 24 = 66 kg

26. (*c*) Length of tangent = $\sqrt{10^2-6^2}$ = 8 cm

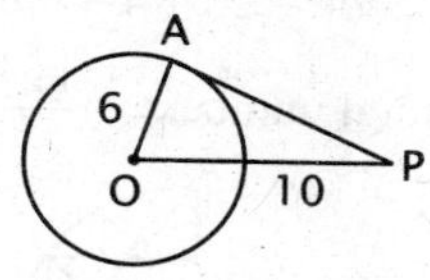

27. (*a*) G.E. = $(-4)^3 - 3(-4)^2 + 3(-4) + 3(-2) + 3(-2)^2 + (-2)^3 = -64 - 48 - 12 - 6 + 12 - 8 = -126$

28. (*b*) $\frac{x}{10} = \cos 30^\circ = \frac{\sqrt{3}}{2} \Rightarrow x = \frac{\sqrt{3}}{2}\times 10 = 5\sqrt{3}$ = 5 × 1.732 = 8.66

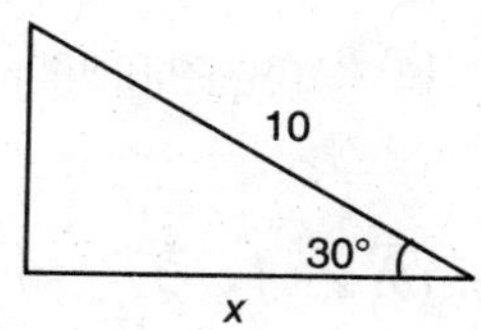

29. (*b*) $\frac{30\times 45}{30+45} = 18$

30. (*d*) $\frac{C}{A} = \frac{2\pi r}{\pi r^2} = \frac{2}{r} \Rightarrow rC = 2A$

31. (*a*) tan 4° tan 43° tan 47° tan 86° = tan 4° tan 43° cot (90° – 47°) cot (90° – 86°)
= tan 4° tan 43° cot 43° cot 4° = (tan 4° × cot 4°) × (tan 43° × cot 43°) = 1 × 1 = 1

32. (*c*) If we put $\theta = 0°$ in equation, $x\cos\theta - \sin\theta = 1$, we get $x = 1$

Now, we put this value in $x^2 + (1+x^2)\sin\theta$, we get $1^2 + (1+1^2)\times 0 = 1$

Alt: If we put $\theta = 30°$ in equation, $x\cos\theta - \sin\theta = 1$, we get

$x\times\frac{\sqrt{3}}{2} - \frac{1}{2} = 1 \Rightarrow \frac{x\sqrt{3}}{2} = \frac{3}{2} \Rightarrow x = \frac{3}{2}\times\frac{2}{\sqrt{3}} = \frac{1}{\sqrt{3}}$

Now, we put this value in $x^2 + (1+x^2)\sin\theta$, we get

$\left(\frac{1}{\sqrt{3}}\right)^2 + \left[1+\left(\frac{1}{\sqrt{3}}\right)^2\right]\times\frac{1}{2} = \frac{1}{3}+\frac{4}{3}\times\frac{1}{2} = \frac{1}{3}+\frac{2}{3} = 1$

33. (*b*) No need to solve. Option (*b*) satisfies this equation.

34. (*d*) Sum of digits of 2055 = 2 + 0 + 5 + 5 = 12 or 1 + 2 = 3

The number is divisible by 27 (i.e. a multiple of 9) if sum of digits is divisible by 9.

$\Rightarrow$ Sum of digits of numbers to be added should be 9 – 3 = 6 i.e. option (d) as 2 + 4 = 6

35. (*b*) Equivalent discount = $10 + 20 - \frac{10\times 20}{100} = 10 + 20 - 2 = 28\%$

36. **(c)** Circle I: Equilateral $\Delta \Rightarrow a = r$

Circle II: Right angle Δ

$\Rightarrow b = \text{Hypotenuse} = \sqrt{2}r = \sqrt{2}a$

Circle I

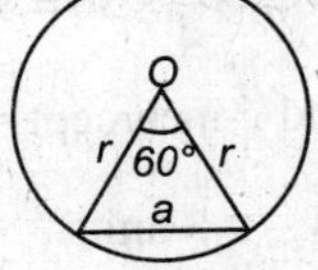

Circle II

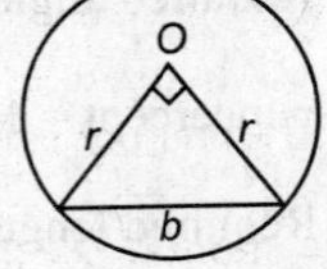

37. **(a)** $(180° - x) = 3(90° - x) \Rightarrow 180° - x = 270° - 3x \Rightarrow 2x = 90° \Rightarrow x = 45°$

38. **(b)** Out of the given choices, 16 is divisble by 4. Let's try this value.

$\frac{16}{4} + \frac{61-16}{9} = 4 + 5 =$ 9 hours. Satisfies the condition and hence the presumption is correct.

39. **(b)** $1 \times 100 \times \frac{4}{25} = 16$ years

40. **(a)** $4 \times \text{Side} = 20 \Rightarrow \text{Side} = 20 \div 4 = 5$ cm $\Rightarrow$ Volume $= a^3 = 5^3 = 125$

41. **(b)** Average of n natural numbers $= \frac{\text{First term + Last Term}}{2} = \frac{(n+1)}{2}$

42. **(a)** Diagonal $= \sqrt{2}x$

$\Rightarrow r = \sqrt{2}x$

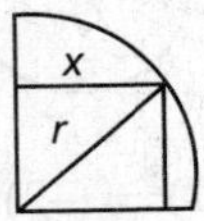

43. **(a)** Required distance $= \sqrt{12^2 + 5^2} = 13$ m

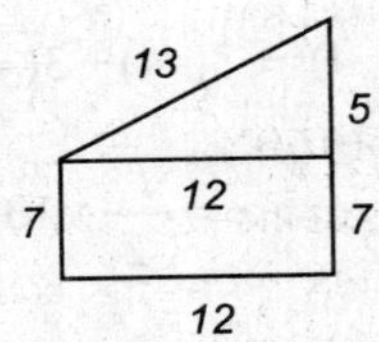

44. **(b)** $\angle P + \angle Q = 180° \Rightarrow 4x + x = 180° \Rightarrow x = 36° \Rightarrow \angle R = \angle P = 4x = 144°$

45. **(d)** Population 3 years ago $= 9261 \div \left(\frac{21}{20}\right)^3 = 9261 \times \frac{20}{21} \times \frac{20}{21} \times \frac{20}{21} = 8000$

46. **(a)** G.E. $= \frac{\frac{1}{2}}{\frac{3}{4}} + \frac{\frac{1}{4}}{\frac{1}{2}} - \frac{\frac{1}{3}}{1} - \frac{\frac{1}{4}}{3} = \frac{1}{2} \times \frac{4}{3} + \frac{1}{4} \times 2 - \frac{1}{3} - \frac{1}{4} \times \frac{1}{3} = \frac{2}{3} + \frac{1}{2} - \frac{1}{3} - \frac{1}{12} = \frac{1}{3} + \frac{5}{12} = \frac{9}{12} = \frac{3}{4}$

47. **(a)** G.E. $= \left(\sqrt{7} + \sqrt{6}\right) - \left(\sqrt{6} + \sqrt{5}\right) + \left(\sqrt{5} + 2\right) - \left(\sqrt{8} + \sqrt{7}\right) + \left(3 + \sqrt{8}\right) = 2 + 3 = 5$

48. **(a)** If $x < 1$, then $x > x^n$, where n is a natural number greater than 1.

The maximum value of $\sin^4 \theta + \cos^4 \theta$ is $= 1$

Alt: $\sin^4 \theta + \cos^4 \theta = \sin^4 \theta + (1 - \sin^2 \theta)^2 = 1 + 0 = 1$ (when $\theta = 90°$)

49. **(c)** Formula: $\frac{xy}{y - x}$

50. **(c)** First discount = ₹720 – 10% = ₹720 – ₹72 = ₹648

Second discount = ₹648 – ₹550.80 = ₹97.20

Rate of second discount $= \frac{97.20}{648} \times 100 = 15\%$

CGL CHSL 2015 (HELD ON 1/11/2015)

Directions: Read it carefully and answer Five questions.

Total cost of production of a firm is ₹250 lakhs. Following pie-chart shows the percentage of costs of production in different purposes.

Letters A, B, C, D represents the following :

A = Cost of raw materials

B = Cost of packing materials

C = Cost of labour

D = Maintenance cost

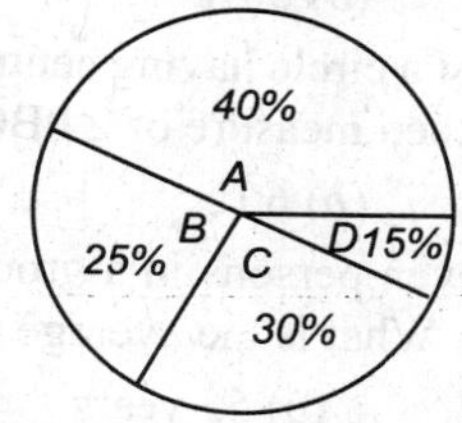

1. If the cost of production doubles in a period of 3 years, then the corresponding maintenance cost in rupees will be

(*a*) ₹75 lakhs (*b*) ₹25 lakhs (*c*) ₹12.5 lakhs (*d*) ₹125 lakhs

2. Cost of packing materials and raw materials together amounts to

(*a*) ₹175 lakhs (*b*) ₹13.75 lakhs (*c*) ₹8.75 lakhs (*d*) ₹162.5 lakhs

3. If the production increases to five times of the present cost, then the percentage increase of the cost is

(*a*) 400% (*b*) 50% (*c*) 300% (*d*) 500%

4. If the total maintenace cost increases from ₹12.5 lakhs to ₹50 lakhs, then the percentage increase of the maintenance cost is

(*a*) 75% (*b*) 300% (*c*) 25% (*d*) 400%

5. If the packing cost increases by 2%, then the new packing cost will be

(*a*) ₹ 5 lakhs (*b*) ₹ 63.75 lakhs (*c*) ₹ 1.25 lakhs (*d*) ₹ 62.5 lakhs

6. One-fifth of half of a number is 20. Then 20% of that number is

(*a*) 40 (*b*) 80 (*c*) 60 (*d*) 20

Directions: Study the following histogram and answer the four questions :

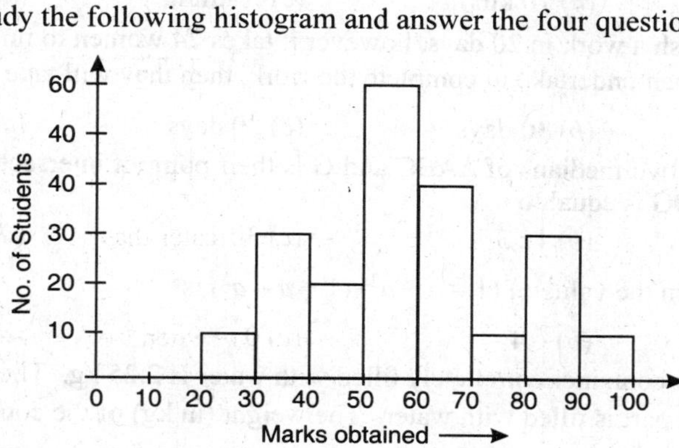

7. The percentage of students securing marks less than 50 is

(*a*) $28\frac{4}{7}\%$ (*b*) $27\frac{5}{7}\%$ (*c*) $23\frac{13}{21}\%$ (*d*) $47\frac{13}{21}\%$

8. The number of students securing marks in the range 90 – 100 is

(*a*) 40 (*b*) 30 (*c*) 20 (*d*) 10

9. The range of marks obtained by maximum number of students is

(*a*) 50 – 60 (*b*) 60 – 70 (*c*) 80 – 90 (*d*) 30 – 40

10. The total number of students on whom this survey was made is

(*a*) 190 (*b*) 210 (*c*) 200 (*d*) 220

11. A discount series of 10%, 20% and 40% is equal to a single discount of

(*a*) 50% (*b*) 60% (*c*) 60.28% (*d*) 56.8%

12. AB is a diameter of a circle having centre at O. P is a point on the circumference of the circle. If ∠POA = 120°, then measure of ∠PBO is

(*a*) 70° (*b*) 60° (*c*) 68° (*d*) 75°

13. Average age of seven persons in a group is 30 years, the average age of five persons of this group is 31 years. What is the average age of the other two persons in the group ?

(*a*) 26 years (*b*) 55 years (*c*) None of these (*d*) 15 years

14. The LCM of two numbers is 12 times their HCF. The sum of the HCF and LCM is 403. If one of the number is 93, then the other is

(*a*) 120 (*b*) 124 (*c*) 116 (*d*) 112

15. In 2 kg mixture of copper and alminium, 30% is copper. How much almunium powder should be added to the mixture so that the proportion of copper becomes 20% ?

(*a*) 1000 grams (*b*) 800 grams (*c*) 900 grams (*d*) 1200 grams

16. A man spends his three months income in four months time. If his monthly income is ₹1,000, then his annual savings is

(*a*) ₹9,000 (*b*) ₹4,000 (*c*) ₹6,000 (*d*) ₹3,000

17. If the angles of a triangle are in the ratio 2 : 3 : 5, then the measure of the least angle of the triangle is

(*a*) 90° (*b*) 36° (*c*) 20° (*d*) 18°

18. A man rows to a place 60 km distant and back in 13 hours 30 minutes. He finds that he can row 5 km with the stream in the same time as he can row 4 km against the stream. Find the rate of the stream.

(*a*) 8 km/hr (*b*) 10 km/hr (*c*) ½ km/hr (*d*) 1 km/hr

19. 15 men can finish a work in 20 days, however it takes 24 women to finish it in 20 days. If 10 men and 8 women undertake to complete the work, then they will take

(*a*) 10 days (*b*) 30 days (*c*) 20 days (*d*) 15 days

20. BE, CF are the two medians of ΔABC and G is their point of intersection. EF cuts AG at O. Ratio of AO : OG is equal to

(*a*) 2 : 3 (*b*) 1 : 3 (*c*) 3 : 1 (*d*) 1 : 2

21. If $a+\frac{1}{a}=1$, then the value of $(1-a+a^2)(1+a-a^2)$ is

(*a*) 1 (*b*) – 4 (*c*) 0 (*d*) 4

22. The weight of a container completely filled with water is 2.25 kg. The container weighs 0.77 kg when its 0.2 part is filled with water. The weight (in kg) of the container when 0.4 part of it is filled with water, is

(*a*) 0.74 (*b*) 0.40 (*c*) 1.14 (*d*) 1.88

23. The time required for a sum of money to amount to four times itself at 15% simple interest p.a. will be

(*a*) 15 years (*b*) 30 years (*c*) 20 years (*d*) 10 years

24. ABC is a triangle in which ∠A = 90°. Let P be any point on side AC. If BC = 10 cm, AC = 8 cm and BP = 9 cm, then AP =

(*a*) $2\sqrt{3}$ (*b*) $3\sqrt{5}$ (*c*) $3\sqrt{3}$ (*d*) $5\sqrt{5}$

25. The value of $\sec^4 A\,(1 - \sin^4 A) - 2\tan^2 A$ is

(*a*) 0 (*b*) 1 (*c*) 2 (*d*) ½

26. A man rows 12 km in 5 hours against the stream, the speed of current being 4 kmph. What time will be taken by him to row 15 km with the stream ?

(*a*) 1 hour $25\frac{7}{13}$ min (*b*) 1 hour $26\frac{7}{13}$ min (*c*) 1 hour $24\frac{7}{13}$ (*d*) 1 hour $27\frac{7}{13}$ min

27. Which of the following is true ?

(*a*) $\frac{7}{24} > \frac{1}{3} > \frac{3}{8} > \frac{5}{12}$ (*b*) $0.5 < \frac{2}{3} < \frac{3}{4} < \left(\frac{16}{25}\right)^{0.5}$ (*c*) $\frac{1}{2} > \frac{2}{3} > \frac{3}{4} > \frac{4}{5}$ (*d*) $0 > \frac{7}{17} > \frac{3}{7} > \frac{3}{5}$

28. If $x = a\sec\theta + b\tan\theta$ and $y = a\tan\theta + b\sec\theta$, then the value of $\frac{x^2 - y^2}{a^2 - b^2}$ is

(*a*) 1 (*b*) 2 (*c*) ½ (*d*) 0

29. A vendor purchased 40 dozen bananas for ₹250. Out of these 30 bananas were rotten and could not be sold. At what rate per dozen should he sell remaining bananas to make a profit of 20%?

(*a*) ₹6 (*b*) ₹8 (*c*) ₹10 (*d*) ₹12

30. Arvind deposited a sum of money with a bank on 1st January, 2012 at 8% simple interest per annum. He received ₹3,144 on 7th August, 2012. The money he deposied with the bank was

(*a*) ₹2,500 (*b*) ₹3,000 (*c*) ₹3,100 (*d*) ₹3,080

31. Base of a prism of height 10 cm is square. Total surface area of the prism is 192 sq. cm. The volume of the prism is

(*a*) 90 cm³ (*b*) 120 cm³ (*c*) 160 cm³ (*d*) 640 cm³

32. ABCD is a cyclic quadrilateral, AB is the diameter of the circle. If ∠ACD = 50°, the measure of ∠BAD is

(*a*) 50° (*b*) 130° (*c*) 140° (*d*) 40°

33. If $x = 5$, then the value of the expression $x^2 - 2 + \frac{1}{x^2}$ is

(*a*) $\frac{24}{25}$ (*b*) $\frac{625}{24}$ (*c*) $\frac{576}{25}$ (*d*) $\frac{24}{5}$

34. The radii of a sphere and a right circular cylinder are 3 cm each. If their volumes are equal, then curved surface area of the cylinder is

(*a*) 74 (3/7) cm² (*b*) 75 (3/7) cm² (*c*) 72 (3/7) cm² (*d*) 65 (3/7) cm²

35. The average marks of 14 students was 71. It was later found that the marks of one of the students has been wrongly entered as 42 instead of 56 and of another as 74 instead of 32. What is the correct average ?

(*a*) 68 (*b*) 67 (*c*) 71 (*d*) 69

36. $(x^3 + y^6)(x^3 - y^6)$ is equal to

(*a*) $x^6 + y^{12}$ (*b*) $x^6 - y^{12}$ (*c*) $x^9 - y^{36}$ (*d*) $x^9 + y^{36}$

37. Simply: $\dfrac{\sqrt[4]{0.0625}+\sqrt[3]{0.008}+\sqrt{0.09}-1}{\sqrt[3]{62.5\sqrt[5]{32}}}$

(*a*) 2.4 (*b*) 0 (*c*) 1.25 (*d*) 1/5

38. A low land, 48 m long and 31.5 m broad is raised to 6.5 dm. For this, earth is removed from a cuboidal hole, 27 m long and 18.2 m broad, dug by the side of the land. The depth of the hole will be

(*a*) 2 m (*b*) 2.5 m (*c*) 2.2 m (*d*) 3 m

39. The value of (cos 0° + sin 45° + sin 30°) (sin 90° – cos 45° + cos 60°) is

(*a*) 5/4 (*b*) 3/4 (*c*) 3/2 (*d*) 7/4

40. A can do half of a piece of work in 1 day, whereas B can do full. B can do half the work as C in 1 day. The ratio of their efficiences of work is

(*a*) 2 : 4 : 1 (*b*) 2 : 1 : 4 (*c*) 4 : 2 : 1 (*d*) 1 : 2 : 4

41. If 5 tan A = 4, then the value of $\dfrac{5\sin A-3\cos A}{\sin A+2\cos A}$

(*a*) 5/6 (*b*) 7/9 (*c*) 9/14 (*d*) 5/14

42. If $2\sin^2\theta-3\sin\theta+1=0$, θ being positive acute angle, then the values of θ are

(*a*) 60°, 45° (*b*) 30°, 90° (*c*) 60°, 55° (*d*) 45°, 50°

43. The length of two parallel chords of a circle of radius 5 cm are 6 cm and 8 cm in the same side of the centre. The distance between them is

(*a*) 1.5 cm (*b*) 3 cm (*c*) 1 cm (*d*) 2 cm

44. Divide 27 into two parts so that 5 times the first and 11 times the second together equals to 195. Then ratio of the first and second part is

(*a*) 2 : 7 (*b*) 3 : 2 (*c*) 17 : 10 (*d*) 5 : 4

45. After deducting a commission of 5%, a T.V. set costs ₹9,595. Its marked price is

(*a*) ₹10,000 (*b*) ₹10,074.75 (*c*) ₹10,100 (*d*) ₹12,000

46. Ram, Rahim and Robert started a partnership business investing ₹30,000, ₹50,000 and ₹40,000 respectively. If they made an annual profit of ₹18,504, then the share of Rahim is

(*a*) ₹7,170 (*b*) ₹7,510 (*c*) ₹7,710 (*d*) ₹7,010

47. A right prism has triangular base. If *v* be the number of vertices, *e* be the number of edges and *f* be the number of faces of the prism. The value of $\dfrac{v+e-f}{2}$ is

(*a*) 10 (*b*) 4 (*c*) 2 (*d*) 5

48. The sum of $\dfrac{1}{x+y}$ and $\dfrac{1}{x-y}$ is

(*a*) $\dfrac{-2y}{x^2-y^2}$ (*b*) $\dfrac{2x}{y^2-x^2}$ (*c*) $\dfrac{2x}{x^2-y^2}$ (*d*) $\dfrac{2y}{x^2-y^2}$

49. A bag contains coins of ₹1, 50 paise and 25 paise in the ratio 2 : 3 : 5. If the total value of these coins is ₹228, then the number of 50 paise coins in that bag was

(*a*) 144 (*b*) 124 (*c*) 112 (*d*) 96

50. If $x=2$, $y=1$ and $z=-3$, then $x^3+y^3+z^3-3xyz$ is

(*a*) 0 (*b*) 8 (*c*) 6 (*d*) 2

ANSWERS AND HINTS

1. **(a)** $2 \times (250 \times 15\%) = 75$ lakhs
2. **(d)** $250 \times (25 + 40)\% = 162.5$ lakhs
3. **(a)** Increase = 500% – 100% = 400%
4. **(b)** $12.5 : 50 = 1 : 4 \Rightarrow$ Increase $= \frac{3}{1} \times 100 = 300\%$
5. **(b)** $\left(250 + \frac{25}{100}\right) \times \frac{102}{100} = 63.75$ lakhs
6. **(a)** $\frac{1}{5} = 20\% \Rightarrow$ Half = 20 $\Rightarrow$ Full = 40
7. **(a)** Total students = (10 + 30 + 20) + (60 + 40 + 10 + 30 + 10) = 210

   $\Rightarrow \% = \frac{60}{210} \times 100 = \frac{200}{7} = 28\frac{4}{7}\%$
8. **(d)** 10
9. **(a)** Maximum students 60 in range of 50–60
10. **(b)** Total students = 210 (Refer to sol. 7 above)
11. **(d)** 20% and 40% = 20 + 40 – 8 = 52%   52% and 10% = 52 + 10 – 5.2 = 56.8%
12. **(b)** $\angle BOP = 180° - 120° = 60°$

    OP = OB = radii

    In ΔPOB, $\angle O = \angle B = 60°$ each

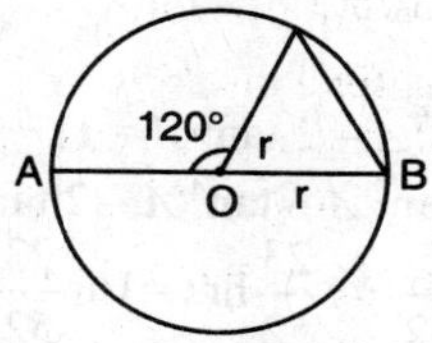

13. **(c)** $\frac{7 \times 30 - 5 \times 31}{2} = \frac{210 - 155}{2} = \frac{55}{2} = 27.5$
14. **(b)** $x + 12x = 403 \Rightarrow 13x = 403 \Rightarrow x = 31 \Rightarrow$ HCF = 31, LCM = $12x = 12 \times 31 = 372$

    $\Rightarrow$ Second number $= \frac{31 \times 372}{93} = 124$
15. **(a)** Copper = 30% of old mixture = 600 g = 20% of new mixture

    $\Rightarrow$ New mixture = 600g × 5 = 3 kg $\Rightarrow$ Alminium added = 3 – 2 = 1 kg.
16. **(d)** Savings = 3 months income = ₹3000
17. **(b)** Smallest angle $= \frac{2}{10} \times 180° = 36°$
18. **(c)** Average speed $= \frac{60}{13.5} = \frac{120}{27} = \frac{40}{9}$

    Average speed $= \frac{2 \times 5 \times 4}{9} = \frac{40}{9}$

    Rate of stream $= \frac{5-4}{2} = \frac{1}{2}$
19. **(c)** 15 men can do in 20 days $\Rightarrow$ 10 men can do in 30 days.

    24 women can do in 20 days $\Rightarrow$ 8 women can do in 60 days.

    $\Rightarrow$ Required number of days $= \frac{30 \times 60}{90} = 20$ days

20. (*c*) Let AD = 6, then

AD : GD = 2 : 1 = 4 : 2

AG : OD = 1 : 1 = 3 : 3

OG = AG – AO = 4 – 3 = 1

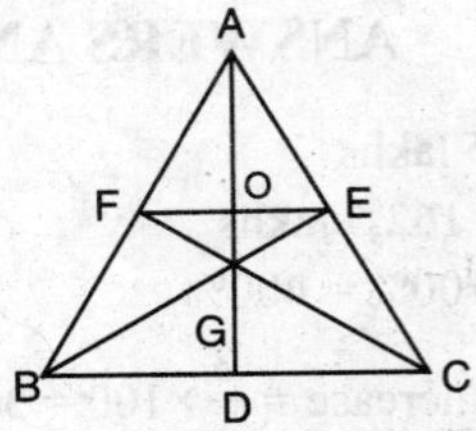

21. (*c*) $a+\frac{1}{a}=1 \Rightarrow a^2+1=a \Rightarrow a^2+1-a=0 \Rightarrow (1-a+a^2)(1+a-a^2)=0$

22. (*c*) 1 Water + 1 Container = 2.25 ⇒ 0.2 Water + 1 Container = 0.77

⇒ 1 Water + 5 Container = 3.85 ⇒ 4 Container = 1.6 ⇒ 1 Container = 0.4

⇒ 0.2 Water + 1 Container = 0.77 ⇒ 0.2 Water = 0.77 – 0.4 = 0.37

⇒ 0.4 Water = 0.74 ⇒ 0.4 Water + 1 Container = 0.74 + 0.4 = 1.14 kg

23. (*c*) $\frac{3\times100}{15}$ = 20 years

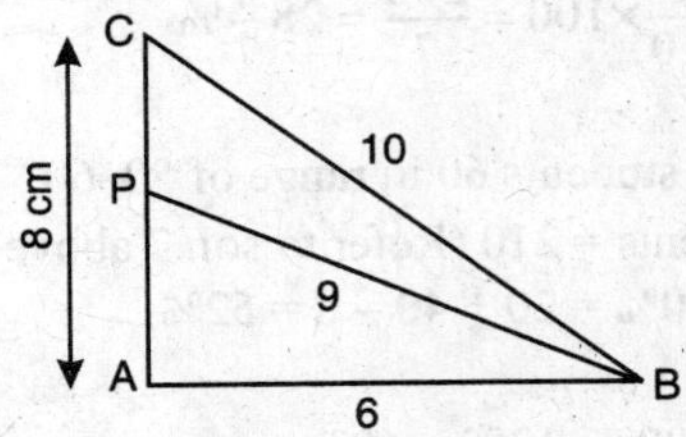

24. (*b*) $AB^2 = 10^2 - 8^2 = 6^2$

$\Rightarrow AP^2 = 9^2 - 6^2 = 45$

$\Rightarrow AP = \sqrt{45} = 3\sqrt{5}$

25. (*b*) $= \sec^4 A - \tan^4 A - 2\tan^2 A = (1+\tan^2 A)^2 - \tan^4 A - 2\tan^2 A$
$= 1 + \tan^4 A + 2\tan^2 A - \tan^4 A - 2\tan^2 A = 1$

26. (*b*) Time $= 15\times\frac{5}{52} = 1\frac{23}{52}$ hrs $= 1$ hr $\frac{23}{52}\times 60$ min $= 1$ hr $\frac{23\times15}{13}$ min $= 1$ hr $26\frac{7}{13}$

27. (*b*)

28. (*a*) $x^2 - y^2 = (a\sec\theta + b\tan\theta)^2 - (a\tan\theta + b\sec\theta)^2$
$= a^2(\sec^2\theta - \tan^2\theta) + b^2(\tan^2\theta - \sec^2\theta) = a^2 - b^2$

29. (*b*) Bananas sold = 480 – 30 = 450 ⇒ Total selling price = 250 + 20% = 300

⇒ Price per dozen = $\frac{300}{450}\times 12$ = ₹8

30. (*b*) Annual interest on ₹3000 = 240. Total days = 219 ⇒ Interest = $240\times\frac{219}{365} = 144$

31. (*c*) $2\times x^2 + 4x\times 10 = 192 \Rightarrow x^2 + 20x = 96 \Rightarrow x = 4 \Rightarrow V = x^2\times h = 16\times 10 = 160$

32. (*d*) ∠C = 50° + 90° = 140°

∠A = 180° – 140° = 40°

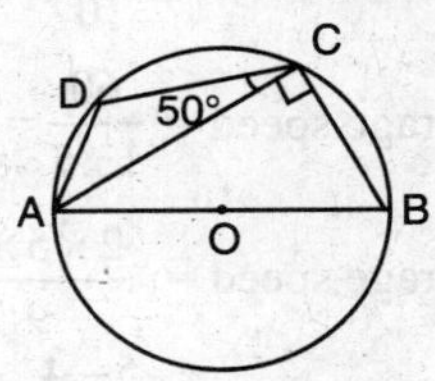

33. (*c*) G.E. $= 25 - 2 + \frac{1}{25} = 23\frac{1}{25} = \frac{576}{25}$

34. (*b*) $\frac{4}{3}\pi r^3 = \pi r^2 h \Rightarrow h = 4 \Rightarrow 2\pi rh = 2\times\frac{22}{7}\times 3\times 4 = 24\times 3\frac{1}{7} = 72 + \frac{24}{7} = 75\frac{3}{7}$

35. (*d*) Difference = (56 + 32) – (42 + 74) = – 28 ⇒ Reduction in average = 28 ÷ 14 = 2

New average = 71 – 2 = 69

36. (*b*) Formula: $(a+b)(a-b) = (a^2 - b^2)$

37. (*b*) Numerator = 0.5 + 0.2 + 0.3 – 1 = 0

38. **(a)** $\dfrac{48\times31.5\times0.65}{27\times18.2}=2$

39. **(d)** G.E. $=\left(1+\dfrac{1}{\sqrt{2}}+\dfrac{1}{2}\right)\left(1-\dfrac{1}{\sqrt{2}}+\dfrac{1}{2}\right)=\left(\dfrac{3}{2}+\dfrac{1}{\sqrt{2}}\right)\left(\dfrac{3}{2}-\dfrac{1}{\sqrt{2}}\right)=\dfrac{9}{4}-\dfrac{1}{2}=\dfrac{7}{4}$

40. **(d)** A = 2 days, B = 1 day ⇒ Ratio of work = 1 : 2 ⇒ C's work = 2B = 4 ⇒ Ratio = 1 : 2 : 4

41. **(d)** G.E. $=\dfrac{5\tan A-3}{\tan A+2}=\dfrac{4-3}{\frac{4}{5}+2}=\dfrac{1}{\frac{14}{5}}=\dfrac{5}{14}$

42. **(b)** $\theta=30°$ and $90°$ satisfies the equation.

43. **(c)** BQ $=\dfrac{1}{2}\times$ AB = 3 cm

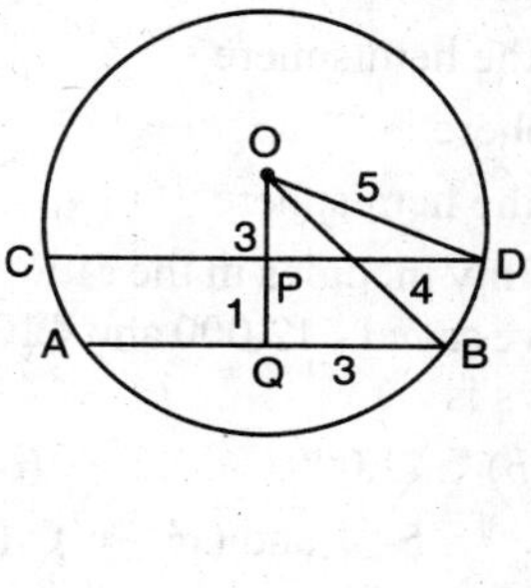

DP $=\dfrac{1}{2}\times$ CD = 4 cm

OP $=\sqrt{5^2-4^2}$ = 3 cm

OQ $=\sqrt{5^2-3^3}$ = 4 cm

⇒ PQ = 4 – 3 = 1 cm

44. **(c)** 27 × 5 = 135; 195 – 135 = 60; 60 ÷ (11 – 5) = 10 ⇒ I = 27 – 10 = 17

45. **(c)** $9595\times\dfrac{100}{95}=10100$

46. **(c)** Ratio of capitals = 3 : 5 : 4 ⇒ Rahim's share = $\dfrac{5}{12}\times18504=7710$

47. **(d)** Vetices = 6, Edges = 9, Faces = 5 ⇒ $\dfrac{v+e-f}{2}=\dfrac{6+9-5}{2}=5$

48. **(c)** $\dfrac{1}{x+y}+\dfrac{1}{x-y}=\dfrac{(x-y)+(x+y)}{x^2-y^2}=\dfrac{2x}{x^2-y^2}$

49. **(a)** Ratio of denominations = 100 : 50 : 25 = 4 : 2 : 1; Ratio of coins = 2 : 3 : 5

Ratio of values = (4 × 2) : (2 × 3) : (1 × 5) = 8 : 6 : 5

Value of 50 paise coins $=\dfrac{6}{19}\times228=72$ ⇒ Coins = 72 ÷ 0.50 = 144

50. **(a)** Here, $x+y+z=0$ ⇒ Given expression = 0

SSC CGL TIER II 2015

1. In a school there were 1554 students and the ratio of the number of the boys and girls was 4 : 3. After few days, 30 girls joined the school but few boys left: as a result the ratio of the boys and girls became 7 : 6. The number of boys who left the school is

   (*a*) 86 (*b* 76 (*c*) 74 (*d*) 84

2. If A : B = 2 : 3 and B : C = 3 : 7, then A + B : B + C : C + A is

   (*a*) 4 : 10 : 9 (*b* 5 : 10 : 9 (*c*) 5 : 8 : 9 (*d*) 4 : 8 : 9

3. If a hemisphere is melted and four sphere of equal volume are made, the reaius of each sphere will be equal to

   (*a*) 1/4 of the radius of the hemisphere
   (*b*) 1/2 of the radius of the hemisphere
   (*c*) radius of the hemisphere
   (*d*) 1/6 of the radius of the hemisphere

4. A and B have their monthly incomes in the ratio 8 : 5, while their monthly expenditures are in the ratio 5 : 3. If they have saved ₹ 12,000 and ₹ 10,000 monthly respectively, then the difference in their monthly incomes is

   (*a*) ₹ 44,000 (*b*) ₹ 42,000 (*c*) ₹ 52,000 (*d*) ₹ 46,000

5. If $(x^3 - y^3) : (x^2 + xy + y^2) = 5 : 1$ and $(x^2 - y^2) : (x - y) = 7 : 1$, then the ratio $2x : 3y$ equals

   (*a*) 2 : 3 (*b*) 4 :3 (*c*) 4 : 1 (*d*) 3 : 2

6. The marked price of a tape recorder is ₹ 12,600. A festival discount of 5% is allowed on it. Further for cash payment, a second discount of 2% is given. The cash payment, in rupees, is to be made for buying it is

   (*a*) 11,730.60 (*b*) 11,370.60 (*c*) 11,703.60 (*d*) 11,073.60

7. Articles are marked at a price which gives a profit 25%. After allowing a certain discount the profit reduces to 12½%. The discount percent is

   (*a*) 12½% (*b*) 12% (*c*) 10% (*d*) 11.1%

8. If $\tan\theta - \cot\theta = 0$ and θ is positive acute angle, then the value of $\dfrac{\tan(\theta+15°)}{\tan(\theta-15°)}$

   (*a*) 3 (*b*) $\dfrac{1}{3}$ (*c*) $\dfrac{1}{\sqrt{3}}$ (*d*) $\sqrt{3}$

9. A number when divided by 361 gives a remainder 47. If the same number is divided by 19, the remainder obtained is

   (*a*) 8 (*b*) 3 (*c*) 1 (*d*) 9

10. In an examination average marks obtained by the girls of a class is 85 and the average marks obtained by the boys of the same class is 87. If the girls and boys are in the ratio 4 : 5, average marks of the whole class (approx) is closest to

    (*a*) 86.4 (*b*) 86.5 (*c*) 85.9 (*d*) 86.1

11. Water tax is increased by 20% but its consumption is decreased by 20%. Then the increase or decrease in the expenditure of the money is

    (*a*) 4% increase (*b*) No change (*c*) 4% decrease (*d*) 5% decrease

12. A man sells an articles at 5% above its cost price. If he had bought it at 5% less than what he had paid for it and sold at ₹ 2 less, he would have gained 10%. The cost price of the article is

    (*a*) ₹ 400 (*b*) ₹ 200 (*c*) ₹ 300 (*d*) ₹ 100

13. If 90 men can do a certain job in 16 days, working 12 hours/day, then the part of the that work which can be completed by 70 men in 24 days, working 8 hours/day is

(a) $\frac{2}{3}$ (b) $\frac{5}{8}$ (c) $\frac{7}{9}$ (d) $\frac{1}{3}$

14. If $(3x - 2y) : (2x + 3y) = 5 : 6$, then one of the value of $\left(\frac{\sqrt[3]{x}+\sqrt[3]{y}}{\sqrt[3]{x}-\sqrt[3]{y}}\right)^2$ is

(a) $\frac{1}{5}$ (b) $\frac{1}{25}$ (c) 5 (d) 25

15. If $\sec\theta - \tan\theta = \frac{1}{\sqrt{3}}$, the value of $\sec\theta : \tan\theta$ is

(a) $\frac{4}{\sqrt{3}}$ (b) $\frac{2}{3}$ (c) $\frac{1}{\sqrt{3}}$ (d) $\frac{2}{\sqrt{3}}$

16. A dealer fixed the price of an article 40% above the cost of production. While selling it he allows a discount of 20% and makes a profit of ₹ 48. The cost of production (in ₹) of the articles is

(a) 320 (b) 360 (c) 400 (d) 420

17. AB and CD are two parallel chords of a circle of lengths 10 cm and 4 cm respectively. If the chords are on the same side of the centre and the distance between them is 3 cm, then the diameter of the circle is

(a) $\sqrt{21}$ cm (b) $\sqrt{29}$ cm (c) $2\sqrt{21}$ cm (d) $2\sqrt{29}$ cm

18. If $a + b = 1$, find the value of $a^3 + b^3 - ab - (a^2 - b^2)^2$

(a) 2 (b) –1 (c) 1 (d) 0

19. If 64 buckets of water are removed from a cubical shaped water tank completely filled with water, 1/3 of the tank remains filled with water. The length of each side of the tank is 1.2 m. Assuming that all buckets are of the same measure, then the volume (in litres) of water contained by each bucket is

(a) 18 (b) 15 (c) 16 (d) 12

20. Given that the ratio of altitudes of two triangles is 4 : 5, ratio of their areas is 3 : 2. The ratio of their corresponding bases is

(a) 8 : 15 (b) 8 : 5 (c) 5 : 8 (d) 15 : 8

21. Pipe A can fill an empty tank in 6 hours and pipe B in 8 hours. If both the pipes are opened and after 2 hours pipe A is closed, how much time B will take to fill the remaining tank?

(a) $7\frac{1}{2}$ hours (b) $2\frac{1}{3}$ hours (c) $3\frac{1}{3}$ hours (d) $2\frac{2}{5}$ hour

22. P and Q together can do a job in 6 days. Q and R can finish the same job in 60/7 days. P started the work and worked for 3 days. Q and R continued for 6 days. Then the difference of days in which R and P can complete the job is

(a) 15 (b) 8 (c) 12 (d) 10

23. There is a wooden sphere of radius $6\sqrt{3}$ cm. The surface area of the largest possible cube cut out from the sphere will be

(a) 864 cm^3 (b) $646\sqrt{3}$ cm^2 (c) $464\sqrt{3}$ cm^2 (d) 462 cm^2

24. The value of (cosec a – sin a) (sec a – cos a) (tan a + cot a)

(*a*) 2 (*b*) 1 (*c*) 4 (*d*) 6

25. A manufacture fixes selling price at 33% over the cost of production. If cost of production goes up by 12% and manufacture raises his selling price by 10% his percentage profit is

(*a*) $28\frac{3}{8}\%$ (*b*) $36\frac{5}{9}\%$ (*c*) $30\frac{5}{8}\%$ (*d*) 35%

Directions : The following graph shows production (in thousands) of two types (P and Q) of vehicles by a factory over the years 2009 to 2014. Study the graph and answer five questions:

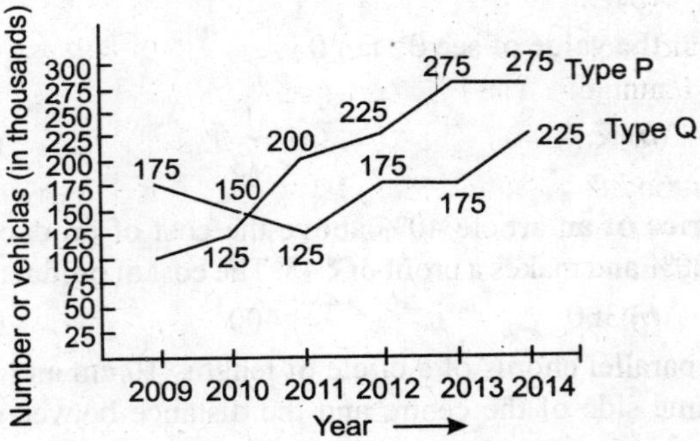

26. In how many of the given years, was the production of Type P vehicles of the company more than the average production of this type vehicles in the given years?

(*a*) 4 (*b*) 2 (*c*) 5 (*d*) 3

27. The production of Type Q vehicles in 2010 was approximately what percent of Type p vehicles in 2014?

(*a*) 75 (*b*) 54.5 (*c*) 45.5 (*d*) 60

28. The ratio of total production of Type P vehicles to total production of Type Q vehicles over the years is

(*a*) 48:41 (*b*) 8:5 (*c*) 5:8 (*d*) 41:48

29. The total production of Type P vehicles in the years 2009 and 2011 is what percent of total production of Type Q vehicles in 2010 and 2014?

(*a*) 75 (*b*) 69.25 (*c*) 80 (*d*) 81.25

30. Approximate percentage decrease in production of Type Q vehicles from 2010 to 2011 is

(*a*) 16.7 (*b*) 10.1 (*c*) 14.3 (*d*) 12.5

31. The radii or two solid iron spheres are 1 cm and 6 cm respectively. A hollow sphere is made by melting the two spheres. If the external radius of the hollow sphere is 9 cm, then its thickness (in cm) is

(*a*) 2 (*b*) 0.5 (*c*) 1 (*d*) 1.5

32. If $x = a \sin\theta - b\cos\theta$, $y = a\cos\theta + b\sin\theta$, then which of the following is true?

(*a*) $\frac{x^2}{y^2}+\frac{a^2}{b^2}=1$ (*b*) $x^2+y^2=a^2+b^2$ (*c*) $\frac{x^2}{a^2}+\frac{y^2}{b^2}=1$ (*d*) $x^2+y^2=a^2-b^2$

33. A plane divides a right circular cone into two parts of equal volume. If the plane is parallel to the base, then the ratio, in which the height of the cone is divided, is

(*a*) $1:\sqrt[3]{2}$ (*b*) $1:\sqrt[3]{2}+1$ (*c*) $1:\sqrt[3]{2}-1$ (*d*) $1:\sqrt{2}$

34. If $3(a^2 + b^2 + c^2) = (a + b + c)^2$, then the relation between a, b and c is

(*a*) $a \neq b = c$ (*b*) $a \neq b \neq c$ (*c*) $a = b = c$ (*d*) $a = b \neq c$

35. The value of $\dfrac{(0.67\times0.67\times0.67)-(0.33\times0.33\times0.33)}{(0.67\times0.67)+(0.67\times0.33)+(0.33\times0.33)}$

(*a*) 1.1 (*b*) 0.34 (*c*) 11 (*d*) 3.4

36. There is a number consisting to two digits, the digit in the units place is twice that in the tens place and if 2 be subtracted from the sum of the digits, the difference is equal to 1/6 of the number. The number is

(*a*) 24 (*b*) 26 (*c*) 25 (*d*) 23

37. A sum of money is paid back in two annual statements of ₹ 17,640 each, allowing 5% compound interest compounded annually. The sum borrowed was

(*a*) ₹ 32,200 (*b*) ₹ 32,800 (*c*) ₹ 32,400 (*d*) ₹ 32,00

38. Two blends of a commodity costing ₹ 35 and ₹ 40 per kg respectively are mixed in the ratio 2 : 3 by weight. If fifth of the mixture is sold at ₹ 46 per kg and the remaining at the rate of ₹ 55 per kg, the profit percent is

(*a*) 50 (*b*) 40 (*c*) 20 (*d*) 30

39. 300 grams of sugar solution has 40% of sugar in it. How much sugar should be added to make it 50% in the solution?

(*a*) 60 gm (*b*) 10 gms (*c*) 80 gms (*d*) 40 gms

40. If 60% of A = 30% of B, B = 40% of C and C = x% of A, then value of x is

(*a*) 800 (*b*) 500 (*c*) 300 (*d*) 200

41. The diameter of each wheel of a car is 70 cm. If each wheel rotates 400 times per minute, the the speed of the car (in kr/hr) is $\left(\text{Takes } \pi = \dfrac{22}{7}\right)$

(*a*) 5.28 (*b*) 528 (*c*) 52.8 (*d*) 0.528

42. A cylinder with base radius 8 cm and height 2 cm is melted to form a cone of height 6 cm. The radius of the cone will be

(*a*) 4 cm (*b*) 6 cm (*c*) 5 cm (*d*) 8 cm

43. If $5\cos\theta + 12\sin\theta = 13$, $0° < \theta < 90°$, then the value of $\sin\theta$ is

(*a*) $\dfrac{6}{13}$ (*b*) $\dfrac{12}{13}$ (*c*) $\dfrac{5}{13}$ (*d*) $-\dfrac{12}{13}$

44. If a shopkeeper wants to give 20% discount on a toy, he has to sell it for ₹ 300. If he sells its at ₹ 405 then his gain percent is

(*a*) 8% (*b*) 5% (*c*) 4% (*d*) 6%

45. Let x be the least number, which when divided by 5, 6, 7 and 8 leaves a remainder 3 in each case but when dividend by 9 leaves no remainder. The sum of digits of x is

(*a*) 21 (*b*) 22 (*c*) 24 (*d*) 18

46. The interior angle of a regular polygon exceeds its exterior angle by 108°. The number of sides of the polygon is

(*a*) 14 (*b*) 12 (*c*) 10 (*d*) 16

47. In triangle ABC, DE || BC where D is a point on AB and E is a point on AC. DE divides the area of Δ ABC into two equal parts. Then DB : AB is equal to

(*a*) $\sqrt{2}:(\sqrt{2}-1)$ (*b*) $\sqrt{2}:(\sqrt{2}+1)$ (*c*) $(\sqrt{2}-1):\sqrt{2}$ (*d*) $(\sqrt{2}+1):\sqrt{2}$

48. Let $x = \dfrac{\sqrt{13}+\sqrt{11}}{\sqrt{13}-\sqrt{11}}$ and $y = \dfrac{1}{x}$, then the value of $3x^2 - 5xy + 3y^2$ is

(*a*) 1717 (*b*) 1177 (*c*) 1771 (*d*) 1171

49. ABCD is a cyclic quadrilateral. AB and DC when produced meet at P, if PA = 8 cm, PB = 6 cm. PC = 4 cm, then the length (in cm) of PD is

(*a*) 6 (*b*) 8 (*c*) 12 (*d*) 10

50. The unit digit in the product $(2467)^{153} \times (341)^{72}$ is

(*a*) 1 (*b*) 9 (*c*) 3 (*d*) 7

51. A and B can do a piece of work in 30 and 36 days respectively. They began the work together but A leaves after some days and B finished the remaining work in 25 days. After how many days did A leave?

(*a*) 6 days (*b*) 11 days (*c*) 10 days (*d*) 5 days

52. Three glasses of equal volume contains acid mix with water. The ratio of acid and water are 2 : 3, 3 : 4 and 4 : 5 respectively. Contents of these glasses are poured in a large vessel. The ratio of acid and was in the large vessel is

(*a*) 407 : 560 (*b*) 411 : 540 (*c*) 417 : 564 (*d*) 40 : 544

53. A boat moves downstream at the rate of 1 km in $7\frac{1}{2}$ minutes and upstream a the rate of 5 km an hour. What is the speed (in km/hour) of the boat in the still water?

(*a*) 8 (*b*) 4 (*c*) $6\frac{1}{2}$ (*d*) $\frac{1}{2}$

54. The average age of 30 students of a class is 14 years 4 months. After admission of 5 new students in the class the average becomes 13 years 9 months. The youngest one of the five new students is 9 years 11 months old. The average age of the remaining 4 new students is

(*a*) 11 years 2 months (*b*) 13 years 6 months (*c*) 10 years 4 months (*d*) 12 years 4 months

55. Ram sold two horses at the same price. In one he gets a profit of 10% and in the other he gets a loss of 10%. Then Ram gets

(*a*) 2% loss (*b*) 1% profit (*c*) no loss or profit (*d*) 10% loss

56. If tan A = n tan B and sin A = m sin B, then the value of $\cos^2 A$ is

(*a*) $\dfrac{m^2+1}{n^2-1}$ (*b*) $\dfrac{m^2+1}{n^2+1}$ (*c*) $\dfrac{m^2-1}{n^2+1}$ (*d*) $\dfrac{m^2-1}{n^2-1}$

57. If $\left(\dfrac{p^{-1}q^2}{p^3q^{-2}}\right)^{\frac{1}{3}} \div \left(\dfrac{p^6q^{-3}}{p^{-2}q^3}\right) = p^aq^b$ then the value of a + b, where p and q are different positive primes, is

(*a*) 2 (*b*) 1 (*c*) 0 (*d*) –1

58. The value of $4 - \dfrac{5}{1+\dfrac{1}{3+\dfrac{1}{2+\frac{1}{4}}}}$

(*a*) $\frac{1}{64}$ (*b*) $\frac{1}{32}$ (*c*) $\frac{1}{16}$ (*d*) $\frac{1}{8}$

59. The simple interest on a sum of money is $\frac{8}{25}$ of the sum. If the number of years is numerically half the rate percent per annum, then the rate percent per annum is

(*a*) 5 (*b*) 8 (*c*) $\frac{1}{64}$ (*d*) 4

60. Let x be the smallest number, which when added to 2000 makes the resulting number divisible by 12, 16, 18 and 21. The sum of the digits of *x* is

(*a*) 5 (*b*) 6 (*c*) 4 (*d*) 7

61. If a man walks at the rate of 5 km/hour, he misses a train by 7 minutes. However if he walks at the rate of 6 km/hour, he reaches the station 5 minutes before the arrival of the train. The distance covered by him to reach the station is

(*a*) 4 km (*b*) 7 km (*c*) 6.25 km (*d*) 6 km

62. The portion of a ditch 48 m long, 16.5 m wide and 4 m deep that can be filled with stones and earth available during excavation of a tunnel, cylindrical in shape, of diameter 4 m and length 56 m is $\left(\text{Take } \pi \frac{22}{7}\right)$

(*a*) $\frac{2}{9}$ part (*b*) $\frac{1}{2}$ part (*c*) $\frac{1}{9}$ part (*d*) $\frac{1}{4}$ part

63. The value of cot 41° . cot 42° . cot 43°. cot 44° . cot 45° . cot 46° cot 47° . cot 48° . cot 49°

(*a*) $\frac{\sqrt{3}}{2}$ (*b*) $\frac{1}{\sqrt{2}}$ (*c*) 1 (*d*) 0

64. AD is perpendicular to the internal bisector of ∠ABC of ΔABC. DE is drawn through D and parallel to BC to meet AC at E. If the length of AC is 12 cm, then the length of AE (in cm) is

(*a*) 8 (*b*) 6 (*c*) 3 (*d*) 4

65. The perimeter of a rhombus is 60 cm and one of its diagonal is 24 cm. The area (in sq. cm) of the rhombus is

(*a*) 108 (*b*) 216 (*c*) 432 (*d*) 206

66. A man starts from a place P and reaches the place Q in 7 hours. He travels 1/4th of the distance at 10 km/hour and the remaining distance at 12 km/hour. The distance, in kilometer, between P and Q is

(*a*) 70 (*b*) 72 (*c*) 80 (*d*) 90

67. 60 kg of an alloy A is mixed with 100 kg of alloy B. If alloy A has lead and tin in the ratio 3 : 2 and alloy B has tin and cooper in the ratio 1 : 4, the amount of tin in the new alloy is

(*a*) 80 kg (*b*) 44 kg (*c*) 24 kg (*d*) 53 kg

68. $\frac{6^2+7^2+8^2+9^2+10^2}{\sqrt{7+4\sqrt{3}}-\sqrt{4+2\sqrt{3}}}$ is equal to

(*a*) 305 (*b*) 355 (*c*) 330 (*d*) 366

69. Quadrilateral ABCD is circumscribed about a circle. If the lengths of AB, BC, CD are 7 cm, 8.5 and 9.2 cm respectively, then the length (in cm) of DA is

(*a*) 16.2 (*b*) 7.7 (*c*) 7.2 (*d*) 10.7

70. If sin A + sin 2 A = 1, then the value of $\cos^2 A + \cos^4$ is

(*a*) $1\frac{2}{3}$ (*b*) 2 (*c*) $1\frac{1}{2}$ (*d*) 1

71. A car covers four successive 7 km distances at speeds of 10 km/hour, 20 km/hour, 30 km/hour, 60 km/hour respectively. Its average speed over this distance is

(*a*) 40 km/hour (*b*) 20 km/hour (*c*) 30 km/hour (*d*) 60 km/hour

72. If $x = a^{\frac{1}{2}} + a^{-\frac{1}{2}} y = a^{\frac{1}{2}} - a^{-\frac{1}{2}}$ then value of $(x4 - x^2 y^2 - 1) + (y^4 - x^2 y^2 + 1)$ is

(*a*) 12 (*b*) 13 (*c*) 14 (*d*) 16

73. If $x - \sqrt{3} - \sqrt{2} = 0$ and $y - \sqrt{3} + \sqrt{2} = 0$, then value of $\left(x^3 - 20\sqrt{2}\right) - \left(y^3 + 2\sqrt{2}\right)$

(*a*) 0 (*b*) 2 (*c*) 3 (*d*) 1

74. The average of five consecutive positive integers is *n*. If the next two integers are also included, the average of all these integers will

(*a*) increase by 1 (*b*) increase by 1.5 (*c*) remains the same (*d*) increase by 2

75. A telegraph post is bent at point above the ground due to storm. Its top just touches the ground at a distance of $10\sqrt{3}$ m from its foot and makes an angle of 30° with the horizontal. The height (in metres) of the telegraph post is

(*a*) 20 (*b*) 30 (*c*) 25 (*d*) 24

76. Average of *n* number is *a*. The first number is increased by 2, second one is increased by 4, the third one is increased by 8 and so on. The average of the new numbers is

(*a*) $a + \frac{2^{n+1}}{n}$ (*b*) $a + \frac{2^n - 1}{n}$ (*c*) $a + 2\frac{2^n - 1}{n}$ (*d*) $a + \frac{2^{n+1} - 1}{n}$

77. A man purchases some orange at the rate of 3 for ₹ 40 and the same quantity at 5 for ₹ 60. If he sells all the oranges at the rate of 3 for ₹ 50, find his gain or loss percent (to the nearest integers).

(*a*) 34% loss (*b*) 32% profit (*c*) 31% profit (*d*) 31% loss

78. If $a - \frac{1}{a-3} = 5$, then the value of $(a-3)^3 - \frac{1}{(a-3)^3}$ is

(*a*) 5 (*b*) 14 (*c*) 2 (*d*) 7

79. The greatest number among 3^{50}, 4^{40}, 5^{30} and 6^{20} is

(*a*) 3^{50} (*b*) 4^{40} (*c*) 5^{30} (*d*) 6^{20}

80. If $x^2 + y^2 + z^2 = xy + yz + zx$, then the value of $\frac{3x^4 + 7y^4 + 5z^4}{5x^2y^2 + 7y^2z^2 + 3z^2x^2}$ is

(*a*) 1 (*b*) –1 (*c*) 2 (*d*) 0

81. A, B and C can do a work separately in 16, 32 and 48 days respectively. They started the work together but B leaving off 8 days and C six days before the completion of the work. In what time is the work finished?

(*a*) 14 days (*b*) 12 days (*c*) 9 days (*d*) 10 days

82. The centroid of a ΔABC is G. The area of Δ ABC is 60 cm ^{2}. The area of Δ GBC is

(*a*) 30 cm^2 (*b*) 40 cm^2 (*c*) 10 cm^2 (*d*) 20 cm^2

83. A and B can do a given piece of work in 8 days, B and C can do the same work in 12 days and A, B, C complete it in 6 days. Number of days required to finish the work by A and C is

(*a*) 16 (*b*) 12 (*c*) 8 (*d*) 24

84. Three Science classes A, B and C take a Life science test. The average score of class A is 83. The average score of class B is 76. The average score of class C is 85. The average score of class A and B is 79 and average score of class B and C is 81. Then the average score of classes A, B and C is

(*a*) 8.15 (*b*) 80 (*c*) 80.5 (*d*) 81

85. The area of an isosceles trapezium is 176 cm^2 and the height is 2/11th of the sum of its parallel sides. If the ratio of the length of the parallel side is 4 : 7, then the length of a diagonal (in cm) is

(*a*) 28 (*b*) $2\sqrt{137}$ (*c*) 24 (*d*) $\sqrt{137}$

86. A right prism has a triangular base whose sides are 13 cm, 20 cm and 21 cm. If the altitude of the prism is 9 cm, then its volume is

(*a*) 1134 cm^3 (*b*) 1314 cm^3 (*c*) 1413 cm^3 (*d*) 1143 cm^3

87. In an office, 40% of the staff is female. 70% of the female staff and 50% of the male staff are married. The percentage of the unmarried staff in the office is

(*a*) 54 (*b*) 60 (*c*) 42 (*d*) 64

88. A and B are centres of two circles of radii 11 cm and 6 cm, respectively. PQ is a direct common tangent to the circles. If AB = 13 cm, then length of PQ will be

(*a*) 13 cm (*b*) 12 cm (*c*) 8.5 cm (*d*) 17 cm

89. A sum of ₹ 7,930 is divided into 3 parts and given on loan at 5% simple interest to A, B and C for 2, 3 and 4 years respectively. If the amounts of all three are equal after their respective periods of loan, then the a received a loan of

(*a*) ₹ 3,050 (*b*) ₹ 2,760 (*c*) ₹ 2,800 (*d*) ₹ 2,750

90. Two places P and Q are 162 km apart. A train leaves P for Q and simultaneously another train leaves Q for P. They meet at the end of 6 hours. If the former train travels 8 km/hour faster than the other, the speed of train from Q is

(*a*) $9\frac{1}{2}$ km/hour (*b*) $8\frac{1}{2}$ km/hour (*c*) $10\frac{5}{6}$ km/hour (*d*) $12\frac{5}{6}$ km/hour

91. Base of a right pyramid is a square of side 10 cm. If the height of the pyramid is 12 cm, then its total surface area is

(*a*) 400 cm^2 (*b*) 360 cm^2 (*c*) 260 cm^2 (*d*) 460 cm^2

92. A sum of money placed at compound interest doubles itself in 5 years. It will amount to eight times itself at the same rate of interest in

(*a*) 12 years (*b*) 10 years (*c*) 20 years (*d*) 15 years

93. In trapezium ABCD, AB || CD and AB = 2 CD. Its diagonals interest at O. If the area of Δ AOB = 84 cm^2, then the area of Δ COD is equal to

(*a*) 26 cm^2 (*b*) 72 cm^2 (*c*) 21 cm^2 (*d*) 42 cm^2

94. If 7 sin $^2\theta$ + 3 cos $^2\theta$ = 4, then the value of tan θ is (θ is acute)

(*a*) 1 (*b*) $\frac{1}{\sqrt{3}}$ (*c*) $\sqrt{3}$ (*d*) $\frac{1}{\sqrt{2}}$

95. If $a+\frac{1}{b}=b+\frac{1}{c}=c+\frac{1}{a}$, where $a \neq b \neq c \neq 0$, then the value of $a^2b^2c^2$ is

(*a*) 1 (*b*) –1 (*c*) abc (*d*) 0

96. If O is the circumcentre of a triangle ABC lying inside the triangle, then ∠OBC + ∠BAC is equal to

(*a*) 110° (*b*) 90° (*c*) 120° (*d*) 60°

97. The H.C.F. and L.C.M. of two number are 21 and 84 respectively. If the ratio of the two number is 1 : 4, then the larger of the two numbers is

(*a*) 48 (*b*) 108 (*c*) 84 (*d*) 12

98. There would be a 10% loss, if rice is sold at ₹ 54 per kg. To earn a profit of 20%, the price of rice per kg will be

(*a*) ₹ 65 (*b*) ₹ 70 (*c*) ₹ 63 (*d*) ₹ 72

99. In Δ ABC, ∠ BAC = 90° and AD ⊥ BC. If BD = 3 cm and CD = 4 cm, then the length (in cm) of AD is

(*a*) 5 (*b*) 3.5 (*c*) $2\sqrt{3}$ (*d*) 6

100. The numerical values of the volume and the area of the lateral surface of a right circular cone are equal. If the height of the cone be *h* and radius be r, the value of $\frac{1}{h^2}+\frac{1}{r^2}$.

(*a*) $\frac{3}{1}$ (*b*) $\frac{1}{9}$ (*c*) $\frac{9}{1}$ (*d*) $\frac{1}{3}$

ANSWERS AND HINTS

1. (*b*) Boys = $\frac{4}{7}\times 1554 = 888 \Rightarrow$ Girls = 1554 – 888 = 666

⇒ Increased girls = 666 + 30 = 696 = 6 ratio (new)

⇒ 1 ratio (new) = 116 ⇒ Boys = 7 ratio = 812

⇒ Boys left = 888 – 812 = 76

2. (*b*) A : B : C = 2 : 3 : 7

⇒ (A+ B) : (B + C) : (C + A) = (2 + 3) : (3 + 7) : (7 + 2) = 5 : 10 : 9

3. (*b*) $\frac{2}{3}\pi^3 R = 4\left(\frac{4}{3}\pi r^3\right) \Rightarrow R^3 = 8r^3 = (2r)^3 \Rightarrow R : r = 2:1$

4. (*b*) $\frac{8x-12000}{5x-10000}=\frac{5}{3} \Rightarrow 24x-36000=25x-50000$

$\Rightarrow x = 14000 \Rightarrow$ Difference = $8x - 5x = 3x = 42{,}000$

5. (*c*) $\frac{x^3-y^3}{x^2+xy+y^2}=\frac{5}{1} \Rightarrow x-y=5$

$\frac{x^2-y^2}{x-y}=\frac{7}{1} \Rightarrow x+y=7$

$\Rightarrow x : y = (7+5) : (7-5) = 12 : 2 = 6 : 1$

$\Rightarrow 2x : 3y = 12 : 3 = 4 : 1$

6. (*a*) 12600 – 5% of 12600 = 12600 – 630 = 11970
11970 – 2% of 11970 = 11970 – 239.40 = 11730.60

7. (*c*) CP = 100 ⇒ MP = 100 + 25 = 125
⇒ SP = 100 + 12.5 = 112.5
⇒ Discount = 125 – 112.5 = 12.5 on 125 ⇒ 10%

8. (*a*) $\tan\theta = \cot\theta \Rightarrow \theta = 45° \Rightarrow \frac{\tan(\theta+15°)}{\tan(\theta-15°)} = \frac{\tan 60°}{\tan 30°} = \sqrt{3} \div \frac{1}{\sqrt{3}} = 3$

9. (*d*) Divide 47 by 19 ⇒ Remainder = 9

10. (*d*) Average = $\frac{85\times4+87\times5}{4+5} = \frac{775}{9} = 86.1\%$

Alt : $85 + \frac{2\times5}{9} = 85 + 1.1 = 86.1$

11. (*c*) Decrease = $\frac{20\times20}{100} = 4\%$

12. (*a*) Let CP = 100 ⇒ SP = 105
New CP = 95 ⇒ New SP = 95 + 9.5 = 104.5
⇒ Difference = 105 – 104.5 = 0.5
Actual difference = ₹ 2 ⇒ CP = 4 ×100 = ₹ 400

13. (*c*) $\frac{70\times24\times8}{90\times16\times12} = \frac{7}{9}$

14. (*d*) $6(3x - 2y) = 5(2x + 3y) \Rightarrow 18x - 12y = 10x + 15y$
$\Rightarrow 8x = 27y \Rightarrow x : y = 27 : 8$
$\Rightarrow \sqrt[3]{x} : \sqrt[3]{y} = \sqrt[3]{27} : \sqrt[3]{8} = 3 : 2$
$\Rightarrow \text{G.E.} = \left(\frac{3+2}{3-2}\right)^2 = 5^2 = 25 = 5^2 = 25$

15. (*b*) $(\sec^2\theta - \tan^2\theta) = 1 \Rightarrow (\sec\theta + \tan\theta)(\sec\theta - \tan\theta) = 1$

$\Rightarrow (\sec\theta + \tan\theta) = 1 \div \frac{1}{\sqrt{3}} = \sqrt{3}$

$(\sec\theta + \tan\theta)^2 - (\sec\theta - \tan\theta)^2 = \left(\sqrt{3}\right)^2 - \left(\frac{1}{\sqrt{3}}\right)^2$

$\Rightarrow 4\sec\theta \,.\, \tan\theta = 3 - \frac{1}{3} = \frac{8}{3} \Rightarrow \sec\theta \,.\, \tan\theta = \frac{8}{3}\times\frac{1}{4} = \frac{2}{3}$

16. (*c*) Let CP = 100 ⇒ MP = 140 ⇒ Discount = 28
⇒ SP = 140 – 28 = 112 ⇒ Profit = 12
Actual Profit = 48 ⇒ CP = 400

17. (*d*) $r^2 = 5^2 + x^2 = (x+3)^2 + 2^2$
$\Rightarrow x = 2$
$\Rightarrow r = \sqrt{5^2+2^2} = \sqrt{29}$
$\Rightarrow$ Diameter = $2\sqrt{29}$

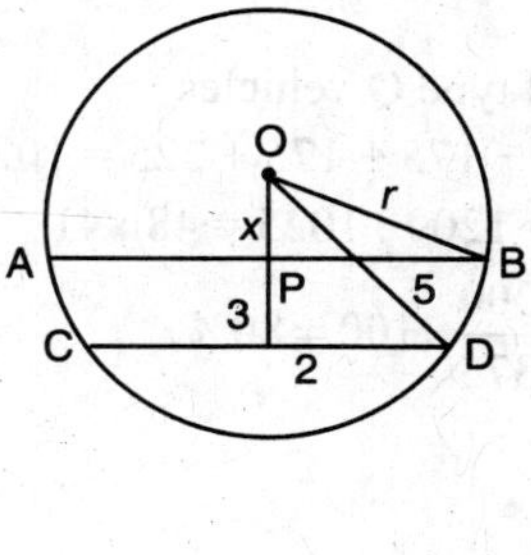

18. (*d*) G. E. $= (a+b)(a^2-ab+b^2) - ab - [(a+b)(a-b)]^2$
$= (a^2 - 2ab + b^2) - (a-b)^2 = (a-b)^2 - (a-b)^2 = 0$

19. (*a*) Volume of tank $= (1.2)^2 m^3 = 1728$ litres $\Rightarrow 64 \times \frac{3}{2}$ buckets = 96 buckets

$\Rightarrow$ 1 bucket $= 1728 \div 96 = 18$ litres

20. (*d*) Required ratio $= \frac{3}{4} : \frac{2}{5} = 15 : 8$

21. (*c*) A's work for 2 hours $= \frac{1}{6} \times 2 = \frac{2}{6} = \frac{1}{3}$ $\Rightarrow$ Remaining work $= 1 - \frac{1}{3} = \frac{2}{3}$

Time taken by $B = \frac{2}{3} \times 8 = \frac{16}{3}$ $\Rightarrow$ Required time $= \frac{16}{3} - 2 = \frac{10}{3} = 3\frac{1}{3}$ hours

22. (*d*) (Q + R)'s work for 6 days $= \frac{7}{60} \times 6 = \frac{7}{10}$

$\Rightarrow$ Remaining work $= 1 - \frac{7}{10} = \frac{3}{10}$ done by P in 3 days

$\Rightarrow$ P alone will take $\frac{10}{3} \times 3 = 10$ days $\Rightarrow$ Q alone will take $= \frac{10 \times 6}{10 - 6} = 15$ days

$\Rightarrow$ R's 1 days work $= \frac{7}{60} - \frac{1}{15} = \frac{3}{60} = \frac{1}{20}$ $\Rightarrow R - P = 20 - 10 = 10$ days

23. (*a*) $r = 6\sqrt{3} \Rightarrow 2r = 12\sqrt{3}$ = Diagonal of cube $\Rightarrow a = 12\sqrt{3} \div \sqrt{3} = 12$
$\Rightarrow$ Surface area $= 6a^2 = 6 \times 12^2 = 864$

24. (*b*) G.E. $= \left(\frac{1}{\sin a} - \sin a\right)\left(\frac{1}{\cos a} - \cos a\right)\left(\frac{\sin a}{\cos a} + \frac{\cos}{\sin a}\right)$

$$= \frac{1-\sin^2 a}{\sin a} \times \frac{1-\cos^2 a}{\cos a} + \frac{\sin^2 a + \cos^2 a}{\sin a \times \cos a} = \frac{\cos^2 a}{\sin a} \times \frac{\sin^2 a}{\cos a} \times \frac{1}{\sin a . \cos a} = 1$$

25. (*c*) Let CP = 100 $\Rightarrow$ SP = 133
New CP = 112 $\Rightarrow$ New SP = 133 + 13.3 = 146.3

New profit = 146.3 – 112 = 34.3 $\Rightarrow$ % profit $= \frac{34.3}{112} \times 100 = \frac{245}{8} = 30\frac{5}{8}\%$

26. (*d*). Total production of type P vehicles
= 100 + 125 + 200 + 225 + 275 + 275 = 1200
$\Rightarrow$ Average $= 120 \div 6 = 200$
In 3 years (2012), 2013, 2014) production was more than 200.

27. (*b*) $\frac{150}{275} \times 100 = 54.5\%$

28. (*a*) Total production of type Q vehicles
= 175 + 150 + 125 + 175 + 175 + 225 = 1025
$\therefore$ Required ratio = 1200 : 1025 = 48 : 41

29. (*c*) $\frac{100+200}{150+225} \times 100 = \frac{300}{375} \times 100 = 80\%$

30. (*a*) Decrease = 150 – 125 = 25 ⇒ % Decrease = $\frac{25}{150}\times100=16.67\%$

31. (*a*) $\frac{4}{3}\pi\left(9^3-x^3\right)=\frac{4}{3}\pi\left(1^3+6^3\right)$

$\Rightarrow 729-x^3=1+216 \quad \Rightarrow x^3=512 \Rightarrow x=8$

⇒ Thickness = 9 – 8 = 1 cm

32. (*b*) $x^2+y^2=a^2\sin^2\theta+b^2\cos^2\theta+a^2\cos^2 a+b^2\sin^2\theta$

$=a^2(\sin^2\theta+\cos^2\theta)+b^2(\sin^2\theta+\cos^2\theta)=a^2+b^2$

33. (*c*) Δ ABC ~ Δ ADE

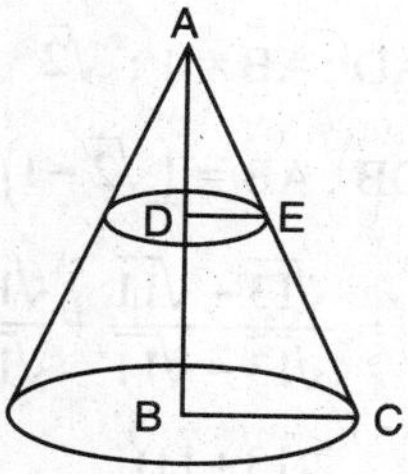

Ratio of volumes of smaller and bigger cones = 1 : 2

⇒ Ratio of radii = $\sqrt[3]{1}:\sqrt[3]{2}=1:\sqrt[3]{2}$

⇒ Ratio of height = $1:\sqrt[3]{2}-1$

34. (*c*) If a = b = c, then $3(3a^2)=(3a)^2 \Rightarrow 9a^2=9a^2$

35. (*b*) G.E. = $\frac{x^3-y^3}{x^2+xy+y^2}=x-y=0.67-0.33=0.34$

36. (*a*) 2 × 2 = 4 ⇒ Option (a)

37. (*b*) Principal of first instalment = 17640 × $\frac{20}{21}$ = 16800

Principal of second instalment = 16800 × $\frac{20}{21}$ = 16000

⇒ Sum borrowed = 16800 + 16000 = 32,800

38. (*b*) CP of 5 kg = 2 × 35 + 3 × 40 = 190

SP fo 5 kg = 1 × 46 + 4 × 55 = 266

⇒ Profit = 266 – 190 = 76 ⇒ % = $\frac{76}{190}\times100=40\%$

39. (*a*) Sugar = 40% of 300 g = 120 g ⇒ Other quantity = 300 – 120 = 180 g

⇒ New sugar = 180 g ⇒ Sugar added = 180 – 120 = 60 g

40. (*b*) A : B = 30 : 60 = 1 : 2 ; B : C = 40 : 100 = 2 :5

⇒ C : A = 5 : 1 ⇒ C = 500% of A

41. (*c*) $2\times\frac{22}{7}\times35\times400\times\frac{1}{100}\times60\times\frac{1}{100}=52.8$ km/kr

42. (*d*) $\frac{1}{3}\pi r^2h=\pi(8)^2 2 \quad \Rightarrow r^2=\frac{3\times8\times8\times2}{6}=64$

$\Rightarrow r=8$ cm

43. (*d*) $5\cos\theta+12\sin\theta=13 \quad \Rightarrow \frac{5}{13}\cos\theta+\frac{12}{13}\sin\theta=1$

$\because \cos^2\theta+\sin^2\theta=1 \quad \Rightarrow \sin\theta=\frac{12}{13}$

44. **(*a*)** $80\% = 300 \Rightarrow 405 = \frac{405}{300} \times 80 = 108\% \Rightarrow$ Profit = 8%

45. **(*d*)** Out of given options, 18 is only multiple of 9.

46. **(c)** Exterior angle = $\frac{180-108}{2} = 36°$

$\Rightarrow$ Number of sides = $\frac{360}{36} = 10$

47. **(*c*)** Area (ΔADE) : Area (ΔBC) = 1: 2

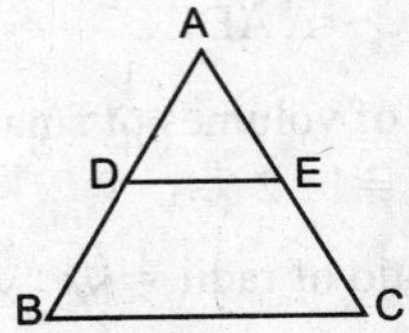

$\Rightarrow$ AD : AB = 1 : $\sqrt{2}$

$\Rightarrow$ DB : AB = $(\sqrt{2}-1) : \sqrt{2}$

48. **(*a*)** $x + y = \frac{\sqrt{13}+\sqrt{11}}{\sqrt{13}-\sqrt{11}} + \frac{\sqrt{13}-\sqrt{11}}{\sqrt{13}+\sqrt{11}}$

$= \frac{2(13+11)}{13-11} = 24; \quad xy = 1$

$3x^2 - 5xy + 3y^2 = 3\,(x+y)^2 - 11 = 3 \times 24^2 - 11 = 1717$

49. **(*c*)** PA. PB = PC. PD

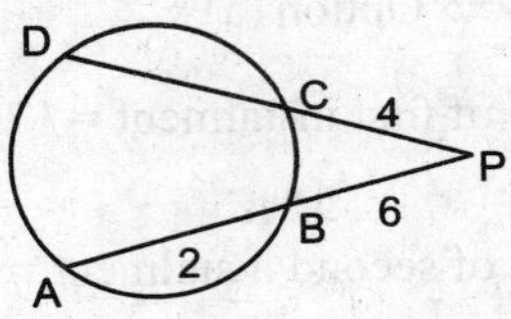

$\Rightarrow$ PD = $\frac{8 \times 6}{4}$ = 12 cm

50. **(*d*)** Unit digit = $7^1 \times 1 = 7$

51. **(*d*)** B's work for 25 days = $\frac{1}{36} \times 25 = \frac{25}{36} \Rightarrow$ Remaining work = $1 - \frac{25}{36} = \frac{11}{36}$

Time taken by (A + B) = $\frac{11}{36} \times \frac{30 \times 36}{66} = 5$ days

52. **(*d*)** Acid = $\frac{2}{5} + \frac{3}{7} + \frac{4}{9} = \frac{126+135+140}{315} = \frac{401}{315} \Rightarrow$ Water = $3 - \frac{401}{315} = \frac{544}{315}$

$\Rightarrow$ Required ratio = 401 : 544

53. **(*c*)** Downstream speed = $1 \div 7\frac{1}{2} = 8$ km/hr

Speed in still water = $\frac{8+5}{2} = \frac{13}{2} = 6\frac{1}{2}$ km/hr

54. **(*c*)** Total age of 5 new students = $5 \times 13\frac{3}{4} - 30 \times \frac{7}{12} = 65 + \frac{15}{4} - \frac{35}{2} = 51\frac{1}{4}$

Total age of 4 new students = $51\frac{1}{4} - 9\frac{11}{12} = 41\frac{4}{12}$

$\Rightarrow$ Average age of 4 new students = $41\frac{4}{12} \div 4 = 10\frac{1}{3}$ = 10 years 4 months

55. (*d*) Net loss = $\frac{(10)^2}{100} = 1\%$

56. (*d*) Let A = 60° and B = 30°, then

$\tan 60° = n \tan 30° \Rightarrow \sqrt{3} = n \times \frac{1}{\sqrt{3}} \Rightarrow n = 3$

$\sin A = m \sin B \Rightarrow \frac{\sqrt{3}}{2} = m \times \frac{1}{2} \Rightarrow m = \sqrt{3}$

$\cos^2 A = \cos^2 60° = \frac{1}{4} = \frac{2}{8} = \frac{3-1}{9-1} = \frac{m^2-1}{n^2-1}$

57. (*) G.E. = $\left(\frac{p^{-4}q^4}{p^8q^{-6}}\right)^{1/3} = \left(p^{-12}q^{10}\right)^{1/3} = p^a \cdot q^b \Rightarrow a + b = \frac{-12+10}{3} = -\frac{2}{3}$

58. (*d*) G.E. = $4 - \frac{5}{1+\frac{1}{3+\frac{4}{9}}} = 4 - \frac{5}{1+\frac{9}{31}} = 4 - 5\times\frac{31}{40} = 4 - \frac{31}{8} = \frac{1}{8}$

59. (*b*) $\frac{\text{S.I.}}{\text{Principal}} = \frac{8}{25} = \frac{32}{100} \Rightarrow 32 = 4\times8$

60. (*d*) LCM fo 12, 16, 18, 21 is multiple of 9 ⇒ sum of digits is divisible by 9
sum of digits of 2000 = 2 ⇒ sum of digits of $x = 9 - 2 = 7$

61. (*d*) $\frac{6\times5}{6-5}\times\frac{12}{60} = 6$ km

62. (*a*) $\frac{\frac{22}{7}\times4\times56}{48\times\frac{33}{2}\times4} = \frac{2}{9}$

63. (*c*) (cot 41° . cot 49°) (cot 42° . cot 48°) (cot 43° . cot 47°) (cot 44° . cot 46°) (cot 45°)
= 1 × 1 × 1 × 1 × 1 = 1

64. (*b*) In ΔABD and ΔFBD,

∠ABD = ∠FBD (Angle bisector)

∠ADB = ∠FDB = 90° (Linear Pair)

BD = BD (common)

⇒ ΔABD ≅ ΔFBD (ASA)

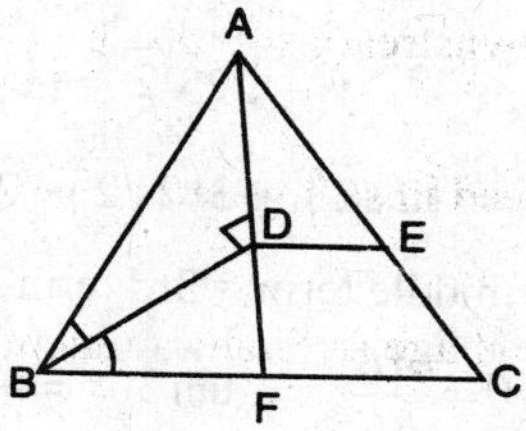

In ΔAFC, DE || FC

⇒ ΔADE ~ ΔAFC

AD = DF ⇒ D is mid-point on AF.

E is mid-poind on AC ⇒ E = $\frac{1}{2} \times 12 = 6$ cm

65. (b) Side of rhombus = 60 ÷ 4 = 15 cm

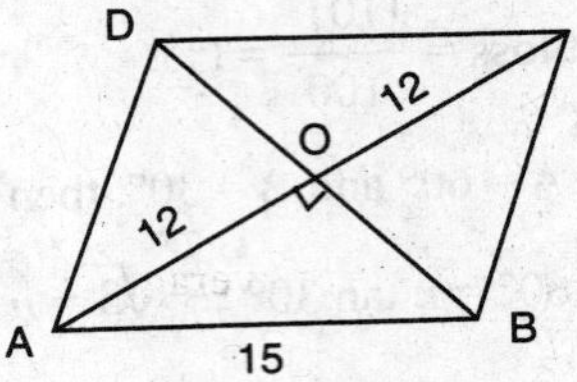

$AD = \frac{1}{2} \times 24 = 12 \text{ cm} = 9 \text{ cm}$

$BO = \sqrt{15^2 - 12^2}$

BD = 2 × 9 cm = 18 cm

$\text{Area} = \frac{1}{2} \times 18 \times 24 = 216 \text{ cm}^2$

66. (c) $\frac{\frac{x}{4}}{10} + \frac{\frac{3x}{4}}{12} = 7 \Rightarrow \frac{x}{40} + \frac{x}{16} = 7 \Rightarrow x = 80 \text{ km}$

Trick: 80 is divisible by 4 and 1/4 is divisible by 10

67. (b) $\text{Tin} = \frac{2}{5} \times 60 + \frac{1}{5} \times 100 = 24 + 20 = 44 \text{ kg}$

68. (c) $\text{G.E.} = \frac{330}{(2+\sqrt{3}) - (1+\sqrt{3})} = 330$

69. (b) (7+ 9.2) – 8.5 = 7.7 cm

70. (d) $\sin A = 1 - \sin^2 = \cos^2 A$

$\Rightarrow \cos^2 A + \cos^4 A = \cos^2 A + \sin^2 A = 1$

71. (b) Let each distance = LCM of 10, 20, 30, 60 = 120 km

$\text{Then total time} = \frac{120}{10} + \frac{120}{20} + \frac{120}{30} + \frac{120}{60} = 12 + 6 + 4 + 2 = 24 \text{ hours}$

$\text{Average speed} = \frac{4 \times 120}{24} = 20 \text{ km/hr}$

72. (d) $x = \sqrt{a} + \frac{1}{\sqrt{a}}; \; y = \sqrt{a} - \frac{1}{\sqrt{a}} \Rightarrow x + y = 2\sqrt{a} \text{ and } x - y = \frac{2}{\sqrt{a}}$

$= \left(x^2 - y^2\right) = \left[(x+y)(x-y)\right]^2 = \left[\left(2\sqrt{a}\right)\left(\frac{2}{\sqrt{a}}\right)^2\right] = 4^2 = 16$

73. (a) $x = \sqrt{3} + \sqrt{2}; y = \sqrt{3} - \sqrt{2} \Rightarrow x - y = 2\sqrt{2}$ and $xy = 1$

$\text{G.E.} = x^3 - y^3 - 22\sqrt{2} = (x-y)^3 + 3xy(x-y) - 22\sqrt{2}$

$= \left(2\sqrt{2}\right)^3 + 3\left(2\sqrt{2}\right) - 22\sqrt{2} = 16\sqrt{2} + 6\sqrt{2} - 22\sqrt{2} = 0$

74. (a) Let middle term = 3rd term = x Than new average = 4th term = x +1

75. (b) $\tan\theta = \frac{AB}{BC} \Rightarrow \tan 30° = \frac{AB}{10\sqrt{3}}$

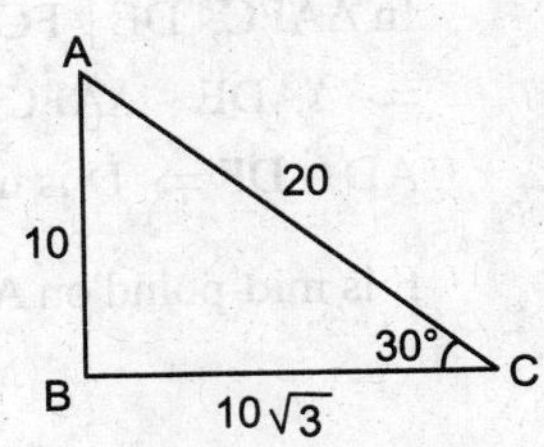

$\Rightarrow \frac{1}{\sqrt{3}} = \frac{AB}{10\sqrt{3}} \Rightarrow AB = 10 \text{ m}$

$\sin\theta = \frac{AB}{AC} \Rightarrow \frac{1}{2} = \frac{10}{AC} \Rightarrow AC = 20 \text{ m}$

⇒ Height of post = 10 + 20 = 30 m

76.(*c*) Increase in Sum = 2 + 4 + 8 + 16 + ...

$= 2^1 + 2^2 + 2^3 + ... + 2^n = \frac{2(2^n - 1)}{2 - 1}$

$\Rightarrow$ Increase in average $= \frac{2(2^n - 1)}{n}$ $\Rightarrow$ New average $= a + \frac{2(2^n - 1)}{n}$

77. (*b*) CP of 3 oranges = 40; CP of 3 other oranges = (60 ÷ 5) × 3 = 36

$\Rightarrow$ CP of 6 oranges = 40 + 36 = 76;

SP of 6 oranges = 50 + 50 = 100

$\Rightarrow$ Profit = 100 – 76 = 24 $\Rightarrow$ Profit $= \frac{24}{76} \times 100 = 32\%$ (Approx)

78. (*b*) $(a-3) - \frac{1}{a-3} = 5 - 3 = 2$

$\Rightarrow (a-3)^3 - \frac{1}{(a-3)^3} = \left[(a-3) - \frac{1}{(a-3)}\right]^3 + 3\left[(a-3) - \frac{1}{a-3}\right]$

$= 2^3 + 3 \times 2 = 14$

79. (*b*) Numbers $= (3^5)^{10}, (4^4)^{10}, (5^3)^{10}, (6^2)^{10} = (243)^{10}, (256)^{10}, (125)^{10}, (36)^{10}$

80. (*a*) $x^2 + y^2 + z^2 = xy + yz + zx \Rightarrow x = y = z$

$\Rightarrow$ G.E. $= \frac{3x^4 + 7x^4 + 5x^4}{5x^4 + 7x^4 + 3x^4} = 1$

81. (*b*) Let total work = LCM of 16, 32, 48 = 96 units

$\Rightarrow$ Their work for one day each = 6, 3, 2 units

Let total days = x. Then $6x + 3(x - 8) + 2(x - 6) = 96$

$11x = 96 + 24 + 12 = 132$ $\Rightarrow x = 132 \div 11 = 12$ days

Alt. Loss of work, due to absence of B = 8 × 3 = 24 units respectively.

Loss of work, due to absence of C = 6 × 2 = 12 units

$\Rightarrow$ Total days = (96 + 24 + 12) ÷ (6 + 3 + 2) = 132 ÷ 11 = 12 days

82. (*d*) Area (ΔGBC) $= \frac{1}{3}$ (ΔABC) $= \frac{1}{3} \times 60 = 20$ cm²

83. (*c*) (A + C) = 2 (A + B + C) – (A + B) – (B + C) $= \frac{1}{3} - \frac{1}{8} - \frac{1}{12} = \frac{1}{8}$

84. (*a*)

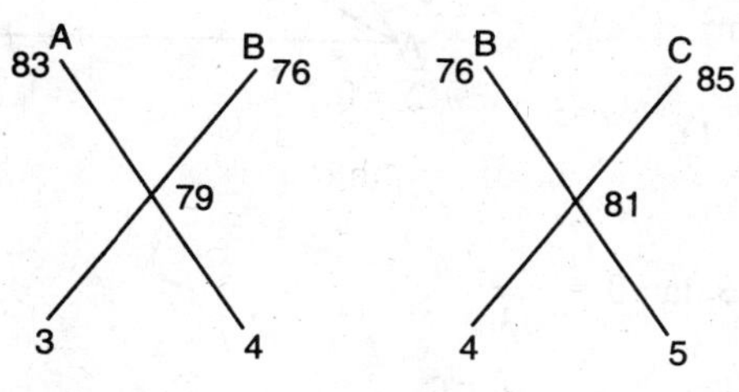

$\Rightarrow$ A : B : C = 3 : 4 : 5

$\Rightarrow$ Average score $= \frac{3 \times 83 + 4 \times 76 + 5 \times 85}{3 + 4 + 5} = \frac{978}{12} = 81.5$

85. (*b*) Area (ABCD)

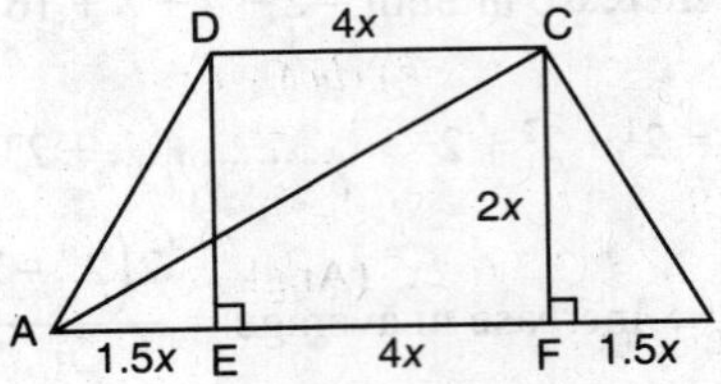

$= \frac{1}{2} \times 11x \times 2x = 176 \Rightarrow x = 4$

In Δ ACF, AF = 22, CF = 6

$AC = \sqrt{22^2 + 8^2} = 2\sqrt{137}$

86. (*a*) Base Area = $\sqrt{27 \times 14 \times 7 \times 6} = 126$

Volume of prism = Base Area × Height = 126 × 9 = 1134 cm²

87. (*c*) Married males = 60% of 50% = 30%

Married females = 40% of 70% = 28%

Married staff = 30 + 28 = 58%

⇒ Unmarried staff = 100 – 58 = 42%

88. (*b*) Tangent = $\sqrt{13^2 - (11-6)^2} = \sqrt{13^2 - 5^2} = 12$ cm

89. (*b*) Ratio of amounts = 110 : 115 : 120 = 22 : 23 : 24

Ratio of principals = $\frac{1}{22} : \frac{1}{23} : \frac{1}{24} = 23 \times 24 : 22 \times 24 : 22 \times 23 = 552, 528, 506$

Loan of A = $\frac{552}{1586} \times 7930 =$ ₹ 2760

90. (*a*) Sum of speeds = 162 ÷ 6 = 27 km/hour

Slower speed = $\frac{1}{2} \times (27 - 8) = \frac{19}{2} = 9\frac{1}{2}$ km/ hr

91. (*b*) slant height of pyramid = $\sqrt{12^2 + (10/2)^2} = 13$ cm

Total Surface Area = $10^2 + 4\left(\frac{1}{2} \times 10 \times 13\right) = 100 + 260 = 360$ cm²

92. (*d*) $2^3 = 8 \Rightarrow$ Time = 5 × 3 = 15 years

93. (*c*) Δ ABO ~ ΔCOD

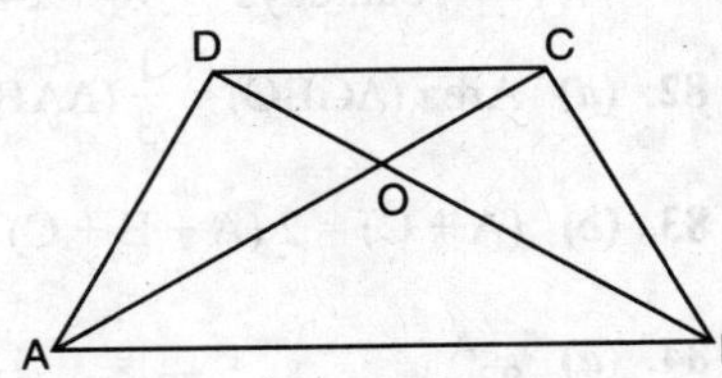

A B : CD = 2:1 (given)

⇒ Area (ΔABO) : Area (ΔCOD) = 4:1

⇒ Area (ΔCOD) = $\frac{1}{4} \times 84 = 21$ cm²

94. (*b*) $7 \tan^2 \theta + 3 = 4 \sec^2 \theta = 4 (\tan^2 \theta + 1)$

$\Rightarrow 7 \tan^2 \theta + 3 = 4 \tan^2 \theta + 4$

$\Rightarrow 3 \tan^2 \theta = 1 \Rightarrow \tan^2 \theta = \frac{1}{3} \Rightarrow \tan \theta = \frac{1}{\sqrt{3}}$

95. (*a*) $a + \frac{1}{b} = b + \frac{1}{c} \Rightarrow a - b = \frac{1}{c} - \frac{1}{b} = \frac{b-c}{bc} \Rightarrow bc = \frac{b-c}{a-b}$

similarly, $ab = \frac{a-b}{c-a}$ and $ca = \frac{c-a}{b-c}$

$\Rightarrow a^2 b^2 c^2 = (ab)(bc)(ca) = \dfrac{a-b}{c-a} \times \dfrac{b-c}{a-b} \times \dfrac{c-a}{b-c} = 1$

96. (*b*) Let $\angle A = x°$

$\Rightarrow \angle BOC = 2x°$ (Angle at centre)

$OB = OC$ (radii)

$\Rightarrow \angle OBC = \angle OCB = y°$

In ΔOBC, $2x + 2y = 180°$

$\Rightarrow x + y = 90°$

$\Rightarrow \angle BAC + \angle OBC = 90°$

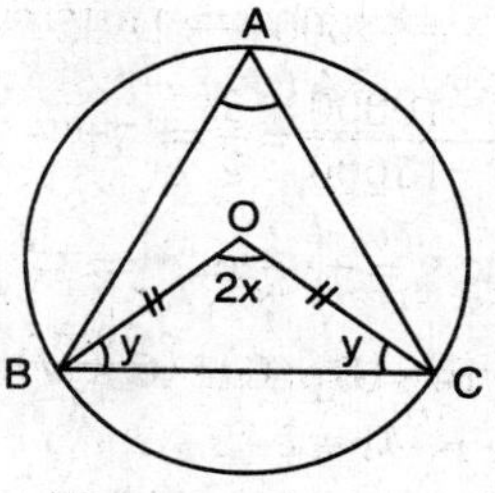

97. (*c*) From options : Numbers are 21 and 84.

98. (*d*) $54 \times \dfrac{120}{90} = ₹\ 72$

99. (*c*) $AC = \sqrt{BD \times CD} = \sqrt{3 \times 4} = \sqrt{12} = 2\sqrt{3}$ cm

100. (*b*) $\dfrac{1}{3}\pi r^2 h = \pi r l$

$\Rightarrow \dfrac{1}{3} rh = \sqrt{r^2 + h^2} \quad \Rightarrow \dfrac{1}{9} r^2 h^2 = r^2 + h^2 \Rightarrow \dfrac{1}{9} = \dfrac{r^2 + h^2}{r^2 h^2} \quad \Rightarrow \dfrac{1}{9} = \dfrac{1}{h^2} + \dfrac{1}{r^2}$